Jane's

SPACE DIRECTORY

Edited by David Baker

Fifteenth Edition

1999-2000

ISBN 0 7106 1919 7

In the USA and its dependencies
Jane's Information Group Inc, 1340 Braddock Place, Suite 300, Alexandria, Virginia 22314-1651, USA

DPA
DIRECTORY & DATABASE PUBLISHERS ASSOCIATION
MEMBER

British Library Cataloguing-in-Publication Data.
A catalogue record for this book is available from the British Library.

Printed and bound in Great Britain by Butler and Tanner Limited, Frome and London

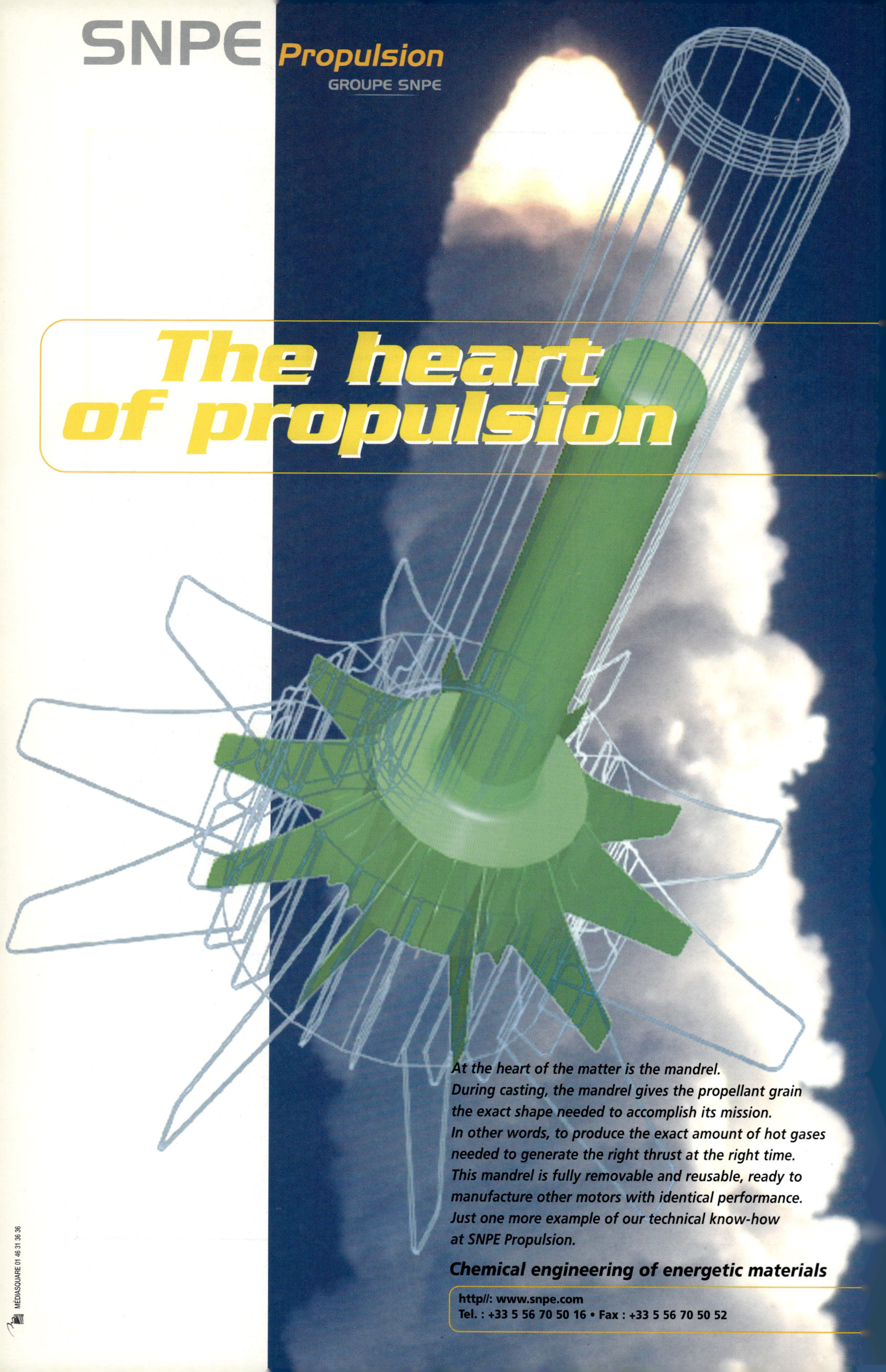
SNPE Propulsion
GROUPE SNPE
The heart of propulsion
At the heart of the matter is the mandrel.
During casting, the mandrel gives the propellant grain the exact shape needed to accomplish its mission.
In other words, to produce the exact amount of hot gases needed to generate the right thrust at the right time.
This mandrel is fully removable and reusable, ready to manufacture other motors with identical performance.
Just one more example of our technical know-how at SNPE Propulsion.
Chemical engineering of energetic materials
http//: www.snpe.com
Tel. : +33 5 56 70 50 16 • Fax : +33 5 56 70 50 52
MÉDIASQUARE 01 46 31 36 36

Contents

How to use *Jane's Space Directory*	[6]
Units, abbreviations and acronyms	[10]
Alphabetical list of advertisers	[12]
Foreword	[15]
Users' Charter	[24]
Government and non-government space programmes	1
Government and non-government institutional organisations	
Military	29
Civil	31
Space industry	
Communication satellites, components and services	63
Earth observation, facilities and equipment	71
Earth observation satellites, components and services	77
General	79
Launchers, propulsion systems and launch site services	105
Microgravity facilities, equipment and services	115
Navigation, search and rescue facilities and equipment	117
Satellite actuation systems and thermal control devices	121
Satellite busses	125
Satellite communications facilities and equipment	127
Satellite guidance, navigation and control components	133
Satellite information storage components	141
Satellite and propulsion support hardware	143
World space centres	
Launch sites	147
Training facilities	165
Telemetry, tracking and command	167
Research and development facilities	171
Space monitoring facilities	177
Launch vehicles (orbital and sub-orbital)	179
Launch vehicle propulsion	247
Civil and commercial satellites, satellite series and satellite constellations	
Communications	293
Data relay	341
Earth observation and meteorological	343
Engineering test	373
Geodetic	381
Microgravity and materials science	383
Navigation	387
Scientific	389
Planetary and space science	419
Human space flight	437
Resupply vehicles	465
Space stations	467
Military satellites, satellite series and satellite constellations	
ASAT	485
Ballistic missile defence	487
Communications	489
Data relay	498
Early warning	499
ELINT	503
FOBS	505
Geodetic	506
Meteorological	507
Minor military	509
Navigation	510
Nuclear surveillance	515
Ocean surveillance	516
Photo reconnaissance	519
Radar calibration	526
Radar imaging	527
Science and engineering test	528

Contractors
Civil and commercial space organisations 535
Military space organisations 543
Communication satellite systems and services 545
Earth observation satellite systems and services 555
Navigation satellite systems and services 561
Civil and commercial ground facilities 563
Space industrial providers 567

Space logs 589

Indexes 629

How To Use *Jane's Space Directory*

The 1999-2000 issue of *Jane's Space Directory* continues the three-year revision begun with the last edition and announced on this page in that edition. Responding to readers' reactions and mindful that this book is only as good as its ability to satisfy the needs of subscribers, we have re-introduced some sections and restructured the general order of content. As we move to what we hope will be the definitive upgrade in timeliness of information, refined presentation and user-friendly content layout, the editor welcomes comment on the way you use the book, what you seek within its pages and how we can improve the product for your needs.

This year the content is structured around the categorisation of functions, beginning with a description of government and non-government space programmes and organisations providing detailed information on relevant activities. This is followed by a directory of the global space industry which serves the aforementioned programmes. These two sections provide the core from which all the technology, research, development and operational activity flows and includes information about trends and objectives, budgets, programme and project prioritisation and about the centres of application. These two sections form the institutional backbone to government and non-government space activity.

The second main section of the book covers civilian operations involving launch sites, launch vehicles, in-orbit satellite activity, planetary and scientific operations and human space flight including space stations and resupply vehicles. The opening part catalogues all known world space centres including launch sites, training and tracking facilities, research and development facilities and space monitoring sites. It continues with a survey and technical description of the world's orbital and sub-orbital launch vehicles followed by a section on propulsion. Civilian satellites follow with information on design types, pedigree, operations history and specifications. Planetary and space science spacecraft follow in a similar format. Human space flight serves to connect major goals outlined in earlier sections with realisation of national or international objectives.

The third main section groups all military operations by programme category and then by country but cross-reference with the launch vehicles section is necessary to determine the characteristics of launchers and propulsion systems where they apply. Categories of military satellites are those exclusively dedicated to military support functions but do not include those commercial or civilian defence-related experiments, conducted by manned vehicles, details of which can be found in the appropriate section.

Finally, the fourth main section is devoted to chronicling the contractors, to providing tables of data and lists of information for reference purposes in a series of space logs and enabling cross-reference through the index.

Since 1994, all new pictures have been dated with the year of publication. All new pictures this year are dated 1999. Some are followed by a seven digit number for ease of identification by our image library.

ADMINISTRATION

Director, Defence Business Unit:	*Alan Condron*
Publisher:	*Karen Heffer*
Managing Editor:	*Simon Michell*
Database Manager/Data Administrator:	*Ruth Simmance*
Team Leader:	*Joanne Fenwick*

EDITORIAL OFFICE

Jane's Information Group Limited, Sentinel House, 163 Brighton Road,
Coulsdon, Surrey CR5 2YH, UK
Tel: (+44 208) 700 37 00 Fax: (+44 208) 700 37 88
e-mail: jsd@janes.co.uk

SALES OFFICE

Jane's Information Group Limited, Sentinel House, 163 Brighton Road,
Coulsdon, Surrey CR5 2YH, UK
Tel: (+44 208) 700 37 00 Fax: (+44 208) 763 10 06
e-mail: info@janes.co.uk

Send enquiries to: *Jo Moon – Senior Sales and Customer Service Manager*
Florina Shons (Continental Europe/Middle East/Africa)
Julie Reeder (UK & Scandinavia)

Send USA enquiries to: *Robert Loughman – Vice-President Product Sales*
Jane's Information Group Inc, 1340 Braddock Place, Suite 300, Alexandria, Virginia 22314-1651, USA
Tel: (+1 703) 683 37 00 Fax: (+1 703) 836 00 29 Telex: 6819193
Tel: (+1 800) 824 07 68 Fax: (+1 800) 836 02 97

Send Asia enquiries to: *David Fisher*
Jane's Information Group Asia, 60 Albert Street, #15-01 Albert Complex, Singapore 189969
Tel: (+65) 336 62 80 Fax: (+65) 336 99 21

Send Australia/New Zealand enquiries to: *Pauline Roberts*
Jane's Information Group, PO Box 3502, Rozelle,
New South Wales 2039, Australia
Tel: (+61 2) 85 87 79 00 Fax: (+61 2) 85 87 79 01

ADVERTISEMENT SALES OFFICES

Australia: *Richard West* (UK Head Office)

Austria: *Richard West* (UK Head Office)

Benelux: *Stephen Judge* (UK Head Office)

Channel Islands: *Stephen Judge* (UK Head Office)

Eastern Europe: *Richard West* (UK Head Office)

Egypt: *Terry Doherty* (UK Head Office)

France: *Patrice Février*
BP 418, 35 avenue MacMahon,
F-75824 Paris Cedex 17
Tel: (+33 1) 45 72 33 11 Fax: (+33 1) 45 72 17 95
e-mail: patrice.fevrier@wanadoo.fr

Germany: *Richard West* (UK Head Office)

Greece: *Stephen Judge* (UK Head Office)

India: *Richard West* (UK Head Office)

Ireland: *Stephen Judge* (UK Head Office)

Israel: *Oreet Ben-Yaacov*
Oreet International Media, 15 Kinneret Street,
IL-51201 Bene Berak
Tel: (+972 3) 570 65 27 Fax: (+972 3) 570 65 26
e-mail: oreetimc@netvision.net.il

Italy and Switzerland: Ediconsult Internazionale Srl
Piazza Fontane Marose 3, I-16123 Genoa, Italy
Tel: (+39 010) 58 36 84 Fax: (+39 010) 56 65 78
e-mail: ediconsult@iol.it

Korea, South: *Young Seoh Chinn*
JES Media International, 2nd Floor ANA Building,
2571 Myungil-Dong, Kangdong-Gu, Seoul 134 070
Tel: (+82 2) 481 34 11 Fax: (+82 2) 481 34 14
e-mail: jesmedia@unitel.co.kr

Middle East: *Terry Doherty* (UK Head Office)

Pakistan: *Richard West* (UK Head Office)

Rest of the World: *Terry Doherty* (UK Head Office)

Russian Federation and Associated States (CIS): *Simon Kay*
33 St John's Street, Crowthorne, Berkshire RG45 7NQ, UK
Tel: (+44 1344) 77 71 23 Fax: (+44 1344) 77 58 85
e-mail: crowkay@msn.com

Saudi Arabia: *Terry Doherty* (UK Head Office)

Scandinavia: *Gillian Thompson*
The Falsten Partnership, 11 Chardmore Road, Stamford Hill,
London N16 6JA, UK
Tel: (+44 181) 806 23 01 Fax: (+ 44 181) 806 81 37
e-mail: falsten@dial.pipex.com

Singapore: *Richard West* (UK Head Office)

South Africa: *Terry Doherty* (UK Head Office)

Spain: *Michael Andrade*
Via Exclusivas SL, Viriato 69SC, 28010 Madrid
Tel: (+34 91) 448 76 22 Fax: (+34 91) 446 01 98
e-mail: via@varenga.com

Thailand: *Richard West* (UK Head Office)

Turkey: *Richard West* (UK Head Office)

United Arab Emirates: *Terry Doherty* (UK Head Office)

UK – South, West, North and Middle England, Wales and Scotland:
Richard West

UK – South East and London: *Terry Doherty*

Advertisement Sales Manager: *Terry Doherty*
Jane's Information Group Limited, Sentinel House, 163 Brighton Road,
Coulsdon, Surrey CR5 2YH, UK
Tel: (+44 208) 700 38 55 Fax: (+44 208) 700 37 44
e-mail: terry.doherty@janes.co.uk

Advertisement Sales Manager: *Richard West*
Jane's Information Group Limited, Sentinel House, 163 Brighton Road,
Coulsdon, Surrey CR5 2YH, UK
Tel: (+44 208) 700 37 39 Fax: (+44 208) 700 37 44
e-mail: richard.west@janes.co.uk

Advertisement Sales Executive: *Stephen Judge*
Jane's Information Group Limited, Sentinel House, 163 Brighton Road,
Coulsdon, Surrey CR5 2YH, UK
Tel: (+44 208) 700 38 53 Fax: (+44 208) 700 37 44
e-mail: stephen.judge@janes.co.uk

USA and Canada: *Ronald R Lichtinger, III,* Advertising Sales Director
Jane's Information Group Inc, 1340 Braddock Place, Suite 300,
Alexandria, Virginia 22314-1651, USA
Tel: (+1 703) 683 37 00 Fax: (+1 703) 836 55 37
e-mail: lichtinger@janes.com

North Eastern USA and East Canada: *Harry Carter*
Jane's Information Group Inc, 1340 Braddock Place, Suite 300,
Alexandria, Virginia 22314-1651, USA
Tel: (+1 703) 683 37 00 Fax: (+1 703) 836 55 37
e-mail: carter@janes.com

South Eastern USA: *Kristin D Schulze*
5370 Eastbay Drive, Suite 104,
Clearwater, Florida 33764, USA
Tel: (+1 727) 524 77 41 Fax: (+1 727) 524 75 62
e-mail: kristin@intnet.net

Western USA and West Canada: *Richard L Ayer*
127 Avenida del Mar, Suite 2A,
San Clemente, California 92672, USA
Tel: (+1 949) 366 84 55 Fax: (+1 949) 366 92 89
e-mail: ayercomm@earthlink.net

Administration:
USA and Canada: *Maureen Nute*
Jane's Information Group Inc, 1340 Braddock Place, Suite 300,
Alexandria, Virginia 22314-1651, USA
Tel: (+1 703) 683 37 00 Fax: (+1 703) 836 00 29
e-mail: nute@janes.com

UK and Rest of World: *Joni Beeden*
Jane's Information Group Limited, Sentinel House, 163 Brighton Road,
Coulsdon, Surrey CR5 2YH, UK
Tel: (+44 208) 700 37 42 Fax: (+44 208) 700 38 59
e-mail: joni.beeden@janes.co.uk

Units, abbreviations and acronyms

The metric system of measurement is employed throughout the Directory. Times and dates are quoted in Greenwich Mean Time (GMT) unless otherwise noted. Shuttle missions, in addition to their official STS designations, are also allocated 'SM' numbers to indicate the actual flight sequence. Abbreviations and acronyms are usually explained in the text; the most common are listed below, along with some unit conversions not given elsewhere.

μg: microgravity
μm: 10^{-6} m (micron)
Å: Angstrom, 10^{-10} m
ACS: Attitude Control System
A/D: Analogue/Digital
AF: Air Force
AFB: Air Force Base
AIT: Assembly, Integration and Testing
AKM: Apogee Kick Motor
AO: Announcement of Opportunity
AOCS: Attitude and Orbit Control System
AS: Air Station
ASEAN: Association of South East Asia Nations
ASIC: Application Specific Integrated Circuit
atm: Earth atmospheric pressure (1.0336 kg/cm^2; 14.7 lb/in^2)
AU: Astronomical Unit (149.598 million km) or ESA Accounting Unit (equivalent to the ECU)
billion: 1,000 million
BOL/EOL: Beginning Of Life/End Of Life
C^3: Command, Control and Communications
C^4: Command, Control, Communications and Computers
CAD: Computer-Aided Design
CCT: Computer-Compatible Tape
CDR: Critical Design Review
cleanliness: the Directory adopts the US system specified by the numerical limit of 0.5 μm particles per ft^3 (0.028 m^3). Thus a class 10,000 environment has fewer than 10,000 particles of 0.5 μm in 1 ft^3 (0.028 m^3). Most satellite and launcher integration and testing facilities are class 100,000. Class 10 and class 100 facilities are used for assembly of sensitive items such as gyros
cm: centimetre (1 inch = 2.54 cm)
CNC: Computer Numerical Control
Conus: Continental United States
COTS: Commercial-Off-The-Shelf
CPV: Common Pressure Vessel
DAMA: Demand Assigned Multiple Access
DBS: Direct Broadcast Satellite
DGPS: Differential GPS
DoF: Degrees of Freedom
DRAM: Dynamic Random Access Memory
DTH: Direct To Home
ECLSS: Environment Control and Life Support System
ECU: European Currency Unit
EGSE: Electrical Ground Support Equipment
EIRP: Equivalent Isotropically Radiated Power
elint: electronic intelligence
EM: Engineering Model
EPC: Electronic Power Conditioner
ESA: European Space Agency
eV: electron Volt
fl: focal length
FM: Flight Model
FoV: Field of View
FSD: Full Scale Development
FWHM: Full Width Half Maximum
FY: Fiscal Year
g: gravitational acceleration (9.806 m/s)
GDP: Gross Domestic Product
GEO: Geostationary Orbit; altitude 35,786 km, 0° inclination, period 23.934 h
GH_2: Gaseous Hydrogen
GIS: Geographic Information System
GMDSS: Global Maritime Distress and Safety System
GMT: Greenwich Mean Time (=UT, Universal Time)
GN&C: Guidance, Navigation and Control
GNP: Gross National Product
GOX: Gaseous Oxygen
GPS: Global Positioning System
GSE: Ground Support Equipment
GSM: Groupe Spéciale Mobile
GSO: Geosynchronous Orbit
GTO: Geostationary Transfer Orbit
HES: Health, Environment and Safety
HGA: High Gain Antenna
HOOD: Hierarchical Object Oriented Design
HPA: High-Power Amplifier
HRVIR: High-Resolution Visible Infra-Red
IC: Integrated Circuit
ICBM: InterContinental Ballistic Missile (range >5,500 km)
IFOG: Interferometric Fibre Optic Gyro
IMU: Inertial Measurement Unit
INS: Inertial Navigation System
INU: Inertial Navigation Unit
IOC: Initial Operational Capability
IRBM: Intermediate Range Ballistic Missile (range 500-5,500 km)
IRU: Inertial Reference Unit
ISDN: Integrated Services Digital Network
km: kilometre (0.6214 mile)
L: Launched
LEO: Low Earth Orbit, typically up to 1,500 km
LEOP: Launch and Early Orbit Phase
LH_2: Liquid Hydrogen
LHCP: Left-Hand Circular Polarisation
LiOH: Lithium Hydroxide
LITVC: Liquid Injection Thrust Vector Control
LNA: Low-Noise Amplifier
LOX: Liquid Oxygen
m: metre (39.37 inch)
MAU: Member Accounting Unit
MEOP: Maximum Expected Operating Pressure
MGSE: Mechanical Ground Support Equipment
micron: 10^{-6} m
MIR: Mid Infra-Red
MMH: Monomethyl Hydrazine
MMIC: Monolithic Microwave Integrated Circuit
MON: Mixed Oxides of Nitrogen
MoU: Memorandum of Understanding
MPD: Magnetoplasmadynamic
MW: Momentum Wheel
N: Newton (0.225 lbf)
NASA: National Aeronautics and Space Administration (US)
Ni/Cd: Nickel Cadmium
Ni/H_2: Nickel Hydrogen

NiMH: Nickel Metal Hydride
NIR: Near Infra-Red
NSSK: North-South Station-Keeping
NTO, N_2O_4: Nitrogen Tetroxide
OBDH: Onboard Data Handling
OEM: Original Equipment Manufacturer
PAN: Panchromatic
PDR: Preliminary Design Review
PI: Principal Investigator
PKM: Perigee Kick Motor
PMD: Propellant Management Device
ppm: parts per million
PROM: Programmable Read-Only Memory
QM: Qualification Model
RAM: Random Access Memory
RCS: Reaction Control System
RDT&E: Research, Development, Test and Evaluation
RF: Radio Frequency
RHCP: Right Hand Circular Polarisation
RLG: Ring Laser Gyro
RLV: Reusable Launch Vehicle
RW: Reaction Wheel
SAR: Synthetic Aperture Radar
SCADA: Supervisory Control and Data Acquisition
SCOE: Special Check-Out Equipment
SEP: Spherical Error Probability
SEU: Single Event Upset
SGLS: US DoD's Space Ground Link Subsystem
sigint: signals intelligence
SL: Sea Level
SLBM: Submarine-Launched Ballistic Missile
SNG: Satellite News Gathering
SOS: Sapphire On Silicon
SSPA: Solid-State Power Amplifier
SSR: Solid-State Recorder
STDN: NASA's Spacecraft Tracking and Data Network
SW: Space Wing
SWIR: Short Wave Infra-Red
t: metric tonne (1,000 kg; 0.984 ton; 1.102 US ton)
TC: Telecommand
TCR: Telemetry, Command and Ranging
TT&C: Tracking, Telemetry and Command
TVC: Thrust Vector Control
TVRO: TV Receive Only
TWTA: Travelling Wave Tube Amplifier
UTM: Universal Transverse Mercator grid projection
VHSIC: Very High-Speed Integrated Circuit
VLBI: Very Long Baseline Interferometry
VNIR: Very Near Infra-Red
VTIR: Visible/Thermal Infra-Red
W: Watt
WWW: World Wide Web (Internet)
XS: multispectral

Alphabetical list of advertisers

A

Alcatel
Alcatel Denmark A/S Lautrupvang 2,
DK-2750 Ballerup, Denmark .... [7]

Alenia Aerospazio
Divisione Spazio, Via Saccomuro 24,
I-00131 Rome, Italy .... *On Front Cover*

Airclaims Ltd
Cardinal Point, Newall Road, Heathrow Airport,
Hounslow, Middlesex TW6 2AS, UK 631

I

Irvin Aerospace Limited
Icknield Way, Letchworth,
Hertfordshire SG6 1EU, UK .... 629 and 633

S

SNPE
12 Quai Henri IV, F-75181 Paris,
Cedex 04, France .... [2]

Reach for the sky
Jane's Strategic Weapon Systems
A comprehensive overview of every weapon system on the market, and those in development. Regularly updated, this key intelligence document includes details of obsolete weapons, the text of major arms-control treaties and country inventories to ensure you stay on top of this fast-moving field.
Jane's Missiles and Rockets
Edited by a former missile engineer, this unique monthly newsletter covers the latest developments in missile and rocket technology, plus procurement and manufacturing issues. Recent exclusives include first details of Pakistan's Ghauri missile last year.
Jane's Land-Based Air Defence
Unrivalled expertise at your fingertips. This crucial reference tool provides you with a complete listing of available systems, and information on those in development to highlight key trends. Country inventories ensure you are aware of the global landscape, and full manufacturer details aid procurement.
Jane's Air-Launched Weapons
A comprehensive survey of the weapons systems available or in development around the globe. Authoritative and impartial, including a detailed analysis table of systems for ease of comparison, and country inventories to ensure awareness of the current strategic landscape.
AVAILABLE ONLINE • CD-ROM • HARDCOPY

The first Ariane 5 lifts off ELA-3 in May 1996 (ESA)

1999/0054379

Foreword

This is a particularly important edition of *Jane's Space Directory*. At the turn of a new millennium the Age of Space has reached a mature and self-sustaining level where growth is solid and the industry sound. No longer is it an outgrowth of the aerospace industry, nor is it an appendage to existing technology endeavours as it was for the first three decades. In the fifth decade of this Age of Space, it has became independent and mature. Nevertheless, government space programmes are in decline and the research investment in applications and exploration is diminishing year by year. While the commercial sector continues to expand, the broad picture in science and engineering is one of uncertainty and faltering commitment. National space agencies have evolved from entrepreneurial, 'seat-of-the-pants' action groups to nervous bureaucracies, unwieldy and overburdened with pleasing their political overlords. Decisions made in the next few months and years will significantly reshape the way both civil and military programmes are managed and there is much to do to restore flexibility and the ability for agencies to operate without micro-managed oversight.

The Market

In terms of numbers and sheer size, the global space industry is healthy and set for burgeoning growth over the next decade but governments are cutting back on space research and technology development while private companies and international corporations are putting more money into space than taxpayers. Recent trends threaten to shake up previous assumptions and entrenched long-term goals are locking big government programmes into inflexible commitments that threaten the traditional, evolutionary, nature of space research and development. Forty years after NASA's first long-range plan, published for internal consumption only, pronounced the goal of building a permanently manned base in space, the International Space Station is finally getting built, albeit 15 years after Ronald Reagan formalised that objective. It should be in full operation by 2004, about the time NASA's last large-scale planetary mission – Cassini – gets to Saturn, but NASA's budget is in decline and most spacefaring nations are cutting back on their investment in basic research. Meanwhile, worldwide revenues from privately owned space-based enterprises continue to expand and give good return on investment while some big players are heading for a showdown with their backers. In every respect this, the fifteenth edition of *Jane's Space Directory*, portends major changes in the industry among the user community as well as in the direction of government commitments.

Over the last decade, space has changed from being a place for high-tech markets with niched applications to a major business initiative with shareholders, investors and developers playing on a global arena where mergers, acquisitions, competition and financial markets jostle for a piece of the action. It is this coterie of well-established controls, regulations and influences that is increasingly responsible for shaping the direction of applications and revenue streams. Interlinked with world economic factors, changes in regional and area markets and technological opportunities, a plethora of services carried by satellites is boosting companies and raising investment opportunities. From the burdensome debt financing and loan restructuring of 1998 to easier money-raising conditions in 1999, entrepreneurial ventures are thick on the ground. Too much so, as witnessed by overlapping services jostling for a piece of the pie and ending up with no place at the table.

Three proposed configurations for the Angara launcher series (Theo Pirard)
1999/0054361

Eutelsat satellite for East-West Communications (Theo Pirard) ***1999***/0054354

Xichang launch site used by CZ-2E and CZ-3 rockets for geosynchronous transfer (Theo Pirard) ***1999***/0054385

NPO Energomash RD-180 engine adopted by Lockheed Martin for the Atlas IIAR, Atlas IIIA and Atlas V series (Theo Pirard) ***1999***/0054362

It is the growing market for consumer and business services that is at the driving end of new opportunities and a wide range of applications, including broadband data markets, analogue and digital satellite TV services, digital radio and mobile communications, that is building demand for satellites and carriers. New applications such as tracking services, desktop teleconferencing, telemedicine, commercial navigation aids and Internet services – to mention but a few – bode well for bus manufacturers, payload technologies and launch service providers. But more of all that later. Space is big business and, perhaps surprisingly, set to grow faster than related industries in aerospace and communications. In both mass market and niche services, past trends show a realistic expectation of growth averaging just over 9 per cent/annum for the foreseeable future.

A Decline in US Funding

At the other end of the spectrum, government-funded space research is in decline at a time when international ventures lock up increasing amounts of money traditionally reserved for new programmes and missions of opportunity. The major problem with most developed countries lies in their social programmes which, without debate, are essential and high-priority services. Nevertheless, it is one in which increasing amounts of government expenditure are tied to mandatory increases for veterans' benefits, health and welfare payments and social investment. In the last 50 years, mandatory payments have taken an increasing toll on discretionary spending so that the amount of a nation's wealth available for non-mandatory programmes has fallen from about 70 per cent to around 35 per cent, even 20 per cent in some countries. This means that developed countries are less and less able to react to funding demands at time of national need or provide surge funds for a national commitment.

It would, for example, be impossible today for the United States to provide the surge funding for a major government initiative such as that made by President Kennedy more than 38 years ago when he authorised the Apollo moon landing programme. NASA's fiscal year 2000 budget is around US$13 billion which, in real terms, when discounted for inflation, is about half what it was at the peak of Apollo development. Yet that peak of what would now amount to an annual budget of US$25 billion reached in fiscal year 1965 was about 4.5 per cent of total federal expenditure that year. It had taken just four years to reach that financial commitment starting from a sum which in today's money would be equal to about US$4.5 billion, or 1 per cent of federal expenditure in that year (1961). In other words, when measured as a percentage of federal outlay, the NASA budget surged by 450 per cent in just four years. Today it would be impossible to increase funding to a major government agency by that same ratio, raising it from US$13 billion to US$58.5 billion in just four years. The ever diminishing reserve of discretionary spending places governments in strait-jackets and, in turn, it is unwise for government agencies to lock a high percentage of their financial resources into mandatory programmes. Yet that is precisely what NASA has done not only to its own budget but, through international co-operation, to those of other, leading, spacefaring nations.

The International Space Station

For the first time, Russia, the United States, Europe, Japan and Canada are teamed on a major space engineering endeavour (the International Space Station – ISS) the benefits and payback from which are impossible to define. Few question the advantages of research in materials processing or the desirability of conducting basic science in microgravity conditions but for more than 15 years the Shuttle has carried experiments and research projects which consistently fail to reveal a commercial imperative for zero-*g* factories. These days it is not enough to say we may never know what we are losing if we fail to explore the possibilities. We are no longer alchemists holding open the promise of precious metals from stone and there is fool's gold aplenty for those with blinkered vision to believe that all research is beneficial. The vast sums of money spent by the world's leading nations on putting people aboard a permanently habitable space station may yet reap rewards to justify the cost. They may also compensate for the lack of progress in other endeavours which that money could support, but it is not clear where that will come from. Surely, it is the purpose of governments to conduct scientific research above and beyond the capacity of commercial enterprise or private organisations. Pursuant to that, is it not important to preserve a balance between competing strands of research and not lose sight of the small gains which can be lost if disproportionately large resources are focused on a single objective?

There is no single justification for the International Space Station and it is as much to do with foreign policy and preserving an operating capability as it is about fundamental scientific research, less still the promise of commercial payback. The history of space station decision-making is littered with shifts in emphasis and changes in its centre of gravity. Back in 1959, when NASA's long-range plan envisaged a permanently manned facility coupled to a common goal of manned circumlunar flight, a space station was confidently expected to emerge by the early 1970s. This plan was effectively carrying out the direction given NASA in the 1958 Space Act to make the US 'pre-eminent' in space. Within two years a presidential decision, augured into reality by powerful political and industry lobbyists, set NASA on a course for the moon, but it was a one-off commitment, with no support for a sustained presence on the moon from the American people or congressional representatives. In making the initial commitment in May 1961, President John F Kennedy effectively adapted the 1958 Space Act to a specific battle within the broader context of the Cold War and a competition with the USSR for the hearts and minds of uncommitted nations around the world. Going to the Moon was as much about reaction to the failed Bay of Pigs fiasco (the attempted invasion of Cuba) and the

Russia's new Meteor 3 m weather satellite (Theo Pirard) ***1999***/0054350

SSTI built UoSat-12 to demonstrate advanced leo communications and Earth observation payloads (SSTL) ***1999***/0054360

flight of Yuri Gagarin (the first to orbit the Earth) as it was about maintaining pre-eminence in space although, to some extent, that was achieved through Apollo. However, after Apollo, that investment did very little to sustain the forward thrust of US 'pre-eminence in space' across the board and, over the next two decades, the US struggled to secure priorities which were both fundable and of value.

Disengaging from the one-off Moon commitment and adapting NASA plans to much reduced budgets, which fell by 62 per cent between 1965 and 1974, US government non-military space programmes went into decline. Ambitious plans for parallel operation of a reusable shuttle and a manned space station beginning in the late 1970s slipped first into sequence and then into oblivion. In January 1972, President Richard Nixon gave approval for development of a reusable shuttle but NASA was to wait a further twelve years before President Reagan approved development of a modular space station. By the time it becomes operational the ISS will have been 20 years in gestation and assembly. By contrast, almost four decades ago it took NASA eight years and two months to mobilise the technology and fulfil the Moon landing goal, raising greater engineering challenges than ISS ever will. There is nothing else that bears greater testimony to the transition of NASA from a pioneering science and engineering organisation to just another government bureaucracy than the dismal saga of its efforts to build a laboratory in space.

To get to this position has taken countless redesigns at great cost, involved the collective support of more than a dozen other countries and required the metamorphosis of its purpose from science facility to a plank in American relations with Russia. Testimony to that is the way Europe, Japan and Canada were central to the station's support until sidelined by a political arm-link between Presidents Clinton and Yeltsin in 1994. The result of this was a commitment for the US taxpayer to subsidise Russian space industries and grow the station concept to twice its original size. There is no doubt that if it had not been an important part of a new deal with Russia, Clinton would have cancelled space station plans when he came to power in 1993. There is a bizarre parallel here with the way the Shuttle was developed when US Air Force support was obtained at the cost of dramatically increasing the size of the orbiter and giving it high cross-range through an expensive delta-wing configuration unnecessary for NASA's original requirement. Now the ISS has been increased to twice its original size to get the Russians on board and justify its existence. Had this not happened, there is little doubt the Clinton administration would have cancelled it, yet NASA persists in referring to it as a science facility.

Space Station Rationales

One outstanding justification for the ISS is to regard it as a vital starting point for the utilisation of space-based technologies in a wide range of futures activity. It is a good place from which to begin the colonisation of the solar system, an important link between manned and unmanned space operations and a useful staging post for in-orbit satellite servicing facilities and the embarkation of planetary probes in a benign environment devoid of the launch hazard. The human exploration of the solar system is a goal capable of mobilising enthusiasm among the young and stimulating the passion for discovery and adventure among all. In-orbit satellite servicing facilities offer financial savings by eliminating the construction and launch of a new satellite bus but the technique requires an in-orbit transfer vehicle capable of retrieving and returning to their respective positions geosynchronous satellites selected for upgrade or refurbishment. It was only in this one sector of activity that econometric studies of manned Shuttle costs, conducted in the early 1970s, displayed savings over expendable launch systems.

Ironically, the in-orbit transfer vehicle (then known as a Space Tug) was abandoned as a cost-saving measure and the primary means by which Shuttle could have saved money was removed! The value of the ISS to assemble and release planetary probes destined for deep-space missions is self-evident, eliminating the risk inherent with unmanned, expendable, launch systems and reducing the necessary energy for escape velocity. Excess lifting capacity on routine ISS servicing missions could accommodate elements of robotic spacecraft dispatched to the environs of the solar system, requiring a delta-velocity little more than one-third that from the surface of the Earth. Cryogenic propellant for these energy-efficient trajectories could be provided by scavenging residual oxygen and hydrogen trapped in the Shuttle's main engine plumbing when it reaches orbit.

To ensure broad utilisation of the ISS, the international partners must think globally and not programmatically. The main failing with the space programme today is that it is divided into separate disciplines, a constricting influence where, for example, remote sensing specialists cannot see across the divide into the world of plasma physicists. Their inability to do this results in failure to appreciate that combining assets on the ISS, in the way discussed in the preceding paragraph, will multiply benefits and enhance each discipline in parallel. Hitherto, advocates of manned or unmanned space flight have fought turf wars to the detriment of the programme itself. In this division they have failed to maximise the opportunity for consensus. It was this very issue which collapsed efforts in the early 1980s to give Britain a bigger stake in space and opened cracks into which politicians drove chisels to ring-fence UK involvement – now less than Brazil spends on space research.

There are lessons here for the way the ISS is being hailed as a

An engineer makes final adjustments to a Globalstar satellite (Theo Pirard) ***1999***/0054356

Model of the GSLV rocket being developed by ISRO for operational duty early in the next century (Theo Pirard) *1999*/0054377

scientific research facility. Sombody, somewhere down the road, is going to seek justification for that claim. But, the real point here is that for the next 30 years NASA and its international partners will spend at least 50 per cent of their discretionary space budgets on one programme. This amounts to more than US$125 billion for NASA, in all probably twice that amount for the collective group of international partners. Even in today's money, you can buy an awful lot of space programme for a quarter of a trillion dollars. Worse still, it will inhibit the onset of new goals for an entire generation of scientists and engineers, including astronauts and cosmonauts, within a diminishing financial resource. The ISS must not be seen as an end in itself, rather a flexible stepping stone to new opportunities.

Decline in US government Programmes:

So why is it that we are seeing the decline of government-funded space activity? The simple fact is that government space programmes carry no votes and not one politician has ever been installed or re-elected on an agenda connected with space. Quite the contrary and at least one Vice-President (Spiro Agnew) has been ridiculed for his support of a manned Mars expedition! In reality there is very little hope for an increase in government spending on space projects over the next decade and signs are that times could get harder. Throughout the past six years the Clinton administration has consistently failed to maintain a level budget for NASA, the amount of money it has been allowed to spend on space (versus aeronautics) falling by almost 20 per cent in real terms. Historically, the Democrats have spent more on space than Republicans and, with a strong Republican Congress, the future looks bleak.

With strong pressure from both houses of Congress for a leaner NASA involved in fewer programmes the agency is unlikely to realise all its published aspirations and several proposed projects have already been cancelled. To many, the human exploration of space is the very embodiment of what the programme is all about yet, for others, the premium for carrying people beyond the atmosphere is not worth the cost. In Fiscal Year 2000 beginning 1 Oct 1999, within a total budget of US$13.5 billion, NASA will spend more than US$5.5 billion on human space flight, less than US$2.2 billion on space science and less than US$1.5 billion on earth and atmospheric science.

Efficiencies and economies implemented by Administrator Dan Goldin have allowed NASA to move forward despite the challenges of a diminished budget. For seven years Goldin has steered NASA through changing times, presiding over a significant increase in programme safety and reliability and implementing cuts in operating costs; when discounted for inflation, Shuttle flight costs have been reduced by about one-third. One of the major accomplishments of the Goldin era has been the successful implementation of the 'better-faster-cheaper' ethic creating low-cost, efficient projects and programmes capable of refining and focusing research into niched areas. One example of this is the restoration of planetary programmes including the Mars frequent-flyer missions with two spacecraft launched at every window little more than two years apart. Invigorated by new and innovative technologies such as micro-miniaturisation and advances in sensor engineering and creative ways of achieving parallel goals at much less cost have proved remarkably successful.

Given its head, a new breed of engineers is revolutionising planetary exploration while the legacy of the 'big is best' era still continues to return solid data. Significant among the latter are the Galileo and Cassini/Huygens missions to Jupiter and Saturn respectively. Begun in 1977, with a planned launch date in 1982, delayed by budget shuffles and changes to the boost stage, Galileo finally got off the ground seven years late. It arrived at Jupiter in December 1995, depositing an instrumented probe into Jupiter's atmosphere before commencing a protracted orbital tour of the Jovian system. For four years Galileo has been returning outstanding images and data from the Jovian environment and the Galilean moons. Launched in October 1997, Cassini/Huygens epitomises the best of international co-operation: a space bus designed and built in the US carrying an advanced probe from Europe to explore the environment of Saturn's moon, Titan.

Despite these achievements, Congress has served notice that it wants a much lower budget for NASA although it is too early yet to say how this will influence spending over the next several years. Suffice to say that, with a new Presidential campaign starting up, social and domestic priorities will once again take precedence. Gallup has conducted a poll which shows that a majority of Americans now believe the space programme is worth the money and should continue but the same majority believes NASA gets enough from the public purse. Paradoxically, the downturn in US government spending on space comes at a time when the nation can afford it and more.

In an unprecedented upturn in growth, projected to remain consistently above 4.4 per cent/annum, the US federal budget is now in surplus for the first time in 29 years (almost US$100 billion for fiscal year 1999). For 1999, US government expenditure is just over US$1.7 trillion versus receipts in excess of US$1.8 trillion. As discussed above, during the height of Apollo, NASA received 4.5 per cent of federal expenditure. Today that allocation would be almost US$77 billion rather than the US$13 billion NASA will actually receive this year. But comparisons like that are deceiving. Apollo was unique and nothing to do with a sustained effort, unsupportable at that level. What is does show is the level of commitment that once existed.

Non-US government Programmes:

Europe

Of sustained scientific value or not, one piece of fallout from the Apollo era was to ignite a commitment to space from other leading nations which today underpins activity in all aspects of space exploration and application. Since the mid-1970s, the European Space Agency (ESA) has been a legitimate member of the global space club with active participation in launchers, human space flight, space science and applications. After a failure on its maiden flight, in the past year the Ariane 5 programme has moved to the operational stage and Arianespace is actively marketing the transition from Ariane 4 series launchers to the new heavyweight. Ariane continues to maintain a healthy segment of the commercial market, launching almost half the world's optional satellite traffic. Ministers have approved an Ariane 5 upgrade programme involving two new cryogenic upper stages for a total investment of US$570 million by 2001. ESA has also committed US$60 million to the Future Launcher Technology Programme (FLTP) in an effort to reach a decision about the kind of system required to significantly lower the cost of space transportation after Ariane 5.

In human space flight, along with NASA, the European Space Agency is moving from periodic visits to zero-*g*, utilising Spacelab, to permanent habitation in space with the Columbus research module, now expected to join the ISS in orbit in February 2004. As a separate commitment, Italy has handed over to NASA the Leonardo multipurpose logistics module for launch in February 2003. Commitment too is the key word in Europe's determination to move ahead with the Galileo navigation satellite system to secure government and commercial custom for what it sees as a quantum step ahead of existing Navstar GPS and Glonass systems. ESA member countries have already pledged the US$483 million believed necessary for the next five years of development. Other areas where Europe is determined to forge ahead, securing a world lead in data collection, extraction, analysis, marketing and sales, include remote sensing and meteorological information. Both areas have seen Europe take strong steps in 1998-99 toward that objective and remote sensing is set to become one of the best growth areas in space applications (see later). Budget allocations for ESA are relatively stable and will probably remain at current levels for the next several years, possibly facing less threat than non-military budgets in the US.

India

India has bucked the downward trend and taken a major step forward with a 50 per cent increase in expenditure over the last two years. This makes sense. India is now reaping the rewards from two decades of sustained development in satellite and launcher technology. Infrastructure is now capable of supporting a strong independent programme and the country will continue to profit from international contact particularly in the area of remote sensing. Typically, India sees the prospect of customers for home-grown satellite launchers too attractive to pass up and is looking to commercialise its PSLV (Polar Satellite Launch Vehicle) in 2000 with infrastructure seemingly in place to support two flights a year. In this sector, India's space programme has come of age and the PSLV, capable of lifting a 1,200 kg satellite to sun-synchronous orbit, could appeal to low-budget earth observation planners. The careful path trodden by India's space planning people in courting support from both East and West over the last 25 years of evolving space-based capability may now bear fruit among a variety of developing countries unhappy with US or European offers for satellites and launch slots.

Proton with Block DM3 launches the Astra 1F satellite on 9 April 1996 (Theo Pirard) ***1999***/0054381

Engineers check out the antennas on DBS-1, North America's first direct-to-home TV satellite (Hughes Space & Communications) ***1999***/0054333

Japan

In a series of progressive steps over the last 40 years Japan has become an economic giant through national goals and trade objectives more reminiscent of a totalitarian state than the democracy it is. In order of commitment, motorcycles, cars and electronics drove the Japanese trade strategy with aerospace an aspiring successor for the 1990s but aviation and space proved harder to apply. Space has become the leading edge of Japan's aerospace ambitions and significant steps made in this last year have moved its launch vehicle programme closer to commercialisation. Many would argue there are too many players on the launcher field already but Japan is convinced it can market the medium-lift H-II and its derivatives for a meaningful return on investment. However, Japan has come late to the commercial launcher market and may have left it too late for a secure slice. Where Japan is committed it is doing well, continuing to expand and improve its applications sectors primarily in remote sensing and meteorology. Its aspirations go further and Japan has made progress toward securing a permanent place in space with the Kibo Japanese Experiment Module now scheduled for attachment to the International Space Station in January 2003. Meanwhile Japan faces financial constraints predicated by the Asian economic crisis of the last year. Despite restrictions on government expenditure, NASDA and ISIS continue to progress with ambitious programmes involving launchers, human space flight and science and applications satellites.

Russia

When compared with former decades, Russian expenditure on space programmes has all but collapsed. Without a vertically tiered market applying technologies to manufactured goods for an internal consumer market or for export, few politicians in Russia see the point of supporting ambitious, ideological, goals which hitherto underpinned a massive space industry. Yet space technology is an important element in acquiring foreign currency and, in itself, provides one of the few export opportunities for Russian industry. The last symbol of Russia's Cold War stance from the Soviet regime, legacy of a commitment made in 1969, the Mir space station was vacated in late August 1999 although, at the time of writing, it was expected to remain in space for several months. It has provided a base from which Russia moves forward as a partner in the International Space Station, effectively marrying Mir-2 with the former US Space Station Freedom concept. Neither would have survived the post-Cold War drawdown had it not been a plank in US foreign policy created by the Clinton administration in 1994. Yet there are other areas of greater financial advantage. Russia has a strong design and engineering base in propulsion with many products eagerly sought by Western companies for their performance, efficiency and

Galaxy VIII-i spacecraft is being prepared for a thermal vacuum test
(Hughes Spacecraft and Communications) ***1999***/0054326

application. Lockheed Martin's use of the RD-180 to power its Atlas 3 launch vehicle is a classic example of the opportunities available to strong Western marketing organisations hooked up to Russian expertise. The joint venture between DaimlerChrysler and the Khrunichev Company under launch service provider Eurockot is a template for alloying complementary strengths for market opportunities and is to be applauded. Quite simply, Russia's enormous technical base in space engineering is an appropriate platform for joint ventures and unique applications. The advantages to US and European organisations are self-evident.

China

Of growing significance, China's space programme has taken several forward steps in the last year and there are indications that a manned space programme is emerging, perhaps to celebrate the 50th anniversary of the revolution which preceded the official beginning of the People's Republic on 1 October 1949. China continues to export technology to friendly states and has conducted a major expansion of its rocket programme. The technology from this has supported an increasingly capable launch industry, offering foreign users a ride into space at competitive rates, as witnessed by the prolific array of orders for Iridium satellite launches by CZ-2C. The first successful flight of the CZ-3B in August 1997 was followed in May 1999 by the first CZ-4B, which successfully placed two satellites into orbit. With a capacity to place 1,500 kg in geosynchronous transfer orbit (GTO), the CZ-4B is the latest in a growing family of launchers with plans under way for the production of a super-heavyweight, capable of placing a 13,000 kg mass in GTO, available after 2005.

China is already co-operating with Brazil on development of a remote sensing satellite system and plans to develop a multispectral system for commercial use from 2003. May 1998 saw the first launch in the Zhongwei (ChinaStar) series of telecommunication satellites for China and southeast Asia and Sinosat 1, first in a new national satellite system, was placed in geosynchronous orbit in July 1998. Recent tightening of export regulations has slowed the transfer of US technology to Chinese organisations. Loral Space and Communications is taking the brunt of a US-China tit-for-tat response as the US State Department refuses to give an export licence to the Chinasat 8 satellite on the pretext that human rights issues remain unresolved. China has forbidden US warships to enter Hong Kong waters as their protest at US anger.

Market Analysis

Because most space-related expenditure is in non-government hands, the opportunity for profit and investment is wide and, as it happens, solid. Total space revenue growth is commendable and healthy, having expanded globally from US$97.6 billion in 1998 to more than US$105 billion in 1999. Of the 1998 sum, 57 per cent is in space infrastructure, 34 per cent is in telecommunications, 4.8 per cent is in the use of space data and assets and 3.8 per cent is in support services. Total worldwide employment on space programmes runs at almost 1.1 million while aggregate growth over the period 1998-2002 is a discounted 9.01 per cent per annum. Year-on growth for this five-year period will see world space revenues total US$577 billion of which US$298 billion will be infrastructure, US$218 billion telecommunications, US$68 billion satellite manufacturing and US$33 billion launch vehicles. GPS revenues are predicted to top almost US$27 billion with almost US$10 billion from remote sensing and data services.

Telecommunications and data handling

According to most analysts the market is buoyant. Broadband services are predicted to accrue revenues in excess of US$65 billion per annum by 2002, US$8 billion of which will go to satellite-driven services increasing to US$16 billion within three years. Internet services are in line to grow from 40 million users to 400 million by 2002. A significant fraction of that expanded traffic will go by satellite. The number of subscribers for direct-broadcast satellite TV services, one of the fastest growth sectors in the industry, is set to expand from 30 million to almost 100 million by 2008 while satellite share increases over cable. Mobile communications services for direct relay or local cellular networks are less predictable, with about 10 million subscribers and US$15 billion in annual revenue possible by 2003, doubling four years after that. Digital audio radio has taken off faster than original predictions, now securing a steady growth indicating a potential for 50 million subscribers generating US$8 billion in revenue. From a start in 1993 VSAT networks are growing fast at an annual rate of 18 per cent and changes in the regulatory structures will help this to expand further in the coming years. Peripheral but fast-growing areas such as desktop videoconferencing, telemedicine, training by satellite and data broadcasting are adding creditable revenue streams.

Expansion and growth are synonymous with a diversity of applications and a strong world market. Consider this. Fewer than 5 per cent of the world's population have a computer, less than 20 per cent have a TV set, over 35 per cent have never used a telephone and 63 per cent do not have access to one. By going direct to a satellite-based communications system, several developing countries have successfully leap-frogged the interim infrastructure developed countries went through in laying down terrestrial communications links. Over the next decade or two the growth in telecommunications and information services is confidently expected to rise to a greater level than any other commercial sector in the history of manufacturing and services. Revenues are set to expand at a commensurate rate making space-based telecommunications and information processing one of the biggest industries ever. While the value of telecommunications satellites launched is expected to remain at around US$11-12 billion per annum the number of units launched is predicted to increase from 130 in 1999 (now slightly modified downward by unprecedented launcher failures) to about 170 within five years. Moreover, the sheer volume of investment capital now sunk into innovative and expensive new-venture systems will preclude many smaller entrepreneurs from reaching the order books with Iridium, ICO, Globalstar and Teledisc alone accounting for almost US$30 billion between them.

Remote sensing

Around for more than 25 years, satellite remote sensing has languished on the sidelines of commercial exploitation, being used more frequently as a means of encouraging international co-operation in government-to-government projects than in the raw commercial environment of investment, sales and revenue potential. In 1998, total raw and processed data markets accrued a mere US$500 million in revenue but this is now growing at a healthy rate and is set to reach US$600 million in 1999 and in excess of US$1 billion by 2002. Set against this commercial revenue stream is the predicted US$17 billion governments and industry will spend in 2000 on obtaining the data from satellites and interpreting it. One of the serious restrictions on remote sensing markets is being demolished, limitations on resolution that governments have attempted to impose to prevent the sale of critically sensitive intelligence information to unfriendly countries and terrorist organisations. There are numerous instances where terrorist groups in countries as far apart as Northern Ireland and Namibia have purchased satellite imagery for target selection. That situation is expected to get worse with the proliferation of high-resolution images but it is an acceptable price to pay for a wide range of systems such as Ikonos with its 1 m resolution and several others with a similar capability.

Global navigation services

The market for the Global Positioning System (GPS) has proved lucrative with projections from several European analysts of a US$50 billion business by 2005, accelerated by a deal between Magellan and Hertz fleet rental for 50,000 in-car navigation units. Within the next two years, integrated multifunctional GPS products will arrive with chips specifically designed to provide position and tracking information integral with cellular and mobile telephones. These will find application in the security industry (integrated chips for tracking stolen vehicles, personnel tracking in secure facilities, and so on) and as integral devices for maintaining location and position information on cars, boats, aeroplanes and trains as well as individuals and animals. The industry will achieve a new dynamism within five years when integrated chips are made cheap enough to add in for the cost of a flashlight and made no bigger than a torch battery.

At this level of affordability the potential user base for GPS systems could expand from the present 2.5 million to a predicted 100 million by 2008. Reacting to major improvements in receiver hardware and software, Europe is investing heavily in the Galileo future navigation system in the belief that this market will become as important as mobile phones to the telecommunications industry. At present, the Russian Glonass system competes to some extent with the US Navstar system and it was this that the Russians wanted Europeans to adopt. However, there are some discrepancies that have driven Europe to a completely new system and this will probably lead to the demise of Glonass over the next decade as it becomes redundant and the user base focuses on the Galileo and Navstar systems.

Launch vehicles

Reusable

The entire launcher industry is going through a transformation and fulfilling the promise that the 1990s would see the emergence of a vibrant commercial endeavour. Unfortunately, year-on-year returns are patchy and the first half of 1999 has seen an unprecedented sequence of failures that blight the strong performance of 1998. Had it not been for the tragic and unforgettable loss of seven astronauts and the Shuttle orbiter Challenger in January 1986 the US may not have had an expendable launch vehicle market at all by now. The post-Challenger reaction to remove all competitive elements from Shuttle launch operations by prohibiting the use of that system for carrying commercial satellites into orbit was a wise and prudent move. It probably saved the Shuttle programme from another disaster, stopped taxpayers having to subsidise commercial operators (paying artificially low launch prices to make the Shuttle competitive with Ariane) and revitalised the US expendables industry which, in every way, has been a good thing.

For the Shuttle it means a more focused remit, less pressure, enhanced safety and a slightly lower launch rate. Significant economies and the best efforts of the United Space Alliance that now effectively runs Shuttle preparation and turnaround have reduced operating costs by 30 per cent. In 1996, NASA effectively froze upgrades and modifications in the belief that a replacement was likely to emerge early in the next century. A year later NASA changed its mind and reinstated earlier plans to spend US$100 million a year on improving the Shuttle and bringing it up to date. Early in 1999, NASA officially endorsed a belief that the Shuttle would remain its only manned launcher for at least the next 13 years. Expectations of an imminent replacement have faded and, at average use rates of up to 10 flights a year, the Shuttle could remain flying for a further three decades. The unique capabilities of the Shuttle and its potential for expanded orbital operations, involving satellite refurbishment, rescue and repair as well as supporting an operations-orientated space station capability, have been discussed above but the Shuttle is destined to see further transitions in its operating mandate.

Mobile communication services provided by Inmarsat P satellites utilising the Hughes HS-601 bus (Hughes Space & Communications) ***1999***/0054339

As conceived in the 1960s, the Shuttle was to have serviced a permanent space station through low-cost reusability but, by the time operations began in the early 1980s, it was hailed as the single successor to expensive, expendable, conventional launch vehicles. It was the race to compete against Ariane that entrenched an on-time/every-time mentality which brought about the Challenger disaster and Shuttle received a new mandate excluding commercial satellites and focusing operations around science missions and high-value, heavyweight, NASA satellites and spacecraft. Now, with the assembly of the International Space Station, the next six years will keep three Shuttle orbiters (Discovery, Atlantis and Endeavour) primarily occupied launching components and elements of that multi-nation facility on more than 40 flights. However, orbiter Columbia is too heavy to carry out space station missions and would be ideal for supplementary objectives. Directives to open the Shuttle once again to a broader range of operations appeared with the 1998 Commercial Space Act signed by President Clinton. Language in the Act opens the door for the Shuttle to be used for selected satellite rescue operations and moves have already been made to return the Shuttle to satellite rescue and re-boost, a role it so ably undertook in the 1980s. Recently, the White House has pushed for amendments to that Act openly endorsing these roles. At the time of writing, Hughes Space & Communications was in discussion with NASA about the possibility of a Shuttle rescue mission to put Orion 3, a communication satellite stranded in low earth orbit by a failed Delta 3 launch, in its intended geostationary transfer orbit.

In the last year significant changes to the strategy for emergent reusable launch vehicles has shifted the emphasis from the pre-operational phase to R&D activities. For example, originally hailed as a technology demonstrator for a reusable single-stage launcher called Venture Star, expected to be operational by 2005, the Lockheed Martin X-33 has been delayed by technical problems and now looks less likely to presage a shift quickly from demonstration to its much larger successor. The reusable X-34 will begin flight trials later in 1999 and, like the X-33, provide atmosphere tests but with a system designed to demonstrate automated landings and an emphasis on operational turnaround. Both systems will systematically explore the radical technologies and meet operational challenges essential for reusable and single-stage-to-orbit launchers. NASA and Boeing have now signed a four-year contract to develop the X-37 which, from 2002, will be carried aboard the Shuttle for demonstration trials with

Solidaridad 1 undergoes vibration tests to simulate launch vehicle dynamics (Hughes Space & Communications) *1999*/0054331

a low-cost system exploring the Mach 10-25 flight regime. This is something the X-33 and the X-34 cannot do and, because of safety factors and operational considerations, something which the Shuttle is unable to explore in the same vigorous and technologically tasking way. The X-37 will be designed to perform aggressive manoeuvres to prove radical and innovative operations techniques for low-cost, low-risk, reusable launch vehicles. At present, most leading pioneers in this field prefer a two-stage system.

Expendable

In the revitalised area of expendables, growth is steady but solid. The Enhanced Expendable Launch Vehicle (EELV) contract awarded by the US Air Force on 16 October 1998 will provide Boeing and Lockheed Martin with a solid order book for continued evolutions of Delta and Atlas. The deal has also been touted as saving the buyer up to 20 per cent over the current Titan 4 that the EELVs are designed to replace but this may be flawed. The Air Force is betting on the Pentagon shifting to smaller and more frequent payloads. However, cost savings are only really made with larger defence satellites of the type now considered too inflexible and costly to risk on single launchers. The same equation is being followed by commercial operators forcing up the payload capability of expendables in the search for reduced costs and lower pricing. With the trend toward launching more than one satellite on each launcher, the growth in satellite traffic over the next 10 years, set to reach 175 per cent of the total launched in the last decade, will increase to more than 1,000 payloads. Because more satellites will ride on fewer rockets, the number of launches over the next decade is predicted to increase by 125 per cent reaching a total of about 820. The strong marketing position of Arianespace has centred on its ability to carry more than one satellite on each launcher but that position will degrade as more powerful Delta 3 and 4, Atlas 3 and Zenit 3SL become available. To counter this, with a GTO capacity of almost 6,300 kg, the Ariane 5 will get a head start on the opposition.

One innovative way of extracting better performance from existing vehicles is to move the launch site for geosynchronous payloads closer to the equator as Sea Launch have done. Comprising a consortium of Boeing Commercial Space, Kvaerner Maritime AS of Norway, RSC Energia of Russia and NPO Yuxhnoe of the Ukraine, Sea Launch consists of the Odyssey launch platform and the Sea Launch Assembly and Command Ship. Launches take place from Odyssey after the two vessels have been moved to a position on the equator. Contrary to persistent media reports that this is the first satellite launcher to fly from a sea-based platform, it follows the precedent set by the San Marco platform used off the coast of Kenya to launch Scout rockets in the 1960s. The system is designed for exclusive use of the Zenit 3SL which consists of the first two stages of the Zenit 2 plus the DM (fourth) stage of the Proton launcher. The high degree of automation in launch and flight sequencing makes this the ideal vehicle for Sea Launch, which successfully achieved its first qualification flight on 27 March 1999. The consortium currently has 15 firm commitments for launches from two customers. The launcher is capable of lifting 4,500 kg to GTO and operators aim to charge no more than US$90 million a shot.

In 1995, to consolidate a firm niche in the expendable launcher market, Lockheed Martin combined with Khrunichev Enterprise and RSC Energia to form International Launch Services, marketing flights of the Atlas and Proton families. Covering the GTO range from 3,000-4,900 kg, ILS launchers have a current backlog of more than US$4 billion in launch fees. To meet the requirements of low earth orbit (LEO) users, DaimlerChrysler Aerospace and the Khrunichev State Research and Production Space Centre are marketing, under the name Eurockot, a satellite launcher based on the two-stage SS-19 *Stilleto* ICBM with a restartable Breeze third stage using storable propellants. To date the SS-19 has achieved 139 out of 142 test flights without failure. Capable of placing 1,900 kg in LEO, Eurockot is flown from the Plesetsk site and a minimum of 45 vehicles have been made available by the authorities. Up to 160 SS-19s are potentially available for conversion into satellite launchers. The small launcher and LEO market is potentially lucrative with mobile telephone applications promising large quantities of relatively small satellites. To fill this market need Europe is developing the FiatAvio Vega for initial launch in 2002 and capable of placing up to 1,000 kg in LEO.

Insurance

The roller-coaster ride between coverage capacity, premium income and claims goes on as major losses hit an otherwise confident industry. Insurance for commercial space projects is now a mature and essential part of managing the not insubstantial risk associated with launch and early life as well as protecting against revenue loss from in-orbit failure. From the heady days of the late 1970s and the over-confidence of the early years of the Shuttle period, space insurance has become a balancing game between premiums and claims. A series of major losses in 1997 totalled US$693 million in claims but premiums of more than US$800 million offset outlays if doing little for profit margins. Last year, against premium income totalling almost US$900 million, claims reached a staggering US$1.76 billion and in the first half of this year claims have already reached a total of US$457 million. This figure would have been considerably higher had the US Air Force insured its US$1.8 billion worth of hardware lost in one Titan failure. Premium rates that had risen to around 20 per cent were coming down to around 15 per cent but these are thought likely to increase slightly in response to events earlier this year. Overall, capacity is high because of substantial profits from earlier dealings and, when compared with other areas of insurance, it is still a good thing to be in.

Military programmes

Major changes are in the offing with a wide range of military satellites and new markets are likely to emerge as former allies move toward greater autonomy and independence from former power blocks. European interest in a new generation of reconnaissance and surveillance satellites, operated by European armed forces, is stronger than ever and the French are quietly leading a campaign to allow the European Space Agency greater freedom to provide services for the exclusive use of the military. In the United States, a debate is now beginning about the proper sizing of major space-based assets of the future and a lobby is developing which favours smaller and more frequently launched satellites. Military communication satellites could disappear over the next decade, their services replaced by leased transponders on civilian satellites. This has a strong precedent in the use of Marisat satellites for US Navy communications during the 1970s and makes good sense. The distinction between military and civilian satellites may gradually erode as increasing use is made of commercial information. The use of space-based assets in Operation Desert Storm and, more recently, in the Balkans, has recruited strong support among all armed services for increasing amounts of intelligence information from satellites.

Acknowledgements

In preparing this, my first, edition of *Jane's Space Directory* I would like to acknowledge the help and assistance of space people around the world who, for the last 35 years, have kept me abreast of developments and reasonably conversant with the technology. My special thanks go to ESA, ISAS, ISRO, NASA, NASDA, the RSA and the USAF Space Command and all the myriad contractors and participants involved in space activities. You all know who you are. Updates to this volume have benefited in several respects from the library of information those and other organisations have provided. It was both a privilege and an honour to have been asked to edit this, the fifteenth, edition of *Jane's Space Directory* and I absolve all those who have provided information of responsibility for any mistakes. They are mine alone.

No book of this size can appear without the tireless efforts of many people. In the front trenches I would particularly like to credit the extraordinary tenacity of David Blake, Diana Burns, Carol Gallagher, Neil Grace, Laura Kew and Sharon Marshall in sifting anomalies, pointing out errors and smoothing the presentation. I should also like to acknowledge the inestimable help of Elizabeth Dawson in collecting material to support the new edition and Katherine Willis for sending through press releases and news items. Sitting almost continuously on my shoulder, team leader Jo Fenwick has cracked the whip when necessary, held things together when they threatened to fall apart and kept the flow going to the finish line.

For her part, thanks to database manager Ruth Simmance and her team for the professional way the information was integrated. Finally, but certainly not least, a special acknowledgement and gratitude to my Managing Editor Simon Michell for commanding a trim ship, always from the bridge and always with sympathy for the many and varied problems that arise in a production of this size. To you all, a heartfelt appreciation. As we move to strengthen and improve *Jane's Space Directory* I look forward to hearing from readers and users. I want to know your thoughts and to hear your suggestions.

David Baker Ph.D
Cambridge, August 1999

Jane's

Users' Charter

This publication is brought to you by Jane's Information Group, a global company with more than 100 years of innovation and an unrivalled reputation for impartiality, accuracy and authority.

Our collection and output of information and images is not dictated by any political or commercial affiliation. Our reportage is undertaken without fear of, or favour from, any government, alliance, state or corporation.

We publish information that is collected overtly from unclassified sources, although much could be regarded as extremely sensitive or not publicly accessible.

Our validation and analysis aims to eradicate misinformation or disinformation as well as factual errors; our objective is always to produce the most accurate and authoritative data.

In the event of any significant inaccuracies, we undertake to draw these to the readers' attention to preserve the highly valued relationship of trust and credibility with our customers worldwide.

If you believe that these policies have been breached by this title, you are invited to contact the editor.

A copy of Jane's Information Group's Code of Conduct for its editorial teams is available from the publisher.

INVESTOR IN PEOPLE

GOVERNMENT AND NON-GOVERNMENT SPACE PROGRAMMES

GOVERNMENT AND NON-GOVERNMENT SPACE PROGRAMMES

ARGENTINA

Argentina's modest space activities emphasise applications and space science, although now the country has broadened the charter to consider developing an indigenous orbital launcher. Until May of 1991 the air force executed overall authority for all space research projects. Then the Argentine government transferred the CNIE National Commission for Space Research into the civil CONAE National Commission for Space Activities, under direct supervision of the country's President. CONAE's strategic programme covering 1995-2006 was approved by President Menem 28 November 1994, projecting expenditure of US$701 million 1995-2006. This figure still reflects significant contributions from unspecified third parties. It is reviewed and extended every two years so that the horizon is always a decade away.

The current plan states these objectives:

Establish a ground station for access to data from international and domestic remote sensing satellites.

Continue the SAC satellite series for remote sensing, carrying national instruments, and science.

Develop the SAOCOM Satellites for Observation and Communications.

Reactivate the old Condor 2 missile facility at Falda del Carmen as the Teófilo Tabanera Space Center, adding laboratories for integrating/testing satellites and payloads. The Mario Gulich Institute for Advanced Space Studies will be located at the centre providing courses tailored to satisfying the needs of CONAE and the space community.

CONAE and its predecessor have worked on the **SAC** Satelite de Aplicaciones Cientificas (Scientific Applications Satellite) project since 1981. Two further SAC missions are planned with NASA, the SAC-C Earth observation (multispectral medium resolution) satellite in December 1999 using a Delta vehicle.

An agreement was signed with Italy in October 1992 to begin discussions on space co-operation, emphasising communications and Earth observation.

Intelsat and Inmarsat are accessed primarily through the Balcarce station near Buenos Aires. CNT signed a 24-year concession in May 1993 with a consortium headed by DASA for the Nahuel S American satellite communications system.

Impsat was formed in 1988 to develop and operate a regional satellite-based network and is now the leading provider of satellite communications in Latin America, operating 2,500 VSATs by 1995. A US$30 million contract was signed February 1995 with Hughes Network Systems for a further 2,000 VSATs and five shared hubs. The Lusat packet radio satellite was provided for Amsat-Argentina and launched by Ariane in January 1990 as one of four Microsats.

Reports emerged during 1989 suggesting that Argentina was co-operating with Iraq and Egypt (Egypt actually withdrew in 1988) on the development of a 900 km range Condor 2 missile that could also form the basis of a space launcher. The single stage Condor 1, capable of lofting a 400 kg payload to 70 km, began development 1977-78 as a sounding rocket with assistance from Germany's MBB. A parallel design entered service 1990 as the Alacran missile, capable of delivering 500 kg over 300 km. The military government approved Condor 2 in late 1982 after the Falklands defeat as a two-stage solid-propellant missile capable of hitting the islands with a 500 kg payload.

Eventually, the air force built a US$400 million installation under a mountain at Falda del Carmen in Cordoba province in accordance with plans prepared by MBB. President Menem announced the project's deactivation in April 1990 as part of defence budget cuts, but apparently all material was handed over to CONAE and the technology was then transferred to Spain's Capricornio launcher programme.

In July 1995, the Asociacion Argentina de Tecnologia Espacial, AATE, obtained a reservation to fly a Getaway Special payload aboard a United States shuttle flight. The formal agreement between NASA and AATE was signed 6 August 1996. The payload called Proyecto PADE (Paquete Argentino de Experimentos) comprises seven educational experiments, designed and manufactured by Argentine universities and scientific institutions.

AUSTRALIA

Australia originally ran its space activities through the NSP, which was established in 1985 to provide a base for the country's participation in the world space industry. Shortly thereafter, the government formed the Australian Space Office in 1987 as the focus for space-related activities.

The Federal government established the Australian Space Council in 1993 as a result of the 1992 review of the National Space Programme by the Curtis committee. The Curtis report recommended an integrated NSP with designated programme funding and the new council. A major task was the formulation of a five-year rolling plan to set new directions for the NSP and determine priorities. Bureaucratically the ASC reported to the Federal Minister for Science & Technology within the Industry, Science & Tourism portfolio. Then, as part of the 1996/1997 Science and Technology budget announced by the Minister for Science, Peter McGauran MP, on 20 August 1996 the Commonwealth Scientific and Industrial Research Organisation (CSIRO), Australia's largest R&D body, got new authority to undertake space projects on behalf of the Australian government. The former Australian Space Office and Australian Space Council were terminated.

The new space plan, the first controlled by the new government, elected 2 March 1996, made a commitment to a viable space research programme and fostering applications. To carry this out ASC recommended the government place priority on the development and demonstration of light launch capabilities, funding the launch of two rockets from Australia within development agreements with private interests with confirmed capital. The launches would create the centrepiece for building other Plan elements. They could be the launch vehicles for carrying Australian-built demonstration satellites, leading to the development of Australian Earth observation and communications payloads. Australia seeks to secure a significant share of Asian space business in launches, small satellites and space-based services, beginning with launches first from Woomera then Darwin and Cape York.

Studies have been undertaken under the Earth observation programme on the need for Australia to develop an Earth observation data network, the development of value-added services and the development of training programmes. Other studies may be undertaken on the development of sensors such as SAR and an imaging spectrometer.

ASO operated with a budget of A$3.2 million 1987-88, A$5.4 million 1988-89, A$4.4 million 1989-90, A$6.299 million 1990-91, A$6.179 million 1991-92, A$6.989 million 1992-93, A$7.653 million 1993-94, A$11.3 million 1994-95 and A$9.0 million 1995-96 (fiscal year begins 1 July).

Foremost among the new space projects the CSIRO plans to introduce are programmes to place one or more micro-satellites into orbit before or during 2001. CSIRO is heading this project with universities, industry and other parties. A proposal to establish a Co-operative Research Centre to conduct this programme is being considered by the Australian government.

A range of other organisations undertakes space-related science and applications programmes, including the Australian Centre for Remote Sensing (ACRES), Commonwealth Industrial & Scientific Organisation (CSIRO), and the Bureau of Meteorology; commercial satellite communications are handled by Telstra and Optus.

Although space activities are necessarily modest, a population of 18 million dispersed over a 7.68 million km^2 land mass, coupled with a dependence on natural resources and agriculture, provides the impetus for development of space applications. Remote sensing plays a significant role in national space activities. Landsat, Spot, NOAA, ERS and JERS data are received by ACRES and the Bureau of Meteorology operates meteorological satellite reception sites at Melbourne, Perth, Darwin and Casey (Antarctica). Some 70 per cent of Australia's overseas telecommunications traffic is carried by Telstra through Intelsat satellites, and Optus's own spacecraft carry a major part of domestic services, including routes to Papua New Guinea, New Zealand and deeper into the southwest Pacific.

Two satellites were launched by Shuttle in 1985, followed by a third in 1987, to improve domestic communications and supply radio/TV to 16 million people over sparsely populated areas. The largest single application is the Homestead and Community Broadcasting Satellite Service (HACBSS), provided by the Australian Broadcasting Corp. By January 1986, 650,000 Australians in remote areas were able to receive radio/TV for the first time. The second of two new generation satellites was launched by China on a CZ-2E from Xichang in August 1994, capable of carrying mobile communications services. L-band satellite phone services were inaugurated 31 August 1994. Before then, only 5 per cent of the country had phone coverage (80 per cent of the population), the remainder relied on UHF radio. A third generation series, Optus C, will begin with the launch of C1 in early 2001 and C2 in 2003.

The Australian Fiducial Network of GPS Reference Stations was established in 1994. There are 14 sites, including a site at Cocos Island (South Pacific), Macquarie Island, Wellington (New Zealand) (South Pacific) and three in Antarctica plus the mainland sites, for satellite integrity monitoring, scientific studies and to provide the precise framework for surveying and mapping. A Norfolk Island (South Pacific) site is imminent. The network also broadcasts error corrections to provide high-accuracy navigation for mobile users.

Space tracking activities are based at the Canberra Deep Space Communication Complex at Tidbinbilla. Tidbinbilla was commissioned in 1965 and employs 150 staff. CDSCC routinely supports about 25 missions, including Voyager and Galileo. The 64 m antenna of the main Deep Space Station, DSS 43, was enlarged to 70 m in 1989 (and is active 80 per cent of the time); the new 34 m DSS 34 became operational in 1996. The two other major stations installed in the Australian Capital Territory during the 1960s, Honeysuckle Creek and Orroral Valley, were closed and their residual functions transferred to CDSCC. Orroral's antenna was donated to the University of Tasmania for radio astronomy, and Honeysuckle Creek's DSS 45 was re-erected at CDSCC to track Earth-orbiting satellites. Stations are also operated for the US agency at Yathagarra in western Australia (Mobile Laser Ranging Facility) and for ESA at Gnangara near Perth. Radio astronomy facilities include the 64 m Parkes telescope and the Australia Telescope at Culgoora, completed in 1988. Parkes was upgraded for linking directly with CDSCC to support Galileo's Jupiter operations in 1996.

A meteorological satellite command/data acquisition station is operated at Perth. Earth station sites are also provided for US military satellites, including the DSP early warning facility at Nurrangar, the signals intelligence satellite station at Pine Gap (near Alice Springs), and the Harold E Holt Naval Communications Station at North West Cape, western Australia, working through DSCS satellites. Harold E Holt remained a Joint Facility until it transferred to Australian control in May 1999, although the US Navy continues to have access. DSCS is used by other military installations, mainly the Australian Defence Signals Directorate in Melbourne. One Ku-band and one L-band transponder are also leased aboard Optus for military communications. DSP will leave Nurrangar in about 2000 as the US consolidates all data reception/processing in a single domestic site. An Australian imaging surveillance system is projected for 2004, possibly using commercial imagery or a purchased dedicated satellite, to cover Australia and the surrounding seas.

AUSTRIA

Austria began participating in the European Space Agency's activities in 1975 and on 1 January 1987 became a full member. ESA accounts for almost all of the national space expenditure. The 1999 contribution is €30.3 million, 1.3 per cent of members' contributions. Former years' contributions were: 1996 €31.4 million; 1995 30.6; 1994 26.7; 1993 25.9; 1992 25.4; 1991 18.8; 1990 16.2; 1989 12.5. It had contributed a total of Sch3.340 million to the agency by September 1996, receiving Sch2.562 million in contracts in return. Its contribution to ESA's optional activities is frozen 1994-97 at Sch250 million annually. Austria's principal interests within ESA are Artemis, DRS (1.5 per cent), ERS 1/2 (0.73/1 per cent), Ariane 5 (0.4 per cent), Envisat (1.0 per cent), Meteosat Second Generation (2.3 per cent), EOPP (2-17 per cent) and FESTIP (2.31 per cent). The Space Research Institute contributed two experiments to the four-satellite Cluster project.

Austria solidified its growing programme by the signing in November 1988 of the protocol for the 'AustroMir' mission. The neo-commercial agreement with Glavkosmos called for a payment of Sch85 million (about US$6 million) plus research equipment provided for the mission. Total cost to Austria was Sch200 million. Joanneum Research in Graz provided management, after the selection of 14 experiments (10 biomedical, three technological and one remote sensing using Mir's MKF-6MA multispectral camera). The science research required 42 h of in-orbit time.

Austrian Franz Viehböck flew the successful Soyuz-TM 13 mission 2-10 October 1991.

Previous Soviet collaboration includes magnetometers carried by the Venera 13/14 and Vega 1/2 Venus/Halley's Comet probes, and similarly modest contributions to the international Phobos and Interball projects. Two ion emitter modules from Joanneum Research and the Research Centre Siebersdorf are flying on Japan's Geotail to cancel electrostatic charging; the design was tested during AustroMir. Interball, ESA's Cluster and Germany's Equator-S carry further modules. Manufacturing and testing support have been provided to Interball's two Magion Czech sub-satellites since 1993. France's Mars balloon includes an Austrian magnetometer, should Russian funding permit the flight.

Austria participates in Intelsat (holding a 0.41 per cent investment share in 1996) and Eutelsat (0.61 per cent 1996). In January 1994 it became EUMETSAT's 17th member through its National Meteorological Service (Zentralanstalt für Meteorologie und Geodynamik).

BELGIUM

Belgium is a federal state comprising the Flanders, Walloon and Brussels economic regions and the Flemish, French and German cultural communities. Space activities at the national level are funded and managed by the federal Ministry of Science Policy through the SSTC Services Fédéraux des Affaires Scientifiques, Techniques et Culturelles (Office for Science, Technology and Cultural Affairs). The three major regions participate separately. For example, the Walloon region contributed to France's Ramses electrophoresis experiment flown on Shuttle in 1994.

National space activities are almost all centred on the European Space Agency, to which it contributes more than 90 per cent of its space budget. The 1999 contribution is €122.1 million. Space expenditure totals about 0.08 per cent of GNP. Belgium is the fifth largest contributor, after France, Germany, Italy and UK, to ESA overall (5.3 per cent of Member State income for 1999) and about the fourth largest contributor to major programmes: 2.9 per cent science 1996-2000, 6 per cent Ariane 5 (0.2 per cent Ariane 5E upgrade), 5 per cent Columbus, 4.0 per cent DRS, 4.5 per cent Envisat.

One of ESA's major ground stations is located at Redu in Belgium, providing TT&C facilities for Eutelsat's satellites. Belgium is leading the proposal for the IRIS (Intercontinental Retrieval of Information via Satellite) satellite system for commercial messaging and localisation. The LLMS Little LEO Messaging System development project is funded within ESA's communications programme. SAIT Systems is system designer and plans to establish a subsidiary for operating IRIS; Alcatel Bell Telephone is building the payload. The first launch will be of an integral package on Russia's Resurs-O1 in 1997, followed by a free-flyer later in 1997.

There is increasing emphasis on remote sensing observations and applications. Belgium is a 4 per cent shareholder in France's Spot remote sensing satellite system and will contribute 4 per cent of the cost for Spot 5 and another 10 per cent to the development of the 'Vegetation' instruments for Spot 4/5. Belgium is a candidate for Vegetation's data processing centre. Mir flies the major MIRAS spectrometer.

Belgium also provides 4 per cent to ESA's Metop costs and 2.7 per cent to the EUMETSATs, 3 per cent to ESA's Meteosat Second Generation costs and 2.7 per cent to the Meteosat Transition Programme.

The first Belgian in space was Dr Dirk Frimout, flying as a Payload Specialist aboard Shuttle STS-45 in March 1992. Physician Marianne Merchez was selected in May 1992 by ESA as one of six new astronauts.

Belgacom manages national participation in Intelsat, Inmarsat and Eutelsat. Belgacom operates the V-STAR 1.2 to 2.4 m VSAT service, operating from hubs in Liedekerke via Eutelsat for Europe and in Lessive via Intelsat for Africa, the Middle East and North and South America.

Belgian Space Expenditure (BFr million)*

Year	ESA	National	Total
1989	3,453.0	368.0	3,821.0
1990	3,928.0	347.6	4,275.6
1991	4,111.7	380.0	4,491.7
1992	4,959.0	408.6	5,367.6
1993	5,750.4	430.3	6,180.7
1994	5,906.2	456.8	6,363.0
1995	6,107.5	450?	6,557?
1996	6,180.8	450?	6,630?

*OSTC declines to provide current National funding figures.

BRAZIL

As a developing country with a land area of 8.512 million km^2, Brazil's space activities emphasise the development of remote sensing and communications programmes in conjunction with a move towards an independent launch capability. Brazil desires an indigenous domestic and commercial satellite infrastructure after the year 2000. The federal government passed the law February 1994 creating the Brazilian Space Agency (AEB) as the successor to the inter-ministerial COBAE Commission for Space Activities. The civil AEB, reporting directly to the President, is now in charge of co-ordinating and planning space programmes. The existing governmental space organisations remain linked to their ministries, but they now function as the operational arms of AEB.

On 8 December 1994, the government approved the revision of the PNDAE National Policy for the Development of Space Activities proposed by the AEB's General Council, setting the main objectives for Brazil's space activities. It promotes the development of technologies and capabilities relevant to Brazil's problems, and the preparation of industry to become competitive in the space goods and services markets. The Brazilian government is particularly interested in space surveillance of the Amazon, patrolling the land borders and the Atlantic coastal area, cataloguing and monitoring natural resources, planning and inspecting land use, forecasting agricultural harvests, weather and climate.

The ambitious MECB (Missão Espacial Completa Brasileira) 'All-Brazilian Space Programme' began with government approval in 1978. It began in earnest in 1980 to design, build and launch four satellites, two 115 kg SCD (Satelites de Coleta de Dados) satellites to relay environmental information from ground data collection platforms, and two 170 kg SSR (Satelites de Sensoriamento Remoto) to provide 200 m resolution Earth imagery from CCD cameras. A third SCD was added in February 1994 and a fourth in 1995.

MECB's charter required the development of the VLS orbital launcher, similar in performance to India's ASLV (first launched 1987), and associated launch sites and tracking/data network. Commercial US Pegasus launched the first data relay satellite February 1993. The first imaging craft is planned for 2001 on VLS, followed by a second satellite in 2002. Total MECB cost was estimated at more than US$1 billion at 1988 rates, divided between US$280 million for the satellites and more than US$700 million for the launcher which is now under test.

In February 1988, the National Institute for Space Research, the national agency responsible for applications satellites, committed US$45 million to the joint development of the China/Brazil Earth Resources Satellite (CBERS), carrying sensors to return 20 to 260 m resolution imagery. Brazil will pay US$15 million for two CZ-4 launches, according to the contract signed November 1993, in return for which China guarantees a 100 per cent return on the commercialisation of Brazilian high-tech products in China, such as aerospace items, through a trade agreement. CBERS-1 is scheduled for launch in July 1999 and Brazil will build, integrate and test CBERS 2 for launch in the year 2000.

INPE's 1996 CBERS budget is US$17 million; 1995 24; 1994 10; 1993 21. Brazil will provide the HSB humidity sensor for NASA's EOS PM 1 satellite. CNES and AEB agreed in May 1996 to launch a joint 80 kg science satellite in 1998.

Technology transfer constraints on European and US companies helped to slow launcher development. Arianespace's successful bid for Brasilsat second generation launch services during 1989 was reported to include proposals for transferring satellite thruster, gyros, other satellite technology and Ariane Viking rocket motor technology. However, the Missile Technology Control Regime signed by the G7 group of the world's seven most industrialised nations prevented technology transfer. As a result, VLS underwent significant redesign. Brazil became part of the MTCR in October 1995, agreeing not to develop missiles capable of delivering 500 kg warheads over 300 km, or exporting technology to others. The first orbital launch attempt of VLS-1 failed on 2 November 1997 when the SCD2A satellite was destroyed.

INPE is responsible for the satellites and their ground system, projecting a cost of US$295 million at 1988 rates. US$190 million had been expended to end-1995 (US$150 million by December 1991 when SCD 1 was completed; US$3 million in 1994 plus US$5 million in 1995). The Satellite Control Centre was inaugurated September 1989 at INPE's São José dos Campos site. Cuiaba is the primary reception station for the data collection and remote sensing satellites, while INPE's Cachoeira Paulista site 80 km northeast of São Paulo handles the environmental data processing and dissemination.

BULGARIA

The Bulgarian Aerospace Agency (BASA) was formed in December 1993 as the national body for the development and application of aerospace research and technology. Bulgaria was one of the most active Interkosmos members, and the only one to have undertaken two manned visiting missions to USSR space stations. The Space Research Institute of the Academy of Sciences is a BASA member and acts as the national centre for space science activities. Twelve ionospheric, geodetic and upper atmosphere experiments were contributed to Interkosmos 22 (Bulgaria 1300), the Fregat TV/spectrometer was provided for Phobos, and scientists are participating in the Aktivny, Apex, Interball, Spektrum-X/γ and Mars-96 projects.

The first space station mission was aborted in April 1979 when Soyuz 33, carrying researcher Georgi Ivanov (real surname was Kakalov, a name which was deemed inappropriate for a cosmonaut), suffered a main engine failure before it could dock with Salyut 7. Ivanov's back-up, Alexander Alexandrov, flew to Mir aboard Soyuz-TM 5 in June 1988 for the Shipka 47-experiment research programme. The CCD-based Rozhen telescope and camera system was demonstrated successfully and left in position for future crews; a version of it was to fly as Spektrum-X's star tracker (subsequently cancelled) and it is offered for other collaborative projects. SRI also provided the Spectrum-256 imaging spectrometer, the Zora equipment for cosmonaut psycho/physiological monitoring and the Svet greenhouse, still operational aboard Mir. The new generation Svet 2 is being developed, and Zora's Neurolab-B medical system successor will be launched to Mir in 1996. SRI signed a distribution agreement with Spot Image in April 1993.

Bulgaria is a member of the Intersputnik, Inmarsat and Eutelsat (joining 1995) satellite communications organisations and joined INTELSAT on the 15 May 1996.

CANADA

Canada's space programme is governed by the Long Term Space Plan. A new LTSP was released 3 June 1994 covering 1995-2004, adding (after a 12 per cent cut in 1995) C$670 million to the C$1.7 billion remaining from the previous plan and focusing on Earth observations, Space Station, science and communications. Many feared that election of a new government in October 1993 and a large national deficit would lead to a reduction in Canada's Space Station commitment and loss of manned flights. But the Mobile Servicing System will be completed and there are sufficient funds for one astronaut a year on Shuttle.

Manned funding is C$498 million 1995-2004. The C$637 million for Earth observation is the single largest element, including C$240 million for a second Radarsat scheduled for launch in March 2001. The C$271 million for space science includes new 250 kg SciSat satellites, a cooperative venture with the US and currently in the process of mission selection; the first was due to be selected in 1996. The C$334 million for communications includes C$160 million for an advanced satellite to demonstrate Ka-band high-rate links for fixed and mobile terminals.

The Communications Research Centre is looking for a private partner to at least equal that investment. Adding in C$271 million for technology, C$236 million for infrastructure and a C$124 million reserve brings the total for 1995-2004 to C$2,371 million.

The Canadian Space Agency oversees the Space Station programme, space science, space technology development, the astronaut programme and Radarsat. It is also responsible for overseeing national space policy.

Industry sales, 70 per cent of them export, topped C$353 million in 1990, about the same in 1991. Sales of goods and services in 1995 reached C$1 billion, while employing more than 5,000. The Space Station represents Canada's largest single expenditure, with the Mobile Servicing System accounting for most. The two primary satellite programs initiated under the previous LTSP are Radarsat 1 (remote sensing) and MSAT (mobile communications). Canada signed a 10-year renewal of an ESA agreement as an associate member in May 1989, contributing about 4 per cent (C$6 million) of the general ESA budget and an additional C$7.4 million for specific European programmes such as ERS. Since 1991, it has contributed a half share on the GNP scale employed to calculate the contributions of full member states. The latest accord, signed 21 March 1991, called for C$130 million over the following five years, with specific interest in ERS 2, Envisat, the ASTP Advanced Systems & Technology Programme and the Artemis Data Relay & Technology Mission. 1996's contribution was 0.7 per cent of the total from member states, or C$32.8 million (1995 29.4; 1994 25.2; 1993 30.6).

SovCan Star Satellite Communications Inc was formed in 1990, 50 per cent owned by Canada's COM DEV, Spar Aerospace, Canadian Satellite Communications and General Discovery, and 50 per cent by Russia's NPO PM and Intersputnik. The goal is to build five GEO telecommunications satellites, with NPO PM providing the launcher and bus and Canada the payload. The first is planned for 1999 launched by Proton following integration in Canada's David Florida Laboratory.

Canada was one of the four founding members of the Cospas-Sarsat satellite-based search and rescue system in 1979. Teleglobe Canada represents it in Intelsat and Inmarsat.

Telesat Canada, jointly owned by the government and telephone/telegraph companies (the government's 53.7 per cent interest was sold to the Alouette Telecommunications Inc consortium for C$154.8 million in March 1992), became the world's first operators of domestic satellites with the launch of Anik A1 in 1972. With Anik C2 launched by Shuttle STS-7 in 1983, Telesat Canada pioneered five-channel TV direct to US subscribers' homes in November 1983. TMI Communications began operating the MSAT system following April 1996's launch. Canada is a prime mover for development of land mobile systems because of its geography.

Years of close collaboration with NASA, and the development at a cost of C$100 million of the Remote Manipulator System for the Space Shuttle, made Canada the largest national contributor to the Space Transportation System outside the USA. In return, in April 1980 NASA ordered three more RMS sets for the next three Orbiters at a cost of US$63 million.

Canada was an observer member of ESA from 1975, and in 1978 signed a co-operation agreement under which it would pay 1 per cent of the net fixed common costs of ESA's general budget. This agreement was renewed in 1988 for a further 10 years, with Canada's contribution to ESA's general budget growing to about 4 per cent in 1990. During 1981-85 Canada spent a total of C$476 million on space activities, including five Anik launches. The hybrid Anik E appeared on Ariane in April and September 1991 and will serve well into the next century. The next generation Anik will begin with the launch of F1 in 2000, an HS-702 bus carrying 36 C-band and 48 Ku-band transponders, followed by F2 in 2002.

In December 1987, Canada became the first foreign partner to reach agreement with NASA and State Dept officials on the text of agreements covering participation in Space Station. Canada is providing the Mobile Servicing System to play a critical role in Space Station assembly, maintenance and servicing. Spar was awarded the prime C$195 million second phase contract in March 1991 (the first phase was worth C$50 million). Canada will have access to 3 per cent of station facilities, which includes instrument space, power, data handling and astronaut servicing time. The Space Vision System is being developed in co-operation with NASA for aiding astronaut RMS control by providing synthetic visual clues. Canada has interests in materials and life sciences projects within the station, in addition to attached payloads for materials, Earth observations, astronomy and space physics.

Canada was among a number of countries to sign collaborative space agreements with the Soviet Union in the 1987/88 period. After a four-day meeting in Moscow between representatives of the National Research Council of Canada and the Soviet Academy of Sciences, the parties signed a joint protocol. Canada agreed to participate in the international Interball magnetospheric research project, and to undertake joint research in solar/terrestrial physics, high-energy astrophysics, sub-mm/mm astronomy, radio astronomy, materials processing, space biology/medicine. A University of Calgary ultra-violet auroral imager is aboard Interball as part of this agreement, following up versions on Sweden's Viking/Freja satellites. Two experiments, including a radiation detector, flew in September 1989 aboard the Cosmos 2044 biosatellite.

The MIM Microgravity Vibration Isolation Mount 6-DOF magnetic levitator for isolating experiments from onboard disturbances is flying on Mir's Spektr module, joined in 1996 by the NFE Nuclear Fragmentation Experiment for measuring neutron and charged particle-induced events inside Mir. A CSA agreement with the Russian Space Agency on space co-operation was signed 21 May 1993. A joint Space Panel with Japan was established in 1989 to co-ordinate co-operative activities, and established working groups in remote sensing, materials science and space science. A Thermal Plasma Analyser, similar to a Freja instrument, is carried by Japan's 1998 Mars orbiter Planet-B. COM DEV is prime under CSA's C$3.7 million contract.

NASA recognised Canada's contributions to the Shuttle programme by providing three flights for Payload Specialists. The first, Marc Garneau, a naval communications and electronic warfare expert, flew on STS-41G in October 1984. The other two flights, which aimed to test the SVS Space Vision System to increase the usefulness of the RMS, and study the disorientation experienced when waking from sleep in zero-*g*, were scheduled for 1986, but postponed indefinitely after the Challenger accident. The National Research Council had reduced 4,300 applicants to a team of six. In January 1989, NASA announced the selection of Drs Roberta Bondar and Kenneth Money as candidates from Canada's cadre of six astronauts for the single life sciences PS post aboard the first Shuttle International Microgravity Laboratory mission in February 1992. Bondar was selected in January 1990 to fly the mission; Bondar and Money left the corps in 1992. Steve MacLean tested SVS aboard STS-52 in October 1992. Bob Thirsk flew as a PS aboard STS-78's Life & Microgravity Spacelab in June 1996. The sixth member of the original group, Bjarni Tryggvason, flew as PS aboard STS-85's earth observation mission in August 1997. A further four astronauts were selected mid-1992 from 5,330 applicants to cope with Shuttle and Space Station demands: Major Chris Hadfield (AF pilot), AF Captain Michael McKay (electrical/computer engineer), Julie Payette (computer engineer) and Dafydd Rhys Williams (physician). Garneau and Hadfield joined NASA's Shuttle Mission Specialist training group in August 1992; Williams joined 1995's MS group. They are paid by CSA but are part of NASA's corps. Hadfield flew as an MS on the second Mir docking mission, STS-74, in November 1995; space vision experiments were again conducted. Garneau flew on STS-77 in May 1996 and Williams flew aboard STS-90 on a 15-day 21-hour mission in April 1998. Payette was launched aboard STS-96 in May 1999.

Fiscal Year	1995-96	1996-97	1997-98
Earth Observation	58.857	86.117	80.017
Manned Flight	139.494	85.900	40.485
Space Science	32.803	36.330	36.845
Communications	8.778	17.495	33.495
Space Technology	50.688	47.576	42.654
Space Awareness	1.432	1.417	1.014
David Florida	7.996	8.505	10.138
Central Services	22.910	14.896	15.003
Contingency	0	13.000	12.800
Total	**322.958**	**311.236**	**272.451**

All figures are C$ million.
CSA accounts for about 70 per cent of all federal government space expenditure.

CHILE

Space affairs are conducted under the auspices of the CAE Committee for Space Matters, created in 1980, as part of the Ministry of Defence. The Committee proposes to the President specific space budgets and projects. The Space Division of the Fuerza Aérea de Chile (FACH, Chilean Air Force) launched the 50 kg FASat-Alfa, as a passenger payload in 1995, into polar orbit aboard a Ukrainian/Russian Tsyklon.

A £3 million contract was signed with the UK's Surrey Satellite Technology Ltd in May 1994 for the satellite, ground station and on-the-job training for eight FACH engineers (launch cost was additional). The satellite was built at Surrey by a Chilean team working alongside SSTL staff. The programme includes sending students to gain Master's Degrees on Surrey University's satellite engineering course. Unfortunately, a pyrotechnic problem meant that FASat-Alfa could not be separated from Sich. FACH thus signed a £1.3 million contract with SSTL for the identical FASat-Bravo launched in July 1998. Educational activities using the satellite will help to create space awareness in the country. FACH plans the similar FASat-Beta (built in Chile). Two others will follow to form a constellation providing LEO communications services. Work is also under way on the FASat-Gamma mini-satellite for remote sensing from 2003. The expertise derived from these projects is expected to lead to a GEO satellite around 2010.

The CEE Centro de Estudios Espaciales was established in 1959 to provide TT&C services to NASA from the Peldehue station some 40 km north of Santiago. The station still operates. CEE also designs/builds GOES WEFAX/VISSR stations for the military, but they might also become commercially available. A GPS station is operated as part of the Topex-Poseidon project. It offers a certified Cospas/Sarsat terminal. Chile was the first South American country to establish a Cospas/Sarsat ground station. The country does not possess Landsat or Spot ground stations but Landsat coverage is available via the Brazil, Argentina and Ecuador facilities.

In conjunction with CEE, Chile's Amsat-CE amateur satellite organisation, formed in April 1992, is working on the Cesar 1 satellite at the University of La Frontera in Temuco for amateur digital voice links. The 10 kg Microsat design is being adapted under agreement with Amsat North America and could include a GPS receiver. Uplink will be in L-band and downlink in S-band. FACH supplied SSTL software for Cesar 1.

Chile's signatory agency to Intelsat and Inmarsat is Entel (Empressa Nacional de Telecomunicaciones), with 1.51 per cent and 0.065 per cent interests respectively. The private CTC Compañia de Telefonos contracted with Scientific-Atlanta of the US in 1989 to build a US$29 million domestic telephone and data network, creating the world's largest digital satellite communications network. Scientific-Atlanta installed two 6 m Earth stations, five 7 m regional stations, an 11 m hub at Santiago, 35 4.5 m remote stations and more than 100 1.8 m VSAT stations for interactive data transmission. Chilesat and Satel Comunicaciones SA (a partnership of Entel and a Comsat subsidiary) provide domestic and international Intelsat services from their respective Intelsat A stations at Melipilla and Santiago.

Following a March 1985 agreement between Chile's Director General of Civil Aviation and NASA, a Shuttle contingency landing site was built on Easter Island, situated 3,700 km due west of Chile's Pacific coast and suitable for an aborted Shuttle launch from Vandenberg AFB in California. As a result of this agreement, there was a possibility of a Chilean national flying on the US Shuttle. FACH, in 1993, expressed interest in a US or Russian manned flight as part of an increased involvement in space development; there is no plan at the moment for such a flight.

CHINA, PEOPLE'S REPUBLIC

China's growing space programme aims in the near-term to operate polar and geostationary meteorological satellites, polar Landsat-type Earth resources satellites, geostationary communications satellites and recoverable Earth resources/microgravity capsules, in addition to marketing its range of CZ/Long March launchers. The most powerful of its launchers provides a capability equivalent to Europe's Ariane 4 and Russia's Proton. Longer term ambitions have been acknowledged to include manned flights and space stations by the middle of next century. Future CZ versions are being studied to provide a LEO capability of at least 20 tonnes within the next decade.

The space agency proposed a manned project within the 1996-2000 five-year plan but this has been delayed. The two-man capsule would be launched on a stretched CZ-2E from Jiuquan. Discussions have been held with a view to acquiring Russian Soyuz technology, particularly in life support and cosmonaut training. Studies have been reported in recent years but the programme now has high level backing. A 10-year shuttle programme was claimed in late 1992 to be under way, with work beginning on a new launch site near Jiuquan.

Studies are under way of a GMSIS Global Mobile Satellite Information System which would provide personal hand-held communications via 18 to 24 satellites in medium orbits. Three or four of the 500 kg satellites could be delivered by a single CZ-3A. Also under study is the SSRSS Small Satellite Remote Sensing System to use frequent revisits for monitoring environmental changes, particularly natural disasters. Seven satellites in 773 km near-polar orbits would observe a particular site twice daily. Several of the 250 kg satellites could be launched by one CZ-4. China is offering this as a collaborative project for the Asia-Pacific region. The Twin Star proposal would use two satellites some 40° apart in the GEO arc to provide positioning, messaging and timing services, after the fashion of OmniTracs. A lunar orbiter was mentioned in 1995 as possible for about 2000.

Estimating China's space expenditure is difficult and figures have only recently begun to appear. An official report in September 1994 claimed that almost ¥15 billion of space goods were produced in 1993, a 35 per cent increase over 1992. Sales increased 32 per cent to ¥15.3 billion (about US$1.7 billion), for a profit of ¥540 million. A civil annual budget of ¥1.4 to 1.5 billion (US$168-180 million) has been claimed for R&D and rocket and satellite production, but excluding launch operations and tracking, which are performed by a separate organisation (see below). By contrast, one US analysis in 1993 estimated US$1.35 billion annual expenditure on all programmes. With the return of Hong Kong to Chinese rule in July 1997, the communications satellite companies in the former-British colony automatically became Chinese.

The Commission of Science, Technology & Industry ultimately directs most civil and military space programmes for national defence. Civil space programmes fall under the aegis of the China Aerospace Corp and the China National Space Administration, the latter being the facet for foreign contact. Both were formed in June 1993 from their dissolved predecessor Ministry of Aerospace Industry, created in 1988 from the merging of the Ministry of Astronautics Industry and the Ministry of Aeronautics Industry. Commercial activities are directed through CASC's China Great Wall Industry Corporation. The Ministry of Posts & Telecommunications is the national signatory to Intelsat, and the Beijing Marine Communications & Navigation Co to Inmarsat. MPT's 'Chinasat' China Telecommunications Broadcast Satellite Co is the state operator of telecommunications satellites.

The principal space and military launcher manufacturing facility is the Beijing Wan Yuan Industry Corporation, founded in 1957 and now called the China Academy of Launch Vehicle Technology (CALT) near the town of Nan Yuan some 15 km south of the capital. Here, the smaller CZ-2 family and cryogenic upper stages are manufactured in a complex of 13 research institutes and seven factories. The 27,000 employees include 9,000 engineers/technicians. Its Shanxi Liquid Rocket Engine Co designs/builds the hypergolic engines. Production of the CZ-3 launchers' lower stages and of the entire CZ-4/2D is undertaken at the Shanghai Bureau of Astronautics (SBA), founded in 1961 and now also known as the Shanghai Academy of Space Technology (SAST), in 17 institutes and 11 factories. It employs 30,000, including 8,000 technical staff. Spacecraft production is the responsibility of the Chinese Academy of Space Technology. Founded in February 1968, its 14 research institutes and factories employ 11,000, including 6,000 technicians. SBA, CAST and CALT all report to CASC. The Shanghai Institute of Satellite Engineering (part of CAST until 1993, when it came under SBA/SAST) builds the FY meteorological satellites with funds and specifications from the State Meteorology Administration and the Shanghai Institute of Technical Physics provides the imaging system. The Xian Institute of Space Radio Technology is similarly providing the 20 m resolution CCD imaging sensors for the CBERS remote sensing satellite. The Institute of Space Medico-Engineering (ISME), founded in April 1968, studies principally the medical, physiological and psychological aspects of manned space flight, including the development of life support systems.

The China Satellite Launch and TT&C General organisation, with more than 20,000 personnel, operates the launch sites and is responsible for all TT&C services through to the conclusion of a mission. It operates under the Commission of Science, Technology & Industry for National Defence. China maintains three orbital launch sites, the most recent identified in 1988. Jiuquan in the Gobi desert was the starting point for China's first satellite and has two pads for CZ-1/2 launches into 57 to 70° orbits. Xichang became operational in 1984 for GEO missions with the CZ-3 and a third site at Taiyuan south of Beijing is utilised for Sun-synchronous meteorological and Earth resources launches with the CZ-4.

Launch and satellite TT&C is provided by six fixed, three mobile and three ship ground stations, headquartered at Xian City in Shaanxi Province. Ground stations at Beijing, Guangzhou and Urumqi feed NOAA and FY meteorological satellite data to the National Satellite Meteorological Centre in the capital. A Landsat Earth resources station has been operating since December 1986, provided by Hughes STX of the US, with the dish at Miyun 100 km northeast of Beijing and the processing facilities in northwest Beijing itself. Facilities were upgraded in 1993 to add Landsat 6 ETM, ERS and JERS 1.

China adopted a 'Twelve year Development Plan of Science and Technology' incorporating rocket propulsion in 1958. In 1965, the Academy of Sciences formulated a satellite programme that resulted in the first successful launch in 1970. China was the fifth nation after the USSR, USA, France and Japan to achieve national launcher capability. The ability to recover payloads from orbit was added in November 1975 (FSW capsules are now offered commercially) and the then only cryogenic upper stage outside of the US and European Space Agency carried the first Chinese GEO communications satellite in 1984.

Development of China's long-range military rockets and space launchers is largely credited to Dr Chien Hsue-shen, expelled from the US in August 1955 as a result of Senator Joseph McCarthy's anti-Communist campaigns. Chien had taken a PhD at Toronto University and then worked as a research engineer at the California Institute of Technology until ordered to return to mainland China. He visited the US in November 1972 heading a team of Chinese scientists. Ren Xinmin, chairman of the Ministry of Astronautics' scientific committee in the 1980s, also returned to China after completing his studies abroad and participated in developing plans for the Earth resources, meteorological and communications programmes now under way.

China began offering commercial launches following the successful positioning of its first GEO satellite in 1984. The issue of a *Long March 3 User's Manual* in 1985 began the appearance of detailed information on most of the launcher family. GWIC indicated in 1986 that up to a dozen departures a year could be handled from the two operational sites, but that at least half would be required for domestic purposes. Although none of the early agreements was converted to a firm contract, they did demonstrate China's willingness to provide undertakings, in order to satisfy western concerns over the potential for technology transfer, that customers' spacecraft would be admitted and processed without inspection. Upgraded handling and pad facilities were introduced to meet international standards, in addition to construction of a second Xichang pad for the new CZ-2E. The first firm launch contract was signed in November 1988 with Hughes for Australia's Optus B 1/2 satellites. AsiaSat became the first to depart, aboard a CZ-3 in April 1990 at a cost of US$30 million. These successes and international concern over CZ pricing policies produced an agreement in December 1988 that China would carry no more than nine foreign satellites by end-1994 and at costs compatible with the international launcher market. It has been reported that China has consistently underbid US counterparts by at least 30 per cent, although comparisons are difficult. A new agreement was reached in January 1995, allowing 11 further GEO satellites by 31 December 2001 (in addition to those signed up under the previous accord). The total may be increased if the annual global requirement is at least 20 satellites, or if western vehicles cannot accommodate the market. Prices must be comparable with the international market and certainly within 15 per cent. LEO launches must also be on a cost par, although no percentage limit is given. There has been notable success in signing up commercial payloads, including Intelsat, Globalstar, Mabuhay, APStar and EchoStar, although mission failures have cast doubts on reliability.

China's first 24-transponder DFH-3 three-axis satellite at last appeared in 1994 to help satisfy the country's burgeoning telecommunications demands, although it had to be abandoned in orbit after its leaking attitude control system exhausted the propellant. Its replacement was launched in May 1997. The first successful demonstration communications satellite was placed in GEO during 1984, a second in 1986, followed by three DFH-2A versions in 1988/90. DASA and CASC established the EurasSpace GmbH joint venture in 1994 to help produce China's domestic satellites and to bid for international contracts. A Local User Terminal and Mission Control Centre, supplied by France's CEIS TM under a US$1.5 million contract, became operational in 1995 in Beijing for the Cospas-Sarsat international search/rescue system.

Collaboration with Brazil has developed over recent years: joint development of the US$150 million CBERS (China/Brazil Earth Resources Satellite) was finalised in 1988. The potential exists for construction of a Brazilian CZ launch site near the equator for GEO missions. A similar Indonesian facility, possibly on Biak Island, was considered in 1987.

Discussions began in 1994 with Canada on the Radarsat programme as microwave imaging would be useful over China's persistently cloudy south. Meteorological satellites have proved difficult to establish: the first two FY-1 polar craft had only short lives and the first FY-2 GEO imager was lost in 1994's ground accident. The replacement was launched in June 1997. As a result, China has continued to rely on NOAA and Japanese GMS satellites and its own ground radar. The country's vast expanse and its susceptibility to natural disasters provides an impetus to spaceborne monitoring. China is estimated to lose ¥50 billion annually from natural disasters, a cost expected to double by 2000.

There were rumours of a Chinese launch attempt in November 1969, but these appear to have related to posters appearing anticipating China's first satellite launch rather than actually celebrating a launch attempt. The first Chinese satellite was Dong Fang Hong, launched on 24 April 1970. While the satellite became infamous for its playing of the Chinese anthem 'The East is Red' (after which the satellite was named), examination of the payload suggests that it carried communications transponders similar to the United States Telstar 1 and 2. This raises the possibility that it might have had a limited communications repeater role.

Shi Jian 1 was a science payload (carrying a magnetometer, cosmic ray and solar X-ray detectors) launched a year later using a modified version of the DFH-1 satellite: this time solar cells were carried and thermal control was achieved by using louvres on the satellite. The triple payload which comprised Shi Jian 2, 2A and 2B was a science mission, the latter satellite being a balloon attached to a heavy mass which quickly decayed from orbit. Shi Jian 3 is a designator used for an abandoned early-1980s meteorological satellite programme and was never used for an orbital mission. Shi Jian 4 – along with KF 1 – was a science payload for the maiden flight of the CZ-3A launch vehicle: KF 1 (the acronym has never been explained by the Chinese) appears to have been an active satellite, but its purpose has never been explained. It is believed that a Shi Jian 5 satellite might be being developed, but few details are available: the satellite is said to be intended to prove a design of a planned satellite but which could have other applications.

The series of Ji Shu Shiyan Weixing satellites has never been discussed in Chinese literature, but it is thought that they comprise a series of Elint satellites, which were never followed up as an operational programme. Since Mao Zedong primarily backed the Shanghai space activities, which were responsible for this programme, it is possible that when he died so did the political support for this programme.

The most numerous satellites in the Chinese programme have been the Fanhui Shi Weixing recoverable satellites, with three main series having flown. The FSW-0 had 10 launches, one launch failure and nine recoveries (lifetimes three days initially, then five days). FSW-1 had five launches, no launch failures, four recoveries (lifetimes typically eight days) and FSW-2 had three launches, no launch failures, three recoveries (lifetimes 15-16 days). The series of successful consecutive recoveries starting in 1975 and ending in 1992 is remarkable, with China being the third nation to demonstrate this ability.

The FSW satellites were originally developed for a photoreconnaissance role, but after a break in launches from 1978-1982 the satellites were modified for a dual-role photoreconnaissance and remote sensing role. They have also been used to carry both domestic and foreign commercial microgravity and biomedical experiments. The final FSW-0 satellite, launched in 1987, carried experiments for MATRA Espace, the second FSW-1 (1988) carried COSIMA 1 for INTOSPACE in the (then) Federal Republic of Germany. COSIMA 2 had been planned for an FSW-1 in August 1989, but the flight was cancelled – in part for political reasons as a reaction against the events in Tiananmen Square in June that year: the experiment actually flew on a Russian Resurs-F1 satellite, launched in September 1989. No details appear to be available, but the fourth FSW-1 is said to have carried a small Japanese 710°C semiconductor furnace.

The next group of applications satellites to begin flying was for communications. The first launch in January 1984 failed to place its DFH-2 satellite into geosynchronous orbit due to a malfunction of the new liquid oxygen/liquid hydrogen third stage of the launch vehicle. But only three months later a test satellite was placed into geosynchronous orbit and it continued to operate over 125° east until June 1988 when it was boosted into a new orbit slightly higher than the geosynchronous orbit band. A third satellite was

List of Chinese Satellite Launches, 1970-1996 (including known test flights)

Launch Date		Satellite	Mass kg	Vehicle	Site	Orbit Class
10 January 1970				CZ-1	J	Test flight
24 April 1970		DFH 1	173	CZ-1	J	E-LEO
3 March 1971		SJ 1	221	CZ-1	J	E-LEO
10 August 1972				FB-1	J	Test flight
18 September 1973		JSSW	1,100 ?	FB-1	J	Failure
14 July 1974		JSSW	1,100 ?	FB-1	J	Failure
4 November 1974		FSW-0	1,800 ?	CZ-2A	J	Failure
26 July 1975		JSSW 1	1,107	FB-1	J	LEO
26 November 1975		FSW-0 1	1,790	CZ-2C	J	LEO (3)
16 December 1975		JSSW 2	1,109	FB-1	J	LEO
30 August 1976		JSSW 3	1,108	FB-1	J	E-LEO
10 November 1976		JSSW	1,100 ?	FB-1	J	Failure
7 December 1976		FSW-0 2	1,812	CZ-2C	J	LEO (3)
14 September 1977				FB-1	J	Test flight
26 January 1978		FSW-0 3	1,810	CZ-2C	J	LEO (3)
16 April 1978				FB-1	J	Test flight
28 July 1979		SJ	770 ?	FB-1	J	Failure
19 September 1981		SJ 2	257	FB-1	J	E-LEO
		SJ 2A	483			E-LEO
		SJ 2B	28			E-LEO
9 September 1982		FSW-0 4	1,783	CZ-2C	J	LEO (5)
19 August 1983		FSW-0 5	1,842	CZ-2C	J	LEO (5)
29 January 1984		SW (DFH-2)	1,000	CZ-3	XC	GEO fail
8 April 1984		STTW A (DFH-2)	1,000	CZ-3	XC	GEO
12 September 1984		FSW-0 6	1,809	CZ-2C	J	LEO (5)
21 October 1985		FSW-0 7	1,820	CZ-2C	J	LEO (5)
1 February 1986		STTW 1 (DFH-2)	1,000	CZ-3	XC	GEO
6 October 1986		FSW-0 8	1,820	CZ-2C	J	LEO (5)
5 August 1987		FSW-0 9	1,820	CZ-2C	J	LEO (5)
9 September 1987		FSW-1 1	2,100	CZ-2C	J	LEO (8)
7 March 1988		STTW 2 (DFH-2A)	1,025	CZ-3	XC	GEO
5 August 1988		FSW-1 2	2,100	CZ-2C	XC	LEO (8)
6 September 1988		FY-1 1	757	CZ-4A	T	SSO
22 December 1988		STTW 3 (DFH-2A)	1,025	CZ-3	XC	GEO
4 February 1990		STTW 4 (DFH-2A)	1,024	CZ-3	XC	GEO
7 April 1990	*	AsiaSat 1 (Hong Kong)	1,442	CZ-3	XC	GEO
16 July 1990	*	Badr 1 (Pakistan)	52	CZ-2E/PKM	XC	LEO
		AUSSAT-B (model only)	2,700 ?		XC	GTO fail
3 September 1990		FY-1 2	881	CZ-4A	T	SSO
		QQW 1	4 ?			SSO
		QQW 2	4 ?			SSO
5 October 1990		FSW-1 3	2,080	CZ-2C	J	LEO (8)
28 December 1991		STTW 5	1,025 ?	CZ-3	XC	GEO fail
9 August 1992		FSW-2 1	2,400	CZ-2D	J	LEO (16)
13 August 1992	*	Optus-B 1 (Australia)	3,000 ?	CZ-2E/STAR-63F	XC	GEO
6 October 1992	*	Freja (Sweden)	259	CZ-2C	J	E-LEO
		FSW-1 4	2,100			LEO (7)
21 December 1992	*	Optus-B 2 (Australia)	3,000 ?	CZ-2E/STAR-63F	XC	GEO fail (satellite exploded)
8 October 1993		FSW-15	2,099	CZ-2C	J	LEO
8 February 1994		SJ 4	410	CZ-3A	XC	GTO
3 July 1994		FSW-2 2	2,600 ?	CZ-2D	J	LEO (15)
21 July 1994	*	APStar 1	1,383	CZ-3	XC	GEO
27 August 1994	*	OPTUS-B 3	3,000 ?	CZ-2E/STAR-63F	XC	GEO
29 November 1994		Zhongxing 6	2,230	CZ-3A	XC	GEO
25 January 1995	*	APStar 2 (Hong Kong)	3,000 ?	CZ-2E/STAR-63F	XC	Failure (satellite exploded ?)
29 May 1995			?	CZ-1D	J	Failure
25 November 1995	*	AsiaSat 2 (Hong Kong)	3,379	CZ-2E/EPKM	XC	GEO
28 December 1995	*	EchoStar 1 (USA)	3,288	CZ-2E/EPKM	XC	GEO
14 February 1996	*	INTELSAT 708 (USA)	4,576	CZ-3B	XC	Failure
4 July 1996	*	APStar 1A (Hong Kong)	1,385 ?	CZ-3	XC	GEO
18 August 1996		Zhongxing 7	1,400 ?	CZ-3	XC	GEO fail
20 October 1996		FSW-2 3	2,600 ?	CZ-2D	J	LEO (15)
11 May 1997		Zhongxing 8	2,230 ?	CZ-3A	XC	GEO
10 June 1997		Feng Yun-2 1R	1,380	CZ-3	XC	GEO
19 Aug 1997		Mabuhay Ayilu 2	2,560	CZ-3B	XC	GEO
1 Sept 1997		Iridium mFS-1	650	CZ-2C/5D	T	LEO
		Iridium mFS-2	650			LEO
22 Oct 1997		APStar 2R	3,747	CZ-3B	XC	GEO
30 Mar 1998		Iridium 51	657	CZ-2C/5D	T	LEO
		Iridium 61	657			LEO
2 May 1998		Iridium 69	657	CZ-2C/5D	T	LEO
		Iridium 71	657			LEO
30 May 1998		Zhongwei 1	2,984	CZ-3B	XC	GEO
18 July 1998		SINOSAT 1	2,840	CZ-3B	XC	GEO
19 Aug 1998		Iridium 3	657	CZ-2C/5D	T	LEO
		Iridium 76	657			LEO

Sub-orbital test flights are shown as 'Test flight'. Abbreviations of satellite names are as follows. DFH – Dong Fang Hong (The East is Red). SJ – Shi Jian (Practice). JSSW – Ji Shu Shiyan Weixing (technical experimental satellite). FSW – Fanhui Shi Weixing (recoverable experimental satellite). SW – Shiyan Weixing (experimental satellite). STTW A – Shiyan Tongbu Tongxin Weixing – (experimental stationary communications satellite). STTW 1-5 – Shiyong Tongbu Tongxin Weixing (operational stationary communications satellite). FY – Feng Yun (wind and cloud). QQW – Qi Qui Weixing (atmospheric observation satellite). An asterisk indicates a non-Chinese satellite, with the owning country shown under the satellite's name. Launch sites are J – Jiuquan, XC – Xi Chang and T – T'ai Yuan Orbits are indicated in general terms: LEO – low Earth orbit, E-LEO – eccentric low Earth orbit, GEO – geosynchronous Earth orbit, GTO – geosynchronous transfer orbit and SSO – Sun-synchronous orbit. 'Failure' indicates that nothing reached orbit: 'GEO fail' indicates a mission intended to reach GEO which failed after the initial orbital injection. For the recoverable FSW satellites the lifetime in days is shown in parentheses following 'LEO'.

launched in February 1986 and this entered geosynchronous orbit over 103° east where it operated until around September 1990: at this time the satellite might have simply run out of station-keeping propellant and 'died' on station.

In 1988, an improved series of communications satellite was introduced, DFH-2A. The launches in March and December 1988 and February 1990 placed satellites in orbit over 87.5° east, 110.5° east and 98° east respectively: all three satellites were still operating as of April 1997. A launch in December 1991 left the final DFH-2A stranded in a low orbit after the launch vehicle's third stage failed once more.

The Chinese have developed, in collaboration with Germany, the new DFH-3 series of communications satellites. The first launch (Zhongxing 6) in November 1994 was initially successful, but problems with the satellite's attitude control system meant that by the time a geosynchronous drift orbit was reached there was not sufficient propellant left to reach a geosynchronous orbit and perform the necessary station-keeping manoeuvres. As a result the satellite was abandoned.

As a stop-gap measure the Chinese purchased a Hughes HS-276 satellite and launched it in August 1996 as Zhongxing 7: however a third re-ignition failure of the CZ-3 launch vehicle's third stage left the satellite stranded in a low orbit.

The third series of Chinese applications satellites has been used for meteorology and remote sensing. Two Feng Yun-1 satellites have been launched to Sun-synchronous orbit, but each suffered in-orbit problems. The first, launched in 1988, only operated for 39 days. The second, launched in September 1990, suffered a loss of attitude control in February 1991, but this was solved after the Chinese spent 50 days trouble-shooting the problem. Later, radiation problems again meant that control of the satellite was lost.

The FY-2 series of meteorological satellites is intended for launch to geosynchronous orbit. The first was lost in April 1994 when there was an explosion during its propellant loading before being mated with the launch vehicle: a replacement flew in June 1997. The accompanying table lists all of the Chinese launches through to the end of March 1999.

Starting in the mid-1980s, the Chinese started to market the CZ-3 as a commercial launch vehicle, which could be used by foreign customers. While companies showed an interest in the Chinese offers, political technology-transfer reasons meant that it was some years before any satellites were ready to be launched. Even when commercial clients appeared, two of the companies were based within Hong Kong.

The first commercial flight was AsiaSat 1 in April 1990, the satellite being successfully launched. The major Chinese success was in winning the launch contract for the two AUSSAT-B satellites to be flown for AUSSAT in Australia: these satellites were the latest-generation Hughes HS-601 satellites. By the time that the launches took place the AUSSAT company had become Optus. The first Optus-B launch in August 1992 was successful, but the malfunctions which surrounded the second Optus-B launch in December that year have never been explained although the Chinese state that their launch vehicle performed flawlessly. However, the Australians stuck with the Chinese and the replacement Optus-B 3 satellite was successfully launched in August 1994. The previous month APStar 1 had been successfully launched.

In 1995, the Chinese suffered a major failure when the CZ-2E launch vehicle with APStar 2 (a further HS-601 satellite) exploded only seconds after launch: as with Optus-B 2 the western media blamed the Chinese launch vehicle for the failure. At the end of the year the first two flights of the CZ-2E with the Chinese new third stage EPKM successfully placed commercial payloads into orbit: AsiaSat 2 in November and EchoStar 1 in December.

Just as the Chinese commercial launch programme was recovering, the maiden flight of the CZ-3 in January 1984 failed when the third-stage engine malfunctioned: it was supposed to re-ignite to take the satellite out of a low Earth orbit and into a geosynchronous transfer orbit. The chamber pressure rose to 90 per cent of the normal value, maintained this for 3 seconds and then declined to zero with a resulting loss of thrust. The problem had its roots in the weightless behaviour of the propellants (it was the first time that a Chinese rocket engine was planned to shut down and then re-ignite in orbit) and modifications were made to the next flight vehicle, successfully launched the following April.

Following a string of successful launches the eighth CZ-3 launch vehicle resulted in a failure in December 1991. As with the maiden flight, the problem appeared when the time came to re-ignite the third stage. 58 seconds after re-ignition the chamber pressure began to drop gradually and failed completely 77 seconds later. Post- failure analysis showed that the problem had been due to leakage of the helium pressurant supply.

The eleventh CZ-3 was the next one to suffer a third-stage failure. 261 seconds after the re-start of the engine what was described as an 'abnormal phenomena' appeared in the engine control gas line system and this caused the engine's thrust to drop, resulting in shut-down some 48 seconds early. After a failure of the first orbited launch attempt with the CZ-3B in February 1996 the second succeeded in August 1997.

Turning to the maiden flight of the CZ-2E vehicle, this carried two payloads. The basic launch vehicle worked as planned and the Pakistani Badr 1 was deployed after parking orbit injection. However, problems arose with the new PKM third-stage which was intended to place a mockup of the AUSSAT-B (later renamed Optus-B) satellite into a 24°, 340 to 25,740 km orbit. Although the PKM/AUSSAT-B model assembly reached parking orbit it was never tracked in orbit by US sensors and it seems possible that when the time came for the PKM burn the assembly was wrongly-oriented and the rocket stage brought itself and the payload out of orbit. Chinese literature has stated that the failure was due to an attitude control and orientation problem with the second stage of the CZ-2E, which would suggest that the scenario just presented is a reasonable one.

The next two failures of the CZ-2E have never been explained in public to everyone's satisfaction, with the Chinese and Hughes (who built the satellites) initially blaming each other. More recently the western media has been happy to place the blame on the Chinese launch vehicles. These two failures involved Optus-B 2 and APStar 2, both Hughes HS-601 satellites launched on a standard two-stage CZ-2E but with a United States STAR-63F third stage added as the perigee kick motor.

Video coverage shows that about 70 seconds after the launch of Optus-B 2 a cloud of gas could be seen emerging from the payload shroud, although the launch vehicle continued in flight. The CZ-2E successfully launched the STAR-63F/Optus-B assembly into LEO – except that after orbital injection no contact would be made with the satellite. After the launch, when recovering the payload shroud, downrange debris from the outer parts of the satellite were also found and, it was therefore obvious that the satellite had exploded within the payload shroud at the time that the gas cloud was seen shortly after launch. The Chinese themselves have said in their official User's Manual for the CZ-3A/3B/3C launch vehicles that, in the case of the Optus-B 2 launch 'the launch vehicle performed well in all aspects, although the spacecraft mission was not completed due to an explosion in the spacecraft'.

Although actual details are sparse, China is planning a new launch vehicle or family of launch vehicles. Since the mid-1980s it has been known that the Chinese were working on a launch vehicle which would be in a similar class to the United States Saturn-1/1B, although no design details have been officially published. In 1992, a paper entitled 'A Modular Space Transportation System' was presented at the International Astronautical Federation. It described a new two-stage launch vehicle using liquid oxygen and kerosene in the first stage and liquid oxygen/liquid hydrogen in the second stage. This basic vehicle would be capable of placing 11 tonnes into a 60°, 300 to 500 km orbit.

The two-stage launch vehicle was described as the basis for a new family of launch vehicles, based upon the clustering concept. The baseline vehicle supplemented by two strap-on boosters only slightly modified from the first stage would be capable of putting 22 tonnes into LEO, while a third modification would be capable of orbiting 38 tonnes (see above illustration). The largest variant would have six strap-on boosters clustered and would be capable of placing 70 tonnes into LEO.

It is not known whether the 1992 Chinese paper represents 'official' thinking or whether it is simply an intellectual exercise. Whichever, it certainly indicates a possible road which the Chinese could follow.

To date the Chinese have not used LOX/kerosene on any of their launch vehicles or missiles and it is therefore interesting that they have shown an interest in purchasing Russian engines using these propellants. Originally the Chinese wanted to buy an RD-171 engine (used on the Zenit first stage) for examination but the Russians refused the sale. The Chinese were able to purchase some (the number is not known – it might only be one) RD-120 class engines which are used on the second stage of Zenit. It is known that a modification of the RD-120, the RD-120K will be used on either the strap-on boosters or the core of the new Russian Soyuz-2 vehicle and therefore the engine has been modified for first stage applications. If the 1992 studies have any official basis, then one might speculate that four uprated RD-120K class engines – designed and manufactured in China – could be clustered for the baseline launch vehicle's first stage.

In 1992, the Chinese stated that a new launch complex is being built as part of the Jiuquan launch site and images from photoreconnaissance satellites show that its size indicates that it is to support far larger launch vehicles than those currently operating.

The only reason for developing launch vehicles, which are destined to place payloads of 20 tonnes or more into LEO, is in support of a manned space programme. Such a programme has been in an 'on-off-on' situation since the mid-1970s as the Chinese government placed a higher financial priority in the development of 'practical' satellites which could be used for reconnaissance, remote sensing and communications. At the end of the 1970s photographs of apparent Chinese astronaut trainees appeared in the press and similarly different photographs from later experienced were published in 1988.

The Chinese have a programme designated 'Project 921' for the launch of a manned spacecraft using a Chinese launch vehicle. It is not known whether the Chinese will first launch a Soyuz-class spacecraft or whether they will try to go straight to a 'shuttle' – the latter term is usually used by the Chinese in their recent discussions of a manned flight. They have said that an unmanned test flight of a future manned vehicle is planned for 1999 with a manned flight – possibly involving two astronauts – coming the following year.

In March 1995, ITAR-TASS stated that the Chinese were interested in purchasing hardware which could be used in a manned programme. The areas of interest were said to be "systems of emergency rescue and thermal control, as well as other 'joints and units' including a docking system". A complete Soyuz life-support system was said to have been purchased along with a spacesuit. It was stated that the Chinese were planning a manned mission around 2000-2002 with an orbital station planned for around 2015.

Visitors to the Russian cosmonaut training centre at Zvezdny Gorodok in July 1996 were told that a group of Chinese astronauts would be arriving in October to undergo basic training for space missions. The number was initially hinted to be up to 10, but in December 1996 only two Chinese trainees were named by the Russians: Wu Tse and Li Tsinlung, said to have been selected for the Chinese national astronaut team in March 1995. There were other Chinese at Zvezdny Gorodok, but it was not clear whether these were also astronaut candidates or simply trainers. While it was thought that a Russian-Chinese manned flight to Mir might be in the planning stage, it would appear that the Chinese simply want their astronauts to go through basic training and then return to China for final training using Chinese hardware for a two-manned flight in 1999-2000. Of course, simply the experiences of using Russian equipment and going through the training process will be of major benefit to the Chinese when they start to train their own astronauts from the initial selection all the way through to orbital flights.

CZECH REPUBLIC

The organisation of space activities became unclear when Czechoslovakia split to become the Czech and Slovak Republics on 1 January 1993. Funding difficulties, uncertainty over Interkosmos projects and preliminary discussions with the European Space Agency all contributed to the lack of focus of future plans. The Czech Republic signed a draft co-operative agreement with ESA in November 1995.

In May 1992, The Ministry of Economy, the national signatory to Inmarsat and Sprava radiokomunikaci Praha became the Intelsat signatory. The Czech Republic are now debating Eutelsat membership and the country is also an Intersputnik member (operating two Earth stations). Czech scientists continue to make significant contributions to international space science missions. Recent involvement includes instrumentation for Phobos and Magion sub-satellites for the Aktivny, APEX and Interball projects. Under the agreement providing for Interkosmos member countries to participate in manned space flights, military pilot Vladimir Remek became the first non-US/Soviet citizen to enter space. As a researcher on the Soyuz 28 mission in 1978 he spent a week aboard Salyut 6.

In addition to the Magion satellites, Czech instruments for Interkosmos have included a laser reflector for geodetic ranging measurements and a micrometeoroid detector. Prognoz was provided with fields and particles instrumentation and Prognoz 10 in 1985 carried a six-channel Czech radiometer. Czechoslovakia also contributed the steerable camera platforms to the Vega probes that intercepted Halley's Comet in 1986. A derived version, the ASP-GM, was provided for mounting instruments on Mir's Kvant 2 module, docked to the space station from December 1989. The development of a micro-accelerometer, possibly for an aeronomy Magion, continues. It was tested successfully aboard Mir in 1992 and returned to Earth.

DENMARK

In 1996 the Danish Research Advisory Board for Space was established following new Danish legislation on research advisory services. The main task of the Danish Advisory Board for Space is to provide the Minister for Research and Information Technology with independent and qualified advice within the space research area, while improving the possibilities of creating cohesion in Danish space research. As part of its task, the Board follows all developments within the field of space research.

Denmark has been a member of ESA since the Agency's creation. The country's contribution of ESA's budget is about 75 to 80 per cent of the domestic annual space budget and in 1999 accounts for 1.1 per cent of the total ESA budget, its latest annual contribution being €24.8 million. The national space budget for 1997 was DKr241 million, 1996 – DKr246.8 million, 1995 – DKr231 million, 1994 – DKr253.3 million, 1993 – DKr201.2 million and 1992 DKr230 million. Around DKr20 million is spent on non-ESA activities.

Denmark participates through the Space Research Institute in Russia's Granat and Spektrum-X astronomical missions. Ørsted is the first national satellite. The country also participates in Intelsat, Inmarsat and Eutelsat through its Post & Telegraph Administration.

The country finished the Ørsted in 1995 and stored it as the US schedule slipped.

FINLAND

Space activities began in the early 1980s, when Finland put instruments aboard the Soviet Phobos Mars probes, handled by the Geophysics Dept of the Finnish Meteorological Institute. Space activities now come under the principal authority of the Ministry of Trade & Industry (KTM). TEKES acts as the co-ordinator and main funding agency.

The Finnish Space Committee was established by the government in 1985 as a policy advisory body and transferred from the Ministry of Communications to KTM in 1989. The Ministry of Education supports space science through the Academy of Finland, and CIS collaboration takes place through a special commission supported by the Academy of Finland. The Finnish Space Committee presented a National Strategy for Space Research and Development in Finland in 1994 and this was updated in 1996 to cover the years 1996-1998. It made several recommendations. First, Finnish space activities should concentrate on ESA programmes, with special emphasis on the sectors, which would best impact Finnish industries. Second, Finland should undertake bilateral programmes as well as promoting national activities. Third, in the areas of remote sensing and satellite geodesy, Finland should place emphasis on the development of operational applications. In space science, Finland should focus on the scientific utilisation of instruments, with Finland continuing to participate in the development of instruments for ESA scientific satellites. Finally, in the fields of satellite communications and navigation, the emphasis should be on industry and the development of key technologies, primarily within ESA programmes to ensure and promote profitability in the field, to expand the markets and to enhance the effectiveness of all Finnish industries.

Two technology programmes began in 1996, jointly funded at FMk8 million per annum: Space 2000 and the RSTP Remote Sensing Technology Programme. Space 2000 aims at creating a network of companies and institutes capable of handling large space projects. Software, structures, electronics and reliability engineering are being emphasised. RSTP will last for five years, covering: remote sensing technology and research projects of international significance; innovative projects demonstrating applications to end-users; implementation of operational remote sensing methods; commercial applications.

Finland contributed the ERNE particle analyser to ESA's Soho and the University of Oulo is co-investigator in the EFW Electric Fields and Waves instrument on the failed Cluster (also on NASA's Polar). Sodankylä Geophysics Observatory has provided several magnetometers for NASA sounding rockets. It has industrial involvement in the ESA's Huygens portion of Cassini; the VTT Technical Research Centre, the University of Oulu and FMI are involved in both elements. Finnish scientists are collaborating on the Russian-led Spektrum-X (X-ray detectors, developed from versions provided for Mars Observer), the failed Mars-96 (FMI's Metegg meteorology sensors), Radioastron (22 GHz receivers) and Interball missions, and contributed the Aspera plasma spectrometer to Phobos. There is involvement with Sweden's Freja and Odin.

FMI is the national signatory to EUMETSAT, with Finland's contribution paid via FMI's budget. Data from polar meteorological satellites have been used since 1969 by FMI: a Primary Data User Station for Meteosat was installed in 1986 and a receiving station for full resolution digital NOAA data was installed during 1987. The National Land Survey (Maanmittauslaitos) is the National Point of Contact for ESA/ESRIN. Finland is involved in the GOMOS ozone monitor for ESA's Envisat 1 polar platform (1.20 per cent interest); Metop 1 platform involvement is 0.70 per cent. It is contributing 1.35 per cent to Meteosat Second Generation and 1.82 per cent to ERS 2 phase E. ERS provides ice data for the Baltic Sea. The Finnish Geodetic Institute participates in the CIGNET Co-operative International GPS Network and the DORIS/Spot programmes.

Outside of ESA, Finland's major space involvement has been in the Scandinavian Tele-X direct broadcast satellite, for which the Nokia group provided the ground control centre. There is a 0.40 per cent commitment to ESA's Data Relay Satellite programme, plus participation in ARTES 1 (1.71 per cent + 1.60 per cent extension), ARTES 5 (4.05 per cent), ARTES 7 (1.50 per cent) and ASTP 4 (1.90 per cent). Telecom Finland is the national signatory to Intelsat, Eutelsat and Inmarsat. Sweden's Telia AB and Telecom Finland International jointly own/operate the Tanum Teleport, on Sweden's west coast, with five Intelsat antennas. Inmarsat communications are routed through Norway's Eik station.

The Space Committee comprises a chairman, vice chairman and 10 members, with the executive working group as the sole subordinated body. The chairman is the national representative on ESA's Council. A new Committee was appointed on 1 April 1995 to serve into 1998.

1996 funding for space R&D was FMk158 million, including contributions to ESA. It is projected to remain flat through 1999. The country became an Associate Member of ESA for a five-year period from 1 January 1987 (extended in 1991 to end-1994), contributing 0.81 per cent of the General Budget and expressing the intention of eventually becoming a full member. The agreement for full membership from 1 January 1995 was signed at the ESA Council Meeting on 22 March 1994 and ratified by Finland's parliament in December 1994. It began participating in the agency's science programme on a full GNP level in 1990 and in 1996 contributed AU17.5 million (FMk10.0 million; 0.68 per cent) to ESA's total budget. In 1999 it contributed €14 million, or 0.6 per cent of ESA's budget. 1988 FMk2.2 million (0.16 per cent); 1989 2.9 (0.2 per cent), 1990 5.9 (0.28 per cent), 1991 7.0 (0.28 per cent), 1992 8.2 (0.30 per cent), 1993 9.5 (0.30 per cent), 1994 11 (0.38 per cent), 1995 14.5 (0.54 per cent).

Finland's space budget (all figures in millions of Finnish marks)

Year	1995	1996	1997	1998
ESA fee[1]	21	18	12	12
ESA programmes				
Science Programme	38	34	23	23
Earth Observation	22	23	27	32
Telecommunications	12	15	18	18
Technology (GSTP)	2	10	10	10
National and Bilateral				
TEKES	32	20	30	25
Academy of Finland	15	10	12	12
Other[2]	65	58	74	69
Total expenditure	**160**	**158**	**165**	**165**

[1] The ESA membership fee is paid by the Ministry of Trade and Industry. The membership fee includes the CSG Kourou contribution paid by TEKES (1997 FMk2.1 million). From 1997 for the following three year period the scale of contribution for mandatory activities will be 1.14% (previously 1.73%). The budget figures for 1995 and 1996 are actual figures: new budget estimates are given in 1993 economic conditions. ESA's official conversion rate for 1997 is 1 AU = FMk5.76. [2] The Ministry of Communications budget allocation to the Finnish Meteorological Institute including EUMETSAT contributions and other related budgets (Technical Research Centre of Finland).

FRANCE

France is Western Europe's major space power and, with Germany and Italy, provides the principal impetus behind ESA. Its expansion into all areas of space activity now includes strong military surveillance interests, exemplified by the launch of Europe's first independent reconnaissance satellite, Helios 1A, in 1995.

France was the third nation after the USSR and the US to achieve national launcher capability and now possesses a significant telecommunications and Earth observation satellite manufacturing capacity, principally through Aerospatiale and Matra Marconi Space. As elsewhere, the French space industry is undergoing consolidation to meet the new financial climate and manufacturing over-capacity. MMS in 1994 acquired the space interests of British Aerospace. CNES (Centre National d'Etudes Spatiales) is the agency responsible for prosecuting French space activities and has been instrumental in establishing a number of companies and GIEs (Groupements d'Intérêt Economique) to commercialise and promote these activities, including Arianespace (Ariane launcher family) and Spot Image (Earth observation).

December 1995's Franco-German summit in Baden-Baden between Chancellor Helmut Kohl and President Jacques Chirac yielded an accord on military satellite programmes previously propounded by France. It opened the door for the long-planned merger of the Aerospatiale and DASA satellite divisions. Aerospatiale Chairman Louis Gallois and DASA President Manfred Bischoff signed an MoU on 18 December 1995 that produced the European Space Industries in 1996. Headquartered in Munich, ESI is the largest European space company (with subsidiaries ESI-France and ESI-Deutschland) and the world's fifth argest, with 2,600 personnel and more than US$1 billion annual sales. Originally, it was planned that France and Germany would contribute equally overall to the FFr11.5 billion Helios 2 optical/IR and FFr13 billion Horus radar satellite projects, while expressing the hope that they will become pan-European. France will consolidate its state-controlled aerospace companies, leading to extensive industry privatisation and additional European alliances.

In April 1998, Germany formally ended its participation in the Helios programme, prompting French reluctance to continue in the Horus programme. In the wake of this blow, France remains committed to the Helios.

France is responsible for the successful series of Ariane commercial launchers, providing 46 per cent of the funding for the €6.5 billion Ariane 5/5E development programmes, in return for prime contractorship. Until Russia became part of the international Space Station project and NASA began flying joint Shuttle-Mir missions, France was the leading collaborator with Russia. The country provided an extensive range of instrumentation to Earth-orbiting and deep space vehicles, including instruments for Mars-96 and the primary Sigma X/gamma-ray telescope of the Granat observatory. Annual agency-level meetings are held alternately in France/Russia to review current co-operation and consider proposals for future missions.

French AF pilot Jean-Loup Chrétien flew twice to Soviet space stations, in December 1988 performing the first non-US/USSR EVA. As part of the accord on long-term co-operative activities signed with Glavcosmos in Moscow December 1989, CNES/Licensintorg formalised an agreement for a French cosmonaut to spend 12 days aboard Mir in July-August 1992. Lt Col Michel Tognini's 'Antares' visit cost France FFr72.3 million, less than the quoted rate because of the close collaborative ties. CNES/NPO Energia signed a memorandum on 29 July 1992 for four further missions, in 1993, 1996, 1998 and 2000 to exploit Mir's facilities. Lt Col Jean-Pierre Haignere's 'Altair' visit used Soyuz-TM 17 in July 1993. Altair back-up Dr Claudie André-Deshays flew August 1996's 'Cassiopeia' mission. FFr100 million was invested in the experiments programme. FFr126 million was paid jointly for the 1993/96 missions, plus equipment which remained aboard Mir for Leopold Eyharts' planned visit in August 1997 on Soyuz-TM 27 (Pegasus mission). That mission was delayed until January 1998 by problems aboard Mir. CNES astronaut Jean-Francois Clervoy was selected in May 1992 by ESA as one of six new astronauts; he flew as a NASA Mission Specialist on STS-66 in November 1994. Dr Jean-Jacques Favier (materials science) and Michel Viso (veterinary science) were selected as CNES astronauts in 1985, followed by test pilot Philippe Perrin in 1990. Favier flew as a payload specialist aboard STS-78's Life and Microgravity Spacelab in June 1996 and Viso was due on 1998's Neurolab Shuttle. Chrétien and Tognini qualified as NASA Mission Specialists in 1995; Chrétien flew aboard the STS-86 mission in September 1997 and Tognini was scheduled to fly on the STS-93 mission in July 1999. France also pioneered the flight of µg payloads aboard China's recoverable spacecraft.

The space industry accounted for FFr18.3 billion of all military and civil aerospace sales of FFr100.6 billion in 1995 (17.5/105.5 1994; 16.7/110 1993). Comparison with 1989's 11.4/112.3 highlights the increasing importance of space. 12,000 are employed full time on space activities, including 8,700 in industry. 1996's military space budget was proposed at

CNES Budget Authority (FFr millions)

	ESA	Total
1984	1,793.0	4,610.1
1985	1,740.2	4,858.3
1986	2,209.8	5,686.9
1987	2,436.2	6,076.1
1988	2,657.9	6,660.5
1989	3,243.0	8,323.7
1990	3,614.1	9,528.1
1991	4,151.2	10,054.3
1992	4,725.0	10,623.1
1993	4,947.0	11,016.0
1994	4,860.0	11,774.7
1995	4,776.3	11,782.5
1996*	5,190.0	12,182.0

* provisional. Note: the contribution to ESA grew by 55% in real terms 1987-92. It is capped at the current real level until 2000 in order to maintain national activities.

FFr4.6 billion; 1995's was FFr4.1 billion, 1994 4.1, 1993 3.9. While 1992's overall military budget remained static at about FFr103 billion, its space element rose from 1991's FFr3 billion (itself a 50 per cent increase over 1989) to FFr3.5 billion for 1992; 1995's was then projected at FFr5.5 billion and 2000's at FFr8 billion. It was announced 26 February 1993 that CNES had responsibility for design and development of military space projects, under the Ministry of Defence.

GABON

The Republic of Gabon is served by the 'Equasat' C-band domestic satellite communications network provided on a turnkey basis by Scientific-Atlanta of the US. A 72 MHz hemispheric transponder was purchased aboard Intelsat's 1° west satellite in October 1986. Elements of this initial network include an 18 m International IDR Intermediate Data Rate gateway station in Franceville, a 15 m hub station in Libreville, eight 11 m stations, two 7 m stations and a single 7 m transportable. These stations provide >400 telephony channels and two TV channels. The IDR link from Franceville to Paris provides additional international connectivity. Completed in 1995, the system covers 36 stations: nine for telephony and TV distribution, two for TV receive and telephony, and 25 TVRO. These support two 17.5 MHz TV, two FM radio and about 370 telephone channels. The country is a signatory to Intelsat and Inmarsat through its TIG Télécommunications Internationales Gabonaises.

GERMANY

The federal government has been supporting space research and technology since 1962. In February 1976, the government decided that there would be no further national or even bilateral research satellite projects such as Symphonie. Consequently, two-thirds of the 1976-79 space budget of DM2.2 billion would be allocated to ESA. However, activities such as the Franco-German TDF/TV-Sat direct broadcast satellite programme were subsequently undertaken. While France focused on the Ariane launcher, West Germany became the largest contributor (53.3 per cent) to Spacelab, developed by prime contractor ERNO and companies from 10 European countries for the US Space Shuttle. In addition, DM40 million was expended over the eight years to 1984 on contributions by Dornier and MBB to Ariane. 1983's budget totalled DM770 million.

The DARA German Space Agency, created in 1989, is in full charge of implementing the space programme, except for the hypersonic technology programme, which remains embedded within the BMFT Ministry for Research & Technology's aeronautics activities. Federal expenditure over 1985-96 totalled DM18.9 billion. 1999 federal funding for ESA space activities amounts to €570.1 million. The federal states, universities, industry and the German Aerospace Establishment (DLR) made minor additional contributions. Germany is the largest contributor to ESA's mandatory programme (25 per cent in 1999) and second only to France in the total budget, including the optional activities. National budgetary restrictions arising from the worldwide recession and unification mean that Germany has reduced its ESA payments and cut back on national activities.

At the end of the Cold War, West Germany had to fold a modest but vigorous East German space programme into its own. The space involvement of the former East Germany was restricted to co-operation with the USSR and other eastern European countries through its membership of Interkosmos, established in 1967. Beginning with a UV photometer aboard Interkosmos 1 in 1969, about 175 items of scientific equipment from institutes and industry flew on more than 40 Soviet spacecraft. The MKF-6 multispectral film camera, first flown aboard Soyuz 22 in 1976, was a permanent fixture on the Salyut 6/7 space stations and was attached to Mir in December 1989 as part of Kvant 2. The camera was operated aboard Salyut 6 in August 1978 by East German guest cosmonaut Sigmund Jähn, launched in Soyuz 31. His flight covered more than 20 experiments in remote sensing, materials science and biomedical research. IR Fourier spectrometers performed thermal mapping of Venus from Venera 15/16 and similar instruments flew on three Meteor Earth weather satellites. Image processing and magnetometers were contributed to Vega and Phobos and VNIR spectrometers to two Interkosmos satellites plus Salyut 7 and Mir for atmospheric remote sensing. Plasma probes were developed for several Interkosmos missions in addition to geophysical and meteorological rockets.

Federal Ministries' expenditure (DM millions)

	ESA	Nat'l	BMV	BMPT	Total
1985	414.1	416.4	36.5	545	1,412.0
1986	559.2	358.3	43.9	480	1,441.4
1987	639.6	418.8	35.0	385	1,478.4
1988	644.1	462.4	31.5	323	1,461.0
1989	712.9	506.5	17.4	570	1,806.8
1990	839.0	576.5	25.0	404	1,844.5
1991	964.0	580.0	30.0	-	1,574.0
1992	1,133.0	604.0	30.0	-	1,767.0
1993	1,202.0	580.0	30.0	-	1,812.0
1994	1,092.0	478.0	30.0	-	1,600.0
1995	1,072.0	300.0	30.0	-	1,402.0
1996	1,003.0	300.0	-	-	1,303.0

BMFT = Bundesministerium für Forschung und Technologie (Federal Ministry for Research & Technology); its figures exclude the German National Hypersonic Technology Programme. BMV = Bundesministerium für Verkehr (Federal Ministry for Transportation); mainly contributions to EUMETSAT. BMPT = Bundesministerium für Post und Telekommunikation (Federal Ministry for Postal Services & Telecommunications); expenditures for satellite broadcasting

East German participation was principally through the Institut für Kosmosforschung (IKF, Space Research Institute), which specialised in spectrometric remote sensing, opto-electronic systems, extraterrestrial physics and the development and testing of space research equipment. It was absorbed by DLR in 1991, which continues the international projects. At that time, a programme of planned IKF international co-operation existed. It included Mir Priroda module (MOS Multispectral Opto-electronic Scanner), Mars-96 (WAOSS Wide-Angle Opto-electronic Stereo Scanner, Orbiter Magnetometer, IR spectrometer, image processing; balloon magnetometer), Regatta (magnetospheric plasma diagnostics), Sodart (Spektrum-X/v high resolution X-ray spectroscopy), APEX Active Plasma Experiment mission (launch preparation, mission assistance, telemetry acquisition), Cassini (cosmic dust analyser; data compression), Okean (data acquisition from Okean satellites), Coronas solar observations (main station for telemetry data acquisition) and Beacon (satellite beacon network).

Germany, in 1994, made its first strong declaration of commitment to military space activities. December 1995's Franco-German summit in Baden-Baden between Chancellor Helmut Kohl and President Jacques Chirac yielded an accord on military satellite programmes previously propounded by France. It opened the door for the long-planned merger of the Aerospatiale and DASA satellite divisions. Aerospatiale Chairman Louis Gallois and DASA President Manfred Bischoff signed an MoU 18 December 1995 that formed the European Space Industries in 1996.

Germany shocked its French partners in April 1998 when it withdrew from the Helios spy satellite programme. France had the role of prime contractor for the Helios and Germany the Horus as part of their joint military space programme. France expressed reservations about continuing in the Horus programme, but Germany has reasserted its commitment to the project.

German involvement in the Helios 2 programme appeared to be in doubt at the beginning of December 1996. Since the satellite was not capable of seeing through cloud cover, Germany saw it as being of limited use and with German budget limitations, funding for the programme appeared to be uncertain.

Two major ESA centres are located in Germany. The ESOC European Space Operations Centre for TT&C of satellites during launch and orbital operations is housed at Darmstadt, heading a network of seven stations worldwide. The European Astronauts Centre was established in 1990 in Cologne as the home base of ESA's astronauts. EAC hosts the nearby CTC Crew Training Complex, where the bulk of Columbus training will be done. Germany's commitment to manned flight is reflected by the GSOC German Space Operations Centre satellite and manned mission control facility at Oberpfaffenhofen, providing the first manned capability outside of the US/Russia and already experienced with the Spacelab D1/D2 and EuroMir 95 missions. It was recently expanded at a cost of DM533 million (national/ESA funding) to cope with European space station operations. EUMETSAT headquarters are sited in Darmstadt and, until December 1995, Meteosat activities were controlled through ESOC's Odenwald station; EUMETSAT then assumed control via its new Fucino station in Italy.

Manned and microgravity programmes have been emphasised since the early 1970s. It has extensive manufacturing and manned flight experience with the Spacelab vehicle and the Spacelab D1/D2 missions, devoted primarily to life sciences and other μg activities. COF's ECU658.5 million fixed price construction contract was signed by ESA and DASA 28 March 1996. Test pilot Maj Klaus-Dietrich Flade flew March 1992's eight-day Soyuz-TM 14 mission to Mir, with physicist Reinhold Ewald as back-up, conducting 14 experiments emphasising biomedical research. DARA provided programme management, while DLR was responsible for the astronauts. A DARA/RKA agreement was signed 9 December 1995 and Ewald flew to Mir aboard TM25 in February 1997. National payload specialists flew the Spacelab D1 (Reinhard Furrer and Ernst Messerschmid. Furrer died on 9 September 1995 in a plane crash) and D2 (Ulrich Walter and Hans William Schlegel) missions. AF test pilot Thomas Reiter was selected in May 1992 by ESA as one of six new astronauts, flying ESA's EuroMir-95 Soyuz-TM 22 mission September 1995 – February 1996. Dr Ulf Merbold has been an ESA astronaut since 1977, flying Spacelab missions 1983/92 and Soyuz-TM 20 to Mir in October 1994 for EuroMir-94.

The former DASA provided ESA's Eureca free-flyer and operates the Texus, Mini-Texus and Maxus sounding rocket μg programmes. The recoverable SPAS platform, flown five times on Shuttle by end-1995, is available for μg payloads but it is now principally the Astro-SPAS astronomical observatory. The Express project in January 1995 launched Germany's reusable capsule on Japan's M-3SII but it was delivered into an untenable orbit; the cost renders future missions unlikely. Instead, Germany continues to lead Europe's utilisation of Russia's Photon μg craft. DASA Jena-Optronik and Kayser-Threde have developed the piggyback MIRCA Micro Re-entry Capsule which was due to debut on Photon in late 1996 for re-entry investigations.

In satellite construction, MBB was one of five Eurosatellite partners that produced the TV-Sat, TDF and Tele-X direct broadcast satellites and developed the national DFS-Kopernikus craft with Dornier. Dornier was also prime for ESA's Cluster satellites (Germany provided two instruments), following its lead role for the ERS 1/2 microwave remote sensing satellites, the Ulysses solar probe and the international Rosat X-ray observatory.

After termination of the former East Germany's membership of Interkosmos, Germany included those projects in its co-operative programme with Russia. Including Mars-96 (involvement in 14 experiments), the programme covers about 40 space science and 10 Earth observation projects. In 1996, DASA split into two new organisations: Dornier Satellitensysteme GmbH and Daimler-Benz Aerospace-Space Infrastructure Division.

To meet space requirements, the federal cabinet decided in April 1989 to create a new structure, broken into three organisations. These are:

A Cabinet Committee on Space Activities, chaired by the Federal Chancellor and with the Federal Minister for Research & Technology taking the lead rôle; A State Secretaries' Committee on Space, chaired by the Secretary of State in the Federal Ministry for Research & Technology; DARA GmbH, the German Space Agency.

The Cabinet Committee is responsible for advising the Federal government on space policy decisions, in addition to fundamental strategic and planning aspects and on the allocation of the space budget. Funds are channelled to DARA through the budgets of the responsible ministries. The committee comprises: Federal Chancellor (chairman), head of the Federal Chancellery, Foreign Minister, Minister of Finance, Minister of Economics, Defence Minister, Minister of Transport, Minister for Research & Technology and Minister of Posts & Telecommunications.

The State Secretaries' Committee on Space prepares the decisions to be taken by the Cabinet Committee, co-ordinates the relevant Federal ministries and defines goals and commissions for DARA. Its members are the Secretaries of State of the ministries represented on the Cabinet Committee. DARA is the central management organisation for German space activities, headed by a Director General and two Managing Directors. Since June 1990 it has been legally responsible for drafting space planning and issuing the programmes, managing them and awarding contracts to science and industry. DARA also represents German interests at international level, most notably ESA.

The increasingly ambitious nature of future projects, with a return to national satellites and the selection in December 1982 of two national astronauts, reflected a growing commitment to space. The Spacelab D1 mission, costing DM565 million, was devoted to materials/life sciences; Spacelab D2 in 1993 cost DM890 million. Microgravity research in the Texus sounding rocket programme began in 1978. DARA, in 1994, withdrew from funding Maxus, preferring the cheaper Texus instead. Launched science missions include ESA's Exosat (MBB was prime), NASA's Galileo mission to Jupiter (MBB provided the propulsion module), AMPTE, Hipparcos, Hubble, ERS and Rosat (Dornier prime). Earth observation instruments include the SCIAMACHY spectrometer for ESA's Envisat, the Millimetre Atmospheric Sounder flying on Atlas Shuttle missions and the major XSAR X-band synthetic aperture radar as part of Shuttle's Space Radar Lab.

GREECE

The first national involvement in space activities was initiated on 27 September 1989 by the signing of an MoU in Athens between Moscow's Institute of Space Research (IKI) and the Crete Research Centre covering participation in the Mars-96 project. The government and ESA signed a co-operative agreement in July 1994. Greece is a member of Intelsat, Inmarsat and Eutelsat through the signatory Hellenic Telecommunications Organisation (OTE). An Inmarsat A/C/B/M station covering the Indian and East Atlantic Oceans is operated at Thermopylae. ESA is providing consultancy to the Ministry of Transport and Communications on the proposed Hellas-Sat national telecom satellite system.

HUNGARY

The Hungarian Space Office has managed all national space activities under the Hungarian Space Board since January 1992. Membership of ESA is a possibility for the future; formal meetings were held with the European Space Agency in July 1990. Hungary is a member of Intersputnik (single Earth station operating via the 14° west Gorizont) and joined Eutelsat in 1993 and Intelsat in January 1994 (two 18 m Intelsat A stations). Israel Aircraft Industries plans a second Amos satellite by 1999, owned by Magyarsat Ltd in Hungary, a 50/50 joint venture by IAI and Antenna Hungaria, the country's public broadcasting monopoly and the second largest telecommunications service provider. For this CERES Central European Communications Satellite to proceed, it required a commitment from the Hungarian government to guarantee two transponders, which was agreed 30 April 1995. The US$150 million financing was expected to be acquired in 1996, although Antenna Hungaria's privatisation could affect the project.

As part of the Interkosmos organisation, Hungary has contributed to several Russian-led science missions. Hungary provided elements of Vega's TV system and the controlling electronics for the Phobos long-lived lander, in addition to involvement in Aktivny, Apex and Interball. Lt Col Bertalan Farkas of the Hungarian Air Force flew to the Salyut 6 space station aboard Soyuz 39 in May 1980 under Interkosmos; radiation dosimeters carried on the mission were subsequently adopted for Salyut 7 and Mir space station monitoring. A space co-operation agreement was signed with India on 27 October 1995, emphasising Earth observation and space physics.

INDIA

India, with more than three-quarters of its 850 million population dependent on agriculture, concentrates its space development activities on applications satellites and autonomy in launch systems. The two prime aims involve operational space-based remote sensing and communications systems: meteorological and Earth resources images are routinely returned from space and 701 TV and 182 radio stations are linked into the space network. The US-built Insat 1 geostationary communications and meteorological satellite family was superseded by indigenous Insat 2s in 1992 and the new IRS 1C, launched in December 1995, is the world's most advanced civil remote sensing satellite.

India was the seventh nation to achieve orbit capability, in July 1980. By late 1985, Insat 1B, which was launched in August 1983, had extended TV coverage from 20 to 70 per cent of the population, with the goal of 90 per cent now satisfied. Following the successful experiments with educational TV from NASA's ATS 6 in 1975, more than 35,000 of India's 520,000 villages have been equipped to receive Insat services. Professor Rao, who became chairman of ISRO in October 1984, commented at the time that space technology had given India the opportunity to convert backwardness into an asset: developing countries could bypass the intermediate technology stage and leapfrog into a high technology era.

Like France, India has benefited from simultaneous co-operation with the CIS/USSR, ESA and US. After the loss of Insat 1A, initial problems with Insat 1B were overcome, and this Ford-built series should have been completed with the Shuttle launch of Insat 1C, accompanied by an ISRO scientist or engineer, in mid-1986. Thus, India would have joined France as the only country to have had specialists aboard both USSR and US spacecraft, but 1986's Shuttle accident required a switch to Ariane. Maj Rakesh Sharma of the Indian Air Force, after two years training, took part in the successful seven-day Soyuz-T11/Salyut 7 mission in April 1984. ISRO and the Russian Space Agency signed a general agreement continuing space co-operation on 23 December 1994.

Two commercial agreements reached in 1995 demonstrate that India has advanced to the front rank of the world's space nations. While development of the US Landsat remote sensing series has stumbled along, EOSAT plans to market imagery and data from the 14 satellites that India expects to fly by 2005. Over 1983-2000 there will have been 12 satellites launched towards polar orbit. They are not only placing India in a strong position to satisfy its own applications needs, but also to dominate a significant portion of the global commercial market. The IRS series is now also moving from land applications to environmental monitoring, with an emphasis on oceanography.

The second notable development, India's agreement with Intelsat meant the European consortium will lease 11 transponders on Insat 2E for 10 years beginning in early 1998 at a cost of around US$100 million. It was noted in February 1996 that the system already generates revenues of Rs4 billion annually.

Launcher and propulsion programmes represent half of space expenditure, which runs at some US$300 million. The PSLV Polar Satellite Launch Vehicle, fully successful on its second attempt in October 1994, provides a capability of placing one tonne-class IRS satellites into Sun-synchronous orbits and is now offered commercially through Antrix Corp. PSLV offers only 450 kg GTO capability but an upgrading into the Geostationary SLV is under way to satisfy 2.5 tonne-class GTO requirements by 2000. Notably, GSLV incorporates a cryogenic KVD-1 engine acquired from Russia. The same engine will fly on Proton. India's first attempt at firing a 10 kN LOX/LH_2 engine ended with an explosion at ignition in July 1993. LH_2 became available in the country only after a US company built a plant at end-1992. The first seven GSLVs will use the Russian engine to allow time for India's own to be developed by 2003-2005. GSLV will also create a planetary capability: Mars, Mercury and Venus missions are under consideration.

1988 saw the successful orbiting of IRS 1A under a semi-commercial agreement with Glavcosmos – the first of its kind – and an agreement was reached in November 1988 for IRS 1B to be carried by another Vostok in August 1991 at a cost of about Rs220 million. The agreement of January 1991 added IRS 1C in December 1995 for Rs500 million. Insat 1D departed 12 June 1990 aboard a Delta 2.

HECL Hughes Escorts Communications Ltd, a subsidiary of Hughes Network Systems of the US, began providing India's first private satellite shared hub service in January 1995 using Insat 2 leased capacity. VSAT-based wide area networks in India are expected to grow rapidly. Afro-Asian Satellite Communications Ltd (ASC), part of India's Essel Group, signed a contract with Hughes Space & Communications International in January 1995 for two satellites to provide GSM-compatible mobile telephone links from 1998, initially covering an area bounded by India, Middle East, Singapore and Moscow. The value was not disclosed but system cost has been estimated at US$700 million. First launch is planned for 2001.

ISRO and Malaysia's Maxstar consortium signed an agreement in January 1995 for development of two 50 kg satellites for launches in 1997 and 2000 as part of Malaysia's space capability development programme.

India's budget summary (Rs million)

	85-86	86-87	87-88	88-89	89-90	90-91	91-92	92-93	93-94	*94-95	*95-96
Rocket development	1192.3	1319.0	1470.1	1604.4	1474.1	1812.9	2057.9	2680.4	2967.2	3678.2	3868.2
Satellite development	380.1	611.8	880.9	1387.4	1389.2	1180.0	1432.4	645.2	326.8	358.5	327.5
Insat/IRS ops**	357.1	722.8	559.8	622.4	481.7	216.0	330.8	819.1	2481.9	2701.7	3925.7
Space applications	242.1	308.4	406.0	438.8	450.7	444.5	493.2	508.8	658.6	649.3	610.6
Space sciences	49.5	50.6	67.1	76.7	84.8	94.7	140.4	132.7	136.2	146.2	141.9
Administration	69.8	87.3	85.7	93.7	105.1	114.2	116.2	122.9	382.7	216.1	293.7
Totals	**2291.0**	**3099.9**	**3469.6**	**4223.4**	**3985.6**	**3862.2**	**4301.0**	**4909.2**	**6953.4**	**7750.0**	**9167.6**

Includes National Remote Sensing Agency and Physical Research Lab expenditure.
*1995-96 figures are estimates; 1994-95 revised.
**IRS included from 1992-93.

INDONESIA

Indonesia is undertaking a modest programme of remote sensing and atmospheric probing, but its emphasis is on satellite-based communications linking the country's 6,064 inhabited islands, with their 182 million population, spanning some 5,000 km along the equator. The first launch in July 1976 resulted in the system, operated by Perumtel (now PT Telkom), being declared operational the following month. The second generation was introduced in June 1983 by Palapa B1 and the Palapa C third generation appeared in 1996 to expand services to China, India, Japan and Pakistan, for example. CI launched 1 February on a US Atlas-2AS and C2 16 May on an Ariane-44L. The private PT Satelindo was set up in 1993 to finance and operate the Palapa C system; it is also expected to take over the Palapa Bs, leaving PT Telkom with the terrestrial tele-communications network. The private Indostar was due to launch its first DBS satellite in first quarter 1997 to provide up to 40 TV channels; CD-quality will be added to Indostar 2. National TV cannot currently be received by 32 per cent (57 million) and radio by 20 per cent (36 million). Indonesia is leading the US$650 million Garuda/ACeS Asia Cellular Satellite System project to provide personal mobile services throughout Asia.

The LAPAN space agency maintains ground stations for reception of Earth resources and meteorological satellite imagery. Indonesia was the 11th country to build a Landsat station, in 1981. A Cospas-Sarsat satellite-based search/rescue system Local User Terminal (from Canada's CAL Corp) was installed in Ambon and one with an Operations Control Console in Jakarta in 1991. In conjunction with the Netherlands' NIVR, LAPAN studied the TERS Tropical Earth Resources Satellite to provide equatorial multispectral coverage from a 1,680 km, 10° inclination orbit, employing a cloud sensor to direct the CCD-based imager towards clear regions. An equatorial launch site for handling China's CZ-3 was considered in 1987, possibly located on Biak Island.

INTERNATIONAL

European Space Agency

Eleven states created ESA in 1975 to replace the European Launcher Development Organisation (ELDO) and European Space Research Organisation (ESRO). The states were Belgium, Denmark, France, West Germany, Ireland, Italy, Netherlands, Spain, Sweden, Switzerland and the UK. Austria and Norway, with Finland as an associate member (full member from January 1995) and Canada as a co-operating state subsequently joined the charter members. Member states contribute to the mandatory general and science budgets (€2,553.2

million approved for 1996-2000 (1995 rates)) on a gross national product basis, and for the remainder receive contracts in proportion to their funding.

ESA's future was mapped out in November 1987 when the ministers responsible for space matters in its 13 member states met in The Hague, Netherlands and approved programmes into the next century. Total budget for this 1988-2000 long-term plan was estimated in 1987 at AU32 million, but the member states suggested a pruning of 15 to 20 per cent at Germany's instigation. Essentially all programmes were approved, with the Columbus space station element and the Hermes manned spaceplane being phased so that after a Phase 1 of three years the member states could review their status.

At the ministerial meeting in The Hague in November 1987 the Hermes spaceplane was approved along with Columbus and Ariane-5. Hermes was the most technically-demanding of the three and a two-phase approach was adopted: Phase 1 began in January 1988 and it was to have ended in December 1989 but was delayed until the end of 1992. At the end of Phase 1 ESA's planning was revised and called for Hermes' maiden flight in 2002, followed by its manned debut in 2003 and its first operational servicing mission in 2004.

Programme cost was estimated at AU7,320 million (1990 rates; AU6,222 million 1986 rates), a 40.5 per cent increase over the AU4,429 million (1986 rates) envelope established at The Hague ministerial meeting in 1987. 17.5 per cent of the increase was credited to technical changes and 23 per cent to the four-year programme stretch-out. AU1,167 million had been expended to end-1992 (1991 rate). However, ESA's need to reduce funding commitments produced November 1992's Granada decision to cut Hermes back to a technology programme, principally on joint ESA/Russian spaceplane studies. The funded 1993-95 reorientation period began studying three strategic options: co-operation with Russia, co-operation with the US and an autonomous European scenario. These covered system studies primarily directed towards definition of an ESA/Russian vehicle and development of critical technologies based on the Hermes definition. A detailed definition study of an Assured Crew Return Vehicle for Space Station, detailed definition studies and pre-development of an Automated Transfer Vehicle (ATV) also entered the mix of options.

These were reworked as MSTP (Manned Space Transportation Programme) and approved by ESA's Council in February 1994. Development of a joint spaceplane with Russia was no longer an option. Full development required approval by the ministerial meeting in October 1995 when ATV was fully approved and CTV began a three-year study. ATV could perform logistics and propellant resupply to ISS beginning February 2003. It will also perform the re-boost function, to raise the orbit of ISS, which is continuously decreased by the residual atmospheric drag. The October 1995 ministerial meeting approved €667 million (1995 rate) for the ATV for 1996-2004. The November 1991 Munich ministerial meeting reconsidered the plan, accepting a revised 1992-2005 plan as a strategic framework for ESA's activities. Cost of the 14-year plan was projected at AU39.1 billion (US$45.0 billion) at 1990 rates. The expected Phase 2 approval for Hermes (launch delayed to 2002), Columbus (its free-flyer element's launch delayed to 2003) and DRS was deferred a year, to a third ministerial meeting in Spain. The Munich meeting did approve POEM 1 start (to be reviewed in Spain) and a 5 per cent reduction in 1992's budget. 1992's Granada meeting approved the 1993-2000 long-term programme, costing AU22.7 billion at 1991 rates (reduced from Munich's planned AU25.8 billion for the same period).

DRS was approved for 1999 but its role had to be redefined now that major user programmes had gone. The Columbus Orbital Facility (COF) module was approved, but with a 5 per cent cost reduction and 5 per cent financial under-subscription. The Columbus Free-Flyer module was cancelled. However, continuing financial pressures and the major reorientation of the international space station again required significant changes in February 1994 of ESA's long-term plan.

Hermes was reduced to a technology programme, principally based on joint ESA/Russian spaceplane studies which showed a questionable utility of space planes, plus studies for a Space Station rescue vehicle. Hermes became the Manned Space Transportation Programme (MSTP) comprising the Crew Transfer Vehicle (CTV), Automated Transfer Vehicle (ATV), External Robotic Arm (ERA) and EVA 2000 EVA suit. MSTP/Columbus since then were jointly managed under the Manned Spaceflight and Microgravity Directorate.

These programmes were funded only for 1994-95 (ERA was fully approved September 1994); the objective being to make a single programme out of the two proposals and then submit the single programme in 1995 at the ministerial meeting in October 1995. The resulting merged programme was called the European Participation in the International Space Station, wherein COF and ATV were fully approved and CTV began a three-year study phase. COF has been significantly reduced until it is about the size of Spacelab. The Toulouse meeting approved €2,651.2 million (1995 rate) for the whole programme in 1996-2004, with Germany contributing 41.0 per cent, France 27.6 per cent, Italy 18.9 per cent, Belgium 3.0 per cent, Switzerland 2.5 per cent, Spain 2.0 per cent, Denmark 1.17 per cent, Netherlands 0.94 per cent and Norway 0.45 per cent. ESA will have access to 5.3 per cent of ISS resources in exchange for ATV services including re-boost.

ESA permanent staff totals 1,820 (as of 1 January 1997), with 349 based at the Paris headquarters (including Washington, Brussels, Moscow, Kourou and Toulouse). Three centres provide support. The ESTEC European Space Research & Technology Centre in Netherlands designs, develops and tests spacecraft and payloads (1,078 personnel). The ESOC European Space Operations Centre, staff 269, in Darmstadt, Germany is responsible for spacecraft operations. And the ESRIN European Space Research Institute at Frascati, Italy with a staff of 147 houses ESA's Information Retrieval Service and the system for collecting, pre-processing and distributing data from remote sensing satellites. The European Astronauts Centre, with 15 personnel, in Cologne is a centre of expertise for ESA's manned space activities and acts as the home base for all ESA astronauts.

The ESA programme plan provides the following status:

Science

Annual funding will remain at a constant €347 million from 1996 to 2000, yielding a reduction in real terms of 15 per cent. This requires some of the later missions to be delayed and some to be placed on indefinite hold. The Science Programme Committee in February 1996 ordered a 10 per cent reduction in the budgets of new missions.

Horizon 2000 comprises four major Cornerstone missions interspersed with smaller projects into the next century.

The loss of the Cluster payloads on the Ariane 501 mission meant ESA needed supplemental funding for a replacement mission launched in 2000. The X-ray astronomy cornerstone is the X-ray Multi-Mirror mission for 1999. The primordial body's cornerstone (Rosetta) was selected in 1993 to fly in 2003, leaving the sub-millimeter astronomy cornerstone (FIRST Far Infra-red/Sub-mm Space Telescope) as the fourth, in 2007. Horizon 2000 also includes medium and small space science projects selected as funding allows. The Giotto Comet Halley fly-by probe of 1986 intercepted a second comet in July 1992 and is now dormant.

ESA shut down the International Ultra-violet Explorer (IUE) in 1996 to save money. The Hipparcos astrometry satellite was orbited in 1989 (mission completed 1993), followed in 1990 by the Hubble Space Telescope (15 per cent contribution to NASA project, covering solar array and Faint Object Camera) and the Ulysses solar polar probe. The Infra-red Space Observatory departed in November 1995, the Cassini Saturn orbiter collaboration with NASA for which ESA is providing the Huygens Titan atmospheric entry probe, the Integral (International Gamma-Ray Laboratory) in 2001 and the M3/M4 medium missions for 2004/2007. Budget cuts mean that M3 is deferred a year while its cost is studied and M4 is shelved for the moment.

Preparatory work began in 1993 on the follow-up Horizon 2000 Plus programme to cover new missions 2007-2016. This new science long-term plan was endorsed by October 1995's ministerial meeting, but full funding is not required to begin until 2000. The call for mission concepts was made 29 June 1993 and 110 were received by the 15 October 1993 deadline. the special Survey Committee presented its report October 1994. With funding remaining at 1995's level, two Cornerstone missions (maximum ECU625 million; 1993 rate) will be complemented by four medium (ECU345 million) plus small missions and Space Station (possibly a high-energy astrophysics instrument). One Cornerstone will establish a Mercury orbiter for an intensive geophysical and magnetospheric survey. The second will be an interferometric observatory, either astrometric or IR imaging. The astrometric would improve on Hipparcos by a factor of 100, to 10 μarcsec, to ascertain the distances, motions and luminosity of tens of millions of Milky Way stars, and the mass distribution of nearby galaxies. The IR would search for Jupiter-like planets and brown dwarf companions of other stars.

ESA participation in Hubble should continue, currently planned through 2002. Annual funding increases of 5 per cent for 2001-2005 are required (but look unlikely in the new financial climate) for a third Cornerstone and augmented technology development. The gravitational waves Cornerstone (the LISA Large Interferometer Space Antenna was studied for M3 but proved far too expensive) in 2017 would explore the very early phases of the Universe and observe massive black holes and their coalescence. Technology studies could lead to X-ray and IR observatories beyond 2017.

Earth observation

Meteosat operations continue with development/procurement of the MSG second generation under way. ESA has budget authority for ERS 1/2 microwave all-weather remote sensing satellites, the first Polar Platform, Envisat 1, remains under development and the second PPF, Metop 1, was approved in 1996 in reduced form.

Microgravity

ESA facilities aboard Columbus, Eureca (stored), Spacelab, Mir, Photon, Bion and sounding rockets such as Texus and Maxus.

ESA's microgravity programme had performed 343 experiments through August 1995, including 183 on manned missions. The first phase of the materials science, fluid physics and life sciences μg research programme ran from January 1982 to 1985 at a cost of AU44 million (1993 rates, AU28 million 1983 rate) and comprised two principal elements: Spacelab experimentation and a sounding rocket programme. Three multi-user facilities were provided for Spacelab: the Biorack unit for investigating bacteria, cells, insects and plants (flown Spacelab D1), the Fluid Physics Module to study hydrodynamic phenomena of floating liquid zones (flown Spacelab 1/D1) and the Sled facility to investigate the human's vestibular system (flown D1). Eureca was added to ESA's overall programme in 1983 as a self-contained programme element and contributed significantly to μg research (see the separate entry above).

The current Phase 2 (known as EMIR 1 since 1993) began in 1985 to accomplish several goals. First, it was meant to maximise use of Spacelab and its multi-user facilities. Secondly, it needed to develop and fly the new Eureca payload facilities and exploit facilities developed nationally for the Spacelab 1/D1 and for the sounding rocket programme. Finally, it had to design, develop and launch new multi-user facilities for all μg disciplines; and provide flight opportunities for existing or new individual experiments. It therefore re-flew Biorack (Spacelab IML-1/2) and developed three fluid physics facilities (Advanced Fluid Physics Module – flown on Spacelab D2, The Bubble, Drop and Particle Unit – flown IML-2/LMS and the Critical Point Facility – flown IML-1/2). The Anthrorack human physiology module (flown Spacelab D2), the Advanced Protein Crystallisation Facility (flown Spacehab 1/IML-2/LMS, STS-72) and the materials science Advanced Gradient Heating Facility (flown LMS Spacelab); further flight was expected on 1998's Neurolab Spacelab. Total Phase 2 cost envelope is AU450.6 million at 1983 rates.

Apart from the short duration flight opportunities on sounding rockets, all other elements required flights of either Spacelab or Eureca. However, Challenger's loss in January 1986 halted all Shuttle flights for more than two years. As a result, Phase 2, conceived as a four-year programme, was stretched into 1997 to cover the delayed Shuttle activities. This extension was performed in two steps: the first was approved in 1988 and the second in 1991. Besides increasing the short duration opportunities, in order to maintain programme momentum, using drop towers, aircraft flights, Get Away Specials and sounding rockets, it emphasised two further broad objectives. The first further utilises Spacelab. Secondly, it sought to develop other research outlets independent of Shuttle/Spacelab: small payloads are flying on other retrievable carriers (such as Biobox and Biopan on Russian capsules) and the EuroMir missions. For example, the Fluid Physics Facility was planned for Photon in 1998 and Biorack was due to fly three times on Shuttle/Mir (STS-76, March 1996; STS-81 December 1996; STS-84, May 1997). In May 1993, the Microgravity Programme Board approved re-orientation of the extended Phase 2 activities towards basic research under the EMIR 1 European Microgravity Research Programme. EMIR 2 is essentially a utilisation period covering 1996-2001, costing AU146 million (1995 rates) and approved July 1996.

Phase A studies for Columbus payloads began 1988/89: Biolab, Fluid Science Laboratory, High Temperature Materials Processing Laboratory and an Anthrolab. Protein and vapour crystal growth units, a Combustion Facility and a Biotechnology Facility were also earmarked for further study. It is now planned to develop/build the multi-user facilities for Space Station under a programme separate from EMIR starting 1997: MFC Microgravity Facilities for Columbus. €206.7 million (1995 rate) was approved October 1995 for 1997-2003. Planned are Biolab, Fluid Science Laboratory, Physiology Modules, New Physiology Facility, Low Gradient Furnace, and Solidification and Quench Furnace. Low Gradient Furnace will fly aboard ISS's US Laboratory module; the others on ESA's COF.

Communications

ESA plans to maintain one Marecs as Inmarsat back-up plus one for experiments, keep three ECS satellites in service with Eutelsat and press ahead with Artemis to demonstrate future data relay, land mobile and other innovative communications technologies.

Navigation
The GNSS Geostationary Navigation Satellite Service will be provided through Inmarsat 3's navigation transponders. Preparatory studies for the GNSS2 second-generation system (2005) are under way. ESA/EU/Eurocontrol are co-operating on the €150 million 1995-2000 programme.

Space transportation
ESA and CNES, with Aerospatiale as industrial architect, were responsible for developing the Ariane 1-4 range of orbital launchers. The vehicles are now procured, marketed and launched by Arianespace. The six Ariane 4 versions will be phased out in 2000.

Ariane 4 became Europe's primary launcher in 1989. ESA is developing Ariane 5 under a €6,572 million (1995 rate, total including preparatory programme) programme to expand capacity. A further €1,713.4 million was approved in October 1995, principally to develop the more capable 5E version. The Kourou base in French Guiana will continue as the launch site. ESA began FESTIP studies in 1994 of future reusable systems.

Manned space transportation
ESA's Manned Space Transportation Programme is developing support systems for Space Station, including the ATV Automated Transfer Vehicle and ERA External Robotic Arm.

The CTV is intended for ISS support and possibly crew rescue. Delivered by an Ariane-5/ATV, the 10 tonne ballistic vehicle could carry a four person crew plus 400 kg in the 51.6°, 400 km orbit. The first manned launch is due around 2002 and the cost is more than €2.5 billion. The October 1995 ministerial meeting approved €52.3 million (1995 rate) for the 1996-1998 definition study, with full approval to be discussed at a further ministerial meeting in late 1997. The 2,665 kg, 2.8 m diameter, 2.04 m high Atmospheric Re-entry Demonstrator (ARD) will fly on Ariane 503, splashing down close to 5° north, 148° west in the Pacific after one circuit and an apogee of 875 km. It is ESA's first re-entry craft, a 0.7-scale Apollo capsule (L/D = 0.3) with 200 sensors to monitor aerodynamic, aerothermodyanamic, thermal protection and blackout conditions. A Sextant Avionique NSS100-P GPS receiver is included to improve landing accuracy and study signal penetration during blackout (as will a TDRS S-band link). Protected by Aerospatiale's Aleastrasyl heatshield and Norcoat 622-50FI on the aft body. The electrical and control systems (including 7 × 400 N hydrazine thrusters) were derived from Ariane 5. Poser comes from two 28 V Spot 4 Ni/Cd batteries plus one pyrotechnics battery. Two 1 Gbit SSRs. The €30 million contract was signed with Aerospatiale on 30 September 1994; some 25 per cent of the contract value went to Belgium (SABCA supplied the aluminium structure).

As part of the 'early deliveries', ESA will also develop a robotic arm, ERA (European Robotic Arm). ERA is a derivative of the HERA (Hermes Robotic Arm). It will be used on the Russian segment of ISS to support assembly and servicing tasks such as the installation and exchange of external equipment and also external inspection. The 10 m long symmetrical arm will have two 'wrists' and be attached at either end, thus being able to relocate itself, operating with either wrist acting as the manipulator's base. The €145 million project was approved in 1994 and has Fokker space of the Netherlands (which is the major contributor to the project) leading an international consortium from Belgium, Denmark, Germany, Italy, Netherlands, Sweden and Switzerland.

Columbus
Orbital Facility to be attached to Space Station in 2003 by Shuttle. €1,234.1 million allocated 1996-2004, plus €206.7 million 1996-2003 for µg equipment.

On 11 and 12 May 1999, the Council of Ministers met in Brussels to confirm the main objectives of the ESA Long Term Plan for the period 1999-2006, incorporating new programmes spanning the fields of space science, access to space (launchers), space applications (Earth observation, telecommunications, navigation and multimedia), exploitation of the International Space Station and advanced technology. Optional programmes included the following:

On top of these optional programmes, the Ministers approved the Level of Resources for the period 1999-2003 amounting to €2.17 billion of which €1.85 billion is for Scientific Programme and the remaining amount for the General Budget. The Level of Resources sees a constant level for the Mandatory Activities (General Budget and Science) at €531.2 million per annum based on the 1998 budget.

Science Programme
Based on the long-term plans Horizon 2000 and Horizon 2000 Plus. The programme was drafted in a consensual manner with the involvement of the whole European community.

Four ESA space science missions are in orbit in 1999:
- Ulysses, launched 1990, exploring the heliosphere;
- the Hubble Space Telescope, launched in 1990 a joint ESA-NASA mission;
- SOHO (part of Cornerstone), the Solar & Heliospheric Observatory launched 1995;
- Huygens/Cassini mission to Saturn and Pluto launched 1997 in collaboration with NASA.

Five spacecraft are under development:
- XMM, X-ray Multi-mirror Mission, the second cornerstone of Horizons 2000 for launch 2000;
- Cluster II, part of first Cornerstone, four-satellite magnetospheric mission launched 2000;
- INTEGRAL, International Gamma Ray Laboratory, to be launched 2001;
- Rosetta, third Cornerstone, for analysis of in-situ comet, to be launched 2003;
- FIRST/Planck, combination fourth Cornerstone and smaller mission, submillimetre launch 2007.

Additional missions under study:
- SMART-1, advanced technology Moon mission for launch 2001;
- MiniSTEP, mission to test Equivalence Principle, launched 2003 in collaboration with NASA;
- four candidates for launch of next Cornerstone 10 years from now: Mercury mission, advanced interferometric astrometry mission (GAIA), interferometric mission to find terrestrial extra-solar planets (IRSI), mission for detection of gravitational waves (LISA).

At its November 1998 meeting the Science Planning Committee worked on two priorities defined by the scientific community:
- Mars Express (to be launched 2003) unanimously approved at a cost of €150 million under the condition that adequate resources are available and that there is no impact on already approved missions.
- the decision was taken to ensure the presence of the European astronomical community on board the HST mission, the plan being to link the HST and Next Generation Space Telescope (NGST) in negotiations with NASA. Collaboration in these two missions should not cost more than €175 million.

The overall science budget requirements for 1993-2003 amount to €1.85 billion, contributions shared on basis of GNP of each member state.

Earth Observation programme
Better known now as the Living Planet Programme, earth observation programmes comprise two main components:
- Earth Explorer Component, including the definition, development, launch and operations of Earth Explorer (Core and Opportunity) missions covering Earth's interior, oceans, atmosphere, cryosphere and land surface;
- Development and Exploitation Component including instrument pre-development for Earth Explorer and Earth Watch; definition of Earth Watch type missions and the preparation of detailed programme proposals for optional Earth Watch type programmes; mission exploitation/market development.

The Living Planet Programme is an optional programme implemented in five-year successive periods, its activities starting under the Earth Observation Preparatory Programme (EOPP) funded in June 1998 as a transitional measure with the following commitment:
- continuation of ERS operations beyond January for three years;
- one Earth Explorer Core Mission;
- two Earth Explorer Opportunity Missions.

The start of the Earth Watch missions includes instrument development, technology development and market development. A new decision will be made in 2001 to include the development of a second Earth Explorer Core Mission in order to accommodate the objectives of the first five years of the Earth Observation Envelope Programme.

Telecommunication Programmes
The new ESA Telecommunications Programme is based on a portfolio of activities that reflect the need to adapt swiftly to market conditions, the main instrument being the ARTES (Advanced Research in Telecommunications Systems) programme consisting of several elements covering different activities and recently extended to include the period 2001-2005. ARTES 1 is intended to carry out preparatory activities in order to identify new mission opportunities and to identify and study major opportunities for partnership programmes. ARTES element 3 is directly aimed at supporting privately led European Multimedia Satellite Systems to offer industry support in the start-up phase of new projects with which there is difficulty in finding finance from the market.

Satellite Navigation Programme
The Global Navigation Satellite System (GNSS) will become a central feature of all European transport systems, and the European Union, in close co-operation with ESA, has defined a strategy which includes:
- the implementation of a navigation satellite constellation and associated ground infrastructure providing worldwide coverage compatible with GPS;
- international co-operation with the US to ensure system compatibility and with Russia to benefit from GLONASS know-how;
- an approach in phases including a definition phase until 2000, a development/validation phase financed mainly through public funding and completed by 2006 and an operational phase involving private funding leading to full operational capability of the Galileo system by early 2008.

ESA's GalileoSat Programme involves the definition, development and validation phases and if private funding is forthcoming this will be added to the €500 million equally divided between the EU and ESA.

Manned Space Flight & Microgravity Programmes
Exploitation of the International Space Station is aimed at developing European operational capabilities for long-term manned space exploration, to build up the necessary know-how to master operations with complex manned outposts in space and to encourage support of the ISS by the European user community. Technical content includes operations and maintenance of the ESA/ISS elements and ground segments (Columbus pressurised module, Automated Transfer Vehicle, data management system. European robotic arm and microgravity facilities for Columbus as well as respective ATV and Columbus control centres. Other commitments include the fulfilment of ISS common-systems operation, including Ariane 5/ATV transportation, refuel and reboost services to the ISS and support and co-ordination of ISS utilisation by European users (payload integration, logistics, data transmission and astronauts).

Exploitation of the ISS is divided into steps: Step 1, early activities in 2000-2001 for €344 million including ATV preparation and development, Ariane 5 in support of a first operational ATV launch in 2004, NASA reimbursable services in the form of payments for ISS utilisation and bartering these services in exchange for development of the Crew Return Vehicle (CRV); Step 2, (2002-2004) €336 million completing the Toulouse commitment.

The microgravity research programme aims to continue with the technological development of microgravity research, implement microgravity missions involving sounding rockets, parabolic flights and Spacehab missions, support early utilisation of the ISS and continue to upgrade existing facilities and to develop new ones. The future microgravity programme is structured in two slices; slice 1 (2000-2003) with €98 million covering precursor missions such as those with sounding rockets and Spacehab flights, early ISS utilisation and new microgravity facilities; slice 2 (2002-2004) with €145 million for decision in 2001. Further details can be found in the section on human space flight.

Launcher Programmes
The launcher development programme aims to acquire new types of launch vehicle to orbit larger telecommunications satellites, satellite constellations and small payloads. Strategy is based around improvements to Ariane 5, to develop complementary launch capabilities and to prepare the technology for future reusable launchers to maintain Europe's strong lead in launch services for global customers.

	million Euro (€)
1. Navigation/GalileoSat (€40 million until end-2000 + €460 million until end-2005)	500
2. Telecommunications/Preliminary studies – ARTES 1 – (2000-2005)	50
3. Telecommunications/multimedia – ARTES 3 – (1999-2002)	309
4. Earth Observation/Living Planet programme (1999-2002)	759
5. International Space Station/Exploitation (2000-2001)	344
6. Microgravity follow-on Programme (2000-2003)	98.4
7. Ariane 5 Plus (step 2 until end-2001)	462
8. Vega (step 2 until end-2002)	317
9. Ariane 5 ARTA (extension of two years – 2001 & 2002)	161
10. Ariane 5 Infrastructure (extension of one year – 2001)	26
11. Future Launchers Technology (until end-2001)	70
12. CSG Korou (extension of one year – 2001)	87

ESA's budget by %

By country	**1991**	**1992**	**1993**	**1994**	**1995**	**1996**	**1997**	**1999**
Austria	0.76	0.93	0.92	1.05	1.14	1.04	1.10	1.30
Belgium	4.04	4.25	4.49	4.90	5.90	5.51	5.70	5.30
Denmark	0.80	0.80	0.85	0.85	0.94	0.92	1.00	1.10
Finland	0.28	0.30	0.30	0.42	0.54	0.60	0.50	0.60
France	24.58	25.66	27.48	26.07	23.92	26.59	26.80	29.00
Germany	20.53	20.54	21.09	19.10	19.01	20.11	20.10	25.00
Ireland	0.19	0.21	0.18	0.20	0.19	0.20	0.20	0.30
Italy	15.07	14.52	15.96	14.61	14.37	14.23	11.90	13.70
Netherlands	2.19	2.15	2.54	2.58	2.97	3.56	3.50	3.60
Norway	0.65	0.70	0.70	0.77	0.80	0.83	0.80	0.90
Spain	3.95	3.99	3.96	4.04	3.93	4.10	3.70	4.50
Sweden	1.87	1.89	2.08	2.29	2.29	2.39	2.20	2.60
Switzerland	1.96	2.02	2.13	2.38	2.62	2.61	2.70	3.30
United Kingdom	5.12	5.04	5.84	6.50	6.47	6.35	6.30	8.00
Canada	0.48	0.59	0.70	0.66	0.67	0.66	0.50	0.60
Other income	17.51	16.42	10.78	13.57	14.36	10.29	13.00	0.20

By activity	**1991**	**1992**	**1993**	**1994**	**1995**	**1996**	**1997**	**1999**
General Budget	5.94	5.86	5.96	6.72	6.35	6.22	7.10	6.30
Associated to GB	3.19	3.41	3.34	4.03	4.03	3.96	4.40	4.10
Science	9.62	9.74	10.44	13.08	12.80	12.85	14.50	13.40
Earth Observation	7.43	6.81	9.65	15.96	18.63	18.59	18.80	24.30
Microgravity	2.26	2.49	3.12	3.56	3.47	3.30	4.10	3.70
Telecommunications	9.25	8.64	9.48	10.63	10.86	10.81	7.70	7.80
Manned[1]	12.16	12.16	14.06	5.14	10.04	9.74	13.30	16.40
Launchers[2]	47.23	46.94	40.54	37.73	29.90	32.04	25.70	18.40
TDP 1/2 & Prodex[3]	0.73	0.73	0.50	0.83	1.16	1.19	1.50	1.50
Third Party Programmes	2.18	3.22	2.91	2.37	2.76	1.22	2.70	3.90
Transformation Programme	-	-	-	-	-	-	0.20	0.10

* 1997 total AU2,417.3 million from Members AU2,104.1 million. 1996's budget totalled AU2,699.5 million, including AU2,294.7 million from Member States. ESA's financial year runs 1 January to 31 December.
For accounting purposes, the agency employs the Accounting Unit, which is taken as the average value of each national currency, based on the ECU (€), in June of the previous year. 1 AU = US$1.33 for 1996 figures (1.18 1995; 1.19 1994; 1.30 1993; 1.15 1992; 1.22 1991).
The columns may not total 100 per cent due to rounding.

Notes [1] 'Space Stations & Platforms' until 1995; [2] 'Space Transportation Systems' until 1995; [3] this element includes Hipparcos operations in 1991-92, and GSTP from 1994. Previous budget totals were: 1983 977, 1984 1,124,1985 1,139, 1986 1,457, 1987 1,605, 1988 1,882, 1989 2,030.5, 1990 2,070.4, 1991 2,552, 1992 2,731.6, 1993 2,817.3, 1994 2,531.3, 1995 2,679.4 MAU.

The Ariane 5 Plus programme aims to increase the performance of the Ariane 5 in a stepped sequence. This includes the Ariane 5V version in 2001, the Ariane 5C-A programme providing a nine-tonne GTO capability in 2002 followed by Ariane 5C-B with an 11-tonne GTO capability in 2006, thus doubling the present Ariane 5 lift capability. The programme has a total budget of €1.064 billion broken down into three steps. The first step for €108 million was decided in June 1998 and is ongoing, the second step is funded at €462 million to cover development until the end of 2001 and the third step involving €484 million covers the period 2002-2006.

In the Vega small launcher development programme with a total multi-year funding of €370 million agreed in June 1998, ESA is developing a low-cost launcher for payloads around one tonne into LEO. In the Future Launchers Technology Programme ESA is analysing a variety of optional launch vehicle concepts for cost-effective and reusable space transportation with a decision required around 2007 for implementation of a development programme. Additional activity is proceeding on maintaining and improving the level of reliability for Ariane 5 with a view to developing increased efficiency, recovery and reuse of solid-propellant boosters and enhanced mathematical models.

Inter-Agency Solar Terrestrial Physics Programme

The Inter-Agency Solar Terrestrial Physics Programme is a 12-satellite project involving NASA, ESA, ISAS and Russia's Institute of Space Research. IASTP is undertaking detailed investigation of the Sun, Earth's space environment and Sun-Earth interaction. The Inter-Agency Consultative Group (IACG), selected Solar Terrestrial Science as the next discipline area for multilateral co-operation following the highly successful Comet Halley collaboration. IASTP, started in 1977 with planned completion in 2001, involves more than 100 universities, research laboratories and major contractors in 16 countries.

ESA's Soho and Cluster contributions, pursued in collaboration with NASA as the STSP Solar-Terrestrial Science Programme, comprise five satellites launched in 1995/6. NASA includes its involvement in Soho, Cluster and Geotail under the ISTP International Solar Terrestrial Physics programme. IKI planned to fly two Regatta satellites 1993-96, the second closely complementing Cluster, but severe financial constraints forced indefinite postponement. The four-satellite Interball is contributing, however. The US Global Geospace Science (GGS) element is designed to measure the terrestrial portion of the solar-terrestrial connection. Its Wind (launched November 1994) and Polar (February 1996) are complemented by ISAS' Geotail (July 1992). Wind is using lunar swing-bys to hold its apogee on Earth's dayside to survey the upstream region out to 250 Earth radii. After two years, it will enter a small halo orbit around the Sun-Earth L1 Lagrangian point to act as the solar wind monitor for the other craft. Germany is planning to co-operate with China in development of a Solar Space Telescope for launch in 2003.

From these measurements, models will trace the energetic plasma particles and fields in the solar wind through acceleration and storage in the Earth's magnetosphere, to their ultimate deposition in the atmosphere. In the polar regions, this energy deposition equals the solar radiant energy input, and therefore has a significant influence on the atmospheric energy balance. The results should make it possible to predict disruptive effects on the ground and in space.

NASA budgeted US$18.6 million in FY88 to start development of Wind/Polar (rising to 64.4 FY89, 57.6 FY90, 96.6 FY91, 75.3 FY92, 72.6 FY93, 21.6 FY94, 40.0 FY95, 0 FY96) and provide instruments for Geotail, Soho and Cluster. Russia's KONUS v-burst detector was delivered 1992 to fly aboard Wind, the first CIS instrument on a US satellite. GE Astro Space was awarded the prime Polar/Wind contracts in January 1989, ending at US$175 million. Total NASA development cost for GGS, Geotail, Soho and Cluster is estimated at US$732 million. Wind/Polar design/construction difficulties resulted in launch delays, leaving Geotail as the first in orbit.

Interball

Interball is a 20-nation collaborative project to investigate the interaction of solar wind with Earth's magnetosphere and its acceleration down into the auroral oval. Satellites based on the new Prognoz-M2 design are operating in two distinct orbits, one monitoring the magnetotail processes in conjunction with its S2-T sub-satellite, and the second at a higher inclination for auroral observations with its S2-A companion. The Czech Institute of Atmospheric Physics' Upper Atmosphere Dept provides the Magion sub-satellites. The lower craft includes a Canadian UV imager first flown aboard Sweden's Viking auroral research satellite in 1986. Relikt 2, should it appear, would provide additional fields/particles data deep in the magnetotail from its halo orbit around the L2 libration point 1.5 million km on the anti-Sun side from Earth. Interball is part of the Inter-Agency Consultative Group plan.

Interkosmos

Interkosmos, formally the Council for International Co-operation in the Studies and Uses of Outer Space, was responsible for collaborative science projects among its 10 member states and with other nations. Based at the Academy of Sciences, it was closed in 1994 and the projects continued under IKI, IZMIRAN and other institutes. Current activities include the Aktivny, APEX, Interball and Relikt science satellite projects. There was no joint Interkosmos fund: Russia offered its technology and launchers free of charge and each country bore its own costs. No new missions have recently been announced and existing projects continue to suffer delays.

Interkosmos was created in April 1967 by the Soviet-affiliated countries of CMEA (Committee for Mutual Economic Assistance): Bulgaria, Cuba, Czechoslovakia, GDR, Hungary, Mongolia, Poland, Romania and USSR; Vietnam was added in 1979. Areas of interest covered space physics, meteorology, communications, biology and medicine, with Earth resources added in 1975. Specific emphasis has been made of fields/particles studies of solar radiation and its interaction with Earth's magnetosphere. The first launch, Cosmos 261 in 1968 from Plesetsk, studied air density and polar auroras. Each member financed its own contributions, with the CIS/USSR supplying the basic satellite and the launch; scientific results were the common property of all Interkosmos members.

The first programme ran 1967-80, with Czechoslovakia a noted contributor to every unmanned mission. This culminated in its provision on Interkosmos 18 of the Magion sub-satellite, the first non-Soviet satellite in the series, Magion 2 in 1989-90 and #3 in 1991 (#4/5 are flying within Interball). The second programme, for 1980-85, saw a move towards Earth surface studies, with the ocean dynamics

RFAS/USSR-led International Earth-orbit Science Missions

Project	Launch Date	Spacecraft	Orbit	Objective
Aktivny-IK	September 1989	AUOS parent + Magion sub-satellite	Low Earth Orbit	Study the effects of LLF waves on the magnetosphere; Interkosmos project; parent still operational
Granat	December 1989	Astron-type	Prognoz-type	Study of gamma/X-ray radiation with France/Denmark; still operational
Gamma	July 1990	Progress-type	Low Earth Orbit	Study of gamma/X-ray radiation in co-operation with France
APEX	December 1991	AUOS parent + Magion sub-satellite	Low Earth orbit	Electron and plasma probing of magnetosphere; Interkosmos project; parent still operational
Coronas-I	March 1994	AUOS-SM	Low Earth Orbit	Helioseismology and solar activity; Interkosmos project, still operational
Interball	August 1995 August 1996	Prognoz-M2+ Magion sub-satellites	Prognoz + intermediate Molniya-type	Study of magnetosphere and plasmasphere in 14-nation co-operation; one Prognoz/Magion pair in each orbit type
Priroda	April 1996	Mir module	Low Earth Orbit	Mir remote sensing module, including France's Alissa radar
Bion 11	December 1996	Vostok-type	Low Earth Orbit	Life sciences research on living organisms, including 2 monkeys; NASA involvement
Relikt 2	?	Prognoz-M2	L2 halo orbit	Study of Universe's microwave background
Coronas-F	1997?	AUOS-SM	Prognoz-type	Helioseismology and solar activity; Interkosmos project
Bion 12	1998	Vostok-type	Low Earth Orbit	Life sciences research on living organisms, including 2 monkeys; NASA involvement
Spektrum-X/γ	1998	Spektrum-type	Prognoz	High-energy astronomical observations
Radioastron (Spektrum-R)	2000?	Spektrum-type	Elliptical 24h	VLBI study of cm-radio emissions with 10 m diameter radio telescope in conjunction with 70 m ground-based telescopes
Spektrum-UV	2002?	Spektrum-type	Elliptical synchronous?	Study of ultraviolet sources with 170 cm diameter telescope

investigations of #20/21. By October 1984, when the 17th Interkosmos conference was held in Berlin, 11 high-altitude Vertikal research rockets and 22 Interkosmos satellites had been launched, the last in August 1981.

The outline of the programme issued for 1986-90 noted there would be economic use of the results already achieved, but made no mention of any new launches. However, as a member of the Inter-Agency Consultative Group, Interkosmos was closely involved in the Vega missions and the international collaboration on the 1985/86 studies of Halley's Comet, and a major contributor to 1988/89's Phobos mission. In April 1985, Prognoz 10 (Intershock), one of the solar radiation series, was co-designated Interkosmos 23. Aktivny, APEX and Interball subsequently appeared as major projects and in October 1987 Interkosmos announced proposals for two Coronas observatories that would monitor the Sun during its period of maximum activity in the early 1990s. The SAS aeronomy satellites were described in 1989 (but later cancelled) and agreement was near on the Regatta contribution (now postponed indefinitely) to the Inter-Agency Solar-Terrestrial Physics programme. During 1988, a second Bulgarian cosmonaut, Alexander Alexandrov, made a Soyuz flight, apparently in compensation for the failed Soyuz 33 docking with Salyut 6 in 1979.

With the exception of Interkosmos 6, a Vostok-type, and #15, an AUOS Automatic Unified Orbital Station prototype, #1-16 were light research satellites. Weighing 320 to 375 kg, they were based around a standard ellipsoidal pressurised shell. Czech scientists made contributions to all these satellites and provided the orientation system for the more advanced 422 kg AUOS, introduced in operational form with Interkosmos 17. Features included onboard data processing and provision for a sub-satellite. APEX is flying the last AUOS-Z model (Z=Zemlia for Earth, denoting three-axis pointing about z-axis as vertical). Coronas-I introduced the AUOS-SM in 1994; the Prognoz-M2 debuted on Interball in 1995.

Orbiting Satellites Carrying Amateur Radio (OSCAR)

The Orbiting Satellites Carrying Amateur Radio (Oscar) series of small satellites was initiated for radio amateurs to experience satellite tracking and participate in radio propagation experiments. The World Administrative Radio Conference allocated frequencies for the Amateur Satellite Service, including 29 MHz (10 m), 145 MHz (2 m), 435 MHz (70 cm), 1,270 MHz (24 cm) and 2,400 MHz (13 cm). Transmitting low-powered signals, initially battery-operated and offering short lives, the satellites have become increasingly sophisticated. They have served school science groups, provided emergency communications for disaster relief, acted as technology demonstrators and transmitted Earth imagery. The UK's UoSAT series alone has involved hundreds of schools and thousands of groups worldwide in activities using simple antennas, receivers and personal computers. The work ranges from telemetry transmission only, through amateur radio communications, to advanced experiments such as testing indium phosphide solar cells and transputers, and monitoring radiation effects on electronics – satellite engineering for a fraction of the time and cost of the more-advanced conventional satellites. Somewhat confusingly, US military Transit satellites also carry Oscar designations.

IRAN

Iran is a signatory to the Intelsat and Inmarsat organisations, and plans its own domestic satellite communications system. An Iran Space Administration was believed to be in the process of formation, but little has been heard recently.

IRAQ

Little is known of Iraq's space activities but it was reported before 1990's invasion of Kuwait that a satellite launch was expected by the end of August 1990. The 24 m 48 tonne Tamouz 1 vehicle launched from the Al Anbar facility 80 km west of Baghdad on 5 December 1989 was claimed by the Iraqis as a satellite launcher test. Only the first of the three stages was live, reportedly attaining 12 km altitude. Stage 1 appears to comprise five clustered Al Abbas versions of the SS-1 Scud missile, topped by a single Al Abbas stage 2 and a short stage 3. A range of 2,000 km was claimed for the Al Aabed (The Worshipper) missile version. A 1989 report suggested that Brazil was to provide an SCD environmental platform data collection satellite. Technical assistance from China and France was also mentioned. Close ties with the Soviet Union suggested involvement. It is possible that plans revealed during 1990 to build a 1 m-bore 'supergun' under Project Babylon by the Brussels-based Space Research Corp included the firing of small payloads into low Earth orbit.

Iraq is a signatory to Intelsat and Inmarsat.

IRELAND

Ireland's national space activities are managed by the Department of Enterprise & Employment, and directed almost exclusively through ESA. Forbairt manages the technological aspects of the Irish programme. Contributions to ESA were €5.7 million in 1999 and €5.0 million (US$5.9 million) for each of 1995 and 1996, €5.1 million (US$6.1 million) in 1994, Ir£5.2 million (€5.0 million) 1993, 4.4 (5.8) 1992, 3.6 (4.8) 1991, 3.3 (4.2) 1990, 3.0 (3.9) 1989, 1.7 (2.4) 1988 and 1987, and 1.6 (2.2) 1986, amounting to 0.2 per cent of ESA's budget, the smallest contribution of any member. 0.2 per cent is contributed to the Ariane 5 programme.

Ireland's first ESA facility, the Microelectronics Technology Support Laboratory, was established in 1988 at the National Microelectronics Research Centre in Cork. MTSL provides support in semiconductor technologies, particularly microwave, opto-electronic, power, digital and analogue functions. Ireland is the sixth country to host an ESA facility. Space Technology (Ireland) Ltd, was established in 1987 following the successful provision of space radiation monitors to the Giotto, Vega and Phobos international missions.

Telecom Eireann is the national signatory to Intelsat and Eutelsat. A commercial satellite communications system was announced in March 1986 by Atlantic Satellites Ltd, 80 per cent owned by Hughes Communications Inc, to use the 31° west slot allocated to Ireland. HCI later withdrew and the project fell into abeyance.

ISRAEL

Israel became the eighth nation to launch an indigenous satellite when a three-stage Shavit solid-propellant booster placed the Ofeq 1 demonstration satellite into LEO from the Palmachim Air Force Base south of Tel Aviv. The launcher is believed to have been developed from the Jericho 2 missile. The satellite was developed by Israel Aircraft Industries under the auspices of the Israeli space agency as a precursor to more sophisticated science and communications carriers.

US President Clinton in January 1994 decided to allow Israeli companies to compete in the US on civil space activities. Reconnaissance platforms are also a possibility. The defence minister hinted in March 1991 at such a project and there have been rumours that at least one has flown. Reconnaissance satellites can be expected to play a major role in the Middle East strategic balance in the next decade. Since the surprise attacks that began the 1973 Yom Kippur War, Israeli intelligence analysts have given high priority to developing an independent source of space-based intelligence. It was charged that immediately before that war the US withheld critical satellite intelligence information on Arab offensive military formations. After the 1990-91 Gulf War, defence minister Moshe Arens explicitly and publicly declared Israel's intention of launching an indigenous reconnaissance satellite.

The Technion Institute of Technology, has proposed three or four small satellites at 7,000 km for continuous early warning coverage, although the Knesset claimed little official interest. Ofeq 3 is primarily an imaging satellite, with UV/visible sensors. If its technical capabilities are similar to those planned for South Africa's Greensat, it is providing 2 m resolution. There has clearly been strong collaboration between Israel and South Africa on launchers and satellites.

The partnership between IAI and Core Software Technology of Pasadena, California, planned a commercial 250 kg remote sensing satellite for launch by Russian Molniya in July 1997 to return 2 m-resolution imagery. CST is responsible for the ground segment. The satellite is based on Ofeq 3. The first improved version was planned for 1998, offering 1 m PAN plus 5 m XS. At least seven regional partners are each required to invest US$60 to 80 million, the cost of flying one satellite. Each would operate a dedicated satellite within its region, with IAI/CST controlling it beyond.

ISA chairman Professor Yuval Ne'eman noted in 1990 that the Ofeq programme began in 1983 and had then cost US$150 million; a principal aim was development of a national space infrastructure. However, funding tightened after 1990 and officials in 1994 said that an Ofeq 4 was not yet financed and there was little support for the launcher programme. The government required commercial potential to be demonstrated first. The US$200 million commercial domestic/subregional Amos 1 telecommunications satellite developed by IAI appeared in 1996 on Ariane. A second satellite is planned in partnership with Hungary in 1999.

Shavit was proposed unsuccessfully in 1990 and 1994 as NASA's Meteor and Ultralight launcher, respectively. IAI is working on the NEXT commercial version of Shavit (Shavit 2), stretching stages 1/2 and adding a bipropellant stage 4. Development depends on commercial commitments, but a test launch is planned for late 2000 followed by the more powerful Shavit 3 in 2001. Unconfirmed reports suggest that a 4,800 km range Jericho 3 is under development, based on NEXT. Early in 1999 the Israeli government let it be known that a Defence Ministry Committee has backed development of a heavy lift satellite launcher. Called Star-460, this vehicle would be capable of lifting 28 tonnes to low

Earth orbit and 12 tonnes to geosynchronous orbit, more than any other launcher currently in operational use.

MBT Systems & Space Technology, part of IAI's Electronics Division, is Israel's major space centre, responsible for developing Ofeq and Amos. Its Space Technology Center includes 3,400 m^2 of satellite integration/test facilities and a 10 m S/X-band ground station. The station's Image Processing Center handles the received Spot imagery (and presumably Ofeq 3); upgrading to handle other satellites is planned. Raw/processed data are transferred to ISA for marketing by ICTAF.

Elop, Israel's largest electro-optics company, has developed the Earth Resources Monitoring System for small satellites as part of a major company effort to lead large scale high-tech R&D programmes for remote sensing, navigation sensors (TechSat) and astronomy (see TAUVEX). The pushbroom linear CCD camera would provide 16 m resolution from 400 km but increased resolution is possible. A similar camera is aboard Ofeq 3.

ITALY

The Italian government established a National Space Programme (PSN) in 1979 based on a plan that was to be updated every five years. The Interministerial Committee for Economic Planning (CIPE) initially entrusted the overall management and administration of the programme to the CNR and then to ASI. Early in 1984, Italy joined West Germany in proposing that ESA should authorise development of the Columbus space station facility, operating either independently or with the US, but leading to a manned free-flying European laboratory by 1999. Columbus draws upon the technology developed by Aeritalia in building the Spacelab pressurised modules and by West Germany with Spacelab, Eureca and SPAS. Italy was instrumental in the October 1991 Munich ESA ministerial meeting in calling for the Hermes and Columbus programmes to be stretched out to reduce annual funding requirements.

The creation of the ASI national space agency in 1988 to manage the National Space Plan was accompanied by allocation of an Lit800 billion annual budget for the first three years, more than doubling 1985's Lit362.6 billion space funding. The next five-year plan was submitted in December 1990 and approved in July 1991 by the CIPE Interministerial Committee for Economic Planning, which directs government spending, totalling Lit6,260 billion. More than Lit1,300 billion for each year 1992-94, following Lit1,029 billion 1990 and Lit1,211 billion 1991. In fact, because of government financial difficulties, only Lit800 billion (US$470 million at 1993 rates) was provided for each of 1992-94, with Lit850 billion planned for 1995. Acquiring hard figures is difficult in the current uncertain climate. The previous plan concluded with Lit772 billion in 1988 and Lit742 billion in 1989.

Italy is ESA's third largest contributor, accounting for 13.7 per cent of member states' 1999 contributions and encompassing a 15.0 per cent interest in Ariane 5 (plus 7 per cent in Ariane 5 future developments), 19 per cent in Columbus, 11 per cent in Envisat and 45 per cent in the Data Relay Satellite. Expenditure was previously divided almost equally between national/bilateral programmes, but the current plan has increased ESA's element to about two-thirds. Italy's financial problems meant that ESA had to take out a Lit189 billion loan on its behalf to help pay its ESA contributions 1994-96. In 1996, the Italian civil space budget was equivalent to US$600 million.

Industry is dominated by the Space Division of Alenia Aerospazio (previously Alenia Spazio SpA which is now incorporated within Finmeccanica). The significant national interest in Columbus arises from Aeritalia's expertise with pressurised modules for manned spacecraft, beginning with Spacelab, while Selenia headed communications programmes such as Italsat. Italy was a major driving force behind Olympus, contributing 31.5 per cent, and is taking the lead role in DRS. National propulsion capabilities are represented by BPD, responsible for the IRIS upper stage motor, the Ariane 4 solid propellant strap-ons and development in co-operation with France's SEP of Ariane 5's large boosters.

The last five-year plan approved development of the San Marco Scout launcher and BPD was awarded ASI's industrialization contract in 1991 aiming for a 1995 maiden flight. However, the University of Rome proposed a separate San Marco Scout programme closely based on the US Scout, prompting legal wrangling. BPD is continuing development of the current Vega design using internal funding plus Ministry of Research support for development of the main Zefiro solid motor. Should government support ever be provided, the qualification flight would be in the next decade. The San Marco site is now regarded as too costly to maintain, and Vega's principal sites would be Kourou and Vandenberg AFB using transportable equipment. Discussions are under way for commonality with Spain's Capricornio and with CNES's ESL. Development cost would be about US$200 million.

The five-year plan also approved a new contribution to Space Station outside of Italy's ESA involvement. NASA/ASI signed an MoU 6 December 1991 for three Alenia Mini Pressurised Logistics Modules to deliver Space Station supplies aboard Shuttle. Cost is about US$300 million. The agreement also covers a Mini-Lab, currently under definition. MPLM will also form the basis for ESA's Columbus module. In another bilateral project with NASA, Italy is providing a significant contribution to the Cassini Saturn orbiter. Alenia Spazio is prime for the high gain antenna, the radar mapper's RF element and parts of the RF subsystem, and Officine Galileo for the VIMS Visual IR Mapping Spectrometer visible channel and stellar reference unit. Italy might also build a Ka-band station on its soil to enhance gravitational wave experiments.

An Italian plasma physics specialist, Dr Franco Malerba (back-up Dr Umberto Guidoni) flew Shuttle's Tethered Satellite mission in August 1992. Guidoni flew the repeat TSS 1R aboard STS-75 in February 1996. AF test pilot Maurizio Cheli was selected in May 1992 by ESA as one of six new astronauts and also debuted on STS-75. A mission to Mir has been discussed with Russia. ASI proposed a corps of 10 to 15 astronauts to satisfy these and future collaborative ventures, working from the National Centre for Space Training & Experimentation, but such a group is now unlikely because of financial constraints.

Telespazio was responsible for developing the Argo system for the Civil Defence Department to provide national emergency communications and datalinks through Eutelsat within 4 h of a major disaster. An inter-service study for a PSM Piano Spaziale Militare was completed in 1990 but funding restrictions render its future uncertain. In addition to participation in France's Helios observation satellite and the proposed Sicral military/civil emergency communications satellite (more likely now a package aboard a civil satellite), it recommended a Helios successor, an electronic intelligence satellite, a navigation system, and communications and data relay satellites, all with international co-operation.

JAPAN

Japan was the fourth country after the USSR, USA and France to achieve national satellite launcher capability and the third into GEO. It now stands only third to the USA and Russia in space expenditure, at about US$2.5 billion state spending, or 0.04 per cent of GDP. The Space Activities Commission in January 1996 released a new 15-year plan, proposing to more than double spending to ¥7,000 billion (US$68 billion) over 1996-2010, emphasising Earth observation, Space Station, launcher and spaceplane activities, and increased deep space ventures. The influential Federation of Economic Organisations in mid-1993 recommended more than doubling such spending to ¥500 billion annually. In the light of significant financial difficulties these plans were deemed unrealistic and Japan's space budget is unlikely to increase.

The space programme has grown significantly since the early 1970s, aiming for independent launcher and applications satellite capabilities from modest beginnings. The H2, for example, became the first wholly-indigenous NASDA launcher when it debuted successfully in 1994 and an important member of the international launcher scene. The Improved H2 is under development for a 2000 debut, principally to reduce production cost by 30 per cent. Development of the Upgraded H2A began FY96 to debut 2001, as a family capable of delivering 4 tonnes GEO and 20 tonnes LEO. Unit cost will be ¥9 billion. That last capability will allow it to fly Space Station logistics missions and carry the HOPE spaceplane.

Japan is aiming ultimately for a national manned space presence, gaining experience via Spacelab and JEM. These also support the growing µg interest, which includes the Space Flyer Unit, USERS and Spacelab. A national space station and a lunar base have been identified as specific goals for the next century, supported by LEO servicing and space transfer vehicles.

Until 1989's launches of Superbird (Space Communications Corp) and JCSat (Japan Communications Satellite Co, now Japan Satellite Systems), telecommunications satellites had been state-owned and operated through the Telecommunications Advancement Organisation of Japan (TAO). The CS-3/BS-3 series and, now, NTT's NStar provide national telecommunications and direct TV services. US pressure helped to open the follow-on programmes to foreign bidders, resulting in both NStar and the BSAT DTH satellites of Broadcasting Satellite System Corp being built in the US. NASDA continues to develop and operate the ETS series of technology demonstration satellites.

Earth observations are well established in Japan: GMS provides routine global meteorological imaging from GEO, the MOS Marine Observation Satellite system was operational 1987-96 and was joined by the synthetic aperture radar of JERS in 1992. The Science & Technology Agency has undertaken an extensive study of remote sensing requirements through 2010 to improve environmental monitoring and research, while complementing other nations' programmes. The inclusion of military imaging satellites in the new National Defence Outline was under consideration in 1994, with reported support from the Defence Agency. The US has traditionally served Japan's intelligence requirements for satellite imagery, with the drawback of limited access.

Although it operates from a modest budget, ISAS maintains an impressive science and launcher programme. The two Comet Halley probes of 1985-6 demonstrated a deep space capability and in March 1990 ISAS established the first lunar orbiter since 1976. The agency was granted permission during 1989 to develop a larger M-5 all-solid launcher which made its first flight in February 1997 carrying the Haruka radio telescope that will test techniques for deploying large space structures before performing a radio astronomy

Space budget summary (¥ billion)

JFY	NASDA	ISAS	Total
1974	47.56	5.08	57.47
1975	61.42	6.54	77.04
1976	72.65	7.14	88.01
1977	79.17	8.42	94.91
1978	79.41	9.23	95.96
1979	81.84	10.28	99.46
1980	83.73	10.37	102.01
1981	85.36	11.24	105.00
1982	86.38	12.93	108.47
1983	86.07	15.18	113.39
1984	84.36	15.83	112.95
1985	88.86	10.96	112.55
1986	90.66	12.38	117.38
1987	92.648	11.825	121.924
1988	96.534	19.790	141.781
1989	106.76	20.79	154.99
1990	116.293	18.022	162.160
1991	128.313	20.841	177.658
1992	140.789	20.87	189.66
1993	152.184	20.646	201.700
1994	164.348	21.414	217.502
1995	172.398	21.414	228.812
1996	182.2	22.072	231.245
1997	168.5	n/a	n/a

April 1996 exchange rate was ¥110 per US$. Figures are for Japan's Fiscal Year of 1 April – 31 March. They exclude private sector spending on production items such as telecommunications satellites and industry-funded R&D. The total does not include non-government contributions, such as from NTT, to NASDA's overall budget since 1979. Some three-quarters of government funding is routed through the Science & Technology Agency, with the remainder through the ministries of Education, International Trade & Industry, Transport, and Post and Telecommunications. ISAS budget for 1997 was not available as this went to press.

mission. The second M-5 carried the Mars orbiter Nozomi launched in July 1998 and future vehicles are assigned to carry Lunar-A, a combined moon orbiter/ penetrometer mission. NASDA has also developed the all-solid vehicle, the J1, to carry LEO payloads too small for the expensive H2.

Independent of the national programme, journalist Toyohiro Akiyama visited the Soviet Mir space station during December 1990 under a commercial agreement between Glavcosmos/NPO Energia and Japan's TBS television network.

KOREA, SOUTH

South Korea seeks to build its second generation satellite system with a large proportion of the technology from domestic sources. The country's programme to develop space technology began with the launches of two small satellites in 1992/3. An agreement worth £2.7 million (US$4.6 million; actual cost was less than £2.4 million) was signed in 1990 by KAIST with the UK's University of Surrey and associated Surrey Satellite Technology Ltd covering five years training, one Kitsat micro-satellite, parts for a second and a ground station. Students took the university's MSc in Satellite Communications & Spacecraft Engineering and the UoSAT 5 project was used for training. The second satellite was built in Korea at a total project cost of less than US$5 million. The indigenous Kitsat 3 is now under construction and Kitsat 4 is in the planning stages.

China and South Korea expected to reach an agreement in 1994 on a US$24 million project to launch the CKETS Chinese-Korean Engineering Test Satellite medium-class LEO satellite in 1998 to provide communications and data collection from environmental platforms, but the project was cancelled. Instead, KARI selected TRW Space & Electronics Group in 1994 in competition with Lockheed and Matra Marconi to jointly develop the Korea Multipurpose Satellite for launch in 1999 carrying science and communications payloads.

KARI launched the first indigenous sounding rocket (KSR 1, the Korean Sounding Rocket 1) on 4 June 1993 to 39 km altitude to study the ozone layer. The one-stage 6.7 m long 41.3 cm diameter 1.1 tonne rocket was unguided. KSR 2, with the same payload, attained 55 km on 1 September 1993. Development of a two-stage vehicle is under way. This has been delayed due to financial restrictions on government spending.

LUXEMBOURG

Luxembourg's space activities are focused on the Astra TV distribution satellite system that began operational broadcasting to Western Europe in February 1989 using a bought-in satellite. Six were orbited by mid-1996, with four more ordered. The government is also a signatory to Intelsat and Eutelsat, but not Inmarsat. Although there is no national space programme and limited industrial capacity, membership of ESA was expressed during 1988 as a possibility for the future.

MALAYSIA

The government, in 1994, identified three principal strategies for developing national space technology capabilities; involvement of the private sector in foreign space programmes, establishing formal links between Malaysian institutes with foreign space agencies and fostering closer regional governmental relationships. The country is in the process of formulating a more comprehensive national space policy and programmes. National telecommunications policy is viewed as a catalyst to Malaysia achieving the status of an industrialised nation by 2020.

The telecommunications company Binariang launched the MEASAT 1 (Malaysian-East Asian Satellite) telecommunications satellite to provide domestic and regional services. MEASAT 1 was launched on 12 January 1996 by Ariane-44L, MEASAT 2 followed on 13 November 1996. Binariang signed a MoU with Arianespace in January 1995 for launching a 50 kg Malaysian educational and science micro-satellite. This is being developed with India's assistance, under an agreement signed on 3 August 1995 between the Maxstar (Malaysian Space and Telecommunications Research) consortium and ISRO. The payload includes amateur communications (voice/data), a CCD Earth observation camera and a GPS receiver. The cost for two or three ground stations will be M$45 million to M$50 million, funded by the government and private sector. The consortium includes universities, government agencies and private companies such as Binariang and Telekom Malaysia.

The MACRES Malaysian Centre for Remote Sensing is being upgraded and operates a major ground receiving station. Meteorological satellite data are already received.

MEXICO

Eighteen million out of a population of 75 million had no access to telephones or TV in 1982 when Mexico ordered two Hughes HS-376 satellites to provide national telephone, TV, data and facsimile services. To launch the satellites, Mexico agreed with NASA to use the Shuttle if Mexican payload specialist, Rodolfo Neri Vela, could fly aboard the Shuttle for the second launch in November 1985. Morelos 2, launched in November 1985 aboard STS-61B, had to be placed in a drift orbit moving over two years to its final position because financing problems and the September 1985 earthquake delayed completion of the ground segment. A request for proposals for a replacement was released in 1996. Hughes also won the contract in 1991 for the two second generation Solidaridad satellites, based on its HS-601 platform and adding L-band transponders for mobile communications. First launch was made in November 1993 aboard Ariane 4; #2 appeared in October 1994.

Telecomunicaciones de Mexico (Telecomm Mexico) was created in January 1990 to assume responsibility for the country's telecommunications systems; SCT now functions as a regulating body. Telecomm was privatised in 1996. Mexico is also a member of Intelsat and joined Inmarsat in January 1994, with Telecomm as signatory.

As part of its winning Solidaridad bid, Arianespace included the future launch of a Mexican micro-satellite and training for technicians with the French/German space agencies. The separate 12 kg Unamsat 1 was built by the UNAM Autonomous University of Mexico using Amsat-NA's Microsat design, but was lost in the 28 March 1995 launch failure of Russia's Start from Plesetsk. The main mission was meteor sounding by detecting echoes from 60 W 1 to 10 ms pulses transmitted every 1 to 10 seconds on 40.097 MHz. Amateur store/forward links were to be provided on 2 m/70 cm up/down. A replacement is expected.

NETHERLANDS

The Netherlands' space activities were undertaken principally within Europe's ESRO and ELDO organisations until 1970. In 1969, however, the government decided to fund the Astronomical Netherlands Satellite (ANS) at a cost of NLG76 million, with NASA help. Instrumentation was also provided to OGO 5, Solar Max, ESRO 2/4, TD 1, COS B, ISEE 2, Exosat and Ulysses. In 1976, the government approved the pioneering Dutch/US/UK IRAS infra-red radiation observatory satellite (launched 1983), but then withdrew from large national space activities in favour of increased ESA participation. In 1999, the Netherlands' participation in ESA was 3.6 per cent, accounting for 69 per cent of its national space budget. As a full ESA member, the Netherlands contributes to the Mandatory Programme on a GNP basis (about 4.5 per cent; for 1972-91 it was 2.74 per cent).

Some 80 per cent of the Netherlands' 1999 space budget was devoted to ESA, largely via the ministries of Economic Affairs (60 per cent) and Education and Science (20 per cent). The contribution is relatively modest, but the country hosts the agency's primary ESTEC technical R&D centre at Noordwijk. The Interdepartmental Committee on Space Research and Technology is responsible for formulating space policy. Space activity management is exercised by the NIVR, Netherlands Agency for Aerospace Programmes, a semi-governmental institute. Space science is controlled through SRON Space Research Organisation Netherlands. Industrial co-operation is formalised through the NISO Netherlands Industrial Space Organisation.

The first Dutch satellite was the ANS astronomical platform launched in 1974, followed by significant participation in IRAS and a 10 per cent interest in the SAX satellite. The SAX bilateral MoU, worth NLG65 million to Dutch companies, was signed on 29 June 1990 and accounts for the attitude control system and X-ray telescope; Fokker Space also provided the solar array. SAX was launched on 30 April 1996 and is operating successfully. The government has declared Earth observation to be a top priority; the SCIAMACHY instrument is being developed with Germany for Envisat under a MoU signed in 1994.

ESA, in 1994, approved the €180 million ERA European Robotic Arm for installation on Russia's segment of Space Station. The Netherlands has a 59 per cent share, making it the largest Dutch space project during 1995-2000. Fokker Space leads the industrial consortium.

Fokker Space represents the Netherlands principal industrial space capability. It is 50 per cent owned by Fokker NV and 50 per cent by Ultra Centrifuge Nederland NV. Its solar arrays are included on Inmarsat 2, Telecom 2, SAX, Soho, Hispasat, Artemis, Hot Bird 2-4, DRS and Envisat and its RARA retractable/ retrievable array flew on Eureca. It provided ISO's attitude/orbit control subsystem and was responsible for the integration and testing of the ERS 1/2 payloads.

Through the Dutch Telecommunications Administration and the Dutch Meteorological Institute, the Netherlands is a member of Intelsat (1.4 per cent), Eutelsat (8.4 per cent), Inmarsat (2.8 per cent) and EUMETSAT (4 per cent).

NEW ZEALAND

There is no single space agency or co-ordinating group in New Zealand. Telecom Corp of New Zealand Ltd is the Intelsat and Inmarsat signatory, operating Intelsat stations at Warkworth, Wellington, Rangiora, Chatham Islands and Scott Base (Antarctica). It also leases capacity on Australia's Optus. MetService in Wellington receives and distributes NOAA and GMS satellite imagery and products. Landcare and its Dept of Scientific & Industrial Research predecessor has undertaken Earth resources satellite data processing since 1973. It maintains a large archive of satellite data of New Zealand and holds licences to distribute Spot and Landsat data. A Cospas/Sarsat ground station (CAL Corp's ATLUT-320) was installed in Wellington in late 1991 for the Ministry of Transport.

NIGERIA

Nigeria has considered development of a national communications satellite system, with three spacecraft located at 20/16/14° east employing C-band frequencies. Meanwhile, Nigeria leases two transponders from Intelsat, with 26 ground stations, for telecommunications and TV services. Studies were also

Netherlands space budget (million guilders)

	1990	1991	1992	1993	1994	1995	1996	1997
ESA contribution	145	146	153	160	150	175	200	190
NIVR technology programme	13	13	13	14	14	14	14	15
SRON space research	19	19	20	20	20	20	20	20
SAX & SCIAMACHY*	10	12	14	28	34	34	21	15

ESA Participation	
General & Science	4.5%
Telecommunications	2%
Micro-*g* & Earth Observation	4%
Ariane 5	2.1%
Manned Systems	7%
ERA	59%

* SCIAMACHY funding began 1992; SAX funding until 1996

made for the establishment of a Landsat and Spot data receiving station. In February 1988, Nigeria became the 54th member of Inmarsat; Nitel is also the national Intelsat signatory.

NORWAY

Norway became an associate member of ESA from 1981, in 1985 sharing a 1.8 per cent contribution to the general budget with Austria, Canada, Finland, Portugal and Turkey. 1985 government direct space expenditure totalled NKr39.4 million, almost equally divided between ESA and national activities. It became a full member from January 1987, contributing 0.7 per cent of the general budget and established NSC as the national space agency five months later. Direct space expenditure more than doubled from its 1986 level of NKr50.6 million (27 to ESA).

The public sector funding of space activities in 1996 amounted to NKr461 million (1995 – NKr430 million). Of this, ESA programmes accounted for 43 per cent, the balance being divided between universities/colleges, Telenor Research, Norwegian Space Centre national programmes, administration, Norwegian Defence Research Establishment and the Research Council of Norway. Space-related sales totaled NKr3.2 billion in 1996, NKr2.6 billion in 1995, and NKr2.7 billion in 1994. More than a third of government space expenditure is devoted to ESA programmes, for which Norway's €22.2 million represents some 0.9 per cent of members 1996 contributions. The Norwegian Space Centre's 1997 budget is NKr304 million (1996 – NKr310 million) which includes the ESA funding. Contracts in return from ESA totalled NKr1,320 million by the end of 1996.

The Andøya Rocket Range is a major European facility, launching primarily sounding rockets undertaking scientific investigations at high latitudes. The commercial Polar Satellite Launch Service is planned for delivering 250 kg satellites into polar orbit, but a suitable launch vehicle remains to be found. The Tromsø Satellite Station reflects strong Norwegian interest in Earth observation and was upgraded in 1991 as a near-real-time (30 minutes) processing and distribution centre for ESA's ERS radar satellite; Radarsat and JERS 1 were added later. The Swedish Space Corp acquired a half interest in 1995.

The country's principal space emphasis is on communications: its status as Inmarsat's fourth largest member, with 6.9 per cent in 1996, complements the size of its maritime fleet and the importance of the major Earth station at Eik. Norwegian firms are among international leaders in supplying coastal Earth stations and ship terminals. Telenor Satellite Services is the national signatory to Intelsat and Eutelsat and is among the largest VSAT and gateway providers in Europe. In addition, Telenor (Norwegian Telecom until January 1995) 'through its subsidiary Telenor Satellite Services' is a member of the Skyphone consortium providing aeronautical communications services through the Inmarsat system. NT bought the UK Marcopolo 2 satellite in 1992 and moved it to 1° west as Thor 1, to provide direct-to-home services in conjunction with an Intelsat at 1° west and Tele-X at 5° east; Germany's TV-Sat 2 has been leased at 1° west since March 1995. 35 channels were planned by 1998 by investing NKr1.5 billion in two new satellites. Hughes built Thor 2 for launch on 21 May 1997 and the larger Thor 2B contract went to Hughes.

The navigation and positioning sector was identified in 1992 as a new focus. The Norwegian Mapping Authority in 1995 completed a network of 11 GPS reference stations providing 5 m navigation accuracy to mobile users. Comparing the known locations with GPS-derived positions allows GPS error corrections to be broadcast on AM/FM radio to users. Norway participates in the European tri-partite project EC-NOS.

The Norwegian Meteorological Institute is the EUMETSAT signatory.

The NDRE Norwegian Defence Research Establishment is leading work on the national NISSE Norwegian Ionospheric Small Satellite Experiment satellite on behalf of the space physics community. Funds come from the Research Council of Norway. Further development is pending due to lack of national funding. Cost would be NKr83 million, including launch.

The national plan approved by Parliament as a step towards ESA membership proposed that by 2000 Norway should equal an average small ESA country in space activity. This required funding of NKr220 million annually (1987 rates); 70 per cent would be channelled through ESA, and 30 per cent through national programmes. Special projects such as Tele-X, the direct broadcast satellite in which Norway originally shared a 15 per cent interest with Sweden (82 per cent) and Finland (3 per cent), would be treated on a case-by-case basis. At the November 1987 ESA ministers' meeting Norway indicated that it would make a 0.4 per cent contribution to each of the Ariane 5 (increased in 1991 to 0.6 per cent) and Columbus projects. It subsequently committed 0.2 per cent to Hermes but withdrew from the re-oriented programme starting 1993. The first ESA scientific project in which Norway could participate fully was Cluster/Soho, followed by Cassini/Huygens. Norway is fully participating in the replacement Cluster-2 programme following the loss of the original satellites aboard Ariane 501. Before that, its ESA activities had concentrated on remote sensing, contributing to Operational Meteosat, ERS and EOPP. ERS and Polar Platform are seen as important steps towards future operational remote sensing systems and it is intended to participate directly in both their utilisation and their ground segments.

Summary of Norway's ESA Contributions 1991-1997 (in millions of NKr)

Year	1991	1992	1993	1994	1995	1996	1997
Science	29.4	33.4	36.9	39.3	41.4	41.5	45.0
Telecommunications	23.8	23.5	27.4	23.0	22.1	21.2	14.4
Earth observation	19.4	24.8	30.1	38.2	43.8	48.3	49.7
Int'l Space Station	9.1	9.6	9.9	4.5	9.2	6.7	8.3
Space transportation	36.9	37.0	38.4	46.0	34.7	40.0	27.5
Microgravity	1.5	1.8	2.4	2.5	2.6	2.5	0.9
General budget	26.4	32.9	39.3	20.9	25.2	23.9	30.4
Other	-	-	-	10.5	12.1	13.9	14.5

PAKISTAN

The development of a space infrastructure has started with the design and development of low-cost Badr ('Full Moon') LEO satellites for communications, remote sensing and science applications. Badr 2 is part of a programme to develop a remote sensing satellite system, although the cost could require an international partner. Remote sensing applications occupy a central position in Suparco's activities. The agency's Islamabad station began reception of Landsat, Spot and NOAA imagery in 1988.

Suparco has also studied a 'Pakstar' Radio Determination Satellite Service based on Geostar's principles. TV and telecommunications capacity is leased on AsiaSat 1. A Local User Terminal & Mission Control Centre for the Cospas/Sarsat distress system was commissioned during 1990 at Lahore. The country is a member of Inmarsat, Intelsat, ICAO, Cospar/ICSU, Cospas/Sarsat, the UN Committee on the Peaceful Uses of Outer Space (Copuos), the ITU, IAF and WMO.

Another long range goal is to develop a small satellite launcher, but western technology transfer restrictions are slowing progress.

The government sold one of its GEO slots to Alcatel Espace for a commercial telecommunications satellite. Pakistan has withdrawn from a domestic system and even the Nahuel approach of licensing the location while acting as an anchor for national use. There are currently no communications services for more than 70 per cent of the rural and remote population. Out of a country of more than 100 million people, three quarters of the 2 million working telephones serve 15 major cities.

POLAND

The country is a member of Inmarsat and Intersputnik, and became Eutelsat's 27th member in 1990 and Intelsat's 130th in January 1994. Polish Telecom is the national signatory. Inmarsat Atlantic and Indian ocean Earth stations operate at Psary, 200 km south of Warsaw, and a 13 m station become operational in 1993 for digital telephony as the Eutelsat system's 19th such station.

The Polspace consortium was formed in 1993 by research centres and businesses to promote and participate in space projects. Poland was a leading member of the Russian-led Interkosmos space science organisation, participating in the Phobos, Vega, Aktivny, Apex, Gamma, Interball, Biokosmos, Mars-96 and Interkosmos missions. Polish researcher Miroslav Hermaszewski visited the Salyut 6 space station in June-July 1978 under the Interkosmos program as part of the Soyuz 30 crew. Discussions began with ESA in 1990 on closer co-operation, with a view to possible membership. A general co-operative agreement was signed 28 January 1994. The Academy of Science's Committee for Space Research co-ordinates space science activities for execution by the Space Research Centre.

PORTUGAL

The first Portuguese satellite, PoSAT 1, was launched in 1993. A Portuguese team, within a technology transfer programme between the UK and Portugal, built it. The primary objective is to stimulate a national space industry and generate a nucleus of engineers with first-hand expertise for possible future satellite projects. The PoSAT consortium is led by INETI National Institute of Engineering & Industrial Technology and includes EFACEC, Alcatel Portugal, OGMA Oficinas Gerais de Material Aeronautico, CPRM Companhia Portuguesa de Rádio Marconi, IST Instituto Superior Técnico, UBI Universidade de Beira Interior and CEDINTEC. Overall cost was about £5 million, funded by the Ministry of Industry through the PEDIP programme. Each consortium member supported its own manpower. The programme included on the job training for 11 engineers at Surrey Satellite Technology Ltd in the UK, who also provided the ground station in Portugal.

ROMANIA

A member of Interkosmos and Intersputnik since their inceptions, Romania, in 1990, joined Eutelsat (28th member state), Intelsat (119th) and Inmarsat (60th) through its Ministry of Communications signatory. It was the first East European country to be a signatory to Intelsat. Its initial use of the Eutelsat system covers international telephony links with six other member countries; an Earth station built by Satellite Transmission Systems of Hauppauge, New York at a cost of about US$3 million became operational in early 1991 near Bucharest. The country has no Intersputnik Earth stations and it declared in 1990 that it would not take part in the organisation's future programmes (although it remains a member). Membership of the European Space Agency has also been mentioned as a possibility; a basic co-operative agreement was signed on 11 December 1992.

Space contributions have been modest and all have been made through Interkosmos. Participation in the Biokosmos, Apex, Interball and Interkosmos 6/17/18/20/21 science missions has been acknowledged but was at a low level. Romania was the ninth/last country to fly a cosmonaut to a Soviet space station under the 1976 Interkosmos agreement: Dumitru Prunariu spent a week aboard Salyut 6 in May 1981 and performed a brief programme of experiments, primarily biomedical studies.

RUSSIAN FEDERATION

Although launches have fallen to about a third of their Soviet levels, the up to 180 satellites operating at any one time is unchanged and reflects improving longevity. Much of the apparent robustness can be attributed to the momentum remaining from the Soviet era, but that is now clearly dissipating. Civil and military agencies chronically complain of inadequate budget allocations, being paid late and only partially. Western currency from commercial contracts now plays a crucial role. Visiting missions to the Mir space station in 1995 brought in some Rb350 billion, highlighting why Russia has decided to try and keep Mir operational through to at least 2000. The income is essential for paying for launch vehicles, which are now hand built to specific order. Momentum was being maintained partly by drawing launchers from the strategic stockpile, but without replacing them. Likewise, much of the rocket engine production capacity has been shut down or, in some cases, dismantled. Those launcher reserves were apparently exhausted in October 1995, forcing Soyuz-TM 23 to be delayed from December 1995 to February 1996 while its rocket was built.

At least the high-profile manned programme has the power and the currency to buy vehicles on a reasonably regular basis, but less glamorous projects such as the Photon microgravity and Resurs-F observation missions struggle for funding. February 1995's Photon, for example, was postponed from September 1994 because of problems in paying the rocket supplier and the VKS Military Space Forces for launch services. Photon and its Bion life sciences equivalent would probably not survive without European and US involvement. There was only one Resurs-F mission in 1994-95 because of the high price of launchers and none in 1996. The improved F1M has been ready since 1994 but there is no flight funding. Science missions are particularly vulnerable: no new programmes have started in recent years, or are likely to, and it has been a struggle to maintain operations of some already flying. The loss of Mars 96 has only made matters worse. French payments were required to keep the Granat observatory in service and Franco-German contributions were crucial for Mars 96. Future planetary involvement may depend on NASA missions.

A 12-point space pact was signed in Minsk, Belarus 30 December 1991 by heads of all members of the new Commonwealth of Independent States (CIS), except for Ukraine and Moldova. It proposed the creation of an ESA-like structure, financed proportionately depending on level of participation, while allowing independent space programmes and agencies. An Interstate Space Council would control CIS activities, while military space operations continue under central control of the joint strategic forces. The Prime Ministers voted on 13 November 1992 to form the council, although its value is not easy to determine. Russia's determination to go it alone does sideline it, though. Most importantly, the pact agreed that Baikonur/Tyuratam would provide free access for other states. Although Russia holds up to 90 per cent of space industry capability, the loss of the Baikonur launch site in Kazakhstan would have been a major blow.

Russia is determined to eliminate its dependence on former USSR countries for any space technologies, materials and capabilities. Major areas include Ukrainian electronics for launchers and spacecraft and the Yevpatoria ground station. Proposals for a new cosmodrome at the decommissioned missile site at Svobodny-18 near the Chinese border, were widely brandished, probably mainly as a strong negotiating ploy against Kazakhstan, although President Yeltsin signed a decree in March 1996 approving its creation. The cost would be high but Kazakhstan could not then dictate terms. First orbital launch from Svobodny came in March 1997. For example, helicopter crews for recovering manned Soyuz and unmanned capsules cannot leave their base on the border until the Kazakh government has approved their flight plans. They also have to carry cash to pay for refuelling.

A positive note among all the negative developments is the completion of the Glonass satellite navigation system. Under deployment since 1982, the launch of triple satellites in December 1995 brought the system up to its full strength of 24 operating satellites for the first time. It provides better accuracy for civil users than GPS (50 m *vs.* 100 m), and combined receivers are now beginning to appear. Russia hoped to create a unified navigation network of differential stations providing at least 10 m accuracy nationwide by 2000 but prospects for keeping Glonass at this level looked bleak in early 1999.

The federal civil budget for 1996 was Rb2,219.5 billion, including Rb719.9 billion for Baikonur, although the RKA Russian Space Agency did not receive it all. RKA head Yuri Koptev claimed in March 1995 that RKA received only 51 per cent of 1994's allocated budget. 1995's budget was Rb1,895 billion after RKA and VKS had each requested about Rb2,900 billion. VKS claimed in January 1996 that it received only half of that request. RKK Energia, Russia's largest space company and responsible for the manned programme, complained in February 1995 that it had been allocated only 2 per cent of the required resources but the following month was guaranteed the funding by Parliament. It is doubtful whether manned activities would survive without integration into international missions. RKA took the lead role in developing a seven-year civil space plan adopted by Parliament in December 1993. This emphasises expansion of telecommunications, such as a 20-fold increase in telephone lines, Earth observation, upgrading the Proton and Soyuz launch vehicles and, above all, continuing the manned programme with Mir and involvement in the international Space Station. Koptev has even floated the idea of Russia becoming an associate member of ESA. Nonetheless, it is still an impressive achievement when the civil space agency's budget is less than that of India's.

Russia's fight to enter the commercial launch market lasted 14 years before Proton successfully dispatched Luxembourg's Astra 1F in April 1996. Inmarsat was the first to sign up to Proton, in 1992 and Proton launched Inmarsat-3 in September 1996 and once that credibility barrier had been breached and Lockheed Martin gave its stamp of approval by jointly marketing the vehicle, there has been marked success. The first accord with the US stipulated that only eight GEO/GTO commercial launches could be conducted through 2000 (limited to two in any 12 months). Although not specifically a Proton agreement, in practice it covered principally that vehicle's GEO version. Part of the deal required prices within 7.5 per cent of the western equivalents. The agreement excluded Proton's earlier Inmarsat contract; LEO and other non-GEO/GTO launches would be dealt with on a case-by-case basis. A more relaxed agreement was signed on 30 January 1996, allowing 16 launches (plus another four should market conditions allow) within 15 per cent of western prices.

Khrunichev is commercially the most successful Russian space company, its employees earning the highest rates in the industry (Rb800,000 monthly, as opposed to Rb230,000 in the TsUP manned control centre). Its resources are allowing it to develop the Breeze-M storable stage and a cryogenic stage for Proton to handle the next generation of large satellites. In fact, Khrunichev plans to begin flight-testing the new Angara vehicle in about 2002 to replace Proton. Not only would it improve Proton's cost and performance, it would free Russia of Proton's dependence on Kazakhstan. Khrunichev's smaller Rockot is being marketed jointly with DASA and can be expected to be successful with smaller payloads Payloads are also appearing on AKO Polyot's Cosmos vehicle and a sea-launched version of Zenit is being developed with Boeing.

While there were calls during 1990 for a strong central space agency, the power of Glavcosmos was declining. Organisations such as RKK Energia were unleashed to offer their particular expertise directly; Energia first took over responsibility from Glavcosmos for commercial manned missions and in late 1991 assumed management of Mir, reporting to the Russian Ministry of Industry. Meanwhile, the Council of Ministers' main space planner had thrown doubt on a Mir successor, condemning national space stations as too expensive. The Russian Federation established its own Ministry of Communications, Information and Space (under Minister Vladimir Bulgak) in 1990 and paid the Ministry of Defence for launching the first of three dedicated Gorizont communications satellites. The Interkosmos international science collaboration programme, based at the Academy, ended in 1994 and the projects were distributed among institutes such as IKI.

Order was slow to emerge following 1991's collapse of the USSR and the subsequent struggle to establish a formal structure inevitably left the space program in an organisational limbo. The Soviet Ministry of Machine Building (MOM), for more than a quarter of a century the principal governmental body with direct oversight of space and missile activities, was disbanded after August 1991's failed coup. The RKA, created by President Yeltsin's decree in February 1992, rapidly established itself under Koptev, formerly MOM deputy minister, as the principal civil authority. Russia's parliament formalised its existence in 1993.

Even before 1991's events, Western perception of the Soviet space programme had undergone a remarkable transformation. Contrasting with the traditional view of a large and vigorous monolithic program, public criticism of space expenditure appeared and the concerted response revealed that Soviet space planners were having to justify and fight for their funding. As part of the reaction, detailed budget figures were released for the first time. They revealed that 1989's total space budget amounted to Rb6.9 billion, or almost 1.5 per cent of the state total. Koptev in late 1992 noted that 1989 represents the peak of funding. More than half, Rb3.9 billion, was allocated to military activities, Rb1.3 billion to development of reusable space systems (principally Energia/Buran) and Rb1.7 billion to the remaining civil programme of applications satellites and manned missions. 1988's figure for the last category was Rb1.3 billion. 1990's budget reportedly fell to Rb6.3 billion (including Rb220 million for the manned programme), and 1991's was Rb5.8 billion. The Academy of Sciences 1991 allocation was Rb300 million. Officials in April 1991 revealed that Rb31.6 billion had been expended 1986-1990, with Rb34.5-36 billion (including Rb15 billion for civil) requested for 1991-95.

Other quoted figures included Rb1,471 million for the whole Mir space station programme up to April 1989 and Rb14 billion over the 13 years of the Energia/Buran programmes. Meteorological satellites were claimed to return Rb500-700 million on their investment annually, supported in 1988 by Rb500 million from communications satellites. Overall, it was estimated that such projects with economic applications had returned some Rb12 billion in 33 years.

Planned investment for the 1991-95 period was some Rb6 billion, comprising Rb1 billion for development of new commercial satellites, Rb2 billion for science research and Rb3 billion for manned flights and new hardware.

The military has publicly criticised the power RKA has acquired, complaining that civil programmes are diverting funding from military efforts. 1993's civil appropriation in December 1993 was Rb164.8 billion (pegged to July 1993's rate, about US$130 million), only half of the 1980s' level after allowing for inflation. That sum breaks down as Rb126 billion for R&D, Rb7.8 billion for maintaining Baikonur, Rb21 billion for purchasing spacecraft production and Rb10 billion for capital construction. 1994's simultaneous allocation was Rb274.8 billion (July 1993 rate), breaking down as 167, 16, 81.8, 10, respectively.

1995's draft allocation to begin 1 April was reported in March 1995 to be Rb2,970 billion and Rb2,889 billion for military (inflation makes it difficult to compare on a year by year basis). Koptev has said that he wants civil budgets to outstrip the military's. Both have suffered, most dramatically from inflation, since the Rb6.9 billion peak in 1989. Civil space funding fell 10 to 12 per cent over 1990-92, while military development funding dropped 30 per cent and production funding 45 per cent for the same period.

Mir's visiting missions are offered through RKK Energia (through Glavcosmos until 1990) at a cost of US$12 million to US$15 million. Austria in November 1988 became the first to sign an agreement, flying in October 1991. Austria paid Russia US$6 million plus the contribution of research apparatus. That mission was preceded by Japanese and UK visits, and it was followed (July 1992) by France's third trip to a Russian station and Germany's first (March 1992). CNES/Energia signed a memorandum on 29 July 1992 for four further missions, in July 1993, August 1996 (Cassiopeia), 1998 and 2000 to exploit Mir's facilities. FFr126 million was paid jointly for the 1993/96 missions, plus equipment. The 2000 Mir visit cannot now occur. 2000's visit may, in fact, be made to Russia's portion of ISS. RKA/DARA signed an agreement in December 1995 for a second German visit, aboard TM25 in February 1997. ESA's two visits were the 30-day EuroMir-94 from October 1994 and the six-month EuroMir-95 from September 1995, paying €45 million for the pair. NASA pays more than US$300 million for its Shuttle visits.

Once it became clear in the 1980s that technology transfer restrictions imposed by the US had effectively strangled Soviet attempts at launcher marketing, the USSR began more vigorously pushing its microgravity, remote sensing, communications and manned space facilities. Commercial accommodation aboard the Photon μg and Resurs-F Earth resources craft are regularly leased. The drive for hard currency is providing an international impetus towards higher resolution Earth imaging, a move the US military has traditionally resisted. Some material has been available since the late 1980s, for example through Soyuzkarta, but its use has been hampered by a rigid and unresponsive system. But in 1993 Sovinformsputnik (whose dominant shareholder is the Central specialised Design Bureau, builders of Resurs-F and military systems) became licensed to sell data from some film cameras of military satellites. Likewise, Priroda in 1993 established the WorldMap consortium with the UK's JEBCO Information Services to sell digitised versions of its Resurs-F imagery. Its archives, holding two million negatives, are being digitised and added to a database to facilitate user searches. Priroda/WorldMap advertised three 1994 Resurs-F flights in advance and invited requests for imaging targets. However, no missions appeared because of a dearth of launch vehicles. Both organisations can provide 2 m resolution (and greater if allowed), exceeding western commercial capability. Priroda plans to distribute current digital 3 m resolution panchromatic imagery from a military satellite, presumably the 5th generation photo-reconnaissance design.

Mars-96's November 1996 departure could be the last launch of a Russian planetary probe for many years. Any talk of manned Mars expeditions is long ended. Funding for science missions has become so tight that external contributions have been required. CNES in early 1992 began contributing FFr1.5 million every six months to keep the Yevpatoria station open to control the joint Granat astronomical satellite.

Participation in the International Space Station should ensure the survival of Russia's manned programme and is providing a major source of hard currency. The international partners on 7 December 1993, formally invited Russia to join the project. The intention is to integrate Russian core elements and fly modified Soyuz-TM craft as attached rescue vehicles and uprated Progress as ferries. Russia will be paid upwards of US$600 million for its contributions, including the US$190 million agreed with Khrunichev in February 1995 for the Salyut. NASA Administrator Dan Goldin and RKA General Director Yuri Koptev signed an additional US$400 million USFY94-97 protocol on 16 December 1993 (with the final agreement signed 21 June 1994). It includes US$305 million for up to 10 Shuttle Mir 1 visits encompassing six NASA astronauts spending a total of 24 months aboard. The visits exist primarily for life/μg sciences research, crew exchanges, cargo delivery and experiment return.

RFAS/USSR PAYLOADS 1957-1998

	57-61	*62*	*63*	*64*	*65*	*66*	*67*	*68*	*69*	*70*	*71*	*72*	*73*	*74*	*75*	*76*	*77*	*78*	*79*	*80*	*81*	*82*	*83*	*84*	*85*	*86*	*87*	*88*	*89*	*90*	*91*	*92*	*93*	*94*	*95*	*96*	*97*	*98*	*Total*
1 Sputnik	4	-	-	-	-	-	-	-	-	-	-	-	-	-	-	-	-	-	-	-	-	-	-	-	-	-	-	-	-	-	-	-	-	-	-	-	-	-	4
2 Luna (Lunik)	3	-	2	-	4	5	-	1	1	2	2	1	1	2	-	1	-	-	-	-	-	-	-	-	-	-	-	-	-	-	-	-	-	-	-	-	-	-	25
3 Vostok/khod	7	2	2	1	1	-	-	-	-	-	-	-	-	-	-	-	-	-	-	-	-	-	-	-	-	-	-	-	-	-	-	-	-	-	-	-	-	-	13
4 Cosmos	-	12	12	27	52	34	61	64	55	72	81	72	85	74	85	101	86	96	79	88	94	97	94	94	99	96	97	79	68	66	54	55	38	38	21	10	10	17	2,363
5 Venera	2	3	-	-	3	-	2	-	2	2	-	2	-	-	2	-	-	2	-	-	2	-	2	2	-	-	-	-	-	-	-	-	-	-	-	-	-	-	26
6 Mars	2	3	-	-	-	-	-	-	-	-	2	-	4	-	-	-	-	-	-	-	-	-	-	-	-	-	-	-	-	-	-	-	-	-	-	1	-	-	12
7 Polyot	-	-	1	1	-	-	-	-	-	-	-	-	-	-	-	-	-	-	-	-	-	-	-	-	-	-	-	-	-	-	-	-	-	-	-	-	-	-	2
8 Elektron	-	-	-	4	-	-	-	-	-	-	-	-	-	-	-	-	-	-	-	-	-	-	-	-	-	-	-	-	-	-	-	-	-	-	-	-	-	-	4
9 Zond	-	-	-	2	1	-	-	3	1	1	-	-	-	-	-	-	-	-	-	-	-	-	-	-	-	-	-	-	-	-	-	-	-	-	-	-	-	-	8
10 Molniya	-	-	-	-	2	2	3	3	2	5	3	6	8	7	10	7	6	6	5	4	8	5	7	4	8	7	1	7	4	6	5	4	5	2	1	2	1	2	158
11 Proton	-	-	-	-	2	1	-	1	-	-	-	-	-	-	-	-	-	-	-	-	-	-	-	-	-	-	-	-	-	-	-	-	-	-	-	-	-	-	4
12 Soyuz	-	-	-	-	-	-	1	2	5	1	2	-	2	3	4	3	3	5	4	6	3	3	2	3	2	2	3	3	1	3	2	2	2	3	2	2	1	2	82
13 Meteor	-	-	-	-	-	-	-	-	2	4	4	3	2	5	4	3	4	-	3	2	2	2	1	1	3	1	2	2	2	2	2	-	1	1	-	-	-	-	58
14 Intercosmos	-	-	-	-	-	-	-	-	2	2	1	3	2	2	2	2	1	2	2	-	2	-	-	-	-	-	-	-	2	-	1	-	-	-	2	-	-	-	28
15 No desig	-	-	-	-	-	2	-	-	-	-	-	-	-	-	-	-	-	-	-	-	-	-	-	-	-	-	-	-	-	-	-	-	-	-	-	-	-	-	2
16 Salyut	-	-	-	-	-	-	-	-	-	-	1	-	1	2	-	1	1	-	-	-	-	1	-	-	-	-	-	-	-	-	-	-	-	-	-	-	-	-	7
17 Aureole	-	-	-	-	-	-	-	-	-	-	1	-	1	-	-	-	-	-	-	-	1	-	-	-	-	-	-	-	-	-	-	-	-	-	-	-	-	-	3
18 Prognoz	-	-	-	-	-	-	-	-	-	-	-	2	1	-	1	1	1	1	-	1	-	-	1	-	1	-	-	-	-	-	-	-	-	-	-	-	-	-	10
19 For other countries	-	-	-	-	-	-	-	-	-	-	-	1	-	-	2	-	1	-	1	-	1	-	-	-	-	-	-	1	-	-	2	-	1	1	8	4	19	15	57
20 Raduga	-	-	-	-	-	-	-	-	-	-	-	-	-	-	1	1	1	1	1	2	3	1	2	2	2	2	2	1	3	3	2	-	2	3	-	1	-	-	36
21 Ekran	-	-	-	-	-	-	-	-	-	-	-	-	-	-	-	1	1	-	2	2	1	2	2	2	1	1	2	2	-	-	-	1	-	-	-	-	-	-	20
22 Progress	-	-	-	-	-	-	-	-	-	-	-	-	-	-	-	-	-	4	3	4	1	4	2	5	1	2	7	6	4	4	5	5	5	5	5	3	4	3	82
23 Radio	-	-	-	-	-	-	-	-	-	-	-	-	-	-	-	-	-	2	-	-	6	-	-	-	-	-	-	-	-	-	-	-	-	1	-	-	-	-	9
24 Gorizont	-	-	-	-	-	-	-	-	-	-	-	-	-	-	-	-	-	1	2	1	-	2	2	2	1	2	1	2	3	3	2	3	2	1	-	2	-	-	32
25 Iskra	-	-	-	-	-	-	-	-	-	-	-	-	-	-	-	-	-	-	-	-	1	2	-	-	-	-	-	-	-	-	-	-	-	-	-	-	-	-	3
26 Astron	-	-	-	-	-	-	-	-	-	-	-	-	-	-	-	-	-	-	-	-	-	-	1	-	-	-	-	-	-	-	-	-	-	-	-	-	-	-	1
27 Mir/module	-	-	-	-	-	-	-	-	-	-	-	-	-	-	-	-	-	-	-	-	-	-	-	-	-	1	1	-	1	1	-	-	-	-	1	1	-	-	6
28 Phobos	-	-	-	-	-	-	-	-	-	-	-	-	-	-	-	-	-	-	-	-	-	-	-	-	-	-	-	2	-	-	-	-	-	-	-	2	-	-	4
29 Photon	-	-	-	-	-	-	-	-	-	-	-	-	-	-	-	-	-	-	-	-	-	-	-	-	-	-	-	1	1	1	1	1	-	1	1	-	1	-	8
30 Okean	-	-	-	-	-	-	-	-	-	-	-	-	-	-	-	-	-	-	-	-	-	-	-	-	-	-	-	1	-	1	1	-	-	1	1	-	-	-	5
31 Buran	-	-	-	-	-	-	-	-	-	-	-	-	-	-	-	-	-	-	-	-	-	-	-	-	-	-	-	1	-	-	-	-	-	-	-	-	-	-	1
32 Resurs-F	-	-	-	-	-	-	-	-	-	-	-	-	-	-	-	-	-	-	-	-	-	-	-	-	-	-	-	-	5	4	4	3	3	-	1	-	1	-	21
33 Pion	-	-	-	-	-	-	-	-	-	-	-	-	-	-	-	-	-	-	-	-	-	-	-	-	-	-	-	-	4	-	-	2	-	-	-	-	-	-	6
34 Nadezhda	-	-	-	-	-	-	-	-	-	-	-	-	-	-	-	-	-	-	-	-	-	-	-	-	-	-	-	-	1	1	1	-	-	1	-	-	-	-	4
35 Granat	-	-	-	-	-	-	-	-	-	-	-	-	-	-	-	-	-	-	-	-	-	-	-	-	-	-	-	-	1	-	-	-	-	-	-	-	-	-	1
36 Gamma	-	-	-	-	-	-	-	-	-	-	-	-	-	-	-	-	-	-	-	-	-	-	-	-	-	-	-	-	-	1	-	-	-	-	-	-	-	-	1
37 Informator	-	-	-	-	-	-	-	-	-	-	-	-	-	-	-	-	-	-	-	-	-	-	-	-	-	-	-	-	-	-	1	-	-	-	-	-	-	-	1
38 Almaz	-	-	-	-	-	-	-	-	-	-	-	-	-	-	-	-	-	-	-	-	-	-	-	-	-	-	-	-	-	-	1	-	-	-	-	-	-	-	1
39 Mak	-	-	-	-	-	-	-	-	-	-	-	-	-	-	-	-	-	-	-	-	-	-	-	-	-	-	-	-	-	-	1	1	-	-	-	-	-	-	2
40 Resurs 500	-	-	-	-	-	-	-	-	-	-	-	-	-	-	-	-	-	-	-	-	-	-	-	-	-	-	-	-	-	-	-	1	-	-	-	-	-	-	1
41 Gals	-	-	-	-	-	-	-	-	-	-	-	-	-	-	-	-	-	-	-	-	-	-	-	-	-	-	-	-	-	-	-	-	-	1	1	-	-	-	2
42 GEO-IK	-	-	-	-	-	-	-	-	-	-	-	-	-	-	-	-	-	-	-	-	-	-	-	-	-	-	-	-	-	-	-	-	-	1	-	-	-	-	1
43 Express	-	-	-	-	-	-	-	-	-	-	-	-	-	-	-	-	-	-	-	-	-	-	-	-	-	-	-	-	-	-	-	-	-	1	-	1	-	-	2
44 Coronas	-	-	-	-	-	-	-	-	-	-	-	-	-	-	-	-	-	-	-	-	-	-	-	-	-	-	-	-	-	-	-	-	-	1	-	-	-	-	1
45 Luch	-	-	-	-	-	-	-	-	-	-	-	-	-	-	-	-	-	-	-	-	-	-	-	-	-	-	-	-	-	-	-	-	-	1	1	-	-	-	2
46 Resurs-O	-	-	-	-	-	-	-	-	-	-	-	-	-	-	-	-	-	-	-	-	-	-	-	-	-	-	-	-	-	-	-	-	-	1	-	-	-	1	2
47 Elektro	-	-	-	-	-	-	-	-	-	-	-	-	-	-	-	-	-	-	-	-	-	-	-	-	-	-	-	-	-	-	-	-	-	1	-	-	-	-	1
48 Tsikada	-	-	-	-	-	-	-	-	-	-	-	-	-	-	-	-	-	-	-	-	-	-	-	-	-	-	-	-	-	-	-	-	-	-	1	-	-	-	1
49 EKA	-	-	-	-	-	-	-	-	-	-	-	-	-	-	-	-	-	-	-	-	-	-	-	-	-	-	-	-	-	-	-	-	-	-	1	-	-	-	1
50 Gonets-10	-	-	-	-	-	-	-	-	-	-	-	-	-	-	-	-	-	-	-	-	-	-	-	-	-	-	-	-	-	-	-	-	-	-	-	3	-	-	3
51 Bion	-	-	-	-	-	-	-	-	-	-	-	-	-	-	-	-	-	-	-	-	-	-	-	-	-	-	-	-	-	-	-	-	-	-	-	1	-	-	1
52 Bankir	-	-	-	-	-	-	-	-	-	-	-	-	-	-	-	-	-	-	-	-	-	-	-	-	-	-	-	-	-	-	-	-	-	-	-	-	1	-	1
53 Tubsat	-	-	-	-	-	-	-	-	-	-	-	-	-	-	-	-	-	-	-	-	-	-	-	-	-	-	-	-	-	-	-	-	-	-	-	-	-	2	2
54 ISS module	-	-	-	-	-	-	-	-	-	-	-	-	-	-	-	-	-	-	-	-	-	-	-	-	-	-	-	-	-	-	-	-	-	-	-	-	-	1	1
Totals	**18**	**20**	**17**	**35**	**65**	**44**	**67**	**74**	**70**	**89**	**97**	**90**	**107**	**95**	**111**	**121**	**105**	**120**	**102**	**110**	**125**	**119**	**116**	**115**	**118**	**114**	**116**	**108**	**100**	**96**	**85**	**78**	**59**	**65**	**47**	**33**	**38**	**43**	**3,132**

SOUTH AFRICA

The country's overt involvement in space has been modest, principally comprising the Satellite Applications Centre. Recent years saw attempts to find commercial applications of defence technologies, but ultimately without success. Denel (Pty) Ltd, formed in 1992 to take on the mantle of South Africa's defence manufacturing from Armscor and expand into the commercial sector, began promoting industrial space capabilities. Although the company is government-funded, projects must show commercial promise for them to be continued.

The government agency charged with developing plans, CSIR, considered proposals in 1988 for a national space programme emphasising communications and remote sensing, but including the possibility of developing an indigenous orbital launcher. The Minister of Trade and Industries announced the formation of a South African Space Committee in 1990. Its brief included the monitoring of trends in space technology applications and encouraging appropriate national utilisation. Committee representation covered CSIR, Dept of Posts and Telecommunications, the Weather Bureau, Dept of Trade and Industries, Industrial Development Corp, and private and official broadcast agencies and academic interests. However, the Committee was disbanded in 1991. The government in 1993 passed legislation to establish a Council for Space Affairs to manage and control national space activities, reporting to the Dept of Trade and Industries, but it did not appear.

Denel's Houwteq division, specialising in missiles and high tech systems, spearheaded the design of the Greensat Earth observation satellite. However, Denel announced in October 1994 that Greensat was being cancelled because of the lack of interest and that Houwteq's space activities had ended.

Denel had planned a three-stage all-solid launcher for delivering small payloads to LEO from 1995, but announced in June 1993, after expending US$55 million, that it would be uneconomic. Termination removed a major obstacle to South Africa's membership of the MTCR Missile Technology Control Regime, which became effective in 1995. Denel had two motors which would still be static tested. Three successful test flights were made from the Overberg Test Range (part of Denel) near Bredasdorp north of Cape Town: June 1989's flew only stage 1 and stages 1/2 were tested July 1989 and November 1990. 1989's flight was originally associated with Israel's Jericho 2 missile or Shavit space launcher; there seems to be some collaboration. Israel was apparently closely linked with South Africa's space efforts.

Of the Denel group, Somchem in Somerset West near Cape Town specialises in propellants and propulsion systems, Kentron in guided weapons and systems, and Eloptro in electro-optical systems.

Telkom SA is the national signatory to Intelsat and Inmarsat (joined April 1994). Studies have been made of a domestic 'Syncom' system. The Satellite Applications Centre (part of the Council for Scientific & Industrial Research, now known only as CSIR) receives, processes and markets ERS, Landsat, Spot and NOAA data and provides TT&C services to foreign space agencies. For example, it currently operates as part of France's 2 GHz TT&C network.

SPAIN

Although the majority of Spain's space activities continue to be routed through ESA, there is a growing national programme. Spain launched a domestic telecommunications satellite service in 1992, Ariane launched a micro-satellite in 1995, a small platform flew in 1997 and a micro-launcher is being developed.

Hispasat SA was created in 1989 to operate the domestic satellite system. The satellites also provide secure government military links at X-band, complementing Spain's 7 per cent participation in France's Helios 1 reconnaissance satellite project.

The nine-nation Western European Union inaugurated a satellite data interpretation and training centre in 1993 at Spain's Torrejon Air Base. Minisat has been developed as a national entry into the space market and studies began in 1991 of the three-stage Capricornio, capable of placing about 100 kg in orbit from Spain. Pedro Duque was selected in May 1992 by ESA as Spain's first astronaut; he was back-up for October 1994's Mir visit. Discussions for a national visiting mission to Russia's Mir space station were conducted in 1989, but no accord was reached.

1999 ESA contribution is €102.1 million or 4.5 per cent of the agency's total budget. 1995 3.9 per cent (AU105.4 million), 1994 4.7 per cent (AU102.2 million), 1993 4.0 per cent (AU97.8 million), 1992 4.0 per cent (AU113.1 million), 1991 4.1 per cent (AU96.3 million); 1990 4.0 per cent (AU71.6 million); 1989 3.8 per cent (AU57.3 million). Space expenditure more than doubled over 1984-87 to about Pta6.5 billion; by 1991 the ESA contribution was Pta12,680 million. Involvement in ESA programmes has been modest, taking small but significant roles in Ariane 5, Envisat, Polar Platform, Data Relay Satellite; it is now contributing 2 per cent to the Columbus Orbital Facility and the Automated Transfer Vehicle.

National space planning and co-ordination is undertaken by the Ministry of Industry & Energy's Centre for the Development of Industrial Technology (CDTI), which in 1986 replaced the Ministry of Defence's Comisión Nacional de Investigación del Espacio (CONIE). INTA (see below) provides R&D and its INSA subsidiary operates ESA and NASA tracking stations at Villafranca (IUE control centre) and Madrid.

Madrid's 70 m, two 34 m and 26 m antennas are part of NASA's Deep Space Network for controlling spacecraft such as Voyager and Pioneer. A renewed space co-operation agreement with the US was signed 11 July 1991, including the continued use of emergency Shuttle landing sites (Moron and Zaragoza). The Instituto Nacional de Meteorología is the national signatory to EUMETSAT, and Telefónica to Inmarsat, Eutelsat and Intelsat.

SWEDEN

Sweden has been an ESA member since 1975, accounting for 2.6 per cent of members' contributions in 1999, 2.7 per cent in 1996, 2.7 per cent 1995, 2.6 per cent 1994, 2.3 per cent 1993, 2.2 per cent 1992, 2.3 per cent 1991, 2.3 per cent 1990, 2.5 per cent 1989, 2.5 per cent 1988 and 2.2 per cent 1987. Bilateral co-operation includes France (Spot) and Germany (Sanger). An MoU was signed with Germany in 1990 agreeing to share hypersonic information; Sweden expended SKr30 million to end-1992 on hypersonic work in Germany's Phase 1 programme. It has also made contributions to CIS/USSR space projects, including Prognoz and Interkosmos. Astrid 1 was the first western satellite launched on a Cosmos-3M. Physicist Dr Christer Fuglesang was selected in May 1992 by ESA as one of six new astronauts: he was back-up for Mir's Soyuz-TM 22 mission in September 1995.

Sweden joined the German Texus μg sounding rocket programme in 1977 and 10 years later began its own Maser, offering one launch annually. The former DASA and the Swedish Space Corp developed the Maxus Castor 4-based vehicle to extend μg time to more than 14 minutes. Sweden's guidance control system ensures the vehicle and payload descend within Esrange. The maiden launch failed in May 1991; number 2 was successful 8 November 1992 and number 3 28 November 1995; number 4 was planned for April 1998.

In April 1979, the government embarked upon a national space programme to enable its industries to be more competitive in the international space communications market. The scientific Viking and the Tele-X direct broadcast satellite were the first results. Telia AB (previously Swedish Telecom), the national signatory to Intelsat, Inmarsat and Eutelsat, operates as a private company but is wholly government-owned, under the Ministry for Transport & Communications.

Sweden's space budget comprises two primary elements: national and ESA/bilateral. The 1997 budget total was SKr548 million. Within its SKr573 million ESA contribution for 18 months in 1995 and 96 (559 1994/95; 556 1993/94; 469 1992/93; 414 1991/92; 418 1990/91; 280 1989/90), it participates in the mandatory basic and scientific program, Ariane 5, telecommunications, remote sensing (including ERS) and μg activities. National interests cover space science, remote sensing, Esrange management and industrial development. Scientific research is traditionally focused on satellite and sounding rocket magnetospheric/ionospheric investigations, including modest but important collaboration on CIS missions.

The other national activity aims to enhance industry's capabilities through satellite projects such as Viking, Freja and Odin. Saab Ericsson Space and Volvo represent the primary national industrial space capability. Three DBS channels from Sweden's Tele-X satellite were augmented in 1994 with the purchase of the UK's Marcopolo 1. Tele-X's 1997 Sirius replacement was contracted in 1995.

Sweden's Space Budget (in millions of SKr)

National Activities	**1994/95**	**1995/96**	**1997**
Management	13	20	14
Basic activities (Esrange)	13	22	15
Space science (including μg)	51	72	46
Remote sensing	18	26	18
Industrial development	47	47	28
National total	142	187	121
International Activities	**1994/95**	**1995/96**	**1997**
Basic activities	64	89	62
Space science	99	129	82
Space transportation	160	235	146
Manned programme	14	40	14
Telecommunications	61	61	22
Remote sensing	238	226	184
Microgravity	38	49	38
International total	674	829	548
Total	**816**	**1,016**	**669**

The annual accounting period was originally 1 July to the following 30 June, but it was decided that the accounting period should be adjusted to begin on 1 January: as a result the 1995/96 figures cover the 18 months period of 1 July 1995 to 31 December 1996. The totals for earlier years are 1993/94 – SKr733 million, 1992/93 – SKr662 million, 1991-92 – SKr589 million, 1990/91 – SKr561 million.

SWITZERLAND

Switzerland's space involvement is almost totally routed through ESA – accounting for 2.8 per cent of the members' contributions. Although modest, funding for ESA more than tripled during the period 1984 to 1994 and has remained almost stable since then: SFr27.5 million in 1984, growing to SFr114 million in 1994, SFr110.8 million in 1995, SFr111.7 million in 1996, SFr113.4 million in 1997 and SFr347 million planned for 1998-2000.

The Federal Council (government) determines the national space policy under the advice of a 20-member Federal Space Affairs Commission (CFAS). Science projects fall under the Federal Office of Education and Science in the Federal Department of the Interior and are carried out by the universities and federal institutes of technology with funding from the Swiss Academy of Sciences' Space Research Commission, the Swiss National Science Foundation and the various Cantons. Switzerland is a signatory to Intelsat, Eutelsat and Inmarsat through the PTT General Directorate of Posts, Telephones and Telecommunications: it is also a member of Eumetsat.

Since 1988, Switzerland has been contributing 4 per cent to all of the new Earth observation and microgravity development programmes. These programmes, together with the mandatory ESA science programme, give Swiss scientists the opportunity when selected to fly experiments on ESA missions which are mainly funded through the ESA Prodex (SFr7 million per year). In the field of launchers Switzerland participates with 2 per cent of the main Ariane-5 development, 2.5 per cent to the Ariane-5 infrastructure programme, 0.65 per cent for Ariane-5 research and technology and 0.38 per cent at the Ariane-5 evolution programme. At the ESA ministerial meeting in October 1995 Switzerland agreed to contribute 2.5 per cent to the European participation in the International Space Station programme, with its main interest being the Automated Transfer Vehicle. Swiss participation in telecommunications development programmes has been limited to a 2 per cent share in the Artemis programme, but following a decision in 1995, future funding of technology and telecommunications programmes has been stabilised at SFr9 million.

No national programme is being pursued. However, the Bern-based International Space Science Institute, co-funded by Switzerland and ESA, was created in 1995 to fully exploit space missions in a multi-disciplinary environment – in particular, the solar-terrestrial physics missions presently co-ordinated by ESA, ISAS, IKI and NASA. The Observatory of Geneva holds the recently opened ESA Integral Science Data, which will serve as an access point for the astronomical community.

More than 30 Swiss companies are directly involved in space; products and on-going developments include satellite structures, Ariane fairings, MGSE, pointing mechanisms, cryogenic valves, advanced motors, sensors, bioreactors, atomic clocks, electronic units, EGSE and dedicated space science instruments. ESA has nominated the Observatory of Neuchâtel as its external laboratory for time and frequency. Swiss scientists are actively involved in Ulysses, HST, ISO, Soho and XMM. There is particular interest in ERS SAR activities, as well as data interpretation from Spot, Landsat and NOAA. EuroMir, Shuttle, Spacelaboratory and sounding rockets are employed for life sciences research.

Claude Nicollier, from Vevey, flew as the first non-American Mission Specialist aboard 1992's Shuttle mission STS-46. He was responsible for deploying ESA's Eureca platform. He flew again on STS-61 (December 1993 – first Hubble servicing mission) and STS-75 (February 1996 – second Tethered Satellite mission).

SYRIA

The country is a member of Intersputnik, joining in October 1991, and Intelsat but not of Inmarsat or Eutelsat. On 22 July 1987, Lt. Col. Mohammed Faris, 36, a Syrian Air Force pilot, represented his country on the Soviets' 12th international mission to a space station. Launched on Soyuz-TM 3, Faris was aboard Mir for six days. In addition to participating in materials processing experiments, he conducted photographic surveys of Syria to help the search for water and mineral resources.

TAIWAN

Taiwan authorised an NT$13.6 billion (about US$540 million) 15-year programme in December 1990 to begin development of space technologies. The premier also appointed a five-member Space Programme Planning Committee, which identified feasible goals and developed a long-range strategic plan by August 1991. The National Space Programme Office, under the auspices of the National Science Council, was established in 1992. The principal goal is to develop an indigenous space technology capability. The programme is creating a cadre of 300 personnel capable of developing future space missions.

The basic plan includes:

the purchase and operation of a remote sensing satellite ground station (operations began 1993);

the establishment of a small satellite development and operations capability, constructing three Rocsat satellites for launch at five-year intervals by existing foreign vehicles.

TRW was awarded the Rocsat 1 bus development contract. It is working alongside Taiwanese companies to help develop critical technologies and transfer capabilities: Taiwan is performing the systems engineering, integration and testing, while TRW will design, manufacture, integrate and test the bus. A team led by AlliedSignal Technical Services Group was awarded a contract in January 1995 worth potentially US$31.6 million for the ground system. France's Intespace was awarded a 1993 contract to assist in building an integration/test centre, including space simulation, vibration, acoustic and metrology facilities.

The Directorate General of Telecommunications and Singapore Telecom in March 1996 jointly ordered the ST-1 satellite from MMS to provide regional broadcasting and telecommunications services from Singapore's 88° east in 1998. The country is not a member of Intelsat or Inmarsat.

THAILAND

Shinawatra Computer & Communications Group, the largest group of Thai companies operating telecommunications and broadcasting service concessions, was selected in 1991 for a 30-year satellite concession to lease transponders to government and private sector users, with excess capacity offered to neighbouring countries. The Hughes HS-376 Thaicom 1 was launched successfully 18 December 1993; number 2 followed 8 October 1994. The larger Thaicom 3 was launched 16 April 1997 by an Ariane-4, to be followed by the duplicate number 4. The first pair offers up to 25 DTH channels; their successors' high power Ku-band payload will be principally devoted to DTH.

The Asian Broadcasting and Communications Network Co Ltd has a 30-year concession from Laos to operate the L-Star DTH and telecommunications services system. Launch is projected for December 1997. Thailand joined Inmarsat in January 1995, through its signatory Post & Telegraph Dept. The first Inmarsat land Earth station is expected at Nonthaburi. The country has been a member of Intelsat since 1966. Shinawatra Satellite and the Communications Authority of Thailand are jointly investing 5 per cent (US$65.4 million) in ICO Global Communications. Population is about 55 million.

The Thai MicroSatellite Co signed a £3 million contract 28 November 1995 with the UK's Surrey Satellite Technology Ltd for a 50 kg micro-satellite. The 18-month technology transfer programme involves a team of Thai engineers working alongside SSTL staff as TMSAT is built. It will carry Earth imaging cameras and communications payloads, controlled from a station installed by SSTL in Bangkok. Its GPS receiver is being developed by SSTL under ESA contract to provide position and attitude data.

China's Xinhua news agency claimed November 1993 that Thailand's Ministry of Transport & Communications planned to establish a satellite technology centre with Chinese assistance. It further claimed there was already an agreement for help in building and launching two small weather and science satellites. The Thailand Remote Sensing Centre operates a Landsat/Spot/MOS/ERS/JERS receiving station in Bangkok. In a region plagued by dense cloud cover up to six months annually, there is interest in microwave data from satellites such as ERS.

On 31 March 1997, Radio Thailand carried a report that there were plans to develop military satellites. It was stated that the Military Satellite for Defence Project was important because the cost would not be too great and it was essential for modern warfare. Two systems were noted as forming the project: one would be a satellite to be used for telecommunications and command, while the other would be for military surveillance and intelligence purposes. No schedules, budget levels, and so on, were given in the broadcast, but three steps in the programme were planned: establishment of ground stations, acquisition of high resolution surveillance satellites and finally the acquisition of a telecommunications satellite. It was stated that the first stage would not require a large budget, while the second and third stages would depend upon the country's economic situation. These plans have been temporarily suspended due to economic difficulties.

TURKEY

Turkey began operating a domestic satellite system in 1994 for TV distribution (including Europe), data and voice services. Its first satellite was lost in Ariane's January 1994 failure but the second was successful in August 1994. Insurance funded 1996's replacement launched on an Ariane-4, 9 July 1996. A joint venture with France's Aerospatiale will provide a second generation satellite before the end of the century.

The country is a member of Intelsat, Eutelsat and Inmarsat, operating two stations at Ata to provide Inmarsat coverage of the Indian Ocean and Atlantic Ocean regions. A ground station for Landsat 6 imagery was planned near Ankara, but none has been built since the launch failure.

UKRAINE

Ukraine, the second major space force to emerge from the break-up of the Soviet Union, is pursuing an independent policy. The future of its major space enterprise, NPO Yuzhnoye, was uncertain. After it appeared the state considered halting production of the Zenit orbital launcher. As commercial markets appeared in the early 1990s, though, the pressure relaxed and now the Zenit's future is assured. The first commercial success came with the May 1995 selection by Space Systems/Loral for three 1998 launches to carry 36 US Globalstar satellites. Boeing, too, issued a vote of confidence in the Zenit when it selected the booster for the Sea Launch venture. This projects calls for three-stage Zenits to be shipped to the US, assembled and then towed into international waters for near-Equator launches. The world's largest commercial telecommunications satellite manufacturer, Hughes Space & Communications Co, placed a firm order in December 1995 with Sea Launch for 10 launches beginning in 1998. At the end of 1998, however, political forces associated with technology transfer questions have prevented Sea Launch from inaugurating its service.

The Ukraine and US signed a launch agreement on 14 December 1995 that will allow Ukraine to launch 16 western satellites through to 2000. Five are allowed on purely Ukrainian vehicles (six if the market grows) and 11 on joint US/Ukrainian ventures, effectively Sea Launch.

The future of Ukraine's Tsyklon launcher is less assured – its launch rate is falling under funding limitations and the loss of suppliers for the ageing avionics. Tsyklon production appears to have halted; it has been suggested that some of Russia's Tsyklon-M strategic stockpile may be converted into the three-stage version. 1995's sole Tsyklon delivered the national Sich ocean radar satellite, controlled by Russia's Military Space Forces until Yevpatoria took it over from October 1995. A Ukrainian spaceport is not likely because of the investment required, although polar orbits would be accessible for southern departures from the Black Sea coast. The new generation Sich/Okean is Ukraine's major national satellite programme, expanding to provide extensive remote sensing capabilities. 1998's Sich 2 will include a synthetic aperture radar with 10 m resolution.

Ukraine President Leonid Kravchuk announced on 2 March 1992 the creation of the Ukrainian Space Agency. The Commission for Space Research was also established, under the Academy of Sciences. Presidents Leonid Kuchma and Bill Clinton establishing direct contact between their space agencies for the first time signed a five-year agreement November 1994. Ukrainian cosmonaut Leonid Kadenyuk flew on Shuttle STS-87 in November 1997 as a consequence. Kazakhstan issued a decree the previous September setting up the Kazakhstan Space Research Agency; cosmonaut Toktar Aubakirov became its General Director in 1993 (when it became the National Aerospace Agency). He stepped down in May 1994 after his election as a parliamentary deputy. Alsultan Kalybayev replaced him. Azerbaijan formed its own ANAKA aerospace agency in February 1992. The Uzbek State Agency for Space Research (Uzbekkosmos) followed in March 1993, headed by Kamol Muminov. Tajikistan approved a plan in January 1994 to set up a space research centre under its Academy of Sciences.

UNITED KINGDOM

The United Kingdom's space programme is brought together by BNSC, the British National Space Centre. Two-thirds of the UK's steady-state £200 million annual space budget is devoted to ESA programmes, with the balance supporting national R&D activity through the universities, research centres and companies.

Historically, the UK was the sixth nation to launch its own satellite into orbit when Prospero flew aboard the fourth Black Arrow launch vehicle (designated R-3) on 28 October 1971: however, this was the final Black Arrow launch to be funded – planned further launches had been cancelled in July 1971. The first two Black Arrow launches had been sub-orbital missions (R-0 on 28 June 1969 failed, R-1 on 4 March 1970 with the X-1 payload was successful), while the third had been the first orbital attempt (2 September 1970 with the Orba/X-2 payload) failed to reach orbit. A sad comment on the programme was made in the United States Library of Congress report 'Worldwide Space Activities' in the introduction to the Black Arrow discussion: 'The only continuous thread which runs through the history of the British space programme has been its lack of continuity'.

MMS continues to market the successful series of Skylark sounding rockets, first launched in 1957 and the backbone of European Texus microgravity launches. MMS has worked with ESA partners on possible two-stage to orbit systems using the Russian Antonov An-225 and interim Hotol spaceplane, with results showing a medium term feasibility. UK interest in reusable launchers is now led by Reaction Engines, with its Skylon vehicle using a derivative of the Hotol dual-mode engine with a new airframe design: one of the people behind Reaction Engines is Alan Bond, who was responsible for the original Hotol concept.

The first Briton in space, food scientist Helen Sharman, flew to Russia's Mir orbital station aboard Soyuz-TM 12 in May 1991 under the commercial Project Juno. British-born Michael Foale (now joint US/UK nationality), a career astronaut rather than a one-off commercial flier, was selected in 1989 as a NASA mission specialist: his first flight was aboard STS-45 in 1992 and when flying STS-63 in 1995 he became the first British-born person to perform an EVA. He flew to Mir in May 1997 aboard the STS-84 mission and returned to Earth in October 1997 aboard STS-86.

By early 1999, the UK had flown 22 indigenously-built national satellites: one launched by Black Arrow, 13 by the US and seven by Ariane. The total includes the five UoSATs of the University of Surrey. These are in addition to craft produced for international and commercial programmes, such as Giotto and Inmarsat. Prospero remains the only UK satellite launched on a UK vehicle. Like the original US/USSR programmes, the national space effort depended initially on the conversion of a ballistic missile into a launcher. This was the Hawker Siddeley Blue Streak, finally cancelled as a missile in 1960 though continued until 1973 as first stage for ELDO's Europa. A more modest attempt at a national launch vehicle was announced in September 1964, developing a small three-stage vehicle from the successful Westland Black Knight research rocket. Construction of three Black Arrows was ordered in March 1967. Success in October 1971 with only the second complete vehicle was contrasted by the programme's previous cancellation in July 1971. The final Black Arrow now resides in a museum. With Prospero, the UK became the sixth nation to achieve orbital capability – after the USSR, USA, France, Japan and China.

Following up its success with the Ariel science series, Britain provided the ground control for the three-nation Infra-red Astronomical Satellite, launched January 1983. RAL also provided the UKs satellite to 1984's AMPTE investigation of the magnetosphere.

By early 1984, NASA had agreed to carry a UK payload specialist on each of the Skynet 4 Shuttle launches due in 1986. Sq-Ldr Nigel Wood was chosen, from four candidates who were given six months' training at an 'expenses only' cost of about US$150,000 each, as prime candidate for the first mission. *Challenger*'s loss in January 1986 resulted in the Skynets transferring to expendable launchers and the team disbanding. NASA's overtures about its readiness to carry a British PS in 1985 met with little response, and the opportunity to have a British scientist accompany the ambitious solar X-ray telescope aboard the Spacelab 2 mission in July was missed. The agreement for the commercially-sponsored non-government Project Juno visiting mission to the Soviet Mir space station, including 26 UK experiments, was signed with the Soviets 29 June 1989. Sponsorship target was £16 million, about £9 million of which was to be paid to the Soviets. The City of London Moscow Narodny Bank (the only wholly owned Soviet state bank outside of the USSR) managed the project through its Antequera Ltd company. Scientist Helen Sharman and Army Air Corps pilot Major Tim Mace, selected from 13,000 applicants, began training in Star Town in November 1989. Juno was re-oriented during 1990 when it became clear that sponsorship was insufficient; the Soviets agreed to continue the flight at below commercial rates and to provide the experiments.

UK space priorities were re-cast in 1996, following long consultation between the government, industry and academia. In a new 'Forward Plan' for space, announced by the then Space Minister, Ian Taylor MP in July 1996, some new aims were set for BNSC and for the UK in space. These aims built on past UK activities to provide a coherent programme for the future and BNSC is charged with updating them annually. These aims are:

Foster the development of a competitive Earth/observation industry.

Develop supportive technology and R&D programmes driven by future needs.

Maintain continuity of global datasets to meteorology and climate monitoring.

Establish a recognised means for procuring Earth observation services, and microgravity.

To exploit, sustain and develop UK science expertise and the links and mutual support which science, Earth observation and industrial applications share.

Support a strong ESA science programme.

Pursue opportunities for bilateral activity and a national 'small missions' programme; and maintain interest in ESA microgravity options.

Encourage a new European approach to spectrum authorising and regulations.

Promote international standards and industrial access.

Take significant roles in Europa's GNSS (Global Navigation Satellite System) programme.

Aid competitiveness through strong partnerships with industry (including R&D and technology demonstrators).

Urge the commercialising of international satellite organisations; and help the development of multimedia services by seeking government uses for them.

Help Europe optimise use of its own launcher services and those provided elsewhere and seek ways of reducing costs while stimulating availability.

Encourage military/civil space synergy; support UK industrial competitiveness.

Utilise and develop technology and technology transfer.

Encourage synergy between ESA (as a provider) and the European Union (as a user) of space services and facilities.

Encourage more co-funding with industry; and heighten awareness and applications of space in society.

Recent BNSC studies have indicated that the UK has a thriving space industry sector with more than 400 organisations, a turnover of £750 million, a workforce of some 6,500 and markets throughout the world.

BNSC itself was established in 1985 to bring together the space interests of government department and

UK civil space budget (£ million; ESA/national)

	1992	**1993**	**1994**	**1995**	**1996**	**1997**	**1998**
Science/μg	28/14	30.05/14.90	35.31/12.11	39.32/13.21	41.61/12.11	41.69/9.51	41.73/7.61
Earth Observation	42/43	45.73/41.14	47.91/48.26	48.58/45.80	53.93/47.84	59.11/48.36	64.72/47.35
Telecoms	18/1	15.62/0.76	16.00/1.32	11.66/4.00	10.55/3.94	6.35/3.30	3.10/3.17
Technology	– /4	– /5.15	– /4.45	0.26/0.31	0.26/0.73	0.26/0.73	– /0.73
Transportation	1/ –	1.58/ –	4.31/0.19	4.63/0.11	5.14/0.19	5.55/0.19	6.61/0.19
ESA General	12/ –	15.90/ –	12.72/ –	16.85/ –	16.29/ –	16.68/ –	16.63/ –
Other national	–	– /0.48	– /0.24	– /0.26	– /0.24	– /0.24	– /0.24
Total	**101/62**	**108.88/62.43**	**116.25/66.57**	**121.30/63.69**	**127.78/65.05**	**129.64/62.33**	**132.79/59.29**

1984 94; 1985 112; 1986 116; 1987 121.1; 1988 132.9; 1989 143; 1990 148; 1991 158. Financial year begins April.

research councils, optimise value for money and provide the UK with a coherent position in European and World space activities. The government through the BNSC partnership, initially focused the UK space effort in satellite communications.

With major participation through BNSC in infrastructure projects, the government enabled and encouraged industrial investment and a sound commercial position for the UK in telecommunications before, in the late 1980s, moving to meet the challenge of Earth observation.

UK contributions to international Earth observation programmes – especially through ESA – are considerable. They include two major instruments on both ESA's ERS (European Remote Sensing) satellites, a 22.5 per cent stake in the next-generation Polar Platform and a 17.5 per cent stake in ENVISAT and its ground segment. ENVISAT is the large Earth observation satellite due for launch by ESA in 2000. The stake in ENVISAT is worth up to £250 million in the UK over 10 years in the form of support of industrial involvement in the platform's development, and in instrument development. The UK provides two key instruments – ASAR (Advanced Synthetic Aperture Radar) and AATSR (Advanced Along-Track Scanning Radiometer).

The ERS missions (ERS 2 is still active) have justified the UK's funding share of £130 million, returning a steady stream of data about the Earth's oceans, ice, land and atmosphere. Key instruments – AMI (SAR) and ATSR, both forerunners of ENVISAT instruments – were provided from the UK. ERS has enabled scientists to derive a series of new technologies. In 1995, when both ERS 1 and ERS 2 were in operation in tandem, joint interpretation of the data led to interferometric techniques capable of detecting surface movements of around 2 cm. At this level of resolution scientists can identify areas of subsidence and ground heave, thus making it possible to predict and monitor the effects of eruptions, earthquakes and flooding with unprecedented accuracy.

ERS data are acquired, processed and distributed by the £20 million Earth Observation Data Centre at Farnborough, one of four ESA Processing and Archiving Facilities (PAF). UK PAF operations are managed under ESA contract by the National Remote Sensing Centre Ltd.

The UK is providing instruments for other international Earth observing missions: HIRDLS (High Resolution Dynamic Limb Sounder) on NASA's EOS and AMSU (Advanced Microwave Humidity Sounder) on later NOAAs. BNSC's partner NERC is part-funding a Global Earth Radiation budget instrument for EUMETSAT's second generation METEOSAT.

Communications remain important commercially to the UK, which is committed to four key ESA programmes. The programmes are the DRTM (Data Relay and Technology Mission, 6.7 per cent to Artemis and 0.5 per cent to the Data Relay Satellite), EMS (European Mobile System) on ITALSAT 2, ARTES Element 4 (competitively improvement programme) and RTES Element 9 (GNSS).

Two Hughes-built satellites, Marcopolos 1 and 2, were launched for the commercial British Satellite Broadcasting (BSB) company and became operational in 1990, providing direct broadcast channels. In late 1990, BSB merged with Sky Television (to become BSkyB). As a result in December 1993 Marcopolo 1 was sold to Nordiska Satellit AB in Sweden who operated it under the new name Sirius 1, while in 1992, Norwegian Telecom (now Telenor Satellite Services AS) had purchased Marcopolo 2 and operated it re-named to Thor 1.

Industrial involvement in satellite navigation has been supported with good effect – Racal's Global Skyfix network now has some 60 reference points worldwide, manned around the clock, providing positioning information to a wide spectrum of users. The UK is a strong advocate through ESA of the EGNOS civil overlay satellite communications positioning system and the planned GNSS programme as a whole.

The UK's Skynet satellite communications system remains the leading one outside the USA and CIS committed to military communications. The UK with British Telecom as a signatory participates in Inmarsat (9.36 per cent), Intelsat (9.19 per cent) and Eutelsat (21.67 per cent) and the UK hosts the headquarters for the Maritime Communications Agency.

While UK science ambitions suffered a serious setback, in June 1996, with the uninsured loss of the Cluster solar monitoring missions on Ariane 501, current missions with UK participation such as ESA's Ulysses, ISO and SOHO, the ESA-NASA Hubble Space Telescope and the US/German/UK ROSAT and future missions with a UK contribution such as Russia's Spektrum-RG, ESA's Rosetta cometary probe and the NASA-ESA Cassini-Huygens mission to Saturn proceeded according to plan. BNSC worked to ensure ESA's smoothest adoption of a new space science strategy, Horizon 2000+, which commences in 2006.

The loss of Ariane 501, carrying Cluster, was immediately subjected to an ESA investigation, from which it was clear that no UK components were at fault. Two UK companies contribute to Ariane 5 and the UK has a commitment of £2.8 million over three years to the new launch vehicle's infrastructure programme. Proving flights are expected to resume in the third or fourth quarter of 1997. The UK continues its parallel commitment to the highly-successful Ariane 4 launch vehicle family, with various companies supplying equipment for the vehicle: MMS supplies the SPELDA payload units and GEC-Marconi Avionics the ring laser guidance units.

The telecommunications expertise developed by British Aerospace and Matra Marconi Space UK has provided the greatest return on space investment. BAe/MMS France in the Satcom International partnership jointly developed the successful Eurostar platform. Marconi Space and Matra Espace merged in 1990 to create Matra Marconi Space to compete in the changing European market against strong aggregations such as Deutsche Aerospace. MMS acquired BAe Space Systems in 1994 for £56 million, at the time, creating Europe's largest space hardware group and the world's third (after Hughes and Lockheed Martin), with annual sales of more than US$1 billion. BAe was once Europe's leader in telecommunications satellite manufacturing and perhaps a factor in its decline was the government's late 1980s switch from satellite communications to remote sensing. BAe acted as prime contractor for the £34 million Giotto probe, which encountered Halley's Comet in March 1986, and ESA's experimental Olympus communications satellite. The company also provided 20 Spacelab pallets in return for Britain's 6 per cent contribution to Spacelab.

In response to a forecast that a world remote sensing market worth £250 million a year would develop by 1990, Britain invested £40 million 1983-86 in a joint science and industry programme covering data acquisition, processing, dissemination, interpretation and forecasting. The RAE (now DRA) National Remote Sensing Centre at Farnborough was established in 1980 and joined a decade later by the Earth Observation Data Centre, initially to process, archive and distribute ERS data. MMS (ex-BAe) now leads ESA's Polar Platform.

UNITED STATES OF AMERICA

NASA hoped that the recommendations of 1990's Advisory Committee on the Future of US Space Policy (generally known as the Augustine Committee after its Chairman) would set the tone for the decade. It concluded that the agency's budget should grow 10 per cent annually in real terms, attaining US$30 billion (FY90 rates) by 2000. This is equivalent to 0.4 per cent GDP, or about half of peak spending during the 1960s. It quickly became clear, however, that civil and military budgets would remain flat at best for the foreseeable future, together accounting for 24 per cent of the federal budget. In fact, NASA's FY95 appropriation, at US$14.46 billion, reflected the first reduction in two decades when inflation is taken into account. FY96 fell to US$13.8 billion, but that budget was not settled well into 1996 because of disputes between Congress and the White House. 1995's proposed plan through FY99 would see a continuing decline, providing only US$12.6 billion in FY99 (FY94 rate) and saving US$5 billion over those years. There was even grimmer news for NASA when the FY97 US$13.8 billion budget was released in March 1996. President Clinton plans to take out another US$3.3 billion as part of his drive to balance the federal budget by 2002, leaving NASA with only US$11.6 billion by Fiscal 2000. The conclusion is that NASA must reinvent itself, divesting itself of responsibility for operations and returning to cutting-edge projects.

US federal space expenditure of some US$30 billion annually was, until the early 1990s, challenged only by that of the USSR, which undertook 5 to 10 times more launches annually. The Department of Defense and the civil National Aeronautics & Space Administration together account for about 99 per cent, with military expenditure standing at around half. Commercial space sales have been increasing at 20 per cent annually during the 1990s, largely because of satellite communications. The US government projects that commercial sales will surpass military and government activities by 2010, partly because of the downturn in those latter areas. Satellite communications continue to represent by far the largest element, accounting for 65 per cent. Space export earnings fell to US$590 million in 1995 from US$750 million in 1994 (US$554 million in 1993). Employment in the missile/space sector is projected to fall to 153,000 by the end of 1996, from 157,000 in 1995, 165,000 in 1994 and 176,000 in 1993.

The major manned programme for the next five years will be the low Earth orbit International Space Station, costing more than US$30 billion to establish and with significant international contributions. The arrival of a new administration and concern over costs resulted in President Clinton in February 1993 ordering yet another restructuring to cut cost to completion to US$9 billion (US$19 billion was achieved; US$9 billion has already been expended). Annual spending is capped at US$2.1 billion. In 1998, the US Congress continues to examine the cost overruns on the station, and a major report released in May indicated that NASA had seriously underestimated the cost to completion of the Station.

Station *Freedom*, as such, ceased to exist in 1993. The programme and its management were overhauled and by the end of 1993, Russia had agreed to become a fully-fledged international partner. NASA Administrator Dan Goldin and Russian Space Agency director general Yuri Koptev signed a US$400 million USFY94-97 final agreement on 21 June 1994. It included US$305 million for up to 10 Shuttle Mir visits encompassing six NASA astronauts spending a total of 24 months aboard the Mir Station, primarily for life and μg sciences research, crew exchanges, cargo delivery and experiment return. Starting with STS-76/SMM-03 a long-term US presence was established on Mir when Shannon Lucid was left on board the station when the shuttle returned to Earth. Her stay on Mir ended when STS-79/SMM-04 docked to leave John Blaha in her place. STS-81/SMM-05 (January 1997) returned Blaha to Earth and Jerry Linenger, who was in turn replaced by Michael Foale in May 1997, took his place. It was courtesy of these residences on Mir that, on 22 March 1997 NASA was able to claim that there had been a continuous American presence in orbit for a full year. By contrast, the Russians far surpassed this mark since there had been a permanent Russian presence in orbit since September 1989. In fact, apart from a short break during April-September 1989, the Russians celebrated ten years of residencies abroad Mir in February 1997. Additional shuttle flights to Mir included STS-86 in September 1997 leaving astronaut Daniel Wolfe and returning (in early October) with Foale, STS-89 in January 1998 leaving astronaut Andy Thomas and STS-91 in June 1998 returning him to Earth. These flights concluded the Shuttle-Mir programme of visits.

President Bush announced on 20 July 1989, the 20th anniversary of the first manned lunar landing, that the next major US goal after Space Station would be a permanent presence on the Moon before beginning manned Mars exploration in 2019. This US$400 billion 30-year Space Exploration Initiative was supported in spirit by the Augustine review, but it depended on funding availability and was not driven by dates. Congress repeatedly denied funding and NASA closed its Office of Exploration in early 1993. The prospect of manned missions to the Moon and beyond has receded well into the next century.

Solar system exploration has arguably suffered most from launcher and budgetary difficulties. The continual drain to fund the space station has turned the tide against the funding of deep space missions. It is now impossible to fund ventures akin to the highly successful Magellan Venus radar mapper and the Galileo Jupiter orbiter. CRAF was sacrificed to save 1997's associated Cassini Saturn orbiter but, at a development cost of US$1.4 billion, that mission survived because it was so far down the road. Termination would have been a major blow to ESA's enthusiasm for joint ventures with NASA. The European agency suffered grievously a decade ago when NASA unilaterally cancelled its half of the Ulysses mission.

Instead, the emphasis will be on frequent and affordable science missions with highly focused objectives: numerous micro-spacecraft carrying advanced miniaturised instruments returning a continuous information flow. NASA unveiled the New Millennium programme in its FY96 budget request for the identification, development and flight validation of key technologies so that 21st Century science missions can take advantage of them without the risks inherent in their first use. Spending will be within US$50 million each year. The 486 kg spacecraft Deep Space 1 was launched in October 1998 to investigate an asteroid and comet. There are also Earth observing equivalents, demonstrating new technologies for Mission to Planet Earth.

NASA's Discovery series of rapid turnaround missions, each costing less than US$150 million, began with the NEAR asteroid mission departure in February 1996 and the Mars Pathfinder lander in December 1996. Pathfinder and Surveyor's 1998 and 2001 follow-up landers are all that remains of the MESUR mission that was to build a network of 12 small surface stations on Mars between 1999 and 2003, but NASA is committed to a Mars launch at every planetary opportunity (25 months).

NASA advertised that the Hubble Space Telescope would provide a tenfold increase in observing power for optical astronomers after a four year launch delay, but embarrassing shortcomings lay unrevealed until it attained orbit. Then the director of the Hubble science programme Eric Chaisson revealed in his book *The Hubble Wars*, that NASA intentionally misled the public on the true performance of the space telescope. The duplicity including duping the prestigious and respected journal *Scientific American*, which published NASA's claims almost *verbatim* without a standard peer review.

The next 'Great Observatory', the Compton Gamma Ray Observatory, followed in April 1991 but the third, the 15 tonne Advanced X-ray Astrophysics Facility, was delayed beyond its planned launch date and is now scheduled for Shuttle flight STS-93 in July 1999.

The military reversed its earlier course and shifted payloads from Shuttle to expendable launchers after the Challenger disaster of January 1986. The military decision provided the stimulus for vigorous commercial marketing of Atlas, Delta and Titan launchers. These vehicles, with their origins in the 1950s, complement the revised Shuttle programme, but will probably be dropped in favour of the EELV early in the next century.

Commercial enterprises have found survival more difficult, but OSC's Pegasus and Lockheed Martin's LMLV have emerged as strong contenders. Lockheed Martin has also been a prime mover in the recent international marketing success of Russia's Proton. Many critics including the US's own Defense Intelligence Agency have stated the US launch capability is out of date, inefficient and expensive. Numerous studies over recent years have recommended radical programmes, including the NLS National Launch System, the air force's Spacelifter and the more exotic single stage to orbit plans of the National AeroSpace Plane and Delta Clipper. DoD, in 1993, opted for gradual updating of existing vehicles, while NASA remained interested in SSTO, but the White House OSTP Office of Science & Technology Policy in September 1993 began the studies anew. The three broad options covered by OSTP were updating existing vehicles, developing a new family of launchers based on current technology, and developing SSTO. NASA envisaged a fully recoverable manned Shuttle 2 by 2010, capable of delivering 11 tonnes to Space Station. The approach is similar to the original Shuttle requirements, providing full re-usability, carrying modest payloads on brief missions and offering rapid turnaround as a Space Station support vehicle.

The Space Transportation Policy, signed by President Clinton 5 August 1994, set the pattern for US launch vehicle development well into the next century. It gave the DoD responsibility for upgrading existing medium and heavy ELVs and NASA the role of developing a reusable next generation after 2000. After several years of deliberation the DoD announced in October 1998 that it had awarded contracts to Boeing Expendable Launch Systems and Lockheed Martin Astronautics for a modern and flexible series of launches in the Evolved Expendable Launch Vehicle (EELV) programme. This will build an existing Delta Atlas and Titan to develop new variants supposedly capable of cutting launch costs by 25-50 per cent over the next 20 years.

NASA, in January 1995, invited industry to submit competing designs for two reusable launch vehicles. The goal is for a payload cost of US$500/kg instead of the current US$900-3,600/kg. Seven proposals were evaluated for the two experimental vehicles: the X-33 single-stage-to-orbit (SSTO) reusable launch vehicle (RLV) leading to development of a definitive vehicle which would eventually replace the Shuttle and on which NASA was committed to spending US$660 million; and the X-34, a much smaller launcher, partially reusable, on which NASA would spend about US$70 million. NASA selected an Orbital Sciences Corporation (OSC)/ Rockwell team to develop the two-stage X-34 aiming for a suborbital flight in 1997 and orbital flights beginning a year later. The team formed American Space Lines as an 'Arianespace' look-alike for commercial development of the small launcher market but in early 1996 the team pulled out believing the costs could not produce a viable commercial return. NASA re-contracted with OSC for a single-stage suborbital demonstrator, still designated X-34. Three X-34 test vehicles are scheduled to support flight trials planned to begin in August 1999 with the winged vehicle launched from the under-fuselage cradle of a converted L-1011 operated by OSC. Capable of a maximum speed of Mach 8 the X-34 will demonstrate all-weather, pilotless landings as a technology precursor to fully reusable launch systems of the future. X-33 finalists McDonnell Douglas, Rockwell and Lockheed competed for the precursor RLV work, the former submitting its DC-X Delta Clipper vertical-take-off-and-landing (VTOL) vehicle with which it had been conducting flight trials as a private venture. In July 1996, Lockheed Martin

US space budget summary (then-year $ millions)

FY Authority	NASA	DoD	DoC	DoE	Total
61	926.0	813.9	–	67.7	1,808.2
62	1,796.8	1,298.2	50.7	147.8	3,294.8
63	3,626.0	1,549.9	43.2	213.9	5,434.5
64	5,016.3	1,599.3	2.8	210.0	6,831.4
65	5,137.6	1,573.9	12.2	228.6	6,955.5
66	5,064.5	1,688.8	26.5	186.8	6,969.8
67	4,830.2	1,663.6	29.3	183.6	6,709.5
68	4,430.0	1,921.8	28.1	145.1	6,528.9
69	3,882.0	2,013.0	20.0	118.0	5,975.8
70	3,547.0	1,678.4	8.0	102.8	5,340.5
71	3,101.3	1,512.3	27.4	94.8	4,740.9
72	3,071.0	1,407.0	31.3	55.2	4,574.7
73	3,093.2	1,623.0	39.7	54.2	4,824.9
74	2,758.5	1,766.0	60.2	41.7	4,640.3
75	2,915.3	1,892.4	64.4	29.6	4,914.3
76	3,225.4	1,983.3	71.5	23.3	5,319.9
TQ	849.2	460.4	22.2	4.6	1,340.5
77	3,440.2	2,411.9	90.8	21.7	5,982.8
78	3,662.9	2,738.3	102.8	34.4	6,518.2
79	4,030.4	3,035.6	98.4	58.6	7,243.5
80	4,680.4	3,848.4	92.6	39.6	8,688.8
81	4,992.4	4,827.7	87.0	40.5	9,977.8
82	5,527.6	6,678.7	144.5	60.6	12,440.7
83	6,327.9	9,018.9	177.8	38.9	15,588.5
84	6,648.3	10,194.9	236.0	34.1	17,135.7
85	6,924.9	12,767.9	422.9	34.0	20,166.5
86	7,165.0	14,126.0	308.9	34.6	21,659.4
87	9,809.0	16,286.8	277.9	47.6	26,447.9
88	8,302.4	17,678.7	351.5	240.8	26,606.7
89	10,097.5	17,906.1	301.3	97.2	28,443.3
90	12,141.6	15,616.0	243.0	78.6	28,140.0
91	13,036.0	14,181.0	251.0	251.0	27,779.0
92	13,199.0	15,023.0	327.0	223.0	28,845.0
93	13,064.0	14,106.0	324.0	165.0	27,729.0
94	13,022.0	13,166.0	312.0	74.0	26,649.0
95	12,543.0	10,644.0	352.0	60.0	23,676.0

FY Outlay	NASA	DoD	DoC	DoE	Total
61	694	710	–	64	1,468
62	1,226	1,029	1	130	2,387
63	2,517	1,368	12	181	4,079
64	4,131	1,564	12	220	5,930
65	5,035	1,592	24	232	6,886
66	5,858	1,637	28	188	7,719
67	5,337	1,673	39	184	7,237
68	4,595	1,890	29	147	6,667
69	4,078	2,095	31	118	6,326
70	3,565	1,756	24	103	5,453
71	3,171	1,693	30	97	4,999
72	3,195	1,470	37	60	4,772
73	3,069	1,557	29	51	4,719
74	2,960	1,777	64	39	4,854
75	2,951	1,831	64	34	4,891
76	3,336	1,864	71	26	5,314
TQ	869	458	23	8	1,361
77	3,600	1,833	87	22	5,559
78	3,582	2,457	101	29	6,188
79	3,743.9	2,891.8	97.4	54.7	6,808.3
80	4,340.1	3,162.3	88.7	48.8	7,667.7
81	4,877.1	4,130.5	81.0	46.9	9,165.5
82	5,463.3	4771.5	142.4	59.5	10,466.2
83	6,100.9	6,246.7	178.0	39.6	12,590.4
84	6,461.4	8,000.2	208.7	33.4	14,726.1
85	6,607.4	10,441.3	155.4	34.0	17,254.8
86	6,756.0	11,448.5	316.9	34.7	18,581.0
87	7,254.0	14,264.3	261.9	37.4	21,843.9
88	8,450.5	14,397.4	333.9	199.1	23,414.1
89	10,195	14,504	306	97	25,143
90	12,292	12,962	279	79	25,671
91	13,351	14,432	266	251	28,360
92	12,838	14,437	298	223	27,865
93	13,092	13,779	295	165	27,398
94	12,363	10,973	297	83.0	23,790
95	12,593	11,494.0	330	70.0	25,564.0

These figures are taken from NASA's *Aeronautics & Space Report of the President*. 'Outlay' represents actual spending; Fiscal Year (FY) runs 1 October – 30 September. TQ = transitional quarter. NASA's exclude non-space activities. DoD = Dept of Defense; DoE = Dept of Energy; DoC = Dept of Commerce. The total includes relevant expenditures of the departments of Interior, Agriculture and others.

X-33, a suborbital horizontal-take-off-and-landing precursor to a full-scale Shuttle replacement called Venture Star which was to rely on composites and integrated design for lightness and a linear aerospike engine for propulsion. After several delays the X-33 is expected to begin flight trials in 2000 or 2001 and demonstrate RLV technologies to a maximum speed of Mach 13.5.

The three principal communications satellite manufacturers, Hughes (the world's largest manufacturer, with about a third of the market), Lockheed Martin and Space Systems/Loral, hold relatively healthy order books. The US accounted for more than 70 per cent (worth more than US$10 billion) of the satellites constructed in the first half of the 1990s, with an increasing proportion of non-domestic contracts. Commercial geostationary telecommunications systems have, of course, been long established, but low Earth orbit exploitation is imminent. The 66-satellite Iridium constellation should go online in late 1998. It has been projected that mobile satellite communications alone will be worth US$20 billion annually worldwide by 2000, including US$11 billion for the satellites, ground segment and launches. The current satellite communications market is some US$6.5 billion worldwide.

Military space programmes accounted for about US$13 billion of FY99's US$263 billion defence budget, sustaining steady growth from US$10.6 billion in 1995. The US military satellite communications network is extensive, employing DSCS, Fleetsatcom and Leasat craft plus the recent additions of UFO and Milstar. Much of the general traffic is being shifted to commercial transponders on long term leases as a means of reducing the estimated US$30 billion cost of replacing the current military systems.

The DoD's Navstar Global Positioning System (GPS) is providing a base for commercial receiver systems. They initially offered 16 m accuracy navigation information to civil users but the signals have been deliberately degraded since mid-1991. Nevertheless, it is a rapidly growing market and the constellation of 24 satellites attained full strength in 1994. The US GPS Industry Council projects a worldwide market for GPS receivers of US$8.47 billion by 2000, from US$867 million in 1994.

The US has long maintained the world's most comprehensive military surveillance and monitoring capability, despite the far higher USSR/CIS launch rates. The LEO and geosynchronous networks of imaging reconnaissance, electronic intelligence and early warning satellites were particularly affected by the launcher problems of 1986-87, but the first US imaging radar satellite was finally launched by the Shuttle in late 1988 and several reconnaissance satellites have departed since. However, it seems that such programmes have suffered deep cuts in recent years, reflected in the decreased requirements for USAF Titan 4 launches through the 1990s. Current expenditure is estimated at around US$5 billion annually, although outgoing CIA Director James Woolsey in January 1995 noted to a Senate committee that a radical restructuring would cut the number of future intelligence satellites by almost half. The four major companies building these systems (Lockheed Martin, Rockwell, TRW and Hughes) are competing for fewer contracts and having to diversify.

The Strategic Defense Initiative in 1993 experienced a fundamental restructuring. The Defense Secretary in May 1993 declared an end to the Star Wars era and renamed SDIO as the Ballistic Missile Defense Organisation. Highest priority is now Theater Missile Defense, in recognition of DoD's focus on regional conflicts and experience in Desert Storm.

The US maintains a double system of civil meteorological satellites providing coverage from both GEO and polar craft. Ideally, two of each type are in service at any one time but the GEO system operated a single vehicle since 1989 until a replacement from the next generation could be launched in April 1994. These GOES satellites complement the similar European Meteosat, Japanese GMS and CIS Elektro craft for synoptic coverage. The civil polar satellites are matched by the DMSP for military applications; their merging is planned to orbit the first satellite in about 2006, halving the operational fleet to two and saving at least US$1 billion over 15 years.

Commercialisation of the Landsat remote sensing satellite system has been far from an unqualified success, partly because of uncertain funding for continued operations with the current Landsat 4/5 and for fabrication of successors Landsat 6/7. Landsat 6's loss in 1993 was a serious blow and Landsat 7, which became entangled in protracted wrangling with the military, was not launched until 1999. Instead, image distributors are increasingly looking towards Indian and Russian sources. The first private commercial system licensed by the Dept of Commerce is EarthWatch's system offering 3 m resolution.

NASA budget summary US$ millions (FY98 is request)

	FY97	**FY98**
Human Space Flight	5,674.8	5,326.5
Science, Aeronautics and Technology	5,453.1	5,642.0
Mission Support	2,564.3	2,513.2
Total Budget Authority	13,709.2	13,500.0
Human Space Flight	5,674.8	5,326.5
Space Station	2,148.6	2,121.3
Development	1,766.3	1,386.1
Operations	177.6	490.1
Research Programme	204.7	245.1
US/Russian Co-Operative Programme		100.0
Space Shuttle	3,150.9	2,977.8
Shuttle Operations	2,514.9	2,494.4
Safety/Performance Upgrades	636.0	483.4
Payload and Utilisation Operations	275.3	227.4
Spacelab	50.3	14.2
Payload Processing And Support	41.7	51.6
Advanced Projects	34.7	58.7
Engineering And Technical Base	148.6	102.9
Science, Aeronautics and Technology	5,453.1	5,642.0
Space Science	1,969.3	2,043.8
Advanced X-Ray Astrophysics Facility	178.6	92.2
Cassini	89.6	9.0
Relativity Mission (Gp-B)	59.6	45.6
Space Infra-Red Telescope Facility (Sirtf)		81.4
Thermosphere, Ionosphere, Mesosphere Energetics And Dynamics (TIMED)	18.2	48.2
Payloads	16.9	12.3
Explorer Development	125.0	142.7
Mars Surveyor Programme	90.0	139.7
Discovery	76.8	106.5
New Millennium	48.6	75.7
Advanced Space Technology	132.0	151.2
Mission Operations and Data Analysis	583.3	507.4
Supporting Research and Technology	246.0	311.2
Suborbital Programme	64.1	84.4
Life and Microgravity Sciences and Applications	243.7	214.2
Life Sciences	97.4	85.5
Microgravity Science Research	105.3	101.4
Aerospace Medicine	3.8	7.5
Shuttle/Spacelab Payload Mission Management and Integration	24.2	6.9
Space Product Development	13.0	12.9
Mission To Planet Earth	1,361.6	1,417.3
Earth Observing System (Eos)	586.7	679.7
Eos Data Information System	254.6	244.7
Earth Probes Development	57.2	40.7
Science	295.4	271.1
Operations, Data Retrieval And Storage	78.0	54.2
Global Observations To Benefit The Environment (GLOBE)	5.0	5.0
Launch Services	84.7	121.9
Aeronautics And Space Transportation Technology	1,339.5	1,469.5
Aeronautical Research And Technology	844.2	920.1
Research and Technology Base	404.2	418.3
Focused Programmes	440.0	501.8
High Performance Computing and Communications	23.3	45.7
High Speed Research	243.1	245.0
Advanced Subsonic Technology	173.6	211.1
Advanced Space Transportation	336.7	396.6
Reusable Launch Vehicle	283.5	353.5
Advanced Space Transportation Technology	53.2	43.1
Commercial Technology Programme	158.6	152.8
Commercial Technology Programmes	25.8	20.0
Technology Transfer Agents	7.8	7.8
Small Business Innovative Research Programmes	125.0	125.0
Mission Communication Services	418.6	400.8
Ground Network	245.6	224.7
Mission Control And Data Systems	147.1	145.0
Space Network Customer Services	25.9	31.1
Academic Programmes	120.4	96.4
Education Programmes	65.6	55.5
Minority University Research And Education	54.8	40.9
Mission Support	2,564.3	2,513.2
Safety, Reliability And Quality Assurance	38.8	37.8
Space Communication Services	277.7	245.7
Space Network	185.1	161.2
Telecommunications	92.6	84.5
Research And Programme Management	2,092.5	2,070.3
Construction Of Facilities	155.3	159.4
Inspector General	**17.0**	18.3

The military's exclusive right to high resolution imagery was removed in March 1994 down to at least 1 m by the Clinton administration. The Dept of Commerce, as the licencing agency, retains the power to order shut down in times of crisis. The same decision also places tight restrictions on the export of remote sensing technology. DoC estimated in 1994 that the US$400 million annual market for remote sensing data would grow to US$2 billion by 2000.

President Clinton signed the executive order in February 1995 for the public to be given access by August 1996 at the National Archives to more than 800,000 images from some reconnaissance satellites flown 1960-72. Impetus has come from US companies looking for new sources of revenue in the post-Cold War world and from Russian sales of digitised film imagery down to 2 m from both civil and military missions.

NASA's Upper Atmosphere Research Satellite began operating in 1991 as the first Mission to Planet Earth satellite and the largest ever flown for atmospheric research. One goal is a better understanding of stratospheric ozone depletion. The US$21 billion EOS Earth Observing System programme running through 2019 is a major inter-agency effort to understand global environmental interactions, humanity's influence and develop techniques for predicting changes.

GOVERNMENT AND NON-GOVERNMENT INSTITUTIONAL ORGANISATIONS

Military
Civil

MILITARY

POLAND

Military Institute of Aviation Medicine

The institute is researching into astronaut fitness during long flights, μg effects on muscles and inner organs, and the influence of gravity changes on mammalian reproductive systems.

UNITED STATES OF AMERICA

11th Space Wing

The 11th SWS at Falcon AFB, Colorado, operates the ALERT Attack and Launch Early Reporting to Theatre system.

21st Space Wing

The 21st Space Wing was activated 15 May 1992, formed from the assets of 1st Space Wing and 3rd Space Support Wing, to operate and maintain a system of space/land-based sensors to detect/track ballistic missile launches, detect space launches, and provide data on foreign ballistic missile nuclear detonations. Space-based early warning is provided by the DSP satellites, the catalyst of the US early warning system as it is the first to detect missile launches. 21st SW DSP squadrons send crucial launch and nuclear detonation reports to the NORAD and USSPACECOM centres at Cheyenne Mountain Air Station.

The 5th Space Warning Squadron at Woomera Air Station in Australia is unique among wing units because it is assigned with the Australian 1st Joint Communications Squadron at the Joint Defense Facility, Nurrangar for DSP.

The wing's ground-based sensors until April 1995 comprised BMEWS and SLBM. The two BMEWS radar units are the 12th SWS at Thule AB in Greenland and the 13th SWS at Clear AS in Alaska. SLBM warning system units are the 6th SWS at Cape Cod AS, Massachusetts, 7th SWS at Beale AFB, California. These sensors also provide essential tracking reports on space objects. The 73rd Space Group at Falcon AFB was absorbed 29 April 1995, along with its eight space tracking squadrons.

As host wing for the Peterson Complex, the 21st provides complete base support services for Peterson AFB and Cheyenne Mountain AS, plus some support functions for Falcon AFB. Personnel total 4,127 military, 846 AF civilian and 3,300 contractor employees.

30th Space Wing

The Western Space and Missile Center in 1991 became the 30th Space Wing. Now elements of AF Space Command, the east and west launch ranges and bases were managed by AF Systems Commands' Space Systems Div until 1 October 1990. 30SW comprises four groups: 30th Operations Group, 30th Logistics Group, 30th Medical Group and 30th Support Group. 30OG operates the Western Range. Its 2nd Space Launch Squadron is responsible for small/medium launch operations (including Atlas II, Delta 2, Pegasus and Taurus 2) and its 4th SLS for heavy lift vehicles (Titan 2 and Titan 4). The 576th Flight Test Squadron handles ballistic operations.

45th Space Wing

The 45th Space Wing (45 SW) designation in November 1991 replaced the Eastern Space and Missile Center organisation, established 1 October 1979. Its origins date back to 1950 with the Army/Air Force establishment of the Joint Long Range Proving Ground. The Air Force assumed full control in 1951, redesignating it the Air Force Missile Test Center. It was renamed the Air Force Eastern Test Range in 1964. A unit of Air Force Space Command, it is the host organisation for Cape Canaveral AS, Patrick AFB and the two island stations (Ascension and Antigua) of the Eastern Range (ER). The eastern and western launch ranges and bases were managed by AF Systems Commands' Space Systems Division until 1 October 1990. The Wing comprises four major Groups: Operations; Support; Logistics; and Medical.

The Operations Group provides operational control over all aspects of DoD space launches. The 1st Space Launch Squadron was activated 1 October 1990 to assume all responsibility for Delta operations; the 3rd Space Launch Squadron activated 3 August 1992 to take over Atlas work; the 5th SLS activated 14 April 1994 to oversee all Titan operations. Each squadron exercises field operations management/control of all DoD satellites launched from CCAS.

A single story 11,800 m² Range Operations Control Center became operational in March 1995 as the ER control facility under a US$134 million Harris Corporation contract. The goal is to allow flights every 12 hours. Operationally, it serves the ER by processing worldwide radar metric data, generating inter-range vectors to aid in acquisition, calculating state vectors for navigation uplink, displaying telemetry and orbital navigation information and functioning as LRCC Lead Range Control Center to co-ordinate radar tracking and retrieval.

50th Space Wing

50th Space Wing is a component of Air Force Space Command at 18 km-distant Peterson AFB. The wing was originally established 8 July 1985 as the 2nd SW and then redesignated 30 January 1992. The facility at Falcon was built under the air force programme known as the CSOC Consolidated Space Operations Center. Its mission is to command and control operational DoD satellites and to manage the worldwide AFSCN AF Satellite Control Network.

Previously, the Air Force Space and Missile Systems Center's Consolidated Space Test Center (CSTC) in California was the prime control facility. Its satellites include DSP, DMSP, GPS, DSCS, NATO 3, NATO 4/Skynet 4, Fltsatcom, IUS, UHF Follow-On and Milstar.

The wing has 3,300 military and civil personnel assigned to its 45 units at 16 locations worldwide. It comprises four groups: 50th Operations, 50th Logistics and 50th Support Groups at Falcon AFB, plus the 750th Space Group at Onizuka AS, California. The 50th Operations Group commands/controls the satellites, trains space operations crews and provides operational support and evaluation for managing the satellite operations centres and ground stations. The group's seven Space Operations Squadrons are:

1st SOPS support during launch early orbit, mission and anomaly phases of GPS and DSP. This includes routinely operating older, more troublesome, satellites and eventually de-orbiting them. Once satellites are checked out, day to day control is turned over to the other SOPS.

2nd SOPS day to day GPS command/control. The Navstar GPS Master Control Station took command of the first GPS operational satellite following launch 14 February 1989. Det 1 at Cape Canaveral AFS operates a ground antenna for prelaunch compatibility testing and on-orbit control.

3rd SOPS day to day command/control of 14 communications satellites, including DSCS 3, Fltsatcom and UFO (Fltsatcom transferred to NAVSOC at Point Mugu on 19 June 1996, to be followed by UFO in 1999). The operators also provide LEOP support, trend analysis and anomaly resolution, which they also do for Milstar (before handing over to 4th SOPS). A total of 20 DSCS, FLTSAT and NATO 3 satellites was transferred 11 July 1991 from CSTC to what was then CSOC.

4th SOPS responsible for day to day operations of Milstar, including satellite command/control, communications management and ground segment maintenance.

5th SOPS at Onizuka AS in California to plan/conduct launch and on-orbit operations for several DoD, allied and commercial missions, including IUS for NASA/DoD, NATO 4/Skynet 4 and DSCS 3. It also supports Shuttle and Titan 4 missions. Responsible for day to day operations of NATO 3 and DSCS 3. Staffed by Lockheed Martin personnel.

6th SOPS located at Offutt AFB in Nebraska for day to day command/control of DMSP (will relocate to Falcon). CSOC assumed DMSP control 1 April 1987. Detachment 1 at Fairchild AFB in Washington is responsible for emergency command/control antenna operations; it will close September 1997 and Vandenberg will provide TT&C station.

7th SOPS the first reserve unit in AFSPC. It provides launch, early orbit, on-orbit and emergency support for DSP and GPS in conjunction with 1st SOPS. Staffed by experienced operators, many also with civilian satellite jobs.

750th Space Group

The 750th Space Group at Onizuka AS is responsible for the daily operation of AFSCN, which AFSPC took over in late 1987. The network comprises nine tracking stations: 23rd SOPS, New Boston AFS (New Hampshire), Det 1 Vandenberg AFB (California), Det 3 Thule AFB (Greenland), Det 4 Mahe Island (Seychelles; planned to close 1997), Det 5 Andersen AFB (Guam), Det 6 Kaena Pointt (Hawaii), 22nd SOPS Colorado Tracking Station (Falcon AFB), Det 2 Diego Garcia (Chagos Archipelago), OL-AE Oakhanger (UK). Two resource control complexes, the 21st SOPS at Onizuka AS and 22nd SOPS at Falcon AFB, schedule satellite contacts for the tracking stations.

Some 400 contacts of 518 minutes each are scheduled daily for about 90 satellites. Camp Parks at Pleasanton in California is responsible for satellite on-orbit testing. The Transportable Vehicle Checkout Facility at Cape Canaveral AS is responsible for pre-launch compatibility testing. AFSCN was recently upgraded, automating many operations. Satellites were controlled from the 1960s under the CDS Current Data System from CSTC, using a 26 m roll of graph paper specifying the contact schedule of each ground station over seven days. The computerised CCS Command and Control System took over at 50th SW and CSTC in spring 1992.

US Army, Space and Strategic Defense Command (USASDC)

USASDC develops space products for army users and supports theatre missile defence projects for BMDO. FY95 budget was US$363.2 million. The US Army Space Command is its operational arm. ARSPACE interfaces with industry and other DoD agencies to identify and demonstrate space technologies useful to Army warfighters through the ASEDP Army Space Exploitation and Demonstration Program.

Army Space Command (forward) is the operational arm of the US Army Space and Strategic Defense Command. Command of the DSCS System Operations Centers and management of the joint tactical use of DSCS satellites is a primary ARSPACE mission. The Joint Staff assigned ARSPACE the operation of five worldwide DSCSOCs, three worldwide AN/MSQ-114s and three worldwide Regional Space Support Centres.

ARSPACE operation of the JTAGS Joint Tactical Ground Stations in Europe and Korea provides theatre Commanders with a transportable system for processing/disseminating near-real-time warning of theatre ballistic missile launches. Using DSP data, JTAGS can process in stereo, improving the sensor's capability. JTAGS ties directly to theatre communications systems to transmit launch point information, impact area prediction, time of flight and positional information. ARSPACE also provides oversight of space surveillance operations at the US Army Kwajalein Atoll. ARSST Army Space Support Teams have deployed with units such as the 10th Mountain Division worldwide, providing weather, terrain analysis, mapping, imagery, communications and early warning information.

US Space Command

The unified USSPACECOM, comprising the Army, Naval and AF Space Commands, was authorised by President Reagan in November 1984 and activated in September 1985. Its singular responsibility is to ensure the US has continuous access to and the use of space. Army, Naval and AF space operations had previously functioned separately since the DoD's space programme began in the 1950s. Staff totals about 647 military and 139 civilian.

AFSPC, formed September 1982, is the largest component with 21,095 military and 4,772 civilian personnel. The FY97 budget was US$1.7 billion; FY96 budget US$1.6 billion; FY95 US$2.84 billion, FY94 US$2.8 billion, FY93 US$1.97 billion, FY92 US$1.5 billion. The Army Space Agency was activated August 1986 and redesignated Army Space Command (ARSPACE) April 1988. It employs 380 military and 89 civilian personnel. FY95 budget was US$47.8 million. Naval Space Command, formed in 1983, is headquartered at Dahlgren, Virginia and is staffed with 408 military and 280 civilian personnel.

AFSPC provides forces for: strategic and tactical ballistic missile warning, space control, satellite operations, DoD satellite launchings and, since 1 July 1993, ICBM missile operations. The 14th AF, activated 1 July 1993, is the day-to-day manager of AFSPC's space forces. The 20th AF has daily operational control over ICBMs. AFSPC's 50th Space Wing (which in 1992 took over the CSOC Consolidated Space Operations Center) at Falcon AFB is the focal point for daily command/control of more than 90 operational satellites.

The nine ground stations of the AF Satellite Control Network (AFSCN) make an average of 11,485 monthly satellite contacts. Responsibility for the space launch bases at Cape Canaveral (45th SW) and Vandenberg (30th Space Wing) was transferred from Air Force Systems Command to Space Command in October 1990. Its Space Launch Squadrons are responsible for Delta, Titan, Atlas and IUS launch operations. Air Force Space Command assumed AFSATCOM responsibility 1 June 1992 from US Strategic Command at Offutt AFB, Nebraska.

Through its component commands, US Space Command operates a worldwide network of missile warning sensors providing tactical warning and attack assessment of ICBM and SLBM launches posing potential threats against North America, and provides tactical warning of ballistic missile attacks to US Commanders worldwide. These are the responsibility of 21st Space Wing (formed May 1992 from 1st Space Wing and 3rd Space Support Wing), which is also the host wing for the Peterson complex. Data are relayed via satellite almost instantaneously to four primary locations: the command's Missile Warning Center (Cheyenne Mountain AS), Strategic Command HQ (Offutt AFB, Nebraska), National Military Command Center (Washington DC) and Alternate NMCC (Maryland). The data are also employed by the NORAD/USSPACECOM Command Center in Cheyenne Mountain Operations Center. NORAD is a bi-national US-Canadian command charged with safeguarding the sovereign airspace of the two countries. The missile warning sensors are also used for space surveillance. Together with dedicated space surveillance sensors (operated by the 21st Space Wing at Peterson AFB, NAVSPACECOM at Dahlgren and contributing ARSPACE and civil assets), they form the Space Surveillance Network, which provides timely and accurate detection, tracking and identification of space objects and events. The network makes 50,000 to 80,000 daily observations for the command's Space Control Center in Cheyenne Mountain and maintains a catalogue of more than 8,000 orbiting objects.

CIVIL

ARGENTINA

Asociacion Argentina de Tecnologia Espacial (AATE)

AATE obtained a reservation for a Getaway Special aboard a US shuttle flight in the second half of 1997.

Centro de Investigaciones Aplicades

This was originally the Instituto de Investigaciones Tecnológicas de la Fuerza Aérea and was part of the air force. In 1991, it was renamed Centro de Investigaciones Aplicades (Centre for Applied Research) and is part of the Instituto Universitario Aeronautico (University Institute of Aeronautics). Its responsibilities are to co-ordinate the research, projects and developments in the fields of space and aeronautical work. It was responsible for Victor 1 (MuSAT before launch) which was launched by the Russians aboard a Molniya-M in August 1996.

Comision Nacional de Actividades Espaciales (CONAE)

The CNIE Comision Nacional de Investigaciones Espaciales (National Commission for Space Research) was transferred from the air force in May 1991 to become the civil CONAE, under the direct supervision of Argentina's President. CNIE directed Argentina's space science activities from the 1960s. It began with development of a two-stage Castor sounding rocket able to carry small payloads up to 500 km, and in 1981 embarked on the SAC science satellite project. The San Miguel Centre houses the SAC-B Mission Operations Control Centre and TT&C station.

Comision Nacional de Telecomunicaciones

CNT is the national signatory to Intelsat and Inmarsat. Intelsat services are routed through the Balcarce ground stations, also expected to provide Inmarsat services. CNT provides oversight of telecom satellite operations.

Instituto de Astronomia y Fisica del Espacio

The IAFE Institute of Astronomy and Space Physics is an institute of Argentina's National Research and Technology Council (CONICET) of SECYT. It is the Principal Investigator institute of SAC-B's HXRS instrument.

AUSTRALIA

Aussat

Aussat, Australia's national satellite operator, was established by the government in late 1981 to own/operate the Aussat system. The company was legislated under the 1984 Aussat Act and was 75 per cent owned by the Commonwealth of Australia, with the remaining portion allocated to the Australian Telecommunications Commission (Telecom Australia). The government announced plans in November 1990 to de-regulate the telecommunications industry, selling Aussat to establish a second carrier to compete with the merged OTC/Telecom Australia. It immediately removed Telecom Australia's 25 per cent and lifted all legislative restrictions imposed upon Aussat's operation. The successful bidder was announced 19 November 1991: Optus Communications Pty Ltd., a 51 per cent Australian-owned company with BellSouth of Atlanta and Cable and Wireless of the UK holding the remainder, acquired Aussat for A$800 million. The government paid out Aussat's debt and terminated the leverage lease financing arrangements for the satellites at a total cost of about A$740 million. The new ownership arrangements came into operation 31 January 1992.

The Australian element, Optus Pty Ltd, has shareholders Mayne Nickless, AMP, National Mutual and other institutional investors whose interests are managed by the AIDC Telecommunications Fund Management Pty Ltd.

Australian Centre for Remote Sensing

ACRES is a unit within the Australian Surveying and Land Information Group (AUSLIG). It is the data acquisition, archiving, processing and distribution facility for Landsat, Spot and NOAA satellite imagery. ERS/JERSS data have also been acquired, archived and processed since 1992. ACRES ground station coverage includes all of Australia, Papua New Guinea and parts of Indonesia. ACRES also operates the Tasmanian Earth Resources Satellite (TERSS), which extends coverage across the Southern Ocean to Antarctica (including New Zealand).

Processed data are used to generate remote sensing products on either magnetic or photographic media. These products are used for agriculture monitoring, mapping, land use assessment, environmental impact monitoring, valuable resource exploration, and regional and urban planning. A digital quick-look on-line browse facility was introduced in 1996.

Centre for Remote Sensing and GIS

Founded in 1981, at the University of New South Wales, the centre is Australia's pre-eminent remote sensing and GIS educational institution. Facilities include a dedicated image analysis laboratory for undergraduate and masters students, and an advanced research laboratory.

Commonwealth GPS Group

The GPS Group was established in September 1992 by the Commonwealth Spatial Data Committee as a forum for Commonwealth government civil and defence agencies. Issues cover co-ordination and standardisation of base stations, development of a consistent Commonwealth approach to commercial opportunities and co-ordination of integrity monitoring.

CSIRO Earth Observation Centre

The centre, collocated with COSSA, was established by CSIRO in 1995 to centralise its collection of remote sensing data; data product management, including core-level processing. It performs strategic research for advancing the scientific base of remote sensing.

CSIRO Office of Space Science and Applications (OSSA)

CSIRO is Australia's primary body for industrial and scientific R&D and is responsible through its OSSA for space science activities. Its annual budget is about A$700 million, of which space-related expenditure is about A$27 million (funding 240 full-time-equivalent staff out of 7,100). The 35 divisions include radio physics, atmosphere, oceans, applied physics and materials science. The collocated Earth Observation Centre was established in 1995 to centralise its remote sensing activities.

CSIRO hosts the Australia Telescope National Facility, which operates a synthesised array of six 22 m antennas at the Paul Wild Solar Observatory (Narrabri, New South Wales), a 22 m antenna at Coonabarabran (New South Wales) and the Australian National Radio Astronomy Observatory at Parkes, New South Wales. Parkes' 64 m dish is used for supporting deep space missions such as NASA's Voyager. ATNF budget is A$12.5 million; 140 staff. It will play a critical role in southern hemisphere observations for Russia's Radioastron and Japan's VSOP radio astronomy satellites.

CSIRO had its role within Australian space re-defined as a result of the major re-organisation ordered in August 1996. With the demise of the Australian Space Office and Australian Space Council, CSIRO has taken over the responsibility of undertaking space projects on behalf of the Australian government. It is leading the project for the planned FEDSAT programme, intended to launch a micro-satellite into orbit by 2001. At the same time it was also announced that the activities of the NASA tracking station at Tidbinbilla would be transferred to CSIRO as a result of CSIRO's experience in radio astronomy and space physics.

OSSA identifies and co-ordinates space-related R&D in more than 20 of CSIRO's divisions. It provides international liaison for CSIRO space interests, manages multi-division projects, provides technical assistance to users of space-related technologies, negotiates commercial agreements and joint ventures, and operates an airborne data collection and processing facility.

Land Information Centre

A division of the Department of Conservation and Land Management LIC's Remote Sensing Unit provides image processing and analysis services, particularly to New South Wales state government agencies.

Remote Sensing Applications Centre (RSAC)

As part of the Department of Land Administration (DOLA), RSAC is the major government distributor of Landsat, Spot and airborne scanner digital and photographic data in Western Australia. Services include full image processing, image writing, geo-rectification to AMG (Australian Map Grid) and bureau facilities.

Satellite Communications Research Centre (SCRC)

The Australian Space Office established SCRC (as the Australian Space Centre for Signal Processing) to undertake strategic R&D of space-related subsystems, including: mobile terminal modems for the Mobilesat L-band service, supported by Optus. It also performs high-speed satellite communication at up to 155.52 Mbit/s, supported by Intelsat; and remote-control and telemetry systems. The 155.52 Mbit/s encoder/decoder was developed for use via Intelsat to deliver such services as up to 10 HDTV channels (or 50 regular channels) bundled for simultaneous transmission over a single transponder. It will also likely be used for Asynchronous Transfer Mode transmissions for high-speed datalinks. Satellite demonstrations were performed in 1995.

Space Centre for Satellite Navigation

This Space Industry Development Centre, based at the Queensland University of Technology, is a joint initiative of the Australian Space Office, QUT School of Electrical and Electronic Systems Engineering, and QUT School of Surveying. In association with industry, it develops space-related products/services, in particular GPS-related.

Space Policy Unit (SPU)

The Space Policy Unit (SPU) was established as part of the re-organisation of the Australian space effort in 1996 and it operates within the Department of Industry, Science and Tourism to maintain an overview of national and international space activities and to set policy in the national interest. The SPU provides advice on space policy and facilitates commercial space activities and interaction with other governments beyond the role and responsibilities of individual organisations currently involved in space research.

Telstra Corp Ltd

Telstra is Australia's government-owned domestic and international telecommunications company, formed in 1992 by the merger of OTC and Telecom Australia. It is Australia's signatory to Intelsat (2.82 per cent in 1996) and Inmarsat (1.55 per cent in 1996). With 1994-95 revenues of A$14.08 billion (1993-94: A$13.4 billion; 1992-93: A$12.6 billion) on assets of A$24.08 billion, Telstra is ranked in the world's top 20 telecommunications companies. Its Sydney International Telecommunications Centre is a major satellite Earth station, with a number of antennas: SYD1A 18 m limited motion, 183° east, primarily for cable restoration, SYD2A 32 m full motion, 174° east, services include IDR, IBS, SCPC, DAMA and itinerate video, SYD3A 18 m limited motion, 180° east, primarily for full time video leases and itinerate video, SYD4A 18 m limited motion, 177° east, IDR, IBS, itinerate video; PAS 2 9– 1 m limited motion, PAS 2, video leases and pay TV video for domestic distribution, AsiaSat 2 9.1 m limited motion for pay TV domestic distribution. Telstra's Perth ITC (Western Australia) is a major satellite communication and tracking centre. It encompasses 32 m Intelsat relocated from Ceduna SA; 7.5 m submarine cable restoration; two 18 m Intelsat and a 27.5 m relocated from Moree NSW (OTC's station opened there in 1969); two 18 m Inmarsat; 20 m TT&C for Intelsat satellites over the Indian and Pacific oceans; four 8 m Intelsat ranging; 7.5 m linking to the Intelsat control centre in Washington DC. A 15 m S/X-band full track antenna, relocated from Carnarvon (Western Australia) in 1987, provides LEOP services for ESA. Telstra is a member of the Satellite Aircom consortium providing aeronautical communications services through the Inmarsat system.

AUSTRIA

Austrian Space Agency

The Österreichische Gesellschaft für Weltraumfragen was founded 1972 under the aegis of the Federal Ministry for Science and Research to co-ordinate the national space programme. Its annual budget is about Sch9 million and staff total is six. ASA's principal associate is the Federal Ministry of Science, Transport and Art; others are the Research Centre Seibersdorf, Joanneum Research, Siemens AG, Schrack Aerospace, Austrian Aerospace Co ORS, Beckel Geospace, Steyr-Daimler-Puch, the Federal Economic Chamber and the City of Graz.

Austrospace

Austrospace was created in 1991 to promote the views of its space industry members.

Joanneum research

Joanneum is one of ASA's supporting centres and was responsible for managing 1991's AustroMir mission. Its Institute of Applied Systems Technology undertakes satellite-based communications R&D. Other specialisations include: remote sensing software and analysis; computer vision systems for autonomous navigation; low temperature materials testing.

Space Research Institute (SRI)

SRI is Austria's principal space science research facility, with emphasis on near-Earth auroral, ionospheric and magnetospheric investigations (including rocket, balloon and radar probing) and deep space fields/particles monitoring with magnetometers and particle detectors aboard ESA Space lab and Huygens and Soviet/RFAS Venera, Vega, Phobos, Interball and Mars-96/98 craft. The Institute provided two experiments for ESA's Cluster mission. Graz also provides one of the world's reference geodesy stations. Its laser unit provides ±8 mm satellite ranging accuracy.

BELGIUM

Belgacom

Belgacom manages national participation in Intelsat, Inmarsat and Eutelsat. Eutelsat's space segment is controlled from ESA's station at Redu in Belgium. Belgacom operates the V-STAR 1.2 to 2.4 m VSAT service, operating hubs in Liedekerke via Eutelsat for Europe and in Lessive via Intelsat for Africa, Middle East and North/South America.

Belgian Institute of Space Aeronomy (BISA)

BISA's space physics and aeronomy research includes solar wind interaction with Earth, terrestrial and planetary magnetospheres, and the physics/chemistry of ionosphere and upper atmospheres. Its infra-red grille spectrometer flew on the Spacelab 1 (1983) and Atlas 1 (1992) missions to observe trace constituents in the high atmosphere. In its improved MIRAS version, it was launched to Mir in 1995 aboard the Spektr module and attached externally by EVA in July 1995.

BISA contributed to SPICAM (atmospheric spectrometer) and MAREMF (magnetometer) on Russia's Mars-96. Work is under way on instrumentation for Space Station.

Centre Spatial de Liège

CSL (IAL Space until April 1992) is an autonomous unit within the University of Liège and, along with Intespace (France) and IABG (Germany), is one of the Co-ordinated European Test Facilities group under ESTEC. The test facilities are based around a 480 m^2 class 10,000 cleanroom and include three thermal vacuum optical test facilities. The 150 m^3 5 m diameter horizontal 33 tonne Focal 5 chamber provides a 10^{-9} atmosphere vacuum and 10 temperature settings over 268 to +70°C and optical bench vibration levels of 10^5 *g* on a 350 tonne seismic block. There is also an associated 30 m^2 class 100 laminar flow clean zone. In the same area is the Focal 2 chamber, with identical performance but 21 m^3/2 m diameter. The 1.2 m^3 Focal 1.5 provides a similar environment but with a 1.5 m vertical diameter and 80 m^2 class 10,000 cleanroom and 4 m^2 class 100 laminar flow clean zone. CSL undertakes testing/calibration of camera tubes, image intensifiers, CCDs, photomultipliers and photon counting systems, in addition to the design/manufacture of meteorological instruments.

Office for Science, Technology and Cultural Affairs (OSTC)

OSTC (formerly the Science Policy Office), under the authority of the Minister of Science Policy, manages Belgium's space programme, representing the national interest within ESA/non-ESA projects such as Spot.

Royal Meteorological Institute of Belgium (RMIB)

RMIB represents Belgium in EUMETSAT and houses the national Space Remote Operations Center, from where scientists can control their experiments aboard Space Shuttle. RMIB's Solcon instrument has flown four times on Shuttle (Spacelab 1, Atlas 1-3) to measure the solar constant, as did its SOVA on ESA's Eureca 1992-93. The Virgo absolute radiometer is now flying on ESA's Soho. Negotiations are under way for Solcon to be included in NASA's Hitchhiker programme, and SOVA (as SOVIM) is under consideration for ESA's Solar payload on Space Station. RMIB is responsible for the optics element of the GERBE Global Earth Radiation Budget Experiment for the Meteosat Second Generation.

BRAZIL

Agencia Espacial Brasileira (AEB)

AEB became operational in March 1994 to co-ordinate and plan Brazil's space programme, replacing the inter-ministerial COBAE Commission for Space Activities. Brazil's President signed a decree 5 December 1991 nominating a commission representing relevant ministries to discuss the creation of the agency.

Andrade Gutierrez Quimica Ltda

AGQuimica's plant at Jacarei-São Paulo can manufacture up to 100 tonnes of ammonium perchlorate annually to US DoD MIL-A-192B standards for solid rocket propellants. Applications include Brazil's Sonda sounding rockets and VLS orbital launcher. The Ministry of Aeronautics' Centro Tecnico Aerospacial initiated research on AP production technology in 1973, and in late 1984, the Andrade Gutierrez Group was selected for the industrial phase. AGQuimica was founded in 1985. Plant start-up was achieved in late 1986 and during June 1987 the product was certified as compatible with international standards.

Centro Tecnico Aeroespacial (CTA)

The Institute of Aeronautics and Space (previously the Institute of Space Activities) is one of the five technical divisions of the Ministry of Aeronautics' CTA (Aerospace Technical Centre), which has 5,000 staff. IAE is responsible for developing the Sonda sounding rockets and VLS orbital launcher for the national space programme, and for operating the Alcantara equatorial launch site.

The Ministry created the GETEPE Executive Group for Space Studies and Projects in June 1964, replaced in 1971 by the Institute of Space Activities.

Empresa Brasileira de Telecomunicacões

Embratel, with a staffing level of 12,000, operates the national Brasilsat satellite system and works through the Intelsat system via its Tangua ground station for international links. It is also Brazil's national Inmarsat signatory.

Embratel was created in 1965 as the operational arm of the Ministry of Communication's Telecomunicacoes Brasileiras (Telebras) component, and is responsible for the domestic Brazilian communications network and links through the Intelsat and Inmarsat systems. Embratel began developing the ground element for a satellite-based domestic system by employing a single leased Intelsat transponder in 1974, rising to seven by 1982, with 21 Earth stations providing telephony, telex and three TV channels.

In June 1982, Embratel awarded Canada's Spar Aerospace a US$125 million contract for two C-band Hughes HS-376 satellites similar to the Canadian Anik Ds, built under license, to create the national Sistema Brasilero de Telecomunicacoes por Satelite (SBTS, or Brasilsat). Attempts to procure a domestic system were initiated during 1975, but were cancelled after two years because of economic considerations. Ariane successfully launched BRASILSAT A1/A2 in 1985-86, working by end-1987 through about 100 Earth stations and with A2 acting as an in-orbit spare.

The ground segment comprises 53 principal stations (one TV distribution, 35 basic telecommunications, seven TVRO and ten 48/64 kbits/s private data) and more than 11,000 private dishes. 1988 use was 54 per cent of capacity, compared to 1984's projection of 83 per cent. With the system fully operational, Brazil's position as Intelsat's 8th largest user at 3.04 per cent in 1985 fell to 15th with 1.24 per cent in 1986. Transponder leasing to neighbouring countries was considered. There was a possibility of a Brasilsat A3 being commissioned for an early-1990s launch to 61° west to extend system lifetime to century's end. However, in early 1989, RfPs were released for a second-generation system offering C- + X-band links on each of two 28-transponder satellites. Hughes was awarded the contract in April 1990. B3 was ordered December 1995 under a US$70 million agreement. Deleting the X-band payload allows it to be optimised for C-band.

The satellites are controlled from the Satellite System Operational Centre (SSOC), located in Guaratiba near Rio de Janeiro, comprising the Spacecraft Control Centre (SCC) and the Communications Operations Control Centre (COCC). SCC provides TT&C commands through 14.2 m and 6.0 m antennas, and the COCC primary traffic routing through a 16.2 m dish.

Instituto de Aeronautica e Espaco (IAE)

IAE is responsible for developing the Sonda sounding rockets and VLS orbital launcher for the national space programme.

Instituto Nacional de Pesquisas Espaciais (INPE)

The INPE National Institute for Space Research was created in 1971 under the Ministry of Science and Technology to succeed the National Space Activities Commission (created 1961) and be responsible for the development of the ground and space segments of applications satellite programmes. As such, it is developing the five MECB satellites and the Brazilian contribution to the joint CBERS Sino-Brazil remote sensing satellite. Its 1996 budget is US$100 million (1995 US$80 million; 1994 US$40 million), excluding personnel costs. Personnel total 1,500. INPE has its headquarters at São José dos Campos (23° south/46° west, where the CRC Satellite Tracking and Control Centre is sited), with additional facilities in Cachoeira Paulista (23° south/45° west, 80 km north of São Paulo, where the mission centre for SCD satellite data collection is located), Cuiaba (15° south/56° west, where it operates Brazil's Spot, Landsat and ERS station), Natal (5° south/35° west; where the owner Ministry of Aeronautics operates a small sounding rocket launching station) and Fortaleza (3° south/38° west).

INPE's interests include spacecraft integration/testing, propulsion (including bipropellant and monopropellant catalytic thrusters) space materials, solar cells, sensors, computer sciences, space and atmospheric sciences, meteorology and remote sensing.

INPE's 10,000 m² LTI Laboratory for Test and Integration opened in 1988 as the largest such facility in the southern hemisphere: 1,600 m² work hall, 450 m² integration room, control room, 13 and 80 kN shakers, 22 m³ thermal vacuum chamber and electromagnetic test chambers; Intespace will complete an acoustic chamber in 1997. CBERS 2 will be assembled here, and Brasilsat B final assembly/testing was undertaken beginning March 1994. All system qualification testing of Argentina's SAC-B satellite was undertaken here.

INPE also operates Tiros/GOES ground stations, in addition to the Local User Terminal in São Paulo for the international Cospas-Sarsat distress system. A 14.2 m VLBI antenna began operating 1993 at Eusebio in northeastern Brazil to link South America geodetically with other continents as part of the International Interferometry Survey, co-ordinated by the International Earth Rotation Service. It also has a GPS station to compare results with VLBI-derived parameters.

BULGARIA

Navigation Maritime Bulgare Ltd

The company is Bulgaria's Inmarsat signatory.

CANADA

Alliance for Marine Remote Sensing (AMRS)

AMRS is a not for profit association designed to develop and promote marine applications of remote sensing technologies. It publishes the *Backscatter* newsletter three times yearly.

Canada Centre For Remote Sensing (CCRS)

Central organisation for the development and promotion of remote sensing technology and applications, including reception and processing of Spot, Landsat, NOAA, MOS, ERS, JERS and Radarsat data.

The CCRS was established in 1971 with a mandate to improve remote sensing technology, facilitate the acquisition and dissemination of remotely sensed data and to work with Canadian industry involved in remote sensing. It has general responsibility for remote sensing research and development within the government. CCRS employs a total of 125 government staff, with a further 100 on contract from industry working on various developmental and/or technology transfer projects.

The centre has full-remote sensing satellite reception capabilities, via ground stations at Prince Albert (Saskatchewan) and Gatineau (Quebec), including Landsat, Spot, NOAA, MOS 1, ERS 1/2, JERS and Radarsat, under the auspices of its Data Acquisition Division. The Methods and Systems Division is developing methods to correct and produce satellite and other data, as well as the international standards associated with such products. It also contains a group focusing on the development of image processing algorithms, and methods to incorporate artificial intelligence and expert systems into image interpretation. The Applications Division is active in developing applications of remote sensing in agriculture, forestry, geology, ice, hydrology, cartography and environmental monitoring. The division is also working on GIS and remote sensing for regional resource management. The Center's major current focus is in radar technology and applications, for ERS 1/2, Radarsat and other planned radar satellites.

The Centre has a number of state-of-the-art systems for the collection, production, correction and analysis of data from ground measurement systems, as well as data from a wide variety of airborne and spaceborne remote sensors. Geocoding is available to superimpose satellite data on to map projections of the National Topographic System of Canada. Geocoding also facilitates comparison of images of the same site on different dates, possibly from different satellites or in combination with aircraft information.

Canadian Space Agency

CSA is responsible for the management, planning and policy development of the Canadian space programme, including co-ordination of the space activities of other agencies of the federal government. It was created in March 1989, drawing together the space activities of the Ministry of State for Science, the Dept of Communications, the Dept of Energy, Mines and Resources, and the National Research Council. It accounts for some 70 per cent of federal space expenditure. The President reports to the Minister of State for Science. Parliament, on 14 December 1990, approved legislation creating the agency, allowing it to present its own budget and appropriate its own funds.

CSA relocated to the new C$70 million 26,600 m² facility at the 0.41 km² St Hubert site in 1993 where it maintains a personnel total of 300. The David Florida Laboratory, a portion of the Space Science Programme and a liaison office remain in Ottawa. The Space Operations Support Center has been operational for supporting space station operations for over a year at St Hubert.

CSA's programmes branch develops new programmes, passing them to the Engineering branch for implementation (for example, Engineering now has Radarsat and Space Station). The Operations branch handles on-orbit operations.

Provincial Remote Sensing Office

PRSO, established in 1973, is part of the Information Resources Division of the Ontario Ministry of Natural Resources, a provincial government agency. PRSO works with government, industry and end-user client groups to facilitate the development/application of remote sensing techniques for natural resources management. Data from air/space platforms are used to update natural resource inventories, conduct environmental studies, geological exploration and mapping.

The centre operates an aircraft modularly equipped for imaging sensors and electronic navigation. Experience includes analogue/digital analysis for processing imaging sensor technologies; aerial photography, thermal infra-red scanning, SAR, multispectral scanners and imaging spectrographs. The technology transfer programme includes training for domestic and foreign scientists. PRSO also provides information on services available from province's remote sensing companies. A new function is the development of corporate data standards by the Data Standards Secretariat.

Teleglobe Canada

Teleglobe Canada, a member of Teleglobe Inc, is the national signatory to Intelsat and Inmarsat as the country's international telecom carrier via satellite and undersea cable. It is the world's fifth largest international communications service provider. It is a member of the Satellite Aircom consortium providing aeronautical voice/data services through the Inmarsat system.

Teleglobe Inc and Orbital Sciences Corp are jointly developing the Orbcomm satellite system. Teleglobe is providing C$95 million. TRW/Teleglobe announced the Odyssey Worldwide Services joint venture November 1994 to build/operate the Odyssey system. Teleglobe will market the services and provide C$60 million equity.

Telesat Canada

This private company, founded in 1969, was 53.7 per cent owned by the Canadian government until 1992 and the remainder by a consortium of nine Canadian common (telephone) carriers (including 24.6 per cent by Bell Canada). The government then decided to sell off its share. Alouette Telecommunications Inc, a consortium formed primarily by the members of Telecom Canada (now called Stentor Canadian Network Management), the association of Canada's major telecommunications companies, was successful with its C$154.8 million bid in March 1992.

1994 reflected net earnings of C$9.0 million, following 1993's C$27.0 million and 1992's C$9.7 million losses. 1994 operating revenue was C$208.4 million (1993 C$200.0 million; 1992: C$195.8 million; 1991: C$190.0 million; 1990: C$176.9 million; 1989: C$146.5 million). Canadian broadcasting business has consistently contributed about half of total revenues; voice, data and image services have remained at about 20 per cent, carrier services 20 per cent and international consulting 10 per cent. A network of 3,500 Earth stations was being maintained at end-1995. Employees totalled 416, principally in Ottawa. Telesat installed 2,730 VSATs by end-1995.

Telesat Canada formerly owned 80 per cent of TMI. Following 1993's restructuring, it retains an interest through preferred shares in a company with an interest in TMI Communications.

Telesat switched control of its five Aniks in 1989 to the new Satellite Control Center, designed and built over four years to handle the more complex Anik Es and MSAT using the SCS-1000 computer system. Telesat will also provide the two Canadian TT&C stations to control Iridium's satellites. Telesat is the owner and operator of Canada's national Anik domestic satellite communication system. The government's controlling interest was sold off in 1992.

CHINA, PEOPLE'S REPUBLIC

Beijing Marine Communications and Navigation Co

In addition to being China's Inmarsat signatory, MCN also leases/sells Inmarsat A/C equipment. Some 80 per cent of sales are to state ministries. Rental customers are mainly tourist groups and foreign oil and mineral extraction companies performing exploratory work.

China Telecommunications Broadcast Satellite Co

China Telecommunications Broadcast Satellite Co or 'Chinasat' is the state satellite operator and an affiliate of the MPT Ministry of Posts and Telecommunications. Capacity is also provided by AsiaSat, which is partly owned by China.

Official estimates claimed more than 40,000 TVROs by 1993. The government in October 1993 announced restrictions on the manufacture, sale and installation of terminals, but it remained to be seen how strictly the regulations would be enforced in such a huge market. A 1994 estimate claimed 500,000 home dishes. There were 2,600 VSATs in operation by 1992, expected to rise to 8,000 by 2002. Hughes Network Systems in 1993 provided 300 VSATs to the Bank of China, working through AsiaSat 1. MPT lost its monopoly in January 1994 and may now be broken up.

MPT is China's signatory to Intelsat, and the Beijing Marine Communications and Navigation Co to Inmarsat.

Preparations for the STTW-T1 included a 1983 agreement to employ Italy's Sirio 1 for communications trials, and a US$20 million March 1984 contract to Spar Aerospace for 26 Earth stations. 15 were in operation by October 1986 at Lhasa, Tibet, Guangzhou and Hohhot, linking off-shore oil/gas production facilities by voice/data. Spar received a further US$36 million contract mid-1991 for 13 m Intelsat A stations at 10 new locations and enhancements at seven existing sites. Spar signed a letter of intent 8 November 1994 with CASC, China National Railway Signal and Communications Corp, and Space Wit Communications Co Ltd to develop Earth stations and terminal networks for voice/data. A new Shanghai station opened August 1992 to provide European links via Indian Ocean satellites. A new Intelsat TT&C and Monitoring station was inaugurated at Beijing in October 1992 as part of the organisation's worldwide network. Scientific-Atlanta in September 1995 was awarded a US$3.7 million contract by MPT for an emergency satellite communication network to support relief operations. The initial phase of the Skylinx DDS Digital DAMA telephony network became operational in early 1996 using AsiaSat 2 Ku-band capacity. It employs a 4.5 m master station in Shanghai working with 3.6 m stations in seven provinces and 37 mobile 1.2/1.8 m VSATs. The service will eventually be expanded to include a 3.6 m station in all 22 provinces, with at least five mobile units per province. A US$9.9 million contract for the Ministry of Foreign Trade and Economic Co-operation installed hubs in Beijing and Shenzhen working to almost 60 VSATs in trading locations for financial and customs communications.

Studies are under way of a GMSIS Global Mobile Satellite Information System, which would provide personal hand-held communications via 18 to 24 satellites in medium orbits. Three or four of the 500 kg satellites could be delivered by a single CZ3A. The Twin Star proposal would use two satellites some 40° apart in the GEO arc to provide positioning, messaging and timing services, after the fashion of OmniTracs and EutelTracs.

Chinese Society of Astronautics (CSA)

CSA was established in October 1979 to promote the development of space science, technology and professional training.

CZECH REPUBLIC

Commission on Space Activity

The 25-man commission was created in 1990 within the Academy of Science to make recommendations on space activities. It officially ceased to exist 31 December 1992 but in practice remains the point of contact for international co-operation.

DENMARK

Danish Research Advisory Board for Space

The Danish Research Advisory Board for Space provides the Minister for Research and Information Technology with independent and qualified advice within the space research area.

Dansk Rumforkningsinstitut

The Danish Space Research Institute (Dansk Rumforkningsinstitut) is a government-funded research organisation, founded in 1966, under the direct supervision of the Research Directorate of the Ministry of Education and Research. Its principal areas of interest are space astronomy, cosmic rays and space plasma physics. Four wide angle x-ray transient and gamma burst detectors which form the Watch all-sky monitor are flying on the CIS Granat satellite and a single Watch unit was provided to ESA for Eureca's 1992 mission. Two large area x-ray telescopes, four imaging detectors and a Bragg crystal spectrometer (Xspect) are being produced as part of the RFAS/Danish Sodart telescope for the Spektrum-X satellite in 1998. Ground support equipment was supplied for the Isophot instrument on ESA's Infra-red Space Observatory in co-operation with Heidelberg's Max Planck Institute of Astronomy, and DSRI is a member of the Hipparcos Northern Data Consortium. The Swedish Viking magnetospheric mission carried a wave experiment built at DSRI; current activities include participation in the Wave and Magnetometer experiments on ESA's Cluster. A series of plasma physics sounding rocket flights in co-operation with Scandinavian countries and NASA is also under way. DSRI is part of Denmark's Ørsted satellite project.

DSRI receives a line item budget voted each year by parliament as part of the national budget. This includes salary, operating and development costs. Additional support is obtained from the Danish Space Board on a project by project basis. DRSI is headed by a Director, who reports to the Research Directorate on financial and personnel matters and to an external Institute Board for science/programme affairs. Otherwise, the Institute is autonomous. 1996's budget is DKr26.1 million; 1995 DKr19.3 million; 1994 DKr19.4 million; 1993 DKr24 million; 1992 DKr22 million; 1991 DKr22.544 million; 1990 DKr21.844 million; 1989 DKr21.335 million. This supports a fixed staff of 40, including 18 physicists and engineers. Included in the total is funding from the Danish Space Board to support seven engineers, scientists and technicians, and a development programme for thin foil x-ray optics.

ECUADOR

Centro de Levantamientos Integrados de Recursos Naturales por Sensores Remotos (CLIRSEN)

The CLIRSEN Centro de Levantamientos Integrados de Recursos Naturales por Sensores Remotos (Centre for Integrated Surveys of Natural Resources through Remote Sensors) was created in December 1977 to be responsible for remote sensing and the cataloguing of natural resources. It operates the Cotopaxi ground station, some 60 km south of Quito, which began taking imagery from Landsat in June 1991 as the system's 18th station; Spot and ERS were added later. The 2,500 km^2 coverage of 25 countries stretches from Mexico's Yucatan Peninsula to Antofagasta in Chile. Products include 23 × 21 cm film, 72 × 97 cm paper, quick look, microfiche (84 scenes each) and digital tapes (6,250/1,600 bpi). Processing levels: bulk, georeferenced, geocoded and user-specified highlighting.

FINLAND

Academy of Finland

The Academy is the central government agency for science policy, planning and administration, responsible to the Ministry of Education. Formed in 1970, it comprises four research councils (Natural Sciences and Technology, Environment and Natural Resources, Culture and Society and Research Council for Health) and the board. The annual budget is about FMk517 million (ECU89 million), of which some FMk10 million (ECU1.7 million) is spent on space research.

Finnish Geodetic Institute

Space-related activities are carried by two of the Institute's four departments. The Department of Geodesy operates a satellite laser ranger at the Metsähovi Space Geodetic Station and permanent GPS stations for the International Geodynamics Service. The Dept of Photogrammetry and Remote Sensing studies/processes remote sensing data from aircraft and satellites. The annual budget is ECU2.9 million, of which ECU850,000 is devoted to remote sensing, space geodesy and geodetic astronomy.

Finnish Meteorological Institute

FMI is Finland's signatory to EUMETSAT with the national contribution paid from FMI's budget. The total staff is 600. The Operations Division undertakes the reception and processing of Meteosat and NOAA data.

The Geophysical Research Division of FMI (FMI/GEO) has undertaken research into magnetospheric and space plasma physics for many years and the present research areas also include planetary atmospheres, the solar wind and stratospheric ozone. FMI/GEOs principal interests concern the scientific results achieved by space missions, but it is actively contributing both to the hardware and to the software of space instruments and to ground support equipment.

It is currently participating in ESA's SOHO, Huygens, ENVISAT 1 and Rosetta, plus the NASA Mars surveyor missions. For the Rosetta cometary mission FMI/GEO will participate in instruments studying the dust and plasma environment of the comet, and its internal structure and composition. FMI/GEO will also manufacture a solid-state mass memory to be accommodated on a small spacecraft to land on the comet's surface as part of Regatta.

In the lost Mars 8 (Mars-96) mission, FMI/GEO designed and manufactured the central data and operations control system as well as the meteorological instruments for an autonomous lander. FMI/GEO devotes about 40 of its 50 man-years to space research.

Finnish Space Committee

The Space Committee comprises a chairman, vice-chairman and 10 members, with the executive working group as the sole subordinated body. The chairman is the national representative on ESA's Council. A new committee was appointed 1 April 1995 to serve into 1998.

Helsinki University of Technology, Space Technology Laboratory

HUT is Finland's oldest and largest technical university. STL main research area is airborne and microwave remote sensing, including instrumentation, measurements and development of geophysical inversion algorithms. It also runs space technology courses and proposes the 50 kg HUTSat.

National Land Survey of Finland (MAANMITT AUSLAITOS)

SIC imports satellite images and offers related services: initial processing and distribution of RS data, archives, technical services, user studies and short tailor-made courses. A major task is to support Finnish industry and the research sector in the use of satellite imagery. It is the National Point of Contact for ESA/ESRIN. SIC is distributor of satellite data products in Finland under agreement with Sweden's SSC Satellitbild and Eurimage. National Land Survey staff 2,160; 1995 operational expenditure FMk469 million (ECU81.6 million). SIC was established in April 1995 and has a staff of nine.

Technology Development Centre (TEKES)

TEKES was established in 1983 by the Ministry of Trade and Industry to co-ordinate and finance applied technical and industrial research and development. It is responsible for implementing, co-ordinating and financing national space projects and Finnish participation in international programmes, including ESA. The 1997 budget is FMk110 million: 1996 – FMk104 million, 1995 – FMk106 million; 1994 – FMk95 million and 1993 – FMk85 million.

FRANCE

Centre National d'Études Spatiales

CNES was established in 1962 with a staff of 27 to execute the national space programme. Organized since April 1993 under the regulating authority of three ministries (Defence; Industry, Posts, Telecommunications and Foreign Trade; Higher Education and Research), CNES is headed by a President and a Director General. It has an 18-member Conseil d'Administration, with six members elected by agency employees. In 1993 it was also allocated responsibility for design/development of military space projects, under the Ministry of Defense. Personnel total 2,483, with 9 per cent located at headquarters in Paris, 10 per cent in Evry, 68 per cent in Toulouse and 13 per cent at the CSG Ariane launch site.

CNES was instrumental in creating eight subsidiaries and four GIE Groupements d'Intérêt Economique to exploit and market its space activities. Arianespace produces and markets Ariane launchers and in conjunction with a US subsidiary, Arianespace Inc, S3R offers insurance as an Arianespace subsidiary. Intespace performs space environment tests and maintains facilities for the tests. Spot Image promotes commercialisation of Spot satellite data, in conjunction with US subsidiary Spot Image Corp, Australian Spot Imaging Service, Swedish SSC Satellitbild and Singapore's Spot Asia CLS. For commercialisation of Argos location and data collection services CNES uses US subsidiary Service Argos Inc and North American CLS. Novespace promotes space technology transfer, including microgravity applications. SCOT Conseil is an Earth observation consultancy. SIMKO provides urban development of Kourou town. MEDES-IMPS researches space medicine and physiology. Satel Conseil is a space telecommunications consultancy. (GIE) Prospace promotes French space sector products and services. (GIE) GDTA gives remote sensing training and (GIE) DERSI develops relations between French and Russian space industries.

CNES does not maintain in-house space science laboratories but supports a number of principal French research facilities. For example, a CESR/SAp collaboration provided the Sigma telescope for Russia's Granat observatory through CNES. The laboratories are: Institut d'Astrophysique Spatiale (IAS, Institute of Space Astrophysics at Verrières-le-Buisson), Laboratoire de Météorologie Dynamique (LMD, Dynamic Meteorology Lab, Paris and Palaiseau), Département de Recherche Spatiale de l'Observatoire de Meudon (DESPA, Meudon Observatory Space Research Dept), Service d'Astrophysique du Département de Physique Générale du CEA (astrophysics section of CEA atomic energy authority, at Saclay), Laboratoire d'Astronomie Spatiale (LAS, Space Astronomy Lab, Marseille), Centre de Recherche en Physique de l'Environnement terrestre et planétaire (CRPE, Centre for Research into the Physics of the Terrestrial and Planetary Environment, Issy-les-Moulineaux), Centre d'Etude Spatiale des Rayonnements (CESR, Centre for Space Investigation of Radiation, Toulouse), Groupe de Recherches de Géodésie Spatiale (GRGS, Space Geodesy Research Group, Grasse and Toulouse), Laboratoire de Physique et Chimie de l'Environnement (LPCE, Environmental Physics and Chemistry Lab, Orléans-la-Source), Laboratoire d'Etude de la Solidification et Cristallogénèse (SESC, Solidification and Crystallogenesis Study Lab, Grenoble), Laboratoire de Physiologie Neurosensorielle (Lab of Neurophysiology, Paris), Laboratoire de Biophysique Médicale (Lab of Medical Biophysics, Tours), Service d'Aéronomie CNFS (Verrières-le-Buisson), Service de Physique de l'Etat Condensé (CEA lab, Gif sur Yvette), Institut de Physique du Globe (IPG, Global Physics Institute, Paris), Laboratoire d'Etudes et de Recherches en Télédétection (LERTS, Toulouse), Laboratoire de Biologie Médicale (Toulouse), Laboratoire de Cytologie et Morphogénèse Expérimentale (Paris), Laboratoire de Biophysique Appliquée (Fontenay-aux-Roses) and Centre d'Etudes et de Recherche de Médecine Aérospatiale (CERMA, Brétigny-sur-Orge). CESR, LPCE and CRPE contributed instruments to the Russian-led Interball; and LMD provides the Scarab Earth radiation budget radiometer for various satellites.

Centre Spatial d'Evry

Evry is CNES' launcher development centre, responsible for the Ariane series for the European Space Agency.

Centre Spatial de Toulouse (CST)

CST at Toulouse is the agency's principal engineering and technology facility and exploits the operational systems. CST also manages the balloon launching sites at Aire-sur-l'Adour and Gap-Tallard. The centre comprises 20 laboratories, including the spacecraft test facility managed by CNES subsidiary Intespace. The operation centres located at Toulouse include: two satellite launch/early orbital phase control rooms (second opened 1991), Sarsat-Cospas control/mission centre; the Telecom 1/2 dedicated control centre; the TDF 1/2 dedicated control centre; the Spot 1-3 control centre and the Spot imagery rectification centre; the Helios 1/Spot 4 Fresnel building; the computer centre; and the 2 GHz network control centre, with stations at Issus Aussaguel (15 km southwest of Toulouse), Kourou, Hartebeesthoek (South Africa) and Kerguelen Island (3,000 km southeast of South Africa).

Etat Major des Armées

The Armed Forces Headquarters is responsible for developing France's multiyear military space plan.

Geosys

Geosys was established in 1987. It is a member of the SGS group (Société Générale de Surveillance) and the Caisse des Dépôts et Consignations (Deposit and Consignment Office). Personnel total 30. It compiles regular crop inventories and prepares yield forecasts using remote sensing data, including Spot, NOAA AVHHR and Landsat TM. It also provides geographic information to the agribusiness and environment sector. It develops/sells GIS handling remote sensing images and vector cartographic data.

Groupement pour le Développement de la Télédetection Aérospatiale (GDTA)

The Groupement pour le Développement de la Télédetection Aérospatiale is a GIE (Groupement d'Intérêt Economique) of CNES, the Institut Géographique National (IGN), Bureau de Recherches Géologiques et Minières (BRGM) and Institut Francais de Recherche pour l'Exploitation de la Mer (IFREMER). The principal aim is to promote the development of remote sensing activities and the use of satellite imagery (in particular, Spot products). Personnel total 33; annual turnover is about FFr25 million. GDTA also has access to the resources and personnel of its four member organisations.

GDTA offers remote sensing training, from introductory to advanced levels. It provides technology transfer services abroad for the French Co-operation and Foreign Affairs Ministries and, in addition to remote sensing training, it provides advisory services for establishing remote sensing centres. GDTA also organises airborne campaigns to test sensors planned for eventual satellite applications. It is currently working on high-resolution European imaging radars.

Météo-France

Coming under the authority of the Ministry of Equipment, Transportation and Sea and responsible for providing space-based meteorological services, Météo-France acted as the national representative in establishing EUMETSAT and in the implementation of the Meteosat operational satellite system. The Ministry's 1995 Meteosat funding was FFr216.7 million.

Its dedicated Centre de Météorologie Spatiale in Lannion receives, processes and relays meteorological satellite data.

Ministère de la Défense

This ministry is the customer through the Délégation Générale pour l'Armement (DGA, military procurement agency) and CNES for the Syracuse military communication element of the Telecom satellite system and the Helios reconnaissance satellite.

Services de Consultance en Observation de la Terre (SCOT CONSEIL)

SCOT (Services de Consultance en Observation de la Terre) was established in 1987 as a CNES affiliate to act as a consultancy and promoter of satellite-based remote sensing data, drawing particularly on Spot-derived imagery. Staff totals 32, including eight engineers. French banks account for 25 per cent shareholding. The company provides a spectrum of services, from user advice on specifications for Earth stations, processing centres and software to overseeing complete remote sensing programmes. It heads a team of contractors selected by the Commission of the European Communities to develop an operational system to produce rapid estimates of areas under cultivation and potential yields of major annual crops at European scale.

Société Nationale des Poudres et Explosifs (SNPE)

The Société Nationale des Poudres et Explosifs SNPE Group, created as a state-owned company in 1971 and now with 5,000 personnel and FFr4.4 billion 1994 turnover (FFr100 million profit), manufactures liquid rocket propellants, with emphasis on UDMH, and develops/produces solid propellants. Along with BPD Difesa e Spazio it participates in the Regulus partnership. The company provides the propellant for Ariane 5's P230 solid boosters using facilities at Kourou constructed by subsidiary SNPE Ingénierie. SNPE maintains facilities at Saint-Médard en Jalles near Bordeaux (solid propellant grains), Toulouse (UDMH, AP), Toulon (pyrotechnics) and Le Bouchet near Paris (HQ and R&D). The Toulouse site manages the production of ammonium perchlorate with a capacity of 6,000 tonnes annually to ensure Ariane 5 supplies. SNPE formed a second partnership with BPD, Société Européenne de perchlorate d'ammonium (EUPERA; SNPE 66 per cent, BPD 34 per cent), to handle production. Toulouse also provides the UDMH for Ariane 4 and the MMH for Ariane 5's upper stage. Subsidiary Pyromeca manufactures detonating cords and flexible linear cutting charges and ultra-fast pyrotechnic valves for Ariane 4 release from the launch pad.

In cryogenic propulsion, SNPE provides the main igniter and turbine starter grain for Ariane 4's HM-7B stage 3 engine and the shaped ducts for the Ariane 5 Vulcain igniter and starter.

Télédiffusion de France (TDF)

TDF, created in 1975 as a limited company with the French government as majority shareholder (and a France Telecom subsidiary since 1988), operates 10,000 TV transmitters/repeaters, 1,500 radio transmitters and 10,500 km of multi-channel microwave links, handling broadcasting for all public/private French TV channels and the principal public/private radio networks. Personnel total 4,000, including 2,500 engineers and technical staff. 1993's sales were FFr4.0 billion. 1992's sales FFr3.8 billion; profit FFr140 million. 1990's estimated FFr200 million loss was largely as a result of its satellite problems.

CNES/TDF undertook a study of a national DBS system, following the signing of the WARC '77 agreements, to provide high quality TV and CD-quality radio. In October 1979, France/Germany agreed to joint development of respective national DBS systems, signing the accord April 1980. The Eurosatellite consortium was awarded the space segment contract in 1984, basing the design on Aerospatiale's Spacebus 300 platform. Aerospatiale acted as prime for TDF 1/2 and MBB for TV-Sat 1/2. The decision to build TDF 2 was taken by the French government in December 1984. There was concern in 1986 that the programme would be cancelled because of developing competition from medium-power systems such as Astra. By then FFr1.5 billion of the FFr3.5 billion allocated to TDF had been expended and the French Prime Minister ruled that it should continue in order to promote D2-MAC as the standard for HDTV in the 1990s. D2-MAC Paquet specifications were adopted by France/Germany in June 1985. FFr550 million was paid to Arianespace for the launches.

TDF underwent minor modifications after TV-Sat 1 was written off when one solar array failed to deploy in November 1987 and thruster overheating was discovered during orbital manoeuvres. TDF was to provide five TV channels from 230 W TWTAs to 45 cm diameter receivers but at the time of launch none had been allocated to users. The failures of two tubes aboard each satellite reduced the number of available transponders from 12 to eight, prompting the 1990 decision to provide only four TV and three digital radio channels in order to maintain full redundancy. At the same time, it was decided there would be no TDF 3 and that the future system would lease Eutelsat from the mid-1990s. By 1992, it was reported that only some 35,000 locations were equipped with TDF dishes. Annual transponder leasing cost was reduced from FFr80 million to FFr35 million. System cost will probably not be recouped.

GERMANY

Aachen Centre for Solidification in Space

The Aachen Centre for Solidification in Space was established at the town's Institute of Technology in 1986 to provide a link between μg technology and potential users. Research topics include containerless processing, metal-ceramic composites, high-temperature superconductors, directional solidification, model systems and numerical simulation. Equipment includes Bridgman-type furnaces for ground-based research accompanying space experiments, containerless processing facilities (EM levitation and a droptube), devices for corrosion investigations and wettability experiments, test setups for determination of mechanical and magnetic properties and microstructure characterisation (SEM, EDX, DTA/DSC, and so on).

Bundesminister für Forschung

The Federal Ministry for Research and Technology is responsible for Germany's space research policy. Following the restructuring of national space administration, responsibility for planning, programming, management and external representation was transferred to DARA.

Deutsche Agentur für Raumfahrtangele genheiten

DARA was founded and began operating in July 1989. Its legal status is that of a private company with limited liability (GmbH) owned and financed by the Federal Government. Personnel total 265. The Management Board comprises three Managing Directors, each responsible for a specific sector covering individual departments: scientific utilisation, Earth observation, application utilisation, orbital infrastructure, transport systems, ground infrastructure and administration. A Supervisory Board of members from Federal Ministries, science and industry provides guidance on policy. An Advisory Board of high-ranking representatives from science and industry provides scientific and economic input. There are also several Advisory Committees on the different fields of science and technology.

Since June 1990, when the 'Law Governing the Transfer of Responsibilities with Regard to Space Activities' was adopted by the German parliament, DARA has had government authority for the space programme, particularly in awarding contracts and representing Germany at the international level. For the first time, domestic space legislation stipulates an obligation for German authorities to transfer responsibilities to a single central body. This applies primarily to BMFT, but also to other governmental authorities with vested interests in space. DARA is in full charge of implementing the German Space Programme, except for the hypersonic technology programme, which remains embedded within BMFT's aeronautics activities. DARA has represented Germany since 1 December 1989 in the various bodies of ESA (Council, Committee and Programme Boards) and other international organisations, in addition to being responsible for multilateral/bilateral agreements.

Deutsche Forschungsanstalt für Luft und Raumfahrt

In close co-operation with DARA, the German Aerospace Research Establishment's responsibility is to execute the research element of the German space programme. It was formed in 1969 and employs a total staff of 4,500, including 1,600 scientists in five research areas: Telecommunications Technology and Remote Sensing, Materials and Structures, Energetics, Fluid Mechanics, and Flight Mechanics/Guidance and Control. The annual budget is about DM700 million. Research centres are located in Berlin, Braunschweig, Göttingen, Köln, Lampoldshausen, Oberpfaffenhofen and Stuttgart, with local branches in Hamburg, Bonn, Trauen, Neustrelitz and Weilheim.

Test facilities include subsonic-to-hypersonic wind tunnels, rocket propulsion test stands, spacecraft TT&C stations and space simulators. Space mission operations, particularly those for Spacelab, Columbus and Eureca, are centred at the Köln-Porz and Oberpfaffenhofen sites. The Crew Training Complex (for ESA's astronaut cadre) and Microgravity User Support Centre in Köln-Porz, and the German Space Operations Centre (GSOC), Manned Space Laboratories Control Centre, User Data Centre and Automation In Orbit Centre at Oberpfaffenhofen have performed manned/unmanned missions for more than 20 years.

The DFD German Remote Sensing Data Centre is the national PAF Processing and Archiving Facility for ERS and other satellites. Köln is the home base for federal astronauts and seat of the EAC European Astronaut Centre of ESA. The Materials and Structures department includes the Institute of Space Simulation, which operates the Microgravity User Support Centre and was most recently primarily concerned with preparing users, astronaut crew and equipment for Spacelab D2. Under the Department of Energetics, the Space Propulsion Division at Lampoldshausen operates test facilities for rocket motors such as Ariane 5's Vulcain. A new stand was completed in 1995 under a bilateral programme with France.

Deutsche Telekom AG

Under the aegis of the German Federal PTT (BMPT), the German Telecommunication Co is the national signatory to Intelsat, Inmarsat and Eutelsat and operator of the DFS-Kopernikus national telecom satellites. It is in the process of privatisation and is moving away from satellite ownership; there may be no DFS follow-on and the sole TV-Sat national DTH satellite was leased off to Norway in 1995.

Deutsche Telekom became a stock corporation on 1 January 1995 and was listed on the stock markets in Frankfurt, New York and Tokyo on 18 November 1996. It is the national signatory to Intelsat, Inmarsat and Eutelsat and operator of the DFS-Kopernikus domestic telecom satellites. A 16.67 per cent interest in Luxembourg's SES was acquired in 1994. As part of the privatisation process, the responsibility for providing mobile services (including Inmarsat) passed to DeTeMobil. The subsidiary agreed in March 1995 to pay US$566 million for a 25 per cent interest in Indonesia's PT Satelindo.

A Franco-German agreement on joint DBS development complying with WARC '77 recommendations was signed in April 1980: Aerospatiale acted as prime for France's TDF 1/2 and MBB for the equivalent TV-Sat 1/2. The system adopted D2-MAC to promote the standard for high definition TV in the 1990s. The sole operational satellite has been leased to Norway since 1995. It commissioned three DFS satellites in 1984 for telecom services. DT operates three large Earth station complexes: Usingen (DFS, Intelsat), Raisting (Inmarsat, Intelsat) and Fuchsstadt (Intelsat). Following 1990's unification, Inmarsat A stations have been contributing to telecommunications services for the former GDR. Licensed by DT, they operate through Raistang. There were 1.8 million telephones in East Germany, compared with 29 million in West Germany in 1989.

DLR Microgravity User Support Centre (MUSC)

MUSC, based at the German Aerospace Research Establishment's Köln centre, supports the preparation, operation and utilisation of materials and life science experiments on Spacelab, Texus, Mir and the International Space Station. It provides user information by seminars, databases and a library, technical support for experiment and experiment facility preparation, operational support for interactive experiment operation (telescience), and scientific support by consultancy and measurement/diagnosis equipment for quick-look analysis.

Facilities: ground versions of Spacelab's Anthrorack, Biolabor, Holop, MEDEA and TEMPUS units, Eureca's core payload and Mir's CSK/TITUS furnace. There is also an experiment control room for remote-control operations.

Institut für Raumfahrtsysteme

The Space Systems Institute (IRS) was founded in 1970 and plasma thruster development has always been a principal interest. Facilities allow MW-class testing at 0.5-2 g/s propellant flow rates (argon) under selected 10^{-6}-10^{-2} atmosphere conditions. 0.5-1 MW MPD stationary thrusters of different geometries have been developed, mainly under AFOSR grants. Based on this experience, IRS has designed and built a range of thermal arcjet thrusters. Current arcjet projects include: Atos 600 to 700 W ammonia, Artus 2 1 to 2 kW NSSK hydrazine (with DASA; DARA contract); 10 kW and 100 kW devices. Atos is a simplified version of Artus 2 for flight on Amsat-Deutschland's amateur satellite aboard Ariane 502.

Four plasma wind tunnels are used to investigate re-entry vehicle thermal protection materials and to validate aerothermodynamic CFD codes. Two have MPD plasma generators (0.10 to 1 MW) and are especially suited for high specific enthalpy (up to 150 MJ/kg) and low total pressure conditions. One tunnel has an inductive plasma generator, specially used for catalycity investigations, and the fourth has a thermal plasma generator for higher total pressure and low specific enthalpy areas.

IRS' Mission and System Analysis division studies development and numerical simulation and design tools for space transportation systems, mission and system optimisation, and performance assessment of air-breathing launchers. The Space Technology and Utilization division encompasses space station design, numerical flow field and simulation methods, and space systems safety.

ZARM

ZARM (Zentrum für angewandte Raumfahrttechnologie und Mikrogravitation) was established in September 1985 in the Faculty for Production Technology at the University of Bremen. μg research interests include fluid mechanics, combustion and thermodynamics, and technical assistance in its industrial exploitation. ZARM's DM30 million 146 m Drop Tower Bremen was commissioned in 1990, providing 10^{-5} *g* for up to 5 seconds. ZARM-FAB mbH (ZARM Fallturm Betriebsgesellschaft mbH: ZARM-Drop Tower Operation and Service Co) co-ordinates drop tower operations. ZARM-Lab GmbH provides technology/science support for users of Columbus and other flight facilities. ZARM-Förderverein eV (ZARM Promoting Association) is a non-profit organisation to support space and μg activities.

A 0.25 m hypersonic wind tunnel (ZARM-HHK) became operational in 1993 within the Hypersonic Technology research division. ZARM was also responsible for the university's Bremsat Shuttle satellite.

The 3.5 m diameter 110 m long drop shaft provides a 4.74 second free-fall through a 10^{-5} atmosphere vacuum (pumping requires 1½ h). The pressurised capsule is decelerated over 6 m by styropore granules. A larger capsule became operational in 1995.

Capsule accommodation: 700 mm diameter, 1,200 mm high; 150 kg mass; 28 V 10 A (100 A short term) power; 1.6 Mbit/s telemetry; peak deceleration 350 m/s² (design must allow for safety factor of 2).

HUNGARY

Hungarian Satellite

HUNSAT is the signatory to Intersputnik, Intelsat and Eutelsat. Two 18 Intelsat A stations are operated: Balaton-2 (Atlantic services) and Balaton-3 (Indian Ocean).

Hungarian Space Office

The new space research structure was formally created 1 January 1992 on the advice of the Scientific Council for Space Research. Before then, the Interkosmos Council managed participation in Interkosmos and related space science projects, such as Phobos and Interball.

KFKI Research Institute for Particle and Nuclear Physics (RIPNP)

The five institutes previously part of the KFKI (Kozponti Fizikai Kutato Intezete, or Central Research Institute for Physics) became independent legal entities on 1 January 1992 belonging to the Academy of Sciences. RIPNP designed the TV camera system and two plasma instruments (Tünde and Plazmag) for Vega, was involved in five plasma detectors on Phobos (in addition to its lander's central control/data acquisition computer), and dosimeters on Salyut 6 (1980) and Challenger (1984). Current projects include Spektrum-X/γ's (now Spektrum-RG) data acquisition system/onboard science computer, Kvant 2's Microsvit image processing system and some onboard software for the CAPS particle analyser and MAG magnetometer on Cassini and co-investigator for the Rapid and Lion particle detectors of ESA's Cluster and Soho, respectively. The Institute is participating in ESA's Rosetta mission to Comet Wirtanen it will develop the on-board electronics and ground support equipment of the RPC plasma experiment on the orbiter. It is also planned that the Institute will produce the onboard computer and software for the lander.

Remote Sensing Centre of the Institute for Geodesy, Cartography and Remote Sensing (FÖMI RSC)

FÖMI RSC has two principal interests: R&D of remote sensing techniques for environmental protection and agriculture; distribution, processing and utilisation of remote sensing data for users. Services cover soil mapping, monitoring of crop cultivation, updating of topographical maps using Spot and Cosmos imagery, and digitally enhanced space photomaps from Spot data.

The centre is participating in the CORINE Community-wide Co-ordination of Information on the Environment programme. ESA is co-operating in upgrading facilities for ERS and NOAA AVHRR data pre-processing.

INDIA

Indian National Satellite System

The Insat system uniquely provides geostationary platforms for simultaneous domestic communications and Earth observation functions. The four first generation Insat 1 satellites were all US-built, but the advanced Insat 2 family is indigenously produced. The Insat system is a joint venture of India's Department of Space, Dept of Telecommunications, India Meteorological Dept, All India Radio and Doordarshan. Overall co-ordination and management rests with the inter-ministerial Insat Coordination Committee (ICC); the DOS has direct responsibility for establishing and operating the space segment. It was noted in February 1996 that the system generates revenues of Rs4 billion annually.

The satellites are handled from the Insat Master Control Facility (MCF) at Hassan, Karnataka through two Satellite Control Earth Stations (one with a 14 m fully steerable antenna and the other with a 7.5 m limited steering dish), one additional 14 m fully steerable antenna and an Insat 1 Satellite Control Centre (SCC) with associated TT&C equipment, on-orbit checkout equipment, computer facilities and auxiliary power services. MCF was upgraded with the addition of the Insat 2 SCC and two associated 11 m Satellite Control Earth Stations.

Indian Space Research Organisation (ISRO)

ISRO was established in 1969 as India's primary space R&D organisation, responsible for developing launcher and propulsion systems, launch sites, satellites and their tracking networks. Personnel totals rose from 13,488 in 1986 to 16,800 in 1996, accommodated in eight main centres and units:

Vikram Sarabhai Space Centre (VSSC). ISRO's single largest facility (5,600 personnel), near Trivandrum, providing the technology base for launcher and propulsion development.

Liquid Propulsion Systems Centre (LPSC). Development wings in Bangalore and Trivandrum are supported by major test facilities at Mahendragiri for a wide spectrum of liquid motors, from reaction control system thrusters to the 720 kN Vikas and cryogenic engines. Personnel total 1,450.

ISRO Satellite Centre (ISAC). ISRO's lead centre for the design, fabrication and testing of science, technology and applications satellites. Staff strength is 2,400.

ISRO Inertial Systems Unit (IISU). Provides inertial systems/components for satellites and launchers.

SHAR Centre. ISRO's orbital launch site and largest solid motor production and test facility with 2,400 employees.

ISRO Telemetry, Tracking and Command Network (Istrac). Headquartered in Bangalore, Istrac operates a network of ground stations to provide TTC support for launcher and satellite operations. Personnel total is 460.

Space Applications Centre (SAC). Located at Ahmedabad, SAC is ISRO's applications R&D centre, including communications, remote sensing and geodesy. For example, it developed IRS 1C's cameras. 2,150.

Development and Educational Communications Unit (DECU) at Ahmedabad.

Insat Master Control Facility (MCF) at Hassan, 180 km from Bangalore. 295 personnel.

National Remote Sensing Agency Ground facilities for the reception, processing and dissemination of IRS, Landsat and ERS data, as well as the development of remote sensing techniques and applications.

National Remote Sensing Agency (NRSA)

The NRSA was created in 1974 as a focus for remote sensing activities and since 1979 has provided India's primary Earth station, at Shadnagar 55 km south of Hyderabad, for reception of NOAA, Landsat, ERS and IRS imagery. It also operates the IIRS training centre. The NRSA Data Centre (NDC) in Hyderabad disseminates photographic and digital products from IRS, Landsat, ERS and NOAA, in addition to maintaining archives of acquired data. There are more than 500 data users, concerned with vegetation, waste land, ground water and forest mapping in addition to agricultural applications.

The agency is a contributing element to the National Natural Resources Management System (NNRMS), which includes five Regional Remote Sensing Service Centers (RRSSC) at Bangalore, Dehra Dun, Nagpur, Jodhpur and Kharagpur. The first, at Bangalore, was inaugurated in August 1987. Each RRSSC provides facilities for digital analysis of satellite data, complemented by photo processing and visual interpretation laboratories. The system is configured around a VAX 11/780 with VIPS 32 image analysis software (provided by France's SEP) and three Numelec

PC 2001 image displays. The system can handle digital classifications for the equivalent of 1 1/2 LISS 2 (IRS) scenes each day, or three Landsat MSS, or three-quarters of a Landsat TM, or three Spot images. Cost/frame to users of processed IRS data is about Rs2,300 for LISS 1 on 24 cm false colour composite paper print.

Physical Research Laboratory (PRL)

India's principal centre for space science activities, covering ground-based telescopes and satellite/sounding rocket instrumentation.

Space Commission/ Department of Space

The Dept of Space (DOS) implements the policies framed by the Space Commission through ISRO, the National Remote Sensing Agency, the Physical Research Laboratory and other agencies.

Videsh Sanchar Nigam Ltd (VSNL)

VSNL is India's national signatory to Intelsat and Inmarsat.

VSCC Liquid Propulsion Systems Centre (LPSC)

LPSC is responsible for the development of launcher liquid and cryogenic propulsion stages and auxiliary propulsion for launch vehicles and spacecraft. It is growing in importance as main liquid engines are introduced to India's orbital launchers and with the increasing size of indigenous satellites. Current main projects cover the 720 kN Vikas PSLV stage 2 engine, PSLV's stage 4 dual 7.5 kN system, Insat 2's 440 N Liquid Apogee Motor and unified network of 22 N ACS thrusters, and the 680 kN Vikas GSLV strap-on. Seven Russian 76 kN cryogenic engines are being provided for the initial GSLVs beginning early in the new century, but India's own engine will complete development in 2003-5.

LPSC's test facilities are sited southeast of Trivandrum at Mahendragiri. Vikas' Principal Test Stand (PTS) was commissioned during 1987 and used in January 1988 for the engine's first full-duration 150-second firing. Altitude facilities are also available for PSLV's 7.5 kN motor, Insat 2's LAM and smaller thrusters. Cryogenic engine and stage facilities were commissioned in 1997.

VSCC Solid Propulsion Group

India's first 75 mm diameter solid motor was produced in 1967, followed by a 125 to 560 mm range for sounding rocket applications and, in the early 1970s, motors for the SLV-3 satellite launcher. Work on the most powerful so far began in 1984: the 2.8 m diameter 3,500 kN thrust solid powers PSLV's first stage. Although the Solid Propulsion Group is an element of VSSC, the primary Solid Propellant Space Booster Plant (SPROB) is sited at SHAR Centre on Sriharikota Island, along with the Vehicle Assembly Static Test and Evaluation Complex (VAST) for solids. At VSSC the motor cases are produced in the Mechanical Engineering Facility and the propellant binders in the centre's Propellant Fuel Complex.

INDONESIA

National Institute of Aeronautics and Space (LAPAN)

LAPAN (Lembaga Penerbangan dan Antariksa Nasional), founded in 1963, is a non-departmental government agency and is directly responsible to the President. Its 1991/92 budget was Rp10,000 million (about US$5 million) and its 1986 budget was Rp6,500 million (about US$4.06 million). Its chief activities are remote sensing and probing of the upper atmosphere by balloon and rocket sondes. It also performs telecommunications research, notably on signal attenuation caused by heavy tropical rainfall. First and second stage solid propellant sounding rockets designed and built by LAPAN are launched up to 100 km from the range at Pameungpeuk some 150 km south of Bandung. The agency maintains Earth stations at Jakarta and Biak Island (Irian Jaya) for reception of GMS, NOAA and Landsat MSS/TM imagery. The Multimission Remote Sensing Satellite Ground Station was commissioned in October 1993 at Pare-Pare on Sulawesi island to receive Spot, ERS and Landsat 5 (Landsat 6 ETM was also to have been included); Japan's JERS was added December 1995.

PT Indosat

The Indonesian Satellite Corp is the national signatory to Intelsat, which the country joined in 1964 as one of the first members, and to Inmarsat, which it joined in October 1986 as the 48th member, with a 0.385 per cent interest. The government in 1994 sold 35 per cent of its interest. Indosat is responsible for overseas communications links, accessing both the Indian and Pacific satellites through its Jatiluhur gateway. It secured a competitively-bid contract in 1983 to provide TT&C services to Intelsat, previously handled by Australia's OTC from its Carnarvon station. Apart from satellite links, Indosat is responsible for national participation in international submarine cables in the region.

INTERNATIONAL

Arabsat

The Arab Satellite Communications Organization was established in 1976 to meet the increasing communications needs of its members. There are 21 members (with percentage shares): Algeria (1.72), Bahrain (2.45), Egypt (1.59), Iraq (1.90), Djibouti (0.12), Jordan (4.05), Kuwait (14.59), Lebanon (3.83), Libya (11.28), Mauritania (0.27), Morocco (0.61), Oman (1.23), Palestine (0.25), Qatar (9.81), Saudi Arabia (36.66), Somalia (0.24), Sudan (0.27), Syria (2.08), Tunisia (0.74), United Arab Emirates (4.66), Yemen (1.65). The first two satellites were launched in 1985 (both removed from service in 1992), and the third/last first generation in 1992, extending operations until the second generation could appear in July 1996, launched by Ariane 44L.

The original operational plan was to carry all traffic on one satellite and to hold the second as an in-orbit spare, but increasing demand required distribution between both. At the beginning of 1992, C-band transponder usage was: nine providing more than 3,500 regional telephone channels, two regional TV, 3¼ Saudi Arabia telecommunications, one Morocco telecommunications, one Oman telecommunications, one Mauritania telecommunications, one ASBU Arab States Broadcasting Union TV, one CNN (US) and 12 India telecommunications. India leased its 12 on 1B for two years beginning 1 October 1989, extending the lease (then on 1C) in 1991 until September 1992. Saudi Arabia leases one S-band.

Argos

Argos is a joint programme of CNES, NASA and NOAA, operated under an MoU signed in 1974 and extended in 1986. System operation and promotion is the responsibility of CLS, established in April 1986 as a subsidiary of CNES (55 per cent), IFREMER (French Marine Institute; 15 per cent) and a pool of banks. CNES operates the French Global Processing Centre (FRGPS) in Toulouse and handles all user relations outside North America. US subsidiary Service Argos Inc is responsible for all US/Canadian users; it operates the identical USGPC, which provides redundancy. North American CLS is concerned with developing new services and products. The Australasian regional processing centre began operations in December 1988.

The system consists of independent user Platform Transmitter Terminals (PTTs, down to 25 g and the size of a matchbox), two NOAA polar satellites with Argos Data Collection and Location System packages to receive PTT messages on a random access basis for separation, time coding, formatting and retransmission to ground stations, and the ground stations and two Global Processing Centers in Toulouse and Landover, where data are retrieved, processed and distributed to users. Location of PTTs can be determined to within typically 150 m, if required, by Doppler techniques.

The PTTs transmit data from up to 32 sensors encoded in 8-bit words on 401.650 MHz with typically 200 mW power at intervals of 60 to 120 seconds for location PTTs and >200 s for data collection-only platforms. Message length is 360 to 920 ms and contains the PTT identification number.

The principal Argos system processes and disseminates space and terrestrial environmental data received by dedicated packages aboard US NOAA polar meteorological satellite from fixed/mobile platforms anywhere in the world. More than 5,000 transmitters are operating; the past five years has seen a 15 per cent growth in activity. More than 20,000 transmitters have used the system.

The NOAA Argos packages receive all messages within a 5,000 km diameter visibility circle at any instant; four can be processed simultaneously by NOAA A-J and eight will be handled by the third generation system aboard NOAA K and its successors from 1996. NOAA 11 can handle data at 960 bit/s, NOAA J 1,200 bit/s and NOAA K-N 2,560 bit/s. NOAA K-N increase bandwidth from 24 kHz to 80 kHz. The data are formatted/stored, then dumped each time the satellite moves within reach of one of the three ground stations. VHF/S-band transmitters also perform real-time relay for any user station within the visibility circle. Current onboard packages can process up to 1,400 data-only PTTs, or 415 requiring the location service, but from NOAA K on the capacity will be quadrupled. The Argos 3 generation will be defined by end-1996 for post-2000 satellites. Japan's 1999 ADEOS 2 will carry an Argos package allowing users to command their PTTs.

Under an agreement signed in April 1990 with EUMETSAT, CLS manages all non-meteorological applications using the Meteosat data collection function. The MAEDS Multisatellite Applications Extended Dissemination Service adds Meteosat and GOES to the existing Argos system to provide high data collection capacity (649 byte messages up to 24 times daily), real-time access to data, and continuous coverage from the Pacific to the Urals. The first MAEDS element became operational in October 1990, followed in 1991 by the GOES satellites covering America and the Pacific. CLS also operates the Doris location system and is involved in the Starsys system.

Global Processing Centers NOAA's tape recorders are read out every 100 minutes over one of the Gilmore Creek (Alaska), Wallops Island (Virginia) and Lannion (France) stations. All of the satellite's data are routed to NOAA's National Environment Satellite, Data and Information Service (NESDIS) in Maryland, where the Argos component is extracted and relayed to CLS in Toulouse and Landover. Here, PTT locations are determined and the sensor data processed. About two-thirds of the processed messages are provided to users within three hours of the uplink. Location is derived from measuring the Doppler shift in the carrier frequency of received messages during the 10 minute overflight. An accuracy of 150 m is typical but for PTTs at fixed positions or on slow-moving carriers 100 m is possible. The transmit frequency must remain stable throughout the pass. Each frequency measurement yields a set of possible positions and with a known elevation only four messages are required on each pass to generate a unique solution. The satellites' own positions are known to within 300 m at any instant from a reference system of 11 beacons worldwide at precisely known geodesic positions.

ASETA

ASETA (Asociacion de Empresas Estatales de Telecomunicaciones del Acuerdo Subregional Andino) comprises the five telecom enterprises of the Andean Group: ENTEL-SAM (Bolivia), TELECOM (Colombia), EMETEL SA (Ecuador), Telefonica del Peru SA (Peru) and CANTV (Venezuela). The Group represents a population of 100 million over 4.8 million km². ASETA was established in 1974 to develop telecom services facilitating regional integration of the Andean countries. It is now developing the Regional Telecommunications Master Plan in co-operation with the ITU, principally focusing on establishing and exploiting the SAT Andean Telecommunication System. SAT's main aim is building the Corredor Andino Digital, the Andean information and telecom highway comprising terrestrial, satellite and submarine route through the five countries. The fibre optic terrestrial route is under construction by each of the five enterprises in their own territories. The submarine route will be developed through the Panamerican Pacific Fiber Optic System Agreement, in which the five enterprises will invest about 40 per cent of the total cost. The satellite route is about to be defined, proceeding with the Condor project under the Andean Enterprise Group. A private operator owned by the five countries, with ASETA member participation, is foreseen.

Until then, Intelsat in 1991 offered eight 36 MHz leases until 2010, in addition to the 11 36 MHz units already in use. Each country contracts directly with Intelsat; the initial leases began operating in May 1995 on Intelsat 705 at 50° west.

ASETA earlier planned a dedicated satellite system. Respective ministers signed the agreement creating the OATS Andean Satellite Telecommunications Organization serving the interests of Bolivia, Colombia, Ecuador, Peru and Venezuela in November 1988. Peru, Colombia and Venezuela were each to provide 28 per cent of the US$900 million Condor program (US$220 million space segment), with Ecuador and Bolivia contributing 7.7 per cent. Four GEO slots at 106°, 72°, 77.5° and 89° west were allocated by the ITU. A two-satellite system was projected for 1993-94. All five, with a total population of 85 million covering 4.74 million km², operate Intelsat A stations and had been considering a regional satellite system since 1974. Chile withdrew from ASETA in 1977. Three participants had to ratify the OATS agreement for it to be activated, but economic constraints instead dictated service leasing.

Committee on Earth Observation Satellites (CEOS)

CEOS was created at the suggestion of the G7 Economic Summit in 1984 and now encompasses all of the world's civil space agencies responsible for Earth observation space programmes, along with agencies that receive and process data acquired remotely from space. It provides a forum to exchange information on programmes, plans and requirements. 1996 activities include: completion of an initial assessment of overlaps and gaps in current/planned space missions to meet observing requirements of the major international operational and research programmes in environment, climate and global change; development of remote sensing training materials for use in developing countries; consideration of what role CEOS should play in defining/implementing a long term global observing system for the next decade and beyond.

COSPAR

The Committee on Space Research was established by the International Council of Scientific Unions (ICSU) in October 1958 to continue the co-operative programmes of rocket and satellite research undertaken during the International Geophysical Year of 1957-58. Membership comprises National Academies of Science or equivalent, and International Scientific Unions adhering to ICSU. Operating under ICSU rules, COSPAR is a non-political organisation for promoting international co-operation through meetings and publication of their proceedings. Publications include: *Advances in Space Research* (the Committee's flagship journal in which the proceedings of the Biennial COSPAR Scientific Assemblies are published – two volumes, 24 issues per year), the trimester *COSPAR Information Bulletin*, the irregular *COSPAR Colloquia Series*, and the *Directory of Organizations and Associates*, every two years.

Cospas-Sarsat

Cospas-Sarsat is an international satellite-based system for search/rescue, established in 1979 by Canada, France, US and USSR (membership now assumed by Russia), based on the detection of distress beacons by four polar satellites. It is capable of locating an activated beacon to within 2 km and was credited with helping to save more than 5,500 lives by June1995 in 1,800 S&R events. The system was governed by an MoU of October 1984 until the International Cospas-Sarsat Program Agreement was signed by the four states on 1 July 1988 in Paris.

Distress signals are relayed by the satellites to Local User Terminals for processing to determine beacon location. The information is passed to a Mission Control Centre to alert the rescue authorities. Some 550,000 121.5 MHz beacons are operational (more than 600,000 projected by 2000), but these are restricted to real-time relay by the satellite and therefore require an LUT to be within range (about 2,500 km) for position determination to within 10 to15 km. The 406.025 MHz units include user identification codes in the message but they still operate on the Doppler shift concept: this provides two locations for each distress signal, an ambiguity resolved by allowing for the Earth's rotation.

If the beacon's frequency stability is sufficient, as with the 406 MHz devices, the true solution is determined on a single satellite pass. The 406 MHz information is not only relayed in real-time but also time-tagged and stored for dumping as each LUT comes into view. This frequency therefore provides a global service with an average waiting time of 44 min, with 94 per cent detected in more than 90 minutes (results from 1990 exercise). All satellite downlinks operate at 1.5445 GHz. By end-1995, more than 119,000 406 MHz beacons had been produced and distributed in more than 130 countries.

The 406 MHz beacons emit a 5 W RF burst of about 500 m/s duration every 50 seconds. The improved frequency stability 10^{-9}/min (a factor of two relaxation was permitted until 1991) assures more precise location accuracy (90 per cent within 5 km), while the high peak power increases the probability of detection on a single pass to more than 98 per cent. Typical delay before detection is 1 to 2 hours near the equator (less at the poles).

Each satellite can handle 90 simultaneous signals. The digital encoded message conveys the country of origin and identification of the vessel or aircraft. Position information from systems such as GPS and Glonass will be encoded in the messages of new 406 MHz beacons from January 1997. This will be particularly useful for the emerging geostationary satellite relays, which at the moment cannot provide the position element. Depending on the beacon type (maritime, airborne or land), it can be activated manually or automatically by shock or immersion. An optional homing device is usually collocated with the 406 MHz beacon for the SAR services but it is not a system specification. Beacons are required to function within a –20/55°C thermal range for a minimum 24 hours and automatically limit any inadvertent continuous transmission to more than 45 seconds.

Mission Control Centres have been established in each country operating at least one LUT to disseminate information to the appropriate Rescue Co-ordination Centres. By end-1995 there were 33 LUTs in 21 countries, with others planned.

The USSR began deploying the space segment with the launch of Cosmos 1383 in 1982. Designated Cospas 1 ('Space System for Search of Vessels in Distress'), the 121.5 MHz band remained operational until Mar 1988, with 406 MHz utilized primarily for interference monitoring. Cosmos 1447 (1983), Cosmos 1574 (1984), Nadezhda 1 (1989), Nadezhda 2 (1990), Nadezhda 3 (1991) and Nadezhda 4 (1994, in the same orbit as Cospas 5 as a spare) adopted the roles of Cospas 2-7, with one more satellite ready for launch as required. The other relays to bring the system up to its full operational complement of four are provided by packages on the US NOAA meteorological satellites. Sarsat 1 (S&R Satellite-aided Tracking) was implemented with NOAA 8 in 1983. NOAA 9-11 brought the system to full strength, with further NOAAs planned for launch as required for meteorological operations. NOAA 14 (Sarsat 6) appeared December 1994 (#12 does not carry the Sarsat package; #13 failed 1993); NOAA 15 is planned for 1997.

An operational geostationary system is under consideration following successful tests of a relay aboard NOAA's GOES 7 meteorological satellite, launched May 1986 to 75° west GEO. The NASA/CNES/DOC demonstration, working through the Washington, Lannion and Ottawa stations, showed that most beacons would be detected within 10 minutes. GEO satellites, however, cannot pinpoint a source but must rely on position information encoded in distress signals. GOES 7 detected its first real mayday 15 June 1988, 75 minutes before the LEO satellites responded. 406 MHz GEO relays are included aboard NOAA's three GOES– Next (first, GOES 8, in April 1994; GOES 9 May 1995) and India's Insat 2 satellites (2A 1992; 2B 1993; 2E planned 1997). Russia's Luch 1 (December 1994) carries a 406 MHz repeater (1989's Cosmos 2054 Luch carried an experimental relay). Japan's GMS-5 began providing an experimental relay in 1995; it is possible that MT-Sat will provide an operational service from 1999 but Japan has yet to decide upon this. A payload will be flown on the Meteosat Second Generation (MSG) geosynchronous-orbit meteorological satellite to be operated by EUMETSAT from 2000. A 406 MHz GEOSAR demonstration and evaluation is under way. Inmarsat also introduced the Inmarsat E distress service in 1992, using beacons operating at 1.6 GHz L-band through its GEO satellites.

Eucosat

Eucosat (European Control by Satellite) is a non-profit organisation promoting joint civil/military use of satellites for treaty compliance, crisis management and environmental monitoring. Created in 1990, its corporate members include DASA, Aerospatiale, CNES, Matra Marconi Space, Alcatel Espace, Alenia Spazio,

Cospas-Sarsat satellite launches

	Launch	Orbit	Payload
C1383[1]	29 June 1982	989 × 1,028 km, 83°	Cospas 1
C1447[2]	24 March 1983	959 × 1,013 km, 83°	Cospas 2
NOAA 8[3]	28 March 1983	803 × 825 km, 98.7°	Sarsat 1
C1574[4]	21 June 1984	965 × 1,005 km, 83°	Cospas 3
NOAA 9[5]	12 December 1984	841 × 862 km, 99°	Sarsat 2
NOAA 10[6]	17 September 1986	808 × 826 km, 98.7°	Sarsat3
NOAA 11	24 September 1988	849 × 865 km, 98.9°	Sarsat 4
Nadezhda 1[7]	4 July 1989	960 × 1,014 km, 83.0°	Cospas 4
Nadezhda 2[12]	27 February 1990	956 × 1,021 km, 83.0°	Cospas 5
Nadezhda 3[8]	12 March 1991	958 × 1,018 km, 82.9°	Cospas 6
NOAA 13[9]	9 August 1993	850 × 863 km, 98.9°	Sarsat 5
Nadezhda 4[10]	14 July 1994	954 × 1,005 km, 82.9°	Cospas 7
NOAA 14[11]	30 December 1994	848 × 863 km, 98.9°	Sarsat 6
NOAA 15	13 May 1998	808 × 823 km, 98.7°	Sarsat 7

GOES 7 (May 1986), 8 (April 1994) and (9 May 1995), Insat 2A (July 1992) and 2B (July 1993), Cosmos 2054 Luch (December 1989), Luch 1 (December 1994) and GMS 5 (March 1995) carried 406 MHz relays into GEO and remain in experimental use.

Notes:
[1] decommissioned March 1988
[2] decommissioned December 1989
[3] decommissioned December 1985
[4] decommissioned June 1990
[5] 406 MHz local/global processor mode operational, global mode reactivated October 1991
[6] 406 MHz global/local modes not operating since September 1988
[7] limited availability in Southern Hemisphere
[8] started S hemisphere operations December 1991
[9] satellite contact lost 21 August 1993 while SAR testing under way
[10] acting as backup to Nadezhda 3 in same orbit
[11] global mode failed soon after launch
[12] decommissioned February 1996

OHB-System, Fokker Space and Systems, and National Remote Sensing Centre Ltd (UK).

European Centre for Space Law (ECSL)

ECSL was established in 1989 at the initiative and under the auspices of ESA. 450 people are involved. Its main objective is to develop/improve the knowledge in Europe of the law of space activities. It also promotes European activities beyond Europe and building a unique position for Europe in space law practice, teaching and publications. ECSL has set up, with ESA's support, the ESALEX space law database, containing basic texts of space law, ESA basic texts, statutes of other international organisations in full text, as well as bibliographical files. ECLS also initiates research: the first concerned the legal protection of remote sensing data; the second concerns intellectual property rights in outer space. A one-day forum is held annually plus the ECSL Summer Course on Space Law and Policy. It also organises, with IISL, the Space Law Moot Court Competition. The *ECSL News* is issued quarterly.

European Commission

The Commission is the executive arm of the European Union. It is Europe's largest purchaser of satellite imagery as a result of the Union's agricultural and environmental interests. Its 1995-98 budget for Earth observation-related activities is ECU270 million; it spends some ECU15 million annually on satellite imagery. It is providing half of the development funding for Spot 4's Vegetation instrument.

Eurospace

Founded in 1961, Eurospace is the Organisation of the European Space Industry. Its members comprise the 60 major industrial space companies from 13 countries in Western Europe. The association is responsible for the promotion of European Space activity. It is the spokesman of industry to the European Space Agency (ESA), in particular with regard to future space programmes and industrial policy matters. Its relationship with the European Union is continuously developing, in particular with the General Directorates directly or indirectly involved in space applications and support to industry.

EUMETSAT

The EUMETSAT Convention came into force 19 June 1986, with the organisation assuming overall and financial control of the MOP Meteosat Operational Program on 12 January 1987. A Council providing one seat per Member State, with four subsidiary bodies controls EUMETSAT: Administrative and Finance Group; Scientific and Technical Group; Policy Advisory Committee; Working Group on Charging Policy. The Council represents the National Meteorological Services of 17 European states. Contributions to programmes beginning with MTP are based on GNP (percentage): Austria 2.23, Belgium 2.70, Denmark 1.76, Finland 1.84, France 16.78, Germany 22.29, Greece 0.96, Ireland 0.54, Italy 15.46, Netherlands 4.03, Norway 1.47, Portugal 0.86, Spain 6.96, Sweden 3.20, Switzerland 3.33, Turkey 1.50 and UK 14.09.

The Meteosat Operational Program, MOP, began 23 November 1983 under ESA's auspices and ended 30 November 1995. Overall responsibility was transferred to EUMETSAT in January 1987 and in-orbit control delegated to ESA. Total cost for the three-satellite series was estimated in 1982 terms at ECU378 million (revised to ECU721 million, or US$901 million, in 1995 terms), funded by Member States (percentage): Belgium 4.4, Denmark 0.58, Finland 0.35, France 25.60, Germany 26.39, Greece 0.30, Ireland 0.11, Italy 12.00, Netherlands 3.00, Norway 0.50, Portugal 0.30, Spain 5.24, Sweden 0.93, Switzerland 3.03, Turkey 0.50 and UK 16.76 (0.01 per cent not covered).

Aerospatiale was awarded an ECU139.1 million contract in May 1984 for three flight models and one spare. MOP's three primary missions were: Earth imaging, dissemination of image and other meteorological data, and data collection/distribution; with two secondary objectives of meteorological processing and data archiving/retrieval. Real-time image data (1,686.833 MHz) and up to 66 channels of DCP data (1,675.181 to 1,675.381 MHz) are transmitted by Meteosat at 333 kbit/s to EUMETSAT's Primary Ground Station in Fucino for relay to the Mission Control Centre in Darmstadt.

The images are processed into a range of formats (for example, Europe–only or 24 frames covering most of the full disc), with lat/long grids and coastlines added for transmission up to Meteosat (channel 1 2,101.5 MHz; channel 2 2,105.0 MHz) and relay to users on two channels). Channel 1 (1,691.0 MHz) operates on conventional analogue WEFAX, compatible with other GEO meteorological satellites and NOAA's Automatic Picture Transmission service, and available to simpler Secondary Data User Stations. Channel 2 (1,694.5 MHz) provides high resolution digital transmissions to Primary Data User Stations. Some of NOAA's GOES images are also available via Meteosat using the Lannion, northwest France relay station. Three other channels (1,695.725, 1,695.756, 1,695.787 MHz) are primarily concerned with the relaying of digitised facsimile and selected conventional met observations to Africa, as the Meteorological Data Dissemination (MDD) service. In 1996 DCP messages received by Meteosat totalled 3,311,644, of which 1.1 million bulletins were distributed to meteorological services around the world.

The MARF Meteorological Archive and Retrieval Facility is the single repository for all Meteosat image data and derived products acquired since 1978. Live data are continuously archived and historic data are being copied. All data are recorded on 6.6 Gbyte optical disks. Online access the MARF's catalogue is available. During MOP, ESOC produced more than 1.1 million images and gathered some 40,000 tapes' worth of data, some dating back to Meteosat 1. This valuable archive is being transferred to MARF.

A new ground segment was developed by EUMETSAT to take full control of all satellites from 1 December 1995. It includes the MCC Mission Control Centre in Darmstadt and Primary/Secondary Ground Stations, replacing the system previously operated from ESOC. The PGS Primary Ground Station is at Fucino, Italy and includes facilities for operational support of two satellites. A high speed link connecting PGS/MCC transmits data, telemetry and satellite commands. The back-up station is at Weilheim, Germany.

The MCC is located at the new ECU32 million EUMETSAT HQ totalling 5,000m² in Darmstadt, opened in June 1995. The centre comprises four main facilities, each being developed by industry under direct contracts from EUMETSAT. The Core Facility is the MCC's central element and includes mission management, satellite control, ground segment control and pre-processing of satellite data. It also manages the dissemination of data and meteorological products to the user community. The MPEF Meteorological Products Extraction Facility completes full data processing and generates a range of meteorological products for the end-users. The MARF Meteorological Archive and Retrieval Facility allows users to access historical data from the satellites. The USDF User Station Display Facility provides real-time visibility of the end products to operations staff within the MCC.

A technical means of controlling access to Meteosat HRI High Resolution Image data was implemented 4 September 1995. This encrypts the data normally received by PDUSs; analogue (WEFAX) transmissions are not encrypted. Over the past decade there has been a rapid increase in the value-added commercial activity by the private sector, based on meteorological forecasts and data provided by meteorological services, which have also been faced with increasing pressure to recover their costs by charging for services. Additionally, they bear the cost of establishing and maintaining the space and ground observing systems. Exchange mechanisms are being established for meteorological satellite data, taking into account their ownership, value and the benefit generated for the user. As a first step, all users of PDUS were requested to register their use. Regular test transmission of encrypted data began in 1994. Decryption requires an MKU Meteosat Key Unit from EUMETSAT and some adaptation to the PDUS for MKU interfacing. However, there is a range of exceptions. For example, HRI images at 00/06/12/18 GMT remain unencrypted, and the 3-hourly encrypted data are free to all National Meteorological Services. In general, the NMS of countries with a GNP per capita greater than US$2,000 have free access to hourly and half-hourly data for internal use; wealthier countries pay according to their GNPs. Educational and science programs have free access to all data. At the beginning of 1996 there were more than 300 PDUSs and almost 2,000 SDUSs.

An independent study published in 1993 concluded that five principal European industries (energy, agriculture and fisheries, construction and manufacturing, transport and services) benefited by ECU130 million annually as a result of Meteosat's products.

Meteosat Second Generation's mission is defined as:
1. Basic multispectral imagery. Improved spatial, temporal and spectral resolution, for nowcasting and shortrange forecasting.
2. High resolution imagery. AVHRR-type spatial resolution in visible band, for mesoscale convective monitoring over Europe.
3. Air mass analysis. Water/CO_2 absorption channels for nowcasting.
4. Meteorological product extraction. Improved wind and temperature data.
5. Support of climate and environment monitoring.
6. Continuity of data collection/dissemination.
7. Secondary mission for the relay of SAR distress signals and observation of the global Earth radiation budget.

EUMETSAT budget (ECU million)	**1994**	**1995**	**1996**	**1997**
MOP	30.5	27.3	–	–
MTP	59.4	55.4	31.0	34.916
MSG	42.0	57.2	86.6	132.507
EPS polar	13.8	11.3	10.8	63.657
General Budget	15.0	15.6	22.4	12.738
Total	160.7	166.8	150.8	243.818

Previous years: 1993 - 131.916; 1992 - 91.720; 1991 - 64.847; 1990 - 50.841; 1989 - 42.8; 1988 - 59.25; 1987 - 62.763. Staff total 141 by end-1996.

Eutelsat

The European Telecommunications Satellite Organization, with 245 staff, is an inter-governmental entity established to operate Europe's regional satellite system for 47 member states. 1996 profit ECU80 million; 1995's profit was ECU67 million (1994 51; 1993 33.6; 1992 30; 1991 21.7; 1990 14.1; 1989: 13.9; 1988: 3.3) on income from satellite operations of 1996 income ECU292 million (1995 275; 1994 260; 1993 244; 1992 231; 1991 183; 1990 122; 1989 100; 1988: 70). 70 per cent of revenue is generated by audiovisual services (TV and radio channel delivery to cable headends and domestic satellite reception systems, SNG, outside broadcasts, European Broadcasting Union programme exchanges) and the rest by telephony, business and land mobile services. 1991 saw the commercial introduction of the EutelTracs two-way messaging and position-reporting land mobile service through the Alcatel-Qualcomm joint venture.

Eutelsat tariffs are designed to cover the satellite transponder alone and do not include charges for uplinks and other services on the ground. The ECU2.8 million annual tariff for a pre-emptible narrowband transponder is based on a three-year contract. For a non pre-emptible service over three years, the charge is ECU3.8 million annually, falling to ECU3.4 million for a five year contract for at least five transponders. The five-year tariff is fixed for the first three years and then revised based on the consumer price index for all European Union members (if the result is a tariff increase, the signatory has the right to terminate the agreement). Other options, for example, provide for one year leases and temporary leases.

Inmarsat

The International Mobile Satellite Organization is a 79 national shareholder, commercial co-operative operating a satellite system to provide telephone, telex, data and fax services to the shipping, aviation, offshore, land mobile and remote industries. It also provides the secretariat for the Cospas/Sarsat search/rescue system. More than 70,000 Inmarsat terminals have been commissioned since global operations began 1 February 1982. 1995 revenue was US$340.2 million (1994 US$334.1 million; 1993 US$349 million).

Most current users operate Inmarsat A ship Earth stations (SES) or their transportable derivatives. The maritime terminals feature driven 0.85 to 1.2 m parabolic dishes housed in radomes. The Inmarsat B system is the updated digital equivalent of Inmarsat-A and is expected to dominate the market for this class of terminal in the future. The Inmarsat C range of terminals, typically about 30 cm high, 20 cm and, handle data at 600 bit/s. Three types of aircraft terminals are available:

Aero- C for low rate store/forward; Aero-L for flight deck and operations datalinks; Aero-H high rate data and one to six voice channels. More than 700 aeronautical terminals had been commissioned by January 1996. Inmarsat 3's advent will allow Aero-I (Intermediate gain) to be introduced in 1997, offering 4.8 kbit/s telephony, 2.4 kbit/s fax/data within the satellites' spot beams (1,200 or 600 bit/s global). Four-channel terminals will weigh only 50 kg, costing US$50,000-100,000. The first voice test, from the ground, was made October 1995 using Inmarsat 2-F4 AOR-W, and prototype flight trials began February 1996. The Inmarsat E service was introduced in February 1992 for maritime distress beacons, although the first was not available until end-1992. Using GPS to generate position data, the beacon transmits a coded 1.6 MHz message for routeing to the appropriate rescue co-ordination centre.

Inmarsat B/M systems became available on a global basis in 1993. B offers services similar to A (which it will eventually replace) but its all-digital nature makes more efficient use of the space segment, offering fax rates up to 9.6 kbit/s and data up to 16 kbit/s. M lightweight (briefcase-sized) mobile terminals offer 4.2 kbit/s telephony and 2.4 kbit/s data/fax services. Mini-M terminals working through Inmarsat 3 satellites are expected in September 1996. A global personal messager, Inmarsat D, will also be fully operational by end-1996.

Land Earth Stations (LESs; also known as Coast Earth Stations in maritime circles and as Ground Earth Stations for aeronautical applications) link the satellites and international telecommunications networks, and are usually owned/operated by their host Signatories. Stations communicate with the satellites in four ocean regions over C-band while mobile users operate in the L-band. Until 1990's Inmarsat 2 debut, the organisation operated with leased satellites, listed in the separate table. The second generation will be followed by the imminent Inmarsat 3 series, providing multiple spot beams.

Inmarsat's constitution also covers the provision of navigation services, and could lead to a global civil system. Inmarsat 3's transponder is the first step, augmenting GPS with ground-derived integrity information, additional ranging signals and wide-area differential corrections. The Inmarsat 4 GEO series, expected to appear 2000+, could include the more capable Navigation Lightsat Payload. It would be much simpler than the GPS satellite payload as it would not have to conform to military standards for radiation hardening or encrypted coding. The Inmarsat 3 integrity monitoring network would allow a reduction in clock stability and data storage. The next step would add these payloads to intermediate orbit satellites (as was planned on the ICO satellites, but Inmarsat's Council in March 1996 rejected the lease proposal). 15 of these teamed with six to eight GEO satellites, although still intended primarily to augment GPS, could provide a 2D service should GPS be removed. A full civil system would require another 15 dedicated satellites in intermediate paths.

Intelsat

The International Telecommunications Satellite Organization is a not-for-profit commercial co-operative of 142 member nations or signatories, established 20 August 1964. These signatories may be either the governments themselves, or telecommunications entities designated by the governments to take all financial and operational responsibilities. Together, they own/operate a global system of communications satellites serving the entire world. Through its network of 24 satellites in the Atlantic, Pacific and Indian oceans and Asian Pacific regions, Intelsat provides international telecom services to more than 180 countries, territories and dependencies, and domestic telecom services to 40 nations. More than 900 Earth station antennas link users via 2,700 pathways. Intelsat carries more than half of all international telephone calls and almost all transoceanic TV. International financial networks, multinational corporations, international news services, and TV/radio broadcasters rely on the system for day-to-day global communications. Full-time Intelsat international traffic exceeds 120,000 derived channels of full-time service. In addition, the system carries more than 60,000 h of occasional use television annually, and 100 full-time leases, three of which are designated to support United Nations peacekeeping operations. Revenue is projected to grow to US$1.28 billion by 2002. 1997 revenue US$962 million; 1996 revenue US$911 million; 1995 revenue was US$805 million; 1994 US$706 million; 1993 US$658 million; 1992 US$622 million; 1991 US$563.4 million; 1990 US$498.6 million; 1989 US$614 million; 1988 US$614 million; 1987 US$519 million. Full-time channel use has grown from 4,258 in 1970, through 13,368 in 1975, to 133,000 in 1992 and 121,000 in 1993.

Ownership/investment is shared among the members according to their respective use of the system. Investment shares determine each signatory's percentage of the total contribution needed to finance capital expenditures. Revenues derive primarily from utilisation charges and, after deduction of operating costs, are redistributed to the signatories in proportion to their investment share as repayment of capital and compensation for use of capital. Non-member users pay the same rates because non-discriminatory access is a basic tenet of the organisation. All net revenues are reinvested into the organisation.

Intelsat operates under a four-tier structure. The Assembly of Parties comprises representatives of all of the governments who signed the Intelsat Agreement that entered into force 12 February 1973. This group considers resolutions and recommendations (essentially from the Board of Governors) affecting the long-term objectives. The Meeting of Signatories is composed of representatives of Signatories to the Operating Agreement who meet to consider issues related to the financial, technical and operational aspects of the system.

Both the Assembly of Parties and the Meeting of Signatories operate under a one nation/one vote rule, where an affirmative vote by at least two-thirds of the parties present is required on substantive issues and a majority vote is required on procedural matters. Representation on the Board of Governors, which functions as the managing directorate, is determined according to a standard minimum investment share established by the Meeting of Signatories. The Board of Governors, composed of representatives of Signatories whose investment shares either individually or as a group meet or exceed the minimum share requirement (currently 1.90598 per cent), is responsible for all major decisions regarding the establishment and operation of the Intelsat system. 27 members of the Board of Governors represent 107 of the Signatories.

Four countries joined Intelsat in 1996: Tajikistan (22 February), Bosnia and Herzegovina (6 March), Bulgaria (15 May) and Equatorial Guinea (11 December).

Earth station antennas accessing Intelsat's satellites are owned/operated by the host telecom organisations; 14 categories of Earth station standards are used. Co-ordination/control of services are the responsibility of the Satellite Control Center (SCC) and the Intelsat Operations Center (IOC) at the organisation's HQ. The IOC is the focal point for the communications networks and handles more than 100 TV transmission, antenna verification and other tests daily. The SCC provides TT&C services and pointing data for transmissions to Earth stations. Six TT&C stations totalling 54 antennas are operated around the world, collocated with Std A stations at Fucino (Italy), Paumalu (Hawaii), Perth (Australia), Raisting (Germany), Beijing (China) and Clarksburg (Maryland, US). A major upgrading of the system was completed October 1992. To cope with the Intelsat 7 generation and its successors, two antennas were added to Fucino and one each to Raisting and Beijing.

International Space University (ISU)

ISU is an international, interdisciplinary educational institution providing courses and qualifications for the space sector, principally through a 10-week summer session and the Master of Space Studies, an 11-month interdisciplinary postgraduate degree course comprising three modules which may be taken over three academic years. The SSP is hosted by an institution in a different country each year. The MSS is conducted at the Strasbourg Central Campus and in industry, government, university and research institutions around the world for a 12-week professional placement period. In both, all major space-related disciplines, technical and non-technical, and their interactions are studied. Students from many countries and cultures work together to produce a Team Project. The MSS also includes an individual project. Fees for: the SSP FFr70,000 (all include food accommodation, transport during the session); MSS: FFr130,000 tuition fees (not inclusive of living expenses).

International Telecommunications Union (ITU)

The ITU, comprising 182 member nations, is the international organisation responsible for the regulation, development and planning of telecom systems worldwide and for establishing equipment and systems operating standards. The Union consists of eight organs, including the Radio Regulations Board (formerly the International Frequency Registration Board until March 1993), which maintains a Master Register of more than 1 million radio frequency assignments.

Frequency band allocations for satellite communication systems were established initially by 1963's Extraordinary Administrative Radio Conference. The regulations for space radiocommunications were revised in 1971 by the World Administrative Radio Conference for Space Telecommunications, allocating frequency bands for the first time to the broadcasting services. This was followed in 1977 by the World Broadcasting Satellite Administrative Radio Conference, in 1979 by the World Administrative Radio Conference, which completely revised the Radio Regulations, in 1983 by the Regional Administrative Radio Conference for the Planning of the Broadcasting Satellite Service in the Americas, in 1988 by the World Administrative Radio Conference on the Use of the Geostationary Satellite Orbit and the Planning of the Space Services Utilising It, and in 1992 by a WARC for dealing with frequency allocations in certain parts of the spectrum (WARC 92), which partially revised the table of frequency allocations to accommodate growing demands for new space services, satellite HDTV and satellite sound broadcasting. WRC 95 considered MSS frequencies. WRC 97 will probably address BSS for Europe, Africa and Asia Pacific, science and Earth observation satellites, inter-satellite links and MSS.

Intersputnik

Intersputnik was established in 1971 to offer satellite-based TV, radio, telephony and datalinks. Membership is Russian Federation, Afghanistan, Bulgaria, Hungary, North Vietnam, Germany, Yemen, North Korea, Cuba, Laos, Mongolia, Nicaragua, Poland, Romania, Czech Republic, Syria, Georgia, Belarus, Kazakhstan, Turkmenistan, Kyrgisia and Tajikistan. A Board composed of one representative from each member governs the organisation; sessions are held at least annually. The Directorate is a permanent executive and administrative body, headed by a Director General. Financial activities are controlled by an Auditing Committee elected for a three-year term. On joining Intersputnik, a member makes a minimum 1 per cent contribution to the Statutory Fund; profits are distributed in proportion to these contributions.

As of 1 January 1996, Intersputnik provides its services through 26 36 MHz C-band transponders on Express 1 at 14° west and eight Gorizont satellites at 40° east, 70° east, 80° east, 85° east, 96.5° east, 103° east, 130° east and 142.5° east, plus two 27 MHz Ku-band transponders on Gals 1 at 70° east. Four transponders (#7/8/10 at 14° west + #8 at 80° east) provide voice/data using digital IDR technology. The others are used for TV, three of which (#7/9 14° west + #10 80° east) are for occasional use. The even-numbered transponders (except for #6) provide minimum 31.0 dBW EIRP hemispheric coverage, the odd numbers are global, at 28.0 dBW minimum. 14°W/80°E are the principal Intersputnik locations. Services will gradually be shifted to the Express satellites as they appear. Intersputnik planned to order two Spacebus 3000 satellites in 1996 to expand its services from 1998.

C-band Earth stations in the system utilise 12 m diameter main dishes of more than 31 dB/K G/T at 3.800 GHz 5° elevation. Transmitter output power is 1 to 3 kW. The stations operate at 3.700 to 4.150 GHz rx, 6.025-6.475 GHz transmit. Ku-band 10 m dishes provide 38.0 dB/K. The network comprises 42 standard Earth stations. TV can be carried in PAL, SECAM and NTSC formats. 6.5 to 7.5 m C-band and Ku-band stations also operate. 2.0 to 3.0 and 3.5 to 4.5 m are used for telephony, fax, telex, data, videoconferencing (3.5-4.5 m only) and TV/radio reception.

Bulgaria, Hungary, German Democratic Republic, Cuba, Mongolia, Poland, Romania, Czechoslovakia and USSR signed the Intersputnik agreement 15 November 1971. The first installations to enter service were those of the USSR and Cuba's Caribe ground station, used for TV transmissions via Molniya 1 of Leonid Brezhnev's 1973 Cuban visit. The system formally became operational in 1974; by 1980 there were eight stations and 11 member countries. By end-1984, the organisation was claiming that 60 per cent of all TV transmissions between member states was being relayed on its channels. Germany's DBP Telkom inherited the GDR membership in German unification.

RASCOM

RASCOM (Regional African Satellite Communications) held its first Assembly of Parties in September 1993 and the meeting of Board of Directors at its Abidjan HQ. A meeting of all the RASCOM countries in October 1993 was held at Intelsat HQ to discuss the pooling of transponders. The agreed upon pooling implementation began January 1994. It was decided in October 1994 to consolidate all services on Intelsat 804 at 21.5° west in 1996. RASCOM's ultimate objective is a dedicated regional satellite system.

The International Telecommunications Union began a feasibility study in March 1987 of a RASCOM system. The first global study of this type in Africa, it analysed the continent's telecom requirements and the terrestrial and satellite options to meet them in both urban and rural/isolated areas. The final report was delivered in late 1990 and considered by a meeting of ministers in Nigeria in February 1991. It was decided to proceed to a transitional stage during which the ITU would be the executing agency under the supervision of a Committee of experts from 15 African countries. The report on this transitional stage, including all elements for the setting up of the RASCOM operating organisation, was submitted to the second Conference of Telecommunication Ministers (25-27 May 1992, Abidjan. The RASCOM Convention was signed at Abidjan by 41 countries (minimum investment is US$50,000), envisaging the initial pooling of the 14 36 MHz Intelsat domestic transponders as a first step towards the ultimate objective of a dedicated African satellite. 35 African countries, UNDP, ITU, the government of Italy and the UNCTADA II Resource Mobilization Committee contributed/pledged towards financing the project activities.

One option proposed for the dedicated system calls for three satellites, including one spare. Cost would be US$1.3 billion: US$500 million for the satellites and US$800 million for the 50 countries' ground segment.

IRAN

Telecommunications Company of Iran

The Telecommunications Company of Iran (TCI) is the national signatory to Intelsat and Inmarsat. Plans for a domestic satellite system under the Shah's regime were subsequently revived, with four slots for Zohreh (Venus) satellites at 34° east, 26° east, 41° east and 47° east confirmed at 1988's ITU meeting (Iran now also holds slots at 59/61.5° east).

An RfP was expected in 1987 to be released by end-1988 for two 24-transponder C-band satellites for launch by Ariane. System cost was estimated in 1987 to be US$600 million. The Soviet Union was reported at one stage to be interested in providing both the satellites and associated launch services. The contract was expected to favour European companies as a US satellite would be unlikely to receive an export license from the State Department. Specifications were presented in March 1989 to invited companies in Rome, calling for two satellites (one a ground spare) featuring 20 Ku-band transponders each (six as back-up) with 54 MHz channel bandwidth. Inclusion of up to 1,500 1.8 m ground terminals was expected in the US$700-800 million project. Bids were due in by end-September 1989. Two responses were reported but rejected.

A further request was issued in early 1991. The Alcatel Espace/Aerospatiale team was announced as winner June 1992 of the US$350 million contract, required to deliver two 1,850 kg satellites with 10-year lives to 26/34° east for TCI in mid/end-1995 by Ariane. However, the contract still has not been signed, reportedly because of disagreement within Iran on whether such a system is necessary. 14 Ku-band 14.0 to 14.5/10.95 to 11.7 GHz up/down transponders would provide EIRP >50 dBW, allowing small stations for TVRO and rural telephone and business services. Four would be switchable to a steerable spot antenna. The L-band payload would provide 100 voice channels of Inmarsat-standard for mobile users. An X-band payload might be included for government/military links. Two TT&C stations would each work via two 7.5 m antennas; one station would include the 5.5 m In-Orbit Test antenna for monitoring and calibrating the payload.

TCI's Inmarsat A station at Boumehen in southwest Iran, working through Inmarsat's Indian Ocean satellite, added Inmarsat C services in 1994.

IRELAND

Team Aer Lingus

Aer Lingus, wholly owned by the Irish government, provides propellant filters for Ariane 4's stage 1 tank, telemetry equipment containers for Ariane 4's stage 3 and supports for the Ariane 5 Vulcain main engine. Space personnel totals about 12, with an annual space revenue of about I£1.2 million.

ISRAEL

Asher Space Research Institute

The institute was inaugurated in 1986 to provide expertise in space sciences and engineering, including astrophysics, spacecraft propulsion, control, materials, structural design and remote sensing. It employs 28 full time engineers and 25 academic members drawn from the aerospace engineering and physics faculties. Its TechSat 1 satellite was launched in March 1995.

Interdisciplinary Center for Technological Analysis and Forecasting (ICTAF)

The Space and Remote Sensing Division at Tel-Aviv University was established in 1985; it has conducted technology surveys in several space-related fields, such as materials, solar cells, bondings and coating, but its principal emphasis is on promoting remote sensing applications within Israel. ICTAF is the national distributor for EOSAT, Spot Image and Eurimage products. It was appointed in 1991 as distributor for Israel Space Agency's receiving station. It is also involved in several applications projects: limnology, urban studies, geology and agriculture, soil and forestry mapping. Experience includes thermal mapping for industrial applications.

It is the executor of a NASA/ISA information exchange agreement under ISA contract.

Israel Military Industries Ltd (IMI)

IMI is a government-owned company comprising a HQ group and four operating groups of 10 divisions and 18 plants. There are 4,000 personnel. The Rocket Systems Division (RSD) of the Systems Group develops/manufactures solid rocket motors, weapon systems and metal/composite material products. For example, it provides the propulsion systems for the Gabriel surface-to-surface and Shafrir air-to-air missiles. In the space field, it developed/produces the stage 1/2 motors for the Shavit/NEXT launchers. These are HTPB cylindrical grains in filament wound composite cases. Steering options are jet vane, LITVC and flexseal. The motors are marketed as the ASTM family in the US by Atlantic Research Corporation. RSD is deeply involved in both Israel's space programme and the SD10 Arrow. The propulsion systems and the pyrotechnics for these programmes are IMI developed and produced.

Israel Space Agency (ISA)

ISA was established in 1983 within the Ministry of Science and Technology to direct national space research and R&D activities. The Ofeq programme is conducted under its auspices.

National Committee for Space Research

An element of the Israeli Academy of Sciences and Humanities formed in 1960, it advises the government, ISA and universities on space-related research projects.

Remote Sensing Laboratory

Established in 1988, at Ben-Gurion University of the Negev, this BGU/IDR research facility specialises in the analysis of arid/semi-arid regions using satellite data. Facilities cover an ERDAS IMAGINE image processor, an ARC/INFO GIS system and a NOAA/AVHRR receiving station. Current activities include: software development for automatic examination of faults, cracks and dykes in satellite images (such features often indicate water). Automatic extraction of surface features such as drainage networks, lineaments and roads from satellite images and DEMs; monitoring environmental trends in the Negev from satellite imagery; studying how dust and aerosols degrade satellite images; looking for water stress in plants before it is apparent on the ground.

ITALY

Agenzia Spaziale Italiana (ASI)

ASI was formally established 25 August 1988 under MURST to manage the PSN national space plan (Piano Spaziale Nazionale), previously directed from within the National Research Council (CNR, Consiglio Nazionale delle Ricerche). ASI is responsible for Italy's ESA involvement and manages the four principal national satellite projects: Italsat, Tethered Satellite, SAX and Lageos 2.

The agency's annual budget is described at the beginning of the national entry. The Cabinet elects chairman of the Board, proposed by the MURST Minister. The DG is appointed by MURST after consultations with the Board of Directors. Board Members, elected by the Cabinet, sit for five years.

The CGS at Matera is part of the worldwide integrated geodetic network. Returns are regularly made using the principal laser ranging satellites, Lageos 1/2, Starlette, Ajisai and Stella. AlliedSignal Technical Services Corp of the US installed a Mobile Laser Ranging System in 1994. The station includes a 20 m VLBI dish. ASI's Milo base in Sicily (38°01′ north/12°35′ east) launches stratospheric balloons along the Mediterranean to Spain's west coast.

Directorate of International Affairs

MURST is the parent body of the ASI space agency. Its Office for Space Affairs co-ordinates the government's space interests.

MARS Centre

The Microgravity Advanced Research and Support Centre was created in 1988 by Alenia and the University of Naples as a user support and operations centre for space experimentation, with specific expertise in fluid physics. 1993-95 turnovers were each Lit5 billion; 30 personnel. The Centre hosts a fluid physics lab for experiment preparations, and a telescience room for the remote control and operation of space experiments. For example, the Telescience Control Room controlled an experiment during 1996's Life and Microgravity Sciences Spacelab mission. MARS publishes the *Microgravity Quarterly* and *Space Technology* journals.

Nuova Telespazio

Nuova Telespazio, 50/50 owned by Telecom Italia and STET, is Italy's international satellite communication carrier under the Ministry of Post and Telecommunications. Telecom Italia is the national signatory to Intelsat, Inmarsat and Eutelsat. Staff total 900. 1995's Lit396.4 billion sales yielded a Lit5 billion profit. Its primary Earth station is the Piero Fanti Space Centre at Fucino, complemented by the Lario facility adjacent to Lake Como and the Scanzano site near Palermo that became operational during 1989. Fucino is a major European TT&C facility: its antennas include 17/27.4/32 m Intelsat Atlantic, 29.5 m Intelsat Indian, 13 m Inmarsat Atlantic, 4/8 m Eutelsat, 11 m Olympus, 10 m Marisat, 7.5 m LASSO, 10 m Landsat and 7.7/9/13.4/16.5/32 m in-orbit test dishes. Lario maintains two 32 m dishes for servicing Intelsat Atlantic/Indian Ocean satellites and a 17 m antenna. Telespazio also receives, processes and distributes Landsat, Spot, ERS and MOS data from Fucino as Italy's National Remote Sensing Centre and the national point of contact for ESA's ESRIN Earth observation data, and operates the Matera laser station for geodetic ranging measurements with the Lageos satellite. The company acquired a 2.3 per cent interest in Spot Image in 1990.

Telespazio was responsible for developing the Argo system for the Civil Defense Department to provide emergency communications and datalinks through Eutelsat within 4 hours of a major disaster. Working through a 9 m Fucino antenna, Rome's Civil Defense HQ control centre has 12 2.2 m Ku-band mobile stations ready on lorries and helicopters to provide telephone, fax and TV links. Three helicopters are equipped for airborne TV. 110 1.8 to 2.5 m fixed terminals provide environmental data (seismometric, volcanic and hydrological, for example) to help disaster forecasting.

In addition to being the national signatory to Intelsat, Inmarsat and Eutelsat and Italy's exclusive carrier for space communications, Nuova Telespazio also receives, processes and distributes Landsat, Spot, ERS and MOS imagery via its major Fucino facility as Italy's National Remote Sensing Centre and the national point of contact for ESA/ESRIN.

JAPAN

Earth Observation Centre (EOC)

NASDA's 115,000 m² EOC was established in October 1978 to receive and process Landsat, Spot and MOS data for distribution as CCT and prints through the Remote Sensing Technology Centre (RESTEC). It is part of NASDA's Office of Earth Observation Systems and it performs MOS/JERS 1/ADEOS mission management, archiving and processing, plus Landsat, Spot and ERS archiving/processing. EOC's two 10 m parabolic dishes channel data to high density digital tape recorders, with associated quick-look units. It can handle 30 SAR and 30 OPS scenes daily from JERS 1.

Institute for Unmanned Space Experiment Free Flyer (USEF)

USEF was established in 1986 to promote development and utilisation of unmanned space systems. The SFU Space Flyer Unit carried nine experiments on its first flight before retrieval by NASA's Shuttle in January 1996. The Express and Space Robotics projects began in 1992. The Users project began in 1995.

USEF is a non-profit organisation under the direction of the Ministry of International Trade and Industry and endowed by 13 companies: Fujitsu, Hitachi, IHI, Kawasaki Heavy Industries, MELCO, Mitsubishi Heavy Industries, Mitsubishi Precision, NEC, Nippon Electronics Development, Denso, Nissan Motors, Sharp and Toshiba.

Institute of Space and Astronautical Science (ISAS)

ISAS was established 14 April 1981 by reorganisation of the Institute of Space and Aeronautical Science, at the University of Tokyo. It was responsible for launching Japan's first satellite in 1970. The university began developing the all-solid M orbital launcher in April 1963 after opening the Kagoshima Space Centre in 1962. The original ISAS was not formed within the University until 1 April 1964. ISAS is one of the National Inter-University Research Institutes answerable to Mombusho (Ministry of Education, Science and Culture) and is responsible for Japan's science satellites, launching every one to two years. It was restricted to all-solid vehicles of 1.4 m diameter but permission for the 2.5 m M-5 was received in 1989 for modest deep space missions from 1996. The Sakigake and Suisei Halley's Comet probes of 1985-86 provided Japan's first experience beyond Earth orbit.

ISAS is headed by a Director-General advised on policy and administrative matters by a 19-member Board of Councillors and on research planning by a 21-member (11 from within ISAS itself) Advisory Council for Research and Management. Mombusho appoints all members. ISAS' organisational structure incorporates two Technical Divisions (Space Operation and Engineering Support) and nine Research Divisions responsible for 49 Research Sections. The Office of Project Co-ordination oversees all project activities. The Research Divisions provide the core of each Project Team; the Research Sections concentrate on specific fields and are headed by a Professor, an Associate Professor and two Research Associates. Personnel total 421, of which graduate and research students account for 115.

ISAS operates seven facilities:
1) Kagoshima Space Centre (KSC), initial use 1962, launch centre for sounding rockets and M orbital vehicles; 349 launches to end-1995. There is also a 20 m space tracking dish at Kagoshima.
2) Noshiro Testing Centre (NTC), established 1961 and located on Asanai Beach for static firing of large solids and research/static tests of a LOX/LH_2 engine and a turbo-ramjet.
3) Usuda Deep Space Centre (UDSC), 64 m S/X-band deep space TT&C station. Constructed for Comet Halley probes of 1985-86, began operations October 1984.
4) Space Utilization Research Centre (SURC), established 1988 at Sagamihara campus.
5) Centre for Planning and Information Systems (PLAIN), established 1993 at Sagamihara campus.
6) Centre for Advanced Spacecraft Technology, established 1995 at Sagamihara campus.
7) Sanriku Balloon Centre, launch site for 500 kg payloads on 200,000 m³ helium balloons to 40 km altitude over the Pacific.

Japan Meteorological Agency (JMA)

JMA is responsible, under the Ministry of Transport, for providing space-based meteorological services and undertaking sounding rocket investigations. The Meteorological Satellite Centre, established April 1977 near Tokyo and attached to JMA, undertakes satellite control, image data acquisition through its Command and Data Acquisition Station (CDAS) at Hatoyama 35 km northwest of Tokyo, and data analysis for the GMS satellites. NASDA procured/launched the satellites under JMA funding and requirements. Raw imagery, data collection platform signals and spacecraft telemetry are received on the two CDAS 18 m Cassegrain dishes; they are then relayed by 2.0 GHz microwave link to the Data Processing Centre (DPC). The DPC maintains four mainframe computers: two M-360R (24 Mbit each) for spacecraft control/operations and two M-380S (32 Mbit each) for image processing and subsequent extraction of parameters such as sea temperature and cloud top height. The smaller machines can assume some of the 380's processing duties in the event of failure. The Stretched-VISSR and WEFAX data are passed to the CDAS for uplinking and broadcast to users: Medium Data Utilization Stations (MDUS) can receive S-VISSR but the Smaller Data Utilization Stations (SDUS) receive the lower-resolution (7 to 8.5 km) WEFAX. Two TARS Turn-Around Ranging Stations are operated on Ishigaki Island (Japan) and Crib Point (Australia) in conjunction with CDAS for satellite orbit determination and prediction.

Japan Microgravity Centre (JAMIC)

JAMIC was established in March 1989 with a capital of ¥2.6 billion to provide μg research and development facilities; it opened in late 1991. The site includes a 710 m drop shaft in a former mine, providing more than 10 seconds of μg conditions of less than 10^{-5} *g* using a standard 7.85 m long 1.8 m diameter capsule. Gas jets compensate for air drag and the payload is carried in a module surrounded by a vacuum jacket. The 490 m free-fall zone is followed by an air drag braking distance of 200 m completed by a mechanical braking system. If the system fails, the bottom 20 m houses shock absorbing material for emergency braking. JAMIC also provides facilities for experiment preparation/analysis. The experiments and administration buildings total 2,494 m².

Japan Space Utilisation Promotion Centre (JSUP)

JSUP was founded in February 1986 to promote industrial exploitation of the space environment. The Science and Technology Agency (STA) and the Ministry of International Trade and Industry (MITI) support it. 1997 revenue ¥4.3 billion. Utilisation promotion includes aircraft parabolic flights and the μg drop test facilities in Hokkaido and Gifu (MGLAB: Microgravity Laboratory of Japan).

Microgravity Laboratory of Japan (MGLAB)

MGLAB began operations in 1993, providing a 1.50 m diameter drop tube with 100 m free-drop zone. The 900 mm diameter 2.280 m capsule is decelerated over 50 m and provides two payload racks to users.

Ministry of International Trade and Industry (MITI)

MITI is Japan's third largest government funding agency for space activities, accounting for ¥11.017 billion in FY96 (¥12.754 billion FY95; ¥12.007 billion FY94; ¥13.987 billion FY93), after the Science and Technology Agency and the Ministry of Education. It is leading the USERS Unmanned Space Experiment Re-entry System project and provided most of the Space Flyer Unit's funding. It was also instrumental in establishing the USEF Institute for the Unmanned Space Experiment Free Flyer. MITI's Agency of Industrial Science and Technology (AIST) operates several institutes with space interests: ETL Electrotechnical Lab (advanced technologies, such as robotics, space power and remote sensing), MEL Mechanical Engineering Lab (R&D such as attitude control for large space structures) and GSJ Geological Survey of Japan (geological applications for remote sensing).

Ministry of Posts and Telecommunications – Communications Policy Bureau

The Ministry of Posts and Telecommunications' Communications Policy Bureau is responsible for space-based communications systems, supervising NASDA and the Telecommunications Advancement Organisation of Japan, the latter established in August 1979 to operate national communications satellites. The Telecommunications Bureau supervises the Nippon Telegraph and Telephone Corp and the Kokusai Denshin Denwa Co, respectively the provider of domestic satellite communication links and the Japanese signatory to the Intelsat and Inmarsat operating agreements.

Ministry of Posts and Telecommunications

Until 1989's launches of Superbird (Space Communications Corporation) and JCSat (Japan Communications Satellite Company), Japanese telecom satellites were state-owned and operated through the Telecommunications Advancement Organisation of Japan (TAO). National space agency NASDA was responsible for procurement and launch of the CS and BS series before handing them over to TAO for operational use. The Japan Broadcasting Corp (NHK: Nippon Hoso Kyokai) and industry-owned Japan Satellite Broadcasting Inc (JSB, created 1984 by the Ministry of Posts and Telecommunications) broadcast through BS-3 transponders.

As a result of US pressure, the 'BS-4' generation was opened to foreign bidders. Hughes Space and Communications was selected in December 1993 by the Broadcasting Satellite System Corp of NHK, JSB and others to negotiate for two BSATs to be launched 1997/98. Nippon Telegraph and Telephone provided communications services through CS-3, but NTT's own two NStars were built by SS/L and launched in 1995, the open market competition resulting from US pressure against a government-funded CS-4 closed to foreign bidders. Satellite Japan Corp (Sajac) planned to launch its Sajac Hughes HS-601 satellites in 1994 but financial problems forced it to merge with JCSat in 1993, creating Japan Satellite Systems (JSAT); JCSat 3 began commercial services in November 1995 and JCSat 4 was ordered in February 1996 and launched in February 1997.

NASDA continues to develop and operate the ETS series of technology demonstration satellites. ETS 6 appeared in 1994 incorporating inter-satellite data relay and mobile communications packages, but a propulsion problem left it in a low orbit and allowed only a truncated programme. A Data Relay and Test Satellite system will build on ETS 6 and COMETS (1997) experience. The OICETS Optical Inter-orbit Communications Engineering Test Satellite is planned to test laser links with ESA's Artemis. The Ministry of Transport plans two GEO MT-Sats to augment GPS signals and air traffic control links for improving aircraft navigation and control; they will also replace the GMS meteorological satellite. See the Earth Observation section for further information. ETS 8 in 2002 will demonstrate a cellular phone system that can cope with Japan's mountainous geography.

MPT's Communications Policy Bureau is responsible for Japan's R&D and utilisation policy in space communications in CRL is a leading research laboratory in the field of satellite communications and broadcasting technologies.

National Aerospace Laboratory (NAL)

The NAL is Japan's national aerospace technology research centre; it was founded in 1958. It provides the lead on aerospace plane activities. Research staff total 334, working to a FY95 budget of ¥12.445 billion (FY94 ¥11.948 billion; FY93 ¥11.5 billion; FY92 ¥10.8 billion); its FY95 space budget is ¥4.000 billion (FY94 ¥4.058 billion; FY93 ¥4.258 billion; FY92 ¥3.25 billion; FY91 ¥2.89 billion).

National Space Development Agency of Japan (NASDA)

The National Space Development Agency of Japan (NASDA) was established in October 1969 to act as a nucleus for implementing space development and promoting space utilisation exclusively for peaceful purposes. It is responsible for the following tasks based upon the Japanese Space Development Programme: development of satellites (including space experiments and space station operations) and launch vehicles, and their launching and tracking; and development of methods, facilities and equipment required for satellite development.

NASDA's first Engineering and Test Satellite was successfully launched in September 1975 using an N1 rocket, based upon the McDonnell Douglas Delta vehicle first stage and built under license. This was followed by the N2 vehicle and later by the H1, the latter being retired in 1992. These three launch vehicles used United States Castor strap-on boosters and Japanese second stages. The first NASDA launch vehicle to be based upon wholly-Japanese stages is the H2, first flown in 1994. The all-solid propellant J1 vehicle was first flown in February 1996 and carries small payloads.

NASDA's personnel totals have increased only slightly to 1,019 from the 1982 level of 904 (1969 – 151), and they are accommodated in six principal facilities:

1) Headquarters: accounting for about half of the total employees.
2) Tsukuba Space Center: established in 1972, it now employs approximately one-third of NASDA personnel.
3) Tanegashima Space Center: NASDA's largest facility.
4) Earth Observation Center (EOC), set-up in 1978 as a NASDA field centre to establish and develop satellite remote sensing technology. EOC receives and processes data from Earth-observing satellites for a variety of applications and research.
5) Katsuura, Masuda and Okinawa Tracking and Data Acquisition Stations: used for feeding data to the Tsukuba Tracking and Control Center. Co-operation with other domestic and foreign stations are arranged as necessary for launch and flight support.
6) Kakuda Propulsion Center: located 270 km northeast of Tokyo.

Remote Sensing Technology Centre (RESTEC)

The Remote Sensing Technology Centre of Japan disseminates remote sensing data received/processed by NASDA's Earth Observation Centre. Radarsat data were added in May 1996. RESTEC was established in July 1975 to conduct R&D into remote sensing techniques and applications, and operates FACOM M1400/10 K-650, VAX 4000 and Sun 4/630 computer systems.

Science and Technology Agency (STA)

In addition to being the SAC secretariat, the STA's research and development bureau is responsible for planning basic policies on space development, international co-operation in space utilisation and acting as NASDA's parent body. It is also responsible for the National Aerospace Laboratory (NAL), where research on basic space technology is undertaken. As the central space body, it was responsible for ¥231 billion FY96 funding; it is responsible for ¥244 billion for FY97 funding.

Space Activities Commission (SAC)

The SAC, composed of four commissioners and chaired by the Minister for Science and Technology, was established in May 1968 to centralise space activities under a systematic programme. The 'Outline of Japan's Space Development Policy' was issued in March 1978 as a guideline for space developments over the next 15 years; a review is produced about every five years (the latest in January 1996). The STA performs the Commission's secretariat functions.

Space Activities Promotion Council (SAPC)

The SAPC was formed in 1968 as part of Keidanren (Federation of Economic Organizations) to promote space activities. Membership stands at 96 companies and trade associations; most major space-related Japanese companies are represented.

Telecommunications Advancement Organisation of Japan

The Telecommunications Satellite Corp of Japan (TSCJ) was established in August 1979 to operate the CS and BS satellite series procured and launched by NASDA. The name changed to TAO in October 1992 as it introduced new services such as R&D. Japan's first two operational communications satellites, CS-2a/2b, each with six Ka-band and two C-band transponders, were launched in 1983 principally for maintaining communications with the more scattered islands of the Japanese archipelago. Both were still operational when they were replaced by CS-3a/3b in 1988. NTT's NStar took over in 1995. The BS-2 series introduced DBS services in 1984 using 100+W transponders but two of BS-2a's three amplifiers failed soon after launch on 23 January 1984. BS-2b brought the system to full capacity in 1986 but a BS-2x was purchased from GE Astro Space by NHK for an Ariane launch to ensure system continuity before the first of the BS-3 series could be launched in 1990. However, BS-2x was lost in the Ariane failure of 22 February 1990; the replacement BS-3H was then lost in its April 1991 launch. Both would have been controlled through TSCJ. BS-3a, launched 28 August 1990, replaced BS-2b but it operates on marginal power. 3b completed the system in 1991, but NHK procured BS-3N in late 1992 (launched July 1994) to guarantee services. The 'BS-4' generation will be operated by B-SAT through its BSAT 1a satellites.

KOREA, SOUTH

Korea Telecom (KT)

KT was 80 per cent government-owned in 1995, expected to fall to 51 per cent in 1996. It began operating the Kumsan coast Earth station in March 1991 for Pacific Inmarsat A services; Indian Ocean services were added 1993. Inmarsat C was added to both in 1994. Korea is a leading maritime nation, with 960 ocean-going ships and 550 deep sea fishing vessels carrying 130 Inmarsat terminals. KT is the Inmarsat and Intelsat signatory.

LIBYA

Libyan Centre for Remote Sensing and Space Science (LCRSSS)

LCRSSS, administered under the Secretariat of Scientific Research, signed a data distribution agreement in 1994 with Spot Image. The two also completed a study using Spot imagery on the impact of urban growth on agricultural land around Tripoli.

MEXICO

Secrétaria de Comunicaciones y Transportes (SCT)

Until 1990, SCT was responsible through the Departamento Espacial of the DGT Dirección General de Telecomunicaciones for Mexico's telecom systems. Telecomm, an independent non-stock company owned by the government, was created in January 1990 to assume control, with SCT operating as a regulating body. It is expected to be privatised in 1997. It is

Mexico's Intelsat (Earth stations at Tulancingo and Cuncun) and Inmarsat signatory.

By 1994, SCT was operating 254 Earth stations, including 185 for TV distribution, 11 for transmitting TV, telephony and data, 15 data, 9 telegraphy, 26 rural telephony, 8 mobile and 186 VSATs.

MOROCCO

Centre Royal de Télédétection Spatiale du Maroc (CRTS)

CRTS was established in 1989 to promote, develop and co-ordinate satellite-based remote sensing for the country. It has specific responsibility for the archiving and distribution of satellite image data, and has distribution agreements with Spot Image for Spot, and Eurimage for Landsat, NOAA and ERS. Morocco's CNTS Comité National de Télédétection Spatiale was set up in 1993, including representatives of ministerial departments involved in remote sensing and GIS.

The Technical University of Berlin is providing its Tubsat C micro-satellite platform for a space mission carrying an Earth imaging camera provided commercially by the UK's Rutherford Appleton Lab. The CCD camera, returning 250 m resolution from 800 km, was developed for Pakistan's Badr 2 satellite.

NETHERLANDS

National Aerospace Laboratory (NLR)

The Nationaal Lucht-en Ruimtevaartlaboratorium (NLR) is the central national institute for aerospace research, employing 900. 70 per cent of its funding is provided through research contracts. It is primarily concerned with aviation (maintaining nine wind tunnels covering M0-6, two laboratory aircraft and research flight simulators), but it also provides expertise in attitude control, spacecraft structures, data processing, thermal control, robotics and telescience. It operates an NEC SX-3 supercomputer for computational fluid dynamics and structural calculations. NLR supports users of spacecraft and remote sensing satellites. NLR also distributes Spot and Landsat imagery as National Point of Contact.

Netherlands Agency for Aerospace Programmes

The Nederlands Instituut voor Vliegtuigontwikkeling en Ruimtevaart (NIVR) founded in 1946, is a semi-governmental non-profit agency with the statutory objective of promoting industrial aerospace activities, including space since 1969. Following on from ANS and IRAS, the current major bilateral programmes are SAX and SCIAMACHY. NIVR is responsible for SAX programme management, co-operating with SRON and Italy's ASI space agency. SCIAMACHY, which will measure a large number of trace gases and constituents in the stratosphere and higher troposphere from ESA's Envisat, is a joint project with Germany's DARA, with a contribution from Belgium. Dutch interest totals NLG90 million.

NIVR is conducting a multiyear space technology programme (NRT) to stimulate national technology developments. This includes the solar array development efforts of Fokker Space. Another example is the technological study by Bradford Engineering which led to the development of different types of gloveboxes for use in Spacelab, Mir, space shuttle and International Space Station.

Netherlands Industrial Space Organisation (NISO)

Initiated in 1986, NISO was officially established in December 1989. It comprises Dutch industries, research organisations and users in the space field with the aims of: defining common interests and strategies; promoting industrial involvement; stimulating the use/application of space facilities and products by industrial enterprises and institutes; broadening the social base of space; and promoting technology spin-offs. Membership is 24.

Netherlands Remote Sensing Board

Beleidscommissie Remote Sensing (BCRS) is the national co-ordinating body for national remote sensing activities. The BCRS Program Bureau is located at the Survey Department of the Ministry of Transport, Public Works and Water Management at Delft.

Observatory for Satellite Geodesy

Operated by the Faculty of Geodesy at the Delft University of Technology.

Prins Maurits Laboratory (PML) TNO

PML is one of the three defence research institutes of the Netherlands Organization for Applied Scientific Research (TNO). PML has 280 employees in two research and development departments that includes propulsion technology. It developed the gas generator igniter, thrust chamber igniter and turbine pump starter for Ariane 5's Vulcain cryogenic engine, in addition to undertaking research and development of high-performance solid-propellants based on phase-stabilised ammonium nitrate and hydrazinium nitroformate for future spacecraft. Facilities include an indoor test stand for solids, a connected pipe facility for ramjets, mixers for propellant and ramjet fuel manufacturing, and equipment for determining the chemical, mechanical and burning properties of propellants.

PTT Telecom Netherlands

PTT Telecom Netherlands is the national signatory to Intelsat, Inmarsat and Eutelsat. Its principal Earth station is at Burum in the north, with smaller stations at Amsterdam, Rotterdam and Nederhorst den Berg. In addition to the antennas listed below, two Inmarsat A/C land Earth stations were installed at Burum to support mobile communications over the Atlantic/Indian regions from 1991.

The dates given below indicate year of installation. Burum 1 28.5 m, 4/6 GHz Intelsat A, AOR, 1973; Burum 2 32.0 m 4/6 GHz Intelsat A, AOR, 1978; Burum 3/4 32.0 m 4/6 GHz Intelsat A, IOR, 1984; Burum 6/7 5.0 m ECS-SMS/IBS, 1988; Burum 8 9.0 m 11/14 GHz ECS 1988. Amsterdam 3 9.0 m 12/14 GHz ECS-SMS/IBS, 1988; Rotterdam 1 6 m 12/14 GHz SMS, 1985; Nederhorst den Berg 1 11 m 11/14 GHz Eutelsat, 1984; Nederhorst den Berg 2 4.2 m 11/14 GHz Eutelsat, 1989.

Space Research Organisation Netherlands

The Stichting Ruimteonderzoek Nederland (SRON) was founded 10 June 1983 within the Netherlands Organization for Scientific Research (NWO) in The Hague, by which it is financed, to manage/execute space science projects. SRON comprises two laboratories, in Utrecht and Groningen, with specific interest in x/γ-ray and IR/sub-mm astronomy. Current and recent projects include detectors for NASA's Compton Observatory and AXAF, the COMIS Coded Mask Imaging Spectrometer for Mir's Kvant 1, two wide-field cameras for SAX, the short wavelength spectrometer for ESA's ISO and a reflection grating spectrometer for ESA's XMM cornerstone mission. SRON also manages Dutch μg, radiation, exobiology and Earth-oriented research. Some 10 per cent of its G20 million annual budget derives from commissioned research.

Technisch Physische Dienst TNO-TU DELFT

TPD, the TNO Institute of Applied Physics, is an element of Netherlands Organization for Applied Scientific Research (TNO). Space personnel total 35; space turnover was NLG15 million in both 1995 and 1994. TPD began space instrumentation research and development in 1964 with the S59 UV stellar spectrometer for Europe's TD-1A astronomy satellite. It now specialises in attitude sensors, science instrumentation and Earth observation sensors. For example, the department provided Giotto's star mapper and Eureca's Sun acquisition sensors. Science instruments cover the wavelength range from hard X-ray up to the far-IR/mm. Recent projects include Hipparcos' modulation grid and refocusing mechanism, and ISO's cryogenic Short Wavelength Spectrometer. Remote sensing activities are concentrating on instruments for Meteosat Second Generation, Envisat and Metop, including optical systems, detection techniques, focal plane layout and precision mechanisms. Examples are the MIPAS cryogenic focal plane assembly (4 to 15 μm) and SCIAMACHY's optical bench. It was intensively involved in the development of the GOME Global Ozone Monitoring Experiment for ERS 2. TPD supplied the onboard calibration unit and performed the whole instrument's ground calibration under ESA contract. For that, TPD developed a calibration facility unique in Europe; it will also be used for the calibration of SCIAMACHY and other atmospheric science instruments, such as MERIS.

NORWAY

Norsk Romsenter

The Norwegian Space Centre is an independent organisation established by the Ministry of Industry 5 June 1987 as the immediate successor to NTNF, Space Activity Division. The 1997 budget is NKr304 million: 1996 – NKr270 million, 1995 – NKr285 million, 1994 – NKr276 million, 1993 – NKr272 million, 1992 NKr251 million, 1991 – NKr225 million and 1990 – NKr221 million. Including daughter companies there are 70 personnel. Its mandate includes compiling and submitting to the Ministry recommendations for co-ordinated long-term programmes. In particular, it represents Norway in ESA and co-ordinates the space-related activities of Norwegian companies, universities and research institutes. The Andoya Rocket Range and Tromso Satellite Station are both limited companies, with the Norwegian Space Centre having 100 per cent and 50 per cent interests.

Norwegian Meteorological Institute

The DNMI is the national signatory to EUMETSAT. It provides operational meteorological services to national users, including reception, processing and dissemination of NOAA and Meteosat imagery/data products.

Telenor Satellite Services AS

TNSS, a Telenor subsidiary, is the national signatory to Intelsat, Inmarsat and Eutelsat. The country is Inmarsat's fourth largest member, with 6.8 per cent in 1996, complementing the size of its maritime fleet and emphasising the importance of the major Earth station at Eik. TNSS is among the largest VSAT and gateway providers in Europe, with an installed base of 1,000 terminals by end-1995. Telenor is a member of the Skyphone consortium providing aeronautical communications services through the Inmarsat system.

PAKISTAN

Space and Atmospheric Research Centre

Located near the University of Karachi on a 0.2 km^2 site, Sparcent is Suparco's main establishment and encompasses Satellite Communications, Remote Sensing Applications, Agriculture and Landuse Applications, Remote Sensing Operations, Space Science and Ionospheric Research divisions. The campus houses a tracking and telemetry facility, which served as the prime TT&C station for Badr 1, an online information retrieval facility, Local User Terminal data collection platforms operating via the Argos system, an APT station and a NOAA TOVS acquisition and processing ground station to extract vertical atmospheric profiles for 45 parameters (such as water content and T) up to 35 km. Suparco operates:

The Remote Sensing Applications Centre (Resacent) functions under the Remote Sensing Applications Division as the national centre responsible for remote sensing data acquisition and applications co-ordination.

Suparco Plant, Karachi: manufacture of propellant and sounding rockets. Static test facilities to handle full size motors are maintained.

Flight Test Range, Miani Beach: Suparco's 200 hectare launch site 50 km northwest of Karachi at about 24.5° north/66.5° east overlooking the Arabian Sea equipped with modern facilities. Several Shahpar 7 m long 50 cm diameter sounding rockets are launched annually, capable of delivering 55 kg to 450 km.

Instrumentation Laboratories, Karachi: provision of flight, tracking, telemetry and sensing instrumentation for sounding rockets.

Satellite Ground Station, Islamabad: acquisition of Landsat TM/MSS, Spot HRV (10 m Cassegrain dual feed) and NOAA AVHRR (2.5 m parabolic) imagery; operational since May 1989. Facilities include data reception, acquisition and pre-processing incorporating high density digital tape recorders and two VAX 11/780 computers, 5210 FPS Array Processors and I^{2}S Image Terminals. EOSAT signed an agreement in 1991 for marketing rights to the Landsat data. SGS covers about 20 countries, including most of Central Asia. In addition to standard products, CD-ROMs will soon be available.

The Pakistan Meteorological Dept operates four NOAA APT stations at Karachi, Islamabad, Lahore and Peshawar.

Space Applications and Research Centre (Sparc), Lahore: established near the University of the Punjab, this Centre will develop into one of Suparco's main facilities for satellite development and space applications. It is currently engaged in satellite tracking, telemetry and reception, processing and analysis of science data. A Cospas/Sarsat Local User Terminal and Mission Control Centre were commissioned in April 1990.

Computer Centre, Karachi: computing facilities in support of Suparco's activities.

Aerospace Institute, Islamabad: the institute aims to satisfy Suparco's growing need for qualified manpower in space science and technology. In addition to regular postgraduate courses, it provides short preparatory courses for scientists and engineers travelling abroad for advanced training. Construction completed in 1995.

Space and Upper Atmosphere (SUPARCO)

Suparco is the national space agency for execution of space technology, science and applications research programmes. It was established in 1961 under the chairmanship of Professor Abdus Salam and re-organised in 1981 as an autonomous commission under the aegis of the 10-man Space Research Council, which is responsible for formulating national space policy, headed by the Prime Minister. The Executive Committee of the SRC, headed by the Federal Minister for Finance, monitors Suparco on behalf of the SRC. Suparco is headquartered at Karachi.

POLAND

Copernicus Astronomical Centre

The centre specialises in astrophysics studies. It is currently participating in ESA's Integral mission.

Institute of Aviation

The institute provides R&D in space science instrumentation, including spectrum analysers, telemetry systems and GPS receivers. It has contributed instruments to Interkosmos 19, Vega and Phobos and most recently spectrum analysers for the ASPI and POLRAD instruments aboard Interball.

Institute of Geodesy and Cartography

The only national state-owned geodetic research institute, a principal aim is development/promotion of space technology for geodesy and remote sensing. 40 of the 120 staff are engaged in space research. The geodetic observatory in Borowa Gora is constantly monitored by GPS to detect its geodynamic behaviour within research programs such as the International Geodynamic Service and EUROPROBE. The institute is developing systems to update topographic maps using satellite imagery and to incorporate remote sensing data in GIS. This includes interpretation of satellite images to monitor lake water pollution, forests, crop forecasting, etc. It is a Spot Image distributor.

Institute of Meteorology and Water Management

The Institute's satellite centre handles NOAA and Meteosat digital and analogue data. Processing of NOAA TOVS data for atmosphere temperature, moisture and ozone vertical profiles and of NOAA AVHRR for sea surface temperature, soil temperature and moisture, cloud classification, Baltic Sea ice, and vegetation index changes.

Polspace Ltd

Polspace represents almost the entire national potential in space-related technology. Membership includes SRC, Institute of Aviation, Warsaw University of Technology, Institute of Geodesy and Cartography, Institute of Telecommunications, Telecommunications Research Institute, Institute of Automation and Measurements, Institute of Precise Mechanics, Machine Tools Research and Design Centre, Polish Aviation Factory PZL-Mielec, Diginet Ltd, MAG Ltd, Milemi Ltd and NAVI Ltd.

Space Research Centre

The SRC was established in 1977 to conduct and co-ordinate national space research. 1995 budget was Zl 53.1 billion (1994 Zl 41.4 billion) with a staff total of about 145. Principal interests are solar system and space physics, geodynamics and satellite geodesy and remote sensing. Current projects are Interball (plasma waves), Coronas (solar X-ray spectroscopy), Mars-96 (plasma waves, atmosphere infra-red spectrometry), Relikt 2 (neutral ions, plasma waves), Cassini (Titan's surface temperature), ISTOK 1 (atmosphere infra-red spectrometry from Mir's Priroda module) and VIZIR (advanced Mir TV).

Telekomunikacja Polska SA

Polish Telecom is the national signatory to Inmarsat, Eutelsat, Intelsat and Intersputnik.

PORTUGAL

Instituto Nacional de Engenharia e Tecnologia Industrial (INETI)

INETI operates within the Ministry of Industry and has a staff of about 1,400, specialising in biotechnology, materials, energy, opto-electronics and computer sciences. Space activities include remote sensing and cartography (which have also involved the Institute of Information Technology). INETI has also acted as a defence research lab in opto-electronics and space.

RUSSIAN FEDERATION

Association for the Advancement of Space Science and Technology (ASCONT)

ASCONT was founded in 1992 by IKI, Lavotchkin, NIIKP, Lunokhod and NII Radio to promote Russia's science and technology capabilities.

AKO Polyot

Polyot designs and builds the Cosmos 3M satellite launcher (marketed by the Cosmos USA joint venture). It has built the Glonass, Tsikada, Cospas, Parus, Strela, Gonets and Informator satellites in conjunction with designer NPO PM. It built the RD-170 engine for NPO Energomash, but production has halted and the plant is now dismantled.

Polyot was created in 1941 to build aircraft. It employed 20,000 as of 1994. Production of the An-74 cargo aircraft resulted in the PO designation changing to AKO.

Astelit

NPO Astra and Italy's Nuova Telespazio are working together in the Astelit joint venture, founded in February 1992, and providing Russia with reliable links into the international telephone network using a Raduga satellite.

Central Specialised Design Bureau, Progress Division

CSDB (TsSKB) was founded in 1959 as a division of Korolev's OKB-1 Moscow design bureau to provide design support for launcher manufacturing at the 'Progress' factory, where the SL-4 Soyuz and SL-6 Molniya vehicles and satellites are now built. It is the parent of KB Photon. Employees, including KB Photon, total about 5,000. Its future appears assured as a principal supplier of launchers and satellites to the Russian Ministry of Defence.

Centre for Space Biotechnology

The Institute of Biorganic Chemistry, a part of the Academy of Sciences research microgravity and provides equipment for research.

ESKOS SA

ESKOS was registered in July 1992 in Paris to promote the products and services of Russia's space industry. Major shareholders include the COSMOS concern (leading Russian space enterprises and institutes) of the Russian Space Agency and the COSMOS International Centre for Advanced Studies based at the Moscow Aviation Inst. ESKOS signed a co-operation agreement with the Russian Space Agency 10 May 1993.

Gascom

RKK Energia and Gaszprom, Russia's state natural gas monopoly (94 per cent of Russia's production, 22 per cent of world's), created Gascom in 1992 to develop the Yamal telecom satellite system. Gascom will be the owner, operator and service provider. Gasprom will use the satellites to communicate with its gas fields, particularly in Siberia. Gasprom were already using 30 stations operating through Gorizont by 1995, with another 75 over 1995-96. The system and satellite design are also being offered commercially. Energia, Gascom and Loral Corp formed a joint venture 9 February 1996. Some of the communications payload is being bought from Space Systems/Loral, and Gascom are jointly marketing the design with SS/L. RKK Energia's advanced design includes Xe ion thrusters, CPV Ni/H_2 battery and unpressurised compartments.

Glavcosmos

Glavcosmos USSR was established in 1985 under the Ministry of Machine Building (MOM) as the prime authority for implementing co-operative agreements, including commercial utilisation of Soviet systems and flying foreign cosmonauts aboard Soviet spacecraft. Responsibility for manned missions was subsequently transferred to RKK Energia. It acts as the technical and management organisation for commercial space systems.

Specific deals are negotiated with the Licentsintorg foreign trade agency. Glavcosmos (dropping the USSR in late 1991) now comes under the Department of Machine Building within the Russian Ministry of Industry. Following the creation of RKA, it now appears more concerned with converting military technology to civil applications. It is responsible for the sale of the cryogenic engine to India.

Institute of Biomedical Problems (IBMP)

IBMP, founded in 1963, is the Russian Academy of Science's premier life sciences research centre. It is responsible for cosmonaut medical care under the Principal Medical Commission, chaired by Grigoriev and the head of the Russian AF's medical section. Poliakov is head of the physician-cosmonauts group, which is based at the institute. He set a space duration record on his return to Earth in March 1995.

IBMP also heads the international Bion biocosmos missions, maintaining a control room on the premises.

Izmiran

The Institute of Terrestrial Magnetism, Ionosphere and Radio Wave Propagation of the Academy of Sciences is Russia's principal investigator of the magnetosphere, ionosphere, Sun and Sun-Earth phenomena. It is operating the Coronas-I solar observatory and plans Coronas-F for 1997/98. Izmiran also provides instruments, particularly fields and particles measurement devices for satellites.

KB Arsenal

Collocated with PO Arsenal, historically a major producer of artillery, KB Arsenal provides the bus for the military Eorsats to TsNPO Kometa. They now offer science and commercial satellite design.

KB Khimautomatiki

The bureau of Semyin A Kosberg (1903-65) was formed in 1941, and began investigating liquid propellant rocket propulsion for aircraft in the early 1950s. It later became responsible for some Soviet space launcher upper stages. In 1974 it was reorganised as KB Khimautomatiki.

Responsible for Energia's RD-0120 cryogenic core engine, Proton's stage 2/3 engines and Soyuz/Molniya stage 2 engines. Formed from old Kosberg upper stage propulsion bureau.

KB Khimmach

A bureau headed by Alexei M Isayev was established in 1944 for research into storable liquid propellant rocket engines. Whereas Glushko's GDL concentrated on major launcher engines and Kosberg's bureau on upper stage propulsion systems, Isayev's produced smaller spacecraft engines, including manned, Earth orbit and deep space systems. A number of designations have been adopted but 'KTDU' (Korrektyroushchaya-Tormorznaya Dvigatelnaya Ustanovka, or Corrective-Braking Rocket Motor) is most frequent, with the others variations on this theme.

Information released in early 1991 indicated the bureau is part of NPO Soyuz in Kaliningrad. It is responsible for the 73.58 kN cryogenic engine of the new Proton KM version, originally designed for the L-3M manned lunar mission, and recently sold to India for its GSLV vehicle. A newly-developed DMT-600 600 N NTO/UDMH thruster was referred to in 1994. Khimmach developed all of the engines for the Heavy Cosmos TKS transport craft.

Starting in 1971 and at the instigation of Isayev, the bureau started the development of low-thrust engines, which could be used for attitude control of spacecraft. A series of 11 engines with thrusts up to 2,206 kN (225 kg) and eight with thrusts up to 49 kN (5 kg) were developed. They are capable of burns lasting from hundredths of a second through to hundreds and thousands of seconds.

KB Kolomna

Kolomna specialises in surface-air missile warheads and anti-tank guided missiles, but also offers the solid-propellant Sphere geophysical research rocket to deliver 230 kg payloads up to 330 km. Sphere employs guidance systems recovered from SS-23 missiles.

KB Machinostroenye

V P Makeyev (1925-1985) founded the bureau in 1955 as an offshoot of Korolev's bureau. It developed the SS-N-6 Zyb, SS-N-8 Vysota, SS-N-18 Volna, SS-N-20 and SS-N-23 Shetal SLBMs. It is developing the Surf satellite launcher with the Sea Launch Investors group of the US, by attaching an SS-N-20 stage 1 to an SS-N-23.

The Makeyev bureau proposes to offer satellite launch services using its converted SLBMs. A Volna was launched 7 June 1995 from the 'Kalmar' Delta 4 boat submerged in the Barents Sea on a 20 minute suborbital µg flight carrying a 120 kg experiment from Germany's ZARM (electrically-driven thermal convection in a fluid shell between two concentric spheres). Recovery was 5,600 km downrange on Kamchatka. There were apparently three SLBM demonstrations 1991-93. A Shtil 2N orbital demonstration launch was reportedly planned carrying a 100 kg Izmiran satellite.

KB Photon

KB Photon designs and builds a range of spacecraft derived from the original Vostok as a result of taking over Korolev's Zenit photo-reconnaissance satellite programme in 1964. This covers the Resurs-F Earth observation film camera satellite, the Photon µg vehicle, the international Bion biocosmos, and the Zenit Gen 3, Yantar Gen 4 and Kuban Gen 6 photo-reconnaissance. Kuban derivatives will apparently replace Resurs-F, Photon and Bion. It operates under the Central specialised Design Bureau. The satellites are built at the Progress factory in Samara.

KB Salyut

KB Salyut's history can be traced back to 1951 as Vladimir Myasischev's (died 14 October 1978) OKB-23 bureau to develop strategic bombers. The first was the M4 Bison, still in service. A modification has transported Buran and Energia's core. In 1959, Vladimir Chelomei took over part of the aircraft bureau (including the Khrunichev factory) to develop ICBMs. It produced the SS-19 and Proton. Chelomei's OKB-52 was responsible for designing the Almaz military space station and its three-man ferry craft. It performed the detailed design work on Salyut and Mir under NPO Energia's general direction; it is now building Alpha's FGB and Service Module. It provided the re-entry vehicle for the German-Japanese Express µg project. Dmitri A Polyukhin was KB Salyut's General Designer until his death 7 September 1993.

Keldysh Institute of Applied Mathematics

IPM was founded in 1953 to perform mathematical research and modelling for the nuclear and rocket industries. Its Ballistic Centre provides navigational computing expertise for manned and unmanned spacecraft.

Keldysh Research Centre

The Keldysh Research Centre – previously known as RNII, NII-1 and then NIITP – was established in 1933 as the first institute for research and development of liquid and solid propellant rocket engines. It currently develops, manufactures and tests advanced prototype of rocket engines, space power systems, high-energy beam generators and particle accelerators of various types. Its products have been used in the development of launch vehicles, spacecraft and orbital stations. Before and during the Second World War the centre concentrated on ground-based and air-launched weapons, the major achievement at that time being the development and manufacture of the first batch of Katyusha missiles, for which the centre received the Order of the Red Star.

During its history the centre boasted having M V Keldysh, S P Korolyov and V P Glushko working within its organisation. As of January 1999 the centre had about 1,000 employees. The sources of finance for the centre are given as: Russian Space Agency – 52.7 per cent, exports – 18.8 per cent, other contracts for weapons and military equipment – 8.6 per cent, various other contracts – 8.4 per cent, Ministry of Defence – 7.5 per cent, other contract for the federal space programme – 3.4 per cent, Ministry of Defence industry 0.6 per cent.

Khrunichev State Research and Production Space Centre

The history of GKNPTs Khrunichev began in 1916 when the factory was established in Fili for building cars by the Russian-Baltic Carriage Factory Joint Stock Co. It began aircraft construction in the 1920s, initially leased to Junkers. Serial Proton production has been undertaken since the 1960s, followed by the SS-19 missile (now available as the Rockot satellite launcher) and space station modules (Salyut, Mir, Almaz, Kvant, Kristall, Priroda, Spektr, FGB and Alpha Service Module). The modular Angara launcher is under development here. President Yeltsin issued a decree 7 June 1993 merging Khrunichev and KB Salyut (Proton's design bureau) to resolve the question over authority for Proton. Lockheed-Khrunichev-Energia (LKE) International was formed in 1993 for international marketing, and International Launch Services in 1994.

Koskon

Koskon (Space Conversion) is the creation of a 15-member consortium involving the former Soviet PTT, Metallurgy, Electronics and Health Ministries, and the

Moscow Aviation Institute; industrial contractor is AKO Polyot. Up to 45 860 kg satellites would be placed in four 1,100 km 83° orbits, offering 4.8 kbit/s real-time and store/forward voice/data traffic to fixed and mobile terminals. User up: 320.8322 and 1,642.5 to 1,634.4 MHz; down: 290-320 and 1,555 to 1,555.9 MHz. Intersatellite links would be at S-band (2,056 to 2,060 and 2,2402,244 MHz). Feeder links would be at C-band. Polyot's Informator 1 was a Koskon demonstrator.

Lebedev Physics Institute Astro Space Centre

The Astro Space Centre is leading the Spektrum-R/Radioastron VLBI project.

Lunokhod

The Innovative Enterprise Lunokhod was founded in 1991 to promote NPOL's expertise.

Ministry for Posts and Telecommunications

The Ministry for Posts and Telecommunications is responsible for space-based communications services (joining Eutelsat in July 1994). The Russian Federation established its own Ministry of Communications, Information and Space in 1990, and paid the Ministry of Defence for launching the first of three dedicated Gorizont communications satellites (the third appeared April 1992).

Morsviasputnik

Morsviasputnik is the national Russian signatory to Inmarsat, accounting for about 4 per cent.

Moscow Aviation Institute (MAI)

MAI was founded in 1930 on the basis of the air mechanics faculty of Moscow's Higher Technical School. It educates specialists for the aerospace industry in 10 faculties: aircraft construction, engines, automatic flight control systems, radio electronics, business/management, aerospace, robotic and intelligent systems for aerospace and military craft, applied mathematics, applied mechanics, humanitarian. Affiliates are located in Arzamas, Zhukovsky, Kaliningrad, Tyuratam and Moscow's radio higher school. There are about 2,000 faculty members, 6,500 other staff and more than 14,000 students. Its extensive museum includes flight-standard spacecraft, including the N1 manned lunar lander.

Moscow Institute of Thermal Technology (MITT)

MITT was founded in 1948 by Alexander Nadiradze's (1914-87) design bureau and was responsible for the design and development of the RS-12/SS-13 silo-based, RS-14/SS-16, RSD-10/SS-20 and RS-12M/SS-25 road mobile and RS-22/SS-24 rail mobile solid propellant ICBMs. Its Science and Technical Centre Complex was created in 1991 to develop the Start orbital launchers based on the SS-20/SS-25; the Start-1/Start test launches were made 25 March 1993/28 March 1995 from Plesetsk.

N D Kuznetsov Company

Nikolai Dmitrievich Kuznetsov (24 June 1911-30 July 1995) established an aircraft engine design and development bureau during the 1940s in Kuibyschev (now Samara). Information finally released in 1989 revealed that the bureau had been assigned the responsibility for the design and development for engines for Korolyov's four-stage N-1 launch vehicle. Previously the bureau had developed engines planned for use on Korolyov's R-9 missile, but politics ensured that the missile used other engines.

In July 1993, Aerojet signed a teaming agreement with N D Kuznetsov SSC to market these engines within the United States. Two NK-33 engines were delivered to Aerojet in 1995. One of them was used by Aerojet for five firings which totalled 410 seconds operating time within a 57 to 114 per cent thrust range which followed the proposed Atlas-2AR launch profile. The engines were under consideration to power the first stage of the Atlas-2AR, but they lost out to the Energomash RD-180 engine.

The United States Kistler Aerospace Corporation is planning to use NK-33 engines on its reusable K-1 launch vehicles: three would be on the first stage and one on the second stage.

NII Argon

Argon designs onboard computers, including those carried by the manned Soyuz and Mir.

NII Grafit

Grafit works on applied research and development of graphite materials including carbon-carbon composites, leading to brakes, nose and leading edges for Buran.

NII Khimicheskogo Machinistroenye

NIIKhM was founded in 1948 as NII-229 and provides Russia's largest rocket engine test facilities, covering 25 km^2 and capable of testing complete stages up to 9 m diameter and 12 MN. The major test stands are, Stand V1: 100 kN cryogenic; V2: 2 positions for 100 to 2,000 kN cryogenic (such as RD-0120); V3: stages to 490 kN. Stand 102: 11.8 MN LOX/kerosene. Six thermal vacuum chambers of 0.3 to 100 m^3 can simulate missions of up to nine months for 5 N to 6 kN engines. 900 m^3 and 8,300 m^3 chambers test complete satellites. Employees total 5,000. SEP heads a European team that began testing the RD-0120 at the site in 1995.

NII Kosmisheskovo Priborostroenye

NIIKP is responsible for Glonass, Tsikada, GEO-IK, remote sensing and meteorological satellite payloads, including the MSU imagers aboard Resurs-O. It provided the 60 laser reflectors for Germany's GFZ 1 laser geodetic satellite.

NII MACHINOSTROENYE R&DIME (Research and Development Institute of Mechanical Engineering)

NII MACHINOSTROENYE R&DIME (Research and Development Institute of Mechanical Engineering) was established in 1958 as part of NII Thermal Processes, but became independent in the early 1980s. Development and production of 0.8 to 400 N spacecraft thrusters (implying connection with old Isayev bureau), including Soyuz, Progress and Mir. Development and small batch production of engines up to 6 kN can be accommodated. Test facilities can handle hydrogen (up to 3 kN), methane and kerosene fuels. The Elektro Earth observation satellite carries 16 DEN-16 electrothermal thrusters. Note: LTRE means Low Thrust Rocket Engine, or RDMT in Russian (Raketnay Dvigatel Maloi Tyagi).

NII Parachutostroenye

The NIIPS Institute of Parachute Construction provides recovery parachutes for Soyuz, Buran, Energia boosters and Progress return capsule. It is also supplying the recovery parachutes for ESA's Ariane 5.

NII PM (Prikladnoi Mekhaniki)

NII Applied Mechanics is a major designer and developer of launcher and spacecraft navigation instruments, including laser gyros.

NII Radio

NIIR is the main research centre of the Ministry of Communications.

NII THERMAL PROCESSES NII TP

NII THERMAL PROCESSES NII TP performs research in all forms of launcher and space propulsion, including ramjet, electric and nuclear. BMDO took delivery in October 1993 of a T-100 1.5 kW Hall-type Stationary Plasma Thruster for evaluation. N_2O_4/UDMH thrusters include 10 N and 1,000 N models. NII TP developed the Phobos spacecraft's network of hydrazine thrusters for attitude control and minor orbital modifications. 28 (24 x 50 N, 4 x 10 N) were mounted on four spherical tanks. A set of 0.5 N hydrazine microthrusters of an unusual design was also included to avoid the need for separate N_2 cold gas jets; the propellant was burned internally and cooled inside an expansion tank before passing to the thrusters.

NPO AP

NPO Automatics and Instrument Engineering was created in 1978 as part of NII AP to develop control, navigation and guidance systems for launchers and spacecraft. The institute was founded in 1946 by Nikolai A Pilyugin (1908-82; Lapigin succeeded him) and provided the flight control systems for the Vostok launcher family, Proton (including Block D/DM), N1, Zenit and Buran.

NPO Astra

Responsible for most of the previous Soviet satellite communication ground segment, including the Moskva and Orbita systems. Astra and Italy's Nuova Telespazio are working together in the Astelit joint venture, providing Russia with reliable links into the international telephone network using a Raduga satellite from early 1992.

NPO Cryogenmash

Design and manufacture of ground-based cryogenic equipment; supply of cryogenic fuel to space programme.

NPO Elas

Elas builds communications and imaging sensor payloads. It is providing the payloads for the Kupon and Courier communications satellites; it is probably responsible for the military Geizer satellite's payload.

NPO Energomash

The Gas Dynamics Laboratory OKB DL was created 1928 in Leningrad (St Petersburg) for research into and production of liquid propellant and electric rocket engines. Valentin P Glushko was director until his death in January 1989. It became responsible for the majority of space launcher lower stage engines, notably the Vostok and Proton families. GDL merged with Sergei Korolev's bureau in May 1974 to form NPO Energia (now RKK Energia). It left Energia after Glushko's death, becoming NPO Energomash under Viktor Radovski. Energomash now reports directly to RKA, employing about 9,000. Two test stands can handle 10 MN engines. An agreement was signed with P&W in October 1992 providing exclusive US marketing rights on Energomash products and technology.

NPO Geofizika

Geofizika is Russia's primary designer and producer of spaceborne attitude control sensors, optical instrumentation, spectrophotometers, radiometers, star/Sun/Earth simulators, star simulator calibration, astronaut training simulators and high precision optical angle encoders, plus test and calibration equipment.

NPO Iskra

Iskra ('Spark') was created in 1978 as part of KB Makayev, a leading developer of solid propellant missiles. It specialises in solid motors and gas generators; for example, it is responsible for the Soyuz launch escape tower.

NPO Istochnik

Istochnik supplies batteries for space applications.

NPO Istok

Istok is Russia's only producer of TWTA travelling wave tube assemblies for communications satellites. It also makes solar panels.

NPO Kompozit

Advanced composite material research in Russia, including Buran thermal protection, takes place at Kompozit. Other products include microgravity research equipment, including semiconductor and optical materials units. Spacecraft and launcher instrumentation.

NPO Kraznaya Zvezda (Red Star)

Red Star specialises in nuclear power systems. It developed the 2 kW Bouk thermionic reactor for the Rorsat programme, and the twice-flown (Kosmos 1818/1867) Topaz 1 5 to 6 kW unit.

NPO Kvant

Kvant builds solar cells and arrays for space applications.

NPO Lavotchkin

NPOL specialises in planetary and deep space craft, science satellites, including Luna, Venera, Zond 1-3, Mars, Phobos, Spektrum, Astron and Granat. The Babakin centre, which emerged from Korolev's OKB-1, is solely responsible for planetary probes and science satellites until its associated NPO Lavotchkin was identified openly in 1989. Georgi Babakin (1914-1971) established his own OKB in 1965 and at the same time took over the satellite arm of OKB Lavotchkin (founded 1937 as an aircraft manufacturer). Lunar and deep space work was transferred from Korolev's OKB-1 in 1965. Sergei Kryukov was chief designer August 1971 to December 1977. Vyacheslav M Kovtunenko then took over (died 11 July 1995, aged 73). Babakin is now directly funded by RKA.

Lavotchkin is providing the Kupon telecom satellite for Global Information Systems. It provides the bus for the Oko and Prognoz early warning satellites to TsNPO Kometa, and is proposing Fregat as a launcher upper stage.

NPO Machinostroenye

Vladimir Chelomei's bureau was divided after his death in December 1984 into KB Salyut and NPO Machinostroenye. Almaz radar satellites, and probably others, are built at the Khrunichev factory. It operates the Almaz foreign trade company to market Almaz data.

NPO Molniya

Molniya studied the Buran aerodynamic design, flew BOR Cosmos subscale missions, and conducted hypersonic vehicle research. Dr Gleb E Lozino-Lozinsky was Director until February 1993.

NPO Orion

Orion is a major designer and developer of detectors, focal plane arrays, cryogenic systems, thermal imagers and processing electronics.

NPO Planeta

Formed in 1989 under what is now Roskomgidromet, the Federal Service for Hydrometeorology and Environmental Monitoring, to manage Earth observations and resources programmes, emphasising meteorological applications.

NPO PM (Prikladnoi Mekhaniki)

NPO Applied Mechanics was responsible for more than 829 satellites by the end of 1995 (13 in 1995), including the Express, Gals, Ekran, Gorizont, Raduga and most other major telecom satellites. Its other responsibilities included the Strela 3 military sextets, Glonass and Cosmos navigation and GEO-IK geodetic satellites (the NIIKP Institute of Space Device Engineering provides the payloads). It remains a state enterprise, reporting to RKA, and has not yet decided if it will privatise. Employees total 10,000. Its business has been divided equally between civil and military, but the military element is now sharply declining. As a result, it is moving into other areas, such as ground antennas.

Since January 1994, NPO-PM controls the Gals and Express satellites itself through its Persei subsidiary, instead of relying on the military. TT&C stations in Krasnoyarsk and Gouss Khroustalni allow control over 20° west-154° east GEO. Its Mercury subsidiary offers transponder leasing. NPO PM began as a design bureau under Korolev in 1959, specialising in missiles, before becoming independent in the mid-1960s under Mikhail F Reshetnev (died 26 January 1996). It developed Yangel's R14/SS-5 Skean missile into the Cosmos space launcher. NPO PM made a commercial breakthrough in July 1995 when Eutelsat selected it to provide 1998's Sesat.

NPO Precision Instruments

NPO TP specialises in manned and unmanned spacecraft payloads and electronics, including control systems, telemetry, reception stations and communications. It developed the Salyut/Mir Igla and Kurs rendezvous and docking systems and the Vostok telecommand system. It is part of the Smolsat consortium developing the Gonets satellite system.

NPO Radio

Payloads for Ekran, Gals, Gelikon and Raduga telecom satellites. Part of the Informcosmos consortium.

NPO Saturn

The Lyulka bureau specialises in military aircraft engines. It also provides turbomachinery and gas generators for space applications, including Buran's gas generators. The 400 kN LOX/LH_2 D57 engine was developed in the 1960s to replace the lunar N1 launcher's stage 3 engines. Lyulka teamed with Aerojet Propulsion in 1993 to improve and market the engine.

NPO SATURN/LYULKA's cryogenic engine, which has been described in the 1990s as the D-57, would appear to be the 11D57 gimballing engine, which was developed during the early 1970s for use on the Block S of the modified N-1 launch vehicle, which would have been flown for the dual-launch L-3M manned lunar missions. The bureau was also tasked to develop the 11D54 fixed-chamber cryogenic engine with the same thrust level of 40 tonnes (390 kN). This engine would have been used in a cluster of six to eight on a modified Block V of the N-1 vehicle.

NPO Scientific Centre (NPO-NTs)

NTs is an affiliate of the electronics industry ministry. Electronic components and μg research equipment, including Mir's Gallar semiconductor furnace.

NPO Soyuz

Soyuz operates a major production and R&D facility for double base and composite solid propellants. It is closely connected with STC Complex, indicating that it provides the propellant for the Start launcher.

NPO Start

Start is a commercial effort to build rocket motor test equipment and stands.

NPO Stekloplastik

Supplier of the composite materials employed on Buran.

NPO Tekhnomash

Tekhnomash is a commercial concern that provides space engineering research, development and consultancy. Branches in Omsk, Tomsk, Krasnoyarsk, Kuibyshev (Samara), Dniepropetrovsk, Kharkov, Kiev, St Petersburg, Orenburg and Urga. It has 8,500 employees about half of whom are graduate engineers.

NPO Tekhnologiya

A supplier of advanced materials for Buran.

NPO Trud

The former Kuznetsov aircraft engine bureau developed engines for the abandoned N1 lunar launcher. They are now offered commercially. Aerojet markets them in the US. Nikolai D Kuznetsov was General Director until 1994, aged 83.

NPO Vega-M

Headed by the MNIIP Moscow Science and Research Institute of Instrument Engineering, Vega provides space SAR radars, including those of the Almaz satellites.

OKB Fakel

Fakel is a propulsion laboratory which was established in the Academy of Sciences in 1955 and re-organised in 1971 as OKB Fakel ('Torch'), specialising in spacecraft attitude control thrusters, ion engines and plasma sources. The first EPS flew in 1972 aboard a Meteor-1 satellite and further models was subsequently flown on the later Meteor generation satellites, including the Meteor-Priroda series. Research and development work on Stationary Plasma Thrusters (SPT) started at Fakel in 1964. Starting in 1982, EPS units based upon STP-70 and the K-10 hydrazine thermal-catalytic thrusters (TCT) have flown on communications satellites such as Geizer and Luch (including flights within the Cosmos programme). In 1994, a third generation EPS was introduced, SPT-100 aboard Gals and Ekspress satellites. In 1992, Fakel together with Space Systems/ Loral and the Moscow Scientific Research Institute of Applied Mechanics and Electrodynamics (RIAME) established a joint venture – International Space Technology Inc (ISTI) – for promoting, marketing and selling the EPS outside Russia. Subsequently, SEP and ARC joined the company. In 1996 the SPT-100 was certified in accordance with western standards: 7,440 hours of firing tests were successfully conducted at Fakel and 5,000 hours were accomplished at the Jet Propulsion Laboratory in Pasadena.

Its SPT-70 and SPT-100 (Stationary Plasma Thruster) Hall electric thrusters have flown on more than 50 Meteor polar meteorological satellites since 1972 to provide orbit control. The numerical designators indicate beam diameter in mm. SPTs employ a DC-gas discharge in an annular chamber, in which a radial magnetic field traps the electrons. These Hall currents ionise the Xe propellant, the ions of which are accelerated inside a quasi-neutral plasma, without grids, and by the discharge voltage itself. The advantages are their rugged simplicity (no grid system or high voltage supply), but they provide lower efficiencies, lower exhaust velocities and increased Xe consumption than ion thrusters such as the UK-10.

OKB Moskovskaya Energetitsheski Institut

This institute builds solid-state recorders for space applications.

PO Mayak

Mayak produces radioisotopes for all applications, including space and nuclear weapons. It is a potential supplier of material for future US RTGs. Users included the Lunokhod lunar rovers.

PO Samara Frunze Engine Building

One of Russia's principal aerospace engine production plants, Samara builds the Soyuz/Molniya RD-107/108 engines for NPO Energomash.

PPO Motorostroitel

One of Russia's leading aircraft engine and gearbox plants, Perm Motorostroitel builds Proton's RD-253 stage 1 engines for NPO Energomash.

Priroda

State Centre Priroda is the primary civil remote sensing acquisition and processing organisation in Russia, under the administration of Roskartografia, the Federal Service of Geodesy and Cartography. It operates a large cartographic division that employs satellite data to create cartographic and GIS products. New imagery is collected during Resurs-F missions. It created the WorldMap çonsortium in 1993 to market digitised versions of its archive of two million film images.

RD&PE Zvezda

Previously known as the Zvezda bureau, RD&PE Zvezda is a joint-stock company. It has specialised in the life support systems for all manned spacecraft as well as EVA and pressure suits for space missions dating back to Vostok 1 and the SK-1 suit. It was responsible for the Krechet 'Gyrfalcon' suit planned for use on the lunar surface in the late 1960s/early 1970s. From this evolved the Orlan family of EVA suits which have seen service on the Salyut orbital stations and currently Mir. The bureau also produced unflown manned manoeuvring units for use in the mid-1960s and the Ikar 'space chair' tested outside Mir. Among its other activities, it designed Buran ejection seats, the design of which was submitted for possible use aboard Hermes. Dornier/Zvezda led ESA's EVA 2000 studies based upon Mir's Orlan-DMA EVA suit, updated with western technology. Zvezda suits will fly on board the International Space Station.

RKK Energia

RKK Energia is Russia's largest space company, responsible for the design, development and manufacture of manned and man-related vehicles: Soyuz, Progress, Mir, Buran, and Energia. Its personnel total 22,000, with 1,000 permanently sited at Baikonur. President Yeltsin signed an order in early 1994 to sell 49 per cent of the company with the government is retaining its 51 per cent interest for at least 3 years. The name became RKK Energia, or Energia Space Rocket Corp.

Energia's predecessor, NPO Energia was formed May 1974 by merger of Sergei Korolev's bureau and the Gas Dynamics Laboratory OKB of Valentin Glushko. Korolev's OKB had been renamed Central Construction Bureau of Experimental Machine Building (TsKBEM) after his death in January 1966. GDL left Energia after Glushko's death in January 1989, becoming NPO Energomash. Mir management and foreign cosmonaut missions are handled by Energia, reporting to the Russian Ministry of Industry.

Soyuz and Progress craft are assembled in the KIS Integration and Test Facility, and shipped to Baikonur in rail containers. KIS is also used for Proton stage 4. As of September 1995, KIS also housed Buran, Mir core, Kvant 1 and Kristall high fidelity mockups, and a Space Station ACRV Soyuz model.

Russian Academy of Sciences (RAN)

RAN established a Space Council in March 1992 to direct fundamental space research, select missions for proposal to RKA and organise international missions. The council comprises eight sections: plasma physics and Sun/Earth interactions, solar system studies, astrophysics, life sciences, materials, legal and social, Earth sciences and international co-operation (Interkosmos until 1994).

Russian Space Agency

RKA was created by President Yeltsin's decree in February 1992 to draw up and manage Russia's civil space programme. Its director, Koptev was formerly MOM deputy minister and an NPO Lavotchkin engineer. It has 300 HQ employees. The nine principal divisions are: state programmes, manned projects and launch facilities, implementation of state programmes, science and commercial satellites, international; ground infrastructure, external relations and legal affairs and resources and business affairs. All rocket engine organisations, such as Energomash and NII TP, now come under RKA.

SKAS (Moscow Institute of Radio Links)

SKAS builds communications payloads and ground terminals.

Smolsat

The Smolsat consortium includes NPO PM, NPO Precision Instruments, Selkhoz Bank and Moscow's Soyuzmedinform Programme Management (connected with the Ministry of Health). It plans the Gonets satellite communication constellation working through hand-held terminals.

Sovinformsputnik

Companies and organisations responsible for military Earth observation satellites formed Sovinformsputnik in 1991. The majority shareholder is the Central specialised Design Bureau, builders of Resurs-F and the Zenit, Yantar, Kometa and Kuban satellites. Other shareholders include the Progress factory, Priroda and the Krasnogorski Zavod joint stock company. Sovinformsputnik has an exclusive right to distribute products/services from these military film satellites; their archives stretch back to the first Soviet photo-reconnaissance satellites in 1962. The main archive dates back to 1981. The principal sources are the complementary KVR-1000 and TK-350 camera systems flown on the Kometa topographic mapping satellites.

Sovzond

The Sovzond consortium of NPP VNIIEM, NPO Planeta and RNIIKP was created in December 1992 to market remote sensing expertise. An agreement with Sweden's SSC Satellitbild is allowing reception, processing and marketing of imagery from November 1994's Resurs-O1 directly at Kiruna.

Soyuzkarta

Soyuzkarta was established in 1987 as a Soviet foreign trade association affiliate of Priroda to market aerial/ space-based photographic, cartographic and geodetic products. In 1992, it was transformed into a joint stock company; Kosmokarta is the space division. 5 m resolution colour film imagery is available from the KFA-1000 camera, 6 to 8 m b/w spectral from the MK-4 and 15 to 30 m colour film from the Kate-200, carried by Mir and Resurs-F. This imagery is now available in digitised form direct from Priroda and its WorldMap consortium. Lower altitude coverage is provided by An-30 and Tu-134 aircraft.

Space Research Institute (IKI)

IKI, part of the Russian Academy of Sciences since its creation in 1965, is Russia's premier space science centre and leads missions such as Spektrum and Relikt 2. Bureaux in Frunze and Tarusa support it.

Splav Technical Centre

The sole Russian centre dedicated to μg research and payloads, it provides the Zona, Konstanta and Splav furnaces for manned Mir and unmanned Photon missions.

TsAGI

The Central Aerohydrodynamics Institute, was formed in 1918 in Moscow. It maintains aircraft and satellite test facilities employing 6,000 technical staff out of 12,000 total personnel. TsAGI worked with British Aerospace on interim Hotol research. In 1993, it worked with ESA on development of a Buran real-time entry flight simulator. Buran pilots trained on TsAGI's 6-DOF simulator.

TsNII Mashinostroenye

TsNIIMash central research establishment of Russia's Ministry of General Machine Building (previously the USSR's MOM) is now part of RKA. The TsUP Space Operations Centre is one of its departments. Originally an artillery factory, it became NII-88 (Science Research Institute 88) in May 1946 and, with the help of interned German V2 specialists, under Korolev developed the first Soviet rockets. Korolev's OKB-1 branched off from NII-88 in August 1956. Utkin was General Director of NPO Yuzhnoye until he moved to TsNIIMash in 1990. TsNIIMash-Export was created in 1991 to market TsNIIMash capabilities. It is working with the US BMDO on TAL (thruster with anode layer) D-55 1.5 kW and TAL D-100 5 kW Hall-type Stationary Plasma Thrusters.

TsNPO Kometa

Kometa is responsible for several high value tactical and strategic space systems, including the ASAT anti-satellite, Oko/Prognoz early warning, Eorsat and Rorsat.

Ural Electrochemical Integrated Plant (UEIP)

UEIP is a major supplier of enriched uranium for nuclear power plants. Although apparently no Russian fuel cell has flown in space, UEIP has developed at least two. The Volna cell, employing circulating KOH as electrolyte, was developed for the manned lunar programme. The Photon fuel cell, immobilising the KOH in an asbestos matrix, was developed and qualified for future Buran missions. ESTEC ran a two-week continuous test campaign on Photon in 1992, attaining 12 kW at 400 A. Photon comprises eight 32-cell stacks connected in parallel, with 176 cm^2 active cell area. UEIP's fuel cell interests now focus on large plants for terrestrial applications.

VNII Elektromekaniki

Founded in 1941, VNIIEM is responsible for the Elektro, Resurs-O and Meteor Earth observation satellites. It contributes instruments to other projects, including Almaz and Mir.

VNII TransMash

TransMash specialises in the design and fabrication of transport vehicles. In the space field, it is developing the Mars-98 rover (Lavotchkin integrates the payload) and was responsible for the Lunokhods, small Mars 3 rover and the Phobos hopper. Employees total 2,000.

SINGAPORE

Centre for Remote Imaging, Sensing and Processing (CRISP)

Funded by the National Science and Technology Board, and located at the National University of Singapore, CRISP was set up in late 1992 and its ground station became operational in September 1995 to receive Spot and ERS data. Radarsat was added January 1996. The 12.8 m antenna (1°17'30" north/103°47'00" east, in campus, 2 km from the processing site) and related ground station facilities were built under a US$7.3 million contract by US companies Datron and International Imaging Systems Inc.

The data processing facility features Silicon Graphics computers running under Unix. The main computer can process 10 ERS SER and 25 Spot scenes in an 8 hour shift. Ocer 1996-97, CRISP is concentrating its research effort on forestry, oceanography and advanced data processing techniques (stereo, SAR interferometry, radargrammetry and optical/radar fusion).

SOUTH AFRICA

Satellite Applications Centre (SAC)

As the leading Earth observation centre in southern Africa, the SAC receives, processes, archives and distributes data and map format products from LANDSAT, SPOT, NOAA and ERS payloads. During 1996, SAC signed an agreement with Core Software Technology in the US, under the terms of which SAC will receive and distribute high-resolution images from future EROS satellites. Starting 1997, SAC also distributes data from Canada's RADARSAT. Further negotiations are under way with several agencies to broaden the range of SAC's products so as to ensure its customers an appropriate choice of spectral and spatial characteristics. The Satellite Applications Centre (SAC) operates as part of CSIR, South Africa, and has two main areas of business:

Earth observation: the reception, processing, archiving and commercialisation of image data from a range of remote sensing satellites.

Tracking, telemetry and command (TT&C) services to international satellites operators and launch agencies.

Located at 25.88° south, 27.70° east, SAC's footprint reaches northwards to 5° south, and it is therefore ideally located to serve any clients with an interest in and over the southern African sub- continent.

The infrastructure in place includes extensive buildings and secure power, reception and transmission facilities covering the L-, S- and X- bands and state-of-the-art communications links and powerful hardware and software for data processing. A team of 60 engineers, technicians and support personnel staffs the Centre. Turnover of the SAC is expected to be in excess of US$5 million during the FY97-98.

The SAC archive contains digital data from 1972 and it remains up-to-date with the daily addition of data. Training opportunities are organised twice a year and are typically designed and presented jointly by SAC and selected partners.

The first SAC TT&C services were for NASA during the 1960s through to the mid-1970s At present SAC's services are available on a contract basis to launch agencies and the owners and operators of satellites. These services are rendered by an experienced team 24 hours a day, throughout the year. In this context SAC is part of the 2 GHz network of the French National Space Agency, CNES.

University of Stellenbosch

Sunsat was announced June 1991 to promote development of industrial and educational space engineering. It is part of a programme to establish an internationally recognised satellite engineering activity at Stellenbosch, started with the Grinaker company's sponsorship of a Chair of Satellite Systems. The control centre will be sited at the University over the five-year life. Stellenbosch already has an Oscar ground station and a 4 m antenna. Students are supported by major electronics companies: Altech and Siemens of Johannesburg, Grinaker Electronics of Pretoria, Plessey SA of Cape Town, Analysis Management and Services of Stellenbosch, and First National Bank of Johannesburg. Cost is estimated at US$800,000 and NASA provides the launch free in exchange for the GPS receiver and laser reflector payloads.

Sunsat was originally planned for Ariane's Helios IA launch but the US$400,000 cost was prohibitive. Instead, it replaces the balance mass for Denmark's Ørsted on Delta's Argos launch. Breadboard prototypes were completed January 1992 to March 1993; the engineering was completed April 1994.

SPAIN

Centro para el Desarollo Tecnologico Industrial (CDTI)

The Ministry of Industry and Energy created the Centre for the Development of Industrial Technology (Centro para el Desarollo Tecnologico Industrial). To implement April 1986's 'Law of Science' policy, which identified space as one of 14 areas requiring improved official support, the military-based CONIE was dissolved by a law decree and the official Spanish delegation to ESA was entrusted to CDTI. It drew together 45 companies and 11 research institutes to compete for ESA and other space-related business.

Instituto Nacional de Técnica Aeroespacial (INTA)

The defence ministry established the Instituto Nacional de Técnica Aeroespacial (INTA) in 1942. It remains an autonomous R&D body reporting to the Ministry of Defence. Personnel total about 1,500 in four divisions, namely, Materials and Structures, Aerodynamics, Energy and Propulsion, and Avionics. It also operates the Aeronautical/Space Documentation and Information Centre (CIDAE). In the space field, it provides management and technical expertise and is currently managing the Minisat and Capricornio programmes. It was responsible for Spain's first satellite (Intasat), conducts INTA 100/300 sounding rocket firings, provides technical management for Spain's Helios contribution, developed S/Ku-band TT&C antennas used on satellites such as Olympus, Hipparcos, Telecom 2, Eureca and Hispasat, and operates the ESA Spasolab Laboratory for testing and certifying photovoltaic cells. INTA's INSA subsidiary operates and maintains the Villafranca tracking sites for ESA and the Madrid site for NASA. Its own 10 m S-band antenna at Maspalomas receives Landsat and Spot imagery, which it distributes as ESRIN's National Point of Contact.

Telefónica de Espana SA

The Compania Telefónica Nacional de España is the national signatory to Intelsat, Eutelsat and Inmarsat. It also holds 25 per cent of Hispasat SA, operators of the national satellite system. Telefónica's Eutelsat Earth station at Guadalajara, east of Madrid, operates an 18 m Ku-band antenna for handling TDMA and TV traffic. Telefonica Sistemas de Satelites was created as a subsidiary to provide business communications.

SWEDEN

Centre for Environmental Satellite Data

MDC is owned by SSC to produce, archive and distribute environmental databases for government and research. The MDC is contracted by the EEA European Environment Agency to lead the ETC/LC European Topic Centre on Land Cover; the ETC/LC is a consortium of 16 European organisations.

Swedish Institute of Space Physics

The Institut för Rymdfysik (IRF) has participated in more than 40 rocket and 22 satellite investigations between

1964 and 1996, primarily targeted at magnetosphere and ionosphere studies. It provided the payload for Astrid 1, Promics particle detectors for Prognoz 7/8, hot plasma and low frequency wave experiments aboard Viking and Freja, the Aspera hot ion composition spectrometers aboard both Phobos, the Promics 3 hot plasma detectors for both Interball orbiters, the Electric Fields and Waves instrument for Cluster, Mars-96's Aspera-C ion/neutral particle imager; planned is an ion spectrograph on Japan's Planet-B Mars orbiter. Headquartered at Kiruna, the institute includes two divisions at Umeå (specialising separately in plasma physics and propagation of mechanical waves) and Uppsala. The Kiruna/Uppsala divisions focus on upper atmosphere, ionosphere and magnetosphere studies.

Swedish National Space Board (Rymdstyrelsen)

The SNSB was created in 1972 under the Ministry of Industry as the central governmental agency responsible for all publicly-funded national and international space programmes, including co-ordination of remote sensing activities. Three advisory committees cover science, remote sensing and industrial policy. Executive management of space projects, including the Esrange sounding rocket range and satellite ground facilities, is assigned to the SSC state-owned limited corporation.

Swedish Space Corporation (SSC)

The Swedish Space Corporation is a government-owned limited company under the Ministry of Industry and Trade, responsible for the technical implementation of Sweden's space and remote sensing programmes, including feasibility studies, operational applications, systems engineering and research and development. The SSC is part-owner of the Nordic Satellite Company, NSAB, which provides capacity for broadcasting and other television and radio broadcasting, as well as the transmission of data using the SIRIUS satellite system (comprising the Sirius 1 and Tele-X satellites located at 5° east). The SSC is responsible for the operation and procurement of NSAB satellites. It also operates the Esrange sounding rocket range and satellite control and data reception facilities. It acquired a 50 per cent interest in Norway's Tromsø station in 1995: The total number of employees is 315. The 1995 turnover was SKr307 million (1994 – SKr350 million, 1993 – SKr311 million, 1992 – SKr432 million).

SSC offers µg flight services within the Maser (Material Science Experiment Rockets) programme, in addition to design/manufacture of specific experiment modules. By using an improved Black Brant 9, 7 minutes µg can be provided for a 375 kg experiment module. Skylark 7 can provide 6 minutes for 370 kg. Launches are made almost annually. Modules are interchangeable between Maser, Texus and MiniTexus with slight modifications. Maser 1 flew at Esrange 19 March 1987; #2 29 February 1988; #3 10 April 1989; #4 29 March 1990; #5 9 April 1992; #6 4 November 1993 (Skylark 7); #7 (Skylark 7) 3 May 1996, with ESA as the principal customer.

SSC and DASA/ERNO jointly developed the Maxus suborbital vehicle to provide 14 minutes µg.

Telia AB

Telia AB (previously Swedish Telecom), the national signatory to Intelsat, Inmarsat and Eutelsat, operates as a private company but is wholly government-owned, under the Ministry for Transport and Communications. Eutelsat is accessed through the station at Ågesta 10 km south of Stockholm. Telia AB and Telecom Finland International jointly own and operate the Tanum Teleport, on Sweden's west coast, with five Intelsat antennas. Inmarsat communications are routed through Norway's Eik station.

SWITZERLAND

International Space Science Institute

ISSI began operating in 1995 with a nucleus of six to eight staff and 10 to 30 research fellows, research associates and visiting scientists.

TAIWAN

Centre for Space and Remote Sensing Research

CSRSR, part of the National Central University, signed a contract 27 April 1993 with Spot Image and began receiving Spot imagery directly through the new station, 50 km west of Taipei, on the Chung-Li university campus (first image 25 May; contract came into formal effect July).

It signed a contract 15 December 1994 with EOSAT to receive and process Landsat 5 data. It signed an agreement 10 May 1994 with ESA to receive and process ERS SAR data. Annual operating budget is FFr24 million; 48 staff operate it 12 hours daily to generate standard products. The centre's Meteorological Satellite Laboratory receives and processes GMS and NOAA AVHRR images. Other facilities, such as digital image processing, digital photogrammetry and GIS are used to generate value-added products. CSRSR has been a commercial Spot distributor since 1988.

THAILAND

Post and Telegraph Department

Thailand joined Inmarsat in January 1995, through its signatory Post and Telegraph Dept. The first Inmarsat land Earth station is expected to open in 1995 at Nonthaburi.

Thai Remote Sensing Center (TRSC)

The TRSC is a division of the National Research Council and has operated the Thailand Landsat Station (TLS) since November 1981. The facility was upgraded to receive Landsat TM and Spot HRV in 1987 with the assistance of the Canadian International Development Agency, added MOS capability in 1988, and ERS/JERS in 1993.

TONGA

Friendly Islands Satellite Communications (Tongasat)

Tongasat is an agency of the government of Tonga that handles the co-ordination/licensing of Tongan satellite systems. A licence agreement with Rimsat Ltd of the US allocated slots at 130/134/142.5° east for 30 years. Tonga also holds positions at 70/83.3/170.75° east.

TURKEY

Turkish Telekom

Turkey has been a signatory to Intelsat since 1968 and in November 1989 became the 58th member of Inmarsat. The Ata station 37 km south of Ankara became operational in July 1989 to handle Inmarsat A communications; it added Inmarsat C in 1994. COMSAT Corp and IDB both operate Inmarsat A stations in Anatolia for the Indian Ocean region.

UKRAINE

Marlin Yug Ltd

Marlin Yug produces environmental units returning data through the international Argos satellite system. The Berkut balloon unit allows the paths of stratospheric balloons to be tracked. A diving drifter transmits ocean vertical profile parameters and is tracked for surface currents. Ground tests conducted for VEGA and Mars-96 mission.

NPO Khartron/ Elektropribor

With facilities in Kiev, Moscow and Zhaporje, Khartron supplies onboard spacecraft computers, including Mir and Buran and it provides Zenit SL-16 and Cosmos 3 SL-8 avionics.

NPO Musson

Musson handles all maritime and airborne satellite communications terminals, including Inmarsat A maritime.

NPO Yuzhnoye

Based at the Yuzhmash factory, Yuzhnoye is the world's largest integrated missile production facility. NPO Yuzhnoye (formerly OKB-586) is a major launcher, missile and satellite design, development and manufacture bureau. It is responsible for the SL-11/14 Tsyklon and SL-16 Zenit space launchers, the SS-4 Sandal (basis of SL-7), SS-5 Skean (basis of SL-8 Cosmos), SS-9 Scarp (basis of SL-11/14 Tsyklon), SS-18 Satan (now offered as a satellite launcher) and SS-N-5 Sark missiles, the Okean-O/Sich satellites, Tselina global Elints, and about 400 Cosmos and Interkosmos satellites (including those based on the AUOS bus family). It also probably handles the military Rorsats. Employees totalled about 50,000 (including 14,000 on R&D), with 6,000 involved in space work by 1991. Facilities total about 185,000 m^2, including titanium forging and propellant tank fabrication (also undertaken for other organisations).

Yuzhmash was producing automobiles and aircraft at the end of the Second World War. In 1951, Stalin allocated it to Sergei Korolev. In about 1954, Korolev's deputy Mikhail K Yangel (25 October 1911 – 25 October 1971) assumed control for production of the SS-6 Sapwood, which formed the basis of the current SL-4 Soyuz and SL-6 Molniya launchers (later switched to the Central specialised Design Bureau). OKB Yangel was also responsible for the manned lunar lander's propulsion system (led by Boris Gubanov, who later became responsible for the Energia launcher), and proposed the R56 manned lunar launcher. NPO Energia allocated it in 1974. These were developed in parallel as the SL-16 Zenit 2. NPO Yuzhnoye was formed out of OKB Yangel in 1974.

PO Monolith

Monolith builds control systems for rockets.

Ukrainian Space Agency

The agency and commission were established in 1992 to manage national space activities.

UNITED KINGDOM

British Association of Remote Sensing Companies (BARSC)

BARSC was established in 1985 as the trade association of UK remote sensing companies. It aims to ensure that the interests of its members are represented on national, international and government committees and to provide a forum for discussion within UK government departments, such as the British National Space Centre, the Department of the Environment and the Ministry of Defence. For example, BARSC plays a prominent role in the UK's all-party Parliamentary Space Committee. There are 30 member companies. It holds general meetings two to three times annually, in addition to one-day workshops.

British National Space Centre (BNSC)

BNSC, created in 1985, focuses the civil space interests of the Department of Trade and Industry (its parent body), the Ministry of Defence, the Foreign and Commonwealth Office, the Cabinet Office, the Particle Physics and Astronomy Research Council, the Natural Environment Research Council and The Meteorological Office. A staff of 235 operates from the London HQ and technical centres such as the Central Laboratory of the Research Councils.

British Telecommunications (BT)

BT has significant investment in three satellite consortia: INTELSAT, EUTELSAT and Inmarsat. Each consortium was established through an inter-governmental treaty as a co-operative but with a requirement to function under accepted commercial principles. BT is designated by the UK government to act as the signatory to each consortium and as a consequence, BT represents UK interests at consortia management board meetings.

INTELSAT and EUTELSAT were established respectively with a view to providing public telecommunications globally, and within Europe. Inmarsat was set up to serve the maritime community, a mission which was subsequently expanded to include aeronautical and land mobile services.

The UK's investment in the satellite consortia, including BT's, is as follows: EUTELSAT – 22.6 per cent (largest investor), INTELSAT – 9.2 per cent (second largest investor) and Inmarsat – 9.3 per cent (second largest investor).

BT provides a wide range of satellite services, accessing satellite consortia capacity and private satellite systems such as PanAmSat, Orion and Astra SES to provide:

- Public Switched Telecommunications Network Services including voice, data and facsimile.
- Multimedia including Turbo-Internet.
- Global mobile satellite services including the Skyphone aeronautical system, maritime communications and land mobile services. BT recently introduced the 'Mobiq' terminal which is smaller than a laptop PC and can provide global direct-dial phone fax and data.
- Entertainment TV and radio. Contribution and distribution TV services as well as CD quality radio.
- Satellite newsgathering and special events coverage.
- In the UK, BT operates earth stations at Goonhilly, Madley, London Teleport and Aberdeen. BT also operates earth station facilities in several European countries providing uplink services for TV broadcasters. BT Broadcast Services operates a fleet of transportable earth stations, which provide uplink services for special event coverage (for example, major sporting events). A BT earth station was scheduled to open in New Zealand in spring 1997, providing a range of BT Inmarsat services via the Inmarsat satellite system.

Central Laboratory of the Research Councils

The Central Laboratory is the focus for the UK's space science projects and directs both ESA and bilateral activities, such as the current Spectrum-X astronomical satellite collaboration with Russia. It was created 1 April 1995 as an independent entity from the merged Daresbury and Rutherford Appleton Laboratories, reporting to the Office of Science and Technology. Its space funding is from PPARC.

DERA – Space Department

DERA is a government-owned UK research and development organisation providing independent services to civil and military organisations and industries world-wide. DERA draws upon well-equipped test and production facilities, and state-of-the-art satellite control and ground receiving stations. DERA is involved in most space activities ranging from concept and feasibility studies through to project management, design and hardware construction. Space Department also supports the satellite communications work of the Strategic Communications and Networks Department at DERA Defford.

Activities include analysis of future satellite and ground system requirements, procurement support, post launch mission analysis and resolution of spacecraft operating anomalies. Experience in the UK Skynet military communications programme is used to support overseas customers in the initial studies and procurement of new civil and military satellite systems.

In the last two decades DERA has had a responsibility to develop remote sensing applications. Capabilities include SAR and visible/infra-red instrument design, and research into the exploitation of satellite remote sensing data for both civil and military use. Two ground stations for operational reception of meteorological data and high resolution radar data and near real-time delivery of application-specific products support the activities. DERA teams with industry to promote and develop commercially self-sustaining satellite remote sensing markets. Examples of collaboration includes Radarsolutions, a consortium to receive and distribute RADARSAT data across the British Isles and Denmark.

DERA's research, development and consultancy is offered on all aspects of space technology, including spacecraft subsystems, attitude and orbit control systems, power systems, onboard data handling, materials and structures, thermal modelling and space debris. In addition, DERA is an acknowledged leader in the field of space environmental effects and ion propulsion. DERA has worked on Ion Thruster technology for more than 20 years, leading the development of the T5 Thruster which is employed in the UK10 Ion Propulsion system.

DERA also maintains a comprehensive range of test facilities available to industry and academia for qualification testing of space hardware, instruments, components and satellites.

DERA runs a micro-satellite development programme offering a complete service to organisations wishing to take advantage of the comprehensive DERA space capability. This includes complete systems, satellite design and predictive modelling, comprehensive testing, manufacture and the complete provision of ground segment facilities.

Facilities

West Freugh satellite ground station
Acquisition of radar data
ERS-1,2; RADARSAT
Reception downlink at X band

Lasham satellite ground station
Acquisition of meteorological data:
NOAA, Meteosat, GOES
COSPAS/SARSAT data dissemination
L&S band TT&C

Image Data Facility
Image and data transcription facilities

Space Test Facility
Thermal vacuum chambers
Solar simulator
Vibration testing
Mass properties facility
Solar cell measurement and testing.

Earth Observation Data Centre (EODC)

EODC archives, processes and distributes data and products from ERS and other remote sensing satellites. One of four ESA off-line ERS Processing and Archiving Facilities (PAFs) was established at EODC. National Remote Sensing Centre Ltd manages its operations under contract to ESA.

Ministry of Defence, Meteorological Office

An Executive Agency within the Ministry of Defence, the department undertakes R&D into the use of satellite instruments for meteorological studies and forecasting of weather systems. The work falls into two principal categories: instrument development and imagery. The department built the SSU Stratospheric Sounding Unit currently flying on NOAA, the Focal Plane Assembly for ERS' Along– Track Scanning Radiometer and the AMSU-B humidity sounders for NOAA K-M.

Meteorologically useful products are derived from AVHRR and Meteosat data for research and operational purposes. TOVS sounding data have been used for many years and R&D work is under way for ATOVS. Images are employed for studies of atmospheric systems and for short-period weather forecasting.

Remote sensing consultancy services are offered through Metstar Consultants, a Met Office business unit. Instrument thermal vacuum calibration is undertaken for industry and space agencies. Testing is planned of MMS, MIMR, MWR; SSU, AMSU-B, MIR, MARSS, AMSU-A have been tested.

Particle Physics and Astronomy Research Centre (PPARC)

PPARC is the primary funding body for space science in the UK (mainly conducted in the universities and CCLRC), the second largest partner in BNSC and provider of much of the UK's subscription to ESA's science budget. FY95 space science budget was £51.7 million, including £38.5 million to ESA's science budget. FY95 total budget was £196.37 million; FY94 was £187.4 million. A constant £196.37 million was planned for each of FY96-FY98. As a result of 1993's White Paper on Science and Technology, the SERC Science and Engineering Research Council, created in 1965, was replaced 1 April 1994 by PPARC and the EPRSC Engineering and Physical Sciences Research Council (EPRSC has no space interests). SERC's Earth observation interests transferred to NERC.

Queen Mary and Westfield College

QMW's Compact Antenna Test Range was opened in 1991 to measure the radiation characteristics of spacecraft antennas up to 200 GHz; it is Europe's second such facility. Single offset antenna 5.4 m focal length, 8-20, 26-40, 90, 150 and 180 GHz; test antenna mounted in class 10,000 tent on rail-mounted turntable, capacity 1 tonne, 360° azimuth motion, ±90° elevation. Test and measurement equipment includes HP Series 9000 model 310 desktop computer and model 375 UNIX workstation connected to an Ethernet LAN for fast pattern processing and efficient file transfer; receiver SA 1771 or HP 8510B. CATR was designed specifically for RF testing with instrument contractor Matra Marconi Space of the UK's AMSU-B Advanced Microwave Sounding Unit contribution to NOAA's Tiros-Next series. Testing of an

EM and three FMs was successfully completed summer 1993.

A new far-field mm-wave anechoic chamber for feed measurements became operational in early 1995. The receiving system is based on a LHe-cooled InSb detector providing 100 to 600 GHz coverage. The CATR and QMW's 6 m far-field Anechoic Chamber are available for commercial use.

Royal Aircraft Establishment, Culham

The RAE (now DRA) provides: programme technical lead, thruster design/manufacture, testing, electronics and hollow cathode development. Culham: thruster testing (including diagnostic life tests), plasma physics, ion beam extraction modelling, UK-25 development/testing. Matra Marconi Space (MMS): development of the UK-10 system for Artemis, including the power conditioning and control equipment and the propellant supply/monitoring equipment and the manufacture and qualification of the flight hardware for Artemis. An element of AEA Technology, Culham houses Europe's largest Xe propellant ion thruster test chamber: 5.8 m long, 1.3 m diameter, equipped with helium cryopumps for handling the inert gas. During 1996, MMS commissioned a new test facility, capable of handling two 25 mN thrusters simultaneously.

Ion thruster development in the UK began in the late 1960s at RAE/Culham, concentrating on a 10 cm diameter thruster with a nominal 10 mN thrust and mercury propellant. The T4A engineering model achieved very high efficiency, stable operation and a long operational lifetime. The propellant was changed to Xe in 1985 and new versions of the T5 flight model thruster have been manufactured/tested in collaboration between DRA, Culham and Matra Marconi Space. 0.2-70 mN thrust has been demonstrated; qualification testing at 18 mN is under way at MMS and DRA, specifically for operational NSSK application on ESA's Artemis. This will be extended to 25 mN during 1997.

A major experimental programme was completed in 1995 at the Aerospace Corp in Los Angeles, using USAF funding and a thruster supplied by DRA. This concentrated on a detailed characterisation of the ion beam. A lower level effort continues at Aerospace funded by DRA, MMS and the USAF.

The thruster comprises a cylindrical discharge chamber closed at one end by a soft iron backplate and with a set of closely-spaced grids at the other. A magnetic field is generated by six equispaced peripheral solenoids. The field lines inside the chamber link an inner cylindrical soft iron pole and a larger diameter outer pole. Propellant gas is introduced through the axial hollow cathode and a bypass distributor on the backplate. A DC discharge is set up between the cathode and the cylindrical anode. This ionises the gas, the efficiency being enhanced by the magnetic field and the correct design of the inner pole/baffle disc arrangement. The positive ions are extracted and accelerated by a high electric field between the grids, attaining typically 30 to 60 km/s. A triple grid design is used to minimise damage caused by the impact of charge-exchange ions. Electrons from an external cathode neutralise the ion beam's charge, and an earthed screen to prevent those electrons from reaching other parts of the device surrounds the whole thruster.

MMS UK is responsible for the electronics and propellant feed systems, and commercial exploitation of the fully qualified operational system. Philips Components Ltd previously manufactured the hollow cathode and neutraliser; the work has now been transferred to DRA. DRA/Culham in 1994 produced a low cost version of the complete system for experimental applications. This was accepted for flight on The Johns Hopkins University's NEPSTP Nuclear Electric Propulsion Space Test Program, since cancelled, which was intended to test the Russian Topaz 2 nuclear reactor. Other flight opportunities are being sought.

ESA's Artemis will carry two 18 mN versions. Another application may be ESA's Gravity Explorer Mission; DRA/Dornier studied the possibility in 1995/6 with encouraging results. DRA's STRV-1A launched in June 1994, included an experiment to allow the hollow cathode assembly to demonstrate spacecraft electrostatic discharging. Although the cathode assembly operated correctly, charging was not observed. The nominal operating parameters given below are in 18/25 mN order.

Design of a laboratory model 25 cm thruster began in early 1986 using scaling laws formulated during the UK-10 programme. Culham testing began October 1986; nominal thrust was 200 mN (highest obtained was 316 mN). Following the successful conclusion of the lab model work, an engineering model was manufactured at the beginning of 1989 and completed a comprehensive test/evaluation programme in 1992. Included were studies of life-limiting factors. Investigation of other propellants was completed in 1995 at Southampton University (although an ideal inert propellant, xenon is costly). Development of a power conditioning system was undertaken at Birmingham University and high current hollow cathodes were developed by DRA/Philips Components. Further work on the ESA-XX hybrid design, using features from this thruster and the German RIT-35, is under way with ESA funding. The first prototype of this thruster, intended for interplanetary missions, has been successfully tested at the University of Giessen in Germany.

Satellite Navigation Group

The Satellite Navigation Group (previously UKCSG) was established by RIN in 1988, and represents UK interests on the US Civil GPS Interface Committee. Membership is 250 corporate/individual members. Services include the organisation of meetings and conferences and the publication of a newsletter. It also acts as a UK node for dissemination of material from the US Coast Guard Navigation Information Service.

University of Dundee Centre for Coastal Zone Research

The centre was established in 1995 jointly by Dundee University's Dept of Applied Physics and Electronic and Mechanical Engineering and Cray's Space and Defense Division. Projected revenue after four years is about £6 million. The main focus is research that directly involves operational end-users of information services, and emphasises the integration of satellite data with information derived from GIS and models of the physical, chemical and biological processes in the coastal regions. Cray will lead development of the operational systems arising from the research.

The initial programme includes work analysing methods of automatically detecting and classifying algal blooms and monitoring coastal ships. Future activities will cover the potential users of satellite data for flood monitoring and related disaster monitoring applications, and methods of using SAR interferometry for monitoring small scale changes in coastal defence systems.

University of Dundee, Satellite Station

University of Dundee operates the Satellite Station, a NOAA HRPT data acquisition, archiving and distribution service funded by the Natural Environment Research Council. Data set consists of daily recordings from NOAA since October 1978. More recently, the Station was upgraded to take SeaWiFS data from SeaStar. specialised ground stations are manufactured for use with GMS S-VISSR, NOAA HRPT and Meteosat PDUS/WEFAX/MDD/DCP.

UNITED STATES OF AMERICA

Center for Commercial Space Communications

CCSC was established in 1991 by Virginia Tech, partly under initial funding from the state's Center for Innovative Technology, as an interdisciplinary university centre to develop, manage and undertake research in commercial aspects of space communications. LEO satellites are of particular interest. CCSC is associated with Virginia Tech's Satellite Communications Group, which has operated for two decades and received more than US$6 million in federal/corporate funding for space communications research. It has designed/built receivers for NASA's ACTS programme.

Center for Satellite and Hybrid Communication Networks

The Center at the University of Maryland Institute for Systems Research was established in October 1991 as one of NASA's Centers for the Commercial Development of Space. Its primary goal is development and commercialisation of advanced space-based communications, emphasising hybrid satellite/terrestrial networks. Specialist areas cover teleconferencing, DBS, HDTV, database access, supercomputing linkage, satellite digital cellular networks, narrowband mobile networks and personal communications.

Civil GPS Services

NIS (until 1995 the GPS Information Center) is operated by the US Coast Guard within the Dept of Transportation as part of the Civil GPS Service to provide basic information and as a point of contact for the civil GPS user. It has a manned watch 24 hours a day. The Bulletin Board is a constantly updated computer service that can be accessed via a modem. The Internet service has been available since July 1995. The other three CGS elements are: CGS Interface Committee (identifies civil GPS technical needs); Differential GPS (operational in 1996, providing 10 m accuracy to US maritime users); PPS Program Office (under formation, allowing qualified civil users access to the PPS signal). NIS gathers, processes and disseminates timely GPS, DGPS, Omega and Loran-C status and other navigation information to users.

Communications Satellite Corporation (COMSAT)

COMSAT is the US signatory to Intelsat (19.1 per cent contribution) and Inmarsat (23.0 per cent). Units include COMSAT General, operators of Comstar satellites, and COMSAT Labs. COMSAT Corp was created in 1963 following passage of the Communications Satellite Act, signed into law by President Kennedy in late 1962. COMSAT acted as technical manager for Intelsat until 1979, forming COMSAT Laboratories in 1967 for technology R&D. COMSAT General was established in 1973 to branch out into domestic satellite communication.

The shareholder-owned Communications Satellite Corp was established in February 1963 to carry out a US government mandate for a global commercial satellite communication system in co-operation with other countries.

The World Systems division provides satellite communication services to US international carriers and other customers through Intelsat. COMSAT Mobile Communications provides satellite and land Earth station services for shipping, offshore oil platforms and international land mobile users through Inmarsat. As part of the Skyways Alliance, it is providing aeronautical communications through the Inmarsat system.

The COMSAT General Corporation subsidiary was established in 1973 as a satellite operator: it manages the five satellites of the Marisat, Comstar and SBS 2 networks. COMSAT Video Enterprises owns and operates a satellite-based network distributing entertainment and videoconference services to the US lodging industry. COMSAT Laboratories perform communications R&D for its parent and other organisations, such as advanced technologies for NASA's ACTS, low-cost antennas, and advanced transmission systems. It developed ACTS' NASA ground station and the master control station. The Laboratory employs some 320 of COMSAT's 1,500 personnel. Radiation Systems Inc was acquired in 1994 and merged with COMSAT Technology Services to create COMSAT RSI, able to provide more than 80 per cent of complete satellite communication ground systems.

The Corporation operates Earth stations at Santa Paula, California and Southbury, Connecticut, each with two 12.8 m dishes and a 10.4 m non-tracking TT&C antenna. 1996 revenues were US$1.0 billion, returning US$8.6 million in net income. 1995 figures were US$862.9 million/US$37.8 million (strategic restructuring cost US$22 million). 1994 US$826.9/77.6 million; 1993 US$754.3/84.4 million; 1992 US$564/42.9 million; 1991 US$523/44.8 million; 1990 US$457 million (US$16.3 million loss).

Department of Commerce, Office of Space Commerce

The Office of Space Commerce was established 2 December 1988 to co-ordinate space-related issues and programmes within the Department. OSC's Director is responsible for advising the Secretary on all space issues.

Department of Defense, Defense Advanced Research Projects Agency (DARPA)

DARPA was established in February 1958, partially as a response to Sputnik, and operates as the Dept of Defense's central R&D organisation for maintaining technical development and proof of concept demonstrations, pursuing them through other DoD and US government agencies. For example, it funded development of the Pegasus and Taurus launch vehicles, several small satellites and numerous advanced technologies for satellite subsystems and payloads.

However, DARPA in 1994 withdrew from space activities. The Office of Director is supported by a number of Special Assistants, five management offices and six programme offices: Defense Sciences; Electronic Technology; Information Systems; Information Technology; Sensor Technology; and Tactical Technology.

Department of Defense, Defense Information Systems Agency (DISA)

DISA is a Department of Defense (DoD) combat support agency whose main objective is to anticipate and respond to the needs of its customers, the warfighters, by providing them with seamless, end-to-end, innovative and integrated information services, which provide a fused picture of the battlefield. It is responsible for planning, developing and supporting command, control, communications, computers and intelligence (C^4I) and information systems that serve the needs of the National Command Authorities (NCA) under all conditions of peace and war. It provides guidance and support on technical and operational C^3 and information systems issues and coordinates DOD planning and policy for the integration of C^4I systems and the insertion of C^4I for the Warrior (C^4IFTW) leading edge technologies into the Defense Information Infrastructure (DII).

Department of Defense, National Imaging and Mapping Agency (NIMA)

The NIMA National Imaging and Mapping Agency was created 1 October 1996 to co-ordinate imagery collection, processing, exploitation, analysis and distribution, absorbing NPIC, CIO and the Defense Mapping Agency. NRO underwent internal reorganisation in 1992, apparently opting for multi-agency staffing of its various branches so that, for example, the CIA and USAF no longer compete on imaging satellite projects.

Department of Defense, National Reconnaissance Office

NRO was formally established 25 August 1960, although its existence was not officially acknowledged by DoD until 18 September 1992, to co-ordinate overhead reconnaissance operations, including those from satellites. A US$5 billion to US$7 billion annual budget has been estimated, although outgoing CIA Director James Woolsey in January 1995 noted to a Senate committee that a radical restructuring would cut the number of future intelligence satellites by almost half. About 4,000 staff. The US$100 million HQ was completed in 1996 near Dulles airport; NRO was previously housed in the Pentagon.

It receives its budget through the National Reconnaissance Program portion of the National Foreign Intelligence Program. The office reports to the Secretary of Defense and the Director of Central Intelligence controls allocation of NRO assets. Until 1992, the inter-agency Comirex Committee on Imagery Requirements and Exploitation met daily to review requests and direct targeting. That role was assumed by the DoD Central Imagery Office, created in June 1992. Imagery is relayed to Fort Belvoir in Virginia, from where it is passed to the National Photographic Interpretation Centre in Washington DC (established 1961 and run by the CIA for all the intelligence community) and other users.

Previous NRO Directors were: Joseph V Charyk (6 September 1961 -1 March 1963), Brockway McMillan (1 March 1963 - 1 October 1965), Alexander H Flax (1 October 1965 - 11 March 1969), John L McLucas (17 March 1969 - 20 December 1973), James W Plummer (21 December 1973 - 28 June 1976), Thomas C Reed (9 August 1976 - 7 April 1977), Hans Mark (3 August 1977 - 8 October 1979), Robert J Hermann (8 October 1979 - 2 August 1981), Edward C Aldridge Jr. (3 August 1981 - 16 December 1988), Martin C Faga (26 September 1989 - 5 March 1993), Jimmie Hill (acting Director), Jeffrey K Harris (19 May 1994 February 1996).

Department of Defense, Office of Space Acquisition and Technology

Military space policy and acquisition responsibilities were consolidated in December 1994 into a single new 'space architect office': Space Acquisition and Technology Programs (SA&TP), headed by a Deputy Defense Undersecretary (DUSD-Space, Robert Davis). Air Force Major Gen Robert Dickman was appointed as the Space Architect 15 September 1995. The new Office has authority over all military space activities and organisations, with a goal of integrating space capabilities into a unified approach. The Joint Space Management Board was activated December 1995 to ensure an integrated approach for the first time between DoD/NRO. JSMB is spending 1996 creating the National Strategic Space Master Plan to cover US$130 billion of spending over 10 years.

Military activities were previously drawn together in 1985 by the activation of a unified US Space Command at Peterson AFB in Colorado. Collocated are the USAF Space Command and US Army Space Command; the Naval Space Command is sited separately in Dahlgren, Virginia. Space Command provides missile and satellite tracking information, operational launch services at Canaveral and Vandenberg and command/control of operational DoD satellites. Management, operation, TT&C and execution of R&D space systems is allocated to the AF Space and Missiles Systems Center in California.

Department of Energy, Office of Space and Defense Power Systems

DoE is developing the radioisotope thermoelectric generators (RTGs) for the Cassini missions (US$45 million FY95). Funding cuts mean that it might be the last unless NASA or other agency provides support; FY96's request excluded all space reactor work. DoE's Office of Energy Research is supporting the Alpha Magnetic Spectrometer for Space Station. The Office of Intelligence and National Security also provides space sensors for treaty verification, nuclear proliferation detection and environmental protection; it funded the Alexis project. See the Los Alamos National Laboratory entry in the World Space Centres section.

Department of the Navy, Office of Naval Research

NRL was established in July 1923 as the Naval Experimental and Research Laboratory. Now an element of ONR, it is the navy's corporate laboratory. Staff total about 3,500, mostly civilian, in Washington DC, Stennis Space Center (Bay St Louis, Missouri), Monterey (California) and Orlando (Florida). NRL's Naval Center for Space Technology (NCST) maintains space technology expertise through two departments: Space Systems Development and Spacecraft Engineering. The Remote Sensing Division (RSD) and the Space Science Division (SSD) come under NRL's Ocean and Atmospheric Science and Technology Directorate.

NRL directed the Vanguard programme and since 1960 has developed more than 80 satellites, most recently the Clementine technology demonstrator lunar/asteroid probe for BMDO. Clementine 2 is now in the design phase for a 1998 launch to three asteroids. It has flown numerous payloads and experiments, such as the Hercules unit carried by STS-53 in December 1992 to allow astronauts to photograph surface features while automatically recording positions within 2 km. It flew again on STS56 in April 1993 and on STS-70 in July 1995; no other flights are yet manifested. The HTSSE High Temperature Superconducting Space Experiment is aboard STP's Argos satellite.

NCST facilities include cleanrooms, anechoic RF chambers, shock/vibration chambers, an acoustic reverberation chamber, thermal/vacuum chambers and long term testing of satellite clock time/frequency standards.

RSD undertakes a broad programme in sensing applications over frequencies from ultra-violet to radio. Sensor systems include RAR Real Aperture Radar, scatterometers, lidars, optical/radio interferometers and passive microwave imagers. Its middle atmosphere sensors include the POAM Polar Ozone and Aerosol Monitor on Spot 3, the MAS Millimetre-wave Atmospheric Sounder on Shuttle Atlas missions (three flown), and a suite of water vapour and ozone monitors as part of the Network for Detection of Stratospheric Change. RSD's facilities include digital image processing, a tactical environmental visualisation centre, an aerosol measurement facility, 25.6/25.9 m radio antennas at Maryland Point Observatory and an optical interferometer at Mount Wilson Observatory for monitoring background environmental emissions at high angular resolution.

SSD maintains facilities for designing, constructing, assembling, calibrating and analyzing space experiments, principally upper atmosphere UV sensing, solar atmosphere spectrometry and celestial radiation over UV to cosmic rays. For example, it provided the SUSIM Solar UV Spectral Irradiance Monitor for UARS and Atlas, the RAIDS Remote Atmospheric and Ionospheric Detection System to measure airglow, the OSSE Oriented Scintillation Spectrometer Experiment gamma instrument on NASA's Compton Observatory, solar x-ray measurements using a Bragg crystal on Japan's Yohkoh, and the LASCO Large Angle Spectrometric Coronograph on Soho. SSD provided three of the main instruments (USA, HIRAAS, GIMI) on Argos.

Department of Transportation, Office of Space Transportation

The Commercial Space Launch Act of 1984 and Executive Order 12465 gave DoT the authority to regulate US commercial ELV activities. Orbital/suborbital firings are listed in OCST's quarterly Commercial Launch Manifest. The first licensed orbital launch was of the BSB satellite in August 1989 by Delta. Licenses are granted based on launch providers presenting evidence they are in compliance with all safety regulations and other requirements, and have sufficient insurance or financial resources to cover any probable losses from a launch mishap. The office is also responsible for the regulation of any future commercial launch sites and to oversee commercial re-entry vehicles. It is also the compliance monitor for the Chinese and Russian commercial launch agreements.

The COMSTAC Commercial Space Transportation's Advisory Committee provides policy guidance. OCST, previously reporting directly to the Office of the Secretary, transferred intact 16 November 1995 to the Federal Aviation Administration. 28 staff. FY96 budget request US$6.5 million; FY95 budget was US$6.0 million.

Federal Communications Commission (FCC)

The FCC is the regulatory body for interstate and international communications by radio, TV, wire, satellite and cable. Satellite proposals are submitted to the commission for licensing consideration. The International Bureau has oversight of all satellite services. Five Commissioners, appointed for five-year terms direct the FCC. There are six operating bureaus: Mass Media, Common Carrier, Wireless Telecommunications, Compliance and Information, International, and Cable Services.

Florida Spaceport Authority

Created in 1989 by Florida's Governor and Legislature, and governed by a board of supervisors, the Spaceport Authority aims to foster development of Florida's space enterprise, including industry, research and education. The Authority is authorised to issue industrial development bonds and its legislation provides statewide tax incentives for space business, and tax exemptions for launch vehicles, payloads and propellants. It has financed the development of the US$30 million Apollo-Saturn 5 Center museum for NASA, and built a US$35 million 6,094 m² Titan 4 solid rocket motor warehouse/support complex for the air force in northeast Florida, both completed in 1996.

Canaveral's Launch Complexes 46 and 20 are being modified by the Authority for suborbital rockets and small orbital vehicles beginning 1997, including AF Minuteman and those employing Thiokol's Castor 120 solid motor, such as Taurus and LMLV.

Geosat Committee

The Committee was formed in 1976 as a non-profit educational organisation dedicated to the civil use of remote sensing technology for geological and environmental applications. Geosat has more than 40 corporate members and 16 alumni/consultant chapter members, including Spot Image, MacDonald Dettwiler, EOSAT, Eurimage, ARCO, Earthsat, Amoco and Texaco. Research projects have been conducted with NASA, US Geological Survey, Australia, France, Germany, Japan and UK.

National Aeronautics and Space Administration, Ames Research Center

Ames was founded in 1939 as an aircraft research laboratory and became part of NASA in 1958. Its major research activities focus on computational and experimental aerodynamics, hypersonic aircraft, aeronautical and space human factors, life sciences, Earth environment, space science, solar system exploration and IR astronomy. Personnel total about 3,400, about half contractor employees. Ames manages Pioneer 10/11 and was responsible for Galileo's Jupiter atmosphere entry probe. It also leads NASA's life sciences co-operation on the international Russian Biocosmos missions. It headed the agency's SETI programme until Congress terminated activities in late 1993. FY96 budget is about US$625 million, FY95 US$624.7 million, FY94 US$679.9 million, FY93 US$686 million, FY92 US$646 million, FY91 US$608 million, FY90 US$559 million.

National Aeronautics and Space Administration, Dryden Flight Research Center

DFRF merged in 1981 with NASA Ames but again became a fully independent centre 1 October 1994. FY97 budget was US$146 million. Some 500 civil service and 500 contractors are employed. It is located 130 km north of Los Angeles on the edge of the 114 km² Rogers Dry Lake at the south end of an 800 km high-speed flight corridor at the USAF Edwards AFB. It has 51,800 km² of restricted airspace to conduct aeronautical research with high performance aircraft such as the SR-71, F- 16XL and the X-36 tail-less fighter.

Dryden is involved in the development of the X-33 and X-34 reusable launchers. The Apollo Lunar Landing Research Vehicle and Lifting Body family operated out of Dryden in the 1960s-70s. The 8 to17 km runways on the dry lakebed and Edwards' main 4,600 m concrete runway are now used as Shuttle landing sites when Cape Canaveral conditions are unacceptable, additionally supporting the ferry flights to NASA's Florida launch site.

The OSC/Hercules Pegasus small orbital launch missions were based initially at the site, released over the Pacific from the same USAF/Boeing B-52 employed for X15 hypersonic research flights up to 1968.

National Aeronautics and Space Administration, Goddard Space Flight Center

GSFC was established in May 1959 around a core of 157 transferred from NRL's Vanguard team. It is the only US national facility that can develop, fabricate, test, launch and analyse data from its own space science missions. FY98 budget is US$3,472 million; for FY97 it was US$3,495 million; FY93-96 budgets each about US$2.5 billion; FY92 US$2,459 million; FY91 US$2,379 million; FY90 US$1,849 million. Total workforce is 11,752, including 3,415 civil servants and 8,337 contract personnel, planned to be 7,000 by 2000. Goddard has managed the development of 160 satellites for NASA and NOAA, including the Explorer series, COBE, Compton Observatory, Solar Max, ERBS, Spartan and UARS. Hubble Space Telescope science operations are controlled by the ST ScI at The Johns Hopkins University. GSFC directs NASA's Delta launch activities.

Goddard directs the operation of NASA's Spaceflight Tracking and Data Network. STDN's Space Network (SN) segment can provide 85 per cent coverage for satellites orbiting up to 1,200 km using six TDRS Tracking and Data Relay Satellites. Each TDRS can simultaneously handle multiple users at up to 300 Mbit/s. A network of fully automated computer-controlled equipment and 18.3 m ground antennas controls the TDRS constellation under Goddard's Network Control Center. The White Sands Complex in New Mexico comprises two terminals, each responsible for three TDRS. STDN's Ground Network (GN) comprises three facilities with S-band (9 m) and C- band tracking systems. Located at Ponce de Leon and Merritt Island (Florida) and Bermuda Island (UK), they are used primarily by GSFC to support all Kennedy Space Center launch operations, including Shuttle and expendables. Three further tracking stations in Canberra (Australia), Goldstone (California) and Madrid (Spain) are operated separately by JPL. Although closely aligned with Goddard's STDN for supplementing Shuttle and selected Earth satellites, these Deep Space Network (DSN) stations are principally dedicated to JPL managed projects such as Galileo and Voyager.

The 8,082 m² US$16 million Spacecraft Systems Development and Integration Facility opened in June 1990 for handling Shuttle payloads of up to 27 tonnes. The 1,161 m² class 10,000 laminar flow cleanroom is one of the world's largest and can accommodate two full size Shuttle payloads (up to 27,216 kg, 4.572 m diameter) simultaneously. Other facilities include: Large Area Pulsed Solar Simulator (test complete panels), Space Simulation Test Facility, Vibration Facility, Battery Test Facility, High Voltage Test Facility, Magnetic Field Component Test Facility (calibrate/align magnetometers), Spacecraft Magnetic Test Facility (entire craft and sounding rockets), High Capacity Centrifuge, Acoustic Facility, Electromagnetic Interference Facility, Static/Dynamic Balance Facility, Optical Thin Film Deposition Facility, Material Properties and Analysis Laboratories.

NASA's Goddard runs the Getaway Special Program (GAS) which accommodates small, autonomous payloads in Shuttle's cargo bay utilising standardised hardware. Hitchhiker offers two carriers providing capabilities beyond GAS. Hitchhiker Junior entered service 1995, using HH avionics and GAS cans but offering satellite ejection (previously from GAS) and limited Orbiter power and services such as pointing.

The canister is an aluminium cylinder provided with a standard experiment mounting plate which, while it may not be altered, does carry provisions for a variety of attachments. GAS can be evacuated and/or pressurised and includes an insulated exterior on the bottom/sides for passive thermal control (an insulated top end cap is available). Operations are independent of the Shuttle Orbiter other than three on/off controls activated by the crew. The experimenter is responsible for providing electrical power, heating/cooling and data acquisition systems. A Motorised Door Assembly is available (satellite ejection no longer is).

A GAS Bridge capable of holding 12 canisters was first flown aboard 61C in January 1986; second flight was STS-40 in June 1991. Payload access is not available following installation in the Orbiter at KSC some two to three months before launch. NASA classifies payloads within three categories: educational (class 1), foreign/commercial (2) and government (3). Within each group, payloads are flown on a first-come-first-served basis. The agency must be provided with detailed information no later than 18 months before proposed launch to ensure adherence to safety regulations. Shipping to KSC is required three to four months before launch. 53 GAS payloads had been flown on the first 24 Shuttle missions before *Challenger's* loss in January 1986, including 13 on the 61C mission alone. The next mission was not until 1991's STS-40, carrying the GAS Bridge with 12 cans. STS- 42 added 10 in January 1992; STS-45 flew #78 March 1992, STS-47 #79-87 September 1992, STS-57 #88-97 June 1993, STS-60 #98 101 February 1994, STS-59 #102-104, STS-64 September 1994 (10 GAS on Bridge), STS 68 September 1994 (3 GAS), STS-67 March 1995 (2 GAS), STS-69 September 1995 (4 GAS), STS-72 January 1996 (5 GAS), STS-76 March 1996 (1 GAS), STS-77 May 1996 (12 GAS on Bridge); STS-80 November 1996 (1 GAS); STS-85 August 1997 (2 GAS); STS-87 November 1997 (1 GAS); STS-89 January 1998 (4 GAS); STS-90 April 1998 (3 GAS); STS-91 June 1998 (8 GAS); STS-95 October 1998 (1 GAS); fewer opportunities are now available as the Space Station era begins.

The cost is US$3,000, US$5,000, US$10,000 US Educational; US$8,000, US$14,000, US$27,000 others; for the three size categories listed below; US$500 non-refundable earnest money required.

The Hitchhiker HH-G/HH-M comprise mechanical mounting structures and an avionics unit connecting to the Orbiter's electrical systems, providing up to six payloads with power, real-time telemetry and crew/ground commands. The canisters are mechanically very similar to GAS cans. HH-G is a side-mount version carrying up to three canisters or 63.5 × 99 cm vertical mounting plates holding up to 272 kg in total. It is normally installed in KSC's OPF at the forward starboard side. The HHM bridge is installed either in the OPF or on the pad, and can accommodate 907 kg in up to 11 canisters or 71 × 71 cm plates (six on each side, one reserved for avionics); four 71 × 91.4 cm top plates are also available. Both HH versions offer two 0.142 m³ 50.2 cm diameter 71 cm high canister types: sealed with dry N_2 or air, 90.7 kg capacity; with a Hitchhiker Motorized Door Assembly, 77.1 kg capacity. Customer hardware must be delivered about 6 months before flight; launch occurs typically 8 weeks after Orbiter integration. Cost is $1.108 million at FY90 rates. First HH mission was HH-G1 on STS-61C in January 1986; #2 (STP 1) STS-39 April 1991; #3 STS-52 October 1992; #4 STS-53 December 1992; #5 (DXS) STS- 54 January 1993; #6 STS-60 February 1994; #7 STS-69 September 1995; #8 STS-85 August 1997; #9 STS-95 October 1998. 36 Hitchhiker experiments have flown to date.

Another programme Hitchhiker Junior (HH-J) was introduced in 1995 as an expanded GAS. Up to 100 W can be provided from the Orbiter, totalling up to 4 per mission. The HH avionics report T/P, battery voltage/current, door status and commanded relay status to a laptop computer in the cabin. Real-time ground command or data are not provided. HH-J can eject a satellite up to 68 kg 48 cm diameter at 0.6-1.2 m/s from a 23.8 cm diameter marmon plate interface.

National Aeronautics and Space Administration, Headquarters

NASA was formally established 1 October 1958 to plan and execute the US civil space programme. It comprises seven principal offices and about a dozen major centres and facilities employing 22,400 civil servants (goal is 17,000 by 1999), including 1,430 headquarters staff (700 goal by October 1997). FY96's budget is US$13.8 billion. The White House in January 1995 directed NASA to cut the budget to US$12.6 billion (FY94 rate) by 2000, saving US$5 billion over those years. President Clinton's March 1996 FY97 request plans were to take out another US$3.3 billion, leaving NASA with only US$11.6 billion by FY2000.

Headquarters offices are:

NASA Office of Space Flight which is responsible for development of Space Station and large propulsion systems; operation of Shuttle, Spacelab and Station through its major Kennedy (as the US manned launch site), Marshall, Johnson and Stennis centres.

Office of Space Science, which conducts, unmanned space activities directed at planetary and astronomical investigations and the science elements of manned missions, through the Jet Propulsion Laboratory and Goddard centres.

Office of Aeronautics exercising aeronautical research and development programmes and directs the Ames, Langley and Lewis research centres.

Office of Space Communications which is responsible for all launcher, satellite and deep space probe tracking, and NASA's communications/data systems.

Office of Life and Microgravity Sciences and Applications, newly formed in 1993 to strengthen this area of research for Space Station and Shuttle.

Office of Mission to Planet Earth (MTPE), also formed in 1993 encompassing missions such as Earth Observing System, Earth Probes and UARS to study changes in the global environment.

Office of Space Access and Technology, pursues innovative technologies in partnership with industry for future space activities, such as reusable launchers.

National Aeronautics and Space Administration, Jet Propulsion Laboratory

JPL is a government-owned facility operated by the California Institute of Technology under contract since 1959 to NASA. The space agency provides about 90 per cent of the budget, with the remainder from the departments of Defense and Energy. Work force totals 5,800, expected to fall to 4,500 by 2000, mostly from the reduction in Galileo and Cassini activities. FY95 total operating budget was US$1.25 billion; FY94 US$1.1 billion. JPL is responsible for most of NASA's deep space missions, including Voyager, Galileo, Cassini, Mars Surveyor and Mars Pathfinder. It manages NASA's portion of the joint Ulysses solar probe, the US/French Topex/Poseidon oceanographic satellite project, the SIR Spaceborne Imaging Radar (Shuttle flights 1994) and developed the Wide Field/Planetary Camera for Hubble.

JPL developed and manages NASA's worldwide Deep Space Network for providing links with spacecraft above 10,000 km. DSN comprises three complexes near Canberra (Australia), Goldstone (California, opened 1958) and Madrid (Spain). There is also a launch support facility at NASA Kennedy, and network control and spacecraft compatibility test facilities at JPL. DSN can supplement Goddard's tracking system for Shuttle and selected Earth satellites. The typical DSN Deep Space Communications Complex, operated by the host government under agreement, comprises: 70 m antenna (S/X-band downlink; 20 kW and 400 kW S-band uplink); 34 m standard antenna (full downlink; 20 kW S band uplink); 34 m high efficiency system (full downlink; X-band uplink) and 26 m antenna (S-band up/down). Goldstone also has a prototype 34 m beam waveguide X-band antenna; the other locations will have it from 1996. To boost data reception rates, the Goldstone complex can be arrayed with the 27 25 m Very Large Array dishes of the US National Radio Astronomy Observatory in New Mexico, and Canberra linked with the 64 m Parkes radio telescope.

Spacecraft assembly/test facilities include an 890 m² high bay area, class 100,000, and a 16.8 m very high bay class 10,000 area of 390 m². A flight system testbed opened in 1993 for development-phase integration/testing of small spacecraft prototypes and subsystems, in a 150 m² lab near the high bays. Environmental test facilities provide vibration and spin balancing. The 6.1 m diameter × 7.63 m (working volume) space/solar simulator was built in 1961 and reopened in early 1994 after refurbishment. It provides 5×10^{-6} torr, -196°C to +93°C, solar intensity to 2.7 Suns (5.6 m diameter beam) or 12 Suns (2.2 m beam). A 3 m simulator, added in 1965, and a 93 m² class 10,000 assembly room supplement it.

Mission data facilities include the MIPL Multimission Image Processing Laboratory based on the digital image processing technology begun at JPL in the early 1960s. The Planetary Data System archives and distributes (principally via CDROM) digital data from planetary missions, ground observations and lab measurements. There are more than 190 titles since inception in 1991.

National Aeronautics and Space Administration, John H. Glenn Research Center (formerly Lewis Research Center)

Founded in 1941, the John H Glenn Research Center (formerly LeRC) is NASA's lead centre for research, technology and development of aircraft propulsion, space propulsion, space power and satellite communication. As such, it oversees the ACTS Advanced Communications Technology Satellite project by operating the Master Control Station. It is the lead centre for the propulsion technology development for NASA's High Speed Research Program. The Microgravity Materials Science Laboratory is housed at the centre for qualifying space experiments, with facilities including a 130 m (1 second) μg drop tower, furnaces and a crystal growth system. The centre manages NASA's Atlas and Centaur launch activities. Staff total 2,400 plus 2,065 contractors. FY96 budget US$914 million; FY95 US$844 million; FY94 US$955 million; FY93 US$1,060 million; FY92 US$1,021 million.

In 1999 the Lewis Research Center was renamed the John H. Glenn Research Center in honour of the first American to orbit the earth.

National Aeronautics and Space Administration, Johnson Space Center

JSC was established in 1961 as NASA's centre for design, development and testing of manned spacecraft; it is currently responsible for the Shuttle programme and was named in August 1993 as the host centre for the revised Space Station programme. Staff comprises 3,200 civil servants and 12,500 contractors. The centre is divided into directorates responsible for specific functions: Administration, Flight Crew Operations, Mission Operations, Engineering, Space/Life Sciences, and Information Systems. JSC also manages the White Sands Test Facility in New Mexico for Shuttle propulsion, power system and materials testing.

Most of JSC's 100 buildings on the 6.5 km² Houston site area provide office space and laboratories. Life sciences, planetary and Earth sciences, artificial intelligence and lunar sample analysis are pursued in the 16 research facilities. The Space Environment Simulation Laboratory houses thermal-vacuum chambers, complementing the Vibration/Acoustic Test Facility and the Anechoic Chamber Test Facility. Four major software laboratories are operated to handle Shuttle computer codes. specialised training facilities include the Shuttle simulators, Shuttle Orbiter trainer in the Mockup and Integration Lab, Precision Air Bearing Facility and Space Station mockups (buildings 9/9A) and the Weightless Environment Training Facility.

Mission Control Center in building 30 has controlled all NASA manned flights since Gemini 4 in 1965. The two identical Flight Control Rooms (FCR, pronounced Fickers) are on floors 2/3, each with 204 stations for controlling a Shuttle mission. The first floor is completely filled by mainframe computers, maintained by an 80-strong staff during missions. Any one of the five main computers can support an FCR. Should a catastrophic failure shut down MCC, KSC's Launch Control Center can bring Shuttle back. However, these old elements are being phased out.

The US$250 million new FCR, built in 1994 in a new wing of the MCC, was first used for Spacelab payload operations during STS-71 in July 1995 and then for some Shuttle control during STS-70 the same month. It controlled its first ascent and entry for STS-75 in February 1996 but it will not fully replace the old system until early 1997. Built by Loral, it replaces NASA's obsolete mainframes with 197 commercially available Unix workstations interconnected on a LAN Local Area Network; most of the software is also off the shelf. The 38 km of fibre optic cable, providing a 100 Mbit/s capacity, is the world's largest fibre data distributed interface network. The workstations and 180 fewer people will control combined Shuttle and Station operations.

When Space Station is complete in 2002, the 9,500 m² five-storey SSCC Space Station Control Center next door will be the focal point of Station operations. Apollo 11's old FCR will be preserved as an historic site.

Shuttle and Mission Simulator, building 5, houses three high fidelity simulators. The motion base flight deck simulator uses hydraulic actuators for 6-DOF to simulate launch, landing and other dynamic manoeuvres. A mid-deck might be added for entire crews to train together. Integrated simulations with MCC begin some 10 weeks before a flight, concluding with a fifth ascent/entry session two days before the crew flies to KSC. The two fixed base simulators provide on-orbit operations training. The detailed aft flight decks covers activities such as RMS, EVA, rendezvous and payload deployment. Experiments, meals and housekeeping chores are practiced in the mid-deck. There is also a Spacelab simulator.

Mockup and Integration Lab houses a high-fidelity Orbiter forward section (flight deck + mid-deck) which can be rotated into launch attitude. Astronauts use it to learn the cabin's layout and engineers make fit-checks of new equipment. A separate cargo bay mockup allows fit-checking of payload mockups. The Manipulator Development Facility trains Mission Specialist for RMS operations.

Weightless Environment Training Facility, building 29. The 7.6 m deep WET-F 9.8 × 24 m tank can accommodate a full scale Orbiter payload bay and working RMS for crews to simulate zero-*g* by neutral buoyancy. A 31 × 62 m 12.2 m-deep pool for Space Station training from 1997 is being built adjacent to Ellington Field.

Lunar Sample Building, #31N, holds most of Apollo's samples.

National Aeronautics and Space Administration, Langley Research Center

LaRC is primarily a research centre for advanced aerospace technology. Major research fields include aerodynamics, materials, structures, flight controls, information systems, acoustics, aeroelasticity, atmospheric sciences and non-destructive evaluation. 40 per cent of the work supports space activities, including technology for advanced space transportation, large space structures and the Earth Observing System as part of NASA's Mission to Planet Earth. Langley covers 3.2 km² in its West area, plus 0.08 km² on its East side, has more than 220 buildings and a replacement value on today's market of US$2.1 billion. LaRC employs 2,500 civil servants and 1,750 contract support personnel. FY95 budget was US$643.7 million; FY94 US$713.3 million.

The centre was established in 1917 as the first research laboratory of the National Advisory Committee for Aeronautics. It became one of the four original NASA facilities (with Ames, Edwards, and Lewis) when the agency was created in 1958. The US manned space programme began there in 1959 as the Space Task Group, which completed its transfer to new dedicated facilities in Houston in mid-1962. LaRC was responsible for the highly successful Lunar Orbiters and Viking Mars orbiters/landers.

Langley's LITE Lidar In-Space Technology Experiment during STS-64 in September 1994 detected stratospheric and tropospheric aerosols, probed the planetary boundary layer and measured cloud top heights. The 1 m diameter telescope with three-wavelength Nd: YAG laser is a testbed for future operational spaceborne lidars. No further Shuttle flights are planned, but a smaller satellite version is being designed. The LASE Lidar Atmospheric Sensing Experiment, the first autonomous aircraft-based lidar to measure water vapour, completed final validation studies aboard NASA's ER 2 in September 1995.

The centre manages the HALOE Halogen Occultation Experiment aboard UARS to monitor the vertical distributions of ozone and key trace gases. The MAPS Measurement of Air Pollution from Satellites instrument has already flown four times on Shuttle to measure the global distribution of carbon monoxide in the free troposphere; it will operate for one year aboard Mir from June 1996. Langley's Earth Radiation Budget Experiment continues operations on the ERBS satellite (which also carries LaRC's SAGE instrument) and NOAA 9/10. A set of instruments is being developed for EOS missions. CERES, the ERBE follow-on, flew on TRMM in 1997 and SAGE III on Russia's Meteor3M in 1998.

The remote sensing office specialises in developing new technologies for future systems, including solid-state lasers, detectors, lightweight optics, deployable telescopes, miniature sensors and onboard data processing. LaRC leads NASA in developing high energy, high efficiency, long-life solid-state lasers for advanced lidar and other applications requiring stable laser sources. Together with NASA Stennis, it is exploring commercial markets for remote sensing technology. LaRC is an EOS Distributed Data Active Archive Center, which carries data from 18 major NASA atmospheric field missions and satellites.

LaRC is providing several experiments for Russia's Mir station. The MEEP Mir Environmental Effects Payload was attached to the docking module by STS-76 in March 1996, where it will remain until late 1997 collecting micrometeoroid particles and the effects of strikes on Station Alpha materials. The Enhanced Dynamic Load Sensors set up has carried Mir's Priroda module to measure the effects of crew movement. The MIDAS Materials in Devices As Superconductors experiment will stay aboard for several months after delivery by STS-79 in August 1996. Four circuit boards, including one with Russian samples, will look at the effects on the electrical and magnetic properties of high temperature superconductor materials.

LaRC has the lead in developing the composite primary structures and the advanced composite thermal protection systems for X-33 and X-34. It is helping the industry teams to develop reusable cryogenic tanks and advanced propulsion, in addition to vehicle systems analysis, aerodynamic and aerothermodynamic testing/analysis, and flight control development. It is performing wind tunnel testing of an X-34 model mated to a Boeing 737.

National Aeronautics and Space Administration, Marshall Space Flight Center

Marshall is one of the largest of NASA's 10 field centres, accounting for about a quarter of the agency's budget. FY96 budget was US$2,495 million. Employees total approximately 2,900 plus 3,000 contractors; the goal for the year 2000 is 2,500 plus 1,900. It is the lead NASA centre for research, technology maturation, design, development and integration of space transportation and propulsion systems. This includes both reusable and expendable launch vehicles, as well as vehicles for orbital transfer and deep space missions. The centre has an Agency-assigned Center of Excellence role for space propulsion.

Marshall is also lead centre for NASA's Microgravity Research Program with responsibilities for planning and direction; resource management; assessment; and definition, development, integration and flight of microgravity science and space processing experiments and facility payloads; as well as programme outreach and education activities.

Marshall developed and continues to provide the Space Shuttle's main engines, solid rocket boosters and external tank.

Other key programme assignments include the Advanced Space Transportation Program which focuses on reducing the cost of Earth-to-orbit transportation and the Reusable Launch Vehicle Technology Program which is demonstrating technologies needed for commercial development of a low-cost, next-generation launch system. Included in the programme are the X-33 and X-34 technology demonstrators.

Marshall provides support to the International Space Station Program office with engineering personnel and facilities to develop pressurised modules, support equipment and assigned payloads. Marshall has lead responsibility for the Advanced X-ray Astrophysics Facility for development, launch and mission operations phases of the X-Ray Observatory program and for archiving the mission and science data. Marshall manages the Gravity Probe-B and other smaller scientific payloads and instruments and conducts research in high-energy astrophysics, solar magnetic fields, and low- energy space plasma physics. Marshall also supports the Mission to Planet Earth Program through research at the Global Hydrology and Climate Center in Huntsville. Since the Astro-1 mission in 1990, Spacelab science activities have been controlled from Marshall's Spacelab Mission Operations Control Center.

Marshall was established in 1960 around a nucleus headed by Dr Wernher von Braun, the team which, as part of the US Army Ballistic Missile Agency, orbited Explorer 1. Marshall became NASA's lead centre for Apollo's Saturn launchers and developed the Skylab space station. It is located on a 7.3 km² tract within the US Army's Redstone Arsenal in Huntsville, Alabama. Also operated by Marshall is the Michoud Assembly Facility in Louisiana, occupying a 3.37 km² site where Shuttle external tanks are manufactured and shipped to Florida by barge.

National Aeronautics and Space Administration, Stennis Space Center

Previously known as the National Space Technology Laboratory, SSC is NASA's primary large propulsion test facility, including development testing of the Space Shuttle SSME main engines. The NASA-related workforce is 1,775. Stennis is also NASA's lead centre for the commercialisation of remote sensing technology. Twenty-one other federal and state agencies occupy the site, including the Naval Meteorology and Oceanographic Command, bringing the workforce to 3,745.

Total land area is 582 km², of which 54.8 km² constitute the operational base and the remainder is held as an acoustic buffer zone. The site was selected in 1961 to accommodate Apollo's F1 and J2 engines and Saturn 5 stages, testing a total of 27 stages. All main Shuttle engines undergo acceptance testing at Stennis before flight. The first SSME test was made during June 1975. Three stands handle single SSME trials. A1 evaluates sea level performance. A2 simulates firing up to 16.5 km altitude, operating with the engine 18° from the vertical to mimic the most extreme flight/gimballing combination. A1/A2 were both built in the 1960s for Saturn 5 stage 2 testing and can each handle 4.89 MN, 10.07 m diameter. Stand B1 is modelled after A2 but the firing angle is vertical. The other side of the B structure is the Main Propulsion Test Article facility where three-engine SSME cluster testing is performed; it originally handled Saturn 5's stage 1 and as such can cope with 48.9 MN, 10.07 m diameter.

The three-cell E-1 Test Facility can handle engines using LH_2 or hydrocarbons, solid or hybrid, high flow rates, high pressure cryogenics and ultra-high pressure gases. The state of the art complex is being used for testing hardware for future launch vehicles. Lockheed Martin from NASA Michoud began seven months of advanced hybrid testing in December 1996. Construction of the H-1 Shuttle Advanced Solid Rocket Motor static test stand was started in early 1991; it was 80 per cent complete when construction ended January 1994 as a result of ASRM's cancellation. It is retained for future use.

The E-2 Test Facility supports testing for hypersonic vehicles, using 10 small H_2/O_2 gas generators to supply 1,900°C gas to test panels. The Diagnostic Testbed Facility is for subscale turbopumps and liquid engines, providing a dual-position stand for development of non-intrusive, remotely sensed exhaust plume diagnostics technology.

NASA's SSC operates an Earth Observation Research Office and a Remote Sensing Applications Program Office. The SRSC is one of NASA's Centers for Commercial Development of Space, providing commercial technology applications for developing satellite remote sensing, image processing and GIS. The Resource21 affiliate company is studying a four-satellite constellation beginning 1999 for crop monitoring in conjunction with Boeing (space segment), GDE Systems and several agribusinesses.

The centre houses an 890 m² Sensor Development Laboratory for the design and R&D of multispectral scanners, and is a prime site for defining the remote sensing instrumentation requirements for the Space Station. The Laboratory was responsible for developing an off-the-shelf thermal imager for ice and leaking cryogen detection during Shuttle pad operations. The US Navy Oceanographic Office's Satellite Processing System is based at Stennis; it processes real-time Tiros data for routeing to NOAA, US Navy and the USAF Global Weather Center.

National Aeronautics and Space Administration, Wallops Flight Facility

NASA's sounding rocket programme is based at the Wallops Flight Facility in Virginia using vehicles from commercial sources or developed by the agency. Some 25 launches are made each year from a range of sites, including Wallops itself, Poker Flat and White Sands. NASA currently employs 15 sounding rocket types (Aries use ended 1993), including several utilising Bristol Aerospace's Black Brant 5 motors. Extensive use is made of military surplus motors and all are unguided. Saab's S19 Boost Guidance System can be added to the Nike Black Brant and Black Brant 5/9/10: the gyro platform controls aerodynamic canards, decreasing impact dispersion by a factor of 5/10, allowing higher apogees and launches in higher winds.

Now a part of Goddard and on Virginia's east coast, Wallops became the third US orbital site on 16 February 1961 with the launch of the Explorer 9 balloon by the all-solid Scout. 19 vehicles achieved orbit from Wallops by the end of 1995 (plus three failures), the most recent in 1985. Although the facility is still available, Scout, its only orbital vehicle, retired in 1994. Orbital activity resumed October 1995, but the commercial Conestoga failed. Wallops also provided control for the B52/Pegasus launch of Brazil's SCD February 1993 (the B-52 took off from Florida). One air-launched Pegasus carried from Wallops by Lockheed L-1011: SAC-B/HETE (Argentina/NASA; November 1996).

Standard orbital inclination is 37.7°. Wallops' major activity now is as the base for NASA's sounding rocket programme. The first firing made from Wallops was a Tiamat on 4 July 1945 and since then more than 14,000 rockets have been dispatched. Current vehicles include Super Arcas, Black Brant, Taurus-Tomahawk, Taurus-Orion and Terrier-Malemute. Some 30 launches are made annually (32 in 1995, 33 in 1994, 26 in 1993; 16 planned for 1996), including Wallops-managed firings from the Poker Flat Research Range in Alaska and the White Sands Missile Range in New Mexico. 1995's schedule included six from Woomera. 1994's had more than 30 from Brazil. Wallops' own total was 807 firings by end-1996.

The Wallops Orbital Tracking Station was established in 1986 to provide NASA tracking support for projects such as IUE, IMP 8, Nimbus 7 and COBE. High-speed data transfer to Goddard at Greenbelt is provided by a satellite link. WFF utilises telemetry receiving and command antennas ranging over 2.4 to 18 m for satellite and sounding rocket support. The radars include an 8.8 m parabolic FPS 6 radar that can track objects out to 60,000 km with a precision of ±3 m and 9 cm/s range-rate; peak output 2.5 MW. Wallops also houses NOAA's Command/Data Acquisition Station of nine 7.3 to 26 m antennas to track, command and take data from the NOAA and GOES meteorological satellites.

National Oceanic and Atmospheric Administration

Operators of the NOAA polar and GOES geostationary meteorological satellite systems, and federal overseer until October 1992 of the commercial Landsat 4/5 remote sensing satellite system (a role it will repeat for Landsat 7). As an element of the Department of Commerce, NOAA is the US signatory to the Cospas-Sarsat agreement. Its Space Environment Laboratory in Boulder, Colorado provides real-time monitoring and forecasting services of solar disturbances and their effects on Earth's environment.

Satellite data are used to monitor weather conditions, provide data for issuing warnings of severe weather, prepare charts and coastal maps, and improve assessment and conservation of marine life. They are also employed to assess the impact of natural factors and human activities on global food and fuel supplies and on environmental quality. NOAA agencies participating directly in space projects include NESDIS, the Systems Acquisition Office (SAO), the National Weather Service, the National Marine Fisheries Service, the Office of Oceanic and Atmospheric Research, and the National Ocean Service. In co-ordination and co-operation with other appropriate organisations, SAO has 'responsibility for designated major systems acquisition', such as the GOES I-M series. NOAA provides government authority for the Landsat 4/5 Earth resources satellites, now operated commercially by EOSAT. Landsat 7 operations will come under NOAA.

NOAA, National Centers for Environmental Prediction, Space Environment Center

NESDIS manages the NOAA polar-orbiting and GOES geostationary meteorological satellite systems, gathering, archiving and distributing environmental data. NESDIS' FY92 budget was US$343.9 million; US$329.5 million FY93; US$349.5 million FY94; US$387.4 million FY95; US$472 million FY96 and US$532 million requested FY97. NOAA was established within the Department of Commerce in 1970.

SEC provides real-time space environment monitoring and forecasting services, develops techniques for forecasting solar disturbances and their effects on the Earth's environment, and researches in solar-terrestrial physics. The data and forecast products are of interest to satellite operators, HF radio users, those concerned with radiation damage to passengers in high-flying aircraft or spacecraft, radio navigation users and operators of electric power distribution networks.

SEC receives some 1,400 data streams daily on the state of the space environment from space platforms, such as Tiros and GOES, and ground-based observatories, such as the National Solar Observatory, USGS's Intermagnet Network and USAF's SEON. The more important databases and indices are available for public access by computer networks.

Office of Science and Technology Policy

The Senior Interagency Group (SIG) on Space was replaced in 1989 by the National Space Council to co-ordinate and monitor national space policies and strategies. NSC was chaired by the Vice-President and included 10 cabinet-level officials: NASA Administrator, director of Office of Management and Budget, Secretaries of State, Defense, Commerce, Energy, Transportation and Treasury, CIA director, presidential Chief of Staff, and the President's assistants for National Security Affairs and for Science and Technology. The Executive Secretary led an 11-member policy staff. The new President replaced NSC in February 1993 with a science and technology council chaired by Vice-President Al Gore.

Ohio State University Center for Mapping

The Center, established in 1986, is a NASA commercial space centre, concentrating on real-time satellite mapping, which makes use of data as soon as they are transmitted from satellites. R&D projects concentrate on commercial products for mapping, remote sensing and GIS user communities. The Center's current research project is the development of a fully-digital, real-time Airborne Integrated Mapping System (AIMS). It combines positioning sensors, GPS and INS, with digital imagery. AIMS has many applications from traditional large-scale topographic mapping to military reconnaissance, determining the extent of an oil spill, corridor surveys such as rail and highway construction and management, drug interdiction efforts and emergency response deployment. Earlier research included the development of the GPSVan, which can map highway and transportation infrastructure at normal traffic speed.

Pennsylvania State University, Center for Cell Research

The CCR was established in 1987 as one of NASA's CCDS Centers for the Commercial Development of Space, focusing on commercial product oriented projects: physiological testing, bioseparations and illumination. NASA withdrew funding in December 1993, but the Center continues operations. Space-based physiological testing involves space-flown animals, tissues and/or cells as pharmaceutical test subjects or in discovery research for health problems that occur on Earth. CCR flew the first commercial proprietary physiological test aboard Shuttle in October 1990.

The CCR bioseparation programmes provide continuous flow electrophoresis and aqueous two-phase partitioning. The USCEPS United States Commercial Electrophoresis Program in Space is the first to couple industry access to space-based systems with ground-based testing and product trial separations. This service enables industrial clients to evaluate, before flight, how much improvement going into space will produce. The team began testing an electrophoresis unit in 1995; was certified in 1996 and became available to commercial interests in 1997.

United States Geological Service, Eros Data Center

The US Geological Survey's Earth Resources Observation System Data Center is responsible for managing a global archive to ensure long-term preservation of satellite remote sensing data. EDC will manage land processes data for the EOS programme a 5,600 m² extension was completed in April 1996 to handle EOS processing/distribution facilities. It is the world's single largest repository of remote sensing data, with more than three million images. EDC maintains the national archive of Landsat imagery; its name derives from the Department of Interior's plan for an EROS Earth Resources Observation Satellite, which evolved into Landsat. A field site for the USGS National Mapping Program, EDC also manages other forms of Earth data, including more than nine million aerial mapping photographs covering the US.

United States GPS Industry Council

USGIC represents the interests of GPS manufacturers and users. It projects the worldwide market for GPS receivers will grow to more than US$8 billion by 2000 from 1994's US$867 million.

University of Alabama, Center for Macromolecular Crystallography

The CMC was established in 1985 as one of NASA's CSC Commercial Space Centers, focusing on space-based growth of biological crystals and their applications. The centre's Protein Crystal Growth experiments have made numerous Shuttle flights.

University of Alabama, Consortium for Materials Development in Space

UA Huntsville's Consortium for Materials Development in Space was established in 1985 as a Center for the Commercial Development of Space (CCDS) through NASA's Office of Commercial Programs to focus on investigations in space to develop new materials and processes. This approach includes commercial materials development that benefits from the space environment, commercial applications of the physical chemistry occurring at the surface of a new material, and how materials are transported to it, and frequent space experiments.

UAH flew six experiments under a US$1 million contract aboard Space Services Inc's first Starfire 1 sub-orbital rocket 29 March 1989 as the Consort 1 mission. Each was developed by one or more principal investigators from the University, industry or both: de-mixing of immiscible polymers, powdered metal sintering, electrodeposition, convective mixing of elastomer modified epoxy resins, foam formation, and operation of the Materials Dispersion Apparatus. Consort 2, funded by a US$1.1 million grant from the Office of Commercial Programs, failed 15 November 1989 because of stage 1 guidance problems. The 455 kg payload was recovered intact and was reflown as Consort 3 on 17 May 1990, with 10 of the 12 experiments successful. Consort 4 flew successfully 16 November 1991 carrying 12 experiments; NASA provided the funding of about US$1.9 million. EER Systems' Space Services Div was awarded a contract worth up to US$9.8 million for up to seven further Consorts through 1996. Consort 5 flew 10 September 1992 but stage 2's burn ended 3 seconds early when the casing burnt through and µg data were not obtained because of the attitude perturbations. Consort 6 flew successfully 19 February 1993 with seven experiments, attaining 301 km. The US$2 million funding was provided from the insurance following C5's partial failure. Conquest 1, the first of a new series, was launched 3 April 1996 carrying eight experiments on Starfire at White Sands Missile Range: providing 7 minutes µg for the 293 kg payload, attaining 320 km. One experiment planned to extrude a 6 m foam beam into space, but it instead formed a large ball. Conquest 2 is not yet planned.

In co-operation with industrial partners the CMDS flew materials and product development experiments on STS- 79 with the Extreme Temperature Translating Furnace in Spacehab 5, and on STS-80 with biological materials in the Shuttle mid-deck lockers. Shannon Lucid using the Russian Optizone furnace processed metal sintering samples aboard the Mir complex.

University of Colorado, Bioserve Space Technologies

The University of Colorado, Department of Aerospace Engineering runs BioServe after NASA established it in October 1987 as one of NASA's CCDS Centers for the Commercial Development of Space. It focuses on space-based bioprocessing and biomedical testing, CELSS technologies and apparatus development. Almost 50 private sector concerns participate in the CCDS. It is responsible for fabricating and qualifying its own apparatus. Development efforts focus on providing a suite of generic life sciences space facilities, including fluids mixing, micro-organisms culture and plant growth: CGBA Commercial Generic Bioprocessing Apparatus (flown STS-50 June 1992, STS-53 January 1993, STS-57 June 1993, STS-60 February 1994, STS-62 March 1994, STS-63 February 1995, STS-69 September 1995, STS-73 October 1995), AGBA Autonomous Generic Bioprocessing Apparatus, A-MASS Animal Module for Autonomous Space Support and P- MASS Plant Module for Autonomous Space Support.

University of Colorado, National Snow and Ice Data Center

The centre archives and distributes analysis of snow and ice data from satellites, including NOAA and DMSP, and maintains an archive of satellite and *in situ* data.

University of Houston, Space Vacuum Epitaxy Center (SVEC)

SVEC was established in 1985 as one of NASA's CCDS Centers for the Commercial Development of Space, focusing on space-based thin-film growth using Molecular Beam Epitaxy (MBE) and Chemical Beam Epitaxy (CBE). More than 60 personnel are located at SVEC within 1,000 m² of laboratory space, including class 10/1,000/10,000 cleanrooms.

University of New Mexico, Earth Data Analysis Center

EDAC was established in 1964 to transfer aerospace technology. Since 1974, it has focused on observations using spatial and spectral analysis, GIS development and GPS applications. EDAC has four service areas: Image Processing; GIS Development; Geographic and Electronic Data Services; Training. These are supported by collocated organisations, including the NM Natural Heritage Program, National Park Service (Southwest GIS Center), US Fish and Wildlife Service, US Geological Survey Geographical Names Information System, and Autometric Inc. Information services include access through Internet and others to provide metadata and digital image transfers. Through strategic partnerships with industry and government, EDAC employs state of the practice hardware and software systems for data processing, modelling, visualisation and environmental simulations.

University of Tennessee Center for Space Transportation

The CSTAR Center for Space Transportation and Applied Research at The University of Tennessee Space Institute headed the original three-flight programmes of the Comet retrievable payload sponsored by NASA. Westinghouse Electric Corporation provided the service module and systems engineering, EER Systems the Conestoga launcher (contract worth about US$45 million), Space Industries (contract US$15 million) the recovery capsule, orbital operations and payload integration. EER assumed responsibility for the programme in late 1994. NASA's FY94 budget allocated US$14.5 million, leaving a shortfall of US$6 million required for the launch. Commercial customers provided the remainder, so NASA's US$14.5 million was released by the 28 March 1995 contract, which allocated the agency half of the capacity. FY93 US$23 million; FY92 US$18 million; FY91 US$10.5 million. The US$85.1 million contract signed in January 1990 called for three Comet flights: #2 September 1995 and #3 March 1997. NASA modified it to a single mission costing US$65.8 million, after projecting that three missions would cost more than US$158 million. That contract did not cover the companies' investments, but the commercial Westar (Westinghouse Space Transportation and Recovery) service was also offered, charging $35 million for each mission or about US$200,000/kg payload. Westinghouse projected 10 annual missions by 2000. Westinghouse, France's Novespace and Arianespace in May 1993 began marketing the service to European users for launch on Ariane. Cost would be US$68,000/kg payload. A suitable Ariane slot would be available about once annually from 1995. However, Westinghouse withdrew from Comet/Westar in 1994.

VENEZUELA

Centre for Digital Image Processing (CPDI)

The CPDI section of Venezuela's Fundacion Instituto de Ingenieria was established in August 1983 and was appointed in 1988 as national representative for distribution of EOSAT's Landsat products/services. More than 50 trained professionals provide advisory and technical services, and conduct studies in remote sensing applications, GIS and software development for special applications.

SPACE INDUSTRY

Communication satellites, components and services
Earth observation facilities and equipment
Earth observation satellites, components and services
General
Launchers, propulsion systems and launch site services
Microgravity facilities, equipment and services
Navigation, search and rescue facilities and equipment
Satellite actuation systems and thermal control devices
Satellite busses
Satellite communications facilities and equipment
Satellite guidance, navigation and control components
Satellite information storage components
Satellite and propulsion support hardware

COMMUNICATION SATELLITES, COMPONENTS AND SERVICES

AUSTRALIA

Satellite Aircom Consortium

MITEC emerged from an R&D Centre of the University of Queensland Electrical Engineering Department in 1987 with 15 staff and an A$700,000 turnover; staffing level is 190 and 1994 turnover A$21 million. It is listed on the Australian Stock Exchange with an issued capital of A$18 million. Major space interests cover satellite and terrestrial communications systems and microwave components. It has designed and manufactured space-qualified transmit and receive equipment for Optus B (three 30 GHz beacons) and Russia's RadioAstron (L-band cryogenic amplifiers and converters). For ground station applications, it produces a range of frequency conversion (up/down) equipment for L, C, X and Ku bands. SSPAs operate up to 70 W Ku band. Low-noise L to Ku amplifiers provide 0.6 to 1.8 dB noise figures. Planar antennas are manufactured for mobile applications and low-noise converters for weather satellite ground stations. MITEC also produced equipment for CSIRO to receive signals from Japan's MOS satellites in Australia.

BELGIUM

Sirius Communications NV

Sirius Communications is producing state-of-the-art digital chips and development systems for CMDA/spread spectrum-based satellite communications. The company supplies its new transceiver chips for a target market in the VSAT and portable voice/data satellite terminals for communications, positioning and messaging communication with its new transceiver chips. About 75 per cent of the company's current revenue comes from satellite communication hardware and services sales and it is involved in W-CDMA (UMTS) developments.

Sirius also provides baseband ICs for broadband CDMA-based satellite links including Internet connections. The chips support both LEO and GEO applications while supplying a worldwide customer base. For specific applications Sirius offers a custom design service for CDMA and spread spectrum users.

One of Sirius' principal products is the ASTRA chip. The Astra is a 3.3V 100-PQFP full-duplex CDMA/spread spectrum transceiver chip. It supports two-parallel QPSK or a four-parallel BPSK channel communication, with spreading code lengths up to 1,023 bits. The spreading codes are fully flexible and all variants of PSK modulation are supported. Data rates from a 50 bit/s up to 150 kbit/s are possible. The Astra has on-chip Direct Digital Synthesis up- and down-conversion. The on-ship receiver has a dual demodulator architecture for operation in S-CDMA (Synchronous CDMA) and QS-CDMA (Quasi-Synchronous CDMA) satellite networks.

The ASTRA Development System including graphical receiver analysis software allows field trials early in the product development cycle. The development boards serve as design examples, which can be customised and miniaturised in the terminal product. FES functions can be executed on plug-in FPGA based 'Gemini' boards available from Sirius.

The DIRAC chip combines a flexible CDMA/spread spectrum receiver chain with an on-chip 32-bit ARM microcontroller core. Sirius Communications works with the customer from the design phases through to the silicon etch for individual or volume manufacture.

INDIA

Afro-Asian Satellite Communications Ltd (ASC)

ASC and Hughes Space & Communications Company announced a contract 19 January 1995 for two Agrani satellites to provide GSM-compatible mobile telephone links from 1998, covering at least an area bounded by India, Middle East, Singapore and Moscow. ASC is part of India's Essel Group. The turnkey contract with Hughes Space & Communications International called for the first launch December 1997, followed by the second six months later (possibly covering Africa). The value was not disclosed but system cost has been estimated at US$700 million. HSC is providing the satellite, Hughes Network Systems the four gateways and the network control centres (the primary gateways) in Bombay and Gibraltar. Likely slots are 46° east and 53° east, provided by Gibraltar. Each HS-601 HP will be able to handle up to 16,000 simultaneous calls using about 300 L-band spot beams and 30 kHz channels. 10 to15 W SSPAs will generate approximately 1 kW RF power. Each spot can be adjusted from 30 to 50 dBW. The dual transmit/receive antenna is 12 m in diameter. The Si solar wings will provide 5 kW; design life is 12 years. Unusually, it will be allowed to drift ±6° north/south as mobile users do not require standard NSSK. Several delays postponed the launch of Agrani 1 to 2000, the satellite now being built by Lockheed Martin Telecommunications.

INDONESIA

PT MediaCitra Indostar

Indostar will provide the world's first DBS dedicated to radio/TV for a single nation. The first of four satellites will provide five S-band transponders for digital TV, using 8:1 compression for 49 channels. The second will add L-band for CD-quality radio. The US$100 million contract for the first satellite and its launch was signed with CTA International 8 December 1993. Funding is provided by: 40 per cent private consortium, 30 per cent PT Amcol Graha Ltd (a Jakarta electronics company) and 30 per cent PT Bimantra Citra. The Indostar 1 DBS system costs US$271 million. PT MediaCitra Indostar is the satellite company of the Indovision group. Indovision is a multichannel pay television company. National TV cannot be received by 32 per cent (57 million) and radio by 20 per cent (36 million). The commercial Indostar system will be received via US$100 hand-held radios, US$100 analogue receivers and US$500 digital satellite decoders with satellite dishes less than 1 m in diameter. It is designed to guarantee quality reception within Indonesia, whatever the weather conditions. Indostar 1 was launched to a slot at 96.7° east.

PT Pasifik Satelit Nusantara

Palapa B1, retired by PT Telkom at 118° east, was sold to PSN in 1991 to provide Pacific Rim services from mid-1992. PSN's teleport at Bataan opened in early 1994. Still operated by PT Telkom, PSN's major shareholder, B1's inclination was allowed to increase to extend its life for three years. Projected revenue for that period was US$3 million. It began drifting in October 1995. B2P was replaced by C1 in February 1996 and transferred to PSN. It is expected the two other Palapa Bs (and then the Palapa Cs) will be sold to PSN as they near the end of their lives. PSN already owns and operates the six extended C-band transponders aboard each Palapa C.

PSN is leading the US$650 million ACeS Asean Cellular Satellite System with the PLDT Philippines Long Distance Telephone Co and Jasmine International Public Company Ltd (Thailand) which planned to develop a cellular phone system by early 1998. An agreement was signed with Lockheed Martin in May 1995 for exclusive negotiations; the contract was announced 6 July 1995. The three partners announced in May 1995 that the financing is in place: they are committing US$150 million and the rest will be provided by debt financing. The Garuda 1 satellite (named after a mythological Indonesian bird) will be able to handle 10,000 simultaneous calls from the 2 million subscribers via the 150-beam phased array antenna.

PSN's shareholders are PT Telkom (40 per cent), PT Electrindo Nusantara (30 per cent), Hughes Aircraft (10 per cent), Telesat Canada (10 per cent) and private investors.

PT Satelindo

PT Satelit Palapa Indonesia was created 29 Jan 1993 to finance and operate the US$240 million new Palapa C generation. Shareholders were PT Bimagraha Telekomindo (60 per cent), PT Telkom (30 per cent) and PT Indosat (10 per cent) until the DeTeMobil subsidiary of Deutsche Telekom agreed in March 1995 to pay US$566 million for a 25 per cent interest. PT Satelindo will also take over the Palapa Bs from PT Telkom. PT Telkom released the Palapa C RfP 10 August 1992, with bids due 10 October 1992. Hughes' HS-601 was selected and the contract signed 15 April 1993. The winner was required to return at least 20 per cent of the value to Indonesian industry. Satelindo has until the end of 1999 to exercise its option for C3.

PT TELKOM (PT Telekomunikasi Indonesia)

Perumtel (Perum Telekomunikasi) became a Limited Company on 24 September 1991, changing its name to PT Telkom. It originally operated the whole satellite network and continues to be responsible for three Palapa Bs.

INTERNATIONAL

ICO Global Communications

ICO was constituted in January 1995 to provide global hand-held telephone services using a medium orbit satellite system. Four million subscribers are projected by 2010. It will begin full operations in late 2000. First launch was scheduled for second half of 1998, with services beginning in late 1999. Cost is projected at some US$3 billion. The 47 investors are providing US$1.5 billion total capitalisation, including Inmarsat's own US$150 million. At least 70 per cent will be owned by Inmarsat and its signatories. (ICO allows current Inmarsat signatories to invest as desired, instead of in proportion to their Inmarsat interests; maximum is US$103 million.) The rest will be financed by outside strategic investors and bank debt.

ICO targets hand-held phones retailing at US$1,500 for dual-mode models (accessing cellular when in range) and US$1,000 for satellite only. Usage cost will be average US$2 per minute. It will also provide fax, 4.8 kbit/s data and alerting services. The satellites will relay calls to SAN Satellite Access Nodes within their view. SANS are interconnected terrestrially to form the ICONet, linked through gateways owned and operated by third parties to the public switched networks. Each satellite will provide 4,500 simultaneous calls or equivalent via 163 beams. A 53 kg navigation payload provided by Hughes under a US$105 million contract was planned to augment GPS but Inmarsat in March 1996 voted against its inclusion.

ICO requires a constellation of 10 satellites plus two spares in two planes at 10,355 km. An order for 12 Hughes HS601 satellites, worth US$1.4 billion, was announced 20 July 1995; the final contract was signed 5 October 1995. An additional US$925 million launch services contract was signed 7 December. Hughes Electronics Corp also became a strategic investor (US$93.8 million). This will be the first HS-601 application in medium orbits; the design will employ GaAs cells, direct injection by the launch vehicle, an enhanced thermal radiator and a modified attitude control system.

Orion Atlantic LP

Orion Atlantic has established a high-power Ku-band trans-Atlantic system optimised for small diameter VSATs and rooftop-to-rooftop communications supporting integrated voice, data and video applications. The structure of what is now known as Orion Atlantic LP in the project was revealed in January 1992. Seven companies joined Orion Network Systems to provide US$90 million equity for shares totalling 66.66 per cent: British Aerospace (UK; 25 per cent), COM DEV (Canada; 4.17 per cent), Martin Marietta (US; 8.33 per cent), Kingston Communications (UK; 4.17 per cent), Matra Hachette (France; 8.33 per cent), Nissho Iwai (Japan; 8.33 per cent), STET (Italy; 8.33 per cent). In addition to the equity investment, an international syndicate of banks, led by Chase Manhattan, provided the partnership a credit facility of US$250 million.

Orion Network Systems (ONS) was formed in 1987 as a telecommunications carrier providing international transmission facilities and services to corporations and US inter-exchange carriers. The OrionNet subsidiary provides digital private line and switched transmission services from its US gateway locations to overseas locations. ONS's subsidiary Orion Satellite Corp is the General Partner and management arm of Orion Atlantic. Its Orion Atlantic Network Services and Orion Atlantic Satellite Services carry out the marketing and sales activities for the partnership.

BAe and Orion Satellite Corporation signed a US$360 million contract on 1 September 1989 for the in-orbit delivery of two satellites. It marked the first US commercial sale by Europe. Construction began 7 January 1992. Under the turnkey contract, Martin Marietta Commercial Launch Services provided launch and insurance services. The FCC granted final authority to proceed 28 June 1991. The all-Ku Orion 2 was ordered by Orion Atlantic 24 October 1995 for 1998 in-orbit delivery. Total cost, including launch and insurance, is US$265 million, underwritten by bank commitments. The Eurostar 2000+ design is similar to Orion 1 but the higher power transponders will allow additional services such as DTH. The coverage will overlap with Orion 1's in eastern US and Europe. Onboard switching will allow beam connectivity reconfiguration according to customer demand.

The company (ONS, its subsidiaries and Orion Atlantic) has returned annual losses through 1999.

ISRAEL

Israel Aircraft Industries Ltd

IAI began the Amos programme in 1990, with full go-ahead formally in January 1992, to develop, build, operate and market a medium-class GEO satellite, principally for subregional TV broadcasting. IAI retains ownership of Amos 1 and SpaceCom was created in 1993 to sell capacity. SpaceCom's equal partners are IAI, Gilat Communication Engineering, General Satellite Services Co and Mer Services Group Ltd. Amos is built from flight-proven hardware to minimise risk and only the FM was assembled. Cost through launch, plus operating costs over 10 years, is projected at US$190 million to US$200 million; adding interest yields a total cost of US$270 million. US$100 million was provided by Israeli banks, guaranteed by the government. The project began when the government guaranteed three transponders over the 10-year life (replacing the three previously leased from Intelsat). A Portugal feed was included early in the programme under a letter of intent, but it is unlikely to be used. Transponder cost is about US$3 million annually non-preemptible. It was hoped that Arab users would be signed up before launch as the majority of the 250 million people within the 700,000 km^2 Middle East beam are Arab-speakers.

IAI plans the launch of the Amos 2/CERES Central European Regional Satellite in 2001-2002. For the project to proceed, it required a commitment from the Hungarian government to guarantee two transponders, agreed 30 April 1995. The US$150 million financing was acquired in 1996. The satellite is owned by Magyarsat in Hungary, a 50/50 joint venture by IAI and Antenna Hungaria, and collocated with Amos 1 with eight active 72 MHz transponders (plus two back-ups). The Israel beam will provide Amos 1 back-up (there will be no Portugal beam). IAI concluded other contracts in 1996.

IAI is marketing the approximately 1 tonne design for domestic and subregional communications. The modular approach allows integration of other payloads, and the low mass offers more launch slots.

JAPAN

Broadcasting Satellite System Corporation

B-SAT was incorporated 13 April 1993 as a consortium of NHK, WOWOW, five other broadcasters and eight banks to procure and operate the 'BS-4' generation of broadcasting satellites for launches in 1997-98. The government is expected to allocate four DBS channels, including HDTV. Hughes Space and Communications International Inc was selected in December 1993 (signed June 1994) for on-orbit delivery, in competition with Space Systems/Loral and Martin Marietta.

Japan Satellite Systems, Inc

Japan Communications Satellite Company Inc (JCSat) was formed in April 1985 to provide the first commercial Japanese satellite communication system. Satellite Japan Corp (Sajac, licensed in May 1991 to launch two satellites in 1994/5) was forced by financing problems to merge with JCSat in 1993, creating JSAT. Hughes Communications sold its 30 per cent equity stake in JCSat and Sajac's Nissho Iwai Corp and Sumitomo Corp major shareholders joined JCSat's Mitsui and ITOCHU. Interests are now: ITOCHU (28.5 per cent), Mitsui (24.5 per cent), Sumitomo (23.5 per cent), Nissho Iwai (23.5 per cent). ¥17 billion revenue was projected for the year ending 31 March 1996 and ¥20 billion for the following year, but has yet to return a profit.

About 65 per cent of the 64 transponders on the first two satellites are leased; there are 55,000 subscribers to the five TV channels. In addition to network and cable TV distribution, Japan Satellite Communications Network Corp (JSNet) employs JCSat to provide VSAT business data services. Three transponders are used for DTH video services and two carry 12 stereo channels of high-quality Digital Audio Broadcast programming. Other full period lease customers provide time-shared and other value-added services for third parties. Annual transponder charge is ¥300 million to ¥650 million, depending on leasing conditions for a full period, full bandwidth transponder; it also leases partial transponders on a full period, power and bandwidth basis. From autumn 1992, JCSat has provided a full transponder time-share service (15-minute increments).

JSAT started digital multichannel broadcasting named 'PerfecTV!' with more than 70 television channels and 100 radio channels in October 1996 using JCSAT 3. In December 1996, there were about 100,000 subscribers to PerfecTV! and more than 300,000 were expected by the end of FY96 (March 1997).

Kokusai Denshin Denwa Company Ltd

KDD is the Japanese international telecommunications carrier and the signatory to Intelsat and Inmarsat, operating through the Ibaraki and Yamaguchi Earth stations. With COMSAT of the US, it is providing aeronautical communications through the Inmarsat system.

Nippon Hoso Kyokai (NHK)

NHK broadcasts two DBS channels through BS-3a and procured BS-3N to guarantee services. It is a major shareholder in Broadcasting Satellite System Corporation.

Space Communications Corporation of Japan

SCC is wholly Japanese-owned by the Tokyo-based Mitsubishi Electric Corp (MELCO, 18.94 per cent), its general trading affiliate Mitsubishi Corp (28.41 per cent), Mitsubishi Heavy Industries (10.0 per cent), The Mitsubishi Bank (5.0 per cent), Mitsubishi Trust and Banking Corp (5.0 per cent), Mitsubishi Estate Co (5.0 per cent), The Tokyo Marine and Fire Insurance Co (5.0 per cent), The Meiji Mutual Insurance Co (5.0 per cent), Kirin Brewery Co (5.0 per cent), and 20 other companies of the Mitsubishi Group (12.65 per cent); share capital totals ¥60 billion. Revenue for the year ended 31 March 1995 was ¥16.4 billion; it was hoped the company would break even in 1995. SCC was established in March 1985 and provides Japan's second commercial satellite communication system (see JSAT); personnel total about 200. Two Superbird satellites were planned, built by Space Systems/Loral and based on the FS-1300 platform. Programme cost was ¥70 billion (US$636 million), covering two satellites, launch services and insurance at premiums of about 20 per cent. Launch of Superbird A 5 June 1989 by Ariane to 158° east was successful but most of its stationkeeping oxidiser was lost in December 1990 and commercial operations were ended; some customers were transferred to the rival JCSat. The insurance claim was reportedly US$170 million. Superbird B, delayed from a December 1989 launch because of transponder problems, was lost in the Ariane 4 accident of 22 February 1990. Insurance cover was US$94.3 million. Launch contract for the replacement Superbird B1 was signed with Arianespace in November 1990; SS/L delivered the satellite within 24 months. Superbird C was ordered from Hughes in March 1995 to expand coverage to the Asia Pacific region, including HDTV and 150 Mbit/s data and is located at 144° east with Superbird B1 at 162° east and Superbird A at 158° east. Cost with launch and insurance is about ¥35 billion (US$412 million).

Telecommunications Advancement Organisation of Japan

The Telecommunications Satellite Corporation of Japan (TSCJ) was established in August 1979 to operate the CS and BS satellite series procured and launched by NASDA. The name changed to TAO in October 1992 as it introduced new services such as R&D. Japan's first two operational communications satellites, CS-2a and 2b, each with six Ka-band and two C-band transponders, were launched in 1983 principally for maintaining communications with the more scattered islands of the Japanese archipelago. Both were still operational when they were replaced by CS-3a and 3b in 1988. NTT's NStar took over in 1995. The BS-2 series introduced DBS services in 1984 using 100+W transponders but two of BS-2a's three amplifiers failed soon after launch on 23 January 1984. BS-2b brought the system to full capacity in 1986 but a BS-2x was purchased from GE Astro Space by NHK for an Ariane launch to ensure system continuity before the first of the BS-3 series could be launched in 1990. However, BS-2x was lost in the Ariane failure of 22 February 1990; the replacement BS-3H was then lost in its April 1991 launch. Both would have been controlled through TSCJ. BS-3a, launched 28 August 1990, replaced BS-2b but it operates on marginal power. 3b completed the system in 1991 but NHK procured BS-3N in late 1992 (launched July 1994) to guarantee services. BSAT 2a is scheduled for launch in fourth quarter 2000 followed by BSAT 2b in second quarter 2001.

LUXEMBOURG

Astra/Société Européenne des Satellites (SES)

SES was incorporated as a private company in March 1985 to establish a medium-power satellite system for TV distribution from one of the GEO slots assigned to the Grand Duchy of Luxembourg. It operates under a franchise extending to 2010 from the Duchy (which retains a 20 per cent interest). Germany's Deutsche Telekom acquired a 16.67 per cent interest in June 1994. By end-1995 the system was reaching 61.33 million homes (39.60 million cable; 21.73 million SMATV/DTH). 1995 profit was Lfr2,800 million from sales of Lfr10,300 million.

A contract was signed with RCA-Astro Electronics in October 1985 for a single series 4000 satellite operating in the FSS range. Europe's first private satellite began operational programme transmissions 1 February 1989. SES then acquired Satcom K3, previously owned by Crimson Satellite Associates. The 5000-class satellite was in an advanced stage of completion and required only modifications to its transmission characteristics. Transmitting on adjacent frequencies and collocated with Astra 1A from 1991, it doubled the number of channels to 32. Rather than using its more powerful tubes to increase EIRP, the 52 dBW coverage is extended. Hughes was awarded the contract of about US$428 million in December 1990 for two further satellites, based on the HS-601 platform, and their launches on Ariane 4 for collocation at 19.2° east in a 140 km cube with 1A and 1B. 1C operates below 1A's frequencies and acts primarily as its back-up, replacing transponders if they fail. 1D acts as 1B and 1C's back-up but its frequencies at the very bottom of the FSS range require receiver retrofitting. It can also switch to BSS frequencies that can be employed for HDTV. Astra 1E was approved in 1992 to provide digital DTH programming. An ECU220 million loan facility was contracted in February 1993 to finance it, bringing total SES capital investment to ECU1 billion. Astra 1F was approved in 1993, also for 19.2° east and digital DTH; the enhanced 1G was ordered January 1995 and the identical 1H in July 1995 to back up 1E to 1G. Astra 1H is scheduled for launch in 1999 followed by Astra 1K in 2000 and Astra 1L in 2002-2003. An ECU175 million loan facility was contracted September 1995; ECU360 million was added July 1996. Total investment through 1996 was ECU1,250 million. The first two satellites of the second-generation, replacing the first and expanding into other business areas from the new 28.2° east slot were ordered in June 1996 and are scheduled for launch in 1999 and 2001. Also, Ka-band services will be introduced by 1H.

Luxembourg has filed for eight further 10.70 to 12.75 GHz slots: 24.2, 26.2, 28.2, 31.5, 35.5, 37.5, 41.2, 43.2° east.

Astra's Betzdorf control facility ***1996***

MALAYSIA

Binariang SDN BHD

Binariang signed an MoU in November 1991 with Hughes for two HS-376 satellites to provide domestic and regional services. The first MEASAT (Malaysian-East Asian Satellite) launch was then planned for 1994 aboard Ariane. Before MEASAT appeared, Malaysia spent some US$30 million annually on Intelsat and Palapa leases. A contract was not signed with Hughes Communications International Inc until 17 May 1994 following a four-month international competition. The agreement covered one satellite to be launched by late 1995, an option for a second (authorised January 1995), the ground control station and training for the Malaysian operators. This HS-376 is the first to carry GaAs solar cells (increased output), a lightweight shaped antenna (improves gain and eliminates multiple feedhorns) and a bipropellant stationkeeping propulsion system. Binariang also holds a licence for international gateway services. MEASAT 1 was launched by Ariane 44L in January 1996 followed by MEASAT 2 in November 1996 on another Ariane 44L.

NORWAY

Telenor Satellite Services AS

TNSS, a Telenor subsidiary, is the national signatory to Intelsat, Inmarsat and Eutelsat. The country is Inmarsat's fourth largest member, with 6.8 per cent in 1996, complementing the size of its maritime fleet and emphasising the importance of the major Earth station at Eik. TNSS is among the largest VSAT and gateway providers in Europe, with an installed base of 1,000 terminals by end-1995. Telenor is a member of the Skyphone consortium providing aeronautical communications services through the Inmarsat system.

Telenor, known as Norwegian Telecom until 1 January 1995, bought the UK's Marcopolo 2 satellite in 1992 for a reported £25 million to provide DTH services from 0.8° west to a potential six million homes. Germany's TV-Sat 2 was added on lease in March 1995 at 0.6° west for the rest of its life. In addition, Telenor leases transponders on an Intelsat at 1° west, employing the Scandinavian spot beam at 50 W per channel. A Nordic channel was previously provided from Tele-X at 5° east. Approximately 20 TV channels (analogue and digital) are currently transmitted from TNSS's 1° west transponders. 35 channels were planned by 1998 by investing NKr1.5 billion in two new satellites. The Thor 2A contract with Hughes was announced in November 1995, with 15 transponders and the larger Thor 2B contract is imminent.

SINGAPORE

Government Sponsored Organisations

Nanyan Technological University and the UK's University of Surrey signed an MoU in September 1995 for joint research and training in satellite engineering communications satellites and, specifically, LEO technology. This will enable NTU to help Singapore develop its own LEO satellite industry. NTU's Merlion initiative is acquiring LEO engineering and applications expertise, at first operating a Surrey-provided ground station and, in March 1996, signing a £1.35 million contract for Surrey's UoSAT 12 to carry NTU's communications payload, funded by the National Science and Technology Board of Singapore. NTU's package comprises: bent pipe L/S-band up/down transponder; L/S-band processing transponder; propagation beacons; processor for various modulation/coding methods and protocols. The ground station will be upgraded to operate at these frequencies. Experiments will include real-time mobile communications and satellite linking of vehicles in GPS-based fleet tracking/control. It is proposed as a follow-up to fly a satellite in equatorial LEO; five would provide continuous coverage for Singapore.

Singapore Telecommunications Ltd

MMS was selected by Singapore Telecom and Taiwan's Directorate General of Telecommunications in March 1996 for the ST-1 satellite to be delivered in-orbit within 24 months and ground stations in Singapore and Taipei. Total project cost is about US$240 million. The Eurostar 3000, approximately 3,000 kg and approximately 12-year life, will be positioned at Singapore's 88° east slot to provide regional broadcast and telecom services using 14 C-band and 16 Ku-band (about eight serving India) transponders. The Ariane first quarter 1998 launch contract was signed 30 July 1996. Singapore Telecom also has a 14 per cent interest in AMSC. ST-1 was eventually launched in August 1998 and will be followed by ST-2 in 2002-2003.

SWEDEN

Swedish Space Corporation

The Nordic countries of Denmark, Finland, Iceland, Norway and Sweden began studying a joint regional satellite system in the mid-1970s. From the beginning, TV direct broadcasting via satellite was regarded as a major role for such a system. In 1977, the World Administrative Radio Conference on broadcast satellite services allocated eight channels to the Nordic group as an entity (excluding Iceland), plus another 17 to the individual countries (including five for Iceland). What became the Tele-X project assumed its current form in 1982, when it was envisaged as providing both DBS and business communications services within Scandinavia, and also as a means of stimulating the space engineering capabilities of these countries. The 'X' represented Experimental and it was planned to have only one satellite in orbit. It was to provide two high-power DBS TV channels (plus one back-up, which might be used operationally) and two plus one video/data channels.

By then, Denmark and Iceland had withdrawn from the project, though Denmark had earlier expressed interest in a beam covering southeast Greenland. An agreement between Sweden and Norway was concluded in 1983, giving them 85 per cent and 15 per cent interests, respectively, in NSAB (Nordiska Satellit AB). 3 per cent of Sweden's share of cost/work in the project (but not in NSAB) was later assigned to Finland in a separate bilateral agreement.

Taking advantage of the Franco-German TDF/TV-Sat programme, initiated in 1980, the Swedish Space Corp ordered a similar satellite based on the Spacebus 300 platform. The contract was signed with both the Eurosatellite consortium and one of its members, Aerospatiale, specifying the latter as prime contractor. NSAB was to be proprietor of the system, while SSC was to be the procurement and executive agency. After launch, ownership and operational responsibility was to pass to a PTT-owned consortium, Notelsat (50/50 per cent Sweden/Norway). Shortly before launch, Norway withdrew from NSAB, trading its 15 per cent share for a lifetime lease of one channel. NSAB is now a 50/50 partnership between SSC and Teracom, a major distributor of TV/radio programming. SSC became wholly responsible for operating the system. System cost is SKr1.25 billion at fixed 1982 rates, now about

Tele-X is based on the Spacebus 300 platform (Aerospatiale) ***1994***

SKr1.8 billion, or US$280 million. All three DBS TV channels are occupied: the Norwegian public service transmits in D-MAC, and TV4 (from September 1990) and TV5 Nordic (from November 1991) operate under PAL.

THAILAND

Asian Broadcasting and Communications Network Company Ltd

ABCN was established in 1994 as a subsidiary of Thailand's The M Group Plc to develop joint ventures with regional governments for satellite systems. A contract was signed November 1995 with the Lao People's Democratic Republic to jointly manage and operate a system from January 1998 at 116° east; the concession extends 30 years. ABCN holds 80 per cent of the shares in the LaoStar Company set up with Laos.

Shinawatra Satellite Public Company Ltd

The government invited tenders in 1987 for a domestic satellite system; the winner would receive a 30 year concession to lease transponders to government and private sector users, with excess capacity offered to neighbouring countries. Thailand's population is about 55 million. Shinawatra Computer & Communications Group (SC&C; chairman Dr Taksin Shinawatra), the largest group of Thai companies operating telecommunications and broadcasting service concessions, was selected in 1991. It has an eight-year monopoly to supply satellite capacity to domestic users (and was required to pay existing users of other satellites to transfer to Thaicom). It began formal negotiations June 1991 with Hughes, signing the contract 8 October 1991 at around US$100 million for two satellites, ground equipment and training. Deliveries were due within 24 to 28 months. The first launch contract was signed with Arianespace 6 December 1991. As lighter versions of Hughes' HS-376 bus, Thaicom was the second to come within Ariane's 1 tonne SDS (Spelda Dedicated Satellite) class. Shinawatra signed a letter of intent 1 November 1993 to negotiate with Hughes for a third satellite but Aerospatiale's Spacebus 3000 platform was selected December 1994 for the more powerful Thaicom 3; negotiations for Thaicom 4 continued. Total cost for Thaicom 3 on station is estimated at US$240 million. 1994 revenue was Bt371.60 million, for a Bt156.50 million loss.

Thailand previously leased 2¼ transponders aboard AsiaSat 1 for telecom services. Thaicom 1 was intended to be placed at 101° east and Thaicom 2 at 78.5° east but an agreement was reached with AsiaSat September 1993 to co-locate at the more westerly slot to avoid interference with AsiaSat 2 at 100.5° east. Thaicom 3 is co-located with Thaicom 2 at 78.5° east.

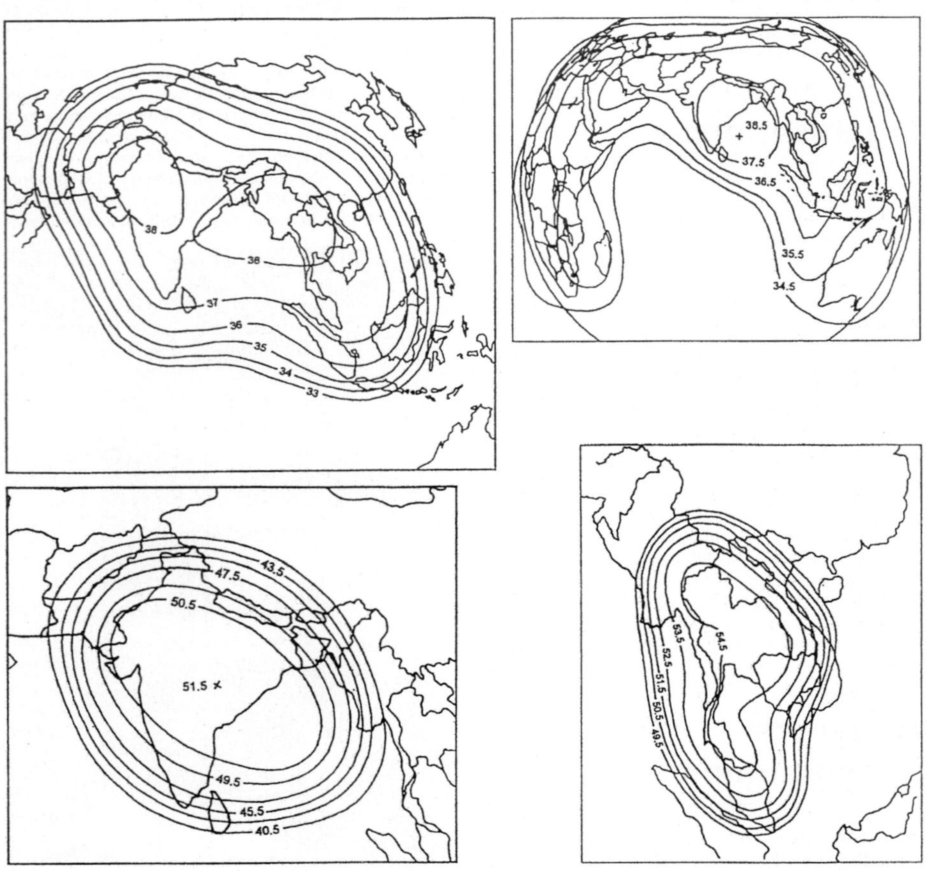

Thaicom 3 footprints (dBW). Top left: C-band India/Indochina; top right: C-band hemispherical; bottom left: Ku-band steerable (India); bottom right: Ku-band Thailand (Shinawatra Satellite Public Co Ltd) ***1996***

UNITED STATES OF AMERICA

American Mobile Satellite Corporation

AMSC was formed in May 1988 by a consortium of telecom companies to launch and operate a mobile satellite system. Four companies own 61 per cent: AT&T Wireless Services (13 per cent), Hughes Communications, Inc (27 per cent), Mtel Corp (7 per cent) and Singapore Telecom (14 per cent). US$236 million was raised by December 1993's stock offering to complete the US$605 million financing. 1995 operating revenues were US$6.87 million (1994 US$3.66 million), for US$66.9 million net loss (1994 US$21.1 million loss). The FCC licensed AMSC in May 1989 to provide voice and data Skycell communications to mobile users on land (Conus, Hawaii, Alaska, Puerto Rico and Virgin Islands), up to 320 km out at sea and in the air. AMSC has an agreement with Canada's TMI Communications to operate compatible systems and use each other's satellite as a spare. The companies issued a joint RfP 7 July 1989 for one satellite each to provide mobile communications for North America from 106.5° west (TMI) and 101° west (AMSC) GEO. The Hughes/Spar Aerospace team was announced as winner in December 1990, with Hughes acting as prime for AMSC and Spar for TMI. Cost was about US$100 million per satellite. Joint operation reduced start-up costs (by about US$200 million), ensured a high degree of mutual back-up and protection, each allowing users to roam anywhere over North America with continuous coverage. Each provides up to 2,000 simultaneous voice channels through L-band spot beams. Westinghouse Electronic Systems Group was awarded a US$80 million contract in May 1992 to integrate a ground segment network and to develop and manufacture satellite phones for both systems. Mitsubishi Electric Corporation is also manufacturing phones for the AMSC system. Combined manufacturing capacity was expected to be 10,000 monthly by mid-1996.

The initial voice service began January 1996, using US$3,000 transportable satellite phone. The full range of Skycell voice products was introduced through 1996: land mobile January 1996, US$1,800 to US$2,300 (dual-mode service 1 February); aeronautical February, US$15,000; maritime March, US$4,000 to US$6,000: nationwide voice dispatch, June.

AMSC subsidiary American Mobile Radio Corporation filed an FCC application 15 December 1992 to provide 55 radio broadcast channels (11 CD-quality) from two satellites at 99° and 103° west. Hughes Space and Communications has been selected to build the US$528 million system's satellites, although no contract has yet been signed. AMSC filed in 1994 for two PCSAT GEO satellites to be launched about 1998, operating in the 2 GHz band set aside by the FCC for emerging technologies. AMSC subsidiary Personal Satellite Communications Corporation would operate the US$872 million system to provide mobile phone, paging and wideband data services.

AT&T Alascom

AT&T Alascom, a subsidiary of AT&T, employs the Aurora satellite to carry domestic telephone, the Alaskan Television Network and emergency messages. Aurora 1 was launched in 1982 as RCA Satcom 5, the first RCA (now Lockheed Martin Astro Space) 'Advanced Satcom' C-band spacecraft to employ SSPAs in its 24 active transponders. AT&T Alascom bought the satellite for US$88 million and leased back four transponders to RCA; GE Americom provided TT&C services. GE's Satcom 1R at 139° west and then C-1 at 137° west acted as in-orbit back-ups. Customers had to be moved to C-1 in late 1990 when Aurora suffered service interruptions because of attitude control problems, forcing retirement in 1991 (the satellite is still held inclined 4.0° at 105° west. Aurora 2 (Satcom C5) was launched in 1991 to 139° west as replacement, incorporating 24 active 11W C-band SSPAs. GE Americom continues to provide operational control. Although it employs basically the same GE 3000 bus, it offers a lifetime of 12 years, an improved redundancy scheme allowing instantaneous payload reconfiguration, Ni/H_2 batteries, larger hydrazine tanks and electro-thermal thrusters. AT&T Alascom signed a US$100 million agreement with GE Americom in spring 1990 to procure two-thirds of the satellite's capacity. Since then, AT&T Alascom has bought two more transponders, leaving GE Americom with a quarter ownership. Telstar 6 will replace it at 129° west in 1999.

Constellation Communications, Inc

A January 1995 Memorandum of Understanding between CCI (Bell Atlantic is the major shareholder) and Brazil's Telebras could develop the ECCO satellite constellation to provide voice, data, position location and other mobile services for rural and remote areas poorly covered by terrestrial systems. Eleven 425 kg low-cost satellites in 2,000 km, 0° orbits (plus a spare) would yield continuous coverage between 23° north/south. System cost: less than US$500 million. Later, five satellites (plus a spare) in each of seven 62° planes would extend coverage to 71° north/south.

DirecTV, Inc

DirecTV was established by Hughes Electronics Corporation in 1992 to operate a domestic US DTH system using Ku-band HS-601s; it became an independent subsidiary in November 1993. DBS 1 was the 100th Hughes-designed telecom satellite to be launched. A 3,720 m² programming uplink facility was added to Hughes Communications Inc's Castle Rock (Colorado) station, working via 13 m antennas, capable of delivering 216 simultaneous channels. DirecTV has an FCC licence to co-locate up to four HS-601 satellites around 101° west offering a total of 27 transponders.

The demise of the Sky Cable consortium of Hughes, Cablevision Systems, NBC and The News Corporation was followed in June 1991 by HCI's creation of a Direct Broadcast Services business unit. Hughes unveiled its DirecTV direct-to-home distribution system at the same time and simultaneously sold five of the 16 transponders aboard the first satellite to the USSB United States Satellite Broadcasting Inc subsidiary of Hubbard Broadcasting for US$100 million plus. The channels are BSS 24, 26, 28, 30, 32. DirecTV owns all 16 on DBS 2. The two satellites will deliver 175 channels. Each transponder can provide four video channels using digital compression, or eight films. The venture requires three million subscribers to recover its US$750 million cost; it projects 10 million by 2000. (one million was achieved as at 2 November 1995).

DirecTV can provide 175 channels directly to 45 cm home antennas (Hughes) ***1994***

EchoStar Satellite Corporation

EchoStar Satellite Corporation signed a contract with Martin Marietta Astro Space in October 1992 for one 7000-series satellite, plus options on six others, with launch operations and support for 100 US DTH channels via 45 cm dishes. One million receivers are projected after one year. EchoStar is 100 per cent owned by EchoStar Communications Corporation, which had 1995 revenues of US$163.9 million (1994: US$190.6 million), for a US$11 million loss. The company is programming some of its own capacity and leases some channels on a non-common carrier basis. EchoStar's own 5,580 m^2 programme uplink facility is in Cheyenne, Wyoming.

The second satellite could use one transponder to provide 30 radio broadcast channels. Sister company Sky Highway Radio Corporation filed an FCC application 15 December 1992. EchoStar has an 80 per cent interest in the E-Sat Inc joint venture with DBS Industries Inc to launch six LEO data relay satellites serving mainly electric and gas utilities. Cost is estimated at less than US$70 million.

Directsat merged with EchoStar in January 1994 and has FCC approval to merge their DBS licences for 119° west. Directsat had been allocated 10 BSS channels at 119° west by the FCC in December 1993. EchoStar made the US$52.3 million winning bid at the FCC's 26 January 1996 auction for the 24-channel slot at 148° west. EchoStar filed an application before the FCC's 29 September 1995 deadline to build a Ka-band broadband system to deliver high-rate data and video services via small terminals.

Ellipsat International Inc

Ellipsat was licensed by the FCC in August 1992 to launch four experimental Ellipso satellites. However, it was not one of the three Big LEO mobile satellite systems awarded licences 31 January 1995. It was granted a further year to file an amended application, although it was appealing against the decision. The system would provide CDMA hand-held communications using 10 1.6 MHz channels in 1.610 to 1.6266 MHz up and 2.4835 to 2.500 MHz down. Construction, launch and one year of operation is projected at US$564 million for a 16-satellite system. The satellites would be simple repeaters as all call setup/processing would be performed on the ground and terrestrial networks would provide the long-distance links. Ellipsat International and Ellipsat Corporation are subsidiaries of Mobile Communications Holdings Inc (MCHI) for the international and domestic markets, respectively. A strategic alliance was formed in 1992 with Fairchild Space & Defense Corporation (Fairchild Space is now part of Orbital Sciences Corporation), which would also provide the bus. However, it was reported in 1996 that an agreement was expected with Rockwell, after OSC withdrew 1995. Other earlier strategic partners include Harris Corporation (payload), Westinghouse Electric Corporation (ground system), InterDigital Communications Corporation (terminal transmit technology) and Barclays de Zoete Wedd (financial advisors). MBT Systems and Space Technology, a subsidiary of Israel Aircraft Industries' Electronics Division, would provide some hardware and act as Ellipsat's local service provider. In addition to these shareholders, financial support would be provided by Arianespace, Cable and Wireless Spectrum Network Systems and Satellite Transmission Systems. Ellipsat estimates that 350,000 subscribers would be required to break even, paying a US$50 monthly fee and US$0.50 per minute for calls.

Four satellites would be sufficient for daylight coverage of the northern hemisphere. The 520 × 7,800 km, 3 hour, 116.6° Sun-synchronous 'Borealis' orbit would follow the Sun and hence daylight demand. 24 hour northern coverage would be provided by eight satellites. Eight satellites in the 'Concordia' 4,000 × 7,800 km equatorial element would add coverage down to 40° south.

GE Americom

GE American Communications Inc, a subsidiary of GE Capital Services, operates 12 domestic satellites (as of January 1996) servicing cable TV, broadcast radio/TV, government and business information, and maintains a supporting network of Earth stations, central terminal offices and five TT&C facilities. The company also installs, operates and maintains dedicated Earth stations for government customers such as NASA, NOAA and DoD. It acquired GTE Spacenet in 1994 (then operating eight satellites), making it the nation's largest commercial operator and a major VSAT company. GE Americom filed an application before the FCC's 29 September 1995 deadline to build a Ka-band broadband system to deliver high-rate data and video services via small home terminals. GE's system would use nine satellites in five GEO slots. It has also applied for a second round Little LEO licence to operate a constellation of 24 small LEO satellites to provide data relay services in four planes at 800 km 98°. This is in addition to its 80 per cent stake in the earlier Starsys system, which was licensed November 1995.

RCA Americom was created in 1976 as a wholly owned RCA subsidiary to own and operate the parent company's domestic system. Following GE's 1987 acquisition, GE Americom began operating as part of GE's Communications & Services Group.

Globalstar Limited Partnership

Loral Corporation and Qualcomm Inc in 1991 formed LQSS Loral Qualcomm Satellite Services Inc. Loral owns directly and indirectly approximately 34 per cent of the outstanding equity of Globalstar. The company filed an FCC application in June 1991 for the US$2 billion Globalstar mobile communications system. The licence was awarded 31 January 1995. The name was changed to Globalstar LP in March 1994 to reflect expansion of the partnership: Loral, Qualcomm, AirTouch Communications, Alcatel, Finmeccanica, DACOM DASA, France Telecom, Hyundai, Space Systems/Loral and Vodafone. SS/L's fixed-price contract for 56 satellites is US$1.3 billion; each will cost US$13.1 million plus US$1.6 million in-orbit incentives.

Globalstar is a worldwide, low-earth orbit satellite services system that will provide low-cost, high-quality telephony and other digital telecommunications services, such as data transmission, paging, facsimile and position location. Globalstar will provide access to existing terrestrial cellular telephone services as it extends, enhances, and operates with public land mobile networks, public switched telephone networks and government and private networks. Users of Globalstar will make or receive calls using hand-held or vehicle-mounted terminals similar to today's cellular phones that will be able to switch from conventional cellular telephony to satellite telephony as required. In remote areas with little or no existing wireline telephony, subscribers will communicate through fixed-site telephones, similar to phone booths or ordinary wireline telephones.

Globalstar's designated service providers, a worldwide network of regional and local telecommunications service providers that includes its strategic partners, will sell access to the service and seek to obtain all necessary local regulatory approvals. Currently, Globalstar has service provider agreements in more than 100 countries, accounting for approximately 88 per cent of the world's population. The service providers – AirTouch Communications, Dacom/Hyundai, France Telecom/Alcatel, Vodafone, Elsag Bailey and Loral – each have the exclusive right to offer Globalstar service in its operating areas. The countries in which Globalstar has most recently established service provider agreements include Brazil, Canada, China, India, Mexico. The first cluster of four Globalstar satellites was launched by Delta 2 in February 1998 followed by a second cluster of four in April 1998. Twelve satellites were lost in launch vehicle failure.

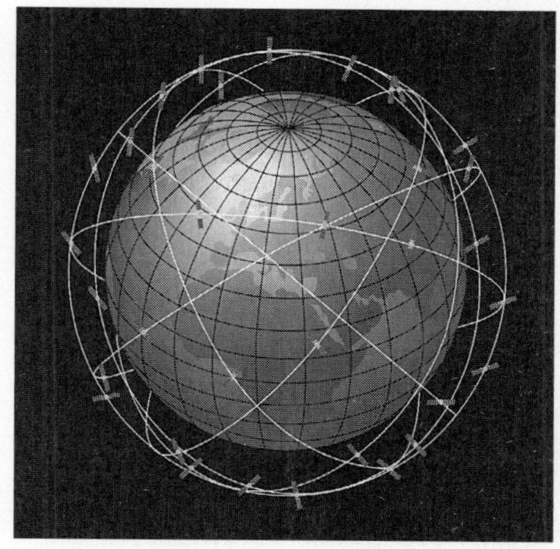

The planned 48-satellite Globalstar constellation, which was to be complete by early 1999 (Globalstar) ***1997***/0005570

Hughes Communications, Inc

HCI is a wholly owned subsidiary of Hughes Electronics Corporation. It operates the world's largest fleet of privately owned commercial communications satellites.

HCI sold off its 30 per cent interest in the Japan Communications Satellite Company (JCSat, now Japan Satellite Systems) in June 1993. The two 32-transponder Ku-band satellites were provided by HCI, which also built JCSat's two satellite control stations. DirecTV, Inc was established by HCI in 1992 to operate a domestic US DBS system using three Ku-band HS-601s (first launch was successful December 1993); DirecTV became an independent Hughes Electronics subsidiary in November 1993. The company also holds a 27 per cent interest in the AMSC American Mobile Satellite Corporation, whose first satellite appeared April 1995.

Current services include radio and cable TV distribution, customised voice, video and data communications, and end-to-end private business network services. Video services provide for cable TV and network broadcasts via Galaxy/SBS, Video Time-sharing Services (VTS) for occasional users, and end-to-end International Television Services (ITS) for direct, broadcast-quality audio and video. Business communications cover data, digital voice and teleconferencing applications; International Business Services (IBS) for direct, high-speed all-digital satellite links connecting customer premises with major foreign business centres using Intelsat and Galaxy Earth stations; Integrated Satellite Business Network (ISBN) for private two-way satellite transmissions for data, video and voice traffic between a hub facility (shared or dedicated) and personal Earth stations (VSATs of 1.8 to 2.4 m diameter or less) at business locations. HCI also provides the Earth stations for integration with the space segment.

Hughes Space and Communications Company

HSC, created in August 1970 around the Space Systems Division, has been responsible for 178 launched satellites and probes (as of May 1999), including more than 40 per cent of all the world's commercial communications satellites. Its 100th telecommunications satellite was launched in December 1993 and its 100th commercial telecommunications satellite in December 1994. 1991 to 96 sales each totalled about US$2 billion per annum (1989: US$1.7 billion; 1988 US$1.4 billion); it has 7,200 employees. Hughes' first major space projects were the Surveyor lunar soft lander and the Syncom geosynchronous communications demonstrator. The company was responsible for Early Bird (Intelsat 1) and the four Intelsat 2s. Its USAF Tacsat tactical communications craft in 1969 demonstrated technology for future generations, introduced to the international market by Intelsat 4/4A. HS 333 became the first generation of domestic telecommunications satellites, with new technology shaping the beam footprint to the host country's requirements. Anik A1 in 1972 was the first domestic GEO satellite. That HS 333 was joined by Canada's Anik A2/A3, Westar 1-3 and Palapa A1/A2; Westar 3 was the last to leave service, in 1990.

Hughes also provided the three Marisats to COMSAT General for the first maritime service in 1976. 22 years later they were still in service. It then provided the payload to British Aerospace for the second-generation Inmarsat. The Comstar design for COMSAT introduced the use of dual polarization to double capacity, preceding development of the most successful commercial satellite yet: the HS 376, first launched in November 1980. By February 1997, 50 had been ordered for domestic and regional systems on five continents. The advent of the Shuttle resulted in the wide-body communications satellite: five Leasat HS 381 models were built to provide US Navy links. A blending of the wide-body approach with the deployable functions of the 376 led to the development, in the early 1980s, of the international Intelsat 6 (five ordered) and the third-generation HS 393 domestic platform (three flown). Australia's Optus B1 introduced the modular three-axis HS 601 design in August 1992. The first HS 601 HP (high performance) version was ordered in January 1995 for Luxembourg's Astra 1G, offering increased power, enlarged payload and ion propulsion. The first 601 HP launch should be PAS 5. HS 702 was unveiled in October 1995, the mass savings increasing the transponder number by 50 per cent. 1998's Galaxy 1 will be the first user.

In addition, Hughes was responsible for five ATS Applications Technology Satellites, numbers 1 and 3 of which remain functional after more than 28 years. It also built 1975's OSO 8 solar observatory, the five Pioneer Venus probes, Magellan's radar payload, Galileo's atmospheric probe, the second-generation GOES and Japan's five HS 378 GMS meteorological satellites. Further information is included in the appropriate sections of *Jane's Space Directory*. HSC's Springback 20.4 kg 4.9 × 6.7 m graphite mesh antennas were introduced on AMSC 1 in 1995. The antennas are rolled into a 5 m-high cone for launch, taking advantage of the normally unused space in the forward fairing. It is assembled on a large graphite mandrel and then baked. Hughes also designs, develops, builds and commissions telemetry, command and satellite control ground terminal facilities on a worldwide basis.

Facilities: HSC has been delivering one satellite per month (on average) for more than three years. As of May 1999 there are 37 satellites ordered to be built and launched before 2001. The smaller spinner requires less than 18 months to build. In 1992, Hughes began streamlining operations so that even an HS 601 or HS 702 from initial contract to launch site delivery requires 24 months or less. Spacecraft are assembled into a full-up configuration and tested for flight readiness in the company's Systems Integration Activity area. The Systems Test and Engineering Laboratory undertakes systems level integration and testing.

Hughes 376

Until HS 601, the HS 376 was numerically the most successful commercial communications satellite, with 54 ordered for 19 domestic and regional communications systems since it was introduced in 1980 to replace the HS-333. As of February 1997, 35 of the 53 orbited since SBS 1 remained operational. The total includes the two Canadian Anik Ds and the Brazilian Brasilsat-A series, built by Spar Aerospace employing licensed HS 376 technology. With a mission life of up to 12 years, it can accommodate up to 24 transponders. Mexico's Morelos was the first to provide for the simultaneous use of C and Ku-band channels, and was the first Hughes model to incorporate a planar array, for receiving the four Ku-band channels. The UK's BSB satellites, numbers 31 and 32 ordered, were the first to use the 376 for DBS applications. They carried a larger solar drum for increased power and unusually created the 110 W DBS output by linking paired 55 W transponders. They were later sold in-orbit to companies in Sweden and Norway. Malaysia's MEASATs are the first to employ GaAs cells and a shaped antenna to eliminate multiple feedhorns.

The enhanced 376W wide body version almost doubles solar power generation and expands the propulsion capability by adopting some HS-393 features. Hughes was awarded a US$175 million contract in April 1990 for the first two 376Ws as Brasilsat's second generation: Brasilsat B3 was ordered in 1995. The first 376L lightweight version, with 439 kg dry mass, was ordered by Thailand in 1991; the two-satellite contract was worth about US$100 million.

Total ordered: 54 by May 1999 (SBS 1-5, Optus A1-3, Anik C1- 3, Morelos 1/2, BSB 1/2, Telstar 301-303, AsiaSat 1 (Westar 6), Palapa B1/B2P/B2R/B4, Galaxy 1-3/1R/1RR/5/6/9, Westar 4/5, Brasilsat B1/B2/B3/B4, Thaicom 1/2, APStar 1, MEASAT 1/2, BSAT 1A/1B, APStar 1/1A, Chinasat 7, Thor 2A (Brasilsat A1/A2, Anik D1/D2 built by Spar under licence)
Total launched: 53 by May 1999 (BSAT 1A/1B, Brasilsat B3, Thor 2A; Galaxy 1R lost in launch failure)
Operating: 35 as of May 1999
Launch mass: about 1,200 kg (excluding perigee stage); 376W 1,750 kg; 376L 1,080 kg
On-orbit mass: about 700 kg BOL; 376W 1,052 kg BOL; 376L 630 kg BOL
Deployed dimensions: 376/376L 2.16 m diameter, 6.6 m height (2.84 m stowed); 376W 3.65 m diameter, 8.3 m height (3.43 m stowed)
Communications section reliability: more than 0.97 at 10 years
Communications payload mass: about 160 kg
Mission life: up to 12 years, depending on application
Structure: cylindrical conventional design in three sections: bus (including AOCS and so on), solar array and despun payload/antenna platform. The antenna typically incorporates a single despun 1.83 m diameter reflector. The lower telescoping concentric solar array is stowed around the upper body for launch
Power: Si cells mounted on cylindrical surface providing 19.7 mW/cm², 1,000 W BOL, 900 W EOL (376W: 1,400 W BOL; 376L: 800/670 W BOL/EOL) at solstice; two Ni/Cd or NiH_2 batteries (376L: 52 Ah) provide 100 per cent eclipse protection. BSB's bottom drum was lengthened 25 cm for 1,100 W BOL. MEASAT is the first to employ GaAs, providing 1.2 kW to the payload
Pointing: antenna accuracy of 0.05° in azimuth and elevation via RF beacon
AOCS: spin-stabilised at 50 rpm by four 22.2 N mono thrusters (two axial plus two radial) supplied with up to 210 kg hydrazine from four peripheral tanks; GEO insertion provided by solid propellant Thiokol Star 30 AKM. MEASAT uses bipropellant HS 601 N_2O_4/MMH attitude thrusters plus Star 30. The HS 376W adopts HS 393 elements: a central Kaiser Marquardt R-4D 490 liquid thruster for GEO insertion plus four 22.2 N ARC attitude thrusters
Thermal systems: quartz mirror radiator band encircling upper body provides primary heat rejection path; area 4.0 m²

Hughes 601

HS 601 is the company's first three-axis design, accommodating up to three bands simultaneously for high-power services such as land mobile, large area DTH and HDTV. The first commercial contract was for Australia's Optus B; Hughes invested US$100 million in platform development. 10 are for USN under a US$1.8 billion contract, including launches. The Solidaridad contract is worth US$183.47 million. The ¥30 billion JCSat 3 contract was signed 29 October 1993. JCSat ordered numbers 4, 5 and 6 in 1996 and 1997. Modular construction reduces usual production time by about a year. The control system incorporates MIL STD 1750A microprocessor (16k RAM). As part of its on-orbit delivery packages, Hughes paid US$128.8 million in early 1992 to cover eight satellites for approximately US$800 million insurance through brokers Marsh and McLennan Aviation and C T Bowring: two Optus, three UFO, three Galaxy. Luxembourg's SES in January 1995 ordered the first HS 601 HP (high performance) version for its Astra 1G. It subsequently ordered two more in 1996 – Astros 1H and 2A. The twin four-panel GaAs wings provide 8 kW BOL, the enlarged payload can carry 32 active 100 W TWTAs, the XIPS xenon ion propulsion system provides NSSK and lightweight contoured reflectors reduces mass. NASA on 23 February 1995 selected medium power HS 601s for its three TDRS H-J under a US$481.6 million fixed-price contract. First delivery is due 1999. Japan's Space Communications Corp ordered the 42nd HS 601 2 March 1995 for its Superbird C, with Mitsubishi Electric providing the cell panels for the solar wings. PanAmSat ordered an HS 601 HP 10 March 1995 for PAS 5 due for launch in 1997. An order of 12 ICO Global Communications satellites, worth approximately US$1.3 billion, was announced 20 July 1995; the final contract was signed 6 October 1995. This makes the HS 601 the most numerous model with 65 ordered as of February 1997. An additional US$925 million launch services contract for the ICO satellites was signed 8 December. Hughes Electronics Corporation also became a strategic investor, with a US$93.8 million investment. This will be the first HS 601 application in intermediate orbits; the design will employ GaAs cells, a simplified orbit adjust system (the launcher will be responsible for direct placement), an enhanced thermal radiator and a modified attitude control system.

Total ordered: 73 through May 1999 (Optus 3; UHF/FO 10; PanAmSat 14; DirecTV 4; SES 7; AMS/TMI 2; Mexico 3; PT Satelindo 2; APT 2; AST 2; SpaceCom 2; Loral 1; NASA 3; NOAA 2; Japan Sat 4; ICO 12)
Total launched: 48 through 4 May 1999
Number operating: 40 through to May 1999 (Optus B2, PAS 3, APStar 2 lost during launch; UFO 1 stranded in low orbit by launcher; AsiaSat 3; Galaxy 10; Orion 3)
On-orbit mass: 1,582 kg on-station BOL (Optus B), 1,925 kg ICO
Deployed dimensions: 0.227 m³ body, solar arrays span 18.3-26 m depending on three- or four-panel wings (satellite envelope 229 × 254 × 254 cm at launch)
Communication section reliability: not available
Communication payload mass: not available
Communication payload power: 3-4.8 kW RF
Design life: 10-15 years, depending on applications
Structure: conventional aluminum structure; payload mounted on north/south panels and Earth-facing floor; east/west faces carry hardpoints for antenna mountings, fixed antenna mountings on Earth-facing wall
Power: twin solar wings of three or four 2.16 × 2.54 m panels carrying K4¾ large area Si cells on Kevlar substrate generate up to 6 kW. HS-601 HP employs twin four-panel GaAs wings for 8 kW BOL. NiH_2 batteries provide 100 per cent eclipse protection
Pointing: 0.025°
AOCS: provided by unified 12 × 22 North Atlantic Research Corporation and single 490 N Kaiser Marquardt R-4D bipropellant thrusters incorporating onboard control processor, Sun and Barnes Earth sensors and two 61N ms two-axis gimballed momentum bias wheels. 1,658 kg NTO/MMH in four spheres. Solar tacking techniques reduce conventional Δ V requirements. Spin-stabilised in transfer orbit. When required, Thiokol's 4.7 tonne Star 63 provides perigee kick. HS-601 HP carries Hughes' XIPS xenon ion propulsion subsystems for NSSK. Comprises four 13 cm 18 mN thrusters (two primary plus two back-up), power supply (439 W input each thruster off 29 to 34 V DC bus) and a propellant storage/control unit. Total XIPS package mass 68 kg, for a 400 kg saving. In normal NSSK operation, the thrusters fire for 2 to 3 hours daily.

Hughes 702

HS 702 performance spans and extends HS 601's range. A payload module tailored to customer preference, and able to carry 90 high-power transponders, (72 plus spares) mounts to a common bus module by four bolts and six electrical connectors. Production cycle is about 24 months.

Total ordered: 9 through March 1999 (Galaxy II, 3-C and PAS-1R; XM-1 and XM-2; Anik F1; Spaceway (3))
Total launched: 0 through February 1996
Number operating: 0 through May 1999 (first launch due in late 1999)
On-orbit mass: up to 5,200 kg at launch
Deployed dimensions: solar arrays span 40 m
Communication section reliability: not available
Communication payload mass: up to 1,200 kg
Communication payload power: 10 kW RF
Design life: 15 years
Structure: conventional aluminum structure; payload module bolted to common bus. East/west faces carry hardpoints for antennas up to 272 cm, fixed-antenna mountings on Earth-facing wall
Power: twin four-panel GaAs wings for 10 kW BOL, growth to 15 kW planned. NiH_2 batteries provide 100 per cent eclipse protection
Pointing: 0.025° being achieved
AOCS: provided by bipropellant (1,700 kg NTO/MMH) thrusters incorporating onboard control processor, Earth sensors and MWs. Hughes' XIPS xenon ion propulsion subsystem for NSSK: four 25 cm thrusters (two primary plus two back-up) consuming only 5 kg per year. XIPS can also perform final orbit insertion.

Iridium, Inc

Motorola unveiled its Iridium mobile communications system proposal in June 1990 after 2½ years of internal work. Iridium planned a system of 66 small LEO satellites by end-1998 providing global pocket mobile digital 4.8 kbit/s telephone services via onboard switching and inter-satellite links for US$3 per minute to the user. Handsets will retail at about US$3,000. Cost was projected at US$3.45 billion for establishing the operational system of 66 satellites and 14 in-orbit spares by 1998, followed by another US$2.8 billion over the next five years for operations and maintenance. 800,000 users are required within five years for viability. Motorola projects 1.824 million by 2001, broken down as 42 per cent pager, 42 per cent portable phone and 7 per cent mobile. After 10 years, in 2006, it projects 3.224 million. Viability is driven by the terminal.

Iridium, Inc was incorporated 14 June 1991 to own and operate the system, with Motorola eventually holding 15 per cent. Owner organisations are: Motorola, Lockheed Martin, Iridium Africa Corporation, Iridium Canada Inc, Iridium India Telecom Private Ltd, Iridium Middle East Corporation, Iridium SudAmerica Corporation, Nippon Iridium Corporation, Korea Mobile

Telecommunications Corporation, Raytheon, Sprint Corporation, STET (Italy), Khrunichev (Russia), China Great Wall Industry Corporation, Pacific Electric Wire & Cable Co (Taiwan), Thai Satellite Telecommunications Corporation and Vebacom GmbH (Germany). A total of US$1.888 billion was raised by early 1996; the remainder will come from debt financing.

The proposal was submitted to the FCC 3 December 1990 and an experimental licence was awarded August 1992 for five satellites. The full licence was awarded 31 January 1995. Launches were to have begun by end-1996. Lockheed Space Systems Division (now Lockheed Martin) was awarded the US$700 million contract in August 1993 for 120 spacecraft by 2003. 79 are required for the initial system; the remaining 41 will be launched as required. The bus is built in Nashua, New Hampshire at Lockheed Martin Sanders and integrated with Motorola's communications module in Motorola's final assembly facility in Chandler. The production line system there can handle five satellites simultaneously; the goal is for final assembly to require 21 days. Satellite CDR was held June 1992. The power and control systems and structure are similar to Lockheed's venerable Agena upper stage. Volume production allows use of GaAs solar arrays. In February 1993, Motorola announced that it had signed a contract (subject to US government approval) with Khrunichev Enterprises for three Proton launches of seven satellites each 1996 to 98; Khrunichev invested US$40 million in return. A similar agreement was signed with GWIC April 1993 for 10 launches of two satellites each 1996 to 2002 on CZ-2C. Eight Delta 2s were ordered to handle five satellites each. For replacements, five Delta 7420 launches were ordered December 1995, each handling three satellites. Each satellite will de-orbit at EOL to avoid pollution.

Each satellite operates 48 L-band cells 670 km across using FDMA/TDMA. Calls will be routed to the Earth gateways and among the space segment at Ka-band. The user's 0.7 kg 0.6 W handset will automatically select a terrestrial cellular network if available and cheaper. The network will maintain a log of each handset's last known position: each device will determine its location via GPS or Doppler shift and relay it to a visible satellite for passing on to the nearest gateway. Each gateway will comprise three terminals dispersed sufficiently to avoid weather problems affecting links. Scientific-Atlanta is providing 14 terminals under US$22.6 million contracts for the System Control Segment.

The first cluster of five Iridium satellites was launched by Delta 2 in May 1997 and it took less than two years to orbit 66 operational satellites for the prime constellation. Seven were lost in launch vehicle failures but by February 1999 there were a total of 79 satellites in orbit of which 13 were spares. Predicted subscriber levels were far higher than actually achieved, however, and by 1 April 1999, only 10,294 subscribers had signed up to the Iridium system, far short of the 500,000-600,000 needed to begin paying debt on the US$4.85 billion programme. In the first three months of 1999 the company announced revenue of US$1.4 million and a net loss of US$505 million bringing down stock from US$72 in 1998 to US$15 by late April 1999, casting doubts on the viability of high-end satellite telephone services. Analysts did, however, point to the high pricing of Iridium, which (with 1.5 billion minutes of capacity per year) has to charge US$3-8 a minute for calls. It needs to recoup at the rate of US$1.4 a minute just to break even. Competing systems such as Globalstar, with 7 billion minutes of capacity can wholesale at 21 cents while ICO with 5 billion minutes can do it for 23 cents. Iridium's telephone handset price of US$2,700 is set to plummet during 1999, a year in which the company's still expects to sign up a total 500,000 subscribers.

Iridium's network of 66 satellites will provide global pocket mobile telephone services via the L/S-band phased array antennas. Iridium is the 77th element, reflecting the original constellation size ***1995***

Loral Space and Communications

Loral Space & Communications, with headquarters in New York City, was spun off from Loral Corporation in April 1996 as part of the sale of its defense electronics and systems integration businesses to Lockheed Martin Corporation.

On a pro forma basis, combining the results of the operations under its management, the company's revenue for what was a nine-month fiscal year, ending 31 December 1996, was US$950 million. Effective 1 January 1997, Loral will report on a 12-month calendar year basis.

In March 1997, Loral completed its acquisition of Skynet, a leading domestic satellite service provider, from AT&T. Loral Skynet advances Loral's strategy of becoming a global provider of satellite-based communications services and will complement its existing satellite manufacturing capabilities.

Loral Skynet owns 60 major US Earth stations and operates a US$100 million satellite monitor and control facility in Hawley, Pennsylvania.

MCI/News Corporation

MCI Communications Corporation acquired the rights to the 110° west DBS slot at the FCC's 25 January 1996 auction with a bid of US$682.5 million. MCI, America's second largest long-distance phone carrier and News Corporation will use the slot in an equal partnership to deliver more than 100 DTH and data channels via 45 cm dishes. The partners awarded Space Systems and Loral a US$400 million contract in March 1996 for the on-orbit delivery of two satellites, each with 32 Ku-band transponders: fourth quarter 1997 on Proton and second quarter 1998 on Atlas 2AS.

Orbcomm Global LP

OSC's Orbital Communications Corporation subsidiary, created at the beginning of 1990, filed an FCC application 28 February 1990 for a constellation of LEO satellites for two-way communications and geolocation services. Orbcomm will provide low-cost alphanumeric data communications and position determination (375 m) for emergency assistance, data acquisition and messaging services using pocket portable and mobile subscriber terminals. The system is optimised for shorter messages of 6 to 250 characters. OSC and Canada's Teleglobe Inc signed an agreement 26 July 1993 for the joint financing and operation of the system. Teleglobe provided US$10 million of the US$55 million cost of the two first phase satellites. For the second phase of 34 more (including eight ground spare), it has increased its equity share in Orbcomm Global LP to 50 per cent. Teleglobe is contributing US$85 million and OSC US$75 million.

The system comprised 28 43 kg satellites by 1998 at 775 km altitude; 24 at 45° inclination (eight in each of three planes) and two at 70° for high-latitude coverage. Total cost is projected at US$220 million, with four launches aboard OSC's own Pegasus and one on Taurus, plus a set of eight ground spares intended for Pegasus. The first two satellites, in polar orbit, appeared successfully 3 April 1995. Both soon developed receiver problems. Number 2's problems in this respect were solved by a self-activated reset 13 May (subsequent satellites carry a separate receiver to allow reset by ground command). Number 1's subscriber receiver was freed 18 May by sending commands correcting an apparent software blockage; the subscriber transmitter was restored 13 June by commanding battery drain, which triggered a computer reset. Number 2 relayed the system's inaugural message 25 May to the Arcade gateway from a production prototype Panasonic personal communicator at Dulles. Both completed acceptance tests 7 July 1995. A message from an Elisra EL-2000 Data Communicator near the Dead Sea in Israel was successfully routed 25 July 1995 to Dulles via the Georgia gateway. The satellites began providing an initial service 1 February 1996, by which time approximately 300,000 messages had already been sent.

Each satellite carries a GPS receiver for the 375 m position-fixing service using Doppler. OSC's own Magellan Systems Corporation will build hybrid GPS/Orbcomm communicator units from late 1996.

Orbcomm was granted an experimental licence April 1992 for two satellites, plus 1,000 user terminals, to demonstrate the system and initiate revenue services. The full licence was granted October 1994, the first for an LEO mobile constellation. The first two satellites provide the polar coverage for the operational system. The 14.5 kg CDS Capabilities Demonstration Satellite was launched 9 February 1993 into 750 km 25° by Pegasus as a pathfinder to demonstrate services via prototype hand-held communicators built by various electronics manufacturers. Planned for a six-month life, it operated for two years.

Orbcomm seeks initially to serve the US market, where it foresees a total of 10 to 20 million users, beginning with 150,000 the first full year of constellation operations and rising to 5.2 million after seven years. Emergency services are expected to provide 75 to 85 per cent of the market. The company does not seek to offer cross-border services, but it is licensing affiliated companies and PTTs in other countries. Three grades of terminals are envisaged: US$100, 5 W power, signalling and data transmit only; US$150 adding position-fixing; US$400 alphanumeric messaging. The first type requires a US$30 to US$50 annual service charge, with the others supported by US$35 monthly plus usage-linked charges.

Each satellite carries 17 data processors and seven antennas, allowing the relay of 50,000 messages hourly in each direction, or more than five million two-way daily for the 28-satellite system. Four regional gateways will cover the US: Oscilla Georgia, Arcade New York, St Johns Arizona and Washington State, with the Network Control Center in Dulles Virginia backed up in Boulder Colorado. The orbits provide 70° north/south coverage with 5,000 km diameter footprint, allowing users to access the system for more than 95 per cent of the time. Polar region accessibility will be 14 minutes every half hour.

The first launch of the operational Orbcomm satellites took place in December 1997 when a cluster of eight was carried into orbit by a Pegasus XL. By February 1999, 24 satellites had been placed in orbit and the system had three sets of eight satellites in each orbit plane, thus completing the first operational phase.

PanAmSat Corporation

On 16 May 1997, the merger between PanAmSat Corporation and the Galaxy Satellite Services was completed: the newly formed company would retain the name PanAmSat Corporation. The merger had been announced on 20 September 1996 by Hughes Electronics Corporation and PanAmSat Corporation, with the PanAmSat stockholders receiving an aggregate of US$1.5 billion in cash and 28.5 per cent of the new company.

The original PanAmSat had been established in 1984; 1995's revenue was US$116.2 million (1994: US$63.7 million). The company was granted an FCC licence in 1985 and took over the partially-completed ASC 3 satellite, surplus to Contel ASC requirements, to provide the first private international satellite communication links outside of Intelsat via PanAmSat's Miami International Gateway at Homestead, south of the city. PAS 1 was launched by the maiden Ariane 4 demonstration mission at a reduced fee of US$8 million; insurance coverage was US$40 million at a 25 per cent premium. Prevented by the FCC from connecting with public switched networks, the service is intended primarily for TV programming of US or Mexican origin, and business data, video and voice traffic. PAS 2 appeared in July 1994 to serve the Asia-Pacific region, and by November 1995 had 17 full-time programmers. Hundreds of customers worldwide use the first two satellites; broadcast services represent 67 per cent of revenue. There are also teleports in Napa California, Homestead Florida and Atlanta Georgia.

PanAmSat is the first private sector operator of a global system, with satellites over the Pacific, Atlantic and Indian Ocean regions. Hughes was selected in 1991 for a contract worth about US$300 million to provide four of its HS-601 models for PAS 2 to 4; the fourth was to be held as a ground spare. Each carries cross-strapped C and Ku-band 54 MHz transponders optimised for TV programmers and corporate telecommunications. The launch contract went to Ariane in June 1992, in return for which Credit Lyonnais arranged French export loans covering 85 per cent of the US$240 million launch costs. To finance the US$700 million expansion, 50 per cent of PanAmSat's equity was acquired by Mexican media company Televisa SA in December 1992 for US$200 million. US$440 million was raised from a bond sale, completed August 1993. The 2,985 kg PAS 3, intended for 43° west to provide DTH services to Latin America from June 1995, was lost in Ariane's 1 December 1994 failure. Insurance was reportedly US$214 million at 17.5 per cent. The ground spare was thus converted for launch as PAS 3R to begin the Latin America service in first quarter 1996. PAS 3R was launched by Ariane 44L in January 1996. PAS 5 was ordered from Hughes Space & Communications in March 1995 and PAS 6 from Space Systems/Loral in

August 1994 to expand the DTH Latin America capacity to approximately 120 digital channels per customer (approximately 500 throughout Latin America). PAS 5 was launched by Proton from Baikonur in August 1997, PAS 6 on an Ariane 44P in the same month followed by Galaxy 8-I in December that year. US$262 million was raised April 1995 in a preferred stock offering to finance PAS 5/6. PAS 7/8 were ordered from SS/L in April 1996. FCC filings were made in September 1995 for the PAS 10/11 Atlantic Ka-band satellites, costing US$409 million through the first year.

In 1998, PanAmSat announced its intention to double its worldwide transmission capabilities by the end of 1999 with the launch of nine additional satellites, launches in 1998 included (in order) Galaxy 10, PAS 7, PAS 8 and PAS 6B. Launches planned for late 1999 included Galaxy 11, PAS 1R, Galaxy 10R, PAS 9 and Galaxy 4R.

Space Systems/Loral

Space Systems/Loral (SS/L) was formed out of the Loral Corporation Space Division when the parent company took over Ford Aerospace Corporation in an acquisition worth US$715 million. Since then, Space Systems/Loral has become the world's fastest growing commercial producer of communications and weather satellites and key subsystems by answering customers' needs for rapid delivery of reliable, state-of-the-art spacecraft. The company is a full service manufacturer that develops, produces and delivers satellites, payloads and complete satellite systems, while offering risk management, insurance, mission control and launch services around the world. Global markets range from broadcast and cable television to cellular mobile communications and direct-to-home TV, on to video conferencing, environmental monitoring, educational and scientific data networks and global positioning.

SS/L has a funded backlog of US$1.46 billion, with US$1.9 billion in orders for the 12 months ended December 31, 1996. Employees total 3,500. Loral Space & Communications Ltd, SS/L's parent company, made a strategic decision in 1996 to increase its ownership in SS/L to 100 per cent by acquiring the 18.3 per cent interest held by the Lehman Partnerships in August 1996, and in February 1997, by acquiring the remaining 49 per cent minority interest held by SS/L's four European aerospace partners – Aerospatiale, Alcatel, DASA and Finmeccanica. SS/L maintains a close working relationship in the marketing and manufacture of satellites with these four companies.

SS/L offers satellites based on its standard three-axis spacecraft. By early 1997, the number of satellites in this family of spacecraft that have been built or ordered totalled 59, including 22 satellites in production. The company has supplied this type of spacecraft to many significant customers including Intelsat, Nippon Telegraph & Telephone, NASA, TCI, PanAmSat, the Japanese Ministry of Transport, APT Satellite, Skynet, Mabuhay Philippines Satellite, MCI, Asia Broadcasting and Communications Network, Pasifik Satelit Nusantara and Chinasat. In February 1997, the satellites built by SS/L accumulated more than 600 years of service in space. This landmark of longevity and reliability represents the combined success of 82 communications and weather satellites built during the past three and a half decades. A new, more advanced bus, offering lower mass, increased power and improved reliability is being designed and is expected to be in production by 1998.

Loral applied to the FCC in May 1995 for the CyberStar satellite to cover North America from 1998. This broadband data communications satellite would use Ka-band to provide business and consumer users with high-quality video and high-speed data services via small, low-cost antennas. Later in 1995, Loral amended its application, increasing the number of satellites for worldwide coverage. Loral also received orbital assignments from the FCC for this international coverage. Space Systems/Loral currently is designing and developing this system to provide service by 1999.

Space Systems/Loral is the prime contractor for the space segment for the Globalstar 48 low-earth orbit satellite-based system. The digital telecommunications system that SS/L and its partners are constructing is designed to enable low-cost, high-quality wireless voice telephony and data services such as paging, facsimile and position location in virtually every populated area of the world, including areas currently underserved or not served by existing wireline and cellular telecommunications systems.

Space Systems/Loral and Russia's OKB Fakel in 1992 established the International Space Technology, Inc. (ISTI) joint venture to market electric thrusters. Using technology licensed by Trimble, SS/L is developing the GPS *Tensor* TM solid-state, lightweight GPS spacecraft attitude sensor. It is a spin-off of work performed with Stanford University researchers on NASA's Gravity Probe B. SS/L is subcontracted by Rocketdyne Division of Rockwell International Corporation to build Space Station Alpha's Ni/H_2 battery system.

The company features the FS-1300 platform.

Total ordered: 40 (including Superbird, GOES, Intelsat VII, N-Star, Tempo, PanAmSat, MTSAT, APStar, TELSTAR, Mabuhay, Sky, L-Star, M²A. Chinasat)
Total launched: 18 (including Superbird, GOES, Intelsat VII/VIIA, N-Star, Tempo)
Total currently operating: 15
Launch mass: up to 4,500 kg, depending on payload: EOL mass up to 1,900 kg
Deployed dimensions: 25.7 m solar array span, 7.5 m high
Availability: production cycle will be 18 months by 1997
Platform reliability: better than 0.875 EOL
Communications payload power: up to 12,000 W, depending on payload mass
In-orbit lifetime: designed for more than 15 years
Power: 100% eclipse capability
Pointing: antenna pointing accuracy is less than 0.10°
Attitude control: a microprocessor-based momentum bias system controls the satellite and auxiliary functions. The processor can be pre-programmed with more than a month of manoeuvres; execution is by either automatic or manual control
Propulsion: a single-integrated bipropellant subsystem provides both apogee (400 N) and on-station manoeuvring
Thermal: a passive thermal subsystem includes heat pipes, and heaters for some components. Large N/S surfaces covered with internal heat pipes allow maximum payload mounting flexibility.
Telemetry, Tracking and Control: custom capability tailored to each programme in any frequency band

Tempo DBS, Inc

Tempo awarded Space Systems/Loral a US$400 million contract July 1994 for two high-power DTH satellites and their launches. Tempo is a subsidiary of Tele-Communications, Inc, the largest cable TV operator in the US. The payload carries 34 Ku-band transponders and is switchable to 200 W transponders, optimised for 120 to 150 digitally compressed channels to home dishes. PrimeStar, owned by six cable TV companies (including 22 per cent by TCI) and GE Americom, lease capacity on the satellites, although operation remains with Tempo. It already provided DTH programming on Americom's FSS Satcom K-1. Launch date has not yet been announced.

Worldspace, Inc

WorldSpace's subsidiary AfriSpace filed an FCC request in July 1990 to launch a single AfriStar satellite to provide digital audio broadcasting to the African-Arabian region. The licence was awarded 21 June 1991. The plan now is for three satellites, each beaming up to 288 radio channels to Africa, Middle East, Asia, the Caribbean and South and Central America. The CaribSpace licence was granted by Trinidad & Tobago in December 1992 for AmeriStar, and in September 1995 Australia notified the AsiaStar satellite for AsiaSpace. A contract was signed 21 January 1995 with Alcatel Espace to build, launch and insure the three satellites (plus a ground spare) for US$600 million. Alcatel Espace signed the Ariane launch contract 18 March 1996. More than US$750 million financing was secured, with investment banker Morgan Stanley of New York serving as financial advisor, by a group of international investors.

The WorldSpace receiver, cost US$150 to US$200 initially, will have several stationary and mobile models, including solar powered and serial data ports for faxcasting. Sales are projected at 20 million annually within five years of the full service. The radio chipset will also be integrated onto computer interface cards for data downloads and audio programmes via PC.

AsiaStar is scheduled for launch by Ariane 5 in October 1999.

EARTH OBSERVATION FACILITIES AND EQUIPMENT

ARGENTINA

Aeroterra SA

Aeroterra has been involved in remote sensing since 1973 for the evaluation, cataloguing and development of natural resources. Thematics maps drawing on Landsat, Skylab and Spot data cover approximately 7,000,000 km^2 of Argentina, Bolivia, Brazil, Paraguay and Uruguay. It is a representative of EOSAT, ESRI and ERDAS and provides systems and services in GIS, image processing, consulting, training and education. Its Training Centre opened in 1992 for remote sensing and GIS courses.

AUSTRALIA

Earth Resource Mapping Pty Ltd (ERM)

ERM specialises in integrated mapping software. Its latest ER Mapper is designed principally to register, process and merge remotely sensed geoscience data along with geographic and other grid overlay information of the user's choice. Development was backed by Sun Microsystems and Hitachi Data Systems as a standard package requiring no customising or programming knowledge. It can handle complete datasets such as full Landsat TM and Spot scenes, and large magnetic datasets. The latest release includes 3-D visualisation and hardcopy production and direct editing of ARC/INFO coverages.

Geoimage Pty Ltd

An independent image-processing and remote sensing consultancy specialising in the sales and processing of satellite and other raster data for the mineral and petroleum exploration industries and land management and environmental disciplines.

Moonraker Technology Pty Ltd

MRT offers products/services for oceanographic and environmental science researchers, including the MRT-112 drifting buoy for use with the Argos satellite data relay system. The streamlined cylindrical hull fully encloses the device, apart from the antenna, and provides standard sea surface temperature, battery level and drogue loss indication sensors, with air temperature and atmospheric pressure sensors as options. Activation is effected by removal of a magnet; the MRT-201 hand-held set confirms operation before deployment. Battery packs for 6- and 12-month lives. Other products include marine animal trackers rated to 1,500 m depth.

Spot Imaging Services Pty Ltd (SIS)

SIS was formed in 1988 as a subsidiary of France's Spot Image to market Spot data in Australia and the South Pacific region. 1996 revenues were approximately US$900,000. Derived products based on Spot images, such as Digital Terrain Models, can be provided in addition to other services such as image processing for geological, cartographic and civil engineering applications. Regional imagery is received directly at the Australian Centre for Remote Sensing.

AUSTRIA

Geospace

Geospace is the authorised distributor of Spot Image, Eurimage, EOSAT, EUROMAP and Radarsat data for Austria. It also represents the ESA Earthnet National Point of Contact and distributes ERS data. It provides consulting services, evaluation of remote sensing data and satellite-based cartography and is equipped for all image processing and production of colour transparencies from processed data on colour laser plotter. The company has published atlases of Austria and Germany based on Landsat TM and Spot imagery at 1:200,000 and 1:220,000, respectively. It is leading the ISY Global Satellite Image Mapping Project.

BELGIUM

CLEO

The Cluster for Earth Observation (CLEO) was created in 1995 to promote the value-added capabilities of Alcatel Bell, SAIT Systems and Trasys. Trasys Space is responsible for management operations at the image-processing and archiving centre for Spot 4's Vegetation instrument. Trasys signed a Radarsat data distribution agreement April 1996.

Eurosense

Walphot's satellite remote sensing activities include digital image treatment and interpretation for Earth sciences, forestry, agriculture and environment, in addition to thematic map production.

CANADA

Array Systems Computing Inc (ASC)

ASC specialises in remote sensing, defence and security software development and systems integration. In remote sensing, it focuses on interpreting and processing captured satellite data in real time.

Radarsat Fastscan Processor, FASTSCAN, processes real-time data and produces real-time imagery for viewing on the Moving Window Display. It supports full resolution ScanSAR processing, fast Data Quality Reports and Catalogues, comprehensive data products and a Raw Data Interface. Radarsat Image Data Calibration Workstations provide operational radiometric and geometric calibration and verification methodologies, to assess and maintain end-to-end SAR image quality. Meteorological Applications Processing System (MAPS), ground stations receive, process and distribute HRPT data from polar and GEO satellites. It combines real-time data reception with on-the-fly image production. Automatic or on-demand data distribution can be made to remote sites.

Barringer Research Ltd

Development of techniques and instrumentation for remote sensing and pollution monitoring. These include GASPEC, a gas filter spectrometer used by two NASA programmes on Space Shuttle, and laser fluorosensors for oceanography and maritime pollution measurements.

Dendron Resource Surveys Inc

Dendron is a forest resources consulting company specialising in the application of remote sensing technologies, including airborne and spaceborne image analysis. The company provides GIS services for the input, manipulation and output of large sets of geographically referenced data.

Eidetic Digital Imaging Ltd

The company was established in 1985 to develop modestly priced, stand-alone, PC-based, digital image analysis systems for remotely sensed data obtained from both satellite and airborne sensors. The EARTHSCOPE system is a new Windows™ version of the original DOS-based RSVGA system. The software consists of a complete set of programs for performing enhancements, classifications, geometric corrections, hard copy output, TIFF and BMP output, false- and true-colour 24-bit image display, and utility functions. The generic TEX1 tape reading software is used for transcribing Landsat data on Exabyte tapes to the hard disk. RSVGA and TEX1 are distributed worldwide by EOSAT. The PEDAGeOG program has been developed for high school and college instruction in processing satellite images. A new Windows™ version of PEDAGeOG is now available. It is the software component of Geo/SAT, a joint project between Eidetic, EOSAT and high school and university teachers in the US to develop remote sensing educational materials. The other components are Scholar series data sets from EOSAT and Teacher's Manuals.

Gregory Geoscience Ltd

Incorporated in 1973, Gregory Geoscience specialises in remote sensing analysis. It has developed the Procom-2 system for optical transfer and integration of diverse data sets from satellite and airborne imagery to a common map scale which is then digitally recorded and entered into a GIS, image analysis or computer mapping system. Company capabilities include very high-resolution scanning and digital analysis of land use change, resource mapping and environmental monitoring. Activities have included the extensive use of satellite and airborne imagery for detecting changes on 1:50,000 maps, mapping peat bogs and wetlands from Landsat, and undertaking a national inventory of mine wastes from satellite coverage. Gregory Geoscience also has expertise in the integration of geographic databases with municipal applications, soil analysis mapping, environmental risk to biodiversity mapping and analysis.

Horler Information Inc

Founded in 1984, HII is a consulting company specialising in the technologies and applications of remote sensing and GIS. Recent projects are the development of an Earth observation strategy for the Canadian Space Agency and an assessment of the geological applications of Radarsat.

Intermap Technologies Corporation

Intermap specialises in airborne SAR and other remote sensing acquisition, processing and analysis services, in addition to GIS, mapping and digital ortho-imaging.

Macdonald Dettwiler

MDA specialises in computer-based systems for aerospace, defence and resource management applications, exporting 75 per cent of its products. In November 1995 it became part of the Orbital Sciences Corporation group. In the year ending 31 March 1995, revenues were C$110 million.

MDA's product line includes: space-qualified software; MIL-SPEC (including DOD-STD-2167A) systems and software; turnkey remote sensing satellite ground stations; image mapping system for generating maps from digital images; optical/radar image analysis systems for space/airborne sensor data; meteorological data analysis systems. For Canada's Space Station Mobile Servicing System, it is contributing to development of software, data processing subsystems and artificial intelligence applications.

MDA is a leading supplier of ground receiving/processing systems for remote sensing satellites, working as prime contractor on 18 turnkey satellite ground stations worldwide and as major subcontractor on 10. It provides systems for all optical and SAR missions. Its SAR team was the first to digitally process SAR data and to develop a commercial digital processor, and is the world's leading supplier of spaceborne SAR processing systems.

RADARSAT International, Inc

RADARSAT International Inc (RSI) is a private company, majority Canadian owned, established in 1989 to market, process and distribute the data products from Canada's RADARSAT. RADARSAT is Canada's first earth observation satellite and the world's first radar satellite totally dedicated to operational applications. This first of the RADARSAT series of three satellites was successfully launched 4 November 1995. A world leader in providing information solutions from space, RSI has the exclusive commercial licence to distribute RADARSAT data worldwide, and additionally distributes ERS data in North America, and SPOT, LANDSAT and JERS-1 data in Canada.

Satlantic

Analysis systems for SAR; receiving, processing, analysis stations for NOAA, SeaWiFS, Radarsat, Spot; marine Inmarsat C/M systems.

Seimac Limited

Seimacs Satellite Asset Tracker Argos PTT provides intermodal container tracking, automatic utility meter reading and environmental applications such as oceanographic research. Positioning accuracy with GPS is less than 100 m (typically 20 m) and without less than 1 km (typically 350 m). Antenna 3 × 13 × 38 cm, electronics 13 × 18 × 19 cm; rechargeable battery (45 day capacity) 8 × 18 × 19 cm; operating tolerance −30 to +60°C. Seimac's 3.2 × 5 × 13 cm 250 g Smart Cat III Argos PTT is small enough for animal tracking. It is commonly used in compact air-deployable buoys and underwater mooring markers. It includes complete data acquisition and logging capability and an 8-bit analogue-to-digital converter.

The Bercha Group

The Bercha Group of companies specialises in the acquisition, processing and analysis of remote sensing data, including satellite and airborne data. Capabilities include digital elevation modelling and specialised SAR processing. Registered Spot distributor.

CZECH REPUBLIC

GISAT

GISAT was founded as a remote sensing and GIS company in 1990. It is a Spot Image, EOSAT and Eurimage distributor within the Czech Republic. It is involved in the CORINE Land Cover and the agricultural MARS programmes of the European Community.

DENMARK

Kampsax Geoplan

The company specialises in digital and analytical mapping from remote sensing images and aerial photography. It is a distributor of Spot and Landsat data products in Denmark under agreement with SSC Satellitbild and Eurimage, respectively.

ECUADOR

CLIRSEN

The CLIRSEN Centro de Levantamientos Integrados de Recursos Naturales por Sensores Remotos (Centre for Integrated Surveys of Natural Resources through Remote Sensors) was created in December 1977 to be responsible for remote sensing and the cataloguing of natural resources. It operates the Cotopaxi ground station, some 60 km south of Quito, which began taking imagery from Landsat in June 1991 as the system's 18th station; Spot and ERS were added later. The 2,500 km^2 coverage of 25 countries stretches from Mexico's Yucatan Peninsula to Antofagasta in Chile. Products include 23 × 21 cm film, 72 × 97 cm paper, quick look, microfiche (84 scenes each) and digital tapes (6,250/1,600 bpi). Processing levels: bulk, georeferenced, geocoded and user-specified highlighting.

EGYPT

Eight M Ltd

Eight M Ltd specialises in environmental and remote sensing services, offering state-of-the-art digital image processing and GIS projects, consultancies and workshops. It represents EOSAT (Landsat products), Eurimage, Earth Resources Data Analysis Systems (ERDAS; image processing, GIS software packages) and Cirrus laser film writers.

FRANCE

Fleximage

Fleximage was established in 1989 to specialise in computer-aided photo-interpretation and geographic databases. Aerospatiale holds 78 per cent. Personnel total 25; turnover is FFr13 million annually. The extraction of thematic maps from satellite imagery is based on the OCAPI (Operational Computer-Aided Photo-Interpretation) software, initially developed for the military to exploit Spot data.

Geosys

Geosys was established in 1987; it is a member of the SGS group (Société Générale de Surveillance) and the Caisse des Dépôts et Consignations (Deposit & Consignment Office). Personnel total 30. It compiles regular crop inventories and prepares yield forecasts using remote sensing data, including Spot, NOAA AVHHR and Landsat TM. It also provides geographic information to the agribusiness and environment sector. It develops and sells GIS handling remote sensing images and vector cartographic data.

ISTAR

Imagerie Stéréo Appliquée au Relief was established in 1988 to market Digital Terrain Models and orthophotos computed by an automatic relief plotting software, for use with Spot stereopairs. DTMs are generated with a 20 m grid cell and an average altimetric accuracy of 10 m. The company offers VUE3D software products for interpreting DTMs and orthophotos; they can create composite images providing 3-D views and calculate successive image sequences for animation purposes. ISTAR is also a consultant in satellite imagery, geometry and relief mapping utilising different sensors.

Spot Image acquired a 20 per cent interest in March 1995. Thomson-CSF is an 18 per cent shareholder.

Matra Cap Systèmes

Matra Cap Systèmes is the European leader in Communication, Control, Command and Intelligence systems and the world leader in Earth observation ground stations. Matra Cap Systèmes offers multisatellite Earth observation ground stations for SPOT, HELIOS, LANDSAT, ERS, RADARSAT, CBERS, either fixed (in Toulouse, Sweden, US, Pakistan, Brazil) or mobile (US Air Force, IOSAT Canada). The company is also working on several demonstration Pilot projects using Earth observation for EEC (image data servers, agriculture calamities information server). Several R&D projects are run by the company for the European Space Agency in new archiving technologies, information client/server software, SAR interferometry applications, etc. Products are offered by Matra Cap Systèmes for image processing and applications development (Multiscope) and digital cartographic systems (Prodigeo, T10) which are used in several remote sensing and mapping centres in the world.

Spot Image SA

Spot Image, with CNES and Matra as the majority shareholders, was established in 1982 to distribute Earth remote sensing data and their products from the Spot satellites, the first of which was launched by Ariane into polar orbit in February 1986, followed by Spot 2 in January 1990 and Spot 3 in September 1993. The two main image receiving stations (Toulouse and Kiruna) are designed to receive/store 250,000 images annually.

GERMANY

Energia Deutschland GmbH

The company is a partnership between Germany's Kayser-Threde and Russia's RKK Energia, principally to exploit an exclusive right to distribute Earth observation data from Mir. A 120 × 120 km 7.5 m resolution colour image from Mir's two KFA-1000 cameras (Priroda 5 package) costs DM1,900.

GAF mbH

The Company for Applied Remote Sensing (Gesellschaft für Angewandte Fernerkundung mbH) specialises in multidisciplinary applications of remote sensing and image processing techniques. It is a distributor of

Radarsat, EOSAT (Landsat, IRS 1C), Spot Image (Spot), Soyuzkarta (KFA-1000) satellite data products. It directly receives, processes and archives IRS 1C data via DLR's Neustrelitz ground station under a contract signed with EOSAT 1 February 1996. GAF and Spain's Minas de Almadèn y Arrayanes SA created Télédéteccion Aplicada SA (TdA) in 1990 as a joint venture specialising in remote sensing services in Spain and Latin America.

VCS Nachrichtentechnik GmbH

VCS has operated in the space ground system market for 16 years, specialising in satellite ground stations, image workstations, satellite communication, meteorological applications and task force engineering and consulting. ESA and EUMETSAT contracts include development of the whole ground infrastructure for the Meteosat MDD mission, GOES/Meteosat relay and projects on data encryption and compression and significant parts of the core mission control centre for the Meteosat Transition Programme.

INDIA

Speck Systems PVT Ltd

Manufacturers and suppliers of digital image processing, GIS and photogrammetry systems; digital film recorders and scanners. Production job services in topographic mapping, map digitisation, image analysis and GIS.

INDONESIA

PT Indica Dharma Consulting Services

Established in 1991, the company distributes data from the Thailand Remote Sensing Centre, LAPAN's Pare-Pare station, Australia's ACRES and FR mapper software distributor. The focus is on the mapping of forestry concession areas, land use, coastal zones, plus training.

INTERNATIONAL

CLS Argos

Collecte Localisation Satellites Argos (CLS Argos) is operating/developing several satellite-based location and data collection systems. The principal Argos system processes and disseminates space and terrestrial environmental data received by dedicated packages aboard US NOAA polar meteorological satellites from fixed/mobile platforms anywhere in the world. Approximately 5,000 transmitters are operating. The past five years have seen a 15 per cent growth in activity. (approximately 20,000 transmitters have used the system). Argos is a joint programme of CNES, NASA and NOAA, operated under a Memorandum of Understanding signed in 1974 and extended in 1986. System operation and promotion is the responsibility of CLS, established in April 1986 as a subsidiary of CNES (55 per cent), IFREMER (French Marine Institute; 15 per cent) and a pool of banks. CNES operates the French Global Processing Centre (FRGPS) in Toulouse and handles all user relations outside North America. US subsidiary Service Argos Inc is responsible for all US/Canadian users; it operates the identical USGPC, which provides redundancy. North American CLS is concerned with developing new services and products. The Australasian regional processing centre began operations in December 1988.

The system consists of independent user Platform Transmitter Terminals (PTTs, down to 25 g and the size of a matchbox), two NOAA polar satellites with Argos Data Collection & Location System packages to receive PTT messages on a random access basis for separation, time coding, formatting and retransmission to ground stations, and the ground stations and two Global Processing Centres in Toulouse and Landover, where data are retrieved, processed and distributed to users. Location of PTTs can be determined to within typically 150 m, if required, by Doppler techniques.

The PTTs transmit data from up to 32 sensors encoded in 8-bit words on 401.650 MHz with typically 200 mW power at intervals of 60 to 120 seconds for location PTTs and approximately 200 seconds for data collection-only platforms. Message length is 360 to 920 ms and contains the PTT identification number.

The NOAA Argos packages receive all messages within a 5,000 km diameter visibility circle at any instant; four can be processed simultaneously by NOAA A-J and eight will be handled by the third-generation system aboard NOAA K and its successors from 1996. NOAA 11 can handle data at 960 bit/s, NOAA J 1,200 bit/s and NOAA K-N 2,560 bit/s. NOAA K-N increases bandwidth from 24 to 80 kHz. The data are formatted/stored, then dumped each time the satellite moves within reach of one of the three ground stations. VHF/S-band transmitters also perform real-time relay for any user station within the visibility circle. Current onboard packages can process up to 1,400 data-only PTTs, or 415 requiring the location service, but from NOAA K on the capacity will be quadrupled. The Argos 3 generation will be defined by end-1996 for post-2000 satellites. Japan's 1999 ADEOS 2 will carry an Argos package allowing users to command their PTTs.

Under an agreement signed in April 1990 with EUMETSAT, CLS manages all non-meteorological applications using the Meteosat data collection function. The MAEDS Multisatellite Applications Extended Dissemination Service adds Meteosat and GOES to the existing Argos system to provide high data collection capacity (649 byte messages up to 24 times daily), real-time access to data, and continuous coverage from the Pacific to the Urals. The first MAEDS element became operational in October 1990, followed in 1991 by the GOES satellites covering America and the Pacific. CLS also operates the Doris location system and is involved in the Starsys system.

Global Processing Centers NOAA's tape recorders are read out every 100 minutes over one of the Gilmore Creek (Alaska), Wallops Island (Virginia) and Lannion (France) stations. All of the satellite's data are routed to NOAA's National Environment Satellite, Data and Information Service (NESDIS) in Maryland, where the Argos component is extracted and relayed to CLS in Toulouse and Landover. Here, PTT locations are determined and the sensor data processed. About two-thirds of the processed messages are provided to users within 3 hours of the uplink. Location is derived from measuring the Doppler shift in the carrier frequency of received messages during the 10-minute overflight. An accuracy of 150 m is typical but for PTTs at fixed positions or on slow-moving carriers 100 m is possible. The transmit frequency must remain stable throughout the pass. Each frequency measurement yields a set of possible positions and with a known elevation only four messages are required on each pass to generate a unique solution. The satellites' own positions are known to within 300 m at any instant from a reference system of 11 beacons worldwide at precisely known geodesic positions.

Eurimage S.c.r.l.

Established in 1989 to supply remote sensing products and services, Eurimage's shareholders are: British Aerospace (UK), Dornier (Germany), SSC Satellitbild (Sweden) and Telespazio (Italy). Data are distributed from Landsat, MOS, NOAA/AVHRR, JERS and ERS. 1995 revenue was ECU5.4 million (1994 ECU5 15 million, 1993 ECU5 million, 1992 ECU3.3 million). Products cover digital and photographic formats of large areas or small regions: raw data, standard products, geocoded and enhanced products. The rush service offers near-realtime cloud cover assessment (within a few hours of acquisition) and enables the final processed product to be dispatched to the end user within three days.

RESURS-F WorldMap joint venture

The WorldMap consortium was created in 1993 by JEBCO, Priroda and Rosvneshgeo. Priroda is Russia's primary non-military remote sensing acquisition and processing organisation. It also operates a large cartographic division that employs satellite data to create cartographic and GIS products. Priroda is under the administration of Roskartografia, the Federal Service of Geodesy & Cartography. Rosvneshgeo is a private Russian company affiliated with Roskomnedra, the Russian State Committee of Geology in charge of mineral and hydrocarbon exploration.

Priroda collects new imagery each year using the Resurs-F spacecraft. Each F1 satellite carries two KFA-1000 plus three Kate-200 cameras; F1M three KFA-3000 plus one Kate-200; F2 one MK-4; F3 two KFA-3000s. Priroda's archives contain some two million XS/PAN scenes at 2 to 20 m resolution covering most of the globe. The data collected for the KFA-1000, MK-4 and Kate-200 film camera systems are now available digitally in a bit sequential raster format. Some 500 scenes monthly are being digitised by two Vexcel VX3000 enhanced scanners linked to Sun Microsystems SPARC stations and output on to 8 mm tape. Cost per archive scene to users is: US$3,000 KFA-1000; US$4,000 KFA-3000; US$3,000/US$4,500 1/3-bands MK-4; US$1,600/2,400 1/3-band Kate-200. Prices drop rapidly for large purchases. JEBCO is the sole distributor of digital data derived from the original Resurs-F satellite negatives. The consortium accepts requests for targets to be included in upcoming flights. Priroda plans to distribute digital 3 m resolution panchromatic imagery from a military satellite, presumably the fifth-generation photoreconnaissance model, which had Cosmos 2320 in operation at the beginning of 1996.

Sovinformsputnik (whose dominant shareholder is the Central Specialized Design Bureau, builders of Resurs-F) is licensed to sell data from the KVR-1000, TK-350 and DD-5 camera systems of military satellites. These negatives are also being digitised and Priroda has a sales agreement with Sovinformsputnik. Orders can be directed through the WorldMap consortium.

The KFA-1000 has a focal length 1,000 mm, 0.57 to 0.68, 0.67 to 0.81 μm, 5 m resolution, 80 km swath width (160 km by two cameras, 240 km by three), frame 300 × 300 mm (optimal enlargement × 10), altitude 220 to 280 km scales 1:220,000 to 1:280,000, stereo overlap 60 per cent, 1,800 frames per camera

The MK-4 has a focal length 300 mm, four filters for usually 0.635 to 0.690, 0.810 to 0.900, 0.515 to 0.565, 0.580 to 0.800 μm; 0.460 to 0.505, 0.400 to 0.700 μm also available. From 220 to 270 km altitude, 5 to 8 m resolution, 130 to 160 km frame width, scale 1:730,000 to 1:900,000, stereo overlap 60 per cent, frame size 18 × 18 cm, 2,500 frames per camera. Film improvement planned for F2M in 1997 to increase resolution 1.4 times

The Kate-200 has a focal length 200 mm, 0.5 to 0.6, 0.6 to 0.7, 0.7 to 0.85 μm, 15 to 30 m resolution, 180 km swath width, 18 × 18 cm frame size, overlap 20, 60 or 80 per cent, 2,500 frames per camera

The KFA-3000 is a high-resolution camera flying aboard Resurs-F3/F1M. Focal length 3000 mm, panchromatic, 2 to 3 m resolution (1.6 m from 180 km), frame 300 × 300 mm, 21 × 21 km, scale 1:70,000

The KVR-1000 is a panoramic low-angle camera flying aboard Kometa since 1983, usually in conjunction with TK-350. Focal length 1,000 mm, 0.56 to 0.8 μm, 75 cm resolution (but available commercially only down to 2 m), frame 18 × 72 cm (37 × 165 km; cut into three), scale 1:220,000 but can be enlarged to 1:10,000 without loss of detail. Images nested within TK-350's

The TK-350 is a topographic camera flying aboard Kometa since 1983. Focal length 350 mm, 0.50 to 0.68 μm, 10 m resolution, frame 30 × 45 cm, 200 × 300 km, scale 1:660,000 (can be enlarged to 1:50,000 without significant loss of detail), 60 to 80 per cent stereo overlap

The DD-5 is a high-resolution data degraded to 2 m resolution in the digitising process. DD-5 (DD means digital data) does not refer to a specific camera system but is a catch-all designation for digitised data from 'conversion' military satellites.

ITALY

Alenia Aerospazio

As co-contractor with Dornier, it developed and integrated the RF subsystem for the German-Italian X-SAR, flown twice in 1994 as part of the Space Radar Laboratory. It developed the RF project for the 12 m planar antenna and was responsible for its integration and testing. It was also responsible for the design and development of the ground segment Mission Planning and Operation System (MPOS). Alenia Spazio led the design and development of ERS 1 and 2's Radar Altimeter and is studying a second-generation instrument for Envisat. It is developing ESA's MIMR Multifrequency Imaging Microwave Radiometer, with 1.6 m antenna, planned for Metop 1.

JAPAN

NEC

NEC's space sensors include the MESSR (MultiSpectral Electronic Self- Scanning Radiometer) flown on MOS, the OPS (Optical Sensor) on JERS 1, the OCTS (Ocean Colour Temperature Scanner) on ADEOS and the VNIR (Visible and Near-Infra-Red) Radiometer of the ASTER (Advanced Spaceborne Thermal Emission and Reflection) Radiometer for NASA's EOS AM1. See the project entries in the space segment section for instrument details. NEC also supplies the ground systems for receiving/processing Earth observation satellite data.

Nissho Iwai Corporation

Nissho Iwai Corporation is a 'Sogo Shosha' trading company, with 180 offices worldwide. It is a partner in the JCSat Japan Satellite Systems company, owner and operator of telecommunications satellites. The Space Project Office was established in 1985 and now distributes Spot and Landsat data to Japanese users. It also offers Terra-Mar's Image Processing and GIS workstations.

KENYA

Teleos International BV

Teleos is a partnership between EOSAT and Nuova Telespazio of Rome, Italy, to provide East and South Africa with remote sensing and GIS applications services in co-operation with Nairobi's Regional Centre for Services in Surveying, Mapping & Remote Sensing (RCSSMRS). Teleos' portable ground station was installed in the grounds of RCSSMRS in September 1994, initially receiving Landsat 5 TM imagery and upgraded in 1995 to add Spot, ERS and IRS.

KOREA, SOUTH

Chang Woo, Inc

The company is a national provider of remote sensing satellite data products: LANDSAT, ERS and Russian KVR-1000 (under agreement with EOSAT), Spot and ERS (Spot Image), Spot, ERS and JERS (Korea Advanced Institute of Science and Technology, a Korean ground station), RADARSAT (Radarsat International) and Russian Resurs-F (Soyuzkarta). It will provide turnkey direct reception ground stations for local markets under agreements with MacDonald Dettwiler and Associates (MDA, an Orbital Sciences Corporation company) for future high-resolution Earth observation satellites like EarthWatch's EarlyBird and QuickBird and OrbImage's OrbView. It has provided a direct reception ground station for Spot, ERS and JERS and has been undertaking a contract for OrbView, RADARSAT and IRS-1C downlink to Korea. It provides ground station facilities for GMS, NOAA and SeaStar and can also provide uplink ground stations or customised facilities for the ground links to Earth observation satellites.

NEW ZEALAND

Meteorological Service of New Zealand Ltd

MetService receives and processes NOAA HRPT and GMS S-VISSR data streams. Image data are rectified on to standard map projections, Argos transmitters located. Sea surface temperature is retrieved from NOAA AVHRR radiances.

Sirtrack Limited

A wholly owned subsidiary of Landcare, Sirtrack packages satellite transmitters working through the Argos system for wildlife and marine applications. For example, the Prestel unit is used to locate lost seismic streamers: if the water pressure exceeds a preset limit (indicating sinking) over a set period, the transmitter package is released. Argos can output the location directly to the host vessel via Inmarsat.

NORWAY

NORUT Information Technology Ltd

NORUT Informasjonsteknologi is a Norwegian institute for research into applied information technology and telematics applications. NORUT's main activities are satellite remote sensing, image processing, GIS, high-speed data communication, electronic data interchange, and object-oriented methods for data modelling and system development. 1996 space turnover was NKr9 million; 35 staff.

SPAIN

AURENSA

Auxiliar de Recursos y Energia SA (AURENSA) is an Earth resources company divided into: remote sensing; geological engineering; drilling engineering. It has represented Spot Image interests in Spain since 1988.

UNITED KINGDOM

Earth Observation Sciences Ltd

EOS was established in 1985 and now has a staff of 65 providing expertise in the science, technology and applications of Earth observation and environmental data management: systems design/development, consultancy, data management, and data and information processing. It is involved in ERS, EOSDIS, ADEOS and CEO.

EDS Ltd

The company's space business included a £2.5 million contract to supply the SAR processor for the UK Earth Observation Data Centre to process ERS data. In addition, the PulSAR high-performance SAR processing workbench has been developed to provide high-quality SAR imagery with rapid throughput on low-cost hardware. It supplied the SARCALQ image QA and calibration system for ESA's Earthnet and is undertaking projects in the interpretation of Earth observation data using neural nets, the development of integrated GIS, and the development of a system for determining terrain heights from SAR imagery.

Feedback Instruments Ltd

A member of The Feedback Group, the company produces meteorological satellite reception and processing systems. The self-contained Weather Satellite Receiver (WSR) 524PC, 528, 538, 548 provide reception from polar/GEO satellites operating on APT, WEFAX and PDUS modes, covering Meteosat, Tiros and GMS. Each utilises an IBM PC-compatible computer to control reception and process the images.

Hunting Technical Services Ltd

Studies and analyses of airborne, satellite optical and microwave remote sensing data for assessment of all renewable and non-renewable natural resources. Colour Fire 240 precision filmwriter for high-quality photographic hardcopy production. Data analysis using ERDAS image processing and ARC/INFO GIS systems; data acquisition and procurement; consultancy and training.

Matra Marconi Space

MMS Bristol (UK) is providing instruments for several Earth observation satellites. A £9.5 million contract was awarded by the UK's Meteorological Office in March 1989 for development and fabrication of the Advanced Microwave Sounding Unit-B (AMSU-B) for NOAA's Tiros-Next series, due for first launch in late 1996, to return atmospheric humidity information. The company designed and built UARS' ISAMS Improved Stratosphere and Mesospheric Sounder for Oxford University/Rutherford Appleton Laboratory to measure gas species and aerosol levels in the upper atmosphere. The company provided the Digital Electronics Unit for the ATSR Along-Track Scanning Radiometer for determining sea surface temperatures and atmospheric water vapour by infra-red radiometry on ERS 1. It developed ATSR 2 for ERS 2 and is now developing Advanced ATSR for Envisat 1 (delivery was due 1996).

MMS offers a Scanning Mechanism for high-accuracy platform pointing of remote sensing instruments. It comprises a scan platform mounted on bearings within a structural housing and driven through a gearing system by an electric motor. A closed-loop control system works off an Inductosyn position pick-off attached to the rotating platform, using an electronic processor to drive a pre-determined scan pattern. The mechanism can incorporate momentum compensation and bearing offload during launch if required, and offers both direct drive and gear drive versions. It employs dry lubrication throughout and offers extended operational lifetime (approximately 100 million revolutions). It is used on AMSU-B and derivatives are baselined for Metop 1's MIMR and Envisat 1's GOMOS steering front assembly.

Size: 300 mm diameter, 180 mm high, 6 kg
Performance (5 kg load mass, 1 A current): 0.01° dynamic accuracy (3s) Earth viewing, 0.005° positional knowledge, 360° scan range
Power: 20 W
Momentum compensation: better than 1%
Life: 100 million revolutions.

Nigel Press Associates Ltd

NPA has been operating as a remote sensing consultancy since 1972, concentrating on the digital and photographic enhancement of satellite imagery, particularly for petroleum exploration. It is leading a consortium developing new applications of ERS data for offshore oil exploration. The NPA Data Centre holds one of the largest independent collections of Landsat imagery outside of the USA. NPA, designated an official UK Spot image distributor, offers production and interpretive facilities using in-house equipment for computer and photographic image processing.

Space Technology Systems Ltd

STS was established in 1958 and now specialises in systems for receiving and processing data and imagery from Earth observation satellites, and meteorological and oceanographic instruments, data buoys and data collection platforms using satellite telemetry links.

Timestep Weather Systems Ltd

Timestep's PROsat II PC-based systems receive imagery from polar and GEO meteorological satellites. The GEO version provides 1,000-image colour animation, 3-D display, real-time display and autoschedule to save image. The NOAA version, with 90 cm dish, saves the complete pass in full resolution, temperature readout with no calibration needed, latitude and long gridding, country and state outline and multi-satellite prediction software. The company also offers Landsat data of England and Wales, together with low-cost 24-bit multispectral image processors.

UNITED STATES OF AMERICA

Aerojet

The plant specialises in electro-optic sensor systems. It was selected for a US$45 million contract by NASA Goddard in September 1993 for the AMSU Advanced Microwave Sounding Unit for NASA's EOS-PM1; US$82.4 million was added November 1995 for four further units (one will fly on Europe's Metop 1). For the new USAF DMSP D3 meteorological satellites, it was awarded a US$62.1 million contract in 1989 for three SSMIS Special Sensor Microwave Imager Sounders. A US$7.9 million option was exercised in January 1990 for a fourth sensor. The 57 kg, 70 W sensor will provide 19 channels over 19 to 183 GHz for observations from 30 km to ground level, including rainfall to an accuracy of 5 mm/h and ocean wind speeds to within 2 m/s. Temperature profiles up to 73 km will also be generated.

The plant also provides the imaging systems for the DSP early warning satellites. Each of the infra-red array's 2,000 PbS cells views a region on Earth less than 3 km across. The last two spacecraft of the second generation were retrofitted with generation 3 sensors, incorporating 6,000 elements.

Alden Electronics, Inc

Alden Electronics specialises in communication systems and weather data display terminals for transmitting, recording and display of meteorological data. The WeatherWorks 2000 provides weather information in the form of NEXRAD, DIFAX, GRIB, satellite and lightning images or text. Operating on a 486 PC with high-resolution graphics option for pan, zoom, loop, rotation, colour and storage functions. Images can be captured and saved in GIF or TIFF formats. The APTS/WEFAX System 6A is a 486 PC-based system to acquire, process, store and display images from polar and GEO satellites. The 6A, running on UNIX, allows simultaneous ingest and data manipulation, image rotation, time lapse zooming, automatic gridding with latitude and longitude insertion, geopolitical boundary generation and superimposition.

EarthSatellite Corporation

EarthSat, founded 1969, utilises remote sensing data from satellites and aircraft and provides value-added services for satellite imagery. It provides image processing and interpretation services and environmental, geologic and GIS consulting. EarthSat is a distributor for all civil satellites, including Spot, Landsat, Russian, Radarsat, ERSI. JERS, AVHRR and GOES.

Intergraph Corporation

Intergraph provides interactive computer graphics systems for GIS, mapping and Earth sciences applications. Image Station Imager (ISI) is a workstation software product providing multispectral digital image enhancement and analysis. MGE Base Imager and MGE Advanced Imager are Windows-based software products providing multispectral digital imaging enhancement and analysis.

Lockheed Martin IR Imaging Systems

Formerly a Loral company, IR Imaging Systems is developing the Atmospheric IR Sounder (AIRS) for JPL under a US$145 million contract for the National Center for Atmospheric Research. IR Imaging Systems also participated in developing the imager and sounder instruments for GOES-Next, for which Space Systems/Loral was prime.

Microwave Telemetry, Inc

The availability of new power supply components and batteries, developed for the fast emerging hand-held computer and personal phone market, has made it possible for the company to produce a new version of the PTT 100. The 20 g version or PICO PTT is making it possible to track birds weighing less than 500 g. By using a single lithium battery and new power supply circuits, it was possible to reduce the size and weight of the PTT to less than 20 g. This version however still retains all of the features of its larger brothers, such as the multiseason duty cycle timer, temperature, battery voltage and activity sensors and is configured as a backpack for harness mounting on many avian species.

Santa Barbara Remote Sensing

Hughes SBRS specialises in remote sensing instruments for Earth, meteorological and planetary surface observations. Its engineering organisation was restructured in 1995 to be an integral part of the Hughes Electro-Optical Systems engineering centre in Los Angeles, reporting to the Space Electro-Optics Business Unit. By January 1997, 69 instruments designed and developed by SBRS since its founding 30 years ago had been launched. There were no failures, on average operating for 2.5 times their design lives, totalling 289 instrument-years. It provided Galileo's Photopolarimeter Radiometer, the Thematic Mapper and Multispectral Scanners for Landsat 4/5 and the Enhanced Thematic Mapper for Landsat 6, in addition to the main instruments for GOES and GMS. It built the Thermal Emission Spectrometer for NASA's Mars Observer. TES 2, completed in 1996, is aboard Mars Global Surveyor. The Visible Infra-Red Scanner (VIRS) delivered to NASA in 1996 for the Tropical Rainfall Measurement System (TRMM) is undergoing its integration and test cycle. Scheduled for completion and delivery in 1997 are the MODIS instrument for the EOS AM platform and the Enhanced Thematic Mapper Plus for LandSat 7. Awaiting a Pegasus launch is SeaWifs instrument which will perform ocean colour science aboard the SeaStar platfor.

Seaspace Corporation

SeaSpace provides complete reception and processing systems for full resolution data from all sensors on polar and GEO satellites: NOAA HRPT (1 km AVHRR, TOVS, Argos), SeaStar (1 km SeaWiFS), DMSP RTD (0.5 km OLS, SSM/I, SSM/T-1, T-2), GOES GVAR (imager, sounder), GMS SVISSR imager, Meteosat PDUS imager. TeraScan software includes 340 UNIX-based functions for multisensor data fusion and analysis.

Space Instruments, Inc

SI specialises in design and studies of visible, infra-red and microwave sensors, their in-flight calibration, data handling, mission planning, control and detailed algorithms for signal processing. It is currently building the Infra-red Spectral Imaging Radiometer (ISIR) for NASA Goddard and the Total and UV Irradiance Radiometer (TUVIR) for the Naval Research Laboratory (to fly on MightySat 2 in 1998). SI also manufactures a line of high-precision cavity radiometers for use as absolute radiometric standards.

Space Liaison and Imaging Corporation

SLIC is a technical and marketing consulting firm, with offices in San Diego and Moscow. It represents Sovinformsputnik in the worldwide distribution of high-resolution satellite imagery from Russian military satellites.

Spot Image Corporation

The company is the exclusive US distributor for Spot products and services. 1995 sales were US$9.5 million; 1994's US$7.7 million revenue yielded a US$590,000 profit; 1993 revenue US$4.7 million. Spot's standard product covers 60 × 60 km; scenes are available on CCT, 8 mm tape, QIC, 4 mm and CD-ROM or as photographic products (prints or transparencies) at a variety of scales and processing levels. A computerised catalogue maintains a record of all imagery. The corporation's services include image processing and production, applications and technical support.

Telonics Inc

Telonics specialises in the design and manufacture of electronic environmental monitoring equipment, including radio telemetry for the wildlife field and related areas of environmental research and remote sensing systems. These include Earth stations with 1.2 to 1.73 m antenna/positioner systems for NOAA AVHRR/HRPT and TIP/TOVS, SeaWiFs, D-band Local User Terminals for Argos downlink processing, certified miniaturised satellite transmitters (PTTs) for oceanographic (low-cost drifters, temperature, depth profile chains and Argos PTTs with GPS positioning capability) and animal (birds; land and marine animals) research. Certified GOES Data Collection Platforms are built for stand-alone operation and for interfacing to dataloggers. Other products include PC-based orbital prediction and graphic display software with multiple propagator models, 1.0/1.2/1.4/1.8/2.4 m tracking antenna receiving systems for D/S/L- band applications, including NOAA HRPT receiving systems and SGLS full duplex command and control systems.

The ST-10 Argos PTT is designed for smaller mammals and birds. One bird version is less than 75 g. The electronics are available in two versions: 4.6 × 4.1 × 1.15 cm flat and 2.4 cm diameter, 4.6 cm high cylindrical. The more powerful ST-9 PTT provides 1 W transmit.

Terra-Mar

Specialist in developing software and integration of computer systems for GIS and raster image processing systems. Terra-Mar is co-operating with EOSAT in a study of the economics of GIS for resource management and environmental applications. It is also a Landsat and Spot satellite data product distributor.

Wirin & Associates

The company specialises in the distribution of Russian satellite imagery, specifically 2 m resolution optical (KVR- 1000), 10 m resolution stereo pairs (TK-350), KFA-1000 and MK- 4.

ZIMBABWE

Realtime Computers (PVT) Ltd

Realtime in 1990 became a sales representative for Landsat's data products.

EARTH OBSERVATION SATELLITES, COMPONENTS AND SERVICES

UNITED STATES OF AMERICA

Orbital Sciences Corporation / Orbimage

Orbital Sciences Corporation's Orbimage subsidiary was formed in 1992 to build and operate OSC's commercial Earth observation system, and to market the company's remote sensing products and services.

Orbital Sciences Corporation's Eyeglass International was formed in 1994 to build and operate the Eyeglass satellite system and to market imagery products and services. Orbimage in 1995 decided instead to pursue the smaller, lower cost OrbView satellite, derived from its MicroLab design. The cost of deploying one satellite and establishing the ground system is less than US$100 million. OSC's MacDonald Dettwiler Associates is managing the ground segment requirements. Saudi Arabia's EIRAD Co signed June 1994 to become a major investor (operating a receive station); DoC approval was granted 5 June 1995. EIRAD took a 20 per cent stake in Orbimage; US regulations impose a 25 per cent limit. OrbView 1 was launched in 1995 followed by OrbView 2 in 1997 and a second-generation system is planned. These consist of OrbView 3 scheduled for launch in late 1999 followed by OrbView 4 in 2000.

Resolution: 1 to 2 m PAN (4 to 8 km swath), 4 m XS (8 km swath)
Launch: planned for fourth quarter 1997 by Pegasus XL
Lifetime: 3 years
Mass: 146 kg
Orbit: planned 460 km polar, 10.30 equatorial crossing, revisit less than 3 days
Spacecraft: 96.5 cm diameter; disc about 60 cm high. Although there will be some onboard storage, most imagery will be downlinked in real time
Sensor: bands (μm): PAN 0.50-0.90; XS 0.45-0.52, 0.52- 0.60, 0.63-0.69, 0.76-0.90.

Space Imaging, Inc

Space Imaging's satellite system was initiated by Lockheed Corporation as the Commercial Remote Sensing System in April 1991. n November 1996, Space Imaging acquired EOSAT (Earth Observation Satellite Co), which in 1995 experienced a US$3 million loss on revenues of US$30 million. Founded in 1985 to take over the commercial marketing of Landsat imagery, and, eventually, the satellites themselves, EOSAT had a worldwide distribution network and access to imagery from several leading space-based imaging systems. These included the Indian government's Remote Sensing System (IRS), the Canadian Space Agency's Radarsat, the Landsat series satellites, Japan's JERS satellite system and the European Space Agency's ERS synthetic-aperture satellite. India's IRS-1C has a 5 m resolution capability, currently among the highest available from any commercial system.

Space Imaging aspires to becoming the world's leading distributor of Earth observation and remote sensing imagery and to revolutionise the way people use imagery to conduct business in the 21st century. It gathers, processes and produces information about the Earth's natural and man-made features with an unequalled combination of resolution, accuracy, information, content, speed of delivery and price. Space Imaging clients are in a wide variety of markets including environmental monitoring, mapping agricultural monitoring, urban planning, oil and gas exploration, disaster assessment, forestry management, telecommunications, real estate, insurance, entertainment and media. The market for high (1 m) resolution imagery worldwide has been estimated at US$3 billion to US$5 billion over the next five to six years.

Space Imaging's imagery and information products are collected in a digital archive known as the CARTERRA, which houses space-based and aerial imagery from multiple sources. Users can enter the CARTERRA Archive using a standard Web browser, search for Ikonos, Landsat and IRS imagery in a point-and-click environment and browse the imagery and accompanying geographic information available online. The catalogue is updated with new imagery several times each day.

Lockheed was granted a licence 22 April 1994 by the Department of Commerce to operate a private, 1 m resolution, Earth imaging system under Presidential Directive 23 March 1994. Lockheed formed Space Imaging in May 1994 to handle the project; it became an incorporated company in December 1994. The cost for two satellites in orbit, two launchers, a primary ground station, two supporting stations, all image processing hard/software and five years of operation was projected at about US$500 million; about US$275 million had been raised by January 1996. Annual revenue is projected at US$500 million. Lockheed Martin is committing US$150 million, Raytheon's E-Systems Inc US$100 million and Mitsubishi Corporation, as a regional partner, US$25 million.

The Ikonos satellite will routinely collect 1 m PAN and 4 m XS digital imagery, plus 1 m PAN-sharpened, or colour enhanced, imagery. Mapping North America's largest 100 cities is an early priority. Products include: radiometrically corrected imagery, geometrically corrected imagery, ortho-rectified imagery, digital terrain models, and multispectral ratio images. Space Imaging is establishing a network of global regional affiliates that will take customer orders, task the satellite to collect requested imagery, and market imagery in their regions. The ground segment covers a primary station in Thornton and two support stations in Alaska and southeast US. The regional affiliates will also operate their own stations. Tasking commands can be modified up to 10 minutes before satellite contact, accommodating last-minute priorities or changing weather conditions. Each station can download images in real time while in contact. Satellite PDR was completed 16 June 1995. A failure involving the Lockheed Martin Astronautics Athena II launcher during the launch attempt on 27 April 1999, destroyed the Ikonos 1 satellite when it failed to reach orbit. Space Imaging plans to launch a replacement on another Athena II but no earlier than the end of 1999.

Resolution: 1 m Pan, 4 m XS. It will be able to image between ±85° latitude and register any point within the image to less than 12 m horizontally. With inclusion of ground control points, products will have less than 2 m horizontal accuracy
Design life: 7 years
Mass: less than 900 kg at launch, including 50 kg hydrazine
Orbit: 680 km circular in 98.20° Sun-synchronous, descending node between 10.00 and 14.00 local time, providing same-scene access every three days or fewer at better than 1 m ground sample distance
Spacecraft: 1.5 m diameter hexagonal bus, 2 m height. High-accuracy 3-axis pointing control and knowledge by CT-601 star trackers, rate gyros, thermal stability and refined scanning laws. Hydrazine thrusters. Slew rate sufficient for both wide-area monoscopic and same-pass stereo collecting. Two-axis high-gain X-band antenna downlinks wideband imagery at 320 Mbit/s; stored in 64 Gbit SSR. 1,500 W from three deployed GaAs/Ge panels plus bus panel; 50 Ah Ni/H_2 battery
Sensor: totally reflecting optical telescope with 0.7 m aperture primary mirror mated to a digital focal plane array, which can generate 6,500 lines/s PAN. 11 km nadir swath. Spectral ranges (μm): Band 1 0.45-0.52, Band 2 0.52-0.60; Band 3 0.63-0.69; Band 4 0.76-0.90; PAN 0.50- 0.90. PAN utilises 12 μm-pixel pushbroom linear CCDs; XS 48 μm-pixel four pushbroom linear CCDs. Low-loss data compression on original 11-bit data.

SPACE INDUSTRY – GENERAL

AUSTRALIA

Asia Pacific Aerospace Consultants Pty Ltd (APAC)

APAC provides space consultancy services for the Asia-Pacific region.

Auspace Limited

Auspace was established in 1983, and in 1986 it became a subsidiary of Matra Marconi Space, with capabilities and experience in satellite system engineering and R&D in space systems, ground communications satellites and electro-optics.

Current and recent projects include: the design and building of the Endeavour experimental ultra-violet imaging system to space-qualify a large format photo counting array detector for use in satellite astronomy. The focal plane array for the ETS 2 along-track scanning radiometer; major contractor for Envisat's Advanced Along-Track Scanning Radiometer (infra-red/visible focal plane array, foreoptics, instrument electronics unit and signal pre-amplifier unit); an airborne prototype of a passive spaceborne atmospheric pressure scanner for CSIRO; a high-performance multispectral scanner for airborne applications; conceptual study of the ARIES satellite which will provide hyperspectral imaging of the Earth in the visible and short-wave infra-red regions of the spectrum for applications in the mining, environmental and agricultural areas.

In 1994, Auspace sold its GPS business to SIGTEC.

Australian Space Insurance Group (ASIG)

ASIG provides capacity for the worldwide space risk insurance market.

British Aerospace Australia (BAeA)

British Aerospace Australia is one of Australia's leading defence and aerospace companies – having supported Australia's Defence Forces for almost 45 years.

With 1,800 employees located at the company's 25 sites around Australia and overseas – British Aerospace Australia is the country's second largest defence employer.

The company's strong commitment to meeting Australian Defence Force (ADF) requirements has led to a broad range of capabilities in the support of land, sea and air defence systems including: aircraft systems integration, military vehicles, missiles, decoys and targets, electronic warfare, electro-optics and communication, command and control.

Under contract to the Australian Space Office, British Aerospace Australia manages and operates the Moblas 5 facility located at Yaragadee in Western Australia. Staffed by five technicians, the station provides satellite laser ranging data for investigation of earth geodynamics.

British Aerospace Australia has managed and operated the Canberra Deep Space Communication Complex (CDSCC) on a turnkey basis for the Australian Space Office since 1990.

The company employs some 140 engineers, technicians and administrative personnel who carry out the entire CDSCC operation. Their technical activities provide a link between the ground control and processing environment and various Earth orbital and deep space spacecraft. In addition British Aerospace Australia provides a broad range of support functions.

On behalf of the Australian Surveying and Land Information Group (AUSLIG), British Aerospace Australia provides data acquisition, processing and distribution services for the Australian Centre for Remote Sensing (ACRES). Located at the Data Acquisition Facility in Alice Springs and the Data Processing Facility in Canberra, British Aerospace Australia employs approximately 35 personnel. These personnel are responsible for the operation of a complex manufacturing system which generates highly customised photographic imagery and digital data products by using data received from orbiting remote sensing satellites. These products are sold through a network of Australian distributors and satellite operations outside Australia.

Electro Optic Systems Pty Ltd (EOS)

EOS specialises in satellite laser ranging and tracking. It has developed high-power eye-safe lasers for communications links to satellites.

Hawker de Havilland Ltd

HDH's space activities focus on structural, mechanical and thermal systems. It has studied small satellites and worked with BAeA on feasibility studies for the Southern Launch Vehicle.

KEL Aerospace Pty Ltd

KEL specialises in HF radar, advanced HF components, electrical transient monitoring, science instrumentation and commercial information systems. Its ERS Frame Synchroniser is a combined high/low rate frame format synchroniser for data processing ground stations.

AUSTRIA

ACT Hochleistungskunst-stofftechnik GmbH

ACT provides development, manufacture and assembly of lightweight structures and components, mainly using composites. It provided the composite struts for ISO's sunshade support, the first satellite composite hardware designed and manufactured nationally. It provided the solar array yoke for Orion and Tempo.

Andritz AG

Andritz fabricates the rings for attaching Ariane 5's solid boosters to the cryogenic core.

Austrian Research Centre Seibersdorf

With a staff of 538, Seibersdorf is Austria's largest non-university research centre. The federal government holds 50.5 per cent of the shares. 1995's income for contract-based research was Sch337 million (1994 – Sch320 million; 1993 – Sch287 million). Activities are focused in five divisions: instrumentation and information technology; process and environmental technologies; engineering; life sciences; systems research (technology, economy and environment). Seibersdorf holds a long-term contract as an ESTEC test house for space materials, qualifying advanced materials and developing space-specific tests, for example, mass loss, tribology and cold welding in the space environment. Seibersdorf further participates with ESA material technology programmes like FESTIP or ALSCAP. Seibersdorf provided support for ESA's Giotto and developed an ion emitter for controlling electrostatic charging of Geotail, Interball and Cluster. Other work includes a micro-analysis station for materials testing in space, investigation of μgravity effects on the human immune system and genetic material.

Böhler Edelstahl GmbH

Böhler is one of the world's largest producers of special steel, including materials for satellite components since 1982. The P802 Invar and P812 Super-Invar steels were employed in the Rosat x-ray astronomy satellite's mirror system to ensure accurate alignment and dimensional stability. Nickel-based alloys are supplied for Ariane.

Fischer Advanced Composite Components – FACC

Fischer Advanced Composite Components (FACC) was established in 1986 to centralise Fischer GmbH's composite component activities. An annual research and development budget of US$2 million supports production of components for space applications.

Österreichische Raumfahrt-Und Systemtechnik GmbH (ORS)

ORS is owned by Saab Ericsson Space (74.9 per cent) and Dornier (25.1 per cent). 1996 sales were Sch98 million (1995 Sch130 million); 61 staff. It is Austria's largest space company. 1994 sales were Sch123 million; 1993 and 1992 each Sch130 million. Flight and ground support equipment covers ERS, ISO, Ariane 5, Columbus, Soho, Cluster, Huygens and Artemis. It designs, develops and manufactures spacecraft structures, mechanisms and thermal hardware and MGSE. For Italsat, ORS supplied the MGSE and a full-size engineering structure manufactured from aluminium honeycomb. It is developing MGSE and thermal hardware for Envisat 1 instruments. The company provided ISO's aluminium honeycomb sunshield and its support.

Other fields concern mainly ECLSS (hygiene, food, water management) and space biology (design of biological and μgravity payloads, utilisation studies). In addition, ORS is engaged in space robotics (end-effectors, sensors) and the design of advanced composite material and high-accuracy structures.

Plansee AG

Plansee is a leading producer of refractory metals (tungsten, molybdenum, tantalum, niobium, chromium and their alloys, heavy metals, tungsten-copper alloys, copper-chromium). It is a manufacturer of components for high-temperature applications, and offers a range of products and services for space industries.

Schrack Aerospace GmbH

Saab Ericsson Space holds an 80 per cent interest and Ericsson Austria 20 per cent. 1996 sales were Sch85 million (1995 Sch85 million); and it employs 60 staff. Schrack Aerospace is a manufacturer of electronics equipment and subsystems for communications, Earth observation and science satellites. It specialises in the design/development of onboard Earth observation instruments, subsystems for optical intersatellite links and satellite control/monitoring electronics. In the ground segment, it is developing satellite checkout equipment.

Siemens AG Österreich

Siemens Austria specialises in communications and data technology, power engineering and automation, and programme and system development. Space activities include onboard processing, application of expert systems and object-oriented software for ground systems, telescience software and data processing, EGSE and software development for ESRIN, ESOC and ESTEC.

Steyr-Daimler-Puch Fahrzeugtechnik AGd CoKG (SDPF)

SDPF developed and produces mechanical components for Ariane 5, such as cryogenic fuel and pressurisation lines, and booster attachment elements. Machining of fuel valve casings and pressurisation manifolds. Moulding of hard foam insulation for feed lines. Manufacture of XMM mirror shell containers. Participation in ESA's FESTIP study and the Austro/Swiss ALPSAT project.

GERMANY

Anite Systems GmbH

Previously Cray Systems GmbH, the company is a subsidiary of Anite Systems Ltd of the UK. It specialises in the design and development of software and systems for ground segments.

Autoflug GmbH & Co

Development and manufacture of parachute recovery systems and shock absorbing landing bags, flotation and location systems for launch vehicles and components.

Bank Reuschel & Co

Bank Reuschel provides space project financing support. Involvement includes Eutelsat, EUMETSAT and Ariane 5.

Carl Zeiss

Carl Zeiss specialises, in the space field, in high-quality optical systems such as telescopes and microscopes. It will provide Abrixas's x-ray optics system. Hipparcos ROSAT and SELEX and ESTEC's large space simulator employ the company's optics. 1994 space turnover was DM30 million.

Computer Anwendung Für Management GmbH

A subsidiary of France's CISI, CAM's space expertise includes control centre software for controlling satellites such as TV-Sat, DFS Kopernikus and Eutelsat, computer facilities for the Columbus payload operations centre, and development of image processing systems for Earth observation satellites.

DaimlerChrysler Aerospace AG (DASA)

DaimlerChrysler created DASA (known until 1 January 1995 as Deutsche Aerospace) in 1989 from the country's four major aerospace companies: MBB, Dornier, MTU and TST. Their resources were reorganised in 1991 into four operating groups: space, aircraft, propulsion and defence/civil applications. MBB/TST were absorbed into DASA in 1992; Dornier retains its identity. 1995's DM15 billion revenue produced a DM4.3 billion loss (1994: DM438 million loss on DM17.39 billion. 1993: DM694 million loss on DM18.6 billion). Workforce totals 75,000, from 86,000 in 1994. The Space Systems Group's 1995 revenue totalled DM1.9 billion (1994 DM1.354 billion; 1993 DM1.4 billion; 1992 DM1.91 billion); personnel total 4,200. SSG comprises two divisions: Satellites Systems (Dornier) and Space Infrastructure; they are specified in greater detail below. DASA in 1992 took a 12.25 per cent interest in Space Systems/Loral of the US. Following reorganisation in 1996 the DASA Space Systems Group was split into Dornier Satellitensysteme GmbH and DaimlerChrysler Aerospace – Space Infrastructure Division.

The Space Infrastructure Division is responsible for infrastructure programmes such as Columbus; DASA is leader (the ECU658.5 million fixed price construction contract was signed with ESA 28 March 1996). The Bremen division was prime contractor for Spacelab, including design, development, integration, qualification and operational support, and for ESA's Eureca retrievable μg carrier. It also conducts the Texus and Express μg research programmes. It is a major contributor to the Ariane programme, including prime for stage 2 integration. The company is prime for the Astro-SPAS reusable platform.

The former Space Transportation Systems and Propulsion division was divided in 1993 into two centres of excellence, one for the Satellite Systems division and the other for the Space Infrastructure division. Strategic responsibility for satellite systems was transferred in 1991 by DASA to Dornier, while production of subsystems remains in Ottobrunn. The former MBB was prime for Exosat and provided Galileo's propulsion module. It undertook 40 per cent of the work on Rosat, provided the Imaging Compton Telescope and part of the Energetic Gamma Ray Experimental Telescope for NASA's Compton Observatory, and developed the Infrared Space Observatory's payload module. Extensive communications satellite involvement includes DFS (platform, integration/testing), TV-Sat (system management, integration/test, AOCS, propulsion, antenna systems, TT&C), TDF (AOCS, propulsion), and Eutelsat 2 and Turksat (AOCS, TT&C, solar array); the Spacebus platform is offered in association with partner Aerospatiale. Earth stations have been provided to Spain, Belgium, Australia and Italy, in addition to a Landsat terminal in Italy and the ERS antenna in Antarctica. Far field and near field ranges are operated for the testing of antenna systems. The division works with France's SEP on Ariane engine development. The division also develops and manufactures opto-electronic sensors and cameras, such as the MOMS-02 package flying on Russia's Priroda Mir module.

Dornier Satellitensysteme GmbH

Dornier Satellitensysteme GmbH was founded in 1995 as a subsidiary of Dornier GmbH, a DaimlerChrysler Aerospace AG company, by integrating the satellite activities. The company is involved with the development, manufacturing and marketing of satellite systems, instruments, subsystems and components for a wide field of applications in the area of scientific space exploration, Earth observation, satellite communications and navigation as well as military satellite systems.

Drägerwerk AG

Drägerwerk specialises in gaseous control systems and gas sensors. These include systems for Spacelab and Columbus ECLSS and for onboard scientific experiments. Equipment developed, manufactured and qualified covers vessels, valves, gas and air filters, servicing units, carbon monoxide removal systems, contamination monitoring and control systems, and water servicing units.

Elbe Space & Technology Dresden GmbH & Co KG (ESTD)

ESTD was formed in 1992 to market RFAS products and services, finance enhancement of Russian systems for the western market and act as a contractor in western projects. Its STIR (Science and Technology International of Russia) subsidiary is headquartered in Moscow. ESTD offers satellite-based navigation and communications systems such as Elekon, GPS/Glonass receivers, Start launch services, IL-76 parabolic flights and software development.

Elektronik-Und Luftfahrtgeräte GmbH

Elekluft, wholly owned by DaimlerChrysler Aerospace, provides system support for ESOC and EUMETSAT and general support for ESTEC, DLR and the Ariane Kourou base. Space personnel total 80.

ESG Elektronik System GmbH

Principal space activities cover system design and analysis, system simulation, information management, logistics, operations support, artificial intelligence and expert systems, and product/quality assurance. Contracts have included Anthrorack failure analysis for Spacelab D2 and logistic concepts for Columbus.

Kayser-Threde GmbH (KT)

KT provides expertise in the design, build and integration of space experiment payloads, covering sounding rockets, Spacelab, SPAS, Photon, Shuttle and re-entry capsules. 1995 sales were DM51.5 million (DM80 million including subsidiaries); staff totals 200. Experience covers: 43 sounding rocket payloads (7 Aries, 33 Texus, 3 Maxus), vestibular experiments on Spacelab 1 and Mir-92, venous pressure measurements on Spacelab 1/D1, high-precision thermostat on Eureca 1, Spacelab D1/D2, Cryostat protein growth facility on IML-1, main contractor for Astro-SPAS' Orfeus telescope and Holop holographic lab for Spacelab D2. It was system integrator for all German experiments on Mir-92. The Energia Deutschland GmbH partnership with NPO Energia exploits an exclusive right to distribute Earth observation data from Mir.

KT built Temisat (plus a spare) under a US$10 million contract to Telespazio for piggyback launch on 31 August 1993 with a Meteor 2 aboard a Tsyklon, demonstrating relay from environmental data collection platforms. Temisat was the first of KT's Blackbird family of micro-satellites, although no others are under development. KT also acted as launch agent for the Technical University of Berlin's Tubsat B in January 1994. It built the GFZ 1 laser geodetic satellite for Germany's National Centre of Earth Sciences for an April 1995 launch on a Russian Progress Mir station ferry.

KT developed a range of DGPS stations for DARA and is involved in the pre-development of an advanced combined GPS/Glonass receiver for ESA. The company has been marketing Glonass receivers since June 1991. It provided the MOMSNAV orbit/attitude determination package for Germany's MOMS-02P imaging system on Mir's Priroda module. This uses two GPS receivers and a high-precision gyro package for 2 m/3 arcsec accuracy. A PRARE Precise Range & Range Rate Experiment package was provided under DARA contract for January 1994's Meteor 3 meteorological satellites. KT developed the transputer-based multiprocessor for Soho's SUMER instrument and the camera control processor for Mars-96's High Resolution Stereo Camera.

KT developed the Biopan biological experiment carrier for the exterior of Photon and Resurs descent capsules under ESA contract. The hinged lid on the 37 cm diameter 22 cm high pan-shaped container opens through 180° to expose samples mounted on the lid and fixed bottom plate, totaling 910 cm^2 for experiments. Its own 2 Mbyte microcontroller commands the power supply for the hinge drive and heaters (maintaining 0-37°C) and performs temperature sensor switching.

Solar aspect angle and 0.25-4.5 μm ultra-violet radiation are also monitored. 3.8 cm thick layer of glass fibre and phenol provides re-entry protection. The test flight was made aboard Photon 5 in October 1992. ESA has approved two further flights; first was June 1994 on Photon 6 carrying six experiments, using the refurbished ground qualification unit. The next was in September 1996.

The company signed an agreement with Glavcosmos in December 1987 for options on Photon and Resurs missions, acting as marketing agent. Four missions flew 1989-91, carrying Intospace's Cosima 2-4, Casimir and LZZ experiments. A further contract was signed in autumn 1991 for up to six flights. The next crystal growth experiments flew on Photon 6 in June 1994. Photon 7 in February 1995 carried seven KT materials processing experiments under DARA contract in the Zona 4M and Konstanta 2M furnaces. MIRKA (Micro Re-entry Capsule) was launched aboard the Photon 8 satellite in October 1997. The 150 kg MIRKA was mounted on the outside of the Photon descent capsule where it remained until after the main spacecraft's de-orbit manoeuvre had taken place. It came down 110 km southeast of Orsk on 23 October 1997.

Kayser-Threde was awarded the contract from DARA to be responsible for the logistical and technical preparation for the MIR-97 visit of Reinhold Ewald to Mir – a role which the company had played for the MIR-92 mission. Progress-M33 launched in November 1996 carried 158 kg of German scientific equipment to Mir in advance of the Soyuz-TM 25 launch. Kayser-Threde were responsible for a MIR-97 experiment involving real-time transmission between Mir and the German control centre at Oberpfaffenhofen.

Krupp VDM GmbH

High-performance aerospace materials: nickel-based alloys, cobalt alloys and special stainless steels. It provides the cobalt-based Conicro 5010W-alloy 25 for the nozzle extension of Ariane 5's upper stage engine. Ariane 4 carries VDM steel propellant tanks.

MAN Gutehoffnungshutte AG

MAN Gutehoffnungshutte AG (MAN GHH) is a subsidiary of the MAN Group and specialises in the design and building of ground installations for the aerospace industry, parabolic antennas and mobile bridge systems. The company is also active as a consultancy within these fields. Within the space-related fields, MAN GHH has been involved with: rocket launching facilities, test stands for rocket motors, buildings for satellite preparation, research and communications antennas and ground stations for telecommunications and research.

In 1976, MAN GHH provided and installed the launch table for the first Ariane launch complex, ELA 1 at Kourou. In 1981, the company was commissioned to build the main components for ELA 2, to be used by the Ariane-3 and Ariane-4 launch vehicles. In 1988, MAN GHH was appointed the general contractor to build the third launch pad at Kourou, ELA 3 for Ariane-5. Also at Kourou, MAN GHH was responsible for the satellite preparation buildings S1B and S3B, as well as the satellite encapsulation bay which is part of the final assembly building of Ariane-5.

MAN GHH has also planned and built stands for the testing of both solid and liquid propulsion systems, including the equipment required for the test operations such as holding devices, cranes, propellant supply systems and safety systems.

The Vulcain engines for Ariane-3 and Ariane-4 are tested on the PF 20 test stand, built by MAN GHH and located at the SEP test site in Vernon, France. At Kourou MAN GHH constructed the BEAP test stand for the solid-propellant boosters for Ariane-5.

MAN GHH has supplied fixed, transportable and mobile ground stations for satellite communications, TT&C and geodesy.

Munich Reinsurance Company

Munich Reinsurance (Münchener Rückversicherungs-Gesellschaft) is one of the largest risk carriers for commercial space projects, providing 15 to 20 per cent of the world's coverage. A team of specialists, set up at the end of the 1970s, concentrates exclusively on the insurance and reinsurance of space risks. The company is represented in 32 countries with 48 offices.

Satellite Operational Services GmbH (SOS)

SOS, founded in 1983 and now with 50 staff, provides expertise in mission operations for manned and unmanned programmes, space flight dynamics and ground station operations. The company participated in Spacelab D1/D2, Helios, TV-Sat, DFS Kopernikus, Eutelsat, Galileo, Rosat, Giotto, IRS and AMPTE. Its current principal customer is the DLR Aerospace Research Establishment at Oberpfaffenhofen. SCS' capabilities include defining and developing satellite operational requirements, training spacecraft operators, control centre software definition and acceptance testing, satellite in-orbit testing, spacecraft command operations, communications networking, ground data/ tracking station operations, orbit and attitude determination, and manoeuvre planning.

W L Gore & Associates GmbH

Gore manufactures space-qualified wires, cables, waveguides and data lines. These include wires and cables with/without drainwire qualified to Spacelab SLP 2110 and ESA/SCC detail specification 3901/009, 017, 018, 019, 021. Applications include ERS, SAX, ISO, PPF/Envisat, XMM, GlobalStar and ERA. Also included are high-voltage cables for TWTs and dielectric waveguides 26.5-110 GHz.

Zeppelin Technologie GmbH

Zeppelin manufactures propellant and pressure tanks for all kinds of space applications, as well as individual parts like bulkheads, parts, flanges and pipes.

INTERNATIONAL

Matra Marconi Space

Matra Hachette SA and GEC Marconi plc in 1990 merged the activities of their respective Matra Espace and Marconi Space Systems companies under Matra Marconi Space. Matra Hachette holds 51 per cent and GEC Marconi 49 per cent. Matra brought assets worth £10.7 million to the partnership and GEC £8.7 million; total workforce was 3,200 and 1993 turnover was FFr5,167 million. MMS acquired British Aerospace Space Systems in 1994 for £56 million, creating Europe's largest space company. BAeSS 1993 sales were £151.6 million, for a £5 million profit. At the time of the purchase, its backlog stood at £512 million (MMS' own was FFr11.5 billion) and the workforce at 1,050. MMS' 1995 turnover was FFr6.7 billion (plus FFr9 billion backlog); 4,400 personnel. MMS is Europe's first fully integrated space company and its largest spacecraft manufacturer. With five major UK/France sites, it can undertake all major applications of space technology: Earth observation, science, civil/military communications, and space transportation and manned space systems.

MMS France is concerned with the design, development and manufacture, as prime contractor, of satellites and spacecraft, platforms, payloads and onboard electronic equipment. The company is prime contractor for Telecom 1/2, Hispasat, Nilesat, Hot Bird 2-5, Spot 1-4, Helios, Soho and Ariane's equipment bay; it is the main subcontractor for Eutelsat 2, Inmarsat 3 and Koreasat. MMS will also play a major role in the Columbus data management and AOCS.

MMS provides attitude control, navigation and guidance equipment for telecom, Earth observation and science satellites, launch vehicles and sounding rockets. Other onboard equipment includes stabilisation systems, Sun sensors, data processors and gyro packages. In the ground-based area, it provides control equipment, Earth stations, automatic test equipment, simulators and image processing systems. It holds 35 per cent interest in Spot Image and 45 per cent in NRSC Ltd.

MMS UK specialises in the design and manufacture of payloads for communications, remote sensing and science applications, together with the associated ground support and control equipment, and user terminals. Expertise also encompasses structures, thermal control, power, propulsion, AOCS, TT&C and mechanisms. Extensive facilities include a fully equipped manufacturing complex, comprehensive AIT resources and high-technology research and development facilities. These include a fully-screened indoor near/far-field antenna test range, one of Europe's largest.

Marconi was prime contractor for Skynet 2 and Ariel 5/6, and provided much of the onboard equipment for the highly successful Marecs, Exosat and IUE satellites. Recent work includes the communications payloads on Skynet 4 and NATO 4, the active microwave system on ERS 1/2 and onboard electronics equipment of the Meteosat series, including the attitude measurement system, AOCS, imaging electronics and communications packages. BAe participated in approximately 100 satellite projects over three decades. It developed Ariane's Spelda carrier, Spacelab's pallets and the Skylark sounding rocket (see the Launchers section). It primed most of Europe's science satellites, including Giotto, provided Hubble's solar arrays and Photon Detector Assembly. Bristol continues as ESA's prime for Polar Platform (first launch as Envisat 1 in 1998). It developed the ISAMS multichannel infra-red instrument for NASA's Upper Atmosphere Research Satellite and the passive microwave AMSU-B sensor for NOAA's Tiros, in addition to providing the Digital Electronics Unit for the UK's Along-Track Scanning Radiometer on ERS.

The company's products span radio, radar and analogue/digital processing applications for civil/ military satellite systems, covering space bands less than 0.2-183 GHz, and extending to optical frequencies for intersatellite links. The subsystems encompass: UHF/SHF/EHF transponders, transmitters, frequency converters, receivers, filters and multiplexers, beacons, local oscillators, launcher and satellite command equipment, modulators/demodulators, encoders/ decoders, meteorological radiometers, SARs, TT&C systems, onboard processors and memories, optical communications systems

Earth Station terminals are available in 1 to 19 m diameter antenna sizes. Military stations include: fixed stations with 7 to 13 m antennas, 6.4 m transportable stations, 1.7/4.5 m vehicle-mounted terminals, 1.06/ 1.2/1.8 m SCOT ship terminals with fully stabilised antennas (more than 70 ordered; the MoD ordered an Enhanced SCOT in March 1990), MASTER aircraft terminals, 12 kg Manpack man-portable SHF terminal and TT&C stations for satellite station keeping. Civil stations include: Inmarsat transportable, Apollo, TVRO stations (fixed and for ships), TV news gathering, meteorological data receiving, satellite measuring stations (antennas up to 19 m), and aircraft data relay. The company provides a turnkey service.

Ground Support Equipment. Up to 45 GHz, covering high power amplifiers, receivers, frequency converters, modems, TDMA systems, filters and multiplexers, and processors

Satellite Ground Control Facilities. MMS UK offers equipment for the monitoring/control of civil/military systems, including launcher command, TT&C links, security/safety systems and satellite simulators. Most recently, it provided Skynet 4's Satellite Operational Facility at RAF Oakhanger.

Research/Technology. Current programmes include onboard processing, Ni/H_2 batteries, ion propulsion, GaAs solar cells, phased array antennas, laser communications, and space-based surveillance systems.

In Toulouse the company performs commercial satellite integration: 1,332 m^2 10 m high High Bay and 1,190 m^2 3 m high Low Bay, both class 100,000; two 100 m^2 class 100 laminar flow areas. Alignment area on seismic block with computer-driven table (accurate to 1 arcsec). It also conducts Ariane vehicle equipment bay integration: three 400 m^2 areas of 3/6/12 m high; class 100,000 available; two seismic blocks for inertial platform calibration and military satellite integration: three 400 m^2 areas of 3/9/12 m high; class 100,000 specified (10,000 measured); 120 m^2 class 100 10 m high laminar flow area.

The Toulouse antenna range has a 400 m slant range, class 100,000 under 14 m radome, 8 tonne positioner, 70 m tower for transmit/rx measurements.

The company's Portsmouth facilities include two principal class 100,000 cleanroom complexes. Building 43's 41 × 30 m includes a 32 × 16 m satellite assembly/integration area with 10 m hook height/3,000 kg crane capacity. Accessed through a 5 × 6 m high door, the conditions are controlled to within 20±2°C and 35 to 60 per cent relative humidity. A 10 m^2 seismic block provides a position for optical measurements on spacecraft critical reference axes. The building also houses a reception area for loading/unloading, clean store, EGSE room, EMC chamber suite, thermal chamber, mass properties area and antenna test range, all maintained to class 100,000 conditions.

The 11 × 13 m (7.3 m high) EMC facility of five rooms, with 5 × 5 m access door, provides better than 80 dB magnetic screening 14 kHz-10 MHz, and better than 120 dB electric field screening 14 kHz-10 MHz, 115 dB 0.10-1 GHz and 126 dB 1-40 GHz.

The 5 × 5 × 5 m thermal chamber can handle heat dissipation of up to 5 kW from equipment under test and control conditions over −40/+110°C with 2°C accuracy. Data recording is performed on 16 channels.

The mass properties measurement area provides mass determination up to 5,000 kg (or 500 kg in 100 gm increments), moment of inertia measurement up to 3,000 kg using Schenck M1, M4 and M6 machines, centre of gravity location for subjects up to 4,000 kg using Schenck WS12 and WM50/6 machines; and spin balancing using Schenck E5 (capacity 500 kg, up to 300 rpm) and E6 (400 kg, 300 rpm) units.

Building 43 also houses Portsmouth's large Near-Field Far-Field antenna test range. A monolithic foundation supports both the elevation/roll-over-azimuth antenna and test positioner and the source/probe tower, track mounted for variable length. Loads up to 4,000 kg can be accommodated. The 15 × 15 × 31 m chamber offer a 5 m diameter quiet zone, 0.25-60 GHz, better than −40 dB reflectivity at 1 GHz rising to better than −55 dB at 7 GHz, 100 dB magnetic screening at 200 kHz (110 dB at 1 MHz) and 130 dB electrical screening 1 kHz-50 MHz (110 dB over 0.050-1 GHz, 100 dB over 16-18 GHz, 80 dB over 18-60 GHz). Positioner tower capacity 1,600 kg, positional accuracy 1 mm and pointing accuracy 0.03°. The range operates primarily in the spherical near-field mode. The radiated field is sampled in the antenna near-field and the far-field patterns calculated. Determination of gain, radiation patterns, cross polar response and boresight location are undertaken.

Building 48 provides a class 100,000 35 × 12.5 m assembly/integration area with optical block and houses a clean store and EGSE operating room. In addition, the building includes 250 m² class 10,000 and 1,300 m² class 100,000 manufacturing facilities with class 100 laminar flow workstations. Other Portsmouth facilities cover three thermal-vacuum chambers in building 23A, each offering a 1,250 × 1,250 mm cylindrical internal working volume and less than 10^{-5} torr over −65/+100°C. An LN_2 turbomolecular pump can achieve 10^{-6} torr within 4 hours. The shrouds can also be operated at cryogenic temps with LN_2 (−196°C). 16 auxiliary T-measuring channels are recorded on a data logger for each chamber. A range of climatic chambers is available, providing up to −70/+175°C.

Building 17 houses seven vibrators installed in three rooms and controlled from a common position. Drop, centrifuge and shock test equipment is also housed in the building. The four principal vibrators are: Ling 980 (5-2,000 Hz, 150.52 kN sine peak, ±12 mm max displacement), Ling 964 (5-2,000 Hz, 75.26 kN, ±12 mm), Derritron VP2500 (5-2,100 Hz, 111.2 kN, ±12 mm) and Derritron VP1200 (5-3,000 Hz, 55.6 kN, ±12 mm). The 1.2 m centrifuge achieves 60 *g* on 50 kg test items (up to 0.5 × 0.5 × 0.5 m) and is equipped with 60 sliprings for functional testing. The Portsmouth site does not offer acoustic test facilities but the high intensity acoustic unit in GEC's Mechanical Engineering Lab at Whetstone has been extended for spacecraft testing.

Building 36's Hydrazine Propulsion and Test Facility houses a 3.4 m diameter × 6.7 m long 55 m³ vacuum chamber with associated pumps and exhaust scrubber, providing an initial chamber pressure of 10^{-2} torr for hot firing tests. T controlled to −15/+60°C. The unit is connected to a hydrazine cart (weighing accuracy 50 gm, pressure up to 35 atmosphere) and an oxidiser cart handling nitrogen tetroxide, nitric oxide or MON. Exhaust is scrubbed by recirculated dilute hydrochloric acid to remove ammonia and unburned hydrazine. Chamber instrumentation includes P/T sensors, flowmeters and thrust transducers associated with a 24-channel data acquisition system. The control system permits thruster pulse duration programming. The remaining 200 m² of building 36's class 100,000 cleanroom is used for propulsion subsystem assembly and test, including a class 100 area for detailed assembly work and automatic particle counting. An x-ray room caters for pipework joint inspection by weld radiography, and flow rate/pressure drop tests using GN_2 are conducted in the area. A 1.5 × 1.5 m vacuum chamber for propane tests is also available.

The Stevenage facilities covering 9 hectares, has approximately 6,000 m² manufacturing, assembly and test floor area, and a new three-storey office building with 7,000 m² for management, admin, design and engineering staff.

The Satellite Assembly Hall measures 30 × 24 m, 11 m high, maintained to Fed-Std-209B class 100,000, air-conditioned to 21±3°C, 35 to 55 per cent relative humidity. Access by 7 × 5 m roller shutter doors. Crane with 10 tonne capacity and 9.7 m hook height, movable over 90 per cent floor area. Two 7.6 × 7.6 m areas can be lowered to provide 16.7 m crane height. Four assembly bays capable of simultaneous accommodation of four satellites and associated checkout equipment. Adjacent 5 × 35 m EGSE control area.

The Microwave/EMC Test Chamber. 12 m high, 10 × 14 m, accessed directly from the Assembly Hall by airlock and protected corridor, and two adjacent 4 m high 6 × 8 m screened rooms. All controlled to class 100,000. Anechoic lining ensures a return loss of greater than 30 dB at 500 MHz and greater than 50 dB at 5 GHz, and can handle an input power of 0.3 W/m². A central ceiling access panel allows antennas to be mounted above the spacecraft to excite payload transponders. Chamber screening effectiveness is better than 80 dB in the H field (14 kHz-10 MHz) and better than 115 dB in the E field (14 kHz-40 GHz). T controlled to 5 to 50°C (±3°C); relative humidity 30 to 60 per cent.

Further test facilities include a 7.7 × 18.3 m, 7.3 m high Antenna Anechoic/EMC Chamber, with cylindrical near/far-field range capabilities, and a 2.4 × 2.4 m High Power RF Vacuum Chamber, with up to 3.5 kW at 8 to 18 GHz and 600 W L-band. The Environmental Test Area contains six Climatic Chambers, six Thermal Vacuum Chambers and 40/160 kN shakers. An ultrasound non-destructive test unit allows the bonding of panels and the largest (4 m diameter) honeycomb panels for Ariane 4's Spelda, to be checked. Manufacturing areas include a 150 m² class 100 horizontal laminar flow Precision Products Clean Room with associated cleaning/test facilities and panel bonding, antenna lay-up, satellite assembly, electronics assembly and Spelda assembly areas totalling 4,700 m². A six-axis CNC carbon fibre filament winding machine capable of 2.4 m diameter × 6 m and its adjacent curing facility (three autoclaves; largest 3.5 m diameter × 5.5 m) and four ovens.

The Thermal Blanket and Secondary Surface Mirror Assembly Facility is a 160 m² controlled to Std-209B class 100,000, 21±5°C, less than 50 per cent relative humidity conditions. Complete blanket manufacturing facility for spacecraft applications

In Bristol the Halley Building (20X) houses the majority of the executive, project management, engineering and drawing office staff on four floors totalling 4,800 m² and accommodating 340. Building 20Y houses the site's manufacturing and AIT facilities, including a single story AIT Assembly Hall with pit and lift, capable of accommodating the largest spacecraft currently envisaged in Europe (equivalent to full STS cargo bay), together with an associated two-storey manufacturing, test and office facility.

The Main Assembly Hall, on the ground floor has a working area 589 m², access 5 × 6 m high door from haulingway (airlock), head room 12.5 m to crane hook (10 tonne limit and full traverse facilities). 10 × 10 m hydraulic lift platform capable of lowering a six-tonne load 7.93 m below normal floor level. Whole area protects against RF interference up to 60 dB, 0.1-18 GHz. Air conditioning 20±3°C, 50±5 per cent relative humidity, class 100,000 with cross flow ventilation (not laminar) at eight changes per hour. Floor is conductive and walls/ceiling are RF shielded with cleanroom linings.

The Detail Manufacturing/Assembly Area is 562 m², air conditioned (warm air only) 20±5°C with a workshop area of 116 m² for producing spacecraft detail parts to flight standard. Equipped with small lathe (Harison M250), guillotine, bench drill, bench drill with compound table, bench folder, finger press, 2 × finishers, grinder, large folder, fly press with associated equipment and various power tools such as jigsaw. The RCS assembly/test area: 116 m², controlled to class 100,000 Fed-Std-209B, air conditioned to 20±3°C, 50 per cent ±5 per cent RH, warning system with interlocks for identification of x-ray in use. The Ultrasonic cleaning facility: includes de-ionised water plant for water rinse and acid holding tank for the passivation and pickling of stainless steels. The paint spray facility with DeVilbis Turbo-Dynaclean Spray Booth (24 m²) with 2.5 m wide double door access to booth and curing oven.

Spacelink Europe

Spacelink was established in 1991 by JRA (UK), Novespace (France), MST Aerospace (Germany) and D'Appolonia (Italy) to co-ordinate ESA's technology spin-off activities. JRA covers UK/Scandinavia and acts as ESA's programme co-ordinator. The network of technology transfer agents spans 13 European countries and Canada. Spacelink produces the annual TEST (Transferable European Space Technologies) catalogue of technologies from space programmes that are available for licence or further development for non-space uses. More than 550 inter-company introductions have been made, the majority of them transnational. Approximately 100 negotiations between space/non-space companies are underway at any one time.

ISRAEL

Amsat-Israel

Amsat-IL is collaborating on the TechSat project, which includes an amateur digital packet communications unit.

Elbit Ltd

Elbit began its space participation in September 1989 with the award by General Dynamics Space Systems of a contract for avionics boxes for the Atlas launch vehicle. It provided TechSat 1's OBC. It is one of Israel's largest electronics systems houses.

ELTA Electronics Industries Ltd

ELTA is part of Israeli Aircraft Industries' Electronics Division and as such participates in the development and fabrication of Israeli satellites. Products cover military electronics systems, radars, active/passive early warning, secure communications, computers, signal processing, microwave components, tubes and antennas, high-voltage power supplies and microelectronics.

Israel Aircraft Industries Ltd (IAI)/ MBT Systems & Space Technology (MBT)

An element of IAI's Electronics Division, MBT is Israel's major space centre. It was responsible for developing Israel's first satellite, Ofeq, its all-solid Shavit launcher, and the Amos commercial telecom satellite. MBT may provide some Ellipsat hardware and act as the local service provider.

MBT's Space Technology Centre includes the Satellite Integration and Test Centre, an S/X-band TT&C station (10 m S/X-band antenna and 2 m S-band back-up), laboratories and fabrication facilities. The ground station's Image Processing Centre handles the received Spot imagery; upgrading to handle other satellites is planned. The satellite centre covers 3,400 m², including 2,000 m² class 100,000 and a class 10,000 30 m² portable laminar flow booth. Class 100 can also be provided.

Part of IAI's Electronics Division, MLM System Engineering and Integration specialises in telemetry systems, communications, C³I and image processing. It is leading the NEXT project, to develop a commercial derivative of Israel's Shavit satellite launcher. MLM provided the solar panels for Technion's TechSat.

An element of IAI's Electronics Division, TAMAM specialises in inertial navigation and reference systems, inertial sensors and stabilised platforms.

Soreq Nuclear Research Centre (NRC)

NRC's multidisciplinary space technology group focuses on the evaluation of space environment effects on materials, electronic components and systems. It has participated in all Israeli space activities: Ofeq, Amos, TechSat and Tauvex. The four branches emphasise electronics, materials, calculation of space hazards, and reliability assessment.

Tadiran Spectralink Ltd

Tadiran is Israel's largest electronics company. In the space field, it provides end-to-end TT&C systems qualified in the Ofeq programme. It provided receivers and transmitters for TechSat. Elisra is a subsidiary.

ITALY

Alenia Spazio SpA

Alenia Spazio SpA, a FINMECCANICA company, is Italy's major space company, born from the merger of two major national space industries: Selenia Spazio SpA and Aeritalia – Space Systems group. 1995 turnover was Lit740 billion (1994 Lit774 billion; 1993 Lit536 billion). It accounts for 70 per cent of Italy's space activities in terms of systems, technology and production, employing 1,900 in seven plants in Rome, Turin, L'Aquila, Milan and Naples. It also co-ordinates the operations of subsidiaries Laben, Proel Tecnologie, SSI Space Software Italia and Space Controls Alenia Honeywell; see their separate entries.

Company expertise covers complete space systems, satellites for telecommunications, remote sensing, meteorology and science applications, manned systems and space infrastructures, transport, launch and re-entry systems, telemetry/control stations and specialist space software. In telecommunications, it is prime for the national Italsat and for ESA's Artemis and Data Relay Satellite. It will integrate Globalstar's 56 satellites and is contributing to the construction of Argentina's Nahuelsat.

It was prime contractor for the Lageos 2 geodetic satellite launched October 1992 by Shuttle. An Italian industrial team led by the company developed the IRIS (Italian Research Interim Stage), a low-cost spinning solid upper stage for deploying lightweight satellites from Shuttle. First use was with Lageos 2. Other Shuttle launches of ASI programmes with Alenia Spazio as prime are X-SAR (X-band Synthetic Aperture Radar) for Earth observation (two flights in 1994) and the Tethered Satellite System for electrodynamic experiments (responsible for the satellite and the Core Equipment experiment). It was prime for the SAX x-ray satellite. In addition to X-SAR, it is developing other Earth observation instruments.

Following its role as a major industrial contractor for Spacelab (for which it built the pressurised module), Alenia Spazio is involved in the Columbus Orbital Facility, with particular responsibility for development and integration of the thermal-mechanical system. It is also responsible for the structure and thermal control. It similarly built Spacehab's pressurised modules. It will build three Mini Pressurised Logistics Modules to deliver Space Station supplies aboard Shuttle; the PDR was completed in October 1993. Significant roles in ESA programmes include: ERS 1/2, Meteosat (development and manufacture of onboard electronics systems), Hipparcos, Eureca, Envisat and Cassini-Huygens. It is undertaking studies on behalf of ASI for the Carina µg capsule (see the Microgravity section).

Assicurazioni Generali SpA

The company provides satellite, launch and space risk assessment and insurance. It is the world's largest space insurance underwriter, with US$110 million underwriting capacity.

Cap Gemini SpA

Onboard software, control centres, data communications, test systems and data processing.

Carlo Gavazzi Space SpA (CGS)

CGS specialises in facilities and payloads for µg applications, onboard instrumentation control electronics, micro-satellites, ground support equipment and ground segment. Principal projects include: EGSE for Cassini, Polar Platform and ERS; Tethered Satellite reflight control units; Spacelab (bubble drop/particle unit); µg instruments for Maxus and Neurolab; feasibility studies for Columbus Anthrolab, Fluid Science and Biology labs; onboard digital image processors for Columbus and Maxus. It is a member of the Europayload Consortium (see Microgravity section). 1995 sales: Lit10 billion; 50 staff.

CISI AID

The company is a subsidiary of France's CISI Ingénierie, created in 1987 to focus on the software capabilities of the CISI group. It specialises in the space field in science payload data processing, mission analysis, orbit determination, software for satellite control centres, telemetry analysis, remote sensing and image processing.

Dataspazio SpA

Dataspazio was established in 1988 by Telespazio (49 per cent) and Datamat (51 per cent) to design/develop systems and software for space applications: satellite Operational Control Centres (OCCs), data processing facilities, satellite simulators and flight dynamics models and software. National programmes include OCCs for Sirio 1/2, Italsat and SAX, and the satellite simulator for SAX. European programmes include the payload data handling subsystem for polar platforms, Olympus data gathering centre, and Columbus user information systems and crew training centre.

Elettronica SpA

Space activities cover microwave/mm-wave technologies, analogue/digital electronics, infra-red sensors and optical instrumentation for the space segment, and small/mobile stations, GPS receivers, HPAs and high power/low noise solid-state components for the ground segment.

FIAR SpA

Fabbrica Italiana Apparecchiature Radioelettriche (FIAR) is primarily concerned with the design, development and production of onboard electronic equipment and small satellite communication terminals. For satellites, FIAR provides full power systems (including batteries), regulators and ancillary equipment, as well as designing and supplying GaAs solar arrays. RF equipment, such as telecom transponders, TWTAs (such as the 135 W 12 GHz version for DBS satellites and 80 W 12 GHz for FSS), SSPAs, frequency generators, oscillators and microwave components. In the Earth station field, FIAR manufactures Prodat, a portable L-band terminal for messaging and position reporting.

FIAR and COMAU formed Tecnospazio in 1986 for space automation and robotics activities.

INTECS Sistemi SpA

INTECS specialises in software technologies, such as artificial intelligence, test/control systems, command/control systems, GIS and real-time embedded systems. 1996 turnover was Lit26 billion; staff total 200.

Laben SpA

Laben, a company of the Alenia Finmeccanica Group, specialises in electronic systems for space applications and is a leading member of the Star Consortium for ESA programmes. 1995 revenue was Lit80 billion; 1994 Lit90 billion; 1993 Lit91 billion; 1992: Lit77 billion; 1991: Lit65 billion; 1990 Lit54 billion; 1989 Lit48.5 billion. Personnel total 380. The company's space activities include the design of onboard computers on chips, electronic data handling, control systems and ground checkout equipment for spacecraft and ground stations. It also provides instrumentation for science, remote sensing and µg satellites. AOCS sensors/actuators and GPS receivers for attitude/orbit determination. The company maintains extensive manufacturing facilities; its thick film hybrid circuit manufacturing procedures are certified by ESA.

Laben Proel Tecnologie Division

Previously Proel Tecnologie, the company merged with Laben SpA in December 1995. It specialises in electron generation and acceleration as well as plasma sources for space, scientific and industrial applications. Activities cover the electron gun assembly for the TSS tethered satellite, the modulated EGA for TSS-3, ion thruster neutralisers for Artemis, plasma contactors and ion propulsion systems. Personnel total 20.

Microtecnica-Torino SpA (MT)

MT was established in 1929 and is now a business unit of UTC Hamilton Standard. 1994 space revenue Lit10.5 billion (1993: Lit16 billion; 1992: Lit15 billion); staffing level is 850. Its main business areas include environmental and thermal control, drive/actuation, power control and generation, and electronic control equipment. Projects include the spin table for Italy's IRIS upper stage, Ariane 5's Vulcain main engine check and helium control valves boxes, Ariane 5's main stage pressurisation system and helium control valve boxes, Columbus main active thermal control system (ATCS) equipment/GSE, and Mini Pressurised Logistics Module main ATCS and ECS equipment. Contributions since 1974, when the company became active in the space field, include Spacelab, Eureca (ACTS elements) and Lageos 2 (structure).

Space Software Italia (SSI)

SSI was established in 1988 for the design/development of system and equipment software. Alenia Spazio owns 92.4 per cent and Alenia-Finmeccanica 7.6 per cent. 1995 revenue was Lit12 billion (1994 Lit10 billion; 1993 Lit21.5 billion; 1992 Lit21.5 billion; 1991 Lit10.2 billion; 1990 Lit5.2 billion); 75 employees. Specialisations cover onboard systems (system and mission management, data processing and communications), mission control and payload operation control systems (mission management and analysis, data processing, mission planning, management and analysis, system simulation, C^3 data processing and test/validation) and services such as logistics support. Space contracts include Columbus advanced flight software prototyping and Central Mission Control Centre detailed definition; XSAR Mission Planning & Operation Software; SAX software ICD for all spacecraft subsystems and payload; Envisat 1 Radar Altimeter EGSE software design/development.

Tecnodata Italia Srl

The main space specialisations are computer services, communications services (VSAT provision and operation; ground station operations, maintenance, scheduling) and software development, including remote sensing and satellite simulation. Principal contracts include software services for ESRIN, ESOC and ESTEC; maintenance and operation of ESOC's Control Centre and Odenwald ground station; software maintenance and systems support for ESA's IRS; ESTEC satellite ground station operations, maintenance and scheduling. 1995 turnover was Lit20 billion (1994: Lit15 billion); 150 personnel.

Tecnospazio SpA

Tecnospazio was formed in 1987 by Comau and FIAR to pursue space automation and robotics activities; it also employs the expertise of those companies. 1995 space revenue was Lit5.5 billion; 1994 Lit5 billion, 1993 Lit5 billion, 1992 Lit6.4 billion, 1991 Lit5 billion, 1990 Lit3.3 billion, 1989 Lit1.6 billion. In 1995, it participated in the Robotic Joint Control experiment during EuroMir 95.

VITROCISET SpA

VITROCISET specialises in integrated logistics support, launch campaigns/operations, software engineering (space/ground), information systems/databases, and remote sensing and image processing. For example, it provides operation, maintenance, technical assistance and engineering support at ESOC and ESTEC, operation and maintenance services at Kourou, and software engineering support at ESOC and ESRIN. Recent activities include design/development of Helios ground segment elements, integrated logistic support for Italy's Helios facilities, design/development of cartography image handling systems, software design/development of Artemis' Control Centre for LEOP and routine operations, and software design/development of generic spacecraft Integrated Central System (SCOS II). Major national contracts include maintenance/operation of ASI's Trapani-Milo stratospheric balloon launch base, and technical assistance to National EGSE for TSS, IRIS-Lageos, SAX, Italsat F2. 1994 space revenue was approximately Lit50 billion (1993: Lit50 billion).

JAPAN

Fuji Heavy Industries Limited (FHI)

FHI is participating in the development of Japan's HOPE unmanned shuttle and the initial design of a horizontal takeoff and landing spaceplane. It built the two ALFLEX models.

Fujitsu Limited

The company produces communications satellites equipment, including small Earth stations, spaceborne electronics such as telemetry systems and MOS-1's visible/thermal infra-red radiometer, and data processing systems such as Yohkoh's onboard data processing computer. It undertook biological and materials processing experiments aboard Mir in early 1992; its equipment was launched aboard Progress-M 11 on 25 January and returned aboard Soyuz-TM 13 on 25 March. It is a major shareholder of the H2 launcher Rocket Systems Corporation.

High-Reliability Components Corporation

High-Reliability Components Corporation was established in 1988 to provide testing, inspection, centralised procurement, distribution and storage of space components. Main activities are QA, vendor quality surveillance, radiation testing, failure analysis, preparation of specifications and selection of space components. The company has been involved with the H-2, JEMLISS, ADEUS, COMETS, ETS-VII, ADEOS 2, OICETS and DRTS programmes. Facilities include the Tsukuba Test Laboratory and the Minami-suna Storage Facility.

IHI Company Ltd

Ishikawajima-Harima Heavy Industries provides the LOX and LH_2 turbopumps for Japan's H2 launcher's cryogenic engines, in addition to stage 2's hydrazine Reaction Control System modules. Other propulsion interests include scramjet and turboramjet systems.

In materials processing, IHI developed systems and elements for the First Material Processing Test (FMPT) for the Shuttle Spacelab-J mission (flown September 1992), including an acoustic levitation furnace, continuous heating furnace and a gradient heating furnace. In addition, the company is actively involved in the development of some instruments for the USERS Unmanned Space Experiment Re-entry System. It is responsible for JEM system engineering and integration support.

ITOCHU Corporation

Formerly C Itoh & Co, ITOCHU is one of Japan's leading sogo shosha (general trading companies), overseeing 790 consolidated subsidiaries and associated companies. In the space field, it provides Earth stations, launch services and satellite control systems. It holds a 28 per cent interest in Japan Satellite Systems (JSAT), which now operates three telecom satellites; see the Communications section for further information.

Japan Aircraft Manufacturing Co Ltd (NIPPI)

NIPPI's space-related structures include: M launch vehicle tail fuselage and fins; H2 booster nosecones and aft skirts; satellite extendible masts (including Akebono and Geotail science booms); extensor for extending motor nozzle; paddles, antennas and satellite structures.

Japan Manned Space Systems (JAMSS) Corporation

JAMSS was created in 1990 by 14 companies, headed by MHI, IHI, NEC and JGC, to support NASDA for safety and product assurance oversight. It also supports NASDA for JEM utilisation, operation and astronaut training.

Kawasaki Heavy Industries Ltd (KHI)

KHI was involved in the construction of launch facilities and the Experimental Geodetic Satellite for the H1 launcher; and launch facilities and the large payload fairing for the H2. It has been participating in system development of the berthing mechanism, airlock and environmental control systems for Space Station's Japanese Experiment Module (JEM) since 1985. It has been involved with the HOPE project and spaceplane research since 1986. It is system integrator of the ALFLEX Auto Landing Flight Experiment and rear body integrator for the HYFLEX Hypersonic Flight Experiment. Both are key steps in developing HOPE.

Space robotics and docking mechanisms with advanced proximity sensors are being studied on ground testbeds for future in-orbit servicing. KHI is developing and producing the docking systems for 1997's ETS-7. In 1994, it completed the WETS Weightless Environment Training System neutral buoyancy facility at NASDA's Tsukuba Space Center for crew basic training, procedure development for on-orbit rack replacement and JEM system development. It was contracted in 1994 for a crew Isolation Chamber at Tsukuba.

Kyokuto Boeki Kaisha, Ltd (KBK)

KBK handles rocket test stands, load cells, flexures, space retroreflectors, lasers and GPS receivers. FY96 annual turnover was ¥182.2 million (FY95 – ¥176.5 million, FY94 – ¥170.6 million), employees 594 (1995 – 621).

Mitsubishi Corporation

The Mitsubishi Corporation comprises a number of companies engaged in space activities. The prime manufacturers are Mitsubishi Electric Corporation, Mitsubishi Heavy Industries Ltd, Mitsubishi Precision Co Ltd and Mitsubishi Space Software Co Ltd.

The first satellite in which the Mitsubishi Electric Corp was involved was the CS domestic satellite communications vehicle in 1977. CS-2a/b were launched in 1983, with CS-3a/b following in 1988. Mitsubishi Heavy Industries has provided airframes, main engines, guidance/control equipment, satellite fairings and vehicle integration for the N and H launchers. Mitsubishi Precision Co was involved in the guidance/control equipment for H1/H2, and Mitsubishi Space Software Co has been responsible for the software.

Mitsubishi Electric Corporation (MELCO)

MELCO provides telecom and Earth observation satellites, remote sensing systems and satellite communication systems. It acted as prime for Japan's CS telecom satellites, the ETS-2/4/5 engineering test satellites, the JERS 1 synthetic aperture radar Earth observation satellite and the ISAS/USEF Space Flyer Unit. It is working with Space Systems/Loral and Aerospatiale as contractors for the Intelsat 7 series and with Lockheed Martin Astro Space on Inmarsat 3. Its xenon ion thruster and inter-satellite communications system flew on ETS-6. MELCO owns the Space Communications Corp of Japan, operators of the Superbird communications satellite system. Subsystem expertise covers AOCS (electronics, wheel drive assembly, Earth/Sun sensors, nutation dampers), communications/data handling (transponders, telemetry/command units, central processing unit), electrical power (battery assemblies, converters, shunt, control units), antennas (shaped beam, multibeam, phased array and omni antennas, feeds, multiplexers), solar arrays (GFRP, KFRP, CFRP substrates, semi-rigid/rigid, solar cells modules, GaAs cells), propulsion (mono hydrazine thrusters, ion and MPD arc jet engines), thermal (louvre, heat pipe, shield, multilayer insulation) and structures.

MELCO provided the microwave scanning radiometer for the MOS satellites and developed the SAR for JERS 1. The company also offers turnkey Earth stations, small Earth stations for SNG, VSAT and USAT, and was responsible for ISAS' 64 m Usuda deep space tracking facility. It is a major supplier of mobile Earth terminals for the US MSAT system.

The class IB spacecraft manufacture/test facility at Kamakura can fabricate/test subsystems and components, and integrate/test the entire spacecraft system in one building. The Satellite Integration and Test Facility includes a centre of gravity machine, mass measuring machine, dynamic balancing machine and a data acquisition system. The Subsystem Test Facility houses an anechoic chamber, antenna range test tower, EMC test system, Sun/Earth simulators, three-axis air bearing and gimballed servo tables, infra-red reflectometer and three-axis co-ordinates measuring machine. The environment test facility accommodates a space simulation chamber, spin table, vibration machine, rotary accelerator, shock test machine and outgassing measuring machine. The manufacturing facility allows fabrication of solar cell modules and heat pipes.

Mitsubishi Precision Co Ltd (MPC)

MPC specialises in inertial reference systems, RW/MW assemblies, and electronics. IRUs, RWAs and MWAs for ETS-5/6/7, JERS-1, SFU, ADEOS-1/2, COMETS and OICETS. ING and RWA for M-3S and M-5. E-PKG and DIU for H1/2 and JEM.

Mitsubishi Space Software Co Ltd

A subsidiary of Mitsubishi Heavy Industries, the company is responsible for the software of the H2 launcher.

Mitsui & Co Ltd

Hardware, software and services for satellites, launchers and Space Station. The company holds a 24.5 per cent interest in Japan Satellite Systems (covered in the Communications section), which began operating its first satellite in 1988.

NEC Corporation

NEC is a major contractor for satellites, rocket borne equipment and ground systems. It has acted as prime for major applications satellites: meteorological (GMS series), Earth observation (MOS series), direct broadcast (BS-3 series) and amateur radio communications (JAS series). Its role as system integrator for Japan's science/test engineering satellites includes: Ohsumi, ETS 1, MS-T1-5, Exos A-D, Astro A-E, Planet-A/B, Muses-A/B, Geotail, Solar-A, Debut, COMETS, OICETS and Lunar-A, totalling 43 satellites by end-1996. In addition, more than 100 other satellites employ NEC's highly reliable bus equipment, communications and special-purpose payloads and other space-related products.

NEC's Space Systems Development Division makes mission payloads including communications transponders, communications antennas and space sensors; TT&C subsystems, including transponders, omni antennas, databus and data processors; AOCS, including attitude sensors/control electronics; power subsystems, including solar array paddles; structural subsystems; thermal control subsystems. Space Station equipment, μg experimental equipment and Get Away Special. Rocket borne equipment, including telemetry transmitters, radar transponders and command destruct receivers. Ground systems, including TT&C, remote sensing stations and meteorological satellites data receive stations.

The Guidance/Electro-Optical makes guidance computers for H2 launch vehicle; SAR high-speed image processing system.

The Overseas Microwave/Communications satellites Systems makes Intelsat/Inmarsat/domsat Earth stations and VSAT/MSAT terminals.

Nippon Sanso Corporation

Space simulation chambers, propellant supply systems for cryogenic engines, hydrogen liquefiers, cryogenic storage tanks, and high-vacuum equipment. It provided Japan's largest solar simulation chamber, the 13 m diameter unit at NASDA's Tsukuba centre. Also Japan's largest is the 1,200 litre/hr liquid hydrogen production facility it built for NASDA. It manufactured the μg drop capsule for the Japan Microgravity Centre.

OKI Electric Industry Co Ltd

Satellite Earth station and command, control and communications systems for launch vehicles.

Sumitomo Heavy Industries Ltd

Rocket launch structures and ground base transportation systems; cryogenic technology, including helium liquefaction plants. In July 1994, Sumitomo Corporation acquired a 20 per cent interest in Hamilton Standard Space Systems.

Taisei Corporation

The Space Technology Section specialises in facilities for space-related ground experiments and the planning and construction of launch bases.

The Japan Amateur Radio League, Inc (JARL)

JARL developed three Fuji Oscar amateur radio satellites, launched piggyback by NASDA's H1/H2 vehicles.

Toshiba Corporation

Spacecraft systems and components (command receivers, star sensors, attitude control electronics, Ni/H_2 batteries), manipulator subsystem for Space Station Japanese Experiment Module, satellite ground systems (Inmarsat ship Earth stations, tracking/control equipment), satellite systems software and high-speed 32-bit Image Processor System. The company was prime contractor for NASDA's ETS 6 and was prime for BS-2. It is prime for NASDA's Precipitation Radar on the US/Japan TRMM mission.

LIECHTENSTEIN

Balzers AG

Balzers provides vacuum pumps/gauges, valves, components, residual gas analysers and leak detectors; vacuum coating systems and space simulation chambers; thin film components for optronics and microelectronics; wear-resistant and tribological surface coatings. For example, it gold-coated Rosat's x-ray mirrors. The company has 1,400 employees.

NETHERLANDS

Brandt Fijnmechanische Industrie BV (BFI)

BFI specialises in low-quantity, high-precision machining of components and tools in steels, titanium, aluminium, super-alloys and ceramics. Space involvement includes Exosat, Solar Max, IRAS, Hipparcos and Ariane.

CCM

Centrum voor Constructie en Mechatronica (CCM) specialises in technical development projects. It designed, developed and built Spacelab biological experiment units that were later modified for sounding rocket application. CCM has designed a protein crystallisation module with Fokker Space and Comprimo to accommodate 400 samples.

DEP BV (Delft Electronische Producten)

DEP/Scientific specialises in the design, development and manufacture of customised high-performance low light level imagers and detectors. The range includes image intensifiers, intensified CCDs, (imaging) photon counters, hybrid photodiodes and related electronics. Detection capabilities can reach down to photon-counting level. The company provided VUV 1,200-1,800 Å auroral imagers for the Canadian contribution to Freja and Interball; Solar Blind Imaging Photon Detectors for Israel's TAUVEX UV telescope aboard Russia's Spektrum-X/γ also developing the Imaging Photon Detectors for XMM's optical monitor (in co-operation with University College London).

DINFA BV

Manufacture of high-precision parts and assembly of components, instruments and special tooling. A member of the Netherlands Aerospace Group.

Dreves Engineering BV

Space mechanisms and structures. Project involvement includes a drill study for ESA's Rosetta comet nucleus sample return mission, design and development of the gearbox for the Hera robot arm joints, and a feasibility study of Fokker's Star mast deployment system.

Drukker International BV

Drukker is a diamond company that has specialised since 1967 in science and technology applications. It provided the diamond windows for the infra-red instruments carried by the Pioneer-Venus and Galileo atmospheric probes, the thin diamond bolometer substrate for COBE's FIRAS instrument, and the diamond substrate for a Nimbus 6 long wavelength pass filter. It also supplies diamond anvils for high-pressure research and heat sinks as semiconductor laser substrates or for high power microwave devices.

Fokker Space BV

Fokker Space was created in January 1996 from Fokker Space & Systems, with equal shareholders Fokker NV (part of DASA since 1992) and the state-owned Ultra Centrifuge Nederland NV. The 1996 bankruptcy of Fokker NV does not affect Fokker Space; Ultra Centrifuge has the first refusal on the shares. Fokker Space is the largest Dutch space company; 1995 revenue was NLG120 million (1994: NLG120 million; 1993: NLG130 million; 1992: NLG120 million); workforce 450 (expected to increase to 500). Fokker developed Ariane 4's carbon fibre interstage 2/3 and stage 3's engine frame; 50 of both units are under order to Aerospatiale. For DASA/ERNO, it is building 96 Ariane 4 booster nosecones. It was contracted for six test and three flight models of Ariane 5's 1.6 tonne aluminium engine thrust frame.

Fokker has been producing solar arrays since 1970. Fokker also designed Ulysses' thermal control system. Attitude control subsystems were provided for IRAS, ISO and SAX. For Envisat, it is providing the solar array, the hold down mechanisms for the Advanced Synthetic Aperture Radar and the SCIAMACHY instrument. It was responsible for ERS 1/2 AIT. The company is a Centre of Excellence of Europe's Robotics, it is prime for ERA (European Robotic Arm) and is participating in the ERA and Eurosim simulators.

Genius Klinkenberg BV

Space-related activities include design/manufacturer of Ariane 5 ground equipment and transport systems, and manufacture of aluminium parts for launchers and satellites.

Hymec BV

Hymec was founded in 1982 to design/produce thick film hybrid circuits. The 1,000 m² class 100,000 + 200 m² class 10,000 cleanrooms and class 100 laminar flow bench production facility is certified by ESA for the supply of thick film hybrid microcircuits; contributions have been made to SAX, Soho, Cluster, Ulysses, Sciamachy. Production technology includes plated-through and multilayer screening (up to eight conductor layers), abrasive and laser resistor trimming (passive/active), chip/wire-bonding, reflow soldering, packaging (hermetic, ceramic or conformal coating) and high-reliability screening (MIL-STD-883). The facility is equipped to produce power-hybrids on aluminium nitride and direct bonded copper.

ICT Aerospace Group

The 15 employees of the Aerospace Group operate within ICT Automatisering Deventer BV. As a systems house, ICT's main space activities cover EGSE, languages and development environment, human computer interfacing, telescience/μg utilisation and simulation, remote sensing, radar system development and transputer technology.

Mechanical Engineering & Consulting (MECON) BV

MECON provides feasibility studies, project management, technical design and detail engineering. Space involvement includes the glovebox camera housing for Bradford Engineering and a science instrument accommodation study for ESA/TNO-TPD. Staff totals more than 20.

Netherlands Aerospace Group (NAG)

NAG is an association of some 40 Dutch aerospace companies; it was established in 1976 as a result of the industrial alignment for the F16 project.

Origin/Nieuwegein BV

Information handling and automation consultancy, project management and software. Experience includes spacecraft command/control, simulators, database management systems, science data processing and man/machine interface techniques.

Polymarin BV

Design, development and fabrication of high-performance components of reinforced plastics, including the nosecones for Ariane 4's liquid strap-on boosters. Studies have also been made of a carbon fibre satellite antenna and for Ariane 5's booster recovery system.

Ramaer Connection Technology BV

Printed circuit board production for aerospace and high-technology applications.

RDM Technology BV

RDM's space interests include machining of frame body rings and final machining of assembled structures for Ariane 4/5.

Satellite Services BV

Telemetry simulators and data transfer interfaces; telemetry, telecommand and analysis systems.

Stork Engineers & Contractors BV

Stork Engineers & Contractors BV (SE&C; formerly Stork Comprimo BV), the main operating company of the Stork Engineering & Contracting Group, serves its clients worldwide with its technological capabilities, design and construction services, backed by over 70 years of experience. Services include consultancy, feasibility studies, project management, management consultancy, conceptual and basic design engineering, procurement, construction, HES (health, environment and safety) engineering, quality and product assurance, commissioning and start-up and operator training. The SE&C Group employs over 3,000 personnel in more than 40 subsidiary offices worldwide.

Regarding space technology, SE&C contributed to the design and construction of ESTEC's Large Space Simulator, acoustic chamber and hydraulic shaker building. R&D is performed on life support systems for short and long duration manned flights and in microgravity research (biological and materials research, space medicine, experiment hardware development, analytical/sensor technology). Current and recent projects include membrane bioreactor systems, biological air filtration units, membrane CO_2 absorption technology, sensor and microsystems technology, a facility for protein crystallisation and a High Performance Capillary Electrophoresis (HPCE) facility for space applications (prototypes have been tested in the Crew Work Station Testbed and the Columbus Automated Testbed at ESTEC); a new HPCE facility based on microsystems technology is under development. SE&C is management consultant to ESA's RADIUS project on zeolite research in space, executed by an international consortium of universities and private industries; responsible for project management, quality assurance, FMECA and safety assessment of biological experiments in the EUROMIR-95 and EUROMIR-E missions; responsible for reliability and safety assessment of ERA's mechanical joints; and participant in the Dutch Utilisation Centre, an infrastructure aimed at microgravity research user support.

Stork Product Engineering BV

Applied R&D for robotics and mechanisms, structural parts, thermal protection systems, propulsion technology, and electronic test/simulation equipment. Personnel total of 70.

NORWAY

AME Space A/S

A subsidiary of Alcatel STK, AME Space designs, develops and produces electronics hardware for space applications. 1996 space turnover was NKr80 million (1995 NKr50 million; 1994 NKr50 million; 1993 NKr60 million; 1992 NKr5 million; 1991 NKr25 million; 1990 NKr15 million); staff total 90. Capabilities include space-qualified thin film hybrid components, space-qualified Surface Acoustic Wave (SAW) components (including bandpass filters, chirp filters and delay lines for applications over 10 MHz to 1 GHz), and onboard signal processing equipment for advanced payloads of communications and remote sensing satellites. Recent contracts include: SAW filter modules for ICO, 900 Globalstar filter modules, IF channel filters for Artemis, SAW bandpass filters for Hot Bird, new generation chirp generator for the Advanced Radar Altimeter on Envisat 1, and development/supply of Ariane 5 electronics (such as solid state power controllers, hybrid circuits for attitude control system, and sensor interface units for cryogenic tank level monitoring).

The company has a class 100 cleanroom for SAW device manufacturing, class 1000-10,000 for thin film hybrids, class 100,000 for PCBs. Total cleanrooms cover 350 m^2.

CAP Gemini Norway A/S

Part of the CAP Gemini Group, the company specialises in: test/checkout systems such as EGSE, SCOE and simulators; space communication networks and data dissemination systems; telemetry/telecommand systems based on Packet TM/TC and CCSDS standards; Earth observation systems (including GIS); onboard data management systems, in particular onboard software; testing/verification of critical software. 1996 space turnover was NKr25 million; space personnel 45.

Det Norske Veritas Industry A/S

DNV Industry's Space Department specialises in R&D for product assurance, structural analysis, noise/vibration analysis, fracture/fatigue analysis, RAMSES analysis, thermal analysis, weight optimisation, human engineering and expert systems. Space contracts include a safety study of the Andøya Rocket Range and a study of in-orbit non-destructive testing. 1994 space/defence turnover was NKr37 million (1993: NKr33 million).

EIDEL (Eidsvoll Electronics A/S)

EIDEL was established in 1967 for electronics R&D and production. 1995 space revenue was NKr6 million (1994 NKr3 million; 1993 NKr5 million; 1992 NKr5 million). Its major activity is based on PCM (Pulse Code Modulation) telemetry encoders, decoders and simulators. EIDEL has developed a HAVE QUICK compatible time distribution and communication control system.

ESTEC has contracted an Ion Propulsion Diagnostic Package for Artemis, comprising sensors on the satellite's outer surface to monitor possible interactions from the ion thrusters. Metallic deposition and charged particles are the main concern. IPDP's general core can be configured for other sensor types and satellites.

Kongsberg Gruppen A/S

The Kongsberg Group is Norway's largest defence products manufacturer, established as an independent company in 1987; 51 per cent is owned by the Ministry of Industry. Its space-related activities are managed by the Space and Avionics Dept of Kongsberg Aerospace: structures and mechanisms; SCOE and simulators; electro-optical systems; ground stations for Earth observation satellites (Kongsberg Spacetec). Principal space contracts cover development of Ariane 5's booster attach/separation system; SCOEs and simulators for Soho, Huygens and Ariane 5; MIPAS optical differential system.

The group's 1995 space turnover was NKr120 million; space personnel 100. Kongsberg Spacetec is a subsidiary of Kongsberg Aerospace.

Konsberg Informasionskontroll A/S or IK is a subsidiary of Kongsberg Defense. Its 1994 space turnover was NKr15 million out of a total NKr40 million. Principal space capabilities cover system development and integration, software engineering computer systems, signal processing and electronics. Main products include software and computer systems for processing and operational analysis of Earth observation data, and EGSE. Contracts include SAR processing systems for ERS and Radarsat, a system for automatic ship detection in SAR images, VICOS Verification Integration Check-Out Software for Columbus Ground System Software, and EGSE (hard/software) for Soho.

Spacetec is an information technology company, a subsidiary of Kongsberg Aerospace, providing system engineering services in electronics, computers and software. 1996 profit was NKr2.2 million on sales of NKr37 million; 50 staff. Located in the Tromsø Satellite Station, it maintains the station under contract to the Norwegian Space Centre. It offers Earth observation ground stations, meteorological systems for Meteosat PDUS/DCP and NOAA AVHRR systems, and value added products and services based on satellite data, including vegetation maps and pollution monitoring.

Nordic VLSI

Nordic VLSI designs ASICs and ASIC-based electronic systems. With Matra Espace, it has designed a radiation tolerant signal processing video front end ASIC for space applications.

Norsk Elektro Optikk A/S

The company, formed by engineers from the Norwegian Defense Research Establishment, provides expertise in electro-optics, laser technology, spectroscopy and fibre optic applications. Space experience includes development of the optical path difference measurement sensor for the MIPAS instrument on Envisat 1. 1994 space turnover was NKr4 million, out of NKr16.5 million; 16 staff.

Norwegian Industrial Forum For Space Activities (NIFRO)

NIFRO is a grouping of R&D and manufacturing companies and research institutes promoting members' participation in space activities. Areas of interest include: ground segment and support data systems, space electronics, materials technology, manufacturing techniques, space transportation technology, remote sensing, telecommunications and quality assurance/control.

Prototech A/S

Prototech was established as a wholly owned limited company in 1988 by the Christian Michelsen Institute (CMI), restructuring its Mechanical Engineering Section. 1994 space turnover was NKr5 million out of NKr18.5 million; 10 space staff out of 34. It specialises in space science and technology, including rocket payload instruments and deployment systems, satellite components/systems, high-pressure gas supply systems for materials science experiments, and custom-designed scientific instrumentation for research institutes and universities. Prototech and its predecessor have designed/built more than 90 payloads for Norway's own sounding rocket programme. The company was responsible for the design, construction, integration and testing of payloads for Hero, Poleward-Leap, Mapwine, Bugatti, Lomas, Madame, Maimik mother/daughter, Mac-Epsilon, Need and Pulsar II. Payloads have been launched from Andøya in northern Norway, Huelva in Spain and Søndre Strømsfjord in Greenland.

Main space contracts include the gas storage module for Spacelab's Advanced Gradient Heating Facility, design/manufacture of a highly serviceable gas supply system for a Columbus high-temperature materials processing laboratory, engineering/flight models of electronic boxes for an ERS 1/2 altimeter chirp generator, as were similar structures for Soho/Cluster, Radarsat and Artemis, a fluid physics experiment for Texus, linear drive mechanisms for Tiros, and design of sensor/electronic boxes for the IMPS experiments of the Norwegian Defense Research Establishment. Company facilities were inherited from CMI: NC/CNC milling machines and benches, 2D/3D CAD/CAM systems, a measurement laboratory, a rocket payload testing laboratory, vibration and environmental test facilities, and high-precision jig-boring machine/tools; a cleanroom. All equipment/facilities of CMI are available to Prototech.

Raufoss Technology A/S

RA is a 53 per cent government-owned company mainly involved in pyrotechnics and solid propellants, lightweight metals and composite materials. Space-related contracts include latch valves, antenna pointing and solar array mechanisms for Tele-X, and the separation motors for Ariane 5's P230 boosters. Space turnover is NKr40 million; 45 staff.

SINTEF Group

The SINTEF Group is a non-profit research organisation working in close co-operation with the Norwegian Institute of Technology (NTH). The staff of 2,069 undertakes 3,000 projects annually; 1994 space turnover was NKr1.7 million out of NKr1,482 million. Space-related expertise covers human engineering and ergonomics, technical operations and logistics, materials and mechanical engineering, and human safety and reliability. The Neutral Buoyancy Facility (Ocean Basin) at MARINTEK is available for simulation and astronaut training, such as EVA, docking, rendezvous and logistics support.

SINTEF DELAB is an R&D institute specialising in information technologies based on computer science and telematics, telecommunications, acoustics and solid-state electronics. Some 20 of the 150 staff work in the space branch, although space-related technical work is also undertaken at other branches. The personnel at DELAB and the Dept of Electrical Engineering & Computer Science, totalling more than 350, make this the largest information technology research centre in Scandinavia. Principal specialisations include: space/ground electronics applying SAW, digital and electrical signal processing; space/ground antennas (particularly reflector antennas and feed systems); distributed databases; knowledge-based systems; Si and GaAs custom-designed ICs; labs for solid-state technology, surface analysis, verification, validation and testing.

Ticon Plastindustrier A/S

Design, development and production of reinforced plastic and composites for radomes sandwich panels and maritime navigation aids. Three production halls totalling 5,000 m^2. Employees total 45. Docking facilities for vessels up to 1,000 tonnes.

SOUTH AFRICA

Denel (Pty) Ltd

Denel was formed in 1992 to take on the mantle of South Africa's defence manufacturing from Armscor and expand into the commercial sector. Of the group, Somchem in Somerset West near Cape Town specialises in propellants and propulsion systems, Kentron in guided weapons and systems, Eloptro in electro-optical systems, and OTR (Overberg Test Range) flight tests and launch services.

SPAIN

Canava Electronica SA

The company specialises in printed circuit boards. Experience includes Hispasat, Shuttle payloads and ground stations.

Construcciones Aeronauticas SA (CASA)

The Space Division of CASA, Spain's leading space sector company, began operations in 1966. Space Division 1995 turnover was Pta8,786 million, divided as ESA 51.5 per cent, Arianespace 39 per cent and others 9.4 per cent; staff is 378. Current activities include: launchers (Ariane 4/5), science satellites (MINISAT, XMM, Soho, Huygens), Earth observation satellites (Envisat, Helios II, Meteosat Second Generation), telecommunications satellites (Artemis, Hotbird, WorldStar, Sesat, Hispasat), research and technology (dual gridded antenna, CFRP rings, ASYRIO), robotics (ROPODA, SRT) and simulations software (Space Vision Device, VISTA). Specialisations include: launcher structures; satellite service module structural, thermal and harness subsystems and integration; spacecraft platforms; high-stability structures for precision optical systems; multibeam reflector antennas from L- to Ka-bands and feed systems; phased array antennas (L- to C-bands); waveguide antennas, mechanisms and associated electronics; power distribution networks; robotics and 3-D simulation software and virtual reality.

For Ariane 5, CASA developed the inner structures of the upper composite: payload adaptors, upper stage propulsive structure and equipment bay. As an Arianespace shareholder, the Space Division produces elements for each Ariane 4, including the equipment bay structure, stage 1 and intertank skirts, and most of the payload adaptors. CASA provided ERS' wind scatterometer antenna and its radar altimeter reflector.

For national entities such as INTA, CASA has developed the Minisat platform for small missions. For SEPI, CASA has developed the Teleoperated Mobile Robot for use in nuclear power plants. It develops/manufactures elements of the Helios military satellites for the Ministry of Defense. International participation include the European Community's ADONNIS, PROMENVIR, PAGE-IN and MULTICUBE, aimed at developing new technologies.

CASA's Space Division is sited in the Barajas Centre, covering 23,250 m^2 of manufacturing and assembly area. Facilities include 700 m^2 of class 100,000 cleanrooms, including 300 m^2 for satellite integration and 100 m^2 for electronics and harness activities, in conjunction with two class 10,000 laminar flow cabins. Test facilities include type LD2 200 N (modal survey), LDS 18 kN and LE 90 kN shakers, a −60/120°C, 10^{-3} atmospheres Sapratin thermal chamber, a −70/125°C Tenney thermal chamber and optical measurement system chamber.

Cray Sistemas SA

A subsidiary company of Cray Systems Ltd of the UK, specialists in the design and development of software and systems for ground segments.

CRISA

CRISA specialises in flight electronics, ground support equipment and software engineering, emphasising control and processing units, power units, communications networks, onboard software and video communications. Shareholders are 50 per cent Matra Marconi Space France and 50 per cent Abengoa; turnover Pta1,900 million, staff 169. Processing units: instrument control units for ASCAT, MIPAS and GOMOS on Envisat and Metop; ASCAT digital processing unit; Columbus vital telecommand unit. Power electronics: Hispasat, Soho and Artemis payload converters; Soho and Ariane 5 electronic unit converters; primary power systems such as battery regulator unit, batteries charge/discharge units, shunt regulators, power distribution units and pyro units. Control electronics: Ariane 5 sequential electronics, Soho antenna pointing mechanism electronics, ISO valve actuation unit. Software engineering: audio/video communications (Central Payload Video System based on 4 Gbit/s data transmission technology); optical communications; control centres; ground stations (baseband equipment, station management, video/audio, monitoring and control functions). EGSE: power (Helios, Spot 4), simulation subsystem (OCOEs), flight unit test benches (Ariane 5, Artemis) and RF (ERS, Hispasat).

GMV, SA

Founded in 1984, GMV provides engineering services and turnkey systems development, maintenance and operations. 1996 turnover was projected at approximately Pta1,035 million and 1995's approximately Pta860 million; 1994 Pta929 million; 1993 Pta1,030 million; 1992 Pta820 million; 1991 Pta540 million. The Technical Department comprises four divisions: Flight Dynamics and Avionics; Operations; Simulation; and Software Engineering. GMV was designated a Centre of Excellence in Orbital Mechanics by ESA in 1989. It is participating in most major ESA and Spanish national programmes, and is providing engineering services and ground control, and data processing systems for other space customers, including Eutelsat, EUMETSAT, CNES and WEU.

Iberica del Espacio SA

Iberica del Espacio SA was created in September 1989 as a 50/50 subsidiary by France's SEP and Spain's Empresarios Agrupados, a grouping of EPTISA, TRSA and GHESA, specialising in defence and engineering for fossil and nuclear energy production. The company has a total of 50 personnel, with 44 engineers; 1993-95 revenues were each about US$5 million.

Expertise covers flight components and propulsion, including significant participation in Vulcain engineering; ground segment (design, supply, installation and commissioning); services and engineering analysis (including RAMS Reliability, Availability, Maintainability and Safety studies), and flight data analysis.

Indra Espacio SA

Principal space specialisations are: spacecraft control systems (control centres, TT&C stations, OBDH, EGSE, in-orbit test equipment); satellite communications (ground stations, VSATs, network management systems); Earth observation systems (data reception stations, image processing centres, archiving/retrieval systems). 1996 sales were Pta2,900 million; 78 personnel. Products include ground stations for meteorological satellite data reception for the Spanish Meteorological Institute. The complete satellite control system was supplied for the Hispasat system, and TT&C stations were provided for ESA at Maspalomas, Kourou, Villafrance, Malindi, Redu and Michelstadt. INDRA participated in the Helios Image Reception Centre and Image Processing Centre.

INSA

Ingeniería y Servicios Aeroespaciales SA was set up in July 1992 as a government company under INTA as Spain's major space services company. 1996 sales were Pta3,080 million, personnel 309 (1995 Pta2,770 million, personnel 326). In the field of aerospace engineering INSA's activity is oriented towards the development of Earth observation services (for example, the FUEGO programme under contract with EC Forestry Observation for ESA). The company also performs specialised technical and managerial services in the development of aerospace systems (ESA's SMOL, INTA's Capricorno launcher). Business development is oriented to provide commercial services based upon space platforms.

NTE SA

NTE's space activities are focused on μgravity, life sciences and life support, with emphasis in thermal design, signal processing electronics, power electronics, image processing, telescience and biology payload hard/software (particularly freezers and coolers).

Sener Ingenieria y Sistemas, SA (SIS)

SIS specialises in space mechanisms, structures and associated electronics. 1994 space revenue was Pta1,800 million; 1993 Pta2,250 million; 1992: Pta2,205 million; 1991: Pta2,400 million; 1990: Pta1,800 million. Activities include the optical path and refocusing mechanism of HST's Faint Object Camera, Ulysses' mechanisms subsystems, mechanisms and structures on Columbus (mainly airlocks and docking equipment), Soho antenna pointing mechanism, Cluster mechanisms and booms, GOMOS instrument electronics, Minisat attitude control system, Helios baffle mechanisms and image processing software, and Meteosat Second Generation's SEVIRI calibration unit, mechanism and pyros subsystem. SIS is also involved in Artemis's Silex payload and Envisat's MERIS and MIPAS instruments. It has produced a series of launcher facilities, such as Ariane's umbilical tower and arms, and ground instrumentation, and provided the Spacelab ground integration system and the planar scanner for ERS.

Tecsidel SA

Tecsidel is part of the GSI Group of France and more directly of the Advanced Technology Division, specialising in information processing systems, data communications networks and embedded systems. Graduate personnel total 38. Space involvement includes studies for ESA on vision processing for in-orbit servicing, Columbus software replaceable units and operating system, and architecture for a distributed real-time operating system.

Universidad Politecnica de Madrid

CIDE/UPM is a centre grouping several laboratories of the Faculties of Aeronautical Engineering, Telecommunications Engineering and Naval Engineering. Space activities include small satellite design/construction, spacecraft thermal control and structural analysis, real-time signal processing and communications, μg fluid sciences/materials, and hypersonic aerodynamics. Its UPM-Sat 1 micro-satellite was launched in July 1995.

SWEDEN

ABB HAFO AB

Mechanical and electronic design and construction, spacecraft structures, sensor technology, microwave antennas.

FFV Aerotech AB

FFV Aerotech, a member of the Celsius Group, comprises three units: avionics, aircraft systems and subsidiaries FFV Test Systems and FFV AvioComp. Employees total 2,000. It concentrates on aviation maintenance and consulting, but in 1989 it was awarded contracts for AIT of Sweden's Freja scientific satellite, which was launched October 1992 by China's CZ-2C. Development and manufacture of Freja's 450 MHz uplink antennas, EGSE, flight harness and thermal blankets were also undertaken by the Avionics Division. Within the Group, composite structure items for space applications are developed and manufactured at FFV Applied Composites AB in Linköping.

Saab Ericsson Space

Saab Ericsson (SE) was founded 1 January 1992 by the merger of Saab Space AB and the Space Department of Ericsson Radar Electronics AB. Saab Combitech AB owns 60 per cent and Ericsson Microwave Systems 40 per cent. 1994's sales were SKr312 million (1993: SKr302 million; 1992: SKr287 million); employees total 400. The company supplies launcher and spacecraft onboard subsystems and equipment, including fault-tolerant computers, data handling systems, antenna systems, microwave electronics, satellite separation systems and guidance systems. Each Ariane 5 carries two new-generation parallel guidance and navigation computers; the equivalent continue to be supplied for Ariane 4. The Columbus Polar Platform employs SE's Cassegrain antenna (including DRS pointing control electronics) and an OBC similar to the Spot 4 fault tolerant concept. For Envisat, SE will deliver control units for two instruments, central RF electronics and a radar signal processing subsystem for the ASAR radar. Meteosat Second Generation will carry a data handling system, UFH receiver/transmitter and TT&C antennas. Thaicom 3, Agila, Nahuel, Sirius 2 and Eutelsat W24 will have SE's data handling systems. Earlier deliveries include computers and microwave antennas for ERS 1/2, computers for Spot 1-3, data handling system, antennas, RF distribution network and central software for Soho, and attitude control computer, antennas and cover release mechanism for ISO. Data handling systems and antennas were supplied for Eutelsat 2, Arabsat 2 and Turksat.

Technology Trade International (TTI)

Initially, TTI's business was to locate and develop possibilities of industrial cooperation between the RFAS space industry and companies/industry with related activities in the rest of the world.

Over the years several important agreements of co-operation and worldwide representation have been reached with major space technology companies of the RFAS. In particular TTI is the representative of the Cosmos Group on the international market. TTI markets the products and services of the Cosos Group and its associated companies including NPO Energomash and Central Specialized Design Bureau (CSDB).

TTI is also the international business representative for STC COMPLEX, the company converting missile technology into the low orbit launch vehicles START and START-1.

Recently TTI has taken additional steps to consolidate the international space co-operation through agreements with interested parties all over the world and in particular in Europe, Asia and Australia. This has among other things resulted in development of new software for improvement of existing Russian satellite hardware.

To adequately meet the demands of the challenging period ahead TTI is enhancing its resources in the space technology field to better serve its customers and partners and also provides advisory services in the satellite and launch vehicle fields.

SWITZERLAND

Aerospace Engineering Office (AEO)

AEO provides engineering consulting services to the aerospace industry, including studies of software and light structures and finite element analyses. Its activities have included static tests of the Ariane fairing, and OTS/ECS solar panel structural analysis. AEO also developed a kinematic model under NASA contract of the Shuttle remote manipulator.

APCO Technologies SA

APCO Technologies specialises in the design and manufacture of metallic flight structures, flight mechanisms, MGSE and special electromechanical test equipment. Recent space work includes XMM Mirror Module support structures, ISO payload module outer vessel, Huygens back cover, Ariane 5 fairing containers, MGSE for Envisat's ASAR, GOMOS and MERIS instruments, and the two main elevator platforms of Kourou's ELA-3 BAF for Ariane 5 final assembly, including the lift system.

Compagnie Industrielle Radioelectrique SA (CIR)

CIR belongs to France's Sextant Avionique. Space turnover is about SFr20 million; personnel total 100. CIR has been active since the 1960s providing TC/TM systems, satellite timing systems and EGSE test/simulation systems. It has developed the image processing system for Envisat's MERIS instrument and for Meteosat Second Generation. Contracts include EGSE equipment for Hipparcos, ECS, Spacelab, ERS and Columbus; time bases for satellite tracking stations; timing equipment for Vulcain engine test facilities; and Anthrorack's conditioning amplifier subsystem.

Contraves

Contraves' Space Division is part of Oerlikon-Bührle Holding Ltd and is responsible for that organisation's space activities. It is Switzerland's largest space-related company, with 1995 space turnover of US$65 million (1994 US$60 million, 1993 US$50 million, 1992 US$50 million). The company provided the spacecraft structures for ESRO 1, GEOS, Giotto, ISEE 2, Exosat, Giotto and Ulysses, in addition to deployment mechanisms for scientific instruments, antennas and solar arrays. Other space interests include the production of Ariane 4/5 payload fairings, thermal control and EODP. It also provides fairings for Lockheed Martin's commercial Titan 3.

Centre Suisse d'Electronique et de Microtechnique SA (CSEM)

CSEM provides research and development and small quantity production in systems design and engineering (including wavefront engineering), instrumentation (based on microsystems), data communications, automation and robotics subsystems, smart structures, optical systems and tribological coatings. 1996 space turnover was about US$4 million. Work for ESA includes a micro-accelerometer (Olympus, Spacelab D2, Shuttle GAS June 1991), lubrication for Exosat, Meteosat and Hipparcos, hand controllers for Hermes, mechanisms and subsystems for optical calibration of Earth observation instruments, and XMM satellite as well as opto-mechanical systems for laser pointing.

Leica AG

Optics and sensors, particularly infra-red detectors, for the space and remote sensing industries.

Mecanex SA

Mecanex is a private company founded in 1959, specialising in the development and manufacture of electromechanical devices, high-performance mechanisms and micro-systems. In the space field, these include slipring brush assemblies for signal/power transmission in solar array drive mechanisms (TV-Sat, Tele-X, TDF, Olympus, Spot, ERS, Eutelsat 2, Turksat, Arabsat, Nahuel), Coarse Pointing Mechanism Assembly for SILEX (Spot 4, Artemis, DRS), Space Bioreactor (Spacelab IML-2), valves and filters mechanisms (Huygens, ACP-Cassini), and actuator units for ERA's end effector.

Oscilloquartz SA

Manufacturer of ultra-stable and very low noise frequency sources, from quartz crystal oscillators to cesium frequency standards. Oscilloquartz also designs/supplies complete integrated systems for the generation and distribution of precise frequency and time references required for satellite communications, time transfer and satellite tracking applications, and equipment for synchronisation of telecommunications networks.

RST Raumfahrt Systemtechnik AG

Provides consultancy, engineering services and space products: space systems, space-related ground systems for Earth observation, navigation and communications. In 1994 it introduced a gyro-stabilised (less than 40 arcmin accuracy elevation/azimuth) 1.2 m antenna for shipboard reception of 1 to 13 GHz satellite TV. In 1996 a new forward-looking CW radar was introduced.

Swiss Aircraft & Systems Company

The company (previously Fabrique Federale d'Avions/ Swiss Federal Aircraft Factory) has provided structural elements, instrumentation and integration of Ariane payload fairings, active cooling systems for Spacelab experiments, structural elements for ESA's Biorack, and satellite containers.

Vibro-Meter SA

Pressure and vibration sensors for space applications. Sensors for Ariane 5's Vulcain engine monitor absolute vibration and dynamic pressure of the engine's gas generator and output nozzle; displacement, relative and absolute vibration, rotational speed and dynamic pressure of Vulcain's turbopumps; absolute vibration of the solid boosters. Sensor tolerance is −253 to 780°C. Shock accelerometers are also provided for Ariane 5 pyrotechnics. High-resolution standard proximity sensor for ESA is under development. Ariane-4 displacement measuring system for the LOX turbopump.

THAILAND

Thai Micro-Satellite Co Ltd (TMSC)

TMSC signed a £3 million contract on 28 November 1995 with the UK's Surrey Satellite Technology Ltd. for a 50 kg micro-satellite to be launched in 1998. The 18-month technology transfer programme involves a team of Thai engineers working alongside SSTL staff as TMSAT is built. It will carry Earth-imaging cameras and communications payloads, controlled from a station installed by SSTL in Bangkok. Its GPS receiver is being developed by SSTL under an ESA contract to provide position and attitude data.

UNITED KINGDOM

Advanced Products (Seals & Gaskets) Ltd

Advanced Products is a subsidiary of the Advanced Products Company of Connecticut and specialises in the manufacture of seals for extreme environmental conditions, including rocket engines, fuel systems, and cryogenic and vacuum systems.

AEA Technology

AEA Technology is a science and engineering services business that solves technical, safety and environmental problems. Turnover in the financial year ending 31 March 1995 was £257 million, with a staff of 4,100, including approximately 2,000 graduate and technically qualified scientists and engineers. During 1995, legislation was passed by Parliament for the privatisation of AEA Technology, the separate commercial division of the UK Atomic Energy Authority. Space activities are focused in the Space and Defence Systems Dept, based at the Culham and Risley sites, although some space work is undertaken at Harwell. Principal areas of expertise cover: RF technology, electromagnetics/dynamics, radar cross-section modelling, antennas, high-power microwaves and tube technology; all aspects of space tribology and mechanism performance, including design, component procurement, lubrication, assembly and functional testing; space and defence equipment, specifically relating to space science, environmental monitoring and Earth observation; space propulsion, environmental simulation, ESD and space concepts. Working with its National Environment Technology Centre, it is increasingly looking to combine its expertise as a potential environmental data user with its Earth observation satellite instrumentation and data processing capabilities.

In antenna systems the company's capabilities range from writing specialist EM design codes and graphical user interfaces to establish design tools for customers, through system level work to solve antenna interpretability issues, to undertaking design, manufacture and test of flight hardware. The contract for design/manufacture of the Skynet 4 Stage 2 UHF antenna subsystem was awarded. The in-house CLASP code was used to predict antenna performance, both in isolation and installed on Skynet, followed by validation using actual tests.

AEA Technology is a major supplier of space-qualified infra-red calibration sources. Units are now being built for Envisat 1's MIPAS and AATSR instruments and for Meteosat Second Generation. Their heritage is the ATSR black bodies, designed/built by Mullard Space Science Laboratory, on ERS 1/2.

A 200 kV D-injector, beam transport system, diagnostics and control systems were designed/ delivered to Grumman Aerospace for the Continuous Wave Deuterium Demonstrator.

Expertise for developing new electrochemical systems ranges from theoretical chemistry and computer modelling to space qualification. Battery/cell development includes a second-generation nickel oxide/hydrogen battery, a solid-state, high-energy rechargeable lithium polymer battery, and a sodium metal halide battery. It can supply specially designed batteries using high-performance Li-ion rechargeable technology.

The design/development of composites for space structures include the space-qualified compact lightweight deployable boom. The development of a low-density metal matrix composite (HIVOL), with thermal conductivity close to aluminium but with a low expansion coefficient matching that of semiconductor devices, has significant applications for space power supply and focal plane instrumentation.

A range of accelerators and irradiation cells assesses the effects of radiation on electronic components and spacecraft materials. A range of large high-pumping-speed plasma facilities, the largest of which is unique in Europe, simulates the space environment up to 2,000 km for investigations such as plasma turbulence and noise generation, and the effects of gas emissions and effluent releases into space.

Multipaction is an RF breakdown phenomenon that can cause problems in high-power communications or remote sensing payloads. AEA Technology has exploited its computational physics expertise to establish a unique suite of multipaction prediction software. A dedicated team provides a routine consultancy service; it also develops new test methods for multicarrier payloads and investigates related effects such as PIM passive intermodulation.

Neural network techniques are under investigation for Earth observation data processing. They have already been applied to processing ERS 1 wind scatterometer data.

Agema Infra-Red Systems Ltd

The Thermovision 900 12-bit thermal infra-red scanning camera is being used by ESA for studying the thermal behaviour of heated liquids in μgravity, at temperatures ranging from zero to 150°C, in its Liquid Structure Facility.

Alcoa International Ltd

High-technology sheets, plates, extrusions and forgings in aluminium, titanium and other metals for the aerospace industry.

Analytical Graphics

A UK-based subsidiary of the United States Analytical Graphics, established in 1996. Markets AG's software packages for orbital and satellite analysis to UK/ European clients, involved with education facilities.

Anvil Technology Ltd

Anvil specialises in software development for spacecraft control, to ESA Software Engineering Standard PSS-05-0.

Argos Electronics Ltd

The company is engaged in printed circuit manufacture, including double sided PTH, multilayer, FR4 and polyamide, to full ESA standards.

Asdor Filters

Asdor specialises in the design and manufacture of filtration systems.

ASTOS

The Association of Specialist Technical Organisations for Space (ASTOS) was formed in 1988 by companies providing specialist manufacturing skills and services to the space industry. It represents the interests of smaller specialist companies and presents their views to all bodies responsible for formulating European space policy. Membership comprises Aegis Systems, Aetheric Engineering, Anvil Technology, Flow Line Communications, Go-Sat International, Moreton Hall Associates, Space Innovations, Set Resource, Spur Electron, Storm Integration, TRL Technology and Westover Flight Technology.

Avica Equipment

Specialists in high-performance ducting and ducting components for space vehicles, including gimbal joints, sliding joints, flanges/clamps, metallic hoses and thermal insulation. Avica provides the fuel feed lines and exhaust ducts for Ariane 5's Vulcain engines. It provided the titanium flexible hoses for Cassini's hydrazine system.

BAJ Coatings Limited

A Meggitt Aerospace company, BAJ provides high-performance surface engineered coatings, applied by the Tribomet, HVOF and plasma spraying processes, to resist wear, corrosion, erosion and oxidation at elevated temperatures.

BP Chemicals Ltd

BP's Aerospace Composites group was formed from the amalgamation of several existing composite companies. It offers design, development manufacturing and repair capability in composite components.

C T Bowring Space Projects Ltd

Part of Marsh & McLennan Space Projects (see US Space Industry), the company provides space insurance brokerage services, develops risk management and insurance programmes for risks of pre-launch, launch, in-orbit operations, manufacturers, product exposures and third-party liability.

Commercial Space Technologies Ltd (CST)

CST is an independent UK technical consultancy, founded in 1983, comprising individual technical specialists and support staff. Work has been undertaken for clients in launchers, propulsion, spacecraft, communications, remote sensing, μgravity and biotechnology. Current specialisations include the procurement of launcher and communication satellite services and brokerage of 'piggy-back' launch services. Direct contacts exist with CIS manufacturers, research institutes, government organisations and universities.

Cossor Electronics Limited

Receivers and other electronic subsystems for ground terminals and space research centres.

Cray Systems Ltd

Cray Systems, a wholly-owned subsidiary of Cray Electronics (Holdings) plc, is a software and systems house with an annual turnover of approximately £75 million generated by 720 staff. Of this, space projects employ 135 and contribute revenue of £9.4 million in the following strategic areas: spacecraft control centres; spacecraft simulators; mission planning/evaluation tools; ground station systems; terrestrial communications systems; Earth observation data processing centres; information systems; software products and system studies.

As prime contractor, Cray Systems has designed/developed software systems for 10 spacecraft control centres over the past decade. Five are operational for Ulysses, ECS, Rosat, DFS and ERS. In collaboration with Germany's DLR, it has developed an Advanced Generic Control Centre. It has been used to date for seven satellites and has been installed for Deutsche Telekom for operation of their telecommunication satellites. Cray Systems designed/developed, on behalf of ESA, the £3 million ERS Mission Management and Control Centre – the largest ESA spacecraft control centre to date. It has recently developed the control centres for ESA's Pastel and Cluster missions, and is now developing the requirements for ESA's Envisat control centre.

The company was awarded the development contracts for eight of the last 11 simulators commissioned by ESA, including Giotto, Meteosat P2, Hipparcos, ERS, ISO and Cluster. It also provides spacecraft simulation consultancy to prime contractors in Italy and Canada.

Software systems and engineering services have been provided for ESA's Estrack packet-switching network since 1976. The company's software systems have been installed at all ESA ground stations. A new generation of station computer, the Monitor and Control Module and Front End Controller, is under development.

Since 1988, Cray Systems has provided engineering services for ESANET, covering strategic planning, product assessment and procurement, installation and pre-operational support for new facilities.

A £4.5 million contract for the Western European Union for a turnkey image-processing facility has been completed. The Dundee Centre for Coastal Zone Research has been established with Dundee University.

Software engineering services have been provided to ESRIN's Information Systems Division since 1993, particularly on the CUIS and Cluster projects, as part of a five-year frame contract.

Satcom Earth station products have been developed for Inmarsat, including: Inmarsat M/B Control/Signalling Component, Production/Factory Test Unit, and Earth Station Test Set for Inmarsat type approval. Cray also provides CLEO (for spacecraft monitoring/control), MMOrbit and VISIM (for spacecraft simulation and operations visualisation and analysis).

EASAMS Limited

Part of GEC-Marconi S3I, EASAMS specialises in business systems, defence consultancy, software and computer services. It has provided onboard guidance and control software for all Ariane versions to date, including: flight software; studies, mission planning and algorithm development; ground integration support software for the launcher electrical systems; launcher simulation for real-time testing, studies, and post-flight analysis of telemetry; and integrated vehicle checkout software at Kourou. EASAMS is contracted to provide individual Flight Programs for each Ariane.

EDS Ltd

The EDS Group employs 8,000 people in the UK. Its space business centres on consultancy, systems development, systems integration, systems management and process management. This included a £2.5 million contract to supply the synthetic-aperture radar processor for the UK Earth Observation Data Centre to process ERS data. In addition, the PulSAR high performance SAR processing workbench has been developed to provide high-quality SAR imagery with rapid throughput on low-cost hardware. ESA projects include: supply of the SARCALQ image QA and calibration system for Earthnet; development of expert systems demonstrators and assessment of multimedia systems for the agency's Information Retrieval Service; and leading an international consortium which is investigating the application of formal methods to space software development for ESTEC. Other projects include: interpretation of Earth observation data using neural nets; development of integrated GIS; development of a system for determining terrain heights from SAR imagery; assessment of radiation effects on spaceborne instruments; simulation of space-based instruments and systems; and finite element analysis for space-based structures.

Electron Tubes Ltd

Established in 1953, Electron Tubes is a major manufacturer of very-low-light-level photodetectors, particularly for space activities. The management buy-out from THORN EMI was completed in January 1994. Current and recent space projects include: image intensifiers for the Faint Object Camera in NASA's Hubble Space Telescope; photomultipliers for the NASA Compton Observatory's Compton Imaging Telescope, Burst/Transient Source experiment and the Gamma Ray Telescope; photomultipliers for ESA's Hipparcos astrometry project; photomultipliers for Ulysses (taking a Cerenkov detector into solar polar orbit); and photomultipliers for Italy's SAX astronomical x-ray detector. Space-related facilities include simulation thermal, vacuum and vibration test equipment, and expertise in encapsulation techniques.

Empcom Systems Ltd (ESL)

ESL provides consultancy, contract negotiation and subcontractor monitoring services to the European space and telecommunications industries.

Encore Computer (UK) Ltd

Part of the Encore Computer Corporation, the company supplies mission-critical computer systems. The Infinity R/T real-time and fault-tolerant systems are used for simulation, control and data-acquisition systems. The Infinity SP Enterprise-wide Storage Systems provide high-capacity storage and sharing of data between mainframe, UNIX and LAN systems.

ESYS Limited

ESYS provides consultancy and technical services to Europe's space market. Key business areas include Earth observation (focusing on the needs for increased operational and applications development), modelling and design of telecommunications systems and future space systems, including small satellites and launch systems.

GEC-Marconi Materials Technology Limited (GMMT)

The company's Caswell facility specialises in the manufacture of GaAs, opto-electronic and pyro-electric sensor products. The GaAs range includes MMIC, WLAN, SART, amplifiers and switches, discrete FETs and a foundry service offering F20 FET and F40 HEMT technologies. The F20 MMIC process is undergoing Capability Approval for ESA. Opto-electronic capacity includes DFB lasers, OEICs, specialist LEDs and GaAs modulators, operating at DC-50 GHz. The company's Towcester facility specialises in radar-absorbent materials and radar-transparent products for endo/exo-atmospheric applications (including anechoic chambers and screened rooms), equipment testing and microwave antenna beam shaping. The transparent high-performance low-loss resin systems (RP13) and lightweight foam materials (P10 range) are designed specifically for radomes.

GEC Plessey Semiconductors

The company was formed in 1990 by the merger of Marconi Electronic Devices and Plessey Semiconductors. It is Europe's largest manufacturer of integrated circuits for space and defence applications. It specialises in VLSI CMOS Silicon on Sapphire (SOS) for high performance and radiation hardness. The SOS range covers advanced monolithic MIL-STD-1750A microprocessors and peripherals, static RAMS to 256 kbit, CMOS logic and MIL-STD-1553B databus products. Semi-custom gate arrays and standard cells to 40k gates are available. Self-certification to MIL-STD-883; ESA 9000 Capability Approval achieved 1993. At the same Lincoln facility, GPS manufactures thick/thin-film hybrid circuits to space quality levels.

Graseby Dynamics Ltd

Graseby Dynamics is a research and development and production specialist in trace-vapour detection and monitoring, using gas chromatography and ion-mobility spectrometry. Space capabilities include the detection and monitoring of hydrazine, total hydrocarbons and volatile organic compounds for flight and ground operations. The company's hydrazine monitor/data-logger was used on Shuttle STS-37/49 and was planned for reflight in 1997. Graseby developed air-quality monitors for Shuttle (completed 1996) and Space Station (completed 1997) installation.

Hunterskil Howard plc

Space project consultants, including µgravity, component engineering and RF systems.

IGG Component Technology Ltd

IGG is the oldest independent European company devoted solely to parts engineering and procurement for the space and defence industries. The service includes procurement of individual line items to total project requirements, specification writing, part type reduction, evaluation, constructional analysis, destructive physical analysis and screening. Project involvement includes most ESA craft and satellites such as Insat 2, Telecom 2, Intelsat 7 and Italsat. Facilities include a Class 10,000 cleanroom, automatic thermal shock chamber and full screening facilities.

IGG's Spartan Microwave specialises in the design/manufacture of multichip microwave components using GaAs and/or Si technology for space and defence applications.

IMI Titanium Ltd

IMI Titanium is Europe's largest producer of semi-finished titanium mill products. Applications development embraces hafnium, niobium, superconductors, zirconium and its alloys. The company holds approvals on testing/analysis from standards authorities such as the UK Ministry of Defence and the Civil Aviation Authority. Production facilities encompass computer monitored vacuum arc melting furnaces capable of producing ingots of up to 17 tonnes, hydraulic forging presses and machining, sawing and cutting equipment. Mechanical, chemical, ultra-sonic, corrosion testing and metallography are conducted throughout the manufacturing operations.

International Space Brokers Ltd

A wholly owned subsidiary of Crawley Warren & Co Ltd, International Space Brokers Ltd are insurance brokers specialising in all aspects of space insurance and risk management, in association with Le Blanc de Nicolay (France) and International Space Brokers Inc (US).

Irvin Group Ltd

Previously Irvin Aerospace Ltd, the Irvin Group comprises: Irvin Aerospace Inc; Irvin Aerospace Ltd; Irvin Aerospace SpA; and Irvin Aerospace Canada Ltd. Each has a development and manufacturing organisation responsible for a wide range of parachutes and associated equipment. The Irvin Group is a leader in parachute design/manufacture for space projects; for example, recovery systems for heavy vehicles such as boosters and retarders for slow descents and landings on planets and moons. Completed projects include: Ariane 1 Stage 1 recovery parachutes, Pioneer-Venus descent parachute, Atlas MA3 main thrust stage recovery parachutes, and Shuttle landing-brake parachute. IRVIN was involved in the Cassini-Huygens Titan probe deployment parachute and a comprehensive software development tool for ESA to evaluate space-related retarder systems. The Group has one of the world's largest research and development divisions for parachute work, with specialist expertise in materials and the prediction and analysis of system performance using advanced computer modelling techniques. Recent research programmes include the behaviour and performance of materials in the space environment.

J S Chinn Engineering Co Ltd

Manufacturer of satellite ground support equipment, transportation packages, and special access and lifting equipment. The company provided the Skynet 4 integration trolley. Metal spinning and flow forming up to a diameter of 2.4 m; maximum mass for single unit: 20 tonnes.

Johnson & Higgins Ltd

Space insurance brokers specialising in pre-launch and launch insurance, satellite in-orbit and legal liability insurance.

JRA Aerospace & Technology Ltd

JRA was established in 1988 to provide marketing and management support within the UK/European aerospace and defence community. As a founder member of the Spacelink Europe consortium, JRA co-ordinates the activities of ESA's technology spin-off programme (while also covering UK/Scandinavia) through a network of 'technology transfer' companies spanning 13 European countries. The company has a 24-hour turnaround interrogation service for all major technical, patent and commercial UK/European databases.

Kalgard (UK) Ltd

The company is a licensee of the Kalgard Company of South Africa. Facilities are available for the application of military standard, resin-based coatings: corrosion-resistant (aluminium-filled resin and aluminium-filled PTFE); dry film lubricants (MOS_2 loaded); high-temperature-resistant coatings; forming and forging lubricants.

KG Coating Limited

KG Coating (formerly Scanwel Scientific) is a specialist manufacturer of advanced technology metal deposition equipment. Applications range from large-scale protective coating of structural components to special multilayer conductive and insulating coatings for pressure and temperature sensing devices. Ion Vapour Deposition services are available for aluminium and cadmium: For aluminium, services include either barrel coating or jigging of components to MIL-C-83488, DEF STAN 03-28/1; capacity 99 cm (length) × 35 cm (diameter); For cadmium, either barrel coating or the individual jigging of components is available to MIL-C-8837, DTD 940, DEF STAN 03-28/2; capacity 61 cm (length) × 30 cm (diameter).

Kinesix (Europe) Ltd

Founded in January 1994, the company manages all aspects of the pan-European markets of 'Sammi', with technical support offices in London, Paris and Rome. 'Sammi' is an X-Windows-based tool-set that enables users to build graphical user interfaces for real-time distributed applications. It is designed to run in high-performance mission-critical environments, where the real-time display of high data volumes is essential. 'Sammi' is used at all NASA sites in the US and has been chosen by ESA for Space Station onboard and ground control systems. It is used extensively by defence contractors, including IBM/Loral, for satellite and telemetry systems.

Ling Dynamic Systems

The company specialises in manufacturing, installing and commissioning vibration test systems for analysis and test applications. The force range is 8 N-300 kN. The company is a leading supplier of vibration test systems for satellite integration facilities.

Logica UK Ltd

Logica supplies computer-based systems for all aspects of space-related activity and undertakes associated studies and consultancy.

Onboard software: design/development for Huygens; definition of Rosetta autonomy requirements; support for ESA Polar Platform; software environment for ESA 31750 and 32-bit processors.

LRE Relays and Electronics Ltd

Part of the L I Group, the company provides relays and contactors for power switching in aerospace, military and industrial applications.

Marshall Aerospace

Marshall is a leading European aeronautical engineering organisation. It has manufactured a number of instrumentation systems for space application, including the Space Sled used aboard Spacelab. In 1993, it completed the conversion, for Orbital Sciences Corporation, of an ex-Air Canada Lockheed L-1011 Tristar to 'belly-carry' the air-launched Pegasus XL orbital vehicle.

Martin-Baker Aircraft Company Ltd

Founded in 1929, the privately owned Martin-Baker specialises in aircraft ejection seats and helicopter crashworthy seats. The company has produced 68,000 ejection seats that have saved a total of over 6,570 aircrew lives in 89 air forces. Latest developments include ejection seats for EF 2000, Pilatus MKII (JPATS) and the new Joint Strike Fighter programme. Martin-Baker crashworthy seats have been selected for both the military and the civil market that includes programmes such as the Eurocopter Tiger and Sikorsky S-92 Helibus. As well as a wide range of safety equipment that includes automatic lifejacket and liferaft inflation units, the company has designed, developed and produced the Descent Control System for the Huygens space probe that will soft-land on Titan, one of Saturn's moons.

Mitsubishi Electric UK Ltd (MEUK)

MEUK was established in 1978 as a wholly owned subsidiary of Japan's Mitsubishi Electric Corporation (MELCO). The Electronics Division was created in 1981 to offer expertise in spacecraft, satellite communications, sensor and image-processing systems, drawing on the extensive space activities of its parent company.

Molniya Space Consultancy

The consultancy specialises in information on the Chinese and RFAS space programmes. Additional expertise covers worldwide launch vehicles, technology and space history. It compiles the monthly *Worldwide Satellite Launches* newsletter.

Moreton Hall Associates (MHA)

Moreton Hall Associates (MHA) provides space consultancy services, including system engineering, insurance, tender preparation, launch and satellite statistics, procurement and illustration services. It is a member of the ASTOS Association of Specialist Technical Organisations for Space.

Normalair-Garrett Ltd

Normalair-Garrett Ltd (NGL), part of Westland Technologies Ltd, designs and manufactures precision products, including environmental control and life-support systems, pneumatic and hydraulic solenoid valves, investment castings, actuators, severe environment recorders and electronic controllers.

Parker Hannifin PLC

The Parker Hannifin Seal Group is part of Parker Hannifin Corporation, the world's largest manufacturer of hydraulic and pneumatic equipment, with more than 70 years experience. Worldwide personnel number approximately 30,000, of which around 5,000 are in Europe and 1,500 in the UK. The Seal Group was established in Europe in 1968 and, with the acquisition of the German-based Pradifa Company in 1979, became a major European supplier of high-quality O-rings and packings. For Spacelab, Spacehab and COF, Gask-O-Seals of up to 4 m diameter are installed between the elements of the pressurised laboratory. For Ariane 4, O-rings of almost 4 m inner diameter are applied to seal the 8,200 litre water tanks used for cooling the first stage gas generators. These special-compound seals are manufactured as endless units, with no joints or connection spots, for very narrow tolerances. For Ariane 5, Parker Pradifa worked with MAN Technologie GmbH to research different sealing compounds on the housings for the solid-propellant P230 boosters.

Plowshare Technology Ltd

Plowshare is the European marketing representative for the Cosmos USA joint venture between AKO Polyot and Assured Space Access, which offers Polyot's Cosmos 3 satellite launcher.

Quality CNC Engineering Ltd

A subcontract engineering company specialising in high-precision components using DNC/CNC. Examples of precision machining from solid raw materials, to flight qualification, include antenna arrays, filter bodies and LNA/SSPA housings.

Queensgate Instruments Ltd

The company is engaged in the development/manufacture of space-qualified piezoelectric translators and high-precision capacitance sensors for positioning and position measurement to less than 10^{-9} m. It also manufactures high-stability Fabry-Perot etalons (including cryogenic) for remote sensing applications. Space experience includes: high-stiffness piezoelectric translators for several pointing mechanisms in Japan's OICETS; development of the Silex Point Ahead Assembly; a scanning interferometer mirror for the WINDII instrument on UARS; mirror control actuators on Soho's Luminosity Oscillations Imager; and capacitance displacement sensors for Space Station.

Rose Bearings Limited

The company is a manufacturer of precision rod end and spherical bearings for the aerospace industry, supplying spherical bearings and self-lubricating bushes for Ariane 4/5. Rose became a wholly owned subsidiary of the MINEBEA Company Ltd in 1988, with associated companies in the US and Japan.

Rosemount Limited

Rosemount is a subsidiary company of Rosemount in the USA. Its instrumentation includes pressure sensors and air data probes on the Space Shuttle, and surface and immersion temperature sensors for launchers and satellites.

Science Systems (Space) Ltd

Formed in 1980, Science Systems offers software development and consultancy services. Staffing is 250, with 1996 turnover being £14.5 million (1995: £13.5 million). The company builds the ENVISAT platform and payload simulator in the UK and is engaged in specification and development of the ENVISAT PDS Centre Monitoring and Control and Common Services Facilities, specification and design of the Master Control Centre for the EGNOS satellite navigation system and development of a Small Low-cost Integrated Kernel (SLIK) for small satellite control.

The company offers a set of real-time software packages, including the Kernel control system – an integrated program suite providing satellite monitoring and control. Used on Skynet 4, the Hispasat Spacecraft Control Centre and the Meteosat Transition Programme. Facilities include telemetry reception and monitoring, telecommand generation and orbit control.

Sedgwick Space Services

Sedgwick provides all aspects of space insurance brokering, specialising in technical, legal, insurance and risk management on behalf of manufacturers, owners and operators. It provides the design and implementation of insurance, self-insurance and captive programmes in respect of build, ground property, pre-launch and in-orbit operations.

Serco Europe Limited

Part of Serco Group plc, the company provides management, technical services and consultancy in the space field. It is a leading service contractor, with 600 staff supporting ESA, EUMETSAT and the UK Ministry of Defence. Areas of support include project management, operations planning, spacecraft control, data communications, computer operations, ground station engineering, documentation and space applications software support.

Smith System Engineering Ltd

Smith System Engineering is an independent company of consulting system engineers, offering expertise in: remote sensing; environmental issues; onboard data processing; advanced computing and specialist information technology; C^3; radar, sonar and electro-optics. Staff level is 2,140, employed in Guildford (UK) and in offices in Brussels and Tokyo. Key space projects include: a review of satellite Earth observation data for climate change studies within the UK; a three-year project on the extraction of useful mesoscale ocean features from microwave satellite data; and a two-year project to demonstrate the utility of satellite data in coastal zone management.

Space Innovations Ltd (SIL)

Space Innovations Ltd (SIL) was established in 1983 and was previously Satellites International Ltd. Capabilities range from feasibility studies and research and development, to build, production and systems integration of space and space/ground equipment. Products include: small satellite busses; onboard data-handling systems; low-cost Sun/Earth sensors; star mappers; magnetorquers; magnetometers and associated attitude-control systems; power conditioning systems and batteries; radiation dose measuring instrumentation (its modular dosimeter operated on Meteosat 3 from 1988); space/ground data-compression systems; EGSE; radiometer instruments; telemetry systems and S-band transmitters/receivers and antennas for space/ground applications; and variable-rate/spread spectrum modems applicable to LEO and deep space missions. SIL is involved in: Soho; Meteosat; ERS; STRV-1a/b; DRS; Artemis Envisat; Ørsted (S-band transceiver) and Badr-B (S-band transceiver, OBDH, magnetometer, magnetorquers, power conditioning, batteries, attitude sensing system).

SIL now offers two families of semi-standardised platforms: MicroSIL for 40 to 80 kg micro-satellites, and MiniSIL for 100 to 350 kg mini-satellites.

MicroSIL is available in three basic forms: MicroSIL-G gravity-gradient stabilised; MicroSIL-S spin-stabilised; MicroSIL-3 three-axis stabilised. Subsystems are densely packed within 170 × 470 × 470 mm on an equipment shelf.

Specifications

MicroSIL

Size: 490 × 490 × 510 mm; 40-80 kg
Payload (kg, for 55 kg total): 13 -G; 18 -S; 14 -3
OBDH: DHS-848B data-handling system, dual T805 or SPARC processors, 1 Mbit program ROM, 48 Mbit SRAM data memory; memory expandable in 32 Mbit modules
Attitude control: two DSS-256 2-axis digital Sun sensors with 0.5° accuracy (options to 0.125°). MFM-3L 3-axis fluxgate magnetometer ±50,000 nT. Control by two MTR-25 magnetorquers. MicroSIL-G nadir pointing ±4° using 6 m boom with 4 kg tip mass; MicroSIL-S 1-10 rpm spin, 0.5° knowledge; MicroSIL-3 pointing within 0.5° using redundant MWs
Communications: two STX-91B 1W S-band transmitters; BPSK at 1 kbit/s to 1 Mbit/s
Power: 20 37-cell strings of 20 × 40 mm GaAs cells + 2 × 4 Ah Ni/Cd batteries. Provides 15-20 W for -G in sunlight (orbit dependent), 20 W for -S/-3

MiniSIL

There are two basic types: MiniSIL-P for narrow fairing Pegasus-class vehicles; MiniSIL-L for larger fairings such as LMLV, Delta and Ariane. Each is then available in a double-height version (2P/2L) and in both spin and three-axis designs. They employ a simple primary structure of a thrust tube with 937 mm V-band interfaces fore and aft. The tube is stiffened by longitudinal stringers and shear panels, to which the eight vertical sides are attached, with GaAs cells.
Size (P/2P/L/2L): 1.10 × 0.64 m/1.10 × 1.28 m/1.60 × 0.64 m/1.60 × 1.28 m; 100-150 kg/150-200 kg/150-250 kg/200-350 kg
Payload (P/2P/L/2L): 40-55 kg/65-85 kg/65-120 kg/90-180 kg
OBDH: DHS-8240B data-handling system, triple T805 or SPARC processors, 3 Mbit program ROM, 240 Mbit SRAM data memory; optional mass memory
Attitude control: six DSS-256 2-axis digital Sun sensors with 0.5° accuracy (options to 0.125°), MFM-3L 3-axis fluxgate magnetometer ±50,000 nT + optional RLG pack, GPS, static Earth horizon sensor or star mapper. Control by six MTR-35 magnetorquers; optional RWs + cold gas thrusters
Communications: two STX-91B 1W S-band transmitters; BPSK at 1 kbit/s to 1 Mbit/s
Power: N × 37-cell strings of 20 × 40 mm GaAs cells + 7 Ah Ni/Cd battery (redundant optional). Provides up to 55W/135W/80W/185 W (P/2P/L/2L).

Spacesense Consultants

Spacesense offers worldwide information, marketing data/analysis, and advisory services to organisations active in space technology and its application. The Spacesense Environmental Research Group (SERG) was set up in 1994 by Imperial College, NRSC Ltd, Phoenix Systems and Spacesense for satellite environmental monitoring consultancy, research and development.

Spectrogon UK Ltd

Spectrogon UK Ltd is a wholly owned subsidiary of Spectrogon AB, Taby, Sweden. Spectrogon UK Ltd has purchased the Diffraction Grating Department of Tayside Optical Technology Ltd, and therefore inherits the equipment and expertise to manufacture high-quality Ion-Etched Diffraction Gratings. Space Optics previously manufactured by this group include the variable line density master grating for ESA's XMM project. Key elements of technical expertise are: holographic optical elements (particularly diffraction gratings); ion etching (or milling) of holographic patterns to give high damage thresholds; and hydrocarbon-free metal coatings suitable for UHV applications. The Spectrogon group also produces Replica Holographic Gratings and Optical Interference Filters for a range of wavelengths from ultra-violet to infra-red.

Spur Electron Limited

Spur Electron offers a total parts procurement service in addition to specialist engineering support, software development and technical publishing expertise. It also specialises in product assurance activities, and parts, materials and process support for space hardware components. Current activities include Columbus (product assurance database, central parts procurement database); Houwteq Greensat (total parts procurement activity); ESA/ESTEC (radiation effects database); UK Defence Research Agency (data validation unit support; STRV satellite procurement support) and UK MoD (reliability study of generic components).

Surrey Satellite Technology Ltd (SSTL)

SSTL was formed in 1985 by the University of Surrey to support technology transfer to industry and commercial development of research activities, in addition to undertaking industrial contracts. The company's expertise includes designing, fabricating and operating low-cost (around £2 million each) micro-satellites: UoSAT 1-12, Kitsat 1/2, S80/T, PoSAT 1, HealthSat 2, Cerise, FASat satellites, Thai-Phutt, TiungSat-1, SNAP-1, PicoSAT, Clementine and Tsinghua-1 launched 1984-99. SSTL is located within the purpose-built 1,500 m^2 Surrey Space Centre, operating as an autonomous business with management and administration separate from the University. There are three varieties of cleanrooms (which can accommodate AIT of five micro-satellites simultaneously): 25 m^2 class 10,000 instrument assembly, 80 m^2 flight subsystem/payload assembly and 80 m^2 flight model AIT. The 300 m^2 R&D laboratories cover RF and ground station support, OBDH and digital systems, power and sensors, and an MCAD/ECAD design office. SSTL's Mission Operations & Control Groundstation was improved and relocated to CSER in March 1994. The three independent tracking antennas (VHF, UHF, S-band) frequently track as many as eight satellites, supporting more than 110 daily passes.

Research studies cover: power systems; OBCs and data-handling networks; modulation/demodulation schemes; RF transmitters/receivers; signal processing; attitude determination; stabilisation and control; imaging; radiation effects on components; failure-resilient architectures; VLSI subsystem design and store/forward techniques from LEO satellites. CSER is developing a 350 N hybrid motor. Using hydrogen peroxide as oxidiser and a solid fuel, it was successfully tested at Royal Ordnance's Westcott facility in February 1995. Under an ESA contract, SSTL is developing a GPS receiver for providing micro-satellite position and attitude; it will be carried by FASat-Bravo and TMSAT.

The modular multipurpose bus forming the core of the UoSat-12 mini-satellite provides S-band communications and enhanced OBDH for 180 kg payloads. The first will carry RWs, a 40 m resolution Earth camera and propulsion system for orbit control.

SSTL contracts have included supply of UoSAT ground stations to South Korea, Pakistan and Portugal; ESTEC spacecraft studies; mission studies for the Swedish Space Corporation; licensing of UoSAT ground station software; manufacture of flight hardware and launch and integration services on UoSAT missions; and supply of platforms for integration of dedicated payloads for commercial customers Matra and Aerospatiale (S80/T and Cerise). SSTL signed an agreement with Arianespace in August 1988 to act as the prime customer for the Ariane Structure for Auxiliary Payloads (ASAP) project. Of the six spacecraft slots on the first ASAP mission in January 1990, two were filled by UoSAT D/E, and four by micro-satellites from Amsat of North America. As a result, SSTL undertook a number of contracts for integration and launch services for experiments from ESTEC, RAE and Volunteers in Technical Assistance (VITA, US). In July 1990, SSTL signed a further Ariane agreement for UoSAT 5 launch aboard ERS 1's vehicle in 1991. The SSTL-built Kitsat 1 was launched in 1992 (SSTL supplied parts for the South Koreans to build 1993's Kitsat 2 based on SSTL's design), PoSAT 1 in 1993 and Healthsat 2 in 1993. A £3 million contract was signed in May 1994 with the Chilean Air Force for the FASat-Alfa satellite, ground station and training. The £1.3 million contract was signed in January 1996 for the identical FASat-Bravo. A £3 million contract was signed in November 1995 with the Thai MicroSatellite Company for the first national Thai micro-satellite, TMSAT, in 1997. At the beginning of 1996, contracts in progress totalled £11 million. A £1.35 million contract was signed in March 1996 for a communications payload aboard UoSAT 2. Photographs of SSTL-built satellites are included in the South Korea, Portugal, UK, and Oscar entries.

The £6 million demonstration mission of SSTL's multipurpose MiniBus platform will be carried free on the fourth and last test launch of Russia's Rockot. MiniBus can support 50 to 180 kg LEO payloads, including SSTL standard module boxes and a payload bay for heavier instruments or those with nadir-pointing apertures. Payload data will be distributed via the Internet.

During 1996, SSTL became part of a team selected by ESA to conduct a Lunar orbiter Mission study including the Technical University of Munich and the Swedish Institute of Space Physics. Known as the Lunar Academic Research Satellite (LunARSat) the study sought to encourage young pupils in schools and universities in a belief that they could participate in useful science. In October 1998, SSTL announced that it had broken into China's tightly controlled internal satellite industry with formation of a collaborative venture company in Beijing to develop micro-satellites. The first £3 million contract had been signed between SSTL and Tsinghua for a 50 kg micro-satellite to be called Tsinghua-1, the demonstrator for a constellation of seven micro-satellites which will provide daily worldwide high-resolution imagery for disaster monitoring. Launch is planned for 2000. Tsinghua-1 is scheduled for launch on a Long March rocket at the end of 1999.

Specifications

UoSat Bus

Configuration: octagonal prism, 1.1 m (diameter) × 80 cm (height), accommodating standard SSTL module boxes

Mass: 300 kg on-orbit (180 kg payload)

Power supply: 9 GaAs panels provide 150 W orbit average processed power; 21 Ah Ni/Cd battery supports eclipse operations

OBCs: 186 and 386 computer operating under a multitasking system provide data handling. Specialist computing resources provided by TMS320C32 and Transputer-T800 systems. Standard SSTL TM/TC protocol. Inherent in the store/forward architecture is a 72 h autonomy period with automatic fault reporting and data downloading

Attitude control: 3-axis by MWs, magnetorquers and cold gas thrusters

Communications: data rates of 9.6-76.8 kbit/s on 400 MHz using redundant UHF and 1 Mbit/s on redundant S-band. 9.6 kbit/s VHF command uplink

TT&C: by SSTL's Mission Control and Operations Centre

MiniSat Bus

Total mass: 50 kg, including launcher fittings

Configuration: 600 × 345 × 345 mm box-shaped body with solar panels on four sides, 6 m boom deployed from top face for gravity gradient stabilisation. Each housekeeping system or payload is housed in a standard module box, mass-produced by a computer-controlled process. The boxes are then stacked to form the main spacecraft structure; their number can be varied to meet mission requirements

Power supply: 4 GaAs panels totalling 4,992 cm^2 providing 18 W (orbit average processed); 6 Ah Ni/Cd battery supporting eclipse operations

OBC/Data Handling: the 80C186 payload experiment computer is used for high-level data manipulation. 80C188 version provides secondary OBC. On UoSAT 5 the payload Transputers are employed for high-performance data processing and attitude control

Attitude control: Earth pointing is maintained by combination of passive gravity gradient stabilisation and active magnetorquing; Sun/Earth horizon sensors and 3-axis fluxgate magnetometer with ±2 nT resolution over ±64 μT range provide attitude information. Gravity gradient supplied by 3 kg tip mass on 6 m boom deployed from top face, magnetorquing by six 150-turn aluminium wire coils (one on each spacecraft facet)

Communications: data rate of 9.6 or 38.4 kbit/s, 3-channel receiver (2 communications, 1 command), redundant transmitter switchable to 1/2.5/10 W RF output, 4-element turnstile VHF antenna, one monopole UHF antenna.

Systems Engineering & Assessment Ltd (SEA)

Systems Engineering & Assessment Ltd (SEA) is an independent space systems software/hardware development company with professional staff of more than 80. It provides consultancy services on feasibility, systems studies and soft/hardware engineering skills. Remote sensing expertise emphasises active sensors such as SAR. For satellite communications and satellite navigation, it is active in bespoke ground segment systems and hardware design/development. SEA develops and markets menu-driven software products, including the Space Systems Analysis Toolset, a modular suite of Earth observation mission-analysis software tools, SAR processing solutions and NextPerf, the SAR Performance Analysis and Prediction Tool for the next generation of advanced SARs.

T S Space Systems

This contract research and development company specialises in designing, building and operating: space-simulation equipment for testing satellite components/materials and ion thrusters; solar cell characterisation equipment; spectral response; close-match simulators; proton irradiation chambers. Expertise includes high-vacuum equipment and cryogenics (pumping/shrouds, computer control/data acquisition).

UK Industrial Space Committee (UKISC)

UKISC, the national trade association of the UK space industry, is sponsored jointly by the Society of British Aerospace Companies (SBAS) and the Federation of the Electronics Industry (FEI). UKISC represents the collective interests of member companies in increasing space and space-related business and their share of the market. Membership stands at 23 companies, representing over three-quarters of the total turnover and employees of the British space industry, and includes Matra Marconi Space, BT, Logica and Vega.

VEGA Space Systems Engineering Ltd

VEGA is an independent systems engineering and consultancy group with approximately 290 engineers. 1996 turnover was £16 million (profit £3.5 million).

VEGA specialises in spacecraft operations from the design study phase, through mission planning, to orbital operations. Since forming in 1978, it has been involved in operations/planning for all of ESA's science and applications missions, and has supported Inmarsat 2/3, Eutelsat 2, Skynet 4, NATO 4, Italsat, SAX and Tethered Satellite. It also produces operations-related software such as spacecraft simulators, data-analysis software and training packages.

Willis Corroon Inspace

Operating within the Willis Corroon Group's Aerospace division, Willis Corroon Inspace provides full space insurance services, including launch, in-orbit, products liability and third-party legal liability. Customers include AsiaSat, NHK and NASDA.

UNITED STATES OF AMERICA

Advanced Products Company

Seals for extreme environmental conditions, including rocket engines, fuel systems and cryogenic and vacuum systems.

AEC-ABLE Engineering Company Inc

Founded in 1975, ABLE provides deployable structures and special-purpose space mechanisms. In addition to deployable coilable and articulated lattice masts and tubular columns, ABLE provides: composite structures; honeycomb panels; wire deployers; solar array deployment and tracking/drive mechanisms; payload separation devices; brushless DC motors; linear/rotary drive systems; hinge devices; structural and fluid quick-connect; atomic oxygen protection and thermal control blankets. ABLE is providing the bearing assembly and solar array deployer for Space Station. The failed Meteor 1 satellite in 1995 carried ABLE's SCARLET solar wing, with four concentrator panels and two Si planar panels. ABLE offers a range of solar arrays, including:

(a) The UltraFlex Solar Array which provides 142 W/kg, 10.6 kg at 1,500 W BOL with Si; 166 W/kg, 9.0 kg at 1,500 W BOL with GaAs/Ge cells. 0.0794 m^3 stowed volume for 1,500 W wing.
(b) The PUMA Solar Array which provides 47 W/kg for 1,300 W BOL; self-synchronising deployment, high stiffness. Due on MSTI 3.

The deployable masts and columns include:

(a) The CoilABLE Boom which is a triangular cross section with three continuous longerons, or rectangular cross section with four, for use with any of three deployment systems: nut; lanyard; carousel. Booms up to 45 m have been deployed in space. A Nut-Deployed CoilABLE boom was first used to deploy the large solar array on Shuttle STS-41D in September 1984. Others include LACE, UARS and Galileo.
(b) The ABLE Articulated Retractable Mast or AARM employing four articulated longerons for greater bending strength and stiffness than above. Deployed by Nut Deployer. The FASTMast is the strongest version, stowing in 2 to 5 per cent of its deployed length. Side length 18 to 274 cm. Italy's TSS Tethered Satellite employs FASTMast, as does ISSA for solar array deployment.
(c) The Tubular Masts which complement the lattice structures consist of a line of tubular masts of 0.6 to 10 cm diameter, including BESTMast and STALK. Flights include six on Transit (SOOS) and a pair for Cornell University's WISP II.

AeroAstro Corporation

AeroAstro was formed in 1988 to design and build small low-cost space systems and components. 1994 revenue was US$3.5 million. There are about 40 personnel. Its first major contract was spacecraft development, test and payload integration and ground station for the Los Alamos National Laboratory Alexis satellite. MIT's 1996 HETE satellite and ground station was developed under a similar contract, awarded in February 1991. The company is building Boston University's Terriers satellite to study the upper atmosphere and ionosphere. AeroAstro is planning a low-cost small satellite launcher.

AeroAstro provided the 1 Gbit solid-state digital mass memory system for the NASA/Los Alamos MOXE x-ray detector on Russia's Spektrum-X satellite in 1998. The same mission carries a 0.4 Gbit memory for the Canadian Space Agency's Euvita experiment.

The company also designs and fabricates spacecraft RF equipment, torque coils, Sun sensors (0.1° accuracy over 10°), solar panels, integrated control systems, radiation-hardened flight computers, software, ground stations and other elements for small satellite projects. It provided Ni/Cd batteries for Sweden's 1992 Freja satellite under a US$340,000 contract.

Co-developed with the USAF, AeroAstro's Bitsy microsatellite provides an autonomous three-axis satellite weighing only 1 kg. Its simple, single circuit board design is easily tailored to specific mission requirements, such as store/forward, asset tracking, and hosting remote sensing instruments. Cost is US$100,000 for the basic satellite, or less than US$1 million for a turnkey system, including launch. Bitsy can be operated by the customer's PC or Macintosh-based ground station, or via AeroAstro on Internet. The debut Bitsy will determine its attitude using a star camera, image the Earth and transmit the compressed image.

Availability: 6 months after order
Mass: 1.0 kg
Size: 5 × 15 × 15 cm with attached solar panel
AOCS: 0.5 mrad (2s) using star camera, 50 mrad (2s) attitude control by cold gas thrusters, 1 mrad/s stability (target, orbit dependent)
Power: 2 to 10 W (increase optional), 8 V unregulated; 4 Wh lithium ion battery
Communications: 8 Mbit minimum, up to 1 Gbit, 5 Mbit/s payload raw data max output; UHF or S-band, 2 kbit/s up/down (10 to 100 kbit/s down option)

The Aerospace Corporation

The Aerospace Corporation was founded in 1960 and operates as a private, non-profit federally funded research and development centre dedicated to national security needs, specialising in space systems and related technologies, and providing general system engineering and integration. Aerospace provides technical oversight for the national security programme, including launchers and satellites such as Milstar, DSCS, GPS, DMSP, DSP and future missile launch detection. For example, it wrote the DSCS II programme plan and provided technical oversight from 1965's study contract through the continuing operations phase. The organisation is funded from more than 60 programme budgets. The 2,800 personnel include 64 per cent technical staff.

Aerospace Industries Association of America Inc (AIA)

AIA is the trade association representing the nation's manufacturers of aircraft, missiles and spacecraft, their propulsion, navigation/guidance systems, and other systems and components.

AlliedSignal Aerospace Company

ASAC's principal interests cover auxiliary power units, environmental control systems, launch vehicle and satellite systems (through Aerospace Equipment Systems, see below), propulsion systems, engine controls, avionics, guidance systems, secure communications and technical services.

AlliedSignal Aerospace Company/Environmental Control Systems (ECS)

Previously named AiResearch, Aerospace Equipment Systems, the company is involved in environmental control systems. AlliedSignal's heat exchangers, pumps, valves, sensors, fans controllers, blowers, smoke detectors and actuators are being combined into Space Station systems. These include the CO_2 Removal Assembly and the Solar Dynamic Power Array.

AlliedSignal Aerospace Company/Power Management and Generation Systems (PMGS)

PMGS specialises in the design, development and manufacture of components and systems for launch vehicles and satellites. These include: valves; actuators; turbopumps; turboalternators; cold gas and advanced monopropellant propulsion; propellant management systems; flight control actuation; power generation; dynamic power systems and attitude control systems. Current programmes include: Atlas; Centaur RL10A-4/E-1; LLV; Vega; Star 37/48; X33; EELV; Castor 4B/120; Gravity Probe B and military projects such as Theater Missile Defense. The NASA Solar Dynamic Ground Test Demonstrator and the NASA/RKA Flight Test Demonstrator are covered in the Thermal and Power industry section.

AlliedSignal Aerospace Company/Technical Services Corporation

The corporation is a technical services unit of AlliedSignal Aerospace, providing systems management for space tracking and data acquisition, systems engineering/design, laser tracking systems, VLBI systems and communication equipment operations. It operates/maintains part or all of three key NASA Goddard networks: the Ground Network, Space Network (including the White Sands TDRSS ground terminal) and the NASCOM NASA Communications Network. In addition, it operates/maintains GSFC's five Payload Operations Control Centers and provides: engineering support to the Data Technology Division; computer operations and maintenance to the Flight Dynamics Facility; and all elements of support to the Information Processing Division in the data reduction of spacecraft telemetry data and delivery to the principal investigators. For NASA's Crustal Dynamics Program, it provides technical support for the Satellite Laser Ranging element, including operation/maintenance of fixed/transportable laser ranging stations, engineering development to improve the accuracy of ranging observations, and precision satellite orbit determination. It is designing, installing and implementing a state-of-the-art satellite laser ranging system for the Italian space agency at the Matera Geophysical Observatory. The VLBI effort includes the design, construction, transportation and operation of hydrogen maser frequency standards, cryogenically cooled low-noise microwave amplifiers and microwave water-vapour radiometers.

At NASA Johnson, Technical Services Corporation is a member of the STSOC Space Transportation Systems Operations Contract team. Functions cover project management, maintenance/operations, flight preparation, production and mission operations involving the Mission Control Center, Shuttle Mission Simulator, Shuttle Avionics Integration Laboratory, Software Production Facility, Central Computing Facility, and the Mockup and Integration Laboratory.

For JPL, the corporation operates/maintains the Deep Space Network elements in Pasadena and Goldstone, providing tracking, data acquisition and network support for manned and unmanned missions.

AMP

AMP specialises in manufacturing electrical and electronic interconnection systems and components for the aerospace industry. Products include packaging systems, cable and board assemblies, shell circular metal, hermetic, rack and panel, fibre optic, coaxial and printed circuit connectors, in addition to solderless terminals, switches and EMP/EMI filters.

Analytical Graphics, Inc

Developers of the Satellite Tool Kit family, an interactive graphical software package to access, manage, display and manipulate data. Aerospace systems can be analysed graphically and numerically. Data are generated using the provided Two-Body Keplerian, Space Command SGP4/SDP4 (which accepts two-line mean element sets) J2 or J4 propagators. Programmer's Library provides integration of STK functions into an existing environment. Visualisation Option permits analysis in 3-D.

ANSER, Inc

ANSER is a not-for-profit public service research institute, founded in 1958, providing research and analysis in strategic and long-range planning, operational requirements, acquisition planning, systems engineering and analysis, and programme management. It operates in six business areas: aerospace systems; defence acquisition; information technology; military operations; special operations; international co-operation. ANSER's Center for International Aerospace Cooperation was formed in 1994 to promote the development and operation of international space systems. It publishes a weekly report on current Russian aerospace technologies, programme funding, political and economic issues, and institutional changes.

Applied Technologies Associates

Systems and software engineering, command/control engineering, automatic data processing, and continuous support for the USAF Satellite Control Network. The company's Real-Time Orbit Determination system provides real-time satellite orbit information autonomously to the user.

ARAC

ARAC, a former NASA Industrial Applications Centre, focuses on the utilisation of existing technology for resolving industrial technological needs. It accesses government laboratories, industry, universities and research facilities and hundreds of technical databases. Twenty in-house technical staff analyse and apply information and prepare a proprietary report according to a client-defined technological requirement. It is the US 'Technomart' co-ordinator. A 'Technomart' event takes place in Seoul, South Korea, every year. ARAC is also involved in meeting management, marketing and design, and is a consultant in technology transfer.

Ardak Corporation

Ardak provides tailored analyses of US government space markets, including launch, ground and customer-specified market niches. Analyses of companies providing space-related products/services are performed, especially competitions and teaming arrangements, merger/acquisition services, and evaluation of independent research and development programmes.

Arrowhead Products

Arrowhead Products designs, develops and manufactures pneumatic and cryogenic ducting systems, metal bellows and other flexible components, elastomeric and composite structures, and integral thermal insulation for engines and spacecraft. Experience includes: Shuttle Orbiter LOX/LH_2 main propulsion feedlines and manifolds; Shuttle External Tank LOX/LH_2 feedlines, LH_2 recirculation line and GOX/GH_2 pressurisation system ducting; argon-jacketed, vacuum-jacketed and polyurethane foam insulation for cryogenic systems to –251°C, as on Shuttle ET.

Astro Aerospace Corporation

A wholly owned subsidiary of Canada's Spar Aerospace, Astro Aerospace specialises in spacecraft deployable systems, including the three-longeron Astromast folding truss, STEM (Storable Tubular Extendable Member), telescopic deployable booms, specialised solar array developments, large deployable mesh antennas (6 m diameter Astromesh) and deployable support structures for radar systems. Examples are Astromasts for Mars Observer instrument booms, the Milstar solar array deployment booms, Hubble Space Telescope solar array deployment booms, the Astro Edge solar arrays for NASA's Clark satellite, and the support structures for Seasat and Radarsat SARs. Astro is designing and building the Mobile Transporter for Space Station. Personnel total about 120.

Autometric, Inc

The Omni space system planning software of Autometric's Space Technology and Applications Division provides interactive 4-D whole-Earth modelling and simulation.

Avibank Manufacturing, Inc

Avibank was founded in the 1940s and specialises in fasteners, devices and multicomponent fastening systems. Products range from quick-release pins and accessories to special bolts, latches and keepers, adjustable-diameter fasteners, hold-open rods, electronic hardware, panel fasteners, grounding hardware and threaded inserts. Ball-lok quick-release pins are provided for space missions. The Truss-lok mechanical joint mechanism has been developed for assembling erectable space structures and attaching equipment to Space Station's truss

Ball Aerospace & Technologies Corporation (BATC)

BATC reported 1998 sales of US$363 million. The company was founded in 1956 and comprises the aerospace systems and telecommunications products divisions. It was prime contractor for NASA's ERBS satellite, SDI's Relay Mirror Experiment, and the CRRES science satellite. Ball was awarded the US$46 million contract in 1992 to build the Geosat Follow-On radar altimeter satellite for the US Navy. It is providing technical assistance to Spain's small satellite programme under a four-year agreement signed in October 1992. Its EarthWatch affiliate plans to return 1 and 3 m imagery from QuickBird and EarlyBird satellites.

BDM International, Inc

A subsidiary of BDM International, BDM Federal, provides space systems engineering and services in military space and C[4], Ballistic Missile Defence, national security and civil space applications. It supports NASA's RSPAC Remote Sensing Public Access Center. 1994 revenue US$775 million.

Bergen Cable Technologies Inc

Mechanical cables and assemblies for the aerospace industry. Users include Space Shuttle.

BFGoodrich Aerospace

The Military Aircraft section of the Aircraft Wheels/Brakes unit is responsible for the Space Shuttle's wheels and brakes. The new carbon-carbon brakes were introduced on Discovery in April 1990. The Carbon Products unit manufactures the materials. Super-Temp develops high-temperature composites for space, aviation, electronics and nuclear applications. Aircraft Integrated Systems provides measurement, sensing and control systems for manned spacecraft, launchers and strategic missiles. Shuttle carries three systems: point level sensors to monitor pre-launch loading/unloading operations; propellant measurement during orbital manoeuvring and re-entry; cryogenics measurement in orbit. Strategic missile sensors measure the burn rate of solid-motor insulation. On expendable launchers, it provides multi-channel excitation and signal conditioning for instrumentation sensors and destruct control systems. Lighting Systems provided the approach systems for the nine prime and back-up Shuttle landing sites, plus other airborne lighting equipment. Rosemount Aerospace is a division of BFGoodrich Aerospace.

Boeing Aerospace Operations, Inc (BAO)

BAO's 4,460 m² Flight Equipment Processing Facility near NASA, Johnson, houses a laboratory dedicated to processing and providing astronaut equipment for each Shuttle mission. EVA suits, photographic/TV equipment (including 16/35/70 mm film cameras), crew communications items, tools, batteries, food, clothes and personal items are stored, assembled, tested and readied for flight. Equipment is packaged and shipped to the Cape in isopods for installation in the Orbiter; items are checked and refurbished after the flight.

Boeing Commercial Space Company (BCSC)

BCSC was formed in July 1987 to pursue commercial space ventures. Its flagship programme is the Sea Launch venture to launch Zenit GTO payloads from international waters. BCSC applied for a licence in January 1996 to operate the Resource 21 four-satellite constellation, beginning 1999, for crop monitoring. Other activities include development of a vapour-transport crystal-growth furnace for Shuttle flights under a Joint Endeavour Agreement with NASA, and participation in NASA Centers for the Commercial Development of Space. The furnace flew first aboard STS-52 Oct 1992.

Boeing Space Systems Group

Boeing signed an agreement on 1 August 1996 to acquire Rockwell's space and defence businesses, which would be renamed Boeing North American Inc, part of Boeing's Defense & Space Group. 1995's combined space revenue was approximately US$3.5 billion. The acquisition would provide Boeing with Rockwell's Shuttle operations and satellite manufacturing expertise (particularly GPS satellite contracts), and Rocketdyne's engine production.

Boeing's Missiles/Space Division, created in May 1991, operates within the Space Systems Group, formally established to centralise all space activities. Boeing Defense & Space Group was named on 17 August 1993 as prime contractor for the revised Space Station. The contract was signed with NASA on 13 January 1995 at US$5.63 billion. Other current projects include the Inertial Upper Stage, covered in detail in the Launchers/Propulsion section.

Booz, Allen & Hamilton Inc

Booz Allen's Space Systems Division, with a dedicated staff of 120 scientists and engineers, consolidates the firm's work in civil space engineering and operations. Services span concept definition, development, production, operation, support and data exploitation. Major programmes supported include Space Station (as part of teams performing systems engineering and integration, and training systems development), AF Space Command and NASA's Space Network.

Brush Wellman Applications Development Center

Design, development and fabrication of spacecraft structures, including solar array booms, antenna booms, central cylinders, optical support structures, thrust cone adaptors, radiators and heat sinks. Antenna struts were provided for Galileo, sunshields and radiators for Voyager and the solar array support structure for Magellan.

California Space Institute

Research and development for space science programmes.

Calspan SRL Corporation

SIII was formed in 1982 to design, market and operate commercial space-based facilities and to provide commercial services to government and industrial users. It merged with Arvin Industries' Calspan R&DTE subsidiary in 1992; Arvin took a 70 per cent interest. The creation of Space Business International was announced in June 1993 by SIII, OHB-System GmbH and Novespace to co-operate on space contracts.

Celestis Inc

Celestis Inc is a company formed to provide a route to permit the launching of a person's ashes into orbit. In 1997, the cost for flying each person's phial of ashes (the mass of each set of remains was limited to 7 g) was US$4,800. The first launch took place on 21 April 1997 with 24 phials being attached to the third stage of the Pegasus-XL launch vehicle, entering a 151°, 554 to 582 km orbit. Among those carried into orbit were Gene Roddenberry ('Star Trek' TV series creator), Timothy Leary (1960s 'icon'), Gerard K O'Neill (who established the modern concept of orbiting space colonies), Todd Hawley (founder of the International Space University) and Krafft Ehricke (engineer with the German V-2 and American Apollo programmes). Further launches aboard Pegasus-XL and Taurus vehicles have been purchased by Celestis.

Center for EUV Astrophysics (CEA)

CEA's core function is carrying out the science programme of the EUVE (Extreme Ultra-Violet Explorer), launched in June 1992. The payload was designed and built at CEA and at UCB's Space Sciences Laboratory and is now commanded from the Science Center.

Cincinnati Electronics Corporation (CE)

CE's main space product lines are the design/manufacture of spacecraft TT&C command systems, high-rate data transmitters, range safety flight termination systems for launchers, and TDRSS telemetry transmitters. The company also provides command encoders, message-format generators, frequency synthesisers and automated test equipment. Spacecraft command systems are offered in SGLS, STDN and TDRSS formats. Range safety flight-termination systems are available in both the standard IRIG 3/4 tone formats as well as the secure, Hi-Alphabet, format. CE is the leading producer of command destruct receivers, with equipment qualified for Shuttle, Atlas, Delta, Titan, Centaur and LMLV-1, plus numerous tactical missiles and ICBMs such as Minuteman and Peacekeeper.

Circle Seal Controls, Inc

Development/manufacture of manifold systems, control valves, check valves, solenoid valves, regulators, flow limiters, relief valves and shutoff valves for use in rockets, spacecraft and space-related programmes. Experience covers Shuttle, Titan 2 and Apollo.

Computer Sciences Corporation (CSC)

CSC provides tracking, data acquisition, processing, analysis and software support to US space programmes. The company's Applied Technology Division was selected in May 1994 for a US$1 billion eight-year contract to provide computer support to NASA Marshall. FY93 saw a US$566 million, six-year USAF contract for range technical services and a US$200 million 7½ year contract to operate/maintain NASA Langley's Central Scientific Computer Complex. Space Software Italia was established by CSC/Aeritalia.

Conatec, Incorporated

Conatec was established in 1984 to provide space engineering and technical services. Support has been provided to Texus, TDRSS and BMDO's target programmes.

COSMIC

NASA's Software Technology Transfer Center distributes computer programs originally developed by NASA and its contractors for the US space programme.

CTA Space Systems (CTA/SS)

A world leader in manufacturing small satellite systems, CTA Space Systems (CTA/SS) has developed, manufactured, fielded and operated nationally important surveillance and communications systems since 1980. Since 1984, the company has designed and built 27 satellites, and has an additional seven satellites in various stages of development. The company's spacecraft have varied widely in capability and have ranged in weight from 150 to 2,700 lbs. CTA/SS satellites currently under construction include the NASA Small Spacecraft Technology Initiative (SSTI) 'Clark' spacecraft; INDOSTAR, a private commercial geostationary direct broadcast mini-satellite; the EarthWatch (formerly World View) EarlyBird commercial remote sensing spacecraft and several DOD experimental mini-satellites. CTA/SS also builds satellite command and control ground systems, meteorological satellite data ingest systems, airborne and air-dropped remote sensor systems, and communications systems.

DSI and DSI's sister company, International Technologies Inc, were acquired by CTA/SS in mid-1992. 1992 and 1991 revenues were each about US$25 million, with 200 personnel. DSI built 27 satellites, launched between 1990 and 1994; six more were under construction for 1996/1997 launches, including NASA's Clark and EarthWatch's Early Bird. The company's first satellite, Global Low Orbiting Message Relay (GLOMR) was released from a Shuttle GAS canister in 1985. The first Pegasus air-launched booster in April 1990 released a DSI satellite for US Navy communications relay. Three 68 kg Maestro satellites, collectively the USAF P87-2 Stacksat, were launched by Atlas 28E on 11 April 1990. This US$11 million Multiple Autonomous Experimental Spacecraft for Telecommunications, Recording and Observations contract started in August 1987 as an Office of Naval Research and DoD Space Test Program project. The company received a US$7.4 million contract in August 1988 from DARPA for seven mini-satellites. The 21.8 kg micro-satellites were carried by the second Pegasus on 17 July 1991, to demonstrate UHF tactical voice/data links. Two 68 kg MACSAT Multiple Access Communications Satellites were launched on 9 May 1990 aboard a single Scout. 1991's Shuttle STS-39 observed propellant samples released from DSI's three 80 kg CRO subsatellites. DSI built the 90 kg ISES Independent Space Experiment System launched by Scout on 25 June 1991 for communications experiments. The Radcal radar calibration satellite was launched by Scout on 29 June 1993. STP's STEP M1/M3 were lost in the June 1994 and June 1995 launch failures, and GEMstar 1 on 15 August 1995 on LMLV 1. REX 2 was launched on 9 March 1996 for STP. DSI built Meteor's service module under contract to Westinghouse.

CTA Incorporated

Formed in 1979, CTA Incorporated has a current staff of over 1,200 employees and 1996 revenue of US$180 million. CTA specialises in the engineering, design and development of complex space and communications systems, and information technology applications and services. Their principal efforts are directed toward the innovative applications of new technologies in software, computer and information sciences and the design and development of small satellites, and associated ground-based command, control, communications and intelligence systems.

CTA provides skilled information technology services to commercial clients and state/federal government agencies. CTA's information technology business base includes engineering and acquisition support contracts which address programmes and systems through the entire life cycle of requirements: specifications, design, development, integration, test, training, and maintenance. CTA's government clients include the US Air Force, NASA, NOAA, DOJ, Treasury, the US Navy and the FAA. Their activities include C³I, avionics, space systems, air traffic control, and automated enterprise management technology applications. CTA participates in the systems engineering of major large-scale systems such as the Air Force Integrated Tactical Warning/Attack Assessment (ITW/AA) System, NASA's Earth Observing Systems, the FAA Advanced Automation System, and the independent test of such systems as the Air Force Peace Shield Air Defense System, the NASA Hubble Space Telescope Ground System and the FAA Advanced Air Traffic Control System.

CTA performs the operation and maintenance of GSA's Information management systems providing support for a wide variety of governmental users. This work includes state-of-the-art Information Technology services including client-server implementations, network engineering and development of relational base applications. CTA has completed a Year 2000 date conversion for the Bureau of Veteran Affairs, Austin Center and has also been selected by Nebraska and Kansas to perform Year 2000 conversions. CTA performs avionics integration for the Naval Air Warfare Center (NAWC) at China Lake and provided avionics software support for Operation Desert Storm. CTA develops range instrumentation for NAWC test ranges under the Range Instrumentation Development contract. CTA has a growing commercial IT business. Customers include Reynolds Metals, United Services Automobile Association and Applied Materials.

CTA International (CTAI), Inc markets, engineers and integrates turnkey communications and broadcast satellite systems for International markets. CTAI is under contract to deliver INDOSTAR-1™, an innovative digital direct-to-home television broadcasting satellite and ground system, to PT MediaCitra Indostar (MCI) and to operate and maintain the satellite broadcast system.

CTA Special Payloads Group supports spacecraft system design, hardware, software, test, and flight operations on Goddard Space Flight Center's Small Explorer Program, Space Shuttle Small Payloads Project, and Spartan Shuttle Missions.

Dallas Remote Imaging Group (DRIG)

DRIG is interested in satellite tracking, telemetry decoding, observation of new launches and digital processing of satellite imagery. The bulletin board makes available satellite information and images from the Okean and Sich satellites and from other observation satellites. There are approximately 15,000 members worldwide, with 100 actively tracking satellites daily.

Decca Valves Corporation

As the fluid control division of Puroflow Corp since 1980, Decca provides hydraulic swivels and titanium control valves for space applications.

Derlan Inc

Derlan provides machined components, fabricated systems and launch vehicle ground-handling systems.

Deskin Research Group, Inc (DRG)

DRG performs complete programme systems engineering, spacecraft production and ground station command/control, from initial design through launch integration and on-orbit operations. It also offers ground antennas up to 5.0 m in diameter, including the 3.0 m Isohex antenna, software engineering services and programming, and specialised communications systems, including LPI and one-of-a-kind packages for tailored applications.

For the Forte satellite, Deskin designed/produced the 7 Ah Ni/Cd battery, tailored to minimise its current-induced magnetic moment. It also provided the attitude-control electronics, flight computer and GPS elements.

Draper Laboratory

The Draper Laboratory continues to build its reputation as a nationally recognised space avionics design and development organisation for manned and unmanned space programmes. This is based upon years of experience in overall guidance, navigation and control system design and supporting sensor, actuator and data-management system design. Draper continues to explore and develop new designs for advanced inertial systems, attitude control and energy storage, and fibre optic gyro designs in support of advanced space systems. Draper has become the expert in on-orbit flight control and powered flight guidance design. Currently it is involved in support of manned and unmanned systems for both commercial and government programmes.

Dynamic Engineering Inc (DEI)

DEI provides design, analysis and manufacturing support for research and development hardware: mechanical design (100 per cent CAD); electronic design, engineering static, dynamic and thermal analyses, structural dynamics and aeroelasticity; conventional and CNC milling, turning, wire EDM; MIL-Q-9858/MIL-I-45208 Quality Assurance programme, control system hard/software development. Staff total 200.

Eagle Aerospace, Inc

Aerospace systems engineering and programme management. Recent activities include Space Station development and telecommunications.

Eastman Kodak Company

Supplier of image capture, storage, processing and printing components and subsystems, employing photographic, optical, electro-optical and electronic imaging technologies. Develops/manufactures custom imaging products and subsystems, including high-performance electronic cameras, imaging workstations, photographic processing equipment and large ultra-lightweight space-qualified optical components and subsystems. This unit is a major subcontractor to TRW for integration of NASA's AXAF telescope system. It is providing the camera payload for Space Imaging's commercial remote sensing satellite. Other current space projects include NASA's Far-Ultra-violet Spectroscopic Explorer (FUSE), a sub-arcsec star-sensing telescope and a custom digital camera now flying on Space Shuttle. It was also a bidder for the Hubble Space Telescope's mirrors and, in fact, cut/polished the back-up primary mirror. The company is believed to provide optical systems for reconnaissance satellites. It also provided the imaging packages for NASA's five Lunar Orbiters in the 1960s.

Eaton Corporation

Eaton's Pressure Sensors Division supplies four types of pressure transducers to the Space Shuttle programme: 41SG51 hydrogen/oxygen fuel-cell supply tanks, 27.2 to 82 atmospheres, –87 to +107°C 48 mV full scale output; 41SG78 Kaiser Marquardt thruster (44 per Orbiter) chamber pressure, 13.6 atmospheres, –1 to +149°C, 30 mV full scale output; 41SG85 LH_2/LOX main engines (19 per engine), 20.4 to 646 atmospheres, –54 to +74°C; 41SG275 main engine hydrogen supply, 340 atmospheres –107 to +38°C. The division is fully certified to NHB5300.

EG&G Florida, Inc

The company was established in 1982 when NASA awarded KSC's base operations a 10-year US$1.6 billion contract. NASA announced on 1 October 1993 that EG&G Florida had won the second 10-year US$1.7 billion contract. It provides operations and maintenance, engineering, computer services, astronaut rescue teams, medical services, security and other support services. EG&G developed a liquid-air pack for the astronaut rescue team and a state-of-the-art meteorological system for the Shuttle Landing Facility. It also prepared design and modification packages for the structures and support systems at KSC.

Encore Computer Corporation

Originally Systems Engineering Laboratories (SEL) and then Gould/Computer Systems Inc, Encore is a major computer supplier for range/tracking operations.

Engineered Systems Co (ESCO)

ESCO was created by the merger of All American Engineering and the Advanced Development and Engineering Center. AAE was selected in 1987 to provide up to seven Space Shuttle Orbiter Arresting Systems (SOAS) for vehicle braking in the event of a Shuttle landing on an emergency airfield. Three were installed in 1988 in time for flight programme resumption, a fourth was ordered in October 1988, and a fifth in March 1989 for installation at Hickam AFB, Hawaii. SOAS is a transportable arresting gear designed to capture an Orbiter should a landing on a short runway be necessary. The vehicle would engage a net (height: 13.7 m) across the runway, pulling a nylon tape from each Model 44 Water Twister energy-absorber tape drum and generating 0.75 *g* (maximum) braking. The 36-element net is a scaled-up version of the company's design used with military aircraft and would stop an Orbiter within 250 m without structural damage. The technique was demonstrated by towing Enterprise into a net at Washington DC's Dulles Airport.

Entran Devices, Inc

Founded in 1970, Entran specialises in miniature and compact static/dynamic measurement devices. In addition to extensive utilisation on missile and aircraft projects, the company's devices have been employed aboard Shuttle, Intelsat 5, Spacelab and Ariane.

Essex Corporation

Essex provides patented ImSyn™ processes for the reconstruction of SAR images, onboard spacecraft/other platforms and as workstation accelerators. Essex supplies custom design and development of trainers and mockups, provides quality training to mission specialists, logistics support and technology publications support.

Fiberite, Inc

Fiberite specialises in glass and graphite fibre combined with epoxy and cyanate resins for the aerospace industry, including the Shuttle Orbiter's main doors and OMS pods, the Hubble Space Telescope's main support structure, and rocket-motor exit cones (including IUS, Ariane and Shuttle).

Furon

Advanced composite components for aerospace and defence applications. Furon has provided components for classified satellite programmes and insulation material for the Atlas launch vehicle. Capabilities include twin sheet and pressure forming of thermoplastic materials, compression moulding, oven and autoclave cure of structural composites, and mechanical assembly.

G&H Technology, Inc

Non-explosive actuators for space applications, including solar array, antenna, cable tension and connector releases and experiment jettison.

General Research Corporation

GRC specialises in decision support and information systems, including space systems and operation support, experiment design, sensor performance modelling/analysis, cost engineering and estimation, systems engineering, advanced database technology, and prototyping and evaluation of distributed real-time simulations. GRC developed the Integrated Technology Assessment System, a regional database of international spacecraft components. Variations are used by the Space Station programme and the National Facilities Study.

Gulton Data Systems

Gulton provides spacecraft and launch-vehicle command, processing and telemetry systems. Recent products include control systems for Space Station. Comprehensive facilities include design, manufacture and environmental testing.

Harris Corporation

Harris Corporation focuses on four major businesses: electronic systems, communications, semiconductors and office equipment. GASD/HSSC are part of Harris Electronic Systems Sector. Space activities emphasise spacecraft data processing and satellite communications, microwave remote sensing, and tracking/control systems. GASD has provided major processing subsystems for Shuttle and the Hubble Space Telescope. It also provided the deployable antennas for TDRS and Galileo. It is developing the internal audio communications and video distribution systems for Space Station under contract to Boeing, and is working with Rocketdyne on solar concentrators for future Space Station power systems.

HSSC, a separate subsidiary, is designing and implementing the Core Electronics System, Kennedy Space Center's next-generation launch-processing and support equipment that will perform checks on Space Station modules and other payloads.

The Semiconductor Sector manufactures radiation-hardened and high-performance analogue/digital signal processing integrated circuits as well as discrete power and power integrated circuit chips for space/military applications. Landsat 6, for example, carried Harris chips in 14 subsystems.

Hernandez Engineering Inc

HEI specialises in manned space engineering and technical services, including mission operations, payload integration, safety, reliability, quality assurance, technical information and training analysis/implementation. Contracts are held with NASA Goddard, Johnson and Kennedy, DLR, ESTEC and Alenia.

Honeywell Inc

Space Systems' main activities concentrate on: inertial systems; onboard processors; altitude and control; control-moment gyroscopes; reaction wheel and momentum wheel assemblies; communications/data handling; and isolation/pointing. This includes the development and spacecraft application of fibre optic rotation sensors to supplement the presently manufactured ring laser and dry tuned gyro inertial measurement units. Integrated satellite attitude-control systems and their components are also available, including star trackers, reaction wheels and control-moment gyroscopic effectors, phased array momentum control systems and their operational software.

Data-management systems, in selectable configurations, employing fibre-optic networks, flat-panel displays and signal processors, are ready for spacecraft incorporation and are being provided for Space Station. Integrated components are also available to include optical mass storage, upgraded Shuttle standard multiplexers/demultiplexers, network and bus interface units and onboard data processors.

Honeywell, under contract to the USAF Phillips Laboratory, developed the GVSC Generic VHSIC Spaceborne Computer, which has a high-performance general purpose 1750A CPU, a local-memory controller and 256 + 8 static RAM. A 32-bit processor and single-board computer is in production and has been ordered for various military and commercial space platforms. All elements are radiation-hardened.

Advanced payload pointing systems with magnetic and passive isolation have been developed to attain sub-arcsecond accuracy with low bandwidth isolation. To support a wide range of satellite pointing applications, high-gain modular antenna pointing systems, solar array drives with integrated power transfer, gimbal actuators and spin/de-spin payload control systems are being manufactured. On Shuttle, Honeywell provides the main engine controllers, hand controllers, flight control system, radar altimeter and autoland software. It is subcontracted (US$70 million) by Rockwell for the Crew Multifunction Electronic Display System (MEDS 'glass cockpit') to replace the Orbiter's old cockpit displays by 1999.

Hughes Danbury Optical Systems Inc

HDOS designs and manufactures precision optics and electro-optical systems for defence and science space applications. Products include ground- and space-based telescopes, science and remote sensing instruments, star trackers, and optical components and systems. It also specialises in surveillance and directed energy systems, with emphasis on visible light sensors, optical systems and multispectral sensors. The company manufactured the Optical Telescope Assembly of the Hubble Space Telescope, including the 2.4 m main and 30 cm secondary mirrors, and the three Fine Guidance Sensors. In 1995, it completed the eight grazing incidence optics for AXAF. The Special Optics Facility, completed in September 1986, supports design, manufacture and complete environmental test capabilities up to the 4 m class, in optics of any geometry.

HDOS completed the acquisition of Itek Optical Systems from Litton Industries in February 1996. Itek company designed/developed large, precision lightweight optical systems for space deployment, adaptive optics and optics for very large astronomical instruments. It produced high-resolution film cameras and real-time electro-optical sensors, with 15 mm to 350 cm focal lengths, for reconnaissance and surveillance missions. Equipment was also produced for processing the electro-optical imagery. Itek built NASA's Large Format Camera, which flew aboard Shuttle in 1984 and returned approximately 2,200 large film frames. These capabilities will be continued in HDOS's main Danbury, Connecticut plant, as well as in facilities located in Lexington, Massachusetts and San Jose, California.

Hughes STX Corporation

Hughes STX Corporation (HSTX) is a wholly owned subsidiary of Hughes Aircraft Company and operates as part of Hughes Technical Services Company. Its primary focus is science, engineering and software support services in: astronomy, Earth and atmospheric sciences; science data management, satellite ground data systems; aerospace engineering and computer networking. For example, HSTX developed the current Landsat ground processing facility for EOSAT. Until 1995 it was co-owner of Starsys Global Positioning Corporation.

ILC Dover, Inc (ILC)

ILC specialises in the development of high-technology soft goods materials and the design of inflatable structures for unique applications. It has designed and manufactured space suits for Apollo, Skylab and Shuttle, and expects to provide suits for Space Station. It has also provided Mars Pathfinder's lander airbag system.

Inframetrics, Inc

The company's radiometer flew aboard Shuttle STS-50 in June 1992 to map oil surface temperature flows as part of a microgravity experiment.

Interferometrics Inc

The company was founded in 1982 for research, scientific investigation, technical analysis and the design/construction of hardware for NASA, DoD, the radio-astronomy community and commercial space telecommunications companies. More than half of the engineering staff have advanced degrees and 80 per cent of the science staff have PhDs. From June 1988 to August 1989, it provided engineering support to the Radio Amateur Satellite Corp (Amsat) for the development of a micro-satellite in return for the commercial rights to this 'Eyesat' technology. The first four were launched in January 1990 aboard Ariane. The fifth, the 10.5 kg Eyesat 1, was launched on 25 September 1993 with France's Spot 3 into a 791 x 806 km, 98.6° orbit to demonstrate worldwide commercial digital store/forward communications. The company is funding the project itself, controlling the satellite from its Chantilly facility.

The Eyesat design is configured as an expandable modular stack and can support a range of LEO missions, from store/forward to specialised science applications, and be launched as a secondary payload on most vehicles. The specially developed general purpose electronic bus interfaces with the command and mission payload equipment, controlled by a 16-bit microprocessor with 18 Mbyte of program/data storage memory. Advanced solar cells and load-side power management optimises power usage. The RF transmission system allows the use of discrete output levels under computer control.

Specifications

Configuration: 23 cm cube carrying no deployables. Each major subsystem is housed in one of five aluminium-frame modules bolted together

Power: 2 x 4 cm Si cells on each face provide 5-7 W orbit average, supported by 6.5 Ah Ni/Cd battery

AOCS: passive magnetic stabilisation with solar spin; magnets maintain alignment with local geomagnetic field

Computer: CMOS V-53 microprocessor, 16 Mbyte RAM data memory, 2 Mbyte RAM EDAC program memory

Communications: primary commercial payload comprises two 300-9,600 baud UHF transmitters, output of each controllable to over 4 W; omni 4-element turnstile antenna. 5-channel VHF receiver, 300-9,600 baud; omni whip antenna. Frequencies not available. Experimental amateur payload: one 300-9,600 baud UHF transmit (436.7975 MHz), output controllable to 0.5 or 1.5 W; omni whip antenna. 2-channel VHF rx (145.85 MHz), 300-9,600 baud; omni whip antenna. It may be configured for a single analogue channel.

International Space Brokers (ISB)

ISB are insurance brokers specialising in all aspects of pre-launch, launch, liability and in-orbit coverage, in association with International Space Brokers Ltd (UK) and Le Blanc de Nicolay (France).

International Technology Underwriters

Underwrites insurance for the builders, launchers and users of satellites. Policies indemnify against physical loss or damage to the vehicle and satellite during launch and satellite operations. Coverage can apply to: launch-vehicle performance, satellite initial operations, launch-vehicle and satellite initial operations and satellite on-orbit operations.

Intraspace Corporation

Intraspace, founded in 1986, was selected in 1993 by Continental Satellite Corporation for a contract worth US$250 million for two DBS satellites.

ISSI

ISSI (previously International Space Systems, Inc) is a space technology R&D company, providing support in mission planning, mission operations and systems analysis. Project participation covers Spacelab, Shuttle, Galileo and Israel's Ofeq satellite. Software packages and services include: Mission Planning and Orbit Definition and Design (mission planning; orbital analysis and requirements definition; orbit lifetime forecasting and impact predictions; launch-vehicle performance and dispersion analysis), Mission Control and Operations (mission control centre design and requirements definition; flight/mission operations; expert systems software for orbital analysis/control; expert systems software for timeline generation; simulation/analysis software), Satellite System/Subsystem Design and Analysis (payload systems design and tradeoff analysis; systems engineering; satellite systems design/evaluation; satellite guidance/control design/analysis; electrical power system design/analysis; communications balance analysis; simulation/analysis software), and Integration and Testing Equipment and Support (systems integration support/testing; automated test equipment).

ITT Defense & Electronics, Inc

The Aerospace and Communications Division provides voice and data communications, voice recognition, and meteorological and navigation spaceborne payloads. It is responsible for the payload of GPS Navstar Block 2R under contract to Lockheed Martin Astro Space; the Imaging Radiometer and Vertical Sounder instruments of GOES 8 to 12 for Space Systems/Loral; AVHRR radiometers and HIRS sounder for NOAA polar.

Johnson & Higgins

Established in 1981 to provide launch and transponder insurance packages and risk management expertise.

Kaiser Electro-Optics, Inc (KEO)

KEO specialises in refractive and reflective optical and electro-optical assemblies for space, surface, airborne and underwater military and commercial applications. Expertise covers production, assembly and testing of spherical, flat, cylindrical and aspheric optics and systems for visible to infra-red wavelengths. The company delivered in 1984 the first of 12 Cassegrain telescopes for DMSP's Operational Linescan System imagers. The 20 cm f/1.0 primary and hyperboloid secondary nickel-coated beryllium mirrors were mounted in a beryllium structure. Eight 15 cm aperture telescopes (f/0.5 parabola primary and hyperbolic secondary) have been delivered for NOAA's High Resolution Infra-Red Sounder, in addition to 20 cm Cassegrain systems for NOAA's Advanced Very High Resolution Radiometer.

Kaman Aerospace Corporation

Kaman specialises in ground/space-based optical systems. It developed the US$86 million Wavefront Control Experiment for SDI's Starlab mission, designed to demonstrate image-distortion compensation techniques employing a flexible mirror (the mission was cancelled). It has developed the 'Pamela' segmented-optics mirror system for DARPA, designed/developed a tracking/pointing system for Lawrence Livermore National Laboratory's KESTREL programme, and completed design efforts on NASA's Space Debris Warning Sensor system. EODC is participating in USAF Philips Laboratory's High Altitude Balloon Experiment (HABE), incorporating a precision pointing/tracking system for detecting/tracking ICBMs. EODC has developed several versions of the Magic Lantern lidar for detecting, classifying and locating floating/submerged objects for naval defence.

Kaman Sciences Corporation

A wholly owned subsidiary of Kaman Corporation, it provides services and products in information technology, consulting and technical assistance, military system effects and analysis, and hardware design, test and evaluation. Specific expertise includes software development, operations and maintenance; surveillance and intelligence systems O&M; modelling, simulation and virtual systems development; information network security; information analysis centres; facilities management; astrodynamics and space mission analysis; open source information analysis; EME assistance; advanced weapon system performance assessments and threat modelling; systems testing; photonics and smart sensor systems.

Kaman Sciences is the prime contractor for Cheyenne Mountain Complex software support, warning systems software support and NORAD/US Space Command/AF Command C^4 support services programmes. It has established the Space Environment and Orbital Debris Resource Center to promote the study of space environmental issues. The Center includes 15 million satellite orbital element data sets dating from the early 1960s. Kaman Sciences also develops specialised software such as the commercial Kaman Satellite Analysis Tool (KSAT), a high-fidelity 3-D animated visual modelling tool for the analysis of orbiting bodies.

Kamatics Corporation

The company is a wholly owned subsidiary of Kaman Corporation. It develops and manufactures self-lubricated slide bearings used on a range of space vehicles.

Ketema Programmed Composites

Specialists in design/fabrication of precision, dimensionally stable, high-strength, lightweight composite components for space applications, including solar-array substrates, bus structures, reflectors, tubes and struts.

KKI

Founded in 1984, KKI develops satellite simulation software: Orion for Windows, Orbit II Plus for Macintosh and PCOrbit for DOS. They support space-based wide area surveillance, mission planning, visibility studies, communications studies, constellation design, ASAT or threat studies, and production support.

KPMG Peat Marwick

KPMG Peat Marwick has provided support to the space industry since 1966 and in 1986 formalised a specialised Commercial Space Group. It provides business advisory services to governments and aerospace companies related to telecommunications, space transportation/infrastructure, space-related research and services. Capabilities cover strategic planning, market research, international joint ventures, technology commercialisation, productivity improvements and cost analysis. The company operates 650 offices in 132 countries.

L'Garde, Inc

L'Garde specialises in lightweight space structures. It produces a variety of space targets and RV decoys. It built the inflatable parabolic antenna (diameter: 14 m) flown as a Spartan payload on Shuttle STS-77 in May 1996. It produced the world's first inflatably deployed and supported photovoltaic array, and is developing technology for power densities of 150 W/kg for more than 1 kW.

Launchspace, Inc

Launchspace provides courses and seminars for the professional space community.

Lockheed Martin Corporation

The world's two largest space companies formally merged on 15 March 1995. Combined space sales in 1993 were US$6.7 billion, with 33,000 space employees. The companies' overall 1994 sales totalled US$22.9 billion, with 170,000 employees. In 1998, sales were US$26.26 billion compared with US$28.07 billion for 1997. Excluding the sale of divested businesses, 1997 net sales would have been US$26.14 billion. At the end of 1998, sales backlog was US$45.3 billion. Employing more than 165,000 people worldwide, L M Co is rated 32nd in the Fortune 500 of the world's largest corporations. The former Lockheed Missiles and Space Company, Martin Marietta Astronautics and Martin Marietta Astro Space now come under the Space and Strategic Missiles Sector; Lockheed Engineering and Sciences Company, Martin Marietta Space Systems and Lockheed Space Operations Company under the Information and Technology Services Sector; and Sanders under the Electronics Sector. Two of the four satellite plants will eventually be closed: most of the East Windsor, New Jersey and Valley Forge, Pennsylvania capabilities are transferring to Sunnyvale, California, although existing commercial, Landsat 7 and EOS AM 1 contracts were completed there. Denver, Colorado, will retain its military and interplanetary projects. All space launcher activities are being consolidated in Denver at Lockheed Martin Astronautics.

Lockheed Martin in April 1996 acquired most of Loral Corporation for US$9.1 billion and took a 20 per cent interest in the remaining Loral Space and Communications Ltd. Total annual sales are expected to be US$30 billion. The former Loral companies are initially grouped into a sixth sector, Tactical Systems.

At the time of the Lockheed-Martin merger, Martin Marietta's Space Group comprised Astronautics, Manned Space Systems and Astro Space. The Martin Marietta Corporation was established in 1961 by the merger of the Glenn L Martin Company and the America-Marietta Corporation. General Electric's aerospace businesses were combined in April 1993 with Martin Marietta for US$3.05 billion, establishing the world's largest aerospace electronics company, employing 97,000. GE Astro Space Division became Martin Marietta Astro Space. Martin Marietta's 1992 revenue was US$5,954 million; GE aerospace's US$5,278 million. 1994 sales totalled US$9.87 billion (1993: US$9.44 billion) for US$636 million profit (1993: US$450 million). Space Group employed some 16,000; 1994 sales were US$3.49 billion (1993: US$3.44 billion) for US$338.4 million operating profit (1993: US$249.3 million). As of May 1999, Lockheed Martin Corporation was divided into five sectors: Space and Strategic Missiles; Electronics; Information and Services; Aeronautics; Energy and Other.

Lockheed Martin Federal Systems

IBM Federal Systems was acquired by Loral in March 1994 for US$1.5 billion. 1993 sales were US$2.2 billion. FSG provides systems integration and specialised information handling and control systems, such as space vehicle guidance systems. In August 1993, NASA extended the contract for developing and maintaining Shuttle's primary flight and avionics software for a further two years, with options for a further eight years, potentially worth US$357 million. A five-year, US$150 million contract was awarded in July 1995 to provide software and engineering support for DSP ground stations. A five-year, US$400 million July 1995 contract will replace the GPS mainframe computers at Falcon AFB with workstations ready for controlling the GPS 2R generation from 1996. The company provided Mars Pathfinder's radiation-hardened flight computer and memory.

Lockheed Martin Missiles & Space Company

Lockheed Missiles & Space Company became the Lockheed Martin Missiles & Space Company in March 1995, part of Lockheed Martin's Space & Strategic Missiles sector. LMSC's Astronautics division was folded into the Space Systems and Missile Systems divisions at the end of 1990. LMSC was systems integrator and main contractor for NASA's Hubble Space Telescope, responsible to NASA for the structure and support; it designed the shroud that encloses the telescope assembly and the scientific instruments. It is building the Iridium satellites for the LEO system. It developed and manufactures the silica fibre insulation tiles for the fleet of Shuttle Orbiters. A range of solar-array designs has been produced for NASA, including an experimental 32 m panel tested aboard Shuttle.

The Research and Development Division has designed, built and flown more than 25 space science payloads, including three on Solar Max, particle spectrometers on CRRES, a soft x-ray telescope on Japan's Solar-A and two major instruments on UARS. For IASTP, the Research and Development Division was involved in Soho and Polar satellite instruments. It is the programme integrator for, and is building, the TRACE Small Explorer and is providing the cryogenic cooler for the WIRE Small Explorer.

In military work, Lockheed has been responsible for more satellites than any other company in the West. For example, it is prime contractor for the Milstar satellite to provide highly jam-resistant strategic and tactical communications. The company is a major BMDO/SDI contractor; it was awarded the US$350 million to US$400 million development contract in November 1985 for the Exoatmospheric Re-entry Interceptor System. It is competing for the follow-on early warning system to replace DSP. LMSC was also responsible for the US Air Force's Agena, the 'most-flown' western upper stage. LMMSC comprises Lockheed Martin Astronautics (including International Launch Services, a joint venture with Khrunichev Energia International), Lockheed Martin Michoud Space Systems, Lockheed Martin Missiles and Space (incorporating Communications and Power, Advanced Technology Center and the Commercial Satellite Center), Space Imaging Inc and United Space Alliance (a joint venture with Boeing).

Lockheed Martin Astronautics

Astronautics is one of the largest elements of Lockheed Martin's Space and Strategic Missiles sector and is headquartered at the Deer Creek facility about 20 miles from Denver, Colorado. It designs, develops, tests and manufactures a variety of advanced technology systems for space and defence among which are planetary spacecraft and other space systems, space launch systems and ground systems. The facility houses some of the most sophisticated production, space simulation and test facilities in the world, including 70 modern laboratories as well as computerised engineering design, manufacturing and administrative buildings. Astronautics also operates the Titan, Atlas and Athena launch vehicle programmes at both Vandenberg Air Force Base, California, and Cape Canaveral, Florida. In January 1999, it employed almost 10,000 people (7,100 at Deer Creek) and had an order backlog of US$8.8 billion. Astronautics is involved in two major core businesses: Space Transportation and Space Systems.

Space Transportation includes Space Launch Systems which is responsible for the Titan II and Titan IV, the Athena I and II, the MultiService Launch System (MSLS) and the Target Delivery System. Advanced Launch Systems is responsible for Atlas and Centaur launchers including development of the Atlas V in co-operation with the USAF Evolved Expendable Launch Vehicle (EELV) programme and Reusable Launch Systems which includes the Crew Return Vehicle (CRV) for use at the International Space Station as an emergency lifeboat.

Space Systems includes Flight Systems, which is responsible for the Mars Global Surveyor spacecraft, the first of a new generation of robot explorers in the Mars Surveyor programme. This also includes Mars Climate Orbiter and Mars Polar Lander which were launched in December 1998 and January 1999 respectively. The company is also building two spacecraft, an orbiter and a lander, for launch to Mars in 2001 and is studying designs and building one lander for each of two Surveyor programme missions in 2003 and 2005. Flight Systems also built the propulsion module for NASA's Cassini spacecraft launched to Saturn in October 1997 for a rendezvous with the planet in June 2004. It is also an industrial partner on Stardust which was launched in February 1999 with the objective of collecting, and returning to Earth in January 2006, sample material from the comet Wild 2. Flight Systems is an industrial partner with JPL in the Genesis programme and after launch in January 2001 will gather substantial quantities of charged particles from the solar wind for return to earth in August 2003 with which scientists will sample primordial material released from the Sun. Flight Systems is also a partner on teams selected by NASA in 1998 to study Aladdin, a spacecraft to sample the Mars moons Phobos and Deimos, and VESPER, which would study weather and climate patterns in the atmosphere on Venus.

Special Programs and Defense Systems includes various classified space and ground support systems.

Lockheed Martin Tactical Systems Sector

The former Loral Corporation elements acquired in 1996 are initially grouped under Lockheed Martin's sixth business sector of Tactical Systems, comprising 30 operating units and facilities employing 35,000. 1995 sales totalled US$6.2 billion. Loral in 1992 acquired the missile division of LTV Aerospace & Defense, builder of the Scout orbital launcher, in a deal worth US$260 million. IBM Federal Systems was acquired in March 1994 for US$1.5 billion. Loral was responsible for building NASA's new Mission Control Center in Houston for Space Station, in addition to modernising the existing Shuttle facility. On Space Station, it is handling systems engineering and integration for propulsion, communications, tracking and information, plus the integration/validation of the flight software and attached payload integration. Ford (then Loral, now Lockheed Martin) has been prime contractor since 1983 for development of a centralised C^3I capability for NORAD's Space Defense Operations Center. It has also provided approximately 200 large satellite ground terminals around the world. The company operated the USAF Satellite Control Network from the 1950s until September 1995 (when Lockheed Martin took over the new contract). The former Loral AeroSys, now Lockheed Martin Space Mission Systems, provides satellite ground systems and operations, including control centres for Hubble, NOAA and commercial Japanese satellites. Expertise covers information processing and data storage systems, integration and engineering services for space/ground support systems, and space segment engineering and operations, including launch, on-orbit support and payload support.

Loral Space & Communications Ltd

Loral Space & Communications, with headquarters in New York City, was 'spun off' from Loral Corporation in April 1996 as part of the sale of its defence electronics and systems integration businesses to Lockheed Martin Corporation.

On a pro forma basis, combining the results of the operations under its management, the company's revenue for what was a nine-month fiscal year, ending 31 December 1996, was US$950 million. Effective 1 January 1997 Loral reported on a 12-month calendar year basis. First quarter 1999 revenue from its operating segments totalled US$415 million, an increase of 15 per cent over first quarter 1998 earnings of US$362 million.

Space Systems/Loral (SS/L), a wholly owned subsidiary of Loral, is one of the world's leading manufacturers of space systems. Globalstar, a system of low-earth-orbiting (LEO) satellites supports a digital telephone service to hand-held and fixed terminals worldwide. Loral and its partners act as Globalstar service providers in Brazil, Canada and Mexico, and with Qualcomm, will hold the exclusive rights to provide an in-flight phone service using Globalstar in the United States.

Loral completed its acquisition of Skynet, a leading domestic satellite services provider, from AT&T in March 1997. Loral Skynet advances Loral's strategy of becoming a global provider of satellite-based communications services, and will complement its existing satellite manufacturing capabilities.

Loral is pursuing additional satellite-based communications service opportunities, including CyberStar, designed to provide interactive, broadband, multimedia data transmission. Loral is also developing a location information service that is designed to improve the accuracy of GPS receivers, and is pursuing opportunities in partnership with others to offer domestic and international direct-to-home services.

In March1998, Loral completed the acquisition of Orion Network Systems, an Internet access, corporate data networking and satellite services company. Loral and its partner Telefonica Autrey, S.A. de C.V. own a 75 per cent share in Satélites Mexicanos, S.A. de C.V. (SatMex) consisting of three operating satellites: Morelos II, Solidaridad 1 and 2 and SatMex 5 soon to be launched.

By 2002, Loral expects to have in place a robust communications network of more than 65 in-orbit geostationary and low-earth orbit satellites providing mobile and fixed telephone services, Internet access, high-speed video and data transmissions, direct-to-home broadcasting and broadband on demand.

Lord Mechanical Products

Isolators, couplings and elastomeric bearings for mounting vibration-, noise- and shock-sensitive equipment. Shuttle navigation units incorporate the company's gyro-suspension system.

Malin Space Science Systems (MSSS)

Company founder, Dr Malin, was principal investigator for the Mars Observer Camera, lost in August 1993's failure, and is principal investigator for the replacement camera on 1996's Mars Global Surveyor, and on new cameras for 1998's Mars Surveyor orbiter and lander. MSSS is involved in developing new cameras for a range of potential missions, and in developing data processing and archiving technologies.

Marsh & McLennan Space Projects

Provides space insurance brokerage services, develops risk management and insurance programmes for risks of pre-launch, launch, in-orbit operations, manufacturers, product exposures and third-party liability.

McDonnell Douglas Aerospace

There has been a merger between Boeing and McDonnell Douglas. The company's revenue from space, missile and electronics activities in 1995 and 1994 was each US$1.9 billion (1993: US$2.6 billion), for a 1995 profit of US$198 million (1994 US$262 million; 1993 US$338 million). MDA is Product Group 1 contractor for Space Station, responsible for building the integrated truss, mobile transporter, data-handling systems, crew health-care and monitoring equipment. MDA plays a multiple role with the production of Shuttle hardware, flight training and mission programming and payload integration. This includes building the aft/fore skirts, frustrum and other structures for the solid-fuel rocket boosters, and the aft propulsion system incorporating an orbital manoeuvring system and a reaction control system. MDA also checks and installs all military Shuttle payloads. Spacehab's Shuttle pressurised module was fabricated by the company.

McDonnell Douglas builds the Delta launch vehicle, with approximately 230 satellites placed in orbit over 35 years. MDA has worked on an SSTO single-stage-to-orbit launcher for BMDO; the experimental DC-X one-third scale version of this Delta Clipper underwent flight testing in 1993.

Metalmart, Inc

Supplier of speciality metals and alloys to the aerospace industry, including high-temperature materials for the Space Shuttle Orbiter. Metalmart was incorporated in 1974 and now employs 42 personnel, with an annual sales volume of approximately US$15 million. The company is the world's largest stocking distributor of sheet and plate magnesium, currently supplying Hughes, Lockheed Martin Astro Space and other satellite manufacturers.

Microcosm, Inc

Microcosm is developing a pressure-fed LOX/kerosene 22.2 kN engine with a claimed manufacturing cost of less than US$5,000 each, excluding the injector. There are only 38 parts, full duration 200 seconds, ablative film-cooled; performance is comparable to solid rockets, although key specifications remain confidential. A 200-second run was successful on 20 November 1995. Assisted by USAF Phillips Laboratory funding (US$3.1 million through September 1996). Microcosm proposes the Scorpius family of vehicles based on the same engine; SR-S sounding rocket (724 kg GLOW, 50 kg to 225 km; US$99,000 recurring cost); SR-1 sounding rocket (400 kg to 190 km; US$295,000); Mini-Lift (100 kg into LEO; US$875,000); Liberty Light Lift (71.7 tonnes GLOW, four stages, seven propulsion pods totalling 49 engines, 1 tonne into LEO; US$1.8 million); Exodus Medium Lift (6,800 kg LEO; US$8.5 million). Non-recurring development cost of the vehicles through Liberty is less than US$30 million (FY96).

MTS Systems Corporation

MTS, founded in 1966, provides aerospace testing needs, including: full scale structural; assembly and substructure; component; materials properties determination; strength/durability of parts; quality control.

NERAC, Incorporated

NERAC is a not-for-profit problem-solving and technology-transfer centre serving US industry since 1966. Resources include a large staff of technical specialists; technical/scientific resources providing global coverage of historic/current developments; US government technology and procurement opportunities; US and European patents; Reference Plus expert and consultant matching; and document retrieval.

Northrop Grumman Corporation (Electronics & Systems Integration Division)

The Division develops/produces lightweight space structures; advanced space engineering services; space power systems; space nuclear thermal propulsion; space-based sensors; high-energy particle beam systems for space/terrestrial applications; military/commercial test equipment; avionics; electronic subsystems including hybrid and MMIC components; special purpose computers and software.

Northrop Grumman Corporation (Technical Services Division)

The Division is involved in operations and maintenance, training support and engineering services for space systems. Until October 1995, it participated in the Shuttle Processing Contract as part of the Lockheed Martin Space Operations consortium; it was responsible for the management of a network of some 300 launch-processing system computers at Kennedy Space Center.

Northrop Grumman Corporation

Westinghouse Electric Corporation's Electronic Systems unit was acquired by Northrop Grumman in 1996 for US$3 billion and renamed the Electronic Sensors and Systems Division. Principal interests include radars and electronics for combat aircraft and battlespace management systems, as well as for military space and underseas systems. A leading supplier of air traffic control radars to the Federal Aviation Administration and to countries in Europe, the Middle East, Africa, Asia and South America. Produces nationwide air defence systems, tactical communication equipment, anti-submarine warfare systems, undersea vehicles, mine countermeasures, marine propulsion, and shipboard instrumentation.

Oceaneering Space Systems (OSS)

OSS is one of the application arms of Oceaneering International, the world's largest publicly owned underwater services company. In May 1993 it acquired ILC Space Systems. Expertise covers thermal and acoustic protection, EVA astronauts' tools and hardware, robotics, life-support systems and neutral-buoyancy training. The range includes thermal insulation blankets, ablative cork insulation, power tools, foot-restraint working platforms, lights, tethers and other tools and containers. OSS is designing the nitrox breathing system for the new buoyancy facility at NASA Johnson.

OSS is contracted for the design, fabrication, test and integration of Space Station's food and science refrigerators and freezers (0.57 m^3 each, −15 to +4°C). Work on other galley components, including oven, drink dispenser, trash compactor and wardroom, is planned. OSS is furnishing Station's EVA support tools and containers. These include a stowage unit mounted on the airlock which will house about 75 subassemblies such as electromechanical tools and crew translation and worksite restraint equipment, much of which is to be robotically compatible. OSS is responsible for the contract from concept design to flight qualification.

OSS occupies a 5,900 m^2 building adjacent to NASA Johnson. It houses a 507 m^2 precision machine shop for fabrication of prototype items and space hardware, capable of handling materials ranging from aluminium and exotic alloys to plastics such as Lexan, Teflon and Vespel. Soft goods are produced in two areas: a 176 m^2 area is oriented towards Shuttle items, fabric bags, containers and other items using conventional materials; the 502 m^2 area is isolated to handle more unusual materials such as quartz, ceramic and carbon cloth. OSS maintains a laminar flow tunnel (14.6 m along flow, 7.3 m wide) for work requiring strict cleanliness. There is access to a 4.6 m (depth) × 7.3 m (diameter) neutral-buoyancy facility at Oceaneering's Park Ten Facilities, 68 km NW. OSS also supports EVA and suit testing in Johnson's neutral-buoyancy tank and KC-135. At NASA Marshall, OSS provides neutral-buoyancy simulations, including Shuttle suit-support and general underwater operations.

Optical Coating Laboratory, Inc (OCLI)

OCLI, founded in 1949, is the world's largest independent producer of thin-film-coated optical products. Thin-film-coating deposition methods such as plasma plating and MetaMode (Metal Mode Reactive Sputtering) have been developed from OCLI's theoretical research. It has provided anti-glare coatings for the windows of every manned US spacecraft, in addition to coatings for optical instrumentation on Voyager, GOES and UARS. It has coated AXAF's eight x-ray mirrors. OCLI also provides solar-cell covers, thermal-control mirrors and instrumentation filters.

Orbital Sciences Corporation (OSC)

OSC recorded revenues of US$364.32 million for 1995, for a net loss of US$4.848 million, and was among NASA's 20 largest prime contractors. Firm contract backlog was US$540 million as of 31 March 1996. It acquired Fairchild Space & Defense Corporation from Matra in August 1994, Magellan Systems Corporation in December 1994, Perkin-Elmer's Applied Science Operation in September 1993 and MacDonald Dettwiler & Associates in November 1995. OSC was established in 1982 to develop/market Space Shuttle upper stages, resulting in the Transfer Orbit Stage. The Pegasus air-launched small orbital delivery system was developed in partnership with Hercules Aerospace, and OSC was contracted in 1989 by ARPA for the Taurus rapid-response orbital launcher based on Pegasus. Space Data Corporation of Tempe, Arizona became a subsidiary in November 1988, extending OSC's capabilities to a wide range of sounding rockets: it has manufactured/launched approximately 600 suborbital boosters in 35 different configurations, in addition to producing up to 1,000 meteorological rockets annually.

OSC's Orbital Communications Corp (Orbcomm) subsidiary plans a constellation of LEO satellites for two-way commercial communications services. First launch was successful in April 1995. By early 1999, 28 Orbcomm satellites were in orbit and during the first quarter total subscriber units in service doubled from 32,000 to 63,000 units. The service is on schedule to commence in late 1999.

The Orbital Imaging Corporation (Orbimage) is building the OrbView Earth-imaging system to provide 1 m resolution from 1997. OSC was selected by NASA in 1991 to deliver multispectral ocean colour data to NASA investigators for five years as part of the agency's Earth Probes programme. This is the first time the US government has purchased global environmental data from a privately designed/operated remote sensing satellite. Orbimage's SeaStar satellite, based on its PegaStar bus, carried the SeaWiFS Sea-viewing Wide Field Sensor in late 1996.

In early 1999, OSC formed the Orbital Navigation Company (Orbnav) to develop, operate and market satellite aided automotive guidance and related value-added information services for the rental car, private car, commercial vehicle and emergency vehicles markets. Orbnav plans to leverage Orbcomm's satellite data network and is currently focusing on the rental car market through a joint venture with the Hertz Corporation.

OSC has developed/integrated several Shuttle payloads, including high-precision two-axis pointing/tracking experiments such as the USAF Cirris 1A. It has also designed, built and installed more than 60 major telemetry/tracking stations in 10 countries.

The OSC Pomona division specialises in imagers and sounders for infra-red to x-ray remote sensing. It is fabricating four Total Ozone Mapping Spectrometers under a US$20 million contract for NASA. To ensure continuity of ozone data, the agency negotiated for the engineering model of Nimbus 7's TOMS to be launched on a CIS/USSR Meteor in 1991 and new models were planned for NASA's own Earth Probe in 1996, Japan's ADEOS in 1996 and Russia's Meteor 3M in 2000.

Picolab
OSC offers PicoLab for ultra-small orbital missions, including instrument development, component proof of concept and space and Earth sciences. Weighing up to 22 kg, PicoLab is carried in the avionics section of OSC's Pegasus or Taurus launchers and can be ejected or remain attached. The firm fixed price cost is quoted at less than US$3 million, which includes satellite, launch and one year of operations. Cost can be reduced to US$1 million if the customer provides its own control/data collection. Launch would be between 9 and 18 months after contract signing. OSC is targeting industry, universities and NASA field centres.
Design life: 3 years
Size: 20.3 cm (diameter), 15 cm (height), 45.7 cm solar array span; 22 kg, including 2.7 kg payload (4,900 cm^3)
Attitude control options: ±0.5° Sun-pointing or inertial spin; ±5° nadir pointing; ±20° passive magnetic
Power: 10-25 Wh orbit average to spacecraft (5-10 W payload), 3.3-15 V regulated bus; deployable or body solar arrays
Payload data: more than 10 Mbyte per day per station, 64 Mbyte (expandable) storage, 32 bit microprocessor payload interface

MicroLab
MicroLab employs the MicroStar platform design developed for the Orbcomm satellites. MicroLab 1 was launched successfully in April 1995 carrying a NASA lightning detector. Total contract was less than US$7 million.
Design life: 3 years
Internal size: 96.5 cm (diameter), 30.5 cm (height), 45 kg payload (0.17 m^3)
Attitude control options: nadir pointing ±0.5° or Sun-pointing, spin-stabilized to ±2°
Power: 95 Wh orbit average to payload, 28 V unregulated bus; two deployable articulated solar discs
Payload data: 2.0 Mbit/s max, 512 Mbit (expandable) storage

PegaStar
The SeaStar and APEX satellites both employ OSC's PegaStar bus. It incorporates launcher/spacecraft avionics functions into a single vehicle, yielding greater power/mass for mission experiments. Cost less than US$20 million; lead time 18 to 24 months. Payload accommodation up to 91 kg/0.454 m^3. The all-aluminum primary structure mates directly to the Pegasus stage 3 motor, and houses launch-vehicle guidance, telemetry/control systems as well as spacecraft housekeeping functions. The power system is shared by the satellite/launcher, incorporating two 6 Ah Ni/H_2 high-energy density batteries (10 Ahr on SeaStar). Four 55.9 × 152.4 cm rigid solar arrays provide 500 W peak BOL (300 W to payload). ACS is provided by a momentum-biased 3-axis system of magnetorquers, magnetometer, Sun sensor and constant-speed momentum wheel to provide Sun pointing to 0.5° in two axes. Control about the Sun line is held to 5.0° with very low angular rates. Four 22.2 N hydrazine thrusters provide initial orbit insertion, drag make-up and precise orbit maintenance; hydrazine capacity 68 kg. A GPS receiver obtains position information to within 100 m independent of ground systems. The multi-processor satellite computer controls all housekeeping functions; an advanced 80 Mbyte SSR stores up to 24 h of experiment data. The telemetry transponder provides data transmit (2.6 Mbit/s; 1.25 Gbit storage).

Oremet Titanium

Titanium sponge, ingots, mill products and castings. Applications include Shuttle propellant tank ring.

Precision Rolled Products, Inc (PRP)

PRP, founded in 1974 and acquired by Germany's Krupp VDM in 1990, is a speciality mill with integrated facilities to hot roll, thermal treat and finish bar products in titanium, titanium alloys, high-T cobalt and nickel-base alloys, and stainless steels for the aerospace industries.

Princeton Synergetics, Inc (PSI)

PSI specialises in policy and economic issues related to space commercial development, transportation and insurance. Analytical tools include simulation models that evaluate launcher and satellite attributes in terms of telecommunications satellite business financial performance; a computer-aided small satellite design tool aimed at mission planners, satellite designers and technology programme planners; a computer model for comparing advanced space transportation architectures explicitly taking into account uncertainty and risk. PSI has supported the Department of Transportation in its setting of space insurance requirements and has supported space insurance companies in establishing space insurance programmes.

Questar Corporation

Questar, incorporated in 1951, manufactures space-qualified telescopes, bore-sighted scopes for optical tracking, long-distance microscopes and non-contact measurement systems for laboratory applications, in addition to a range of catadioptric telescopes. All optics are diffraction limited and hand crafted. The company manufactured the first telescope to fly in space (Gemini, 1965) and is currently involved in several space projects.

R-Cubed Composites Inc

Composite structures for spacecraft and launch vehicles, including Orbcomm, CBERS, Pegasus, Taurus, Clementine, Clark and MicroLab: customers include Hughes, TRW, Space Systems/Loral and Motorola.

Rosemount Aerospace Inc

Now a division of BFGoodrich, Rosemount Aerospace was established in 1956. The company designs, develops and produces temperature, pressure and air data sensors for measuring critical flight and engine control system parameters. Development/testing facilities include sub/supersonic wind tunnels, icing and flow tunnels, and extensive environmental facilities. Rosemount sensors have been used in every US manned space programme, including temperature sensors in Apollo's lunar surface science packages. Shuttle sensors monitor and control main engine functions such as cryogenic and hydraulic temperature, hot gas temperature, and speed/flow rates. Approximately 10,000 sensors were supplied for onboard and ground support systems. Specifications below for the Space Shuttle Main Engine temperature sensors illustrate the operating environment. In orbit, flow sensors monitor the operation of the Shuttle fuel cells; during entry, two air data probes measure the Orbiter's angle of attack, total temperature and airspeed. The measurements are made from 30 km altitude (and at speeds in excess of M4.0) to touchdown.

Scaled Composites, Inc

Scaled Composites provides complete composite aerospace vehicle design, manufacture and test services. It builds the fins and wing for Orbital Sciences Corporation's Pegasus launcher, and provided the aeroshell for the DC-X demonstrator. It is currently working on the K-1 vehicle for Kistler Aerospace Corporation and the HMX Roton.

Schonstedt Instrument Company

The company's magnetometers have flown on several hundred satellites and planetary probes as spacecraft attitude-control sensors and local magnetic field measuring instruments. They were added to the Hubble Space Telescope during its December 1993 in-orbit servicing. Magnetometers for rocket aspect measurement include: SAM-73 tri-axis fluxgate sensor, SAM-72 dual-axis fluxgate sensor and SAM-71 single-axis fluxgate sensor; RAM-5C, S29-S30 single-axis fluxgate sensor.

Schuller International

Acoustic and thermal insulating material for aerospace applications, including Space Shuttle tiles.

Science Applications International Corporation (SAIC)

SAIC is an employee-owned company specialising in computer systems integration, training-device systems, electronic warfare and countermeasures, systems modelling/simulation, electro-optical and signal processing, military space sciences and technology, aeronautical engineering, astronomy and astrophysics, software development and systems integration. 1996 revenue was US$2.4 billion (1995: US$1.9 billion, 1994: US$1.7 billion).

Sedgwick Space Services

Sedgwick provides all aspects of space insurance brokering, specialising in technical, legal, insurance and risk management on behalf of manufacturers, owners and operators. It provides the design and implementation of insurance, self-insurance and captive programmes in respect of build, ground property, pre-launch and in-orbit operations.

SETI Institute

The non-profit institute is making microwave observations looking for extra-terrestrial life.

Space Analysis & Research, Inc (SAR)

SAR offers satellite and launch database software systems for IBM-type PCs. SPACE 2000 contains all past launches and attempts plus plans through 2012. Includes all 25,000+ catalogued objects detected by US Space Command, orbital element sets for all orbiting payloads and basic orbital data for the approximately 6,000 pieces of debris on orbit.

Space Applications Corporation

Space Applications specialises in systems engineering, spacecraft command/control, space-qualified hardware and mission planning tools. It developed the station acquisition, orbit/attitude control, command generation and status telemetry monitoring software for GPS and CRRES. It provided experiment control software for the ultra-violet plume instrument flying on SDI's LACE satellite and the Drop Physics Module for Shuttle's US Microgravity Laboratory. The company is also providing systems engineering and technical assistance to the AF Satellite Control Network.

Space Commerce Corporation (SCC)

SCC was formed in 1986 to market Soviet launcher vehicles and space services. It now acts as a consultant to high-technology firms doing business in the Russian Federation.

Space Electronics Inc (SEI)

SEI manufactures Rad-Pak™ space-qualified radiation-hardened microelectronic devices for space applications. 1996 revenues were US$10.6 million. 1995 sales were US$4.0 million (revenue US$3.75 million for US$325,000 profit), with a backlog of approximately US$6 million. Experience includes ISTP, TOMS, Geotail, Mars-96, SeaStar, DMSP and Cassini. Globalstar contracts total US$1.7 million. A new family of microcircuits includes 1M EEPROM (128k × 8), 256K EEPROM (32k × 8), 1 Mbit SRAM (128k × 8), Microprocessor (80386RP + 80186RP) and DRAM (4 Mbit). Devices survive a minimum of 100 krad (Si) in most orbits and have SEU tolerance more than 43 MeV/mg/cm^2 and single event latchup tolerance approximately 120 MeV/mg/cm^2.

Space Industries, Inc (SII)

SII was founded in 1982 by Dr Max Faget, the chief engineer responsible for the designs of Mercury, Apollo and Shuttle. In March 1996, SII was purchased by GB Tech Inc and is now 'Space Industries Inc, a division of GB Tech Inc'. The company is responsible for the design/fabrication of the Wake Shield Facility, the Meteor capsule and several thermally controlled experiment modules. WSF successfully grew seven thin film substitutes on STS-80, demonstrating its capabilities as an ultra-vacuum research platform. SII has developed the 7 kg, less than US$1 million Nanosat for a range of applications, including communications and ISS inspection. The 58 × 25 × 36 cm satellite could manoeuvre around ISS for 12 hours using cold gas thrusters and miniature gyroscopes controlled from a joystick terminal.

Space Machine Advisors, Inc

The company provides risk management analysis, insurance and financial services to the space and satellite communications industries.

Space Policy Institute

Established in July 1987, the Institute conducts research on space policy issues, organises seminars, symposia/conferences, and offers graduate courses on space policy. It focuses on US space policy issues and competitive and co-operative interactions with other countries.

Space Studies Institute (SSI)

SSI is a non-profit organisation for conducting and sponsoring research on the use of space resources. Current projects include the design of solar-power satellites built from lunar materials. The Institute publishes a bimonthly newsletter.

Sparta Inc

High-technology company focusing on US military space contracts.

Specialty Steel & Forge

Distributor of high-temperature metals to the aerospace industry, including stainless steel, aluminium, titanium, nickel, copper, brass, glass sealing alloys and tungsten alloys. Users include the Space Shuttle.

Spectrum Astro, Inc

Spectrum Astro provides research, design, development and manufacture of low-cost, advanced technology space systems. 1995 sales were US$7 million and the company has 60 staff. The six product lines are: research and development for advanced space systems; development of rapid prototype advanced space systems; production of multimission spacecraft; design/manufacture of high-reliability space-qualified disk memory systems, space electronics and complete subsystems, and EGSE.

Spectrum Astro was awarded a contract in 1991 to design, fabricate, integrate, test and deliver on-orbit the MSTI Miniature Sensor Technology Integration satellites for BMDO/Phillips Laboratory. MSTI 1 was launched successfully in November 1992 and MSTI on 2 May 1994, both providing five months of mission data, meeting all mission objectives. MSTI 3 departed in May 1996. JPL's first New Millennium bus contract was awarded to Spectrum Astro in September 1995. The bus for Deep Space 1 was delivered in February 1997 for launch on the satellite in October 1998.

The company has developed, space-qualified and flown a low-cost, low-mass 8.7 Gbit Magnetic Disk Data Mass Memory unit for ARPA's Advanced Space Technology Program (see the Recorders & Memories section). Carried by MSTI 3 and STEP M3.

The SA line of multimission small platforms offers standard payload interfaces and high payload mass fraction. The SA-2 first flew for MSTI 2. The SA-1000 would typically carry a small communications payload in GEO. No customers have yet been announced.

Specifications

The specifications below are provided in SA-2/SA-3/SA-1000 order:
Orbit: LEO Sun-oriented/LEO all inclinations/GEO
Schedule: 12/18/24 months
Design life: 3/5/10 years
Dry bus mass: 113/181/295 kg
Payload mass: 181/181/363 kg
Payload power: 190/760/1,200 W
Pointing/control: 0.25°/0.1°/0.1°; 3-axis

Taber Industries

Stainless steel bonded strain gauge pressure transducers covering from 0 to 2,040 atmospheres and from −210 to +190°C. Users include Insat, Intelsat, Arabsat and Tethered Satellite.

Teledyne Brown Engineering (TBE)

TBE specialises in mission analysis and planning, hard/software development and payload integration. It is the Systems Engineering and Technical Assistance contractor for the US Army Space & Strategic Defense Command and is the Payload Integration Contractor for Spacelab. Current space-qualified hardware emphasis is on materials-processing systems for Shuttle and Space Station: TBE's high-temperature Crystal Growth Furnace, zeolite CGF and Protein Crystal Growth apparatus flew on Shuttle USML 1/2 in 1992/95.

For the US Army, TBE integrates BMDO experiments and provides technical and programme support for the development of exoatmospheric surveillance sensors, ground-based interceptors, airborne sensors, the Ground-Based Radar and BM/C³ elements.

Teledyne Electronic Systems (TES)

TES is the consolidation of the former Teledyne Systems Co, Teledyne Lewisburg, Teledyne Electronics and Teledyne Ryan Electronics. It specialises in redundant, fault-tolerant and radiation-hardened electronic systems such as launch computers, on-orbit mission computers (such as Milstar's), navigation systems (inertial, Doppler, GPS), flight-control systems/sensors, telemetry, remote multiplexers, inertial attitude control/pointing, data links, encryption and cryptographic modules, signal conditioning and interface electronics. TES is a JPL support contractor for space electronics.

Telegenix/Grim Corporation

Specialists in electronic switching systems, technical control equipment, intelligent illuminated graphic displays and electronic contract manufacturing.

Textron Inc

Textron was responsible for the design, development and production of Peacekeeper's re-entry system and expanded its manufacturing facilities for producing the Mk21 Re-entry Vehicle. In 1991, the Electronic Systems business line was formed, including specialised surveillance systems and communications antennas. The Aerostructures Division designs/manufactures structures for spacecraft: it was responsible for the intertank structure of Shuttle's External Tank, and for the nose cones and aft skirts of Titan 4's solid boosters. TSM is a developer/manufacturer of high-strength, lightweight advanced composite and fire-protection materials. Composite materials include boron and carbon fibre, silicon carbide reinforced aluminium, titanium and ceramics, and carbon-carbon for structures, turbine engines and rocket motors. Carbon-carbon throat inserts were built for Trident and MX. The company manufactures 32 tonnes of carbon-carbon annually. TSM developed a silicon carbide-reinforced titanium composite for NASP/X30's skin and internal support.

Thomas Electronics Inc

Thomas Electronics furnishes special-purpose CRTs for the aerospace industry, including the Shuttle Orbiter's set of triple displays (these were due to be replaced by flat LCD screens by 1998).

TRW Space & Electronics Group

TRW's space, defence and information business achieved 1998 sales of US$5.3 billion. The work backlog (without options) in January 1999 stood at about US$6 billion. TRW's S&EG came into being on 1 January 1993 as a merging of the Space & Technology and Electronic Systems groups; personnel total 7,000. It has built approximately 185 satellites, built/integrated approximately 130 payloads, developed approximately 200 space instruments and integrated some 550 experiments into spacecraft. Military programmes include DSP, DSCS II, Fltsatcom, Milstar's payload and many covert activities such as KH11 DSCS covers 16 satellites, launched from 1971 and designed to operate into the 1990s. UHF military communications satellites are provided by the US Navy's Fltsatcom; TRW was the lead contractor for NASA's TDRS satellites, the seventh and last of which was launched in July 1995.

TRW provided three High Energy Astronomical Observatories in the 1970s to investigate x-ray sources; these were followed by the Compton Observatory in 1991. NASA selected the company in August 1988 as AXAF prime contractor. Launch is scheduled for 1999. TRW signed a US$5.48 million contract on 17 April 1990 for the US Air Force's STEP Space Test Experiment Platform. STEP is designed to produce satellites of less than 450 kg, with lives of one to three years using small ground stations. Options provide for a further 11 satellites; the first three are to be followed by five more (the fourth was contracted mid-1992 for US$14.5 million). The platform can provide 300 W power and be three-axis or gravity gradient stabilised. Debut was on Taurus in March 1994 with the TAOS Technology for Autonomous Operational Survivability payload but STEP-2 was lost in a launch vehicle failure.

TRW is marketing this Eagle/STEP design: it was selected in April 1994 for Taiwan's Rocsat 1 (US$61 million contract) and in September 1994 for South Korea's Kompsat (US$75 million contract signed in March 1995). TRW was selected in June 1991 for NASA's US$29.3 million 1994 TOMS Earth Probes contract. TRW built the spacecraft and integrated the instrument. NASA awarded the US$398.7 million contract on 15 September 1995 for the EOS Earth Observing System Common Spacecraft, with its AB1200 bus derived from Odyssey's AB940 bus. The initial contract covers PM 1 and Chem 1; two further options for AM 2/PM 2 could take the total to US$668.5 million. The launch of EOS PM is scheduled for 2000.

In the ground terminal field, the GEODSS Ground-based Electro-Optical Deep Space Surveillance system was developed by TRW Systems Integration Group for USAF. It designs/develops space communications and processing payloads; programmes include Milstar, TDRSS, DSP and SMTS. The division also handles design, development and production of the ground data-handling systems for these payloads.

TRW makes satellite bus systems for the USAF Step programme. Each system comprises three modules: the standardised self-sustaining core, and the payload and propulsion modules that can be tailored to mission requirements. Each module is a 12-sided prism (diameter: 96.5 cm) of monocoque aluminium construction.

The primary or Core Module has a mass of 91.6 kg and a height of 29 cm.
Attitude control: 3-axis (gravity gradient or spin available); scan wheels provide horizon scanners and RWs for pitch bias momentum control system. 3-axis magnetometer + magnetorquers unload wheels and control yaw. 0.5° knowledge/control; 1°/s drift rate.
Power: six deployable panels + body-mounted cells provide 110 W orbit average (75 W for payload). Articulated array offers 350 W orbit average. Five Ni/Cd batteries provide 330 Wh storage. 18 VDC bus.
Communications: UHF BPSK or FSK rx, 10 W transmit, 1/125 kbit/s up/down; 5 W S-band transponder, 1/256 kbit/s up/down.
Digital Processing System: 80C86 digital processor/controller, 200 kips; 512 kbyte RAM + 512 kbyte EPROM; 8.2 Mbyte data storage, 256 kbyte command storage, mass memory options

The Payload Module has a height of 36.8 cm, and a payload volume 0.255 m³, body solar cells add 15 W orbit average. Modules are stackable to accommodate long items such as telescopes

The Propulsion Module has a height of 58.4 cm, monopropellant hydrazine (Olin Aerospace 178 N MR-107 + four 5 N MRE-1), body-mounted solar cells.

Unisys Space & Information Systems (USIS)

USIS, with 1,700 personnel, supplies software and systems engineering, products and services for spacecraft and operations. It manages, modifies and maintains Shuttle ground-based software, which totals approximately 19 million lines of computer code. The software encompasses 15 programming languages, including Fortran, Assembler, PL1 and DB2. Eighteen operating systems are used, along with equipment produced by eight manufacturers, including Unisys, IBM, Perkin-Elmer and SEL. The software runs on almost 300 computers in 13 separate facilities connected through LANs.

USIS has software and systems engineering responsibilities for Space Station operations, including software development, integration, testing and sustaining engineering, and systems requirements development, design support and performance analysis. It performs sustaining software engineering, testing, analysis and systems support for the Control Center Complex for Shuttle and Space Station at Johnson. At Goddard, it has evaluated the safety, reliability and quality of electronic, mechanical and other spacecraft components since 1965. Manned space activities are performed at the primary Houston facility.

United Technologies Corporation/Hamilton Standard Space Systems International, Inc (SSI)

SSI's space interests focus on environmental control, thermal control and life support systems. Japan's Sumitomo Corporation acquired a 20 per cent interest in July 1994. Hamilton Standard provided Apollo's lunar suit life-support system and the Lunar Module's environmental and thermal control system; it continues to manufacture environmental control systems for the Space Shuttle Orbiter and provide the Shuttle EMU space suit. The current EMU contract calls for strengthening the design to extend its life, reduce on-orbit maintenance and improve its capability for Space Station assembly. SSI and Russian suit maker Zvezda are collaborating on development of common components between the EMU and Russia's Orlan suit for Space Station.

SSI provides seven major Orbiter systems: Regenerable CO_2 Removal System; Atmosphere Revitalisation System (provides shirtsleeve environment, avionics bays cooling air and a water coolant loop to collect and transport cabin heat to the external Freon Coolant Loop); FCL (collects heat from the ARS and other equipment in the Orbiter mid-body/aft areas and delivers it to one of three heat rejection devices); Flash Evaporator System (rejects all FCL heat during ascent/entry when Orbiter approximately 30 km; supplements radiators on-orbit and rejects excess water collected onboard); Water Spray Boiler (cools the hydraulic fluid and lubricating oil during launch/descent); Ambient Temperature Catalytic Oxidiser (converts CO to CO_2); Waste Collector Subsystem (collects/stores crew solid/liquid body wastes).

For Space Station, SSI is providing: temperature- and humidity-control systems to circulate air between the modules, remove airborne particulates and cool/dehumidify cabin air; similar components also circulate air to cool avionics and other heat loads; waste water processor, storing it for drinking; active thermal control systems, including pumps, heat exchangers and cold plates for the external loop (as part of Rocketdyne's team). Elements are being provided for the modules of Japan, ESA and Italy.

SSI's Hamilton Standard Management Services Inc subsidiary in Houston provides engineering support at NASA Johnson for its EMU and Shuttle hardware, and furnishes EVA tools and training aids. A separate Space and Sea Systems Group focuses on military and commercial applications of proton exchange membrane SPE electrolysers, undersea naval auxiliary equipment, and aircraft cabin oxygen generation and humidification equipment.

Valcor Engineering Corporation

Fuel, hydraulic, cryogenic, pneumatic and special purpose valves for the aerospace industry.

Veda Incorporated

Specialising in engineering and technical services, Veda has 1,400 employees in offices throughout the US. Space activities focus on analysis, engineering and management support: satellite communications; constellation design, analysis, optimisation; TT&C of aerospace systems; satellite communications system engineering and integration; communications network scheduling, simulation, control; spacecraft design; GPS applications and analysis; COTS telemetry systems; launch-vehicle operations analysis and support; economic analysis and market assessment; LEO + MEO satellite systems engineering; legal technical assistance; advanced engineering seminars.

Wear-Cote International Inc

Electroless coatings for ferrous/non-ferrous metals. The company's electroless nickel is employed in Shuttle ignition and mechanical systems.

Wesgo Metals

The Metals Division of Wesgo Inc provides high-purity low vapour pressure brazing alloys. Applications include rocket engines, vacuum tubes and lasers.

Willis Corroon Inspace

Operating within the Willis Corroon Group's Aerospace division, Willis Corroon Inspace provides full space insurance services, including launch, in-orbit, product liability and third-party legal liability.

Wyle Laboratories

Specialists in engineering, testing and support services, small payload integration and small experiment payload engineering.

LAUNCHERS, PROPULSION SYSTEMS AND LAUNCH SITE SERVICES

CANADA

Bristol Aerospace Limited

In addition to the supply of basic rocket vehicle hardware, Bristol provides a range of self-contained vehicle/payload support systems, including thrust termination, parachute recovery, payload separation, despin, telemetry systems and fairing deployment. Two launcher types are usually employed: a rail-type typically providing 6 to 8 m of guided rail travel and a 30 m travel tower. In cases where the normal dispersion is unacceptable, the Saab/SES S19 boost guidance system is available; Saab's Spinrac attitude control system provides attitude control for a spinning exoatmospheric stage.

CHINA, PEOPLE'S REPUBLIC

Government sponsored organisations

The China Great Wall Industry Corp is the foreign trade company, under the aegis of the China Aerospace Corp, responsible for launch services marketing, commercial negotiations and contract execution. National launches are charged to the government at about 60 per cent of the commercial rate; some 30 per cent of GWIC's profits are paid in tax. The China Satellite Launch and TT&C General organisation, with a workforce of more than 20,000, operates the launch sites and is responsible for all TT&C services through to the conclusion of a mission. It operates under the Commission of Science, Technology & Industry for National Defence but for commercial activities it is responsible to GWIC. China's primary space and military launcher manufacturing facility is the China Academy of Launch Vehicle Technology (CALT, formerly the Beijing Wan Yuan Industry Corporation) and near the town of Nan Yuan about 15 km south of the capital. Here, the CZ-2, DF-5 ICBM (a CZ-2 variant) vehicles and the CZ-3 cryogenic upper stages are manufactured in a complex of 13 institutes and seven factories. CALT is the technical co-ordinator for GWIC and provides on-site technical supervision of launcher procedures. The cryogenic engines are built here, but the other stages' engines are built by its Shanxi Liquid Rocket Engine Company. The plant also operates large stands for vibration and thermal testing and has a large anechoic chamber for antenna development. Some of the launcher production is conducted at the Shanghai Academy of Space Technology (SAST, also known as the SBA Shanghai Bureau of Astronautics) with a 30,000 workforce (8,000 engineering/technical personnel) manning 17 research laboratories and 11 factories: CZ-3 first two stages, CZ-3 cryogenic third-stage attitude control system, and all the launchers' guidance systems. It is responsible for the entire CZ-4/2D vehicles. The Chinese Academy of Space Technology in Beijing is responsible for sounding rockets but is primarily known for the recoverable FSW platforms now offered for μg/Earth resources activities.

Space launcher production capacity for the two manufacturing sites is assessed to be 10 to 12 vehicles annually. In June 1989, GWIC and Avibras of Brazil established International Satellite Communications Ltd (Inscom) to market their combined launcher, spacecraft and tracking expertise, but the company fell into abeyance.

FRANCE

Fortech Airforge

L'Air Liquide, with more than 400 plants around the world, specialises in the production, handling and utilisation of gases and cryogenic liquids. Its contribution to Ariane 4 includes stage 3's tanks and ground equipment for handling cryogenic liquids. Arianespace signed a FFr400 million joint contract with L'Air Liquide and Aerospatiale in 1989 for 50 Ariane 4 stage 3 tanks in the P9 production lot. The tanks, being delivered 1991-98, are manufactured by the company at its Sassenage facility and then integrated by Aerospatiale at Les Mureaux. The two companies established the Cryospace GIE (55 per cent L'Air Liquide, 45 per cent Aerospatiale) in 1988 to develop/fabricate the main tanks for Ariane 5's principal cryogenic stage. The first tank completed main structural tests in November 1991. L'Air Liquide was contracted in 1990 to build a LH_2 production plant, with an annual 10^6 litre capacity, at Kourou to meet Ariane 4/5 requirements. It entered service in 1991.

Pyrospace

The company was established in 1989 by Aerospatiale and SNPE of France and OEA of the US to develop/manufacture pyrotechnic components and systems. For Ariane 4/5, Pyrospace provides stage separation, destruct, strap-on ignition/separation, fairing separation and non-contaminating dual-satellite separation. For satellites, it offers open/close of hydraulic and propellant valves, and solar array and antenna deployment by separation nuts and rod cutters.

Société Européenne de Propulsion (SEP)

The principal space business of SEP, a subsidiary of SNECMA, is the design, development and construction of propulsion systems for launch vehicles and satellites, including Ariane's Viking, HM-7, Vulcain and P230 (partnered with BPD in Europropulsion), motors. In satellite work, SEP produces Mage apogee kick motors, hydrazine thrusters for attitude control, bipropellant liquid engines and electric/ion engines. The company also produces solar array drive mechanisms, μg furnaces, surface tension tanks, magnetic bearings, transducers and flow regulators. In 1995, personnel totalled 3,500, with a turnover of FFr5,400 million generating a FFr150 million profit. Aerospatiale in January 1994 sold its 13.6 per cent interest for FFr178.9 million. SEP and Spain's Empresarios Agrupados established Iberespacio in October 1989 to undertake joint development of space propulsion systems. SEP and its SNECMA parent established Hyperspace in January 1990 to co-ordinate propulsion activities for hypersonic aircraft and launchers. Other interests include: Arianespace 8 per cent, Europropulsion (50/50 with BPD), G2P 75 per cent (large solid-propulsion systems for military applications), Spot Image 12 per cent, Carbone-Industrie 100 per cent (carbon brakes), TECHLAM 50 per cent (laminated elastomeric components) and S2M 54 per cent (magnetic bearings).

Test/Assembly Facilities: SEP maintains a 4,500 m² Ariane propulsion assembly hall containing 14 Viking engine assembly stands, five for HM-7 and nine for stage 1 propulsion systems. There is also a 400 m² class 100,000 cleanroom for cryogenic engine component assembly. TDF propulsion was installed in an integration building provided with two 300 m² class 100,000 rooms. The Vulcain Engine Assembly Building can produce eight engines annually. The 1,220 m² class 100,000 cleanroom accommodates 12 assembly cells and 25 class 100 laminar flow booths/hoods for storage before assembly. Also in the building are a 700 m² warehouse, a 1,400 m² engineering and mechanical area and 1,150 m² of office space.

Engine test facilities are concentrated at the 1.5 km² Vernon site but some are located at the 24,300 m² Villaroche area. For storable propellant motors, there are four component test stands (PF1-3, A48, F22) and one engine stand (PF2). Cryogenic testing is undertaken on three stands: the two of Vernon's PF41 (one with altitude simulation) and one at Villaroche (with altitude simulation). Eleven existing cryogenic component stands were joined in 1988 by PF52 for Ariane 5 Vulcain gas generator and turbopump development. The PF50 stand for complete Vulcain engine tests was inaugurated September 1990 at Vernon. SEP inaugurated its 11,000 m² production plant for Ariane 5's P230 nozzles in October 1990 at Haillan. A FFr180 million investment, it will fabricate 10 nozzles annually by year 2000. Assembly space is 650 m², plus 825 m² class 100,000 cleanrooms.

Société Nationale des Poudres et Explosifs (SNPE)

The Société Nationale des Poudres et Explosifs (SNPE) Group, created as a state-owned company in 1971 and now with 5,000 personnel and FFr4.4 billion 1994 turnover (FFr100 million profit), manufactures liquid rocket propellants, with emphasis on UDMH, and develops/produces solid propellants. Along with BPD Difesa e Spazio in the Regulus partnership, the company provides the propellant for Ariane 5's P230 solid boosters using facilities at Kourou constructed by subsidiary SNPE Ingénierie. SNPE maintains facilities at Saint-Médard en Jalles near Bordeaux (solid propellant grains), Toulouse (UDMH, AP), Toulon (pyrotechnics) and Le Bouchet near Paris (HQ and R&D). The Toulouse site manages the production of ammonium perchlorate with a capacity of 6,000 tonnes annually to ensure Ariane 5 supplies. SNPE formed a second partnership with BPD, Société Européenne de perchlorate d'ammonium (EUPERA; SNPE 66 per cent, BPD 34 per cent), to handle production. Toulouse also provides the UDMH for Ariane 4 and the MMH for Ariane 5's upper stage. Subsidiary Pyromeca manufactures detonating cords and flexible linear cutting charges and ultra-fast pyrotechnic valves for Ariane 4 release from the launch pad. In cryogenic propulsion, SNPE provides the main igniter and turbine starter grain for Ariane 4's HM-7B stage 3 engine and the shaped ducts for the Ariane 5 Vulcain igniter and starter.

The propulsion package was designed/produced for the 153/227 kg Eclipse T1/T2 supersonic rocket and autonomous launch ramp in partnership with CAC Systèmes. Applications include meteorological sounding and μg research. T1/T2 capacities are 20/25 kg to 12/72 km.

STARSEM SA

STARSEM is a joint-venture company, set up to market the Soyuz family of launch vehicles on a commercial basis. The partners are Aerospatiale and Arianespace in France, the Russian Space Agency (RSA) and the Samara Space Centre (comprising the TsSKB design and engineering bureau and the Progress production facilities) which produces the launch vehicles. The agreement for the creation of STARSEM was signed by the partners on 17 July 1996 in Moscow and on 6 August the statutes of the company were signed in Paris. The company is a French-registered corporation, set up with FFr500,000. The backing for the project is split with Aerospatiale having 35 per cent, Samara and RSA 25 per cent each and Arianespace 15 per cent.

The company proposes to market three basic launch vehicles: the current Soyuz-U, the current Molniya-M and a new Soyuz-U/Ikar version which adds a new upper stage to the existing Soyuz-U.

GERMANY

DaimlerChrysler Aerospace AG

Previously DASA, DaimlerChrysler Aerospace AG (DBAA) is the main contractor for Ariane's second stage and the PAL liquid strap-on boosters, and provides the thrust chamber for the HM-7B third-stage engine; it similarly provides the injection head, combustion chamber and expansion bell for Ariane 5's Vulcain first-stage engine, under a contract signed with SEP in April 1989. During 1991-98, DBAA is providing 50 second stages and 120 PALs under a DM1,600 million contract, in addition to 50 HM-7B thrust chambers under a DM52 million award. Another 15 stage 2s and 26 PALs have been ordered for 1998-99. The company leads Ariane 5's EPS upper stage, the first development of an Ariane stage outside of France. Fourteen will be built between 1997 and 2002. DBAA, as MBB, also built the P3.2 test stand for Vulcain chamber trials. Its other engines include storable bipropellant (4, 10, 400, 27,500 N), hydrazine monopropellant (0.02-350 N) and electric (10-15 and 150-200 mN).

As MBB, the company began development of the world's first 50 kN topping cycle engine in 1958 and began testing at 85 atmosphere chamber pressure in 1963. In 1968, MBB's 180 kN LOX/LH_2 engine attained the existing chamber pressure record of 282 atmosphere, leading to NASA and Rocketdyne acquiring a licence for utilising the technology in the Space Shuttle Main Engine. The current 10 N and 400 N NTO/MMH thrusters are combined to form a Unified Propulsion System (12 × 10 N + 1 × 400 N), employed by the Galileo Jupiter orbiter and the TV-Sat/TDF, Tele-X, DFS Kopernikus, Eutelsat 2, Turksat and Amos telecom satellites. Bipropellant 4 N thrusters are currently under development and qualification.

DBAA's Lampoldshausen and Ottobrunn sites are equipped with test stands for high-pressure thrust chambers up to 1 MN (P3.2 stand for Vulcain) and for gas generators and turbopumps with supply pressures up to 800 atmosphere, stands for LH_2 ramjets, sea level and vacuum stands for 4 N to 50 kN bipropellant and 0.5-500 N monopropellant engines (Trauen site), and a surface tension tank propellant behaviour simulation facility.

MAN Technologie AG

Studies with MT acting as ESA's prime contractor include structural components such as apogee boost motor casings, CFRP shells for Rosat, Italsat high-pressure gas tanks, the CFRP structure for Orfeus, Silex structures, and radio astronomy high-precision reflectors.

For Ariane 4, MT Space Systems provides the stage 1 thrust frame, the load-bearing gas generator water coolant tank (fibre composite/aluminium, 30 per cent/150 kg less than conventional design), Viking turbopumps and gas generators, aft/forward booster adaptors, and booster separation and release mechanisms. In December 1996, MT took over the production of the second-stage main tank.

For Ariane 5, MT builds the flow-turned steel booster casings in Augsburg, the main tank's front skirt, the GAT/GAM high-pressure vessels for TVC of Vulcain and booster nozzles, Vulcain's gimbal joint, and the steering mechanism's heat shields. In December 1996, MT took over the production of the main tank bulkheads. MT has production facilities in Augsburg, Munich and Oberpfaffenhofen.

MT played a leading role in the development of Hermes' thermal protection system, drawing on its expertise in ceramic composites, high-temperature materials technology and complex system modelling and testing. MT also developed methods and tooling for manufacturing the hot structural elements. The experience is reflected in its development of Sänger's hypersonic intake ramp model.

MT upgraded and extended its know-how in the fields of development and the production of components incorporated into space transportation systems, in order to qualify itself to be a 'general contractor' for entire systems. The company being commissioned with the overall responsibility for the co-ordination of the configuring of Germany's portion of work to be carried out on the European Crew Transport Vehicle's (CTV) systems represents an initial step towards that goal. MT was also able to secure orders for key components (structures, tanks) of the Automated Transfer Vehicle (ATV) project.

MT has delivered approximately 700 Viking turbopumps and associated gas generators at a production rate of 60 units annually. The two single-stage centrifugal pumps use eight-blade steel wheels to move 270 kg/s propellant while raising its pressure from 5 to 57 bar. The Curtis velocity-compounded impulse turbine incorporates two rows of rotating blades to drive the shaft from the gas generator's water-cooled exhaust. The generator's spherical chamber is cooled by eight water jets facing the propellant's 40 injection ports. The turbopumps also serve as part of the thrust structure.

INDIA

Alfa-Laval (India) Ltd

In the space field, the company undertakes precision fabrication of liquid engine components.

Larson & Toubro Ltd

Founded in 1949, Larson & Toubro specialises in precision fabricated and machined components. It provides the motor casings in 15CDV6, MDN250 and T16AL4V, fuel tanks and gas bottles for India's PSLV and GSLV satellite launch vehicles.

INTERNATIONAL

Europropulsion

Europropulsion was established in 1985 by France's SEP and Italy's BPD to bid for solid motor development contracts within Europe's civil space programme. Its principal activity is Ariane 5's P230 strap-on boosters. The FFr3.8 billion fixed price contract, following two preliminary awards, was signed with CNES in late 1989, covering development/qualification of the motors, including 10 ground tests to mid-1995. Individually, BPD provides the lead for segment and igniter development/integration, SEP contributes the nozzle assembly and insulation liner, and sub-contractors MAN Technologie provide the segment casings, SNPE propellant development and Regulus (the SNPE-BPD partnership) for the two large segment casting operations at Kourou. The small forward segment is completed by BPD in Italy. The EUPERA SNPE/BPD partnership produces the ammonium perchlorate at SNPE's Toulouse site (the AP for 501/503 was provided by Kerr McGee Corporation of the US and for 502 by SNPE; future division remains to be decided). SEP inaugurated its nozzle production plant (UTB: Usine Tuyère Booster) in October 1990 at Haillan. A FFr180 million investment, it will fabricate 10 nozzles annually by 2000. MAN commissioned its production centre in Augsburg during September 1988: this 5,400 m^2 facility includes an 1,800 m^2 air conditioned area for final machining of the clevis/tang connection between the booster segments. The 500 tonne counter-roller flow-forming machine rolls the initial 40 mm thick, 1 m high, 3 m diameter steel cylinders into 8 mm thick 3.5 m high segments before heat treatment produces an ultimate strength of 1,500 N/mm^2.

The first demonstration motor, a reduced scale version loaded with 15 tonnes of propellant, was fired successfully in December 1989. The first full scale P230 was successfully tested 16 February 1993. Seven full scale tests were then planned: B1 with battleship (thick) casings, five (M1-M5) development and two (Q1-Q2) qualification. The last two tested the entire stage. M1, the second firing, was successful 25 June 1993. This was the first in-flight configuration, confirming thrust profile, intersegment seals and nozzle gimballing. M2 was planned for mid-November 1993 but voids were found in the motor's propellant and that motor was shelved (it is expected to be fired after an ageing programme). The third firing, M3, was successful 20 June 1994. M4, with a flight-type actuation system and lighter thermal protection, was successful 30 September 1994. Flight-standard M5 was successful 15 December 1994. Q1 was successful 10 March 1995, Q2 21 July 1995.

International Space Technology, Inc (ISTI)

Space Systems/Loral and Russia's OKB Fakel and RIAME Research Institute of Applied Mechanics & Electrodynamics formed the ISTI joint venture in 1992 to market Fakel's SPT-100 0.08 N Hall xenon stationary plasma thrusters using western electronics. They provide a 500 kg mass saving over a conventional system on a 3.5 tonne satellite over a 15-year life. The venture was later joined by SEP and Atlantic Research Corp. ISTI's thrusters became available in 1995. Fakel has built approximately 100 of the devices, which have flown on approximately 50 Meteor polar meteorological satellites since 1972 to provide orbit control. Gals 1, launched January 1994, is carrying a set of eight to provide the first NSSK on a Russian satellite. SEP joined ISTI in 1993, receiving commercialisation rights in Europe in exchange for the other partners to have equivalent rights in North America and Russia on the Mk 2 version that SEP is developing. This will increase SI from 1,650 to 2,200 seconds. Development began in May 1993 at SEP Villaroche. France's 1999 Stentor satellite was to be equipped with them, but technical problems mean the satellite will be equipped with only slightly modified versions.

Lockheed-Khrunichev-Energia International, Inc (LKE) / International Launch Services (ILS)

After the merger of Lockheed and Martin Marietta in 1995, ILS was created to handle the joint commercial marketing of Proton and Atlas for the Lockheed Khrunichev Energia venture and Lockheed Martin Commercial Launch Services (LMCLS). LKE was formed in early 1993 as a joint venture to market Russia's Proton launch vehicle and associated launch services. LMCLS had been doing business since 1988 as the commercial marketing company for the Atlas launch vehicle.

The ILS services cover spacecraft integration, Proton and Atlas supply, mission management, insurance brokering, launch site support, post-mission support and customer support.

Regulus

Regulus is a partnership between France's SNPE (40 per cent) and Italy's BPD Difesa e Spazio (60 per cent), formed to operate Kourou's Ariane 5 booster propellant facility Usine Propergols en Guyane (UPG). The plant, comprising 40 buildings totalling 26,000 m^2 on a 300 hectare site, was commissioned 24 October 1991. Cost was AU180 million (about FFr1,250 million). Each strap-on contains 237 tonnes of HPBT solid in three sections. Regulus provides the propellant required for the mid and aft units, while the forward element is shipped from BPD. When the facility became fully operational in 1995, its 150 personnel were able to support eight flights a year, requiring 3,300 tonnes of propellant for 32 segments. The first segment was poured November 1991 for the B1 battleship motor, tested 16 February 1993.

Sea Launch Co, LDC

The Sea Launch joint venture was created in 1994 for three-stage Zenits to be shipped to the US, assembled and then transported into international waters for near-Equator launches. The consortium comprises Boeing Commercial Space Co, Norway's Kvaerner group, NPO Yuzhnoye and RKK Energia. Hughes Space and Communications Company, placed a firm order December 1995 for 10 launches. With options, the contract could be worth around US$1 billion. In early 1997, there were 13 firm orders from Hughes and five from Loval, with Hughes holding additional options. The first launch with a demonstration payload took place in March 1999. Kvaerner Rosenberg converted a 30,000 tonne oil drilling rig into the ocean-going launch platform in Stavanger under a US$78 million contract; Kvaerner Govan in Scotland built the 34,000-tonne 200 m Assembly & Command Ship (ACS) for US$93 million.

ISRAEL

Rafael Manor

Rafael's Propulsion Dept (RPD) specialises in the R&D and production of solid, liquid and air-breathing propulsion systems. It developed Shavit's AUS-51 solid motor (marketed in the US by Atlantic Research Corp) and Ofeq's hydrazine AOCS. This module provides the means for despin, final orbit positioning, station-keeping and MW unloading. It comprises hydrazine thrusters, tank, latch valves, pressure transducers, filters, thermal control and fill/drain valves.

Israel Military Industries Ltd (IMI)

IMI is a government-owned company comprising an HQ group and four operating groups of 10 divisions and 18 plants. There are 4,000 personnel. The Rocket Systems Division (RSD) of the Systems Group develops/manufactures solid rocket motors, weapon systems and metal/composite material products. For example, it provides the propulsion systems for the Gabriel sea to sea and Shafrir air-to-air missiles. In the space field, it developed/produces the stage 1/2 motors for the Shavit/NEXT launchers. These are HTPB cylindrical grains in filament wound composite cases. Steering options are jet vane, LITVC and flexseal. The motors are marketed as the ASTM family in the US by Atlantic Research Corporation. RSD is deeply involved in both Israel's space programme and the SD10 Arrow. The propulsion systems and the pyrotechnics for these programmes are IMI developed and produced.

Rotem Industries Ltd

Rotem is the commercial arm of Israel's Nuclear Research Centre-Negev (NRCN). It is the official hydrazine supplier to the Israel Space Agency, IAI, Rafael and other major users. A full range of aniline-free hydrazine-based fuels for thrusters and gas generators is offered: space grade (more than 99 per cent), monopropellant (more than 98.5 per cent), standard (more than 98 per cent), H92, H80, H70 and blends/derivatives.

Rotem also specialises in vacuum/pressure vessels, space simulators (up to 5 m diameter), vacuum chambers (to 10^{-10} atmosphere), cryogenic thermal systems and thermal shrouds.

ITALY

Alenia Aerospazio

IRIS (Italian Research Interim Stage) is Italy's most powerful upper stage and the first designed for operations from a dedicated cradle in NASA's Shuttle cargo bay. Alenia Aerospazio is prime contractor; BPD is responsible for the IRIS Spinning Stage (ISS) portion of the assembly. The spinning, solid propellant stage is optimised for injecting 900 kg science and small telecom satellites into GTO from the standard Shuttle parking orbit, with a maximum 25 per cent propellant offloading available for 600 kg payloads. IRIS' only scheduled mission, however, was targeted at a lower orbit. The Lageos 2 geodetic satellite, also with Alenia Spazio as prime, was placed in 5,610 × 5,950 km, 52.8° this way in 1992. No further applications have been manifested but it is available for other vehicles. It could also be employed as a planetary kick stage with a launcher such as Atlas Centaur.

The Shuttle ASE is now stored in Turin.

The IRIS system is designed for a Shuttle load factor of 0.125 (⅛ of cargo bay volume), with the ASE – essentially a cradle providing structural/thermal protection, power/command links and spinup motors – designed for a minimum of five missions. The ASE interfaces with the Shuttle via four lateral and one keel fittings. The 4.5 m diameter, 2.1 m deep cradle restrains the IRIS Spinning Stage (ISS) with two 'grabber' devices on the Payload Attach Fitting and releases them for the 45 to 100 rpm spinup with redundant electric motors. All the deployment electronics are mounted in the ASE, with the computer responsible for commanding spinup and severing of the base clamp by explosive bolts for four springs to impart 0.4 m/s. An aluminium platform around the motor's dome end supports stage subsystems such as avionics/telemetry systems. Separation switches trigger the stage sequence. 128 channels provide telemetry at 2 kbit/s on 2.2270 GHz through the 5 W transmitter for 50 minutes. BPD's Nutation Control System uses N_2 for controlling coning during coast. All coast operations are commanded by two redundant electrical sequencing systems, finally releasing a yo-yo tumble mass to avoid payload contact. Motor ignition occurs about 45 minutes after deployment some 50 km from Shuttle.

Payload dynamic envelope is a 1,900 mm diameter, 2,455 mm high cylinder. The Payload Attach Fitting 937.62 mm diameter upper flange mates with the payload's fitting with a two-sector clamp band, separated pyrotechnically; four springs provide 0.5 m/s separation rate. Two 37-pin flyaway pins provide electrical interfaces at the clamp band level.

BPD Difesa e Spazio

BPD Difesa e Spazio, wholly part of FiatAvio since January 1996, is involved with propulsion systems, including the study and production of solid, liquid and gaseous propellants, as well as studies on magneto-plasma dynamic systems. This involvement includes the design, development and manufacture of apogee, stage separation and space launcher motors, strap-on boosters, orbital transfer systems and attitude control equipment. 1995's space revenue was projected at Lit280 billion (1994 Lit262 billion; 1993 Lit323 billion). It is Europe's seventh largest space company.

BPD has provided bi/monopropellant and gaseous attitude control systems for satellites such as Italsat 1/2, Lageos 2, SAX and Artemis. It provided the propulsion systems for Sirio and GEOS 1/2, and the motors for the European Mage family of apogee motors for which it is responsible for the propulsion and thermal protection, as well as overall integration/testing This motor was selected for ESA satellites such as Giotto, Telecom, ECS 2-5, Meteosat, Hipparcos and Lageos 2. BPD also provided the motor for Italy's IRIS upper stage. A large propulsion unit, known as ALFA, was also developed and, as a result of this programme, BPD designed and produced the Ariane 3/4 solid propellant strap-on boosters, and is a partner with SEP in Europropulsion, responsible for developing Ariane 5's P 230 solid boosters. Regulus is a partnership between BPD and France's SNPE formed to operate Kourou's Ariane 5 booster propellant facility. Each strap-on contains 237 tonnes of HPBT solid (68 per cent AP, 18 per cent aluminium, 14 per cent binder) in three sections. Regulus provides the propellant for the mid/aft units, while the forward element's 23 tonnes is shipped from BPD. The facility became fully operational in 1995.

FiatAvio SpA

FiatAvio entered the space propulsion field in 1985 as an extension of its primary aero gas turbine business. Under SEP contract, it designed, developed and manufactures Vulcain's LOX turbopump. Preliminary development of Vulcain Mk 2's pump, increasing power from 3 to 5 MW, was underway for testing in 1998. Test facilities accommodate specialised testing of critical cryogenic components. It is collaborating with DASA and Techspace Aero on ESA's FESTIP studies, and it is performing R&T studies for advanced cryogenic turbomachinery under the Euro/Russian RECORD program led by SEP. Previous activities include studies of component development for 30 to 80 kN liquid engines for orbit propulsion modules and launcher upper stage applications. FiatAvio is a stock holder of Intospace and Arianespace, and since January 1996 has held 100 per cent of BPD Difesa e Spazio.

JAPAN

Daicel Chemical Industries Ltd

Daicel makes rocket motor and propellant products. Employees total 300.

IHI Company Ltd

In space propulsion, IHI provides the cryogenic turbopumps for the H2's stage 1/2 cryogenic engines, in addition to H2's hydrazine Reaction Control System modules. It previously provided the pumps and gas generator for H1's stage 2 engine. ISAS, IHI, KHI and MHI are pursuing development of the ATREX-500 air turboramjet, which could power the flyback booster of a two-stage spaceplane. The hydrogen expander ATREX would take the vehicle to M6/30 km. The 5.4 kN ATREX-500 ¼-scale version, 2.2 m long and fan inlet 30 cm diameter, totalled 1,400 seconds in 40 static sea level runs during 1990-95 at ISAS' Noshiro Testing Centre. Efforts now centre on a 1.4 tonne flying testbed launched from Japan and parachuting into the sea. A jet engine will take the FTB to M0.5 from a horizontal rail for the 370 kg ATREX-500 with 190 kg of LH_2 to attain 25 to 35 km altitude 200 km downrange. The initial metal turbo machinery will achieve M4.5, with M6 targeted for the carbon-carbon composite design. Goals for the composite ATREX are 17.0 to 18.9 kN and 3,150 to 3,300 seconds SL static.

IHI's Rocket Test Centre was established in October 1975 at the Aioi Plant as a cryogenic test centre of IHI's Research Institute. It was incorporated in July 1980 into the Space Development Division for testing the cryogenic engine and bipropellant apogee engines.

IHI developed the 2 kN bipropellant apogee engine under NASDA contract (in co-operation with NAL) for ETS 6 and COMET. IHI is independently developing a 10 kN CUS Cryogenic Upper Stage LOX/LH_2 expander cycle engine for upper stage or orbital transfer vehicle applications. SI 471 seconds, chamber pressure 28 atmospheres, LOX turbopump 43,000 rpm, hydrogen turbopump 85,000 rpm.

UPS provides GEO insertion and attitude/orbit control for 2 tonne-class satellites; it is derived from the LAPS of ETS 6. It carries a single 1.7 kN bipropellant Apogee Kick Engine plus 4 × 50 N + redundant 8 × 0.86 N hydrazine catalytic thrusters. COMETS's flight model was shipped March 1996.

Iwatsu Electric Company, Ltd

Iwatsu performs rocket motor check out and hot firing test, control and telemetry/data acquisition systems. The company provided H1's propulsion system console and its engine data acquisition system.

Mitsubishi Heavy Industries Ltd

Mitsubishi provides Japan's primary capability in liquid rocket engines, most recently developing the cryogenic oxygen/hydrogen LE-7 stage 1 and LE-5A stage 2 engines for NASDA's H2 launcher. The company has prime industrial responsibility for H2 (as it did for H1). The 10-year US$108 million LE-5 development programme produced Japan's first operational LOX/LH_2 engine, providing H1's stage 2 with a restart capability. The technology formed the basis for developing the larger and more advanced LE-7 and the LE-5A direct outgrowth.

MHI also fabricated Rocketdyne MB-3 engines under licence for the H1 and produced the indigenous LE-3 storable propellant engine for the seven N1 rockets. Two advanced smaller engines, the 9.8 kN RE6 and 2.94 kN RE10-300, have been tested for possible applications in future orbit transfer vehicles.

MHI is testing Liquefied Air Cycle Engine (LACE) hardware, in which LH_2 is used to liquefy atmospheric oxygen, using the LE-5 as a demonstrator. Mitsubishi's Tashiro Test Field includes stands for LE-5/5A/5B engine and stage sea level firings, a short-duration LE-7 stand and an H2 stage 1 battleship tower (full-duration firings of flight-type engines and stages are undertaken at Tanegashima's H2 launch complex). MHI completed a 105 m long 30 m wide 20 m high assembly/test facility at its Nagoya plant for the H2 in May 1988, allowing stages to be shipped from Nagoya bay to the launch site. MHI is supplying stage 1/2 LOX tanks for McDonnell Douglas' Delta 3. MHI is also providing the injectors for Pratt & Whitney's RL-10B-2 engine.

Nippon Propellant Industry Company Ltd

Nippon does R&D and production of anhydrous hydrazine, MMH, UDMH and hydrazine decomposition catalyser for launcher and spacecraft.

Nissan Motor Company Ltd

Nissan's Aerospace Division is Japan's primary developer/manufacturer of solid propellant rocket motors and launch vehicles. These include the M-5 satellite launcher, the MT-135P, S-310, S-520 and K-9M sounding rockets, boosters for the H2, and several kick stages. The company is also involved in propulsion studies for Japan's spaceplane and is responsible for the experiment logistics module exposed section of the Japanese Experimental Module contribution to NASA's Space Station.

Manufacture/assembly is undertaken at Nissan's Ogikubo plant (planned to move to Tomioka City); the Taketoyo Test Facility opened in 1980 for vacuum firings of spinning upper stage motors. Taketoyo's Spin Static Firing Test Stand provides a 39 to 78 mbar vacuum for 1.8 m diameter motors of 0 to 147 kN at spin rates up to 150 rpm. The company provided test equipment for the H2 solid booster static stand at NASDA's Tanegashima site. R&D activities are conducted at the Kawagoe plant, including research into liquid bipropellant satellite thruster systems, ram propulsion and gimballing TVC. The company is also a major supplier of pyrotechnic devices for space applications.

Nissan developed the four-segment solid strap-ons for Japan's H2 launcher, with four demonstration firings during 1988-91 at NASDA's Tanegashima Space Centre. First test was conducted April 1988, prototype number 1 was fired June 1989 and prototype number 2 13 December 1989; the qualification model firing was achieved 29 May 1991. The units represent the country's largest and most powerful indigenously developed solids. The improved H2's SRBs will be reduced to three segments, debuting in 1999. Upgraded H2A, from 2001, will use a monolithic composite case design.

NOF Corporation

NOF has participated in solid rocket propellant production since 1954 and is now the sole supplier to NASDA/ISAS. Plant for satisfying H2 booster requirements was installed 1987. The company also manufactures pyrotechnics.

Rocket Systems Corporation

RSC was incorporated on 5 July 1990 by 75 Japanese aerospace companies, insurance organisations, banks and trading companies. Its original business was to block buy H2 launchers from the suppliers for selling on to NASDA for national missions. With H2's major contractors, RSC has expanded to cover production management and launch preparation activities at Tanegashima, in addition to marketing H2 commercially. RSC also offers the TR-1A suborbital μgravity vehicle. H2 prime contractor Mitsubishi Heavy Industries leads the corporation, with Nissan Motor Co, IHI, NEC, Kawasaki Heavy Industries, Fujitsu Ltd, Japan Aviation Electronics Industry and Toshiba Corp. RSC has concluded launch services agreements with two satellite suppliers: both were in 1996, the first being with Hughes Space and Communications for 10 launches during 2000-05 (plus additional optional launches) and the other is with Space Systems/Loral for 10 launches during 2000-05.

Showa Denko K.K.

SDK manufactures LH_2 at its Oita Petrochemical Complex for NASDA's H2 launcher, and in 1986 formed a joint company, Pacific Hydrogen Co Ltd. (PHC), with L'Air Liquide SA of France and Teisan Kabushiki Kaisha to create a 7.4 million litre annual production capacity. Hydrogen gas of 96 per cent purity is generated as a by-product at an adjacent ethylene plant and purified to 99.9999 per cent or better by pressure swing and cryogenic adsorption techniques. The gas is then liquefied at −253°C by Claude's process by passing it through Freon/LN_2 coolers and into a cold box comprising a multi-stage catalyst bed and coolers. PHC operates 11,000 and 18,000 litre insulated vehicles for shipping.

NETHERLANDS

Aerospace Propulsion Products BV

APP is owned equally by TNO and Stork. APP, Stork Product Engineering BV and TNO 1986-95 developed Vulcain's starter and igniters. APP is responsible for complete production as a subcontractor to Stork. Following five years of R&D, the feasibility of using Hydrazinium NitroFormate (HNF) in high performance solid propellants (7 per cent greater than current AP/Al/HTPB) was demonstrated in 1995. APP, under ESA contract, is preparing for the industrialisation of HNF propellants. APP's pilot plant produces 5 kg batches.

RUSSIAN FEDERATION

Eurockot Launch Services GmbH

The commercial Rockot launcher, derived from the RS-18 2-stage ICBM, is marketed by Eurockot, the 10-man Khrunichev/DASA 49/51 per cent joint venture announced 23 March 1995. Eurockot is responsible for all commercial activities, acting as the customer's prime contractor, but Khrunichev develops and produces all of the hardware. The first revenue launch is planned for the fourth quarter of 1998 from Plesetsk, using a modified Cosmos pad, now designated exclusively for Rockot. Launches will be from a tube, replicating a silo, on stage 1 skids jettisoned pyrotechnically.

There have been three demonstration launches from Tyuratam, all silo-based. The first two (20 November 1990; 20 December 1991) were suborbital, demonstrating basic vehicle performance and KB Salyut's new Breeze stage 3 (inactive on the first ascent). The first orbital flight was 26 December 1994, into 1,875 × 2,253 km 64.80° carrying the 70 kg RS-15 Radio-Rosto amateur relay. Commercial Rockot carries a large new fairing developed by Khrunichev. Breeze was originally designed to dispense multiple communications satellites into low orbits in the event of nuclear conflict. The RS-18s would have resided in silos awaiting launch on demand. The derived Breeze-M is being developed as Proton's new stage 4.

KB Khimautomatiki

The bureau of Semyin A Kosberg (1903-65), formed 1941, began investigating liquid propellant rocket propulsion for aircraft in the early 1950s and later became responsible for some Soviet space launcher upper stages. In 1974 it was reorganised as KB Khimautomatiki.

KB Khimmach

A bureau headed by Alexei M Isayev was established in 1943 for research into storable liquid propellant rocket engines. Whereas Glushko's GDL concentrated on major launcher engines and Kosberg's bureau on upper stage propulsion systems, Isayev's produced smaller spacecraft engines, including manned, Earth orbit and deep space systems. A number of designations have been adopted but 'KTDU' (Korrektyroushchaya-Tormorznaya Dvigatelnaya Ustanovka, which means "Corrective-Braking Rocket Motor") is most frequent. Information released in early 1991 indicated the bureau is part of NPO Soyuz in Kaliningrad. It is responsible for the 73.58 kN cryogenic engine of the new Proton KM version, originally designed for the L-3M manned lunar mission, and recently sold to India for its GSLV vehicle. A newly-developed DMT-600 600 N NTO/UDMH thruster was referred to in 1994. Khimmach developed all of the engines for the Heavy Cosmos TKS transport craft.

Starting in 1971 and at the instigation of Isayev, the bureau started the development of low-thrust engines which could be used for attitude control of spacecraft. A series of eleven engines with thrusts up to 2,206 kN (225 kg) and eight with thrusts up to 49 kN (5 kg) were developed. They are capable of burns lasting from hundredths of a second through to hundreds and thousands of seconds.

Korolev Design Bureau

Korolev's bureau developed Proton's Block D/DM and Molniya's Block L engines. Mikhail Melnikov designed the Proton engine.

N D Kuznetsov Company SSC

Nikolai Dmitrievich Kuznetsov (24 June 1911-30 July 1995) established an aircraft engine design and development bureau during the 1940s in Kuibyschev (now Samara). Information finally released in 1989 revealed that the bureau had been assigned the responsibility for the design and development for engines for Korolyov's four-stage N-1 launch vehicle. Previously the bureau had developed engines planned for use on Korolyov's R-9 missile, but politics ensured that the missile used other engines.

In July 1993, Aerojet signed a teaming agreement with N D Kuznetsov SSC to market these engines within the United States. Two NK-33 engines were delivered to Aerojet in 1995. One of them was used by Aerojet for five firings which totalled 410 seconds operating time within a 57 to 114 per cent thrust range which followed the proposed Atlas-2AR launch profile. The engines were under consideration to power the first stage of the Atlas-2AR, but they lost out to the Energomash RD-180 engine.

The United States Kistler Aerospace Corporation is planning to use NK-33 engines on its reusable K-1 launch vehicles: three would be on the first stage and one on the second stage.

NII Machinostroenye R&DIME

Research & Development Institute of Mechanical Engineering was established in 1958 as part of NII Thermal Processes, but became independent in the early 1980s. Development and production of 0.8-400 N spacecraft thrusters (implying connection with old Isayev bureau), including Soyuz, Progress and Mir. Development and small batch production of engines up to 6 kN can be accommodated. Test facilities can handle hydrogen (up to 3 kN), methane and kerosene fuels. The Elektro Earth observation satellite carries 16 DEN-16 electrothermal thrusters. For the thrusters specified below, LTRE means Low Thrust Rocket Engine, or RDMT in Russian (Raketnay Dvigatel Maloi Tyagi).

NII Thermal Processes

NII TP performs research in all forms of launcher and space propulsion, including ramjet, electric and nuclear. BMDO took delivery in October 1993 of a T-100 1.5 kW Hall-type Stationary Plasma Thruster for evaluation. N_2O_4/UDMH thrusters include 10 and 1,000 N models. NII TP developed the Phobos spacecraft's network of hydrazine thrusters for attitude control and minor orbital modifications. 28 (24 × 50 N, 4 × 10 N) were mounted on four spherical tanks. A set of 0.5 N hydrazine microthrusters of an unusual design was also included to avoid the need for separate N_2 cold gas jets; the propellant was burned internally and cooled inside an expansion tank before passing to the thrusters.

NPO Energomash

The history of NPO Energomash can be traced back to 15 May 1929 when the Group for the Development of Rocket Engines was organised at the Gas Dynamics Laboratory (GDL) in Leningrad (now St Petersburg). For more than 60 years the organisation repeatedly changed its name and location, but remained at the forefront of rocket engine design. The founder and the permanent head of the GDL as Academician Valentin P Glushko, considered to be the initiator of the Russian rocket engine industry.

In May 1974, the GDL merged with Sergei Korolyov's design bureau, the new organisation being re-named NPO Energiya, with Glushko as its head. Following Glushko's death in January 1989, the former GDL left Energiya to become NPO Energomash, continuing to concentrate on the development of rocket engines: the bureau was headed by Vitali P Radovsky during 1974-1994. The current general director and chief designer is Boris I Katorgin.

NPO Energomash incorporates the design bureau, plant and test facilities, with the capability to test engines with thrusts of up to 10 MN either as components or complete. An agreement was signed with Pratt and Whitney in October 1992 providing exclusive US marketing rights for Energomash's LOX/kerosene and tri-propellant products and technology.

NPO Saturn

The cryogenic engine which has been described in the 1990s as the D-57 would appear to be the 11D57 gimballing engine which was developed during the early 1970s for use on the Block S of the modified N-1 launch vehicle which would have been flown for the dual-launch L-3M manned lunar missions. The bureau was also tasked to develop the 11D54 fixed-chamber cryogenic engine with the same thrust level of 40 tonnes (390 kN). This engine would have been used in a cluster of six to eight on a modified Block V of the N-1 vehicle.

OKB Fakel

Fakel specialises in spacecraft attitude control thrusters, ion engines and plasma sources. Its SPT-70 and SPT-100 (Stationary Plasma Thruster) Hall electric thrusters have flown on approximately 50 Meteor polar meteorological satellites since 1972 to provide orbit control. The numerical designators indicate beam diameter in mm. SPTs employ a DC-gas discharge in an annular chamber, in which a radial magnetic field traps the electrons. These Hall currents ionise the Xe propellant, the ions of which are accelerated inside a quasi-neutral plasma, without grids, and by the discharge voltage itself. The advantages are their rugged simplicity (no grid system or high voltage supply), but they provide lower efficiencies, lower exhaust velocities and increased Xe consumption than ion thrusters such as the UK-10.

SWEDEN

Volvo

Volvo specialises in the design, development and production of rocket engine and hypersonic propulsion systems for space applications. 1996 turnover SKr237 million; 1995 space business totalled SKr232 million (1994 SKr181 million; 1993 SKr142 million; 1992 SKr124 million; 1991 SKr221 million; 1990 SKr150 million). Space personnel total 150. Current products include the combustion chambers/nozzles for Ariane 4's Viking 4/5 engines, and production of the turbines (subcontracted by FiatAvio for the LOX turbine and by SEP for the LH_2 turbine) and nozzle extension for Ariane 5's cryogenic Vulcain. Volvo is currently developing the new nozzle and Lux turbine for the Vulcain 2, in co-operation with SEP/FIAT. The company also has an agreement to deliver nozzle skirts for the LE-7B engine to be used on the Japanese H2A launcher. Research activities include combustion, hypersonics and turbopumps and nozzles for storable propellant upper-stage engines and MTU.

UKRAINE

NPO Yuzhnoye

Yuzhnoye produces verniers for some of its own vehicles (Zenit stage 2 and Tsyklon stages 1-3) and the 11D25 main engine for Tsyklon's stage 3.

UNITED KINGDOM

Air Products PLC

A wholly owned subsidiary of Air Products and Chemicals of the US the company provides liquid oxygen/hydrogen and other gases, plus cryogenic engineering expertise.

Dowty Aerospace Wolverhampton

DAW is a member of the Dowty Aerospace Division of the TI Group, which comprises more than 50 operations worldwide. The Space Projects department manufacturers diaphragm, capillary and pressurant tanks.

DAW's space manufacturing facilities were opened in 1989 to provide a fabrication capacity for tanks up to 120 × 150 cm long. The 400 m² class 100,000 cleanroom, with 150 m² test area, includes a computer-controlled inert gas welder for welding titanium tank shells, computer-controlled resistance welder for fabricating internal components, and orbital tube welder for titanium and stainless steel piping. Test facilities include ultra-sonic cleaning, bubble testing, helium leak testing, pressure and expulsion testing, cryogenic proof pressure testing, dye penetrant and radiographic inspection.

The capillary tanks are surface tension propellant management device tanks and are under contract for MMS, Hughes HS-601 and BPD.

DAW manufactures diaphragm tanks for mono hydrazine systems. ESA's Infra-red Space Observatory carries them and Orbital Sciences Corp contracted in 1991 for the 96-litre titanium alloy elastomeric diaphragm hydrazine AOCS tanks for SeaStar. In addition, DOW provided 40 titanium alloy centrifuge tanks for Cluster.

DAW is under contract by MMS UK to supply Xe tanks for the UK ion thruster system on ESA's Artemis. Net volume is 30 litres and it can store 44 kg of Xe at 125 atmosphere.

The company is teamed with Vacco Industries to manufacture titanium etched disc filters, latch valves fill drain valves and non-return valves for the European market.

Fairey Microfiltrex Ltd

The company designs, develops and manufactures filtration equipment for fuel, hydraulic, gas and oil systems. Fairey Microflex has expertise in spacecraft propulsion systems and has supplied the filter element assemblies for the hydrazine thruster reaction control systems of ESA's ISO and Soho satellites. Ariane 4 and 5 also utilise the company's filters.

Reaction Engines Ltd

Reaction Engines is designing the Skylon SSTO spaceplane to deliver up to 12 tonnes into LEO. Reaction Engines is conducting supportive investigations in conjunction with Bristol University and AEA Technology Harwell to investigate some of the more technically contentious features. Project funding is sought; Reaction Engines has brought together an interim UK working group before the formation of an industrial consortium, which will eventually be expanded to include international partners.

Rotadata Limited

Rotadata has an active programme of developing measuring techniques in turbomachinery and aerospace development hardware. Following the design of a number of fully-instrumented full-scale model turbines for the Space Shuttle Main Engine turbopump, the company has developed the following new products, both hardware and software, for applications to aerospace turbomachinery development and monitoring.

Rotodata new product line features a new high-performance pyrometer sensor with a better sensitivity and a high-frequency bandwidth for turbine blade surface temperature mapping at full rotation speed. They make turnkey digital and analogue telemetry systems for rotor instrumentation, and high-temperature strain-gauging for turbine blades and discs. In other NDT applications, they make miniature capacitance probes for blade tip clearance and rotor orbit monitoring and an optical smoke density meter for engine emissions.

The company is also working on a novel CO_2 sensing unit for cabin condition monitoring. Meanwhile, the traditional Rotadata activities in turbomachinery instrumentation, aero-thermal rakes and probes, specialist positioning systems and high quality instrumented test hardware continues to provide a core business from an increasing range of customers.

UNITED STATES OF AMERICA

Aerojet

A segment of GenCorp of Akron, Ohio, Aerojet was founded in 1942 by astronautics pioneer Dr Theodore von Karman and initially developed Jet-Assist Take-Off (JATO) rockets for aircraft. It was the first US company to produce storable bipropellant rockets. The propulsion plant, as Aerojet TechSystems (formerly Liquid Rocket Co), provided liquid engines for Aerobee, Bomarc, Nike, Vanguard, Delta, Atlas Able, Titan 1 and Apollo, and continues to manufacture Titan, Delta and Shuttle engines. Aerojet, Rocketdyne and P&W formed the Space Transportation Propulsion Team in 1990 for work on the STME Space Transportation Main Engine.

Aerojet manufactured the solid motors for MX, Minuteman, Small ICBM, Standard Missile (rolling out the 10,000th motor in June 1989) and Hawk, and successfully bid in partnership as Aerojet Space Boosters (becoming Aerojet ASRM Division) with Lockheed for NASA's Shuttle Advanced Solid Rocket Motor contract. Congress cancelled ASRM in October 1993. Propulsion activities are performed at the 52.6 km² Sacramento plant, founded 1951 and capable of testing cryogenic and storable propellant motors up to 6,670 kN. The Mach 8/30 km-altitude hypersonic Hytest facility was added for National AeroSpace Plane engines and components tests of up to 35 seconds, and the company has access to the Hypulse M25 facility at subsidiary General Applied Sciences Laboratories Inc (Ronkonkoma, New York), which it acquired in 1989.

Aerojet teamed with Lyulka in 1993 to improve and market the 400 kN LOX/LH_2 D57 engine, originally developed in the 1960s for the improved N1-L3M manned lunar mission. It is a candidate for an SSTO demonstrator. A teaming agreement was signed July 1993 with Kuznetsov/NPO Trud to use the NK series exN1 engines and their technology in the US market. Aerojet made five tests totalling 410 seconds with an NK-33 in October and November 1995. If it had been selected for a US vehicle as the AJ26-NK33A (it lost to the RD-180 for Lockheed Martin's new Atlas 2AR), a US production line would have been established. Unit cost would have been about US$4 million.

A similar agreement was signed October 1994 with KB Khimautomatiki for the RD0120, principally developing it into a tripropellant engine. Work is being performed under a US$17.2 million NASA Marshall contract.

Air Products & Chemicals, Inc

Provision of LH_2/LOX propellants and LN_2/LHe for propulsion systems. Air Products Europe offers a similar service.

Alliant Techsystems, Inc

Alliant Techsystems acquired Hercules Aerospace Company in March 1995 for US$296 million. Hercules was selected in October 1987 as the propulsion contractor for Titan's Solid Rocket Motor Upgrade programme, potentially worth US$725 million for the development, qualification and production of 15 sets of Titan 4 solid boosters. Lightweight graphite composite materials replaced the steel used in current motor cases, and a high-performance propellant similar to that developed for the Delta 2 strap-on is employed. PQM-1 (Preliminary Qualification Motor) was the first full-scale test, but the motor failed after a few seconds at Edwards AFB 1 April 1991. The added PQM-1 firing was successful 12 June 1992. The QM-2 third test, run at 2.5°C, was successful 21 February 1993. QM-3 at the 41°C upper limit was successful 2 June. The final QM-4, 12 September 1993, would allow flight introduction late 1995, although debut was expected fourth quarter 1996. Delivery of the first flight set was made first quarter 1994. Lockheed Martin awarded the company a contract for Titan launch vehicles support which extends to 2003.

The company was awarded a McDonnell Douglas contract in 1987 to develop the stretched solid rocket GEM Graphite Epoxy strap-on Motors for Delta 2; they first flew November 1990. The initial contract for 144 motors (16 flight sets) was followed in 1991 by a second for 117 (13 sets) beginning production in 1993, and a third in June 1995 for 144 (16 sets). GEM production rate is six per month. For Delta 2, from 1996, the expansion ratio for the airlit nozzles will be increased from 10.6 to 16.3 by lengthening the nozzle by 30 cm and increasing its diameter by 19.5 cm. The GEMVN version incorporates a vectorable nozzle; the qualification firing was made May 1994. The TVC system is provided by AlliedSignal and the nozzle by BP-Hitco as part of the strategic partnership. Delta 3's nine strap-ons will be lengthened by 1.22 m and their diameters increased to 116.8 cm. Three of the six ground lit motors will carry gimballed nozzles for TVC, and the three airlit will improve performance with extended nozzles. The first of 16 sets ordered June 1995 were delivered late 1997.

Alliant produces Pegasus' Orion solid motors, derived from GEM, and the fairing. Taurus also uses the Orion motors. Hercules also designs/fabricates composite structures: spars, struts and optical benches used in satellites.

Ardé Inc

Ardé was formed in 1952 by engineers from the Rocket Group of M W Kellog. In the space field it specialises in lightweight pressure gas storage vessels and propellant tanks. Programmes include Shuttle/Centaur, Mars Observer, Atlas 2 and GOES helium pressurant tanks, GOES satellite and Shuttle SRB TVC auxiliary power hydrazine tanks, and Shuttle EMU backpack breathing oxygen tank. The product lines include cryoformed 301/304L pressurant and propellant tanks, prestressed composite fibre over-wrapped metal tanks, and ring stabilised and apex rolling diaphragm expulsion devices.

Cryoformed 301/304L Tanks Under-sized 'preform' tanks are conventionally fabricated from austenitic stainless steel and then immersed/filled with LN_2 before pressurisation to a level about equal to its room temperature design burst pressure. The membranes/welds are stressed beyond the 195°C yield and the vessel plastically stretches: 7 per cent in spheres and 15 per cent for cylinders. The temperature causes a transformation to martensite steel in both parent metal/welds with an attendant 5:1 increase in room temperature yield strength. approximately 3,000 flight tanks using this process have been delivered over 20 years. The Shuttle EMU backpack O_2 tank is a 36.3 cm long, 13 cm diameter cylinder pressurised to 71.4 atmospheres; mass 0.9 kg.

PSC Pressurant Tanks Prestressed composite fibre over-wrapped pressurant (PSC) tanks provide pressure vessels up to 50 per cent lighter than equivalent all titanium tanks. Kevlar or graphite fibre is wrapped on a metal liner and immersed in LN_2 as in the cryoforming process to stress both materials.

Assured Space Access, Inc

Assured Space Access and Russia's AKO Polyot signed an agreement 21 September 1995 to establish the Cosmos USA joint venture for exclusive marketing of Polyot's Cosmos-3M small satellite launcher outside of the Russian Federation and the former Soviet states. In addition to marketing, Assured Space Access serves as the integration services agent between Polyot and commercial customers: support with export licensing and certifications; integration, transport and logistics planning; interface requirements and documentation; provision of cleanrooms, integration area, crew accommodation; satellite integration and launch support.

A request for an exemption from the US Government (USG) policy prohibiting the foreign launch of USG payloads made by NASA was rejected. As a result the Cosmos-3M was ineligible to win the Med-Lite competition. Since the launch vehicle had passed NASA's technical scrutiny it would appear that this was a purely political decision, not based upon the launch vehicle's merit.

Atlantic Research Corporation

Atlantic Research Corporation (ARC), founded in 1949, is a manufacturer of solid propulsion motors and gas generators, and in 1987 expanded into liquid propulsion with the acquisition of Bell Aerospace Textron. The company's Virginia and Arkansas facilities take solid motor concepts from design through to high-volume production; the company's annual propellant production capacity is 9,000 tonnes. ARC produces propulsion units for missiles such as Stinger, Tomahawk, Peacekeeper and Trident 2, but also undertakes space-related projects: the company teamed with Hercules in December 1987 to bid (unsuccessfully) for NASA's Shuttle Advanced Solid Rocket Motor. Israel's 1.3 m and AUS51 'Marble' motor, flown as Shavit's stage 3, is marketed by ARC/Rafael. ARC markets the ASTM family of Shavit and NEXT launcher stage 1/2 motors in the US under agreement with TAAS Israel Industries Ltd. ARC also manufactures rocket motor components, including igniters, initiators, casings and nozzles using advanced materials and filament winding/braiding techniques. November 1987's acquisition of Bell Aerospace Textron added liquid bipropellant attitude control and apogee satellite propulsion systems, and their associated tanks and valves. ARC's 22 N (5 lbf) bipropellant thruster was developed for Intelsat 6's attitude control system and has demonstrated 400,000 pulses, 32 hours aggregate firing and 748 kg total throughput. This thruster is carried by Optus B, UFO and Intelsat 7. As Bell, ARC/LP developed the Agena propulsion system that flew on approximately 370 missions. The engine remains available for 31.1120 kN thrust.

Boeing Space Systems

Boeing signed an agreement 1 August 1996 to acquire Rockwell's space and defence businesses, which was renamed Boeing North American Inc, part of Boeing's Space and Communications Systems group. 1995's combined space revenue was approximately US$3.5 billion.

The former Rockwell's Space Systems Division was one of Rockwell's Aerospace and Defense businesses, along with the former Rocketdyne and Autonetics and Missile Systems. Space Systems designs, develops, builds and integrates space vehicles and supporting hardware. As prime contractor for NASA's Shuttle Orbiters, it also maintains their technical integrity and configuration; assists NASA in the integration of Shuttle system elements; and provides support services, including mission planning and simulation, flight design, mission data production, flight crew/controller training, ground facility engineering and operations support, payload integration, launch operations support, flight analysis, direct mission support, Orbiter turnaround assistance, logistics and engineering support.

Space Systems supports Orbiter logistics operations at KSC, in addition to Orbiter processing and cargo/system integration for NASA Johnson. Also at Houston, the Rockwell Space Operations Company, a subsidiary, provides further Shuttle and advanced programme services. The Florida Operations encompass all work associated with the purchase of Orbiter and Ground Support Equipment (GSE) spares and repair of existing Orbiter and GSE hardware. The NASA Shuttle Logistics Depot is certified to perform Orbiter component overhaul and repair on approximately 3,200 of the Shuttle's 4,000 line replaceable units; 14,864 m² was added to the 10,219 m² facility in 1990. In October 1991, NASA extended Rockwell's logistics contract to 1994 with a US$453 million award, followed by a US$600 million five-year award in October 1994 (with options totalling US$800 million through 2004).

The Orbiters' forward/aft fuselages, crew compartment, forward RCS, vertical stabiliser and secondary structures were fabricated at SSD's 223,605 m² Downey facility. The Orbiters are assembled or modified at the 10,350 m² Palmdale, California site, near Edwards AFB, the only facility capable of handling a complete vehicle. All manufacturing was shifted by 1995 from Downey to Palmdale to save NASA US$20 million annually.

McDonnell Douglas now subsumed into Boeing performs Delta marketing/sales and contracting.

Space Systems is also responsible for Sea Launch, the Inertial Upper Stage, and studies on future space transportation systems.

Boeing Rocketdyne

Rocketdyne was a division of Rockwell International and in December 1996 was included in the Rockwell merger with Boeing. It is devoted primarily to the design and manufacture of rocket engines – it is the Western world's largest producer, operating from its Canoga Park engineering and 16 km distant 10.5 km² Santa Susana test facilities. 1995 revenue was US$880 million; 5,400 personnel. It was established as a separate division of North American Aviation Inc on 8 November 1955 and following its first major propulsion system for the Navaho missile has delivered around 3,500 engines for Jupiter, Redstone, Thor, Delta, Saturn, Atlas and Shuttle. It currently provides the power plants for Atlas and Delta, and continues to test, upgrade and refurbish Shuttle's main engines, still the only large reusable systems in service. It designed and fabricated Peacekeeper's stage 4, its first integration of a full stage.

Rocketdyne was selected in December 1987 to develop Space Station's silicon photovoltaic solar arrays, batteries and power management/distribution. 1994 Station revenue was US$282 million; 1993 US$340 million.

Rockwell's Shuttle Operations Co, established in 1985, was renamed the Space Operations Co in 1989 to reflect a broader range of involvement in space activities, including Space Station. A further change of name took place following the Boeing merger with Rockwell in late 1996. RSOC, a Rockwell Space Systems Division subsidiary, manages Shuttle mission planning, flight design, mission data production, flight crew and flight controller training, ground facility engineering and operations support, and direct mission support. NASA's STS Operations Contract (STSOC), worth US$685 million over four years, was awarded to RSOC in September 1985, effective 1 January 1986. A US$2.3 billion extension through December 2000 was exercised March 1991, bringing total contract value to US$4.8 billion. RSOC became responsible for managing 22 Shuttle operations functions previously performed by 17 contractors.

RSOC was selected in 1990 for Johnson's Operations Support Contract, beginning April 1990. The 10-year cost plus award fee contract is valued at about US$814 million.

E'Prime Aerospace Corporation

EPAC is offering its Eagle S-series solid propellant orbital launchers employing stages derived from Peacekeeper motors. Improvements include changing from Kevlar casings to graphite. The primary launch site would be at the Devil's Ash Pit on the UK's Ascension Island (polar missions from Vandenberg AFB), with the vehicle ejected by high-pressure steam from its canister before stage 1 ignition. The Eagle and Eagle S-I carry a bipropellant PostBoost Vehicle (PBV) derived from Peacekeeper's stage 4. The S II would carry the USTM (Unified Satellite Transfer Module) or the PASS (Payload Assist SubSystem). USTM is planned as a generic satellite bus containing the essential subsystems and a payload compartment for a customer payload (usually, the electronics and antennas that are integrated into the bus). USTM contains a gimballed bipropellant low-thrust engine for apogee or other orbital manoeuvreing. PASS, attached to USTM and later separated, is a gimballed high-thrust bipropellant engine module for perigee or other orbital manoeuvring. The larger SIII/S-IV would be launched in the future from Cape Canaveral.

The Eagle/PBV combination would deliver 1,360 kg into LEO for US$10 million, the Eagle SI/PBV 2,722 kg LEO for US$20 million and the Eagle SII/PASS 1,716 kg GTO for US$30 million.

The company's debut Loft 1 was launched 17 November 1988 from Canaveral's complex 47 to 4.6 km altitude carrying a 15.5 kg educational payload.

EER Systems Inc

Space Services Inc (SSI) was acquired in December 1990 by EER Systems. SSI designed the Conestoga solid propellant family of orbital launchers capable of handling up to 2.1 tonnes into LEO. It was selected in early 1991 to provide Conestoga launch services for the NASA sponsored Meteor μgravity carrier programme. A preliminary demonstration launch using a single Minuteman 1 second-stage motor was conducted in September 1982. In March 1989, SSI fired its first commercial vehicle: a two-stage Starfire sounding rocket for the University of Alabama under a US$1 million contract carrying a μgravity payload.

SSI's first launcher, Percheron, with a liquid first-stage, exploded on its debut from Matagorda Island off the Texas coast 12 August 1981, and the company switched its efforts to the all solid Conestoga designs. Conestoga 1, using a single 206.8 kN thrust Minuteman 1 second-stage motor, was successfully fired as a demonstration on a 10.5 minute, 313 km altitude suborbital trajectory 9 September 1982 from Matagorda. The current Conestoga orbital versions are based on Thiokol's Castor 4A/4B solids with Star injection stages. The 1679 was selected for BMDO's MSTI-5, although no funding was provided. The debut (1620) October 1995 launch self-destructed at 45 seconds after faulty navigation data produced extensive nozzle movements and depleted the TVC hydraulic fluid.

Hughes Research Laboratories

Hughes has developed an 18 mN Xenon Ion Propulsion Subsystem (XIPS) for NSSK of three-axis or spin-stabilised telecom satellites. It was to debut on Hughes' own Galaxy 3R satellite in October 1995; Astra 1G and ASC 1/2 are to use it. XIPS comprises four 13 cm thrusters (two primary + two back-up), power supply (439 W input each thruster off 29 to 34 V DC bus) and a propellant storage/control unit. Total XIPS package mass is 68 kg, for a saving of 400 kg on an HS-601 satellite. 12,000 hours of operating time was accumulated by early 1995. In normal NSSK operation, the thrusters will fire for 2 to 3 hours daily.

Xe is fed through a heated (1,000°C) hollow cathode inside the thrust chamber. A ring cusp magnetic field confines the plasma and a three-grid ion optics assembly extracts the thrust beam as 3,125 beamlets. Xe is stored at 74.8 atmospheres and reduced to 0.68 atmospheres for feeding into the cathode.

The 13 cm XIPS was developed from work on a 25 cm diameter 63.5 mN thruster that provided SI 2,800 seconds at 1.3 kW input. It was also operated at 179 mN, 4,000 seconds, 4.5 kW.

ICI Explosives

ICI designs, develops and manufactures a wide range of explosively actuated aerospace, ordnance and commercial products, including: actuators/motors, cutters, detonators/explosive leads, gas generators, igniters, switches, electric frangible link assemblies and pyrotechnic actuated valves.

Kelly Space and Technology

Kelly Space and Technology Inc (KST) is a technology development company which was incorporated in April 1993 under the laws of the State of California. The company's current focus is the development of the 'Eclipse' family of low-cost, reusable, commercial orbital and suborbital space launch vehicles which are intended to make access to space both affordable and routine. KST has focused initial development activities on solving this problem by offering what could become a significantly more operationally efficient launch system at substantially lower cost, with plans to reduce costs even further by incrementally incorporating new technology into improved operational vehicles as it becomes available. The Eclipse launch vehicles are designed to address this need in the near-term by combining the best of available, proven technology to mitigate development risk and improve operational flexibility while defining an implementation plan that minimises non-recurring investment to ensure sound financial performance.

The Eclipse concept utilises a Boeing 747 to tow the Eclipse winged launch vehicle from a conventional runway to the launch altitude of approximately 12 km. At this altitude, the Eclipse vehicle's rocket engine is ignited, the tow line is released and the vehicle climbs to the payload separation altitude of approximately 120 km. Following ejection from the Eclipse vehicle, the upper stages are ignited to deliver the payload to the specified destination while the Eclipse vehicle descends as a glider. For final descent from approximately 9 km, the Eclipse deploys two turboprop engines located aft to support powered approach and landing.

Importantly, KST has mitigated its development risk by employing only existing, proven technology in its launch vehicle configurations. No technological advances or breakthroughs are required.

The Eclipse tow launch concept is currently in a flight demonstration programme being conducted by KST under a United States Air Force Phase II Small Business Innovation Research (SBIR) contract. Under this cost-share contract KST will demonstrate the key features of the Eclipse launch concept using a QF-106 jet powered supersonic aircraft towed behind a C141A aircraft. The USAF has provided the Company with two QF 106 aircraft under the SBIR contract to serve as scaled representatives of the Eclipse launch vehicle. The USAF Flight Test Centre at Edwards AFB is supporting flight test operations by providing a C141A aircraft for use as the tow plane. Flight tests are being supported by and conducted at NASA's Dryden Flight Research Center (DFRC) at Edwards Air Force Base. Flight test activities continue using the QF-106 aircraft throughout the remaining useful life of the aircraft.

The Eclipse family of reusable launch vehicles consists of three basic, sequentially developed, flight vehicle configurations: Eclipse Sprint, Eclipse Express and Eclipse Astroliner. Each vehicle supports the development of the next and is placed into commercial use following testing to service incrementally larger segments of the space launch market. Adopting this approach enables each vehicle to serve initially as a test vehicle and to then generate revenues concurrent with the development of the next vehicle.

The first vehicle in the family, a small launch vehicle named Eclipse Sprint, will be constructed to serve both as a development test bed for the Eclipse Astroliner and as a commercial suborbital launch vehicle to generate early revenue. The Eclipse Sprint vehicle will be used to develop flight operations procedures to ensure safety, evaluate and demonstrate various recovery options, gain operational flight experience with the Eclipse tow launch method, support test activities of the Eclipse Astroliner, and generate early revenues. Eclipse Space Lines Inc, will be established and licensed by KST to commercially provide Eclipse Sprint suborbital launch services. The Eclipse Sprint vehicle, a small-scale rocket powered glider which is representative of the Eclipse Astroliner, will support successive Eclipse development activities and provide suborbital launches to service approximately half of the existing sounding rocket market. Sounding rockets provide payloads with up to 20 minutes of exposure to the space environment supporting microgravity experiments, Earth observation, atmospheric sampling and numerous US Department of Defense applications. The Eclipse Sprint service offers significant price advantages with reliability not provided by any of the expendable rockets currently servicing this market.

The Eclipse Express is a follow-on and enhanced version of the Eclipse Sprint vehicle which incorporates upper stages that could be ejected with a payload to service the balance of the existing suborbital market as well as being capable of launching small satellite payloads (under 90 kg) into low-earth orbit. Development of the Eclipse Express will be initiated as a follow-on to Eclipse Sprint operations in order to escalate the development of the Eclipse Astroliner. The Eclipse Sprint vehicle will serve as the test vehicle to support Eclipse Astroliner development activities. As currently envisioned, Eclipse Express would be developed under the same financing approach and business model as the Eclipse Sprint and is expected to be deployed before flight testing of the Eclipse Astroliner.

The flagship of KST's family of launch vehicles will be the Eclipse Astroliner. This vehicle will be capable of launching payloads of up to 1,600 kg into a 465 km polar orbit or up to 2,300 kg into a 465 km equatorial orbit. Commercial operations are projected to commence around mid-2000. The Eclipse Astroliner recoverable flight vehicle consists of the airframe, propulsion systems, thermal protection systems, avionics packages (including software), attitude control system, ordnance initiation system, and landing gear. The Eclipse Astroliner airframe will be a low to mid-wing delta monoplane with a single vertical stabiliser. The airframe will include an articulated nose door similar to a 747 Freighter or C-5 aircraft for both ground loading of the payload and payload deployment at altitude. Access doors servicing the payload and all other units requiring inspection and maintenance are also incorporated into the airframe.

Kaiser Marquardt

The Marquardt company was formed in 1944 to undertake R&D leading to the first US subsonic ramjet in 1945. Its principal engineering business continues to be advanced aerospace propulsion and the supply of ram air turbine power systems. The company has been developing and manufacturing bipropellant thrusters since 1959, notably the attitude control thrusters for Apollo spacecraft and the Space Shuttle Orbiter. Precision bipropellant engines are available for 0.004, 0.02, 0.06, 0.11, 0.22, 0.44, 0.89, 4 kN sizes, and requests up to 8 kN can be accommodated. The monopropellant hydrazine thruster line of UTC's Hamilton Standard Division was acquired in 1993.

Kistler Aerospace Corporation

Kistler Aerospace Corporation is proposing the development of a two-stage, fully reusable launch vehicle, designated the K-1. The vehicle will use Russian engines designed and built by NPO Trud (previously the Kuznetsov design bureau) for the aborted Soviet manned lunar programme.

The company completed the phase 1 preliminary design in September 1995 and has secured the right of first refusal for all NK33 and NK43 engines which were built for the N-1 manned lunar landing vehicle. Phase 2 development and testing was completed in 1996 and Kistler is currently undertaking phase 3 development and testing, scheduled for 1997-1999. Phase 3 calls for completion of the detailed design of the launch vehicle, the establishment of an advisory group comprising contractors and Kistler supporters, finalisation of the contract team, fund raising, assembly and test of the first K-1 vehicle, conduction of flight tests and the initial commercial operations.

Each of the two stages of the K-1 vehicle would be recovered using a combination of parachutes and airbags. Ideally, both stages would automatically return to the launch site, the second stage after initially entering orbit with the payload.

When complete a fleet of five K-1 vehicles is proposed, with each vehicle designed for 100 flights. Each vehicle would have a turnaround time of nine days, with each stage being returned to the launch site within 24 hours of launch. The launch vehicle would be assembled horizontally, transported and erected on a launch stand using a wheeled mobile launcher. A Memorandum of Understanding with the State of Nevada to establish a launch site was signed as part of the phase 1 work for the vehicle.

Kistler is forecasting a price of US$1,700 per lb (or about US$770 per kg) to a 28.5°/km LEO.

Lockheed Martin Astro Space

The USAF Space & Missiles Systems Center began the IABS Integrated Apogee Boost Subsystem programme in 1988 after Atlas 2 was selected for the 10 DSCS 3B military communication satellites. Cost of 10 stages and their launch vehicle integration is about US$200 million; unit cost is about US$15 million. DSCS contractor LMAS was awarded the contract, with the first stage delivered during 1991.

IABS is used to inject DSCS into a near GSO orbit from GTO, permitting the use of a smaller launcher. Payload capacity is 1,180 kg. Spin stabilisation and commands are provided by the payload. For DSCS, the main 60-minute burn is made at apogee 5 into a 0.4° intermediate 'burn short' circular orbit; an 80-second trim at apogee 7 establishes a sub-synchronous drift orbit. The payload separates for its own thrusters to achieve GEO; IABS remains in the drift path. The pair can remain in transfer orbit for up to 12 days for mission flexibility. The peripheral solar array generates 240 W and DSCS's stowed array adds 190 W (there are no batteries; IABS employs DSCS's).

Lockheed Martin Astronautics

Lockheed Martin Astronautics is one of four operating elements of Lockheed Martin's Space and Strategic Missiles sector. It already encompasses Titan and Atlas production and operations, but the former Lockheed space launcher activities of LKE (Proton) and Lockheed Missiles and Space Company (LMSC) are also consolidated here. Denver is also a centre of satellite production, particularly classified satellites such as Lacrosse.

The Astronautics element of the Martin Marietta Corporation was established in 1956 to develop and produce the Titan 1 ICBM. Headquarters for Astronautics is the Deer Creek facility, 32 km southwest of downtown Denver in the Rocky Mountain foothills. Acquired in 1987, it provides 66,140 m² of office space on a 4 km² site in Deer Creek Canyon. The main Astronautics facility is located on 20.5 km² 8 km south of HQ and houses advanced production, space simulation and test facilities. Included in the US$550 million plant are approximately 70 laboratories, as well as computerised engineering design, manufacturing and administration facilities. Astronautics also operates the USAF Titan and Atlas launch facilities at Vandenberg AFB and Cape Canaveral.

NASA programmes include Viking, Magellan, Tethered Satellite and Shuttle's Manned Maneuvring Unit. It is believed to be prime for the USAF Lacrosse imaging radar spacecraft. The Transfer Orbit Stage was built under contract to Orbital Sciences. A US$30 million five-year contract was awarded in May 1992 to convert Minuteman 2 ICBMs into suborbital launchers for the MultiService Launch Systems Programme. A three-stage orbital version, costing US$4 million to convert, was due to debut in 1997/98. The company acquired a 15 per cent interest in Canada's Radarsat International June 1994 for US$3 million, giving it an exclusive licence for marketing Radarsat data to the US Federal government.

Lockheed Martin Space Launch Systems

Lockheed Missiles & Space Co in May 1993 announced plans for a family of all-solid LLV Lockheed Launch Vehicles to deliver payloads ranging from 800 to 3,200 kg to LEO for US$15 million to US$26 million. LMLV 1 employs a Castor 120 stage 1 and Orbus 21D stage 2. LMLV 2 inserts a second Castor 120 between the two, while LMLV 3 would add at least two Castor 4A strap-ons (3/4/6 strap-ons also possible). All versions carry a hydrazine orbit adjust module, which also houses the vehicles' avionics. CDR was conducted April 1994; PDR July 1993. Transfer orbits and Earth escape could be provided by a Thiokol Star stage 4 with TVC. The Castor 4A could also be replaced by the 4AXL version.

The programme moved from LMSC in Sunnyvale to Denver in late 1995, although the vehicles (now named LMLV) continue temporarily to be built at the Californian site. LMSC and CTA Inc signed a less than US$5 million contract 18 February 1994 for the launch of CTA's 113 kg GEMStar 1 into a 480 km polar orbit from VAFB on the debut LMLV 1. Launch was from an adaptor on one of the solid booster mounts of the Shuttle SLC-6 complex. This flight used an existing Orbus 21 modified to 21D; propellant was offloaded for another mission not flown. The vehicle was destroyed by range safety after 159 seconds, when separate stage 1 TVC and IMU problems left it uncontrollable.

LMSC began investigating the possibilities of converting excess missiles in 1987 (it was responsible for the Polaris, Poseidon and Trident SLBMs) but dropped the approach in 1990 when studies showed the vehicle would not be reliable enough (90 to 95 per cent) for the commercial market. Lockheed Corporation approved the LLV development programme in January 1993. The philosophy is to use existing or imminent hardware, aluminium structures rather than composite (lower cost), higher margins in lieu of acceptance tests and not to push performance at the expense of reliability or lower cost.

Lockheed Martin Manned Space Systems

Manned Space Systems was formed in 1973 to design, assemble and test external tanks for NASA's Space Shuttle. It is part of Lockheed Martin's Information and Technology Services Sector. Staff total 2,200. A US$1,797 million contract extension in August 1989 added a further 60 tanks to the 59 already under order. The first was completed in late 1991; the 60th is due in 2000. The first tank, the main propulsion test article, was delivered to NASA 7 September 1977. Manned Space Systems is located at NASA's 3.37 km² Michoud Assembly Facility, which features one of the world's largest manufacturing plants (174,000 m²) under one roof and a port with deep water access for barging ETs to Florida. Three ocean-going barges are available for the five-day journeys. Manned Space Systems is working under a US$172.5 million contract to develop the Super Lightweight Tank, using the 2195 Weldalite aluminium lithium alloy instead of 2219 aluminium, saving 3,400 kg. The total saving will reach 3,600 kg by fine tuning the foam insulation's spray-on process and optimising other tank structures. First was ET-96 for first Space Station assembly mission. A 12.2 m-long full diameter version, essentially a segment of the hydrogen tank with a LOX tank dome at one end, was delivered to NASA Marshall 1 February 1996 for six months' dynamic and pressure testing to verify the structural design. The alloy is 30 per cent stronger and 5 per cent less dense than the current material and comprises 1 per cent Li, 4 per cent Cu, 0.4 per cent Ag, 0.4 per cent Mg and 94.2 per cent Al.

Manned Space Systems studies include advanced and reusable launch vehicles, hybrid propulsion systems, on-orbit use of ETs, and the development of related composite materials technologies. Lightweight pressurant tanks for the propulsion system of Lockheed Martin Astro Space's A2100 satellites are in production.

Microcosm

Microcosm is developing a pressure fed LOX/kerosene 22.2 kN engine with a claimed manufacturing cost of less than US$5,000 each, excluding the injector. There are only 38 parts, full duration 200 seconds, ablative film cooled; performance is comparable to solid rockets, although key specifications remain confidential. A 200-second run was successful 20 November 1995. Assisted by USAF Phillips Lab funding (US$3.1 million through September 1996), Microcosm proposes the Scorpius family of vehicles based on the same engine: SR-S sounding rocket (723 kg GLOW, 50 kg to 225 km; US$99,000 recurring cost); SR-1 sounding rocket (400 kg to 190 km; US$295,000); Mini-Lift (100 kg into LEO; US$700,000); Liberty Light Lift (71.7 tonne GLOW, 4 stages, 7 propulsion pods totalling 49 engines, 1 tonne into LEO; US$1.8 million); Exodus Medium Lift (6800 kg LEO; US$8.5 million). Non-recurring development cost of the range through Liberty is less than US$30 million (FY96).

Olin Aerospace Company

OAC, previously the Rocket Research Company, specialises in monopropellant hydrazine engines and gas generators for spacecraft and upper stage applications; approximately 9,300 assemblies and 100 propulsion systems have been produced since the first system was qualified in 1964. There are 600 personnel at the Redmond and Moses Lake sites. The MR508 hydrazine arcjet was qualified in 1991 and Telstar 4 in 1993 became the first commercial satellite to carry an arcjet for station-keeping. The model is superseded by the MR-509/510. OAC has delivered approximately 250 gas generators for Shuttle's Auxiliary Power Unit.

Facilities: OAC's 0.29 km² site at the Grant County Airport in Moses Lake houses 2,840 m² of solid propellant manufacturing plant, a hazardous device assembly area, storage magazines, a short-range flight test facility, a sea level large liquid engine test complex, ballistic test bays and chemistry laboratory.

Olin Chemicals

Olin is the world's largest supplier of hydrazine and hydrazine-based propellants for space propulsion applications, including anhydrous hydrazine, MMH and UDMH. Users include Titan, Delta, Shuttle and most commercial/military spacecraft. In addition, hydrazine and hydrazine blends are used in auxiliary power units such as Shuttle's APU. An AF contract worth US$92 million over five years was awarded January 1995.

Orbital Sciences Corporation

OSC was established in 1982 to develop Space Shuttle upper stages, resulting in the solid propellant Transfer Orbit Stage. In June 1988, OSC announced a partnership with motor manufacturer Hercules Aerospace (now Alliant Techsystems) to develop the Pegasus air-launched small orbital delivery system. OSC was contracted in 1989 by DARPA for the Taurus rapid response orbital launcher based on Pegasus. Space Data Corp of Tempe, Arizona became a subsidiary in November 1988, extending OSC's capabilities to a wide range of sounding rockets: it has manufactured/launched approximately 600 suborbital boosters in 35 different configurations, in addition to producing up to 1,000 meteorological rockets annually. The division has participated in approximately 60 Minuteman 1 re-entry tests since 1971. A series of winged air-launched suborbital vehicles was announced September 1992. OSC's L-1011 aircraft can carry three 1- or 2-stage suborbital vehicles derived from the company's other carriers, totalling 11,340 kg with payloads up to 1,130 kg. The OSC/Rockwell team was selected by NASA in March 1995 to develop the semi reusable X34 launcher, the same month the OSC/McDonnell Douglas team was awarded NASA's MedLite launch contract, involving Taurus for the smaller satellites. At this date the X-34 was a two-stage, small, reusable winged launcher, partially reusable, on which NASA would spend about US$70 million. Moving with uncharacteristic haste, NASA selected an Orbital Sciences Corporation (OSC)/Rockwell team to develop the two-stage X-34 aiming for a suborbital flight in 1997 and orbital flights beginning a year later. The team formed American Space Lines as an 'Arianespace' lookalike for commercial development but in early 1996 it pulled out, believing the costs could not produce a viable commercial return. NASA re-contracted with OSC for a single-stage suborbital demonstrator, still designated X-34. Three X-34 test vehicles are scheduled to support flight trials planned to begin in August 1999 with the winged vehicle launched from the under-fuselage cradle of a converted L-1011 operated by OSC. Capable of a maximum speed of Mach 8 the X-34 will demonstrate all-weather, pilotless landings as a technology precursor to fully reusable launch systems of the future.

Para-Flite Incorporated

Para-Flite develops/manufactures 0.5 to 280 m² ram-air gliding parachutes for aerospace recovery systems, including orbital capsules and sounding rocket payloads. Providing a lift-to-drag ratio of three, the parachute carries an Airborne Guidance Unit to process signals from a ground-based transmitter (GPS is an option) to manipulate the control lines and home in on the target. The automatic mode provides 60 m accuracy, complemented by 15 m for manual control. Payloads can be recovered from suborbital rocket and balloon carriers: 100 to 2,500 kg capacity systems are available off-the-shelf.

Pressure Systems, Inc

PSI is the world's largest independent manufacturer of titanium pressure vessels for launchers and spacecraft; it now ships 200 to 300 annually, worth about US$25 million in total. Approximately 3,800 have been delivered, without flight failure, including approximately 600 propellant management device tanks of approximately 20 designs and more than 700 elastomeric diaphragm tanks of approximately 15 designs. PSI acquired Programmed Composites Inc in June 1995, allowing it to offer tanks integrated with satellite core structures, in addition to precision reflectors, solar array substrates and antenna support structures. PSI tanks provide the propellant for most commercial communications satellites, including Hughes HS 601, Loral FS1300 and Lockheed Martin A2100. PSI tanks are also used on Milstar, Cassini, Soho, NEAR, Tirus, Inmarsat, Intelsat, ETS 7, TRMM and Axaf. PSI is completing a US$4 million contract from Lockheed Martin for 75 hydrazine tanks for Iridium satellites. In 1996, PSI delivered the 1000th upper stage high pressure helium tank and the 100th upper stage propellant tank for the Lockheed Martin Atlas Centaur launch vehicle family.

Sandia National Laboratories

Sandia was established in 1949 and is now operated for the US Dept of Energy by Lockheed Martin's Sandia Corp. The Laboratories are active in defence and energy national security programmes, with principal emphasis on nuclear weapons R&D. Personnel total 8,400, including the associated facilities at Livermore in California. The HQ and main laboratory are located on Kirtland AFB on the southeast edge of Albuquerque. Aerospace interests include flight control/telemetry systems, robotics, materials, and computerised and physical laboratory testing of space debris impact on satellites (the Hypervelocity Launcher gas gun accelerates particles up to 12 km/s). It also develops/ operates suborbital launch vehicles for the DoE, the military and other users. Tonopah Test Range at the north end of Nellis AFB, Nevada houses test facilities for small rockets.

Sandia operates the Kauai Test Facility in Hawaii on the Navy's Pacific Missile Range Facility, maintaining its own launch, handling and launch control equipment. For example, the Strypi family of solid propellant rockets has been employed in re-entry research programmes and as targets for orbital sensors. Strypi also launches the Laboratory's SWERVE Sandia Winged Energetic Re-entry Vehicle Experiment as an M14 hypersonic testbed.

The STARS strategic target system has been developed to extend these capabilities, particularly for BMDO demonstrations, and could be modified for modest orbital missions by adding a stage 4. STARS is based on Lockheed's two-stage solid propellant Polaris A3 SLBM motors carrying the new UTC/CSD Orbus 1 solid as stage 3, featuring a flex nozzle for steering. The configuration is capable of throwing 295 kg some 4,255 km downrange at a unit flight cost of about US$5 million. On a typical target mission, STARS is launched from Kauai to impact in the Kwajalein Atoll. First launch was projected for 1991, but the project was stalled by environmental study demands. Two are planned to be launched for BMDO's MSX satellite to observe.

Space Vector Corporation

Founded in 1969, SVC is a wholly owned subsidiary of Precision Standard Inc and specialises in the design and manufacture of spacecraft and launcher structures, inertial guidance and control systems, telemetry subsystems and launch support. The Aries suborbital vehicle, based on government furnished Minuteman booster motors, was developed at SVC primarily for experiments of up to 2,270 kg and altitudes up to 480 km. The company serves as booster integrator, performs all design analyses and provides the electronic interfaces, ground support equipment and experiment modules. Approximately 40 Aries-class vehicles have been launched. An orbital version has been proposed, capable of delivering 226 kg into LEO. Aries is now changing from Minuteman stage 2 M56 motors for its first stage, to the more powerful Minuteman 2 stage 2 SR19. This model will be used for BMDO's HERA TMD, the GBI development test vehicle and Altair. In addition, SVC has provided attitude control systems to approximately 130 smaller rockets and approximately 600 Miniature Inertial Digital Attitude System (MIDAS) platforms. The strapdown Advanced Guidance System (AGS) is capable of ±0.25° accuracy or better. The Magnetic Control System (MCS) is a gyro-less system that senses Earth's magnetic field. SVC also provides 0.4111 N cold gas thruster packages for sounding rockets and satellites.

Israel's Shavit was offered in 1994 for NASA's ultralight launcher competition (won by Pegasus). Team leader SVC would have integrated and launched the vehicle.

SSI, Inc

SSI is a partnership between ITT Federal Services Corp and California Commercial Spaceport Inc to provide commercial launch services for small to medium polar vehicles from VAFB beginning in 1997. Typical users would be LMLV, Taurus and Delta-Lite. SSI is building a new pad and is leasing the US$250 million mothballed Payload Preparation Room at SLC-6, renovated as the IPF Integrated Processing Facility (GEMstar 1 and STEP M0 has been processed by end 1995). Annual capacity will be 24 launches, using the Integrate-Transfer-Launch (ITL) approach, allowing the use of simple stationary pad structures. The 25-year lease was signed by WCSC/AF for the undeveloped 0.10 km² site near SLC6; the annual fee is US$70,000 plus about US$10,000 monthly for operations costs and utilities. The government is providing US$10 million, mostly as AF grant money, and ITT is committing US$30 million. WCSC was formed as a non-profit corporation in May 1992 to pursue commercial opportunities at VAFB. It is designed by the State Legislature as The California Spaceport Authority. Subsidiary CCSI was established in August 1993 as a for-profit company to meet grant requirements for matching funds from space companies.

The SLF Spaceport Launch Facility operates a single pad at 34°34′35″ north/120°37′53″ west (92 m altitude) between two SCF Stack & Checkout Facilities on an east-west axis. Most vehicles will depart due south (180° azimuth), so will not overfly the SCFs. 168 to 220° azimuths are available without impacting the local community or off-shore oil rigs; 150° is possible. A vehicle is stacked on an MLP Mobile Launch Platform and transferred launch-ready to the pad on launch day. SCF-E is intended for Delta 2 and Delta-Lite (but can handle Castor 120); SCF-W is optimised for Castor 120-based vehicles. IPF handles booster and payload processing, fairing/storage and payload encapsulation, and houses three launch control rooms in the 12.2 × 42.7 m Launch Control Center (LCC). The processing areas include a 9 × 30 m airlock, three 10.7 × 13.4 m cells and 9 × 45 m high bay, with 47 m high 68-tonne crane.

TRW

TRW develops and manufactures propulsion components, engines and systems for spacecraft attitude and velocity control, orbit insertion, manoeuvring and adjust, tactical missile steering, low-cost orbit transfer stages and low-cost engines for Earth to orbit systems. Since the company produced the first space operated monopropellant hydrazine system for 1960's Able lunar probe, it has provided approximately 2,000 mono/bipropellant thrusters and reaction control systems. The bipropellant Lunar Module Descent Engine was responsible for landing 12 astronauts on the Moon 1969-72, and Apollo 13's rescue, before it was converted into a fixed thrust model for Delta's second stage; 84 were built for the lunar programme and 75 flew on Delta from 1974. It remains under consideration for future upper stages. A further derivative was to power later versions of NASA's Orbital Manoeuvring Vehicle (cancelled). Propulsion research covers low-thrust monopropellant, bipropellant, colloid, ion, radioisotope, MPD, electrothermal and low-cost engines.

Propulsion test facilities are centred at TRW's 8 km² Capistrano Test Site (CTS) 108 km south of Redondo Beach, equipped for thermal vacuum, vibration, shock, cold flow calibration, pressure/leak, functional and electrical, and accurate hot fire performance and life testing. Engine designs burning storable, cryogenic, gelled and high energy bipropellants can be evaluated under sea level and simulated altitude conditions. Continuous operation, steam exhaust pumped altitude chambers can test high-expansion ratio 450-670 N engines at ambient atmospheric pressures equivalent to an altitude of 46 km. The HEPTS High-Energy Propellant Test Stand has four 222 kN test positions, two of which simulate altitude firings. Cold flow and hot fire tests of low-thrust hydrazine, ion/plasma electric and resistojet gas thrusters are undertaken at Redondo Beach.

United Technologies

Solid propellant programmes are run by CSD in San Jose. CSD specialises in the design, development and production of solid propellant, hybrid and ramjet propulsion systems for space, strategic, tactical and launch booster systems. Its Orbus upper stage family includes Orbus 6/6E/21 (IUS), Orbus 7S (JCSat), Orbus 21S (Intelsat 6), Orbus 21 (TOS), Orbus 21D (LMLV launcher) and Orbus 1 (STARS and Starbird upper stage). The numerical designator corresponds to propellant mass to the nearest 1,000 lb, S denotes spin stabilisation and E the use of an extending exit nozzle (EEC) system. CSD also manufactures the segmented SRMs for Titan 4 and the Space Shuttle SRB separation motors. It was responsible for the Algol stage 1 motor on Scout, which retired in 1994.

CSD's Coyote facility in San Jose includes the world's largest solid rocket vertical test stand: 29 m high, it can handle 26.7 MN thrust motors and was most recently used for qualification firings of the 7 segment USAF Titan 4 strap-on motor. Small/medium– sized horizontal and vertical test stands are controlled from a central complex and structural test facilities can apply omni-axial simulated flight loads of up to 4.9 MN to motor cases and attach skirts. CSD's integral rocket/ramjet test complex can simulate booster, transition and ramjet operations in a single sequence under sea level to high altitude conditions up to M6.0. The Coyote site also houses a 5,400-tonne annual capacity solid propellant mixing facility comprising two 2,270 litre, one 2,840 litre and one 1,510 litre mixers.

United Technologies Pratt & Whitney

UTC combined the space propulsion activities of CSD, USBI and P&W in 1990 within P&W's Government Engines & Space Propulsion unit based at West Palm Beach in Florida. USBI is under contract to NASA Marshall to provide assembly, test and refurbishment of the non-motor segments of Shuttle's solid rocket boosters. NASA extended the contract in 1994 for recovering and refurbishing boosters to September 1997, worth US$1.8 billion. The previous US$1 billion award in January 1989 ran through September 1994. It was extended in 1996 through September 1999.

Universal Propulsion Company, Inc

The company makes rocket motors, escape systems, rocket catapults, initiators/gas generators, electro-explosive devices, safe/arm units and pyrotechnic simulators for space and aircraft.

Western Electrochemical Company

Pacific Engineering & Production satisfied half (18,000 tonnes) of US ammonium perchlorate solid propellant oxidiser requirements until explosions destroyed its plant in May 1988. Users included Delta, Titan and Shuttle boosters; each Shuttle launch consumes 771 tonnes. A new facility operated by Western Electrochemical (WECCO), a wholly owned subsidiary of American Pacific Corp, a Las Vegas based manufacturer of specialty chemicals, was completed during 1989, offering an initial annual production rate of 14,000 tonnes capable of rising to 18,000 tonnes with only minor capital additions. The 33-building complex was constructed at a cost of US$92 million and employs a staff of 175. The batch process under automated control can produce AP to custom chemical characteristics and physical distributions, as well as standard materials in large homogeneous lots. High purity AP is used in low smoke reduced signature rocket motors.

MICROGRAVITY FACILITIES, EQUIPMENT AND SERVICES

FRANCE

Novespace

Novespace was established in July 1986 by CNES and eight banks as a consultancy for commercial utilisation of space technology, focusing on μgravity applications and technology transfer. 1995 revenue was FFr17 million (profit FFr1.3 million); 1994 FFr16.8 million (profit FFr1.6 million), 1993 FFr16 million (profit FFr0.89 million); 1992 FFr16.5 million; 1991 FFr14.5 million; 1990 FFr9.5 million; 1989 FFr6 million. The company offers μgravity campaigns aboard an Airbus A300 aircraft flying parabolic curves. Each week-long campaign costs about US$28,000 for a 150 kg experiment and two people flying for three days. Each day provides 40 parabolas of about 25 seconds μgravity. Testing by CNES began in January 1989 with a Caravelle; Novespace flew more than 40 campaigns by end-1995. The A300 took over in June 1996.

The creation of Space Business International was announced in June 1993 by Novespace, OHB-System GmbH and Space Industries International, Inc to co-operate on space contracts. Novespace is a founder member of the Spacelink Europe consortium, a network of space technology transfer companies spanning 13 European countries.

GERMANY

DaimlerChrysler Aerospace AG

DaimlerChrysler extensive μgravity flight facilities range from the 4 minutes provided by the Mini-Texus sounding rocket vehicle, through the MAUS NASA GAS-based canisters, to SPAS and Eureca. The company manages the Texus and Maxus commercial programmes. Mini-Texus, Texus and MAUS can accommodate small payloads for μgravity levels up to 10^{-5} *g* whereas free-flying SPAS and Eureca offer conditions enhanced by one to two magnitudes. SPAS can be deployed from the Shuttle cargo bay as a free-flyer but its autonomous lifetime is constrained to hours/days by battery supplies. Eureca is based on a double-SPAS structure, but solar arrays permit a typically six month active life followed by a period of dormancy up to three months before retrieval. The larger Maxus rocket was introduced in 1991 in co-operation with Sweden to provide about 14 minutes μgravity. The Express capsule flew 15 January 1995 but failed because of a launcher problem. No further missions are planned.

Leybold DurferriT AG

Leybold produced high-vacuum furnaces for Spacelab D1/D2; the Isothermal Heating Facility is a multi-purpose system for isothermal and temperature gradient experiments. The company also produced TEM high-vacuum furnaces for dedicated experiments on the Eureca free-flying platform.

OHB-System GmbH

Orbital- und Hydrotechnologie Bremen-System GmbH, founded in 1958, undertakes system engineering, hardware development/production and project management for space and environmental programmes. 1995 sales were DM41 million. Principal activities are: design/development of mechanical and electronic equipment; μgravity systems, experiment facilities/equipment; manned/unmanned space systems and mission operations, re-entry technology and aerodynamics/aerothermodynamics; small satellites. Its μgravity activities include industrial lead in the Mikroba programme, parabolic flights in chartered aircraft (NASA's KC-135 and CNES' Caravelle), collaboration in the construction and utilisation of ZARM's drop tower at Bremen University (providing the drop capsule), development of the Holop holographic camera for Germany's Spacelab D2 mission and the High Speed Centrifuge for ESA's Anthrorack. The company also developed the electronic control unit for Intospace's Cosima crystal growth facility, first flown aboard a recoverable Chinese platform in August 1988.

OHB manufactures the flight harness for the BPDU Spacelab rack, it developed the video control for IML 2's NIZemi, and it was responsible for the design of ergometry, blood and urine monitoring experiments flown on Germany's Mir-92 mission. On Spacelab D2, OHB provided the Gravitational Biology element of incubators and centrifuges. OHB is MGSE prime contractor for ESA's Polar Platform, it was awarded the Phase A study for Columbus' Antrolab facility and within the Europayload Consortium, it is working on the Fluid Science Laboratory and Automated Bio Laboratory for Columbus.

OHB developed Bremsat for the University of Bremen and its SAFIR-R1 demonstration package was launched November 1994 attached to a Russian Resurs-O1, preceding 1997's SAFIR 2 small satellite to provide global two-way data services. Belgium plans the derivative IRIS (Intercontinental Retrieval of Information via Satellite) system from 1997; OHB-System has overall responsibility for the space segment. If the project is approved, it will provide the bus for the German-Israeli David Earth observation satellite.

The creation of Space Business International was announced in June 1993 by OHB, Space Industries International, Inc and Novespace to co-operate on space contracts.

ITALY

Alenia Spazio

The company was awarded the 18-month Lit8 billion Phase B work in mid-1990 on a small recoverable capsule for μgravity applications. Work was completed October 1991; the 15-month Phase B2 definitive design phase was authorised January 1993 and concluded April 1994. As Phase C/D requires 32 months, first launch will be no earlier than 1998. Carina (Capsula di Rientro Non Abitata, unmanned re-entry capsule) would be launched into 360 km 2.9° by Vega from San Marco (Pegasus is a possible back-up), carrying more than 160 kg of payload for five days. The 1.067 m diameter, 1.248 m high Apollo-type REM capsule (13° wall slope) permits access up to one day before launch and samples recovery 6 hours after splashdown. Service Module: 1.244 m diameter, 60.95 cm high. Re-entry is initiated typically by a 760-second burn from four 20 N (four back-up) monohydrazine thrusters. The 28.7 kg thermal shielding has 25 mm base and 15 mm wall thickness. Recovery under a 2.1 m conical drogue and then a 13.9 m main parachute, is made following a 7 m/s splashdown. Two flights annually will be available. The first payload has already been selected by ASI: SCGE Superconductive Crystal Growth Experiment, growth of YBa2Cu3O7-x; MMSFI Marangoni Migration and Solidification Front Interaction to analyse the inclusion, coalescence and migration of bubbles/drops inside liquid and solid matrices; MBI Marangoni Benard Instability in Two Immiscible Liquid Layers.

Specifications
Size: 1.244 m diameter, 1.888 m high
Mass: 599 kg (165 kg payload, 142 kg REM, 292 kg SM)
Microgravity levels: 10^{-5} *g* for 0-100 Hz, 10^{-3} *g* above 100 Hz (re-entry loads less than 9 *g*)
Mission duration: five days
Payload volume: 0.3 m^3
Power supply: eight lithium batteries provide 180 W mean to the payload
Telemetry: 100 kbit/s (6 kbit/s payload) and 2 kbit/s real time; command uplink 2 kbit/s. 40 Mbit mass memory.

JAPAN

Space Technology Corporation

STC provides expertise on μgravity materials experiments from experience gained on Spacelab (such as D2), sounding rocket and dropshaft experiments.

NETHERLANDS

Bradford Engineering

Bradford Engineering is a manufacturer of glovebox systems for space applications. A glovebox provides two levels of environmental containment in the Shuttle and International Space Station, permitting the performance of experiments which might contaminate the living environment. A video link and two-way audio ground link give ground-based scientists the ability to follow the experiments which an astronaut is conducting in orbit. The Bradford glovebox has flown on the USML-1 and 2 Spacelab missions. An improved design, the Middeck Glovebox (MGBX) has flown on the Shuttle middeck and is installed in the Priroda module which is part of the Mir Complex.

Present developments include the Microgravity Science GloveBox (MSG) and the Biological GloveBox (BGB) for ISS, as well as the Universal Transportable GloveBox (UTGB), a flexible low-cost glovebox design which can be carried by individual astronauts.

Bradford is additionally active in equipment such as valves, video systems, cooling loops and software. Many of these components are derived from proven glovebox designs. The Avionics Air Assembly (AAA) air-to-water heat exchanger will provide 500 to 1,500 W cooling power for ISS racks. Control valves from Bradford Engineering were selected by Alenia for use in the water-cooling system of the MPLM logistics module and the Columbus Orbital Facility. This valve is safety critical and therefore required an extensive testing and qualification programme. The company also developed application and control and monitoring software for microgravity science equipment.

CCM designed, developed and built Spacelab biological experiment units that were later modified for sounding rocket application. CCM has also designed a protein crystallisation module in co-operation with Fokker/Stork Comprimo to accommodate 400 samples.

NORWAY

Prototech A/S

Prototech provides sounding rocket payload design, development, testing and integration, and is involved in production of μgravity flight facilities. The group's debut in manned space technology came with the design/production of a sensor and electronics box for a Spacelab 1 plasma experiment. It has produced a series of biological cell chambers and, for ESA, a gas supply module and associated ground handling equipment for a μgravity furnace. Prototech is developing the first Norwegian fluid physics experiment to be flown on Texus.

SWEDEN

Swedish Space Corporation (SSC)

SSC offers μgravity flight services within the Maser (Material Science Experiment Rockets) programme, in addition to design/manufacture of specific experiment modules. By using an improved Black Brant 9, 7-minute μgravity can be provided for a 375 kg experiment module. Skylark 7 can provide 6-minute μgravity for 370 kg. Launches are made almost annually. Modules are interchangeable between Maser, Texus and MiniTexus with slight modifications. Maser number 1 flew at Esrange 19 March 1987; number 2: 29 February 1988; number 3: 10 April 1989; number 4: 29 March 1990; number 5: 9 April 1992; number 6: 4 November 1993 (Skylark 7); number 7 (Skylark 7) 3 May 1996, with ESA as the principal customer. Maser 8 is planned for November 1997.

UNITED STATES OF AMERICA

Instrumentation Technology Associates, Inc

ITA provides commercial Shuttle flight opportunities, fabricates and leases space hardware and offers support services (payload development, integration, consulting and engineering services) for μgravity experiment activities. It provides turnkey leasing of its MDA Automated Lab, Liquids Mixing Apparatus, CMIX Flights and ISEM-G GAS canister modules for Shuttle, sounding rocket and recoverable capsule flights.

Materials Dispersion Apparatus (MDA) is designed for Shuttle, sounding rocket, satellite recovery capsule and Space Station applications for mixing of up to 100 samples of two-three fluids in space at precisely timed intervals. MDA weighs 1.82 kg and occupies less than 2,000 cm^2, operating on 16 VDC. Up to six units can fit into NASA's R/IM Refrigerator/Incubator Module or Commercial R/IM within a single Shuttle mid-deck locker. MDA's first flight was made 29 March 1989 aboard the Consort 1 sounding rocket mission. The second flight ended with launcher failure in November 1989, and during the Consort 3 mission of 17 May 1990, one of two MDAs malfunctioned. The problem was solved and the unit requalified. MDAs flew on the Joust 1 (vehicle failed 18 June 1991) and Consort 4 (16 November 1991) and number 5 (10 September 1992) sounding rockets. An integrated package of four flew on STS-37, 43, 52, 56. MDAs flew on STS-67 in March 1995 and STS-69 August 1995.

The MDA was designed to accommodate a large number of data sets in small volumes. Principal experiments are biotechnology, focusing on protein crystal growth. Experiments in bioprocessing, thin-film membrane casting, zeolite crystal growth, microencapsulation and cell research have also been performed. Two blocks of inert material, each with an equal number of sample test wells in the upper and lower halves, are held together with a sealing mechanism in an aerospace housing. The wells are misaligned at launch, thus separating the fluids to be mixed. After μgravity has been achieved, the blocks are aligned by a motor-cam mechanism. There is an option to mix a third or fourth fluid to fix the process while in μgravity or before re-entry. Each MDA can have up to 100 test cells, and each well can accommodate samples in the 0.125 to 0.5 μlitre range.

The LMA (Liquids Mixing Apparatus) complements MDA by allowing the mixing of larger volumes (up to 6 ml) at variable rates, from laminar to highly turbulent. Each vial can also be loaded individually. A vial comprises a clear plastic tube for the mixing of two or three fluids with the actuation of a plunger. 16 LMA vials flew for the first time as part of CMIX-3 on STS 67 in March 1995. Each was manually operated on that flight but on subsequent missions, beginning 1996, all vials will be actuated simultaneously.

The CMIX (Commercial MDA ITA Experiments) commercial agreement offers six Shuttle flights of 4-6 MDA units in a T-controlled mid-deck locker. Individual samples or the whole package may be leased. CMIX-1 flew STS-52 October 1992, CMIX-2 April 1993, CMIX-3 on STS-67 in March 1995, CMIX-4 on STS-69 in August 1995; numbers 5/6 are planned for 1996-97.

The ISEM-G (ITA Standardised Experiment Module) for NASA's Get-Away Special programme provides an aluminium structure with mounting hard points, support avionics and house-keeping instrumentation for user payloads. A customer is awaited. ISEM-G fits in the 0.07/0.14 m^3 canisters. The standard support equipment in the lower section comprises a power supply, recorder, sequencer, and temperature, pressure and μgravity accelerometer instrumentation. Up to 10 experiments can be accommodated. ISEM-G can also be flown on Black Brant sounding rockets.

Specifications

The specifications below are given in 0.07/0.14 m^3 GAS version order:
ISEM mass: 13.6/27.2 kg
Experiment mass: 31.7/63.5 kg
Experiment volume: 0.037/0.1 m^3
Power: 1.2 by lead acid cells at 5/16/24 V.

NAVIGATION, SEARCH AND RESCUE FACILITIES AND EQUIPMENT

CANADA

CAL Corporation

CAL is a supplier of Sarsat satellite-aided search/rescue ground terminals, providing 75 per cent of the world's total. It received a C$1.4 million contract in late 1989 for an Advanced Technology Local User Terminal (ATLUT) and two Operations Control Consoles for Hong Kong's Kai Tak airport, CAL's 17th for the system. In the first quarter 1991, Japan acquired two ATLUTs co-located at Yokohama, and Indonesia installed one ATLUT in Ambon and one with an Operations Control Console in Jakarta. New Zealand installed an ATLUT-320 at Wellington in autumn 1991. Singapore began operations mid-1992 with a dual ATLUT-320 and three OCC workstations under a C$2 million contract.

Canadian Marconi Corporation

In the space field, CMC designs and manufactures GPS navigation sensors and receivers and airborne satellite communications antennas.

The 12-channel CMA-3012 GPS sensor unit is the first designed for certification under anticipated rules that will allow GPS as a sole means of airline navigation. The CMA-3112 is a single card engine derived from the 3012; 116 cm^2, 1.1 kg. The ALLSTAR mobile is a self-contained 12-channel all-in-view GPS receiver packaged in a ruggedised enclosure for applications such as automatic vehicle location. The ALLSTAR base version is designed for differential GPS applications, such as mapping and site positioning.

FRANCE

Informatique Electronique Sécurité Maritime

SERPE-IESM offers several versions of its Ka and 406 Cospas/Sarsat beacon. 406 m: pocket-sized Personal Locator Beacon, targeted at pilots or military forces on special assignment. 406 ATP: aircraft, activated by a *g* switch or manually. 406S: hand-held for pleasure crafts, small fishing vessels and life rafts. 406FH: float-free for large fishing and commercial vessels, housed in a case containing the hydrostatic release mechanism which ejects and activates the beacon automatically at 4 m depth. IESM also manufactures Argos transmitters and Local User Terminals, and the Gonio 400 direction-finder to pinpoint and identify Sarsat and Argos beacons.

Sextant Avionique

The company made its first deliveries in 1992 of its Topstar 200 GPS receiver for civil aircraft. The 3 kg unit combines GPS with Omega. The 10-channel Topstar 100 GPS receiver is available in both P-code (100-P) and C/A-code (100-S) military forms. Europe's first satellite-borne GPS receiver, the eight-channel C/A Topstar 300 is carried by the US HETE satellite under CNES contract. A Topstar 100-P will be carried by ESA's ARD Atmospheric Re-entry Demonstrator in 1997. Several GPS space application studies are underway with CNES and ESA.

GERMANY

Man Technologie AG

MT acted as prime contractors for the ESA NAVSAT Phase B1 study and for the Design and Development of an Experimental Health/Integrity Monitoring Unit for the GPS and GLONASS Satellite Navigation Systems. Within ESA's Global Navigation Satellite System (GNSS) 2 Mission Analysis study and the European Geostationary Navigation Overlay System (EGNOS) Project, MT is co-operating with Thomson-CSF (France) in the international teams. In co-operation with 35 Navigation, USA, MT developed and marketed the world's first commercially available integrated GPS/GLONASS receiver, the GNSS200/300. Within the German Luftfahrtforschungsprogramm, MT is developing in co-operation with Litef GmbH, Freiburg, a combined airborne navigation system called GLOGINAV. Various national projects funded by DARA cover new technological developments and different applications.

JAPAN

Japan Radio Company Ltd

The JLR-6800 GPS Ship Navigator is a single channel multiplexing receiver that can maintain constant track of up to eight satellites. It can be integrated with other systems such as Loran C, and provides fully automatic satellite selection, reception and position fixing under microprocessor control. It displays position data, waypoints, bearing, range and time to go, course deviation, route tracking, clock and stopwatch. JRC is providing its JLR-3300 8-channel land mobile GPS receivers to the car industry. JRC has built a prototype of a wristwatch-size GPS receiver, the Micro GPS-A. The 50 × 56 mm, 28 mm-high unit houses an eight-channel receiver in a case that also acts as the radome for its dielectric patch antenna. A lithium battery provides up to 7 hours continuous use, or five days for receiving every 10 minutes. Another receiver, 54 × 86 mm, 5 mm thick, is designed for insertion in compact portable personal computers with PCMCIA type 2 slots.

JRC's JQE-3A Cospas/Sarsat EPIRB can be provided with two mounting brackets: float-free with hydrostatic release (to satisfy Category 1 requirements), or standard with quick release for manual deployment (Category 2). Activation can be via remote control or sea water contact. The 406 MHz transmission includes a digital message bearing the vessel's identity and nationality, and a separate 121.5 MHz homing signal provides location information for nearby rescue services. A 0.75 candela strobe light flashing 20 to 30 times per minute is included as a visual aid.

NORWAY

Jotron Electronics AS

Jotron and Total Marine Norsk A/S jointly developed the Tron 30S Mk2 Sarsat/Cospas beacon for maritime applications in two basic models. The FB3 version is mounted in a hydrostatic float-free release bracket, while its associated MB3 is designed for attachment to a bulkhead without the automatic release feature. The free versions are released at a preset water depth, typically at 2 to 3 m, and their mass distributions cause them to invert, activating a mercury trigger switch in conjunction with a seawater contact switch. Options include a flashlight and homing transmit on 121.5 MHz.

Kongsberg Navigation AS

Kongsberg is a supplier of precision position and navigation products and services. GPS activities are based on differential applications: DiffStar includes 15 reference stations that provide high-precision services through a range of receivers in the Barents, Norwegian and North Seas. Each reference station houses a GPS receiver with 12 channels and a real-time computer reading pseudo-range and ephemeris from the receiver. It calculates the difference between the measured and theoretical pseudo-range for transmission to users' GPS systems on 300 kHz/2 MHz bands.

UNITED KINGDOM

GEC-Marconi Radar and Defence Systems

The division developed the world's first miniature P-code GPS receiver and has established a range of GPS user equipment for military aircraft and marine applications. Current products include the five-channel PA9052E Nu airborne (fixed- and rotary-wing aircraft) and the PA9150 naval receiver series. Both provide the full 16 m accuracy and can be fitted with a module to ensure the equipment remains unclassified. A range of antenna subsystems is also available.

PA9052E Nu GPS Receiver is a five-channel receiver providing either stand-alone operation or it can be incorporated into an aircraft's integrated navigation system. Size/mass are designed for high dynamic and rotary-wing aircraft where space is at a premium. It provides a navigation solution at 10 Hz and a separate navigation filter to estimate INS errors. It supports dual redundant 1553, ARINC 429/575 and PTTI interfaces, but alternative interfaces can be provided. In service with RAF Tornado, Harrier and Nimrod aircraft.

PA9150 Series GPS receivers were designed for the marine environment but employ GPS modules common with the airborne equipment. The built-in Control/Display Unit provides navigation displays, including route planning. Interfaces permit integration into platforms with either Synchro or digital interfaces. The back-up battery allows operations through power outages of up to 15 minutes. The unit also supports remote Control/Display Units. Selected for Royal Navy, Royal Netherlands Navy, Turkish Navy, Royal Norwegian Navy ('Oslo' class frigates) and Royal Australian Navy ANZAC frigates.

Lokata Limited

Lokata produces three versions of a 406 MHz Emergency Position Indicating Radio Beacon (EPIRB) for use on marine craft, designed for compatibility with the Cospas-Sarsat international satellite system. Approximately 7,000 are in use. The 406H/406MH activate automatically when the vessel sinks by using a hydrostatic release mechanism. The 406M, designed for life rafts, has its own mounting bracket and is activated manually, as is the 406P personal version. All 406 MHz beacons transmit the user's identity to the satellite, which then relays the information and the location of the beacon to within 2 to 5 km to the ground stations. Once alerted, the search and rescue authorities use the 121.5 MHz transmissions from the beacon to home in on the vessel in distress.

Navstar Systems Ltd

Nova's RT260M is a Cospas-Sarsat EPIRB. Stable 5 W RF on 406.025 MHz ensures global satellite location; it also transmits at 121. 5MHz for location by aircraft. Frequency transmission is simultaneous and signals are encoded with a vessel's individual call sign or MMSI number to ensure recognition of multiple alerts.

Satellite Navigation Group

The Satellite Navigation Group (previously UKCSG) was established by RIN in 1988, and represents UK interests on the US Civil GPS Interface Committee. Membership is 250 corporate/individual members. Services include the organisation of meetings and conferences and the publication of a newsletter. It also acts as a UK node for dissemination of material from the US Coast Guard Navigation Information Service.

UNITED STATES OF AMERICA

3S Navigation

3S offers a family of Glonass and integrated Glonass/ GPS receivers.

The R-100 line of receiver products is designed to provide the scientific community with the highest possible precision measurements in the areas of time transfer, Ionospheric measurement and precision navigation using a combined solution of the GPS L1 C/A-code and GLONASS L1/L2 C/A and P- code signals.

The GNSS-300 is a 12-channel, broadband receiver with the capability of dynamically configuring any channel to utilise the GPS L1 C/A-code or GLONASS L1 C/A-code. Applications include precision navigation in harsh environments such as cities and precision survey in remote regions. The GNSS-300 can also be configured as a differential correction reference station producing standard RTCM-SC-104 correction messages for both the GPS and GLONASS, or accepting these same messages for sub-metre positioning accuracy in real time.

Allen Osborne Associates, Inc

GPS receivers cover geodesy/survey, precise time/ frequency, differential, ionospheric calibration, spaceborne and SA/A-S capability. AOA's TurboStar receiver has been supplied for Orbital Sciences Corp's MicroLab 1 satellite and for Denmark's Ørsted. For MicroLab, tracking of dual-frequency carrier phase and amplitude at 50 Hz from selected satellites are yielding atmospheric temperature and moisture content. TurboStar allows both code (C/A, P) and P-codeless (cross correlation) operation. It automatically acquires and tracks up to eight satellites at power-up without initialisation.

Ashtech, Inc

Ashtech manufactures and markets approximately 20 types of GPS receiver systems and related subsystems for precision geodetic survey, GIS mapping, and land, marine and aerial navigation. Its major surveying markets include: land boundary and cadastral, including real-time static, rapid-static and kinematic; precise geodetic; hydrographic and bathymetric. The product line includes the all-in-view Z-12 receiver and the less costly Z-12 Field Surveyor, both based upon Ashtech's Z- Tracking technology that mitigates the effects of anti-spoofing. Available in 1996 is a new line of GIS field acquisition tools, featuring Ashtech's Super C/A receiver capable of sub-m accuracy in real time or through the use of Windows-based GIS post-processing software. The company also produces the GG24 receiver, the world's first single-board receiver to integrate the GPS and GLONASS satellite-based navigation technologies.

CAST

CAST's GPST GPS evaluation system features include a 10-channel signal generator (L1, L2, C/A, P-code), standard avionics busses, and fully integrated jamming capability with up to eight jammers of six types. Options include dual navigation points (differential or relative), externally generated trajectory, and SA/AS.

Datum Inc

Datum's dual-channel GPS receivers primarily provide time transfer and optional frequency measurement information, accessing to UTC through the GPS master clock system. The 9390-5500 is designed for mounting in a standard 48 cm rack; the 9390-5700 is the airborne/portable version. Timing precision less than 1 μs (100 ns with respect to GPS tracking intervals if reference position accurate to 10 m). Position performance less than 25 m with at least 5 minutes of averaging. Only one visible satellite is required for time/ frequency monitoring providing that the receiver's location is known. Otherwise, the sets themselves provide automatic positioning through four Navstars. Initial data entry of geodetic co-ordinates is not required because worldwide geographical positioning is automatically performed by an 'anywhere' algorithm. When interfaced with an external printer, the 9390 generates hardcopy reports detailing satellite position, time/frequency, navigation data and elevation plot. The 9390-52000 was introduced in 1993. Timing accuracy is less than 300 ns under current selective availability. Position determination less than 25 m.

Del Norte Technology, Inc

Del Norte's GPS receivers incorporate a PC processor and mass memory to provide logging and interfacing capabilities. The 1009+ 12-channel C/A receiver can operate as a geodetic receiver with post-processing software, as a differential reference receiver or as a differential mobile receiver. The 4012 12-channel C/A receiver operates as a differential mobile unit and, like the 1009+, can accept differential corrections in the RTCM104 format. Both units are RTK/OTF capable.

Garmin International, Inc

The GPS 10 receiver module was introduced in 1992 for embedding in other products. The GPS 150 was introduced in 1993 and GPS 155 in 1994 for general aviation panel mounting; the GPS 95 AVD is a hand-held aviation receiver with moving map display. Garmin's GPS 100 Euro personal navigator provides 250 waypoints and a 10-route storage capability.

The FCC, in November 1995, approved the Starsys application (initially filed 4 May 1990) to build, launch and operate a system of 24 satellites in 1,000 km 53° orbits to provide two-way positioning and messaging services for remote user terminals at VHF/UHF using Spread Spectrum Multiple Access (SSMA). GE Americom at the same time acquired an 80 per cent interest in Starsys, with North American CLS retaining 20 per cent (Hughes STX sold its 5 per cent share to NACLS). GE Americom continues to pursue its separate and later 24-satellite system. Building and launching two Starsys satellites in late 1997 will cost GE Americom US$46 million, including one year of operations. The full system will cost US$150 million to US$190 million, depending on the launchers. Alcatel Espace was selected as the satellite manufacturer.

Capacity would be approximately 50,000 messages hourly at each ground station. Position determination is based on the Argos Doppler system plus ranging for 100 m accuracy. The FCC issued an experimental licence in April 1992. The company demonstrated the spread spectrum messaging concept in the allocated frequency bands 1992-93 using a fixed station working with leased capacity on CNES's S80/T satellite. Starsys has entered into a number of Early Entry Programme contracts with companies around the world for demonstration and testing of satellite positioning/ messaging using the Argos transponder aboard Tiros.

WARC 92 in March 1992 allocated VHF/UHF frequencies for the Little LEO mobile satellite service; the FCC followed suit domestically in January 1993. The forward link will be FDMA: 149.9-150.05 MHz up and 400.15-401 MHz down. The return link will be SSMA: 148.0-148.905 MHz from terminal to satellite and 137-138 MHz down to main station. The initial core of the system was to have been CNES' TAOS satellites, which the French agency decided in March 1994 not to pursue.

Honeywell Inc

The company, a wholly owned subsidiary of Figgie International Inc, was primarily a military contractor after its founding in 1956. It is now increasingly targeting commercial markets, with its GPS receiver family and the range of bandwidth on demand satellite communication terminals. Interstate introduced the SCS 2400 GPS simulator in 1995, generating 24 independent RF channels of L1/L2 GPS signals that can be routed to two RF ports. The signals simulate ground, air/space vehicles in motion with up to 30 *g* acceleration. Multipath, wide area augmentation, common GPS jammers, GPS ground transmitters, ionosphere, S/A and anti-spoofing effects can be simulated. Interstate is jointly developing with Airport Systems International a differential GPS system for airports to transmit corrections to aircraft on final approach.

The SEM-E size P(Y) and C/A code GPS five-channel continuous-tracking receiver is designed for military environments and is packaged for embedded applications. Dual antennas minimise losses of signal and allow for simultaneous tracking of satellites and pseudo-satellites. A factory-installed satellite almanac minimises acquisition time by eliminating the sky search. An option obtains stabilised fixes in less than 5 seconds.

B2 bombers were fitted in 1996 with 900 kg Mk 84 bombs carrying a Northrop Grumman tail kit incorporating Interstate Electronics' GPS receivers for 6 m CEP accuracy.

Leica, Inc

Magnavox's commercial navigation satellite division was sold to Leica in 1994. It provides complete turnkey DGPS system solutions, including communication links and application software.

Litton Aero Products

Litton's LTN-2001 is an eight-channel, third-generation C/A GPS receiver, featuring continuous tracking for every satellite in view and advanced integrity monitoring. A contract was signed with Boeing in January 1992 to provide the system for all of Boeing's production aircraft. First hardware was delivered late 1992 for flight testing. The LTN-92 ring laser gyro inertial navigation system also offers GPS operations with the addition of a single- card plug-in module. The LTN-311 Omega/VLF navigation system can be enhanced with a GPS package. The LTN-101 Flagship GPS inertial system was certified in 1996 as a primary means of navigation; its inertial system detects GPS errors before they can significantly affect navigation. The LTN-700 single channel C/A navigation set was one of the first designed specifically for commercial aviation applications.

Magellan Systems Corporation

A pioneer in developing positioning and navigation systems using the United States government's Global Positioning System (GPS), Magellan brought to market the world's first hand-held commercial GPS receiver. Focusing on expanding markets and high-volume production, Magellan's product line offering has swelled from a single GPS receiver to over 30 satellite navigation and communications products and systems. Since 1989, the company has shipped more than 1,000,000 GPS receivers while annual sales have grown from zero to nearly US$100 million.

Key worldwide markets include aviation, geo-sciences, marine, mapping, outdoor recreation, survey, vehicle tracking and navigation, wireless communications and systems integration. Magellan's products are sold throughout the United States and in more than 70 foreign countries. Foreign sales of the company's communications and navigation products account for about 35 per cent of Magellan's total annual sales.

Incorporated in 1986 as a venture capital startup, Magellan began operations in 1987 with five employees. Today, the company employs more than 350 people at its headquarters in San Dimas, California and assembly facility in Mexicali, Mexico. In 1994, Magellan merged with Orbital Sciences Corporation (NASDAQ: ORBI), a diversified commercial space products company with interests in low-cost launch vehicles, small satellites, space sensors, avionics, satellite tracking systems, Earth observation systems and related services and products. Magellan Systems operates as an independent subsidiary within OSC's Communications and Information Systems Group.

Marine users, including commercial and recreational boaters and sailors, represent a key market for GPS receiver products. The world leader in portable GPS for the marine market, Magellan's product line offers boaters a combination of differential GPS, full graphics plotter capability and portability. Products include the panel-mount NAV 6500™ GPS/chartplotter; the hand-held NAV 6000™ chartplotter; the hand-held GPS 2000 XL™, GPS 3000 XL™, NAV DLX-10™ and Meridian XL™ receivers; the low-cost NAV 1200XL™ panel-mount receiver, the GPS Sensor™ and the Differential Beacon Receiver and DBR-2.

Used in geology, geography, GIS data collection, survey, mapping, oil and gas exploration, forestry, utility management, wildlife studies, public safety, environmental remediation and other professional applications, Magellan's ProMARK X-CM™ with centimetre accuracy offers outstanding 10-channel GPS performance, unlimited data collection and multiple differential GPS approaches. Magellan's professional product line also includes the hand-held ProMARK X, FieldPRO V™, the MBS-2™ base station, and the DBR-2 DGPS reference receiver. Magellan's SkyNav 5000™ panel-mounted GPS receiver and its three portable GPS receivers, the hand-held SkyBlazer XL™ with moving map display, the lower-cost SkyBlazer LT™ and the EC-10X with terrain cartography and aviation database, are targeted to the general aviation market to meet the position, navigation, flight planning and operations needs of pilots around the world. In 1995, Magellan introduced the CNS-12™ integrated two-way digital communications, GPS navigation and automatic dependent surveillance avionics unit.

Magellan developed advanced GPS receiver modules adaptable to a wide range of OEM markets. The 10-channel AIV-10™ is used by systems integrators in applications ranging from vehicle tracking and navigation to avionics instrumentation and marine autopilots.

In 1995, Magellan introduced the world's first hand-held GPS priced under US$200. The GPS 2000 XL™, as well as the GPS 4000 XL™ and Trailblazer XL™ receivers, is designed specifically for the outdoor recreation market, including hunting, hiking, fishing, snowmobiling, camping, skiing, off-road driving and many other activities.

In 1996, Magellan introduced the GSC 100, the world's first hand-held global satellite communicator, offering worldwide two-way, e-mail messaging via the ORBCOMM low-Earth-orbit satellite network. With the Magellan handset, the user can send and receive digital text messages in an e-mail format to and from any Internet e-mail address or other ORBCOMM communicator product anywhere in the world.

Magellan also offers the microCOM-M, the world's smallest, lightest and lowest priced satellite communications telephone for global voice, fax and data communication via the Inmarsat-M satellite system. About the size of a notebook computer, and weighing just 5.5 lb, it will slip comfortably into a briefcase or carry-on.

Micrologic

Micrologic specialises in consumer navigation electronics, using GPS and Loran. The company was acquired in 1995 by Singapore's Vikay Industrial Ltd, a leader in OEM navigation electronics. The Millennium set of GPS products was due to be introduced in late 1996: ML-150 hand-held; ML-250 fixed mount; ML-350 fixed GPS and chartplotter.

The US$1,895 Admiral maritime receiver continuously tracks up to five satellites and offers a 128 × 160-pixel high resolution graphic display. 250 waypoints can be programmed (100 protected), nine routes with 20 waypoints each.

The US$1,195 hand-held Supersport waterproof receiver continuously tracks up to five satellites. 133 map datums are set in the memory and 250 waypoints can be programmed (with 100 protected by a locking system).

Motorola Inc

Motorola is one of the 100 largest US companies and a leading world electronics supplier. It introduced the six-channel hand-held GPS receiver in 1992 and the eight-channel Oncore in 1994. The Government Space Systems Operations has developed space-qualified GPS receivers for LEO autonomous satellite navigation. The Viceroy is a low-power 12-channel two-antenna C/A code receiver provided for SeaStar, MOMSNAV, MSTI, Forte and Meteor. Monarch is a 12 to 24 channel one to four antenna C/A, P-Y code receiver capable of anti-spoof/selective availability, radiation hardened to 100 krad (Si). Technologies include ASIC, MMIC and VHSIC. The receiver plays an important role in NASA's Explorer Platform and the Topex mission.

Sokkia Corporation

A subsidiary of Japan's Sokkia Co Ltd, the company is a supplier of precision products for surveying, mapping, measurement and positioning. These include electronic field books and total stations, GPS/GIS systems, software and 3-D positioning systems. The SGS Sokkia Global Surveying GPS system comprises an eight-channel receiver, antenna and post-processor to determine 3-D positions with static surveying accuracy of 5 mm ±2 PPM and kinematic surveying accurate to 2 cm ±2 PPM. Data are stored on removable 256 kbyte cards, each holding up to 9 hours. Battery allows up to 4 hours continuous use.

Sony Electronics Inc

Sony's first GPS product, the four-channel Pyxis IPS-360, receiver was launched in 1992. Designed for outdoor sports enthusiasts, it offers a range of modes in addition to the standard latitude, longitude and altitude display: navigation (distance/direction to destination, with current direction/speed), track (current direction, altitude relative to start, and distance travelled) and mark (store details of 50 waypoints and 50 markpoints in memory). Cost is about US$1,150/£685. The IPS-760 was introduced in 1993 as the first portable GPS system incorporating Arnav's Aviation Database on removable InfoCards. The eight-channel receiver offers 1,000 waypoint storage. In Japan, the NVX-F10 car navigation system is produced at 12,000 units monthly. The US$2,000 system was introduced June 1993 and comprises a GPS receiver, a CD-ROM, small GPS antenna and 10 cm colour LCD monitor.

Stanford Telecom

Stanford Telecom produces GPS satellite signal generators to simulate the exact signal environment experienced by high performance aircraft during manoeuvres. It has developed a high-performance GPS receiver embedded in a Doppler Navigator for US Navy and Marine Corps helicopter applications under subcontract to Teledyne Electronics Systems.

Tecom Industries, Inc

TECOM provides a range of GPS antennas that includes Glonass and space qualification. One version of the 401163 is optimised for L1/L2, and another at L1/Glonass frequencies. The 401163-5 is fully qualified for space applications. The 401226 houses the basic 401200 volute L1 antenna and can house a GPS receiver. The range of military GPS antennas provides gains of more than −2.5 to more than −3.0.

Universal Navigation Corporation

The GPS-950 five-channel receiver interfaces with UNS Flight Management and Navigation Management Systems. It is housed in a 1 MCU box weighing 1.50 kg. The UNS-1M NMS, available from January 1994, incorporates a GPS receiver. The company introduced its UNS 764-1 combined two-channel GPS/Omega/VLF aeronautical civil receiver in 1990. Priced at US$37,000, series production is undertaken by Canadian Marconi.

Welnavigate, Inc

Welnavigate offers GPS signal test generators; approximately 300 have been sold. Current models are Orbiter I/II, GS 100 and GS 500. The Orbiter C/A dynamic simulator provides five channels (up to 16 available) of GPS RF, an 80486 desktop computer, VGA video card and Multisynch Colour Monitor. Version I emulates plane, boat or car movements; II mimics missile or LEO satellite motion. I: US$81,000; II: US$127,000. A constellation of up to 32 satellites can be specified, along with data messages, parity, health, power levels, visibility, ephemeris errors and ionosphere/troposphere effects. Precomputed receiver vehicle motion is supplemented by selectable manoeuvres (turn, climb, accelerate), automatically controlled jerk, and vehicle silhouette masking and signal outage. It tests receiver sensitivity, tracking performance with variable south/north, dynamic tracking capability, antenna and preamp performance, and data demodulation and decoding.

SATELLITE ACTUATION SYSTEMS AND THERMAL CONTROL DEVICES

CANADA

Novatronics Inc

Novatronics Inc is engaged in the design and development of custom components and systems for the sensing and actuation of rotary and linear motion. These include brushless DC and stepper motors, linear actuators, brakes and clutches, solenoids, RVDTs and LVDTs, and electronic units to integrate them with other systems. Novatronics is working with Spar on the development and manufacture of motor modules and brakes for the Space Station and Shuttle manipulator systems.

DENMARK

Alcatel Danmark A/S

Alcatel Danmark Space's main product areas are: electronic power conditioners for SSPAs; AC/DC converters for equipment; power distribution units and solid-state power switches; servo amplifiers; magnetics components. It has for more than 21 years supplied custom-designed power electronics for Ariane and satellites such as GEOS, OTS, ECS, Marecs, Viking, Telecom 1/2, Hipparcos, Eureca, Skynet 4, ERS, Soho, Cluster, Spot 4, Inmarsat 3, EMS, Artemis, MSG, Globalstar, Thaicom, MT-SAT, WorldStar, Astra 2B and ICO.

FRANCE

Aerospatiale

Aerospatiale has developed a range of rigid solar arrays optimised for 1 to 10 kW outputs. The first generation GSR-1 is employed on Arabsat 1, Italsat 1, Telecom 1, TDF/TV-Sat and Tele-X. The GSR-3 array offers 40 to 45 W/kg and 115 W/m² (12-year EOL). The arrays are flying on Spacebus 2000/3000, such as Italsat 2 (5.3 kW) and Turksat (3 kW). The GSR-3 on Spot 4/Helios 1 use ADELE frictionless hinges. Up to seven panels, each a maximum 3 × 6 m, can be accommodated in a variety of 2-D/3-D configurations. Standard or ultra-thin (50 μm) cells can be employed, up to 6 × 8 cm. The panels comprise an aluminium honeycomb sandwich faced by 2-D carbon skins and insulated by a single Kapton sheet. Final panel mass density is 1.050 kg/m². The GSR-4 family is under qualification in 1996. The panels of co-cured aluminium sandwich with 50 μm Kapton reinforced carbon fibre face skins (max density 1 kg/m²) can host advanced Si and GaAs cells. ILSA is a deployable flexible Kapton array for 4 to 7 kW missions such as Spot 1 to 3 and ERS 1/2. A 55 kW system has been studied for ESA.

The company operates an 875 m² class 100,000 solar array integration area at La Frayère near its major Cannes facility, housing a 35 m³ vacuum chamber, a Spectrolab flasher for solar cell testing and two deployment jigs.

Alcatel/SAFT

Alcatel SAFT's Space Department was established in 1963 and has more than 25 years' experience in providing space-qualified batteries and cells. Gates' aerospace battery element was acquired at end-1993; the manufacturing facilities completed transfer to Poitiers in April 1995. SAFT's principal types are: nickel cadmium, silver zinc, nickel hydrogen and lithium couples. The company provides Ariane's 6/10 Ah primary and rechargeable silver zinc batteries. 250 Ah 38 kg $LiSOCl_2$ lithium thionyl chloride batteries have been developed for Titan's Centaur; first launch was May 1995. SAFT's energy density is up to 500 Wh/kg, with a high rate capability of 1 kW/kg. The large high-rate cell is being produced with capacities covering 5 to 250 Ah.

SAFT offers a range of hermetically sealed prismatic Ni/Cd cells/batteries designed specifically for space applications. Capacities cover 7 to 50 Ah, providing energy density of 40 Wh/kg and GEO life expectancy of more than 15 years. The VOS range of cell designs can be accommodated to match mission specifications; SAFT batteries are employed by Meteosat, Spot, Olympus, IRS and GPS; Radarsat carries three 50 Ah batteries. Ni/Cd batteries for launcher and micro-satellite applications: 1 to 6 Ah VRE cylindrical cells. VRE cells have been carried by Ariane from its first flight.

A 23-cell 60 Ah battery is under development for Artemis and DRS.

Cell rated capacity (Ah): 40-100 (VHS-CM series), 35-200 (AN series)
Depth of discharge: 70 per cent for 15 years (1,500 cycles) in GEO, 40% for 5-6 years (30,000 cycles) in LEO
Energy density: 50-60 Wh/kg
Cell design: cylindrical Inconel 718 metal case containing a stack of electrochemically impregnated nickel (positive) electrodes. Two insulated feed throughs are located on one hemispherical end. Minimum burst pressure is 190 atmospheres; maximum pressure is 75 atmospheres when 100 per cent charged.
Battery construction: the modular architecture can accommodate all space platform requirements, allowing extension of the standard configuration with minimised battery height.

GERMANY

DaimlerChrysler Aerospace AG

DASA's Solar Generator Department in Munich (see the General and Launcher/Propulsion sections for company information) was responsible for designing, developing and manufacturing the solar arrays for Intelsat 6 and the Space Systems/Loral FS-1300 satellite communication bus. Intelsat's 10.66 kW is generated at 64 W/kg and could be further improved by replacing the Kevlar face sheets with carbon fibres. Japan's Superbirds, based on SS/L's FS-1300, are the first to use DASA's Mark 4 40 W/kg array, featuring superlight open net substrates with multi-insulation. The motor-controlled deployment mechanism can also be used for retraction. 22 satellites were launched 1990-95 with DASA arrays (Intelsat 6 F1/3/5, Inmarsat 2 F1-4, Superbird A1/B/B1, JCSat 2, DFS 2/3, Eutelsat 2 F1-6, Meteosat 5/6, Italsat 1). A DASA-led consortium under ESA contract in 1994 completed development of 25.1 per cent-efficiency 4 × 6 cm Si cells for use under low-intensity conditions. This is the highest Si efficiency attained without optical concentration devices. The cells are intended for deep space missions such as ESA's Rosetta comet rendezvous. Improvements are being studied to raise efficiency to the 27 to 28 per cent at −150°C (illumination 0.03 solar constant) required by Rosetta. CISE developed 23 per cent efficiency GaAs cells under subcontract to DASA.

The former Telefunken Systemtechnik, part of DASA since 1992, provides electrical power systems, including solar cells and complete solar arrays (rollout, flexible foldout, rigid deployable and body-mounted), power conditioning, storage/distribution, in addition to spacecraft power/signal harnesses, static converters/inverters, and data-handling equipment.

Friwo Silberkraft

Silberkraft designs, develops and manufactures silver zinc, nickel cadmium and lithium batteries for space applications. Lithium primary cells covering 0.35 to 36 Ah (cell voltage 3 V) are available offering shelf lives up to 10 years. Vehicle users include Helios, AMPTE, MAUS, Mikroba, Cosima, SPAS, GAS, Texus, ASTRO-SPAS, Navex, Falke, IBBS.

Linde AG

Refrigerator and cryostat systems for handling oxygen, nitrogen, helium, hydrogen and neon. The company participated in developing the cryogenic cooling system for ESA's Infra-red Space Observatory, and provided the project's cryogenic ground support equipment. It is involved in developing a freezer system for Space Station.

Teldix GmbH

Teldix, a member of Litton Industries Inc (USA), was founded in 1960 as an affiliate of Telefunken GmbH and the Bendix Corporation. The company specialises in high-precision devices for satellite attitude-control systems, such as momentum/reaction wheels, solar array drive assemblies and antenna pointing mechanisms.

Teldix developed its Solar Array Drive Assembly system for West Germany's DFS Kopernikus but it is suitable for pivoting the solar arrays of any high-power three-axis satellite; it is used on China's DFH-3. It was qualified under the DFS programme for 15-year missions. The modular design has three subassemblies: bearing unit; redundant drive motor with sensors; slipring unit. Control is by redundant electronic units.

Specifications
Electrical transmission (A): 2 × 22.5, 8 × 4, 3 × 1.3, 12 × 0.25
Size: 22.5 cm diameter, 13.0 cm height, less than 5.6 kg
Stepwidth: 0.0375°
Torque: more than 3 Nm hold, more than 0.7 Nm stepping
Power consumption: less than 3.3 W
Lifetime: more than 10 years

ITALY

FIAR SpA

FIAR provides spacecraft full power systems (including batteries), regulators and ancillary equipment, as well as designing and supplying GaAs solar arrays. For Argentina's SAC-B, it provided four deployable carbon fibre panels carrying a total of 1,032 30 × 40 cm GaAs cells, generating 205 W EOL at 28 V. For France's Arsene, it provided six panels with a total of 986 2 × 4 cm GaAs cells. Arsene was the first satellite to be powered completely by European GaAs cells.

JAPAN

Sharp Corporation

Sharp has provided silicon solar cells for 60 satellite programmes since 1971, including Inmarsat, Intelsat and ETS 6 (50 μm).
Cell thickness: 50-200 μm
Cell size: 2 × 4 to 4 × 6 cm
Conversion efficiency: 14.8% (typical) for 100 μm BSFR cells; 17.0% (typical) for 100 μm NRS/BSF cells
Cell type: bare cell, glassed cell and CIC

Silicon diodes
Thickness: 300 μm maximum
Size: 2 × 2 cm
Absolute max rating: V_R 140 V, I_F 3 A, T_{op} −160/150°C

NETHERLANDS

Fokker Space BV

Fokker Space has been producing solar arrays since 1970, beginning with the ESRO 4 science satellite. Its Advanced Rigid Array MkII, designed for 2 to 4 kW at 40 W/kg after 10 years with panels up to 2 × 3.5 m, was selected for Inmarsat 2, Telecom 2, SAX, Soho, Hispasat and Artemis; first launch was in 1990 aboard Inmarsat. It was further developed (and completed qualification June 1988) into the Retractable And Retrievable Array (RARA) for ESA's Eureca platform (first flown 1992-3), providing 5 kW and 10-year life over five mission cycles. A power range of 2 to 8 kW can be accommodated by the design. Work on the ARA MkIII began in 1992, offering up to 7 kW (12 year EOL) for Si cells and up to 10 kW using GaAs. It can accommodate up to five panels of 7 m² each. Eutelsat's Hot Bird 2-4 will employ it, providing 5.5 kW after 12 years. Fokker is developing the 'Flatpack' array for Envisat, providing more than 6.6 kW after four years from 70 m² of Si cells on one wing comprising 14 1 × 5 m panels. The company is also providing NASA's AXAF array.

SWITZERLAND

Cryophysics SA

Cryophysics was established in 1967 to provide state-of-the-art cryogenic equipment and instrumentation to European industry and research laboratories. Its activities were broadened in the 1970s to include high-power carbon dioxide lasers. Product lines include helium liquefiers, cryogenic sensors and temperature controllers, super-insulation, closed cycle cryogenic refrigerators and cryogenic storage vessels. The Barber-Nichols cryogenic pumps are the industry standard for the efficient transfer of liquid cryogens, including LHe, LN_2 and LH_2.

ETEL SA

Études Électromécaniques SA (ETEL) specialises in the development and manufacture of high-performance electric motors and controllers: brushless torque ring motors (PM synchronous), brushless high and very high speed motors (up to 200,000 rpm), stepper (200 and 400 steps), limited angle motors, reluctant motors, linear motors (voice-coil, moving magnet), linear or rotative geared actuators and EM actuators. All can be delivered frameless or with housing, bearings, position sensors, brake and associated electronics. Examples include all motor brakes, resolvers and associated electronics for ESA's External Robot Arm; brushless motors for driving Telecom 2's optical sensors (1 W consumption, 60 to 600 rpm, stability better than 1:1000 over 10 years); electromagnets for opening ISO's container.

The 2,400 m² facility, built in 1995, includes a class 100 laminar air flow unit in a laboratory cleanroom equipped for electronic testing. 1996 turnover was more than SFr14 million, with space accounting for 40 per cent; personnel total 100.

Reusser AG

Space-related activities cover design/fabrication of cooling systems for space applications: water/freon/ammonia pumps of 2 to 800 W input. The company's pump package was used to cool experimental furnaces flown on the Spacelab 1/D1/D2 and Eureca 1 missions.

UNITED KINGDOM

Control Techniques Dynamics Ltd

The company provides high precision electric motors for space applications: solar array deployment drives, instrument positioning drives, shutter drives and antenna deployment drives. Users include Eureca, Inmarsat and ERS.

EEV Ltd

A subsidiary of GEC plc, EEV manufactures specialist electronic components and subsystems based on solid state and vacuum tube technology.

EEV develops and manufactures advanced high-performance GaAs cells and complete panels. The MGA series of GaAs/GaAs cells offer BOL efficiencies up to 20 per cent AMO (25C) and standard devices up to 16 cm². The MGE series of GaAs/Ge cells offer lower mass and BOL efficiencies up to 18.5 per cent AMO (25C), with device sizes up to 36 cm². EEV offer a full design, solar cell/panel manufacturing service for panels up to 2 m² within a fully integrated facility. Flight proven since 1991 (UoSat 5: Europe's first GaAs array), the panels have a 100 per cent reliability record. Missions include SSTL's UoSat 4/5, Chile's FASats, Pakistan's Badr-B, UK's STRV 1, Thailand's TMSat. Brazil's SACI-1, Malaysia's MYSAT. Ultrathin (5 μm) GaAs and cascade (23 per cent AMO efficient) cells are under development.

A dedicated CCD facility produces a range of custom and standard Si CCDs tailored for optical, ultra-violet and x-ray applications. Almost any size of linear or 2-D device can be manufactured, subject only to the limitation of the 125 mm Si wafer. EEV is ESA's supplier of choice: Envisat (CCD25-20 for MERIS optical imaging spectroscopy; CCD26-10 for GOMOS ultra-violet/optical spectroscopy) and XMM (CCD15-30 for RGS high resolution x-ray spectroscopy; CCD22-20 for EPIC x-ray imaging spectroscopy). Devices can be supplied to agreed specifications and screened/qualified to the required levels. Standard CCDs have been supplied for micro-satellite missions such as UoSat and FASat.

Matra Marconi Space

The former British Aerospace Space Systems element designs, develops and manufactures spacecraft power and active/passive thermal control systems, and offers cryogenic coolers derived from a unit developed for NASA's Upper Atmosphere Research Satellite. The company has experience in rigid, advanced rigid, flexible fold-out and roll-out retractable solar arrays, including the twin 16 m² silicon wings for NASA's Hubble Space Telescope. Studies are underway to evaluate the application of InP and GaAs cells in the development of flat and concentrator arrays. BAe developed an 8 kg array regulator introduced aboard France's Telecom 2 series. Using a high-speed switching technique, sections of unregulated solar array are switched in and out of circuit to match demands for regulated power within 1 per cent and with a 98 per cent efficiency. In addition, BAe developed a Solar Array Simulator.

BAe's Ni/Cd batteries are all designed/tested for GEO and LEO operational conditions. The electrical designs require that all of the cells used in assembling a flight battery are selected from one cell manufactured lot. Thermal control is achieved using secondary surface mirrors or a radiator, and a (resistive) heater chain. The battery is mounted on the spacecraft using standoffs providing electrical and thermal isolation.

The cell mechanical design requires a hermetically sealed, prismatic, thin-wall case providing a high strength to mass ratio. Cells and divider plates are sandwiched between end plates using tie bars, and the assembly bolted to a baseplate. The cells are electrically connected in series, but each cell case is isolated.

Nickel cadmium batteries
Model 3686 Nickel cadmium battery
Rated capacity: 32 Ah at 32.0 V average, 3.5 A max discharge current
Mass: 30 kg (26 cells)
Size: 336 × 425 × 202 mm
Environmental: tested to 20 *g* at 30 Hz for 60 s, −5/25°C

Model 3600 Nickel cadmium battery
Rated discharge: 35 Ah at 17.0 V average, 15.0 A max discharge current
Mass: 17 kg (16 cells)
Size: 183 × 439 × 128 mm
Environmental: tested to 17 *g* at 30 Hz for 60 s, −5/15°C

HST solar array
For design and construction of the HST Solar Array, BAe was awarded a £9.5 million contract in June 1988 for a replacement high-power version of the Hubble Space Telescope's solar array. The two wings deliver 5.1 kW BOL, reducing to 4.5 kW after two years, at 34 V when operating at 90°C. The array is qualified for LEO applications over 400-600 km, encountering 30,000 thermal fatigue cycles (−90 to +100°C) during a five year mission. The replacement array was installed December 1993.
Size: 2.6 × 12.0 m each wing deployed, 4.5 × 0.65 × 0.5 m stowed
Mass (2 wings): 87.2 kg four Si cell blankets, 115.1 kg for two Secondary Deployment Mechanisms, 81.8 kg for two Primary Deployment/Solar Array Drive Mechanism and Jettison Adaptors, 10.6 kg for two primary electrical harnesses, 14.4 kg for drive electronics. Total mass for two arrays 332.6 kg.

Solar array simulator
The solar array power simulator comprises up to 38 modules, each with variable control capable of outputting up to 150 V/10 A. Modules are designed as either a complete racked system, with controller and mains distribution functions available or in a stand-alone mode. The master cabinet houses eight array simulator modules and can accommodate three slave cabinets of 10 simulator modules each.

Solar array drive subsystems
MMS manufactures low, medium and high-power solar array drive mechanisms and their associated drive electronics and array Sun sensors. The medium power drive (carried by Inmarsat 2, Hispasat, Orion 1 and Telecom 2) employs indirect drive, with rigidly pre-loaded large diameter ball bearings capable of withstanding all launch and array deployment loads. The high-power version (employed by Olympus) uses two redundant stepper motors driving through a reduction gear. Both operate in open-loop mode and require no launch lock. The drive electronics provide pointing control for two prime and two redundant drive motors, including a differential or common offset pointing facility for solar sailing. The 75 g (52 × 23 × 35 mm) Sun sensor provides 10mV/degree error signals for Sun angles less than 4°. The low power range BAPTA (Bearing And Power Transfer Assembly) permits up to 0.8 kW power transfer and is incorporated on Telecom 1, ECS and ECS derivatives.

Specifications

High power drive mechanism
Drive motors: two stepper motors driving through a 30:1 reduction gear, generating an 11.8 Nm torque per motor at the array flange
Sliprings: 26 or 52 power sliprings each rated at 8.0 A; 50 signal sliprings each rated 0.5 A; 18 pyrotechnic sliprings each rated at 10 A for 100 ms
Friction torque: 3.2 Nm (52 sliprings) to 2.0 Nm (26 sliprings)
Power consumption: 7.5 W for Sun tracking and acquisition
Power transfer: 2.4-10.2 kW per mechanism
Position setting: two orthogonal axes to ±0.05°
Size: 288 mm diameter × 250 long max; 9.8 kg (52 sliprings) to 8.7 kg (26 sliprings)
Drive electronics: 50 VDC, 23 W (worst analytical case), 8.5 kg, 337 × 222 × 169 mm
Medium power drive mechanism
Drive motor: one stepper motor with dual stack (one redundant) driving through a 60:1 reduction gear, generating a 15.6 Nm torque at the array flange
Sliprings: 12, 18, 24, 30 or 36 power sliprings each rated at 8.0 A; 30 signal sliprings each rated 0.5 A
Friction torque: 2.4 Nm (36 sliprings) to 1.8 Nm (12 sliprings)
Power consumption: 0.07 W for Sun tracking at 1 rev/day; 6 W max for Sun acquisition at 96 rev/day
Power transfer: 0.8-5.6 kW per mechanism
Position setting: two orthogonal axes to ±0.05°
Size: 230 mm diameter × 265 long max; 7.2 kg (36 sliprings) to 5.1 kg (12 sliprings)
Drive electronics: as high power model.

BAPTA
The BAPTA is a low power mechanism using direct drive DC torque motors operating in closed-loop mode in conjunction with solar array Sun sensors. A pyrotechnic device offloads launch and deployment forces. The control electronics provides pointing control for two units and can operate in Sun acquisition, normal Sun tracking, inhibit and differential or offset pointing modes. Sun tracking error is less than 2°.

Drive motors: two DC brushed torque motors each producing 0.71 Nm at stall
Sliprings: 13 power sliprings rated at 3.5 A each, 19 signal sliprings at 0.1 A
Friction torque: 0.16 Nm
Power consumption: 0.04 W for Sun tracking at 1 rev/day; 0.85 W for Sun acquisition at 1°/s
Power transfer: up to 0.8 kW per mechanism
Size: 171 mm diameter × 267 mm long; 4.2 kg
Control electronics: −50 VDC, +5 VDC and ±15 Vdc; 2.5 W typical, 262 × 171 × 89 mm, 2.2 kg

Cryogenic cooler
Oxford University and Rutherford Appleton Laboratory developed a Stirling Cycle 80 K cooler for the Improved Stratospheric And Mesospheric Sounder (ISAMS) on NASA's Upper Atmosphere Research Satellite (UARS), and a design variant is incorporated in the Along-Track Scanning Radiometer on ESA's ERS. MMS offers a production version of the cooler for Earth observation, upper atmosphere measurements, and short wavelength/IR/radio sensor applications. The cooler operates on the Stirling Cycle principle and comprises compressor and displacer units with drive/control electronics. The compressor sets up the pressure cycle and heat is shuttled from the cold tip of the displacer by a regenerator operating in a precise phase relationship with the pressure cycle. Long life is attained from the low dynamic stress levels and the absence of contacting sliding surfaces. A two-stage 20 K Stirling Cycle cooler has been developed and qualified under ESA contract. MMS has a separate ESA contract to build another 20 K cooler together with an additional Joule-Thompson expansion stage for achieving down to 4 K.

Of the first commercial batch of six 80 K coolers, one was delivered to NASA's Jet Propulsion Laboratory and two to the US Naval Research Laboratory in January 1990; a fourth was delivered to TRW; Goddard took delivery of the fifth for evaluation in late 1990; the sixth was ESA's pre-qualification cooler. Of the second batch, three have been delivered to Japan and two to Canada; one remains at MMS on life test. Of the third batch, two have been sold to Toshiba (for Japan's ADEOS satellite) and four to Rockwell (for Brilliant Eyes). MMS is now building the Mk2 cooler (increased heat lift and power efficiency), a 50-80K cooler based on an ESA contract to optimise the 80K cooler. An initial batch of eight is being built. The first two were ordered at £460,000 by Com Development Atlantic for the MOPITT instrument, to fly on EOS-AM1 in 1998. Other orders are for Envisat 1: three for AATSR and four for MIPAS. The 4K cooler was delivered in mid-1994 to ESA for evaluation.

General Specification for 80 K cooler
Cold tip temperature: 80 K standard (60 K possible)
Cooling power at 80 K: 800 mW
Lifetime: 10 years
Size (mm): 200 long, 120 diameter; 3.2 kg compressor
210 long, 78 diameter; 1.08 kg displacer
290 × 205 × 105; 3.9 kg electronics
Power consumption: 40 W (28 VDC, adaptable to other voltages)
Operating temperature tolerance: −20/40°C
Vibration (non-operating): 20-80 Hz increasing at 3 dB/octave; 80-350 Hz level at 0.24 g^2/Hz; 350-2,000 Hz decreasing at 3 dB/octave.

General Specification for 50-80 K cooler
Cold tip temperature: 50-80 K nominal
Cooling power (mW): 800/55 K, 1,100/65 K, 1,750/80 K
Lifetime: 10 years
Size (mm): 200 long, 120 diameter compressor; 3.24 kg typical
230 long, 78 diameter displacer; 1.09 kg typical
Power consumption: 50 W
Operating temperature tolerance: −10/40°C
Vibration: 0.22 N rms back to back compressor; 0.044 N rms back to back displacer

General Specification for 20-50 K cooler
Cold tip temperature: 20-50 K nominal
Cooling power (mW): 60/20 K, 300/30 K
Lifetime: 10 years
Size (mm): 165 long, 140 diameter compressor; 3.5 kg each
210 long, 75 diameter displacer; 1.3 kg
Power consumption: 37 W
Operating temperature tolerance: −10/40°C
Vibration: 0.22 N rms back to back compressor; 0.044 N rms back to back displacer

Pilkington Space Technology

PST represents the civil business of Pilkington Optronics, a joint venture company owned equally by Pilkington plc and Thomson-CSF. PST specialises in thin space-qualified glass used for solar cell covers and second surface mirrors. More than 200 satellites, including Landsat, Hubble, GOES, Giotto, Intelsat 5-8, NStar and Cluster carry PST products. Facilities include a custom-designed glass melting furnace, advanced vacuum coating plant, automatic cleaning lines and an accelerated UV machine. The coverglasses and SSMs are manufactured from borosilicate-based glasses carrying a nominal 5 per cent cerium dioxide content to provide radiation stability by preventing the formation of colour centres under proton/electron irradiation. The glasses are available at 0.05 to 0.5 mm thickness and with a variety of coatings.

Saft Nife Limited

The company manufacturer of nickel cadmium, silver zinc, lithium and thermal batteries for the space industry.

Space Innovations Ltd

It produces the NC-X series of Ni/Cd batteries for LEO eclipse operations. Nominal operating voltage 20 V and, operated to 20 per cent depth of discharge, are suitable for LEO missions of up to two years with no reconditioning. Nominal capacities are 4/7/10 Ah, and can be paralleled. 16-cell 4 Ah pack less than 3.8 kg, 7 Ah less than 5.7 kg, 10 Ah less than 85 kg. Pakistan's Badr-B carries a 4 Ah pack. SIL also produces the 30 to 300 W PCS-28R power conditioning system which generates a fully regulated +28 V power bus.

UNITED STATES OF AMERICA

Eagle-Picher Industries, Inc (EPI)

EPI began battery development in 1920 and established the Electronics Division in 1940. 1995 sales were about US$800 million. The division accounts for 95 per cent of batteries used on NASA manned missions, 85 per cent in US missiles/weapons and 90 per cent worldwide of satellite Ni/H_2 batteries. Ni/H_2 batteries aboard more than 60 LEO/GEO satellites have demonstrated more than 170 million cell-hour in-space without failure. Satellites include Eutelsat 2, Milstar, Hubble, Telecom 2 and Iridium. 320 Ni/H_2 cells are being supplied to Space Systems/Loral for Space Station. Ground testing has exceeded 110,000 cycles at depths of discharge up to 80 per cent. Production capacity is 3,000 cells annually in the 4,600 m² dedicated facility supported by 300 personnel. Some 90 per cent of materials and components involved is produced inhouse. Test facilities can accommodate more than 1,500 cells simultaneously.

Also available are ready to mount spacecraft power subsystems, including solar arrays, interconnecting cabling, conditioning electronics and mechanical/thermal support structures and enclosures.

The company is developing three Lithium Ion cell sizes: 0.50; 10; 50. Advantages include 125 Wh/kg, conventional rectangular battery packaging, and 3.6 V/cell for lower cost.

Engineered Magnetics, Inc (EM)

EM, founded in 1955, designs, develops and manufactures custom power conditioning and conversion equipment for space applications, specialising in switching power converters. Programmes include Landsat, ERBS, GRO, EUVE, Topex, UARS and DMSP. Products cover voltage/power regulators, DC/DC converters, DC current sensors, battery chargers and Ah meters. The electronics assembly area of the 9,300 m² plant is certified for ESD; facilities cover solar array simulators, solar simulator (space chamber), vacuum chamber and vibration tables/chambers.

Their DC current sensors are a family of non-contact current sensor products for space applications using the saturable reactor principle. Modules cover full scale ranges over 1-250 A unipolar and to 65 A bipolar. Isolated output is linearly proportional to the sensed current. Optional configurations and multiple units can be accommodated.

General Atomics

General Atomics makes space nuclear power programmes, electromagnetic launchers, neutral particle beam devices, robotics, unmanned air vehicles, image intensifiers, power conditioning and advanced survivability materials. GA is a system contractor for the design, development and testing of thermionic power systems. GA develops active optics and Millimetre wave imaging systems for surveillance. Under a USAF contract, GA is designing, building and testing a prototype system for class 1.1 solid rocket motor propellant disposal using cryogenic washout for propellant removal and supercritical water oxidation for disposal. The only current methods are open burning or detonation.

John Crane Belfab

The company designs and manufactures welded metal bellows products for missiles, space vehicles and satellites, including reservoirs, accumulators, volume compensators, temperature/pressure sensors and mechanical shaft seals. It has provided several types of accumulators for Space Station. The Urine Water Recovery System version, tested aboard Shuttle, incorporates 20 cm diameter metal bellows with more than 100 individually welded stainless steel diaphragms. Another, 40.6 cm diameter and 61 cm long, welded with Inconel diaphragms, is used in the Ultra Pure Water System.

Lockheed Martin Missiles & Space Company

In a 5,700 m² facility (recently expanded from 3,700 m² to accommodate Space Station work), using an automated weld process, Lockheed fabricates and tests solar arrays with flexible or rigid substrates. Over the past 27 years, approximately 1,100 m² of array surface area has been space qualified. Based on technology that was flight proven on the Shuttle Solar Array Flight Experiment, solar arrays are being built for Space Station.
Cell type: silicon
Cell size: 2 × 4 cm to 8 × 8 cm
Power to area ratio: 120 W/m² (rigid)
Power to weight ratio: 63 W/kg (flexible).

Cryo coolers
The coolers are based on the Reverse Split Stirling Cycle, featuring: compressor and displacer moving components located by low axial and high radial rate suspension springs to maintain alignment within the gas clearance seals; linear motor-driven displacer for precise phase control with the compressor for optimum thermal performance and minimum vibration; helium working gas; low exported vibration (less than 0.18 N from a pair of compressors and 0.004 N for a pair of displacers) by the electronic controller precisely slaving the moving masses together.
Cold tip temperature: 50-120 K
Cold fingers diameters (mm): 7, 10, 12
Cooling power at 65 K: ranging 0.5-4 W
Target life: 90,000 h running, 1,000 on-off operation cycles
Connecting pipe length: 20-600 mm
Environmental tolerance: −50/70°C

Lockheed Martin Vought Systems

The former Loral Vought supplies the reinforced carbon-carbon thermal protection system for Shuttle Orbiter's nose cap, wing leading edge and chin panel. It also supplies the Orbiter's heat rejection radiators and fluid thermal controls. The division is developing two types of large deployed radiators for Space Station.

Loral Corporation acquired the missile division of LTV Aerospace & Defense in 1992 in a deal worth US$260 million. This ended a lengthy bidding competition for LTV's aircraft and missile businesses, put up for sale as a means of helping to lift LTV Corp from bankruptcy. Lockheed Martin acquired most of Loral's businesses in April 1996. LTV's best-known space involvement was the all-solid Scout orbital launcher. It completed its service in mid-1992 as NASA's smallest launcher and the last departed in 1994 for DoD.

Schaeffer Magnetics Inc

Schaeffer Magnetics was established in 1967 to produce electric motors, actuators and actuation systems for space applications. Employees total 115. The company provided the solar array drive system for the NASA/Fairchild Explorer platform, the Magellan Venus orbiter and Iridium. It also supplied the biaxial drive for Telecom 2's spot beam antenna, the shutter motor and filter wheel assemblies for the Soft X-ray Telescope on Japan's Yohkoh solar observatory, the 8 hp 800 rpm brushless DC motor for the Tethered Satellite's deployment system, the scan mirror assembly for Mars Observer's Pressure Modulator IR Radiometer instrument, the solar array drive and antenna positioning mechanism on Japan's Muses-B, and the biaxial antenna pointing mechanisms (using the type 55 drive) for ADEOS and COMETS. Its 6.35 kg scan/motor encoder will drive the 4.3 kg beryllium mirror of the MODIS instrument for NASA's EOS. The 16-pole two-phase brushless DC motor includes a 14-bit optical encoder with 11 μrad accuracy. Schaeffer is also introducing a family of reaction and momentum wheels and, later, control moment gyros. A class 100 cleanroom became operational in 1990 for assembly, cleaning, inspection and lubrication of RW bearings.

The Hubble Space Telescope carries 57 actuators, for unlatching and opening the main aperture door (type 5 actuator), unlatching and deploying the high-gain antennas, unlatching its solar panels, positioning the secondary mirror, figure controlling its primary mirror, selecting the Fine Guidance System optical filters, selecting its Wide Field Planetary Camera optical filters, and operating its tape recorders. Actuators are carried by Pioneer 10/11 and TDRS' antenna gimbal drives.

The Rotary Incremental Actuator can also be supplied with brushless DC motors.

Specifications

Types: TY-1, -2, -3, -5, -7
Output step angles: 0.06, 0.02, 0.01, 0.0075, 0.004°/step
Harmonic drive ratios: 60, 100, 160, 200
Output step rates: 31.25, 9.00, 3.75, 2.25, 0.80°/s
Power: 5, 8, 10, 12, 16 W; units can be wound for any voltage or current
Output capabilities: 1.5, 6.8, 45.2, 108.5, 361.5 Nm; estimated minima, higher capability exists
Powered holding torques: 1.5, 16.9, 45.2, 108.5, 361.5 Nm; adjustable to requirements
Unpowered holding torques: 0.7, 5.7, 19.8, 33.9, 163.8 Nm, adjustable to requirements
Torsional stiffness: 0.34, 0.68, 4.52, 11.29, 42.93 kNm
Shaft load capabilities
axial: 4.9, 7.8, 11.1, 13.3, 53.4 kN
transverse: 4.9, 7.1, 9.3, 11.1, 41.4 kN
moment: 61, 106, 203, 298, 1,628 Nm
Total actuator mass: 0.50, 0.91, 1.50, 2.0, 7.7 kg.

Spectrolab Inc

A Hughes Electronics company, Spectrolab in the space field specialises in solar cells, panels and arrays, and provides solar simulator test equipment. Spectrosun solar simulators provide up to 2,000 solar constants (pulsed) and 200 continuous: X-10 1 kW Xe; X-25 2.5 kW Xe; X-200 for large areas (hexagonal beam 127 cm diameter at 3.6 m distance) at 2 solar constants; the LAPSS (Large Areas Pulsed Solar Simulator) can illuminate and characterise large arrays in 1 ms. Spectrolab also produces the SCATS Solar Cell Automatic Tester Sorter for production line testing of up to 2000 cells/h. When used in a semiautomatic cell test system, the operator task is limited to handling the cell.

The K7700B Silicon Solar cell represents the company's capabilities in space power production.

K7700B
Cell thickness: 62 μm
Cell size: up to 8 × 8 cm
Efficiency: 14.6% minimum average
Jsc: 42.5 mA/cm²; Jmp: 40.2 mA/cm²
Vmp: 0.490 V;
Voc: 0.600 V
Pmp: 19.7 mW/cm²
Cff: 0.77
Mass (mg/cm², bare): 24

GaAs/Ge
Cell thickness: 175 μm standard; 90 μm available
Cell size: up to 6 × 6 cm
Efficiency: 18.5% minimum average
Jsc: 29.7 mA/cm²; Jmp: 28.1 mA/cm²
Vmp: 0.89 V;
Voc: 1.020 V
Pmp: 25.0 mW/cm²
Cff: 0.82
Mass (mg/cm², bare): 100 for 175 μm, 55 for 175 μm
Cell thickness: 175 μm standard, 90 μm available
Cell size: up to 6 × 6 cm
Efficiency: 19% minimum
Jsc: 30 mA/cm², Jmp 28.4 mA/cm²
Vmp: 0.905 V
Voc: 1.020 V
Pmp: 25.7 mW/cm²
Cff: 0.84
Mass (mg/cm²): 100 for 175 μm (bare)
55 for 90 μm (bare)

GaAs/GaAs Concentrator Cells
Cell thickness: 200 μm
Cell size: application dependent
Efficiency: 21.5% minimum average
Jsc: 650 mA/cm²; Jmp: 630 mA/cm²
Vmp: 0.925 V;
Voc: 1.080 V
Pmp: 582 mW/cm²
Cff: 0.83
Mass (mg/cm², bare): 115

Sundstrand Aerospace

Sundstrand Aerospace specialises in electrical generating systems, actuation systems/components, and environmental control systems, secondary and emergency power systems, and auxiliary and missile power units (such as Shuttle's APUs). Westinghouse's Electrical Systems Division was acquired by Sundstrand in 1992. Westinghouse in the 1960s entered the space market as the developer/manufacturer of Power Static Inverters for Apollo's Command Module. This led to its selection in the 1970s to develop both the inverter and a family of solid state power controllers (RPCs – Remote Power Controllers) for the Space Shuttle. Each Orbiter employs approximately 600, in three package sizes and six current ratings up to 20 A, working at 28 V DC. Current projects include design/development of candidate load conditioners (converters and inverters) both AC/DC, Remote Power Controllers both AC/DC, transformers and related components for the Space Station testbed programmes.

Sundstrand's Mechanical Systems specialises in the development and manufacture of lubrication systems for space applications; thermal hot gas systems for shutdown/startup in space; and other long life reusable equipment (seven APUs per shipset) for Shuttle Orbiter and SRBs; missile hydraulic power units.

Sundstrand's Power Systems Division specialises in small gas turbine engines, generator sets, Auxiliary Power Units and pneumatic system components.

Tayco Engineering, Inc

Tayco develops and manufactures cryogenic Platinum Resistance Temperature Detectors (PRTDs) for space applications. Users include Shuttle, Delta and satellites. The model 2921 was originally developed for Delta's propulsion system, designed for a −260/260°C range, 150 ms response in flowing liquid and able to cope with high levels of vibration and mechanical shock.

Tecstar Inc

Allied Solar Energy Corp changed its name to Tecstar in February 1996 to reflect its expanding interests. With the March 1996 acquisition of Honeywell Electro Components of Durham, North Carolina, it can offer solar arrays with deployment and motion mechanisms. 1995 revenue was about US$40 million; 1996's is estimated at US$80 million to US$100 million. ASEC began manufacturing silicon solar cells in 1958, with its 5 per cent-efficient cells used to power Vanguard 1. Its cells will help to power Space Station. Key products include space solar cell assemblies, panels and array subsystems of Si (13.8 to 17 per cent efficiency) and GaAs (18.5 to 20 per cent). 21 to 22 per cent multijunction cascade cells are produced for panels and arrays, with development work under USAF contract directed at 24 to 26 per cent. Electromechanical products include solar array drive and solar power transfer assemblies, reaction wheels, and rotary and linear actuators, components and systems. It operates the world's largest cell production facility.

Previously affiliated with Honeywell Satellite Systems Operation, Durham Electro Components was acquired by Tecstar Inc in March 1996. It designs and manufactures electromechanical and electromagnetic motion control devices and control electronics for spacecraft. These include AC, DC, brushless DC and stepper motors or eddy current dampers, coupled with gearheads, differentials and output shafts for solar array deployment/stowage drive units, solar array drive assemblies, boom/mast deployment drive units, as well as motors, clutches and brakes for the Shuttle RMS. The facility also produces motors, tachometers and resolvers for control moment gyros and reaction wheels. A space products catalogue is available.

TRW Space & Electronics Group

TRW offers a range of cryocoolers (30 to 300 K), including two innovations: a miniaturised 65K 0.25 W Stirling cooler, and a line of pulse tube coolers. Flexure bearings compressors and passive pulse tube cold heads promise 10-year lives. Life tests of both types are into their third year. Pulse tubes increase reliability and producibility by replacing the Stirling cooler moving displacer piston with a fixed geometry and pretuned orifice to control cold-end mass flow. Helium is the working gas. Efficiencies comparable to Stirling coolers have been demonstrated down to 30 K. Vibration output of the compressor is controlled to very low levels 0 to 1,000 Hz at the fundamental drive frequency (40 to 70 Hz) and its harmonics using either a driven balancer 180° out of phase or an opposed piston. All compression and balancer components are arranged symmetrically along a single axis in a hermetically sealed pressurised metal housing. Cooler mass range 1.4 to 22 kg, with 0.4 to 10 W capacity at 80 K. Maximum input 15 to 250 W.

TRW has designed and delivered four different flight qualified pulse tube cooler designs under AF Phillips Lab contracts: 1 W 35 K; 2 W 60 K; 0.3 W 35 K; miniature (two are cooling IR instruments on NASA's Lewis satellite). The flight cooler was delivered in 1996 for the Multispectral Thermal Imaging satellite. A miniature Stirling cooled payload, including a high temperature superconducting device, was delivered to NRL in 1994 for the HTSSE experiment on 1997's AF Argos satellite. NASA's Earth Observing System AIRS redundant cooler system (1.5 W at 55 K) was completed in 1996.

Yardney Technical Products, Inc

YTP, established in 1944 as Yardney Electric Corporation, is a supplier of batteries for space applications. Current and recent applications include Inertial Upper Stage, Delta, Atlas, Shuttle, Landsat, Space Suit Life Support System, Giotto, IRIS, Tethered Satellite and Cluster. The company has approximately 23,250 m² of floor space for the design, development, qualification and manufacture of space-qualified hardware. Facilities include a class 300,000 cleanroom, several Dry Room facilities and a complete Battery Test Laboratory.

YTP has developed and qualified 1-400 Ah 6 to 31 V space batteries. At battery level, energy density yields vary from 60 Wh/kg up to 130 Wh/kg.

Silver cadmium cells are space-qualified silver cadmium cells of 3/5/10/16 Ah and providing 45 to 70 Wh/kg are available as are lithium thionyl chloride batteries of 250 Ah, and providing 200 to 300 Wh/kg at battery level.

YTP is developing prismatic Li-ion cells for 3/4/5/20/100 Ah space batteries. This new technology provides 115 Wh/kg, 250 Wh/l and 1,200 cycles.

SATELLITE BUSSES

UNITED KINGDOM

Matra Marconi Space

MMS and its predecessor companies (British Aerospace Space Systems and Matra Espace) had been prime on 14 GEO communications programmes by early 1996, totalling 38 satellites. Beginning September 1977, 29 were launched and 26 achieved operational orbit, the remainder suffering from launch failures. Accumulated in-orbit time was 156 per year by January 1996, with 21 satellites still operational. BAe provided ESA's OTS satellite and derived from it three maritime Marecs, the ECS family of five European Communications Satellites (operated by Eutelsat) and the six Skynet 4s for the UK Ministry of Defence and two NATO 4s. BAe/Matra in the Satcom International partnership jointly developed the Eurostar platform from ECS as a medium-class communications satellite, with 2½ times the power of ECS. For example, BAe was prime contractor for four Inmarsat 2s based on Eurostar and Matra similarly built four Telecom 2s and two Hispasats. BAe was also prime contractor for ESA's Olympus technology demonstration satellite.

BAe was appointed prime contractor for ESA's technology demonstration Orbital Test Satellite (OTS), which formed the basis for the European Communications Satellite (ECS). The system busses are represented by the ECS 5 and Skynet 4A-C satellites specifications shown below.

Total ordered: 21 by March 1996
Total launched: 18 by March 1996
Operating: 11 as of January 1996
Future launches: Skynet 4D/4E/4F
Launch mass: 1,172 kg (Skynet 4 1,433 kg)
On-orbit mass: 680/576 kg BOL/EOL ECS (Skynet 4 790 kg BOL)
Deployed dimensions: body 2.1 m high, 1.9 m deep and 1.4 m wide; 13.8 m solar array span (17 m for Skynet 4)
Power: 1,260/1,000 W BOL/EOL derived from twin 24 kg 3-segment AEG BSR silicon wings, each panel 122 cm wide, 173 cm long; two SAFT 24 Ah Ni/Cd batteries (Skynet 4 34 Ah)
In-orbit lifetime: 7 years
Structure: communications module is U-shaped unit sitting on octagonal service module; major structural elements are predominantly aluminium alloy honeycomb with aluminium or CFRP skins. Payload components are mounted on both sides of the upper floor and on the sidewalls. The central thrust cone supporting the 91.4 cm diameter solid AKM is CFRP honeycomb with aluminium support struts
Pointing: 0.07° roll/pitch, 0.35° yaw (Skynet 4); antenna accuracy 0.2° half-cone (ECS), 0.1° (Skynet)
AOCS: momentum biased three-axis control and ±0.1°E-W stationkeeping. Two Teldix 25 Nms fixed MWs (and one 16 Nms bi-directional transverse MW for redundancy) provide pitch/roll from redundant Galileo IRES. Yaw controlled during manoeuvres by rate integrating gyro. 20 redundant 0.7/2 N hydrazine thrusters drawing on four N_2-pressurised tanks totalling 118 kg propellant (Skynet 0.7/2/5/20 N, two tanks carry 71 kg).
Thermal control: heaters plus 2.4 m² passive radiators, including secondary surface mirrors and 10 to 15 layer Kapton blankets

The Eurostar Platform Eurostar was developed 1983-86 directly from the ECS family by BAe/Matra under the Satcom International partnership as a medium-class communications platform originally offered in three models: 1000 (1,500 kg GTO mass), 2000 (2,200 kg) and 3000 (3,000 kg). It has since evolved into a general range with payload capability up to 550 kg/5.5 kW (Eurostar 2000+), increasing to 650 kg/6.5 kW with the 2000+i ion propulsion option. Eurostar 3000 is being developed to be available from early 1997 for, initially, at least 850 kg/10 kW.

BAe was awarded a £150 million contract in April 1985 for three Inmarsat 2s (1000), with options for a further six. The fourth was approved in March 1988. Matra was appointed prime contractor in 1988 for three Telecom 2s, in June 1989 for two Hispasats, and in June 1989 for two Locstars (later cancelled). Telecom 2D was ordered December 1993. BAe/Orion Satellite Corp signed a contract 1 September 1989 for the in-orbit delivery of one satellite (2000), marking the first US commercial sale by Europe. Under the turnkey contract, Lockheed Martin provided Atlas launch and insurance services. Eurostar 2000+ was selected in March 1994 for Eutelsat's Hot Bird 2. The first was exercised December 1994 for HB3; HB4 was ordered in July 1995 and Hot Bird 5 in March 1996. Orion Atlantic signed the contract 24 October 1995 for Orion 2 (2000+; 475 kg/5 kW), with an Orion 3 option. Nilesat (2000) was ordered in October 1995 for in-orbit delivery; the contract includes the complete ground segment and parts for a second satellite. Alcatel Espace chose 2000+ in December 1995 for the three WorldSpace satellites. Singapore Telecom selected Eurostar in March 1996 for ST-1. SES selected Eurostar for Astra 2A in June 1996. In January 1997, Intelsat procured a Eurostar for Intelsat K-TV, the first time that Intelsat has selected a European company as prime contractor for one of its satellites.

Total ordered: 23 to February 1997
Total launched: 12 to February 1997 (4 Inmarsat 2, 4 Telecom 2, 2 Hispasat, Orion 1)
Operating: 12 to February 1997
Launch mass: 1,190 kg Inmarsat 2, 2,145 kg Telecom 2, 2,350 kg Orion 1, 2,910 kg Hot Bird 2, 3,200 kg Orion 2
On-orbit mass: 730 kg Inmarsat 2, 1,100 kg Telecom 2, 1,450 kg Orion 1, 1,600 kg Hot Bird 2, 2,000 kg Astra 2B
Deployed dimensions: Inmarsat 2 body 2.557 m high, 1.586 m long, 1.48 m wide, solar wing span 15.23 m; Telecom 2 2.1 × 1.7 × 1.9 m, span 22.02 m; Orion 1 2.1 × 2.5 × 1.7 m, span 21.6 m
Platform reliability: 0.87 after 10 years (0.99 to station acquisition, typically 700 h after launch)
Payload mass: up to at least 550 kg; 650 kg (2000+); 850 kg (3000). Hispasat payload mass 250 kg, Orion 1 370 kg, Telecom 2 390 kg, Hot Bird 270 kg, Orion 2 475 kg.
Payload power: up to 5,500 W (2000+); 6,500 W (2000+i); 10 kW (3000). Telecom 2 2,600 W, Hispasat 2,250 W, Orion 1 2,350 W, Hot Bird 2 4,200 W, Orion 2 5,000 W, Astra 2B 7,000 W, Intelsat K-TV 7,600 W
In-orbit lifetime: up to 15 years, depending on tank selection
Structure: 1.4/1.6/2.2 m high carbon fibre/aluminium honeycomb thrust cone with shear walls separating the propellant tanks. The payload mounted on north/south panels and Earth-facing floor. East/west faces carry hardpoints for antenna mountings. Structure designed for all launcher load cases.
Power: 1 to 7 kW from twin fold-out wings of 3 to 5 panels each; each panel up to 7.5 m² Si or GaAs cells. Eclipse power from 160 Ah Ni/H_2 batteries (Ni/Cd on Inmarsat 2). Single or dual bus systems, regulated at 42.5 V sunlight (50 V option), regulated at battery voltage in eclipse.
Pointing: antenna pointing accuracy typically 0.1° half-cone; bias correction provides greater accuracy
AOCS: automatic momentum-biased three-axis control avoids the need for ground intervention. The fixed MWs provide pitch control, with solar sailing controlling yaw/roll using data from IRES.
Propellants: NTO/MMH helium-pressurised in two sets of twin tanks
Thrusters: Kaiser Marquardt R-4D 490 N liquid apogee motor, two sets of seven DASA platinum-rhodium 10 N thrusters (2000+; sets of six on other models)
Thermal systems: standard heat pipes, heaters and thermostats. Radiator area on north/south faces.

UNITED STATES OF AMERICA

Lockheed Martin Astro Space

The former General Electric Astro Space Division in 1993 became part of Martin Marietta's Space Group and in March 1995 part of Lockheed Martin. GE Astro Space was part of RCA Aerospace & Defense until it was acquired in 1986 by General Electric, which already operated its own space facility at Valley Forge. The East Windsor and Valley Forge production complexes house more than 40 environmental test chambers and 10 class 10,000 integration bays. Workforce is about 5,000. Lockheed Martin announced in June 1995 that both will close after most of the work is transferred to Sunnyvale, California, although existing commercial, Landsat 7 and EOS AM 1 contracts will be completed first. Sunnyvale, with four times the area, will become responsible for commercial, military and civil projects, including DSCS, GPS, DMSP and NOAA. The new commercial production line will build two A2100 satellite communication platforms on spec to demonstrate the capability. Some 550 jobs will move to Sunnyvale, plus 1,100 at a new East Pennsylvania facility, building satellite communication payloads and power elements.

Together, the two existing sites are responsible for 220 launched satellites through March 1996, and are working on a backlog of 62. Their record covers Tiros (38 launched over 35 years), DMSP (31 in 29 years), 14 Earth resources/observation (Nimbus 1-7, Landsat 1-6), 14 planned DSCS-3, NASA's Upper Atmosphere Research Satellite, Mars Observer, Dynamics Explorer, Transit, Nova, Polar, Wind, EOS and GPS' Block 2R satellite series. Astro provided the Viking Lander communications subsystems, the TV cameras used by Apollo on the Moon and Shuttle's closed circuit TV system. Valley Forge was selected by the Dept of Energy in mid-1990 to provide the RTGs for the Cassini Saturn mission. The 10 DSCS-3 departing Atlas 2s are delivered into GEO by the company's IABS Integrated Apogee Boost Subsystem, the first of which was delivered in early 1991.

Communications satellites form the largest single space element, beginning in 1958 with RCA providing the payload carried into orbit aboard an Atlas to transmit a taped message from President Eisenhower. The three-axis Satcom platform was introduced in 1975 and now operates with the 3000, 4000 and 5000 series (designating the approximate launch mass in pounds); the 5000 class debuted in 1991, the 7000 class in 1993 and the A2100 in 1996.

The 3000 series is the original member of the Satcom family and is second only to Hughes' HS-376 in its numerical popularity. Satcom 1 introduced the model in December 1975.

Total ordered: 34 through March 1996 (Aurora 1/2, PAS 1, ASC 1/2, Anik B, BSE, BS-2a/b, BS-3a/b, BS-2x, BS-3H, Satcom 1-3R, Satcom 1-4, Satcom C1/C3/C4, GStar 1-4, Spacenet 1– 3, Spacenet 3R, Koreasat 1/2, BS-3N)
Total launched: 34 through March 1996
Operating: 18 through March 1996
Launch mass: about 1,290 kg (about 545 kg dry)
On-orbit mass: about 760 kg
Deployed dimensions: 163 × 132 × 99 cm bus and 14.33 m solar array span
Communications section reliability: not available
Communications payload mass: not available
Payload power: not available
Design life: 10 years minimum
Structure: electronic units, batteries, propulsion and AOCS equipment mounted on six honeycomb structural pallets or panels. Transponders and housekeeping components mounted on four panels, two each on the north/south sides. Additional housekeeping equipment mounted on anti-Earth base panel.
Power: twin Si wings of three panels each provide 1,400/1,350 W BOL/EOL supported by 3 × 35 Ah Ni/H_2 batteries.
Pointing: 0.05° roll, 0.05° pitch, 0.10° yaw
AOCS: three-axis by 6,000 rpm MW (pitch) and magnetorquing (roll/yaw) controlled by Earth Sensor Assembly error signals. Propulsion subsystem is conventional 41 blowdown mono hydrazine type: four surface tension tanks (181 kg hydrazine) manifolded to two redundant sets of eight thrusters (six Olin Aerospace MR-103C catalytic and two MR-501 NSSK electro-thermal). GEO insertion by Thiokol Star 30B/C/E solid AKM
Thermal systems: passive and heaters.

The 4000 class was introduced with the launch of General Electric Americom's Satcom Ku-2 in November 1985 by Shuttle. NASA's ACTS is based on the platform, which was also selected in 1990 for Inmarsat's third generation spacecraft (Matra Marconi Space UK is responsible for the payload). The DoD's Navstar GPS Block 2R navigation satellites is derived from the 4000 series; GE was awarded the initial US$93.1 million contract in 1989, including options worth up to US$916 million for 20 to 26 satellites (21 are baselined 1996-2002).

Total ordered: 30 (Satcom Ku-1/2, Astra 1A, ACTS, Inmarsat 3 F1-5, 21 GPS 2R)
Total launched: five through April 1996 (Inmarsat 3 F2-F5, GPS 2R remain)
Operating: five
Launch mass: 1,820 kg
On-orbit mass: 1,050 kg BOL
Deployed dimensions: 170 × 213 × 152 cm body, 19.3 m span across solar arrays, total height 305.8 cm at launch
Communications section reliability: not available

Communications payload mass: not available
Payload power: not available
Design life: 10 years minimum
Structure: equipment is mounted on eight aluminium honeycomb structural pallets or panels; transponders and some housekeeping units mounted on north/south faces; antenna and reflectors, including the primary Kevlar honeycomb reflectors, and Earth sensors on Earth-facing panel.
Power: twin four-panel solar wings using N-on-P Si cells provide 3,576/2,309 W BOL/EOL; 3 × 50 Ah Ni/H_2 batteries provide 100 per cent eclipse protection
Pointing: not available
AOCS: three-axis by MW, magnetorquing and 16 blowdown mono hydrazine thrusters divided into two redundant sets of eight: six 0.89 N MR-103 Olin Aerospace catalytic Reaction Engine Assemblies (REAs) accompanied by two 0.18-0.33 N Electrothermal Hydrazine Thrusters (EHT) for increased performance NSSK, drawing on four tanks. GEO insertion by Thiokol 37F solid AKM.
Thermal systems: passive and heaters

The first 5000 satellite, Anik E2, was launched in 1991. GE provided the bus to Canada's Spar Aerospace for Anik E1/2; the stored Satcom Ku-3/4 are flying as Astra 1B and Intelsat K, respectively.
Total ordered: four (Astra 1B, Intelsat K, Anik E1/2)
Total launched: four
Operating: four
Launch mass: 2,580 kg (Astra 1B)
On-orbit mass: about 1,540 kg BOL
Deployed dimensions: 2.84 × 2.26 × 2.18 m body, 24.3 m span across solar wings, total height 2.56 m at launch (3.30 m deployed)
Communications section reliability: not available
Communications payload mass: not available
Payload power: not available
Design life: 10 years minimum
Structure: equipment is mounted on eight aluminium honeycomb structural pallets or panels; transponders and some housekeeping units mounted on north/south faces; Kevlar honeycomb antenna reflector assembly deployed from west side, Earth sensors on Earth-facing panel.
Power: twin four-panel solar wings using N-on-P Si back surface reflector cells provide 4,850/3,460 W BOL/EOL. During transfer orbit operations, the wings remain stowed against the north/south faces, the outer panels providing housekeeping power. Four 50 Ah Ni/H_2 batteries provide 100 per cent eclipse protection.
Pointing: not available
AOCS: three-axis by MW, magnetorquing and 20 Olin Aerospace blowdown mono hydrazine thrusters, divided into two redundant sets of 10: eight catalytic Reaction Engine Assemblies (REAs) accompanied by two 0.36-0.50 N MR-502 Improved Electrothermal Hydrazine Thrusters (IEHT) for increased performance NSSK. GEO insertion by dual- redundant 500 N Royal Ordnance Leros (used on Astra 1B) or 454 N TRW (used on Anik E & Intelsat K) bipropellant motors. After insertion, the bipropellant manifolds are isolated from the RCS system. Four hydrazine spheres, single N_2O_4 central cylinder.
Thermal systems: passive with heater-augmentation. Transponder panel equipment cooled by ammonia heat pipes and radiators.

AT&T requested bids in 1988 for the largest US domestic communications satellites to date: two hybrid Telstar 4s plus a ground spare. Astro was awarded the contract in July 1989; first launch was December 1993. The 7000 was selected September 1992 by Intelsat for two satellites, worth US$165 million, and in October 1992 for AsiaSat 2 (with option for a second for a potential contract of US$133 million). Intelsat 806 was ordered March 1994 for US$72.3 million, plus US$10 million in-orbit performance incentive.
Total ordered: 12 (Telstar 401/402/402R, Intelsat 801-806, AsiaSat 2, EchoStar 1/2)
Total launched: five through March 1996
Total operating: four (Telstar 402 failed minutes after launch)
Launch mass: 3,375 kg (Telstar 4); payload 420 kg
On-orbit mass: not available
Deployed dimensions: 2.16 × 2.46 × 3.15 m body, 24.3 m solar array span
Communications section reliability: not available
Communications payload mass: not available
Payload power: not available
Design life: 12 years
Structure: as 5000
Power: twin four-panel solar wings totalling 48.6 m^2 using N-on-P Si BSR cells providing 7.2 kW BOL. During transfer orbit operations, the wings remain stowed against the N/S faces, the outer panels providing housekeeping power. Two 50 Ah Ni/H_2 batteries provide 100% eclipse protection.
Pointing: ±0.1° pitch/roll, ±0.25° yaw accuracy; ±6.0° pitch, ±2.0° offset pointing control
AOCS: three-axis by MW, magnetorquing and monopropellant hydrazine thrusters. NSSK provided by two Olin Aerospace 0.20-0.23 N 1.8 kW hydrazine MR-508 arcjets. GEO insertion provided by dual-redundant 500 N Royal Ordnance Leros 1 or 454 N TRW bipropellant motors. After insertion, the bipropellant manifolds are isolated from the RCS system. Four hydrazine spheres, single NTO central cylinder.
Thermal systems: 112-312 W passive with heater-augmentation. Transponder panel equipment cooled by ammonia heat pipes and radiators.

The new A2100 platform is part of GE Americom's fleet, carrying 24 C- + 24 Ku-band transponders. Full details have yet to be released. It is expected that the A2100 will supersede the 7000 series as it can carry the same payload for 25 per cent lower total mass and has a shorter production cycle. Lockheed Martin announced 6 July 1995 the contract for two Garuda satellites for the ACeS Asean Cellular Satellite System. It will employ the A2100AX platform, the largest variant in the family.
Mass: 2,585 kg at launch, 1,575 kg on-station BOL.
Total ordered: seven (GE 1/2, Garuda 1/2, DBSC 1/2, ChinaStar 1)
Total launched: 0 through May 1996
Total operating: 0 through May 1996
Design life: 15 years
AOCS: three-axis by MWs, arcjets and hydrazine thrusters (redundant sets)
Power: twin four-panel solar wings, 5,000 W EOL, 2 × 100 Ah Ni/H_2 batteries

SATELLITE COMMUNICATIONS FACILITIES AND EQUIPMENT

BELGIUM

Alcatel Bell NV

Bell Telephone is active in the spaceborne and Earth station fields. Since 1987 it has been part of the Alcatel NV group; personnel total 5,700 (128 space). 1995 space revenue was ECU20 million. Spaceborne applications include equipment for Intelsat 7, ECS 1-5, Marecs, Giotto, Telecom 2, Spacelab, Envisat, Artemis and AC&S. It is providing the communications payload for LLHS. Equipment/subsystems supplied cover satellite electronics, power conditioning systems, Ka-band systems, LNAS. The company also provided the cooler/freezer and the incubator electronics for Biorack.

Complete turnkey capabilities are available for satellite control Earth stations; Alcatel Bell has delivered approximately 10 currently operational terminals. Examples are the three S-band terminals installed at Toulouse, Kourou and Pretoria as part of CNES' 'Reseau'. Kiruna's Tele-X and Michelstadt's Meteosat and Hipparcos terminals were also provided by the company. It also provides receiving equipment for Earth observation satellites (for example, Landsat, Spot and ERS). Alcatel Bell is also providing onboard switching control systems and ATM switching for satellites.

Newtec CY NV

Newtec CY NV includes research, development and manufacturing of various types of digital satellite modulators and demodulators (IDR, DVB, SDH with QPSK, 8PSK, 16QAM, and so on) for television broadcast and PTT operators. Built-in and stand-alone up/down converters are also available from 50 MHz to 30 GHz. The company is also a manufacturer of C- Ku- and Ka-band Earth stations (Mesh, VSAT-DAMA, and so on). Equipment has been supplied to all major broadcasting stations and PTT administrations.

SAIT Systems SA

Part of the SAIT RadioHolland Group and specialising in telecom satellite control and Earth stations, including data processing subsystems. Personnel total about 800. The company has provided Intelsat A, IBS/SMS and ECS Earth stations and mobile communications systems. SAIT Systems is system designer for the IRIS (Intercontinental Retrieval of Information via Satellite) satellite system for commercial messaging/localisation. Debut was planned for 1997. SAIT Marine offers Inmarsat A/B/C/M terminals, maritime e-mail systems, GMDSS consoles and IRCS (Integrated Radio Communications System).

TSAT manufactures TSAT 2000 which is a low-cost full duplex system for 300-2400 bit/s data transfer via satellite. A common hub station controls remote terminals in a closed star network.

INTERNATIONAL

Alcatel Qualcomm

Qualcomm Inc of the US began development of the OmniTracs mobile communications system in 1985 to provide low-cost two-way satellite communications to mobile units via existing Ku-band transponders. The service is directed at trucking fleets and other mobile vehicles. The EutelTracs European version, operating through commercial Eutelsat transponders, is marketed by the Alcatel Qualcomm joint venture, formed in March 1990 as part of the Alcatel Radio-communication, Space & Defense Group; Qualcomm Inc holds 49 per cent. It is commercialised throughout the coverage zone by a network of 18 national service providers. The QASPR automatic satellite position reporting service provides 200 m accuracy covering the whole of western and eastern Europe, the Mediterranean Basin and the Middle East using two Eutelsat satellites. In fact, accuracy has been 80 m since September 1993 when messaging was transferred from Eutelsat 1-F2 at 1° east to 1- F4 at 25.5° east; positioning remains on 2-F4 at 7° east. As Eutelsat operates at Ku-band, the antenna locks on to the first satellite, rotates to the second (Ranger) and back - the sequence requiring about 10 seconds. The Ranger provides information on the time difference, precise satellite positions and the terrain altitude model stored at the hub. Computed position accuracy depends on this model, timing errors and the satellites' orbital separation.

Individual customer dispatch facilities connect via regional centres to the Network Management Facility (9 m antenna) at Rambouillet in France to send or receive messages of up to 1,900 characters to any terminal in their fleets. A message is coded and dispatched via satellite to the prescribed vehicle or group of vehicles. The Facility waits for acknowledgement that it was received correctly; if not, it automatically repeats the procedure according to a system-wide retransmission algorithm. The driver's unit can retain the last 99 messages or 256 lines.

Testing within Europe began in June 1989 through Eutelsat, using the UK's Goonhilly as the hub. Commercial operations began 1 January 1991, offering a system capacity of 45,000 users within the Eutelsat region. The primary market is the road transport industry, with others such as non-civil, tourism and emergency links. Projected market size is about 80,000 for the first 10 years; there were 5,500 active terminals by end-1994 and 10,000 by end-1995 (11,000 delivered for 18 service providers). 10,000 were required to break even. 19,000 are projected by end-1996 for 24 service providers.

The Mobile Communications Terminal (MCT) comprises three units: Outdoor Unit, Communication Unit and Display Unit. The Outdoor Unit contains the antenna assembly and front end electronics. Including electronics, it is 28.95 cm diameter, 17.0 cm high and weighs 3.94 kg. The Communications Unit contains an analogue section, digital electronics and the QASPR receiver, all on four circuit card assemblies. Size: 32.4 × 23.4 × 11.4 cm, 4.07 kg. The Display Unit consists of a 40 character by four line display with keys for pre-programmed user functions.

SOUTH AFRICA

Grinaker Telecom

GEL introduced the Thinroute M Inmarsat M package in 1994 for communications from remote areas. It comprises a conventional NERA Inmarsat M satellite phone with a specially developed 2.3 m antenna. The 60 kg modular parabolic reflector is designed for fixed/semi-fixed external applications, connected by a single coaxial cable to the satellite phone inside a building. An upgrade supports 16 simultaneous Inmarsat M voice, fax and data channels.

SPAIN

Alcatel Espacio SA

Alcatel Espacio was created in 1988 as an Alcatel Standard Eléctrica subsidiary to specialise in spaceborne communications equipment and subsystems: RF passive equipment (microwave filters and diplexers, Ku-band input multiplexer); RF active (TT&C S-band transponder); digital electronics (antenna pointing system, data handling, baseband). Project involvement includes TT&C S-band transponders (Hispasat, Helios 1/2, Spot 4/5, Soho, Cluster, Telecom 2D, Hot Bird 3/4, XMM, WorldStar); filters, diplexers and imuxes (Turksat, Amos, Artemis, Envisat, Arabsat 2, Telecom 2D, Globalstar, Mabuhay, MT-Sat); antenna pointing system (Polar Platform, Envisat), data handling (Spot 4 remote terminal unit, Helios 2/Spot 5 standard Module de Gestion, Soho common data processing unit; XMM data bus unit, Sesat communication controller unit); baseband equipment (transmit interface ASIC for onboard processing switch, Artemis optical BPSK modulator, MSG raw data modulator, Skyplex). Personnel total 110; 1995 turnover was Pta1,750 million (1994 Pta1,550 million; 1993 Pta1,450 million; 1992 Pta1,400 million; 1991 Pta1,300 million). The company operates a 7,300 m² facility, including a 1,000 m² class 100,000, 10,000 and 100 cleanroom.

Telefonica Sistemas De Satelites SA

A subsidiary of Telefónica de España, TSS was created to provide satellite business telecommunications. As Spanish service provider, it specialises in satellite communication provision, development, integration, space segment access and provision, installation, operation and maintenance of VSAT networks and business TV as main market segments. Coverage is Spain and Latin America; the merger with Unisource Satellite Services provides access to the whole of Europe. TSS's centre in Madrid operates approximately 21 VSAT networks using Hispasat, Eutelsat and Intelsat. It also offers Inmarsat services.

TAIWAN

Jonsa

Jonsa was formed in 1988 and now manufactures approximately 90,000 C/Ku-band satellite TV reception dishes monthly in its 8,200 m² facility. The 0.35 to 4.2 m offset/prime focus dishes offer ground, wall and polar mounts.

UNITED KINGDOM

Andrew Limited

Andrew Limited builds Earth station antenna systems, ranging from 1.8 m VSATs to 9.3 m Intelsat B.

COM DEV Europe Ltd

The company was established in 1985 as a wholly owned subsidiary of COM DEV Canada to design, develop, manufacture and test multiplexers, filters, receivers, amplifiers, ferrite devices and antenna beam-formers. Project involvement includes GE 1/2, Telecom 2, Artemis, Cassini and Intelsat 7/8. Test facilities include RF benches of up to 40 GHz and thermal/vacuum chambers.

GEC-Marconi Communications Ltd (GMCL)

GMCL is a Management Company of GEC-Marconi Ltd and designs/manufactures systems and equipment for satellite communication ground stations; tropospheric scatter communications; sound broadcasting and TV; line transmission systems; civil, military and naval communications in most frequency bands; mobile radio communications, including fixed/mobile; cellular car phone systems; network management; digital data systems; LF to VHF antennas; and computer software management. It employs 1,300 at its main plant in Chelmsford. All manufacturing, testing, design, development systems engineering and supporting activities are carried out at this site.

GMCL is structured on a divisional basis; the Civil Communications Division specialises in tropospheric scatter radio; specialised supervisory systems for monitoring, control and management of communication networks; satellite stations for Intelsat, Eutelsat, Inmarsat & domsat; Inmarsat ship/land terminals are handled by Marconi Marine.

Marconi entered the space communications business in the mid-1960s with the design, manufacture and installation of an Earth station for Cable & Wireless on Ascension Island in connection with the early NASA space programme. Since then, it has been involved in building all classes of satellite Earth stations, ranging from Intelsat A/B C-band and Intelsat C and Eutelsat Ku-band stations to the transportable/portable Earth stations for SNG and portable communications. The product range was expanded in 1990 to include stations for offshore platform deployment and in 1991 to include standard EBU Earth stations. The Systems Engineering Dept supported by the GEC-Marconi Research Centre undertakes system studies, planning and link analysis as well as the system design of Earth stations. Marconi's equipment includes modems and frequency conversion equipment for digital communications, allowing the upgrade of existing Earth stations from analogue to cost-effective digital operation.

Go-Sat International

Go-Sat International (GSI) was formed in 1992 to bring together the partner's knowledge and expertise in the satellite arena, including many years of board level involvement with the international satellite consortia. GSI provides broad-based independent management consultancy services mainly, but not exclusively, related to the field of satellite telecommunications.

Kudos Technology Ltd

Kudos Technology provides consultancy in space/ground-segment satellite communication systems and supplies ground terminals (including Inmarsat portable terminals) and subsystems for large Earth stations.

M/A-Com Limited

Microwave amplifiers, switches, phase shifters and microwave semiconductors for space applications. Project-qualified for approximately 20 payloads, including Skynet, Italsat, Telecom 2 and Eutelsat. Semiconductors ESA qualified to ESA/SCC 5010.

Marconi Marine

A division of GEC-Marconi Communications, Marconi Marine designs/produces the Oceanray family of satellite communication stations, marine/land mobile satellite communication systems and peripherals and data transmission, facsimile, telex and video pictures (via Code C compression) over satellite.

Satpax Marconi Marine's Satpax TP-01 Inmarsat A terminal provides telephone, telex, fax and data services using equipment transported in two cases and operational within 20 to 25 minutes. The 46 kg single-case (34 × 54 × 63 cm) Satpax 2 was introduced in 1992.

Marine Technology International Ltd

STC International Marine's communications satellites and marine radio business was acquired by Mobile Telesystems Inc of the USA in January 1993 as its marine specialist arm.

Multipoint Communications Ltd

Founded in 1982, Multipoint specialises in fixed, mobile and flyaway Earth stations for C/X/Ku-band TV, data and voice, along with SCPC and MCPC, MESH or STAR networks. The SE9530/9730 Mobile Earth Station M2400 is a 1.8/2.4 m fast-action station for fitting on a transit van or Land Rover. The 1.5/2.4 m SE9510/9710 Ranger is a 318 kg flyaway diamond-antenna station for TV, data and voice. The basic terminal can carry 300 W HPAs for TV or 20 W SSPAs for data. For voice, eight channels are offered in a 128 kbit/s data stream.

Racal Antennas Ltd

Earth terminals for all applications, including Inmarsat C and airborne.

Racal Avionics Ltd

The division is the group's centre of expertise for military satellite communication and Earth observation systems. The leading UK contractor for military ground stations, Siemens Plessey Systems supplies turnkey networks, large anchor stations, network control centres, spacecraft operations facilities, medium/small fixed terminals, medium/small transportable terminals, man-pack terminals and Earth observation ground stations. Studies are underway in connection with Skynet 5, the UK's next generation of military satellites.

Acting as prime contractor and system design authority, it supplied the £40 million FSC 646 satellite anchor station at RAF Oakhanger to the UK MoD as the hub/central control point of the UK Military Satellite Communications System (UKMSCS). The station was handed over in 1986 but in 1990 the company was awarded a further £10 million contract to provide significant expansion. It has now implemented the £60 million expansion and diversification: a second large anchor station, a new Satellite Network Control Centre, improvements to Oakhanger and another existing large ground station and provision of two large transportable terminals with network access points. These facilities are linked by terrestrial bearer systems to provide an integrated, survivable ground segment network transparent to the users.

The company supplies full hemispheric coverage ground stations for polar orbiting Earth observation satellites such as ERS, Landsat and Spot. For example, it constructed the UK's 13 m ERS 1/2 station at West Freugh in Scotland.

The Transportable Rapid Deployment Steerable Mount Antenna (Tradesman) is a 7 m diameter transportable antenna that provides high RF gain to satellite ground terminals requiring a rapid deployment capability. The X-band version, supplied to the UK MoD for the TSC 666 Transportable Satellite Ground Terminals, is designed for high-capacity, multicarrier terminals that handle a variety of communications circuits and are equipped with anti-jam facilities. The S-band version, used for the FSC 671 Satellite Ground Terminal, is for TT&C of satellites on-station or in drift or transfer orbits. A fully transportable TT&C station is also being supplied to NATO.

Provost is a self-contained medium TSGT for rapid-deployment SHF satellite communications. It comprises a single 8.5 m trailer package with an electronics cabin and a 4.5 m Cassegrain antenna with two hinged sections on the main reflector folding inwards for transportation. It can be set up by a two-man crew and left unattended for supervision by a remote Monitoring/Control computer.

The X-band Flyaway Terminal is a 1.8 m antenna military terminal packaged à la civil SNG equipment, transported in rugged transit cases for an aircraft hold or small van. It can be deployed by one man in minutes, operated in a stand-alone configuration or for user circuits and networks.

Satpac is a man-portable terminal that can be carried in a small suitcase or rucksacks. They employ DAMA demand assignment multiple access to minimise usage of satellite transponder power/bandwidth, allowing hundreds to operate within a network. Satpac offers secure voice and data traffic over all current military SHF satellites.

EOSGT can directly acquire data from all current X/S-band observation satellites, capable of tracking down to 500 km altitude and receiving up to 144 Mbit/s. Initially designed for use with ERS, Spot and Landsat, the modular approach allows for future expansion to Radarsat, Envisat and US and Japanese Earth observation systems. A terminal was acquired by the Royal Aerospace Establishment (now part of the Defence Research Agency) as the UK's primary data reception station, at West Freugh. The high gain antenna offers a figure of merit better than 35 dB/K at X-band for high-quality data reception (BER at 1 in 10^{-5} for data recorded from ERS at 105 Mbit/s). Single channel monopulse tracking ensures rapid satellite acquisition near the horizon, and the 7° antenna tilt axis eliminates the keyhole problem with zenith passes.

Dual-band operation is achieved by the use of an S-band feed at the prime focus and an X-band feed at the vertex. The integrated control/monitoring subsystem provides full control from a single console.

The Management of Satellite and Terrestrial Resources & Operations system provides management of complete satellite communication networks. A version has been supplied to the RAF for managing the diversified UKMSCS ground segment. The System Management Facility is the central point of management and enables system, terrestrial network and satellite access using resource and connectivity databases; the Access Monitoring & Policing Subsystem conducts automatic spectrum monitoring by scanning the downlink and comparing the carriers with the SMF access plan; the Satellite Network Expert helps to identify failures and anomalies.

The Siemens Integrated Monitor And Control System, although originally designed for satellite ground stations, can be applied in any installation where control/monitoring of diverse equipment is required from a central location. Based on COTS hard/software, it offers a graphical user interface presenting the operator with mimic diagrams grouping functionally related information. The Windows operating environment allows users to monitor important system data, such as alarms, while simultaneously performing lower level tasks.

Siemens Plessey Systems was recently awarded a contract by the NATO CIS Agency, won under international competition, to supply a Transportable Satellite Ground Terminal (TSGT) for telemetry and command of NATO spacecraft. The terminal design is based on a similar system already successfully in service within the UK SATCOM Anchor Station System, for which the company was also the prime contractor.

Operating in the 2 GHz S-band, the TSGT will provide a secure and reliable RF link to ensure the health and integrity of the spacecraft and to command the satellite in response to changing operational requirements. The electronics equipment is contained in a standard ACE II equipment shelter and the antenna is the established Tradesman design as supplied to the Royal Air Force. Full remote control and status monitoring to allow unmanned operation is provided by SPS's proprietary SIMACS system.

The MATELO project upgrades the UK component of the NATO Maritime Air Telecommunications Organisation facility, which is part of a NATO-wide distributed communications network providing telegraph and voice circuits between UK command centres and NATO aircraft, in support of maritime operations within the Eastern Atlantic region.

The upgrade is being achieved by expanding an existing RAF system, the Strike Command Integrated Communications System (STCICS) through the provision of new terminal and control equipment and by connecting existing MATELO HF radio facilities (suitably modernised) into the expanded STCICS. The resulting system combines the capacity to accommodate a major increase in traffic with improved operational flexibility for command staff.

As prime contractor for MATELO, SPS has managed the whole project, including the civil works. The project team has been responsible for the system design and integration, including interfaces with existing systems (principally STCICS and UNITER). Work has been carried out at nine sites, as far apart as Cornwall and the North of Scotland, with civil works at two of those sites. One of the key requirements has been the maintenance of full operational capabilities during the implementation programme.

For those elements of the system located at RAF St Mawgan and RAF St Eval (both Cornwall), SPS was responsible for the design, supply, installation and commissioning. On both these sites new buildings have been constructed and at RAF St Eval a new antenna system has been erected. At the other sites – RNWS Penhale Sands, RAF Bampton Castle, RAF Kinloss, RAF Chelveston, RAF Milltown, Northwood and Pitreavie Castle – SPS was responsible for design and supply, including the provision of installation design packages.

SPS delivered the first Containerised Relocatable Satellite Terminal (CREST) in the autumn of 1994. CREST was developed by Siemens Plessey Systems to cater for those satellite communication users who need low-cost, high-capacity, high-reliability ground terminals which can be easily 'relocated' from one operating site to another if requirements or priorities change.

Set-up and tear-down takes two or three days, but CREST has the advantage that it can be shipped complete as standard commercial freight. In addition, the use of Commercial-Off-The-Shelf (COTS) equipment throughout means that the costs of initial purchase and through-life support are minimised.

Individual CREST components are selected for their reliability, thus achieving a high system availability. This, combined with a comprehensive graphics-based control and monitor system, based on SIMACS, allows worry-free, unmanned operation.

CREST provides a flexible solution to most communication needs – each system can be configured to match individual customer requirements. Systems are available in C, X and Ku-band configurations and have been delivered with communications capacities ranging from 64 kbit/s to 8 Mbit/s.

The Joint Tactical Information Distribution System (JTIDS) is a secure, jam-resistant, extended line-of-sight RF communications bearer for Link 16, providing integrated digital and voice communications, navigation and identification capabilities. Communications survival is ensured by the use of frequency hopping and spread spectrum techniques.

Siemens Plessey Systems has been executing the Royal Navy JTIDS Ship System (RNJSS) contract since June 1995. The contract is for Full Development and Production, including installation into Type 42 Destroyers and CVSs (Carrier Versatile Support). The order is for 12 complete systems and modification kits for three JTID S terminals bought by MOD under the US Foreign Military Sales programme. These terminals are currently installed in HMS Illustrious and at the Land-Based Test Site situated at Portsdown.

RNJSS consists of two major elements – the JTIDS Class 2 Terminal and the Network Control Initialisation and Data Preparation Sub-System (NCIDPSS) – plus two antennas and audio ancillaries.

The Class 2 Terminal comprises JTIDS equipment, currently in service with the US Navy, adapted to meet Royal Navy requirements. The terminals, modification kits and spares are being procured from Rockwell Collins Avionics and Communications. The first Terminal is due for delivery to Christchurch in March 1997, whereupon it will form the initial building block of the Integration and Test system. The remainder were delivered on a monthly basis until completion in 1998.

The NCIDPSS provides the JTIDS network management facility. Its major function is to initialise, monitor and control the JTIDS Ship System, to manage the JTIDS/Link 16 Network, to prepare initialisation data for the Aircraft Initialisation Peripheral in the CVS and to supply initialisation data parameters to the Data Link Processing System for Link 16 and multi-link operations. The NCIDPSS is a new design and represents the main development item in the RNJSS project. To date the majority of software modules have been completed and integration has commenced. The prototype hardware has completed environmental and EMC testing and is now undergoing informal TEMPEST testing.

UNITED STATES OF AMERICA

Andrew Corporation

The parent company of Canada's Andrew Antennas Co Ltd; annual sales through September 1995 were US$626.5 million for a US$67.8 million profit (FY94 US$558.5 million/US$44.4 million).

Antenna Technology Communications Inc

ATCI manufactures the Simulsat Multibeam Antenna, a semi-parabolic antenna capable of receiving 35+ satellites simultaneously within a 70° view arc at C/Ku-band. Available in 3/5/7 m equivalent sizes. The company also provides new/refurbished 1.8 to 32 m parabolic antennas, satellite head end electronics and maintenance.

Astroguide International

The company was created in 1990 to develop/market three satellite products: the Trax II satellite tracking system, a refrigeration system for cooling focal point electronics of satellite dishes (40°C differential), and the Quad Trax VI motor (designed to eliminate counting errors). Trax was originally designed to track the Gorizont 3 Statsionar 12 satellite in its inclined orbit for customers in the Middle East receiving CNN broadcasts. It continuously measures the signal strength by interfacing with the automatic gain control or tuning meter and moves the antenna to pick up the strongest available signal. Each search is triggered by the drop in signal strength at a user-set value (strength is displayed on a range of 1-250). The IIC is a rack-mounted commercial version.

AT&T Tridom

The Aegis Network Management System provides real-time event-driven telemetry for navigating, monitoring and troubleshooting large complex satellite-based networks. Aegis can detect, isolate and assess network problems, control growth or configuration changes, manage capacity and control access. The company has provided approximately 8,000 of its Clearlink VSAT terminals since 1986. Clearlink products are manufactured at Tridom's HQ and at the AT&T facility in Merrimack Valley, New Jersey. The Clearlink System 400 third generation, typically 1.8 m diameter, features a software-based architecture that can be programmed for a range of satellite transmission formats. It provides complete Ku/C-band data, voice and video. VSATs transmit data to the hub at 64 kbit/s and receive transmissions from the hub at 128/256/512 kbit/s. Outgoing data are transmitted in a first come, first served continuous time division multiplexed fashion. Access to the inbound channels is shared by multiple VSATs using a multiple access technique such as Aloha, fixed frame TDMA, reservation TDMA or fixed assignment. Addition of the Clearlink Voice Link Module provides two full-duplex digital voice channels.

The hub station (3.5, 4.5, 6.1 m diameter) acts as a central satellite access and switching node, providing connections to the Clearlink Network Control Computer and to Host Interfaces that connect host computers to the network. Hubs can be shared or private; AT&T Tridom operates shared hubs in the USA. All hub equipment is fault tolerant, with automatic switchover to prevent interruption in the event of a failure.

Aydin Corporation

Telecommunication, telemetry and avionics equipment for satellites, spacecraft and missiles. Aydin produces satellite uplink HPAs, low noise GaAs FET amplifiers, up/down convertors, continuous and burst-type satellite modems, and TDMA systems. Turnkey Earth stations and integrated transmission networks at C/Ku-band have been provided to Belgium, Chile, China and South Korea, in addition to US common carriers. The company's 1000 series GPS amplifier is a linear power UHF unit employed with the navigation satellite tracking system. The SP-100 Microprocessor Based Data Reformatter is used in missiles and standoff weapons to provide GPS tracking information to communications downlinks. Aydin's airborne measurement system is installed in the redesigned Shuttle solid booster. Employees total 1,300, including 500 engineers and computer specialists.

Ball Aerospace & Technologies Group

BATG reported 1996 sales of US$362 million, with an end-1996 backlog of US$326 million. The company was founded in 1956 and comprises the Aerospace Division and the Telecommunications Products Division. The latter division produces antennas and video products to customers in the scientific, military and aviation communities. Ball Telecommunications Products has been involved in the aviation industry since the early 1970s when engineers designed the first-of-its-kind airborne antenna for the FAA: this was the precursor to the Airlink satellite communications system used by more than 20 airlines around the world, providing telephone and fax services to airline passengers. Ball-built antennas that can access the GPS provide navigation information, returning accurate location data to military pilots. Ball antennas are also found on many space missions: the Manpack antenna is launched on each Shuttle mission, for example. Flown twice aboard the Shuttle in 1994, the Spaceborne Imaging Radar C returned radar images of the Earth and is planned for a third launch in 2000 on the Shuttle Radar Topographic Mapper.

California Microwave, Inc

Founded in 1968, the company specialises in satellite and wireless communications, including satellite Earth stations. 1995 sales were US$467.9 million (1994: US$369.0 million). Satellite Transmission Systems, Government Communications Systems, EFData and Sunnyvale products are business units with space-related products.

Center for Satellite & Hybrid Communication Network

The centre was established in October 1991 as one of NASA's Centers for the Commercial Development of Space. Its primary goal is development and commercialisation of advanced space-based communications, emphasising hybrid satellite/terrestrial networks. Specialist areas cover teleconferencing, DBS, HDTV, database access, supercomputing linkage, satellite digital cellular networks, narrowband mobile networks and personal communications.

Collins Air Transport Division (CATD)

CATD's SAT-906 six-channel Satcom system has replaced the SAT-900 as the standard Collins ARINC 741 offering for commercial aircraft communications satellites; the first entered service December 1993. Since then, approximately 400 have been delivered. Approximately 130 of the low data rate SAT-900 were delivered; all have been upgraded to the 906. Operating through Inmarsat, SAT-906 comprises three line replaceable units: satellite data unit, high-power amplifier and RF unit. Any channel can support cabin or cockpit voice, fax or data transmission. The modular channel architecture allows the loss of one channel without total system failure.

Collins Avionics & Communications Division (CACD)

CACD introduced its two Inmarsat M satellite phones in 1993: SecSat provides an STU-111 phone to encrypt all voice, fax and data communications; ExecSat operates without this. The STU-111 is the US government standard phone encryption device; several hundred thousand are in use by government agencies worldwide. Each satellite phone is packaged in a 20.3 × 45.7 × 33 cm briefcase; total mass 11.4 kg. The flat antenna is removed and aimed at the satellite; a dial tone is heard in the handset when the satellite is acquired.

Communications & Power Industries, Inc (CPI)

CPI, formerly Varian Electron Device Business, designs/manufactures klystrons, travelling-wave tubes, magnetrons, cross-field amplifiers, gyrotrons, mm-wave tubes, power-grid tubes, microwave amplifiers, power supplies and radar components. CPI's Satcom Division is a market leader in high-power uplink amplifiers, with approximately 10,000 amplifiers in more than 151 countries. It offers L/C/X/Ku/Ka bands up to 10 kW. Accessory products include computer and remote control, output switching and power-combining networks.

COMSAT RSI

Radiation Systems Inc, founded in 1960, was acquired by COMSAT in 1994 and merged with COMSAT Technology Services to create COMSAT RSI, able to provide approximately 80 per cent of complete satellite communication ground systems. Turnkey systems range up to 32 m.

Comsearch

A member of The Allen Group, Comsearch provides engineering services and software to the telecommunications industry, including mobile, microwave and PCS.

ComStream

ComStream was founded in 1984 by management and engineers from Com Corporation's communications division as a developer/manufacturer of products for digital satellite data networks, from modems and protection switches to fully integrated Earth stations. In December 1992 it became a wholly owned subsidiary of Canada's Spar Aerospace. 1995 revenue was C$250.7 million, for C$10.5 million operating income. The company is supplying approximately 10,000 receive-only VSATs for use in the Reuters Information Services data-broadcast network worldwide. Latest product is Spar's VSAT Plus II C- or Ku-band multiservice system for low-traffic-volume locations on private or public networks. Data rate 1 to 15 Mbit/s; for example, four voice-encoding rates up to 64 kbit/s and data transmit/videoconferencing up to T1/E1. It does not require a hub because of its distributed network control architecture. Others are the DT7000 Earth station (1.2 to 2.4 m antenna system for two-way Ku- or C-band services) and the IntelliCast 501 75 cm to 1.8 m Ku-or C-band rx-only terminal. The DT5000 offers higher-power RF units or C-band. Modems available are: CP101 continuous or burst mode for transmission and burst mode for receiving; CM701 IDR- IBS- SMS-compatible variable-rate modem with electronic selection of data rates in a 4.8 to 2,048 kbit/s range, digitally tunable in 1 Hz steps. Protection switches, high-performance coding and Doppler buffers are also marketed. IntelliCast 501 Data Broadcast Receiver: IntelliCast is an integrated rx-only Earth station operating at C- or Ku-band and data rates over 9.6 to 2,048 kbit/s. Variable symbol rates, programmable in 1 bit/s steps. The terminal comprises an antenna, low-noise block down converter and the receiver indoor unit housing the demodulator, RF converter and power supply. Applications include online document or image retrieval, database transfer and real-time news or financial information networks.

Comtech Antenna Systems Inc

A division of Comtech Systems Inc, the company manufactures a range of satellite communications antennas. This includes the 3.0/3.8 m C-band transmit/rx systems and 2.4 × 5.5 m Offsat transportable antennas (designed specifically to meet the FCC specifications for 2° spacing). The 3.8 m inclined-orbit antenna uses a dual-axis polar mount to follow satellites straying beyond GEO. The 2.4 m Ku-band is designed for the VSAT market. The 7.3 m Earth station features a dual-axis polar mount. The 1.8 m C- or Ku-band flyaway terminal can be packed in three cases totalling 150 kg and shipped as air baggage.

Cynetics Corporation

Communications systems engineering and small-satellite, off-the-shelf digital and analogue transponders. The CTL 57 mode L digital transponder was provided for SEDSat 1, 0.83 kg, 3 to 7 W out, 400 to 450 MHz. The CTA 1 mode A linear transponder translates a 60 kHz bandwidth at 146 MHz down to 29 MHz; three-year life, 10 W out, 0.85 kg.

Datron Systems Inc

Datron provides satellite communication antenna systems for shipborne, ground-based and tracking requirements; its image-processing division in San Jose was incorporated into Simi Valley in July 1996. It delivered 13 m Multi-Satellite Data-Acquisition Systems to the Swedish Space Corporation and the National Central University of Taiwan and, in 1995, completed the 12.8 m antenna and related complex for the Center for Remote Imaging, Sensing and Processing at the National University of Singapore.

Datron has installed X/S-band remote sensing data-reception terminals in the US, UK, Norway, Sweden, Kenya, Thailand, Taiwan, Indonesia, Australia and South Africa to handle Landsat, Spot, ERS, JERS, MOS and Radarsat imagery, supporting 55 to 105 Mbit/s. The 9.2 m dual-band satellite-tracking antenna system comprises an antenna, series 8400 pedestal, RF feeds, antenna control unit, data receivers and associated patch facilities. The system is typically configured to receive 2.2 to 2.3 GHz (Landsat, Spot, DMSP) and 1.67 to 1.72 GHz (NOAA, Meteosat, GOES, GMS).

Datron's Metrack-8 was the first system specifically designed for use with polar-orbiting meteorological satellites; 16 were installed in 12 countries. The Mk II provides simultaneous reception at LH/RHCP over 1.540 to 1.710 GHz. It allows autotrack of NOAA's HRPT downlink, in addition to the Sarsat search/rescue NOAA packages. Orbital parameters are entered for any number of satellites and prediction software determines satellite positions for the daily tracking schedule. The 70 kg positioner is a two-axis elevation over azimuth configuration mounted on a 1.2 m riser base.

Decom Systems Inc

Mobile/fixed telemetry ground equipment for flight testing and satellite communication signal acquisition and processing. Complete front-end telemetry acquisition systems, including bit synchronisers, modems and decommutators; high-performance real-time telemetry data processing and information distribution in highly integrated packages.

Deskin Research Group, Inc (DRG)

DRG offers ground antennas of up to 5.0 m in diameter, including the 1.8/3.0 m alti/azimuth Isohex antenna, a prime focus antenna comprising seven hexagonal (4.76 mm thick) reinforced glass fibre panels. Pointing is manual. Specifications are provided in C/X/Ku-band order for 1.8; 3.0 m.

Dorne & Margolin Inc

Design, development of antennas for communications and navigation: GPS, Satcom (UHF, Ku, Inmarsat) telemetry, DF and EW for 1.5 to 18 GHz. Capabilities include base and mobile antennas for wireless markets.

E-Systems Inc (ECI Division)

The company was acquired by SSE Telecom in 1996 for US$6.2 million. It specialises in high-speed digital modems and other Earth station products, accounting for more than 15,000 installations.

ElectroMagnetic Sciences, Inc (EMS)

EMS's 1995 sales were US$129 million (profit US$2.3 million) generated by three operating companies: EMS Technologies Inc; LXE Inc, CAL Corporation. There are 900 employees. EMS has been involved in more than 60 space projects over 26 years. It specialises in space-qualified beam-forming networks for satellite communications antennas, such as those used on NASA's ACTS vehicle. EMS owns 74 per cent of Canada's CAL Corporation.

ESSCO Collins Limited

ESSCO was founded in 1961 to develop ground-based radomes and precision antenna systems. The precision antenna systems are generally used at 30 to 300 GHz for applications in which precision reflector-surface accuracy, pointing and tracking are required, ranging from mm-wave radar to radio astronomy. Reflector sizes range from 6.1 to 26 m diameter and systems are radome-enclosed. The company was responsible for the first large radio telescope in the southern hemisphere, a 13.7 m antenna in Brazil. The radomes include sandwich, metal and dielectric space frame, solid laminate and air supported. Approximately 3,000 are operational worldwide. Applications include 2-D/3-D radar, air traffic control systems, communications satellites and weather radar. Sizes from 1 m to more than 49 m. ESSCO has supplied radomes for major military and commercial programmes worldwide.

ESSCO's Corporate HQ is on a 142,000 m^2 site, including a 610 m test range, 5,580 m^2 of manufacturing space; an Irish facility has 2,800 m^2.

FEI Microwave, Inc

FEI Microwave designs/builds spaceborne microwave assemblies, including receivers, frequency converters and solid-state switch matrix products, utilising qualified MIC/MMIC technology.

Glocom, Inc

Glocom in 1994 introduced the system's third Inmarsat B satellite phone, the GP3000, supporting 16 kbit/s voice, 9.6 kbit/s fax and data services. Requires 160 W in transmit mode. Prices start at about US$24,000. Glocom's Inmarsat M Global Phone 2000t and 2000r models were type-approved September 1993. The 2000 t is an 8 kg self-contained briefcase model, while the 6 kg 2000r can remain outdoors for more rugged duty. The user is guided in pointing the built-in patch antenna towards the satellite by software, a signal level indicator and multilingual graphic and synthesised voice instructions. Ports allow addition of optional cordless phone and Gp 3 fax.

Frequencies (GHz): 1.6265 to 1.6605 transmit; 1.5250 to 1.5590 rx

EIRP: 19 or 25 dBW (nominal) per carrier

G/T: -12 dB/K minimum, 14.5 dB gain

Data rate: 6.4 kbit/s full duplex voice with IMBE voice coded; 2.4 kbit/s for fax and data

Power: 96 W transmit; 37 W rx; 90 to 250 VAC or 10 to 16 VDC

Size: 457 × 340 × 127 mm, 8 kg (2000 t)

Harris Corporation

GCSD is responsible for a range of satellite communication systems, Earth stations, and transmission and ground terminal products. Earth stations include jam-resistant secure terminals such as the AN/GSC-49, tactical SHF equipment and 2.4 m mobile terminals. More than 4,200 terminals have been produced over 30 years. Probably the most important project is TDRSS ground station at White Sands, New Mexico. Harris is prime contractor for this station, which operates in conjunction with the TDRS satellites to provide the prime datalink for Shuttle, Spacelab and other space payloads. Harris developed a 1.544 Mbit/s prototype terminal for NASA's Advanced Communications Technology Satellite (ACTS).

Hughes Electron Dynamics Division

Hughes EDD is the only US supplier of space-qualified TWT/TWTAs. More than 6,800 units have been delivered, recording 150 million hours in space. Available designs cover 1.5 to 65 GHz at output power from 10 W to greater than 1 kW. EDD's EPCs have accumulated more than 88 million hours operating in orbit. Linearised TWTAs are also available.

Hughes Network Systems Inc (HNS)

HNS was founded as Digital Communications Corporation in 1971, became COM Telecommunications Inc in 1978 and was acquired by Hughes Aircraft Company in 1987. The company is a leader in the VSAT industry, claiming 65 per cent of the world market and delivering more than 130,000 terminals through 1996. 1996 revenue was about US$1.06 billion, of which about half was from communications satellites. A US$30 million contract was signed February 1995 with Argentina's Impsat for 2,000 VSATs and five shared hubs. 6,000 units are being provided to Ford Motor Company and 2,000 to Best Western hotels. A contract was won in 1990 to supply 10,000 1.8 m units for General Motors' Pulsat Network. In 1993, it provided Personal Earth Station and Telephony Earth Station VSATs at 300 locations for the People's Bank of China, working through AsiaSat 1. Shanghai-Hughes Network Systems Ltd (SHNS) was created in 1994 as a joint venture with two major Chinese telecommunications companies, primarily for manufacturing/installing VSATs in China.

The company also provides satellite trunking, modem and telephony products such as the Model 4100 digital modem, digital SCPC, Telephony Earth Stations and videoconferencing with its InTELEconference product. Subsidiary Hughes Escorts Communications Ltd (HECL) began providing India's first private satellite shared hub service in January 1995, using Insat 2 leased capacity. The UK subsidiary HNS Ltd provides the Inmarsat C 3100 land mobile station.

Integral Systems, Inc

Integral Systems Inc offers satellite ground command and control systems. The EPOCH 2000 software can handle as many satellites of different types as required, control ground communications networks and be used in satellite integration and testing. It is used by China's DFH-3 telecommunications satellite system.

Keiser Engineering, Inc

The company was established in 1975 to perform feasibility studies and engineering analyses for satellite telecommunications, including voice, video and data. Included are modulation and multiplexing techniques, (FDMA, TDMA, CDMA), error-correction coding, interference analyses, antenna design and propagation studies.

LNR Communications Inc

LNR develops/manufactures communications satellites equipment, specialising in Earth stations, LNAs, frequency converters and solid-state TWT drivers. LNR's amplifiers are used by the TDRS space/ground segments and by Shuttle. The Tramp-T trailerised clamshell Earth station provides C-, X- and/or Ku-band links via a 2.2 m antenna (1.2/1.8 m options). Meets Intelsat IESS 308/309 and DSCS/NATO requirements. The FTSAT Flyaway Triband Satellite Terminal is being produced for the US Army under a US$27 million 1995 contract to provide C/X/Ku-band links via DSCS, NATO, Intelsat, Eutelsat and PanAmSat. The DVF series of digital flyaway stations provides SNG. Options: 1.8/2.4 m diameter, C/Ku-band, 20 to 400 W amplifiers, 220 to 325 kg, digital video encoders for multirate selectable to 3.072 Mbit/s or full broadcast quality at 8.448 Mbit/s.

Lockheed Martin Conic-Terracom

Formerly a division of Loral Corporation, supplying satellite transponders, transmitters, receivers and RF power amplifiers for projects such as Radarsat, Rosat, Landsat, STEP, Tiros, Clementine, GGS, Olympus, Lewis and Clark. The CXS-600B 5 W transponder provides TT&C with STDN and DSN ground stations. It demodulates the PCM/PSK telecommand signal and outputs command data and bit timing. For transmission, it receives data from a telemetry encoder and modulates it on an internal subcarrier and/or directly on the S-band downlink. MIL-STD- 2000; designed to withstand 10 years in GEO environment.
Size: 11.2 × 18 × 14.7 cm; 3.9 kg max
Power: 34 W at 5 W; 24 W at 3 W
Rx/Transmit: 2.025 to 2.120/2.200 to 2.300 GHz

Lockheed Martin Missiles & Space Company, Inc (LMMSC)

LMMSC designs, builds and tests antennas and microwave devices for space vehicles. Antennas range from simple omniconical spirals to broadband multimode systems incorporating integrated beam-forming networks and low-noise amplifiers. It has designed, fabricated and delivered more than 45 deployable antenna reflectors in 30 years, the largest being the (9.1 m aperture) reflector for NASA's ATS 6 in the mid-1970s. The Wrap-Rib™ reflector design is space-qualified over 1 to 10 m apertures and 0.15 to 12 GHz. It features low stowed volume, lightweight and proven reliability, never having had a deployment failure.

LMMSC is also designing, building and testing a line of lightweight phased-array antennas using LMSC T/R modules. Arrays being developed include capacitive-fed, probe-fed and aperture-coupled arrays with steerable, shaped and multiple beams, using printed feed lines with built-in beam shaping.

MCL Incorporated

MCL was founded in 1961 as a design and manufacturing specialist in coaxial and waveguide components, and now operates with 138 personnel in a 5500 m² facility. 1995 revenue was approximately US$29 million, 75 per cent attributable to satellite communication transmitters. MCL produces satellite communication amplifiers (TWT & Klystron), airborne TWT amplifiers, SNG transportable amplifiers, coaxial cavities and power systems. The satellite communication amplifiers range from 50 W to approximately 3,000 W CW, covering L- to Ka-band and DBS. Approximately 4,000 Earth station amplifiers are in service worldwide. Coaxial cavities and power systems range from 10 W to approximately 500 kW over 0.010 to 18 GHz.

Microdyne Corporation

Microdyne ATD is involved in the manufacture, design and integration of aerospace telemetry RF acquisition reception, processing and display. Recent new products include the 700-MR receiver, 1620-PC diversity combiner, WCS-CS cosecant squared tracking antenna system, VXI-2000 multibus Quick Lock system and the TSS-2000 Telemetry Signal RF Simulator.

Miteq, Inc

Subsystems and components for satellite communication ground segments, including up/down converters, frequency translators, integrated RF front ends, FM modulators and demodulators, RF amplifiers (1 MHz to 50 GHz), signal converters/low-noise front ends to 50 GHz, and oscillators/frequency sources.

Mobile Telesystems Inc

Mobile Telesystems Inc (MTI), leading supplier of Inmarsat communications terminals, introduces a new line of state-of-the-art video transmission products suitable for wireless news gathering, wireless cable TV and distance-learning applications.

MTI's Multichannel Multipoint Distribution System (MMDS) broadcasts TV programmes at microwave frequencies from a central point, or head end, to small receiving antennas on the subscriber's roof. MTI's 8 × 8 in receiver/down converter is the smallest in the industry.

MTI's V-Link is an Audio/Video Microwave Transmission System, designed to relay live TV programmes from the camera, outside broadcast car or studio to relaying systems such as SNG, microwave or cable/fibre systems, or directly transmit TV programmes to remote TV stations or CATV head-end systems for broadcasting or distributing to subscribers. V-Link's compact size and wireless transmission capabilities make it a cost-effective solution to traditional time-consuming installations of heavy coaxial cables.

MTI's CompaLink Radio System is a compact, economical, and high-performance digital microwave radio for point-to-point applications in fast-growing wireless rural telephony, rural telephone corridor networks, remote cellular links, wireless local-loop backbones, video teleconferencing, and telemedicine. Due to the ruggedness feature, the CompaLink has the capability to withstand severe operating conditions in most climates.

Motorola Space & Systems Technology Group (MSST)

The division specialises in the design, development and production of tactical and secure communications equipment and systems. The LST-5D Tactical DAMA Satcom/LOS Transceiver is 5.3 kg, 11.4 cm high, 15.2 cm wide, 24.9 cm deep, and integrates into man-pack, vehicular, airborne and fixed-station units. Embedded DAMA, COMSEC, TRANSEC. Nine programmable transmit/rx pairs. Narrow/wideband dedicated satellite communication (MIL-STD-188-181); 5 kHz DAMA satellite communication (MIL-STD-188-182); 25 kHz DAMA satellite communication (MIL-STD-188-1832). The transceiver is packaged in the LSSC Series UHF Satcom/LOS Terminals, which add a 9 dB fold-out antenna, battery/charger, interconnect wiring, data port, and voice/data interface module.

Raytheon Company

The Satellite Terminal Systems Directorate of the Equipment Division develops and manufactures UHF/SHF/EHF airborne, ship– and ground-based communications terminals and peripheral equipment for military applications. It was awarded a US$122.6 million contract in November 1994 for the main mission antennas and transceiver modules for the 75 Iridium satellite, following a US$94 million award for five prototypes. A family of AF/USN EHF and UHF/EHF terminals has been developed for Milstar and other EHF-capable satellites. Other products include high-power amplifiers, modems and terminals to operate with SHF satellites.

Raytheon E-Systems Richardson Division

Formerly Electrospace Systems, Inc, Raytheon E-Systems Richardson designs, integrates, and installs ground, airborne, and ship SATCOM systems. Its AN/USC-54 VME Integrated Communications System (VICS) is the only full-duplex transceiver in the world to receive Demand Assigned Multiple Access (DAMA) certification from the US Defense Information Systems Agency. Its SHF antenna system supports ship links via DSCS, NATO 4, and Skynet. Its modular ground-based LF/VLF solid-state transmitters provide power savings with efficiencies in excess of 85 per cent. Its lightweight 2.4 m 'flyaway' antenna is DISA/DSCS certified in X-band and meets Intelsat E1 certification requirements for Ku-band and Standard G certification for C-band. Its HMMWV-transportable 6.1 m triband SATCOM antenna system provides a rugged, stable platform for mobile communications requirements. All antennas operate with Raytheon E-Systems' state-of-the-art 93C-30 antenna positioning controller, which provides high-speed processing control based on the industry's premier trajectory-estimation algorithm.

Scientific-Atlanta, Inc / Network Systems Group

The company is a leading manufacturer of satellite communication Earth stations, with approximately 46,000 systems in 135 countries. The product line ranges from telephony and TV Earth stations to encrypted and digital transmission systems, including Intelsat A-Z gateway stations, VSAT and mobile. It also provides the system engineering, installation and training for turnkey systems. SkyRelay is a VSAT network designed specifically for distributed computing environments. The remote VSATs feature Ethernet and Token Ring LAN support, providing bridging/routeing capabilities. The Skylinx.DDS (Digital DAMA System) family of terminals provides full-mesh single-hop connectivity for voice/data at 9.6 to 2,048 kbit/s. For thin-route sites, the μDAMA provides one to two channels. The μMCPC terminal provides point-to-point connectivity. The company was awarded a US$3.7 million contract by China's Ministry of Posts and Telecommunications in September 1995 for an emergency satellite communication network to support relief operations. The initial phase of the Skylinx.DDS network became operational in early 1996 using AsiaSat 2 Ku-band capacity. It employs a 4.5 m master station in Shanghai, working with 3.6 m stations in seven provinces and 37 mobile 1.2/1.8 m VSATs. The service will eventually be expanded to include a 3.6 m station in all 22 provinces, with at least five mobile units per province. Skylinx.DDS is also being used by Teleport-TP to create Russia's largest private VSAT network. The first phase, completed April 1996, links 30 remote sites to the 18 m Intelsat A hub in Moscow, working through Intelsat 704 at 66° east. On completion, there will be 125 remote sites by the end of this century.

The Model 9821 MariStar-M (for maritime users) and Model 9826 TerraStar-M (land users) terminals provide global voice (6.4 kbit/s), fax/data (2.4 kbit/s) Inmarsat M services. MariStar-M comprises an above-deck radome with a 20 kg, 66 cm diameter, stabilised antenna and a small below-deck electronics/ communications unit. The Model 9820 MariStar Multi-M delivers 3 to16 channels. The 5.2 kg Model 9805 TerraStar-C terminal and 5.5 kg Model 9801 MariStar-C Satellite Terminal (for maritime users) provide global data/telex Inmarsat C services, including data reporting and automatic polling of terminal position. The optional Model 9803 GPS Receiver (2 kg, C/A code, three-channel, eight-satellite, one-second update rate, 25 m SEP without Selective Availability) provides position determination through the same antenna. The Model 9806 TerraStar- C Portable totals 23 kg in a 356 × 572 × 243 mm flight case, including a Ni/Cd pack providing four-hour operation. EIRP 12 dBW min, G/T −23 dB/K at 5° elevation.

Sea Tel, Inc

Sea Tel is the world's largest manufacturer of stabilised antenna platforms, with approximately 19,000 delivered since 1979. Antenna sizes cover 3.3 to 3.66 m with accompanying radomes of 4.27 m diameter. Applications include Inmarsat A/B/M, marine digital satellite TV and TVRO systems, C/Ku-band VSAT high-speed data and multichannel voice for oilfield and cruise-ship communications, and spread-spectrum systems.

SierraCom

Formerly Lucas Espec Division and now part of Sierra Networks Inc, SierraCom designs/manufactures VSAT terminals/antennas, SSPAs, microwave components and subsystems for communications satellites.

Space Communications Technology Center

SCTC was established in November 1991 as one of NASA's Centers for the Commercial Development of Space. It focuses on developing critical technologies for commercially viable digital transmission systems for video, audio and data by satellite where a large commercial market can be shown to exist. Another focus is in digital signal processing, HDTV signal processing and RF propagation studies for LEO satellite and Ka-band transmission using ACTS.

SSE Technologies Inc

SSE Technologies manufactures digital modems and C/Ku/X-band transceivers utilised in VSATs, hubs or point-to-point applications. Products include the Ku-band ASAT 1214 (SSPA up to 25 W; TWT 75 to 300 W), C-band ASAT N406 (SSPA 60 W; TWT 75 to 400 W); X-band 5 to 40 W SSPA; up to 300 W TWT. The S-series offers complete RF terminals in one-piece outdoor units for communications satellites in rural and remote areas: C-band ASAT S406 (SSPA up to 25 W); Ku-band ASAT S1214 (SSPA up to 16 W). The C-STAR/K-STAR transceivers incorporate MMIC technology: Ku-band 2 W or 4 W SSPA built-in (external booster for up to 60 W); C-band 2 W or 5 W SSPA built-in (external boost up to 25 W).

SSE Telecom is the parent company of SSE Technologies and in 1996 acquired Fairchild Data. It had an 80 per cent interest in the Directsat Corporation (DSAT). Directsat merged with Echostar Satellite Corporation in January 1994 and merged their DBS licences for 119° west. Directsat had been allocated 10 BSS channels at 119° west by the FCC in December 1993. SSE Telecom has an 8 per cent interest in Direct Broadcast Satellite Corporation (DBSC), which has 11 channels at 61.5° west and 11 at 175° west for western USA.

ST Systron Donner Corporation

ST Systron Donner, part of Signal Technology, specialises in instrumentation and GaAs FET HPAs. It has provided the first 150 W 1.61 to 1.626 GHz SSPA for Iridium; efficiency is at least 23 per cent. In 1993, it delivered a 100 W C-band unit for US Naval links and in 1994 a 125 W X-band unit.

Stanford Telecommunications Inc

The company develops/manufactures Earth terminal and satellite network control systems and VLSI integrated circuits. Revenue for FY95 was US$114.4 million, 44 per cent was DoD (FY94: US$98.1 million, 47 per cent DoD; FY93: US$93 million, 53 per cent DoD). Major products include: GPS satellite signal generators to simulate the exact signal environment experienced by high-performance aircraft during manoeuvres; anti-jamming satellite-communication equipment for worldwide secure communications; TDMA modems; tracking/ranging receivers for automated remote tracking stations; automated satellite signal-monitoring systems for optimum employment of communication satellites' resources; multipoint communication network systems for real-time, high-speed, subscriber-to-subscriber network controlled communications; and semi-custom high-speed digital signal-processing integrated circuits for forward-error correction, digital synthesis and spread-spectrum applications.

The company has developed the AN/FSC-111 ICBM SHF Satellite Terminal (ISST) for USAF's satellite communication system. It continues to produce the AFSATCOM AN/FSC-97 Single Channel Transponder Injection System, the uplink transmitting and satellite control system for AFSATCOM payloads now in orbit. A Milstar test terminal was developed and delivered to Lockheed to support the satellite's factory testing; a second terminal undertakes on-orbit tests. TT&C subsystems for USAF's Automated Remote Tracking Stations were delivered under subcontract to Loral Space and Range Systems. They service almost all military satellites, tracking the range, receive telemetry and transmit commands. A Transportable Ground Station was developed under internal R&D funds to incorporate state-of-the-art subsystems. In September 1992, it was awarded a five-month contract with four one-year options totalling US$37.8 million by the US Army for support to the DSCS Operations Control System at Colorado Springs. It provides continued on-site operations, maintenance and training, plus maintenance support at the other DSCS centres around the world. It is designing/building Intelsat's new Communications System Monitor for high-speed measurements on satellite downlink signals, including signal power, carrier frequency and bandwidths.

TECOM Industries, Inc

TECOM designs and manufactures antennas and computer-controlled electromechanical positioners for communications, aerospace, navigation, surveillance and command control. These include Inmarsat aeronautical antenna systems, American Mobile Satellite Corporation (AMSC) land mobile antennas, and GPS antennas.

The new T-4100 low-gain blade antenna supports aeronautical data services. An intermediate-gain antenna system is being developed from the T-4000 for the new Aero I service of the Inmarsat 3 satellites.

TelSys Inc (TSI)

TSI develops and manufactures ground-station communication systems for both traditional telemetry processing and bridging/switching functions to interconnect local/wide-area networks to space-ground communications networks.

TIW Systems, Inc

TIW Systems is a privately held company founded in 1976. It is a turnkey supplier of antenna systems and Earth stations for communications satellites, tracking/ telemetry, radio astronomy and optical astronomy. Its stations are used in the Intelsat and Inmarsat networks, and its antennas by COMSAT, Teleglobe, Telespazio, Indosat, British Telecom, Embratel, GE, Hughes Communications, SES, Lockheed, JPL and NASA.

In 1996, TIW Systems completed the Uplink Antennas for MEASAT Broadcast Network System (MBNS) Direct-To-Home (DTH) digital video and audio system. Antennas were installed for EchoStar's DISH Direct Broadcast services as well as DirecTV Latin American DBS services. Installation was completed on the JPL Deep Space Network antenna system. An installation of the Missile Engagement Simulation Area (MESA) for the US Naval Weapons Center was also completed.

Trimble Navigation

Trimble has introduced the Galaxy combined Inmarsat C/GPS transceiver.

W L Pritchard & Co, Inc

Consultants providing intersystem co-ordination and engineering; economic, competitor and market research for communications satellites, radio, fibre optics and private telecom networks.

SATELLITE GUIDANCE, NAVIGATION AND CONTROL COMPONENTS

FRANCE

Aerospatiale

Aerospatiale has developed a family of magnetically suspended MW/RWs, accumulating 400,000 hours on Spot 1-3 and ERS 1. A new generation is qualified for Helios 1, Spot 4 and 5 and Envisat. Specifications in Spot 1-3 and Helios 1 order are:
Angular momentum: 15/40 Nms
Torque: 0.2/0.45 Nm
Diameter: 350/345 mm
Mass: 8.5/16 kg
Maximum power: 65/160 W
Speed: 2,450/2,500 rpm
Induced vibrations: 5/0.5 μm
Life: approximately 7/5 years

SAGEM SA

SAGEM, one of Europe's principal inertial systems manufacturers, comprises three divisions: Defence and Security; Terminals and Telecommunications; and Electronics and Industry. Research and Development, prototype design and system integration are concentrated in four sites (Argenteuil, Eragny, Saint-Christophe, Cergy), with four manufacturing centres (Montluçon, Saint-Etienne-du-Rouvray, Fougères and Poitiers). Total floor area of 200,000 m² of class 10-10,000 cleanrooms. Principal specialisations for satellites and launchers cover gyros, linear and rotating mechanisms (with associated electronics), bubble memories, optics, microelectronics and navigation/guidance systems. The linear and rotating mechanisms and the mechanical systems used in telescopes and optical instruments, mirror refocusing, optical systems and structures for spectrometers include: Spacelab spectrometer, mirror scanning system for Metop's IASI instrument, Mir MIRAS spectrometer, and Helios refocusing mechanism.

Regys 3S, the newest member of the Regys family of dry-tuned two-axis inertial reference units developed for space applications, it incorporates the Gildas 3 gyro. A fully autonomous unit, it provides error-compensated angular rates in digital form ready for AOCS computer processing. No thermal control is needed as there is built-in compensation for thermal-induced inertial errors. The separate gyro and electronics units are connected by a flexible harness; the electronics units can be stacked. Regys 35 has been selected for the following programmes: ODIN, XMM, Integral, Stentor and Proteus.
Size: 103 × 96 × 95 mm sensor unit; 283 × 164 × 45 mm electronic unit; less than 2.3 kg total
Reliability: 0.99 for a three-axis redundant configuration on a 15-year telecom application
Power requirement: 15 W maximum, at 22-51 V DC
Output interfaces: OBDH/RU, 1553, MACS (optional)
Range (°/s): ±2 fine, ±40 coarse
Resolution: less than 0.1 arcsec
Bandwidth: more than 5 Hz (up to 20 Hz by software adaptation)
Scale factor stability: 10^{-3} BOL; 10^{-2} EOL
g-insensitive drift stability: 0.15° max over 1 h, with ΔT±2°C
Environmental tolerance: −10/50°C; vibration 22 *g* rms 0-2 kHz random, 20 *g* 0-100 Hz 0-peak; radiation 15 year GEO

The Regys 10 rate gyro package is a digital-output package for sensing satellite angular rates about two orthogonal axes, incorporating a Gildas 1/4 dynamically tuned gyro adapted from an air navigation system, and mounted with a shock absorber in a cradle that can be adjusted for required axis orientation. Regys 10 is used by Spot 4, Helios 1 and Envisat, and the newer version aboard Spot 5 and Helios 2. The computed angular displacement is delivered on request at least every 200 ms. If the measured rate is approximately 2°/s, the OBC can order the gyropackage to operate at quarter speed in security mode and provide coarse data (0.400 arcsec/impulse in contrast to 0.0288 arcsec/impulse in fine mode).
Size: 16 × 16 × 16 cm; 2.5 kg
Power requirement: 21 W null speed, 29 W at 8°/s
MTBF: 200,000 h
Range (°/s): ±0.6 fine, ±2 coarse, ±8 security
Scale factor (arcsec/impulse): 0.0288 fine, 0.0972 coarse, 0.400 security
Scale factor stability: 10^{-4} short-term; 10^{-3} long-term
Environmental tolerance: 0/50°C operation, −20/50°C transient; 40 *g*/4 ms shock: 100 krad radiation

SAGEM began developing a space-qualified brushless DC motor under ESA/CNES contracts in 1975, producing a high resolution unit offering 1,200 steps/rev for application in: stepper mode for solar array drive mechanisms on LEO/GEO spacecraft, synchronous mode for de-spinning onboard structures (as on Giotto), precision positioning mode for mirror pointing (as on Spot) and mini-stepping mode for manipulator articulation. 100 μrad accuracy has been demonstrated for the stepper version, with a 6 per cent maximum variation at 0.01 rpm in the synchronous variant. SAGEM's 18EM brushless DC motor drives the rotating mirror of Sodern's STD 12 Earth horizon sensor aboard Spot.

Sydem is a motorised unfolding and pointing system designed to actuate and precisely position various appendages including large antennae on satellites. Typical use includes also single-shot movement with final locking.

Sydem (Type 42) is qualified for space applications, and represents the top of the Sydem range. A smaller and lighter Sydem (Type 32) supports less demanding applications.

Sextant Avionique

It provides a range of gyroscopes, accelerometers, onboard digital computers and data handling equipment for missiles and space launchers. The company is contracted to provide 87 Quasar ring laser gyro systems for Ariane 4's strapdown inertial system; 72 had been manufactured and 65 launched by end-1995. This comprises an Inertial Measurement Unit of three 33 cm-path length laser gyros and three accelerometers, linked to a Sextant series 7800-based processing unit. The system was qualified by July 1986 and flown aboard the first Ariane 4 in June 1988. The unit is now radiation hardened. Sextant also provides Ariane 5's RLG inertial reference system. This features an improved RLG, a 68020/68882 processing unit, radiation hardened memories, new algorithms and reduced power.
Mass: 20 kg, size 390 × 230 × 210 mm
Power requirement: 35 W at 28 V
Range: 100°/s, 40 *g* shock
Bias drift: 0.03°/h
Drift at 14 g rms vibration: 0.9°/h
Random drift: 0.005°/h½

SODERN

The Société Anonyme d'Etudes et Réalisations Nucléaires (SODERN) specialises in instrumentation for particle radiation detection, Sun/Earth/star sensors for satellite attitude control systems, and spaceborne rendezvous/docking cameras. Personnel total about 300. Approximately 100 SODERN attitude sensors have been operated in space since the first launch of a SODERN sensor in 1974 (Symphonie). Sensor development began with the STR 01 Earth horizon-crossing and STA 01 Earth static sensing devices for the GEO Symphonies, launched 1974/75. All the Earth-sensing devices operate at 14 to 16 μm IR. An improved STA 03 followed on OTS and Marecs, and four STR 03s on each Meteosat. A slightly improved STA 03A is employed by the ECS/Eutelsats. 35 STR 04 horizon-crossing sensors were provided for Intelsat 5 and the self-redundant STA 04 was introduced on Telecom 1. The STD 12 two-axis conical scan sensor was developed for Spot and ERS, and has been further developed as the STD 15/16 for operations in all orbits from low altitude to GEO. The SED 04 star tracker was employed by Spacelab 2's Instrument Pointing System. In 1985 CNES awarded SODERN the contract for developing the SED 12 CCD star tracker for the Franco-Soviet Granat X/γ observatory. SODERN also developed the Earth-imaging cameras for Spot 1-4 and Helios, and Envisat's MERIS spectrometer and is now developing the Spot 5 and Helios II cameras.

SED 04 star/sun tracker
The SED 04 was developed to update the reference gyros of ESA's Instrument Pointing System, first flown aboard Spacelab 2 in 1985.

SED 12 CCD star tracker
The SED 12 was designed and built to provide attitude data to CNES's control system for the French Sigma telescope aboard the Soviet Granat astronomical observatory. The SED 12D model is used by France's Helios 1 reconnaissance satellite and has been adapted for Helios 2, Envisat and Sweden's Odin. The tracker is based on Thomson's TH7861 CCD area array, each element measuring 23 × 23 μm and providing 40 mV/lx sensitivity. Acquisition probability is more than 0.997 for stars of magnitude +8 to −1. Simultaneous tracking of five stars over the 7.5 × 10° rectangular Field Of View (FOV) is possible. From an initial standby mode, the sensor can operate in search mode for up to 30 stars over the whole FOV (requiring minimum detectable magnitude, number of stars to be tracked after search, and apparent star motion), or in tracking mode for a maximum of five local fields (requiring co-ordinates of five stars, expected magnitude and motion). For both modes, the sensor outputs the located stellar co-ordinates, magnitudes, time delay between charge collection and read-out, detector/optical head/electronics temperatures.
Field Of View: 7.5 × 10° rectangular
Accuracy (arcsec): 3 after one observation, 0.5 after filtering/calibration (1 σ); 17 on two axes after one observation (3 σ), 5 after filtering/calibration
Acquisition time: less than 8 seconds overall field; less than 1.5 seconds in 0.25° square field. Probability of single star acquisition more than 0.997
Maximum star rate: 0.1°/s for magnitude brighter than 9, 1.0°/s for magnitude brighter than 3
Size: 141 mm diameter, 230 mm length/2.5 kg optical head without baffle; 271 × 222 × 217/7.2 kg electronics unit
Input power: 2 W optical unit with cooler off; 2-8 W optical unit with cooler on; 40 W electronics unit typical; 20-50 V DC (PWM 1525)
Temperature range: −20 to +50°C
Memory capacity: 40-64 kbyte ROM; 16-32 kbyte RAM depending on mission
Failure rate: 3,800 × 10^{-9}/h

SED 15 CCD star tracker
The SED 15 CCD Star Tracker is under development for GEO satellites to provide pitch/roll data. One version will track only the Pole Star. 30 arcsec accuracy in 5 × 7° FOV. By adding barycentric signal processing, the accuracy holds for a larger Field of View (FoV), such as 25 × 35°.

STD 15 scanning earth sensor
The STD 15 scanning earth sensor derived from the STD 12 of Spot and ERS 1/2. Its first flight was aboard Telecom 2 in December 1991; 12 are now flying on Telecom 2A/2B, Hispasat 1A/1B and Hot Bird 2. The sensor has been selected by NPO PM (Russia) for a new telecommunications platform. The sensor permits the attitude control subsystem to generate roll/pitch angular deviation data by using the four Earth-space/space-Earth transitions of the two scan traces (generated by the rotating mirror and two fixed mirrors); an optical encoder measures the angular position of each transition with respect to the spacecraft reference axes. Data rate 1.25 Hz in normal mode, 5 Hz in stationkeeping mode.
Field Of View: 1.5° roll/12° pitch; 14° roll/30° pitch in acquisition mode. Instantaneous FOV 1.8 × 1.8°
Accuracy: 0.025° 1 second
Mass: 3.4 kg
Input power: 7.5 W at 22 to 50 V DC
Temperature range: −20− to +50°C operating; −40 to +60°C storage
Vibration tolerance: 20 *g* rms over 20 to 2,000 Hz
Lifetime: 15 years; 0.9 × 10^{-6}/h failure rate

STD 16 scanning earth sensor
STD 16 scanning earth sensor is Sodern's latest sensor, designed for LEO satellites: Helios 1, Spot 4, Envisat, ADEOS and ETS 7. Operation principle is as STD 15. Nine were supplied to Matra for Spot 4/Helios and Envisat: six three were supplied 1992-93 to Toshiba for ADEOS and ETS 7.
Field Of View: effective scan for one trace is 152°
Altitude range: 200 to 2,000 km
Accuracy: 0.045° bias, 0.03° noise; 3 seconds
Mass: 3.5 kg; length 386 mm; height 175 mm

Input power: 7.5 W at 22 to 52 V DC
Temperature range: −20 to +50°C; −40 to +65°C storage
Failure rate: 1.3 x 10^{-6}/h
Lifetime: 10 years

DTA 01 high-resolution imager
Spot 1-3 incorporate Sodern's DTA 01 high-resolution visible push broom imaging unit. Each of the four spectral bands employs four Thomson TH7801A 1728-element CCD arrays (Spot 1's utilised Fairchild's 122 DC detectors). PAN passes directly through the prisms; its spectral range is determined by absorption and interference filters on the CCDs. Green band B1 is reflected off to the side by a dichroic mirror; red band B2 and near-IR B3 are separated by a second mirror. The rigid primary titanium structure holds the beam splitter and the CCD assembly of four butted lines; each line of four CCDs is straight to within 2 μm. A mechanically isolated secondary structure carries the front-end electronics. Resistance heaters help to maintain CCD temperature to within 2°C.
Spectral response (μm, at half-ht): 0.495-0.580 B1; 0.610-0.665 B2; 0.775-0.785 B3, 0.490-0.715 PAN
Number of pixels: 3,000 (1,326 μm) B1-3; 6,000 (1,313 μm) PAN
Resolution (from 800 km): 20 m B1-3; 10 m PAN
Typical peak beam splitter transmittance: 0.8 B1; 0.75 B2; 0.8 B3; 0.85 PAN
Saturation spectral irradiance (W/m²/μm): 0.004 B1; 0.005 B2; 0.003 B3; 0.006 PAN
S/N for max/min illuminance: 700/90 B1; 650/40 B2; 800/50 B3; 500/70 PAN
Response uniformity: 5% B1-3; 3% PAN
Size: 211 mm wide, mounting flange 255 mm diameter; 2.5 kg
Power consumption: 2.7 W
Thermal: 18 to 22°C operating, −20 to +60°C storage.

DTA 03 high-resolution imager
The DTA 03 high-resolution imager, derived from DTA 01 for Spot 4, enhanced by a 1.6 μm SWIR channel and by using the B2 channel (instead of PAN) for 10 m resolution. The SWIR scanline is formed by 3,000 30 × 30 μm InGaAs Thomson detectors in two staggered rows of 1,500 each with a centre to centre pitch of 26 μm. The three B channels use TH7811 detectors, derived from the TH7801. Short-term stability is 0.01°C at 5°C for SWIR and 0.1°C at 20°C for the CCDs.
Spectral response (μm, at half-ht): 0.495-0.580 B1; 0.610-0.665 B2; 0.775-0.785 B3, SWIR 1.54-1.75
Typical peak beam splitter transmittance: 0.8 B1; 0.7 B2; 0.8 B3; 0.7 SWIR
S/N for max/min illuminance: 900/250 B1; 350/60 B2; 1,000/100 B3; 650/3 SWIR
Response uniformity: 4% B1/2; 4% B3; 10% SWIR
Size: 200 × 250 mm, mounting flange 250 mm diameter; 7.4 kg
Power consumption: 10.5 W
Thermal: 18 to 22°C operating, −15 to +50°C storage.

DTA 02/04 very high-resolution imagers
SODERN developed the first prototype of the imager (DTA 02) for the Samro military satellite. When the programme re-started as Helios, the company became responsible for the whole DTA 04 detection chain, including both focal plane and video electronics. An improved new detection chain DTA 06 is being developed for the new-generation Helios II. The programme is classified.

DTA 05 multispectral cameras
The DTA 05 multispectral cameras were developed for Spot 4's Vegetation payload, this imaging subassembly comprises separate focal plane detection units: 1,728-pixel CCDs for 0.49, 0.58, 0.64, 0.83 μm visible and a 1,500-pixel AsGaAs hybrid detector for the 1.6 μm SWIR channel. The front-end electronics are linked to a common video electronic unit, each CCD being read sequentially in series. Each detection unit is mounted on optics (developed by SODERN subsidiary Cerco) with an interposed dedicated interference filter. Optics are telecentric, ±50.5° FOV with registration less than 0.3 pixel regardless of incident angle. Global MTF 0.3 along track. 23 W power consumption.

DTA 10 spectro imager
The DTA 10 spectro imager was developed for Envisat 1's MERIS. Five identical cameras each analyse a 1,200 km swath (300 m resolution) with a spectrometer. Up to 15 bands can be programmed in position and width within the swath. Optics developed by Cerco.
Spectral range: 400 to 1,050 nm
Spectral resolution: 2.5 nm
Radiometric accuracy: less than 2%

Other imagers
A new imager is being developed for Spot 5. The detection unit is built around a five spectral band beam splitter. Four bands are fitted with a Thomson TH7834 12,000 element array and the fifth band (SWIR) is fitted with a 30,000 element GoAo detector line. The PA channel is designed to achieve a ground resolution of 5 m.

ALICE BOE/BOS
The ALICE BOE/BOS was developed for the ALICE 1 and ALICE 2 instruments and used aboard the Mir Complex for microgravity experiments, the ALICE BOE and ALICE BOS achieve the optical diagnosis of cells which are filled with liquid subjected to critical conditions. Cells are temperature-regulated by high-precision thermostats and are analysed during the physical phase transitions by the optic-electronic devices of BOE and BOS. The ALICE optical capabilities include: microscopic inspection at various magnifications, interferometry, optical transmission, index gradient visualisation and optical diffusion. The total volume is 0.15 m³ and the mass is 55 kg.

POLDER-BOM
The POLDER-BOM was developed for the POLDER instrument aboard Japan's ADEOS 1 and 2 spacecraft, the optical-mechanical sub-assembly POLDER-BOM comprises a wide angle lens, a rotating filter wheel and a CCD focal plane. The optics have a 114° field of view, low distortion and constant illumination over the local plane. The filters and polarisers achieve spectral analysis from violet to the near-IR in different directions of polarisation. Eight spectral channels are available, three of which have polarisers.
Spatial resolution: 244 × 276 pixels
Temperature: 15 to 20°C operating
−20 to +60°C non-operating.

SIS 01 very broad band 1-axis seismometer
The SIS 01 very broad band 1-axis seismometer, a vertical-axis seismometer, was developed for the Russian Mars 8 (Mars-96) mission for operations on the Martian surface.
Band pass: 0.03 to 8 Hz
Acceleration resolution: 10^{-8} m/s²
Power consumption: 60 mW
Dimensions: 90 × 90 × 90 mm
Mass: 350 g

SIS 02 very broad band 3-axis seismometer
The SIS 02 very broad band 3-axis seismometer is an improved-technology development of SIS 01 with three seismic masses on a pyramid corner. Developed for planetary seismology.
Band pass: 0 to 5 Hz
Acceleration resolution: 2 × 10^{-8} m/s²
Power consumption: 1 mW
Dimensions: sphere, diameter 22 cm
Mass: less than 1.6 kg

GERMANY

Daimler Jena-Optronik GmbH

DJO (51 per cent DASA; 49 per cent Carl Zeiss) was founded in January 1992 from the Space Research and Development elements of Jenoptik; the Scientific and Technical Equipment Development Division was established in 1980, building on the space research and development team formed in 1973. The company specialises in the design, development and manufacture of: space optical and opto-electronic systems and components; systems, subsystems and components for science observations, measurement, pointing and tracking; precision mechanisms; optical and opto-electronic calibration and test equipment.

All space involvement before German unification in 1990 was with Soviet organisations. For example, the division was responsible for the science instrumentation on the Soyuz 22 manned mission. More recently, its six-channel MKF-6MA multispectral camera and Astro 1 star sensor package (developed 1984-87) were launched to the Mir space station aboard the Kvant 2 module. The improved Astro 1M was contracted for the Spektrum-X/γ astronomy mission but was cancelled because of Russian funding problems.

DJO is providing PSS, the Precision Sun Sensor for ESA's Artemis. A fully redundant system comprises six optical heads and the redundant electronics unit (two identical channels, each operating with three heads).

Teldix GmbH

Teldix specialises in flywheel actuators for satellite attitude control systems, solar array drive assemblies antenna pointing mechanisms, systems for optical satellite communications (SILEX) and other mechanisms. Company details and solar array drive specifications are provided in the Thermal and Power section. Teldix has provided wheels and drive electronics for the attitude control systems of nearly 100 satellites accumulating more than 700 years of orbital operations without failure. By December 1996, Teldix had 206 stabilisation wheels aboard 95 satellites. The company has been developing a new class of Large Momentum and Reaction Wheels under ESA contract to satisfy the greater momentum capacity and output torque requirements of future large European vehicles.

INDIA

ISRO Inertial Systems Unit

ISRO's IISU specialises in inertial systems and components for launch vehicles and spacecraft.

Laboratory for Electro-Optic Systems (LEOS)

LEOS provides India's capabilities in star, Earth and Sun sensors, responsible for the units on Aryabata, Bhaskara, Apple, IRS, SROSS and Insat 2. ISAC has also provided tri-axial magnetometers (range ±45,000 γ, accuracy 500 γ) for every Indian satellite.

ISRAEL

Elop Electro-Optics Industries Ltd

Elop's Remote Sensing Operation is principally concerned with airborne and ground electro-optical observation systems, warning systems and optical processing. Civil Research and Development activities cover remote sensing, astronomy, Earth resources monitoring and electro-optical navigation sensors. It developed Israel's TAUVEX UV telescope in co-operation with Tel Aviv University for Russia's Spektrum-X/γ and developed the static horizon sensor and low resolution camera for Technion's TechSat 1. It was also probably responsible for Ofeq 3's camera. Elop will provide the camera for the German-Israeli David Earth observation satellite.

The compact static horizon sensor, using state of the art Si-monolithic non-cooled line of 32 14-16 μm infra-red detectors, was developed for Israel's TechSat 1 but is suitable for all small LEO satellites. The signals from the four identical optical units are analysed by the integral computer to locate the radiance profile in the horizon's vicinity. This is compared to the memory data and the angle to the horizon in each of four directions (hence the nadir angle) is computed. It can also compute the angle rate of change.
Accuracy: 0.1-0.15° (1 σ)
Field of view: 14.6° total
Size: 110 × 120 × 120 mm; 1.25 kg
Power requirement: less than 1 W; 11 to 14.5 V DC or 5 V DC regulated

The 6.5 kg earth resources monitoring system is a pushbroom linear CCD 0.5 to 0.90 μm visible/near infra-red 300 mm fl f/5 15° FOV camera. It provides 16 to 25 m resolution (40.6 μrad) in a 100 km swath from 400 km aboard a small satellite. It is currently a PAN instrument but it can be converted to XS if required. Resolution can also be improved. Ofeq 3's camera is similar.

TAMAM Precision Instrument Industries

An element of IAI's Electronics Division, TAMAM specialises in inertial navigations/reference systems, inertial sensors and stabilised platforms.

TAMAM's MW 1243 momentum wheel is flying on the small HETE satellite (the first was lost on TechSat in Mar 1995). The electronics are accommodated within the wheel case. The balanced flywheel is driven by an ironless armature brushless DC motor.
Torque: 20 mN-m maximum
Momentum: 4 Nms maximum
Size: 255 mm diameter, 60 mm height; 2.5 kg
Power consumption: 14 W at 3,500 rpm (20 mN-m); 4 W at 1,700 rpm (10 mN-m)
The MW 2427 is designed for medium and large satellites. It is flying on Ofeq 3.
Torque: 200 mN-m maximum
Momentum: 16 Nms maximum
Size: 355 mm diameter, 95 mm height; 5.5 kg
Power consumption: 3 W at no torque; 60 W at 2,000 rpm (200 mN-m); 13 W at 750 rpm (100 mN-m)

The Coarse Rate Gyro model 2424 provides analogue information of the three mutually orthogonal angular rates sensed by three microminiature rate gyros. It also provides back-up to the main stabilisation system. It has flown on Ofeq 1-3.
Dynamic range: 200°/s X/Y, 600°/s Z
Threshold/resolution: 0.01°/s max X/Y, 0.03°/s max Z
Size: 60 × 70 × 90 mm; 0.580 kg
Power consumption: 50 W max start, 8 W max run

The Inertial Spacecraft Stabilisation Unit is a fully redundant strapdown attitude reference unit for satellite stabilisation and pointing. No mission has yet been claimed. It provides compensated angular incremental pulses and instantaneous digital angular velocity. Accuracy better than 1 mrad in three axes can be achieved when updated by Earth horizon, Sun sensor and magnetometer. The main unit comprises three dual axes thermally modelled dry tuned gyros in orthogonal configuration on four shock absorbers. The electronics unit is separate.
Attitude accuracy: 0.8 mrad
Resolution: 1.6 μrad/pulse
Dynamic range: 2°/s fine, 8°/s coarse
Size: 8 × 13 × 18.9 cm sensor unit, 15 × 14.8 × 22.2 cm electronics unit; 6.6 kg

The model 2425 fluxgate magnetometer is designed for small satellites; it was carried by TechSat 1. It comprises a sensor of coils wound on high-permeability ferromagnetic cores, and an electronic unit with an oscillator that excites the sensors and three control capture loops that process the sensor signals and provide the output analogue signals.
Dynamic range: ±0.6 gauss
Linearity: ±1% of full scale
Scale factor: 10 V/gauss
Size: 27 × 53 × 90 mm; 0.200 kg
Power consumption: 0.9 W maximum

ITALY

Alenia Difesa

The Optics and Space Division (DIOS) was established in 1983, with a staff of 250 by end-1995, to complement the activities of the Military Systems and Environment divisions. It is concerned principally with onboard equipment and instrumentation for spacecraft, including electro-optical sensors for attitude determination (IR Earth, Sun and star sensors), CCD video systems, optical sensors for robotic rendezvous and docking, remote sensing electro-optical instruments (visible and near-infra-red), and mechanisms and components for optical payloads (lightweight mirrors, baffles and optics). DIOS has provided more than 250 attitude sensors for Eureca, Turksat, Soho, SAX, Inmarsat, Telecom, Hipparcos, Olympus, Eutelsat, Italsat, Tethered Satellite, ISO, DFH 3, Cassini, Skynet, Orion, Hispasat, Arabsat 2 and Nahuelsat. The light baffle and aluminium mirror for Giotto's multicolour camera were produced by DIOS, as were the CDD monitoring cameras of Eureca's Protein Crystallisation Facility. Other instruments have been delivered or are under development for astrophysics research and Earth observation projects. UVCS/Soho precision mechanisms for Soho's three-channel UV coronograph spectrometer. JET-X + XMM Optical Monitor ancillary telescopes to detect faint stars in visible/near-UV close to the optical axis. Cassini VIMS camera visible channel on the orbiter; Huygen's Atmospheric Structure Experiment to probe Titan's atmosphere; optica heads for star reference unit. IASI IR atmospheric sounding interferometer for Metop 1 polar satellite for temperature and humidity soundings and trace gases. GOME (Global Ozone Monitoring Experiment) for ERS 2. MIPAS (Michelson Interferometer for Passive Atmospheric Sounding) for Envisat 1. Sounding rocket μg: MITE (Measuring Interfacial Tension Experiment); INEX MAM (Interactive Experiment for Marangoni Migration).

The MSS, Modular Star Sensor, designed for three-axis satellites such as ISO, Soho and SAX, is a two-axis star tracker incorporating a catadioptric head providing attitude control data in addition to the precise pointing of astronomical instrumentation. In the latter role, it also provides magnitude measurements of stars brighter than 9^{mv}. It first flew aboard STS-52 in October 1992 for ESA's In-Orbit Technology Demonstration Programme. With a few modifications (10 × 10° FOV), it will fly on Cassini.
Field of view: 4 × 3°
Accuracy: 6 arcsec uncalibrated
Size: 137 × 134 × 227 mm/3.5 kg (sensor head)
137 × 228 × 187 mm/2.5 kg (electronics unit)
Detector: silicon CCD matrix of 288 × 384 pixels
Input power: 10.5 W

The IFPSS, Integrated Focal Plane Star Sensor is a focal plane star tracker developed for ISO but is suitable for operation in any IR astronomical satellite with provision for cooling down to 2 K.
Accuracy: 1 arcsec rms
Size: 70 × 70 × 30 mm/1 kg (optical head at cryogenic temperatures)
200 × 150 × 70 mm/1.5 kg (electronics unit at −30 to +60°C)

The FDSS, Fine Digital Sun Sensor, is a two-axis wide angle static precise Sun sensor for application aboard three-axis satellites such as Eureca, Italsat and TSS.
Number of optical heads: two or more
Field of view: ±45°
Accuracy: 0.02° rms
Size: 170 × 94 × 88 mm/1.0 kg (optical head)
204 × 152 × 121 mm/2.7 kg (electronics)
Input power: 5 W

The CASS, Coarse Analogue Sun Sensor is a two-axis sensor designed for coarse determination of Sun angle aboard three-axis satellites such as Italsat. The sensor head incorporates four redundant Si cells assembled on a truncated pyramid with light baffle.
Field of view: 2π steradians
Accuracy: 0.1° rms
Size: 86 × 86 × 53 mm/0.230 kg

The ESS, Earth Sun Elevation Sensor, incorporates two 15 μm IR telescopes and two visible light solar split sensors to measure Earth/Sun elevation angles for spin-stabilised satellites. Users include ECS, Hipparcos, Giotto and Inmarsat.
Field of view: ±55° Sun
Accuracy: 0.03° both Earth/Sun, rms
Size: 166 × 150 × 127 mm/1.4 kg total package
Input power: 0.7 W

LACES, the Low Altitude Conical Earth Sensor is a two-axis IR (15 μm thermistor bolometer) Earth horizon sensor designed for pitch/roll determination of three-axis LEO satellites. Users include Eureca and TSS; it also flew October 1992 aboard STS-52 with MSS.
Number of sensors: 2 optical heads with conical scanning equipment
Field of view: 360° pitch, 90° roll
Accuracy: 0.1° rms
Size: 170 × 100 × 116 mm/1.55 kg (optical head)
204 × 172 × 107 mm/2.5 kg (electronics)
Input power: 6 W

The IRES, Infra-red Earth Sensor, is a two-axis IR (15 μm thermistor bolometer) Earth horizon sensor designed for pitch/roll determination of three-axis GEO satellites. Users include ECS, Inmarsat Olympus, Skynet, Eutelsat, DFH-3, Turksat, Italsat, Orion, Arabsat 2 and Nahuelsat.
Field of view: ±11° pitch, ±15° roll
Accuracy: 0.02° rms
Size: 137 × 190 × 158 mm/3 kg integrated unit
Input power: 2.2 W

IRES with new electronics is replacing the original, improving accuracy to less than 0.02°rms and failure rate to 1,700 × 10^{-9}/h, while reducing cost and mass. Qualified for Sinosat and Eutelsat 3.

The CCD onboard camera is a 288 × 384 pixel CCD TV-standard camera (625 lines/50 Hz) is designed for space operations such as rendezvous/docking, robotic vision, ERA teleoperations and routine monitoring. The optics, processing electronics and LED illumination system are integrated in a single housing.
Field of view: 30 × 40°, with 1.8 × 10^{-13} W lower detection limit
Size: 106 × 100 × 95 mm/1.5 kg
Power input: 2 W camera head, 10 W processing electronics and illumination system.

JAPAN

Japan Aviation Electronics Industry Ltd

The company provides the strapdown RLG-based inertial measurement unit for Japan's H2 launcher, and the strapdown FOG-based inertial sensor package for NASDA's TR-1A suborbital rocket. It also provides liquid propellant gauges for H2.

NEC

The ESG-100A/110A/120A Earth sensors generates roll/pitch errors for geosynchronous satellites, employing an infra-red telescope assembly with two pyroelectric detectors and a scanning mirror with an optical encoder of 0.0025° resolution. Users: ETS 6 (1994), EchoStar (1995), Intelsat 7A/8 (1996), PAS (1996), Tempo (1997), COMETS (1997).
Instantaneous field of view: 1.3° diameter
Accuracy: ±0.03° bias error
Output: 8 Hz
Failure rate: 619 × 10^{-9}/h
Size: 122 × 157 × 124 mm; 1.95 kg
Power consumption: 5.0 W

Following the successful flights of the microprocessor-controlled STT-100B star trackers flown on Astro-C (1987) and Solar-A (1991), Muses-B (1996) will carry a new version, the STT-200A utilising a 510 × 1,000-pixel CCD image sensor for satellite attitude reference.
Field of view: 64°-sq (focal length 61.5 mm, effective aperture 48 mm)
Position accuracy: 16 arcsec max (static/random 3 σ)
Sensitivity: better than +6 magnitude
Exposure time: 62.5 ms-2 s (selectable)
Size: optics 24 × 10 cm diameter, 3.5 kg (excluding hood); electronics 9 × 20 × 27 cm, 3.1 kg plus 9 × 26 × 29 cm, 4 kg
Input power: 16 W (27 W max with cooling device)

SWEDEN

CelsiusTech Electronics AB

The company specialises in optics and electronics. Space activities include a yaw attitude sensor, optical processing, spectrometers and optical subsystems.

UNITED KINGDOM

GEC-Marconi Radar & Defence Systems Ltd (Navigation & Electro-Optic Systems Division)

NESD manufactures a range of inertial navigation, guidance, control and stabilisation systems for air, land, sea, subsea and space applications. At its Silverknowes and Granton sites in Edinburgh, NESD employs about 1,200 personnel. Of these, approximately 300 are qualified engineers and scientists engaged in the research, design, development and manufacture of inertial systems and components.

Over the past 25 years, NESD has become a specialised supplier of gyroscopes, gyro packages and inertial guidance systems for satellites, experimental payloads and launch vehicles. NESD's first space-related equipment were the SF500 and FE 610 Attitude Reference Units for ELDO's Europa launcher, 1966-69. These were followed by the FE 650 system for the four UK national Black Arrow launchers of 1969-71. Both systems incorporated the Type 125 floated rate integrating gyro. This gyro has since been the mainstay of the company's space units. In 1974, NESD was selected to supply Ariane's Inertial Sensing System. A similar system was built under licence in Japan for NASDA's H1 launch vehicle. A dozen 125s have been running continuously since 1976 as part of an open-ended life test programme; there have been no failures and all remain within specifications.

The 125 gyro has also been employed in the series of gyro package sensors flown on Miranda, IRAS, Exosat, Olympus, Rosat and Spacelab's IPS Instrument Pointing System. Closed loop stabilisation and pointing accuracies of less than 1 arcsec and open loop drifts of less than 2 arcsec/h have been achieved. Further packages have been supplied for NStar, Tempo, Artemis, ISO, Superbird and Intelsat 7 communications satellites. Named DIRA, this gyro package had received 63 orders by the end of 1996. NESD is now developing a smaller, low-cost system, DIRA Mk 2, which will be used on scientific, communications and observation missions well into the next century.

NESD continues to supply Ariane 4 with strapdown RLG guidance systems using the Type 162 32 cm-path ring laser gyro and has been requested to look at providing a new guidance system for Ariane-5 based on the new Type 164 24-cm ring laser gyro. In addition to the gyro package and launcher GNU, NESD developed the dynamic monitoring accelerometer package for ESA's Cluster mission, with a specified accuracy of 0.05°.

Assembly is conducted in the 490 m^2 category B cleanroom at Silverknowes and in the 52 m^2 class A 100,000 cleanroom; commissioning of all systems takes place in the latter.
Path length: 32 cm
Mass: 0.5 kg
Power: 4 W
Maximum rate: 400°/s
Scale factor: 0.5 arcsec/pulse
Bias drift: 0.004°/h
Random drift: 0.003°/$h^{1/2}$
Scale factor stability: 3 PPM

The company also makes the Type 125 Floated Rate Integrating Gyro:
Mass: 0.45 kg
Size: 83 mm long, 53 mm diameter
Scale factor: 42.24°/h/ma
Maximum rate: 5000°/h
Vertical short term drift: 0.0007°/h
Vertical day to day drift: 0.0033°/h

It also makes a standard generic gyro package:
Resolution: 0.5 arcsec (ISO), 0.002 arcsec (IPS)
Maximum rate: 2°/s
Drift stability: 0.0005°/h in zero-g orbit
Scale factor: 0.1% or better
Noise: 0.2 arcsec/s rms (0-2 Hz) or better
Bandwidth: 2 Hz-10 Hz
Power: 5.5 W per channel
Heater power: 1 W at 50°C, 12 W at −20°C

The DMA Package has the following performance:
Accelerometer measuring range: ±20 mg
Accelerometer measuring resolution: 0.01 mg
Accelerometer scale factor: 222 mV/mg nominal
Temperature sensitivity of scale factor: 210 PPM/°C
Bandwidth: 22 Hz
Input power: ±13.9 V less than 0.65 W

Sira Limited

Sira specialises in the design and manufacture of space instrumentation, from the study phase through to the manufacture, integration and test of flight hardware. Principal areas of interest are CCD-based star trackers and other sensors, visible and infra-red optical systems for remote sensing, data handling electronics, x-ray detectors and µgravity payloads. CCD detector testing covers electro-optical performance, radiation damage and up-screening of commercial parts for space applications.

Sira supplied the CCD star trackers for the Rosat and Astro-SPAS satellites (in partnership with MBB) and for the Rosat wide field camera, the Exobiology and Radiation Assembly core payload for Eureca, and the Digital Electronics command/data handling computer for Compton Observatory's Comptel instrument. Sira undertook the initial design of the MERIS Medium Resolution Imaging Spectrometer for ESA's Envisat 1. This also uses a CCD array detector. It is developing four modules for Envisat's GOMOS (Global Ozone Monitoring by Occultation of Stars) instrument: the SATU star acquisition/tracking unit CCD star sensor to hold the star image on the entrance slits of the two CCD spectrometers and the fast photometer detection module for 1 kHz two-band scintillation monitoring of the star image. It is undertaking the pre-Phase A feasibility study for the PRISM (Process Research by an Imaging Space Mission) sensor with DASA/MMS under ESA contract. It combines the features of two earlier instruments, HRTIR (High-Resolution Thermal Infra-Red) and HRIS (High-Resolution Imaging Spectrometer), to provide four-band sensing over 0.450 to 2.35 µm, with 100 to 150 Å and 50 m resolution.

Sira provided an experimental infra-red telescope for launch on STRU-2 in the second half of 1997. The 0.2 m diameter telescope will be capable of imaging a 15 km swath of the Earth's surface with a ground-level resolution of 15 to 30 m. It will be capable of detecting aircraft and other targets of potential military interest.

Sira is fabricating the Acquisition and Tracking Sensor Detection Units, using a CCD array, for the highly accurate tracking of an incoming communications laser beam between satellites. It will fly on ESA's Artemis, Spot 4 and DRS; Matra Espace is Silex prime.
Resolution: 0.12 µm, equivalent to 2.9 µrad
Field of view: 7,440 × 7,440 µrad
Incident laser power: 25 to 1,500 pW
Size: 115 × 80 × 65 mm; 0.422 kg
Input power: 1.0 W

Space Innovations Ltd

SIL makes a low-cost attitude control system that is based on the SACE-MDS modular electronics unit, which can drive and interface with various combinations of sensors and actuators.

The unit is called the SACE-MDS Attitude Control Electronics. Its baseline performance can support two rod magnetorquers, two two-axis DSS and one three-axis magnetometer.
Size: 110 × 170 × 195 mm; 2.5 kg
Power consumption: less than 2.5 W (magnetorquers off)
Environmental tolerance: −40 to +50°C, more than 25 krad, EMC MIL-STD-461C, 25 *g* sine 25-80 Hz.

Space Innovations makes the SSM-20/2 Star Mapper, a high-performance three-axis star mapper designed for LEO small satellites but with capabilities and radiation tolerance (approximately 20 krad) suitable for GEO missions. 576 × 770-element passively cooled CCD. The FOV guarantees at least three stars brighter than 6.0.
Accuracy: 20 arcsec (3s)
Field of view: 12 × 16°
Power requirement: 6 W
Size: 100 × 100 × 120 mm 1.5 kg optical head unit; 40 × 120 × 160 mm 2.2 kg processing unit.

In addition it manufactures the DSS-256 Digital Sun Sensor. The sensor head utilises a pair of orthogonal photodiode arrays behind thin slits. It converts the signal into a 10-bit digital word corresponding to the solar aspect angle. DSS is suitable for gravity gradient, three-axis and spin (up to 10 rpm).
Accuracy: 0.5° up to 5 rpm; 1° 5-10 rpm
Field of view: ±60° azimuth, ±60° elevation
Power requirement: less than 0.8 W total
Size: 38 × 51 × 78 mm 0.150 kg sensor head;
18 × 118 × 159 mm 0.180 kg interface card.

One other Earth/Sun sensor is known as the FES-140 Fan Beam Earth Sensor/FSS-140 Sun Sensor. The configuration is typically two FSS and one FES on each of two triplets, with additional FSS at 90° to detect coning.
Attitude reconstruction accuracy: better than ±0.5°
Field of view: 140 × 1°
Range: 250 to 120,000 km orbits, 1-60 rpm spin
Power requirement: less than 0.7 W
Size: FES-140 22 × 44 × 52 mm 0.85 kg;
FSS-140 16 × 44 × 52 mm 0.70 kg

The MFM-3L Fluxgate three-axis Magnetometer has a configuration featuring a cylindrical head with three orthogonal and one redundant sensitive axes. It can be supplied with or without boom.
Resolution: 250 nT
Dynamic range: ±50,000 nT low gain, ±2,500 nT high gain
Orbital range: 300 to 10,000 km
Power requirement: less than 1 W total
Size: 110 × 50 mm diameter 0.40 kg (no boom), interface card 18 × 118 × 159 mm 0.180 kg

The MTR-25 Rod Magnetorquer is suitable for any orbit with perigee less than 1,000 km. Larger units are available.
Magnetic dipole moment: 2.2 × 10^{-5} Wbm
Drive current: 0.1 A
Total power dissipation: less than 1.0 W per rod from +5 V
Size: rod 30 × 60 × 270 mm 0.400 kg; interface card 18 × 118 × 159 mm 0.180 kg

UNITED STATES OF AMERICA

Adcole Aerospace Products

Adcole Corporation's Aerospace Division specialises in spacecraft Sun attitude sensors, including units for spin-stabilised vehicles, two-axis digital sensors, analogue sensors and gating/switching units. The company's devices have seen service aboard approximately 400 satellites, including Tiros, DMSP, GPS, Magellan, COBE, Compton Observatory, Radarsat, Inmarsat 3, Meteor, Amos, XTE, Intelsat 7, MOS, Insat, UARS, EUVE, IUE, Landsat, Satcom, Geotail, Mars Observer, ACTS, GOES, Wind, Polar, Topex, Soho, ISO, Milstar and numerous classified US programmes; they have been selected for GPS 2R, TRMM, SAC-B, Cassini and ACE.

The company's popular Sun angle sensor, the Model 17061, was flown aboard the USAF Block 5 DMSP and NOAA Tiros meteorological satellites, consists of a single-axis Sun sensor head and a package of signal processing electronics. The head incorporates two reticles for Sun angle sensing: a binary code reticle for coarse angle determination and a periodic pattern version for fine.
Max number of sensors: 1
Field of view: ±50°
Accuracy: 0.05°
Size: 109 × 64 × 28 mm/0.322 kg (sensor)
102 × 86 × 51 mm/0.517 kg (electronics)
Input power: +5 V DC at 65 mW; +28 VDC at 0.7 W
Operating range: −5 to +45°C
Acceleration tolerance: 25 *g* (40 *g* shock, 1 ms)

The Model 17140/28520 is a single-axis analogue Sun sensor used for solar array pointing, instrument platform pointing, and initial stabilisation from spinning to three-axis stabilised. Programme involvement includes GOES, Intelsat 5/7, Arabsat, Superbird, Insat, MOS, ETS 5, SCS.
Max number of sensors: 3 (6 for redundant electronics)
Field of view: ±10 to 60° acquisition, ±½ to 20° linear range, ±8-60° insensitive axis
Null accuracy: ±0.01 to 1.5°
Size: varies with specific model
Input power: varies with specific model
Operating range: −40 to +60°C
Acceleration: typically 20 *g* rms

The Model 18656 is a high resolution (0.1°) Sun angle spin sensor system. Users include Geotail, Wind, Polar, GPS 2R, Lockheed Martin Satcom series and classified satellites.
Max number of sensors: 2
Field of view: 128° (each sensor)
Accuracy: 0.1° (incidence is less than or equal to 40°); 0.25° (incidence is greater than or equal to 40°)
Size: 66 × 33 × 25 mm/0.109 kg (sensor)
104 × 58 × 94 mm/0.726 kg (electronics)
Input power: +28 V DC at 400 mW
Operating range: −20 to +60°C

The Model 20910: two-axis fine digital sensor system designed originally for NASA's Solar Max and comprising a two-axis sensor head with processing electronics for coarse and fine Sun angle sensing. Users include Topex, Radarsat, JERS 1, ISO and EUVE.
Max number of sensors: 1
Field of view: 64° each axis
Accuracy: 0.017°
Size: 84 × 109 × 24 mm/0.317 kg (sensor)
198 × 114 × 51 mm/1.044 kg (electronics)
Input power: +28±7 V DC; 2.8 W
Operating range: −10 to +50°C
Acceleration tolerance: 25 *g* (50 to 200 *g* for 200 to 600 Hz shock)

The Model 21730/20850 is a two-axis precision digital sun sensor employed as the primary pointing sensor for Solar Max. This Fine Pointing Sun Sensor (FPSS NASA Standard) operated successfully throughout Solar Max's life (1980-89). The redundant system comprises two sensors co-aligned on a mounting bracket and two electronics units. A similar sensor, offering greater resolution, was supplied for ESA's Soho.
Sensors/electronics: 1
Field of view: ±4° acquisition, ±½° linear
Accuracy: 5 arcsec
Resolution: 0.11 arcsec (0.025 arcsec Soho)
Size: 123 × 84 × 36 mm/0.5 kg (sensor)
165 × 90 × 120 mm/1.0 kg (electronics)
Input power: 28 V DC
Operating range: −10 to +50°C
Random vibration: 14.6 *g* rms

The Model 27290/27460 is a high-accuracy wide field of view digital Sun sensor used for yaw

measurement. Autonomous yaw update can be achieved by using a Sun crossing pulse generated when the Sun is in the yaw plane. Designed and qualified for Milstar; also flown on SDS and classified programmes. Similar units flown on NOAA Tiros and USAF DMSP.
Sensors/electronics: 1
Field of view: ±50° measurement direction, ±50° insensitive axis
Accuracy: ±0.017°
Size: 168 × 87 × 26 mm/0.8 kg (sensor)
152 × 152 × 62 mm/1.2 kg (electronics)
Input power: 28 V DC
Operating range: −40 to +60°C
Random vibration: 18.6 *g* rms

AlliedSignal Aerospace Company (Government Electronic Systems)

GES designs, develops and produces gyroscopic equipment, accelerometers, digital computers and other electronic devices. These include spacecraft pointing and control equipment for Hubble Space Telescope, star scanners such as that for Galileo, and star sensor assemblies designed for spacecraft attitude control using very faint stars. Attitude control equipment using an RW/MW is used aboard satellites such as GPS and various orbiting observatories and technology satellites. It is developing a new generation of angular rate sensors. Utilising lasers, fibre optics and solid-state technology, these gyroscopes will be smaller, less expensive and more reliable than their spinning mass counterparts. It has developed a low-cost MW assembly and a three-axis RLG assembly for Iridium. Other efforts involve the application of advanced technologies such as charge coupled devices and Alexandrite lasers.

Engineering and manufacturing facilities and field support offices include the primary facilities in Teterboro, New Jersey and major operations in Boyne City, Michigan; Mishawaka, Indiana; Westlake Village, California; Orlando, Florida. GES also has one subsidiary company, the Cheshire Corporation in Cheshire, Connecticut, a manufacturer of precision rate gyros.

The Trihex single-fault tolerant flight control and inertial measurement unit debuted aboard Delta 2 in December 1995. Six RL-20 ring laser gyros and six accelerometers are contained in a triply redundant modular architecture. Each lane contains two sensor channels, a 1750A processor and I/O interfaces. Two lanes have a 1553B bus for vehicle data and the third incorporates RS-422 for GSE.

GES builds single- and double-gimbal CMGs covering a stored angular momentum range of 7 to 4,750 Nms and a torque output range of up to 4,600 Nm. Reliability of these units is indicated by accumulated life testing approximately 400,000 h. GES is under contract to build five double-gimbal CMGs for Space Station.

Recent spacecraft using GES reaction wheels include Insat, DSP and GPS. Over the past three decades, the Division has delivered approximately 600 RWs with a cumulative operating time approximately 7 million hours in orbit. The small wheels range over 0.4 to 136 Nms with torque over 0.028 to 0.353 Nm. The large MWs range over 814 to 2,440 Nms with 0.177 Nm torque.

GES provided the precision rate sensor system for Hubble's Pointing Control System, capable of measuring spacecraft motion to a resolution of 0.00025 arcsec. This gyroscopic inertial instrument utilises a hydrodynamic gas bearing to achieve spacecraft pointing to 0.007 arcsec and support mission life requirements of 15 years.

Innovations include on-orbit performance optimisation over a wide focus range, and 'a priori' star crossing location and identification to minimise hardware channels.

Galileo's star scanner provides absolute position information to the attitude control system. Designed to operate in Jupiter's intense radiation fields, which comprise primarily high-energy electrons and protons, the optics and photomultiplier tube are configured to minimise fluorescence and Cerenkov effects and the electronic components to minimise ionisation and displacement damage. The scanner can detect the 200 brightest stars at a 20°/s scan rate. The focal plane's V slit modulates the star signal and the clock and cone angles of each star are uniquely determined from the time relationships between each star crossing.
Field of view: 10°
Accuracy: 60 arcsec two-axis position (1s)
Sensitivity: 200 brightest stars
Power consumption: 3 W
Size: 368 × 321 × 251 mm without sunshade, 482 diameter × 975 mm with sunshade; 20 kg
Inputs: commands for power, threshold, channel; 400 Hz AC 28 V DC power
Outputs: star transit pulse; star intensity signal
Operating range: −10 to +50°C.

Ball Aerospace Systems Group

BECD was created in 1988 to provide electro-optical, cryogenic and tactical subsystems. The hardware includes science instruments such as HST's second-generation ST imaging spectrograph, Near IR camera and advanced camera, plus the COSTAR Corrective Optics Space Telescope Axial Replacement; multi-spectral sensors and imaging; attitude sensors; pointing, tracking and electromechanical subsystems; and stored cryogens and closed-cycle refrigerators. Ball provides the Shuttle Orbiter's power reactant storage assembly pairs of cryogenic LOX/LH_2 tanks. It is building three SAGE Stratospheric Aerosol and Gas Experiment sensors under a US$20 million contract for NASA; the first is due in 1998 aboard Russia's Meteor-3M. It provided the antenna, star tracker and instrument integration for NASA's SWAS Small Explorer, plus a star tracker for the similar WIRE satellite.

Ball has extensive experience in attitude sensors for rockets, satellites and ground-based systems; approximately 2,000 star trackers and scanners have been delivered and flown approximately 300 times. Recent programmes include the NASA standard fixed head star tracker flown on Landsat 4/5, HST, UARS and Compton Observatory; the solid-state star scanners flown on Pioneer-Venus, Magellan and the Inertial Upper Stage; the Shuttle Star Tracker, flown on every Orbiter and now adapted as the CT-611 to charge-coupled device (CCD) technology. The Division has completed research, development and engineering for CCD and charge-injection device (CID) sensors for space target acquisition, tracking and vehicle range/attitude determination. The technology applies also to ground-based astronomic position determination. Completed hardware includes the Retroreflector Field Tracker and the Rendezvous/Docking Tracker (two flights each on Columbia and Endeavour), and RME's Coarse Acquisition Sensor.

BECD operates 190 m² of production rooms and 95 m² of alignment and test areas dedicated to attitude sensors. A high-accuracy test facility is contained in a class 10,000 cleanroom for testing star sensors to accuracies less than 0.03 arcsec. Thermal vacuum test facilities provide 10^{-8} torr and environmental vibration testing is available with up to 133 kN shakers.

Ball makes the CS-203 solid-state star scanner.
Field of view: 4.6° elevation, 5.8° azimuth
Sensitivity: +1.6 to −1.4 magnitude at 1° spin
Spin rate: 0.2-1.0°/s
Accuracy: 18 arcsec
Power consumption: 7 W
Size: 16.8 × 18.0 × 24.1 cm (no shade); 5.5 kg

The first production units of Ball's CT-601 High Accuracy Star Tracker are flying on BMDO's MSX and NASA's SWAS Explorer satellites. It will be used by Space Imaging's satellite.
Field of view: 8 × 8°
Accuracy: 3 arcsec (1σ)
Sensitivity: +1 to +6 magnitude
Numbers of stars tracked: five
Power consumption: 12 W, 22-34 V DC
Size: 17.8 cm diameter, 29.5 cm long; 8.2 kg (no shade)
Output: MIL-STD-1553
Environment: −30 to +50°C

The CT-611 Space Shuttle star tracker is the world's only man-rated reusable general purpose attitude sensor. It emulates the original image dissector tube tracker but has 420 parts in contrast to 1,100. NASA has certified it for 100 missions. CCD is 512 × 512 pixels.
Field of view: 10 × 10°
Accuracy: 15 arcsec (1σ)
Sensitivity: −1 to +3.9 magnitude
Numbers of stars tracked: one
Power consumption: 12.5 W, 22-34 V DC
Size: 21.2 × 16.8 × 28.6; 8.26 kg (no shade)
Output: Manchester biphase
Environment: 10 to 60°C

The CT-621 Wide Field Star Tracker FOV is enlarged at the cost of high-end performance. It has the same electrical/mechanical configuration as the CT-601. The wider FOV improves roll determination by allowing greater separation of tracked stars.
Field of view: 20 × 20°
Accuracy: 11 arcsec (1σ)
Sensitivity: +0.1 to +4.5 magnitude
Numbers of stars tracked: five
Power consumption: 12 W, 22-34 V DC
Size: 17.8 cm diameter, 21.6 cm long; 5.4 kg (no shade)
Output: MIL-STD-1553
Environment: −30 to +50°C

The CT-631 Lightweight star tracker repackages the basic CT-600 electronic design with wide field lens in to a small unit intended primarily for the miniature satellite market.
Field of view: 20 × 20°
Accuracy: 20 arcsec (1σ)
Sensitivity: +0.1 to +4.5 magnitude
Numbers of stars tracked: five
Power consumption: 12 W, 22-34 V DC
Size: 12.7 cm diameter, 12.7 cm long; 2.3 kg (no shade)
Output: MIL-STD-1553
Environment: −30 to +50°C

Barnes Engineering Division

A division of EDO Corporation, Barnes Engineering was founded in 1952 to develop and produce IR detectors, instruments and systems. It has delivered approximately 1,200 mission-critical systems, including approximately 1,000 infra-red Earth sensors and 104 science payloads. Space systems currently in production or operating on-orbit include: scanning Earth sensors, miniature static Earth sensors, GEO static Earth sensors and infra-red detectors. Current programmes include Earth sensors for Hughes (HS-601), Lockheed Martin Astro Space (DMSP, GPS, Landsat, Tiros) and miniature Earth sensors for Lockheed Martin (Iridium), OSC (Orbcomm) and Space Systems/Loral (Globalstar).

The company's Static GEO Earth Sensor, model 13-405 is in production for Hughes' HS-601 + CTA's Indostar, combining horizon crossing indicators used for attitude control during the spin-stabilised transfer orbit, with a static sensor suitable for 15-year life. First used with GPS.
Accuracy: ±0.1° (3σ)
Reliability: 0.998 5 years, 0.996 10 years, 0.994 15 years
Detector: thermopile
Size: 16.0 cm diameter, 37.3 cm long; 4.72 kg
Power consumption: 3.2 W
Redundancy: full internal, transfer orbit and operational static sensors

The single-axis miniature static earth sensor measures one attitude component of a three-axis satellite with respect to Earth's IR horizon. Multiple sensors can be configured for high accuracy over a limited manoeuvring range. It employs the same large-area thermopile IR detector technology as classic GEO static sensors, but with order of magnitude improvements in mass, power, size and cost. There are two standard versions: CHS coarse horizon sensor; FHS fine horizon sensor. A US$15 million contract was awarded in 1993 by Lockheed for Iridium, in 1994 for approximately US$1 million by OSC for Orbcomm, and in 1995 for US$10.5 million (including the Sun sensor specified below) by DASA for Globalstar.
Accuracy: CHS 1.0° (3σ) over 22° operating range; FHS 0.1° (3σ) over 4° (accuracy 0.08° at optimum point)
Detector: IR thermopile
Size (cm): CHS 4.1 diameter 5.6 cylinder, 0.11 kg; FHS 16.5 diameter 8.1 cylinder, 0.28 kg
Power consumption: CHS less than 30 mW; FHS less than 0.30 W

The analogue sun sensor assembly, model 13-517 assembly provides medium accuracy Sun position sensing for three-axis satellites in any orbit. 13-517 provides dual-axis Sun tracking coverage over a central 128° square region and single axis over 168°. Fully redundant and self-calibration maintains accuracy. A contract was awarded in 1995 by DASA for Globalstar.
Accuracy: 0.3° (3σ EOL)
Reliability (15 years): 0.997 non-redundant, 0.99999 redundant
Size (cm): 13.0 × 6.1 × 4.8; 0.43 kg
Power consumption: 0.3 W satellite conditioned

Condor Pacific Industries, Inc

Established in 1964, Condor Pacific manufactures rate gyros, tuned gyros, solid-state resonant gyros and the world's smallest multisensor (two gyros and two accelerometers). The T-100 tuned gyro is used in inertial reference units for space applications. Users include Hughes' HS-601 series (Aussat, Galaxy, Astra, MSAT, UFO, DBS, Solidaridad). Specifications are given in gimballed, strapdown and high-rate strapdown order (series 1/2/4):
Mass: 100 g for flange mounted
Dynamic range: ±20/±100/±300°/s continuous
Random drift: 0.01 to 0.05/0.02 to 0.05/0.03 to 0.1°/h 1s
G sensitive drift: 5/5/10°/h/*g*
Non-G sensitive drift: 20/25/30°/h
Linearity: 0.05/0.05/0.05%
Run power: 2/2/2 W
Typical run up time: 10/10/10 s

Vibration: MIL-STD-810
Operating temp: 0 to 85/−54 to +85/−54 to +85°C
Shock hard mounted: 50/300/300 *g*

Delco Systems Operations

Delco Systems Operations is part of the Delco Electronics Corporation, a subsidiary of GM Hughes Electronics Corporation, and responsible for the company's space activities. Delco's Magic 3 series of guidance computers is employed in the Delta and Titan family guidance systems and was adapted for the USAF/Boeing Inertial Upper Stage. The M352 Delta Inertial Guidance System (DIGS) computer has been in service since 1975 and will continue into the second half of the 1990s aboard Delta 2. Delco's Universal Space Guidance System has provided Titan 3 control since the early 1970s and was integrated into the Titan 2 ICBM in the Rivet Hawk vehicle update of 1975. An improved USGS is carried by the current Titan 2-4 space launchers.

Delco's Space Inertial Reference Unit is a highly accurate fault-tolerant three-axis inertial reference. The Hemispherical Resonator Gyro (82 cm³, 0.14 kg) has the highest performance to size ratio of state-of-the-art gyros. The system contains redundant sensors and electronics that enable continued performance to specification after any combination of one gyro, one processor, one power supply or one accelerometer failure. SIRU will fly on Cassini, NEAR, Clark and a new Martin Marietta telecom satellite.
Size: 20.3 × 9.5 × 30.2 cm; less than 5 kg
Power requirement: 21 W
Bias stability: less than 0.015°/h 3σ
Angle random walk: less than 0.001°/h½ 3σ
Scale factor stability: 100 PPM 3σ
MTBF: more than 6 million hours

The M362S derivative of the Magic 3 series is utilised as the onboard computer for the USAF/Boeing IUS Space Shuttle and Titan 3/4 upper stage. Each stage carries two computers for redundancy. The memory subsystem offers 66,536 16-bit words of high-speed CMOS RAM, electrically alterable and non-volatile via a self-contained backup battery, producing a central processor computational performance more than 10^6 operations/s. A memory error detection and correction system is used, correcting single-bit errors in real-time without affecting throughput. The unit is sealed in a dry nitrogen atmosphere and passively cooled; a single cover provides access to the 25 modules for individual replacements.
Size: 36.6 × 36.6 × 15.2 cm; 23 kg
Power requirement: 180 W at 28 V DC

Honeywell Inc

Space Systems' RWs have been provided for Landsat 7, Topex, Explorer platform, HST, Milstar, TDRSS, Magellan, Compton Observatory, Globalstar; spin bearing and scan drive assemblies for Galileo, and IMU and star sensors for DMSP. The GG1308 ring laser gyro flew on Clementine in 1994.

The Miniature Inertial Measurement Unit is being supplied for the Aerospatiale Spacebus and a variety of other commercial, military and scientific spacecraft. The electronics are hardened to approximately 100 krad.
Range: ±375°/s, ±25 *g*
Size: 0.041 m³/3.45 kg
Power: 33 W
Computer: Honeywell 1750A Generic VISIC
Accelerometers: three Sundstrand QA3000
Gyros: three Honeywell GG1320C RLG
Reliability: more than 173,000 h MTBF
Life: 15 years

Honeywell's Space Systems operation has qualified the Fault Tolerant Inertial Guidance and Navigation System for the next generation of launchers and upper stages. It provides detection and isolation of up to two gyro and accelerometer failures, and improved accuracy using the output of all sensors to compute body rates. Processing and redundancy management is performed by a dual self-checking pair of 1750A radiation-hardened processors arrangement to provide protection against single event upsets and hardware processor failures. With two accelerometer and two gyro failures, the IMU provides 0.999999 reliability. It will be adopted by the USAF/Boeing IUS beginning with IUS-27 from 1997/98.
Size: 0.047 m³/33 kg
Power: 135 W
Accelerometers: five Sundstrand QA3000
Gyros: five Honeywell GG1320 RLG

Honeywell's IFOG Interferometric Fibre Optic Gyro was provided for NASA's XTE, the first IFOG used on a NASA satellite.
Performance: 1-10°/h low, 0.01-1°/h medium, 0.001-0.01°/h high, 0.0001-0.0001°/h very high

The Laser Inertial Navigation Unit was introduced on Centaur in December 1991.
Size: 50.8 × 30.5 × 22.9 cm; 31.75 kg
Power: 105 W average
Accelerometers: three Sundstrand QA3000
Gyros: three Honeywell GG1342 RLG
Processors: two Performance Series Corporation 1750A 1 MIPS
Reliability: 25,000 hours MTBF ground benign.

Control Moment Gyroscopes from Honeywell's Phoenix operation features high momentum-to-weight ratios, low-torque ripple and emitted vibration, long life and proven performance.

Honeywell's Arizona facility provides six different versions of reaction/momentum wheel assemblies for commercial, military, scientific and special purpose applications ranging from LEO to GEO. Masses range from 3.8 to 19.7 kg: power requirements 14 to 44 V.

Hughes Danbury Optical Systems

The HD-1003 miniature star tracker started deliveries in 1996 for NASA and DoD programmes; earlier models were ordered for Topex/Poseidon. The all-solid-state HD-1003 employs a frame transfer 512 × 1024 CCD to obtain the required accuracy, sensitivity and stability over the FOV. Lower cost models are available for missions requiring less pointing accuracy; other specifications remain as below.
Field of view: 8 × 8° or 20 × 20°
Accuracy: up to 2.5 arcsec rms. Accuracy per star at 6.0_{mv}: 6 arcsec pitch/yaw, 40 arcsec roll
Update rate: 10 Hz
Number of stars tracked: 6 maximum
Size: 38 × 15.7 × 11.2 cm; 3.3 kg (with light shade)
Detector: CCD matrix of 512 × 1,024 pixels
Input power: 11 W
MTBF: 10^6 hours
Environment: −20 to +60°C, 12 *g*.

Humphrey Incorporated

The company was founded in 1946 and incorporated in 1951 to design and manufacture precision electromechanical instruments for defence and aerospace applications. Product lines include gyroscopes, rate sensors, accelerometers, magnetometers and switches. The RT02-0800 series three-axis solid-state rate sensor consists of three single axis rate transducers packaged with electronics in a 6.6 × 5.8 cm unit weighing 0.4 kg. Rates up to 5,000°/s can be accommodated; life is 10,000 hours minimum.

ITHACO Incorporated

ITHACO designs, manufactures and sells attitude determination and control components and systems for spacecraft. Their Earth horizon sensors, electromagnetic torquers (TORQRODS™), reaction and momentum wheels, magnetometers, Sun sensors, and associated electronics have been used on over 125 US satellites and on many international spacecraft. Specialised test equipment and post-delivery support are also offered to meet customer specific needs.

The T-Scanwheel is a momentum/reaction wheel with an integral high accuracy scanning Earth sensor. The wheel portion provides both angular momentum and control torque, while the sensor portion provides precise two-axis attitude information. The wheel consists of a precision balanced aluminium flywheel driven by an ironless armature brushless DC motor and suspended by a pair of angular contact ball bearings. The T-Scanwheel optics consist of a scan mirror coupled to the flywheel shaft, a parabolic objective mirror, a 14.25 to 15.65 μm spectral filter and an infra-red detector. The scan mirror reflects radiation causing the Field Of View (FOV) of the T-Scanwheel to sweep a 45° half apex angle cone in space. The signals produced by the alternating space-Earth and Earth-space transitions are amplified and sent to a separate electronics box for processing into phase and chord or pitch and roll attitude information. The T-Scanwheel is available in two sizes, A and B, with the following specifications:
Operating speed (rpm): 500-3,000
Angular momentum (Nms): 2.25/9.75 at 3,000 rpm
Available reaction torque (mN-m): more than 13/31
Mass (kg): T-Scanwheel; 3.2/5.8; electronics 2.6
Steady state power: less than 9/15 at 3,000 rpm

The T-Wheel series of fully capable momentum wheels and bi-directional reaction wheels is built specifically for attitude control of small- and medium-sized spacecraft to provide both reaction torque and angular momentum storage for either three-axis reaction control systems or momentum bias stabilised systems. The T-Wheel is available in three sizes, A, B and E. The Type A and B designs can be augmented with horizon sensor optics to provide both attitude determination and attitude control (see T-Scanwheel). The Type E design provides a significant increase in momentum storage capability and is primarily targeted for medium to large satellites. The following specifications are given for the Type A/B/E wheels, respectively:
Angular momentum (Nms): 4/16.6/50
Available reaction torque (mN-m): more than 20/40/300
Mass (kg): T-Wheel 2.55/5.1/10.5; electronics 0.9/0.9/3.6
Steady state power at max speed (W): less than 9/17/40
Peak power at max speed (W): less than 25/50/280

The Conical Earth Sensor is a scanning horizon sensor used to gather attitude information for Earth orbiting satellites. The most common application utilises one or more CESs to provide pitch and roll attitude data for three-axis stabilised Earth oriented satellites. The CES can also support autonomous navigation and attitude determination for spin-stabilised satellites. The CES can be used from low Earth orbits to above geosynchronous orbits, as well as during transfer orbits. The sensor head is usually mounted with its spin axis located in either the satellite roll/yaw plane or the pitch/yaw plane, while the associated electronics is conveniently mounted inside the satellite. Output signals can be presented in either an analog or a digital format.
Accuracy (deg): less than 0.10° post-processed
Mass (kg): SH 1.1; SE 1.9
Power (W): 8

The ITHACO Torqrod is an electromagnet designed to provide complete momentum management on an Earth orbiting satellite (even at geosynchronous). Typically, one Torqrod is aligned with each axis of the satellite. Torqrods should be mounted away from any instruments sensitive to magnetic fields and should also be separated from one another in order to avoid cross coupling effects. Numerous Torqrod sizes have been developed, with the following range of specifications:
Dipole magnitude (Am²): 12-790
Length (cm): 16-55.1
Mass (kg): 0.9 to 30
Power (W) at max linear voltage: 0.7 to 8.8

The ITHACO two-axis (IM-102) and three-axis (IM-103 and IM-203) magnetometers produce continuous vector measurements of the Earth's magnetic field as the satellite orbits the Earth. These measurements are used by the satellite's magnetic attitude control system to generate the control commands that stabilise the satellite attitude and/or spin rate. The magnetometer design incorporates both the sense coils and the signal processing electronics into a single box. Both the IM-102 and the IM-103 require regulated input voltages, while the IM-203 incorporates a DC/DC converter and can accommodate an unregulated input voltage. The following specifications are for the IM-102, IM-103, and IM-203 units, respectively:
Field measurement range (gauss): ±0.6/±0.6/±1.0
Mass (g): less than 220/227/500
Volume (cm³): 165/233/419
Input voltage (V): +5/±15/+28±6

The Horizon Crossing Indicator is intended for use on a spinning spacecraft for the purpose of providing attitude and timing information. While a single HCI is sufficient for a circular orbit, either two HCIs or the use of a single Steerable HCI is effective for an elliptical orbit. The sensor head contains infra-red optics which respond to the infra-red energy reflected from the Earth as compared to cold space. The sensor electronics converts the detected signals into pulses that are accurately referenced to the Field Of View (FOV) and the Earth's infra-red horizon.
Accuracy (deg): less than 0.10° (post-processed)
Mass (kg): HCI – 0.65; SHCI – 2.2
Power (W): HCI – 0.7; SHCI – 2.5 static/10.0 stepped

Kearfott Guidance & Navigation Corporation

Kearfott is a subsidiary of the Astronautics Corp of America, specialising in space-qualified inertial guidance and inertial reference systems. Employees total 1,850. Principal space hardware covers one-, two- and three-axis attitude control and pointing systems, star trackers, rotary components and cryogenic systems. In its New Jersey facility, Kearfott also produces a line of gimballed and strapdown GNC systems, using both ring lasers and dynamically tuned gyros (DTGs), test equipment and inertial sensors. In North Carolina and Mexico, it designs and manufactures precision rotary components, actuators and electronics.

Kearfott initially furnished floated rate integrating gyros to Vela Hotel, OAO, Mariner, Nimbus 2-4, Surveyor, Viking, Skylab and a three-axis rate system (GRA) for a classified satellite. These programmes involved approximately 170 gyros, accruing approximately 290,000 hours in space without failure.

In the early 1970s, Kearfott developed the first of the SKIRU (Space-qualified Kearfott Inertial reference Unit) three-axis attitude reference systems employing the Gyroflex DTG, which has 6.2 million hours of data on 82 life test units. SKIRU is used on Galileo, DSCS III, Voyager, TDRS, Milstar, XTE, TRMM, EOS AM 1 and AXAF-I, plus classified satellites. 70 SKIRUs have been delivered. Some systems have 45,000 to 110,000 hours in space without failures; total is 1.7 million operating hours. SKIRU has also been developed into a fine pointing system, utilising the same basic components with temperature control and circuit modifications to permit very accurate space pointing over long periods. SKIRU I to V are two-gyro single-string units; the D-II is a three-gyro internally redundant unit that is interchangeable with NASA's DRIRU II Dry Gyro Inertial Reference Unit.

Performance: 0.002-0.009°/h, 3σ, 8 h
Size (cm): 17.8 × 22.8 × 9.5 IV; 17.8 × 22.8 × 14.2 IV-H; 17.8 × 22.8 × 15.0 V; 17.8 × 35.5 × 17.8 D-II
Mass (kg): 3.2 IV; 5.0 IV-H; 5.0 V; 12.7 D-II
Power (max quiescent; W): 25 (typical 17) IV; 25 (typical 22.5) IV-H; 26 (typical 17) V; 26 D-II

TARA, the Two-Axis Rate Assembly utilises an IRU comprising a two-axis CONEX MOD I/S DTG. 15,000 CONEX gyros have been built. TARA's modular concept allows two packages to be used for three-axis control, while three would provide complete redundancy. It was developed and qualified by early 1993 under a company-funded programme. Users are TOMS-EP, Lewis, MSTI 2/3, Rocsat 1 and Komsat. The MOD I/S CONEX is also being delivered as part of a three-axis IRU for Hughes' HS-601.

Performance: 0.05°/h
Size (cm): 16.0 × 14.3 × 8.6
Mass: 1.8 kg
Power (max quiescent; W): 11 (typical 7) operating from unregulated DC input.

Litton Applied Technology

Applied Technology, a division (ATD) of Litton Industries, was founded in 1959 to develop and produce microwave amplifiers, reconnaissance and surveillance receivers and countermeasure jammers. The Applied Technology Advanced Computer (ATAC) was developed in the 1970s; the company now offers space processor designs with ATAC, 1750A and RISC architectures to optimise radiation hardening and fault tolerance with maximum thermal survivability. ATAC is now performing attitude and articulation control functions for NASA's Galileo Jupiter orbiter; it was also carried by the Magellan Venus orbiter.

Litton's Guidance and Control Systems division develops and produces navigation and guidance systems for space launchers, missiles and tactical aircraft. Some 2,000 are employed in the division's southern California locations near Los Angeles, and another 900 in its main manufacturing facility in Salt Lake City, Utah.

Litton's first space launcher inertial guidance system, the LN-100L, debuted on Lockheed Martin's LMLV in August 1995. A product line derivative of the LN-100 strengthened for the space launch environment, it comprises three Zero-Lock laser gyros, an A-4 triad of miniature accelerometers, sensor electronics, a digital processor and two power supply modules.

Performance: 1.48 km/h, 0.76 m/s rms, 0.05° azimuth rms, 0.015° pitch/roll rms
Size: 17.8 × 17.8 × 24.9 cm; 8.4 kg
Power: 27 W running; 27 V DC
Operating ranges: 20 *g* all axes, unlimited attitude, approximately 400°/s three-axes, approximately 1,500°/s² three-axes
Outputs: two dual MIL-STD-1553B, three RS-485/RS-422, one bi-directional SDLC, one input-only SDLC
Navigation processor: 32-bit 80960, Ada
Operational service life: 10,000 h or 15 years
Environmental tolerance: MIL-E-5400, sea level to vacuum, −54 to +71°C, vibration (random) 22.6 *g* rms endurance and 8.1 *g* rms performance, vibration (sine) ±5 *g* 5 to 2,000 Hz, shock 21 *g* 40 ms handling and 1,130 *g* pyro, acoustic 140 dB
Zero-Lock laser gyro: 18 cm non-dithered, 0.75 arcsec/pulse resolution (0.003 enhanced), 0.003°/h bias repeatability, 0.2 PPM scale factor stability, less than 0.0015°/h½
A-4 Accelerometer Triad: 25 μg bias repeatability, 120 PPM scale factor, 2 arcsec alignment

Lockheed Martin Missiles and Space Company

LMMSC is the largest producer of US Earth sensors. In 1964, it designed the first sensor for spin-stabilised satellites to use the 15 μm filter (flown 1965). Approximately 250 sensors for 12 different types of spinning spacecraft (0.16 to 180 rpm) have since been delivered. Conceptual work on a sensor for three-axis satellites began in 1968, incorporating an oscillating mirror approach now used on all LMMSC scanning sensors. In 1970, the concept of a single mirror with two detectors was proposed, permitting the simultaneous determination of pitch and roll for GEO satellites. The NOHS and PESA sensors (see below) are current examples.

Lockheed began development of the Wide Angle Sun Sensors (WASS) in 1974. Users include HST. The Intermediate Sun Sensor (ISS) is used for high-accuracy Sun-pointing experiments.

The Null Operating Horizon Sensor is a scanning mirror Earth sensor for application aboard three-axis stabilised GEO spacecraft, the NOHS provides pitch and roll information. Versions have flown on Intelsat, GOES and DSCS 2. An upgraded version, the VLSIC Earth Sensor, combines the NOHS optical system with new VLSIC electronics for reductions in size, mass and input power.

Instantaneous field of view: 1.65° square
Accuracy: ±0.04° (3s)
Range: ±6.4° pitch, ±2.24° roll
Size: 10.4 × 13.8 × 16.5 cm (11.5 cm VLSIC)
Mass: 1.6 kg (1.36 kg VLSIC)
Input power: 4 W (3 W VLSIC)
Output rate: 4/s
Output signals: serial digital 0.01° LSB
Failure rate: 0.8 × 10⁻⁶/h.

The Precision Earth Sensor Assembly, or PESA, is designed to handle a large dynamic range in both vehicle attitude and orbit altitude. Its three detectors compute the Earth's position in parallel to generate a serial-digital output word containing the information from all three, linear to within 0.01° over up to ±16°.

Instantaneous field of view: 1.10° square
Accuracy: ±0.03° (3s)
Range: ±16°
Size: 11.4 × 17.0 × 22.9 cm/3.37 kg
Input power: 8 W
Output rate: 1/s
Output signals: serial digital 0.01° LSB
Failure rate: 2.5 × 10⁻⁶/h

The Dual-Horizon Sensor Assembly contains two identical sensors in a single housing for application on spinning satellites. Fixed level threshold detectors are used for high accuracy in roll at GEO altitude. A bi-level signal proportional to Earth's chordwidth crossed by the field of view is available as an output. Amplifier outputs providing radiometric information are also available and can be externally processed for improved accuracy.

Instantaneous field of view: 1.65° square
Accuracy: ±0.05° (3s)
Size: 7.6 × 8.3 × 12.1 cm/0.68 kg
Input power: 2.5 W
Output rate: one per revolution
Output signals: 5 V sq wave width of Earth
Failure rate: 0.3 × 10⁻⁶/h (× 2).

Wide Angle Sun Sensor is available in several configurations; the detectors may be wired as a single-axis or two-axis control sensor or as a hemispherical sensor. The 1.8 mm Si photo detectors are mounted on a pyramid, sized to accommodate up to three detectors per face for full redundancy, and the pyramid enclosed under a fused silica dome. Each detector output is a cosine function peaking at 620 μA when the Sun is normal to the face. WASS is qualified for a −100 to +85°C thermal operating range and the new design is nuclear hardened.

Accuracy: up to ½° at null
Range: 2π steradian
Size: 6.1 × 5.1 × 4.8 cm; electronics 6.6 × 5.8 × 5.1 cm
Mass: 0.155 kg; electronics 0.132 kg
Input power: none for head, 70 mW electronics
Output: 620 μA peak/detector; 0-5.2 V electronics.

Lucas Control System Products

Schaevitz designs, develops and manufactures acceleration, pressure and displacement sensors to military aircraft and space standards. The company's linear and angular servo accelerometers include the LSXH, designed specifically for flight control applications such as satellite nutation sensing systems.

Type: linear
Size: 44.4 × 31.8 × 39.6 mm/0.114 kg
Range: ±5 *g*
Scale factor: 1 V/*g*±1%
Scale factor stability: 500 PPM over 60 days
Bias: 5 mg
Natural frequency: 35 Hz
Linearity: 0.10% over full range
Temperature: −55 to +71°C operating, −65 to +90°C storage
Shock survival: 1,000 *g* for 0.5 ms

Meda, Incorporated

Meda offers its TAM-1C three-axis magnetometer for attitude control systems and science instrumentation. The series meets MIL-STD-97 5G Class S (Class B also available).

Servo Corporation of America

Founded in 1946, Servo was a pioneer of IR technology, now covering satellite attitude determination, Earth resource measurements, gas analysis, imaging and laser power monitoring. Its bolometers and pyroelectric detectors form the sensing element of horizon sensors for most telecom satellites; they have served on hundreds of satellites with no known failures. Servo also builds its own Earth sensors for stabilised and spinning satellites.

Servo's Horizon Sensor and IR Detection Facility comprises research and development laboratories, environmental test facilities, Earth sensor test facility, manufacturing facilities and cleanrooms, including dedicated equipment for all thin film, ceramic thick film and wet chemistry processing of infra-red detection cells, fabrication, testing and environmental screening of infra-red systems and special optics. Specialised equipment includes vacuum evaporator stations, sputtering stations, electron beam equipment, leak detectors, humidity chambers, environmental chambers, spectrophotometers, a full range of electro-optic test equipment, multispindle polishing machines, optical grinder, sand blasters, lens generators, specialised microscopes and a JEOL scanning electron microscope.

The MiDES, Mini Dual Earth Sensor is a state-of-the-art Earth horizon sensor designed to provide pitch and roll information for LEO satellites, this being accomplished by two pairs of pyrotechnic arrays positioned 90° apart in the image plane. Each array is spatially separated into two eight-element columns to provide for Sun/Moon rejection. An algorithm calculates the position of the horizon from the voltage obtained from the pixels which subtend the horizon gradient (typically two) and the pixels that look at space and Earth. The lifetime is improved by there being no motor but a low-power tuning fork like a shutter mechanism. The MiDES can also be adapted for GEP operations. The MiDES first test flight will be aboard the NASA/DARA CRISTA SPAS 2 to be flown as part of the STS-85 mission.

Accuracy: 0.06° (3 sigma)
Field of view: ±5.5°
Dimensions: 13.3 × 13.3 × 13.0 cm
Mass: 1 kg
Spectral band: 14-16 micron
Power input: 28 V DC, +7/−6 V DC
Power consumption: less than 800 mW
Signal output: RS-232, -422 or -485 digital data stream of 8-bit words representing 32 pixels voltages plus temp
Environment: −20 to +60°C, 19 g rms, 50 krads

The Model 110 Horizon Crossing Indicator is a state-of-the-art component for use in attitude determination of micro spinning satellites with short lives. The low-power electronics and pyroelectric detector make it self-powered. It will operate for a minimum three years, with a five-year shelf life.

Spin rate: 1-50 rpm
Mass: 0.120 kg
Power consumption: 100 μW

The Series 13000 Broadband Radiometer Sensor/Telescope is used as the primary sensing element in an Earth resource monitoring system requiring uncooled operation, broad spectral sensitivity and large radiance dynamic range. It was planned to fly as part of the CERES instrument on NASA's EOS AM 1 in 1998 and PM 1 in 2000 and NASA/NASDA's TRMM in 1997. Optional filters provide specific selectivity up to 150 μm.

Spectral response: 0.3-50 μm
Detector: uncooled thermistor bolometer
Time constant: 9 ms maximum.

The Servo Wide Angle Earth Sensor (SWAES) is a 120 × 10° field of view Wide Angle Earth Sensor which has been developed for use on missions which are prohibited from viewing the Earth with the payload sensor. Developed especially for NASA's WIRE (Wide-Field Infra-Red Explorer) mission, the WAES consists of two sensor heads and an electronic control unit. On the WIRE mission the sensor heads are arranged back-to-back 180° apart so that they can monitor one axis of rotation, while a Sun sensor is used for the other axis. This is of extreme importance to the mission because the telescope features a detector which is cooled by a solid hydrogen cryogen. Accidental viewing of the Earth or Sun would cause the cryogen to boil off and terminate the mission.

Accuracy: ±1.5° (exclusive of radiance)
Mass: 1 kg
Power: less than 1 W

Smiths Industries Aerospace

Smiths Industries is a UK company, emphasising aerospace electronics, medical systems and specialised industrial products, deriving about half of its revenue from US activities. 1998 aerospace turnover was US$764 million versus US$683 million for 1997, US$622m million for 1996 and US$591 million for 1995. The company provides six Space Shuttle elements: engine interface unit, mission/event timers, FM signal processor, ground command interface logic controller, and attitude director indicator. Its fibre optic gyro inertial measurement unit (0.4 kg, 10.7 × 7.1 × 6.6 cm) is used by Boeing's LEAP BMDO demonstration kinetic kill vehicle.

Systron Donner

Systron Donner comprises three divisions, of which the Inertial Division is the most space-orientated. It provides precision inertial systems for space purposes, including inertial angle sensors, accelerometers, gyros, rate sensors and vibration transducers. Equipment has been provided for Shuttle, Landsat, Meteosat, Intelsat, Galaxy and several communications satellites. The division and Rockwell Autonetics in 1994 began producing the tactical-grade DQI digital quartz inertial measurement unit. It is the first solid-state inertial assembly in Rockwell's family of Miniature Integrated GPS and Inertial Navigation System Tactical Systems (MIGITS).

The Edcliff Division provided pressure transducers for the Shuttle and several NASA satellites.

SATELLITE INFORMATION STORAGE COMPONENTS

FRANCE

Alcatel Espace

Alcatel Espace and IBM Montpelier Technologies were selected by CNES in February 1995 to provide solid-state recorders for Spot 5 and Helios 2. Capacity of each is nearly 100 Gbit, enough for 200 Spot images. For Helios 2, Alcatel Espace is also responsible for the image telemetry subsystem.

Matra Marconi Space

MMS is developing mass memories for remote sensing and science satellites. Multichip module technology enables 1 Gbit/kg capacity. Reed Solomon encoders, decoders and data compression circuits have been built using ASIC technology, allowing further improvement in capacity and bit error rate.

Schlumberger Industries

The division is a leader in high-data-rate mass storage magnetic recording. Product lines cover ground station and spaceborne systems, complemented by telemetry capabilities. The DV 6000 series of rotary head digital ground recorders (MIL-STD-2179) include the DV 6410 designed for harsh environments. It has flown on Shuttle for the SIR-C and ATMOS science missions.

Schlumberger's principal spaceborne magnetic tape recorder is the high capacity (50 Mbit/s, 120 Gbit) heavy duty (2,200 h/five year on-orbit) digital recorder for Spot 4/Helios 1. It comprises two modules: a sealed tape transport module (70 × 60 × 41 cm, 95 kg) and an electronics module containing an error detection and correction system (50 × 42 × 20 cm, 26 kg). Power requirement is 175 W record; 260 W replay; 12 W standby. Bit error rate less than 10^{-7} (typically 10^{-8}).

Designed for unattended operation in automatic system environments, the 3801 TT&C digital processor unit is the key communication unit for ground stations operating satellites, from integration, through the positioning phase and in-orbit operation. Contracts have been won for Astra and Nilesat TT&C. The 3301 CCSDS Mux/Demux is a modular stand-alone 3U high chassis housing up to seven Euronorn format VME Cards. It handles CCSDS formats at 3 Mbit/s with Reed Solomon encoders and decoders and/or convolution coders and Viterbi decoders.

GERMANY

Racal-Heim GmbH

In 1996, Racal Recorders Ltd. in Southampton (UK) acquired the German company Joseph Heim kg in Germany, and as a result the instrumentation activities were renamed Racal-Heim GmbH and were relocated to Germany as shown above. The new company manufactures instrumentation recorders for transport, aerospace, science and research and development applications. Storeplex and 'A' series portable recorders offer bandwidth up to 2.56 MHz analogue and 51.2 Mbit/s digital. The rugged 'D' series records both analogue and digital signals (including ARINC and 1553 busses) up to 12 Mbit/s total bandwidth. Storehouse is a modern IRIG recorder compatible with both 2 and 4 MHz standards.

UNITED KINGDOM

Ampex (GB) Ltd

The UK office of Ampex Corporation Inc designs and manufactures advanced data recording and processing systems for mobile and fixed installations. Users include satellite ground receiving stations such as Helios. The DCRsi transverse digital recording unit is a widely used rotary cassette system, with approximately 1,300 units installed. It stores 50 Gbyte on a single tape cassette, and can sustain 0 to 240 Mbit/s continuous transfer or 0 to 300 Mbit/s burst. Ampex also offers a range of high-capacity (12,800 Gbytes) D-2 technology-based mass storage devices with a maximum data rate of 160 Mbit/s.

Penny and Giles Data Systems Ltd

A member of the Bowthorpe Group, the company specialises in advanced high-density digital and analogue magnetic tape recording systems for Earth resources satellite ground stations, radio astronomy, pre-launch check-out, meteorology, range telemetry, rocket motor development and general instrumentation data acquisition. Principal space contracts include HDDRs for Landsat, ERS 1/2, IRS 1C, Radarsat, Spot, MOS 1, JERS 1 and Spacelab. Recorders have also been supplied to SEP, DLR and DASA for Ariane Vulcain engine development, to CNES for Ariane 4 engine pre-launch testing and to ISRO for range telemetry and satellite development.

The Pegasus cassette-based recorder, introduced in 1992, is capable of recording rates approximately 100 Mbit/s on a standard D-1(L) cassette. Used for IRS 1C, ERS 1/2, Landsat, Spot and Radarsat.

Input/output data rate: 100 kbit/s to 108 Mbit/s continuous, variable or burst; 120 Mbit/s burst for 750 ms (repetition not less than 1 s)
Recording duration: 30 minutes at 100 Mbit/s; pro rata other rates
Storage capacity: 22 Gbytes per cassette
Interfaces: serial, parallel

The Sirius laboratory recorder is capable of recording 42 4 MHz channels simultaneously or a single 144 Mbit/s serial digital stream; alternatively, multiple streams of lower rates. Four were provided to China for recording Landsat imagery.

The VLBI is under a user agreement with the US National Radio Astronomy Observatory. Penny & Giles offers a range of VLBI-compatible recorders, reproducers, data acquisition racks and headstacks. VLBI recorders have been selected for radio telescopes in Chile and Poland and at VLBI correlator sites in China, Germany and Netherlands.

SATELLITE AND PROPULSION SUPPORT HARDWARE

UNITED KINGDOM

Dowty Aerospace Wolverhampton

Dowty Aerospace Wolverhampton is a member of the Dowty Aerospace Division of the TI Group, which comprises more than 50 operations worldwide. The Space Projects department manufacturers diaphragm, capillary and pressurant tanks.

DAW's space manufacturing facilities were opened in 1989 to provide a fabrication capacity for tanks up to 120 by 150 cm long. The 400 m^2 class 100,000 clean room, with 150 m^2 test area, includes a computer-controlled inert gas welder for welding titanium tank shells, computer-controlled resistance welder for fabricating internal components, and orbital tube welder for titanium and stainless steel piping. Test facilities include ultra-sonic cleaning, bubble testing, helium leak testing, pressure and expulsion testing, cryogenic proof pressure testing, dye penetrant and radiographic inspection.

The capillary tanks are surface tension propellant management devices. Tanks are under contract for MMS, Hughes HS-601 and BPD.

DAW manufactures diaphragm tanks for mono hydrazine systems. ESA's Infra-red Space Observatory carries them and Orbital Sciences Corporation contracted in 1991 for the 96-litre titanium alloy elastomeric diaphragm hydrazine AOCS tanks for SeaStar.

In addition DAW provided 40 titanium alloy centrifuge tanks for Cluster.

DAW is under contract by MMS UK to supply Xe tanks for the UK ion thruster system on ESA's Artemis. Net volume 30 litres; can store 44 kg of Xe at 125 atmospheres.

The company is teamed with Vacco Industries to manufacture titanium etched disc filters, latch valves, fill drain valves and non-return valves for the European market.

Fairey Microfiltrex Ltd

The company designs, develops and manufactures filtration equipment for fuel, hydraulic, gas and oil systems. Fairey Microflex has expertise in spacecraft propulsion systems and has supplied the filter element assemblies for the hydrazine thruster reaction control systems of ESA's ISO and Soho satellites. Ariane 4 and 5 also utilise the company's filters.

Firth-Rixson Plc

Firth makes open and closed die forgings, rolled rings and superalloys for aerospace applications. Products include steel, nickel and titanium alloy closed die forgings for missile and rocket motor casings.

Rotadata Limited

Rotadata has an active programme of developing measuring techniques in turbomachinery and aerospace development hardware. Following the design of a number of fully instrumented full-scale model turbines for the Space Shuttle main engine turbopump, the company has developed a number of new products, both hardware and software, for applications to aerospace turbomachinery development and monitoring.

Rotodata new product line features a new high-performance pyrometer sensor with a better sensitivity and a high-frequency bandwidth for turbine blade surface temperature mapping at full-rotation speed. The company makes turnkey digital and analogue telemetry systems for rotor instrumentation, and high-temperature strain-gauging for turbine blades and discs. In other NDT applications the company manufactures miniature capacitance probes for blade tip clearance and rotor orbit monitoring and an optical smoke density meter for engine emissions

The company is also working on a novel CO_2 sensing unit for cabin condition monitoring. Meanwhile, the traditional Rotadata activities in turbomachinery instrumentation, aero-thermal rakes and probes, specialist positioning systems and high-quality instrumented test hardware continues to provide a core business from an increasing range of customers.

UNITED STATES OF AMERICA

Ardé Inc

Ardé was formed in 1952 by engineers from the Rocket Group of M W Kellog. In the space field it specialises in lightweight pressure gas storage vessels and propellant tanks. Programmes include Shuttle/Centaur, Mars Observer, Atlas 2 and GOES helium pressurant tanks, GOES satellite and Shuttle SRB TVC auxiliary power hydrazine tanks, and Shuttle EMU backpack breathing oxygen tank. The product lines include cryoformed 301/304L pressurant and propellant tanks, prestressed composite fibre over-wrapped metal tanks, and ring stabilised and apex rolling diaphragm expulsion devices.

Cryoformed 301/304L Tanks: Under-sized 'preform' tanks are conventionally fabricated from austenitic stainless steel and then immersed and filled with LN_2 before pressurisation to a level about equal to its room temperature design burst pressure. The membranes and welds are stressed beyond the 195°C yield and the vessel plastically stretches: 7 per cent in spheres and 15 per cent for cylinders. The temperature causes a transformation to martensite steel in both parent metal and welds with an attendant 5:1 increase in room temperature yield strength. Approximately 3,000 flight tanks using this process have been delivered over 20 years. The Shuttle EMU backpack O_2 tank is a 36.3 cm long, 13 cm diameter cylinder pressurised to 714 atmosphere; mass 0.9 kg.

PSC Pressurant Tanks: Pre-stressed composite fibre over-wrapped pressurant (PSC) tanks provide pressure vessels up to 50 per cent lighter than equivalent all titanium tanks. Kevlar or graphite fibre is wrapped on a metal liner and immersed in LN_2 as in the cryoforming process to stress both materials.

Brunswick Defense

Brunswick Defense has three major product lines for space applications: fluid control devices; thin foil extendible booms; fire detection and suppression systems. The Lincoln facility produces composite filament wound pressure vessels for life support systems and control mechanisms. Each Space Shuttle carries 23 He, O_2 and N_2 vessels.

Brunswick Corp acquired Wintec Aerospace in 1972 as specialists in filters and propellant management systems, including high-performance ultra-clean wire mesh filters; large screen propellant management assemblies; rocket engine tubular heat exchangers; Shuttle bay vents; Shuttle shock and vibration mounts; and high-capacity GSE cryo filters. Wintec has a design inventory of approximately 1,500 filter part numbers. The filters are available in all stainless and/or stainless screen with Ti body details, are manufactured in a cleanroom and many are processed to 50 PPM or better. Wintec developed the fabrication design technology of a number of Propellant Management Devices (PMDs), some of which required the joining of stainless screen to larger Ti frames; notably for the Shuttle OMS tank and Peacekeeper missile PMDs. Others include the Harpoon missile, Shuttle RCS tanks, a new satellite design and several developmental units, including a toroidal design.

EG&G Wright Components, Inc (WCI)

WCI designs, develops and manufactures solenoid and pressure operated valves, mechanical valves, relief valves and manifolds for aerospace applications, including rocket propulsion systems. The major space products are hydrazine thruster valves 0.45 to 778 N. WCI has 75 valves on Space Shuttle, ranging from hydraulic valves controlling the landing gear and main engine gimballing, to helium systems for purging oxygen lines. Shuttle's Auxiliary Power Units use hydrazine and relief valves.

Kaiser Marquardt

The Marquardt company was formed in 1944 to undertake research and development leading to the first US subsonic ramjet in 1945. Its principal engineering business continues to be advanced aerospace propulsion and the supply of ram-air turbine power systems. The company has been developing and manufacturing bipropellant thrusters since 1959, notably the attitude control thrusters for Apollo spacecraft and the Space Shuttle Orbiter. Precision bipropellant engines are available for 0.004, 0.02, 0.06, 0.11, 0.22, 0.44, 0.89, 4 kN sizes, and requests up to 8 kN can be accommodated. The monopropellant hydrazine thruster line of UTC's Hamilton Standard Division was acquired in 1993.

Moog Inc

Moog is a worldwide manufacturer of precision fluid control components and systems. SPD specialises in mono/bipropellant valves, hot/cold gas thrusters, regulators, latch valves and electric propulsion system components. It developed the Xe propellant feed system for the SPT-100 plasma thruster. Moog's Model 51172 latch/thruster combination bipropellant torque motor valve is used in the Tempo satellite. It is based on the Model 5289 thruster/thruster valve used on Aerojet's Satprop, for which approximately 150 have been provided. Propulsion components have been provided for Eurostar, Spacebus, Italsat, Cluster, HS601, Hot Bird, Cassini, Artemis and Ariane 5. Cold gas thrust control is provided for Pegasus, Meteor and SAFER.

Para-Flite Incorporated

Para-Flite develops and manufactures 0.5 to 280 m^2 ram-air gliding parachutes for aerospace recovery systems, including orbital capsules and sounding rocket payloads. Providing a lift-to-drag ratio of 3, the parachute carries an Airborne Guidance Unit to process signals from a ground-based transmitter (GPS is an option) to manipulate the control lines and home in on the target. The automatic mode provides 60 m accuracy, complemented by 15 m for manual control. Payloads can be recovered from suborbital rocket and balloon carriers: 100 to 2500 kg capacity systems are available off-the-shelf.

Parker Bertea Aerospace

Parker Bertea Aerospace, part of the Parker Hannifin Corporation, designs, manufactures and services fluid systems, components and related electronic control systems for aerospace applications. Personnel total 5,500. The space-related products of its divisions are specified below.

Air and Space. This division specialises in the design and manufacture of valves and coupling devices to handle gaseous, liquid and cryogenic propellants, including hydrogen, nitrogen, oxygen, helium, NTO and MMH. Such systems fly on Shuttle, RL10 engine, Peacekeeper and Pegasus.

Parker's high-pressure helium regulator is used for ullage pressure control in a lightweight 680 atmosphere blowdown propellant feed system for missile and spacecraft propulsion systems. The regulator controls ullage pressure within ±10 per cent with a 10:1 variation in inlet pressure and from lockup to rated flow. It is a welded assembly constructed entirely from titanium alloys. The basic regulator design can be used for a variety of fluids and gases.

The company's cold gas thruster module is designed to provide three-axis control for small launch vehicles; it is flown on Pegasus. The module comprises three cold gas thrusters operated by integral independent solenoid valves. The central thruster generates 111.2 N and the two lateral thrusters 55.6 N when operated with a 136 atmosphere N_2 gas supply. SI of each is 68 seconds. The solenoid valves are pilot operated and have a response time of more than 5 ms at maximum gas inlet pressure.

Pressure Systems, Inc

PSI, is the world's largest independent manufacturer of titanium pressure vessels for launchers and spacecraft: it now ships 200/300 annually, worth about US$25 million. Approximately 3,800 have been delivered, without flight failure, including about 600 propellant management device tanks of approximately 20 designs and more than 700 elastomeric diaphragm tanks of approximately 15 designs. PSI acquired Programmed Composites Inc in June 1995, allowing it to offer tanks integrated with satellite core structures, in addition to precision reflectors, solar array substrates and antenna support structures. PSI tanks provide the propellant for most commercial communications satellites, including Hughes HS 601, Loral FS1300 and Lockheed Martin A2100. PSI tanks are also used on Milstar, Cassini, Soho, NEAR, Tirus, Inmarsat, Intelsat, ETS 7, TRMM and Axaf. PSI is completing a US$4 million contract from Lockheed Martin for 75 hydrazine tanks for Iridium satellites. In 1996, PSI delivered the 1000th upper-stage high-pressure helium tank and the 100th upper-stage propellant tank for the Lockheed Martin Atlas Centaur launch vehicle family.

Puroflow Corporation

Surface tension devices for spacecraft propellant tanks. Filters and strainers for propellants, pneumatics, hydraulics and cryogenic systems. Users include Shuttle, Titan 3, GOES, Milstar and Eurostar. An ultra-violet system is used to treat the deluge water on the Shuttle launch pad.

WORLD SPACE CENTRES

Launch sites
Training facilities
Telemetry, tracking and command
Research and development facilities
Space monitoring facilities

LAUNCH SITES

AUSTRALIA

Cape York

Location: 12.25°S/143°E

Background
The Australian government began a feasibility study in October 1986 into the possibility of an international equatorial launch site on the Queensland coast, offering launches over the Pacific. By August 1987, five specialist studies had reported favourably on the project, which would be operated by private enterprise. Two commercial consortia then undertook studies. The Australian Spaceport Group included Martin Marietta and Aussat, and focused on the peninsula's western mining town of Weipa. The group concluded in 1989 that the space market did not require such a facility. However, the Cape York Space Agency submitted a proposal to the Queensland and federal government in October 1989 to develop a 607 km² site at Temple Bay, on the eastern side, as the Cape York Space Port.

Capital investment would be about US$470 million. The goal was to operate Ukraine-supplied Zenith launchers using local crews, trained initially by the former Soviets. Full commercial operations would fly five missions annually. Russian satellites would be barred from the site. Facilities would include a 30 × 60 × 10 m processing building and horizontal transporter, and a 50 × 50 m concrete pad with 70 m service tower and support services. The project would require the building of a town to accommodate some 700 people and a runway to handle Boeing 747F and equivalent aircraft.

The government endorsed the proposal in December 1989, subject to a full study of environmental impacts and agreement with local Aboriginal communities. CYSA planned to contract United Technologies' USBI to manage and operate the site (a Memorandum of Understanding was signed June 1989). USBI received a US Technical Assistance Agreement export licence in summer 1990 covering the initial design/development stage. A further licence would have been required for construction. CYSA, owned by Essington Developments Ltd, was wound up in October 1991. Space Transportation Systems Ltd was awarded the mandate to develop the spaceport and Kaiser Engineers Australia was to be project engineers.

1992's INSP report recommended that no further resources should be invested unless financial backers were positively in place by end-1992. STS withdrew in 1992 to concentrate on studying a Proton launch site in Papua New Guinea. The Euro-Pacific Capital Group consortium, based in Sydney, was reported in 1993 to be investigating Cape York's viability.

Darwin

Location: 12.28°S/130.5°E

Background
An Australian joint venture in 1993 signed an agreement with AeroAstro of the US for two launches of small communications satellites from Australia. KITComm Pty Ltd, an Australian-owned joint venture between Kennett International Technology Pty Ltd of Queensland, and several private investors, would team with AeroAstro to provide low-cost launches from a new site near Darwin. The commercial facility would be regionally focused to cater for the growing satellite requirements of Asian countries. AeroAstro planned to launch its PA-2 liquid rocket, capable of carrying 230 kg into orbit for US$6 million, from the site.

Woomera

Location: 31.1°S/136.8°E

Current status
The Australian Space Council is studying whether to establish a new space launch facility first at Woomera and then at Darwin under its five year rolling plan. Its goal is to make Australia a significant launch site for Asian small satellites and space-based services. However, the government has yet formally to endorse the launch programme.

Background
About 430 km north of Adelaide and running eastwards for 2,000 km across the desert, the Woomera range in South Australia was established jointly by the UK and Australia in 1946. Australian hopes that what was originally a test range for ballistic missiles and sounding rockets, and for target practice at pilotless aircraft, would have a future as a satellite launch centre, were not fulfilled.

During the period when the UK's Blue Streak was under development as the main stage of ELDO's Europa rocket, a township of 4,500, with more than 500 houses supplied with power and water across 160 km of desert, was created. However, Woomera was not suitable for equatorial launches and France preferred to develop its own centre at Kourou. After successive cuts in its contributions to the range, the UK announced there would be no more work for Woomera after 1976. NASA also decided to close down its Carnarvon tracking station, which had played a major part in manned and unmanned space flights. The small Wresat test satellite, launched into polar orbit by a US Redstone in November 1967, and the UK's Prospero in October 1971 remain the only two satellites orbited from Woomera. The range reopened in August 1987 for the launch of a West German-funded Skylark sounding rocket to undertake about 5 minutes of observations of Supernova 1987A.

The South Australian government has long wanted to revitalise Woomera as a commercial site for aircraft and aerospace systems testing, but rejected initial proposals in 1990 for relying excessively on the use of the Australian armed forces. Woomera's advantages include polar orbit access, a sparsely inhabited downrange area, an existing infrastructure and largely cloud-free weather.

Woomera has been proposed in the past for redevelopment and refurbishment to handle the Southern Launch Vehicle, capable of delivering 800 kg into LEO. BAe, Auspace and Hawker de Havilland, supported by an A$1.25 million ASO contract, completed preliminary feasibility studies. However, the joint venture was terminated in 1993.

Germany's Express orbital capsule was to be recovered at Woomera 21 January 1995 following ascent from Japan, but was lost soon after launch. NASA Goddard launched six Black Brant 9 sounding rockets October/November 1995 to study the Large Magellanic Cloud. Japan's 760 kg Alflex spaceplane automatic approach/landing experiment was released from a helicopter on a series of tests using the 45 × 2,600 m runway in 1996 to demonstrate GN&C for automatic landings.

BRAZIL

Alcantara

Location: 2°17'S/44°23'W

Background
The Ministry of Aeronautics runs the larger of Brazil's two launch sites, Alcantara. The CLA Alcantara Launch Center was expanded on a 520 km² site on the Atlantic coast outside Sao Luis to handle the VLS orbital launcher. CSA also built pads for the Sonda 3 and Sonda 4 sounding rockets, meteorological rockets and other science vehicles (NASA, for example, launched four Nike Orions here in 1994). The position near the equator offers 25 per cent greater initial launch energy due to the earth's rotation compared to Cape Canaveral. Press reports in mid-1990 indicated that discussions were under way for commercial Proton operations.

Brazilian government and military officials held a formal opening in February 1990.

Barreira do Inferno (Barrier of Hell)

Location: 5°55'S/35°10'W

Background
A smaller site than Alcantara near Natal, established by the Minister of Aeronautics in 1964 for Sonda 1, BdI used until the late 1980s for Sonda (the first Sonda 4 departed from here in November 1984) and international vehicles such as Skylark. All launches have now transferred to CLA. It was upgraded in 1979 to provide radar, telemetry and real-time data processing services for Ariane launches from French Guiana. Similar support is provided for sub-orbital and orbital launches from other sites.

CANADA

Churchill

Location: 58.46°N/94.1°W

Current status
SpacePort Canada's Orbital Complex number 2, was due to be operational in 1999, and will include the following new facilities: launch site, mission control centre, vehicle assembly building, launch vehicle storage building, payload preparation facility and instrumentation sites; supporting infrastructure includes power, base time, and communications; SpacePort operations and data processing systems; emergency services.

Background
Akjuit signed a 30-year lease with the Manitoba government on 18 July 1994 for an 80 km² site, including the former Churchill Research Range, for the development of SpacePort Canada, a commercial launching facility for sub-orbital and LEO payloads. Akjuit estimates that the first phase of expansion leading to orbital missions will cost C$100 million. The C$550,000 refurbishment of existing sounding rocket facilities was completed in September 1994, this including an upgrade to the universal launcher. The launcher handles vehicles up to 18 m long and with a mass of up to 4,530 kg. The enclosed launch bay is temperature controlled.

In the past, Nike Tomahawk, Black Brant 5B/5C/8/9/10, Taurus Orion and Nike Orion, have all flown from Churchill. The hazardous assembly building, payload assembly area, blockhouse and above-ground tunnels link all operations.

The launch site (Flat Pad) includes a concrete foundation with embedded anchor hooks for installing launch stands for the launching of Russian Start and Start-1 vehicles, plus a movable launch building. The mission control centre will be equipped with separate consoles for the payload owner, launch vehicle provider, senior director, operations director, range system control, mission flight control, range safety control and launch control. GPS information as well as computer simulation and base time will be available at the control centre, which will also act as the hub for the operations and data processing system. A separate room for customer (payload owner) use will also be available. The centre will have a dual-processing capability and thus two launch vehicles can be processed in parallel.

The vehicle assembly building is designed for the horizontal processing of two Start-size launch vehicles in parallel. It has a climate-controlled, 10,000 cleanliness-class assembly room and separate rooms for two users' technical and launch vehicle teams. It includes mechanical, electrical and boiler rooms and an independent back-up power generator, as well as a control room. Two 20 tonne bridge cranes are available for operations.

The payload preparation facility will provide the opportunity for the parallel preparation of up to four satellites up to 150 kg mass under climate-controlled conditions and with a cleanliness class not less than 100,000. The facility will have air lock access, climate-controlled storage for satellites and special storage for satellite propellants. Separate rooms for the customer's team, as well as separate storage for the customer's equipment, will be available.

Instrumentation includes: Churchill site (IS-1), comprising two identical tracking receive systems with S-band antennas and telemetry receive equipment and one UHF command-destruct system; Chesterfield Inlet (IS-2), comprising two identical tracking higher sensitivity receive systems with S-band antennas and telemetry receive equipment and one UHF command-destruct system; P-band telemetry receiving systems will be available for the Russian launch vehicles.

The mission operations and data processing system provides instantaneous impact predictions, flight termination criteria and instrumentation status displays. It supports data types from sources including TT&C instrumentation, facilities, utility systems and range systems. It uses graphics-based software architecture, which permits decisions to be made quickly and accurately. The architecture is based upon the distributed client/server model.

SpacePort Canada will be connected with the North American power network through the special 148 kV line; a back-up standby power plant will be constructed.

CHINA, PEOPLE'S REPUBLIC

Long March 4 at the Taiyuan launch site used for Sun-synchronous missions (Theo Pirard) ***1999***/0054384

Hainan/Harbin

Location: 19°N/109.5°E

Background

A sounding rocket site on the west coast of Hainan Island was inaugurated in late 1988 with the launches of four Zhinui 1 ('Weaver Girl') vehicles over Beibuwan Bay. A Zhinui 3 attained 120 km in January 1991. In the previous Five Year Plan, the Chinese began to upgrade the facilities at Hainan to handle 300 km-altitude vehicles. The Harbin site in northeast Manchuria was established in the late 1960s to provide a 3,500 km land range for missile tests, including the CSS- 2 and CSS- 3.

Jiuquan satellite launch centre

Location: 40.6°N/99.9°E

Background

The Chinese built Jiuquan in the early 1960s in the Gobi deserts 1,600 km west of Beijing near the Soviet border, north of Jiuquan city in Kansu province. Hundreds of missile tests have been reported, including into the Lop Nor nuclear test site. Formerly known in the West as Shuang Cheng Tzu (still used as the US Space Command designator), this was China's first launch site but it is limited to southeastern launches into 57 to 70° orbits to avoid over-flying Russia and Mongolia. Sounding rockets also utilise the site. It consists of two pads 416 m apart, with a shared mobile service tower that can be employed separately for CZ-2/FB-1 and CZ-1 space vehicles. In the early 1980s, the latter added the capability for CZ-2C missions. CZ-2D flew from this pad as well, beginning in 1992.

The site's base area (called Dong Feng, 'East Wind') houses the Huxi Xincun range control centre, power station and residential accommodation. The technical centre, where vehicles are assembled and checked out, is 18 km north. Supporting facilities BL 1 and BL 2 can each integrate two launchers and several satellites simultaneously. BL 1, an enormous 140 m long building has an 18 × 90 m test and assembly hall, where the launch vehicles are delivered on rail from the connected transfer hall. There are 25 test rooms down each side of the main hall, including a class 100,000 satellite assembly and test room. The pads are another 30 km further north.

Within the chamber, 41.6 m towers share a 55 m mobile service tower on 17 m span rails for preparation and delivery of the launch vehicle. The mobile tower's maximum speed is about 15 m/min. It has 15/5 tonne-capacity main/auxiliary cranes. To prepare the satellite, commercial users have a class 100,000 clean room which employs a 6 × 6.5 m 13.1 m high cabin to protect the vehicle's upper section. Temperature can be maintained at 20 ±5°C, with 50 to 70 per cent humidity.

A launch control centre is 200 m from pad 1. Both road and rail systems link the facility to the nearest airport, 94 km to the south. A 270 km line connects to the main railway network so satellites can be delivered for launch from virtually any point in China. Jiuquan will continue to be used primarily for recoverable Earth observation and µgravity missions but, because of its geographical constraints, greater commercial activities are focused on the other two launch bases.

If a manned project develops out of the 1996-2000 Five Year Plan, a new CZ-2E pad would be built at Jiuquan. Construction continues at the site possibly for a launch complex for vehicles larger than those flown to date.

A 10-year shuttle programme was claimed in late 1992 to be under way, with work beginning on a new launch site near Jiuquan.

Jiuquan provides two pads for CZ-1/2 launchers but the site's location limits missions to 57 to 70°

Taiyuan satellite launch centre

Location: 37.5°N/112.6°E

Background

Taiyuan was inaugurated in September 1988 for launches by CZ-4 into polar orbits for remote sensing, meteorological and reconnaissance missions. The second launch was made during September 1990. There is a single pad. It was initially operated for missile testing as an extension of Jiuquan for larger vehicles, including the DF-5. The first photographs of the site were released in 1990. CZ-2C launches began with a load of iridium satellites. US Space Command still calls the site Wuzhai.

The China Academy of Launch Vehicle Technology (CALT, launcher manufacturer) provides on-site supervision of launcher procedures.

Xichang satellite launch centre

Location: 28.25°N/102.0°E

Background

Xichang was selected from a shortlist of 16 sites, of 81 surveyed, for a more favourable GEO mission base than Jiuquan. Construction work in the canyon site at the foot of Mt Liang Shan, 65 km north of Xichang city, began in 1978 and resulted in the first launch in January 1984. The site hosted its first commercial mission when it launched the Hong Kong AsiaSat 1 communications satellite on a CZ-3 in April 1990, followed by Australia's two Optus Bs in 1992.

The local population has traditionally been allowed to live close to the pads. When the maiden CZ-3B crashed on 14 February 1996, into a hillside 1.5 km from the newer complex, the official death toll was six plus 57 injured, and 250 homes were damaged or destroyed. When the CZ-2E of 25 January 1995 exploded at 51 seconds, the debris killed six and injured 23 others in a village 7 km downrange.

As it stands now, the dry season, running from October to May, provides the most suitable time for launch campaigns, but the site enjoys 320 days of sunshine annually. Annual temperature range is −10 to +33°C, because the site sits at 1,839 m above sea level. Xichang's main airport, upgraded to handle B-747 and C-130s, is 50 km south of Xichang city.

The single CZ-3 pad can support up to five missions annually. It has a fixed 77 m high/11-level gantry. In order to set the launch azimuth a technician hand cranks a firing table. The firing room is buried in the hillside 60 m away. Originally, the Chinese stacked vehicles on the pad and mated the payload in the open air but, as

Xichang: the main launcher processing facility can handle three vehicles simultaneously. A CZ-2E requires all three tracks because of the strap-ons

Xichang launch site used by CZ-2E and CZ-3 rockets for geosynchronous transfer (Theo Pirard) ***1999***/0054385

part of the CZ-3 commercial marketing, a stage 3/payload class 100,000 clean room was added.

Vehicle stacking requires about two days, followed by 14 days of checkout with a launch rehearsal on the day 15 aiming for the actual launch on the day 20. Stage 1/2 UDMH loading from underground reservoirs begins 16 hours before launch, followed by N_2O_4. Stage 3's cryogenic loading about 5½ hours before launch. 36 days are allowed for launcher and payload checkout.

The second complex is sited 300 m distant for CZ-2E/3A/3B vehicles with add-on boosters. It was first used on 16 July 1990. The pad is at 28.25° north/102.02° east, 1,826 m above sea level. Like the original pad, this one has a fixed 74 m 1,025 tonne tower plus 98 m 4,300 tonne mobile tower. Just before launch the ground crew withdraws the tower 130 m.

Payloads are encapsulated within their fairings in the class 100,000 preparation facilities and delivered to the pad for vehicle mating. Vehicle erection begins 11 weeks before launch, the encapsulated payload is mated at six weeks, launch rehearsal concluded at three weeks and fuelling completed at one week.

The Mission Command and Control Centre (MCCC) lies 7 km southeast of the pads and was upgraded 1994-95. At this site, a communications centre includes an 11 m Intelsat B Earth station. Vehicle/payload processing facilities are 2.2 km from the pads in the technical centre.

Launcher stages are received by rail directly inside a 31 × 14 m transit hall and transferred to an associated 92 × 28 m assembly room capable of processing three complete vehicles. The launchers are assembled, checked out and then broken up for transfer by road for stacking in stages on the pad. The Non-hazardous Operations Building (BS2) houses the main 756 m² 42 × 18 × 18 m satellite test hall, class 100,000, climate-controlled (20 to 25°C, humidity 35 to 55 per cent) facility. Personnel access it by an airlock with 5.4 m wide 12.5 m high door. The building includes offices, test rooms and storage.

The associated Hazard Operations & Fuelling Building (BS3) is a 324 m² spacecraft fuelling and assembly hall, with a class 100,000 clean room. The door is 5.4 m wide 13 m high. Solid kick motors are installed in here and the payload encapsulated in its fairing.

After the stack begins its preparation, the Checkout/Preparation Building (BM) is used to inspect the solid motors. It is a 108 m² 12 × 9 × 9.5 m hall (door 3.6 m wide 4.2 m high). Next door sits the SPM x-ray Building (BMX), a solid motor test facility with an x-ray hall sized 125 m² 12.5 × 10 × 15.4 m high (the door is 3.2 m wide 4.5 m high). In the same building the Cold soak chamber, 9.6 m² 3.3 × 3.0 × 4.0 m, door 3.2 m wide, 3.5 m high, provides 0 to 15°C.

CZ-2E ready for launch at Xichang. The older pad is in the background ***1995***

Xichang's Mission Command and Control Centre ***1995***

FRANCE

Kourou (Centre Spatial Guyanais–CSG)

Location: 5° 14'N; 52° 46'W

Background

Although owned by CNES, the CSG is made available to ESA/Arianespace under a governmental agreement that guarantees access to the ESA-owned Ariane launch facilities. CSG is the most favourable major site for GEO launches, its near equatorial position providing a 15 per cent payload advantage over Cape Canaveral for eastward launches. Guiana's coastline permits launches into both equatorial and polar/Sun-synchronous paths, with inclinations up to 100.5° possible. The annual

Ariane 5 emerges from the BAF final assembly building (ESA) ***1996***

CSG's Vertical Assembly Building can stack one Ariane 4 on a mobile launch platform in a month. The assembled payloads/fairing are integrated on the pad, as are the solid propellant boosters

operating cost for the site is FFr1 billion, with France, including its contribution through ESA, paying about 70 per cent. Operating costs have continued to go down since 1991 and now are about 15 per cent less than they were that year. CSG staff total 1,400.

CSG extends more than 30 km along the coast from Kourou town to Sinnamary, with the Ariane facilities about 18 km northwest of Kourou. CSG's Technical Centre, close to Kourou, incorporates spacecraft reception and checkout facilities (able to handle five satellites in overlapping campaigns) and the Jupiter range control centre. The 2,600 m² Jupiter 2 handled its first campaign in January 1996 and will be used for all future launches. Its predecessor, Jupiter, had been operational since 1968. Radar and telemetry stations are sited at Montagne des Pères and Montabo, respectively, 10 and 60 km southeast of the Technical Centre, with downrange stations for easterly trajectories in Natal (Brazil), Ascension Island and near Libreville in Gabon.

Since CSG became operational in April 1968, 497 launches of sounding rockets and orbital vehicles have been made through 1995. The site has also witnessed more than 60 balloon flights (the last in 1980).

CSG's meteorological conditions reflect its tropical location. The mean daytime temperature is 26.4°C, with maximums of 34.0°C, and minimums of 18.1°C. The windspeed has an annual mean of 3.3 m/s, a maximum of 23 m/s and the rainfall is 2.9 m annually, 438 h/year over 235 days.

Ariane 4 launch requirements preclude launches if storms exist within 10 km at T−10 minutes or if convective clouds exist at 5,000 to 6,500 m altitude at T-0. These criteria have been used since flight 19.

ESA's Ariane Launch Base comprises two elements:

Ariane is transported 950 m by rail on its mobile launch platform; the central turntable (with cover off, at right) permits two to pass. The launch control centre (CDL-2) is the low building to the VAB's left in the background (Arianespace)

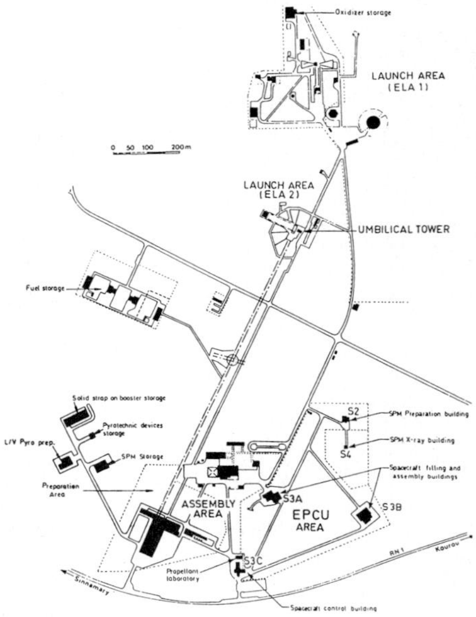

Kourou ELA-2 launch complex layout

the ELA (Ensemble de Lancement) Ariane launch areas, including facilities for launcher final assembly/checkout, and the EPCU Ensemble de Préparation Charges Utiles for satellite processing.

The major installations are:

ELA-1: Pad 5°14′09″N; 52°46′29″W

This pad was responsible for the first Ariane launches from 1979. ELA-1 was closed down in December 1989 and handed back to ESA. The 70 m service tower was lowered to the ground in June 1991.

Vehicles were stacked on the pad, limiting campaigns to two month intervals. Platform 8 in the service tower provided a class 100,000 100 m² clean room for payload mating/checkout. The 34 m diameter protected CDL-1 (Centre de Lancement) blockhouse was sited 300 m from the pad to handle pre-launch and countdown operations.

ELA-2: Pad 5°13′56″N; 52°46′32″W

Vehicles are stacked up to equipment bay level on one of two mobile platforms in the Vertical Assembly Building, 950 m by rail from the launch stand, permitting monthly departures of Ariane 4s. The CDL-2 control centre is adjacent to the VAB. Payloads are mated on the pad within the 86 m, 5,500 tonne service tower's class 100,000 clean room (PFCU Plateforme Charge Utile, platform 16). ESA/Arianespace considered making ELA-3 compatible with Ariane 4 but instead hardened ELA-2, which CNES guarantees to restore in nine to 10 months in the event of a launch accident.

ELA-3: Pad 5°14′21″N; 52°46′06″W

Construction of the Ariane 5 facilities began in 1988 to allow up to eight departures annually (the current plan is for five), including two a month apart. Each launcher campaign will last 22 days. The new pad cost the GSC FFr5,000 million, with a similar amount invested in upgrading other industrial facilities in Guiana. The simplified pad concept deletes the requirement for large

Ariane 5 pad. The 90 m tower holds 1,500 m³ of water for trench cooling and sound suppression. Four pylons provide lightning protection (ESA) ***1996***

cryogenic arms on the umbilical tower by feeding the propellants from below the mobile launch table. The 600 ha ELA-3 includes the Batiment d'Intégration Lanceur (BIL) launcher integration building where the core is erected on the mobile platform and the boosters added, and the Batiment d'Assemblage Final (BAF) assembly building for completing the vehicle.

To the south is the Batiment d'Intégration Propulseur (BIP) integration hall for the P230 boosters. Solid propellant production and loading for the two large segments is undertaken at the 300 ha Usine Propergols en Guyane (UPG) plant, managed by Regulus, a BPD-SNPE Franco-Italian partnership, and commissioned 24 October 1991 at a cost of FFr1,250 million (ECU180 million). The 40 buildings total 26,000 m². UPG made the first casting, unsuccessfully, in November 1991. The forward segment is shipped from BPD in Italy.

The filled segments are taken by road into the 45 m high BIP and stacked as a single P230 on an 80 tonne 8 × 8 m 5 m high rail transporter. From BIP, it can be moved 1 km south to BEAP or to BIL 2.8 km north. P230's seven firing tests were performed 1993-95 on BEAP's 50 m vertical stand over the 200 m long, 60 m deep 35 m wide pit cut into the granite bedrock. No further tests are planned, but BEAP is maintained for possible future development tests and could even be converted to a small satellite launcher pad.

While the P230s are being stacked in BIP, the core is stacked on the 900 tonne 21 × 25 m 7 m high mobile platform in the BIL's 30 × 43 m 58 m high integration hall. The platform can move at a maximum 4 km/h on

Ariane 5 emerges from the BAF final assembly building (ESA) ***1999***/0054383

Launcher Integration Building of the Preparation and Launch Control Centre (ESA) ***1999***/0054386

the 16 m gauge track. At this stage, the platform has only the 30 m integral lower umbilical mast. The BIL is 31 × 127 m overall, including the 26 × 83 m storage hall. This hall is only partly clad, on the rain-prone side, and used to store stage 1/2 in their containers. Stage 1 is rotated to the vertical in the erection hall.

Each P230 is rolled in to the integration hall and attached, remaining on its own pallet throughout the campaign. With the boosters now providing rigidity, the upper stage and vehicle equipment bay can be added. Meanwhile, final spacecraft preparation and encapsulation takes place in the upper portion of the 85 × 52 m 83 m high BAF. The upper payload is mated with the Speltra in the 30 × 30 m class 100,000 encapsulation area and the fairing fitted. The stack is rolled in to BAF's launcher integration hall and the lower payload attached. The upper portion then fits over. The platform's upper mast is added, batteries fitted, the upper stage fuelled under remote control, the hydrazine attitude control system loaded manually and the countdown started. CDL-3, located near BIL, houses two control rooms in separate wings for conducting simultaneous campaigns. The 1½ to 2 hour rollout to the pad 2.8 km to the north begins 8 hours before launch through BAF's 4-section 24 × 62 m door. The pad is designed to be operational again six months after an explosion.

ELA-4

A fourth site for Ariane 5's successor will not be required for 15 to 20 years but potential locations have been considered because of the expected growth of nearby Kourou town.

Payload Preparation Complex (EPCU, Ensemble de Préparation Charges Utiles)

Payload preparations are conducted in three phases, 1) satellite processing/checkout, 2) hazardous operations, and 3) integration with the launcher on the pad. Buildings S1A and S1B, in the CSG Technical Centre, provide clean room facilities for satellite preparation. S2 and S4, near the ELAs, are designed for solid kick motor preparation and X-ray operations. S3A/3B, also close to the two ELAs, are assigned to payload propellant filling operations, final integration, assembly of the Spelda dual launch carrier, and encapsulation in the Ariane 4 fairing. S3C is used for monitoring the hazardous operations. Ariane 5 payloads continue to use these facilities, although final encapsulation is performed in the BAF.

S1A/B

S1A, the older of the two satellite preparation/checkout buildings, provides a 488 m² class 100,000 area accessed by a 4.5 m wide, 10 m long, 10 m high airlock for simultaneous processing of two Ariane 3 class satellites in three bays. S1B was added to the north for handling two Ariane 4 class craft simultaneously. It provides a class 100,000 800 m² clean area accommodation with a total area of 300 m² for three satellite checkout stations, and air-conditioned 340 m² and clean 60 m² storage rooms, in addition to offices and meeting rooms for two satellite teams. Payloads pass from this Phase 1 processing to Phase 2 in buildings S3A/B.

S2/4

S2 is the restricted-access Solid Propellant Motor (SPM) preparation building for processing motors and pyrotechnic devices before integration with payloads in buildings S3A/B. Two halls of 97 and 48 m² include mechanical test equipment for Mage-type motors and electrical systems for Mage and Star 30/37 safe/arm devices. Building S4 is used for cold soaking and x-ray inspection of SPMs up to 3,000 kg 2 m diameter, with the 75 m² x-ray hall including a Linatron 2000 source (5.5/8/10 MeV) linked to an automatic, 32-position rotating table. The 9 m², 4 m high cold soak room allows motors to be held at 0 to 15°C.

S3A/B

S3A was designed for handling Ariane 3-class payloads and Syldas but it continues in use for smaller Ariane 4 craft. Hydrazine propellant tank filling, integration of ordnance, balancing and loading into the pad transfer containers for two satellites are undertaken in class 100,000 facilities. These are a 185 m² assembly hall, equipped with filling positions over a drain and a

MAN tractor pulls the Ariane 5 launch table from the Final Assembly Building (ESA) ***1999***/0054382

Schenck-type E-50S balancing machine, and a 172 m² Sylda assembly room. For Ariane 3 launches, the satellites were then encapsulated and transferred to the pad for the fairing to be added. On Ariane 4 campaigns, the payload is transferred into the newer S3B. This includes a 320 m² assembly and filling room equipped with a weighing machine (5,000 kg±1 per cent, resolution 1 kg), and a 400 m² encapsulation room. Encapsulation inside Spelda and the fairing is completed before pad transfer. Ariane 5 payloads are transferred to the BAF for their final handling.

INDIA

Balasore Rocket Launching Station

Location: 21°25'N/87°00'E.

Background

Balasore has been considered for due south departures into polar paths, enhancing PSLV's Sun-synchronous capability by 500 kg. ISRO currently uses it for RH-200 sounding rockets.

Shar Centre

Location: 13°41'N/80°14'E

Background

India's orbital launch site is housed at ISRO's Shar Centre on Sriharikota Island, but the first rocket firings were made from the Thumba Equatorial Rocket Launching Station in 1963. Together with the northerly Balasore Rocket Launching Station, the facilities are operated under the ISRO Range Complex (IREX) headquartered at Shar. Shar launches the full range of Indian vehicles from sub-orbital up to the PSLV but range safety considerations limit Shar to less than 140° launch azimuths.

Shar, encompassing the Sriharikota Island on the east

PSLV's Mobile Service Tower provides payload clean room facilities. At bottom is the Checkout Terminal Room for vehicle testing before the more distant Launch Control Centre takes over (ISRO)

coast of Andhra Pradesh, is ISRO's satellite launching base and additionally provides launch facilities for the full range of Rohini sounding rockets. 2,380 staff support the site. The Vehicle Assembly, Static Test and Evaluation Complex (VAST, previously STEX) and the Solid Propellant Space Booster Plant (SPROB) are located at Shar for casting and testing PSLV solid motors.

The ASLV orbital launcher was integrated vertically, beginning with motor and subassembly preparations in the Vehicle Integration Building (VIB) and completed on the pad within the 40 m tall Mobile Service Structure. The PSLV launch complex was commissioned in 1990. It has a 3,000 tonne 76.5 m high Mobile Service Tower (MST) which provides the SP-3 payload clean room. The MST is rolled back 180 m for launch. Shar also has a 52 m umbilical tower, an open steel structure next to the launch pedestal. Modifications are under way to allow the complex to handle GSLV. The Launch Control Centre is 5.6 km distant and the CTR Checkout Terminal Room is next to the pad.

Liquid stage fuelling is done remotely from the LCC. All solid motors are processed in the SMPF Solid Motor Preparation Facility. As a launch campaign nears completion, stage 1's segments are stacked on the pad. Satellite preparation/integration is undertaken in three SP clean room facilities. SP-1 at the LCC has three bays: class 100,000 17 × 15 × 8 m high bay, materials airlock bay and dynamic balancing room. It provides satellite inspection, subsystem tests, RCS leak tests and solar panel deployment tests. SP-2 next to the pad is used for RCS filling and checking compatibility with PSLV's adaptor. SP-3 is part of MST for integrating payloads with the launcher.

The VAST was originally designed to handle motors up to 1 m diameter, 10 m length and 980 kN. ISRO modified it for PSLV for 2.8 m diameter and 4.9 MN levels (first PSLV stage 1 test was made in 1989). The horizontal firing stands provide 1-/6-component facilities. The vibration platform was upgraded from 13.5 tonnes to a 30 tonne dual shaker operating over 5-2,000 Hz with maximum acceleration of 100 *g*. Specimens up to 2 tonnes can be shock tested to 80 *g* by free-fall to simulate motor ignition shocks, and upper stage motors are fired in a 50 km-equivalent high-altitude test facility. The 25 km^2 SPROB was commissioned in March 1977 for SLV-3's motors but it was then augmented to cope with PSLV's 2.8 m diameter 25 tonne segments, requiring a 2.5 tonne vertical mixer and 15 MeV linear accelerator for the radiographic facility in addition to the previous 8 MeV machine.

Thumba Equatorial Launching Station

Location: 8°32'N/76°52'E.

Background
TELS supports RH-200/300 sounding rockets. It was the site of India's first sub-orbital launch in 1963.

INDONESIA

Paneungpeuk

Location: 1.0°S/136°E

Background
Indonesia launches its one- and two-stage solid propellant sounding rockets by the LAPAN national space agency for upper atmospheric research from the Paneungpeuk range some 150 km from Bandung. China considered contracting with Indonesia to build an equatorial launch site nearby for China's CZ-3 during 1987. One possible location under consideration was Biak Island.

ISRAEL

Palmachim air force base

Location: 31.52°N/34.45°e

Background
Israel launches its Ofeq orbital shots from an area near or within Israel's Palmachim air force base south of Tel Aviv, and near the town of Yavne. The Israelis inaugurated it as an orbital site with the launch of the Ofeq 1 satellite 19 September 1988. The facilities are classified although they are reportedly visible from the coast road. Located at the eastern end of the Mediterranean, the site is restricted to retrograde departures for range safety considerations (although the trajectory still passes over southern Europe). The second orbital launch was made in April 1990 and the third in April 1995.

ITALY

San Marco

Location: 2.9°S/40.3°E

Background
Italy's offshore orbital launch platform was last employed in March 1988 – its first firing in 12 years.

INTERNATIONAL

Introduction

Number of orbital launches by site

Country	Launch site	Latitude	Longitude	1957-85	1986	1987	1988	1989	1990	1991	1992	1993	1994	1995	1996	1997	1998	1957-98
RFAS/USSR	Tyuratam (Baikonur)	45.6°N	63.4°E	693	35	47	43	29	27	22	21	21	30	19	15	18	16	1,036
	Barent Sea	69.3°N	35.3°E	0	0	0	0	0	0	0	0	0	0	0	0	0	1	1
	Kapustin Yar	48.4°N	45.8°E	82	0	1	0	0	0	0	0	0	0	0	0	0	0	83
	Plesetsk	62.8°N	40.1°E	1056	56	47	47	45	48	37	33	26	18	13	10	9	7	1,452
	Svobodny	51.37°N	128.3°E	0	0	0	0	0	0	0	0	0	0	0	*0	2	0	2
USA	Cape Canaveral/ KSC (ER)	28.5°N	81.0°W	370	3	3	4	16	23	12	24	20	19	23	24	24	23	588
	EdwardsAFB/ B52	35°N	118°W	0	0	0	0	0	1	1	0	1	2	0	-	1	0	6
	Vandenberg AFB (WR)	34.4°N	120.35°W	467	3	5	7	2	3	5	4	2	5	4	8	11	11	537
	Wallops Island	37.8°N	75.5°W	19	0	0	0	0	0	0	0	0	0	0	1	2	2	24
France	Hammaguir	31.0°N	8.0°W	4	0	0	0	0	0	0	0	0	0	0	0	0	0	4
Italy	San Marco platform	2.9°S	40.3°E	8	0	0	1	0	0	0	0	0	0	0	0	0	0	9
Australia/UK	Woomera	31.1°S	136.8°E	2	0	0	0	0	0	0	0	0	0	0	0	0	1	3
Japan	Kagoshima	31.2°N	131.1°E	16	0	1	0	1	1	1	0	1	0	1	0	1	0	23
	Tanegashima	30.4°N	131.0°E	13	2	2	2	1	2	1	1	0	2	1	1	1	1	30
France/Europe	Kourou	5.2°N	52.8°W	18	2	2	7	7	5	8	7	7	6	11	10	12	11	113
China	Jiuquan	40.6°N	99.9°E	13	1	2	1	0	1	0	2	1	1	0	1	0	0	23
	Taiyuan/Wuzhai	37.5°N	112.6°E	0	0	0	1	0	1	0	0	0	0	0	0	2	4	8
	Xichang	28.25°N	102.0°E	2	1	0	2	0	3	1	2	0	4	2	2	4	2	25
India	SHAR, Sriharikota	13.9°N	80.4°E	3	0	0	0	0	0	0	1	0	2	0	1	1	0	8
Israel	Palmachim/Yavne	31.5°N	34.5°E	0	0	0	1	0	1	0	0	0	0	1	0	0	0	3
Brazil	Alcantara	2.3°S	44.4°W	0	0	0	0	0	0	0	0	0	0	0	0	0	0	0
Spain	Torrejon AB	40.488°N	3.457°W	0	0	0	0	0	0	0	0	0	0	0	**0	0	0	0
South Africa	OTB	34.583°S	20.317°E	0	0	0	0	0	0	0	0	0	0	0	0	0	0	0
			Total Launches	**2,766**	**103**	**110**	**116**	**101**	**116**	**88**	**95**	**79**	**89**	**75**	**73**	**88**	**79**	**3,978**

*First launch 4 March 1997
**First launch 21 April 1997

The Italian Vega (previously called San Marco Scout) would use the site, although the vehicle's future is highly uncertain. San Marco is owned and operated by the Italian government. It comprises two platforms in Formosa Bay 4.8 km off the coast of Kenya. The San Marco platform provides the launch pad, while the Santa Rita platform houses blockhouse facilities. The range is maintained in operational condition and the telemetry and tracking station still works under ESA contract. A new MoU was signed with Kenya in March 1995.

The range became operational in 1966 and by mid-1976 eight satellites had been launched, four Italian, one UK and three US. The launch platform has 20 steel legs sunk into the sandy seabed, and is linked by 23 cables to the control platform, a modified oil rig, 920 m distant. Explorer 42, the small astronomy satellite launched by an Italian crew on 12 December 1970, was the first US satellite launched by a foreign country. Because of the position of this equatorial site, the Scout put the 195 kg Explorer 43 into orbit, which it could not have done, from a more northerly US location.

In addition to the Scout facilities, there is a ramp for sounding rockets such as US Nike Apache and Nike Tomahawk. The last launch from San Marco was in 1988.

JAPAN

Kagoshima

Location: 31.2°N/131.1°E

Background

ISAS' assigned launch site, Kagoshima, is primarily a sounding rocket facility, but handles a science satellite every one to two years. It was restricted, by government directive, to all-solid vehicles of 1.4 m diameter but approval for the 2.5 m diameter M-5 was made in 1989 to undertake modest deep space missions. The Sakigake and Suisei Halley's Comet probes of 1985-8 provided Japan's first experience of beyond Earth orbit.

Japan's first six satellites were launched from these levelled hilltops facing the Pacific Ocean at Uchinoura on the southern tip of Kyushu Island. By the end of 1995, orbital launchings totalled 21 successful, with 349 launches of all types. Construction of a sounding rocket site began in February 1962, with extensions for science satellite launchings by M rockets completed in 1966. The first Lambda 4 was fired in 1964 but it was another six years before the first successful orbital launch was achieved. Kagoshima's M-3SII all-solid ISAS/Nissan orbital launcher retired in January 1995. The S-310, S-520 and K-9M sounding rockets and the MT-135 meteorological rocket are currently supported. Annual launch rate is typically three to four for the larger sounding rockets.

A satellite processing facility is linked with the M pad and assembly room, and a sounding rocket payload assembly centre with the covered pad. Facilities have been upgraded to support the new M-5 orbital vehicle.

Departures from this site are normally restricted to 45 day January-February/August-September 'launch seasons' because of range safety procedures and the influential fishing industry lobby.

In addition to Usuda's 60 m station, ISAS operates a 20 m dish at Kagoshima built originally for lunar fly-by communications. It provides S-band uplink and S/X-band downlink.

ISAS' seven other principal facilities are the Kagoshima Space Centre, Noshiro Testing Centre (established 1961 on Asanai Beach for solid/liquid engine static testing), Usuda Deep Space Centre (64 m S/X band deep space tracking and TT&C station), Sanriku Balloon Centre (500 kg payloads on 200,000 m³ helium balloons up to 40 km altitude), the Space Utilisation Research Centre (established 1988 on the new campus), the Centre for Planning & Information Systems (established 1993), and the Centre for Advanced Spacecraft Technology (1995).

Tanegashima

Location: 30°24'N; 130°58'E.

Background

Departures from both sites are normally restricted to 45-day January-February/August-September 'launch seasons' because of range safety procedures and the influential fishing industry lobby (March 1995's H2 launch was a unique exception). The solid strap-ons of NASDA's H1 impacted some 19 to 29 km out, but H2's fall beyond the 190 km range safety limit, provided the potential for year-round firings (although launch failures would still be potentially hazardous). Rocket Systems Corporation is negotiating to increase launch opportunities for commercial H2. NASDA orbital launch facilities are centred on Tanegashima Island, with the Tsukuba Tracking & Control Centre acting as the primary satellite command facility. Subordinate stations are located at Katsuura, Okinawa and Masuda, and facilities on Ogasawara and Christmas Island provide downrange launch tracking. The Kakuda Propulsion Centre is responsible for testing high-performance propulsion systems, but NASDA's major technical facility is the Tsukuba Space Centre. New facilities include a second NDTF Non-Destructive Test Facility for computed tomography examination of SRB segments. Kick motors are still tested in H1's NDTF-1.

NASDA's TNSC orbital launch centre is sited on the southeast tip of 8.6 km² Tanegashima Island, 1,000 km southwest of Tokyo. The more northerly Osaki Launch Site includes the H2/J1 pads and liquid engine static test facilities, while the Takesaki Launch Site handles sounding rockets and provides facilities for H2 solid booster static firings and the H2 Range Control Centre.

H2's 150,000 m² Yoshinobu complex was built from 1985 to 1992 for launching the larger vehicle from the headland west of J1/H1's site. H2 is integrated in the Vehicle Assembly Building on a mobile platform. A single platform is adequate for two annual launches, but increasing the rate to four would require a second platform. A new pad will be completed in 1999 for the Upgraded H2. LE-7 stage 1 engines are static fired in a dedicated facility off to one side, and the strap-on boosters have the uprated N/H1 solid static site at the Takesaki Launch Site.

The pad at 30°23'45"N; 130°58 22"E handled its last H1 launch in February 1992 and was then modified for the heavier J1, requiring a larger crane, adapted access platforms and elimination of one umbilical tower. J1 stands on a 6 m stool on the launch deck to minimise the changes needed for the Mobile Service Tower. The complex was adapted for H1 by adding cryogenic facilities and a second, taller, umbilical tower to the existing N1/2 complex. The 6.4 m high 12 m wide launch deck was serviced by the two umbilical towers 43 and 49 m high (43 m removed for J1) and the 67 m 2,650 tonne MST, rolled back 100 m before launch.

Rocket Assembly Buildings 1/2 received H1's first two stages for processing before erection on the pad. MB-3 engines and components could be static tested north of the complex. Solids and pyrotechnics are received in the Solid Motor Test Building, with NDTF-1 for ultra-sonic and X-ray inspection of solid motors, and the Spin Test Building used for integration of H1's stage 3 with its spin table. Payloads are processed in the class 100,000 STA-1 in preparation for mating with their upper stage in the Third Stage & Spacecraft Assembly (TSA) building. The assembly was then shipped to the Mobile Service Tower in a protected container for mating with stage 2.

J1 countdowns are still controlled from the Osaki Block House, a reinforced concrete partially submerged structure 170 m from the pad. Control after lift-off, and until the payload separates, is the responsibility of the more distant Takesaki Range Control Centre.

Now satellites are processed in the new class 100,000 3,930 m² STA-2 Spacecraft Test & Assembly building and then moved to H1's modified class 100,000 4,010 m² SFA Spacecraft & Fairing Assembly building, where propellants are loaded, pyrotechnics and solid motors inserted, and the fairing is added. The complete payload assembly is then transferred to the pad to be attached to the waiting H2.

H2's stages and SRBs are received in the VAB's 32 × 46 m 19.1 m high low bay before hoisting on to the mobile launcher platform in the 32 × 27 m 66 m high high bay in solids/stage 1/stage 2 order. The 800 tonne 18 × 22 m 4 m high ML is powered by electric motors on four dollies at 0 to 0.48 km/h for the 500 m rail transfer to the pad. ML provides LOX/LH_2 to stage 1 and is protected by a water deluge system. The exhaust opening is 4 × 9 m.

The PST Pad Service Tower comprises a fixed tower plus two rotating service structures. The 8 × 15 m 67 m high 1,500 tonne tower provides 10 floors, H2 umbilicals and payload assembly installation. The 2,500 tonne main RSS provides 12 floors, the top two adjustable, and 20 tonne crane. The 800 m² launch blockhouse, submerged 4.8 m, is next to the VAB. Control after lift-off, and up to about 30 minutes later when the payload separates, is the responsibility of the distant new Takesaki Range Control Centre, built because the existing Osaki centre is within the launch danger zone.

NORWAY

Andøya Rocket Range

Location: 69°17'N/16°01'E

Background

The first sounding rocket, a Nike Cajun, was launched

NASDA's Takesaki Range Control Centre (top) assumes responsibility for H2 after lift-off, taking over from Yoshinobu's blockhouse (bottom) (NASDA)

Orbital launch facilities at NASDA's Tanegashima site. The static test stand at centre right is used for LE-7 firings. H2 is stacked on its mobile platform in the Vehicle Assembly Building at right and moved the 500 m by rail to the pad. At top centre is the J1 pad. Beyond it, the old Osaki Range Control Centre stands unused as it is too close for H2. Launcher and payload processing buildings are at top right (NASDA) ***1996***

from Andøya on 18 August 1962. ARR is now an operational unit within the Norwegian Space Centre, with 27 staff. There were 619 launchings of all types to the end of 1995 (13 in 1995, 19 in 1994. 17 were planned for 1996: 13 Viper 3A, 3 Taurus Orion, 1 Nike Orion), primarily to investigate the upper atmosphere in the polar region and to study ionospheric and magnetospheric processes at high latitudes.

The range has been supported since 1972 by ESA through the Special Project Agreement, under which it is maintained by and made available to some ESA states. It is also operated for commercial and bilateral projects. Vehicles currently handled include Viper, Dart, Loki, Nike Orion, Orion, Skylark and Black Brant. There are eight pads, including three universal launchers (two with 3 tonne and one with 20 tonne load limits) and a small rail launcher (over-slung type). The new universal launcher 3 (U3) with 20.5 m rail length and 20 tonne safety weight limit became available from late 1993.

A vehicle is prepared horizontally in the protective 36 m long 5.5 m wide 5.2 m high concrete housing and raised shortly before launch. At launch, the launch elevation can be set from 65 to 90° (nominally 85°), the azimuth 260 to 20° (nominally 340°). Four-stage rockets up to 23 m long, 1.2 m diameter, 520 kN thrust (Black Brant 12 class) can be launched into an impact area about 1,900 km downrange.

U3 can also handle the launching of small satellites up to 250 kg into high-inclination orbits. If ESA ever exercised the option this site would be the only satellite pad in Europe. The Polar Satellite Launch Service has passed PDR and is searching for a suitable vehicle. Several US and Russian launchers have been studied, but none satisfies the PSLS cost-performance requirement. PSLS seeks the capacity to put 250 kg into a 750 km orbit. NSC's studies showed that it could capture three orbital launches annually, with a capacity of six. PSLS will be limited to vehicles with propellant totalling less than 40 tonnes because of ground safety considerations.

The 272 m^2 Main Assembly Hall accommodates vehicles up to 15 tonnes.

The User Science Operations Centre is located 300 m northeast of the control centre with two laboratories and a recording room. Payload recovery is performed routinely from the Norwegian Sea using ships or helicopters. New tracking facilities using radar and slant range systems are designed to improve impact prediction to 500 m for sea-recovered capsules.

A downrange telemetry station and a sounding rocket mobile launch facility will be operational at Ny- Ålesund (79° north) on Svalbard after 1998. Climatically, Svalbard is an arctic desert, warmed by the Gulf Stream (mean monthly winter low of −14°C) with 200 to 300 mm annual precipitation. Longyearbyen is the administrative centre, with 1,000 residents and an international-standard airport. The position of the Svalbard Satellite Station (SvalSat) on the Plataberget close to Longyearbyen will allow it to cover every orbit of any polar satellite. Data would be relayed to the mainland via Telenor's terminal at Isfjord Radio.

There is a 20 m VLBI antenna to measure continental drift, and the Norwegian Mapping Authority also operates GPS/Glonass receivers as part of the systems' core network to determine precisely the satellites' orbits. The SvalRak launch facility allows 3 tonne rockets to climb parallel and perpendicular to the magnetic field, as well as directly into the polar cusp and cleft regions.

ARR handles balloons of up to 50,000 m^3. A mobile tracking telemetry unit allows flight times of about 72 h. The balloons drift east in winter and west in summer. An increase in environmental research campaigns (including aircraft, long-duration balloons of up to 600,000 m^3, sounding rockets and ground-based observations) is planned for the updated range. The ALOMAR Arctic Lidar Observatory for Middle Atmosphere Research became operational in July 1994. For the first time, data from several lidars are combined for routine atmospheric observations, including composition, over 10 to 100 km. One lidar is designed for ozone monitoring.

The Cosmonaut Yuri Gagarin *was the flagship of the Soviet tracking fleet, sporting pairs of 25 m diameter 'Ship Shell' and 'Ship Bowl' dishes. It was the only ship to carry the 240 tonne larger antenna* (Jane's Soviet Intelligence Review/LL van Ginderen)

RUSSIAN FEDERATION

Baikonur/Tyuratam

Location: 45.6°N/63.4°E

Background

All manned flights and planetary missions are launched from Tyuratam. The site is prohibited from using due-east azimuth launches (the most efficient) because the lower stages would impact in China. Lieut Gen Alexei Shumilin, appointed in 1992, commands the facility. Interestingly, he is an officer of the Kazakhstan armed forces, overseeing a facility developed by the Soviet Rocket Forces. VKS personnel still totalled 14,500 by 1995, but all civil launch operations are being transferred to the RKA. Some 700 RKA staff, in blue uniforms, were already in position for the Soyuz-TM 22 launch in September 1995. 300 had transferred from VKS.

The Baikonur Kosmodrome offers nine launch complexes, with 15 pads, to accommodate space and missile activities. It is the only Kosmodrome supporting Proton, Zenit and Energia. Energia's three pads include the original facilities of the ill-fated N1 lunar launcher, which failed in four attempts between 1969 and 1972.

The Kosmodrome was built by a team led by V P Barmin, to house a site for missile and rocket tests which began in the area in the early 1950s. As a historical site, the first pad built launched both Sputnik 1 and Yuri Gagarin. Western media were allowed, in May 1988, to observe the departure of a mapping satellite from this facility. It is now designated as the Gagarin Launching Complex.

A second pad with manned launch capability was completed in 1965 about 30 km from the first. It is now used exclusively for military vehicles.

A Japanese astronomer pinpointed the site's location in 1957 as being near Tyuratam in Kazakhstan, east of the Aral Sea and about 370 km southwest of the real Baikonur. However, the Soviets continued for at least 17 years to give its latitude/longitude as that of the town of Baikonur. Kazakhstan finally named it after Tyuratam in 1992, although locals still refer to it as Baikonur, as does the entrance sign at the main gate.

Not until Soyuz-T 6 in June 1982 was a Western journalist admitted, but US reconnaissance satellites provided detailed information for the military, while Landsat and Spot images brought the sprawl of pads into the public domain in 1974.

Leninsk, 2,100 km from Moscow and founded 5 May 1955, has an official population of 60,000, down from the 100,000 peak of the mid-1980s, although the recent influx of Kazakh peasants probably yields a much higher total. It grew up alongside the launch facilities as the administrative centre of the 1,560 km^2 kosmodrome. 46,011 km^2 is allocated for downrange stage impacts. Cosmonauts are accommodated in a hotel complex with swimming pool and sports centre. RKK Energia earlier maintained a permanent staff of 2,000 at Tyuratam, reduced by 1993 to 800 to 1,000 (mainly for Mir and Proton) because of the loss of Buran and Energia work.

Tyuratam's largest known accident came on 24 October 1960, when at least 165 were killed by the on-pad explosion of a developmental SS-7 ICBM. Although the West had known for years that a major catastrophe had occurred, the USSR did not acknowledge it until 1989 nor give the death toll until the 30th anniversary. Marshall M I Nedelin, Cdr of the Strategic Rocket Forces, was under pressure from Khruschev to hold to schedule when a fault prevented launch. Rather than drain the propellants, Nedelin saved time by ordering engineers to work on the vehicle as it stood. Stage 2 ignited and detonated stage 1. SS-7's designer, Mikhail Yangel, narrowly escaped death. A Moscow TV programme 26 February 1995, claimed that 91 deaths were caused by a 1969 launch failure.

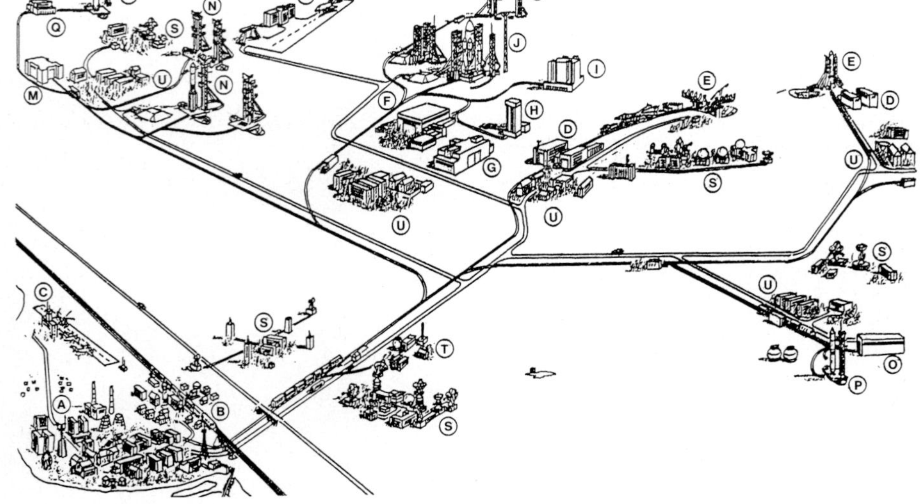

Baikonur Kosmodrome launch and processing facilities. A: Leninsk. B: Tyuratam. C: Krainiy airport. D: Soyuz launcher/payload processing (older is site 2; newer is site 49). E: Soyuz pads; Gagarin pad (often referred to as Pad 1) is site 2; newer is site 31. F: Energia processing building, building 112 site 130 (ex-N1). G: Buran processing building, site 254. H: Energia mechanical test stand. I: Buran/Energia fuelling building. J: Energia/Buran pads (2, site 110; ex-N1). K: Energia pad; complex 113 (but site 250 also referred to). L: Buran Yubileini landing strip (site 251). M: Proton processing building, newer is site 92. N: Proton pads, 81L/R plus 200L/R. O: Zenit processing building, site 42. P: Zenit pads, 45L/45R. Q: Tsyklon processing building. R: Tsyklon pad, 90L/90R. S: launch tracking system. T: KAZ LOX/LN$_2$ plant, site 3. U: technical facilities **1994**

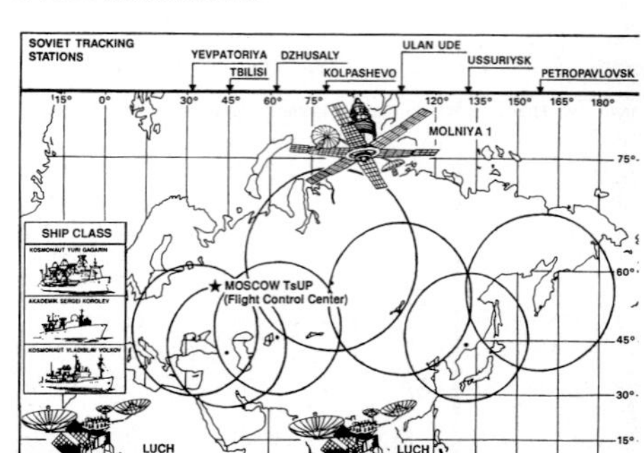

The network for controlling manned missions is augmented by Luch satellites at the central and western SDRN geostationary positions. Molniya provides links from the ships and remote land sites to TsUP (ships now not operational) (Teledyne Brown Engineering)

Tyuratam's Gagarin pad has supported more than 350 orbital launches. Vehicles are moved to the pad horizontally by rail – the transporter is visible at right

The Minsk agreement of 30 December 1991 declared that Baikonur would provide free access for other states. Although Russia holds up to 90 per cent of space industry capability, Baikonur's loss would have been a major blow. Under the Russian/Kazakh agreement of 25 May 1992, Russia pays 94 per cent of operating costs and Kazakhstan 6 per cent.

Before the USSR's break-up, Tyuratam cost Rb3 billion annually to run. Disagreements over Baikonur have continued to rumble on, with the military complaining of lack of investment. Conscripts from Kazakh battalions have rioted because of the poor conditions. Russian premier Viktor Chernomyrdin signed an MoU December 1993 that Russia would rent some of the 4.5 million ha currently occupied by Baikonur. The MoU also called for gradual demilitarisation; there were 28,000 servicemen based there (it was agreed later that 16,000 would remain for space activities). Presidents Yeltsin and Nazarbayev agreed 28 March 1994 that Russia has sovereignty and jurisdiction over Tyuratam and Leninsk and their personnel for 20 years (with a further 10-year option) for the equivalent of US$115 million annually. The final agreement was signed 10 December 1994. Leninsk formally returned to Russian jurisdiction 1 August 1995. That US$115 million might not involve actual payments, but is likely credited to the Kazakh national debit to Russia.

The climate can be extreme, +60°C in summer, while suffering violent snowstorms and –40°C frosts in winter.

Soyuz and Progress spacecraft and their launchers are integrated in the MIK building. Following mating, the vehicle is transferred on a rail transporter and erector to the pad, 5 km distant. The cosmonauts suit up in a small room in the rear of MIK and ride to the pad in a bus. Launch operations are controlled from a blockhouse 500 m from the pad; a public viewing stand is only 300 m further out.

Northwest of the Soyuz buildings is the Energia and Buran area, originally built for processing the N1. Energia was integrated horizontally in the 40 m high 274 × 160 m old N1 Building 112 after the core was shipped in by a Mya 201M aircraft. The payload was added and the combination lifted on to the transporter/erector. The stack moved to a separate building for Buran fuelling.

In September 1995, a fully integrated Energia and two cores sat in residence. The adjacent Buran building will be upgraded for processing payloads to Western standards. Only the flown Buran remained in residence by September 1995, whereas in October 1994 three other models were also nearby. Since then, Mir's Spektr module became the first to be processed for launch since Buran. Space Station elements are also processed here.

There are two Energia launch complexes 20 km apart: the static test stand (number 113) used for the maiden launch and a pair originally built for N1 capable of handling the Shuttle (number 110). Energia's static stand was purpose-built because there was no existing N1 equivalent – the lunar vehicle was considered too large for static tests.

Some 12 km further northwest is the 4,500 m long 84 m wide Shuttle Yubileini landing strip on a northeast-southwest orientation. It supported jet-powered test and training flights and provided a microwave system for fully automatic landings. Operations are controlled from a nearby building. The runway fell into disrepair after Buran's 1988 landing, but a German company repaved the 52 × 3,200 m central portion in 1995, under contract to LKE, to receive commercial Proton payloads.

To the west of the Shuttle strip is the Proton preparation and launch area. The building (number 92) can handle up to six launchers simultaneously. Rail transfer of the fully assembled vehicle to the pad requires about 7 hours, followed by five days' preparations before launch. Most launches are currently from pad 81L, but will eventually switch to the refurbished 81R. 200R has been reactivated for commercial and Space Station launches (and transferred from VKS to RKA control). 200L is certified for only a handful more launches before it must be refurbished. It will then transfer to RKA. Number 81 is the original complex. Number 200 was built in the 1970s several km to the east with pads 600 m apart. By the time of Granat's launch on 1 December 1989, number 200 had serviced 114 launches.

Zenit's pad 45R was severely damaged 4 October 1990 when a vehicle exploded within seconds of departure; 45L was inaugurated with August 1991's departure. Together, they would be capable of supporting 15 missions annually. 45R was so severely damaged, collapsing the launch table, that it was not repaired until the late 1990s.

Kapustin Yar

Location: 48.4°N, 46.5°E

Background

Known as the Volgograd Station to its personnel, this was the first rocket development centre, hosting its first launch in 1947. It became a source of renewed interest when it was used in June 1982 for Kosmos 1374, the first test of the subscale spaceplane flights. Its two launches in 1984 included the fourth spaceplane test. Kapustin Yar in later years handled only occasional missions, possibly for radar calibration. There has been no launch since 1987, leaving the total at 83. It is no longer under VKS control but there are still some missile testing activities. Kosmos sub-orbital launches continue and AKO Polyot offers commercial orbital launches, partly using Kosmos USA processing facilities transported from Plesetsk. Its first orbital launch was Kosmos 1 in 1962. In 1975, a Landsat 2 image revealed publicly that its facilities extend over a 96 × 72 km area northeast of the River Volga and 965 km southeast of Moscow.

During its early years, Kapustin Yar tested captured V2 missiles and conducted sounding rocket experiments carrying dogs and other animals up to 500 km. By 1980, there had been 70 orbital launches, mostly small Kosmos science satellites but, as work was switched to Plesetsk, the annual launch rate fell to an average of only one. The last launch from Kapustin Yar took place in 1987.

Plesetsk

Location: 62.8°N/40.1°E

Background

Plesetsk was, for a long time, by far the world's busiest spaceport but has now been overtaken by Baikonur. Orbital launches have declined for several years, falling from a steady 62 per year in 1983-85 to 56 in 1986, 47 in 1987, 47 in 1988, 45 in 1989, 48 in 1990, 37 in 1991, 33 in 1992, 26 in 1993, 18 in 1994, 13 in 1995, 10 in 1996, 9 in 1997 and 7 in 1998. The decline has accelerated as programmes transfer to newer vehicles at Tyuratam. Plesetsk's total record of 1,452 vehicles by the end of 1998 is 416 higher than Baikonur's.

The site's 1,762 km^2 base is located 170 km south of Archangel, which enables communications and spy satellites to be placed in polar (but never retrograde) and highly elliptical orbits. Range safety restrictions now limit most flights to 62.8°, 67.1°, 73-74°, 82-83°.

Plesetsk's existence was established publicly by Geoffrey Perry, of Kettering Grammar School (England), following the first orbital launch Kosmos 112 on 17 March 1966 from pad 1. It was officially acknowledged only in 1983. Originally intended as an ICBM site, construction of the first of several pads for the F7/A class vehicle began in April 1957, and was completed in December 1959. The pad and ICBM began active duty in January 1960. Three other pads were active by July 1961.

Landsat imagery taken in 1973 showed Plesetsk to be about 100 km long, housing at least four complexes and defended by surface-to-air missiles. Mirny has grown alongside to house staff and their families; population is 40,000.

There are nine operational pads: three for Cosmos at complex 133 (one converted for Rockot), four Soyuz/Molniya (number 16 converted for Russia, number 41 being converted for Rus, 2 at number 43) and two Tsyklon number 32 (at 62°54′26″ north/40°47′23″ east). Complex 35 has two Zenit pads and vehicle/spacecraft processing facilities now under construction for polar missions. The new Angara will also use the complex. One pad was retired in 1989 after the launch of the Cosmos 2044 international biosatellite.

Climatic conditions are extreme ranging from –47 to +34°C, with up to 85 cm of snow cover.

The most powerful launcher available at Plesetsk is still the Molniya. Tsyklon is processed horizontally in a six-storey building, where stage 3 is fuelled, and not delivered to one of the two pads 40 km away by train until about T–3 h.

Two major accidents were acknowledged in 1989. Nine technicians and soldiers were killed on 26 June 1973 (SL8 Cosmos) and a further 50 on 18 March 1980 (SL-3 Vostok with Meteor). On both occasions a vehicle exploded during fuelling operations. Collection of the hundreds of strap-ons and first stages discarded over the area started in 1991 in response to complaints from local residents over environmental contamination; compensation was also claimed. The junk is now processed to recover useful materials.

Svobodny

Location: 50°00′-51°40′N/128°00′-128°30′

Background

Proposals for a new Kosmodrome at the decommissioned missile site at Svobodny, near the Chinese border, surfaced during Russia's negotiations with Kazakhstan over Baikonur's future. They were widely perceived as a negotiating ploy, but President Yeltsin signed a decree on 1 March 1996 approving the Kosmodrome's creation. Start used it first, employing its own transportable launcher. Rockot will follow, employing one of five silos. The largest investment will be a two-pad Angara complex. 60 SS-11s were silo based at the site. The maiden orbital launch for Svobodny took place on 4 March 1997 when Zeya was placed into orbit by a Start-1 flown from a mobile launch platform. A second Start-1 flight on 24 December 1997 carried the Early Bird 1 commercial satellite for EarthWatch Inc. No further launches had taken place by the end of April 1999.

SOUTH AFRICA

Overberg Toetsbaan (Overberg test range)

Location: 34°35′S/20°19′E

Background

OTB is a 43,000 ha site along 70 km of southeast coastline of the Western Cape. Its position near Africa's southernmost tip is ideal for launches into Sun-synchronous, polar and Earth-synchronous orbits down to about 38° inclination. The site is a division of Denel, with a permanent staff of 175. Construction of the test range and launch facility began in 1983, allowing the first simple tests in 1987. A first stage test of South Africa's planned orbital vehicle was launched in June 1989, followed by stage 1 and 2 in July 1989 and November 1990, leading to the facility's formal qualification by the end of March 1991. RSA cancelled its satellite launch programme here in 1994, but OTB's core capability is maintained through regular use on a wide variety of flight tests. South Africa joined the MTCR as a full member in 1995, allowing OTB to be used by other parties subject to non-proliferation controls.

The range, close to the villages of Bredasdorp and Waenhuiskrans, is divided into east/west sectors by the De Hoop Nature Reserve, with most of the instrumentation and infrastructure in the western half. An adjacent air force airfield, part of the SAAF Test Flight and Development Centre, can accommodate a Boeing 747 and is available for OTB clients. A good quality highway between Cape Town and Bredasdorp allows the 200 km to be covered by car in 2 hours. OTB's own accommodation caters to 100 people.

OTB's central control room (Denel-OTB) ***1996***

The area has a mild Mediterranean climate with a temperature of +5/25°C, 40 cm annual rainfall (mostly May to September) and 80 per cent humidity).

OTB's preparation and launch facilities are:

Launch Preparation Complex

The LPC handles final preparation/checkout and mating of satellite and launcher. The preparation control centre for monitoring this work is in an LPC basement area. LPC comprises:

S1

Spacecraft Launch Preparation for final satellite preparation, mating with the final stages, fairing encapsulation, EM integration with the Satellite Ground Station via S-band. 12 × 17 m 13 m high class 100,000, 5 tonne overhead crane with 11.4 m free hook height. A 3.5 × 5.5 m stabilised column provides satellite alignment. 12 to 20°C clean N_2 at 10 m^3/min provides satellite cooling after encapsulation. EGSE can be housed in S1 or C1/LCC.

S2

Hydrazine Filling is a class 100,000 for fuelling satellite and launcher final stage from adjacent S3 bulk container. Water drenching and neutralisation for spillage. 4 × 10.5 m, 7 m high.

S4

Spacecraft Long-Term Storage, a 4 × 10.5 m, 7 m high class 100,000, for storing fully integrated satellite/final stages for fast reaction launches. At least two 2 × 2 m, 4 m high satellites can be stored simultaneously.

L1

Launch Vehicle Preparation made up of two sections, a 21 m and a 14 m high high bay, both serviced by a 90 tonne crane, 11 m high hook. In high bay, first stages can be mechanically and electrically integrated horizontally on mobile launch platform, and elevated to vertical for system functional testing.

C1

Launch Preparation Control Centre for controlling and monitoring satellite and vehicle preparations, and the first few seconds of flight. The total launch operation is controlled from the Test Range control centre. 5.6 × 10.5 m, 4 m high.

A1

Hydrazine Storage (storage and chemical analysis).

A2

Pyrotechnics Magazine.

A3

System Magazine (solid motor and fully assembled stage storage).

A4

Flammable Liquid Store.

Launch is made up of three main areas:

FIC

Final Integration Centre, the final integration of first stages with final stages/satellite before pad transfer. Fibre optic and RF links with LCC. Cooling N_2 for payload cooling. 20 × 50 m; 10 tonne overhead crane.

Launch Pad

A launch platform is towed by road from FIC. Fibre optic and RF links with LCC. Cooling N_2 for payload cooling. Launch rehearsal verifies system functionality and countdown procedures.

Range Instrumentation

This sits apart from the L1 fixed lateral site in the eastern sector (fixed telemetry station, Doppler receiver, tracking radar) and similar western sites (with two radars).

OTB has numerous measurements sites spread over 20 × 60 km. Sites are linked with the command and control centre by fibre optic and microwave. Some instruments can be moved outside the Range for a 300 km baseline. A fully self-contained telemetry station with stabilised antennas can be deployed on land or onboard ship. All instrumentation is linked to a central computer in the Range Control centre, and can be slaved to the tracked object's state vector, computed by the central computer and based on prioritised pointing and direction information from the instruments and telemetry. Instrumentation includes, three tracking radars, four mobile cinetheodolites, one mobile optical high-speed cameras track mount (with film development laboratory and film/video readers, two telemetry stations, one downrange telemetry station (ship or land), four Doppler (two mobile); Satellite Ground Station (10 m S/X band, 2 m S-band TT&C), Command Destruct System; electronic sky screens.

There is also a fully equipped meteorological station, with rocket sondes that can probe up to 65 km.

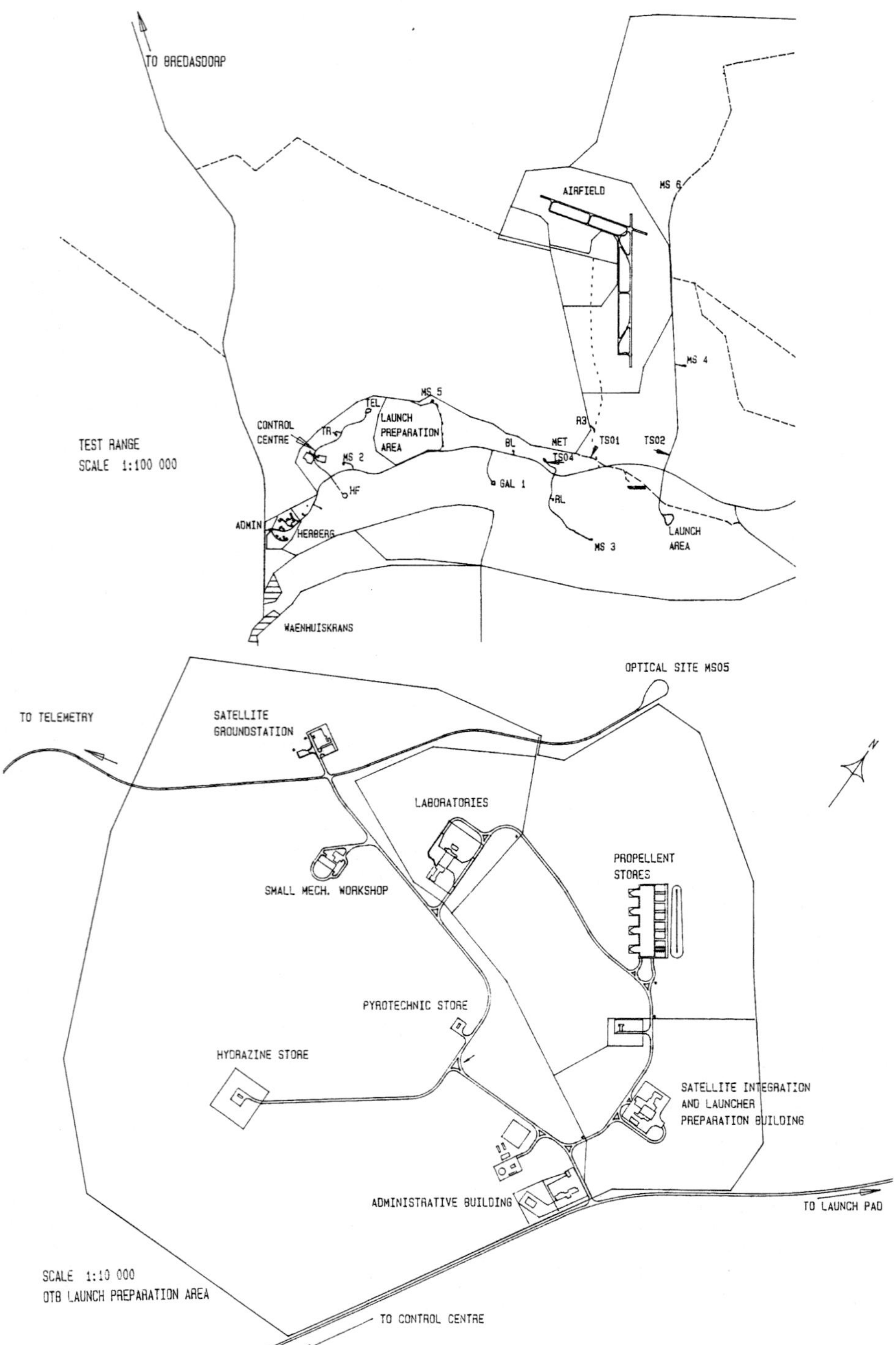

OTB layout (Denel-OTB) ***1996***

SWEDEN

Esrange

Location: 67° 56′ N/21° 04′ E

Background

Esrange is the Swedish Space Corporation's sounding rocket and satellite operations range north of the Arctic Circle. More than 400 launches were made here between 1966 and 1995. Esrange includes seven permanent launch facilities able to handle most types of sounding rocket (including Terrier, Black Brant and Skylark 12). The seventh became operational in 1991 to handle the large Maxus vehicle. With altitudes of up to 900 km possible, the Swedish German Microgravity Sounding Rocket Programme uses it most frequently to provide ESA members with µgravity conditions in preparation for Spacelab missions. Sweden has been studying the commercial use of Esrange for orbiting small polar satellites.

Esrange has a 75 × 120 km impact area available for land recovery. Payload and control facilities have been upgraded, partly to accommodate the Maxus long-duration sounding rocket, based on Thiokol's Castor 4B and developed jointly by SSC/DASA. The 23 m high Maxus building has a 15 m-high umbilical mast. A 145 m^2 storage building can house two Castors at 20±2°C.

Stratospheric balloons of up to 10^6 m^3 are launched to 40 km and regularly recovered within Sweden, Finland and Russia.

Since 1978, facilities for receiving, processing and disseminating data from remote sensing satellites have been used for Landsat, and have been extended to handle data from France's Spot, Japan's MOS 1 and JERS 1, and Russia's Resurs O1. ESA's station dedicated to ERS is located at Salmijärvi near the range. Esrange is contracted for operation/maintenance.

UNITED STATES OF AMERICA

Cape Canaveral Air Station

Location: 28°26′48.73″N/80°33′54.58″W

Current status

In early 1998, industry spokesmen and US congressmen called on the United States air force to invest in a new launch infrastructure to bring Cape Canaveral up to the same standards as the US' major space competitors. By historical accident the air force has authority over Cape Canaveral and Vandenberg. A spokesman for the air force said, in March 1998, that the air force would not commit any additional funds to upgrades at this time.

Background

The 45th Space Wing (45SW) designation in November 1991 replaced the Eastern Space & Missile Center organisation, established 1 October 1979. Its origins date back to September 1948 when the Banana River Naval Air Station was transferred to the USAF as a base for operations as a joint-service missile range. The Headquarters opened on 10 June 1949 but joint operations proved unwieldy and it was replaced by the air force's Long Range Proving Ground Division on 16 May 1950. On 30 June 1951, in keeping with policy to designate intermediate headquarters as 'centres', the long Range Proving Ground Division became the air force Missile Test Center. It was renamed the air force Eastern Test Range in 1964. A unit of air force Space Command, it is the host organisation for Cape Canaveral AS, Patrick AFB and the two island stations (Ascension and Antigua) of the Eastern Range (ER). The eastern and western launch ranges and bases were managed by air force Systems Commands' Space Systems Division until 1 October 1990. The Wing comprises four major Groups: Operations; Support; Logistics; Medical.

CCAS encompasses active Titan, Atlas and Delta complexes in addition to providing support facilities for the military, NASA and commercial organisations. Military, civilian and contractor personnel at Patrick and the Cape total about 12,000 and of these, up to 4,000 are involved in processing vehicles and satellites. 455 space launches had been made by the end of 1995. No other site has been used for US manned missions.

The annual launch rate is now limited to 25 to 30 because, despite upgrading, the Cape is based on 1950s concepts. The pads are recognised as the choke points, with missions requiring an average of 26 to 40 days processing before launch.

Launch Complex 17 (Delta) and LC 36 (Atlas Centaur), pads, formerly under NASA control, were turned over to the air force following the 1987 announcement of a mixed fleet policy and 1988's revised national space policy endorsing the growth of a commercial ELV industry. The last NASA-managed Atlas Centaur launch took place from LC 36 in September 1989 and of a Delta in November 1989 from Vandenberg. NASA continues to have oversight responsibility for ELVs carrying agency payloads, but now contracts with either USAF or vehicle manufacturer for launch services.

Extensive support is provided for Shuttle launch operations by the military. The USAF Eastern Range, of which the Cape is part, provides weather forecasting, range safety, tracking and preparation of DoD payloads. USAF also operates Titan pad 41 on a 1.44 km² patch inside the KSC boundary. KSC operates a Vandenberg Launch Site Resident Office as the interface with USAF for base support of all NASA payloads flying from the west coast.

The Operations Group provides operational control over all aspects of DoD space launches. The 1st Space Launch Squadron (SLS) was activated 1 October 1990 to assume all responsibility for Delta operations. The 3rd SLS activated 3 August 1992 to take over Atlas work. The 5th SLS activated 14 April 1994 to oversee all Titan operations. Each Squadron exercises field operations management/control of all DoD satellites launched from CCAS.

A single storey 11,800 m² Range Operations Control Center became operational in March 1995 as the ER control facility under a US$134 million Harris Corporation contract. The goal is to allow flights every 12 hours. Operationally, it serves the ER by processing worldwide radar metric data, generating inter-range vectors to aid in acquisition, calculating state vectors for navigation uplink, displaying telemetry and orbital navigation information, and functioning as LRCC Lead Range Control Center to co-ordinate radar tracking and retrieval. Facilities cover communications, range control and scheduling, data processing, range safety and weather information.

CCAS offers five principal hangars for non-hazardous spacecraft processing. Hangar AO with a 14 × 53 3 m class 100,000 high bay clean room capable of handling two payloads simultaneously. Hangar AE with class 10,000 facilities and the Delta mission director's centre. Hangar AM with a 4.4 × 5 m clean room and solarium, the SPIF Shuttle Payload Integration Facility, and the SAB Satellite Assembly Building. Loading of payload propellants and pyrotechnics can be performed in three facilities, Explosive Safe Area 60A and KSC's Spacecraft Assembly and Encapsulation Facility number 2, both of class 100,000 standard, or KSC's Cargo Hazardous Servicing Facility Complex, built to handle Shuttle-sized elements.

Principal current launch activities are Titan operations at pads 40/41, Delta from Space Launch Complex 17 (A/B) and Atlas Centaur from SLC 36 (A/B). The orbital inclination is limited without doglegging to 57° by range safety considerations. Titan pad 41 is actually north of the CCAS boundary on a 1.44 km⁻² site allocated to the air force. The ER's tracking network stretches 16,000 km over the Antigua and Ascension Island stations and into the Indian Ocean, where it meets the Western Range. Mainland Florida stations are maintained at the CCAS, Jonathan Dickinson and Malabar annexes, augmented by the fleet of ARIA advanced range instrumentation aircraft routinely scheduled from the 452nd Test Squadron/ENO Edwards AFB. In addition, ER can use instrumentation operated by NASA on Bermuda and Wallops Island. Another station is available at Argentina in Newfoundland for high inclination launches on northeastern flights of about 37°.

Cape Canaveral's Titan operations. Core stacking is performed in the Vertical Integration Building in the centre, followed by strap-on mating in the next building along the causeway. Further along is the new Solid Motor Assembly and Readiness Facility for processing the strap-ons. Straight on is pad 41, off to the right is pad 40 (USAF)

The status of the launch complexes is:

1-4 Inactive. Snark, Bomarc, Bumper, Redstone, Jason, X-17, Polaris launches; 'Fat Albert' tethered aerostat tests and radar testing of air defence equipment to be used at other locations.
3 First Cape launch, Bumper, 24 July 1950.
5-6 Dismantled. First US manned launches: Mercury Redstone 3/4 (Shepard/Grissom suborbital flights, May/July 1961); Jupiter C, Explorer, Pioneer. Part of CCAS air force Space Museum since January 1964.
9/10 Dismantled. Used for Navaho, Minuteman 1-3 and Pershing 1A and later renumbered 31/32 Challenger debris stored here from 1987.
11 Dismantled.Used for Atlas.
12 Dismantled. Used for Atlas and Atlas Agena: ATS, Mariner, OAO, OGO, Pioneer, and Ranger.
13 Atlas and Atlas Agena: Mariner, Lunar Orbiter and DoD missions.
14 Dismantled. Mercury Atlas 6-9 manned orbital flights, starting with John Glenn, February 1962. Demolished in 1976 because of dangerous state from salt corrosion.
15 Dismantled. Titan 1/2.
16 Inactive. Previously Titan 1/2 and Pershing 1/1A/2.
17 Active. Delta Pads A/B (A: 28°26′48.73″ north 80°33′54.58″ west; B: 28°26′44.12″ north 80°33′57.78″ west), providing an annual capacity of 12. Plans had been made to close them down at end-1986, but selection of Delta 2 to launch GPS satellites ensured their survival. One has been modified to accommodate the cryogenic Delta 3 from 1998. Delta stage 1/2 checkout typically takes place over 14-7 weeks before launch in USAF Hangar M and the Delta Mission Checkout Facility. Stage 1/interstage erection occurs five to six weeks before launch, followed immediately by strap-on and stage 2 attachment; simulated launch/flight in second/third week; stage 3 and mated payload during second; final week sees fairing installation and stage 2 propellant loading. GPS satellites and stage 3 are processed at Area 59. Commercial payloads meet the stage 3 assembly for the first time in the Payload Spin Test Facility, 2.8 km west of complex 17, where it is mated/balanced at 0 to 300 rpm.
18 Dismantled. Vanguard, Blue Scout, Thor.
19 Dismantled. Titan 1/2; all 10 Gemini Titan 2 manned flights 1965-66.
20 Dismantled Titan 1/2 pad; modified for Titan 3A testing 1964/65. Renovated to launch Starbird suborbital vehicles as targets for SDI's Starlab Shuttle mission (cancelled). Demonstration Starbird launched 18 December 1990 – the first launch since the 1960s. The unsuccessful Joust 1 materials processing flight for the University of Alabama in Huntsville's Consortium for Materials Development in Space was launched 18 June 1991 by Orbital Sciences Corporation. Successful Aries Red Tigress launches were made by OSC for SDIO 20 August 1991, 14 October 1991, 28 May 1993. Florida Spaceport Authority is modifying the complex for commercial operations from 1997.
21-22 Dismantled. Mace and Bull Goose.
25 Inactive. Trident 1, Poseidon and Polaris.
26 Deactivated. Now at now USAF Space Museum. Two pads launched 24 Jupiter IRBMs, two Jupiter bio monkey flights, two Redstone IRBMs, two Jupiter C (Explorer 1/2), and six Juno 2 (Explorer 8/11
29 Deactivated. Polaris and UK's Chevaline.
31/32 See 9/10.
34 Dismantled. Apollo Saturn 1/1B. Apollo 7 October 1968 was that programme's only (and NASA's last manned) launch from CCAFS.
36 Active. Atlas Centaur. From 1985 only Pad B remained for Atlas launches. Pad A was converted at a cost of US$20 million into a Shuttle Centaur Checkout Facility in time for the 1986 launches of Ulysses and Galileo – delayed by the Shuttle accident. The centre would then have handled the common Centaur stage being developed for NASA/DoD Shuttle operations but abandoned in June 1986. In October 1987, USAF and General Dynamics signed a commercialisation agreement, supplementing an earlier agreement with NASA as owner of the complex, allowing the company to use Complex 36. Both commercial and government-sponsored Atlas Centaur launches are involved, with General Dynamics reimbursing the government for the direct cost of goods and services provided. Pad 36A is used exclusively for USAF Atlas 2 launches, supporting five annually (six under surge conditions). As part of the modifications, all Shuttle Centaur structures were

removed from 36A. 36B was modified by the end of 1990 to accommodate all new Atlas 2 types. The last NASA-managed launch was September 1989 departure of AC 68 with the last FltSatCom.

37 Inactive. Pads A/B used for Apollo Saturn 1/1B and Pegasus 1-3.

40 Titan 4, previously Titan 3C/34D and commercial Titan 3. The Intelsat 6 launch of June 1990 was the pad's 35th. Martin Marietta and engineering contractor Bechtel National Inc then closed it down for a US$340 million rebuild. It was ready to handle September 1992's Mars Observer launch, after which work was completed for it to accommodate all Titan 4 models. The essentially new complex has a 15-year life. The 5,000 tonne 79.3 m high Mobile Service Tower is encased by 7,000 m^2 of shielding against electromagnetic interference of the vehicle and payload. The major 8 × 37 m vehicle and 5.2 × 26.8 m crane doors are sealed by inflatable bladders. The new 55 m umbilical tower and the rest of the facility can withstand 185 km/h winds. Four 90 m towers with strung cables between were added in 1993 around both pads as lightning protection.

Titan's core is stacked on its launch platform at the Vertical Integration Building. The VIB houses four high bays, with Bay 1 dedicated to commercial Titan operations, Bay 2 to USAF Titan and Bay 3 for Centaur. The new LOCC Launch Operations Control Centre replaced the Launch Control Centre in the VIB in 1993 in a different area. The stack is then moved 1.4 km by rail to the Solid Motor Assembly Building where the strap-ons are stacked and attached before moving the final 4.2 km to pad 40. The SMAB also houses the USAF Spacecraft Processing Integration Facility, a secure equivalent of NASA's Vertical Processing Facility for preparing classified payloads. The new 5,600 m^2 SMARF Solid Motor Assembly and Readiness Facility for simultaneous processing of up to six sets of Titan 4 boosters was completed in 1994. The SMAB is also used for IUS stage assembly/processing. Construction began January 1993 of a separate Centaur Processing Facility to decouple Centaur pad checkout from other Titan 4 elements.

41 Formerly used for military Titan 3C and NASA Titan 3E Centaur launches such as Viking and Voyager, subsequently modified for Titan 4 (first launch June 1989).

42 Once proposed as a second Titan 4 pad on the same 1.44 km^2 USAF island, as pad 41 within KSC.

43 Deactivated. Used for weather rockets.

45 Deactivated. Used to launch ROLAND mobile missiles

46 USN Trident 2 but being modified by Florida Spaceport Authority under US$4.89 million in Federal grants and US$2 million in industry matching funds to accommodate small orbital vehicles, initially those employing Thiokol's Castor 120 solid motor. Modifications include a mobile access tower and adaptable launch mount. Naval Ordnance Test Unit retains Trident launch capability.

47 Active: sounding rocket launches.

Kennedy Space Center (KSC)

Location: 28°36'33"N/80°36'15"W

Background

KSC is NASA's site for processing, launching and landing the Space Shuttle and its payloads, including Space Station components. Located on Merritt Island adjacent to the USAF launch facilities of the Cape Canaveral Air Station, Kennedy was originally built to support the Apollo lunar landing programme of the 1960s. After the last Apollo lunar launch in 1972, Launch Complex 39 supported Skylab in 1973-74, Apollo-Soyuz in 1975 and Shuttle from the late 1970s. Shuttle mission STS-96 in May 1999 brought the number of manned launches to 108 from pads 39A/B. The installation covers 567 km^2. All non-operational areas are part of the Merritt Island National Wildlife Refuge. KSC extends some 55 km north to south and 16 km across its widest point.

Land acquisition to build the centre began in 1962, and clearing for the Vehicle Assembly Building began in November 1962. The VAB was structurally completed in mid-1965. The first LC39 launch was Apollo 4 in November 1967. Shuttle processing and launch facilities are in the LC 39 area. About 8 km to the south, office buildings, payload processing facilities (including the new Space Station Processing Facility) and flight crew quarters are in the Industrial Area. NASA also uses some buildings on neighbouring Cape Canaveral AS. The USAF Titan pad 41 lies within a 1.44 km^2 island assigned to USAF inside KSC.

About 14,000 personnel work at Kennedy. 1,925 are civil service employees, but contractors employ the majority. There are four major contracts: Lockheed Martin Space Operations Co was the Shuttle Processing Contractor (SPC) during 1983 to 1996, McDonnell Douglas Aerospace handles payload ground operations, and Rockwell supports logistics related to the spaceplane, including spare parts. EG&G holds the base operations contract.

Columbia occupies pad 39A in the foreground during STS-35 preparations, while Discovery is on 39B for STS-41. The white Payload Changeout Room in the Rotating Service Structure is used for installing vertically processed payloads on the pad. A total of 82 Shuttle launches had been made by February 1997 (NASA)

In October 1996, United States Alliance officially began work on the space flight operations contract – a consolidation of Shuttle processing contracts at KSC.

Today, final Shuttle assembly takes place in the VAB and the stack is then transported either 5,535 m east to pad 39A or 6,828 m northeast to 39B. LC 39's design reversed the fixed launch concept of earlier missions, under which assembly, checkout and launch were conducted at the pad. Kennedy's first director, Dr Kurt Debus, developed the mobile launch concept. The Shuttle processing and launch facilities are:

Vehicle Assembly Building (VAB)

160 m high, 218 m long, 158 m wide, divided into four 160 m high bays and a 64 m low bay, transected by a north to south transfer aisle. One of the world's largest buildings by volume, it covers 32,400 m^2. A maintenance shop for Shuttle main engines and holding areas for the SRB forward assembly and aft skirt is in the low bay. External Tanks are stored and checked out in high bays 2/4, with 4 also used for payload canister handling. high bays 1/3, facing east, are used for assembly of the three main Shuttle elements (two SRBs, ET, Orbiter) on the Mobile Launch Platform. The stack is transported from here to the pad.

Launch Control Center (LCC)

A four-storey building adjacent to the VAB on the east side, housing four firing rooms. Firing rooms 1/3 provide full control of launch and orbiter operations; number 3 was previously secured for classified DoD departures. Number 2 is usually used for software development and testing. Number 4 is only a partial firing room, primarily used as an engineering analysis and support area for launch and checkout operations. Plans are currently under way to turn number 4 into a full-up firing room, with construction beginning in late 1997. Computer and software equipment comprising the LPS Launch Processing System is also housed in the LCC. Designed at Kennedy for the Shuttle programme, the LPS automatically controls many of the procedures formerly performed mechanically, while also requiring fewer personnel (about 90). Shuttle checkout, countdown and launch are conducted from the LCC using the LPS. A countdown begins typically about three days before scheduled lift-off, at the T−43 h mark and including an average of 28 hours of built-in holds. As soon as the two solids ignite at T−0, control of the mission shifts from Kennedy to Mission Control at Johnson Space Center in Houston. KSC again assumes responsibility after the crew exits (usually within an hour of touchdown) and ground cooling has been established.

Mobile Launcher Platform

Final Shuttle assembly, transport to the pad and launch are conducted on the MLP. A two-storey steel structure, 7.6 m high, 49 m long, 41 m wide, 4,190 tonnes. MLP-3, previously used for Apollo 11, manned Skylab and ASTP, was added to the existing pair in late 1989 and first used by STS-32.

Crawler-transporter

Two crawlers, each weighing 2,700 tonnes, were in service by early 1967 for the Apollo lunar programme. About 40 m long, 35 m wide and height adjustable to 6 to 8 m, with an operator control cab at each end. There are eight tracks, each with 57 0.9 tonne shoes. Two 2,750 hp diesel engines drive four 1 MW DC generators powering 26 traction motors. Two 750 kW AC generators, driven by two 1,065 hp diesels, provide jacking, steering, lighting and ventilation. Two 150 kW AC generators provide MLP power. Maximum speed is about 3.2 km/h unloaded, 1.5 km/h loaded, consuming about 350 litres of diesel oil per km. It requires between 6 and 8 hours to move Shuttle from the VAB to 'hard down' at the pad. Vehicle apex maintained vertical to within ±10 arcmin. Crawler 2 passed its 1,000 mile (1,609 km) mark, 22 April 1990 carrying Columbia to 39A for STS-35.

Launch Pads

39A/B identical roughly octagonal design. Central structure on hard stand is the 106 m Fixed Service Structure, including 24.4 m lightning mast. FSS has three service arms, the ET gaseous oxygen vent arm (with beanie hood), the Orbiter access arm, the ET hydrogen vent umbilical and intertank access arm. Attached to the FSS is the Rotating Service Structure, which rotates 120° on a track to encase the Orbiter so that its Payload Changeout Room fits flush with the cargo bay for vertical payload installation. All manned flights after Apollo 7 have been launched from complex 39, and all except Apollo 10 during the lunar programme from 39A. The first 24 Shuttles also went from 39A; the fateful Challenger 51L mission was the first to use 39B. During the post-Challenger hiatus, 39B underwent extensive weather protection modifications. Others included safety improvements to the crew's emergency escape system. The Orbiter access arm was covered with solid panels for fire protection and a water spray system was incorporated. Two slidewire baskets were added to the existing five, along with brakes; and improvements were made to the emergency shelter bunker near the end of the slidewires.

Modifications were made to improve clearance between the Shuttle as it lifts off and the ET hydrogen vent umbilical as it retracts. Environmental controls were improved in the Payload Changeout Room. Also added were an SRB joint heater umbilical to keep booster field joints at 24°C, a system to recirculate pad water services to prevent them from freezing, and installation of debris traps in the ground interfaces between the Orbiter and LOX/LH_2 servicing systems on the mobile launcher platform. The Apollo-era elevators were also replaced. Pad B was used for the new series of launches while 39A underwent identical modifications. Pad A returned to service in January 1990 for STS-32. Pad B was again removed from service after STS-40 in June 1991 to undergo US$3.3 million of modifications by Lockheed

Space Operations Corportation to prepare it for Endeavour's roll-out in March 1992. Modifications to both Shuttle pads are scheduled periodically.

Orbiter Processing Facility
The original OPF is capable of handling two Orbiters in identical high bays 29 m high, 60 m long, 46 m wide. A third bay came online in September 1991 in the converted Orbiter Modification & Refurbishment Facility, north of the OPF. The bulk of Orbiter processing between missions is performed in the OPF. Equipped with movable work platforms that completely envelop an Orbiter, they are equipped to fuel the Auxiliary Power Units and thrusters with hypergolic propellants, in addition to draining them after missions. Horizontal payloads such as Spacelab are installed in the Orbiter within the OPF (vertical payloads are installed on the pad). Before transfer to the VAB for final assembly with the ET/SRBs, the Orbiter is weighed and its centre of gravity located. Conversion of the OMRF into OPF bay 3 cost US$85 million; Discovery was the first to use it, after STS-48 in 1991. Some of the equipment was transferred from the mothballed Shuttle facilities at Vandenberg AFB, including US$40 million of ground support equipment and US$6 million to US$7 million of unique work platforms.

Thermal Protection System Facility
Two storey 4,088 m² opposite the original OPF for final manufacture and repair of thermal tiles, and gap fillers.

Vehicle Assembly Building and Shuttle landing strip (background). The Launch Control Center is the low building angled at the VAB's base. Note Mobile Launcher Platform 3 at right, awaiting modification from its Apollo configuration (NASA)

Rotation/Processing Facility
One of several facilities devoted to SRB processing. Includes four buildings north of the VAB. New and reloaded SRB segments shipped by manufacturer Thiokol in Utah are received here. Recovered casings, refurbished at Hangar AF, are installed on rail cars for shipping to Utah. Inspection, rotation and aft booster build-up occur in the processing building. Completed aft skirt assemblies, brought from the ARF (see below), are integrated here with aft segments. Two nearby surge buildings store up to eight segments each (two flight sets). Final assembly and stacking into complete SRBs is performed in the VAB.

SRB Assembly & Refurbishment Facility
Managed by NASA Marshall for refurbishment and subassembly of inert SRB hardware, including the forward and aft skirt assemblies, frustrum, thrust vector controls, recovery systems and electronic controls takes place in this seven-building complex. The three-storey Manufacturing Building includes an ordnance area for installing separation devices; 24.4 × 61 m high bay; class 100,000 area for building up aft skirt elements; two 13.6 tonne bridge cranes and three of the world's largest overhead gantry industrial robots for painting and spraying on insulation. A hot fire test stand on the southeast corner of the 178,000 m² site supports testing of the SRB's hydrazine TVC system. Completed aft skirt assemblies are transported to the RPF for integration with the motor segments.

Shuttle Landing Facility
4.8 km northwest of the VAB on a northwest/southeast alignment. Paved runway is 4,572 m long (305 m overrun at each end), 91 m wide (plus 15 m asphalt shoulders) and 41 cm thick at the centre (38 cm at sides). The 61 cm centre to edge slope facilitates drainage. Post-Challenger, a section 1,067 m long at each end was ground down to create a smoother surface without cross grooves to reduce Orbiter tyre wear. The 6.3 mm grooves provide a skid-resistant surface and rapid water runoff. The centre's surface was also smoothed off (but without reducing the grooves) in September 1994. The reduced tyre wear might allow the current crosswind constraint of 15 kt/h (27.8 km/h) to be increased to 20 kt/h (37.1 km/h), cutting the number of launch scrubs. STS-63 in February 1995 was the first to use it. The strip is considered as two runways: Runway 15 with approach from northwest; Runway 33 from southeast. The TACAN system provides range/bearing measurements from 44.2 km altitude; the Microwave Scanning Beam Landing System provides more precise slant range, azimuth and elevation from 5.5 to 6.1 km. A 149 × 168 m aircraft parking apron is at the southeast end.

On the northeast corner of the ramp is the Mate/Demate Device that attaches the Orbiter to or lifts it from the Shuttle Carrier Aircraft during ferry operations. There are also movable platforms for access to some Orbiter components. The device can lift 104.3 tonnes and withstand winds up to 200 km/h. A similar unit at NASA's Dryden Flight Research Facility supports Shuttle landings in California. Five missions ended on the Kennedy runway to April 1985 (STS-51D) but future landings were deferred because of tyre, braking and weather problems; STS-38 became the 6th in November 1990. Upgraded main landing gear, carbon brakes,

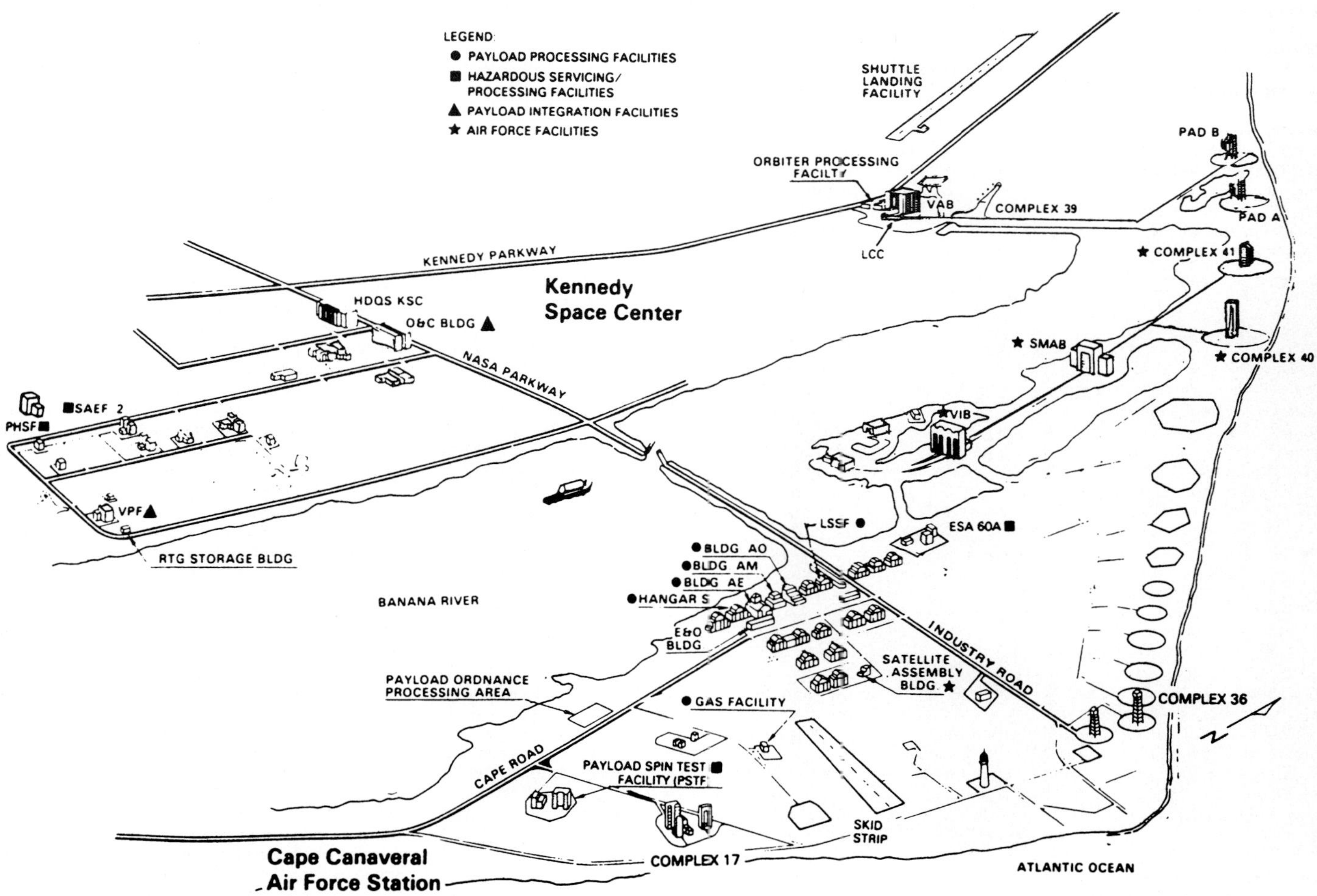

Principal launch and payload handling facilities at KSC and CCAFS (NASA)

additional nosewheel steering and improved tyres (plus later drag 'chutes), meant that KSC in 1991 was again designated a primary landing site, with equal priority to Edwards depending on weather. Through February 1997, 35 missions landed at KSC.

Logistics Facility
21,351 m² main building south of the VAB. Houses some 160,000 Shuttle parts under an automated storage/retrieval system. The separate NASA Shuttle Logistics Depot maintains and repairs 95 per cent of Shuttle's 4,000 Line Replaceable Units and parts.

Operations Support Building
Six-storey building, 27,700 m², completed mid-1990, southwest of VAB, housing technical documentation centre, library and photo analysis area.

Shuttle Launch Processing

A KSC-led team prepares a returning Orbiter for turnaround whether it lands at KSC or EAFB. It is returned from Edwards atop the Shuttle Carrier Aircraft (a modified 747) to the Shuttle Landing Facility, where it is removed using the Mate/De-mate Device. Processing for the next flight then begins in the OPF, including ordnance safing, payload removal, payload bay reconfiguration, removal and installation of the three main engines, horizontal payload insertion and any required maintenance. During this work, the SRBs are stacked in the VAB atop the MLP and the ET attached. The Orbiter is then transferred on a transporter, lifted vertically and mated to the ET. Crawler moves the stack to the pad, where connections and hookups are made and checkout begins. Equipment in a room below the pad hard stand provides the links to the Launch Control Centre. If the payload is not installed in the OPF, it is inserted on the pad.

Processing Control Center
The three-storey 9,200 m² PCC between the OPF and Operations Support Building, began operations in 1992 for Orbiter testing (three control rooms monitor Shuttle processing in the three OPF high bays), launch team training and LPS maintenance.

Shuttle Payload Processing

Shuttle payloads can be processed horizontally or vertically. Facilities are located in the Industrial Area and across the river on Cape Canaveral AS. Horizontal payloads such as Spacelab are received, assembled and tested in the Operations & Checkout Building, the Industrial Area's largest building. They are then transferred to the OPF for installation in the cargo bay; returned payloads follow a reverse procedure. Astronaut crew quarters are also housed in the O&C. Vertically processed payloads are handled on the Cape in:

Hangar L
Live specimens are received here, also a life sciences research facility.

Hangar S
Satellite reception and storage; plus storage of SCAPE Self-Contained Atmospheric Protective Ensemble suits for ground crews.

Buildings AE/AM/AO
In the CCAFS Industrial Area contain high bay clean rooms for processing deployable satellites. The management team in AE's Mission Directors Center oversees launches of NASA, USAF and commercial payloads on Delta and Atlas. AE includes Launch Vehicle Data Center and Telemetry Facility.

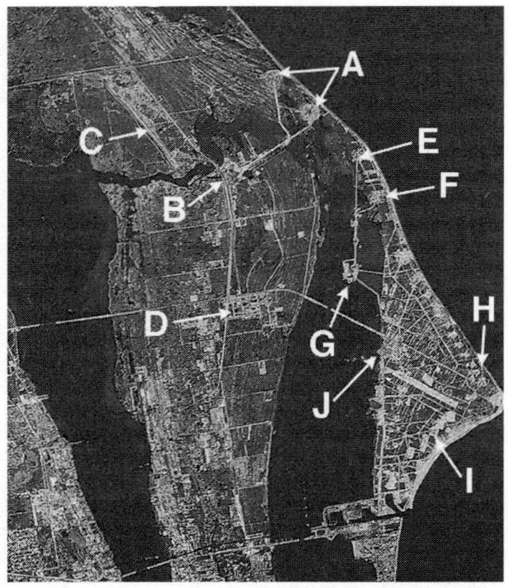

Landsat image of Cape Canaveral. A: Shuttle pads; B: VAB; C: Shuttle runway; D: NASA HQ and industrial area; E: Titan complex 41; F: Titan complex 40; G: Titan Vertical Integration Building; H: Atlas complex 36; I: Delta complex 17; J: CCAS industrial area ***1995***

Vertical Processing Facility
Processing of vertical payloads, including upper stages, takes place in the Vertical Processing Facility south of the O&C in the KSC Industrial Area. Contains Cargo Integration Test Equipment (CITE) for verification of orbiter/payload interfaces. 32 m high bay and airlock and single storey support facilities along the sides of the high bay. The entire facility is a class 100,000 clean room.

Spacecraft Assembly & Encapsulation Facility 2
SAEF 2, also in the southern part of the Industrial Area, is used for assembly, testing, encapsulation and sterilisation of spacecraft, particularly large payloads.

Payload Hazardous Servicing Facility
A nearby class 5,000 facility opened in 1986 for large payload processing, integration of solid motors and fuelling operations. The west end of the 21.3 × 33.6 m 28.7 m high service bay has a 6.1 × 12.2 m sloping area for fuelling. PHSF operations are handled from the separate (for safety) Facility Control Building, with two 112 m² control rooms.

High Energy Radiographic Facility (HERF)
Operations on Cape Canaveral AS began in early 1990 for non-destructive Linatron testing of kick and other large motors, including those on upper stages. Tests are performed in a 16.5 × 19.2 m high bay with wall/door of reinforced concrete 203 cm thick.

The 5,132 acre launch area contains five launchers rated from 1 to 20 tons, two movable launcher enclosures and separate facilities for payload assembly, rocket motor assembly, motor storage, scientific support, telemetry, radar and data communication (Credit Evelyn Trabant/Poker Flats) ***1997***/0008102

The Canister Rotation Facility
With 670 m² high bay was officially activated December 1992 for maintaining Shuttle's two environmentally controlled 19.8 m long payload canisters. Previously, rotation had to be performed in the VAB.

At the Cape, Hangar AF receives recovered SRBs for cleaning and disassembly (some components are reconditioned here) and Building E&O is the headquarters for operations related to ELVs.

Space Station Processing Facility
In order to prepare for the coming Space Station era, NASA has built the Space Station Processing Facility, a three-storey 42,500 m² facility east of the O&C which receives, processes and checks out Space Station flight hardware as it arrives at KSC. In fact, SSPF is a generic facility capable of handling any Shuttle payload. The 5,860 m² class 100,000 clean room includes a high bay and an intermediate bay, accessed by a 465 m² airlock. Eight 13.7 × 27.8 m areas each support a complete Shuttle payload. Full-scale work began in April 1991 and initial occupancy began April 1994. Total cost will be about US$400 million. Personnel total 1,040. The operational readiness date is June 1997, but the first processed payload was STS-74's Mir Docking Module, arriving in June 1995.

Kodiak Launch Complex (KLC)

Background

AADC was established in July 1991 to promote development of the aerospace industry in Alaska. As the spaceport authority for the State of Alaska, it is constructing the commercial US$18 million to US$20 million KLC Kodiak Launch Complex on Narrow Cape, Kodiak Island on 12.55 km² of state-owned land, provided free for 30 years. It will be the only non-federally run commercial launch range in the US.

KLC will target launches of up to 3.6 tonne payloads into LEO ±40° polar orbits, taking advantage of the wide launch azimuth. AADC completed the design and siting of KLC's facilities in early 1995 and broke ground in April 1996, aiming for an initial operating capability some time before the end of the century. It is still looking for strategic partners. The design represents the state-of-the-art in launch facilities: all-weather, flexible and adaptable to all current small vehicles. Facilities will comprise Launch Control & Management Centre; Payload Processing Facility; Integration & Processing Facility; Spacecraft & Assemblies Transfer Facility. One pad and service structure is planned but the site allows for two additional pads.

Poker Flat Research Range (PFRR)

Location: 65° 07'N, 147° 28'W

Background

Poker Flat Research Range (PFRR), primarily a sounding rocket launch facility dedicated to auroral and middle to upper atmospheric research, is 50 km northeast of Fairbanks. Operated by the Geophysical Institute, it is the world's only university-owned launch range. It is also the only high-latitude and auroral zone rocket launch facility on US soil. PFRR's central complex lies on 21 km², on a site which includes telemetry, optical observatory, five pads, blockhouse, rocket storage/assembly buildings, maintenance and communication facilities, a payload assembly building, radar and the range office.

With flight zones covering over 67,000 km² of northern Alaska, it is the largest land-based launch facility in the world and, as such, permits extensive payload recovery programmes along with established down range instrument sites.

Poker Flat is funded through contracts with DoD, NASA, NOAA and the National Science Foundation. NASA currently funds the US$1.1 million annual operation cost. Five employees work year-round directing design and construction of new facilities, co-ordinating science projects, maintaining the physical plant, obtaining the various waivers, approvals and agreements necessary for each launch, and providing launch services. Additional employees are added during major campaigns.

Although NASA launch campaigns only occur every other year on odd-numbered years, an average of two to 10 major sounding rockets are launched annually. Since construction of the facility began in the Summer of 1968, 270 major rocket experiments were conducted through 1997. Nike Tomahawk made the most flights

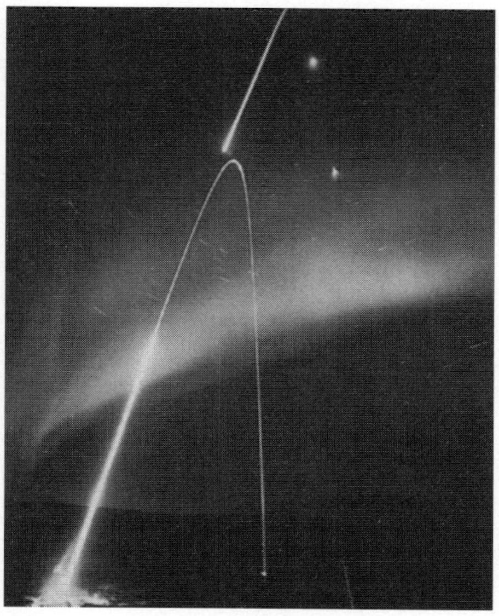

Black Brant 10 launch from Poker Flat March 1984. Two barium releases are visible at 800 and 1,000 km (apogee) (Technology International Corp)

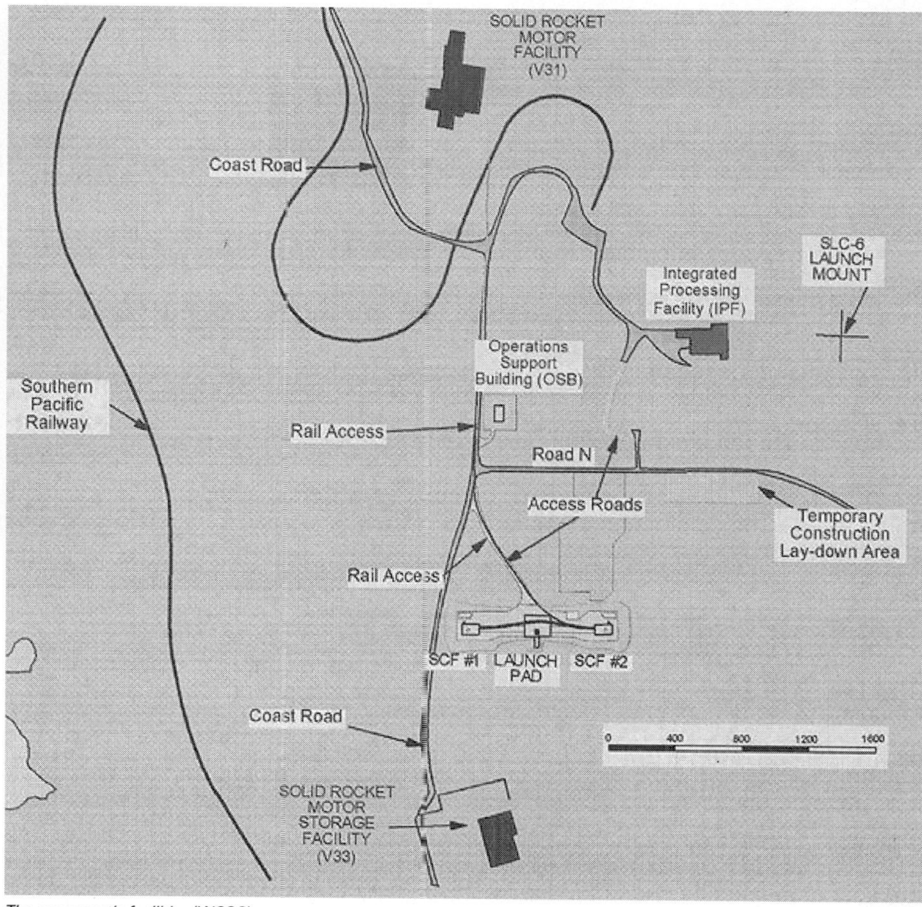

The spaceport's facilities (WCSC) ***1995***

from here, with 64 launches, but the larger and more complex payloads of recent years have led to the use of larger motors and combinations as large as the Aries and Sargent. A Black Brant 12 established a Poker Flat altitude record of 1,422 km on 2 April 1993.

Most missions require simultaneous ground-based data from a worldwide network of stations or are in co-ordination with satellite experiments. The T. Neil Davis Science Operations Center (SOC) houses magnetometers and riometer data displays, all-sky and narrow field auroral TV cameras, meridian scanning photometers and other optical and RF observing instruments for auroral and upper atmospheric research. A large worldwide data display is available at the SOC for launch team use. A major international co-operative effort with the Japanese Communications Research Laboratory is under way. A 256-element imaging riometer is operational and design for large-scale MF radar has begun.

The range operates five launch pads. Pads 1 and 2 each have a 3,400 kg MRL rail launcher. Pad 3 has a 9,000 kg rail launcher. Pad 4 has a 16,000 kg Athena launcher. Pad 5 has a 1,800 kg twin boom launcher and is currently dedicated to the NASA-sponsored Student Rocket Programme, which is affiliated with the Space Grant Programme at the university. A static test stand for cold weather testing of rocket motors is in the design stage in this programme. De-activated Pad 6 was employed for meteorological rockets.

Range telemetry support is provided by two S-band autotrack systems, incorporating 2.44 and 4.9 m dishes, provided by NASA and located in a 15.3 m dome at the top of the hill overlooking the complex. A new telemetry building served one 10 and two 8 m S-band antennas for rocket and satellite telemetry. A NASA C-band radar unit was established at Poker Flat in 1983 and was replaced with a new C-band system in 1995.

Authorisation for overflight, landing and recovery of rockets and payloads on 67,000 km² of land, roughly northeast of the range head to the coast of the Beaufort Sea, is granted by the Bureau of Land Management, the US Fish and Wildlife Service, the State of Alaska Dept of Natural Resources (Div of Lands), the Venetie Tribal government, the Traditional Councils of Arctic Village and Venetie and Doyon, Ltd (the Interior Alaska Regional Native Corporation). The Federal Aviation Administration approves rocket flight zones and co-ordinates air space use during launches and the Alaska Department of Transportation and Public Facilities authorises range personnel to stop road traffic on the Steese Highway during launches. In addition to the normal NOTAMS procedure, each launch is co-ordinated with NORAD.

Improvements to launch facilities and research infrastructure are continuing. Major additions completed in 1996 include satellite and rocket telemetry antennas, a refurbished Athena launcher, two launcher enclosures and road improvements. In a separate initiative, the State of Alaska commissioned a feasibility study through the Alaska Industrial Development and Export Authority (AIDEA) on the market potential for commercial launch activities, and established the Alaska Aerospace Development Corporation (AADC) in July 1991 to encourage commercial space applications. AADC won a US$1.1 million federal grant in 1993 to design an all-weather orbital launch facility to handle payloads of up to 3.6 tonnes. The initial site study of environment, construction and safety led AADC to choose a site at Cape Chiniak on Kodiak Island. Orbital launches from there would be directed over the Pacific into 0 to ±40° polar orbits.

Spaceport West

Location: 34.4°N/120.35°W

Background

SSI is a partnership between ITT Federal Services Corporation and California Commercial Spaceport Incorporated to provide commercial launch services for small to medium polar vehicles from VAFB. Typical users would be LMLV, Taurus and Delta-Lite. SSI is building a new pad and is leasing the US$250 million mothballed Payload Preparation Room at SLC-6, renovated as the IPF Integrated Processing Facility (GEMstar 1 and STEP M0 has been processed by end-1995).

Annual capacity will be 24 launches, using the ITL integrate-transfer-launch approach, allowing the use of simple stationary pad structures. The 25-year lease was signed by WCSC/AF for the undeveloped 0.10 km² site near SLC6; the annual fee is US$70,000 plus about US$10,000 monthly for operations costs and utilities. The government is providing US$10 million, mostly as AF grant money, and ITT is committing US$30 million. WCSC was formed as a non-profit corporation in May 1992 to pursue commercial opportunities at VAFB. It is designated by the State Legislature as the California Spaceport Authority. Subsidiary CCSI was established in August 1993 as a for-profit company to meet grant requirements for matching funds from space companies.

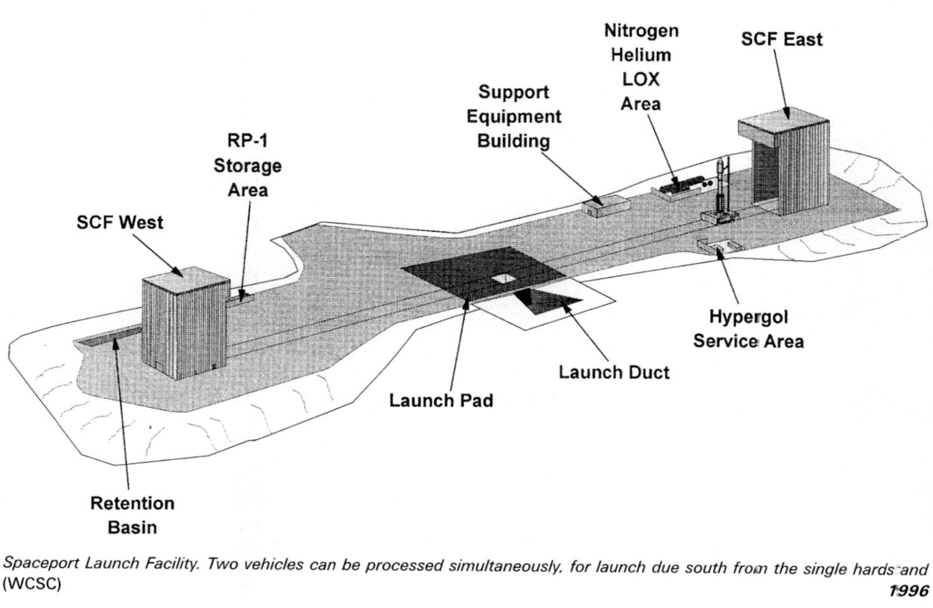

Spaceport Launch Facility. Two vehicles can be processed simultaneously, for launch due south from the single hardstand (WCSC) ***1996***

The spaceport's Integrated Processing Facility was SLC-6's payload preparation facility. The Shuttle pad is off to the right and the spaceport's to the left (WCSC) ***1995***

The SLF Spaceport Launch Facility operates a single pad at 34°34′35″ north/120°37′53″ west (92 m altitude) between two SCF Stack & Checkout Facilities on an east-west axis. Most vehicles will depart due south (180° azimuth), so will not overfly the SCFs. 168 to 220° azimuths are available without impacting the local community or off-shore oil rigs. 150° is possible.

A vehicle is stacked on an MLP Mobile Launch Platform and transferred launch-ready to the pad on launch day. SCF-E is intended for Delta 2 and Delta-Lite (but can handle Castor 120). SCF-W is optimised for Castor 120-based vehicles. IPF handles booster and payload processing, fairing/storage and payload encapsulation, and houses three launch control rooms in the 12.2 × 42.7 m Launch Control Center (LCC). The processing areas include a 9 × 30 m airlock, three 10.7 × 13.4 m cells and 9 × 45 m high bay, with 47 m high 68 tonne crane.

Vandenberg

Location: 34.4°N/120.35°W

Background

Vandenberg is responsible for missile and space launches on the west coast and operates the Western Range tracking network extending into the Indian Ocean, where it meets the ER system. Vandenberg provides the US with access to polar orbits using due south launches and was to have provided a base for Shuttle departures on high-inclination missions. 1995's four orbital missions brought the total to 512. Delta missions are supported by the Space Launch Complex (SLC)-2W pad, Titan by SLC-4 and Atlas by SLC-3. Scout retired with its last launch May 1994 from SLC5. The Taurus maiden flight, March 1994, used pad 576E for its transportable launch equipment.

The Western Space & Missile Center in 1991 became the 30th Space Wing. Now elements of air force Space Command, the east and west launch ranges and bases were managed by AF Systems Commands' Space Systems Division. The 30th Space Wing is made up of four groups: 30th Operations Group, 30th Logistics Group, 30th Medical Group and 30th Support Group. 30OG operates the Western Range. Its 2nd Space Launch Squadron is responsible for small/medium launch operations (including Atlas II, Delta 2, Pegasus and Taurus 2) and its 4th SLS for heavy lift vehicles (Titan 2 and Titan 4). The 576th Flight Test Squadron handles ballistic operations.

Vandenberg is also an important base for development and operational ICBM flight testing. Above-ground pads and underground silos support Peacekeeper and Minuteman launches. General Dynamics signed agreements with NASA/USAF by October 1987 allowing up to eight Atlas launches annually from Cape Canaveral, on a cost reimbursable basis. GD has refurbished surplus Atlas ICBMs at Vandenberg and launched them as space vehicles since 1970; the last departed March 1995.

The 399 km^2 base was created along 40 km of coastal scrub, similar to the Canaveral area. The 1,100 buildings house 53 host organisations, 37 associate units and 49 contractor organisations, supporting more than 13,000 military, civil and contractor personnel and service families. From here are launched classified satellites and missiles are test-fired under operational conditions over the Western Range, extending 8,000 km across the Pacific and into the Indian Ocean.

From its southern tip, Point Arguello, spacecraft are launched into polar orbits for both NASA and DoD.

The major facilities are:

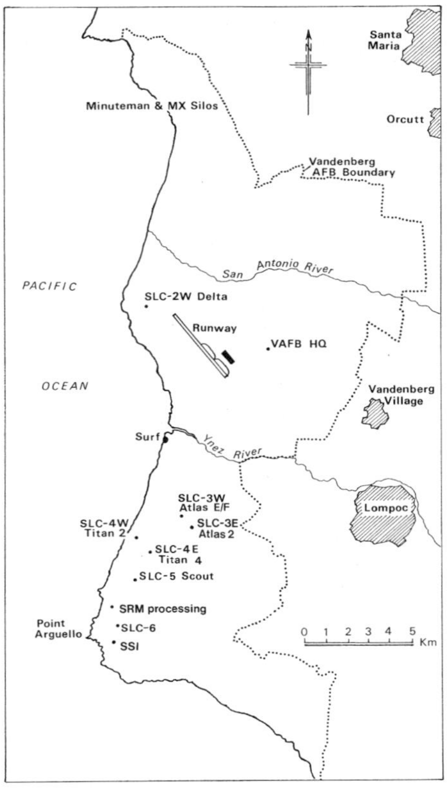

Vandenberg provides access to polar orbits. The northern area is generally employed for missile development and operational test launches; most are de-activated but Minuteman and Peacekeeper ICBM silos and pads remain in use. Apart from the SLC-2W Delta pad, the orbital pads are clustered in the Point Arguello area: SLC-3E/3W Atlas, SLC-4W Titan 2 (previously Titan 3B), SLC-4E Titan 4 NUS (previously for Titan 34D) and SLC-5 Scout (retired 1994). SLC-6 was converted for Shuttle. SLC-7, south of SLC-6, was to have been a second Shuttle pad and was then proposed for Titan 4. Spaceport Systems International's pad is being built at 34.576° north/120.631° west south of SLC-6 for commercial operations starting in 1997 ***1996***

SLC-2

Built in 1958 as part of a group of seven launch installations for the Thor IRBM. Pad 2W completed upgrading in 1995 to accommodate Delta 2 (Radarsat was the first launch). The mobile service tower was raised hydraulically to insert a 3.7 m spacer to handle the longer Delta. Before that, NASA's COBE in November 1989 was its first use since 1984. The second Delta pad, 2E, is no longer operational. 2W can support Delta operations up to the 7920-10 model; it can cope with six launches annually. The 54.3 m gantry supports nine working levels, including a white room enclosing Levels 4-6. Payloads are delivered to the Spacecraft Laboratory in Vandenberg's Building 836, housing assembly areas, laboratory areas, clean rooms, an RF screen room, computer facility and the T/M Doppler station. The 404 m^2 Spacecraft Laboratory 1 includes a 6.1 × 7.5 m curtain clean room providing class 100,000 conditions. A 4.5 tonne bridge crane is used. The 521 m^2 Laboratory 2 includes a 9.2 × 5.1 × 5.1 m class 100,000 clean room and a smaller clean room in the high bay area. Laboratory 3, 284 m^2, accommodates a curtain clean room similar to that in Laboratory 1. Following processing, payloads are transported to the Delta spin test facility some 3 km from SLC-2. The 10.7 × 18.6 × 13.7 m building houses a Trebel dynamic balancing machine and necessary equipment for build-up, alignment and balancing of the upper stage solid motors and spacecraft. Payload propellants are loaded in this facility. Launch operations are conducted from 2W's nearby blockhouse and the Mission Director's Center in Building 840.

SLC-3

SLC-3E became the nation's newest launch complex when it attained initial launch capability in October 1996 following a three year, US$300 million modification project to launch the Atlas II family of launch vehicles. To reach this milestone, Air Force Material Command and Lockheed Martin Astronautics designed and built a state-of-the-art mobile service tower and umbilical tower while transferring countdown operations and blockhouse functions to building 8510 on north VAFB. SLC3E will provide customers with access to sub-synchronous and polar orbits. The 2nd Space Launch Squadron operates SLC-3E. The last launch from SLC-3W was in March 1995. Currently kept in caretaker status by the 2nd Space Launch Squadron. SLC-3W's future missions are yet to be determined.

SLC-4

The two Titan pads, SLC-4E/4W, suffered US$50 to 60 million of damage when a Titan 34D exploded 8 seconds after lift-off on 18 April 1986. Debris damaged ground support equipment and ignited brush fires up to 1 km away. Following this failure, SLC-4's blockhouse no longer conducts the entire launch: it now takes the countdown from 54 hours to within 8.5 hours, when it is vacated. A remote facility in north Vandenberg takes over. Titan 2 launches continue to be fully handled from the blockhouse's other firing room. The first Titan 2 ICBM refurbished as a space launcher was mounted on the repaired SLC-4W in November 1987 and launched the following Sep. 4W, previously devoted to Titan 3B, is now used exclusively by Titan 2. 4E was refurbished from its Titan 34D configuration at a cost of US$235 million and declared operational for Titan 4/NUS in May 1990; it can handle two missions annually. The first Titan 4, on the pad since June 1990, departed 8 March 1991. Martin Marietta was awarded a US$10 million contract in 1990 to install a sound suppression system on 4E as protection against the solid motors' ignitions. The system dumps water into the exhausts for 2 seconds before and 7 seconds after ignition to reduce the pressure wave.

SLC-6

Lockheed Martin's LMLV 1 vehicle was the first launch from the pad, in August 1995. It sits on an adaptor on one of the pad's Shuttle SRB mounts, allowing it to use an existing flame duct. Lockheed Space Operations Corporation was awarded the contract in July 1990 to modify the existing SLC-6 as the TCLC Titan 4 Centaur Launch complex, aiming for a first launch in 1996. However, reduction in projected Titan launches by early 1991 again brought into question the need for converting SLC-6 at all. Like SLC-4E, it would also have been able to handle the Titan 4 NUS version. Projected cost was US$314 million for the integrated collection of structures, systems and equipment required for direct ground processing and launch capability. Design work began 1991, with site modifications planned for 1992-94 and testing for 1995. The conversion was cancelled 1991.

SLC-6 was created in the 1960s for the Manned Orbiting Laboratory. When MOL was cancelled in 1969, the mothballed facility consisted of the pad, mobile service tower, flame duct and launch control centre. In 1975, on the basis that using the existing facilities could save US$100 million, it was decided to make it the West

SLC-4 and SLC-3 complexes. From left: SLC-4W (Titan 2), SLC-4E (Titan 4), SLC-3E (being refurbished for Atlas 2 family from 1998) and SLC-3W (Atlas E/F) (USAF)

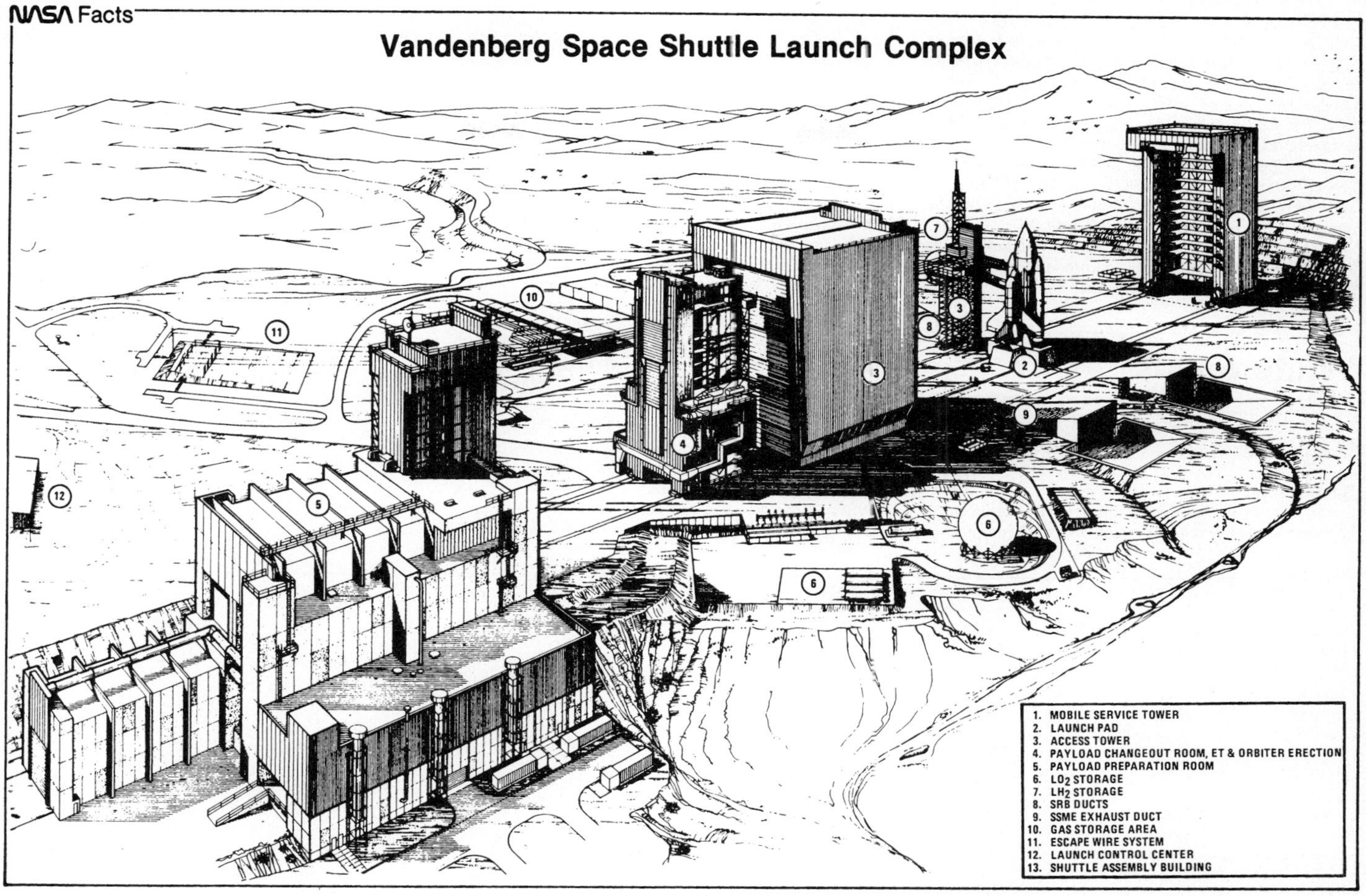

SLC-6 was created in the 1960s for the Manned Orbiting Laboratory, before it was converted for Shuttle operations (shown here). Lockheed Martin's small LMLV 1 uses one of the SRB mounts. The Payload Preparation Room (5) is being leased to Spaceport Systems International (qv) as a processing facility and launch control centre

Coast Shuttle launch site, and construction began in 1979. Up to 20 flights annually were projected at one stage. US$3.2 billion was expended on expanding the runway to 4,500 m, on building the Orbiter maintenance/checkout facility, and building accommodation for support personnel at north Vandenberg. The funds included modifications to SLC-6 proper, which took in the launch control centre (only 347 m from the pad), payload preparation/changeout facilities, Shuttle Assembly Building (SAB), access tower, launch mount, mobile service tower and three exhaust ducts at south Vandenberg.

Construction delays were compounded when it was found that high winds and low temperatures would not permit the planned assembly of SRB/ET/Orbiter on an unprotected pad. At the end of 1981, a 71 m high movable assembly building to protect the Shuttle cluster and technicians during launch preparations was approved. Lengthening of the existing runway, 25.7 km north of the pad, was completed in 1983. The Orbiter was carried from there to the pad on a 76-wheel transporter, a journey requiring 3 to 4 hours. Shuttle processing contrasted with Kennedy's: instead of the cluster being moved to the pad after assembly, stacking took place on the pad under the protection of three mobile buildings. The first step was for the service tower, using a 200 tonne crane, to mount the SRBs; two cranes would then position the ET between. Then the 4,000 tonne SAB would move in from the opposite direction to mate with the service tower and enclose the assembly area.

With a 125 tonne crane in the roof of the SAB, the Orbiter was to be lifted into position and attached to the SRBs. The third mobile building, an 18 storey Payload Changeout Room, would then move through the SAB and transfer satellites and other payloads to the Orbiter. These payloads would come from another, stationary building, the Payload Preparation Room (this building is being leased to Spaceport Systems International to support its commercial launch services from a new pad south of SLC-6). After fuelling, the three buildings would move back on their tracks for the launch. To protect the pad, 1 million gallons of cooling water would flood into the three exhaust ducts in 23 seconds. At lift-off, control was to pass from Vandenberg to Houston for NASA flights or to the new Support Operations Center at Cape Canaveral AS for DoD missions. The complex was placed in an operational caretaker status in 1986-87 to save US$1 billion following the Challenger accident. The mothballing process was completed by the end of 1989. By end-1986, 1,700 Shuttle contractor employees had been laid off. In December 1986, a General Accounting Office report to Congress estimated a cost of US$200 million in FY89 to maintain SLC-6 in operational status, capable of full reactivation in 18 months. Congress rejected the USAF claim for a new SLC-7 for Titan 4 Centaur and directed instead that SLC-6 should be modified; cost was estimated in 1990 at US$700 million. Before Challenger, subject only to a successful Flight Readiness Firing with Discovery in January 1986, Vandenberg was to provide the Shuttle with a polar orbit launch and landing site for use by NASA and USAF from July 1986. That first DoD flight was to carry the US$500 million Teal Ruby satellite.

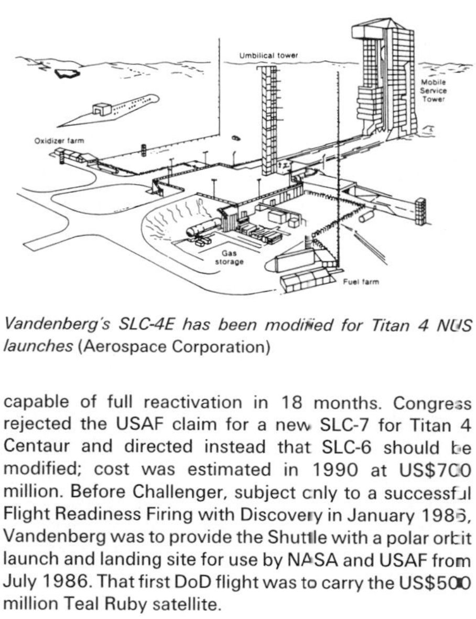

Vandenberg's SLC-4E has been modified for Titan 4 NUS launches (Aerospace Corporation)

Wallops Flight Facility

Location: 37.85°N, 75.46° W

Current status

NASA maintained the small Wallops centre in for Scout vehicles, although only one was launched from 1979 until Scout's 1994 retirement and it is now primarily a sounding rocket base. The maiden Conestoga failed from there in October 1995.

Background

Now administered by NASA Goddard, Wallops Flight Facility became the fifth space port and the third US orbital site when NASA launched Explorer 9 from this island on Virginia's east coast on16 February 1961. In all, 19 vehicles entered space from this point through the end of 1985, when NASA and the air force largely suspended orbital operations. Although NASA actively maintains the service buildings for sounding rocket tests and other classified experiments, Wallops' space launch history ended with the retirement of the Scout vehicle in 1994. A commercial consortium, EER Systems tried to revive the site with a launch of their Conestoga in October 1995, but the vehicle failed and the programme's future remains uncertain. As a flight control centre, Wallops provided orbital support and control for the B-52/Pegasus launch of Brazil's SCD February 1993. The B-52 took off from Florida.

Wallops served NASA and its predecessor NACA (the National Advisory Committee on Aeronautics) for many years before emerging as a spaceport. NACA launched its first sounding rocket, a Tiamat on 4 July 1945, inaugurating the site as a missile research facility. Since 1945, NACA, NASA, the air force, the navy, SDIO and many foreign governments have launched more than 800 sounding rockets from a pad on the far end of the island. Even today more than 30 launches a year occur at Wallops using vehicles as diverse as Super Areas, Black Brant, Taurus-Tomahawk, Taurus-Orion and Terrier-Malemute.

NASA and the air force built extensive facilities on the island over the years and the active maintenance of these facilities reflects a commitment on the part of the US government to maintaining this infrastructure. The buildings, which are still operational, include:

Launch Areas

Launch Area 1 for heavy-lift sub-orbital rockets. Launch Area 2 and associated blockhouse for most sounding rocket launches. Wallops used this pad for the maiden 1945 sounding rocket firing. Launch Area 3 for the Scout orbital launch vehicle. Launch Area 4 for classified payloads and special projects. Launch Area 5 most recently used for US Navy Vandal missile tests. Launch Area Oar for EER System, which built and owns the pad.

Block Houses

Blockhouse 2 to service Launch Area 2. Blockhouse 3, serving all pads.

Workshops and preparation facilities

Scout Assembly Shop which NASA and the air force used to prepare all Scouts in the Scout Assembly Shop. When the Scout was ready the ground crew rolled back the shelter building on rails and raised the launch rail into the firing position.

Assembly Shop 1 for sounding rocket assembly and checkout.

Spin Balance Facility for vertical/horizontal spin testing of payloads and motors up to 13.6 tonnes. A Trebel FVD-3000 machine spins the test object to test and modify its dynamic balance. Typically 32 motors, six satellites and three sounding rocket payloads pass through the test procedure annually, including 80 per cent of AKM and PKM motors for GEO satellites.

Wallops Orbital Tracking Station established by NASA in 1986 to provide tracking support for projects such as IUE, IMP 8, Nimbus 7 and COBE. NASA performs high-speed data transfer to Goddard at Greenbelt through a satellite link located at Wallops. To assist NASA, WFF has installed telemetry receiving and command antennas ranging from 2.4 to 8 m. Wallops' tracking radars include 8.8 m parabolic FPS 6 radar that can track objects out to 60,000 km with a precision of ±3 m and with a line of sight velocity resolution of 9 cm/s. The antenna radiates a peak output 2.5 MW.

NOAA's Command/Data Acquisition Station of nine 7.3 to 26 m antennas to track, command and take data from the NOAA and GOES satellites.

White Sands

Location: 32.23°N/106.28°W

Background

The 8,100 km^2 White Sands is operated by the US Army and was the site of the first major US rocket firings after the Second World War before the move was made to Florida's larger range. It is also the site of the first atomic explosion; the Trinity Site is on the northern part of the range. White Sands is still a major sounding rocket firing base, however, and supports BMDO flight-testing.

Some 26 km northeast of Las Cruces, White Sands Missile Range has two separate propulsion test sites, designated areas 300 and 400, established for trials of Apollo CSM/LM main engines/thrusters. Area 300 began testing the CSM's main engine in September 1964, while area 400 accommodated hundreds of LM ascent/descent engine firings. There are seven test stands, four altitude chambers and three open. The largest vacuum stand is 302, 10 m diameter 17.7 m high; it was employed for Viking's lander thrusters. Stand 405 is an altitude facility for spinning solid motors up to 125 rpm, on cylinders as large as122 cm diameter. It produces 111.2 kN thrust. Stand 401 accommodates cryogenic testing; the others utilise hypergolic propellants. Space Station propulsion module test article firings began November 1992 (The Space Station's redesign no longer requires this element). Materials testing expanded after 1967's Apollo 1 fire, emphasising flammability work.

WSTF also provides precision cleaning facilities, including a class 100 clean room for spacecraft parts. For example, it ensured that Viking's soil sampler and the tools used in the Apollo Lunar Receiving Laboratory at JSC were free of terrestrial contaminants.

White Sands Space Harbor provides the third Shuttle landing site in the US after Edwards and Florida; the third mission landed on the Northrup Strip, consisting of two 11 km long gypsum-sand runways 1,200 m above sea level, in March 1982 after the primary Edwards site became waterlogged.

TRAINING FACILITIES

EUROPEAN SPACE AGENCY

European Astronauts Centre (EAC)

Background
EAC was founded in 1990, to provide expertise for ESA's manned space activities and act as the astronauts' home base. It is responsible for all astronaut-related matters, such as selection, medical surveillance, support and training. Since its foundation, ESA astronauts have flown on STS-42/Soyuz-TM 20 (Ulf Merbold), STS- 46/ 61/75 (Claude Nicollier), STS66 (Jean-Francois Clervoy), Soyuz-TM 22 (Thomas Reiter) and STS-75 (Maurizio Cheli). Following 1992's selection of six new astronaut candidates, EAC organised their basic training leading to their qualification as European astronauts in December 1993. EAC provided payload training to the astronauts and cosmonauts involved in the EUROMIR 94 and 95 missions. Six ESA astronauts are currently in active status and training with ESA's partners under EAC co-ordination, preparing for the International Space Station.

RUSSIAN FEDERATION

Zvezdny Gorodok (Star City)

Background
The Gagarin Cosmonaut Training Centre (TsPK) was established at Star City 30 km northeast of Moscow in 1960 and trained some 300 in its first 35 years, including back-ups and unflown programmes. RKA/VKS jointly fund and manage TsPK now, assuming control from the air force. The Director and most section heads are veteran cosmonauts. Previous Director Gen Vladimir Shatalov retired in 1991. He succeeded Gen Georgi Beregovoi. Alexei Leonov was head of training until he, too, stepped down in 1991. Star City has a population of 5,000. Only 10 to 20 experienced cosmonauts are active at any one time, plus an equal number of rookies (a total of 'about 30' was claimed in September 1995). Of these, six two-man crews (usually comprising one experienced man and one rookie) train for missions. Three organisations are responsible for crew selections: the air force, RKK Energia and the Ministry of Health. The Buran pilots were from VKS. Rookies normally fly five to 10 years after the start of one and a half to two years of basic training. Every cosmonaut has to undergo a week's medical examination every year to retain flight status. They are kept in a pool until assignment to a mission. Added to these pairings are specialists as necessary.

Facilities include classrooms, laboratories, sports complex, centrifuges and thermal chambers. A high fidelity mockup of Mir and its modules provide equipment and procedures training. Two Soyuz-TM and one Soyuz-T simulators are housed in a separate room. One TM offers full simulation, while the other is a visual docking simulator (another is reportedly in the Mir hall). A 23 m diameter 12 m deep water tank containing spacecraft mockups allows EVA training. Shuttle training facilities are housed in a separate building; a Shuttle simulator is based at the TsAGI Central Aero-Hydrodynamics Institute. The Swedish-built 18 m 3-DOF centrifuge became operational in 1979, providing a maximum 30 *g* and 300 tonne rotating mass. The cabin can simulate different altitudes, temperature and air mixtures. Training requires a cosmonaut holding a dead-man's handle to undergo an initial 3 *g* for 14 seconds, rest for 10 minutes, 5 *g* for 30 seconds, break for one day, 4 *g* for 40 seconds, rest for 10 minutes, then conclude with 6 *g* for 30 seconds. There is also a 7 m centrifuge. A modified Il-76 aircraft is available for parabolic flight training, and is equipped with an airlock to provide EVA rehearsals. It is also available for commercial μgravity flights in a joint venture with Germany's Intospace and other western companies. First demonstration flight was 25 November 1992.

UNITED STATES OF AMERICA

White Sands Space Harbor

Background
Part of WSTF, WS Space Harbor is one of the primary training areas for Shuttle pilots. NASA selected the site in early 1976 within the US Army's White Sands Missile Range and the original 3 km Northrup gypsum lakebed strip was lengthened to 4.6 km in time for the first Shuttle Training Aircraft flight in August 1976. Altitude above sea level is about 1,200 m. About 10 STA training sorties are made weekly; the aircraft are housed in the NASA hangar at El Paso International Airport, Texas. A second runway was added in 1979 and both lengthened to 10,680 m, with landing aids, to allow WSSH to serve as an alternate Shuttle landing site. Columbia landed there in March 1982 after rain prevented use of the primary Edwards AFB facility. A full set of convoy equipment for safing the Shuttle is kept at WSSH. The runways were laser-levelled and widened to 275 m during the post- Challenger standdown. A duplicate of the trans-Atlantic abort runway at Ben Guerir (Morocco) was added in 1989 for TAL practice.

TELEMETRY, TRACKING AND COMMAND

AUSTRALIA

Woomera Tracking and Telemetry Station

Background
The United States Air Force's 21st Space Wing's 5th Space Warning Squadron works with its Australian counterpart at the Woomera Air Station in Australia. The Australian 1st Joint Communications Squadron at the Joint Defence Facility, Nurrangar handles DSP downlinked information for the eastern hemisphere.

CHINA, PEOPLE'S REPUBLIC

XSSC-Xian Satellite Control Centre

Background
China's launch and satellite TT&C network is centred on the XSSC Xian Satellite Control Centre in Shaanxi province, controlling six fixed, three mobile and three ship (Yuanwang 1-3) stations. Yuanwang 3 was delivered in March 1995. The fixed sites are Weinan (near Xian, and the lead site), Min-Xi (Fujian province), Xiamen, Changchun (Jilin), Karshi (Xinjiang) and Nanning (Guanxi), with the first three providing GEO control. XSSC issues de-orbit commands to FSW recoverable satellites through the two mobile stations, while the third locates the re-entry modules during descent. XSSC's hardware covers three NCI 2780, two VAX 8700 and several VAX-II computers.

GERMANY

German Space Operations Centre (GSOC)

Background
Operated by DLR, GSOC is responsible for all operational tasks of the national space programme and provided the first manned mission control outside of the US/USSR. It first controlled the German funded Spacelab D1 (1985) and then Mir-92, D2 (1993) and ESA's EuroMir-95 (1995-96). It may operate as the Columbus Orbital Facility Control Centre. GSOC also handles unmanned spacecraft, specifically LEOP. TV-Sat 1's rescue attempt was mounted from here and initial TT&C was provided for the operational TV-Sat and DFS satellites. All Eutelsat 2s were positioned by GSOC, which also supported Giotto through its Weilheim station. Others include AMPTE and Rosat. The centre also supports mobile rocket launches for high altitude investigations.

INDIA

ISTRAC-ISRO Telemetry Tracking and Command network

Background
Istrac operates a network of TT&C stations for ISRO's launch vehicle and satellite activities. The facility employs 465. Stations are sited at Trivandrum (8 m), Sriharikota (10 m), Carnicobar (8 m, principally for launcher downrange TT&C), Lucknow (10 m) and Bangalore (10 m), which also accommodates the network's Spacecraft Control Facility. Its two 10 m Cassegrain dishes (20.5 dB/K, 2.200– 2.300/2.025-2.120 GHz down/up link, 9°/s slew rate, 0.8° beamwidth).

ISTRAC has two VAX 11/985 computers to currently support the IRS remote sensing satellite (imagery is routed via the National Remote Sensing Agency's Hyderabad station) with a back-up VHF command system. Insat is controlled from its own dedicated station at Hassan.

Three satellites can be controlled simultaneously, one undergoing orbital checkout and two operational. SSC has a main control room, mission analysis room, two dedicated control rooms and a data control room, the first two operating until orbital checkout is completed and the satellite is handed over to a dedicated room. The data control room acts as a node for all the network stations and SCC.

The network also operates an S-band receive-only TT&C station near Port Louis, Mauritius. Optical/laser tracking is performed at the Satellite Tracking & Ranging Station (STARS) established at Kavalur in co-operation with Russia.

INTERNATIONAL

European Space Operations Centre (ESOC)

Background
ESOC, with a permanent staff of 272, and a total staff of about 700, provides control of ESA spacecraft during launch and orbital operations through an operations control centre comprising a main control room (for LEOP), dedicated control rooms and computer facilities. The centre became operational in May 1968 for the control of ESRO 2B.

ESOC controls the Estrack network of seven ground stations: Perth (Australia), Kiruna (Sweden), Malindi (Kenya), Kourou (French Guiana), Redu (Belgium), Villafranca (Spain), Odenwald (Germany). Satellites are assigned to individual stations for operational support. Redu has full autonomy for ECS communications satellite control on behalf of Eutelsat and Villafranca for the ISO mission. The Ground Systems Engineering Department (GSED) is responsible for all engineering functions related to the implementation of the ground systems for operations; all operational launches related to the support of missions entrusted by ESO, comprising spacecraft and ground operations are consolidated in the Mission Operations Department (MOD).

The European Space Operations Centre of the European Space Agency was created in 1967 in order to operate the satellites launched by the Agency. Its expertise encompasses operations for scientific, remote sensing, telecommunications and microgravity satellites. ESOC also serves as the operational computer centre of ESA. Nearly 700 staff including 272 permanent employees, are in charge of operations, software development, ground stations and mission preparation (ESA)
1998/0019685

Bangalore's Istrac complex. Centre: main control centre; right: TT&C station. The twin 10 m dishes are controlling the IRS remote sensing satellites (ISRO)

ESOC is also at the forefront of research into orbital debris, maintaining the DISCOS (Database and Information System Characterising Objects in Space). Created in 1990, DISCOS holds more than 2.1 million elements and 25,000 more are added monthly.

Facilities include:

Redu

The site (50.0° north/5.1° east) of some 19 ha, in a valley outside the town of Redu, houses the station control room, as the network interface with ESOC's OCC; the main equipment room housing the telemetry, telecommand and ranging baseband racks, and the station computers. The ECS control room and computer room; and the test and monitor room. The SHF facilities include four fixed in-orbit test antennas (TMS3A, 3B, 3C, 7) for measuring telecom satellites' performances before handing them over for routine operations, four fixed telemetry (TM1, 2, 3, 6), one ranging (RG1), one satellite multi- services terminal and a radiometer. A new S-band 15 m antenna became operational in mid-1995 to support two Cluster satellites. Staffing is about 50, mostly from the Belgian company CISET, responsible to ESA for maintenance/operations.

Villafranca

ESA's Villafranca del Castillo satellite tracking station (40.4° north/3.9° west), Vilspa, occupies a 10,000 m^2 site some 30 km northwest of Madrid, and was inaugurated 12 May 1978 for IUE. The principal building comprises the main equipment room (telemetry, command/ranging baseband racks, the station computers, timing system and network communications interface with ESOC's OCC), the computer room (IUE real-time operations and image processing), IUE control room and observatory, the payload test lab (communications testing), and the Villafranca computing centre (computers and terminals for IUE image de-archiving, data processing/editing). A new operations building accommodates ISO's control centre and science centre; the infra-red observatory is operated from Villafranca. The antenna farm covers: two 15 m and one transportable 5.5 m S-band antennas, one 12 m C-band, one 4 m L-band, one 3 m Ku-band, one VHF uplink and nine crossed yagi arrays. The majority of the more than 70 staff are employees of the Spanish INTA. Vilspa has been providing 16 hours of real-time support to IUE daily since 1 October 1995. For the 18 years before that it was 8 hours by Vilspa and 16 hours by NASA Goddard. The site also provides continuous TT&C services on behalf of ESOC for the Marecs B2 satellite over the Atlantic.

Kourou

The Diane S-band station (5.2° north/52.8° west), 27 km from Kourou and also known as Kourou 93, provides telemetry, tracking/command during LEOP of ESA missions. Odenwald uses a transponder for four-way Meteosat ranging measurements. VITROCISET is responsible for maintenance/operations.

Malindi

The station (2.9° south/40.1° east) is at Italy's San Marco Scout base camp in Formosa Bay. The S-band front end belongs to the University of Rome, while ESA provides the baseband equipment.

Perth

The station (31.8° south/115.8° east) is on the Perth International Telecommunications Centre complex of Telstra Corp, which provides management. The S/X-band antenna supports LEOP of ESA missions and, until mission end in 1993, routine Hipparcos operations.

Kiruna

The 15 m S/X- band antenna (67.9° north/21.0° east) at Salmijärvi primarily supports ERS 1/2 TT&C and data reception under Swedish Space Corp management. The station is remotely controlled from ESOC's ERS Mission Management and Control Centre.

Odenwald

(49.7° north/8.9° east). The main function is the support of two Cluster satellites (the other two to be handled at Redu). A VHF command facility provides Redu's ECS back-up. Thorn Security provides maintenance/ operations support.

JAPAN

Tsukuba Space Centre

Background

Located in Tsukuba Science City, 60 km northeast of Tokyo, TKSC is NASDA's largest technical facility, with seven sections responsible for launcher and satellite R&D, design and testing. The Tsukuba Tracking and Control Centre provides Japan's satellite command node.

The Tracking and Control Centre is the command hub for all Japanese-launched satellites, including those of ISAS and commercial organisations. The Okinawa Tracking and Data Acquisition Station (26°30′ north; 127°54′ east) provide support to the Masuda TDA Station (30°33′ north; 131°01′ east) at Tanegashima, and the Katsuura TDA (35°13′ north; 140°18′ east) south of Tsukuba. NASA, CNES and ESA support is provided as required. The Ogasawara (27°04′ north; 142°13′ east) and Christmas Island (2°03′ north; 157°27′ west) stations provide downrange launch tracking and command relay for Tanegashima's Takesaki Range Control Centre, with Christmas Island employed principally in GEO launches.

Japan's JEM Space Station contribution will be controlled and supported from the SSIP Space Station Integration & Promotion facility at Tsukuba, completed in 1997 . It provides the SEL, Space Experiment Lab, user experiment preparation/testing building, the SST Space Station Test building for JEM testing, training, payload integration and operations simulation, the ATF Astronaut Training Facility for crew operations, WET Weightless Environment Test building (16 m diameter 11 m deep water tank) for procedure verification and crew training. It acts as the SSOF Space Station Operation Facility for JEM operation.

NORWAY

Tromsø Satellite Station (TSS)

Background

TSS, established in 1967, is Norway's national receiving station for data from polar satellites, emphasising near-realtime data processing and distribution. Its northerly position provides coverage for 10 out of 14 revs for polar satellites. A fourth antenna, for SAR reception, was added in 1994 to the 10 m L/S/X– band, 2.4 m and 3 m L-band dishes. TSS became a shareholding company in 1995, equally owned by NSC and the Swedish Space Corporation. 1995 turnover was NKr21 million (1994 18.5; 1993 16.5; 1992 15.2). The number of staff is only 26.

The station's previous main activity was reception and archiving of NOAA data as part of ESA's Earthnet but it completed upgrading in 1991 as the national facility for near-realtime processing and distribution of ERS SAR data. The CESAR fast processor generates a SAR full resolution, a SAR low resolution and a SAR low-resolution calibration image in less than 8 minutes, covering 100 × 100 km. TSS offers the same products from Japan's JERS 1; the 75 × 75 km images are available some 20 minutes after data acquisition. Radarsat was added in 1996. Since summer 1994, TSS has been operating a pre-operational pilot service for near-realtime detection of oil spills using ERS SAR data. NOAA data have been received since 1973; quick look, raw data sets, geo-corrected data sets, SST and weather images are available for near-realtime delivery via Internet or ISDN.

TSS operates a Cospas/Sarsat Local User Terminal in direct contact with the Main Rescue Co-ordination Centre in Bodø. A new LUT was installed in 1993. TSS also provides telemetry support for Andøya launches.

RUSSIAN FEDERATION

Support ships

Background

As its manned spacecraft are within direct visibility of its territory for only 9 hours daily, the RFAS/USSR traditionally compensated for its lack of land bases with a flotilla of support ships and, increasingly, relay satellites. However, it was noted in early 1992 that Mir mission control was operating without satellite relays because of the cost, restricting contact to 15 hours daily. It also appeared that the ships were returning to port. Indeed, 1993-95 saw no ship activity at all – indicating this element had been retired. Some have been working as cruise ships and making promotional tours. Operation of many of the space surveillance and TT&C sites is now erratic, with facilities dropping out and then returning to service. The relay satellites continue to be used only when necessary because of the large fees – the manned and communications programmes come under different ministries.

The Soviets and their Russian successors put more than 10 ships into service from 1960, operating under the flag of the USSR Academy of Sciences. The flagship, *Kosmonaut Yuri Gagarin* (45,000 tonnes), carried 75 antennas, including two 25 m 'Ship Shell' and two 12 m 'Ship Bowl' dishes. Communications could be maintained simultaneously with two or more satellites, manned craft or orbital stations. The flagship was supported by the *Akademik Sergei Korolev* (21,465 tonnes) and four 9,000 tonne-class Kosmonaut ships: *Pavel Belyayev, Georgi Dobrovolski, Viktor Patsayev* and *Vladislav Volkov*. The *Kosmonaut Vladimir Komarov* was transferred to Leningrad for conversion to an ecological data collection platform. It is claimed that *Komarov, Gagarin* and *Korolev* sit in Odessa, too expensive for Russia to hire from Ukraine. They may even be up for sale, possibly with China interested. The *Morzhovets, Borovichi, Kegostrov* and *Nevel*, former timber carriers, were taken out of service by mid-1989 and sold to India for scrap. The *Marshal Nedelin* was completed in 1983 but was rarely mentioned in relation to the space programme, although she did support Buran's maiden flight in November 1988. Sister ship *Marshal Krylov* was completed in 1989. On 12 April 1988 the keel of the *Akademik Nikolay Pilyugin*, lead ship of yet another large class of space support ship, was laid in Leningrad.

Western navies monitored the fleet's activities because their positioning provides advance warning of manned and other launchings. Molniya relay communications to TsUP.

Russia fully intended to mothball the entire fleet as geosynchronous relays, particularly SDRN, entered full operation but both have suffered from financial cuts.

TsDUC Russian deep space network

Background

High Earth orbit and deep space missions are controlled by the TsDUC Long Range Space Communications System centred at Yevpatoria in the Ukraine. The Soviet Academy of Sciences and VKS controlled the centre until 1991's independence. It has the responsibility for controlling missions such as Mars, Granat, Interball, Spektrum, Phobos, Astron, Vega and Venera. It took over full control of the Ukraine's Sich 1 11 October 1995 after working in conjunction with VKS. For deep space missions in very high orbits and planetary probes, a special six-station network (Medvezhi Ozera, Pushino, Simeiz, Ulan Ude, Ussurisk and Yevpatoria) of 22 to 70 m radio-telescopes is employed.

This Crimean site was selected by Korolev and Keldysh in 1957 and equipped with eight 16 m antennas mounted on battleship gun turrets. It now operates 70 and 32 m dishes and is complemented by 70 and 32 m antennas at Ussurisk and a 25 m system at Ulan Ude. The 70 m RT-70 radio telescope was being built on the Suffa Plateau, but its status is unclear. This core system also calls on radio telescopes at other sites: 64 m at Medvezhi Ozera (Moscow Power Engineering Institute), 22 m at Simeiz (Crimean Astrophysical Observatory) and 22 m at Pushchino (Lebedev Institute of Physics). Six 32 m telescopes were under construction for VLBI work from the mid-1990s, but have probably been abandoned.

It was claimed in early 1994 that TsDUC would that year become part of the ground system for more 'profitable' satellites such as Molniya, Meteor and Gorizont. This is reflected by its control of Sich 1. During the previous 18 months it had been used for only for Granat and as Mir/Soyuz back-up. CNES has been making payments since 1992 to keep it open for Granat operations.

TsUP-Tsentr Upravlenyia Polyotom

Background

To maintain control of the some 150 operational satellites, the Russian Military Space Forces (VKS) operate a far-flung network of about a dozen sites across the former USSR. The HQ of this KIK (komandno-izmeritelnyy kompleks) command and control system is at Golitsyno-2 in the Moscow region. Political problems associated with operating non-Russian stations prompted the first deployment in January 1994 of a mobile station (eight vehicles and 20 officers). Mir operations are controlled primarily using a ground network of seven stations (Dzhusaly, Kolpashevo, Petropavlovsk, Tbilisi, Ulan Ude, Ussurisk and Yevpatoria) co-ordinated by the FCC Flight Control Centre at Kaliningrad outside of Moscow.

The manned and unmanned Flight Control Centre (abbreviated in Russian to TsUP, pronounced 'soop'), 10 km northeast of Moscow, began construction in 1970. 1973's Soyuz 12 was its first manned mission; previous missions were controlled from Yevpatoria. TsUP is manned constantly by 2,000 staff working in three shifts, earning an average salary of Rb230,000 monthly in mid-1995 (about US$50 at the official exchange rate). TsNII Mashinostroenye manages it, where it is located. The main control room is responsible for controlling Mir and its supporting Soyuz and Progress missions. Ten PS-2000 and three Elbrus mainframe computers handle the more demanding missions. There are 287 Mir controllers divided into five shifts (including one reserve), each lasting 24 h. A similar, but independent, control room for Buran was used for the first time in 1988 to support the maiden flight. It is used for Mir support when the main room is undergoing modifications and will be updated as the main Space Station control room.

The centre can handle 10 spacecraft simultaneously, assuming control from Tyuratam at launcher separation. A smaller control room handles Soyuz and Progress flights independent of Mir. Another room accommodates weekly two-way TV broadcasts with cosmonauts' families.

UNITED STATES OF AMERICA

Army Space Command ground stations

Background

Army Space Command (forward) is the operational arm of the US Army Space & Strategic defence Command. Command of the DSCS System Operations Centres and management of the joint tactical use of DSCS satellites is a primary ARSPACE mission. The Joint Staff assigned ARSPACE the operation of five worldwide DSCSOCs, three worldwide AN/MSQ-114s and three worldwide Regional Space Support Centres. ARSPACE operation of the JTAGS Joint Tactical Ground Stations in Europe and Korea provides theatre commanders with a transportable system for processing/disseminating near-realtime warning of theatre ballistic missile launches.

Using DSP data, JTAGS can process in stereo, improving the sensor's capability. JTAGS ties directly to theatre communications systems to transmit launch point information, impact area prediction, time of flight and positional information. ARSPACE also provides oversight of space surveillance operations at the US Army Kwajalein Atoll. ARSST Army Space Support Teams have deployed with units such as the 10th Mountain Division worldwide, providing weather, terrain analysis, mapping, imagery, communications and early warning information.

Deep Space Network (DSN)

Background

The NASA Deep Space Network – or DSN – refers to three antennas located roughly equidistant around the globe that receives downlinked information from interplanetary spacecraft and performs radio and radar astronomy observations. The three active locations for DSN nodes are Goldstone, in California's Mojave Desert; near Madrid, Spain; and near Canberra, Australia. As a two way communications system the DSN is the largest and most sensitive scientific instrument in the world.

DSN draws its heritage from a JPL programme begun in January 1958. JPL, then under contract to the US Army, travelled to Nigeria, Singapore, and California to place mobile tracking and communications dishes in anticipation of the army-launched Explorer 1, the first successful US satellite. NASA, which received an official charter on 1 October 1958, thus consolidated the separately developing space-exploration programmes then under way in the three branches of the US military.

On 3 December 1958, NASA accepted a transfer of JPL from the army, thus inheriting the precursor network. At the same time, NASA turned to JPL to design and carry out the lunar and planetary exploration programmes using remotely controlled spacecraft. As a means to communicate with these new satellites, NASA conceived of a Deep Space Network, initially under a separately managed programme. This new autonomous network would avoid the need for each flight project to reinvent a specialised space communications network. As an autonomous entity the Deep Space Network standardised communications to and from satellites regardless of the builder.

DSN researches and investigates new communications theory and as a result has set a standard for low-noise receivers, tracking, telemetry and command systems, digital signal processing, and deep space navigation.

All DSN antennas are steerable, high-gain, parabolic reflector antennas. They specifically: acquire spacecraft telemetry; transmit commands to spacecraft; locate spacecraft and measure their flight characteristics such as velocity; perform very-long-baseline interferometry observations; measure variations in radio waves for radio science experiments; gather science data: monitor and control the performance of the network; and the JPL continues to manage the facility for NASA.

Eastern Test Range

Background

This stretch of ocean and islands from Florida to the South Atlantic had its origins in 1950, when the army and air force jointly established the Joint Long Range Proving Ground. The air force assumed full control in 1951, redesignating it the Air Force Missile Test Center. In October 1952, the Air Research and Development Command needed to extend the Eastern Range's length to 5,000 miles in order to be able to communicate with the new SNARK and NAVAHO missile, over its entire flight path. The army and air force tentatively selected Antigua, St Lucia, Fernando de Noronha and Ascension as new tracking hosts, and the Eisenhower administration signed agreements with the governments of United Kingdom, St Lucia, Brazil and Ascension to secure their co-operation.

By the end of 1954, Cape Canaveral had three tracking stations up and running at the Jupiter Auxiliary Air Force Base and Grand Bahama Island. The air force was also building smaller installations capable of tracking satellites on the islands of Eleuthera, San Salvador, Mayaguana and Grand Turk. That same year the Eisenhower administration entered into agreements with the Dominican Republic and Puerto Rico and had already broken ground on still more tracking stations on those islands. In all, nine tracking stations spread out along 1,000 miles of the Atlantic, entered service before the end of the year. To communicate with this far-flung network, the US set up five separate government contracts, to lay submarine cable. One-by-one the installations entered operation. The last submarine cable linking Florida to Puerto Rico sent its first message in August 1954.

In December 1954, Mayaguana opened for operations followed by Eleuthera, San Salvador and Grand Turk between 18 July 1955, and the end of August 1955. The stations in the Dominican Republic and Puerto Rico became operational on 5 December 1956. The St Lucia site was activated on 5 December 1956, and Antigua and Ascension were ready for operations in October 1957. The Fernando de Noronha station was activated off the coast of Brazil on 18 September 1958.

A coverage gap still opened up between Antigua and Ascension though and the navy provided 12 small telemetry ships downrange in 1957 and 1958. With the ships in place, the air force fired its first 5,000 mile test, a SNARK test flight, on 31 October 1957.

In arguments between themselves and with the civilian government, the US military services finally wrested control of the new space exploration programme from the civilians associated with the International Geophysical Year. In part, the military argued that it alone had the tracking stations to support the mission, whereas civilians did not. In secret memos, the air force cited the ETR as the principal reason it should host all space launches. In fact, the ETR did not cover enough of the globe to be useful to IGY and the new NASA had to place portable antennae around the world. But eventually, after the launch of Sputnik the National Science Foundation had to capitulate and launch its satellites from Cape Canaveral. This ensured the future of Canaveral as America's space port.

All of the Eastern Range stations stared out with single-point radars. These radars came directly from the Second World War vintage SCR-584 radar systems. The so-called MOD I detected winged craft such as the Snark the easiest and so became the most economical solution for winged missile requirements. By the end of 1957, the ranges had optical systems, infra-red tracking equipment, cinetheodolites and ribbon-framed cameras. The CZR-1 ribbon-framed cameras covered the missile during the first 1,000 ft of flight, and cinetheodolites followed the vehicle out to a distance of 20 miles. Wild BC-4 ballistic cameras captured optical data beyond the tracking radars' beamwidths.

Ballistic missiles forced the ETR to upgrade its tracking systems in the 1950s and 1960s. An AZUSA continuous wave tracking system began operation at Cape Canaveral in the mid-1950s and on Grand Bahama in the early 1960s. The AN/FPS-16 radar system went into service at Cape Canaveral, Grand Bahama, San Salvador, Ascension and East Island between 1958 and 1961, followed in the early 1960s, by the MISTRAM (Missile Trajectory Measurement) system at Valkaria, Florida and Eleuthera Island. This new system tracked Minuteman's throughout their flight.

Pershing missiles required an even better system, the UDOP (Ultra-High Frequency Doppler) trajectory measuring system and an AN/MPS-25 radar placed at Grand Bahama's Cay and Carter Cay.

As this system reached out to cover the entire Atlantic, the navy's role diminished. The navy retired its Eastern Range's FS-type telemetry ships in 1960.

Alongside the ground-based systems, the air force routinely supported missile launches with specially instrumented aircraft for communications and telemetry support. The 6550th Operations Squadron preceded important missile tests with the flight of a C-54 aircraft in the 1950s. Eleven C-130s acting in the same role replaced the C-54s in 1961 and 1962. At the dawn of the Apollo programme the ETR introduced the more capable Advanced Range Instrumentation Aircraft (ARIA). Eight C-135A aircraft for the ARIA programme joined the C-130s in January 1966. Each ARIA carried a steerable telemetry antenna.

Half a dozen range stations, no longer had a purpose after the last MERCURY mission in May 1963, and the air force retired them. Mayaguez station shut down in August 1961, and the East Island Annex in 1963. The Dominican Republic reabsorbed the Dominican station on 6 November 1962. Project FRESH LOOK eliminated the MOD II radar network in 1965. San Salvador had a caretaker staff only after March 1965, finally closing for good on 31 January 1970. St Lucia reverted to its original owners in early December 1967. Mayaguana facilities were abandoned on 16 June 1970, except for a still active runway. The Missile Guidance Annex on Fernando de Noronha was returned to Brazil on 14 January 1969.

When the Shuttle Challenger exploded on 28 January 1986, it carried a TDRS satellite meant to speed the replacement of ground installations. Congress directed a national space recovery effort to consider a revitalisation of America's unmanned space vehicle industry. It included an effort to modernise the Eastern Range with a new Range Operations Control Center (ROCC), fibre optic communications, consolidated instrumentation facilities on Antigua and Ascension and radar, telemetry and optics improvements. The USNS Redstone was deactivated on 6 August 1993, but a new range site was completed in Argentina, Newfoundland in June 1993 to support northbound flights of the TITAN IV from Cape Canaveral. Operational testing for the ROCC continued in 1993 and 1994, and the ROCC was declared operational on 1 March 1995.

Space Surveillance Network/NORAD

Background

NORAD is a bi-national US-Canadian command charged with safeguarding the sovereign airspace of the two countries. Through its component commands, US Space Command operates a worldwide network of missile warning sensors providing tactical warning and attack assessment of ICBM and SLBM launches posing potential threats against North America, and provides tactical warning of ballistic missile attacks to US Commanders worldwide. The missile warning sensors are also used for space surveillance.

Together with dedicated space surveillance sensors (operated by the 21st Space Wing at Peterson AFB, NAVSPACECOM at Dahlgren and contributing ARSPACE and civil assets), they form the Space Surveillance Network, which provides timely and accurate detection, tracking and identification of space objects and events. The network makes 50,000 to 80,000 daily observations for the command's Space Control Center in Cheyenne Mountain and maintains a catalogue of more than 8,000 orbiting objects.

Data are relayed via satellite almost instantaneously to four primary locations: the command's Missile

Warning Center (Cheyenne Mountain AS), Strategic Command HQ (Offutt AFB, Nebraska), National Military Command Center (Washington DC) and Alternate NMCC (Maryland).

The 21st Space Wing's ground-based sensors until April 1995 comprised BMEWS and SLBM. The two BMEWS radar units are the 12th SWS at Thule AB in Greenland and the 13th SWS at Clear AS in Alaska. SLBM warning system units are the 6th SWS at Cape Cod AS, Massachusetts, 7th SWS at Beale AFB, California. These sensors also provide essential tracking reports on space objects. The 73rd Space Group at Falcon AFB was absorbed 29 April 1995, along with its eight space tracking squadrons. The 11th SWS at Falcon AFB, Colorado, operates the ALERT Attack & Launch Early Reporting to Theater system.

As host wing for the Peterson Complex, the 21st provides complete base support services for Peterson AFB and Cheyenne Mountain AS, plus some support functions for Falcon AFB. Personnel total 4,127 military, 846 air force civilian and 3,300 contractor employees.

The NORAD/USSPACECOM Command Center in Cheyenne Mountain Operations Center also receives the data.

RESEARCH AND DEVELOPMENT FACILITIES

BELGIUM

Centre Spatial de Liège (CSL)

Background
CSL (IAL Space until April 1992) is an autonomous unit within the University of Liège and, along with Intespace (France) and IABG (Germany), is one of the Co-ordinated European Test Facilities group under ESTEC. The test facilities are based around a 480 m² class 10,000 clean room and include three thermal vacuum optical test facilities. The 150 m³ 5 m diameter horizontal 33 tonne Focal 5 chamber provides a 10^{-9} atmosphere vacuum and 10 temperature settings over −268°C to +70°C and optical bench vibration levels of 10^5 *g* on a 350 tonne seismic block. There is also an associated 30 m² class 100 laminar flow clean zone. In the same area is the Focal 2 chamber, with identical performance but 21 m³/2 m diameter. The 1.2 m³ Focal 1.5 provides a similar environment but with a 1.5 m vertical diameter and 80 m² class 10,000 cleanroom and 4 m² class 100 laminar flow clean zone. CSL undertakes testing/calibration of camera tubes, image intensifiers, CCDs, photomultipliers and photon counting systems, in addition to the design/manufacture of meteorological instruments.

CANADA

Canadian Space Agency – David Florida Laboratory

Background
Operated on government funding and collocated with the Communications Research Center (CRC), CSA's DFL is the national facility for spacecraft AIT; its facilities are available on a fee for service basis. Built in the early 1970s for the Canadian-US Hermes Communications Technology Satellite project, it has since supported Anik C-E, Shuttle remote manipulator, Brasilsat, Olympus, MSAT and Radarsat development, and will play an integral role in the next Anik generation and Canada's contribution to Space Station.

The current activity centres on the testing of Canada's contribution to the International Space Station programme – the Mobile Servicing System (MSS).

Facilities include:

Clean rooms
Integration and assembly areas comprise three temperature (22±3°C) and humidity (45±5%) controlled clean rooms: Bay 1 440 m²; Bay 2 560 m²; Bay 3 1,100 m². All three high bays provide class 100,000 conditions, with class 10,000 available when required, and are traversed by travelling bridge cranes with hook heights of 9, 11 and 16 m, respectively. A 6 × 12 m low bay is used for subsystem assembly/checkout. The clean rooms house spin balance and mass property-measuring equipment. The MRC Mk VII-6 Vertical Axis Measurement System (VAMS) measures static/dynamic imbalances and moments of inertia for 25 to 2,300 kg payloads spun at up to 200 rpm, and calculates the required balance masses and their locations. A Horizontal Axis Measurement System (HAMS) provides complementary data, for payloads of up to 2,700 kg, 3 m diameter. Both systems can be housed within DFL's 7 × 10 m vacuum chamber.

Thermal Vacuum Test Facility
Four principal chambers of 1, 2.5, 3 and 7 m diameter exist. PC-based data processing system.

7 × 10 m vertical chamber: 1.3 × 10^{-6}-10^{-7} mbar, 1.1 solar constants/50 per cent shroud area (maximum load 260 kW) using infra-red lamps, 6.7 × 10.7 m test volume within LN_2 shrouds, accessible via 7 m diameter removable top lid and 2.2 and 1.5 m man-doors. Hardpoints permit 27 tonnes mounted on floor, 2,200 kg from lid, or 13,600 kg from 36 hardpoints in six rings on chamber walls.

3 × 9 m vertical chamber: 1.3 × 10^{-6}-10^{-7} mbar, LN_2 temperature range plus thermal cycling with infra-red lamp system, side- loading 1.8 × 2 1 m door.

2.5 × 2.5 m horizontal chamber: 1.3 × 10^{-6}-10^{-7} mbar, LN_2 or ±150°C GN_2 temperature range.

1 × 1 m horizontal chamber: 1.3 × 10^{-6}-10^{-7} mbar, LN_2 or ±150°C GN_2 temperature range.

Ambient pressure thermal chamber: thermal cycling and/or temp soak, and (modified) vibration or RF testing at controlled temperatures. Can be configured to different volumes, −150 to +149°C, ±1°C.

Vibration Test Facility
UD 4000: 178 kN sinusoidal, 165 kN random, up to 183 cm² slip plate, 122 cm diameter head expander.
MB C-150: 76 kN sinusoidal, 65 kN random, up to 152 cm² slip plate, 81 cm diameter head expander.

Ling A395: 27 kN sinusoidal, 23 kN random, up to 61 cm² slip plate and 61 cm diameter head expander.

Static Load Test Facility
The new SLT facility, first used in 1992 by the Radarsat bus structural model, incorporates a 12 channel Cyber Fatigue Master 7000 Digital Control System combined with a new data processing/acquisition system accommodating 200 strain gauge channels and up to 40 LVDTs.

RF Test Facility
Comprises four anechoic chambers, a rooftop range, an EMC facility and PIM (Passive Intermodulation Measurement) facility. The chambers and rooftop turntable provide outdoor ranges for automatic antenna measurements; the three-axis positioners and sources are controlled via a fibre optic link between the 62 m tower and control room. DFL has added a spherical near field antenna measurement facility for performing near field to far field transformations, and is developing an RCS radar cross section measurement capability. DFL is designated by Inmarsat as both an AATH Authorised Antenna Test House, authorised to conduct antenna tests to Inmarsat specifications, as well as a Designated Inmarsat Representative (DIR), providing the DFL representative with the authority to approve antennas on Inmarsat's behalf.

Chamber 1: 24 × 12 × 20 m high for half its area, with 9,000 kg gantry crane access, reflection co-efficient −30 to −55 dB over 0.250-20 GHz. A two-axis positioner has total vertical load of 18,145 kg and 0.05° pointing accuracy, 5 m tower for short (9 and 400 m) outdoor range tests.

Chamber 2: 6.1 × 6.1 × 6.7 m, reflection coefficient −30 to 55 dB over 0.500-20 GHz, magnetic/electric field shielding 90/100 dB in 200 kHz-15 GHz range. A three-axis positioner provides 0.05° accuracy for 4,536 kg loads, 3 m pyramidal horn with 1.8 × 1.8 m aperture for 6 m indoor range tests.

Chamber 3: 6.1 × 4.3 × 2.9 m, shielded, reflection coefficient better than −40 dB, 2-100 GHz. Azimuth positioner accuracy +1±0.005°. R&D into RCS measurements, near field systems and EHF-bands FF testing.

Rooftop Range: 0.250-18 GHz, 6 × 6 m platform, two-axis positioning of 13,959 kg loads, source inclination 5° for 400 m outdoor range, <1° for 16 km long range to Kingsmere.

EMC Facility: 4.8 × 5.5 × 3.6 m shielded room is lined with high performance anechoic material to provide electric/magnetic field shielding of 100 dB over 200 kHz-15 GHz; 2 × 3 × 2 m Faraday cage; 3 × 2.5 × 2.4 m anteroom. The EMC facility is equipped to test to MIL-STD-461A/B/C/D/462/462D RTCA/DO-160A/B/C. Emission susceptibility up to 40 GHz.

PIM Facility: the passive intermodulation measurement facility evaluates the magnitude of intermodulation products generated by the components and subsystems of satellite transponders. The three DFL-designed systems operate in three separate bands (275-330, 862-873, 1,511-1,578 MHz) and measure down to −146 dBm.

HF Anechoic Chamber: 7.6 × 3.35 × 3.35 m high, reflection co-efficient −30 to −50 dB over 1-40 GHz, two-axis high speed/accuracy, total vertical load 30 kg, <0.005° pointing accuracy. Currently set up for spherical near field measurements and automated gain extrapolation.

To complement and support classical vibration and acoustic testing, DFL is developing modal analysis through both conventional and innovative modal test techniques. A 366 tonne seismic mass has been built as a modal analysis testbed; a new LMS vibration control/modal analysis system capable of supporting swept sine, random or shock control and analysis has been procured.

DENMARK

Danish Space Research Institute (DSRI)

Background
The Danish Space Research Institute (Dansk Rumforkningsinstitut) is a government-funded research organisation, founded in 1966, under the direct supervision of the Research Directorate of the Ministry of Education & Research. Its principal areas of interest are space astronomy, cosmic rays and space plasma physics. Four wide angle x-ray transient and gamma burst detectors, which form the Watch all-sky monitor, are flying on the RFAS Granat satellite. A single Watch unit was provided to ESA for Eureca's 1992 mission. Two large area x-ray telescopes, four imaging detectors and a Bragg crystal spectrometer (Xspect) are being produced as part of the CIS/Danish Sodart telescope for the Spektrum-X satellite in 1998. Ground support equipment was supplied for the Isophot instrument on ESA's Infra-red Space Observatory in co-operation with Heidelberg's Max Planck Institute of Astronomy, and DSRI is a member of the Hipparcos Northern Data Consortium. The Swedish Viking magnetospheric mission carried a wave experiment built at DSRI; current activities include participation in the Wave and Magnetometer experiments on ESA's Cluster. A series of plasma physics sounding rocket flights in co-operation with Scandinavian countries and NASA is also under way. DSRI is part of Denmark's Ørsted satellite project.

DSRI receives a line item budget voted each year by Parliament as part of the national budget. This includes salary, operating and development costs. Additional support is obtained from the Danish Space Board on a project by project basis. DSRI is headed by a Director, who reports to the Research Directorate on financial and personnel matters and to an external Institute Board for science/programme affairs. Otherwise, the Institute is autonomous. 1996's budget is DKr26.1 million; 1995 DKr19.3 million; 1994 DKr19.4 million; 1993 DKr24 million; 1992 DKr22 million; 1991 DKr22.544 million; 1990 DKr21.844 million; 1989 DKr21.335 million. This supports a fixed staff of 40, including 18 physicists and engineers. Included in the total is funding from the Danish Space Board to support seven engineers, scientists and technicians, and a development programme for thin foil x-ray optics. Facilities include state-of-the-art equipment for determining x-ray optical, scattering and diffraction properties of surfaces, crystals and gratings (transmission/reflection) over a wide range of energies. A high-power rotating anode x-ray tube was installed most recently. Technical personnel were trained/certified by ESA/NASA for flight hardware fabrication; the mechanical workshop is equipped for precision manufacturing of materials such as glass fibre epoxy composites and titanium. CAD/CAM facilities are maintained, electronic circuits can be designed/manufactured and, when required, circuits can be assembled/tested under clean room conditions. At other institutes, DSRI has access to vibration tables, a small thermal vacuum chamber, shock testing equipment and units for determining centre of gravity and moment of inertia.

FRANCE

Intespace

Background
Intespace (Ingéniére Tests en Environnement Spatial) has operated CNES's Toulouse test facilities since 1983 (taking over from Sopemea), incorporating 20,000 m² of

MPLM during acoustic testing at Intespace ***1997***/0008101

test halls, laboratories and clean rooms able to handle Ariane 4/5-class 3/4 tonne satellites. 1995 sales were FFr120 million (1994 FFr119 million; profit FFr3.6 million), 1993 FFr108 million, 1992 FFr105 million, 1991 FFr94 million. The facility employs 150 personnel.

CNES/Sopemea hold 35 per cent each, Aerospatiale/ Matra Marconi 9 per cent each, Alcatel Espace 3 per cent and others 9 per cent. It is part of the Co-ordinated European Test Facilities group, along with IABG (Germany) and Centre Spatial de Liège (Belgium), organised by ESTEC. Intespace also supervised the installation of new facilities, opened June 1988, at Brazil's INPE, plus those of Taiwan's NSPO and South Korea's KARI. The 1996 principal test programmes included SILEX (LEO and GEO); Hot Bird, MPLM, Telecom 2, Mars-96, Ariane 4 and Ariane 5.

Three 280 m2 preparation/integration halls support Intespace's own key facilities, each with an access airlock and meeting standard cleanroom requirements. Facilities include:

Thermal chambers

600 m^3 6 m diameter with 1,600 W/m^2 3.8 m diameter solar simulator for objects to 2.5 to 4 m, –170/120°C; 30 m^3 with 1 m diameter solar simulator; also 1, 1.5, 8 m^3 chambers with controlled humidity and –170/ 120°C temperature range.

Thermal vacuum

SIMMER (Simulateur de l'Environnement) became operational in 1992 (first used by Hispasat 1B) for 5 tonne (vertical), 2.5 (horizontal) subjects. Primary pumps 2.2 × 10^{-2} atmospheres in 2 hours; cryogenic/ turbomolecular pumps 4 × 10^{-6} atmospheres in 10 hours. 10 m diameter 13 m long, –173/120°C temperature range, 400 temperature measurements, 200 electrical measurements. Other chambers: 0.25, 3, 20 m^3 volume, temperature –170 to +120°C, vacuum down to 10^{-9} atmospheres.

Other capabilities

Acoustic: 156 dB, 1,100 m^3 chamber
Salt spray chamber: 1.2 m^3 volume
Electrical: 45 + 1,700 m^3 Faraday cages; susceptibility analyses over 20 Hz-40 GHz in 17 × 10 × 8 m anechoic chamber; impulse magnet charger and magnetic measurement facility rated at 0-5,000 gauss
Mechanical: 27, 80 (× 2), 150, 170, 300 kN shakers operating over 5-2,000 Hz and driving a 5 × 5 m block under a 13 m ceiling. 10,000 *g*/kg centrifuge; shock simulator (400 *g* for 4 ms); 8 × 10 × 13 m high 1,100 m^3/156 dB and 1 m^3/173 dB acoustic chambers; mass properties (moments of inertia, mass, centre of gravity) for articles up to 2 t in 150 m^2 central hall.
Compact payload test range: the FFr65 million Mistral range is planned to enter operation in 1998.

Office National d'Etudes et de Recherches Aérospatiales (ONERA)

Background

The Office National d'Etudes et de Recherches Aérospatiales was founded in 1946 as a national aeronautical research centre under the General Delegate for Armament (DGA). 2,200 staff are employed at eight major centres, specialising in aerodynamics, fluid mechanics, flight mechanics, structures, materials, optics, acoustics, electronics, radars, computer sciences, and ground, wind tunnel and flight testing. The Châtillon site houses HQ, operational departments and laboratories, Chalais-Meudon research wind tunnels, Palaiseau specialises in energetics, Lille in flight mechanics and Toulouse accommodates the CERT Center d'Etudes et de Recherches de Toulouse. Large test facilities are operated at two principal centres: Modane-Avrieux and le Fauga-Mauzac. Modane-Avrieux maintains the S1MA continuous sonic wind tunnel (8 m diameter, 14 m long), the S2MA continuous transonic/ supersonic (1.8 × 1.75 m), the S3MA transonic/ supersonic blowdown (0.8 m high test sections), the S4MA hypersonic (M12.0) and R4.3 transonic/ supersonic cascade tunnel. The S4B 15 mbar pressure chamber (2.5 m diameter, 10 m long), is primarily used today for the calibration of nacelles and air-powered engine simulators. Fauga-Mauzac operates the subsonic F1 (pressurised) and F2, and the high enthalpy F4 wind tunnels. A new laboratory was created in 1992 at l'Ecole de l'Air de Salon-de-Provence for common research such as flight mechanics and optronics.

GERMANY

Industrieanlagen-Betriebsgesellschaft GmbH

Background

IABG was established in 1961 to provide aircraft and spacecraft test capabilities; it is now an element of the Co-ordinated European Space Test Facilities group, along with Intespace (France), Centre Spatial de Liège (Belgium) and ESTEC. For space, IABG offers thermal vacuum, thermo-mechanical, thermo-environmental, vibration, acoustic, modal, static, magnetic and EMC testing on launcher and satellite components and assemblies. It has tested more than 100 satellites and major satellite subsystems; recent activities include Huygens, Cassini, Artemis, ISO, Cluster, Soho, Freja and Ariane 5 stages and major components. Facilities include:

Electromagnetic (EMC) Test Facility

The facility comprises a shielded test chamber, fitted with RF absorbers and providing class 100,000 conditions, and EMC measuring equipment. For emission testing, a computer-controlled 10 Hz to 40 GHz spectrum surveillance system, current probes and antennas are employed. For susceptibility measurements, high-power broadband amplifiers provide up to 600 V/m field strength from 10 kHz to 18 GHz. Special pulse generators and current probes allow various susceptibility tests, including electrical tests for power supply simulation.

Usable dimensions: 10.5 m length, 7 m width, and 8 m height.

Access door: 4 m width, 6 m height.

Shielding attenuation: H-field >60dB for 10kHz, >100 dB above 1 MHz; E-field and plane waves 100 dB for 1 MHz to 10 GHz.

Internal reflectivity: >20 dB above 100 MHz, >40 dB for 300 MHz to 10 GHz.

Vibration Test Facility

A 300 kN vibration system can operate horizontally or vertically under class 100,000 conditions, powered by four coupled electrodynamic shakers. Specimen response information is fed into a 360-channel digital data acquisition system. Sensors include 500 accelerometers with graded performance ranges. There are also several smaller single-shaker facilities.

Mounting table: 3 × 3 m (80 × 80 mm M10 hole pattern).

Maximum specimen mass: 15,000 kg.

Maximum acceleration: 15 *g* without load, 4.5 *g* with 5.0 t load.

Maximum displacement: ±25 mm.

Frequency ranges: 4-2,000 Hz sine low level, 4-300 Hz sine high level, 10-2,000 Hz random high level.

Modal Test Facilities

The systems provide tuned sinusoidal and non-tuned broadband excitation. The largest is housed in a transportable container; the two smaller ones are easily transported. The software undertakes test control, data acquisition/processing, modal analyses and result presentation. A separate hall is available for testing a larger range of structures, under clean room conditions if needed.

Excitation: sine, random 8 channels.

Vibration exciters: 26, 10-7,000 N.

Measurement: 400 accelerometers, plus force and displacement transducers.

Data acquisition: 882 channels (in 7 blocks) up to 2 kHz; 384 channels (in 8 blocks) up to 4(20) kHz; 16 channels up to 25 kHz

Acoustic Test Facility

Acoustic environments are simulated by electro-pneumatic noise generators that can reproduce overall sound pressure levels and spectra of operational fields. The total acoustic power available is 80 kW AC. The recording system includes 20 microphone channels, 128 accelerometer channels and 24 strain gauge channels. The facility comprises several reverberation chambers, control/computer rooms, air supply systems (5 kg/s) and 9.6 m height × 15.4 × 11.5 m preparation hall. The 800 m^3 chamber is an irregular pentagon 5.7 m high, 13.0 m wide, 10.8 m deep and can create 150 dB levels in class 100,000 conditions. The 206 m^3 chamber, 4.7 m high, 8.0 m wide, 5.5 m deep, can attain a maximum of 162 dB. The Progressive Wave Tube provides up to 170 dB in a 0.8 × 1.2 m test section. A thermoacoustic facility can simulate combined environments such as surface temperatures up to 1,300°C on a specimen plus acoustic excitation up to 158 dB.

Magnetic Field Simulation

The facility consists of a square, triaxial coil system with four coils per axis and with an edge length of 15 m. The test volume measures 10 × 10 × 10 m with free access of 4 × 4 m. It is used for measuring the magnetic cleanliness of objects, recording magnetic moments and eddy current fields, attitude control testing of magnetically stabilised spacecraft, calibrating magnetometers, and magnetising or demagnetising objects to determine permanent, remnant and induced fields. DC field values up to 75,000 nT can be produced, corresponding to Earth's magnetic field level at the poles. The limit for uniform AC fields is 25 kHz. Hall probes, fluxgate magnetometers and search coil magnetometers are available; for precise measurements, proton spin magnetometers and optically-pumped magnetometers are employed.

Thermal-Vacuum Facilities

A 6.8 m diameter chamber (6.2 m diameter test volume) providing more than 10^{-8}atm vacuum conditions and a 3.6 m diameter or 3 × 4.5 m rectangular solar beam for test objects up to 2.5 tonnes, 4 m diameter/5 m high was introduced in 1983. The solar beam, with ±2° collimation angle, provides up to 1.4 solar constants. The cryogenic shroud creates 100 to 385 K conditions within the 13 m long chamber. Test specimens can be mounted on a two-axis motion simulator, providing up to 10 rpm continuous spin and ±200° attitude excursions and levelling capability.

Other Thermal-Vacuum Chambers

These chambers are used for subsystem and component trials. 3 m chamber: 3.2 m diameter, 3.8 m long shroud, 10^{-8}atm, 100 to 400 K, infra-red radiators; 1.3 m chamber: 1.35 m diameter, 2.2 m high shroud, 10^{-8}atm, 100 to 400 K. Clean room conditions from class 100,000 to 100.

Thermo-Mechanical Vacuum Facility

This is a 1.5 m diameter cylindrical vacuum chamber, with two-zone heating by graphite elements up to 1,600°C, mechanical loading by vacuum-tight feed-through and hot load introduction. Displacement measurement in hot zones relative to fixation points. 600 × 600 × 100 mm maximum sample.

Thermo-Mechanical Facilities

Mechanical loading is applied to shingle-type test samples (up to 350 × 350 mm) by pressure difference in a cylindrical pressure chamber. Seven infra-red modules with total 252 kW$_e$. Maximum temperature is 1,450°C applied thermal loading. Maximum pressure difference is 500 mbar. Reaction forces, temperature, pressure and displacements can be measured and visualised online. In addition, cover furnaces with resistance heaters up to 1,600°C in air or inert gas atmosphere; metallic and carbon heater up to 2,000°C in vacuum or inert gas available. Samples can be loaded and measured by extensometer and/or displacement by LVDT.

Thermo-Environmental Test Facility

A 1.5 m diameter cylinder of nominal 1.5 m length, extendible by modules. Houses a water-cooled reference system for, for example, displacement measurements. Several heating systems can be installed; maximum sample temperature =1,800°C. Different atmospheres and pressure profiles can be simulated. Mechanical loads can be applied by a hydraulic external loading device.

INDIA

ISRO satellite centre

Background

ISAC is the lead centre for ISRO's satellite development and construction. It grew from the Indian Scientific Satellite Project established in 1972 to build the Aryabhata satellite to a current staff of 2,400 responsible for the design, fabrication and testing of the IRS, SROSS and Insat 2 series.

ISAC is organised into five main operating groups: Spacecraft Electronics (including responsibility for spacecraft integration and ground testing), Mechanical Systems, Control, Mission, and Technical Support Services. Three independent divisions cover: Quality Assurance, Technical Physics (rocket and satellite science instrumentation), and Programme Planning & Evaluation. The PP&E division provides technical and managerial planning, resource allocation and progress monitoring on each project. Facilities include:

Thermal-Vacuum Chambers

The US$25 million, 9 m diameter 9 m high Large Space Simulation Chamber (LSSC) was commissioned in 1991 for testing Insat 2-class spacecraft. The 7 m diameter auxiliary chamber houses 12 20 kW Xenon lamps (11 used at any one time) to produce a 3.5 m diameter 1.7 kW/m^2 beam for solar simulation. The pumping system achieves 10^{-9} atmosphere in 3 to 4 hours. Two LN_2 tanks each store 1,000 m^3.

4 m diameter chamber, 10^{-8} atmosphere, walls cooled by LN_2. No solar simulation is available. Previously the main test chamber, used for IRS.

2 and 1.2 m diameter chambers for instrument package testing to 10^8 atm/±100°C. A −10/50°C chamber provides thermal cycling testing for instrumentation.

Mechanism Laboratory 40 × 12 × 24 m high, for boom and solar panel deployment testing. A second facility provides horizontal deployment across a water tank.

Motion Simulation Laboratory, 2 m diameter Earth and 25 mm diameter Sun simulators in conjunction with three-axis servotable or three-axis air table for sensor and AOCS testing. SROSS and IRS are the most recent subjects. A moving Earth simulator for GEO/GTO satellite systems has been added, in addition to a thermal chamber for infra-red Earth sensors.

Vibration Facility

Two Ling Dynamics Systems (UK) tables accommodate SROSS and IRS/Insat 2, respectively, with both providing 2.5 cm peak-to-peak sinusoidal (100 *g*) and 30 *g* random.

Antenna and EMC

EMC testing is conducted in a 12 × 40 m anechoic chamber. Satellite integration is conducted in 12 × 12 m class 10,000 (for SROSS-class satellites) or 55 × 14 m class 100,000 (for IRS/Insat 2 class) cleanrooms.

Space Applications Centre (SAC)

Background

ISRO's prime centre for space applications research, employing 2,140 (including DECU), specialising in communications, satellite remote sensing, meteorology and geodesy. The centre is organised into SCA Satellite Communications Area, Remote Sensing Area and Microwave Remote Sensing Programme. The Insat and IRS payloads are developed by SAC. It also supplied extended C-band Earth station equipment to the Insat Master Control Facility.

Vikram Sarabhai Space Centre (VSSC)

Background

VSSC, located around the village of Thumba near Trivandrum and with a personnel total of 5,600 is ISRO's largest centre. It provides the development base for the country's indigenous launchers: ASLV, PSLV, GSLV and sounding rockets such as Rohini, along with the solid motor programme (Trivandrum's Liquid Propulsion Systems Centre operates as a separate entity).

Supporting the programmes are specialised R and D groups in: Avionics and Mission Dynamics, Solid Propulsion, Propellants and Chemicals, MMS (Materials and Mechanical Systems), Systems Reliability, Computers and Information Systems, and Programme Planning and Management. Extension facilities are located at Valiamala for the PSLV launcher project, at Vattiyoorkavu for composite development, and an experimental ammonium perchlorate plant at Alwaye.

Views of ESA's main establishments: HQ, ESTEC, ESOC, ESRIN ***1998***/0019686

INTERNATIONAL

European Space Research Institute (ESRIN)

Background

ESRIN is one of four ESA establishments and operates with a permanent staff of 140. Its main focus is Earth observation satellite data handling, including management of the ground segment to acquire, pre-process and archive data, and handling distribution, either directly or via a distributor. ESRIN handles missions for non-ESA clients as well including Landsat, Tiros, MOS, JERS 1 and Spot but the principal emphasis is on ERS and preparations for Envisat.

The ESRIN ERS Central Facility (EECF) maintains constant links with the ERS Mission Management and Control Centre at ESOC. ESOC executes the mission operation plan prepared by the EECF, with ERS ground stations for the scheduling of near– realtime distribution of Fast Delivery products. The Processing and Archiving Facilities handle data product orders, providing users with access to the online worldwide catalogue of ERS data. To this end, ESRIN is responsible for stations in Europe and Canada, ERS PAFs in Italy, Germany, France and UK, and has contracts with worldwide national stations for ERS data acquisition.

ESRIN also operates the ESA-IRS Information Retrieval Service which provides online access to bibliographic and factual databases covering most fields of science and technology. The ESIS European Space Information System caters for the astronomy and space physics community.

European Space Research and Technology Centre

Background

ESTEC is ESA's largest single establishment, with 1,108 employees. It provides project management for science, communications, earth observation, µgravity and space station programmes, executes the space science programme, performs future satellite programme studies, and undertakes design, development and testing of components and complete space vehicles. Half of the on-site staff belongs to the spacecraft project teams reporting to the Programme Directors at ESA Headquarters; the majority of the remainder belong to the specialised technical divisions based at ESTEC.

Capital investment is about US$55 million for buildings and US$150 million for the technical facilities. The development laboratories include Mechanical Systems, Propulsion, RF Systems, European Space Battery Test Centre, Materials/Processes, Components, Onboard Data Processing, Simulation and Electrical Facilities for Automated and Manned Missions, including a robotics testbed. The Satellite Communications Building supports special telecom services such as videoconferencing, mobile links and other in-orbit testing. The Fuel Cell Test Facility is an annexe to the Space Battery Centre.

The largest laboratory, the ESTEC Test Centre, operates a wide range of environmental test facilities. Working with Intespace (Toulouse), IABG (Ottobrunn) and Centre Spatial de Liège, ESTEC heads the Co-ordinated European Test Facilities to ensure that national facilities can be used for agency projects. All test areas are air-conditioned (19 to 23°C; 40 to 60 per cent RH), class 100,000.

ESTEC's test facilities include:

Large Space Simulator (LSS)

Europe's largest high-performance solar/vacuum simulator became operational in 1986 and was inaugurated January 1987 for Ariane 4-class payload testing. The first major test was on Alenia's IRIS upper stage test model. The LSS consists of a main 10 m diameter chamber with a removable lid for ease of access and a 5 m side port, with an auxiliary chamber containing the collimating mirror. Payloads can be supported on a vibration-isolated (≤10^{-2} *g*) 3.2 × 3.2 m platform, or suspended from the upper volume, or mounted on a two-axis motion simulator. An array of 19 unfiltered 25 kW high-pressure Xenon lamps provides a 6 m diameter collimated beam, with a maximum 1.3 solar constant intensity. The LSS is depressurised using both the test centre's central pumping system and a dedicated high-vacuum system of turbo-molecular pumps and a liquid helium cryo-pump to attain 3 × 10^{-7} mbar in about 10 hours. Both chambers incorporate stainless steel shrouds operating on LN_2 or GN_2.

Chamber capacity: 10 m diameter × 15 m high, auxiliary chamber 8 to 11.5 m diameter × 14.5 m long. The contractor is BSL.

Vacuum conditions: 3 × 10^{-7} mbar in 10 hours; repressurised by GN_2 to 100 mbar followed by filtered air over 4-24 hours.

Illumination: 1 solar constant (1,360 W/m²) by 12 of 19 × 25 kW Xe lamps. 32 kW lamps can be used if needed. 6 m diameter horizontal beam with ±4% uniformity; 7.2 m collimation mirror consists of 121 hexagonal segments, collimation angle 1.9°. Contractor Carl Zeiss.

Shroud temperature: <100K by LN_2 circulation, 150-350K by GN_2. Cool/warm-up period ±2 hours. Shrouds supplied by Leybold-Heraeus.

Motion simulator: spinbox (1-6 rev/min + 1-24 rev/day), turntable (1-24 rev/day). The contractor is SIGRI.

Thermal Facilities

HBF3 thermal vacuum chamber. The chamber is employed for subsystem vacuum temperature cycling in a 3 m diameter usable test volume. The facility is equipped with a demountable infra-red rig for solar panel testing, comprising a LN_2 shroud system with two compartments 3.85 m long, 2 m deep each fitted with 38 500 W infra-red lamps. Maximum temperature range on carbon fibre-backed solar arrays is 170/+90°C. The basic chamber incorporates five shrouds temperature controlled by LN_2 or GN_2. Loading is via removable top lid.

Vacuum conditions: 2-6 × 10^{-6} mbar in 5 h (0.02 mbar in 1 hour) from atmospheric pressure.

Shroud temperature: <100K by LN_2 circulation, 200-373 K by GN_2. 2 K/min cooling; 3 K/min heating.

Small thermal vacuum chambers: VTC 1.5 usable volume 1.5 m diameter × 2.5 m high, up to 500 kg test subjects, 1.3 × 10^{6} mbar vacuum, <100 and 125-425 K using LN_2/GN_2. Corona horizontal cylinder for up to 400 kg masses in 1.8 m diameter × 3.2 m long test volume. The LN_2 shroud with two 1.21 m deep, 2.9 m long compartments each equipped with 32 500 W infra-red lamps for solar panels testing, featuring rapid de-pressurisation (down to 1.3 × 10^{-4} mbar in 7 min), down to 6.6 × 10^{-6} mbar, 170/+110°C achievable on carbon fibre-backed solar arrays (the chamber can also be used for outgassing tests); Ultra-High Vacuum Chamber UHV 0.5 0.120 m³ test volume in 45 cm diameter × 76 cm high cylinder, attaining <1.3 × 10^{-10} mbar with bake-out facility up to 523 K; Accelerated Thermal Cycling Chamber ATC II for rapid automatic thermal cycling of small lightweight items at atmospheric pressure, T range 93-403 K using dry N_2 in 48 cm high × 50 cm long × 25 cm wide usable volume, typical cycle +100 to −100°C and back in 3 minutes.

Electrodynamic Shakers

Located in the same complex as the Large Space Simulator, the centre is equipped with a 280 kN multishaker and a 70 kN shaker. The larger system incorporates two Ling Dynamics Systems 984 LS vibrators, each capable of 144 kN thrust. The 70 kN system uses one Ling Dynamics 964 LS exciter. All systems generate 52,000 Hz sinusoidal or random vibrations in horizontal and vertical modes. The data acquisition system allows for 250 channels, processed on-line.

HYDRA Hydraulic Shaker the 5.5 m diameter 22 tonne octagonal table (flush with the floor) of the 6-DOF shaker is driven by four actuators vertically and two in each lateral direction. Each actuator has a ±70 mm stroke, max 0.8 m/s piston velocity and 630 kN force rating, providing about 5 *g* for a 5 tonne test payload and 3.5 *g* for a 15 tonne mass in the vertical axis. Frequency range 0.1-100 Hz. In addition to the traditional sine dwell and sine sweep tests, HYDRA can provide transient excitations in 6-DOF. The 6-DOF allows the specimen to be sine tested along the vertical/lateral axes with one single set-up. Adding transients was a major HYDRA objective: multidirectional transients at the launcher/spacecraft interface produces a more realistic structural response. The shaker is supplied by Mannesmann-Rexroth.

Large European Acoustic Facility (LEAF)

The 154.5 dB 1624 m³ LEAF, the largest in Europe, became operational in 1990. It provides an internal height of 16.4 m, allowing testing of Ariane 4/5 payloads. The noise generation system, employing pressurised N_2, comprises four horns with cut-off frequencies of 25, 35, 80 and 160 Hz. The overall noise level may be increased to 158.5 dB in the future.

Chamber size: 1624 m³, 9 × 11 × 16.4 m (W × L × H), accessed by 7 × 16.4 m (W × H) door.

Cleanliness level: class 100,000.

Temperature range: 20 ±2°C during test.

Suspension points: 9 of 80 kN capacity, 35 of 15 kN; max crane load 160 kN.

Overall sound pressure level range: 125-154.5 dBL.

Specified octave band pressure levels in empty chamber at 154.5 dBL: 136.5 dBL at 31.5 Hz centre frequency, 141.5 at 63, 147.5 at 125, 150.5 at 250, 147.5 at 500, 144.5 at 1,000, 137.5 at 2,000, 131.5 at 4,000, 125.5 at 8,000.

Field homogeneity in test volume: ±2 dBL.

Control tolerance of sound field: ±1.5 dBL overall.

Noise measurement/data acquisition: 16 microphones, 250 accelerometers, and 50 strain gauges.

Compact Payload Test Range (CPTR)

The CPTR became operational in mid-1992 primarily for measuring the electrical performance characteristics of Ariane 4-class RF-radiating payloads in their operational configurations. It comprises a shielded anechoic chamber with a feed scanner room and control room together with a test preparation area. The PWZ Plane Wave Zone test volume is obtained using two large reflectors in an offset Cassegrain configuration. Payloads of up to 5 tonnes are located at the PWZ centre by means of a positioner providing movement in azimuth, elevation and polar. Test data can be presented in a wide range of formats, including 2-D radiation patterns, contour plots and projections in 3-D. System performance parameters can also be measured. By locating feed horns at different positions in the focal plane (in the scanner room), the direction of the electrical boresight can be changed. In this way, the performance of the satellite transmit /rx from stations at different locations is verified. Similarly, Earth coverage contours or in-orbit reconfiguration can be verified. As the chamber is shielded, it can undertake a range of electromagnetic compatibility measurements.

Internal size: 10.9 × 9.6 × 24.5 m.

Frequency range: 1.5-40 GHz.

Plane wave zone: 7 × 5 × 5 m.

PWZ performance: ±0.2 dB amplitude ripple, ±4° phase ripple, <0.4 dB taper, <-40 dB cross polar.

Reflectors: 9.2 × 8.0 m subreflector, 10.2 × 7.6 m main reflector, less than 100 μm peak-to-peak surface accuracy.

Environmental: class 100,000, 20±2°C, 50±10% humidity.

EMC/ESD Facility

The electromagnetic compatibility/electrostatic discharge facility comprises a shielded anechoic chamber and two operating rooms, including a complete range of automated test equipment. The 20 Hz 40 GHz range satisfies all test requirements for science satellites (low magnetic) and communications satellites (high RF power). All emission and susceptibility measurements are fully automated with on-line data reduction narrow/broadband identification. Data output is corrected for probe antenna factors and compensated for resonance effects.

Test area: 6 × 6 × 4.5 m, accessed by 3.5 × 4.5 m door.

Environmental: class 10,000, 20 to 24°C, 610 to 775 mm pressure, 40 to 55 per cent humidity.

JAPAN

Tsukuba Space Centre

Background

Located in Tsukuba Science City 60 km northeast of Tokyo, TKSC is NASDA's largest technical facility, with seven sections responsible for launcher and satellite R&D, design and testing. The Tsukuba Tracking and Control Centre provides Japan's satellite command node.

TKSC offers a 25 m high 8.5 m diameter thermal vacuum test chamber, a radiometer test laboratory, ion engine test facility, two anechoic chambers (the larger 22 × 25 × 39 m), vibration platforms for 13.6/7.9 tonne payloads, a 5.5 × 3.5 m surface plate for testing satellite separation and solar array deployment, a spacecraft magnetic test facility, and a four-chamber acoustic facility. A 10,950 m² Spacecraft Integration & Test building was completed August 1988 for handling/testing H2-class 2 tonne payloads. 6,000 m² of SITE offers class 10,000 conditions.

Facilities include:

Acoustic chamber

9 × 10.5 × 17 m, 1,607 m³, compressed air generates max 151 dB (31.5 Hz-8 kHz). Concrete walls are 50 cm thick.

Solar simulation chamber

Internally 13 m diameter, 16 m long, <10^{-7} torr, 6 m diameter solar simulator using Xe lamps. 1,150 measuring channels. Evaporated nitrogen is re-liquefied and recycled.

Vibration test

3 × 3 m table to handle 4.5 tonne payloads. 318 kN vertical and 238 kN horizontal, sinusoidal 5 to 100 Hz and random 5 to 200 Hz.

Mass characteristics measurement chamber

One facility for centre of gravity and moment of inertia determination tests payloads up to 6 t/98 kNm². Second facility for aligning satellite sensors and other devices; payload up to 5 tonnes, 10 m high; accuracy ±0.5 mm/±20 arcsec.

UNITED KINGDOM

Central Laboratory of the Research Councils

Background

The Central Laboratory was created 1 April 1995 as an independent entity from the merged Daresbury and Rutherford Appleton Labs, reporting to the Office of Science & Technology. It is operated by the Council for the Central Laboratory of the Research Councils (abbreviated to CCLRC). Its space funding is mostly from PPARC and NERC but some 20 per cent comes from commercial contracts. RAL's Space Science Department is a partner member of the British National Space Centre. RAL is the focus for the UK's space science projects and directs both ESA and bilateral activities, such as the Spectrum-X astronomical satellite collaboration with the Soviets. It hosts 200 personnel.

Centralised facilities for university research groups include a satellite operations centre, data processing facilities, and the Starlink data analysis network, the British Atmospheric Data Centre and assembly/testing of space payloads. Projects include:

Badr-B

CCD camera for cloud monitoring. Chilton will act as back-up to Lahore ground station. The same camera is used for Morocco.

Cassini

Participation in Saturn orbiter's dust analyser and plasma electron spectrometer, and Titan lander's surface science package.

Cluster

Participated in the development of the four satellites' 16 plasma sensors.

SOHO

Led development of Coronal Diagnostic Spectrometer.

ISO

RAL managed development of the Long Wavelength Spectrometer.

Spektrum-X

Participation in the JET-X grazing incidence multimirror x-ray telescope.

IUE

Managed UK participation until shutdown in 1996.

Rosat

Participated in development of the Wide Field Camera.

Minisat

Built star tracker for Legri gamma telescope.

UARS

Joint developer of the Improved Stratospheric & Mesospheric Sounder, incorporating a Stirling cycle cooler now available from Matra Marconi Space.

ERS 1/2
ATSR Along-Track Scanning Radiometer developed by RAL-led consortium. Advanced ATSR calibration carried out at RAL. The centre provides data processing for all three.

IRAS
Controlled from RAL's Satellite Control Centre, via the 12 m antenna. The IRAS database is available at RAL for the UK astronomical community.

Ginga
Leicester University/RAL provided the Large Area Counter to Japan's X-ray satellite (re-entered 1991).

Solar Max
Major collaboration in the X-ray Polychromator for studying solar flares.

Spacelab 2
CHASE Coronal Helium Abundance Spectrometer Experiment developed by RAL/Mullard Space Science Laboratory.

UKS
RAL and MSSL satellite and sub-satellite.

Yohkoh
Developed Bragg Crystal Spectrometer with MSSL and US Naval Research Laboratory.

Rosetta
Management and a system engineering for sample analysis instrument on the lander and mother spacecraft to visit comet Wirtanen.

EOS
Management of High-Resolution Dynamics Limb Sounder to measure global temperature and atmospheric chemical species.

Polar
RAL provided hardware on two instruments to study Earth's magnetosphere

Facilities include:

Clean rooms
A 13 × 8 m room containing 4 × 3 m Class 100 tunnel and three class 100 laminar flow benches; a 90m ^{2} room housing 7.5 × 3.5 m Class 100 working area in form of vertical laminar flow downdraft units

Thermal vacuum
5.5 × 3.0 m diameter tank with cryopumps down to 10^{-10} atmosphere −190/100°C, plus 1.7 × 1.0 m diameter with turbomolecular pump down to 10^{-10} atmosphere, −170/100°C. Bakeout Tank: 90 × 60 cm diameter with LN_2 traps on diffusion pumps down to 10^{-10} atmosphere. Ambient temperature to +150°C.

Vibration facility
Electromagnetic vibrator with 90 × 90 cm slip table capable of vibrating a 35 kg mass to ESA/NASA specs in three-axes. Max displacement ±19 mm, max rate 1.78 m/s. The facility has a cryostat to fit on the vibrator for testing 30 kg objects at down to 4 K. A second electromagnetic vibrator can test a 5 kg mass to ESA/NASA specs. Maximum displacement ±22 mm, max rate 1.2 m/s.

UNITED STATES OF AMERICA

Applied Physics Laboratory of The Johns Hopkins University

Background
Founded in 1942, APL is a not-for-profit R&D laboratory and independent division of The Johns Hopkins University. It operates primarily under a contract with the US Naval Sea Systems Command. The DOD sponsors some three-quarters of APL's effort. Full-time staff total 2,700, of which 400 work within the Space Department's two branches, named Engineering and Technology and Science and Analysis.

APL is a major centre for satellite development, construction and operation. Through 1996, it had built 57 satellites (45 of them 50-180 kg). The Laboratory gained early experience by putting instruments on captured German V2 rockets after World War II. Based on its analysis of Sputnik 1's Doppler signals, APL conceived and built the Transit satellite navigation system for USN. It built and operated Geosat, provided the altimeters on Geos-C, Geosat, Seasat and Topex, and led the AMPTE project for NASA. Topex also includes APL's laser retro-reflectors and frequency reference unit, the most stable oscillator ever orbited. Earlier satellites include Dodge, Anna, Triad, Traac, Magsat, Nova and three small astronomy satellites, Injun, Hilat, Polar Bear. The HUT Hopkins UV Telescope (a joint project with the Department of Physics & Astronomy) flew on Shuttle's 1990 Astro 1 mission and March 1995's Astro 2. Voyager, Galileo and Ulysses carry particle radiation detectors. Japan's Geotail includes the EPIC Energetic Particles and Ion Composition experiment. Sweden's Freja and NASA's UARS carry APL magnetic field instruments.

Current projects include the NEAR Near-Earth Asteroid Rendezvous craft (launched February 1996), Mid-course Space Experiment (MSX, launched April 1996), the ACE Advanced Composition Explorer, the TIMED spacecraft and the Magnetospheric Imaging Instrument for Cassini. Facilities include:

Richard B Kershner Space Integration and Test Facility
The test facility, 7,350 m², was dedicated in 1983. It houses seven clean rooms: three class 100,000 (139 + 2 × 93 m²); two 93 m² class 10,000; two 22 m² class 100 rooms are suitable for assembling small space instruments; they can also be held sterile for assembling implantable biomedical devices.

The Space Simulation Laboratory
The SSL includes two 2.4 m diameter 3 m high vertical thermal vacuum chambers (−196/+121°C; 10^{-7} torr cryopump and diffusion pump), a 0.91 m diameter 0.91 m long space simulation chamber (−78/+126°C; 10^{-8} torr) and a 1.52 × 1.83 × 2.13 m T/humidity/altitude chamber (−100/+125°C; 35 to 95 per cent humidity; 0 to 61 km altitude), in its 447 m² area.

The Vibration Test Laboratory
A facility that offers: two Unholtz-Dickie shakers (T4000: 178 kN, 152 × 152 cm Team bearing-line slip table; T1000: 89 kN, 4.4 cm peak-to-peak displacement, 61 × 61 cm combined slip table assembly) and an MRC mass properties machine (2,270 kg capacity).

The Solar Array Laboratory and Simulator Control Room
A Large Area Pulsed Solar Simulator (LAPSS) characterises and calibrates solar panel and coupon performance using a Xe source. LAPSS can calibrate panels up to 1.83 m across in 1 ms. Smaller laboratories serve the Satellite Reliability Group, where new parts are inspected and radiation tested.

Satellite Communications Facility
Originally developed as a dedicated Transit station, the 623 m² facility has been upgraded for S/L/X-band station for a wide range of NASA and DoD satellites. It is used for continuous MSX operations (a 450 m² addition houses a new command/control centre and a data-processing centre for MSX). Antennas: 18.3 m parabolic, 2.2 to 2.3 GHz autotrack feed, receive G/T 23 dB/K, simultaneous RH/LHCP. A 5 m parabolic, 2.2 to 2.3 GHz + 1.650 to 1.750 GHz feeds with selectable RH/LHCP, rx G/T 14 dB/K at 2.2 GHz; 10 m parabolic X/S/L-band simultaneous coverage, fully operational SGLS site with 0.2/2 kW transmit, associated 1 m acquisition antenna. Real-time computers steer the dishes and redundant back-up control and configure the 10 m dish.

Time and Frequency Standards Laboratory
Supports APL projects requiring highly stable time (1 pps), frequency (1, 5, 10 MHz) and time code signals (IRIG-A/B). Other activities include measurement of frequency standards stability, design of atomic hydrogen masers, spaceborne oscillators and synthesisers, and superconducting oscillators.

Anechoic Chamber
7.62 × 7.62 m internal, accessed by 1.83 × 2.44 m door, attenuates outside interference by >70 dB over 0.1 to 10 GHz.

Arnold Engineering Development Center (AEDC)

Background
AEDC operates the largest and most advanced complex of aerospace simulation test facilities in the world, with some 40 units for simulating conditions up to 1,600 km altitude and speeds up to 32,000 km/h. Construction of the J6 Large Rocket Test Facility, costing more than US$200 million, began in FY89 as a national site for handling large solid propellant rocket motors at simulated 30 km altitudes from late 1994.

Los Alamos National Laboratory (LANL)

Background
LANL, founded in 1943, is a multidisciplinary research facility operated by the University of California for the Department of Energy, undertaking research in nuclear and non-nuclear defence programmes, nuclear safeguards/security, biomedical sciences, computational sciences, materials science, environmental cleanup and basic science. Its first director was J Robert Oppenheimer, who helped to found the Laboratory to develop the first atomic weapons. It encompasses 111 km² of mesas and canyons in northern New Mexico, has an annual budget of about US$1 billion and employs 7,000. It houses a number of interdisciplinary research centres, including the Center for Materials Science, the Advanced Computing Laboratory, the Center for Nonlinear Studies, the Center for Human Genome Studies, a branch of the Institute for Geophysics & Planetary Physics, the Center for International Security Affairs, and the Exploratory Research & Development Center.

Los Alamos continues to have responsibility for monitoring outer space for nuclear explosions, employing its detectors aboard the DSP early warning and lower altitude GPS navigation satellites. DSP also carries LANL's Magnetic Plasma Analyser and Synchronous Orbit Particle Analyser to monitor spacecraft charging and radiation environment. Its Alexis satellite was launched in 1993 to conduct an all-sky survey of the soft x-ray background glow. LANL/Sandia National Laboratory launched their FORTE Fast On-orbit Recording of Transient Events satellite on Pegasus to demonstrate technology to look for pulses from low-technology nuclear explosions.

LANL provided the x-ray detectors for the HETE mission, three instruments on NASA's Polar, four imaging spectrometers on ESA's Cluster, two γ-ray detectors on Mars-96 and a γ-ray spectrometer on Rosetta's comet lander. It is collaborating with Finland's VTT to provide Cassini's Ion Beam Spectrometer. LANL operates the only US facility capable of making Pu-238 RTGs; Cassini has three, Galileo two and Ulysses one. The laboratory provided technical direction for SDI's BEAR (Beam Experiments Aboard a Rocket) flight of July 1989, which fired the first neutral particle beam in space.

Naval Research Laboratory/Office of Naval Research

Background
NRL was established in July 1923 as the Naval Experimental & Research Laboratory. Now an element of ONR, it is the navy's corporate laboratory. Staff total about 3,500, mostly civilian, in Washington DC, Stennis Space Center (Bay St Louis, Missouri), Monterey (California) and Orlando (Florida). NRL's Naval Center for Space Technology (NCST) maintains space technology expertise through two departments: Space Systems Development and Spacecraft Engineering. The Remote Sensing Division (RSD) and the Space Science Division (SSD) come under NRL's Ocean & Atmospheric Science & Technology Directorate.

NRL directed the Vanguard programme and since 1960 has developed more than 80 satellites, most recently the Clementine technology demonstrator lunar/asteroid probe in conjunction with the Lawrence Livermore National Laboratory. It has also flown numerous payloads and experiments, such as the Hercules unit carried by STS-53 in December 1992 to allow astronauts to photograph surface features while automatically recording positions within 2 km. It flew again on STS-56 in April 1993 and on STS-70 in July 1995. Another NRL project, the HTSSE High Temperature Superconducting Space Experiment flew aboard STP's Argos satellite.

The RSD undertakes a broad programme in sensing applications over frequencies from UV to radio. Sensor systems include RAR Real Aperture Radar, scatterometers, lidars, optical/radio interferometers and passive microwave imagers. Its middle atmosphere sensors include the POAM Polar Ozone & Aerosol Monitor on Spot 3, the MAS Millimetre-wave

Atmospheric Sounder on Shuttle Atlas missions (three flown), and a suite of water vapour and ozone monitors as part of the Network for Detection of Stratospheric Change.

RSD's facilities include digital image processing, a tactical environmental visualisation centre, an aerosol measurement facility, 25.6/25.9 m radio antennas at Maryland Point Observatory and an optical interferometer at Mount Wilson Observatory for monitoring background environmental emissions at high angular resolution.

SSD maintains facilities for designing, constructing, assembling, calibrating and analysing space experiments, principally upper atmosphere ultra-violet sensing, solar atmosphere spectrometry and celestial radiation over ultra-violet to cosmic rays. For example, it provided the SUSIM Solar ultra-violet Spectral Irradiance Monitor for UARS and Atlas, the RAIDS Remote Atmospheric & Ionospheric Detection System to measure airglow, the OSSE Oriented Scintillation Spectrometer Experiment gamma instrument on NASA's Compton Observatory, solar x-ray measurements using a Bragg crystal on Japan's Yohkoh, and the LASCO Large Angle Spectrometric Coronograph on Soho. SSD provided three of the main instruments (USA, HIRAAS, and GIMI) on Argos.

Phillips Laboratory

Background

An element of the AF Space & Missile Systems Center, the Phillips Laboratory was founded in December 1990 as part of the AF's reorganisation of the 14 existing laboratories and research centres into four 'super laboratories'. Phillips studies military space, missile, directed energy, propulsion and geophysics R&D. It replaced the USAF Space Technology Center (created October 1982) at Kirtland, the USAF Weapons Laboratory (Kirtland), AF Geophysics Lab (Hanscom) and AF Astronautics Lab (Edwards). In fiscal year 1996, the air force provided a budget request of US$510 million; FY95 US$629 million; FY94 US$552 million; FY93 US$756 million. Personnel at Kirtland total 576 civilian and 442 military. Hanscom has 311 civilians and 66 military. Edwards employs 184 civilian and 99 military.

The laboratory's seven technical directorates are:

Advanced Weapons and Survivability

AWS performs and directs R&D of technologies for directed energy weapons (microwave, laser and particle beam) and related advanced weapon concepts, space systems survivability and EM effects. It measures and models directed energy effects on air and space systems and determines lethality to threat targets, develops hardening countermeasures and transitions survivability technology to product and logistic centres. High-energy density plasma and radiation source research and pulsed power development are performed at the Shiva Star facility, a 10 MJ 3×10^{12} W fast capacitor bank, one of the nation's largest.

Lasers and Imaging

This new optical observatory began operating 1994 as part of the Starfire Optical Range. The 3.5 m telescope, the DoD's largest and 12th in the US, augments the existing 1.5 m telescope and 1 m laser beam director for research into space object tracking and atmospheric compensation techniques. This directorate also operates the AF Maui Optical Station in Hawaii, devoted to satellite and rocket tracking and imaging. Its existing 1.6 m and 2×1.2 m telescopes was to be enhanced with a US$120 million 3.67 m telescope for completion in early 1997 and fully operational in spring 1998. The directorate also operates the NC-135 Argus aircraft as a flying laboratory for space imaging work. The former AFWL handles the directed energy weapon research for BMDO, which included the Alpha hydrogen fluoride chemical laser. It was also responsible for BMDO's Relay Mirror Experiment.

Space and Missiles Technology

SMT focuses on spacecraft structures and controls, power/thermal management, sensors, communications, electronics, satellite control and simulation, orbital dynamics, ballistic missile technologies and reusable launch vehicles. Its Thermionic System Evaluation Test is testing an unfuelled Topaz 2 Russian nuclear reactor. The Space Nuclear Thermal Propulsion Program aimed at developing an advanced nuclear rocket engine with twice the specific impulse of the best current liquid engines, but it was cancelled May 1993 on cost grounds, after expending US$55.5 million FY93 plus US$65 million FY92.

Space Experiments

This branch conducts balloon and sounding rocket payload integration, and designs, integrates and tests whole spacecraft systems. Its MightySat programme will provide low cost and regular access to space for testing new technologies. It is responsible for the TAOS mission of the Space Test Program, and BMDO's HABE programme supports ATP systems for boost phase DEW missile defence.

Geophysics Directorate

The centre for air force research into the near-Earth space environment, the geophysics directorate includes ionospheric, meteorological, optical backgrounds and Earth dynamics measurements in its charter. All research is related to military space applications, such as radar propagation, crustal motion effects on ballistic missiles, and vehicle IR signatures against natural backgrounds. Its projects include MSX.

Propulsion

This directorate undertakes R&D of advanced solid, liquid, hybrid and electric missile and space propulsion on a 170 km^2 site in Edwards AFB's northeast corner. Facilities include test stands accommodating engines up to 45 MN thrust and altitude chambers in which attitude control thrusters can fire up to 7 hours continuously. The National Hover Test Facility to support BMDO was inaugurated in November 1988 by the first hovering trials of a space-based interceptor demonstration model.

Airborne Laser

The ABL programme is developing an airborne demonstrator of a high energy laser for defence against theatre missiles in the boost phase. Two contractor teams: Boeing (lead)/TRW/Lockheed and Rockwell (lead)/Hughes/E-Systems. From the two concept designs produced by the current Phase I, one team will be selected to develop a demonstrator aircraft, culminating in a lethality demonstration against boosting missiles after 2000.

SPACE MONITORING FACILITIES

RUSSIAN FEDERATION

The SSS Space Surveillance System

Background

The USSR, during 1969-70, officially established the SKKP System for Monitoring Outer Space using the then recently deployed first generation BMEWS radars, known in the USSR as Dnepr. In 1970, the Soviet satellite catalogue was more than 250, or about 10 per cent of the US catalogue. Today, Russia's SSS Space Surveillance System relies primarily on Dnepr and Darayl-UM radars operating in the VHF range near 150 MHz at eight sites. The Russian sites are Irkutsk, Murmansk and Pechora in Russia, Sevastopol and Uzhgorod in Ukraine, Balkhash in Kazakhstan, Mingechaur in Azerbaijan, and Riga in Latvia (Riga's Darayl-UM was scheduled for 1995 demolition. Dnepr should continue through 1998). Another site at Baranovichi was under construction, but its status is unclear. Operation of many of the space surveillance and TT&C sites is now erratic, with facilities dropping out and then returning to service.

C-band and higher frequency radars are beginning to be acknowledged. Observations are also provided to the SSS by two ABMD Anti-Ballistic Missile defence radars in the Moscow region near Sofrino and Chekhov operating at UHF near 400 MHz. This combined network generates some 50,000 observations daily to maintain a catalogue of nearly 5,000 objects, most in LEO. Owing to geographical limitations, Earth satellites at low inclinations are difficult or impossible to track. As a whole, the SSS radar sensors appear to be limited to a range of about 4,000 km.

Also supporting the SSS are more than 20 optical and electro-optical facilities at 14 locations: Irkutsk, Kourovka, Krasnodar, Uzhno-Sakhalinsk and Zvenigorod in Russia, Kiev, Odessa, Simeiz and Uzhgorod in Ukraine, Alma Ata in Kazakhstan, Burokan in Armenia, Abastumani in Georgia, Dushanbe (Hissar) in Tadjikistan, and Ashgabad in Turkmenia. Seven of these sites are supported jointly by SSS and the Russian Academy of Sciences (RAN), while five are run independently by SSS and two by RAN. A wide range of Soviet and German instruments of optical diameters 45-100 cm are employed, including the AZT-8, AZT-14, VAU, SBG and Zeiss-series telescopes. As in the US, these optical and electro-optical facilities perform the majority of deep space observations, especially in the eastern hemisphere GEO arc. RAN previously maintained sites in Bolivia, Chile, Ecuador and Egypt.

In addition to generating positional and ephemeris data, the SSS maintains a catalogue of satellite characteristics such as size and stability. Papers published by RAN scientists reveal an ability to discern the configuration (shape and materials) of high altitude satellites through an inverse photometric process by extended observations at multiple wavelengths, typically U/B/V/R. One of the principal instruments involved in this research is at Simeiz. The 101.6 cm diameter Zeiss-1000 with 2.5 arcmin FOV and about 17.5 limiting magnitude is augmented with a PG-3M two-stage cascade image tube with a 256 × 256 CCD.

UNITED STATES OF AMERICA

Ground-based Electro-Optical Deep Space Surveillance (GEODSS)

Background

A more recent monitoring system than the Space Surveillance Network, GEODSS is also operational. Three sites around the world each house three telescopes on rapid-motion mounts positionable to within 1.5 arcsec. The air force claims they are able to identify football-sized objects in orbits between 5,500 to 37,000 km. They supersede the network of Baker-Nunn film cameras in place since 1958 (of which the remaining example at St Margarets, New Brunswick, Canada was closed April 1992). Data from GEODSS's TV image are relayed to SCC's computers. If a satellite is not detected immediately at its expected co-ordinates, a search for it is automatically triggered on-site. The sites, at Socorro (New Mexico), Choe Jong San (South Korea) and Maui (Hawaii), were operational by 1983. A fourth, on Diego Garcia in the Indian Ocean, was completed in February 1987. A fifth was originally planned for Portugal, but cancelled. The Korean site closed in 1993 to save money (the weather made it a poor site). Each site has three observatories: two with a 102 cm 2° FOV Cassegrain to look for faint (to +16th magnitude) slow objects at high altitude and the third with a 38 cm 6° FOV Schmidt to search lower altitudes for faster objects (Diego Garcia has three Cassegrains and no Schmidt). MIT Lincoln Laboratories continue to operate the GEODSS prototype for R&D. The air force is interested in flying a similar system in space, believing an operational system would cost no more than operating an international ground system, nor would cloud cover be a constraint.

Maui Space Surveillance Site

Background

The USAF's Phillips Laboratory's Lasers & Imaging Directorate operates several facilities for testing advanced imaging and acquisition, pointing and tracking techniques. The Maui Space Surveillance Site at 3,000 m on the summit of Mount Haleakala comprises GEODSS, the AMOS AF Maui Optical Station and the MOTIF Maui Optical Tracking & Identification Facility. MOTIF's paired 1.2 m telescopes became dedicated SSN sensors in October 1979 and provide satellite imaging, tracking and infra-red signature data down to +19th magnitude. AMOS's adjacent 1.6 m telescope with deformable mirror coupled to a 6,940 Å laser radar is primarily for research but also contributes to SSN. Rockwell Power Systems under a five year contract operate MOTIF and AMOS. Under development is the AMOS Daylight Optical Near Infra-red System (ADONIS), which will extend imaging capabilities to 24 hours daily. A major addition to AMOS is the AEOS (Advanced Electro-Optical System): the US$120 million 3.67 m telescope was completed in early 1997 and fully operational by Spring 1998.

NAVSPASUR

Background

The Space Surveillance Network includes Naval Space Command's electronic fence, known until 1993 as NAVSPASUR. It is a series of 216.980 MHz transmitters and receivers, operational since 1961, forming an electronic fence across the southern US, stretching 4,800 km from Georgia to California, 1,600 km off each coast and 24,100 km into space. Three transmitters are sited at Lake Kickapoo (Texas; primary transmitter, 766.8 kW, array 3,269 m long), Gila River (Arizona; 40.5 kW; 497 m) and Jordan Lake (Alabama; 38.4 kW; 314 m). They work in conjunction with six receivers at San Diego (California), Elephant Butte (New Mexico), Silver Lake (Mississippi), Red River (Arkansas), Tattnall and Hawkinsville (Georgia). The Hawkinsville and Elephant Butte sites operate 25 to 366 m arrays for high altitude satellites; the others use 12 to 122 m systems for lower objects. The data are processed at NAVSPACOM headquarters in Dahlgren (Virginia), which since 1984 has also served as the Alternate Space Control Center.

Starfire Optical Range

Background

A new optical observatory began operating 1994 at Kirtland AFB in New Mexico as part of the Phillips Laboratory's Starfire Optical Range. The 3.5 m telescope, DoD's largest and 12th largest in the US, joined an existing 1.5 m telescope and 1 m laser director for research into space object tracking and atmospheric compensation techniques. The directorate also operates the NC-135 Argus aircraft as a flying laboratory for space imaging research. The 1.5 m mirror carries 341 actuators 7 mm apart on its 1 mm-thick glass. The system samples atmospheric distortion 1,700 times/s by observing laser scattering off molecules in the lower atmosphere, and the continuously adjusts the deformable mirror. In 1993, it produced the first important astronomical observation via adaptive optics: ionised hydrogen clouds around Orion nebula stars.

The Maui Space Surveillance Site on Mount Haleakala is devoted to satellite and rocket tracking and imaging. MOTIF (top left) uses paired 1.2 m telescopes for satellite imaging, tracking and IR signatures. AMOS's 1.6 m telescope (top right) is coupled to a laser radar. GEODSS is housed in the three identical domes. The two small central domes illuminate targets with lasers for night imaging. The AEOS Advanced Electro-Optical System 3.67 m telescope with adaptive optics will be added to AMOS (Phillips Laboratory)

USSPACECOM Space Surveillance Network

Background

Currently, NORAD, under the control of the US Space Command's Space Control Center and Missile Warning Center inside the hollowed-out Cheyenne Mountain, monitor all space debris and active satellites. The 15 steel buildings house a series of control rooms and offices, cushioned on steel springs inside vast chambers cut into the granite mountain. More than 1,100 personnel (US Army, Navy, Marine Corps, Air Force, Canadian Forces and civil technicians) keep the complex permanently operating. When the site was chosen in the 1950s, it was believed to be impregnable to a nuclear attack.

The centre issues warnings when objects appear likely to pass within 5 km radially, 25 km downtrack or 5 km out of orbital plane of the Shuttle. The Shuttle makes avoidance manoeuvres whenever the values are 2/5/2 km, respectively. Collision avoidance information is increasingly important. Facilities include:

HAX Haystack Auxiliary

The HAX Haystack Auxiliary 16.7 GHz 13 m dish, also known as the NEAR Near-Earth Assessment Radar, became operational in 1993, although it cannot detect low inclination objects. However, it should handle the new 51.6° Station orbit well. The main 36.6 m diameter 400 kW 10 GHz Haystack Long Range Imaging Radar has been used periodically since 1990 to observe LEO debris, capable of seeing objects down to 1 cm at 1,000 km. One discovery is that Russia's retired nuclear-powered Rorsats appear to be leaking liquid metal coolant; some 70,000 droplets were found around 900 to 1,000 km. Six solid spheres (two 15 cm diameter, 5 kg, aluminium; two 10 cm diameter, 4.3 kg, stainless steel and two 5 cm diameter, 0.53 kg, stainless steel) were released from STS-60 in February 1994 to help calibrate radar and optical tracking systems. The primary objective was to fine tune the Haystack Radar. STS-63 in February 1995 deployed a second array. This array had a 5 cm 0.53 kg polished stainless steel, 10 cm 4.24 kg white aluminium and 15 cm 5.0 kg black aluminium spheres, as well as three platinum iridium alloy dipoles. Two of the dipoles were 1.5 g, and 13.3 cm long, the other one was 0.5 g and 4.42 cm long. All were 1.0 mm diameter).

BMEWS

Development of US space surveillance, operated by the US Air Force with help from the UK, Canada and other countries, began in the late 1950s to cover both missile firings and satellite launches. As the USSR developed ICBMs, the West responded with the BMEWS Ballistic Missile Early Warning System. It extended the then existing range of tracking radars from a few hundred km to about 5,000 km. The most probable flight path of a Russian ICBM attack is over the Arctic, so three BMEWS radar stations were built at Thule in northern Greenland, at Clear in Alaska and at Fylingdales, northern England. Thule and Clear had detection radars the size of football fields to pick up a missile as it rises above the horizon. Trackers (Thule and Clear each had 25.6 m dishes; Fylingdales had three but no detection radar).

The BMEWS system has undergone a major update. The first site, Thule, which covers the northern perimeter, was given a new US$80 million phased-array antenna. This is a 50 m high, 122 m long structure, providing two phased-array faces accommodating 2,560 active elements and 1,024 additional elements per face. It began test transmissions in 1987 and was completed in summer 1988. A similar array was constructed at Fylingdales Moor in Yorkshire, England, operated by the RAF. The array replaced the familiar giant 'golf ball' radomes that dominated the landscape of this area of northern England (the radomes were dismantled June 1994). Raytheon completed the new US$200 million system in 1992. Unlike Thule, it has three 25.5 m diameter active faces, providing 360° coverage and improving detection of SLBMs. It can track 800 objects simultaneously. The warning time is about the same as under the old system but radar resolution is much improved. Upgrading the Clear facility is planned but not yet funded. It is often the first to pick up a Plesetsk launch.

Passive Space Surveillance System

Air Force Space Command has recently deployed a new sensor network: the PASS Passive Space Surveillance System using satellite RF emissions to compute angle of arrival observations. The air force deployed tow versions by 1994, the DSTS Deep Space Tracking System and LASS Low Altitude Space Surveillance. A third, CROSS Combined RF-Optical Space Surveillance, became operational August 1991 in San Vito, Italy using an optical telescope and an RF receive antenna but was closed in 1993. The LASS sites in Edzell (Scotland), San Vito (closed 1993) and Osan (Korea), which became operational May 1992, provide RF surveillance of near-Earth satellites. DSTS sites at Feltwell (England), Grifiss (New York) and Misawa (Japan) provide worldwide coverage of geosynchronous satellites.

PAVE PAWS

Detection and warning of SLBMs started in 1968 with an interim system based on eight radar sites located on the Atlantic, Pacific and Gulf coasts. They could detect the launch of a missile from a patrolling submarine and then track its course. It was replaced in 1980 by the USAF/Raytheon Pave Paws, a pair of solid-state 420 to 450 MHz (carriers 422/435/439 MHz) 31 m diameter radar systems on the US east/west coasts. With no mechanical parts to limit the speed of its radar scan, it embodies phased-array technology – thousands of small radar antennas co-ordinated by two large computers – allowing it to detect, track and predict the impact point and numbers of missiles, the technology now extended to BMEWS. Until 1995, four Pave Paws sites provide continuous overlapping coverage from the northeastern US, across the southern periphery and up to the northwest. All four sites also fed display data on the position and velocity of satellites to Space Command. They have achieved 99 per cent reliability. Additional automated data processing equipment and security improvements became operational in 1992. The southeast and southwest sites at Robins AFB in Georgia and Eldorado AS in Texas became operational in 1987 (but were de-activated in 1995), joining the two original operational sites at Cape Cod AS in Massachusetts and Beale AFB in California.

It was the development of Pave Paws that led to the abandonment of Over-The-Horizon Radar (OTHR), which went into operation briefly in March 1968. This was to meet the threat of the fractional orbit bombardment system (FOBS), Soviet missiles launched into orbit via the South Pole, thus avoiding detection by BMEWS. They would travel three quarters of the way around the world before re-entering to strike their target. OTHR was not a conventional radar system, consisting instead of a series of transmitters generating a continuous signal that reflected back and forth repeatedly between the ionosphere and Earth's surface until received at another station several hundred km away. A missile penetrating the ionosphere disturbed the signal and was thus detected. One of these stations operated at Orfordness on England's east coast for several years until it was closed down in 1974. The problem was that the ionosphere was not always suitably positioned – such as at night, when looking north. One solution proposed space mirrors for heating the ionosphere, making it conform to military requirements.

Cobra Dane

Another phased-array radar with greater sensitivity also performed dual missile warning and space surveillance until 1994. The Cobra Dane radar (AN/FPS-108, 1,175 to 1,375 MHz) on Shemya Island (52.7° north/174.1° east) at the western edge of the Aleutian Islands served from 1977 as a primary source of data on Soviet ICBM and SLBM tests impacting near Kamchatka Peninsula. Its location ensured that it was often the first US radar to track Soviet space launches. A major upgrade to the system was completed in 1993; testing was still under way in early 1994 – but it was closed down 1 April 1994. The new Cobra Dane radar could track more than 100 objects simultaneously, identify targets more rapidly, provide more accurate measurements and maintain a significantly larger (12,000 *vs.* 5,000) catalogue. Cost of the Raytheon upgrade was more than US$58 million.

Other Phased-Array Radars

SSN made the first use of the Eglin AFB phased-array radar, consisting of two fixed electronic arrays set in a 10-storey high bank of concrete sloped at 45° to the horizon providing 120 to 240° azimuth coverage. One array, of 5,184 transmitter modules, emits an electronically steered beam that sweeps the area of coverage in ms; the other array, 4,660 modules in a hexagonal arrangement, receives the signal as it bounces back from an orbiting satellite. Florida was chosen for the first of these arrays because most space objects pass within its coverage daily. The Eglin radar system was to be shut down when the Robins Pave Paws site received a power upgrade. However, concerns during 1988 about the Robins upgrade led to an indefinite postponement. Transmitter/receiver upgrades were completed in 1995 at Eglin, where 140 civilian and more than 30 military personnel maintain the facility. Eglin's primary mission was shifted in 1987 to space surveillance. Since the mid-1970s it has provided most of SSN's radar cross-section data. The second oldest phased-array system, the PARCS Perimeter Acquisition Radar Characterisation System, was originally constructed to support the US Safeguard ABM system in North Dakota in the 1970s. Although it is one of the most capable sensors in the Space Surveillance Network, the air force is now considering its closure.

The UHF 26 m radar at Pirinclik in Turkey, operational since 1969, provides early tracking of missiles and space launches from Tyuratam and surveillance of GEO satellites over the eastern hemisphere. C-band radars at Kaena Point (Oahu), Antigua and Ascension Island also provide important early observations: Tyuratam's manned, GEO and deep space missions pass directly over Ascension during the first orbit. Kaena Point sees Tyuratam's high-inclination missions during the first half rev; Antigua tracks most LEO satellites in orbit before Conus sensors.

LAUNCH VEHICLES (ORBITAL AND SUB-ORBITAL)

LAUNCH VEHICLES (ORBITAL AND SUB-ORBITAL)

BRAZIL

Sonda

Background

Sonda 1 was a two-stage vehicle first launched 1964, and flown more than 200 times before retirement. Sonda 2's development then took over in 1966 and has flown about 60 times, offering 44 kg capacity to 80 km. Single stage, 300 mm diameter, 4.1 m long, 360 kg launch mass. Three versions are still employed, principally for technology-proving flights. Sonda 3 development began 1969 to provide a two-stage vehicle capable of delivering 50 kg to 500 km, providing three-axis payload control and sea recovery. The S3 basic version handles 50 to 80 kg to high altitudes using the Sonda 2 as the S20 stage 2. A reduced Sonda two-stage 2 (S23) is carried for S3-M1 missions with 130 to 160 kg payloads destined for lower altitudes. (The M2 version, using the smaller S24, has been used once, 23 August 1979.)

Preliminary studies of a Sonda 4, began in 1974 and produced the decision to launch five prototypes for vehicle qualification and as technology demonstrators for the VLS (Velculo Lancador de Satelites) satellite launcher, which incorporates clustered Sonda 4 stage 1 motors. Four launches have been made: 21 November 1984, 19 November 1985, 8 October 1987, and 28 April 1989. Stage 2 failed to separate in 1987's flight. The programme has developed 300M steel for the motor casings, digital control system (although Sonda has so far carried an analogue system), payload fairing ejection and three-axis control (including liquid injection TVC and movable nozzle TVC). MBB provided technical assistance in developing the LITVC system. #4 demonstrated movable nozzle TVC for the first time on stage 2.

IAE, under the aegis of the Ministry of Aeronautics, has developed the Veiculo Lancador de Satelites. Originally intended to launch the MECB satellites. But the programme ran afoul of the Missile Technology Control Regime signed by the G-7 group of the world's seven most industrialised nations. Many components such as the liquid roll control package planned forward of stage 2 and replaced by solid motors (but is now again liquid), had to be reduced to less capable technology.

Brazil had major problems acquiring the inertial guidance platform and the vehicle has yet to be launched. The four flights of the original MECB were planned in 1989 to begin 1992, followed by the second in 1993 and the remaining two during 1994. This never materialised but the first launch attempt for VLS-1 took place on 2 November 1997. It failed to reach orbit. The 50 tonne four-stage all-solid vehicle stands 19 m high, capable of placing 200 kg into a 750 km circular 25° orbit or a 450 km Sun-synchronous orbit.

The strap-ons and two of the three core stages are derived from Sonda 4's stage 1. The four strap-ons ignite on the pad, employing nozzle flexure for steering. The core stage 1 is nested within the strap-on cluster and ignites at 20 km altitude. A liquid stage 1 is under consideration as a future upgrading. Stage 2 is a shortened version of the S-43 motor. The stage 3 orbit injection motor is newly developed and spin-stabilised with a fixed nozzle. By end-1995, two of four planned strap-on static firings had been made, two of three stage 1, four of five stage 2 and four of six stage 3. The 6.6 tonne 9.52 m 'VS-40' vehicle was launched 2 April 1993 on a 24 minute, 1,248 km altitude flight testing VLS's stages 2/3. A 9.76 tonne 12.0 m VS-43 vehicle using VLS stage 1/3 could deliver 1.2 m diameter 200 to 500 kg suborbital payloads to 1,000-2,000 km.

A ⅓ VLS model was launched May 1989 to demonstrate strap-on separation (IAE/CTA)

Sonda 3/3-M1
First launch: 26 February 1976
Number launched: 28 (2 failures: 31 October/14 November 1983) to end-1996; M1 4 flights, successful
Number of stages: 2 (solid propellant)
Overall length: 8.0 m both versions
Principal diameter: 557 mm
Launch mass: 1,521 kg (M1: 1,527 kg), excluding payload
Typical performance: 60 kg to 600 km (M1: 140 kg to 275 km).

Sonda 3/3-M1 stage 1
Overall length: 373 cm
Principal diameter: 557 mm
Stage mass: 1,205 kg both versions
Propellant mass: 860 kg both versions
Average thrust: 102 kN sea level both versions
Burn time: 24 s.

Sonda 3/3-M1 stage 2
Designation: S20 (M1: S23)
Overall length: 290 cm (M1: 160 cm), excluding payload
Principal diameter: 300 mm both versions
Stage mass: 316 kg (M1: 182 kg)
Propellant mass: 229 kg (M1: 113 kg)
Average thrust: 33 kN vacuum (M1: 18 kN)
Burn time: 15 s (M1: 15 s).

Sonda 4
First launch: 21 November 1984
Number of launches: 4, to end-1995 (8 November 1987 stage separation failure)
Number of stages: 2 (solid propellant)
Overall length: 11.0 m
Principal diameter: 1,008 mm
Launch mass: 6,800 kg, excluding payload
Typical performance: 500 kg to 650 km
Guidance: the inertial platform and control system are housed in the equipment bay atop stage 2. Three-axis control during stage 1 burn is provided by LITVC (2.5° max vector deflection; 81.6 atmospheres injection by N_2; 9 ms valve response time; Three valves per quadrant) for pitch/yaw and thrusters for roll.

Sonda 4 stage 1
Overall length: about 537 cm
Principal diameter: 1,008 mm

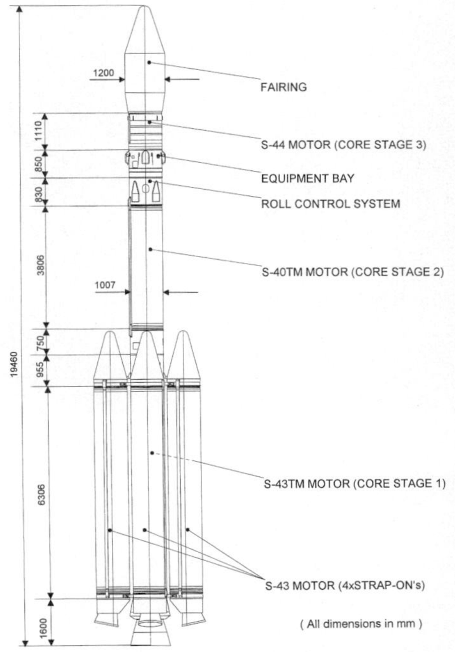

BRAZILIAN VLS SATELLITE LAUNCHER

VLS principal features (IAE/CTA) ***1997***/0003408

Stage mass: 5,670 kg
Propellant mass: 4,220 kg
Average thrust: 203 kN sea level
Burn time: about 60 s

Sonda 4 stage 2
Overall length: about 325 cm, excluding payload
Principal diameter: 555 mm
Stage mass: 1,130 kg
Propellant mass: 869 kg
Average thrust: 95 kN vacuum
Burn time: 28 s.

VLS
First launch: 1997 (fail)
Launch site: Alcantara
Principal uses: delivery of small payloads to LEO
Schedule of missions: SCD 2A fourth quarter 1997?; first half 1998 SCD 3; SSR 1 1999; SSR 2 2000
Typical performance: 200 kg into 750 km, 25°
Number of stages: 3 + 4 strap-ons (all solid)
Overall length: 19.46 m
Principal diameter: 1.00 m
Launch mass: 50 t
Guidance: inertial, mounted in equipment bay below stage 3
Launch sequence (min:s):

0.00	strap-ons ignite
0.55	stage 1 ignites
1.07	strap-ons burn out/separate
1.58	stage 1 burnout/separation and stage 2 ignition
2.07	heatshield separates
3.00	stage 2 burnout
3.08	stage 2 separates
7.12	equipment bay separates/stage 3 Ignites
8.23	stage 3 burnout/separate

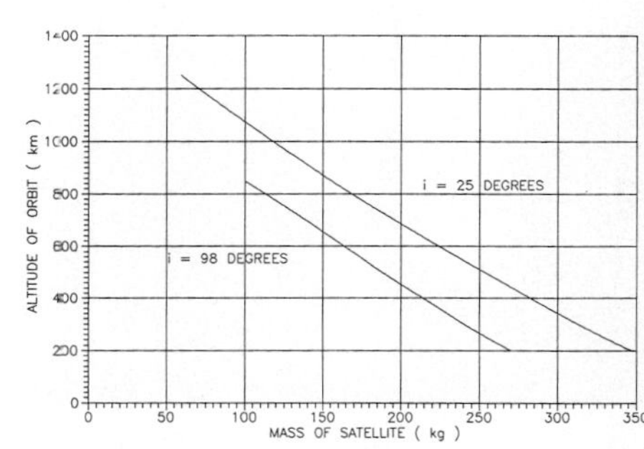

VLS circular orbit performance (IAE/CTA)

Specifications
VLS motor

Motor	S-43	S-43TM	S-40TM	S-44
Motor mass (kg)	8,210	8,400	5,340	917
Propellant mass (kg)	7,180	7,180	4,450	810
Burn time (web, s)	58.9	58.9	56.4	67.9
Action time (s)	62.0	62.0	62.0	72.0
Average pressure (atmospheres)	56.2	55.3	57.2	39.5
Average thrust (kN, vacuum)	309.0	327.0	212.5	33.9
Total vacuum imp (MNs)	18.42	19.52	12.00	2.31
SI (s, vacuum)	260.0*	277.0	274.9	281.6
Expansion ratio	12.82	37.68	25.97	66.0
Nozzle exit diameter (mm)	700	1,200	800	602

* sea level SI 230.5 s. S-44 employs Kevlar epoxy casing; others are steel

The VS-40 tested VLS stages 2/3 (IAE/CTA) ***1995***

VLS strap-ons
Designation: S-43
Overall length: 8.92 m
Principal diameter: 1,006.6 mm
Stage mass: 8,550 kg each (7,180 kg propellant + consumables)
Average thrust: 309 kN vacuum each
Burn time: 58.9 s (62 s. action time)
Attitude control: movable nozzles provide up to 3.0° TVC for three-axis control from four motors
Separation: each strap-on is attached by an aft spherical thrust transmission pin and paired forward/aft arm mounts. Following burnout, the arms are severed pyrotechnically and internal gas-pressurised actuators provide separation velocity.

VLS stage 1
Designation: S-43TM (similar to strap-on but nozzle reconfigured optimised for altitude operations)
Overall length: 8.86 m
Principal diameter: 1,006.6 mm
Stage mass: 8,720 kg (7,180 kg. propellant plus consumables)
Average thrust: 327 kN vacuum
Burn time: 59 s (62 s. action time)
Attitude control: nozzle flexure up to 3° provides pitch/ yaw TVC; roll managed by solid thrusters

VLS stage 2
Designation: S-40TM
Overall length: 6.24 m
Principal diameter: 1,006.6 mm
Stage mass: 5,660 kg (4,450 kg propellant plus consumables)
Average thrust: 212.5 kN vacuum
Burn time: 56.4 s (62 s. action time)
Attitude control: nozzle flexure up to 3° provides pitch/ yaw TVC; roll managed by liquid thrusters

Scale model of the VLS launcher with four solid propellant strap-on booster rockets (Theo Parard) ***1999***/0054363

End view of Sonda 4 rocket motor propellant grain, for which AGQuimica manufactures the ammonium perchlorate

VLS stage 3
Designation: S-44
Overall length: 1.75 m
Principal diameter: 1,006.6 mm
Stage mass: 1,025 kg (810 kg propellant plus consumables)
Average thrust: 33.9 kN vacuum
Burn time: 67.9 s (72 s. action time)
Attitude control: spin initiated by solids before separation of equipment bay
Heatshield/payload accommodation:
The payload is mounted on its adaptor by a V-clamp and separated by springs following pyrotechnic initiation. The two heatshield aluminium halves provide an internal diameter of 1,180 mm (external 1,200 mm) and cylindrical payload section 1,180 mm high. Mass <150 kg. Pyrotechnic clamps and springs separate the fairing after stage 2 ignition.

CANADA

Black Brant

Background

Bristol Aerospace's solid propellant Black Brant launcher is offered in five basic versions (5, 8, 9, 10, and 12). Payloads of 70 to 850 kg can be delivered to 150 to 1,500 km from conventional boom rails or three to four fin towers, with the BB10 providing 18 minutes of μgravity conditions for 100 kg. >800 vehicles of all types have flown since 1962, of which some 280 (BB6/7) were meteorological rockets. The single stage BB5 forms the basis of all currently available versions, with the addition of upper stages and/or US Army surplus boost motors. Combination with Nike/Terrier boosters, respectively, creates the BB8/9 models. NASA continues to be the major user and the vehicle also flies under the CSA μgravity materials research programme. With the closure of the Churchill Research Range in 1984, the Canadian programme was interrupted but one or two missions are now conducted annually from US/European ranges. The BB9 also forms the basis for EER Systems Space Services Division's Starfire vehicle, first launched March 1989.

In addition to the supply of basic rocket vehicle hardware, Bristol provides a range of self-contained vehicle/payload support systems, including thrust termination, parachute recovery, payload separation, de-spin, telemetry systems and fairing deployment. Two launcher types are usually employed: a rail-type typically providing 6 to 8 m of guided rail travel and a 30 m travel tower. In cases where the normal dispersion is unacceptable, the Saab/SES S19 boost guidance system is available; Saab's Spinrac attitude control system provides attitude control for a spinning exoatmospheric stage.

A 16.0 kg and 22.6 cm high igniter housing module incorporating de-spin and payload separation systems caps the stage 1 motor. The forward end provides a separation interface, comprising a deploying manacle ring with associated pyrotechnically actuated release guns and springs to generate 0.5 to 5 m/s separation. A yo-yo system can accommodate the standard range and profiles of despin requirements. The housing also incorporates the deployment logic for each system, including battery power, sequencing timers and safety altitude switches. The payload can provide the latter functions.

A thrust termination system manufactured by Explosive Technology Inc is provided as a complete assembly with the exception of the range authority-supplied command receiver. Two 'paddles' housing the shaped pyrotechnic charge are attached to the motor's forward end, but all other components, including explosive transfer lines, safe/arm assembly, batteries and command receiver, are mounted in the igniter housing. If activated, two holes 180° apart are cut in the motor but the payload is unaffected and events continue as pre-programmed.

Specifications

Black Brant 5
Description: The single stage BB5, which forms the basis for all BB variants, is available in 3/4-fin versions, designated the 5B/5C, respectively.
First launch: June 1965.
Number operational launches: 139 to end-1995.
Success rate: 97% to end-1995.
Number of stages: 1 (solid).
Overall length: 5.29 m + payload bay/nosecone.
Principal diameter: 43.8 cm (151.4 cm fin circle diameter).
Launch mass: 1,263 kg + payload/nosecone.
Typical performance: 200 kg to 290 km.
Payload recovery system: two basic (forward/aft-mounted) designs are available. The forward version is housed in a 3:1 ogive stainless steel fairing, the forward section of which is deployed soon after burnout to expose the recovery section heatshield. The aft-mounted recovery system is housed in a cylindrical aluminium bay immediately forward of the payload separation module. Both are self-contained with the necessary deployment logic equipment. The system is automatically armed by the closure of timers and altitude switches on the upleg phase. The heatshield is deployed, beginning the parachute deployment sequence, at about 5 km during re-entry. Both systems are available with either an 8.5 m circular main canopy or a 14 m paraform type. Each recovery system is available in two versions. The forward type comes as a 44.1 kg/1.32 m long unit qualified to 340 kg/320 km apogee or a 50.9 kg/1.38 m long unit qualified to 450 kg/500 km. The aft cylindrical type comes as a 33.2 kg/40 cm long/43.8 cm diameter unit qualified to 340 kg/320 km or a 44.1 kg/52 cm long/44 cm diameter unit qualified to 450 kg/500 km. Two new recovery systems based on the existing forward-mounted hardware have been introduced. The HARS High-Altitude Recovery System extends the recoverable mass/apogee to 725 kg/1,000 km; it flew successfully on the first three Maxus flights, in May 1991 (150 km), November 1992 (718 kg/715 km) and November 1995 (505 kg/706 km). The second system is a water recovery version (WRSA) of the standard 450 kg/500 km system. It uses an integral parachute/ flotation system and can accommodate both floating and sinking payloads. First operational flight was made successfully 9 December 1992. Two (successful) flights were made through 1995; no others are yet planned.

Black Brant 5 stage 1
Motor: Bristol Aerospace 26 KS 20000.
Overall length: 5.29 m.
Principal diameter: 43.8 cm.

Black Brant 12 departure from Wallops

Stage mass: 1,265 kg (propellant 1,000 kg).
Average thrust: 76.87 kN sea level.
Burn time: 33.0 s.
Payload environment/accommodation:
A range of deployable and fixed nose fairings are available in conical (5:1) or ogival (3:1 or 4.25:1) fineness ratios. BB9/10 utilise lightweight 1.5 mm aluminium skins, the others 1.5 mm stainless steel. Deployable fairings are available in split clamshell design with a horizontal (for low spin rates) or vertical (high spin) hinge. Forward-ejecting fairings provide typical separation rates of 5 m/s and a lateral impulse can be provided to prevent vehicle re-contact or ensure a clear field of view for instrumentation.
Peak longitudinal acceleration: 13 *g*/181 kg payload.
Shock: instrument qualification to 100 *g*/11 ms.
Vibration: <10 *g* at any frequency.
Temperature: 50°C maximum

Black Brant 8
Description: BB8 was developed in 1974 as the first upgrade of the BB5, with the US Army surplus Nike booster providing a 40% apogee performance increase. Otherwise, vehicle configuration is identical.
First launch: December 1975.
Number operational launches: 104 to end-1995.
Success rate: 99% to end-1995.
Number of stages: 2 (solid).
Overall length: 8.995 m + payload/nosecone.
Principal diameter: 43.8 cm.
Launch mass: 1,850 kg + payload.
Typical performance: 200 kg to 400 km.
Payload recovery system: as BB5

Black Brant 8 stage 1
Motor: Nike M5E1.
Overall length: 370.4 cm.
Principal diameter: 41.9 cm.
Stage mass: 601 kg (propellant 361 kg).
Thrust: 195.6 kN average, sea level.
Burn time: 3.35 s.
Comment: At burnout, the two stages immediately separate by differential drag, augmented by a drag ring at the Nike's forward end. Stage 2 coasts for 5 s before ignition at 8.5 s/2.0 km.

Black Brant 8 stage 2
Configuration: Same as BB5, and the payload accommodation is identical to BB5's. Peak longitudinal acceleration is 10.3 *g* for 435 kg payload. Other payload environment specifications are the same as for BB5.

Black Brant 9
Description: The BB9 adds a US Army surplus Terrier booster to the BB5 to augment apogee performance by 75%. BB9/10 are also available with the Thiokol TX664 booster; 'mod 1' is added to the BB designation. It provides about 30% additional performance than the Terrier but is identical in geometry.
First launch: 15 March 1982.
Number operational launches: 109 (+ 12 mod 1) to end-1995.
Success rate: 99% (+ 92% mod 1) to end-1995.
Number of stages: 2 (solid).
Overall length: 9.56 m + payload/nosecone.
Principal diameter: 43.8 cm (159.6 cm fin circle diameter).
Launch mass: 2,128 kg + payload.
Typical performance: 200 kg to 475 km.
Payload recovery system: as BB5

Black Brant 9 stage 1
Motor: US Army Terrier Mk 12 mod 1.
Overall length: 4.27 m.

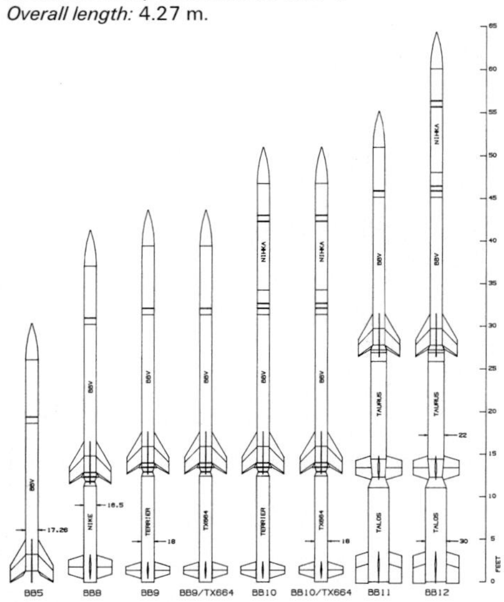

Bristol Aerospace's family of Black Brant rockets. Dimensions are in feet and inches ***1995***

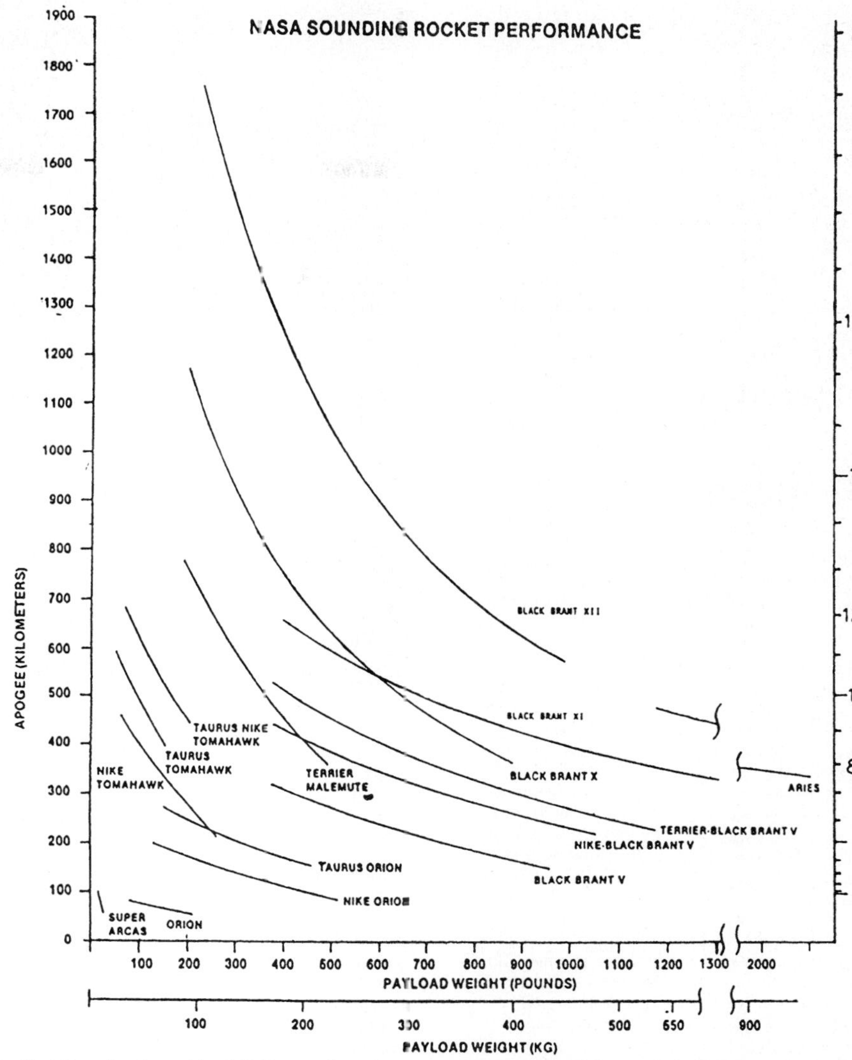

The Black Brant family provides NASA's sounding rocket programme, based at Wallops, with its highest performance vehicles (NASA)

Principal diameter: 45.7 cm.
Stage mass: 878.0 kg (propellant 534 kg).
Thrust: 257.7 kN average, sea level.
Burn time: 4.4 s.
Comment: Staging occurs at burnout using the same interface hardware as BB8. In this case, a 7.5 s coast precedes stage 2 ignition at 11.9 s.

Black Brant 9 stage 2
Configuration: Same as BB5's. Payload accommodation is the same as for BB5. Peak longitudinal acceleration is 13.4 *g* for 431 kg payload. Other payload environment specifications are the same as for BB5.

Black Brant 10
Description: BB10 was introduced in 1981 by adding a new Bristol Aerospace Nihka upper stage to the BB9 configuration. The flight sequence is as for BB9 until stage 2 burnout: the vehicle coasts for 33 s until the stage 2/3 separation mechanism is activated and the finless stage 3 ignites at 82 s/85 km. The payload fairing can be deployed before stage 3 ignition for a 4-8% apogee performance increase, typically at 60 s when dynamic pressure has fallen below 480 N/m². The Thiokol TX664 booster replaces the Terrier in the mod 1; see the BB9 for comments.
First launch: 14 August 1981.
Number operational launches: 30 (+ 4 mod 1) to end-1995.
Success rate: 93% (+ 100% mod 1) to end-1995.
Number of stages: 3 (solid).
Overall length: 11.88 m + payload/nosecone.
Principal diameter: 43.8 cm.
Launch mass: 2,560 kg + payload.
Typical performance: 200 kg to 750 km, 400 kg to 400 km.
Payload recovery system: high altitude version of the forward mount system

Black Brant 10 stage 1
Configuration: Same as BB9 stage 1

Black Brant 10 stage 2
Configuration: Same as for BB5

Black Brant 10 stage 3
Motor: solid propellant Bristol Aerospace Nihka 17 KS 12000, developed specifically for BB10.
Overall length: 1.92 m.
Principal diameter: 43.8 cm.
Stage mass: 402.1 kg (propellant 322 kg).
Thrust: 50.41 kN average, vacuum.
Burn time: 17.8 s.
Attitude control: none
Payload accommodation: Same as for BB5. Peak longitudinal acceleration 22 *g* for 136 kg payload.

Black Brant 12
Description: Bristol Aerospace and NASA Wallops completed studies in 1987 for a vehicle capable of delivering a 140 kg payload to 1,500 km altitude to satisfy plasma physics requirements for investigating ion acceleration over 900-2,000 km. A four-stage vehicle was selected, produced by replacing the BB10's Terrier booster with US Army surplus Talos and Taurus motors. Two demonstration flights were made from Wallops Flight Facility and the first operational flight from Poker Flat (Alaska) in March 1990. On the first demonstration flight in September 1988 the Nihka stage 4 suffered premature loss of thrust. The second vehicle, fired 5 December 1989, qualified the re-design and used a new spin motor to reduce vehicle dispersion. Of the six BB12s flown to end-1995, the one failure was due to Talos explosion at ignition. A BB11 3-stage version, without the Nihka stage 4, is available; its debut launch of 1 February 1990 carried a 547 kg plasma experiment to an altitude of 375 km from Poker Flat.
First launch: 30 September 1988.
Number launched: 2 BB11 + 7 BB12 to end-1995.
Success rate: 100% BB11 + 86% BB12 to end-1995.
Number of stages: 4 (solid).
Overall length: 16.11 m + payload.
Principal diameter: 43.8 cm (upper stages).
Launch mass: 5,253 kg + payload mass.
Typical performance: 140 kg to 1,500 km, 300 kg to 800 km.
Payload recovery system: as BB10

Black Brant 12 stage 1
Motor: US Army Talos.
Overall length: 3.51 m.
Principal diameter: 79.0 cm.
Stage mass: 2,053 kg (propellant 1,285 kg).
Thrust: 489.1 kN average, sea level.
Burn time: 5.42 s.
Separation: drag

Black Brant 12 stage 2
Motor: Taurus.
Overall length: 4.18 m.
Principal diameter: 58.1 cm.
Stage mass: 1361 kg (propellant 755 kg).
Thrust: 490.7 kN average, sea level.
Burn time: 3.30 s.
Separation: spring

Black Brant 12 stage 3/4
Configuration: Same as for BB10 stages 2/3, respectively.

CHINA, PEOPLE'S REPUBLIC

Long March

Lift-off of LM-2E from Xichang launch site (Theo Pirard) ***1999***/0054375

Background

Rocket propulsion was included as one of the key technologies in the 'Twelve Year Development Plan of Science and Technology' approved in 1958. This plan led to the launches of the first indigenously developed sounding rocket on 19 February 1960, and first successful ballistic missile on 29 June 1964 (failure on the first test 21 March 1962 led to a major redesign). The initial CZ-1 space launcher, derived from the CSS-3 MRBM, utilised nitric acid/UDMH but each subsequent operational and planned stage 1/2 has used N_2O_4 as oxidiser. Development of a LOX/LH_2 cryogenic upper stage for GEO missions began in 1977, leading with its launch as CZ-3 stage 3 in January 1984 to China becoming only the third such user after the US/ESA. The China Academy of Launch Vehicle Technology (CALT, formerly the Beijing Wan Yuan Industry Corporation) near the town of Nan Yuan 15 km south of the capital develops/builds the cryogenic engines and its Shanxi Liquid Rocket Engine Co the storable engines; SLREC also handles solid motors. The Beijing Rocket Test Centre 50 km southwest of Beijing maintains five major test stands. No 1 handles altitude testing of spacecraft thrusters of up to 490 N, No 2 provides single-engine cryogenic facilities, with No 4 accepting a complete H-8 cryogenic stage, and No 5 is devoted to hydrazine engines. Other test facilities are operated at the launch sites and SLREC probably has stands for the storable engines.

Fewer details are known of Chinese thrusters and solid propellant motors. The CZ-1D stage 3 is a solid of 625 kg propellant and satellites such as FSW incorporate solid kick motors. CZ-2E's 1.7 m diameter EPKM solid stage 3 debuted in 1995. CZ-2C's 1.2 m diameter CPKM is imminent. Sounding rockets with solids are also flying. Solid propellant ICBM/IRBMs are under development. Hydrazine thrusters of unspecified nature provide three-axis control for the H-8 cryogenic stage. However, CZ-2E's stage 2 employs a hydrazine system acknowledged as derived from H-8: 4 × 45 N pitch thrusters, 2 × 45 N yaw + 4 × 11 N roll. The FSW recoverable capsule is also three-axis controlled by cold gas, and the CZ-1D solid stage is spin-stabilised by N_2 thrusters.

China's space launchers draw on a small range of main engines, the majority utilising hypergolic NTO/UDMH although more recently two cryogenic upper stages were added. No LOX/kerosene engines as used extensively elsewhere are known, reflecting the space variants' parallel development alongside storable propellant long range missiles. The engines carry YF designations, indicating Yei-ti Fa-dong-ji (liquid state engine).

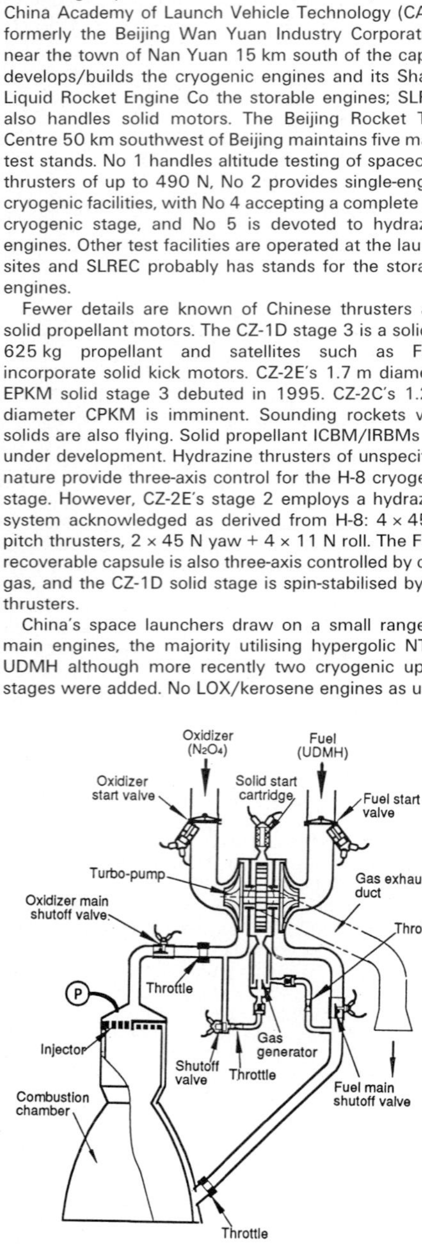

YF-20/21/22 schematic

YF-20/21 and YF-20B/21B

The YF-20/21 is the backbone of the Chinese launcher range, clustered in fours (creating the YF-21/21B assembly) for every first stage of the CZ-2 to -4 families and modified into the high altitude YF-22 for the second stages. In the L-140 stage 1s (the 140 indicating 140 tonne propellant loading) it operates at 696 kN thrust, but the stretched L-180 stages and strap-ons of the more powerful vehicles run it at 740 kN. Chinese statements suggest that still higher levels are possible. The LB-40 strap-ons of the CZ-2E/3B/3C each incorporate a single YF-20B.

Development of the cryogenic YF-73 began in 1976 for the restartable H-8 third stage, the first use of LH_2 propulsion technology outside of USA/Europe. Principal use is delivery of payloads into GTO. Firings of flight-type engines began 1981. After completion of acceptance testing, the first flight engine was delivered 4th quarter 1983. By end-1983, about 120 tests totalled more than 30,000 seconds. The four chambers, gimballed ±24° for attitude control, and are serviced by a single 37,000 rpm turbopump. A pair of high pressure nitrogen spheres initiates the two starts. The valves, pumps, gearbox, gas generator and propellant lines carry 10 to 20 mm of polyurethane foam to minimise heat transfer. The engines are manufactured at CALT. There have been two flight failures. The first was January 1984's debut mission when the re-start after coast attained 90 per cent thrust and then rapidly decayed after a few seconds. Cause was attributed to propellant flow fluctuations following the μgravity coast. Similarly, the second (December 1991) suffered a pressure drop 58 seconds into the second burn and shut down after a further 77 seconds. Cause was believed to be helium supply failure to control valves.

Development of the cryogenic restartable YF-75 drew extensively on YF-73 experience to create the H-18 stage, comparable to Centaur D. Its employment in pairs aboard the CZ-3A vehicles provides a GTO capability equivalent to the lower range of Ariane 4. Aboard CZ-3B, it surpasses Ariane 4 and almost equals Proton. The engines are manufactured at CALT.

The solid motor carried by the CZ-1D appears to be related to the GF-02 solid flown on the original CZ-1 vehicle. Work began in 1965 at the Solid Fuel Engine Research Academy under deputy head Yang Nansheng. The first spin firing took place 26 January 1968 but the motor exploded after 30 seconds, causing some fatalities. During 1968-70 a total of 19 tests, including five altitude simulations, was conducted successfully.

The HEXI Corp's indigenous EPKM FG-46 commercial perigee kick stage debuted for AsiaSat 2 in November 1995, providing 3,460 kg into GTO. An early version was carried on the first 2E. Nine test firings included two failures at the nominal 40 rpm spin rate. 31 May 1993, test M1, prototype/design qualification. 24 October 1993, M2, prototype/structural qualification. 31 January 1994, C1, 40 rpm spin (fail). 5 May 1994, C2, altitude simulation. 6 July 1994, C3, 40 rpm spin (fail). 30 October 1994, C5, 40 rpm spin. 8 November 1994, C4, altitude simulation. 16 January 1995, C6, 40 rpm spin. 20 March 1995, Z1, 40 rpm spin/acceptance.

Specifications

YF-2A
Applications: cluster of four for CZ-1 family stage 1 and CSS-3 IRBM
First flown: June 1964?
Oxidiser: nitric acid
Fuel: UDMH
Propellant flow rate: 115 kg/s?
Cycle: gas generator
Thrust: 275.3 kN sea level
Specific impulse: 241 s sea level
Burn time: about 130 s operational

YF-3
Applications: CZ-1 family stage 2, CSS-3 IRBM
First flown: June 1964?
Oxidiser: nitrogen tetroxide
Fuel: UDMH
Propellant flow rate: 50 kg/s?
Thrust: 294.2 kN vacuum
Specific impulse: 287 s vacuum
Burn time: about 130 s operational

YF-20
Applications: CZ-2C/3 stage 1 for YF-20/21; CZ-2E/3B/3C strap-on for YF-20B; CZ-2D/2E/3A/3B/3C/4 for YF-21B
First flown: 5 November 1974 (first CZ-2 flight)?
Oxidiser: nitrogen tetroxide
Fuel: UDMH
Propellant flow rate: 270 kg/s?
Thrust: 696.2 kN sea level for YF-20/21 of L-140 stage 1 of CZ-2C/3; 740.35 kN, E 12.69, exit area 7,728.6 cm² for YF-20B of LB-40 strap-on and YF-20B/21B of L-180 stage 1 of CZ-2E/3A/3B/3C/4.
Specific impulse: 259 s SL YF-20/21; 260.66 s SL YF-20B/21B
Mixture ratio: 2.10 YF-20/21; 2.12 YF-20B/21B
Burn time: 125-160 s operational

YF-22/22B and YF-24/24B
Descriptions: The YF-22/22B is a high altitude derivative of the YF-20. It is flown singly with four gimballed YF-23/23B verniers, the assemblies designated 24/24B, providing steering on the L-35 stage 2. YF-22B/23B use larger expansion ratios for improved SI
Applications: CZ-2C/3 stage 2 for YF-24; CZ-2D/2E/3A/3B/3C/4 stage 2 for YF-24B
First flown: 5 November 1974 (first CZ-2 launch)?
Oxidiser: nitrogen tetroxide
Fuel: UDMH
Propellant flow rate: 270 kg/s?
Thrust: 719.8 kN vacuum for YF-22, 740.4 kN for YF-22B, but 845.3 kN attainable
Specific impulse: 289 s vacuum YF-22; 298 s vacuum YF-22B
Mixture ratio: 2.181 YF-22 and -22B
Burn time: 300 s CZ-2E; 110 s others

YF-23/23B
Four gimballed YF-23/23Bs provide steering for the fixed single YF-22/22B on the second stages
Applications: as YF-22/22B

CZ-3's H-8 third stage is powered by the 4-chamber cryogenic YF-73

First flown: as YF-22/22B
Oxidiser: nitrogen tetroxide
Fuel: UDMH
Propellant flow rate: 16 kg/s for all four chambers with single pump?
Thrust: 46.1 kN YF-23 and 47.1 kN YF-23B total (4 chambers) vacuum
Specific impulse: 281.7 s YF-23 and 289 s YF-23B vacuum
Mixture ratio: 1.57 YF-23/23B
Burn time: up to 300 s. 136 s for CZ-4

YF-40
Description: The engine appears to have been uprated for CZ-4B. Where known, the different specifications are provided in 4A/4B order
Applications: flown paired on CZ-4 stage 3; uprated for 4B?
First flown: 6 September 1988
Mass: each 70/83 kg dry
Oxidiser: nitrogen tetroxide
Fuel: UDMH
Thrust: 98.1/101.03 kN vacuum for pair
Specific impulse: 305/303 s
Mixture ratio: 2.14
Chamber pressure: 45.39 atmospheres
Expansion ratio: 55
Exit diameter: 630 mm
Burn time: 321/412 s (single/single or dual burn)
Nozzle materials: niobium alloy, radiation cooled

YF-73
Application: H-8 third stage of CZ-3
First flown: 29 January 1984 (failed to restart)
Number flown: 9 sets, to end-1995 (2 failures)
Engine cycle: gas generator
Dry mass: 236 kg
Length: 1,438 mm
Maximum diameter: 2,220 mm
Oxidiser: liquid oxygen at 9.086 kg/s
Fuel: liquid hydrogen at 1.817 kg/s
Thrust: 44.147 kN total (4 chambers) vacuum
Specific impulse: 420 s vacuum
Mixture ratio (O/F): 5.0
Chamber pressure: 25.96 atmospheres
Expansion ratio: 40:1

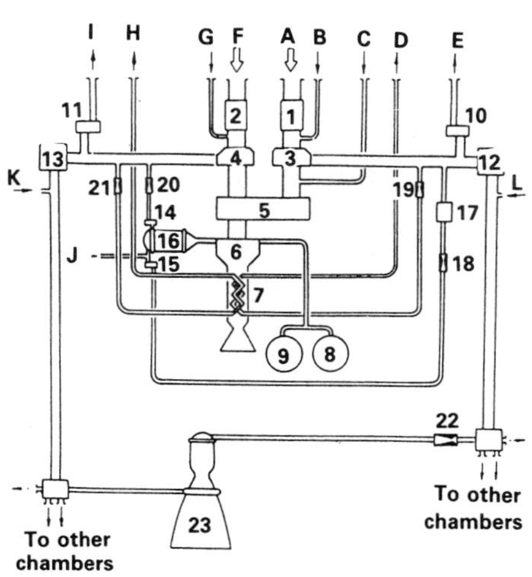

YF-73 schematic. A) LOX inlet, B/C) purge, D) LOX tank pressurisation, E) bleed, F) liquid hydrogen inlet, G) purge, H) LH2 pressurisation, I) bleed, J-L) purge; 1/2) isolation valves, 3) LOX pump, 4) LH2 pump, 5) gearbox, 6) turbine, 7) heat exchanger, 8/9) start tank, 10/11) bleed valve, 12/13) main valve, 14/15) gas generator valve, 16) gas generator, 17) pressure stabiliser, 18-22) venturi, 23) combustion chamber (one of four)

Long March family of launchers: (left to right) LM-3B, LM-2E, LM-3A, LM-2C, LM-4 (Theo Pirard) ***1999***/0054376

Chamber materials: stainless steel, regeneratively cooled by supercritical hydrogen
Injector: brazed stainless steel, 38 coaxial elements
LOX/LH$_2$ pumps: 37,000 rpm single stage centrifugal pump with helical inducer. LH$_2$ and turbine share common shaft; LOX shaft is driven via reduction gears. Gears cooled by hydrogen.
Burn time: two burns of 500 s + 300 s

YF-75
Application: H-18 stage 3 of CZ-3A/3B/3C
First flight: 8 February 1994
Number flown: 2 sets to end-1995
Oxidiser: liquid oxygen
Fuel: liquid hydrogen
Propellant flow rate: 19 kg/s each chamber?
Thrust: 78.4 kN vacuum each chamber
Specific impulse: 442 s vacuum
Mixture ratio: 5.0
Chamber pressure: 37.15 atmospheres
Expansion ratio: 80:1
Burn time: about 470 s total, dual start

GF-02/CPKM/EPKM/FG-46
First flown: 28 November 1995
Number flown: 2, to end-1995
Mass: 6,001 kg (5,444 kg propellant, 529 kg burnout, 28 kg ablated)
Length: 2.5 m
Diameter: 1,700 mm
Propellant: HTPB
Burn time: 87 s
Thrust: 210 kN max vacuum
Specific impulse: 292 s vacuum
Total impulse: 15.5881 MNs vacuum
Pressure: 6.0 MPa max
Casing materials: glass fibre
Nozzle: carbon-carbon

INDIA

General History

Current status

India has achieved its long sought goal of launcher autonomy. In 1998, it completed the GSLV launch programme providing it with a full suite of launch vehicles capable of LEO, Sun-synchronous and geosynchronous orbits. After the detonation of a new nuclear device in June 1998, the western economies imposed sanctions on India, but currently, every part on the Indian space launch vehicles is or can be indigenously produced.

Background

India has undergone a gradual evolution from the all-solid SLV-3 of 1980-83 to today's geo capability. The first all-India launch had the capacity to only place a 42 kg satellite in LEO. The ASLV expanded this to 150 kg for the Augmented SLV largely by adding strap-ons derived from the existing stage 1. ASLV retired in

India's orbital launch record

	Date	Vehicle	Payload
1*	10 August 1979	SLV-3	Rohini (35 kg)
2	18 July 1980	SLV-3	Rohini 1 (35 kg)
3	31 May 1981	SLV-3	Rohini 2 (32 kg)
4	17 April 1983	SLV-3	Rohini 3 (41.5 kg)
5*	24 March 1987	ASLV-D1	SROSS 1 (150 kg)
6*	13 July 1988	ASLV-D2	SROSS 2 (150 kg)
7	20 May 1992	ASLV-D3	SROSS C (106 kg)
8*	20 September 1993	PSLV-D1	IRS 1E (846 kg)
9	4 May 1994	ASLV-D4	SROSS C2 (113 kg)
10	15 October 1994	PSLV-D2	IRS P2 (804 kg)
11	21 March 1996	PSLV-D3	IRS P3 (930 kg)
12*	28 September 1997	PSLV-D4	IRS P4 (1,250 kg)

* indicates vehicle failure (SROSS C entered low orbit because stage 4 spun up to 80 rpm instead of 180 rpm). All launches made from Sriharikota.

1994 after four launches. Six of those same boosters fly on the new Polar SLV, allowing India to launch its own 1 tonne-class IRS remote sensing satellites into Sun-synchronous orbits. PSLV has been offered commercially through Antrix Corp since March 1996's flight. The Geosynchronous SLV builds upon the PSLV by replacing the six solid strap-ons with four liquids and substituting a cryogenic stage for the two upper stages. This provides a 2.5 tonne GTO capacity, freeing India from dependence on foreign vehicles for Insat 2-class satellites. PSLV's pad is being modified to accept the new vehicle.

GSLV

Background

ISRO is developing an Ariane 4-class Geosynchronous SLV with a 3.4 m diameter fairing, capable of handling 2.5 tonne Insat 2-class satellites. To develop the plans, some 500 configurations were reduced to four candidates during 1986-88, divided equally between solid/liquid cores and both with solid/liquid strap-ons. GSLV replaces PSLV's six solid strap-ons with four liquids powered by Vikas engines (although the initial version will use PSLV strap-ons) and substitute a cryogenic stage for the two upper stages. The vehicle is comparable to Ariane 44L except for the lower performance solid stage 1.

Glavkosmos signed a Rs2,350 million agreement with ISRO in January 1991 to supply India with the cryogenic engine. This KVD-1 76 kN KB Khimmach engine was originally developed to power Russia's N-1 manned booster for the L-3M programme in the 1970s. The contract called for two engines plus the technology transfer required for subsequent Indian production. However, US pressure under the Missile Technology Control Regime forced the Russians to formally terminate the agreement in August 1993. ISRO noted that two years had been lost on the GSLV programme and forced a decision on the development of an indigenous cryogenic engine. Later in the year, the Indian government approved plans for a domestic water-cooled LOX/GH_2 subscale engine which ISRO tested successfully 21 July 1989. Critics of international control regimes point out that the US action to stop the Russian sales actually created another source of such engines, namely, India, and that a resentful India may not be restrained in efforts to sell its technology overseas.

The course of engine development was not easy for India. The contractor LPSC's first attempt at firing a 10 kN LOX/LH_2 engine ended with an explosion at ignition in early July 1993. LH_2 became available in the country only after a US company built a plant at LPSC's Mahendragiri site at end-1992.

Programme has been restructured to include basic GSLV with 1,500 kg GTO compatibility, GSLV Mk2 (KVD-1) with 2,500 kg GTO compatibility and GSLV Mk 3 (C-20) with 3,000/3,500 kg GTO capability. Flight dates are uncertain.

The GSLV utilises Insat 2's 119 kg (dry mass) unified bipropellant liquid propulsion system made up of a single 440 N LAM Liquid Apogee Motor and two redundant networks of 8 × 22 N RCS thrusters for final orbital placement. The qualification LAM was tested for a total 9,550 seconds, including a single burn of 3,000 seconds. RCS testing accumulated 30,500 seconds including a single continuous burn of 10,000 seconds. Pulsed firing mode logged >250,000 pulses, each component being tested over three life cycles. The engines, pressure regulator, check valve, fill, drain and vent valve, pyro valve and pressure transducer were developed indigenously; the tanks, gas bottles, filters and latch valves were procured from foreign suppliers. The system carries about 450 welded joints.

Model of the GSLV rocket being developed by ISRO for operational duty early in the next century (Theo Pirard) ***1999***/0054377

Ground test version of the Vikas engine to be used on the strap-on boosters for the GSLV; the original Vikas variant flew on the second stage of the PSLV (Liquid Propulsion Systems Centre) ***1997***/0003431

Specifications

GSLV

First launch: first quarter 1998
Launch site: SHAR Centre (Sriharikota)
Principal uses: medium-class GTO
Cost: not available
Availability: to be announced
Schedule of missions: to be announced
Performance: 2,500 kg into GTO (1,500 kg for initial version)
Number of stages: 3 (1 solid/2 liquid) + 4 liquid strap-ons (6 PSLV solids initially)
Overall length: 50.9 m
Launch mass: 402 t
Guidance: similar to PSLV

GSLV liquid strap-ons

Comment: PSLV's Vikas engine is being adapted for GSLV's strap-ons. A successful 200 s run 24 July 1995 tested the new indigenously-developed silica-phenolic throat, designed to withstand the operational 1,300°C
Designation: L40
Overall length: 19.7 m
Principal diameter: 2.1 m
Total mass: 4 × 45.5 t
Propellant: 4 × 40 t UDMH/NTO
Thrust: 4 × 735 kN vacuum average
Burn time: 158 s
Attitude control: gimballed ±5° for 3-axis

GSLV stage 1

Designation: S125
Other specifications: same as PSLV

GSLV stage 2

Designation: L37.5
Other specifications: same as PSLV

GSLV stage 3

Designation: CS
Overall length: 8.72 m
Principal diameter: 2.80 m
Total mass: 14.60 t
Propellant: 12.50 t LOX/LH_2
Thrust: 76 kN vacuum average (450 s Isp), including two gimballed LOX/GH_2 verniers fed by main chamber pumps
Burn time: 800 s
Attitude control: gimballed, auxiliary and cold gas
Payload fairing: 3.4 m and, 7.8 m long, aluminium

LAM

First flight: July 1992 (Insat 2A)
Dry mass: 4 kg
Length: 570 mm
Maximum diameter: 276 mm
Mounting: fixed
Engine cycle: pressure-fed in regulated mode
Propellants: MMH/NTO from tanks at 16 atmosphere
Mixture ratio (O/F): 1.65
Thrust: 440 N vacuum nominal
Specific impulse: 310 s vacuum minimum
Expansion ratio: 160
Chamber pressure: 6.9 atmosphere
Injector: coaxial titanium welded to chamber
Combustion chamber: silicide-coated columbium alloy radiatively cooled
Burn time: qualified to 3,000 s continuous; minimum impulse bit not available

Insat 2 RCS Thruster

First flight: July 1992 (Insat 2A)
Dry mass: 850 g
Length: 249 mm
Maximum diameter: 60 mm
Mounting: fixed
Engine cycle: pressure-fed in blowdown mode
Propellants: MMH/NTO from tanks at 16 atmosphere
Mixture ratio (O/F): 1.65
Thrust: 22 N vacuum nominal
Specific impulse: 285 s vacuum minimum
Expansion ratio: 100
Chamber pressure: 6.9 atmosphere
Throat temperature: 1,200°C
Injector: coaxial titanium welded to chamber
Combustion chamber: silicide-coated columbium alloy radiatively cooled
Burn time: qualified to 1,000 s continuous; minimum impulse bit not available

PSLV

Background

India's Polar Satellite Launch Vehicle (PSLV) represents the first stage in acquiring launcher autonomy for applications satellites. Sized for placing 1 tonne IRS Earth resources satellites in 817 km Sun-synchronous orbits from Sriharikota. The Indians have paid Rs4

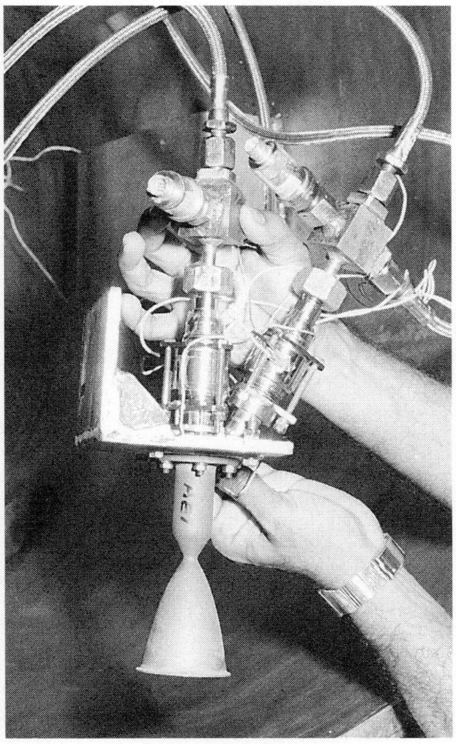

The 22 N RCS thruster carried on INSAT 2 (Liquid Propulsion Systems Centre) ***1997***/0003444

The liquid propellant engine, two of which are carried on the fourth stage of the PSLV (Liquid Propulsion Systems Centre) ***1997***/0003446

The 440 N LAM carried on INSAT 2 (Liquid Propulsion Systems Centre) ***1997***/0003445

155 million, to develop the PSLV through its first two flights and the ground infrastructure. Each launcher itself costs about Rs450 million. The government, in October 1994, approved six further tests known as PSLV C1-C6, (C = continuation) vehicles, after the first three development flights named PSLV D1-D3, (D = development). Some estimates state the total cost for all six flights is Rs4,000 billion. The C version should improve performance to 1,300 kg. IRS-1D (1,200 kg) will use the PSLV.

PSLV employs an unusual combination of liquid and solid stages, using the first Indian liquid systems for stages 2 to 4 and clustering six ASLV strap-ons around the solid first stage. Stage 2's single Vikas engine is based on SEP's Viking from Ariane. In order to prepare for the eventual programme, the Indian Space Research Organisation began long-duration testing on the engine in January 1988 at LPSC's Mahendragiri facility.

In practice, Sriharikota has a launch constraint on azimuths of less than 140°, forcing PSLV to execute a 55° yaw manoeuvre after 100 seconds and stage 1 separation to reach a Sun-synchronous path. Without this constraint the PSLV could launch 1.6 tonne into Sun-synchronous orbit.

PSLV-D1 performed well but an uncorrected pitch problem after stage 2 separation left it at only 340 km altitude after stage 3's burn instead of the planned 414 km. Stage 4 had insufficient reserve to attain a useful orbit. Investigators later found stage 2 engine nulling occurring about 3.7 seconds before stage 3 ignition and as a consequence, the stages recontacted because two small solid retros failed, producing a pitch disturbance that exceeded the preset limit. PSLV could have removed it but a software error in the control loop prevented action. D2, with modified software, was successful, followed by D3 in March 1996. D4 launched in September 1997 placed the IRS satellite in an eccentric orbit instead of the planned circular orbit due to a fourth stage malfunction.

Specifications

PSLV

First launch: 20 September 1993
Number launched: 4
Launch sites: SHAR Centre (Sriharikota)
Principal uses: 1 t-class remote sensing payloads into Sun-synchronous orbit, 3 t-class into LEO
Vehicle success rate: 67 per cent
Cost: Rs450 million each D1-D3, Rs650 million C1-C3
Availability: launches projected every year
Performance:
Sun-synchronous: 1,000 kg into 817 km, 99.1° from SHAR (1.6 t without range safety constraints)
GTO: 450 kg
LEO: 3,000 kg into 400 km
Number of stages: 4 (2 solid/2 liquid) plus 6 solid strap-ons
Overall length: 44.2 m
Fairing diameter: 3.2 m (5.1 m base circle with strap-ons)
Launch mass: 283 t
Guidance: Inertial Guidance System located in vehicle equipment bay surrounding stage 4 base. Redundant strapdown inertial navigation system (Resins, incorporating three dry-tuned gyros + four servo accelerometers) feeds the navigation processor, which produces navigation data every 500 ms for the guidance & control processor to issue steering commands at similar intervals. Open loop guidance is employed during stage 1 burn before switching to closed loop from stage 2 onwards. Target 3Σ injection accuracy for Sun-synchronous payloads ±35 km altitude, ±0.2 inc.
Launch sequence (typical IRS mission, min/s):

00:00	stage 1 ignite
00:01	2 solids ignite
00:30	4 solids ignite at 3 km
00:54	2 solids burnout at 17 km
01:15	2 solids sep at 24 km
01:19	4 solids burnout at 36 km
01:30	4 solids sep at 38 km
01:45	stage 2 ullage motors ignition, stage 1 sep + stage 1 retro motors ignition
01:45	stage 2 ignition at 48 km
02:32	fairing sep at 105 km
02:37	closed loop guidance begins
04:21	stage 2 shutdown, sep, retro ignition, stage 3 ignition at 232 km
05:33	stage 3 burnout at 350 km
05:07	stage 3 sep at 405 km
10:17	stage 4 ignition at 700 km
17:06	stage 4 shutdown at 817 km

PSLV solid strap-ons.
Designation: PSOM (PSLV strap-on motors).
Comment: The segmented motors are almost identical to the ASLV strap-ons. Two are ignited on the pad, followed 30 s later by the remaining four, with separation occurring at 73/90 s respectively. Nozzles are canted 9°. One ground-lit and one air-lit strap-on include TVC secondary injection to augment roll control.
Overall length: 11 m
Principal diameter: 1.0 m
Propellant: 8,628 kg of HTPB solid
Average thrust: about 440 kN vacuum each, 662 kN max
Burn time: 47 s
Attitude control: exhaust secondary injection
Separation: pyrotechnic nuts and springs

PSLV stage 1
Designation: PS1
Overall length: 20.3 m
Principal diameter: 2.8 m
Total mass: 156.0 t
Propellant: 129.0 t of HTPB solid in five maraging steel segments, each 3.4 m long and cast individually (centre three are interchangeable); segments held by 144 pins in tongue/groove joint.
Thrust: max 4,600 kN vacuum, 3,500 kN thrust at launch
Burn time: 103 s
Attitude control: pitch/yaw provided by secondary injection TVC of strontium perchlorate from two 70 cm diameter 1,650 litre steel tanks through quadrants of six valves from a total of 24 into stage 1 nozzle (35% of distance from throat to exit). Each quadrant can generate a 200 kN side force; valve flow rate is 12 litres maximum. Roll control provided by two continuous-burn swivelling 6.4 kN vacuum NTO/MMH thrusters below SITVC tanks.
Separation: flexible linear shaped charge and eight 48.5 kN mean thrust 1,414 × 209 mm diameter retros, each burning 26.8 kg propellant in 1.3 s

PSLV stage 2
Comment: The integrated system was first tested at Mahendragiri 21 March 1990. Final qualification test October 1992 for 150 s. Four 15.3 kN solid motors burn

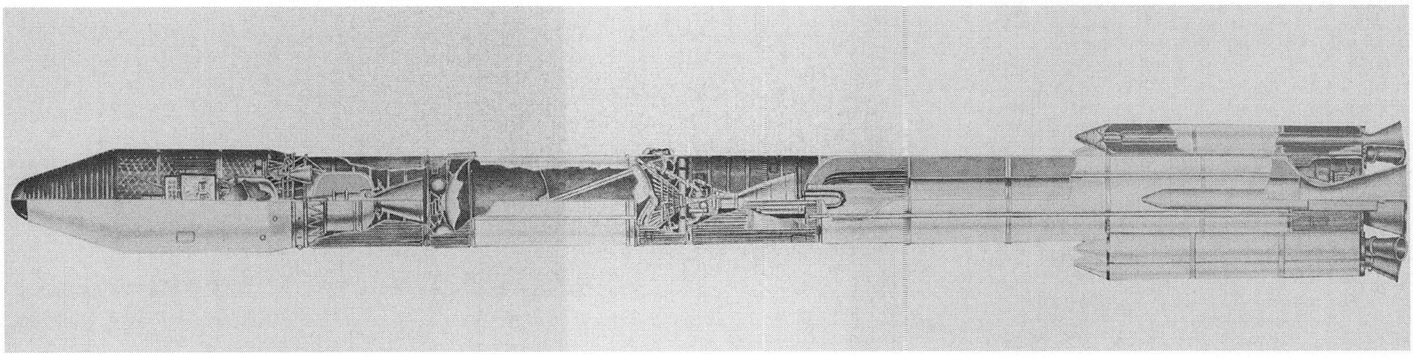

PSLV provides a 1 t Sun-synchronous capacity (ISRO)

for 5.5 s at stage 1 separation to provide ullage for stage 2.
Designation: PS2/L37.5
Engine: Vikas open gas generator cycle engine (based on SEP Viking), gimballed for pitch/yaw control, single start, film + radiatively cooled, pumps driven by single 9,400 rpm shaft powered by gas generator.
Overall length: 11.5 m
Principal diameter: 2.8 m
Dry mass: 5.3 t
Propellant: 37.5 t of UDMH/NTO
Stage thrust: 725 kN vacuum
Burn time: 149 s
Attitude control: Vikas gimballing for pitch/yaw; roll control provided by two on/off 300 N nozzles drawing hot gas from Vikas gas generator
Separation: at 249 s/243 km, with four 22.6 kN 1.6 s retros providing separation from third stage

PSLV stage 3
Comment: Stage 3 is suspended inside a 2.8 m diameter shroud from the fourth stage skirt
Designation: PS3
Overall length: 354.1 cm
Principal diameter: 2.0 m
Propellant: 7,260 kg of HTPB in Kevlar casing
Thrust: max 386 kN vacuum
Burn time: 73.1 s (78.3 s action time)
Attitude control: pitch/yaw by ±2° nozzle flexure (qualified to ±3°) controlled by two actuators; roll control from the 50 N roll thrusters on stage 4.
Separation: ball release and springs

PSLV stage 4
Comment: The integrated stage was first tested at Mahendragiri on 27 September 1989, beginning a programme including two full duration trials and a demonstration of re-start capability.
Designation: PS4/L2
Overall length: 2.65 m
Principal diameter: 1.335 m
Dry mass: 920 kg
Propellant: 2 t MMH/NTO
Average thrust: 14 kN
Burn time: 425 s (D2: 397 s)
Attitude control: six 50 N thrusters in two blocks drawing propellants from stage 4's tanks provide 3-axis control during coast period; main engine gimballing employed during burn.
Separation: Merman clamp and springs

PSLV stage 4 engines
Description: two ISRO 7.5 kN pressure-fed bipropellant engines, gimballed up to ±3° in orthogonal axes for 3-axis control, 8.37 atmosphere chamber pressure, 60:1 expansion ratio. The titanium alloy tanks are pressurised at 19.2 atmospheres by helium stored at 300 atmospheres in five titanium spheres. Tank incorporates surface tension-type propellant acquisition system to ensure supply under adverse acceleration conditions.
First flight: 20 September 1993
Dry mass: 28 kg
Length: 1.1 m
Maximum diameter: 0.63 m
Mounting: gimballed up to ±3° in two orthogonal planes by a closed loop servo control system to provide stage pitch, yaw and roll control during thrust phase.
Engine cycle: pressure-fed
Propellants: MMH/NTO from tanks at 19.2 atmosphere
Mixture ratio (O/F): 1.4
Thrust: each 7,500 N vacuum nominal
Specific impulse: 308 s vacuum minimum
Expansion ratio: 60
Chamber pressure: 8.37 atmosphere
Injector: AISI 304 stainless steel; 45 sets of triple elements
Combustion chamber: AISI 304 stainless steel film cooled along single helical grooving
Nozzle: silicide-coated columbium alloy radiatively cooled
Burn time: 425 s mission nominal
Qualification time: 530 s (continuous)

PSLV L37.5 Vikas engine
Comment: India's first large liquid propellant rocket engine is derived from Viking technology acquired from SEP. The engine achieved its first full-duration 150 s firing in January 1988 on the principal test stand at the Liquid Propulsion Test Facility, Mahendragiri, and completed the first integrated stage test 21 March 1990 (qualification completed October 1992). A 686 kN version will be used for GSLV's strap-ons. A successful 200 s run 24 July 1995 tested the new indigenously-developed silica-phenolic throat, designed to withstand the operational 1,300°C.
First flown: 20 September 1993
Number flown: 4 to May 1999
Dry mass: 876 kg
Length: 3.509 m
Mounting: gimballed for pitch/yaw control
Engine cycle: gas generator (9400 rpm turbine)
Propellants: UDMH/NTO
Mixture ratio (O/F): 1.86
Thrust: 72 kN vacuum (686 kN for GSLV)
Specific impulse: 295 s vacuum
Time to full thrust: 2.4s
Expansion ratio: 31 (13.88 C-SU strap-on)
Chamber pressure: 51.9 atmospheres
Injector: 216 like on like doublets set in six rows for each propellant in light alloy annular injector
Chamber cooling: UDMH film supplied through additional channels on the injector's lower section
Igniter: hypergolic
Nozzle materials: cobalt alloy with SEPHEN (phenolic resin/silica fibre) throat
Burn time: 160 s nominal mission

Payload fairing/accommodation
Hindustan Aeronautics Ltd. builds 3.2 m diameter aluminium fairing, 8.3 m long. The 4.5 m long isogrid central cylindrical section is topped by a nose with a 70 cm radius cap/20° sloping wall; the boat-tail narrows along a 14° slope to 2.8 m diameter. A Merman clamp, explosive bolt cutters and zip cord, effects separation. 937 mm payload adapter using 12.2 kN Merman clamp; 4 springs provide 80 cm/s max separation velocity.
Acceleration: 6.0 *g* max, during stage 3
Acoustic: payloads qualified to 146 dB overall for two min, acceptance testing to 142 dB for one min

PSLV stage 1 motor
PSLV's first stage is the most powerful model produced by ISRO; it was successfully static fired for the first time 21 October 1989. The second was successfully fired 23 March 1991.
Application: PSLV stage 1
First flight: 20 September 1993
Number flown: 4 to May 1999
Mass (kg): 145,000
Length (cm): 2,034.5
Diameter (cm): 280.4
Propellant: type: HTPB, segment-cast with three central segments interchangeable

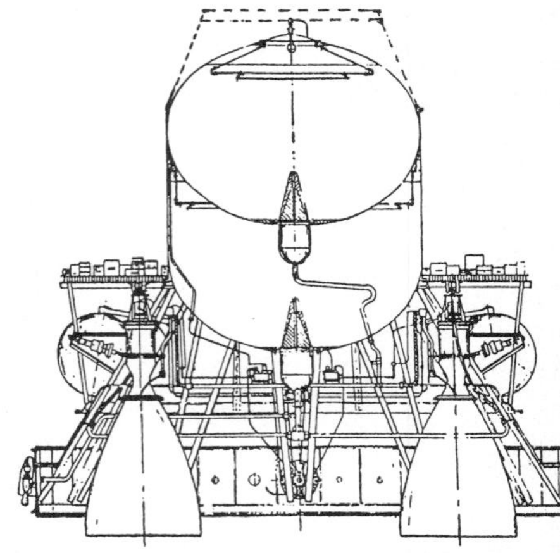

PSLV's stage 4 incorporates two 7.5 kN engines and, for coast attitude control, six 50 N thrusters. Surrounding the tank is PSLV's vehicle equipment bay. On top is IRS 1E (ISRO)

shape: forward segment deep 10-fin to burn 22 t loading in 20 s to generate high launch thrust; remainder tubular with average 120 cm diameter bore.
mass fraction: 0.890 (129 t loading)
Burn time (s): 100
Thrust (kN, vacuum): 3,463 mean, 4,600 max at 17 s
Specific impulse (s, vacuum): 265
Total impulse (kNs, vacuum): 326,000
Pressure (atmosphere): 58 max at 17 s
Nozzle: throat diameter (cm): 80
length (cm): 2,034.8
exit diameter (cm): 227
expansion ratio: 8
materials: carbon phenolic throat insert, tape-wound carbon phenolic on divergent cone changing to silica phenolic below TVC inlets.
Casing materials: five segments of M250 maraging

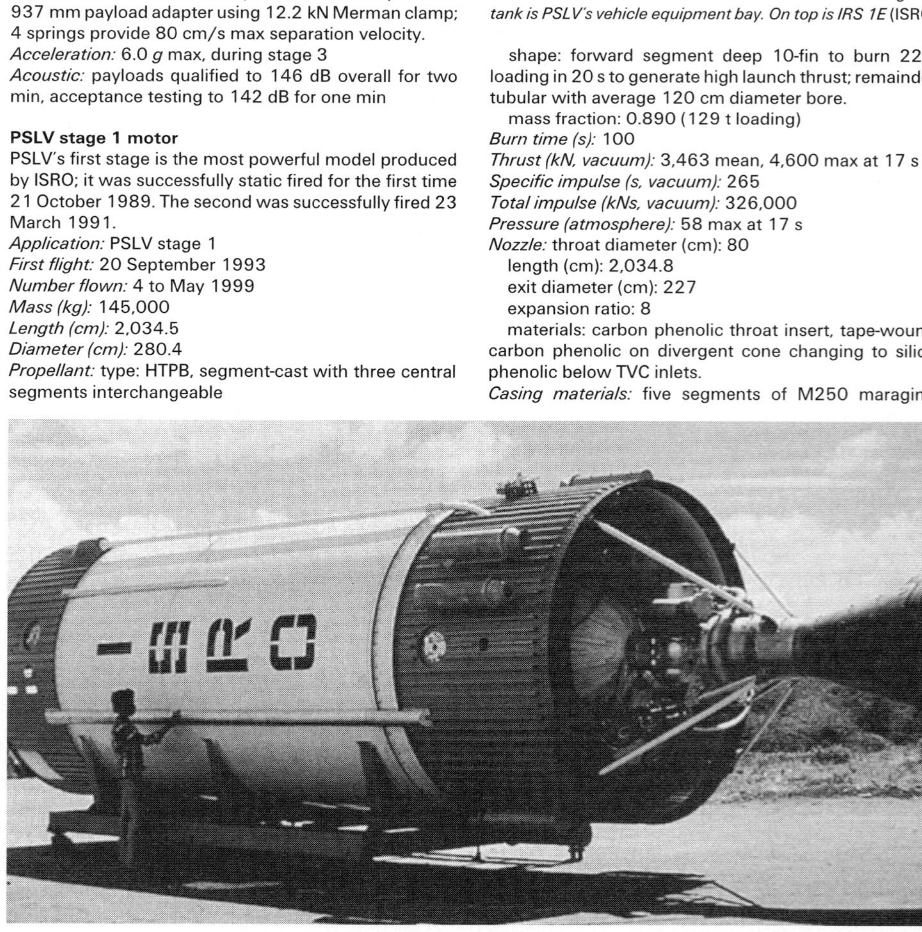

PSLV stage 2: ISRO developed the Vikas engine from Ariane's Viking to power the first large Indian liquid stage. The toroidal water coolant tank is visible inside the base skirt, with solid ullage and retro motors on the exterior (ISRO)

steel, 3.4 m long. Joints tongue/groove with two O-rings and 144 pins. Insulated by nitrile rubber, with silicon potting compound on joints.
Igniter type: pyrogen head-end 30 cm diameter, 125 cm long containing Pedpro 2661 HTPB and its own 2 kg-charge igniter

PSLV stage 3 motor
Comment: This is ISRO's most advanced solid propellant motor, designed for a 0.915 mass fraction and the first to adopt a submerged flex nozzle design for TVC. Off-loading is also possible. Test #5 made 29 July 1991, #6 17 October 1991. Tests 8, 9 and 10 were for qualification.
Application: PSLV stage 3 (PS3)
First flight: 20 September 1993
Number flown: 4 to May 1999
Mass (kg): 7,975
Length (cm): 244.20 (casing 208.35)
Diameter (cm): 198.8 (202.5 skirt)
Propellant: type: HTPB casebound, 86% solid
shape: slotted
mass fraction: 0.910 (7,260 kg standard loading)
Burn time (s): 73.1; 78.1 action time
Thrust (kN, vacuum): 243.2 mean, 386 max
Specific impulse (s, vacuum): 293
Total impulse (kNs, vacuum): 20,850
Pressure (atmosphere): 57.4 max
Nozzle: the 19.6 per cent submerged contoured nozzle is provided with a ±3° (±2° flight) conical flex bearing. The elastomer, based on natural rubber of nominal 3 mm thickness is bonded with steel reinforcements of 4.5 mm thick run. Each seal consists of six reinforcements and seven elastomer pads. The seal positioned beyond the throat housing is driven by two 11.8 kN electromechanical actuators 90° apart; a thermal boot protects the bearing.
throat diameter (cm): 19.4
length (cm): 180.2
mass (kg): 280
exit diameter (cm): 141.4
expansion ratio: 52.3
materials: graphite throat insert, carbon and silica phenolic composites for nozzle.
Casing materials: Kevlar filament-wound, EPDM insulation and aluminium forgings for skirt and end-opening fittings; total mass 340 kg. Nozzle opening 750 mm diameter, 200 m head end.
Igniter type: pyrogen head-end

PSLV strap-on motor
Comment: The same basic motor, derived from the SLV-3's stage 1, was employed for ASLV's stage 1 and strap-on, and adopted as PSLV's initial strap-on. The strap-ons are essentially identical to the ASLV stage 1 version, with the principal exception of canted nozzles. For PSLV, the secondary injection TVC systems are deleted on one ground-lit and three air-lit strap-ons. The specifications below are for ASLV's stage 1 model; differences for the strap-ons are noted where applicable.
Application: ASLV stage 1 + strap-ons, PSLV strap-ons
First flown: 24 March 1987, ASLV-D1 launch
Number flown: 24 PSLV strap-ons, 4 ASLV stage 1, 8 ASLV strap-ons to May 1999
Mass: 10.4 t
Length (cm): 998.99
Diameter (cm): 100
Propellant: type: AP/18%Al/HTPB case-bonded
shape: star
mass fraction: 0.853 (8.9 t loading; 8.63 t for strap-on)
Burn time (s): 46 action time (49.95 action strap-on)
Ignition delay: about 200 ms strap-on
Thrust (kN, vacuum): 502.6 mean, 643 max (strap-on: 440 mean, 580 max)
Specific impulse (s, vacuum): 259 (252 strap-on)
Total impulse (kNs, vacuum): 22,600 (21,450 strap-on)
Pressure (atmosphere): 43.5 max (40.2 max strap-on)
Nozzle: secondary injection ports for TVC; 9°-canted nozzles
expansion ratio: 6.7
materials: graphite throat, carbon phenolic composite for convergent section, tape-wound carbon and then silica phenolic composites for divergent.

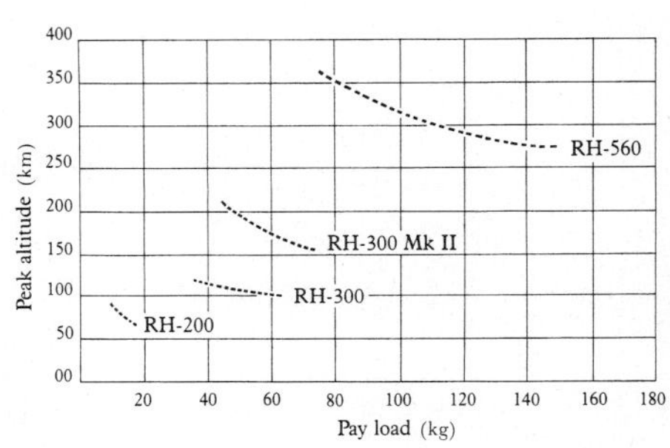

Rohini sounding rocket performance curves. The improved RH-560 will deliver 100 kg to 500 km

Casing materials: 3.5 mm thick low carbon steel with nitrile rubber insulation. Three-segment casing with tongue/groove joints.
Igniter type: pyrogen head-end

Rohini Sounding Rockets (RSRs)

Background
India's space programme began in November 1963 with the launch of a two-stage US Nike-Apache sounding rocket from Thumba, near Trivandrum. Production of indigenous Rohini Sounding Rockets (RSRs) at VSSC created the core of expertise for an orbital launcher programme, beginning with the 75 mm diameter RH-75. Four Rohini models remain in service, launching at an average of one a week for meteorology (RH-200), middle atmosphere (RH-300) and ionosphere (RH-560) investigations from SHAR, Thumba and Balasore. The Indians favour the RH-200s, launching about 40 during 1987. These rockets carry radar chaff out of Balasore to test and calibrate Indian Meteorological Department equipment.

The RSR programme is sited at the Vikram Sarabhai Space Center, which includes a Rocket Propellant Plant and Rocket Fabrication Facility.

Specifications
Rohini 200
Applications: meteorological sonde up to 80km
Number launched: 0 in 1995, 18 in 1994, six in 1993
Number of stages: 2 (solid)
Overall length: 3.60 m
Principal diameter: 207 mm
Launch mass: 108 kg
Typical performance: 10 kg to 80 km

RH-200 stage 1
Motor: solid from VSSC
Overall length: 1.48 m
Principal diameter: 0.21 m
Average thrust: 16.89 kN sea level
Burn time: 5.7 s
Attitude control system: spin at max 270 rpm

RH-200 stage 2
Motor: RH-125 solid from VSSC
Overall length: 2.12 m
Principal diameter: 0.12 m
Average thrust: 10.49 kN sea level
Burn time: 2.9 s action time

Rohini 300
Applications: middle atmosphere probing above 100 km
Number launched: 8 to May 1999
Number of stages: 1 (solid)
Overall length: 4.14-6.84 m depending on payload
Principal diameter: 305 mm

The Rohini 300 provides access to the middle atmosphere for Indian scientists (ISRO)

Launch mass: 370 kg
Typical performance: 50 kg to 140 km

RH-300 stage 1
Motor: solid from VSSC
Overall length: 2.844 m
Principal diameter: 305 mm
Propellant mass: 240 kg
Average thrust: 37.94 kN sea level
Burn time: 14.5 s

Rohini 300 MkII
Comment: The RH-300 Mk II was introduced during 1987 for investigations of the lower ionosphere, beyond the RH-300's capabilities. Major modifications included stretching motor length 1.2 m, reducing casing thickness by 0.4 mm from 2.0 mm, and adopting case-bonded HTPB propellant for high volumetric loading and burning characteristics resembling those of a two stage system. The resulting propellant mass increase was 100 kg.
Applications: upper middle atmosphere, lower ionosphere probing.
First launch: 8 June 1987
Number launched: 6 to end-1995
Number of stages: 1 (solid)
Overall length: 5.39-5.89 m depending on payload.
Principal diameter: 304.7 mm
Launch mass: 504 kg
Typical performance: 60 kg to 160 km

RH-300 MkII stage 1
Motor: solid from VSSC
Overall length: 3.4 m
Principal diameter: 304.7 mm
Propellant mass: 330 kg of HTPB propellant
Average thrust: 39.14 kN sea level
Burn time: 20.2 s

Rohini 560
Comment: The RH-560 is India's largest sounding rocket, although a 560/300 MkII is under development to deliver 100 kg to 500 km. A static test of an RH-560 with case-bonded HTPB was successful in 1995, and flew in 1995.
Applications: ionospheric and astronomical payloads
Number of stages: 2 (solid)
Number launched: 33 to end-1995 (1995:1; 1994: 0; 1993: 1)
Overall length: 8.376-9.176 m depending on payload
Principal diameter: 561 mm
Launch mass: 1,350 kg
Typical performance: 100 kg to 350 km

RH-560 stage 1
Motor: solid from VSSC
Overall length: 3.29 m
Principal diameter: 0.56 m
Average thrust: 75.97 kN sea level
Burn time: 19.3 s

RH-560 stage 2
Motor: solid from VSSC (based on RH-300)
Overall length: 3.78 m
Principal diameter: 0.30 m
Average thrust: 37.94 kN sea level
Burn time: 14.5 s action time

INTERNATIONAL

Ariane

Background
France designed and optimised Ariane as a direct ascent GTO launcher, beginning with the Ariane 1 variant in 1979. ESA and CNES, with Aerospatiale as the industrial architect, then further developed the Ariane 1 to 4 range of orbital launchers. Arianespace now procures, markets and launches the vehicles. The company plans to phase out the six Ariane 4 versions in 2000 but ESA's Ariane 5 development programme has been quite successful. It was approved in 1987, to produce a vehicle offering 60 per cent additional GTO capacity at significant cost reduction and able to carry a manned spaceplane. Following the development flight programme, Ariane 5 will be assigned to Arianespace for commercial launches, beginning in late 1998. The cost to customers for Ariane 5 is 20 per cent less expensive than a launch using the most powerful Ariane 4, except that Ariane 5 provides a free relaunch in the event of failure.

Kourou can launch up to eight annual departures from the dedicated ELA-3 pad, but only five are now projected. Arianespace made its first order for 14 commercial vehicles on 12 June 1995 for launch as:

503-505 in 1997, 506-509 in 1998, 510-515 in 1999. Following the loss of Ariane 501 this was revised to 503-506 in 1998, 507-511 in 1999 and 512-518 in 2000.

Ariane 3 introduced the Sylda (Système de Lancement Double Ariane) for dual launches, encapsulating the lower satellite and mounting the forward payload on its front face. Ariane 4 introduced British Aerospace's (now Matra Marconi Space) larger aluminium honeycomb Spelda (Structure Porteuse pour Lancements Doubles Ariane) as part of the fairing assembly. The lower 3.97 m diameter, 2.78 m high cylinder (Short Spelda) is attached to the VEB by 180 bolts and topped by either an 80 cm high cylinder, with a 1 m high frustum on top presenting a 1,920 mm diameter interface to the upper payload. Short Spelda masses 408 kg and extends the fairing by 280 cm; it provides a 34 m^3 payload envelope and can carry a 2,800 kg forward load. Separation is achieved by pyrotechnic line charge around the circumference, and six springs. The 458 kg Long Spelda adds a further 1 m cylindrical section, provides a 42 m^3 payload envelope and can carry a 1,900 kg upper load. Arianespace began offering an SDS Spelda Dedicated Satellite service in 1990, available from 1992 (Insat 2B in July 1993 was the first; using the 3,080 mm-long Mini-Spelda to carry 24 m^3 payloads of 400 to1,100 kg). This 337 kg unit can handle a 3,200 kg forward load. The 350 kg Mini-Spelda+300 version became available in 1996, stretching Mini-Spelda's cylinder by 300 mm for 26 m^3 payloads. It can handle a 3,200 kg forward load.

The development of a larger lift Ariane variant to become the standard vehicle through the mid-1990s, was formally approved by ESA's Council in January 1982, and management responsibility assigned to CNES. Arianespace completed technical and scheduling negotiations in 1988 with contractors for a single purchase of 50 Ariane 4 vehicles (increased in March 1995 to 55, and in February 1996 to 75) as the 'P9' final batch, for first launch mid-1992. The contracts were signed 15 February 1989, totalling about FFr18 billion; the large order yielded a 20 per cent saving in production costs. 1996's purchase of another 10 is worth about FFr5 billion. P9 introduced a stretched stage 3 (H10+) on V50 in April 1992, carrying an additional 340 kg of propellants and increasing burn time by 30 seconds, to accommodate satellite growth. The modification and weight-saving measures improved GTO capability by 130 kg, increasing 44L's to 4,460 kg. The H10-3 modification, introduced on V70 in December 1994, increased that to 4,720 kg. Other weight-saving, improved engine calibration and new trajectory optimisation, yielded 4,901 kg from V93.

Percentage interests in the Ariane 4 development programme were: France 62.60, Germany 16.28, Italy 5.87, UK 4.64, Belgium 4.05, Spain 1.77, Switzerland 1.55, Sweden 1.15, Netherlands 1.76, Denmark 0.17, and Ireland 0.07. Development programme cost was ECU650 million, including launcher development (ECU 340 million), ELA-2 construction (ECU210 million) and V22 demonstration flight (ECU100 million). Ariane 4 was designed primarily to handle Intelsat 6 and to cater for IUS-class payloads. Mixing pairs of solid and/or liquid strap-on boosters creates six variants; in order of increasing performance, they are:

Ariane 40	no strap-ons
Ariane 42P	two solids
Ariane 44P	four solids
Ariane 42L	two liquids
Ariane 44LP	two solids/two liquids
Ariane 44L	four liquids

While the Ariane 2/3 launch vehicle variants were based upon the proven technology of Ariane 1 and Ariane 4 was similarly based upon the experience and hardware of the earlier versions, Ariane 5 marked the introduction of a totally new launch vehicle design and associated technology.

Ariane 5 is optimised for dual GTO launches, but will not be as attractive for polar missions, as there will be fewer Earth observation satellites that can be paired with a primary payload. Ariane 4 can be better matched with single polar payloads. Arianespace projects that, for GTO, 3 to 3.5 tonne satellites will dominate in coming years. Ariane 5's current 5.9 tonne dual capacity will be upgraded by 2003 to allow easier mixing of payloads.

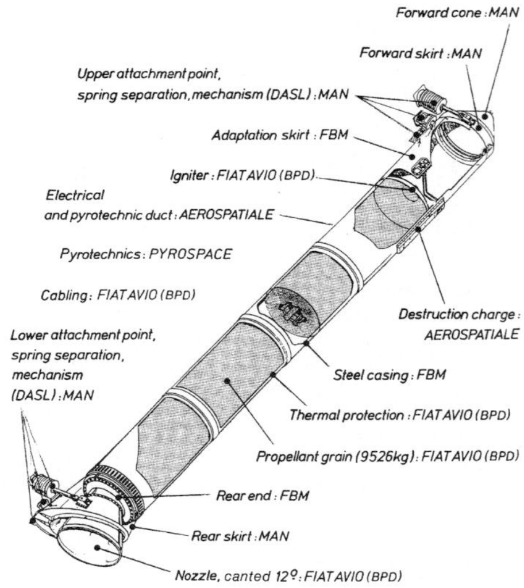

Ariane 4's PAP solid propellant strap-ons are stretched versions of those carried by Ariane 3: 126 should be used operationally within the Ariane 4 programme (Arianespace) ***1997***/0003432

Three principal changes achieved the 7.4 tonne GTO capacity. The dual payload carrier will be reduced by 400 kg, translating into a direct 400 kg GTO increase. The booster casings will be welded instead of bolted, yielding a 1.8 tonne saving for a 150 kg GTO increase. The principal increase (800 kg), and the most costly, accrues from improving the core stage. Vulcain's mixture ratio will be increased 5.3 to 6.2 by increasing the LOX flow rate by 18 per cent, accommodated by lowering the core's common bulkhead by 70 cm to raise propellant mass from 155 to 172 tonne. Vulcain thrust will be raised to 1,350 kN vacuum, achieved by a 10 per cent wider throat (expansion ratio from 45 to 61.7), 10 per cent chamber pressure increase, the higher mixture ratio, and extending the nozzle, its lower part using a simple film-cooled sheet. The launcher's structures are sized to accommodate such upgrades. Should the ECU167 million be unblocked, stage 2 improvements would add another 400 kg GTO. The propellant load would be increased to 15 tonne and thrust to 35 kN.

Other options studied include using two current Vulcains and stretching the core to accommodate 210 tonnes of propellants. This would increase LEO capacity by 3.5 tonnes, but hardly improve GTO performance because staging would no longer be optimised. Other studied enhancements include replacing EPS with a cryogenic upper stage (H10) and adding a short segment to the booster to create a P270 strap-on. A heavy lift version for 35 tonnes into a trans-lunar path would employ four P230s, a 7.5 m diameter core carrying 620 tonne propellants for five Vulcain MkIIs, and a cryogenic upper stage with 80 tonne propellants and a single restartable Vulcain derivative.

On 6 June 1996, the first Ariane 5 (Ariane 501, Volume 88) was launched from Kourou. About 40 seconds after the start of the flight sequence and at an altitude of 3.7 km the launch vehicle veered off its flight path, broke up and exploded with the total loss of the four Cluster scientific satellites. Within a few days an independent inquiry board was established to determine the cause of the launch failure. They investigated whether the qualification and acceptance tests were appropriate in relation to the failure, and recommended corrective action to remove the cause of the 'anomaly', and other possible weaknesses of the system found to be at fault. The report was published on 19 July 1996.

The countdown on 6 June had been trouble-free, with the exception of a hold at HO–7 minutes because of weather problems (HO is the time of the command for the main cryogenic engine ignition). When the conditions improved the launch sequence resumed at

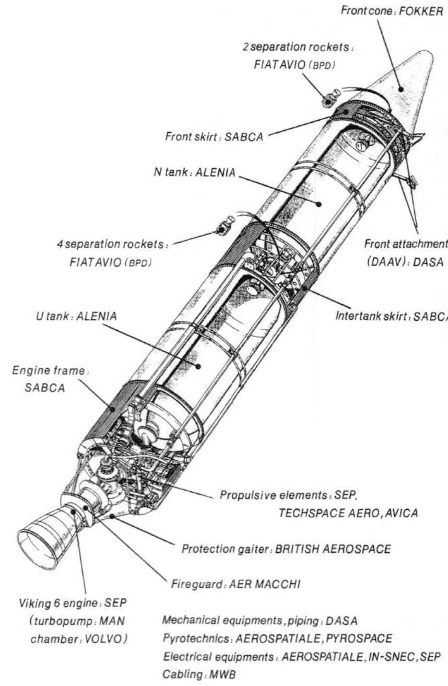

Ariane 4's PAL liquid boosters are of comparable size and performance to stage 2: 194 should be used operationally within the Ariane-4 programme (Arianespace) ***1997***/0003406

12 hours 33 minutes 59 seconds UT, with ignition of the first stage core and two solid booster engines being nominal. The vehicle performed a nominal flight until HO+36 seconds when there was a failure of the back-up inertial reference system, immediately followed by a failure of the active inertial reference system. The nozzles of the two solid-propellant boosters swivelled to their extreme positions. An instant later the Vulcain engine's nozzle did the same, causing the launch vehicle to abruptly veer off-course. This triggered (correctly) the self-destruction of the launch vehicle by the rupture of links between the strap-on boosters and the core stage. Thus, the origin of the failure was quickly narrowed down to the flight control system and specifically the inertial reference system, which had ceased operating simultaneously at HO+36.7 seconds.

The inquiry board's report found that at HO+22 seconds, variations of 10 Hz frequency started to appear in the hydraulic pressure of the actuators which commanded the nozzle of the main engine. Although this was considered to be significant and was not fully explained in the report, it was not considered to be relevant to the failure of the launch.

At HO+36.7 seconds the computer within the back-up inertial reference system – which was operating on stand-by for guidance and attitude control – became inoperative, this being caused by an internal variable related to the horizontal velocity of the launch vehicle exceeding a pre-programmed limit. Approximately 0.05 seconds later the active inertial reference system (identical to the back-up system in terms of hardware and software) failed for the same reason. With neither system operating, the correct guidance and attitude information could no longer be obtained. The fate of the mission thus became inevitable.

As a result of its failure, the active inertial guidance reference system transmitted what was essentially diagnostic information to the launch vehicle's main computer, where it was interpreted as actual flight data and used for flight control calculations. On this basis, the main computer commanded the booster nozzles, and slightly later, the core's main engine nozzle, to make a major correction for what was an imaginary deviation in the vehicle's attitude. The rapid change of attitude caused the launch vehicle to disintegrate at HO+39 seconds due to excessive aerodynamic forces. The

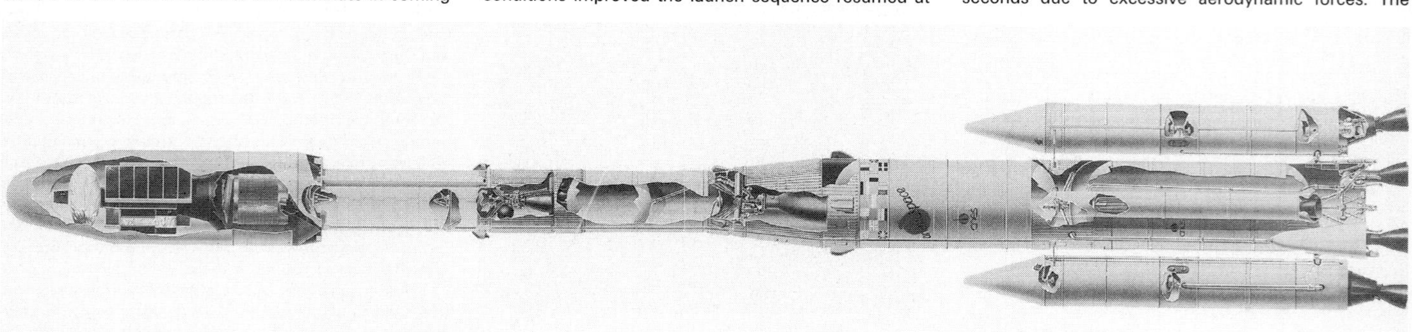

The 44LP is Ariane 4's most powerful version

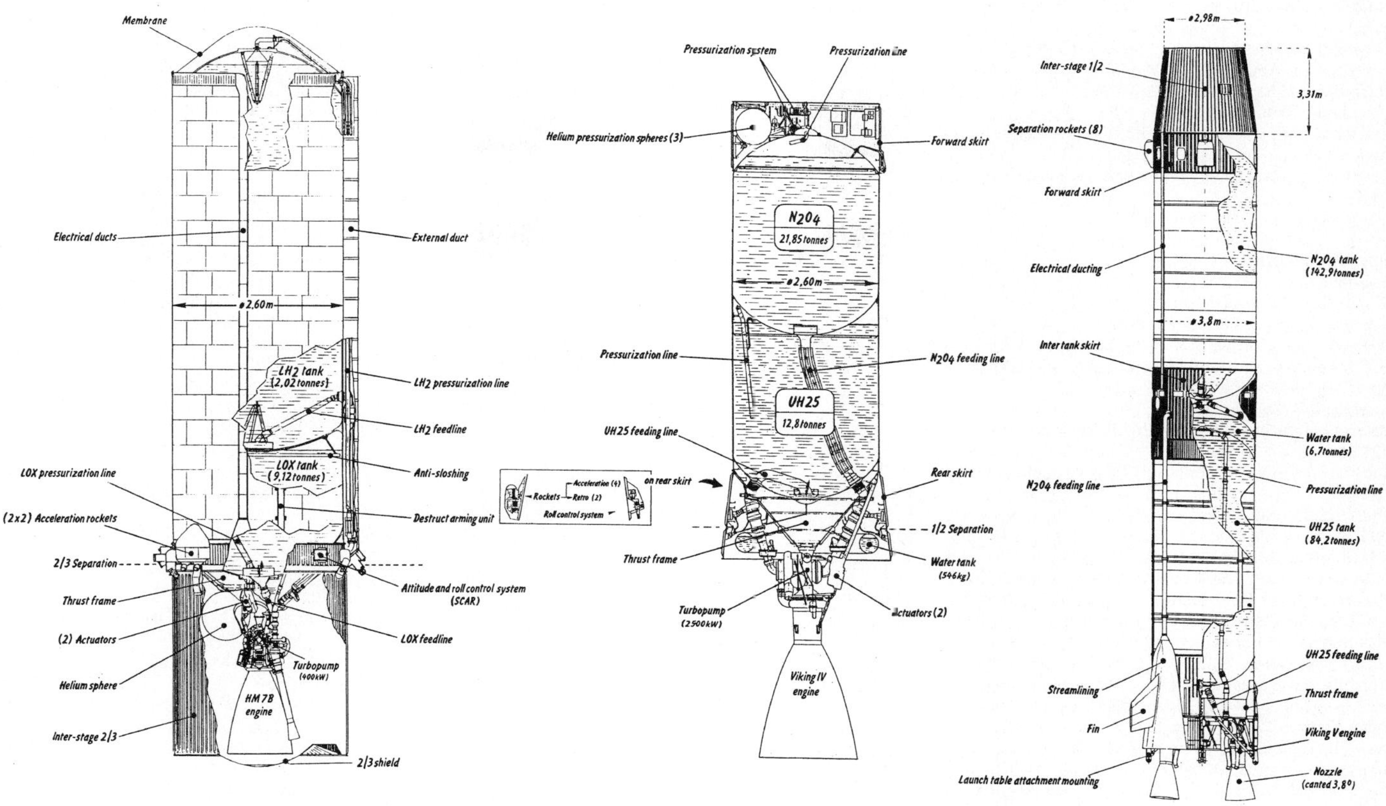

Ariane 4 stages 1-3 principal features. The H10+ stage 3 version is shown. The four $2m^2$ stage 1 fins are carried on the 42L and 44L versions only (Arianespace)

vehicle's destruction was automatically initiated at an altitude of about 4 km and at a distance of 1 km from the launch pad.

The Ariane 5 inertial reference system was essentially common to a system which is used for Ariane 4. The part of the software which caused the interruption in the inertial system computers is used before launch to align the inertial reference system and for Ariane 4 is also used to permit a rapid re-alignment of the system in the case of a late hold in the countdown. This re-alignment function does not serve any useful purpose when included in the Ariane 5 software, but it was retained for commonality reasons, and as on Ariane 4 flights it would be allowed to operate for approximately 40 seconds after launch.

During the design of the inertial reference system used for both Ariane 4 and Ariane 5, Arianespace made a decision not to protect the inertial system computer from being made inoperative by an excessive value of the variable related to the horizontal velocity. This projection was provided for several other variables of the alignment software. When this decision was made it had not been realised which values this variable might assume when the alignment software was allowed to operate after the launch of an Ariane 5.

For Ariane 4 flights the horizontal-velocity variable cannot reach (with adequate operating margins) the level which is pre-set in the software. However, Ariane 5 has a high initial acceleration and a trajectory which leads to a high horizontal velocity component which is five times the value for Ariane 4: when the limiting value of the variable was reached during the Ariane 501 launch the inertial system computers ceased operations.

Since it was designed for Ariane 4 launches, the specifications for the inertial reference system and the tests performed at an equipment level did not include simulated Ariane 5 trajectory data. Thus the re-alignment function was not tested under Ariane 5 flight conditions. As a result the design error was not discovered until after the loss of Ariane 501. Post-flight simulations were carried out on a computer running the inertial reference system and with a simulated environment which included the actual trajectory data from Ariane 501. These simulations faithfully reproduced the chain of events leading to the failures of the inertial reference systems.

As a result of its deliberations, the inquiry board made a series of recommendations which included switching-off the inertial reference system immediately after launch: more generally, no software should be run in-flight unless it is specifically required. All the flight software should be reviewed to ensure that all assumptions which were made when the code was written, are validated against the restrictions of the actual launch vehicle and also verification should be made of any internal or communication variables in the software.

The second launch of Ariane 5 carried MAQSAT H and MAQSAT B, instrumented platforms to measure the launch and ascent environment. The two were successfully placed in orbit on 31 October 1997. The third flight (Ariane 503) successfully carried the Atmospheric Re-entry Demonstrator (ARD) to a ballistic re-entry and MAQSAT 3 (a mockup of the Eutelsat W communications satellite) to GTO on 21 October 1998. The first Ariane commercial flight (Ariane 504) is scheduled for the second half of 1999.

The costs for building Ariane are divided among France 46.2 per cent, Germany 22.0 per cent, Italy 15.0 per cent, Belgium 6.0 per cent, Spain 3.0 per cent, Netherlands 2.1 per cent, Sweden 2.0 per cent, Switzerland 2.0 per cent, Norway 0.6 per cent, Austria 0.4 per cent, Denmark 0.2 per cent, Ireland 0.2 per cent. Following go-ahead in January 1985 for the Ariane 5 Preparatory Programme, Arianespace received AU718 million (AU780 million 1995 terms) to proceed. Then October 1995's ESA Ministerial conference approved ECU1,026.2 million over 1995-2003 for the Ariane 5E upgrade. Of this, ECU167 million is blocked, intended to increase stage 2 propellant load to 15 tonnes and its thrust to 35 kN. The same Ministerial meeting also approved ECU351.5 million 1996-2000 for ARTA Ariane Research & Technology Accompaniment, to maintain the vehicle's overall reliability, plus ECU335.7 million 1996 to 2000 to maintain ESA's launch and production facilities.

ESA plans to make its own orbital transfer vehicle to fit atop the Ariane 5 for Space Station operations. Arianespace is awaiting advances in the Space Station programme before expending funds to man rate the Ariane 5. During its flight programme the core stage had 0.99 reliability, an order of magnitude greater than for Ariane 3/4. The unmanned vehicle's overall reliability target is 0.985. During 1988, Aerospatiale/BAe and MBB/Matra teams undertook preliminary studies of an Ariane 5 Transfer Vehicle, then aiming for a 1996 initial operating capability, based on enhanced versions of Ariane's EPS upper stage, Vehicle Equipment Bay and hardware from other projects as much as possible to minimise development costs. The baseline

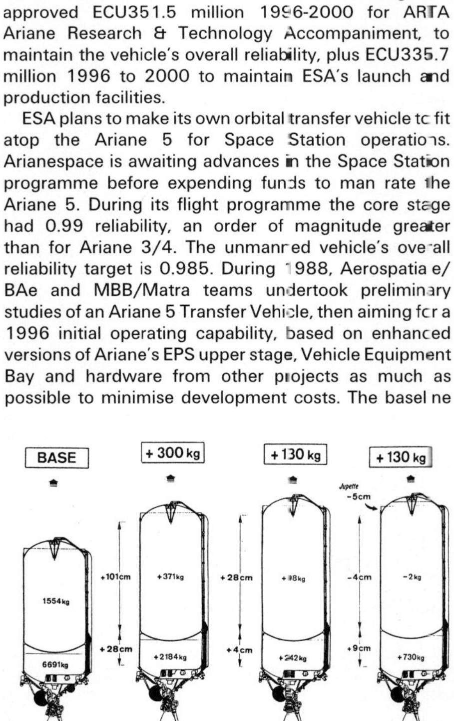

Ariane stage 3 evolution (Arianespace)

ATV is intended to be expendable; the cost effectiveness of a reusable version has been shown to be less attractive.

ESA projects one or two flights annually with about 16 tonnes of cargo in logistics carriers. For Space Station missions, ATV would execute an automatic approach on a collision avoidance trajectory. Safety and contamination regulations may demand that only cold-gas thrusters are used for the final approach and grappling by Canada's Shuttle-type Remote Manipulator System. Dwell time at the Station could be up to six months.

The Aerospatiale/MMS (ex-British Aerospace) proposal considerably modifies Ariane's EPS/VEB into a compact 3.936 m diameter (4.57 m diameter envelope), 2.47 m high stage of 2,998 kg dry mass (plus 2,983 kg propellant). The propulsion module incorporates gimballed twin 500 N bipropellant thrusters drawing on 2,400 kg MMH/NTO in four 108.5 cm diameter He-pressurised tanks, supported and backed up by four 4 × 250 N hydrazine thruster clusters. Proximity operations switch to four clusters of 20 N nitrogen jets.

The MBB/Matra studies show an ATV encompassing the VEB and elements of the EPS, maintaining the basic Ariane 5 upper stage configuration as much as possible. The VEB guides Ariane during ascent and, after ATV separation, becomes an integral part of ATV. Dry mass is 3,903 kg with 3,086 kg propellants. The 5.4 m diameter ATV is powered by a single 27.3 kN bipropellant gimballed engine, supported by 4 × 4 350 N hydrazine thrusters and 4 × 4 10 N cold-gas jets for proximity operations.

ESA began the AU37.7 million (1994 rate) Future European Space Transportation (FESTEP) Investigation Programme in February 1994 to define which reusable launchers are relevant to Europe and to develop the required technology. Contributions are 33.5 per cent Germany, 15 per cent Italy, 6 per cent Belgium, 5 per cent Spain, 4 per cent Netherlands, 3.35 per cent Sweden, 2.3 per cent Austria and 1.6 per cent Norway, totalling 70.75 per cent. DASA led the system studies (AU13 million), beginning October 1994. Five technology studies (AU8 million) were placed in first half 1995 on aerothermodynamics, materials, structures, propulsion and heat management. Proposals for FESTIP's next slices, covering larger scale work and experimental vehicles, were discussed in 1996. The continuation of the programme is under request. Concepts which were defined included single-and two-stage launch vehicles, essentially using rocket propulsion and being either partially or fully reusable. The design work is managed by an integrated system concept located at DASA Ottobrunn. Technologies being investigated are in the fields of aero-thermodynamics, materials, structures, propulsion and heat management.

For either concept, manned safety rules require ATV to back off automatically if it suffers major failures,

Ariane family statistics
First launch: 24 December 1979
Number launched: 115 to end of April 1999
Launch sites: Kourou ELA-1 (Ariane 1 to 3; site closed December 1989) and ELA-2 (Ariane 2 to 4); ELA-3 completed 1995 for Ariane 5
Vehicle career success rate: 93.04% through end of April 1999
Success rate, past 25 launches: 100% to the end of April 1999
(these figures exclude Ariane 501)

Specifications

Ariane 4
First launch: 15 June 1988
Number launched: 63 to end 1996
Principal uses: delivery of 2,100-4,720 kg payloads into GTO for apogee kick motor insertion into GEO and using the H-10+ third stage
Launch sites: Kourou ELA-2
Vehicle career success rate: 96.5% through end April 1999
Success rate, past 25 launches: 100% through end April 1999
Cost: US$40-50 million for 1,200-1,600 kg (A40), US$55-65 million for 2,000-2,500 kg (A42P), US$65-80 million for 2,500-3,000 kg (A44P), US$90-110 million for >3,000 kg (A42L and above), US$1 million for ASAP platform, US$19.5-26 million for 400-1,100 kg SDS Spelda Dedicated Satellites using the Mini-Spelda. (European governments typically pay a 15-20% surcharge, but ESA in October 1995 agreed to drop the requirement.) Users are separately offered a FFr2.5 million package of services, including payload transport from Cayenne airport to Kourou and the use of the integration facilities. The payment plan typically calls for a down payment on contract signature of 10-15% of the total price, followed by quarterly instalments of 10% and four payments of 5% in the last eight months before launch. Contract signature is preferred 36 months before launch. A reflight is guaranteed (paid for by insurance) within nine months of request in the event of vehicle failure. If a satellite fails after <27 months in orbit a new launch can be made within 9-12 months of request. Third party liability is covered by a mutual non-recourse clause and insurance subscribed by Arianespace for damages up to FFr400 million (the French government covers higher claims).
Availability: 12 launches per year.
Performance:
dedicated 7° GTO: 1,900 kg (A40; H10+ 2,020; H10-3 2,105), 2,600 (A42P; H10+ 2,740; H10-3 2,930), 3,000 (A44P; H10+ 3,290; H10-3 3,465), 3,200 (A42L; H10+ 3,350; H10-3 3,480), 3,700 (A44LP; H10+ 4,030; H10-3 4,220), 4,200 (A44L; H10+ 4,460; H10-3 4,720).
Sun-synch: 2,740 (only used by A40).
Number of stages: 3 plus up to 4 strap-ons.
Overall length: 56.35-60.13 m depending on fairing/payloads (minimum length uses short fairing; maximum uses medium fairing plus short Spelda).
Principal diameter: 380.0 cm.
Launch thrust (kN): 2,720 (A40), 3,944 (A42P), 4,060 (A42L), 5,140 (A44P), 5,270 (A44LP), 5,400 (A44L).
Ignition mass (H10+/10-3): 245 t (A40), 324 t (A42P), 356 t (A44P), 363 t (A42L), 421 t (A44LP), and 484 t (A44L).
Guidance: Matra Marconi Space inertial in stage 3 Vehicle Equipment Bay, with Sextant laser gyro providing attitude data (GEC-Marconi gyro back-up) from stage 1 separation. Saab Ericsson Space provide the computers. Some 600 parameters are telemetered on 2.1 GHz. Ariane follows pre-programmed attitude profile during first stage burn. Typical GTO standard deviation values are 50 km apogee, 1 km perigee, 0.018° inclination

Ariane 4 solid strap-ons
Stretched versions of Ariane 3's PAP (Propulseur d'Appoint Poudre) strap-ons are carried by the 42P/44P/44LP variants, integrated on the pad. Nozzle canted 12°. The solid pair ignites at launch, 4.2 s after the main engines. Two spring pairs (66 kN forward, 59 kN aft) ensure 6°/s rotation away from base at 5 m/s separation rate. Separation in pairs 89 s after launch (42P), 70 + 70.3 s (PAP 1/3 + 2/4, 44P) and 67 s (44LP).
Contractor: Aerospatiale (stage), Fiat Avio (BPD (propulsion), MAN (mechanical equipment)
Length: 1,204.8 cm
Principal diameter: 107.1 cm
Mass at ignition: 12,660 kg each
Propellant mass: each 9,500 kg of CTPB 1613
Average thrust: 650 kN each, sea level (625 kN at H0)
Burn time: 236 s
Casing material: AISI 4130 steel, 5 mm thick

Ariane 4 liquid strap-ons
Pairs of PAL (Propulseur d'Appoint Liquide) strap-ons, comparable in size/performance to stage 2, are carried by 42L/44LP/44L (the last two pairs). Each comprises a

Ariane launch record

	Launch	Type	Payload
L01	24 December 1979	1	CAT (monitoring)
L02*	23 May 1980	1	CAT, Firewheel, Amsat 3A
L03	19 June 1981	1	CAT, Apple, Meteosat 2
L04	20 December 1981	1	Marecs A
L5*	10 September 1982	1	Marecs B, Sirio 2
L6	16 June 1983	1	ECS 1, Oscar 10
L7	19 October 1983	1	Intelsat 507
V8	5 March 1984	1	Intelsat 508
V9	22 May 1984	1	Spacenet 1
V10	4 August 1984	3	ECS 2, Telecom 1A
V11	10 November 1984	3	Spacenet 2, Marecs B2
V12	8 February 1985	3	Arabsat 1A, Brasilsat 1
V13	8 May 1985	3	GStar 1, Telecom 1B
V14	2 July 1985	1	Giotto
V15*	12 September 1985	3	Spacenet 3, ECS 3
V16	22 February 1986	1	Spot 1, Viking
V17	28 March 1986	3	GStar 2, Brasilsat 2
V18*	31 May 1986	2	Intelsat 514
V19	16 September 1987	3	Aussat/Optus A3, ECS 4
V20	21 November 1987	2	TV-Sat 1
V21	11 March 1988	3	Spacenet 3R, Telecom 1C
V23	17 May 1988	2	Intelsat 513
V22	15 June 1988	44LP	Meteosat 3, PAS 1, Oscar 13
V24	21 July 1988	3	ECS 5, Insat 1C
V25	8 September 1988	3	GStar 3, SBS 5
V26	28 October 1988	2	TDF 1
V27	11 December 1988	44LP	Skynet 4B, Astra 1A
V28	27 January 1989	2	Intelsat 515
V29	6 March 1989	44LP	JCSat-1, Meteosat 4
V30	2 April 1989	2	Tele-X
V31	5 June 1989	44L	Superbird 1, DFS 1
V32	11 July 1989	3	Olympus 1
V33	8 August 1989	44LP	Hipparcos, TV-Sat 2
V34	27 October 1989	44L	Intelsat 602
V35	22 January 1990	40	Spot 2, UoSAT 3 & 4, Microsats 1-4
V36*	22 February 1990	44L	Superbird B, BS-2X
V37	24 July 1990	44L	TDF 2, DFS 2
V38	30 August 1990	44LP	Eutelsat 2 F1, Skynet 4C
V39	12 October 1990	44L	SBS 6, Galaxy 6
V40	20 November 1990	42P	Satcom C1, GStar 4
V41	15 January 1991	44L	Italsat 1, Eutelsat 2 F2
V42	2 March 1991	44LP	Astra 1B, Meteosat 5
V43	4 April 1991	44P	Anik E2
V44	17 July 1991	40	ERS 1, UoSAT 5, SARA, Orbcomm-X, Tubsat A
V45	14 August 1991	44L	Intelsat 605
V46	26 September 1991	44P	Anik E1
V47	29 October 1991	44L	Intelsat 601
V48	16 December 1991	44L	Inmarsat 2 F3, Telecom 2A
V49	26 February 1992	44L	Arabsat 1C, Superbird B1
V50	15 April 1992	44L+	Telecom 2B, Inmarsat 2 F4
V51	9 July 1992	44L	Insat 2A, Eutelsat 2 F4
V52	10 August 1992	42P	Topex-Poseidon, S80/T, Kitsat 1
V53	10 September 1992	44LP+	Hispasat 1A, Satcom C3
V54[1]	28 October 1992	42P+	Galaxy 7
V55	1 December 1992	42P+	Superbird A1
V56	12 May 1993	42L	Astra 1C, Arsene
V57[1]	25 June 1993	42P+	Galaxy 4
V58	22 July 1993	44L+	Hispasat 1B, Insat 2B
V59	26 September 1993	40	Spot 3, Stella, Kitsat 2, Itamsat PoSAT 1, Healthsat 2, Eyesat
V60	22 October 1993	44LP	Intelsat 701
V61	20 November 1993	44LP+	Solidaridad 1, Meteosat 6
V62	18 December 1993	44L+	DBS 1, Thaicom 1
V63*	24 January 1994	44LP+	Turksat 1, Eutelsat 2 F5
V64	17 June 1994	44LP+	Intelsat 702, STRV 1A/1B
V65	8 July 1994	44L+	PAS 2, BS 3N
V66	10 August 1994	44LP+	Brasilsat B1, Turksat 1B
V67	9 September 1994	42L+	Telstar 402
V68	8 October 1994	44L+	Solidaridad 2, Thaicom 2
V69[1]	1 November 1994	42P+	Astra 1D
V70[1]*	1 December 1994	42P/3+	PAS 3
V71	28 March 1995	44LP+	Hot Bird 1, Brasilsat B2
V72	21 April 1995	40+	ERS 2
V73	17 May 1995	44LP/3+	Intelsat 706
V74	10 June 1995	42P/3+	DBS 3
V75	7 July 1995	40/3+	Helios 1A, Cerise, UPM-Sat 1
V76	3 August 1995	42L/3+	PAS 4
V77	29 August 1995	44P/3+	NStar a
V78	24 September 1995	42L/3+	Telstar 402R
V79	19 October 1995	42L/3+	Astra 1E
V80	17 November 1995	44P/3+	ISO
V81	6 December 1995	44L/3+	Telecom 2C/Insat 2C
V82	12 January 1996	44L/3+	PAS 3R, Measat 1
V83	5 February 1996	44P/3+	NStar b
V84	14 March 1996	44LP/3+	Intelsat 707
V85	20 April 1996	42P/3+	MSAT 1
V86	16 May 1996	44L/3+	Palapa C2, Amos 1
V88*	4 June 1996	5	Cluster
V87	13 June 1996	44P/3+	Intelsat 708
V89	9 July 1996	44L/3+	Arabsat 2A, Turksat 2C
V90	8 August 1996	44L/3+	Italsat 2, Telecom 2D
V91	11 September 1996	42P/3+	Echostar 2
V92	13 November 1996	44L/3+	Arabsat 2A, Measat 2
V93	30 January 1997	44L/3+	GE 2, Nahuel 1A
V94	1 March 1997	44P/3+	Intelsat 801

Ariane launch record (continued)

	Launch	Type	Payload
V95	16 April 1997	44L/3+	Thaicom 3, B-Sat 1A
V97	3 June 1997		Immarsat-3 F4, Insat 2D
V96	25 June 1997		Intelsat 802
V98	8 August 1997		PAS-6
V99	2 September 1997		Hot Bird-3, Meteosat-7
V100	24 September 1997		Intelsat 803
V102	12 November 1997		Sirius-2, Cakrawarta-1
V103	2 December 1997		JCSAT-5, Equator-S
V104	22 December 1997		Intelsat 804
V105	4 February 1998		Brasilsat B3, Inmarsat-3 F5
V106	27 February 1998		Hot Bird-4
V107	24 March 1998		Spot 4
V108	29 April 1998		Nilesat 101, BSAT-1B
V109	25 August 1998		ST-1
V110	16 September 1998		PAS-7
V111	5 October 1998		Eutelsat W2, Sirius 3
V113	28 October 1998		GE 5, AfriStar
V114	6 December 1998		SATMEX 5
V115	22 December 1998		PAS-6B
V116	26 February 1999		Arabsat 3A, Skynet 4E
V117	2 April 1999		Insat 2E

* indicates launcher failure (L02: stage 1 Viking combustion instability. L5: HM7 turbopump lubrication. V15/V18: stage 3 igniter. V36: stage 1 engine cooling. V63: stage 3 LOX pump bearing. V70: blockage in LOX stream to stage 3 gas generator). Arianespace assumed operational responsibility from flight ('vol') 9, following ESA's L01-04 development and L5-8 promotional phases. Notes: + indicates use of H10+ stage 3, +3 = H10+3 stage 3; 1: use of PVA Perigee Velocity Augmentation. 1993 Ariane contracts (chronological order): Inmarsat 3-F3, DBS 2, Telstar 4C2, Telecom 2C/2D, PanAmSat 2-4, Meteosat 7, GE 1, Palapa C1 (swapped with Telstar 402R from Atlas), Eutelsat 2-F5, Amos, and Intelsat 709/801/802. 1994: Intelsat 803/804, Insat 2C/2D, Arabsat 2A, Skynet 4E, MEASAT 1, BSAT 1a/1b, Turksat 1C, Nahuel 1, Inmarsat 3-F5, PAS 5. 1995: PAS 3R, Hot Bird 3, Telstar 402R, Indostar, MEASAT 2, Thaicom 3, Helios 1B, Meteosat 8, Sirius 2, Hot Bird 4, Eutelsat 3-F1/F2, Brasilsat B3. 1996: Arabsat 2B, PAS 7, Nilesat, Insat 2E, Afristar, Asiastar, Caribstar, L-Star 1/2, EchoStar 2, Lovalsat 1998 satellites, 3 Astra, ST1, Jesat 5, Skynet 4F, Sieral, Envisat 1.

fixed Viking 6 engine drawing on propellants from identical 2.1 mm-thick stainless steel tanks (weighing 830 kg each), with water coolant provided via stage 1's intertank skirt. The PALs are pyrotechnically released at 149.1 s/37.5 km after burnout at 143.6 s and separated by six solid BPD motors identical to the stage 1 retrorockets.
Contractor: DASA (prime, integration), Alenia (main tanks), Sabca (thrust frame, intertank, forward skirts), Fokker (nose cone).
Engine: single SEP Viking 6 canted at fixed 10°
Length: 1,900 cm
Principal diameter: 221.6 cm (internal 217.4 cm)
Dry mass: 4,540 kg
Oxidiser: nitrogen tetroxide, in forward tank pressurised at 4.45 atmospheres.
Fuel: UH25 (UDMH+25% hydrazine hydrate), in aft tank pressurised at 4.45 atmospheres.
Propellant mass: 39,000 kg (A44L/44LP), 37,000 kg (A42L).
Average thrust: 739 kN each vacuum (670 kN at launch).
Burn time: 140 s (igniting with main engines, 3.4 s before launch)

Ariane 4 stage 1

The Ariane 3 stage was stretched 6.7 m to increase propellant capacity from 144 t to 227 t and the four Viking 5 engines qualified to 300 s to accommodate the burn increase from 138 to 205 s. Fins are carried only on the 42L/44L variants. The two 10.09 mm long, 3,800 mm diameter identical 15CDV6 steel tanks are separated by a 2,688 mm high interstage and an 6,736 l water tank 730 mm high with a GFRP forward dome and sharing the UH25 tank's forward bulkhead. The water coolant is also routed to the PAL liquid strap-ons, when carried. The stage's 1,500 mm high forward skirt supports the eight Fiat Avio/BPD separation retrorockets. Stage 1 + 4 PALs (i.e. 44L configuration) provide about 3,040 m/s.
Designation: L220
Contractor: Aerospatiale (prime): Man (thrust frame, water tank), Fokker (interstage), CASA (inter tank, forward skirt), SABCA (streamlining, fins)
Engine: four gimballed SEP Viking 5, canted at fixed 3.8°. For each, a 35 atmosphere gas generator drives a single 10,000 rpm turbopump to deliver propellants and water coolant in addition to maintaining tank pressures at 5 atmospheres. UH25/NTO delivery pressures are 67/69 atmospheres.
Overall length: 28.388 m (including 3.31 m interstage 1/2)
Principal diameter: 380 cm
Dry mass: 17,560 kg
Oxidiser: nitrogen tetroxide, in forward tank
Fuel: UH25 (UDMH + 25% hydrazine hydrate) in aft tank
Propellant mass: 167.5 t (A40), 217.2 t (A42P), 201 t (A42L), 227.1 t (A44P/44LP/44L).
Stage thrust: 2,718 kN sea level
Burn time (s): 150 (A40), 196 (A42P), 181 (A42L), 205 (A44P/44LP/44L)
Attitude control: engine gimballing
Separation: stage 1/2 separation is commanded by the VEB when stage 1 thrust decays to 50%. The four stage 2 acceleration solids fire, followed 2 s later by the interstage explosive cord + stage 1's eight Fiat Avio/BPD solid retros. Stage 2's Viking 4 ignites after a further 1 s, its solids burn out 4 s later and are jettisoned with a 4 s delay.

Ariane 4 stage 2

Essentially the same as Ariane 3's stage 2 but structurally stiffened. The aluminium alloy propellant tanks form a cylinder, height 6,515 mm, with hemispherical bulkheads, divided into two vessels by a common hemispherical bulkhead with its concave face forward. The NTO feed line passes through the lower tank. The thrust frame includes a 188 mm high cylindrical section and a 1,350 mm high cone bearing the gimbal unit mounting flange. The frame connects to the stage 1/2 interstage conical skirt via the aft conical skirt, height 1,570 mm. This aft skirt incorporates the Viking 4 engine's toroidal water coolant tank (overall diameter 2,240 mm, tube diameter 340 mm). The 1,245 mm high forward skirt connects to the carbon fibre stage 2/3 interstage and houses three 300 atmosphere helium spheres for 3.5 atmosphere tank pressurisation.
Designation: L33
Contractor: DASA (prime, integration), DASA/Dornier (tanks), Zeppelin (water tank), Aljo (thrust frame, skirts).
Engine: single SEP Viking 4 (chamber pressure 58.5 atmospheres), gimballed for pitch/yaw control
Overall length: 1,161 cm
Principal diameter: 260.0 cm
Oxidiser: nitrogen tetroxide
Fuel: UH25 (UDMH + 25% hydrazine hydrate)
Propellant mass: 34.6 t
Dry mass: 3.4 t
Average thrust: 794 kN vacuum
Burn time: 125 s
Attitude control system: pitch/yaw by engine gimballing; roll by two sets of 50 N thrusters mounted on rear skirt fed from Viking's gas generator. Guidance begins about 10 s after stage 1 separation (stage 1 executes pre-programmed profile).
Separation: at about 344 s, 147.4 km, 5,379 m/s, assisted by two solid deceleration rockets; sequence begins when stage 2 has added required ΔV.

Ariane 4 stage 3

Essentially the same as Ariane 3's stage 3 to provide direct burn into GTO (although sub-GTO first used on V54), but structurally stiffened. The associated aft carbon fibre interstage skirt, developed by Fokker Space, is 53 kg lighter than its aluminium predecessor, adding 12.5 kg to GTO capability. The cryogenic propellants are housed in 7020 aluminium alloy tanks separated by a phenolic honeycomb bulkhead under vacuum; external surfaces are treated with thermal insulation. The tank

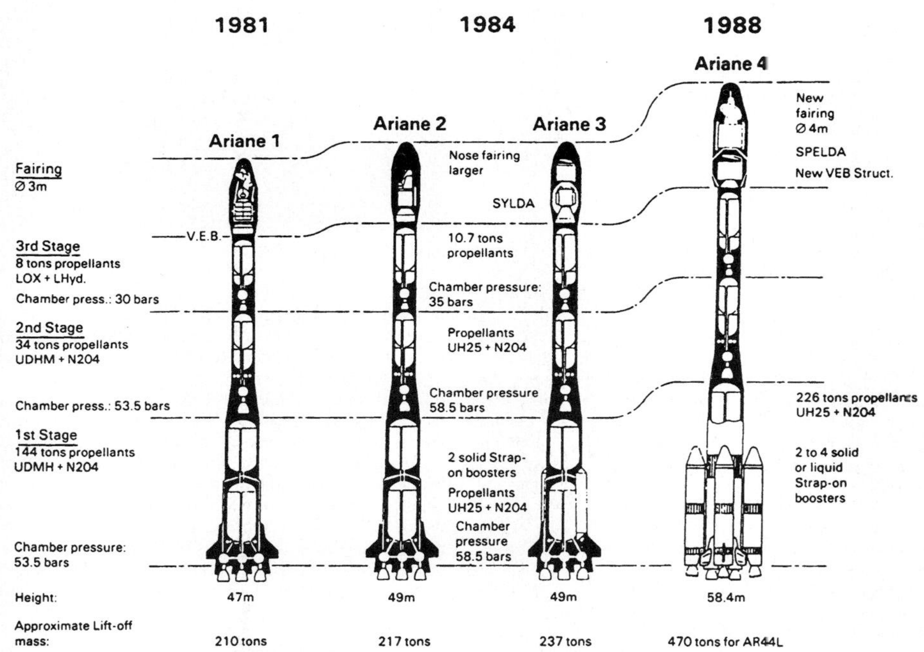

124 Ariane 1-4s have been ordered in 10 batches: 4 development (Ar 1 1979-81), 6 P1 (Ar 1 1982-84), 4 P2 (1 × Ar 1+3 × Ar 2/3 1984-85), 5 P3 (Ar 2/3 1985-87), 7 P4 (Ar 2/3 1987-88), 3 P5 (Ar 4 1988-89), 6 P6 (2 × Ar 2/3+4 × Ar 4 1989), 6 P7 (Ar 4 1989-90), 8 P8 (Ar 4 1990-91), 75 P9 (Ar 4 1992-2000). Totals are 11 Ar 1, 17 Ar 2/3, 96 Ar 4

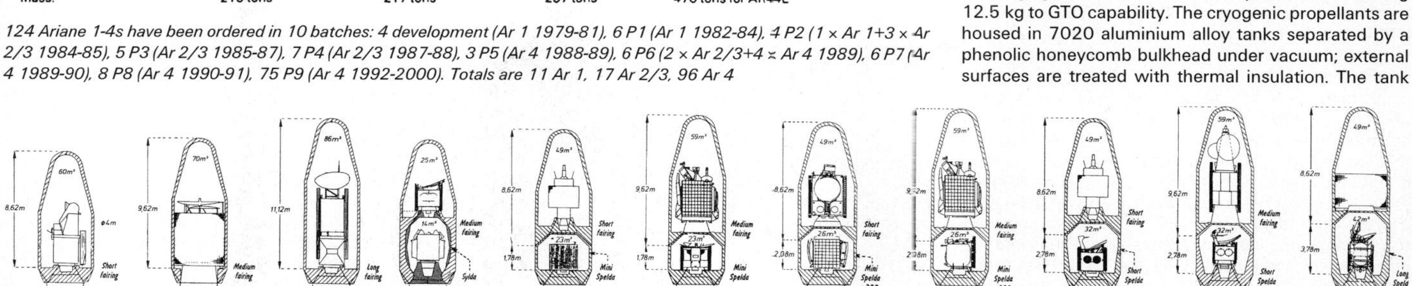

Ariane 4 offers two basic Oerlikon-Contraves fairings: 8.62 m and 9.62 m long, but an 11.12 m version is available on request. Sylda and Spelda encapsulations provide a dual-launch capability. Mini-Spelda debuted in 1993. The Mini-Spelda+300 version became available in 1996, stretching Mini-Spelda by 300 mm (Arianespace) ***1996***

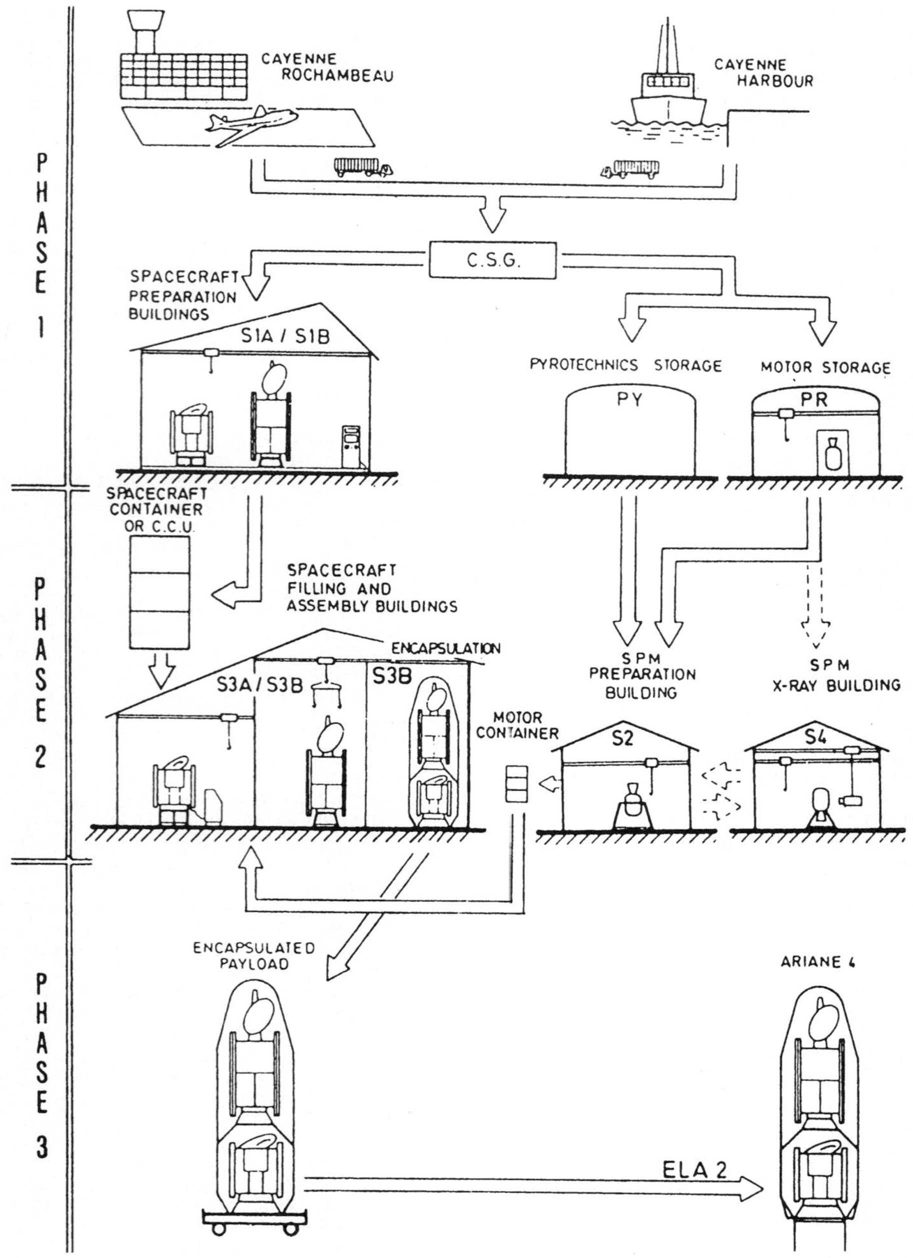

Typical Ariane 4 payload processing flow at Kourou

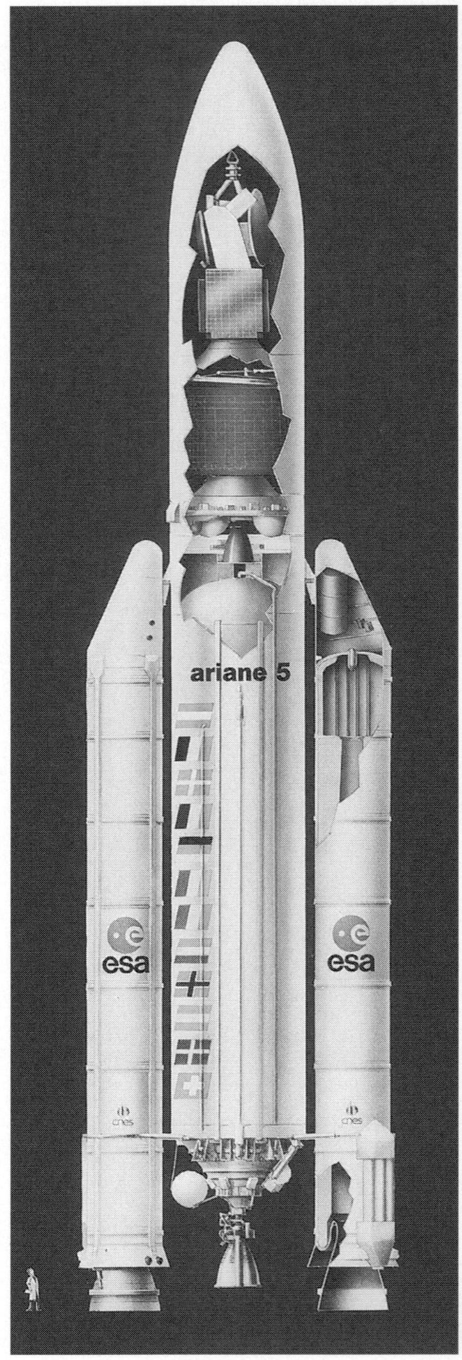

Ariane 5 in dual-launch Speltra configuration (ESA) ***1997***

assembly is extended by a 451 mm high forward skirt connecting to the equipment bay, and at the rear by a 288 mm high skirt that provides a mounting for the thrust frame, consisting of a 1,184 mm high cone and 499 mm high cylindrical section. This cylinder carries the acceleration rockets and pyrotechnic cutting system for stage 2/3 separation. The stage 2/3 skirt, 2,730 mm high, is sealed at its lower end to minimise heat exchange. The P9 Ariane production batch introduced a stretched stage 3 (H10+) on V50 in 1992, carrying an additional 340 kg of propellants (242 kg LOX + 98 kg LH_2) in stretched tanks (4 cm LOX + 28 cm LH_2) and increasing burn time by 30 s, to accommodate satellite growth. Dry mass was reduced by 30 kg by reducing wall thickness to 1.5 mm, 1.4 mm with internal integral longitudinal stiffeners. The modification and vehicle weight-saving measures improved GTO capability by 200 kg (stage 3 contributing 130 kg), increasing 44L's to 4,460 kg. The last of the old tanks flew on V60 in October 1993. PVA Perigee Velocity Augmentation was used for the first time on V54. Stage 3 established a sub-GTO with 27,720 km apogee, allowing the payload itself to establish GTO. The H10-3 modification was introduced on V70 in December 1994, improving 44L capacity to 4,720 kg. This is achieved by reducing the thermally stratified unused LH_2, increasing the flow rate and moving the internal bulkhead slightly (see diagram) to carry more LOX (increased mixture ratio). These measures together make another 730 kg LOX available. The safety margin of unused propellants was previously 180-200 kg and once was 250 kg. The performance of the turbopump exhaust heater was improved to maintain the same margin on the He pressurisation system. V93 introduced an increase to 4,900 kg GTO by further weight saving, better engine calibration and new trajectory optimisation. Specifications are given in H10/H10+/H10-3 order.

Designation: H10/H10+/H10-3
Contractor: Aerospatiale (prime, integration), Air Liquide (tanks), Fokker (thrust frame, interstage 2/3).
Engine: single cryogenic open-cycle SEP HM-7B, gimballed for pitch/yaw control. The LOX/LH_2 gas generator drives a turbopump including two 60,000 rpm LH_2/13,000 rpm LOX shafts. LOX is routed directly to the injectors; LH_2 is heated to 150 K as it cools the engine nozzle. This gas also pressurises the fuel tank; before ignition both tanks are pressurised by helium from a single 90K/200 atmosphere sphere. Chamber pressure is 36.2 atmospheres (37.2 atmospheres H10-3).
Overall length: 1,073.4/1,105.4/1,105.4 cm
Principal diameter: 260.0 cm
Oxidiser: liquid oxygen
Fuel: liquid hydrogen
Propellant mass: 10.80/11.14/11.86 t
Dry mass: 1.2t/1.24/1.24 t, excluding interstage 2/3
Mixture ratio: 4.76/4.66/5.04

Ariane 5's Vehicle Equipment Bay (MMS) ***1996***

Average thrust: 63/63.2/64.8 kN vacuum
Burn time: 720/750/780 s, until desired GTO has been achieved in single direct ascent burn beginning about 347 s after launch.
Attitude control system: the Roll & Attitude Control System (SCAR) maintains roll control during main engine burn (engine gimballing provides pitch/yaw) and three-axis following cut-off. SCAR consists of two sets of diametrically opposed triple thrusters fed with GH_2; total impulse >15 kNs. Two aft-facing nozzles depressurise the LOX tank after cut-off and are sequenced as part of the attitude control system and collision avoidance manoeuvre following payload separation. Spin stabilisation of about 10 rpm is provided for payload deployment.

Ariane 4 Vehicle Equipment Bay

Matra Marconi Space's 520 kg 1,037 mm high VEB performs guidance (Sextant or GEC-Marconi laser gyros), data processing, sequencing, telemetry, tracking and destruction function, as on all Ariane variants, but was completely reconfigured for Ariane 4. A digital flight control system replaced Ariane 2/3's analogue equipment. There are four principal elements fabricated from aluminium honeycomb with carbon fibre facing (provided by CASA). They are the internal cone, providing 1,920 mm diameter interface to the payload, the external 1,000 mm high cone mating to either the 3,936 mm diameter fairing or Spelda at its forward face, a horizontal annular plate carrying the electronics, and 12 removable carbon fibre panels enclosing the equipment compartment while providing simple access on the pad.

Ariane 5's L9 EPS stage (ESA) ***1995***

Payload fairing/carriers

Three basic lengths of Oerlikon-Contraves (prime contractor) 25 mm-thick aluminium alloy honeycomb two-piece fairings are available. They are 8.6 m (type 01, 740 kg, 60 m³), 9.6 m (type 02, 800 kg, 70 m³) and 11.1 m (type 03, 86 m³, available on special request). Ariane 4 is the first fairing to provide a standard 4 m diameter, with 3,650 mm payload envelope. Pyrotechnic cord and piston separate the fairing halves when the heating flux has reduced to 1,135 W/m², at about 285 s/115.3 km during stage 2's burn. The base clamp band is released first. Integration is the responsibility of SF Emmen. The following standard payload adaptors are available (numerical designation indicates diameter in mm at satellite interface):

937 truncated cone providing four springs for 0.5 m/s separation. Eight versions: 937/937A, 1.5 t, 48 kg carbon fibre; 937B, 2.3 t, 60 kg carbon fibre; 937C, 1.4 t, 54 kg metallic; 937V, 1.8 t, 75 kg carbon fibre; 937VB, 2.3 t, 80 kg carbon fibre, 937VD, 2.4 t, 88 kg, metallic; 937D, 3.2 t, 90 kg, metallic.

1194 in four versions: 1194, 2.6 t; 1194A, 3.5 t, 45 kg carbon fibre; 1194B, 2.8 t, 60 kg metallic truncated cone; 1194V, 3.5 t, 72 kg carbon fibre.

1666A, 2.8 t, 50 kg aluminium alloy truncated cone, 4-8 actuators provide 0.5 m/s separation.

1920, adapts 1666A for HS-601 satellites. Three versions, 400 mm height (22 kg), 500 mm height (25 kg), 550 mm heightt (26 kg). The prime contractors for the payload adapters are Aerospatiale (937/937A, 937C, 937D, 937VD, 1194A, 1194B, 1920), CASA (937B, 937V, 937VB, 1194V, 1666A) and Matra Marconi Space (1194).

Acceleration load: 4.0 *g* at end stage 1 burn, 4.1 *g* at end stage 2, 1.6 *g* end stage 3 (for A40); 7.3 *g* rms max vibration.

Acoustic load: 142 dB integrated at launch + transonic

Thermal load: max 500 W/m² radiated by fairing/Spelda

The 60 kg ASAP Ariane Structure for Auxiliary Payloads developed by Matra Marconi Space, was introduced in January 1990 to carry up to six small (<50 kg) piggyback payloads. (Up to a total of 200 kg on a 40 mm-thick aluminium honeycomb circular plate with an outer diameter 290 cm, and an inner 210 cm extending into the base of the payload fairing volume below the primary satellite). ASAP is designed for missions with a single primary payload, usually found on polar launches. Each payload and its separation system are contained within a 350 × 350 × 450 mm volume. First flight was V35.

Vehicle/payload processing

The Operations Directorate is responsible for the vehicle stages from the time they leave the factory until launch. Launch campaigns now require typically 27 working days and some 400 people for the vehicle, payloads and site. New procedures introduced in 1992 (with the P9 batch) reduced the number of launch operations from 1,200 to 700. Kourou's ELA-2 complex is designed to accommodate up to 10 Ariane 4 departures annually. Stages are shipped down the Seine from Aerospatiale typically nine weeks before launch from Les Mureaux to Le Havre, then to Cayenne (Kourou's Pariacabo harbour from 1995) for road transfer to CSG. Propellants (except LOX/LH_2) are also shipped in. Erection of the three stages and liquid boosters in the assembly building requires about three weeks before transfer to the pad, two weeks before launch and a 38 h countdown distributed over two days. The solid strap-ons are added on the pad. Payloads are received and undergo the first processing phase in building S1, while kick motors are handled in S4 (x-ray) and S2. Spin balancing and hydrazine filling are conducted in S3A/B, completed by encapsulation and transfer to ELA-2 typically six days before launch, for mating the following day. Stages 1/2 + PALs are fuelled the day before launch and final countdown begins 16 hours 40 minutes before ignition (times are referenced to ignition; lift-off occurs 4.4 seconds later). Service tower removal at 5:25 before launch, stage 3 filling start at 3:35 apogee kick motor armings at 50/45 minutes, and finally initiation of the automatic launch sequence 6 minutes before ignition. Ariane switches to internal batteries at 60 seconds and the inertial platform is released at 9 seconds. The two computers in the Ariane Launch Center confirm that all liquid propellant engines are burning satisfactorily 4.1 seconds after ignition, and then fire the solids and release the hold-down arms. If any hold in the synchronised sequences is called up to 5 seconds before ignition, Ariane automatically resets to its 6 minute configuration.

Ariane 5

Launch sites: Kourou, French Guiana
First launch: 4 June 1996
Number of launches: 3, through May 1999
Success rate: 66.66%, through May 1999
Performance: 6,800 kg single or 5,970 kg double payload version into 7° GTO. 10,000 kg into 800 km 98.6° Sun-synchronous. 18 t into 70 × 300 km, 51.6° (Space Station transfer). 4,450 kg dedicated into lunar transfer orbit, shared GTO mission could deliver 400 kg lunar orbiter + 200 kg-payload lunar lander.
Number of stages: 2 + 2 strap-ons
Overall length: 45.71-51.37 m
Principal diameter: 5.40 m
Launch mass: 746 t
Guidance: provided by the Vehicle Equipment Bay encircling stage 2. Accuracy (1 Σ): 26 km semi-major axis, 0.02° inclination GTO; 4 km, 0.04° 800 km 98.6° Sun-synchronous

Ariane 5 boosters

Designation: P230 (EAP Etage Accélérator à Poudre)
Contractor: Aerospatiale (stage integrator), Europropulsion (motors), Parachutostroenye (recovery parachutes).
Overall length: 31.16 m
Principal diameter: 3.049 m
Propellant type: PCA + HTPB + aluminium solid
Propellant mass: 237.7 t each
Empty mass: 40 t
Thrust: 5,250 kN sea level (5,904 kN vacuum) at launch
Burn time: 132 s

The boosters from two launches annually will be recovered from the Atlantic 450 km downrange following separation at about 55 km altitude for inspection to check design/manufacturing margins. Six chutes (auxiliary, three drogues, secondary, main) for 27 m/s impact. Originally sized to carry 170 t of propellant each, the boosters were upgraded to cope with satellite growth (including Hermes) and a demand for uprated GTO performance over initial requirements. First static test 16 February 1993; last two of the seven used flight-type motors (#6 first quarter was successful 10 March 1995; number 7 second quarter 21 July 1995). See the Europropulsion entry for motor details.

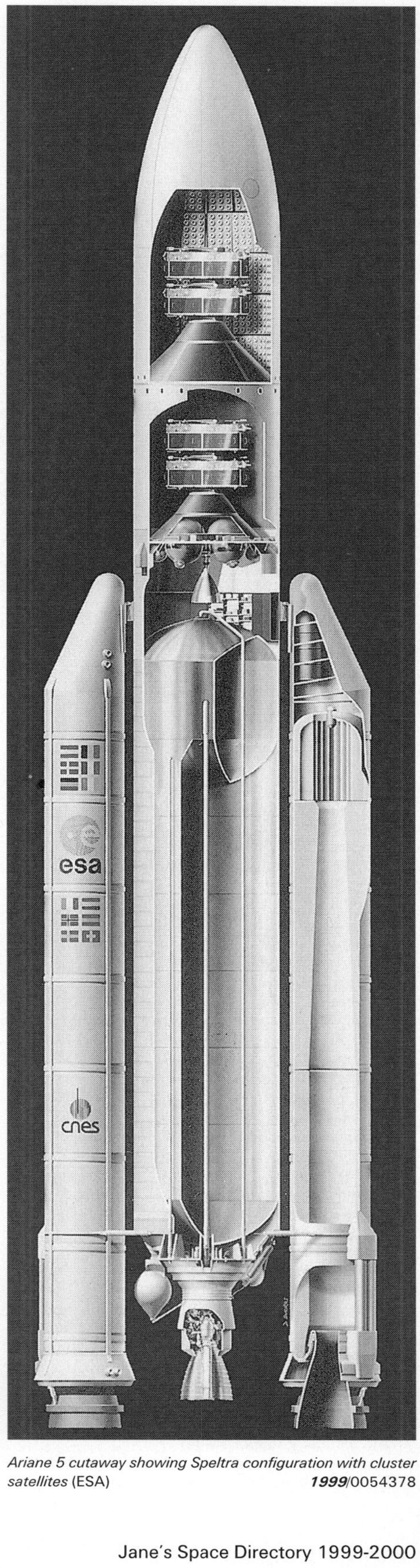

Ariane 5 cutaway showing Speltra configuration with cluster satellites (ESA) ***1999***/0054378

Ariane 42P launches the Hot Bird 4 satellite in February 1998 (ESA) ***1999***/0054364

The first Ariane 5 lifts off ELA-3 in May 1996 (ESA) ***1999***/0054379

Ariane 5 stage 1 (core)

The core stage delivers the upper composite into 50 × 1,300 km, 7° on GTO missions (which ARD will employ for its re-entry mission).

Designation: H155 (EPC Etage Principal Cryotechnique)
Contractors: Aerospatiale (stage integrator), SEP (engine), Fokker (thrust frame), MAN (forward skirt), Cryospace (tanks), SABCA (engine actuation system).
Engine: single cryogenic gas generator cycle SEP Vulcain, began stage-level qualification firings 17 November 1994 using a battleship stage (concluded 27 January 1995); flight-type stage testing May 1995 – January 1996. Thrust structure 1.6 t, 3.6 m ht, 5.40 m diameter. 501 carried Vulcain M14, 502 M16, 503 M17 (M15 used for tests).
Overall length: 30.7 m
Principal diameter: 5.40 m
Oxidiser: 130.60 t max of liquid oxygen in 120 m^3 forward tank, pressurised to 3.5 atmosphere by helium (140 kg in 830 mm diameter 400 atmosphere spheres)
Fuel: 25.60 t max of LH_2 in 390 m^3 aft tank, pressurised to 2.5 atmosphere by GH_2 from Vulcain
Dry mass: 12.6 t; launch mass 167.5 t
Average thrust: 1,145 kN in vacuum, 900 kN at launch
Burn time: 580 s
Attitude control system: Vulcain gimballed ±6° for pitch/yaw

Stage 1 will decay naturally from its low perigee. The core, originally sized to carry 120 t of cryogens, was upgraded because of satellite growth (including Hermes') and a demand for uprated GTO performance. PDR was conducted March 1990. As the aluminium 2219 5 t tank is <2 mm thick. The tank is first riveted to the forward skirt from which it is suspended. Then it is placed on the thrust frame. A 2 cm layer of expanded polyurethane insulates the tank. 18.5 cm diameter lines carry the propellants to the engine. On the launch pad the two boosters at the level of the forward skirt support the vehicle, and the EPC is suspended. The thrust from the boosters is transferred to the vehicle via the skirt.

Ariane 5 stage 2 (upper stage)

An upgrading of stage 2 from L5 to L7 and then L9.7 EPS was necessary to optimise the configuration and accommodate project payload growth. Reliability goal is 0.997. A design requirement calls for the stage to accommodate 60-day standbys after propellant loading (precluding cryogenics). PDR was conducted October 1989, CDR February 1995. Stage safing at mission's end will prevent explosive creation of debris. First complete stage testing began September 1994; the first 1,075 s long duration burn was successful 5 October 1994. Qualification using a flight-type stage at Lampoldshausen was conducted 18 July – 20 September 1995 with two 1,100 s and nine 10-100 s tests simulating the various loading, pressurisation and burnout configurations.

Designation: L9.7 (EPS Etage à Propergols Stockables).
Contractor: DASA (stage integrator, propulsion); subcontracts to CASA (structure), Zeppelin (tanks).
Engine: single DASA 29 kN Aestus storable propellant, open cycle, pressure-fed, reignitable, gimballed ±6° for pitch/yaw control.
Overall length: 3.3 m
Principal diameter: 3.936 m
Propellants: 6.5 t N_2O_4 + 3.2 t MMH in four 1.410 m diameter tanks at 20 atmosphere, two 0.3 m^3 helium pressurant 830 mm diameter spheres (400 bar).
Dry mass: 1,190 kg
Thrust: 29 kN vacuum
Burn time: 1,100 s.

Ariane 5 VEB

Matra's Vehicle Equipment Bay commands missions from its position encircling stage 2 and, unlike its predecessor Ariane 4 VEB, incorporates an attitude control system, providing roll control during stage 1/2 burns and three-axis control after burnout. The 5.4 m diameter, 1.56 m high, 1,400 kg unit carries 70 kg of hydrazine.

Payload fairing/carriers

Contraves provides two standard 5.40 m diameter fairings to accommodate 4.57 m diameter payloads: 12.70 m/2.3 t short and 17 m long units. The ogive fairing jettisons at about 191 s/106 km during stage 1 burn when aerodynamic heating has reduced to 1,135 W/m^2. Dornier's 880 kg Speltra (Structure Porteuse Externe de Lancements Triples Ariane), the equivalent of Ariane 4's Spelda, increases the length of the upper composite by 5.66 m and permits the launch of double payloads. Dornier was awarded the contract in early 1990; the first Speltra was delivered first quarter 1995. Standard payload adaptors are available with interface diameter 937; 1,194; 1,666; 2,624 mm.

Acceleration load: 4.2 *g* at end P230 burn, 3.4 *g* at end EPC burn, 0.4 *g* at end EPS burn; lateral acceleration does not exceed 0.25 *g*.
Acoustic load: 142 dB at launch/transonic
Thermal load: max 1,000 W/m^2 radiated by fairing + VEB

Ariane 5 vehicle processing

Kourou's ELA-3 and associated processing areas were constructed as dedicated Ariane 5 facilities to permit up to eight launches annually. Unlike Ariane 4, the payload assembly is integrated with the vehicle before they are transported to the pad only 8 h before launch, in order to minimise pad operations (similar to CIS launch operations). A launch campaign covers 21 days; the payload is mated six days before launch. The simplified pad concept deletes the requirement for large cryogenic arms on the umbilical tower by feeding the propellants from below the mobile launch table. It also reduces vulnerability to pad and launch accidents. There are four principal buildings in the preparation zone. The BIP Batiment d'Intégration Propulseur integration hall for the P230s to be assembled/checked out, the BIL Batiment d'Intégration Lanceur launcher integration building where the core is erected on the mobile platform and the boosters added, the BAF Batiment d'Assemblage Final assembly building where the payload composite is assembled and erected, the stage 2 tanks filled and the final electrical checkout conducted, and the Launch Center (CDL-3) for launch operations with two vehicles simultaneously. The pad was also used for stage 1 static firings. LOX/LH_2 production and mixing/casting P230's two large segments are performed on-site.

Ariane 5 P230

First flight: 4 June 1996 (February 1993 first static firing, July 1995 end of ground qualification).
Mass: 268.7 t (237.7 t propellant): 106.9 t aft, 107.4 t centre, 23.4 t forward; nozzle 6 t; internal insulation 4.3 t; casing 20 t; igniter 0.3 t; 0.4 t miscellaneous).
Length: 26.774 m (stage 31.16 m)
Diameter: 3.049 m
Propellant: type: HTPB 68% ammonium perchlorate, 18% aluminium, 14% PBHT liner.
shape: main section cylindrical, forward segment 15 rectangular slots mass fraction: 0.884 for 237.7 t loading.
Burn time: 132 s
Thrust: 5250 kN sea level at launch (5904 kN vacuum); 4931 kN average vacuum, 6637 kN max vacuum.
Specific impulse: 271 s vacuum
Total impulse: better than 630 MNs
Maximum pressure: 59.9 atmosphere
Nozzle: mass: 6 t
throat diameter: 90.0 cm
length: 3.420 m (2.22 m exposed)
exit diameter: 2.826 m (initial expansion ratio 9.6)
materials: carbon-carbon throat; nozzle flexbearing provides 6° deflection (6.6° max). Employs two actuators 90° apart with ±145 mm strokes. Actuator system mass 150 kg, hydraulic 1100 l, 345/270 atmosphere initial/end pressure.

Ariane 501 on its roll-out from the Launcher Integration Building showing the enclosed aft area of the first stage (ESA) ***1999***/0054380

Casing materials: 8 mm-thick D6AC steel heat-treated at 14,800 atmosphere ultimate strength; thermal insulation EPDM/Silica (GSM55) or Kevlar (EG2). Six 3.351 m-long segments are assembled into two motor sections of three segments each with two factory joints and 1.005 m-long aft dome. Tipped by 3.398 m-long forward section.
Igniter type: pyrogen head-end (25 kg composite propellant).

Rockot

Background

Eurockot markets the commercial Rockot launcher, derived from the RS-18 two-stage ICBM, a 10-man Khrunichev/DASA joint venture announced 23 March 1995. Eurockot is responsible for all commercial activities, acting as the customer's prime contractor, but Khrunichev develops and produces all of the hardware. The first revenue launch was planned for the fourth quarter of 1998 from Plesetsk, using a modified Kosmos pad now designated exclusively for Rockot. Launches will be from a tube, replicating a silo.

To prepare, Eurockot has performed three demonstration launches from Tyuratam, all silo-based. The first two (20 November 1990 and 20 December 1991) were suborbital, demonstrating basic vehicle performance and KB Salyut's new Breeze stage 3 (inactive on the first ascent). The first orbital flight was 26 December 1994, into 1,875 × 2,253 km 64.80° carrying the 70 kg RS-15 Radio-Rosto amateur relay. Commercial Rockot carries a large new fairing developed by Khrunichev.

Development of the silo-launched missile began in 1965 and it was first deployed operationally in 1974. The latest Mod 3 entered service in 1979. At the height of the Cold War, the Soviets deployed a peak 360 SS-18s in silos by 1982. Some would be retired under the START 2 treaty. In 1991, Russia and other countries had 300 silos at four operational sites, Kozelsk (60 silos), Tatischevo (110), Khemilnitsky (90, Ukraine) and Pervomaysk (40, Ukraine). In December 1994, the operational stockpile had dwindled to 290 missiles and Khrunichev had 15 at its factory available for conversion as of October 1995.

Military secrecy still surrounds this space launcher. It is known that stage 1 carries four engines, but it is not known whether the RD-0233 or the RD-0234 carries the tank pressurant system: there will be one engine of the type which carries the pressurant system and three of the other engine.

Soviet nuclear planners originally built the Breeze to dispense multiple communications satellites into low orbits in the event of nuclear conflict. The RS-18s would have resided in silos awaiting launch on demand. A derivative, the Breeze-M is being developed as Proton's new stage 4.

Breeze's restart capability, plus 7 hour flight life, allows payloads to be dispensed into different orbits. A single ignition is used for orbits up to 400 km. Breeze performs collision avoidance and can also de-orbit. During coast, it holds three-axis (1 to 10° accuracy) in

Rockot will depart from Plesetsk using a tube to replicate a silo launch (Theo Pirard/Space Information Centre) ***1996***

any selected attitude within ±120° of Sun axis. It can provide 5 rev/minute about the longitudinal axis to the payload on release. Three burns are required for Sun-synchronous missions: 1, to attain 200 × 700 km, 63°; 2, circularise at 700 km, 63°; 3, plane change to 700 km, 98°.

No designators are available for the main or vernier engine systems used on the Breeze (Briz) third stage on Rockot. It has been indicated that the main engine has flown on Phobos, VEGA, Venera and Mars spacecraft, suggesting that it is (or derived from) the Isayev bureau's (now Khimmash) KTDU-425. As a comparison with the third stage main engine data the KTDU-425A thrust is variable within the range 9.86 to 18.89 kN and the specific impulse within the range 293 to 315 seconds.

Eurockot is developing an 80 kg double launch system (DOLASY). Payload integration and checkout is performed in a 180 m² class 100,000 facility at 18 to 25°C and 30 to 70 per cent humidity. After encapsulation in the adjacent 90 m² clean room, air conditioning maintains the payload at 10 to 25°C, 15 to 60 per cent humidity, class 100,000. Air conditioning is interrupted only during two periods each of 1 hour max: mounting on the pad and final assembly of the 'silo' container. There is no access after payload encapsulation; the fairing is not radio transparent. Customer accommodation is provided in Mirny, some 42 km from the launch site, during campaigns. The airport is 35 km from the complex. A campaign requires 30 days: stage 1/2 assembly/testing L-30 to L-22; Breeze testing/fuelling L-21 to L-13; satellite fuelling L-13/12; upper composite integration L-11 to L-8; stage 1/2 transfer to pad L-7; transfer upper composite to pad L-6; launch preparations L-5 to L-1; fuelling/launch L-0.

Specifications

Rockot

First orbital launch: 26 December 1994
Number launched: 1, through May 1999 (plus 2 sub-orbital)
Launch sites: Tyuratam (retired?), Plesetsk planned available 1999
Principal uses: small-medium payloads (up to about 1,900 kg) into intermediate orbits
Vehicle success rate: 100% through end May 1999.
Past 25 launches: last 25 RS-18 100% successful
Missile career: 141 successes from 144 launches
Future commercial missions: none announced
Availability: 5/yr (up to 12/yr) on 20-month cycle between contract signature and launch
Cost: on application to Eurorockot
Insurance: information not available
Number of stages: 3 (liquid)
Overall length: 28.496 m
Principal diameter: 2.500 m
Launch mass: 107 t
Guidance: inertial, carried in compartment attached to front end of stage 3, provides three-axis control until payload separation commanded. Accuracy for 300 km 82° orbit ±3 km altitude, ±0.03° inclination; 700 km 63° orbit ±14 km altitude, ±0.05° inc
Launch sequence (min:s) for 700 km, 63° (some values have been estimated)

0.00	launch
0.52	Q_{max}, 605 m/s, 11 km altitude
2.01	stage 1 shutdown & separation (stage 2 verniers provide hot sep)
2.27	stage 2 main ignition
3.06	fairing separation, 3.74 km/s, 119 km
4.44	stage 2 main shutdown
5.04	stage 2 verniers shutdown and separation
08.40	Breeze ignition 1, max altitude (257 km)
14.06	Breeze shutdown 1, orbit 200 × 700 km
63.50	Breeze ignition 2 (circularisation)
64.26	Breeze shutdown 2, orbit 700 × 700 km

Rockot stage 1

Bureau: Khrunichev/KB Salyut
Engines: RD-0233 and RD-0234
Overall length: 17.2 m
Diameter: 2.5 m
Oxidiser: nitrogen tetroxide
Fuel: UDMH
Dry mass: not available
Propellant mass: 81 t
Total thrust: 2,030 kN (vacuum), 1,834 kN (sl)
Specific impulse: 310 s (vacuum), 285 s (sl)
Burn time: 121 s
Comment: Once the first stage engines have ignited they cannot be shut down: there are no brackets to hold the launch vehicle onto its launch table and no engine parameter verification before launch. The first stage carries four solid-propellant retro-rockets for separation from the second stage. The propellant tanks have a common bulkhead and tank pressurisation is achieved by means of hot gas from one of the main engines.

Rockot stage 2

Bureau: Khrunichev/KB Salyut
Engines: RD-0235 (main engine), RD-0236 (verniers)
Overall length: 3.9 m
Diameter: 2.5 m
Oxidiser: nitrogen tetroxide
Fuel: UDMH
Dry mass: not available
Propellant mass: 15 t

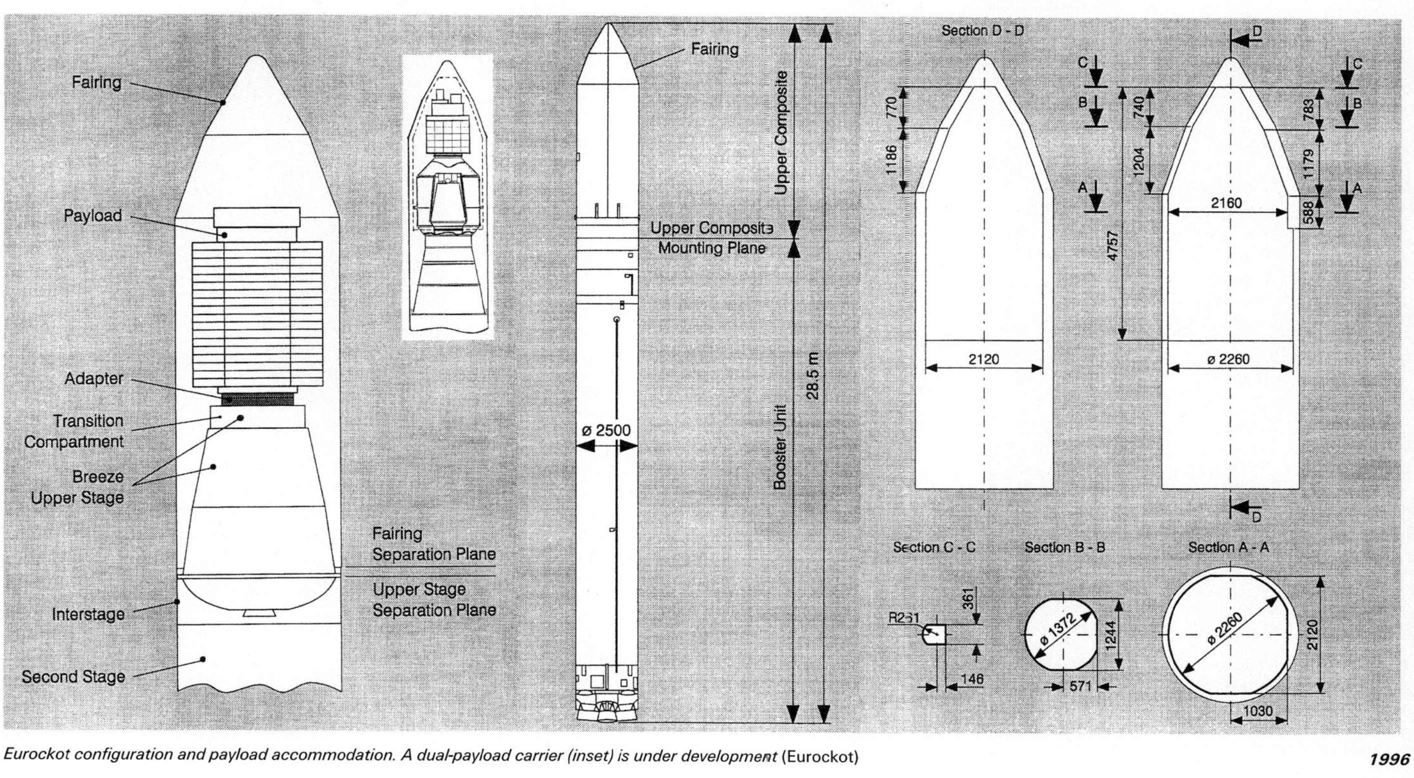

Eurockot configuration and payload accommodation. A dual-payload carrier (inset) is under development (Eurockot) ***1996***

Rockot's new stage 3 was first displayed at 1995's Paris Air Show (Ted Hooton) ***1995***

Total thrust: RD-0235 – 235 kN (vacuum), RD-0236 – 14.7 kN (vacuum)
Specific impulse: RD-0235 – 320 s (vacuum)
Burn time: RD-0235 – 158 s, RD-0236 – 183 s
Comment: Separation of the first and second stages is a 'hot' one since the RD-0236 vernier engines are ignited just before separation: the exhaust gases are diverted using special hatches within the first stage.

Rockot stage 3 (Breeze)
Bureau: Khrunichev/KB Salyut
Engines: KTDU-425? (main engine) ? (verniers)
Overall length: 3.38 m
Diameter: 2.28 m (maximum)
Oxidiser: nitrogen tetroxide
Fuel: UDMH
Dry mass: =~3.1 t
Propellant mass: 3.5 t
Total thrust: main engine – 20 kN (vacuum), verniers – 3.9 kN (vacuum)
Specific impulse: main engine – 325.5 s (vacuum), verniers – 275 s (vacuum)
Burn time: main engine – 564 total (max six burns)
In addition, the third stage carries 12 attitude-control thrusters: each has a thrust of 0.13 kN and a specific impulse of 270 s.

Payload processing/accommodation
The 8.5 m long upper composite includes the fairing, Breeze and interstage. The 7.9 m long, 2.5 m diameter cylindrical/biconic two-piece composite fairing, developed specifically for Rockot, provides a payload

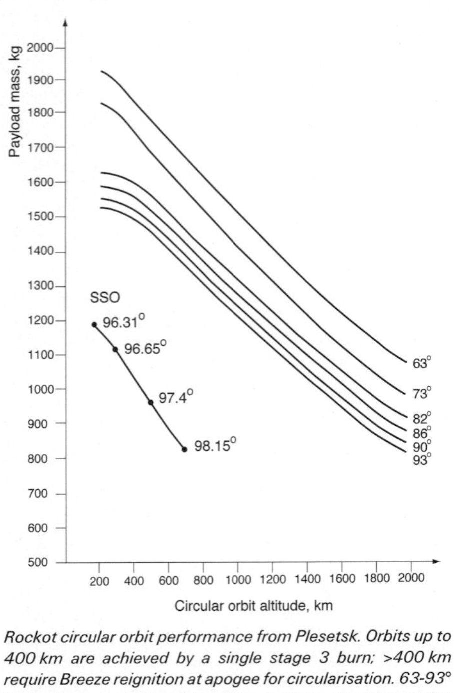

Rockot circular orbit performance from Plesetsk. Orbits up to 400 km are achieved by a single stage 3 burn; >400 km require Breeze reignition at apogee for circularisation. 63-93° inclinations are allowable without plane changes during launch (Khrunichev/DASA) ***1995***

Rockot is launched by ejecting it from a tube simulating the silo from which the SS-18 is usually fired (Theo Pirard) ***1999***/0054371

envelope of 4.757 m long 2.260 m diameter. Bolted payload adaptor is 591 or 937 mm diameter. The separation system comprises pyrolocks (four for 937 mm, three for 591 mm), three springs, guide pins (eight for 937 mm, six for 591 mm) and two separation sensors. The upper ring of each adaptor remains bolted to the payload: 15 kg for 937 mm, 12 kg 591 mm. After the pyrolocks are cut by pyrotechnics, the springs impart a minimum 0.5 m/s. A Western standard system using a 937 mm Marman clamp band is also available. The adaptors carry up to eight 50-pin payload electrical connectors, relaying up to 20 A total 28 V DC during launch preparation. 100 lines allow payload checkout until launch. During launch, Rockot's batteries in the transition section between Breeze and the adapter can provide up to 5 A totalling 15 Ah over 7 hours. Up to 50 lines can be provided for Rockot's computer to issue payload commands. 32 payload telemetry channels are available.
Acceleration load: longitudinal static 7.2 *g* at end of stage 1 burn, 3.0 *g* at end of stage 2 burn, 1.6 *g* max at end of Breeze burn (depends on payload mass); vibration 7.5 *g* rms long/lat.
Acoustic load: 142 dB max integrated level during lift-off and transonic
Thermal load: <500 W/m² at any point of the fairing; <1,135 W/m² after fairing jettison.

IRAQ

Tamouz 1

Background
The Tamouz 1 vehicle launched from the Al Anbar facility 80 km west of Baghdad in 1989 was claimed by the Iraqis as a satellite launcher test. Only the first of the three stages was live, reportedly attaining 12 km altitude. A range of 2,000 km was claimed for the Al Aabed (The Worshipper) missile version, apparently with a 750 kg payload. It is possible that plans revealed in 1990 to build a 1 m-bore 'supergun' under Project Babylon by the Brussels-based Space Research Corp included the firing of small payloads into low Earth orbit.

Specifications
Tamouz 1
First launch: 5 December 1989
Number launched: 1 to end-1996 (two Al Aabed missile versions were apparently launched previously).
Launch site: Al Anbar
Performance: about 150 kg into LEO
Number of stages: 3 (liquid)
Overall length: 24.4 m
Launch mass: about 48 t
Guidance: inertial

Tamouz 1 stage 1
Comment: The stage appears to be a cluster of five units derived from the Al Abbas version of the SS-1 Scud missile.
Length: about 11 m
Principal diameter: about 2.3 m
Propellants: liquid
Stage thrust: about 686 kN sea level

Tamouz 1 stage 2
Comment: The stage appears to be a single unit derived from the Al Abbas version of the SS-1 Scud missile
Length: about 9 m
Principal diameter: 90 cm
Propellants: liquid

Tamouz 1 stage 3
Comment: Little is known of the third stage, but length is about 3.5 m and it is believed to utilise liquid propellants

ISRAEL

Shavit/NEXT

Background
The Shavit ('Comet') all-solid orbital launcher, built by Israel Aircraft Industries, is believed to be derived from the two-stage Jericho 2 missile, first fired into the Mediterranean in 1986. A full range firing of 1,400 km is believed to have taken place from South Africa's Overberg site near Cape Town in June 1989, followed on 14 September 1989 by a 1,300 km test into the Mediterranean. The Jericho 1 500 km-range missile is reported to have been developed in the 1960s with French assistance, based upon the Dassault MD-600. The Israelis launched the second Shavit carrying Ofeq 2 on 3 April 1990. Ofeq 3 appeared 5 April 1995 and Ofeq 4 failed in 1998. It has been claimed there was a failed launch between Ofeq 2/3.

Delta Research Inc of Huntsville, Alabama joined forces with IAI to propose Shavit in 1990 for NASA's Meteor payload carrier. It was then offered in 1994 for NASA's ultra-light launcher competition (won by Pegasus). Team leader Space Vector would have integrated/launched the vehicle. Atlantic Research Corp would have loaded the stage 1/2 casings provided by IAI. Manufacturer Rafael/ARC already markets stage 3's AUS-51 motor. TRW in 1995 approached the Shavit manufacturer for the launch of its TSX 5. Israel is pressing for exemption from the White House rule that US vehicles must be used for government payloads from US soil.

IAI is working on the NEXT commercial version of

The Shavit satellite launch vehicle flanked on either side by missiles with propulsion systems developed by Israel Military Industries (Israel Military Industries) ***1997***/0003371

Shavit, stretching stages 1/2 and adding a bipropellant stage 4. Development depends on commercial commitments. Unconfirmed reports suggest that a 4,800 km-range Jericho 3 is under development, in common with NEXT.

Specifications

Shavit
First launch: 19 September 1988 (Ofeq 1 test satellite)
Number launched: 4, through May 1999 (a 4th possibly failed)
Launch sites: Palmachim Air Force Base
Vehicle success rate: 75% (four launches; but see above)
Principal uses: small payloads into LEO
Performance: 156 kg into 248 × 1,170 km, 142.9° orbit
Number of stages: three solid propellant
Overall length: 17.7 m
Principal diameter: 1.352 m
Launch mass: 22-23 t
Guidance: inertial, mounted on stage 2. Redundant; MIL-STD-1553B bus. Flight angle commanded as function of time (zeroed for stage 1 separation)

Shavit stage 1
Motor: TAAS Israel Industries Ltd. PBAN solid with filament wound graphite epoxy casing. SI 240 s sea level, expansion ratio 9
Overall length: 5.25 m
Principal diameter: 1.352 m
Stage mass: unavailable (9.1 t propellant)
Average thrust: 610 kN vacuum
Burn time: 43 s
Attitude control: four jet vanes (jettisoned after vertical phase) plus four air vanes
Staging: pyrotechnic separation of external V-clamp holding the two bulkheads with trapezoid teeth

Shavit stage 2
Motor: similar to stage 1 but with 23.4 expansion ratio for altitude performance
Overall length: 5.676 m
Principal diameter: 1.352 m
Stage mass: unavailable (9.1 t propellant)
Average thrust: 564 kN vacuum
Burn time: 52 s
Attitude control: pitch/yaw by LITVC modules. Four 0.8 kg modules each injects 0.35 kg/s (at 51.0 atmosphere) strontium perchlorate through a single orifice; max operating pressure 85.0 atmosphere)
Staging: as stage 1, hot separation

Shavit stage 3
Motor: Rafael AUS-51 'Marble'
Overall length: about 2.1 m
Principal diameter: 1.3 m
Stage mass: about 2 t (1,895 kg propellant)
Average thrust: 55
Burn time: 92.5 s
Attitude control: spin-stabilised, spun up/down by Rafael's ST-200N thrusters. NEXT's AUS-51 will employ LITVC. Rafael's ACT-25N thrusters provide attitude/nutation control

NEXT
Comment: IAI is working on the NEXT commercial version of Shavit, stretching stages 1/2 and adding a bipropellant stage 4. Development depends on commercial commitments. All propulsion systems were expected to be qualified by end-1995 but no test flight appeared.
First launch: 2000?
Principal uses: small payloads into LEO
Performance: see diagrams
Number of stages: three solid-propellant plus one liquid
Overall length: about 21 m
Principal diameter: 1.352 m
Launch mass: about 31 t
Guidance: gimballed platform on stage 2, succeeded by GPS receiver on stage 4. MIL-STD-1553B bus

NEXT stage 1
Motor: as Shavit, but lengthened about 20%. A flexseal nozzle has been developed that might be employed by future NEXT.
Overall length: about 6.8 m
Principal diameter: 1.352 m
Stage mass: 14.3 t (12.87 t propellant)
Average thrust: 637 kN vacuum
Burn time: 51 s
Attitude control: as Shavit
Staging: as Shavit

NEXT stage 2
Motor: as Shavit, but lengthened about 20%
Overall length: about 7.3 m
Principal diameter: 1.352 m
Stage mass: 14.4 t (12.87 t propellant)
Average thrust: 705 kN vacuum
Burn time: 60 s?
Attitude control: as Shavit
Staging: as stage 1, hot separation

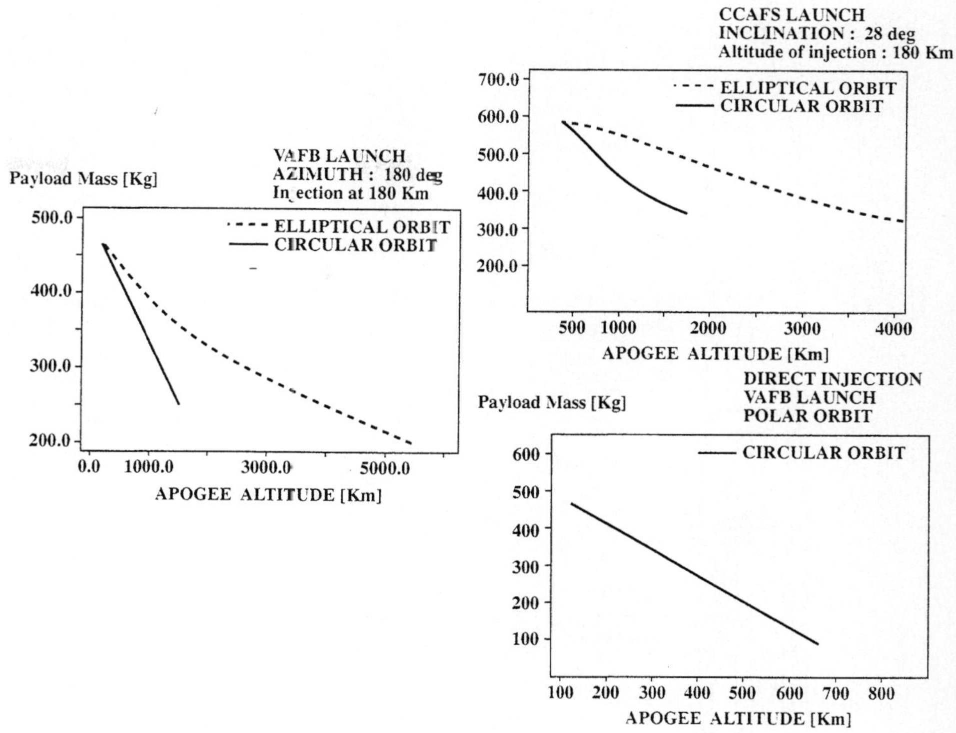

NEXT payload performance (IAI) ***1995***

NEXT stage 3
Same as Shavit, but three-axis attitude control by LITVC modules as used on stage 2

NEXT stage 4
Designation: BPM Bipropellant Module
Engine: IAI bipropellant (see Propulsion section for specifications)
Overall length: unavailable
Principal diameter: 1.35 m
Dry mass: 60 kg; structure is composite
Propellants: 120 kg MMH/N_2O_4
Stage thrust: 3,000 N
Burn time: 120+30 s 2-phase circularisation
Attitude control: two clusters of three thrusters for three-axis control; 6 × 20 N or 6 × 150 N depending on mission requirements
Separation: unavailable

ITALY

Vega/San Marco Scout

Background

The Italian government accepted proposals from industrial concerns in 1988 to upgrade Scouts, with plans to launch them from San Marco. Among the various proposals was one from BPD to double the Scout's base performance. Italian government approval followed, in July 1991, of the country's 1990-94 space plan which allowed ASI to award BPD the Lit5,120 billion industrialisation contract on 3 December 1991. ASI's plan included little US involvement by making all stages indigenously produced. Versions without, or with two or four strap-ons might allow the Italians to modularise the design for economic efficiency.

BPD's offered its Zefiro (Zephyr) motor, developed from the company's Ariane experience, with thrust vector control. It conducted four successful static tests during 1991, but the first flight test ended by detonation from range safety 10 seconds into the planned 40 second burn on 18 March 1992 from the Sardinian military range. Ground qualification continued until final completion in mid-1995 and two qualification flights at the end of the year.

A design rework in 1994 dropped the strap-on approach and increased Zefiro's diameter from 1.35 to 1.90 m. The two base designs are now called Vega K0 and Vega K3. A K0 minus stage 1 could substitute for Capricornio (150 kg, US$9 million) and an enlarged K4 could become the European Small Launcher (1,100 kg, US$18 million).

Italy's 1990-94 five-year plan approved development of the San Marco Scout launcher for operations from Italy's equatorial platform off the Kenyan coast. Payloads were to include recoverable μgravity carriers such as the Carina capsule. The Italians expected to make their first launch in 1995. However, the University of Rome proposed a separate San Marco Scout programme closely based on the US Scout, prompting legal wrangling. BPD is continuing Vega development with internal funding plus Ministry of Research funding for Zefiro development. Should government support be provided, BPD will press on with its programme to the qualification flight.

In the meantime, Italy's San Marco site has become too expensive for the Italian government to maintain, and Vega's principal sites would be Kourou and Vandenberg AFB using transportable equipment to

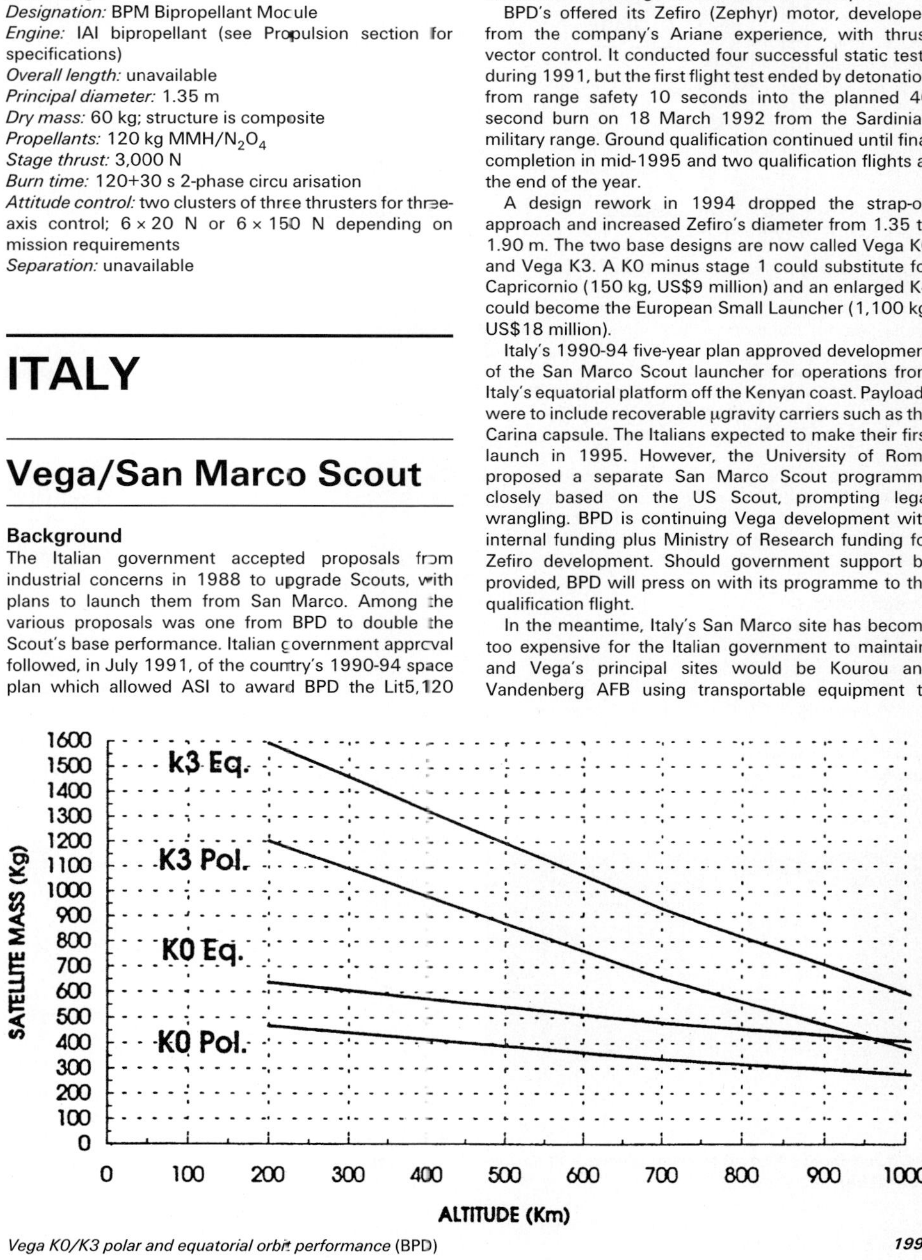

Vega K0/K3 polar and equatorial orbit performance (BPD) ***1996***

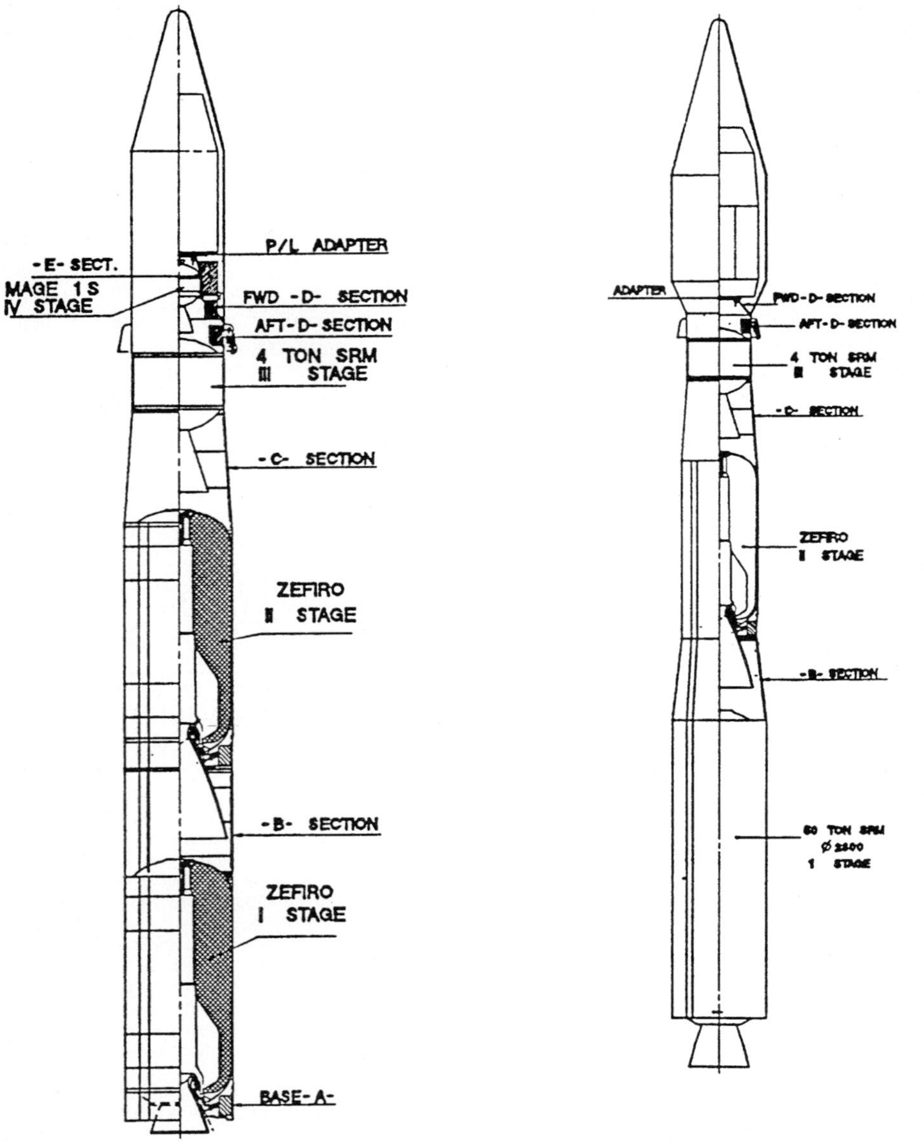

Vega K0 (left) and K3 configurations (BPD) ***1996***

RD-861 motor proposed by Fiat Avio to equip second stage of the Vega Launch vehicle (Theo Pirard) ***1999***/0054372

support campaigns of 7 to 15 days. Discussions are under way for commonality with Spain's Capricornio and with Europe's ESL at a total development cost of US$200 million. The vehicle family must be commercially viable, addressing the 200 to 600 kg market for 200 to 1,000 km orbits. The configurations are not optimised but yield lower development and recurring costs.

In May 1999, the European Space Agency approved US$340 million (€317 million) for development but only after successful completion of additional technical, financial and economic assessments in September 1999.

Specifications

Vega/San Marcos

First launch: 1999?
Launch sites: Kourou, VAFB
Principal uses: 300-700 kg payloads into LEO
Performance: K0/K3 340/660 kg into 700 km 90°; see diagram
Cost: US$12 million K0; US$16 million K3
Number of stages: 4 K0; 3 K3
Overall length: 21 m K0; 25 m K3
Principal diameter: 1.896 m K0; =~2.3 m K3

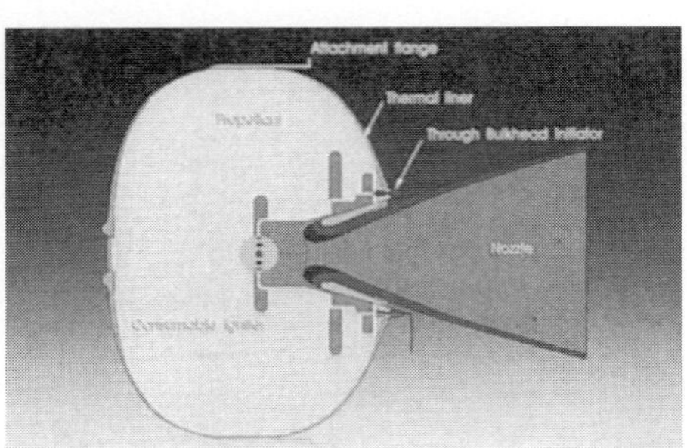

BPD's End Burning Motor is an improved version of the IRIS motor (ESA)

Launch mass: 42 t K0; 80 t K3
Guidance: inertial above stage 3. Accuracy ±20 km altitude, ±0.05° inc. Stage 1/2 TVC for pitch/yaw, stage 3 ACS for roll control during stage 1/2 and three-axis during coast and stage 3 burn; stage 4 ACS for three-axis stage 4, orbit trim and payload release. MIL-STD-1773 optical bus.

Vega K0 stage 1

Motor: Zefiro (with SL nozzle), finocyl with nine-point aft star, 16% Al HTPB-1614, filament wound casing with EPDM (EG8/EG2) insulation, 0.915 mass fraction
Overall length: 4.886 m
Principal diameter: 1.896 m
Stage mass: 18 t (16.5 t propellant)
Average thrust: 568 kN vacuum; max 850 kN (41.5 MNs total impulse)
Burn time: 73 s
Attitude control: nozzle flexure ±6.5° provides pitch/yaw TVC; roll by stage 3 ACS

Vega K3 stage 1

Motor: P50 solid from ESL, or equivalent such as Castor 120

Vega K0/K3 stage 2

Designation: Zefiro with altitude nozzle
Overall length: 4.886 m
Principal diameter: 1.896 m
Stage mass: 18 t (16.5 t propellant)
Average thrust: 568 kN vacuum; max 850 kN (41.5 MNs total impulse)
Burn time: 73 s
Attitude control: nozzle flexure ±6.5° provides pitch/yaw TVC; roll by stage 3 ACS

Vega K0/K3 stage 3

Comment: Stage 3 would be a bought-in motor of about 4 t propellant, such as Thiokol's Star 63F. If none available, the smaller IRIS could be used
Designation: IRIS
Overall length: 194 cm
Principal diameter: 158 cm
Stage mass: 1,950 kg (1,574 kg HTPB-1813)
Thrust: 66 kN peak vacuum
Burn time: 78 s
Attitude control: three-axis by stage 3 ACS

Vega K0 stage 4

Designation: Mage 1S
Overall length: 129 cm
Principal diameter: 77 cm
Motor mass: 448 kg (410 kg CTPB-1612)
Average thrust: 23.7 kN vacuum
Burn time: 50 s
Attitude control: three-axis by stage 4 ACS, also providing final orbit correction

Heatshield/payload accommodation

K0 provides 4 m^3 in a 1.65 m diameter fairing, with 590 mm payload adaptor. K3's 2.0/2.4 m diameter accommodates 590/937 mm adaptors. Static acceleration limited to 8 *g*, random vibration less than 6 *g* rms at payload interface, acoustic <144 dB overall.

End burning motor

Comment: BPD has developed the end burning motor (actually spherical burning) as an improved version of the IRIS motor capable of injecting a 2,195 kg payload into GEO from LEO. It was successfully ground tested 4 November 1994; further qualification depends on finding a customer. An innovative feature is the consumable igniter bonded to the grain head-end, replacing current heavy integral devices. After pyrotechnic ignition, 10 equatorial nozzles spray combustion products and the motor burning consumes all the epoxy and most of the carbon fibre. The remainder of the soft fibre cloth is exhausted. MAN provides the motor casing.
Mass: 1,754.1 kg (1,612.7 kg propellant, 141.4 kg empty)
Length: 1,736 mm
Maximum diameter: 1,351 mm
Propellant
  type: composite 11HTPB-18Al-71AP
  shape: head-end web with circular port and radial slots
  loading fraction: 0.94
  mass fraction: 0.9265
Burn time: 83.2 s (25°C)
Thrust: 55.9 kN (vacuum average, 25°C)
Specific impulse: 293.9 s vacuum
Total impulse: 4.65 MNs
Pressure: 3.64 MPa (25°C); MEOP 5.39 MPa (40°C)
Nozzle: throat diameter: 95 mm
  expansion ratio: 60:1
  length: 1,015 mm (of which 300 mm submerged)
  exit diameter: 736 mm internal, 765 mm external
  materials: carbon-phenolic and silica-phenolic with carbon-carbon throat insert
  mass: 58.2 kg
Casing: epoxy-aramide fibre wound, 79 kg (including thermal protection)
Igniter type: consumable pyrogen head-end, 0.56 kg
Motor mount: single composite adaptor flange

JAPAN

H1/H2

Background

Japan's Space Activities Commission gave the go-ahead, in February 1984, for the development of a large all-Japanese H2 capable of placing 2 tonnes in GEO and a small spaceplane in LEO. Work had already started on the current H1 in 1979 as an interim step. N1/2's McDonnell Douglas stage 1 with Rocketdyne's MB-3 engine transferred directly to the programme, but the cryogenic stage 2, inertial guidance and solid upper stages were all-Japanese. Two demonstration missions flew 1986/87, with seven operational departures (five GEO) following. H2 completed the transition by adopting indigenous solid strap-ons and large cryogenic first stage, creating a capability comparable to that of Ariane 4 and Titan 3 but with a significantly lower launch mass because of the more advanced stages. Development of the Improved H2 and the Upgraded H2 (H2A) is under way.

The N1/2 launchers provided the initial expertise but they depended on McDonnell Douglas Delta stages built under a 1969 licence that prohibited orbiting of third party payloads without US approval. The interim H1 introduced Mitsubishi's cryogenic stage 2 and all-Japanese inertial guidance; H2 completed the transition in 1994 by adopting a large cryogenic first stage and solid strap-ons.

To enhance the cost efficiency of the H2, the Improved H2 and Upgraded H2 (H2A) programmes are under way. The Improved H2 will debut in 1999 with a production cost reduced by about 30 per cent. The H2A development began FY96 and should debut about 2000, giving a launch vehicle family with a payload capability of up to 15 tonnes/3 tonnes (LEO/GEO) for the H2A. A growth potential of up to 20 tonnes/4 tonnes (LEO/GEO) is under consideration in the design.

These increased capabilities will allow it to fly logistics missions to the International Space Station and carry the HOPE spaceplane.

Introduction of the H2 in 1994 provided Japan with its first large indigenous launcher, capable of placing 4-tonne payloads in GTO, and dispatching deep space probes. Freed from US licensing restrictions, it is offered commercially but cost renders it a difficult sell. Unit cost is ¥19 billion, but the Improved H2 is under development for debut in 1999, principally to reduce cost by 30 per cent. It will provide *direct* GEO injection. The H2A plans to debut 2000, as a family capable of delivering up to 4 tonnes GEO and 20 tonnes LEO. Unit cost will be about ¥8.5 billion. That last capability will allow it to fly Space Station logistics missions and carry the HOPE spaceplane.

An initial batch of six H2s was built, the first two considered to be prototypes.

The development of the Improved H2 began in 1995, principally to reduce unit cost and increase reliability to >0.96. GTO capacity will be 4 tonnes. First flight will carry MTSAT in 1999. Stage 1 will be little changed. Payloads will be attached to the adaptor before reaching the pad. Stage 2 will be extensively reworked, principally separating the main tanks and using the improved LE-5A engine. The current common LOX/LH_2 bulkhead is expensive and thermal leakage limits coast duration.

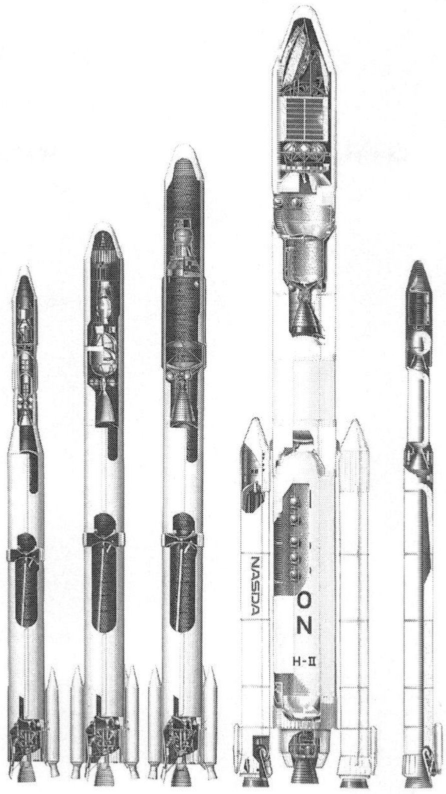

Comparison of NASDA's family of launchers (from left): N1, N2, H1 (essentially the N2 with improved upper stages), H2 and J1 (NASDA) ***1995***

Improved H2 will use tanks separated by a truss structure, reducing cost and allowing a 5 hour long coast. LH_2 tank 4.0 m diameter, 43.9 m³, 2.3 atmosphere; LOX tank 3.2 m diameter, 12.7 m³, 2.75 atmosphere. Propellant mass 16.6 tonnes, thrust increased to 137.3 kN. Electromechanical actuators replace hydraulic TVC. The ambient He bottles will be changed from titanium to composite. Captive Firing Test of the first Flight Model Stage 2 came in 1998.

Development of the H2A began FY96 to debut 2001, 2000 as a family which eventually will cover the launch capability from the existing H2 to 15 tonnes to LEO, 3 tonnes to GEO, with a growth potential up to 20 tonnes to LEO, 4 tonnes to GEO being considered in the design. Unit cost will be less than ¥8.5 billion for the 2-tonne GEO configuration. These increased launching capabilities will allow it to fly Space Station logistics missions using the HTV H2 Transfer Vehicle and to carry the HOPE spaceplane. Improved H2's SRBs will be replaced by monolithic composite designs. Liquid boosters, identical to stage 1 with two LE-7A engines on each, will provide LEO/GTO/GEO capabilities of 15/6/3 tonnes (one booster) and 20/8/4 tonnes (two boosters) respectively. Other improvements include propellant cross-feeds, an intelligent electrical system, automatic checkout functions and simplified launch operations. The new pad will be completed in 1999. Hughes Space and Communications and Space Systems/Loral have

Fourth H2 launch in August 1996 carrying ADEOS and JAS-2 into orbit (NASDA) ***1997***/0003372

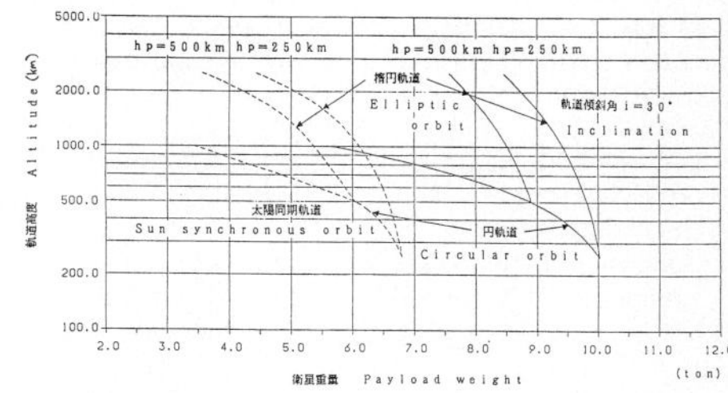

H2 performance from Tanegashima (NASDA)

each reached agreements (in 1996) with RSC for 10 launches each during the period 2000-2005.

Nissan developed the four-segment solid strap-ons for Japan's H2 launcher, with four demonstration firings during 1988-91 at NASDA's Tanegashima Space Center. The first test was conducted April 1988, prototype number 1 was fired June 1989 and prototype number 2 13 December 1989, with the qualification model firing following on 29 May 1991. The units represent the country's largest and most powerful indigenously-developed solids. The Improved H2's SRBs will be reduced to three segments, debuting in 1999. The upgraded H2A, from 2001, will use a monolithic composite case design.

Cost to first launch was estimated at ¥200 billion (then about US$800 million), with 23 per cent allocated to LE-7 development and 20 per cent for launch site construction. A 1993 estimate was ¥270 billion, including facilities and the first two launches. ¥35.5 billion was allocated in FY89 for vehicle development, followed by ¥37.80 billion, plus ¥10 billion supplemental for the previous year, in FY90, ¥32.407 billion FY91, ¥28.13 billion FY92, ¥16.93 billion FY93, ¥1.656 billion FY94 request. LE-7 development had cost about ¥50 billion by the beginning of 1990. In 1989-90, the Japanese deferred the first launch into early 1993. Because of continuing, the country made a decision in July 1990 that LE-7's target performance would be reduced to 92 per cent, a level already almost achieved in tests. Overall H2 capability would not be affected, however, because SRB and stage 2 performance exceeded target. Further LE-7 problems in 1992 deferred first launch to early 1994.

Tanegashima can handle four flights annually, but acceleration to six would require a second pad.

Specifications

H2

First launch: 3 February 1994
Number launched: 6 to May 1999
Industrial integrator: Mitsubishi Heavy Industries, Nagoya
Launch sites: Yoshinobu Launch Complex, Tanegashima
Principal uses: delivery of 4 t-class payloads into GTO
Vehicle success rate: 100% for 6 launches
Cost: ¥19 billion; Improved H2 goal is ¥14 billion
Availability: fishing/range safety requirements currently

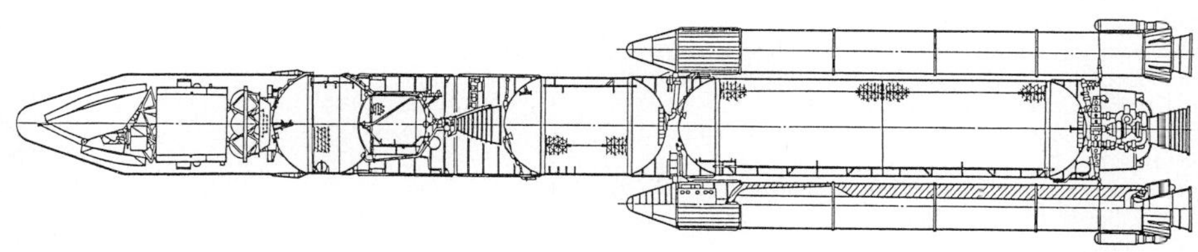

H2 launch vehicle, shown with ETS 6 payload in a 4S single fairing. H2 can add two small strap-ons to stage 1 (NASDA)

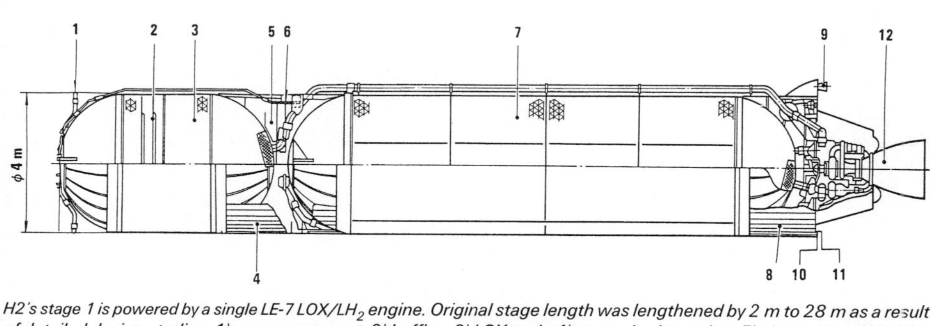

H2's stage 1 is powered by a single LE-7 LOX/LH_2 engine. Original stage length was lengthened by 2 m to 28 m as a result of detailed design studies. 1) oxygen gas vent, 2) baffles, 3) LOX tank, 4) centre body section, 5) electronics, 6) H_2 gas vent, 7 LH_2 tank, 8) engine section, 9) auxiliary engine, 10) umbilical connector, 11) LOX/LH_2 fill port, 12) LE-7 (NASDA)

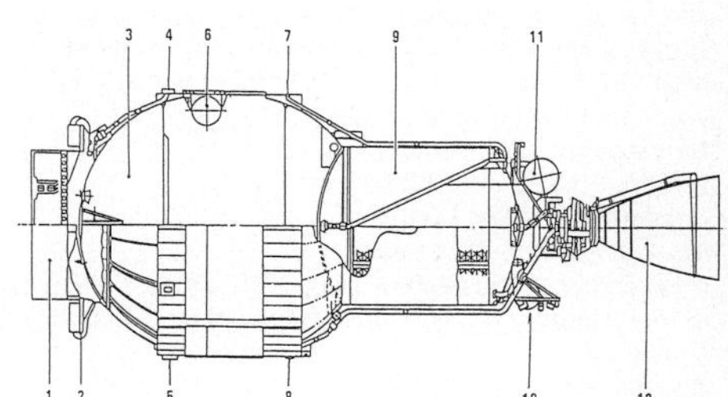

H1's stage 2 was enlarged for H2, with the LE-5 engine modified for increased thrust and improved re-start characteristics. 1) payload attach fitting, 2) guidance section, 3) LH_2 tank, 4) H_2 gas vent port, 5) umbilical connector, 6) cryogenic helium bottle, 7) LOX fill port, 8) LH_2 fill port, 9) LOX tank, 10) RCS thrusters, 11) ambient helium bottle, 12) LE-5A engine

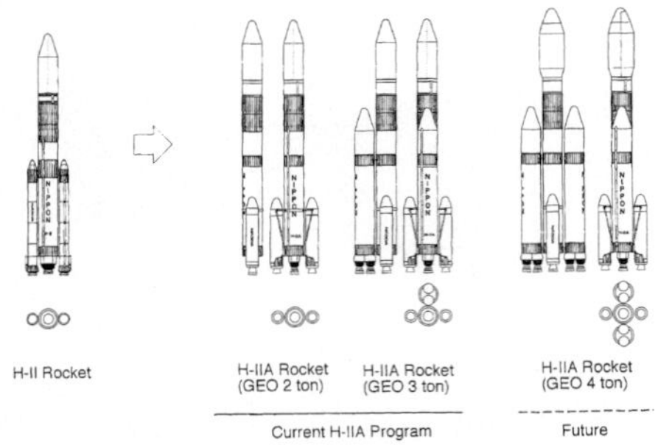

H2 rocket together with the planned H2A launch vehicle family (NASDA) ***1997***/0003447

constrain launches to 45-day periods in January/February and August/September each year. With second mobile platform, plan to have 4/year capability starting 1999.

Performance:

LEO: 10,000 kg into 250km circular 30° orbit
GTO: 4,000 kg
GEO: 2,000 kg
Sun-synchronous: 4,300 kg into 800 km
Escape: 2.8 t Moon, 2 t Venus, 1.7-2.2 t Mars (for 3,650 m/s transfer injection), 700 kg Mercury, 400 kg Jupiter

Number of stages: 2+2 strap-ons
Overall length: 50 m with 4S fairing; 51.1 m with 5/4D fairing
Principal diameter: 4.0 m
Launch mass: 260 t plus payload for 4S fairing; 278.9 t plus payload for 5/4D fairing and SSBs
Guidance: strapped-down inertial mounted at top of stage 2. NEC inertial guidance computer, 14.2 kg, 45 W consumption, mission reliability >0.999, 32 kword main memory capacity (16-bit words), qualified to 12.8 *g* rms random vibration. Three Japan Aviation Electronics laser gyros and accelerometers.
Launch sequence (min:s) for GTO single fairing/LEO+GTO dual fairing + SSBs:

-00:06/-00:06	stage 1 ignition
00:00/00:00	SRBs ignite/lift-off
/00:10	SSBs ignite
/01:16	SSBs burnout
/01:30	SSBs separation
01:34/01:34	SRBs burnout
01:39/01:37	SRBs jettison
03:46/03:39	fairing/upper fairing jettison
05:46/05:47	stage 1 shutdown
05:56/05:56	stage 1 separation
06:02/06:02	stage 2 ignition 1
12:45/12:20	stage 2 shutdown
/13:16	LEO payload separation
/16:11	lower fairing separation
24:44/25:03	stage 2 ignition 2
28:01/27:02	stage 2 shutdown
28:21/27:52	GTO payload separation

Schedule of future missions:

MTSAT	1999
ARTEMIS/DASH/ETS-VIII 1/6 ANTENNA	2000
MDS-1/DRTS-W	2000
ADEOS-II	2000
OICETS	2000
MDS-2	2001
HTV#1	2002
HTV#2	2003
SELENE	2003
HTV#3	2003
HOPE-X	2003

H2 SRB solid strap-ons

Comment: The first full-scale static test was undertaken at Tanegashima 15 April 1988 (including 5° nozzle deflection), followed by prototype model (PM) firings June and 13 December 1989 at Tanegashima. The single qualification model (QM) run was conducted 29 May 1991. The first, successful, booster separation test was conducted in July 1988.
Motor contractor: Nissan Motor Co
Length: 23.362 m (19.26 m motor)
Principal diameter: 1.81 m
Mass at ignition: 70.4 t each
Propellant: 59.15 t each of 14% HTPB/ 68%AP/18%Al, in four-segment NT-150 steel case held by 108 bolts at each joint.
Average thrust: 1,560 kN each average sea level
Burn time: 94 s
Steering: TVC by ±5° deflection of flexible nozzle joint provides vehicle three-axis control

H2 SSB solid strap-ons

Comment: The twin Solid Sub-Boosters are carried by stage 1 only when required by mission performance. One flight so far: #3.

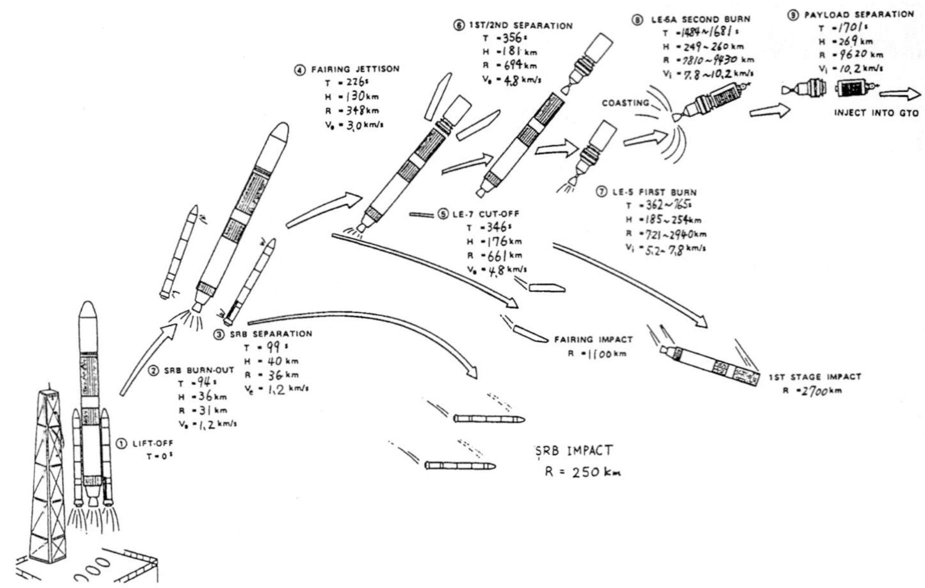

Typical H2 GTO flight sequence (NASDA)

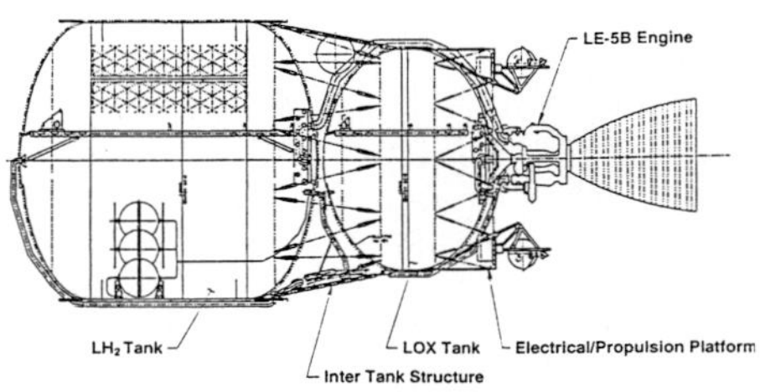

Improved H2's stage 2 will separate the cryogenic tanks (NASDA) ***1996***

Motor contractor: Nissan Motor Co
Length: 9.053 m
Principal diameter: 1.125 m
Mass at ignition: 10.5 t each
Propellant: 8.4 t each of 14% HTPB/ 68%AP/18%Al in steel case.
Average thrust: 34.5 kN each average sea level (SI 267 s vacuum)
Burn time: 66 s
Steering: none, fixed nozzle canted 10° towards H2's centre of gravity.

H2 stage 1

Comment: The stage 1 Captive Firing Test series began 23 February 1993 with a 10 s run on the Tanegashima pad. The full 350 s was achieved 15 June 1993. The last run, 60 s, was performed 8 July 1993. Battleship tests were held at Mitsubishi's Tashiro site October 1990 to May 1992. The bracketed values below indicate goals, reduced during 1990 because of LE-7 development problems.
Contractor: Mitsubishi Heavy Industries
Engine: single Mitsubishi LE-7 staged combustion cycle, single start, liquid oxygen/hydrogen, regenerative cooling, 54:1 (60) expansion ratio, 6.0:1 mixture ratio, gimballed for pitch/yaw control.
Overall length: 28.0 m
Principal diameter: 4.0 m
Total stage mass: 98.1 t.
Oxidiser: 73.6 t LOX in forward tank; both tanks of semi-monocoque 2219 aluminium alloy, external polyisocyanurate foam thermal insulation. Pressurised at 4.26 atmospheres by helium from spheres immersed in LH_2 tank.
Fuel: 12.6 t LH_2 in aft tank, pressurised at 3.39 atmospheres by GH_2 tapped off from LE-7 nozzle extension.
Stage thrust: 843.4 kN (911.96 kN) sea level; 1,078 kN (1,180 kN) vacuum.
Burn time: 346 s
Attitude control system: two 1,500 N base auxiliary engines bleed GH_2 from LE-7 for roll control following SRB separation

H2 stage 2

Comment: An enlarged variant of H1's stage 2. It originally selected the same LE-5 engine, but NASDA decided to increase thrust from 102.96 kN to 121.5 kN (vacuum SI increase of 3 s to 452 s, chamber pressure from 37 to 39 atmospheres) by switching from the gas generator cycle to the hydrogen bleed cycle, tapping gaseous fuel from the nozzle skirt to drive the turbines. The design also provides improved re-start characteristics. LE-5A prototype firings concluded May 1988 after achieving a total run time of 2,800 s. Stage 2 battleship firing tests were held January to March 1990 at Mitsubishi's Tashiro test site, followed by flight-type stage testing August to October 1991. The Improved H2 will significantly redesign stage 2.
Contractor: Mitsubishi Heavy Industries
Engine: single Mitsubishi LE-5A hydrogen bleed cycle, dual start, LOX/LH_2, regenerative cooling, 130:1 expansion ratio, 5.0:1 mixture ratio, gimballed 3.5° square for pitch/yaw control.
Overall length: 10.6 m
Principal diameter: 4.0 m

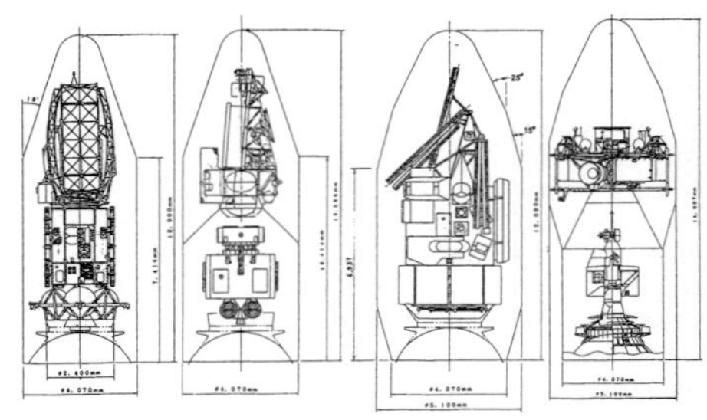

H2 payload accommodation (scales differ). From left: type 4S (single 4.1 m), 4/4D (dual 4.1 m), 5S (single 5.1 m), 5/4D (5.1 + 4.1 m). The 5.1 m fairing handles 4.6 m diameter payloads. 5/4D first flew on mission 3 (NASDA)

N, H and J series orbital launch history

	Date	Vehicle	Payload
1	9 September 1975	N1	ETS-1
2	29 February 1976	N1	Ume-1
3	23 February 1977	N1	ETS-2
4	16 February 1978	N1	Ume-2
5*	6 February 1979	N1	Ayame-1
6	22 February 1980	N1	Ayame-2
7	11 February 1981	N2	ETS-4
8	11 August 1981	N2	GMS-2
9	3 September 1982	N1	ETS-3
10	4 February 1983	N2	CS-2a
11	6 August 1983	N2	CS-2b
12	23 January 1984	N2	BS-2a
13	3 August 1984	N2	GMS-3
14	12 February 1986	N2	BS-2b
15	13 August 1986	H1	EGS/Fuji-1
16	19 February 1987	N2	MOS-1
17	27 August 1987	H1	ETS-5
18	19 February 1988	H1	CS-3a
19	16 September 1988	H1	CS-3b
20	5 September 1989	H1	GMS-4
21	7 February 1990	H1	MOS-1b/Fuji-2/DEBUT
22	28 August 1990	H1	BS-3a
23	25 August 1991	H1	BS-3b
24	11 February 1992	H1	JERS-1
25	3 February 1994	H2	OREX/VEP
26	28 August 1994	H2	ETS-6
27	18 March 1995	H2	SFU, GMS-5
28[1]	11 February 1996	J1[1]	Hyflex
29	17 August 1996	H2	ADEOS
30	27 November 1997	H2	ETS-8, TRMM
31	21 February 1998	H2	Kakehashi

* indicates vehicle failure. All launches made from Tanegashima. *N/H Breakdown* (launch total for each type, followed by number of failures; through May 1996): N1 7(1); N2 8(0); H1 9(0); H2 3(0); J1 1(0).
1 J1 #1 was suborbital.

H2A will have new monolithic composite strap-on boosters and greatly improved payload capability (Theo Pirard) ***1999***/0054367

Total stage mass: 19.7 t
Oxidiser: 13.9 t of liquid oxygen in aft tank
Fuel: 2.8t of liquid hydrogen in forward tank
Stage thrust: 121.5 kN vacuum, but 5% idle mode provides precise orbit control + de-orbit capability.
Burn time: 609 s max; two burns (403 s to establish LEO parking orbit + 197 s for 28.5° GTO) for typical GEO missions.
Attitude control system: LE-5A gimballing provides pitch/yaw control during burn; two IHI hydrazine thruster modules (each 4 × 50 N + 2 × 18 N) provide roll control during burn and three-axis during coast

Payload fairing/attach fitting
Comment: Kawasaki's standard 4.1 m diameter, 12.0 m high 1.4 t aluminium honeycomb core/skin 4S fairing provides a 3.7 m diameter envelope for single payloads. The 4/4D is stretched to 15.0 m for dual main payloads, with an 11 m aluminium upper section and 4 m CFRP lower portion; it will be used for ETS 7/TRMM. A 5.1 × 12.0 m version (5S) offers accommodation for Shuttle-class (4.6 m diameter) payloads; ADEOS used it. Flight #3 was the first to carry the '5/4D' 14.1 m-long dual fairing, with 5.1 m diameter 9.6 m-long aluminium upper portion and 4.1 m diameter 4.5 m-long CFRP lower. A Mild Detonating Fuse effects separation of the clamshell halves and springs when aerodynamic heating has fallen to 1,135 W/m² during stage 1 burn.
Acoustic loads: 141 dB peak during lift-off and transonic flight
Vibration: sinusoidal 2 *g* longitudinal, 1 *g* lateral
Acceleration: peak 4 *g* longitudinal at stage 1 shutdown, design requirement for 5 *g*
Thermal environment: 500 W/m² radiation from inner wall
Pad environmental control: class 100,000, 15-25°C air conditioning

HOPE (H2 orbiting plane)

Background

NASDA initiated studies of the HOPE unmanned reusable spaceplane in 1986, intending to provide servicing missions from 1998 to Japan's Experimental Module attached to the Space Station. Both HOPE and the Space Station have been considerably delayed. From their work, the Space Activities Commission in mid-1993 released a report recommending development of a winged reusable vehicle using H2. However, given the problems with the Shuttle and the Soviet equivalent, the Commission recommended that Japan hold back on a full commitment to the programme until after demonstration of critical technologies.

Japan seemed especially interested in understanding hypersonic re-entry and building a guidance and control system so the spaceplane could have autonomous operation and not require the same expensive overhead associated with training a crew. These technologies were actively pursued through three experimental vehicles. The OREX re-entry vehicle carried by H2 number 1 in February 1994 tested thermal protection and navigation. J1 number 1 in February 1996 carried the Hyflex hypersonic flight experiment spaceplane model on a suborbital path. The Alflex automatic approach/landing experiment was released and landed successfully 13 times from a helicopter July-August 1996 to demonstrate automatic GN&C beginning at 1,500 m altitude and 2,700 m out from Australia's 2,600 m Woomera runway. Japan now appears to be able to operate a spaceplane completely autonomously. It is only the second country after Russia to be able to accomplish this feat, a skill the United States has not yet attained for its Shuttle.

The 760 kg 6.11 m length, 3.79 m span Alflex is a 37 per cent scale version of HOPE's FY92 04C double delta design. It includes an IMU, differential GPS, radio altimeter and microwave landing system. Fuji Heavy Industries built two models. The full scale HOPE-X demonstrator, approved FY96 and started Phase C, HOPE-X is planned for launch by a single-stage H2A in 2000 to a 120 to 200 km orbit, completing one circuit of the Earth. HOPE-X will re-enter, glide and land automatically within 2 hours. HOPE-XA is planned for a brief LEO mission. FY97 funding is ¥11 billion (NASDA 10.5 plus NAL 0.5), FY96 was ¥9.5 billion (8.5 + 1.0). HOPE will be modified from HOPE-X and increase the ISS cargo capacity from 3 to 5 tonnes utilising standard 500 kg, 1 × 1 × 1.8 m ISS racks. Earth-return capacity would be 5 tonnes (for wing loading of 250 kg/m²).

HOPE's own propulsion system circularises its path at Space Station altitude and then the craft performs an automatic rendezvous/docking after two days. After exchanging cargoes and undocking, it would re-enter within one day and execute an automatic 300 km/h touchdown on a 3,000 m runway. Crossrange capability is projected at 1,500 km, with maximum time on-orbit about 100 h. In addition to Station and platform servicing, HOPE would act as a Japanese technology demonstrator for future spaceplane development. Principal contractors are Mitsubishi, Kawasaki and Fuji.

The HOPE now awaits the development of the more powerful H2A launcher. HOPE test flights would begin in the early 2000s; technology development has been approved. Future versions might be manned.

Specifications

HOPE-X
Length (m) × width (m) × height (m): Approx. 16 × 10 × 5
Total weight (kg): Approx. 14,000 (at launch) /9900 (at re-entry)
Launcher: Single Stage type H-11A Rocket (Improved)
Landing: Automatic landing on 1800 m class runway
Shape: Delta wing with tip fins

Alflex in 1996 demonstrated HOPE's landing sequence (NASDA/NAL) ***1997***/0003449

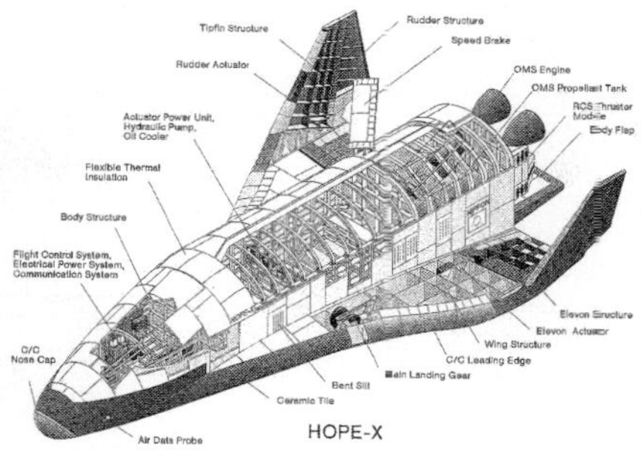

HOPE is planned to deliver 3 tonnes of cargo to Space Station from about 2010 (NASDA) ***1997***/0003454

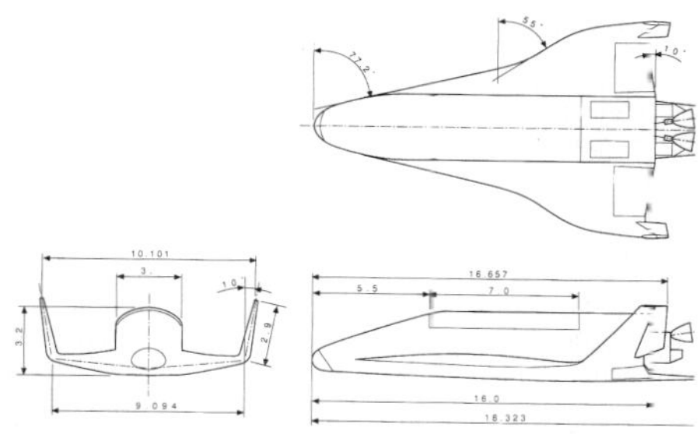

HOPE-X's double delta design provides lower mass and better controllability in all flight regimes (NASDA) ***1997***/0003452

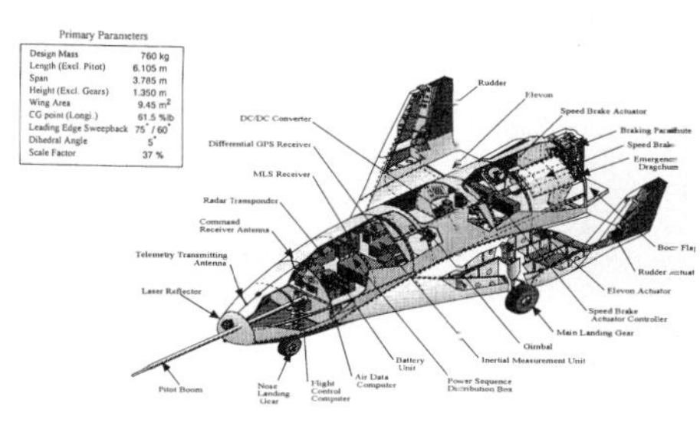

Alflex in 1996 demonstrated HOPE's landing sequence (NASDA/NAL) ***1997***/0003450

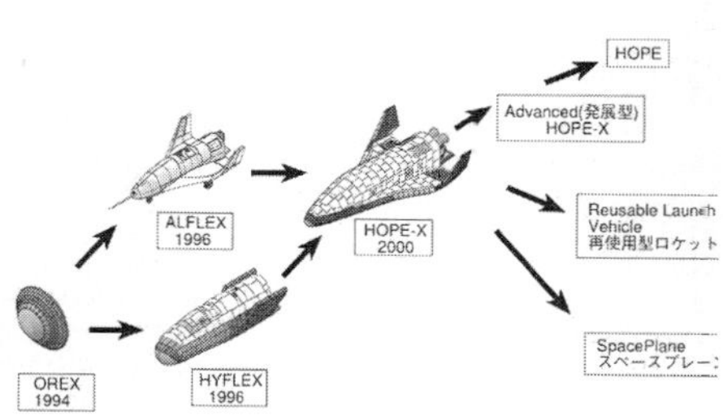

Hope families ***1997***/0003451

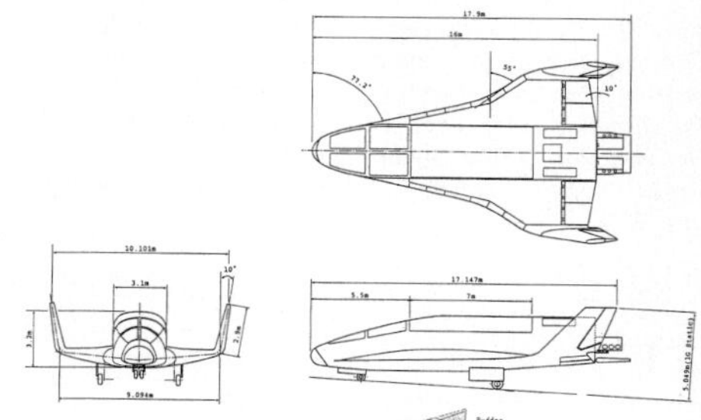

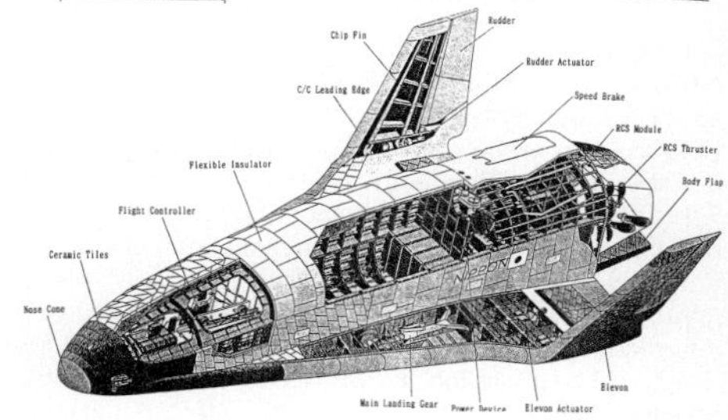

HOPE is proposed as a reusable unmanned Space Station ferry. Tests would begin in the early 2000's (NASDA) ***1995***

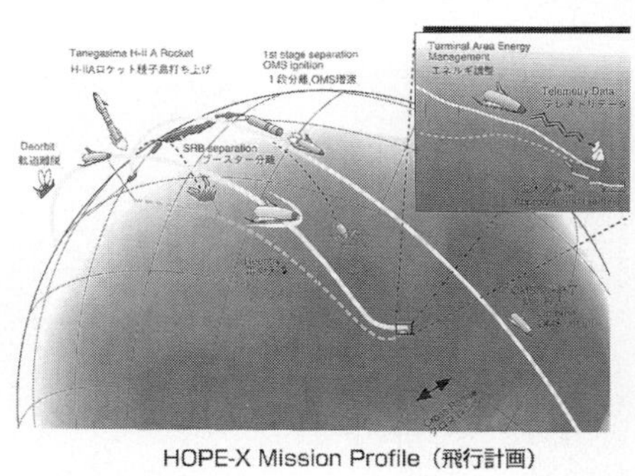

Hope-X mission profile ***1997***/0003453

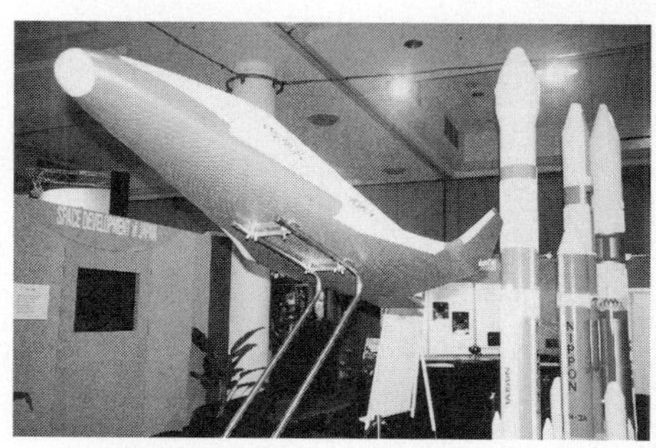

Scale model of HOPE spaceplane shows double-delta wing platform and simulated thermal tiles (Theo Pirard) ***1999***/0054365

Structure: Al alloy and carbon/polyimide composite
Thermal protection system: carbon/carbon composite, ceramic tile and flexible thermal insulator
Propulsion system: Orbital Manoeuvring System (OMS) and Reaction Control System (RCS)
Attitude control method: Aerodynamic control surface and RCS

J1

Background

NASDA has, unusually, developed the all-solid J1, capable of delivering 900 kg into LEO. Stage 1 employs H2's strap-on, while stages 2 and 3 use M-3SII's second and third stages. The debut two-stage suborbital test in February 1996 was successful, but J1's manifest remains sparse – only one other launch is scheduled.

The Space Activities Commission in July 1991 approved development of the ¥11.30 billion all-solid J1 intermediate launcher. ¥434 million was authorised for FY92, ¥3,215 million FY93, ¥2,909 million FY94, ¥4,767 million FY95, ¥0 FY96. NASDA sought minimum development cost and risk by adapting existing hardware. Tanegashima's old H1 pad is utilised. Two strap-ons derived from the TR-1 would increase LEO capacity to about 1 tonne; using two H2 SRBs would yield 2 tonnes.

Specifications

J1
First launch: 11 February 1996 (2-stage suborbital)
Number launched: 1 by end of April 1999
Launch site: Tanegashima
Principal uses: delivery of intermediate payloads to LEO
Vehicle success rate: 100 per cent
Cost: about ¥5 billion, excluding development cost (not planned commercially).
Availability: potentially two annually. Fishing/range safety requirements constrict launches to January-February/August-September annually.
Performance: 870 kg into 250 km 30° circular
Number of stages: 3
Overall length: 33.110 m
Principal diameter: 1.81 m
Launch mass: 87,700 kg plus payload
Guidance: inertial/radio, largely derived from M-3SII. The B2CNE stage 2 control electronics unit is mounted on stage 2's forward end, and includes the RIG Rate Integrating Gyro on the SFAP Spin Free Analytic Platform for pitch/yaw/roll data. Ground commands are received by B2CNE should the Nogi Radar Station on the Tanegashima Islands 115 km south of Kyushu detect deviation from the programmed path. B2CNE routes radio commands and attitude error signals to stage 1's B1CNE, which also uses a stage 1 rate gyro package.

J1 stage 1
Motor contractor: Nissan Motor Co
Length: 20.992 m
Principal diameter: 1.8094 m
Mass at ignition: 70.9 t
Propellant: 59.15 t of 14% HTPB/ 68%AP/18%Al, in 4-segment NT-150 steel case held by 108 bolts at each joint

J1 debut was successful in 1996 (NASDA) ***1996***

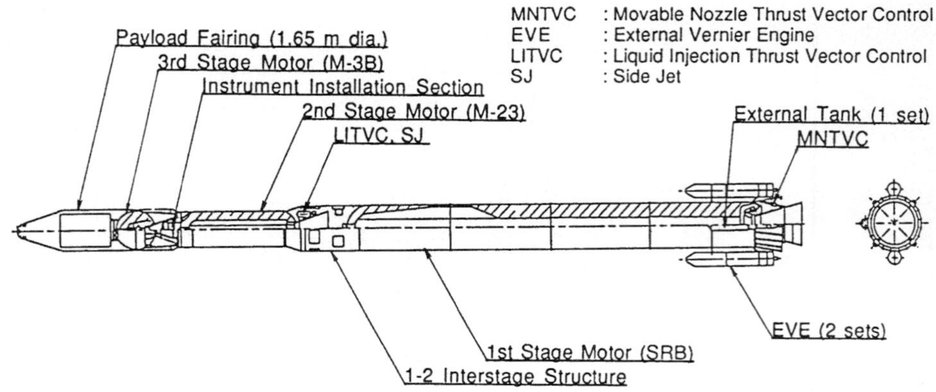

J1 principal features (NASDA) ***1995***

Average thrust: 1,560 kN average sea level
Burn time: 89 s
Steering: TVC by ±5° deflection of flexible nozzle joint provides vehicle pitch/yaw control. Two aft External Vernier Engine pods ignite 5 s before launch to provide roll control during burn, then three-axis during coast (previously planned solid motors on fins for roll control, as on M-3SII). EVE 4.177 m long, 56 cm diameter, carrying 110 kg N_2O_4/hydrazine. 3.43-2.45 kN vacuum blowdown (from hydraulic accumulator), gimballed ±45° pitch/roll, ±11° yaw.

J1 stage 2
Designation: M-23
Contractor: Nissan Motor Co
Overall length: 6.3 m
Principal diameter: 140.7 cm
Mass at ignition: 12.7 t (excluding fairing)
Propellant: 10.4 t of solid seven-star grain HTPB composite in steel case
Average thrust: 524 kN vacuum
Burn time: 55 s
Attitude control system: pitch/yaw control during burn maintained by LITVC, injecting 40 litres of $NaClO_4$ solution into nozzle at eight points. Roll control during burn and three-axis control in preparation for stage 3 spin-up/separation provided by four clusters of 4 × 150 N Side Jet (SJ) thrusters at stage 2's base drawing on 40 litres hydrazine

J1 stage 3
Designation: M-3B
Contractor: Nissan Motor Co
Overall length: 2.7 m
Principal diameter: 149.5 cm

L and M series orbital launch history

	Date	Vehicle	Payload
1*	26 September 1966	L-4S	Test satellite
2*	20 December 1966	L-4S	Test satellite
3*	13 April 1967	L-4S	Test satellite
4*	22 September 1969	L-4S	Test satellite
5	11 February 1970	L-4S	Ohsumi
6*	25 September 1970	M-4S	Science sat 1
7	16 February 1971	M-4S	Tansei 1
8	28 September 1971	M-4S	Shinsei
9	19 August 1972	M-4S	Denpa
10	16 February 1974	M-3C	Tansei 2
11	24 February 1975	M-3C	Taiyo
12*	4 February 1976	M-3C	Corsa A
13	19 February 1977	M-3H	Tansei 3
14	4 February 1978	M-3H	Kyokko
15	16 September 1978	M-3H	Jikiken
16	21 February 1979	M-3C	Hakucho
17	17 February 1980	M-3S	Tansei 4
18	21 February 1981	M-3S	Hinotori
19	20 February 1983	M-3S	Tenma
20	14 February 1984	M-3S	Ohzora
21	8 January 1985	M-3SII	Sakigake
22	18 August 1985	M-3SII	Suisei
23	5 February 1987	M-3SII	Ginga
24	21 February 1989	M-3SII	Akebono
25	24 January 1990	M-3SII	Hiten/ Hagomoro
26	30 August 1991	M-3SII	Yohkoh
27	20 February 1993	M-3SII	Asuka
28*	15 January 1995	M-3SII	Express
29	12 February 1997	M-5	Hanika
30	3 July 1998	M-5	Nozomi

* indicates vehicle failure. All launches made from Kagoshima. *M-3S Breakdown* Launch total for each type, followed by number of failures. M-3S: 4(0); M-3SII: 8(1); M-5 0(0).

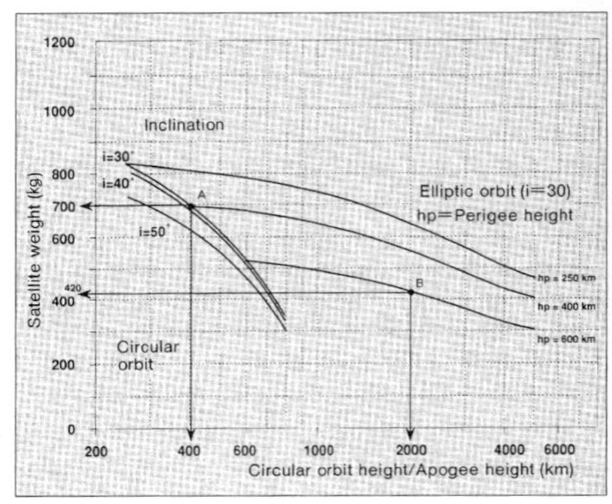

J1 payload capability from Tanegashima ***1995***

Mass at ignition: 3.59 t
Propellant: 3.28 t of seven-star HTPB composite solid in welded titanium alloy casing
Average thrust: 132 kN vacuum
Burn time: 71 s
Attitude control system: upper stages spin induced by stage 2 attitude control system before separation

Payload fairing
The 500 kg glass fibre honeycomb 2 cm-thick sandwich clamshell fairing, as used by the M-3SII, is pyrotechnically separated after stage 2 burnout. Nissan's 6.86 m long, 1.65 m diameter structure, which also covers stage 3, provides a payload envelope of about 3.0 m long, 1.4 m diameter. Peak vehicle acceleration is 7 *g* longitudinal (2 *g* at launch).

M launch vehicles

Background

ISAS has maintained its own stable of orbital and suborbital rockets for science missions since the late 1950s. All solid-fuelled, these have been integrated by ISAS with Nissan Motor as prime industrial contractor. The largest sounding rocket, L-3H, was augmented by a kick stage to create the L-4S orbital vehicle as a technology demonstrator for the M-4S launcher, flying Japan's first satellite in 1970 after four failures. The M-3SII model was responsible for launching two Halley's comet probes in 1985 – the first deep spacecraft dispatched by all-solid rockets. It retired January 1995 after its eighth (and first unsuccessful) mission.

The division of launcher responsibilities between ISAS/NASDA limited the scientific agency to rockets with main diameters no greater than 1.4 m, but studies were initiated in 1987 of the 2.5 m diameter M-5 design that will almost triple LEO capacity and permit a wider range of deep space missions. Development was approved during 1989, then aiming for first launch in 1995.

Studies were undertaken during 1987-88 of a large successor to the M-3SII, capable of placing 2.2 tonnes in LEO and dispatching 520 kg to the Moon. Development cost was estimated at US$133 million, with a US$36 million unit cost. ISAS estimated in 1990 a unit mission cost of ¥5.85 billion.

After the feasibility studies showed promise for the rocket, the Japanese government approved the M-5 in 1989. A September 1994 test launch of an ST-735-2 vehicle succeeded in demonstrating stage 2's 'fire-in-the-hole' separation technique. Then ISAS/Nissan began studies in 1990 of an air-launched three-stage winged version, capable of delivering 1,270 kg into a 250 km circular orbit at a cost of ¥360,000/kg. The 17 m-long 51.85 tonne vehicle would be released from the back of a 747-200 at an altitude of 10 km and speed of

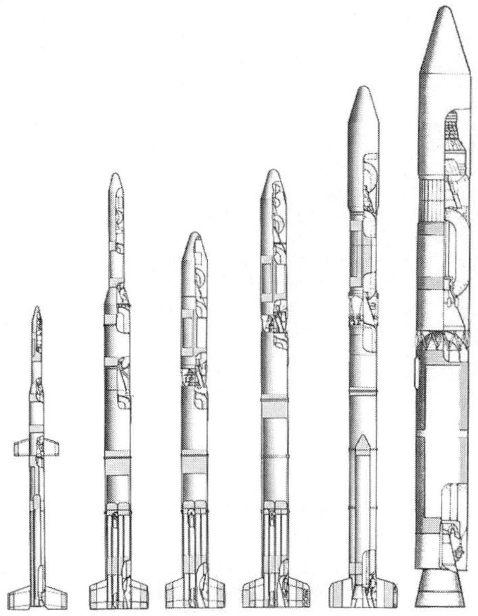

ISAS orbital launchers (from left): L-4S, M-4S, M-3C, M-3S, M-3SII, M-5 ***1995***

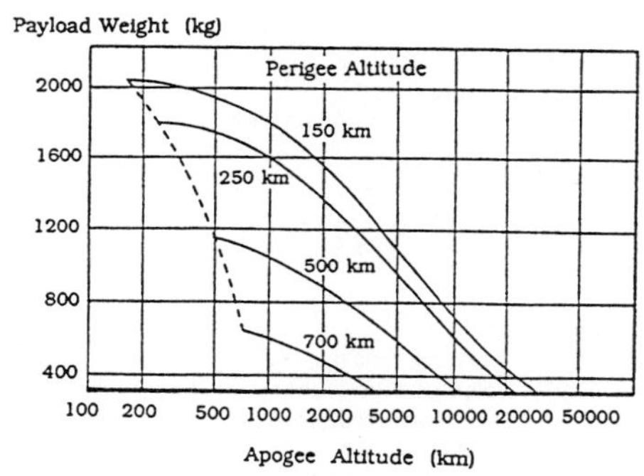

M-5 performance from Kagoshima (ISAS)

200 m/s. This vehicle derives its stage 1 from M-5's stage 2 M-24, its stage 2 from M-5's stage 3 M-34, and its stage 3 from M-3SII's M-3B stage 3. The advantages would be a 2.5 per cent payload ratio (1.5 per cent for M-5) and relaxation of the fishing season launch constraints. The first M5 was launched on 12 February 1997 followed by the second on 3 July 1998.

Specifications

M-5

Launch site: Kagoshima
Performance: 200 × 200 km, 31°: 2,000 kg
500 × 500 km, 31°: 1,200 kg
Sun-synchronous: 800 kg
GTO: 800 kg
Lunar probe: 520 kg
Mars/Venus, C3 9.0 km²/s²: 450 kg
Number of stages: 3 plus optional stage 4
Overall length: 31.2 m
Principal diameter: 2.50 m
Launch mass: 128.4 t (excluding payload)
Guidance: active radio guidance with inertial using fibre optic gyros and microprocessor as back-up on early flights. Inertial system may be used later. Other electronics, such as telemetry, ranging and command systems transferred from M-3SII, are also in a bay at top of stage 3

M-5 stage 1

Comment: Prototype stage 1 was successfully fired 21 June 1994. A second firing, using a thinner nozzle, was successful June 1995.
Designation: M-14
Contractor: Nissan Motor Co
Overall length: 13.65 m
Principal diameter: 2.50 m
Mass at ignition: 82.5 t
Propellant: 71.7 t solid propellant in HT-230M maraging steel case
Average thrust: 4,119 kN sea level

Preparing for the first M-5 (ISAS) ***1996***

Burn time: 47.1 s (effective)
Attitude control system: movable nozzle, actuated by gas generator

M-5 stage 2

Comment: Stage 1 separates at 70 s and stage 2 ignites simultaneously (fire-in-the-hole) to reduce gravity losses. The lattice structure of the 1,762 kg 3.84 m high interstage vents the exhaust and the three 120° supporting panels are released by FLSC Flexible Linear Shaped Charges and pushed out by springs from top. Stage 2 incorporates an extending nozzle, increasing expansion ratio from 27.9 to 37.1 and Isp vacuum from 288 s.
Designation: M-24
Contractor: Nissan Motor Co
Overall length: 6.71 m
Principal diameter: 2.51 m
Mass at ignition: 34.3 t
Propellant: 31.1 t solid in HT-230M maraging steel case
Average thrust: 1,206 kN vacuum
Burn time: 70.2 s (effective)
Attitude control system: liquid injection TVC

M-5 stage 3

Comment: 904 kg 2.45 m high aluminium interstage. Extending nozzle saves 300 kg and 1 m length.
Designation: M-34
Contractor: Nissan Motor Co
Overall length: 3.55 m
Principal diameter: 2.2 m
Mass at ignition: 11.1 t
Propellant: 10.1 t solid in carbon fibre 12-layer filament wound case. Expansion ratio 96.0, sp 301 s vacuum.
Average thrust: 289 kN vacuum
Burn time: 103.5 s (effective)
Attitude control system: movable nozzle (electrical actuation)

Payload accommodation

715 kg clamshell fairing of CFRP skins on honeycomb sandwich, length about 9 m, 2.50 m diameter. Outer surface covered by cork for thermal protection

N1/2

Background

NASDA began developing the Delta-based N1 orbital launcher in 1970 under US license, flying seven missions 1975-1982. The 130 kg GEO capability was increased to 350 kg for the eight N2s launched to February 1987.

IAS and NASDA continue the operation of two distinct launcher types. ISAS remains under constraint by a 1970 ruling limiting their missions to science objectives, using only solid propellant vehicles, while NASDA provides the national capability for launching applications satellites aboard liquid vehicles of increasing capabilities. Until 1990, ISAS was limited to making vehicles with base diameters no greater than 1.4 m but development of the 2.5 m diameter M-5 was approved in 1990. NASDA, on the other hand had the charter to build an indigenous vehicle for 2-tonne GEO applications, with a view to commercialisation.

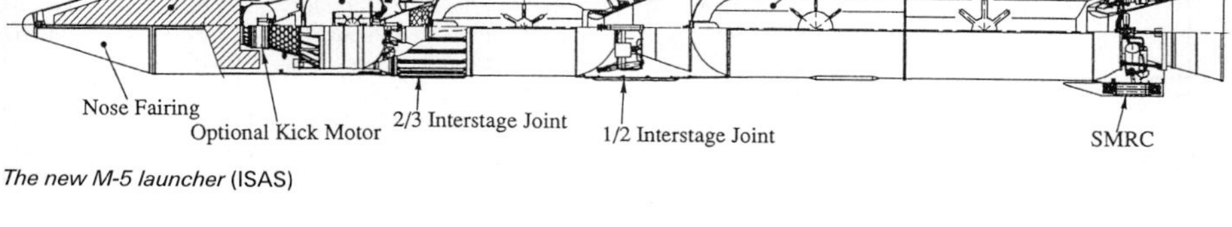

The new M-5 launcher (ISAS)

Models of the M5 and J1 launch vehicles (Theo Pirard)
1999/0054366

S-520 sounding rocket

Background

Nissan's S-520 is the largest of ISAS's range of sounding rockets, capable of delivering 250 kg to 400 km. In addition to space science missions, it is employed for recoverable µgravity experiments. The single solid propellant motor, with steel casing, is supported by four aluminium honeycomb tail fins with niobium/titanium alloy leading edges. ISAS is studying a three-stage version for orbiting ultra-small satellites.

Specifications

S-520

First launch: 18 January 1980 (first recoverable flight 5 September 1981)
Number launched: 24 to end-April 1999 (20 Kagoshima, 4 Andøya)
Success rate: 100% to end-April 1999
Launch site: Kagoshima (Japan), Andøya (Norway)
Performance: 150 kg to 350 km, using 82° launch angle
Number of stages: 1 (solid)
Overall length: 931.0 cm
Principal diameter: 524 mm
Launch mass: 2,285 kg
Guidance: fin spin stabilisation, but de-spin system included.

S-520 stage 1

Contractor: Nissan Motor Co
Overall length: 6,053 mm
Principal diameter: 524 mm (179 cm across four fins)
Thrust: average 143.17 kN vacuum
Burn time: 28.7 s effective

Payload accommodation

GFRP clamshell ogive nose, providing 494 mm payload base diameter, narrowing to 156 mm after 1,252 mm length. The support section below houses telemetry, timing, ignition relay, radar transponder and power supply equipment. Immediately below are the attitude control and recovery sections.

RUSSIAN FEDERATION

Angara

Background

Khrunichev/KB Salyut is developing Angara to replace Proton early in the next century. Khrunichev's design won out in a competition by RKA over a concept forwarded by Energia-M and others. It significantly improves Proton performance but relies on existing or imminent technology, drawing on Zenit for stage 1,

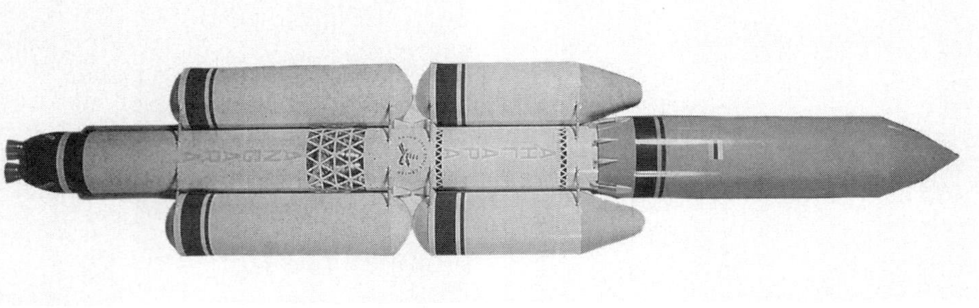

Angara employs an unusual suspended tank configuration for both main stages. Stage 1's outer tanks carry LOX, stage 2's LH_2 (Khrunichev) ***1996***

Energia for stage 2 and the Breeze-M and cryogenic stages already under development for Proton KM. It will employ Zenit's Plesetsk pad, freeing Russia of Proton's reliance on Kazakhstan. A fly-back stage 1 is possible, as is replacing the stage with an Antonov An-225 to deliver 6 tonnes into 200 km 51° orbit for US$1,500/kg.

Specifications

Angara
First launch: planned about 2005
Launch sites: Plesetsk Zenit complex 35
Contractors: Khrunichev/KB Salyut (prime), NPO AP (GN&C), NPO Energomash (main engines)
Principal uses: geostationary, escape and large LEO
Performance (from Plesetsk): 26.0 t into 200 km 63° two-stage; 3.5 t GEO using Breeze-M; 4.5 t GEO using Cryogenic. See diagram for further information.
Number of stages: two for LEO; three for higher orbits
Overall length: 55.5 m (5.0 m diameter fairing); 52.4 m (4.35 m diameter fairing).
Principal diameter: 4.10 m core
Launch mass (with payload): about 640 t
Launch thrust: 7,622 kN sea level
Guidance: unknown

Angara stage 1
Engines: RD-174 from NPO Energomash, four gimballed chambers.
Overall length: 14.5 m
Diameter: 4.10 m core, 3.90 m suspended tanks
Thrust: 7,622 kN sea level
Burn time: not available
Dry mass: not available
Oxidiser: LOX in two suspended tanks
Fuel: kerosene in core

Angara stage 2
Engines: NPO Energomash RD-0120 single fixed reignitable chamber with verniers
Overall length: 9.5 m
Diameter: 4.10 m core, 3.90 m suspended tanks
Thrust: not available
Burn time: not available
Dry mass: not available
Oxidiser: LOX in central core
Fuel: LH_2 in two suspended tanks

Three proposed configurations for the Angara launcher series (Theo Pirard) ***1999****/0054361*

Angara stage 3
Configuration: Angara will employ the same Breeze-M and cryogenic stage 3 as Proton KM's stage 4. See Proton's entry for further details.
Payload accommodation: two fairings are planned: 5.00 m diameter, 22.50 m long and 4.35 m diameter, 19.42 m long.

Dnepr-1 (SS-18K)

Background

When NPO Yuzhnoye started to market this launch vehicle – based upon the R-36M2 missile – they took the NATO designator 'SS-18' and added a 'K' to indicate a space application. It remained the SS-18K until 1996 when the name Dnepr started to be applied to the vehicle. Yuzhnoye offered the booster for the cost of US$20 million to US$40 million in mid-1992. Tsyklon's S5M stage 3 or Lavotchkin's Fregat stage could be adapted for orbital injection/circularisation.

Development of the silo-launched missile began in 1964 and it was first deployed in 1974. Different modifications carried single or 10 warheads. Yuzhnoye is particularly interested in launching μgravity payloads of up to 700 kg, providing up to 2,000 hours of 10^{-6} *g* μgravity on orbital missions and 2 hours on ballistic flights.

Members of the design and launch team claimed the missile achieved a 97% success rate on 157 launch attempts between 1973 and April 1999, and the backlog can be gauged from the US Dept of Defense which notes 308 missiles (204 in Russia, 104 in Kazakhstan) remain extant. Production ended in 1991 under the first phase of the START arms control agreement. By May 1999, 174 rounds were due to be destroyed by 2002 if START-2 is ratified and 154 missiles can be converted into satellite launchers. US Defense Secretary William Perry, confirmed in May 1996, that China had been seeking to obtain SS-18 technology from Russia and possibly Ukraine.

The first Dnepr was launched successfully from a silo at Baikonur on 21 April 1999, carrying the UK satellite UoSAT-12

Sources originally reported that the RS-36M2 missile used the RD-251 engine on its first stage, but it is now known that this is the designator applied to the first stage of Tsyklon. It is known that the first stage uses an engine assembly (four chambers, but not known if it is a single engine, two or four engines) from the GDL/Energomash, but the designator is not known: based upon the propellants it should be in the RD-200 series. A thrust of 1,128 kN was reported for the first stage engine assembly, but this figure was linked to the now-discredited report that the RD-251 was used on the first

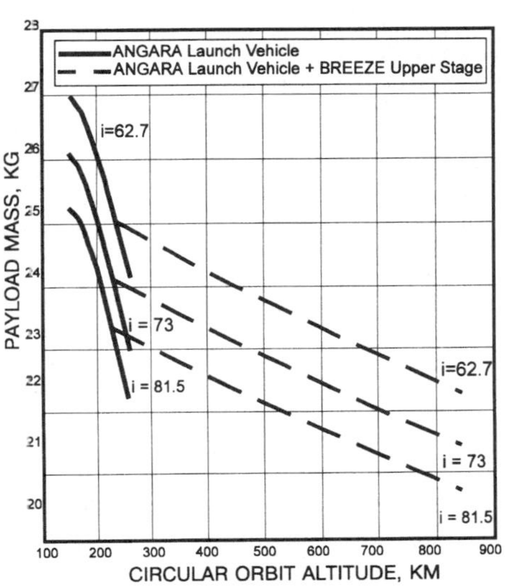

Angara performance from Plesetsk for the 2-stage (solid lines) and Breeze-M (dashed) versions (Khrunichev) ***1996***

stage. *Stage 2:* Original reports indicated that the RD-252 engine was carried by the R-36M2 second stage. But it is now known from recently-released Russian information that a single RD-0256 is carried as the main engine and the RD-0257 is the four-chamber vernier engine system. These engines were designed by KB Khimavtomatiki, originally the Kosberg bureau. A thrust of 755 kN was reported for the second stage main engine, but this figure was linked to the now-discredited report that the RD-252 was used on the second stage.

The payload occupies the volume previously allocated to the warhead re-entry vehicle/s. The 5.25 m long 3.0 m diameter fairing provides a 21.0 m^3 payload envelope; 8.25 m long fairing is possible. Maximum static launch acceleration is 7.5 *g* during powered flight; 1.3 *g* lateral. Advanced versions are in development capable of placing 3,000 kg in sun-synchronous orbit.

Specifications

Dniepr
First launch: –
Launch sites: Tyuratam silo (10)
Principal uses: delivery of medium-class payloads into LEO
Typical performance: 4,400 kg into 200 km circular at 65°; 4,100 kg into 200 km circular at 90°; 3,700 kg into 200 × 1,000 km, 65°; 3,500 kg into 200 × 1,000 km, 90°. Injection accuracy: ±3.5 km altitude, ±2.5 s period, ±2.0 arcmin inc.
Number of stages: three (liquid)
Overall length: 34.30 m (37.3 m with extended payload fairing)
Principal diameter: 3.00 m
Launch mass: about 210 t
Guidance: computer-controlled inertial

Kosmos 3M

Background

For many years the smallest vehicle in the operational inventory, Kosmos originated as the R14 (8K65)/SS-5 Skean medium range missile, developed by the Yangel bureau (now NPO Yuzhnoye). NPO-PM made it a space launcher, because of Yuzhnoye's heavy workload. In its first incarnation the Soviets called it Kosmos and then the current Kosmos 3M (11K65M). It is now the responsibility of AKO Polyot in Omsk. It remains as the only vehicle to be fired from all three Soviet sites. Its debut space launch in August 1964 orbited the first Soviet triple satellite payload and it is now employed principally for placing small military craft into medium Earth orbits. About half of all launches have performed suborbital hypersonic and orbital entry tests. The Russians continue suborbital flights from Kapustin Yar. Orbital launch frequency has decreased as new generations of large satellites have transferred to Tsyklon. The precipitous decline in the launch rate will only get worse as the VKS Military Space Forces switch to Rockot. At this point, commercial success for the launcher is critical for Polyot's survival. Kosmos was one of only two bidding for NASA's 1995 Med-Lite launch contract (awarded to McDonnell Douglas/Orbital Sciences Corp).

Kosmos released its first commercial payloads during the 24 January 1995 launch. They were Sweden's 28 kg Astrid 1 science satellite and Final Analysis Incorporated's 114 kg store/forward Faisat 1. FAI planned a second satellite in 1996 as a prelude to 26 data messaging satellites. Polyot planned to provide the launch in return for being the Russian service provider, but in April 1997, the satellite was not carried by the planned launch vehicle, apparently because paperwork was incomplete. The 5 September 1996 flight had carried EUNAMSAT 2, a Mexican satellite which replaced an earlier launch in 1995, where the satellites were lost during the failure of the first Start mission. Germany's DLR-Tubsat and the similar Tubsat being built for Morocco may also fly on Kosmos. The German government is expected to buy a launch, through OHB-Systems, for its Abrixas astronomy satellite. The Belgium-led IRIS is a likely payload as well. Polyot and Assured Space Access Inc of the US signed an agreement 21 September 1995 to set up the Kosmos USA joint venture for the global marketing of Kosmos 3 outside of the former Soviet Union. UK's Plowshare Technology Ltd acts as the European representative. Commercial contacts are now made via Kosmos USA, although pre-existing agreements with Final Analysis and OHB-System are still being executed directly by Polyot.

The 11K55 vehicle was designed with state approval in the 1980s by Yuzhnoye/Polyot to replace Kosmos and Tsyklon, but development was prevented by the Soviet Union's collapse. Stage 1 would have employed RD-120.

The improved Vzlet version is under study, increasing performance by 15 to 20 per cent, increasing the precision of the final orbit to ±2.5 km by including

Kosmos is still Russia's smallest operational orbital launcher. One of stage 2's external tanks can be seen at right

Glonass receivers in the updated guidance system, and adding the capability of dispensing multiple payloads into different orbits. Stage 2 would be burned to depletion and de-orbited. Funding problems may delay the upgrade or cancel it altogether. Polyot does not have commercial or state support for the work but Germany may become involved. The Kosmos USA agreement does not include Vzlet.

Specifications

Kosmos 3M

First orbital launch: August 1964 (Kosmos 38-40)
Number launched: 417 attained orbit by end-1998. 735 launches in total 1970-98, including 345 suborbital re-entry tests.
Launch sites: Plesetsk, Kapustin Yar (KY currently suborbital only, but orbital available).
Principal uses: small military store/dump communications, navigation and unknown minor missions at medium altitudes; Interkosmos science satellites; commercial.
Vehicle success rate: 94.4 per cent orbital attempts; 23 orbital failures: 23 October 1964, 16 November 1966, 26 June 1967, 27 September 1967 (pad explosion), 4 June and 15 June 1968, 27 December 1969, 27 June 1970, 23 December 70, 22 July 71, 17 October 72, 25 May 73, 3 June 75, 19 December 75, 29 November 77, 20 December 78 (Kosmos 1064 attained orbit but stage 2 failed to circularise), 4 March 82, 18 June 82 (Kosmos 1380 attained orbit but stage 2 failed to circularise), 30 August 82, 24 November 82, 25 January 83, 23 October 85, 25 June 91, 6 October 95 (stage 2 sticking fuel valve left Kosmos 2321 in low orbit, although still claimed as usable). Nine were killed 26 June 1973 at Plesetsk when Kosmos exploded during fuelling operations.
Cost: about US$10 million for complete capacity; US$6,500/kg for secondary payloads (Sweden was charged US$5,000/kg for Astrid).
Availability: main payloads can be accommodated with three months' notice. Capacity of Polyot and facilities is 30 launches annually (four to six military launches planned for 1996).
Performance: 1,400 kg into 180 km circular, but principally used for about 1 t payloads into 800-1,500 km circular orbits. Accuracy into 1,000 km ±40 km perigee/apogee, +0.04/−0.08° inc.
Number of stages: 2

Kosmos launchers stockpiled in Omsk. The vehicle stands on the four protruding feet, which also house graphite jet vanes for steering; the vanes are installed on the pad (Plowshare Technology Ltd) ***1995***

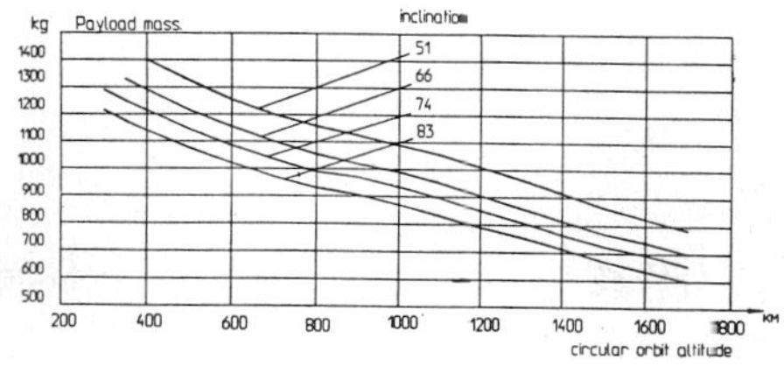

Kosmos 3M circular orbit performance from Plesetsk. The curves indicate different inclinations. Precision into a 1,000 km circular path is ±40 km; ±30 km for 1,600 km

Overall length: 32.4 m
Principal diameter: 2.40 m
Launch mass: 109 t
Launch thrust: 1,726 kN
Guidance: inertial, on stage 2 forward end, by Khartron

Kosmos 3M stage 1

Engines: RD-216 assembly (four fixed chambers) from NPO Energomash with storable propellants. Steering by graphite vane inserted into each exhaust (installed at end of pad preparations).
Overall length: 22.5 m
Diameter: 2.40 m
Thrust: 1,726 kN
Burn time: about 130 s
Oxidiser: N_2O_4 in NO_2
Fuel: UDMH (aft tank)
Propellant mass: 82.0 t; dry mass 5.30 t
Separation: cold separation, assisted by solid braking rockets

Kosmos 3M stage 2

Configuration: Stage 2 adds two side tanks for missions into 1,000-1,500 km circular orbits, using four smaller thrusters for the climb to the required altitude. The combined UDMH/N_2O_4 external tanks run the length of stage 2. They are not carried on the suborbital version, which also ignites stage 2 only once.
Engine: KB Khimmach 11D49: pump-fed fixed single chamber + four steering thrusters (used for steering, orbit trims, manoeuvres for multiple deployments, climb to high orbits).
Overall length: 4.205 m (6.585 m from nozzle to payload mount).
Diameter: 2.40 m (3.2 m across external tanks)
Thrust: 157.3 kN main chamber plus 4 × 1.4-1.8 kN steering/adjust thrusters (each with a pressure-fed 93 N thruster for three-axis coast control and ullage).
Propellants: as stage 1, carried in spherical tanks
Burn time: 325-335 s burn 1 plus 2-8 s burn with 2 main engine
Propellant mass: 18.7 t; dry mass 1,434 kg. Fuel in aft tank.

Fairing/payload environment

The 2.40 m diameter 5.720 m long 40° apex coni-cylindrical fairing provides 12.58 m^3 payload volume, with four access doors in the 1.809 m high cylindrical section and seven in the conical. The vehicle is integrated horizontally and the payload attached in the same building. Access provides two 3 × 3 m class 100,000 cleanrooms in the building for satellite integration/checkout, in addition to the building's own 28 m-long satellite area. Launcher preparation up to pad arrival requires 105 men over 34-36 h; pad activities 120-135 men for 8-10 h; launch is by a crew of 20-25. Fuelling is completed 4 h before launch and air conditioning to the payload fairing ends at T-1 h. Launches are made within −40/50°C and 20 m/s wind surface conditions. Kosmos provides 29.5±2 V DC to the payload. The fairing halves separate mechanically at about 75 km. Payload is separated by four pyropushers within 20 s of shutdown; stage 2 ignites a small lateral solid 1.25 s after separation for collision avoidance. Payload angular disturbance at separation is ±3.0°/s pitch/yaw + ±0.5°/s roll for 1,000 kg.
Mechanical loads: 1.0-6.5 *g* longitudinal, <1.3 *g* lateral
Acoustic load: 140 dB rms max
Thermal: 280°C conical and 180°C cylindrical internal wall maximum temperatures

Molniya-M

Background

Molniya is the most powerful of the Sapwood-based space launchers. It was employed until 1972 for planetary missions, but is now principally used for placing Molniya communications and early warning satellites into highly eccentric Earth orbital paths. The Soviets and their Russian successors have not yet used it for a GEO mission, but the first Sun-synchronous mission came in December 1995.

Russians have been rumoured to be upgrading the new derivative, the Rus, but have not yet released figures on its performance.

The 8K78 Molniya first flew successfully in February 1961 carrying the Venera 1 Venus probe. The original 8K78 model employed the RD-0107 engine to power stage 2. This was replaced by the RD-0110 for the current 8K78M, first flown 19 February 1964 (failed to reach orbit). Before that an earlier variant attempted to put two unsuccessful Mars probes on the way to the Red Planet on 10 and 14 October 1960. In general, the Molniya itself is a capable launcher, but the Block L family of escape ('e') stages has proved to be one of the more unreliable, occasionally stranding payloads in the LEO parking orbit. The Kosmos 2084 early warning mission of June 1990 is the most recent example. NPO Lavotchkin proposes replacing Block L with its more powerful and flexible Fregat stage, based on its Phobos Mars stage.

The Soviets used various design bureaus as competing agencies and the Molniya 8K78 variant had at least one rival. The Korolyov bureau proposed an 8K77 launch vehicle using a Block Zh second stage instead of the Block I. A version designated 8K76 was also discussed, this being the 8K77 without the Block L class final stage.

Commercial customers, such as India's IRS 1C use the Molniya as well. The Indian launch occurred in December 1995, accompanied by BMDO's Skipper. Molniya's first Sun-synchronous mission uniquely used Block L for the orbital injection, the previous stages all being suborbital. In mid-1996, Starem, a new joint venture company, opened in France to market the Molniya-M launch vehicle on a commercial basis.

Molniya and Molniya-M annual launch rates

1960: 2	1967: 6	1974: 7
1961: 2	1968: 6	1975: 12
1962: 6	1969: 4	1976: 11
1963: 6	1970: 7	1977: 10
1964: 8	1971: 3	1978: 9
1965: 12	1972: 11	1979: 7
1966: 10	1973: 10	1980: 12

Subsequent missions are covered in greater detail in the table overleaf.

The Molniya-M is employed primarily for Cosmos early warning and Molniya communications satellites

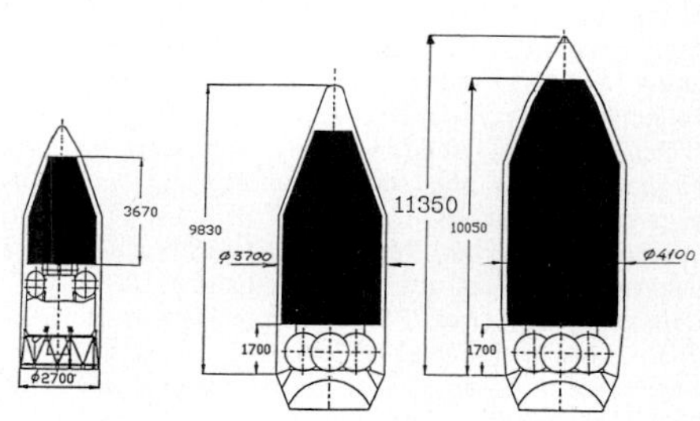

The current Molniya payload envelope and Block L configuration are shown at left. To the right are NPO Lavotchkin's additional proposals for the Rus upgrade: replacing Block L with Fregat and using Proton's larger fairings

Molniya-M Launches 1981-1998

	Date	Site	Payload
1	9 January 1981	Plesetsk	Molniya 3-14
2	30 January 1981	Plesetsk	Molniya 1-49
3	19 February 1981	Plesetsk	C1247 early warning
4	24 March 1981	Plesetsk	Molniya 3-15
5	31 March 1981	Plesetsk	C1261 early warning
6	9 June 1981	Plesetsk	Molniya 3-16
7	19 June 1981	Plesetsk	C1278 early warning
8	24 June 1981	Plesetsk	Molniya 1-50
9	4 August 1981	Plesetsk	C1285 early warning
10*	11 September 1981	Plesetsk	Molniya failure
11	17 October 1981	Plesetsk	Molniya 3-17
12	31 October 1981	Plesetsk	C1317 early warning
13	17 November 1981	Plesetsk	Molniya 1-51
14	23 December 1981	Tyuratam	Molniya 1-52
15	26 February 1982	Plesetsk	Molniya 1-53
16	3 March 1982	Plesetsk	C1341 early warning
17	24 March 1982	Plesetsk	Molniya 3-18
18	7 April 1982	Plesetsk	C1348 early warning
19	20 May 1982	Plesetsk	C1367 early warning
20	28 May 1982	Plesetsk	Molniya 1-54
21	25 June 1982	Plesetsk	C1382 early warning
22	21 July 1982	Tyuratam	Molniya 1-55
23	27 August 1982	Plesetsk	Molniya 3-19
24	22 September 1982	Plesetsk	C1409 early warning
25*	8 December 1982	Tyuratam	Molniya failure
26	11 March 1983	Plesetsk	Molniya 3-20
27	16 March 1983	Plesetsk	Molniya 1-56
28	2 April 1983	Tyuratam	Molniya 1-57
29	25 April 1983	Plesetsk	C1456 early warning
30	1 July 1983	Tyuratam	Prognoz 9 science
31	8 July 1983	Plesetsk	C1481 early warning
32	19 July 1983	Tyuratam	Molniya 1-58
33	30 August 1983	Plesetsk	Molniya 3-21
34	23 November 1983	Plesetsk	Molniya 1-59
35	21 December 1983	Plesetsk	Molniya 3-22
36	28 December 1983	Plesetsk	C1518 early warning
37	6 May 1984	Plesetsk	C1541 early warning
38	16 March 1984	Plesetsk	Molniya 1-60
39	4 April 1984	Plesetsk	C1547 early warning
40	6 June 1984	Plesetsk	C1569 early warning
41	3 July 1984	Plesetsk	C1581 early warning
42	2 August 1984	Plesetsk	C1586 early warning
43	10 August 1984	Plesetsk	Molniya 1-61
44	24 August 1984	Plesetsk	Molniya 1-62
45	7 September 1984	Plesetsk	C1596 early warning
46	4 October 1984	Plesetsk	C1604 early warning
47	14 December 1984	Plesetsk	Molniya 1-63
48	16 January 1985	Plesetsk	Molniya 3-23
49	26 April 1985	Tyuratam	Prognoz 10
50	29 May 1985	Plesetsk	Molniya 3-24
51	11 June 1985	Plesetsk	C1658 early warning
52	18 June 1985	Plesetsk	C1661 early warning
53	17 July 1985	Plesetsk	Molniya 3-25
54	12 August 1985	Plesetsk	C1675 early warning
55	22 August 1985	Plesetsk	Molniya 1-64
56	24 September 1985	Plesetsk	C1684 early warning
57	30 September 1985	Plesetsk	C1687 early warning
58	3 October 1985	Plesetsk	Molniya 3-26
59	22 October 1985	Plesetsk	C1698 early warning
60	23 October 1985	Tyuratam	Molniya 1-65
61	28 October 1985	Plesetsk	Molniya 1-66
62	9 November 1985	Plesetsk	C1701 early warning
63	24 December 1985	Plesetsk	Molniya 3-27
64	1 February 1986	Plesetsk	C1729 early warning
65	18 April 1986	Plesetsk	Molniya 3-28
66	19 June 1986	Plesetsk	Molniya 3-29
67	5 July 1986	Plesetsk	C1761 early warning
68	30 July 1986	Plesetsk	Molniya 1-67
69	28 August 1986	Plesetsk	C1774 early warning
70	5 September 1986	Plesetsk	Molniya 1-68
71*	3 October 1986	Plesetsk	early warning
72	15 October 1986	Plesetsk	C1785 early warning
73	20 October 1986	Plesetsk	Molniya 3-30
74	15 November 1986	Plesetsk	Molniya 1-69
75	20 November 1986	Plesetsk	C1793 early warning
76	12 December 1986	Plesetsk	C1806 early warning
77	26 December 1986	Plesetsk	Molniya 1-70
78	22 January 1987	Plesetsk	Molniya 3-31
79	4 June 1987	Plesetsk	C1849 early warning
80	12 June 1987	Plesetsk	C1851 early warning
81	21 December 1987	Plesetsk	C1903 early warning
82	26 February 1988	Plesetsk	C1922 early warning
83	11 March 1988	Tyuratam	Molniya 1-71
84	17 March 1988	Plesetsk	Molniya 1-72
85	26 May 1988	Plesetsk	Molniya 3-32
86	12 August 1988	Plesetsk	Molniya 1-73
87	30 August 1988	Plesetsk	C1966 early warning
88	29 September 1988	Plesetsk	Molniya 3-33
89	3 October 1988	Plesetsk	C1974 early warning
90	25 October 1988	Plesetsk	C1977 early warning
91	22 December 1988	Plesetsk	Molniya 3-34
92	28 December 1988	Plesetsk	Molniya 1-74
93	14 February 1989	Plesetsk	C2001 early warning
94	15 February 1989	Tyuratam	Molniya 1-75
95	8 June 1989	Plesetsk	Molniya 3-35
96	27 September 1989	Plesetsk	Molniya 1-76
97	23 November 1989	Plesetsk	C2050 early warning
98	28 November 1989	Plesetsk	Molniya 3-36
99	23 January 1990	Plesetsk	Molniya 3-37
100	27 March 1990	Plesetsk	C2063 early warning
101	26 April 1990	Plesetsk	Molniya 1-77
102	28 April 1990	Plesetsk	C2076 early warning
103	13 June 1990	Plesetsk	Molniya 3-38
104*	21 June 1990	Plesetsk	C2084 early warning
105	25 July 1990	Plesetsk	C2087 early warning
106	10 August 1990	Plesetsk	Molniya 1-78
107	28 August 1990	Plesetsk	C2097 early warning
108	20 September 1990	Plesetsk	Molniya 3-39
109	20 November 1990	Plesetsk	C2105 early warning
110	23 November 1990	Plesetsk	Molniya 1-79
111	15 February 1991	Plesetsk	Molniya 1-80
112	22 March 1991	Plesetsk	Molniya 3-40
113	18 June 1991	Plesetsk	Molniya 1-81
114	1 August 1991	Plesetsk	Molniya 1-82
115	17 September 1991	Plesetsk	Molniya 3-41
116	24 January 1992	Plesetsk	C2176 early warning
117	4 March 1992	Plesetsk	Molniya 1-83
118	8 July 1992	Plesetsk	C2196 early warning
119	6 August 1992	Plesetsk	Molniya 1-84
120	14 October 1992	Plesetsk	Molniya 3-42
121	21 October 1992	Plesetsk	C2217 early warning
122	25 November 1992	Plesetsk	C2222 early warning
123	2 December 1992	Plesetsk	Molniya 3-43
124	13 January 1993	Plesetsk	Molniya 1-85
125	26 January 1993	Plesetsk	C2232 early warning
126	6 April 1993	Plesetsk	C2241 early warning
127	21 April 1993	Plesetsk	Molniya 3-44
128	26 May 1993	Plesetsk	Molniya 1-86
129	4 August 1993	Plesetsk	Molniya 3-45
130	10 August 1993	Plesetsk	C2261 early warning
131	22 December 1993	Plesetsk	Molniya 1-87
132	5 August 1994	Plesetsk	C2286 early warning
133	23 August 1994	Plesetsk	Molniya 3-46
134	14 December 1994	Plesetsk	Molniya 1-88
135	24 May 1995	Plesetsk	C2312 early warning
136	2 August 1995	Plesetsk	Interball 1 science
137	9 August 1995	Plesetsk	Molniya 3-47
138	28 December 1995	Tyuratam	IRS 1C/Skipper
139	14 August 1996	Plesetsk	Molniya 1-89
140	29 August 1996	Plesetsk	Victor 1 (Argentina) plus Maglen 5 (Czech Republic) plus Interball 2 science
141	24 October 1996	Plesetsk	Molniya 3-48
142	9 April 1997	Plesetsk	C2340 OKO earlywarning
143	14 May 1997	Plesetsk	C2342 OKO earlywarning
144	24 September 1997	Plesetsk	Molniya-1 90
145	7 May 1998	Plesetsk	C2351 OKO earlywarning
146	1 July 1998	Plesetsk	Molniya-3 49
147	28 September 1998	Plesetsk	Molniya-1 91

* indicates launcher failure (all those shown attained orbit but Block L failed from LEO)

Specifications

Molniya/Molniya-M

First launch: 10 October 1960 (Mars probe).
Number launched: 305 to end-1998, 290 attaining LEO and 272 successfully departing LEO.
Launch sites: Tyuratam, Plesetsk.
Principal uses: Kosmos early warning, Molniya communications and science satellites, all in highly eccentric Earth orbits.
Vehicle success rate: 89% to end-1998 (includes nine failures to LEO + 24 upper stage failures in LEO).
Commercial launches: IRS 1C/Skipper December 1995, Victor 1 (Argentina) August 1996.
Cost: about US$15 million for commercial launches. India paid Rs500 million for IRS 1C.
Performance
Molniya-type (400 × 40,000 km, 63-65°): 2,000 kg
Sun-synchronous, 800 km: 1,500 kg
Prognoz-type (400 × 200,000 km, 65°): 900 kg
Lunar delivery: 1,620 kg
Venus delivery: 1,180 kg
Mars delivery: 950 kg
Comment: The corresponding version of Rus could deliver 2,100 kg into 200 × 36,000 km, 51.6° orbit from Tyuratam. Replacing Block L by Fregat would increase that to 2,700 kg.
Number of stages: 3 + 4 strap-ons
Overall length: 43.44 m
Principal diameter: as SL-4
Launch thrust: as SL-4
Launch mass: 306 t (305.481 t for Interball 1)

Molniya-M stage 3

Description: The 329 kg solid propellant ullage motor package (BOZ Blok Obespecheniya Zapuska: launching provision block) fires to settle the stage's propellants for ignition and is then jettisoned. It has been speculated that this stage might have been scaled up to create the Block D for N1 and Proton. OkO launches employ the Block 2BL version; Molniya satellites Block ML 2BL-SM2 was referred to for Interball 1. There are believed to be others.
Designation: 11D33, Block L family
Engine: Korolev bureau
Stage length: 2.64 m.
Stage diameter: 2.41 m (suspended inside shroud)
Oxidiser: liquid oxygen (2.66 t for Interball 1)
Fuel: kerosene (1.11 t for Interball 1)
Propellant mass: 3.77 t for Interball 1 (2BL dry mass 897 kg); excludes ullage package.
Thrust: 67 kN vacuum
Burn time: about 200 s
Attitude control: unknown
Fairing: The payload and Block L are protected by an 825 kg 7.90 m long 2.70 m diameter fairing, providing a payload envelope of 3.670 m long 2.65 m diameter.

Proton-K

Background

Proton entered the stable of Russian launch vehicles in 1965, by Chelomei's OKB-52 bureau, as a two-stage vehicle capable of delivering 12.2 tonnes into LEO. Almost immediately, three and four stage versions of the vehicle made the original obsolete. The Proton has the distinction of being the Soviet's first space launcher that did not begin as a military ballistic missile. Although the design originated in 1962 as the UR500 ICBM, the military need for that calibre of ICBM diminished as international targeting doctrines changed and the Soviet dropped plans for the missile. A two-stage test version left over from the early missile programme served as the testbed for the new space launch programme.

Chelomei used the original Proton project in his bid to become the father of the Soviet Moon programme. He aimed to use it to launch a one-man circumlunar flight using the original two-stage launcher, thus stealing some of Korolev's fame. This programme, called LK-1, received official approval in early 1965, but on 25 December 1965 Korolev wrested control back from Chelomei, adding it to his own manned landing programme. Slippage of the whole lunar programme prompted the addition of N1's Block D stage to Proton. This was the UR500K-L1 project and was publicly designated Zond. L1 was cancelled in early 1969 and the effort transferred to the Almaz and Salyut space station projects. KB Salyut divided off from Chelomei's bureau following the designer's death in late 1984.

KB Salyut (the design authority) and Khrunichev (the builder) in recent years promoted Proton separately, disagreeing over rights to the vehicle. President Yeltsin issued a decree 7 June 1993 merging them as the 'Khrunichev State Research & Production Space Center', which Lockheed joined as a minority shareholder to form Lockheed-Khrunichev-Energia (LKE) International in early 1993. Because of proliferation concerns, the US and Russia reached a draft accord in May 1993 limiting total sales of the Proton to eight GEO/GTO commercial launches through 2000 (further limited to two in any 12 months). Although not specifically a Proton agreement, in practice it covers principally that vehicle's GEO version. Part of the deal required prices within 7.5 per cent of the western equivalents, indicating the US concerns had more to do with launcher sales than arms control. The agreement excludes Proton's earlier Inmarsat contract. LEO and other non-GEO/GTO launches are dealt with on a case-by-case basis. A more relaxed agreement was signed 30 January 1996, allowing 16 launches (plus another four should market conditions allow) within 15 per cent of western prices.

Confusingly, the Russians refer to both three- and four-stage versions of this launch vehicle as the Proton-K. This is done because the basic launch vehicle is considered to be the Khrunichev three-stage Proton-K and if required one of Energiya's Block D or Block DM variant fourth stages can be added for missions beyond low Earth orbit.

The Proton's original fourth stage started life as the fifth stage of the N-1 manned lunar vehicle developed in the late 1960s. Two basic types of fourth stage are flown on the Proton-K: the Block D family and the Block DM family.

Launches using the Block D family employ the Block D to perform initial orbital injection, the first three stages of the vehicle being sub-orbital. The Block D then performs a second burn to launch the spacecraft out of a low parking orbit. The Block D family does not carry its own command and control system and it relies upon the payload to relay commands for engine firing, separation, and so on. These upper stages have been used exclusively for deep space missions. Different versions have appeared over the years: the original Block D (11S824), lunar and planetary missions during 1967-1983, the Block D-1 (11S824M) for Astron, Venera

Proton stage 1 production at Khrunichev's Moscow factory **1996**

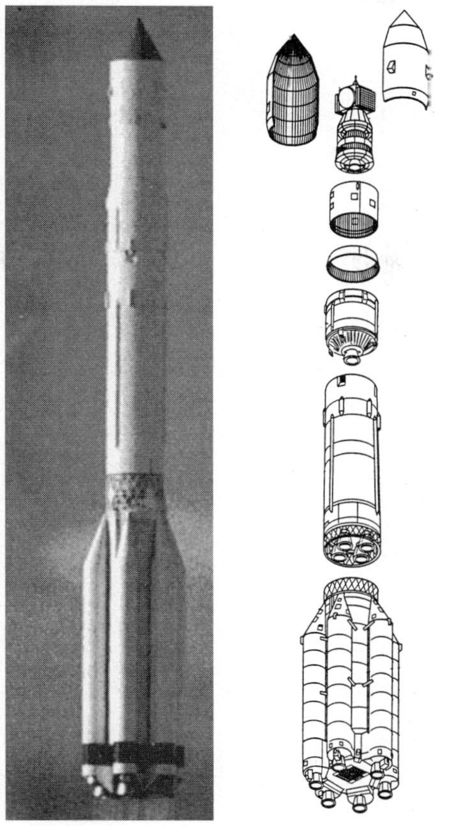

The Proton launcher (LKE International/C P Vick) **1994**

11-16, VEGA 1-2 and Granat, and the Block D-2 (11S824F) used for Phobos 1-2 and Mars 8 (=Mars-96). Actual differences between these upper stages have not been described in Russian literature.

Proton's first three stages are assembled by Khrunichev's factory close to Moscow. The production rate declined to eight annually in 1989 from the previous 12. The factory can handle 16 per year and requires orders for six to remain a viable commercial entity. Only the propellant tanks are welded at Khrunichev; the vehicle itself is assembled at Baikonur over 1½ months as components arrive at the main 300 × 400 m, 20 m high assembly hall. The Block D/DM stage 4 is managed and built by RKK Energia for integration with the vehicle at Baikonur. This stage was designed by Korolev's OKB for the N1 lunar launcher.

Until the Vega missions of December 1984, western observers had never seen the launcher in any detail. Then it became apparent that six cylinders, each with one engine, were clustered around a central propellant reservoir. It had been believed previously that these were strap-ons separated at altitude for core engines to continue burning. The unique configuration was adopted to allow rail transport to Baikonur, where the outer tanks could be attached. Stage 1's descent is now controlled to minimise the pollution problem in Kazakhstan. Impact dispersion is limited for a team to clean up the area.

In the past, vehicles up to seven years old have been launched.

Proton's future employment for GEO payloads and planetary missions requires enhancements. The Russian Space Agency announced in late 1993 that it is funding improvements to both Proton and Soyuz. Proton's stage 1 engines will be upgraded (as they were in 1986 and possibly in 1991), Zenit's avionics will be adopted, a 5 m diameter payload fairing will be added, and the interstages strengthened to cope with the heavier loads. These improvements will provide 22.3 tonnes LEO and almost 3-tonne GEO capacities. The coming generations of larger payloads will be beyond this GEO capability. However, the Proton KM (M means 'modernised') is planned for about 2000 to handle 3.2 tonne in GEO using Khrunichev's own Breeze-M (adapted from Rockot), and a new cryogenic stage 4 post-2001 will provide 4.2 tonnes GEO. A new stage 2 has also been mooted, powered by a single RD-0120 LOX/LH_2 engine. Addition of NPO Lavotchkin's Fregat as a fifth stage under the existing fairing has been studied to increase Proton K GEO capacity to 3.3 tonnes. Fregat has flown three as the injection module for 1988's Phobos missions and on Mars-96.

The possibility of flying non-Soviet payloads first arose with an offer to ESA for launching the Marecs maritime satellite, followed by a more vigourous campaign in 1983 to launch the Inmarsat 2 generation. Commercial promotion developed during 1986-87 but, despite the lower rates and assurances that customer representatives can accompany payloads at all times to avoid technology transfer problems, no customers were

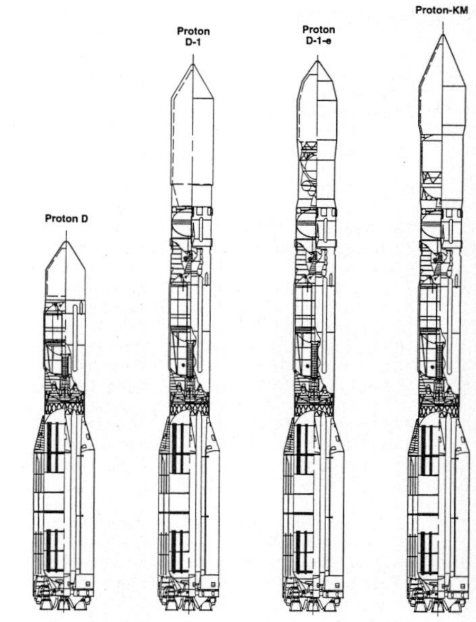

The Proton family (LKE International) **1994**

found. The Soviets claimed in 1990 that firm contracts had been signed in late 1987 for two US telecom satellite launches 1989-90, but the deals were blocked by the refusal of US export licences. It was reported in mid-1990 that discussions were under way with the Brazilian government for commercial operations from the Alcantara site. Zenit replaced Proton in plans for Australia's Cape York spaceport. Australia's Space Transportation Systems Ltd has studied a commercial spaceport based on Proton from the Papua New Guinea island of Manus or Emirau. Requiring US$900 million, the near-equatorial location would improve GEO performance to 4.6 tonnes without KM. Lockheed Martin/ILS completed a study in July 1996 of an alternative launch site for Proton. Candidates included Florida, Brazil and Australia.

The first marketing success came in November 1992 when Inmarsat selected Proton and KB Salyut for a US$36 million contract to launch an Inmarsat 3; the contract was signed 27 April 1993. (Inmarsat has noted its total cost will be closer to US$50 million.) The success built on the Bush administration's June 1992 approval for one Proton launch of a US-built satellite. Motorola, in February 1993, announced that it had signed a contract (subject to US government approval) with Khrunichev Enterprises for three Proton launches of seven satellites each. In return, Motorola expected Khrunichev to invest US$40 million in Iridium Hughes Space & Communications Co, and in February 1995, signed a general launch services agreement for an unspecified number. Proton is baselined for the 2001 launch of ESA's Integral satellite. Launch would be provided in return for observing time with the γ-ray observatory.

The Block DM family of upper stages have been used for launching GLONASS payloads. The Soviets used the original Block DM (11S86) for launches to geosynchronous orbit during 1974-1990 until its last flight for Gorizont 20. The Block DM-2 (11S861) put all GLONASS satellites into position, delivered Kosmos 1700 and the majority of GEO missions starting with Kosmos 1961 (in 1988), and put the Elint launches of Kosmos 1603/1656 into position. The Block DM-2M (11S861-01) has seen limited use in the launches of geosynchronous communications satellites. The commercial launch of Astra-1F introduced a new version to the Block DM family: the Block DM-3.

The three-stage Proton-K places the fourth stage and payload assembly into a low parking orbit and then the fourth stage ignites for the first time in LEO to enter a transfer orbit and then a second time to circularise the orbit with an orbital inclination of around 47 to 48°. At the first pass through apogee the SOZ motors would fire to settle the remaining propellant in the main tanks and they would separate as the main engine ignited to circularise the orbit at geosynchronous altitude and reduce the inclination to 1.5° or less. The Block DM-3 performed the first burn to enter a geosynchronous transfer orbit but retaining the original 51.6° inclination: at the apogee pass the second burn changed the orbit to 11,970 to 35,940 km, but inclined at 6.95°. In addition, no SOZ motors appear on radar screens after the satellite reaches its final position suggesting that they might have remained attached to the Block DM-3.

All of the Block D and DM variants discard two ullage motors as they start their final main engine burns. The ullage motors (designated SOZ by the Russians) fire briefly just before the final burn of the fourth-stage engine to settle the propellants in the rocket stages tanks.

There have been two propellant versions used within

Proton launch history (all launches made from Tyuratam)

	Date	Stages	Payload
1	16 July 1965	2	Proton 1
2	2 November 1965	2	Proton 2
3*	24 March 1966	2	Proton
4	6 July 1966	2	Proton 3
5	10 March 1967	4	Kosmos 146/L1-P
6	8 April 1967	4	Kosmos 154/L1-P
7*	28 September 1967	4	Zond lunar
8*	22 November 1967	4	Zond lunar
9	2 March 1968	4	Zond 4 lunar
10*	23 April 1968	4	Zond lunar
11	14 September 1968	4	Zond 5 lunar
12	10 November 1968	4	Zond 6 lunar
13	16 November 1968	3	Proton 4
14*	20 January 1969	4	Zond lunar
15*	19 February 1969	4	Luna-1969A/Lunokhod
16*	27 March 1969	4	Mars-1969A
17*	2 April 1969	4	Mars-1969B
18*	14 June 1969	4	Luna-1969B
19	13 July 1969	4	Luna 15
20	7 August 1969	4	Zond 7 lunar
21*	23 September 1969	4	C300 (lunar?)
22*	22 October 1969	4	C305 (lunar?)
23*	16 November 1969	4	Kosmos/L1-E
24*	6 February 1970	4	Luna 1970A
25	12 September 1970	4	Luna 16
26	20 October 1970	4	Zond 8
27	10 November 1970	4	Luna 17
28	2 December 1970	4	Kosmos 382/L1-E
29	19 April 1971	3	Salyut 1
30*	10 May 1971	4	C419 (Mars)
31	19 May 1971	4	Mars 2
32	28 May 1971	4	Mars 3
33	2 September 1971	4	Luna 18
34	28 September 1971	4	Luna 19
35	14 February 1972	4	Luna 20
36*	29 July 1972	3	Salyut (DOS 2)
37	8 January 1973	4	Luna 21
38	3 April 1973	3	Salyut 2
39	11 May 1973	3	C557/Salyut
40	21 July 1973	4	Mars 4
41	25 July 1973	4	Mars 5
42	5 August 1973	4	Mars 6
43	9 August 1973	4	Mars 7
44	26 March 1974	4	Kosmos 637
45	29 May 1974	4	Luna 22
46	24 June 1974	3	Salyut 3
47	29 July 1974	4	Molniya-1S
48	28 October 1974	4	Luna 23
49	26 December 1974	3	Salyut 4
50	6 June 1975	4	Venera 9
51	14 June 1975	4	Venera 10
52	8 October 1975	4	Kosmos 775
53*	16 October 1975	4	Luna-1975A
54	22 December 1975	4	Raduga 1
55	22 June 1976	3	Salyut 5
56	9 August 1976	4	Luna 24
57	11 September 1976	4	Raduga 2
58	26 October 1976	4	Ekran 1
59	15 December 1976	3	Kosmos 881-882
60	17 July 1977	3	Kosmos 929
61	23 July 1977	4	Raduga 3
62*	4 August 1977	3	Dual Kosmos
63	20 September 1977	4	Ekran 2
64	29 September 1977	3	Salyut 6
65	30 March 1978	3	Kosmos 997-998
66*	27 May 1978	4	Ekran
67	18 July 1978	4	Raduga 4
68*	17 August 1978	4	Ekran
69	9 September 1978	4	Venera 11
70	14 September 1978	4	Venera 12
71*	17 October 1978	4	Ekran
72*	19 December 1978	4	Gorizont 1
73*	21 February 1979	4	Ekran 3
74	25 April 1979	4	Raduga 5
75	22 May 1979	3	Kosmos 1100-01
76	5 July 1979	4	Gorizont 2
77	3 October 1979	4	Ekran 4
78	28 December 1979	4	Gorizont 3
79	2 February 1980	4	Raduga 6
80	14 June 1980	4	Gorizont 4
81	15 July 1980	4	Ekran 5
82	5 October 1980	4	Raduga 7
83	26 December 1980	4	Ekran 6
84	18 March 1981	4	Raduga 8
85	25 April 1981	3	Kosmos 1267
86	26 June 1981	4	Ekran 7
87	30 July 1981	4	Raduga 9
88	9 October 1981	4	Raduga 10
89	30 October 1981	4	Venera 13
90	4 November 1981	4	Venera 14
91	5 February 1982	4	Ekran 8
92	15 March 1982	4	Gorizont 5
93	19 April 1982	3	Salyut 7
94	17 May 1982	4	Kosmos 1366
95*	23 July 1982	4	Ekran
96	16 September 1982	4	Ekran 9
97	12 October 1982	4	Kosmos 1413-15
98	20 October 1982	4	Gorizont 6
99	26 November 1982	4	Raduga 11
100*	24 December 1982	4	Raduga
101	2 March 1983	3	Kosmos 1443
102	12 March 1983	4	Ekran 10
103	23 March 1983	4	Astron 1
104	8 April 1983	4	Raduga 12
105	2 June 1983	4	Venera 15
106	6 June 1983	4	Venera 16
107	1 July 1983	4	Gorizont 7
108	10 August 1983	4	Kosmos 1490-92
109	25 August 1983	4	Raduga 13
110	29 September 1983	4	Ekran 11
111	30 November 1983	4	Gorizont 8
112	29 December 1983	4	Kosmos 1519-21
113	15 February 1984	4	Raduga 14
114	2 March 1984	4	Kosmos 1540
115	16 March 1984	4	Ekran 12
116	29 March 1984	4	Kosmos 1546
117	22 April 1984	4	Gorizont 9
118	19 May 1984	4	Kosmos 1554-56
119	22 June 1984	4	Raduga 15
120	1 August 1984	4	Gorizont 10
121	24 August 1984	4	Ekran 13
122	4 September 1984	4	Kosmos 1593-95
123	28 September 1984	4	C1603/Tselina 2
124	15 December 1984	4	Vega 1
125	21 December 1984	4	Vega 2
126	18 January 1985	4	Gorizont 11
127	21 February 1985	4	Kosmos 1629
128	22 March 1985	4	Ekran 14
129	17 May 1985	4	Kosmos 1650-52
130	30 May 1985	4	C1656/Tselina 2
131	8 August 1985	4	Raduga 16
132	27 September 1985	3	Kosmos 1686
133	25 October 1985	4	Kosmos 1700
134	15 November 1985	4	Raduga 17
135	24 December 1985	4	Kosmos 1710-12
136	17 January 1986	4	Raduga 18
137	19 February 1986	3	Mir
138	4 April 1986	4	Kosmos 1738
139	24 May 1986	4	Ekran 15
140	10 June 1986	4	Gorizont 12
141	16 September 1986	4	Kosmos 1778-80
142	25 October 1986	4	Raduga 19
143	18 November 1986	4	Gorizont 13
144*	29 December 1986	3	Almaz
145*	30 January 1987	4	Kosmos 1817
146	19 March 1987	4	Raduga 20
147	31 March 1987	3	Kvant 1
148*	24 April 1987	4	Kosmos 1838-40
149	11 May 1987	4	Gorizont 14
150	25 July 1987	3	Kosmos 1870
151	3 September 1987	4	Ekran 16
152	16 September 1987	4	Kosmos 1883-85
153	1 October 1987	4	Kosmos 1888
154	28 October 1987	4	Kosmos 1894
155	26 November 1987	4	Kosmos 1897
156	10 December 1987	4	Raduga 21
157	27 December 1987	4	Ekran 17
158*	18 January 1988	4	Gorizont
159*	17 February 1988	4	Kosmos 1917-19
160	31 March 1988	4	Gorizont 15
161	26 April 1988	4	Kosmos 1914
162	6 May 1988	4	Ekran 18
163	21 May 1988	4	Kosmos 1946-48
164	7 July 1988	4	Phobos 1
165	12 July 1988	4	Phobos 2
166	1 August 1988	4	Kosmos 1961
167	18 August 1988	4	Gorizont 16
168	16 September 1988	4	Kosmos 1970-72
169	20 October 1988	4	Raduga 22
170	10 December 1988	4	Ekran 19
171	10 January 1989	4	Kosmos 1987-89
172	26 January 1989	4	Gorizont 17
173	14 April 1989	4	Raduga 23
174	31 May 1989	4	Kosmos 2022-24
175	21 June 1989	4	Raduga-1 1 (24)
176	5 July 1989	4	Gorizont 18
177	28 September 1989	4	Gorizont 19
178	26 November 1989	3	Kvant 2
179	1 December 1989	4	Granat
180	15 December 1989	4	Raduga 25
181	27 December 1989	4	Kosmos 2054
182	15 February 1990	4	Raduga 26
183	19 May 1990	4	Kosmos 2079-81
184	31 May 1990	3	Kristall
185	20 June 1990	4	Gorizont 20
186	18 July 1990	4	Kosmos 2085
187*	9 August 1990	4	Ekran
188	3 November 1990	4	Gorizont 21
189	23 November 1990	4	Gorizont 22
190	8 December 1990	4	Kosmos 2109-11
191	20 December 1990	4	Raduga 27
192	27 December 1990	4	Raduga-1 2 (28)
193	14 February 1991	4	Kosmos 2133
194	28 February 1991	4	Raduga 29
195	31 March 1991	3	Almaz 1
196	4 April 1991	4	Kosmos 2139-41
197	1 July 1991	4	Gorizont 23
198	13 September 1991	4	Kosmos 2155
199	23 October 1991	4	Gorizont 24
200	22 November 1991	4	Kosmos 2172
201	19 December 1991	4	Raduga 30
202	29 January 1992	4	Kosmos 2177-79
203	2 April 1992	4	Gorizont 25
204	14 July 1992	4	Gorizont 26
205	30 July 1992	4	Kosmos 2204-06
206	10 September 1992	4	Kosmos 2209
207	30 October 1992	4	Ekran 20
208	27 November 1992	4	Gorizont 27
209	17 December 1992	4	Kosmos 2224
210	17 February 1993	4	Kosmos 2234-36
211	25 March 1993	4	Raduga 31
212*	27 May 1993	4	Gorizont
213	30 September 1993	4	Raduga 32
214	28 October 1993	4	Gorizont 28
215	18 November 1993	4	Gorizont 29
216	20 January 1994	4	Gals 1
217	5 February 1994	4	Raduga-1 3 (33)
218	18 February 1994	4	Raduga 34
219	11 April 1994	4	Kosmos 2275-77
220	20 May 1994	4	Gorizont 30
221	6 July 1994	4	Kosmos 2282
222	11 August 1994	4	Kosmos 2287-89
223	21 September 1994	4	Kosmos 2291
224	13 October 1994	4	Ekspress 1
225	31 October 1994	4	Elektro 1
226	20 November 1994	4	Kosmos 2294-96
227	16 December 1994	4	Luch 1
228	28 December 1994	4	Raduga 35
229	7 March 1995	4	Kosmos 2307-09
230	20 May 1995	3	Spektr
231	24 July 1995	4	Kosmos 2316-18
232	30 August 1995	4	Kosmos 2319
233	11 October 1995	4	Luch-1 1
234	17 November 1995	4	Gals 2
235	14 December 1995	4	Kosmos 2323-25
236	25 January 1996	4	Gorizont 31
237*	19 February 1996	4	Raduga 36
238	8 April 1996	4	Astra 1F
239	23 April 1996	3	Priroda
240	25 May 1996	4	Gorizont 32
241	6 September 1996	4	Inmarsat-3 2
242	26 September 1996	4	Ekspress 2
243	16 November 1996	4	Mars 8 (Mars-96)
244	24 May 1997	4	Telstar 5
245	6 June 1997	4	Kosmos 2344
246	18 June 1997	4	Iridium (× 7)
247	14 August 1997	4	Kosmos 2345
248	28 August 1997	4	PAS-5
249	14 September 1997	4	Iridium (× 7)
250	12 November 1997	4	Kupon 1
251	2 December 1997	4	Astra 1G
252	24 December 1997	4	AsiaSat 3
253	7 April 1998	4	Iridium (× 7)
254	29 April 1998	4	Kosmos 2350
255	7 May 1998	4	EchoStar 4
256	30 August 1998	4	Astra 2A
257	4 November 1998	4	PAS-8
258	20 November 1998	3	ISS (Zarya)
259	30 December 1998	4	Kosmos 2362, 2363, 2364
260	15 February 1999	4	Telstar 6
261	28 February 1999	4	Raduga 1-4
262	21 March 1999	4	AsiaSat 3S

* indicates launcher failure. February 1996 stage 4 failed to reignite for GSO injection. The review board concluded that the triethyl aluminium ignition hypergol had not reached the gas generator or main chamber because of reduced pressure from a leaking joint in the pipe from the TEA bottle. The fitting nut had a broken lockwire, probably from poor installation, allowing it to back off during burn 1. Corrective actions include adding a second lockwire. May 1993 failure due to stage 2/3 engine burn-throughs caused by contaminated UDMH. 1990 stage 2 engine failure believed to be due to a worker leaving a rag in the vehicle. Stage 3 failure January 1988. Proton breakdown launch total for each type, followed by number of failures in ( ): two-stages 4(1), four-stages 213(28), three-stages 27(3): these give success rates of 75%, 87% and 89% respectively. Four stage failures include Block D missions where the failure was with the payload (for example Mars 8). Two-stages 4(1)/75%, four-stages 209(27)/87%, three-stages 27(3)/88.9%.

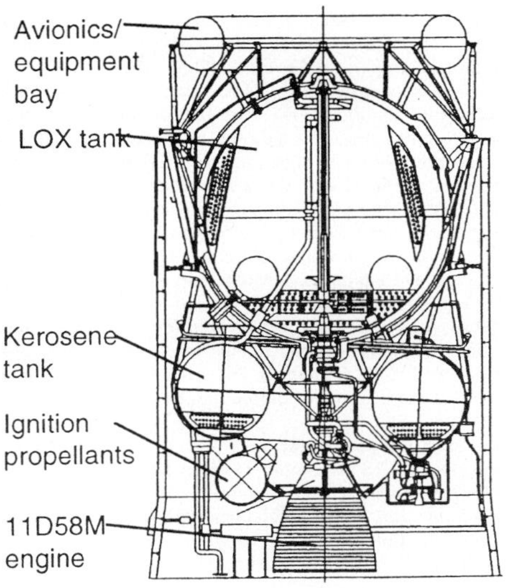

Block DM carries the vehicle's avionics in its forward torus (Boeing) ***1996***

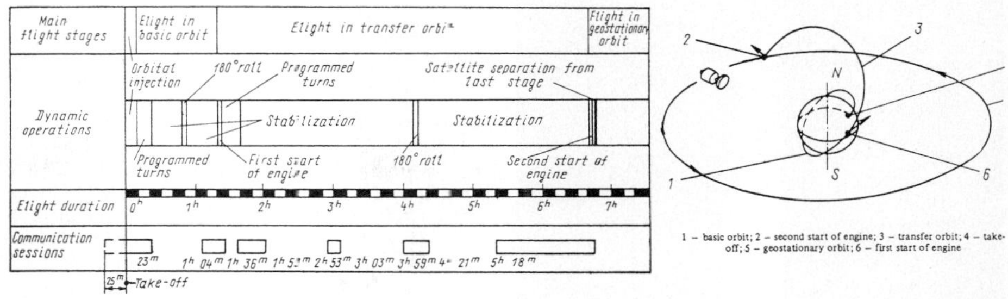

Standard Proton GEO flight profile. 1988 introduced a different technique in which the vehicle remained in its parking orbit for a further 6 hours ***1989***

the Block DM (and presumably Block D) flight programme. Originally, the Soviets used a standard LOX/kerosene combination, but starting in the 1980s – possibly with the introduction of the Block DM-2 stage – the higher-performance LOX/SYNTIN combination was introduced. It seems likely that the later Block D variants also used the new propellant combination. In September 1996, the Russians halted any further production of SYNTIN. While the Russians have said that they have sufficient SYNTIN to meet their commercial needs with the Proton-K, it remains to be seen for how long the domestic programme will continue before the Block DM family have to revert to using LOX/kerosene. This problem will be alleviated when the Proton-KM starts to fly with its LOX/hydrogen fourth stage and when the Briz-M replaces the existing upper stage.

The significantly uprated Proton KM (M = modernised) will debut about 2000, replacing Energia's Block D with Khrunichev's Breeze-M for a 3.2 tonne GEO capacity. It had earlier been expected that a cryogenic stage 4 would appear by 1995 for 4.5 tonne GEO, combined with improved stage 3 performance. That is now projected for post-2000, providing 4.2 tonne GEO, 6.4/7.9 tonnes 7/28° GTO, 7.9 tonne Earth escape. The cryogenic stage has been under development since 1988/89 using the KVD-1 engine designed for the N1-L3M manned lunar lander project. India's GSLV will also use this engine and the same stage is planned for Angara.

KB Salyut's Breeze-M, and its 50-hour active life, coupled with multiple restarts, will allow Proton to dispense satellites into different orbits. Its compactness leaves 98 m³ payload volume under the largest current fairing and 150 m³ with the new hammerhead design. Breeze-M uniquely adopts a core surrounded by a jettisonable neo-toroidal module of NTO/UDMH tanks with a common bulkhead. The core is based on Rockot's flight-proven Breeze stage, with the fixed main chamber and verniers to provide ullage and three-axis coast during coast and burn. Breeze's core houses Proton's GN&C equipment.

Typical GEO/GTO missions will jettison the external tank, but it can remain attached for missions to 15,000 to 25,000 km orbits. Breeze-M can fly with the core only for LEO missions.

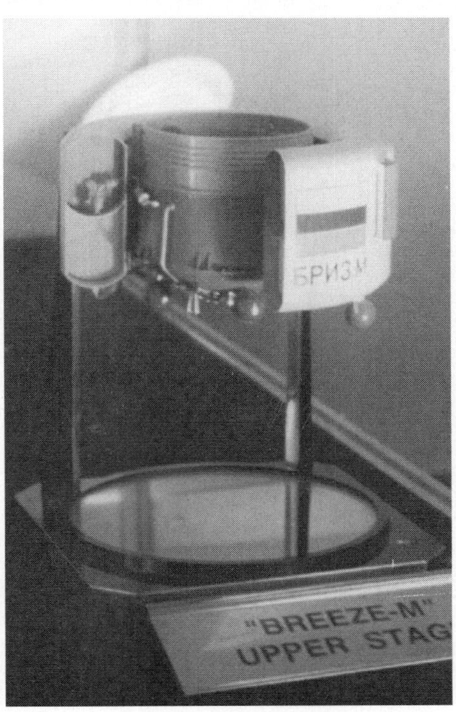

Proton's Breeze-M stage 4 (Khrunichev/KB Salyut) ***1996***

Specifications

Proton launcher
First launch: 16 July 1965 (two-stage version)
Number launched: 262 to end-April-1999
Launch sites: Baikonur

Proton K (four stages)
First launch: 10 March 1967
Number launched: 229 to end-April 1999
Launch sites: Baikonur
Principal uses: geostationary, escape and Glonass navigation satellite missions
Vehicle success rate: 86.8% to end-1996. *Past 25 launches:* 88.0% to end-1996. *Past 50 launches:* 92.0% to end-1996
Availability: two-year lead time
Cost: ILS/LKE declined to provide current commercial costs. US$56 million quoted in 1992 for dedicated (but Inmarsat's 1992 contract is for about US$36 million). About US$40 million quoted October 1994, with Rb14-15 billion for internal flights. Quoted 1989: US$35 million for dedicated GEO (US$28 million GTO), US$12 million dual launch, US$8 million triple.
Insurance: arranged through Ingosstrakh, 12% premium for reflight; commercial broker quotes available through LKE for specific missions.
Performance: GTO (48°): 4,800 kg
LEO: not used for LEO missions, but vehicle 20 t in 200 km parking orbit before GTO injection
GEO: 2,600 kg using Block DM-2M. 2,400 kg using Block DM and Syntin. 3,200 kg using Breeze-M.
Sun-synchronous: 2,800 kg (not yet flown)
Lunar delivery: 5,700 kg
Venus delivery: 5,300 kg
Mars delivery: 4,600 kg
Number of stages: 4
Overall length: SL-12 57.07 m, SL-13 57.76 m (42.340 m without stage 4/payload)
Principal diameter: 7.400 m stage 1, 4.100 m upper stages
Launch mass: 690 t
Guidance: inertial, carried in stage 4, provides three-axis control by means unknown until payload separation commanded. GEO injection accuracy ±1°, ±20 min orbital period. Analogue control system carried by Block DM or by payload for Block D. GN&C systems provided by NPO AP. Stage 1 conversion to closed loop guidance will add 500 kg LEO by improving efficiency and reducing propellant reserve.

Proton K stage 1
Bureau: Khrunichev/KB Salyut
Engines: single storable liquid non-restartable NPO Energomash RD-253 carried in six cylindrical UDMH side modules around central oxidiser tank.
Overall length: 21.1 m
Principal diameter: 7.4 m overall, 1.6 m cylinders around 4.1 m core
Oxidiser: nitrogen tetroxide in core tank
Fuel: UDMH in side tanks
Propellant mass: 420 t; 31.0 t dry

Comparison of Proton-K and Proton-M four-stage vehicles

	Proton-K	Proton-M
Payload fairing:		
Usable diameter	3.9 m	3.9 m
Usable length	7.3 m	10.0 m
Fairing mass	2,000 kg	2,300 kg
Payload capabilities:		
LEO (three-stage version)	20,100 kg	22,000 kg
GSO (four-stage version)	2,100 kg	3,000 kg
Length	60 m	58 m
Launch mass	688,000 kg	691,790 kg
Stage 1:		
Length	21.18 m	21.18 m
Diameter *	7.4 m	7.4 m
Dry mass	31,000 kg	30,600 kg
Propellant mass	419,410 kg	419,410 kg
Total thrust (vacuum)	10,000 kN	10,497 kN
Specific impulse	316 s	316 s
Stage 2:		
Length	17.05 m	17.05 m
Diameter	4.10 m	4.10 m
Dry mass	11,700 kg	11,400 kg
Propellant mass	156,113 kg	156,113 kg
Total thrust (vacuum)	2,325 kN	2,325 kN
Specific impulse	326.5 s	326.5 s
Stage 3:		
Length	4.11 m	4.11 m
Diameter	4.10 m	4.10 m
Dry mass	4,185 kg	3,700 kg
Propellant mass	46,562 kg	46,562 kg
Total thrust (vacuum)	613 kN	613 kN
Specific impulse	326.5 s	326.5 s
Stage 4:	Block DM-3	Briz-M
Length	6.28 m	2.61 m
Diameter	3.7 m	4.0 m
Dry mass **	3,130 kg	2,370 kg
Propellant mass	15,050 kg	19,800 kg
Total thrust (vacuum)	88 kN	19.6 kN
Specific impulse	361.0 s	325.5 s

These figures have been extracted from an ILS brochure issued in 1997. * Diameter measured across a pair of external propellant tanks and the central core propellant tank. ** For the Block DM-3 this figure includes the fourth stage shroud discarded in LEO and the two ullage motors. The vehicles use nitrogen tetroxide throughout other than the Block DM-3 which uses LOX/SYNTIN. Some of the data quoted by ILS differs from those available from other Russian literature: for example it is known that some domestic Proton-K LEO payloads have been in excess of 20,600 kg mass and some GSO payloads are around 2,500 kg.

Proton with Block DM3 launches the Astra 1F satellite on 9 April 1996 (Theo Pirard) ***1999***/0054381

Stage thrust: 8.8 MN at sea level, 9.8 MN at altitude
Burn time: 130 s

Proton K stage 2
Description: The original stage 2 was about 8.5-9 m long, 6 t dry, 80 t propellant. It was then stretched for the later versions. Chelomei's one-man LK-1 circumlunar mission would have apparently employed the original short version.
Bureau: Khrunichev/KB Salyut
Engines: gimballed liquid chambers, developed by KB Khimautomatiki. Three are RD-0210 and one is RD-0211 version (with gas generator).
Overall length: 14.56 m
Principal diameter: 4.15 m
Oxidiser: nitrogen tetroxide
Fuel: UDMH
Stage thrust: 2,376 kN vacuum
Burn time: up to 300 s
Propellant mass: 156 t, 11.7 t dry

Proton K stage 3
Comment: Stage 3 injects the payload and stage 4 into a 200 km circular, 51.6° parking orbit about 10 min after launch.
Bureau: Khrunichev/KB Salyut
Engine: single fixed RD-0212 version of RD-0210 with four gimballed verniers for steering
Overall length: 6.52 m
Principal diameter: 4.15 m
Oxidiser: nitrogen tetroxide
Fuel: UDMH
Stage thrust: 593.6 kN main chamber plus 4 × 7.875 kN verniers
Burn time: about 250 s
Propellant mass: 46.6 t; 4,185 kg dry

Proton K stage 4 Block DM
Designation: Block D (11S824, first flown 1967), Block 'D-1' [?] (11S824M, 1978), Block D-2 (11S824F, 1988), Block DM (11S86, 1974), Block DM-2 (11S861, 1982), Block DM-2M (11S861-01, 1994), Block DM-3 (?, 1996): possibly a Block DM-4 in 1997.
Bureau: RKK Energia; control system by NPO AP
Engine: 58/58M restartable single chamber developed by Korolev bureau for D/DM.
Overall length: 5.366/6.218 m D/DM-2
Principal diameter: 3.70 m
Stage mass: 17.3 t D/18.46 t DM-2; dry mass 2.5 t D/3.37 t DM-2 (dry masses include the 800 kg casing + SOZ).
Oxidiser: 10,610 kg liquid oxygen in forward sphere
Fuel: 4,330 kg kerosene or syntin in aft torus
Stage thrust (vacuum): 85 kN kerosene; 83.5 kN Syntin.
Burn time: about 600 s total over two burns
Comment: Stage 4/Block D/DM-2 ignites about 80 min after stage 3 separation, making a maximum 450 s burn for injection into a 36,000 km, 48° GTO. About 400 min after launch, it ignites for a maximum of 230 s at first apogee to slot the payload directly into GEO – no satellite apogee boost motor is required. (DM injected Astra 1F into 12,100 × 36,000 km, 7.0° GTO). D/DM-2 carries two 110-120 kg (56 kg dry) 60 × 100 cm NTO/UDMH 'SOZ' thruster packages for three-axis control during coast and to settle the stage's propellants after coasting. Each houses five thrusters: 2 × 22 N pitch/roll, 1 × 44 N yaw and 2 × 11 N ullage. The ullage motors ignite about 300 s before main stage burn starts and are jettisoned 5 s after. About 12 of these packages have broken up years after launch. Vehicle three-axis control is lost at the instant of shutdown and the stage control system orders payload spring separation (at 1.5±0.3 m/s) within 15 s. By this time stage tumbling is typically 2°/s. Combined with separation disturbances, the payload's attitude control system has to cancel 2.5°/s rotation about the longitudinal axis and 3°/s about lateral axes. All Soviet satellites were inserted into GEO close to 90° east for 14 years, sometimes requiring up to several weeks to drift to their operating positions, but in April 1988 Kosmos 1940 was held in its parking orbit for an additional 6 h and then slotted directly into 15° west. This pattern followed with Kosmos 2155 in 1991, Kosmos 2209 (20° west) in 1992 and Kosmos 2282 (25° west) in 1994.

Fairing: The spacecraft is protected until 351 s after launch at an altitude of 150 km. Class 100,000 air conditioning on the pad through a 45 cm² hatch provides 10-25°C (30°C possible for up to three days), 30-60% humidity inside the fairing until it is disconnected 2 h before launch. There is also a 4-5 h interruption while the vehicle is erected on the pad. Power and communications links are terminated 90 min before launch; telemetry contact is not available until stage 4 of the flight profile. Surface wind limit 15-18 m/s, depending on fairing type.

Acceleration load: about 4 *g* longitudinal ±1 *g* lateral at 5.7 Hz during ascent, 1.2 ±0.8 *g* longitudinal at 10-15 Hz/±3.0 *g* lateral at 5-7 Hz at lift-off; 1.25 *g* ±4.75 *g* longitudinal at 10-15 Hz/±1.5 *g* lateral at 5-7 Hz during separation.
Acoustic load: 144 dB rms
Temperature: 130°C sidewall max after 200 s

Payload processing
The customer is responsible for ensuring that a payload is protected against −40 to 40°C conditions for the air transfer to Baikonur and the subsequent 100 km truck journey. A commercial payload undergoes initial processing in the 270 m², class 100,000 20±3°C Satellite Preparation Room 119, with the adjacent 100 m² class 10,000 room 120 for electrical tests. It is then transferred into a separate 100 m² 20±5°C facility for hydrazine and nitrogen loading. The 500 m² General Assembly Room number 100 at 15-30°C, 40 to 60% humidity is then used for the horizontal attachment to the spacecraft adaptor, to stage 4 and fairing enclosure. Subsystems testing and storage rooms are also available. Stage 4 is then filled with propellants in a separate area. The old Buran building at site 254 is expected to be converted for processing. The assembly then transfers to the Proton building 70 km distant for vehicle integration; the building can handle up to six launchers simultaneously. The fully assembled Proton then undertakes a 7 h journey to the pad, where five days are normally allocated to preparations before launch. All payload checkout procedures must be completed 8 h before launch; contact is severed at T-90 min. Two complexes several km apart and each with two pads are available, although not all pads are in service simultaneously. Most launches are currently from pad 81L, but will switch to the refurbished 81R in 1996. 200R has been reactivated for commercial and Space Station launches (it has also been transferred from VKS to RKA control); 200L is certified for only a handful more launches before it must be refurbished.

Proton K (three-stages)
First launch: 16 November 1968 (Proton 4 satellite)
Number launched: 28 to end-April 1999
Principal uses: large LEO spacecraft, principally space station modules.
Cost: LKE declined to provide current commercial costs: US$20 million quoted 1987
Vehicle success rate: 89% to end-1996
Performance: 20-21 t LEO (51.6°), other missions not flown.
Number of stages: 3
Overall length: 44.3 m + payload

Proton KM Breeze-M stage 4
Designation: Breeze-M
Bureau: Khrunichev/KB Salyut
Engine: single fixed restartable pump-fed engine from KB Khimmach, plus three-axis control provided by verniers

Proton KM cryogenic stage 4
Designation: KVRB Kislorodno-Vodourodny Razgonny Block (oxygen/hydrogen upper stage).
Bureau: Khrunichev (stage), KB Khimmach (engine).
Engine: single chamber cryogenic KVD-1M (with two verniers) capable of five burns and 7.5 h coast.
Overall length: 8.6 m
Principal diameter: 4.0 m
Oxidiser: liquid oxygen
Fuel: liquid hydrogen
Dry mass: 3.4 t
Propellant mass: up to 19 t
Stage thrust: 73.58 kN vacuum (69.66 kN main chamber plus 2 × 1.96 kN LOX/GH_2 two-axis verniers)
Burn time: 450 s total. First burn starts at 650 s.

Forward section of a completed three-stage Proton at the Khrunichev factory (Theo Pirard) ***1999***/0054370

Soviet era heavy lift launch vehicles

Background

In the early 1960s, there were three contenders for the launch vehicle to place men on the moon: Korolyov's OKB-1 proposed the N-1, Chelomei's OKB-52 proposed the UR-700 and Yangel's OKB-586 proposed the R-56.

The Yangel R-56 proposal formed part of a plan to develop three related missiles and space launch vehicles. The first, the R-36, was developed both as a missile and as the Tsyklon launch vehicle family. The second, the R-46 would have been capable of launching a 50 MT warhead, but it was not developed as either a weapon or a launch vehicle, and the third, the R-56 could have been the basis of a manned lunar programme Considering just sub-orbital (missile?) trajectories, the R-36 could launch a payload of 5 tonnes, the R-46 15 tonnes and the R-56 50 tonnes.

R-56's launch mass would have been about 1,400 tonnes and two different payload capabilities have been quoted, namely, 30 tonnes to an unspecified orbit or 40 tonnes to a 200 km 'polar' orbit. These figures indicate that any manned lunar landing programme based upon this vehicle would have required multiple launches and rendezvous and docking of several lunar spacecraft components in Earth orbit. It is therefore little surprise that the R-56 was the first of the lunar launch vehicles to be cancelled, this taking place on 19 June 1964.

Originally, the N-1, the second contender would have been just one of a series of launch vehicles with common stages. The N-11 was planned using Blocks B, V and G: this would have had a launch mass of 700 tonnes and could have placed 20 tonnes into a 300 km orbit. The N-111 would have used the N-1 Blocks V and G plus the second stage of the R-9A missile: this would have had a launch mass of 200 tonnes and could have put 5 tonnes to a 300 km orbit. However, these plans were not developed and only the N-1 reached flight status.

The N-1 vehicle was developed and launched on four occasions during 1969-1972. None of the launches placed a payload into Earth orbit. The configuration of the planned N-1/L-3 stack was for the first three stages of the N-1 (Blocks A, V and G) to be used to place the fourth stage (Block G) and lunar complex into a low Earth orbit. The Block G would perform trans-lunar injection and be discarded. The Block D (also flown as the fourth stage of the Proton-K launch vehicle) would perform lunar orbit injection and then the main descent burn to the lunar surface with the lunar lander (LK), while the lunar Soyuz (LOK) command craft would remain in lunar orbit. The four stages of the N-1 and the Block D would all have used LOX/kerosene propellants.

Korolyov looked towards future versions of the N-1 launch vehicle, using liquid hydrogen as a fuel rather than kerosene. New LOX/LH Blocks S and R would replace the developed Blocks G and D. In support of these plans, OKB-2 (Isayev, now KM Khimmash) developed the 11D56 engine with a thrust of 7.5 tonnes for the Block R and OKB-165 (Lyulka, now NPO Saturn) developed the 11D54 (fixed chamber) and 11D57 (gimballing chamber), both with a thrust of 40 tonnes. It was planned that six to eight 11D54 engines would be used on a modified Block V of the N-1, while a single 11D57 would be used on the Block S.

The Block R which would perform lunar orbit injection and descent to the Moon would have a dry mass of 4.3 tonnes and would carry up to 18.7 tonne of propellant: the diameter would have been 4.1 m and the length 8.7 m.

The first test-firing of the 11D56 closed-cycle engine was carried out in June 1967: at around the same time Kuznetsov's OKB-276 was conducting development of the NK-15V LOX/hydrogen engine with a thrust of 200 tonnes, this being planned for the second stage of the improved N-1 launch vehicle. This new design would make the N-1 appear more like the United States Saturn-5. The US vehicle used LOX/kerosene (RP-1) in the first stage and LOX/hydrogen in the second and third stages. The new N-1, on the other hand, would have had LOX/kerosene in the first stage and LOX/hydrogen in its second, third and fourth stages (the 'fifth' stage, whether Block D or Block R, is considered to be part of the payload rather than the launch vehicle).

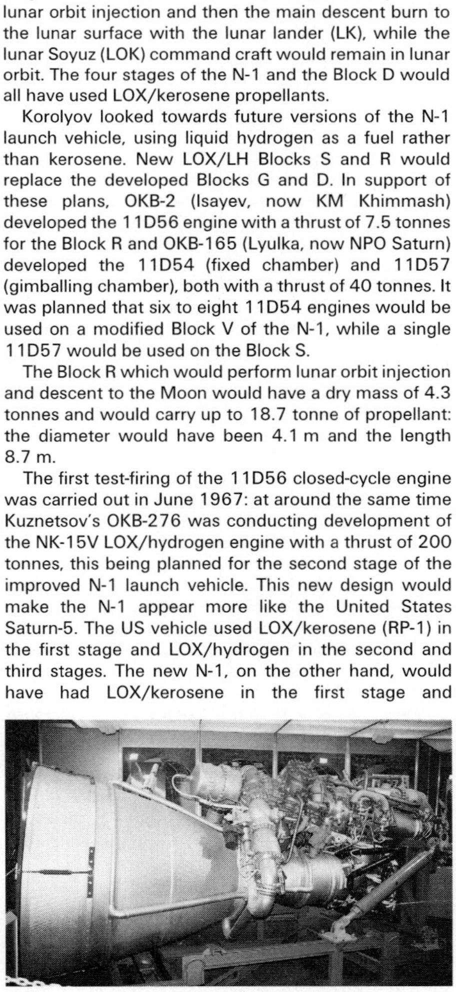

NK-33 engine from the N-1 heavy lift vehicle programme procured by Aerojet (Theo Pirard) ***1999***/0054373

In May 1970, the Politburo decided to develop a single new stage, designated Block Sr, which would combine the functions of the Block S and Block R. The Block Sr would have used 11D56M engines having a diameter of 5.2 m, a length of 16.5 m, a dry mass of 1.5 tonnes and propellant mass 66.4 tonnes.

The Block Sr would have been used on the dual-launch N-1/L-3M manned lunar missions, but the N-1 programme was cancelled in 1974 when Valentin Glushko became head of the combined Korolyov bureau and GDL (NPO Energiya). Despite this, five Block R stages were built and testing of the technology continued through 1977: the live firing of a Block R came on 12 October 1976.

Finally, while the Korolyov bureau was refining the designs of the N-1, OKB-52 was working on its UR-700 launch vehicle – again seen as part of a family of vehicles which would have included the UR-200 missile (cancelled in 1964) and the UR-500 missile. The UR-500 eventually turned into the Proton-K launch vehicle. The overall design of the UR-700 was finally revealed in a paper presented at the 1996 IAF Congress, although there are still many unknown details concerning its staging technique, and so on.

Chelomei's UR-700 would have had a launch mass of 4,823 tonnes and would have placed 151 tonnes into a 200 km orbit: the payload injected to a trans-lunar orbit would have been 50 tonnes. The design of the UR-700 appears to be unique. To reach Earth orbit three pairs of strap-on boosters (first stage) would have surrounded a two-stage central core (stages two and three). Each of the two core stages would have comprised three independent sets of propellant tanks and engines. The strap-on boosters would have each carried a single RD-270 engine, as would each second-stage module: the third-stage modules would each have carried an altitude-ignition version of the RD-253 engine used on the first stage of the Proton. The requirement for railway transportation meant that each strap-on or core unit was limited to a maximum diameter of 4.1 m.

The unusual aspect of the launch vehicle is that each of the strap-on boosters had three propellant tanks, and for each pair the upper propellant tanks were of different sizes. It would appear that each pair of strap-on boosters carried either an extra tank of nitrogen tetroxide or UDMH atop the tanks used by that strap-on, and this extra propellant load would be fed into the adjacent main central core to 'top-up' the fuel being used in flight. With all nine RD-270 engines firing at launch (six on the first stage strap-ons, three on the second-stage core), the total thrust generated would have been approximately 56,450 kN (9 × 6,272 kN or 5,755 tonnes) sea level, 60,420 kN (9 × 6,713 kN, 6,160 tonnes). The second-stage core would have burned for longer than the strap-ons and by the time that the fuel transfer from each pair of strap-ons to the adjacent second-stage core unit was complete, the strap-ons would be ready to separate after burning of their dedicated propellant.

The third-stage engines would have each had a thrust of 1,716 kN (175 tonnes) and would have burned through to Earth orbit injection. A sketch showing the LK-700 manned lunar landing complex attached to its trans-lunar stage was published in the December 1991 issue of the now-defunct Znanye journal. That article suggested that the stage might be based upon the original short second stage of the Proton vehicle (used on the two-stage variant in 1965-1966).

The high-thrust RD-270 engines were not developed in time to save the UR-700 lunar programme, which was finally passed over in favour of Korolyov's N-1/L-3 programme in 1966. However, some of the engines were subsequently built and tested.

This was not the end of the UR-700 proposal. Although he had lost the chance to fly men to the Moon, Chelomei proposed an even more powerful version of the UR-700, the UR-700M, for a manned Mars expedition. The problems with the RD-270 meant that this engine had to be dropped from the project and in its place each first-stage strap-on and second-stage core module would now have a cluster of four RD-253 engines. By 1968, these engines were flying on the Proton launch vehicle's first stage with a reasonable reliability. This would have meant 36 engines firing at launch – more than the 30 engines on the Block A of the much-criticised N-1 vehicle. Also, a fourth stage using the GDL's RD-410 nuclear motor would have been added, presumably to perform the heliocentric orbit injection from an Earth parking orbit. The UR-700M could have placed 240 tonnes into Earth orbit. When Chelomei was making this proposal – presumably in parallel with Vasili Mishin putting forward the Korolyov bureau's plans for a manned Mars mission based upon the uprated N-1 launch vehicle – the politicians thought that such a programme was too expensive and they had no interest in supporting it. As a result all work on the UR-700 class launch vehicle was stopped around early 1970.

When Glushko took over as head of the new NPO Energia he brought with him plans for another family of launch vehicles which would give the Soviet Union both a launcher for a space shuttle orbiter (Buran) and launch vehicles with different orbital capabilities. The vehicles would all be based around a central core having a different number of strap-on boosters. The launch vehicles which Glushko proposed were Groza (thunderstorm) with two strap-ons, which would become Energia-M, and Grom (thunder) with four strap-ons eventually turning into Energia, and Vulkan (volcano).

Of these three, Vulkan was the real monster. It would have eight strap-on boosters surrounding a modified Energia central core and a second core stage mounted on top. It was intended for a manned lunar mission.

Specifications

Payload mass: 170 t (LEO), 28 t (GEO)
Maximum diameter: 16.5 m (over strap-ons)
Length: 88 m
Strap-on engines:
8 × RD-170: 8 × 7,255 kN = 58,040 kN (sl)
8 × 7,905 kN = 63,240 kN (vacuum)
Core engines:
4 × RD-0120: 4 × 1,865 kN = 7,460 kN (vacuum)
Specific impulse: 351 s (s1), 452 s (vacuum)
1 × 11D57M: 1 × 410 kN 410 kN (vacuum)
Specific impulse: 460 s (vacuum)
(The original data are given as tonnes for thrusts: in the conversions the results are rounded to the nearest 5 kN. The payload mass includes the mass of the dry Block V.)

The oft-rumoured six strap-on version of Energia does not appear to have been part of Glushko's original plans, and it is unclear at what point it appeared as a design study. Development of the Vulkan launch vehicle ended soon after the proposal, although into the early 1990s the Russians were still talking about an eight strap-on variant of Energia as a distant concept. The Energia-M and standard four-strap-on Energia were cancelled in 1993 due to a lack of funding, although should funding become available once more it is possible that the vehicle might be revived.

Soviet launch vehicles

Background

Historically, the general Soviet philosophy was to cluster numbers of relatively modest, low chamber pressure open cycle engines, mass-produced with few modifications in their decades-long careers. Most are assigned RD (Raketnay Dvigatel, rocket motor) designations. NPO Energomash is the major centre for rocket engines; KB Khimautomatiki, was identified in 1990 as responsible for Energia's core engines, and emerged from the Kosberg bureau, which concentrated on upper stage propulsion, as does KB Khimmach of the old Isayev bureau. NII Machinostroenye undertakes more basic research but does offer spacecraft thrusters. The Kuznetsov (late NPO Trud, now RD Kuznetsov Co SSC) aircraft engine bureau developed engines for the abandoned N1 lunar launcher for Korolev. Korolev's bureau itself developed Proton's Block D/DM and Molniya's Block L engines. The Lyulka aircraft engine bureau (now part of NPO Saturn) developed the LOX/ LH_2 D-57 to replace the lunar N1 launcher's stage 3 engines.

The history of NPO Energomash can be traced back to 15 May 1929 when the Group for the Development of Rocket Engines was organised at the Gas Dynamics Laboratory (GDL) in Leningrad (now St Petersburg). For more than 60 years the organisation repeatedly changed its name and location, but remained at the forefront of rocket engine design. The founder and the permanent head of the GDL was academician Valentin P Glushko, considered to be the initiator of the Russian rocket engine industry.

In May 1974, the GDL merged with Sergei Korolyov's design bureau, the new organisation being re-named NPO Energia, with Glushko as its head. Following Glushko's death in January 1989, the former GDL left Energia to become NPO Energomash, continuing to concentrate on the development of rocket engines: the bureau was headed by Vitali P Radovsky during 1974-1994. The current general director and chief designer is Boris I Katorgin.

NPO Energomash incorporates the design bureau, plant and test facilities, with the capability to test engines with thrusts of up to 10 MN either as components or complete. An agreement was signed with Pratt and Whitney in October 1992 providing exclusive US marketing rights for Energomash's LOX/kerosene and tri-propellant products and technology.

1996 released the designations of many previously-unidentified engines used on missiles and launch vehicles, and these are included under the appropriate design bureaus. For easy cross-referencing between a launch vehicle and the propulsion systems, please refer to the table which accompanies this entry.

Rocket engines used on selected RFAS missiles and launch vehicles

Vehicle	Designator	Stage 0	Stage 1	Stage 2	Stage 3	Stage 4
1) Selected missiles (development related to early launch vehicles)						
R-9	8K75 (original)		4 × NK-9	1 × NK-19		
	(developed)		1 × RD-111	1 × RD-0106		
R-9M	8K77 (not developed)		4 × NK-9	4 × SI-5400		
R-16	8K64		1 × RD-218	1 × RD-219		
GR-1	8K713		4 × NK-9	1 × NK-19	1 × 8D726	
UR-200	8K81		? × RD-0203	1 × RD-0206		
			? × RD-0204	1 × RD-0207		
2) Existing and retired launch vehicles						
Sputnik	8K71PS	4 × RD-107	1 × RD-108			
Sputnik	8A91	4 × RD-107	1 × RD-108			
Vostok-L	8K72	4 × RD-107	1 × RD-108	1 × RD-0105		
Vostok	8K72K	4 × RD-107	1 × RD-108	1 × RD-0109		
Vostok-2	8A92	4 × RD-107	1 × RD-108	1 × RD-0109		
Meteor	8A92M	4 × RD-107	1 × RD-108	1 × RD-0109		
Voskhod	11A57	4 × RD-107	1 × RD-108	1 × RD-0107		
Soyuz	11A511	4 × RD-107	1 × RD-108	1 × RD-0108		
Soyuz-L	11A511L	4 × RD-107	1 × RD-108	1 × RD-0108		
Soyuz-M	11A511M	4 × RD-107	1 × RD-108	1 × RD-0108		
Soyuz-U	11A511U	4 × RD-107	1 × RD-108	1 × RD-0110		
Soyuz-U2	11A511U2	4 × RD-107	1 × RD-118	1 × RD-0110		
-	11A59	4 × RD-107	1 × RD-108	1 × RD-0110		
Soyuz-2		4 × RD-120K ?	1 × RD-120K ?	1 × RD-0124		
Molniya	8K78	4 × RD-107	1 × RD-108	1 × RD-0107	1 × SI-5400	
Molniya-M	8K78M	4 × RD-107	1 × RD-108	1 × RD-0110	1 × 11D33	
Kosmos	63S1		1 × RD-214	1 × RD-119		
Kosmos-2	63S1M		1 × RD-214	1 × RD-119		
Kosmos-2M	11K63		1 × RD-214	1 × RD-119		
Kosmos-1	65S1		1 × RD-216	1 × 11D49 ?		
Kosmos-3	11K65		1 × RD-216	1 × 11D49 ?		
Kosmos-3M	11K65M		1 × RD-216	1 × 11D49		
Proton	8K82		6 × RD-253	3 × RD-0210		
				1 × RD-0211		
-	11A510	4 × RD-107	1 × RD-108			
R-36-0	8K69		1 × RD-218	1 × RD-219		
Tsyklon-2	11K67		1 × RD-251	1 × RD-252		
Tsyklon-M	11K69		1 × RD-251	1 × RD-252		
Proton-K	8K82M		6 × RD-253	3 × RD-0210	1 × RD-0213	1 × RD-54
				1 × RD-0211	1 × RD-0214	
Proton-K	8K82M		6 × RD-253	3 × RD-0210	1 × RD-0213	
				1 × RD-0211	1 × RD-0214	
Tsyklon	11K68		1 × RD-251	1 × RD-252	1 × RD-861	
N-1	11A52		30 × NK-33	8 × NK-43	4 × NK-39	1 × NK-31
Zenit-2	11K77		1 × RD-171	1 × RD-120		
Energiya	11K25	4 × RD-170	1 × RD-0120			
Rockot			? × RD-0233	1 × RD-0235		
			? × RD-0234	1 × RD-0236		
3) Future launch vehicles						
Angara			1 × RD-174	1 × RD-0120		
Dnepr			?	1 × RD-0256		
				1 × RD-0257		
Riksha			6 × RD-169	1 × RD-185		
Shtil			? × RD-0244	?		
			? × RD-0245	?		

Some launch vehicles use clusters of slightly different engines on the same stage. For example, the Proton family uses three RD-0210 and one RD-0211 on the second stage, the only difference being that the RD-0211 provides the tank pressurisation: the assembly is called the RD-0210. Vernier engines usually have separate designators and these are indicated where known: in such cases there is one main engine and one set of verniers. In the case of the Proton-K third stage it is known that the RD-0213/RD-0214 assembly is designated RD-0212.

RD-704 operating data

Operating Mode	Mode 1	Mode 2
Fuel	kerosene/liquid hydrogen	liquid hydrogen
Oxidiser	LOX	LOX
Thrust: vacuum	1,960 kN	781 kN
sea level	1,720 kN	n/a
Specific impulse: vacuum	407 (401) s	452 (450) s
sea level	356 (351) s	n/a
Mixture ratio (O/F)	4.38	6.0
Chamber pressure	290 atmosphere	120 atmosphere
Area ration	74	
Mass	2,417 kg	
Single bell diameter	1.8 m	
Length	3.8 m	

n/a – not applicable. For the specific impulse values the nominal value is quoted, followed by a worst case value in parentheses

Soyuz-U

Background

The Vostok, Soyuz and Molniya family of launch vehicles all derived from Sergei Korolyov's R-7 (8K71) ICBM. Called SS-6/Sapwood in the West: the vehicle was first launched on 15 May 1957. In the 8K71PS version – little modified from the missile – it launched the first two Sputniks, while the further modified 8A91 was used for the Sputnik 3 launch. The Vostok and different Soyuz variants have carried all Soviet and Russian manned missions. They are now the responsibility of the Central specialised Design Bureau in Samara. CSDB was founded in 1959 as a division of Korolyov's OKB-1 Moscow design bureau to provide design support and launch vehicle manufacturing at the Samara 'Progress' factory where the vehicles are now built.

The launch vehicles that evolved from the R-7 missile are all built around the standard 'core plus four strap-ons' concept. Soviet era commentators designate the strap-ons as Block B, V, G and D while the core is Block A. Vostok added a stage designated Block E atop the core.

As the Soyuz family matured, the Soviets considered two variants to carry on the design. The first, the 8K72 incorporated the Kosberg RD-0105 engine into the Block E. The second, the 8K73 would have used Glushko's RD-0109 on that stage. The 8K72 eventually won out and the Soviets launched the initial Luna missions during 1958-1960 with it. Some Russian literature has called this the Vostok-L. Later, an improved Block E returned the RD-0109 engine to the first stage to give the 8K72K vehicle, used for the manned Vostok flights. Further improvements gave the Vostok-2 8A92, used within the photo reconnaissance programme and a final variant was called either Vostok-2M or Meteor (since it was used for launching Elint and Meteor satellites), the 8A92M.

The next variant to appear introduced two new upper stages: the Block I replaced the Block E and final Block L stage was added: that in turn spawned the Molniya launch vehicle. The Molniya carried the RD-0107 in its Block I, and this vehicle without the Block L was flown as the Voskhod launch vehicle. Western observers erroneously grouped the Voskhod launches with the Soyuz launch vehicles since they were very similar vehicles and flew the same types of missions. Although the Molniya-M launch vehicle had switched to an improved Block I engine compared with the original Molniya, there is no evidence that a similar switch took place with the Voskhod vehicle's Block I.

The true Soyuz launch vehicle, 11A511, came along in 1966, when the Soviet Union launched Kosmos 133 (the first Soyuz spacecraft launch). This was an improved version of the Voskhod vehicle, carrying the RD-0108 engine on the Block I. There were two minor variants of the Soyuz which flew a few missions, namely the Soyuz-L (three test flights of the lunar module and the Soyuz-M (eight flights of photo reconnaissance satellites).

The Soyuz-U, 11A511U, replaced the older engine design with the RD-0110 Block I engine in May 1973 and put Kosmos 559 in orbit. In a design sense the U directly replaced and retired the Voskhod vehicle for good. The Voskhod saw its last mission in June 1976.

Summary Vostok, Soyuz and Molniya launch record summary

Launch vehicle		First launch	Final launch	Successes	Failures	Total
Vostok-L	8K72	23 September 1958 *	26 April 1962	7	7	14
Vostok	8K72K	28 July 1960*	16 June 1963	10	2	12
Vostok-2	8A92	1 June 1962 *	12 May 1967	45	5	50
Molniya	8A92M	28 February 1967	29 August 1991	87	2	89
			Totals	149	16	165
(No name)	11A59	1 November 1963	20 July 1966	2	0	2
Voskhod	11A57	16 November 1963	29 June 1976	293	13	306
(No name)	11A510	27 November 1965	20 July 1966	2	0	2
Soyuz	11A511	28 November 1966	24 May 1975	28	2	30
Soyuz-L	1A511L	24 November 1970	12 August 1971	3	0	3
Soyuz-M	11A511M	27 December 1971	31 March 1976	8	0	8
Soyuz-U	11A511U	18 May 1973	(operational)	634	20	654
Soyuz-U2	11A511U2	28 December 1982	29 September 1995	66	0	66
			Totals	1,036	35	1,071

The 11A59 launches were the core and strap-ons plus the Polyot satellites as upper stages: the 11A510 used a Tsyklon upper stage as a technology test. Final launches were: Voskhod – June 1976, Soyuz-L – August 1971, Soyuz-M – March 1976 and Soyuz-U2 – September 1995. The number of Soyuz-U launches is through to the end of March 1997. * These maiden flights were launch vehicle failures.

SL-4/A-2 launch of a manned Soyuz-TM spacecraft. Icing reveals liquid oxygen tank locations

Starting with launches in the ASTP programme and then the more general Soyuz programme which gave the world its first dedicated resupply vessel, the unmanned Soyuz 20, the Soyuz-U became the workhorse for manned and unmanned missions through to 1985. With the recent retirement of the Soyuz-U2 vehicle, Soyuz U has resumed its former role as the backbone of the Russian manned missions.

The Russian Space Agency announced in 1993 that it is funding a major upgrade of the Soyuz launch vehicle to create the Rus or Soyuz-2 vehicle. RSA has scheduled the first flight of Soyuz-2 in 2000.

Voskhod and Soyuz variants orbital launches annual launch rates

1963: 1	1975: 41	1987: 43
1964: 5	1976: 38	1988: 42
1965: 13	1977: 37	1989: 38
1966: 15	1978: 45	1990: 30
1967: 22	1979: 45	1991: 24
1968: 34	1980: 45	1992: 24
1969: 37	1981: 42	1993: 17
1970: 31	1982: 43	1994: 15
1971: 33	1983: 43	1995: 12
1972: 31	1984: 44	1996: 2
1973: 40	1985: 40	1997: 10
1974: 36	1986: 36	1998: 8

Soyuz-U2 launch record summary

Programme	Flight Period	Number of launches
5th Gen P-R	1982-1995	20
6th Gen P-R	1989-1993	5
Soyuz-TM	1986-1995	22
Progress-M	1989-1994	19

Gen = generation, P-R = photo-reconnaissance

Specifications

Soyuz-U

Launch sites: Tyuratam, Plesetsk
Principal uses: photo-reconnaissance, manned Soyuz, Progress, remote sensing, Bion, other science; all LEO
Vehicle success rate: 96.8% (end of April 1999)
Commercial launches: 0
Cost: US$15 million quoted 1989 (during June 1988 Bulgarian cosmonaut mission to Mir space station it was reported that Soyuz-U launcher hardware cost was R 2 million; in November 1992 it was quoted at R 56 million, in March 1995 at R 12 billion), in October 1995 at R 15 billion.
Performance: 6,855 kg into 220 km, 51.6°
Number of stages: 2 plus 4 strap-ons
Overall length (m): 45.22 unmanned (with 10.14 m-long fairing)
Principal diameter: 10.3 m across fins
Launch mass: 309.7 tonnes
Launch thrust (kN, vacuum): 5,932
Guidance: inertial. NPO AP is responsible for the GN&C systems

Soyuz-U strap-ons

Number of strap-ons: 4
Designations: individually Blocks B, V, G, D
Engines: RD-107 from NPO Energomash, each with four fixed chambers and two gimballed verniers
Length: 19.8 m
Diameter: 2.68 m at base
U: 1,013 SL; 1,239 vacuum
U2: 1,030 SL; 1,242 vacuum
Burn time: about 120 s
Oxidiser: 27.8 t liquid oxygen
Fuel: 11.8 t kerosene
Propellant mass: 39.6 t each strap-on; dry mass 3.770 t

Soyuz-U core stage 1

The strap-ons separate at 120 s but the hammerhead core continues for a further 190 s
Designation: Block A
Engines: single RD-108 from NPO Energomash, based on the RD-107 but with four gimballed verniers for steering
Overall length: 28.75 m
Diameter: 2.95 m maximum at strap-on attach points on LOX tank; 2.15 m lower kerosene tank
Thrust (kN, vacuum): 978 U
Burn time: 310 s
Oxidiser: liquid oxygen
Fuel: kerosene
Propellant mass: about 94 1/2 t; dry mass 6.8 t

Soyuz-U stage 2

Designation: Block I
Engines: RD-0110 from KB Khimautomatiki, incorporating four fixed chambers and four gimballed verniers for steering
Overall length: 8.1 m
Diameter: 2.66 m
Oxidiser: liquid oxygen
Fuel: kerosene
Propellant mass: about 23 t; 2 t dry mass (excluding payload adaptors, and so on)
Thrust: 298 kN vacuum
Burn time: about 245 s
Fairing: A 4.5 t fairing protects the payload until 152-171 s after launch (77-93 km), depending on ascent profile. The final stage's control system commands shutdown at orbit injection about 9 min after launch, and 3.3 s later orders payload release for spring separation at about 1 m/s. Attitude control is not available following shutdown and stage rotation can be up to 3.5°/s at payload separation. The stage executes a collision avoidance manoeuvre 0.7 s after the separation command. Unmanned payloads employ the 10.14 m-long two-piece fairings, consisting of a 6.72 m-long cylindrical section and providing a 9 m long 2.65 m and payload envelope.
Acceleration loads: 4.5 *g* longitudinal, 1.5 *g* lateral; payloads rated for 10 *g* longitudinal, 2 *g* lateral for 10 min
Acoustic: 144 dB for 60 s, payloads rated for 148 dB for 120 s
Vibration: 0.5 *g* 10-40 Hz actual
Thermal environment: fairing inner wall reaches maximum 57°C before separation at about 170 s

Vehicle/payload processing

On the pad, air conditioning provides a 15-25°C environment under the fairing until services are withdrawn 40 minutes before launch, with a 0-30°C variation around the payload possible. All payload checkout operations must be completed 40 minutes before launch, with ground test systems disconnected by 30 minutes later, and no payload telemetry is available to avoid interference with the launcher until the final stage's telemetry is terminated 35 seconds after separation.

RFAS/Soviet launchers are transported horizontally to the pad by rail and erected for firing. Manned SL-4 shrouds carry four panels deployed on aborts to decelerate the escaping Soyuz craft

Fregat

Comment: NPO Lavotchkin's stage has flown three times on Proton as the injection module for 1988's Phobos missions and for Mars-96
Engine: NPOL single chamber; up to 20 ignitions. Four clusters of 22.2 N hydrazine thrusters (SI 225 s) provides attitude control/ullage
Overall length: 1.500 m
Principal diameter: 3.350 m
Stage mass: up to 6,530 kg (up to 5,350 kg main engine propellant plus 85 kg hydrazine for RCS)
Oxidiser: N_2O_4 in two neo-spherical tanks
Fuel: UDMH in two neo-spherical tanks
Stage thrust (vacuum): 19.6 kN main (SI 327 s)
Burn time: above values indicate 6.1 kg/s propellant flow rate, leading to 877 s total burn time

Soyuz-U2

Background

Soyuz-U2 was an uprated version of the Soyuz-U, flying with improved control systems. The Soviets, by introducing LOX/SYNTIN on the central core stage, rather than LOX/kerosene improved the Soyuz performance by a factor of 25 per cent. There are reports that an RD-118 engine replaced the RD-108 second stage of the Soyuz-U and other vehicles within the family.

On 5 September 1996, ITAR-TASS announced that because of the halt in the production of Syntin fuel, for budgetary reasons the Soyuz-U2 had been retired from use. As a result, launches of Soyuz-TM spacecraft to Mir cannot reach the altitude at which Mir operates and it is now flying around 10 km lower to compensate, so that a three-person crew can still reach the station using the Soyuz-U launch vehicle.

The final launch of the venerable vehicle put Kosmos 2320 in orbit on September 1995. There appear to have been 66 orbital launches of the Soyuz-U2 and Russian literature states that there were no launch failures.

Soyuz-2

Background

Soyuz-2 replaces the existing Soyuz-U launch vehicle. As a commercial vehicle Russian literature first referred to it as the 'Rus' and occasionally as Soyuz-M, a name which would cause confusion with the Soyuz-M variant flown in the 1970s.

The new launch vehicle will retain the concept of the existing and well-proven Soyuz design by keeping the four strap-on boosters surrounding a central core with an orbital stage on top. The strap-on booster and central core will both use engines derived from the RD-120 which has flown on the second stage of the Zenit-2 launch vehicle.

A 1991 paper from NPO Energomash described the development of different variants of the RD-120 to be used on the first stages of launch vehicles. While the literature is sketchy and possible unreliable, it is possible that the RD-120.01-2 variant forms the basis of the strap-on boosters engine, while the RD-120.01-1 might be used on the core. Other references state specifically that the RD-120K engine is used on the Soyuz-2.

A new payload fairing with a diameter of 3.3 m will be introduced to complement the existing 3.0 m diameter fairings used on Progress-M and Soyuz-TM missions. The payload capability to LEO will be around 8.2 tonnes.

Soyuz/Ikar

Background

In mid 1996, a new joint venture company called STARSEM offered to market a modified booster based on the Soyuz-U design. Based in France, the company added an Ikar fourth stage. The Ikar stage is described as 'an orbital module which has already been used successfully for more than 30 missions'. From a line drawing included on the STARSEM brochure, the Ikar appears to be the propulsion system used on the fifth and possibly sixth and seventh generation Zenit photo reconnaissance satellites. If this identification is accurate then the Ikar stage is capable of performing many small orbital corrections or a smaller number of larger ones, limited in total magnitude by the propellant supply.

For the Soyuz-U/Ikar launch vehicle the mechanical interfaces (which are not specified), together with equipment such as the infra-red Earth sensor had to be redesigned, necessitating a new flight control programme. In this new design, controllers keep the Ikar stage under ground control during its operations. A computer generates commands for the manoeuvres on the ground, transmits them to the stage and receives updates on the attitude and position.

RD-0124M developed as the upper stage engine for the Soyuz-2 (Theo Pirard) ***1999***/0054369

The addition of Ikar to the Soyuz-U permit a much greater flexibility to the variety of orbits available to commercial users. Historically, the Soyuz class of launch vehicles simply placed payloads into low Earth orbits (perigee 170 to 200 km, apogee typically 300 to 400 km, rarely higher), but Ikar will bring higher orbits within range. If used like the Molniya-M fourth stage with a single burn then Ikar can reach higher-apogee orbits than can the Soyuz-U. With its restart capability Ikar can also circularise the orbit. STARSEM literature states that 2.5 tonnes can be placed in a 51.8°, 1,400 km orbit or 2.35 ° to a 90°, 1,400 km orbit.

Space Clipper (SS-24)

Background

In 1991, NPO Yuzhnoye proposed conversion of the SS-24/RS-22 Scalpel MIRV ICBM into a medium-class LEO vehicle air-launched from an Antonov An-124. By 1994, sources estimated the entire programme would cost the Russians US$220 million for system development and US$11 million per launch. Operationally the Russians can fly the vehicle to the customer's payload rather than vice versa. This neatly sidestepped objections of US export control restrictions regarding shipping US technology satellites into the former Soviet Union. Since the original proposal this concern has diminished however and the company is looking for western investment to develop a canister-based service. User cost would be about US$23 million to US$25 million. The Talisman launcher, using SS-24 technology, is being jointly studied with Dassault Aviation.

SS-24 development began in 1971 as an answer to the US MX. The ballistic missile achieved the first test flight October 1982 from Plesetsk. Deployment of the rail mobile version began in 1987, followed by the silo model two years later. It was announced in August 1990 that missile production would end in 1991, and the Russians would retire all 92 silo/rail-based missiles under the START 2 treaty.

To make the Space Clipper a viable space launch vehicle designers must add a new control system, stage 4, fairing and attitude control thruster module. Stage 1 yaw/pitch control is provided by nozzle flexure and stage 1/2 roll control is from fairing fins. The new thruster module, probably using elements of the SS-18/24 NTO/UDMH units, would provide the remaining control. The 105 tonne vehicle would be ejected from its launch canister by a cold gas system before ignition. Stage 4 could be liquid or solid. A liquid stage would be more suited to circular LEO orbits, delivering 2,500 kg into 200 km circular (accuracy 10 km altitude, 0.05° inclination). A solid stage is preferable for elliptical paths: 1,000 kg into 150 × 36,000 km (accuracy ±1,000 to 2,500 km, ±0.2 to 0.6°).

Even more remarkable than flying the missile to the payload, the Space Clipper can be launched directly from its transport vehicle. The launcher is rear-loaded and checked aboard the An-124 before flying to the customer's airfield. Once at the delivery site a customer can use the onboard clean room by loading the payload through the front hatch. Then the Antonov flies to the release area, while the launcher's control/guidance system is initialised, and the rear doors are opened. At an altitude of 10 to 11 km and speed of 800 to 850 km/h, a pilot parachute extracts the vehicle, which ignites some 3 to 5 seconds later.

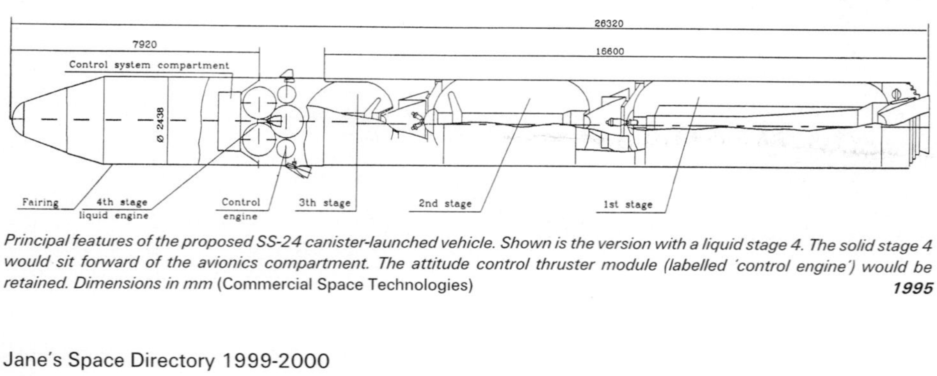

Principal features of the proposed SS-24 canister-launched vehicle. Shown is the version with a liquid stage 4. The solid stage 4 would sit forward of the avionics compartment. The attitude control thruster module (labelled 'control engine') would be retained. Dimensions in mm (Commercial Space Technologies) ***1995***

Specifications

Space Clipper

Typical performance: 570 kg (1234) or 330 kg (2234) into 400 × 40,000 km 65°; 820 kg (1234), 530 kg (2234) or 160 kg (0234) into 150 × 20,000 km 63.5°; 450 kg (1234), 200 kg (2234) or 60 kg (0234) on escape trajectories.
Number of stages: 3 or 4 (solid)
Overall length: 21.00 m (1234/1235), 20.50 m (2235), 14.30 m (0235)
Principal diameter: 2.40 m
Thrust (kN, vacuum): 1: 2,059; 2: 1,049; 3: 206; 4: 59; 5: 20
Launch mass: 64 t (1234/1235), 52 t (2234/2235), 32 t (0234/0235)
Guidance: computer-controlled inertial. Control can be provided for 60 min post-boost. Injection accuracy: ±3.5 km altitude, ±2 s period, ±2 arcmin inc.
Fairing/payload accommodation: the two-piece aluminium alloy fairing separates at about 100 km altitude. Two versions available: payload envelope 2.140 m diameter, 3.590 m long for 1235/2235/0235 and 2.150 m diameter, 4.790 m long for 1234/2234/0234. Maximum static acceleration 10 *g* during powered flight. 25 ±5°C maintained within the fairing during air transport.

Start-1/Start

Background

The first demonstration launch of the four-stage Start 1 was made from Plesetsk 25 March 1993, delivering a 260 kg communications payload into 695 × 966 km, 75.8°. The second Start-1 launch trailed along on 4 March 1997, inaugurating the Svobodny launch site. Its sister version, the first five-stage Start was lost 28 March 1995. Stage 4 shut down 12 seconds early and stage 5 did not ignite. Israel's TechSat, Mexico's Unamsat and STC Complex's own test payload were lost in the failure. Start-1 is a converted three-stage solid propellant road mobile SS-25 Sickle (RS-12M Topol) ICBM with an

Start installed in its launch canister before transfer to its launch stand (Theo Pirard/Space Information Centre) ***1996***

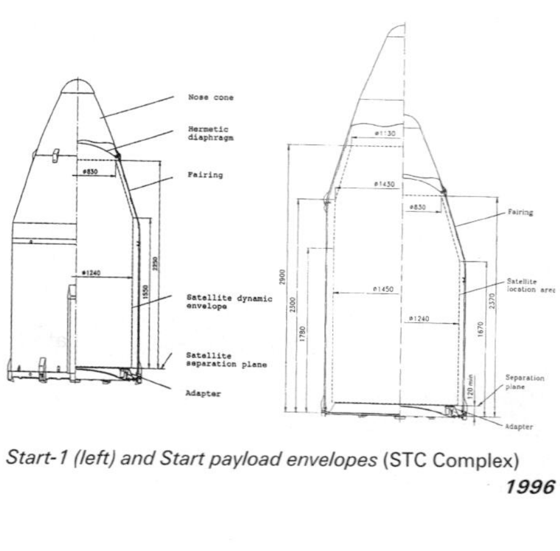

Start-1 (left) and Start payload envelopes (STC Complex) ***1996***

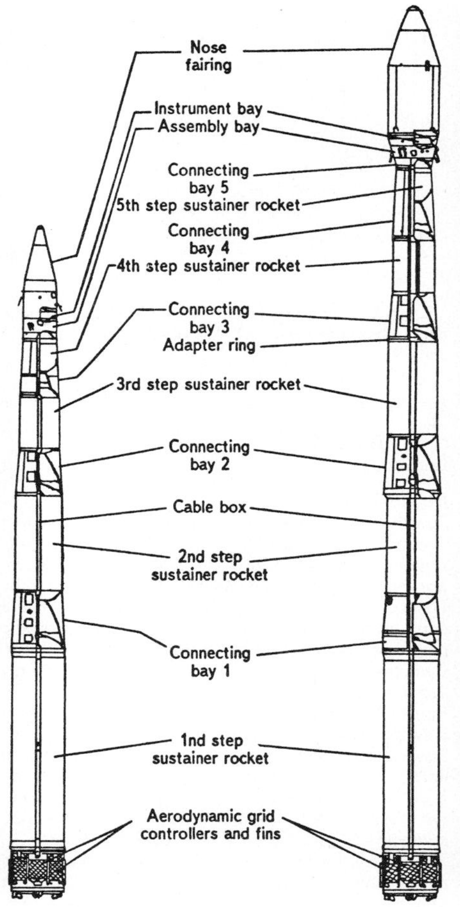

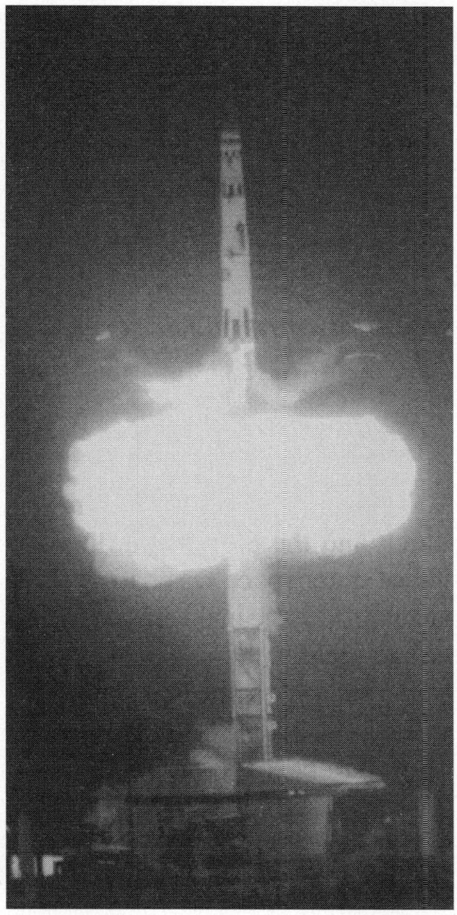

Start 1 (left) and Start principal features. At right is the Start 1 demonstration launch of March 1993. The vehicle is ejected from its fixed canister by gas pressure before stage 1 ignition. Base grids pop out for stability

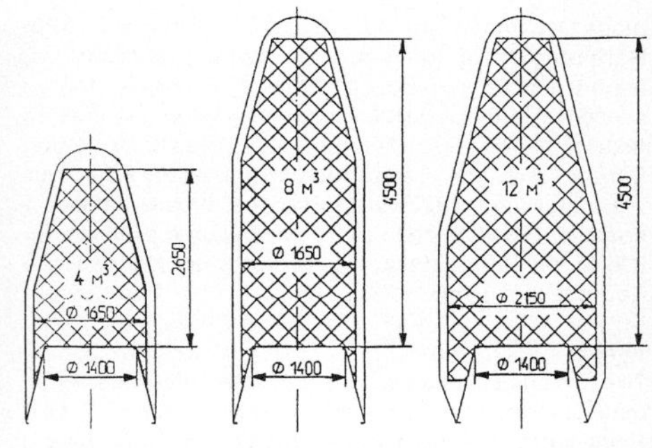

Surf's three payload fairing options; all are watertight (Glavcosmos) ***1996***

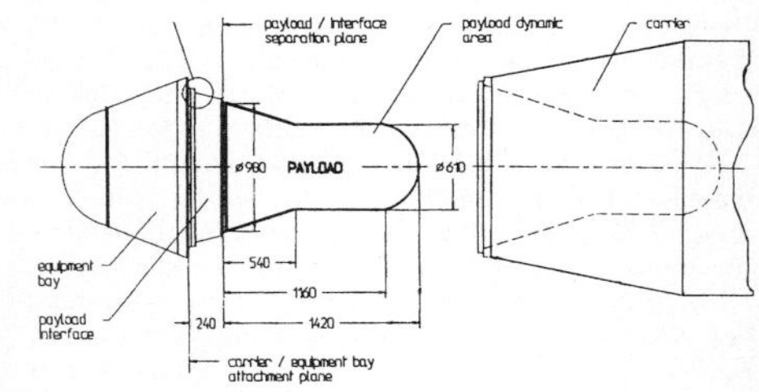

Vysota payload accommodation. The payload hangs below the nose in the space previously occupied by the warhead re-entry vehicle

added solid stage 4 which apparently repeats stage 2. A Start launch costs about US$6 million, although costs of up to US$10 million have also been claimed. 1995's launch was free to the users.

Some elements from the related two-stage SS-20 IRBM are included. Both missiles came from the Nadiradze bureau and built at the Votkinsk plant, Udmurt. Unusually, the project is performed under the aegis of the Strategic Rocket Forces. STC Complex provides customer contact. The SS-25 began development in 1971 and had its first test flights in 1982 (the first, 27 October 1982, was unsuccessful). Testing completed with the launch of 23 December 1987. It is capable of delivering a single 1 tonne 550 kT nuclear warhead over 10,500 km. It entered service in 1985 and in 1991 there were 288 missiles deployed at nine sites. The current road mobile form is expected to be replaced by the Topol-M silo-based version. It is heavier and has a larger-diameter stage 1; first launch was 20 December 1994. It will be deployed in former SS-18 silos.

All stages burn out fully (about 1 minute each). The 10 to 20 second coast after stage 1 burnout reduces dynamic pressure so that stage 2 is controllable after separation. The base aerodynamic fins and exhaust vanes provide control during stage 1 coast. N_2 jets in stage 5's aft section (stage 4 on Start-1) are used for roll control during stage 1 and three-axis during subsequent burns. There are no coast periods for the intermediate stages, but one of several 100 seconds before last stage firing, while the N_2 jets provide control.

EarthWatch planned to launch its EarlyBird Earth observation satellite on Start in 1996 but the flight was delayed. South Africa's Houwteq, in 1994, selected it for the Greensat, before the project was cancelled. Another proposal is for Russia's own Courier system of LEO relay satellites in a joint venture with NPO ELAS.

Specifications

Start 1/Start

First launch: 25 March 1993/28 March 1995
Number of launches: 2/1, through May 1998
Success rate: 100/0%, through May 1998
Number of stages: 4/5 (plus each can carry 70/80 kg post-boost stage)
Launch mass: 47/60 t
Length: 22.7/28.8 m
Maximum diameter: 1.8/1.8 m
Guidance: inertial; avionics in sealed compartment atop last stage
Performance: with post-boost stages 420/645 kg into 300 × 300 km, 90°; 300/530 kg into 500 × 500 km, 90°; 110/275 kg into 1,000 × 1,000 km, 90°. Accuracy 5 km apogee ±3' inclination with post-boost stage (solid propellant gas generator); 60-140 km apogee uncertainty without. Performance values increased by 70/80 kg without post-boost. Start-1's performance is reduced by 50 kg if larger fairing used.
Payload accommodation: Launcher is assembled horizontally in an 18 × 18.5 60 m class 100,000 building. Payload attachment requires 3 h, fairing/canister installation 5 h, tests 10 h and final launch operations 0.5 h. The campaign requires two to three days from when the payload is ready. The facility, canister and launch site maintain the payload at 10.5-26°C, 25-70% humidity. The payload can be accessed until 2-3 h before launch. The vehicle is transported horizontally to the 5 × 8 m launch box and raised. Start provides payload power and issues commands. Telemetry transmit: 203.27, 219.70, 75.67 MHz. On release, payload motion does not exceed 2°/s pitch/yaw + 1.5°/s roll with post-boost stage; 4°/s + 3.5°/s without.
Acceleration (g): 2.8/2.3 ejection, 0.15/3.3 stage 1, 6.5/4.4 stage 2, 6.5/6.1 stage 3, 9.0/6.3 stage 4, -/8.0 stage 5.
Acoustic load: ≤135 dB over 50-4,000Hz

Surf (Priboi)

Background

The Makeyev bureau proposes to offer satellite launch services using its converted SLBMs. The bureau was founded by V P Makeyev (1925-1985) in 1955 as an offshoot of Korolev's bureau and developed the RSM-25 Zyb (SS-N-6 Serb), RSM-40 Vysota (SS-N-8 Sawfly), RSM-50 Volna (SS-N-18 Stingray), RSM-52 Rif (SS-N-20 Sturgeon) and RSM-54 Shtil (SS-N-23 Skiff) SLBMs. A Volna was launched 7 June 1995 from the 'Kalmar' Delta 4 boat submerged in the Barents Sea on a 20 minute suborbital μgravity flight carrying a 120 kg experiment from Germany's ZARM (electrically-driven thermal convection in a fluid shell between two concentric spheres). Recovery was 5,600 km downrange on Kamchatka. There were apparently three SLBM demonstrations 1991-93. A Shtil 2N orbital demonstration launch was reportedly planned for 1996 carrying a 100 kg Izmiran satellite, but the first known launch did not take place until 7 July 1998 when TUBSAT N and TUBSAT N1 were launched from a submerged submarine in the Barents Sea.

Surf

The 104 tonne Surf satellite launcher will attach a Rif solid propellant 2.4 m diameter stage 1 to two stages of the 1.90 m diameter liquid propellant Shtil ('Calm Sea') and adding new stages 4/5. Length is 33.9/34.8/34.8 m depending on the 4/8/12 m³ fairing. Water-based launches are planned but land ascents are possible using the Antonov An-225 as transporter. Up to three vehicles could be assembled and transported on an *Ivan Rogov*-class ship and released aftwards to float vertically in the sea for launch. Commercial launches would cost <US$11 million. Capacities into circular orbits are (kg, 0°/90° inclination): 2,400/1,840 200 km; 2,200/1,650 400 km; 2,020/1,500 600 km; 1,850/1,360 800 km; 1,700/1,230 1,000 km; 1,570/1,110 1,200 km; 1,460/1,000 1,400 km; 1,360/910 1,600 km; 1,270/830 1,800 km; 1,200/770 2,000 km; 170/-6,500 km.

Shtil

The individual vehicles are also offered commercially. The Shtil is available in three versions: -1N 430 kg into 200 km, 185 kg into 700 km, 40 tonne launch mass, 1.5 m³ payload volume using original warhead vehicle; -2N 265 kg into 200 km, 2 m³ payload volume; -3N 670 kg into 200 km, 410 kg into 700 km, 46 tonne launch mass, 3.6 m³ payload volume. SLBM testing began in 1983 and Shtil entered service in 1986; there were 112 aboard Delta 4 boats in 1991. Inertial guidance plus stellar reference update; 8,300 km range.

Zyb

The Zyb is offered with two types of re-entry capsules for suborbital microgravity missions: 450 kg Sprint capsule (100 kg payload, original warhead vehicle) and 650 kg Efir (65 to 80 kg payload). A Zyb/Sprint test launch was

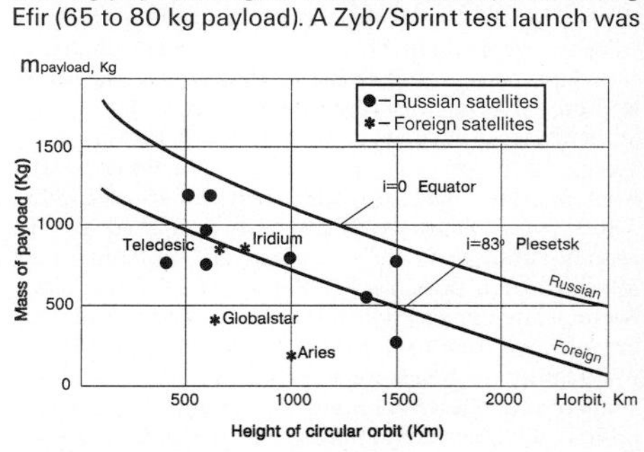

Riksha payload performance ***1996***

Riksha would be powered by liquid natural gas ***1996***

made December 1991 and a Zyb/Efir December 1992. Development of the two-stage liquid propellant Zyb began in 1958 and the 2,400 km-range missile entered service in 1967. A 3,000 km-range version began flight tests October 1972. Launch mass 14,200 kg, length 9.65 m, diameter 1.5 m, inertial guidance. 600 Zybs were built and 192 were in service aboard Yankee 1 boats in 1991. A total of 419 will become available for use as they are expected to be removed from service from the mid-1990s.

Vysota
The two-stage liquid propellant Vysota ('Altitude') began development in 1961 and entered service in 1971; there were 244 operational missiles in 1991 plus a further 177 in storage. It seems likely that Vysota and its boats will be decommissioned to meet the requirements for reductions in SLBM numbers under the START 1/2 agreements. In December 1994, it is believed that two Delta 1 boats had been taken out of service, reducing the number of operational missiles to about 220. Typical space performance would be 130 kg into 200 to 250 km circular at 0 to 25°; 130 kg into 170 to 220 × 1,000 to 1,500 km 0 to 25°. Length 14.2 m, diameter 1.80 m, launch mass 33 tonnes, burn time about 230 seconds. The payload occupies the volume previously allocated to the warhead re-entry vehicle, hanging inside the carrier below the nose equipment bay. Separation of the forward assembly is achieved by residual air pressure in the carrier. Temperature is maintained at 15 to 30°C; 200 W is available to the payload. Maximum static launch acceleration is 10.6 *g* during powered flight; 13.5 *g* for 0.1 second at stage separation.

Volna
The three-stage liquid propellant Volna ('Wave') began development in 1968; there were 224 missiles in service in 1991 aboard Delta 3 boats and a further 100 in storage (no reduction in operational numbers had been reported as of December 1994). Typical orbital performance is as Vysota but with a larger volume. Length 15.6 m, diameter 1.80 m, launch mass 35.3 tonnes.

Rikscha
The new Riksha is proposed, requiring US135 million and four years to develop, and capable of delivering 300 to 1,700 kg into orbit for US$10 to 11 million. New Energomash's LOX/methane engines would power the 64 tonne 24.5 m 2.4 m diameter two-stage vehicle. Six RD-169s would be clustered for stage 1, and a single RD-185 altitude version would power stage 2 (see Propulsion for further details). NPO Energomash, but now comes under the KompoMach consortium with Makeyev as lead initiated the project.

Tsyklon-2

Background
In the early 1960s, the Yangel design bureau (now Yuzhnoye) proposed a family of launch vehicles to be called Tsyklon. These would be based upon the bureau's R-16 and R-36 missiles and topped by a common S5M third stage for orbital missions. The Tsyklon-1 would become the R-16 missile with the S5M added (launch mass 145.4 tonnes, payload to a polar 1,000 km orbit =~700 kg). The Tsyklon-2 would simply be a modified two-stage R-36 missile (launch mass 179 tonnes, 1.5 tonnes to a polar 200 km orbit). And the Tsyklon-3 would be the Tsyklon-2 with the S5M stage added (launch mass 185.5 tonnes, 2.5 tonnes to polar 1,000 km orbit). Of these three options, the Soviets backed away from the Tsyklon-1 and never developed it.

The Soviets began their first series of test launches of the R-36 missile in either late 1965 or early 1966. Reports differ because the vehicles involved in various launch failures have never been positively identified. For years, rumours of a R-36-0 Fractional Orbit Bombardment System (FOBS) test on 16 December 1965 and 5 February 1966 have circulated. On 16 March 1966, an R-36-0 exploded on the launch pad, followed by a possible FOBS test on 20 May 1966. Sorting out the payload for these vehicles is difficult because, by their very nature, the FOBS launches had the dummy warheads recalled after slightly less than a single orbit. Since the difference between a successful partial orbit and a failed full orbit is unclear to observers the exact purpose of the launch in some cases depends on the interpretation of the observer. Since the FOBS programme was looked upon as an extension of the ICBM programme the Soviets might have considered that flights which entered orbit but were recalled after less than one orbit were not 'true' satellites to be announced publicly and registered via the United Nations.

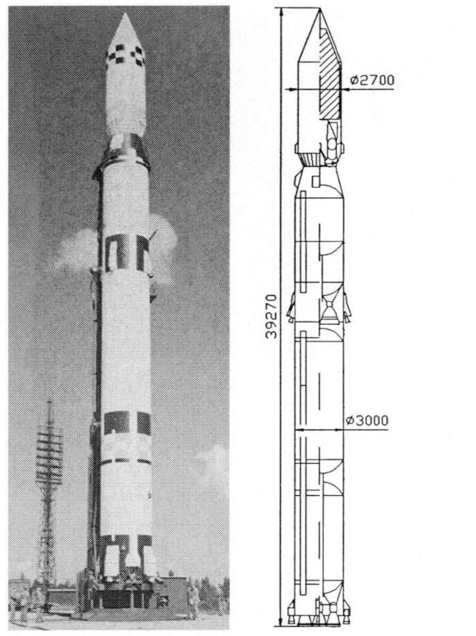

Tsyklon is based on Tsyklon M **1996**

Two unannounced orbital launches on 17 September and 2 November 1966 are particularly curious. On each occasion the dummy FOBS exploded shortly after orbital injection, and since there was operating or intact payload in orbit, the Soviets might not have thought that they should register the launches. In reality, the debris was tracked in the West and the launches appeared in all of the western launch catalogues - public and otherwise. This seems to have brought a change of heart within the Soviet bureaucracy, since the next launch in the series - Kosmos 139 on 25 January 1967 - was announced by TASS.

On 22 March 1967, a FOBS failed to reach orbit, and then a series of generally successful tests of the system took place during May 1967 to August 1971. Kosmos 160 (17 May 1967) seems to have been a payload failure since it apparently decayed from orbit rather than being de-orbited. Following the August 1971 launch the FOBS system was considered operational, although the system was subsequently dismantled under arms reductions treaties.

The Russians seem to be in two minds concerning the R-36-0 launches since the flights were of little-modified R-36 missiles and were part of the ICBM programme: therefore sometimes they are considered to be part of the two-stage Tsyklon launch programme and at other times they are not.

Launches of the dedicated space version of the R-36, the Tsyklon-2 vehicle started in October 1967, but this variant had a short period of flights and was replaced by the launch vehicle referred to as both Tsyklon-2A and Tsyklon-M. For the now-retired RORSAT launches the payload used its own propulsion system to reach orbit, the two Tsyklon rocket stages being sub-orbital. While the spacecraft propulsion systems have failed either on their way to orbit or in orbit, the actual Tsyklon-M launch vehicle has never failed in flight.

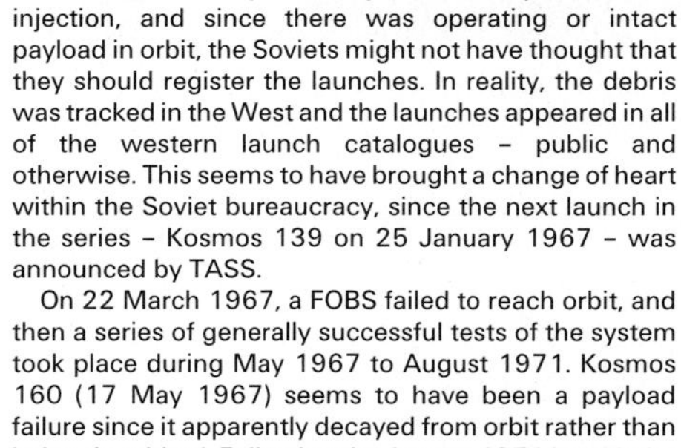
Tsyklon is based on the SS-9 Scarp ICBM, with an unusual six-chamber first stage configuration

Summary R-36-0 and two-stage Tsyklon launch record

Launch vehicle		First Launch	Final Launch	Successes	Failures	Total
R-36-0	8K69	16? December 1965 **	August 1971	22	2	24
Tsyklon-2	11K67	27 October 1967	25 January 1969**	7	1/1	8
Tsyklon-M *	11K69	6 August 1969	(operational)	100	0/5	100

* Also sometimes called the Tsyklon-2A. ** Launch vehicle failure. Tsyklon-2 and Tsyklon-M failures are shown as X/Y where X is the number of launch vehicle failures and Y the number of payload propulsion system failures.

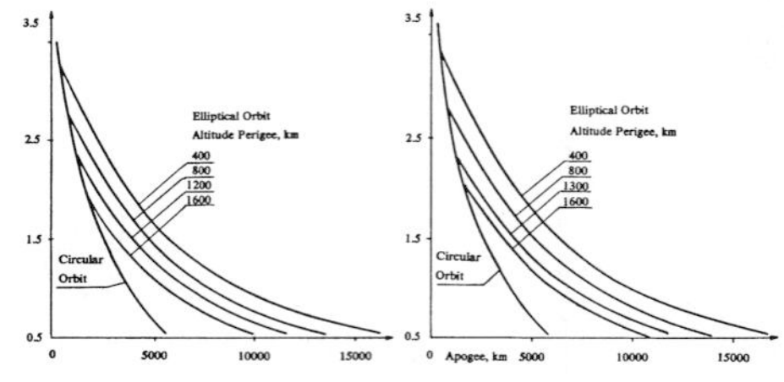

Tsyklon performance (t) into 82.5° (left) and 73.5° (right) **1996**

Launches of the Tsyklon-M are still continuing, although only the EORSAT programme is now using the vehicle.

Tsyklon 3

Background
The three-stage variant of Tsyklon - referred to as either Tsyklon-3 or simply Tsyklon, was part of the original 'Tsyklon family' proposed in the early 1960s. It took over from the Meteor launch vehicle (a modification of the Vostok launcher) to accomplish the launches of Meteor and Elint satellites. Two Meteor launches carried paying piggyback payloads, Italy's Temisat (August 1993) and Germany's Tubsat B (January 1994). The UK's Surrey Satellite Technology Ltd used Tsyklon in August 1995 for Chile's 50 kg FASat-Alfa and was charged a fee of only US$400,000. FASat-Bravo will also use Tsyklon. However, the launch rate is falling because of funding limitations and the loss of suppliers for the ageing avionics. Tsyklon production appears to have halted, but the Russians have hinted that some of Russia's R-36 strategic stockpile may be converted into Tsyklon's.

Tsyklon is the three-stage version and was first offered as the commercial Tsyklon launcher in October 1987 at Sputnik's 30th anniversary forum in Moscow. It supports a range of medium military and civil missions but its single most numerous payloads are electronic intelligence spacecraft. It is launched only from Plesetsk, but Tyuratam's Tsyklon 2M facilities should be compatible.

Specifications
Tsyklon
First launch: 24 June 1977 (Kosmos 921)
Number launched: 117 to March 99 (113 attaining orbit)
Launch sites: Plesetsk
Principal uses: electronic intelligence, Meteor, oceanographic, geodetic, remote sensing
Vehicle success rate: 96.6% to April 1999
Cost: US$10 million quoted in 1989
Performance: see diagrams
Number of stages: 3
Overall length: 39.270 m
Principal diameter: 3.000 m
Launch mass: 185.5 t plus
Launch thrust: 2,747 kN SL
Guidance: inertial; accuracy into 1,500 km orbit ±25 km altitude, 12 s period, 5 arcmin inclination; 600 km orbit ±15 km, 5 s, 3 arcmin

Tsyklon stage 1
Engines: Tsyklon-M uses the six-chamber RD-251 engine from NPO Energomash and four gimballed NPO Yuzhnoye verniers: there are some reports that the three-stage Tsyklon uses a modification of this assembly which is designated RD-261.
Overall length: 19.38 m
Diameter: 3.000 m
Thrust: 2,364 kN SL (2,643 kN vacuum); verniers 296 kN SL (334 kN vacuum)
Mass: 127.0 t at launch; 6,320 kg dry
Burn time: 120 s (280 s quoted for combined stage 1/2 burn times)
Oxidiser: nitrogen tetroxide; 84,860 kg consumed. 60.15 m^3 tank pressurised to 2.5 atmospheres
Fuel: UDMH; 33,810 kg consumed. 44 m^3 tank pressurised to 2.1 atmospheres
Propellant mass: 120.7 t

Tsyklon stage 2
Engines: Tsyklon-M uses the two-chamber RD-252 engine from NPO Energomash: there are reports that the three-stage Tsyklon uses a modification of this assembly which is designated RD-262
Overall length: 10.9 m
Diameter: 3.000 m
Thrust: 940.5 kN vacuum
Mass: 53.3 t at launch; 3,460 kg dry
Burn time: 158 s
Oxidiser: nitrogen tetroxide; 34,820 kg consumed. 25.00 m^3 tank pressurised to 3.7 atmospheres
Fuel: UDMH; 13,640 kg consumed. 18.14 m^3 tank

Tsyklon-M launches (from 1983)

	Date	Payload
1	7 May 1983	Kosmos 1461
2	29 October 1983	Kosmos 1507
3	30 May 1984	Kosmos 1567
4	29 June 1984	Kosmos 1579
5	7 August 1984	Kosmos 1588
6	31 October 1984	Kosmos 1607
7	23 January 1985	Kosmos 1625
8	18 April 1985	Kosmos 1646
9	1 August 1985	Kosmos 1670
10	23 August 1985	Kosmos 1677
11	19 September 1985	Kosmos 1682
12	27 February 1986	Kosmos 1735
13	21 March 1986	Kosmos 1736
14	25 March 1986	Kosmos 1737
15	4 August 1986	Kosmos 1769
16	20 August 1986	Kosmos 1771
17	1 February 1987	Kosmos 1818
18	8 April 1987	Kosmos 1834
19	18 June 1987	Kosmos 1860
20	10 July 1987	Kosmos 1867
21	10 October 1987	Kosmos 1890
22	12 December 1987	Kosmos 1900
23	14 March 1988	Kosmos 1932
24	28 May 1988	Kosmos 1949
25	18 November 1988	Kosmos 1979
26	24 April 1989	Kosmos 2033
27	27 September 1989	Kosmos 2046
28	24 November 1989	Kosmos 2051
29	14 March 1990	Kosmos 2060
30	23 August 1990	Kosmos 2096
31	14 November 1990	Kosmos 2103
32	4 December 1990	Kosmos 2107
33	18 January 1991	Kosmos 2122
34	30 March 1993	Kosmos 2238
35	28 April 1993	Kosmos 2244
36	7 July 1993	Kosmos 2258
37	17 September 1993	Kosmos 2264
38	2 November 1994	Kosmos 2293
39	8 June 1995	Kosmos 2313
40	20 December 1995	Kosmos 2326
41	11 December 1996	Kosmos 2335
42	9 December 1997	Kosmos 2347

All launches from Tyuratam; all of ocean surveillance satellites. Failures 16 March 1966, 18 September 1966 (in orbit), 2 November 1966 (in orbit), 22 March 1967, 25 January 1969, 25 April 1973.

Tsyklon launches

	Date	Payload
1	24 June 1977	C921
2	24 September 1977	C956
3	27 December 1977	C972
4	28 June 1978	C1025 electronic intelligence
5	26 October 1978	C1045/RS 1/2
6	12 February 1979	C1076 ocean
7	23 January 1980	C1151 ocean
8*	23 January 1981	GEO-IK
9	24 August 1981	C1300 ocean
10	21 September 1981	Aureole 3
11	30 September 1981	C1312 GEO-IK
12	3 December 1981	C1328 ocean
13	25 March 1982	Meteor 2-8
14	10 June 1982	C1378 ocean
15	16 September 1982	C1408 ocean
16	24 September 1982	C1410 GEO-IK
17	23 April 1983	C1455 electronic intelligence
18	23 June 1983	C1470 electronic intelligence
19	28 September 1983	C1500 ocean
20	24 November 1983	C1510 GEO-IK
21	15 December 1983	C1515 electronic intelligence
22	8 February 1984	C1536 electronic intelligence
23	15 March 1984	C1544 electronic intelligence
24	5 July 1984	Meteor 2-11
25	8 August 1984	C1589 GEO-IK
26	28 September 1984	C1602 ocean
27	18 October 1984	C1606 electronic intelligence
28*	27 November 1984	C1612/Meteor 3
29	15 January 1985	C1617-22
30	24 January 1985	C1626 electronic intelligence
31	6 February 1985	Meteor 2-12
31	5 March 1985	C1633 electronic intelligence
33	14 June 1985	C1660 GEO-IK
34	8 July 1985	C1666 electronic intelligence
35	8 August 1985	C1674 electronic intelligence
36	9 October 1985	C1690-95
37	24 October 1985	Meteor 3-1
38	22 November 1985	C1703 electronic intelligence
39	12 December 1985	C1707 electronic intelligence
40	26 December 1985	Meteor 2-13
41	17 January 1986	C1726 electronic intelligence
42	11 February 1986	C1732 GEO-IK
43	19 February 1986	C1733 electronic intelligence
44	15 May 1986	C1743 electronic intelligence
45	27 May 1986	Meteor 2-14
46	12 June 1986	C1758 electronic intelligence
47	28 July 1986	C1766 ocean
48	30 September 1986	C1782 electronic intelligence
49*	15 October 1986	Sextuplet communication satellites
50	2 December 1986	C1803 GEO-IK
51	10 December 1986	C1805 electronic intelligence
52	18 December 1986	C1809 science
53	5 January 1987	Meteor 2-15
54	14 January 1987	C1812 electronic intelligence
55	20 February 1987	C1823 GEO-IK
56	3 March 1987	C1825 electronic intelligence
57	13 March 1987	C1827-32
58	27 April 1987	C1842 electronic intelligence
59	1 July 1987	C1862 electronic intelligence
60	16 July 1987	C1869 ocean
61	18 August 1987	Meteor 2-16
62	7 September 1987	C1875-80
63	20 October 1987	C1892 electronic intelligence
64	6 January 1988	C1908 electronic intelligence
65	15 January 1988	C1909-14 communication satellites
66	30 January 1988	Meteor 2-17
67	15 March 1988	C1933 electronic intelligence
68	30 May 1988	C1950 GEO-IK
69	14 June 1988	C1953 electronic intelligence
70	5 July 1988	Okean 1 ocean
71	26 July 1988	Meteor 3-2
72	11 October 1988	C1975 electronic intelligence
73	23 December 1988	C1985
74	10 February 1989	C1994-99
75	28 February 1989	Meteor 2-18
76*	9 June 1989	Okean
77	28 August 1989	C2037 GEO-IK
78	14 September 1989	C2038-43
79	28 September 1989	Interkosmos 24
80	24 October 1989	Meteor 3-03
81	26 December 1989	C2053
82	30 January 1990	C2058 electronic intelligence
83	28 February 1990	Okean 2 ocean
84	27 June 1990	Meteor 2-19
85	30 July 1990	C2088 GEO-IK
86	8 August 1990	C2090-95
87	28 September 1990	Meteor 2-20
88	28 November 1990	C2106 electronic intelligence
89	22 December 1990	C2114-19
90	24 April 1991	Meteor 3-4
91	16 May 1991	C2143-48
92	4 June 1991	Okean 3
93	13 June 1991	C2151 electronic intelligence
94	15 August 1991	Meteor 3-5
95	28 September 1991	C2157-62
96	12 November 1991	C2165-70
97	18 December 1991	Interkosmos 3/Magion 3
98	13 July 1992	C2197-2202
99	20 October 1992	C2211-16
100	24 November 1992	C2221 electronic intelligence
101	22 December 1992	C2226 GEO-IK
102	25 December 1992	C2228 electronic intelligence
103	16 April 1993	C2242 electronic intelligence
104	11 May 1993	C2245-50
105	4 June 1993	C2252-57
106	31 August 1993	Meteor 2-21/Temisat
107	25 January 1994	Meteor 3-6/Tubsat
108	12 February 1994	C2268-2273
109	2 March 1994	Coronas-I
110*	25 May 1994	electronic intelligence?
111	11 October 1994	Okean 4
112	29 November 1994	GEO-IK 1
113	26 December 1994	C2299-2304
114	31 August 1995	Sich 1/FASat-Alfa
115	19 February 1996	C2328-30/Gonets
116	14 February 1997	C2337-9/Gonets
117	15 June 1998	C2352-7/Strela-3

* indicates vehicle failure. All launches from Plesetsk

pressurised to 2.2 atmospheres
Propellant mass: 49.8 t

Tsyklon stage 3
Designation: S5M
Engines: NPO Yuzhnoye RD-861 single restart embedded in toroidal tanks. Three-axis control during burns by eight fixed nozzles (6 x 20 N + 2 x 98 N) fed by main engine gas generator. Three-axis coast control by 10 thrusters of separate system.
Overall length: 2.58 m
Principal diameter: 2.25 m
Thrust: 78.90 main kN vacuum
Mass: 4.63 t at launch; 1,407 kg dry
Burn time: 125 s in two burns (dual burn used for orbits >250 km)
Oxidiser: nitrogen tetroxide; 2,030 kg consumed. 1.6 m^3 tank pressurised to 7.5 atmospheres
Fuel: UDMH; 970 kg consumed. 1.5 m^3 tank pressurised to 4.2 atmospheres
Propellant mass: 3.2 t

Fairing
Length: 10.0 m.
Diameter: 2.700 m; payload volume 19 m^3 (15 m^3 in 2.4 m diameter cylindrical section).
The fairing can be installed up to 5 h before launch. Peak loads: 4.5/10 *g* stage 1/2 longitudinal, 1.5 *g* lateral

Vehicle/Payload Processing
Plesetsk vehicles are processed horizontally in a six-storey building, where stage 3 is fuelled, and not delivered to one of the two pads 40 km distant until about T-3 h by train. Processing takes 38 hours. Once at the pad, all operations are performed remotely. Propellant loading through the base requires about 1 hour from T-75 min. The complex can handle three launches monthly. A launch can be made 72 hours after request if Tsyklon is specially stored and the payload/s ready.

Zenit-2/Zenit-3

Background
When first proposed, Zenit was planned to be one member of a family of launch vehicles based upon the commonality of rocket stages. The 11K55 and 11K66 vehicles would have been used for small satellite launches, the 11K77 (Zenit-2 itself) for medium satellites and the 11K37 was offered as a variable-configuration assembly for the launching of heavy payloads within the range 30 to 60 tonnes.

The family of vehicles would all have used the same first stage as a building block. Presumably the 11K55 and 11K66 vehicles would have used second stages (different second stages for the different designators) which were smaller than the existing Zenit-2 second stage. The 11K37 would have used clusters of two, three or four Zenit first stages to give the proposed range of payload performances. Of these options only the 11K77 vehicle was actually built, and this was the first new Soviet launcher developed since the unsuccessful N-1 of 1969-1972. It provides a payload capability midway between those of Proton and Soyuz.

Zenit-2/3 and the Energia Zenit-1 strap-on are the responsibility of NPO Yuzhnoye in Dniepropetrovsk, and its chief designer Dr Yuri A Smetanin. In the cavernous integration hall at Yuzhnoyes, workers can assemble about 20 first stages simultaneously. Titanium forging and tank production is performed on-site. The Soviets only reluctantly released information about the programme because of its military pedigree even as the country tried to market the launcher commercially. The first brief glimpse came in Energia footage released in 1988, and a relatively detailed description appeared during the second 'Progress in Space Transportation' conference held in Bonn, Germany, May 1989 to support commercial marketing. Further information, particularly on the three-stage version, was provided at

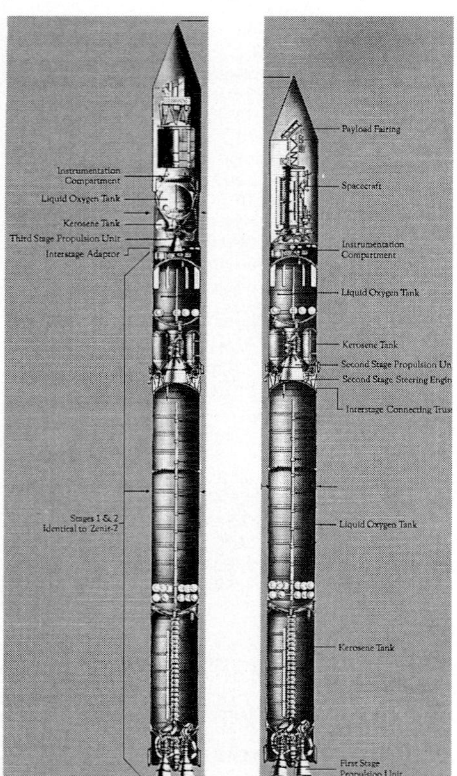

Zenit 2/3-stage versions. Zenit 2 is depicted with an Okean ocean radar satellite

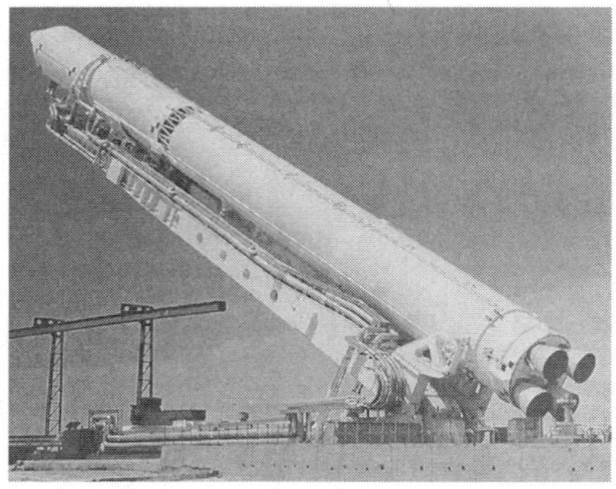
Zenit 2 roll out to one of the two Tyuratam pads

the Montreux commercial space conference in February 1990.

The Soviets claimed a manned capability (stage 1 is man-rated as Energia's strap-on), for the booster and stated a plan for carrying a two-man spaceplane called Uragan (Hurricane) to intercept US Shuttle missions out of Vandenberg. The project was cancelled in 1987 after west coast Shuttle launches were abandoned in the wake of *Challenger's* loss. Zenit was also originally intended to replace Soyuz manned launches.

A new cryogenic stage 3 was designed for Zenit 3 but discarded in favour of the existing Proton Block DM (which will be used by Sea Launch's Zenit 3SL version). It seems that Yuzhnoye has continued working on its own stage 3, although its status is unclear. The Zenit M3 would provide 900 kg GEO and 5.16 tonne GTO capacity with the 23.25 kN NTO/UDMH stage 3. The single chamber (SI 325 seconds, 1,600 seconds total burn), gimballed for pitch/yaw control, would be embedded in a single 2.72 m diameter spherical tank, with 10 × 29 N verniers for roll control, three-axis coast control and ullage. The Zenit M4 would add a 2.6 tonne NTO/UDMH apogee stage for 1.7-tonne GEO capacity from Baikonur. An air-launched version was described in 1994 for 1998 first test: released from an Antonov 225, it could handle 9-tonne LEO and 1-tonne GEO. Yuzhnoye began consideration of a Heavy Zenit in early 1989. Capable of delivering 25 tonnes to LEO, it was intended to supplant Proton by adopting the Proton approach of clustering Zenit stage 1's around a central tank. The Zenit's first stage appeared in parallel with Energia's strap-ons (designated Zenit-1), although Yuzhnoye's original proposal was based on the SS-18 ICBM.

The Ukrainian Space Agency said in September 1994 that an agreement had been reached with Russia to use Zenit for Space Station resupply (a 1996 Progress launch to Mir was expected to demonstrate the capability but it did not take place). At the same time, the Russian military indicated a requirement for two to three vehicles annually through 2000. Zenit's first commercial success came with its May 1995 selection by Space Systems/Loral for three launches to carry 36 Globalstar satellites. Zenit's survival thus appears assured.

The Soviets used the vehicle in the early programme to launch Tselina-2 heavy electronic intelligence satellites.

The Ukrainian literature issued in support of marketing the Zenit-2 has claimed that all four 1985 launches succeeded, although in reality three of the four were failures. At the time Western observers thought that the launches in April and June 1985 were simply sub-orbital test flights, and it was decided by the Ukrainians to continue this story. In reality both had been orbital attempts with Tselina-2 test satellites, confirmation coming from Russian sources in 1995. Additionally, Kosmos 1714 in December 1985 was stranded in an unplanned eccentric orbit following the failure of the payload shroud to separate and the explosion of a second stage engine.

By implication, these launches were finally confirmed as having been failures by ITAR-TASS when announcing the launch failure on 20 May 1997. The news agency stated that this had been the seventh launch vehicle failure in the Zenit-2's history – a figure which can only be reconciled if the three 1985 launches were also failures.

Specifications
Zenit 2
First launch: 13 April 1985
Number launched: 31 to May 1999
Launch sites: Tyuratam (two complex 45 pads, capable of supporting 15/year; pad 45R severely damaged October 1990 and still not in service end-1995). Plesetsk complex 35 was expected to be operational about 1997-8.
Contractors: NPO Yuzhnoye (prime), NPO AP (GN&C),

Zenit-2 flight history

Launch	Date	Payload	Mass kg	Inclination degrees	Altitude km	Comments
1985	13 April	Kosmos	3,250 ?	Launch failure		Elint test: no details of failure
	21 June	Kosmos	3,250 ?	64.41	197-340	Elint test: only debris reached orbit: vernier engine on second stage exploded 511 s after launch
	22 October	Kosmos 1697	3,250 ?	70.99	850-854	Elint
	28 December	Kosmos 1714	3,250 ?	70.99	163-853	Elint: payload shroud did not separate: second stage engine exploded preventing orbital circularisation
1986	10 July	Kosmos 1767	10,000 ?	64.88	196-207	Purpose unknown
	22 October	Kosmos 1786	500 ?	64.88	190-2,564	Atmospheric density sphere
1987	14 February	Kosmos 1820	10,000 ?	64.83	180- 252	Purpose unknown
	18 March	Kosmos 1833	3,250 ?	71.01	849-852	Elint
	13 May	Kosmos 1844	3,250 ?	71.01	849-853	Elint
	1 August	Kosmos 1871	11,000 ?	97.03	179-199	First Zenit Sun-synchronous attempt
	28 August	Kosmos 1873	11,000 ?	64.84	177-255	Purpose unknown
1988	15 May	Kosmos 1943	3,250 ?	71.01	849-851	Elint
	23 November	Kosmos 1980	3,250 ?	71.00	849-854	Elint
1990	22 May	Kosmos 2082	3,250 ?	71.00	849-855	Elint
	4 October	Kosmos	3,250 ?	Launch failure		Elint: first stage exploded
1991	30 August	Kosmos	3,250 ?	Launch failure		Elint: second stage failure
1992	5 February	Kosmos	3,250 ?	Launch failure		Elint: second stage failure
	17 November	Kosmos 2219	3,250 ?	71.01	849-855	Elint
	25 December	Kosmos 2227	3,250 ?	71.02	849-854	Elint: second stage exploded in orbit after satellite deployment
1993	26 March	Kosmos 2237	3,250 ?	71.02	849-853	Elint: second stage exploded in orbit after satellite deployment
	16 September	Kosmos 2263	3,250 ?	71.00	849-855	Elint
1994	23 April	Kosmos 2278	3,250 ?	71.01	849-855	Elint
	26 August	Kosmos 2290	10,600 ?	64.81	212-293	Photo-reconnaissance
	4 November	Resurs-01 1	1,900 ?	98.05	661-663	Remote sensing
	24 November	Kosmos 2297	3,250 ?	71.00	849-854	Elint
1995	31 October	Kosmos 2322	3,250 ?	71.02	849-852	Elint
1996	4 September	Kosmos 2333	3,250 ?	71.01	849-852	Elint
1997	20 May	Kosmos	3,250 ?	Launch failure		Elint; first stage exploded 48 s after launch
1998	10 July	Resurs-02/01	2,800?	98.79		remote sensing
		Fasat-Bravo	50?	98.79		Chilean Air Force
		Tinsat	50?	98.79		Thai remote sensing
		Techsat 1B	50?	98.79		remote sensing
		Westpac	24?	98.79		Geodetic
		Safir 2	60	98.79		Relay
	28 July	Kosmos 2360	3,250?	71.02		Elint
	9 September	Globalstar × 12	12 × 450			ComSatellite failure

The 1985 launch vehicle failures are not acknowledged as such in Russian/Ukrainian marketing literature, although when the May 1997 launch failure was reported, the ITAR-TASS announcement confirmed that this had been the seventh failure of the launch vehicle in 28 flights. In turn this confirmed that the 1985 missions were indeed failures.

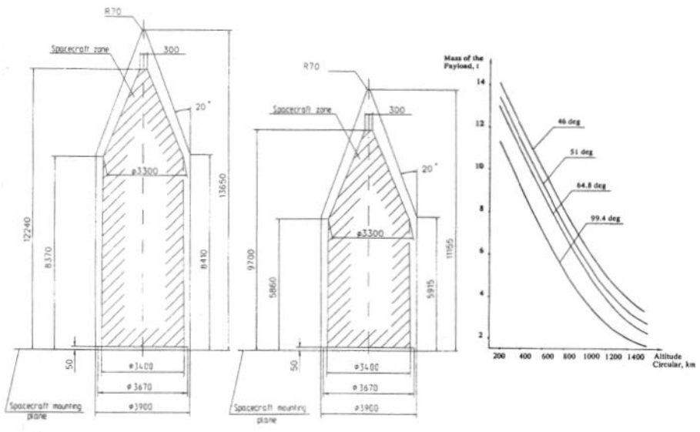

Zenit 2 fairings and performance ***1996***

NPO Elektropribor (avionics), PO Komunar (inertial guidance package), NPO Energomash (main engines).
Principal uses: Tselina 2 electronic intelligence, Resurs-O, Kuban; potential: commercial, Advanced Okean, super-Progress, manned to LEO.
Cost: Rb23 billion quoted October 1995
Vehicle success rate: 71% to May 1999
Performance: Zenit 2 13,740 kg into 200 km 51.6°, 11,380 kg into 200 km 99°, both from Tyuratam; 15.7 t into 200 km 12° for equatorial site. Injection accuracy: ±3.5 km altitude, ±2 arcmin inclination, ±2.5 s period. Zenit 3 capabilities from Tyuratam: 5,370 kg GTO, 1,600 kg GEO
Number of stages: 2 (3 for Zenit 3)
Overall length: 57.00 m (61.40 m) including 13.65 m payload fairing; 11.155 m fairing available
Principal diameter: 3.90 m
Launch mass: 449 t (466 t) without payload
Launch thrust: 7,259 kN (7,622 kN) sea level
Guidance: inertial. 1 Mbit/s telemetry provides information on 1,000 parameters during launch. Receives navigation update just before launch via laser optical receiver package mounted on side; package is then blown free.
Comment: Zenit 2 stage 1 Stage 1 and SL-17 Energia's strap-ons were developed in parallel. Both stages are aluminium with integrally machined stiffeners.
Engines: RD-171 from NPO Energomash, four gimballed chambers. RD-172 will be used on Zenit 3 (5% thrust increase).
Overall length: 32.94 m
Diameter: 3.90 m
Thrust: 7,911 kN vacuum, 7,259 kN sea level
Burn time: about 144 s (146 s Resurs-O1)
Dry mass: 28,080 kg
Propellants: 318,800 kg of liquid oxygen (210.55 m³ forward tank) and kerosene (107.31 m³ aft)
Separation: four retro solids at aft end
Zenit 2 stage 2 *Engines:* NPO Energomash RD-120 single fixed reignitable chamber with four NPO Yuzhnoye verniers
Overall length: 11.50 m
Diameter: 3.90 m
Thrust: 833.5 kN main + total 78.4 kN verniers, vacuum
Burn time: about 300 s total, verniers can burn for up to 1,100 s (Resurs-O1: 285 s main, 499 s verniers)
Dry mass: 8,300 kg
Propellants: 80,600 kg of liquid oxygen (53.43 m³ forward tank) and kerosene (27.87 m³ aft torus)
Separation: four retro solids at aft end

Vehicle/payload processing
Zenit is assembled horizontally, with the payload integrated on stage 2 before stage 1/2 mating. Assembly of the vehicle alone requires 80 h, increasing to 116 h with the payload. The vehicle can be stored for a year. Transfer to pad (45L/45R) is by rail; erection/launch processing is highly automated, requiring 21-80 h between initial integration and launch. Umbilicals retract 12 min before launch, returning to their protected positions. At 4 min, the train returns to the integration building. 4 m³/s of water is sprayed into the flame pit beginning at 15 s to reduce acoustic loads and temperature. The pad is automatically reconfigured within 5 h for the next launch. Zenit stages 1/2 separate at 150 s (Resurs-O1: 145 s) and the payload shroud during stage 2 burn between 160-1,240 s (Resurs-O1: 160 s). Payload volume is 90 m³ for the 13.65 m long shroud, reduced to 43-50 m³ for three-stage missions. An 11.155 m shroud is also available. Payload separation is achieved at 450-1,250 s, depending on the mission (Resurs-O1: 640 s). Acoustic load 140 dB.

Zenit-3SL

Background
Boeing initiated the Sea Launch joint venture in 1994 for three-stage Zenits to be shipped to the US, assembled and then transported into international waters for near-Equator launches. The consortium comprises Boeing Commercial Space Co, Norway's Kvaerner group, NPO Yuzhnoye and RKK Energia. Hughes Space & Communications Co, placed a firm order December

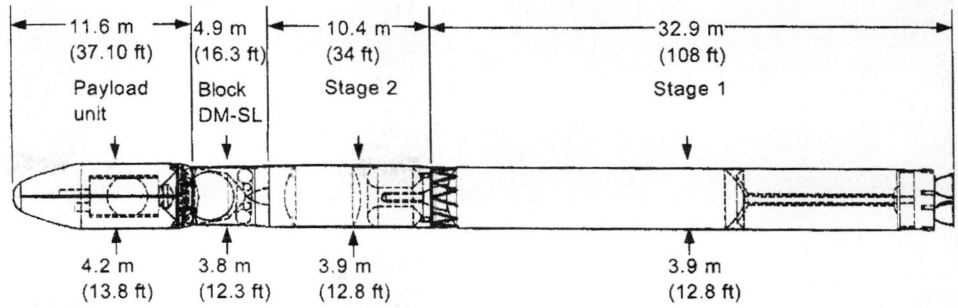

Sea Launch's Zenit 3SL configuration (Boeing) ***1996***

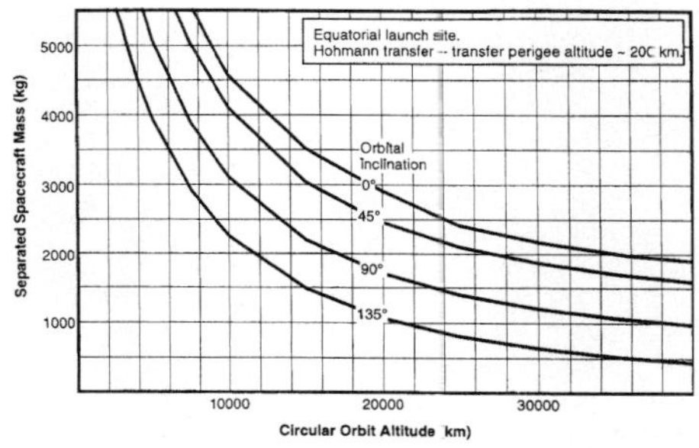

Zenit 3SL capability to circular orbit (Boeing) ***1996***

1995 for 10 launches beginning in 1998. With options, the contract could be worth around US$1 billion. In early 1997, there were 13 firm orders from Hughes and five from Loval, with Hughes holding additional options. Kvaerner Rosenberg is converting a 28,000 tonne oil drilling rig into the ocean-going launch platform in Stavanger under a US$78 million contract; Kvaerner Govan in Scotland is building the 30,000 tonne 200 m support/control ACS Assembly & Command Ship for US$93 million. The first Zenit-3 was successfully launched on 28 March 1999 carrying DEMOSAT, a dummy manufactured by Boeing to simulate the mass of a Hughes HS-702 satellite.

Specifications
Zenit 3SL
First launch: 28 March 1999
Number launched: 1
Launch sites: typically 0°north/152°west, south of Hawaii.
Contractors: NPO Yuzhnoye (stage 1/2), RKK Energia (Block DM, vehicle assembly, range services), Kvaerner (vessels, sea-going operations), Boeing Commercial Space Co (project manager, mission operations, spacecraft processing, Home Port, fairing, adaptor).
Principal uses: commercial GEO launches
Cost: US$90-100 million
Availability: 18 months after contract signature (12 months for repeat satellites).
Vehicle success rate: 100% (of 1)
Performance: 5.25 t GTO, 2.0 t GEO. GTO accuracy: 5/40 perigee/apogee altitude, 0.20° inclination.
Number of stages: 3
Overall length: 59.8 m with 11.390 m fairing
Principal diameter: 3.90 m
Launch mass: 470 t with 5t payload
Launch thrust: 7,370 kN sea level
Guidance: as Zenit 2
Launch sequence (typical GTO mission, min:s):

00:00	stage 1 ignite/lift-off
00:08	begin pitch over
01:04	max Q (5,300 kgf/m²)
01:48	max acceleration (3.8 g)
01:49	begin throttle to 75%
02:09	begin throttle to 50%
02:21	stage 2 vernier ignite
02:23	stage 1 shutdown
02:26	stage 1 separation
02:31	stage 2 main engine ignite
03:37	fairing jettison (117 km, <1,135 W/m²)
07:09	stage 2 begin throttle to 85%
07:29	stage 2 main engine shutdown
08:44	stage 2 vernier shutdown/separation
08:49	stage 3 casing jettison
08:54	stage 3 ignition 1
12:46	stage 3 shutdown 1 (LEO coast)
42:46	stage 3 ignition 2
49:02	stage 3 shutdown 2/GTO insertion
49:17	payload separation

Zenit 3SL stage 1 As Zenit 2
Zenit 3SL stage 2 As Zenit 2
Zenit 3SL stage 3 *Engine:* 11D58M restartable single chamber developed by Korolev bureau.
Overall length: 6.2 m
Diameter: 3.70 m
Thrust: 85 kN vacuum
Burn time: about 600 s total over two burns
Dry mass: 2.4 t + 800 kg casing + 200 kg residual propellants
Propellants: 14.6 t liquid oxygen in forward sphere and kerosene in aft torus

Vehicle/payload processing
The vehicle stages and payload are delivered to Sea Launch's Home Port in Long Beach, California. The stages are loaded into the ACS Assembly & Command Ship via the aft ramp for assembling on the main deck (nine days). ACS can handle three vehicles in parallel. Block DM is handled in the bow, where its kerosene is loaded (LOX is loaded on-pad). Meanwhile, the payload is processed and encapsulated in Boeing's two-piece composite fairing in land facilities. The bolted or Marmon clamp adaptors are 937, 1194 and 1666 types. The 3,000 m² Payload Processing Facility includes two 15 × 20 m class 100,000 cells for fuelling and adding ordnance. The encapsulated unit is loaded on to the ACS for integration with Zenit (three days). The completed vehicle is then loaded into the Launch Platform's horizontal hangar (two days), where it remains for the 11-day journey to the launch site. There, the LP semi-submerges 22.5 m and Zenit is erected for launch. Fuelling begins at T–3 hours. Launch operations are controlled from the Launch Control Center aboard ACS, 5 km distant. Water deluge into the flame bucket protects the LP and reduces acoustic effects. ACS provides tracking and telemetry for 410 s, then the Selena-M Russian tracking ship until LEO parking and then Russian Altair satellites (Russian ground stations also available). Launch can occur with waves up to 2.5 m high.
Acoustic loads: 142 dB overall
Maximum internal fairing T: 65°C

SPAIN

Capricornio

Background
INTA's began its own programme to develop an orbital launcher by building upon its own small INTA 100 and 300 sounding rockets. The agency calls the result of its evolutionary programme the Capricornio's. Its first and second stages are expected to fly as the 9.2 m Argos sounding rocket in the near future.

Specifications
Capricornio
First launch: planned 1998
Launch sites: Isla de El Hierro, Canary Islands
Principal uses: 60-140 kg payloads into LEO
Cost: unavailable
Number of stages: 3 (solid)
Overall length: 18.25 m
Principal diameter: 1.0 m
Launch mass: 15,035 kg

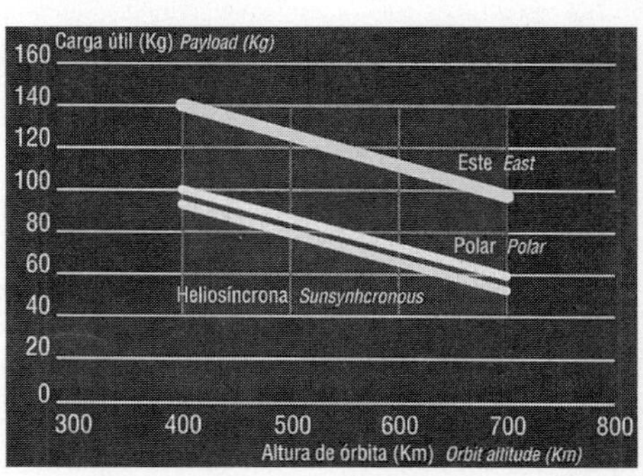

Capricornio performance from the Canary Islands (27° north) (INTA) ***1996***

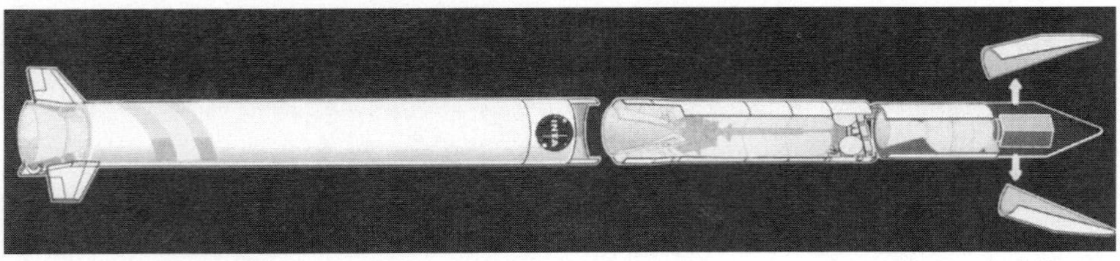

Capricornio will employ one US and two Spanish solid motors (INTA) ***1996***

Guidance: inertial above stage 3. Accuracy ±50 km altitude, ±1° inclination.

Capricornio stage 1
Motor: Thiokol Castor 4B
Overall length: 8.990 m
Principal diameter: 1.019 m
Stage mass: 11,674 kg (9,974 kg propellant)
Average thrust: 429 kN average (26.142 MNs total impulse)
Burn time: 61 s
Attitude control: ±6° provides pitch/yaw TVC; roll by stage 3 ACS.

Capricornio stage 2
Designation: Deneb-F
Overall length: 3.000 m
Principal diameter: 0.830 m
Stage mass: 2,549 kg (motor 2,436 kg; 2,124 kg propellant).
Average thrust: 167.9 kN average (5.976 MNs total impulse)
Burn time: 35.6 s
Attitude control: nozzle pitch/yaw TVC; roll by stage 3 ACS?

Capricornio stage 3
Designation: Mizar-B
Overall length: 2.104 m
Principal diameter: 0.830 m
Stage mass: 717 kg (motor 686 kg, propellant 607 kg)
Thrust: 50.29 kN average (1.700 MNs total impulse)
Burn time: 33.8 s
Attitude control: unknown.

Heatshield/payload accommodation
0.55 m³ in a 25 kg 1.722 m-long 83 cm diameter fairing, able to accommodate two satellites.

UNITED KINGDOM

HOTOL/SKYLON

Background

The unmanned Hotol Horizontal Take-Off and Landing concept emerged from British Aerospace in September 1984, with the goal of reducing launch costs of LEO payloads by 80 per cent (700 to 1,000 ECU/kg) of current Shuttle levels. BAe and the UK government put up £3 million for a two-year proof-of-concept study. It concluded in September 1987, and the UK government declined to fund the submitted three year feasibility study proposal, believing Hotol development cost would be about £4.5 billion. Rolls-Royce exercised its option in April 1988 to buy designer Alan Bond's patent on the RB-545 dual-mode engine but halted work after government funding ran out. Some basic information was released during 1988 and the patent was declassified during 1991, allowing a detailed description to be published in 1993. Bond's Sabre engine design has not been patented in order to prevent its classification. It differs by driving the air compressors from a separate power loop rather than using the hydrogen stream directly.

The Hotol SSTO fully reusable concept relied upon a dual mode engine, using atmospheric oxygen and onboard LH_2 for take-off, and onboard LOX/LH_2 as it left

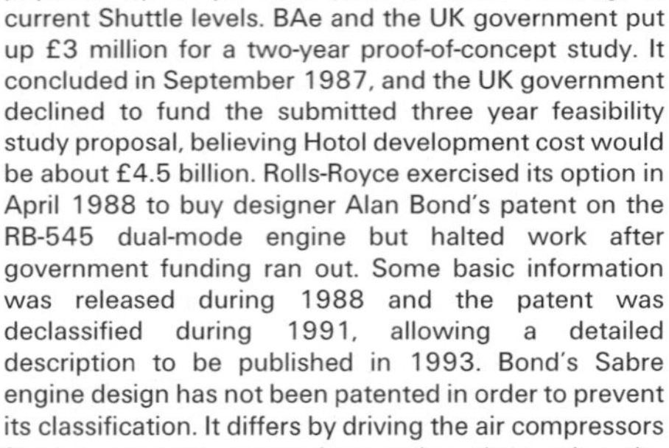

Skylon configuration C1 (Reaction Engines) ***1996***

Capricornio's launch pad will be built on the Isla de El Hierro (Theo Pirard/Space Information Centre) ***1996***

the atmosphere. The air is cooled by two heat exchangers to just above its condensation point using LH_2 and then compressed to 150 atmospheres (reaching 750 K) to feed into the rocket chamber, where it burns at 100 atmospheres and 2,700 K with hydrogen to generate 340 kN thrust. Air-breathing continues up to M5.5/26 km, for about 10 minutes, when the engine would switch to pure rocket mode using onboard oxygen, providing 735 kN thrust from 3,400 K and 150 atmosphere chamber conditions (SI 459 s). The engine also carries a small ramjet to utilise the excess air gathered at low speeds (M0.5 to 4.5). The 250 tonne vehicle could deliver 7 tonnes into LEO. BAe studied an Interim Hotol with CIS partners, powered by four RD-0120 engines and released from the back of an An-225. The full Hotol and interim Hotol entries are on pp246-247 of 1994-95's edition.

Building on Bond's design, Reaction Engines is designing the Skylon SSTO spaceplane to deliver up to 12 tonnes into LEO. Like Hotol, Skylon would be unpiloted (although eventually capable of carrying passengers) and it would take off and land horizontally using conventional runways on its own undercarriage. The airframe is designed for 200 flights. Aircraft-type refuelling and checkout procedures would minimise turnaround time. Two twin-chambered SABRE (Synergetic Air-Breathing and Rocket Engine) engines are mounted on each wingtip in a circular nacelle. These dual-mode engines would employ an air-breathing hydrogen-fuelled mode from a standing start before switching to the closed cycle LOX/LH_2 rocket mode at about M5/26 km. Skylon is designed for maximum abort capability, flyable with one nacelle shut down. Projected launch cost, including amortised purchase cost, insurance, fuel, maintenance, range activity and spaceport handling charges, is estimated at about US$10 million.

Reaction Engines is conducting supportive investigations in conjunction with Bristol University and AEA Technology Harwell to investigate some of the more technically contentious features. Project funding is sought; Reaction Engines has brought together an interim UK working group before the formation of an industrial consortium, which will eventually be expanded to include international partners.

In rocket mode, each Sabre engine driving the vehicle provides 729.2 kN thrust at 459 seconds SI and 145 bar chamber pressure. The air entry temperature into the engine in air-breathing mode is up to 1,000°C at high Mach numbers, so it is cooled before compression by passing it through a closed-cycle helium precooler and using the cryogenic hydrogen fuel flow as the heat sink. After compression, the airflow is divided between the main chamber and the fuel-rich preburner. The preburner's hot gas, en route to the main chamber, passes through a heat exchanger where it raises the helium's temperature to drive the compressor. The helium is then cooled by the hydrogen heat exchanger: the warmed hydrogen drives the hydrogen pump and the helium's circulator; the cold helium passes to the precooler to cool the incoming air, completing the cycle. Owing to the thermodynamics, the fuel flow for this type of engine is greater than stoichiometric but the installed SI is still three to six times better than a pure rocket engine. The airflow is drawn into the engine via an axi-symmetric intake which, at supersonic speeds, contains one oblique and one normal shock wave. To minimise spill drag, the centrebody translates to maintain the shock on the lip and finally closes altogether to seal the nacelle for re-entry. The intake operates at maximum capture during air-breathing, but as the core engine operates at an almost constant air mass flow the excess flow at moderate Mach numbers is diverted down a spill duct where it is heated by combustion with part of the hydrogen flow and expelled through a separate nozzle.

The fuselage and wing load-bearing structure is made from CFRP; the 2014-T6 aluminium propellant tank is suspended within, free to move under thermal and pressurisation displacements. The aeroshell is made from hot-pressed SiC fibre reinforced glass-ceramic and carries only aerodynamic pressure loads, transmitted to the fuselage structure through flexible suspension points. The 0.5 mm aeroshell is corrugated for stiffness. It is free to move under thermal expansion during the latter stages of the aerodynamic ascent and re-entry. It reaches a maximum 855 K during ascent, and the trajectory is controlled during re-entry via active feedback of measured skin temperatures to keep it down to 1,100 K.

On-orbit the main tanks are vented and allowed to warm to ambient. AOCS is provided by the LOX/LH_2 OMS/RCS system, which can remain operational for seven days. The RCS employs gaseous propellants, which are also supplied to the fuel cells and APU turbines.

Specifications

Skylon
Fuselage length: 83 m
Wing span: 25 m
Mass: 275 t gross take-off, 220 t propellant (ascent: 150.2 t LOX + 66.8 t LH_2; 2.36 t OMS/RCS), 55 t maximum landing mass
Performance: 12 t into 300 km 0°; 10.5 t 460 km 0°; 9.5 t 460 km 28.5° from equatorial site
Payload bay: 4.6 m diameter, 12.319 m long (compatible with ISO standard containers)

Skylark sounding rocket

Background

The Royal Aircraft Establishment originally developed Skylark for the International Geophysical Year of 1957-58, and consequently it is therefore the longest-running current space programme in the World. It proved so successful that, following the first launch on 13 February 1957, its launch total for all models was 429 by the end of 1995. The first parachute recovery occurred in 1961. In 1964, the 100 total was passed. The 200 mark was achieved five years later, followed by 300 in 1972 as the launch rate began to moderate. Skylark 7 was introduced in late 1974 with launch 335 and Skylark 12 in mid-1976.

The Skylark 12 is the most powerful vehicle in the range, capable of delivering 100 kg to 1,000 km

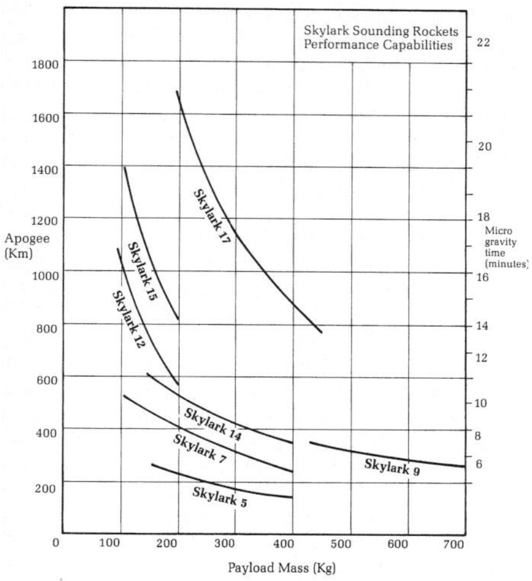

Skylark family apogee and microgravity performances, for 88° launch angle. Skylark 5, 7 and 12 are available, the rest remain as studies (Matra Marconi Space)

MMS, the prime industrial contractor, offers Skylark launchers in three versions (5/7/12 variants) although other motor combinations are available on request, such as the four Skylark 2s flown 1988-89. The launch rate is maintained primarily by Germany's Texus programme and Sweden's Maser with the Skylark 7 from Kiruna, attaining apogees of about 250 km with 370 kg payloads to create µgravity conditions for 6 minutes.

The Skylark 7 consists of a Raven XI solid boosted by a Goldfinch IIA; the 12 model adds a Cuckoo IV upper stage for lofting a 100 kg payload above 1,000 km.

A cast interstage adaptor is bolted to the front end of the Goldfinch steel casing and interfaces with the Raven motor spacer bay by dowel pins to permit clean drag separation. This adaptor houses a spin-up system of six Imp-18 motors armed by a lanyard at the 'instant of move' and triggered by microswitches leaving the launcher rail. A transient roll rate of about 250°/s is achieved in 0.5 seconds.

Skylark 5 is the smallest in the range, comprising a single Raven XI capable of delivering a 250 kg payload to 200 km. A Skylark 17 model was proposed unsuccessfully in 1988 for ESA's long duration sounding rocket programme, with the Stonechat-Mage 2 combination designed to provide 15 minutes of µgravity conditions for a 350 kg payload.

UK	266
ESRO	82
Germany	72
NASA	4
Sweden	6
UK/Argentina	2

Specifications

Skylark 7

First launch: November 1974
Number launched: 57 to end-1995
Success rate: 96.5% to end-1995
Success rate for last 25 launches: 92% to end-1995
Number of stages: 2 (solid)
Overall length: 15.3 m + payload bay/nosecone as required (<=5.5 m).
Principal diameter: 43.8 cm (1.96 m fin circle diameter)
Launch mass: 1,700 kg plus payload/nosecone
Typical performance: 100 kg to 460 km; 300 kg to 270 km; 400 kg to 210 km (250 kg to 210 km for Skylark 5) for standard launches at 5° from vertical.
Payload recovery system: a standard 36.1 cm-long Skylark recovery parachute bay is fitted at the rear of the payload. A pair of barometric switches initiates a pyrotechnic sequence at about 4 km altitude to deploy a stabilising drogue, followed 5 s later by the 7 m diameter main canopy for a 13 m/s final descent rate. Internally mounted equipment is generally recovered undamaged and can be reflown after refurbishment.

Skylark 7 stage 1
Motor: Goldfinch IIA (Royal Ordnance)
Overall length: 249 cm
Principal diameter: 43.8 cm
Total mass: 420 kg (propellant 305±5 kg)
Peak thrust, sea level: 191 kN (total impulse 701 kNs)
Burn time: 3.7 s

Skylark 7 stage 2 (stage 1 for Skylark 5)
Motor: Raven XI (Royal Ordnance)
Overall length: 545 cm
Principal diameter: 43.8 cm
Total mass: 1,280 kg (propellant 990 kg)
Peak thrust, sea level: 105 kN (total impulse 2,502 kNs)
Burn time: 39 s
Comment: A three-blade fin unit is attached to the bottom end with the fin cant angle of 0.40° providing roll control. For single rail launcher systems, the rear launch shoe is positioned on the fin barrel, with the second on the motor head end ring. A 23 cm-long spacer bay is bolted to the barrel's rear face to protect the nozzle during Goldfinch separation. The self-contained Raven Ignition Unit includes time delay switches, inertia switches, redundant batteries and arming sockets. An acceleration-triggered time-delay unit at 6 or 12 s initiates the igniter sequence after launch (the latter for lower payload masses to avoid excessive heating). A yo-yo despin system is carried above this unit to cancel the fin-induced roll for attitude stabilisation to begin. A manacle ring and three separation plungers are fired by a gas pressure source at a programmed time to release a 300 kg payload at 2 m/s.

Payload environment/accommodation
Peak longitudinal acceleration: 10.5 *g* for 300 kg, 13.5 *g* for 100 kg
Shock: 20 *g* for 15 ms
Vibration: the complete payload assembly is tested at 5.2 *g* rms over 20-2,000 Hz for 2 min (longitudinal), and 5.6 *g* for lateral.
Temperature: 157°C max skin for parallel bay; 253°C for nosecone
Comment: Standard magnesium alloy body sections are available in 8.3, 20.3, 30.5, 40.6, 50.8 cm lengths clamped together by quick release manacle rings. Hatches/access holes can be cut in the skin where required providing adequate reinforcement is added. Located at each body section junction is either a sealing ring or a bulkhead for mounting experiments. The whole or part of the payload can be sealed at ground pressure by fitting rubber O-rings in the body junctions; venting is also possible. Generally, the experiments are housed in the forward end, with the user-provided telemetry and housekeeping instruments, tracking aids and attitude determination systems occupying the remainder of the volume. A range of stainless steel/glass fibre nosecones is available: split versions released pneumatically are employed when the payload protrudes into the nose space, forward-ejecting (3 m/s) for non-intrusive payloads, and fixed for payloads with no exposure requirements (such as µgravity payloads). Glass fibre cones are available for experiments with RF or magnetometer elements.

Skylark 12

Skylark 12 is basically a Skylark 7 with a Cuckoo IV third stage motor, almost doubling the peak apogee

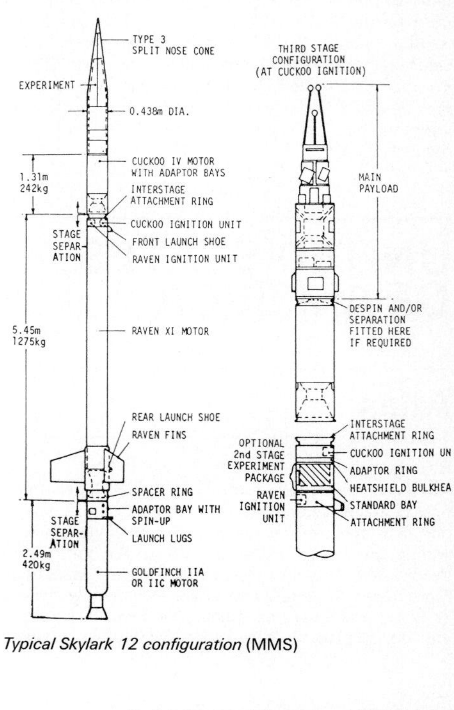

Typical Skylark 12 configuration (MMS)

Skylark launches from 1981

No	Type	Date	Site	Payload/Customer
383	7	April 1981	K	Texus 3B/DFVLR
384	7	May 1981	K	Texus 4/DFVLR
385	7	December 1981	K	Energy/DFVLR
386	7	April 1982	K	Texus 5/DFVLR
387	7	May 1982	K	Texus 6/DFVLR
388	12	November 1982	A	Caesar/DFVLR
389	7	March 1983	K	Texus 7/DFVLR
390	7	April 1983	K	Texus 8/DFVLR
391	6	November 1983	K	Mapwine/DFVLR
392	7	March 1984	K	Texus 9/DFVLR
393	7	April 1984	K	Texus 10/DFVLR
394	12	March 1985	N	Interzodiac 1/DFVLR
395	7	April 1985	K	Texus 11/DFVLR
396	7	April 1985	K	Texus 12/DFVLR
397	7	April 1986	K	Texus 13/DFVLR
398	7	April 1986	K	Texus 14/DFVLR
399	7	April 1987	K	Texus 14B/DFVLR
400	7*	April 1987	K	Texus 15/DFVLR
401	7	August 1987	W	Super Nova/DFVLR
402	7*	November 1987	K	Texus 16/DFVLR
403	7	May 1988	K	Test flight/BAe
404	12	September 1988	N	Interzodiac 2/DFVLR
405	7	November 1988	K	Texus 19/MBB-ERNO
406	2	November 1988	A	Rose 1/DFVLR
407	7	November 1988	K	Texus 20/MBB-ERNO
408	2	December 1988	A	Rose 2/DFVLR
409	2	January 1989	K	Rose 3/DFVLR
410	2	February 1989	K	Rose 4/DFVLR
411	7	April 1989	K	Texus 21/MBB-ERNO
412	7	April 1989	K	Texus 22/MBB-ERNO
413	7	November 1989	K	Texus 23/MBB-ERNO
414	7	December 1989	K	Texus 24/MBB-ERNO
415	6	March 1990	K	SISSI 1/DLR
416	7	May 1990	K	Texus 25/MBB-ERNO
417	7	May 1990	K	Texus 26/MBB-ERNO
418	6	July 1990	K	SISSI 2/DLR
419	7	August 1990	K	SISSI 3/DLR
420	7	November 1990	K	Texus 27/MBB-ERNO
421	12	November 1990	K	Maxus test/SSC & MBB
422	7	March 1991	K	SISSI 4/DLR
423	7	November 1991	K	Texus 28/MBB-ERNO
424	7	November 1992	K	Texus 29/DASA-ERNO
425	7	May 1993	K	Texus 30/DASA-ERNO
426	7	November 1993	K	Maser 6/SSC
427	7	November 1993	K	Texus 31/DASA-ERNO
428	7	May 1994	K	Texus 32/DASA-ERNO
429	7	November 1994	K	Texus 33/DASA-ERNO
430	7	March 1996	K	Texus 34/DASA-ERNO
431	7	May 1996	K	Maser 7/SSC
432	7	November 1996	K	Texus 35/DASA-ERNO

* indicates vehicle failure. Texus 14 (April 86) was rated as a partial success because the payload failed to achieve µg conditions. Texus 7 (March 83) suffered a nozzle burn-through and was rated as partially successful. Texus 17/18 flew aboard Black Brant. Skylark 2 comprised Raven VIII plus Cuckoo I for the four Rose (Rocket Scatterometer Experiment) firings of 1988/89. SISSI = Spectroscopic Infra-red Structure Signatures Investigation. Skylark 6 was essentially identical to Skylark 7. Key: A=Andoya, K=Kiruna, N=Natal, W=Woomera

performance. It is designed for a third-stage payload range of 100-200 kg, which includes everything forward of the Cuckoo motor adaptor ring. The first two operational launches occurred from Andøya in November and December 1976. Since then, its primary employment has been for science payloads requiring high altitudes.
First launch: November 1976, from Andøya, Norway.
Number launched: 15 to end-1995
Vehicle success rate: 100% to end-1995
Number of stages: 3 (solid)
Overall length: 9.25 m + payload/nosecone
Principal diameter: 43.8 cm (1.96 m diameter fin circle).
Launch mass: 1,937 kg + payload/nosecone
Performance: 100 kg to 1,030 km, 200 kg to 575 km.
Payload recovery system: as Skylark 7
Typical flight sequence (time in s):

0	Goldfinch ignition
0.5	Spin-up
6	Raven ignition
45	Raven all-burnt
60	ejection of nose-cone and experiment hatches, under control of payload timer
70	stage 3 separation, ignition at 70.4 s
85.4	Cuckoo all-burnt
96	earliest time for payload separation, boom deployment and so on

Skylark 12 stage 1
as Skylark 7

Skylark 12 stage 2
as Skylark 7, but with fins canted 0.5° for roll and the availability of a small experiment bay atop the Raven (see payload accommodation below).

Skylark 12 stage 3
Motor: Cuckoo IV (Royal Ordnance)
Overall length: 131 cm
Principal diameter: 43.8 cm
Mass: 242 kg (186 kg propellant)
Peak thrust, vacuum: 44 kN (total impulse 452 kNs)
Burn time: 15 s at 20°C
Comment: Magnesium alloy adaptor bays are bolted to each end of the motor and provide the standard manacle feature; the short 78 mm forward ring can accommodate a de-spin or payload separation system described in the Skylark 7 entry. At 70 s, the Cuckoo ignition unit at Raven's forward end actuates two pistons that push the third stage away at 0.2 m/s; 87 mm separation triggers the firing pulse from the Raven along a flying lead.

Payload environment/accommodation
Peak longitudinal acceleration: 40 *g* for 100 kg, 30 *g* for 150 kg
Peak lateral acceleration: 1.5 *g*
Shock: 20 *g* for 15 ms
Vibration: the complete payload assembly is tested for 2 min at 5.2 *g* rms over 20-2,000 Hz longitudinal, 5.6 *g* for lateral.
Spin: 230 rpm final
Temperature: 283°C on nosecone, 182°C on side wall for 150 kg payload.
Comment: A 30 cm-long, 40 kg experiment bay can be carried at the Raven's forward end, with a solid bulkhead as a heatshield against the Cuckoo efflux. The magnesium body sections for the main section are described in the Skylark 7 entry. The recommended nosecone for Skylark 12 is the stainless steel type 3 split version since it has to be ejected without inducing significant pitch rate before Cuckoo burn. Both halves incorporate a sealed pneumatic system with the gas provided by a Conax cylinder fitted with a pyrotechnic valve; the half cones are held by a tip latch and an aft manacle body ring.

UNITED STATES OF AMERICA

Atlas

Background
Atlas was developed as the first US ICBM, becoming operational in September 1959. A total of 159 was at one time deployed across the US before the last was withdrawn from service in 1965. As they were deactivated, they were refurbished as space launchers. More than 500 in the Atlas family have been launched. They are responsible for orbiting the first US astronaut and dispatching the first successful probes to Mars,

Atlas 2AR will provide a 3.8 tonne GTO capacity ***1996***

Venus, Mercury and Jupiter. Of the 140 Atlas E/F types originally stored at Norton AFB, the last departed in March 1995. A new build Atlas J was studied to complement the refurbished Titan 2, providing a 3,400 kg capacity (with spacecraft kick motor) into polar orbit. Studies were also made of using the Atlas 2 family for Space Station resupply.

The vehicle's future depends on the Atlas Centaur combination, first launched in 1962. In January 1987, it lost out to the Delta as the USAF Medium Launch Vehicle for the GPS Navstar satellites. General Dynamics in 1987 committed US$100 million of company funding to 18 new commercial Atlas Centaur vehicles providing a 2,335 kg GTO capability. The company later increased the plan to US$400 million/62 vehicles. Eutelsat was the first to select this Atlas 1, for one of its second generation satellites in January 1990 (it actually flew on the first of the Atlas 2 family, December 1991). In October 1987, NOAA booked three GOES at a fixed price of about US$200 million for delivery into GTO and a free reflight in the event of a launcher failure. The last Atlas 1 departed in 1997.

Atlas became the air force's choice in May 1988 as the USAF MLV-2 to handle DSCS3, beginning in 1992. 62 vehicles are being built to support projected launch requirements through 2000. Another eight Atlas 2AS were added in 1995 for launch through 2002. Martin reportedly charges the USAF US$45 million per launch. Atlas 2 achieved the required DSCS 3 capability by stretching the booster and upper stage tanks by 2.75 m and 92 cm, respectively, uprating the Atlas booster powerplant from 1,668 kN to 1,841 kN and incorporating new lightweight and low cost avionics. Lockheed Martin's IABS Integrated Apogee Boost Subsystem provides insertion of DSCS into near GEO. This new Atlas 2 (the previous 18 Atlas Centaurs were retrospectively renamed Atlas 1) provides 2,812 kg GTO capability. The last Atlas 2 was launched on 16 March 1998.

The MLV2 success permitted GD to offer a derived Atlas 2A commercial version (3,045 kg GTO) from 1992. The 2A is an enhanced version of the 2 model for commercial and government applications. The major improvement is the incorporation of P&W's RL10A-4 engines on the Centaur stage. A 508 mm long all welded columbium (niobium) extension skirt is deployed following Centaur separation and provides a 6.5 second SI increase to 449.5 second vacuum (thrust 185 kN total vacuum). There is no thrust compartment length increase required and no additional mass. The Atlas 1/2 avionics is further updated by replacing the sequence control unit and the servo inverter unit with a remote-control unit, providing 128 channels of solid state switching for vehicle/spacecraft sequencing instead of the current electromechanical relays.

GD began offering a 2AS version in 1993, uprated with the inclusion of four solid propellant jettisonable Castor 4A boosters. Two are ignited on the pad and jettisoned at 88 seconds after their 54 second burn. The second pair ignite at 57.5 seconds for jettison at 114 seconds. The Atlas structure required strengthening to accommodate the boosters. 22 are being built (eight were added in 1995 as insurance against 2AR problems). Intelsat's three launches are contracted at US$93 million to US$98 million each.

The 2AS was conceived to bid for the Intelsat 7 contract; the original intention was to propose the 2A but Intelsat's mass growth dictated the addition of strap-ons. Intelsat awarded a US$9.5 million contract in 1991 to pay for enhancing Atlas 2AS capacity into GTO by 121 kg, the additional satellite propellant providing a 16 year life. Atlas is not able to handle the heavier Intelsat 7A. The **Block 1** upgrade became available in 1994 for 2AS/2A, able to improve GTO capacity by 8 per cent. In addition to software improvements, the two air-lit Castor 4As on the 2AS replace their 11° canted nozzles with 7°, the Centaur's RL10A-4 is upgraded to the 10A41 with thrust increase per engine of 6.67 to 99.19 kN. First flight was January 1995 for Intelsat 704. The **Block 2** upgrade is planned to handle larger payloads such as EOS AM 1 into intermediate LEOs. The fairing will be stretched 90 cm, and the payload adaptor and Centaur equipment module strengthened.

Lockheed Martin, on 7 November 1995, announced development of the 2AR (R means re-engined) to replace all previous models and, potentially, as the basis of a common core family, for handling medium to heavy payloads. The radical redesign will reduce cost by more than 20 per cent, principally by adopting Russia's RD180 main engine and a single engine Centaur. It will carry 3,810 kg GTO.

Initially, the company will build 18, using Russian built engines. US built engines will then become available (US law requires US engines for government payloads). The two-engine Centaur may be offered for high inclination or multiple satellite LEO missions.

Systems integration and final assembly of the Atlas boosters and Centaur was transferred to Denver in 1995 when GD sold its space division to Martin Marietta for US$208.5 million. Denver's existing Waterton facility accommodates seven vehicles, with four being actively

Final Atlas assembly is now undertaken in Denver's Waterton facility

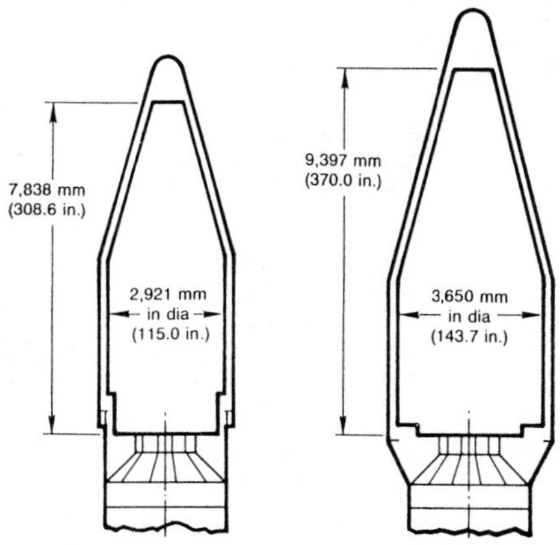

Atlas payload fairings (Lockheed Martin)

worked on at any one time. A new class 100,000 3,720 m² Final Assembly Building now handles Centaur. Tanks continue fabrication at the Naval In Service Engineering West facility (formerly Air Force Plant 19) in San Diego (26 annual capacity) and fairings, thrust structures, adaptors and other components continue to be provided from a company facility in Harlingen, Texas.

Martin can launch up to four vehicles annually (five under surge conditions) from each of Canaveral's two complex 36 pads. 36A is used for USAF launches but open slots are available for commercial flights. 36B is dedicated to commercial operations. Work is under way on converting VAFB's SLC-3E to handle the Atlas 2 family.

On 8 April 1998, Lockheed Martin announced changes to the designation system whereby Atlas 2AR would become Atlas 3A and a variant known as the 2ARC would become the Atlas 3B. The Atlas 3B incorporates two RL-10A cryogenic engines to power a stretched version of Centaur and extends the lift capability to 4,500 kg to GTO, an increase of more than 770 kg over the Atlas 2AS, the most powerful launch vehicle in the Atlas family presently flying. Atlas 3A was scheduled for first flight in the second half of 1999 followed by the maiden flight of Atlas 3B a year later. In October 1998, Lockheed was awarded contracts to complete development of a new family of launchers under the USAF Evolved Expendable Launch Vehicle (EELV) programme. The air force envisions that these vehicles will eventually replace current Atlas series vehicles through utilisation of the new Common Core Booster (CCB) stage and reduce launch costs by up to 25 per cent. The CBC is essentially the Atlas 3A first stage, 3.8 m in diameter and 27 m tall with a single Russian built RD-180 engine and equipped with fixtures for solid rocket boosters that could be attached to increase lifting capacity.

On 29 July 1998, an RD-180 was test fired in a prototype Atlas 3A for the first time for 10 seconds at the NASA Marshall Space Flight Center Advanced Engine Test Facility. This was the first time a Russian rocket motor had been fired at a US government test facility. The advanced Engine Test Facility had previously been used to test Shuttle Orbiter main engines and the Saturn V F-1 first-stage engines before that. The RD-180 delivers 3,825.28 kN giving the Atlas 3 series more than twice the lift-off thrust of the original Atlas space launch vehicle and at the date of this test nine development engines had been successfully tested for more than 10,000 seconds. In flight, the RD-180 will operate for about 186 seconds in providing first-stage energy.

On 2 February 1999, Lockheed Martin named this new family Atlas 5, the baseline intermediate class vehicle being able to lift payloads weighing more than 5,000 kg to GTO. The common Centaur upper stage has the performance option of one or two RL10 engines and the new Centaur will be flight proven on Atlas 3 more than a year before the first use on Atlas 5. By carrying up to five Aerojet solid rocket boosters attached to the CCB, each producing 1,110 kN thrust, the most powerful commercial variant of Atlas 5 will be able to carry payloads weighing up to 8,200 kg to GTO. Developed for the USAF EELV requirement, the Atlas 5 Heavy configuration will comprise three CCB stages placed side by side with the common Centaur upper stage placed atop the centre CCB and encapsulated in a 5.4 m diameter Contraves composite payload fairing. The Atlas 5 Heavy will be capable of placing a payload in excess of 13,000 kg to GTO or directly injecting a payload of 6,500 kg into geostationary orbit. As of May 1999, Lockheed Martin has a backlog of 30 launch orders for Atlas 3A and 3B.

Specifications

Atlas 1

First launch: 25 July 1990 (AC-69 CRRES)
Number launched: 10 to April 1999
Launch sites: Canaveral pad 36B

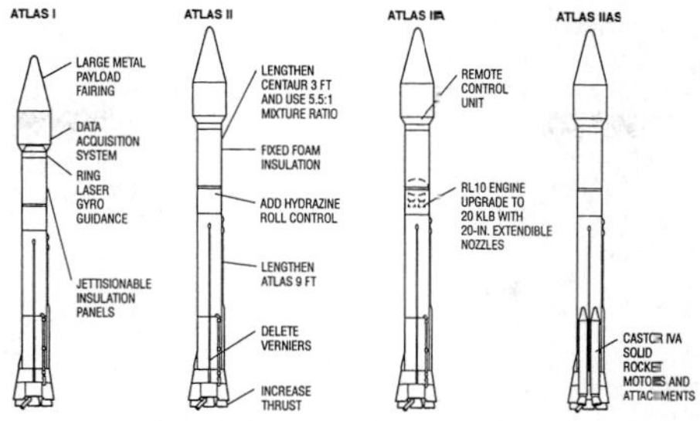

The Atlas Centaur family (Lockheed Martin)

Principal uses: medium-class telecom and meteorological satellite payloads into GTO.
Vehicle success rate: 82% to April 1999
Performance
LEO (185 × 185 km, 28.5°): 5,900 kg medium fairing/5,700 kg large.
GTO (167 × 35,788 km, 27.0°): 2,375 kg medium fairing/2,255 kg large; 3Σ errors of 106 km apogee, 2.4 km perigee and 0.02° inc
Earth escape: 1,520 kg medium fairing, 1,400 kg large
Number of stages: 2 1/2 (booster engines burn in parallel)
Overall length: 42.0 m with medium fairing, 43.9 m with large
Principal diameter: 3.05 m
Launch mass: 163,900 kg with medium fairing, 164,290 kg with large
Launch thrust: 1,953 kN sea level
Guidance: Honeywell's Inertial Navigation Unit mounted on Centaur's forward equipment module performs the inertial guidance and attitude control computations for both Atlas and Centaur. Some initial Atlas 1s retained the existing Honeywell inertial unit and Teledyne flight control computer but subsequent vehicles incorporate a Honeywell ring laser gyro INU + Gulton digital data acquisition unit, saving 36 kg and enhancing reliability.

Atlas stage 1

Comment: Atlas uniquely incorporates two booster engines fired in parallel with the central sustainer until the base section is jettisoned by the release of pneumatically- actuated latches at about T+156 s/5.5 *g* longitudinal acceleration. The propellant tanks are thin-walled (0.361.04 mm), fully monocoque stainless steel separated by an ellipsoidal bulkhead. Structural integrity is maintained in flight by the pressurisation system and on the ground by either internal pressure (N_2 at 0.330.68 atmosphere) or application of mechanical stretch. The tank design is driven primarily by the 'max a Q condition', the angle of attack due to jet stream winds near the trajectory's maximum dynamic pressure. Atlas separation is ensured by eight solid propellant retros around the base firing angled at 40° to the vertical to prevent spacecraft contamination. An external Atlas pod houses range safety, propellant utilisation, pneumatics and instrumentation.
Engines: Rocketdyne MA-5 propulsion system of two booster, one sustainer and two vernier single start liquid bipropellant engines. Each booster yields 839.5 kN sea level, the sustainer 269 kN sea level, and the verniers 2,975 N each. The engines are gimballed hydraulically to provide three-axis control during Atlas burn.
Overall length: 22.16 m
Principal diameter: 3.05 m
Propellant: 137.53 t LOX (forward tank) + RP1
Burn time: 156 s boosters, 266 s sustainer

Interstage adaptor

Mass: 477 kg
Dimensions: 3.96 m long, 3.05 m diameter
Comment: The ISA supports Centaur until separation at about 268 s is effected by a flexible linear shaped charge around the forward circumference. Construction is an aluminium skin/stringer and frame.

Fairing/payload accommodation

Two fairing designs are available for spacecraft protection during ascent: 4.19 m diameter 12.22 m long 2,005 kg, or 3.30 m diameter 10.36 m long 1,375 kg mass. Usable diameters are 365.0 cm and 292.1 cm, respectively. Both employ an aluminium skin/stringer/frame structure and non-contaminating pyro separation bolts for jettison in halves at about 205 s before sustainer engine cut-off when the heating rate has reduced to 1,135 W/m². Four 762 mm2 doors in the aft section provide payload access. The highest steady state longitudinal acceleration occurs at Atlas booster engine cut-off. It is typically 5.5 *g* but can be reduced with some loss of performance. Laterally, the highest steady state acceleration is 0.4 *g* caused by the vehicle reacting to winds. During the assembled payload/fairing transport to the pad for vehicle mating, air conditioning can provide a 15-25°C environment around the spacecraft. On the pad, air with a 2°C dew point is used until 2 h before launch, when a switch to N_2 is made. Gas is provided to class 10,000 (5,000 available) with 1,029°C inlet temperature. *Acceleration load:* 5.5 *g* maximum longitudinal, 0.4 *g* lateral *Acoustic:* max 138.9 dB overall.

Atlas provides five basic payload adaptors compatible with the Ariane 937B (Atlas type A), 937A (A1), 1194A (B, B1) and 1666A (D, Hughes HS 601) interfaces, and two (C1, C2) providing bolted interfaces for satellite adapters or mission peculiar requirements. The numerical designators indicating diameter in millimetres. Separation is effected by springs following V-clamp release commanded by Centaur's guidance system. The upper stage's sequence control unit can also issue up to 10 control commands to the spacecraft.

Comparison of current and future launch vehicle variants

	Atlas 2A	Atlas 2AS	Atlas 2AR	Atlas 2ARS
Usable diameter	3.65 m	3.65 m	3.65m	3.65 m
Usable length	9.39 m	9.39 m	10.3 m	10.3 m
Fairing mass	2,076 kg	2,076 kg	2,215 kg	2,215 kg
Payload capabilities:				
LEO at 28.5°	7,316 kg	8,618 kg	8,641 kg	9,095 kg
LEO at 90°	6,[illegible]92 kg	7,212 kg	7,121 kg	7,398 kg
GTO	3,066 kg	3,719 kg	4,037 kg	4,264 kg
Overall vehicle:				
Length	47.5 m	47.5 m	52.8 m	52.8 m
Launch mass	185,427 kg	233,750 kg	214,535 kg	238,044 kg
Stage 1: strap-on boosters		2 pairs		1 pair
Motor		Castor 4A		Castor 4A
Length		9.9 m		9.9 m
Diameter		1.0 m		1.0 m
Dry mass ground ignited		3,101 kg		3,101 kg
Dry mass altitude ignited		3,224 kg		3,244 kg
Propellant mass		10,200 kg		10,200 kg
Thrust (sl)		433 kN		433 kN
Specific impulse (sl)		229 s		229 s
Stage 1: Atlas				
Motor	MA-5A	MA-5A	RD 180	RD180
Length	24.9 m	24.9 m	29.0 m	29.0 m
Diameter	3.05 m	3.05 m	3.05 m	3.05 m
Dry mass	7,151 kg	8,856 kg	11,613 kg	11,613 kg
Propellant mass	156,260 kg	156,260 kg	183,530 kg	183,530 kg
Booster thrust (sl)	1,843 kN	1,843 kN	3,826 kN	3,826 kN
Sustainer thrust (sl)	269 kN	269 kN		
Booster specific impulse (sl)	261.8 s	261.8 s	311.3 s	311.3 s
Sustainer specific impulse (sl)	220.4 s	220.4 s		
Stage 2: Centaur				
Motor	2 × RL-10A-41	2 × RL10A41	1 × RL-10A4 1	1 × RL-10A 41
Length	10.1 m	10.1 m	10.1 m	10.1 m
Diameter	3.05 m	3.05 m	3.05 m	3.05m
Dry mass	1,735 kg	1,735 kg	1,680 kg	1,680kg
Propellant mass	17,060 kg	17,060 kg	16,946 kg	16,946 kg
Total thrust (vacuum)	198.4 kN	198.4 kN	99.2 kN	99.2 kN
Specific impulse (sl)	450.5 s	450.5 s	450.5 s	450.5 s

These figures have been extracted from an ILS brochure issued in 1997

NPO Energomash RD-180 engine adopted by Lockheed Martin for the Atlas IIAR, Atlas IIIA and Atlas V series
(Theo Pirard) ***1999***/0054362

Payload Processing
Cape payload processing facilities, including those of Astrotech Space Operations in nearby Titusville, are described under the McDonnell Douglas Delta entry. Each pad at complex 36 includes a 64 m Mobile Service Tower, 50 m Umbilical Tower and a Blockhouse for launch operations. The Atlas is erected on the pad typically 71 days before launch, followed by Centaur (69 days), its insulation panels and forward equipment module (45 days). Payload separation tests and terminal countdown demonstration are held 40 days and 29 days before launch, respectively. The spacecraft is encapsulated at one of the remote safe facilities and the assembly installed atop Centaur at 13-8 days. Atlas' RP1 fuel is loaded at 3 days. A countdown typically requires 910 h, including two 30 min and 5 min built-in holds. Centaur's LOX is loaded at T-75 min, followed by Atlas' LOX at 55 min and Centaur's LH_2 at 43 min. The booster and its upper stage switch to internal power at 4 min and 2 min, respectively.

Atlas 2
First launch: 7 December 1991
Number launched: 10 to April 1999
Launch sites: Canaveral pads 36A/B, VAFB SLC 3E from 1998 for access to Sun-synchronous, polar and 63.4° orbits for military and Earth observation satellites.
Principal uses: delivery of DSCS– 3 satellites into GTO; future: high inclination missions.
Vehicle success rate: 100% to April 1999
Schedule of further missions:
Performance:
LEO (185 × 185 km, 27.0° Canaveral): 6,780 kg medium fairing, 6,580 1kg large
LEO (185 × 185 km, 90.0° VAFB): 5,510 kg large
GTO (160 × 35,788 km, 28.5°): 2,950 kg medium fairing, 2,810 kg large
Earth escape: 2,000 kg large fairing
Number of stages: Same as Atlas 1
Overall length: 46.8 m with medium fairing; 47.4 m with large fairing
Principal diameter: 3.05 m
Launch mass: 187,170 kg with medium fairing, 187,560 kg with large
Launch thrust: 2,159 kN sea level
Guidance: see Atlas 1

Atlas 2 stage 1
Comment: To complement the higher performance MA5A engines, stage 1 tanks were stretched 1,702 mm (LOX) plus 1,041 mm (RP-1), adding 2,585 kNs total impulse.
Engines: Rocketdyne MA-5A single start liquid bipropellant consisting of two booster engines (1,841 kN total SL) and one sustainer (269 kN sea level). The sustainer remains unchanged from that of Atlas 1 but the booster portion now uses the Delta RS-27 engine for an SI increase from 259.1 s to 263.1 s sea level. The side mounted verniers have been deleted; hydrazine thruster modules on the interstage adaptor have replaced their roll control function.
Overall length: 24.9 m
Principal diameter: 3.05 m
Propellant: 156,260 kg of LOX/RP1
Stage thrust: 2,159 kN sea level
Burn time: 169 s boosters, 277 s sustainer

Atlas launch history, from 1980

	Launch	Atlas/Centaur	Site	Payload
1980				
453	17 January	5029D/3220	36A	Fltsatcom 3. AC 49
454	9 February	35F	3E	GPS Navstar 5
455	3 March	67F	3W	NOSS 3
456	26 April	34F	3E	GPS Navstar 6
457*	29 May	19F	3W	NOAAB
458	30 October	5037D/3228	36A	Fltsatcom 4. AC 57
459	6 December	5034D/3225	36B	Intelsat 502. AC 54
460*	8 December	68E	3W	NOSS 4
1981				
461	21 February	5023D/3213	36A	Comstar 1D. AC 42
462	23 May	5036D/3227	36B	Intelsat 501. AC56
463	23 June	87F	3W	NOAA 7
464	6 August	5039D/3230	36A	Fltsatcom 5. AC59
465	15 December	5035D/3226	36B	Intelsat 503. AC55
466*	18 December	76E	3E	GPS Navstar7
1982				
467	4 March	5038D/3229	36A	Intelsat 504. AC 58
468	28 September	5040D/3231	36B	Intelsat 505. AC 60
469	20 December	60E	3W	DMSP F6
1983				
470	9 February	6001H	3E	NOSS 5
471	28 March	73E	3W	NOAA 8
472	19 May	5041D/3232	36A	Intelsat 506. AC 61
473	10 June	6002H	3E	NOSS 6
474	14 July	75E	3W	GPS Navstar 8
475	17 November	58E	3W	DMSP F7
1984				
476	5 February	6003H	3E	NOSS 7
477*	9 June	5042G/3233	36B	Intelsat 509. AC 62
478	13 June	42E	3W	GPSNavstar 9
479	8 September	14E	3W	GPS Navstar 10
480	12 December	39E	3W	NOAA9
1985				
481	12 March	41E	3W	Geosat
482	22 March	5043G/3234	36B	Intelsat 510. AC 63
483	29 June	5044G/3236	36B	Intelsat 511. AC 64
484	28 September	5045G/3237	36B	Intelsat 512. AC 65
485	8 October	55E	3W	GPS Navstar 11
1986				
486	9 February	6004H	3E	NOSS 8
487	17 September	52E	3W	NOAA 10
488	4 December	5046G/3238	36B	Fltsatcom 7. AC66
1987				
489*	26 March	5048G/3239	36B	Fltsatcom 6. AC67
490	15 May	6005H	3E	NOSS 9
491	19 June	59E	3W	DMSP F8
1988				
492	2 February	54E	3W	DMSP F9
493	24 September	63E	3W	NOAA 11
1989				
494	25 September	5047G/3240	36B	Fltsatcom 8. AC 68
1990				
495	11 April	28E	3W	Stacksat
496	25 July	5049/3241	36B	CRRES. AC69; A1
497	1 December	61E	3W	DMSP F10
1991				
498*	18 April	5050/3242	36B	BS-3H. AC70; A1
499	14 May	50E	3W	NOAA12
500	28 November	53E	3W	DMSP F11
501	7 December	8102/202	36B	Eutelsat 2 F3. AC102; A2
1992				
502	11 February	8101/203	36B	DSCS 3B14. AC101; A2
503	14 March	5052/3244	36B	Galaxy 5. AC 72; A1
504	10 June	8105/205	36B	Intelsat K. AC105; A2A
505	2 July	8103/201	36A	DSCS 3B12. AC-103; A2
506*	22 August	5051/3243	36B	Galaxy 1R. AC71; A1
1993				
507*	25 March	5054/3246	36B	UFO 1. AC 74; A1
508	19 July	8104/204	36A	DSCS 3B9. AC 104; A2
509	9 August	34E	3W	NOAA 13
510	3 September	5055/3247	36B	UFO 2. AC75; A1
511	28 November	8106/206	36A	DSCS 3B10. AC106; A2
512	16 December	8201/601	36B	Telstar 401. AC108; A2AS
1994				
513	13 April	5053/3245	36B	GOES8. AC 73; A1
514	24 June	5056/3248	36B	UFO 3. AC 76; A1
515	3 August	8107/207	36A	DBS 2. AC-107; A2A
516	29 August	20E	3W	DMSP F12
517	6 October	8202/602	36B	Intelsat 703. AC-111; A2AS
518	29 November	8109/208	36A	Orion 1. AC 110; A2A
519	30 December	11E	3W	NOAA14
1995				
520	10 January	8203/603	36B	Intelsat 704. AC113; A2AS
521	29 January	8110/210	36A	UFO 4. AC112; A2
522	22 March	8204/604	36B	Intelsat 705. AC 115; A2AS
523	24 March	45E	3W	DMSP F13
524	7 April	8111/211	36A	AMSC 1. AC 114; A2A
525	23 May	5057/3249	36B	GOES 9. AC77; A1
526	31 May	8112/212	36A	UFO 5. AC 116; A2
527	31 July		36B	DSCS 3B-7. AC-118; A2A
528	29 August	8205/605	36B	JCSat 3. AC 117; A2AS
529	22 October	8114/214	36A	UFO 6. AC119; A2
530	2 December	8206/606	36B	Soho.AC-121; A2AS
531	15 December	8115/215	36A	Galaxy 3R. AC 120; A2A
1996				
532	1 February		36B	Palapa C1. AC126; A2AS
533	3 April		36A	Inmarsat 3F1. AC 122; A2A

Atlas launch history, from 1980 (continued)

	Launch	Atlas/Centaur	Site	Payload
534	30 April		36B	SAX. AC-78; A1
535	25 July		36B	UFO7. AC125; A2
536	8 September		36B	GE 1, AC123, Atlas 2A
537	21 November		36B	Hot Bird 2, AC 124, Atlas 2A
538	18 December		36A	INMARSAT-3 F3, AC-129, Atlas 2A
1997				
539	17 February		36B	JCSAT 4, AC-127, Atlas 2AS
540	8 March		36A	Tempo 2, AC 128, Atlas 2A
541	25 April		36B	GOES 10, AC-79, Atlas 1 (final Atlas1 launch)
542	28 July		36B	Superbird C
543	4 September		36A	GE 3
544	4 October		36B	EchoStar 3
545	25 October		36A	USA 134
546	8 December		36B	Galaxy 8I
1998				
547	29 January		36A	USA 137
548	28 February		36B	Intelsat 806
549	16 March		36A	UFO 8
550	18 June		36A	Intelsat 805
551	9 October		36B	Hot Bird 5
552	20 October		36A	UFO 9
1999				
553	16 February		36A	JC-SAT 5
554	12 April		36A	Eutelsat

* denotes launch failure (June 1984 due to Centaur leak after abnormal separation shock; March 1987 due to lightning strike; April 1991 and August 1992 due to Centaur engine failures; March 1993 due to Atlas booster low thrust, payload attained inadequate low orbit). 36A/B = Canaveral launch; 3E/W = Vandenberg. AC is Atlas Centaur designator (Atlas 2 series began at #101). Launch number at left includes all Atlas vehicles flown. A1= Atlas 1; A2= Atlas 2; A2A= Atlas 2A; A2AS= Atlas 2AS; A2AR= Atlas 2AR

Interstage Assembly
The ISA is similar to that on Atlas 1 but two hydrazine thruster modules similar to those on Centaur provide roll control.

Atlas 2 stage 2
Comment: Centaur's LOX/LH_2 tanks are stretched 338/577 mm to increase propellant loading by 2,990 kg. Foam panels bonded to the tanks. replace the four jettisonable insulation panels of Atlas 1.
Engines: as Atlas 1 but oxidiser/fuel ratio increased from 5.0 to 5.5
Overall length: 10.1 m
Principal diameter: 3.05 m
Propellant: 16,780 kg LOX (aft) + LH_2
Stage thrust: 146.8 kN vacuum
Burn time: typically as Atlas 1

Fairing/payload accommodation
Same as Atlas 1.

Atlas 2A
First launch: 10 June 1992 (Intelsat K)
Number launched: 10 to end-1996
Vehicle success rate: 100%
Performance
LEO (185 × 185 km, 28.5° Canaveral): 7,316 kg large.
LEO (185 × 185 km, 90.0° VAFB): 6,190 kg large.
GTO (167 × 35,788 km, 27.0°): 3,040 kg medium, 2,900 kg large; Block 1 upgrade: 3,180 kg medium fairing, 3,066 kg large.
Earth escape: 2,160 kg large fairing

Atlas 2AS
First launch: 15 December 1993
Number launched: 7 to end 1996
Vehicle success rate: 100% to end 1996
Performance
LEO (185 × 185 km, 28.5° Canaveral): 8,610 kg large fairing.
LEO (185 × 185 km, 90.0° VAFB): 7,210 kg large fairing.
GTO (167 × 35,788 km, 27.0°): 3,730 kg medium, 3.620 kg large; Block 1 enhancement: 3,810 kg medium, 3,700 kg large.
Earth escape: 2,680 kg large fairing

Castor 4A strap-ons
Motor contractor: Thiokol, Huntsville division. TX7803 ground-ignited (quads 1+3), TX7804 air-ignited (quads 2+4) to minimise umbilical tower plume impingement.
Length: 11.28 m
Diameter: 1,016 mm
Average thrust: each 433.6 kN sea level
Burn time: 53 s
Mass at ignition: 11,649 kg (ground), 11,743 kg (air).
Burnout mass: 1,452 kg (ground), 1,529 kg (air).
Propellant: 10,200 kg of HTPB solid.

Atlas 2AR
First launch: planned December 1998
Launch sites: Canaveral pad 36B, Vandenberg SLC3E.
Principal uses: medium-class telecom and meteorological satellite payloads into GTO.

Performance
LEO (185 × 185 km, 28.5°): unavailable.
GTO (167 × 35,788 km, 27.0°): 3,810 kg (4,080 kg with two Castor 4Bs); 3Σ errors unavailable.
Availability: typically four launches/year per pad. Launch typically 24-30 months following contract.
Cost: about US$45 million into GTO
Number of stages: 2
Guidance: Honeywell's Inertial Navigation Unit mounted on Centaur's forward equipment module performs the inertial guidance and attitude control computations for both Atlas/Centaur.
Launch sequence (s) into GTO 27.0°:

00.0	stage 1 ignite (75 per cent thrust)
04.8	stage 1 throttle up to 85 per cent, 30 m
12.9	start pitch programme, 290 m
24.4	ramp to zero angle of attack
32.4	zero angle attack, 2.44 km
44.2	pre-max Q throttle down to 61 per cent
67.6	post-max Q throttle up to 85 per cent
147.9	max 5.0 *g*, begin throttle down
184.9	stage 1cut-off/separation (49 per cent throttle)
199.9	Centaur ignition1
209.9	jettison payload fairing
760.2	Centaur cut-off 1
1,449.9	Centaur ignition 2
1,625.3	Centaur cut-off 2

Atlas 2AR stage 1
Comment: The two-chamber throttleable RD-180 eliminates the booster section and provides a more benign environment. Max thrust 85%; 49% at cut-off. The LOX tank is extended by 2.60 m to accommodate the change in mixture ratio..
Engines: NPO Energomash/P&W RD180 two chamber engine throttleable 40–105%, nominal 3,594 kN SL/3,923 kN vacuum. The chambers are gimballed hydraulically to provide three-axis control during Atlas burn.
Burn time: 185 s

Interstage adapter
Comment: The 4.42 m long, 3.05 m diameter ISA supports Centaur until separation at 185 s is effected by a flexible linear shaped charge around the forward circumference. Construction is an aluminium skin/stringer and frame.

Fairing/payload accommodation
Comment: 2AR will adopt the Block 2 fairing upgrade, to be introduced on 2AS. Other specifications as for earlier models.
Acceleration load: 5.0 *g* maximum longitudinal, 0.4 *g* lateral.
Acoustic: unavailable
Max Q: 427 N/cm²

Centaur upper stage

Background
GD began development of the world's first LOX/LH_2 upper stage in the late 1950s. Through 1995 it was launched 120 (including nine R&D) times with seven operational flight failures. Seven out of 97 operational Atlas Centaur launches failed, but none of the 14 operational Titan Centaurs did. The stage continues in service with the Atlas 1/2 vehicles.

A Centaur G prime derivative debuted February 1994 on Titan 4. NASA purchased the Centaur G and the G prime for applications aboard the Shuttle by increasing LH_2 tank diameter to 4.32 m. The longer G prime version can accommodate almost 50 per cent greater propellant loading than the Atlas 1 version. The USAF's G model was 3 m shorter to provide maximum payload length within the cargo bay but could still deliver 4,500 kg to GEO.

By late 1985, there were seven orders, four G variants for classified payloads and three of the longer G prime for Galileo, Ulysses and Magellan. The G version development cost US$280 million, with NASA adding a further US$88 million for modifications to the *Challenger* and *Atlantis* Shuttle orbiters, and US$80 million for changes to the Shuttle launch pad. Among the modifications, the air force changed the old Atlas Centaur pad 36A for the new version of the upper stage. Orbiter modifications included dump valves to permit jettison of 20 tonnes of cryogenic propellants in a few seconds in an abort.

Centaur G and G prime were cancelled in June 1986 as Shuttle stages because of increased safety concerns over carrying a liquid vehicle inside a manned spaceplane. Development and cancellation costs totalled an estimated US$1 billion. Those built were scrapped.

Centaur G prime was then adapted for employment with Titan 4, creating the West's most powerful launcher combination and providing the air force with the required 4.5 tonne GEO capability (5.7 tonnes with upgraded Titan boosters). The first Centaur for Titan was delivered to the Cape in December 1990 under a US$1.3 billion contract with Martin Marietta that calls for 15 such stages over five years. This model will continue to use the RL10A33A engine. A single engine version employing the RL10C is under consideration to increase capacity (545 kg more into GSO) and improve

Final Centaur assembly is now undertaken in a new building in Denver. Two Titan 4-type stages are at left and two Atlas-type at right. The tanks are either pressurised or mechanically tensioned to prevent the thin-wall structures from collapsing (Lockheed Martin)

reliability and reduce cost; such an Atlas stage would cost about US$105 million to develop.

The upper stage Atlas version carries LOX/LH_2 separated by a double wall vacuum insulated intermediate bulkhead. External insulation on Atlas 1 is provided by glass fibre honeycomb panels jettisoned 25 seconds after booster cut-off. Like Atlas, Centaur is constructed of thin-wall fully monocoque stainless steel, stabilised only by helium pressurisation. An advanced avionics programme for both commercial and Titan Centaurs has replaced the Teledyne digital computer and Honeywell's three-axis inertial platform with a Honeywell INU incorporating RLGs and state of the art processors.

Specifications

Centaur

Engines: two P&W RL10A-3-3A cryogenic multiple start engines of 146.8 kN total vacuum thrust gimballed for three-axis control; 12 × 27 N (Olin Aerospace MR-106) hydrazine thrusters provide three-axis control during coast, spin-up for payload separation and collision avoidance manoeuvre.
Overall length: 9.15 m
Principal diameter: 3.05 m
Propellant: 13,790 kg LOX (aft) + LH_2
Stage thrust: 146.8 kN vacuum
Comment: The single engine configuration, pneumatic system self-test capability and replacing the hydraulic engine gimbal actuators with an electromechanical system reduces cost and increases reliability (reduces part count by 4,200).
Burn time: two burns of 560.3 s + 175.4 s.

Conestoga

Background

SSI designed the Conestoga solid propellant family of orbital launchers capable of handling up to 2.1 tonnes into LEO. It was selected in early 1991 to provide Conestoga launch services for the NASA sponsored Meteor μgravity carrier programme. A preliminary demonstration launch using a single Minuteman 1 second stage motor was conducted in September 1982. In March 1989, SSI fired its first commercial vehicle, a two-stage Starfire sounding rocket for the University of Alabama under a US$1 million contract carrying a μg payload.

SSI's first launcher, Percheron, with a liquid first stage, exploded on its debut from Matagorda Island off the Texas coast 12 August 1981, and the company switched its efforts to the all solid Conestoga designs. Conestoga 1, using a single 206.8 kN thrust Minuteman 1 second stage motor, was successfully fired as a demonstration on a 10.5 min, 313 km altitude suborbital trajectory 9 September 1982 from Matagorda. The current Conestoga orbital versions are based on Thiokol's Castor 4A/4B solids with Star injection stages. The 1679 was selected for BMDO's MSTI-5, although no funding was provided. The debut (1620) October 1995 launch self-destructed at 45 seconds after faulty navigation data produced extensive nozzle movements and depleted the TVC hydraulic fluid.

Specifications

Conestoga 1620

First launch: 23 October 1995 from Wallops Island
Number launched: 1, to April 1999
Launch sites: Wallops Island
Principal applications: small payloads to LEO
Vehicle success rate: 0%.
Schedule of future missions: none known
Performance: 945 kg into 463 km, 40°
Availability: 20 month baseline
Cost: about US$18 million (65% is cost of solids)
Number of stages: 4
Overall length: 15.7 m with 4.88 m long fairing
Principal diameter: 3.57 cm
Launch mass: 87,407 kg
Avionics/guidance: all configurations use TVC for three-axis control through the stage 1/2 phases. The six motor cluster uses only four motors for control authority; the other two have fixed (noncanted) nozzles. TVC provides pitch/yaw control during core burn; RCS system in Payload Attach Fitting provides roll during stage 3 and three-axis during stage 4 coast/burn. The H-MACS Hydrazine Manoeuvring & Attitude Control System can be carried on top of Star for three-axis control and orbit trim. The avionics module is housed in the interstage structure above the core. The Mars guidance system (two tuned rotor gyros + three accelerometers; Guidance Dynamics Corp) interfaces with the 1553 avionics bus to provide all TVC, sequencing, staging and prelaunch monitoring/control. Injection accuracy: ± 37 km/±0.20° 3Σ (±18.5 km/±0.05° 3Σ with H MACS)

Each Centaur upper stage requires paired RL10s, designated C-1 and C-2 (Lockheed Martin)

Performance:

Conestoga 1229 295 kg to 180 km, 90°; 363 kg to 180 km, 38°. Two stage 1 Castors strapped to stage 2 core Castor. Injection by Star 48V stage 3.

Conestoga 1620 1,000 kg to 180 km, 90°; 1,180 kg to 180 km, 38°. Six Castors (4 stage 1 + 2 stage 2) clustered around stage 3 core Castor. Injection by Star 48V stage 4.

Conestoga 1679 1,250 kg to 180 km, 90°; 1,500 kg to 180 km, 38°. As 1620 but Star 63 stage 4.

Launch sequence (s) into 463 km 40°:

00.0	stage 1 ignite
56.4	stage 1 4A sep 16.35 km
59.4	stage 2 ignite 18.01 km
66.4	stage 1 4B sep 22.03 km
119.6	stage 3 ignite 64.58 km
125.6	stage 2 sep 70.93 km
147.6	fairing sep 96.80 km
185.6	stage 3 sep 157.87 km
475.7	stage 4 ignite 460.3 km
562.0	stage 4 burnout 463.0 km
742.0	payload sep 463.0 km

Conestoga 1620 stage 1

Motor: two Castor 4A + two Castor 4B (Thiokol solids) 4Bs provide 6° blowdown TVC
Stage length: 9.14 m
Thrust: 1,884 kN avg
Burn time: 52/61 s Castor 4A/4B
Mass at ignition: 47,137 kg
Propellant mass: 40,164 kg

Conestoga 1620 stage 2

Motor: two Castor 4B (as stage 1), 6° blowdown TVC
Stage length: 9.14 m
Thrust: 869.8 kN average vacuum, 1,049.7 kN peak
Burn time: 61 s.
Mass at ignition: 23,764 kg
Propellant mass: 19,949 kg

Conestoga 1620 stage 3

Motor: one Castor 4B, 6° blowdown TVC; as stage 1 but with increased expansion ratio (12.97) for altitude operation.
Stage length: 9.24 m
Stage diameter: 101.6 cm
Thrust: 434.9 kN average vacuum, 524.9 kN peak.
Burn time: 61 s
Mass at ignition: 11,914 kg
Propellant mass: 9,974 kg

Conestoga 1620 stage 4

Motor: Thiokol Star 48V, with 4° pitch/yaw TVC (Meteor cold gas thrusters provide roll).
Stage length: 2.03 m
Stage diameter: 124.5 cm
Thrust: 66 kN average vacuum
Burn time: 52 s
Mass at ignition: unavailable
Propellant mass: 9,974 kg

Payload accommodation

1.829 m diameter composite fairing provides lengths of 4.27 6.56 m. Separation system derives from Delta technology.
Acceleration: 11 *g* longitudinal
Acoustic: 128.5 dB for acceptance tests

DC-XA Delta Clipper

Background

McDonnell Douglas developed the DC-X vehicle for BMDO's Single Stage Rocket Technology programme, making a total of five flights at White Sands Missile Range by mid-1994. All flights in this first series ended in failure. Three further flights costing US$5 million to US$6 million were successful in May-July 1995 to complete the series.

McDonnell Douglas Space Systems was awarded BMDO's 24-month US$58.9 million Phase 2 contract 16 August 1991 for the design of an SSTO to provide a launch and rapid replacement capability for Brilliant Eyes and Brilliant Pebbles. The experimental DC-X ⅓ scale 11.9 m high, 3.7 m base diameter version (7.2 tonnes dry, 18.88 tonnes at launch, 1.5 tonne LH_2, 7.3 tonne LOX) of their Delta Clipper was demonstrated in summer 1993 at White Sands. Four P&W RL10A5 provide propulsion oxygen/hydrogen engines derived from the RL10A-3 proven by long service on the Centaur upper stage. P&W's US$12 million contract called for four engines plus spares for a fifth. Each 143 kg engine, incorporating a conical nozzle for low altitude operations, can be throttled 30 100 per cent (19.5-65.7 kN) and gimballed ±8°. Four Aerojet gaseous O_2/H_2 1.78 kN thrusters provide RCS.

Phase 1 evaluated three basic launch/recovery concepts: vertical take-off/landing; vertical take-off/ horizontal landing; horizontal take-off/landing. By eliminating the dead weight of wings, vertical/vertical is the lightest and most cost effective design. Even without wings, cross-range would be more than 2,970 km. Very high controllability from four body flaps, GPS navigation and powered landings (at 15 to 20 per cent thrust) would enable it to operate in rain and high winds that would force a diversion for Space Shuttle. Flight control would be totally autonomous and handled by onboard computers. The crew module would be removed for unmanned missions. Airline type operations were planned:

The first ascent (18 August 1993) was a hop to 80 m, demonstrating throttling but remaining vertical throughout. The second (11 September) attained 100 m and the third (30 September) 370 m. Two flights were planned 20/23 October to attain 1.3 km plus various manoeuvres and demonstrate a three-day turnaround, but an engine problem delayed the first flight and funding ended 22 October.

NASA, in early 1994, provided US$1 million to prepare DC-XA for further tests, and DoD in May 1994 released another US$5.1 million for five flights over June to July 1994. Flight 4 attained 790 m on a 135.9 second venture 20 June 1994, for the first time with a full propellant load. Flight 5, intended to last for 136 seconds on 27 June 1994, was aborted after an explosion at ignition blew a 1 × 4 m hole in the composite aeroshell. Controller Pete Conrad triggered the automatic abort sequence at 17 seconds and DC- X landed safely at 78 seconds. The cause was identified as vehicle-vented GH_2 collecting in a service trench.

Flight 6, the first of the new series, came 16 May 1995: 124 seconds, 1,300 m altitude, 15° angle of attack, landed 340 m downrange. Firsts included GPS data for navigation and differential throttling for primary attitude control. Number 7 on 12 June, 132 seconds, 1,900 m altitude, 070° AOA, 590 m downrange. The gaseous attitude thrusters were used for the first time. Number 8: 7 July, 124 seconds, 2,500 m altitude. Pitched to 10° to show that an SSTO could enter the atmosphere nose first and then flip over for landing. Faulty radar altitude readings produced a heavy landing, cracking the aeroshell and cancelling flight 9.

As a result of White House policy decisions on the future route for US launchers, DC-X was transferred to NASA after flight 8.

Following BMDO's Phase 2, the full size 38.7 m- high 458 tonne (36 tonne dry) DC-Y prototype was proposed. Payload cost for the operational version would be around US$250 2,500/kg, depending on flight rates. The goal was for 4,500 kg into 740 km polar and 10,000 kg LEO.

McDonnell Douglas predicted that less than 100 people could maintain all operational and ground support needs for a fleet of six, each with a three-day turnaround. Cargo containers would be loaded à la commercial airfreight into a 4.57 × 4.57 × 6.7 m bay. A major factor in reducing ground processing is the lack of Shuttle type thermal tiles; a maximum 1,430°C around the carbon/carbon nose is predicted, some 440°C below maximum protective capability. The flaps and the lower temperature nose areas would be protected by a carbon/silicon carbide composite. The underbody and other medium temperature areas would carry Inconel 718 metal multiple layers. The aeroshell would be a graphite epoxy/aluminium honeycomb sandwich supported by a P75/P100 graphite/epoxy truss. The

LOX tank would be an aluminium/copper/lithium alloy (Weldalite 049); LH_2 woven P75 graphite cloth and toughened epoxy. DC–Y would carry eight bell nozzle engines, four with extending nozzles. They would fire at 85 per cent for the 1.3 *g* launch, reserving 100 per cent for engine-out situations and 110 per cent in more extreme cases. The goal was for 200 flights between tear-down inspection.

Delta-1

Background

In 1959, NASA's Goddard Space Flight Center contracted the Douglas Aircraft Company to develop an interim space launcher from the company's Thor IRBM and Vanguard upper stages programmes. Douglas had been working on the Thor since 1954 for the air force. NASA envisioned a vehicle capable of putting about 45 kg into GTO. As the space race matured, Delta's growth potential and the need for a medium launcher for communications, meteorological and science satellites paced a continuous set of improvements to the original vehicle.

Strap-ons were first added in 1964, and in 1965 the booster propellant capacity was increased, a larger fairing was introduced and both upper stages were upgraded. The number of strap-ons was increased to six in 1970 and to nine in 1972, at the same time as Aerojet's Titan Transtage engine was adopted for stage 2. Further improvements were made in 1973 with the introduction of Rocketdyne's RS-27 stage 1 engine and in 1975 with the introduction of Thiokol's Castor 4 strap-ons to replace the Castor 2s. These latter two configurations, when using TRW's TR201 engine, derived from the Apollo Lunar Module's descent engine, were designated the Delta 2914 and 3914, respectively. These and their variants were responsible for the bulk of launches after 1975

Delta-2

Background

NASA ordered Delta's Huntington Beach production line to close down in 1984 but Shuttle's failure in January 1986 resulted in its reactivation eight months later to produce three further vehicles (#182 to 184) under a US$130 million contract. Delta's subsequent emergence as a commercial vehicle resulted from its selection as USAF's Medium Launch Vehicle (MLV) in January 1987 under an initial US$316.5 million contract. The basic buy of seven eventually expanded to 20 when the air force exercised two options in February 1988 (seven vehicles) and December 1988 (six vehicles), totalling US$680 million, to launch the network of GPS Navstars offloaded from Shuttle.

Delta became associated with civil science again when the US$140.6 million contract for the Geotail, Wind and Polar science satellites was signed with NASA

Delta launches (from 1978)

No	Payload	Date	Pad	Type
138	IUE	26 January 1978	17A	2914
139	Landsat 3/Oscar	5 March 1978	2W	2910
140	BSE	7 April 1978	17B	2914
141	OTS 2	11 May 1978	17A	3914
142	GOES 3	16 June 1978	17B	2914
143	Geos 2	14 July 1978	17A	2914
144	ISEE3	12 August 1978	17B	2914
145	Nimbus 7	24 October 1978	2W	2910
146	NATO 3C	19 November 1978	17B	2914
147	Anik 4	16 December 1978	17A	3914
148	Scatha	30 January 1979	17B	2914
149	Westar 3	10 August 1979	17A	2914
150	RCA Satcom 3	7 December 1979	17A	3914
151	Solar Max	14 February 1980	17A	3910
152	GOES 4	9 September 1980	17A	3914
153	SBS 1	15 November 1980	17A	3910/PAM
154	GOES 5	22 May 1981	17A	3914
155	Dyn Exp 1/2	3 August 1981	2W	3913
156	SBS 2	24 September 1981	17A	3910/PAM
157	SME + Uosat 1	6 October 1981	2W	2310
158	RCA Satcom 3R	20 November 1981	17A	3910/PAM
159	RCA Satcom 4	16 January 1982	17A	3910/PAM
160	Westar 4	26 February 1982	17A	3910/PAM
161	Insat 1A	10 April 1982	17A	3910/PAM
162	Westar 5	9 June 1982	17A	3910/PAM
163	Landsat 4	16 July 1982	2W	3920
164	Anik D1	26 August 1982	17B	3920/PAM
165	RCA Satcom 5	28 October 1982	17B	3924
166	IRAS	26 January 1983	2W	3910
167	RCA Satcom 6	11 April 1983	17B	3924
168	GOES 6	28 April 1983	17A	3914
169	Exosat	26 May 1983	2W	3914
170	Galaxy 1	28 June 1983	17B	3920/PAM
171	Telstar 3A	28 July 1983	17A	3920/PAM
172	RCA Satcom 7	8 September 1983	17B	3924
173	Galaxy 2	22 September 1983	17A	3920/PAM
174	Landsat 5 + UoSat 2	1 March 1984	2W	3920
175	AMPTE	16 August 1984	17A	3924
176	Galaxy 3	21 September 1984	17B	3920/PAM
177	NATO 3D	13 November 1984	17A	3914
178*	GOES G	3 May 1986	17A	3914
180	DoD (SDI)	5 September 1986	17B	3920
179	GOES 7	26 February 1987	17A	3924
182	Palapa B2P	20 March 1987	17B	3920/PAM
181	DoD (SDI)	8 February 1988	17B	3910
184	Navstar 21	14 February 1989	17A	6925
183	Delta Star	24 March 1989	17B	3920
185	Navstar 22	10 June 1989	17A	6925
186	Navstar 23	18 August 1989	17A	6925
187	Marcopolo 1	27 August 1989	17B	4925
188	Navstar 24	21 October 1989	17A	6925
189	COBE	18 November 1989	2W	5920
190	Navstar 25	11 December 1989	17B	6925
191	Navstar 26	24 January 1990	17A	6925
192	RME/LACE	14 February 1990	17B	69208
193	Navstar 27	26 March 1990	17A	6925
194	Palapa B2R	13 April 1990	17B	69258
195	Rosat	1 June 1990	17A	692010
196	Insat 1D	12 June 1990	17B	4925
197	Navstar 2-8	2 August 1990	17A	6925
198	Marcopolo 2	18 August 1990	17B	6925
199	Navstar 29	1 October 1990	17A	6925
200	Inmarsat 2 F1	30 October 1990	17B	6925
201	Navstar 210	26 November 1990	17A	7925
202	NATO 4A	8 January 1991	17B	7925
203	Inmarsat 2 F2	8 March 1991	17B	6925
204	ASC 2	13 April 1991	17B	7925
205	Aurora 2	29 May 1991	17B	7925

No	Payload	Date	Pad	Type
206	Navstar 211 +Losat	4 July 1991	17A	7925
207	Navstar 2-12	23 February 1992	17B	7925
208	Navstar 2 13	10 April 1992	17B	7925
209	Palapa B4	14 May 1992	17B	7925 8
210	EUVE	7 June 1992	17A	6920-10
211	Navstar 2 14	7 July 1992	17B	7925
212	Geotail	24 July 1992	17A	6925
213	Satcom C4	31 August 1992	17B	7925
214	Navstar 2 15	9 September 1992	17A	7925
215	DFS 3	12 October 1992	17B	7925
216	Navstar 2-16	22 November 1992	17A	7925
217	Navstar 2 17	18 December 1992	17B	7925
218	Navstar 218	3 February 1993	17A	7925
219	Navstar 2-19/SEDS	30 March 1993	17A	7925
220	Navstar 220	12 May 1993	17A	7925
221	Navstar 2-21	26 June 1993	17A	7925
222	Navstar 222	30 August 1993	17B	7925
223	Navstar 2 23	26 October 1993	17B	7925
224	NATO 4B	8 December 1993	17A	7925
225	Galaxy 1RR	19 February 1994	17B	79258
226	Navstar 224/SEDS	10 March 1994	17A	7925
227	Wind	1 November 1994	17B	792510
228[1]	Koreasat 1	5 August 1995	17B	7925
229	Radarsat	4 November 1995	2W	792010
230	XTE	30 December 1995	17A	792010
231	Koreasat 2	14 January 1996	17B	7925
232	NEAR	17 February 1996	17	7925
233	Polar	24 February 1996	2W	7925 10
234	Navstar 225	28 March 1996	17	7925
235	MSX	24 April 1996	2W	7920 10
236	Galaxy 9	24 May 1996	17B	7925
237	Navstar 226	16 July 1996	17	7925
238	Navstar 227	12 September 1996	17	7925
239	Mars Global Surveyor	7 November 1996	17	7925
240	Mars Pathfinder	4 December 1996	17	7925
241*	Navstar 2R1	17 January 1997	17	7925
242	Iridium 4 8	5 May 1997	17	7920
243	Thor 2A	20 May 1997	17	7925
244	Iridium (× 5)	9 July 1997	17	7920
245	Navstar 29	23 July 1997	17	7295
246	Iridium (× 5)	21 August 1997	17	7920
247	ACE	25 August 1997	17	7920
248	Iridium (× 5)	26 September 1997	17	7920
249	Navstar 38	6 November 1997	17	7925
250	Iridium (× 5)	9 November 1997	17	7920
251	Iridium (× 5)	20 December 1997	17	7920
252	Skynet 4D	10 January 1998	17	7925
253	Globalstar (× 4)	14 February 1998	17	7420
254	Iridium (× 5)	18 February 1998	17	7920
255	Iridium (× 5)	30 March 1998	17	7920
256	Globalstar (× 4)	24 April 1998	17	7420
257	Iridium (× 5)	17 May 1998	17	7920
258	Thor 3	15 June 1998	17	7925
259*	Galaxy 10	27 August 1998	17	Delta 3
260	Iridium (× 5)	8 September 1998	17	7920
261	Deep Space-1/ SEDSAT	24 October 1998	17	7326
262	Iridium (× 5)	6 November 1998	17	7920
263	Bonum 1	23 November 1998	17	7925
264	Mars Climate Orbiter	11 December 1998	17	7424
265	Mars Polar Lander	27 January 1998	17	7425
266	Stardust	7 February 1999	17	7426
267	Argos/Orsted/ Sunsat	23 February 1999	17	7920
268	Landsat 7	15 April 1999	17	7920
269*	Orion 3	4 May 1999	17	Delta 3

* indicates launcher failure. Note 1: Koreasat 1's Delta did not achieve GTO because number 6 GEM failed to seperate; the satellite's own station-keeping propellant was required to reach final GEO. The pad designator indicates launch site; 17A/B = Cape Canaveral; 2E/W = WTR. The full list to the end of 1977 is included on p 276 of the 1990/91 edition.

Delta 7925 GPS launch from Canaveral's pad 17A **1995**

in December 1990. It included VAFB pad modifications and pre-priced options for 12 further vehicles to mid-1998. The previous 3920/PAM models could handle 1,284 kg to GTO, but the Delta 2 6925 raised that capacity to 1,447 kg, and the 7925 to 1,841 kg.

The most recent NASA/USAF ELV reimbursement agreement of October 1987 sets the cost to NASA for a Delta at US$25.6 million cost (FY89 rate), and in 1990 the USAF agreed to pay US$37 million plus US$4 million launch services.

USAF Delta 2 requirements called for a switch to the more capable 7925 following their first nine launches to cope with the heavier GPS Navstar Block 2A models. The 7925 similarly incorporates a 2.9 m diameter fairing and 3.6 m longer first stage but adds graphite epoxy strap-ons and uprated stage 1 thrust. 6925 made its last flight with Geotail in 1992. 7925 was selected April 1993 for USAF's MLV3 contract. 25 are required (+11 options) to launch the GPS Block 2R replenishment architecture from 1997-2002.

Launches are required within 60 days of request. GPS performance is improved to 2,141 kg and GTO to 1,882 kg, principally by increasing the expansion ratios of the three air-lit strap-ons. MLV-3 employs the triply redundant RIFCA Redundant Inertial Flight Control Assembly and the improved FTS Flight Termination. The fairing is modified so that the satellite, stage 3 and fairing can be transported to the pad as a single unit. The 7920 series differs from the 6920 principally with the adoption of the 12:1 expansion ratio RS27A (also known as RS-2701C) stage 1 engine and more powerful solid strap-ons. The 7925A additional option (not exercised) is to enlarge the strap-ons to 116.8 cm diameter. This would improve GPS performance to 2,254 kg and GTO to 1,987 kg.

Delta 7320, 7326, 7420, 7425, 7426 series

MDA/OSC teamed to bid for NASA's Med-Lite contract, providing a performance in the current gap below Delta at half the cost (US$25 million). The team was selected March 1995, and the contract awarded February 1996 for five firm launches from 1998. Two vehicles were to have satisfied the requirements: MDA's DeltaLite and OSC's Taurus, but MDA now has no plans to develop Delta Lite. That would have comprised Castor 120 stages 1/2, the current Delta 2 stage 2 as stage 3 and a Star 37 kick motor. This combination would deliver 1,930 kg into 185 km, 28.7°; 1,463 kg 185 km, 90°; 650 kg GTO; 331 kg C^3 10 km^2/s^2. With two Castor 4B strap-ons, these capacities would be 2,540/1,968/837/440 kg, respectively. Taurus equivalent capacities would be: 1,770/1,435/445/230 kg and 2,180/1,750/640/275 kg. Instead, the Delta 7325 will be used at a special discount. Equivalent capacity is: 2,865/2,095/1,004/594 kg. It was reported in February 1996 that this vehicle, although double the cost, will replace Delta-Lite and the excess capacity be offered for other payloads.

Delta 4925/5920

Based on the 3920, these models were produced as interim models to help ease the RS27 first stage engine shortage until the full Delta 2 manufacturing capacity became available. The 4925 was essentially a 3920/PAM but with uprated Castor 4A strap-ons (as used on Delta 2) compensating for an old 765.06 kN Thor IRBM Rocketdyne MB-3 Block 3 stage 1 engine from storage at Norton AFB, California. Orbital performance was thus similar to that of the 3920/PAM. The 5920, flown only with NASA's Cosmic Background Explorer in November 1989, was a 3920 (that is retaining the RS 27) but with Castor 4A strap-ons providing 1,406/3,849 kg GTO/LEO capacities. Only two 4925s were flown, carrying BSB 1 and Insat 1D, because only two MB3 engines were available, built in 1978 from USAF spares. As the engine required RJ-1 fuel, no longer produced in the US, it had to be imported from Japan, where it remained in use aboard NASDA's H1 vehicle.

Company sources claim a production capacity of 12 to 18 per year is possible from their Huntington Beach (manufacture) and Pueblo, Colorado (assembly; 250 work force) facilities, although the rate is 12 initially. Eight sales appear to be the average at least for the near future.

Delta 2 had problems with the launch of Koreasat 1: on this flight one of the GEM solid propellant boosters failed to separate. Orbit was reached, but it was lower than planned and the satellite had to use its own propellant to reach geosynchronous orbit thus reducing its operational life.

The first complete loss of a Delta-2 came on 17 January 1997 when a 7925 was destroyed seconds after launch, this being due to a malfunction of one of the strap-on boosters. The payload – the first Navstar2R satellite was destroyed.

Specifications

Delta 2/Delta 7925/7920

First launch: 26 November 1990
Number launched: 58 to April 1999
Launch sites: Canaveral pads 17A/B, VAFB SLC2W.
Principal uses: launches of GPS Navstar Block 2A; medium class telecom satellites into GTO.
Vehicle success rate: 96.55%
Performance:
GTO: 1,842 kg (1,882 kg from December 1995)
GPS: 2,090 kg (2,141 kg from December 1995)
LEO: 5,039 kg (28.7°) 7920 two-stage
Sun-synchronous: 3,175 kg, 830 km 7920 VAFB
Molniya-type (63.4°, 370 × 40,094 km): 1,275 kg
Lunar delivery: 1,240 kg
Availability: typically 30 months from contract signature to launch.
Number of stages: 3 plus 9 strap-ons (7920 excludes stage 3)
Overall length: 38.41 m
Principal diameter: 2.44 m
Launch mass: 231,870 kg
Launch thrust: 3,110 kN (six solids fire at launch)
Guidance: the digital inertial guidance system is mounted inside a cylinder at stage 2's forward end, controlling the vehicle during stage 1/2 and commanding spinup/separation of stage 3 in addition to triggering its fuze based sequencing system. The system is a strapped down all-inertial unit incorporating a Delco guidance computer and a Delta Redundant Inertial Measurement System (DRIMS) containing three gyros, four accelerometers and conditioning electronics. The computer also issues preprogrammed sequence commands and provides attitude control. GTO perigee altitude error is typically <5.6 km + 0.6° inclination. Typical 3Σ errors for 2-stage missions into 1851,850 km circular orbits are 2 km/0.05°. The triply redundant RIFCA Redundant Inertial Flight Control Assembly replaced DRIMS from the XTE launch of December 1995. The single fault tolerant RIFCA comprises six RL 20 AlliedSignal ring laser gyros and six QA3000 Sundstrand accelerometers in a triply redundant modular architecture. Each lane contains two sensor channels, a 1750A processor and I/O interfaces. Two lanes have a 1553B bus, for vehicle data, and the third incorporates RS422 for GSE. There are two separate power sources. The data are continuously processed, updating the required course throughout ascent.

Alliant GEM solid strap-ons

Comment: Hercules' Aerospace Group (now Alliant Techsystems) of Magna, Utah was contracted in February 1987 to provide 144 (16 flight sets) higher performance, lightweight Graphite Epoxy Motors for Delta 7925. A second contract was awarded 1991 for a further 117 (13 sets), to begin production 1993. A third award, for 16, was made June 1995. For MLV3, the expansion ratio for the air-lit nozzles is increased from 10.6 to 16.3 by lengthening the nozzle by 30 cm and increasing diameter by 19.5 cm (debut was on XTE December 1995).
Length: 12.96 m
Diameter: 101.6 cm
Average thrust: each 446.02 kN SL; 499.18 kN vacuum
Burn time: 63.3 s
Mass at ignition: 13,232 kg each for six ignited on pad; 13,061 kg each for three air-lit
Propellant mass: 11,765 kg each
Propellant type: solid
Burn sequence: six GEMs ignite at launch, followed by the remaining three 2.5 s after the others burnout at 63 s. The original six are spring separated (under command of a sequencer between stage 1's tanks) in two symmetrical sets at 66/67 s; the final three burnout at 129 s and separate 3 s later.

Delta 7920/7925 stage 1

Comment: Incorporating uprated strap-ons, which provide the majority of the launch thrust, permitted increasing RS 27A main engine expansion ratio from 8:1 for improved altitude performance.
Engines: Rocketdyne 12:1 expansion ratio RS 27 single start liquid bipropellant of 890 kN sea level thrust (1,084.8 kN vacuum), hydraulically gimballed for pitch/yaw control; two Rocketdyne 4.45 kN LR101-NA11 verniers provide roll control.
Overall length: 26.047 m
Diameter: 2.44 m
Stage mass: 101,718kg
Oxidiser: liquid oxygen
Fuel: RP-1
Propellant quantity (usable): 96,033 kg
Oxidiser tank length: 13.00 m, extended 2.23 m over 3920 version
Fuel tank length: 8.38 m, extended 1.43 m over 3920 version
Stage thrust: 890 kN SL + two 4.45 kN verniers
Burn time: 260.5 s
Comment: The two isogrid tanks are separated by a 77 cm centre body section that houses control electronics, ordnance sequencing and telemetry

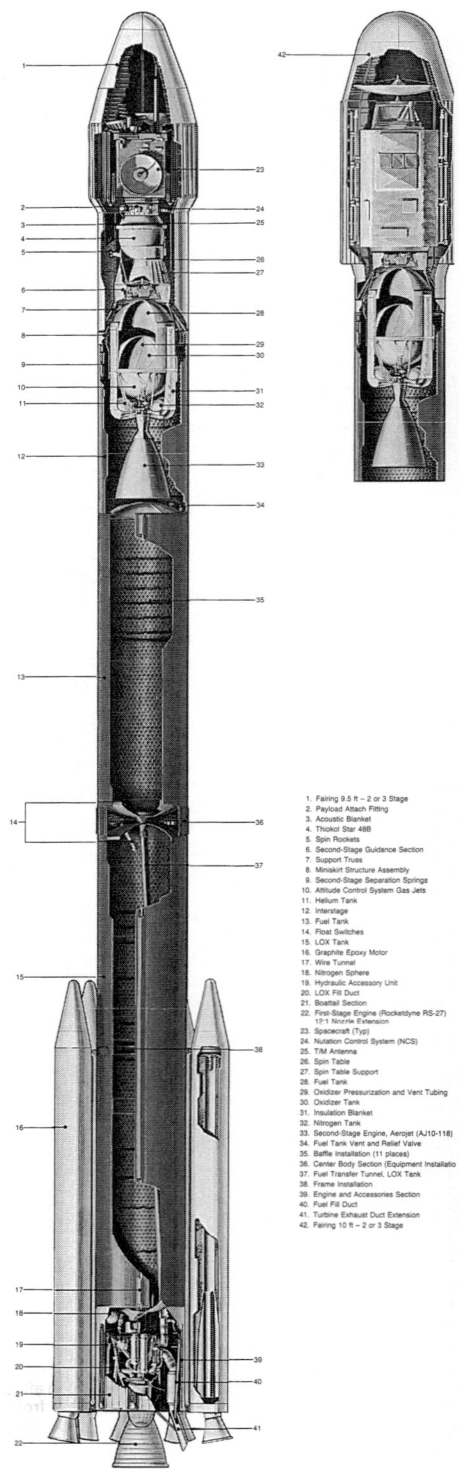

Delta 7925/7920 principal elements (MDA)

Hoisting Delta stage 2 at pad 17A (NASA) ***1995***

Delta launch manifest (as of January 1999)

Year/Quarter	Launch Site	Customer	Payload	Version
1999				
2nd	Cape Canaveral Air Station, Florida	Space Systems/ Loral	Globalstar-3	7420
2nd	Vandenburg Air Force Base, California	Motorola-12	Iridium	7920
2nd	Cape Canaveral Air Station, Florida	NASA	FUSE	7320
3rd	Cape Canaveral Air Station, Florida	Space Systems/ Loral	Globalstar-4	7420
3rd	Cape Canaveral Air Station, Florida	Hughes Space & Communications, Intl, Inc.	ICO-4	Delta III
3rd	Cape Canaveral Air Station, Florida	Space Systems/ Loral	Globalstar-5	7420
3rd	Vandenberg Air Force Base, California	Motorola-13	Iridium	7420
3rd	Vandenberg Air Force Base, California	Space Systems/ Loral	Globalstar-6	7920
4th	Vandenberg Air Force Base, California	Space Systems/ Loral	Globalstar-7	7920
4th	Cape Canaveral Air Station, Florida	Hughes Space & Communications, Intl, Inc.	ICO-5	Delta III
4th	Vandenberg Air Force Base, California	NASA	EO-1/SAC-C	7320
4th	Cape Canaveral Air Station, Florida	Space Systems/ Loral	Globalstar-8	7420
2000				
1st	Cape Canaveral Air Station, Florida	Space Systems/ Loral	Globalstar-9	7420
1st	Cape Canaveral Air Station, Florida	NASA	IMAGE	7326
1st	Cape Canaveral Air Station, Florida	Hughes Space & Communications, Intl, Inc.	ICO-7	Delta III
1st	Cape Canaveral Air Station, Florida	TRW	SBIRS-LOW	7420
2nd	Vandenberg Air Force Base, California	NASA	Jason/Timed	7920
2nd	Cape Canaveral Air Station, Florida	Hughes Space & Communications, Intl, Inc.	ICO-10	Delta III
3rd	Cape Canaveral Air Station, Florida	Hughes Space & Communications, Intl, Inc.	ICO-12	Delta III
4th	Cape Canaveral Air Station, Florida	Hughes Space & Communications, Intl, Inc.	Hughes 6	Delta III
4th	Vandenberg Air Force Base, California	NASA	Gravity Probe-B	7920
4th	Cape Canaveral Air Station, Florida		MAP	7420
4th	Vandenberg Air Force Base, California		EOS-PM	
2001				
1st	Cape Canaveral Air Station, Florida	NASA	GEOLITE	792X
1st	Cape Canaveral Air Station, Florida	Hughes Space & Communications, Intl, Inc.	Hughes 7	Delta III
1st	Vandenberg Air Force Base, California	NASA	Mars Orbiter	792X
2nd	Cape Canaveral Air Station, Florida	NASA	Mars Lander	742X
2nd	Cape Canaveral Air Station, Florida	Space Systems/ Loral-1	TBA*	Delta III
3rd	Cape Canaveral Air Station, Florida	Hughes Space & Communications, Intl, Inc.	Hughes 8	Delta III
4th	Cape Canaveral Air Station, Florida	Space Systems/ Loral-2	TBA*	Delta III
4th	Cape Canaveral Air Station, Florida	NASA	SIRTF	7920H
2002				
1st	Cape Canaveral Air Station, Florida	Hughes Space & Communications, Intl, Inc.	Hughes 9	Delta III
2nd	Cape Canaveral Air Station, Florida	Space Systems/ Loral-3	TBA*	Delta III
3rd	Cape Canaveral Air Station, Florida	Hughes Space & Communications, Intl, Inc.	Hughes 10	Delta III

equipment. The package receives commands from RIFCA and drives servo amplifiers for engine gimbal; DC batteries in the centre section provide power. A rate gyro was added forward of this section to ensure adequate stability margins with the extended tanks and larger fairing.

Interstage
The 4.72 m-long isogrid interstage extends from the top of stage 1 to stage 2's miniskirt, encircling most of the upper stage until explosive bolt detonation 8 s after stage 1 burnout. Six spring driven separation rods at the forward end then separate the stages and stage 2 ignition occurs 5 s later.

Delta 7920/7925 stage 2
Engine: pressure-fed restartable 65:1 Aerojet AJ10 118K, developed for the USAF Improved Trans-stage Injector programme
Overall length: 5.89 m.
Diameter: 1.70 m, suspended by 2.44 m diameter mini skirt and support truss.
Stage mass: 6,930 kg.
Oxidiser: nitrogen tetroxide.
Fuel: Aerozine-50.
Propellant mass (usable): 6,006 kg.
Thrust: 43.37 kN vacuum.
Burn time: 432 s (restartable).
Attitude control system: the nitrogen cold gas jet Redundant Attitude Control System (RACS) provides three-axis control during coast periods and roll control during powered flight; hydraulic gimballing of the main engine provides pitch/yaw control under the command of RIFCA at the forward end. Three helium spheres provide tank pressurisation. Four aft isogrid panels provide equipment attach points and tank fill ports; the gas spheres are filled through the miniskirt on pad.
Burn time: On a typical 7925 GTO mission, stage 2 will burn for 343 s to enter a low circular orbit, coast under control of RACS for 620 s and then burn near the equator for a final 70 s. For the two-stage 7920, the burns are 395/25 s, respectively.

Delta 7925 stage 3
Designation: PAM-D
Motor: Thiokol Star 48B solid
Stage length: 2.04 m
Diameter: 1.24 m
Propellant: 1,756-2,025 kg solid, depending on mission requirements.
Thrust: 66.4 kN vacuum
Burn time: 87.1 s
Mass: 2,141 kg before ignition; 132 kg after burnout.
Comment: Delta 2's third stage is derived from components and concepts used on the PAM-D and the USAF SGS-2 upper stage. The Star 48B is base supported by a spin table that mates to the top of the second stage guidance section. Spin rockets fire typically 50 s after stage 2 burnout to create a stabilising spin of 30,110 rpm for stage 3 firing. Four or eight spin rockets can be used in 334/823/934 N thrust versions, firing for 1 s. Stage 2's RIFCA triggers the stage 3 fuze based sequencing system, which produces spin-up, pyrotechnic/spring separation 2 s later and ignition after a further 38 s. A Nutation Control System derived from SGS2 and mounted above the PAM suppresses coning during motor burn and post-burn phases; a single axis rate gyro senses the motion and fires a counteracting hydrazine 178 N thruster system (2.7 kg propellant loading). Stage 3 can also house a 3.6 kg C-band radar transponder tracking system to aid early orbit determination; if integrated with the telemetry unit it adds 2.3 kg. The latter returns spacecraft environment data such as vibration, acceleration transients, shock, velocity increment, indication of separation and temperatures.

Three stage missions adopt the standard 3712 Payload Attach Fitting at stage 3's forward end, capable of holding 3,084 kg. The spacecraft is secured by means of a two-piece V-block clamp assembly, held by two instrumented studs. Two ordnance cutters sever the studs for four springs to effect a 0.6-2.4 m/s separation rate. A yo-yo weight tumble system despins and imparts a coning motion to the expended motor 2 s after separation to modify its momentum vector and prevent recontact with the spacecraft. On Delta 6920 missions, a 6019 Payload Attach Fitting was used designed originally for NASA's Multimission Modular Spacecraft and capable of holding 2,177 kg with three explosive nuts and a secondary latching system. Stage 2's nitrogen gas thrusters ensure a separation speed of 31 cm/s.

Fairing
The spacecraft is protected by an 848.8 cm long, 839 kg aluminium fairing 243.8 cm diameter at the base and 289.6 cm at the widest portion. Essentially it

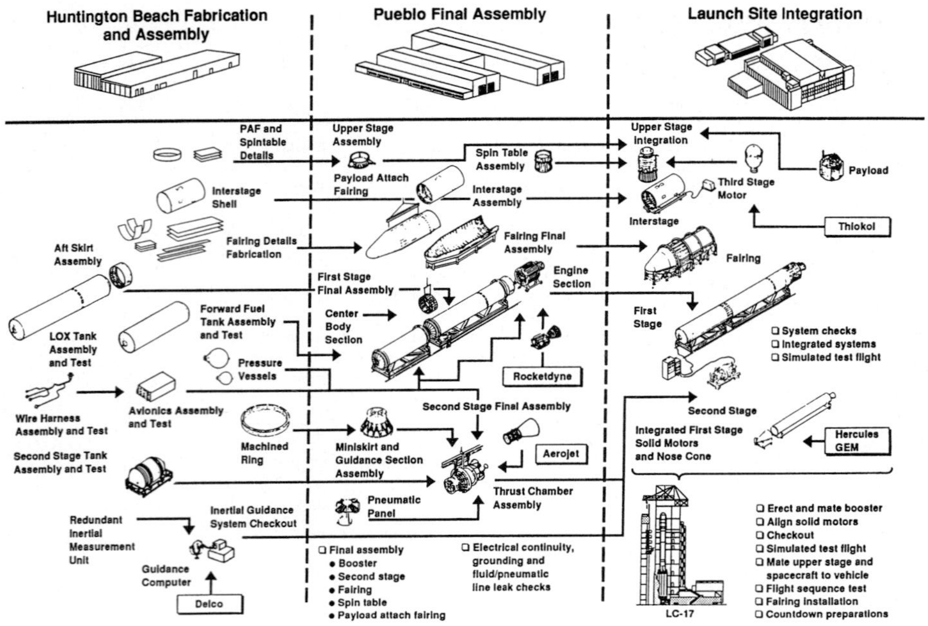

Delta 2 production flow (MDA)

consists of a Titan 3C section atop the previous Delta fairing cylinder. The halves are pyrotechnically separated during second stage burning at a height of about 111 km where aerodynamic heating has reduced to 1,135 W/m². However, 243.8/304.8 cm diameter versions are available; in these cases, 8/10 are added to the normal Delta 4-digit vehicle configuration designations, respectively. A composite 304.8 cm diameter 8.89 m long fairing has been developed Iridium. Air conditioning on the pad can maintain 14 26.7°C, max humidity 50%. Four standard fairing doors provide limited access on the pad and others can be provided. Payload closeout and fairing installation occurs typically three days before launch. An umbilical to stage 2 provides 64 hardwire payload links via the fairing.

Acceleration load: 6.25 *g* maximum at end of stage 1 burn for heaviest GTO payloads; 4.24 *g* at stage 3 burnout; 1.31 *g* at launch.

Acoustic load: maximum about 130 dB at launch/ transonic; payloads subjected to 140 dB for 60 s for flight acceptance test.

Temperature: 33°C max on fairing inner wall after 300 s.

Vehicle/Payload processing

Delta stage 1/2 checkout typically takes place over 13 to 7 weeks before launch in Hangar M of the Cape Canaveral Air Station (CCAS). Stage 1/interstage erection followed by strap-on attachment occurs in the sixth week before launch, with stage 2 mating in the fifth, checkout over the fourth to second week, stage 3 with attached spacecraft mating seven days before, followed by ordnance installations. The final week sees fairing installation, mission rehearsal and stage 2 propellant loading. Launch is typically scheduled for a Thursday. MDA provides transport/handling for payloads between local airports and processing facilities. CCAS offers three principal hangars for non-hazardous spacecraft processing. Hangar AO has a 14 × 53.3 m class 100,000 high bay clean room capable of handling two payloads simultaneously. Hangar AE has class 1,000 facilities and the Delta launch control centre, and Hangar AM with a 4.4 × 5 m clean room and solarium. Loading of payload propellants and pyrotechnics can be performed in three facilities: Explosive Safe Area 60A and KSC's Spacecraft Assembly & Encapsulation Facility #2, both of class 100,000 standard, or KSC's Cargo Hazardous Servicing Facility Complex, built to handle Shuttle sized elements. The spacecraft then meets the stage 3 assembly for the first time in the Payload Spin Test Facility 2.8 km west of complex 17, where it is mated and balanced at 300 rpm.

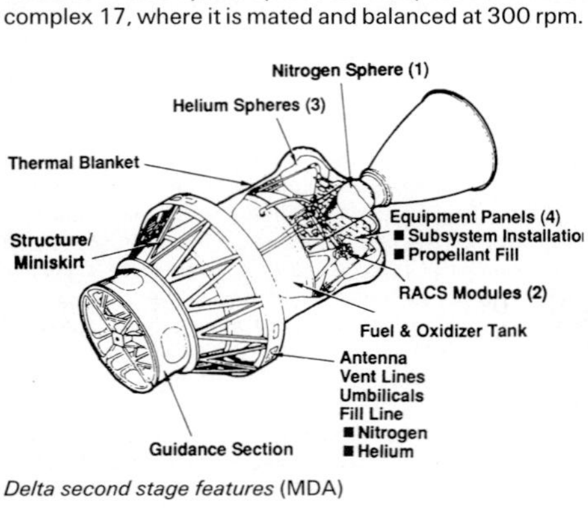

Delta second stage features (MDA)

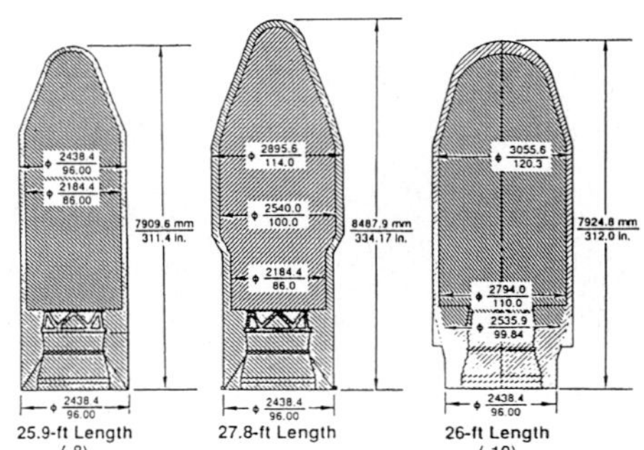

Delta 2 fairing options. A composite 3 m version is also now available

The combination is then ready for transport to the pad in MDA's 6.1 m high, 3.05 m diameter N_2 purged handling can. Alternatively, Astrotech Space Operations offer equivalent commercial processing facilities at their Titusville site. Building 1 provides three 18.3 × 12.2 m class 100,000 high bays for non-hazardous processing, building 2 handles the hazardous processing of PAM, payload fuelling & ordnance loading, mating and spin balancing.

Delta-3

Background

Delta-3 will provide a low Earth orbit payload capability of 8,350 kg and a geosynchronous transfer orbit capability of 3,800 kg, twice the payload of the Delta-2. To meet the specification Delta-3 has a larger composite fairing to house bigger payloads, a larger and more powerful strap-on solid rocket motors, a new cryogenically propelled single engine upper stage.

Both Delta-2 and Delta-3 use the same first-stage engine, as well as similar avionics systems, launch operations infrastructure and liquid oxygen tanks to minimise cost and maximise flexibility. The first stage is powered by a Rocketdyne RS 27A main engine and two Rocketdyne vernier engines to control roll during main engine burn and attitude control between main engine cut-off and second-stage separation. The diameter of the booster fuel tank was increased from Delta-2 to reduce length and improve control margins.

Nine Alliant Techsystems built strap-on solid rocket motors supplement the first stage: these are directly evolved from the Delta-2's proven motors, but provide 25 per cent more thrust. Three of the nine are equipped with thrust vector control to further improve vehicle manoeuvrability and control. The second stage carries a Pratt & Whitney RLB– 10-2 engine which is derived from the RL10 engine in use for more than three decades aboard the Centaur stage used on Atlas and Titan launch vehicles. The RLB 102 also incorporates a larger exit cone for increased specific impulse and payload capability: it relies heavily on the flight-proven RL10 engine components, with operability improvements for increased reliability.

Mitsubishi Heavy Industries will supply 16 sets of stage 1 fuel and stage 2 LH2 tanks, delivery starting in 1996. These will only be used for commercial launches. The Delta-3 will use the SLC 17B at Cape Canaveral. Workers will modify the white room platform locations (part of the 10-storey mobile service tower used to protect the vehicle during processing), add a ground liquid hydrogen propellant tank and increase the overhead crane capability for Delta-3. The solid rocket motor storage area, Delta checkout facility, and high pressure test facility at Cape Canaveral will also be modified to assemble and check out the Delta 3. Vehicle control and launch operations will be conducted from the operations building.

Delta launcher family record

First launch: 13 May 1960

Number launched: 268 to end-April 1999

Launch sites: Canaveral pads 17A/B, Vandenberg pad 2W

Vehicle success rate, career: 95.5% excluding Delta 3: 94.8% including 2 Delta 3 failures.

Vehicle success rate, past 25 launches: 100% excluding Delta 3; 92% including 2 Delta 3 failures; 98% in past 50 launches.

Delta-4

Background

The Delta-4 series of medium-weight and heavyweight launchers will become the mainstay of the Delta family and provide a wide range of optional payload capabilities for military and commercial users. In a development cycle that began in 1995, the US Air Force began a programme leading to procurement of a series of satellite launchers in the Evolved Expendable Launch Vehicle (EELV) programme. During the first phase four competitors completed a 15-month contract to validate low-cost concepts and in December 1996 two contractors were selected to participate in the second phase known as the Pre-Engineering, Manufacturing and Development (Pre-EMD) phase under a fixed-price US$60 million contract for each company. In November 1997, the US Air Force announced that it would introduce competition into the EELV programme by sustaining a dual-source procurement strategy.

On 16 October 1998, the US Air Force announced that it was to buy 19 Boeing Delta-4 launchers for the EELV programme at a cost of US$1.38 billion. The

External Tank production in NASA's Michoud Assembly facility (Lockheed Martin Manned Space Systems) **1996**

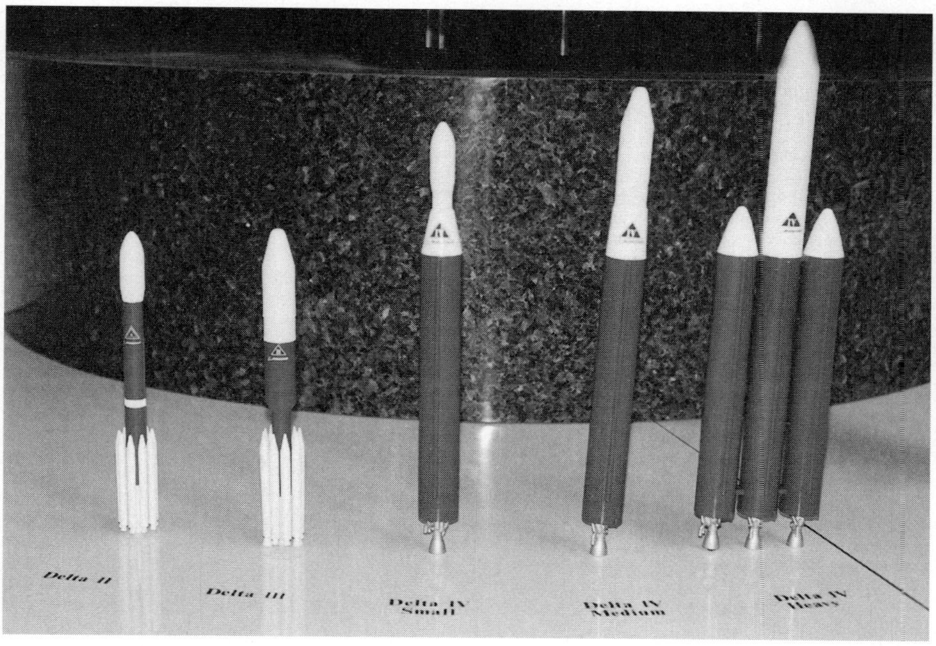

Boeing Delta-4 family (Theo Pirard) **1999**/0054374

TDRS using an IUS for geo injection, departing from Shuttle (NASA) **1996**

Delta-4 family includes five launch vehicles with payload capabilities ranging from 4,173 to 13,154 kg: medium, heavy and three variants of the medium vehicle collectively known as the medium-plus variants. All vehicles will use the Common Booster Core (CBC) powered by the cryogenic liquid hydrogen/liquid oxygen fuelled Boeing (Rocketdyne) RS-68 engine delivering 2,900 kN thrust. The RS-68 is designed for low-cost production and has 80 per cent fewer moving parts than the Shuttle Main Engine. Modified Delta 2 and Delta 3 upper stages will be added to the CBC to complete each vehicle.

Members of the Delta-4 family include the medium which carries the cryogenic second stage of the Delta 3 and that launcher's composite 4 m diameter fairing to provide a 4,200 kg GTO capability. Medium-plus members also available for commercial operations include the 4,2 with two solid rocket motors, a 4 m payload fairing and a GTO capability of 5,800 kg. The 5,2 also has two solid rocket boosters but a 5 m fairing currently manufactured by Boeing for the Titan 4 and a GTO capability of 4,800 kg. The 5,4 has the same payload shroud but four solid propellant rockets and can lift 6,700 kg to GTO. The Delta-4 heavy links three of the CBCs together in parallel and carries an enlarged Delta second stage for increased propellant. With a 5 m diameter fairing it can lift 13,200 kg to GTO.

Boeing selected Decatur, Alabama, as the home of the Delta-4 CBC manufacturing programme and construction of the 139,350 m² facility began in November 1997. The CBC cores are 37.92 m in length with a diameter of 4.85 m – roughly the size of a Boeing wide-body airliner fuselage. Completed CBCs will be moved to docks on the Tennessee River using a specially designed transporter, loaded onto a cargo vessel and delivered to launch sites at Cape Canaveral Air Force Station (CCAFS), Florida, and Vandenberg Air Force Base, California. Employment at the Decatur facility began with 100 people in 1998, will grow to almost 600 by the end of 1999, increasing to 2,000 by 2004. The RS-68 engines will be manufactured at the Boeing (Rocketdyne) Canoga Park facility. Start-up for production commenced early 1999 with the first Delta-4 scheduled to leave the facility in March 2000. First launch is scheduled for 2001.

In July 1998, Boeing unveiled plans for a major expansion of launch capabilities at SLC-37, CCAFS, to accommodate Delta-4. A new launch pad, mobile service tower, fixed umbilical tower, 6,967 m² horizontal integration facility as well as support and test facilities will be provided. In addition, Boeing plans to modify SLC-6 at Vandenberg AFB to handle west coast launches.

IUS

Background

Boeing's two-stage all-solid Inertial Upper Stage is designed to deliver payloads of more than 2 tonnes directly into GEO from LEO following launch by either Shuttle or expendable Titan. The USAF Space & Missile Systems Center IUS Office in Los Angeles maintains overall responsibility for development, procurement and operations, allocating vehicles to NASA Marshall for civil use. Most of the manufacturing is conducted at Boeing's Kent Space Center south of Seattle, but the motors themselves come from United Technologies/Chem cal Systems Division of California.

NASA and the USAF conceived of the Interim Upper Stage as a stop-gap solution while they awaited the introduction of a reusable orbital transfer vehicle, now not likely until next century. Boeing's built their concept around their successful dual-stage solid propellant Burner 2, but growth of mass requirements into GEO dictated a more powerful vehicle. The company was awarded the development contract in August 1976 following a 2½ year competition, leading to the final qualification in 1981. The first, US$700 million, contract called for eight stages, with a US$277.3 million option for a further six to be delivered by April 1985 subsequently exercised. A third batch (vehicles 15 to 19) was contracted at a 32 per cent lower unit cost. By September 1984, the USAF programme was estimated to have cost US$1.7 billion.

IUS, qualified to Mil-Std 1540A standards, was designed as the most reliable upper stage ever built (98 per cent). Centaur's 4,540 kg GEO performance delivered special payloads for an increasing number of demanding military and NASA planetary missions from Shuttle's cargo bay, but the June 1986 decision to cancel the highly energetic liquid Centaur stage from the manned spaceplane on safety grounds ensured further IUS employment. The Magellan, Ulysses and Galileo deep space probes were consequently switched to Shuttle/IUS deployment. The stage, now renamed the Inertial Upper Stage, is also offered for the Titan 4 and commercial Titan 3 vehicles.

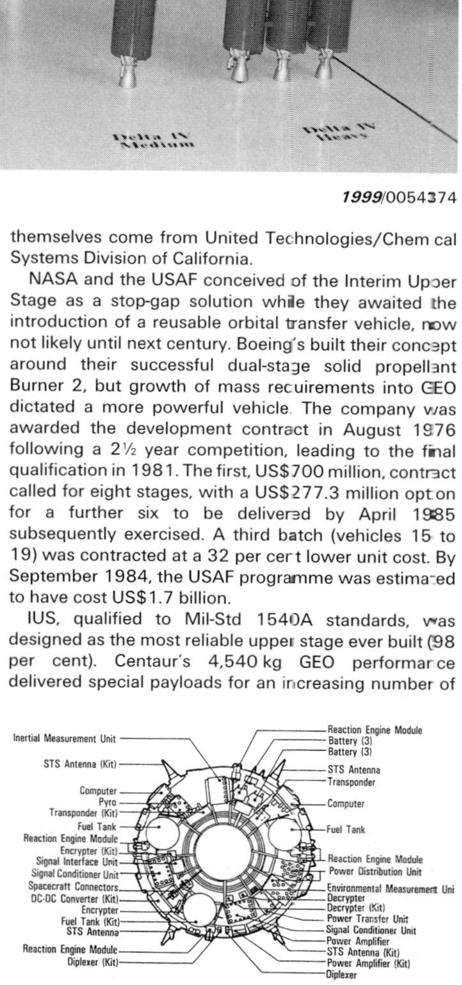

IUS Equipment Support Section in Shuttle configuration (NASA)

Boeing completed studies in 1993 on an IUS upgrade that could be available 1999. The second stage replaced the solid with storable liquid propellants and improved avionics to reduce mass and cost. Previous studies considered increasing GEO capacity to 3,765 kg by using Aerojet's Transtar engine on a liquid stage 2. Other possibilities included adding a third stage (a PAM was for Ulysses) and enlarging stage 2.

Originally, the unit cost was US$60 million to US$70 million, depending on mission unique requirements, within production lots of threes. The air force ordered 23 IUSs in the initial buy and the last was delivered in mid1991. A follow-on contract delivered one in September 1994 to carry TDRS 7 in 1995. The first serial buy also included two for USAF Titan 4, but they were cancelled. NASA, in May 1994, decided to order another under a US$25 million contract to carry its AXAF astronomy satellite in 1998. Shuttle/IUS flight turnaround is eight months.

Specifications

Inertial Upper Stage

First flight: 30 October 1982
Number flown: 21 to April 1999
Vehicles: Shuttle, Titan 34D, Titan 4; Titan 3
Success rate: 95% through April 1999 (TDRS1 payload on failed stage 2 was able to attain operating orbit)
Number of stages: 2 solid.
Overall length: 5.17 m
Principal diameter: 2.9 m

Inertial Upper Stage flight record

	Launched	Vehicle	Payload
1	30 October 1982	T 34D	DSCS II-15/III1
2*	4 April 1983	STS 6	TDRS 1 communication satellites
3	24 January 1985	STS 51C	Magnum electronic intelligence
4	3 October 1985	STS 51J	DSCS III 2/3
5*	28 January 1986	STS51L	TDRS 2 communication satellites
6	29 September 1988	STS 26	TDRS 3comms
7	13 March 1989	STS29	TDRS 4 communication satellites
8	4 May 1989	STS 30	Magellan probe
9	14 June 1989	T4-1	DSP early warning
10	4 September 1989	T 34D	DSCS II 16/III-4
11	18 October 1989	STS34	Galileo probe
12	23 November 1989	STS33	Magnum electronic intelligence
13	6 October 1990	STS41	Ulysses probe
14	13 November 1990	T4-3	DSP early warning
15	2 August 1991	STS43	TDRS 5 communication satellites
16	24 November 1991	STS 44	DSP 16 early warning
17	13 January 1993	STS54	TDRS 6 communication satellites
18	22 December 1994	T411	DSP 17 early warning
19	13 July 1995	STS70	TDRS 7 communication satellites
20	24 February 1997	T4-20	DSP 18 early warning
21	9 April 1999	T4B	DSP 19 early warning

T = Titan. STS = Shuttle mission. * stage 2's TVC system failed on second mission but payload separated and attained final GEO with its own thrusters; TDRS 2 and its IUS were lost in the Shuttle Challenger explosion. DSP-19 lost in failure of SRM-2 to ignite following successful Titan mainstage launch. Originally, Boeing did not acknowledge all of the IUS missions, but the WWW site for the IUS put together by Boeing now lists all 20 missions. STS-38, in November 1990, has been claimed as a Magnum flight but a second IUS only 2 days after DSP's is unlikely.

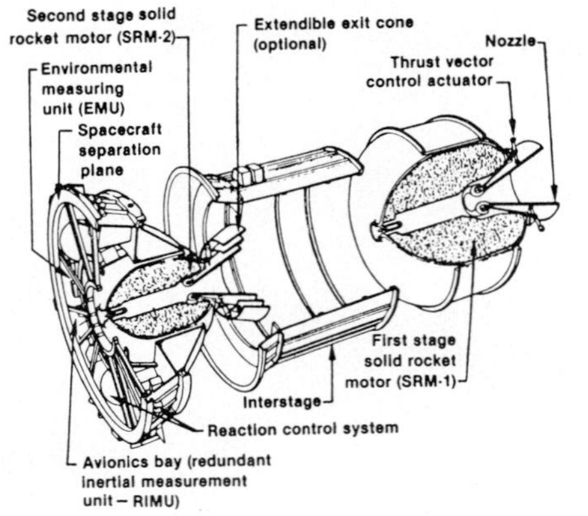

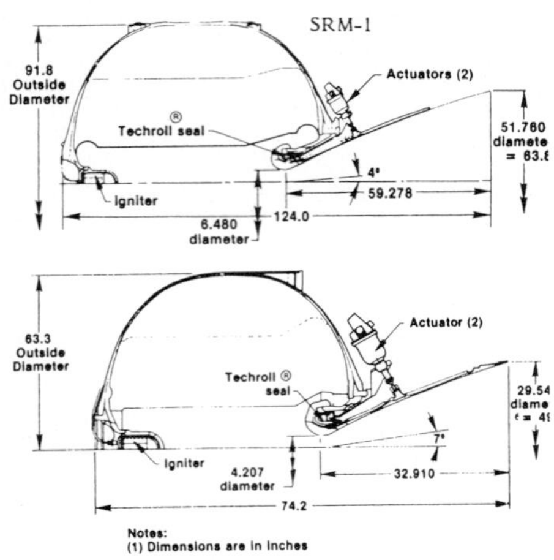

The Inertial Upper Stage's two solid propellant motors (UTC Chemical Systems Division)

Mass (excluding payload): 14,760 kg (Shuttle version), 13,100 kg (Titan 34D)
Performance: (motors can be offloaded 0-50%, vehicle sized for up to 3,600 kg payloads, but 7,200 kg possible with modifications)
GEO: 2,268 kg (Shuttle), 1,817 kg (Titan 34D), 2,364 kg (Titan 4 with CSD SRM strap-ons), 2,860 kg (Titan 4B with Hercules SRMU strap-ons)
GTO: 4,944 kg (using payload for apogee kick into GEO)
Polar (96°, 24 h): 1,860 kg from 57° Cape Canaveral orbit
Molniya (925 × 39,450 km, 63°, 12 h): 3,645 kg (T4 SRM via 185 km, 55°), 3,866 kg (T4 SRM via 185 km, 57°), 4,189 kg (T4 SRM via 185 km, 63°), 4,933 kg (T4 SRMU via 185 km, 55°), 5,253 kg (T4B SRMU via 185 km, 57°).
Escape: 820 kg for C^3 of 60 km²/s², 3,550 kg for 7.0 km²/s².

IUS Motors
Description: IUS incorporates two UTC Chemical Systems Div solid propellant motors. On a typical GEO mission, the larger SRM1 performs GTO injection, followed by SRM2 insertion into GEO. The motors are designed from a common approach, with Kevlar filament wound cases, 3-D carbon/carbon integral throats and 2-D carbon-carbon exit cone nozzles. Thrust vector pitch/yaw control is provided by the nozzles moving on a Techroll joint pressurised by silicon oil under the action of electromechanical actuators. SRM1 can gimbal 4°, SRM2 7°. The Techroll joint failed on the second mission near the end of the SRM2 burn but the payload was able to separate and attain GEO with its own thrusters. SRM2 can be flown without its Extendible Exit Cone (EEC), resulting in a reduced specific impulse.

IUS SRM1 stage 1
Motor length: 3.52 m
Motor/stage diameter: 2.34 m
Propellant loading: 4,853-9,709 kg
Thrust: 185.1 kN average vacuum
Specific impulse: 295.5 s vacuum
Burn time: 152 s 100% loading

IUS SRM2 stage 2
Motor length: 2.08 m without nozzle extension
Motor diameter: 1.61 m
Propellant loading: 1,361-2,749 kg
Thrust: 78.41 kN average vacuum
Specific impulse: 289.1 s (without EEC), 303.5 s (with EEC), vacuum
Burn time: 103.35 s 100% loading

IUS interstage and ESS
Description: An aluminium skin stringer interstage connects the SRM1 to the Equipment Support System (ESS), which houses SRM2, most of the avionics and the payload interfaces (3.05 m diameter payload mounting ring and electrical connectors). The eight longerons on the aluminium skin stringer ESS assembly provide primary load paths and the spacecraft attach points.

Reaction control subsystem
Description: Six modules each with two Kaiser Marquardt 133 N hydrazine thrusters provide roll control during main motor burns, three-axis control during coast periods, vernier orbit corrections after SRM2 burnout, and collision avoidance manoeuvres after payload separation. The configuration provides dual-nozzle redundancy and there are no forward facing thrusters to avoid payload contamination. Two tanks hold a total of 109 kg of hydrazine but one and three tank versions are possible. RCS burns to depletion after payload release in order to remove IUS as far as possible from GEO.

Avionics
Description: IUS is completely autonomous once separated from its launcher. Avionics are redundant throughout; the A side is normally in control but the B side performs the same functions and issues the same commands in parallel and takes over should A fail. This redundancy saved three missions through TDRS 6, on IUS1/6/7. IUS1 lost its A computer memory before SRM2 ignition because of a cosmic ray upset, an IUS6 thruster stuck on, and IUS7's SRM2 failed to arm. On IUS13 (TDRS 6), side B's power was lost. The avionics consist of five subsystems: TT&C, Guidance & Navigation, Data Management, TVC, and Electrical Power. Telemetry is provided at 16/64 kbit/s rates, allowing 1/4 kbit/s, respectively, from the payload. Eight primary and eight back-up discrete commands can be sent to the payload. The Guidance & Navigation Subsystem (GNS) provides measurements of angular rates and linear accelerations to the two Delco M362S 65,536, 16 bit word computers of the Data Management Subsystem from a redundant strapdown inertial measurement unit (RIMU) and a star scanner. The RIMU incorporates five Hamilton Standard RI1010 rate integrating gyros and five Kearfott 2401 accelerometers. IUS uses an explicit guidance algorithm (gamma guidance) to generate thrust steering commands, SRM ignition time and RCS commands. The vehicle is oriented to a thrust attitude before each burn based on normal performance of the remaining propulsion stages. Stage 1 carries four batteries: two 100 Ah + one 140 Ah for the avionics, plus one 100 Ah for the spacecraft. Five 13 Ah (one for the payload) provide power for stage 2 operations. All are AgZn 28±4 V batteries. The current Delco computers and Hamilton Standard RIMU may be replaced by a new Honeywell fault tolerant system.

IUS/payload deployment
Description: On Shuttle missions, the IUS/Payload is spring released from the ASE tilted at 58°. The RCS does not activate until 10 min later at a separation of about 60 m. SRM1 ignition does not occur for a further 50 min until the Orbiter is some 18 km distant. The times in the following schedule are from Orbiter/IUS deployment (min:s):

10:00	Activate RCS three-axis control
34:00	Star scanner attitude update
47:10	manoeuvre to SRM1 burn attitude; data to 64 kbit/s
59:48	SRM1 ignition
61:39	RCS corrects SRM-1 errors
66:39	RCS manoeuvre to coast attitude, with roll for thermal control
370:55	switch to stage 2 batteries
371:12	separate stage 1, extend SRM2 nozzle 7 s later
371:47	manoeuvre to SRM2 burn attitude; data to 64 kbit/s
373:47	SRM2 burn
379:14	RCS corrects SRM2 errors
389:14	Separate payload
394:56	Collision avoidance manoeuvre by IUS

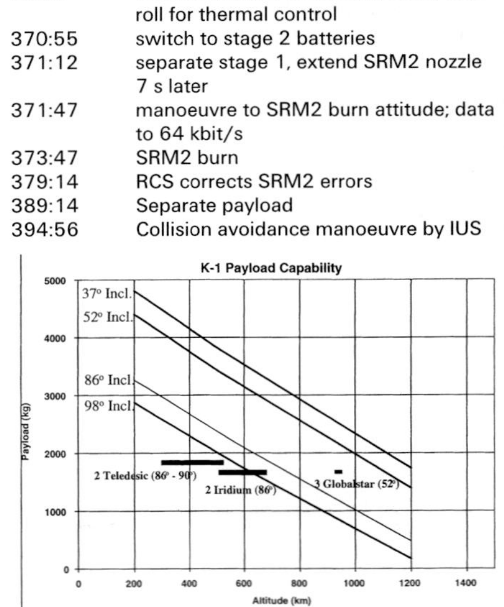

K-1 payload capacity **1997**/0003456

IUS/payload processing
The IUS, without motors, is fabricated and tested at the Boeing Space Center and shipped to the Cape, where CSD's motors are added. Processing/checkout are performed in the east low bay of the USAF Solid Motor Assembly Building (SMAB) at Cape Canaveral, followed by spacecraft mating in the Shuttle Payload Integration Facility (SPIF) in the SMAB west bay. For NASA missions, integration is performed in the Vertical Processing Facility. In both cases, the assembly is transferred to pad 39 in the MultiMission Support Equipment canister. For Titan missions, the IUS is processed as for the Shuttle, but the payload is mated on the pad following Titan/IUS integration.

ATHENA (LMLV)

Background
LMSC began investigating the possibilities of converting excess missiles in 1987 (it was responsible for the Polaris, Poseidon and Trident SLBMs) but dropped the approach in 1990 when studies showed the vehicle would not be reliable enough (90 to 95 per cent) for the commercial market. Lockheed Corp approved the LLV development programme in January 1993. The philosophy is to use existing or imminent hardware, aluminium structures rather than composite (lower cost), higher margins in lieu of acceptance tests and not to push performance at the expense of reliability or lower cost.

Lockheed Launch Vehicles should deliver 800 to 3,200 kg to LEO for US$15 million to US$26 million. LMLV 1 employs a Castor 120 stage 1 and Orbus 21D stage 2. LMLV 2 inserts a second Castor 120 between the two, while LMLV 3 would add at least two Castor 4A strap-ons (3/4/6 strap-ons also possible). All versions carry a hydrazine orbit adjust module, which also houses the vehicles' avionics. Transfer orbits and Earth escape could be provided by a Thiokol Star stage 4 with TVC. The Castor 4A could also be replaced by the 4AXL version.

The programme moved from LMSC in Sunnyvale to Denver in late 1995, although the vehicles (now named LMLV) continue temporarily to be built at the Californian site. LMSC and CTA Inc signed a roughly US$5 million contract 18 February 1994 for the launch of CTA's 113 kg GEMStar 1 into a 480 km polar orbit from VAFB on the debut LMLV 1, 15 August 1995. Launch occurred from an adaptor on one of the solid booster mounts of the Shuttle SLC-6 complex. This flight used an existing Orbus 21 modified to 21D. The vehicle was destroyed by range safety after 159 seconds, when separate stage 1 TVC and IMU problems left it uncontrollable. Second launch carried the satellite Lewis into orbit on 23 August 1997. Before the launch of the first LMLV-2, the vehicle programme was renamed Athena. Athena 2 made its maiden flight with the launch of Lunar Prospector on 7 January 1998. A third Athena 1 launch carried the ROCSAT satellite for Taiwan into orbit on 27 January 1999.

Specifications
LMLV 1 Athena 1
First launch: 15 August 1995 from SLC6 Vandenberg AFB
Number launched: 2, to end 1998

LMLV 1 debut, August 1995 **1996**

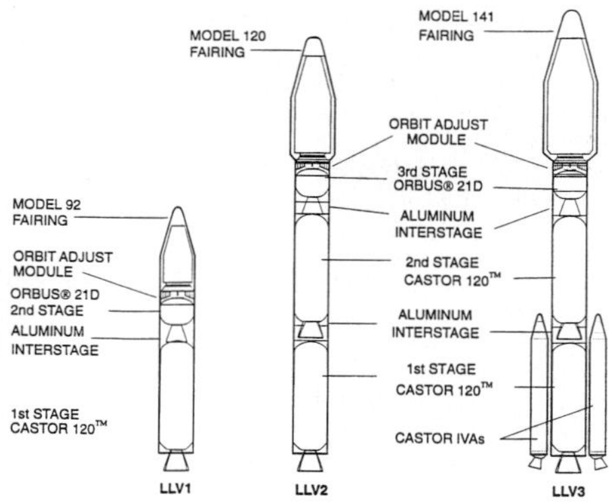

LMLV family. The market requirements for an LMLV 3 are under evaluation. The LMLV 3 strap-ons would not separate from stage 1 ***1995***

Launch sites: SLC-6 VAFB; planned LC46 Cape Canaveral.
Principal applications: small payloads to LEO
Vehicle success rate: 50%
Availability: current production capacity can support 12 flights annually (increase to 24 if required). Typical mission flown 18 months after receiving order
Cost: US$16 million (excluding range + insurance)
Number of stages: 2
Overall length: 18.9 m with 2.34 m Model 92 fairing
Principal diameter: 236.3 cm
Launch mass: 66,345 kg
Launch thrust: 1,780 kN
Avionics/guidance: flight electronics, batteries, telemetry, inertial navigation unit and attitude control carried in OAM. Autopilot is hosted in Lockheed RISC 3000-based avionics processor and uses a Litton LN-100L IMU with software modified for LMLV to provide incremental velocity and angle data to Lockheed's navigation software. Strapdown IMU uses three non-dithered Litton Zero-lock 18 cm path length RLGs and three A-4 accelerometers. Avionics performance is monitored by the avionics processor and formatted for downlink. Castor 120 provides ±5.5° movable nozzle TVC pitch/yaw control (enabled at T-7 s but restricted until pad cleared). Orbus 21D provides ±4° movable nozzle TVC pitch/yaw control. OAM provides roll control during burns, three-axis during coast and orbit adjust. Accuracy: 3Σ ±5.4 km at 1,100 km circular altitude, 0.06° inc.

Launch sequence (s) into 463 km 28.5°:

00.0	stage 1 ignite
85.9	stage 1 burnout 57.1 km/3.33 km/s
132.0	fairing separation 109.1 km/3.18 km/s
137.0	stage 1 sep/stage 2 ignition 113.9km/3.16 km/s
291.0	stage 2 burnout 208.6km/7.54 km/s
517.0	OAM ignition 329.6 km/7.39 km/s
1,885.0	OAM burnout 463 km/7.64km/s

LMLV 1 stage 1
Motor: Thiokol Castor 120
Stage length: 9.02 m
Diameter: 236.3 cm
Thrust: 1,780 kN at launch, 1,554.8 kN average vacuum, 2.00 MN peak vacuum at 20 s
Burn time: 85.9 s
Mass at ignition: 52,852 kg
Propellant mass: 48,809 kg
Separation: hot separation; linear shaped charge separates the interstage's aluminium skin. Interstage is 184.7 cm high

LMLV 1 stage 2
Comment: The first launch employed an Orbus previously assigned to another mission and offloaded.
Motor: Orbus 21D employing carbon phenolic nozzle instead of previous carbon/carbon because of US$700,000 lower cost
Stage length: 317 cm
Diameter: 233.7 cm
Thrust: 192.4 kN average vacuum, 267.8 kN peak
Burn time: 154 s
Mass at ignition: 10,644 kg
Propellant mass: 9,779 kg

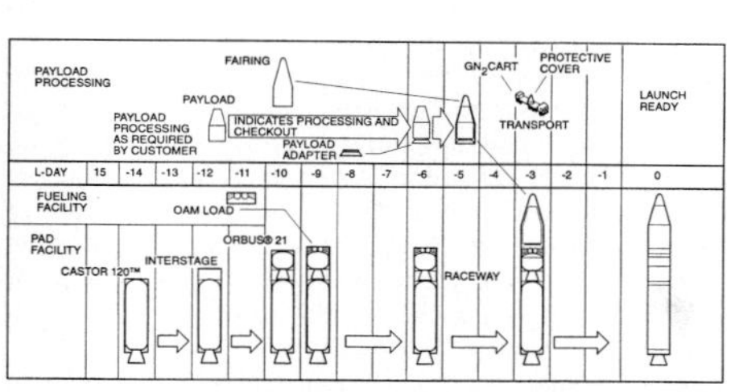

LMLV 1 processing flow (Lockheed Martin) ***1996***

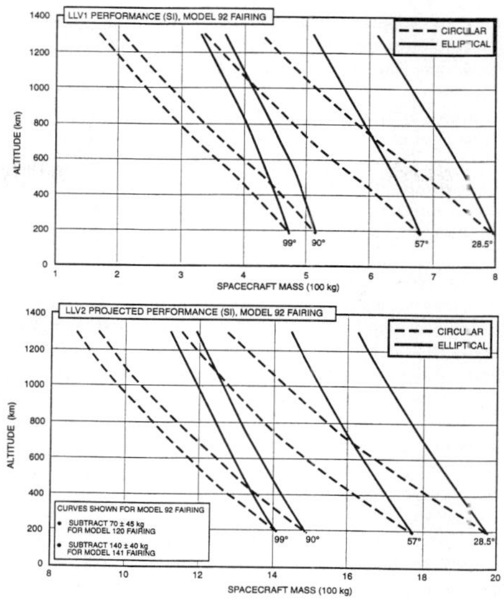

LMLV 1/2 orbital performance ***1995***

Separation: clean Zip tube fractures the OAM separation ring.

LMLV 1 OAM orbit adjust module
Olin Aerospace integrates the ACS into OAM for each vehicle. OAM can carry 2, 4 or 6 hydrazine aluminium tanks for 118/236/354 kg loading (total 607/653/818 kg). LMLV 1/2 typically carry 236 kg (354 kg on flight 1). 10 Olin MR-107 blowdown thrusters: 4 × 222 N axial for orbit trims + 6 × 111 N for three-axis control. Tanks pressurised to 31.3 atmospheres by N_2. Total impulse 254.4 kNs. Elastomeric bladders. Activated 1 s after stage 1 ignition. After payload release, the axial thrusters fire for separation and to deplete the tanks

Fairing and payload accommodation
The Model 92 233.7 cm diameter 792 kg two-piece clamshell fairing provides a 10.6 m³ 198.4 cm diameter × 429 cm long payload envelope. LMLV 2/3 can also employ: Model 120 305 cm diameter, providing 29.5 m³ 274.3 × 710.1 cm envelope; Model 141 358 cm diameter, providing 56.3 m³ 327.7 × 872 cm. A Zip tube cuts the base and springs separate the halves. 74 copper conductors + two fibre optic cables are available for payload links via the T-0 umbilical. 10 analogue + 10 discrete monitors, plus five signal conditioned continuity loops and five separation indicators, are standard via telemetry. The avionics computer can issue eight discrete payload commands during ascent.
Acceleration load: 8 *g* longitudinal (stage 1/2), ±2.5 *g* lateral.
Acoustic load: 133.5 dB overall
Maximum internal fairing temp: 107°C
Recommended acceptance vibration: unavailable.

Vehicle and payload processing
LMLV 1 assembly/checkout requires about 14 days by a crew of 2,025 (LMLV 2 18 days; LMLV 3 25 days). OAM is checked out at the factory and delivered directly to the hydrazine loading facility. It arrives at the pad on T-9 days. The payload is mounted to its adaptor (mating to LMLV's 167.6 cm diameter bolt circle) and the fairing added before transfer to the pad three-days before launch. Launch control is provided from a 12 m long van 2,700 m from the pad connected via two-way fibre optics in the T-0 flyaway umbilical.

LMLV 2 Athena 2
First launch: 7 January 1998
Number launched: 1
Launch sites: as LMLV 1
Schedule of future missions: Lunar Prospector October 1997, Space Imaging December 1997.
Performance: see diagram. Also, 450 kg into lunar transfer.

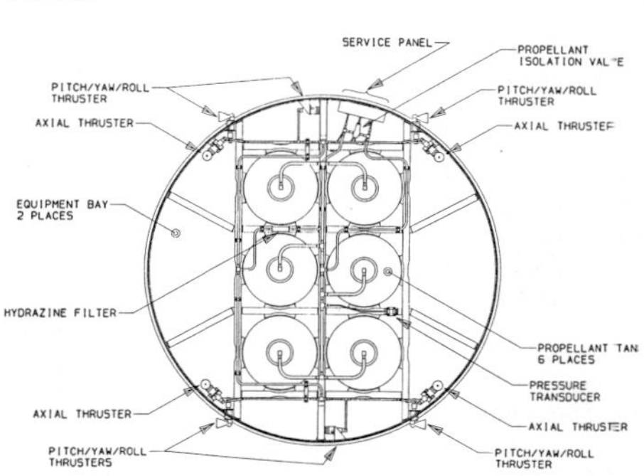

OAM configuration (Lockheed Martin) ***1996***

Cost: US$21 million
Number of stages: 3
Overall length: 28.2 m with 2.34 m fairing; 30.2 m with 2.95 m fairing; 31.7 m with 3.58 m fairing.
Launch mass: 120,150 kg.

LMLV 2 stage 1
Same as LMLV 1.

LMLV 2 stage 2
Same as LMLV 1 stage 1 but expansion ratio increased from 17 to 24 to improve altitude performance.
Thrust: 1,606 kN average vacuum
Mass at ignition: 53,020 kg.

Fairing and payload accommodation
Acceleration load: 8 *g* longitudinal stage 1/2, ±2.5 *g* lateral.
Acoustic load: to be determined
Maximum internal fairing temp: to be determined
Recommended acceptance vibration: unavailable.

Orion and Orion derivatives (sounding rockets)

Background
The Orion is a single-stage unguided rocket, incorporating a surplus US Army motor, optimised for 39 to 68 kg payloads. An Orion version with 20 per cent increased total impulse was tested successfully 5 April 1994 as the second stage on a Terrier. The test included a GPS receiver, attitude/rate gyros and a new water recovery system.

Specifications
Orion
First launch: May 1974
Number launched: 33 to end-1996
Success rate: 100%
Number of stages: 1 (solid)
Overall length: 534 cm
Principal diameter: 35.6 cm
Launch mass: 422.3 kg plus payload
Typical performance: 39 kg/88 km, 68 kg/71 km
Payload recovery system: payload can be separated if required.
Payload environment: separable clamshell nose accommodates standard 35.6 cm diameter, 183-254 cm length.
Guidance: unguided, three fins provide 4 rev/s roll stabilisation at burnout.

Orion Derivative
Terrier-(Improved) Orion
Comment: A new two-stage vehicle employing surplus military motors. Stage 2 employs an enhanced version of the basic Orion. Performance is similar to Taurus Orion, but the environment is more benign, allowing bulbous (42.8 cm diameter) payloads.
First launch: 5 April 1994
Number launched: 3 to end-1996
Success rate: 100%
Number of stages: 2 (solid)
Overall length: 10.6 m (6.6 m without payload)
Principal diameter: 45.7 cm
Launch mass: 1,316 kg plus payload
Typical performance: 290 kg to 118 km
Payload recovery system: IRMA Ignition and Recovery Modular Assembly
Payload environment: 35.6 cm or 42.8 cm diameter payloads with various deployable nosecones. 19 *g* peak for 290 kg payload.
Guidance: unguided, fin spin-stabilised. Spin motors. Burnout roll rate about 6.5 Hz.

Terrier (Improved) Orion stage 1
Motor: US Army Aerojet solid, surplus
Overall length: 268.1 cm
Principal diameter: 35.6 cm
Stage mass: 422.3 kg
Thrust: 7.74 kN average sea level
Burn time: 32 s

Terrier-(Improved) Orion stage 2
Motor: improved XM22E8 Orion
Overall length: 268 cm
Principal diameter: 35.56 cm
Stage mass: 438 kg (293 kg propellant)
Thrust: unavailable
Burn time: 25.4 s (T+15 s ignition typical).

Nike Orion
Comment: A boosted version of the basic Orion vehicle, extending payload performance to >200 kg
First launch: September 1975

Number launched: 103 to end-1996
Success rate: 97%
Number of stages: 2 (solid)
Overall length: 880.9 cm
Launch mass: 1,021.8 kg plus payload
Typical performance: 68 kg/190 km, 204 kg/90 km
Payload recovery system: Wallops' 35.56 cm (14 inch) recovery system
Payload environment: as for Orion
Guidance: unguided, fin spin stabilisation

Nike Orion stage 1
Specifications as for Nike Tomahawk.

Nike Orion stage 2
Specifications as for the Orion vehicle, but the motor ignites 9 s after lift-off and the four fins are canted to generate 4 rev/s at burnout.

Nike Tomahawk
First launch: March 1965
Number launched: 230 to end-1996
Success rate: 97%
Number of stages: 2 (solid)
Overall length: 7.3 m
Principal diameter: 22.9 cm
Launch mass: 885 kg
Typical performance: 45 kg/370 km, 113 kg/220 km
Payload recovery system: non-standard
Payload environment: normally 22.9 cm diameter under 3:1 tangent ogive nose; a 30 cm diameter version is available but requires the larger fins. Payload length 183-305 cm. Clamshell nose is separable; payload can be separated if required. De-spin module is standard.
Guidance: unguided, fin spin-stabilised

Nike Tomahawk stage 1
Motor: Nike M5-E1 solid (US Army Allegheny Ballistics Laboratory)
Overall length: 345 cm
Principal diameter: 41.9 cm
Stage mass: 597 kg (342 kg propellant)
Thrust: 190.29 kN average sea level, total impulse 651.8 kNs at sea level
Burn time: 3.2 s
Comment: An interstage adaptor is bolted to Nike's forward face, its conical portion slip fitting into stage 2's nozzle, allowing drag separation. Four fins, each 0.23 m^2, are canted to generate 2 rev/s at Nike burnout.

Nike Tomahawk stage 2
Motor: Thiokol TE-M-416 solid
Overall length: 361 cm
Principal diameter: 22.9 cm
Stage mass: motor 245 kg (175 kg propellant)
Thrust: 48.9 kN average sea level, total impulse 416.6 kNs sea level
Burn time: 9 s, ignited 12 s after launch
Comment: Fin and shroud assembly mass 22 kg. Four fins, each 0.13 m^2, are canted to generate 6 rev/s at burnout. 0.17 m^2 fins are available for increased aerodynamic stability.

Taurus Orion
First launch: July 1977
Number launched: 53 to end-1996
Success rate: 96%
Number of stages: 2 (solid)
Overall length: 986.8 cm
Principal diameter: 57.8 cm
Launch mass: 1,800 kg plus payload
Typical performance: 68 kg/260 km, 227 kg/140 km
Payload recovery system: Wallops' 35.56 cm (14 inch) recovery system)
Payload environment: standard payload 35.6 cm diameter, length 183-380 cm. Separable clamshell nose; payload can be separated.
Guidance: unguided, four-stage 10.45 m^2 Ajax fins are canted to generate 2 rev/s spin at burnout

Taurus Orion stage 1
Motor: US Army surplus solid (Allegheny Ballistics Laboratory)
Overall length: 4.5 m
Principal diameter: 57.8 cm
Stage mass: 1,363 kg
Thrust: 457 kN average sea level
Burn time: 3.5 s

Taurus Orion stage 2
Specifications as for Nike Orion.

Taurus Tomahawk
First launch: October 1978
Number launched: 14 to end-1995
Success rate: 100%
Number of stages: 2 (solid)
Overall length: 821 cm
Principal diameter: 22.75 cm
Launch mass: 1,621 kg plus payload
Typical performance: 27 kg/590 km, 59 kg/490 km
Payload recovery system: non-standard
Payload environment: as Nike Tomahawk
Guidance: unguided, fin spin stabilisation

Taurus Tomahawk stage 1
Specifications as for Taurus Orion stage 1.

Taurus Tomahawk stage 2
Specifications as for Nike Tomahawk stage 2, but ignition is 18 s after launch.

Taurus Nike Tomahawk
A boosted version of the Nike Tomahawk. Stage 2 hardware includes an interstage locking device that holds the Nike/Tomahawk motors together during the coast before Nike ignition. Standard stage 3 hardware includes a Tomahawk Firing De-spin Module.
First launch: September 1983
Number launched: 15 to end-1996
Success rate: 100%
Number of stages: 3 (solid)
Overall length: 11.64 m
Principal diameter: 22.86 cm (Tomahawk)
Launch mass: 2,229.9 kg plus payload
Typical performance: 32 kg/700 km, 125 kg/400 km
Payload recovery system: non standard
Payload environment: typically 22.9 cm diameter, length 137 366 cm under 3:1 ogive cone. 30 cm diameter is possible but special analysis is required because of high dynamic load factors.
Guidance: unguided, fin spin stabilisation

Taurus stage 1
Specifications as for Taurus Orion stage 1 but fins canted for 1 rev/s spin at stage 1 burnout.

Nike stage 2
Specifications as for Nike Tomahawk stage 1 but ignition is made 11 s after launch.

Tomahawk stage 3
Specifications as for Nike Tomahawk stage 2 but ignition after 18 s and action time is 11.5 s.

PA-X sounding rocket

Background
AeroAstro's PacAstro subsidiary is developing the PA-X suborbital rocket for USAF Phillips Laboratory, primarily as a target for ballistic missile defence testing. PA-X is also available commercially for US$1 million. The two-stage PA-Y version is under development for handling micro-satellites, possibly beginning as early as 1998. Funding for the US$1,020 million development cost was under negotiation at the beginning of 1996. The future PA-1 (suborbital) and PA-2 (orbital) vehicles would employ Russia's RD-120M (stage 1) and 11D58M (stage 2).

Specifications
PA-X
Launch sites: compatible with Vandenberg AFB, White Sands Missile Range, NASA Wallops
Principal applications: small suborbital payloads
Performance: 45 kg to 64 km at 90° elevation
Cost: US$1 million
Overall length: 9.1 m
Principal diameter: 91 cm
Launch mass: 2,400 kg
Thrust: 62.3 kN SL/68.9 kN vacuum
Engine: PacAstro's PA-X engine. Fixed, liquid propellant, 10.2 atmosphere chamber pressure, He pressure-fed stainless steel injector, graphite over-wrapped chamber, ablative cooling, restartable, throttleable.
Fuel: 435 kg RP-1 kerosene
Oxidiser: 1,010 kg liquid oxygen
Burn time: 54 s
Payload accommodation: conical envelope 122 cm high, 91 cm diameter at bottom, 61 cm diameter at top. 15 *g* max acceleration. Payload integration and final assembly performed on unfuelled rocket.

PA-Y
First launch: planned for fourth quarter 1998 from site TBD
Launch sites: compatible with Vandenberg AFB, White Sands Missile Range, NASA Wallops
Principal applications: small orbital payloads
Performance: 20 kg into 100 × 280 km
Cost: US$3-4 million
Overall length: 11.1 m
Principal diameter: 1.22 m
Launch mass: 3,035 kg (800 kg dry)
Launch thrust: 72.1 kN SL/83.4 kN vacuum
Guidance: inertial

PA-Y stage 1
Engine: PaCAstro's enhanced PAX
Overall length: 7.1 m
Principal diameter: 1.22 m
Fuel: 580 kg RP-1 kerosene
Oxidiser: 1,345 kg liquid oxygen
Dry mass: 725 kg
Thrust: 72.1 kN SL/83.4 kN vacuum
Burn time: 72.5 s

PA-Y stage 2
Engine: PacAstro's PA-Y-2 engine. Restartable and throttleable.
Overall length: 4.0 m
Principal diameter: 0.91 m
Fuel: 128 kg RP-1 kerosene
Oxidiser: 298 kg liquid oxygen
Dry mass: 73 kg
Thrust: 14.7 kN vacuum
Burn time: 94 s

PA-Y stage 3 options
Comment: PA-Y delivers the payload into an elliptical orbit, but AeroAstro's miniature cold gas package or the user's own kick stage can circularise. 54 m/s delta-V is required at 280 km.

Payload accommodation
Conical envelope 61 cm high, 61 cm diameter at bottom, 46 cm diameter at top. 15 *g* max acceleration. Payload integration and final assembly performed on unfuelled rocket.

Payload Assist Module (PAM) (upper stage)

Background
As the Delta family grew and evolved, NASA sought ways to put ever larger payloads into orbit. One chief means was the introduction of the new PAM, or Payload Assist Module, a spinning upper stage based around Thiokol's Star 48 motor. The 3920/PAM configuration first flew in 1982 and incorporated a larger capacity stage 2 with an Aerojet AJ10 118K engine to create a 1,284 kg GTO capacity.

MDA's Payload Assist Module was first conceived to promote the transition of payloads from the company's Delta launcher to the Shuttle. It could fly as a conventional third stage on the expendable, or be released into LEO with its payload from a dedicated cradle in the Orbiter. The Shuttle's difficulties and resultant offloading of payloads meant that PAM's future lies with Delta; it is not now considered as a separate stage. 18 PAM-Ds and two PAM-D2s flew from Shuttle, one PAM-D2 from Titan 3 and the one-off PAM-S from Shuttle with Ulysses solar probe. The company also provided the SGS-2 stage for initial Navstar GPS launches, and the Payload Assist Stage for SDI Delta launches.

Pegasus

Background
OSC was established in 1982 to develop Space Shuttle upper stages, resulting in the solid propellant Transfer Orbit Stage. In June 1988, OSC announced a partnership with motor manufacturer Hercules Aerospace (now Alliant Techsystems) to develop the Pegasus air-launched small orbital delivery system. The 28-month development schedule was then already under way as a private venture, aiming for a first launch in July 1989 carrying a DARPA Glomar data relay and NASA Pegsat chemical release satellites.

Pegasus is capable of placing a 201 kg payload into a 460 km polar orbit (289 kg equatorial) following aircraft release at 11.9 km/Mach 0.8. Six launches (at 12.2 km) utilised the USAF-owned B-52 NASA-operated aircraft employed for X-15 hypersonic drops in the 1960s. A Lockheed L1011 Tristar, named 'Stargazer', was acquired by OSC from Air Canada in 1992 on a 10-year lease as the belly carrier beginning with STEP M1 in 1994. It is now leased from the owner, ITT Financial. Stargazer can handle 35.5 tonne vehicles and operate from most major runways.

The base vehicle consists of three graphite epoxy composite case Alliant solid motors of conservative design, a fixed delta planform composite wing, an aft skirt assembly including three composite control fins, an avionics section forward of the stage 3 motor, and a two-piece composite payload fairing. Some 94 per cent of the structural mass is graphite epoxy, 5 per cent aluminium, 1 per cent titanium. Air-launching at 11.9 km altitude reduces ΔV requirements by 10 to 15 per cent.

Work began on the 18 tonne, 15 m long, 6.7 m wingspan Pegasus in 1987 with US$40 to 50 million

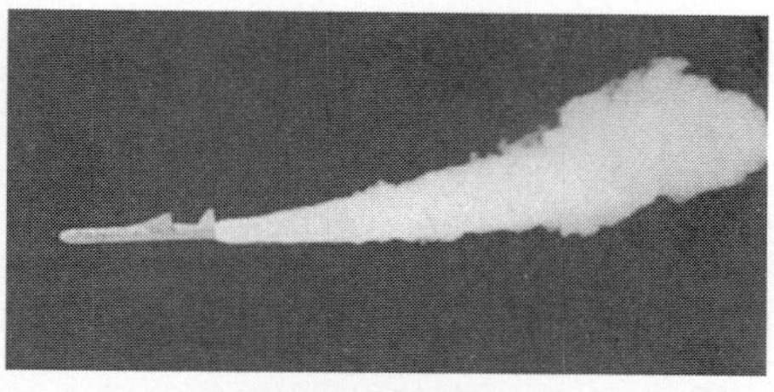

Pegasus stage 1 ignites 5 seconds after release (OSC)

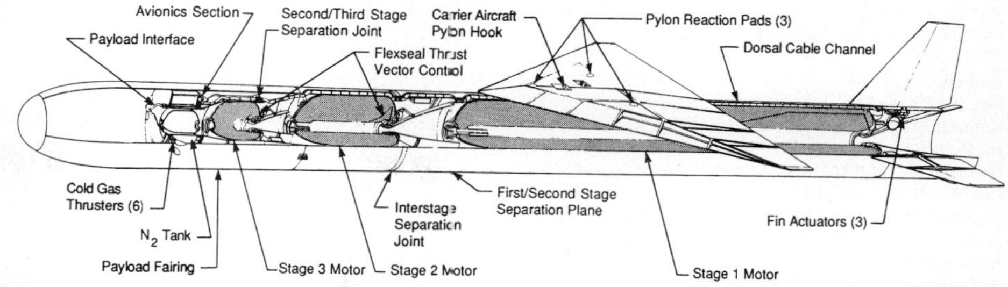

Pegasus features all-composite wing and solid rocket motor structures. The HAPS stage 4 replaces the N_2 tank shown here (OSC/Alliant)

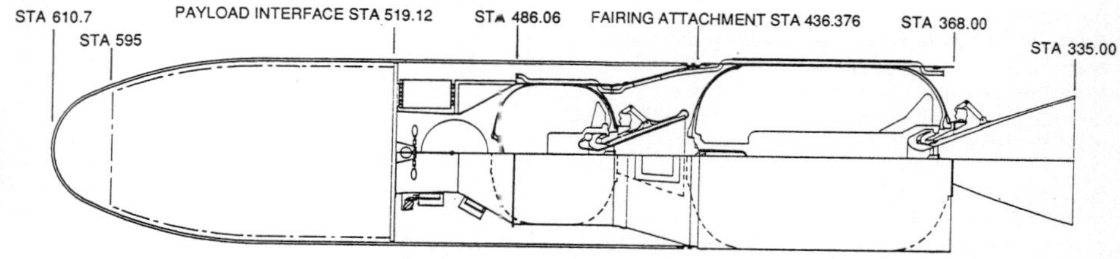

Pegasus payload accommodation. Station positions (standard Pegasus) are given in inches. Length is reduced 10 cm with HAPS (OSC/Alliant)

private funding divided equally between the partners. Commercial launch costs were estimated at US$10 to 12 million, requiring 16 to 18 missions in 2 to 2½ years to recoup costs. DARPA contracted for six launches (responsibility from number 3 was transferred April 1993 to the AF Space & Missile Systems Center), paying US$6 million for the first and US$7.3 million each for numbers 2/3. The increased cost reflects inclusion of the liquid HAPS Hydrazine Auxiliary Propulsion System stage 4, added to cope with increased mass of flight number 2's Microsats and to improve injection accuracy. This variant, known as Pegasus XL, lengthened version improves performance into LEO by about 10 per cent.

NASA Goddard selected Pegasus in March 1991 to provide Small Expendable Launch Vehicle Services for its Small Explorer science satellite programme. The total contract, worth some US$80 million, covers seven firm and three optional launches from 1993. It was selected April 1991 for one firm (late 1992) and 39 optional launches under the AFSLV Air Force Small Launch Vehicle programme, bringing orders to 15 firm and about 45 optional. BMDO, in July 1992, awarded a US$14.7 million contract for one launch (plus nine options) for its MSTI 3 satellite. NASA selected Pegasus as its Ultralight launcher in November 1994, with two firm (USRA in 1997) and eight optional launches. Spain's INTA selected it in October 1994 for Minisat 01, which will be the first west European orbital launch (from Spain's Torrejon AB). Contract value is US$10.2 million.

Specifications

Pegasus

First launch: 5 April 1990
Number launched: 26 through April 1999
Launch sites: now hosted by L-1011 operating out of Vandenberg AFB, NASA Wallops and Cape Canaveral (Spain's Minisat will be from Torrejon AB near Madrid)
Principal applications: small payloads into LEO
Vehicle success rate: 88%
Performance: 3-stage typically 200 kg (XL: 279 kg) into 463 km polar, 288 kg (XL: 382 kg) into 463 km equatorial (payload fraction 2.2%). For 740 km polar orbit, accuracy is ±0.2° inclination, ±85 km 3Σ deviation from circular; with HAPS: ±0.1°/±19 km. 165 kg GTO capacity projected with XL + stage 4
Availability: 1 flight/45 days projected
Cost: US$53,000/kg or US$12 million for 227 kg payload
Number of stages: 3
Overall length: 15.512 m; XL: 16.8 m
Principal diameter: 127.1 cm (wingspan 670.6 cm)
Launch mass: 18,518 kg; XL: 22,583 kg
Guidance: autonomous autopilot forward of stage 3, with Oettle & Reicher's 4.46 kg 68430-based flight computer (previously AlTech's 8.2 kg 5E 102 MS 68020 based) containing mission specific mission data load (MDL). Inertial/attitude data are provided from 5.8 kg Litton LR81 Inertial Measurement Unit (IMU); GN&C performance is monitored by the autopilot processor and processor data are formatted within the flight computer for direct downlink via the 56 kbit/s S-band vehicle telemetry system. Flight number 2 demonstrated a Trimble GPS receiver for navigation. It is now standard; IMU is retained for attitude data. XL carries upgraded avionics for increased payload mass/dynamic volume.

Pegasus stage 1

Stage 1 incorporates a Scaled Composites Inc graphite composite wing with 45° sweptback leading edge and 670.6 cm span. Subsonic L/D is 4.0. The airfoil is a double wedge with 2.5 cm radius leading edge. The wing thickness is truncated to 20.3 cm, with upper/lower parallel surfaces facilitating attachment to the motor case wing saddle. Three active aft fins provide control during stage 1 burn as part of an aluminium assembly attached to the motor's skirt extension. Control is augmented at end of stage 1 burn by firing small fin-mounted motors.
Motor: Alliant Techsystems Orion 50S, stretched by 137 cm for XL. All three motors use IM7/55A graphite epoxy composite cases with aramid filled EPDM rubber internal insulator and integral skirts. Each nozzle consists of a carbon phenolic exit cone with 3D carbon/carbon integral throat/entry. Propellant is HTPB with 88% solids, grain designed for low burn rates. The igniter is mounted on the forward dome. A flight termination charge is mounted on the aft dome of stage 1 and forward dome of stage 2 for range and aircraft safety requirements. Stage 1 employs a fixed nozzle.
Stage length: 9.39 m; XL: 10.76 m
Diameter: 127.1 cm
Thrust: 486.696 kN average vacuum, 583.773 kN max vacuum attained about mid-way through burn (total impulse 35.05 MNs); XL: 619.9 kN average vacuum
Burn time: 72.3 s; XL: 64.3 s
Mass at ignition: 14,020 kg; XL: 17,157 kg
Propellant mass: 12,152 kg; XL: 15,051 kg
Separation: the forward skirt, which also serves as interstage adaptor, incorporates two linear-shaped charges.
Burn sequence: vehicle is released from L 1011, activating sequencer and autopilot. At 2 s, the arming sequence begins for stage 1 arming/ignition at 3/5 s, respectively. Burnout occurs 76 84 s.

Pegasus stage 2

Motor: Alliant Techsystems Orion 50 (50XL stretched 45.0 cm). See stage 1 for comments. Stage 2 employs a silicon elastomer flexseal nozzle and Parker Bertea Aerospace electromechanical actuators for TVC
Stage length: 383.7 cm; XL: 428.7 cm
Diameter: 127.1 cm
Thrust: 113.874 kN (160.136 kN) average vacuum, 131.667 kN (195.266 kN) max vacuum. Total impulse 8.67 MNs (11.22 MNs)
Burn time: 75.9 s; XL: 70.1 s
Mass at ignition: 3,370 kg; XL: 4,314 kg
Propellant mass: 3,024 kg; XL: 3,919 kg
Attitude control: motor nozzle flexing provides pitch/yaw control during powered flight; roll control and three-axis control during coast provided by six N_2 thrusters in two Parker Bertea Aerospace modules on stage 3.

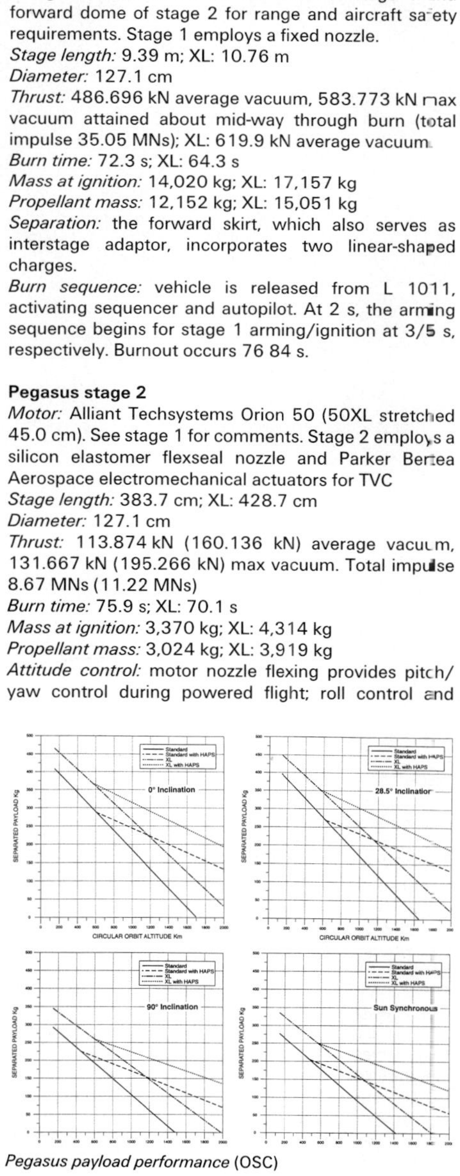

Pegasus payload performance (OSC)

Pegasus stage 3

Motor: Alliant Techsystems Orion 38. See stage 1/2 for comments. Stage 3 also employs flexing for pitch/yaw control during burn. Igniter is toroidal. Head-end grain design maximises propellant density. XL uses the same stage 3 as the standard Pegasus.
Stage length: 208 cm
Diameter: 96.5 cm
Thrust: 34.314 kN average vacuum, 38.686 kN max vacuum. Total impulse 2.18 MNs.
Burn time: 63.8 s
Mass at ignition: 893 kg
Propellant mass: 771 kg
Attitude control: motor nozzle flexing provides pitch/yaw control during powered flight; roll control and

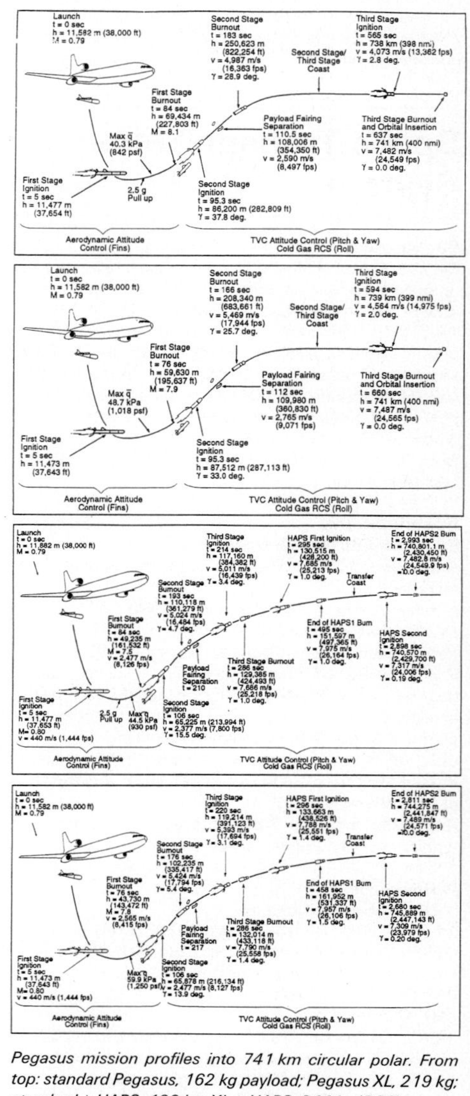

Pegasus mission profiles into 741 km circular polar. From top: standard Pegasus, 162 kg payload; Pegasus XL, 219 kg; standard + HAPS, 190 kg; XL + HAPS, 244 kg (OSC)

three-axis control during coast provided by six N_2 thrusters in two Parker Bertea Aerospace modules.

Pegasus stage 4

The HAPS Hydrazine Auxiliary Propulsion System is an optional stage carried to improve injection accuracy and performance by about 36 kg for 720 km, 82°. It was ordered by DARPA to cope with the Microsat mass growth (launch number 2, July 1991). It occupies the payload adaptor interior, replacing the N_2 tank of the 3-stage version. The attitude jets are supplied instead from two high-pressure (395 atmosphere) N_2 tanks on the adaptor's outer surface. On typical missions, the first burn establishes an elliptical orbit, followed by circularisation.

Thrusters: three fixed blowdown monopropellant Olin Aerospace MR-107s
Propellant: 72.6 kg of hydrazine stored in central sphere
Stage thrust: 3 × 222 N vacuum
Burn time: typically 2 burns of 131 and 110 s
Attitude control: nitrogen jets and differential firing of main thrusters

Fairing and Payload Accommodation

Alliant's 127 kg 2-piece composite fairing, which also covers stage 3 and the avionics packages (and stage 4 when carried), is jettisoned during stage 2 burn. The payload may remain attached to stage 3 or be separated by clamp, following pointing to ±2° accuracy or spin up. The payload can be provided with 60 V DC (up to 140 W) from the carrier aircraft until release, and up to 16 discrete commands can be issued by Pegasus' computer. Telemetry at 3 kbit/s from the vehicle's total 56 kbit/s downlink is available to the payload. On the ground, air conditioning maintains 21±5°C.

Payload volume: 1.17 m diameter, 2.13 m long for 3 stage; 1.76 or 1.79 m long when HAPS carried
Acceleration load: 8 *g* longitudinal max at end of stage 1 burn (lighter payloads can exceed 8 *g* during stage 3 burn).

Vehicle and Payload Processing

Vehicle and payload processing are performed in two 15 × 36 m bays of the OSC Vehicle Assembly Building at VAFB, controlled at 21±5°C and humidity of 40±10%. The stages and other subassemblies arrive typically 4 months before launch. Payload mating at T12 days is in a 3.66 × 7.32 m class 100,000 soft-walled clean room. The N_2 ACS tank is then filled and the fairing attached. All processing is undertaken horizontally. Pegasus is attached to the carrier aircraft at T−1 day. Takeoff is 1 hour before release.

Pegasus launches

	Date	Type	Carrier/Base	Payloads
1	5 April 1990	std	B-52/DFRC	192 kg Pegsat + 68 kg SECS
2	17 July 1991	HAPS	B52/DFRC	7 × 21.8 kg Microsats
3	9 February 1993	std	B 52/KSC	115 kg SCD 1 + 14.5 kg CDS
4	25 April 1993	std	B52/DFRC	109 kg Alexis
5	19 May 1994	HAPS	B52/DFRC	180 kg STEP M2
6*	27 June 1994	XL	L 1011/VAFB	348 kg STEP M1
7	3 August 1994	std	B52/DFRC	261 kg APEX
8	3 April 1995	std	L1011/VAFB	2 × 40 kg Orbcomm 1/2 + 68 kg MicroLab 1
9*	22 June 1995	XL	L1011/VAFB	268 kg STEP M3
10	9 March 1996	XL	L 1011/VAFB	113 kg REX 2
11	17 May 1996	std	L1011/VAFB	211 kg MSTI 3
12	2 July 1996	XL	L-1011/VAFB	294 kg TOMSEP
13	21 August 1996	XL	L1011/VAFB	180 kg FAST
14 !	4 November 1996	XL	L-1011/WI	2200 kg SAC B+NETE
15	21 April 1997	XL	L1011/C1	209 kg MINISAT 1/Celestis
16	1 August 1997	XL	L-1011/VAFB	309 kg Seastar
17	29 August 1997	XL	L-1011/VAFB	215 kg FORTE
18	22 October 1997	XL	L-1011/WI	395 kg STEP M4
19	23 December 1997	XL	L-1011/WI	8 × 43 kg Orbcomm
20	26 February 1998	XL	L-1011/VAFB	132 kg SNOE; 45 kg Teledisc T-1
21	2 April 1998	XL	L-1011/VAFB	250 kg TRACE
22	2 August 1998	HAPS	L-1011/WI	8 × 43 kg Orbcomm
23	23 September 1998	HAPS	L-1011/WI	8 × 43 kg Orbcomm
24	23 October 1998	Std	L-1011/K5C	540 kg STEX
25	6 December 1998	XL	L-1011/VAFB	283 kg SWAS
26	5 March 1999	XL	L-1011/VAFB	270 kg WIRE

*launcher failure. DFRC = Dryden Flight Research Center; KSC=Kennedy Space Center; VAFB=Vandenberg Air Force Base; WI = Wallops Island. 1996 Minisat launch is planned from Spain's Torrejon Air Base, near Madrid. Number 6 failure: the review board concluded the guidance software was unable to cope with unpredicted aerodynamic responses on the first XL vehicle. The design was then tested in wind tunnels to develop an aerodynamic model, rather than previous reliance only on computer modelling. Number 9 failure: stage 1/2 interstage remained attached during stage 2 burn (one of three separation skids improperly installed) satellite failed to separate from Pegasus final stage. Flight 14 from Wallops Island, 15 from Canary Islands.

Pegasus XL with the Celistis payload launched 21 April 1997 (Theo Pirard) ***1999***/0054368

Space Transportation System

Background

NASA's Space Transportation System relies for its propulsion on four elements: two Solid Rocket Boosters (SRB), the Orbiter main engines (SSME), an External Tank (ET), and small manoeuvring engines on the Shuttle Orbiter (OMS). At an altitude of about 43 km, after some 2 minutes, the boosters separate for recovery and refurbishment. The Orbiter continues under SSME power until about 8 minutes 50 seconds after launch, when the ET separates for destructive re-entry. One or two Orbital Manoeuvring System firings establish operational orbit.

Specifications

Space Shuttle Launch Operations

First launch: 12 April 1981 from Kennedy Space Center complex 39A
Number launches: 93 to the end of 1998
Launch sites: KSC pads 39A/39B
Vehicle success rate: 98.9% to December 1998
Principal uses: US manned capability to beyond 2010 (four reusable Orbiter fleet), 25 t payload delivery to LEO, satellite retrieval/in situ repair, short duration science platform, Space Station assembly/servicing.
Cost: restricted commercial access, STS27 military mission cost to NASA in December 1988 was reported at US$375 million.
Availability: typically 7-8 flights/year manifested but with restricted commercial access.

Performance

LEO (204 km, 28.45°)
21,140 kg OV-102, 24,950 kg OV-103/104/105; each additional 1 km altitude reduces capacity by 25 kg, each crew member beyond 5 persons reduces capacity by 230 kg into basic orbit.
LEO (204 km, 57°)
14,800 kg OV-102, 18,600 kg OV-103/104/105 (57° is usually the highest inclination flown from Florida, but dog-legging provides access to higher angles)
LEO (204 km, 98° Sun-synchronous VAFB)
13,426 kg OV-103/104/105, assuming use of Advanced Solid Rocket Motors. Neither VAFB nor ASRM will be used.
Molniya (925 × 39,450 km, 63°)
3,563 kg using IUS via 222 km, 57° parking orbit.
Crossrange: 2,040 km max; 1,020-1,300 km used operationally by NASA.
Number of stages: 1 plus 2 strap-ons plus payload injection stages
Overall length: 56.14 m (Orbiter 37.24 m)
Launch mass: about 2,040 t
Landing mass: 104,328 kg max, defined by abort requirements
Launch thrust: 34,677 kN (three SSMEs firing at 104%/1,754 kN each, two SRBs at 14,680 kN each; SSMEs throttle down to 65% by 35 s and up to 104% by 65 s).

Solid Rocket Boosters (SRB)

Description: Thiokol was selected as Shuttle SRM prime in 1974. NASA announced July 1991 a US$2.6 billion Buy 3 contract extension to Thiokol for 67 further flight pairs through 1999/STS105, plus eight ground test motors. Another extension is expected for launches through 2012. The SRM is cast in four segments in Utah and then transported by rail to Kennedy Space Center, Florida, where they are stacked vertically in pairs either side of the External Tank on the mobile launch platform in the Vehicle Assembly Building. The boosters provide the majority of launch thrust, with TVC provided by a flexible bearing nozzle driven by two hydraulic actuators. After burnout at about 2 minutes, the boosters are separated pyrotechnically and fall into the Atlantic for recovery. The motors are cleaned, disassembled and returned to Utah for refurbishment and reloading. The segments are designed for 20 flights; nine is currently the maximum. The configuration underwent extensive reworking and testing following Challenger's loss, and the new booster was redesignated RSRM (Redesigned Solid Rocket Motor, changed by NASA in 1995 to Reusable SRM).

A High-Performance Motor (HPM) design for an improved SRM incorporates lightweight steel cylindrical case segments, a higher expansion nozzle, and altered motor thrust characteristics to increase Shuttle payloads by >1,540 kg. This design was static tested in 1982 and went into service on STS-8 in August 1983. Graphite epoxy composite cylindrical SRM case segments were also developed to increase Shuttle capacity by 2 t; the configuration was shelved following Challenger's loss. Thiokol was subcontracted to Aerojet to develop/produce the nozzle for the Advanced SRM that was to enter Shuttle service in about 1998. ASRM was cancelled in 1993.

Mass: 569,636 kg.
Dimensions: Length 45.46 m, diameter 3.7 m
Propellant configuration: 504 t of ammonium perchlorate/aluminium powder/iron oxide/binding agent in four segments with 11 point star, with max

thrust at launch, reducing by one-third after 50 s (maximum dynamic pressure period).

Propellant Data
Type: TP-H1148 HB Polymer, 16% Al, 70% AP shape: 11-point star forward; double taper cylinder two centre segments; triple taper cylinder aft.
Mass fraction: 0.883 (501,746 kg)
Burn time (s): 123.7 action, 111.4 web
Thrust (vacuum): 11.52 MN average, 14.77 MN max
Specific impulse: 268.0 s vacuum
Total impulse: 296.3 MNs vacuum
Pressure (atmosphere): 62.1 max (MEOP 69.1); 45.0 avg
Throat diameter: 136.8 cm
Length: 454.0 cm
Exit diameter: 380.1 cm (expansion ratio 7.72)
Materials: D6AC steel for fixed housing, throat and forward exit cone housings; aluminium for nose inlet and aft exit cone housing
Casing materials: D6AC steel, 1.165 cm minimum thickness
Igniter: forward end internally mounted. 60.7 kg of TP H1178 case bonded 40-point star.
Control: Nozzle gimbals 8° for steering
Design lifetime: 20 missions (current max 9)
Thrust: 29.36 MN at sea level (70% of total lift-off thrust)
Comment: SRB separation occurs at about 124 s, 45 km altitude (triggered when pressure falls to 3.4 atmospheres), landing speed <100 km/h under 3 × 41 m diameter parachutes.
Cost: US$64million/pair (1983), US$25million per reflight.

Space Shuttle Orbiter-Main Engines (SSME)
Description: Three LOX/LH_2 staged combustion Rocketdyne SSMEs fed by External Tank
Burn time: typically 520 s
Design life: 55 missions or 7½ h at 1,668 kN sea level (specific impulse 363 s)/2,090 kN in vacuo (SI 455 s).
Performance: Typically throttled over 65 104% rated power setting (109% was to have been used for Vandenberg polar missions, but is now available for transatlantic aborts). Constant 104% setting, extending lifetime, would have been possible with Advanced Solid Rocket Boosters providing required thrust profile (cancelled 1993). Pump and other improvements will allow 106% for Space Station assembly.
Control: The engines are individually gimballed hydraulically by two APU powered actuators at up to ±10.5° pitch and ±8.5° yaw; each engine was originally controlled in 20 ms steps by a 97 kg top-mounted package housing two redundant Honeywell 360 16.4 kbit-memory computers. The controllers are commanded by the Orbiter's General Purpose Computers, or by the crew in emergencies. Block 2 Honeywell controllers are now installed (introduced on OV105 and retrofitted to the others), providing four times the memory using the first space-qualified 68000 series microprocessors. They also allowed elimination of a separate 13.6 kg vibration monitoring box.
Pump performance: The incoming oxidiser is boosted by the first turbopump from 7 to 29 atmospheres before the High-Pressure Oxidiser Turbopump increases it to 500 atmospheres in two stages. This pump has suffered durability problems, with some engines requiring overhaul after three missions (33 min). Pratt & Whitney, one of the original SSME bidders, was awarded a US$188 million contract, as part of a 10-year US$1 billion Phase II SSME improvement programme beginning in 1983. In 1986 to develop a new power head containing liquid oxygen and hydrogen high-pressure turbopumps and a single pre-burner chamber (the Rocketdyne design incorporates a pre-burner for each). The goal is for 7½ h firing (55 missions) between major overhauls, but problems increased development cost to US$504 million. NASA thus halted fuel pump work from December 1991 in order to focus on the more critical oxygen unit (it restarted work May 1994 for STS86 September 1997 debut once the LOX pump was on schedule for 1995 debut). First flight was aboard STS70 in July 1995 with one pump. As that was successful, all three engines were equipped for STS 77 in May 1996.
Comment: Some of the oxygen turbopump output is routed into a pre-burner where it is mixed with fuel and burned hydrogen-rich at 760°C. The hydrogen half of the system likewise employs a low-pressure pump (2 to 18.7 atmosphere), high-pressure pump (18.7 to 442 atmosphere) and a pre-burner. Some GH_2 is directed from the high-pressure unit as a coolant through the 390 main chamber wall channels and the 1,080 tubes composing the 2.87 m long, 2.38 m max diameter nozzle. The products of the two pre-burners are fed into the main chamber with additional oxygen through 600 elements in 13 annular rows in a 56 cm diameter injector plate and burned at 3,300°C/204 atmospheres (100% thrust level) in a 6:1 oxygen to hydrogen ratio. Some 770 kg LH_2 + 1,680 kg LOX are trapped in the system at cut-off, producing an 18 cm aft shift in the Orbiter's centre of gravity. Sporadic venting during orbital flight would affect guidance and possibly contaminate the payloads, and the hydrogen could combine during re-entry with atmospheric oxygen to form a potentially explosive mixture. Instead, after ET separation, the LOX is dumped through the SSME nozzles and the LH_2 is released through the starboard T0 umbilical overboard fill/drain valves.

External Tank (ET)
Dimensions: Length 46.88 m (LH_2 tank 29.47 m, LOX tank 16.64 m), diameter 8.40 m.
Dry mass: 35.43 t for early version, reduced to 29.93 t in later models, 26.33 t for aluminium lithium alloy Super LightWeight version, loaded typically 760 t.
Flight pressure: 3.0 atmosphere LH_2, 1.43 atmosphere LOX.
Contractor: Lockheed Martin Manned Space Systems at Michoud, Louisiana
Unit cost: US$30million (US$33.5million for AlLi version).
Capacity: 104,308 kg LH_2 at 253°C in 1,514 m³ aft, 625,850 kg LOX at 183°C in 559 m³ forward aluminium alloy semi-monocoque tanks.
Supply lines: Orbiter's SSMEs, by two 43.2 cm outlets, discarded after about 530 s, 110 km altitude.
Insulation: 25 mm-thick polyurethene foam insulation. Super LightWeight version first flight ET-96 December 1997.

Space Shuttle overall external dimensions

Wing span	23.79 m (78 ft 0.7 in)
Orbiter length	37.24 m (122 ft 2 in)
Orbiter height	17.27 m (56 ft 8 in)
Length overall	56.14 m (184 ft 2.4 in)
Length of external tank	46.88 m (153 ft 9.6 in)
Length of boosters	45.46 m (149 ft 2.0 in)
Height overall	23.35 m (76 ft 1.6 in)

Masses

Shuttle stack	2,040,000 kg (4,500,000 lb)
OV-102 inert	82,288 kg (181,445 lb)
OV-103 inert	78,448 kg (172,978 lb)
OV-104 inert	78,687 kg (173,505 lb)
OV105 inert	79,135 kg (174,493 lb)
Max landing	104,328kg (230,000lb)
External tank (full)	750,975 kg (1,665,600 lb)
Boosters (2), each	589,670 kg (1,300,000 lb)

Thrust

Total, at lift-off	34,677 kN (7,781,000 lb)
Orbiter, SSME (3), each	1754 kN (393,800 lb)
Boosters (2), each	14,680 kN (3,300,000 lb)

Space Shuttle Orbiter-OMS/RCS
Description: The Orbital Manoeuvring System and aft Reaction Control System engines are housed in two 6.45 m long 8.6 t McDonnell Douglas pods mounted on the aft fuselage either side of the tail. The 1.8 t composite (skin panels and frames) and 2,124 aluminium (main structural members) pods are each attached by 11 bolts accessible through 24 doors. Each carries 4.1 t MMH plus a 6.7 t NTO in 2.45 × 1.25 m titanium tanks for OMS, although this supply can also be made available to RCS. Cross lines provide redundancy and balancing for re-entry.
Performance: 26.7 kN/313 s SI vacuum engine
Mass: 118 kg dry mass
Dimensions: 1.96 m long
Operating pressure: 8.5 atmospheres chamber pressure, helium pressure fed
Control: Gimballed ±6% pitch and ±7° yaw
Design lifetime: 10 years/100 mission/1,000 starts/15 h firing duration
Comment: Earlier missions required two OMS burns to attain operational orbit but a direct ascent technique first demonstrated on mission 41C is now employed, omitting the OMS-1 burn following ET separation and relying on the OMS-2 burn at apogee about 45 minutes after launch. The OMS engines are normally fired in pairs for a minimum of 2 s, but for fine orbit adjustments of 1.8 m/s or less one can be gimballed through the Orbiter's centre of mass for solo firing. Some 80-95 m/s capacity is reserved from the about 305 m/s total ΔV available for the 120-170 s re-entry burn. Aerojet was contracted in 1984 to add turbopumps for increased chamber pressure and SI; no fitting schedule has been announced. 12 OMS engines were originally produced under US$140 million contracts, including four spares stored at KSC. A further pair was ordered for Challenger's replacement.

The pods also house part of the Orbiter's RCS of 44 hypergolic thrusters providing attitude control and fine orbit tuning. Each pod incorporates 12 Kaiser Marquardt 3,870 N R40 and two 106 N R1E radiatively and fuel film cooled engines drawing on 675 kg NTO plus 422 kg MMH in titanium spheres. The larger motors, operating with 10.3 atmosphere chamber pressures, are grouped into three up, three down, two aft and four side, rated for 50,000 individual firings totalling 330 minutes in 1,150 s bursts or 80 ms (min) pulses. The vernier pairs, modified from USAF Manned Orbiting Laboratory motors, are capable of 500,000 ignitions over 35 h lifetimes of 125 s bursts and 80 ms (min) pulses. The system can also take propellants from OMS tanks if necessary.

The forward RCS module is removable for maintenance. It houses identical motors and 99 cm titanium spheres but the nozzles are machined flush with the thermal tiles. The 14 R40s are grouped into three forward, three up, two each side and two angled 45° down each side for roll control, as are the twin solo verniers. Propellants are loaded on the pad via servicing panels protected by thermal tiles (forward RCS) or through the T0 panels (OMS pods). Post-mission, the propellants are drained in the Orbiter Processing Facility.

Starfire sounding rocket

Background
SSI's first commercial launch was successfully undertaken 29 March 1989 using a Bristol Aerospace Black Brant 9 boosted by a Thiokol Mark 70 solid motor. SSI calls the configuration 'Starfire'. This first launch carried a 295 kg package under a US$1 million contract from the Consortium for Materials Development in Space at the University of Alabama. Operations were conducted from White Sands' complex 36 under US Navy control. Vehicle guidance is provided by Saab's S19 package, incorporating a Space Vector Corporation gyro platform.

The second flight failed 15 November 1989 because a capacitor broke loose in the guidance system. Number 3 succeeded on 17 May 1990 with a repeat payload from number 2 and Consort 4, launched 16 November 1991, also succeeded. The company was awarded a contract worth up to US$9.8 million for up to seven further Consorts through 1996. Consort 5 flew 10 September 1992 but stage 2's burn ended 3 s early and µg data were not obtained. Consort 6 flew successfully 19 February 1993 with seven experiments, attaining 301 km. The US$2 million funding was provided from the insurance following C5's partial failure. Conquest 1, the first of a new series, flew 3 April 1996 carrying eight experiments.

SSI offers larger Starfire models for longer duration µg missions, but none has flown. Starfire 2 is a one-stage Castor 4B capable of providing 15 minutes µgravity time for 454 kg; Starfire 3 adds two Castor 4B strap-ons to create 20 minutes µgravity for 900 kg payloads.

Specifications
Dimensions: Length: 15.25 m, 46 cm diameter
Mass: 2,234 kg
Performance: Up to 8 min *µg* can be provided for 454 kg payloads.

Super Arcas sounding rocket

Background
Super Arcas has been used since 1965 to carry meteorological sensors up to around 100 km. A high-altitude decelerator slows and stabilises the payload for making measurements in the mesosphere; it also acts as part of the recovery system.

Specifications
First launch: January 1965
Number launched: 201 to end-1996
Success rate: 97%
Number of stages: 1 (solid)
Overall length: 268.4 cm
Principal diameter: 11.4 cm
Launch mass: 37.6 kg
Typical performance: 4.5 kg to 92 km (86° launch)
Payload recovery system: high-altitude radar reflector decelerator; payload section separates 148 s after launch
Payload environment: 7 *g* peak acceleration and 25 revs/s at burnout. Payload diameter 11.4 cm, typical length 100 cm, accommodated under tangent ogive nosecone
Guidance: unguided, fin spin stabilised

Super Arcas stage 1
Motor: Atlantic Research Corporation Marc 60A2
Overall length: 192.5 cm
Principal diameter: 11.4 cm
Stage mass: 37.7 kg
Thrust: 1,445 N average sea level
Burn time: 40.2 s

Taurus

Background

DARPA contracted OSC in 1989 to build the Taurus rapid response orbital launcher using the company's Pegasus as a baseline. OSC had already taken on Space Data Corporation of Tempe, Arizona in November 1988. SDC had launched a wide range of sounding rockets, including more than 600 suborbital boosters in 35 different configurations. The division has participated in more than 60 Minuteman 1 re-entry tests since 1971, and so it is not surprising that the joint corporation announced that it would make a series of winged air-launched suborbital vehicles using some Minuteman technology.

According to the contract, Taurus had to be a self-contained road transportable system capable of rapid response from austere sites. The vehicle and all its support equipment was required to be transported up to 1,600 km before the launch site was established and the vehicle integrated/tested within five days. After this setup, the vehicle can wait for the payload to arrive, from when launch takes a further 72 hours.

The original SSLV requirement was for a 450 kg payload into a 740 km polar orbit, increased to 815 kg in 1991 as a result of a DARPA-funded modification to add a vectorable nozzle to stage 1. The vehicle's three upper stages are derived from Pegasus, supported for the first launch (and any subsequent USAF mission) by a Thiokol Peacekeeper stage 1.

After that successful debut, the commercial Taurus uses Thiokol's Castor 120, based on Peacekeeper stage 1 technology, and a 120 per cent larger fairing. Uprating options are the Taurus XL (same stretched motors as Pegasus XL) and Taurus XLS (two strap-ons from 1998), increasing the 1,400 kg LEO capacity from Canaveral to 1,550 kg (XL) and 1,950 kg (XLS). XLS would carry two Thiokol Castor 4B motors. For transfer orbit or deep space missions, stage 3 can be replaced with a spin-stabilised Thiokol Star 37 for perigee kick, or followed by Star 37. The OSC/McDonnell Douglas team was awarded NASA's MedLite launch contract in March 1995. Taurus XL/XLS, along with optional upper stages, will be used for the smaller payloads in the MedLite spectrum.

Specifications

Taurus

First launch: 13 March 1994 from VAFB
Number launched: 3
Launch sites: number 1 from VAFB 576E; potential Canaveral complex 46 and VAFB SSI; transportability allows launch from almost any range
Principal applications: medium LEO, small GTO or Earth escape payloads
Vehicle success rate: 100%
Performance: typically 800 kg into 695 km polar, 1,300 kg into 400 km at 28.5°. For 645 km polar orbit, accuracy is projected at ±0.2° inclination, ±37 km deviation from circular. See performance diagrams. 470 kg GTO and 320 kg lunar transfer with Star 37 stage 3.
Availability: current production capability can support six flights annually. Typical mission cycle is 18 months after receiving order.
Cost: US$14,000/kg standard; US$13,000/kg XL/XLS
Number of stages: 4 (optional 3/5 stage)
Overall length: 26.72 m launch 1; 27.56 m standard; 29.12 m XL/XLS
Principal diameter: 236.3 cm stage 1 (233.7 cm for launch 1)
Launch mass: 72,576 kg (68,040 kg for launch 1)
Guidance: autonomous autopilot hosted in Oettle & Reicher RCOM06OSC 68020-based flight computer contains the mission specific mission data load (MDL). Inertial/attitude data provided from Litton's LR-81 Inertial Measurement Unit (IMU) aided by position data from Trimble GPS receiver. The autopilot processor monitors GN&C performance and processor data are formatted within the flight computer for direct downlink via the vehicle telemetry system.

Orbital's Taurus rocket on the launch pad for GFO/ORBCOMM mission 9 February 1998 at VAFB ***1998***/0024769

Taurus Strap-ons

Comment: Optional paired strap-ons augment XLS performance. They mount to reinforcing rings on the Castor 120 forward/aft ends using machined ball fittings mated to receptacles.
Motor: Thiokol Castor 4B. Welded 4130 steel case with TI R300 internal insulator. Nozzle features carbon phenolic ablative exit cone with 4130 steel structural shell. Nozzle mounted in a laminated rubber/glass fibre flexseal for ±6° vectoring with AlliedSignal hydraulic cold gas blowdown TVAs. Igniter mounted in forward dome. Redundant conical shaped charges on forward dome allow flight termination.
Stage length: 8.98 m
Diameter: 101.9 cm
Thrust: each 429 kN average vacuum (total impulse 26.1 MN)
Burn time: 60.9 s
Mass at ignition: 15,001 kg each
Propellant mass: 9,990 kg each
Separation: clamp bands retaining ball fittings are pyrotechnically released, with separation force provided by hot gas pushoff thrusters
Flight sequence: TVA enabled after stage 0. Internal count sequencer performs nozzle sweep then commands simultaneous ignition of stage 0/strap-ons. Strap-on nozzles vectored to augment core control authority and provide roll control.

Taurus stage 0

Comment: The first vehicle (and any later USAF missions) utilised a Peacekeeper stage 1. Commercial launches adopt Thiokol's Castor 120 motor. The Peacekeeper specifications are given first, followed by those related to Castor 120.
Motor: Thiokol TU-903 solid, filament wound Kevlar epoxy casing structure of 23 helical layers and 55 hoop plies. Attachment to these skirts provided by aluminium rings. TU903 employs HTPB propellant and a silica-filled EPDM internal insulator. All stages' motors employ 88% solids, with grain designed for low-burn rates. Protection from aerodynamic heating and static charge build-up provided by an external rubber insulating layer impregnated with graphite. Ignition by pyrogen igniter on forward dome. Each stage nozzle consists of a carbon phenolic exit cone with 3-D carbon/carbon integral throat/entry. Stage 0 uses a silicone elastomer flexseal nozzle with hydraulic actuators. Hydraulic power is provided by a hot gas-driven turbopump. Redundant flight termination charges are mounted along the casing.
Stage length: 8.48 m
Diameter: 233.7 cm
Thrust: 2,204.5 kN average vacuum, 2,583.7 kN max vacuum.
Burn time: 56 s
Mass at ignition: 48,960 kg
Propellant mass: 44,660 kg
Separation: aft skirt extension incorporates redundant MDF to sever connection with launch stand. Aluminium interstage uses linear-shaped charge to release stage 1. Interstage vent covers released when bolt cutters separate forward and aft clamp bands.
Flight sequence: stage 0 TVA enabled at T2 s. Internal count sequencer performs nozzle sweep then commands simultaneous release from stand and ignition. Flies preprogrammed pitch profile to limit aerodynamic loading; roll not controlled. Vent panels released just before burnout, relieving pressure for stage

1 ignition (keyed to predetermined thrust tail-off; staging occurs 0.9 s later).
Motor: Thiokol Castor 120 solid. Employs Toray 100068 graphite epoxy wound composite case with an aramid filled EPDM rubber internal insulator. External protection from aerodynamic heating also provided by EPDM rubber layer. The nozzle features a carbon phenolic ablative exit cone with a graphite epoxy structural over-wrap and a 3-D carbon/carbon integral throat/entry. Nozzle mounted in a laminated rubber/glass fibre flexseal for ±5° vectoring. The 88% solids HTPB propellant is machined for fine tailoring of thrust profile. Igniter mounted in the forward dome. Redundant linear shaped charge (LSC) in external raceway allows flight termination.
Stage length: 7.72 m
Diameter: 236.3 cm
Thrust: 1,703 kN average vacuum, 1,942 kN max vacuum attained after 29 s (total impulse 133.9 MN).
Burn time: 83 seconds
Mass at ignition: 53,070 kg
Propellant mass: 48,988 kg
Separation: bolted connection to launch stand removed several hours before launch. Aluminium interstage LSC releases stage 1. Interstage vent covers released when bolt cutters separate forward/aft clamp bands
Flight sequence: stage 0 TVA enabled at T-30 s. Internal terminal count sequencer performs nozzle sweep then commands ignition. Remaining sequence similar to TU 903.

Taurus stage 1
Motor: Alliant Techsystems Orion 50S (50XL stretched by 137 cm for XL/XLS). All three Alliant motors use IM7/55A graphite epoxy wound composite cases with aramid filled EPDM rubber internal insulator and integral skirts. Flight termination charges mounted on the aft domes of stage 1/2 motors for range safety requirements. Stage 1 employs silicon elastomer flexseal nozzle with AlliedSignal hydraulic cold gas blowdown TVAs. Changes from Pegasus include removal of the wing saddle, thickening of the aft skirt and skirt doublers, revision of the aft attach ring and aft skirt for a tapered unit that mates to the interstage, and a vectored nozzle (4°) to replace fins for vehicle control.
Stage length: 763.8 cm (943.9 cm)
Diameter: 127.1 cm
Thrust: 471.511 kN (627.199 kN) average vacuum, 556.028 kN (720.612 kN) max vacuum attained about mid-way through burn. Total impulse 34.10 MNs (43.58 MNs).
Burn time: 72.4 s (69.3 s)
Mass at ignition: 13,270 kg (16,230 kg)
Propellant mass: 12,143 kg (15,072 kg)
Separation: the forward skirt, which also serves as interstage adaptor, incorporates LSC
Flight sequence: stage 1 TVA enabled just before stage 0 burnout. Following release of the interstage vent panels, stage 1 ignites, followed immediately by the LSC firing in the 0/1 interstage. Stage 1 also follows a preprogrammed pitch profile. Nozzle vectoring provides pitch/yaw control during powered flight. Burnout occurs typically 154 s (151 s), followed by staging at 158.2 s (150 s). Springs push the stack away, providing clearance for stage 2 ignition.

Taurus stage 2
Motor: Alliant Techsystems Orion 50 (50XL stretched 45.0 cm for XL/XLS). Changes from Pegasus are reinforcement of the forward/aft skirts and attach rings. Stage 2 also employs silicon elastomer flexseal nozzle and Parker Bertea Aerospace electromechanical actuators for TVC. A graphite epoxy facesheet/syntatic foam core sandwich structure boat-tail transitions from the stage 2 diameter to the 160 cm diameter fairing. A self-contained non-contaminating frangible separation system on the boat-tail's forward end allows stage 2 to be jettisoned. Four springs provide separation for the upper stage/payload.
Stage length: 265.7 cm (310.6 cm)
Diameter: 127.1 cm
Thrust: 113.874 kN (160.136 kN) average vacuum, 131.667 kN (195.266 kN) max vacuum. Total impulse 8.67 MNs (11.22 MNs).
Burn time: 75.9 s (70.1 s)
Mass at ignition: 3,370 kg (4,314 kg)
Propellant mass: 3,024 kg (3,919 kg)
Separation: frangible joint at boat-tail forward end allows separation of avionics skirt/stage 3/payload
Flight sequence: stage 2 ignites at 165 s as the remainder of the 1/2 interstage is jettisoned. Payload fairing jettisoned at 168 s. Stage 2 burnout at 237 s. Roll control during burn and three-axis control during coast by six N_2 thrusters in two Parker Bertea Aerospace modules on avionics section.
Attitude control: as stage 1

Taurus stage 3
Motor: Alliant Techsystems Orion 38 (also for XL/XLS). Stage 3 also employs vectored nozzle for pitch/yaw control during burn. Head end grain with toroidal igniter. Unlike Pegasus, the motor is not a load-carrying member but is cantilevered from its forward attach ring. The annular avionics shelf encircles the motor to mount most of the vehicle's electronics in addition to the N_2 attitude control system. This arrangement allows replacement of the motor with a spinning upper stage, such as Star 37, for perigee kick.
Stage length: 133.9 cm
Diameter: 96.5 cm
Thrust: 34.314 kN average vacuum, 38.686 kN max vacuum. Total impulse 2.18 MNs.
Burn time: 63.8 s
Mass at ignition: 893 kg
Propellant mass: 771 kg
Flight sequence: stage 3 ignites at apogee for orbit circularisation, using nozzle flexing for pitch/yaw control and the N_2 jet system for roll control and three-axis control during coast. Following injection, stage 3 manoeuvres to prescribed payload deployment attitude and provides a separation command to the payload, typically 30 s after burnout

Fairing and payload accommodation
Two bisector fairings are available: 1.6 m and 2.3 m diameter external. Fairing shells are fabricated by Courtaulds Structural Composites, Inc (1.6 m) and R Cubed (2.3 m) from graphite epoxy facesheets and aluminium honeycomb core. An OSC proprietary frangible joint separation system secures the fairing halves at the base and along the longitudinal seams until jettison during stage 2 burn. A pneumatic thruster system ejects the fairing, which is guided by fall-away hinges. Class 10,000 is maintained inside the fairing at all times, beginning from encapsulation until launch On the ground, air conditioning maintains 15.6+4/ 7°C and 2065% humidity. The payload can be provided with 28 V DC (up to 140 W) until T0. 90 passthrough circuits allow payload commanding and monitoring before launch. The Taurus flight computer can issue up to five discrete payload commands and can monitor five discrete talkbacks. RS-422/485 serial data interfaces can also issue commands and accept payload telemetry in the vehicle data stream. Four pyro commands can be issued. Data for 2.6 m fairing in ().
Payload volume: 1.37 m (4.06 m) diameter × 2.3 m (5.84 m) long
Acceleration load: 6.9 *g* longitudinal max at end of stage 1 burn
Acoustic load: 133.5 dB OASPL (XLS: 138 dB OASPL)

Vehicle and Payload Processing
Stages are pre-integrated and tested at a Stage Integration Facility. For early missions, this is the Missile Assembly Building at VAFB. Stage and launch support equipment are transported to the launch site as late as 8 days before launch. The LSV launch support van is positioned outside the caution/hazard corridor and connected to the pad via a fibre optic bundle. Portable generators can power the system if facility power is not available. Once the launch stand is bolted to the concrete pad, stage 0 is mounted and scaffolding erected up to its top. In parallel, the upper stages are processed in an adjacent integration tent. Jumper harnessing between the mated upper stages and the erected stage 0 allows end-to-end integrated system tests and mission simulations.

In parallel with (or before) vehicle integration, the payload is processed and encapsulated in an appropriate facility. OSC provides the payload cone, mounted on a GSE handling adaptor, for spacecraft integration/test. The fairing is then mated to the cone, encapsulating the cargo element. The environmental control system is connected and the cargo element moved to the launch. This can be as late as 72 h before launch. At the launch site, the cargo element is electrically and mechanically mated to the upper stages. This stack is then lifted and mated to stage 0. Final systems tests are completed, followed by alignment of the IMU. Vehicle closeout and final countdown starts at T7 h, controlled from the LSV. Dedicated consoles are provided for payload personnel, range safety personnel and the customer launch director as well as the OSC personnel controlling the launch.

Terrier Malemute sounding rocket

Background
NASA's sounding rocket programme is based at the Wallops Flight Facility in Virginia using vehicles from commercial sources or developed by the agency. Some 25 launches are made each year from a range of sites, including Wallops itself, Poker Flat and White Sands. NASA currently employs 15 sounding rocket types (Aries use ended 1993), including several utilising Bristol Aerospace's Black Brant 5 motors. Extensive use is made of military surplus motors and all are unguided. Saab's S19 Boost Guidance System can be added to the Nike Black Brant and Black Brant 5/9/10. The gyro platform controls aerodynamic canards, decreasing impact dispersion by a factor of 5 to 10, allowing higher apogees and launches in higher winds.

This combination is particularly suited to lower mass payloads; performance reduces appreciably as the weight increases. Bulbous payloads can be accommodated but high-dynamic pressures result in high-aerodynamic heating rates and vehicle structural loads. In general, it is employed for relatively lightweight plasma physics experiments and rail launched.

Specifications
Terrier-Malemute
First launch: June 1975
Number launched: 27 to end-1996
Success rate: 85%
Number of stages: 2 (solid)
Overall length: 7.2 m
Principal diameter: 40.6 cm
Launch mass: 1,478 kg plus payload
Typical performance: 91 kg to 670 km
Payload recovery system: custom built
Payload environment: 26 *g* peak acceleration for 90 kg payload, accommodated typically in 40.6 cm 3:1 ogive nose
Guidance: unguided, spin stabilised by four 0.22 m^2 fins

Terrier-Malemute stage 1
Motor: US Navy Terrier Mk12 mod 1 (Hercules Powder Co and Allegheny Ballistics Laboratory)
Overall length: 4.27 m
Principal diameter: 45.7 cm
Stage mass: 878 kg (534 kg propellant)
Thrust: 257.7 kN average sea level
Burn time: 4.4 s

Terrier-Malemute stage 2
Motor: Thiokol Malemute TU-758
Overall length: motor 3,302 mm
Principal diameter: 40.6 cm
Stage mass: 600 kg (499 kg propellant)
Thrust: 64 kN average SL (total impulse 1,223 kNs SL)
Burn time: 20 s

Titan 2

Background
Martin Marietta began development of the Titan 2 LGM-25C ICBM in 1960, achieving the first of 85 ICBM-related launches in March 1962 and beginning deployment in 1963 in three wings of 18 missiles each at Little Rock AFB (Arkansas), Davis Montham AFB (Arizona) and McConnell AFB (Kansas). The missile carried 5 MT thermonuclear warheads. Over the years, Martin Marietta had delivered 141 vehicles to USAF: 36 Research and Development N types and the rest operational B versions. Production attained a peak rate of 12 per month during the mid 1960s.

The missile was declared operational after 32 launches, but a further 49 test/training launches were made from Vandenberg until the last in June 1976. Twelve flew in the manned Gemini programme of 1964-1966, 10 carrying two-man crews. ICBM deactivation began July 1982 and was completed with the removal of the last missile from its silo at Little Rock 23 June 1987 after its last duty cycle 5 May.

The Titan 2G is the base version of the Space Launch Vehicle. 2B and 2S models were also planned. 2S competed unsuccessfully in 1993 for USAF's MLV3 medium-launch vehicle contract. It was originally planned for BMDO's MSX mission, which switched to Delta. A 2L version was also considered, adding two liquid strap-ons to handle 8,165 kg into 185 × 185 km 28.5° from the Cape.

Martin Marietta received a US$45.2 million letter contract in January 1986 to proceed with the final system design review of the Titan 2 as a space launcher. The US government signed a US$483.7 million contract in October 1986 to refurbish eight ICBMs, with a later exercised option for five more (a 14th was added in November 1987), bringing the total contract value to US$638.7 million to September 1995. The work included modifying the stage 2 forward structure to accommodate McDonnell Douglas' 3.05 m Titan 34D-type payload fairing, refurbishing the Aerojet engines, upgrading the Delco inertial guidance systems (to Titan 4 standard), developing command, destruct and telemetry systems, and modifying Vandenberg's Titan 3B SLC4W pad to accept the smaller vehicle. The first refurbished Titan, designated as the Titan 23G Space Launch Vehicle, was rolled out at the company's Denver plant 3 August 1987 scheduled initially for April 1988 launch. The launch actually occurred the following September.

This first vehicle employed the B56 stage 1 and

B98's stage 2 (Titan 2 was designed to allow stage interchanging). B106 was used for the second mission.

Martin Marietta finished refurbishing the first 14 in December 1990, each vehicle requiring typically 12 months' work. The remainder of the 55 were stored at Norton AFB, California but they began October 1993 moving to an army storage facility near Pueblo, Colorado when Norton shut its doors as part of the Base Closings Act. There are no plans for either Canaveral launches or commercial marketing. The most recent USAF/NASA ELV reimbursement agreement of October 1987 provides for space agency use at a basic US$24.6 million cost (FY89 rate). Launch cost to USAF was reported in 1990 at US$33 million. NOAA was charged US$36.5 million by USAF for 1993's Landsat 6 launch vehicle and services.

The Titan 2 Basic is similar to the 2G configuration but with minimal modification to the ICBM. It incorporates new wiring and avionics, and offers the option of an attitude control system. Instead of the McDonnell Douglas fairing, it uses the existing re-entry vehicle fairing.

The Titan 2 Solid Thrust Augmented version enhances the 2G's performance by adding 28 Castor 4A strap-ons. This requires a stage 1 extension to provide attach points. Alliant's higher performance GEM graphite epoxy motors (now flying on Delta 2) may also be used to increase payload capability.

Specifications

Titan 2 SLV

First orbital launch: 8 April 1964 (Gemini Titan 1)
Number of orbital launches: 17 to end 1998, plus 3 suborbital
Launch sites: Canaveral pad 19 (deactivated), Vandenberg SLC 4W
Principal uses: currently small and medium payloads into low polar orbits; 10 manned Gemini launches 1965/66
Vehicle orbital success rate, career: 100% to end July 1996
Performance: 2,177 kg into 185 km polar, 2,360 kg into 185 km 63.5°; with Star 37 kick stage 3,028 kg into 546 km Sun-synchronous polar; 3,175 kg into 185 km, 28.6° from Canaveral; 1,043 kg into GTO with SSPS upper stage. In addition, Castor 4A strap-ons can be added (Titan 2S) to increase LEO polar orbit capacity to 3.7 t, depending on inclination (3,540 kg 185 km 99°); performance to 1,100 km circular orbits >1.9 t. 2L: 8,165 kg 185 km 28.5°
Number of stages: 2 plus optional kick stages
Overall length: 31.4 m
Principal diameter: 3.05 m
Launch mass: 153.7 t plus payload
Launch thrust: 1,913 kN
Guidance: Delco Electronics Titan 4 digital inertial mounted on stage 2
Launch sequence (min:s):

–00:05	go inertial
00:02	stage 1 ignition signal
00:09	roll to flight azimuth
00:19	end roll
00:30	begin zero lift
01:50	end zero lift
02:37.7	stage 1/2 hot separation (65 km, 2,704 m/s)
03:31	payload fairing separation (119 km, 3,393 m/s)
05:32.8	stage 2 shutdown (189 km,7,785 m/s)
05:49	enable attitude control system
06:18	converge to commanded attitude, begin velocity trim
06:40	end velocity trim

Comment: An Enhanced Attitude Control System, carrying 726 kg hydrazine, became available in 1993 to reach higher orbits and/or maintain control for longer.

Titan 2G stage 1

Engines: refurbished gimballed paired Aerojet LR87AJ5 single-start hypergolic.
Overall length: 21.4 m
Principal diameter: 3.05 m
Stage mass: 4,220 kg dry
Propellant: 118.3 t N_2O_4/Aerozine 50
Stage thrust: 1,913 kN sea level
Burn time: 158 s (154 s Landsat 6)

Titan 2G stage 2

Engines: refurbished Aerojet LR91 AJ5 single-start hypergolic single chamber.
Overall length: 12.2 m
Principal diameter: 3.05 m
Stage mass: 2,860 kg dry
Propellant: 28.44 t N_2O_4/Aerozine 50
Stage thrust: 445 kN vacuum
Burn time: 175 s (182 s Landsat 6)

Fairing

McDonnell Douglas supplies the 3.05 m diameter 6.1-9.2 m Titan 34D-type aluminium skin/stringer payload fairing, providing a 2.83 m diameter, 9.1 m long envelope. A 15.3 m long fairing is feasible. Martin Marietta completed development in 1993 of an aluminium honeycomb/graphite epoxy composite skirt and payload adaptor that will yield an 87 kg payload increase.

Titan 2S

Motor contractor: Thiokol, Huntsville division
Length: 11.16 m
Diameter: 1.01 m
Average thrust: 434 kN sea level
Burn time: 53.1 s
Mass at ignition: 11.6 t
Propellant: TP-H8299 HTPB polymer, 20% aluminium
Propellant mass: 10.1 t

Titan 3

Background

Martin Marietta announced at the UK's Farnborough Air Show in September 1986 the availability from spring 1989 of a commercial version of its Titan 34D USAF launcher. Federal Express made the first reservation that month. Commercial Titan is essentially a 34D with a stretched stage 2 and 4 m diameter fairing to accommodate single or dual Shuttle-class payloads with perigee kick motors to LEO. Vehicles are handled on the same production line as Titan 4, allowing low-volume commercial operations.

Commercial Titan Inc was established in May 1987, but absorbed into Space Launch Systems in summer 1990. In order to launch the missile commercially, Martin concluded an agreement to utilise Canaveral pad 40 with the Air Force in August 1987. GE Astro Space made a 15 satellite block reservation in January 1988 to offer in orbit delivery of their satellites. GE and Martin later mutually terminated the deal.

Emphasis is now placed on flying single large payloads. The contract for launching Mars Observer in 1992 reportedly brought in US$156.6 million, including US$20 million for mission specific modifications. In order to make the vehicle available to a wider commercial base, Martin Marietta once considered making a restartable stage 2 and adopting Titan 4's fairing and uprated strap-ons for Space Station support and assembly.

A Titan 3L version was mooted in 1991, employing two liquid strap-ons to handle 9.1 tonnes LEO at about 70 per cent of the current Titan 3 cost. The Titan 3 family has now been retired.

Titan 3 family record

First launch: 1 September 1964 (18 June 1965 for strap-ons)
Number launched: 156 to end-1998
Launch sites: Canaveral pads 40/41, Vandenberg SLC4E/4W
Vehicle success rate, career: 91.0%
Vehicle success rate, past 25 launches: 84% to end 1996 (four failures in 10 years before most recent launch)

Titan 3 launch history

	Date	Type	Site	Payload
1*	1 September 1964	3A	ETR 20	test; stage 3 failure
2	10 December 1964	3A	ETR 20	test
3	11 February 1965	3A	ETR 20	LES 1
4	6 May 1965	3A	ETR 20	LES 2
5	18 June 1965	3C	ETR 40	test
6*	15 October 1965	3C	ETR 40	OV2 1; stage 3 failure
7	21 December 1965	3C	ETR 41	LES 3/4, OV2-3
8	16 June 1966	3C	ETR 41	IDCSP 1-7, GGTS1
9	29 July 1966	3B	SLC4W	KH 8 close-look
10*	26 August 1966	3C	ETR 41	IDCSP
11	28 September 1966	3B	SLC4W	KH 8 close-look
12	3 November 1966	3C	ETR 40	OV43, 4-1R, 1-6, 4 1T
13	14 December 1966	3B	SLC4W	KH 8 close-look
14	18 January 1967	3C	ETR 41	IDCSP 8 15
15	24 February 1967	3B	SLC4W	KH 8 close-look
16*	26 April 1967	3B	SLC4W	KH 8 close-look
17	28 April 1967	3C	ETR 41	Vela Hotel 7/8, OV5-3/5 4
18	20 June 1967	3B	SLC4W	KH 8 close-look
19	1 July 1967	3C	ETR 41	IDCSP 1618, LES 5, Dodge
20	16 August 1967	3B	SLC4W	KH 8 close-look
21	19 September 1967	3B	SLC4W	KH 8 close-look
22	25 October 1967	3B	SLC4W	KH 8 close-look
23	5 December 1967	3B	SLC4W	KH 8 close-look
24	18 January 1968	3B	SLC4W	KH 8 close-look
25	13 March 1968	3B	SLC4W	KH 8 close-look
26	17 April 1968	3B	SLC4W	KH 8 close-look
27	5 June 1968	3B	SLC4W	KH 8 close-look
28	13 June 1968	3C	ETR 41	IDCSP 19-26
29	6 Aug 1968	3B	SLC4W	KH 8 close-look
30	10 September 1968	3B	SLC4W	KH 8 close-look
31	26 September 1968	3C	ETR 41	LES 6, OV25, 5-2, 54
32	6 November 1968	3B	SLC4W	KH 8 close-look
33	4 December 1968	3B	SLC4W	KH 8 close-look
34	22 January 1969	3B	SLC4W	KH 8 close-look
35	9 February 1969	3C	ETR 41	Tacsat1
36	4 March 1969	3B	SLC4W	KH 8 close-look
37	15 April 1969	3B	SLC 4W	KH 8 close-look
38	23 May 1969	3C	ETR 41	Vela Hotel 9/10, OV5
39	3 June 1969	3B	SLC4W	KH 8 close-look
40	22 August 1969	3B	SLC4W	KH 8 close-look
41	24 October 1969	3B	SLC4W	KH 8 close-look
42	14 January 1970	3B	SLC4W	KH 8 close-look
43	8 April 1970	3C	ETR 40	Vela Hotel 11/12
44	15 April 1970	3B	SLC4W	KH 8 close-look
45	25 June 1970	3B	SLC4W	KH 8 close-look
46	18 August 1970	3B	SLC4W	KH 8 close-look
47	23 October 1970	3B	SLC4W	KH 8 close-look
48	6 November 1970	3C	ETR 40	DSP1 early warning
49	21 January 1971	3B	SLC4W	KH 8 close-look
50	21 March 1971	34B	SLC4W	Jumpseat
51	22 April 1971	3B	SLC4W	KH 8 close-look
52	5 May 1971	3C	ETR 40	DSP2 early warning
53	15 June 1971	3D	SLC4E	KH 9 Big Bird#1
54	12 August 1971	3B	SLC4W	KH 8 close-look
55	23 October 1971	3B	SLC4W	KH 8 close-look
56	3 November 1971	3C	ETR 40	DSCS 1/2
57	20 January 1972	3D	SLC4E	KH 9 Big Bird #2
58*	16 February 1972	34B	SLC4W	Jumpseat
59	1 March 1972	3C	ETR 40	DSP-3 early warning
60	17 March 1972	3B	SLC4W	KH 8 close-look
61*	20 May 1972	3B	SLC4W	KH 8 close-look

Titan 3 launch history (continued)

	Date	Type	Site	Payload
62	7 July 1972	3D	SLC4E	KH 9 Big Bird #3
63	1 September 1972	3B	SLC4W	KH 8 close-look
64	10 October 1972	3D	SLC4E	KH 9 Big Bird #4
65	21 December 1972	3B	SLC4W	KH 8 close-look
66	9 March 1973	3D	SLC4E	KH 9 BigBird #5
67	16 May 1973	3B	SLC4W	KH 8 close-look
68	12 June 1973	3C	ETR 40	DSP-4 early warning
69*	26 June 1973	3B	SLC4W	KH 8 close-look
70	13 July 1973	3D	SLC4E	KH 9 Big Bird #6
71	21 August 1973	34B	SLC4W	Jumpseat
72	27 September 1973	3B	SLC4W	KH 8 close-look
73	10 November 1973	3D	SLC4E	KH 9 Big Bird #7
74	13 December 1973	3C	ETR 40	DSCS II3/4
75*	11 February 1974	3E	ETR 41	test/Viking mass Centaur failure
76	13 February 1974	3B	SLC4W	KH 8 close-look
77	10 April 1974	3D	SLC4E	KH 9 Big Bird #8
78	30 May 1974	3C	ETR 40	ATS 6
79	6 June 1974	3B	SLC4W	KH 8 close-look
80	14 August 1974	3B	SLC4W	KH 8 close-look
81	29 October 1974	3D	SLC4E	KH 9 Big Bird #9
82	10 December 1974	3E	ETR 41	Helios 1
83	10 March 1975	34B	SLC4W	Jumpseat
84	18 April 1975	3B	SLC4W	KH 8 close-look
85*	20 May 1975	3C	ETR 40	DSCS II
86	8 June 1975	3D	SLC4E	KH 9 Big Bird #10
87	20 August 1975	3E	ETR 41	Viking 1
88	9 September 1975	3E	ETR 41	Viking 2
89	9 October 1975	3B	SLC4W	KH 8 close-look
90	4 December 1975	3D	SLC4E	KH 9 Big Bird #11
91	14 December 1975	3C	ETR 40	DSP5 early warning
92	15 January 1976	3E	ETR 41	Helios 2
93	15 March 1976	3C	ETR 40	LES 8/9, Solrad 11A/B
94	22 March 1976	3B	SLC4W	KH 8 close-look
95	2 June 1976	34B	SLC4W	SDS
96	26 June 1976	3C	ETR 40	DSP6 early warning
97	8 July 1976	3D	SLC4E	KH 9 Big Bird #12
98	6 August 1976	34B	SLC4W	SDS
99	15 September 1976	3B	SLC4W	KH8 close-look
100	19 December 1976	3D	SLC4E	KH 11 #1
101	6 February 1977	3C	ETR 40	DSP7 early warning
102	13 March 1977	3B	SLC4W	KH 8 close-look
103	12 May 1977	3C	ETR 40	DSCS II 7/8
104	27 June 1977	3D	SLC4E	KH 9 Big Bird #13
105	20 August 1977	3E	ETR 41	Voyager 2
106	5 September 1977	3E	ETR 41	Voyager 1
107	23 September 1977	3B	SLC4W	KH 8 close-look
108	25 February 1978	34B	SLC4W	Jumpseat
109	16 March 1978	3D	SLC4E	KH 9 Big Bird #14
110*	25 March 1978	3C	ETR 40	DSCS II 9/10
111	10 June 1978	3C	ETR 40	Chalet electronic intelligence
112	14 June 1978	3D	SLC4E	KH 11 #2
113	5 August 1978	34B	SLC4W	SDS
114	14 December 1978	3C	ETR 40	DSCS II 11/12
115	16 March 1979	3D	SLC4E	KH 9 Big Bird #15
116	28 May 1979	3B	SLC4W	KH 8 close-look
117	10 June 1979	3C	ETR 40	DSP8 early warning
118	1 October 1979	3C	ETR 40	Chalet electronic intelligence
119	21 November 1979	3C	ETR 40	DSCS II 13/14
120	7 February 1980	3D	SLC4E	KH 11 #3
121	18 June 1980	3D	SLC4E	KH9 Big Bird #16
122	13 December 1980	34B	SLC4W	Jumpseat
123	28 February 1981	3B	SLC4W	KH 8 close-look?
124	16 March 1981	3C	ETR 40	DSP9 early warning
125	24 April 1981	34B	SLC4W	SDS
126	3 September 1981	3D	SLC4E	KH 11 #4
127	31 October 1981	3C	ETR 40	Chalet electronic intelligence
128	21 January 1982	3B	SLC4W	radar test satellite?
129	6 March 1982	3C	ETR 40	DSP-10 early warning
130	11 May 1982	3D	SLC4E	KH 9 Big Bird #17
131	30 October 1982	34D	ETR 40	DSCS II15 &III1/IUS
132	17 November 1982	3D	SLC4E	KH 11 #5
133	15 April 1983	3B	SLC4W	KH 8 close-look
134	20 June 1983	34D	SLC4E	KH 9 Big Bird #18
135	31 July 1983	34B	SLC4W	SDS or Jumpseat
136	31 January 1984	34D	ETR 40	Chalet?/Trans-stage
137	14 April 1984	34D	ETR 40	DSP 11/Trans-stage
138	17 April 1984	3B	SLC4W	KH 8 close-look
139	25 June 1984	34D	SLC4E	KH 9 Big Bird #19
140	28 August 1984	34B	SLC4W	SDS
141	4 December 1984	34D	SLC4E	KH 11 #6
142	22 December 1984	34D	ETR 40	DSP 12/Trans-stage
143	8 February 1985	34B	SLC4W	SDS or Jumpseat
144*	28 August 1985	34D	SLC4E	KH 11 #7
145*	18 April 1986	34D	SLC4E	KH 9 Big Bird #20
146	11 February 1987	34B	SLC4W	SDS or Jumpseat
147	26 October 1987	34D	SLC4E	KH 11 #8
148	29 November 1987	34D	ETR 40	DSP 13/Transtage
149*	2 September 1988	34D	ETR 40	Chalet (Vortex)/ Trans-stage failure
150	6 November 1988	34D	SLC4E	KH 11?
151	10 May 1989	34D	ETR 40	Chalet (Vortex)/ Trans-stage
152	4 September 1989	34D	ETR 40	DSCS II-16 & III 4/IUS
153	1 January 1990	3	ETR 40	Skynet4A/JCSat 2
154*	14 March 1990	3	ETR 40	Intelsat 603
155	23 June 1990	3	ETR 40	Intelsat 604
156	25 September 1992	3	ETR 40	Mars Observer/TOS

Specifications

Commercial Titan 3

First launch: 1 January 1990
Number launched: 4 to end-1998
Vehicle success rate: 75.0% for commercial flights, 91.6% for series
Launch sites: Canaveral pad 40
Principal uses: delivery of Shuttle class payloads to LEO for perigee kick motor insertion into GTO.
Cost: US$110 million for a dedicated flight; US$100,000 reservation deposit required; reflight insurance offered at 10 per cent premium.
Availability: 33 months after contract go-ahead; potentially 2/3 flights/year.
Schedule of future missions: none

Performance

LEO (148 × 259 km, 28.6°): 14,334 kg dual payload, 14,742 kg single payload.
GTO 26.4°: 1,279 kg with PAM D; 1,851 kg with PAM-D2, dual carrier; 4,944 kg with IUS; 4,990 kg with TOS, single carrier
GEO: not applicable
Space Station transfer orbit: 11,700 kg with no upper stage.
Sun-synchronous: not available
Lunar delivery: 3,400 kg with TOS
Venus delivery (Type II): 2,600 kg with TOS
Mars delivery (Type II): 2,430-2,600 kg with TOS
Number of stages: 2 plus 2 strap-ons
Overall length: 47.3 m dual carrier, 44.06 m single carrier
Principal diameter: 3.05 m
Launch mass: 680 t
Guidance: Delco Systems Operations inertial, incorporated in stage 2 with Olin Aerospace's 178 N MR 107 hydrazine attitude control thrusters. 3Σ injection accuracies: apogee 259±11.1 km, perigee 148+1.9/-0 km, inclination 113.3±3.0°
Launch sequence (min:s)

00:00	solids ignition
00:54	max dynamic pressure
01:48	stage 1 ignition
01:56	solids jettison
04:29	stage 2 ignition
04:30	stage 1separation
04:40	fairing jettison
08:14	stage 2shutdown
08:30	park orbit insertion
08:40	trim burn
60:30	payload1 deploy attitude
63:30	payload 1spin-up
67:16	payload 1 deploy
77:00	avoidance manoeuvre
112:16	payload 1 GTO burn
148:00	eject payload carrier
151:00	payload 2-deploy attitude
152:00	payload 2 spin-up
155:50	payload 2 deploy
188:00	avoidance manoeuvre and burn to depletion
200:50	payload 2 GTO burn.

Titan 3 solid strap-ons

Motor contractor: United Technologies' Chemical Systems Division, San Jose, California. CSD was awarded a $100 million contract in January 1988 for commercial Titan motor supply. Each motor comprises 5½ segments, each joint secured by 237 hand-placed clevis pins. The base segment extension bears the weight of the whole vehicle on the pad.
Length: 27.57 m
Principal diameter: 3.11 m
Mass at ignition: 250.387 t each
Propellant: UTP-30001 B solid
Propellant mass: 210.63 t
Average thrust: 6,227 kN each, vacuum
Burn time: 113.7 s
Steering: secondary injection through 24 nozzles into exhaust of N_2O_4 from 3,630 kg capacity tank (8.5 m long, 1.1 m diameter), although typically less than half consumed. Injection adds about 17.8 kN thrust to each motor.
Separation: jettison at 116 s by eight 20 kN solid motors firing for 1 s.

Titan 3 stage 1

Contractor: Lockheed Martin, Denver
Engines: paired Aerojet gimballed LR87AJ11
Overall length: 24.0 m
Principal diameter: 3.05 m
Oxidiser: 71.7 t nitrogen tetroxide in forward tank, loaded 3 days before launch.
Fuel: 38.0 t Aerozine-50 in aft tank, loaded 2 days before launch.
Stage thrust: 2,340 kN vacuum
Burn time: about 160 s, igniting at 118 s
Attitude control: gimballed main engines for pitch/roll; forward boat-tail carries four external modules, each

with three nozzles (pitch, roll, yaw), providing three-axis control. 102 kg hydrazine, thrust 80-133 N. The thrusters also provide ΔV corrections and payload spin-up before deployment.
Separation: at 270 s.

Titan 3 stage 2
Comment: Stage 2 employs a single burn but a restart capability has been considered for enhancing payload capability to circularised orbits. For example, restart would raise the 11,700 kg injection capacity into transfer orbit for 370 km circular 28.5° by 2,270 kg.
Contractor: Lockheed Martin, Denver
Engines: gimballed Aerojet LR91AJ 11, igniting at 269 s.
Overall length: 9.85 m
Principal diameter: 3.05 m
Oxidiser: 18.3 t nitrogen tetroxide in forward tank, loaded 3 days before launch.
Fuel: 10.3 t Aerozine-50 in aft tank, loaded 2 days before launch.
Thrust: 467.04 kN vacuum
Burn time: 225 s (269-494 s after launch)
Separation: stage 2 can provide spin of up to 2 rpm in addition to optional spin table performance; payload spring separated at 0.6 m/s.

Payload fairing/carrier
Contraves AG of Zurich, Switzerland was awarded the contract in 1987 for the 10.4 m long, 3.95 m internal diameter aluminium honeycomb/graphite epoxy fairing, based on their Ariane 4 design. The standard fairing sits atop the Payload Carrier in dual or single form; Germany's Dornier was awarded an initial US$18 million contract in October 1987 (in competition with Aerospatiale's Ariane Sylda and British Aerospace's Ariane Spelda) for one test, two single and two dual carriers. In the single form, it accommodates the payload and Contraves' fairing, creating a payload unit 12.69 m long. With a dual carrier, the fairing is separated in halves at 280 s during stage 2 burn when atmospheric heating has reduced to 1,135 W/m². The forward payload is deployed 67 minutes after launch, the 5.59 m high aft carrier cover is jettisoned forward as a unit at 148 minutes for payload number 2 deployment 7:50 minutes later. Standard Titan 34D 3.05 m diameter fairings are also available, as are spin tables providing up to 70 rpm.
Acoustic loads: 142 dB peak after 10 s and during transonic flight (54 s)
Vibration: typically 4.2 *g* rms for 60 s
Acceleration, longitudinal: 2.8 *g* strap-on burn peak, 4 *g* stage 1 peak, 2.5 *g* stage 2 peak. Payload design requirement for 6 *g*.
Thermal environment: fairing internal wall surface 100°C peak by 280 s.
Pad payload umbilical: 171 hardwire links to the two payloads are available through a pad umbilical at the top of stage 2, below the aft carrier.
Pad environmental control: 9.4 37°C, 30-50% humidity, class 10,000 filtration air conditioning provided inside fairing.

Vehicle/payload processing
Titan's core is stacked on its launch platform at launch complex 40/41's Vertical Integration Building. The VIB houses four high bays: Bay 1 is dedicated to commercial Titan operations, Bay 2 to USAF Titan and Bay 3 for Centaur. VIB's Launch Control Center was replaced in 1993 by the new Launch Operations Control Center elsewhere. The stack is then moved 1.4 km by rail to the SMAB Solid Motor Assembly Building where the strap-ons are stacked and attached before moving the final 4.2 km to pad 40/41. The new SMARF Solid Motor Assembly & Readiness Facility was added in 1994 to prepare Titan 4's strap-ons.

Payload processing facilities, including those of Astrotech Space Operations in nearby Titusville.

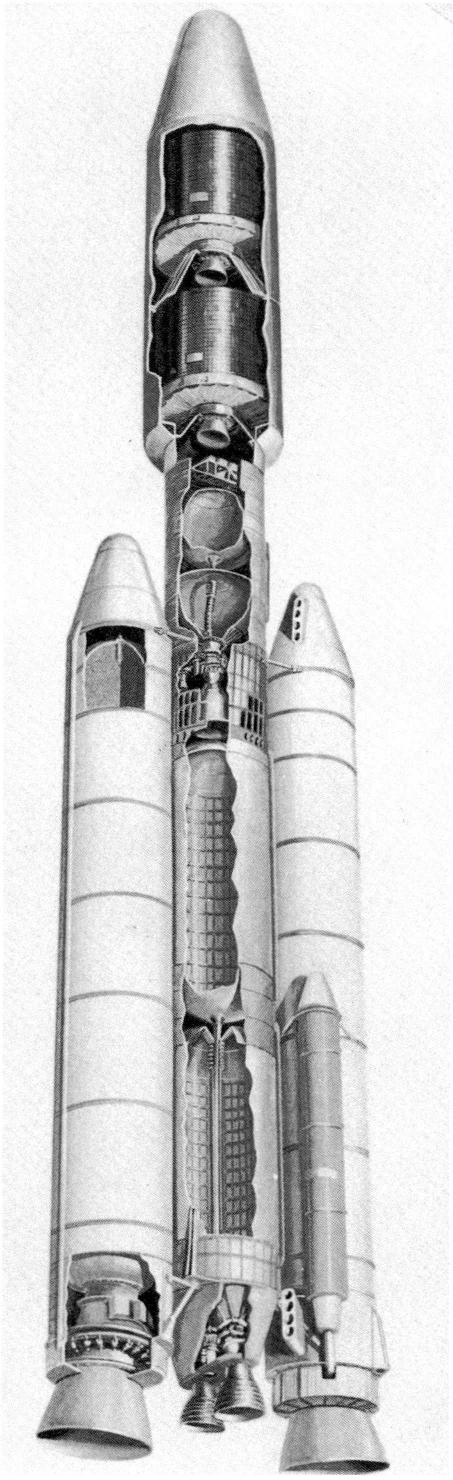

Commercial Titan 3. A second satellite is accommodated in the carrier below the first (Lockheed Martin)

Titan 4

Background
The US Air Force began a programme in March 1984 for a Complementary Expendable Launch Vehicle (CELV) because of Shuttle's failure to provide a 14,515 kg (32,000 lb) capability into polar orbit for the new generation of surveillance platforms and as a hedge against operating difficulties with the manned craft. Martin Marietta's standard Titan 34D launcher of KH11s was upgraded to a 34D7 version by stretching the 5½ segment strap-ons to seven segments, stretching stages 1/2, incorporating a 5.09 m fairing and adding, when required at all, either an IUS or Centaur upper stage (NUS = No Upper Stage).

The company was awarded a US$2.100 billion contract in February 1985 for 10 of these Titan 4s, extended in August 1986 to 23 by a further US$1.970 billion contract. The awards included US$250 million to Aerojet TechSystems for 23 stage 1/2 engine sets (now US$350 million for 41 sets). A US$1.600 billion contract for a further 18, reducing unit cost from US$150 million to US$89 million, was announced in November 1989, with an option for eight more, bringing the total to 41 by September 1995 under Martin Marietta's US$7.8 billion contract. USAF Space Systems Division planned to increase the buy through 1995 to 55 (and 78 through 1997), comprising 21 for USAF payloads and 34 for other users such as the CIA. 30 would have been launched from the Cape.

A 1990 GAO report noted that cost overruns totalled US$298 million, with development problems delaying Initial Launch Capability from October 1988 to February 1989, and the Centaur version's ILC by 14 months to spring 1991. USAF's 1990 projection was for 90-100 launches through FY2000, but by early 1991 programme cutbacks had reduced the plan to 61 launches. Martin Marietta's contracted 41 vehicles were due to be completed in 1995 but that schedule has been stretched for the 41st launch in 2004.

The reduced production rate of five and a half years from 1991, instead of the planned 10, proved too high and vehicles are being stockpiled. A USAF order reducing the rate to three a year was made in mid-1993 (a US$216 million contract increase was awarded December 1994 to maintain production capability during the stretch-out). The purchase of another five is under discussion, the last departing in 2007. The company receives a US$7 million fixed incentive for each successful launch and loses US$45 million for each failure.

As the liquid Centaur upper stage was deleted from Shuttle scheduling on safety grounds in June 1986, NASA requires Titan 4 Centaur for its most demanding deep space missions. The cost of the Cassini mission in 1997 however, stunned NASA and means that one civil use will probably stand as a unique instance.

Space Station assembly or resupply missions could also employ the vehicle, but none has been scheduled. Martin Marietta claimed a Titan 4 cost of US$200 million for NUS and US$250 million for Centaur (1993 rates). In 1996 the figures were reported as US$250 million and US$350 million, respectively.

Air force planners awarded Martin Marietta US$64.3 million contract in February 1990 for modification design work to allow CCAS pad 40 to handle Titan 4 in addition to Titan 3; the conversion was completed in 1992. The new 5,600 m² SMARF Solid Motor Assembly & Readiness Facility for simultaneous processing of up to six sets of Titan 4 boosters began full operations in August 1994. The SMAB is also used for IUS stage assembly/processing.

Construction began January 1993 of a separate Centaur Processing Facility to decouple Centaur pad checkout from other Titan 4 elements. Vandenberg pad SLC4E, declared operational for Titan 4 in May 1990, was refurbished at a cost of US$235 million. Congress rejected the USAF claim for a new SLC7 and directed instead that the SLC6 Shuttle pad should be modified. Cost was estimated in 1990 at US$700 million. USAF announced in September 1990 that work would begin in 1991 for a 1996 first launch. By early 1991, however, the reduction in projected launch rates brought into question the need for SLC6; it was cancelled that year. Titan 4 and IUS authority was transferred from the Systems Command to the Space Command in 1994: 5th Space Launch Squadron for Canaveral operations 4th SLS for VAFB.

The most powerful Titan 4 version, 401 (Centaur), debuted in February 1994 with Milstar. The second departed May 1994, although the original vehicle was stacked on the pad June 1991, with launch planned for mid-1992. However, it was de-stacked after corrosion and water were found in the joints June 1992 (the SRMs are certified for 1 year on the pad). New SRMs and core were built up but, largely because of delays caused by investigations into two Centaur failures on Atlas launches, a third vehicle had to be assembled.

Specifications
Titan 4
First launch: 14 June 1989
Number launched: 26 to end April 1999
Vehicle success rate: 88.4%
Launch sites: Canaveral pad 41 for LEO/GEO missions (plus pad 40 from 1992), Vandenberg SLC4E for NUS version from 1991 (capacity 2/yr). Canaveral pad 41 was used in the 1970s for Titan 3E Centaur D1T launches of deep space probes. A new pad 42 complex was at one stage under consideration for Cape Canaveral.
Principal uses: large military payloads into Sun synchronous orbit from Vandenberg; large military payloads such as electronic intelligence, DSP and Milstar into GEO from Canaveral, BMDO tests in lower orbits (no upper stage), NASA deep space launches.
Availability: no commercialisation plans

Performance
LEO (28.6°): 17,770 kg NUS plus SRM 21,900 kg NUS+SRMU
GEO: 4,545 kg with Centaur plus SRM. 5,773 kg with Centaur plus SRMU: 2,364 kg with IUS plus SRM : 2,860 kg with IUS plus SRMU.
Sun-synchronous (VAFB): 14,090 kg NUS plus SRM

Molniya (12 h, 925 × 39,450 km, 63°) from CC
Deep space: 4,545 kg at 20.5 km²/s² (SRM)
Number of stages: (2 or 3) plus 2 strap-ons
Overall length: 63.14 m for Centaur, 53.99 m for IUS, 53.99/57.04 m for NUS.
Principal diameter: 3.05 m.

Launch mass
Centaur: 868,644 kg (SRM); 939,301 kg (SRMU).
IUS: 910,018 kg (SRM); 924,515 kg (SRMU).
NUS: 906,937 kg (SRM); 910,018 kg (SRMU).
Guidance: GM's Delco Systems Operations digital inertial. 4B's standardised avionics from vehicle #24 will use Honeywell system: 3 Honeywell GG 1342 RLGs, 3 Sundstrand QA 3000 accelerometers, 1750A 1 Mips processors, 25,000 h MTBF, 30 kg, 0.03 m³, 110 W. 38 kg 250 Ah SAFT lithium thionyl chloride battery.

Titan 4 strap-ons (stage 0)
Comment: Titan 4 is initially employing United Technologies Chemical Systems Division's steel casing strap-ons. Lack of capacity to satisfy at least eight

launches per year and the requirements of some missions for higher performance resulted in a US$725 million contract (cost in July 1993 reported at US$1.2 billion) for 15 booster sets being awarded to Hercules Aerospace (now Alliant Techsystems) in October 1987.

The 3-segment filament-wound graphite composite casings and the high-performance HTPB propellant similar to that used in Peacekeeper's stage 1 will increase Titan 4 capacity by typically 27%. Development firings were due to begin March 1990 but were delayed by bonding problems with the second segment, cast December 1989. The new motors were to have debuted on launch 11, in spring 1992, and have taken over from #17. As insurance, USAF authorised long-lead procurement of seven further CSD sets.

On 7 September 1990, a 136 t SRMU practice segment ignited in an accident at the USAF Astronautics Laboratory in California. The unit, incorporating an aft skirt/nozzle, was being lifted from a test stand when its crane overbalanced, killing a nearby worker. The segment was ignited by friction sliding into the flame trench. The first full-scale SRMU firing was planned for the same stand later that year in the schedule to qualify the booster by October 1991. The PQM1 (Preliminary Qualification Motor) exploded after a few seconds 1 April 1991 because of unpredicted internal aerodynamic choking. The added PQM1's firing was delayed to 12 June 1992, largely because the test stand had to be rebuilt. The QM2 third test, run at the 2.5°C lower limit, was successful 21 February 1993. QM3 at the 41°C upper limit was successful 2 June. The final QM4, 12 September 1993, would have allowed flight introduction in 1995, although the initial requirement is for K24 fourth quarter 1996. The first production flight set was delivered first quarter1994.

CSD's seven segment boosters are based on the model developed for the USAF Manned Orbiting Laboratory Titan 3M launcher, cancelled 1969. Two qualification firings only, therefore, were required for Titan 4, accomplished 21 December 1987 and 14 February 1988. Martin Marietta was awarded a US$10 million contract in 1990 to install a sound suppression system on 4E as protection against the solid motors' ignitions. The system dumps water into the exhausts to reduce the pressure wave. The company also received a US$5.3 million contract for upgrading motor storage facilities by April 1991 to handle the SRMU. Titan 4 #7 exploded after 101 seconds when the casing of SRM1's segment 3 burned through. The cause was a slit in the grain from repairs to de-bonded restrictor material on top of the segment. This material retards burning on top. It was expected that compression would close the slit in flight, but tests showed ignition pressure opened a route to the wall.

Contractor: UTC CSD for heavyweight steel casing (D6AC) SRM versions; Alliant for SRMU lightweight filament wound composites.
Length: 34.43 m (CSD), 34.25 m (Alliant)
Diameter: 3.11 m (CSD), 3.20 m (Alliant)
Mass at ignition: 342.8 t (CSD), 349.6 t (Alliant)
Propellant: PBAN (CSD), HTPB (Alliant)
Propellant mass: 295.5 t (CSD), 344.4 t (Alliant)
Average thrust: 7.117 MN each sea level (CSD), 7.562 MN each vacuum (Alliant)
Burn time: 126.5 s (CSD), 145 s (Alliant).

Titan 4 stage 1
Comment: Aerojet was contracted in December 1990 for a modified stage 1 nozzle; it became available in 1994. It required an increased expansion ratio for ignition at greater altitude when Alliant's higher-performance strap-ons are carried. The nozzle exit closure is enlarged and improved to provide better protection during early flight, and a thermal protection clamshell is fitted around the combustion chamber. Projected LEO capacity increase is 136 kg; future modifications under consideration could add a further 318 kg.
Contractor: Lockheed Martin
Engines: paired gimballed Aerojet LR87AJ11A single start pressure fed. First Titan 4 full duration 200 s test firing 20 July 1987.
Overall length: 26.38 m
Principal diameter: 3.05 m
Oxidiser: 111 t of nitrogen tetroxide
Fuel: 59 t of Aerozine-50
Stage thrust: 2,433.8 kN vacuum
Burn time: 186 s (ignites at 116 s)
Separation: stage 2 hot separation

Titan 4 stage 2
Contractor: Lockheed Martin
Engine: gimballed Aerojet LR91 AJ11A single start liquid propellant pressure fed
Overall length: 9.94 m
Principal diameter: 3.05 m
Oxidiser: 24.5 t of nitrogen tetroxide
Fuel: 13.9 t of Aerozine-50
Stage thrust: 472.0 kN vacuum
Burn time: 240 s

Titan 2 space launch history

	Date	Site	Payload
1*	8 April 1964	ETR 19	Gemini 1
2	19 January 1965	ETR 19	Gemini 2
3	23 March 1965	ETR 19	Gemini 3
4	3 June 1965	ETR 19	Gemini4
5	21 August 1965	ETR 19	Gemini 5
6	4 December 1965	ETR 19	Gemini 7
7	15 December 1965	ETR 19	Gemini 6
8	16 March 1965	ETR 19	Gemini 8
9	3 June 1965	ETR 19	Gemini 9
10	18 July 1965	ETR 19	Gemini 10
11	12 September 1965	ETR 19	Gemini 11
12	11 November 1966	ETR 19	Gemini 12
13	5 September1988	SLC4W	electronic intelligence
14	5 September 1989	SLC4W	electronic intelligence
15	25 April 1992	SLC4W	electronic intelligence
16*	5 October 1993	SLC4W	Landsat 6
17	25 January 1994	SLC4W	Clementine1
18	4 April 1997	SLC4W	DMSPSD2 14
19	13 May 1998	SLC4W	NOAA-15

* Gemini 1 was a sub-orbital mission. Landsat 6 mission failed but Titan's suborbital portion was successful

Titan 3 breakdown
Launch total for each type, followed by number of failures in ()/followed by success rate. Upper stage failures included.

3 family: 156(14)/91.0%	*3B:* 54(3)/94.4%
3A: 4(1)/75%	*3C:* 36(4)/88.9%
34B: 14(1)/92.9%	*34D:* 15(3)/80%
3D: 22(0)/100%	*Comm 3:* 4(1)/75%
3E: 7(1)/85.7%	

ETR indicates Canaveral launch; SLCVandenberg.
* indicates launcher or upper stage failure

Titan 4 stage 3
Comment: Centaur and IUS are the principal upper stage choices but a range of spinning solids is possible. Centaur performs three burns to: achieve low Earth parking orbit, establish GTO and inject into GEO. The two-stage IUS is released in LEO and performs a stage 1 burn for GTO, followed by the stage 2 injection into GEO.
Designation: Centaur or IUS (designated Titan 401 when Centaur carried; 402 for IUS)
Contractor: Lockheed Martin (Centaur), Boeing (IUS)
Type: liquid (Centaur), solid (IUS)
Engine: P&W RL10-3-3A (Centaur), CSD Orbus (IUS)
Overall length: 8.94 m (Centaur), 5.00 m (IUS)
Principal diameter: 4.51 m (Centaur), 2.896 m (IUS)
Centaur propellants: 23 t of liquid oxygen/hydrogen
IUS propellant: 9,818 kg stage 1, 2,722 kg stage 2
Stage thrust: 146.8 kN vacuum (Centaur), 202 8 kN stage 1/82.3 kN stage 2 (IUS)
Mass at ignition: 16.2 t (IUS), 26 t (Centaur)
Burn time
IUS: 152 s for stage 1 GTO insertion, 289 s for stage 2 GEO insertion
Centaur: 617 s total
Attitude control: 3-axis both Centaur/IUS

Titan 4 fairings
Contractor: McDonnell Douglas
Construction: isogrid aluminium 3061

Canaveral pad 40 preparations for launching the first Titan 4 Centaur (Lockheed Martin)

Length: 17.08 m NUS, 26.23 m Centaur, 17.06 m IUS (15.25, 20.1, 23.2 m versions also available). 23.2 m version used for first time by Milstar DFS-1.
Diameter: 5.09 m
Mass: 4,033 kg NUS, 6,073 kg Centaur, 4,026 kg IUS
Separation: low-explosive detonating fuse along seams and 12 explosive bolts divides fairing into three sections in 0.2 s

Titan Family Development History
Titan 1 (HGM-25A) Development began in October 1955 as a back-up to Atlas ICBM. First of 67 launches January 1959, entered service 1962 with 54 missiles in

Titan 4 launch history

	Date	Site	Stage	Type	Payload
1	14 June 1989	ETR 41	IUS	402	DSP14
2	8 June 1990	ETR 41	-	405	ocean surveillance?
3	13 November 1990	ETR 41	IUS	402	DSP 15
4	8 March 1991	SLC4E	-	403	Lacrosse?
5	8 November 1991	SLC4E		403	ocean surveillance?
6	28 November 1992	SLC4E		404	Adv KH 11?
7*	2 August 1993	SLC4E		403	ocean surveillance?
8	7 February 1994	ETR 40	Centaur	401	Milstar DFS 1
9	3 May 1994	ETR 41	Centaur	401	Adv Jumpseat?
10	27 August 1994	ETR 41	Centaur	401	Adv Vortex?
11	22 December 1994	ETR 40	IUS	402	DSP 17
12	14 May 1995	ETR 40	Centaur	401	Adv Orion?
13	10 July 1995	ETR 41	Centaur	401	Adv Jumpseat?
14	6 November 1995	ETR 40	Centaur	401	Milstar DFS 2
15	5 December 1995	SLC4E		403	?
16	24 April 1996	ETR 41	Centaur	401	Adv Orion?
17	12 May 1996	SLC4E		?	?
18	3 July 1996	ETR 40	-	405	ocean surveillance
19	20 December 1996	SLC4E		404?	Advanced KH-11?
20	24 February 1997	ETR 40	IUS	402	DSP18 (Titan 4B)
21	15 October 1997	ETR 40	Centaur	401	Cassini/Huygens
22	24 October 1997	SLC4E		403	Lacrosse/Vega
23	8 November 1997	ETR 41	Centaur	401	Trumpet (Elint)
24	9 May 1998	ETR 40	Centaur	401	Advanced Orion (Elint)
25*	12 August 1998	ETR 41	Centaur	401	Mercury (Vortex)
26*	9 April 1999	ETR 41	IUS	402	DSP-19

* vehicle failure (SRM 1 burn through on #7 caused vehicle to explode after 101 s). 401=Centaur, 402=IUS, 403=NUS from VAFB, 404=NUS plus 15.25 m fairing, 405=NUS from Canaveral.

six squadrons; deactivated 1965. Launch mass 99.79 t, length 29.87 m with Avco Mk4 warhead, stage 1: 3.05 m diameter, stage 2: 2.44 m diameter, propellants LOX/RP-1, engines: two Aerojet LR87AJ1s, stage 1 (1,334 kN), single LR87AJ-1, stage 2 (356 kN). The only IRBM/ICBM developed 1950s/early1960s not adapted as space launcher

Titan 2 144 built (36 R&D, 108 operational), first launch 1962, last of 54 deployed ICBMs deactivated May 1987. Modified as space launcher

Titan 3A 4 launched 1964-5 as demonstrations of Titan 3C's core. Essentially a Titan 2 with added Trans-stage upper stage. Length 37.8m

Titan 3B 54 launched, all with KH8 close look reconnaissance satellites from VAFB's SLC4W, 1966-1985. Essentially Titan 2 with added Agena upper stage. A further 14 were launched as the **Titan 34B** (also known as Titan 3B Ascent Agena) 1971-87 from VAFB with stretched stage 1 and 3.05 m diameter payload fairing, carrying the highly classified SDS communications or Jumpseat electronic intelligence into Molniya-type orbits with apogees over the Soviet Union

Titan 3C 36 launched 1966-82 from Florida carrying Vela and DSP early warning, DSCS communications, GEO electronic intelligence and unclassified satellites, including NASA's ATS 6. Titan 3A as core stage with five segment solid strap-ons, 25.9 m long, 3.05 m diameter each generating 5.34 MN. Stage 1's engines ignited at altitude for greater efficiency, producing 2.09 MN

Titan 3D 22 launched from VAFB 1971-83 with KH9 Big Bird and KH11 reconnaissance satellites. Essentially west coast version of the 3C but without Trans-stage.

Titan 34D introduced in 1982, incorporating 34B's stretched stage 1, either Transtage or IUS as upper stage, and adding a half segment to each strapon. number 15/last 34D launched September 1989

Titan 3E Seven launched 1974-77 from Canaveral with Helios, Voyager and Viking. Titan 3D modified for NASA deep space missions: Centaur cryogenic upper stage and 4.27 m diameter fairing

Titan 3M Seven-segment strap-on version of the 3C, intended as the launcher of the USAF Manned Orbiting Laboratory but development ended when MOL was cancelled in 1969. Titan 4 eventually introduced the seven-segment variant

Titan 3 commercial Detailed below; first launch 1 January 1990

Titan 4 In service. Stretched stages1/2, seven-segment strap-ons, IUS, Centaur or no upper stage. First launch June 1989. For the standardised **Titan 4B**, strap-ons replaced from end1996 by the Solid Rocket Motor Upgrade (SRMU) of larger three-segment strap-ons employing filament wound graphite epoxy cases

X-33 VentureStar

Background

The Reusable Launch Vehicle (RLV) programme originated from a NASA study in 1993, which addressed the United States' future launch vehicle requirements. The Access to Space Study concluded that the most beneficial option would be the development and

Comparison of Space Shuttle and VentureStar

	Space Shuttle	VentureStar X-33	VentureStar Full Scale
Launch mass, tonnes	2,050	124	990
Length, m	56	20.5	38.8
Width, m	24	20.8	39.1
Payload to 28.5°, 185 km orbit, tonnes	23	-	>23
Payload bay size, m	4.6 × 18.3	1.5 × 3.1	4.6 × 13.7
Propulsion	3 × LOX/LH2 + 2 SRBs	3 × LOX/LH2 + 2 SRBs	7 × LOX/LH2

deployment of a fully reusable launch vehicle based upon pure rocket propulsion.

The Phase 1 studies during 1995/1996 called for concept verification which was government-led and funded, and industry executed. For Phase 2 from 1996 to 1999 – Lockheed Martin was chosen to develop the X33 flight demonstrator which would be a scaled version of what could become an operational launch vehicle, the full-scale Venture Star. Phase 3, which is 2000 and beyond will see the development and operations of a RLV which will be industry led and investor financed.

The Lockheed Martin consortium which won the contract for the development of the Venture Star vehicles is working in collaboration with AlliedSignal (systems and avionics), Boeing (linear aerospike engine and RCS), Rohr (thermal protection system) and Sverdrup (launch facilities).

The numerical data are taken from Lockheed Martin material, which quoted the original figures in imperial units.

The first sub-orbital flight is planned for late 2000 at the earliest and the vehicle will be capable of flying to a maximum altitude of 80 km and with speeds of up to Mach 13: launch and landing will be from Edwards Air Force Base. NASA has budgeted US$941 million for the project, with Lockheed Martin investing US$200 million of its own money. If the X-33 meets with NASA's approval then Lockheed Martin and its partners will have to decide whether it is economically feasible to develop the estimated US$5 billion fully Reusable Launch Vehicle.

The full-scale VentureStar – should it be developed – would be capable of matching the Shuttle's current capability to LEO, carry about 11 tonnes of payload to the International Space Station or carry a payload of 5.9 tonnes to geosynchronous transfer orbit with a suitable upper stage.

VentureStar ***1997***/0003457

X-34

Background

The Space Transportation Policy, signed by President Clinton 5 August 1994, gave the DoD responsibility for upgrading existing medium and heavy ELVs. Clinton allowed NASA to take the lead after 2000. To respond, NASA, in January 1995, invited industry to submit competing designs for two reusable launch vehicles. The first the X-33, will test the feasibility of replacing Shuttle and the larger ELVs by 2012 with the Reusable Launch Vehicle.

On paper, the 54-tonne 27 m long X-34 should handle satellites of about 1.1 tonnes. Orbital Sciences Corporation and Rockwell were selected March 1995, agreeing to invest US$100 million in addition to NASA's US$70 million through FY99. The two-stage liquid propellant vehicle was to be air-launched from NASA's Boeing 747 and the reusable stage 1 returned to a conventional runway after releasing stage 2 at 100 km altitude and M16. NASA wanted to keep mission costs below about US$3 million and the vehicle would be proposed for its next mission by a 15-man crew in less than three weeks. In the first development of the plane, NASA wanted to use existing engines. Russia's RD 120 was selected in favour of Rocketdyne's RS27. Later advanced, more efficient units that may be supplemented by air breathing engines to increase payload without significantly increasing operating costs would replace these engines.

NASA wanted the X-34 also to demonstrate RLV technologies, including advanced thermal protection, all composite structures, reusable cryogenic tanks and state-of-the-art avionics. Moving with uncharacteristic haste NASA selected an Orbital Sciences Corporation (OSC)/Rockwell team to develop the two-stage X-34 aiming for a suborbital flight in 1997 and orbital flights beginning a year later. The team formed American Space Lines as an 'Arianespace' lookalike for commercial development of the small launcher market but in early 1996 the team pulled out believing the costs could not produce a viable commercial return. NASA re-contracted with OSC for a single-stage suborbital demonstrator, still designated X-34. Three X-34 test vehicles are scheduled to support flight trials planned to begin in August 1999 with the winged vehicle launched from the under-fuselage cradle of a converted L-1011 operated by OSC. Capable of a maximum speed of Mach 8 the X-34 will demonstrate all-weather, pilotless landings as a technology precursor to fully reusable launch systems of the future.

LAUNCH VEHICLE PROPULSION

LAUNCH VEHICLE PROPULSION

INTERNATIONAL

DASA/ERNO and SSC

DASA/ERNO and the Swedish Space Corporation jointly developed the Maxus suborbital vehicle to extend the μg time provided by the Skylark-based Texus and the Terrier Black Brant-based Maser rockets. Thiokol's Castor solid propellant motor was selected following 1987-88 studies. Already qualified as a Delta strap-on booster, it can handle 420 kg useful payload to 850 km altitude, providing >14 min μg. Sweden's Guidance Control System ensures the rocket and payload descend within Esrange with <7 km 1σ dispersion. Laboratories offering basic equipment are provided at the site to assist experimenters' final flight preparations. At least one flight was planned annually, but ESA budget limitations now restrict them to one every 1½-2 yr. A launch campaign typically requires 3 weeks in Apr or Nov/Dec, balancing daylight and the need for frozen lakes to avoid water recovery. User cost is about $10,000/kg. DASA/ERNO has invested about DM20 million (principally in 10 Castors) and SSC about SKr60 million.

The debut launch attained only 200 km altitude instead of the 800 km planned. The flight was smooth to 37 s/21 km but then began pitching; at 46 s the all-ESA payload separated and was recovered. It was later determined that electrical cables of the nozzle's TVC system had burned through; thermal protection was added for future vehicles. ESA paid about DM17 million for the flight and was guaranteed a reflight in the event of failure; insurance was 3%. Maxus 1B was fully successful 8 Nov 1992, providing 12 min 32 s μg time to the seven experiments. Five were controlled by telescience, including one via ESA's Olympus satellite. Maxus 2 was successful 28 Nov 1995, again with an ESA payload (505 kg), attaining 706 km to provide min μg to eight experiments. Maxus 3 is planned for Apr 1998, the gap highlighting the lack of funded payloads. DARA in early 1994 withdrew from Maxus, preferring instead the cheaper Texus.

First launch: 8 May 1991
Number launched: 3
Success rate: 67%
Launch site: Esrange, Sweden
Number of stages: one (solid)
Overall length: 14.877 m min, 15.460 m max
Principal diameter: 1.018 m
Launch mass: 11,800 kg + 420 kg payload + 300 kg service systems
Typical performance: 720 kg total to 850 km, providing 14 min of useful microgravity time
Payload recovery system: flat spin deceleration, subsonic parachute descent (10 m/s landing), helicopter land retrieval

Maxus stage 1

Motor: Thiokol Castor 4B (±6° nozzle TVC)
Overall length: 9.61 m
Principal diameter: 1.018 m
Thrust: 450 kN average
Burn time: 63 s
The motor is topped by an adapter, motor telemetry module and Saab Ericsson's Guidance Control System.

Payload Environment/Accommodation

Total length/mass of all payload modules must be <350 cm/420 kg; 6-8 can be carried. Modules are available on a rental basis, 640 mm ∅, 5 mm aluminium wall (0.5 mm zirconium oxide thermal coating), 640 mm Radax mechanical joints. Access to experiments is possible up to 20 min before launch; the payload is returned to the launch site within an hour of landing. The module below the payload section provides telemetry (2.2-2.3 GHz S-band) + telecommand (450 MHz P-band) to all experiments via a standardised interface. If required, TV signals are transmitted at S-band in PAL B/G using pre-emphasis according to CCIR 405/625. After separation, the cold gas fibre optic gyro Attitude & Rate Control System in the service module above the payload section performs a despin manoeuvre before the μg phase begins at 90 s. If roll rate >1.5°/s or the cone angle >45° (for the RF downlink) during the μg phase, ARCS activates in low thrust mode to minimise experiment disturbance below 2×10^{-4} *g*. At T-891 s, a pyrotechnic valve opens to dump the residual gas through two roll nozzles to create a >60 rpm spin for thermal distribution during re-entry.

Advanced Technology Engine (ATE)

Royal Ordnance (UK), FiatAvio (Italy), SEP (France) and Volvo (Sweden) have been undertaking ESA studies since 1986 on an advanced 20 kN European engine for early next century. Development through ground qualification would require 8-9 yr at a cost of 140 MAU ($190 million); recurring cost is estimated at 1.7 MAU. ATE's outline specification was set at a minimum specific impulse of 345 s vac, 10 restarts/mission and an operational firing time/mission of 2 h, its 4 h qualification life thus permitting its incorporation in re-usable stages. Four principal mission models were considered: re-usable orbital transfer vehicle, Ariane 5 upper stage (10% improvement by replacing DASA's 27.5 kN Aestus), deep space missions (specifically comet sample return) and planetary missions, emphasising Mars and Earth-return capabilities.

ESA selected a staged combustion cycle, resulting in a 74 kg (metallic option) or 58 kg (ceramic option), 1.7 m high, turbopump-fed 9 MPa chamber engine using MON3/MHH propellants. The metallic chamber would be regeneratively cooled, while the ceramic version would be film cooled. Such an engine would provide a 787 kg increase in GTO injection capacity over current 309-316 s SI engines. Used with an Orbital Propulsion Module, ATE yields a >800 kg improvement over the 27.5 kN engine for LEO-GEO or LEO-GTO missions.

In the metallic chamber option, most of the MON3 flows at supercritical pressure through small channels to cool the chamber and nozzle up to a 1:35 expansion ratio. Part of the heated MON3 flows directly to the injector, but most is routed to the pre-burner. Most of the MMH flows to the combustion chamber, where some acts as a film coolant.

Dry mass: 74.2 kg metallic, 57.9 kg SEP ceramic
Length: 1,710 mm metallic, 1,720 mm ceramic
Maximum diameter: 764 mm
Oxidiser: NTO + 3% NO (MON3)
Fuel: MMH
Propellant feed method: 162 kW 76,400 rpm turbopump (Fiat version), 0.4 MPa inlet, 18.1 MPa MMH pump exit, 19.5 MPa NTO pump exit
Thrust: 20 kN vacuum
Specific impulse: 351 metallic, 347 ceramic)
Mixture ratio (O/F): 1.86:1
Expansion ratio: 400 (exit diameter 760 mm)
Expansion bell: high temperature stainless steel (Volvo) or C/SiC Ceramic (SEP)
Chamber pressure: 9 MPa
Chamber materials: Narloy-Z or similar copper alloy, carrying 122 NTO coolant channels or SiC/SiC film cooled. The cool outside of the metallic chamber liner is closed by an electro-deposited metal transition layer and lightweight fibre overwrap
Injector: metallic option employs central pintle or swirl with annular fuel sheet and fuel cooling; ceramic option employs a classical impingement injector with fuel film cooling.

Electric Propulsion

ESA's Artemis mission (see ESA's Communications section entry) will carry an ion propulsion subsystem, conceived and sized to support the full 10 yr of NSSK required for commercial telecommunications missions. It will carry north/south face packages, each with a 15 mN RIT-10 German and an 18 mN UK-10 British thruster. An engineering model of the RIT-10 thruster has already demonstrated more than 2,000 h of operations on the ground.

FEEP The Field Emission Electric Propulsion (FEEP) thruster offers the unique characteristics of extremely low thrust levels (down to 1 mN) and pulsing and any desired frequency. This makes it the only system suitable for ultra-fine AOCS and drag compensation. Several science missions have chosen it as the only propulsion system capable of fulfilling their ultra-fine pointing requirements. In addition, the compact design and high specific impulse of the FEEP system make it a good

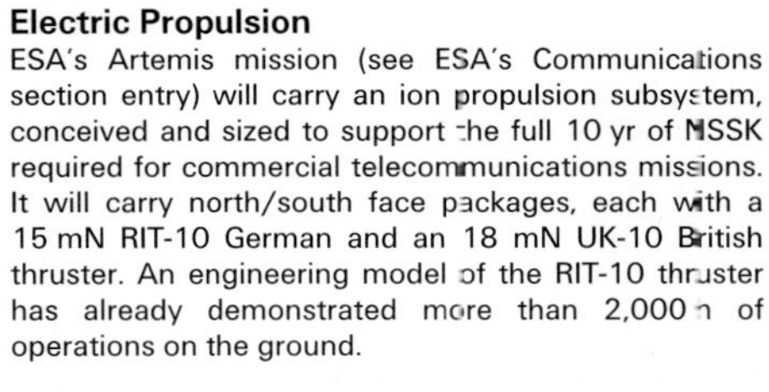

ATE's pre-combustion cycle. Values refer to Royal Ordnance's turbopump (ESA/ESTEC)

candidate for future commercial missions like small satellites and constellations in low Earth orbit. A >7 kV electric field ionises liquid caesium and provides acceleration. ESTEC and CentroSpazio (Italy) are responsible for the testing and development. Current plans foresee a FEEP flight test in 1998 aboard a Shuttle Get Away Special canister.
Mass of thruster module: 450 kg
Mass of PCU: 2,500 g
Thrust: 1 to 1,000 mN
Thrust resolution: 0.1 mN
PCU input power: 9-60 W
Specific impulse 8,000 s

LMIS A space-proven Liquid Metal Ion Source (LMIS) developed by the Austrian Research Centre, Seibersdorf (ARC) and flown on many science missions for charge control, is being investigated by ESA for micro-propulsion applications within the range 1-10 mN. The device uses liquid indium as propellant, flowing on a tungsten needle which is then ionised and accelerated at the needle top by field emission.

ESA-XX Primary electric propulsion, particularly for high delta-V planetary missions, requires thrust levels of 200-300 mN. ESA is supporting the development of the ESA-XX xenon thruster, combining TIT's radio frequency electrodeless discharge, the simple and reliable grid system of the 25 cm 200 mN UK-25 (see the UK propulsion section) and Italy's high power electronics. The thruster was studied jointly by Germany and Russia for interplanetary missions. The self-sustaining RF plasma is ignited by electrons from the neutraliser(s). The ion beam is extracted, accelerated and focused by an inward-dished 9,100-hole grid system, comprising the holder grid (+1.5 to +2.0 kV), the accelerator grid (−0.5 kV) and the (normally) grounded decelerator grid. Prototype tests at Giessen University during 1993-1994 mapped the motor over the 50-200 mN and up to 2.0 kV beam voltage. This prototype is being followed by an advanced breadboard-EM for further parametric testing.
Diameter: 25 cm beam exit; 39 cm thruster
Length: 39 cm
Mass: 9 kg (target mass), 12.6 kg (prototype)
Thrust: 200 mN nominal
Exhaust velocity: 54 km/s (at 2 kV)
Specific impulse: 5,365 s at 200 mN, 2.0 kV
Acceleration voltage: 2.0 kV at >100 mN
Discharge power: 650 W at 200 mN
Power input: 6.2 kW at 200 mN
Specific power consumption: 31 kW at 200 mN
Specific propellant consumption: 19 mg/Ns
Total efficiency: 83% maximum
Lifetime: 15,000 h (target)

EUROPROPULSION

Europropulsion was established in 1985 by France's SEP and Italy's BPD to bid for solid motor development contracts within Europe's civil space programme. Its principal activity is Ariane 5's P230 strap-on boosters. The FFr3.8 billion fixed price contract, following two preliminary awards, was signed with CNES in late 1989, covering dev/qual of the motors, including 10 ground tests to mid-1995. Individually, BPD provides the lead for segment and igniter dev/integration, SEP contributes the nozzle assembly and insulation liner, and sub-contractors MAN Technologie provide the segment casings, SNPE propellant development and Regulus (the SNPE-BPD partnership) for the two large segment casting operations at Kourou. The small forward segment is completed by BPD in Italy. The EUPERA SNPE/BPD partnership produces the ammonium perchlorate at SNPE's Toulouse site (the AP for 501/503 was provided by Kerr McGee Corp of the US and for 502 by SNPE; future division remains to be decided).

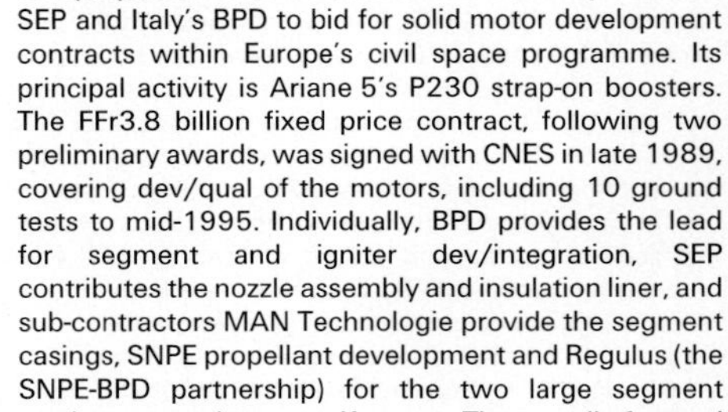

Ariane 5's strap-on booster was developed by the Europropulsion partnership. The masses shown are the segment propellant loadings **1996**

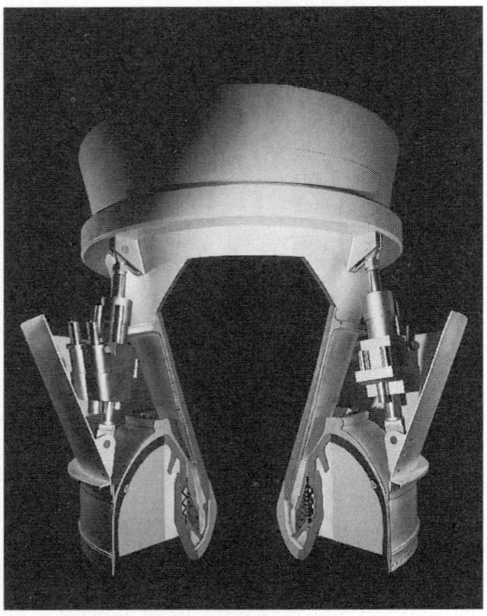

Ariane 5's nozzle flex-bearing provides up to 6.6° thrust vector control (SEP) ***1996***

SEP inaugurated its nozzle production plant (UTB: Usine Tuyère Booster) in Oct 1990 at Haillan. A FFr180 million investment, it will fabricate 10 nozzles annually by year 2000. MAN commissioned its production centre in Augsburg during Sep 1988: this 5,400 m² facility includes an 1,800 m² air conditioned area for final machining of the clevis/tang connection between the booster segments. The 500 t counter-roller flow-forming machine rolls the initial 40 mm thick, 1 m high, 3 m ∅ steel cylinders into 8 mm thick 3.5 m high segments before heat treatment produces an ultimate strength of 1,500 N/mm².

The first demonstration motor, a reduced scale version loaded with 15 t of propellant, was fired sucessfully in Dec 1989. The first full scale P230 was successfully tested 16 Feb 1993. Seven full scale tests were then planned: B1 with battleship (thick) casings, five (M1-M5) development and two (Q1-Q2) qualification. The last two tested the entire stage. M1, the second firing, was successful 25 Jun 1993. This was the first in flight configuration, confirming thrust profile, intersegment seals and nozzle gimballing. M2 was planned for mid-Nov 1993 but voids were found in the motor's propellant and that motor was shelved (it is expected to be fired after an ageing programme). The third firing, M3, was successful 20 Jun 1994. M4, with a flight-type actuation system and lighter thermal protection, was successful 30 Sep 1994. Flight-standard M5 was successful 15 Dec 1994. Q1 was successful 10 Mar 1995, Q2 21 Jul 1995.

Ariane 5 P230 specifications

First flight: 4 Jun 1996 (Feb 1993 first static firing, Jul 1995 end of ground qualification)
Mass: 268.7 t (237.7 t propellant): 106.9 t aft, 107.4 t centre, 23.4 t forward; nozzle 6 t; internal insulation 4.3 t; casing 20 t; igniter 0.3 t; 0.4 t miscellaneous)
Length: 26.774 m (stage 31.16 m)
Diameter: 3.049 m
Propellant
  type: HTPB 68% ammonium perchlorate, 18% aluminium, 14% PBHT liner
  shape: main section cylindrical, forward segment 15 rectangular slots mass fraction: 0.884 for 237.7 t loading
Burn time: 132 s
Thrust: 5250 kN sea level at launch (5904 kN vac); 4931 kN avg vac, 6637 kN max vac

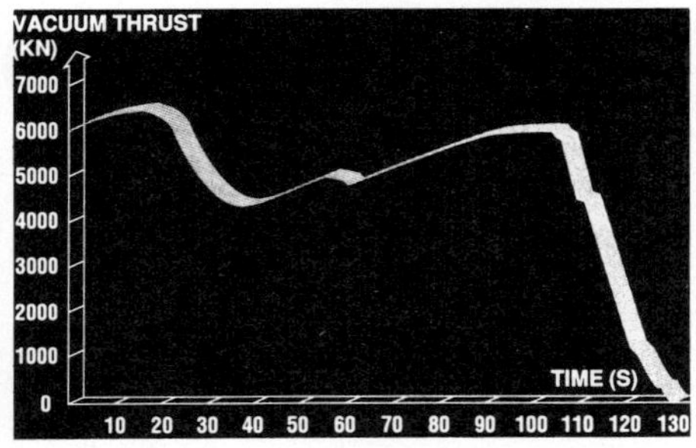

P230 typical thrust curve (Europropulsion)

Specific impulse: 271 s vacuum
Total impulse: better than 630 MNs
Maximum pressure: 59.9 atm
Nozzle
  mass: 6 t
  throat diameter: 90.0 cm
  length: 3.420 m (2.22 m exposed)
  exit diameter: 2.826 m (initial expansion ratio 9.6)
  materials: carbon-carbon throat; nozzle flexbearing provides 6° deflection (6.6° max). Employs two actuators 90° apart with ±145 mm strokes. Actuator system mass 150 kg, hydraulic 1100 l, 345/270 atm initial/end pressure
Casing materials: 8 mm-thick D6AC steel heat-treated at 14,800 atm ultimate strength; thermal insulation EPDM/Silica (GSM55) or Kevlar (EG2). Six 3.351 m-long segments are assembled into two motor sections of three segments each with two factory joints and 1.005 m-long aft dome. Tipped by 3.398 m-long forward section
Igniter type: pyrogen head-end (25 kg composite propellant).

REGULUS

Regulus is a partnership between France's SNPE (40%) and Italy's BPD Difesa e Spazio (60%), formed to operate Kourou's Ariane 5 booster propellant facility (UPG Usine Propergols en Guyane). The plant, comprising 40 buildings totalling 26,000 m² on a 300 hectare site, was commissioned 24 Oct 1991. Cost was 180MAU (about FFr1,250 million). Each strap-on contains 237 t of HPBT solid in three sections. Regulus provides the propellant required for the mid & aft units, while the forward element is shipped from BPD. When the facility became fully operational in 1995, its 150 personnel were able to support eight flights a year, requiring 3,300 t of propellant for 32 segments. The first segment was poured Nov 1991 for the B1 battleship motor, tested 16 Feb 1993.

CHINA

China's space launchers draw on a small range of main engines, the majority utilising hypergolic NTO/UDMH although more recently two cryogenic upper stages were added. No LOX/kerosene engines as used extensively elsewhere are known, reflecting the space variants' parallel development alongside storable propellant long range missiles.

Rocket propulsion was included as one of the key technologies in the 'Twelve Year Development Plan of Science and Technology' approved in 1958, leading to the launches of the first indigenously developed sounding rocket on 19 Feb 1960 and first successful ballistic missile on 29 Jun 1964 (failure on the first test 21 Mar 1962 led to a major redesign). The initial CZ-1 space launcher, derived from the CSS-3 MRBM, utilised nitric acid/UDMH but each subsequent operational and planned stage 1/2 has used N_2O_4 as oxidiser. Development of a LOX/LH_2 cryogenic upper stage for GEO missions began in 1977, leading with its launch as CZ-3 stage 3 in Jan 1984 to China becoming only the third such user after the US/ESA. The **China Academy of Launch Vehicle Technology** (CALT, formerly the Beijing Wan Yuan Industry Corporation) near the town of Nan Yuan 15 km south of the capital develops/builds the cryogenic engines and its Shanxi Liquid Rocket Engine Co the storable engines; SLREC also handles solid motors. The **Beijing Rocket Test Centre** 50 km southwest of Beijing maintains five major test stands: No 1 handles altitude testing of spacecraft thrusters of up to 490 N, No 2 provides single engine cryogenic facilities, with No 4 accepting a complete H-8 cryogenic stage, and No 5 is devoted to hydrazine engines. Other test facilities are operated at the launch sites and SLREC probably has stands for the storable engines.

Fewer details are known of Chinese thrusters and solid propellant motors. The CZ-1D stage 3 is a solid of 625 kg propellant and satellites such as FSW incorporate solid kick motors. CZ-2E's 1.7 m ∅ EPKM solid stage 3 debuted in 1995. CZ-2C's 1.2 m ∅ CPKM is imminent. Sounding rockets with solids are also flying. Solid propellant ICBM/IRBMs are under development. Hydrazine thrusters of unspecified nature provide 3-axis control for the H-8 cryogenic stage. However, CZ-2E's stage 2 employs a hydrazine system acknowledged as derived from H-8: 4 × 45 N pitch thrusters, 2 × 45 N yaw + 4 × 11 N roll. The FSW recoverable capsule is also 3-axis controlled by cold gas, and the CZ-1D solid stage is spin-stabilised by N_2 thrusters.

The engines carry YF designations, indicating Yei-ti Fa-dong-ji (liquid state engine).

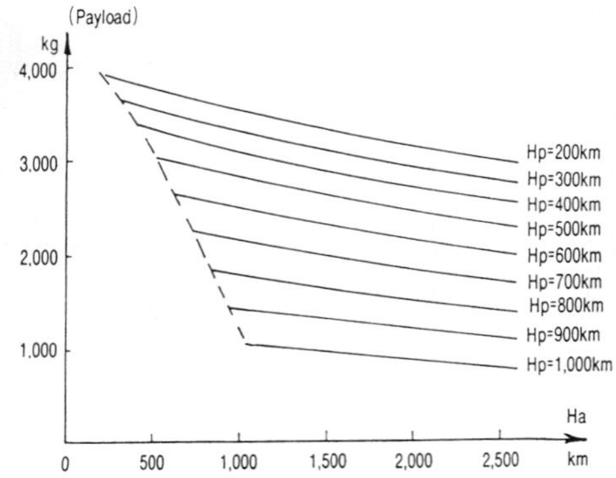

CZ-4A payload capacity into elliptical (solid lines) and Sun-synchronous (dashed) orbits from Taiyuan (GWIC)

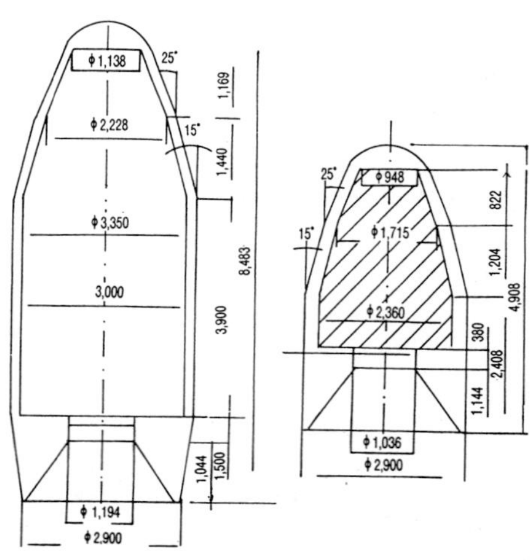

CZ-4 fairing and payload envelopes. A lower section can be added to the A-type at right to carry a secondary payload below (GWIC)

YF-2A

Applications: cluster of four for CZ-1 family stage 1 and CSS-3 IRBM
First flown: Jun 1964?
Oxidiser: nitric acid
Fuel: UDMH
Propellant flow rate: 115 kg/s?
Cycle: gas generator
Thrust: 275.3 kN sea level
Specific impulse: 241 s sea level
Burn time: about 130 s operational

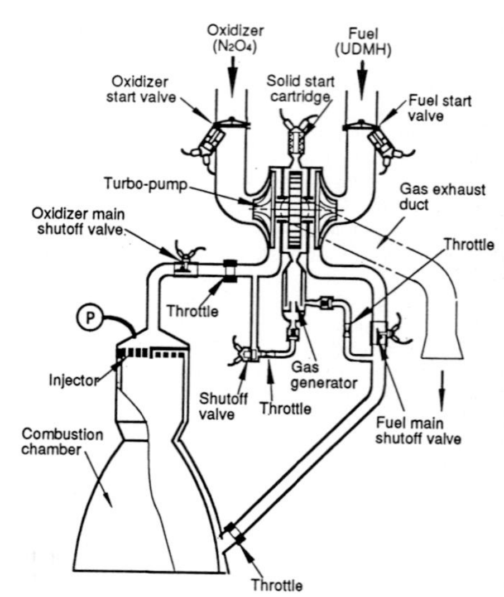

YF-20/21/22 schematic

CZ-3's H-8 third stage is powered by the 4-chamber cryogenic YF-73

YF-3

Applications: CZ-1 family stage 2, CSS-3 IRBM
First flown: Jun 1964?
Oxidiser: nitrogen tetroxide
Fuel: UDMH
Propellant flow rate: 50 kg/s?
Thrust: 294.2 kN vacuum
Specific impulse: 287 s vacuum
Burn time: about 130 s operational

YF-20/21 and YF-20B/21B

The YF-20/21 is the backbone of the Chinese launcher range, clustered in fours (creating the YF-21/21B assembly) for every first stage of the CZ-2 to -4 families and modified into the high altitude YF-22 for the second stages. In the L-140 stage 1s (the 140 indicating 140 t propellant loading) it operates at 696 kN thrust, but the stretched L-180 stages and strap-ons of the more powerful vehicles run it at 740 kN. Chinese statements suggest that still higher levels are possible. The LB-40 strap-ons of the CZ-2E/3B/3C each incorporate a single YF-20B.
Applications: CZ-2C/3 stage 1 for YF-20/21; CZ-2E/3B/3C strap-on for YF-20B; CZ-2D/2E/3A/3B/3C/4 for YF-21B
First flown: 5 Nov 1974 (first CZ-2 flight)?
Oxidiser: nitrogen tetroxide
Fuel: UDMH
Propellant flow rate: 270 kg/s?
Thrust: 696.2 kN sea level for YF-20/21 of L-140 stage 1 of CZ-2C/3; 740.35 kN, E 12.69, exit area 7,728.6 cm^2 for YF-20B of LB-40 strap-on and YF-20B/21B of L-180 stage 1 of CZ-2E/3A/3B/3C/4
Specific impulse: 259 s SL YF-20/21; 260.66 s SL YF-20B/21B
Mixture ratio: 2.10 YF-20/21; 2.12 YF-20B/21B
Burn time: 125-160 s operational

YF-22/22B and YF-24/24B

The YF-22/22B is a high altitude derivative of the YF-20. It is flown singly with four gimballed YF-23/23B verniers, the assemblies designated 24/24B, providing steering on the L-35 stage 2. YF-22B/23B use larger expansion ratios for improved SI.
Applications: CZ-2C/3 stage 2 for YF-24; CZ-2D/2E/3A/3B/3C/4 stage 2 for YF-24B
First flown: 5 Nov 1974 (first CZ-2 launch)?
Oxidiser: nitrogen tetroxide
Fuel: UDMH
Propellant flow rate: 270 kg/s?
Thrust: 719.8 kN vacuum for YF-22, 740.4 kN for YF-22B, but 845.3 kN attainable
Specific impulse: 289 s vacuum YF-22; 298 s vacuum YF-22B
Mixture ratio: 2.181 YF-22 & -22B
Burn time: 300 s CZ-2E; 110 s others

YF-23/23B

Four gimballed YF-23/23Bs provide steering for the fixed single YF-22/22B on the second stages listed above.

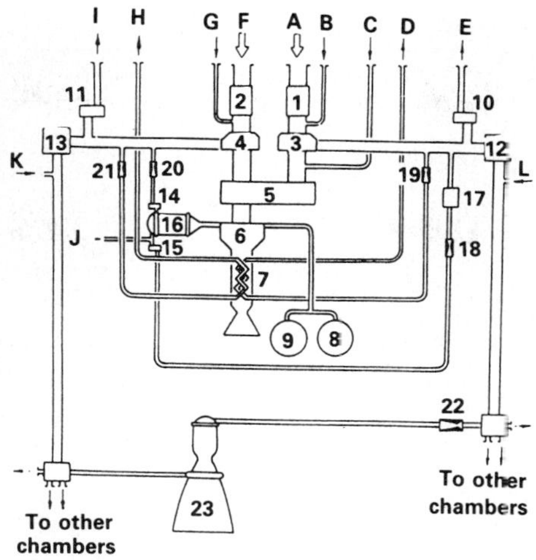

YF-73 schematic. A LOX inlet, B/C purge, D LOX tank pressurisation, E bleed, F liquid hydrogen inlet, G purge, H LH2 pressurisation, I bleed, J-L purge; 1/2 isolation valves, 3 LOX pump, 4 LH2 pump, 5 gearbox, 6 turbine, 7 heat exchanger, 8/9 start tank, 10/11 bleed valve, 12/13 main valve, 14/15 gas generator valve, 16 gas generator, 17 pressure stabiliser, 18-22 venturi, 23 combustion chamber (one of four)

Applications: as YF-22/22B
First flown: as YF-22/22B
Oxidiser: nitrogen tetroxide
Fuel: UDMH
Propellant flow rate: 16 kg/s for all four chambers with single pump?
Thrust: 46.1 kN YF-23 and 47.1 kN YF-23B total (4 chambers) vacuum
Specific impulse: 281.7 s YF-23 and 289 s YF-23B vacuum
Mixture ratio: 1.57 YF-23/23B
Burn time: up to 300 s, 136 s for CZ-4

YF-40

The engine appears to have been uprated for CZ-4B. Where known, the different specifications are provided in 4A/4B order.
Applications: flown paired on CZ-4 stage 3; uprated for 4B?
First flown: 6 Sep 1988
Mass: each 70/83 kg dry
Oxidiser: nitrogen tetroxide
Fuel: UDMH
Thrust: 98.1/101.03 kN vacuum for pair
Specific impulse: 305/303 s
Mixture ratio: 2.14
Chamber pressure: 45.39 atm
Expansion ratio: 55
Exit diameter: 630 mm
Burn time: 321/412 s (single/single or dual burn)
Nozzle materials: niobium alloy, radiation cooled

YF-73

Development of the cryogenic YF-73 began in 1976 for the restartable H-8 third stage, the first use of LH_2 propulsion technology outside of USA/Europe. Principal use is delivery of payloads into GTO. Firings of flight-type engines began 1981. After completion of acceptance testing, the first flight engine was delivered 4th quarter 1983. By end-1983, about 120 tests totalled >30,000 s. The four chambers, gimballed ±24° for attitude control, are serviced by a single 37,000 rpm turbopump. The two starts are initiated by a pair of high pressure nitrogen spheres. The valves, pumps, gearbox, gas generator and propellant lines carry 10-20 mm of polyurethane foam to minimise heat transfer. The engines are manufactured at CALT. There have been two flight failures. The first was Jan 1984's debut mission when the re-start after coast attained 90% thrust and then rapidly decayed after a few seconds. Cause was attributed to propellant flow fluctuations following the μg coast. Similarly, the second (Dec 1991) suffered a pressure drop 58 s into the second burn and shut down after a further 77 s. Cause was believed to be helium supply failure to control valves.
Application: H-8 third stage of CZ-3
First flown: 29 Jan 1984 (failed to restart)
Number flown: 9 sets, to end-1995 (2 failures)
Engine cycle: gas generator
Dry mass: 236 kg
Length: 1,438 mm
Maximum diameter: 2,220 mm
Oxidiser: liquid oxygen at 9.086 kg/s
Fuel: liquid hydrogen at 1.817 kg/s
Thrust: 44.147 kN total (4 chambers) vacuum
Specific impulse: 420 s vacuum
Mixture ratio (O/F): 5.0
Chamber pressure: 25.96 atmospheres
Expansion ratio: 40:1
Chamber materials: stainless steel, regeneratively cooled by supercritical hydrogen
Injector: brazed stainless steel, 38 coaxial elements
LOX/LH_2 pumps: 37,000 rpm single stage centrifugal pump with helical inducer. LH_2 and turbine share common shaft; LOX shaft is driven via reduction gears. Gears cooled by hydrogen
Burn time: two burns of 500 s + 300 s

YF-75

Development of the cryogenic restartable YF-75 drew extensively on YF-73 experience to create the H-18 stage, comparable to Centaur D. Its employment in pairs aboard the CZ-3A vehicles provides a GTO capability equivalent to the lower range of Ariane 4. Aboard CZ-3B, it surpasses Ariane 4 and almost equals Proton. The engines are manufactured at CALT.
Application: H-18 stage 3 of CZ-3A/3B/3C
First flight: 8 Feb 1994
Number flown: 2 sets to end-1995
Oxidiser: liquid oxygen
Fuel: liquid hydrogen
Propellant flow rate: 19 kg/s each chamber?
Thrust: 78.4 kN vacuum each chamber
Specific impulse: 442 s vacuum
Mixture ratio: 5.0
Chamber pressure: 37.15 atm
Expansion ratio: 80:1
Burn time: about 470 s total, dual start.

GF-02/CPKM/EPKM/FG-46

The solid motor carried by the CZ-1D appears to be related to the GF-02 solid flown on the original CZ-1 vehicle. Work began in 1965 at the Solid Fuel Engine Research Academy under deputy head Yang Nansheng. The first spin firing took place 26 Jan 1968 but the motor exploded after 30 s, causing some fatalities. During 1968-70 a total of 19 tests, including five altitude simulations, was conducted successfully.

The HEXI Corp's indigenous EPKM FG-46 commercial perigee kick stage debuted for AsiaSat 2 in Nov 1995, providing 3,460 kg into GTO. An early version was carried on the first 2E. Nine test firings included two failures at the nominal 40 rpm spin rate: 31 May 1993, test M1, prototype/design qualification; 24 Oct 1993, M2, prototype/structural qualification; 31 Jan 1994, C1, 40 rpm spin (fail); 5 May 1994, C2, altitude simulation; 6 Jul 1994, C3, 40 rpm spin (fail); 30 Oct 1994, C5, 40 rpm spin; 8 Nov 1994, C4, altitude simulation; 16 Jan 1995, C6, 40 rpm spin; 20 Mar 1995, Z1, 40 rpm spin/acceptance.

The 1.2 m ∅ CPKM is imminent on the CZ-2C.

EPKM specifications
First flown: 28 Nov 1995
Number flown: 2, to end-1995
Mass: 6,001 kg (5,444 kg propellant, 529 kg burnout, 28 kg ablated)
Length: 2.5 m
Diameter: 1,700 mm
Propellant: HTPB
Burn time: 87 s
Thrust: 210 kN max vacuum
Specific impulse: 292 s vacuum
Total impulse: 15.5881 MNs vacuum
Pressure: 6.0 MPa max
Casing materials: fibre glass
Nozzle: carbon-carbon

CIS/USSR

Historically, the general Soviet philosophy has been to cluster numbers of relatively modest, low chamber pressure open cycle engines, mass-produced with few modifications in their decades-long careers. Most are assigned RD (Raketnay Dvigatel, rocket motor) designations. **NPO Energomash** is the major centre for rocket engines; **KB Khimautomatiki**, was identified in 1990 as responsible for Energia's core engines, and emerged from the **Kosberg** bureau, which concentrated on upper stage propulsion, as does **KB Khimmach** of the old **Isayev** bureau. **NII Machinostroenye** undertakes more basic research but does offer spacecraft thrusters. The **Kuznetsov** (late NPO Trud, now RD Kuznetsov Co SSC) aircraft engine bureau developed engines for the abandoned N1 lunar launcher for Korolev. Korolev's bureau itself developed Proton's Block D/DM and Molniya's Block L engines. The Lyulka aircraft engine bureau (now part of NPO Saturn) developed the LOX/LH_2 D-57 to replace the lunar N1 launcher's stage 3 engines.

In 1996 the designations of many previously-unidentified engines used on missiles and launch vehicles were released, and these are included under the

Rocket Engines Used on Selected CIS Missiles and Launch Vehicles

Vehicle	*Designator*	*Stage 0*	*Stage 1*	*Stage 2*	*Stage 3*	*Stage 4*
1. Selected Missiles (development related to early launch vehicles)						
R-9	8K75 (original)		4 x NK-9	1 x NK-19		
	(developed)		1 x RD-111	1 x RD-0106		
R-9M	8K77 (not developed)		4 x NK-9	4 x SI-5400		
R-16	8K64		1 x RD-218	1 x RD-219		
GR-1	8K713		4 x NK-9	1 x NK-19	1 x 8D726	
UR-200	8K81		? x RD-0203	1 x RD-0206		
			? x RD-0204	1 x RD-0207		
2. Existing and Retired Launch Vehicles						
Sputnik	8K71PS	4 x RD-107	1 x RD-108			
Sputnik	8A91	4 x RD-107	1 x RD-108			
Vostok-L	8K72	4 x RD-107	1 x RD-108	1 x RD-0105		
Vostok	8K72K	4 x RD-107	1 x RD-108	1 x RD-0109		
Vostok-2	8A92	4 x RD-107	1 x RD-108	1 x RD-0109		
Meteor	8A92M	4 x RD-107	1 x RD-108	1 x RD-0109		
Voskhod	11A57	4 x RD-107	1 x RD-108	1 x RD-0107		
Soyuz	11A511	4 x RD-107	1 x RD-108	1 x RD-0108		
Soyuz-L	11A511L	4 x RD-107	1 x RD-108	1 x RD-0108		
Soyuz-M	11A511M	4 x RD-107	1 x RD-108	1 x RD-0108		
Soyuz-U	11A511U	4 x RD-107	1 x RD-108	1 x RD-0110		
Soyuz-U2	11A511U2	4 x RD-107	1 x RD-118	1 x RD-0110		
-	11A59	4 x RD-107	1 x RD-108	1 x RD-0110		
Soyuz-2		4 x RD-120K ?	1 x RD-120K ?	1 x RD-0124		
Molniya	8K78	4 x RD-107	1 x RD-108	1 x RD-0107	1 x SI-5400	
Molniya-M	8K78M	4 x RD-107	1 x RD-108	1 x RD-0110	1 x 11D33	
Cosmos	63S1		1 x RD-214	1 x RD-119		
Cosmos-2	63S1M		1 x RD-214	1 x RD-119		
Cosmos-2M	11K63		1 x RD-214	1 x RD-119		
Cosmos-1	65S1		1 x RD-216	1 x 11D49 ?		
Cosmos-3	11K65		1 x RD-216	1 x 11D49 ?		
Cosmos-3M	11K65M		1 x RD-216	1 x 11D49		
Proton	8K82		6 x RD-253	3 x RD-0210		
				1 x RD-0211		
-	11A510	4 x RD-107	1 x RD-108			
R-36-0	8K69		1 x RD-218	1 x RD-219		
Tsyklon-2	11K67		1 x RD-251	1 x RD-252		
Tsyklon-M	11K69		1 x RD-251	1 x RD-252		
Proton-K	8K82M		6 x RD-253	3 x RD-0210	1 x RD-0213	1 x RD-54
				1 x RD-0211	1 x RD-0214	
Proton-K	8K82M		6 x RD-253	3 x RD-0210	1 x RD-0213	
				1 x RD-0211	1 x RD-0214	
Tsyklon	11K68		1 x RD-251	1 x RD-252	1 x RD-861	
N-1	11A52		30 x NK-33	8 x NK-43	4 x NK-39	1 x NK-31
Zenit-2	11K77		1 x RD-171	1 x RD-120		
Energiya	11K25	4 x RD-170	1 x RD-0120			
Rockot			? x RD-0233	1 x RD-0235		
			? x RD-0234	1 x RD-0236		
3. Future Launch Vehicles						
Angara			1 x RD-174	1 x RD-0120		
Dniepr			?	1 x RD-0256		
				1 x RD-0257		
Riksha			6 x RD-169	1 x RD-185		
Shtil			? x RD-0244	?		
			? x RD-0245	?		

Some launch vehicles use clusters of slightly different engines on the same stage. For example, the Proton family uses three RD-0210 and one RD-0211 on the second stage, the only difference being that the RD-0211 provides the tank pressurisation: the assembly is called the RD-0210. Vernier engines usually have separate designators and these are indicated when known: in such cases there is one main engine and one set of verniers. In the case of the Proton-K third stage it is known that the RD-0213/RD-0214 assembly is designated RD-0212.

appropriate design bureaus. To permit easy cross-referencing between a launch vehicle and the propulsion systems the following table has been prepared.

NPO ENERGOMASH

The history of NPO Energomash can be traced back to 15 May 1929 when the Group for the Development of Rocket Engines was organised at the Gas Dynamics Laboratory (GDL) in Leningrad (now St Petersburg). For more than sixty years the organisation repeatedly changed its name and location, but remained at the forefront of rocket engine design. The founder and the permanent head of the GDL as Academician Valentin P Glushko, considered to be the initiator of the Russian rocket engine industry.

In May 1974, the GDL merged with Sergei Korolyov's design bureau, the new organisation being re-named NPO Energiya, with Glushko as its head. Following Glushko's death in Jan 1989, the former GDL left Energiya to become NPO Energomash, continuing to concentrate on the development of rocket engines: the bureau was headed by Vitali P Radovsky during 1974-1994. The current general director and chief designer is Boris I Katorgin.

NPO Energomash incorporates the design bureau, plant and test facilities, with the capability to test engines with thrusts of up to 10 MN either as components or complete. An agreement was signed with Pratt and Whitney in Oct 1992 providing exclusive US marketing rights for Energomash's LOX/kerosene and tri-propellant products and technology.

RD-100

Developed directly from the V-2/A-4 engine, this was the first large Soviet engine employing a turbopump feed system, powered by catalytically decomposed hydrogen peroxide. Tested May 1948.
Applications: R-1 missile and research rockets
Dry mass: 885 kg
Oxidiser: liquid oxygen
Fuel: 75% ethyl alcohol solution
Feed method: turbopump driven by hydrogen peroxide gas generator
Thrust: 267 kN sea level, 307 kN vacuum
Specific impulse: 203 s sea level, 237 s vacuum
Chamber pressure: 16 atm
Burn time: about 65 s

RD-101

Direct development from the RD-100 and used by the R-2 missile and V-2A research rocket. Solid catalyst in gas generator replaced hydrogen peroxide decomposition.
Thrust: 363 kN sea level, 405 kN vacuum
Specific impulse: 214 s sea level, 242 s vacuum
Chamber pressure: 21 atm
Burn time: 85 s

RD-103

Applications: developed in the early 1950s for the R-5M missile and V-5V research rocket.
Dry mass: 870 kg
Oxidiser: liquid oxygen
Fuel: 92% ethyl alcohol solution
Thrust: 432 kN sea level, 500 kN vacuum
Specific impulse: 220 s sea level, 248 s vacuum
Chamber pressure: 24 atm
Burn time: about 120 s

RD-105

This single-chamber engine was proposed for use on the strap-on boosters for the R-7 early in the missile's development. There were design difficulties during the development of the engine due to burning instabilities in the combustion chamber. It was also realised that the engine would not be powerful enough for the R-7, which also went through major design changes.
Application: Early concept R-7 strap on
Oxidiser: liquid oxygen
Fuel: kerosene
Thrust: 540 kN sea level

RD-106

This single-chamber engine was proposed for use on the central core stage for the R-7 early in the missile's development. There were design difficulties during the development of the engine due to burning instabilities in the combustion chamber. It was also realised that the engine would not be powerful enough for the R-7, which also went through major design changes.
Application: Early concept R-7 central core
Oxidiser: liquid oxygen
Fuel: kerosene
Thrust: 520 kN sea level, 645 kN vacuum

RD-107

The world's most used rocket engine. Those now flying on Soyuz-U are designated 11D511 and those on Molniya-M 11D728. Each of the four strap-ons incorporates a single RD-107 of four fixed combustion chambers, a single turbopump and two ±45° gimballed verniers. These engines and their RD-108 derivative of the launcher's sustainer core were responsible for launching the first satellite, man in space, lunar probes and still account for every manned flight. Developed over 1954-57 by Glushko's GDL, the original versions were uprated after the first Sputnik launches. The RD-107/108 engines are built by the PO Samara Frunze

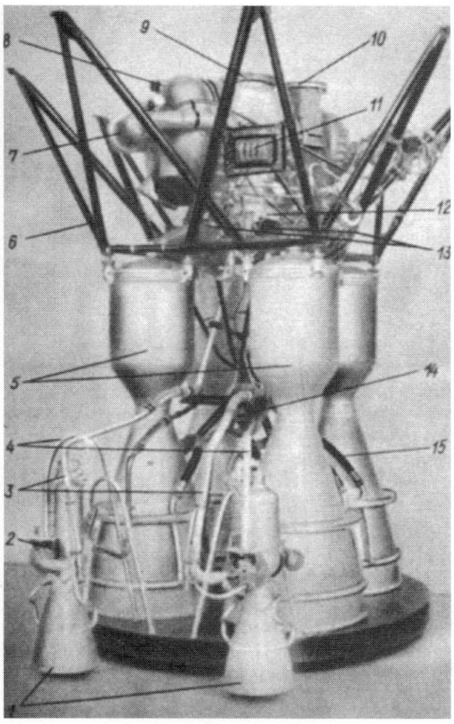

RD-107: the world's most utilised rocket engine, with four main chambers and two steering verniers

Engine Building factory.
Applications: R-7 ICBM, Vostok/Soyuz/Molniya space launcher family
First flown: May 1957
Dry mass: 1,155 kg
Engine cycle: open
Oxidiser: liquid oxygen at 226 kg/s
Fuel: kerosene at 91 kg/s
Feed method: this engine introduced a more efficient turbopump to feed four main chambers, consuming four times the peroxide flow of the RD-100 but yielding ten times the power. The single 8,300 rpm shaft carries a double-sided shrouded impellor for LOX, a single-sided impellor for kerosene, a small centrifugal pump for the two vernier chambers and for feeding LN_2 to the evaporator in the turbine exhaust, and a small pump to feed the H_2O_2 to the gas generator (previously done by compressed air). Turbine power is 3,800 kW. The kerosene is routed from the nozzle base through the cooling channels before reaching the injectors at 210°C
Thrust: 1,000 kN vacuum, 821 kN sea level (verniers about 38 kN each); uprated from original 976 kN vacuum
Specific impulse: 257 s sea level, 314 s vacuum; uprated from original 305 s vacuum
Chamber pressure: 60 atm, gas temperature 3,520°C. For maximum thermal conductivity the inner wall was originally copper, later replaced by a brass-chromium alloy. The outer wall is alloy steel, the two shells joined by both soldering/welding in an inert atmosphere, the joints being on ribs milled on the outer face of the inner shell or to a corrugated plate between the shells. The heads are made from two flat plates welded with concentric tubular rings in which the propellants are premixed before injection through the inner plate. The 337 injector holes are arranged in 10 rings, the outer ones spraying on to the inner wall to hold T at 380°C
Chamber internal diameter: 430 mm, throat 166 mm
Burn time: 120 s, but 140 s possible.

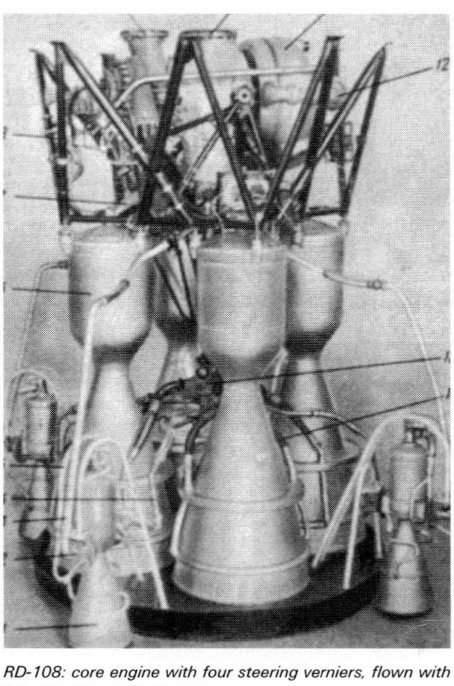

RD-108: core engine with four steering verniers, flown with RD-107 strap-on stage engines

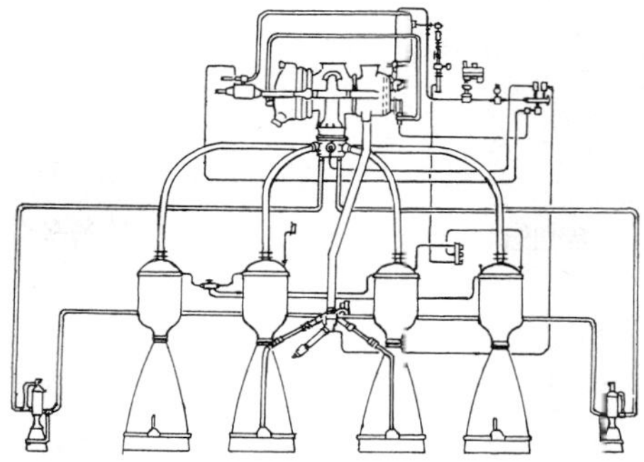

RD-107 propellant flow diagram

RD-108

Developed in parallel during 1954-57 with the RD-107 but with reduced thrust and chamber pressure to permit longer operating time (330 s instead of 120 s) as the core engine for the Vostok and related family of launchers. As only one engine is carried, the number of verniers was increased to four for steering. The current Soyuz-U model is 11D512, for Molniya 11D727.
Applications: R-7 ICBM, Vostok/Soyuz/Molniya launchers
First flown: May 1957
Dry mass: 1,250 kg
Engine cycle: open
Oxidiser: liquid oxygen at 52 kg/s each main chamber
Fuel: kerosene at 21 kg/s each main chamber
Feed method: as RD-107
Thrust: 745 kN sea level, 941 kN vacuum; uprated from original 912 kN vacuum
Specific impulse: 248 s sea level, 315 s vacuum; uprated from original 308 s vacuum. More recent uprating to 323 s is possible
Chamber pressure: 52 atm
Burn time: up to 330 s

RD-109

Little is known about this engine. It was proposed as the orbital stage engine for the original Vostok launch vehicle, but the RD-0105 was chosen in its place. Presumably it was a single-chamber engine, possibly with verniers.
Application: Proposed for early Vostok Block E.
Oxidiser: liquid oxygen
Fuel: kerosene
Thrust: ~100 kN

RD-111

This engine replaced the rival NK-9 as the first stage of Korolev's R-9 missile, first flight 9 Apr 1961 (failure) and entering service 21 Jul 1965.
Applications: first stage R-9 missile
Dry mass: 1,480 kg
Engine cycle: open
Oxidiser: liquid oxygen
Fuel: kerosene
Feed method: turbopump, feeding four gimballed chambers
Thrust: 1,407 kN sea level, 1,628 kN vacuum
Specific impulse: 275 s sea level, 317 s vacuum
Chamber pressure: 80 atm

RD-119

Developed 1958-62 for the Cosmos/Cosmos-2 launcher second stage, the Soviet Union's smallest satellite vehicle and retired in 1977. The single fixed chamber is constructed principally of titanium and unusually burns UDMH as fuel. Steering is effected by routing the turbopump exhaust through four large and four small fixed verniers.
Application: Cosmos satellite launcher stage 2
Dry mass: 168 kg
Engine cycle: open (turbine exhaust utilised for steering)
Oxidiser: liquid oxygen
Fuel: UDMH
Feed method: turbopump
Thrust: 105 kN vacuum
Specific impulse: 352 s vacuum
Chamber pressure: 80 atm
Burn time: 260 s

RD-120/11D123

The single fixed chamber is supported by four NPO Yuzhnoye 'RD-8' (actually the true designator might be in the 'RD-800' series) verniers for steering. Each engine is

GDL's RD-111 four-chamber engine which was used on the first stage of the R-9 missile. (William C Ingalls)
1997/0003410

test-fired at the factory for the intended flight duration. Reliability level is 99.5%. The design minimises the upper section for packaging; on Zenit it protrudes from a toroidal tank. An uprated version, the RD-120.03, has been mooted; its characteristics are included below in (). Two RD-120.01 versions are being considered for stage 1 applications: thrust 715.8/784.5 kN sea level (784.5/853.1 kN vac); SI 302/304 s sea level (329/330 s vac); chamber pressure 160.6/174.2 atm; length 2,430 mm; 1,400 mm ∅; dry mass 1,065/1,079 kg. (It has been claimed the 120.03 and 120.01 versions *have* been built and tested.) The second 120.01 version is now designated the **RD-120M**, with 4.96 expansion ratio, 2.6±10% mixture ratio, −40/+3.5% throttling and ±6° 2-axis gimballing. It was selected for the OSC/Rockwell X-34 before the companies pulled out. It has been suggested for the Rus Soyuz upgrade.

An RD-120 production engine was tested Oct 1995 by Pratt & Whitney in West Palm Beach, Florida, believed to be the first of an in-production and in-service engine outside of Russia. The three <5 s runs showed the engine was compatible with US ground systems; further testing requires a sea level nozzle and a customer.
Application: Zenit stage 2

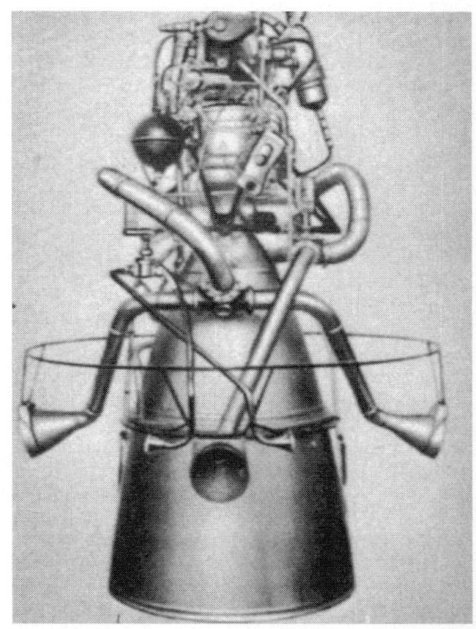

RD-119: single main chamber and eight verniers

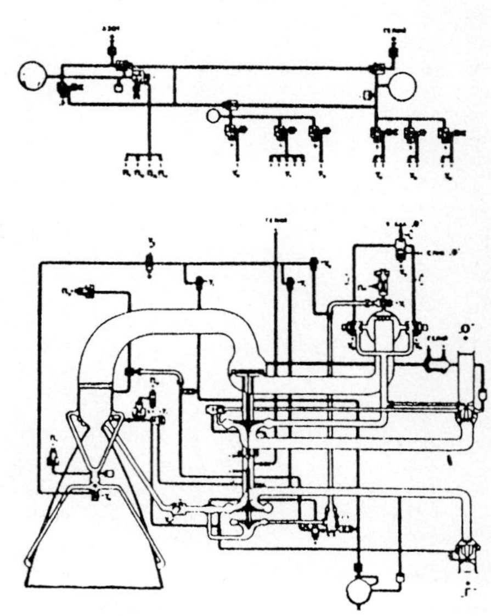

RD-120 engine schematic

The RD-120 powers Zenit's stage 2 (Commercial Space Technologies)

Dry mass: 1,125 kg dry, 1,285 kg with propellant; (1,125 kg)
Length: 3,872 mm (3,872 mm)
Maximum diameter: 1,954 mm (1,954 mm)
Engine cycle: closed gas generator
Oxidiser: liquid oxygen at 175.4 kg/s
Fuel: kerosene at 64.46 kg/s
Feed method: 19,000 rpm turbopump; gas generator 462°C
Thrust: 834 kN vacuum (throttling −8.5/+3.5%); (882.5 kN)
Specific impulse: 350 s vacuum; (353 s)
Expansion ratio: 106:1
Chamber pressure: 160.6 atm (nozzle 0.126 atm); (170.3 atm)
Burn time: 200-315 s

RD-161

Soyuz-2 may use the single chamber engine. The engine gimbals ±6° in two planes and can ignite up to 17 times in flight. It also provides high pressure LOX for attitude control verniers. The specifications are given below in order of with/without the uncooled nozzle extension.
Application: launcher upper stage
Dry mass: 141/119 kg
Length: 2,205/1,700 mm
Maximum diameter: 1,020/780 mm
Engine cycle: closed gas generator
Oxidiser: liquid oxygen
Fuel: kerosene
Mixture ratio (O/F): 2.6
Feed method: turbopump
Thrust: 19.9/19.6 kN vacuum
Specific impulse: 365/360 s vacuum
Expansion ratio: 19.25/18.75
Chamber pressure: 116.1 atm
Burn time: 900 s maximum

The RD-170/171 was publicly displayed for the first time at 1989's Paris Salon ***1997***

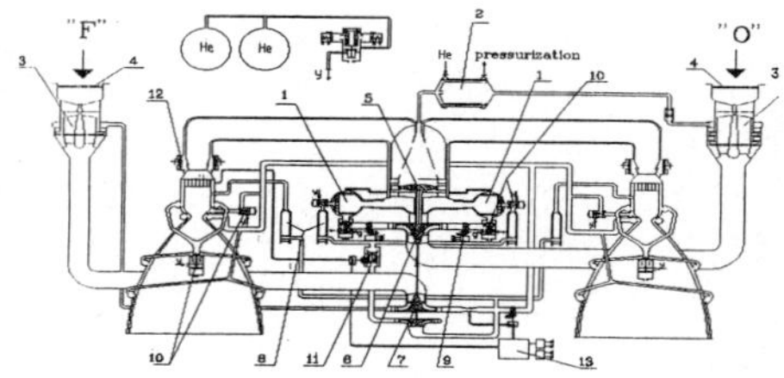

RD-170/171 engine schematic. 1: pre-burners, 2: heat exchanger, 3: booster pumps, 4: filters, 5: turbine, 6: oxidiser pump, 7: fuel pump, 8: start fuel cartridges, 9: throttles, 10: valves, 11: regulator, 12: gimballing unit, 13: command pressure determinator (NPO Energomash via CP Vick)

RD-169/RD-185

The RD-169 is proposed to power KB Machinostroenye's new Rikscha small launcher. Six would be clustered for stage 1 (gimballing 6°), followed by the RD-185 altitude version for stage 2 (gimballing 3°). Specifications are given below in RD-169/185 order.
Dry mass: 215/445 kg
Length: 1.5/3.0 m
Maximum diameter: 0.5/1.5 m
Engine cycle: closed gas generator
Oxidiser: liquid oxygen
Fuel: methane
Mixture ratio (O/F): 3.4/3.4
Feed method: turbopump
Thrust: 147/– kN SL; 167/179 kN vac
Specific impulse: 307/– s SL; 349/376 s vac
Expansion ratio: not available
Chamber pressure: not available
Burn time: not available

RD-170/171/172/191

With the Energia core engine, the most recent Soviet rocket engine development and the highest known chamber pressure. It is also the highest thrust liquid propellant rocket engine ever flown, surpassing Rocketdyne's single chamber F-1. Development began in 1976. The 4-chamber engine (paired chambers) flies in the first stage of the Zenit 2 medium lift launcher, which is also modified as Energia's reusable strap-ons. It can make 10 flights although there is no evidence that it has been re-used. There are only minor differences between Energia's RD-170 (11D520) and Zenit's RD-171 (11D521); for example, 1-axis gimballing for Zenit and 2-axis for Energia. Engines were manufactured from each batch by Polyot in Omsk; one was taken through three life cycles while the remainder underwent a single acceptance test. Polyot production for both RD-170/171 has halted and the plant dismantled, but Energomash itself can build up to 20 annually. The 190 MW 14,500 rpm turbopump of each chamber pair is driven by pre-burners (see diagram). The design permits smooth transition to full thrust over 2 s. At burnout, thrust drops to 70% over 30 s, to 50% over the next 2 s, where it holds for 10 s, and finally shuts down in 0.5 s.

Development of the improved **RD-172** for Zenit 2 was expected to achieve certification in 1996 but there has been no word on this recently. A 5% thrust increase is provided by a chamber pressure of 255.4 atm. The similar **RD-174** version is expected to be available in 1996. The single-chamber **RD-191**, utilising 70% of RD-170 elements, would provide 2,059 kN.
Applications: Zenit stage 1, Energia strap-ons
First flown: 13 Apr 1985
Dry mass: 8,755 kg
Length: 3.56 m
Maximum diameter: 399 cm
Engine cycle: staged combustion
Oxidiser: liquid oxygen at 432 kg/s
Fuel: kerosene at 166.2 kg/s
Thrust: 7,903.77 kN vacuum, 7,256.56 kN SL (but can be throttled back to 4,413 kN vac) for Energia, 7,911 kN vac and 7,259 kN SL for Zenit 2/3
Specific impulse: 308 s SL, 336 s vacuum for Energia; 337 s for Zenit-2 vacuum
Expansion ratio: 36.8
Chamber pressure: 242 atm; 250 atm maximum
Chamber mass: 480kg
Burn time: 140-150 s

RD-180

NPO Energomash is developing RD-170's 2-chamber version in co-operation with Pratt & Whitney for Lockheed Martin's Atlas 2AR (qv), planned for first launch Dec 1998. As only the main turbopump and boost pumps require development (all other components are interchangable with the RD-170), it will be available within 2½ yr. The first 18 vehicles will use Russian-built engines. It is also the baseline engine for Lockheed Martin's proposal for the USAF Evolved Expendable Launch Vehicle (EELV) competition.
Applications: launcher stage 1
First flown: planned Dec 1998

The RD-180 will power Lockheed Martin's Atlas 2AR (P&W)

Dry mass: 4,377 kg
Length: 3.56 m
Maximum diameter: not available
Engine cycle: staged combustion
Oxidiser: liquid oxygen
Fuel: kerosene
Thrust: 3,923 kN vacuum, 3,594 kN sea level, throttleable 40-105%. Atlas 2AR will use it at max 85% (75% at launch)
Specific impulse: 309 s sea level, 337 s vacuum
Expansion ratio: 36.8
Chamber pressure: 242 atm; 250 atm maximum
Chamber mass: 480kg
Burn time: 185 s on Atlas 2AR

RD-190

The RD-190 is a single chamber derivative of the RD-170 engine. No application for the engine is known other than a potential first stage of a launch vehicle. The engine can be throttled within the range 40-100%.
Dry mass: 2,517 kg
Length: 3.7 m
Maximum diameter: 1.8 m
Engine cycle: stages combustion
Oxidiser: liquid oxygen
Fuel: staged combustion
Thrust: 7E1,900 kN sea level, 2,075 kN vacuum
Specific impulse: 311 s sea level, 337 s vacuum

RD-211

Little is known of this engine, which completed development in 1963 with specific impulse of about 314 s. It has been suggested to be an upper stage engine.

RD-214

The first Soviet engine to be identified which uses storable nitric acid as oxidiser for a rapid response time from the R-12 IRBM, the RD-214 carried four fixed chambers and no verniers – steering was effected by refractory vanes protruding into the exhaust.
Applications: Cosmos satellite launcher stage 1, R-12 IRBM stage 1
First orbital flight: Mar 1962
Number flown: 144 on orbital missions
Dry mass: 645 kg
Engine cycle: open
Oxidiser: nitric acid
Fuel: kerosene
Feed method: single turbopump driven by hydrogen peroxide gas generator feeding four fixed chambers
Thrust: 635 kN sea level, 730 kN vacuum
Specific impulse: 230 s sea level, 264 s vacuum
Chamber pressure: 45 atm
Chamber: 480 mm ∅ internal, 176 mm throat ∅
Burn time: 140 s

RD-216

The first Soviet engine to utilise the storable nitric acid/UDMH combination as the first stage propulsion system for the R-14, subsequently adapted for the Cosmos 3 launcher and, until Start, the smallest satellite vehicle in the inventory. RD-216/11D614 comprises four (two pairs) combustion chambers. The engines are built by the Ust-Katavski Machinebuilding Plant at Ust-Katav in the Urals.

The RD-216 assembly powers Polyot's Cosmos 3M stage 1 (AKO Polyot) ***1995***

Applications: Cosmos 3/3M satellite launcher stage 1, R-14 IRBM stage 1
First orbital flight: Aug 1964
Dry mass: 662 kg
Engine cycle: open
Oxidiser: nitric acid/27% N_2O_4
Fuel: UDMH
Feed method: single turbopump feeds each chamber pair
Thrust (4 chambers): 1485.6 kN sea level, 1744.6 kN vacuum
Specific impulse: 248 s SL, 291.3 s vac
Chamber pressure: 75 atm
Burn time: 170 s (130 s on Cosmos 3M)

RD-218

The RD-218 engine was originally reported to have been the first stage of the R-36 missile and the Tsyklon launch vehicles. It is now known that the RD-218 was used on the R-16 missile's first stage. It comprised three engine assemblies, each with two combustion chambers and a turbo-pump assembly. The development period is given as 1958-1961.
Application: R-16 missile first stage
First flown: 1961
Dry mass: 1,960 kg
Length: 2.188 m
Maximum diameter: 2.789 m
Engine cycle: closed
Oxidiser: nitric acid
Fuel: UDMH
Thrust: 2,215 kN sea level, 2,597 kN vacuum
Specific impulse: 247 s sea level, 290 s vacuum
Chamber pressure: 73.5 atm

RD-219

Two-chamber, high-altitude engine developed during 1958-1961 for the R-16 missile second stage (although originally it was reported that it was used on the R-36 missile and Tsyklon launch vehicles).
Application: R-16 second stage
First flown: 1961
Dry mass: 665 kg
Length: 2.030 m
Maximum diameter: 2.162 m
Engine cycle: open
Oxidiser: nitric acid
Fuel: UDMH

RD-218 six-chamber engine which was used on the first stage of the R-16 missile. (William C Ingalls) ***1997****/0003411*

The RD-219 incorporates two thrust chambers fed by a single turbopump

Feed method: single turbopump feeding two chambers
Thrust: 883 kN vacuum
Specific impulse: 293 s vacuum
Chamber pressure: 75 atm
Burn time: 125 s

RD-250

No details are available concerning this engine, other than it was an early version of what became the RD-253, the engine used on the first stage of the Proton-K. Presumably it had a single thrust chamber.
Oxidiser: nitrogen tetroxide?
Fuel: UDMH

RD-251

The RD-251 is used on the first stage of the R-36 missile and the Tsyklon-2 and Tsyklon-M launch vehicles. It was developed in parallel with the RD-218 used on the R-16 missile. Like the RD-218, it comprises three independent two-chamber engines clustered together and operating as a single unit. The development period is given as 1961-1965.
Application: R-36, Tsyklon first stage
First flown: 1965
Dry mass: 1,729 kg
Length: 1.762 m
Maximum diameter: 2.520 m
Engine cycle: open
Oxidiser: nitrogen tetroxide
Fuel: UDMH
Feed method: turbopump
Thrust: 2,364 kN sea level, 2,643 kN vacuum
Specific impulse: 269 s sea level, 301 s vacuum
Chamber pressure: 85 atm

RD-252

Like the RD-251 being developed in parallel with the RD-218, the RD-252 for the second stage of the R-36 missile and Tsyklon launch vehicle was developed in parallel with the RD-219. The engine has two combustion chambers and was developed during the period 1961-1966.
Application: R-36, Tsyklon second stage
First flown: 1965
Dry mass: 715 kg
Length: 2.186 m
Maximum diameter: 2.590 m
Engine cycle: open
Oxidiser: nitrogen tetroxide
Fuel: UDMH
Feed method: single turbopump
Thrust: 940.5 kN vacuum
Specific impulse: 318 s vacuum
Chamber pressure: 91 atm

RD-253/11D43

The first known Soviet closed cycle and N_2O_4/UDMH engine, the gimballed single chamber provides the first stage power for Proton's six clustered external tanks. The engine was displayed at the 1987 Paris Air Show. A modified configuration with improved performance (SI increased by 30 s) was introduced on Proton in 1986. The original version incorporated a high-mounted pump but the new version is side-mounted. The engines are built by the PPO Motorostroitel (=Engine Builder; see their National section entry).
Application: Proton stage 1
First flown: 16 Jul 1965
Dry mass: 1,280 kg
Length: 2.72 m
Maximum diameter: 1.50 m
Engine cycle: closed
Oxidiser: nitrogen tetroxide
Fuel: UDMH
Propellant flow rate: about 528 kg/s; mixture ratio 2.67

RD-253 employs a side-mounted turbopump (Andrew Wilson) ***1996***

Feed method: 18.7 MW turbopump driven by pre-burner gas
Thrust: 1,745 kN vacuum (1,474 kN sea level), uprated from 1,635 kN (1,460 kN sea level). Profile is 40% full thrust in 0.1 s, held at 40% for 2 s, then to 100% in 0.1 s
Specific impulse: 317 s vacuum, 285 s sea level
Chamber pressure: 150 atm
Burn time: 130 s

RD-261 and RD-262

There have been some reports that while the RD-251 and RD-252 engines are used on the first and second stages (respectively) of the Tsyklon-M vehicle, modifications of these engines designated RD-261 and RD-262 (respectively) are carried on the first and second stages of the three-stage Tsyklon. Russian confirmation of these reports has not been found.

RD-270

Although apparently not tested beyond component level, the RD-270 was designed for Chelomei's UR700 lunar vehicle and then rejected for Korolev's N1. It would have been the largest Soviet storable propellant engine, operating with a chamber pressure greater than RD-170's.
Application: launcher stage 1
Dry mass: 4,770 kg (5,603 kg wet)
Length: 4.850 m
Maximum diameter: 3.300 m
Engine cycle: gas-gas closed staged
Oxidiser: nitrogen tetroxide
Fuel: UDMH
Propellant flow rate: unknown
Feed method: turbopump

The RD-251 six-chamber engine which was used on the first stage of the R-36 missile and the two-stage Tsyklon-M launch vehicle. The engine was developed in parallel with the RD-218 used on the related R-16 missile. (William C Ingalls) ***1997****/0003412*

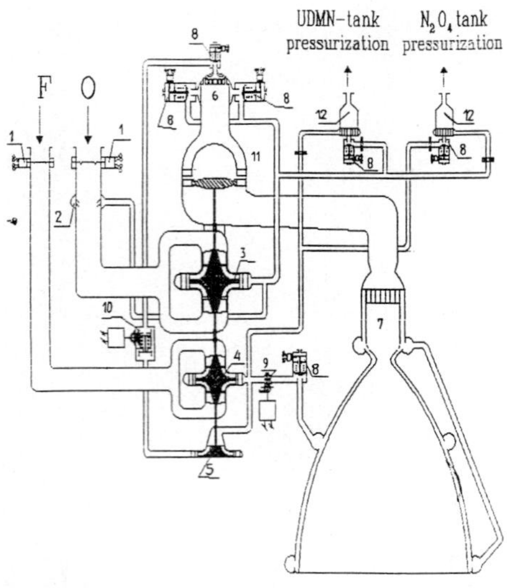

RD-253 engine schematic. 1: *valves,* 2: *ejector,* 3: *oxidiser pump,* 4: *main fuel pump,* 5: *pre-burner pump,* 6: *preburner,* 7: *combustion chamber,* 8: *shutoff valves,* 9: *fuel throttle,* 10: *proportional flow regulator,* 11: *turbine,* 12: *tank pressurisation generators* (NPO Energomash)

Thrust: 6,713 kN vacuum, 6,272 kN sea level
Specific impulse: 322 s vacuum, 301 s sea level
Chamber pressure: 266 atm
Burn time: unknown

RD-301

Developed 1969-75, the RD-301 was intended as Proton's stage 3 engine but flight tests were cancelled when it was decided to be too toxic. It was the first known example of experimentation with high SI exotic propellants, although details on the earlier RD-502 subsequently emerged.
Mass: 183 kg
Length: 1.88 m
Maximum diameter: 98 cm
Engine cycle: open
Oxidiser: fluorine
Fuel: ammonia
Mixture ratio (O/F): 2.7:1
Feed method: 1,265 kW turbopump driving 2-stage oxidiser & single-stage fuel pumps and a turbine
Thrust: 98.1 kN vacuum
Specific impulse: 400 s vacuum
Expansion ratio: 108.7
Chamber pressure: 120 atm
Burn time: 750 s

RD-502

Developed 1960-66, the RD-502 was intended as Proton's stage 3 engine but work was halted when it was decided, like RD-301 before it, to be too toxic.
Mass: 132 kg (140 kg wet)
Length: 2.510 m
Maximum diameter: 1.180 m
Engine cycle: closed staged
Oxidiser: hydrogen peroxide
Fuel: pentaborane
Mixture ratio (O/F): unknown
Feed method: turbopump
Thrust: 98.06 kN vacuum
Specific impulse: 380 s vacuum
Expansion ratio: unknown
Chamber pressure: 150 atm
Burn time: unknown

RD-701

NPO Energomash has been working on this tripropellant RD-170 derivative engine since 1989 to propel a small 22 t shuttle-type vehicle designed by NPO Molniya for release from an An-225 at altitude. For the first 2½ min, it would burn principally kerosene and oxygen (mode 1), using LH_2 for cooling from the separate tank of 205 t of propellants. It would then switch (mode 2) to the less dense LH_2 at altitude for the final 6-8 min to orbit. The required 5 yr engine development would cost >$500 million; international partners are sought. Each engine would comprise two thrust chambers gimballed in 2-axes. Each chamber would employ two turbopumps: one channeling LOX/kerosene into a pre-burner where they are burned oxygen-rich for the product to burn with hydrogen in the main chamber. US marketing is being provided by P&W. NASA awarded P&W a $5.4 million contract in Aug 1994 towards converting the RD-701 into a tripropellant engine for a US reusable launcher. The **RD-704** single chamber derivative is also proposed for US applications.
Mass: 3,990 kg assembly, 1,840 kg each dry
Length: 5.4 m extension down, 3.8 m extension up
Diameter: 2.4 m
Engine cycle: closed staged combustion
Oxidiser: liquid oxygen
Fuel: mode 1: hydrogen/kerosene; mode 2: hydrogen
Mixture ratio (O/F; mass): mode 1 5.27:1 kerosene and 13.2:1 hydrogen; mode 2 6.1:1
Flow rates (each chamber, kg/s): mode 1 388.4 LOX, 29.5 LH_2/73.7 kerosene; mode 2 148.5 LOX/24.7 LH_2
Thrust (each chamber, vac): 1,960 kN mode 1, 785 kN mode 2 (throttle 40-100%)
Specific impulse (s): mode 1 415 vac, 330 SL; mode 2 460 vac
Chamber pressure: mode 1 290 atm; mode 2 122 atm
Expansion ratio: 170 extension down, 70 extension up

KELDYSH RESEARCH CENTER

Some details of engines designed by the Keldysh Research Center are available, although none of the engines described here is believed to have actually flown.

Hybrid Rocket Engines

The Keldysh Research Center is undertaking work on the development of hybrid rocket engines which could be used on lightweight launch vehicles or upper stages of launch vehicles. The advantages of hybrid engines are seen as: simplicity of design and reduction of payload cost by 30%, friendly to the environment during production and when in service, manufacture and service safety, fire and explosion safety preventing the failure of launch vehicles, high reliability, a multiple burn capability while throttling over a wide thrust range and the use of proven designs, advanced technologies and production equipment which reduce the cost of engine production by a factor of 1.5-2 compared with existing engines.

The principles of the design of hybrid engines have been developed at the Center and a model and demonstration prototype with a thrust of 29 kN (Keldysh literature states 3 t) have been developed. The design performances of hybrid engines have been proved by actual performances and the documentation for a modular hybrid engine giving thrusts of 177 kN (18 t) and 294 kN (30 t).
Oxidiser: oxygen, hydrogen peroxide
Fuel: rubber
Specific impulse: 3,140-3,335 kN (first stage), 3,335-3,725 kN (second stage)
Chamber pressure: 20-60 atm
Burn time: 400 s (max)

Ion Thruster for Small Space Vehicles

An ion thruster has been developed for the attitude control of small satellites intended for either remote sensing or communications. Performance data are as follows:
Power input: 50-150 W
Thrust: 2-5 mN
Specific impulse: 3,000-3,500 s

Liquefied Air Cycle Engine (LACE)

LACE is an advanced hybrid engine being studied for an aerospace plane. Advantages of the design are given as: 2.5-4.5 fold improvement of performance compared with liquid propellant engines, specific thrust in the atmosphere is 1.5-2 times higher compared to other air-breathing engines, use of a single-type engine as the propulsion system, maximum possibility of final development using the existing test models and ease of engine integration into spacecraft. When developed using LACE in aerospace planes would permit a payload of 10 t to be launched with a single-stage take-off mass of 340 t: compared with a rocket-based single stage launch vehicle the launch mass would be reduced by a factor of 1.5-2 and the cost of launching each kilogramme into orbit would be reduced by a factor of 15-20 compared with an expendable launch vehicle.

The tripropellant dual mode RD-701 could power a spaceplane. (William C Ingalls) ***1997***/0003413

RD-704 Operating Data

Operating Mode	*Mode 1*	*Mode 2*
Fuel	kerosene/liquid hydrogen	liquid hydrogen
Oxidiser	LOX	LOX
Thrust: vac	1,960 kN	781 kN
sea level	1,720 kN	(N/A)
Specific impulse: vac	407 (401) s	452 (450) s
sea level	356 (351) s	(N/A)
Mixture ratio (O/F)	4.38	6.0
Chamber pressure	290 atm	120 atm
Area ration	74	
Mass	2,417 kg	
Single bell diameter	1.8 m	
length	3.8 m	

N/A – not applicable. For the specific impulse values the nominal value is quoted, followed by a worst case value in parentheses.

The Keldysh Research Center states that concepts and configurations for a LACE have been developed and experimental investigations of the main stages of the operation of the engines have been studied experimentally using sub-scale engines. A sub-scale demonstration LACE is being assembled at the Center's test facility and technical proposals for the LACE and the aerospace plane are being prepared with other specialist Russian design bureaus. The investigations are being undertaken within the framework of a joint programme between the Russian Space Agency and the European Space Agency.

A few numbers have been released concerning LACE. Two modes of operation are noted, with a specific impulse within the range 1,300-2,000 s being attained. The engine would operate in two modes: using liquefied air for Mach 0-6 and in rocket mode for Mach 6-25. In the short term a structural specific mass of 60 kg/t of thrust should be obtainable, reducing to 35 kg/t of thrust in the longer term.

'Rocket Engine for the 21st Century'

A new liquid engine is being studied as the basis for the development of a new generation of 21st century launch systems. The requirements for the launch system are: increased reliability and safety in operation, everything should be reusable and major savings in the cost of putting spacecraft into orbit. The Keldysh Research Center is proposing a new family of liquid propellant rocket engines which would use ecologically-friendly propellants. It is hoped that these proposed engines would provide: reduction of costs as well as ease of development, testing and manufacture of the engine, high energy and mass characteristics, minimum delays between inflight servicing and reuse.

Model prototypes of these engines have undergone successful test firings. The most attractive propellants for the programme appear to be oxygen/methane for the first stage and oxygen/hydrogen for the second stage. The engines could be developed by 2002 (one assumes that finances are the limiting factor) and the Center Claims that there is nothing like their proposed engines being developed elsewhere. However, there are no further details available concerning the engines.

T-100 Electric Thruster

The T-100 electric thruster is described by the Keldysh Research Center as being 'a superb generator of a high-velocity directed flow of xenon plasma', and applications in both the space programme and new technologies within Russian industry are foreseen. The thruster was developed as an actuator for spacecraft manoeuvring systems for orbital transfers, injection of a spacecraft into a planned orbit, attitude control and orbital corrections. The T-100 is described as being highly efficient with a low-level of radiation, a long life and a short time for preparation. It is unclear whether this thruster has actually been flown on spacecraft.
Dry mass: <32 kg
Dimensions: 140 × 135 × 230 mm
Propellant: high-purity xenon
Thrust: 80 mN
Specific impulse: 1,600-1,800 s (literature says 16,000-18,000 m/s)
Efficiency: 50% (minimum)
Propellant flow rate: 5 mg/s
Supply voltage: 300 V AC

Input power: 1.4 kW
Preparation time for operation: <1 s
Burn time: 8,000 h
Start/shutdown cycles: 6,000

KB KHIMAUTOMATIKI

The bureau of Semyin A Kosberg (1903-65), formed 1941, began investigating liquid propellant rocket propulsion for aircraft in the early 1950s and later became responsible for some Soviet space launcher upper stages. In 1974 it was reorganised as KB Khimautomatiki.

RD-0105

Also called the RO-5, this engine was the original one used on the Vostok launch vehicle's Block E. The engine assembly comprised a single main combustion chamber plus four verniers. It has not been used since the failed Luna launches in Apr 1960.
Application: Original Vostok Block E
First flown: 1958
Dry mass: ~125 kg
Feed method: pump
Thrust: 49 kN vacuum
Specific impulse: 316 s vacuum
Chamber pressure: 44 atm
Burn time: 480 s

RD-0106

The RD-0106 was the first engine in the family which would be used on the Block I orbital stages of the Molniya, Soyuz and Voskhod families of launch vehicles. The RD-0106 itself was used on the second stage of the R-9A missile. Assuming that it was like the later variants, it had four main combustion chambers plus four vernier chambers.
Application: R-9A missile
First flown: 1961
Oxidiser: liquid oxygen
Fuel: kerosene

RD-0107/8D715

The RD-0107 was the first development from the RD-0106, and it was used on the Block I of the Molniya and then Voskhod launch vehicles. The engine has also been designated the RO-9. The engine had four main combustion chambers and four vernier chambers.
Application: Molniya, Voskhod Block I
First flown: 1960
Dry mass: 409 kg
Oxidiser: liquid oxygen
Fuel: kerosene
Thrust: 294 kN vacuum
Specific impulse: 326 s vacuum
Chamber pressure: 67 atm
Burn time: 250 s

RD-0108

The RD-0108 was used on the Block I of the original Soyuz, Soyuz-L and Soyuz-M launch vehicles. The performance details are probably similar to those of the RD-0110 which replaced it. Like the earlier versions of the same basic engine, it carried four main combustion chambers and four vernier chambers.
Application: Soyuz, Soyuz-L, Soyuz-M Block I
First flown: 1966
Oxidiser: liquid oxygen
Fuel: kerosene

RD-0109/8D719

This engine replaced the RD-0105 on the orbital stage of the Vostok launch vehicle: it has also been referred to by the RO-7 and RD-448 designators. The designator 11D719 has also been used rather than the 8D719, and it is unclear which one is correct. The engine has a single main combustion chamber plus four vernier chambers.
Application: Vostok Block E
First flown: 1960
Dry mass: 121 kg
Length: 1.535 m
Maximum diameter: 2.1 m?
Oxidiser: liquid oxygen
Fuel: kerosene
Feed method: pump
Thrust: 54.8 kN vacuum
Specific impulse: 324 s vacuum
Chamber pressure: 49 atm
Burn time: 430 s

RD-0110/11D55

The RD-0110 replaced the RD-0107 on the Block I of the Molniya launch vehicle and the RD-0108 on the same stage of original group of Soyuz launch vehicles. The engine has four main combustion chambers plus four vernier chambers.
Application: Molniya-M and Soyuz-U Block I
First flown: 1964
Dry mass: 408 kg
Length: 1.575 m
Maximum diameter: 2.240 m
Oxidiser: liquid oxygen

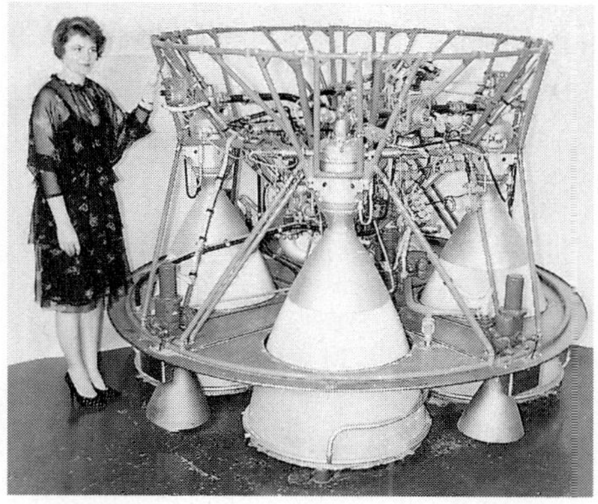

RD-0110 engine used on the Soyuz-U and Molniya-M launch vehicles' orbital stages. (KB Khimavtomatiki) ***1997***/00034⁻5

Fuel: kerosene
Thrust: 298 kN vacuum
Specific impulse: 330 s vacuum
Chamber pressure: 69 atm
Burn time: 240 s

RD-0120/RO-200/11D122

Energia's four single chamber core engines is the first operational Soviet cryogenic system, following space introduction by the US (1962), ESA (1979), China (1984) and Japan (1986). Only the US Shuttle had previously flown a cryogenic first stage. The engine was first displayed in Moscow in Sep 1990. Each chamber gimbals ±11° in two axes for steering. >800 tests exceeding 170,000 s in total had been performed to end-1995 since runs began in 1979. A version capable of 25 flights was proposed for the joint Interim Hotol vehicle (see the UK Launchers section). An agreement was signed Oct 1994 with Aerojet of the US to codevelop and market derivatives in the US; it could be developed into a tripropellant engine for a future reusable vehicle. Aerojet's work is being performed under a $17.2 million NASA Marshall contract. Europe's RECORD (Russia-Europe Cooperation on Rocket Engine Demonstration) project, initiated in 1993 by SEP and its Vulcain industrial partners, began tests at NII Khimmach, Russia's largest test site, in Aug 1995 to help validate modelling for future engines. The ECU20.5 million cost over 3 yr is covered by: 13 European Commission; 6 ESA; 1.5 industry.
Application: Energia core, 4 engines
First flown: 15 May 1987
Number flown: 8 to end-1996
Dry mass: 3,449 kg
Length: 4,549 mm
Maximum diameter: 2,420 mm
Engine cycle: closed
Oxidiser: liquid oxygen at 376.7 kg/s
Fuel: LH_2 at 62.78 kg/s
Mixture ratio (O/F): 6.0:1
Feed method: 33,385 rpm dual-stage turbopump on each chamber. A single pre-burner burns fuel-rich at 527°C/408 atm to drive the single-shaft high pressure turbopump. Some of the pre-burner gas drives the oxygen low pressure pump; the fuel low pressure pump

RD-0109 engine used on the Vostok 8K72K launch vehicle's orbital stage: it was this variant which was used for the six manned launches in the Vostok programme (KB Khimavtomatiki) ***1997***/0003414

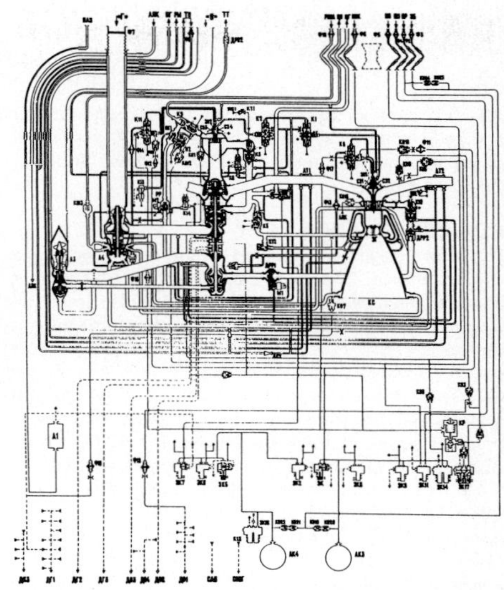

RD-0120 engine schematic (KB Khimautomatiki)

The RD-0120 is Russia's first operational cryogenic engine (KB Khimavtomatiki) ***1997***/0003370

is driven by GH_2 from the main chamber cooling loop. Main LH_2 discharge pressure 448 atm; LOX stage 1 342 atm; LOX stage 2 571 atm
Thrust: 1,961 kN vacuum, 1,451 kN sea level; throttle range 25-114%
Expansion ratio: 85.7:1 (throat ∅ 261 mm, exit ∅ 2,420 mm)
Specific impulse: 455 s vacuum
Chamber pressure: 215.7 atm
Burn time: about 480 s (600 s max operational)

RD-0124

This engine is planned to replace the RD-0110 when the new Soyuz-2 launch vehicle is introduced.
Application: Soyuz-2 Block I
First flown: 1997-1998?
Dry mass: 450 kg
Length: 1.575 m
Maximum diameter: 2.273 m
Oxidiser: liquid oxygen
Fuel: kerosene
Thrust: 294 kN

RD-0124 engine planned to fly on the orbital stage of the new Soyuz-2 ('Rus') launch vehicle: as such, this engine is a replacement for the RD-0110. (KB Khimavtomatiki) ***1997***/0003416

Specific impulse: 359 s
Chamber pressure: 157 atm

RD-0202

RD-0202 is the designator applied to the cluster of four engines on the first stage of the UR-200 missile: it possibly comprises three RD-0203 engines and one RD-0204 engine.

RD-0203/RD-0204

These two engines were used on the first stage of the UR-200 missile. The stage carried four single-chamber engines, and it was thought that one of these carried the tank pressurisation system while the other three did not (hence the two different engine designators for the stage). However, according the KB Khimautomatiki neither of the engines carries a tank pressurisation system, and therefore the difference between them is not known. The Proton-K launch vehicle's RD-0210 and RD-0211 are believed to be altitude-ignition variants of these two engines. The development period is given as 1961-1964.
Application: UR-200 missile first stage
First flown: 1964
Engine cycle: closed
Oxidiser: nitrogen tetroxide
Fuel: UDMH

RD-0205

This designator is applied to the combined RD-0206 main engine and RD-0207 vernier engines on the UR-200 missile's second stage.

RD-0206/RD-0207

The RD-0206 is the main engine carried on the second stage of the UR-200 missile, with the RD-0207 being the vernier engine on the same stage. It is possible that the RD-0206 is a modification for altitude ignition of the RD-0204 and RD-0205 engines. The RD-0206 carries a tank pressurisation system. The development period is given as 1961-1964.
Application: UR-200 second stage
First flown: 1964
Engine cycle: RD-0206 – closed, RD-0207 – open
Oxidiser: nitrogen tetroxide
Fuel: UDMH

RD-0210/8D411

Three RD-0210 engines are carried on the second stage of the Proton-K launch vehicle. They are single-chamber engines.
Application: Proton-K second stage
First flown: 1965
Engine cycle: closed (oxidiser pre-burner gas routed to main chamber after driving turbine)
Oxidiser: nitrogen tetroxide
Fuel: UDMH
Thrust: 594 kN
Specific impulse: 327.4 s
Chamber pressure: 148 atm
Burn time: 210 s

RD-0211/8D412

A single RD-0211 engine is carried on the second stage of the Proton-K launch vehicle. They are single-chamber engines and carry the tank pressurisation system. The performance data are as for the RD-0210.

RD-0213 main engine and RD-0214 vernier system used on the third stage of the Proton-K launch vehicle. (KB Khimavtomatiki) ***1997***/0003419

RD-0216 engine on the first stage of the UR-100 missile. (KB Khimavtomatiki) ***1997***/0003418

RD-0212/8D49

This designator is applied to the RD-0213/RD-0214 engine combination which is used on the third stage of the Proton-K.

RD-0213

The RD-213 is the main engine on the Proton-K third stage, and is a modification of the RD-0210/RD-0211 engines used on the launch vehicle's second stage.
Application: Proton-K third stage main engine
First flown: 1967
Engine cycle: closed
Oxidiser: nitrogen tetroxide
Fuel: UDMH
Thrust: 593.6 kN
Specific impulse: 325.3 s
Chamber pressure: 148 atm
Burn time: 240 s

RD-0214

The RD-0214 is the four-chamber vernier engine assembly carried on the third stage of Proton-K. It carries the tank pressurisation system for the stage.
Application: Proton-K third stage vernier engine
First flown: 1967
Engine cycle: open
Oxidiser: nitrogen tetroxide
Fuel: UDMH
Thrust: 31.5 kN
Burn time: 240 s

RD-0216

Three RD-0216 engines were carried on the first stage of the UR-100 missile. It was developed during 1963-1966. It carries a single main combustion chamber.
Application: UR-100 missile first stage
First flown: 1965
Engine cycle: closed
Oxidiser: nitrogen tetroxide
Fuel: UDMH

RD-0217

A single RD-0216 was used on the first stage of the UR-100 missile and it carried the stage's pressurisation system. It was developed during 1963-1966. It carries a single main combustion chamber.
Application: UR-100 missile first stage
First flown: 1965
Engine cycle: closed
Oxidiser: nitrogen tetroxide
Fuel: UDMH

RD-0225

Developed as the main manoeuvring engine of the military Almaz manned space station. A 3.9 kN Khimmach engine is used by the Heavy Cosmos modules.
Applications: paired as Almaz orbital manoeuvring engine
First flown: 3 Apr 1971, Almaz 1
Engine cycle: pressure fed
Oxidiser: nitrogen tetroxide
Fuel: UDMH
Thrust: 3.9 kN vac
Specific impulse: 291 s vac
Chamber pressure: 8.8 atm
Burn time: ?

RD-0229

A single RD-0229 is used as the main engine on the second stage of the R-36M missile, and it carries the

RD-0229 main engine and RD-0230 vernier engine assembly used on the original versions of the R-36M missile. This assembly was placed by the RD-0256/RD-0257 on the later R-36M2 missiles. (KB Khimavtomatiki) ***1997***/0003417

tank pressurisation system for the stage. The development period is given as 1967-1974.
Application: R-36M missile second stage main engine
Engine cycle: closed
Oxidiser: nitrogen tetroxide
Fuel: UDMH

RD-0230

The RD-0230 is the four-chamber vernier engine system for the R-36M missile. The development period is given as 1967-1974.
Application: R-36M missile second stage vernier engine
Engine cycle: closed
Oxidiser: nitrogen tetroxide
Fuel: UDMH

RD-0231

This engine was developed during the period 1968-1970 and was used on a spacecraft designated 'Granit': however it is not known to which particular spacecraft this name refers – it is certainly not the Granat astrophysical observatory.
Application: 'Granit'
Engine cycle: closed
Oxidiser: nitrogen tetroxide
Fuel: UDMH

RD-0233

First stage engine for the UR-100N missile and the Rockot launch vehicle. It carries the tank pressurisation system for the stage, and therefore one of these engines is used on the stage. It was developed during 1969-1974.

RD-0233 engine on the first stage of the UR-100N missile and the derived Rockot launch vehicle (KB Khimavtomatiki) ***1997***/0003420

RD-0235 main engine and RD-0236 vernier engine assembly used on the UR-100N missile and Rockot satellite launcher. (KB Khimavtomatiki) ***1997***/0003421

Application: UR-100N missile first stage, Rockot first stage
Engine cycle: closed
Oxidiser: nitrogen tetroxide
Fuel: UDMH
Specific impulse: 291 s
Chamber pressure: 20.5 atm

RD-0234

Three of these single-chamber engines are used on the first stage of the UR-100N missile and Rockot launch vehicle: they are almost identical with the RD-0233, except that they lack the tank pressurisation system. They were developed during 1969-1974.
Application: UR-100N missile first stage, Rockot first stage
Engine cycle: closed
Oxidiser: nitrogen tetroxide

RD-0235

The RD-0235 is the main engine for the second stage of the UR-100N missile and Rockot launch vehicle: it is a single-chamber engine, developed during 1969-1974.
Application: UR-100N missile and Rockot second stage main engine
Engine cycle: closed
Oxidiser: nitrogen tetroxide
Fuel: UDMH

RD-0236

The RD-0236 is the four-chamber vernier engine system for the UR-100N and Rockot second stages. It carries the pressurisation system for the propellant tanks. Development period was 1969-1974.
Application: UR-100N missile and Rockot second stage vernier engine
Engine cycle: open
Oxidiser: nitrogen tetroxide
Fuel: UDMH
Specific impulse: 293 s
Chamber pressure: 17.5 atm

RD-0237

The small RD-0237 is carried on the post-boost third stage of the UR-100N missile. It was developed during 1969-1974.
Application: UR-100N third stage
Engine cycle: open
Oxidiser: nitrogen tetroxide
Fuel: UDMH

RD-0244

First stage engine for the RSM-54 submarine-launched ballistic missile, having a single combustion chamber. It carried the propellant tank pressurisation system for the stage. The development period is given as 1977-1985.
Application: RSM-54 SLBM first stage main engine
Engine cycle: closed
Oxidiser: nitrogen tetroxide
Fuel: UDMH
Specific impulse: 301 s
Chamber pressure: 27.5 atm

RD-0245

Four-chamber vernier engine system for the first stage of the RSM-54 SLBM. The development period is given as 1977-1985.
Application: RSM-54 SLBM first stage main engine
Engine cycle: open
Oxidiser: nitrogen tetroxide
Fuel: UDMH

RD-0244 single-chamber main engine and the RD-0245 four-chamber vernier engine assembly, used on the first stage of the RSM-54 submarine-launched ballistic missile. (KB Khimavtomatiki) ***1997***/0003427

RD-0256

This engine replaced the RD-0229 on the R-36M to give the R-36M2 and it carries the pressurisation system for the propellant tanks. The development period is given as 1983-1987.
Application: R-36M2 missile second stage
Engine cycle: closed
Oxidiser: nitrogen tetroxide
Fuel: UDMH

RD-0257

The four-chamber vernier engine system for the R-36M2 missile, replacing the RD-0230 of the earlier R-36M. The development period is given as 1983-1987.
Application: R-36M2 missile second stage vernier engine
Engine cycle: open
Oxidiser: nitrogen tetroxide
Fuel: UDMH

RD-0410

Development of the 35 kN nuclear thermal design began in 1965 as a prototype for manned Mars vehicles. Specific impulse of the 1.6 m ∅, 3.5 m high liquid hydrogen motor was 910 s.

KB KHIMMACH

A bureau headed by Alexei M Isayev was established in 1943 for research into storable liquid propellant rocket engines. Whereas Glushko's GDL concentrated on major launcher engines and Kosberg's bureau on upper stage propulsion systems, Isayev's produced smaller *spacecraft* engines, including manned, Earth orbit and deep space systems. A number of designations have been adopted but 'KTDU' (Korrektyroushchaya-Tormorznaya Dvigatelnaya Ustanovka = Corrective-Braking Rocket Motor) is most frequent, with the others variations on this theme. Information released in early 1991 indicated the bureau is part of NPO Soyuz in Kaliningrad. It is responsible for the 73.58 kN cryogenic engine of the new Proton KM version, originally designed for the L-3M manned lunar mission, and recently sold to India for its GSLV vehicle. A newly-developed DMT-600 600 N NTO/UDMH thruster was referred to in 1994. Khimmach developed all of the engines for the Heavy Cosmos TKS transport craft.

Starting in 1971 and at the instigation of Isayev, the bureau started the development of low-thrust engines which could be used for attitude control of spacecraft. A series eleven engines with thrusts up to 2,206 kN (225 kg) and eight with thrusts up to 49 kN (5 kg) were developed. They are capable of burns lasting from hundredths of a second through to hundreds and thousands of seconds.

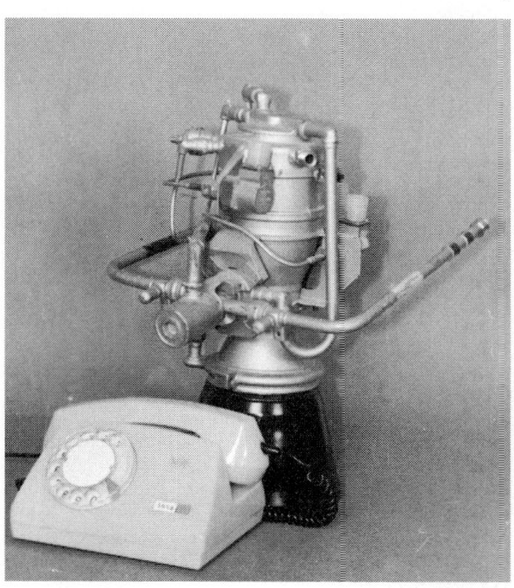

RD-0237 engine used on the post-boost stage of the UR-100N missile. (KB Khimavtomatiki) ***1997***/0003423

RD-0256 main engine and RD-0257 vernier engine assembly which replaced the RD-0229/RD-0230 assembly for the second stage of the R-36M missile. (KB Khimavtomatiki) ***1997***/0003426

11D49

It was confirmed in 1994 by the *Directory* that the bureau is responsible for the Cosmos 3M stage 2 engine. The development period was 1965-1971.
Application: Cosmos 3M stage 2
First flown: May 1967
Dry mass: 225 kg
Length: unknown
Maximum diameter: unknown
Engine cycle: pump-fed
Oxidiser: nitrogen tetroxide
Fuel: UDMH
Mixture ratio: 2.33/0.35
Thrust: 157.3 kN main + 4 × 1.4-1.8 kN steering thrusters (each with a 25 N pressure-fed thruster for 3-axis coast + ullage)
Expansion ratio: main 103.4; steering 12.0
Specific impulse: 303 s vac (176 s steering)
Chamber pressure: 102 atm
Burn time: main chamber about 350 s operationally

DOK-10

Application: Spacecraft thruster
Mass: 0.6 kg
Propellant: hydrazine
Thrust: 10 kN
Specific impulse: 229 s
Chamber pressure: 1.5 atm
Burn durations: 0.05-600 s
Total burn time: 1,500 s
Number of firings: 4,000

Khimmach's 11D49 engine power's Polyot's Cosmos 3M stage 2 (KB Khimmash) ***1997***/0003373

DOK-50
Application: Spacecraft thruster
Mass: 1.1 kg
Propellant: hydrazine
Thrust: 50 N
Specific impulse: 229 s
Chamber pressure: 1.5 atm
Burn durations: 0.05-600 s
Total burn time: 1,500 s
Number of firings: 4,000

DOT-5
Application: Spacecraft thruster
Mass: 0.9 kg
Propellant: hydrazine
Thrust: 5 N
Specific impulse: 230 s
Chamber pressure: 1.2 atm
Burn durations: 0.05-8,000 s
Total burn time: 120,000 s
Number of firings: 55,000

DOT-25
Application: Spacecraft thruster
Mass: 1.3 kg
Propellant: hydrazine
Thrust: 25 kN
Specific impulse: 235 s
Chamber pressure: 1.2 atm
Burn durations: 0.05-6,000 s
Total burn time: 25,000 s
Number of firings: 6,000

DST-25
Application: Spacecraft thruster
Mass: 0.9 kg
Oxidiser: nitrogen tetroxide
Fuel: UDMH
Mixture ratio (O/F): 1.85
Thrust: 25 N
Specific impulse: 285 s
Chamber pressure: 1.5 atm
Burn durations: 0.03-4,000 s
Total burn time: 25,000 s
Number of firings: 300,000

DST-100
Application: Spacecraft thruster
Mass: 1.1 kg
Oxidiser: nitrogen tetroxide
Fuel: UDMH
Mixture ratio (O/F): 1.85
Thrust: 100 N
Specific impulse: 276 s
Chamber pressure: 2.5 atm
Burn durations: 0.05-300 s
Total burn time: 10,000 s
Number of firings: 10,000

DST-100A
Application: Spacecraft thruster
Mass: 1.5 kg
Oxidiser: nitrogen tetroxide
Fuel: UDMH
Mixture ratio (O/F): 1.85
Thrust: 100 N
Specific impulse: 304 s
Chamber pressure: 1.6 atm
Burn durations: 0.05-4,000 s
Total burn time: 50,000 s
Number of firings: 450,000

DST-200
Application: Spacecraft thruster
Mass: 1.3 kg
Oxidiser: nitrogen tetroxide
Fuel: UDMH
Mixture ratio (O/F): 1.85
Thrust: 200 N
Specific impulse: 280 s
Chamber pressure: 2.5 atm
Burn durations: 0.03-300 s
Total burn time: 5,000 s
Number of firings: 10,000

DST-200A
Application: Spacecraft thruster
Mass: 1.7 kg
Oxidiser: nitrogen tetroxide
Fuel: UDMH
Mixture ratio (O/F): 1.85
Thrust: 200 kN
Specific impulse: 300 s
Chamber pressure: 1.6 atm
Burn durations: 0.05-4,000 s
Total burn time: 10,000 s
Number of firings: 100,000

DST-600
This engine can use either UDMH or MMH as the fuel: hence two sets of mixture ratios and specific impulses are given.
Application: Spacecraft thruster
Mass: 4.2 kg
Oxidiser: nitrogen tetroxide
Fuel: UDMH/MMH
Mixture ratio (O/F): 1.85/1.65
Thrust: 25 N
Specific impulse: 301 s/306 s
Chamber pressure: 1.6 atm
Burn durations: 0.05-1,500 s
Total burn time: 5,000 s
Number of firings: 6,000

KDU-414
This engine was used on the first-generation of planetary missions which were launched using the Molniya and Molniya-M vehicles. The final use for a deep space mission was for Venera 8 and the failed Cosmos 472 Venus probe in 1972, but it is still used on Molniya-1 and probably Molniya-3 communications satellites. It has a single fixed combustion chamber.
Applications: Mars 1, Zond 1-3, Venera 2-8, Molniya 1
First flight: Nov 1962 (Mars 1, but earlier, failed, missions likely)
Oxidiser: nitric acid
Fuel: UDMH
Feed method: pressure fed
Thrust: 1.96 kN vacuum
Specific impulse: 271 s vacuum
Chamber pressure: 12 atm
Burn time: 40 s in at least two burns

KRD-61
Apparently designed specifically for the 2.708 km/s ΔV final leg of lunar sample return missions, achieving three successes 1970-1974. Single fixed main chamber with four large verniers providing pitch/yaw control and four small verniers for roll. Single ignition.
Applications: third generation lunar sample return missions (Luna 16/20/24 successful)
First flight: 14 Apr 1969 (failed Luna attempt)
Oxidiser: nitrous oxide
Fuel: UDMH
Feed method: turbo pump
Thrust: 18.8 kN vacuum
Specific impulse: 313 s vacuum
Chamber pressure: 95 atm
Burn time: 53 s

KRD-79
No details are known about this engine other than it is an engine used on Mir core module: it is not known whether it is the main engine or one of the attitude control thrusters.

KRD-442
No details are known about this engine other than it is an engine used on Mir core module: it is not known whether it is the main engine or one of the attitude control thrusters.

KTDU-5A
Retromotor for semi-soft landing or braking into orbit of second generation Luna craft. Single fixed chamber and four peripheral verniers fired for about 43 s to cancel the 2.6 km/s direct arrival speed from Earth. The first successful lunar landing (Luna 9, Feb 1966) and lunar orbiter (Luna 10, Apr 1966) were made possible by this engine.
Applications: spacecraft single burn retro-motor
First flight: lunar landing attempt of Jan 1963?
Oxidiser: nitric acid
Fuel: amine-based
Feed method: pressure fed
Thrust: 45.5 kN vacuum
Specific impulse: 278 s vacuum
Chamber pressure: 64 atm
Burn time: 43 s in two burns

KTDU-35/-53/-66
Principal application of the KTDU-35 system was the orbital manoeuvring and de-orbit system of the original manned Soyuz (1-40) spacecraft. A single multi-start chamber was fed with propellants by a turbopump. Retrofire required a burn of typically 150 s. >4,000 ground/space ignition tests were conducted to qualify the engine. There were also twin nozzles either side providing a single start 4.03 kN backup capability; this was called on during Soyuz 33 when the primary system failed. The **KDTU-53** was essentially the same system used for unmanned Zond lunar round trips but without the backup chambers and with reduced propellant supply. It was further modified as the Salyut 1 space station **KDTU-66** orbital manoeuvring system.
Applications: primary Soyuz/Zond propulsion, Salyut 1 space station orbital adjustment
First flown: Nov 1966 (Soyuz unmanned test)
Engine cycle: open
Oxidiser: nitric acid
Fuel: UDMH
Feed method: turbopump

Soyuz manned spacecraft KTDU-35 propulsion system consists of a single primary central chamber and two reserve chambers. The four smaller associated chambers are part of the separate hydrogen peroxide attitude control network

Thrust: 4.09 kN vacuum primary single nozzle; 4.03 kN vacuum backup twin nozzles
Specific impulse: 281 s vacuum prime, 270 s vacuum backup
Chamber pressure: 40 atm primary
Burn time: up to 1,000 s in multiple burns
Few details are known of the Soyuz-T propulsion system and almost nothing of the current Soyuz-TM model. A switch was made to hypergolic UDMH/N_2O_4 and the attitude control thrusters were integrated with the larger system, providing a second de-orbit backup capability. Thrust of the primary chamber is 3.1 kN vacuum. A space station version was introduced on Salyut 6 for AOCS, with two primary chambers. The current Mir station appears to incorporate a similar system, with a thrust of 5.9 kN.

KTDU-80
No details are known about this engine other than it is the main propulsion system for the Soyuz-TM and Progress-M spacecraft.

KTDU-417
Introduced for the soft-landing stage of the third generation Luna craft (starting in 1969). In this role the engine included a single 11-start main chamber for the greater part of the descent, followed by twin low-thrust chambers for the final approach and landing. Both were variable thrust. The S5.92 engine (qv) appears to be a modernised version of the KTDU-417. (This engine has also been described in Russian literature using the designator KRD-417: it is unclear which is the correct designator.)
Applications: lunar landing and planetary propulsion stage?
First flight: 14 Apr 1969 (failed Luna)
Oxidiser: nitric acid
Fuel: UDMH
Feed method: pressure fed
Thrust: 7.35-18.92 kN vacuum variable prime, 2.06-3.43 kN vacuum variable secondary
Specific impulse: 308-314 s vacuum primary, 249-254 s vacuum secondary
Burn time: 650 s primary, 30 s secondary

KTDU-425A
The KTDU-425 was used as the main propulsion system for the Mars 2/3 missions in 1971: presumably it was also used for the failed Mars probe Cosmos 419 (an intended Mars orbiter) and the two Mars probes launched and lost in 1969. In 1973 it was replaced by the KTDU-425A, first launched on Mars 4. It is the later version which has been described in Russian literature. A single chamber, pump-fed and gimballed.
Applications: Mars 4-7, Venera 9-16, Vega 1-2, Phobos 1-2
First flight: Jul 1973 (Mars 4)
Oxidiser: nitrous oxide
Fuel: amine-based
Feed method: pump
Thrust: 9.86-18.89 kN vacuum variable
Specific impulse: 293-315 s vacuum
Chamber pressure: 95 atm
Burn time: 560 s in up to seven burns

KVD-1/(11)D56
This cryogenic engine was originally developed for the Block R of the abandoned N-1 launch vehicle which would have flown the dual launch L-3M manned lunar missions in the 1970s: the N-1 programme was cancelled in 1974. When the engine was finally revealed in the 1990s it was identified as the KVD-1 or D-56, but clearly it is the same engine as the 11D56 developed in the 1970s. The engine is now to be used on the Proton-M launch vehicle's planned cryogenic fourth stage and it has been sold to India to be used as an upper stage engine on the GSLV.
Applications: upper stages (Proton, Angara, GSLV, possibly Zenit)
First flight: planned 1998 on India's GSLV
Dry mass: 292 kg

The KVD-1 cryogenic engine will fly on India's GSLV

Length: 2.146 m
Diameter: 1.28 m
Engine cycle: closed staged combustion
Oxidiser: liquid oxygen
Fuel: liquid hydrogen
Feed method: turbopump
Thrust: 69.6 kN vacuum; 73.58 kN with two gimballed LOX/GH_2 verniers fed by main chamber pumps; 76 kN quoted for GSLV
Specific impulse: 463 s vacuum
Expansion ratio: unknown
Chamber pressure: 58.1 atm
Burn time: 24,000 s in 6 starts (450 s for Proton)

R2.2000

This engine appears to be a replacement for the KTDU-425A (qv). It can operate in either low or high thrust modes, and thus corresponding low/high thrust mode parameters are shown for some data.
Applications: Mars 8 and Phobos spacecraft main engines
First flown: 1988
Dry mass: 74 kg
Length: 1.025 m
Maximum diameter: 0.920 m
Oxidiser: nitrogen tetroxide
Fuel: UDMH

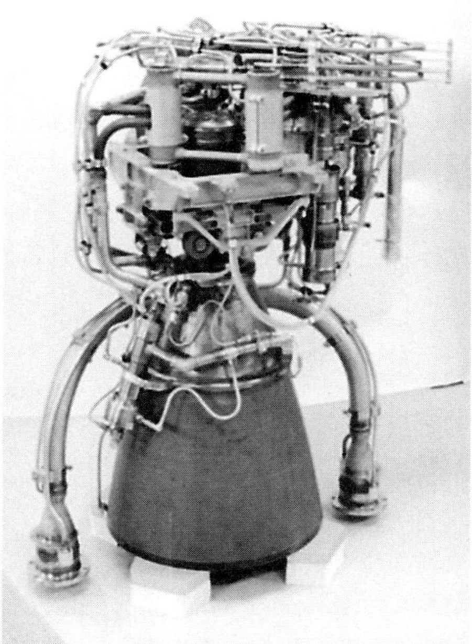

S5.92 engine used as the main propulsion system on the Fregat module used on the Phobos and Mars 8 missions: the same engine might be used on the Breeze third stage of the Rockot launch vehicle. (KB Khimmash) ***1997**/0003429*

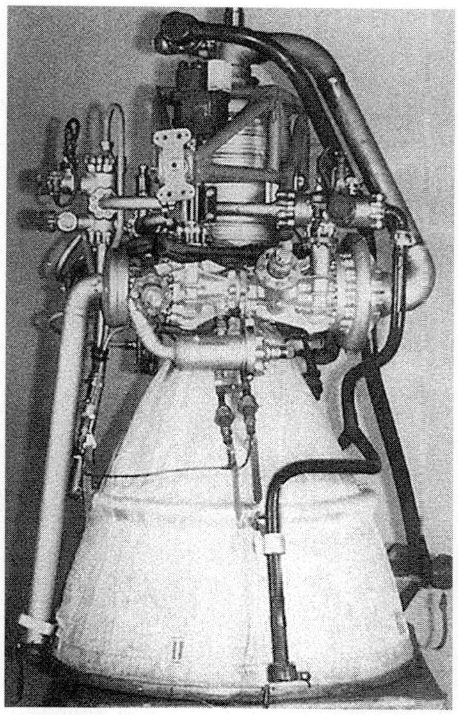

Molniya's Block L engine ***1996***

Mixture ratio (O/F): 2.0/1.95
Thrust: 13.725 kN/19.613 kN
Specific impulse: 316 s/325 s
Burn time: 1,000 s (max 25 burns)

S5.92

This engine has not been described in great detail but could be a modification of the KTDU-417 (qv). It has been stated that it was used on the Phobos 1 and Phobos 2 missions for launching from Earth orbit, course corrections and entry into Mars orbit, and therefore it is clearly the engine used on the Fregat module – also carried by the ill-fated Mars 8 (Mars-96) mission in 1996. It is also possible that this engine (or a modification) is used on the Breeze (Briz) third stage of the Rockot launch vehicle.
Applications: Phobos 1/2, Mars 8 main propulsion modules, Fregat, possibly Breeze on Rockot
First flown: 1988
Dry mass: not known
Length: not known
Maximum diameter: not known
Oxidiser: nitrogen tetroxide
Fuel: UDMH
Mixture ratio (O/F): not known
Feed method: not known
Thrust: 20 kN
Specific impulse: 327 s
Expansion ratio: not known
Chamber pressure: not known
Burn time: 2,000 s (max of 50 burns)

KOROLEV DESIGN BUREAU

Korolev's bureau developed Proton's Block D/DM and Molniya's Block L engines. Mikhail Melnikov designed the Proton engine.

RD-58M/11D58M: Proton stage 4

The engine provides a direct GEO and planetary trajectory injection capability. The stage was originally developed for Korolev's N1 lunar vehicle and then transferred to Proton. There are two versions flying: the standard kerosene version and one using 'sintin' for higher Isp. Sintin is known only as a synthetic hydrocarbon-based fuel. The same engine might power Zenit's stage 3 and possibly Energia's Retro & Correction Stage. Buran utilised two **17D12** sintin versions as its main orbital manoeuvring engines. The configuration was changed only slightly (see photograph); the performance and overall dimensions are as given here for the sintin 11D58M.
Applications: Proton stage 4, Zenit stage 3?, Buran OMS
First flown: 1967
Dry mass: 303 kg
Length: 2.268 m
Diameter: 1.167 m
Oxidiser: liquid oxygen
Fuel: kerosene or sintin
Thrust (vac): 85 kN kerosene; 83.5 kN sintin
Specific impulse (s, vac): 352 kerosene, 362 sintin
Chamber pressure: 76.4 atm kerosene, 78.4 sintin
Burn time: 600 s in multiple burns

C1-5400/11D33: Molniya final stage

Little is known of the propulsion system on the Molniya Block L final stage but it is a fixed single chamber. It was also used as the injection stage for deep space missions until Proton was introduced.
Applications: Molniya stage 3 Block L
First flown: Oct 1960? (as planetary probe escape stage)
Oxidiser: liquid oxygen
Fuel: kerosene
Thrust: 67 kN vacuum
Specific impulse: 340 s vacuum
Burn time: about 200 s

The 11D58M (left) powers Proton's stage 4; its 17D12 modification was used on Buran (Andrew Wilson) ***1996***

N D KUZNETSOV COMPANY SSC

Nikolai Dmitrievich Kuznetsov (24 Jun 1911-30 Jul 1995) established an aircraft engine design and development bureau during the 1940s in Kuibyschev (now Samara). Information finally released in 1989 revealed that the bureau had been assigned the responsibility for the design and development for engines for Korolyov's four-stage N-1 launch vehicle. Previously the bureau had developed engines planned for use on Korolyov's R-9 missile, but politics ensured that the missile used other engines.

In Jul 1993 Aerojet signed a teaming agreement with N D Kuznetsov SSC to market these engines within the United States. Two NK-33 engines were delivered to Aerojet in 1995. One of them was used by Aerojet for five firings which totalled 410 s operating time within a 57-114% thrust range which followed the proposed Atlas-2AR launch profile. The engines were under consideration to power the first stage of the Atlas-2AR, but they lost out to the Energomash RD-180 engine.

The United States Kistler Aerospace Corporation is planning to use NK-33 engines on its reusable K-1 launch vehicles: three would be on the first stage and one on the second stage. See the Kistler entry in the United States Launcher section for further details.

NK-9 and NK-19

These two engines were proposed for use on the Korolyov bureau's R-9 missile: the first stage would have

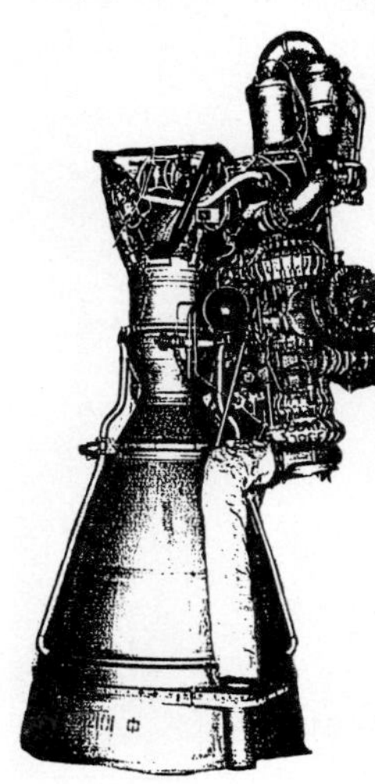

A cluster of 30 of Kuznetsov's 1,510 kN engines powered the lunar N1 stage 1 (NPO Trud) ***1993***

The NK-31 is a modification of the NK-39 and one was carried on the N-1 fourth stage which would have performed trans-lunar injection. (N D Kuznetsov Company SSC) ***1997***/0003423

used four NK-9 engines and the second stage a single NK-19. After political considerations became dominant, the missile used the GDL's RD-111 on the first stage and Kosberg's RD-0106 on the second stage. The N-1's NK-39 and NK-31 engines were developed from the NK-19.

NK-31

Application: N-1 Block G fourth stage (1 engine)
First flown: 20 Feb 1969
Dry mass: 722 kg
Wet mass: 798 kg
Maximum diameter: 1.8 m
Engine cycle: closed-loop (similar to NK-33)
Oxidiser: liquid oxygen
Fuel: kerosene
Feed method: single-shaft turbopump, 26,150 rpm
Thrust: 402 kN vacuum
Specific impulse: 353 s vacuum
Burn time: 1,140 s (530 s for N-1)

NK-33

It is expected that Aerojet modifications will add about 136 kg to the masses quoted below.
Applications: N-1 Block A first stage (30 engines) Kistler K-1 first and second stages
First testbench firing: 12 Feb 1969

Aerojet offers modified NK-33s for future US launch vehicles (Aerojet) ***1996***

The NK-43 engine was a modification of the NK-33, used on the second stage of the N-1 launch vehicle. (N D Kuznetsov Company SSC) ***1997***/0003424

First flown: 20 Feb 1969
Dry mass: 1,236 kg
Wet mass: 1,393 kg
Maximum diameter: 1.5 m
Engine cycle: closed-loop with staged kerosene combustion in oxygen
Oxidiser: liquid oxygen
Fuel: kerosene
Propellant flow rate: 500 kg/s
Mixture ratio (O/F): 2.547
Feed method: two-shaft turbopump, 17,300 rpm
Thrust: 1,510 kN sea level
Specific impulse: 331 s vacuum 297 s sea level
Burn time: 600 s (120-150 s for N-1)

NK-43

Application: N-1 Block B second stage (8 engines)
First testbench firing: 29 Jul 1970
First flown: 20 Feb 1969
Dry mass: 1,359 kg
Wet mass: 1,542 kg
Maximum diameter: 2.5 m
Engine cycle: closed-loop with staged kerosene combustion in oxygen
Oxidiser: liquid oxygen
Fuel: kerosene
Propellant flow rate: 500 kg/s
Mixture ratio (O/F): 2.547
Feed method: two-shaft turbopump, 17,300 rpm
Thrust: 1,757 kN vacuum
Specific impulse: 346 s vac

NK-39

Application: N-1 Block V third stage (4 engines)
First flown: 20 Feb 1969
Dry mass: 584 kg
Wet mass: 635 kg
Maximum diameter: 1.8 m
Engine cycle: closed-loop (similar to NK-33)
Oxidiser: liquid oxygen
Fuel: kerosene
Feed method: single-shaft turbopump, 26,150 rpm
Thrust: 402 kN vacuum
Specific impulse: 353 s vac
Burn time: 1,140 s (380 s for N-1)

NII MACHINOSTROENYE

R&DIME Research & Development Institute of Mechanical Engineering was established in 1958 as part of NII Thermal Processes, but became independent in the early 1980s. Development and production of 0.8-400 N spacecraft thrusters (implying connection with old Isayev bureau), including Soyuz, Progress and Mir. Development and small batch production of engines up to 6 kN can be accommodated. Test facilities can handle hydrogen (up to 3 kN), methane and kerosene fuels. The Elektro Earth observation satellite carries 16 DEN-16 electrothermal thrusters. For the thrusters specified below, LTRE = Low Thrust Rocket Engine, or RDMT in Russian (Raketnay Dvigatel Maloi Tyagi).

Four NK-39 engines were carried on the third stage of the N-1 which would have performed Earth orbit injection. (N D Kuznetsov Company SSC) ***1997***/0003422

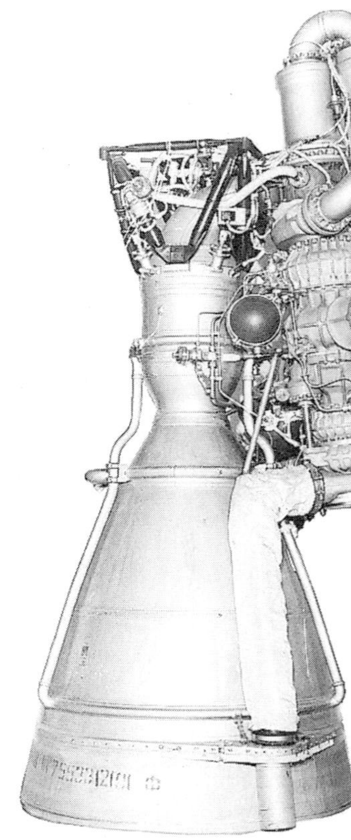

N D Kuznetsov with his NK-33 engine intended for the first stage of the N-1 launch vehicle (bottom) and a full length view of the NK-33 (top): thirty of these engines would have been clustered on the first stage of the N-1. It is now planned to use these engines on the Kistler K-1 launch vehicle. (N D Kuznetsov Company SSC) ***1997***/0003425/0003430

LTRE-400

This 400 N man-rated engine is employed in clusters aboard Mir's Kvant, Kristall, Spektr and Priroda modules. It was developed for the manned Almaz station and the Heavy Cosmos transport craft. It was test-fired by Aerojet in the US in 1994 using MMH, attaining 293 s SI. A redesigned version is estimated to be capable of 310-320 s SI.
Dry mass: 2.5 kg
Length: not available
Max diameter: 15 cm exit ∅
Oxidiser: NTO
Fuel: UDMH
Propellant feed method: pressure
Inlet pressure: not available
Valves: 2 A at 27 V (20-34 V supply)
Thrust: 400 N vacuum
Specific impulse: 255 s vacuum nominal (after 30 s)
Response time: 30 ms to 90% thrust (30 ms decay to 10%)
Mixture ratio (O/F): 1.9:1 (fuel rich); 30-40% of NTO flow used for film cooling. 1,038°C limit; 93°C nominal throat
Expansion ratio: 56:1
Expansion bell: stainless steel, NTO film-cooled
Chamber pressure: 6.5 atm
Chamber materials: stainless steel, NTO film-cooled

LTRE-12

This 12 N thruster is employed in clusters aboard Mir's Kvant, Kristall, Spektr and Priroda modules.
Dry mass: 0.55 kg
Oxidiser: NTO
Fuel: UDMH
Propellant feed method: pressure
Valves: 0.2 A at 27 V (20-34 V supply)
Thrust: 12 N vacuum
Specific impulse: 279 s vacuum nominal
Response time: 15 ms to 90% thrust (10 ms decay to 10%)

LTRE-50

Application: unmanned satellite
Dry mass: 1.0 kg
Oxidiser: NTO
Fuel: UDMH
Propellant feed method: pressure
Valves: 1 A at 27 V (20-34 V supply)
Thrust: 50 N vacuum
Specific impulse: 255 s vacuum nominal
Response time: 30 ms to 90% thrust (30 ms decay to 10%)

LTRE-100/135

This man-rated engine is employed in clusters aboard Mir, Soyuz-T/TM, Progress and unmanned satellites in 98 N or 137 N versions.
Dry mass: 1.2 kg
Oxidiser: NTO
Fuel: UDMH
Propellant feed method: pressure
Valves: 1 A at 27 V (20-34 V supply)
Thrust: 98 N or 137 N vacuum
Specific impulse: 260 s vacuum nominal
Response time: 30 ms to 90% thrust (30 ms decay to 10%)

LTRE-200

Application: Almaz module
Dry mass: 2.0 kg
Oxidiser: NTO
Fuel: UDMH
Propellant feed method: pressure
Valves: 2 A at 27 V (20-34 V supply)
Thrust: 200 N vacuum
Specific impulse: 255 s vacuum nominal
Response time: 30 ms to 90% thrust (30 ms decay to 10%)

LTRE-200K

Application: Buran AOCS
Dry mass: 5.5 kg
Oxidiser: GOX
Fuel: kerosene
Valves: 2.5 A at 27 V (20-34 V supply)
Thrust: 200 N vacuum
Specific impulse: 265 s vacuum nominal
Response time: 50 ms to 90% thrust (60 ms decay to 10%)

NII THERMAL PROCESSES

NII TP performs research in all forms of launcher and space propulsion, including ramjet, electric and nuclear. BMDO took delivery in Oct 1993 of a T-100 1.5 kW Hall-type Stationary Plasma Thruster for evaluation. N_2O_4/UDMH thrusters include 10 N and 1,000 N models. NII TP developed the Phobos spacecraft's network of hydrazine thrusters for attitude control and minor orbital modifications. 28 (24 × 50 N, 4 × 10 N) were mounted on four spherical tanks. A set of 0.5 N hydrazine microthrusters of an unusual design was also included to avoid the need for separate N_2 cold gas jets; the propellant was burned internally and cooled inside an expansion tank before passing to the thrusters.

NPO SATURN/LYULKA

The cryogenic engine which has been described in the 1990s as the D-57 would appear to be the 11D57 gimballing engine which was developed during the early 1970s for use on the Block S of the modified N-1 launch vehicle which would have been flown fo r the dual-launch L-3M manned lunar missions. The bureau was also tasked to develop the 11D54 fixed-chamber cryogenic engine with the same thrust level of 40 t (390 kN). This engine would have been used in a cluster of 6-8 on a modified Block V of the N-1 vehicle.

It has been claimed that the D-57 (ie, 11D57) is the only rocket engine developed by the bureau, suggesting that the 11D54 either had development problems and was abandoned or that the N-1/L-3M programme was simply scrapped before the engine's development was completed.

Development of the engine now called D-57 is said to have started in 1960 (which is too early for the N-1/L-3M programme: perhaps it was designed in the hope of finding a later application?) and work ended in 1977. When production of the engine ended in 1975 – a year after the L-3M programme was terminated – 105 of the engines had been built with more than 53,000 s of testing having been completed in 470 runs, including a single burn of 5,400 s and 22 starts. Only four complete unused engines remain, two used engines (one with the 2-position nozzle), 33 thrust chambers/skirts and other spares. Aerojet teamed with Lyulka in 1993 to improve and market the engine. NASA has studied employing it on an SSTO demonstrator; for example, 1×NK-33 + 2×D-57.
Application: N1-L3M stages 3/4
First flown: -
Dry mass: 840 kg
Length: 3,660 mm
Maximum diameter: 1,860 mm
Mounting: ±2.5° 2-axis gimbal (powered by H_2)
Engine cycle: staged combustion
Oxidiser: liquid oxygen
Fuel: liquid hydrogen
Mixture ratio (O/F): 5.8:1
Feed method: main/boost ungeared pump turbines on separate single shafts are powered by 540°C fuel-rich exhaust; pump outputs 254 atm LOX and 170 atm LH_2. Most of exhaust is routed into main chamber; some provides roll control. GOX injection from chamber LOX cooling
Thrust: 400 kN vacuum; throttle down to 10%
Expansion ratio: 143:1; also version 2-position nozzle providing 170:1 and 88:1 (throat ∅ not available, exit ∅ 1,860 mm)
Specific impulse: 456.5 s vacuum (461 for 170:1 nozzle)
Thrust chamber
materials: non-tubular stainless steel
cooling: hydrogen

Lyulka's only rocket engine is the LOX/LH_2 D-57, developed for the improved N1 manned lunar launcher (SEP

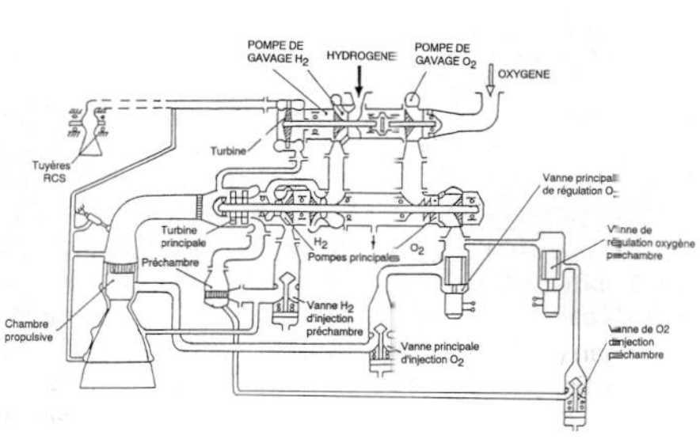

D-57 propellant flow schematic (SEP)

Combustion chamber
pressure: 109 atm (100% thrust)
temperature: not available
materials: milled copper liner
cooling: LOX cooled, providing GOX for injection
ignition: not available
Burn time: 800 s

NPO YUZHNOYE

Yuzhnoye produces verniers for some of its own vehicles (Zenit stage 2 and Tsyklon stages 1-3) and the 11D25 main engine for Tsyklon's stage 3.

RD-800 Series Engine

The engine's four nozzles gimbal ±33° for vehicle steering.
First flown: Apr 1985 (Zenit 2 stage 2)
Cycle: closed
Dry mass: 400 kg (418 kg wet)
Length: 1,000 mm
Maximum diameter: 430 mm
Oxidiser: oxygen
Fuel: kerosene
Thrust: 78.4 kN total four nozzles
Mixture ratio: 2.40:1
Specific impulse: 342.0 s vac
Expansion ratio: 103.7:1 main
Chamber pressure: 75.5 atm main
Burn time: 3,300 s (1,100 s nominal)

RD-861

Each engine includes 6 × 20 N + 2 × 98 N fixed thrusters for steering during the fixed main chamber's burn.
Application: Tsyklon third stage
First flown: Jun 1977
Cycle: open, with pump turbine exhaust routed to steering thrusters
Dry mass: 123 kg (138 kg wet)
Length: 1,555 mm
Maximum diameter: 795 mm
Oxidiser: nitrogen tetroxide
Fuel: UDMH
Thrust: 78.90 kN main vacuum
Mixture ratio: 2.1:1 main, 1.6:1 steering
Specific impulse: 317 s main vac; 230 s steering vac
Expansion ratio: 111.2:1 main
Chamber pressure: 90.5 atm main
Burn time: 118 s

OKB FAKEL

Fakel specialises in spacecraft attitude control thrusters, ion engines and plasma sources. Its SPT-70 and SPT-100 (Stationary Plasma Thruster) Hall electric thrusters have flown on >50 Meteor polar metsats since 1972 to provide orbit control. The numerical designators indicate beam ∅ in mm. SPTs employ a DC-gas discharge in an annular chamber, in which a radial magnetic field traps the electrons. These Hall currents ionise the Xe propellant, the ions of which are accelerated inside a quasi-neutral plasma, without grids, and by the discharge voltage itself. The advantages are their rugged simplicity (no grid system or high voltage supply), but they provide lower efficiencies, lower exhaust velocities and increased Xe consumption than ion thrusters such as the UK-10.

The 0.08 N SPT-100 has been sold to Space Systems/Loral for commercial applications. Thrust range 0.02-0.2 N (nominally 84 mN), SI 1,600 s, 3.2 kg, 1,350 W, 4,000 h life, 48% thrust efficiency. The International Space Technology, Inc (qv) joint venture with SS/L, SEP, ARC and Moscow's RIAME Research Inst of Applied Mechanics & Electrodynamics holds exclusive marketing and distribution rights outside of Russia. ISTI now offers complete propulsion systems. Gals 1, launched Jan 1994, is carrying a set of eight to provide the first Russian NSSK in GEO. SEP is developing a Mk 2 version that will increase SI to 2,200 s. Development began in May 1993 at SEP Villaroche. RIAME/Fakel are investigating SPT thrusters up to 90 cm ∅ and power levels up to 100 kW for primary propulsion. Krypton is a possible propellant.

The SPT-70 offers: thrust 40 mN, SI 1,600 s, 1.5 kg, 640 W, 9,000 h life, 50% thrust efficiency. An improved SPT-50 is under development for the small satellite market, drawing 300 W. Fakel has developed the SPT-140, SPT-200 and SPT-290 for primary propulsion. The specifications below are given in 140/200/290 order:
Diameter: 14/20/29 cm beam exit; 18/26/37 cm thruster
Length: 10/20/23 cm
Mass: 6/12/18 kg
Thrust: 200/500/1,000 mN for 200-600 V discharge voltage
Specific impulse: 1,530-3,160 s
Acceleration voltage: 0.2-0.6 kV discharge
Discharge power: 3-6/6-12/12-25 kW
Power input: 4.5/8.0/18.0 kW
Specific power consumption: 22.5/16.0/18.0 kW/N
Specific propellant consumption: 52.0/51.2/ 50.8 mg/Ns

Total efficiency: -/50-71/50-70%
Lifetime: 6,000/8,000/12,000 h

Plasma Thrusters

In addition to OKB Fakel's SPT-100 (see above), the US Ballistic Missile Defense Organization is evaluating several other Russian plasma thrusters. For example, BMDO took delivery in Oct 1993 of NII TP's T-100 1.5k W Hall-type SPT. See the BMDO entry in the Military section for further details.

FRANCE

SOCIÉTÉ EUROPÉENNE de PROPULSION

The principal space business of SEP, a subsidiary of SNECMA, is the design, development and construction of propulsion systems for launch vehicles and satellites, including Ariane's Viking, HM-7, Vulcain and P230 (partnered with BPD in Europropulsion, qv) motors. In satellite work, SEP produces Mage apogee kick motors, hydrazine thrusters for attitude control, bipropellant liquid engines and electric/ion engines. The company also produces solar array drive mechanisms, μg furnaces, surface tension tanks, magnetic bearings, transducers and flow regulators. In 1995, personnel totalled 3,500, with a turnover of Fr5,400 million generating a Fr150 million profit. 1994: 3,550/4,800/133.5 million profit; 1993: 3,500/4,250/110; 1992: 3,600/4,355/140; 1991: 3,800/4,540/87.2; 1990: 4,050/4,447/103, respectively. Aerospatiale in Jan 1994 sold its 13.6% interest for FFr178.9 million. SEP and Spain's Empresarios Agrupados established **Iberespacio** (qv) in Oct 1989 to undertake joint development of space propulsion systems. SEP and its SNECMA parent established **Hyperspace** in Jan 1990 to co-ordinate propulsion activities for hypersonic aircraft and launchers. Other interests include: Arianespace 8%, Europropulsion (50/50 with BPD), G2P 75% (large solid propulsion systems for military applications), Spot Image 12%, Carbone-Industrie 100% (carbon brakes), TECHLAM 50% (laminated elastomeric components) and S2M 54% (magnetic bearings).

Test/Assembly Facilities: SEP maintains a 4,500 m² Ariane propulsion assembly hall containing 14 Viking engine assembly stands, five for HM-7 and nine for stage 1 propulsion systems. There is also a 400 m² class 100,000 cleanroom for cryogenic engine component assembly. TDF propulsion was installed in an integration building provided with two 300 m² class 100,000 rooms. The Vulcain Engine Assembly Building can produce eight engines annually. The 1,220 m² class 100,000 cleanroom accommodates 12 assembly cells and 25 class 100 laminar flow booths/hoods for storage before assembly. Also in the building are a 700 m² warehouse, a 1,400 m² engineering & mechanical area and 1,150 m² of office space.

Engine test facilities are concentrated at the 1.5k m² Vernon site but some are located at the 24,300 m² Villaroche area. For storable propellant motors, there are four component test stands (PF1-3, A48, F22) and one engine stand (PF2). Cryogenic testing is undertaken on three stands: the two of Vernon's PF41 (one with altitude simulation) and one at Villaroche (with altitude simulation). Eleven existing cryogenic component stands were joined in 1988 by PF52 for Ariane 5 Vulcain gas generator and turbopump development. The PF50 stand for complete Vulcain engine tests was inaugurated Sep 1990 at Vernon. SEP inaugurated its 11,000 m² production plant for Ariane 5's P230 nozzles in Oct 1990 at Haillan. A FFr180 million investment, it will fabricate 10 nozzles annually by year 2000. Assembly space is 650 m², plus 825 m² class 100,000 cleanrooms.

SPT 100 Mk 2

SEP joined the International Space Technology Inc partnership (qv) in 1993. ISTI is marketing OKB Fakel's SPT-100 0.08 N Hall xenon stationary plasma thruster using western electronics. France's Stentor satellite will carry a slightly modified version. SEP began development in May 1993 of a Mk 2 version at Villaroche to increase SI from 1,600 s to 2,200 s. See the ISTI and OKB Fakel entries for further details.

Pyrospace pyromechanisms for space applications. It provides more than 230 for each Ariane 4 ***1995***

Ariane Viking Engines

SEP provides four first stage, one second stage and the required number of liquid strap-on engines for the Ariane launcher (the cryogenic HM-7B is specified separately below). India's Vikas engine is derived from Viking technology. The design of the Viking engine was based on the M40 400 kN feasibility demonstrator. The M55, first fired 8 Apr 1971, was renamed Viking in the tradition of using names beginning with 'V' (for Vernon). At that time it was intended to power ELDO's Europa III, cancelled in 1972. Full development of Viking 2 began 1973 for Ariane; replacement of the conical nozzle with a bell form in 1976 created the Viking 5 and increased sea level thrust from 588.5 kN to 623.7 kN. Three qualification firings were successful in 1979. Viking 3 was a ground test version of the Viking 4 altitude variant of Viking 5. By end-1994, Viking engines had logged 110,400 s of operation, including 68,100 s in flight and 1,261 start-up sequences; a total of 680 units had been fabricated. Almost 800 will have been produced by the end of its career when Ariane 4 is phased out. The initial Ariane 1 vehicle was powered by Viking 5A (611 kN thrust each) and Viking 4A (721 kN); Ariane 2/3 switched to UH25 fuel (75% UDMH/25% hydrazine hydrate) from UDMH and carried the more powerful Viking 5B (752 kN) and Viking 4B (805 kN) models. The current Ariane 4 launcher retains the second stage Viking 4B, but its first stage Viking 5C's firing time is extended by 72 s and up to four Viking 6 engines, derived from Viking 5, can be carried as strap-ons.

Viking 5C & 6 specifications

Applications: Ariane 4 first stage propulsion (four Viking 5C), Ariane 4 PAL liquid strap-on propulsion (single Viking 6, specifications as for Viking 5C but with different layout for PAL accommodation)
First flown: 15 Jun 1988 on Ariane V22-401
Number flown: 212 × Viking 5C, 106 × Viking 6, to end-1995 (delivered: 452 Viking 5 + 123 Viking 6)
Dry mass: 826 kg
Length: 287.3 cm
Maximum diameter: 99.0 cm
Mounting: gimballed for pitch/yaw/roll control
Engine cycle: open gas generator
Oxidiser: NTO at 173.3 kg/s (Viking 6 173.7 kg/s)
Fuel: UH25 at 101.9 kg/s (Viking 6 101.5 kg/s)
Mixture ratio (O/F): 1.70 (Viking 6 1.71)
Turbopump: 10,000 rpm, 2,500 kW power rating. Single-shaft turbopump driven by gas generator consuming 1.2 kg/s propellants. A third pump adds water to limit turbine operating T. A main flow control unit maintains thrust at a pre-determined reference level by adjusting the relative flows of the three liquids entering the gas generator. Another control loop adjusts the thrust chamber pressure by controlling gas production and regulating the turbopump
Thrust: 752 kN vacuum, 678 kN sea level
Specific impulse: 278.4 s vacuum
Expansion ratio: 10.5
Nozzle materials: cobalt alloy with SEPHEN (phenolic resin/silica fibre) throat
Combustion chamber
pressure: 58 atm
temperature: about 3,000°C
injector: 216 like on like doublets set in six rows for each propellant in light alloy annular injector
cooling: UDMH propellant film supplied through additional channels on the injector's lower section
ignition: hypergolic
Burn time: 209 s Viking 5B, 143 s Viking 6

746 Viking engines of all types have been produced through 1995 (SEP)

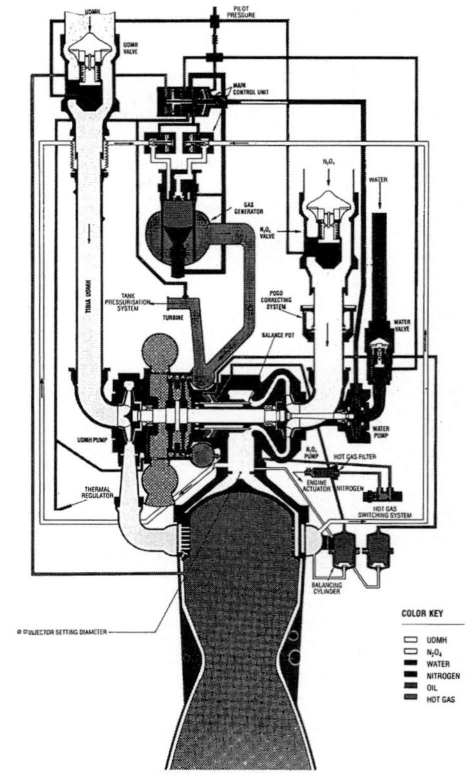

Viking 5 is clustered to power Ariane's first stage

Viking 4B specifications

Ariane's second stage is powered by a single derivative of the Viking 5 engine, optimised for operation at altitude. Gimbal mounting provides pitch/yaw control and roll control is exercised by auxiliary jets feeding from the main engine's gas generator.
Applications: Ariane 2-4 stage 2 engine
First flown: 4 Aug 1984 on Ariane V10
Number flown: 75 to end-1995 (118 delivered)
Dry mass: 826 kg
Length: 350.9 cm
Maximum diameter: 170.0 cm
Mounting: gimballed for pitch/yaw control
Oxidiser: nitrogen tetroxide at 175.0 kg/s
Fuel: UH25 at 103.0 kg/s
Mixture ratio (O/F): 1.70
Turbopump: as Viking 5B/6
Thrust: 805 kN vacuum
Specific impulse: 295.5 s
Expansion ratio: 30.8:1
Chamber pressure: 58.5 atm
Burn time: 125 s

Ariane HM-7B engine

Ariane 4's stage 3 cryogenic engine was developed by SEP from experience with the 40 kN HM-4, tested 1962-69. The HM-7 model was introduced on Ariane 1 in 1979 but subsequent launcher versions have utilised the HM-7B, offering 265 s longer firing duration. Recent enhancments to Ariane's stage 3 have required engine modifications: values are given below for the H10, H10+ (first flown V50, 1992) and H10-3 (debut V70, Dec 1994) stages. Ariane V15 (Sep 1985) and V18 (May 1986) failed because of stage 3 engine ignition

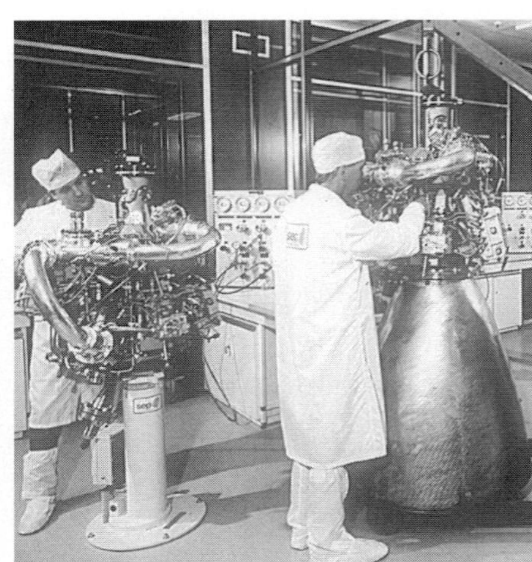

The HM-7 is Europe's first liquid hydrogen/oxygen engine. The 63.8 kN version debuted in Dec 1994. Future developments could include a composite nozzle (SEP) ***1995***

problems. The pyrotechnic igniter was redesigned to expel its hot gases at 45° instead of the previous vertically downwards parallel to the propellant flow. SEP conducted extensive ignition trials with the new design, concluding with nine firings of one engine totalling 3,130 s, including three long-duration runs of about 900 s. Ariane V63 (Jan 1994) was lost because of abnormal heating of the immersed LOX pump bearing. Insufficient cooling and aggravating factors resulted in failure. The solution was installation of a purge line in the bearing cavity to enhance cooling and helium purging, and fitting the bearing with a self-lubricating MoS_2 coating to decrease its sensitivity to possible aggravating factors. V70 failed because stage 3's gas generator was starved of LOX. The engine stabilised at 70% and shut down 40 s early. As a result, a filter was added to the oxygen line upstream of the generator to prevent blockage in future and more stringent contamination prevention measures introduced. By end-1995, HM-7B had made 898 firings totalling 180,225 s (48,240 s in-flight). Firing trials of an HM-7B equipped with a Novoltex-reinforced SEPCarbinox 25 kg composite nozzle were made in 1989 as part of a technology programme. The nozzle does not require active cooling to cope with the 1,800°C, allowing more efficient use of the liquid hydrogen. An **HM-7S** featuring a deployable nozzle has also been considered for a future Ariane 5 upper stage: 62.2 kN, SI 447 s, 4.56 mixture ratio.

Applications: Ariane stage 3 propulsion system
First flown: 4 Aug 1984 on Ariane V10
Number flown: 79 HM-7/7B to end-1995, including five failures (11 HM-7 flown, one failure)
Dry mass: 155 kg
Length: 201.3 cm
Maximum diameter: 99.2 cm
Mounting: gimballed for pitch/yaw control
Oxidiser: liquid oxygen
Fuel: liquid hydrogen
Mixture ratio (O/F): 4.76:1 (gas generator 0.87) for H10 stage, 4.66 H10+, 5.04 H10-3
Turbopump: 60,500 rpm, 380 kW single turbine powered by gas generator requiring 0.25 kg/s propellants. Single-stage hydrogen pump raises pressure from 3 atm to 55 atm; single-stage oxygen pump 2.5-50 atm. Gas generator exhaust temperature 800-900K
Thrust (vac): 62.3 kN H10/H10+, 63.8 kN H10-3
SI (s, vac): 446 H10, 446.1 H10+, 445 H10-3
Expansion ratio: 83.1:1
Cooling method: hydrogen is heated to 150K during cooling of combustion chamber wall along 128 longitudinal channels for injection in gaseous form. The nozzle cooling system employs 242 helical tubes; 0.15 kg/s of hydrogen is dumped overboard
Combustion chamber
- pressure: 35.5 atm
- ignition: pyrotechnic, comprising Isolite 1431 Rauline grain, with redundant initiators, providing two flame jets angled at 45° to propellant flow
- materials: stainless steel

Burn time (s): 735 H10, 760 H10+, 790 H10-3

Vulcain

CNES initiated studies in 1978 for launcher requirements following the Ariane 4 series, while SEP conducted preliminary studies of a 500 kN hydrogen/oxygen engine. During 1979-81, a 900 kN HM-60 engine emerged to power Ariane 5's H-60 second stage employing the gas generator cycle instead of staged combustion because of cost considerations (development cost saving was estimated at about 30% for a 10-15 s performance loss). Subsequently, HM-60 was assigned to Ariane's core stage and uprated to >1 MN (vacuum). Target reliability is 0.9946; programme cost about ECU1 billion. Contractor responsibilities are: SEP prime, gas generator and LH_2 turbopump; DASA thrust chamber/nozzle; Volvo Flygmotor nozzle (DASA subcontract) and pump turbines; FiatAvio LOX turbopump; Techspace Aero and Microtechnica fluid control valves; MAN gimbals.

The preparatory programme ran Oct 1984-87, followed by development 1988-94. Subsystem testing began during 1987. The first engine, the short-nozzle M1, was installed on Vernon's PF50 60 m vertical stand 4 Apr 1990 and performed its first ignition test 5 Jul. M1 achieved 15 runs, including two at full thrust for 8 s. PF50 allows one test per week. DLR's identical P5 stand at Lampoldshausen received the short-nozzle M2 10 Aug. Engine M3 took over on PF50 in early 1991 for flight duration 600 s tests: its first attempt on 29 May was halted after 104 s and on 5 Jun after 185 s as a result of T sensor problems. M3 achieved the first 605 s full duration full-rating run 13 Jun 1991. This was the 42nd Vulcain test and M3's 13th, bringing it to 1153 s. On 4 Jul it ran for 886 s until the stand's propellants were exhausted. A full nozzle was added to M1 (redesignated M1R) for trials on P5, the first 65 s run 2 Jul 1991 leading to the full 605 s later that month. It was then replaced by new engine M4. By end-Aug, the three engines totalled 3,048 s. This first test phase, running Apr 1990 to end-1992, required one engine to achieve 6,000 s over 20 runs; attained were 6,819 s in 23 starts. By end-Jun 1992, 97 tests on the six engines had accumulated 11,329 s, attaining a 1,275 kN max thrust and running for 793 s at the 1,120 kN flight thrust. Seven development engines had been tested by end-1992. Through Jul 1993, 145 tests on nine engines (two revalidated) totalled 29,000 s. 26 were >580 s and 13 >780 s. The longest, 960 s, was achieved by M8 (which completed 20 tests totalling 9,700 s through Jul 1993). By Apr 1994, 180 tests on 13 engines had achieved 44,730 s; M8 totalled 10,000 s. M6-R1 suffered a serious LOX turbopump fire 11 Apr 1994 as a result of impeller friction with the casing (this was an old design; the new version avoids the problem). The stiffer nozzle to prevent buckling was introduced in early 1994. By end-1994, 207 tests on 18 engines had achieved 56,920 s. By end-1995, 275 tests on 18 engines had achieved 85,837 s. Pad firings at Kourou began 17 Nov 1994 using M9 on a battleship stage 1, ending 27 Jan 1995. Qualification firings with a flight-type stage were completed 6 Jan 1996 (the first 680 s firing was achieved 16 Jun 1995). Ariane 5's debut, flight 501, employed engine M14; 502 M16; 503 M17.

Other facilities include the PF52 stand at Vernon (capable of running both turbopumps and gas generator together or independently for 100 s), P5.9 LOX turbopump and gas generator stand at Ottobrun (DASA), and the P3.2 thrust chamber stand at Lampoldhausen. The specifications below are typical of a GTO mission.

Applications: Ariane 5 core stage propulsion
First flight: 4 Jun 1996
Number flown: 1
Dry mass: 1,475 kg
Length: 300 cm
Maximum diameter: 176.0 cm
Mounting: gimballed ±6° for pitch/yaw control
Engine cycle: open gas generator
Propellants: LOX at 225.6 kg/s into chamber, LH_2 at 36.4 kg/s into chamber; total propellant flow 271.5 kg/s includes 9 kg/s for gas generator, attitude control thrusters, nozzle dump cooling and pressurisation
Mixture ratio: 6.2:1 in chamber, 0.94:1 gas generator. Valves are pneumatically operated and provide only two mixture ratio settings
Turbopumps: separate fuel/oxidiser turbopumps. Hydrogen pressure raised to 160 atm by 34,500 rpm 11.9 MW (range 30,340-39,330 rpm, 7.7-16.9 MW) 240 kg 2-stage centrifugal pump; oxygen to 150 atm by 130 kg 13,700 rpm 3.3 MW (range 11,300-14,880 rpm, 1.8-4.3 MW) 1-stage pump
Thrust: 1,145 kN vacuum, 900 kN sea level
Specific impulse: 431.2 s vacuum
Expansion ratio: 45:1
Combustion chamber
- pressure: 110 atm
- cooling: regenerative hydrogen through 360 longitudinal channels in copper-silver-zirconium alloy liner
- materials: stainless steel
- injector: Inconel 718 516-element co-axial
- ignition: pyrotechnic. Start-up and shutdown sequences are controlled by microprocessor-based system incorporated into the engine

Nozzle
- diameter: 176.0 cm (surface area 7.6 m^2)
- mass: 170 kg
- cooling: hydrogen circulated through helical tubes and dumped into exhaust through 228 ports. Capacity about 23 MW. Nozzle is composed of these 456 Inconel 600 helical 4 × 4 mm square tubes (wall thickness 0.4 mm) joined by 2 km of welding. Additional hatbands were added around the lower nozzle following tests, to cope with startup/shutdown transients

Burn time: 580 s (design life: 6,000 s + 20 starts)

Vulcain MkII

Commercial payloads require the upgrading of Ariane 5 performance by 15-20%, a programme approved by ESA in Oct 1995. The Vulcain MkII preparatory phase ran Jan 1995 to Mar 1996 for full development to begin 1996 and qualification in 2002. 120 tests totalling 56,000 s are planned from mid-1998 to end-2001 on seven engines. Vulcain's mixture ratio will be increased to 6.2 by increasing the LOX flow rate by 18%, accommodated by lowering the core's common bulkhead by 70 cm to raise propellant mass from 155 t to 172 t. The other requirement is to improve Vulcain thrust to 1,350 kN vac. This will be achieved by a 10% wider throat (expansion ratio from 45 to 61.7), 5.5% chamber pressure increase, the higher mixture ratio, and extending the nozzle, its lower part using a simple sheet film-cooled by the injected turbine exhaust. Dry mass will increase to 1,850 kg, mainly because of the larger nozzle. The launcher's structures are sized to accommodate such upgrades. The 185 kg LOX pump will increase its flow by 23%, pressure by 19%, power to 5.4 MW and outlet T to 875K. The LOX dome will be redesigned to cope with the increased flow. The thrust chamber will use a 7.1 mixture ratio, 306 kg/s flow and 116 atm pressure. 1.7 kg/s for dump cooling.

Liquid AOCS Thrusters

SEP has developed two sizes of monopropellant hydrazine thrusters for spacecraft attitude and orbit control subsystems. The 3 N and 15 N class thrusters are based on the catalytic decomposition of hydrazine using CNESRO catalyst and are/have been used by D5A, Spot, Geos, Exosat, ERS, Helios and Envisat. The specifications below are in 3/15 N order.

Dry mass: 0.320/0.355 kg
Length: 107.7/145.1 mm
Mounting: fixed
Thrust: 3.5/15.6 N vacuum at max pressure
Inlet pressure range: both 22-5.5 atm
Specific impulse: 230/232 s vacuum
Start temperature: both 180°C
Number of hot pulses: 380,000/116,000

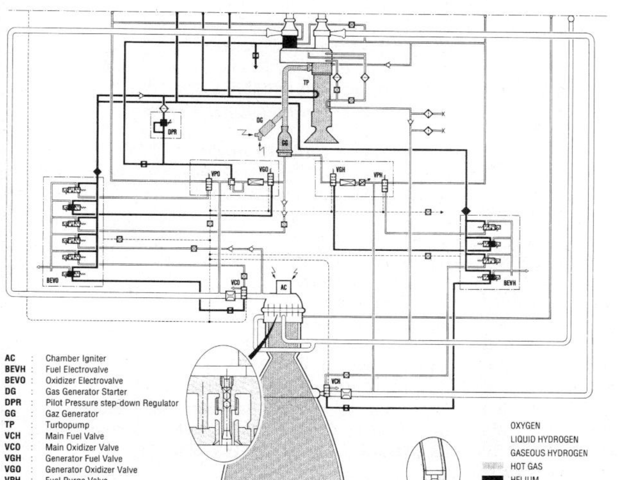

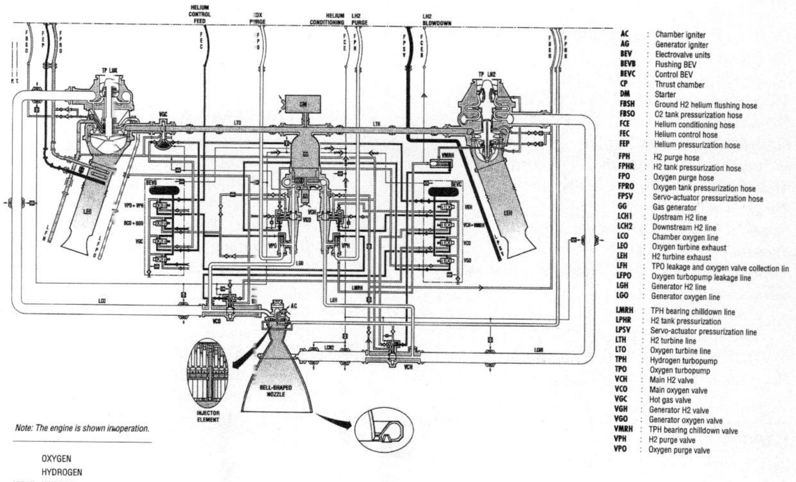

Left: HM-7B engine flow diagram. Hydrogen is tapped off for the attitude control thrusters. Right: Vulcain propellant flow (SEP)

A single Vulcain cryogenic engine (left) powers Ariane 5's first stage. Vulcain MkII (right) is being developed to improve Ariane 5 performance (SEP) **1996**

Number of starts: 9,000/13,500
Total on time: 40/17 h
Total impulse: 520/615 kNs
Valve power supply: both 27 V and 6 W
Heater power supply: 5.2/6 W at 34 V

SEP has developed NTO/UDMH thrusters for military applications which form the basis for space versions suited to Ariane Transfer Vehicle (8 kN + 200 N) and satellite AOCS (20 N) applications.

Mage Solid Propellant Motors

SEP is the industrial lead for ESA's range of three solid propellant apogee kick motors. Mage 1, first launched 1981 on the third Ariane vehicle, is designed for GEO injection of 400-500 kg class satellites. Its 1S derivative carries a greater propellant loading and is intended for spin-stablised platforms. Mage 2 can accommodate propellant loadings of up to 490 kg and is used by 500-680 kg payloads. As lead, SEP is responsible for 55% of the programme, BPD for 35% (propellant, thermal protection, test firings), and Germany's MAN for 10% (structure/casing). See pp242 1995-96 for full details.

Composite Motors

SEP's composite activities include development of carbon-ceramic structures for the Hermes spaceplane and SEPCarbinox and CeraSEP (carbon Novoltex/silicon carbide) composite liquid rocket engine nozzles. Trials of an Ariane HM-7B with a composite nozzle were successfully performed in 1989. Endurance trials were performed on a range of NTO/MMH CeraSEP thrusters into 1989: 5 N (50 h), 200 N (3 h) and 6.2 kN (925 s). Nozzle temperatures of 1,700K were routinely maintained. SEP has been working since 1988 on an 8 kN pressure-fed engine featuring an advanced injector and CeraSEP chamber/bell. Propellants are fed through 509 holes: 89 pentads (central fuel + four surrounding oxidixer holes) + 32 doublets. The pentads intersect for efficient propellant mixing, while the doublets spray a fuel-rich film against the chamber wall. Specifications are:

Applications: launcher upper stages and space vehicle manoeuvring
Dry mass: 20 kg (excluding actuators)
Length: 1.5 m
Maximum diameter: 150 cm
Mounting: gimballed for pitch/yaw control
Oxidiser: NTO at 1.557 kg/s
Fuel: MMH at 0.943 kg/s
Propellant feed pressure: 21.8 atm
Mixture ratio (O/F): 1.65:1
Thrust: 8 kN vacuum
Specific impulse: 320 s vacuum
Expansion ratio: 150:1
Cooling method: spray film and radiative
Combustion chamber
pressure: 14.8 atm
ignition: hypergolic
materials: C/SiC ceramic matrix
Burn time: 1,000 s total

SNPE

Le Bouchet Research Center
9 rue Lavoisier, BP2, F-91710 Vert le Petit.
Tel: (+33 1) 64 99 12 34
Fax: (+33 1) 64 99 13 55
WWW: http://www.snpe.fr
Chairman/CEO: Jean Faure
Advanced Technologies & Propulsion: Jacques Cardin
Space Propulsion Programs: David Quancard
• **Space Propulsion Programs**
BP57, F-33166 Saint-Médard en Jalles Cedex.
Tel: (+33 5) 56 70 50 16
Fax: (+33 5) 56 70 50 52
• **Pyromeca**
Chemin Charles Battezzati, Quartier Lagoubran,
BP2148, F-83063 Toulon.
Tel: (+33 4) 94 22 86 86
Fax: (+33 4) 94 22 86 99

The Société Nationale des Poudres et Explosifs SNPE Group, created as a state-owned company in 1971 and now with 5,000 personnel and FFr4.4 billion 1994 turnover (FFr100 million profit), manufactures liquid rocket propellants, with emphasis on UDMH, and develops/produces solid propellants. Along with BPD Difesa e Spazio in the **Regulus** partnership (see the international propulsion section), the company provides the propellant for Ariane 5's P230 solid boosters using facilities at Kourou constructed by subsidiary SNPE Ingénierie. SNPE maintains facilities at Saint-Médard en Jalles near Bordeaux (solid propellant grains), Toulouse (UDMH, AP), Toulon (pyrotechnics) and Le Bouchet near Paris (HQ and R&D). The Toulouse site manages the production of ammonium perchlorate with a capacity of 6,000 t annually to ensure Ariane 5 supplies. SNPE formed a second partnership with BPD, Société Européenne de perchlorate d'ammonium (EUPERA; SNPE 66%, BPD 34%), to handle production. Toulouse also provides the UDMH for Ariane 4 and the MMH for Ariane 5's upper stage. Subsidiary Pyromeca manufactures detonating cords and flexible linear cutting charges and ultra-fast pyrotechnic valves for Ariane 4 release from the launch pad. See also the Pyrospace entry. In cryogenic propulsion, SNPE provides the main igniter and turbine starter grain for Ariane 4's HM-7B stage 3 engine and the shaped ducts for the Ariane 5 Vulcain igniter and starter.

The propulsion package was designed/produced for the 153/227 kg Eclipse T1/T2 supersonic rocket and autonomous launch ramp in partnership with CAC Systèmes. Applications include meteorological sounding and μg research. T1/T2 capacities are 20/25 kg to 12/72 km.

STARSEM SA

33 rue Fernand Forest
F-95150 Suresenes.
Tel: +33 1 46 25 03 60
Fax: +33 1 45 06 05 78
President: Francois Calaque (Managing Director of Aerospatiale's Space Business Unit)
Director-General: Viktor Kuznetsov (Russian Space Agency)
Commercial Operations: Jean-Charles Vincent (Arianespace)

STARSEM is a joint-venture company, set up to market the Soyuz family of launch vehicles on a commercial basis. The partners are Aerospatiale and Arianespace in France, the Russian Space Agency (RSA) and the Samara Space Center (comprising the TsSKB design and engineering bureau and the Progress production facilities) which produces the launch vehicles. The agreement for the creation of STARSEM was signed by the partners on 17 Jul 1996 in Moscow and on 6 Aug the statutes of the company were signed in Paris. The company is a French-registered corporation, set up with FFr500,000. The backing for the project is split with Aerospatiale having 35%, Samara and RSA 25% each and Arianespace 15%.

The company proposes to market three basic launch vehicles: the current Soyuz-U, the current Molniya-M and a new Soyuz-U/Ikar version which adds a new upper stage to the existing Soyuz-U.

GERMANY

DAIMLER-BENZ AEROSPACE AG

Previously DASA, Daimler-Benz Aerospace AG (DBAA) is the main contractor for Ariane's second stage and the PAL liquid strap-on boosters, and provides the thrust chamber for the HM-7B third stage engine; it similarly provides the injection head, combustion chamber and expansion bell for Ariane 5's Vulcain first stage engine, under a contract signed with SEP in Apr 1989. During 1991-98, DBAA is providing 50 second stages and 120 PALs under a DM1,600 million contract, in addition to 50 HM-7B thrust chambers under a DM52 million award. Another 15 stage 2s and 26 PALs have been ordered for 1998-99. The company leads Ariane 5's EPS upper stage, the first development of an Ariane stage outside of France. 14 will be built 1997-2002. DBAA, as MBB, also built the P3.2 test stand for Vulcain chamber trials. Ariane engine specifications are included in France's SEP company entry. Its other engines include storable bipropellant (4, 10, 400, 27,500 N), hydrazine monopropellant (0.02-350 N) and electric (10-15 and 150-200 mN).

As MBB, the company began development of the world's first 50 kN topping cycle engine in 1958 and began testing at 85 atm chamber pressure in 1963. In 1968, MBB's 180 kN LOX/LH_2 engine attained the existing chamber pressure record of 282 atm, leading to NASA and Rocketdyne acquiring a licence for utilising the technology in the Space Shuttle Main Engine. The current 10 N and 400 N NTO/MMH thrusters are combined to form a Unified Propulsion System (12 × 10 N + 1 × 400 N), employed by the Galileo Jupiter orbiter and the TV-Sat/TDF, Tele-X, DFS Kopernikus, Eutelsat 2, Turksat and Amos telecom satellites. Bipropellant 4 N thrusters are currently under development and qualification.

Facilities: DBAA's Lampoldshausen and Ottobrunn sites are equipped with test stands for high pressure thrust chambers up to 1 MN (P3.2 stand for Vulcain) and for gas generators and turbopumps with supply pressures up to 800 atm, stands for LH_2 ramjets, sea level and vacuum stands for 4 N to 50 kN bipropellant and 0.5-500 N monopropellant engines (Trauen site), and a surface tension tank propellant behaviour simulation facility.

Monopropellant Hydrazine Thrusters

DBAA has developed a 0.5-2,500 N range of monopropellant catalytic hydrazine thrusters (CHTs), together with complete mono propulsion systems. It has produced >550 units for satellites such as Skynet 4, ECS, Telecom, Giotto, Ulysses and Hipparcos.

CHT 0.5 (−1 straight nozzle, −2 90°)

Applications: OTS 2, ECS, Telecom 1, Skynet 4, NATO 4
First flight: May 1978 aboard OTS 2
Number produced: 160 units
Dry mass: 0.19 kg
Length: 113.6 mm
Mounting: fixed
Engine cycle: decomposition of hydrazine over heated catalyst bed of Shell 405 ABSG or KC-12 GA
Thrust: 0.2-0.75 N vacuum (0.5 N nominal)
Specific impulse: 216-227 s vacuum
Chamber pressure: 5.5-22 atm
Power requirement: 5 W valve, 2.5 W heater
Burn time: qualified for 25.2 s single burn, 143 h total
Minimum impulse bit: 0.005-0.015 Ns

CHT 2.0 (−1 straight nozzle, −2 90°)

Applications: OTS 2, ECS, Telecom 1, Giotto, Ulysses, Skynet 4, Meteosat MOP, ISO
First flight: May 1978 aboard OTS 2
Number produced: 211 units
Dry mass: 0.200 kg
Length: 123.9 mm
Mounting: fixed
Engine cycle: see CHT 0.5
Thrust: 0.6-2.0 N vacuum
Specific impulse: 214-227 s vacuum
Chamber pressure: 5.5-22 atm
Power requirement: 5.0 W valve, 1.0 W heater
Burn time: qualified for 36 s single burn, 22.5 h total
Minimum impulse bit: 0.015-0.036 Ns

CHT 5.0

Applications: Skynet 4, NATO 4, Hipparcos
First flight: Dec 1988 aboard Skynet 4B

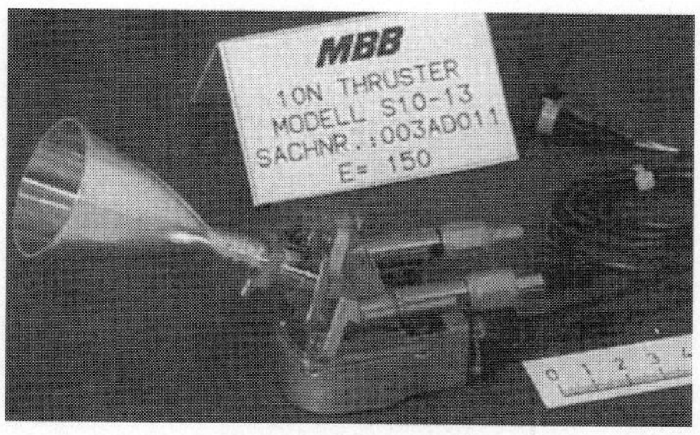

DASA's 10 N bipropellant thruster (DASA)

Number produced: 69 units
Dry mass: 0.220 kg
Length: 129.3 mm
Mounting: fixed
Engine cycle: see CHT 0.5
Thrust: 1.85-6.0 N (5.0 N nominal) vacuum
Specific impulse: 216-228 s vacuum
Chamber pressure: 5.5-22 atm
Power requirement: 5.0 W valve, 3.37 W heater
Burn time: qualified for 3.6 s single burn, 12.5 h total
Minimum impulse bit: 0.038-0.096 Ns

CHT 10.0

Applications: Meteosat MOP series, SAX
First flight: Mar 1989 aboard Meteosat 4
Number produced: 51 units
Dry mass: 0.240 kg
Length: 142.34 mm
Mounting: fixed
Engine cycle: see CHT 0.5
Thrust: 3.0-10.0 N (10.0 nominal) vacuum
Specific impulse: 220-230 s vacuum
Chamber pressure: 5-22 atm
Power requirement: 5.0 W valve, 2.4 W heater
Burn time: qualified for 1.5 s single burn, 3.4 h total
Minimum impulse bit: 0.070-0.190 Ns

CHT 20.0

Applications: Eureca retrievable platform; −1 version utilises constant pressure, −2 is blow down version
First flight: Jul 1992 on Eureca 1
Number produced: 46 units
Dry mass: 0.360 kg
Length: 195.74 mm
Mounting: fixed
Engine cycle: see CHT 0.5
Thrust: 7.2-24.0 N (20 N nominal) vacuum
Specific impulse: 222-235 N vacuum
Chamber pressure: 5-22 atm
Power requirement: 13.0 W valve, ≤6.0 W heater
Burn time: qualified to 3.6 s single burn, 4 h total
Minimum impulse bit: 0.165-0.370 Ns

CHT 400

Applications: Ariane 5 attitude control system
First flight: Jun 1996
Number produced: 53 units to end-1995
Dry mass: 1.8kg
Length: 307 mm
Mounting: fixed
Engine cycle: see CHT 0.5
Thrust: 110-450N (350N nominal) vacuum
Specific impulse: 218-235 s vacuum
Chamber pressure: 5.5-22 atm
Valve power: 70 W
Minimum impulse bit: 5 Ns

Bipropellant (MON/MMH) Thrusters

DBAA has developed a 4-27,500 N range of bipropellant thrusters, together with fill systems and fill/drain valves. 600 units have been produced for satellites such as Galileo, TDF, Telecom 2, Hispasat, Turksat, Amos and Cluster.

Aestus Ariane L9 Engine

DBAA is responsible for development of Ariane 5's L9 EPS upper stage, the first development of an Ariane stage outside of France. Development firings were completed in late 1994, totalling almost 11,000 s on six engines in >800 tests, including a single 1,380 s run. EPS achieved its first nominal duration firing, 1,075 s, 5 Oct 1994. Engine qualification firings started Dec 1994 for completion Mar 1995. Pre-development work began in Jan 1995 on uprating Aestus to 35 kN, including increasing the number of injector elements to about 160, for potential future Ariane 5 upgrades.
First flight: Jun 1996
Dry mass: 111 kg
Length: 2,195 mm
Max diameter: 1,263 mm
Mounting method: gimballed ±6° by electromechanical actuators
Engine cycle: pump-fed (feed pressure 18.0 atm)
Oxidiser: nitrogen tetroxide at 5.89 kg/s
Fuel: monomethyl hydrazine at 2.87 kg/s
Mixture ratio: 2.05:1

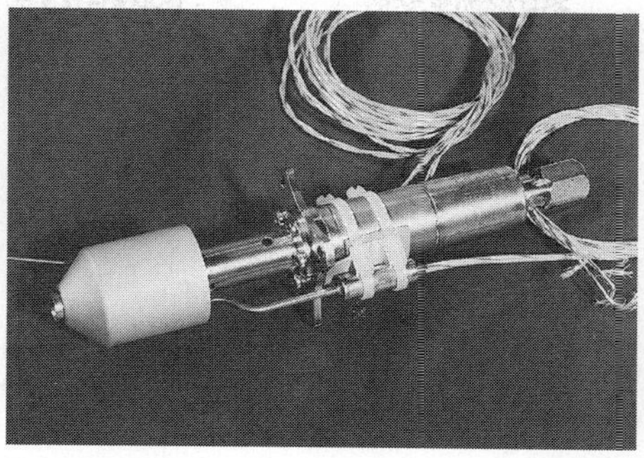

CHT 1.0 N hydrazine thruster for Globalstar **1996**

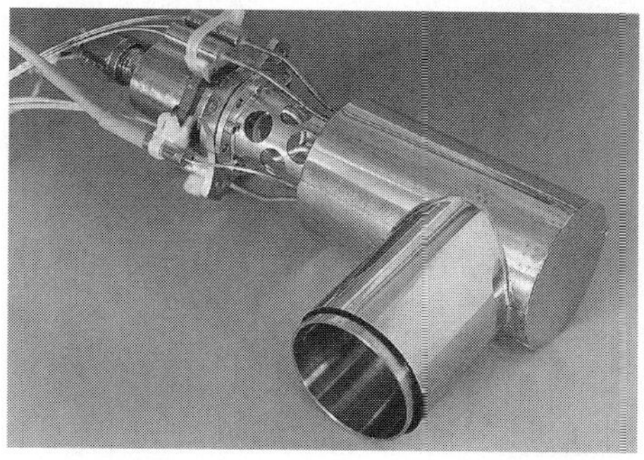

DASA's 20 N hydrazine thruster is available in straight and 90°-canted nozzle configurations (DASA) **1996**

Thrust: 27.5 kN vacuum
Specific impulse: 324 s vacuum (potential 331 s)
Expansion ratio: 83.3:1
Combustion chamber
pressure: 10.42 atm
cooling: regeneratively (fuel) cooled nozzle to 10:1 area ratio
materials: stainless steel inner liner with milled cooling passages and electrolytically deposited nickel layer for outer closure
injector: 132-element coaxial arranged in spirals. Fuel enters injector from the chamber cooling channels. Oxidiser is distributed to elements via manifold at rear
Nozzle
cooling: radiation cooled extension below 10:1 area ratio

S400 Engine

The 400 N bipropellant thruster is incorporated in the DFS, TV-Sat, Tele-X, TDF, Eutelsat 2, Amsat, Turksat, Amos 1, Cluster and Artemis propulsion systems and provides the Galileo Jupiter probe's primary propulsion capability. See also the S10 specifications. DASA has developed a second generation engine under ESA contract to increased SI to 317 s. Performance optimisation is underway for formal qualification by end-1996. Modifications: the pilot operated pneumatic valves were replaced with Moog 53-193 single seat dual coil valves; the combustion chamber was completely redesigned (regenerative cooling loop eliminated, chamber pressure increased, throat section recontoured and made from platinum/rhodium alloy); reduced throat diameter increased expansion ratio to 220 without changing engine length or exit diameter. It has been tested over sea level and vacuum conditions over 1.2-1.8 mixture ratio and 110-145 g/s flow rates. Pulse mode demonstration tests were performed with on-times of 100, 250, 500 and 800 ms and off-times of 1,000 and 100 ms. Specifications are provided in ().
Dry mass: 2.8 kg (3.120 kg)
Length: 531 mm (498 mm)
Max diameter: 248 mm (248 mm)
Mounting method: fixed
Engine cycle: pressure fed, at 18 atm (inlet presure 13.8 atm)
Oxidiser: MON-1 (NTO + 1%NO)
Fuel: monomethyl hydrazine
Mixture ratio: 1.64:1 (1.65:1)
Thrust: 400 N (404 N, flow rate 130 g/s)
Specific impulse: 308 s (317 s)
Expansion ratio: 102/150 (220)
Expansion bell: stainless steel (stainless steel and platinum-rhodium for higher heat loads)
Chamber pressure: 6.9 atm (9.87 atm)
Chamber materials: stainless steel using MMH regenerative cooling (stainless steel upper and platinum-rhodium lower radiative cooling)
Injector: stainless steel swirl injector, double cone

DASA has developed a second generation 400 N MON/MMH bipropellant thruster **1996**

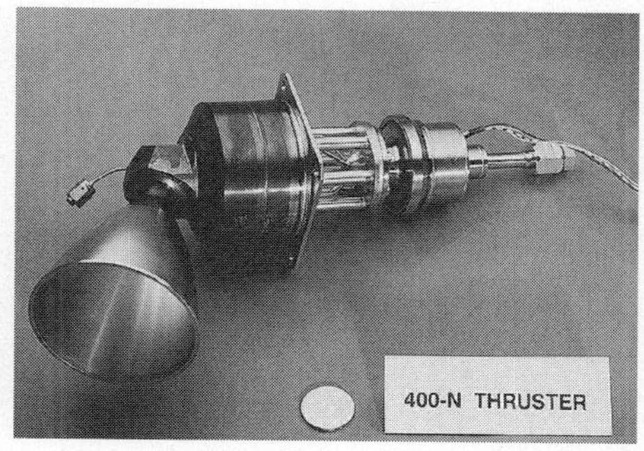

CHT 400 hydrazine thruster for Ariane 5's attitude control system (DASA) **1996**

S10 Thruster

This 10 N (nominal) bipropellant thruster is teamed with the 400 N thruster (see above) in the Galileo, DFS, TV-Sat, TDF, Tele-X, Eutelsat 2, Turksat, Amos 1 and Cluster propulsion systems. >100 were also provided for Inmarsat 2, Telecom 2 and Hispasat. A second generation version, without regenerative cooling, has been qualified; its different specifications are noted in (). Cluster, Amos 1 and Artemis are users.
Dry mass: 0.350 kg
Length: 138 mm
Diameter: 67 mm
Mounting method: fixed
Engine cycle: pressure fed
Oxidiser: MON
Fuel: monomethyl hydrazine

Aestus thrust chamber assembly (DASA)

Mixture ratio: 1.55-1.72 over thrust range
Thrust: 8-11.4 N vacumm
Specific impulse: 287 s (292 s)
Expansion ratio: 90 (150)
Chamber pressure: 18-12 atm over thrust range

S4 Thruster

The S4 will be utilised by the next generation of telecom satellites.
Dry mass: 0.270 kg
Length: 105 mm
Diameter: 29 mm
Mounting method: fixed
Engine cycle: pressure fed, at 5-25 atm
Oxidiser: MON
Fuel: monomethyl hydrazine
Mixture ratio: 1.5-1.85 over thrust range
Thrust: 1.77-5.62 N
Specific impulse: 290 s
Expansion ratio: 190

S3K Engine

S3K is a technology programme that could form the basis for a future orbital transfer vehicle.
Dry mass: 14.5 kg
Length: 1,030 mm
Max diameter: 530 mm
Mounting method: fixed
Engine cycle: pressure fed
Oxidiser: MON 3
Fuel: monomethyl hydrazine
Mixture ratio: 1.6-2.1
Thrust: 3,500 N
Specific impulse: 352 s
Expansion ratio: 125
Chamber pressure: 9-12 atm

INSTITUT FÜR RAUMFAHRTSYSTEME

Universität Stuttgart
Pfaffenwaldring 31, D-70550 Stuttgart.
Tel: (+49 711) 685-23 75
Telex: 7255445 univ
Fax: (+49 711) 685-35 96
WWW: http://www.irs.uni-stuttgart.de
Director: Prof Dr Ernst W Messerschmid
Deputies: Prof Dr-Ing habil Monika Auweter-Kurtz (Space Transportation Technology), Dr Ulrich Schöttle
The Space Systems Inst (IRS) was founded in 1970 and plasma thruster development has always been a principal interest. Facilities allow MW-class testing at 0.5-2 g/s propellant flow rates (argon) under selected 10Д^{6}-10Д^{3} atm conditions. 0.5-1 MW MPD stationary thrusters of different geometries have been developed, mainly under AFOSR grants. Based on this experience, IRS has designed & built a range of thermal arcjet thrusters. Current arcjet projects include: Atos 600-700 W ammonia, Artus 2 1-2 kW NSSK hydrazine (with DASA; DARA contract); 10 kW & 100 kW devices. Atos is a simplified version of Artus 2 for flight on Amsat-Deutschland's amateur satellite aboard Ariane 502.

Four plasma wind tunnels are used to investigate re-entry vehicle thermal protection materials and to validate aerothermodynamic CFD codes. Two have MPD plasma generators (0.10-1 MW) and are especially suited for high specific enthalpy (up to 150 MJ/kg) and low total pressure conditions. One tunnel has an inductive plasma generator, specially used for catalycity investigations, and the fourth has a thermal plasma generator for higher total pressure and low specific enthalpy areas.

IRS' Mission & System Analysis division studies development & numerical simulation and design tools for space transportation systems, mission & system optimisation and performance assessment of air-breathing launchers. The Space Technology & Utilisation division encompasses space station design, numerical flow field & simulation methods, and space systems safety.

DASA's propulsion system for telecommunications satellites incorporates a single 400 N and 14 10 N bipropellant thrusters (DASA)

Artus 2 typical operation specifications
Thrust: 0.188 N
Power level: 1.5 kW at 18 A
Mass flow: 35 mg/s simulated hydrazine (2:1 H_2:N_2)
Thrust efficiency: 33%
Specific impulse: 537 s

MAN TECHNOLOGIE AG
Space Systems Division

Liebigstrasse 5a, D-85757 Karlsfeld.
Postal: Postfach 1347, D-85751 Karlsfeld.
Tel: (+49 813) 18901
Fax: (+49 813) 1891900
Director, Space Systems Division: Klaus Kaiba
1995/96 Space Systems turnover was DM250 million (1994/95: 157), plus DM387 million (1994/5: 146) orders received, including satellite navigation orders. Space projects represent >50% of MT's turnover: system studies, development and manufacture of advanced metallic and composite lightweight structures and mechanisms, thermal protection units, mobile satcoms, satnav and system monitoring (see the Navigation entry and the System Electronics Division entry in the Satcom section). Studies with MT acting as ESA's prime contractor include the definition of the AROC recoverable μg capsule, the NAVSAT phase B1 system study, and the definition and development of an experimental health monitoring unit for GPS and Glonass. Structural components include apogee boost motor casings (see below), CFRP shells for Rosat, Italsat high pressure gas tanks, the CFRP structure for Orfeus, Silex structures, and radio astronomy high precision reflectors.

For Ariane 4, MT Space Systems provides the stage 1 thrust frame, the load-bearing gas generator water coolant tank (fibre composite/aluminium, 30%/150 kg less than conventional design), Viking turbopumps and gas generators (see below), aft/forward booster adaptors, and booster separation and release mechanisms. In Dec 1996 MT took over the production of the second stage main tank. For Ariane 5, MT builds the flow-turned steel booster casings in Augsburg, the main tank's front skirt, the GAT/GAM high pressure vessels for TVC of Vulcain and booster nozzles, Vulcain's gimbal joint, and the steering mechanisms heat shields. In Dec 1996 MT took over the production of the main tank bulkheads. MT has production facilities in Augsburg, Munich and Oberpfaffenhofen.

MT played a leading role in the development of Hermes' thermal protection system, drawing on its expertise in ceramic composites, high temperature materials technology and complex system modelling and testing. MT also developed methods and tooling for manufacturing the hot structural elements. The experience is reflected in its development of Sänger's hypersonic intake ramp model.

MT upgraded and extended its know-how in the fields of development and the production of components incorporated into space transportation systems, in order to qualify itself to be a 'general contractor' for entire systems. The company being commissioned with the overall responsibility for the co-ordination of the configuring of Germany's portion of work to be carried out on the European Crew Transport Vehicle's (CTV) systems represents an initial step towards that goal. MT was also able to secure orders for key components (structures, tanks) of the Automated Transfer Vehicle (ATV) project.

MT has delivered >700 Viking turbopumps and associated gas generators at a production rate of 60 units annually. The two single-stage centrifugal pumps use 8-blade steel wheels to move 270 kg/s propellant while raising its pressure from 5 to 57 bar. The Curtis velocity-compounded impulse turbine incorporates two

IRS's Artus 1.5 kW thermal arcjet suspended from a pendulum thrust balance. Nozzle temperatures reach 1,600K under steady state firing

MAN Technologie manufactures Ariane 5's main tank front skirt ***1996***

rows of rotating blades to drive the shaft from the gas generator's water-cooled exhaust. The generator's spherical chamber is cooled by eight water jets facing the propellants' 40 injection ports. The turbopumps also serve as part of the thrust structure. Key characteristics are:

Viking engine	4B	5C	6
Propellant flow (kg/s)	278.1	276.4	276.4
Turbine speed (rpm)	10,000	10,000	10,000
Turbine power (kW)	2,500	2,500	2,500
Turbopump mass (kg)	331	325	326
Gas generator mass (kg)	28	28	28
Gas flow (kg/s)	6	7	7

MT's apogee boost motor cases consist of a pressure reservoir precision-wound from high strength aramide fibres, with two different pole openings for the nozzle and the cylindrical transition structure for spacecraft mounting. Key casing characteristics are:

	Mage 1	Mage 1S	Mage 2	IRIS
Mass (kg)	9	11.4	13.7	53
Volume (l)	212	260	309	985
Length (mm)	628	728	845	1,020
Diameter (mm)	766	766	766	1,310

INDIA

India's original orbital launchers utilised solid propellant main motors, drawing on liquid systems only for interstage control systems. The PSLV, however, incorporates two major liquid engines. ISRO's Liquid Propulsion Systems Centre with headquarters in Trivandrum and units in Bangalore and Mahendriagiri, is

The Rohini 300 provides access to the middle atmosphere for Indian scientists (ISRO)

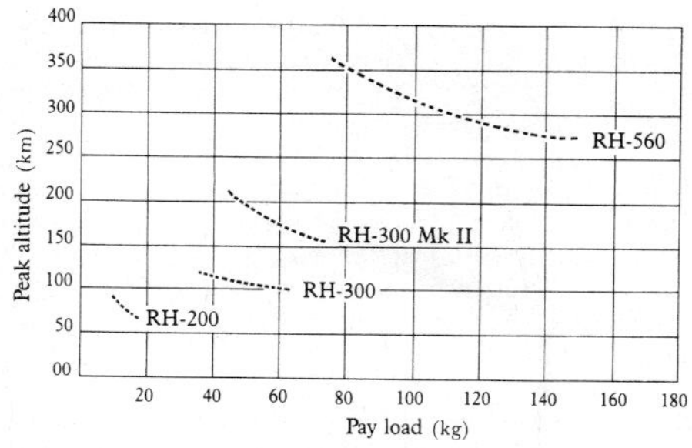

Rohini sounding rocket performance curves. The improved RH-560 will deliver 100 kg to 500 km

responsible for R&D of all Indian liquid motors, from satellite attitude control thrusters to the 725 kN Vikas. Solid motors are developed by the Vikram Sarabhai Space Centre at Trivandrum and the SHAR Centre.

Air-breathing technology for the ABR 200(I) engine was flight tested 14/19 Oct 1992 on sounding rockets launched from Sriharikota.

Liquid Propulsion Systems Centre

Valiamala, Trivandrum 695547.
Tel: (+91 471) 57 22 57
Telex: 0435-396 LPSCIN
Fax: (+91 471) 46 26 86
Director: Mr G Madhavan Nair

LPSC is responsible for the development of launcher liquid and cryogenic propulsion stages and auxiliary propulsion for launch vehicles and spacecraft. It is growing in importance as main liquid engines are introduced to India's orbital launchers and with the increasing size of indigenous satellites. Current main projects cover the 720 kN Vikas PSLV stage 2 engine, PSLV's stage 4 dual 7.5 kN system, Insat 2's 440 N Liquid Apogee Motor and unified network of 22 N ACS thrusters, and the 680 kN Vikas GSLV strap-on. Seven Russian 76 kN cryogenic engines are being provided for the initial GSLVs beginning 1998, but India's own engine will complete development in 1999.

LPSC's test facilities are sited southeast of Trivandrum at Mahendragiri. Vikas' Principal Test Stand (PTS) was commissioned during 1987 and used in Jan 1988 for the engine's first full-duration 150 s firing. Altitude facilities are also available for PSLV's 7.5 kN motor, Insat 2's LAM and smaller thrusters. Cryogenic engine and stage facilities will be commissioned in 1997.

Insat 2 Propulsion System

Insat 2's 119 kg (dry mass) unified bipropellant liquid propulsion system comprises a single 440 N LAM Liquid Apogee Motor and two redundant networks of 8 × 22 N RCS thrusters. The qualification LAM was tested for a total 9,550 s, including a single burn of 3,000 s. RCS testing accumulated 30,500 s including a single continuous burn of 10,000 s. Pulsed firing mode logged >250,000 pulses, each component being tested over three life cycles. The engines, pressure regulator, check valve, fill, drain & vent valve, pyro valve and pressure transducer were developed indigenously; the tanks, gas bottles, filters and latch valves were procured from foreign suppliers. The system carries about 450 welded joints.

LAM Specifications

First flight: Jul 1992 (Insat 2A)
Dry mass: 4 kg
Length: 570 mm
Maximum diameter: 276 mm
Mounting: fixed
Engine cycle: pressure-fed in regulated mode
Propellants: MMH/NTO from tanks at 16 atm
Mixture ratio (O/F): 1.65
Thrust: 440 N vacuum nominal
Specific impulse: 310 s vacuum minimum
Expansion ratio: 160
Chamber pressure: 6.9 atm
Injector: coaxial titanium welded to chamber
Combustion chamber: silicide-coated columbium alloy radiatively cooled
Burn time: qualified to 3,000 s continuous; minimum impulse bit not available

Insat 2 RCS Thruster Specifications

First flight: Jul 1992 (Insat 2A)
Dry mass: 850 g
Length: 249 mm
Maximum diameter: 60 mm
Mounting: fixed
Engine cycle: pressure-fed in blowdown mode
Propellants: MMH/NTO from tanks at 16 atm
Mixture ratio (O/F): 1.65
Thrust: 22 N vacuum nominal
Specific impulse: 285 s vacuum minimum
Expansion ratio: 100
Chamber pressure: 6.9 atm
Throat temperature: 1,200°C
Injector: coaxial titanium welded to chamber
Combustion chamber: silicide-coated columbium alloy radiatively cooled
Burn time: qualified to 1,000 s continuous; minimum impulse bit not available

PSLV stage 4 engines

PSLV's final stage carries two identical gimballed liquid engines to provide guided injection.
First flight: 20 Sep 1993
Dry mass: 28 kg
Length: 1.1 m
Maximum diameter: 0.63 m
Mounting: gimballed up to ±3° in two orthogonal planes by a closed loop servo control system to provide stage pitch, yaw and roll control during thrust phase
Engine cycle: pressure-fed
Propellants: MMH/NTO from tanks at 19.2 atm
Mixture ratio (O/F): 1.4
Thrust: each 7,500 N vacuum nominal
Specific impulse: 308 s vacuum minimum
Expansion ratio: 60
Chamber pressure: 8.37 atm
Injector: AISI 304 stainless steel; 45 sets of triple elements
Combustion chamber: AISI 304 stainless steel film cooled along single helical grooving
Nozzle: silicide-coated columbium alloy radiatively cooled
Burn time: 425 s mission nominal
Qualification time: 530 s (continuous)

PSLV L37.5 Vikas engine

India's first large liquid propellant rocket engine is derived from Viking technology acquired from SEP. The engine achieved its first full-duration 150 s firing in Jan 1988 on the Principal Test Stand at the Liquid Propulsion Test Facility, Mahendragiri, and completed the first integrated stage test 21 Mar 1990 (qualification completed Oct 1992). A 686 kN version will be used for GSLV's strap-ons. A successful 200 s run 24 Jul 1995 tested the new indigenously-developed silica-phenolic throat, designed to withstand the operational 1,300°C.
First flown: 20 Sep 1993
Number flown: 3 to end-1996
Dry mass: 876 kg
Length: 3.509 m
Mounting: gimballed for pitch/yaw control
Engine cycle: gas generator (9400 rpm turbine)
Propellants: UDMH/NTO
Mixture ratio (O/F): 1.86
Thrust: 72 kN vacuum (686 kN for GSLV)
Specific impulse: 295 s vacuum
Time to full thrust: 2.4s
Expansion ratio: 31 (13.88 C-SU strap-on)
Chamber pressure: 51.9 atm
Injector: 216 like on like doublets set in six rows for each propellant in light alloy annular injector
Chamber cooling: UDMH film supplied through additional channels on the injector's lower section
Igniter: hypergolic
Nozzle materials: cobalt alloy with SEPHEN (phenolic resin/silica fibre) throat
Burn time: 160 s nominal mission

Solid Propulsion Group

Vikram Sarabhai Space Centre
Trivandrum 695022.
Tel: (+91 471) 56 24 44/56 25 55
Telex: 0435-201, 0884-202
Deputy Director: R Nagappa

India's first, 75 mm ∅, solid motor was produced in 1967, followed by a 125-560 mm range for sounding rocket applications and, in the early 1970s, motors for the SLV-3 satellite launcher. Work on the most powerful so far began in 1984: the 2.8 m ∅ 3,800 kN thrust solid powers PSLV's first stage. Although the Solid Propulsion Group is an element of VSSC, the primary Solid Propellant Space Booster Plant (SPROB) is sited at SHAR Centre on Sriharikota Island, along with the Vehicle Assembly Static Test & Evaluation Complex (VAST) for solids. At VSSC the motor cases are produced in the Mechanical Engineering Facility and the propellant binders in the Centre's Propellant Fuel Complex.

PSLV stage 1 motor

PSLV's first stage is the most powerful model produced by ISRO; it was successfully static fired for the first time 21 Oct 1989. The second was succesfully fired 23 Mar 1991.
Application: PSLV stage 1
First flight: 20 Sep 1993
Number flown: 3 to end-1996
Mass (kg): 145,000
Length (cm): 2,034.5
Diameter (cm): 280.4
Propellant
type: HTPB, segment-cast with three central segments interchangeable
shape: forward segment deep 10-fin to burn 22 t loading in 20 s to generate high launch thrust; remainder tubular with avg 120 cm ∅ bore
mass fraction: 0.890 (129 t loading)
Burn time (s): 100
Thrust (kN, vacuum): 3,463 mean, 4,600 max at 17 s
Specific impulse (s, vacuum): 265
Total impulse (kNs, vacuum): 326,000
Pressure (atm): 58 max at 17 s
Nozzle
throat diameter (cm): 80
length (cm): 2,034.8
exit diameter (cm): 227
expansion ratio: 8
materials: carbon phenolic throat insert, tape-wound carbon phenolic on divergent cone changing to silica phenolic below TVC inlets
Casing materials: five segments of M250 maraging steel, 3.4 m long. Joints tongue/groove with two O-rings and 144 pins. Insulated by nitrile rubber, with silicon potting compound on joints
Igniter type: pyrogen head-end 30 cm ∅, 125 cm long containing Pedpro 2661 HTPB and its own 2 kg-charge igniter

PSLV stage 3 motor

This is ISRO's most advanced solid propellant motor, designed for a 0.915 mass fraction and the first to adopt a submerged flex nozzle design for TVC. Off-loading is also possible. Test #5 made 29 Jul 1991, #6 17 Oct 1991. Tests 8, 9 and 10 were for qualification.
Application: PSLV stage 3 (PS3)
First flight: 20 Sep 1993
Number flown: 3 to end-1996
Mass (kg): 7,975
Length (cm): 244.20 (casing 208.35)
Diameter (cm): 198.8 (202.5 skirt)
Propellant
type: HTPB casebound, 86% solid
shape: slotted
mass fraction: 0.910 (7,260 kg standard loading)
Burn time (s): 73.1; 78.1 action time
Thrust (kN, vacuum): 243.2 mean, 386 max
Specific impulse (s, vacuum): 293
Total impulse (kNs, vacuum): 20,850
Pressure (atm): 57.4 max
Nozzle: the 19.6% submerged contoured nozzle is provided with a ±3° (±2° flight) conical flex bearing. The elastomer, based on natural rubber of nominal 3 mm thickness is bonded with steel reinforcements of 4.5 mm thick run. Each seal consists of six reinforcements and seven elastomer pads. The seal positioned beyond the throat housing is driven by two 11.8 kN electromechanical actuators 90° apart; a thermal boot protects the bearing
throat diameter (cm): 19.4
length (cm): 180.2
mass (kg): 280
exit diameter (cm): 141.4
expansion ratio: 52.3
materials: graphite throat insert, carbon and silica phenolic composites for nozzle
Casing materials: Kevlar filament-wound, EPDM insulation and aluminium forgings for skirt and end-opening fittings; total mass 340 kg. Nozzle opening 750 mm ∅, 200 m head end
Igniter type: pyrogen head-end

PSLV strap-on motor

The same basic motor, derived from the SLV-3's stage 1, was employed for ASLV's stage 1 and strap-on, and adopted as PSLV's initial strap-on. The strap-ons are essentially identical to the ASLV stage 1 version, with the principal exception of canted nozzles. For PSLV, the secondary injection TVC systems are deleted on one ground-lit and three air-lit strap-ons. The specifications below are for ASLV's stage 1 model; differences for the strap-ons are noted where applicable.
Application: ASLV stage 1 + strap-ons, PSLV strap-ons
First flown: 24 Mar 1987, ASLV-D1 launch
Number flown: 18 PSLV strap-ons, 4 ASLV stage 1, 8 ASLV strap-ons to end-1996
Mass: 10.4 t
Length (cm): 998.99
Diameter (cm): 100
Propellant
type: AP/18%Al/HTPB case-bonded
shape: star
mass fraction: 0.853 (8.9 t loading; 8.63 t for strap-on)
Burn time (s): 46 action time (49.95 action strap-on)
Ignition delay: about 200 ms strap-on
Thrust (kN, vacuum): 502.6 mean, 643 max (strap-on: 440 mean, 580 max)
Specific impulse (s, vacuum): 259 (252 strap-on)
Total impulse (kNs, vacuum): 22,600 (21,450 strap-on)
Pressure (atm): 43.5 max (40.2 max strap-on)
Nozzle: secondary injection ports for TVC; 9°-canted nozzles
expansion ratio: 6.7

materials: graphite throat, carbon phenolic composite for convergent section, tape-wound carbon and then silica phenolic composites for divergent

Casing materials: 3.5 mm thick low carbon steel with nitrile rubber insulation. Three-segment casing with tongue/groove joints

Igniter type: pyrogen head-end.

JAPAN

IHI COMPANY LTD

Space Development Division

As a major Japanese aerospace company, Ishikawajima-Harima Heavy Industries' main details are to be found in the Space Industry section. In space propulsion, IHI provides the cryogenic turbopumps for the H2's stage 1/2 cryogenic engines, in addition to H2's hydrazine Reaction Control System modules. It previously provided the pumps and gas generator for H1's stage 2 engine. ISAS, IHI, KHI and MHI are pursuing development of the ATREX-500 air turboramjet, which could power the flyback booster of a 2-stage spaceplane. The hydrogen expander ATREX would take the vehicle to M6/30 km. The 5.4 kN ATREX-500 ¼-scale version, 2.2 m long and fan inlet 30 cm ∅, totalled 1,400 s in 40 static sea level runs during 1990-95 at ISAS' Noshiro Testing Centre. Efforts now centre on a 1.4 t flying test bed launched from Japan and parachuting into the sea. A jet engine will take the FTB to M0.5 from a horizontal rail for the 370 kg ATREX-500 with 190 kg of LH_2 to attain 25-35 km altitude 200 km downrange. The initial metal turbo machinery will achieve M4.5, with M6 targeted for the carbon-carbon composite design. Goals for the composite ATREX are 17.0-18.9 kN and 3,150-3,300 s SL static.

IHI's Rocket Test Centre was established in Oct 1975 at the Aioi Plant as a cryogenic test centre of IHI's Research Institute. It was incorporated in Jul 1980 into the Space Development Division for testing the cryogenic engine and bipropellant apogee engines described below.

Spacecraft Propulsion

IHI developed the 2 kN bipropellant apogee engine under NASDA contract (in co-operation with NAL) for ETS 6 and COMETS (see below). ETS 6's LAPS Liquid Apogee Propulsion Systems was specified in detail p258 1995-96. IHI is independently developing a 10 kN CUS Cryogenic Upper Stage LOX/LH_2 expander cycle engine for upper stage or orbital transfer vehicle applications. SI 471 s, chamber pressure 28 atm, LOX turbopump 43,000 rpm, hydrogen turbopump 85,000 rpm.

Unified Propulsion System

UPS provides GEO insertion and attitude/orbit control for 2 t-class satellites; it is derived from the LAPS of ETS 6. It carries a single 1.7 kN bipropellant Apogee Kick Engine plus 4 × 50 N + redundant 8 × 0.86 N hydrazine catalytic thrusters. COMETS's flight model was shipped Mar 1996.

Dry mass: UPS total 163 kg, AKE 15.7 kg (including valves)

AKE length: 102.5 cm

AKE maximum diameter: 58.4 cm

Engine cycle: regulated pressure-fed

Oxidiser: nitrogen tetroxide at 0.255 kg/s

Fuel: hydrazine at 0.284 kg/s

Propellant feed: GHe 17 atm in two 82 cm ∅ NTO tanks + 130 cm ∅ N_2H_4 tank; titanium alloy. 223 atm GHe in two 48 cm ∅ CFRP spheres

Thrust: 1,700 N AKE

Specific impulse: 321.4 s AKE

Time to 90% thrust: 1 s max

Mixture ratio (O/F): 0.9

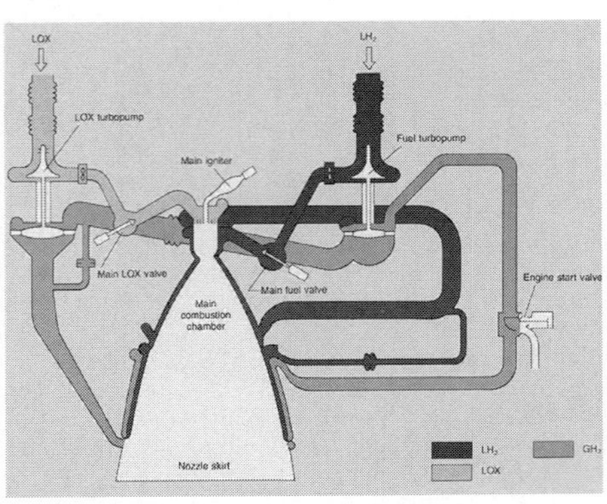

LE-5A nozzle hydrogen bleed cycle **1996**

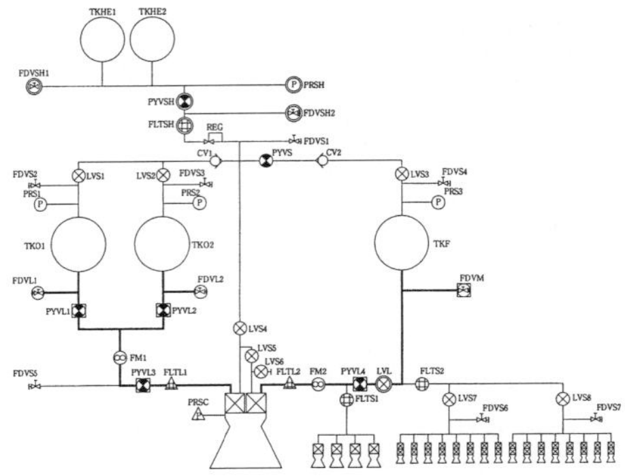

IHI's Unified Propulsion System will provide COMETS with apogee kick and attitude/orbit control **1996**

Expansion ratio: 240 (exit ∅ 584 mm)

Expansion bell
- length: 83.8 cm
- material: columbium (niobium)/silicide coating
- cooling method: radiative

Combustion chamber
- pressure: 8.3 atm
- temperature: 1,770K max wall
- stability: ±5%
- materials: as expansion bell
- cooling method: fuel film + radiative
- ignition method: hypergolic

Duty cycle: AKE 3,500 s total, 10 burns max (typically two required); 50 N thruster 40,000 pulses min total; 1 N thruster 270,000 pulses min.

LE-7/5/5A turbopumps

IHI developed H1's cryogenic turbopumps in co-operation with NASDA/NAL. For LE-7, the LOX/LH_2 pumps provide flow rates of 35.7/211 kg/s at discharge pressures of 266/254 atm working at 42,000 rpm. For LE-5A, the pumps provide 17.4/51.6 l/s at discharge pressures of 56/64 atm working at 16,500/50,800 rpm. In the LE-5, the pumps were driven by combustion gases from the gas generator: LOX/LH_2 burned at 840K/25.4 atm and a mixture ratio of 0.85 to produce an exhaust flow rate of 0.43 kg/s.

Reaction Control Systems

IHI developed hydrazine RCS systems for the ETS 3-6, MOS, CS-3 and JERS satellites and the H1/H2 second stages. It developed an EHT thruster for NASDA (see below) and a magnetoplasmadynamic (MPD) thruster for ISAS' Space Flyer Unit. H2 roll control during burn phases, 3-axis control during coast periods and ullage acceleration is provided by two IHI thruster modules at stage 2's base, either side of the LE-5A engine. Each houses six anhydrous hydrazine thrusters (4 × 50 + 2 × 18 N) pressure-fed at 20 atm. For direct ascent trajectories, without LE-5A re-start, a 15 kg dry mass version carries 6 kg of hydrazine to provide 10.67 kNs total impulse. On re-start missions, the 20 kg (dry mass) module houses 27 kg of propellant for a 52.27 kNs total impulse. ETS 6's blowdown RCS system comprises 16 × 1 N (0.95 N±10%) thrusters + two EHTs (see below). Dry mass 29.6 kg; propellant (hydrazine) 118 kg; SI 221-192 s; inlet pressure 24.6-7.0 kg/cm², 463,000 pulses demonstrated.

IHI's LE-7 turbopumps: the liquid hydrogen unit is at bottom, oxygen at top **1995**

Electrothermal Hydrazine Thrusters

IHI/Toshiba developed an experimental 0.2-0.5N EHT under NASDA contract. Two sets are carried by ETS 6 for NSSK trials; they were tested successfuly over 4.2 h.

Dry mass: 2.0 kg (heater control unit adds 2.85 kg)

Length: 13 cm

Mounting: fixed

Engine cycle: decomposition of hydrazine over a heated catalyst bed. The 600°C gases then pass to the heat exchanger chamber where they are pre-heated by channeling along the wall (which also provides wall cooling) and then heated by a 16.5 A double coil tungsten filament to 1,100°C (low power) or 1,600°C (high power)

Thrust: 0.2-0.5 N vacuum

Specific impulse: goal is 290 s vacuum. 170°C without heating, 240 s low level heating and 280 s high level heating demonstrated. Thermal efficiency 80% demonstrated; 85% goal

Inlet pressure: 6.8-23.8 atm

Power requirement: 300 W low (1,400°C heater), 500 W high (1,900°C heater); 9.8 W valve (plus 4.6 W valve heater), 3.0 W catalyst bed heater, total power 1,132 W under maximum operating conditions

Total impulse: >10^5 Ns demonstrated, 2.7 × 10^5 Ns goal.

MITSUBISHI HEAVY INDUSTRIES LTD

Space Systems Department

Mitsubishi provides Japan's primary capability in liquid rocket engines, most recently developing the cryogenic oxygen/hydrogen LE-7 stage 1 and LE-5A stage 2 engines for NASDA's H2 launcher. The company has prime industrial responsibility for H2 (as it did for H1). The 10 yr $108 million LE-5 development programme produced Japan's first operational LOX/LH_2 engine, providing H1's stage 2 with a restart capability. The technology formed the basis for developing the larger and more advanced LE-7 and the LE-5A direct outgrowth. Nine LE-5s flew, successfully, 1986-92 (see p271 of the 1991-92 edition for detailed information). MHI also fabricated Rocketdyne MB-3 engines under licence for the H1 and produced the indigenous LE-3 storable propellant engine for the seven N1 rockets. Two advanced smaller engines, the 9.8 kN RE6 and 2.94 kN RE10-300, have been tested for possible applications in future orbit transfer vehicles. MHI is testing LACE liquefied air cycle engine hardware, in which LH_2 is used to liquefy atmospheric oxygen, using the LE-5 as a demonstrator. Mitsubishi's Tashiro Test Field includes stands for LE-5/5A/5B engine and stage sea level firings, a short-duration LE-7 stand and an H2 stage 1 battleship tower (full-duration firings of flight-type engines and stages are undertaken at Tanegashima's H2 launch complex). MHI completed a 105 m long 30 m wide 20 m high assembly/test facility at its Nagoya plant for the H2 in May 1988, allowing stages to be shipped from Nagoya bay to the launch site. MHI is supplying stage 1/2 LOX tanks for McDonnell Douglas' Delta 3. MHI is also providing the injectors for Pratt & Whitney's RL-10B-2 engine.

LE-5A

H1's LE-5 was initially intended for direct application aboard H2's stage 2 but NASDA decided at the beginning of the detailed design phase in 1987 to uprate performance and improve re-startability by increasing chamber pressure and throat diameter, and by switching from gas generator to hydrogen bleed cycle, respectively. IHI's turbopumps are powered by tapping GH_2 from the nozzle coolant loop, reducing complexity and mass.

Application: H2 stage 2 main engine

First flight: 3 Feb 1994

Number flown: 4

Dry mass: 244 kg

Length: 2,668 mm

Maximum diameter: 1,625 mm

Engine cycle: nozzle hydrogen bleed

Oxidiser: liquid oxygen at 22.83 kg/s total

Fuel: liquid hydrogen at 5.86 kg/s total

Mixture ratio: 5.0:1 (20 for 5% idle)

LOX turbopump: contractor IHI, 17,363 rpm rotor speed for pump discharge pressure of 57.8 atm

LH_2 turbopump: contractor IHI, 50,534 rpm rotor speed for pump discharge pressure of 63.5 atm

Thrust: 121.5 kN vacuum; 5% idle mode provides precise orbit control and stage de-orbit

Time to full thrust: 9 s

Specific impulse: 452.9 s (>200 s for 5% idle)

Expansion ratio: 130:1

Chamber pressure: 39.3 atm

Cooling method: regenerative hydrogen circulation through tubular walls

Burn time: 400 + 210 s dual burn.

LE-5B

Improved H2 in 1999 will introduce the LE-5B, reducing cost and increasing flexibility, particularly restarts for deploying multiple satellites into different LEOs and direct injection into GEO by a third burn at apogee. The engine cycle will be changed from a nozzle expander bleed to a chamber expander bleed; the heat exchanger function is thus integrated into the chamber cooling function. This reduces nozzle complexity – the skirt can use dump cooling with a simple double wall sheet metal structure. Chamber cooling will be changed from the current brazed tube walls to machined channels for better heat absorption. The injector elements will be reduced from 208 to 127. The engine can operate in throttled condition with throttling valves added. Feasibility test (8 firings, 236.9 s total duration) was conducted in 1995. Phase 1 development test (13 firings, 930.7 s total duration) was conducted in 1996. Phase 2 test (qualification) will be conducted in 1997.
Dry mass: 269 kg (estimated)
Length: 2,625 mm
Maximum diameter: 1,713 mm
Engine cycle: chamber hydrogen bleed
Oxidizer: liquid oxygen at 26.02/16.80 kg/s total
Fuel: liquid hydrogen at 5.20/3.36 kg/s total
Mixture ratio: 5/5:1
LOX turbopump: 17,772/12,897 rpm rotor speed for pump discharge pressure of 54.48/31.10 atm
LH2 turbopump: 51,103/40,535 rpm rotor speed for pump discharge pressure of 67.62/44.61 atm
Thrust: 137.2/88.1 (throttled) kN vacuum; 5% idle mode provides precise orbit control and stage de-orbit
Specific impulse: 450/448 s
Expansion ratio: 110:1
Chamber pressure: 35.26/23.05 atm.

LE-7

Development of Japan's largest and most advanced engine was authorised in 1985 simultaneously with that of its host vehicle, the H2. Like Rocketdyne's Space Shuttle Main Engine, it employs staged combustion in which the propellants are partially burned in the pre-burner, routed to drive the turbopumps and then combined with more oxygen in the main combustion chamber. Complexity is reduced, however, by the lower main chamber pressure and the requirement for only a single start. LE-7 development covered 21 engines, 282 firings and 15,639 s, including 20 mission duty cycles. EG307 flew on H2#1.

Because of the development delays, it was decided during the CDR in Jul 1990 that LE-7's target performance would be reduced to about 90%, a level already achieved in tests. Overall H2 capability would not be affected, however, because SRB and stage 2 performance had exceeded target. NASDA's projected development cost increased from ¥80 billion to ¥100 billion. The specifications provided below reflect the revised performance; the original specifications are retained in parentheses.
Application: H2 first stage
First flight: 3 Feb 1994
Number flown: 4
Dry mass: 1,714 kg (1,560 kg); vacuum thrust:mass 64
Length: 340 cm (350 cm)
Maximum diameter: 180 cm (190 cm)
Mounting method: gimballed to provide ±7.5° pitch/yaw control, hydraulic pump powered by auxiliary turbine mounted on engine
Engine cycle: closed, staged combustion
Oxidiser: liquid oxygen at 25.0 kg/s in pre-burner and 186.6 kg/s in main chamber
Fuel: liquid hydrogen at 35.66 kg/s in pre-burner (resultant fuel-rich gas then burned in main chamber with added oxygen)
Mixture ratio: 0.7:1 in pre-burner; 6.0:1 main chamber, 5.93 overall
LOX turbopump: contractor IHI, 18,285 rpm (20,000 rpm) single-stage centrifugal pump with pre-burner pump; 178.6 atm (210 atm)/253.6 atm (323 atm) main/pre-burner discharge pressures, respectively
LH_2 turbopump: contractor IHI, 41,596 rpm (46,000 rpm) two-stage centrifugal pump; 264 atm (315 atm) discharge pressure
Thrust: 843 kN (910 kN) SL, 1,078 kN (1,180 kN) vac
Specific impulse: 445.6 s (449 s) vacuum
Expansion ratio: 54 (60); exit diameter 173.7 cm
Main combustion chamber: 452-hole coaxial main injector, 130 atm (145 atm) combustion pressure
Pre-burner: 237-hole coaxial injector, with damp-cooled combustion wall; 206 atm (237 atm) pressure, 600°C (698°C) temperature
Cooling method: regenerative hydrogen; LH_2 at 44K in both channel-structured chamber and tubular nozzle extension
Expansion bell length: 216 cm
Burn time: 348 s single start.

1997/0003459

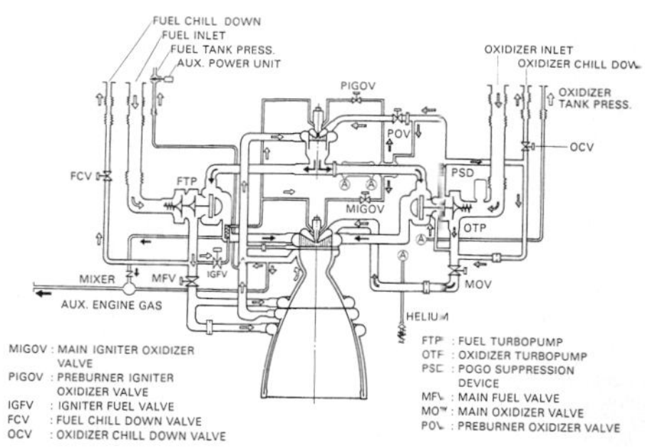

LE-7A

The H2A in 2000 will debut the LE-7A, reducing cost and increasing robustness. For the main injector assembly: reduce and relocate the welding lines, cast the LOX dome, reduce the number of injector elements, remove the hydrogen heat exchanger, reduce the injector ∅ from 1,400 mm to 1,160 mm. The main chamber throat area will be enlarged by up to 10%. The lower nozzle will be changed from regenerative to dump cooling (allowing simple sheet metal construction instead of tubular). For the pre-burner the number of elements is reduced and the LoX dome weld assembly is replaced by a single-cast part. The turbopumps will use more precise cast housings. Throttling is important for reducing maximum aerodynamic pressure on missions such as HOPE. The design will allow longer operation at current thrust or at LE-7's original 1,180 kN original specification. Specifications are given below in that order:
Length: 0.366 m
Max diameter: 0.182 m
Chamber mixture ratio: 6.3
Throttling: 70%
LOX turbopump: 18,000/19,2000 rpm
LH_2 turbopump: 40,600/42,900 rpm
Thrust: 1,078/1,180 kN vac
Specific impulse: 441/442 s vacuum
Chamber pressure: 117/127 atm
Pre-burner temperature: 447/487°C
Burn time: 600/350 s single start.

RE6/RE10

The RE6 cryogenic expander bleed cycle engine has been tested for potential orbit transfer vehicle and launcher stage 3 applications. The lower-thrust RE10 was investigated as a high performance space-storable apogee engine for satellite main propulsion and orbiter manoeuvring systems. Turbopump, pre-burner and main combustion chamber tests were completed since development began in 1986; sea level engine system trials followed. Work ended 1992. See p293 of the 1992-93 edition for details.

NISSAN MOTOR COMPANY LTD

Nissan's Aerospace Div is Japan's primary developer/manufacturer of solid propellant rocket motors and launch vehicles. These include the M-5 satellite launcher, the MT-135P, S-310, S-520 and K-9M sounding rockets, boosters for the H2, and several kick stages. The company is also involved in propulsion studies for Japan's spaceplane and is responsible for the experiment logistics module exposed section of the Japanese Experimental Module contribution to NASA's Space Station.

Manufacture/assembly is undertaken at Nissan's Ogikubo plant (planned to move to Tomioka City); the Taketoyo Test Facility opened in 1980 for vacuum firings of spinning upper stage motors. Taketoyo's Spin Static Firing Test Stand provides a 39-78 mbar vacuum for 1.8 m ∅ motors of 0-147 kN at spin rates up to 150 rpm. The company provided test equipment for the H2 solid booster static stand at NASDA's Tanegashima

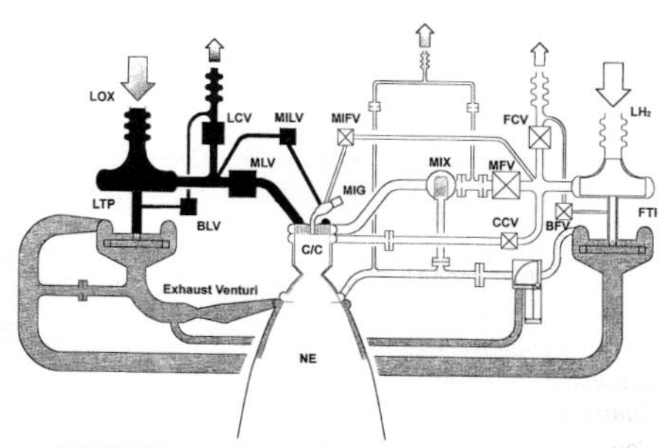

LE-5B's improvements include switching to a chamber hydrogen bleed cycle (IHI) **1997**/0003455

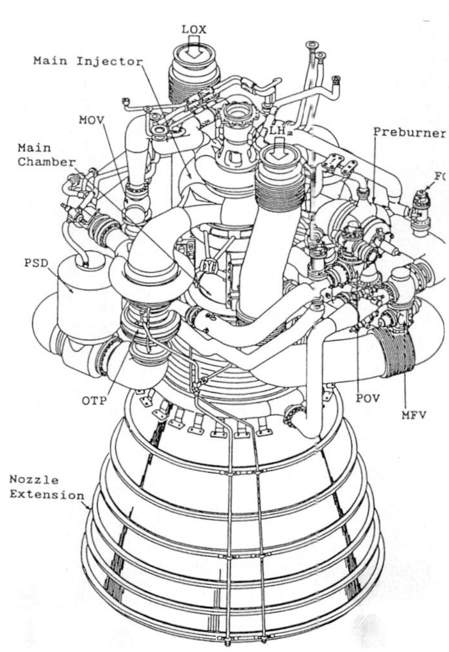

The LE-7 is the first Japanese engine to employ the staged combustion cycle (NASDA)

Nissan developed the TR-1A rocket, providing some 6 min μg (NASDA) **1997**/0003448

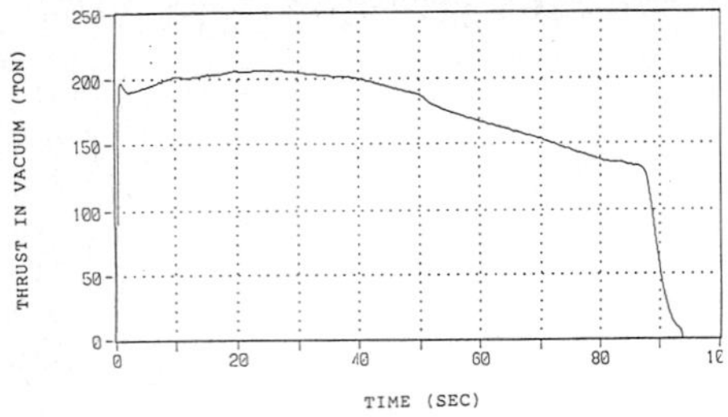

Nissan/NASDA H2 booster thrust curve (thrust given in tonnes)

site. R&D activities are conducted at the Kawagoe plant, including research into liquid bipropellant satellite thruster systems, ram propulsion and gimballing TVC. The company is also a major supplier of pyrotechnic devices for space applications.

H2 Solid Rocket Boosters

Nissan developed the 4-segment solid strap-ons for Japan's H2 launcher, with four demonstration firings during 1988-91 at NASDA's Tanegashima Space Centre. First test was conducted Apr 1988, prototype #1 was fired Jun 1989 & prototype #2 13 Dec 1989; the qualification model firing was achieved 29 May 1991. The units represent the country's largest and most powerful indigenously-developed solids. The Improved H2's SRBs will be reduced to three segments, debuting in 1999. Upgraded H2A, from 2001, will use a monolithic composite case design.

First flight: 3 Feb 1994
Number flown: 3 sets, through May 1996
Mass: 70.4 t
Length: 2,336 cm total booster, 1,926 cm for motor without nose section
Principal diameter: 1,809 cm
Propellant
type: 14HTPB-68AP-18Al
mass: 59.15 t in 4 segments
Burn time: 94 s
Thrust: 1,560 kN, sea level
Specific impulse: 273 s vac
Chamber pressure: 46 atm average
Nozzle: throat 53.4 cm ∅, expansion ratio 10; nozzle gimballed ±5°
Case materials: four 4.0 m-long segments of 4.7 mm-thick NT-150 low carbon steel.

S-520 rocket motor

Nissan produces a range of sounding rockets for ISAS, NASDA, the Japan Meteorological Agency and the National Inst of Polar Research. One of the largest is the S-520 single stage sounding rocket for ISAS (qv). Motor specifications are given below.

First flown: 18 Jan 1980
Length: 6,053 mm
Principal diameter: 524 mm
Propellant
type: polybutadiene composite
mass: 1,610 kg
Burn time: 28.7 s effective, 35.7 s action
Thrust: 143.168 kN avg vac
Specific impulse: 265 s vacuum
Total impulse: 4,109 kNs
Chamber pressure: 55 atm maximum
Nozzle
throat diameter: 161 mm
expansion ratio: 7.0
materials: CFRP/GFRP conical
Case materials: high tension steel
Igniter type: pyrogen head end, SAD ignition with CFRP case.

UNITED KINGDOM

BRITISH AEROSPACE DEFENCE CO LTD

Royal Ordnance RMD provides the principal UK capability in propellants and rocket motors, including double base/composite solids and mono/bipropellant liquid engines. Solid rocket performances cover 0.008-600 s burning time, 0.012-425 kg charge mass and 0.8-200 kN thrust.

Liquid Propellant Engines

The 500 N Leros liquid apogee engine represents the first phase of RO's expansion into civil space propulsion, with the long-term objective of developing a comprehensive range of mono and bipropellant engines. Leros 1 was the first to be qualified; Leros 1b, 2, 20 and 20H have been developed since. RO is improving performance by using advanced materials in thrust chambers for prolonged operation at elevated temperatures.

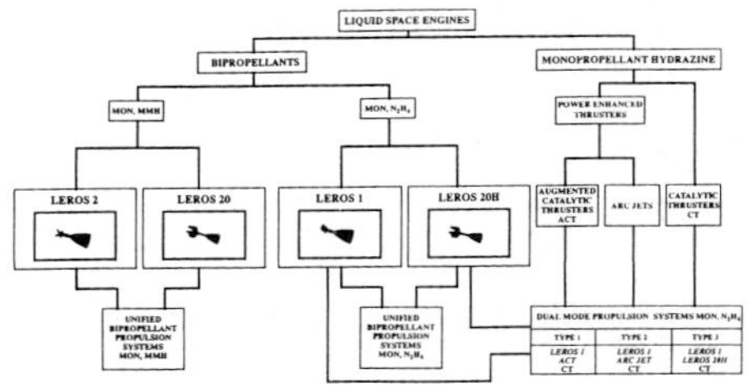

Royal Ordnance's space liquid propulsion family

Olin Aerospace's MR 103C 0.5 N mono thruster is marketed in Europe under licence by RO. It is being supplied to MMS for Skynet 4. RO has a long history of work with catalytic thrusters for attitude control applications or stage attitude control systems. For example, in 1974 RO was appointed as Technical Authority for Chevaline's Hydrazine Actuation System.

Leros 1

RO began development of Leros in 1986 to satisfy the orbital manoeuvring requirements of large telecom satellites using dual mode propulsion, which provides a high performance bipropellant MON/hydrazine liquid apogee engine combined with mono hydrazine thrusters for attitude control and stationkeeping. The design approach was based on proven design features, optimisation of chamber length for smooth combustion over a wide operating envelope, a long combustion chamber to reduce thermal soakback and to ensure a cool running injector assembly and stable film conditions, conventional fabrication and assembly, and in-process injector screening in a slave chamber for acceptable performance.

Leros 1 design verification testing was completed in spring 1989, with the qualification programme at Boeing Tulalip for Martin Marietta Astro Space demonstrating by mid-1990 that the engine can handle a 3 h burn (double the worse-case mission duration identified). Tests covered oxidant depletion, bubble ingestion, multiple hot restarts, hard vacuum ignition and helium-saturated propellants. The 140 starts accumulated 23,000 s and accounted for 3.7 t propellant throughput. Lockheed Martin Astro Space has purchased 29 Leros 1 engines for use in telecom platforms. Leros 1 became the world's first flight-proven dual mode LAE when a pair placed Astra 1B into GEO in Mar 1991. Other users are Telstar 4, AsiaSat 2, Echostar and Intelsat 8/8A.

Applications: dual mode satellite propulsion systems
Dry mass: 4.2 kg
Length: 610 mm
Max diameter: 288 mm (nozzle)
Oxidiser: MON3
Fuel: hydrazine
Propellant feed method: pressure, by helium
Inlet pressure: 15 atm
Valves: Moog model 53-177 torque motor valve, dual coil single seat configuration, based on the model 53-135 unit used on Rocketdyne's Peacekeeper attitude control bipropellant thrusters. >1,500 produced

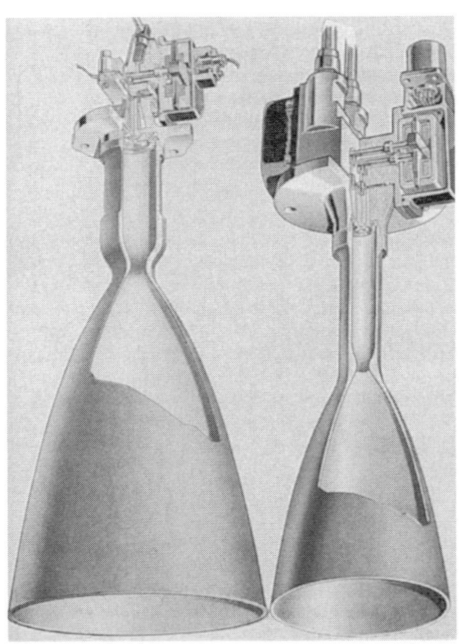

Leros 1 (left) and Leros 20. Not to same scale

Thrust: 500±25 N vacuum
Specific impulse: 314 s vacuum nominal
Response time: <10 ms
Mixture ratio (O/F): 0.8
Expansion ratio: 150:1 with 8.5° exit plane half angle
Expansion bell: disilicide-coated niobium C103, radiatively cooled
Chamber pressure: 7.0 atm
Chamber combustion stability: ±12% above 100 Hz, ±3% below 100 Hz
Chamber materials: disilicide-coated niobium C103, T≤1,360°C
Injector: 6Al/4V titanium alloy, jet impingement type using eight unlike doublets for the core around which are 16 fuel film coolant orifices. A backing plate encloses the injector elements and provides the welded interface with the engine mounting flange and mounting points for the propellant control valves
Duty cycle: unlimited, but qualified to 3 h.

Leros 1b

This improved version delivers 318 s Isp. Design/construction is identical in all respects to Leros 1 with the exception of the propellant valve, which has been replaced by two Moog solenoid valves (model 53-200). Leros 1b was qualified in 1995 for Lockheed Martin and flight engines have been developed.

Leros 1c

Under development for a major US company using high T materials for the thrust chamber. Target vacuum SI is 325 s.

Leros 2

The Leros 2 apogee engine is designed for use in unified MON/MMH bipropellant systems. It differs from Leros 1 only in its injector configuration and combustion chamber geometry. An alternative valve option is available. The engine has completed development and is being used as a springboard for the development of the higher performance (320 s Isp) **Leros 2A** version under ESA contract employing an advanced high temperature material chamber capable of withstanding prolonged operation at up to 2,000°C. Design verification testing is planned for 3Q 1997. **Leros 2B** features conventional columbium/disilicide technology for the thrust chamber and the 300:1 expansion cone. Demonstrated vacuum SI >317 s.

Dry mass: 3.40 kg
Inlet pressure: 16.0 atm
Thrust: 556 N
Mixture ratio (O/F): 1.65
Specific impulse: 312 s.

Leros 20

This second generation bipropellant attitude control thruster was developed for MMH propulsion systems.

Dry mass: 0.73 kg
Length: 202 mm
Max diameter: 60 mm (nozzle)
Oxidiser: MON3 at 4.8 gm/s
Fuel: MMH at 2.9 gm/s
Propellant feed method: pressure, by helium
Inlet pressure: 10-20 atm
Valves: Moog single or dual seat
Thrust: 22 N nominal vacuum
Specific impulse: 294 s vacuum nominal
Mixture ratio (O/F): 1.65
Expansion ratio: 180:1
Expansion bell materials: niobium C103 or titanium
Chamber pressure: 7.5 atm
Chamber materials: niobium C103, T≤1,400°C
Injector: multiple impinging jet element with three unlike doublets and six fuel film cooling orifices
Duty cycle: unlimited
Minimum impulse bit: 25 mNs

The **Leros 20H** MON/hydrazine engine is aimed at dual mode systems under ESA contract: S_I 300 s, mixture ratio 0.7-0.8, minimum impulse bit 15 mNs. Performance characterisation testing is scheduled for 3Q 1996.

Test and Production Facilities

The company most recently commissioned two major high altitude test facilities for its liquid thrusters. Westcott's new facility handles attitude control thrusters, while the Boeing site at Marysville (Washington state) is used for the apogee motors. Westcott's unit provides:

Pumping system: two stage ejector, 52 km at 10 gm/s (airflow) and 0.57 mbar for Leros 20
Vessel volume: 3.5 m³
Propellant feed system: 30 l oxidant tank, 20 l fuel tank, 35 bar working pressure, 4-45°C thermal conditioning, helium saturation
Thrust measurement: single axis, high response, piezo electric transducer
Instrumentation: 64 channels, 120 kHz, analogue recorders, Agema thermal imaging system, Cyclops IR thermometer, CCTV high speed ciné.

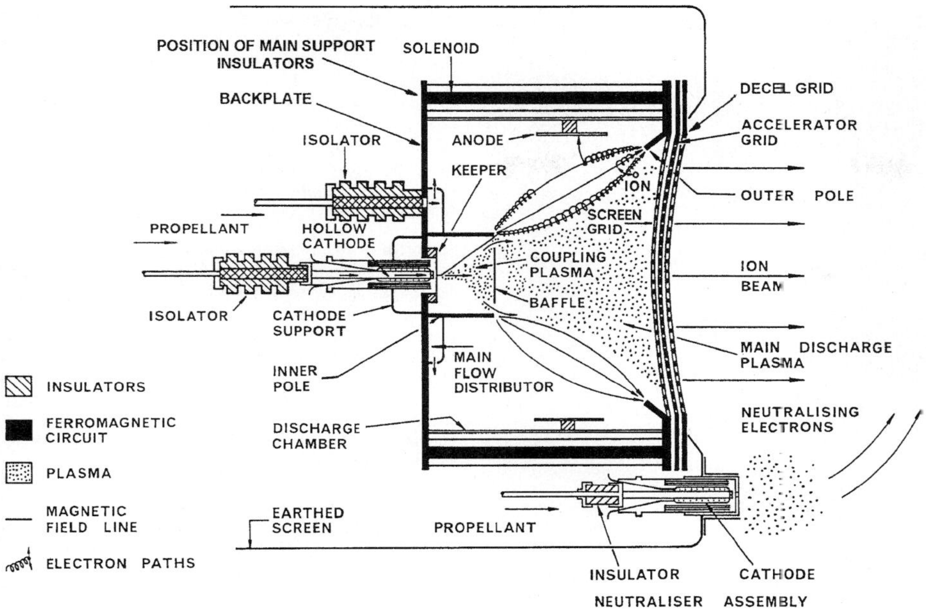

Kaufman thruster principal features, with triple grid (DRA) **1995**

The UK-25 xenon ion thruster was performance-mapped in Culham's vacuum facility (UKAEA Culham)

The Boeing facility provides:
Pumping system: 5-stage steam ejectors with condensors and engine diffuser
Performance: 55 km at 195 gm/s (airflow) and 0.22 mbar for Leros 1
Vessel volume: 285 m^3
Propellant feed system: 450 l oxidant tank, 450 l fuel tank, 20 bar working pressure, ambient to 30°C thermal conditioning
Thrust measurement: six component, multi-axis fast response
Instrumentation: 50 channels, 1 kHz, analogue recorders, Agema thermal imaging system, Cyclops IR thermometer, CCTV high speed ciné.

Westcott provides other firing sites for mono/bi system testing at SL/altitude. The altitude site utilises three interconnected chambers totalling 600 m^3; two positive displacement pumps reduce the pressure in a single chamber to about 4 mbar, and to 6 mbar with all three connected. The 60 l hydrazine expulsion rig can supply propellant at up to 45 atm. The sea level facilities accommodate all storable propellants and thrusts up to 100 kN. Westcott also offers:
Climatic chambers six chambers covering −55/90°C; seven cold chambers for long-term storage testing down to −60°C, with additional hot air and hot water long-term chambers
Vibration two tables, one covering 5-2,000 Hz, 144 kN sine thrust (±25.6 mm amplitude) and the other 10-3,000 Hz, 27 kN sine thrust (±12.7 mm amplitude)
Shock 1,000 kg test items subjected to 250 *g*
Centrifuge two centrifuges for testing solid propellant motors under varying *g* conditions. One provides up to 100 *g* for up to 33.4 kN motors on a 3 m arm, the second 60 *g* for 800 kN motors on a 5 m arm
Dynamic balancing up to 1,000 kg items on 56-300 rpm 1 m ∅ table; minimum measurable out-of-balance 0.3 kg-mm at 270-300rpm.
Westcott's **Liquid Motor Processing Facility** was established to process propellant and gas storage assemblies for Chevaline's post boost propulsion system and has handled >1,000 tanks and gas storage assemblies. The LMPF processes, fills, tests and performs final closure welds on aluminium and stainless steel tanks, with class 100,000 clean rooms. Propellants in use include IRFNA, MAF 1 and 4 and hydrazine.

UK ION PROPULSION

UK ion thruster development began in the 1960s with a 10 mN 10 cm mercury Kaufman thruster at RAE/Culham. The 10 cm work was re-activated in 1985 using Xe because of projected satcom growth in the 1990s. UK-10 in its qualified form provides a thrust of typically 10-25 mN but can be smoothly throttled from 0.2 to 30 mN. Culham has also performed extensive trials on the UK-25 25 cm version, capable of delivering a throttlable 50-300 mN. ESA's Artemis will carry two 18 mN UK-10 versions for NSSK and elements of the UK-25 are being adapted for the ESA-XX (qv) interplanetary thruster. A thruster of intermediate size, in the 50 mN category is also under development by the DRA (ex RAE).

DRA provides: programme technical lead, thruster design/manufacture, testing, electronics and hollow cathode development. Culham: thruster testing (including diagnostic life tests), plasma physics, ion beam extraction modelling, UK-25 development/testing. Matra Marconi Space (MMS): development of the UK-10 system for Artemis, including the power conditioning and control equipment and the propellant supply/monitoring equipment and the manufacture and qualification of the flight hardware for Artemis. An element of AEA Technology, Culham houses Europe's largest Xe propellant ion thruster test chamber: 5.8 m long, 1.3 m ∅, equipped with helium cryopumps for handling the inert gas. During 1996 MMS commissioned a new test facility, capable of handling two 25 mN thrusters simultaneously.

UK-10 ion thruster development in the UK began in the late 1960s at RAE/Culham, concentrating on a 10 cm ∅ thruster with a nominal 10 mN thrust and mercury propellant. The T4A engineering model achieved very high efficiency, stable operation and a long operational lifetime. The propellant was changed to Xe in 1985 and new versions of the T5 flight model thruster have been manufactured/tested in a collaboration between DRA, Culham and Matra Marconi Space. 0.2-70 mN thrust has been demonstrated; qualification testing at 18 mN is underway at MMS and DRA, specifically for operational NSSK application on ESA's Artemis. This will be extended to 25 mN during 1997. A major experimental programme was completed in 1995 at the Aerospace Corp in Los Angeles, using USAF funding and a thruster supplied by DRA. This concentrated on a detailed characterisation of the ion beam. A lower level effort continues at Aerospace funded by DRA, MMS and the USAF. The thruster comprises a cylindrical discharge chamber closed at one end by a soft iron backplate and with a set of closely-spaced grids at the other. A magnetic field is generated by six equispaced peripheral solenoids. The field lines inside the chamber link an inner cylindrical soft iron pole and a larger diameter outer pole. Propellant gas is introduced through the axial hollow cathode and a bypass distributor on the backplate. A DC discharge is set up between the cathode and the cylindrical anode. This ionises the gas, the efficiency being enhanced by the magnetic field and the correct design of the inner pole/baffle disc arrangement. The positive ions are extracted and accelerated by a high electric field between the grids, attaining typically 30-60 km/s. A triple grid design (see diagram) is used to minimise damage caused by the impact of charge-exchange ions. The ion beam's charge is neutralised by electrons from an external cathode, and the whole thruster is surrounded by an earthed screen to prevent those electrons from reaching other parts of the device. MMS UK is responsible for the electronics and propellant feed systems, and commercial exploitation of the fully qualified operational system. Philips Components Ltd previously manufactured the hollow cathode and neutraliser; the work has now been transferred to DRA. DRA/Culham in 1994 produced a low cost version of the complete system for experimental applications. This was accepted for flight on Johns Hopkins Univ's NEPSTP Nuclear Electric Propulsion Space Test Program, since cancelled, which was intended to test the Russian Topaz 2 nuclear reactor. Other flight opportunities are being sought.

DRA's UK-10 thruster under test at the Aerospace Corp. Detailed characterisation of the ion beam was completed in 1995 (The Aerospace Corp) **1997**/0003458

ESA's Artemis will carry two 18 mN versions. Another application may be ESA's Gravity Explorer Mission; DRA/Dornier studied the possibility in 1995/6 with encouraging results. DRA's STRV-1A (see UK National), launched in Jun 1994, included an experiment to allow the hollow cathode assembly to demonstrate spacecraft electrostatic discharging. Although the cathode assembly operated correctly, charging was not observed. The nominal operating parameters given below are in 18/25 mN order.
Thrust: 18/25 mN
Beam voltage: 1,100/1,100 V
Accelerator voltage: −250/−250 V
Decelerator voltage: −50/−50 V
Beam current: 0.329/0.457 A
Exhaust velocity: 40,233/40,233 m/s
Specific impulse: 3,084/3,131 s
Discharge power: 80/106 W
Beam power: 362/503 W
Power/thrust ratio: 26.4/25.7 W/mN
Electrical efficiency: 76/78%
Total efficiency: 57/60%

UK-25 design of a laboratory model 25 cm thruster began in early 1986 using scaling laws formulated during the UK-10 programme. Culham testing began Oct 1986; nominal thrust was 200 mN (highest obtained was 316 mN). Following the successful conclusion of the lab model work, an engineering model was manufactured at the beginning of 1989 and completed a comprehensive test/evaluation programme in 1992. Included were studies of life-limiting factors. Investigation of other propellants was completed in 1995 at Southampton Univ (although an ideal inert propellant, xenon is costly). Development of a power conditioning system was undertaken at Birmingham Univ and high current hollow cathodes were developed by DRA/Philips Components. Further work on the ESA-XX hybrid design, using features from this thruster and the German RIT-35, is underway with ESA funding. The first prototype of this thruster, intended for interplanetary missions, has been successfully tested at the Univ of Giessen in Germany.

USA

AEROJET

A segment of GenCorp of Akron, Ohio, Aerojet was founded in 1942 by astronautics pioneer Dr Theodore von Karman and initially developed JATO jet-assist take-

Aerojet's paired LR-87-AJ-11 engines power the Titan 3/4 first stages

off rockets for aircraft. It was the first US company to produce storable bipropellant rockets. The propulsion plant, as Aerojet TechSystems (formerly Liquid Rocket Co), provided liquid engines for Aerobee, Bomarc, Nike, Vanguard, Delta, Atlas-Able, Titan 1 and Apollo, and continues to manufacture Titan, Delta and Shuttle engines (see individual entries below). Aerojet, Rocketdyne and P&W formed the Space Transportation Propulsion Team in 1990 for work on the STME Space Transportation Main Engine.

Aerojet manufactured the solid motors for MX, Minuteman, Small ICBM, Standard Missile (rolling out the 10,000th motor in Jun 1989) and Hawk, and successfully bid in partnership as Aerojet Space Boosters (becoming Aerojet ASRM Division) with Lockheed for NASA's Shuttle Advanced Solid Rocket Motor contract. Congress cancelled ASRM in Oct 1993. Propulsion activities are performed at the 52.6 km^2 Sacramento plant, founded 1951 and capable of testing cryogenic and storable propellant motors up to 6,670 kN. The Mach 8/30 km-altitude hypersonic Hytest facility was added for National AeroSpace Plane engines and components tests of up to 35 s, and the company has access to the Hypulse M25 facility at subsidiary General Applied Sciences Laboratories Inc (Ronkonkoma, NY), which it acquired in 1989.

Aerojet teamed with Lyulka in 1993 to improve and market the 400 kN LOX/LH_2 D-57 engine, originally developed in the 1960s for the improved N1-L3M manned lunar mission. It is a candidate for an SSTO

Aerojet's LR-91-AJ-11 Titan 3/4 stage 2 engine

demonstrator. A teaming agreement was signed Jul 1993 with Kuznetsov/NPO Trud (qv) to use the NK series ex-N1 engines and their technology in the US market. Aerojet made five tests totalling 410 s with an NK-33 in Oct-Nov 1995. If it had been selected for a US vehicle as the AJ26-NK33A (it lost to the RD-180 for Lockheed Martin's new Atlas 2AR), a US production line would have been established. Unit cost would have been about $4 million. See the CIS Propulsion section for further information. A similar agreement was signed Oct 1994 with KB Khimautomatiki for the RD-0120, principally developing it into a tripropellant engine. Work is being performed under a $17.2 million NASA Marshall contract.

AJ10-118K

This pressure-fed engine, optimised for altitude operation, has flown as the stage 2 propulsion system for McDonnell Douglas' Delta and the related Japanese N vehicle, in addition to flying paired (as the AJ10-138) on Titan 3's Transtage upper stage. Contracts continue through 1998, building about 10 annually.
Applications: Delta stage 2
First flown: Aug 1982, Delta 164
Dry mass: 124.7 kg
Length: 269 cm
Mounting: fixed

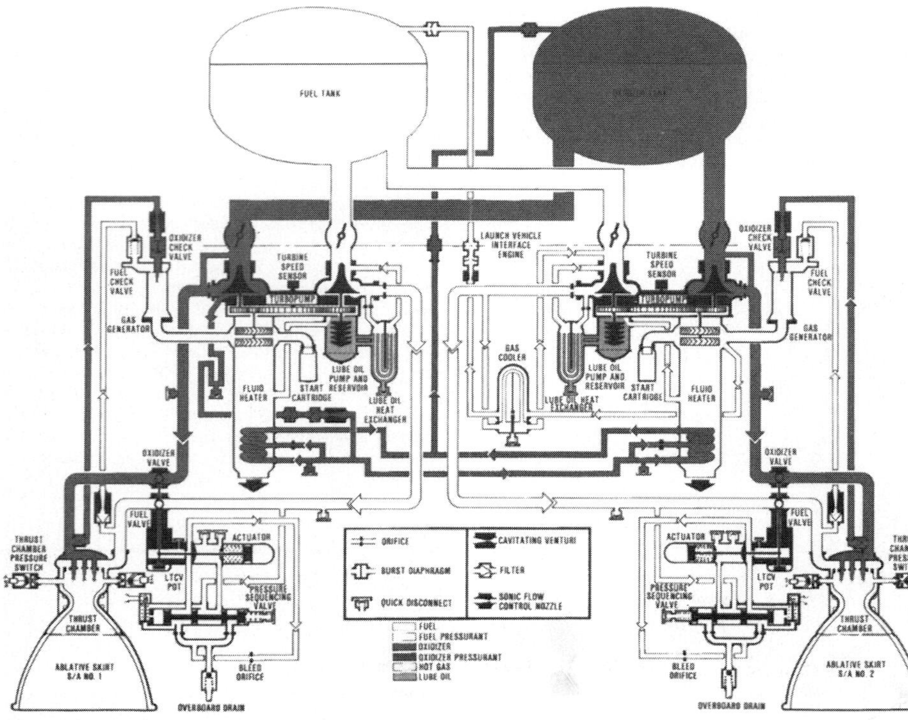

LR-87-AJ-11 schematic (Aerojet)

Engine cycle: pressure-fed
Oxidiser: nitrogen tetroxide at 9.1 kg/s
Fuel: Aerozine-50 at 4.76 kg/s
Mixture ratio: 1.9:1
Thrust: 43.38 kN vacuum
Specific impulse: 320.5 s vacuum
Expansion ratio: 65:1
Combustion chamber pressure: 8.84 atm
Cooling method: ablative chamber, radiative skirt
Burn time: qualified up to 500 s (unlimited starts).

Orbital Maneuvering System (OMS)

NASA's Shuttle Orbiter carries two OMS pods, each housing a single Aerojet OMS engine for orbit insertion, manoeuvring and re-entry initiation. The engines are capable of 100 missions and 500 starts in space. 12 were originally delivered for the Shuttle fleet (including four reserves), with two more added following *Challenger's* loss (#102/112) in 1986 (these were completed as #115/116 in summer 1989 and first flown on STS-50 in Jun 1992). Aerojet was awarded a $3.6 million 38-month contract in Nov 1986 to continue development of an uprated OMS featuring increased chamber pressure through the addition of a pump-feed system. Successful testing beyond 600 s was completed in autumn 1990 at NASA's White Sands altitude simulation facility.

OMS is derived from Aerojet's Apollo Service Propulsion System, designed with high reliability for manned lunar missions. The first OMS prototype and demonstration tests were completed 1972-3 and the company was awarded the full development contract in 1974. The thrust chamber was successfully tested for the first time in 1976 and the first prototype development engine was delivered to NASA in Feb 1977 for extensive White Sands testing to begin the following Aug. A further engine, used to qualify the system as part of the Orbiter's pod, was delivered in Jan 1979 and fired 270 times that year, accumulating 10,817 s. The first two production engines, for *Columbia*, were delivered Mar 1979, and testing of the final qualification engine, delivered Oct 1979, was completed in Jun 1980. In the first 24 Shuttle missions, nine of the 12 engines were carried and fired 254 times for a total of 21,875 s. In the 27 missions following *Challenger* to end-1992, eight engines were used for 377 firings totalling 39,995 s without anomaly.
Applications: Space Shuttle orbit/de-orbit insertion, circularisation
First flown: 12 Apr 1981, Orbiter *Columbia*
Number flown: 14, to end-1995 (103/104 have never flown)
Dry mass: 118 kg
Length: 195.6 cm
Maximum diameter: 116.8 cm
Mounting: gimballed ±7° yaw/±6° pitch by two electromechanical actuators for thrust vector control
Engine cycle: pressure-fed (improvement programme underway for pump-feeding)
Oxidiser: 6,743 kg nitrogen tetroxide in each pod (pods can be cross-linked)
Fuel: 4,087 kg of monomethyl hydrazine in each pod (pods can be cross-linked)
Mixture ratio: 1.65:1
Thrust: 26.7 kN vacuum
Specific impulse: 316 s vacuum
Expansion ratio: 55:1
Combustion chamber pressure: 8.62 atm
Cooling method: fuel regenerative for chamber, radiative for nozzle
Burn time: qualified for 500 starts, 15 h/100 mission life, longest firing 1,250 s, de-orbit burn typically 150-250 s.

Titan Space Launcher Engines

Aerojet has developed and manufactured the engines for four generations of USAF/NASA Titan ICBMs and space launchers since 1958, attaining a peak annual production rate of 270. Each Titan core carries a dual-chamber engine in the first stage and a scaled-down single-chamber version to power stage 2. The company has refurbished 14 sets of Titan 2 engines for orbital launcher applications, and is manufacturing new 11-family engines for Titan 4. The Titan 4 contracts, worth some $350 million, call for 41 USAF and four NASA engine sets.

AJ-5 Titan 2 engines

Aerojet began testing the first stage 1/2 experimental engines in 1960 and completed production of the 474 chambers by end-1967. The ICBMs were overhauled 1974-82 and the company subsequently refurbished 14 sets for the space launcher version. The Gemini-Titan 2 manned version flown 1964-66 utilised man-rated 7-type engines.
Designation: Aerojet LR-87-AJ-5
Configuration: twin fixed motors with individual turbopump assemblies
Applications: Titan 2 stage 1
First flown: 1962 ICBM; Sep 1988 orbital
Dry mass: 1,266 kg for full assembly

Length: 2.3 m
Maximum diameter: 1.1 m
Engine cycle: gas generator
Propellants: hypergolic nitrogen tetroxide and Aerozine-50, delivered at 754 kg/s
Mixture ratio: 1.93:1
Thrust: 1,913 kN sea level
Specific impulse: 259 s sea level
Expansion ratio: 8:1
Combustion chamber pressure: 53.7 atm
Burn time: about 165 s

Designation: Aerojet LR-91-AJ-5
Configuration: scaled-down version of stage 1 engine, featuring fixed single chamber
Applications: Titan 2 stage 2
First flown: as stage 1 engine above
Dry mass: 472 kg
Length: 2.80 m
Maximum diameter: 1.68 m
Engine cycle: gas generator
Propellants: as stage 1, at 146.5 kg/s
Thrust: 444.8 kN vacuum
Specific impulse: 312 s vacuum
Expansion ratio: 49.2:1
Combustion chamber pressure: 56.2 atm
Burn time: about 185 s.

LR-87-AJ-11
Titan 4's first stage (and Titan 3's if it flies again) are powered by paired AJ-11 engines. The 9 model was first tested in 1963 and 476 were produced before being replaced by the 11 in late 1968, first tested earlier that year. Together with its stage 2 derivative, it is the only US engine to be operated on storable N_2O_4/A-50 *and* tested on LOX/RP1. Ablative skirts with 12:1 and 15:1 expansion ratios are available, producing 254 s (sea level) and 302 s (vacuum) SI. USAF awarded a contract to Martin Marietta in 1990 for a modified Titan 4B nozzle. This requires an increased expansion ratio for ignition at greater altitude when Hercules' higher performance strap-ons are carried.
Applications: Titan 3/4 stage 1
First flown: 1968 Titan 3, 1989 Titan 4
Dry mass: 1,874 kg (paired), 758 kg (single)
Length: 3.84 m to top of thrust structure, 3.23 m to top of turbopump assembly
Maximum diameter: 1.6 m
Mounting: gimballed pair
Engine cycle: gas generator
Oxidiser: nitrogen tetroxide at 513 kg/s
Fuel: Aerozine-50 at 268 kg/s
Mixture ratio (O/F): 1.91:1
Thrust: 2,340 kN vacuum paired
Specific impulse: 301 s vacuum
Expansion ratio: 15:1
Combustion chamber pressure: 55 atm
Cooling method: fuel regenerative and ablative skirt
Burn time: about 200 s.

LR-91-AJ-11
Development of the current Titan 3/4 stage 2 engines followed that of the stage 1 version. It has flown without failure on every operational Titan 3 mission.
Applications: Titan 3/4 stage 2
First flown: late 1968 Titan 3, 1989 Titan 4
Dry mass: 589 kg
Length: 281 cm
Maximum diameter: 163 cm (skirt outer diameter)
Mounting: gimballed, turbine exhaust utilised for roll control
Engine cycle: gas generator
Oxidiser: nitrogen tetroxide at 97.0 kg/s
Fuel: Aerozine-50 at 54.7 kg/s
Mixture ratio: 1.86:1
Thrust: 467 kN vacuum
Specific impulse: 316 s vacuum
Expansion ratio: 49.2:1
Combustion chamber pressure: 58.5 atm
Cooling method: fuel regenerative thrust chamber, with separate ablative skirt
Burn time: about 247 s.

Transtar
Transtar was developed as an upper stage engine using injector, chamber and nozzle derived from OMS. However, the propellants are pump-fed, increasing chamber pressure and SI, and permitting use of low-pressure lightweight tankage.
Applications: upper stage
First flight: not flown
Dry mass: 67 kg
Length: 127 cm
Mounting: gimballed ±10° by two electromechanical actuators
Propellants: N_2O_4/MMH
Mixture ratio: 1.8:1
Thrust: 16.680 kN vacuum
Specific impulse: 328 s vacuum
Expansion ratio: 132:1
Combustion chamber pressure: 23.8 atm

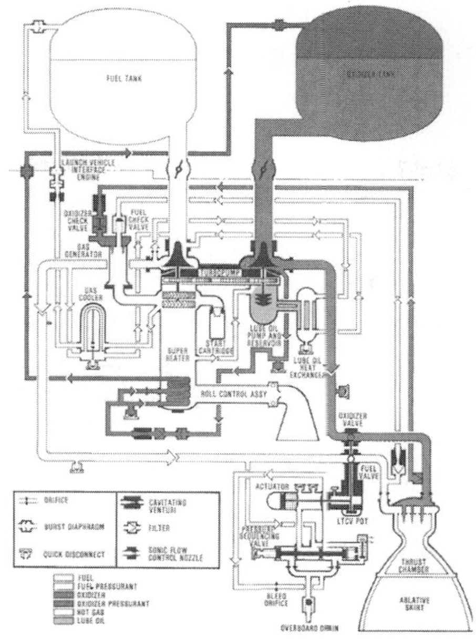

LR-91-AJ-11 schematic (Aerojet)

Cooling method: fuel regenerative for chamber, radiative for extension
Burn time: not available, 15 starts.

Aerojet Satellite Engines
Aerojet also produces 2, 21, 62 and 445 N bipropellant (NTO/MMH) thrusters for satellite orbit-adjust and attitude control systems. Qualification of a 21 N thruster is being completed for a USAF satellite. Principal characteristics are:

Thrust (N, vac)	*2.00*	*21.35*	*62*	*445*
Specific impulse (s, vac)	265	285	287	309
Dry mass (kg)	0.27	0.57	1.13	1.86
Expansion ratio	150	150	75	150
Mixture ratio (O/F)	1.65	1.60	1.65	1.65

ALLIANT TECHSYSTEMS, INC
Alliant Techsystems acquired Hercules Aerospace Co in Mar 1995 for $296 million. Hercules' rocket motor, weapons and ordnance businesses totalled $660 million revenue in 1993, for a $105 million operating profit. Hercules was selected in Oct 1987 as the propulsion contractor for Titan's Solid Rocket Motor Upgrade programme, potentially worth $725 million for the development, qualification and production of 15 sets of Titan 4 solid boosters. Lightweight graphite composite materials replaced the steel used in current motor cases, and a high performance propellant similar to that developed for the Delta 2 strap-on is employed. PQM-1 (Preliminary Qualification Motor) was the first full scale test, but the motor failed after a few seconds at Edwards AFB 1 Apr 1991. The added PQM-1' firing was successful 12 Jun 1992. The QM-2 third test, run at 2.5°C, was successful 21 Feb 1993. QM-3 at the 41°C upper limit was successful 2 Jun. The final QM-4, 12 Sep 1993, would allow flight introduction late 1995, although debut is now expected 4Q 1996. Delivery of the first flight set was made 1Q 1994. Lockheed Martin awarded the company a contract for Tital launch vehicles support which extends to 2003.

The company was awarded a McDonnell Douglas contract in 1987 to develop the stretched solid rocket GEM graphite epoxy strap-on motors for Delta 2; they first flew Nov 1990. The initial contract for 144 motors (16 flight sets) was followed in 1991 by a second for 117 (13 sets) beginning production in 1993, and a third in Jun 1995 for 144 (16 sets). GEM production rate is 6/month. For Delta 2, from 1996, the expansion ratio for the air-lit nozzles will be increased from 10.6 to 16.3 by lengthening the nozzle by 30 cm and increasing its ∅ by 19.5 cm. The GEM-VN version incorporates a vectorable nozzle; the qualification firing was made May 1994. The TVC system is provided by AlliedSignal and the nozzle by BP-Hitco as part of the strategic partnership. Delta 3's nine strap-ons will be lengthened by 1.22 m and their diameters increased to 116.8 cm. Three of the six ground-lit motors will carry gimballed nozzles for TVC, and the three air-lit will improve performance with extended nozzles. The first of 16 sets ordered Jun 1995 will be delivered late 1997.

Alliant produces Pegasus' Orion solid motors, derived from GEM, and the fairing. Taurus also uses the Orion motors. Hercules also designs/fabricates composite structures: spars, struts and optical benches used in satellites.

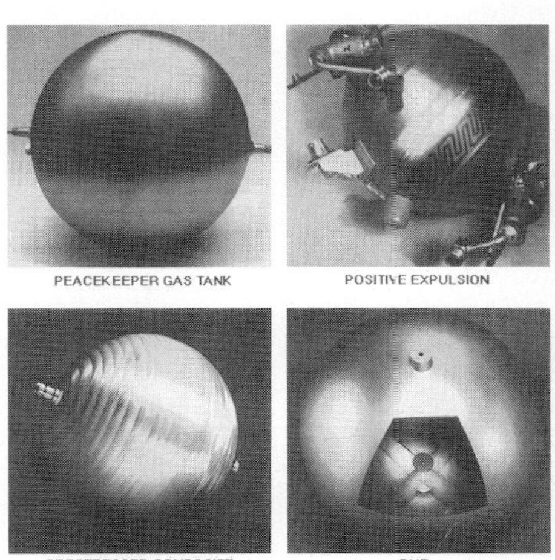

Ardé's cryoformed 301 CRES tankage ***1995***

ATLANTIC RESEARCH CORP
ARC, founded in 1949, is a manufacturer of solid propulsion motors and gas generators, and in 1987 expanded into liquid propulsion with the acquisition of Bell Aerospace Textron. The company's Virginia and Arkansas facilities take solid motor concepts from design through to high volume production; the company's annual propellant production capacity is 9,000 t. ARC produces propulsion units for missiles such as Stinger, Tomahawk, Peacekeeper and Trident 2, but also undertakes space-related projects: the company teamed with Hercules in Dec 1987 to bid (unsuccessfully) for NASA's Shuttle Advanced Solid Rocket Motor. Israel's 1.3 m and AUS-51 'Marble' motor, flown as Shavit's stage 3, is marketed by ARC/Rafael. ARC markets the ASTM family of Shavit and NEXT launcher stage 1/2 motors in the US under agreement with TAAS Israel Industries Ltd.

ARC also manufactures rocket motor components, including igniters, initiators, casings and nozzles using advanced materials and filament winding/braiding techniques. Nov 1987's acquisition of Bell Aerospace Textron added liquid bipropellant attitude control and apogee satellite propulsion systems, and their associated tanks & valves. ARC's 22 N (5 lbf) bipropellant thruster was developed for Intelsat 6's attitude control system and has demonstrated 400,000 pulses, 32 h aggregate firing and 748 kg total throughput. This thruster is carried by Optus B, UFO and Intelsat 7. As Bell, ARC/LP developed the Agena propulsion system that flew on >370 missions. The engine remains available for 31.1-120 kN thrust.

BOEING – SPACE DIVISION
Rocketdyne was a division of Rockwell International and in Dec 1996 was included in the Rockwell merger with Boeing. It is devoted primarily to the design and manufacture of rocket engines – it is the western world's largest producer, operating from its Canoga Park engineering and 16 km-distant 10.5 km² Santa Susana test facilities. 1995 revenue was $880 million; 5,400 personnel. It was established as a separate division of North American Aviation Inc on 8 Nov 1955 and following its first major propulsion system for the Navaho missile has delivered around 3,500 engines for Jupiter, Redstone, Thor, Delta, Saturn, Atlas and Shuttle. It currently provides the powerplants for Atlas and Delta,

International Space Technology's plasma thruster is flight-proven on Russian Meteor satellites (Space Systems/Loral) ***1995***

and continues to test, upgrade and refurbish Shuttle's main engines, still the only large reusable systems in service. It designed and fabricated Peacekeeper's stage 4, its first integration of a full stage.

Rocketdyne was selected in Dec 1987 to develop Space Station's silicon photovoltaic solar arrays, batteries and power management/distribution. 1994 Station revenue was $282 million; 1993 $340 million.

MA-5 Atlas Engine System

The MA-5 propulsion system comprises a YLR89-NA-7 dual-chamber liquid propellant booster engine, a YLR105-NA-7 single-chamber liquid sustainer and two YLR101-NA-15 verniers to control vehicle roll and provide final velocity and directional control following sustainer burnout. The single-start main engines are fed from the same propellant tanks but Atlas uniquely separates the two outer boosters some 172 s into flight, leaving the altitude-configured sustainer to provide propulsion to near-orbital velocity. This system was introduced into space-launcher Atlas versions in 1961 while its direct predecessor continued in service with Atlas E/F ICBMs, subsequently refurbished as orbital launchers and last flown in 1995.

Applications: Atlas 1 propulsion system
First flown: 18 Nov 1961, Atlas 117D
Number flown: 217 MA-5 (to Apr 97), 152 MA-3, 118 MA-2 to end-1995
Dry mass: 1,423 kg for both boosters, 470 kg sustainer
Length: 340 cm booster, 269 cm sustainer (with aspirator)
Max diameter: 119 cm booster, 117 cm sustainer
Mounting: all thrust chambers are gimballed
Engine cycle: gas generator
Oxidiser: liquid oxygen at 458 kg/s for booster, 89 kg/s for sustainer
Fuel: RP-1 hydrocarbon at 203 kg/s for booster, 39 kg/s for sustainer
Mixture ratio: 2.25:1 booster, 2.27:1 sustainer
Oxidiser turbopump: each 1,678 kW, 6,732 rpm, 68 atm discharge pressure for boosters, 846 kW, 10,568 rpm, 73 atm discharge pressure for sustainer
Fuel turbopump: each 1,116 kW, 68 atm discharge pressure for boosters; 508 kW, 73 atm discharge pressure for sustainer
Thrust: 1,882 kN for booster pair vacuum/1,681 kN SL; 374 kN sustainer vacuum/269 kN SL; verniers are 2,975 N each
Specific impulse (s): 292 vac/259 SL for boosters; 309 vac/220 SL for sustainer
Time to full thrust: 2.0 s
Expansion ratio: 8:1 boosters, 25:1 sustainer
Thrust chamber
  length: 249 cm both booster/sustainer
  materials: nickel 200
  cooling method: regenerative by two passes of fuel through 292 tubes in booster chambers, 240 tubes in sustainer
Combustion chamber
  pressure: 44 atm booster/50 atm sustainer, at injector end
  temperature: 3,316°C both booster/sustainer
  materials: nickel 200
  cooling: as thrust chamber
  ignition: hypergolic fluid cartridge enclosed in burst diaphragms
Burn time: 167 s max booster, 368 s max sustainer
Verniers: each 24.1 kg mass, 2.24/2.97 kN SL/vac thrust, 172/231 s SL/vac SI, 1.8 mixture ratio, 5.66 expansion ratio (exit ∅ 9.65 cm), 391 s burn time.

MA-5A/RS-56 Atlas Engine System

The MA-5 system was uprated for Atlas 2 by replacing the booster engines, now designated RS-56BA, with Rocketdyne's RS-27 (qv), providing a sea level SI increase of 4 s. The side-mounted verniers were deleted; their roll control and final adjustment functions were assumed by thruster modules on the vehicle's interstage. Sustainer engine (RS-56SA) specifications are as for the MA-5 except that total oxidiser/fuel flow rates are 90/38 kg/s, respectively; only booster engine specifications are given below. **MA-5B** and **MA-5C** have been studied for future Atlas versions, but lost out to Russia's RD-180 for Atlas 2AR.

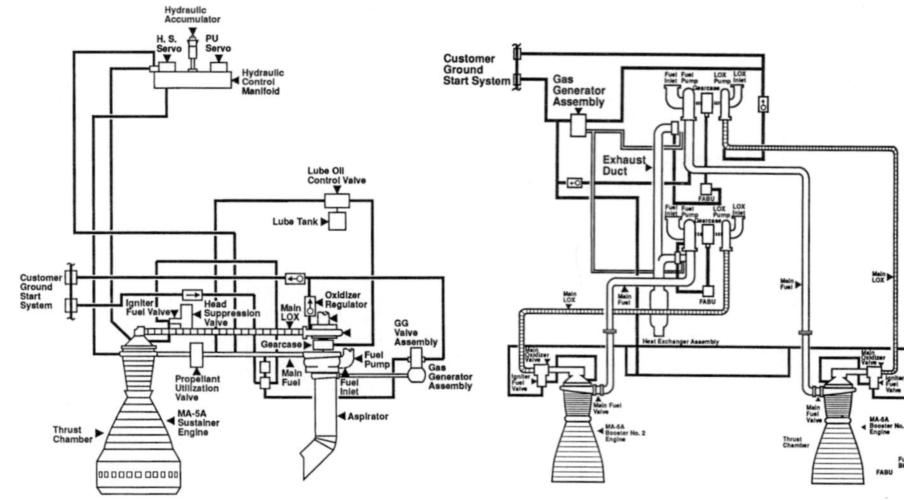

Atlas 2 MA-5A propulsion schematic. Note that Atlas 1/MA-5's integrated thrusters have been deleted

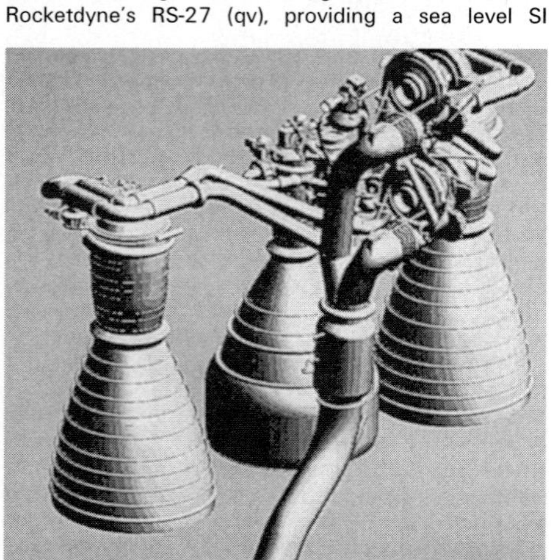

The MA-5A propulsion system for Martin Marietta's Atlas 2 incorporates RS-27 engines from Delta's powerplant as the boosters

Applications: Atlas 2 primary propulsion system
First flown: 7 Dec 1991
Number flown: 26 to end-1996 (49 systems contracted)
Dry mass: 1,610 kg
Length: 343 cm
Maximum diameter: 119 cm
Mounting: all thrust chambers are gimballed
Engine cycle: gas generator
Oxidiser: liquid oxygen pump-fed at 505 kg/s
Fuel: RP-1 hydrocarbon pump-fed at 224 kg/s
Mixture ratio: 2.25:1
Oxidiser turbopump: each 1,903 kW, 6,730 rpm, 70 atm discharge pressure
Fuel turbopump: 1,362 kW, 75 atm discharge pressure
Thrust: total 2,100 kN vacuum, 1,890 kN SL
Specific impulse (s): 295 vac, 263 SL
Time to full thrust: 2.0 s
Expansion ratio: 8:1
Thrust chamber
  length: 249 cm
  materials: 347 CRES austenitic stainless steel
  cooling method: regenerative, two passes of fuel through 292 chamber tubes
Combustion chamber
  pressure: 48 atm at injector end
  temperature: 3,316°C
  materials: 347 CRES austenitic stainless steel
  cooling: as thrust chamber
  ignition: hypergolic fluid chamber enclosed in burst diaphragms
Burn time: 167 s max flight duration.

RS-27/RS2701A, B, RS-27A/RS2701C

The RS-27 powerplant comprises an RS2701A/B main engine and twin LR101-NA-11 verniers. It was introduced in 1974 on McDonnell Douglas' Delta 2000 series launcher, replacing the MB-3, completed its Delta service on the 6000 model in 1992, and continues in service as part of Atlas' MA-5A powerplant. The uprated RS-27A version featuring a 12:1 expansion ratio for improved altitude performance is employed by the 7000 series; its specifications are given below, where different, in parentheses. The engine is a hybrid design utilising the turbopump, turbine, gas generator, valves and thrust chamber of Rocketdyne's H-1 Saturn engine, and the control system, start system and component packaging arrangement of the MB-3.

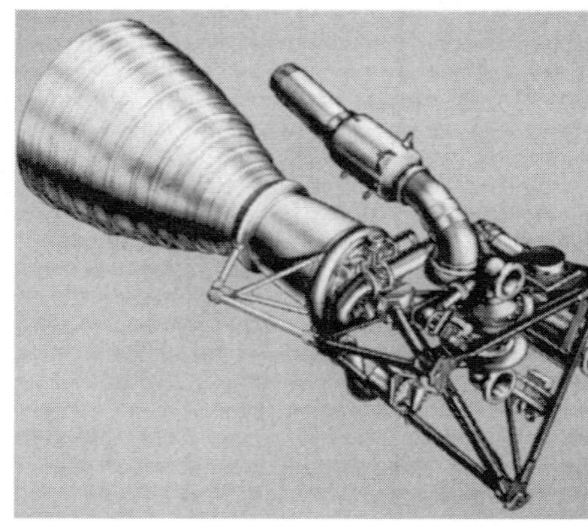

Rocketdyne's RS-27 was in service with Delta 1974-92; it continues aboard Atlas 2

Applications: Delta 6000 series (RS-27A: 7000 series)
First flown: 18 Jan 1974 on Delta 100/Skynet 2A (RS-27A debut 1990)
Dry mass: 1,027 kg (1,091 kg)
Length: 363 cm from top of thrust ring (378 cm)
Maximum diameter: 170 cm envelope
Mounting: gimbal-mounted for pitch/yaw control + gimballed verniers for roll control
Engine cycle: gas generator
Oxidiser: liquid oxygen at 250 kg/s
Fuel: RP-1 hydrocarbon at 111 kg/s
Mixture ratio: 2.245:1
Oxidiser turbopump: 1,900 kW, 6,784 rpm (7,085 rpm at altitude), 70 atm discharge pressure
Fuel turbopump: 1,289 kW, 70 atm discharge pressure
Thrust: 971 kN SL/1032 kN vac (890 kN SL/1,054.2 kN vac)
Specific impulse (S): 264 SL/295 vac (255 SL/302 vac)
Expansion ratio: 8:1 (12:1)
Thrust chamber
  length: 219 cm (234 cm)
  materials: 347 CRES austenitic stainless steel
  cooling: regenerative, two passes of fuel through 292 tubes
Combustion chamber
  pressure: 48 atm at injector end
  temperature: 3,315°C
  materials: 347 CRES austenitic stainless steel

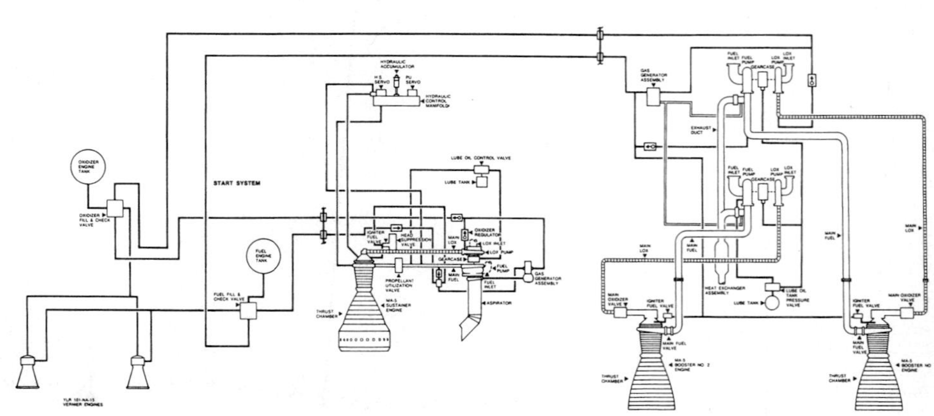

Atlas MA-5 propulsion system schematic (Rocketdyne)

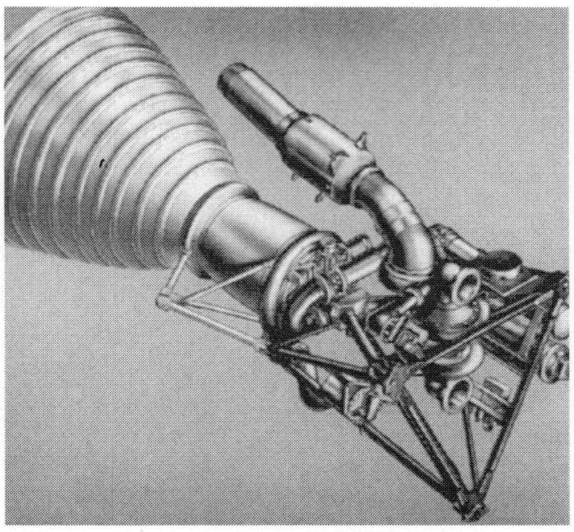

The RS-27A, designed for Delta 7000 vehicles, replaces the 8:1 expansion bell with a 12:1 for improved altitude performance (Rocketdyne)

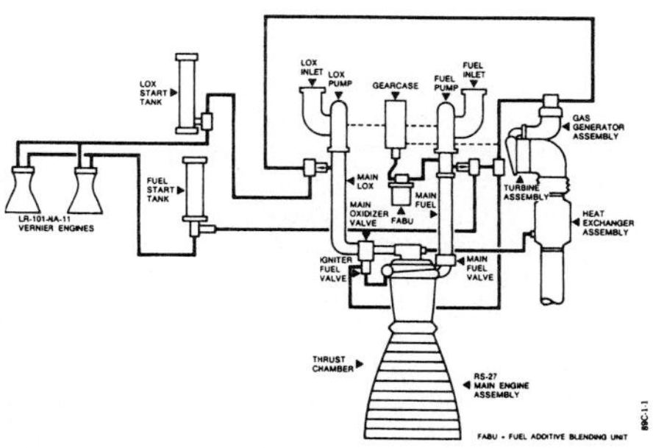

RS-27 engine schematic (Rocketdyne)

cooling: as thrust chamber
ignition: hypergolic fluid cartridge enclosed in burst diaphragms
Burn time: 274 s
Verniers: each LR101-NA-11 21.8 kg mass, 4.63/5.30 kN SL/vac thrust, 209/246 SL/vac SI, 1.8 mixture ratio, 5.6 expansion ratio (9.8 cm exit ∅), 283 burn time.

RS-68

Rocketdyne is working on the 2.89 MN engine for the Delta 4 EELV proposal. Larger than SSME, it would be simpler and cheaper, derived from SSME, STME and J2. Subscale injector tests at NASA Marshall in late 1995 demonstrated >98% combustion efficiency.

Space Shuttle Main Engine (SSME)

Rocketdyne was selected by NASA in Jul 1971 to develop Shuttle's high pressure primary propulsion system. Three oxygen/hydrogen staged combustion SSMEs burn typically for 520 s to propel the vehicle slightly short of orbital velocity. Typically throttled over 65-104% rated power setting, with 109% in abeyance (originally intended for Vandenberg polar missions). Some Space Station assembly missions might require a 106% setting, allowed by the improvements outlined later. Slightly smaller than Rocketdyne's F-1 engine used in the Saturn 5 lunar launcher, it was designed as the first large reusable engine, capable of 55 missions or 7½ h firing time between major overhauls. Flight certification was achieved in Dec 1980, after two

Rocketdyne YLR105-NA-7 sustainer on Atlas Centaur 68. Note the booster unit separation rail (MM)

certification cycles had been completed on each of two engines. Each cycle required a minimum of 13 tests and 5,000 s, including simulation of nominal and abort mission profiles. 40 tests totalling 12,750 s were made on one engine, and 34 tests, totalling 10,650 s, on the other. Another cycle in the Full Power Level certification programme was completed Apr 1983, requiring four test series of 5,000 s each involving engine power levels of 104/105/109/111%. Each flight engine undergoes three tests at NASA's Stennis: 0.5 s ignition, 300 s calibration and 550 s mission duty cycle. The first ascent including 104% power was made by *Challenger* in Apr 1983 for STS-6. The engines are individually gimballed hydraulically by two APU-powered actuators at up to ±10.5° pitch and ±8.5° yaw; each engine is controlled in 20 ms steps by a 97 kg top-mounted package housing two redundant Honeywell 360 16.4kbit-memory computers. The controllers are commanded by the Orbiter's General Purpose Computers, or manually in emergencies. Block 2 Honeywell controllers are installed on OV-105 (and are being retrofitted to the other Orbiters), using the first space-qualified 68000 series microprocessors. They also allow elimination of a separate 13.6 kg vibration monitoring box and add 20 engine parameter monitoring channels to the current 120. The first hot firing with a Block 2 controller was made in Apr 1990. The last of the 30 ordered was delivered Apr 1993.

The incoming oxidiser (propellants are channelled into the Orbiter from the External Tank via 43 cm ∅ pipes and routed to individual engines by 30.5 cm ∅ versions) is boosted by the first turbopump from 7 atm to 29 atm before the High Pressure Oxidiser Turbopump increases it to 500 atm in two stages. This pump has suffered durability problems, with some engines requiring overhaul after three missions (33 min). Pratt & Whitney was awarded a $188 million contract in Aug 1986, as part of NASA's 10 yr $1 billion Phase II SSME improvement programme that began in 1983, to develop a new powerhead containing two LOX and LH_2 high pressure turbopumps and a single pre-burner chamber (the current design incorporates a pre-burner for each). The goal is 7½ h firing (55 missions) between major overhauls. Problems increased turbopump development cost to $504 million, so NASA halted fuel pump work from Dec 1991 in order to focus on the more critical oxygen unit for the Block 1 upgrade. It restarted work May 1994 for 1997 debut as part of the Block 2 upgrade (see below) once the LOX pump was on schedule for its Jul 1995 flight debut. The Block 1 upgrade includes a single coil heat exchanger (eliminating welds) and changing the 3-duct hot gas

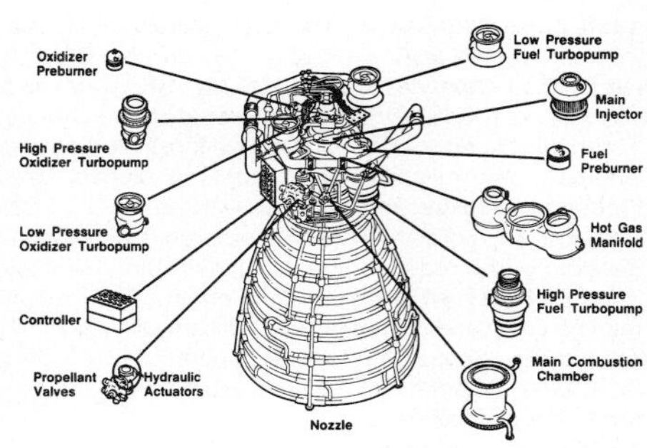

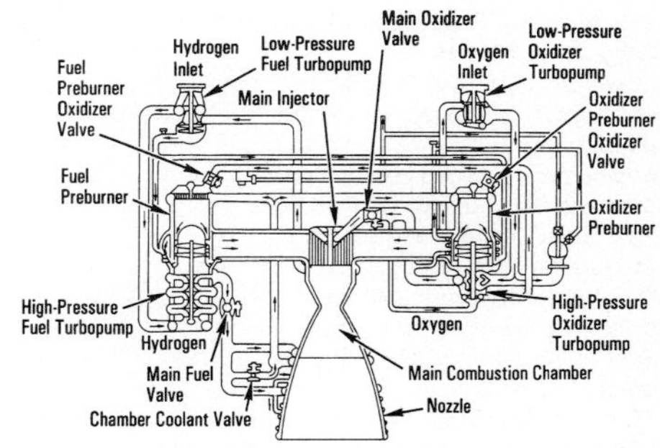

SSME principal features and propellant flow scheme (Rocketdyne)

manifold to a 2-duct design, which allows the fuel pump to run at lower T/P. A single Block 1 upgrade engine (#2036) flew flawlessly on STS-70 in Jul 1995. All three engines were upgraded for STS-77 in May 1996. The pump is now certified for 10 flights between detailed inspections. Advanced fabrication techniques have produced a more reliable engine, with the 157 kg increase offset by a 1 s SI improvement. Testing of the Block 2 upgrade began at NASA Stennis on #0521, once Block 1 certification was completed on #0422/0423 26 May 1995 (0423 ran for a final 761 s; it was later converted to 0523 Block 2). Block 2 includes P&W's fuel pump and a 10% ∅ larger throat combustion chamber that will reduce wear, reducing pump discharge pressure by 4.5-6% and pump operating T by 7-12%. The $90 million large throat contract covers development and eight flight units. All three engines on STS-86 in Sep 1997 will debut Block 2. STS-85 will be the last mission to fly with an old engine; Block 1 or 2 will be used on all subsequent missions.

Some of the oxygen pump output is routed into a pre-burner where it is mixed with fuel and burned hydrogen-rich at 760°C. The hydrogen half of the system likewise employs a low pressure pump (2 to 18.7 atm), high pressure pump (18.7 to 442 atm) and a pre-burner. The pre-burner products are fed into the main chamber with additional oxygen through 600 elements in 13 annular rows in the 56cm ∅ injector plate.

In the 18 months before the resumption of Shuttle missions following *Challenger's* loss in Jan 1986, SSME testing at NASA Stennis accumulated the equivalent of 40 launches. By Jul 1990, an aggressive test campaign covered 293 equivalent ascents, all elements apart from the turbopumps exceeding 55 missions. The programme, including flights, attained 500,132 s 15 Nov 1991 in >1,900 firings, the equivalent of 320 ascents. By end-1993, SSME attained 601,303 s in >2,100 firings. 1993's record was 37,806 s in 88 tests.

SSME static firing at Rocketdyne's Santa Susana Field Laboratory

SSME delivered in 1977 for NASA's three-engine Main Propulsion Test Article. The engine carries a 35:1 expansion ratio bell for ground firings (Rocketdyne)

1994's was 49,181 s in 104 tests. Jan 1995 totalled 7,296 s in 14 tests, including 12 on the Block 1 certification engines (0422/0433). Rocketdyne's contract received a 4 yr $819.8 million extension in Aug 1990. SSME unit cost is up to $40 million. Including test failures, total reliability is 0.9991 in >500,000 s. Production is expected to continue to at least 2010, although no more than the 36 ordered are planned. Because of the reduced Shuttle mission rate, NASA is aiming for only 30 reuses of each engine. They might then be employed in some future unmanned vehicle.
Applications: Space Shuttle Main Engine
First flown: 12 Apr 1981, Orbiter *Columbia*
Number produced: 36 planned
Number flown: 30 to end-1995
Dry mass: 3,177 kg (3,334 kg Block 1)
Length: 4,242 mm
Maximum diameter: 2,388 mm
Mounting: gimballed by hydraulic actuators ±10.5° pitch/yaw
Engine cycle: staged combustion
Oxidiser: liquid oxygen at 408 kg/s
Fuel: liquid hydrogen at 68.0 kg/s
Mixture ratio (O/F): 6:1
Oxidiser turbopump: 261 kg, 17,900 kW, 28,500 rpm, discharge pressure 292.5 atm
Fuel turbopump: 351.1 kg, 46,230 kW, discharge pressure 415 atm
Thrust: 1,668/2,091 kN SL/vac at 100%, throttlable 65-109% in 1% increments. Launch is made at 104%
Specific impulse (s): 452.9 vac (Block 1 goal 453.3)
Time to full thrust: 4.2 s
Expansion ratio: 77.5:1
Thrust chamber
materials: stainless steels
cooling: GH_2 is directed from the high pressure fuel pump through the 1,080 tubes comprising the chamber wall
Combustion chamber
pressure: 204 atm (100% thrust)
temperature: 3,300°C (100% thrust)
materials: NARloy-Z copper alloy and stainless steels
cooling: GH_2 is directed from the high pressure fuel pump through the 390 main chamber wall channels
ignition: augmented spark igniters
Burn time: 520 s typical, 761 s max.

XLR-132

Rocketdyne has completed the advanced technology development phase of the pump-fed high performance upper stage engine. Its performance, mass and extended space storability make it a candidate for perigee/apogee stages, transfer vehicles and lunar and Mars missions. 10-start, 4,000 s demonstration testing was performed on the injector, turbopump, combustion chamber and gas generator. Sea level and simulated altitude testing has been completed in both the breadboard (1991) and flightweight engine programmes (1992), including a 500+ s run in Feb 1992. Current efforts include design studies of engines with thrusts to 111.2 kN, adjustments of the expansion nozzle configuration, and a heavy emphasis on design refinements and producibility to achieve demanding cost goals.
Applications: kick stages, deep space, Space Transfer Vehicle
Dry mass: 54 kg
Length: about 120 cm
Maximum diameter: about 60 cm
Engine cycle: gas generator
Oxidiser: nitrogen tetroxide at 3.3 kg/s
Fuel: monomethyl hydrazine at 1.68 kg/s
Thrust: 16.68 kN vacuum
Specific impulse (s): 340 vacuum minimum
Expansion ratio: 400:1
Thrust chamber
materials: columbium
cooling: radiative
Combustion chamber
pressure: 102 atm at injector end

XLR-132 flightweight engine test (Rocketdyne)

cooling: regenerative by oxidizer
ignition: hypergolic
Burn time: 4,000 s in 10 starts.

HUGHES RESEARCH LABORATORIES

Hughes has developed an 18 mN xenon ion propulsion subsystem (XIPS) for NSSK of 3-axis or spin-stabilised telecom satellites. It will debut on Hughes' own Galaxy 3R satellite in Oct 1995; Astra 1G and ASC 1/2 will use it. XIPS comprises four 13 cm thrusters (two primary + two backup), power supply (439 W input each thruster off 29-34 VDC bus) and a propellant storage/control unit. Total XIPS package mass is 68 kg, for a saving of 400 kg on an HS-601 satellite. 12,000 h of operating time was accumulated by early 1995. In normal NSSK operation, the thrusters will fire for 2-3 h daily.

Xe is fed through a heated (1,000°C) hollow cathode inside the thrust chamber. A ring-cusp magnetic field confines the plasma and a 3-grid ion optics assembly extracts the thrust beam as 3,125 beamlets. Xe is stored at 74.8 atm and reduced to 0.68 atm for feeding into the cathode.

The 13 cm XIPS was developed from work on a 25 cm ∅ 63.5 mN thruster that provided SI 2,800 s at 1.3 kW input. It was also operated at 179 mN, 4,000 s, 4.5 kW. The 25 cm version will be used by the HS-702's improved XIPS.
Thrust: 17.8 mN
Beam voltage: 751 V
Accelerator voltage: −300 V
Beam current: 0.405 A
Accelerator current: 0.97 mA
Exhaust velocity: 30 km/s
Specific impulse: 2,585 s
Beam power: 304 W
Power/thrust ratio: 24.6 W/mN
Electrical efficiency: 88-90%
Total efficiency: 51.3%
Mass: 5.0 kg each thruster (×4); 6.8 kg each power processor (×4), 2.0 kg each Xe tank (×2).

INTERNATIONAL SPACE TECHNOLOGY, INC

Space Systems/Loral and Russia's OKB Fakel and RIAME Research Inst of Applied Mechanics & Electrodynamics formed the ISTI joint venture in 1992 to market Fakel's SPT-100 0.08 N Hall xenon stationary plasma thrusters using western electronics. They provide a 500 kg mass saving over a conventional system on a 3.5 t satellite over a 15 yr life. The venture was later joined by SEP and Atlantic Research Corp. ISTI's thrusters became available in 1995. Fakel has built >100 of the devices, which have flown on >50 Meteor polar metsats since 1972 to provide orbit control. Gals 1, launched Jan 1994, is carrying a set of eight to provide the first NSSK on a Russian satellite. SEP joined ISTI in 1993, receiving commercialisation rights in Europe in exchange for the other partners to have equivalent rights in N America and Russia on the Mk 2 version that SEP is developing. This will increase SI from 1,650 s to 2,200 s. Development began in May 1993 at SEP Villaroche. France's 1999 Stentor satellite was to be equipped with them, but technical problems mean the satellite will be equipped with only slightly modified versions. See the Fakel entry at the end of the CIS propulsion section for other SPT information.

SPT-100 specifications

Thrust: 83 mN
Specific impulse: 1,600 s
Efficiency: 0.48
Electric power: 1,350 W
Mass flow rate: 5.3 mg/s

Five of the Space Shuttle Orbiter reaction control system engines. The smallest is Marquardt's R-1E; the other four are R-40A models in long scarf and short scarf versions (NASA)

SPT 100 MkII under test at SEP ***1995***

Size: 12.5 × 15 22 cm; 3.5 kg
Design cycles: 4,000 *Design total impulse:* 1 MNs (>2 MNs demonstrated).

KAISER MARQUARDT

The Marquardt company was formed in 1944 to undertake R&D leading to the first US subsonic ramjet in 1945. Its principal engineering business continues to be advanced aerospace propulsion and the supply of ram air turbine power systems. The company has been developing and manufacturing bipropellant thrusters since 1959, notably the attitude control thrusters for Apollo spacecraft and the Space Shuttle Orbiter. Precision bipropellant engines are available for 0.004, 0.02, 0.06, 0.11, 0.22, 0.44, 0.89, 4 kN sizes, and requests up to 8 kN can be accommodated. The monopropellant hydrazine thruster line of UTC's Hamilton Standard Div was acquired in 1993; this is described separately at the end of this entry.

Kaiser Marquardt R-1E

This 110 N hypergolic thruster was qualified for the USAF Manned Orbiting Lab and then produced as the Space Shuttle Orbiter's vernier attitude control/orbit adjust thruster. Six are employed in conjunction with 38 3,870 N R-40 thrusters.
Applications: Shuttle Orbiter vernier, satellite orbit adjust
First flown: Apr 1981 on Shuttle *Columbia*
Dry mass: 3.7 kg
Length: 279 mm, 312 mm satellite version
Maximum diameter: 15.2 cm satellite version, 14.0 cm Shuttle version
Mounting: fixed
Engine cycle: pressure-fed at 15 atm (range 6.8-27.2 atm)
Oxidiser: nitrogen tetroxide at 0.0256 kg/s
Fuel: monomethyl hydrazine at 0.0354 kg/s
Mixture ratio (O/F): 1.65 (range 1.0-2.7)
Thrust: 110 N (range 67-155.7 N)
Specific impulse: 280 s vacuum
Expansion ratio: 100:1 standard
Combustion chamber: single chamber, min expansion ratio 26 with orthogonal and scarfed nozzles. Made of silicide-coated columbium (niobium) and insulated for buried installation. Started by electrical signal to on/off solenoid valve. Single doublet injector with hypergolic ignition
Burn time: min 82,000 s demonstrated, 0.89 Ns min impulse bit.

Kaiser Marquardt R-4D

Development of the R-4D began in 1962 as an attitude control thruster for Apollo's Service/Lunar Modules, each employing four quadruple clusters; about 800 were built. The thruster is currently employed in the R-4D-10 version by the US Navy's Leasat, R-4D-11 by Insat 1, Arabsat 1, HS-376W, HS-393, Milstar, Intelsat 6, Italsat, Olympus and Eurostar. Versions are also used on GOES, Superbird, HS-601, Italsat, IABS, Mars Observer and Cassini.
Applications: apogee/perigee manoeuvres, orbit adjust, attitude control
First flown: Feb 1966 (Apollo 201)
Number produced: >900
Dry mass: 3.63 kg
Length: 55.41 cm
Maximum diameter: 27.94 cm (38 cm 11-300 model)
Mounting: fixed
Engine cycle: pressure-fed
Propellants: hypergolic N_2O_4 and MMH or hydrazine, pressure-fed at 0.056 kg/s fuel and 0.091 kg/s oxidiser at 15.2 atm
Mixture ratio: 1.65 (1.0-2.4 range)
Thrust: 490 N vacuum standard
Specific impulse: 311.5 s (164:1) and 315.5 s (300:1) vac
Expansion ratio: 164:1 and 300:1
Chamber pressure: 6.84 atm
Combustion chamber: single chamber/nozzle of welded coated columbium (niobium), radiatively and film-cooled, controlled by on/off solenoid, multiple doublet injector with hypergolic ignition

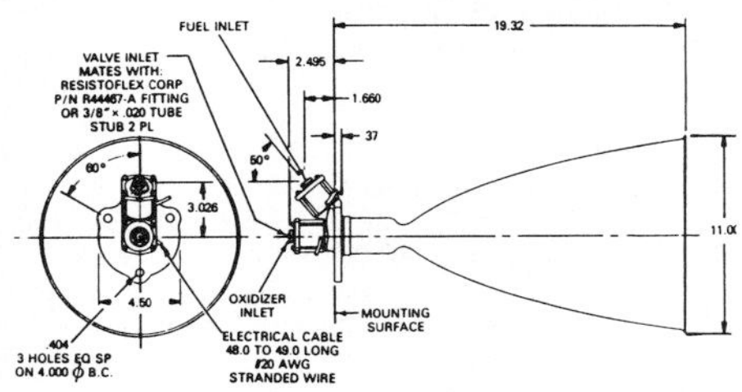

Marquardt's R-4D bipropellant thruster was initially developed for the Apollo manned spacecraft but is now incorporated in satellites for orbital manoeuvres. Dimensions are in inches. The photograph shows the R-4D-11-300 model, which has a 38 cm ∅ nozzle for a 300:1 expansion ratio

Burn time: 45,000 s total demonstrated (including 10,000 s continuous, limited by test facility); 2.67 Ns min impulse bit.

Kaiser Marquardt R-6C

The 6C is a derivative of an engine developed by Marquardt for the Advent communications satellite in the early 1960s; it was qualified in 1981 for Insat 1, Arabsat 1, Olympus and HS-393 satellites. Olympus, for example, employed a network of 22 for attitude control. Milstar also incorporates it. The 22 N thrust is reduced in the R-6C-2.2 version to 10 N.
Applications: satellite AOCS
Dry mass: 0.66 kg
Length: 251.6 mm
Mounting: fixed
Engine cycle: pressure-fed
Oxidiser: N_2O_4 at 0.005 kg/s at 15-17 atm
Fuel: MMH at 0.003 kg/s at 15-17 atm
Mixture ratio (O/F): 1.6±0.1
Thrust: 22 N (range 6.2-32.9 N). R-6C-2.2 version operates at 10 N nominal
Specific impulse: 290 s vacuum at 22 N thrust
Exit diameter: 55.9 mm
Expansion ratio: 100:1
Combustion chamber: single chamber of silicide-coated columbium (niobium). Pressure 6.84 atm. Started by electrical signal to on/off solenoid valve; multiple doublet injector with hypergolic ignition
Burn time: unlimited.

Kaiser Marquardt R-40A

The R-40 was developed for Shuttle Orbiter orbit control: the vehicle carries 38 in long scarf, short scarf or no scarf

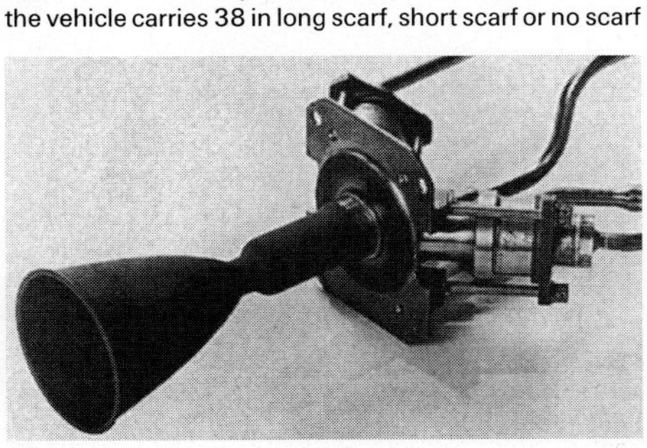

The Marquardt R-6C 22 N bipropellant thruster is utilised by the Milstar military satellites

configurations, depending on location. It has also flown on classified satellites.
Applications: spacecraft orbital manoeuvring, perigee kick engine
First flown: Apr 1981 on Shuttle *Columbia*
Dry mass: 10.25 kg
Length: 55.4 cm short-scarf version, 103.9 cm long-scarf, 88.9 cm for satellite version with 100:1 expansion ratio
Maximum diameter: 30.9 cm long-scarf version, 51.8 cm satellite version of 100:1 expansion ratio
Mounting: fixed
Engine cycle: pressure-fed
Oxidiser: N_2O_4 at 0.838 kg/s at 16.4 atm
Fuel: MMH at 0.526 kg/s at 16.4 atm
Mixture ratio (O/F): 1.6
Thrust: 3,870 N (range 3,114-5,338 N)
Specific impulse: 281 s (306 s for 120 expansion ratio)
Expansion ratio: 20 (40-150 available for satellite version)
Exit diameter: 267 mm (328-635 mm for satellite version)
Chamber pressure: 10.5 atm
Combustion chamber: single chamber of silicide-coated columbium (niobium) with welded-on orthogonal and scarfed nozzle extension in same material. Internal film cooling. Exterior insulated for buried installation. Started by electrical signal to on/off solenoid valve. Multiple doublet injector with hypergolic ignition
Burn time: max 500 s for single burn, qualified to 15,319 s life; 289 Ns min impulse bit.

Kaiser Marquardt R-42

The R-42 was developed for a classified programme. It will be used for Space Station reboost.
Applications: satellite orbit adjust
First flown: classified
Dry mass: <0.45 kg (5.2 kg with series redundant valves)
Length: 787 mm
Maximum diameter: 406 mm
Mounting: fixed
Engine cycle: pressure-fed at 15.0 atm (range 6.8-28.9 atm)
Oxidiser: nitrogen tetroxide
Fuel: monomethyl hydrazine
Mixture ratio (O/F): 1.65 (range 1.0-2.2)
Thrust: 890 N (range 578-1,200 N)
Specific impulse: 303 s vacuum
Expansion ratio: 160:1
Combustion chamber: C-103 silicide-coated columbium (niobium), radiation and fuel film cooling. Started by electrical signal to on/off solenoid valve. Single doublet injector with hypergolic ignition
Burn time: >22,000 s, 15.6 Ns min impulse bit, 20.0 MNs max.

Kaiser Marquardt R-53

Applications: satellite AOCS
First flown: unflown, but development completed
Dry mass: 0.41 kg
Length: 165 mm
Maximum diameter: 51 mm
Mounting: fixed
Engine cycle: pressure-fed at range 6.8-20.4 atm
Oxidiser: nitrogen tetroxide
Fuel: monomethyl hydrazine
Mixture ratio (O/F): 1.65 (range 1.0-2.2)
Thrust: 9 N (range 4.0-12.9 N)
Specific impulse: 295 s vacuum
Expansion ratio: 300:1
Combustion chamber: C-103 silicide-coated columbium (niobium), radiation and fuel film cooling. Started by electrical signal to on/off solenoid valve. Single doublet injector with hypergolic ignition
Burn time: 70,000 s demonstrated, 0.03 Ns min impulse bit.

Monopropellant Thrusters

The hydrazine thruster line of UTC's Hamilton Standard Div was acquired in 1993. Hamilton Standard was the world's largest supplier of catalytic hydrazine thrusters and propulsion systems; it delivered >80 monopropellant systems. It provided the complete propulsion system for the USAF DSCS 3 milcomsat and the reaction control modules for the IUS upper stage. The IUS carries six modules, each with two 133 N hydrazine thrusters, to provide roll control during main motor burns, 3-axis control during coast periods, vernier orbit corrections after stage 2 burnout, and collision avoidance manoeuvres following payload separation. The configuration provides dual-nozzle redundancy and there are no forward-facing thrusters, in order to avoid payload contamination. Two tanks hold a total of 109 kg of hydrazine but 1- and 3-tank versions are available. The low thrust (0.89-26.7 N) engine range includes the KMHS 10 (0.89 N), KMHS 17-6 (2.22 N), KMHS 17-12 (4.45 N), KMHS 16 (22.2 N) and KMHS 39-2 (22.2 N). The higher power motors are used for precession despin and ΔV manoeuvres; the 0.89 N versions are employed for wheel desaturation, N-S/E-W stationkeeping and precise attitude control. The 22.2 N thrusters, first flown 1967, were the first space-qualified catalytic engines. The KMHS 39 is capable of either one million cycles or >20 h continuous operation without thrust degradation. The range of mid-thrust (44.5-556 N) engines includes the KMHS 22-5 (53.4 N), KMHS 22-2 (91.6 N), KMHS 22-16 (133.4 N), KMHS 22-17 (177.9 N) and KMHS 20-4 (556 N). These are used generally for major ΔV manoeuvres and atmospheric drag make-up. The KMHS 20-4 is capable of operating for >4 h in total, delivering 4.45 MNs impulse and consuming >2 t of hydrazine.

KMHS Model 10

Applications: satellite AOCS
First flight: 1971. Topex, IUE, Apple, BS-2
Number produced: 7,200
Dry mass: 0.33 kg
Length: 146 mm
Maximum diameter: 31 mm
Engine cycle: decomposition of hydrazine over heated catalyst bed
Thrust: 0.89 N vac (0.45-1.42 N range)
Specific impulse: 226 s vac
Inlet pressure: 15.0 atm (range 6.0-27.2 atm)
Burn time: 0.01 Ns min impulse bit, 111.2 kNs max
Total pulses: >45,000 cold start, >350,000 hot.

KMHS Model 16/39

Applications: satellite AOCS
First flight: 1971. Skynet 4, Clementine, COBE, CRRES, Topex, TRMM
Number produced: 7,200
Dry mass: 0.51 kg
Length: 254 mm
Maximum diameter: 32 mm
Engine cycle: decomposition of hydrazine over heated catalyst bed
Thrust: 22.2 N vac (7.1-25.4 N range)
Specific impulse: 235 s vac (220 s at 5.1 atm inlet)
Inlet pressure: 3.4-27.2 atm
Burn time: 0.31 Ns min impulse bit, >501 kNs max
Total pulses: >60,000 and 227 kg throughput Model 16; >10^6/520 kg Model 39.

KMHS Model 17

The thruster was originally designed and qualified for 1.78 N at 14.3 atm inlet pressure.
Applications: satellite AOCS
First flight: 1970. DSCS 3, SeaStar, Geosat
Number produced: 7,200
Dry mass: 0.38 kg
Length: 203 mm
Maximum diameter: 32 mm
Engine cycle: decomposition of hydrazine over heated catalyst bed
Thrust: 4.45 N vac (1.56-5.34 N range)
Specific impulse: 230 s vac (220 s at 14.3 atm inlet)
Inlet pressure: 5.1-29.3 atm
Burn time: 0.089 Ns min impulse bit, 311.4 kNs max
Total pulses: >10^6 and 141 kg throughput.

KMHS Model 22

Applications: satellite AOCS
First flight: 1980. IUS, TOS, ASAT, TRS, AXAF
Number produced: >200
Dry mass: 1.02 kg
Length: 305 mm
Maximum diameter: 84 mm
Engine cycle: decomposition of hydrazine over heated catalyst bed
Thrust: 88.9 N vac (22-200 N range)
Specific impulse: 230 s vac (225 s at 6.8 atm inlet)
Inlet pressure: 5.1-30.6 atm
Burn time: 0.89 Ns min impulse bit, >3,496.6 kNs max
Throughput: 771 kg demonstrated.

OLIN AEROSPACE CO

OAC, previously the Rocket Research Co, specialises in monopropellant hydrazine engines and gas generators for spacecraft and upper stage applications; >9,300 assemblies and 100 propulsion systems have been produced since the first system was qualified in 1964. There are 600 personnel at the Redmond and Moses Lake sites. The MR-508 hydrazine arcjet was qualified in 1991 and Telstar 4 in 1993 became the first commercial satellite to carry an arcjet for stationkeeping; see p295 1995-96 for specifications. The model is superseded by the MR-509/510. OAC has delivered >250 gas generators for Shuttle's Auxiliary Power Unit.
Facilities: OAC's 0.29 km^2 site at the Grant County Airport in Moses Lake houses 2,840 m^2 of solid propellant manufacturing plant, a hazardous device assembly area, storage magazines, a short-range flight test facility, a sea level large liquid engine test complex, ballistic test bays and chemistry lab.

MR-50 22.2N

Applications: attitude control for SMS, Viking, Meteosat, GOES, Voyager, GPS, Intelsat 5, Scatha, Lockheed

Magellan thruster module: 2 × 445 N MR-104; 3 × 0.89 N MR-103; 1 × 22.2 N MR-50 (OAC)

Martin 5000, Delta Star, Magellan, Wind/Polar, EOS AM 1
First flown: SMS metsat (May 1974)
Flown: 695 (748 produced) to end-1995
Dry mass (g): 680
Length (cm): 18.3
Max diameter (cm): 6.6
Mounting method: bolted three places
Propellant: hydrazine at 4.67-17.33 g/s (Shell 405 catalytic decomposition)
Feed method: GN_2 or GHe at 4.8-32.7 atm, through 29 W solenoid dual seat valve
Thrust (N): 9.79-38.7 (22.2 nominal)
Specific impulse (s, vacuum): 215-228
Time to full thrust (ms): 150 to 90% PC
Nozzle area ratio: 40:1
Nozzle length (cm): 3.95
Nozzle cooling method: radiative
Combustion chamber
pressure (atm): 2.9-11.3
temperature (°C): 800
cooling method: radiative
Duty cycle (s): 0.020 min to 5,400 max; 471,000 pulses
Total impulse (kNs): 459.4.

MR-103 0.89N

Applications: attitude control thruster for Voyager, GPS, Intelsat 5, Lockheed Martin 3000, 4000, 5000 & 7000, Mars Observer, ACTS, Magellan, Cassini, Iridium
First flown: ATS 6 (May 1974)
Flown: 1,260 (2,705 produced) to end-1995
Dry mass (g): 332
Length (cm): 14.7
Max diameter (cm): 3.43
Mounting method: bolted three places
Propellant: hydrazine at 0.091-0.499 g/s (Shell 405 catalytic decomposition)
Feed method: GN_2 at 4.8-28.6 atm through 9W solenoid dual seat valve
Thrust (N): 0.19-1.12 (0.89 nominal)
Specific impulse (s, vacuum): 206-227
Time to full thrust (ms): 150 to 90% PC
Nozzle area ratio: 100:1
Nozzle length (cm): 0.98
Nozzle cooling method: radiative
Combustion chamber
pressure (atm): 4.35-23.9 atm
temperature (°C): 800
cooling method: radiative
Duty cycle (s): 0.008 min to unlimited max burn; 750,000 pulses
Total impulse (kNs): 158.46.

MR-104 445N

Applications: attitude control and ΔV corrections, Voyager, Magellan, DMSP, Tiros N, Landsat
First flown: Voyager 2 (Aug 1977)
Flown: 68 (101 produced) to end-1995
Dry mass (g): 1,860
Length (cm): 46.0
Max diameter (cm): 15.2
Mounting method: bolted three places
Propellant: hydrazine at 91-290 g/s (Shell 405/ LCH-202 catalytic decomposition)
Feed method: GN_2 or GHe at 6.8-28.6 atm through Wright Components 30 W single seat solenoid valve
Thrust (N): 205-572 (445 nominal)
Specific impulse (s, vacuum): 228-239
Time to full thrust (ms): 50 to 90% PC
Expansion ratio: 53:1 (∅ 15.2 cm)
Nozzle length (cm): 17.8
Nozzle cooling method: radiative
Combustion chamber
pressure (atm): 3.8-19.0
temperature (°C): 800
cooling method: radiative
Duty cycle (s): 0.022-2,000 single firing; 2,654 cumulative; 1,742 pulses
Total impulse (kNs): 693.9.

MR-106 27N

Applications: spacecraft and upper stage attitude control and ΔV corrections, PAM A/S, Radarsat, GPS Block 2R, HAS/Peace Courage, Titan Centaur, Atlas Centaur, A2100, Lunar Prospector, NEAR
First flown: HAS/Peace Courage
Flown: 396 (2,381 produced) end-1995
Dry mass (g): 476
Length (cm): 17.8
Max diameter (cm): 6.4
Mounting method: bolted three places
Propellant: hydrazine at 4.08-11.79 g/s (LCH 227/202 catalytic decomposition)
Feed method: GN_2 or GHe at 6.8-30.6 atm through Wright Components 27 W solenoid valve
Thrust (N): 8.9-26.7 (26.7 nominal)
Specific impulse (s, vacuum): 218-232
Time to full thrust (ms): 200 to 90% PC
Expansion ratio: 61:1
Nozzle length (cm): 4.72
Nozzle cooling method: radiative
Combustion chamber
pressure (atm): 6.12-10.8
temperature (°C): 800
cooling method: radiative
Duty cycle (s): 0.016 min to 2,000 max; 12,397 pulses
Total impulse (kNs): 167.

MR-107 178N

Applications: spacecraft and upper stage attitude control and ΔV corrections, Delta 2, Titan 2, Commercial Titan, PAM D, Small ICBM, HAS/Peace Courage, Atlas roll control module, STEP, Pegasus, LMLV/OAM
First flown: HAS/Peace Courage
Flown: 328 (1,898 produced) to end-1995
Dry mass (g): 885
Length (cm): 21.8
Max diameter (cm): 6.6
Mounting method: bolted three places
Propellant: hydrazine at 24-113 g/s (LCH 207/ LCH 202 catalytic decomposition)
Feed method: GN_2 or GHe at 6.8-34 atm through 50 W solenoid valve
Thrust (N): 51.2-257.9 (178 nominal)
Specific impulse (s, vacuum): 217-236
Time to full thrust (ms): 200 to 90% PC
Expansion ratio: 21:1
Nozzle length (cm): 4.72
Nozzle cooling method: radiative
Combustion chamber
pressure (atm): 1.9-9.5
temperature (°C): 800
cooling method: radiative
Duty cycle (s): 0.016 min to 2,137 max; 7,005 pulses
Total impulse (kNs): 332.3.

MR-111 2.2-4.4N

Applications: attitude control, Intelsat 5, ERBS, ACTS, Radarsat, Lockheed Martin 4000, 5000, 7000, Wind/ Polar, Landsat, Mars Observer, Radarsat, ACE, Skynet 4, MGS, EOS AM 1, MSTI 2/3, Mars Pathfinder, Mars 98, Clementine 1
First flown: Intelsat 5 (Dec 1980)
Flown: 410 (627 produced) to end-1995
Dry mass (g): 345
Length (cm): 16.7
Max diameter (cm): 35.6
Mounting method: bolted three places
Propellant: hydrazine at 0.19-2.4 g/s (Shell 405 catalytic decomposition)
Feed method: GN_2 or He at 4.08-27.2 atm through Wright Components 9 W dual seat solenoid valve
Thrust (N): 0.44-5.34 (2.2-4.4 nominal)
Specific impulse (s, vacuum): 213-229
Time to full thrust (ms): 150 to 90% PC
Expansion ratio: 200:1
Nozzle length (cm): 2.18
Nozzle cooling method: radiative
Combustion chamber
pressure (atm): 3.1-13.9
temperature (°C): 800
cooling method: radiative

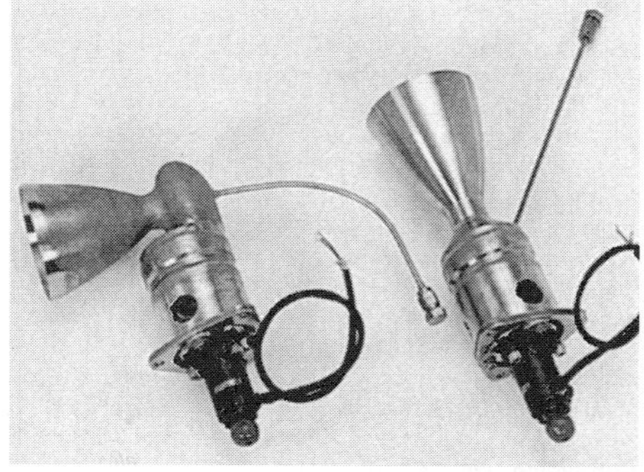

The MR-107 provides Delta, Titan and Atlas attitude control (Olin Aerospace Co)

Duty cycle: 0.020 s min to 15 h max; 420,000 pulses
Total impulse (kNs): 260.21

MR-501 Electrothermal Hydrazine Thruster

Applications: communications satellite NSSK, Lockheed Martin 4000
First flown: Satcom 1R (Apr 1983)
Flown: 72 (92 produced) to end-1995
Dry mass (g): 816
Length (cm): 19.4
Max diameter (cm): 8.84
Mounting method: bolted four places
Propellant: hydrazine at 0.059-0.127 g/s over Shell 405 catalyst and 350-510 W electric heater
Feed method: GN_2 or GHe at 6.8-23.8 atm through Wright Components 9 W dual seat solenoid valve
Thrust (N): 0.18-0.33
Specific impulse (s, vacuum): 280-304
Time to full thrust (ms): not applicable
Expansion ratio: 100:1
Nozzle length (cm): 0.879
Nozzle cooling method: radiative
Combustion chamber
pressure (atm): 3.4-6.5
temperature (°C): 800
cooling method: radiative
Duty cycle: 300 s min to 1.7 h max; 500,000 pulses
Total impulse (kNs): 311.4.

MR-502 Improved EHT (IMPEHT)

Applications: communications satellite NSSK, Lockheed Martin 5000
First flown: Astra 1B (Mar 1991)
Flown: 16 (32 produced) to end-1995
Dry mass (g): 848
Length (cm): 19.4
Max diameter (cm): 8.84
Mounting method: bolted four places
Propellant: hydrazine at 0.119-0.167 g/s over Shell 405 catalyst and 610-839.6 W electric heater
Feed method: GN_2 or GHe at 21.8-23.8 atm through Wright Components 9 W dual seat solenoid valve
Thrust (N): 0.36-0.50
Specific impulse (s, vacuum): 280-304
Time to full thrust (ms): not applicable
Expansion ratio: 100:1
Nozzle length (cm): 0.879
Nozzle cooling method: radiative
Combustion chamber
pressure (atm): 3.4-6.5
temperature (°C): 800
cooling method: radiative
Duty cycle: 300 s min to 2.0 h max; 430,000 pulses
Total impulse (kNs): 525.

MR-509 Low Power Hydrazine Arcjet

Applications: communications satellite NSSK, Lockheed Martin 7000
First flight: AsiaSat 2 Nov 1995
Flown: 8 (36 produced) to end-1995
Dry mass (g): 1,338
Length (cm): 24.4
Max diameter (cm): 9.27
Mounting method: bolted four places
Propellant: hydrazine at 0.068-0.077 g/s over Shell 405 catalyst and 1.8 kW (PCU input) electric arc
Feed method: GN_2 or GHe at 14.9-20.4 atm through Wright Components 9 W dual seat solenoid valve
Thrust (N): 0.21-0.25
Specific impulse (s, vacuum): 511
Time to full thrust (ms): not applicable
Expansion ratio: 100:1

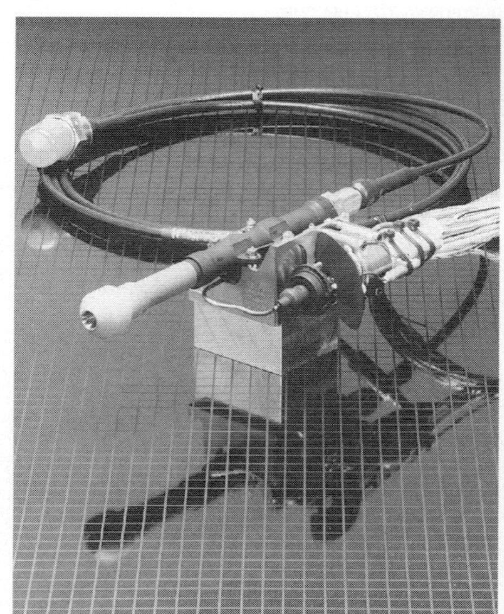

OAC's MR-509 arcjet was introduced on AsiaSat 2 **1996**

Nozzle length (cm): 0.879
Nozzle cooling method: radiative
Combustion chamber
  pressure (atm): 4.1-4.7
  temperature (°C): 800
  cooling method: radiative
Duty cycle: 5 min minimum to 65 h max
Total impulse (kNs): 883.

MR-510 Low Power Hydrazine Arcjet

Applications: communications satellite NSSK, Lockheed Martin A2100
First flight: planned GE 1 1996
Flown: 0 (44 produced) to end-1995
Dry mass (g): 1,338
Length (cm): 24.4
Max diameter (cm): 9.27
Mounting method: bolted four places
Propellant: hydrazine at 0.068-0.077 g/s over Shell 405 catalyst and 2.2 kW (PCU input) electric arc
Feed method: GN_2 or GHe at 14.9-20.4 atm through Wright Components 9 W dual seat solenoid valve
Thrust (N): 0.21-0.25
Specific impulse (s, vacuum): 586
Time to full thrust (ms): not applicable
Expansion ratio: 100:1
Nozzle length (cm): 0.879
Nozzle cooling method: radiative
Combustion chamber
  pressure (atm): 4.1-4.7
  temperature (°C): 800
  cooling method: radiative
Duty cycle: 5 min minimum to 65 h max
Total impulse (kNs): 812.

PARKER BERTEA AEROSPACE

- **Air/Space Division**

18321 Jamboree Road, Irvine, California 92715.
Tel: (+1 714) 833 30 00
Fax: (+1 714) 851 33 41
General Manager: Jim Sabin

- **Control Systems Division**

14300 Alton Parkway, Irvine, California 92718.
Tel: (+1 714) 833 30 00
Telex: 678427
Fax: (+1 714) 586 84 56
VP/General Manager: Robert Barker
VP Marketing: Jim Ryder
1425 West 2675 North, Ogden, Utah 84404.
Tel: (+1 801) 782 31 00
Telex: 388424
Fax: (+1 801) 786 30 45
Plant Manager: Mike Romito

Parker Bertea Aerospace, part of the Parker Hannifin Corp, designs, manufactures and services fluid systems, components and related electronic control systems for aerospace applications. Personnel total 5,500. The space-related products of its divisions are specified below.

Air/Space This division specialises in the design/manufacture of valves and coupling devices to handle gaseous, liquid and cryogenic propellants, including hydrogen, nitrogen, oxygen, helium, NTO and MMH. Such systems fly on Shuttle, RL10 engine, Peacekeeper and Pegasus.

High Pressure Helium Regulator
Applications: ullage pressure control in a lightweight 680 atm blowdown propellant feed systems for missile and spacecraft propulsion systems. The regulator controls ullage pressure within ±10% with a 10:1 variation in inlet pressure and from lockup to rated flow. It is a welded assembly constructed entirely from titanium alloys. The basic regulator design can be used for a variety of fluids/gases.
Inlet pressure: 680-102 atm
Regulated pressure: 87 atm ±10%

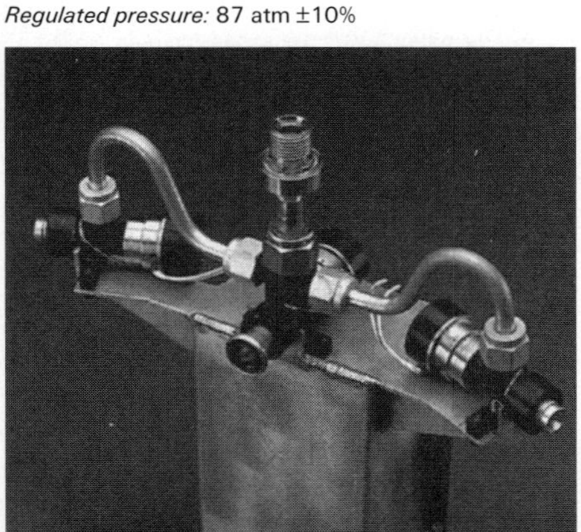

Parker Bertea Aerospace produces the nitrogen thruster packages for Pegasus

Flow rate: 0.16 kg/s helium min at 102 atm inlet
Inlet/outlet ports: 6.35 mm ∅ × 0.89 mm wall tube stubs
Startup transient: 95.8 atm max regulated pressure at 680 atm inlet pressure
Mass: 0.077 kg

Cold Gas Thruster Module
This module is designed to provide 3-axis control for small launch vehicles; it is flown on Pegasus. The module comprises three cold gas thrusters operated by integral independent solenoid valves. The central thruster generates 111.2 N and the two lateral thrusters 55.6 N when operated with a 136 atm N_2 gas supply. SI of each is 68 s. The solenoid valves are pilot operated and have a response time of <5 ms at max gas inlet pressure.
Operating pressure range: 136-13.6 atm
Thrust at 136 atm: 4.5-111.2 N + 8.9-55.6 N
Voltage: 28±4 VDC
Response time: 8 ms max
Mass: 0.086 kg per thruster
Size: 25.4 mm ∅ × 63.5 mm per thruster.

Control Systems The division, with 1,700 employees, designs/manufactures flight control actuators and associated electronics for space applications that provide system functions of proportional control, performance enhancement and availability enhancement utilising available hydromechanical, electrohydraulic, electromechanical and electro-optic technologies. CSD provides the stage 2 3-axis fin control actuator system for Pegasus and the stage 2/3 TVC actuator system.

PRESSURE SYSTEMS, INC

PSI, is the world's largest independent manufacturer of titanium pressure vessels for launchers and spacecraft: it now ships 200-300 annually, worth about $25 million. >3,800 have been delivered, without flight failure, including >600 propellant management device tanks of >20 designs and >700 elastomeric diaphragm tanks of >15 designs. PSI acquired Programmed Composites Inc in Jun 1995, allowing it to offer tanks integrated with satellite core structures, in addition to precision reflectors, solar array substrates and antenna support structures. PSI tanks provide the propellant for most commercial communications satellites, including Hughes HS-601, Loral FS-1300 and Lockheed Martin A2100. PSI tanks are also used on Milstar, Cassini, Soho, NEAR, Tirus, Inmarsat, Intelsat, ETS 7, TRMM and Axaf. PSI is completing a $4 million contract from Lockheed Martin for 75 hydrazine tanks for Iridium satellites. In 1996 PSI delivered the 1000th upper stage high-pressure helium tank and the 100th upper stage propellant tank for the Lockheed Martin Atlas-Centaur launch vehicle family.

PUROFLOW CORP

Surface tension devices for spacecraft propellant tanks. Filters/strainers for propellants, pneumatics, hydraulics and cryogenic systems. Users include Shuttle, Titan 3, GOES, Milstar and Eurostar. A UV system is used to treat the deluge water on the Shuttle launch pad. See also the Decca Valves entry in the Space Industry section.

QUANTIC INDUSTRIES, INC

Quantic designs, develops, manufactures and tests ordnance devices (primarily electrically-initiated), including pyrotechnic initiators, bolts, valves and cutters. The PAM upper stage and Delta launch vehicle carry the company's explosive bolt cutters. It also offers a line of lasers and laser ordnance.
Facilities: manufacturing facilities cover propellant mixing, powder blending, glass-to-metal sealing, protective finishing, inert assembly, welding/soldering and automated explosive loading. The 0.42 km² site houses 26 buildings, providing 5,600 m² under roof for

Through to Feb 1997 Pressure Systems delivered 60 titanium hydrazine tanks for the Iridium satellites. (PSI)
1995

administration, engineering, manufacturing, inspection and testing, in addition to inert storage and explosive magazines.

THIOKOL CORPORATION

Established in 1929, Thiokol Chemical Corp produced and marketed the first synthetic rubber manufactured in the US. In 1943, the discovery by Thiokol of liquid polymer, a new type of synthetic rubber, paved the way for the development of the case-bonded principle of rocket power plant design. The company's polysulphide liquid polymer proved to be the catalyst for the first mass production of efficient solid propellant rocket motors, as well as for the development of *large* solid propellant motors. Thiokol Corp merged with Morton Norwich Products in Sep 1982, but separated off in Jul 1989. It reported $52.3 million net income in FY95 on sales of $956.8 million ending Jun 1995; FY94 60.3, FY93 63.8, FY92 63.0, FY91 53.4. Personnel total >7,500.

Thiokol's Space Operations produces the Space Shuttle Solid Rocket Motor (SRM). FY95 revenues were $467.4 million for an operating income of $60.4 million (FY94 500.8/51.5; FY93 519.3/61.5; FY92 555.6/58.5). Utah DLV Operations (until 1995 Strategic Operations) in a joint venture with Hercules Inc produces stage 1/2 of the USN Trident II D-5 missile. It provided the complete stage 1, ignition systems and flight termination systems for USAF's Peacekeeper. It also produces the Castor 120 and Castor 4 motors, flares and the HARM and Standard Missiles. Elkton produces gas generators, multiple pulse motors and the Star series of spacecraft kick and launcher stage motors. Most Castor motors, long used as Delta strap-on boosters and more recently on Atlas 2AS, were handled by Huntsville. The Castor family has flown on >1,800 launches since 1959. Huntsville also produced Sidewinder, Hellfire and Maverick motors. The facility closed in 1996. Huck produces fasteners for Space Shuttle.

A 10 yr licensing agreement with the UK's Royal Ordnance Rocket Motors Div was announced in Nov 1989 that allows Thiokol to manufacture solid propellant and motor casings developed by RO. These include smokeless flameless elastomer modified cast double base propellant and steel strip laminate cases.

Castor 4 (TX526)

The TX526-0 motor, combined with four Recruit strap-on motors, was the booster for the Athena H launch vehicle. The TX526-2 (with an 11° canted nozzle) and the TX526-3 (7° nozzle) were used as strap-on boosters for Delta, each of which had nine motors. A total of 342 flew on 38 Delta 3000-series missions (#118 to #183); 243 of the 244 ignited on the pad were successful.
First flown: 1971
Number flown: 360 to end-1995 (0 in 1995; 0 in 1994; 0 in 1993; 99.72% success)
Mass: 10,534 kg
Length: 9,072 mm
Diameter: 1,019 mm
Propellant
  type: TP-H8038 PBAA Polymer, 14% aluminium
  shape: radially slotted centre perforate
  mass fraction: 0.880 (9,265kg)
Burn time: 53.9 sec at 27°C
Thrust: 428kN vac at 27°C
Specific impulse: 228.4 s
Total impulse: 23,748 kNs vac at 27°C
Pressure (atm): 45.4 max; 37.0 avg
Nozzle
  throat diameter: 29.54 cm
  length: varies – SL/altitude/canted available
  exit diameter: 81.66 cm
  materials: 4130 steel
Casing materials: 4130 steel.

Castor 4A (TX780)

Castor 4A began development in the early 1980s under a NASA Goddard programme to improve Delta performance by 11% by switching to HTPB propellant. A development motor was fired 25 Feb 1983 and a qualification motor 31 Aug 1983 before the programme was halted when it appeared that Delta would be terminated. Development re-started when Delta 2 was selected as the USAF GPS launcher; three further qualification firings followed. Each vehicle carried nine strap-ons (later Deltas switched to Hercules' motors). The TX780-0 with an 11° canted nozzle is used as the ground ignited booster; the TX780-1 (7° cant) and TX780-2 (11°) are ignited at altitude. The TX780-3/-4 were developed as the ground and altitude ignited boosters, respectively, for Atlas 2AS (first flight Dec 1993; six, successful, by end-1995). The TX780-5 (straight nozzle) powered OSC's Prospector suborbital vehicle and two flew on Conestoga in Oct 1995. **4H** (114.3 cm ∅) and **5** (127.0 cm ∅) versions have been studied. See the separate **4AXL** entry below. The Huntsville plant closed in 1996; Castor 4 production transferred to Utah DLV Operations.

First flown: 14 Feb 1989, Delta 184
Number flown: 211 to end-1995 (100% success)
Mass: 11,578 kg (propellant 10,121 kg) ground-ignited
Length: 8,087/9,199 mm loaded case with/without nozzle
Diameter: 1,016 mm
Propellant
type: TP-H8299 HTPB Polymer (20% aluminium)
shape: cylindrical, 76% web fraction; transitions to four longitudinal slots. Tailored for ignition/burnout thrusts to be similar to Castor 4
mass fraction: 0.874 for standard loading
Burn time: 53.1 s web
Thrust (kN, sea level, 21°C): 476.661 max, 436.736 avg
Specific impulse: 237.6 s sea level 21°C
Total impulse: 23,846 kNs sea level 21°C
Pressure (atm): 46.60 avg, 49.86 max sea level 21°C
Nozzle
throat diameter: 278.6 mm
materials: carbon-phenolic
Casing liner: standard HT polymer
Igniter type: pyrogen, head-end.

Castor 4AXL

The 4AXL version, extended by 2.44 m, was first tested May 1992. Its 30% performance increase would improve performance of vehicles such as Atlas and Conestoga (Martin Marietta included it as part of its unsuccessful bid for the USAF MLV-3 bid, won by Delta in 1993). QM-1 was successfully tested Nov 1992 and QM-2 completed qualification 3 Feb 1993. Thiokol has invested about $12 million in the programme. 4AXL was to have been combined with 4B's TVC system to create the 4BXL stage 1 motor for CTA's ORBEX small launcher.
First flight: -
Number flown: -
Mass: 14,851 kg (propellant 13,128 kg for ground ignited strap-on)
Length: 12,279 mm with nosecone adapter, 13,711 mm with nosecone
Diameter: 1,018.5 mm
Propellant
type: TP-H8299 HTPB Polymer, 20% Al, 68% AP
shape: forward cylindrical perforate, seven aft longitudinal slots
mass fraction: 0.884 ground ignited strap-on
Burn time: 60.1 s
Thrust (kN, vac): 599.81 avg, 700.5 max
Specific impulse: 269.2 s vacuum
Total impulse: 34.679 MNs vacuum
Pressure (atm): 41.70 avg, 54.42 MEOP
Nozzle
throat diameter: 318.8 mm
length: 1,216.8 mm
exit diameter: 937.3 mm
materials: 4130 steel, graphite phenolic throat insert, carbon phenolic exit cone
Casing material: AISI 4130 steel 0.28 mm thick
Igniter type: TX544 (>500 units flown successfully) forward internal pyrogen, 2.45 kg TP-H8027 HTPB propellant, cartridge loaded.

Castor 4B (TX859)

The 4B, which incorporates TVC in the series for the first time, was developed for ESA's Maxus μg carrier and first flown in 1991. Ten were ordered. Five flew on Conestoga's maiden launch, Oct 1995. It is now used by the US Army Theater Missile Defense programme, planned to debut in mid-1996. CTA's cancelled ORBEX was to use the **4BXL** version, which would combine the 4AXL motor with 4B's TVC system (±6°). Thiokol may still qualify it.
First flown: 8 May 1991
Number flown: 8 to end-1995 (100% success: 3 Maxus + 8 Conestoga)
Mass: 11,482 kg
Length: 8,984 mm
Diameter: 1,019 mm
Propellant
type: TP-H8229 HTPB Polymer, 20% aluminium
shape: radially slotted centre perforate
mass fraction: 0.868 (9,969 kg)

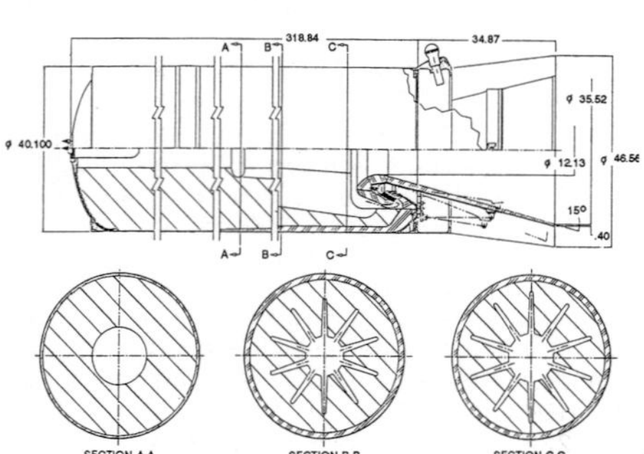

Castor 4B first flew in 1991 on the suborbital Maxus launcher. Dimensions are in inches

Burn time: 60.9 s at 25°C
Thrust: 429 kN vac at 25°C
Specific impulse: 267.4 s vac
Total impulse: 26,141 kNs vac at 25°C
Pressure (atm): 50.6 max; 30.7 avg
Nozzle
throat diameter: 30.84 cm
length: varies – SL/altitude/canted available
exit diameter: 90.2 cm
materials: 4130 steel
Casing materials: 4130 steel.

Castor 120

The motor is similar in size to the Peacekeeper missile stage 1 motor, filling the gap between Castor 4A and the large segmented motors, but using proven technology throughout. Primary goals are >0.999 reliability and 50% cost reduction from Peacekeeper. It represents the upper limit for a monolithic motor that can be handled and transported by commercial methods with few restrictions. Unit cost would be about $4.0 million for volume production, rising to $5.5 million for one annually. Thiokol has invested >$50 million in its development. Its first application was as stage 1 of Lockheed Martin's LMLV small launcher in Aug 1995, but failed because of TVC cable overheating. OSC's Taurus satellite launcher will use it beginning with the second vehicle. Three flights are planned for 1996 (two LMLV + one Taurus); six motors are in production for 1997-98 flights. A potential application is as a strap-on, such as for Delta. It can be used as a stage 1 or 2 or strap-on because both the forward/aft ends can take tension or compression forces. ±5.0° TVC is provided by movable nozzle, but the cold gas blowdown TVC system (provided by AlliedSignal Equipment Systems) can be deleted. The grain can be tailored to reduce thrust during max-q, for high initial thrust or for a regressive thrust to reduce acceleration.

Fabrication of the first 120 (DM120-1) was started in late 1990; first static firing was made successfully 22 Apr 1992. Second test (DM120-2), at the −1°C lower limit, was successful 4 Mar 1993.
First flight: Aug 1995
Number flown: 1 (0% success)
Mass: 53,118 kg (propellant 49,032 kg)
Length: 9,017 mm
Diameter: 2,363 mm
Propellant
type: HTPB
shape: three aft conical slots, one forward slot
mass fraction: 0.923 (49,032 kg)
Burn time: 81 s
Thrust: 1,650 kN vac average
Specific impulse: 280 s vacuum
Total impulse: 133,440 kNs vac
Pressure (atm): 101 max; 78 avg
Nozzle
throat diameter: 368 mm
length: 1,448 mm
exit diameter: 1,600 mm
materials: glass epoxy/polyisoprene flex bearing; 3D carbon-carbon throat; carbon phenolic exit cone; aluminium stationary shell; steel throat support
Casing materials: carbon epoxy, lined with silica-filled EPDM rubber.

Space Shuttle Solid Rocket Motor

Thiokol was selected as Shuttle SRM prime in 1974. NASA announced Jul 1991 a $2.6 billion Buy 3 contract extension to Thiokol for 67 further flight pairs through 1999/STS-105, plus eight ground test motors. Another extension is expected for launches through 2012. The SRM is cast in four segments in Utah and then transported by rail to Kennedy Space Center, Florida, where they are stacked vertically in pairs either side of the External Tank on the mobile launch platform in the Vehicle Assembly Building. The boosters provide the majority of launch thrust, with TVC provided by a flexible bearing nozzle driven by two hydraulic actuators. After burnout at about 2 min, the boosters are separated pyrotechnically and fall into the Atlantic for recovery. The motors are cleaned, disassembled and returned to Utah for refurbishment and reloading. The segments are designed for 20 flights; nine is currently the maximum. The configuration underwent extensive reworking and testing following *Challenger's* loss, and the new booster was re-designated RSRM (Redesigned Solid Rocket Motor, changed by NASA in 1995 to Reusable SRM). Further information is provided within the US National entry.

Thiokol's new Castor 120 fills the gap between the Castor 4 and large segmented motors

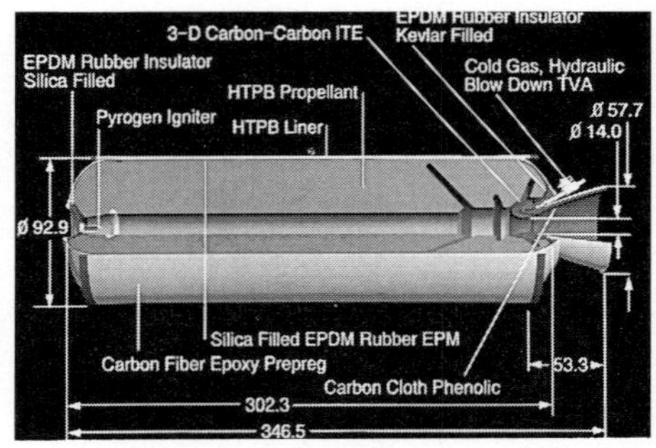

Castor 120. Dimensions are in inches **1994**

A High Performance Motor (HPM) design for an improved SRM incorporates lightweight steel cylindrical case segments, a higher expansion nozzle, and altered motor thrust characteristics increased Shuttle payloads by >1,540 kg. This design was static tested in 1982 and went into service on STS-8 in Aug 1983. Graphite epoxy composite cylindrical SRM case segments were also developed to increase Shuttle capacity by 2 t; the configuration was shelved following *Challenger's* loss. Thiokol was subcontracted to Aerojet to develop/produce the nozzle for the Advanced SRM that was to enter Shuttle service in about 1998. ASRM was cancelled in 1993.

Reusable Solid Rocket Motor
Mass: 569,636 kg
Length: 38.47 m
Diameter: 3,708 mm
Propellant
type: TP-H1148 HB Polymer, 16% Al, 70% AP
shape: 11-point star forward; double taper cylinder two centre segments; triple taper cylinder aft
mass fraction: 0.883 (501,746 kg)
Burn time (s): 123.7 action, 111.4 web
Thrust (vac): 11.52 MN average, 14.77 MN max
Specific impulse: 268.0 s vac
Total impulse: 296.3 MNs vac
Pressure (atm): 62.1 max (MEOP 69.1); 45.0 avg
Nozzle
throat diameter: 136.8 cm
length: 454.0 cm
exit diameter: 380.1 cm (expansion ratio 7.72)
materials: D6AC steel for fixed housing, throat and forward exit cone housings; aluminium for nose inlet and aft exit cone housing
Casing materials: D6AC steel, 1.165 cm minimum thickness
Igniter: forward end internally mounted. 60.7 kg of TP-H1178 case-bonded 40-point star

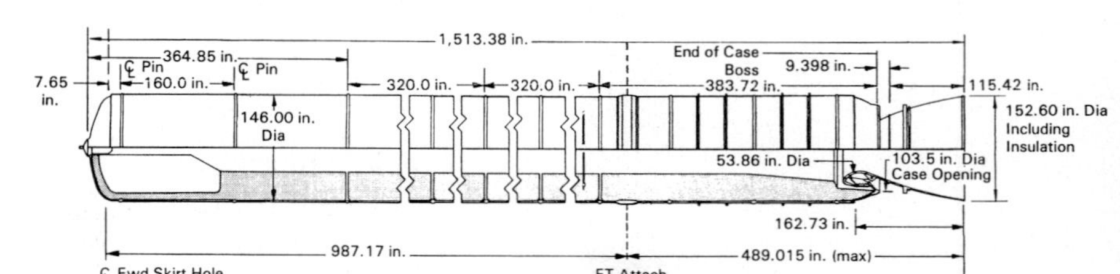

Thiokol's Shuttle Reusable Solid Rocket Motor design **1996**

Star 5CB (TE-M-344-16)
Reduced aluminium HTPB propellant (2.1 kg) minimises contamination when used as Titan 4 stage separation motor. 121 mm main ∅, 341 mm length, 4.5 kg total mass, 5.5 kNs total impulse, 2.77 s action time. Adapted for Mars Pathfinder.

Star 6B (TE-M-790)
Space kick motor, principal ∅ 166 mm, length 403 mm, total mass 10.26 kg (propellant 6.1 kg). Delivers 2.6 kN thrust vac over 6.65 s action time; total impulse 16.4 kNs. Offloading provides impulse tailoring to 7.6 kNs. Sweden's Freja scientific satellite used a 6B to circularise its orbit in Oct 1992.

Star 13/13B (TE-M-458/763)
Braking motor used in NASA's Anchored Interplanetary Monitoring Platform programme. 31 kg charge of AP/Al urethane contained in a spherical case of 6Al-4V titanium, with graphite/vitreous silica phenolic nozzle. Loaded mass 35.65 kg; average thrust 3.8 kN for burn time of 21.8 s. Sweden's Freja scientific satellite used it Oct 1992 to raise apogee. The 13A, with 33 kg propellant, is used to de-orbit Meteor's return capsule. The 13B was developed by adding a 56 mm cylindrical section to the spherical Star 13. Propellant mass 40.86 kg; first flown 1984.

Star 17/17A (TE-M-479/521)
The TE-M-479 is a 442 mm spherical motor developed for NASA's Radio Astronomy Explorer as an AKM. 687 mm long; mass 78.8 kg (propellant 69.7 kg), titanium case. Star 17A, 444 mm ∅, 980 mm long, developed by adding a 175 mm straight section to the spherical 17. 17A mass 126 kg (propellant 112 kg). Circularised orbit for Skynet 1, NATO 1 and IMP-H/J.

Star 27 (TE-M-616)
694 mm ∅, 1303 mm long AKM, 338 kg charge of AP/HTPB/Al in 6Al-4V titanium case and graphite/carbon phenolic nozzle. Mass 365.7 kg; mass fraction 0.925 (propellant mass can be reduced up to 25%). Average thrust 27 kN vac for 34.5 s burn, SI 289.5 s. AKM for Canada's CTS, Japan's GMS/BS, and several USAF GPS and NOAA GOES satellites. GMS 4 in 1989 employed the TE-M-616-5, propellant mass 348 kg, Isp 287.8 s vac. A TVC nozzle was demonstrated on a Star 27 in 1987.

Star 30BP/C/E (TE-M-700-20/18/19)
TE-M-700-20 (STAR-30BP) 762 mm ∅ 1,506 mm long AKM. 505 kg charge of AP/HTPB/Al in 6Al-4V titanium case; nozzle with carbon/carbon throat insert; aft-end igniter. Mass 542.8 kg (mass fraction 0.951); propellant mass can be reduced 20%. Burn 54 s, vacuum total impulse 1,460 kNs, effective Isp 292.0 s. Used on Thaicom. TE-M-700-18 (Star 30C) has shorter exit cone + 126.2 mm longer case for total 1,493 mm length.

Star 48 provides the propulsion for McDonnell Douglas' Payload Assist Module (MDSSC)

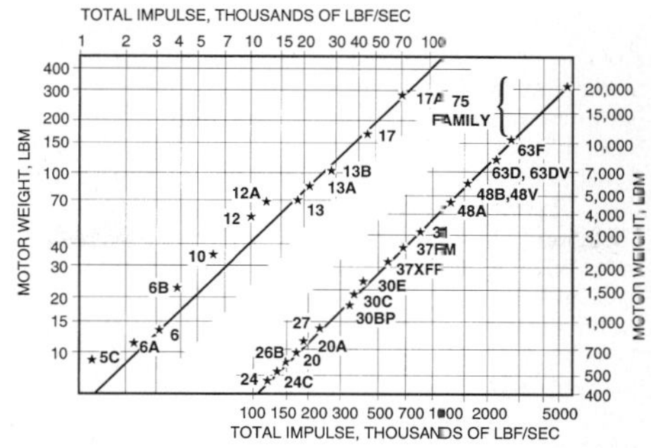

Star family performance. Star 26B and 20A have been deleted (Thiokol) ***1995***

Mass 626.9 kg (propellant 585 kg max). Vacuum total impulse 1647 kNs, effective Isp 284.6 s. Used on Lockheed Martin series 3000 telecom satellites.

TE-M-700-19 (Star 30E) has longer case and expansion cone than Star 30C: 1,683 mm motor length and 668 kg. Vacuum total impulse 1,780 kNs, effective Isp 289.2 s. CTA planned to use it for stage 4 of the ORBEX small orbital launcher. Used by Thaicom and Koreasat (628 kg propellant, mass fraction 0.940).

Star 31 (TE-M-762 Antares 3)
762 mm ∅, 2,873 mm long motor carries 1,287 kg of AP/HTPB/Al in a filament wound Kevlar case and a carbon/phenolic nozzle with a carbon/carbon throat insert. 1,393 kg including external insulation. Average thrust 80 kN over 46 s burn. Qualified as an improved Scout stage 3; first flight late 1979, last 1994.

Star 37FM (TE-M-783)
Developed originally as AKM for USN's Fleetsatcom. Offloading propellant weight by 20% is a standard option. 37FM will be employed by the GPS 2R navsat series. Clementine 1 used it Feb 1994 to enter a lunar transfer orbit.

Mass: 1,148 kg (propellant 1,066 kg, qualified for offloading to 1,024 kg)
Length: 1,676 mm loaded case with nozzle
Diameter: 933 mm
Propellant
 type: AP/HTPB/Al
 mass fraction: 0.929 for standard loading
Burn/action time: 62.7/64.1 s
Thrust (kN, sea level, 21°C): 54.80 max, 47.26 avg
Specific impulse: 289.8 s sea level 21°C
Total impulse: 30,512 kNs sea level 21°C
Pressure (atm): 36.73 avg, 43.67 max sea level 21°C
Nozzle
 throat diameter: unavailable
 materials: the semi-submerged, contour nozzle has a carbon phenolic expansion cone and carbon-carbon throat insert
Casing material: 6Al-4V titanium
Casing liner: unavailable
Igniter type: aft end, initiated from a remotely located Model 2134 safe/arm device.

Star 37XFP (TE-M-714-16/-17)
The GPS 2A navsats use the -16 version for orbit insertion. Propellant offloads to 20% are also produced. It has flown from Atlas Centaur, Ariane, Delta and Shuttle.

Mass: 955.7 kg (propellant 884 kg)
Length: 1,515 mm loaded case with nozzle
Diameter: 933 mm
Propellant
 type: AP/HTPB/Al
 shape: unavailable
 mass fraction: 0.925 for standard loading
Burn/action time: 66/67 s
Thrust (kN, sea level, 21°C): 42.48 max, 37.72 avg
Specific impulse: 290.0 s sea level 13°C
Total impulse: 2,537 kNs vac 13°C
Pressure (atm): 35.58 avg, 39.18 max vac 13°C
Nozzle
 throat diameter: unavailable
 materials: the semi-submerged nozzle has a carbon-carbon throat insert
Casing material: 6Al-4V titanium
Casing liner: unavailable
Igniter type: aft end, initiated remotely.

Star 48B (TE-M-711)
TE-M-711-17 1,245 mm ∅, 2,032 mm long PKM with 2,011 kg charge of AP/HTPB/Al in 6Al-4V titanium case; semi-submerged nozzle. Aft end igniter initiated from a remotely located safe/arm device. 2,137 kg (mass fraction 0.941). Average vac thrust 66 kN over 84 s burn; total impulse 5,647 kNs, effective Isp 286 s. Employed in McDonnell Douglas's PAM for satellite launches from Shuttle and Delta. The TE-M-711-18 has a 203 mm longer exit cone for use in launches from Delta, mass 2,142 kg, delivers effective Isp 292.1 s, total impulse 5,799 kNs.

A Star 63DV was fired Jan 1994 with the TVC system developed for Star 48V ***1995***

Star 48V (TE-M-940-1)
Developed for Conestoga's upper stage, providing a ±4° capability at 30°/s using the same loaded case as the 48B. First test was made Jun 1992; the second and final firing was made 2 Mar 1993. Debut Oct 1995.

Star 63D/F (TU-936/TE-M-963-2)
Perigee kick for the McDonnell Douglas PAM-D2 system. 3,502 kg (propellant 3,254 kg). 1,600 mm ∅, 1,779 mm long, SI 283 s. 63F is an extended version, overall length 2,685 mm, 4,266 kg propellant, SI 298 s. 63F provided Optus B's perigee kick from China's Long March. A Star 63DV was fired 26 Jan 1994 with the TVC system developed for Star 48V.

Star 75 (TE-M-775, 105-KS-45, 220)
A tested demonstration motor as a first step in developing PKMs in the 4,080-7,940 kg propellant range. It features a slotted, centre perforated grain in a graphite epoxy filament wound case, a semi-submerged nozzle with carbon phenolic exit cone, and a consumable wafer-type igniter. Propellant mass fraction 0.9299, burn rate 0.56 cm/s (at 34.0 atm, 15.5°C), length 259.0 cm, 190.5 cm ∅, burn/action time 105/107 s, max thrust 242.83 kN, SI 289.6 s, total impulse 21.330 MNs.

TRW SPACE & ELECTRONICS GROUP
TRW develops and manufactures propulsion components, engines and systems for spacecraft attitude and velocity control, orbit insertion, manoeuvring and adjust, tactical missile steering, low cost orbit transfer stages and low cost engines for Earth to orbit systems. Since the company produced the first space-operated monopropellant hydrazine system for 1960's Able lunar probe, it has provided >2,000 mono/bipropellant thrusters and reaction control systems. The bipropellant Lunar Module Descent Engine was responsible for landing 12 astronauts on the Moon 1969-1972, and Apollo 13's rescue, before it was converted into a fixed thrust model for Delta's second stage; 84 were built for the lunar programme and 75 flew on Delta from 1974. It remains under consideration for future upper stages. A further derivative was to power later versions of NASA's Orbital Maneuvering Vehicle (cancelled). Propulsion research covers low thrust monopropellant, bipropellant, colloid, ion, radioisotope, MPD, electrothermal and low cost engines.

Propulsion test facilities are centred at TRW's 8 km^2 Capistrano Test Site (CTS) 108 km south of Redondo Beach, equipped for thermal vacuum, vibration, shock, cold flow calibration, pressure/leak, functional and electrical, and accurate hot fire performance and life testing. Engine designs burning storable, cryogenic, gelled and high energy bipropellants can be evaluated under sea level and simulated altitude conditions. Continuous operation, steam exhaust-pumped altitude chambers can test high expansion ratio 450-670 N engines at pressures equivalent to 46 km. The HEPTS High Energy Propellant Test Stand has four 222 kN test positions, two of which simulate altitude firings. Cold flow and hot fire tests of low thrust hydrazine, ion/plasma electric and resisto-jet gas thrusters are undertaken at Redondo Beach.

TRW's 454 N DM-LAE dual mode liquid apogee engine is being improved for Odyssey

ESEX Arcjet

TRW has developed a 30 kW_e 2 N ammonia arcjet under contract to the AF Phillips Lab for test aboard the Argos satellite in 1997. Objectives of this ESEX electric space experiment are to measure plume deposition, EM interference, thermal radiation and acceleration. The mission calls for 10 firings of 15 min each at 26kW_e avg power.
Dry mass: unavailable
Fuel: ammonia at 0.25 gm/s steady state
Specific impulse: 800 s
Design life: 1,500 h and 400 on/off cycles.

Secondary Combustion Augmented Thruster

SCAT was qualified in 1995 as a high performance bipropellant thruster for stationkeeping, TVC, attitude control and orbit insertion/change. It uses a gas/gas mixing and combustion process that provides a wide operational range of mixture ratio and thrust for both regulated and deep blowdown systems. It features a standard catalyst decomposition chamber, a low cost regeneratively cooled thrust chamber (no refractory material or oxidation-resistant coatings) and operates near 100% efficiency in the bipropellant mode. It can operate in either mono hydrazine or bipropellant mode and can readily be designed as a bilevel thruster for TVC. Flight unit production is underway (SCAT will be carried on Odyssey), and advanced designs are in test for higher performance orbit change/insertion applications.
Dry mass: 1.8 kg
Length: 30.5 cm
Maximum diameter: 7.9 cm (125:1 nozzle)
Mounting: fixed (four threaded standoffs)
Engine cycle: pressure-fed 8-24 atm. Decomposition of N_2H_4 over Shell 405 catalyst bed mixed with vaporised N_2O_4 (gas-gas mixing)
Oxidiser: N_2O_4
Fuel: N_2H_4
O/F ratio: 0.9 to 2.0:1
Thrust: 20-67 N vacuum
Specific impulse: 287-326 s vacuum
Expansion ratio: 125:1
Chamber pressure: 6.8 atm
Ignition: hypergolic
Burn time: >150,000 s total, >7,000 s single burn demonstrated. >1,362 kg total throughput, >600 cycle life.

Dual Mode-Liquid Apogee Engine

DM-LAE was developed for reduced propulsion system complexity and lighter mass, in which the hydrazine supply is used for both the apogee engine (with N_2O_4) and the mono attitude control thrusters. The coaxial pintle injector is tuned for an optimum combination of performance and operating temperature margin. This basic injector concept began with the design of TRW's Apollo throttleable Lunar Module Descent Engine. Two DM-LAEs were successfully used to inject each of the Canadian Anik E2/E1 telecom satellites into GEO in Apr/Sep 1991. Intelsat K followed in 1992. No further deliveries have been made but an improved version (SI 321 s and 328 s) qualified in 1996. A 330 s model will fly on TRW's Odyssey satellites and the 321 s version is qualified for AXAF-I.
First flight: Apr 1991 (Anik E2)
Number flown: 6 to end-1995
Dry mass: 4.8 kg
Length: 56.05 cm (68.40 cm total)
Maximum diameter: 30.15 cm
Mounting: fixed (3 lugs at 120°)
Engine cycle: pressure-fed (15 atm)
Oxidiser: MON 3; flow rate 0.076 kg/s
Fuel: high purity hydrazine, at 0.071 kg/s
O/F ratio: 1.070:1
Thrust: 454 N vacuum
Specific impulse: 314.5 s vac (320 s + 328 s versions expected to be qualified in 1996)
Expansion ratio: 204:1
Cooling method: radiation
Chamber pressure: 7.0 atm
Ignition: hypergolic, initiated by 28 V electrical signal to on/off bipropellant torque motor valve
Burn time: 24,000 s total, 3,600 s max single burn.

OMV Variable Thrust Engine (VTE)

This deep throttling engine is derived from TRW's man-rated Apollo LMDE. Development began 1986; qualification was due to be completed in 1991 but NASA's Orbital Maneuvering Vehicle was cancelled in 1990. It completed preliminary development and is ready to enter pre-qualification. It was recently considered as a lander engine for precursor planetary probes. A set of four was to provide low acceleration primary propulsion for OMV, using low thrust ramp-up starts to minimise dynamic interaction and propellant sloshing. Thrust modulation of the four provides vehicle TVC.
Applications: OMV primary propulsion; rendezvous, interception and inspection missions
First flight: -
Dry mass: 6.8 kg
Length: 70.05 cm
Maximum diameter: 27.13 cm
Mounting: fixed
Engine cycle: pressure fed (tanks at 19.7 atm)
Oxidiser: N_2O_4; 0.119 kg/s at max thrust
Fuel: MMH; 0.0745 kg/s at max thrust
O/F ratio: 1.64
Thrust: 57.8-578 N vac (10:1 throttle range, 110 N/s throttle rate)
Specific impulse: 280-308 s vacuum
Expansion ratio: 125:1
Cooling method: radiative columbium (niobium) alloy chamber, fuel film barrier; designed to operate buried with full radiation heat shield
Chamber pressure: 0.75-6.8 atm (inlet pressure 19.7 atm)
Ignition: hypergolic
Burn time: 28 h total. 2,400 s max single burn at max thrust, 7,200 s at min thrust; 30 s min single burn at both max/min thrusts.

Delta TR-201

The TR-201 is a fixed thrust version of the LMDE for Delta's stage 2. Multi-start operation is adjustable up to 55.6 kN and propellant throughput up to 7,711 kg, and the engine can be adapted to optional expansion ratio nozzles. The engine made 75 flights during its Delta career over 1974-88. It is under consideration for future upper stages (at one time for PacAstro's PA-2).

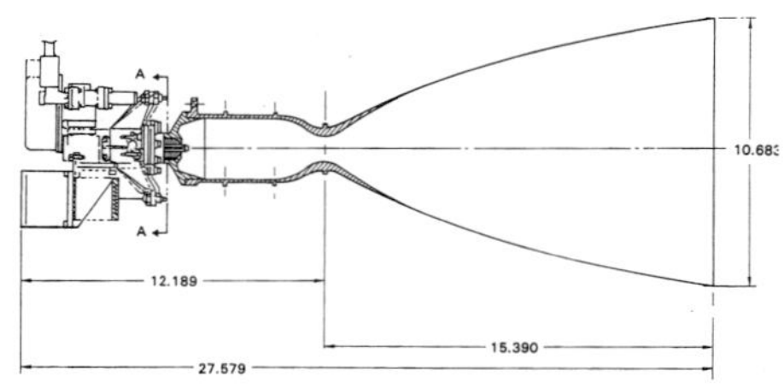

TRW's primary OMV bipropellant thruster, based on Apollo's descent engine, was designed to provide a 10:1 throttling range

Upgraded versions are under development for advanced high performance LOX/LH_2 upper stages. It has been demonstrated at NASA Lewis using LOX/LH_2 at 97% performance efficiency at 89 kN. It is under contract for McDonnell Douglas's EELV proposal, scaled up to 2.89 MN.
First flown: Jan 1974, Delta 100
Number flown: 75
Dry mass: 136 kg
Length: 227.1 cm
Maximum diameter: 148 cm
Mounting: gimbal attachment above injector
Engine cycle: pressure fed (15.5 atm reservoir)
Oxidiser: 50/50 N_2O_4/UDMH at 8.92 kg/s
Fuel: monomethyl hydrazine at 5.62 kg/s
O/F ratio: 1.60
Thrust: 44.035 kN vac
Specific impulse: 303 s vacuum
Expansion ratio: 43:1
Cooling method: quartz phenolic chamber ablation and columbium (niobium) nozzle radiation
Chamber pressure: 7.1 atm
Ignition: hypergolic, started by 28 V electrical signal to on/off solenoid valves
Burn time: 500 s for total of 5 starts; 10-350 s single burn.

Monopropellant Hydrazine Thrusters

TRW produces a family of six RCS thrusters ranging over 0.44-222 N and has provided hydrazine RCS systems

TR-201: Delta's fixed thrust version of the Lunar Module Descent Engine. Upgraded versions are being developed for LOX/LH_2 upper stages

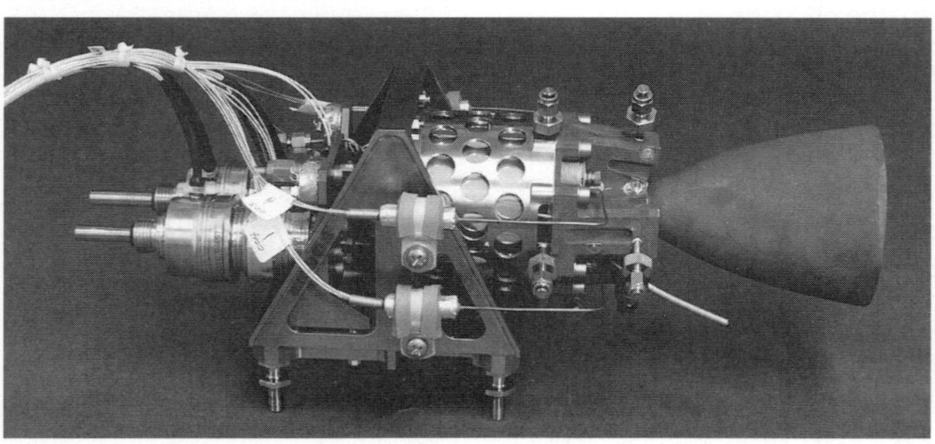

SCAT's gas/gas combustion process allows wide thrust and mixture ratio ranges (TRW) ***1996***

TRW's MRE-1 is NASA's standard 5 N thruster

for DSP, DSCS 2, Mariner, Pioneer, Fleetsatcom, HEAO, ISEE, TDRS and Compton Observatory.

MRE-0.1

Fleetsatcom's RCS incorporates TRW's MRE-0.1 + MRE-1 to manage spin/despin, precession, attitude, positioning, stationkeeping and MW unloading. The pressure-fed system includes sets of redundant thrusters and a complement of latching isolation valves for system redundancy management. Each system comprises 16 MRE-1 + four MRE-0.1 thrusters in dual modules, supplied from 122 kg of hydrazine in two 56 cm ∅ pressure spheres. System dry mass is 30 kg.
Applications: long duration missions requiring continuous precision pointing
First flight: 1978, Fleetsatcom 1
Number produced: 32
Length: 8.9 cm
Engine cycle: decomposition of hydrazine over heated Shell 405 catalyst bed
Thrust: 0.3-0.9 N vac
Specific impulse: 205-220 s vacuum
Total impulse: 71.2 kNs vac from 34 kg throughput
Inlet pressure: 24-79.6 atm
Burn time: 28,800 s maximum single burn
Minimum firing bit: 0.02 s
Total pulses: >370,000
Random vibration: 21 *g* rms.

MRE-1

MRE-1 is NASA's standard 5 N hydrazine thruster and has passed six different qualification programmes, the latest as a low cost version for the STEP/TOMS lightsat class. In addition to its deep space application aboard Pioneer 10/11, it was responsible for saving the TDRS 1 mission by boosting the 2 t satellite up to its operational GEO, consuming 363 kg during 45 h of cumulative firing. TDRS's system employs redundant groups of 12 thrusters, with associated latching isolation valves, drawing on 680 kg of hydrazine in two 102 cm ∅ tanks; dry mass is 86.5 kg.
Applications: satellite attitude and velocity control; Pioneer 10/11, Fltsatcom, HEAO, TDRSS, DoD
First flight: Mar 1972, Pioneer 10
Number produced: >1,500
Mass: 0.82 kg dual thruster module
Length: 15 cm
Engine cycle: decomposition of hydrazine over heated Shell 405 catalyst bed
Thrust: 5 N nominal, range 0.9-5 N vac
Specific impulse: 210-220 s vac steady state, 185-217 s vac pulsed
Total impulse: ≥1,156 kNs vac for 544 kg hydrazine throughput
Inlet pressure: 5.4-38 atm
Burn time: 11,200 s max single burn, >90 h demonstrated total firing time
Minimum firing bit: 0.02 s
Total pulses: 458,000
Random vibration: 21 *g* rms.

MRE-4

Development began in 1967 for Intelsat 3. It was responsible for re-routing NASA's ISEE 3 in 1985 to the first comet intercept. The longest demonstrated firing time is almost 2 h. Nozzle angle is selected to fit the geometry of the host spacecraft.
Applications: spacecraft attitude/velocity control; Intelsat 3, DSCS 2, ISEE 3, DSP, Atmospheric Explorer
First flight: Dec 1968, Intelsat 3-F2
Number produced: >300
Mass: 0.41 kg
Length: 20 cm
Engine cycle: decomposition of hydrazine over heated Shell 405 catalyst bed
Thrust: 2.2-18.2 N vac
Specific impulse: 212-230 s vac steady state, 208-222 s vac pulsed
Total impulse: 512 kNs vac; demonstrated propellant throughput >227 kg
Inlet pressure: 3.4-272 atm
Burn time: 7,080 s maximum single burn.

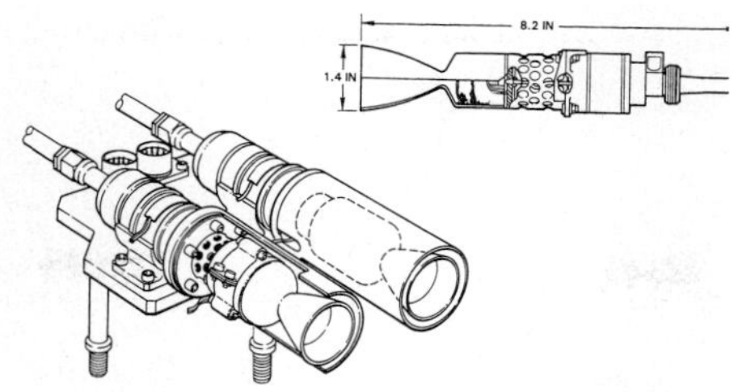

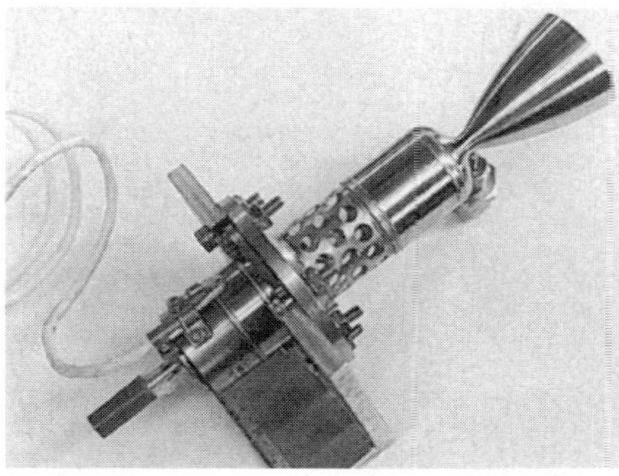

TRW's MRE-5 is used by NASA's Compton Observatory for attitude control

Minimum firing bit: 0.025 s
Total pulses: 507,000
Random vibration: 26.8 *g* rms.

MRE-5/Compton Observatory

Development began in 1975 for USAF programmes; 11 flights units were delivered in 1984 for attitude control of NASA's Compton Observatory. Operational design life is 2 yr, with 450 kg hydrazine throughput and capable of 120 cold starts from <32°C. Carrying 1,920 kg of hydrazine in four tanks and weighing 281 kg dry, it is the largest mono system built, incorporating four dual thruster modules (each a primary + redundant 22 N thruster) and one orbit adjust module of four 445 N thrusters. The MRE-15 (see below) is a scaled-up version for OMV.
Applications: missions requiring duty cycle flexibility, high reliability and long life in steady state/pulse modes
First flight: Apr 1991, Compton Observatory
Number produced: 10 development units, 11 flight units
Mass: 1.13 kg dual thruster module
Length: 21 cm
Engine cycle: decomposition of hydrazine over heated Shell 405 catalyst bed
Thrust: 12-24.5 N vac
Specific impulse: 232-240 s vac steady state
Inlet pressure: 6.8-31.6 atm
Burn time: 0.050 s to several hours
Minimum firing bit: 0.025 s
Total cycles: 50,000
Random vibration: 21 *g* rms.

MRE-15/OMV

The MRE-15 completed development and design verification (pre-qualification) as a scaled-up version of the MRE-5 for use on the NASA/TRW Orbital Maneuvering Vehicle (the programme was cancelled in 1990). 28 were to be clustered in sets of seven in each of four RCS orbital replacement units. Propellant throughput of each thruster was 450 kg of hydrazine. TRW was also to supply the MMH/N_2O_4 propulsion module engines (see below) and the 24 × 22 N nitrogen thrusters. No other applications are yet planned.
Applications: -
Mass: 1.13 kg
Length: 26.9 cm
Engine cycle: decomposition of hydrazine over heated Shell 405 catalyst bed
Thrust: 44.5-89 N vac
Specific impulse: 225 s vac steady state
Inlet pressure: 12-27 atm
Burn time: 0.020 s to steady state
Total cycles: 100,000.

Positive Expulsion Devices

>200 of TRW's elastomeric bladders and diaphragms have flown in hydrazine systems. Large-opening diaphragms and small-opening bladders have been tested under low/moderate accelerations for resistance to slosh/vibration and longterm compatibility with rocket and ramjet propellants. Oxidiser-compatible diaphragms up to 80 cm ∅ have been fabricated and expulsions demonstrated over −40/370°C.

UNITED TECHNOLOGIES

P&W SPO is responsible for UTC's space propulsion activities. Liquid engine programmes are pursued at the West Palm Beach facilities (see the separate entry), while solid programmes are run by CSD in San Jose. CSD specialises in the design, development and production of solid propellant, hybrid and ramjet propulsion systems for space, strategic, tactical and launch booster systems. Its Orbus upper stage family includes Orbus 6/6E/21 (IUS), Orbus 7S (JCSat), Orbus 21S (Intelsat 6), Orbus 21 (TOS), Orbus 21D (LMLV launcher) and Orbus 1 (STARS and Starbird upper stage). The numerical designator corresponds to propellant mass to the nearest 1,000 lb, S denotes spin stabilisation and E the use of an extending exit nozzle (EEC) system. CSD also manufactures the segmented SRMs for Titan 4 and the Space Shuttle SRB separation motors. It was responsible for the Algol stage 1 motor on Scout, which retired in 1994.

CSD's Coyote facility in San Jose includes the world's largest solid rocket vertical test stand: 29 m high, it can handle 26.7 MN thrust motors and was most recently used for qualification firings of the 7-segment USAF Titan 4 strap-on motor. Small/medium-sized horizontal and vertical test stands are controlled from a central complex and structural test facilities can apply omniaxial simulated flight loads of up to 4.9 MN to motor cases and attach skirts. CSD's integral rocket/ramjet test complex can simulate booster, transition and ramjet operations in a single sequence under sea level to high altitude conditions up to M6. The Coyote site also houses a 5,400 t annual capacity solid propellant mixing facility comprising two 2,270 l, one 2,840 l and one 1,510 l mixers.

Orbus

The main Orbus range derives from the two motors developed for the Inertial Upper Stage programme. The US Air Force awarded the Boeing/CSD team a contract in Oct 1976 to develop the IUS for Shuttle and Titan 34D. The stage completed its validation phase 19 Dec 1977 after a successful test firing of the Orbus 21 at the Arnold Engineering Development Center. Key components, such as the advanced flightweight nozzle, proved their survivability at a simulated altitude of 33 km. Full scale development of the IUS propulsion system began 1 Mar 1978 under an award from Boeing Aerospace. The 2-stage IUS incorporates the large Orbus 21 as the first stage SRM-1, topped by the Orbus 6E with EEC extendable exit cone as SRM-2. In May 1978 CSD was contracted by Boeing to develop the TVC electromechanical actuator system with redundant drive motors. The following Dec, CSD was contracted to develop the actuators for raising the stage and attached payload in the Shuttle cargo bay prior to deployment.

The first offloaded Orbus 21, at minimum 50% propellant loading, was fired 13 Aug 1980. By Jan 1983, no failure had occurred in 30 static firings despite the combination of such advanced features as Kevlar fibre case, carbon/carbon integrated throat/entrance, the world's first flightworthy EEC, Techroll seal and lightweight TVA. The first IUS flight placed two DSCS milcom satellites in GEO; the first Shuttle application came aboard STS-6 in Apr 1983 with NASA's TDRS 1 data relay satellite. SRM-1 fired successfully on that occasion, but SRM-2's Techroll joint overheated and failed, leaving the vehicle in an incorrect orbit. See the IUS entry for flight history. On a typical GEO mission, SRM-1 performs GTO injection, followed by SRM-2 insertion after a 6 h coast into GEO. Variations in mission requirements are accommodated by propellant offloading of up to 50%; SRM-2 can also fly without EEC, resulting in reduced specific impulse, or deployed before launch. Orbus 21 was also adapted as the Intelsat 6 perigee motor (21S spinning model) and for Orbital Science Corp's TOS upper stage which has been used

Orbus 21D was qualified as LLV's stage 2 motor by its 30 Jun 1994 firing (CSD) ***1995***

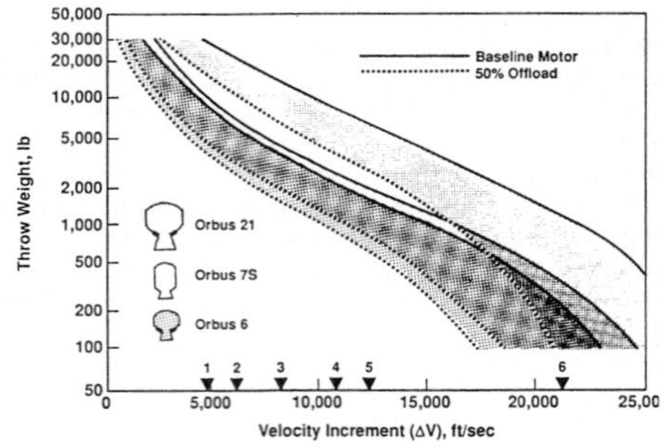

Baseline Orbus motor performances. Numbered triangles indicate 1 apogee burn for circularising 185 km perigee GTO, 2 apogee burn for circularising 185 km perigee GTO with 28.5° plane change, 3 185 km perigee burn for GTO, 4 parabolic escape from 185 km circular orbit, 5 Mars flyby from 185 km circular orbit, 6 Jupiter flyby from 185 km circular orbit (CSD)

Orbus 21 is the most powerful solid developed for in-space applications. Shown is Intelsat 6'2 21S version **1995**

for NASA's ACTS and Mars Observer. All the 'S' models (6, 7, 21) fly with fixed nozzles. The 21D version was test fired 30 Jun 1994 to qualify it as the upper stage of Lockheed's LMLV. Differences are a carbon phenolic exit cone in place of the previous 2-D carbon-carbon (this adds 136 kg inert mass but is $700k cheaper), 4-D throat replacing the 3-D throat, and state of the art TVC actuators (digital *vs* analogue). Specifications provided below are for the IUS and LMLV versions (21D specifications given only where different).

Orbus 21 specifications

First flown: 30 Oct 1982 on Titan 34D IUS
Number flown: 21 on IUS, 2 TOS, 6 Intelsat, 1 LMLV, through Feb 97
Applications: IUS stage 1 motor, TOS motor, Intelsat 6 perigee boost motor, LMLV stage 2
Mass: 10,398 kg fully loaded (10,619 kg 21D); 639 kg burnout (780 kg 21D)
Length: 315.0 cm (317.2 cm 21D)
Maximum diameter: 233.7 cm
Propellant
type: 86% solids HTPB UTP-19360A
shape: based on 52 cm ∅ bore
mass fraction: 0.935 fully loaded (9,709 kg; 0.941 for 21S model without TVC system), 0.47 max 50% offloading (4,853 kg). 21D 0.921 (9779 kg), 0.457 max offloading (4,583 kg)
Burn time: 154 s (146 s 21D)
Thrust (vacuum): 195.7 kN average, 267.77 kN peak (192.4 kN avg 21D)
Total impulse (MNs, vacuum): 28.38 (28.14 21D)
Specific impulse (s, vacuum): 295.5 (293.5 21D)
Pressure (atm): 57.9 atm max
Nozzle
throat diameter (cm): 164.6 mm
length (cm): 150.56 cm
exit diameter (cm): 1,315 mm (63.9 expansion ratio)
materials: single-piece integral throat and entrance of 3-D carbon-carbon (4-D for 21D); exit cone 2-D carbon-carbon (carbon phenolic with graphite/epoxy structural overwrap for 21D)
Casing materials: 353.8 kg, Kevlar/epoxy with silica-filled EPDM
Igniter type: pyrogen head-end, 16.3 kg with 3 nozzles, UTP-1,095 propellant
Thrust vector control: ±4° TVC (up to 20°/s) provided by battery-powered orthogonal electromechanical actuators and Techroll movable nozzle joint pressurised by silicon oil; system mass 22.4 kg.

Orbus 6/6E specifications

First flown: 30 Oct 1982 on Titan 34D
Number flown: 21, on IUS, to Feb 97
Applications: IUS stage 2 motor
Mass: 3,018/2,749 kg fully loaded with/without EEC
Length: 198 cm with EEC stowed, 320 cm EEC deployed
Maximum diameter: 160.0 cm
Propellant
type: 86% solids HTPB UTP-19360A
shape: cylindrical
mass fraction: 0.902/0.914 with/without EEC fully loaded (0.921 without TVC system), 0.45 with max 50% offloading
Burn time: 103 s fully loaded
Thrust (kN, vacuum): 80.95 average, 113.87 peak
Total impulse: 8.14 MNs
Specific impulse (s, vac): 303.5/289.1 with/without EEC
Pressure (atm): 57.1 atm max
Nozzle
throat diameter (cm): 106.9 mm
length (cm): 122/83.6 with/without EEC
exit diameter (mm): 1,438.0 (181.1 expansion ratio) with EEC; exit ∅ 750.3 (47.3 expansion ratio) without EEC
materials: single-piece integral throat and entrance of 3-D carbon-carbon; exit cone/s 2-D carbon-carbon
Casing materials: 90.7 kg Kevlar/epoxy with silica-filled EPDM
Igniter type: 9.5 kg pyrogen head-end, three nozzle, UTP-1095 propellant
Thrust vector control: as Orbus 21 but 7°.

Orbus 6E incorporates an extending nozzle for Shuttle cargo bay compactness, increasing IUS GEO capacity by 100 kg. The motor can fly without the 122 cm extension (CSD)

Orbus 7S specifications

The Orbus 7S was qualified as the PKM for Hughes' HS-393 family of communications satellites. Derived from CSD's Minuteman 3 stage 3 motor, with the thrust termination and vector control systems removed, it could be stretched mid-cylinder and incorporate a lengthened nozzle. An Orbus 7 version is available incorporating a liquid injection TVC system.
First flown: 30 Aug 1984 as part of USN/Hughes Leasat 2 milcomsat
Number flown: 8 to end-1995
Applications: PKM, launcher upper stage

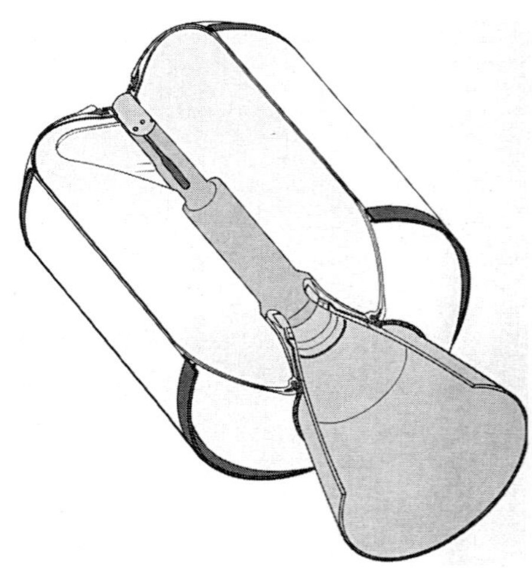

Orbus 7S derives from Minuteman 3's stage 3 motor

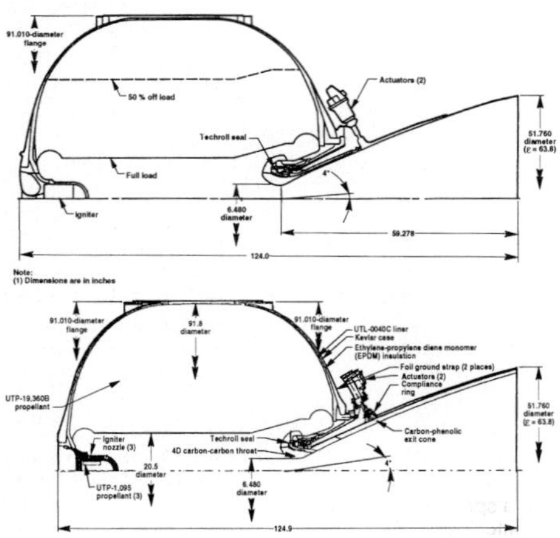

Orbus 21 cutaway principal features: 21 (top) and 21D (bottom) (CSD) **1995**

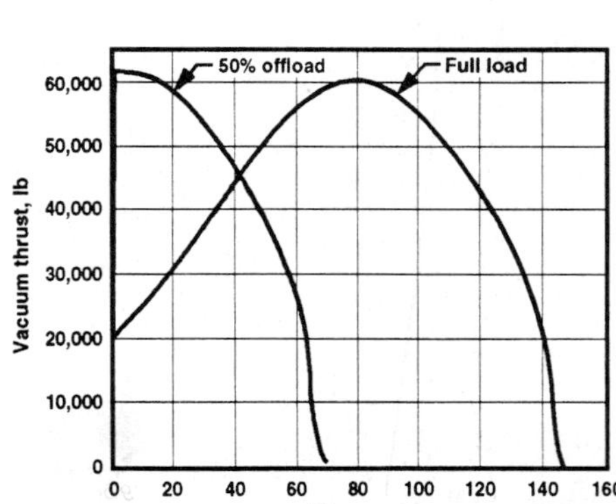

Orbus 21 thrust profile (CSD) **1995**

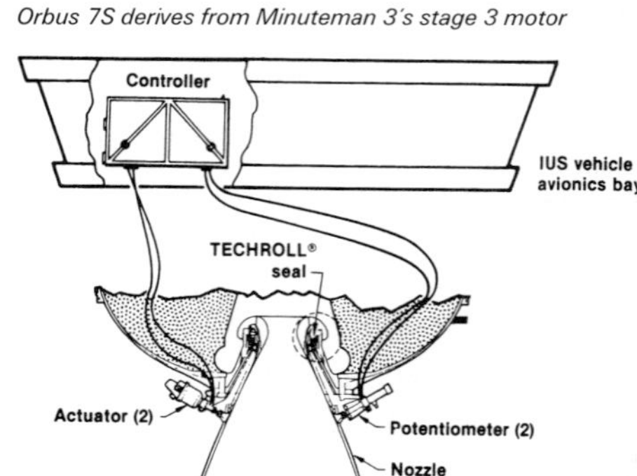

Orbus 21/6E provide steering for the 3-axis Inertial Upper Stage through the Techroll nozzle joint and orthogonal actuators. Orbus 21 produces a maximum deflection of 4°, with 7° for the upper stage Orbus 6E

Mass: 3,547 kg fully loaded, 196 kg burnout
Length: 227.3 cm
Maximum diameter: 132 cm
Propellant
  type: 86% solids CTPB ANB-3066; forward end finocyl
  shape: stepped cylinder
  mass fraction: 0.935
Burn time: 60.1 s
Thrust in vacuum: 153.467 kN average, 191.274 kN max
Total impulse: 12.5 MNs
Specific impulse (s): 281.9
Pressure (atm): 35.3 average, 43.5 max
Nozzle
  throat diameter: 174.7 mm
  length: 90 cm
  exit diameter: 84.76 cm (expansion ratio 23.55)
  materials: tungsten throat insert, carbon phenolic exit cone, total mass 63.2 kg
Casing materials: 93.94 kg S-901 fibreglass/epoxy with silica-filled Buna-N
Igniter type: 4.54 kg head-end pyrogen, 6-nozzle.

Orbus 1

The smallest motor in the series was developed for the upper stages of the Starbird and STARS target vehicles of the US Army Strategic Defense Command. The motor is a complete stage including a flexseal movable nozzle, electromechanical actuators, TVC control electronics and thermal battery, flight termination system and cables. The Orbus 1S spin version is also available.
First flown: 17 Dec 1990, Starbird stage 3/4
Number flown: 3 to end-1995
Applications: upper stage and space motor
Mass: 470.4 kg fully loaded, 53.2 kg burnout
Length: 124.9 cm
Maximum diameter: 69.2 cm
Propellant
  type: 90% solids HTPB
  shape: head-end web
  mass fraction: 0.88 (including TVC/TVA)
Burn time: 39 s
Thrust in vacuum: 30.38 kN average, 36.25 kN max
Total impulse: 1.19 MNs
Specific impulse (s): 293.3
Pressure (atm): 58.4 average, 66.7 max
Nozzle
  throat diameter: 55.1 mm
  length: 525.7 cm
  exit diameter: 375.4 mm (expansion ratio 46.4)
  materials: carbon/carbon throat, carbon phenolic exit cone
Casing materials: T-40 graphite epoxy with silica-filled EPDM insulation
Igniter type: aft-end toroidal pyrogen.

Shuttle Booster Separation Motors

NASA selected CSD in late 1975 to develop and produce solid propellant Shuttle booster separation motors (BSMs) for the first six development missions. Each strap-on fires a cluster of four motors at the forward and aft ends to separate from the vehicle's External Tank and begin the recovery sequence. The motor was accepted by the space agency in Sep 1978 and the last of 104 originally ordered was shipped in early 1980. CSD also provides separation motors for the Titan launcher strap-ons (see below). >1,500 have been produced.
First flown: 12 Apr 1981 on Shuttle mission 1
Number flown: 16 per Shuttle mission (1,280 by end-1996 on 80 missions)
Mass: 73.0 kg
Length: 838 mm
Maximum diameter: 327 mm
Propellant
  type: 86% solids HTPB (UTP-19048)
  shape: 16-point star
  mass fraction: 0.472 (34.5 kg propellant)
Burn time: 0.8 s
Thrust (kN, vac): 82.6 average, 129.5 max
Total impulse: 78.1 kNs
Nozzle: 20° canted; graphite throat insert with steel exit cone
Casing materials: aluminium with silica-filled EPDM insulation
Igniter type: each cluster of four is ignited by firing redundant NSD (NASA Standard Detonators) pressure cartridges into redundant confined detonating fuse manifolds.

Titan segmented boosters

CSD test fired the first large segmented solid propellant motor on 15 Dec 1960 and in May 1962 was contracted to provide Titan 3's strap-on motors. This 120-inch programme (referring to the motor diameter) achieved its first full-scale static firing 20 Jul 1963; 13 others followed by its conclusion on 28 Apr 1965 shortly before Titan 3C's debut. This booster comprised five segments with pinned clevis joints, fixed canted nozzles, LITVC system, aft skirt, nosecone, separation system and destruct systems. Production ended in 1979; the last of 65 missions was in Nov 1982.

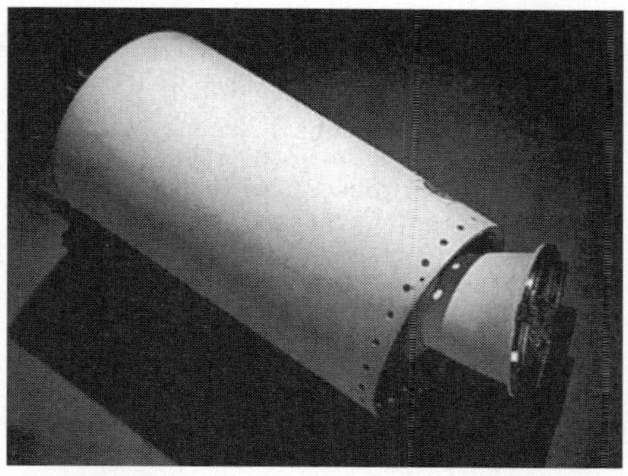

Each Shuttle strap-on carries eight CSD separation motors to initiate the recovery sequence

A '5½-segment' version was created by adding a top segment 60% of normal 320 cm length for Titan 34D; the 15th/last departed in Sep 1989. Lockheed Martin's commercial Titan 3 launcher is derived from this vehicle; further information is provided under that company's entry. Separation of insulation from the steel casing because of a manufacturing/processing anomaly was determined to be the cause of the Titan 34D failure 18 Apr 1986; during the revalidation phase heater strips were also added to the segment joints to avoid *Challenger*-type failures.

A 7-segment version was selected for the USAF Manned Orbiting Laboratory programme and first fired 26 Apr 1969. Three other full-scale trials followed successfully in 1970, but MOL had already been cancelled. The design was selected for the Titan 4 launcher in the 1980s, requiring only two qualification firings, because of the previous development work, on 21 Dec 1987 and 14 Feb 1988. CSD's steel-casing motors are utilised on Titan 4 launches; later missions will employ Alliant Techsystem's lightweight filament-wound graphite composite casing motors, although development problems prompted procurement in early 1990 of long lead items for seven further sets of CSD's design. 20 of the 26 ordered flight sets had been delivered by end-1995. Titan strap-ons provide steering by injection of N_2O_4 into the exhaust from a 3,630 kg-capacity tank; less than half is normally consumed. The injection adds typically 17.8 kN thrust to each motor. CSD also provides the eight separation motors per booster, mounted in clusters of four fore/aft. Each 1.5 m long 15 cm ∅ 38.5 kg motor contains 25 kg of solid propellant and provides 20 kN thrust in the 1 s burn time.

Titan solid rocket motor specifications
The specifications given below apply to the Titan 3 5½-segment SRMs; where Titan 4's 7-segment characteristics differ they are included in parentheses.
First flown: Oct 1982 on Titan 34D, Jan 1990 on commercial Titan 3 (Titan 4 7-segment 1989)
Number flown: 15 sets on Titan 34D + four sets on commercial Titan 3 to end-1995 (15 sets Titan 4)
Mass (t): 238 (316.6)
Length (m): 27.57 (34.43), each segment 328 cm
Maximum diameter (cm): 311 (311)
Propellant
  type: 84% solids PBAN (UTP-3001B)
  shape: forward segment has fins; others are tubular, tapering from 98.3 cm thick to 85.6 cm aft for desirable tail-off characteristics
  mass fraction: 0.885/210.6 t propellant (0.846/268.1 t Titan 4 propellant)

CSD tested this first 7-segment motor in Apr 1969 before the Manned Orbiting Laboratory programme was cancelled. The same motor is now utilised by Titan 4

Burn time (s): 113.7 (119.5)
Thrust (kN): 6,227 (7,117) vacuum average
Specific impulse: 265.2 (272.0) s vacuum
Total impulse (vacuum): 545 MNs (706.8 MNs)
Pressure (atm): 58 max (joints designed for 72.5/1.25 times MEOP)
Casing materials: D6AC high-strength steel <12.7 mm thick. Segments are stacked with clevis joints incorporating 240 pins and single O-ring. Segments are interchangeable. Insulation provided by silica-filled Buna-N rubber.

UNITED TECHNOLOGIES PRATT & WHITNEY

Government Engines & Space Propulsion

PO Box 109600, West Palm Beach,
Florida 33410-9600.
Tel: (+1 407) 796 67 96
Fax: (+1 407) 796 72 58
President, P&W GESP: John A Balaguer
Public Relations: Patrick W Louden

- **P&W Space Propulsion Operations**

PO Box 49028, San Jose, California 95161-9028.
Tel: (+1 408) 779 91 21
Fax: (+1 408) 778 45 99
Executive VP/GM: Douglas A North
VP Liquid Space Programs: Joseph P Zimonis
RL10 Program Director: Peter G Scharf
SSME/ATD Program Manager: John L Price

P&W Space Propulsion Operations, a GESP unit, is responsible for the company's space propulsion activities. Liquid engine programmes are pursued at the West Palm Beach facilities, while solid rocket programmes are run by the Chemical Systems Div in San Jose (see the separate UTC CSD entry). USBI provides Shuttle booster services; see the separate entry. UTC combined the space propulsion activities of CSD, USBI and P&W in 1990 within GESP. P&W has supplied RL10 cryogenic engines for the Centaur upper stage since the early 1960s. In 1986 the company was awarded a NASA contract to develop alternate turbopumps for the Space Shuttle main engine. An agreement was signed with NPO Energomash (qv) in Oct 1992 providing exclusive US marketing rights on the Russian company's products and technology. An RD-120 production engine was tested Oct 1995 in West Palm Beach, believed to be the first test of an in-production and in-service engine outside of Russia. Modified versions are proposed for several US vehicles, including the X34 and Kistler's K1. The three <5 s runs showed the engine was compatible with US ground systems; further testing requires a sea level nozzle and a customer. A major activity continues to adapt variants of the 4-chamber RD-170/171. The RD-180 2-chamber derivative has been selected by Lockheed Martin for the Atlas 2AR; it is being developed by NPO Energomash and provided through a P&W/Energomash joint venture. The first 18 vehicles will use Russian-built engines, but P&W will then begin production (commercial vehicles can employ Russian engines, but US government payloads must use US engines by law).
Test facilities: at peak, the Florida site operated seven RL10 test stands, including one for Centaur-type dual-engine firings. Stand E6 has been refurbished for the new production RL10s and stand E8 accommodates Space Shuttle ATD turbopump work.

P&W RL10

The company's RL10, initially fired in Jul 1959 as the world's first LOX/LH_2 engine and first delivered to NASA in Aug 1960, will continue in paired service for the Centaur upper stage to century's end and could be developed further to produce 156 kN for a future Space Transfer Vehicle. RL10 design work began in Oct 1958, leading to the first Centaur ground test firing in Mar 1962 and the stage's flight debut in Nov 1963. Oct 1966 marked the first completely successful in-space restart of a hydrogen engine. Fluorine and methane have also been used separately in test firings. The current A-3-3A version completed qualification tests in Nov 1981 and first flew on AC-62 in Jun 1984; its development was supported by Intelsat to handle their satellites. By end-1990, 178 RL10s of all types had flown in space, accumulating 290 firings totalling >21 h without a single failure. The first failure occurred 18 Apr 1991 in the AC-70 launch of Japan's BS-3H satellite, although the blame was fixed on pre-launch gantry procedures and not on RL10's operation. General Dynamics' review board decided that turbopump foreign object damage, possibly combined with water/ice contamination, was a likely cause. The Feb 1992 Atlas 2 launch included the 300th space firing. AC-71 on 22 Aug 1992 produced the second failure: icing in the hydrogen pump was identified as the cause for both AC-71/70. The engines were chilled before launch to −256°C by liquid helium from a pad umbilical. It appears that a spring-operated check valve stuck open when the umbilical pulled free, allowing air to be sucked in where the nitrogen could condense and freeze. Subsequent vehicles carry electrically-operated valves and are not as

Pratt & Whitney's RL10 was the first operational liquid oxygen/hydrogen engine and will continue in service next century. Shown is the RL10A-4 (P&W) ***1996***

chilled so that air cannot freeze even if it does penetrate. Before this second failure, RL10 had achieved 190 flights and 312 firings. By end-1995, 246 RL10s of all types had flown in space, accumulating 436 firings totalling >29.3 h. In addition to Centaur applications, clusters of six RL10A-3 powered Saturn 1's S4 stage on six missions during 1964/65. The original RL10A-1 model generated 66.72 kN vacuum thrust with 412 s SI and 20.4 atm chamber pressure. The current RL10 is available in the A-3-3A version offering 5:1 oxidiser: fuel mixture ratio for the Atlas Centaur combination (Atlas 1) and the USAF Titan 4 Centaur. Atlas 2 employs a 5.5:1 mixture ratio for 442.3 s SI and 73.39 kN thrust; Atlas 2A required a significant enhancement to 92.5 kN. Further information on this RL10A-4 model is provided below. Titan 4 will continue to use the A-3-3A, but if its Centaur converts to a single engine then A-4 could be adopted. Further upratings are under consideration, particularly for improving ELV performance and space transfer vehicles. The 93.4 kN **RL10B-2** will emphasise performance improvement by increasing SI to 466.5 s; expansion ratio 285:1. Its selection for Delta 3's stage 2 was announced in Jun 1995; first delivery is due mid-1997. The **RL10C-1**, providing 156 kN using a 147:1 expansion ratio, has been studied by NASA for a single-engine Centaur.

P&W expects to produce 16-24 RL10s annually to end-1999 for the Atlas, Titan 4 and Delta 3 programmes. 26 were manufactured in 1995 (1994: 14; 1993: 7; 1992: 12; 1991: 13; 1990: 18).

P&W was awarded a $12 million contract by McDonnell Douglas Aerospace to provide four **RL10A-5** for BMDO's DC-X SSRT single stage rocket technology suborbital demonstrator. Each 143 kg engine, incorporating a 10° conical nozzle for low-altitude operations, can be throttled 30-100% (19.5-65.7 kN) and gimballed ±8.5°. The settings correspond to 380.5/373 s SI, respectively, and 9.66/32.3 atm chamber pressures. Length 106.9 cm, expansion ratio 4.28:1.

RL10A-3-3A specifications
Applications: Centaur stage of Atlas & Titan
First flown: Nov 1963 (3-3A first flight was Jun 1984)
Number produced: 380 of all types and including ground test engines to end-1995
Number flown: 262 of all types to end-1996 (two inflight failures caused by vehicle pre-flight procedures and not attributed to engine operations)
Dry mass: 138 kg
Length: 178 cm
Maximum diameter: 102 cm
Mounting: ±4° gimbal for pitch/yaw control
Engine cycle: expander
Oxidiser: liquid oxygen at 14.0 kg/s
Fuel: liquid hydrogen at 2.79 kg/s
Mixture ratio (O/F): 5.0:1
Oxidiser pump: 11.3 kg mass, 88 kW power, 13,100 rpm rotor speed producing 45.6 atm discharge pressure
Fuel pump: 34 kg, 32,800 rpm rotor speed, 76.2 atm discharge pressure
Thrust: 73,390 N vacuum
Specific impulse: 444.4 s vacuum
Time to full thrust: typically 2.15 s
Nozzle area ratio: 61:1
Expansion bell
  length: 117 cm
  diameter: 102 cm (expansion ratio 61)
  materials: AISI 347 stainless steel
  cooling method: regenerative by hydrogen

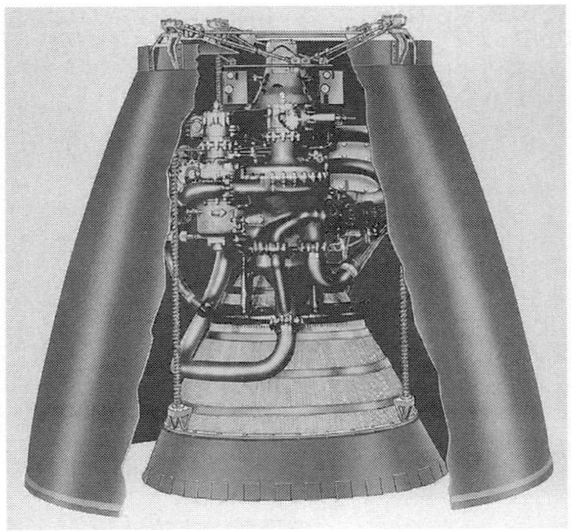

The RL10B-2 will power Delta 3's cryogenic stage (P&W) ***1995***

Combustion chamber
  pressure: 32.2 atm
  temperature: 3,340°C
  stability: ±0.5%
  materials: AISI 347 stainless steel
  cooling: regenerative by hydrogen
  ignition: spark igniter. Qualified for 20 starts (seven made on Helios 2 mission)
Burn time: about 600 s required on Titan 4 Centaur, engine qualified to 4,000 s.

RL10A-4

The commercial Atlas 2A/2AS versions incorporate RL10s improved to 92.5 kN thrust each. Incorporation of the extending nozzle contributes a 6.5 s SI gain without additional thrust compartment length increase. The all-welded columbium (niobium) 508 mm long skirt is electromechanically deployed following Centaur separation from the Atlas booster to create an 84:1 expansion ratio. The A-4 was flight certified in May 1991, followed by delivery in Jun 1991 of the first of 74 ordered by General Dynamics. By end-1992, ground testing accumulated 41,946 s in 275 firings; first flight was made successfully Jun 1992. The **A-4-1** version is available to improve Atlas 2A/2AS performance. First flight was AC-113 10 Jan 1995 to cope with the larger Intelsat 7 series; second AC-115 22 Mar 1995. Thrust is increased by 6.67 kN and SI by 2 s by a modified injector improving propellant mixing in the thrust chamber. The tangential swirl oxidiser injector was developed in the 1960s for the USAF XLR-129 engine. The **RL10E** version is being developed for the single-engine Centaur of Atlas 2AR, due to debut in Dec 1998. It replaces the hydraulic and pneumatic valves with electromagnetic actuators, uses EMAs for gimballing and employs a solid state controller.

Specifications are given below for the RL10A-4 where they differ from the A-3-3A (A-4-1 values are given in parentheses).
First flown: Jun 1992 (Jan 1995)
Number produced/flown: 26/18 by end-1995
Dry mass: 168 kg
Length: 178 cm stored, 229 cm deployed
Maximum diameter: 117 cm

RL10A-4's 508 mm skirt extension improves specific impulse by 6.5 s for the Atlas 2A (P&W)

Oxidiser: LOX at 17.78 kg/s (19.10); initial P 2.9 atm
Fuel: LH_2 at 3.22 kg/s (3.45); initial P 1.95 atm
Mixture ratio (O/F): 5.5:1
Oxidiser pump: 14,300 rpm (14,950) rotor speed producing 56.5 atm (57.6) discharge pressure
Fuel pump: 35,800 rpm (37,400) rotor speed, 92.5 atm (94.6) discharge pressure
Thrust: 92,500 N (99,190) vacuum
Specific impulse: 448.9 s (451.0) vacuum
Nozzle area ratio: 84:1
Expansion bell
  length: 119.4 cm primary + 51 cm extension
  diameter: 102 cm primary; 117 cm extension (expansion ratio 84)
Combustion chamber
  pressure: 39.46 atm (41.5)
  temperature: unavailable
  stability: better than ±0.5%
Burn time: qualified to 3,000 s in multiple starts.

RL10B-2

Delta 3 will employ the B-2 for its cryogenic stage 2. Development began in mid-1995, and both certification and first production delivery are planned for 4Q 1997. The B-2 features a very high area ratio nozzle to produce the highest SI of any operational chemical engine. SEP is developing the 100 kg 3 m-high carbon-carbon extending nozzle. Specifications are given below where they differ from the A-3-3A.
Applications: Delta 3 stage 2
Dry mass: 259±5.5 kg
Length: 415.3 cm extended
Maximum diameter: 222.3 cm
Oxidiser: liquid oxygen at 19.9 kg/s
Fuel: liquid hydrogen at 3.3 kg/s
Mixture ratio (O/F): 6.0:1
Oxidiser pump: 14,676 rpm rotor speed, 55.6 atm discharge pressure
Fuel pump: 36,690 rpm rotor speed, 96.7 atm discharge pressure
Thrust: 105,645±1,955 N vacuum
Specific impulse: 466.5 s vacuum

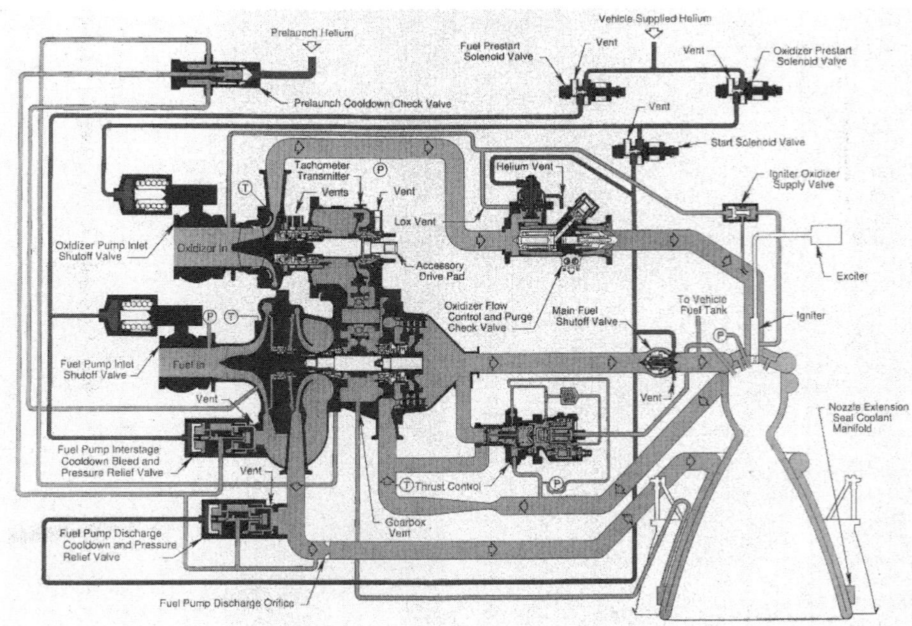

RL10A-4 propellant flow schematic (P&W) ***1994***

Nozzle area ratio: 285:1
Expansion bell
length: 349 cm extended
diameter: 222.3 cm maximum
Combustion chamber
pressure: 42.2 atm
stability: better than ±0.5%
Burn time: 3,500 s in multiple starts.

Shuttle Main Engine Turbopumps

NASA awarded P&W a $188 million contract, subsequently increased by $20 million, in Dec 1986 for new Shuttle SSME high pressure turbopumps as part of the engine's Block 1/2 upgrades. Alternative Turbopump Development (ATD) design was completed in May 1988 for component testing of bearings, seals and turbine blades to begin. Hardware procurement for the initial pumps also began in 1988. Development problems, including casting the LOX main housing in three pieces to eliminate the old welded configuration, coupled with a rotor vibration and bearing wear problem, increased programme cost to $504 million. Testing started in early 1990 at P&W's reactivated and modernised facilities in West Palm Beach. The fuel pump attained rated power by year's end and the first development pump set was delivered in 1991 for engine testing at NASA Stennis. NASA halted fuel pump work from Dec 1991 in order to focus on the more critical oxygen unit. It restarted work May 1994 once the LOX pump was on schedule for its 1995 flight debut. LOX pump certification was completed 15 Mar 1995, testing two pumps >10,000 s each in >20 tests. Block 1 first flight was successful STS-70 in Jul 1995 with one pump. All three engines were equipped for STS-77 in May 1996. The LOX pump completed 235 firings by end-1995 at Stennis, totalling 110,000 s with flight representative hardware. It will be fully retrofitted into the fleet by 1998. The fuel pump will fly as part of the Block 2 upgrade, debuting on STS-86 in Sep 1997 and completing fleet retrofit by 2000.

A separate contract provides for 15 production LOX units in 1995-98 and 15 fuel 1997-2000. The pumps are not intended to increase engine performance but instead offer improved durability of 7½ h firing, or 55 missions, at 109 per cent rated thrust. The design eliminates the 469 welds in Rocketdyne's fuel pump and all but seven of 300 in Rocketdyne's oxidiser pump, and houses fewer rotating parts. Stiffer rotors with robust single row ball and roller bearings are aimed at meeting the long life requirements. The fuel pump will deliver 55,650 l/min at up to 455 atm; 36,300 rpm, 54.66 kW. The LOX side provides 28,000 l/min at 520 atm; 24,000 rpm, 19.46 kW.

UNITED TECHNOLOGIES
USBI Co

UTC combined the space propulsion activities of CSD, USBI and P&W in 1990 within P&W's Government Engines & Space Propulsion unit based at West Palm Beach in Florida. USBI is under contract to NASA Marshall to provide assembly, test and refurbishment of the non-motor segments of Shuttle's solid rocket boosters. NASA extended the contract in 1994 for recovering and refurbishing boosters to Sep 1997, worth $1.8 billion. The previous $1 billion award in Jan 1989 ran through Sep 1994. It was extended in 1996 through Sep 1999.

WESTERN ELECTROCHEMICAL CO

Pacific Engineering & Production satisfied half (18 kt) of US ammonium perchlorate solid propellant oxidiser requirements until explosions destroyed its plant in May 1988. Users included Delta, Titan and Shuttle boosters; each Shuttle launch consumes 771 t. A new facility operated by Western Electrochemical (WECCO), a wholly owned subsidiary of American Pacific Corp, a Las Vegas-based manufacturer of speciality chemicals, was completed during 1989, offering an initial annual production rate of 14 kt capable of rising to 18 kt with only minor capital additions. The 33-building complex was constructed at a cost of $92 million and employs a staff of 175. The batch process under automated control can produce AP to custom chemical characteristics and physical distributions, as well as standard materials in large homogeneous lots. High purity AP is used in low smoke reduced signature rocket motors.

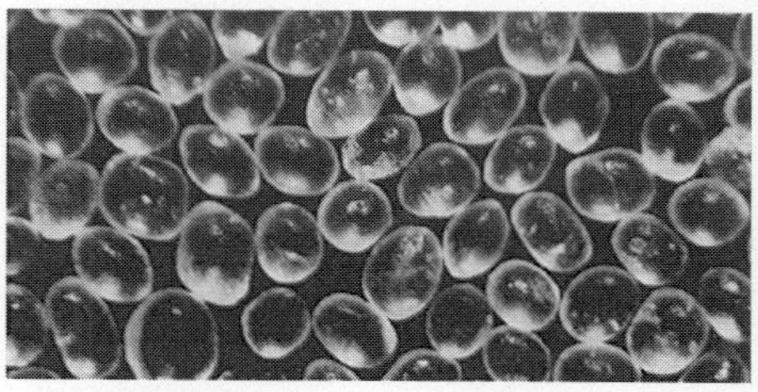

Western Electrochemical's 400 µm-grain ammonium perchlorate ***1996***

CIVIL AND COMMERCIAL SATELLITES, SATELLITE SERIES AND SATELLITE CONSTELLATIONS

Communications
Data Relay
Earth Observation & Meteorological
Engineering Test
Geodetic
Microgravity & Materials Science
Navigation
Scientific

COMMUNICATIONS

ARGENTINA

NAHUEL series

Current status

The status of the Nahuelsat SA satellite for 75.8° west is unclear. DASA signed an agreement 5 May 1995 with China Great Wall Industry Corp for a CZ-3C launch in first quarter 1997, although work had not begun by May 1999.

Background

In 1991, Argentina issued a RfP for a domestic/regional satellite communication system in late 1991. Argentina wanted to operate two Nahuel ('Jaguar' or 'Tiger') satellites at one of two sets of locations 59° west, 71.8° west, 75.8° west, 80° west, 85° west. The country planned to pay about US$250-300 million for the capability and generate US$60-70 million revenue annually. Bolivia, Brazil, Chile and Uruguay all planned to benefit from the system. To establish a framework, the Argentine government organised the Comision Nacional de Telecomunicaciones on 16 December 1992 to negotiate with a consortium headed by DASA, and including Aerospatiale, Alcatel Espace and Alenia Spazio. The agency signed a contract 27 May 1993 for a term of 24 years and five orbital slots, to serve the southern part of the continent for the first time with high performance Ku-band links. NahuelSat SA created by the three European negotiation partners has US$100 million in equity to own/operate the system. The consortium plans to hold about half the shares.

Argentina operates the system as a concession and began service in 1997 following the successful launch of Nahuel 1A. Nahuel provides the equivalent of 36 TV programmes (180 with digital compression), 18,000 duplex telephone calls or 9,000 VSAT interactive connections. Connections require 1-2 m dishes.

Before Nahuel 1A orbited, Argentina used Telesat Canada's Anik C1/C2 satellites (as Nahuel A/B) to provide a Ku-band interim service. Telesat Canada sold the satellites in January 1993 for C$37 million over 3 years to the Paracom SA consortium. Telesat maintained a 10 per cent interest to secure participation in marketing services.

Specifications

Nahuel 1A

Launch: 30 January 1997 by Ariane 4 from Kourou, French Guiana
Location: 71.8° west geostationary
Design life: 12 years
Contractors: Aerospatiale (prime), Alcatel Espace (payload), DASA (AOCS, solar arrays)
Transponders: 18 55 W TWTA 14/12 GHz Ku-band South American beams, bandwidths 54 MHz, EIRP 50 dBW, orthogonal linear polarisation
Principal applications: TV distribution, telephony, VSAT data and business services
Configuration: Spacebus 2000 platform, 3-axis control, 22.4 m span across solar panels, single dual reflector multifeed antenna. GEO location maintained within ±0.05°
Mass: 1,820 kg at launch; about 1,100 kg on-station BOL
AOCS: provided by DASA's 22 × 10 N + 400 N MMH/MON bipropellant system
Power system: twin solar wings providing 2.9 kW equinox EOL.

CANADA

Anik A series

Background

Anik initiated commercial communications over Canada's sprawling 9.22 million km² land mass. The US launched the initial series beginning with Anik A1, 9 November 1972, and followed by Anik A2, on 20 April 1973, Anik A3, 7 May 1975. All went up on Delta launch vehicles from Cape Canaveral. Canada retired A1 at 114° west, in July 1982; Anik A2 at 109° west, in October 1982; Anik A3 at 119° west capable of handling 10 colour TV channels or 9,600 phone circuits on 21 November 1984.

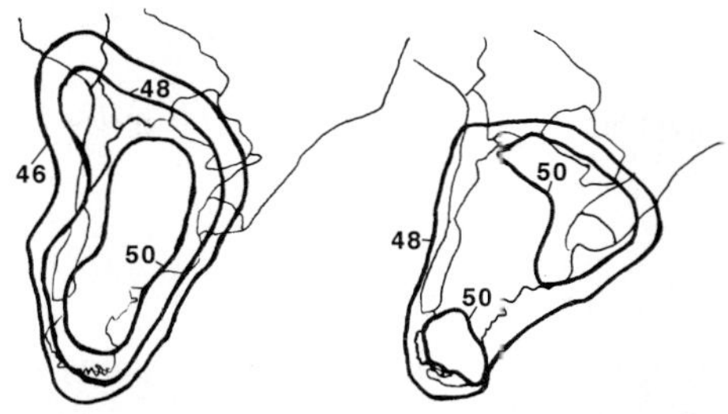

Nahuel EIRP (dbW) contours: horizontal polarisation (right) and vertical (left) ***1995***

Anik B series

Background

Canada upgraded the Anik A to the Anik B, the world's first dual-band (C/Ku-band) satellite, which followed up CTS with experiments in education, telemedicine, conferencing and Inuit broadcasting. Anik B took on the bulk of the C-band East-West traffic from the retired Anik As. It weighed 474 kg on-station and went into orbit on 16 December 1978 aboard a Delta from Cape Canaveral. Canada retired it in December 1986.

Anik C series

Background

Canada purchased Aniks C1-3 from Hughes under a 1978 contract worth C$53.6 million plus C$13.7 million incentive awards. C3 operated in the Canadian orbital slot 117.5° west as the prime satellite until May 1989, with C1 held in orbital storage. The satellite took advantage of pre-1986 Shuttle tariffs. Service switched to C1 because the introduction of Mexico's Morelos required the orbital slot 1° away. C1 remained at 107.3° west, where E2 joined it in 1991. It continued to provide Ku-band services, while E2 employed C-band.

Between 1983 and May 1984, GTE Spacenet leased Anik C2, which in turn leased capacity to United Satellite Communications Corporation to provide DBS service to northern US states. The relatively low-power beam required receiving antennas in the 1.2-1.8 m range, a standard that did not succeed commercially. C2/C3 began inclined orbital operations 20 August/20 July 1991, respectively, to extend their lives. C3 still sees service as a backup to fibre optic providers for the Canadian telephone companies.

C1, with only 4.5 years of station-keeping propellant, began to move from 109.25° west on 29 March 1993 to take up service for Argentina. It arrived at its new 71.8° west slot on 15 April 1993 and began service 1 June 1993. C2 likewise began the move from 109.15° west 5 April 1993 and arrived at its new 75.8° west slot 18 June 1993.

Specifications

Launched: C3 11 November 1982, C2 18 June 1983, C1 13 April 1985, all by Shuttle
Locations: C1: 119° west; C2: 75.8° west (inclination 3.56° end-1995); C3: 114.9° west GEO (inclination 3.68° end-1995)
Design life: 10 years
Contractors: Hughes Space & Communications
Transponders: 16 15 W TWTA 14.003-14.497/11.703 12.197 GHz up/down Ku-band in four overlapping spot beams covering most of populated Canadian landmass, 54 MHz bandwidth, EIRP 46.5 dBW min, linear polarisation
Principal applications: 32 TV channels (1.2 m home antennas) or 21,504 simplex voice links, business communications
Configuration: Hughes HS-376 spin-stabilised bus
Mass: 562.5 kg in orbit (including 99 kg hydrazine)
TT&C: primary from Allan Park (Ontario) with full backup from Edmonton (Alberta) controlled from HQ
Power: 1,135 W from body array, K7 cells; three Ni/Cd batteries.

Anik D/Telstar 301 series

Background

The fourth Anik series Anik E inaugurated the true commercialisation of Canadian services. Canada initially stored D2 in orbit, to benefit from advantageous pre-1986 Shuttle tariffs. It operated at 110.9° west until its services were transferred to nearby E1 in 1991. Then GE Americom bought it and moved it to 82° west in December 1991 as cover for Satcom 4. GE paid C$18 million. Telesat Canada bought it back in April 1993 and put it in an inclined orbit to conserve NSSK fuel. Arabsat (who called it Arabsat 1D) bought it for C$5.5 million, to provide mostly video services from 19 to 20° east beginning in August 1993. Telesat still provided control from Telesat HQ, via a full TT&C facility in Tunisia. D2 left service and exited above GEO in January 1995 as it came to the end of its propellant. D1 left GEO 16 December 1991 after being replaced by E.

Following the Anik E control problems from January 1994, Telesat acquired Telstar 301 from AT&T on 28 March 1994. It arrived at its new 107.1° west slot on 4 April 1994 from AT&T's 96° west to restore services until E2 began commercial services again 1 August 1994. Telesat sold Telestar to Arabsat on 7 June 1994 and began moving it on 3 August to 20° east. It arrived February 1995 and began service as Arabsat 1DR.

Specifications

Launched: D1 26 August 1982 by NASA Delta 3920; D2 9 November 1984 on Shuttle mission 14
Design life: 10 years
Contractors: Spar Aerospace prime, technology licensed from Hughes. Canadian content >50%
Transponders: 24 11 W TWTA 6/4 GHz up/down C-band national and north US beams, 36 MHz bandwidth, EIRP 36 dBW min (33 dBW US coverage), orthogonal polarisation
Principal applications: voice/data services, broadcast distribution (24 TV channels or 21,600 simplex voice circuits)
Configuration: based on Hughes HS-376 spin-stabilised bus
Mass: 633 kg in orbit
TT&C: as Anik C
Power: 1,000 W from body array.

Anik E series

Background

Telesat contracted with Spar Aerospace in November 1986 to build two hybrid Anik Es at costs of C$85.9 million and C$68.1 million, respectively, excluding performance incentive payments. The combination of its

The hybrid Anik E appeared in 1991, the world's most powerful domestic communications satellite ***1996***

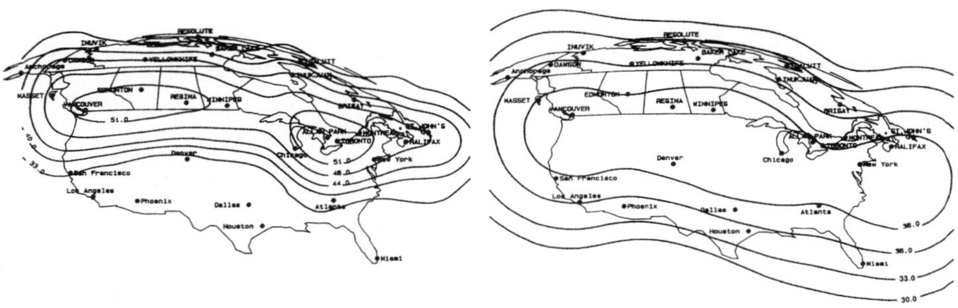

Typical Anik E2 EIRP (dBW) contours. Left: Ku-band national horizontal beam; right: C-band (Telesat) ***1996***

C/Ku-band service requirements within a single system has provided considerable economies; each E replaces one C and one D model. Technical problems during 1989, requiring design changes in the liquid apogee engines and the C-band solid-state power amplifiers, caused delivery and launch delays. E2's service got started 3 months late after its C-band antenna refused to deploy, probably because of thermal snagging. It was finally released 3 July after an unprecedented series of manoeuvres.

E2 provides broadcast services over east/west beams, E1 carries only national beams. E1 suffered a catastrophic failure of a momentum wheel assembly and E2 lost both, probably as a result of electrostatic discharges built up during 10 days of geomagnetic disturbances. E1 returned to service after 7 hours using MWA 2. E2's traffic was switched to E1 and Galaxy 6, and then to Telstar 301. Engineers concentrated on installing software to use thrusters for attitude control, which would reduce E2's life by about 3 months. Full control returned 20 July 1994 and commercial services resumed 1 August.

Specifications

Launched: E2 4 April 1991 by Ariane 44P (entered service 1 September 1991, collocated with C1); E1 26 September 1991 by Ariane 44P
Locations: E2 107.3° west, E1 111.1° west GEO
Design life: 13.5 years (E2 12 years projected)
Contractors: Spar Aerospace (prime, bus from GE Astro-Space); David Florida Lab (final integration/testing)
Transponders: 24 (plus 6 backup, 5-for-4) 11.5 W SSPA 6/4 GHz up/down 36 MHz beams. All-Canada beam, with extended coverage to all of continental USA/Alaska with horizontal polarisation, and to northern states with vertical polarisation. Edge EIRP 35.5 dBW, 37.5 dBW for specified cities. 4-for-2 receiver redundancy

16 (plus 2 backup) 50 W TWTA 14.003 14.497/11.703-12.197 GHz up/down 54 MHz Ku-band beams: National (E1 carries only National), East (vertical polarisation) and West (horizontal). Seven E2 West channels can be switched to the East or National beams; two East-beam channels can also be switched to the National beam. Edge of coverage EIRPs: 44.0 dBW (regional beams), 40.5 dBW (National beam) increasing to 49-44 dBW for specified cities
Configuration: 3-axis GE Satcom 5000 bus. Two large deployable antennas, each generating horizontally/vertically polarised beams. C-band antenna steerable by ground command. Employs GE SCOTS integrated liquid propulsion systems a 454 N TRW bipropellant apogee injection motor sharing hydrazine tanks with mono AOCS thrusters. The two redundant sets of thrusters comprise 12 catalytic versions for attitude control, east/west station-keeping to within ±0.05°, spin-up/nutation control, and four high performance Electrothermal Hydrazine Thrusters for NSSK
Mass: 2,932 kg at launch, 1,781 kg on-station BOL, 1,335 kg (dry)
TT&C: as Anik C
Power: 3,888 W EOL, 4 × 56 Ah Ni/H_2 batteries.

Anik F series

Current status

The next generation Anik F series adopts the Hughes HS702 bus with a payload of 36 C-band and 48 Ku-band transponders. Anik F1 weighs 4,600 kg and is scheduled for launch in first quarter 2000, to be followed into orbit by Anik F2 in 2002. Anik F1 will reside at 107.3° west, with F2 at 118.7° west.

MSAT series

Background

Telesat Mobile Inc began operations in 1988 with the mandate to construct and operate MSAT, Canada's first commercial, mobile-communications system. Telesat Canada (80 per cent) and Japan's ITOCHU Corporation held the only shares. Telesat Mobile Incorporated announced in the spring of 1993 that it had been unable to secure additional financing in the face of cost increases. Telesat Canada was unable to continue funding under the circumstances. Accordingly, on 5 April 1993, the company filed a 'Notice of Intention to Make a Proposal' pursuant to the Bankruptcy and Insolvency Act, in order to provide the time to pursue the possibilities of a plan for reorganisation.

MSAT will provide mobile communications over North America (Spar Aerospace) ***1995***

TMI launched its initial service offerings in spring 1990 through leased capacity on an Inmarsat satellite. Known as KITs (Keep In Touch), these provided full two-way digital messaging, automatic vehicle location and fleet management services for Canadian users throughout North America. The services ceased as a result of the reorganisation.

Specifications

MSAT 1
Launch: 20 April 1996 by Ariane 4 from Kourou, French Guiana
Location: 106.5° west geostationary
Design life: 12 years
Contractors: Hughes Space & Communications Co (management, bus) and Spar Aerospace (payload, integration, testing)
Transponders: 16 (plus 4 backup in 10-for-8 open ring) 35 W SSPA 1.6315-1.6465/1.530-1.545 GHz up/down and 1.6465 1.6605/1.545-1.559 GHz up/down in 2 × 8 matrix feeding 6 elliptical spot beams (4 North America, a Caribbean beam for Puerto Rico, Virgin Islands and Mexico, and Alaska/Hawaii. Aggregate edge of coverage EIRP 56.5-57.3 dBW, circular polarisation. Bandwidth capacity for 2,000 5 kHz channels. Each '35 W SSPA' comprises four 20 W SSPAs in parallel but operated at 35 W for required 16 dB noise/power ratio

1 (plus 2 backup) 110 W TWTA 13.000 13.15 and 13.20-13.250/10.750-10.950 GHz up/down Ku-band 36 dBW feeder/return link in single North American beam
Principal applications: mobile radio, telephone, data, aeronautical, maritime, safety/distress, position location, wide area paging. High quality voice, packet and circuit switched data
Configuration: Hughes' HS-601 3-axis bus, 2.3 m box-shaped body, 3.63 m diameter stowed, 4.44 m stowed ht, twin solar arrays spanning 20.96 m, 18.9 m across deployed antennas. Two 4.9 × 6.7 m 20 kg graphite mesh reflectors deployed from east/west faces provide separate L-band transmission/receive, respectively, illuminated by 23 cup dipoles. Single 76 cm Ku-band shaped reflector for North America coverage.
Mass: 1,710 kg on-station BOL, 2,855 kg at launch (1,330 kg dry)
AOCS: onboard control processor for control of momentum wheel assembly. Unified 12 × 22 + 1 × 490 N bipropellant thrusters (NTO/MMH in four tanks).
Antenna pointing: 0.125° accuracy using Earth sensors
TT&C: two facilities for redundancy and security: Satellite Operations Centre (satellite control) and Network Operations Centre (network control and channel access on demand)
Power system: twin 9 m long 3-panel solar arrays producing 3.6 kW BOL, 3 kW EOL from K4¾ large area silicon cells; 160 Ah Ni/H_2 battery provides 100 per cent eclipse protection
Thermal control: dissipation by north/south radiators, with heat transported by heat pipes.

MSAT 1's two 4.9 × 6.7 m graphite mesh antennas weigh only 20 kg each (Communications Research Centre) ***1996***

CHINA, PEOPLE'S REPUBLIC

Apstar series

Background

The Chinese government, which owns 75 per cent of The Asia Pacific Telecommunications Satellite Company launches and maintains the APstar satellites. APT plans to deliver a three satellite system for the Asia-Pacific region. The consortium signed a contact with Hughes Communications International Incorporated on May 25 1992 to build the APStar 1, and the Macau ground station for US$135 million. Hughes increased the price to US$200 million when APT included the hybrid HS-601 APStar 2 option, primarily for television broadcasting. Apstar also asked Hughes to extend the coverage region to Europe, India and Russia. Apstar changed its plans when Apstar 2 suffered a launch failure on 25 January 1995 and instead substituted the APStar 1A HS-376 in March 1995. Apstar chose the smaller design because Hughes promised it could be ready in 11 months, rather than the HS-601's 18 months. Alpstar 1A was launched by CZ-3 from Xichang launch site, a near identical twin to Apstar 1, on 3 July 1996.

The company originally placed APStar 1 at 131° east for operational service, but Rimsat & Japan (CS-3a) complained of signal interference and consequently APT moved Apstar 1 to 138° east. It presently resides at 134.0° east.

Specifications

APStar 1

Launch: 21 July 1994 by CZ-3 from Xichang, China, People's Republic
Location: 138° east GEO. Initially located at 137° east for testing and was then to move to 131° east for operational service, but complaints from Rimsat & Japan (CS-3a) of signal interference with their nearby satellites resulted in positioning at 138° east, reportedly leased from Tonga
Design life: 12 years
Contractors: Hughes Space & Communications Co (contract with Hughes Communications International, Inc)
Transponders: 24 (plus six backup, 30-for-24) 16 W TWTA 5.850-6.425/3.625-4.200 GHz up/down C-band regional beam, 20 × 36 + 4 × 72 MHz bandwidth, EIRP (dBW) >35 China, 36.0 Hong Kong, 35.9 Philippines, 36.1 Thailand, 36.4 Korea & 35.1 Singapore, orthogonal polarisation
Principal applications: regional telecommunications; VSAT, data & video
Configuration: Hughes' HS-376 platform, cylindrical body 2.16 m diameter total height 6.59 m
Mass: 1,383 kg at launch, 557 kg dry
AOCS: spin-stabilised at 50 rpm by hydrazine thrusters. ±0.05° east/west and north/south station-keeping; antenna pointing accuracy 0.021° pitch & roll & 0.199° yaw
TT&C: through Macau station. Command uplink at 6,422 MHz; telemetry down at 4,198.125 and 4,199.625 MHz.
Power system: solar array mounted on cylindrical body and extension provide 1,070/970 W at solstice/equinox EOL.

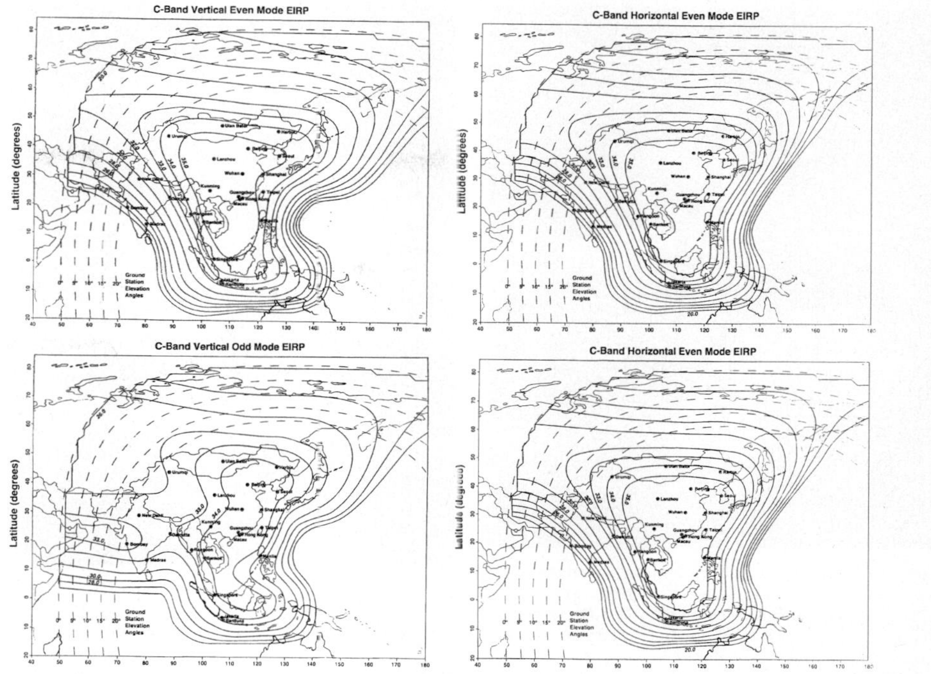

APStar 1A EIRP contours (dBW). Odd-numbered transponders are stretched for Indian coverage (APT) ***1996***

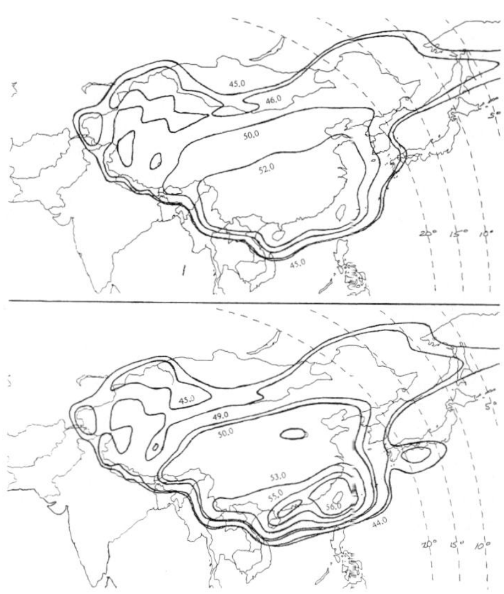
APStar 2R Ku-band EIRP (dBW) coverage; beam 1 top, beam 2 bottom (APT) ***1996***

APStar 1 was joined in 1996 by the near-identical 1A (Hughes) ***1994***

APStar 1A

Launch: 3 July 1996 by CZ-3 from Xichang, China, People's Republic
Location: planned for 76.5° east GEO
Transponders: as APStar 1 but additional feedhorns extend coverage to all India (see EIRP contour diagram) Other specifications as for APStar 1.

APStar 2R

Launch: planned 1997 on CZ-3B from Xichang, China, People's Republic
Location: planned 76.5° east GEO
Design life: 15 years minimum
Contractors: Space Systems/Loral (prime)
Transponders: 28 (plus 12 backup, 36-for-28) 60 W TWTA 5.850-6.425/3.625-4.200 GHz up/down C-band regional beam, EIRP 38 dBW main region (see map), 27 × 36 MHz + 1 × 30 MHz bandwidth, linear orthogonal polarisation

16 (plus 8 backup, 24-for-16) 110 W TWTA 14.00-14.50/12.25-12.75 GHz up/down Ku-band China beams 1&2, EIRP see map, 15 × 54 MHz + 1 × 36 MHz bandwidth, linear orthogonal polarisation. Beam 1 has eight even-number channels; beam 2 eight odd number channels; the four horizontal polarisation beam 2 channels can switch to beam 1
Principal applications: China (Ku) and regional services
Configuration: 3-axis stabilised SS/L FS-1300 2.41 × 2.58 × 2.20 m box-shaped bus with twin solar wings on North/South sides, 27.3 m span
Mass: 3,700 kg at launch, 1,415 kg dry
AOCS: 3-axis bias momentum system; single Marquardt R-4D 490 N apogee thruster and two sets of six 22 N RCS thrusters, common NTO/MMH propellant system

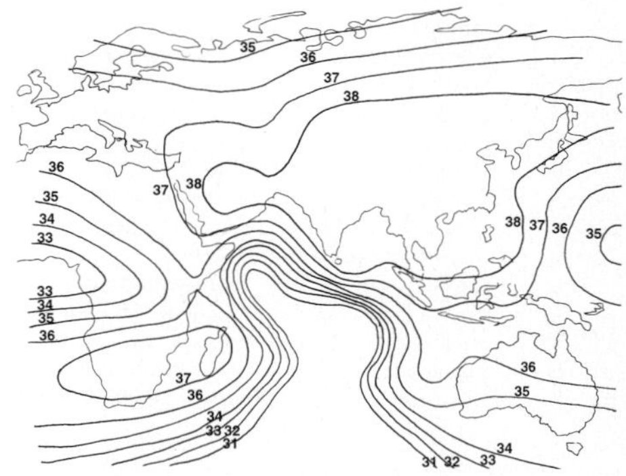
APStar 2R C-band EIRP (dBW) coverage (APT) ***1996***

Antenna pointing: 0.13°
TT&C: as APStar 1/1A
Power system: twin 4-panel GaAs wings providing 8,125 W BOL. Ni/H_2 battery provides 100 per cent eclipse protection.

AsiaSat series

Current status

The Asiasat 3 satellite, a fully functional HS601HP, ended up in an incorrect orbit when the fourth stage of the Proton booster carrying the satellite failed. The final orbit, achieved after launch on 25 December 1997, resembled a GTO rather than a GEO orbit. On 10 April 1998 Hughes began a series of orbital manoeuvres resulting in a translunar flyby in an attempt to reposition the satellite in its correct orbit.

The manoeuvre consisted of a series of satellite thruster firings by 7 May 1998 converting the elliptical orbit of 350 km by 36,000 km into a circumlunar trajectory. This is the first commercial use of the moon and the first lunar mission attempted by a non-government entity.

When owners Asia Satellite Telecommunications Co Ltd of Hong Kong filed an insurance claim of $215 million Hughes Global Services, Inc, obtained title for the salvage operation and renamed the satellite HGS-1. HGS would share any profits with etc insurers.

The first encounter on 13 May carried the satellite within 6,200 km of the lunar far side before looping back to Earth. Engineers adjusted the Earth encounter to effect a second fly by the Moon at a distance of 34,000 km on 6 June. Thrusters were fired at the second Earth fly-by to place HGS-1 in a 46 hour orbit of 36,000 km by 82,000 km. Subsequent thruster firings settled it into a geostationary orbit at 36,000 km.

Background

Lockheed Martin Astro Space, in competition with Hughes and Matra Marconi signed a contract for a hybrid AsiaSat 2 satellite on 26 October 1992. The contract had a potential worth of US$133 million if China exercised all of the options including a second

After the failure of the geosynchronous placement motor on AsiaSat 3, manufacturer Hughes Space and Communications Company chose to salvage the satellite by repositioning it with a lunar gravity assist manoeuvre and onboard station-keeping propellant
(Hughes Space and Communications Company) ***1998***/0022572

satellite purchase. China paid COMSAT to give additional technical support.

The CZ-2E launch vehicle received the nod to put the satellites in space, but China withheld a launch decision until 1995 to allow time to determine the cause of the Telstar 402 (which used the same bus) and CZ- 2E failures to be identified. The first CZ-2E/EPKM launch vehicle combination built by the Chinese put AsiaSat 2 in space on 28 November 1995. It commenced commercial operations in January 1996.

With success assured, the consortium backing AsiaSat took out a US$220 million loan on 16 February 1996 to finance AsiaSat 3, and then signed a contract with Hughes Space & Communications International for the manufacture and on-orbit delivery of an HS-601 HP.

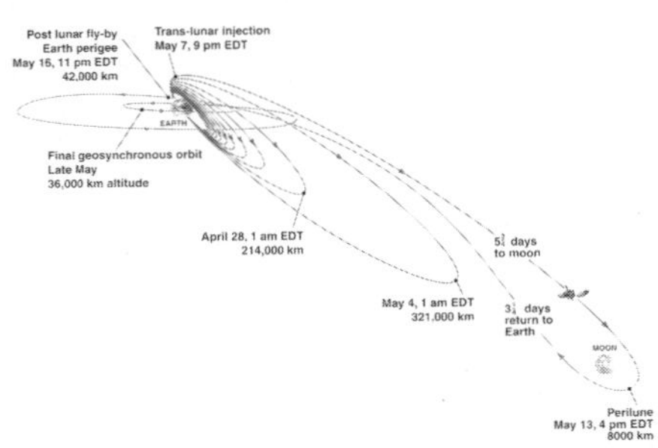

The trajectory for Asiasat 3's Lunar flyby rescue mission required nearly 45 days and used a large fraction of the satellite's station-keeping propellant. All dates shown on the diagram occurred in 1998
(Hughes Space and Communications) ***1998***/0022575

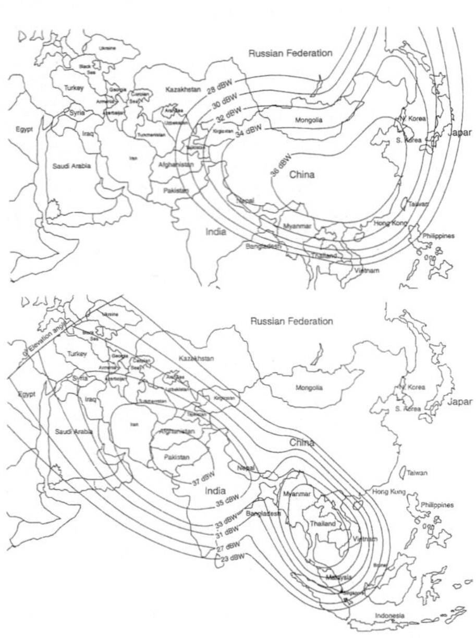

AsiaSat 1 provides two beams. The northern beam covers China, while the southern footprint is optimised for Thailand and Pakistan (AsiaSat) ***1995***

AsiaSat 3 is planned for October 1997 (Hughes) ***1993***

On 22 August 1996, a US$250 million loan facility entered effect with 12 international banks called upon to refinance the existing loans for AsiaSat 1, AsiaSat 2 and AsiaSat 3 as well as to support future capital expenditure.

AsiaSat 3 carries 28 C-band and 16 Ku-band high-powered linearised transponders. The C-band footprint can cover Asia, the Middle East, the CIS and Australasia. There are also two separate footprints providing high-powered Ku-band coverage. In addition, a steerable Ku-band beam provides added market flexibility.

AsiaSat 4 has a launch delivery schedule of 2000 when AsiaSat 1 reaches the end of its planned life. Asia Satellite Telecommunications selected Hughes to build a replacement for the AsiaSat 3 after its launch was unsuccessful on 25 December 1997. The AsiaSat 35 is an exact replica of AsiaSat 3 and was launched by Proton on 22 March 1999 toward its operating slot at 105.5° east. In November 1998, Hughes was contracted to build a backup, designated AsiaSat 35B until its predecessor had been successfully launched subsequent to which the order has been converted to a future AsiaSat requirement.

AsiaSat's ultimate holding company went public in June 1996. Both the Hong Kong and New York Stock Exchange list it with 31 per cent of the company's shares held by the public. The remaining shares reside equally with Cable and Wireless PLC, China International Trust and Investment Corporation (CITIC) and Hutchinson Whampoa.

Specifications

AsiaSat 1

Launch: 7 April 1990 by CZ-3 from Xichang (launched initially as Westar 6 by STS-41B, 3 February 1984; retrieved by STS 51-A in November 1984)
Location: 105.5° east geostationary

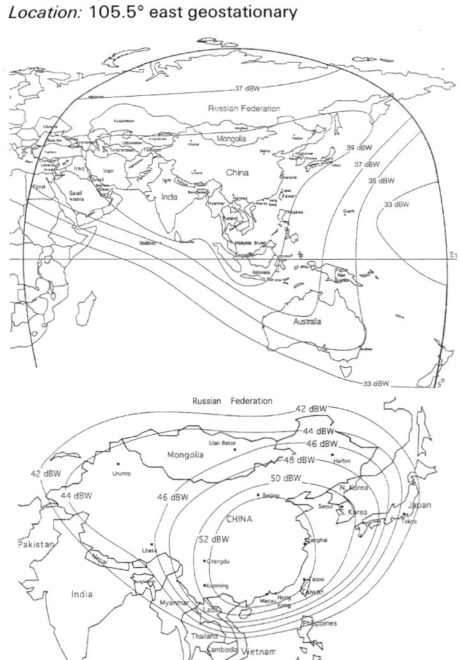

AsiaSat 2 dBW contours: C-band (top) and Ku-band (bottom) (AsiaSat) ***1995***

Design life: 9-10 years
Contractors: Hughes Space & Communications Co
Transponders: 24 (plus six back-up) 8.2 W TWTA 5.925 6.425/3.700-4.200 GHz up/down C-band, 12 each in two beams (see map), 36 MHz bandwidth, China 32-36 dBW, Thailand 35-37 dBW, Pakistan 35-37 dBW, Korea 35 dBW, Japan 30-34 dBW, Singapore 32 dBW, linear polarisation (12 V+12 H)
Principal applications: TV, radio, teleconferencing and VSAT links. TV users include Star TV, China, Mongolia, Myanmar, Pakistan. Each transponder can handle 3,360 telephone channels via a 7 m station. TVRO diameter 2.6 m for 36 dBW
Configuration: Hughes' HS-376 platform, cylindrical body 2.16 m diameter, total height 6.59 m
Mass: 1,280 kg at launch, 654 kg on-orbit BOL
AOCS: spin-stabilised at 50 rpm by hydrazine thrusters
Power system: solar array mounted on cylindrical body and extension provide 900 W at BOL, about 700 W at EOL
TT&C: Stanley Earth Station, Hong Kong.

AsiaSat 2

Launch: 28 November 1995 by CZ-2E from Xichang
Location: 100.5° east GEO (arrived 14 December 1995, entered service mid-January 1996)
Design life: 13 years
Contractors: Lockheed Martin Astro Space
Transponders: 24 (plus 8 backup, 16-for-12) 55 W TWTA 5.845-6.425/3.620-4.200 GHz up/down C-band, 20 × 36 + 4 × 72 MHz bandwidth, 40 dBW EIRP max, linear orthogonal polarisation

9 (plus three backup, 12-for-9) 115 W TWTA 14.000-14.300/12.200-12.500 GHz up/down Ku-band, 54 MHz bandwidth, 53 dBW EIRP max, covering Hong Kong, China, Taiwan, Japan, Korea, linear orthogonal polarisation
Principal applications: TV and radio distribution, telecoms, VSAT
Configuration: series 7000 platform, 3-axis box-shaped bus with twin solar arrays spanning 23 m
Mass: 3,351 kg at launch
AOCS: nominal pointing accuracy ±0.1° pitch/roll, ±0.25° yaw. 1.8 kW hydrazine arcjets provide NSSK, China's EPKM perigee kick motor, Liquid Apogee Engine provides GEO insertion, Reaction Engine Assemblies for attitude control
Power system: twin 4-panel wings totalling 48.6 m^2, providing 4,780 W EOL, supported by Ni/H_2 battery
TT&C: Stanley Earth Station, Hong Kong.

AsiaSat 3

Launch: 25 December 1997 by Proton-K from Baikonur, Kazakhstan. Replacement AsiaSat 35 launched 22 March 1999.
Location: 105.5° east GEO (AsiaSat 35)
Design life: 15 years
Contractors: Hughes Space & Communications
Transponders: 28 (plus 6 back-up, 34-for-28) 55 W TWTA 5.845-6.425/3.620-4.200 GHz up/down C-band, 36 MHz bandwidth, 41 dBW EIRP maximum, coverage as AsiaSat 2, linear orthogonal polarisation

16 (plus 4 backup, 20-for-16) 138 W TWTA 14.000-14.500/12.25/12.75 GHz up/down Ku-band, 54 MHz bandwidth, 55 dBW EIRP max, three beams (North Asia, East Asia, steerable), linear orthogonal polarisation
Principal applications: TV distribution, telecoms, VSAT
Configuration: improved HS-601 HP (High Power) platform, featuring GaAs solar arrays, enlarged payload and lightweight contoured reflectors
Mass: 3,480 kg at launch, 2,560 kg on-station BOL
AOCS: Bipropellant system.
Power system: twin solar wings of four panels providing 8.7 kW BOL (7.8 kW total required); 350 Ah, Ni/H_2 battery provides 100% eclipse protection
TT&C: Stanley Earth Station, Hong Kong.

STTW (Chinasat) series

Current status

Although planned for only 4 years of operations STTW-2, -3 and -4 were still operating on-station in 1998.

Background

MBB/CGWIC signed a DM51.2 million contract on 14 July 1987, effective from 8 March 1988, to develop China's communications and broadcast infrastructure and some satellite subsystems. The Chinese Space Agency CAST acts as prime contractor for both space/ground segments. MBB supported the system definition work and built the satellite antennas and solar array mechanical elements.

The result of this collaboration, the DFH-2A/Chinasat-1 flew successfully three times between 1988 and 1990. Each satellite carries four instruments (electron, proton and solar X-ray detectors and potentiometer) to monitor the GEO radiation environment, the satellite's surface charging and solar activity.

DASA and China Aerospace Corp signed a MoU 16 November 1993 to co-operate in building and launching the next generation satellites. These entities formed the EurasSpace joint venture 8 July 1994, with an initial staff of 12, to help produce China's domestic satellites and to bid for international contracts.

When it was completed, 80 per cent of China's first 3-axis GEO satellite came for indigenous sources. US sanctions did create delays. The satellites incorporate Teldix GmbH's SADA solar array drive assemblies. Officine Galileo provided the IRES IR Earth sensor to determine pitch/roll in GEO. M/A-COM Inc had contracted to supply a transponder but President Bush imposed a ban on export of US components for Chinese spacecraft in the summer of 1991. TIW Systems of the US received a contract in 1993 for the TT&C complex in Beijing.

Specifications

STTW-T1

Launch: 29 January 1984 by first CZ-3 from Xichang, initially into 451 × 309 km parking orbit then after 12 revs boosted to 359 × 6,479 km, 36°. First attempt at placing experimental DFH-1 telecom satellite at 125° east STW-1 but the stage 3 failed to re-ignite for GTO insertion. Satellite kick motor was fired as a demonstration.

STTW-T2

Launch: 8 April 1984 by CZ-3 from Xichang to 125° east GEO (STW-1 position). 420 kg on-station BOL, 915 kg at launch; design life 3 years. Provided two 8 W TWTA 6/4 GHz up/down 20 dBW global beams using horn antenna working to 10 m ground stations. The DFH-1 configuration is similar to STTW 2, but was 3.1 m high and employed a de-spun horn antenna. Operational life apparently ended 1987 when allowed to drift off-station.

STTW-1

Launched: 1 February 1986 by CZ-3 from Xichang to 103° east STW-2 position. DFH-2 design; specifications similar to STTW-2. Broadcast to 8 July 1989; corrections made only occasionally during 1990.

STTW-2/Chinasat-1A

Launched: 7 March 1988 by CZ-3 from Xichang
Location: 87.5° east geostationary (initially 88° east)
Design life: 4 years
Contractors: CAST Chinese Academy of Space Technology
Transponders: four 10 W TWTA 6.050-6.425/3.825-4.200 GHz up/down C-band regional beams using parabolic reflector providing 36 dBW centre, 32 dBW beam edge, 36 MHz bandwidth. Payload can handle 5 TV + 3,000 phone calls simultaneously
Configuration: 2.1 m diameter, 1.6 m height spin-stabilised cylindrical bus, including single de-spun parabolic antenna. Total height 3.68 m. 77.6 kg station-keeping propellant. Antenna pointing accuracy 0.6° North/South, 0.42° East/West
Mass: 441 kg on-station BOL, 1,024 kg at launch
TT&C: provided through XSCC Xian Satellite Control Centre
Power system: 351 W provided by 20,000-cell Si (11% efficiency) array on cylindrical exterior, supported by 15 Ah Ni/Cd batteries
Apogee kick motor: indigenous solid propellant.

STTW-3/Chinasat-1B

Launched: 22 December 1988 by CZ-3 from Xichang
Location: 110.5° east geostationary (initially 111° east)
Other specifications as for STTW-2.

STTW-4/Chinasat-1C

Launched: 4 February 1990 by CZ-3 from Xichang
Location: 98° east geostationary
Other specifications are the same as for STTW-2.

STTW-5

Launched: 28 December 1991 by CZ-3 from Xichang but short stage 3 GTO burn left it in 219 × 2,451 km, 31.1°. Its kick motor was fired 29 December to attain 205 × 35,087 km, 31.6°. 4th/last flight of DFH-2A/Chinasat-I design. Possibly intended as STTW-2 replacement. Satellite was a ground spare upgraded to flight status.

Zhongxing series

Specifications

Zhongxing 5 /Spacenet 1

Launched: 23 May 1984, was owned/operated by GTE at 120° west until it was sold in December 1992 to China and moved to 115.5° east by June 1993. CNPTAC China National Postal & Telecommunications Appliances Corp bought it for use by Chinasat for about US$30 million. See p326 1995-96 for specifications. It exhausted its propellant in second quarter 1996. When DFH-3 1 failed to replace it, Chinasat 7 was ordered from Hughes.

Zhongxing 6/DFH-3/Chinasat-III

Launched: 29 November 1994, abandoned in geosynchronous drift orbit.

Zhongxing 7/ChinaStar 7

Launched: 18 August 1996, left stranded in a 27.2°, 200-17,230 km orbit when the second burn of the C2-3 third stage ended prematurely
Mass: 1,200 kg with propellant, Hughes HS376 satellite bus.

Zhongxing 8

Launched: 11 May 1997 using C2-3A from Xichang
Location: 115.3° east
Design life: >8 years
Transponders: six 16 W TWTA 6.2-6.4/4.0-4.2 GHz up/down C-band regional beams. EIRP >37 dBW (allowing 3 m TVRO antennas)

18 8 W SSPA 6.2-6.4/4.0-4.2 GHz up/down C-band regional beams. 35 dBW for telephony (8,000 duplex calls)
Configuration: 3-axis 2.2 × 1.72 × 2.0 m bus, total height 5.71 m with single 2 m diameter reflector deployed, 18.1 m solar array span. Antenna pointing 0.15° pitch/roll, 0.50° yaw. North South/East West station-keeping 0.10°. Liquid bipropellant motor for GEO insertion
Mass: 1,145 kg on-station BOL, 2,230 kg at launch
TT&C: provided through Beijing complex
Power system: two 3-panel solar wings provide 2,000 W BOL/1,700 W EOL.

FRANCE

STENTOR satellite

Background

France approved the Stentor (Satellite de Télécommunications pour Expérimenter les Nouvelles Technologies en Orbite) on 4 October 1994 (simultaneously with Spot 5) to develop and demonstrate a telecommunications platform, aimed at helping French industry compete with the US and Japan. The French government expected to pay FFr1.8 billion for the satellite, plus FFr800 million for launch operations and ground segment. France cut this by FFr520 million in 1996. CNES heads the project committee, which includes France Telecom and DGA, in close collaboration with three co-contractors.

Specifications

Design life: 9 years
Contractors: Three co-primes: Aerospatiale (platform prime, solar arrays, AOCS), Alcatel Espace (payload prime), MMS (power supply, onboard data management, attitude control, plasma experiments)
Transponders: 1 (plus 1 back-up) 110 W linearised TWTA Ku-band Western Europe beam, 55 dBW EIRP. Phased array antenna with 48 transmission elements (100 W total RF output), reconfigurable coverage and beam hopping

1 L-band providing four European spots, 42 dBW. Ku-band receive

1 20 W TWTA 20.7 GHz Ka-band beacon, 45 dBW France, 44 dBW French Guiana
Principal users: new Ku-band services, L-band air traffic management, Ka-band military link
Configuration: Spacebus platform with Eurostar electronics. NSSK by SEP's modified Russian SPT-100 xenon plasma thrusters (Mk 2 design was to have been employed but will not be ready in time); conventional propellants will provide 3 years life as backup. Deployable radiator
Mass: 2,000 kg at launch
TT&C: via the Toulouse Space Centre working through Ku-band antennas at Aussaguel-Issus near Toulouse
Power system: 2,400 W after 2 years from GaAs/Ge cells and lithium carbon batteries.

TELECOM series

Background

France authorised France Telecom to build the Telecom satellite system in 1979 to link domestic and overseas territories with television, telephone, digital data and military links. To meet these needs, the quasi-public company immediately ordered the Telecom 1 satellite and then later ordered three Telecom 2s in December 1987 from Matra/Alcatel. The Telecom 2 has three times the power of the Telecom 1. Each carries a Syracuse (Système de Radio Communications Utilisant on Satellite) package for secure Ministry of Defence links. The X-band transponders work through three Earth stations: in Brest (8 m dish), Paris (8 and 18 m dishes and the tracking, telemetry and command) and France-Sud (2, 18 m dishes, including back-up tracking, telemetry and command).

Syracuse 2 expands access to 40-90 cm antennas, permitting 75/200 bit/s telegraph; 2-4/16 kbit/s telephone; 75/2,400/ 16,000 bit/s data routeing and a secure backup in case of damage to the primary communication centre.

France Telecom may rely in future on leased Eutelsat coverage under privatisation. It is a member of the Satellite Aircom consortium providing aeronautical communications services through the Inmarsat system.

Specifications

Telecom 1A

Launched: 4 August 1984, was raised above GEO in late 1992 after being replaced at 8° west by 2A. 1B, launched 8 May 1985, lost both attitude control systems 15 January 1988. 1C, launched 11 March 1988, retired from 3° east in February 1996 after being replaced by 2C.

Telecom 2A

Launched: 16 December 1991 by Ariane 44L from Kourou, French Guiana. In-orbit acceptance tests completed 15 March 1992 at 3° east; moved April to 8° west
Location: 8° west geostationary
Design life: 10.25 years minimum
Contractors: Telecom 2A: Matra (platform, integration), Alcatel Espace (payload), Sodern (Earth sensors), Galileo (Sun sensors), SEP/MMS (propellant tanks), Sagem (rate gyros), MBB (thrusters), Fokker (solar arrays), Eagle Picher (batteries), Teldix (reaction wheels), Thomson (TWTs)
Transponders: 10 (plus 4 backup; 6-for-4 + 8-for-6) 8.5 W min SSPA 5.925-6.425/3.700-4.200 GHz up/down C-band Atlantic regional beams, 6 × 50 MHz + 4 × 92 MHz bandwidths, circular polarisation. Semi-global coverage at 32.5 dBW EIRP on channels C1-C4: mainland France, Reunion, Antilles/Guiana, St Pierre/Miquelon. Spot coverage: Antilles/Guiana (39.0 dBW; C5A, C6A, C7A), mainland France (42.4 dBW; C5B, C6B, C7B), St Pierre/Miquelon (39.0 dBW; C5B, C6A, C7A); channels C6A/C7A can be reconfigured onboard between Antilles/Guiana and St Pierre/Miquelon, C5B is divided

11 (plus 4 back-up in 15-for-11) 55 W min TWTA 14.00-14.250/12.50-12.75 GHz up/down Ku-band European regional beams, 36 MHz bandwidth, 52.4 dBW min EIRP central zone, 49.4 dBW min EIRP

Telecom 2C in Matra Marconi Space's Toulouse test facilities ***1996***

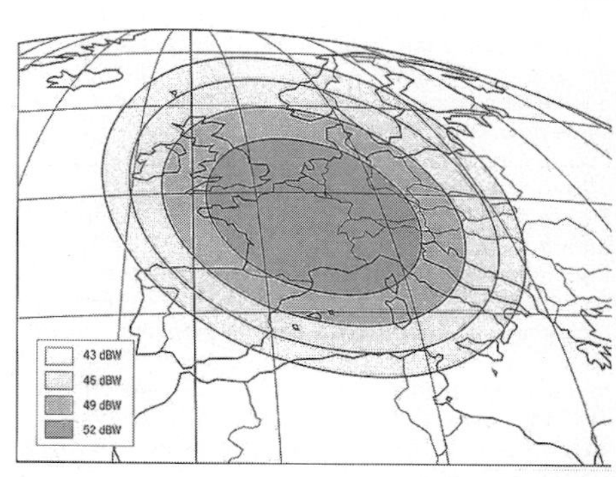

Telecom 2 Ku-band beam coverage ***1995***

surrounding zone, vertical polarisation. This payload also transmits a 12 GHz beacon signal for ground station pointing

3 (plus 3 backup; two groups, 3-for-2 + 3-for-1) 20 W min TWTA 7.900-8.395/7.250-7.745 GHz up/down X-band beams, circular polarisation, channel X1 60 MHz bandwidth central European beam EIRP 44.0 dBW, X4 40 MHz steerable spot EIRP 31.3 dBW, X5 80 MHz global 28.5 dBW

2 (plus 1 backup, 3-for-2) 40 W min TWTA 7.900-8.395/7.250-7.745 GHz up/down X-band beams, circular polarisation, channels X2/X3 40 MHz bandwidth global EIRP 31.3 dBW

Principal users: domestic/overseas telephony, data, radio, TV, military

Configuration: Matra/BAe Eurostar 2000 bus. Four military antennas: LHCP global horn transmitter, RHCP global horn receiver, 2.2 m diameter offset paraboloid 1.65 m fl for European coverage, and a steerable offset paraboloid spot reflector

Mass: 2,275 kg at launch, 1,380 kg on-station BOL, 1,124 kg dry mass (400 kg payload)

Pointing: antenna pointing accuracy typically 0.13°

AOCS: 3-axis, 12 × 10 N DASA hydrazine thrusters + Marquardt R-4D 490 N bipropellant apogee motor in unified bipropellant propulsion system

TT&C: via the Toulouse Space Centre working through antennas at Bercenay-en-Othe near Troyes and Aussaguel-Issus near Toulouse

Power system: twin 4-panel solar arrays, 2.050 m wide, spanning 22.02 m provide 3,770 W equinox EOL with full eclipse battery protection (78 Ah Ni/H_2). 2,600 W payload power requirement.

Telecom 2B

Launched: 15 April 1992 by Ariane 44L from Kourou, French Guiana

Location: 5° west from mid-1992 after commissioning at 3° east

Other specifications as for Telecom 2A.

Telecom 2C

Launched: 6 December 1995 by Ariane 4 from Kourou, French Guiana. Replaced 1C February 1996 after testing at 1° east

Location: 3° east (replacing 1C)

Mass: 2,283 kg at launch, 1,360 kg on-station BOL, 1,120 kg dry mass (400 kg payload)

Other specifications as for Telecom 2A.

Telecom 2D

Launched: 8 August 1996 by Ariane 4 from Kourou, French Guiana

Location: 8° west (collocated with 2A)

The 11 Ku-band transponders use 14.25-14.50/11.45-11.70 GHz up/down. Other specifications as for Telecom 2A.

TELEDIFFUSION DE FRANCE series

Background

TDF, created in 1975 as a limited company with the French government as majority shareholder (and a France Telecom subsidiary since 1988), operates the TDF television repeater satellites.

The Eurosatellite consortium was awarded the space segment contract in 1984, and the company built the original satellites basing TDFs design on Aerospatiale's Spacebus 300 platform. Aerospatiale acted as prime for TDF 1 and 2 and backup for TV-Sat 1 and 2. The French made the decision to build TDF 2 in December 1984. By 1986 the programme verged on cancellation because of developing competition from medium-power systems such as Astra, but by then the I.5 billion francs of the 3.5 billion francs allocated to TDF had been spent. The French government ruled that it should proceed despite the financial difficulties in order to promote D2MAC as the standard for high definition television in the 1990s.

The TDF's under construction underwent minor modifications after TV-Sat 1 failed because one solar array remained stuck in November 1987 and thruster overheating occurred during orbital manoeuvres. At the same time, France decided there would be no TDF 3 and that the future system would lease Eutelsat space. By 1992, only about 35,000 locations had TDF dishes. France cut its annual transponder leasing cost from 80 million francs to 35 million francs. Both TDFs suffered from malfunctions that limited their usefulness. TDF 1's channel 1 transponder failed in August 1989 because of arcing promoted by leaking thruster propellant and TDF shut down channel 17 in September 1990 after 5 months of current variations. TDF 2's two Thomson TWTAs (channels 1/13) failed simultaneously in October 1990, when safety mechanisms irreversibly shut them down.

TDF 1 during final flight readiness review. The solar arrays have yet to be attached (Aerospatiale) ***1996***

Specifications

TDF 1

Launched: 28 October 1988 by Ariane V26 from Kourou, French Guiana

Location: During August-September 1996 TDF 1 was relocated from 19° west to 21-22° west, and then on 4 October the satellite was boosted out of the geosynchronous orbit band, to retirement

Design life: 8 years (sufficient AOCS propellant after orbital positioning for 9 years operations)

Contractors: Eurosatellite (Aerospatiale/MBB prime, Alcatel Espace payload), AEG (three 260 W TWTAs), Thomson Tube Electroniques (three 230 W TWTAs)

Transponders: 5 (plus 1 backup only for channel 9) 230 W min TWTA 17.3-17.7/11.714-12.045 GHz up/down Ku-band elliptical beams providing 64 dBW min EIRP coverage of France, 27 MHz bandwidth, RHCP

Principal applications: direct TV/radio broadcast channels 1/5/9/13/17 and HDTV demonstrations

Configuration: Spacebus 300 platform, body 1.65 × 2.4 × 7.1 m box-shaped, solar array span 19.3 m, 2 m diameter circular receive antenna, 2.4 × 0.9 m elliptical transmission antenna (9-horn feed), both carbon fibre

Mass: 2,136 kg at launch, 1,318 kg on-orbit BOL

AOCS: 3-axis control by 14 × 10 N MBB thrusters of unified bipropellant system incorporating 400 N apogee kick motor

Antenna pointing: ±0.01° driven by RF sensor detecting beacon signal from control centre

TT&C: orbital positioning, station-keeping and performance monitoring by CNES from Toulouse; programme feeds allocated to TDF's Bercenay-sur-Othe station in the Champagne region

Power system: twin 4-panel solar arrays (43,000 cells) providing 4.3/3.3 kW BOL/EOL (8 years).

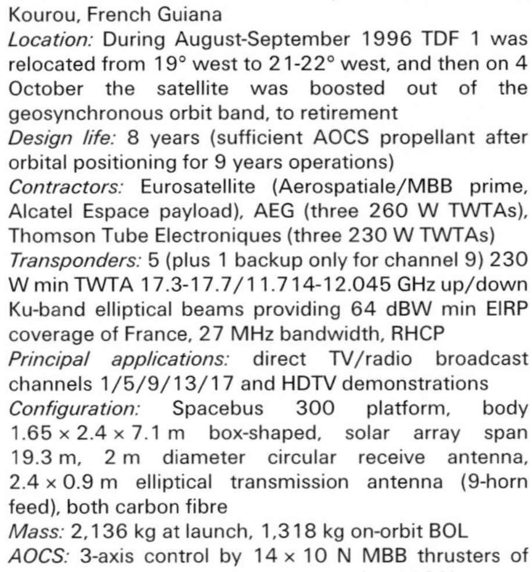

TDF works to 45 cm home receivers in the inner France-centred ellipse

TDF 2

Launched: 24 July 1990 by Ariane V37 from Kourou, French Guiana

Location: 19° west geostationary

Mass: 2,096 kg at launch, 1,255 kg on-station BOL, 1,040 kg dry mass

Other specifications as for TDF 1.

GERMANY

DFS-Kopernikus series

Background

Beginning as an arm of the West German government, Deutsche Telekom became a stock corporation on 1 January 1995 and began public quotations on the stock market during that year. DT commissioned three satellites in December 1983 at DM81.5 million each, plus launch costs and ground segment equipment. DT reportedly paid 250 million marks for its first DFS Ariane launches. In contrast, the 1990 DFS 3 launch contract awarded to McDonnell Douglas cost 75 million marks.

DFS-Kopernikus, which stands for Deutscher Fernmeldesatellit, employs Ku/Ka-bands for television and cable distribution, telephony, voice, facsimile, data and VSAT applications. A ground network of 32 Ku-band and two Ka-band stations handles 7,500 simultaneous telephone conversations and seven television/stereo radio programmes. Licensed by DT, the DFS operate through Raistang and service the former East Germany as well. At the time of reunification there were four million telephones in East Germany, compared with 40 million in West Germany.

In early 1993, DFS 3 replaced the original DFS 1 which entered service over 23.5° east on 5 June1989.

Specifications

DFS 1

Launched: 5 June 1989 by Ariane and positioned at 23.5° east

Mass: the 1,416 kg prototype model

Current Position: moved to 33.5° east as system backup.

DFS 2

Launched: 24 July 1990 aboard Ariane V37 from Kourou, French Guiana

Location: 28.5° east geostationary

Design life: 10 years (Ariane accuracy left sufficient propellant for 11 years); 67% reliability after 7 years

Contractors: GESAT (MBB, ANT, Dornier, Siemens, SEL), ANT (TWTs), AEG (TWTs)

Transponders: Seven (plus three back-ups) 20 W TWTA 14.25-14.5/12.5-12.75 GHz up/down Ku-band providing 49 dBW min EIRP coverage (54.2 dBW max) of Germany, 44 MHz bandwidth, orthogonal polarisation. Each 1 TV or 450 telephone channels

Three (plus two back-ups) 20 W TWTA 14.0-14.25/11.45-11.7 GHz up/down Ku-band providing 49 dBW min EIRP coverage (53.7 dBW max) of Germany, 90 MHz bandwidth, orthogonal polarisation. Each two TV or 900 telephone channels

One (plus one back-up) 20 W TWTA 29.5-29.6/19.7-19.8 GHz up/down experimental Ka-band providing 48 dBW min EIRP (51.9 dBW max), 90 MHz bandwidth, linear polarisation

Principal applications: national telecom systems, TV/

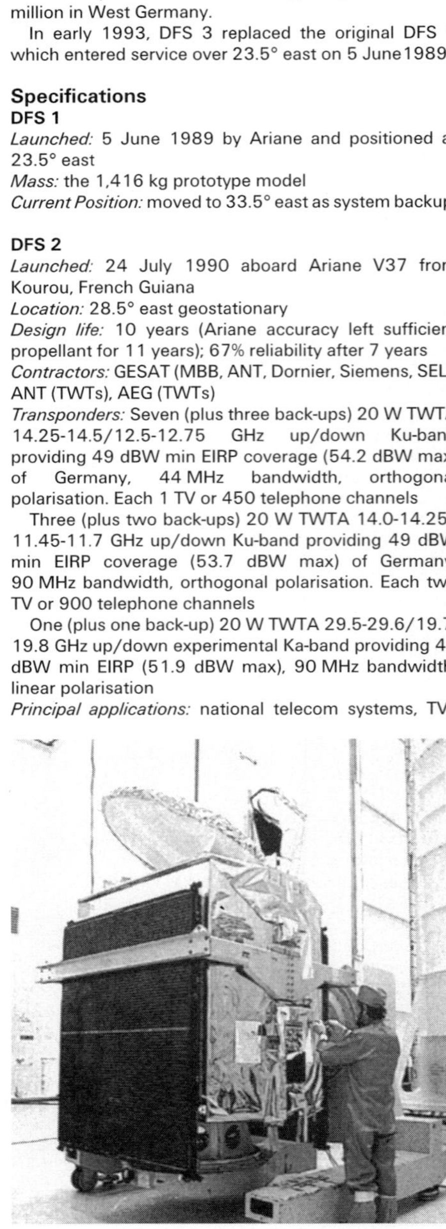

DFS-Kopernikus provides public and business telecom services for Germany; a Ka-band transponder was included for propagation experiments and, ultimately, assignment to commercial operations (DASA)

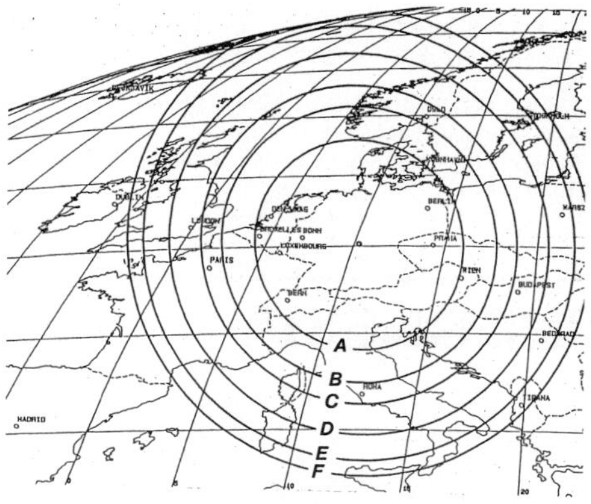

DFS 3 footprint. Antenna diameters (11.5/12.6 GHz): A 1/0.6 m, B 1.3/0.8 m, C 1.5/0.85 m, D 2.2/1.4 m, E 3.35/2.2 m & F 4.35/3.0 m

cable distribution, VSAT and business applications, TV-Sat backup. The narrow bandwidths can handle digital data at 60 Mbit/s, the wideband and Ka-band 140 Mbit/s. DFS 3 dedicated to TV/radio broadcasts; DFS 2 to other services
Configuration: ECS/Spacebus 100 platform, span 15.40 m, height 4.15 m
Mass: 1,418 kg at launch, 850 kg on-station BOL, 645 kg dry
AOCS: 3-axis by 14 × 10 N MBB thrusters plus associated 400 N engine for GEO insertion; GEO position maintained to within 0.07°; antenna pointing accuracy 0.16°
TT&C: initial orbital operations/checkout by German Space Operations Centre at Oberpfaffenhofen (15 m antenna), operational control via DT's Usingen Earth Station (18/9.5/4.5 m)
Power system: twin solar wings provide 1,700 W BOL reducing to 1,580 W EOL; two 35 Ah Ni/Cd batteries provide full eclipse protection.

DFS 3
Launch: 12 October 1992 aboard Delta 7925 from Cape Canaveral
Location: 23.5° east geostationary (replacing DFS 1 early 1993)
Mass: 1,411 kg at launch
Other specifications as for DFS 2.

TV-Sat series

Background

Originally, Deutsche Telekom planned to use TV-Sat to provide Europe's first direct broadcast services, offering four television channels through 35-45 cm dishes, collocated with France's equivalent TDF satellites.

An Ariane launched TV-Sat 1 on 20 November 1987, but one solar panel failed to deploy 25 minutes after launch and could not be shaken free with thrusters firings. DT abandoned the satellite in February 1988; and the insurers paid a US$51 million insurance claim. Initially, DT allocated three channels to private broadcasters, including SAT 1 and RTL, with the fourth used by the ARD association of public broadcasters from 6 pm-1 am for TV and the remaining time for 16 digital sound broadcasting channels.

DT modified TV-Sat 2/TDF to change the thrusters because of overheating discovered on TV-Sat 1. TV-Sat 2 was moved from its 19.2° west slot in February 1995 for leasing Norway's Telenor at 0.6° west in March 1995 for the rest of its life.

INDIA

Insat 1 series

Background

Indian plans for the Insat 1 system originally envisaged two multipurpose satellites, each providing two high-power television broadcast and twelve national coverage transponders, in addition to meteorological services. These two original satellites both ran into difficulties. Orbit inaccuracies forced Insat controllers to use propellant faster than anticipated reducing the satellite's useful life to just 4 months. Then it took Ford and ISRO engineers at India's Hassan Satellite Control Centre several months to deploy the second satellite's solar array. By the end of 1991, 11 transponder channels on Insat 1D, supplemented by two leased on an Indian Ocean Intelsat satellite operated in the Insat system. The Insat system has 140 point-to-point routes with 133 Earth stations of various types.

Insat is controlled through Hassan's Master Control Facility, with the two 14 m antennae joined by two 11 m dishes for Insat 2 (DOS)

India purchased a replacement satellite for the lost Insat 1A but a power system failure from a solar array isolation diode short soon removed half of its capacity. However, the meteorological services, powered through either bus, remained unaffected. Then on 22 November 1989 Insat 1C lost Earth lock and India had to abandon it.

India procured an Insat 1B with expanded battery and propellant capacities and christened it 1D. During payload encapsulation the ground crew damaged the satellite and the repairs delayed the launch. It was repaired at a cost of US$10 million. It also sustained US$150,000 worth of damage during the October 1989 California Earthquake.

The failed 1A/1C reportedly produced insurance payouts of about US$70 million each. The long-lived 1B, launched 30 August 1983 by Shuttle, retired in 1991. It served as a spare from 17 July 1991 and from 1992 was used only for experiments. 2B replaced it at 74° east in 1992 and it retired from 93.5° east when 2B arrived August 1993. NSSK had not been provided since August 1989.

Specifications

Insat 1D
Launched: 12 June 1990 by Delta 6925 from complex 17, Cape Canaveral. Assumed primary role 17 July 1990 from 1B
Location: 83° east geostationary
Mass: 1,292 kg at launch
Design life: 7 years
Contractors: Ford Aerospace (prime)
Transponders: 12 (plus one back-up, for numbers 11/12) 4.5 W TWTA 5.935-6.425/3.710-4.200 GHz up/down C-band all-India beam, 36 MHz bandwidth, 32 dBW min EOL EIRP over primary coverage area, linear polarisation

Two (plus one back-up) 50 W TWTA 5.855-5.935/2.555-2.635 GHz up/down S-band all-India beam, 42 dBW EOL EIRP over primary coverage area linear polarisation
Principal applications: domestic communications on C-band, one S-band channel dedicated to direct TV broadcast and other to five low-level carriers for services such as radio programme distribution, disaster warning, and so on. The Insats also relay 402.75 MHz transmissions from 100 hydrological, meteorological and oceanographic data collection platforms
Spacecraft: 2.18 × 1.42 × 1.55 m box-shaped 3-axis stabilised bus with asymmetrical 5-panel 11.5 m² solar wing providing 1,185 W BOL (930 W EOL) to two 12 Ah Ni/Cd batteries and creating total span of 19.4 m. The asymmetrical solar array configuration provides an unobstructed view into space for the radiative cooler of the VHRR; a solar sail is used for balance. A magnetorquer with current coil around the body provides fine control; a unified bipropellant system is used for orbit transfer, station-keeping and attitude maintenance
Mass: 1,152 kg at launch, 650 kg in-orbit BOL (550 kg dry).

Insat 2 series

Background

The Department of Space forged ahead with the Insat 2 indigenous satellite following approval of funding in April 1985. Initially, the DOS sized the Insat 2 to fit aboard the Space Shuttle, but the continued launch delays associated with the Challenger accident forced a redesign of the satellite so it could fit on the Ariane, and the first satellite appeared on station in 1992.

India is now close to accomplishing its original goal of complete space communication autonomy with the introduction of the Insat 2 communications payload. India had the four first-generation Insat 1 satellites built in the United States, but the advanced Insat 2 family is now indigenously produced. India's Department of Space (DOS), Department of Telecommunications, Meteorological Department, All India Radio and Doordarshan all participate in the Insat system. Overall co-ordination resides in the inter-ministerial Insat Co-ordination Committee (ICC). The DOS establishes and operates the space segment, with the entire system generating revenues of about 4 billion rupees annually.

Insat 2C replaces the meteorological payload with Ku-band and mobile services (DOS) ***1996***

The construction of the Insat 2 Test Spacecraft began with approval in April 1985. India expected to launch the satellites by Ariane in 1992/3, followed by three operational satellites beginning 1994/5 by its own vehicles. The design was sized initially for Shuttle/PAM-D2 deployment. Because of increasing demand, India added to the number of C-band transponders bringing the total to 18 (six at extended C-band) and a 406 MHz transponder adds a distress beacon detection capability. 2C/D had three Ku-band TWTAs and increased C-band power to improve TV coverage and business and remote area services.

A ground station under the direction of the Insat Master Control Facility (MCF) at Hassan Kamataka handles all station-keeping duties through two Satellite Control Earth Stations. One has a 14 m fully steerable antenna and the other a 7.5 m limited steering dish. The facility includes one additional 14 m fully steerable antenna and an Insat 1 Satellite Control Centre (SCC) with associated tracking, telemetry and command equipment, on-orbit checkout equipment, computer facilities and auxiliary power services. The Indian government added an Insat 2 satellites control centre at the Master Control Facility MCF at Hassan, Karnataka for the next generation satellite.

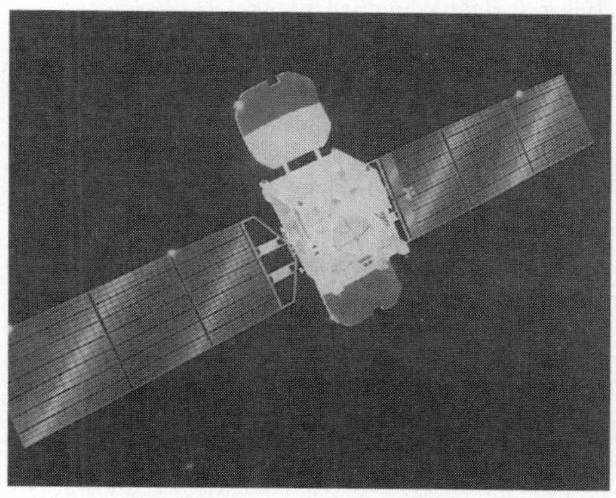

Insat 2: the first Indian-built new generation of multipurpose GEO satellites appeared in 1992. India will be providing transponders to Intelsat from 1998. Shown is the Insat 2C/2D configuration, which does not carry the Earth imager and thus has two solar wings (ISRO) ***1995***

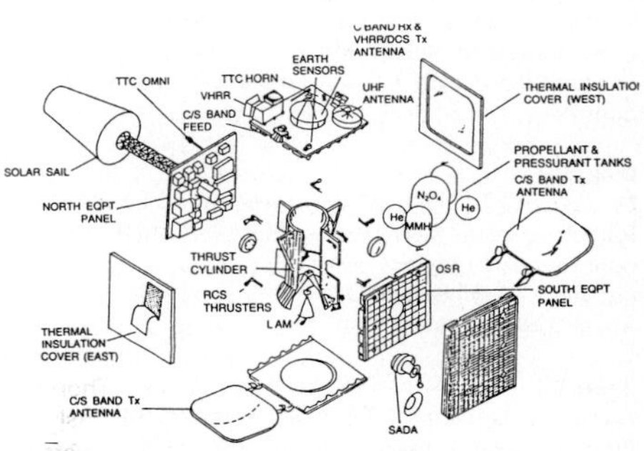

Insat 2A/2B/2E configuration exploded view (DOS)

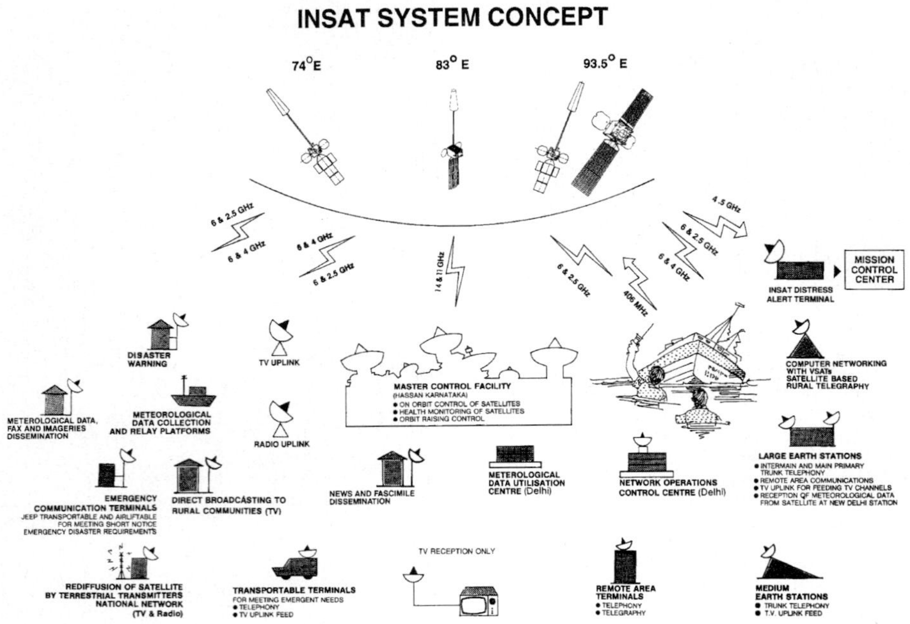

Insat 2 system concept (DOS) ***1995***

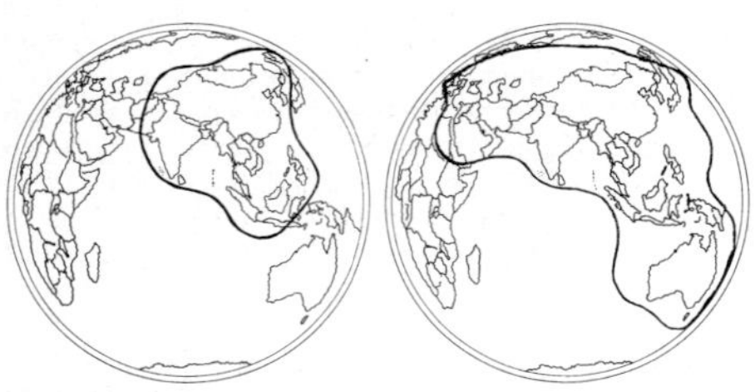

Insat 2E coverage: zonal beam (left) and wide beam (DOS) ***1995***

Specifications

Insat 2A

Launch: 9 July 1992 on Ariane 4 V51 from Kourou, French Guiana (entered service 6 August 1992)
Location: 73.9° east GEO
Design life: 7 years
Contractors: ISRO and Indian companies; HAL Bangalore (structural elements), HAL Hyderabad (C-band SSPAs)
Transponders: 12 (plus one back-up, for numbers 11/12) 4.5 W SSPA 5.930-6.410/3.705-4.185 GHz and six (three back-up; 3-for-2) 6.735-6.975/4.510-4.750 GHz up/down C-band and extended C-band, respectively, all-India beam, 36 MHz bandwidth, 32 dBW EOL EIRP for 16 transponders, 34 dBW for two

Two (plus one back-up, 3-for-2) 50 W TWTA 5.850-5.930/2.550-2.630 GHz up/down S-band all-India BSS beam, 36 MHz bandwidth, 42 dBW EOL EIRP

Single 402.75 MHz Data Collection System transponder

Single 406 MHz Cospas/Sarsat search/rescue package

Principal applications: as for Insat 1
Spacecraft: 193 × 164 × 170 cm box-shaped 3-axis bus around 930 mm diameter central aluminium cylinder with asymmetrical 5-panel solar wing providing 1,400 W at BOL (1,180 W EOL) through 2 × 18 Ah Ni/Cd batteries and creating 23 m span. Reflectors of 1.77 m diameter for S/C-band services located on East/West faces. A 90 cm body-mounted antenna on Earth face performs the C-band receive and DCS/VHRR/S&R data transmit functions
Mass: 1,906 kg at launch, 1,162 kg BOL (911 kg dry); communications payload 179 kg, meteorological 56 kg
AOCS: unified liquid bipropellant 440 N LAM Liquid Apogee Motor and two redundant networks of 8 × 22 N RCS thrusters. Propellant capacity is 1,015 kg. N_2O_4 carried in 520 l central cylinder inside thrust structure, MMH in two 260 l cylinders, pressurised at 16.1 atmospheres by helium stored at 232 atmospheres in two bottles. Latch valves isolate LAM after apogee operations.

Insat 2B

Launched: 22 July 1993 on Ariane 4 V58 from Kourou, French Guiana. Arrived on-station 6 August, declared operational 10 August
Location: 93.5° east GEO
Mass: 1,931 kg at launch.

Insat 2C

Launch: 6 December 1995 on Ariane 4 from Kourou, French Guiana. LAM number 1 firing 01.28-02.32 GMT 8 December raised GTO to 14,420 × 35,790 km, 1.8°; LAM 2 08.32-09.02 GMT 9 December 31,894 × 35,786 km, 0.5°; LAM 3 186 s starting 05.20 GMT 11 December, established GEO. The solar arrays and antennas deployed 12 December
Location: 93.5° east GEO (collocated with 2B)
Transponders: 12 (2 × 50 W TWTA + 7 × 10 W SSPA + 3 × 4 W SSPA; 3-for-2 channel 11/12) 5.930-6.410/3.705-4.185 GHz up/down C-band all-India beam, 36 MHz bandwidth, 36+36+32 dBW edge EIRP

6 (2 × 10 W + 4 × 4 W SSPAs, 3-for-2), 6.735-6.975/4.510-4.750 GHz up/down extended C-band all-India beam, 36 MHz bandwidth, 35+32 dBW edge EIRP

3 (plus one back-up, 4-for-3) 20 W TWTA 14.252-14.493/11.452-11.693 GHz up/down Ku-band all-India beam, 1 × 77 MHz + 2 × 72 MHz bandwidth, 41 dBW edge EIRP

Two (plus one back-up, 3-for-2) 50 W TWTA 5.850-5.930/2.550-2.630 GHz up/down S-band all-India BSS beam, 36 MHz bandwidth, 42 dBW EOL edge EIRP

One 50 W TWTA 2.671-2.690/2.501-2.519 GHz up/down S-band all-India MSS beam, 18 MHz bandwidth, 35 dBW edge EIRP

One (plus one back-up) 4 W SSPA 6.452-6.468/3.681-3.699 GHz up/down C-band all-India MSS link, 18 MHz bandwidth, 30 dBW edge EIRP
Spacecraft: box-shaped 3-axis bus around 896 mm diameter central aluminium cylinder 1.685 m high. Anti-Earth panel carries most of propulsion elements. Twin 3-panel (each panel 1.8 × 2.15 m) solar wings spanning 14.6 m, totalling 23 m², providing 1,620 W EOL (summer solstice; 1,100 W payload required), 2 × 24 Ah Ni/Cd batteries. 1.77 m diameter deployed parabolic reflectors on East/West faces for C-band FSS & S-band BSS transmission; 90 cm Earth face S-band MSS transmission/receive; 70 cm C-band receive; 60 cm Earth face Ku-band transmission/receive; 15 cm Ku-band global horn
Mass: 2,050 kg at launch (747 kg bus, 233 kg payload, 1,070 kg propellant)
AOCS: unified liquid bipropellant system as 2A/2B. Antenna pointing better than ±0.2° pitch/roll, ±0.4° yaw. 3-axis body stabilised by MW/RWs, momentum dumping by four 22 N thrusters on East/West faces.

Insat 2D

Launch: 3 June 1997 on Ariane 4 from Kourou, French Guiana
Location: 74° east geostationary (collocated with 2A)
Mass: 2,050 kg
Transponders: as 2C, but orthogonal polarisation
Other specifications as Insat 2C.

Insat 2E

Launch: 2 April 1999 by Ariane 42P from Kourou
Location: 83° east geostationary
Design life: 12-14 years
Transponders: 16-20 36 MHz up/down C-band Asia & Australia to Western Europe beams

Single 402.75 MHz Data Collection System transponder

Single 406 MHz Cospas/Sarsat search & rescue package

Principal applications: as for Insat 1, plus Intelsat services
Spacecraft: box-shaped 3-axis bus central aluminium cylinder with asymmetrical 4-panel GaAs solar wing on south side providing 2,000 W EOL through two 50 Ah Ni/H_2 batteries and creating 27 m span. C-band transmission reflectors of 1.7/2.0 m diameter on East/West faces; C-band receive antenna on Earth face
Mass: 2,500 kg at launch, 1,100 kg dry
AOCS: unified liquid bipropellant 440 N LAM Liquid Apogee Motor and two redundant networks of 8 × 22 N RCS thrusters
Other specifications as Insat 2A/2B.

INDONESIA

Indostar series

Current status

Indonesia's Indostar (Cakrawarta-1) developed battery charger problems in November 1997 and the operators will silence one of the satellite's five transponder during both the spring and autumn eclipse periods in order to compensate.

Background

Indostar will provide the world's first DBS dedicated to radio/TV for a single nation. The first of four satellites will provide 5 S-band transponders for digital TV, using 8:1 compression for 49 channels. The second will add L-band for CD-quality radio. The US$100 million contract for the first satellite and its launch was signed with CTA International 8 December 1993. Funding is provided by 40 per cent private consortium, 30 per cent PT Amcol Graha Ltd. (a Jakarta electronics company) and 30 per cent PT Bimantra Citra. The Indostar 1 DBS system costs US$271 million. PT MediaCitra Indostar is the satellite company of the Indovision group. Indovision is a multichannel pay television company. National TV cannot be received by 32 per cent (US$57 million) and radio by 20 per cent (US$36 million). The commercial Indostar system will be received via US$100 hand-held radios, US$100 analogue receivers and US$500 digital satellite decoders with satellite dishes less than 1 m diameter. It is designed to guarantee quality reception within Indonesia, whatever the weather conditions.

Specifications

Indostar 1

Launch: 12 November 1997 on Ariane 44L from Kourou, French Guiana
Location: number 1 107.7° east
Design life: 10 years (12 years propellant)
Contractors: CTA International
Transponders: 5 (plus 4 back-up, 5-for-4) 70 W TWTA 8.120-8.270/2.520-2.670 GHz up/down S-band domestic TV 3.6 × 8.35° beams (8:1 compression for 40 channels), 24 MHz bandwidth, EIRP 48-50 dBW min Indonesia, linear orthogonal polarisation

From Indostar 2: 2 TWTA 8.067-8.092/1.467-1.492 GHz up/down L-band domestic radio beams, 300 kHz bandwidth, EIRP 48-50 dBW min Indonesia, linear orthogonal polarisation

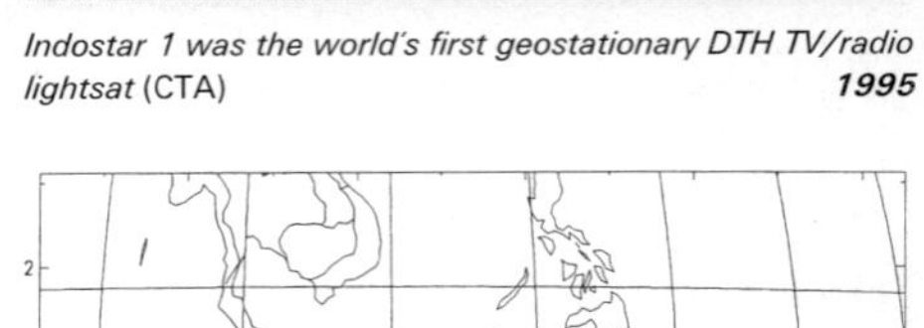

Indostar 1 was the world's first geostationary DTH TV/radio lightsat (CTA) ***1995***

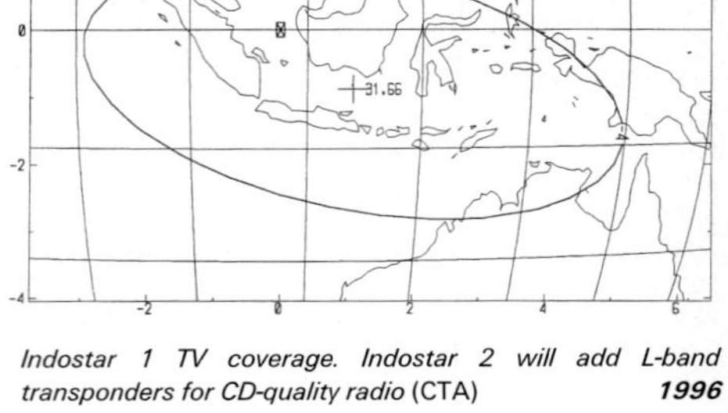

Indostar 1 TV coverage. Indostar 2 will add L-band transponders for CD-quality radio (CTA) ***1996***

Principal applications: direct domestic TV/radio
Configuration: CTA's Star lightsat, bus 165 × 185 × 150 cm. AKM Star 30C
Mass: 1,400 kg at launch; 430 kg on-station BOL
TT&C: through Master Control Station at Jakarta; Backup Control Station at Surabaya
Power system: 1,700 W EOL; two SAFT 52 Ah Ni/H_2 batteries.

Palapa series

Background

Indonesia witnessed the commercially successful domestic satellite systems in Canada (Anik) and US (Westar) for large territorial communication and determined that its 6,064 inhabited islands arcing across 5,000 kms along the equator would benefit from the same type of system. The country built 40 large Earth stations, some 200, 5 m telephony stations and more than 1,000 TVRO stations (television receive only). Indonesia decided to begin its satellite programme with the Palapa ('Fruit of labour'), but did not announce the plans until after negotiations with Hughes had been completed by Perumtel (now PT Telkom). The Palapa became fully operational in August 1976 after the launch of Palapa AI by a Delta on 8 July 1976. Indonesia had a completely functioning network by March 1977 after Palapa 2 completed orbit tests.

The satellites ended their operational lives in June 1985 and January 1988 respectively but in 1979 Indonesia had contracted with Hughes to build a follow-on system for two second-generation satellites at a cost of US$74.5 million. NASA threw in two Shuttle launches of at a cut-rate price of US$18 million. The B1 series Palapa proved a success operating from its June 1983 Shuttle launch continuously until retirement in 1990. PT Pasifik Satelit Nusantara purchased it and wrung three more years of life out of it.

Indonesia originally launched the B2 from the Space Shuttle in February 1984 but the PAM-D GTO insertion stage failed stranding it in a low earth orbit. Indonesia used the US$75 million insurance payment to purchase a replacement of the lost Palapa B1P (P means pengganti or 'replacement' in Indonesian). Shuttle mission 51A recovered it and the insurers sold it in November 1984 to Sattel Technologies Incorporated of Van Nuys, California for about US$18 million. Sattel planned to launch the refurbished satellite by Delta 2 in 1990 and then resell it to Indonesia once in-orbit. PT Telkom retired the Palapa B1 at 118° east and sold it to PSN in 1991 to provide Pacific Rim services. Still operated by PT Telkom, PSN's major shareholder, the company has allowed the inclination to drift in order to extend its life by three years.

Indonesia set up PT Satelit Palapa Indonesia (PT Satelindo) in January 1993 to finance and operate the

Palapa C1 expands coverage to southeast Asia, China and Australasia (Hughes Space & Communications) ***1996***

US$240 million Palapa C programme. Indonesia plans to increase the five telephones per 1,000 population to 200 by 2019 using follow-on systems to the Palapa. The third-generation Palapa C appeared in February 1996, expanding coverage to China, India and Japan. Indonesia created PT Satelindo to finance and operate the new generation. Satelindo plans to develop the Palapa D series but plans have been delayed due to economic difficulties and schedules are undetermined.

Specifications

Palapa B2
Launched: 13 April 1990 by Delta 6925 from Cape Canaveral
Location: 108° east GEO (originally at 118° east it replaced Palapa B1 in 1990). Replaced by C2 in 1996
Mass: 1,240 kg at launch, 652 kg on-station BOL
Other specifications as for Palapa B2P.

Palapa B4
Launched: 14 May 1992 by Delta 7925 from Cape Canaveral
Location: 118° east geostationary (to be replaced by C3)
Mass: 692 kg on station BOL
Other specifications as for Palapa B2P.

Palapa B2P/Palapa Pacific 2
Launched: 20 March 1987 by Delta from Cape Canaveral

Based on the powerful Hughes HS-601 bus, Palapa-C series was introduced in 1996 (Hughes Space & Communications) ***1999***/0054340

The second-generation Palapa satellites will remain in service into the early 2000s (Hughes Space & Communications)

Location: 144° east geostationary from August 1996 (replaced at 113° east 1996 by C1)
Design life: 8 years
Contractors: Hughes Space & Communications Co
Transponders: 24 (plus 6 back-up) 10 W TWTA 5.925-6.425/3.72-4.16 GHz up/down C-band regional beam, 36 MHz bandwidth, EIRP 34 dBW min Indonesia, 32 dBW min ASEAN countries (Philippines, Singapore, Thailand, Malaysia), linear orthogonal polarisation
Principal applications: telecoms, regional/domestic TV
Configuration: Hughes' cylindrical HS-376 bus, 2.16 m diameter, 6.83 m deployed height, spin-stabilised, position/spin by hydrazine thrusters, GEO injection by solid AKM
Mass: 1,200 kg at launch; 628 kg on-station BOL
Pointing: 0.05° accuracy
TT&C: through PT Telkom's Master Control Station at Cibinong, near Jakarta
Power system: 1,062 W BOL/830 W EOL from K7 solar cells on cylindrical body and skirt.

Palapa C1
Launched: 1 February 1996 by Ariane 4 from Kourou, French Guiana
Location: 113° east GEO (replacing B2P). Initially: in March 1996 manoeuvred off-station and re-located over 112° east again manoeuvred off-station August 1996 and the following month re-located over 150° east
Lifetime: 12 years full performance; 14 years basic performance
Transponders: 24 (plus 6 back-up) 21.5 W SSPA 5.925-6.425/3.700-4.200 GHz up/down C-band regional beam, 36 MHz bandwidth, EIRP 37 dBW min Indonesia, ASEAN countries (Philippines, Singapore, Thailand, Malaysia), Vietnam, Brunei, Papua New Guinea, New Zealand/Eastern Australia, linear orthogonal polarisation

6 (plus 2 back-up) 26 W SSPA 6.425-6.665/3.400-3.640 GHz up/down extended C-band beams for PT Pasifik Satelit Nusantara, 36 MHz bandwidth, EIRP 37 dBW min Pakistan, India, Burma, China, People's Republic, Korea, Taiwan, Japan, linear orthogonal polarisation

4 (plus 2 back-up, 6-for-4) 130 W TWTA 13.750-14.490/10.950-11.690 GHz up/down Ku-band regional beams, 72 MHz bandwidth, EIRP 50 dBW min Indonesia, Japan, Korea, Eastern China, Thailand, Vietnam, Singapore, linear orthogonal polarisation. A battery supply problem means this payload cannot operate while C1 is in eclipse.
Principal applications: telecoms, regional/domestic TV
Configuration: Hughes HS-601 platform, 3-axis, solar array span 21.0 m (satellite within 2.7 × 3.1 × 3.5 m envelope at launch). Payload mounted on North/South panels and Earth-facing floor; East/West faces carry hard points for two 216 cm diameter C-band antennas, fixed antenna mountings on Earth-facing wall for 178 cm diameter extended C-band + 152 cm diameter Ku-band; all antennas octagonal dual-gridded with single feed horns
AOCS: 3-axis, unified 13 ARC 22 N and one Marquardt 490 N bipropellant thrusters incorporating onboard

control processor, Sun/Earth sensors, and two 61 Nms 2-axis gimbaled momentum bias wheels. The NTO/MMH is contained in four spheres. Spin-stabilised in transfer orbit
Mass: 2,989 kg at launch, 1,775 kg on-station BOL
TT&C: as Palapa B
Power system: twin solar wings of three 2.16 × 2.54 m panels carrying K4¾large area Si cells on Kevlar substrate provide 3,730 W; EOL 3,400 W. Ni/H_2 batteries provide 100% eclipse protection.

Palapa C2
Launch: 16 May 1996 by Ariane 4 from Kourou, French Guiana
Location: 123° east GEO (replacing B2R). Initially: during June 1996 relocated to ~112° east
Mass: 2,989 kg at launch, 1,803 kg on-station BOL, 1,669 kg dry
Other specifications as for Palapa C1.

INTERNATIONAL

Arabsat 1 series

Background

In May 1981, Aerospatiale and Ford Aerospace received a contract worth US$134.4 million for a regional satellite communication network for the Gulf region. The contract, from the newly formed Arabsat Organisation, asked for development and fabrication of three satellites (one a ground spare) for launch aboard either Ariane or Shuttle. In November 1982, Arabsat and NASA signed an agreement for a Shuttle/PAM launch for US$20.1 million (plus US$6 million for the PAM-D perigee kick stage); an earlier Ariane launch had been negotiated for US$23 million.

The Arabsat 1 system cost about US$500 million in 1986 dollars. The Arab Fund for Economic & Social Development made a loan of US$53 million to the 11 members Arabsat Organisation in 1985 to build Earth stations, in addition to US$123 million it had loaned earlier to 13 members for modernisation of communications systems. Then Arabsat paid Japan's NEC to build the two TT&C stations, while Aerospatiale operated the stations and trained Arabsat personnel over a 2 year period.

At the outset a lack of funding hampered the Arabsat programme. By the beginning of 1989, outside sources estimated the total debt at US$60 million with only 40 per cent utilisation of its total capacity on one active satellite. A drop in oil revenues throughout the Arab world precipitated a debt-reduction plan and gave incentive to Arabsat to increase revenues. The review indicated that the 50 W S-band semi-DBS links were not operational, and some 25 per cent of the utilisation shortfall arose because Egypt did not participate due to its suspension in 1979 from the Arab Telecommunications Union resulting from its dealings with Israel. When Egypt rejoined in April 1989 the debt crisis ended.

Arabsat launched Arabsat 1A on 8 February 1985. One solar panel failed to open during the GTO orbital placement phase but thruster firings successfully freed the unit. About one month later, power to two AOCS gyros failed and consequently NSSK at 19° east had to be exercised manually from Saudi Arabia's Dirabh TT&C station. Arabsat declared the satellite, although still operational an in-orbit spare and claimed US$75 million insurance based on failure to meet specified operational quality. The company allowed the satellite to be used from an inclined orbit to add three years to its seven year lifetime by saving station-keeping propellant. These operations began at the end of 1990, but the fuel ran out September 1991 and the satellite began drifting. Arabsat declared it lost 30 March 1992, and de-activated it 31 July 1992.

Arabsat 1B went up 18 June 1985 by Shuttle to 26.2° east and it began inclined operations in late 1991 to extend useful life by 3 years but, again, there was insufficient fuel and Arabsat took it out of service in October 1992 and deactivated 3 April 1993. Arabsat representative Prince Sultan Salman Al-Saud accompanied the second satellite into orbit.

Aerospatiale/Ford completed the ground spare, 1C, in 1985 and held it in storage at Aerospatiale's facility in Cannes. Intelsat negotiated purchasing it during 1989 for Pacific operations and Arabsat withheld payment of the last 30 per cent of the cost. Repayments began again in June 1990 and China's GWIC received a US$25 million launch contract in January 1990 for departure November 1991. Then Arabsat applied for both US and French export licences because some 60 per cent of the components are US-derived. In March 1991, the US claimed that the licences for the Chinese launch were not complete, Arabsat cancelled the launch for this reason and the Persian Gulf war. Then Arabsat renegotiated the launch with Arianespace instead on 15 May 1991.

Specifications

Arabsat 1C
Launched: 26 February 1992 by Ariane 4 from Kourou, French Guiana
Location: 30.5° east GEO
Design life: 7 years
Contractors: Aerospatiale (prime; structure, thermal control, solar arrays, integration/testing), Ford Aerospace (now SS/L; communications payload, attitude control, propulsion, power conditioning/storage, telemetry, command & ranging), Alenia (antenna reflectors, TT&C transponders), TNT/AEG (solar array electrical network), MBB (propulsion & attitude control integration)
Transponders: 25 8.5 W TWTA 5.925-6.425/3.70-4.20 GHz up/down C-band regional beams, 33 MHz bandwidth, EIRP 31 dBW min, circular polarisation. Two (plus one back-up) paralleled 50 W TWTA 5.925-6.425/2.540-2.655 GHz up/down S-band downlink for semi-DBS broadcasts (1.5 m dishes), 33 MHz bandwidth, 41 dBW min EIRP, linear polarisation
Principal applications: regional TV, telephony, data, fax relay. Each C-band can handle 1,466 SCPC telephone channels or one TV
Configuration: Aerospatiale/MBB Spacebus 100 (now classed as 1000) 3-axis stabilised 2.26 × 1.64 × 1.49 m box-shaped bus with 20.7 m-span twin solar arrays
Mass: 1,310 kg launch, 785 kg on-station BOL, 600 kg dry
TT&C: through Saudi Arabia's Dirabh station, plus back-up station at Eddikhila near Tunis
Power system: 1,400 W BOL/1300W EOL provided by twin silicon solar arrays.

Arabsat 1D-R (ex-Telstar 301)
Launched: 29 July 1983 as Telstar 301 by Delta from Cape Canaveral
Location: 20° east GEO, February 1995-September 1996
Design life: 10 years
Contractors: Arabsat 1D-R: Hughes Space & Communications Co
Transponders: 18 5.5 W SSPAs 5.925-6.425/3.700-4.200 GHz up/down C-band regional & spot beams, 36 MHz bandwidth, 34 dBW min EIRP, linear polarisation. 12 5.5 W TWTAs, other specs as for SSPAs
Principal applications: regional telephony, data, TV broadcast
Configuration: Hughes' HS-376 spin-stabilised bus, 2.2 m diameter, 6.8 m high deployed
Mass: 3,423 kg with PAM-D GTO stage attached, 625 kg on station BOL
TT&C: from Telesat Canada HQ via Tunis station
Power: 800 W BOL/670 W EOL provided by 15,588 silicon cells on cylindrical exterior
Apogee kick motor: Thiokol solid propellant.

Arabsat 2 series

Current status

To satisfy a growing demand, and to guarantee services, Arabsat launched Arabsat 2B in July 1996. Previously, Arabsat intended to hold the satellite as a ground spare, dispatching it at the latest in 1999 before 1C is retired. Arabsat 3A was launched in February 1999 to provide direct TV broadcasting to North Africa and the Near East.

Background

Arabsat requested bids for a successor system to Arabsat 1 in early 1990 and allowed a 15 October 1990 deadline. The Iraqi invasion of Kuwait forced a delay. The final Arabsat 2 RFP went to bidders in 1991 projecting a new launch date. In October of 1992, Arabsat tentatively chose Hughes Space & Communications Company to build the new satellites on a US$258 million bid for two HS- 601 satellites. British Aerospace and Aerospatiale also submitted proposals. However, Hughes and BAe reportedly refused to agree to some of Arabsat's terms and the organisation instead placed the order for US$257.9 million for Aerospatiale's Spacebus 3000 platform on 17 April 1993. The contract included first launch on Ariane 4.

Specifications

Arabsat 2A
Launched: 9 July 1996 by Ariane 4 from Kourou, French Guiana
Location: planned 26° east geostationary
Design life: 12 years (16 years propellant)
Contractors: Arabsat 2A: Aerospatiale (prime; structure, thermal control, solar arrays, integration/testing),

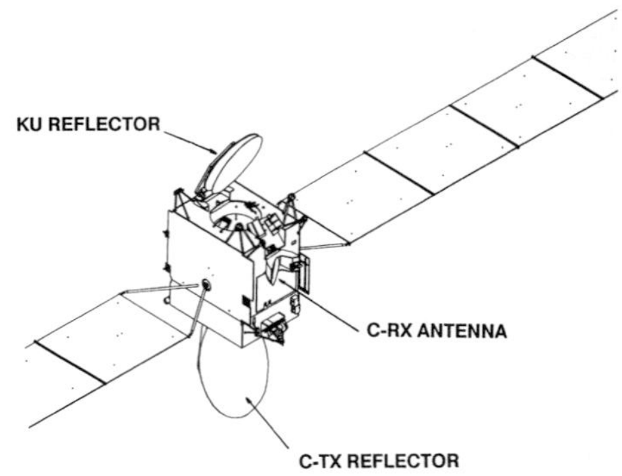

The Arabsat 2 generation became operational in 1996 (Aerospatiale) ***1995***

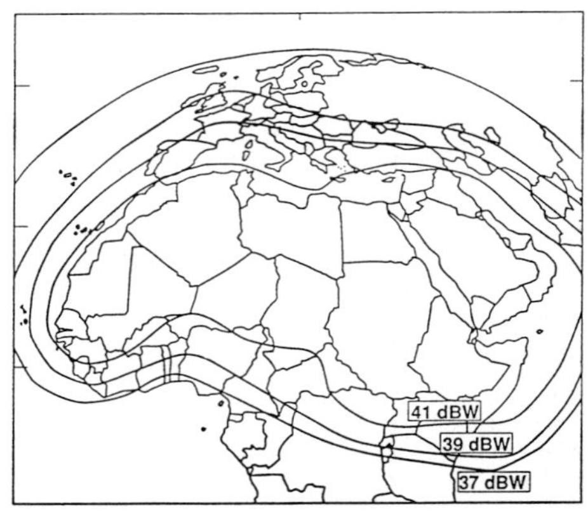

Arabsat 2 high-power C-band EIRP contours (Aerospatiale)

Alcatel Espace (communication payload), Detecon (contract monitoring on behalf of Arabsat), DASA (propulsion/attitude control integration), Alenia (TT&C)
Transponders: 14 (plus 6 back-up: 10-for-7) 15 W SSPA 5.925-6.425/3.70-4.20 GHz up/down C-band regional beams, 12 × 36 + 2 × 54 MHz bandwidths, EIRP 35 dBW min, circular polarisation for telecom services.

8 (plus 4 back-up, 12-for-8) 57 W TWTA 5.925-6.425/3.70-4.20 GHz up/down C-band regional beams, 36 MHz bandwidth, EIRP 41 dBW min, circular polarisation for semi-DBS via 0.8-1.0 cm dishes

12 (plus 5 back-up, 17-for-12) 95 W TWTA 13.75-14.0/12.5-12.7 GHz up/down Ku-band regional beams, 8 × 36 + 4 × 30 MHz bandwidths, 47 dBW min EIRP, linear polarisation
Principal applications: regional TV, telephony, data, fax relay
Configuration: Aerospatiale/DASA Spacebus 3000 3-axis stabilised 1.8 × 2.6 × 2.3 m box-shaped bus (7 m deployed height) with 26.3 m-span twin solar wings. C-band: 2.1 × 2.2 m transmission west panel dish + 60 cm receive Earth face dish; Ku-band: single 1.6 m dual gridded dish on east panel. C-band repeater assembly on N panel; Ku-band on S panel
Mass: 2,617 kg launch, 1,570 kg on-station BOL, 1,108 kg dry, 260 kg payload
TT&C: as Arabsat 1
Power system: 5,074 W EOL provided by twin Si solar wings totalling 46 m². Two 52 Ah Ni/H_2 batteries.

Arabsat 2B
Launch: 13 November 1996 by Ariane 4 from Kourou, French Guiana
Location: 30.5° east geostationary.

Arabsat 3 series

Current status

Alcatel Space Industries was contracted to provide the first of two direct TV broadcast satellites based on the SB-3000B bus. Arabsat 3A was launched on 26 February 1999 for the Arab Satellite Communications Organisation based in Saudi Arabia. A contractor for Arabsat 3B has yet to be determined.

Specifications

Arabsat 3A
Launched: 26 February 1999 by Ariane 44L
Location: 26° east
Design type: 15 year
Transponders: 20 active in Ku-band BSS
Configuration: 3.83 × 3.35 × 2.26 m; solar array span 29 m; onboard power 6,400 W
Mass: 2,708 kg at launch; 1,646 kg BOL; 1,200 kg dry.

Artemis satellite

Background

The ESA Council approved three new experimental payloads in June 1990 and folded them into the ARTEMIS research satellite. The new experiments ESA wants to investigate are the Silex (Semiconductor Laser Intersatellite Link Experiment), SKDR data relay as a DRS precursor, and an L-band mobile channel to demonstrate services for European land vehicles in addition to backing up EMS. Technology experiments. Originally, ESA planned a launch aboard an Ariane 502 in April 1996 but technical problems delayed the launch date to late 1997.

Estimated to cost ECU665 million at 1992 rates, the cost of the Artemis satellite has swollen to about ECU920 million. The cost overrun has prompted NASDA to offer a free H2 launch in return for making Japan a full partner. Italy provided about 45 per cent of the original funding. ESA signed an ECU446 million contract with Alenia Spazio on 4 October 1993. Phase B2 began July 1989.

Specifications

Launch: planned early 2000 by Ariane 5
Location: 16.4° east, then 59° east when DRS appears
Design life: 10 years minimum
Contractors: Alenia Spazio (prime), Matra Espace (Silex prime), BPD/DASA (unified propulsion), DASA (S400 + S10MkII thrusters for AOCS), SAFT (Ni/H_2 batteries), CASA (structure), FIAR (power conditioning/distribution), Galileo (attitude sensors)
Transponders: three (plus one back-up, 4-for-3) feeder link for Silex and SKDR, 27.5-30/18.1-20.2 GHz up/down, EIRP 43 dBW, G/T 0 dB/K, 234 MHz bandwidth, linear vertical polarisation Silex. Data rate 50 Mbit/s, bit error rate $<10^{-6}$. Two optical terminals are being built: the GEO element for Artemis and the LEO unit on Spot 4. Under an agreement with CNES, a pre-operational Silex service will be provided to Spot 4 after the initial experimentation phase. Each terminal employs a 25 cm diameter optical telescope mounted on a coarse pointing mechanism; total 140 kg; pointing accuracy 0.2 µrad (random), 0.8 µrad (static). Optical power source: 8,300 Å GaAlAs semiconductor laser diode with a peak output of 160 mW (60 mW continuous), and a beamwidth of 0.0004°
Receiver: SI-APD silicon avalanche photodiode followed by a low noise trans-impedance amplifier; 1.5 nW useful received power. Sira's CCD acquisition/tracking sensors direct Teldix's fine pointing mechanism of orthogonal mirrors. Phase B concluded October 1990; C/D started mid-1991. A 1 m telescope at the Teide Observatory on Tenerife in the Canary Islands will be used for experiments with Artemis.

Artemis includes large reflectors to provide L-band mobile beams (left reflector) and S/Ka-band inter-satellite data relay (right reflector). Their feeder link antennas are centrally mounted: Ku-band for L-band (top right) and Ka-band (top left) for S/Ka/laser. The Silex laser system is the box unit at bottom right (Alenia)

Technicians at Estec check out the Artemis satellite bus (Theo Pirard) ***1999***/0054353

DRS could add S-band phased arrays, for multiple access, to Artemis's Silex and SKDR payloads. The 48-element receive array is deployed at bottom; the 24-element transmit array is mounted on the Earth face (Alenia)

Comstar/SBS series

Background

Comsat established the Comsat General Corporation subsidiary in 1973 to act as the satellite operator of its growing inventory of orbiting hardware. The subsidiary now manages the five satellites of the Marisat, Comstar and SBS 2 networks. Another subsidiary, Comsat Video Enterprises owns and operates a satellite-based network distributing entertainment and video conference services to the US lodging industry.

Comsat General's 4-satellite Comstar system was the first to be integrated into the US national telephone network, providing voice, data and TV services for the whole of the US, Alaska, Hawaii and Puerto Rico. Hughes received COMSAT's US$65 million contract in September 1973. Comsat immediately leased all of the capacity to AT&T for the 7-year satellite operational lives at US$1.3 million monthly for each. Comstar D1, launched 13 May 1976, took its slot at 76° west next to D2 as their capacity declined. D3 replaced it 29 June 1978. Each satellite provided 18,000 duplex voice channels. COMSAT announced that it planned to move D4 over China in 1993; AsiaSat had an option on D4 but it expired 31 July 1993. D2, launched 22 July 1976, drifting January 1994.

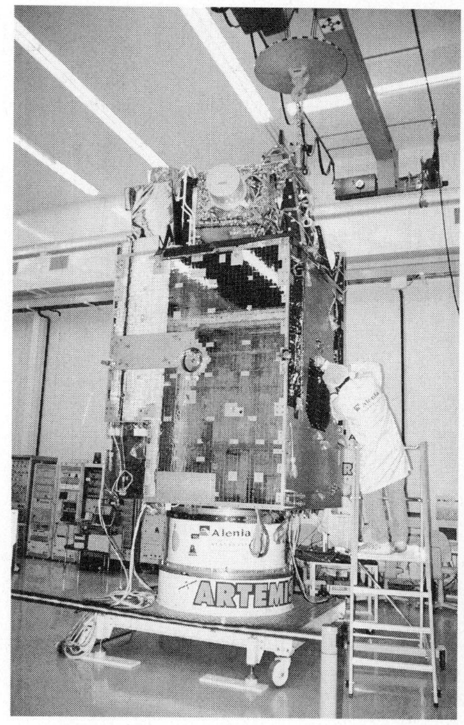

Artemis and Silex (top) (Theo Pirard) ***1999***/ 0054355

Comsat also established a new Satellite Business Systems (SBS) subsidiary in December 1975 with ownership held at 41.3 per cent. IBM holds an equal 41 3 per cent share and Aetna Life & Casualty 17.4 per cent. In April 1984, after the successful launches of three satellites, Comsat sold its interest to its partners. The Federal Communications Commission approved of transfer of SBS to MCI Communications Corporation in November 1985. Comsat then announced the purchase of SBS 1 and 2 from MCI in March 1987.

Then, after Comsat successfully launched three SBS satellites through April 1984, Comsat sold its interest to its partners. Part of the partnership arrangement of commercial owners entailed a transfer of SBS to MCI Communications Corporation, approved by the FCC in November 1985. But SBS 4 and subsequent satellites remained majority partner IBM's wholly-owned Satellite Transponder Leasing Corporation (STLC) subsidiary. SBS 4 began operations in October 1984 to be joined in late 1988 by the last of the series' HS-376-class satellites, SBS-5. The more powerful HS-393-class SBS 6 had a Shuttle launch date of September 1988 when the *Challenger's* loss occurred in January 1986. It eventually departed October 1990 aboard Ariane.

STLC services provided by SBS 4 and 5 included network TV, wideband TDMA (multiple voice and high speed data) and corporate network VSATs for data networks and video broadcasting. HCI reached agreement with IBM for purchase of the SBS system in July 1989.

SBS 6 is the only HS-393 satellite of the SBS series, offering triple the capacity of its predecessors. Galaxy 4 replaced it at 99° west in November 1993. HCI moved SBS 6 to 95° west in March 1994 until Galaxy 3R's December 1995 launch. With the Galaxy in place HCI then moved the SBS satellite to 74° west. Galaxy 7 replaced SBS 4 at 91° west in January 1993. SBS 4 then moved in February 1993 to 77° west pending a decision on its future. NBC bought its entire capacity in September 1993, and began using it on a full-time basis by September 1995. It is now operating in an inclined orbit to extend its life beyond 2000.

Specifications

Comstar D4
Launched: 21 February 1981 by Atlas Centaur from Cape Canaveral complex 36
Location: 76° west geostationary, inclined 8.6°
Design life: 10 years
Contractors: Hughes Space & Communications
Transponders: 24 5/5.5 W TWTA 5.925-6.425/3.70-4.20 GHz up/down C-band CONUS / Hawaii / Alaska / Puerto Rico beams, edge EIRP 33 dBW, 34 MHz bandwidth, orthogonal polarisation. 19.0/28.6 GHz beacons for propagation research
Principal users: national voice, data, TV
Configuration: similar to Intelsat 4A. Spin-stabilised (56 rpm) cylindrical satellite with despun communication platform incorporating two offset

parabolic 127 × 178 cm reflectors each handling 12 channels. Stationkeeping provided by hydrazine thrusters
Mass: 1,516 kg at launch, 911 kg on-station BOL
TT&C: Comsat's Washington facility working through Southbury, CT and Santa Paula, CA stations
Power: 570 W BOL from 17,000 silicon cells on cylindrical surface; two Ni/Cd batteries
Apogee kick motor: Thiokol solid propellant.

SBS 2
Launched: 24 September 1981 by Delta/PAM from Cape Canaveral
Location: 71° west GEO, inclination 6.7°; 97° west until January 1994
Design life: 7 years
Transponders: 10 (plus 4 back-up) 20 W min TWTA 14.025-14.466/11.725-12.166 GHz up/down Ku-band CONUS beams, 43 MHz bandwidth, EIRP 43.7 dBW edge CONUS
Principal users: domestic business voice, video, data, and VSAT services
Configuration: based on Hughes' HS 376 spin-stabilised bus, 2.16 m diameter, 6.6 m height with solar array drum extended (2.8 m stowed). Single antenna comprising two offset grid reflectors with horizontal/vertical feed system on despun platform
Mass: 546 kg on-station BOL
Attitude control: position + 50 rpm spin control maintained by four Hughes 22.2 N hydrazine thrusters
TT&C: through Comsat's Clarkburg, MD facility
Power system: >16,000 Si body cells provide 900 W EOL in conjunction with two Ni/Cd batteries
Apogee kick motor: Thiokol Star 30 solid.

SBS 4
Launched: 30 August 1984 by Shuttle STS-41D/PAM-D from Cape Canaveral
Location: 77° west GEO, inclination 2.4°
Design life: 10 years minimum GEO
Mass: 1,117 kg at launch, 571 kg on-station BOL
Other specifications as for SBS 5 except that SBS 4 cannot double-up its transponders for 40 W output and does not incorporate the four 110 MHz units.

SBS 5
Launched: 8 September 1988 by Ariane V25 from Kourou, French Guiana
Location: 123° west geostationary (to be replaced by Galaxy 10)
Design life: 10 years minimum
Transponders: 10 (plus 4 back-up) 43 MHz and 4 (plus 2 back-up) 110 MHz bandwidth 20 W min TWTA 14.0-14.46/11.72-12.16 GHz up/down Ku-band CONUS plus Alaska and Hawaii spot beams, EIRP 39 dBW edge CONUS (45-46 dBW E/W concentrations), orthogonal polarisation. Four of the 43 MHz and all four of the 110 MHz transponders can be paralleled to provide 40 W output, principally for VSAT services
Principal users: domestic business voice, video, data, and VSAT services
Configuration: based on Hughes' HS-376 spin-stabilised bus, 2.16 m diameter, 6.6 m height with solar array drum extended (2.8 m stowed). Single antenna comprising two offset grid reflectors with horizontal/vertical feed system on despun platform
Mass: 1,241 kg at launch, 725 kg on-station BOL
AOCS: 50 rpm spin control and station-keeping maintained by four Hughes 22.2 N hydrazine thrusters
TT&C: through the Castle Rock, Colorado station
Power system: >16,000 Si cells on cylindrical surface provide 1,200/1,150 W BOL/EOL in conjunction with two Ni/Cd batteries
Apogee kick motor: Thiokol Star 30 solid.

SBS 6
Launched: 12 October 1990 by Ariane 4 from Kourou, French Guiana. Four apogee firings 13/14 October for GEO insertion
Location: 74° west geostationary
Design life: 10 years (hydrazine for 15.6 years remaining after Ariane launch)
Transponders: 19 (plus 11 back-up; 30-for-19 ring redundancy) 41 W min TWTA 14.00-14.49/11.700-12.19 GHz up/down Ku-band CONUS plus switchable Hawaii/Alaska spot beams, 43 MHz bandwidth, CONUS EIRP 49-53 dBW, orthogonal polarisation
Configuration: Hughes HS-393 spin-stabilised bus 3.66 m diameter, 3.6 m stowed height/10.0 m deployed, payload and single 2.4 m diameter antenna mounted on despun platform
Mass: 2,478 kg at launch, 1,514 kg on-station BOL, 1,138 kg dry
AOCS: spin-stabilisation maintained by six 22 N thrusters, GEO insertion and major manoeuvres performed by two Marquardt R-4D 489 N thrusters, both systems drawing on 1,200 kg of NTO/MMH in four tanks each
Antenna pointing: 2-axis RF beacon tracking maintains 0.06°-accuracy pointing, backed up by Earth sensors
TT&C: Same as SBS 5

Power system: 2,300 W equinox BOL/2,200 W solstice EOL provided by Si solar array on cylindrical body (K7 cells) and extension skirt (K4¾ cells) in conjunction with two 48 Ah Ni/H_2 batteries
Thermal control: quartz mirror radiator band encircling upper body provides primary heat rejection path.

ECS series

Background

The first-generation European Communications Satellite system, developed from the successful OTS 2 programme. ESA declared it fully operational in October 1984 and handed ECS-2 over to Eutelsat at that time. ESA then embarked on a programme to procure and launch five ECS for Eutelsat, because by then, the demand for channels had led to three more satellites being ordered. ESA handed over the last of these satellites in-orbit in 1988.

At 1,175 kg launch mass, the ECS satellites are about twice the size of OTS, and in addition to telephone, telex and business traffic the system transfers TV between member countries of the European Broadcasting Union. Under a 10 year agreement, ESA provided the first generation space segment for Eutelsat, which became the owner and manager of each satellite following in-orbit testing. Able to carry 12,000 phone circuits plus two TV channels, they came from the production line of the Mesh consortium, led by British Aerospace.

ESA intended to launch ECS 1/2, by Deltas but eventually transferred them to Ariane. It ordered the first ECS in 1978 at a cost of US$114 million plus Ariane launches at US$26.4 million each. Ariane failures and Marecs electrostatic problems delayed the launches. The ECS 1/2 development phase probably cost ESA about 220 MAU, and ECS 3-5 cost 221 MAU in 1984 rate currencies.

ESA Communications Demonstration Missions

Current status

ESA plans to launch communications demonstration missions in the near future. Plans call for GEO missions, such as EMS European Mobile System, OBP Onboard Processing and the DRTM Data Relay & Technology Mission, comprising the Artemis and DRS Data Relay System satellites. ESA also hopes to stimulate the non-GEO market with Archimedes for European mobile users, including mobile communications, sound broadcasting and navigation.

Eutelsat 1 series

Background

ESA designed, commissioned and financed the first generation of the European Communications Satellite (ECS) and transferred them to Eutelsat, once checked-out in orbit, for commercial exploitation. The international organisation launched the first satellite, known initially as ECS 1 until it began operations as Eutelsat 1-F1, in 1983. ECS 5 (Eutelsat 1-F5) completed the first network in July 1988. Each Eutelsat 1 operates up to 10 Ku-band transponders, simultaneously providing up to five beams.

ESA handed over Eutelsat 1 at 10° east on 12 October 1983 primarily for TV distribution services. Eutelsat moved it to 16° east after the launch of F5 in July 1988. After 2-F3 appeared in December 1991, Eutelsat allowed the satellite to begin a slow drift to 25.5° east. In August 1993, Eutelsat then moved it to 48° east, where it arrived 14 December, to provide CIS services and retain connectivity to the West. Eutelsat moved it again in March 1996 to 36° east, where a quick motor firing boosted it out of the geosynchronous orbit band and into retirement in mid-December 1996. Eutelsat 1 lost part of the solar array capacity after 15 months through short circuits across the Kapton insulating layer between the cells and their grounded panel structure. But, in all, the satellite logged 640,410 channel-hours by 30 June 1993.

Eutelsat 1 F2 added the Satellite Multiservice System to provide companies with access to a European network of dedicated links for high speed data transmission, video conferencing, facsimile and remote printing. It handled 2.4-2,000 kbit/s data. ECS 2 and successors had expected lifetimes of only 4.9 years NSSK because of heavier payload and the Eutelsat retired 1-F2 from service at 1° east 13 September 1993 by raising it 396 km above GEO. In all, it provided 634,969 channel-hours in its life.

ECS 4 and 5 cost US$37 million. ECS 4 became the property of Eutelsat 30 October 1987 after employing the new Arianespace launch-to-separation insurance coverage, on the open market for final deployment. Initially positioned at 10° east, then 13° east for TV distribution, Eutelsat moved it in late 1990 after 2-F1's commissioning to 7° east for fulltime leases, SMS traffic and occasional use, and then in 1992 to 25.5° east. It had provided 480,461 channel-hours by 30 June 1993. F5 completed the first-generation system of four operational satellites in 1988. It operated initially at 10° east and moved in 1991 to 21.5° east to carry telephony and occasional use traffic.

Eutelsat tariffs cover the satellite transponder alone and do not include charges for uplinks and other services on the ground. A lease for an Eutelsat 1 transponder costs a user ECU2.8 million annually for a pre-emptible narrowband transponder when a 3-year contract is signed. Eutelsat also offers a non-pre-emptible service over three years, which costs ECU3.8 million annually, and ECU3.4 million if a user selects a 5-year option for at least five transponders. Eutelsat revises the transponder lease costs based on the consumer price index for all European Union members (if the result is a tariff increase, the Signatory has the right to terminate the agreement).

Specifications

Eutelsat 1-F1 (ECS 1)
Launched: 16 June 1983 by Ariane L6 from Kourou, French Guiana
Location: 36° east geostationary, inclination increased to 5.2° by end-1995
Design life: 9 years, with 74 per cent reliability for 16 channels
Contractors: Satcom International (MMS)
Transponders: ten (plus two back-up) 20 W TWTA 14.00014.500/10.950-11.200 + 11.450-11.700 GHz up/down Ku-band three spot beams (telephony, spots SA Atlantic, SE east, SW west) and one Eurobeam (TV), 72 MHz bandwidth, spot EIRP 40.8-46.3 dBW, Eurobeam EIRP 34.8-41 dBW, linear orthogonal polarisation
Principal applications: leases
Configuration: ECS/OTS platform, 3-axis controlled

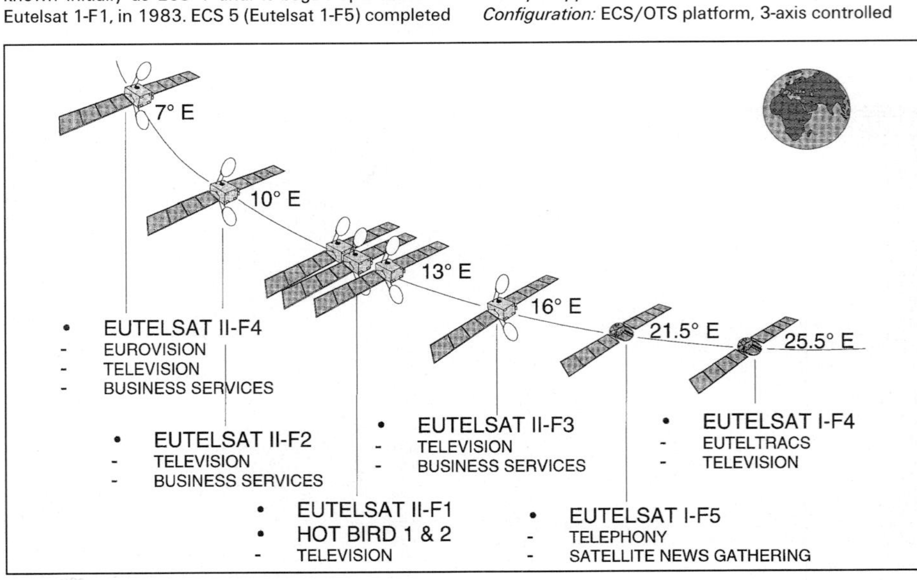

The Eutelsat System in January 1997
1997/0005564

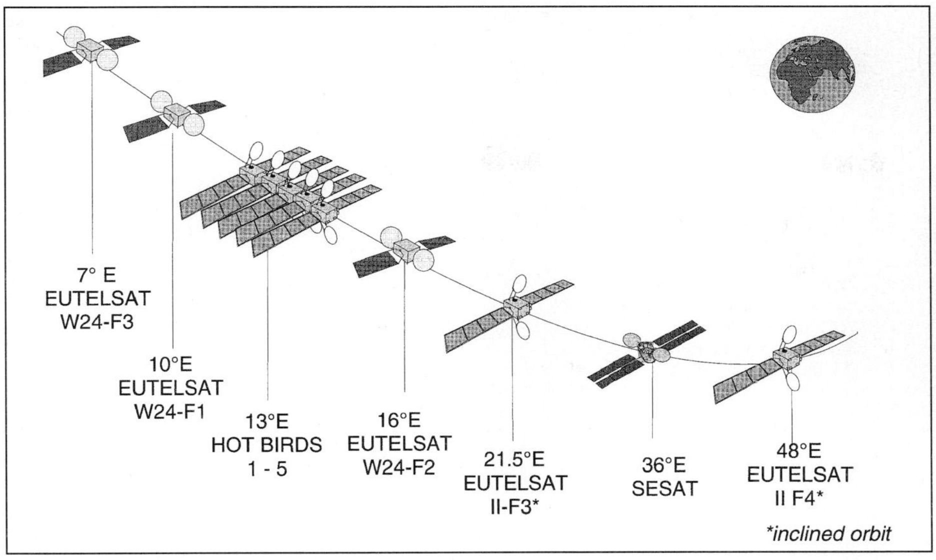

The Eutelsat System in 2001 **1997**/0005565

The Eutelsat 2 series is based on Aerospatiale's Spacebus 2000 platform. Each carries 16 channels working through the East and West 1.6 m reflectors

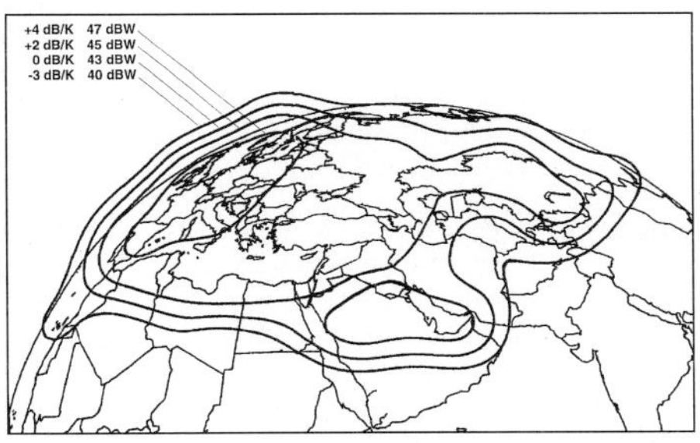

Sesat coverage from 48° east (Eutelsat) **1996**

Mass: 1,185 kg at launch, including 475 kg Mage AKM and 107 kg hydrazine AOCS propellant (2.5 kg remained August 1995), 680 kg on-station BOL
TT&C: launch and on-orbit TT&C performed by ESOC at Darmstadt, Germany; operational TT&C exercised from ESOC's station at Redu, Belgium through 13.5 m antenna
Power system: twin 5.2 × 1.3 m wings spanning 13.8 m providing 1 kW BOL, 900 W (solstice) EOL, supported by 2 × 24 Ah Ni/Cd batteries.

Eutelsat 1-F2 (ECS 2)
Launched: 4 August 1984 by Ariane V10, given to Eutelsat 12 October 1984
Transponders: Same as Eutelsat 1-F1 plus two 20 W TWTA on 12.5 12.75 GHz Ku-band downlink offering 39.8-43.5 dBW EIRP Satellite Multiservices System business links.

Eutelsat 1-F3 (ECS 3)
Destroyed in the Ariane's 12 September 1985 launch failure.

Eutelsat 1-F4 (ECS 4)
Launched: 16 September 1987 by Ariane V19 from Kourou, French Guiana
Location: 25.5° east geostationary, inclination 2.0° (last NSSK made April 1993)
AOCS: 116 kg hydrazine loaded; 14 kg remained as of August 1995
Other specifications as for Eutelsat 1-F2.

Eutelsat 1-F5 (ECS 5)
Launched: 21 July 1988 by Ariane V24 from Kourou, French Guiana
Location: 21.5° east GEO, inclination 1.5°
Mass: 1,185 kg at launch, about 700 kg on-station BOL. 122 kg hydrazine at launch; 22 kg as of August 1995 (mean usage 1.23 kg monthly)
Other specifications as for Eutelsat 1-F2.

Eutelsat 2 series

Eutelsat itself manages the follow-on Eutelsat 2 programme that began with the 2-F1 satellite launched in August 1990. The cost of the six satellites plus launch and insurance ran about ECU1,094 million at 1994 rates. Each F2 offers 16 Ku-band transponders and improved on-board redundancy. When it built its economic model for the second generation, Eutelsat assumed all the traffic carried by the first generation would continue and included sufficient capacity for expansion of services.

The second generation offers 16 simultaneously-operational transponders. 2-F1 has occupied the prime 13° east slot since 1990 to guarantee services without the need for users to re-direct their antennas. Eutelsat contracted with Aerospatiale for three follow-on satellites in May 1986 at the price of ECU225 million. They took out an option on the fourth in June 1987 and planned to launch the new series at intervals of 6 months. Then due to increased demand Eutelsat added a fifth to the overall purchase in March 1989 at a cost of ECU52.8 million, and a sixth asking for delivery during 1990.

2-F6 features with higher power TWTAs in the expansion band frequencies to provide 49 dBW over Europe and to reach to the Urals. Collocation with 2-F1 at 13° east provides up to 39 TV channels from the one slot without receiver modification. Direct reception is possible with 70 cm antennas in all of central/western Europe, and with slightly larger antennas for cable and community reception as far east as Turkey and Ukraine. Commercial services began 29 April 1995. Built by Matra Marconi, the Eutelsat 2-F7/-8/-9/-10 series were launched between November 1996 and October 1998 providing digital and analogue television and multimedia series. The first in a series of widebeam telecommunications satellite built by Alcatel Eutelsat W-F2 was launched by Ariane 44L on 5 October 1998 to 16° east.

Logica built a new TCR system, for Eutelsat 2 including the Satellite Control Centre in Paris and TCR equipment centres in Rambouillet (France) and Sintra (Portugal). Demand exploded when Eastern Europe left the Soviet Union in 1990. Eutelsat met the demand for services to eastern Europe, by stretching 2-F4 and 2-F5's widebeams to the Urals and modifying 2-F6 (Hot Bird 1) by adding radically new and more powerful TWTAs.

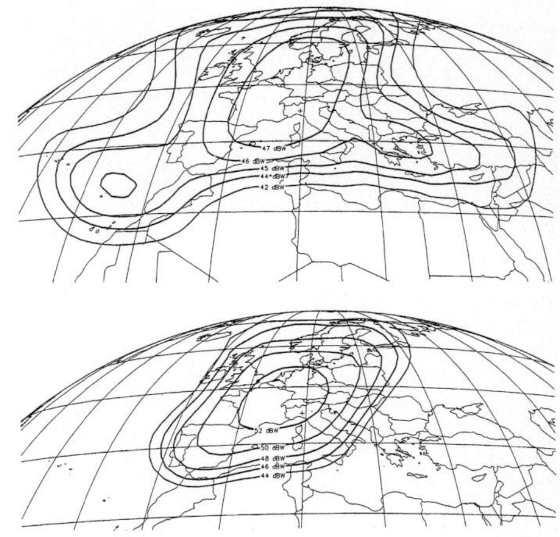

Eutelsat 2-F1 coverage from 13° east. Top: Widebeam; bottom: Superbeam

Specifications

Eutelsat 2-F1
Launch: 30 August 1990 by Ariane V38 44LP from Kourou, French Guiana
Location: 13° east geostationary
Design life: 9 years, reliability 0.74 for 16 channels and 7 years
Contractors: Aerospatiale (prime), Marconi Space Systems (payload), MBB (20% share: AOCS, solar arrays, TT&C equipment) Alcatel Espace, CASA, Aeritalia, ETCA, ERA
Transponders: 16 (plus 8 back-up; 12-for-8 ring redundancy) 50 W TWTA 14.0-14.5/10.95-11.20 + 11.45-11.7 + 12.50 12.75 GHz up/down Ku-band

Eutelsat 2-F4 integration at Aerospatiale's Cannes facility **1994**

Eutelsat satellite for East-West Communications (Theo Pirard) ***1999***/0054354

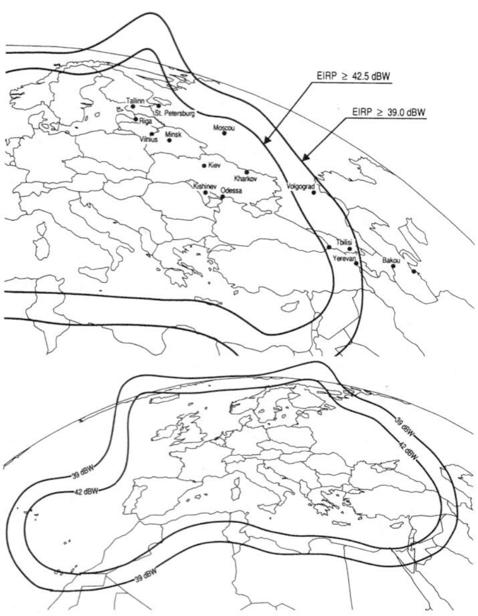

The Widebeam coverage of 2-F4/F5 was modified to reach the Urals (Eutelsat)

regional Superbeam and Widebeam, bandwidths 72 MHz (seven transponders) and 36 MHz (nine), EIRP 44- 52 dBW Superbeam over western Europe + 44-47 dBW pan-European Widebeam, linear orthogonal polarisation
Principal applications: TV/radio distribution to cable networks and domestic satellite systems
Configuration: derived from Spacebus 2000 platform, 3-axis control, 22.4 m span across solar wings, two 1.60 m diameter dual reflector multifeed antennas, one providing transmission/receive, the other transmission only, switchable between Widebeam/Superbeam as required. GEO location maintained within ±0.05°
Mass: 1,878 kg at launch, propellant for 10 years operation; 1,123/866 kg on-station BOL/EOL
AOCS: provided by MBB's 22 × 10 N + 1 × 400 N MMH/MON bipropellant system
TT&C: initially through GSOC German Space Operations Centre at Oberpfaffenhofen, Germany, then via Eutelsat's TCR system, comprising the Satellite Control Centre in Paris and TCR equipment centres in Rambouillet (France) and Sintra (Portugal; also equipped as the remote SCC back-up facility)
Power system: twin solar wings 3,500/3,000 W BOL/EOL.

Eutelsat 2-F2
Launch: 15 January 1991 by Ariane V41 from Kourou, French Guiana
Location: 10° east geostationary
Principal applications: TV/radio distribution, domestic telephony, SMS business traffic
Other specifications as for Eutelsat 2-F1.

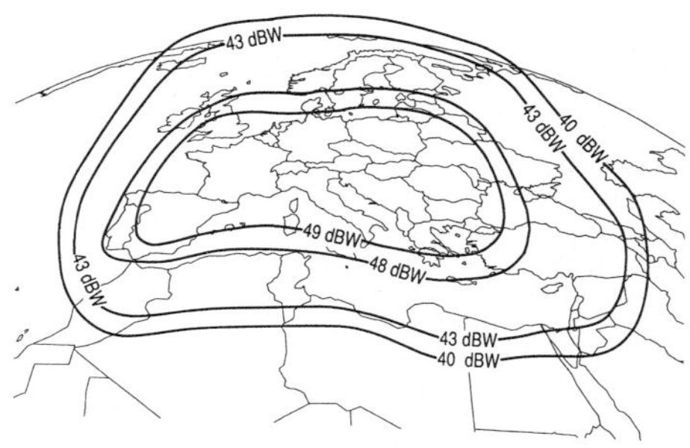

Instead of being divided into Widebeam/Superbeam coverage, 2-F6/Hot Bird 1 offers increased power European beams stretching to the Urals. Its new frequencies and collocation with 2-F1 at 13° east provide up to 39 TV channels from the one slot (Eutelsat) ***1994***

Eutelsat 2-F3
Launch: 7 December 1991 by first Atlas 2 (AC-102) from Cape Canaveral
Location: 16° east geostationary
Principal applications: TV/radio distribution, national/international telecom services, SMS business traffic
Other specifications as for Eutelsat 2-F1.

Eutelsat 2-F4
Launch: 9 July 1992 by Ariane 4 from Kourou, French Guiana
Location: 7° east geostationary
Other specifications as for Eutelsat 2-F1 but Widebeam modified for coverage as far as Urals, providing >42.5 dBW to Moscow and >39.0 dBW to Volgograd.

Eutelsat 2-F5
Lost in Ariane's 24 January 1994 failure. Planned for 36° east, with specifications as 2-F4 (Widebeam modified for coverage to Urals). Launch/satellite fully insured for ECU180 million.

Eutelsat 2-F6/Hot Bird 1
Launch: 28 March 1995 by Ariane 4 from Kourou, French Guiana
Location: 13° east geostationary (collocated with 2-F1)
Design life: 11 years
Transponders: 16 (plus 8 back-up; 12-for-8 ring redundancy) 70 W TWTA 12.895-13.25/11.20-11.55 GHz up/down Ku-band European beams, 36 MHz bandwidths, EIRP 40-49 dBW, linear orthogonal polarisation
Configuration: one transmit 1.60 m diameter dual reflector multifeed antenna and one elliptical receive antenna
Mass: 1,798 kg at launch, 1,078 kg on-station BOL, 840 kg dry
Other specifications as for Eutelsat 2-F1.

Eutelsat 2-F7/Hot Bird 2
Launch: 21 November 1996 by Atlas 2A from CCAFS
Location: 13° east geostationary (colocated with 2-F6)
Design life: 15 years
Transponders: 20 × 115 WTWTA in Ku-band (11.7-12.5 GHz)
Configuration: two 2.3 m transmit antennas for widebeam coverage from Kazakhstan to Persian gulf and 'super beam' coverage for Europe and North Africa.
Mass: 2,923 kg at launch; approx 1,700 kg BOL
Bus: 3.65 × 2.30 × 3.99; span 27.9 m; onboard power 5,500 W (EOL).

Eutelsat 2-F8/Hot Bird 3
Launch: 2 September 1997 by Ariane 44LP
Location: 13° east geostationary
Lifetime: 12 years
Configuration: as Eutelsat 2-F7/Hot Bird 2
Mass: 2,915 kg at launch; 1,715 kg BOL; 1,310 kg dry.

Eutelsat 2-F9/Hot Bird 4
Launch: 27 February 1998 by Ariane 42P
Location: 13° east geostationary
Lifetime: 12 years
Mass: 2,885 kg at launch; 1,770 kg BOL; 1,310 kg dry.

Eutelsat W-F2
Launch: 5 October 1998 by Ariane 44L
Location: 16° east
Transponders: 24 × Ku-band (72 MHz and 36 MHz)
Configuration: 1 fixed widebeam over Europe, North Africa and the Middle East (40-49 dBW), 1 steerable beam positioned over Mauritius and Reunion Island (42-53 dBW)
Mass: 2,950 kg at launch, 1,810 kg BOL: 1,375 kg dry
Bus: 4.6 × 2.5 × 1.8 m; span 29 m; onboard power 5,900 W (EOL).

Eutelsat 2 F10/Hot Bird 5
Launch: 9 October 1998 by Atlas 2A
Location: 13° east.

Inmarsat 1 (Marecs) series

Background

In 1975, the Inter-Government Maritime Consultative Organisation (IMCO) began discussing the need for a global satellite system serving shipping in the same way that Intelsat links countries. This body followed up with an agreement to establish Inmarsat, with headquarters in London in 1979. Several countries contributed to the operating costs on inauguration day including the US 23.37 (per cent), USSR 14.09, UK 9.89, Norway 7.88, and Japan 7.00.

With an initial London-based staff of 65 (now totalling 516), and with 37 member countries, Inmarsat took over responsibility for maritime links from the US-run Marisat on 1 February 1982. By March 1987, Inmarsat membership had grown to 48 countries. More than 5,000 of the world's 74,000 ships greater than 100 tonnes, plus oil rigs and other maritime platforms, have each been equipped with terminals costing about US$25,000 and providing direct access to high-quality telephone, telex and other services. The Inmarsat system includes dial phones and an emergency button giving priority to distress and other urgent messages. Despite the Soviets' large initial stake, only 10 of their ships had been equipped with receiving stations, which either the US or Japan built. In comparison the US had equipped 240 US ships by inauguration day.

Initially, Inmarsat leased Comsat's 10 telephone circuit capacity aboard each of the existing Marisats, stationed over the Atlantic at 15° west, the Indian Ocean at 73° east and the Pacific at 176.5° east. Intelsats 5-F5 to 5-F9 leased maritime packages to Inmarsat as well, each providing 30 telephone circuits to ships and providing connections to Inmarsat shore stations through Intelsat's normal frequencies. But this still left spotty service. ESA built upon its ECS experience to propose and develop three Marecs satellites, a maritime version of ECS, produced by the Mesh consortium.

ESA's Marecs A entered service in 1981, offering an additional 40 telephone circuits over the already saturated Atlantic route at 26° west. Inmarsat leased the Marecs A for an initial five-year period for a total of US$65 million. On 1 May 1982 after problems had been overcome with MARECS A telemetry and command systems, a build up of electrostatic charge occurred on the satellite bus. ESA traced the problem and began modifications to Marecs B and ECS 1, to fix the situation. These modifications delayed the MARECS B's launch by 6 months and the satellite foundered in the 10 September 1982 launch failure of Ariane L-05 due to a stage 3 malfunction. ESA declared its cost as US$37.5 million plus US$26.4 million for the launch. ESA paid 160.5 MAU (1983 rate) to complete Marecs A and 107 MAU (1984 rate) for Marecs B.

In March 1983, the USSR brought an Inmarsat station into use covering both Atlantic and Indian Oceans from

Marecs B2 remains operational with Inmarsat's system ***1995***

Odessa, and the service assisted Soviet space tracking ships. Another major ground station followed at Nakhodka with Soviet electronics and Japanese aerials in 1985, to cover the Pacific/Indian oceans. When the new coast Earth station at Jeddah (Saudi Arabia) opened in January 1987, Inmarsat stations had reached 17, which then grew to 20 by October 1987.

BAe Dynamics, as prime contractor for Marecs, received the go ahead for Marecs B2 in October 1979. An Ariane launched it November 1984 into the 177.5° west slot to cover the Pacific from January 1985 through the present. (Intelsat was covering the Indian Ocean). With Marecs A providing 50 per cent more capacity over the Atlantic than its specified capability, and Marecs B2 underemployed over the Pacific, ESA decided to switch the two satellites in the early months of 1986 to better match their respective capacities. By March 1988, B2 functioned as the prime Atlantic link, with its predecessor relegated to back-up status over the Pacific.

By mid-1985, Inmarsat covered the Atlantic with Marecs A at 26° west (Intelsat MCS as a spare at 18.5° west), the Indian Ocean with Intelsat MCS at 63° east (Intelsat MCS as spare at 60° east), and the Pacific with Marecs B2 at 177.5° east (Marisat F3 as spare at 176.5° east). ESA leased the A/B2 to Inmarsat initially for 5 years for a total of US$65 million.

The system underwent its most fundamental rearrangement in autumn 1990 since becoming operational in 1982, in order to support its expansion into land and aeronautical services. On 29 September 1990, it split into two regions, and Marecs B2 moved west from 26° west to 55.5° west during 5 October – 4 November under the control of ESOC at Darmstadt via the Spanish Villafranca station. The move also closed the coverage gap over parts of North America and the Pacific. This new crowding of GEO slots required Inmarsat to certify precise frequency co-ordination to avoid interference with the four satellites it passed en route – three Intelsats at 27.5/34.5/53° west and a PanAmSat at 45° west. Today the satellite remains active at Inmarsat's Atlantic East 15° west slot.

ESA removed Marecs A from Inmarsat Pacific (178° east) service in mid-1991 after high solar activity degraded its array output and then moved it to 22.5° east (inclined 7.7°) where it served as a testbed for mobile communications experiments. Another section of the solar array stopped functioning in July 1996 and ESA abandoned operations at that time. On 21 August 1996, ESA commanded a series of manoeuvres to boost the satellite above the geosynchronous orbit band to retirement.

Inmarsat dropped the lease on Marecs B2 at the end of 1996. ESA still owns the satellite, and from 1 July 1997 and later, began renting it to Telespazio for back-up of the EMS system on Eutelsat-2 F-2.

Marisat and Marecs continued service even after the Inmarsat 2s became operational. Marisat F1, for instance, provided Inmarsat's first polar links in 1993 from its inclined orbit, and the US National Science Foundation's South Pole Station transmitted data for routing through Comsat's Southbury station.

Any ship can use the Inmarsat 1 service, irrespective of its country, or status as a shareholder. To receive the signal, a ship must have the necessary 1 m diameter parabolic antenna inside a protective radome above deck. Panama, with 130 ships on inauguration day, exemplifies the typical non-shareholding user. Inmarsat adjusts shareholdings periodically according to national use of the system.

As of April 1999 Inmarsat satellites were deployed to cover four regions as follows: Atlantic Ocean Region-West (AOR-W) Inmarsat 2 F-4 at 17° west and Inmarsat 3 F-4 at 54° west with Inmarsat 2 F-2 at 55° west as spare; Atlantic Ocean Region-East (AOR-E) Inmarsat 3 F-2 at 15.5° west; Indian Ocean Region (IOR) Inmarsat 3 F-1 at 64° east with Inmarsat 2 F-3 at 65° east as spare; Pacific Ocean Region (POR) Inmarsat 3 F-3 at 178° east with Inmarsat 2 F-1 at 179° east as spare.

Inmarsat 2 series

Background

Inmarsat requested bids for its first dedicated satellites in August 1983 and the consortium awarded a contract to British Aerospace in March 1985 for three satellites based on the Eurostar 1000 platform, with options for a further six. The Inmarsat board authorised a fourth satellite in March 1988. The overall programme cost about US$675 million, excluding cost of financing and TT&C capital/operating costs. Each satellite provides 125 voice channels forward, and 250 return, five times the previous capacity.

Inmarsat finalised an agreement in October 1988 to sell and lease back its first three satellites from the North Sea Marine Leasing Co (a partnership of the leasing arms of four major UK banks: NatWest, Barclays, Lloyds, Midland (now HSBC)) for 10 years. The consortium insured each satellite for up to US$160 million (including an element of self insurance) to cover launch through the first year's operation.

These satellites mark the first time Inmarsat manage its own satellites from the HQ's US$10 million Satellite Control Centre, working through ground stations in Pennant Pt (Canada), Lake Cowichan (Canada), Fucino (Italy) and Beijing (China).

Specifications

Inmarsat 2-F1

Launch: 23.16 GMT 30 October 1990 by Delta 6925 from Cape Canaveral pad 17B. Three apogee burns 1-4 November provided GEO injection at 44° east; arrived 30° east 9 November, where payload activated 13 November for testing. The final move began 26 November, arriving/entering service 8 December, a week earlier than planned. It became a spare at the same location when 3-F1 became operational

Location: 64.4° east GEO over Indian Ocean, relocated to 179°F November 96

Design life: 10 years minimum; estimated BOL sufficient hydrazine for 14 years

Contractors: BAe (prime; assembly, integration/test; thermal; power subsystems), Matra (AOCS), Fokker (solar arrays), Lockheed (propellant tanks), Marquardt (liquid apogee thruster), Hughes (payload, repeaters and, with Spar, antennas)

Transponders: four (plus two back-up) 45 W TWTA 1.621.64/1.53-1.54 GHz L-band global beams, 5+22.5 kHz bandwidths, 39 dBW min EIRP, circular polarisation, to provide 250 voice circuits shore to ship through the 43 cup-dipole radiating elements

one (plus one back-up) 10 W TWTA 6.42 6.44/3.60-3.62 GHz C-band global beams, 5+22.5 kHz bandwidths, 24 dBW min EIRP, circular polarisation, providing 125 voice circuits ship to shore through seven cup-dipole radiating elements. Each ring is separately excited for beam shaping, providing higher power at edge of coverage

Principal applications: maritime, air and land mobile

Configuration: based on box-shaped Eurostar 1000 bus, 2.557 m high, 1.586 m deep, 1.48 m wide, centred on 1.8 m high thrust cone. Payload and solar arrays mounted on North/South panels and Earth-facing floor. The payload uses separate receive/transmission antennas for both L/C band, all four cup-dipole phased arrays mounted on the Earth-facing wall

Mass: 1,385 kg, 824 kg on-station BOL; 130 kg payload mass, 624 kg dry

AOCS: unified redundant 3-axis control and apogee motor thrusters drawing on 110 kg MMH + 130 kg NTO; Kaiser Marquardt R-4D 490 N AKM and two sets of six MBB 10 N thrusters. Roll/yaw control provided by solar sailing with solar array flaps. Single MW with redundancy

Antenna pointing: typically 0.1° half-cone

TT&C: through Inmarsat's London HQ on C-band (control assumed 10 November 1990 from CNES Toulouse facility)

Power systems: 1,200 W BOL/1,142 W EOL provided by twin Si wings, each comprising three 122 cm wide, 173 cm long panels and spanning 15.23 m. Regulated 42.5 V voltage. Eclipse protection provided by two Ni/Cd batteries.

Inmarsat 2-F2

Launch: 8 March 1991 by Delta 6925 from Cape Canaveral pad 17B. Entered service 13 April

Inmarsat 2-F1 EMC test preparations

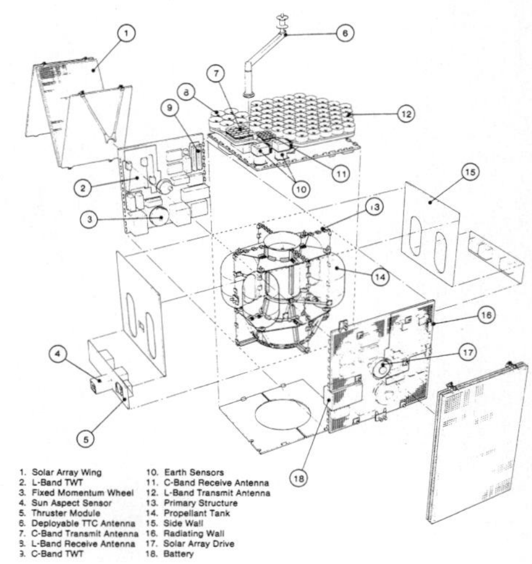

Inmarsat 2 principal features (Inmarsat)

Location: 15.5° west GEO over Atlantic Ocean until December 96, located at 55° west from January 97

Other specifications as for F1.

Inmarsat 2-F3

Launch: 16 December 1991 by Ariane 4 from Kourou. Entered service 19 January 1992

Location: 178.0° east GEO over Pacific Ocean until February 97, relocated to 65° east

Mass: 1,310 kg launch

Other specifications as for F1.

Inmarsat 2-F4

Launch: 15 April 1992 by Ariane 4 from Kourou. Entered service 31 May 1992

Location: 54° west GEO over Atlantic Ocean (tested over 32° west)

Other specifications as for F1.

Inmarsat 3 series

Background

Inmarsat issued a request for a proposal on 2 October 1989 for its third generation of at least three satellites which it then planned to launch in the mid-1990s. The consortium wanted to build a network of global, high-power multiple spot beams for mobile personal communications services, and L-band to L-band connections for direct mobile-to-mobile communications which it expected to be valuable in emergency and search/rescue operations. Responses to the RfP came in during February 1990 from British Aerospace leading a TRW/NEC/Matra team; Martin Marietta Astro Space/MMS UK; Aerospatiale/Alcatel with Ford Aerospace/Mitsubishi; Hughes; and the Indian Space Research Organisation. Inmarsat let the US$320 million contract for four satellites, with options for up to five more, to GE Technical Services Co on behalf of GE Astro Space/Marconi Space on 1 February 1991.

The satellites can reallocate power and bandwidths between the five spots and a single global beam to meet changing traffic requirements. Each satellite provides a capacity 10 times that of Inmarsat 2 and can transmit overlay signals almost identical to those of GPS/Glonass to provide three civil navigation functions: improvement of accuracy, compensation for coverage gaps; and continuously updated, independently monitored integrity monitoring. These functions identify faulty satellites, which is essential for air navigation safety.

COMMUNICATIONS PAYLOAD
1 L-band receiver reflector panel
2 Reflector panel support structure
3 Hinged reflector support strut
4 Strut hinge
5 Launch position shear ties
6 L-band receiver feed assembly with thermal sunshield
7 Gold film multi-layer insulation, applied to exterior of satelite except for North and South radiant panels
8 West feed panel with return combiner
9 East feed panel
10 Feed assembly mounting
11 Feed motor drive electronics
12 L-band output network
13 L-band transmitter feed assembly
14 L-band transmitter reflector
15 C-band receiver antenna
16 C-band transmitter antenna
17 Navigation antenna
18 Graphite/epoxy antenna support structure
19 Pilot tone injection system [PTIS]

NORTH TRANSPONDER PANEL [Z+ Wall]
20 Wall structure [aluminium face panel with honeycomb core]
21 C-band receiver [CMR]
22 Navigation solid state power amplifiers [NAV HPA] [2]
23 L-band solid state power amplifiers [SSPA] [33]
24 Output switches
25 Forward beamforming matrix [BFMJ
26 Power supply electronics [PSE]
27 Power junction assembly [PJA]
28 Return IF processor [RIFP]
29 Forward IF procesor [FIFP]
30 Remote terminal unit [RTU]
31 Frequency generator [FGEN]

SOUTH PAYLOAD AND BATTERY PANELS [Z- wall]
32 Aluminium honeycomb wall structure
33 No 1 beacon transmitter [CBX 1]
34 No 2 beacon transmitter [CBX 2]
35 C-band power amplifiers [CHPA]
36 Power amplifier output switch
37 Power amplifier input switch
3e Battery panels [2]
39 Nickel-Hydrogen batteries [46]
40 Low emmisivity gold plated battery end caps
41 High emmisivity black painted battery centre sec tions
42 Battery pressure monitor [BPM]
43 Aluminium honeycomb wall structure
44 North and South panels optical solar reflector external radiant mirror surfaces
45 Battery control voltage monitor [BVCM]

TELEMETRY TRACKING AND COMMAND SYSTEM
46 TTC Omni-antenna
47 Omni-antenna hinged mounting [ODS]
48 Telecommand lOgic processor [TCP]
49 Central logic extender [CLX]
50 Redundant telemetry module [RTM]
51 Pyrotechnic relay assembly 1 [PRA 1]
52 Pyrotechnic relay assembly 2 [PRA 2]

ATTITUDE AND ORBITAL CONTROL SUB-SYSTEM [AOCS]
53 Roll control torquer
54 Yaw control torquer
55 Roll rate measuring assembly
56 Yaw rate measuring assembly
57 Pitch rate measuring assembly
58 Attitude system processor [ASP]
59 Momentum wheel assembly 1 [MWA 1]
60 Momentum wheel assembly 2 [MWA 2]
61 Horizon sensing assembly [HSA]
62 Earth sensor assembly 1 [ESA 1]
63 Earth sensor assembly 2 [ESA 2]
64 Sun sensor detector 1 [SSD 1]
65 Sun sensor detector 2 [SSD 2]

REACTION CONTROL SUB-SYSTEM
66 Catalytic rocket engine assemblies [REA'S] [12], attitude control
67 Improved performance electrothermal hydrazine thrusters [IMPEHT's] [4], North - South control
68 IMPEHT power electronics
69 Hydrazine propellant tanks [4] with helium pressurant
70 Propellant tank aluminized Kapton multi-layer insulation
71 Hydrazine feed lines to thrusters
72 Propellant fill and drain valves
73 Tank sump, propellant management device [PMD]
74 Tank support struts

BUS MODULE
75 Cylindrical core structure
76 Radial bulkheads/shearwalls
77 Launch vehicle interface/separation ring
78 STAR 37XFP solid propellant apogee kick motor [AKM]
79 Carbon phenolic exhaust cone
80 AKM mounting/thrust ring frame
81 Plume shield

SOLAR ARRAY
82 Single-sided solar panels, inner - mid and outer
83 Outer solar panel provides trickle charge for batteries prior to full deployment
84 6cm x 4cm solar cells
85 Interpanel hinges
86 Hinged berrylium Support boom
87 Solar array boom hinges
88 Solar array launch Position latch
89 Solar array drive motor [SAD]

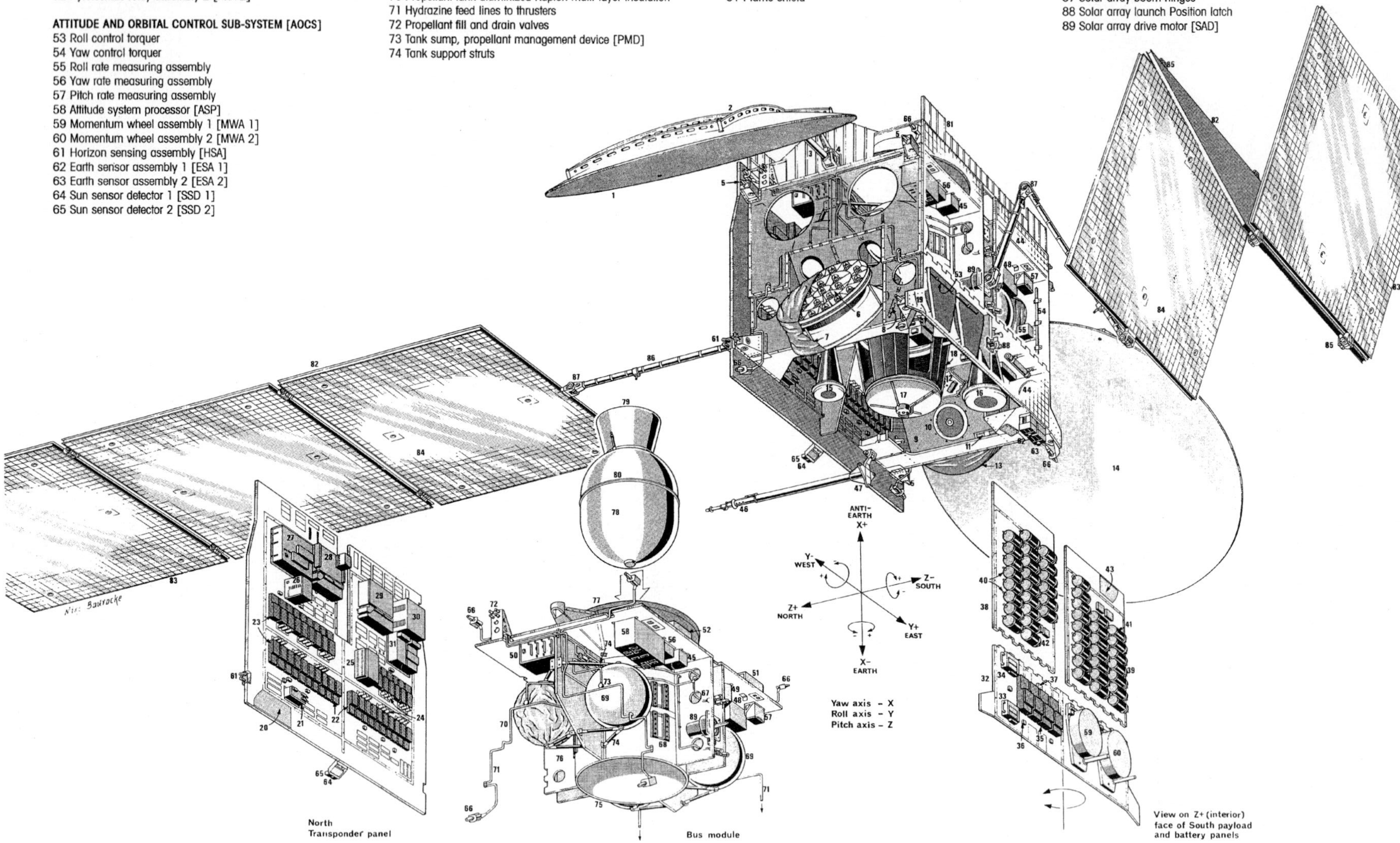

Inmarsat 3 principal features (Inmarsat/Mike Badrocke)

1996

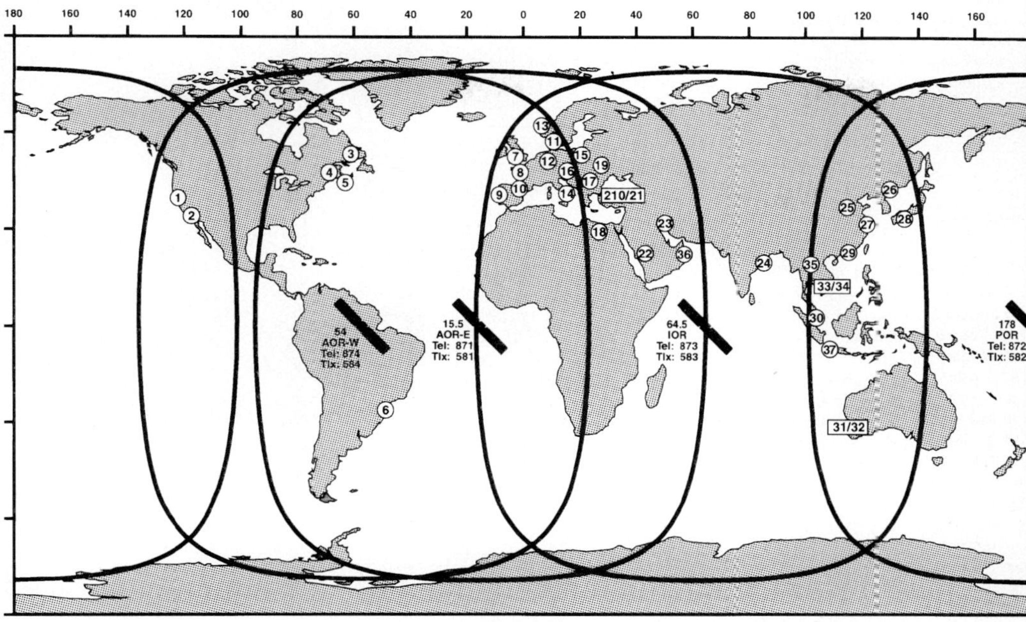

System Availability	Inmarsat-A	Inmarsat-C	Inmarsat-M/B	Inmarsat-Aero
1. Niles Canyon	●			●
2. Santa Paula	●	●	●	●
3. Laurentides			●	●
4. Southbury	●	●	●	●
5. Staten Island	●			
6. Tangua	●	●		
7. Goonhilly	●	●	●	●
8. Pleumeur Bodou	●			
9. Sintra		●		
10. Aussaguel		●	●	●
11. Blaavand		●		
12. Burum	●	●	●	
13. Fucino	●	●	●	
14. Eik	●	●	●	●
15. Psary	●			
16. Raisting	●	●	●	
17. Thermopylae	●	●	●	
18. Maadi	●			
19. Odessa	●			
20. Anatolia	●			
21. Ata	●	●		
22. Jeddah	●			
23. Boumehen	●	●		
24. Arvi	●	●	●	
25. Beijing	●	●	●	
26. Nakhodka	●			
27. Kumsan	●	●		
28. Yamaguchi	●	●	●	●
29. Hong Kong	●		●	
30. Singapore	●	●	●	●
31. Perth	●	●	●	●
32. Gnanagara	●			
33. Comsat Eurasia			●	
34. Kuantan			●	
35. Nonthaburi		●	●	
36. Towi Al Saman			●	
37. Jafilahur			●	

Inmarsat system stations and their capabilities ***1996***

Inmarsat has tentatively allocated its transponders to Comsat for the Pacific and West Atlantic satellites, France Telecom/Comsat for East Atlantic and to Deutsche Telekom for Indian Ocean service. Inmarsat plans to have 1 million terminals in operation by 2000.

Specifications

Inmarsat 3-F1
Launch: 3 April 1996 by Atlas 2A from Cape Canaveral
Location: Initially 28° east for testing, moved to operational location of 64° east in May 1996
Design life: 13 years
Contractors: Lockheed Martin Astro Space (prime), MMS UK (communications payload), Olin Aerospace (hydrazine thrusters)
Transponders: 490 W SSPA matrix 1.63-1.66/1.53-1.56 GHz up/down L-band global + five spot beams, 34 MHz bandwidth, EIRP 48 dBW total (switchable between global/spot in any proportion of total EIRP), circular polarisation

two (plus two back-up) 12 W SSPA 6.43 6.45/3.60-3.63 GHz up/down C-band global beams (feeder for L-band), 29 MHz bandwidth, EIRP 27 dBW in each of two circular polarisations

one (plus one back-up) 20 W SSPA 1,575.42 MHz GPS L1 L-band global beam, 2.2 MHz bandwidth, EIRP 27.5 dBW (translated from C-band uplink). GPS/Glonass overlay signals compensating for coverage gaps and enhancing their accuracy. Also, continuously updated information on these systems' integrity broadcast to identify faulty satellites. A C-to-C communication channel used by the uplink station for more precise signal tracking (less susceptible to ionosphere disturbances than L-band); channel can also be used for precise time and frequency applications
Principal applications: maritime, air and land mobile
Configuration: based on Astro Space boxed-shaped 4000 bus, body 2.1 × 1.8 × 1.7 m, 2.5 m high, 3.2 m radial envelope, 16.7 m span, centred on thrust cone. Payload and solar wings mounted on N/S panels. L band receive/transmission reflectors mounted on E/W panels and fed by cup-shaped radiating elements
Mass: 2,068 kg at launch, 1,100 kg on-station BOL, 190 kg payload, 860 kg dry
AOCS: 3-axis momentum bias stabilisation with MWs and magnetorquers; pivoted MW provides antennas pointing for up to 2.7° inc. Mono hydrazine thrusters, 283 kg capacity, 12 Reaction Engine Assemblies + four Improved Electrothermal Hydrazine Thrusters. GEO insertion by Thiokol Star 37XFP or 37FM
Power system: 2.8 kW EOL provided by twin Si wings (six panels total 30.5 m²) spanning 16.7 m. Eclipse protection by 2 × 62 Ah Ni/H_2 batteries.

Mobile communication services provided by Inmarsat P satellites utilising the Hughes HS-601 bus
(Hughes Space & Communications) ***1999***/0054339

Inmarsat 3-F2
Launch: 6 September 1996 on Proton from Tyuratam
Location: 15.5° west Atlantic Ocean
Other specifications as for F1.

Inmarsat 3-F3
Launch: 18 December 1996 on Atlas-2A from Kourou
Location: Initially tested over 158° east, then relocated to operational 178° east January 1997
Other specifications as for F1.

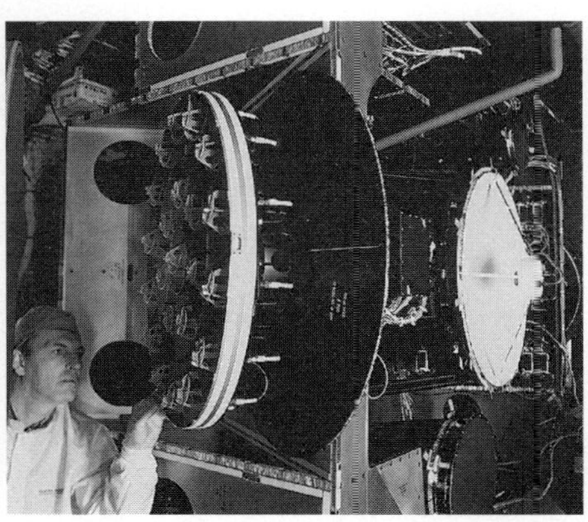

Inmarsat 3 antenna feed assembly. At right is the navigation antenna (MMS) ***1995***

Inmarsat 3-F4
Launch: 3 June 1997 on Ariane-4 from Kourou
Location: 54-55° west Atlantic Ocean
Other specifications as for F1.

Inmarsat 3-F5
Launch: 4 February 1998 on Ariane 4 from Kourou
Location: 28° east
Other specifications as for F1.

Intelsat 1 (Early Bird)

Background

Description: 70 cm diameter, 60 cm height, 68 kg at launch, 38.5 kg after apogee motor burn. Capacity 240 telephone circuits (triple the trans-Atlantic capacity of the time) or one TV channel. The single Early Bird was launched 6 April 1965 and became operational over the Atlantic at 325° east on 28 June 1965. The world's first commercial communications satellite, it made scheduled trans-oceanic TV possible for the first time, its antenna covering North America/Europe with a 6 W transponder, 50 MHz bandwidth. Design life 18 months, but operated for 3½ years; reactivated in 1984 in celebration of Intelsat's 20th anniversary.

Intelsat 2 series

Background

Description: Extended coverage to two-thirds of the world, though only the last two achieved the planned 3-year life. 162 kg at launch, 86 kg after apogee burn. F1, 26 October 1966, failed. F2, launched 11 January 1967 was operational for 2 years over Pacific. F3, launched 22 March 1967, was operational 3½ years over Atlantic. F4, launched 27 September 1967 was operational 3½ years over Pacific.

Intelsat 3 series

Background

Description: Capacity increased to 1,200 circuits or four TV channels, with 5 years design life (not achieved). The satellite weighed 293 kg at launch and 151 kg after apogee burn. F1, 18 September 1969, failed. F2, launched 18 December 1968, was operational 1½ years over Atlantic. F3, launched 5 February 1969, was partially operational 7 years over Indian Ocean. F4, launched 21 May 1969, was operational 3 years over Pacific. F5, 25 July 1969, failed. F6, launched 14 January 1970, was operational 2 years over Atlantic. F7, launched 22 April 1970, was operational 16 years over Atlantic. F8 failed. Starting with this series, satellites were boosted above GEO if possible at EOL.

Intelsat 4 series

Background

Description: Capacity increased to 4,000 circuits and two TV channels, with 7 years design life. 1,414 kg, reducing to 730 kg on station. F2, launched 25 January 1971, was operational 12 years over Atlantic. Raised above GEO and deactivated 1983. F3, 19 December 1971, primary Atlantic 3 years, operational >12 years, placed in orbital storage 1983. F4, 22 January 1972 (first commercial satellite assembled outside US, by British Aircraft Corp at Bristol, UK; provided TV link between US/China for President Nixon's visit to Peking), operational over Pacific >7 years, raised above GEO orbit and deactivated 1983. F5, 13 June 1972, operational over Indian Ocean >7 years, raised above GEO and deactivated 1983. F7, 23 August 1973, major-path satellite Atlantic 2 years, operational >6 years, retired and placed in orbital storage 1983. F8, 21 November 1974, operational Pacific from December 1974, in use for international and domestic lease 1983. F6, 20 February 1975, failed to achieve transfer orbit. F1, 22 May 1975 after 4 years storage, operational Indian Ocean 3 years, moved to Atlantic as major-path satellite 1979, domestic lease service 1983.

Intelsat 4A series

Background

Description: Capacity increased to 6,250 circuits plus two TV channels, with 7 years life. The satellite weighed 1,515 kg at launch and 825 kg after apogee motor burn. F1, launched 25 September 1975, was the primary Atlantic satellite from February 1976. F2, 29 January 1976, major-path Atlantic 2 years from 1976, then leased. F4, launched 26 May 1977, leased for Atlantic until December 1977, then major-path satellite. The last fourth-generation satellite to be removed from service, 17 August 1989. F5, 29 September 1977, launch failure. F3, 6 January 1978, leased over Indian Ocean until 1979, then primary satellite.

Intelsat 5/5A series

Background

Intelsat began its new quasi-public communication service by purchasing the basic Intelsat 5 satellite model (501-509) which it enhanced by the addition of three cross-polarised spot beams at 6/4 GHz for domestic leased services. The last three of the six 5As operate at both 14/11 GHz and 14/12 GHz to facilitate Intelsat Business Services. The organisation ordered nine satellites in total at US$300 million. Launches included: 501 (Atlas Centaur) 23 May 1981; 502 (Atlas Centaur) 6 December 1980; 503 (Atlas Centaur) 15 December 1981; 504 (Atlas Centaur) 4 March 1982; 505 (Atlas Centaur) 28 September 1982; 506 (Atlas Centaur) 19 May 1983; 507 (Ariane) 18 October 1983; 508 (Ariane) 4 March 1984; 509 (Atlas Centaur) 9 June 1984 (launcher failure); 510 (Atlas Centaur) 22 March 1985; 511 (Atlas Centaur) 29 June 1985; 512 (Atlas Centaur) 28 September 1985; 514 (Ariane) 30 May 1986 (launcher failure).

Each satellite has three-axis stabilisation rather than spin stability. They carry 12,250 circuits plus two TV channels, with 7 year life. On station each weighed 1,012 kg on station and had 19.9 m solar wing span. 505-509 carried maritime packages for lease to Inmarsat which carried the equivalent of 30 phone calls in L-band (1.5 to 1.6 GHz).

Intelsat ordered six 5As from Ford Aerospace in January 1981 and contracted four launches to Atlas Centaur and two to Ariane. By the end of the production run Intelsat increased capacity to 15,000 2-way circuits, plus two TV channels and made improvements to the basic Intelsat 5 package through three additional cross polarised spot beams at 6/4 GHz for domestic leased services. The last two Series 5 satellites operate both in the 14/11 and 14/12 GHz bands to provide the Intelsat Business Service.

As of May 1999, Intelsat 505 is operating at 72° east, 510 is operating at 33° east and 511 is operating at 29.5° west. The remaining 11 satellites that successfully reached orbit have been retired. Intelsat 505 has now been operating for more than 16 years.

Specifications

Design life: 7-9 years
Contractors: Ford Aerospace (prime), Aerospatiale (modular structure), Thomson-CSF (11 GHz TWTs), MBB (largest subcontractor, at 10 per cent; AOCS, solar array)
Transponders: 32 (plus 12 back-up) 8.5 W TWTAs 5.9 6.4/3.7-4.2 GHz C-band hemispherical zonal and spot beams, 36/72 MHz bandwidth, EIRP 29 dBW (hemisphere), 29 dBW (zone), 32.5 dBW (spot), dual polarisation

6 (plus 4 back-up) 10 W TWTAs 14.0 14.5/10.9-11.7 GHz Ku-band spot beams, 77 MHz bandwidth, 41-44 dBW EIRP, dual polarisation satellites 505-509 carry maritime packages for lease to Inmarsat, each capable of carrying the equivalent of 30 telephone circuits. L-band: 1.6365-1.644/1.535- 1.5425 GHz up/down, 33 dBW EIRP. C-band: 6.4175-6.425/ 4.1925-4.200 GHz up/down, EIRP 21.0 dBW
Principal applications: Intelsat services
Configuration: 3-axis box-shaped bus with tower and 15.6 m span solar wings; total height 6.4 m. Panels provide 1,300/1,271 W BOL/EOL. Thiokol solid AKM
Mass: 1,998 kg at launch, 1,188 kg on-station BOL
TT&C: through SCC at Intelsat HQ.

Intelsat 6 series

Background

Industry proposals for a new communication satellite with three times the Intelsat 5 capacity, went out to industry in 1980. Many international consortia bid on the request led by Hughes and Ford. Intelsat closed the competition in April 1982 and awarded a US$700 million contract to Hughes for five satellites, with first launch in 1986. The contract called for an option on another six, which would bring the contract to US$1,300 million.

Hughes proposed a spin-stabilised hybrid that offers 33,000 2-way phone circuits and four TV channels. Originally Hughes designed the satellite for launch by Shuttle or Ariane 4; and the first (602) flew aboard an Ariane on 27 October 1989 and entered full service 7 April 1990.

In full service, the five satellites ordered provide the equivalent capacity of up to 120,000 simultaneous 2-way telephone circuits using digital modulation techniques and at least three TV channels utilising 6/4 and 14/11 GHz communication links. Intelsat became the first commercial satellite to employ SS/TDMA Satellite Switched/Time Division Multiple Access, allowing beam interconnections according to traffic requirements. The satellites reuse the C-band frequencies six times through two hemispherical beams and four zone beams with dual circular polarisation. The K-band frequencies undergo reuse twice by spatially isolated east west spots using linear orthogonal polarisation. All provide higher K-band power to promote the use of small Earth stations.

Intelsat launched the first (602) on 27 October 1989 on an Ariane and it entered service 7 April 1990. Titan 3 stranded the second (603) in LEO following launch on 14 March 1990. Shuttle flight STS-49 carried a propulsion module to the stranded satellite and re-boosted it to a correct orbit May 1992, providing a 10 year life. 604 went to space on a Titan 23 June 1990, followed by 605 (14 August 1991 by Ariane) and 601 (29 October 1991 by Ariane). 604's success allowed

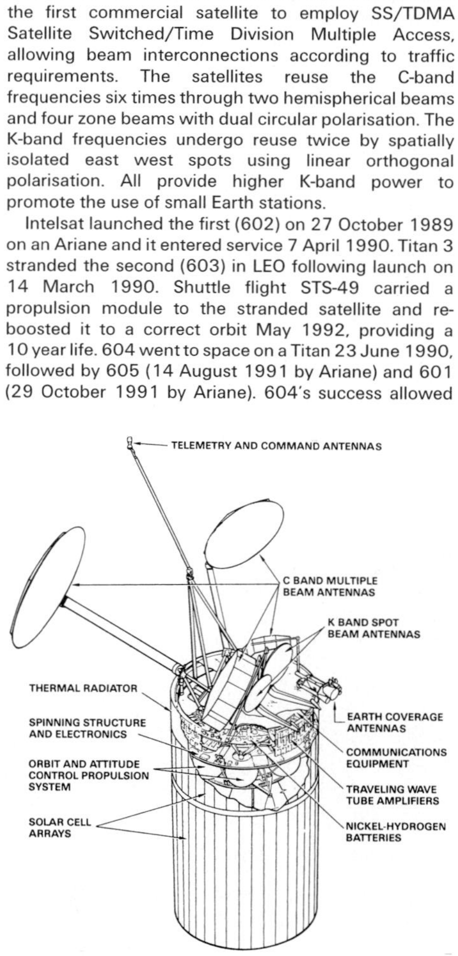

Principal Intelsat 6 features (Intelsat)

Intelsat 6 communications platform
(Hughes Space & Communications)

Intelsat to move six satellites in the second half of 1990, followed by a further six during 1991, in two batches after 605/601 appeared. In May 1999 all five satellites were still operating from the following locations: 601, 325.5° east; 602, 62° east; 603, 335.5° east; 604, 60° east; 605, 332.5° east.

Specifications

Design life: 13 years min 602/3; 15 years 605
Contractors: Hughes Space & Communications Co (prime), British Aerospace (Ku/C-band reflectors, power electronics, Ariane adapter), Pilkington (solar cell covers), Alcatel Espace (Ku/C- band receivers, output multiplexer), Thomson-CSF (Ku-band TWTs), Alenia Spazio (spot antenna, telemetry transmitters/receivers, remote/central telemetry units), AEG (solar cells), MBB/ERNO (solar panels, contract worth about DM50 million), NEC (Ku-band receivers, SSPAs, master oscillators)
Transponders: 8 (plus 3 back-up) 20 W and 2 (plus 1 back-up) 40 W TWTA 14.0-14.5/10.95-11.2 + 11.45-11.70 GHz Ku-band steerable spots, 72 MHz bandwidth, EIRP 45 dBW min, dual polarisation

6 (plus 2 back-up) 16 W TWTA 5.85 6.42/3.62-4.20 GHz C-band hemispherical and zone beams, 72 MHz bandwidth, EIRP 30 dBW min, dual polarisation

32 (plus 10 back-up) 10 W SSPA 5.85 6.42/3.62-4.20 GHz C-band hemispherical and zonal beams, 72 MHz bandwidth, EIRP 30 dBW min, dual polarisation
Principal applications: Intelsat services
Configuration: Hughes' HS-393 platform spin-stabilised, cylindrical body 3.63 m diameter, deployed height 11.84 m (height aboard Ariane 4 6.4 m, height with PKM 8.06 m). Each 3.2 m (30 kg carbon, Kevlar/boron fibre composite) + 2.0 m C-band antenna provides two fixed beams for hemispherical coverage and four isolated beams for zone coverage (these can be reconfigured in orbit to match the requirements of the three ocean regions). Steerable 1.1 m + 1.0 m diameter Ku-band reflectors provide spot coverage
Mass: 4,600 kg at launch (includes adapter), 2,546 kg on station BOL, 1,896 kg dry
Attitude control: 30 rpm spin-stabilised; thrusters fired automatically when attitude excursions exceed pre-set limits
Antenna pointing: ±0.05° maintained in beacon tracking mode, ±0.10° with Earth/Sun sensors
TT&C: through Intelsat's SCC HQ station
Power system: drum-mounted solar arrays provide 2.6 kW BOL/2.252 kW EOL; K7 cells on skirt, K4¾ on fixed body. Two Ni/H_2 batteries provide eclipse protection
Propulsion: perigee kick from Titan 3 or Shuttle provided by 8.76 t UTC/CSD Orbus 6S solid propellant stage. ACS unified with apogee kick: MMH/NTO supply four 22 N radial thrusters for E-W station-keeping + spinup/down, two 22 N axial for NSSK + attitude control, and two 490 N for apogee kick + reorientation manoeuvres.

Intelsat 7 series

Background

The Intelsat Board requested proposals in October 1987 for fixed price bids for 2-3 satellites, with an additional bid for a fourth, for a new generation of Pacific region satellite. Intelsat expected to put the satellites in service in late 1992 or early 1993. Previous satellites had met Atlantic region requirements; which differed from Pacific service. Space Systems/Loral received the contract in June 1988. Matra served as a back-up

The first Intelsat 7 entered service in 1994 (Intelsat)

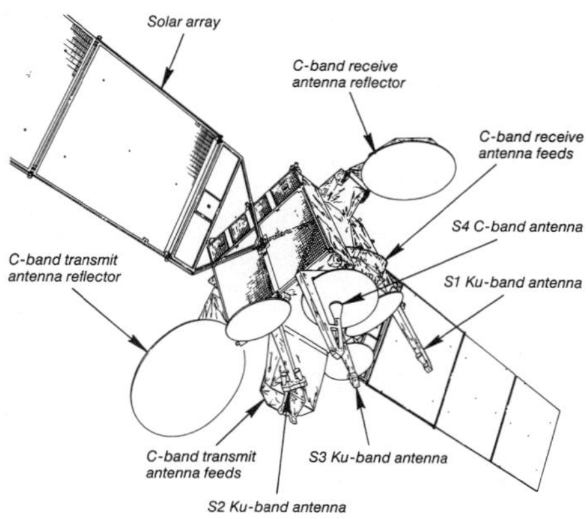

Intelsat 7 configuration (Space Systems/Loral)

source. On 9 September 1988, Intelsat awarded a US$394.28 million contract for five series 7 satellites with the first launches in mid-1992 and early 1993.

The satellites, combining an Intelsat 5-type payload with Ford's FS-1300 bus, offer increased power and coverage for smaller ground stations and flexibility through three independently steerable Ku-band spots and a steerable C-band spot. Intelsat decided in December 1989 to increase the steerability and Ku-band power of F3 and its successors. The satellite operates inverted at 66° east/1° west, simplifying antenna design. General Dynamics' Atlas 2AS has launched three, charging Intelsat US$75 million each. ArianeSpace's Ariane 4 launched two at US$86.5 million each.

Intelsat awarded General Dynamics a US$9.5 million contract in 1991 to pay for enhancing Atlas 2AS capacity into GTO by 121 kg to give the new satellite additional propellant providing a 16 year life. Intelsat decided in March 1993 to order a sixth satellite (709) and completed the order September 1993.

Specifications

Launch schedule: 701 22 October 1993 (Ariane 44LP; entered service 15 January 1994), 702 17 June 1994 (Ariane 44LP), 703 6 October 1994 (Atlas 2AS), 704 10 January 1995 (Atlas 2AS; entered service 25 February), 705 (Atlas 2AS) 22 March 1995 (Atlas 2AS), 709 15 June 1996 (Ariane 44P)
Locations: 701 180° east, 702 177° west, 704 66° east, 705 342° west, 706 307° east, 707 359° east, 709 18° west
Design life: 65% transponder operability after 10.9 years; sufficient propellant for 15 years
Contractors: Space Systems/Loral (prime), Alcatel Espace (payload), NEC (20/30 W SSPAs), MELCO (10/16 W SSPAs), AEG (TWTs)
Transponders: 26 SSPA (10, 16, 20, 30 W) in 7-for-5 rings 5.8-6.5/3.6-4.2 GHz up/down C-band global (26-29 dBW), hemispherical + zonal (33 dBW) and spot (33-36 dBW), 36/72 MHz bandwidth, dual polarisation

5 × 35 W (7-for-5) + 5 × 50 W (8- for-5) TWTA 14.0-14.5/11.0-11.2 + 11.5-11.7 GHz Ku-band spot beams, 44-47.5 dBW EIRP, 6 × 72 MHz + 4 × 112 MHz (allows 155 Mbit/s) bandwidth, dual polarisation
Configuration: box-shaped SS/L FS-1300 bus with Intelsat 5 type communication platform with three steerable Ku-band dishes and one C band, in addition to fixed C-band receive (1.57 m diameter) + transmission (2.44 m diameter)
Mass: 1,450 kg dry mass, 3,650 kg GTO (for Ariane, with 2,160 kg propellant load), typically 1,800 kg on-station BOL
AOCS: 3-axis (from immediately after launcher separation). 12 × 22 N N_2O_4/MMH, 490 N thruster permits Perigee Velocity Augmentation during Atlas orbital insertion, adding 2.1 years to life. Propellant loading 2,160 kg for 92% fill
TT&C: through Intelsat's SCC HQ station
Power system: twin 3-panel solar wings, spanning 21.843 m, providing 3.9 kW after 10.9 years. Supported by four Ni/H_2 batteries totalling 85.5 Ah.

Intelsat 7A series

Background

Intelsat authorised the purchase of two 7A satellites in December 1990 at a total fixed price cost of US$202.4 million (33.3 per cent in-orbit performance incentive to replace the 5As over the Atlantic during 1995 and 1996. Almost identical to the 7 model, they offer increased Ku-band capacity and higher power global transponders. The increased mass requires a more powerful Ariane booster. Each new satellite costs about US$85 million each. Intelsat ordered a third 7A (708) from Space Systems/Loral in September 1992 for US$140 million to keep up with demand. China supplied a US$60 million CZ-3B launcher, but 708 was lost in the February 1996 failure of the Long March.

Specifications

Launch schedule: 706 17 May 1995, 708 14 February 1996 (launch failure), 707 14 March 1996
Locations: 706 307° east, 707 359° east (708 was planned for 50° west)
Transponders: C-band global EIRP 29.0 dBW, C-band spot 36.1 dBW

Ku-band 35 W + 50 W TWTAs replaced by 49 W + 73 W, respectively, and can be paralleled for 49.5 dBW EIRP. Four wideband added for total of 6 × 72 MHz + 8 × 112 MHz
Configuration: body stretched about 45 cm for additional Ku transponders, increased radiator area and stretched propellant tanks
Mass: 1,748 kg dry mass, 4,180 kg GTO (for Ariane 44LP/H10- 3, with 2,820 kg 100% propellant load)
Power system: twin 4-panel solar wings with improved cells, providing 5.3 kW after 10.9 years. Supported by Ni/H_2 batteries totalling 120 Ah.

Intelsat 8 series

Background

Intelsat issued a RfP in December 1991 for three new satellites, with options for a further four if required. Instead of the full purchase, it allocated funds for a third 7A from SS/L, while Lockheed Martin's 7000 platform received the contract for the two new generation 8 satellites at US$165 million. Generation 8 satellites feature a design to meet the growing Pacific region need, using 6-fold C-band and 2-fold expanded C-band frequency reuse.

Intelsat selected Ariane 4 boosters to launch the satellites and announced the decision in December 1992, then it decided to exercise options for three further satellites, including the first 8A (805) landmass satellite. The contract provides the satellites at a favourable price as part of the settlement with Martin Marietta Astro Space over the loss of Intelsat 603.

Specifications

Launch schedule: 801 (Ariane 44LP) 27 February 1997; 802 (Ariane 44LP) 25 June 1997; 803 (Ariane 42L) 23 September 1997; 804 (Ariane 42L) 22 December 1997
Locations: 801, 328.5° east, 802, 174° east, 804, 64° east. The rest are presently inoperable
Design life: 10 years (13-19 years propellant)
Contractors: Lockheed Martin Astro Space
Transponders (801-804): 18 27-38 W variable power SSPA C band, providing 6 global (EIRP 29.0 dBW) + 12 hemispherical (EIRP 34.5-36.0 dBW) bandwidths 34, 36, 41, 72, 77 MHz

20 10-20 W variable power SSPA C- band zonal beams, EIRP 34.5-36.0 dBW, bandwidths 34, 36, 41, 72, 77 MHz

The first Intelsat 8 was launched in 1997 (Intelsat)
1997/0005566

6 43 W TWTA in two Ku-band steerable spots (2.8° circular + 1.9 × 4.3°), EIRP 44/47 dBW inner/outer, 602 MHz bandwidth (as 34, 72, 77, 112 MHz)
Principal applications: Intelsat services
Configuration: 7000 series platform. 3-axis box-shaped 216 × 246 × 315 cm bus with central cylinder
Mass: 3,245 kg at launch, 1,538 kg dry
AOCS: spin-stabilised in transfer orbit; 3-axis on-station using roll/yaw magnetorquers in normal mode, hydrazine thrusters for station-keeping. Redundant 1.8 kW hydrazine arcjets provide NSSK. Pointing: ±0.1° pitch/roll, ±0.25° yaw accuracy; ±6.0° pitch + ±2.0° offset pointing control
TT&C: through Intelsat's SCC HQ station
Power system: twin 4-panel wings (totalling 48.6 m^2) providing 4,800 W BOL. 2 × 45 Ah Ni/H_2 batteries provide 100% eclipse protection.

Intelsat 8A series

Background

Intelsat, in September 1993, exercised its option on the generation 8 contract and purchased the first 8A (805) landmass satellite. It will provide expanded C/Ku-band coverage for Asia-Pacific users, with a minimum 36 dBW EIRP C-band and fully-manoeuvreable Ku beams to cover all of China, Korea and most of Japan. Intelsat ordered 806 in March 1994 for US$72.3 million, plus US$10 million in-orbit performance incentives, to cover the Americas.

Specifications

Launch: 806 28 February 1998 (Atlas-2AS), 805 18 June 1998 (Atlas2AS)
Locations: 805 304.5° east, 806 319.5° east
Transponders: 28 60 W TWTA C-band broad hemispheric beams, 805 EIRP 37.2/36 dBW inner/outer (806 37.5 dBW). 3 130 W TWTA Ku-band steerable shaped beams, 805 EIRP 49/42 dBW inner/outer (806 48 dBW)
Mass: 3,570 kg at launch, 1,595 kg dry
Other specifications as for Intelsat 8.

Intelsat 9 series

Background

In December 1996, Intelsat management chose to procure two follow-on spacecraft (FOS-11) from Space Systems/Loral, which it wanted to operate over the Indian Ocean as replacement for Intelsat 6 satellites. Renamed Intelsat 9, the programme has now been expanded to include five satellites.

Digital voice and data services will be provided in both C-band and Ku-band. Intelsat is investing approximately $1 billion in the first few primary satellites with a fifth under contract for back-up or reassignment.

Key parameters include 32 × 72 MHz and 12 × 36 MHz C-band transponders with uplink frequency of 5.850 to 6.425 GHz and downlink frequency of 3.625 to 4.200 GHz. Typical e.i.r.p range is 43.0-37.0 dBW for the hemi-beam, 44.8-37.0 dBW for the zone beam and 34.0-31.0 dBW for the global beam. In addition, the satellites each have 8 × 72 MHz and up to 8 × 36 MHz Ku-band transponders. Uplink frequencies are in the range 14.00-14.50 GHz with downlink frequencies of 10.95-11.20 GHz and 11.45-11.70 GHz. Typical e.i.r.p. ranges are 52.0-47.0 dBW for

Spacecraft delivery and launch periods:

Spacecraft	Launch Vehicle	Spacecraft Delivery	Launch	Orbit Location
Intelsat 901	Proton/Ariane	July 2000	3rd Qtr 2000	60°E (IOR)
Intelsat 902	Ariane/Proton	November 2000	4th Qtr 2000	62°E (IOR)
Intelsat 903	Ariane	March 2001	1st half 2001	335.5°E (AOR)
Intelsat 904	Ariane	July 2001	2nd half 2001	325.5°E (AOR)
Intelsat 905	Undetermined	November 2001	1st half 2002	Undetermined

Spot 1 and 52.0-47.0 dBW for Spot 2. In all, Intelsat 9 has 60 transponders. Carrying the target satellite capacity in the Intelsat system, Intelsat 9 will have the flexability and power to provide DTH services to 60 cm antennas.

Intelsat K satellite

Background

In 1989, Intelsat purchased the partially completed GE Satcom K4, with extensive payload modifications (coverage, frequencies and switching matrix), to guard against potential transponder shortages. It is Intelsat's first all-Ku satellite and the first the organisation bought from an outside source. It operates in collocation with Intelsat 512 (C-band) over the Atlantic. The satellite cost Intelsat an estimated US$80 million for procurement and modifications, plus US$10-15 million for on-time delivery and in-orbit performance incentives.

Specifications

Launch: 10 June 1992 by Atlas 2A from Cape Canaveral. Commercial service began 20 August 1992
Location: 21.5° west geostationary (collocated with Intelsat 502 into 1994; then 512 from September 1994) now Intelsat 803
Design life: 10 years (10-12 years station-keeping propellant)
Contractors: GE Astro Space Division (through GE Technical Services Co)
Transponders: 16 (2 rings, 11-for-8) 62.5 W TWTA 14.00 14.50/11.45-11.70 (all beams), 11.70-11.95 (North/South America), 12.50 12.75 (Europe) GHz up/down Ku-band beams providing 47 dBW edge (50 dBW centre) over North America + Europe and 45 dBW edge over South America (42.7 dBW North/South America in split mode), 54 MHz bandwidth, orthogonal linear polarisation. In normal mode, North America/Europe covered by 8 transponders each. Eight North America downlinks can be switched between North America/South America downlinks, or operate in a split North America/South America mode. Four of these can also be switched to Europe (making 12 in all). South America coverage is a satellite transmit-only service
Principal applications: primarily video (27 MHz) and business services. The high EIRP/gain are designed for small Earth stations (<1 m for receive only). Italian/German TV/radio began DTH to 66-88 cm dishes in the Americas in 1992
Configuration: GE 5000 series platform with a 3-axis box-shaped bus with central cylinder. 90 × 80 cm fixed reflector dish and tower on Earth face for South America coverage and deployable 2.15 m dual gridded reflector on west face for North America and Western Europe coverage
Mass: 2,836 kg at launch, 1,547 kg BOL on-station, 1,219 kg dry
AOCS: spin-stabilised at 10 rpm in transfer orbit; 3-axis on-station using roll/yaw magnetorquers in normal mode, hydrazine thrusters for station-keeping. Two 445 N dual mode liquid apogee motors for GEO insertion using NTO/MMH. Redundant branches of eight catalytic hydrazine thrusters for AOCS (each 1 × 4.45 N, 5 × 0.89 N, 2 × 22.2 N) plus redundant 2 × 0.44 N EHT for station-keeping
TT&C: through Intelsat's SCC HQ station
Power system: two 15.6 m span solar wings of four panels each (totalling 34.70 m²) provides 4.8 kW BOL/3.5 kW EOL (10 years). 4 × 50 Ah Ni/H_2 batteries provide 100% eclipse protection and EHT firing support.

Marisat series

Background

The United States government decreed in early 1963 that it would form a global commercial satellite communication service in conjunction with other countries. Shares in the new company became available in February 1963 as the public offering of the Communications Satellite Corporation (Comsat). Comsat General introduced pre-Inmarsat maritime service through its three Marisats, positioned over the Atlantic, Pacific and Indian oceans during 1976, and then leased them to Inmarsat when that organisation became active in February 1982. By 1980, Marisat had 300 commercial users, including cargo ships, oil tankers and offshore platforms. The US and UK navies both used the system and Comsat reserved three UHF channels for US government use. Despite their five year design lives, all three Marisats continued into 1995 as Inmarsat spares over the ocean regions. The US National Science Foundation's South Pole Station transmits data for routing through Comsat Southbury station.

Intelsat K is collocated at 21.5° west with Intelsat 512

Specifications

Marisat 1

Launched: 19 February 1976 by Delta from Canaveral
Location: 106° west geostationary, inclined 12.2°
Design life: 5 years
Contractors: Hughes Space & Communications
Transponders: single L-band 1.6385-1.642/1.537-1.541 GHz up/down, EIRP 27 dBW with a 4 MHz bandwidth. Single C-band 6.420-6.424/4.195-4.199 GHz up/down, EIRP 18.8 dBW, 4 MHz bandwidth. Three UHF channels
Principal users: spare in Inmarsat's maritime system
Configuration: spin-stabilised cylindrical platform with despun antenna section; height 231 cm. Solar array of 7,000 Si cells on cylindrical surface provided 330 W BOL
Mass: 655 kg at launch, 362 kg on-station BOL
TT&C: Comsat's Washington facility working through the Southbury, Connecticut and Santa Paula, California stations.

Marisat 2

Launched: 10 June 1976 by Delta from Canaveral
Location: 72.5° east geostationary, inclined 11.4°
Other specifications as for Marisat 1.

Marisat 3

Launched: 14 October 1976 by Delta from Canaveral
Location: 178.6° east geostationary, inclined 12.8°. Manoeuvred out of the geosynchronous orbit band in September 1996 and retired
Other specifications as for Marisat 1.

Olympus satellite

Background

Olympus 1 introduced the world's largest civil communications platform when ESA launched it in 1989. An Ariane put the Olympus in orbit on 12 July 1989 and positioned it at 19° west. During the ESA experimental phase it suffered its most serious in-orbit problems when ground controllers lost control in mid-1991 and the satellite began drifting. A complex recovery programme returned it to 19° west after 77 days, on 13 August 1991. But the rescue drew heavily on propellant reserves, and ESA began to lose hope that it would maintain the projected nominal 5 year mission, in early 1994. When control ceased again on 11/12 August 1993 the few remaining kg of station-keeping propellant proved insufficient for recovery, resumption of services and end of life de-orbiting. ESA decided instead to lower Olympus from GEO and ended the mission formally on 30 August 1993. The satellite nonetheless demonstrated new market applications and helped to establish requirements for the projected Data Relay Satellite (DRS).

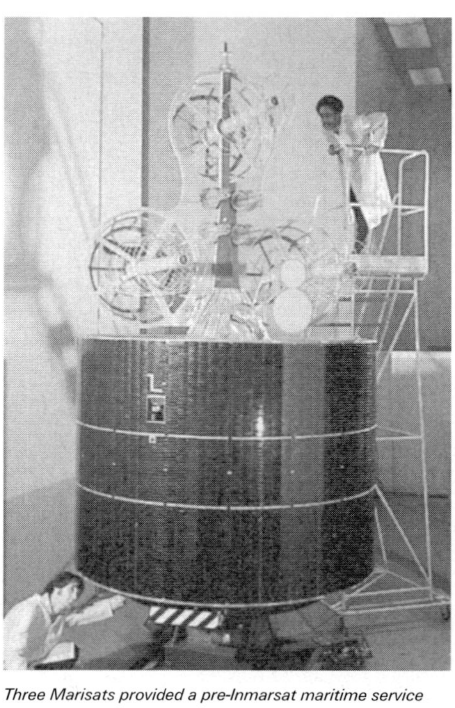

Three Marisats provided a pre-Inmarsat maritime service

Beginning development in 1979 and known until 1983 as L-Sat, the 2,612 kg Olympus, covered direct and semi-direct TV/radio broadcasting, inter-city telephone routing, business communications and mm-wave communications. It carried in total four payloads: a 2 channel 230 W DBS 18/12 GHz a 4-channel 12/14 GHz Specialised Services Payload; the 20/30 GHz advanced experiments; and a 12/20/30 GHz beacon package for propagation studies.

The CCTS established a group of Experimenters for Olympus assisted by Eutelsat, which approved 22 experiments by March 1990 for using the DBS, 20/30 GHz and Specialised Services payloads. Technical institutes across Europe took measurements on the propagation payload and co-ordinated their results. On board the Olympus the IOC (In-Orbit Communications) experiment operated the first Ka-band inter-satellite datalinks, working with Eureca 1 between August 1992 and 24 June 1993. More than 100 organisations in 12 countries used the facility.

Outside auditors estimated the Olympus cost ECU700 million at 1988 rates for the complete programme, including two spacecraft, launcher, insurance, ground segment investments and 5 years of operations. Austria 0.9 per cent, Belgium 3.2 per cent, Canada 11.0 per cent, Denmark 1.2 per cent, Italy 31.5 per cent, Netherlands 10.4 per cent, Spain 2.7 per cent, UK 39.1 per cent all participated in the programme.

OTS satellite

Background

ESA used its initial communications satellite programme, the OTS Orbital Test Satellite technology demonstration mission, to provide the basis for developing the medium-class ECS European Communications Satellite and its Marecs maritime derivative.

The ESA board approved OTS in 1971, but the programme began inauspiciously. OTS 1, intended to be ESA's first communications satellite, disappeared in the 14 September 1977 Delta 3914 explosion, which occurred 54 seconds after lift-off from Cape Canaveral. ESA, had already suffered losses and delays as a result of a series of US launch failures, and to protect itself, it insured the launch for a total of US$29 million. The identical OTS 2 back-up successfully went into space aboard a Delta on 12 May 1978 at 10° east GEO. Its six antennas covered Western Europe, the Middle East, North Africa and Iceland.

Hawker Siddeley (later part of British Aerospace) built the 865 kg (444 kg on-station), 6-sided satellite, and acted as prime contractor for the 10 Mesh countries. The satellite became among the first to operate in the 11/14 GHz Ku-band. This OTS satellite provided four wideband channels able to accommodate up to 7,200 telephone circuits; alternatively, each channel could handle one or two TV transmissions.

After 1 year of non-revenue experimental use, covering telephone, data transmission and TV (including transmissions between Europe, Egypt and Morocco), ESA announced that the commercial potential of OTS had been proved and released the satellite to Eutelsat. Under a joint management structure with ESA, Eutelsat scheduled and controlled all communications traffic through OTS for itself and its customers.

Eutelsat ceased using OTS at the end of 1983. During 1984, ESA conducted a series of end of service tests, involving some risky manoeuvres. The space agency put the satellite in an intentional flat spin and then recovered it, the first time a 3-axis GEO craft had been so recovered. ESA also established a method of station-keeping by solar sailing allowing future spacecraft to tighten control specifications from ±0.1° to within ±0.035° without consuming significantly more propellant.

Solar sailing found an operational application for more than 2 years during normal mode control. In late 1984, OTS then entered hibernation for final studies of long term subsystem degradation. In May 1988, it broadcast celebrations marking its 10th anniversary. The last of its eight TWTAs failed at the end of 1990. ESOC commanded multiple burns during 2/3 January 1991 to raise its orbit 320 km above GEO, leaving it to complete a westward circuit every 90 days.

ISRAEL

Amos series

Current status

IAI plans to launch the Amos 2/CERES Central European Regional Satellite in 2001/2002. For the project to proceed, the Hungarian government agreed to lease two transponders, on 30 April 1995. The satellite will be owned by Magyarsat in Hungary, a 50/50 joint venture by IAI and Antenna Hungaria, and collocated with Amos 1 with eight or sixteen active 72 MHz transponders (plus two back-ups). The Israel beam will provide Amos 1 back-up (there will be no Portugal beam).

Background

Israel Aircraft Industries began the Amos programme in 1990, with the full go-ahead formally approved in January 1992, to develop, build, operate and market a medium-class GEO satellite, principally for sub-regional TV broadcasting. IAI retains ownership of Amos 1. It created a commercial entity named SpaceCom in 1993 to sell capacity. SpaceCom is in equal partnership with IAI, Gilat Communication Engineering, General Satellite Services Co and Mer Services Group Ltd. Industry sources project the cost through launch, plus operating costs over 10 years at US$190-200 million. Israeli banks provided US$100 million, guaranteed by the government. The Israeli government initiated the project by guaranteeing it would use three transponders over the 10 years life and thereby replacing the three previously leased from Intelsat. Transponder cost is about US$3 million annually non-pre-emptible. Israel hoped that Arab users would be signed up before launch since the majority of the 250 million people within the 700,000 km² Middle East beam are Arabic speakers.

Specifications

Amos 1

Launch: 16 May 1996 by Ariane 4 from Kourou, French Guiana
Location: 4° west (Israel also holds registrations at 39° east/1.5° east)
Design life: 10 years
Contractors: IAI (MBT; prime, systems engineering, mission life operations, structure, AOCS, TT&C, data handling, thermal), Alcatel (Ku-band payload), DASA/Dornier (power), DASA/MBB (thrusters), ANT (transponders), CDI (OBCs), Gates (Ni/H_2 battery, integrated by SS/L), Teldix (momentum wheels)
Transponders: 7 (plus 2 back-up in 7-for-2 ring) 36 W TWTA 14.0045-14.4955/10.9545-11.1935 & 11.4565-11.6955 GHz Ku-band up/down in two of three beams using single reflector: Israel beam can take 3-6 transponders, EIRP 55 dBW max, 14.0 dB/K max, -99 to -74 dBW/m²; European beam (centred on Hungary) of 1-4 transponders, one switchable to Portugal. 72 MHz bandwidth, orthogonal polarisation
Applications: projected 70-80% TV distribution; DTH to 50-60 cm dishes; VSAT video, voice and data (including interactive learning) HDTV
Configuration: 1.21 × 1.670 × 1.930 m bus, solar array span 10.550 m. Single 1.73 m diameter deployed Ku-band antenna on west face. 3-axis stabilisation by MWs (no gyros); wheel unloading/station-keeping by redundant sets of 7 × 10 N thrusters. 400 N thruster accomplishes GEO insertion. Control by two 1,750A OBCs. Internal structure + East/West faces are composite; North/South are aluminium honeycomb. Ku-band horn for TT&C
Mass: 996 kg at launch, 479 kg dry, 580 kg on-station BOL
TT&C: from MBT's Yehud complex, Ku-band via two 9 m antennas. Same control room as Ofeq programme
Power system: 1.23 kW EOL from twin Si solar wings supported by 25-cell 50 Ah Ni/H_2 battery.

The completed Amos 1 at IAI/MBT ***1996***

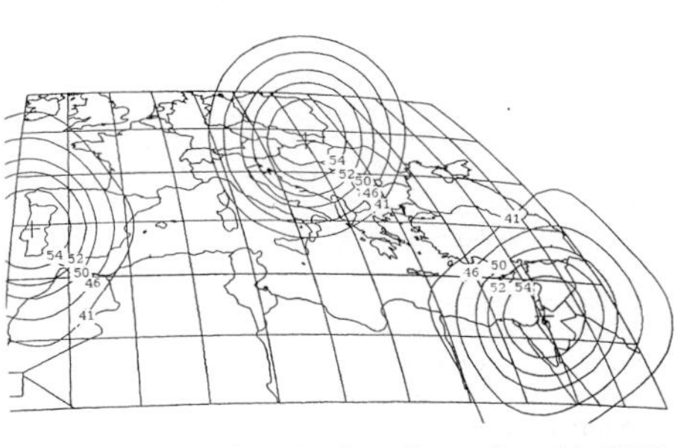

Amos 1 EIRP (dBW) footprints from 4° west. Boresight EIRP is 56 dBW (IAI/MBT) ***1995***

ITALY

Italsat series

Background

Italsat 1 is a Lit545 billion pre-operational programme to investigate the use of Ka-band in domestic voice/datalinks. It features onboard baseband-switched TDMA and makes available up to 12,000 telephone channels. ASI managed the demonstrations over the first two years, before transfer to the PTT. Telespazio provided in-orbit control. The contract included six Earth stations.

Italsat is 77 per cent Italian (Alenia 52 per cent) and the ground segment 100 per cent (Alenia 75 per cent). Alenia served as both prime contractor and systems integrator for a Lit160 billion contract awarded in 1990 for a second satellite. Italsat F2 adds the European Mobile Services (EMS) payload for ESA. Telespazio plans to build about 30 ground stations for the 2-satellite system. Italsat's switching capability means that it can transfer capacity to areas hit by disasters and to holiday regions during the summer. ESA's Artemis will also employ the Italsat platform.

Specifications

Italsat 1

Launched: 15 January 1991 by Ariane 44L from Kourou, French Guiana; spacecraft commissioned March 1991; payload testing completed end-September 1991
Location: 13.2° east geostationary
Design life: 5 years minimum for Ka-band demonstrations (Ariane launch accuracy could allow 7 years life)
Contractors: Alenia Spazio (prime, bus, payload, structure, thermal control), BPD (propulsion), Matra (AOCS), Aerospatiale (solar panels), SS/L (batteries), Fiat (power control/distribution), and Galileo (attitude sensors)
Transponders: 3 (plus 2 back-up) 20 W TWTA 29.5-30/19.7-20.2 GHz up/down Ka-band 1.55 × 1.06° national beams (Global Beam payload), 36 MHz bandwidth, EIRP 46.2 dBW min, 25 Mbit/s data rate, linear polarisation. Primary application business point-to-point

6 (plus 4 back-up) 20 W TWTA 27.7-29.5/18.5-20.5 GHz up/down Ka-band Multibeam payload, 120 MHz bandwidth, EIRP 57.0 dBW min, 147.5 Mbit/s data rate, linear polarisation. TWTAs can be paired to burn through precipitation. Each of two non-steerable 2 m antennas generates three 0.59° spots (360 km diameter), four covering mainland + one each Sicily/Sardinia. 32 ms TDMA bursts handled by the synchronous baseband switching matrix. Capacity 12,000 duplex voice at 32 kbit/s; also analogue video at 2 Mbit/s and digital TV to 70 Mbit/s. Up to 16 ground stations can operate within a beam

1 (plus 1 back-up) 1 W 40/50 GHz Propagation beacon, providing 27 dBW EIRP over 3 × 3° European coverage
Principal applications: demonstration of Ka-band and onboard baseband switching for domestic TV, telephony and datalinks
Configuration: Alenia Spazio's platform, body 3.46 m high, solar array 21 m span, 3-axis stabilisation, Eurostar-type AOCS provided by 16 × 22 N + one Marquardt R-4D 490 N (for apogee insertion and major manoeuvres) thrusters drawing on 915 kg MMH/NTO, two 2 m diameter reflectors for six spot beams and single 64 × 88 cm elliptical reflector for national beam
Mass: 1,865 kg at launch, 890 kg on-station BOL, 950 kg dry mass, 915 kg propellant, 290 kg payload
Antenna pointing: ±0.03° for spot beams, ±0.15° for national beam; maintained by RF auto-tracking
TT&C: through Telespazio's Fucino Earth station. System performance is investigated through stations at CSELT in Turin, CNR's Centre for Satellite Telecommunications Studies in Spino d'Adda, and at the PTT's Higher Institute in Rome. Two beacon stations for antenna pointing control at Cagliari (Sardinia) and Courmayeur (foot of Mt Blanc). Three propagation beacon stations at Milan, Rome and Turin.
Power system: twin solar wings providing 1,600/1,400 W BOL/EOL, in conjunction with two 30 Ah Ni/H_2 batteries.

Italsat 2

Launched: 8 August 1996 by Ariane from Kourou, French Guiana
Location: 13° east
Design life: 8 years minimum
Contractors: as Italsat 1, plus Alenia Spazio is prime for EMS payload
Transponders: Same as Italsat 1, except the propagation beacon is replaced by ESA's 47 kg EMS European Mobile Services payload:

4 (2-for-1 redundancy) 23 W SSPA 1.6315-1.6355, 1.6415-1.6455, 1.6565-1.6605/1.530-1.534, 1.540-1.544, 1.555-1.559 GHz up/down L-band Europe/North Africa/Turkey EMS receive/forward beams, 4 MHz bandwidths (divided into 1 MHz segments), EIRP 41.5 dBW min (42.5 dBW over 95%), RHCP (switchable to LHCP), uses 2 m reflectors of Ka-band Multibeam

1 (2-for-1 redundancy) 4.5 W SSPA 14.236-14.240, 14.246-14.250, 14.241-14.245/12.736-12.740, 12.746-12.750, 12.741-12.745 GHz up/down Ku-band Western Europe EMS feeder/return beams, 4 MHz bandwidths (divided into 1 MHz segments), EIRP 32.5 dBW min, linear polarisation
Principal applications: as Italsat 1 plus mobile service demonstrations
Mass: 1,990 kg at launch, 1,200 kg on-station BOL, 1,025 kg dry mass, 975 kg propellant, 299 kg payload
Power system: as Italsat 1, except 1,880 W BOL
Other specifications as for Italsat 1.

Temisat satellite

Background

Telespazio invested US$31 million in the Telespazio Micro Satellite (Temisat) project to demonstrate a commercial service for pulling data from environmental

Italsat 1 was launched in 1991 to demonstrate Ka-band applications (Alenia Spazio)

platforms over eight 2.4 kbyte/s UHF channels. Germany's Kayser-Threde built the 32 kg 35 cm^3 Temisat and a spare under a US$10 million contract that included a Russian piggyback launch on 31 August 1993. Telespazio designed the satellite to last for three to five years but it failed unexpectedly after a little over a year. Telespazio marketed 1,000 transmitters and 50 Data Collection Centre terminals to Italian customers. In addition to receiving data, the DCCs could also send commands and data to their platforms.

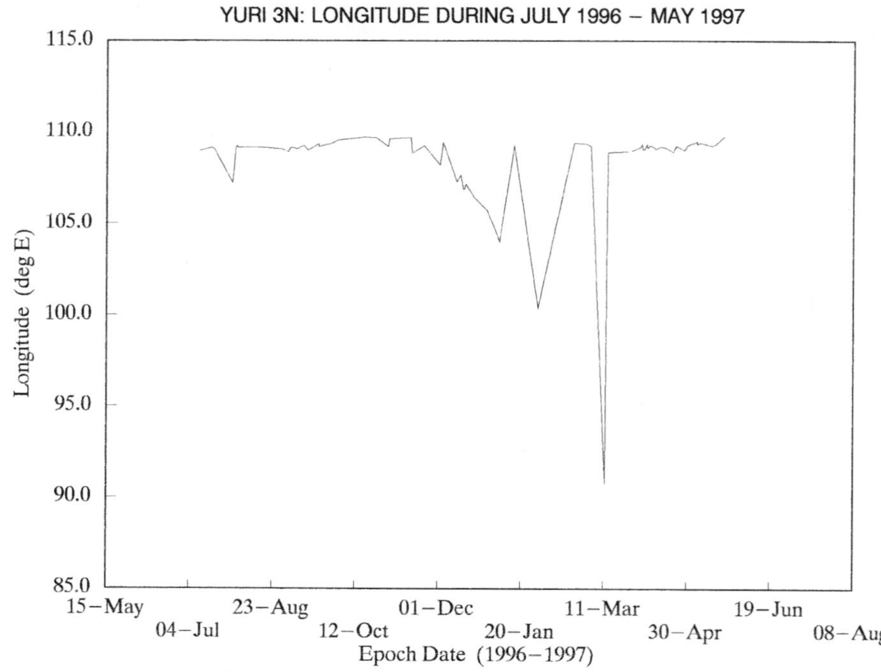

Although operating primarily over 110° east, data from USSPACECOM suggests that from September to December 1996 the Yuri 3N satellite drifted to about 105° east. Some of the USSPACECOM data might be in error (eg, single sets of data for 100° east or 90° east), but this drift appears to have been real. During 1997 the satellite has been shown over 110° east by USSPACECOM ***1997***/0055567

JAPAN

BS series

Current status

Japan took the decision to de-orbit the BS-3A and the Ministry of Posts and Telecommunications began the process beginning 30 April 1998. The satellite's remaining fuel boosted the orbit by 340 km. Bsat 1A replaced BS-3A as prime in August 1997.

Background

Until 1989, Japan owned and operated all of its communications satellites through the Telecommunication Advancement Organisation of Japan (TAO). Organisationally the National Space Development Agency (NASDA) procured and launched (or obtained launch services for) the BS and CS satellites and then handed them over to TAO for operational use.

When it launched the second generation Broadcast Satellites (BS), also known as Yuri, Japan became the first country to provide direct broadcast services to home receivers. Japan investigated the concept for two years with BS-1, launched 8 April 1978 by Delta from Cape Canaveral. The total programme cost the Japanese government 28.8 billion yen. Six years later, Japan launched the first operational satellite, BS-2A, on 23 January 1984 by N-2 from Tanegashima. Two of its three 100 W transponders failed within 3 months and Japanese DBS service then relied on the sole BS-2B. The Japanese removed the BS-2A from GEO in April 1989.

BS-2B brought the system to full capacity in 1986, but NHK purchased a 3000-class GE Astro Space satellite carrying three 200 W transponders, to maintain the system until the BS-3 series became operational in 1991. The total system cost the Japanese US$24.48 million for the satellite, about US$49 million for the launch and insurance.

BS-2X disappeared in the Ariane failure of 22 February 1990 and NHK initially indicated they would not seek a replacement craft because of BS-3's imminent appearance. Nonetheless, after a discussion with GE Astro Space, NHK purchased a second satellite for launch in January 1991. NHK had awarded GE the US$88 million contract (mostly paid for by the BS-2X insurance) to deliver the replacement BS-3H satellite in orbit. The replacement BS-3H was then itself lost in its April 1991 launch. BS-3A, launched 28 August 1990, replaced BS-2B but it operates on marginal power. 3B completed the system in 1991. Again faced with growing demand NHK procured BS-3N in late 1992 (launched July 1994) to guarantee services.

Japan carried on the tradition it pioneered with BS-2's DBS services, on the current generation BS-3 Yuri. NHK announced in late 1991 that it intended to procure a BS-3N for launch Spring 1994. The proportion of indigenous systems and components on the BS-3s has reached 83 per cent. The programme cost Japan about 78.8 billion yen. As it prepared to launch the first in the series NHK reportedly self insured BS-3A. In-orbit testing revealed that one of the four solar array power segments failed, leaving the satellite with only marginal power supply.

BSAT-1A began services, including high-definition TV broadcasting, in 1997
(Hughes Space & Communications International Inc) ***1995***

Specifications

BS 3A/Yuri 3A

Launch: 28 August 1990 by H1 from Tanegashima
Location: 110° east geostationary (arrived end-September 1990; AKM fired 30 August); collocated with BS-3b
Design life: 7 years (specified EOL reliability >0.8 bus, >0.9 broadcast transponder, >0.8 wideband transponder)
Contractors: NEC (prime, communications payload) with GE Astro Space (bus), Nissan (solid AKM)
Transponders: 3 (plus 3 back-up) 120 W TWTA 14.069-14.219/11.766-11.919 GHz up/down Ku-band domestic DBS beams, 27 MHz bandwidth, EIRP 55 dBW for mainland, circular polarisation. Channel allocations (up/down GHz, centre frequencies) number 3: 14.06584/11.76584; number 7 14.14256/11.84256; number 11 14.21928/11.91928

1 (no back-up) 20 W TWTA 14.37/12.64 GHz up/down Ku-band for 60 MHz wideband relay
Principal applications: DBS TV services: two channels (number 7 + 11) NHK (3rd channel, number 3, not used because of power limitation)
Configuration: GE 3000 series, 3-axis stabilised box-shaped 1.3 × 1.6 × 1.6 m bus with twin 3-panel 1.5 × 5 m solar wings and single 11 kg 80 × 170 cm offset parabolic elliptical reflector. In-orbit span 15 m, height 3.2 m
Mass: 1,115 kg at launch, 550 kg on-station BOL
AOCS: bias momentum 3-axis, antenna beam pointing accuracy ±0.1°, ±0.1° East/West + North/South. 4 × EHT + 12 × catalytic REA thrusters 0.39-0.91 N drawing on four hydrazine (170 kg total) surface tension tanks. Pitch control using MWs, roll/yaw using magnetorquers, active nutation damping. Solid AKM
TT&C: through TAO Kimitsu Satellite Control Centre; NHK programming uplinked through 8 m Tokyo antenna
Power system: 1,443/1,093 W BOL/EOL to be provided by twin solar wings (11,515 high voltage + 4,183 low voltage N-on-P 2 × 4 cm BSR) on North/South walls, but one of the four supply segments is not contributing. Supported by 2 × 17 Ah Ni/Cd batteries.

BS 3B/Yuri 3B

Launch: 25 August 1991 by H1 from Tanegashima
Location: 110° east GEO (collocated with BS-3a)
Design life: 7 years, as BS-3a, but insertion precision left 9 years propellant
Transponders: as BS-3a but using channels 5/9/15 (up/down GHz, centre frequencies): 14.1042/11.8042; 14.18092/11.88092; 14.2960/11.0660
Applications: one channel is providing JSB's DBS; one has been demonstrating HDTV 8 h daily since 25 November 1991 to some 400 sets; SDAB digital audio broadcasting is also a user
Other specifications as for BS-3a.

BS-3N/Yuri 3N

Launch: 8 July 1994 by Ariane 4 from Kourou, French Guiana
Location: 109.85° east geostationary, as 3a/b back-up
Transponders: 120 W TWTA, as BS-3a
Applications: as BS-3a/3b
Mass: 1,210 kg at launch, 699 kg on-station BOL, 575 kg dry

BS-3N provides three active 120 W DBS transponders (Martin Marietta Astro Space) ***1995***

Configuration: as BS-3a except 50 cm circular antenna added for 17 GHz uplink
Power system: as BS-3a except Ni/H_2 batteries carried
Other specifications as for BS-3a.

BSAT-1A
Launched: 16 April 1997 by Ariane 4 from Kourou, French Guiana
Location: Initially over 121° east, but planned for 110° east (collocated with 1b)
Design life: 10 years minimum (12.5 years propellant)
Contractors: Hughes Space & Communications
Transponders: 4 (plus 4 back-up, 8-for-4) 106 W TWTAs 17.3-17.8/11.7-12.2 GHz up/down Ku-band domestic beams, 27 MHz bandwidth, >55 dBW EIRP mainland (as BS-3), RHCP
Principal applications: DTH, including high-definition TV
Configuration: Hughes' HS-376 spin-stabilised, 7.97 m high deployed (3.15 m stowed), 2.17 m diameter. Despun single reflector and communications shelf
Mass: 1,250 kg at launch, 720 kg on-station BOL
Pointing: <0.1° accuracy by RF beacon
AOCS: 50 rpm spin-stabilisation and station-keeping by four 22.2 N hydrazine thrusters. GEO insertion provided by solid Thiokol Star 30BP AKM
TT&C: by B-SAT
Power system: K4¾solar cells mounted on cylindrical exterior and deployed drum providing 1,200 W BOL. Two Advanced Ni/Cd batteries provide 100 per cent eclipse protection.

BSAT-1B
Launch: 29 April 1998 by Ariane 44P from Kourou, French Guiana
Location: planned for 110° east (collocated with 1a)
Other specifications as for BSAT 1a.

CS-3/JCSat series

Background

Two Communications Satellite-3 (CS-3), also known as Sakura, superseded CS-2A/B and provide domestic Ka/C-band services, with both C-band area beams and seven of the 10 primary Ka-band spot beams dedicated to NTT operations. Japan estimates the programme cost 66.3 billion yen.

Japan created the JSB Japan Satellite Broadcasting Incorporated in 1984 by the Ministry of Posts & Telecommunications to commercially develop and launch communication satellites. The government ownership of JSB incited US pressure to open more of Japan's satellite purchases to foreign competition. Japan did relent and bids for the 'BS-4' generation resulted in the selection of Hughes Space & Communications as the prime contractor to build the satellites. Until then The Japan Broadcasting Corporation (NHK: Nippon Hoso Kyokai) held a large stake in broadcast through BS-3 transponders.

Then in 1989, the Diet released all restrictions on commercial systems development and several projects resulted. Nippon Telegraph and Telephone provided communications services for the CS series through CS-3, but NTT turned to another manufacturer, Space Systems/Loral (SS/L), to build its own two NStars which it launched in 1995. Satellite Japan Corporation (Sajac) planned to launch its own Sajac Hughes HS-601 satellites in 1994 until financial problems forced it to merge with JCSat in 1993, creating Japan Satellite Systems (JSAT); owner and operator of the JCSat series of satellites. JCSat 3 began commercial services in November 1995 and JSAT ordered JCSat 4 in February 1996.

With no further Japanese government restriction on commercial venture, Broadcasting Satellite System Corporation (B-SAT) incorporated itself on 13 April 1993 as a consortium of NHK, WOWOW, five other broadcasters and eight banks to procure and operate the 'BS-4' generation of broadcasting satellites for launches in 1997 and 1998.

About 65 per cent of the 64 transponders on the JCSat's satellites are now leased; and the company claims there are 55,000 subscribers for five television channels. In addition to network and cable television distribution, Japan Satellite Communications Network Corporation (JSNet) employs JCSat to provide VSAT business data services. Three transponders are used for the video services and two carry 12 stereo channels of high quality Digital Audio Broadcast programming. JCSat charges 300-650 million yen annually to lease a transponder.

Specifications

CS-3A / Sakura 3A
Launched: 19 February 1988 by H1 from Tanegashima
Satellite operated over 132° east until late September 1996: it was then boosted to a retirement orbit
Design life: 7 years
Contractors: Mitsubishi Electric Corp (prime and antenna), NEC (transponders)
Transponders: 2 (plus 1 back-up) 4 W min TWTA 5.925-6.425/3.7-4.2 GHz up/down C-band global beams,

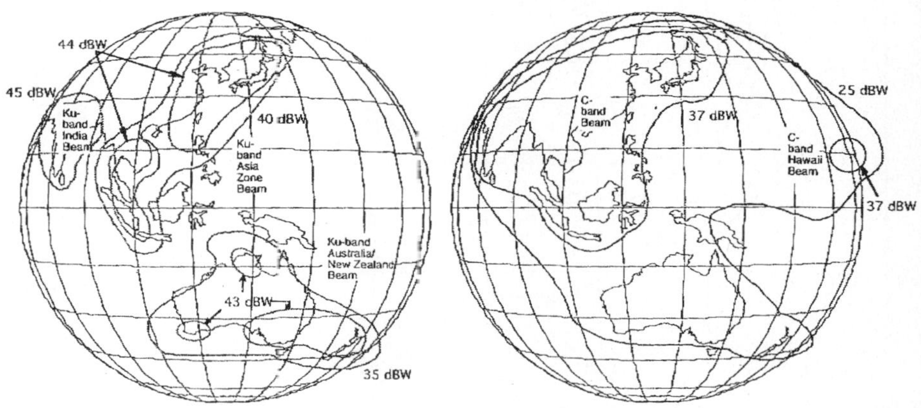

JCSat 3 Ku-band (left) and C-band EIRP footprints (JSAT) **1995**

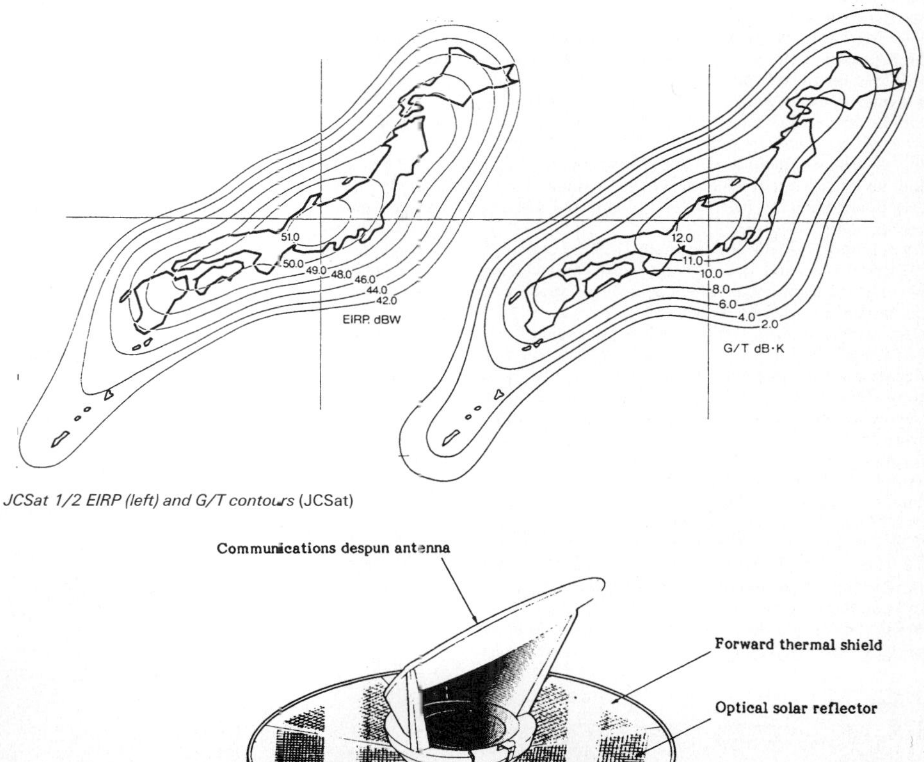

JCSat 1/2 EIRP (left) and G/T contours (JCSat)

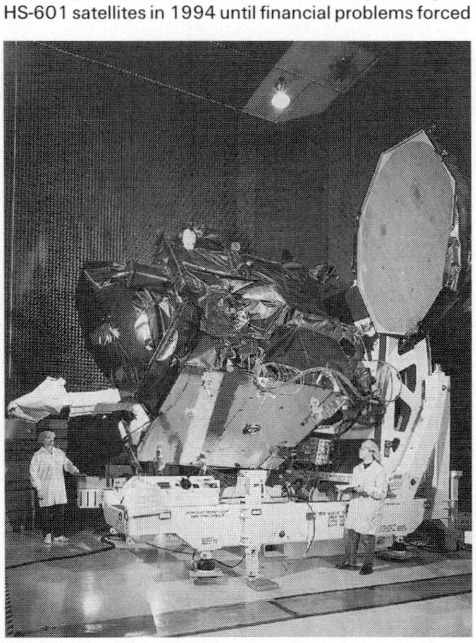

The hybrid JCSat 3 provides Ku/C-band services to the Pacific Rim and Asia (Hughes Space & Communications) **1996**

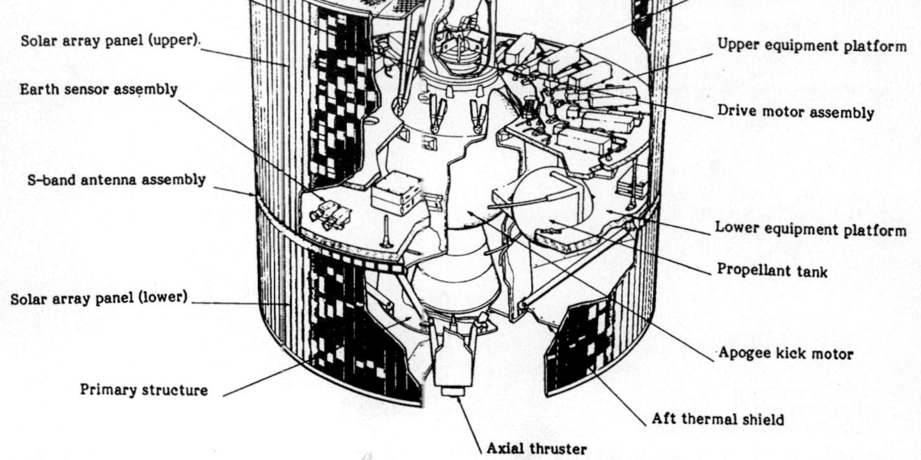

CS-3 has a single de-spun antenna for providing domestic communications links (NASDA)

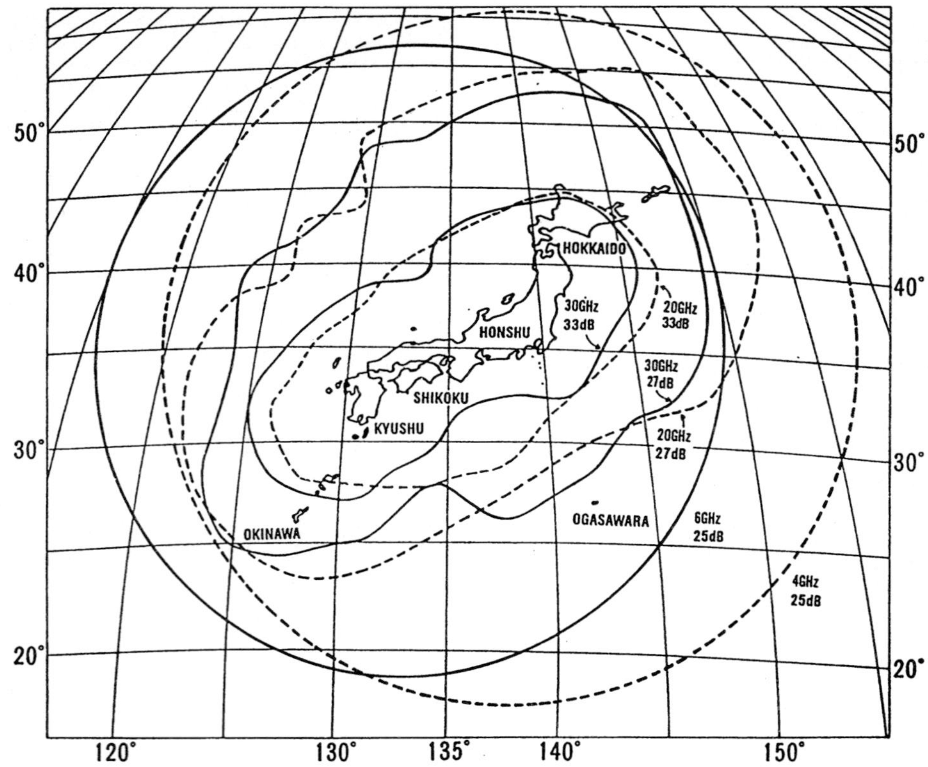

CS-3 coverage: C-band links the remoter communities, the higher-capacity Ka-band the main islands. The solid lines indicate CS-3a coverage from 132° east, the broken CS-3b from 136° east (NASDA)

180 MHz bandwidth, output power >=36.0 dBm, 25 dB min gain, circular polarisation

10 (plus 5 back-up) 4.2 W min TWTA 27.5-29.25/ 17.75-19.45 GHz up/down Ka-band spot beams, 100 MHz bandwidth, output power <=36.2 dBm, 33 dB min gain to main islands, 27 dB min to Okinawa islands, circular polarisation, plus 19.450 GHz beacon

Principal applications/users: NTT (7 Ka + 2 C; routine and emergency telephone, TV and datalinks for major and remote islands), National Police Agency (1 Ka), Construction Ministry (0.4 Ka), other users <0.4 Ka-band transponders

Configuration: 218.4 cm diameter, 242.9 cm high cylindrical bus, and overall height 356 cm including single despun antenna. Monocoque shell supporting two horizontal equipment platforms

Mass: 550 kg BOL on-station, 1,099 kg at launch

AOCS: spin-stabilised at 90 ± 9 rpm, 0.12° BOL spin axis adjustment, ±0.26° measuring accuracy of Earth width from GEO, four 20 N thrusters (2 radial, 2 axial; continuous mode or 90 ms pulses) drawing on 118.5 kg hydrazine in three spheres for attitude control and ±0.05° East/West-North/South station-keeping. GEO insertion by Thiokol Star 30B solid

Antenna pointing: <=0.2° half cone angle accuracy, ±0.03° stepping angle by despin assembly

TT&C: through TAO's Kimitsu Satellite Control Centre; 1-3 W DSN-compatible, near-isotropic S-band transponder up/down-converted to C-band; 125 bit/s 253-command command detector/decoder

Power system: 2-segment GaAs cylindrical solar array (36,621 P-on-N GaAs primary power; 396 N-on-P Si cells for battery charging), 833/750 W autumnal equinox/ summer solstice EOL, two 35 Ah Ni/Cd batteries (55% discharge depth, 616 charge cycles in 7 years), main bus voltage 29.4 ±0.2 VDC

Thermal control: passive, augmented by heaters. Primary heat emitted through solar array and end shields.

CS-3B / Sakura 3B

Launched: 16 September 1988 by H1 from Tanegashima

Location: Initially over 136° east: during April 1997 relocated to 153-154° east

Other specifications as for CS-3a.

JCSat 1

Launched: 6 March 1989 by Ariane 44LP V29 from Kourou, French Guiana. It formally entered service 1 May 1989

Location: 150° east geostationary

Design life: 10 years (launch accuracy left sufficient propellant for 14 years operation, but a leak is expected to end useful life in August 1997)

Undergoing tests in the Hughes Space Environment Chamber, a JCSat Satellite has its extendable solar cell drum retracted (Hughes Space & Communications)
1999/0054338

Transponders: 32 (plus 8 back-up, 5-for-4) 20 W TWTA 14.0-14.5/12.25-12.75 GHz up/down Ku-band domestic Japanese beams, 27 MHz bandwidth, EIRP 50 dBW min (51.8 dBW peak) over most densely populated regions, orthogonal linear polarisation

Principal applications: TV/cable distribution, telephone, business services, and satellite news gathering. Each transponder can carry one high-quality TV channel, a 45 Mbit/s data stream or 250 full duplex telephone circuits

Configuration: Hughes' HS-393 cylindrical platform, 3.66 m diameter, 3.6 m height stowed at launch, 10.0 m height deployed, payload and single 2.4 m diameter antenna, of two orthogonally offset grid reflectors, mounted on despun platform

Mass: 2,280 kg at launch, 1,376 kg on-station BOL, 1,097 kg dry

AOCS: station-keeping + 30 rpm spin-stabilisation maintained by six 22 N thrusters (two axial, four radial); GEO insertion and major manoeuvres performed by two Marquardt R-4D 489 N thrusters, both systems drawing on 1,200 kg NTO/MMH in four tanks each

Antenna pointing: 2-axis RF beacon tracking maintains <0.06°-accuracy pointing, backed up by Earth sensors

TT&C: though the company's Yokohama station with 11 m antenna for transfer orbit operations and two 5.5 m for on-station TT&C. A back-up facility is maintained in Gunma Prefecture

Power system: 2,240 W equinox BOL/1,808 W solstice EOL provided by Si solar array mounted on cylindrical

Engineers check out the payload section of a Hughes JCSat (Hughes Space & Communications) ***1999***/0054337

With nearly 55,740 sq metres of manufacturing space the Hughes Integrated Satellite Facility accommodates a 1,783 cu metre thermal vacuum chamber in which here seen is JCSat-6 (Hughes Space & Communications) ***1999***/0054335

A Hughes HS-393 derived from the Intelsat UI bus
(Hughes Space & Communications) ***1999***/0054336

body (K7 cells) and extension skirt (K4¾cells); 2 × 38.4 Ah Ni/H_2 batteries
Thermal control: quartz mirror radiator band encircling upper body provides primary heat rejection path.

JCSat 2
Launch: 1 January 1990 by Titan 3 from Canaveral
Location: 154° east geostationary
Principal applications: as JCSat 1 plus DTH TV + DAB
Other specifications as for JCSat 1, but Titan launch required perigee kick stage derived from Minuteman stage 3.

JCSat 3
Launch: 29 August 1995 by Atlas 2AS from Cape Canaveral
Location: 128° east geostationary
Design life: 12 years
Transponders: 28 (plus 6 back-up, 17-for-14) 60 W TWTA 13.75-15.0/12.25-12.75 GHz up/down Ku-band beams: Japan (>55 dBW mainland, >51 dBW Okinawa, >46 dBW other), India (45 dBW), Asia zonal (>45 dBW), Australia/New Zealand (>46 dBW). 12 × 36 MHz + 16 × 27 MHz bandwidth (27 MHz switchable to 54 MHz), orthogonal linear polarisation

12 (plus 4 back-up, 16-for-12) 34 W SSPA 6.22-6.49/3.93-4.2 GHz up/down C-band zonal beam covering Russia, Korea, Japan, China, India, South Asia, Indonesia, Hawaii, Australia, New Zealand. 36 MHz bandwidth, orthogonal linear polarisation
Principal applications: as JCSat 1/2
Configuration: Hughes' HS-601 3-axis bus, 26.1 m solar array span
Mass: 3,100 kg at launch, 1,820 kg on-station BOL
Power system: twin solar wings providing 5.2 kW BOL, supported by 200 Ah Ni/H_2 battery.

JCSat 4
Launch: 17 February 1997 by Atlas 2AS from Cape Canaveral
Location: Initially over 150° east, but planned 124° east geostationary
Transponders: 16 60 W TWTA 13.75-14.5/12.25-12.75 GHz up/down Ku-band beams as JCSAT 3 except four 27 MHz transponders can be switched to 120 W/54 MHz

4 (plus 4 back-up, 8-for-4) 90 W TWTA 13.75-14.5/12.25-12.75 GHz up/down Ku-band beams: Japan (>55 dBW mainland, >53 dBW Okinawa), 36 MHz bandwidth, orthogonal linear polarisation.

C-band as JCSat 3
Mass: 3,100 kg at launch, 1,820 kg on-station BOL
Other specifications as for JCSat 3.

JCSat 5
Launch: 2 December 1997 using an Ariane-4 from Kourou
Location: 150° east
Contractors: Hughes Space & Communications Co
Bus: Hughes HS-601
Transponders: Ku-band, 60 W/90 W, 16 × 27 MHz and 16 × 36 MHz (4 switchable to 2 × 72 MHz)
Service area: Japan, Asia and Hawaii.

JCSat 6
Launch: 16 February 1999 aboard an Atlas 2 AS
Location: 124° east
Transponders: Ku-band, 70 W, 32 × 27 MHz
Service area: Japan.

CS-4/NStar series

Background

The CS-4 series dates back to 1984 when two Super CS satellites were envisaged to replace CS-3 with an enlarged capacity of 60-90 multibeam transponders on a 2 tonne platform tailored around the H2 launcher. ¥2.57 billion was allocated to NASDA in FY90 to continue CS-4 studies. In 1989, MPT approved development of two satellites for launches in 1994/96. Transponders working at S/L-band would provide operational mobile services. The project would be closed to foreign bidders because it was designated as R&D. However, although NASDA was to provide 25 per cent of the funding, the domestic telephone monopoly NTT would fund the remainder in return for use of the operational transponders. US pressure resulted in early 1990 with CS-4 being transformed into two distinct projects: NASDA's 25 per cent was applied to the COMETS satellite and NTT acquired a satellite system on the open market.

NTT called for bids 20 June 1991, closing the competition 5 September 1991 SS/L received the contract December 1991, against GE and Hughes competition, for the NStar satellites to be launched 1995 to replace CS-3. The on-orbit delivery contract was worth some US$600 million; the launch contract signing with Arianespace was announced 10 April 1992. Each carries 26 active transponders: six Ka multibeam, five Ka-shaped beam, eight Ku-shaped beam, five C-shaped beam and one S plus its C feeder for mobile links.

Specifications

CS-4A/ NStar A
Launch: 28 August 1995 by Ariane 4 from Kourou, French Guiana
Location: 132° east (replacing Sakura 3A)
Design life: 10 years minimum; sufficient propellant for 15 years
Contractors: Space Systems/Loral (prime), MELCO, NEC, Thomson Tubes Electroniques (TH 3781 12 GHz; TH 3919 20 GHz TWTs), Hughes Electron Dynamics Div, Aerospatiale (reflectors)
Transponders: 5 (plus 2 back-up) 15 W min TWTA 27.5-29.25/17.75-19.45 GHz up/down Ka-band domestic beams, 100 MHz bandwidth, EIRP 44 dBW, circular polarisation

6 (plus 3 back-up) 30 W min TWTA 27.5-29.25/17.75-19.45 GHz up/down Ka-band multibeams, 200 MHz bandwidth, EIRP 51 dBW, circular polarisation, for 1.56 Mbit/s data transmission or broadband ISDN, with onboard switching capacity

8 (plus 3 back-up) 55 W TWTA 13.84-14.56/12.19-12.91 GHz up/down Ku-band domestic beams, EIRP 53 dBW, 54 MHz bandwidth, linear polarised

5 (plus 2 back-up) TWTA 5.925-6.425/3.7-4.2 GHz up/down C-band domestic beams, EIRP 38 dBW Tokyo, 72 MHz bandwidth, circular polarised; plus 6th TWTA as feeder link for S-band (see below)

1 (plus 1 back-up) 250 W TWTA 2.655-2.690/2.500-2.535 GHz S-band domestic beam, EIRP 52 dBW, 15 MHz bandwidth; one C-band feeder link (as other C-band transponders; one of Ku-band payload serves as feeder backup). Principally for maritime links, replacing 200 terrestrial stations along Japan's coast
Principal applications/users: continues CS-3 services, mobile communications, and business services

NStar appeared in August 1995 ***1995***

Configuration: box-shaped FS-1300 bus, two 2.6 × 3.25 m C/S + 2.6 × 4.5 m Ku/C deployable + one 2.1 m reflectors
Mass: 3,410 kg launch, 1,617 kg dry, 2,057 kg on-station BOL
AOCS: 3-axis (from immediately after launcher separation). 12 × 22 N N_2O_4/MMH; 490 N thruster permits Perigee Velocity Augmentation. Propellant loading 2,160 kg for 92 per cent fill
TT&C: through TAO Kimitsu Satellite Control Centre; transponders controlled through NTT's Inuishi and Tokyo stations
Power system: twin 4-panel solar wings, spanning 27 m. Supported by Ni/H_2 batteries.

CS-4B / NStar B
Launch: 5 February 1996 by Ariane 4 from Kourou, French Guiana
Location: 136° east (replacing Sakura 3B)
Mass: 3,420 kg launch, 1,618 kg dry, 2,062 kg on-station BOL
Other specifications as for Nstar A.

ETS series

Current status

The Communications and Broadcasting Engineering Test Satellite (COMETS) failed to achieve the correct orbit after lift-off on 21 February 1998. Japanese investigators traced the failure to a premature shutdown of the H2's, LE-5A second-stage engine. NASDA immediately announced a plan to rescue the satellite and the orbit out of passage through the van Allen belts where it was stranded.

Background

NASDA still conducts communication experiments on its own set of Engineering Test Satellites. NASDA launched four between 1975 and 1982, before privatising its space network. NASDA demonstrated the fruits of its research in August 1987 when it launched the ETS 5, the first Japanese 3-axis stabilised satellite in geosynchronous orbit and the first to use an indigenous apogee kick motor.

ETS 6 appeared in 1994 incorporating inter-satellite data relay and mobile communications packages. NASDA had intended to demonstrate a new ion propulsion engine to maintain north-south station-keeping on this two-tonne platform. The launch vehicle left it short of geosynchronous orbit however after launch in August 1994. The first of three planned firings by the new ETS 2 kN LAPS (Liquid Apogee Propulsion

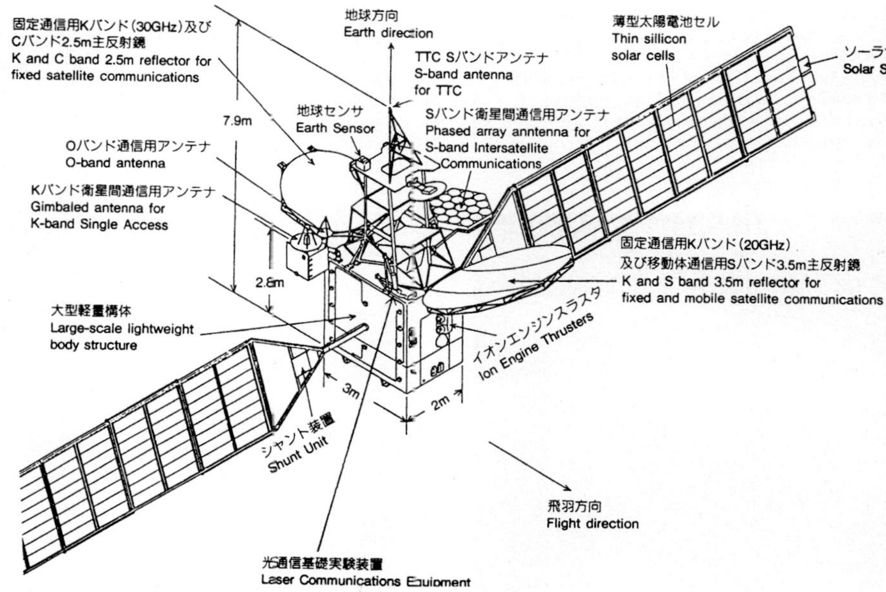

Principal ETS 6 features (NASDA)

COMETS under integration at NEC (NEC) ***1995***

System) ended when controllers found the engine attained only 10 per cent of full thrust, probably because a fuel line valve had failed to open. The tanks ran out of oxidiser supply on the fifth attempt and ground controllers separated the LAPS on 31 August by command via NASA Goldstone, leaving ETS 6 in 7,796 × 38,707 km, 13.1° inclination orbit.

NASDA planned to launch the COMETS Communications & Broadcasting ETS in 1997 to develop new technologies, preceding the operational DRTS Data Relay & Test Satellites in 2000. The OICETS Optical Inter-orbit Communications Engineering Test Satellite should follow in 1998 to test laser links with ESA's Artemis. ETS 8 in 2002 will demonstrate a single-hop cellular phone system and a CD-quality radio broadcast service.

The Ministry of Posts & Telecommunications established the Partners (Pan-Pacific Regional Telecommunications Network Research Satellite) project in 1992, using Kiku 5 to transmit educational teleconferences (64 kbit/s) and material to less developed Asia and Pacific sites, such as Fiji and Papua New Guinea.

Specifications

ETS-5/Kiku 5

Launched: 27 August 1987 by H1 from Tanegashima
Location: 150° east GEO, 3.7° inclination
Design life: 1½ years (hydrazine depletion projected imminent)
Contractors: Mitsubishi Electric, Nissan (AKM)
Transponders: Two 30 W SSPA 1.6445-1.6472/1.543-1.545 GHz up/down L-band beams for 23.2 dBW EIRP Experimental Mobile Satellite System experiment covering North Pacific (Japan-Alaska) and the Philippines/Indonesia/Australia regions for voice/datalinks with small maritime and aircraft terminals

One 6 W SSPA 5.95-5.97/5.22-5.24 GHz up/down C-band beam for NASDA link
Principal applications: spacecraft technology demonstration but experimental mobile communications package included
Configuration: 3-axis box-shaped bus with twin solar wings, GEO insertion by solid AKM
Mass: 1,081 kg at launch, 550 kg on-station BOL
TT&C: through NASDA's Tsukuba station.

ETS-8

Launch: January/February 2002 by H2 from Tanegashima
Locations: 135° east, 129° east or 146° east GEO
Design life: 3 years
Contractors: to be determined
Transponders: S-band for mobile-mobile connections

BSS mobile broadcasting of CD-quality radio
Configuration: to be determined, but with two 10 m-class deployable antenna
AOCS: to be determined
Power: to be determined.

OICETS will demonstrate laser inter-satellite links

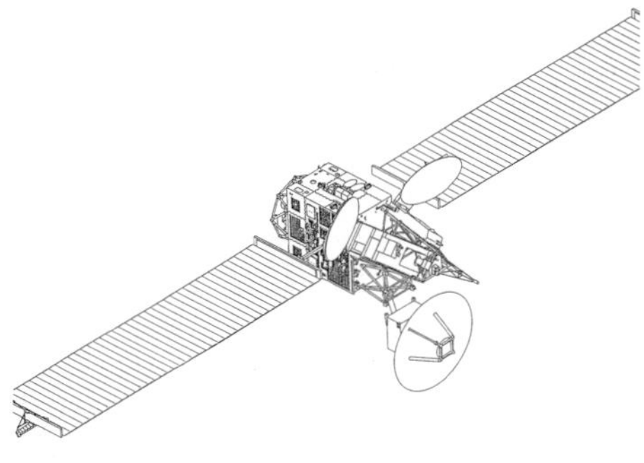

COMETS will demonstrate technology for future data relay, mobile communications and DBS. The largest antenna provides inter-orbit links ***1996***

COMETS

Launch: 21 February 1998 by H2 from Tanegashima
Location: planned 121° east geostationary
Design life: 3 years
Contractors: NEC (system integrator)
Transponders: S-band 2.0-2.1/2.2-2.3 GHz forward/return inter-orbit link, working through 3.6 m antenna (steerable ±10°). Redundant 50 W 20 MHz SSPA/3 W 10 MHz SSPA forward/return, max EIRP 45.0 dBW. Developed by NASDA

Ka-band 23/26 GHz forward/return inter-orbit link (working through same 3.6 m antenna). Redundant 3 W 30 MHz SSPA/30 W 150 MHz TWTA forward/return, EIRP >56.0 dBW. Developed by NASDA

Ka-band 30/20 GHz + O-band 50/40 GHz for high quality mobile communication via 2.0 m feeder link antenna. Developed by Communication Research Lab

2 (plus 1 back-up, 2-for-1) Ka-band 21 GHz 200 W TWTA 120 MHz high quality DBS, covering Japan in two beams (Tokyo area + Kyushu Island) via 2.3 m antenna. Developed by CRL/NASDA
Technical Engineering Data Acquisition: TEDA payload will monitor the space environment and its effect on the spacecraft with Single Event Upset Monitor, Dosimeter, Magnetometer, Potential Monitor, Solar Cell Monitor
Principal applications: development of technology for inter-orbit data relay and tracking services, mobile communications and high definition DBS
Configuration: 2 t-class, similar to ETS 6. Solar array span 33.2 m, bus 1.96 × 2.96 × 2.8 m
Mass: 3.9 t at launch, 2 t on-station BOL
AOCS: highly accurate ACS (<±0.05° pitch/roll, <±0.15° yaw) required to control large antenna pointing. MELCO's 20-30 mN ion thrusters (35 kg Xe fuel) paired on East/West panels, provide NSSK. GEO insertion and attitude/orbit control by UPS Unified Propulsion System derived from ETS 6's LAPS: single 1.7 kN bipropellant Apogee Kick Engine plus 4 × 50 N and redundant 8 × 0.86 N hydrazine catalytic thrusters
Power: twin solar wings 2.93 × 15.6 m (rolled up for apogee motor firings) provide 4.7 kW EOL; Ni/H_2 battery provides 100 per cent eclipse protection.

OICETS

Launch: J1 from Tanegashima
Orbit: planned 550 km circular, 35°
Design life: 1 year mission planned
Contractors: NEC (system integrator)
Transponders: 2.04425/2.22000 GHz S-band forward/return inter-orbit links (COMETS, Artemis); EIRP 4.4 dBW, G/T -30.6 dB/K

2.04425/2.22000 GHz S-band up/down link, EIRP 12.3 dBW, G/T -47.5 dB/K

Laser communications demonstration: 8,010 Å beacon, 8,190 Å receive (2 Mbit/s), 8,470 Å transmit (50 Mbit/s) 200 mW (CW) AlGaAs diode; 26 cm diameter telescope
Principal applications: demonstration of optical inter-satellite communications
Configuration: 3-axis stabilised box-shaped 78 × 110 × 150 cm bus, 9.36 m span across twin solar wings, 2.5 m high from base to top of optical assembly. Coarse pointing assembly provides 0-120° elevation, ±190° azimuth range

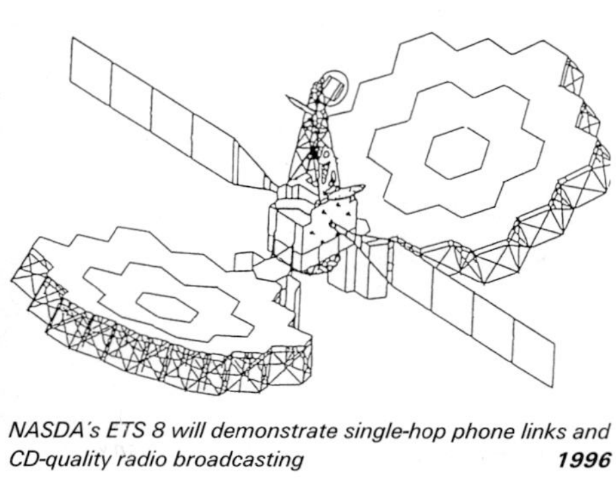

NASDA's ETS 8 will demonstrate single-hop phone links and CD-quality radio broadcasting ***1996***

Mass: about 550 kg in orbit
Power system: twin 3-panel solar wings spanning 9.36 m, 1.75 m wide; 13 Ah NiMH battery.

Superbird series

Background

A Japanese consortium planned to launch two Superbird satellites built by Space Systems/Loral and based on the FS-1300 platform. The Japanese stated the programme cost ¥70 billion (US$636 million), covering two satellites, launch services and insurance at premiums of about 20 per cent. An Ariane launched the Superbird A on 5 June 1989 to 158° east. Most of the satellite's station-keeping oxidiser bled off in December 1990 and commercial operations ended. Some customers transferred to the rival JCSat. Superbird claimed a reported US$170 million for the satellite. Superbird B, delayed from a December 1989 launch because of transponder problems, disappeared in the Ariane 4 accident of 22 February 1990. Insurance covered US$94.3 million. SS/L delivered the replacement Superbird B1 satellite within 24 months. Hughes received the order for Superbird C to expand coverage to the Asia Pacific region, including HDTV and 150 Mbit/s data. The satellite cost about 35 billion yen (US$412 million) with launch insurance.

Specifications

Superbird B1

Launched: 26 February 1992 by Ariane 4 from Kourou, French Guiana; entered service 6 April 1992
Location: 162° east geostationary
Design life: 10 years minimum
Contractors: SS/L (prime), DASA (solar array), Hughes Electron Dynamics Div (EPCs, TWTA), AEG (TWTs), Teldix (BAPTA), Gates (battery cells)
Transponders: 23 (plus 8 back-up, in 3 × 8-for-6 + 1 × 7-for-5) 50 W TWTA 14.00-14.50/12.25-12.75 GHz up/down Ku-band national Japanese beams, min EIRP 47.5 dBW, 36 MHz bandwidth, linear polarised (11 horizontal, 12 vertical)

3 (1 reserved as back-up) 29 W TWTA 28.4-28.8/18.615-18.715, 18.735-18.835 + 18.855-18.955 GHz up/down Ka-band national Japanese beam, min EIRP 43 dBW, 100 MHz bandwidth, LHCP up, RHCP down)
Principal applications: national TV and cable distribution, telephone, business services, and government
Configuration: 3-axis stabilised SS/L FS-1300 2.41 × 2.58 × 2.20 m box-shaped bus with twin solar

Superbird C will provide business and TV services to Japan, south/east Asia and Hawaii. The steerable spot could extend coverage to Australia/New Zealand
(Hughes Space & Communications) ***1995***

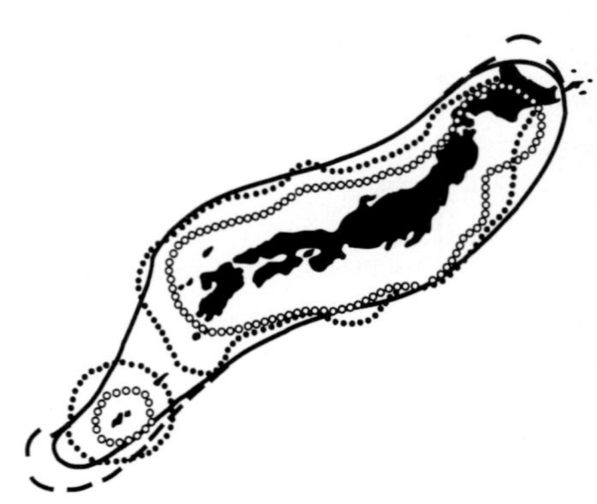

Superbird A/B coverage. Solid line: Ku 47.5 dBW EIRP; dashed: 4 dB/K. Dotted line: Ka 43 dBW; open dotted: 3 dB/K

wings on north/south sides, 20.3 m span; 2.1 m diameter Ku-band dual gridded reflector on West face, 2.0 m Ka-band reflector on East face
Mass: 2,560 kg at launch, 1,532 kg on-station BOL, 1,224 kg dry
AOCS: 3-axis bias momentum system; single Marquardt R-4D 490 N apogee thruster and two sets of 6 × 22 N AOCS thrusters, common NTO/MMH propellant system
Antenna pointing: 0.09° azimuth/0.08° elevation Ku-band; 0.09° azimuth/0.07° elevation Ka-band
TT&C: 14 GHz command frequency, 12 GHz telemetry frequency
Power system: twin 3-panel wings providing 3,984 W BOL summer solstice, 4,460 W BOL autumnal equinox, 3,479 W EOL summer solstice, 3,778 W EOL autumnal equinox. 42 ±0.5 V bus (sunlight), 27.0-42.5 V bus (eclipse). Two 85.5 Ah Ni/H_2 batteries.

Superbird A1
Launch: 1 December 1992 by Ariane 4 from Kourou, French Guiana. Entered service 20 February 1993
Location: 158° east geostationary
Design life: 10 years (propellant for 13 years)
Mass: 2,780 kg at launch, 1,665 kg on-station BOL, 1,255 kg dry (propellant increase over B1 allowed by uprated Ariane stage 3)
Other specifications are the same as for Superbird B1.

Superbird C
Launch: 28 July 1997 on Atlas 2AS from Cape Canaveral
Design life: 13 years
Contractors: Hughes Space & Communications (prime), MELCO (steerable spot)
Transponders: 24 90 W TWTA Ku-band beams, orthogonal linear polarisation
Principal applications: as Superbird B + Asia-Pacific services
Configuration: Hughes' HS-601 3-axis bus, 26.1 m solar array span. Two 216 cm shaped-beam antennas plus steerable spot
Power system: twin solar wings providing >4 kW BOL, supported by Ni/H_2 batteries.

KOREA, SOUTH

Koreasat (Mugunghua) series

Current status
Industry observers expected KT to order a third Koreasat in 1995 for regional services, but after the first experienced problems the decision has been delayed until 1999.

Background
Korea Telekom, formerly owned by the South Korean government, asked satellite builders to supply a US$260 million Koreasat (Mugunghwa) domestic satellite telecommunication system in March 1991. Four teams headed by GE Astro, Space Systems/Loral, Hughes and British Aerospace responded and KT issued a US$145 million contract for two satellites and the ground control system to GE Astro in December 1991, supported by Matra Marconi Space UK (payload), Goldstar Information & Communications of Korea, and Korean Air. Hughes Network Systems of the US provided a hub and network of 70-80 cm VSAT dishes.

At launch, Koreasat 1 had to expend its own onboard propellant to achieve orbit, because the Delta launch vehicle fired incorrectly for geostationary transfer orbit when a strap-on failed to jettison. Koreasat 1 used about half of its station-keeping propellant to achieve final geosynchronous orbit after the apogee kick motor fired. The satellite launch insurer withheld payment of US$104 million because the satellite remains operational.

Specifications
Mugunghua 1 (Koreasat 1)
Launched: 5 August 1995 by Delta 7925 from Cape Canaveral complex 17
Location: 116° east geostationary (with number 2)
Design life: 10 years (+2 years on-orbit storage), but Delta shortfall expected to limit to <5 years
Contractors: Lockheed Martin Astro Space (prime), MMS UK (payload), and Hughes Electron Dynamics Div (TWTAs)
Transponders: 12 (plus 4 back-up, 16-for-12) 14 W TWTA 14.00-14.50/12.25-12.75 GHz up/down Ku-band S Korean FSS 0.86° beam, 36 MHz bandwidth, EIRP peak/edge 52.5/50.4 dBW, orthogonal polarisation

3 (plus 3 back-up, 6-for-3) 120 W TWTA 14.50-14.80/11.70-12.00 GHz up/down Ku-band South Korean DBS 0.86 × 1.06° beams, 27 MHz bandwidth (4 channels per transponder), EIRP peak/edge 62/59.4 dBW, LHCP
Principal applications: domestic telecommunications, VSAT, data, video and DBS (HDTV from 1998)
Configuration: LMAS 3000 3-axis 163 × 142 × 174 cm box-shaped platform with 15.45 m solar array span and 1.52 × 1.83 m diameter reflector. Electronic units, batteries, propulsion and AOCS equipment are mounted on six honeycomb structural pallets or panels. Transponders and housekeeping components are mounted on four panels, two each on North/South sides. Additional housekeeping equipment is mounted on anti-Earth base panel
Mass: 1,459 kg at launch, 833 kg on-station EOL, 641 kg dry, 144 kg payload
AOCS: 3-axis by 6,000 rpm MW (pitch) and magnetorquing (roll/yaw) controlled by Earth Sensor Assembly error signals. Propulsion subsystem is a conventional blow-down monopropellant hydrazine type: four surface tension 50 cm diameter tanks (max capacity 50.3 kg each, initial 26.2 atmosphere; max blow-down 4:1) are manifolded to two redundant sets of eight Olin Aerospace thrusters (six catalytic 0.89 N MR-103C + two NSSK electro-thermal 0.4 N MR-501); 187.5 kg loaded. GEO insertion provided by Thiokol Star 30E solid AKM
TT&C: Yong-In primary, Taejon back-up; 11 m for TT&C, testing and manoeuvres, 6.4 m for on-station TT&C
Power system: twin Si arrays of three panels each provide 1,686/1,533 W BOL/EOL (10 years) supported by 2 × 42.5 Ah Ni/H_2 batteries.

Mugunghua 2 (Koreasat 2)
Launch: 14 January 1996 by Delta 7925
Location: 116° east GEO (with number 1)
Other specifications as for Mugunghua 1.

LUXEMBOURG

Astra series

Current status
Luxembourg has filed for eight further 10.70-12.75 GHz slots: 24.2, 26.2, 28.2, 31.5, 35.5, 37.5, 41.2, 43.2° east.

Background
The Grand Duchy of Luxembourg permitted SES to be incorporated as a private company in March 1985 for the purpose of establishing a medium-power satellite system for television distribution SES now operates under a franchise extending to 2010 from the Duchy, which also holds a 20 per cent interest in the company. The company contracted with RCA-Astro Electronics in October 1986 for a single series 4000 satellite. The resulting Astra 1 launched on 11 December 1988 and

Astra 2A will be deployed at 28.2° east, using ion thrusters for NSSK (Hughes Space & Communications) ***1996***

became Europe's first private satellite when it began operational programming on 1 February 1989, after its orbital trials.

The success of this first trial encouraged SES to expand and it acquired Satcom K3, previously owned by Crimson Satellite Associates. This 5000-class satellite had almost reached the end of the assembly line and the manufacturer needed to make only modest changes to its transmission characteristics to make it acceptable for SES. Rather than using its more powerful tubes to increase EIRP, the 52 dBW extended coverage instead. When it arrived at geosynchronous orbit it doubled the number of channels to 32.

SES continues to build on its transmission infrastructure. It awarded Hughes a US$428 million contract in December 1990 for two further satellites, based on the HS-601 platform. These new satellites, located within a 140 km cube of space, act primarily as back-up in the event of the primary satellite's failure. Astra 1D backs up 1B and 1C. It can also switch to BSS frequencies that can be employed for high definition television. The Duchy approved the purchase of Astra 1E in 1992 to bring digital programming to Europe. SES borrowed ECU220 million in February 1993 to finance it, bringing total SES capital investment to ECU1 billion. Approval for Astra 1F followed in 1993.

Astra 1F approved in 1993 will provide DTH service from 19.2° east and the enhanced 1G, ordered in January 1995 and the identical 1H in July 1995 will back up 1E-1G. SES used an ECU175 million loan contracted September 1995 to purchase the satellites, which it then supplemented with an additional ECU360 million loan in July 1996. The first two satellites of the second generation, replacing the first and expanding into other business areas from the new 28.2° east slot were ordered in June 1996.

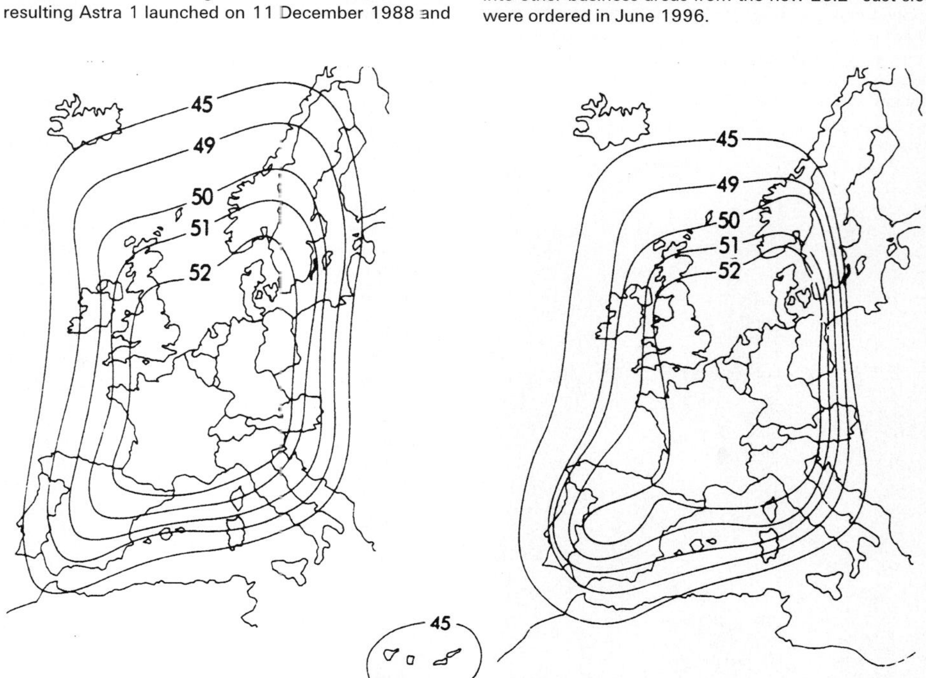

Astra 1B coverage in dBW (left: horizontal mode 2; right: vertical mode 2). 52 dBW is equivalent to 60 cm dish reception; 50 dBW to 85 cm. Astra 1A's lower power provides a smaller central region. Beginning with 1E, the higher power ellipse is stretched eastwards, providing the Moscow region with 75 cm dish coverage (SES)

Astra 1C-1H employ Hughes' HS-601 platform. Shown is Astra 1G/1H, featuring the new lightweight contoured antenna and GaAs solar wings ***1995***

Specifications

Astra 1A

Launched: 11 December 1988 by Ariane 4 V27 from Kourou, French Guiana. To be replaced by Astra 2A in 1999
Location: 19.2° east geostationary (above Zaire)
Design life: 10 years min (projected 12.1 years)
Contractors: GE Astro Space (prime), COM DEV (transponder input-output multiplexers, filters), Parsons (structure), Composite Optics (reflectors, tower), Lockheed Space Systems (Earth sensors), Adcole (Sun sensors)
Transponders: 16 (plus six back-up, 11-for-8) 45 W TWTA 14.25-14.5/11.20-11.45 GHz up/down Ku-band European beams, 26 MHz bandwidth, EIRP 50 dBW min (52 dBW in central region), orthogonal linear polarisation
Principal applications: TV/radio distribution (DTH, cable, SMATV) to Europe
Configuration: GE 4000 platform, 170 × 213 × 152 cm body, 19.3 m solar array span, total height 305.8 cm at launch, 3-axis stabilisation. Equipment mounted on eight aluminium honeycomb structural pallets or panels; transponders and some housekeeping units mounted on North/South faces; antenna/reflectors, including the primary Kevlar honeycomb reflectors, and Earth sensors on Earth-facing panel
Mass: 1,817 kg at launch, 1,043 kg on-station BOL
AOCS: 3-axis by 6,000 rpm MW, magnetorquing and 16 blow-down mono-hydrazine thrusters. Thrusters divided into two redundant sets of eight: six catalytic Reaction Engine Assemblies (REAs) accompanied by two Electrothermal Hydrazine Thrusters (EHT) for increased performance NSSK, drawing on four tanks. Apogee insertion provided by Thiokol's 808 kg Star 37XFP solid
TT&C: through SES' Betzdorf control centre 25 km east of Luxembourg. TT&C routed through 11 m dish and programme uplinking via 9.5 m antenna. Each of the Satellite Control Facility's Hewlett-Packard A900s can process six telemetry streams and command two satellites simultaneously. A 4.5 m transportable Earth station provides backup TT&C

Wrapped ready for launch the Astra 1F satellite is shipped to Russia for a flight on a Proton launcher
(Hughes Space & Communications) ***1999***/0054346

Power system: twin 4-panel solar arrays using N-on-P Si cells provide 2,790/2,309 W BOL/EOL; 3 × 50 Ah Ni/H_2 batteries provide full eclipse protection.

Astra 1B

Launch: 2 March 1991 by Ariane 44L V42 from Kourou, French Guiana. Broadcasting began 15 April
Location: 19.2° east geostationary (above Zaire)
Design life: 10 years min (projected 14 years)
Contractors: GE Astro Space (prime), AEG (TWT), Rocket Research (RCS thrusters), Royal Ordnance (liquid apogee engines), TRW Pressure Systems (tanks), COM DEV (transponder input/output multiplexers, filters, diplexers, switches), Lockheed Space Systems (Earth/horizon sensor assemblies), Program Components (antenna reflectors), Parsons of California (structure), Adcole (Sun sensors), Northrop (rate measuring assembly), Composite Optics (solar array substrates)
Transponders: 16 (plus six back-up, quasi 22-for-16) 60 W TWTA 14.00-14.25/11.450-11.700 GHz up/down Ku-band European beams in 250 MHz band adjacent to Astra 1A, 26 MHz bandwidth, EIRP 50 dBW min (52 dBW in central region), orthogonal linear polarisation
Principal applications: TV/radio distribution (DTH, cable, SMATV) to Europe and the Canary Islands (45 dBW)
Configuration: GE 5000 platform, 2.84 × 2.2 × 2.18 m body, 24.3 m solar array span, total height 2.56 m at launch (3.30 m on-orbit), 3-axis stabilisation. Equipment mounted on eight aluminium honeycomb structural pallets or panels; transponders and some housekeeping units mounted on North/South faces; 213 cm antenna reflector assembly deployed from west side; Earth sensors on Earth-facing panel
Mass: 2,580 kg at launch, 1,537 kg on-station BOL
Attitude control: 3-axis by 6,000 rpm MW, magnetorquing and 20 Olin blow-down mono hydrazine thrusters divided into two redundant sets of 10: eight catalytic REAs accompanied by two Improved EHTs for increased performance NSSK. GEO insertion provided by dual redundant Royal Ordnance Leros 1 500 N bipropellant motors. Dual mode system drawing on four hydrazine spheres + single N_2O_4 central cylinder
TT&C: as Astra 1A
Power system: twin 4-panel solar arrays using N-on-P BSR Si cells providing 4,850/3,700 W BOL/EOL; 4 × 50 Ah Ni/H_2 batteries provide full eclipse protection.

Astra 1C

Launch: 12 May 1993 by Ariane 4 V56 from Kourou, French Guiana. Broadcasting began 1 July
Location: 19.2° east geostationary (above Zaire)
Design life: 15 years min
Contractors: Hughes Space & Communications Co
Transponders: 18 (plus six back-up) 63 W TWTA Ku-band European beams, 2 at 10.90-10.95 GHz +16 at 10.95-11.20 or (to replace Astra 1A) +16 at 11.20-11.45 GHz down, 26 MHz bandwidth, EIRP 50 dBW min (52 dBW in central region), orthogonal linear polarisation
Principal applications: as 1B
Configuration: HS-601 platform, 3-axis, solar array span 21 m (satellite within 229 × 254 × 254 cm envelope at launch). Payload mounted on North/South panels and Earth-facing floor; East/West faces carry hard-points for two dual-gridded antennas, fixed antenna mountings on Earth-facing wall
Mass: 2,790 kg at launch, 1,700 kg on-station BOL, 1,180 kg dry
AOCS: 3-axis, unified 12 ARC 22 N + one Marquardt 490 N bipropellant thrusters incorporating onboard control processor, Sun/Earth sensors + 2 × 61 Nms 2-axis gimbaled momentum bias wheels. 1,658 kg NTO/MMH is contained in four spheres. Spin-stabilised in transfer orbit.
TT&C: as Astra 1A
Power system: twin solar wings of three 2.16 × 2.54 m panels carrying K4¾large area Si cells on Kevlar substrate to satisfy 3.3 kW requirement. Ni/H_2 batteries provide 100 per cent eclipse protection.

Astra 1D

Launch: 1 November 1994 by Ariane 4 from Kourou, French Guiana. Broadcasting began 1 January 1995
Location: 19.2° east (above Zaire)
Transponders: 18 (plus six back-up) 63 W TWTA Ku-band European beams, switchable as 16 at 10.70-10.95 or 10.95-11.20 (1C back-up) or 11.45-11.70 GHz (1B back-up), 26 MHz bandwidth or 18 at BSS 11.70-12.07 GHz (1E back-up) 33 MHz bandwidth, EIRP 50 dBW min (52 dBW in central region), orthogonal polarisation
Mass: 2,924 kg at launch, 1,700 kg on-station BOL, 1,250 kg dry
Other specifications as for Astra 1C.

Astra 1E

Launch: 19 October 1995 by Ariane 4 from Kourou, French Guiana
Location: 19.2° east (above Zaire)
Transponders: 18 (plus six back-up) 85 W TWTA Ku-band European beams, switchable as 18 at BSS 11.70-12.10 GHz 33 MHz bandwidth, or 16 at 10.70-10.95 (Astra 1D back-up) or 10.95-11.20 (1C back-up) or 11.45-11.70 GHz (1B back-up), 26 MHz bandwidth, EIRP 50 dBW min (52 dBW in central region), orthogonal polarisation
Mass: 3,010 kg at launch, 1,803 kg on-station BOL, 1,343 kg dry
Power system: twin solar wings (spanning 26 m) of four 2.16 × 2.54 m panels carrying K4¾ large area Si cells on Kevlar substrate providing 4.7 kW.
Other specifications as for Astra 1C.

Astra 1F

Launch: 8 April 1996 on Proton-K from Tyuratam, Kazakhstan
Location: 19.2° east (above Zaire)
Transponders: 22 (plus two back-up) 82 W TWTA 12.10-12.50 GHz Ku-band European beams (switchable to 18 at 11.70-12.10 GHz as 1E back-up, or 16 at 11.2-11.45 GHz as 1A back-up), 33 MHz bandwidth, EIRP 50 dBW min (52 dBW in central region), orthogonal polarisation
Mass: 3,010 kg at launch, 1,900 kg BOL
Other specifications as for Astra 1E.

Astra 1G

Launch: 2 December 1997 on Proton-K from Baikonur, Kazakhstan
Location: 19.2° east (above Zaire)
Design life: 15 years min
Transponders: 32 100 W TWTA 12.50-12.75 GHz Ku-band European beams (switchable to 18 at 11.70-12.10 GHz as 1E back-up, or 22 at 12.10-12.50 GHz as 1F back-up), 26/33 MHz FSS/BSS bandwidths, EIRP 52 dBW min, orthogonal polarisation
Configuration: improved HS-601 HP (High Power) platform, featuring GaAs solar arrays, enlarged payload and lightweight contoured reflectors
Mass: 3,300 kg at launch, 2,485 kg on-station BOL
AOCS: Hughes' 68 kg XIPS xenon ion propulsion system will provide NSSK: two (+ two back-up) 18 mN 13 cm thrusters
Power system: twin solar wings of four panels providing 8 kW BOL (7,075 W total required) and 100% eclipse protection
Other specifications as for Astra 1E.

Astra 1H

Launch: planned for 1999 on Ariane from Kourou, French Guiana
Location: 19.2° east (above Zaire)
Transponders: Same as Astra 1G, except two 29.50-30.00/18.80-19.30 GHz up/down Ka-band beams will introduce interactive services
Other specifications as for Astra 1G.

Astra 2A

Launch: 30 August 1998 on a Proton-K from Baikonur
Location: 28.2° east
Design life: 15 years min
Other specifications as for Astra 1G.

Astra 2B

Launch: planned for June on an Ariane 44P.
Location: 28.2° east
Design life: 15 years min
Transponders: 30 108 W TWTA 11.70-12.75 GHz Ku-band European beams (16 switchable to steerable beam), EIRP 52 dBW min, orthogonal polarisation
Configuration: MMS Eurostar 2000+i
Mass: 3,200 kg at launch
Power system: twin solar wings providing >6.4 kW BOL; 100% eclipse protection.

MALAYSIA

MEASAT series

Background

Malaysia which had spent US$30 million annually on Intelsat and Palapa leases made a national decision to take over responsibility for its own communications network in 1990. It uses Binariang, a private company to purchase satellites and launch services. Binariang signed a memorandum of understanding with Hughes in November 1991 to buy two HS-376 satellites to provide domestic and regional services. Considering Hughes production schedule at the time, Malaysia scheduled the first MEASAT (Malaysian-East Asian Satellite) launch for 1994 aboard Ariane. But Malaysia did not sign a final contract with Hughes Communications International Incorporated until 17 May 1994 following a 4-month international competition. The agreement specified that Malaysia would launch one satellite by late 1995. The contract also kept an option open for a second satellite,

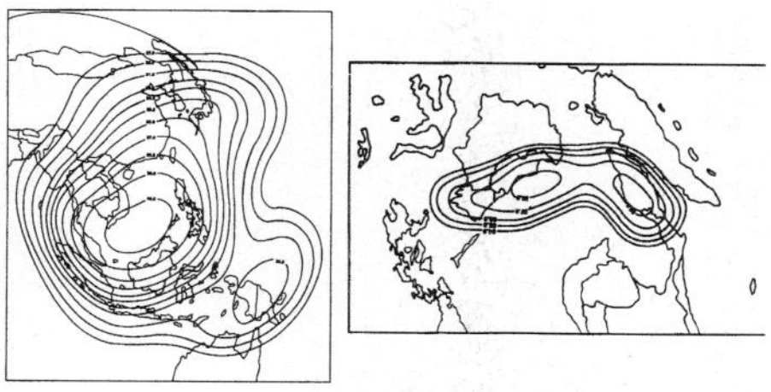

MEASAT coverage from 91.5° east. C-band (left) central EIRP is 40.0 dBW (33 for secondary peak); Ku-bands are both 58.0 dBW. The contours reduce in 1 dBW steps ***1996***

which the Malaysian government authorised in January 1995, as well as the construction of a ground control station and training for the Malaysian operators. This HS-376 is the first to carry GaAs solar cells, a lightweight shaped antenna (improves gain and eliminates multiple feedhorns) and a bipropellant station-keeping propulsion system.

Specifications

MEASAT 1

Launched: 12 January 1996 by Ariane 4 from Kourou, French Guiana
Location: 91.5° east geostationary
Design life: 12 years minimum
Contractors: Hughes Space & Communications (prime)
Transponders: 12 (plus three back-up) 12 W SSPA 5.925-6.425/3.7-4.2 GHz up/down C-band regional beam, EIRP 35 dBW edge (40 dBW centre), 40 MHz bandwidth (36 MHz usable), orthogonal linear polarisation

5 (plus one back-up) 112 W TWTA 14.0-14.25 + 13.75-14.0/10.95-11.2 + 12.2-12.5 GHz up/down Ku-band domestic beam (India and Philippines selectable), EIRP 55 dBW edge (58 dBW centre), 54 MHz bandwidth, orthogonal linear polarisation
Principal users: Ku-band for domestic direct to user services, including DTH TV via 50 cm dishes to Malaysia, Philippines and India; C-band for regional telecom services
Configuration: Hughes' HS-376 spin-stabilised bus, 2.16 m diameter, and 6.6 m height with solar array drum extended (2.8 m stowed). Single-shaped antenna on despun platform
Mass: 1,450 kg at launch, 839 kg on-station BOL, 626 kg dry
AOCS: 50 rpm spin control maintained by four (two radial and two axial) 22.2 N NTO/MMH bipropellant thrusters. GEO insertion provided by Thiokol Star 30 solid
TT&C: Satellite Control Facility on Langkawi Island
Power system: GaAs cells on cylindrical surface provide 1,700 W (1,200 W BOL to payload); 16-cell 142 Ah Ni/H_2 battery.

MEASAT 2

Launch: 13 November 1996 on Ariane 4 from Kourou, French Guiana
Location: initially over 147° east but planned for 91.5° east GEO (collocated with number 1)
Transponders: 3 (plus one back-up) 12 W SSPA 5.925-6.425/3.7-4.2 GHz up/down C-band regional beam, EIRP 35 dBW edge (40 dBW centre), 72 MHz bandwidth, orthogonal linear polarisation

6 65 W TWTA Ku-band domestic, 54 MHz bandwidth, orthogonal linear polarisation

6 100 W TWTA Ku-band domestic, 54 MHz bandwidth, orthogonal linear polarisation
Other specifications as for MEASAT 1.

MEXICO

Morelos series

Current status

Hughes provided the new Morelos 3 (Satmex 5) satellite, initially called Morelos-2-FOS, launched by Ariane 42L on 6 December 1998.

Background

Despite its participation in international communications bodies, Mexico decided in 1982 to develop an independent communications system. During that year, SCT ordered two Hughes satellites to expand domestic television and other services available to the entire population of the country. Before that time, telecommunications only served the 18 million people living in and around Mexico City.

The Mexican government's Departamento Especial of the DGT Direccion General de Telecomunicaciones for

An HS-601 HP, SATMEX 5 relies on Zenon propulsion for North/South station-keeping
(Hughes Space & Communications) ***1999***/0054329

Mexico's telecommunications systems gave authority to SCT, Secretaria de Comunicaciones y Transportes (SCT) to develop and service a communications network. Then in 1990 the Mexican government created Telecomm, an independent non- stock company owned entirely by the federal authority, with SCT operating as a regulating body.

Mexico concluded a US$92 million contract with Hughes which included the control centre at Iztapalapa and the Mexican government selected the name Morelos for its satellites to honour the revolutionary Jose Maria de Morelos y Pavon a hero of the Mexican revolution. The Shuttle launched the first Morelos on 18 June 1985 to 113.5° west. Solidaridad 1 retired the Morelos in March 1995 when it began service. True to its practice of inducing countries to launch satellites on the Shuttle despite the enormous cost, NASA sweetened the pot by allowing Mexican payload specialist Rodolfo Neri Vela, to accompany Morelos 2 when it was deployed. The satellite entered a two-year geosynchronous drift orbit because Hughes had not completed the ground segment to control when it was launched. It entered service May 1989 (requiring Telesat Canada to move Anik C3 from 117.5° west to avoid interference) and it should continue operations into 1998.

The specifications for a new Morelos 3 satellite took more than a year to develop, but the bidding process started on 18 March 1996. Five companies responded initially, but only three presented proposals at the end of May.

Specifications

Morelos 2

Launched: 27 November 1985 by Shuttle mission 61B
Location: 116.5° west geostationary
Design life: 10 years
Contractors: Hughes Space & Communications Co

As replacement for Morelos 2, SATMEX 5 carries 10 × the wattage payload power and 48 C-band and Ku-band transponders (Hughes Space & Communications)
1999/0054328

Transponders: 12 (plus 4 back-up, 14-for-12) 7 W TWTA 5.925-6.425/3.70-4.20 GHz up/down C-band domestic beams, 36 MHz bandwidth, EIRP 36 dBW min, orthogonal polarisation

6 (plus 2 back-up, 8-for-6) 10.5 W TWTA 5.925-6.425/3.70-4.20 GHz up/down C-band domestic beams, 72 MHz bandwidth, EIRP 39 dBW min, orthogonal polarisation

4 (plus 2 back-up, 6-for-4) 20 W TWTA 14.00-14.50/11.70-12.20 GHz up/down Ku-band domestic beams, 108 MHz bandwidth, EIRP 44 dBW min, circular polarisation
Principal applications: domestic TV distribution, telephony and data services
Configuration: Hughes HS-376 cylindrical spin-stabilised bus, 2.16 m diameter 6.60 m high, single 1.8 m diameter dual reflector for C-band and Ku-band transmission, planar array for Ku-band receive (first Hughes civil satellite with planar array); AOCS provided by hydrazine thruster system, GEO insertion by Thiokol solid AKM, GTO insertion by McDonnell Douglas PAM-D
Mass: 1,140 kg launch mass (including 133 kg hydrazine), 645 kg on-station BOL
TT&C: through the Iztapalapa Earth station 10 km South East of Mexico City. Full tracking antenna + two 11 m antennas
Power system: solar array of K-7 cells mounted on cylindrical body and skirt extension provides 950 W BOL/800 W EOL; two Ni/Cd batteries.

Morelos 3

Launch date: 6 December 1998 by Ariane 42L
Location: 243.2° east
Design life: 15 years
Mass: 4,144 kg at launch; 1,950 kg 130L
Power: 10.9 kW beginning of life, 9.2 kW end of life
Transponders: 24 × C-band for continental coverage: bandwidth 36 MHz, TWT 36 W, EIRP 38.0 dBW 24 × Ku-band for regional and continental coverage: bandwidth 36 MHz, TWT 110 and 132.5 W, EIRP - Ku-1 49.0 dBW (EOC), Ku-2 - 46.0 dBW (EOC).

Solidaridad series

Background

In March 1991, Hughes emerged as the leader in the contractor race to provide Mexico with two Solidaridad satellites. Hughes won with its HS-601 platform. Mexico signed the US$183.47 million contract 8 May in Mexico City, calling for first delivery in 28 months. In addition to continuing C/Ku-band services, Solidaridad has an L-band payload providing Mexico with its first domestic mobile links in the Americas. Arianespace signed the dual launch contract 24 September 1991, against Atlas 2A and Long March competition.

Specifications

Solidaridad 1

Launch: 20 November 1993 on Ariane 4 V61 from Kourou, French Guiana
Location: 109.2° west geostationary
Design life: 14 years
Contractors: Hughes Space & Communications C
Transponders: 18 (plus 6 back-up) 10-16 W SSPA 5.925-6.425/3.70-4.20 GHz up/down C-band beams serving Region 1 (Mexico, South West US, Northern Central America), R2 (R1 + Caribbean, Central America, Northern South America), R3 (most of remaining South America except Brazil), 12 × 36 MHz + 6 × 72 MHz

Solidaridad 2 assembly, with the 26-element L-band array prominent (Hughes Space & Communications) ***1995***

Part of the communications payload for Solidarad 2 (Hughes Space & Communications) ***1999***/0054330

bandwidth, R1/R2/R3 EIRP 37.5/37/37 dBW min for 36 MHz, 40.5 dBW min for 72 MHz, orthogonal polarisation. All 72 MHz transmission to R1 only, 4 × 36 MHz switchable between R1/R2, 4 × 36 MHz between R1/R3 and 4 × 36 MHz between R2/R3. R1/R2 feeds use W reflector, R3 E

16 (plus 4 back-up) 42.5 W TWTA 14.00-14.50/ 11.70-12.20 GHz up/down Ku-band beams serving Region 4 (as R1) and R5 (major US cities and small parts of Canada and Cuba), 54 MHz bandwidth, R4/R5 EIRP 47.0/46.4dBW min, orthogonal polarisation. All transmission/receive for R4 but two receive (and one of those two transmission) switchable to R5

One channel using four paralleled 21 W 1.6295-1.6605/1.528-1.559 GHz up/down L-band SSPAs (plus 2 back-up) for mobile links throughout Mexico and up to 370 km offshore, 17 MHz selectable out of available 34 MHz bandwidth, EIRP 45 dBW min, RHCP. 42.5 W 14.248-14.265/11.9515-11.9885 GHz up/ down Ku-band feeder link provides 46.4 dBW min, 17 MHz bandwidth

Principal applications: domestic TV distribution, telephony, data and mobile services; international services

Configuration: Hughes HS-601 platform, 3-axis, solar array span 21.0 m (satellite within 2.7 × 3.1 × 3.5 m envelope at launch). Payload mounted on North/South panels and Earth-facing floor; East/West faces carry hard-points for two dual-gridded antennas, fixed antenna mountings on Earth-facing wall: 1.8 × 2.4 m C-band on West side; 1.8 × 2.4 m Ku-band on East side; 26 L-band cup dipoles on Earth face

AOCS: 3-axis, unified 13 ARC 22 N and one Kaiser Marquardt 490 N bipropellant thrusters incorporating onboard control processor, Sun/Earth sensors and two 61 Nms 2-axis gimbaled momentum bias wheels. The NTO/MMH is contained in four spheres. Spin stabilised in transfer orbit.

Mass: 2,776 kg at launch, 1,671 kg on-station BOL, 1,291 kg dry

Solidarad 1 undergoes vibration tests to simulate launch vehicle dynamics (Hughes Space & Communications) ***1999***/0054331

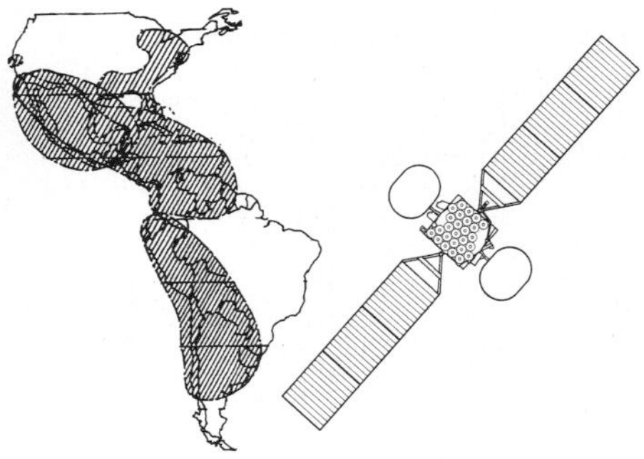

Solidaridad C/Ku-band coverage ***1995***

TT&C: as Morelos

Power system: twin solar wings of three 2.16 × 2.54 m panels carrying K4¾ large area Si cells on Kevlar substrate to satisfy 3.37 kW requirement. 160 Ah Ni/H_2 battery provides 100 per cent eclipse protection.

Solidaridad 2

Launch: 8 October 1994 by Ariane 4 from Kourou, French Guiana

Location: 113.5° west geostationary

Mass: 2,776 kg at launch, 1,640 kg on-station BOL

Other specifications: as for Solidaridad 1.

NORWAY

Thor series

Current status

On 18 May 1997, Norway announced that Hughes Space & Communications would build Thor 3. The satellite will be a high-power HS-376. The agreement contains an option for an additional satellite.

Background

Thor 1 began life as the Marcopolo 2 satellite, part of the UK's British Satellite Broadcasting company. BSB merged into rival Sky Television in November 1990, in preparation to moving all programming to Sky's Astra carrier. In the meantime, Hughes received a follow-up contract for US$300 million contract in 1987 for the on-orbit delivery of two HS-376 satellites to provide five TV channels employing the D-MAC transmission standard. The contract represented Hughes' first sales of the HS-376 models configured for DBS, using paired 55 W TWTAs to achieve the required power levels.

Specifications

Thor 1

Launched: 00.42 GMT 18 August 1990 by Delta 6925 from Cape Canaveral. Satellite separated at 13.10 GMT, followed by AKM firing at 16.17 GMT 20 August for positioning at 50° west for a month of in-orbit testing. Ownership transferred to BSB in mid-December 1990 at 31° west collocated with Marcopolo 1. Transferred over 15 September to 29 October 1992 to 1° west under Norwegian ownership

Location: 0.8° west geostationary

Design life: 10 years minimum

Contractors: Hughes Space & Communications Co

Hughes' HS-376 bus was uprated to meet DTH power requirements. Thor 2A is similar

A future Thor 3 ***1997***/0005568

Transponders: six paired 55 W TWTAs 17.385-17.992/ 11.7715-12.1054 GHz up/down Ku-band Nordic beam (BSB channels 4, 8, 12, 16, 20), 27 MHz bandwidth, 59 dBW min EIRP (63 dBW max), RHCP

Principal applications: DTH via 45 cm dishes

Configuration: Hughes' HS-376 spin-stabilised, 7.2 m high deployed (2.7 m stowed), 2.16 m diameter. Despun single reflector and communications shelf

Mass: 1,250 kg at launch, 660 kg on-station BOL

Pointing: 0.05° accuracy by RF beacon

AOCS: station-keeping and spin-stabilised at 50 rpm by four 22.2 N hydrazine thrusters; GEO insertion by solid Thiokol Star 30B AKM

TT&C: via Nittedal station

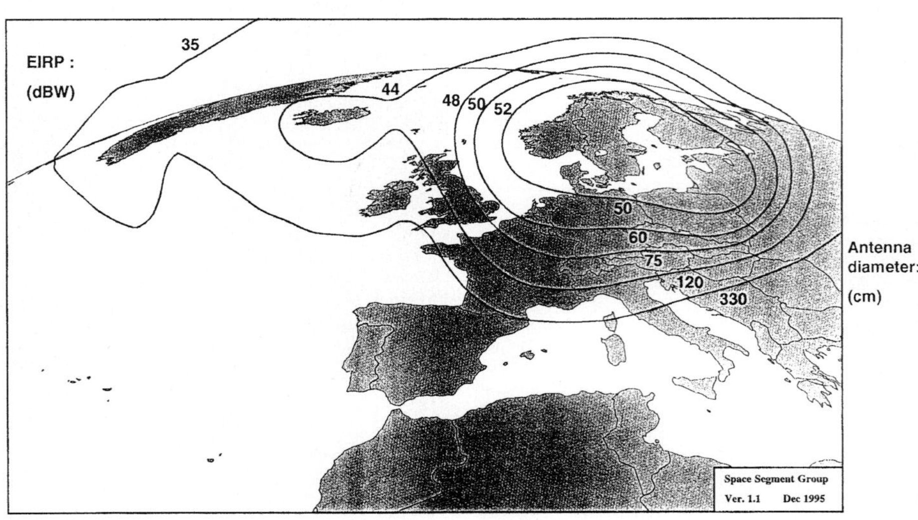

Thor 2A preliminary EIRP and antenna sizes ***1996***

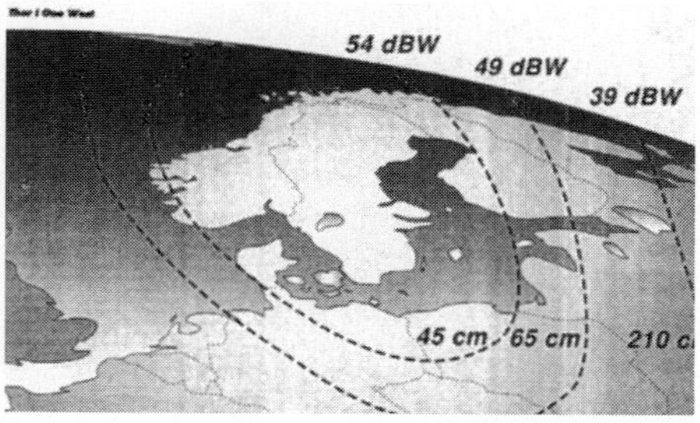

Thor 1 coverage from 1° west

Power system: K7 solar cells mounted on cylindrical exterior and deployed drum providing 1,024 W BOL (drum extended 25 cm over normal 376). Two Ni/Cd batteries provide 100% eclipse protection.

Thor 2
Launch: 20 May 1997 by Delta 7925 from Cape Canaveral
Location: 0.8° west geostationary
Design life: 11 years minimum
Transponders: 15 40 W TWTAs 17/11.2-11.45 GHz up/down Ku-band Nordic beam, 26 MHz bandwidth, 52 dBW min EIRP
Principal applications: DTH via 45 cm dishes
Configuration: Hughes' HS-376HP spin-stabilised, 7.2 m high deployed (2.7 m stowed), 2.16 m diameter. Despun single reflector and communications shelf
Mass: unavailable
AOCS: station-keeping and spin-stabilised at 50 rpm by four 22.2 N hydrazine thrusters; GEO insertion by solid Thiokol Star 30B AKM
TT&C: via Nittedal station
Power system: GaAs solar cells mounted on cylindrical exterior and deployed drum providing 1,200 W BOL for payload. Batteries provide 100 per cent eclipse protection.

Thor 3
Launch: 11 June 1998 by Delta 2
Location: ~1° west
Design life: >11.5 years
Transponders: 14 × 47 W Ku-band.

RUSSIAN FEDERATION

Ekran (Screen) series

Background

Ekran 1, launched on 26 October 1976, brought direct broadcast services to the world for the first time. By March 1996, 20 had been placed into Statsionar-T positions near 99° east over the Indian Ocean. The longevity of the collocated Ekran-M 2, 3, 4 required no further launches 1989-91; M 5 appeared October 1992.

Flight 16 in 1987 introduced the improved Ekran-M, perhaps explaining the reduction in launch rate. Ekran-M4 in orbit since December 1988 performed a small end- of-life manoeuvre in early October 1996 and drifted off-station. This left Ekran M5, launched October 1992, as the sole operating Ekran satellite.

The Ekran satellites, each carry a single operational 200 W transponder. NPO PM has been controlling satellites itself instead of relying on the military since January 1994 through its Persei subsidiary at tracking, telemetry and command stations in Krasnoyarsk and Gauss Khroustalni. To service the system, the Soviet Union and its successor states built more than 5,000 receivers and now this network broadcasts television to some 20 million viewers in Siberia and the far north east..

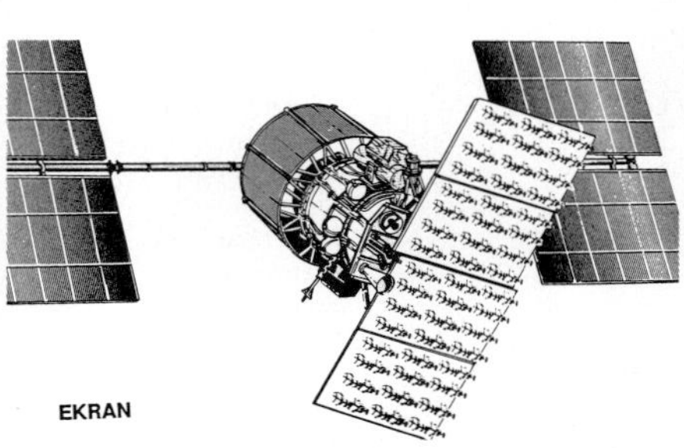

Ekran direct broadcast satellite, first launched 1976. Its 200 W transmitter provides TV coverage to remote Russian territory. Ekran-M has added a smaller panel to each of Ekran's four to boost power to 1.8 kW

Since no other signals interfere with the Ekran in these remote regions the satellite can broadcast directly at UHF (0-7 Ghz). The Russian Federation plans to launch four more Ekrans during 1996 to 1998, but the country plans to eventually phase out the system and replace it with the Gals. However, Ekran satellites may endure beyond their design life because Gals financing appears questionable.

Specifications

Launched: previously typically twice annually by 4-stage Proton-K from Tyuratam, but one 1989-95
Locations: 99° east geostationary
Design life: >3 years (previously 1 year)
Contractors: NPO PM (spacecraft), NPO Radio (payload)
Transponders: one (plus one back-up) 200 W Klystron 6.188 6.212/0.702-0.726 GHz up/down beam covering Siberian region, 24 MHz bandwidth, EIRP 55.5 dBW, circular polarisation
Principal users: direct broadcast TV
Configuration: similar to Gorizont, but planar array carries 96 helical transmit aerials. Solar wings are configured to avoid antenna shadowing
Mass: 1,970 kg (320 kg payload)
AOCS: 3-axis; east-west station-keeping to 0.5° by liquid thrusters
Power: Ekran-M's solar wings were enlarged to provide 1.8 kW.

Elekon series

Background

The GEO-IK Musson geodetic satellite launched 29 November 1994 by Tsyklon from Plesetsk into 1,480 × 1,527 km, 73.61° includes an Elekon payload which Russia intends to use to demonstrate location and data relay for cargo transport services. NPO PM and Germany's Elbe Space & Technology Dresden GmbH, entered into a joint agreement to operate the global store/forward system consisting of L/S/C-band payloads on seven Tsikada navigation satellites

Express series

Background

Gorizont's successor, called the Express, appeared in 1994. InterKosmos operated and marketed the satellite service through its consortium of NPO PM, NIIKP, NPO Radio and the Vostok Bank (along with Gals). Most of the Express payload directly duplicates the Gorizont, although it deletes the L-band Volna service (transferred to Marathon). Express is the second RFAS/Soviet satellite, after Gals, to provide NSSK. Intelsat agreed in March 1993 to lease up to three satellites beginning in mid-1994, but the agreement lapsed in 1994 because InformKosmos required payment in advance. Intelsat only agreed to begin leasing once each satellite was on-station.

Rimsat Ltd, an Asia-Pacific telecom company based in Fort Wayne, Indiana, planned to lease four Express satellites from InformKosmos for the Asia-Pacific region, with first launch in October 1994, but payments reportedly stopped in April 1994.

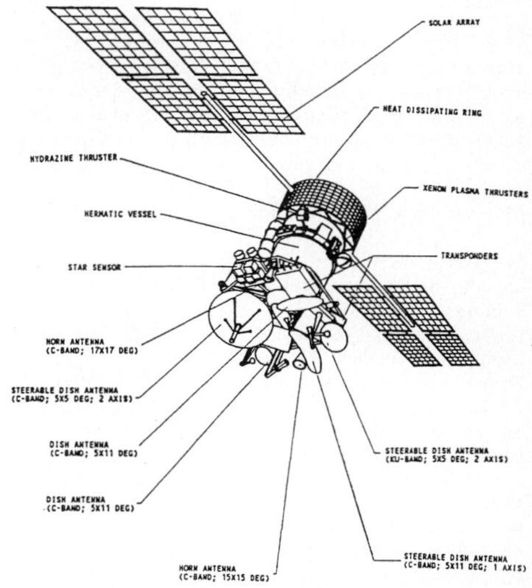

Express is replacing Gorizont (Intersputnik) ***1996***

Specifications

Express 1-2 satellite
Launch: number 1 13 October 1994 by Proton-K from Tyuratam; located over 70° east before moving to 14° west in February 1995 for Intersputnik. Express 2 launched 26 September 1996, located over 80° east and operated by Intersputnik
Locations: initially 14° west and 80° east
Design life: 5-7 years
Contractors: NPO PM (spacecraft), NIIKP (payload)
Transponders: one (plus one back-up) 60-80 W TWTA 6.000/3.675 GHz up/down C-band beam, 40 MHz bandwidth, transponder number 6, switchable to A1 global (EIRP 35.6/32.6 dBW max/min) or A2 steerable (7.3° north, 5° south, 5.5° East/West) 4.6° spot (EIRP 46.1/43.0 dBW max min), RHCP, for single channel TV distribution to 2.5 m Moskva antenna

Nine 10.5 W TWTA 6.050-6.450/3.725 4.125 GHz up/down C-band global, hemispheric, zonal and spot beams, 36 MHz bandwidth, numbers 14/16 jointly switchable to A2 steerable (7.3° north, 5° south, 5.5° east/west) 4.6° spot (EIRP 38.0/35.0 dBW max/min) or A4 steerable 7.5° north-south 4.6 × 10.6° elliptical (EIRP 34.7/31.7 dBW max/min), numbers 8/10 jointly A4 steerable 7.5° north-south 4.6 × 10.6° elliptical (EIRP 39.2/36.2 dBW max/min) or A3 fixed 14.6° quasi-global (EIRP 33.4/30.4 dBW max/min), numbers 7/9/11 jointly A5 fixed 14.6° quasi-global (EIRP 28.9/ 25.9 dBW max/min) or A6 north-south steerable 4.6 × 10.6° elliptical (EIRP 34.7/31.7 dBW max/min), numbers 15/17 as 7/9/11, RHCP transmission.

Two 15 W TWTAs centred at 14.325/11.525 and 14.425/11.625 GHz up/down Ku-band beams, numbers 12/20 working through fully steerable (north-south and east-west) 4.6° spot antenna, 36 MHz bandwidth, EIRP 39.2/36.6 dBW maximum/minimum, orthogonal polarisation (vertical receive and horizontal transmission) for Luch service
Principal users: TV distribution via Moskva system for Russia, telecommunications services
Configuration: 3-axis stabilised, based on pressurised 2 m diameter central pressurised cylinder (maintained at 0-40°C). Seven antennas provide zonal, hemispheric

Express M2000, the most powerful communications satellite designed by Russian telecommunications engineers
(Theo Pirard) ***1999***/0054358

and global beams. Power comes from twin four-panel solar wings spanning 21.0 m, cylindrical base also carries solar cells; 2,400 W EOL. AOCS is provided by liquid hydrazine thrusters (25.5 kg hydrazine); north-south/east-west station-keeping to ±0.2° by eight OKB Fakel SPT-100 plasma thrusters (56 kg Xe loaded). ±0.1° pointing accuracy. Express can operate autonomously for 30 days
Mass: about 2,500 kg at launch (payload 430 kg).

Express-M satellite
Launch: had been planned 1998 by Proton-K from Tyuratam
Locations: initially 40° east, 103° east, 140° east
Design life: 10 years
Transponders: 8 50 W TWTA 14.000-14.250/10.950 11.200 GHz up/down Ku-band 3 × 3° beams, 72 MHz bandwidth, EIRP 46.9 dBW, for 1,024 Mbit/s data services to 1.5-2.5 m dishes
12 (plus 4 back-up, 3 × 4-for-3 + 1 × 2-for-1) 32 W TWTA 14.340-14.500/11.540-11.700 GHz up/down Ku-band 3 × 3° beams, 2 × 36 MHz bandwidth, EIRP 44.9 dBW, for TV distribution and telecom services
Principal users: TV distribution, telecommunications services
Configuration: 3-axis stabilised, based on pressurised 2 m diameter central cylinder. 4 transmission/receive antennas. Twin 6-panel solar wings spanning 26.6 m providing 6 kW (Ni/H_2 batteries) and the base also carries solar cells. AOCS by hydrazine thrusters; north-south/east-west station-keeping to ±0.1°, by Xe ion thrusters (120 kg Xe)
Mass: 2,570 kg at launch.

Gals/Gelikon series

Background
Three Gals satellites, the first of which appeared January 1994 are beginning to replace the Ekran. Gals introduced North-South station-keeping to Russian satellite operations. Russia has updated the satellite to be competitive with the West by building the Gals-R version. Each R derivative carries four or six (reports vary) active transponders. Six Gelikon satellites (possibly also called Gals-RM) working in the FSS range, will complement these satellites. Together the Russians call these three systems the STV-12.

Specifications
Gals series
Launched: number 1 20 January 1994 and number 2 17 November 1995 by Proton from Tyuratam
Locations: number 1 initially 44° east GEO and then 70° east by 13 June 1994. During May-June 1996 number 1 relocated to 35° east, number 2 relocating to this longitude by early July 1996
Design life: 7 years
Contractors: NPO PM (spacecraft), NPO Radio (payload)
Transponders: Two (plus 1-for-1 back-up) 90 W + one (plus 1 for-1 back-up) 45 W TWTA 17.3-18.1/11.72748-12.47550 GHz up/down working through antennas A1 0.9 × 1.2° and A2 1.25 × 2.5°, 27 MHz bandwidth, peak EIRP 57 dBW (56.0 dBW central region), circular polarisation. Both antennas are north-south and east-west steerable and A2 can rotate ±30°
Principal users: direct broadcast TV to 90 cm home (50 cm in 56.0 dBW region) and 1.5-2.5 m professional
Configuration: similar to Express. 21.0 m-span solar array provides 2.40 kW EOL. 100 per cent eclipse protection comes from Ni/H_2 batteries. All four main antennas (A1/A2 transmission + A4/A5 receive) are carbon fibre
Mass: 2,500 kg on-station (375-kg payload)
AOCS: 3-axis; ast-West station-keeping to 0.5° by hydrazine thrusters, NSSK by eight OKB Fakel SPT-100 plasma thrusters
TT&C: C-band via horn antenna. 12 m ground stations at four sites, headed by NPO PM's Krasnoyarsk. Gals allows 30 days' autonomous operations.

Gelikon series
Launch: by Proton from Tyuratam
Locations: planned 23, 44, 74, 110, 140° east

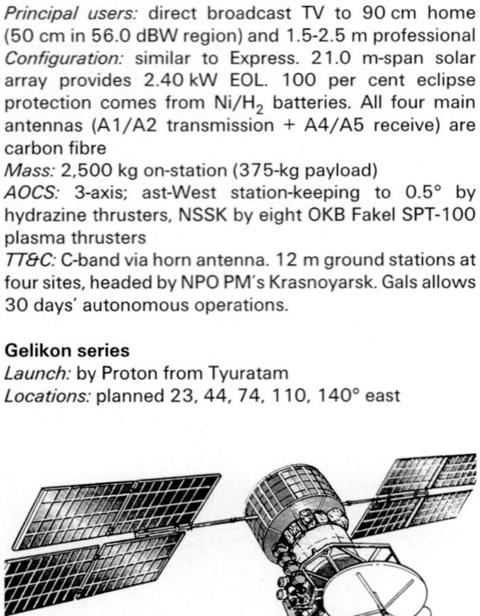

Gals has begun to replace Ekran
(Teledyne Brown Engineering)

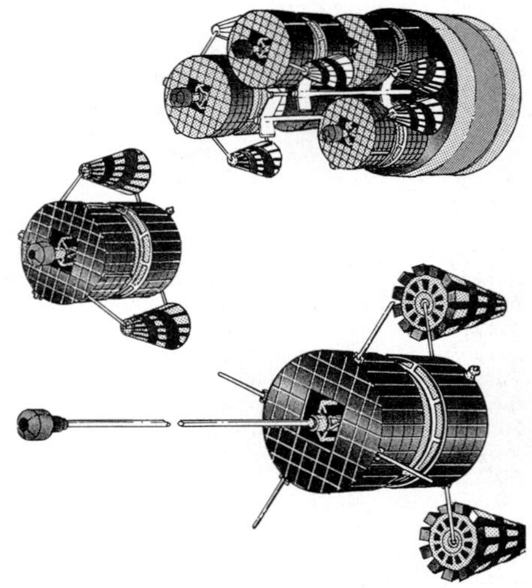

Derivatives of the military sextet store and forward satellites are offered commercially. Flight testing of the Gonets system is underway (Teledyne Brown Engineering)

Design life: 10 years
Transponders: 12 TWTA 17.3-18.1/11.7-12.5 GHz up/down, peak EIRP 55 dBW, circular polarisation
Principal users: direct broadcast TV to 60-90 cm home and 1.5-2.5 m professional
Mass: 2,500 kg on-station
AOCS: 3-axis; north-south/east-west station-keeping to 0.1°.

Gonets-D1 constellation

Background
The Gonets system of small communications satellites come directly from the equivalent military Strela-3 communications satellite programme. Russia flew two Gonets-D demonstrator missions named Cosmos 2199 and Cosmos 2201 as part of the sextet launched on 13 July 1992. The two sextet missions launched on 19 February 1996 and 14 February 1997 each carried three of the military Strela-3 satellites and three Gonets-D1 test satellites.

The Gonets– D1 (Gonets means 'messenger') are cylindrical (covered in solar cells), 0.8 m diameter and 1.6 m long, plus an extended boom for gravity stabilisation and cone– antennae: the mass of each satellite is 225-230 kg. AKO Polyot builds the satellites for NPO PM.

The Smolsat consortium, including NPO PM, NPO Precision Instruments, Selkhoz Bank and Moscow's Soyuzmedinform Program Management (connected with the Ministry of Health), plan two systems. The first will provide e-mail store/forward and real-time (within the 3,000 km diameter footprint) bulk data relay. The second, Gonets-R, will offer mobile phone links.

Network Services International in New York claimed to be the western representative for Gonets and the Russian Space Agency provides state support. The first 36-satellite system was projected to be operational by end-1996, employing six satellites in each of six 82.5° planes at 1,300-1,500 km altitude for 4.8/9.4/64 kbit/s links. 312-315/387-390 MHz up/down 10 W link. Each 230 kg satellite will handle 1.85 Mbit daily, plus 250 Gbit for the 1.6/1.5 GHz up/down L-band, EIRP 10 dBW.

Gorizont (Horizon) series

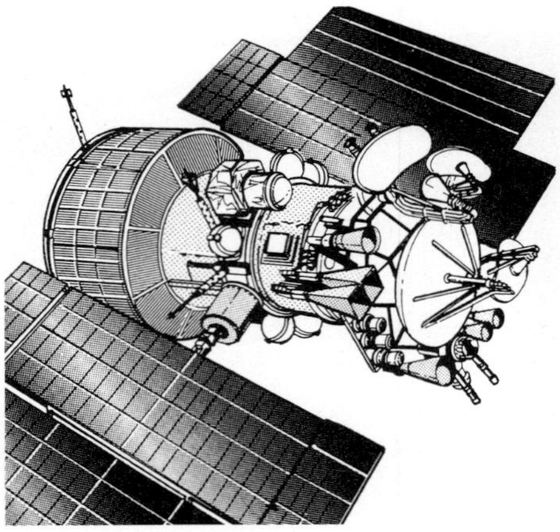

The Gorizont multipurpose communications satellites are available commercially

Background
Gorizont 1 appeared 19 December 1978, although it achieved only 22,553 × 49,023 km, 11.3°. Despite its non-GEO path, it provided some service and was joined by three others for coverage of 1980's Moscow Olympic Games. A single 40 W transponder, similar to its Molniya 3 equivalent, relays TV to more than 1,000 2.5 m diameter dishes in the Moskva system. Gorizont now carries the US/Russia hotline previously supplied by Molniya 3 (Intelsat has also carried it since 1974). The full Gorizont capacity was offered by Glavcosmos at about US$40 million, including launch. Intersputnik provides most of its services but Express will eventually replace them.

Gorizont 22 appeared 23 November 1990, and was positioned at 40° east, the Statsionar 12 location. It was the first of three satellites for the Russian Federation's Ministry of Communications & Information Technology, which paid Rb100 million to the Ministry of Defence for launch services. The satellite provided TV services to 21,000 settlements in European Russia, while the second, Gorizont 23 launched 1 July 1991 to 103° east Statsionar 21, added 3,500 in the Urals and Siberia. The third appeared 2 April 1992 as Gorizont 25, also over 103° east (number 23 apparently failed mid-1992). Gorizont 20 carried a Mayak transmitter, possibly to demonstrate mobile L-band communications. What should have become Gorizont 28 was lost in the Proton launch failure of 27 May 1993. It was to have been the first occupant of the Statsionar 16 145° east slot, from where transponders numbers 9-11 were to have been marketed by SovCan Star and Vista Satellite

Russia and the former Soviet Union launched a total of 32 named Gorizont satellites by end of 1995 into Statsionar and, now, Tonga slots. Russia said January 1996's Gorizont 31 would complete the series, Gorizont 32 appeared in May 1996. Gorizont 29 sits at 130° east (a Tonga slot) and is leased from InformKosmos as Rimsat 1 by Rimsat Ltd. Rimsat initially leased 'Tongasat 1' (Gorizont 17) at 134° east from July 1993, but when the satellite moved January 1995, the lease ended. Gorizont 30 (Rimsat 2) was launched 20 May 1994 towards 142.5° east.

Specifications
Launched: typically twice annually, by Proton-K from Tyuratam
Locations: 40/53/80/90/96.5/103/140° east and 14/11° west geosynchronous (plus Rimsat/Tonga slots), inclined 1.5° BOL
Design life: 3 years
Contractors: NPO PM (spacecraft), NIIKP (payload)
Transponders: One (plus one back-up) 65 W TWTA 6.000/3.675 GHz up/down C-band spot or global beam for transponder number 6, 40 MHz bandwidth, switchable between A3 4.6° spot EIRP 46.1/43.0 dBW max/min or A1 global 35.6/32.6 dBW maximum/minimum, RHCP, for TV distribution to 2.5 m Moskva antenna. The six C-band (also see below) together comprise the Statsionar service.

Operational Gorizont satellites (as of May 1999)

Launch Date	Satellite	Location	Location Name
20 June 1990	Gorizont 20	25° east	
3 November 1990	Gorizont 21	145° east	Statsionar-16
23 November 1990	Gorizont 22	140° east	Statsionar-7
23 October 1991	Gorizont 24	80° east	Statsionar-13
2 April 1992	Gorizont 25	103° east	Statsionar-21
27 November 1992	Gorizont 27	96.3° east	Statsionar-14
28 October 1993	Gorizont 28	90° east	Statsionar-6
18 November 1993	Gorizont 29	161° east	
25 January 1996	Gorizont 31	40.5° east	Statsionar-12
25 May 1996	Gorizont 32	53° east	Statsionar-5

Three Gorizonts are located at positions which have not been registered to the RFAS: Gorizont 19 is close to the Statsionar-2 location (35° east) and Gorizont 20 close to Statsionar-19 (23° east), but there is no RFAS registration close to the Gorizont 29 location.

Five 15 W TWTA 6.050-6.250/3.725- 3.925 GHz up/down C-band global and hemispheric beams, 34 MHz bandwidth, numbers 8/10 on A4 9 × 18° EIRP 31.0/28.0 dBW max/min, numbers 7/9/11 switchable between A6 global 28.5/25.5 dBW max/min and A8 6 × 12° EIRP 34.0 dBW max; RHCP. The six C-band (also see above) together comprise the Statsionar service.

One (plus one backup) 15 W TWTA 14.325/11.525 GHz up/down Ku-band beams, 36 MHz bandwidth, number 12 on A9 4.6° spot EIRP 39.2/36.0 dBW max/min 4.6° Luch service

One (plus one back-up) 1.6377 1.6386/1.5362-1.5371 GHz up/down L-band; transponder cross-strapped with a 15 W C-band transponder for link with Earth station, for Volna maritime service, EIRP 40 dBW

Principal users: TV distribution via Moskva system, telecommunications services, maritime and aeronautical mobile

Configuration: 3-axis stabilised, based on pressurised 2 m diameter central cylinder, about 5 m long. 11 antennas provide zonal, hemispheric and global beams. Twin solar wings, of two major/two minor panels each, spanning 9.460 m generate 1,280 W; cylindrical base also carries solar cells. Liquid bipropellant thrusters provide AOCS. East-West station-keeping to within 0.5° is exercised but north/south excursions are uncontrolled: 2° is possible after 3 years

Mass: about 2,120 kg originally, now about 2,500 kg.

Informatr satellite

Background

The 800 kg Informatr 1 Koskon prototype appeared 29 January 1991. AKO Polyot possibly derived this satellite from a military store/forward design. The press release described it as being launched for the Ministry of Geology to relay data collection platform data, provide communications for survey parties and to support disaster relief operations. Five mobile ground stations worked through the satellite, four in Krasnoyarsk and one in Archangelsk.

Informatr also carried the RS-14 amateur transponders for the Soviet Amsat. Two transponders provided: 435.102-435.022 MH_Z up and 145.852-145.933 MH_Z down, and 435.123-435.043 MH_Z up and 145.866- 145.946 MHz down. Part of this Radio-M1 amateur payload constituted a second version of Amsat Germany's Rudak digital packet communications unit (designated Oscar 21). Amateurs continued to use the satellite until it was deactivated in September 1994 by its VKS Military Space Forces controllers at Golitsyno-2.

Kupon satellite

Current status

Plans to build a system of three Kupon satellites, featuring independently steerable spot beams and full inter-transponder switchability have fallen through.

Background

Lavochkin proposed the Bankir project in 1990 to provide messaging services on 400-800 MHz UHF through 6-8 LEO satellites. Housed on a platform derived from Lavotchkin's Phobos craft, the payload would have been developed by NPO ELAS. Up to 10,000 transmissions of messages up to 400 characters could have been handled daily. The major user would be the Bankir network providing Russian banks with a 40,000-terminal VSAT network with a centralised control centre (the service began in 1993 using existing Potok satellites). Bankir cost US$100 million to establish, and requires 20 per cent of Kupon's capacity.

The company, GIS, became the commercial backer of Kupon in 1991 with shareholders NPO Lavotchkin; NPO ELAS; SovFinTrade, Financial & Trading Co; Pragma Electronics and a consortium of Russian banks. It appears that Russia's Central Bank took control in 1994 after the joint venture failed to raise sufficient capital for the full system.

At the beginning of 1994, the Central Bank of Russia supposedly guaranteed financing of the first three satellites, but failure of the joint venture in 1994 to raise sufficient capital reportedly left only one satellite in prospect. A February 1993 agreement with the Manila based Communication & Broadcast Managers, Inc to set up the Galaxy Satellite & Telecommunications fell through and the project has languished since then.

Specifications

Launches: numbers 1-3 planned 1996-97

Locations: number 1 55° east; number 2 91.75° east, number 3 9.5° west

Design life: 6 years minimum (8 years expected)

Contractors: NPO Lavotchkin (bus); NPO ELAS (payload)

The first Kupon is planned for 1996. Seen at NPO Lavotchkin, the launch configuration shows the radiator at left, solar wing at right and the phased array at top **1996**

Each Molniya 1 carries at least one 40 W X-band transponder for government and military links

Transponders: 16 14.020-14.500/10.960-11.200 and 11.460-11.700 GHz up/down Ku-band spots. Each of four transmission blocks comprises 64 SSPAs and can shape up to four independent spot beams. Each spot is independently steerable ±8.5° and adjustable on either axis from 2 × 2° to 3.5 × 3.5° in 0.5° steps. Each repeater block comprises 6 × 36 MHz 'transponders': four for normal traffic (one per beam) and two either for switching traffic between any beam on any block or as additional transponders in the block. Total power of each block is 20 W and redundancy is provided by the multiple SSPAs (10 per cent failure would provide −0.8 dB loss in overall EIRP). EIRP is adjustable on each block 35 to 50 dBW; equalised power for all beams (2°) would be 39 dBW linearised EIRP (equivalent to 49 dBW saturated), SFD -80 dBW/m² to -95 dBW/m², orthogonal linear polarisation

Principal user: Bankir VSAT network, international diameter domestic VSAT, DBS, cable/TV broadcast, SNG, videoconferencing, point to point

Configuration: 3-axis platform. 4 × transmission and 4 × receive phased array antennas on Earth face, 3-panel solar wings on North/South faces, radiators on East/West faces, electronics housed in central pressurised cylinder

Mass: 2.5 tonnes BOL on-station

TT&C: control centre and ground station both outside of Moscow

Power system: twin 3-panel solar wings.

Molniya (Lightning) constellation

Current status

Eight Molniya 1 and eight Molniya 3 satellites still operate in planes spaced at 45° intervals. Russia has formally ended the Molniya 2 programme. Russia continues to launch Molniyas, but not at the same rate as their original introduction in April 1965. Russia launched only three Molniya types between August 1996 and the end of May 1999.

Background

Sergei Korolev's design bureau first conceived of Molniya in 1962 as a method to cover the USSR's northern expanses from highly inclined orbits. Since boosters had not yet attained the ability to place a satellite in geosynchronous orbit when the USSR launched the first Molniya it put it in a sub-geosynchronous orbit that became a unique signature of the satellite. With a usual perigee over the Southern Hemisphere, each satellite lingers around apogee over the northern territories for about 8 hours each day. Their unusual orbits also avoid the severe launch penalty imposed on geosynchronous satellites from latitudes at Tyuratam and Plesetsk. By coincidence, the 63.4° Molniya inclination also nulls orbital perturbations created by the oblateness of the Earth and so the satellites require much less station-keeping propellant than most other highly elliptical orbits. On the other hand, the orbit passes through the Van Allen radiation belt, which imposes a radiation dosage five to six times the amount experienced in geosynchronous orbit. Hardening to mitigate this radiation dose makes the satellites about 20 per cent heavier than their non-hardened counterparts.

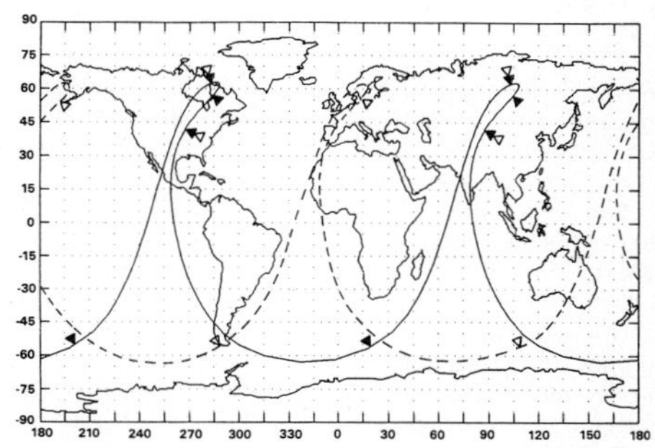

Molniya 1/3 satellites usually follow a common groundtrack taking them over the heart of the Russian landmass. Since 1983, some Molniya 3s have flown with groundtracks (dotted line) between the original constellations
(Teledyne Brown Engineering)

The USSR inaugurated communication service with the first Molniya 1 series satellite, launched 23 April 1965. This Molniya carried the first operational television and telephony links between Moscow and Vladivostok before decaying 16 August 1979. Molniya 1-3, launched almost exactly one year later, added Earth imaging to complement the Meteor series, and exchanged colour television signals with France. Three satellites in a constellation with the same orbital parameters provide 24-hour coverage. Generally however, the USSR placed eight satellites in planes 45° apart to ensure un-interruptible service. The original network of Molniya 1 satellites probably carried military/government links over one transponder.

The Molniya 1 appears to be employed primarily for military and government links through at least a single 40 W X-band transponder. Some early models incorporated a camera to add synoptic Earth views to low-altitude Meteor imagery. Molniya 1-20, launched 4

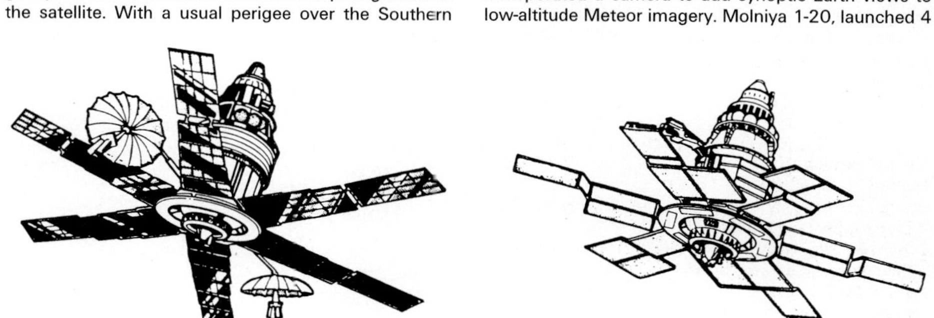

Molniya 3 (right) compared with Molniya 1 (Teledyne Brown Engineering)

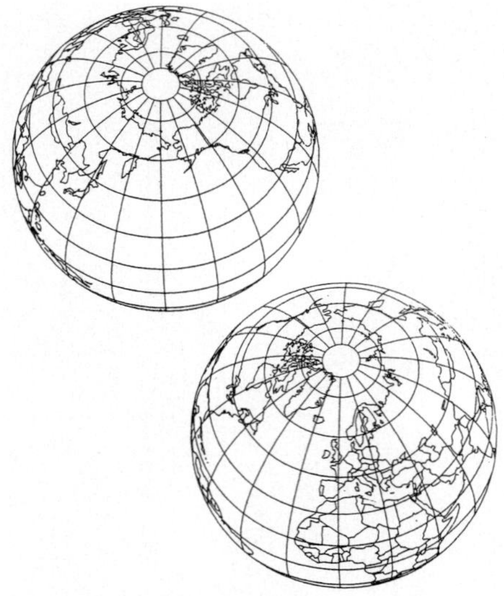

One group of Molniya 3s reaches apogee over the Pacific and East Atlantic, possibly for maritime links. Shown is the vantage point on each of the two daily apogees (Teledyne Brown Engineering)

April 1972 from Plesetsk carried France's piggyback research SRET 1. Molniya 1-30 similarly released SRET 2 on 5 June 1975. Molniya 1S, different from the standard series acted as a geosynchronous demonstration.

Beginning with Molniya 1-11 in 1969, the number above the horizontal at any one time was increased from two to three by reducing the equatorial spacing to 90°. Following the introduction of the Molniya 2 series in 1971 and Molniya 3 in 1974, they were positioned in groups of three employing different frequencies. As traffic increased, the Soviets had to decrease the spacing between Molniya 1s to 45°.

The Molniya constellation achieved great success at bringing communications to remote parts of the USSR. Starting with Molniya 1-11 in 1969, the USSR decreased the orbital stationing to 90° instead of 120° in order to keep at least two satellites above the horizon at all times. The Molniya 2 constellation followed in 1971, and the Molniya 3 constellation in 1974. But the newly increased traffic demands caused the USSR to rely more on geosynchronous satellites and the last Molniya 2 began its service in 1977.

The first Molniya 2, a more capable series employing higher frequencies orbited on 24 November 1971. The 17th and last satellite appeared on 11 February 1977.

From early 1976, Molniya 3 provided one of the two independent satellite-based systems (the other via Intelsat) to maintain the Washington-Moscow hotline. Presidents Nixon and Brezhnev agreed to the hotline concept during the SALT negotiations of 1971. To ensure contact, the United States agreed to continually track the entire Molniya constellation from the US station at Fort Dietrick (Frederick, Maryland). Dietrick tested the circuits hourly. (Traffic from the hotline has now shifted mostly to the Gorizont).

Until 1984, all Molniyas had a common ground track roughly bisecting Asia/North America. In late 1984, Molniya 3-21 manoeuvred to re-establish its ground track over the Pacific and the extreme eastern edge of the Atlantic. During 1985, Molniyas 3-25, 3-26 and 3-27 followed suit. These new satellites had a planar spacing of 90°, but the ascending nodes occurred mid-way between the other Molniya 3s. Western observers speculate that this new constellation might be connected with maritime communications.

The Molniya 3, NPO-PM's newest class of satellites offers three C-band transponders for general telecommunication services, including television distribution over the 100 or more ground stations of the Orbita 2 network. 12 m ground stations in the CIS and affiliated countries such as Cuba and Mongolia.

157 named satellites have been launched in the three series. Molniya 1-85 was at first described as a Molniya -1T, although the Russians subsequently indicated that Molniya-1T satellites had been operating for many years before this. In the same source the Russians referred to Molniya 1-88 as the 48th 1T, but they gave no indication of the difference with the standard design.

Specifications

Molniya 1

Launched: typically one a year by Molniya-M from Plesetsk (very rarely Tyuratam)
Orbit: typically 400 × 40,000 km, 63°; placed in eight orbital planes separated by 45°
Design life: 2 years
Contractors: NPO PM
Transponders: At least one 40 W 1.0/0.8 GHz up/down X-band; possibly total of three transponders
Principal users: military/government
Configuration: pressurised cylinder with domed ends, 1.6 m diameter, 3.4 m long, accommodates electronics and provides mounting for two boom-mounted 90 cm diameter dishes. Six windmill solar wings generate 500-700 W. AOCS provided by liquid bipropellant engine assembly at other end; 3-axis control
Mass: about 1,600 kg.

Molniya 3

Launched: typically now one in two years
Orbit: typically 400 × 40,000 km, 63°; placed in eight orbital planes at 45° intervals
Design life: 4 years?
Transponders: 2 (plus 1 back-up) 30 W TWTA 5.9756.225/3.650– 3.900 GHz up/down C-band global beams, 50 MHz bandwidth, 35 dBW edge EIRP, circular polarisation
Principal users: domestic TV/telecommunications
Configuration: pressurised cylinder with domed ends, 1.6 m diameter, 4.2 m long, accommodates electronics and provides mounting for six 3-panel windmill solar wings spanning 8.13 m and generating 1.2 kW (950 W payload requirement). AOCS provided by 1.96 kN liquid bipropellant engine assembly at other end; 3-axis control
Mass: 1,750 kg (payload 220 kg).

SPAIN

Hispasat series

Current status

The increase in demand on Hispasat's capacity due to digital television and the liberalisation of satellite services has led Hispasat SA to plan the purchase of a third satellite, Hispasat 1C before the first generation satellites require replacement. Hispasat 1C would be located over 30° west (with 1A and 1B) and would provide improved performance and capacity over Europe and the Americas.

Background

The Spanish government originally considered building its own satellite network in 1985, but then rejected the idea. Parliament resurrected again in July 1988 to address the communications needs for the Barcelona Olympic Games and the Columbus quincentenary celebrations. After approval, The Secretaria General de Comunicaciones issued tenders for a two-satellite domestic broadcasting and communications system, plus ground segment. The programme received government approval in April 1989 for two satellites, and a ground spare, to provide three to five Spanish direct broadcast service channels, two North and South America television distribution channels, 8 to 16 television broadcast and communications channels and two government/military X-band channels.

Spain gave the satellite construction contract to the Satcom International partnership of Matra/British Aerospace in June 1989 for Pts19,898 million. The design reflected closely Matra's Telecom 2 because Spain sought to have the satellite in space by 1992. Spain chose an Ariane 4 instead of an Atlas 2 to launch the satellite despite the lower US bid because of the presumed greater stability of the peseta against the franc.

Hispasat 1A became operational in January 1993 despite the fact the 2.2 m direct broadcast satellite antenna pointed slightly too far north, probably because of thermal distortion. The addition of 1B created a total of five operational DBS channels. The Spanish military used the Hispasat for the first time in March 1993 when Loral Western Development Labs installed an SCT-10 anchor terminal in Madrid to provide 12 voice/ data circuits to a transportable unit in Bosnia.

Specifications

Hispasat 1A

Launch: 10 September 1992 by Ariane 4 from Kourou, French Guiana

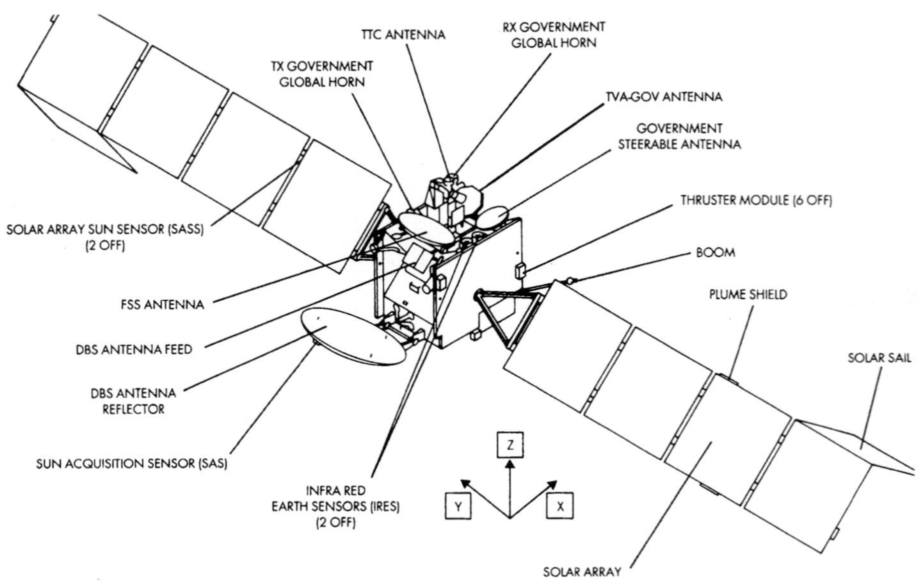

Hispasat configuration: the large dish provides DBS services to 60 cm receivers within Spain

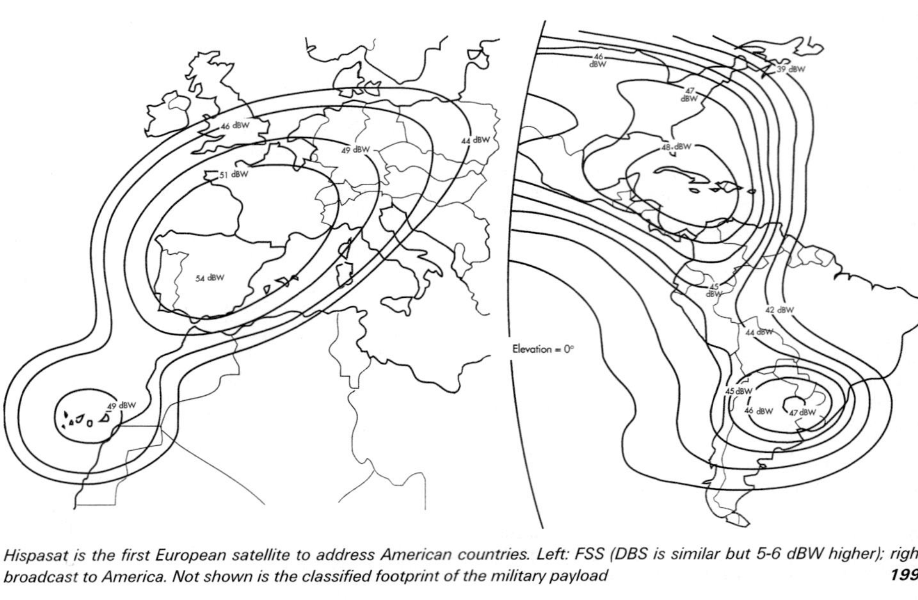

Hispasat is the first European satellite to address American countries. Left: FSS (DBS is similar but 5-6 dBW higher); right: broadcast to America. Not shown is the classified footprint of the military payload ***1996***

Location: 30° west geostationary
Design life: 10 years min, 12 years possible
Contractors: Satcom International (MMS France prime), MMS UK (payload), Thomson Tubes Electroniques (TWTs), ANT (EPCs)
Transponders: three (plus two back-ups shared with TVA) 110 W TWTA 17.314-17.648/12.136-12.470 GHz up/down Ku-band supporting two DBS beams over Spain and the Canary Islands working to 30-60 cm dishes, 27 MHz bandwidth, 55-58 dBW Spain (54-56 dBW Canary Islands) EIRP

One (plus two back-ups shared with DBS) 110 W TWTA 14.249/12.078 GHz (12.015 GHz on 1B) Ku-band supporting TVA Spanish-speaking TV distribution service to North/South America through 1.5-2.5 m stations, 36 MHz bandwidth, 41-47 dBW EIRP

Eight (plus four back-up) 55 W TWTA 14.00-14.50/11.450-11.700 + 12.50-12.750 GHz up/down Ku-band FSS European/Canary Islands beams supporting TV broadcast + telecom services, 4 × 36, 2 × 54, 2 × 72 MHz bandwidth, 49.5-52 dBW

Three (plus one back-up) 40 W TWTA 7-8 GHz X-band Spanish beams for government/military applications, 20-40 MHz bandwidth, EIRP 42 dBW
Principal applications/users: DBS, TV distribution, datalinks, point-to-point applications, rural telephony, and defence. FSS controlled by Telefonica, Retevision and Spanish Post Office; military/government by INTA
Configuration: Matra/BAe Eurostar 2000 bus
Mass: 2,194 kg at launch, 1,325 kg on-station BOL, 1,013 kg dry, 280 kg communications payload
Attitude control: 3-axis employing inertial/Earth sensors in conjunction with hydrazine thrusters
TT&C: identical to Telecom 2 Satellite Control Centre Three dedicated Ku-band + single S-band antennas at the Arganda facility, near Madrid
Power system: twin 4-panel solar wings, 2.050 m wide, spanning 22 m, providing 3,237 W at equinox BOL and full eclipse protection with Ni/H_2 batteries Communications payload requirement 2.50 kW
AOCS: unified 3-axis control and apogee GEO insertion motor using NTO/MMH; two sets of 10 N thrusters and single 490 N motor.

Hispasat 1B
Launch: 22 July 1993 by Ariane 4 V58 from Kourou, French Guiana
Location: 30° west (collocated with Hispasat 1A)
Transponders: Same as 1A but two FSS channels (1 × 54 + 1 × 72 MHz) have switchable uplink to relay TV (TVR) from the Americas to Spain
Mass: 2,210 kg at launch, 1,330 kg on-station BOL, 1,052 kg dry
Other specifications are the same as for Hispasat 1A.

SWEDEN

Tele-X/ Sirius series

Current status
The Swedish Space Corporation signed a contract for a single Sirius 2 large satellite to replace Tele-X on 4 July 1995 with Aerospatiale. The contract included a duplicate payload for Sirius 3 but SSC subsequently went to Hughes Space & Communications for this satellite.

Background
The Nordic countries of Denmark, Finland, Iceland, Norway and Sweden began studying the feasibility of a joint regional satellite system in the mid-1970s. From the beginning, these countries sought television direct broadcasting via satellite for their far northern regions. In 1977, the World Administrative Radio Conference on broadcast satellite services (WARC) allocated eight channels to the Nordic group as an entity (excluding Iceland), plus another 17 to the individual countries (including five for Iceland). What became the Tele-X project assumed its current form in 1982, when the Nordic countries sought to provide both direct broadcast satellite and business communications services within Scandinavia, and also as a means of stimulating the space engineering capabilities of these countries. The 'X' represented Experimental, to acknowledge the developmental nature of this indigenous satellite.

By then, Denmark and Iceland had withdrawn from the project, though Denmark had earlier expressed interest in a beam covering southeast Greenland. Sweden and Norway agreed to continue in a formal agreement signed in 1983, giving them 85 per cent and 15 per cent interest, respectively, in NSAB (Nordiska Satellite AB). Three per cent of Sweden's share of cost/work in the project (but not in NSAB) was allocated to Finland later in a separate bilateral agreement.

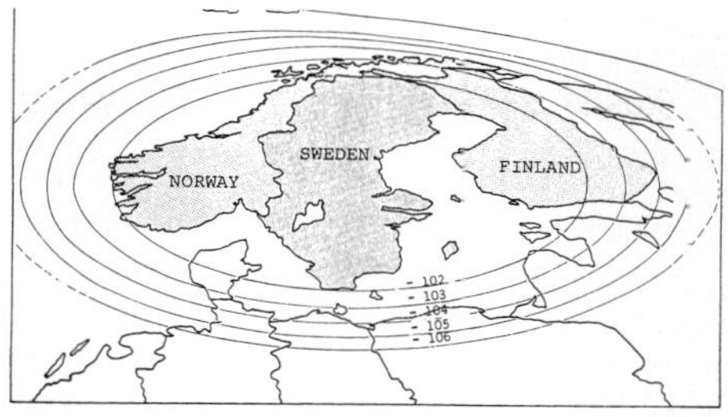

Tele-X transponder coverage; the values represent power density in dBW/m²

Taking advantage of the Franco-German TDF/TV-Sat programme, initiated in 1980, the Swedish Space Corporation (SSC) ordered a similar satellite based on the Spacebus 300 platform. Nordiska Satellite AB (NSAB) planned to be proprietor of the system, while SSC planned to act as the procurement agent and executive agency. Shortly before launch, Norway withdrew from NSAB, trading its 15 per cent share for a lifetime lease of one channel. NSAB now constitutes a 50/50 partnership between SSC and Teracom, a major distributor of television/ radio programming. SSC became wholly responsible for operating the system.

The Sirius system eventually cost Sweden SKr1.25 billion at fixed 1982 rates, or US$280 million. All three direct broadcast satellite television channels are occupied and the Swedes have sought other capacity. NSAB bought the UK's Marcopolo 1 DBS satellite from National Transcommunications Ltd in 1993 for a reported £2.1 million to add five channels. NTL had acquired the satellite from BSkyB. It was renamed Sirius and moved from 3.1° west in January 1994 to begin services at the end of February from 5° east.

Specifications
Tele-X
Launched: 2 April 1989 by Ariane V30 from Kourou
Location: 5° east geostationary
Design life: 6 to 8 years
Contractors: Aerospatiale/Eurosatellite were joint prime contractors. Aerospatiale also handled satellite AIT, thermal control, transmit reflectors/antenna subsystems, and built the solar arrays. Scandinavian industry received 30 per cent of the programme contract value: Saab was associate prime, and was involved in structure, TT&C, antenna integration and OBC work. Ericsson was responsible for the communications payload, receive antennas & beam-tracking receiver. Other Eurosatellite members involved were AEG (solar cells, TWTs, power subsystems), ANT (TWTs), Alcatel Espace (communication module integration, linear amplifiers, other components), ETCA (power system components) & MBB (AOCS). Ground segment contracts went to Elektrisk Bureau (Norway) and Nokia Research Centre, Teleste, Valmet, VTT Technology (Finland)
Transponders: Two (plus one back-up) 220 W (230 to 260 W planned) TWTA 17.7-18.1/11.7715-12.489 GHz up/down Ku-band Nordic TV DBS beams, 27 MHz bandwidth, EIRP 61 dBW max, LH circular polarised

Two (plus one back-up) 220 W (230-260 W planned) TWTA 14-14.25/12.5-12.75 GHz up/down Ku-band Nordic data/video services, 40 & 86 MHz bandwidths, data rates of 0.64/2/34 Mbit/s, EIRP 61 dBW max at beam edge
Configuration: Aerospatiale/MBB Spacebus 300 3-axis satellite with 2.4 × 1.65 × 2.4 m box-shaped bus. Unified liquid bipropellant propulsion system for GEO insertion + AOCS, in conjunction with MWs, IR Earth sensors. RF antenna pointing system. The same deployable receive/transmission antennas are used for each payload, with 0.165°/0.075° pointing accuracies, respectively. Body pointing accuracy is ±0.3°
Mass: 2,130 kg at launch, 1,277 kg BOL on-station
TT&C: Ku-band operational (S-band during transfer phase), via SSC Earth station at Kiruna, Sweden
Power: 3,200 W EOL, generated by twin 4-panel solar wings with 19 m span. One 18 Ah Ni/Cd battery.

Sirius 1
Launched: 27 August 1989 by Delta 4925 from Cape Canaveral pad 17B
Location: 5.2° east geostationary
Design life: 10 years
Contractors: Hughes Space & Communications
Transponders: six paired 55 W TWTAs 17.385-17.992/11.7715-12.1054 GHz up/down Ku-band beam (BSS channels 4, 8, 12, 16, 20), 27 MHz bandwidth, 59 dBW min EIRP (63 dBW max), RHCP
Principal applications: TV direct to home
Configuration: Hughes' HS-376 spin-stabilised, 7.2 m high deployed (2.7 m stowed), 2.16 m diameter. Despun single reflector and communications shelf
Mass: 1,250 kg at launch, about 660 kg on-station BOL
Pointing: 0.05° accuracy by RF beacon
Attitude control: spin stabilised at 50 rpm by four 22.2 N hydrazine thrusters; GEO insertion provided by solid Thiokol Star 30B AKM
Power system: K7 solar cells mounted on cylindrical exterior and deployed drum providing 1,024 W BOL (drum extended 25 cm over normal Hughes 376 configuration). Two Ni/Cd batteries provide 100 per cent eclipse protection.

Tele-X is based on the Spacebus 300 platform (Aerospatiale)

Sirius 2
Launch: 12 November 1997 by Ariane 4 from Kourou, French Guiana
Location: planned 5° east geostationary
Design life: >12 years (15 years propellant)
Contractors: Aerospatiale (prime), Space Systems/Loral (repeater), DASA (AOCS, solar arrays), Saab Ericsson Space (antennas, data handling), Alenia (TCR-RF)
Transponders: 13 57 W TWTA 17.30-18.10/11.70-12.50 GHz up/down Ku-band Scandinavian BSS beams, 33 MHz bandwidth, EIRP 55.5 dBW edge, linear vertical polarised

13 85 W TWTA 17.30-18.10/11.70-12.50 GHz up/down Ku-band European BSS steerable beams, 33 MHz bandwidth, EIRP 49.7 dBW edge, linear horizontal polarised

6 85 W TWTA 14.00-14.25/12.50-12.75 GHz up/down Ku-band Northern/Central European FSS beams, 36 MHz bandwidth, linear polarised
Principal applications: DTH, TV distribution, and data services
Configuration: Spacebus 3000B 3-axis with 1.8 × 2.3 × 2.86 m box-shaped bus. Two 1.8 m fixed dual gridded antennas for FSS and Scandinavian BSS coverage, plus steerable antenna for BSS European beam
Mass: 2,920 kg at launch; 1,760 kg BOL; 1,250 kg dry
Pointing: standard body pointing accuracy 0.1-0.15° following in-orbit calibration. Fine pointing by RF sensors and antenna pointing mechanisms (0.05°)
AOCS: unified bipropellant (NTO/MMH) system of single 400 N and redundant networks of 7 × 10 N thrusters provide GEO injection + 3-axis control. Two titanium propellant tanks, titanium/Kevlar pressurisation tanks. Two 2-axis IRES IR Earth Sensors, two Sun sensors (4 heads each, 60° FOV), two RIGA gyro packages, two 52-68 Nms fixed momentum pitch wheels controlled by digital programmable control electronics. Normal stabilisation is based on pitch by wheel, roll/yaw by Earth sensor using roll/yaw thrusters. Station-keeping by 10 N thrusters; yaw reference by Sun sensors or gyros
TT&C: via SSC Earth station at Kiruna, Sweden
Power system: 5,845 W EOL from twin 4-panel wings spanning 26.3 m. Two 62 Ah Ni/H_2 batteries provide 100% eclipse protection.

Sirius 3
Launch: 5 October 1998 by Ariane 44L from Kourou, French Guiana
Location: 5° east
Design life: 12 years
Contractor: Hughes Space & Communications HS-376
Transponders: 14 Ku-band 11.7-12.5 GHz
Configuration: body diameter 2.16 m, height extended from 3.32 m at launch to 7.76 m in orbit.
Mass: 1,465 kg at launch; 815 kg BOL; 630 kg dry
Power: 1,400 watt.

THAILAND

Thaicom series

Current status

The Asian financial crisis has suspended negotiations for Thaicom 4. Total cost for Thaicom 3 on station is estimated at US$240 million.

Background

Thailand requested that satellite offerers provide them with quotes for a new satellite broadcasting system in 1987 to reach its entire population of 55 million persons. The announcement predated Thailand's eventual entry into Inmarsat in January 1995. The country expected to offer a 30 year concession to lease transponders to government and private sector users, with excess capacity offered to neighbouring countries.

Shinawatra Computer and Communications Group (SC&C), the largest group of Thai companies operating telecommunications and broadcasting service concessions, won an eight year monopoly in 1991 to supply satellite capacity to domestic users. The founder and major shareholder of Shinawatra is Dr Taksin Shinawatra. SCC began formal negotiations June 1991 with Hughes, and signed a contract on 8 October 1991 for about US$100 million to buy two satellites, ground equipment and training.

As a lighter version of Hughes' HS-376 bus, Thaicom was the second to come within Ariane's SDS (Spelda Dedicated Satellite) 1 tonne launch limit. Shinawatra signed a letter of intent on 1 November 1993 to negotiate with Hughes for a third satellite. The total cost for Thaicom 3 on station is estimated at US$240 million.

Specifications

Thaicom 1

Launch: 18 December 1993 by Ariane 4 from Kourou, French Guiana. Handed over by Hughes to Nonthaburi (Bangkok) station 31 December; entered full service February 1994
Location: 78.5° east GEO (collocated with number 2); will move to 120° east after replaced by number 3
Design life: 13.5 years minimum, 15 years goal
Contractors: Hughes Space & Communications (prime), Telespace Ltd. (technical support)
Transponders: 10 (plus 2 back-up, 12-for-10) 11 W SSPA 5.925-6.425/3.700-4.200 GHz up/down C-band Thailand, Indochina, North Pacific rim beam, EIRP 35 dBW edge (37.5 dBW over Thailand), 36 MHz bandwidth, orthogonal linear polarisation

2 (plus 1 back-up, 3-for-2) 47 W TWTA 14.3159-14.4951/12.5679-12.7471 GHz up/down Ku-band Thailand beam, EIRP 51 dBW edge (54 dBW centre), 54 MHz bandwidth, orthogonal linear polarisation
Principal users: domestic phone, TV, cable TV, voice, video, data, VSAT services
Configuration: lightweight version of Hughes' HS-376 spin-stabilised bus, 2.16 m diameter, 6.7 m height with solar array drum extended (2.6 m stowed). Single antenna comprising two offset grid reflectors with horizontal/vertical feed system on despun platform
Mass: 1,080 kg at launch, 629 kg on-station BOL, 436 kg dry
AOCS: 55 rpm spin control/position maintained by four Hughes 22.2 N hydrazine thrusters. GEO insertion provided by Thiokol Star 30BP solid

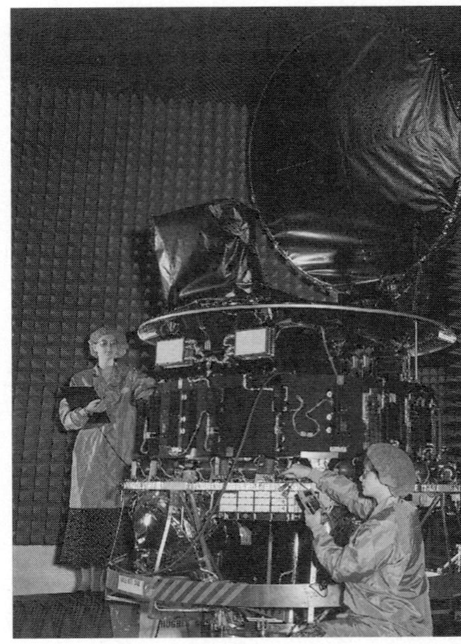

Thaicom 2 anechoic chamber antenna performance tests (Hughes Space & Communications) ***1995***

TT&C: through the Nonthaburi station, 15 km north of Bangkok
Power system: Si cells on cylindrical surface provide 801/671 W BOL/EOL (summer solstice) in conjunction with 51.6 Ah Ni/H_2 battery.

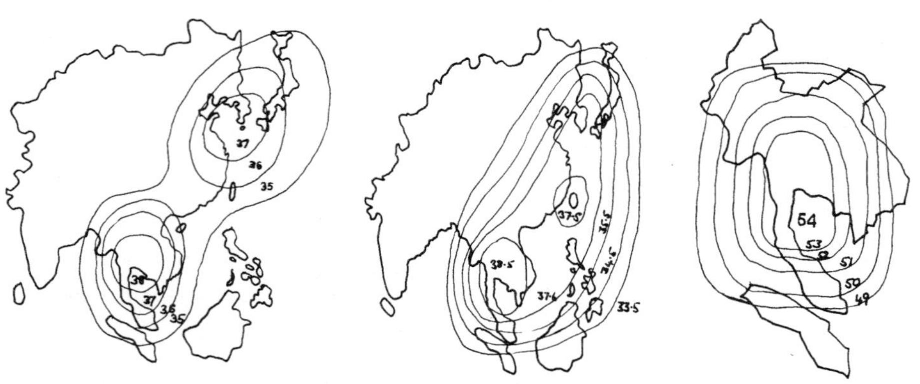

Thaicom 1/2 footprints from 78.5°east. Left: Thaicom 1 C-band; centre: Thaicom 2 C-band; right: Thaicom 1/2 Ku-band (Shinawatra Satellite Public Co Ltd) ***1995***

Thaicom 2

Launch: 8 October 1994 by Ariane 4 from Kourou, French Guiana
Location: 78.5° east GEO (collocated with number 1)
Other specifications as for Thaicom 1, but see footprint diagram.

Thaicom 3

Launch: planned for 16 April 1997 on Ariane 4 from Kourou, French Guiana
Location: 78.5° east GEO (co-located with number 2)
Design life: 14 years (16 years propellant)
Contractors: Aerospatiale (prime; structure, thermal control, solar arrays, integration, testing), Alcatel Espace (communication payload), DASA (AOCS integration)
Transponders: 18 (plus 6 back-up; 24-for-18) 19 W SSPA 5.925-6.725/3.405-3.700 GHz up/down C-band India/Indochina regional beams, 36 MHz bandwidths, EIRP 37 dBW min, orthogonal linear polarisation

6 (plus 3 back-up; 9-for-6) 42 W SSPA 6.425-6.725/3.405-3.700 GHz up/down C-band hemispheric beams, 36 MHz bandwidths, EIRP 34.5 dBW min (38.5 maximum), orthogonal linear polarisation

14 (plus 4 back-up, 18-for-14) 97 W TWTA 14.000-14.500/12.200-12.750 GHz up/down Ku-band domestic DTH beams and one steerable beam (max 7 36 MHz transponders), 2 × 54 MHz + 12 × 36 MHz bandwidth, EIRP 54 dBW min Thailand and 45 dBW India (51.5 maximum), orthogonal linear polarisation
Principal applications: Ku-band principally for DTH

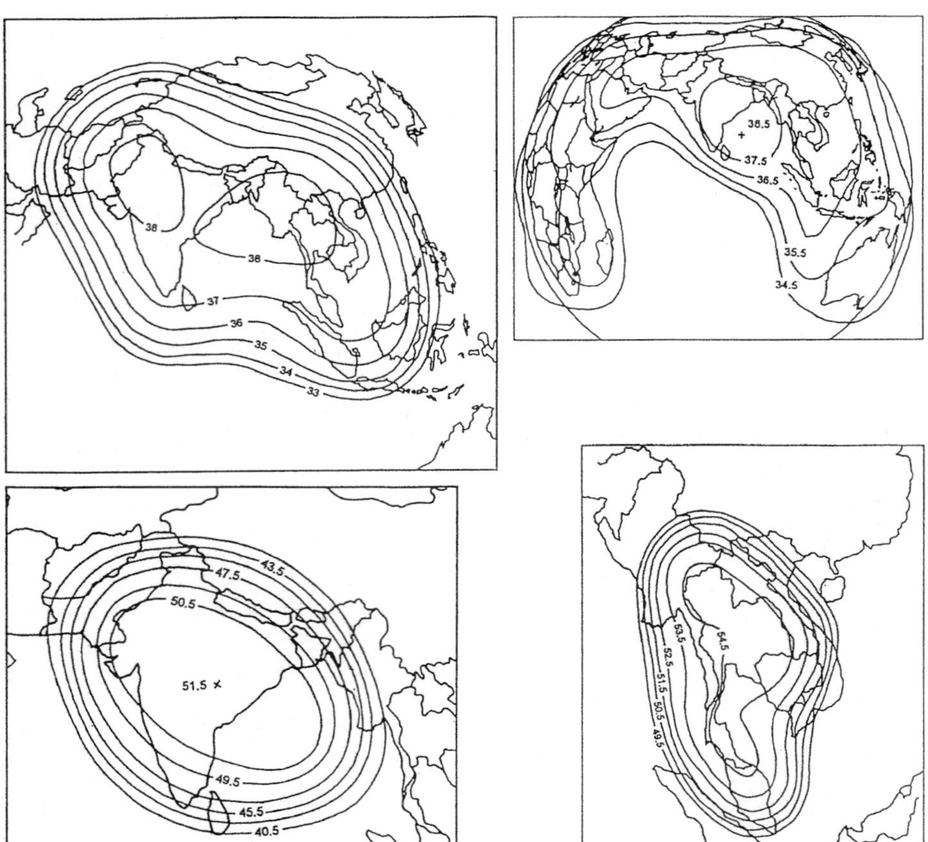

Thaicom 3 footprints (dBW). Top left: C-band India/Indochina; top right: C-band hemispherical; bottom left: Ku-band steerable (India); bottom right: Ku-band Thailand (Shinawatra Satellite Public Co Ltd) ***1996***

Thaicom 1, ANHS-376L gets an antenna beam checkout in an anechoic chamber (Hughes Space and Communications) ***1999***/0054345

Technician checks the feedhorn array for Thaicom 1 (Hughes Space & Communications) ***1999***/0054344

Thailand's first domestic communications satellite, a lightweight HS-376 was launched on 18 December 1993 ***1999***/0054343

Configuration: Aerospatiale/DASA Spacebus 3000 3-axis stabilised 1.8 × 2.6 × 2.3 m box-shaped bus (7 m deployed height) with 25 m-span twin solar wings
Mass: 2,650 kg launch, 2,450 kg on-station BOL, 1,160 kg dry
TT&C: as Thaicom 1/2
Power system: 5,300 W EOL provided by twin Si solar wings totalling 46 m². Two 57 Ah Ni/H_2 batteries.

LStar 1
Launch: Ariane 4 from Kourou, French Guiana
Location: planned 116° east GEO (collocated with number 2)
Design life: 14 years minimum
Contractors: Space Systems/Loral (prime), Telesat Canada (consulting, management services)
Transponders: 32 (48-for-32 ring redundancy) 113 W TWTA 17.32748-17.76862/11.72748-12.16862 GHz up/down Ku-band (paired TWT for high power mode, single TWT for medium power) in three beams: 55.5 dBW min West (India, Pakistan, Nepal, Sri Lanka); 56.8 dBW min Central (Thailand, Laos, Cambodia, Vietnam, Singapore, Western Malaysia); 53.8 dBW min East (Eastern China, Hong Kong, Taiwan and Korea). 33 MHz bandwidth, orthogonal linear polarisation
Principal users: DTH TV, VSAT services
Configuration: 3-axis FS-1300 bus, two 4-panel GaAs solar wings (plus Ni/H_2), Gregorian antenna for each beam. 12 × 22 N NTO/MMH thrusters and Marquardt R-4D 490 N for GEO insertion
Mass: 3,500 kg at launch, 2,132 kg on-station BOL, 1,480 kg dry
TT&C: through the Vientiane, Laos station; Perth back-up.

LStar 2
Launch: Uncertain
Location: planned 116° east GEO (collocated with number 1)
Other specifications as for LStar 1.

TURKEY

Turksat series

Background
Turkey chose to build a completely turnkey satellite operation in order to reach Turkish speaking people in the Central Asian land mass. The government released a request for proposals in June 1989 and received offers from Aerospatiale Hughes/Selenia Spazio, BAe-Matra/ANT and Aerospatiale/MBB by the deadline of December 1989. Aerospatiale signed the US$315 million prime contract, 21 December 1990 for turnkey delivery of two satellites, associated ground stations, insurance, financing and training. It is the first complete system ever delivered on a turnkey basis by a European team. Each satellite carries 16 active Ku-band transponders providing television/radio distribution (also serving the large Turkish populations in Germany/Austria and CIS), data and voice services.

Turkey's first satellite disappeared in the Ariane failure of 24 January 1994. In the contract, Turkey required Aerospatiale to provide on-orbit delivery and the company had taken full insurance, so it provided a replacement in 1996.

Turkish Telecom and Aerospatiale were expected to form the Eurasiasat joint venture in August 1996 before ordering Turksat 2A.

Specifications
Turksat 1B
Launch: 10 August 1994 by Ariane 4 from Kourou, French Guiana
Location: 42° east GEO initially, relocated to 31° east during October 1996.
Design life: 10-13 years (10 years contracted)
Contractors: Aerospatiale (prime), Alcatel Espace (payload), DASA/MBB (AOCS, solar arrays), Thomson Tubes Electroniques (TWTAs), Eagle Picher (battery), SEP (BAPTA), Contraves (structure), Sextant Avionique (power control, distribution unit), ETCA (power conditioning), Teletas (TV receive antennas)
Transponders: 16 (plus 8 back-up; 12-for-8 ring redundancy) 55 W TWTA 14.000-14.500/10.950-11.700 GHz up/down Ku-band Central European, Turkey, Central Asia, bandwidths 6 × 72 MHz + 10 × 36 MHz, EIRP 51 dBW Turkish, 48 dBW Central European, 45 dBW Central Asia, linear polarisation. Up to four channels can be switched to Central Europe and up to two to Asia
Principal applications: TV/radio direct (12 TV + 20 radio to 50 cm antennas), telephony, VSAT data and business services, mobile links
Configuration: Spacebus 2000 platform, 3-axis control, 22.4 m span across solar panels, single dual reflector multifeed antenna. GEO location maintained within ±0.05°
Mass: 1,783 kg at launch; 1,078 kg on-station BOL; dry 827 kg
AOCS: provided by DASA/MBB's 22 × 10 N + 400 N MMH/MON bipropellant system
TT&C: Ankara (prime). Istanbul (back-up)
Power system: twin solar wings providing 2.9 kW equinox EOL.

Turksat 1C
Launch: 9 July 1996 by Ariane 4 from Kourou, French Guiana
Location: initially over 31° east for testing, then manoeuvred to 42° east in September 1996
Mass: 1,743 kg at launch; 1,078 kg on-station BOL; dry 789 kg.

UNITED STATES OF AMERICA

ACTS satellite

Background
NASA Lewis developed and is responsible for the Advanced Communications Technology Satellite which it is using to demonstrate new technologies required for multiple antenna beam communications and satellite switched operation in the 30/20 Ghz Ka-band. More than 100 industry, government and university organisations use ACTS in about 80 different application, technology and propagation experiments. Most applications trials are oriented towards services

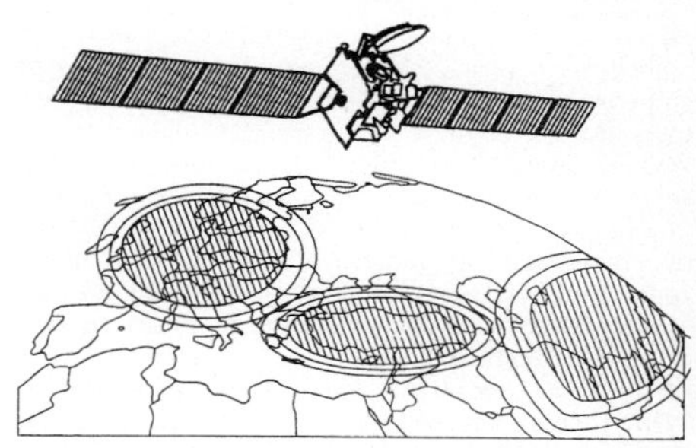

Turksat provides three coverage beams: Central Europe, Turkey and Central Asia (Aerospatiale) ***1994***

with commercial potential: business networks, ISDN, education, DoD tactical reconnaissance imagery, broadcast video, and aeronautical links.

NASA Lewis contracted with RCA Astro-Electronics for US$260 million to build the ACTS in 1984 with the intention of launching the satellite in 1989. Hughes Communications Galaxy Incorporated/Galaxy challenged the government for using public funds to build the ACTS on the grounds that it was planning a similar project using its own corporate funds. The US government put ACTS on hold at the time, but then Congress directed the space agency to revive the satellite. Within six months, the project was again shelved and re-approved – a process repeated several times.

Congress ultimately determined that with the imminent congestion of the spectrum and GEO arc, US leadership in this might erode with respect to Europe and Japan and that NASA needed to reinvigourate its experimental communication work. NASA Lewis resurrected the programme in February 1988, but Congress still demanded that costs should be capped at US$475 million. NASA argued successfully that this ceiling compromised its technology goals, and a limit of US$499 million was agreed to instead.

Motorola is incorporating ACTS-type technology into Iridium and Hughes's Spaceway proposed 48-spot Ka-band video phone satellite system. The US Army used ACTS to provide secure videoconferencing links with Ft Bragg, NC from Haiti following September 1994's occupation.

Specifications
Launch: 12 September 1993 by Shuttle-TOS, ejected 21.13 GMT, TOS fired 21.58 GMT to enter GTO
Location: 100° west geostationary
Design life: minimum goal was 2 years; current station-keeping propellant projection shows operations through May 1998 (longer for inclined orbit)
Contractors: Lockheed Martin Astro Space (prime, system design, bus, payload, integration, flight operations), Motorola (baseband processor), Watkins Johnson (TWTs), Electromagnetic Sciences (beamforming networks), OSC (Transfer Orbit Stage), Comsat Labs (ground system, network operations), Harris (experimenter Earth station), BBN Inc (high data rate experimenter Earth station)
Transponders: 4 Ka-band 19.2-20.2 GHz transmitters and 29.0-30.0 GHz receivers with 46 W TWTAs give 4-for-3 redundancy with stationary beams and 4-for-2 redundancy with the hopping beams. Beam-forming network of 45 feed horns covers about 20% of CONUS area, and provides two 0.3° hopping spotbeams with opposite polarisation, three 0.3° fixed beams and one 1.0° mechanically-steerable beam. EIRP 60 dBW fixed, 59 dBW hopping spots, 53 dBW steerable. Baseband processor stores/switches 64 kbit/s circuits, providing demand-assigned multiple access between all spotbeams. Total throughput 220 Mbit/s. FDM-TDMA access on uplink at 27.5/110 Mbit/s burst rates, TDM only on downlink at 110 Mbit/s. Automatic rain-fade compensation. High speed programmable 3 × 3 switch matrix provides 4 input + 4 output channels of 900 MHz bandwidth (one for redundancy). 20/30 GHz propagation beacons (stability 1.5 dB over 2 years)

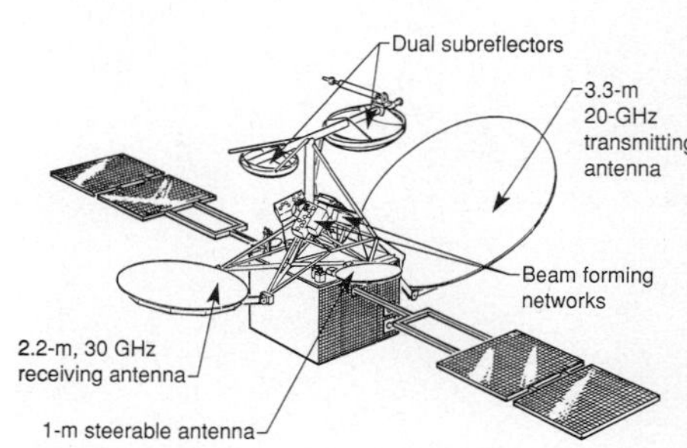

ACTS is demonstrating new Ka-band communications technologies. It is NASA's first experimental communications satellite since 1973

Configuration: modified series 4000 3-axis bus. Two multi-beam offset Cassegrain antennas: receive antennas 2.2 m + 1 m (steerable) diameter; transmission 3.3 m + 1 m (steerable) diameter. Bus 203 cm long, 213 cm wide, 190.5 cm deep; solar array span 14.3 m. Blowdown hydrazine system carrying 263 kg propellant in four tanks supplying 16 thrusters (two redundant half systems) of 0.9/2.2/4.4 N; station-keeping ±0.05°. 3-axis stabilised via Earth/Sun sensors + MW; 0.025° pitch/roll, 0.25° yaw; offset pointing control ±6° pitch ±2° roll
Mass: 2,540 kg at launch, 1,474 kg BOL on-station
TT&C: LMAS's station in E Windsor, NJ working through NASA Lewis ACTS Master Ground Station
Power system: 12.53 m² solar wings providing 1,770 W BOL and 1,418 W after 4 years: 2 × 19 Ah Ni/Cd batteries power only essential functions during eclipses.

AMSC series

Background

In May 1989, the FCC licensed AMSC to provide voice and data Skycell communications to mobile users on land (Conus, Hawaii, Alaska, Puerto Rico and Virgin Islands), up to 320 km out at sea and in the air. AMSC agreed with Canada's TMI Communications to operate compatible systems and use each other's satellite as spares. The companies jointly issued a request on 7 July 1989 for one satellite each to provide mobile communications for North America from 106.5° west (TMI) and 101° west (AMSC) GEO. The Hughes/Spar Aerospace team won the competition in December 1990, with Hughes acting as prime for AMSC and Spar for TMI. Each satellite cost about US$100 million. The two companies claim that their unique joint operation reduced start-up costs by about US$200 million, and ensure a high degree of mutual backup and protection. Users may tap either satellite from anywhere over North America with continuous coverage. Each satellite provides up to 2,000 simultaneous voice channels through L-band spot beams.

Westinghouse Electronic Systems Group manufactures the satellite phones for both systems. Mitsubishi Electric Corp also manufactures phones for the AMSC system. Combined manufacturing capacity reached 10,000 monthly by mid-1996. AMSC began voice service in January 1996, using US$3,000 transportable phones. The full range of Skycell voice products became available in 1996. The products included land mobile phones, US$1,800-US$2,300, aeronautical phones US$15,000; maritime phones US$4,000-US$6,000.

AMSC subsidiary American Mobile Radio Corp filed an FCC application 15 December 1992 to provide 55 radio broadcast channels (11 CD-quality) from two satellites at 99° west and 103° west. Hughes Space & Communications were to build the US$528 million system's satellites, although no contract has yet been signed.

Specifications

AMSC 1
Launch: 7 April 1995 by Atlas 2A from complex 36A, Cape Canaveral
Location: 101° west GEO (62/139° west also authorised)
Design life: 12 years
Contractors: Hughes Space & Communications Co (management/bus) and Spar Aerospace (payload)
Transponders: Forward (Ku/L) 13.023-13.232/1.530-1.559 GHz up/down, 16 35 W 55 PA (4 back-ups in 10-for-8 redundancy scheme) in 2 × 8 matrix feeding 6 elliptical spot beams (4 North America, a Caribbean beam for Puerto Rico, Virgin Islands and Mexico, and Alaska/Hawaii). Aggregate edge of coverage EIRP 56.5-57.3 dBW, circular polarisation. Bandwidth capacity for up to 2,000 5 kHz channels. Each '35 W SSPA' comprises four 20 W SSPAs in parallel but operated at 35 W for required 16 dB noise/power ratio

Reverse (L/Ku) 1.6315-1.6605/10.783-10.932 GHz up/down one (plus two back-up) 110 W TWTA 13.000-13.250/10.750-10.950 GHz up/down Ku-band 36 dBW feeder/return link in single North American beam
Principal applications: mobile radio, telephone, data (gateway to feeder 13.2405-13.2455/10.9405-10.9455 up/down Ku-Ku in single North American beam) aeronautical, maritime, safety and distress, position location, wide area paging. High quality voice, packet and circuit switched data
Configuration: Hughes' HS-601 3-axis bus, 2.3 m box-shaped body, 3.63 m stowed diameter, 4.44 m stowed height, twin solar wings spanning 20.96 m, 18.9 m across deployed graphite mesh antennas. Two Hughes 4.9 × 6.7 m 20 kg Springback mesh reflectors deployed from east/west faces provide separate L-band transmit/receive, respectively, illuminated by 23-cup dipoles. Single 76 cm Ku-band shaped reflector for North America coverage
Mass: 1,650 kg on-station BOL, 2,720 kg at launch (1,270 kg dry)
AOCS: onboard control processor for control of momentum wheel assembly. Unified 12 × 22 N + 1 × 490 N bipropellant thrusters (N_2O_4/MMH in four tanks)
Antenna pointing: 0.125° accuracy using Earth sensors
TT&C: Hughes Communications controls AMSC 1 under a five year contract from its OCC Operations Control Centre in El Segundo via its Spring Creek, New York ground station. AMSC's own Network Operations Centre in Reston, Virginia provides network control/channel access on demand
Power system: twin 9 m long 3-panel solar wings producing 3.6 kW BOL, 3 kW EOL from K4-¾ large area Si cells; 160 Ah Ni/H_2 battery provides 100 per cent eclipse protection
Thermal control: dissipation by north/south radiators, with heat transported by heat pipes.

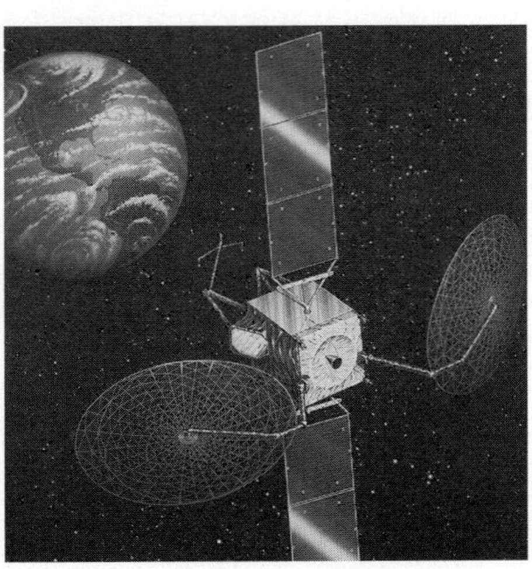

AMSC provides North American mobile services (Hughes Space & Communications) ***1995***

AT&T (Telstar/Skynet) series

Background

AT&T built and operated the world's first commercial satellites, Telstar 1 and 2 in 1962 and 1963. It revived the designation two decades later to commemorate the achievement when the FCC authorised it to proceed with the Telstar 300 series to replace the Comstar satellites.

The first of the series Telstar 301 went into space on 12 July 1982. Following its Anik E2 control problems beginning in January 1994, Telesat Canada decided to acquire 301 from AT&T on 28 March 1994. After purchase, Telesat Canada moved it to its new 107.1 degree west slot on 4 April 1994 to restore services until Anik E2 began commercial operations. Then Telesat Canada sold the Telestar to Arabsat on 7 June 1994 and began moving it to 20° east where it arrived in February 1995 and was designated Arabsat 1DR.

In October 1987, AT&T further requested approval to build a hybrid Telstar 4 offering 48 Ku-/C-band transponders. Space Systems/Loral contracted with AT&T to build the satellite for Skynet and possibly exercise three more satellite options for on-orbit delivery. AT&T planned to launch the Telstar 5 on a Proton. The company paid about $489 million for three satellites. The first of the series orbited in September 1994 by Ariane 4. 10 minutes after separation ground controllers lost contact with the satellite. An investigation concluded that the firing of a pyrovalve to open a propellant line had ignited some hydrazine. Insurance providers paid out a reported $187 million. AT&T quickly substituted 403 as 402R.

A surge in electromagnetic radiation following a major event on the sun's surface on 6 January 1997 resulted in the electronics on Telstar 401 short-circuiting four days later, causing the loss of the satellite. As a result the aging Telstar 303 was relocated to the Telstar 401 longitude as a replacement.

Space Systems / Loral built Telstar 5,6,7,8 and 9, plus a ground spare, for launches between 1997 and 2000. The new satellites target specific markets, whereas the 400s are more flexible. Telstar 6 replaces Aurora 2 for Alaska services from January 1999. Loral Skynet's VoiceSpan system would use 12 satellites in seven GEO slots. As of May 1999, Telstar 5 and 6 had been orbited successfully by Proton K launchers with Telstar 7 and 9 scheduled for launch in June followed by Telstar 8 in 2000.

Specifications

Telstar 302 (3C)
Launched: 30 August 1984 by Shuttle STS-41D
Location: 85° west geostationary; inclined orbit operations began 1 January 1995
Design life: 10 years
Contractors: Hughes Space & Communications
Transponders: 18 5.5 W SSPAs 5.925-6.425/3.700-4.200 GHz up/down C-band CONUS + spot beams, 36 MHz bandwidth, 34 dBW min EIRP, linear polarisation

12 5.5 W TWTAs, other specs as for SSPAs
Principal applications: domestic telephony, data and TV broadcast, including AT&T's range of Skynet business services
Configuration: Hughes' HS-376 spin-stabilised bus, 2.2 m diameter, 6.8 m high deployed
Mass: 3,423 kg with PAM-D GTO stage attached, 625 kg on-station BOL
TT&C: through AT&T's Satellite Management & Control Facility in Hawley, Pennsylvania
Power: 800 W BOL/670 W EOL provided by 15,588 silicon cells on cylindrical exterior
Apogee kick motor: Thiokol solid propellant.

Telstar 303 (3D)
Launched: 17 June 1985 by Shuttle STS-51G
Location: 123° west geostationary, inclined operations began June 1995. Following the loss of Telstar 401 in early 1997, Telstar 303 was manoeuvred off-station from 123° west in January and relocated to 89° west the following month
Other specifications as for Telstar 302.

Telstar 401 (4A)
Launch: 16 December 1993 by Atlas 2AS from Cape Canaveral
Location: 97° west GEO (replacing Telstar 301). Arrived 89° west 24 December 1993 for testing. Services began formally 1 February 1994. Satellite control was lost in January 1997 after damage caused by a solar electromagnetic storm
Design life: 12 years' propellant carried
Contractors: Lockheed Martin Astro Space
Transponders: 24 11-21 W (variable, plus four back-up in 7-for-6) SSPAs 5.925-6.425/3.7-4.2 GHz up/down C-band CONUS and spot beams, 36 MHz bandwidth, EIRP CONUS 37.8/35.0 dBW (in 21/11W order), Alaska 33.8/31.0 dBW, Hawaii 35.8/33.0 dBW, Puerto Rico 33.8/31.0 dBW, linear orthogonal polarisation

24 60-120 W (12 variable 60-120 W, plus 12 back-up in 18-for-12) TWTAs 14/12 GHz up/down Ku-band CONUS and spot beams, 54 MHz bandwidth (eight can be reconfigured to 2 × 27 MHz), EIRP CONUS 44.3/47.1 dBW (in 60/120 W order), Alaska 40.3/43.1 dBW, Hawaii 44.3/47.1 dBW, Puerto Rico 44.3/47.1 dBW, linear polarisation
Principal applications: TV broadcast, data, distance learning
Configuration: MMAS 7000 series 3-axis platform, 246 × 216 × 315 cm bus, 24 m solar array span
Mass: 3,375 kg at launch
AOCS: 3-axis control by momentum bias, thrusters + magnetorquers; NSSK by 1.8 kW hydrazine arcjets. Pointing accuracy ±0.1° pitch/roll, ±0.25° yaw; offset pointing control ±6.0° pitch + ±2.0° roll. Station-keeping ±0.05° lat/long (inclined orbit ops to 5°). GEO injection by dual mode liquid apogee engines
TT&C: as Telstar 3
Power: 7.2/5 kW BOL/EOL provided by Si cells on twin 4-panel solar wings, totalling 48.6 m². 100 per cent eclipse protection by 2 × 50 Ah Ni/Cd batteries.

Telstar 402R (4B)
Launch: 24 September 1995 by Ariane 4 from Kourou, French Guiana. Services began 20 November 1995
Location: 89° west geostationary
Design life: 13 years
Mass: 3,410 kg launch, 2,097 kg on-station BOL, 1,578 kg dry
Other specifications are the same as Telstar 401.

Telstar 5
Launch: 24 May 1997 by Proton K from Baikonur
Location: 263° east
Design life: 12 years
Mass: 3,650 kg at launch; 1,400 kg dry.

Telstar 6
Launch: 15 February 1999
Location: 267° east
Mass: 3,650 kg at launch; 1,400 kg dry.

EchoStar series

Background

When EchoStar Satellite Corporation announced its intentions to enter the direct-to-home broadcast market

EchoStar 1 provides US direct to home TV broadcasts (Lockheed Martin Astro Space)

it became the second large corporation to do so, setting up a competitive market for TV services. EchoStar signed a contract with Martin Marietta Astro Space in October 1992 to build one 7000-series satellite, plus options on six others. EchoStar plans to deliver 100 channels via 45 cm dishes. DirectSat merged with EchoStar in January 1994 and has FCC approval to merge their direct broadcast satellite licences for 119° west. DirectSat had been allocated 10 BSS channels at 119° west by the FCC. EchoStar made the US$52.3 million winning bid at the FCC's 26 January 1996 auction for the 24-channel slot at 148° west.

Specifications

EchoStar 1
Launch: 28 December 1995 by CZ-2E from Xichang; entered service 3 March 1996
Location: 119° west geostationary
Design life: 12 years' propellant carried
Contractors: Lockheed Martin Astro Space
Transponders: 16 130 W TWTAs 14/12.2-12.7 GHz up/down Ku-band CONUS beams (BSS channels, odd numbers 1-21), 24 MHz bandwidth, EIRP CONUS 51 dBW edge, circular polarisation
Principal applications: DTH TV
Configuration: GE 7000 series 3-axis platform
Mass: 3,288 kg at launch
AOCS: station-keeping ±0.5° lat/long. NSSK by 1.8 kW hydrazine arcjets
TT&C: by AT&T Skynet Satellite Services from operations facility at Hawley, Pennsylvania.

EchoStar 2
Launch: 11 September 1996 by Ariane 4 from Kourou, French Guiana
Location: 119° west geostationary
Mass: 2,865 kg at launch
Other specifications as for EchoStar 1.

EchoStar 3
Launch: 4 October 1997 by Atlas 2AS from Cape Canaveral
Location: 298° east geostationary
Other specifications as for EchoStar 1.

EchoStar 4
Launch: planned for 7 May 1998 on Proton-K from Baikonur, Kazakhstan
Location: 212° east geostationary
Local programming spot beams may complement the full CONUS coverage of EchoStar 1-3.

Faisat satellite

Current status

Faisat 2V experienced problems generating sufficient power while in Earth's shadow. Final Analysis Incorporated planned to develop a new software package to mitigate the problem.

Background

Final Analysis Incorporated entered business as a commercial telecommunication provider after modifying a bus originally built for a classified mission for data store/forward. The company received a Department of Commerce export licence 25 March 1994 for the first US commercial launch aboard a Russian vehicle. FAI launched its 114 kg, Faisat 1, as a

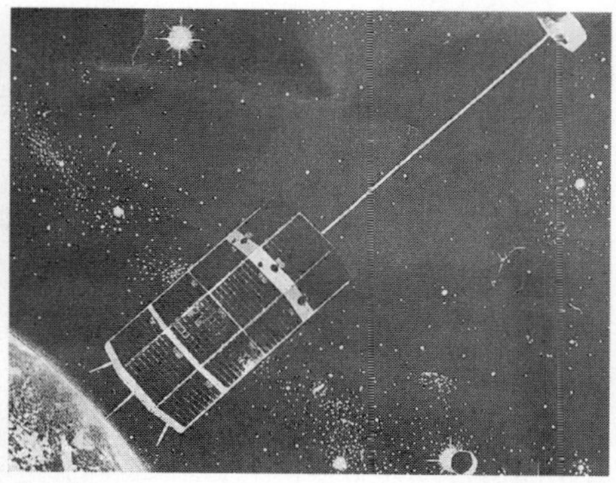

Faisat 1 demonstrated store/forward techniques ***1995***

piggyback payload aboard AKO Polyot's SL-8 Cosmos on 24 January 1995 into 968 × 1,021 km, 82 to 93° inclination orbit. The 46 cm diameter 89 cm high bus is gravity gradient stabilised by a 3 m boom with the assistance of magnetorquers. Faisat is demonstrating data store/forward from interrogated remote terminals in preparation for a 26-satellite constellation. Information is unlinked at 140 MHz/9.6 kbits/s and downlinked at 400 MHz. The company controls the satellite from its Logan, Utah station. Polyot may launch the full constellation in exchange for acting as the system's CIS service provider. FAI is targeting applications such as utility company meter reading, two-way messaging and glacier research.

Globalstar constellation

Current status

Globalstar is on schedule to begin operations with its 48-satellite network. Full 48-satellite coverage will occur in the third quarter of 1999. As of May 1999, Globalstar expected to launch a total of 32 satellites by the end of the year, added to the 20 orbited previously.

Specifications

Launch: 1-4 on Delta 7420 (14 February 1998); 5-8 on Delta 7420, 24 April 1998; 9-12 on Soyuz-Ikar, 9 February 1999; 13-16 on Soyuz-Ikar, 15 March 1999; 17-20 on Soyuz-Ikar, 15 April 1999
Orbits: 1,414 km, 52°; 6 active satellites in 8 planes (plus 1 spare in each plane)
Design life: 7 ½ years (EOL de-orbit)
Contractors: SS/L (prime: satellite and ground/space operations), Loral (operations control centre, TT&C stations), Alcatel Espace (payload modules), Alenia Spazio (payload/platform integration), DASA (AOCS, solar array), Qualcomm (gateways, operations control centres, handsets)
Principal applications: 4.8 kbit/s voice/fax, 9.6 kbit/s data, and position location (300 m)
Transponders: 1.610-1.6265/2.4835-2.5000 GHz user up/down, 5.091-5.250/6.700-7.075 GHz gateway up/down. Overall EIRP 39 dBW
Configuration: 3-axis 0.60 × 1.50 × 1.60 m box-shaped bus. Attitude/orbit determination by onboard GPS receivers using IBM RISC 6000 processors, reducing

The first Globalstar launch is planned for September/October 1997 aboard a Delta 2 (Globalstar) ***1997***/0005569

An engineer makes final adjustments to a Globalstar satellite (Theo Pirard) ***1999***/0054356

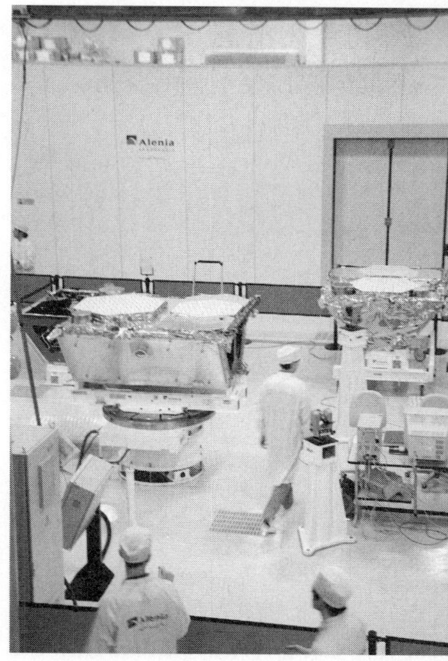

Final assembly of the Globalstar satellites (Theo Pirard) ***1999***/0054357

ground operations and eliminating Earth sensors. 1,200 W from twin solar wings spanning 12 m. AOCS by hydrazine thrusters
Mass: 450 kg at launch, 400 kg dry
TT&C: two SOCC Satellite Operation Control Centres.

Healthsat satellite

Background

SatelLife is a non-profit organisation that began by the initiative of the 1985 Nobel Peace Prize-winning International Physicians for the Prevention of Nuclear War. The project seeks to reduce the isolation of the world's health workers, particularly in developing countries, and link them with information resources. Its HealthNet programme (which began in five East African medical schools) employs the Healthsat 2 digital store/forward micro-satellite to provide e-mail links with Internet. SatelLife originally owned 90 per cent of UOSAT 3 (Healthsat 1), as it does Healthsat 2. UoSAT-builders Surrey Satellite Technology Ltd provided the

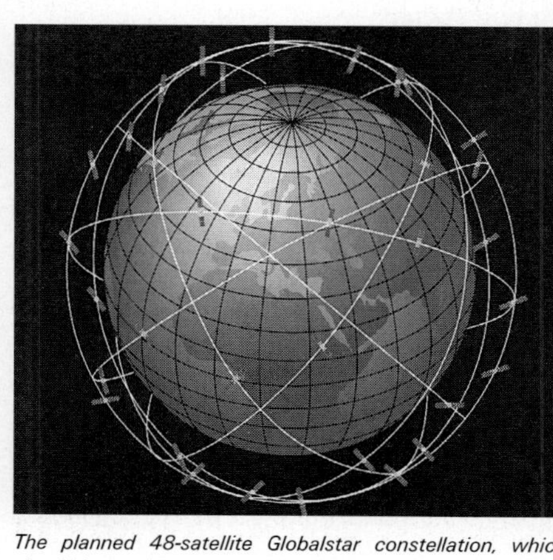

The planned 48-satellite Globalstar constellation, which should be complete by early 1999 (Globalstar) ***1999***/0005570

Final integration of HealthSat 2 ***1995***

Galaxy 3R is providing the first DTH service for Latin America, based on the US DirectTV system (Hughes Space & Communications) ***1995***

first dedicated satellite, Healthsat 2, under a US$1.7 million contract, including launch. HealthNet connects about 4,000 users in 25 developing countries with Internet. The user needs only a small radio, antennas and a computer. Messages are stored in 32 Mbyte memory for later downloading by ground stations.

Specifications

Healthsat 2

Launch: 26 September 1993 as Ariane ASAP passenger
Orbit: 793 × 805 km, 98.68°
Communications: store/forward packet communication. Messages stored in 32 Mbyte memory for later downloading by ground stations. 3-channel VHF receiver (two 9.6 kbit/s communication, one command), redundant UHF transmitters (9.6 or 38.4 kbit/s) switchable to 2/12.5 W RF output, 4-element turnstile VHF antenna, one monopole UHF (400 MHz band) antenna
Configuration: based on SST's UoSAT 5 design. 600 × 345 × 345 mm box-shaped body with solar panels on four sides, 6 m boom deployed from top face for gravity gradient stabilisation. Each housekeeping system or payload is housed in a standard module box, mass-produced by a computer-controlled process. The boxes are then stacked to form the main spacecraft structure. Payload is a dual 'hot' redundant OBC supporting both housekeeping and the store/forward function: two 80C186 processors (+80C188 back-up) and 32 Mbyte message memory
Total mass: 47.5 kg
Attitude control: gravity gradient Earth-pointing by 3 kg tip mass on 6 m boom deployed from top face, magnetorquing by six 150-turn aluminium wire coils (one on each spacecraft face); Sun/Earth horizon sensors and 3-axis fluxgate magnetometer (±2 nT res over ±64 μT) provide attitude information
TT&C: SSTL provides TT&C services. Gateway stations at St. John's, Newfoundland and Boston, Massachusetts
Power supply: four GaAs panels (each 34 W) totalling 1,344 cells provide 20 W (orbit average processed); 6 Ah Ni/Cd batteries.

Hughes (Galaxy/DBS) series

Current status

HCI currently maintains the Galaxy 5 and 6 with 24 C-band transponders each, the hybrid (24C/24Ku) Galaxy 7, and the similar Galaxy 4. Following the loss of Galaxy 10 on 27 August 1998, replacement satellite 10R was scheduled for late 1999 launch. Galaxy 11 also scheduled for launch in this period.

Background

Hughes Electronics Corporation established Hughes Communications Incorporated, or HCI as a wholly owned subsidiary to operate the world's largest fleet of privately-owned commercial communications satellites, known as the Galaxy. HCI first began satellite operations with the launches of its three initial Galaxy satellites during 1983-84. The system provides point-to-point and point-to-multipoint C-band services for users in the continental US, Mexico, Alaska, Hawaii and the Caribbean Basin. Then in 1993, HCI established DirecTV, Inc to operate a domestic US DBS system using three Ku-band HS-601s. DirecTV became an independent Hughes Electronics subsidiary in November 1993.

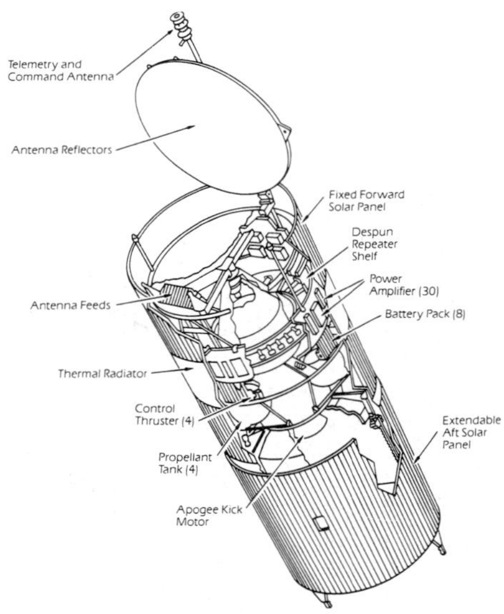

The smaller Galaxies are based on Hughes' HS-376 design, the most commonly used western platform for commercial communications (Hughes Space & Communications)

From the beginning, HCI intended to make cable TV distribution a priority and dedicated Galaxy flights 1RR/5/7 to this application. Galaxy 3R/4/6 provided business users and audio subscribers with a continental United States service. HCI launched Galaxy 1 on 28 June 1983 and replaced it in March 1994 with the higher power, 16 W Galaxy 1RR. The first planned replacement, Galaxy 1R disappeared in a 22 August 1992 launch failure. Galaxy 2 followed on 22 September 1983 and retired from 74° west in May 1994 after replacement by Galaxy 6. Galaxy 3, launched 21 September 1984, left its assigned spot at 93.5° west on 30 September 1995 when 3R replaced it. Then HCI acquired the Westar satellite system and used its geosynchronous locations for its own spacecraft, Galaxy 6/4/5, respectively. This expanded the Galaxy fleet to six satellites.

Galaxies 1RR/5 are stretched HS-376s, originally built as the ground spare for Galaxy 4/7. Their 24 Ku-band transponders provide an initial DTH Latin America service via 60 cm dishes. Although its orbital slot is licensed only for US domestic services, HCI successfully argued that it is only a temporary measure as the Ku-band coverage will be switched back to the US when the dedicated Galaxy 8I appears to provide the full service.

The old C-band Westar 4 slot (temporarily serviced by Galaxy 6) and Ku-band SBS 6 slot at 99° west has been occupied since November 1993 by the hybrid Galaxy 4. HCI decided in late 1989 to expand the Galaxy series with Galaxy 4, at 91° west. Ariane launched it into a sub-GTO (apogee 27,670 km) parking orbit and the Galaxy made two perigee burns to attain GTO. The perigee burns optimised the climb to GEO sparing station-keeping propellant and lengthening the satellite lifetime by about 1 year.

Galaxy 7 joined in 1997 and HCI has scheduled Galaxy 8I (I indicates international) for a 1998 launch to provide dedicated Latin America DTH, replacing the initial service from Galaxy 3R.

Hughes Electronics spun off DirecTV in 1992 after Sky Cable collapsed as a corporate vehicle to operate a domestic US direct-to-home television system using Ku-band transmission from a HS-601 satellite. When it produced its own DBS 1 satellite it became the 100th Hughes-designed telecommunications satellite to be launched. Hughes built a 3,720 m^2 facility for programming uplink facility at Hughes Communications Incorporated's Castle Rock (Colorado) station. The new facility uses 13 m antennas, capable of delivering 216

Galaxy VIII-i spacecraft is being prepared for a thermal vacuum test (Hughes Spacecraft and Communications) ***1999***/0054326

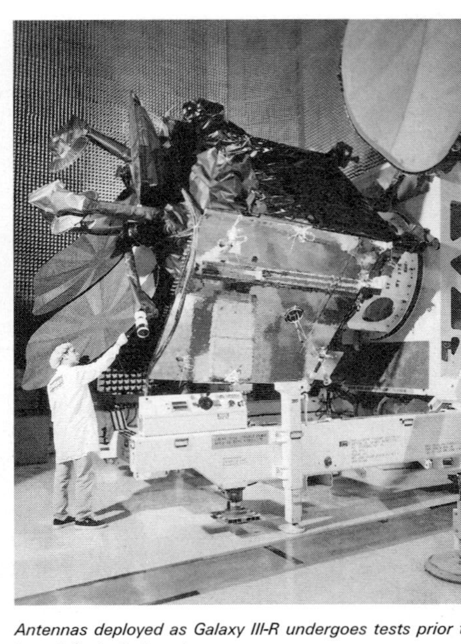

Antennas deployed as Galaxy III-R undergoes tests prior to launch for Latin American (Hughes Space & Communications) ***1999***/0054327

simultaneous channels of television. DirecTV has received an FCC licence to collocate up to four HS-601 satellites over 101° west with a total of 27 transponders.

Hughes sold 5 of the 16 transponders aboard the first satellite to the United States Satellite Broadcasting Incorporated (USSB), a subsidiary of Hubbard Broadcasting for more than US$100 million. DirecTv owns all 16 channels on DBS 2. The two satellites deliver a combined 175 channels of television. Each transponder can provide four video channels using digital compression, or eight films because of the slower motion. In order to turn a profit Hughes estimates the venture requires three million subscribers. It projects 10 million by 2000. One million homes had subscribed by November 1995.

HCI's Galaxy, SBS and Leasat satellites are controlled from the Operations Control Center (OCC) located at its corporate headquarters.

Specifications

DBS 1

Launched: 18 December 1993 by Ariane 4 from Kourou, French Guiana. HSC handed over control of number 1 to HCI 6 January 1994; broadcasting began 17 June
Location: 101.2° west geostationary
Design life: 12-15 years
Contractors: Hughes Space & Communications Co, Hughes Aircraft Company (prime)
Transponders: 16 (plus eight back-up; 12-for-8) 120 W TWTAs (can be paired into 8 × 240 W) 17.3-17.84/12.2-12.74 GHz up/down Ku-band CONUS beams, 27 MHz bandwidth, 48-53 dBW EIRP, 40 Mbit/s, R/LHCP for DBS 1, L/RHCP DBS 2. BSS channels 1-32. Each shaped reflector is fed by a single feed horn; the shape concentrates power on areas such as the SE affected by rain fade
Principal applications: direct broadcast to 45 cm diameter antennas
Configuration: Hughes' HS-601 3-axis bus, 26.2 m solar array span, two 2.14 m transmission reflectors, 102 cm receive reflector, 36 cm beam tracking antenna. Each antenna array is mounted at only one point to minimise thermal distortion to shaped profile
Mass: 2,860 kg at launch, 1,270 kg dry mass, 1,725 kg on-station BOL
TT&C: from HCI's Operations Control Center in El Segundo, CA, via Castle Rock, CO + Spring Creek, NY
Power system: twin solar arrays providing 7,000 W BOL. 100 per cent eclipse protection by Ni/H_2 battery.

DBS 2

Launched: 3 August 1994 on Atlas 2A from Cape Canaveral
Location: 100.8° west GEO, about 300 km from number 1
Other specifications as for DBS 1.

DBS 3

Launched: 10 June 1995 by Ariane 4 from Kourou, French Guiana
Location: 100.8° west GEO, collocated with DBS 2
Mass: 2,934 kg at launch, 1,259 kg dry mass, 1,707 kg on-station BOL
Other specifications as for DBS 1.

Galaxy 1RR

Launch: 19 February 1994 by Delta 7925 from Cape Canaveral
Location: 133° west geostationary (replacing number 1)
Mass: 1,393 kg at launch, 798 kg on-station BOL
Other specifications as for Galaxy 5.

Galaxy 3

Launched: 21 September 1984 by Delta 3920/PAM from Cape Canaveral
Location: drifting
Design life: 10 years, sufficient AOCS propellant for 9 years
Transponders: 24 (plus six back-up; 5-for-4 redundancy) 9 W TWTA 5.925-6.425/3.700-4.200 GHz up/down C-band CONUS/Alaska/Hawaii beams, 36 MHz bandwidth, EIRP 34 dBW over CONUS, orthogonal polarisation
Principal applications/users: cable TV distribution
Configuration: Hughes' HS-376 platform, spin-stabilised at 50 rpm, drum-shaped 2.16 m diameter, 6.6 m deployed height (2.8 m stowed) single 1.83 m diameter reflector (vertical/horizontal polarisation) and communications shelf despun
Mass: 1,200 kg at launch, 654 kg on-station BOL
AOCS: station-keeping and spin-stabilised at 50 rpm by four hydrazine thrusters (136 kg propellant); GEO insertion provided by solid propellant Thiokol Star 30 apogee kick motor
TT&C: stations at Spring Creek (Brooklyn) in New York and Fillmore in California relay signals to the Operations Control Center in El Segundo, California. Users are served by Hughes Communications Network Operations (HCNO), collocated with the OCC
Power system: solar cells mounted on 47 panels on cylindrical surface providing 19.7 mW/cm², 990 W BOL, 900 W EOL; two Ni/Cd batteries provide eclipse protection.

Galaxy 4

Launched: 25 June 1993 by Ariane 42P+ V57 from Kourou, French Guiana
Location: 99° west geostationary (replacing Galaxy 6/SBS 6)
Design life: 15 years, but with sufficient propellant for 13.5 years operations
Transponders: 24 (15-for-12 redundancy) 16 W SSPA 5.925-6.425/3.700-4.200 GHz up/down C-band CONUS/Hawaii/Alaska beams, 36 MHz bandwidth, EIRP 37 dBW CONUS, 30 dBW Alaska/Hawaii, orthogonal polarisation

8 54 MHz + 16 27 MHz (15-for-12 redundancy) 50 W TWTA 14.0-14.46/11.72-12.16 GHz up/down Ku-band CONUS and four of the 54 MHz switchable to Alaska, Hawaii, Virgin Islands and Puerto Rico, EIRP 45 dBW CONUS, 41 dBW Alaska/Hawaii, orthogonal polarisation. Two of the 54 MHz transponders can be cross-strapped Ku-to-C/C-to-Ku
Principal applications: network TV, radio, VSAT, business video and data
Configuration: Hughes' HS-601 3-axis platform
Mass: 2,980 kg at launch, 1,692 kg BOL, 1,320 kg EOL
AOCS: provided by unified 12 × 22 N + 1 × 490 N bipropellant thrusters incorporating onboard control processor, Earth/Sun sensors, MWs. 1,658 kg NTO/MMH in four tanks. Solar sailing techniques reduce conventional δ V requirements
TT&C: stations at Spring Creek (Brooklyn) in New York and Fillmore in California relay signals to the Operations Control Center in Long Beach, California. Users are

Engineers check out the antennas on DBS-1, North America's first direct-to-home TV satellite (Hughes Space & Communications) ***1999***/0054333

An HS6D1 body stabilised satellite was selected for the first Direc-TV satellite (Hughes Space and Communications) ***1999***/0054334

served by Hughes Communications Network Operations (HCNO), collocated with the OCC
Power system: twin solar wings spanning 24.7 m of four 2.16 × 2.54 m panels carrying K4¾ large area Si cells generate up to 6 kW (4.7 kW EOL). Ni/H_2 batteries provide 100 per cent eclipse protection.

Galaxy 5
Launch: 00.06 GMT 14 March 1992 by Atlas 1 from Cape Canaveral
Location: 125° west geostationary (replacing Westar 5)
Design life: 10 years
Transponders: 24 (plus 6 back-up; 30-for-24 ring redundancy) 16 W TWTA 5.925-6.425/3.700-4.200 GHz up/down C-band CONUS/Alaska/Hawaii beams, 24/36 MHz bandwidth, EIRP 40/37.5 dBW peak/min over CONUS, orthogonal polarisation
Configuration: stretched HS-376, 3.10/7.49 m height stowed/deployed
Mass: 1,412 kg at launch, 788 kg dry.

Galaxy 6/Westar 6S
Launch: 12 October 1990 by Ariane 4 from Kourou, French Guiana. GEO insertion 17 October, solar drum and antennas deployed 19 October
Location: in August 1993 it was moved from 99° west to 103° west temporarily; in late 1991 it was positioned at 99° west to replace Westar 4 until Galaxy 4 appeared; previously at 91° west (replacing Westar 3); it moved from 103° west to 74° west in May 1994 to replace the retiring Galaxy 2
Design life: 10 years (hydrazine for 13.9 years remaining after Ariane launch)
Transponders: 24 (plus 6 back-up, in 5-for-4 ring) 10.2 W TWTA 5.925-6.425/3.700-4.200 GHz up/down C-band CONUS, Hawaii, central America, Caribbean beams, 36 MHz bandwidth, EIRP 35-39 dBW CONUS, 28-34 dBW Alaska, 27-30 dBW Hawaii, Caribbean/central America 30-32 dBW, Puerto Rico 28-31 dBW, orthogonal polarisation
Principal applications: system C-band back-up and occasional video services
Configuration: HS-376 bus, solar arrays and attitude control tanks enhanced
Mass: 708 kg on-orbit BOL, 1,212 kg at launch, 509 kg dry mass
Power system: as HS-376 but improved to 869 W EOL.

Galaxy 7
Launch: 28 October 1992 by Ariane 4 from Kourou, French Guiana
Location: 91° west GEO (replacing SBS 4)
Design life: 15 years but with sufficient propellant for 13.5 years operations
Mass: 2,968 kg at launch, 1,680 kg BOL, 1,303 kg dry
Other specifications are as Galaxy 4.

Galaxy 8 (IIIR)
Launch: 15 December 1995 by Atlas 2AS from CCAFS
Location: 95° west
Design life: 12 years
Transponders: 24 × 16 W C-band for cable and broadcast TV; 24 × 63 W Ku-band for data, voice and video
Other specifications as Galaxy 4.

Galaxy 8I
Launch: 8 December 1997 by Atlas 2AS from CCAFS
Location: 280° east
Design life: 12 years
Dedicated Latin American service: (I for International) replacing Galaxy 8 (IIIR).

Galaxy 9
Launch: 24 May 1996 by Delta 7925 from Cape Canaveral
Location: 123° west initially (collocated with SBS 5), then 129° west geostationary
Transponders: 24 (plus 6 back-up; 30-for-24 ring redundancy) 16 W TWTA 5.925-6.425/3.700-4.200 GHz up/down C-band CONUS/Alaska/Hawaii beams, 24/36 MHz bandwidth, EIRP 40/37.5 dBW peak/min over CONUS, orthogonal polarisation
Configuration: stretched HS-376, 3.10/7.49 m height stowed/deployed
Mass: 1,397 kg at launch
Other specifications as for Galaxy 5.

Galaxy 10
Launch: 27 August 1998 on Delta 3 from Cape Canaveral, launch vehicle failure resulted in destruction of the satellite
Location: planned for 237° east geostationary (replacing SBS 5 and Galaxy 9)
Design life: 15 years
Transponders: Same as Galaxy 4 except power increased for 24 × 20 W C-band + 24 × 50 W Ku-band
Configuration: HS-702 platform. GaAs solar arrays span 27 m, providing 10 kW BOL. Ni/H_2 batteries provide 100 per cent eclipse protection. AOCS by bipropellant (1,700 kg NTO/MMH) thrusters incorporating onboard control processor, Earth sensors and MWs. Hughes' XIPS xenon ion propulsion subsystem for NSSK: four 25 cm thrusters (2 primary and 2 backup) consume 5 kg/years
Other specifications as for Galaxy 4
Comments: The contract for this first HS-702 model includes the Galaxy 11 ground spare, which will serve initially as the ground spare but will be launched.

Iridium constellation

Current status
In May 1998, a Delta launch from Vandenberg AFB California completed the 72 satellite (66 prime, 6 spares) Iridium constellation. Iridium planned to begin initial operations as scheduled in the early autumn. Ground observers repeatedly noticed the satellites in the sun's glare near evening and in the early morning.

Background
Motorola unveiled its Iridium mobile communications system proposal in June 1990 after 2½ years of internal planning. Iridium plans a system of 66 small LEO satellites by end-1998 providing global pocket mobile digital 4.8 kbit/s telephone services via onboard switching and inter-satellite links for US$3/min to the user. Iridium plans to sell handsets for about US$3,000. The company planned to spend about US$3.45 billion for establishing the operational system of 66 satellites and 14 in-orbit spares by 1998, followed by another US$2.8 billion over the next 5 years for operations/maintenance. For an economic break even Iridium needs to sign up 800,000 users within 5 years. Motorola business plans project 1.824 million by 2001, broken down as 42 per cent pager, 42 per cent portable phone and 7 per cent mobile. After 10 years, in 2006, it projects 3.224 million.

Motorola submitted its proposal to the FCC 3 December 1990 and the Commission granted an experimental licence in August 1992 for five satellites. A full licence followed on 31 January 1995. Lockheed Space Systems Division (now Lockheed Martin) received a US$700 million contract in August 1993 for 120 Iridium spacecraft to be built through the year 2003. Lockheed builds the satellite bus in Nashua, New Hampshire at Lockheed Martin Sanders and then integrates it with Motorola's communications module in Motorola's final assembly facility in Chandler, Arizona. Chandler's production line can handle five satellites simultaneously and takes 21 days from start to finish of a satellite.

Lockheed drew on its Agena upper stage experience to build the satellites power and control systems. Volume production allows use of GaAs solar arrays. Motorola said in February 1993 that it planned to use the Khrunichev Enterprises Proton booster to put the first three packages of seven satellites in orbit. Khrunichev invested US$40 million in return. Then Motorola made a similar agreement with GWIC April 1993 for 10 launches of two satellites each between 1996-2002 on the CZ-2C. Eight Delta 2 launches with five satellites each, will establish the initial constellation in 1998. As replacements are required, Motorola will use Delta 7420 launches between 1998 and July 2000. To avoid on-orbit debris and pollution, Iridium will de-orbit at the end of its useful lifetime.

Individual satellites broadcast to 48 L-band cells 670 km across using FDMA/TDMA. Calls are routed to the Earth gateways and among the space segment at Ka-band. The user's 0.7 kg 0.6 W handset automatically selects a terrestrial cellular network if one is available and cheaper. Each device has a GPS receiver to maintain a log of each handset's last known position. Each device relays its position to a visible satellite for passing on to the nearest gateway. Each gateway has three terminals

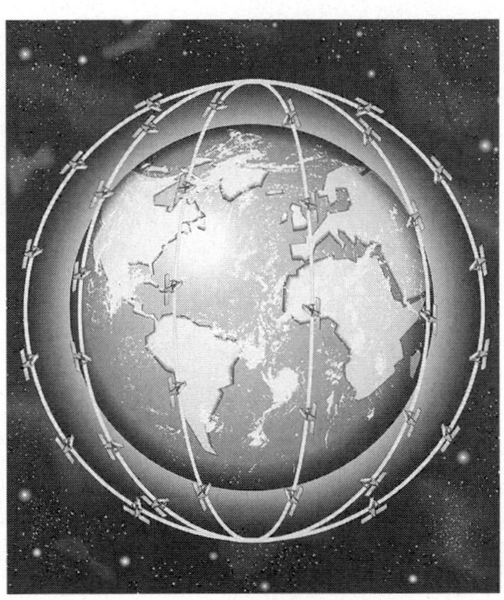

Iridium's network of 66 satellites will provide global pocket mobile telephone services via the L/S-band phased array antennas. Iridium is the 77th element, reflecting the original constellation size ***1995***

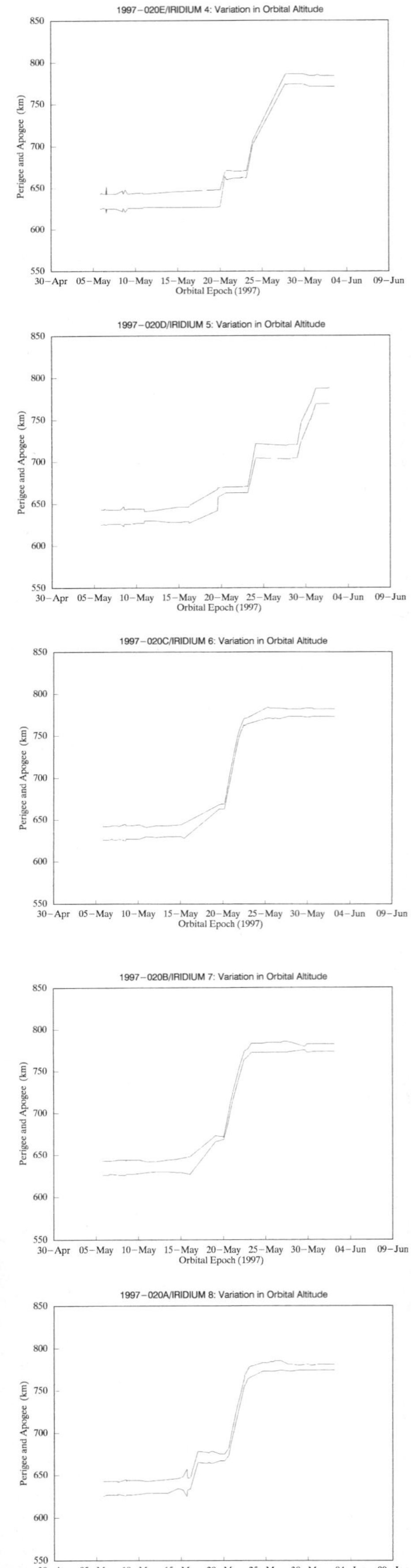

These graphs show the manoeuvres completed during the first month in orbit by the five Iridium satellites which comprised the first launch in the programme. It will be noted that Iridiums 6-8 started their manoeuvres around 16 May and completed them within 10 days. Iridiums 4-5 started their manoeuvres around 20 May with Iridium 4 reaching its planned altitude around 27 May and Iridium 5 in early June (Phillip S Clark) ***1997***/0005571/2/3/4/5

The Iridium LLC filled all of the orbital slots for its constellation in early 1998. Barring any further loss of on-orbit spares, the company planned to offer services beginning in September 1998 (Iridium LLC) ***1998***

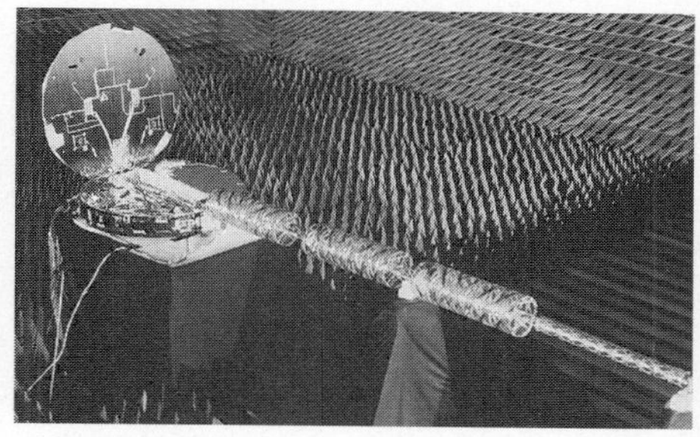

ORBCOMM anechoic chamber testing (OSC) ***1996***

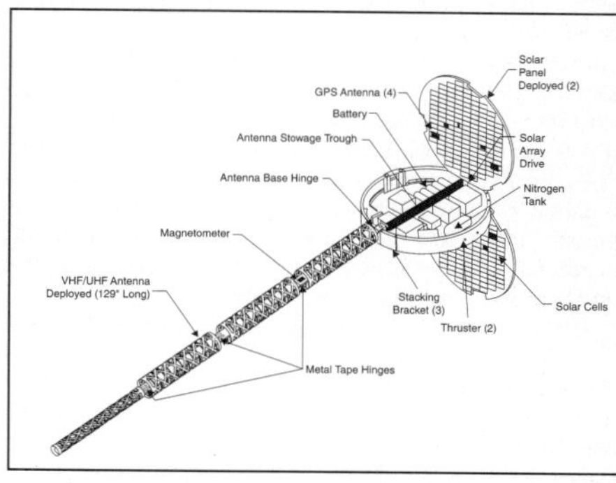

ORBCOMM's design is based on OSC's MicroStar bus

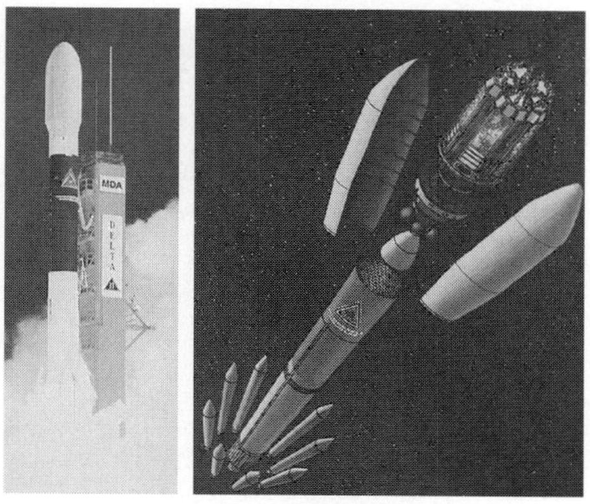

Delta is launching Iridium in clusters of five (McDonnell Douglas Aerospace) ***1995***

dispersed sufficiently to avoid weather problems affecting links.

Specifications

Launches: 5 × Delta 2, 5 May 1997; 7 × Proton K4, 18 June 1997; 5 × Delta 2, 9 July 1997; 5 × Delta 2, 21 August 1997; 7 × Proton K4, 14 September 1997; 5 × Delta 2, 26 September 1997, 5 × Delta 2, 9 November 1997; 5 × Delta 2, 20 December 1997; 5 × Delta 2, 18 February 1998; 2 × CZ-2C/SD, 25 March 1998; 5 × Delta 2, 30 March 1998; 7 × Proton K4, 7 April 1998; 2 × CZ-2C/SD, 2 May 1998; 5 × Delta 2, 17 May 1998; 2 × CZ-2C/SD, 19 August 1998; 5 × Delta 2, 8 September 1998; 5 × Delta 2, 6 November 1998; 2 × CZ-2C/SD, 19 December 1998

Orbits: 780 km circular, 86.4°; 11 active satellites in 6 planes (plus 1 spare in each plane)

Design life: 5-8 years

Contractors: Motorola (payload, integration), Lockheed Space Systems Division (bus), Raytheon (phased array antennas), COM DEV (inter-satellite and feeder link antennas), Olin Aerospace (thrusters), Barnes Engineering Division (miniature infra-red Earth horizon sensors), Tecstar Inc (solar cells), Ithaco, Johnson Controls, Schaeffer Magnetics, Pressure Systems Inc (hydrazine tanks)

Principal applications: voice, fax and data global handheld services

Transponders: three phased array antennas each provide 16 user up/down L/S-band beams from 106 elements at 1.610-1.6265/2.4835-2.500 GHz

Single phased array antenna provides inter-satellite up/down Ka-band beam

Single dish antenna provides feeder link up/down Ka-band beam

Configuration: 3-axis triangular bus, 1 m diameter, accommodating twin solar wings, body-mounted Ka-band inter-satellite antenna, and lower Earth-pointing section with three angled 86 × 188 cm 30 kg L/S-band antennas, each with 106 radiating elements. 115 kg hydrazine (sufficient for >10 years) in single 53.5 cm diameter 71.1 cm height titanium tank

Mass: 690 kg at launch

TT&C: 10 worldwide System Control Segment terminals, controlled by master site at Landsdowne, Virginia (back-up at Nuova Telespazio, Rome, Italy). Motorola's Chandler facility will initially control the first 40 satellites

Power: two deployed 5-panel wings of GaAs cells; 50 Ah battery.

ORBCOMM GLOBAL constellation

Background

OSC's Orbital Communications Corporation subsidiary, created at the beginning of 1990, filed an FCC application 28 February 1990 for a constellation of LEO satellites. OSC planned to provide for two-way communications and geolocation services with low-cost alphanumeric data communications and position determination (375 m) for emergency assistance, data acquisition and messaging services. Circuit design has optimised the system for shorter messages of 6-250 characters. In order to finance the system, OSC and Canada's Teleglobe Inc signed an agreement 26 July 1993. Teleglobe and OSC will jointly operate the system. Teleglobe provided US$10 million of the US$55 million cost of the two first phase satellites. For the second phase, which requires 34 more satellites (including eight ground spare), Teleglobe has increased its equity share in ORBCOMM Global LP to 50 per cent. Teleglobe contributes US$85 million to OSC's US$75 million.

OSC launched the 16.7 kg ORBCOMM-X as an experimental forerunner by Ariane into a polar orbit 17 July 1991. Virginia's Center for Innovative Technology, co-developed the original design and invested US$250,000 as part of its VaStar satellite R&D programme. The state's universities and institutions planned to use it for collecting environmental data from remote platforms. OSC expected ORBCOMM-X to survive for 18 months. OSC's Boulder station made initial contact some 14 hours after launch and verified basic operations, but was unable to establish subsequent links. It abandoned attempts after several weeks. OSC has traced the failure to a software glitch in a power control circuit.

When finished, the ORBCOMM constellation will have 28 43 kg satellites by 1998 at 775 km altitude; 24 at 45° inclination (eight in each of three planes) and two at 70° for high latitude coverage. OSC will conduct four launches aboard its own Pegasus and one on Taurus, plus a set of eight ground spares intended for Pegasus. The first two satellites, in polar orbit, appeared successfully 3 April 1995. Both soon developed receiver problems. ORBCOMM 2's solved its problem by a self-activated reset on 13 May, and ground commands solved ORBCOMM 1's by freeing the receiver 18 May. Apparently battery drain triggered a computer reset.

ORBCOMM 2 relayed the system's inaugural message 25 May to the Arcade gateway from a production prototype Panasonic personal communicator at Dulles airport near Washington, DC. A message from the Elisra EL-2000 Data Communicator near the Dead Sea in Israel routed through the system to Dulles on 25 July 1995 via the Georgia gateway.

ORBCOMM seeks to serve the US market at first, where it foresees a total of 10 million to 20 million users, beginning with 150,000 the first full year of constellation operations and rising to 5.2 million after seven years. Emergency services may provide 75 to 85 per cent of the market. ORBCOMM plans to market three classes of terminals, a 5 W power, signalling and data transmit only, a US$150 position-fixing and a US$400 alphanumeric messaging terminal. The first type requires a US$30-50 annual service charge, with the others supported by US$35 monthly plus usage-linked charges.

ORBCOMM workers prepare a stack of eight LEO satellites that will form part of the ORBCOMM constellation. When complete, the constellation will have 28 of these 43 kg satellites circling the globe at 775 km altitude. Orbital Sciences launched its two latest satellites, the sixth launch in a series of ten, on 10 February 1998, by Taurus rocket (ORBCOMM Global, L.P.) ***1998***/0022576

Each satellite carries 17 data processors and seven antennas, allowing the relay of 50,000 messages hourly in each direction, or more than five million two-way daily messages for the 28-satellite system. Four regional gateways will cover the US from Oscilla, Georgia, Arcade, New York, St Johns, Arizona and Washington State.

Specifications

Launch: 3 April 1995 (numbers 1-2); 8 × Pegasus-XL, 23 December 1997; 2 × Taurus, 10 February 1998; 8 × Pegasus-XL+HAPS, 2 August 1998; 8 × Pegasus-XL+HAPS, 23 September 1998
Orbits: numbers 1-4 736 × 749 km 70.0°; numbers 5-28 775 km, 45°
Design life: 4-6 years
Transponders: 4,800/9,600 bit/s downlink uses 320 kHz in the 137.0-138.0 MHz VHF band (10 channels) and 50 kHz at 400.1 MHz UHF (1 channel, for GPS time data only). EIRP 16.5 dBW at 137 MHz and 19.5 dBW at 400 MHz. For 2,400 bit/s uplinking, a Dynamic Channel Activity Assignment System directs terminal users to a block of 24 out of a possible 760 channels at 148.00-149.90 MHz depending on instantaneous usage by other systems
Principal applications: two-way packet data communications
Mass: 43 kg at launch
Configuration: MicroStar design, allowing eight to be stacked on single Pegasus. Stowed size 104 cm diameter × 16.5 cm. Deployed length 432 cm, width 224 cm (across solar arrays). ±5° nadir pointed by active magnetorquers and gravity gradient; attitude sensing by Earth sensors + magnetometer. 328 cm VHF/UHF antenna deployed by four hinges. Orbital spacing provided by N_2 thrusters
TT&C: via Dulles Satellite Control Centre; 57.6 kbit/s VHF
Power system: twin circular Sun-tracking solar arrays hinge out from main body providing 230 W BOL; 14 V system; 5V regulated bus for electronics. Supported by Ni/H_2 battery.

Orion (Orion Atlantic LP) series

Background

BAe and Orion Satellite Corp signed a US$360 million contract on 1 September 1989 for the in-orbit delivery of two satellites. It marked the first US commercial sale by a European company. Construction began 7 January 1992. Under the turnkey contract, Martin Marietta Commercial Launch Services provided launch and insurance services. The FCC granted final authority to proceed 28 June 1991. Orion cancelled the second satellite, planned for launch early 1996 by Atlas 2A to 47° west, in April 1993, but BAe's remaining contract had a value of US$227 million. Orion 1 is the World's first all-solid-state Ku-band satellite. An agreement was reached in October 1995 with Telecomm Mexico to exchange capacity on their respective satellites, providing Orion with access to Mexico, Central and South America.

Orion ordered the all-Ku Orion 2 on 24 October 1995 for 1998 in-orbit delivery. It will eventually pay US$265 million, including launch and insurance. The Eurostar 2000+ design resembles Orion 1, but Orion 2's higher power transponders allow additional services such as DTH. The coverage overlaps with Orion 1's in the US and Europe. Onboard switching will allow beam connectivity reconfiguration according to customer demand. The contract includes an option for the similar Orion 3, which would be owned and operated by ONS' wholly owned Orion Asia Pacific subsidiary. Total cost, including ground segment, launch, insurance and operations, has been estimated at US$275 million.

Specifications

Orion 1

Launch: 29 November 1994 aboard Atlas 2A from Cape Canaveral
Location: 37.5° west geostationary (arrived 25 December 1994, MMS handed over control 16.00 GMT 22 January 1995 after completing checkout)
Design life: 12 years
Contractors: BAe Space Systems (prime, now part of Matra Marconi Space), Martin Marietta (launch and insurance services), Matra (platform electronics), NEC (communications payload), Lockheed (propellant tanks), Kaiser Marquardt (apogee motor), MBB (10 N thrusters), Fokker (solar arrays), Spectrolab (solar cells), Eagle Picher (Ni/H_2 batteries), COM DEV (output multiplexers) and Telesat Canada (LEOP services)
Transponders: 34 (plus 8 back-up: 5-for-4 54 MHz and 7-for-6 36 MHz) 15 W SSPAs 14.0-14.5/11.45-12.2 and 12.5-12.75 GHz up/down Ku-band Europe and North America spots & regional beams, bandwidth 28 × 54 and 6 × 36 MHz, EIRP 50 dBW (54 dBW max) spots and 45 dBW (48 dBW max) regional. Beam connectivity via switch matrix provides not only trans-Atlantic services but also regional & domestic coverage for Europe and US. 17 transponders allocated to North America (11.7-12.2 GHz down) and 17 to Europe (11.45-11.7 and 12.5-12.75 GHz down). Spots: 3 Europe (North, South and Central) & 4 North America (Northeast, North, South and Central). Regional: 1 North America and 1 Europe
Principal applications: optimised for VSAT applications (1.2 m diameter 2-way; 85 cm diameter receive only). Flexible interconnectivity between high powered spot beams provide for both trans-Atlantic and regional/domestic networking
Mass: 2,361 kg at launch, 1,140 kg dry, payload 330 kg
Configuration: MMS Eurostar 2000 bus. Deployed size

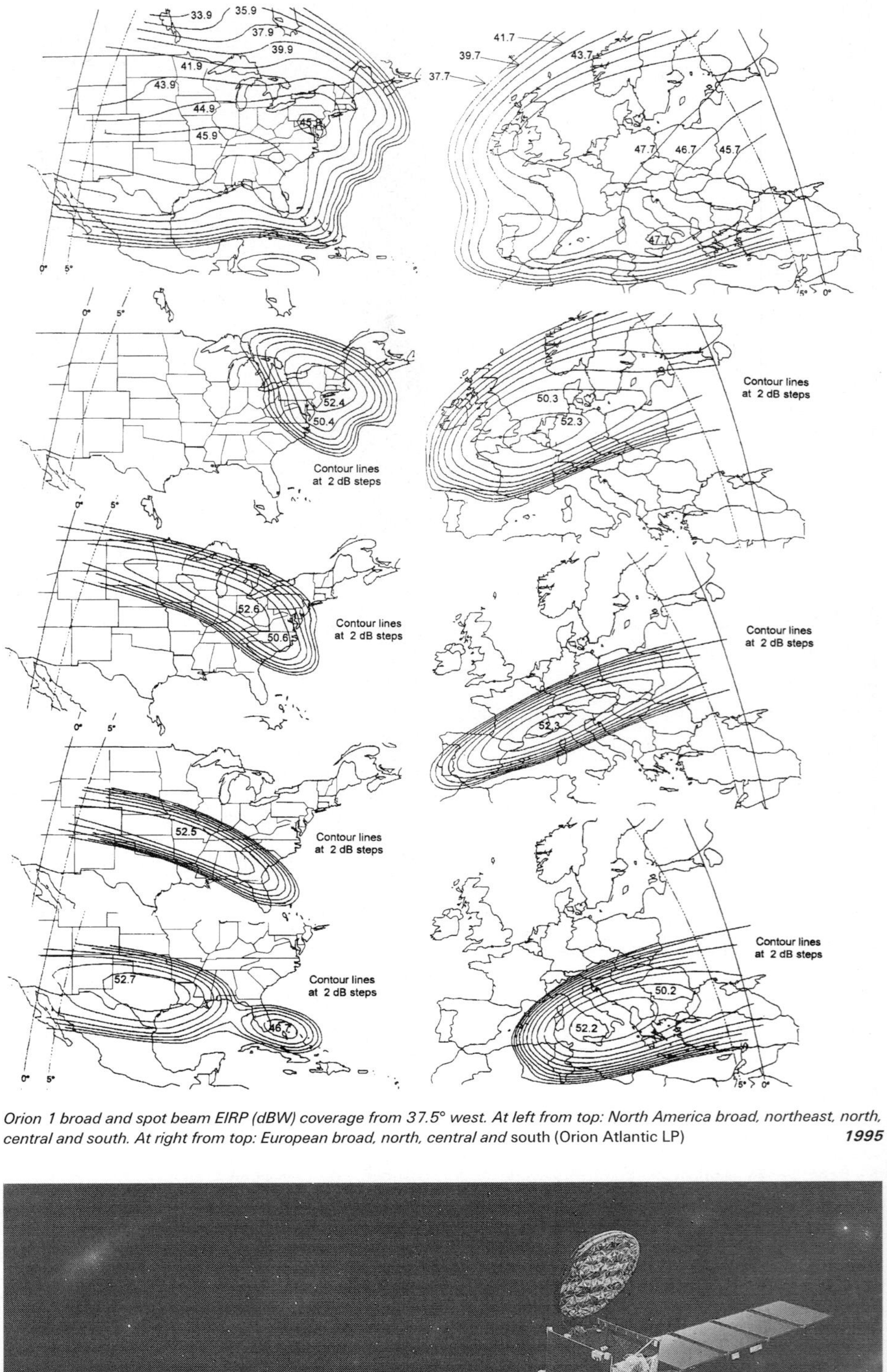

Orion 1 broad and spot beam EIRP (dBW) coverage from 37.5° west. At left from top: North America broad, northeast, north, central and south. At right from top: European broad, north, central and south (Orion Atlantic LP) ***1995***

Orion 1 began providing trans-Atlantic services in January 1995 (Orion Atlantic LP) ***1995***

2.1 × 1.7 × 1.9 m; span 22.0 m. Two 2.3 m antennas on east/west faces
AOCS: 490 N R-4D for apogee raising, 12 × 10N MBB for station-keeping and wheel unloading; NTO/MMH in 4 tanks
TT&C: 15 m fully steerable & two 9 m antennas at Mt Jackson, Virginia
Power system: twin 4-segment Si solar wings providing 3.5 kW EOL, supported by two 78 Ah Ni/H_2 batteries. Payload power is 2,200 W.

Orion 2
Launch: planned for 1999
Location: planned 12° west geostationary
Design life: 13 years
Contractors: Matra Marconi Space (prime), Kaiser Marquardt (apogee motor), DASA (10 N thrusters)
Transponders: 32 TWTA 14.0-14.5/11.45-12.2 & 12.5-12.75 GHz up/down Ku-band Europe & North America spots & regional beams, 54 MHz bandwidth. Beam connectivity via switch matrix provides not only trans-Atlantic services but also regional & domestic coverage for Eastern US & Europe
Principal applications: as Orion 1, plus DTH
Mass: 3,200 kg at launch
Configuration: MMS Eurostar 2000+ bus. Two antennas on East/West faces
AOCS: 490 N R-4D for apogee raising, 12 × 10N DASA for station-keeping and wheel unloading; NTO/MMH in 4 tanks
TT&C: as Orion 1
Power system: twin 4-segment Si solar wings providing 7 kW EOL, supported by two 160 Ah Ni/H_2 batteries.

Orion 3
Launch: 5 May 1999 aboard Delta 3 which failed due to a premature stage 3 shutdown.
Location: projected 139° east GEO (applied for through Marshall Islands)
Design life: 13 years
Transponders: 32 TWTA 14.0-14.5/11.45-12.2 & 12.5-12.75 GHz up/down Ku-band Asia-Pacific spots & regional beams, 54 MHz bandwidth
Principal applications: as Orion 2
Mass: 3,200 kg at launch
Configuration: as Orion 2
AOCS: as Orion 2
Power system: as Orion 2.

PanAmSat series

Current status

After Panamsat 6 suffered electrical failures limiting use on some transponders, Panamsat announced in March 1998 that Hughes would build a replacement satellite named PAS-6B. As of May 1999, the launch schedule comprised the following: Galaxy XI (HS702) 4th Qtr 1999, in service 1st Qtr 2000; Galaxy XR (HS601HP) 1st Qtr 2000, in service 2nd Qtr 2000; Galaxy IVR (HS601HP) 1st Qtr 2000, in service 2nd Qtr 2000; PAS-9 (HS702) 2nd Qtr 2000, in service 3rd Qtr 2000; PAS-9 (HS601HP) 2nd Qtr 2000, in service 3rd Qtr 2000; PAS-10 (also backup for PAS-4) (HS601HP) 2nd Qtr/3rd Qtr 2000; 3rd Qtr/4th Qtr 2000; Galaxy IIIC (HS702) 1st Qtr 2001, in service 2nd Qtr/3rd Qtr 2001.

Background

PanAmSat operates the first global geosynchronous private sector telecommunications system, with satellites over the Pacific, Atlantic and Indian Ocean regions. The company selected Hughes in 1991 to provide four of its HS-601 models for PAS 2-4. PanAmSat paid Hughes about US$300 million for all of the satellites. The fourth satellite remained on the ground as a ground spare. Each bus carries cross-strapped C-/Ku-band 54 MHz transponders optimised for TV programmers and corporate telecommunications. Ariane received the contract to launch the entire series in June 1992, in return for which Credit Lyonnais arranged French export loans covering 85 per cent of the US$240 million launch costs. Then Televisa SA of Mexico stepped up to purchase 50 per cent of PanAmSat's equity, to finance the US$700 million expansion in December 1992, for US$200 million in cash and US$440 million raised from a bond sale.

With financing assured, PanAmSat launched its first satellite PAS 1, on the maiden flight of the Ariane 4 booster. ArianeSpace sold the spot to PanAmSat for the reduced fee of US$8 million. Insurance coverage cost PanAmSat another US$40 million at a 25 per cent premium. PanAmSat after proceeding with its plans, remained prevented by the FCC from connecting with public switched networks. So the service relies on TV programming of US and Mexican origin, and business data traffic. PAS 2 appeared in July 1994 to serve the Asia-Pacific region, and by November 1995 it had 17 full-time programmers. Broadcast sales that year made up 67 per cent of revenue.

PanAmSat is the first private sector operator of a global system (Hughes Space & Communications)

The 2,985 kg PAS 3, intended for 43° west to provide DTH services to Latin America disappeared in Ariane's 1 December 1994 failure. Insurance providers paid an estimated US$214 million at 17.5 per cent. PanAmSat outfitted its ground spare which became PAS 3R, and began the Latin America service in early 1996.

The company announced plans for further satellites and ordered PAS 5 from Hughes Space & Communications in March 1995 and PAS 6 from Space Systems/Loral in August 1994 to expand the DTH Latin America capacity to more than 120 digital channels per customer. PanAmSat raised US$262 million in April 1995 in a preferred stock offering to finance PAS 5/6. SS/L took an order for PAS 7/8 in April 1996, and PanAmSat filed the necessary paperwork with the FCC in September 1995 to ensure a slot for the PAS 10/11 Atlantic Ka-band satellites.

Specifications

PAS 1
Launched: 15 June 1988 by Ariane 401 from Kourou, French Guiana
Location: 45° west geostationary
Design life: 11 years (sufficient propellant for 13¼ years)
Contractors: Lockheed Martin Astro Space
Transponders: 12 (plus 2 back-up) 8.5 W 36 MHz bandwidth SSPA and six (plus 1 back-up) 16.2 W 72 MHz bandwidth TWTA 5.92-6.42/3.7-4.2 GHz up/down C-band global and spot beams for US/Latin America and domestic Latin America; 33-35 dBW EIRP for 36 MHz, 37-40 dBW for 72 MHz, linear orthogonal polarisation
Six (plus 1 back-up) 16.2 W TWTA 14-14.5/11.5-12.2 GHz up/down Ku-band spot beams, 72 MHz bandwidth, 43-45dBW EIRP
Principal applications: TV, voice, data, telex, facsimile
Configuration: Lockheed Martin Astro Space 3000 bus, 1.63 × 1.32 × 1.9 m box-shaped bus, 15.8 m across solar wings. 3-axis stabilised by hydrazine RCS + MW assemblies. GEO injection by solid AKM
Mass: 1,220 kg at launch, 708 kg on-station BOL
TT&C: through Astro Space's Satellite Operations Control Centre at East Windsor, New Jersey. Station-keeping to within 0.1° longitude
Power system: two identical solar wings providing 1,400 W BOL/1,200 W EOL.

PAS 2
Launch: 8 July 1994 by Ariane 4 from Kourou, French Guiana
Location: 169° east geostationary
Design life: 15 years min (Ariane accuracy left propellant sufficient for 16½ years)
Contractors: Hughes Space & Communications Co (prime)
Transponders: 16 (plus 4 back-up) 30 W SSPA two C-band 5.925-6.425/3.70-4.20 GHz up/down Pacific beams (Australasia, Northeast Asia, China; Western US/Hawaii), 54 + 64 MHz bandwidth, EIRP 33-38 dBW for >=2.4m Earth stations, orthogonal linear polarisation. 8 transponders can be cross-strapped C-Ku
16 (plus 4 back-up in 20-for-16) 63 W TWTA three 14.00-14.50/12.250-12.750 GHz up/down Ku-band Pacific beams (Australia/New Zealand; Northeast Asia/China; China/Japan), 54 + 64 MHz bandwidth, EIRP 44-51 dBW for >=90 cm Earth stations, orthogonal linear polarisation. 8 transponders can be cross-strapped Ku-C

Another utilisation of the HS-601HP, this PanAmSat has 24 high-power transponders in C-band and Ku-band
(Hughes Space & Communications) ***1999***/0054341

Principal applications: TV distribution and business communications
Configuration: HS-601 platform, 3-axis, solar array span 26.2 m (satellite within 2.7 × 3.8 × 4 m envelope at launch). Payload mounted on North/South panels and Earth-facing floor; East/West faces carry two dual-gridded rectangular antennas, Earth-facing wall carries two dual-gridded dishes
Mass: 2,920 kg at launch, 1,802 kg on-station BOL, 1,291 kg dry
AOCS: 3-axis, unified 12 ARC 22 N and one Marquardt 490 N bipropellant thrusters incorporating onboard control processor, Sun and Barnes Earth sensors and two 61 Nms 2-axis gimbaled momentum bias wheels. NTO/MMH in four spheres. Spin stabilised in transfer orbit
TT&C: controlled by Hughes Communications Inc from its OCC Operations Control Center in El Segundo, California via its Fillmore, California station
Power system: twin solar wings (spanning 26.2 m) of four 2.16 × 2.54 m panels carrying K4¾ large area Si cells on Kevlar substrate provide 4.225 kW EOL. 100% eclipse protection by Ni/H_2.

PAS 3R
Launch: 12 January 1996 by Ariane 4 from Kourou, French Guiana

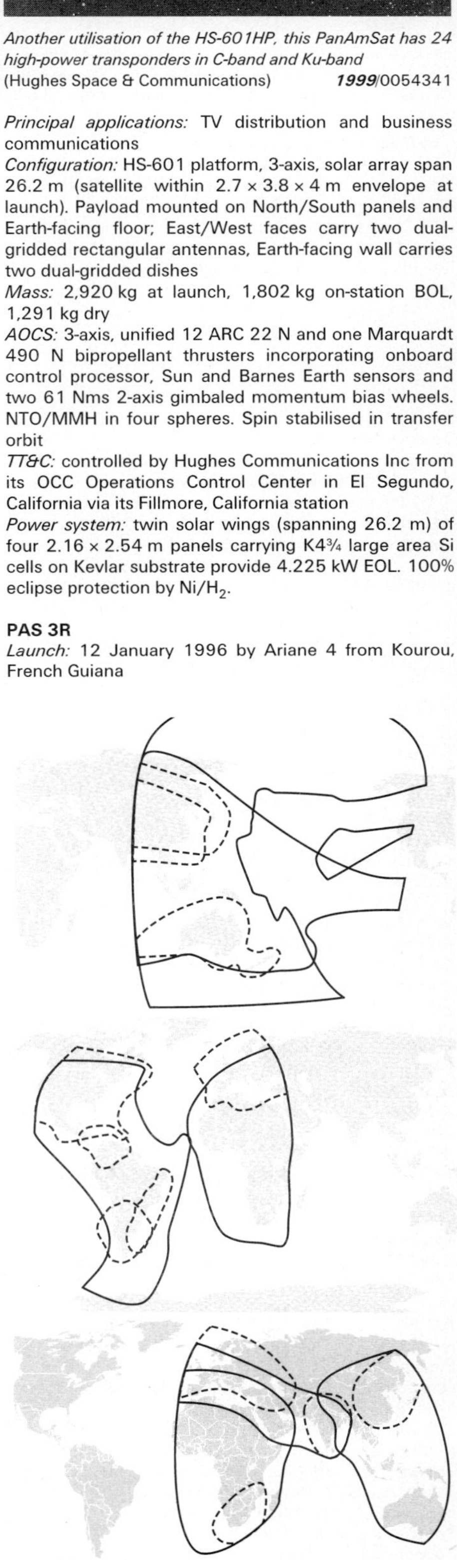

PAS 2-4 footprints (from top). Ku-band is dashed line; C-band is solid

Replacement for PanAmSat 3 lost in an Ariane failure in December 1994, PAS-3RR is made ready for a January 1996 launch (Hughes Space & Communications) ***1999***/0054342

Location: 43° west geostationary
Transponders: Same as PAS 2 but two C-band beams cover Americas and Africa/Europe and five Ku-band US, Canada, Europe and South America
TT&C: controlled by Hughes Communications Inc from its OCC Operations Control Center in El Segundo, California via its Spring Creek, New York station
Mass: 2,918 kg at launch, 1,753 kg on-station BOL, 1,318 kg dry
Other specifications as for PAS 2.

PAS 4
Launch: 3 August 1995 by Ariane 4 from Kourou, French Guiana
Location: 68.5° east geostationary
Transponders: C-band as PAS 2/3 but three beams cover Africa; Middle East/India; Asia
24 (plus 4 back-up) 60 W TWTA four 14.00-14.50/ 12.250-12.750 GHz up/down Ku-band Pacific beams (southern Africa; Northeast Asia; Europe/Eastern Russia; Middle East/India), 16 × 27 + 8 × 54/64 MHz bandwidth, EIRP 44-53 dBW for >=90 cm Earth stations, orthogonal linear polarisation. 8 transponders can be cross-strapped Ku-C
Mass: 3,043 kg at launch, 1,868 kg on-station BOL
Other specifications as for PAS 2.

PAS 5
Launch: 28 August 1997 by Proton-K from Tyuratam, Kazakhstan
Location: 58° west geostationary
Design life: 15 years min
Transponders: 24 (plus 4 back-up in 28-for-24) 50 W TWTA two C-band 5.925-6.425/3.70-4.20 GHz up/ down North/South American beams, 36 MHz bandwidth, orthogonal linear polarisation
24 (plus 4 back-up in 28-for-24) 6 × 60 W + 18 × 110 W TWTA 14.00-14.50/12.250-12.750 GHz up/down Ku-band North/South American beams, 36 MHz bandwidth
Principal applications: TV distribution (including Latin America DTH) and business communications
Configuration: improved HS-601 HP (High Power) platform, featuring GaAs solar arrays, enlarged payload and lightweight contoured reflectors
Mass: 3,720 kg at launch
AOCS: Hughes' 68 kg XIPS xenon ion propulsion system will provide NSSK: two (+2 back-up) 18 mN 13 cm thrusters
TT&C: controlled by Hughes Communications Inc from its OCC Operations Control Center in El Segundo, California via its Fillmore, California station
Power system: twin solar wings of four panels providing 8 kW BOL (7,075 W total required); 100 per cent eclipse protection.

PAS 6
Launch: 8 August 1997 on Ariane 44LP from Kourou, French Guiana
Location: 316.5° east geostationary (collocated with PAS 3R)
Design life: 15 years minimum
Contractors: Space Systems/Loral (prime)
Transponders: 36 (plus 12 back-up, in 12-for-9) 100 W TWTA Ku-band South American beams, 36 MHz bandwidth, linear orthogonal polarisation
Principal applications: DTH Latin America
Configuration: 3-axis stabilised SS/L FS-1300 2.41 × 2.58 × 2.20 m box-shaped bus with twin solar wings on North/South sides, 27.3 m span; 2.4 m diameter Ku-band reflectors on East/West faces plus 1.2 m receive on Earth face
Mass: about 3,420 kg at launch, 1,285 kg dry
AOCS: 3-axis bias momentum system; single Marquardt R-4D 490 N apogee thruster and two sets of six 22 N RCS thrusters, common NTO/MMH propellant system
Antenna pointing: 0.09° azimuth/0.08° elevation Ku-band
TT&C: under development
Power system: twin 4-panel GaAs wings providing about 8 kW BOL. Ni/H_2 battery provides 100 per cent eclipse protection.

PAS 7
Launch: 16 September 1998 on Ariane 44LP from Kourou, French Guiana
Location: 68.5° east geostationary (collocated with PAS 4)
Design life: 15 years minimum
Transponders: 14 (plus 4 back-up in 9-for-7) 50 W TWTA 6.425-6.725/3.40-3.70 GHz up/down C-band beams, 36 MHz bandwidth, maximum EIRP 39 dBW, orthogonal linear polarisation
30 (plus 10 back-up in 40-for-30) 100 W TWTA 13.75-14.25/10.95-11.70 GHz up/down Ku-band, 36 MHz bandwidth, max EIRP 53 dBW
Principal applications: telecom + DTH
Configuration: 3-axis stabilised SS/L FS-1300 2.41 × 2.58 × 2.20 m box-shaped bus with twin solar wings on north/south sides, 27.3 m span; 1.8 m diameter C/Ku-band reflectors on east/west faces plus 1.2 m receive on Earth face
Mass: about 3,838 kg at launch, 2,118 kg BOL, 1,466 kg dry
AOCS: 3-axis bias momentum system; single Marquardt R-4D 490 N apogee thruster and two sets of six 22 N RCS thrusters, common NTO/MMH propellant system
Antenna pointing: 0.09° azimuth/0.08° elevation Ku-band
TT&C: location to be determined
Power system: twin 4-panel GaAs wings providing about 9 kW BOL. Ni/H_2 battery provides 100 per cent eclipse protection.

PAS 8
Launch: 4 November 1998 on Proton-K from Tyuratam, Kazakhstan
Location: 166° east geostationary
Design life: 15 years minimum
Transponders: 24 (plus 4 back-up in 28-for-24) 50 W TWTA 5.925-6.425/3.70-4.20 GHz up/down C-band beams, 36 MHz bandwidth, orthogonal linear polarisation
24 (plus 8 back-up in 32-for-24) 100 W TWTA 13.75-14.00/12.25-12.75 GHz up/down Ku-band, 36 MHz bandwidth
Principal applications: telecom + DTH
Configuration: 3-axis stabilised SS/L FS-1300 2.41 × 2.58 × 2.20 m box-shaped bus with twin solar wings on north/south sides, 27.3 m span; 2.4 m diameter C/Ku-band reflectors on East/West faces plus 1.2 m receive on Earth face
Mass: about 3,800 kg at launch, 2,100 kg BOL; 1,466 kg dry
AOCS: 3-axis bias momentum system; single Marquardt R-4D 490 N apogee thruster and two sets of six 22 N RCS thrusters, common NTO/MMH propellant system
Antenna pointing: 0.09° azimuth/0.08° elevation Ku-band
TT&C: location to be determined
Power system: twin 4-panel GaAs wings providing about 9 kW BOL. Ni/H_2 battery provides 100 per cent eclipse protection.

RCA/GE (Satcom/ Aurora/Gstar) series

Current status

In early 1998, GE decided to redirect the footprint of the GE-1A and GE 2A satellites to provide better coverage of China, Taiwan, and Northeast Asia. This represents a shift in market thinking for Americom which had previously targeted the satellite's footprint to cover more of Southeast Asia.

Background

RCA spun off its wholly-owned subsidiary RCA Americom in 1976 to own and operate the parent company's domestic satellite system known as Satcom. Following GE's 1987 acquisition of RCA, a new corporate entity GE Americom took over the company as part of GE's Communications & Services Group. In the days when it was still RCA, the company launched the original Satcom 1, on 13 December 1975 and began the Satcom network. Satcom 2, launched 26 March 1976, continued the series until Satcom 2R replaced it. Satcom 2R subsequently retired from service 28 February 1995

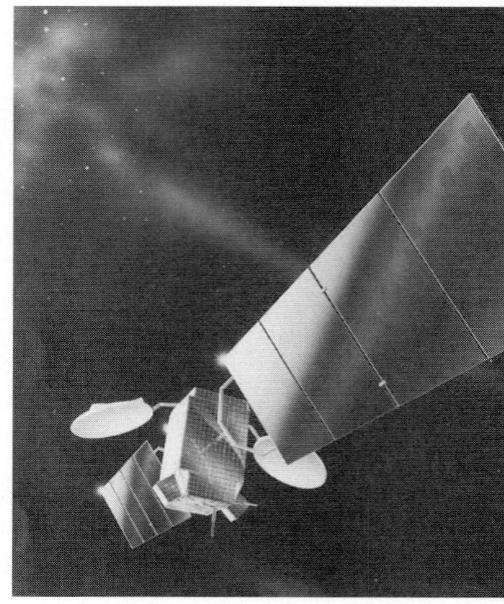

Americom's GE-1 hybrid may debut in 1996 (Lockheed Martin Astro Space) ***1994***

from 72° west. Satcom 3 foundered following a bad firing of its apogee kick motor 7 December 1979; but 3R quickly occupied its slot. Launched 20 November 1981, 3R retired from 131° west in 1991. Satcom 4 and 5 completed the original C-band system in 1982. Then Alascom bought Satcom 5 and renamed it Aurora 1, working until 1991.

Satcom 5 was the first C-band spacecraft to employ solid state power amplifiers in its 24 active transponders. At the time, AT&T Alascom existed as a subsidiary of Pacific Telecom, until AT&T acquired it in 1995. Aurora 1 transferred hands in the acquisition. The Aurora carries domestic telephone, the Alaskan Television Network and emergency messages. As part of the new structure, Alascom bought the satellite for US$88 million and leased back four transponders to RCA.

Aurora served mostly the remote reaches of Alaska, until its customers had to be moved to Satcom C-1 in late 1990, when Aurora suffered service interruptions because of attitude control problems. Eventually Aurora's aging systems forced its retirement in 1991. The satellite now sits in a 40° inclination orbit at 105° west. GE Aerospace launched Aurora 2 (Satcom C5) in 1991 to replace Aurora 1. The new satellite incorporates 24 active 11 W C-band solid-state power amplifiers. GE Americom continues to provide operational control.

Satcom C-1, launched November 1990, began the next generation of communication satellites. Three replacement 3000-class C-band satellites appeared 1990-92. Satcom C-5 replaced C in May 1991. C-4 launched August 1992 to 135° west took traffic from Satcom 4R at 82° west. C-3 appeared in September 1992 to replace the retiring 1R at 131° west.

Americom's Satcom 4, launched 16 January 1982, failed in October 1991, but Telesat Canada's Anik D2 launched November 1984, leased by Americom under a 1990 agreement, stepped in as a temporary measure (designated Satcom 4R) until C-4 appeared and arrived at 82° west in December 1991.

Then the Satcom K generation introduced by GE Americom established a high-power Ku-band data capacity. The Shuttle launched both Satcom K-1 and K-2 in 1985 and 1986, respectively. Crimson Satellite

Satcom C-1 appeared in 1990 as the first of GE Americom's C-band replacements (MM Astro Space)

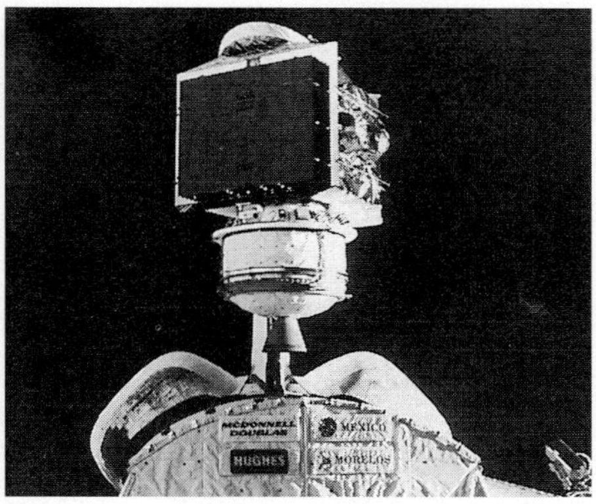

Deployment of Satcom K-2 from Shuttle Columbia's cargo bay, November 1985. The attached PAM-D2 kick motor provided GTO insertion

Associates, a GE Americom/Home Box Office partnership planned to operate Satcoms K-3 and K-4. The company later dissolved, but before that SES purchased K-3 in 1989 for use as Astra 1B. Intelsat bought K-4 the same year to become Intelsat K.

The new GE Americom ordered its first two hybrid satellites from Lockheed Martin Astro Space for debut in 1996, to supply more capacity. Half the payload of Sweden's 1997 Sirius 2 was bought in mid-1995 for the 16 transponders to provide pan-European services.

As a transition, Americom bought all of Spacenet 2's transponders from GTE Spacenet. The WARC originally authorised 85° west for GE-1 (later Gstar-1), where Alaska and Hawaii coverage would have been poor. Americom and Spacenet signed an agreement in 1993 to share the 103° west slot, GStar 1's current position. GE-2, originally slated to remain on the ground as a spare became a firm order when GE's demand grew beyond expectations.

GStar 2 and 4 swapped positions in 1992. Gstar 2 resides next to Hughes' Galaxy 5 for home dish customers. It has combination C-/Ku-antennas and can receive either without repositioning. GStar 2 remained in GEO until June 1995, when GE began operating it in an inclined orbit. After Gstar 3's 55 second GEO insertion burn by the Thiokol Star 30B AKM placed it in a 1.5° inclusive elliptical orbit below GEO, Americom wrote off the satellite as lost. The failure occurred because of a static imbalance in the hydrazine tanks following incorrect ground loading. GStar 3 became a declared US$77 million insurance write-off including US$5 million for Geostar's navigation package.

Controllers, cycling the NSSK thrusters, managed to nurse Gstar 3 all the way to GEO on 15 November 1989 with sufficient propellant for a 7 year life. GTE, the owner, requested FCC permission to operate the satellite from the more accessible 93° west, leaving the planned 125° west to GStar 4. Because GTE wanted to conserve station-keeping propellant, it allowed the satellite to drift in inclination. This inclination required the development and manufacture of a VSAT/TVRO tracking antenna.

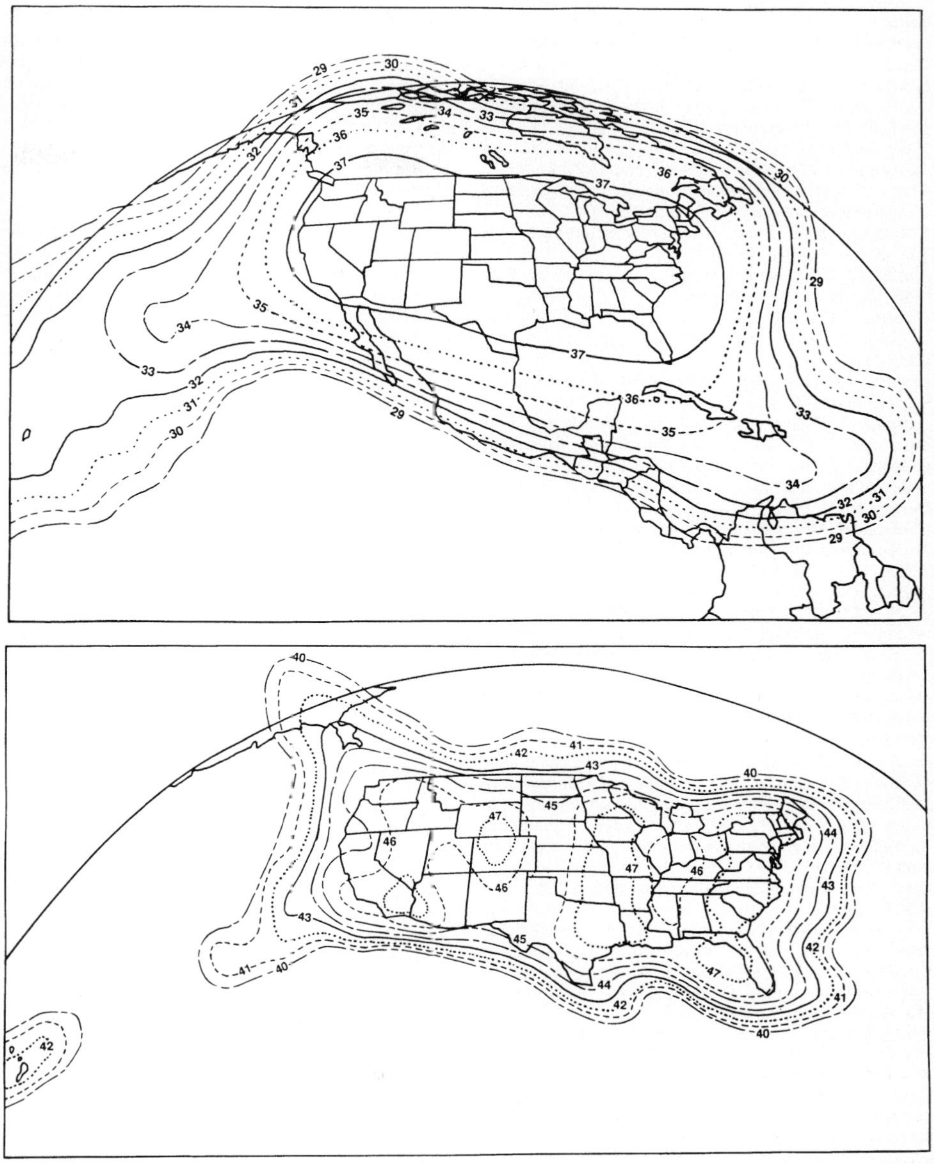

GE-1 EIRP (dBW) performance from 103° west. C-band top; Ku-band bottom (GE Americom) ***1994***

Specifications

Aurora 2

Launched: 29 May 1991 by Delta 7925 from Cape Canaveral
Location: 139° west geostationary
Design life: 12 years
Contractors: GE Astro Space Division
Transponders: 24 (plus 8 back-up in 8-for-6 configuration) 11W SSPA 5.925-6.425/3.70-4.20 GHz up/down C-band CONUS/Alaska/Puerto Rico/Virgin Islands beams, 36 MHz bandwidth, edge EIRP 34-39 dBW Alaska, 34 dBW CONUS, 30 dBW Puerto Rico, 28 dBW Virgin Islands, orthogonal linear polarisation
Principal users: telephony, TV networking, data transmit
Configuration: GE Astro Space 3000-series platform. 3-axis 163 × 132- × 99-cm box-shaped platform with 14.33m solar wing span. Electronics units, batteries, propulsion and AOCS equipment are mounted on six aluminium honeycomb structural pallets or panels. Transponders and housekeeping elements are mounted on four panels, two each on the North/South sides. Additional housekeeping equipment is mounted on the anti-Earth panel
Mass: 1,338 kg launch, 729 kg on-station BOL
AOCS: 3-axis by 6,000 rpm MW (pitch) and magnetorquing (roll/yaw) controlled by Earth Sensor Assembly error signals. About 180 kg hydrazine for 12 Reaction Engine Assemblies + two electrothermal thrusters. Station-keeping accuracy is ±0.1°. GEO insertion by a Thiokol Star 30B solid AKM
Pointing accuracy: 0.1° (±6.0° pitch offset pointing control)
TT&C: through GE Americom's stations at Vernon Valley, NJ + South Mountain, California
Power system: twin Si wings of three panels each, totalling 12.9m², two 40 Ah Ni/H_2 batteries.

GE satellites

Launch: GE-1, 8 September 1996 by Atlas-2A from Cape Canaveral; GE-2, 30 January 1997 by Ariane-4 from Kourou, French Guiana
Location: GE-1 103° west geostationary, GE-2 85° west, GE-3 127° west, GE-4 87° west, GE-5 93° west, GE-6 83° west, and GE-7 58° west
Design life: 15 years
Contractors: Lockheed Martin Astro Space (prime)
Transponders: 24 (plus 8 back-up; 16-for-12) 12-18 W variable SSPA 5.925-6.425/3.7-4.2 GHz up/down C-band CONUS/Alaska/Hawaii beams, 36 MHz bandwidth, edge EIRP 37 dBW CONUS, 29 dBW Alaska, 32 dBW Hawaii, 33 dBW Caribbean, linear orthogonal polarisation

24 (plus 12 back-up; 18-for-12) 60 W TWTA 14.0-14.47/11.73-12.17 GHz up/down Ku-band CONUS + Hawaii beams, 36 MHz bandwidth, edge EIRP 45 dBW CONUS, 42 dBW Hawaii beam, orthogonal polarisation
Principal users: cable, broadcast, education, government and business
Configuration: A2100A platform, 3-axis stabilisation
Mass: 2,585 kg at launch, 1,575 kg on-station BOL,
AOCS: 3-axis by MW, magnetorquing, arcjets and hydrazine thrusters. Arcjets and thrusters divided into two redundant sets
TT&C: as Satcom C-1
Power system: twin 4-panel solar wings providing 5,000 W EOL; 2 × 100 Ah Ni/H_2 batteries.

GStar 1

Launched: 8 May 1985 by Ariane from Kourou, French Guiana
Location: 103° west geostationary
Design life: 10 years
Contractors: RCA/GE Astro Space Div (prime)
Transponders: 14 (plus 5 back-up, 19-for-14) 20 W TWTA 14.0-14.50/11.40-12.20 GHz up/down Ku-band CONUS, half-CONUS and spots, 54 MHz bandwidth, EIRP min 38.5 dBW CONUS, 39 dBW half-CONUS, 47 dBW spot beams, orthogonal polarisation

2 (plus 1 back-up, 3-for-2) 27 W TWTA 14.0-14.5/11.70-12.20 GHz up/down Ku-band CONUS, half-CONUS and spot, 54 MHz bandwidth, EIRP min 44 dBW CONUS, 46 dBW half-CONUS, 50 dBW spot, orthogonal polarisation
Principal applications: domestic telecommunications; VSAT, data and video
Configuration: LMAS 3000 3-axis 163 × 132 × 99 cm box-shaped platform with 14.33 m solar array span, 1.58 m diameter reflector. Electronic units, batteries, propulsion, AOCS equipment mounted on six honeycomb structural pallets or panels. Transponders and housekeeping components mounted on four panels, two each on North/South sides. Additional housekeeping equipment mounted on anti-Earth base panel
Mass: 1,270 kg at launch, 759 kg on-station BOL
AOCS: 3-axis by 6,000 rpm MW (pitch) and magnetorquing (roll/yaw) controlled by Earth Sensor Assembly error signals. Blowdown monopropellant hydrazine propulsion: four surface tension tanks manifolded to two redundant sets of eight thrusters (six catalytic + two electro-thermal). GEO insertion by Thiokol Star 30B solid AKM
TT&C: as Satcom C-1
Power system: twin Si 3-panel wings provide 1,750 W BOL/1,352 W EOL (10 years) supported by 3 × 30 Ah Ni/H_2 batteries.

GStar 2

Launched: 28 March 1986 by Ariane V17 from Kourou, French Guiana
Location: 125° west geosynchronous, inclination 0.6° end-1995
Other specifications as for GStar/Spacenet 1.

GStar 3

Launched: 8 September 1988 by Ariane V25 from Kourou, French Guiana
Location: 93° west geosynchronous inclination (7.0° end-1995)
Design life: 8 years but GEO-insertion failure reduced station-keeping hydrazine to about 7 years useful life
Other specifications as for GStar/Spacenet 1 plus Geostar navigation package.

GStar 4
Launched: 20 November 1990 by Ariane 42P from Kourou, French Guiana
Location: 105° west GEO (swapped 1992 with GStar 2)
Design life: 10 years, but station-keeping propellant sufficient for >13 years projected
Transponders: Same as GStar 1 but no spot beams and additional 27 W transponder replaces one 20 W unit
Principal applications: one-third leased, one-third data/voice services, one-third SNG + business TV
Configuration: as GStar 1 but lighter structure as not intended for Shuttle launch
Mass: 1,295 kg at launch, 741 kg on-station BOL (543 kg dry; 181 kg hydrazine)
Power: as GStar 1.

Satcom C-1
Launched: 20 November 1990 by Ariane 4 from Kourou, French Guiana
Location: 137° west geostationary, providing in-orbit protection for C-band fleet (polarising switching capability allows operations from other locations to restore interruptions in multipoint services)
Design life: 12 years minimum
Contractors: Martin Marietta Astro Space
Transponders: 24 (plus 4 back-up; 7-for-6) 9.6 W SSPA 5.925-6.425/3.7-4.2 GHz up/down C-band CONUS/Alaska/Hawaii beams, 36 MHz bandwidth, edge EIRP 33 dBW CONUS, 32 dBW Alaska, 29 dBW Hawaii, linear orthogonal polarisation (odds horizontal down)
Principal users: cable programming, NBC TV and in-orbit protection
Configuration: RCA Advanced Satcom series platform. 3-axis 163 × 132 × 99 cm box-shaped platform with 14.33 m solar array span. Electronics units, batteries, propulsion, AOCS equipment mounted on six aluminium honeycomb structural pallets or panels. Transponders and housekeeping elements mounted on four panels, two each on North/South sides. Additional housekeeping equipment is mounted on the anti-Earth panel
Mass: 1,169 kg at launch, 682 kg on-station BOL
AOCS: 3-axis by 6,000 rpm MW (pitch) and magnetorquing (roll/yaw) controlled by Earth Sensor Assembly error signals. Blowdown mono hydrazine propulsion: four surface tension tanks manifolded to two sets of seven redundant thrusters. GEO insertion by a Thiokol Star 30B solid AKM
TT&C: through the Astro Satellite Operations Control Center at East Windsor, New Jersey for launch and transfer orbit phases; Americom's own stations at Vernon Valley New Jersey, South Mountain California, Goddard, Maryland, Woodbine, Maryland and Grand Junction, California for operational control
Power system: twin Si wings of three panels each provide about 1,500 W BOL/1,230 W EOL, supported by 3 × 24 Ah Ni/Cd batteries for eclipse operation.

Satcom C-4
Launch: 31 August 1992 by Delta 7925 from Cape Canaveral
Location: 135° west geostationary
Transponders: 24 (plus 8 back-up; 8-for-6) 17.5 W TWTA 5.925-6.425/3.7-4.2 GHz up/down C-band CONUS/Alaska/Hawaii beams, 36 MHz bandwidth, edge EIRP 39 dBW CONUS (40 dBW peak), 27 dBW Alaska, 31 dBW Hawaii, linear orthogonal polarisation (evens horizontal down)
Principal users: cable programming
Mass: 1,402 kg at launch, 791 kg on-station BOL
Power: 1,950 W BOL; 2 × 50 Ah Ni/Cd batteries
Other specifications as for Satcom C-1.

Satcom C-3
Launch: 10 September 1992 by Ariane 4 from Kourou, French Guiana
Location: 131° west geostationary
Mass: 1,375 kg at launch, 789 kg on-station BOL
Other specifications as for Satcom C-4.

Satcom K-1
Launched: 12 January 1986 by Shuttle
Location: 85° west geostationary
Design life: 10 years
Transponders: 16 (plus 4 back-up) 47 W TWTA 14.0-14.47/11.73-12.17 GHz up/down Ku-band CONUS + east/west US beams, 54 MHz bandwidth, edge EIRP 43 dBW CONUS, 48 dBW West beam, orthogonal polarisation
Principal users: PrimeStar Inc, business video, occasional video and data networks
Configuration: MMAS 4000 series platform, 170 × 213 × 152 cm body, 19.3 m span across solar arrays, total height 305.8 cm at launch, 3-axis stabilisation. Equipment mounted on eight aluminium honeycomb structural pallets or panels; transponders and some housekeeping units on north/south faces; antenna/reflectors, including primary Kevlar honeycomb reflectors, and Earth sensors, on Earth-facing panel
Mass: about 1,900 kg at launch, 1,021 kg on-station BOL, about 780 kg EOL
AOCS: 3-axis by MW, magnetorquing and 16 hydrazine thrusters, divided into two redundant sets of eight, drawing on four spherical tanks. GEO injection by Thiokol solid; GTO injection by PAM-D2 solid
TT&C: as Satcom C-1
Power system: twin 4-panel Si solar wings providing 3.7 kW BOL, 2,490 W EOL; 3 × 50 Ah Ni/H_2 batteries.

Satcom K-2
Launched: 27 November 1985 by Shuttle
Location: 81° west geostationary
Principal users: NBC TV network, satellite news gathering, business video and syndication
Mass: 1,926 kg at launch, 996 kg on-station BOL
Other specifications as for Satcom K-1.

Spacenet 2
Launched: 10 November 1984 by Ariane from Kourou, French Guiana
Location: 69° west geostationary
Contractors: Lockheed Martin Astro Space (then RCA Astro-Electronics)
Transponders: 12 (plus 2 back-up, 7-for-6) 8.5 W SSPA 5.925-6.425/3.7-4.2 GHz up/down C-band CONUS, Alaska, Hawaii beams, 36 MHz bandwidth, EIRP min 35-36 dBW CONUS, 25-32 dBW Alaska/Hawaii, horizontal polarisation
6 (plus 1 back-up, 7-for-6) 16 W TWTA 5.925-6.425/3.7-4.2 GHz up/down C-band CONUS, Alaska, Hawaii beams, 72 MHz bandwidth, EIRP min 37-38 dBW min CONUS, 30-35 dBW Alaska/Hawaii, vertical polarisation
6 (plus 1 back-up, 7-for-6) 16 W TWTA 14.0-14.5/11.7-12.2 GHz up/down Ku-band CONUS beams, 72 MHz bandwidth, EIRP min 40 dBW, horizontal polarisation
Mass: 1,195 kg at launch, 692 kg on-station BOL
Power system: 1,530 W BOL/1,250 W EOL
Other specifications as for GStar 1.

Spacenet 3R
Launched: 11 March 1988 by Ariane V21 from Kourou, French Guiana (replacing Spacenet 3 satellite lost in Ariane failure of September 1985)
Location: 87° west geostationary
Transponders: Same as Spacenet 1 but additionally 2-channel 16.5 MHz L-band receive-only transponder providing CONUS and Caribbean/Central American Geostar position-reporting service (inactive since 1991)
Other specifications as for GStar/Spacenet 1.

Spacenet 4 (ASC 2)
Launch: 13 April 1991 by Delta 2 from Cape Canaveral
Location: 101° west geostationary
Design life: 10 years
Transponders: 12 (7-for-6 redundancy) 8.5 W min SSPA 5.925-6.425/3.7-4.2 GHz up/down C-band US beams, 36 MHz bandwidth, EIRP 34-37 dBW continental US, 31-33 dBW Alaska, 28-30 dBW Hawaii, 27-29 dBW Puerto Rico
6 (4-for-3 redundancy) 16.2 W min TWTA 5.925-6.425/3.7-4.2 GHz up/down C-band US beams, 72 MHz bandwidth, EIRP 37-40 dBW continental US, 32-35 dBW Alaska, 31-32 dBW Hawaii, 30-31 dBW Puerto Rico
4 16.2 W min TWTA 14.0-14.5/11.7-12.2 GHz up/down Ku-band US beams, 72 MHz bandwidth, EIRP 38-43 dBW continental US
2 30.0 W min TWTA 14.0-14.5/11.7-12.2 GHz up/down Ku-band US beams, 72 MHz bandwidth, EIRP 38-43 dBW continental US
Principal applications: private C-band data networks
Configuration: LMAS 3000 3-axis satellite. Hybrid C/Ku-band payload with dual antennas
Mass: 728 kg BOL on-station
TT&C: as Spacenet C-1
Power: 1,212 W EOL; 100 per cent eclipse protection.

Western Union (Westar) series

Background

Western Union owned and operated the original Westar franchise and thus became the first US domestic satellite system. The Westar 4 satellite doubled the number of transponder channels available when it introduced its spin-stabilised platform with an orthogonal polarising reflector. The system relayed voice, data, video, facsimile and other services to the continental US, Hawaii, Alaska, Puerto Rico, and the Virgin Islands.

The first Westar satellite is based on Hughes' HS-333 design. Western Union inaugurated the series beginning with Westar's 1 launch 13 April 1974 followed closely by Westar 2 on 10 October 1974 by Delta from Canaveral. Each had a mass of 572 kg at launch, 291 kg in-orbit. Westar 3, launched 10 August 1979 by Delta 2914, operated at 91° west as the system's last HS-333 until it retired 26 January 1990. Westar 4 introduced the larger HS-376 design, almost tripling power from the solar array by adding an extending skirt and doubling the number of transponders. Launched 26 February 1982 by Delta 3910, it operated at 99° west until retirement 5 November 1991.

Western Union's role as a satellite operator ended in February 1988 when Hughes Communication Incorporated bought all outstanding interest in Westar. HCI took over operating Westar 3-5 satellites and ownership of Westar 6S.

Westar 1 ended operational life April 1983. Westar 3, launched 10 August 1979 by Delta 2914, operated at 91° west as the system's last HS-333 until it retired in January 1990.

Westar 4 introduced the larger HS-376 design, almost tripling power from the solar array by adding an extending skirt and doubling the number of transponders. Launched on 26 February 1982 by Delta 3910, it operated at 99° west until HCI retired it in November 1991. Galaxy 6 provided temporary replacement until Galaxy 4 could be launched in 1993. Western Union launched Westar 5 8 June 1982 by Delta 3910 and then retired the satellite from 122.5° west in March 1992 after replacement by Galaxy 5. Westar 6 went into space on 4 February 1984 aboard STS-41B but the PAM-D upper stage malfunctioned and stranded the satellite in a useless 307 × 1,128 km orbit. To ensure that it did not re-enter, pending decisions about its future, Western Union fired its apogee kick motor 12 May 1984 to raise it to 1,060 × 1,666 km. The Shuttle subsequently retrieved it after its hydrazine motors lowered the orbit for rendezvous Shuttle mission 51A in November 1984.

The underwriters, represented by Merrett Syndicates of London paid NASA US$2.75 million for the retrieval, plus US$5 million to Hughes Aircraft for the necessary Shuttle equipment. Pan Am Pacific Satellite Corporation paid US$20 million for the satellite in mid-1986 to provide services to dispersed nations over the Pacific. Westar replaced the satellite with Westar 6S, using the US$110 million insurance, offering greater capacity but had not launched it at the time of purchase by HCI.

DATA RELAY

BRAZIL

SCD-1 satellite

Background
The Brazilian government launched SCD-1 on 9 February 1993 by Pegasus into 725 × 790 km, 24.97° orbit. Little equipment degradation was observed after one year of life. The 115 kg provided real-time relay of data from environmental data collection platforms using the Argos system frequency (401.65 MHz) and 401.62 MHz, which alone will handle 200 platforms. 65 DCPs were in place at end-1995; another 250 will be added during 1996, mainly for monitoring hydrographic basins and for flood control. The 1.70 m high octagonal prism carries University of Sao Paulo solar cells on the eight outer and top faces to generate 60 W average in sunlight, supported by an 8 Ah Ni/Cd battery. S-band/ UHF antennas are mounted on the top and bottom.

SCD-1 monitors the condition of the Amazon River basin and the surrounding rain forests. It also monitors CO_2 and ozone levels, mainly produced from burning vegetation.

The Brazilians spun SCD at release to 120 rpm and thereafter maintained simple attitude control by a magnetorquer. The main requirement for satellite survivability is to hold the bottom face away from the Sun for heat dissipation. The satellite contains 160 telemetry channels.

SCD-1 was ready for flight in 1991, but had to wait until the Brazilian government could select a launcher. OSC's Pegasus won the US$11.5 million contract in October 1991 and it was signed 20 August 1992. The SCD-1 satellite cost US$20 million; 73 per cent of the funding was expended in Brazil.

FRANCE

Eole satellite

Background
France launched the Eole on 16 August 1971 by Scout from Wallops into 664 × 870 km, 50.1° orbit. It weighed 84 kg and the country planned to use it to test a concept for data relay of the satellite interrogation of 500 meteorological balloons drifting in the atmosphere at 12,000 m. It returned data for five months.

GERMANY

SAFIR series

Background
Germany plans to use the SAFIR (Satellite for Information Relay) as a two-way data collection/ distribution system using six small satellites in 640 to 800 km 62 to 98° orbits.

Germany launched the 38.5 kg SAFIR R1 core demonstration package 4 November 1994 attached to Resurs-O1 N3 into 551 × 663 km, 98.05°. The satellites launch was delayed because the lifetime of the previous Resurs-O1's went on longer than expected and deferred departure of its replacement. The satellite body houses the SAFIR electronics module, software and TM/TC box; Resurs provides the power. OHB-System operates the hub station in Bremen. SAFIR interrogates fleet stations and stores the data in its 10 Mbit solid-state memory for later transfer and/or multi-user access. In the operational system, 1.2 kg macro-stations would provide 300 bit/s and 2.4 kbit/s of data transmission.

Co-funded by DARA and OHB-System, the SAFIR 2 which draws on Bremsat technology, will be ejected from Russia's Resurs-O1 N4 Earth observation satellite in 1997 for a DM200,000 fee.

Specifications
Size: 45 × 45 × 50 cm; 60 kg
Power: 25 W orbit average, 50 W peak, from Si cells on five faces; 4 Ah Ni/Cd and 4 Ni/H_2 batteries
Stabilisation: ±5° nadir pointing accuracy using gravity gradient boom and magnetorquer. determination by magnetometer and Sun/star sensors
Data rate: 300 and 2,400 bit/s
Frequencies: users, UHF up/down; TT and C VHF up/ down
Configuration: two modular compartments (lower 30 cm high; upper 20 cm high). The bottom plate carries launcher interface and two communications units. The middle plate houses the batteries, power supply unit, computer and gravity gradient boom interface. Controlled by modified Bremsat 1 OBC, based on Inmos T800 transputer with 10 Mbit mass memory. TT&C on 2 m band.

INTERNATIONAL

Data Relay satellite

Background
In order to reduce some of the costs for building and maintaining a satellite ground station network, ESA has considered a TDRS-like space-based relay satellite network and proposed to run the project in a series of phases. The first, (Phase A1), ran from the middle of 1987 to early 1988 with the work emphasising system design and ground and spaceborne system configuration. This phase confirmed that a DRS using largely available technology could be developed and deployed by 1996. Phase A2 defined the satellite configuration, and Phase B1, lasting from mid-1989 to Spring 1990 developed the designs for the selected satellite and ground segment facilities down to the subsystem and initial equipment levels.

The ESA Council then approved Phase 1 in June 1990. At their Granada meeting in November 1992, ESA also approved Phase 2 full development for 1999 launch but the project has been under review since early 1995, now that ESA has finally dropped Hermes and the Columbus free-flyer will not require it. One other possible concept now under investigation by ESA is a more cost-effective two-satellite relay system with Artemis.

Specifications
Launch: originally planned for 2000 by Ariane from Kourou, French Guiana
Location: planned 44° west GEO
Transponders: Silex/SKDR, except feeder link has four active transponders plus two back-ups. DRS adds an advanced service at S-band using phased arrays for multiple access with 24/48 transmission/receive elements (Artemis can handle only one user satellite at a time) and provides more data channels than on Artemis. However, a second SKDR antenna might replace this S-band phased array
Ground system: Fucino (operational control centre), Darmstadt (mission control centre), Villafranca (S-band). Other specifications as for Artemis.

RUSSIAN FEDERATION

Luch/Altair/SDRN/SSRD series

Current status
Employed positions are 40, 45, 53, 85, 90, 96.5 103 and 140° east; 11, 14, 25 and 170° west.

Background
The Soviet Union gave notice in 1981 to the International Frequency Registration Board that it intended to establish a Satellite Data Relay Network (SDRN), similar to NASA's TDRSS. Soviet officials planned to place satellites at 95° east, 16° west and 160° west between 1983 and 1985. The system would eliminate the Soviet problem of communicating with its spacecraft from the geographically limited Soviet territory. The system would feature an Eastern SDRN for communications with Salyuts and other low-orbiting spacecraft, and using frequencies of 10.82, 11.32, 13.7 and 13.52 GHz for downlink, and 14.62 and 15.05 GHz for uplink.

The first satellite in the series Kosmos 1700, launched 25 October 1985, placed a Luch at 95° east (CSDRN), relayed extensive communications with the new Mir space station in 1986, and subsequently with Soyuz-TM. In October 1986, the satellite began drifting off station. A replacement Kosmos 1897 appeared 26 November 1987. It moved to 12° east in July to August 1988 to support the Buran Shuttle mission, and then returned to 95° east in February 1989. After the satellite drifted off station in early 1993 it appears to have been abandoned.

The WSDRN position at 34.4° east has been occupied by Kosmos 2054 since mid-January 1990 (launched 27 December 1989). With both positions operational, the expanded Mir coverage reduced the requirements for Soviet communications ships. The first named Luch appeared 16 December 1994 for positioning at 95° east, while Kosmos 2054 has exceeded its planned operating lifetime and appears to be ready for retirement. At the end of March 1997, Luch-1 left station and in early May re-located itself close to the Kosmos 2054 longitude.

Excess Luch capacity is being leased on a commercial basis through NPO PM's Mercury subsidiary, particularly for TV in the US and Argentina. Today, Luch (Beam) is both a service and a system. Some Gorizont and Raduga satellites carry Luch transponders, usually as only one of the spacecraft's payloads.

Specifications
Locations: 16° west, 77° east, 95° east, 160° west GEO
Design life: 5 years
Contractors: NPO PM (spacecraft), NPO Radio (payload)
Transponders: TWTA 15.05/13.5 GHz up/down Ku-band 0.5° steerable, EIRP 56.2 dBW
TWTA 14.62/11 GHz up/down Ku-band 1.0° steerable, EIRP 42.3 dBW TV/34.2 dBW voice
TWTA 900/700 MHz up/down UHF 5.0° steerable, EIRP 40.7 dBW
Configuration: 16.0 m-span solar array provides 1.8 kW
Mass: 2,400 kg on-station
AOCS: 3-axis; east-west station-keeping to 0.5° by liquid thrusters.

UNITED STATES OF AMERICA

TDRS constellation

Background
TDRSS provides NASA with communications coverage of satellites and manned vehicles below 3,000 km, permitting links during 85 per cent of each orbit and allowing NASA to close much of its extensive ground system. NASA planned for TDRSS to be the backbone of the Space Shuttle communication system, and provide 80 per cent of the voice datalinks instead of the 15 per cent coverage previously available via ground stations.

When it first proposed the system in December 1979, NASA projected a cost of US$796 million, but by late 1985 costs had exploded to about US$3 billion. Part of the cost overrun resulted form NASA's realisation in 1978, when the satellites were already designed and partially constructed, that they might suffer radio frequency interference from Soviet radars in Eastern Europe. NASA immediately made US$70 million available to correct the problems and harden the satellites. Then NASA entertained a USAF proposal in 1981 to add a Batson 11 security system to protect against Soviet electronic interference. The air force offered the system to NASA for US$500 million for the six satellites. In the end, the air force dropped the plan, but the bureaucratic discussion delayed the programme for several months and further added to the costs.

In the first independent satellite launch from the Shuttle, the Boeing IUS failed and the satellite limped to its correct orbit six months later after 30 separate burns of its control thrusters. At that point the system was two years behind schedule and Spacelab 1, originally

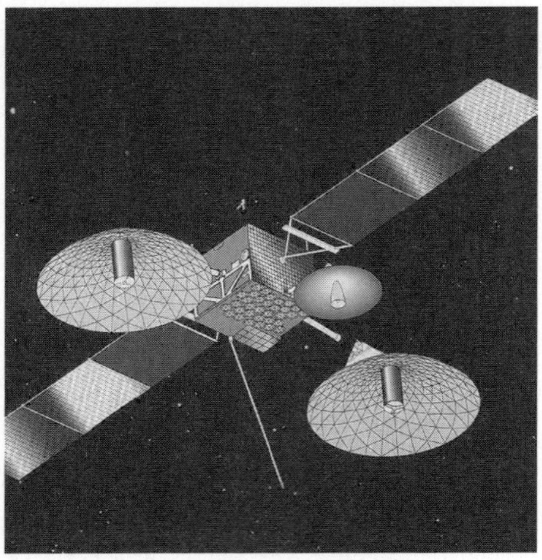

Hughes will build three new TDRS satellites for launches beginning 1999. The composite mesh Springback antennas will handle S/Ka/Ku-band. The S-band phased arrays are mounted directly on the bus. The smaller dish is the Ku-band feeder link (HSC) ***1995***

designed to use both TDRS 1 and 2 for full exploitation, went into space with only the replacement TDRS 1 available. Because of the delays, NASA had to reprogramme the use of data relay for the Landsat Thematic Mapper images and ironically denied the air force the use of the system until full service could be established.

Up and running, TDRS 1 operations led ground controllers to conclude a fundamental flaw in the satellite design caused several timing-oriented problems, which would have been compounded by multiple satellite operations.

Besides technical problems, TDRS suffered from financing problems as well. NASA attempted to spread the cost over 10 years of operation by leasing the satellite to itself instead of paying the purchase price. It entered into its first large-scale joint venture in the form of a leasing contract with Western Union's SpaceCom. Funding problems, including interest charges totalling US$600 million, created another string of delays. TRW originally contracted to build six satellites; two for lease by NASA, one for SpaceCom for commercial Westar users, the fourth as a joint on-orbit spare for NASA/SpaceCom, and two as flight-ready spares. Two months before the TDRS 1 launch, NASA/SpaceCom agreed on cancellation of the commercial service, making it a 100 per cent NASA system, though still owned by SpaceCom. Though increasing NASA's costs by US$216 million, this decision extended TDRSS' life in the relay role to 15 years, allowing numerous ground stations to be closed.

When the Shuttle resumed operations after the Challenger accident, NASA allotted two out of the first three missions to TDRS. TDRS 1 continued operations despite an additional loss of capacity following an antenna failure on 28 November 1986. NASA planned to replace TDRS 2, lost in the Challenger accident, at a cost of about US$250 million compared with the original cost of US$100 million. In the meantime, NASA kept six ground stations open at Guam, Hawaii, Santiago, Ascension Island, Dakar and Yarragadee, which were to have been closed after TDRS 2 became operational. These stations cost NASA US$35 million per month

With much less demand for satellite communications during the Shuttle stand down, an international team used TDRS-1 in July and August 1986 for radio astronomy observations. In the first use of a synchronous satellite in such a role, the Jet Propulsion Laboratory linked TDRS-1 with ground radio telescopes in Australia and Japan in a Very Long Baseline Interferometry (VLBI) technique to accurately locate quasar sources.

Despite deployment difficulties with their 4.9 m antennas, TDRS 3 and 4 brought the system to full operational status in September 1989, when NASA formally accepted TDRS 4. One of TDRS 3's two 4.9 m steerable antennas become locked in LHCP (Left Hand Circular Polarisation) in January 1990, preventing RHCP reception from the Shuttle's RHCP datalinks. NASA decided to shunt TDRS 3 to 174° west, augmenting it with TDRS 1 at 171° west to cover for the lost Ku-band capacity.

NASA stopped using the Shuttle as a launch vehicle, when it decided in 1991 to transfer such payloads to expendable vehicles, rather than using the costly and fragile Shuttle.

The appearance of TDRS 6 in 1993 created for the first time a full system of two operational satellites and a stored spare. The military pays some US$100 million annually for use of the system. NASA's operating costs are about US$65 million annually. The satellites work through the White Sands Complex in New Mexico, reporting to Goddard's Network Control Center (NCC). The station can handle up to 2,400 satellite passes each day at data rates up to 300 Mbit/s.

Congress halted funding for an Advanced TDRS successor in 1993. But NASA still awarded a US$481.6 million fixed price contract to Hughes for three modestly enhanced satellites. Congress has yet to approve the funding.

TDRS 7 was launched in July 1995 as a spare, the last of the first-generation type built by TRW plans.

In February 1995, NASA selected Hughes Space and Communications to build three Tracking and Data Relay Satellites (TDRS-H,-I and J) for launch in 1999 and 2000. They were each to be functionally equivalent to number 7 but adding Ka-band, to help avoid interference from commercial satellites and to offer higher data rates. Based on HSC's HS-601 bus, each will have two 4.58 m Springback steerable single access antennas. These upgrade satellites will provide simultaneous S-/Ku-band or S-/Ka-band operations: 6/0.30 Mbit/s receive/transmit S-band, 300/25 Mbit/s receive/transmission Ku-/Ka-band. An S-band phased-array will receive from five spacecraft.

Specifications

TDRS 1

Launched: 4 April 1983 by Shuttle STS-6/IUS
Location: 49° west GEO (inclination 8.7° January 1996). Replaced as TDRS-E by TDRS 4 in 1989 and moved to 79° west as system spare; in May 1990 transferred to 171° west as cover for lost TDRS 3 Ku-band capacity. Arrived 85° east 7 February 1994 (after number 6 appeared) to provide Compton Observatory link through Australia's Tidbinbilla for relay to White Sands via Intelsat; moved off-station May 1995, stopped 139° west August to December 1995 for Antarctica science operations, before arriving at 49° west June 1996
Design life: 10 years minimum
Contractors: TRW Space & Communications (spacecraft), Harris Electronics Systems Group (ground/space antennas)
Transponders: 2 (plus 4 back-up) 26 W TWTA 2.030-2.113/2.219-2.290 GHz forward/return S-band 2.08° steerable ±22.5° east -west/±31° north-south beams working via 4.9 m antenna, 20/12 MHz forward/return bandwidth, <=0.300/12 Mbit/s forward/return, EIRP >=43.6 dBW min, LH/RHCP selectable

2 (plus 2 back-up) 1.5 W TWTA 13.775/15.0034 GHz forward/return Ku-band 0.28° steerable ±22.5° east-west/±31° north-south beams working via 4.9 m antenna, 50/225 MHz forward/return bandwidth, <=25/12 Mbit/s forward/return, EIRP >=46.5 dBW min, LH/RHCP selectable

30 3.5 W SSPA phased-array for accessing up to 20 satellites simultaneously in 13° beam. 30 elements receive: 2.2875 GHz, 5 MHz bandwidth, LHCP, <=50 kbit/s. 8 elements (plus 4 back-up) also transmission: 2.1064 GHz, 6 MHz bandwidth, LHCP, <=10 kbit/s, 4° beam, EIRP >=34.0 dBW

2 (plus 4 back-up) 25 W TWTA 14.6-15.23/13.4-14.05 GHz up/down Ku-band 0.70° beams linking

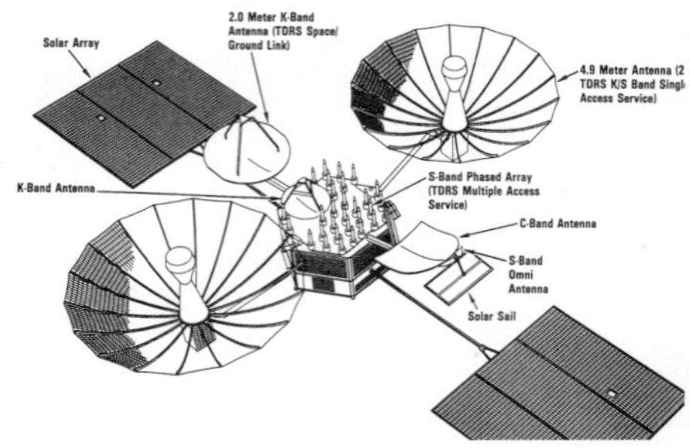

NASA's TDRS system provides communications links for LEO spacecraft over more than 85 per cent of each orbit. The K-band antenna is not used; that and the C-band payload are deleted on TDRS 7 (TRW)

White Sands with TDRS, 650 MHz bandwidth, <=300 Mbit/s, EIRP >=50.8 dBW min, orthogonal linear polarisation

12 5 W TWTA 5.925-6.425/3.700-4.200 GHz C-band up/down beams, 36 MHz bandwidth, EIRP 32 dBW, linear polarisation
Principal users: NASA (including orbiting spacecraft), US government agencies and commercial C-band leasing
Configuration: hexagonal bus/payload modules, 17.4 m span across deployed solar wings
Antennas: two 24.3 kg steerable ±22.5° east-west/±31° north-south 4.9 m S/Ku-band single access molybdenum wire mesh dishes (that is, each accesses one target spacecraft at a time), one 2 m 15.0 kg Ku-band ground link antenna, a 30-element S-band receive phased array for accessing up to 20 satellites and 12 of those elements for S-band transmit, 1 m 5.0 kg K-band dish (not used), 1.2 × 1.6 m 13.6 kg C-band, S-band omni TT&C antenna
Mass: 2,200 kg at launch, 2,120 kg on-station BOL
AOCS: 3-axis, maintained by Earth sensors, RWs + 24 × 4.45 N thrusters (385 kg hydrazine)
TT&C: NASA's White Sands Complex
Power system: two 380 × 380 cm solar wings provide 2.4 kW BOL, 1.85 kW after 10 years (1.7 kW required), supported by 3 × 40Ah Ni/Cd batteries.

TDRS 3

Launched: 29 September 1988 by Shuttle STS-26/IUS
Location: 85° east geostationary
Other specifications as for TDRS 1.

TDRS 4

Launch: 13 March 1989 by Shuttle STS-29/IUS
Location: 41° west GEO TDRS-E (replaced number 1)
Other specifications as for TDRS 1
Comments: In addition to the NASA mission TDRS 4 – as well as TDRS 5 and TDRS 6 – are providing commercial C-band services.

TDRS 5

Launch: 2 August 1991 by Shuttle STS-43/IUS
Location: 174° west GEO TDRS-W (replaced number 3)
Other specifications as for TDRS 4.

TDRS 6

Launch: 13 January 1993 by Shuttle STS-54/IUS
Location: 46° west geostationary (from November 1993)
Other specifications as for TDRS 4.

TDRS 7

Launch: 13 July 1995 by Shuttle STS-70/IUS
Location: 150° west as reserve until May 1996: the satellite was then re-located to 172° west at the end of June 1996
Mass: 2,225 kg at launch
Comments: The K-band antenna and C-band packages were deleted (the C-band antenna was replaced by a larger solar sail); SSPAs replaced the S/Ku-band TWTAs of the single-access system. Other specifications as for TDRS 1.

EARTH OBSERVATION AND METEOROLOGICAL

CANADA

RADARSAT

Background

After a successful launch aboard a McDonnell Douglas Delta II from Vandenberg Launch Facility in California on 4 November 1995, RADARSAT began full operations 28 March 1996.

Designed, constructed, launched, and operated by the Canadian Space Agency (CSA), RADARSAT is the world's most powerful commercial radar remote sensing satellite totally dedicated to operational applications. During its design life of five years, RADARSAT C-Band Synthetic Aperture Radar (SAR) will gather data for numerous global applications including ice surveillance, crop monitoring, disaster assessment, forest applications including clearcut monitoring, mapping, geological exploration, and coastal zone and ocean monitoring. RADARSAT's unique and state-of-the-art characteristics enable RADARSAT International, the Canadian company tasked with international marketing and distribution of RADARSAT data, to customise products to meet specific client requirements. Seven beam modes of 10 to 100 m resolution and 50 to 500 km swath widths combined with 25 beam positions ranging from 10 to 60° incidence angles mean a wide range of products are available for use in environmental monitoring and natural resource mannagement applications worldwide.

In the ScanSAR Wide beam mode, the whole of Canada can be covered every 72 hours and the Arctic daily. Imagery can be provided to high priority users electronically, such as the Canadian Ice Centre to navigate ships through the high Arctic, and within 4 hours of data acquisition. The RADARSAT system is designed to operate with no backlog, so that all imagery received within a 24 hour period can be processed and distributed (or stored) within 24 hours. For most of its

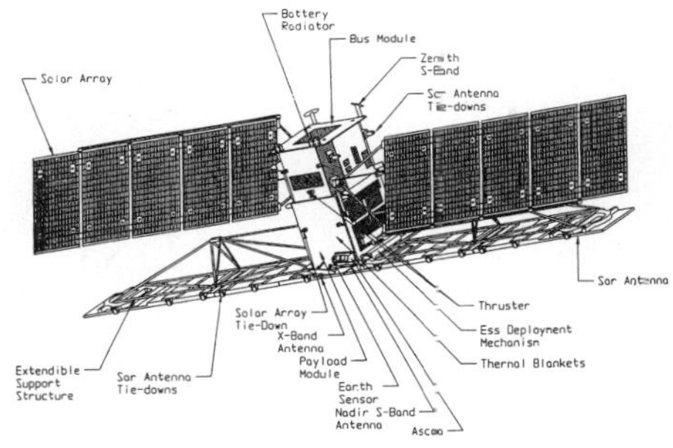

RADARSAT principal features

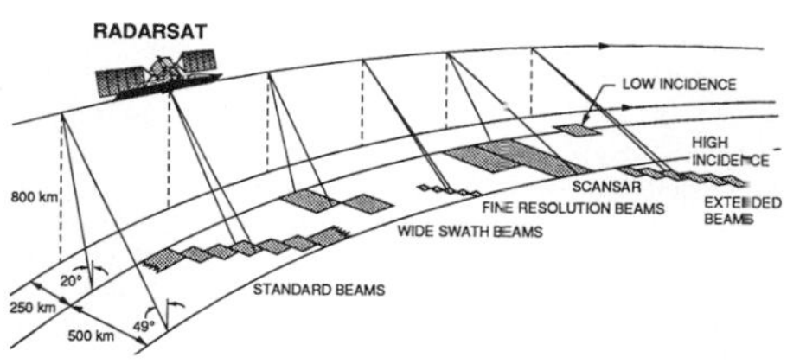

RADARSAT SAR operating modes

life, RADARSAT will fly with the SAR antenna 'looking to the right, providing coverage over the Arctic region. However, the spacecraft will be yawed 180° twice in its lifetime to switch the beam to the left for Antarctica mapping, during maximum and then minimum sea ice cover. The RADARSAT mission control centre is in Saint-Hubert, Quebec, Canada (near Montreal), and the principal data processing centre, operated by RADARSAT International, is in Gatineau, Quebec (near Ottawa, Ontario).

Total cost of the RADARSAT-1 programme reached C$642 million, with the Federal government contributing C$520 million, which included the C$75 million launch, the provinces C$59 million, and the private sector, namely RADARSAT International, C$63 million. NASA provided the launch in exchange for US research access to 15 per cent of RADARSAT SAR on-time and US private sector participation in distribution. The projected operations costs of RADARSAT over five years is expected to be recovered from RADARSAT International by the Canadian Space Agency in the form of 20 per cent royalty payments on the estimated C$250 million revenue.

The next satellite in the series, namely RADARSAT-2, is expected to launch in the 2000-02 timeframe and may be equipped with additional features such as GPS and solid-state tape recorders.

Beam modes

Fine	36-48° incidence angle, 8 m Resolution:	*50 km swath width, 1 × 1 look	5 beam positions
Standard	20-49° incidence angle, 25 m Resolution:	100 km swath width, 1 × 4 looks	7 beam positions
Wide	20-49° incidence angle, 30 m Resolution:	150 km swath width, 1 × 4 looks	3 beam positions
ScanSAR Narrow	20-46° incidence angle, 50 m Resolution:	300 km swath width, 2 × 2 looks	2 beam positions
ScanSAR Wide	20-49° incidence angle, 100 m Resolution:	500 km swath width, 2 × 4 looks	1 beam position
Extended Low	10-23° incidence angle, 35 m Resolution:	170 km swath width, 1 × 4 looks	1 beam position
Extended High	49-59° incidence angle, 25 m Resolution:	75 km swath width, 1 × 4 looks	6 beam positions

*Tape recorded images may vary in size.

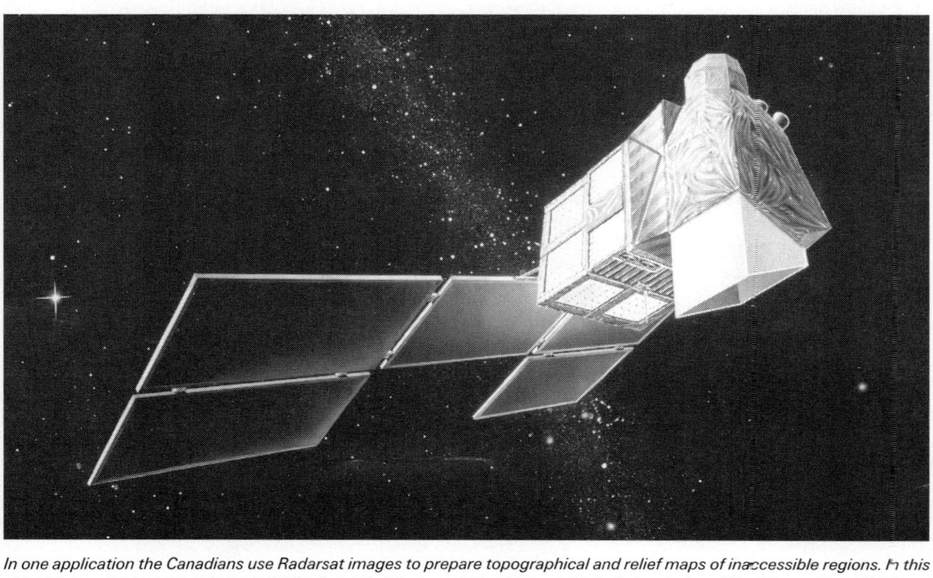

In one application the Canadians use Radarsat images to prepare topographical and relief maps of inaccessible regions. In this view, Radarsat International merged a Radarsat-1, 25 m resolution image with both Landsat and Spot to produce a land use profile of Nagano, Japan prior to the Winter Olympic Games. Normally, Nagano is cloud shrouded and surrounded by mountains making conventional photographic mapping difficult (Radarsat International) ***1998***/0022564

Specifications

RADARSAT 1 Satellite

Contractors: Spar Aerospace (prime), Ball Aerospace Systems Division (bus), CAL Corp. (slotted waveguide panels), Telesat, COM DEV (low-power electronics), SED (satellite operation, special test and ground equipment), MacDonald Dettwiler (mission control and ground processor), Odetics (two digital tape recorders), Dornier/AEG (TWT), SAFT (batteries), Astro Aerospace Corp. (SAR antenna deployment mechanism)
Resolution: variable over 7.6-100 m
Mass: 2,749 kg (dry bus, solar array 1,315 kg, payload 1,366 kg, hydrazine 68 kg), plus 55 kg adapter and 352 kg margin
Orbit: 798 km circular at 98.6° Sun-synchronous, 18.00 ascending node, 24 day repeat cycle to provide optimum revisit opportunities (Arctic regions daily; 49-70° north every 3 days). 14 revisits daily (28 min SAR on-time/orbit), 100.7 min/orbit)
Spacecraft: 3-axis controlled to 0.1° in all axes by reaction wheels, magnetometers, magnetorquers, Earth and Sun sensors. Hydrazine thrusters to maintain orbit altitude within at least 10 km to ensure track repeat. Downlinking on two X-band (8.2 15-8.400 GHz) channels (one for real time, other recorder dumping) at 85-100 Mbit/s to Prince Albert (Saskatchewan) and Gatineau (Quebec) stations in Canada, plus NASA's Fairbanks station in Alaska. Additional international network stations currently receiving RADARSAT data include Tromso Satellite Station (Norway), West Freugh (UK), and Singapore. Rigid Spar array of two 5-panel wings provide 3 kW, supported by three 48 Ah SAFT Ni/Cd batteries. Two tape recorders store 10 min of data each at 85 Mbit/s SAR antenna 1.5 × 15 m
Sensors: 5.3 GHz C-band SAR, HH polarisation, 50-500 km selectable swath width, 10-60° selectable incidence angle. The 300 W (average power; 5 kW peak) TWT is slightly modified from ERS- 1. RF bandwidth 11.6, 17.3 or 30.0 MHz, transmit pulse length 42.0 μs, pulse repetition frequency 1,270-1,390 Hz
Processing Levels: Path Image, Path Image Plus, Map Image, Precision Map Image, Signal Data, Single Look Complex
Derived products: Mosaics, Digital Elevation Models.

CHINA, PEOPLE'S REPUBLIC

Feng Yun (FY)

Background

China launched the first NOAA-compatible FY-1 polar-orbiting meteorological satellite in 1988 by the first CZ-4 from the new Taiyuan polar site. The Chinese began developing the satellite in the late 1970s. Then the country planned to build the FY-2 family and introduce it in April 1994 for GEO observations. The Chinese announced they would launch the FY-2A in April 1994, but the satellite was destroyed in a fire and explosion during final checkout in test hall number 2 at Xichang on 2 April. One worker died later and the fire injured 31 more.

The Shanghai Institute of Satellite Engineering builds the satellites with funds and specification requirements from the State Meteorology Administration. FY-1A, with one year design life, operated for only 39 days but yielded high quality imagery. Its far Infra-Red (IR)

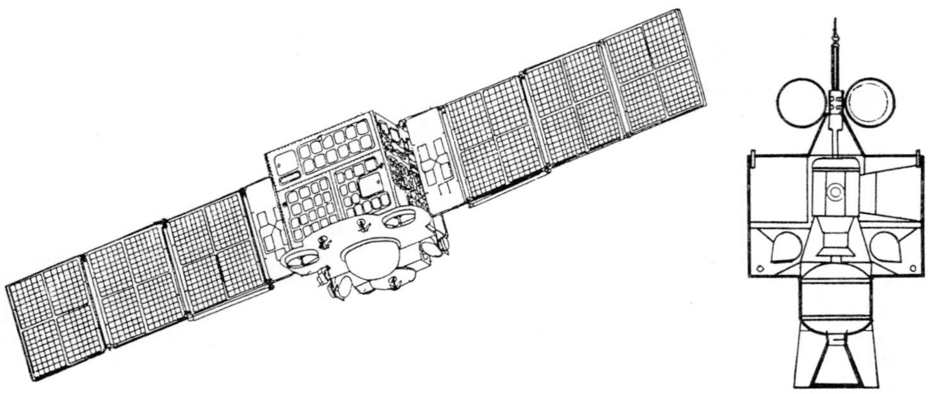

Two FY-1 metsats (left) have been launched into polar orbit. The geostationary FY-2 (right) is planned for 1996. Scales are comparable

channel, however, failed to produce useful information because water vapour contaminated the detectors and radiation cooler, a fault rectified on 1B. The heavier FY-1B appeared a year later, accompanied by two balloons for upper atmosphere density measurements. It lost attitude control February 1991 because of radiation upsets, but was recovered after a 50 day effort. The recovery came in time for it to monitor severe floods and help forecast the weather for the Asian Games. However, radiation damage broke control again. Chinese government officials noted in January 1994 that contact was being maintained but data were unusable. The Chinese used the FY-1A/1B as test satellites, preceding the operational 1C/1D in 1997/98. To ensure that the satellites can survive the radiation hazards of space, the Chinese have tested subassemblies of equipment at their Lop Nor nuclear site.

The first FY-3 second-generation polar satellite will carry a 10-channel scanning radiometer, infra-red and µwave radiometers, space environment monitor and an Earth radiation budget scanning radiometer. The 10-channel scanning radiometer will cover: 0.433-0.453, 0.49-0.51, 0.54-0.56, 0.58-0.68, 0.745-0.785, 0.725-1.1, 1.58-1.64, 3.55-3.95, 10.3-11.3, 11.5-12.5 µm. Later satellites in the series may add solar backscatter UV radiometer, µwave imager, µwave humidity detector, medium resolution imaging spectrometer, altimeter and scatterometer.

Various sources estimated in 1994 that the FY-2 programme cost ¥300 million (US$35 million).

Specifications

Feng Yun ('Wind and Cloud') series

FY-1A

Launched: 6 September 1988 by CZ-4 from Taiyuan (design life one year but failed after 39 days)
Mass: 757 kg at launch
Orbit: 881 × 904 km, 99.13° (afternoon orbit)
Power: 774 W generated BOL.

FY-1B

Launched: 00.53 GMT 3 September 1990 by CZ-4 from Taiyuan
Mass: 881 kg at launch
Contractors: Shanghai Institute of Satellite Engineering (platform), Shanghai Institute of Technical Physics (radiometers)
Resolution: 1.1 km visible high, 4 km infra-red
Orbit: 886 × 900 km, 98.9° (morning orbit), providing seven passes per day over Chinese territory and real-time image coverage of 55-145° east/0-65° north in conjunction with three stations able to receive data within 2,600 km
Spacecraft: body 1.4 × 1.4 × 1.4 m, height 1.76 m. Solar array span 8.6 m. 750 W 42.5 VDC BOL (design goal) power provided by 14,000 2 × 4 cm + 2 × 6 cm Si cells covering 6.8 m^2, supported by two 48 Ah Ni/Cd batteries. 3-axis attitude control by N_2 gas thrusters, RWs (magnetorquer unloading), gyros, infra-red horizon sensors. High-resolution image transmissions made at 6.654 kbit/s in real time on 1.6955/1.7045 GHz, APT on 137.795/137.035 MHz; housekeeping telemetry on 180.0 MHz. Global data stored on tape recorder
Sensors: Two 95 kg 360 rpm scanning 5-channel VHRSR Very High-Resolution Scanning Radiometers operating at (in µm): 0.48-0.53 (ocean colour, land), 0.53-0.58 (ocean colour, land), 0.58-0.68 (daytime cloud, land), 0.725-1.1 (daytime cloud, water, snow, ice, vegetation), 10.5-12.5 (diurnal cloud, land/sea T). 4-24 MeV/n solar/cosmic ray composition monitor.

FY-2

Launch: 10 June 1997 by CZ-3 from Xichang
Mass: 1,200-1,250 kg at launch, 600-700 kg on-station
Contractors: as FY-1
Resolution: 1.25-1.41 km visible, 5-5.76 km infra-red, 5-5.76 km water vapour
Orbit: planned 105° east geostationary
Spacecraft: similar configuration to Meteosat/GMS. Spin-stabilised with despun antenna platform. Control system derived from DFH 2. 2.1 m diameter, 4.5 m height, 280 W generated by drum-mounted solar cells. GEO insertion by solid AKM (separated after insertion)
Sensors: 0.55-0.75 visible, 10.50-12.50 IR, 5.70-7.10 µm water vapour scanning radiometer. Payload includes data collection and image relay systems
TT&C: China's three S/VHF-band stations at Beijing (116°16′ east/40°03′ north), Guangzhou (113°20′ east/23°09′ north) and Urumqi (87°34′ east/43°52′ north) receive real-time high resolution and APT imagery from NOAA and FY polar satellites for relay to the data processing/archiving facility at Beijing's National Satellite Meteorological Centre, part of the State Meteorological Administration. The local station utilises a microwave link to the Centre but the two remote sites communicate via GEO communications satellites through ground stations 10 km distant. The stations receive data when satellites are within 2,600 km, the spread providing pan-China coverage of 55-145° east/0-65° north. Data processing is accomplished by three IBM 4381 model 3s in conjunction with two IBM 7350 image processors and two IBM 5080 graphic processors.

FRANCE

Spot

Current status

The Spot 4 earth observation satellite was successfully launched by an Ariane 40 launcher from the European spaceport in Kourou, French Guiana, on 24 March at 2.46 am French time. Spot 4's HRVIR instruments possess an additional spectral band in the Short-Wave Infra-Red (SWIR). The SWIR band offers better discrimination between different types of crops and plant cover. With Spot 1 and Spot 2 still in service, the enhanced capability of Spot 4, provides commercial stereographic images at the 5 m resolution level. Today, Spot Image holds a 60 per cent share of the market, supported by a worldwide commercial network of three subsidiaries (United States, Australia and Singapore), nearly 90 distributors and 23 receiving stations.

Background

CNES began the Spot programme in 1978 and subsequently established Spot Image to market the data. Spot Image had a start-up capitalisation of US$41.1 million, allocated by public institutions and private companies in Belgium, France and Sweden. Initially the company had several partners. The original percentages were: (current values shown in parentheses), CNES 39.0 (35.32), Institute Geographique National 10.0 (11.27), Matra 8.8 (28.01), Aeropart (5.40), Bureau de Recherches Geologiques et Minieres 7.4 (1.00), Alcatel Espace (3.33), Crédit Lyonnais (1.19), Banque Nationale de Paris (1.19), Société Générale (1.19), Paribas (0.60), Swedish Space Corp 6.0 (5.94), Nuova Telespazio (2.32) and Belgium 4.0 (3.27). Spot Image has three subsidiaries, the Spot Image Corp in the US, Spot Imaging Services in Australia and Spot Asia in Singapore.

As the satellite operator, CNES is at the centre of the organisational structure, directly responsible from Toulouse for orbit maintenance, payload programming, reception and pre-processing. Data downlinks to two SRIS (Station de Reception des Images Spatiales) stations at Aussaguel near Toulouse (SRIS-T) and one at Esrange, near Kiruna, Sweden. These receive both real-time and recorded data as Spot passes over the north polar region, Europe and North Africa within 2,500 km range, in addition to the stored images from other regions. Together, they have a reception capacity of 500,000 images annually. Each SRIS is associated with a pre-processing/archiving centre or CAP (Centre d'Archivage et de Prétraitement), with the equivalent of 700 scenes archived every 24 hours at Toulouse.

Spot Image has responsibility for the Direct Receiving Stations (SDRS) around the world, which receive real-time imagery only. SDRS are located at Prince Albert and Gatineau (Canada), Maspalomas, Spain, station managed by ESA), Cuiaba (Brazil), Lad Krabang (Thailand), Hatoyama (Japan), Islamabad (Pakistan, commissioned 1989), Hartebeesthoek (South Africa, 1989), Riyadh (Saudi Arabia, 1989), Alice Springs (Australia, 1990), Tel Aviv (Israel, 1 February 1991), Cotopaxi (Ecuador, July 1992), Taipeh (Chung-Li University, Taiwan, July 1993), Pare-Pare (Indonesia, October 1993), Fucino (Italy, July 1995) and Singapore (September 1995); the USAF Eagle Vision transportable system began operating in June 1994 (another 20 will be ordered). South Korea was expected to be added in 1996, and Russia's Resurs-O station at Obninsk will be upgraded for Spot 2 imagery during the two warmest months.

CNES categorises processing operations into several levels of sophistication. Level 1 provides basic radiometric and geometric corrections and it does not involve ground control points or satellite attitude restitution data. Level 1A downlinks essentially raw data, apart from the normalisation of CCD detector response in each spectral band. It is useful for stereoplotting and basic radiometric studies. Level 1B gives full radiometric (desmearing) and limited geometric corrections with the pre-processing level for photo interpretation and thematic analysis. Users may also attain stereoscopic pairs at this level. Level 2 gives advanced rectifications according to a given cartographic projection and Level 2A corresponds to Level 2 precision processing but can be implemented without use of map ground control points. The Spot introduced Level 1AP in 1990, which is optimised for photogrammetric applications using analytical stereoplotters. The commerical suppliers provide products in standard Computer-Compatible Tapes, photographic form and CD-ROM.

Spot's 4 millionth scene was archived 19 July 1995. 1995's revenue from data/product sales and receiving station fees totalled FFr207.5 million (US$41.5 million); 1994 220; 1993 178, profit 0.031; 1992 218, profit 1.38; 1991 204; 1990 165; 1989 123; 1988 90; 1987 55 and 1986 15. 1993's decrease reflected lost sales because of Spot 2 recorder problems before Spot 3 became available. Asia/Pacific (largely Japan) accounted for 25.3 per cent of 1995's revenue; Europe 33 per cent, North America 25 per cent, Middle East 6.7 per cent, Latin America 3.8 per cent, Africa 6.3 per cent. Accumulated sales exceeded FFr1.4 billion (US$273 million) by end-1995 (70 per cent from outside Europe); the government had invested FFr8 billion. Europe provided 32 per cent revenue in the first 10 years, followed by North America 22 per cent and Asia/Pacific 20 per cent. The enhanced Spot 4 design was approved in 1989 for launch about 1994 (now late 1997). Revenue was planned to be US$100 million by 1996 in order to fund at least part of the 5 m resolution Spot 5. France expects that Spot Image will pay all operating costs by end of the century.

Spot 3 launch preparation (Arianespace) ***1995***

In this view of the Spot 4 satellite undergoing testing in the Interspace anechoic chamber, the vegetation camera appears as a cylinder in the horizon position, under the twin imaging cameras (Matra Marconi) ***1998***/0038492

Spot 1

Spot 1 was placed in orbital storage 15 January 1991, but with its health monitored at least every 6 revolutions so that it could resume operational service and meet all image quality specifications within two weeks. It became operational again in March 1992 to help Spot 2 meet increasing demand through to late October 1992 during the northern hemisphere's growing season. It provided real-time transmissions (recorders are inoperative; some 60,000 to 80,000 images were received) for Europe/Middle East, while Spot 2 worked in recording mode (and normally for the rest of the world). The two were phased to optimise stereopair acquisition above 41° latitude. It was revived similarly for April-July 1993, at a total cost of FFr7.7 million, requiring a third control centre, as the second is now used for Spot 3. The European Commission this time contributed ECU350,000, in addition to paying for each scene used for its agricultural monitoring work. Spot 1 was again retired after Spot 3 entered service in late 1993, but can still be called upon at a few days' notice. Spot 1 and 2 carry two identical push-broom CCD High Resolution Visible (HRV) imagers, pointed for a total swath width of 117 km with a 3 km overlap. Site revisit period is 26 days but up to 27° off-nadir viewing (creating total swath width of 950 km) via a tilting mirror permits a single area to be imaged on seven successive passes at equatorial latitudes and 11 passes at 45°. Image width is 60 km for nadir viewing, stretching to 80 km at maximum mirror tilt. The capability also

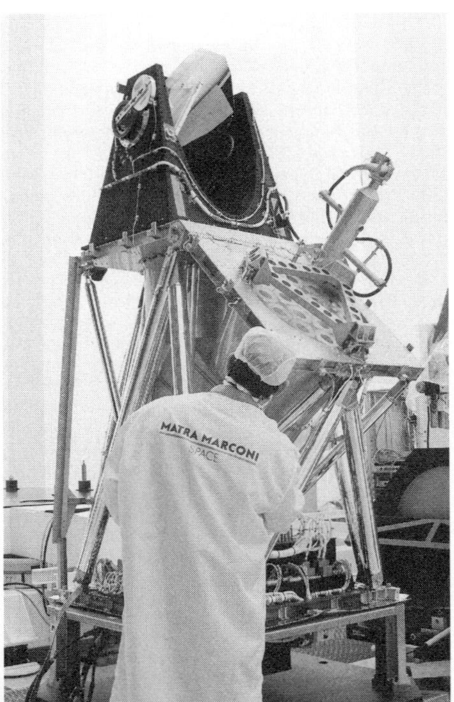

The Spot 4 IR instrument receives a final vibration checkout before launch (Matra Marconi) ***1998***/0038493

Spot 4 is made ready for launch at the Kourou Launch Complex (Theo Pirard) ***1999***/0054347

permits stereo imaging. Through 1995, Spot 1's cameras totalled 2,927 h, returning 1.8 million images.

Spot 2

The satellite returned its first multispectral image, from HRV 2, at 10.18 GMT 23 January 1990, 32 hours after attaining orbit. It arrived in the desired orbit 180° from Spot 1 on 30 January, having expended 12 kg hydrazine. The image quality review of 5 March concluded it to be of commercial quality, and Spot 2 was declared ready for operational exploitation from 21 March 1990. 1.9 million images were returned by end-1995. One tape recorder failed after one year. A second tape recorder in February 1993 began suffering from recording-head wear, leaving spots on imagery. The prime image telemetry channel has failed but the spacecraft is otherwise in perfect order. Spot 3 replaced it as prime satellite November 1993. Spot 2 requires FFr12 million in annual sales to cover operations costs; contracts from the European Commission allowed simultaneous Spot 2/3 operations through 1994-95.

Spot 3

The first image was transmitted within 33 hours, showing the Strait of Bonifacio separating Corsica/ Sardinia. Spot 3 was declared operational 29 November 1993. It expended only 17 kg hydrazine attaining its final orbit, 42 and 208° ahead of Spot 1/2, respectively. It is exhibiting slightly better gain control and dark current characteristics than its predecessors. 686,028 images had been returned by end-1995. Spot 3 cost FFr1.4 billion, including FFr650 million for launch. Spot 3 unexpectedly ceased operating on 14 November 1996.

Spot 4

The French government approved Phase C/D in July 1989 for the expanded Spot 4. It employs the same augmented multimission bus as the Helios 1 reconnaissance satellite, adds a mid-infra-red band to the HRV (now HRVIR) by replacing Sodern's DTA 01 detection unit with the DTA 03 model, incorporating Thomson's 7811 CCDs. Sodern's DTA 05 detection unit gives images at 1 km resolution for monitoring crops, natural vegetation and geosphere/biosphere interactions. The bands are equivalent to HRVIR plus a blue channel for atmospheric corrections. The European Union in January 1994 agreed to contribute half of VGT development costs; Sweden, Belgium (10 per cent) and Italy are also contributing separately. The total VGT programme is expected to cost ECU109.7 million.

Spot 4's Vegetation is an independent payload, having its own recording, transmission and management systems, working to its own ground segment for payload programming, image reception/ processing and product distribution. Local users can acquire 1 km resolution imagery broadcast in real time at 1.704 GHz L-band 1 Mbit/s on their existing metsat receivers. The global data, recorded on the 2.25 Gbit SSR, are downlinked at 3.4 Mbit/s at 8.153 GHz X-band using Spot 4's main antenna. The CPV Centre de Programmation de VGT in Toulouse draws up the daily plan for uplinking by the main Spot 4 control centre. The SRIV Station Reception des Images VGT at Esrange takes 4/5 data dumps daily totalling about 8 Gbit. The CTIV Centre de Traitement des Images VGT in Mol (Belgium) archives/processes the data for a daily synthesis, increasing progressively to a 10-day global synthesis. It can deliver dedicated products within three days if ordered in advance. The QIV Centre Qualité Image VGT in Toulouse is in charge of instrument calibration and image quality.

Spot 5

Definition of Spot 4's successor began in 1993, following an extensive user market survey in 1991-92. 5 m resolution is necessary to address new markets such as civil engineering and an along-track 5 m stereo capability was to generate simultaneous stereo pairs, whereas its predecessors are limited to separate passes. Resolution is legally barred from being less than 5 m. The goal is to provide simple generation of 1/50,000-1/ 100,000 standard maps (1/25,000 in some cases) and digital elevation models. The government approved the FFr6.3 billion programme 4 October 1994 for the first of two launches in 2002 (but available late 1999), employing a bus jointly developed for Helios 2. Sweden and Belgium (4 per cent) are again contributing; Belgium is separately paying 10 per cent towards VGT. The programme was reduced in early 1996 to FFr4.4 billion, cutting Spot 5B and losing the simultaneous stereo imaging by dropping one camera and nadir pointing the remaining two.

Specifications

Spot 1

Launched: 22 February 1986 by Ariane 1 from Kourou, French Guiana (2 year design life; initially retired 31 December 1990)
Mass: 1,830 kg at launch
Contractors: Matra (prime, imaging system), Aerospatiale (structure, thermal control, solar array), Sodern (optical sensors)
Resolution: 20 m multispectral, 10 m panchromatic
Orbit: 824 × 829 km (822 km after 3 years) in 98.7° Sun-synchronous, crossing equator at about 10.30 local time, with 26 day repeat cycle (369 revs). Altitude corrections executed every 2 months and inclination adjustments annually.
Spacecraft: 3.5 m high, 2 m^2, consisting of Multimission Platform (PFM) and forward Spot payload module. 2 year design life. 15.6 m solar wing provided 1,382/ 1,326 W BOL/EOL (equinox values; 36 V) with 3 × 24 Ah Ni/Cd batteries. Panel degradation after 3 year and 18,000 thermal cycles was 4 per cent instead of the projected 21 per cent. 3-axis platform pointing accuracy of 0.1° (0.15° specified) maintained by three MWs unloaded by two magnetic coils. System includes digital Earth sensors, 2 × 2° FOV. orbit control by 8 × 15.6 N hydrazine thrusters (two 75 kg tanks, 130 kg remained after 3 years). Each of the two HRV imagers returns data at 25 Mbit/s on 8.025-8.400 GHz real-time using 20 W Thomson-CSF TWTA
TT&C: Spot Mission Centre is located at Toulouse. 8.307 GHz beacon multiplexed with the transmitted signal to facilitate ground antenna pointing. Communications subsystem mass 240 kg; power consumption 170 W for direct tx, 270 W for recorder playback. The 70 kg Odetics recorder could hold 22 min of data. Recorder number 1 (a qualification model flown to save money) failed in September 1986, number 2 was shut down after problems appeared July 1990, and the prime telemetry TWT failed August 1989. 3.7 kg 2.025-2.120 GHz S-band transponder provides housekeeping telemetry (2,048 bit/s) + TC (2 kbit/s)
Sensors: High Resolution Visible (HRV) Spot was the first satellite to incorporate push-broom sensors, avoiding the need for a mechanical scanning mirror. On Spot 1, each spectral band utilises Fairchild's 122 DC detectors; Spot 2/3 employ four Thomson-CSF TH 7801A 1,728-element CCD arrays, combined in each HRV's Sodern DTA 01 detection unit. The spectral ranges (in μm) of the bands are: Band 1 0.50-0.59, Band 2 0.61-0.68, Band 3 0.79-0.89; panchromatic 0.51– 0.73

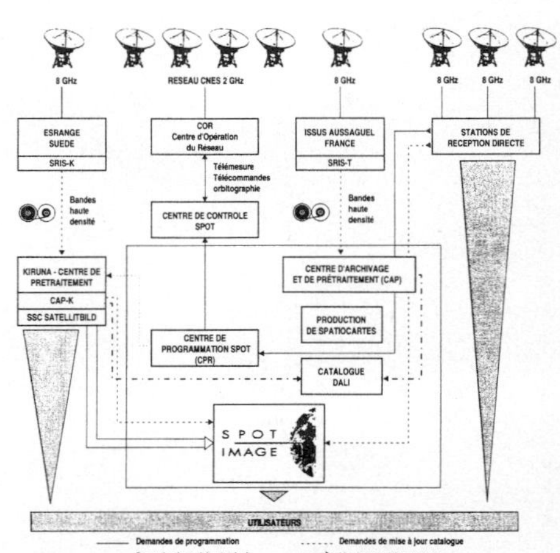

Spot control and data reception network ***1996***

Spot 5 is being developed in parallel with Helios 2 (CNES/D Ducros) ***1995***

Passenger: radar transponder for calibration of Kourou's ground radar and operator training.

Spot 2
Launch: 01.35.27 GMT 22 January 1990 by Ariane 40 from Kourou, French Guiana; declared operational 21 March 1990
Mass: 1,870 kg at launch
Resolution: same as Spot 1
Orbit: 819 × 822 km, 98.7°, initially 180° from Spot 1, now 166° behind Spot 1
Spacecraft: same as Spot 1 but with 18.5 kg 20 W DORIS (Doppler Orbitography and Radiopositioning Integrated by Satellite) unit for accurate orbit determination in preparation for 1992's joint Franco- US Topex mission and hydrazine loading 160 kg
Sensors: Same as Spot 1.

Spot 3
Resolution: same as Spot 2
Launch: 26 September 1993 by Ariane V59 from Kourou, French Guiana
Mass: 1,907 kg launch; dry 1,749 kg; 158 kg hydrazine
Orbit: 816 × 818 km, 98.68°
Spacecraft: as Spot 2, but TWTs and recorders modified slightly for longer lives
Sensors: Same as Spot 2, plus 25 kg POAM II Polar Ozone & Aerosol Measurement instrument carried for USAF Space Test Program.

Spot 4
Launch: 24 March 1998 by Ariane 40 from Kourou, French Guiana
Mass: 2,755 kg launch; dry 2,600 kg: payload 1,400 kg
Orbit: same as Spots 1-3
Contractors: MMS France (prime, assembly, integration, test, onboard data management, AOCS, inertial units, Sun/star sensors), Aerospatiale (VGT, mechanical subsystem, thermal control, solar array), Alenia (structural elements, wheels/magnetorquer drive electronics), Alcatel Espace (transponder), CASA (cabling harness), CRISA (software, EGSE), Sextant (housekeeping unit, bus couplers, decoding/reconfiguration unit), ETCA (electrical power supply), Laben (wheels/magnetorquers drive electronics), Saab (central communications unit), SAFT (batteries), SAGEM (gyroscopes), SEP (propulsion), Sodern (Earth sensors, star sensors, optical head), Schlumberger (tape recorders)
Resolution: HRVIR 20 m multispectral/10 m monospectral; VGT 1 km/4 km
Spacecraft: based on Spot 1-3 design but with additional compartment between platform and payload to accommodate the new tape recorders and passenger experiments. Design life is 5 years (0.8 reliability after four years). New rigid 74 kg 5-panel 1.90 × 2.60 m Aerospatiale GSR-3 solar array will provide 2.5 kW BOL/equinox and 2.2 kW EOL (5 years)/Summer solstice, feeding four SAFT 40 Ah Ni/Cd batteries. 3-axis control to <0.03°, employing three RWs and two magnetorquers, Sodern STD-16 Earth horizon infra-red sensors, Sun sensors and Sagem's Regys 10 gyro orbit and yaw adjustment by four pairs of 15.6 N thrusters; roll/pitch by four pairs of 3.5 N thrusters supplied from four 75 kg hydrazine tanks. Control system based on radiation hardened microprocessor: Central Communication Unit (that is, main computer) incorporates a Fairchild 9450 microprocessor, providing 128,000 words redundant RAM. Telemetry rate at 4,096 bit/s; up 2.025-2.120, down 2.200-2.290 GHz. Two 121 kg 50 Mbit/s 120 Gbit Schlumberger redundant tape recorders store up to 40 min imagery each (175 W required in record mode, 260 W replay, 12 W standby; 70,000 on/off cycle specification). Imagery can be encrypted. VGT has its own recorder and telemetry system: 2.25 Gbit SSR (3 Mbit/s replay at 8.153 GHz); real-time at 1.704 GHz. Spot 4/Helios 1 controlled from the new Fresnel building at Toulouse
Sensors: HRVs paired as in Spot 1-3 are upgraded to HRVIRs (High-Resolution Visible Infra-Red) by the addition of a mid-infra-red band for vegetation water registration; independent and simultaneous observing schedules possible. The new wide-angle VGT is optimised for monitoring of vegetation, crops and geosphere/biosphere interactions

HRVIR Paired instruments with spectral ranges (in μm): Band B1 0.50-0.59; Band B2 0.61-0.68; Band B3 0.79-0.89; SWIR 1.580-1.750; yielding 20 m resolution in spectral mode. In monospectral mode, covering 0.61-0.68, resolution is 10 m (the previous panchromatic band is replaced by sampling the B2 band at 10 m or 20 m resolution). Data rate 25 Mbit/s. Viewing characteristics as Spot 1-3. See Sodern's entry in the Space Industry GN&C section for detailed imager information

VGT bands as HRVIR B2, B3, SWIR but with B1 replaced by B0 0.43-0.47 μm for atmospheric corrections. Swath 2,200 km (101° FOV), allowing daily global coverage
Passenger payloads: DORIS (Doppler Orbitography and Radiopositioning Integrated by Satellite), improved for precise (1 m) onboard real-time autonomous orbit determination and distribution of timing pulse; 10 kg, 14 W

Vega Radar Transponder for calibration of Kourou's ground radars and operator training

PASTEC (Passager Technologique), a group of experiments for studying the orbital environment

PASTEL (Passager Spot de Telecommunication Laser), demonstration of optical 50 Mbit/s datalinks via GEO satellites as part of ESA's Artemis SILEX programme. ESA's Artemis satellite will be used to relay HRVIR images

POAM III Polar Ozone & Aerosol Measurement instrument carried for USAF Space Test Program.

Spot 5
Contractors: MMS France (prime, assembly, integration, test, onboard data management, AOCS, inertial units, Sun/star sensors), Aerospatiale (VGT, mechanical subsystem, thermal control, solar array), ETCA (electrical power supply), Saab (central communications unit), Alcatel Espace/IBM (solid-state recorders)
Resolution: HRG 5 m PAN, 10 m XS, 20 m SWIR; VGT 1 km/4 km
Launch: planned for late 2001 by Ariane 5 from Kourou, French Guiana
Mass: 2,600 kg launch
Orbit: as Spots 1-4
Spacecraft: based on Spot 4/Helios 1 design. Design life 5-7 years. 3-axis control. Two 100 Gbit SSRs. 150 Mbit/s real time/storage
Sensors: HRG three 270 kg High-Resolution Geometry cameras operating independently and simultaneously were planned, but there will now be two nadir-pointing (losing the simultaneous stereo pairs). The PAN band is as Spot 1-3 but at 2.5-5 m resolution; the visible/NIR B1/B2/B3 are as Spot 4, but at 10 m resolution; the 20 m SWIR band for vegetation monitoring is as Spot 4. For three cameras, they would be sighted at nadir, 19.2° fore, 19.2° aft. 60 km nadir swath + ±27° cross-track tilting (as on Spot 1-4) provides swath widths up to 160 km for two cameras operating in conjunction, and 240 km for three. Three operating modes: multi-spectral, 60-240 km swath, 10 m B1/B2/B3 + 20 m SWIR; high resolution, 60-240 km swath, 5 m PAN + 10 m B1/B2/B3; stereo, 60-80 km swath, 5 m PAN + 10 m B3 VGT as Spot 4
Passenger payloads: DORIS as Spot 4, PASTEC as Spot 4.

Spot 5 will provide routine 5 m imaging (CNES/Spot Image/ISTAR) ***1995***

INDIA

Insat

Background
The Indian National Satellite System uniquely provides geostationary platforms for simultaneous Indian domestic communications and Earth observation functions (Japan's 1999 MT-Sat will adopt a similar approach). The first-generation Insat 1 satellites were all US-built, but the advanced Insat 2 family is indigenously produced. 2C/2D replaces the imaging, DCP and DRT payloads with an enhanced communications element. 2E is more like 2A/B, but with expanded imaging capability.

The Insat system is a joint venture of the Department of Space (DOS), the India Meteorological Department (IMD), the Department of Telecommunications (DOT) and All India Radio. The Insat office at Bangalore is responsible for controlling the system through the Master Control Facility (MCF) at Hassan. Meteorological imagery and data are received at the Delhi Earth Station and relayed by microwave to the key Meteorological Data Utilisation Centre (MDUC) at the India Meteorological Department (IMD) headquarters in New Delhi for processing and analysis. 11 VHRR images (each three visible and one infra-red) are acquired each day. 1D returned more than 24,500, 2A more than 11,900 and 2B more than 9,100 images by end of 1995. The processed material is relayed to 22 Secondary Data Utilisation Centres (SDUCs), initially by land lines but converted by 1989 for retransmission through Insat. Information from Data Collection Platforms is also processed at the MDUC, through Insat's Data Relay Transponder.

Insat 1B retired August 1993 after returning more than 36,000 images. 1C failed in November 1989. 1D assumed the prime role in July 1990 from 1B. The five-satellite Insat 2 system began in 1992 with 2A, which took over the prime imaging role from 1D 6 October 1992. Insat 1A was launched 10 April 1982 and positioned at 74° east but it was abandoned 4 September 1982 following exhaustion of its attitude control fuel during eclipse. 1B was launched by Shuttle in August 1983 but it was not until mid-September that Ford and ISRO engineers at India's Hassan Satellite Control Centre succeeded in deploying its solar array. Full operation was achieved in October 1983. 1C was launched 21 July 1988 by Ariane V28. A power system failure (from a solar array isolation diode short) soon removed half of the communications capacity but the meteorological services, powered through either bus, remained unaffected from 93.5° east.

Insat 2C solar array deployment test (Theo Pirard) ***1999***/0054348

However, the satellite lost Earth lock 22 November 1989 and was abandoned. 1C procurement cost was reported at Rs800 million, with Ariane launch cost adding Rs740 million. The failed 1A/1C reportedly produced insurance payouts of about US$70 million each.

Specifications

Insat 1D

Contractors: Ford Aerospace (prime)
Launched: 12 June 1990 by Delta 6925 from complex 17, Cape Canaveral. Entered service 17 July 1990
Orbit: 83° east geostationary
Resolution: 2.75 km visible, 11 km infra-red
Spacecraft: A 2.18 × 1.42 × 1.55 m box-shaped, 3-axis stabilised bus with asymmetrical 5-panel 11.5 m^2 solar wing providing 1,185 W BOL (930 W end) and creating total span of 19.4 m. The asymmetrical solar array configuration provides an unobstructed view into space for the radiative cooler of the VHRR and a solar sail is used for balance. A magnetorquer with current coil around the body provides fine control; a unified bipropellant system is used for orbit transfer, station-keeping and attitude maintenance
Sensors: 2-channel Very High-Resolution Radiometer (VHRR) providing 0.55-0.75 μm visible + 10.5-12.5 μm infra-red images of full Earth disc every 30 minutes. The Insats also relay 402.75 MHz transmissions from up to 100 hydrological, meteorological and oceanographic data collection platforms (system dormant until Insat 2 appeared because of Insat 1 failures).

Insat 2A

Launch: 9 July 1992 by Ariane 4 V51 from Kourou, French Guiana (design life 7 years)
Mass: 1,162 kg BOL, 911 kg dry
Orbit: 74° east geostationary
Contractors: ISRO and Indian companies; ISRO Space Application Centre (VHRR)
Resolution: 2 km visible, 8 km infra-red
Spacecraft: box-shaped 3-axis controlled bus based around 930 mm diameter central cylinder with asymmetrical 5-panel solar wing providing 1,400 W BOL (1,180 W EOL) and creating 23 m span. Integral liquid propellant (NTO/MMH) apogee boost motor and 16 associated attitude control thrusters
Sensors: VHRR similar to Insat 1's but with improved resolution; redundant Data Relay Transponder for Data Collection Platforms. VHRR detectors are redundant linear arrays of four Si photodiodes (visible) and redundant HgCdTe detectors at 105,000 (using passive radiative cooler) for thermal-IR. VHRR operates in 3 modes: full frame 20 × 20° full Earth scanned in 33 min; normal mode 14° north-south 20° east-west scan 23 min; sector scan 4.5° north-south 20° east-west scan 7 min for rapid repetitive coverage during severe weather conditions such as cyclones.

Insat 2B

Launch: 22 July 1993 by Ariane 4 from Kourou, French Guiana. Returned first image 21 July; declared operational 10 August 1993
Orbit: 93.5° east geostationary.

Insat 2E

Launch: 2 April 1999 by Ariane 42P from Kourou
Location: 83° east geostationary
Mass: 2,500 kg at launch
Sensors: 2E will be configured like 2A/2B, but with 5.7-7.1 μm water vapour channel added to VHRR and a new 3-band CCD camera providing 1 km resolution 0.63-0.69 μm vis, 0.77-0.86 μm NIR, 1.55-1.70 μm SWIR.

IRS series

Background

India has built its first domestic dedicated Earth resources satellite programme and named it the Indian Remote Sensing satellite system as an element of the National Natural Resource Management System. Between 1988 and 2000 India will launch 12 polar orbit satellites. These satellites put India in a position to satisfy its own applications needs, but also to dominate a significant portion of the global commercial market.

The country launched the first in the series, the Rs650 million IRS 1A by a Soviet Vostok booster in March 1988, then signed a Rs220 million contract for the identical IRS 1B in November 1988 for an August 1991 departure. IRS 1C's December 1995 was brought about with a contract signed on 18 January 1991.

1C/1D offer improved resolution, stereo viewing and more frequent revisits, making them the world's most advanced civil remote sensing satellites. The 846 kg IRS 1E, 1A's refurbished engineering model, was lost on PSLV-D1 20 September 1993 carrying Germany's MEOSS Monocular Electro-Optical Stereo Scanner (as flown on the failed SROSS 2) and India's LISS 1 camera and CO_2-band Earth radiance monitor. The demonstration IRS P2 launched on PSLV number 2 in October 1994 carries a LISS 2 camera. IRS P3 on PSLV in March 1996 carries a German scanner and an improved wide-field imager. IRS P4 launched in 1997 performs ocean reconnaissance. P5 in 1998 will have a 10 m resolution vegetation camera. IRS P6 in 1999 will provide 2.5 m cartographic stereo images (improved to 1 m in 2002 by the Cartosat 2 follow-on). India has devoted the IRS P7 in 2000 to ocean and fishing applications and IRS P9 in 2001 will address environmental and atmospheric studies.

EOSAT and ISRO signed an agreement 21 October 1993 for the US company to distribute IRS data globally. EOSAT began receiving IRS imagery at its Norman, Oklahoma station on a daily basis in June 1994. EOSAT and the Indian government signed an agreement 2 February 1995 for exclusive worldwide marketing rights for IRS data. EOSAT expects that, within five years, the majority of its new imagery will be from IRS.

All three IRAS 1A imaging systems remained operational from 7 April 1988 until ISRO retired 1A 17 March 1995 from routine service. It remains on standby to supplement the other satellites. The IRAS returned more than 500,000 scenes by the end of 1993.

IRS P2 after being declared operational on 7 November 1994 demonstrated the 1C/1D bus. During the first year, it returned more than 60,000 images of India, in addition to EOSAT imagery received at two US stations. IRAS 1C/1D offers improved spatial and spectral resolution, onboard recording, stereo viewing capability and more frequent revisits. The VNIR resolution is about 20 m, the SWIR about 70 m and the PAN 10 m. SWIR provides data on water stress and pest infestation. The WiFS Wide Field Sensor with 180 m resolution monitors vegetation. All three cameras were returned their first images, 5 January 1996. Outside observers estimate the launch cost India Rs500 million.

IRAS P3 applies its Earth imagers to four major applications: ocean chlorophyll analysis, vegetation assessment, snow studies and geological mapping for identifying prospective mineral sites. ISRO activated the WiFS at 04.40 GMT 22 March 1997, during orbit number 15 and the MOS on 23 March during orbit 29.

India conducts TT&C functions at ISRO's Bangalore station, with imaging downlinked to Shadnager near Hyderabad for distribution by the National Remote Sensing Agency (NRSA). Cost of each LISS 1 (1:250,000) and LISS 2 (1:125,000) photo product is Rs3,000.

Specifications

IRS 1A

Launched: 17 March 1988 by Russian SL-3 Vostok from Tyuratam
Mass: 975 kg in-orbit at beginning of 3 year life
Contractors: ISRO Satellite Centre (prime), Hindustan Aeronautics (structure), and ISRO Space Applications Centre (imaging system)
Resolution: 72.5 m LISS 1, 36.25 m LISS 2
Orbit: 867 × 913 km (aiming for 904 km circular) 99.03° Sun-synchronous, crossing equator on descending node at 10.25 local time (allowed to drift to 10.10 after 2 years), with 22 day repeat cycle (307 revs) and 2,872 km ground track separation at equator
Spacecraft: box-shaped 1.6 × 1.56 × 1.1 m bus with two Sun-tracking solar wings totalling 8.58 m^2, providing 709 W EOL. Two Ni/Cd 40 Ah batteries provide eclipse power. 3-axis control with a 0.3° pitch/roll + 0.5° yaw pointing accuracy provided by zero-momentum RW system utilising Earth/Sun/star sensors + gyros; 1 N hydrazine thrusters (80 kg loaded) for AOCS + momentum dumping. Real time LISS 2A/B data downlinked at 10.4 Mbit/s each to the 10 m dish at Shadnager on 20 W X-band + LISS 1 data on 5 W S-band at 5.2 Mbit/s. No onboard recorder. Satellite control exercised from Bangalore, with Lucknow station providing backup
Sensors: Linear Imaging Self-Scanning (LISS). Three push- broom CCD units operate in four bands compatible with Landsat TM and Spot HRV. Band ranges (μm) and applications: Band 1 0.45-0.52 (coastal environment, soil/vegetation); Band 2 0.52-0.59 (vegetation vigour, rock/soil discrimination, turbidity, bathymetry); Band 3 0.62-0.68 (chlorophyll absorption, plant species); Band 4 0.77-0.86 (delineation of water features, land forms).

LISS 1. Four 2,048-element linear CCD imagers with spectral filters; total unit mass 38.5 kg. Focal length 162.2 cm, FOV 9.4°, generating resolution of 72.5 m over 148 km swath, framing LISS 2 image pairs. Data transmitted on S-band at 5.2 Mbit/s

LISS 2A/B. Eight 2,048 element linear CCD imagers with spectral filters. Focal length 324.4 mm, FOV 4.7° each, 36.25 m resolution over 74 km swath. The two 80.5 kg units are positioned either side of LISS 1 and view either side of the ground track with a 3 km lateral overlap; two pairs thus cover a single LISS 1 frame. Data transmitted on X-band at 10.4 Mbit/s each.

IRS 1B

Launched: 29 August 1991 by Russian SL-3 Vostok from Tyuratam. Declared operational 16 September 1991
Orbit: 857 × 919 km, 99.25° Sun-synchronous
Other specifications as for IRS 1A.

IRS 1C

Launch: 28 December 1995 by Russian SL-6 Molniya
Mass: 1,250 kg in-orbit at beginning of 3 year life
Contractors: ISRO Satellite Centre (prime), Hindustan Aeronautics (structure), ISRO Space Applications Centre (imaging system), Thomson Tubes Electroniques (40 W X-band TWTA), ISRO Inertial Systems Unit for momentum/reaction wheels
Resolution: 23.5/70.5 m LISS 3, 5.8 m PAN, 188 m WiFS
Orbit: 817 km 98.6° Sun-synchronous, crossing equator on descending node at 10.30 local time, with 24-day

IRS 1C undergoes dynamic balancing. The large drum-shaped PAN camera is at top right; WiFS and MOS at top left (IRS) ***1996***

LISS 1 (Linear Imaging Self-Scanning) CCD camera system (ISRO)

repeat cycle (341 revs) for LISS 3 and 5 days for PAN/ WiFS; 2,820 km ground track separation at equator
Spacecraft: box-shaped 1.6 × 1.56 × 1.1 m bus with two Sun-tracking solar wings (each three 1.1 × 1.46 m panels) totalling 9.6 m^2, providing 830 W EOL. Two Ni/Cd 21 Ah batteries provide eclipse power. 3-axis control with a 0.15° pitch/roll + 0.2° yaw pointing accuracy provided by a zero-momentum RW system (four 5 Nm wheels) utilising Earth/Sun/star sensors and gyros. Propulsion system employs 1/11 N hydrazine thrusters for orbit control, attitude manoeuvres and momentum dumping. Lockheed 62 Gbit tape recorder
Communications: satellite control exercised from Bangalore, with Lucknow station providing back-up. Recorded data downlinked on the 40 W 8 GHz X-band link
Sensors: three cameras all utilise push-broom CCD units, continuing and expanding the IRS 1A/B imagers.

LISS 3. Similar to LISS 1/2 but replaces one visible band with Short Wave IR (SWIR). 85 W power operating. Three 6,000-element linear visible CCD imagers with spectral filters for bands 2,3 and 4. 23.5 m resolution, 142 km swath width. Data rate 35.70 Mbit/s. One 2,100-linear SWIR InGaAs CCD imager cooled to −10°C for band 5 (1.55-1.70 μm). 70 m resolution, 148 km swath width. Data rate 2.02 Mbit/s.

PAN. Single-band panchromatic (0.50-0.75 μm), 5.8 m resolution, three linear CCDs with 23.9 km swath widths combined for 70 km width. Three off-axis mirrors with focal plane splitting by isosceles prism. Swath steering of ±398 km by Payload Steering Mechanism with ±26° steerability (entire camera rotates, yielding better S/N + MTF). 0.2°PSM step yields 2.57 km nadir. 55 W power operating. Data rate 84.903 Mbit/s

WiFS. Wide Field Sensor similar to LISS 1 of IRS 1A/B. Four 2,048-element linear CCDs, dual band (0.62-0.68, 0.77- 0.86 μm), focal length 56.420 mm, 188 m resolution, 774 km swath width (yielding 5-day repeat cycle), FOV ±27° (±13.5° each imager), data rate 40.43 Mbit/s.

IRS 1D
Launch: 28 September 1997 by PSLV which suffered fourth stage malfunctions and left ID in an eccentric orbit of 308 × 822 km rather than the planned circular orbit at 820 km
Other specifications as for IRS 1C.

IRS 2A
Launch: planned for 2000 on PSLV from Sriharikota
Sensors: LISS 3. Same as IRS 1C
LISS 4. 3-band, 5-10 m resolution, steerable
WiFS. Same as IRS P5.

IRS P2
Launched: 15 October 1994 on PSLV-D2, from Sriharikota
Mass: 804 kg at launch
Orbit: 817 km 98.7° Sun-synchronous, crossing equator on descending node at 10.40, with 24-day repeat cycle (341 revs). The first IRS with a frozen perigee to minimise scale variations

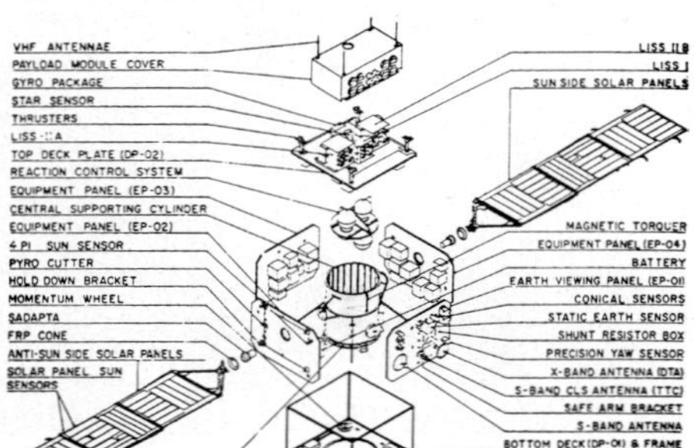

Exploded view of IRS 1A/B (ISRO)

Spacecraft: as 1C, with 6.42 m^2 4-panel wing providing 510 W (with two 21 Ah Ni/Cd batteries). Data rate 2 × 10.4 Mbit/s
Sensors: LISS 2.Single camera of 1A/1B design except each band carries staggered CCDs for one unit to provide same coverage as two LISS 2 previously. 131 km swath width.

IRS P3
Launch: 21 March 1996 by PSLV from Sriharikota
Mass: 922 kg at launch
Orbit: 819 × 821 km, 98.8° Sun-synchronous, crossing equator on descending node at 10.30 local time, with 22 day repeat cycle
Sensors: WiFS. Same as 1C's but with added 1.5-1.7 μm SWIR for determination of vegetation dynamics. In conjunction with 1C, provides 2-3 day coverage of India.

MOS. German (DLR) Modular Opto-electronic Scanner optimised for oceanography. 18-channel VNIR, 248 km swath. MOS-A: 0.7567, 0.7606, 0.7635, 0.7664 μm, resolution 2,520 × 2,520 m; MOS-B: 0.408, 0.443, 0.445 0.485, 0.520, 0.615, 0.650, 0.685, 0.750, 0.815, 0.870, 1.010 μm, resolution 720 × 580 m; MOS-C: 1.600 μm, resolution 720 × 1,000 m.

XAP. X-ray astronomy payload for time variability and spectra of sources and detection of x-ray transients. Three Pointed-mode Proportional Counters 2-20 keV. XSM x-ray Sky Monitor 3-6 keV pinhole camera (PSPC). The Earth imagers have no onboard storage; real-time imagery is transmitted to one Indian and two German stations. XSM operates continuously, but PPC requires P3 to operate in sky-pointing mode, precluding Earth observations.

IRS P4 (Oceansat 1)
Launch: May 1999 on PSLV C1 from Sriharikota
Orbit: planned 720 km circular; 98.28° inclination; Period 99.31 min
Sensors: OCM. 8-band Ocean Colour Monitor. Band 1: 0.402-0.422, 2: 0.433-0.453, 3: 0.490-0.510, 4: 0.500-0.520, 5: 0.555-0.575, 6: 0.655-0.675, 7: 0.745-0.785, 8: 0.845-0.885. IFOV 250 m VNIR, 500 m SWIR. +20/ 0/-20° along track steering, 17.35 Mbit/s, 75 kg.

MSMR (Multi-frequency Scanning Microwave Radiometer) 6.6/10.6/18/21 GHz, 3 dB beam widths 4.2/2.6/1.6/1.4°, <1.0K T resolution, 120/75/45/ 40 km resolution over 1,500 km swath, 86 cm diameter antenna. 5 kbit/s, 76 W, 6 kg
Size: 1.8 m × 1.98 m × 2.57 m; length 11.67 m fully deployed
Mass: 1,050 kg
AOC: 3-axis body-stabilised using reaction wheels, magnetic torques and hydrajure thrusters
Power: 9.6 sq m solar array generating 750 W; two 21 Ah Ni-Cd batteries
Life: 5 years.

IRS P5
Launch: planned for 2000-2001 on PSLV C2 from Sriharikota
Sensors: 3-band XS vegetation camera, 10 m resolution over 40 km swath, across-track steering. Will provide multiple-crop discrimination, species-level discrimination, and so on

WiFS. Improved Wide Field Sensor, 125 m resolution over 750 km swath. Improves frequency/accuracy for assessing mono-crop areas

IRS P6
Launch: planned for 2001-2002 on PSLV C3 from Sriharikota
Sensors: Cartographic PAN camera providing 2.5 m resolution, fore-aft stereo. 2002 Cartosat 2 could improve resolution to 1 m.

IRS P7/Oceansat
Launch: undetermined
Sensors: Ku-band scatterometer; Ku-band altimeter: Microwave radiometers: Thermal-IR radiometer.

IRS P8/Atmos
Launch: undetermined.

INTERNATIONAL

ERS

Background

ESA built the ERS for global measurements of sea wind/ waves, ocean and ice monitoring, coastal studies and a small amount of land imagery. Following ERS 1 approval in October 1981, West Germany, as 24 per cent contributor, provided the instrumentation. The Phase C/D contract, worth about DM840 million, was signed with the Dornier group 29 October 1986 for development, construction, launch preparation and the associated ground station. Under a January 1986 ESA/ NASA agreement, direct readout of ERS 1 SAR data at the Fairbanks, Alaska, station is permitted, and NASA scatterometer and radar imagery are exchanged for other ERS data of interest. The exchange enhances NASA/ESA-supported polar ice research, and complements NASA experimental activities with the Topex/Poseidon (see International) and Shuttle Imaging Radar-C. By permitting NASA direct data readout from ERS, ESA reciprocates earlier NASA provisions allowing European data readout from Seasat and Nimbus 7.

The European Remote Sensing satellite's three primary all-weather instruments provides systematic, repetitive global coverage of ocean, coastal zones and polar ice caps, monitoring wave height/wavelengths, wind speed/direction, precise altitude, ice parameters, sea surface T, cloud top T, cloud cover and atmospheric water vapour content. The Active Microwave Instrument (AMI) can operate as a wind scatterometer or Synthetic Aperture Radar (SAR, with wave scatterometer mode), the Along-Track Scanning Radiometer & Microwave Sounder (ATSR-M) provides the most accurate sea surface T data to date, and the Radar Altimeter (RA) measures large-scale ocean/ice topography and wave heights. The follow-on ERS 2 was approved in June 1990 to provide continuity until Envisat 1 appears in 1999 (see the POEM entry at end). ERS 1/2 operated simultaneously 16 August 1995 to mid-May 1996, the first time that two identical civil SARs worked in tandem. The orbits were carefully phased to provide 1-day revisits, allowing the collection of interferometric SAR image pairs and improving temporal sampling. Although still working perfectly, ERS 1 funding required it to be put on standby from the end of May 1996. Extending ERS 2 operations to 2000 is under discussion.

ESA's Council announced in June 1990 that subscriptions were sufficient to start the ERS-2 programme. Cost is 371 MAU (1989 rate). Dornier was awarded the DM480 million prime contract (DM360 million went to non-German subcontractors). Similar instruments and mission to ERS 1 provides coverage until Envisat 1 becomes available. ERS 2's first SAR image, covering Italy's Campania region, was acquired 2 May 1995. The first ATSR 2 image was acquired 5 May 1995, covering Britain/Ireland. AMI's wind scatterometer mode did not begin operations until 22 November 1995 because of an anomaly. ATSR stopped normal operations 22 December 1995 because of a problem with its scan mirror. The SAR, RA and Microwave Sounder have been operating nominally since September 1995; ATSR, PRARE and GOME since January 1996.

Matra of France (18.31 per cent) supplied a Spot-type bus. The UK (13.34 per cent) provided the AMI lead contractor (Marconi). Other contributors to the 13-nation project include Italy (10.61 per cent), Canada (9.1 per cent), Netherlands (5 per cent), Sweden (3.9 per cent), Belgium (3.72 per cent), Spain (2 per cent), Denmark (1.99 per cent), Switzerland (1.7 per cent), Norway (1.5 per cent). ERS 1 development phase was estimated at 584 MAU at 1984 rates; ESA projected cost to completion was 728 MAU at 1989 rates.

The ERS 1 payload was delivered to Matra at the end of February 1990 for integration with the platform. Following launch, the solar array began deployment within 1 minute of separation, followed by the SAR antenna after 75 minutes and the wind scatterometer three-part antenna after 4 hours. All the instruments were activated in the first two weeks and the first SAR image (of Spitzbergen) was returned to Kiruna 27 July 1991. The operational phase began January 1992. Apart from PRARE's failure, the only significant in-flight anomalies have been some interruption to AMI operations during 1994 caused by recoverable problems in the transmitter, a sudden gain loss in the SAR telemetry downlink requiring a switch to the redundant unit, and failure of ATSR's 3.7 μm channel. All instrument and satellite performances meet or exceed expectations. >100,000 SAR images were returned in the first year.

Specifications

ERS 1
Launch: 17 July 1991 by Ariane 4 from Kourou, French Guiana
Mass: 2,384 kg in-orbit BOL (1,100 kg payload, 317.6 kg hydrazine)
Contractors: Dornier (prime), Matra (bus), Marconi Space Systems (AMI), Alenia Spazio (RA), Laben (payload data handling), Contraves (payload module structure), Fokker (payload module thermal control, harness, integration), Aerospatiale (solar array, laser reflectors), British Aerospace (ATSR), Insitut für Navigation Universität Stuttgart (PRARE)
Resolution: 30 m SAR (100 km swath), 1 km/0.5K ATSR-M, 3 cm Radar Altimeter, 2 m/s wind scatterometer, 5-10 cm PRARE satellite range (failed)
Orbit: controlled within 765-825 km, 98.5° Sun-synchronous for required track repeat period, crossing

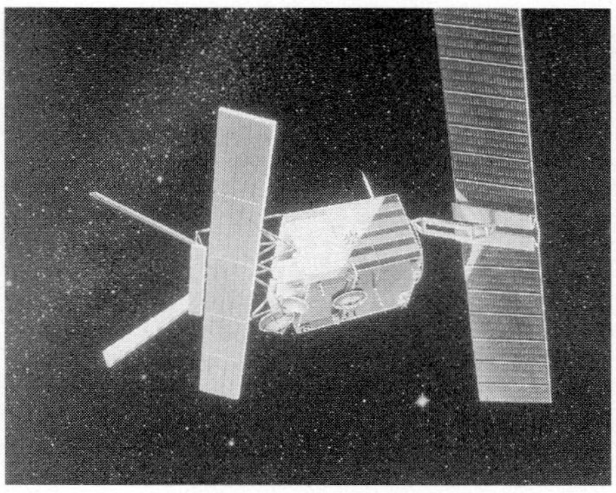

The ERS-1 carriesan ESA SAR, and a NASA scatterometer ***1998***/0038495

equator southbound at 10.30 am local time. 3-day (43 rev) repeat cycle for 3-month commissioning phase, then 35-day cycle alternating with 3-day cycle periods in early 1992 and 1994. From April 1994, a 168 day repeat 'geodetic' orbit has been flown

Spacecraft: 11.8 m high, 11.7 m span on-orbit. Based on Matra's 3-axis Spot bus. Payload electronics housed in compartment mounted on platform. Two 2.4 × 5.8 m solar wings (total of 22,260 cells) to provide 2.2 kW after 2 years, supported by 4 × 24 Ah SAFT Ni/Cd batteries. ESOC Mission Management & Control Centre in Darmstadt controls via Kiruna

Sensors: Active Microwave Instrument (AMI) operates in the SAR mode to produce C-band Earth imagery and separately in the AMI-wind mode as a 3-beam C-band scatterometer. Synthetic Aperture Radar: spatial resolution 30 m; frequency 5.3 GHz (C-band), VV polarisation, pulse length 37.1 μs, data rate 105 Mbit/s; swath width 100 km, with 23° incident angle at mid-swath (up to 35° possible via experimental 'roll-tilt' attitude control system mode). Wave Scatterometer: the programmable wave mode operates every 200-300 km providing a 6 × 6 km image for extraction of information on wave length and direction. Wave length accuracy of ±25% over 100-1,000 m, direction accuracy ±20°. AMI Wind Scatterometer: three antennas (mid 0.35 × 2.3 m, fore/aft 0.25 × 3.6 m) providing fore/mid/aft beams sweep a 500 km swath in 50 km cells at 5.3 GHz, allowing the radar returns to be analysed for surface wind vectors: 4-24 m/s + 0-360° with ±20° accuracy. Incident angle varies over 27-58°. Spatial resolution: 50 km.

Radar Altimeter (RA).The nadir– viewing 13.8 GHz Ku-band 1.3° beamwidth 1.2 m dia antenna altimeter measures, in Ocean Mode (330 MHz bandwidth), wind speed (2 m/s accuracy from measuring backscattering coefficient to 0.5 dB), wave heights of 1-20 m with 0.5 m accuracy over 1.6-2.0 km footprints, and determine altitude to about 5 cm (<1 cm resolution). Ice Mode (82.5 MHz bandwidth) operates with a coarser resolution to determine ice sheet topography, ice type and sea/ice boundaries.

Along-Track Scanning Radiometer & Microwave Sounder (ATSR-M). An experimental passive instrument comprising an advanced four-channel infra-red radiometer (UK-Australia) and a two-channel nadir-viewing microwave sounder (France) for T and water vapour measurement, respectively. ATSR-M IR Radiometer: res 0.5K over 50 × 50 km square, 1 km spatial res; wavelengths 1.6/3.7/11/12 μm; swath width 500 km. Microwave Sounder: passive nadir-viewing radiometer operating at 23.8/36.5 GHz measuring the vertical column water vapour content within a 22 km footprint, providing corrective data for ATSR sea surface T + RA measurements.

Precise Range And Range Rate Experiment (PRARE) for orbit determination (ranging accuracy 3-7 cm) using 8.489 GHz X-band 1 W signals transmitted to a network of 60 cm 2 W mobile ground transponders, with 2.248 GHz 1 W S-band transmissions permitting ionospheric

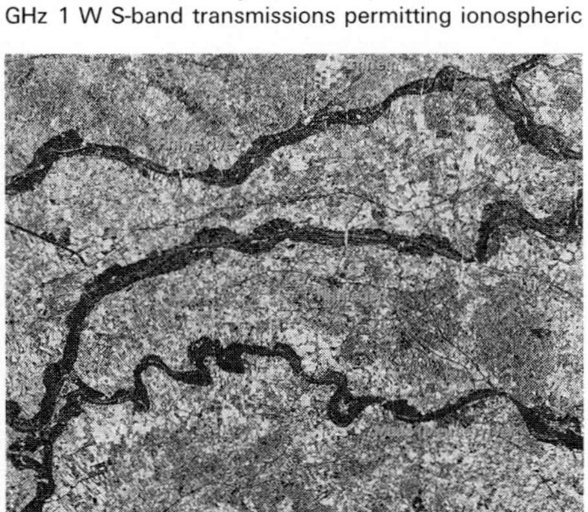

ERS 1 SAR image of flooding in Northern Europe in January 1995 (ESA) ***1995***

ERS 2 final preparations. Prominent is the ATSR (curved apertures) (DASA) ***1996***

corrections. PRARE failed after 3 weeks in orbit, concluded to be destructive proton-induced memory latch-up over the South Atlantic Anomaly. Russia's Meteor 3-06 launched January 1994 is successfully operating a PRARE provided by Kayser-Threde under DARA contract.

Laser Retroreflector permits precise range/orbit determination, although less frequently, and RA calibration.

ERS 2

Launch: 21 April 1995 by Ariane 4 from Kourou, French Guiana

Mass: 2,516 kg at launch, 2,516 kg on-station BOL, 2,110 kg dry

Contractors: as ERS 1 except Officine Galileo (GOME), Schrack (microwave radiometer)

Orbit: similar to ERS 1, phased for 1 or 8 day revisits in turn. 771 × 797 km 98.55° at injection

Spacecraft: Saab developed a new computer memory due to ERS 1 design obsolescence

Sensors: Active Microwave Instrument as ERS 1

Radar Altimeter as ERS 1

Global Ozone Monitoring Experiment (GOME) scanning near-UV/visible spectrometer measuring backscattered Earth radiance in 3,500 channels over four bands, 240– 295, 290-405, 400-605, 590-790 nm (2-4 Å res), to determine ozone/trace gases in tropo/stratosphere. Scan angle and integration variable within ±2-31°/0.1-3,000 s. Internal, Sun, Moon calibration. 52 kg, 30 W operating; 50 × 60 × 70 cm. Each band focused on 1,024-pixel photodiode array cooled to -40±1°C by Peltier coolers. Essentially a scaled-down version of SCIAMACHY

ATSR augmented by 0.55, 0.67, 0.78 μm visible channels to improve monitoring of land applications (vegetation moisture, state, species). Also improvements to mechanical and electronics design.

PRARE redesign of control processor and memory with radiation-tolerant components plus redundant unit, to avoid ERS 1-type loss

Ground segment: ERS are controlled from ESOC at Darmstadt, Germany with ESA ground receiving stations at Salmijärvi, near Kiruna (Sweden, the primary station with 15 m antenna; also used for TT&C), Fucino (Italy), Gatineau (Canada), Maspalomas (Spain), plus national stations at Fairbanks (Alaska), Prince Albert (Canada), West Freugh (UK), Alice Springs (Australia). SAR's 105 Mbit/s image data are returned in real time only, available only when the wave/wind modes are inactive (other data are recorded onboard, thus providing global coverage). ESA's ESRIN ERS Central Facility (EECF) facility at Frascati, Italy is the data management centre and prepares the mission operation plan for ESOC, with processing/archiving facilities at Brest (France), Farnborough (UK), DLR Oberpfaffenhofen (Germany) and Matera (Italy). Some products, such as from the wind scatterometer, are available within 3 hours of observation. Dornier was awarded a DM5 million contract in April 1989 for 20 to 30 mobile PRARE ground stations. Some were deployed in 1994 in support of Meteor 3 and the remainder in 1995.

POEM series

To solidify its civil remote sensing plans ESA developed the Polar Orbit Earth Observation Mission Programme, envisaged as two series (Envisat/Metop) of polar platforms providing data to a wide user spectrum. POEM's bus derives from the ESA designed modular platform built as part of the Columbus programme until it was unified in 1994 with Envisat 1. POEM-1 Phase 1 was approved at 1991's Munich Ministerial Council meeting and the 1,174.5 MAU Ph 2 at Spain in 1992. Envisat 1, the largest and most expensive satellite ever built by Europe, will be launched in 1999 for environmental research, providing continuity for ERS. 40 MAU preliminary work on Metop 1 was approved; the full 760 MAU (1994 rates) was reduced in early 1996 by 45 per cent when it became clear it could not be met. The cost was cutting the programme from 15 years to 10 years and deleting the microwave imager. Developed in co-operation with EUMETSAT for 2001 launch, its climatic research instruments will be accompanied by operational meteorological units. The PPF prime contract with MMS is worth ECU502 million (1988 rate; ECU675 million 1995). Ph C/D contract was signed 24 July 1995 with MMS, including Envisat 1 payload integration. DASA has 30 per cent.

France provides 24 per cent of the funding, UK 19 per cent, Germany 18 per cent, Italy 11 per cent; 5 per cent remains uncovered. The Phase C/D contract, worth ECU675 million (1995 rates; ECU502 million 1988) was signed with MMS 24 July 1995 for Polar Platform development and Envisat 1 integration. MMS Bristol is responsible for platform development and Envisat integration, and MMS Toulouse for the service module.

ESA plans to gather environmental data with the Envisat using eight major instruments to perform atmospheric ozone and turbulence monitoring ***1998***/0038494

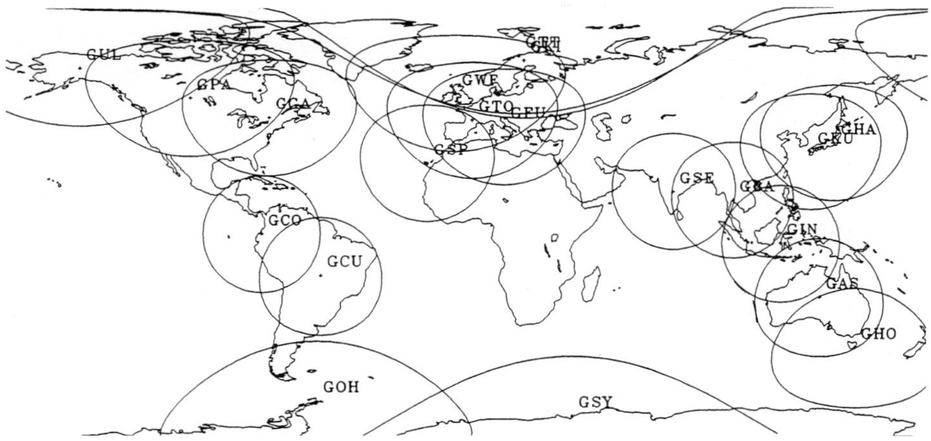

ERS receiving stations. Alice Springs (GAS, Australia), Aussaguel (GTO, France), Bangkok (GBA, Thailand), Cotopaxi (GCO, Ecuador), Cuiaba (GCU, Brazil), Fairbanks (GUL, US), Fucino (GFU, Italy), Gatineau (GCA, Canada), Hatoyama (GHA, Japan), Hobart (GHO, Australia), Hyderabad (GSE, India), Kiruna (GKI, Sweden), Kumamoto (GKU, Japan), Maspalomas (GSP, Spain), O'Higgins (GOH, Antarctica/Germany), Pare-Pare (GIN, Indonesia), Prince Albert (GPA, Canada), Syowa (GSY, Antarctica/Japan), Tromso (GTT, Norway) and West Freugh (GWF, UK) (Eurimage)

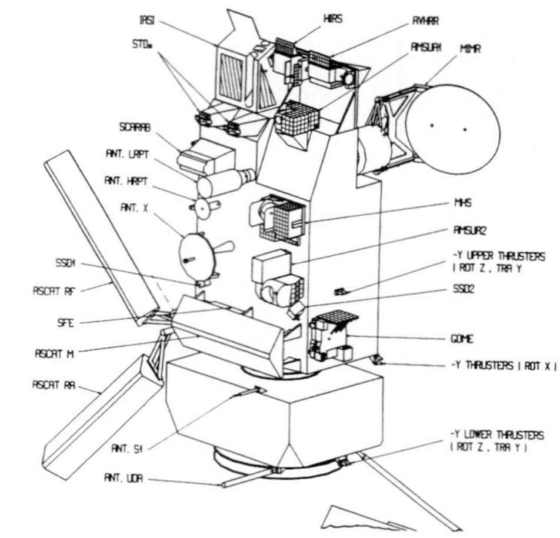

Metop will provide operational meteorology and climate monitoring. Scarab and MIMR have been deleted (ESA) ***1995***

ASAR, RA-2, MERIS, MIPAS, GOMOS, MWR and LRR will be provided by ESA; AATSR (UK/Australia), SCIAMACHY (Germany/Netherlands) and DORIS (France) will be funded nationally. The ECU810 million contract for payload development and system engineering was signed 17 July 1996 with Dornier.

Metop 1 is being jointly developed by ESA and EUMETSAT in the frame of a single space segment as the prototype of a series of operational satellites addressing operational meteorology and climate monitoring missions of EMETSAT. The single space segment, comprises Metop 1, 2 and 3 satellites and instruments like ASCAT, GOME and GRAS. The procurement will be done by a single space segment team led by an ESA project manager assisted by an EUMETSAT project manager. The other instruments will be provided by EUMETSAT, NOAA and the 'Centre National d'Etudes Spatiales' (CNES). EUMETSAT is responsible for launcher procurement, ground segment, satellite operations and data distribution. In addition to complementing the meteorological payload, the climate

ERS orbital configuration (ESA)

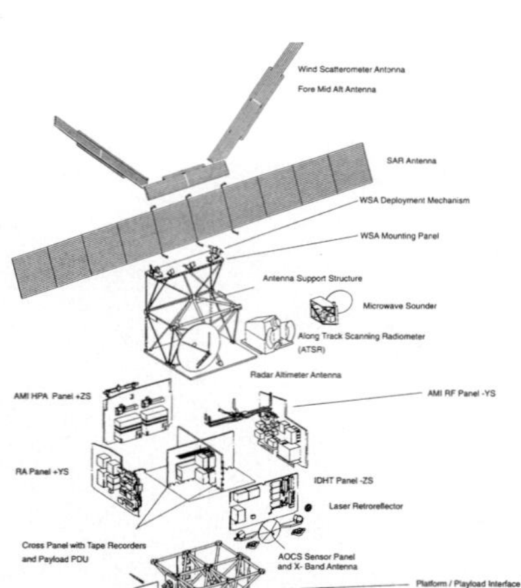

ERS: ESA's ocean and ice monitoring platform

instruments will establish a global ocean, ice and ozone monitoring system. Phase A started October 1993 for completion 1994; Phase B began July 1996, EUMETSAT's cost is projected at ECU 1.5 billion (1994 rate) for two satellite launches, and 14 years of operations: ESA contributes financially, partially to Metop-1, the development of ASCAT, GOME and GRAS. Metop- 2 as planned for 2007 but it is to be available for launch within 18 months of the launch of Metop-1. Metop-3 is envisaged for 2011.

Envisat 1

Launch: planned for May 2000 by Ariane 5 from Kourou, French Guiana

Design life: 5 years

Mass: about 8 t; 1,900 kg payload

Contractors: Dornier (mission prime, MIPAS), MMS (platform, service module, integration, GOMOS, ASAR, AATSR), Alenia Spazio (RA, MWR), Fokker Space (solar array, SCIAMACHY), CASA (service/payload module structures, ASAR panels), Aerospatiale (MERIS)

Orbit: planned 800 km 98.55° morning orbit, crossing equator 10.00 am, 35 day repeat cycle

Spacecraft: 3 axis 10 m long, 3 m wide, comprising the payload module + service module. Single 5 × 14 m 14-panel solar array generates 6.7 kW after 4 years (1.9/4.1 kW average/peak to payload); 8 × 40 Ah Ni/Cd batteries. Data handling: one 100 Mbit/s channel (for ASAR) + 10 × 32 Mbit/s for others. 4 × 30 Gbit tape recorders; record 5 Mbit/s, playback 50 Mbit/s. FOCC Flight Operations Control Centre at ESOC (via Kiruna S-band or DRS); PDCC Payload Data Control Centre at ESRIN. X-band real time and recorded data direct and Ka-band via DRS (steerable 90 cm antenna on 2 m mast); each two channels 50/100 Mbit/s. Hydrazine thrusters (300 kg loading) for attitude control; RWs + magnetorquers for fine control (0.1° 3)

Sensors: Advanced Synthetic Aperture Radar (ASAR) improved version of ERS SAR. 5.331 GHz, 1.3 × 10 m array of 20 66.4 × 99.5 cm radiating panels (each 16 rows of 24 microstrip patches). Imaging mode HH or VV, 29 × 30 m/2.5 dB resolution, 7 swaths 100-56 km at 20-45°, 96.3 Mbit/s, 1,200 W DC power consumption; alternating polarisation mode HH + VV, 29 × 30 m/3.5 dB, 7 swaths 100-56 km at 20-45°, 96.3 Mbit/s, 1,200 W; wide swath mode HH or VV, 150 × 150 m/2.5 dB, >400 km 5 subswaths at 17-42°, 96.8 Mbit/s, 1,200 W; global monitoring mode HH or VV, 1,000 × 1,000 m/1.5 dB, >400 km 5 subswaths at 17- 42°, 0.9 Mbit/s (allowing onboard storage), 750 W; wave mode HH or VV, 30 m/2.0 dB, two vignettes of 5 × 5 km every 100 km in any swath at 20-45°, 0.9 Mbit/s, 520 W. Up to 30 min of high resolution imagery each orbit.

Radar Altimeter (RA-2) fully redundant nadir pointing pulse limited radar using single 1.2 m dish at 13.575 & 3.3 GHz. Derived from ERS RA but improved performance without significant increase in mass and power consumption. 3.3 GHz channel added to correct for ionosphere propagation effects. Fixed pulse repetition frequencies of 1,800/450 Hz are respectively used by the two channels. RA-2 onboard autonomous selection of transmitted bandwidth (CW/20/30/320 MHz) avoids dedicated operational mode commanded from ground and makes possible continuous operation over ocean, ice, land and their boundaries. Altitude accuracy after ionospheric correction improved to <4.5 cm for Significant Wave Height up to 8 m. Operation supported by MWR/LRR/DORIS measurements.

Medium Resolution Imaging Spectrometer (MERIS) is the first programmable imaging spectrometer. VNIR (400-1,050 nm), 250 m/12.5 nm resolution (adjustable as required), swath width 1,130 km at 800 km altitude. Aimed at water quality measurements, such as phytoplankton content, depth and bottom-type classification and monitoring of extended pollution. Secondary goals: atmospheric monitoring and land surfaces processes.

Michelson Interferometer for Passive Atmospheric Sounding (MIPAS) is a Fourier transform spectrometer observing mid-IR 4.15-14.6 μm limb emissions with high spectral resolution (<0.03 cm^{-1}), allowing day/night measurement of trace gases (including the complete nitrogen- oxygen family and several CFCs) in stratosphere and cloud-free troposphere. Global coverage, including poles.

Global Ozone Monitor by Occultation of Stars (GOMOS). Two spectrometers observe setting stars over UV to near-infra-red for 50 m vertical resolution and 0.1% annual variation sensitivity of ozone/related gases. Occultation method is self-calibrating and avoids the long term instrumental drift problems of current sensors. Spectrometer A: 2,500-6,750 Å, res 0.3 nm/pixel; B: 9,260-9,520 Å (H_{2O}) + 7,560-7,730 Å (O_2), res 0.05 nm/pixel. The Fast Photometer Detection Module provides 1 kHz 2-band scintillation monitoring of the star image.

Advanced Along-Track Scanning Radiometer (AATSR). Performance almost identical to ATSR 2 on ERS 2. Provided by UK Department of Environment + Australia.

MicroWave Radiometer (MWR) is a nadir-viewing 23.8/36.5 GHz Dicke radiometer with 3 dB IF bandwidth 600 MHz. ERS's MWR design modified mainly in the mechanical layout and antenna configuration. Radiometric stability <0.5K over 1 year. Periodic onboard calibration by switching receiver input between two references: a horn pointing at the cold sky and a hot radiator at ambient. MWR determines tropospheric columnary water vapour content by measuring the radiation received from Earth's surface to correct RA-2 altitude measurements.

Scanning Imaging Absorption Spectrometer for Atmospheric Chartography (SCIAMACHY) 2,400-23,800 Å grating spectrometer (limb/nadir viewing) for detrimental trace gas measurement in troposphere/stratosphere. resolution 2.4 Å UV and 2.2-14.8 Å vis/infra-red. Swath 1,000 km in nadir mode.

DORIS as Spot 3/4 + Topex/Poseidon for precise (1 m) onboard real time autonomous orbit determination and distribution of timing pulse; 10 kg, 14 W.

LRR Laser Retro Reflector for precise orbit determination

Ground segment: Comprises two major elements: FOS Flight Operations Segment and PDS Payload Data Segment. FOS employs the FOCC Flight Operations Control Centre at ESOC to provide mission and operations planning, and command/control via the primary S-band TT&C station at Kiruna-Salmijärvi. The PDS is composed of the PDCC Payload Data Control Centre at ESRIN (controlling and monitoring all services offered by PDS facilities); two Payload Data Handling Stations at Kiruna-Salmijärvi (X-band) and at ESRIN (via DRS) for acquiring global and regional data and providing near-realtime and FD services; the PAC Processing and Archiving Centres (offline services). The PDS will offer a comprehensive set of user services, with online access to catalogue, browse and ordering services. ESA's products will range from raw reformatted data (level 0) to geophysical products (level 2) and images (from browse to high resolution). PDS development has been awarded to a consortium led by Thomson-CSF.

Metop 1

Launch: planned for mid-2003 by Ariane 5 from Kourou, French Guiana

Design life: 5 years

Mass: 4,813 kg, payload 919 kg

Contractors: Matra Marconi Space/Dornier

Orbit: 830 km Sun-synchronous, 09.30 descending node

Spacecraft: reduced Envisat platform, PPF/Spot avionics, 10- panel solar array feeding 5 × 40 Ah Ni/Cd batteries. Eight paired 15 N hydrazine thrusters drawing on four tanks, fine pointing (±0.15°) by three 40 Nms RWs

Sensors: Advanced Visible/Infra-Red High-Resolution Radiometer (AVHRR/3). Day/night imaging in 6 bands over 0.68-12.50 µm, sea surface T, ice, snow and vegetation cover. 31.3 kg, 29 W, 822.0 kbit/s.

High-Resolution Infra-red Temperature Sounder (HIRS/3). 20 bands over 0.69-14.95 µm, vertical T profile. Moisture content, cloud height, surface albedo. 33.1 kg, 24 W, 2.88 kbit/s.

Advanced Microwave Sounder Unit (AMSU-A1/A2). Two modules covering 23.8 MHz-89 GHz: A1 13 channels + A2 2 channels for T/humidity atmospheric profiles. A1: 53.3 kg, 88.3 W, 2.080 kbit/s; A2: 47.4 kg, 37.25 W, 1,120 bit/s.

Microwave Humidity Sounder (MHS). 90-190 GHz 5-channel self-calibrating for humidity profiling. 66 kg, 100 W, 3.95 kbit/s.

Argos. DCS 2 data collection/location. 47 kg (inc antennas), 27.5 W, 2.56 kbit/s.

IR Atmospheric Sounding Interferometer (IASI). 3.5-15.5 µm Fourier Transform spectrometer. 148 kg, 196 W, 1.5 Mbit/s.

Advanced Wind Scatterometer (ASCAT). 6.255 GHz C-band radar for ocean surface wind vectors, 213 kg, 250 W, 45 kbit/s.

Global Ozone Monitoring Expt (GOME). As ERS 2. 59 kg, 44 W, 50.0 kbit/s for Metop 1, 2.

OMI-Ims considered for Metop-3.

Global Navigation Satellite System Receiver for Atmospheric Sounding (GRAS) Spaceborne GFS/GLONASS receiver for Earth science applications with dual frequency measurement from GPS and GLONASS carrier phase measurements to mm-level precision good quality (<1 m) code phase measurements.

Civil Earth Observation Satellites

Satellite	Launch	Orbit	Range
International			
ERS 1	17 July 1991	780 km, 98.5°	radar/infra-red
ERS 2	21 April 1995	780 km, 98.5°	radar/infra-red
Meteosat 5	2 March 1991	0° GEO	visible/infra-red
Meteosat 6	20 November 1993	10° west GEO	visible/infra-red
Meteosat 7	2 September 1997	0° west GEO	visible/infra-red
Canada			
Radarsat 1	4 November 1995	792 km, 98.6°	radar
China, People's Republic			
FY (Fung Yen) 2	10 June 1997	105° east GEO	visible/infra-red
China/Brazil			
CBERS 1	Third quarter 1999	778 km, 98.5°	visible/thermal infra-red
France			
Spot 1[3]	22 February 1986	825 km, 98.7°	visible/near infra-red
Spot 2	22 January 1990	820 km, 98.7°	visible/near infra-red
Spot 3	26 September 1993	820 km, 98.7°	visible/near infra-red
Spot 4	24 March 1998	820 km, 98.7°	visible/infra-red
India			
Insat 1D	12 June 1990	83° east GEO	visible/infra-red
IRS 1A[4]	17 March 1988	900 km, 99.0°	visible/near infra-red
IRS 1B	29 August 1991	900 km, 99.3°	visible/near infra-red
Insat 2A	9 July 1992	74° east GEO	visible/infra-red
Insat 2B	22 July 1992	93.5° east GEO	visible/infra-red
IRS P2	15 October 1994	817 km, 98.6°	visible/near infra-red
IRS 1C	28 December 1995	817 km, 98.6°	visible/near infra-red
IRS P3	21 March 1996	820 km, 98.8°	visible/near infra-red
IRS 1-D	28 September 1998	300 × 823 km, 98.6°	visible
Japan			
GMS 4	6 September 1989	140° east GEO	visible/infra-red
JERS 1	11 February 92	568 km, 97.7°	visible/infra-red/radar
GMS 5	18 March 1995	140° east GEO	visible/infra-red
MT-Sat	August 1999	140° east GEO	visible/infra-red
RFAS			
Meteor 2-21	31 August 1993	950 km, 82.6°	visible/infra-red
Meteor 3-06	25 January 1994	1,200 km, 82.6°	visible/infra-red
Meteor 3M-2	August 1998	925 km, 82.5°	visible/infra-red
Mir[1]	20 February 1986	375 km, 51.6°	See note 1
Okean 4	11 October 1994	650 km, 82.6°	visible/infra-red/µ/radar
Sich 1	31 August 1995	670 km, 98°	visible/infra-red/µ/radar
Elektro 1	31 October 1994	76° east GEO	visible/infra-red
Resurs-O1 N3	4 November 1994	662 km, 98°	visible/infra-red
Resurs F1-M	10 November 1997		
Resurs-O1 N4	10 July 1998	835 km, 98.8°	visible/infra-red
Resurs-O2 1	10 July 1998	818 km, 98.8°	visible/infra-red radar
United States			
GOES 7	26 February 1987	96° west GEO	visible/infra-red
Landsat 4	16 July 1982	705 km, 98.2°	visible/infra-red
Landsat 5	1 March 1984	705 km, 98.2°	visible/infra-red
NOAA 9	12 December 1984	855 km, 98.9°	visible/infra-red
NOAA 10	17 September 1986	815 km, 98.8°	visible/infra-red
NOAA 12	14 May 1991	816 km, 98.7°	visible/infra-red
GOES 8	13 April 1994	75° west GEO	visible/infra-red
NOAA 14	30 December 1994	855 km, 98.9°	visible/infra-red
Microlab 1 (Orbview 1)	3 April 1995	740 km, 70°	visible/infra-red
GOES 9	23 May 1995	135° west GEO	visible/infra-red
NOAA 15 (K)	13 May 1998	855 km, 98.9°	visible/infra-red
SeaStar (Orbview-2)	1 August 1997	705 km, 98.2°	visible/near infra-red
Lewis	23 August 1997	failed in post launch	
Early Bird	24 December 1997	failed in post launch	
Landsat 7	15 April 1999	705 km, 98.2°	visible/infra-red

Notes. 1: manned space station carrying array of cameras, spectrometers and modules. 2: Resurs-F3 film-return mission expected in 1996. 3: Spot 1 retired 31 December 1990 but revived March-October 1992 and April-July 1993 to return real-time imagery; still held in standby mode. 4: IRS 1A retired from routine service 17 March 1995. Italicised text indicates satellites yet to be launched.

Meteosat series

Background

Meteosat constitutes ESA's contribution to the international World Weather Watch of the Global Atmospheric Research Programme (GARP) which began with Meteosat 1, launched 23 November 1977 by Delta from Cape Canaveral. Although fully operational for only two years of its projected three year planned life, it gathered data until it exhausted its supply of hydrazine propellant in October 1985. During this lifetime, the satellite played a key role in the Global Weather Experiment, a major exercise involving almost all of the 147 member nations of the World Meteorological Organisation. GWE had almost ended when, a day after its second anniversary, Meteosat 1 suffered an onboard radiometer failure. Despite efforts spread over some months, its imaging capacity never reached its full programme again, although its data collection function remained unimpaired. By then, however, ESA had gathered more than 40,000 images and submitted an outline proposal for the five-satellite Meteosat programme for 1984-1994.

Engineers modified M2 after M1's radiometer problems and repaired damage during ground tests that were designed to assess its resistance to the additional vibration involved in an Ariane launch instead of the Delta. The launch occurred on 19 June 1981. Its primary imaging mission successfully started, but ESA ground controllers could not activate the DCP system. Despite the problems the two satellites together provided a full service until the end of 1984, when M1 began to malfunction and its hydrazine approached exhaustion. NOAA then loaned ESA the partially operational GOES 4, moving it to 10° west by June 1985, as a return for M2's help following a GOES failure over the Atlantic. Data collection by GOES 4 continued from 43° west into 1988, with M2 providing the images, until M3 (P2) assumed the relay role. Although M2 had a three year design life, the quality of its imagery continued to be excellent into 1991. As part of the effort to extend its life into 1989 (when M4 became available), ESA performed a small inclination manoeuvre at end of 1986 to hibernate it at 10° west. ESA planned to raise it above GEO, using the 2 kg hydrazine remaining, when M5 was commissioned in 1991. The satellite returned a commemorative photo on its 10th anniversary image 19 June 1991. In all, the satellite returned some 284,000 images during its career. Its retirement move began 2 December 1991, when its thrusters fired for 8 minutes to raise altitude by 140 km. During the third burn sequence two days later, the hydrazine became exhausted and the orbit only increased by 334 km instead of the planned 700 km. The transmitter was shut down 6 December 1991.

ESA built M3/P2 as a qualification model for the pre-operational programme and launched it 15 June 1988 as a back-up to the ageing M2 until the delayed MOP series could begin. Financial constraints dictated that the 12 year old spacecraft could not be upgraded to full flight standards, but its primary role was only to provide the DCP service unavailable from M2. Although its electronically despun antenna experienced 3 dB fluctuations, it operated as prime satellite from August 1988 until M4 appeared. ESA considered loaning M3 to NOAA at 50° west to compensate for the GOES 6 failure of January 1989. Both organisations approved the move June 1989 and completed the transfer from 3° west to 50° west on 4 November 1989. It returned to assume the primary role in January 1990 after the appearance of

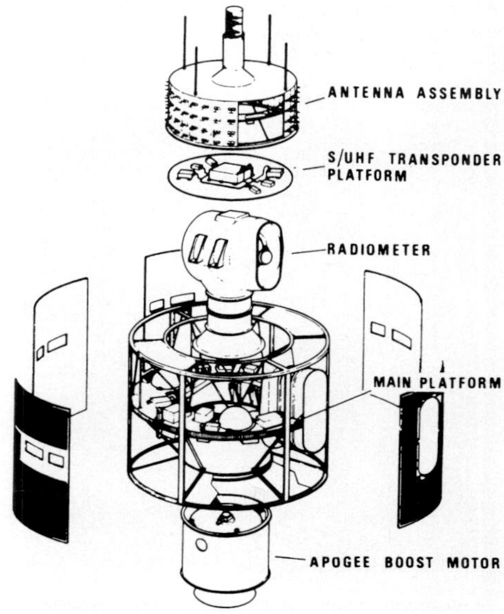

Meteosat principal features

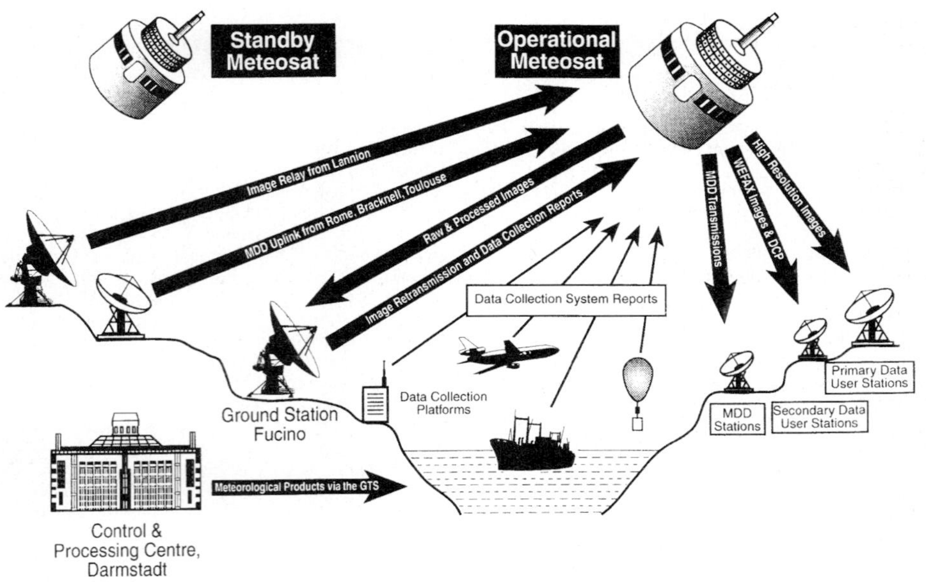

Meteosat system overview (EUMETSAT) ***1996***

Meteosat integration at Aerospatiale

M4's imaging problems. M4 again assumed the prime role in April 1990 after M3 lost dissemination channel 1. M3 moved to 3° west as standby; after Meteosat 5's appearance it began the move again 29 April 1991 to 50° west, arriving 19 July, again to substitute for GOES. Service began 1 August 1991.

Limited visibility to earth stations limited a move further west for Meteosat. After a new relay station came online February 1993 at Wallops Island (Virginia), this problem disappeared and Meteosat began to move again 27 January 1993 to 75° west, arriving 24 February, to provide coverage of the whole continental US. The Wallops station was operated remotely by ESOC. All imagery received there returned to Germany via PanAmSat, then after processing returned to Wallops, where it was uplinked to Meteosat for broadcast to users. M3 moved to 70° west by 14 March 1995, from where it was boosted 21 November 1995 into 36,725 × 36,762 km, 2.82°, using the last 5.1 kg hydrazine.

MOP-1/M4 was launched 6 March 1989 and began service as the prime satellite 19 June 1989. But the satellite began to generate erroneous grey levels in October 1989 and image rectification became impossible. A switch to the second Synchronisation and Imaging Chain (SIC-2) in November 1989 resumed nominal imaging, but it was replaced as prime at 0° by M3 24 January 1990 to allow detailed failure analysis and an intensive test programme. The cause was found to be a malfunction of the DC/DC converters supplying the SICs. All future satellites were modified. Procedures were initiated to switch periodically from one SIC to the other and the image processing software was improved to detect/correct the erroneous grey levels (if not too numerous). Following the failure of an M3 dissemination channel, M4 returned as prime in April 1990. It was then replaced as prime by M5 in February 1994 and held at 8° west until 10 May 1995, when it moved to 8° east before being boosted 9 November 1995 into 36,619 × 36,777 km, 1.50°.

M5 has remained the operational satellite, but the on-board fuel reserves for inclination control have been consumed and thus inclination is increasing at about 0.8°/year. Due to the high inclination and the impact upon user station reception, the operational mission was switched to M6 in February 1997: M6 was relocated from 10° west to 0° in early 1997.

When the EUMETSAT Convention came into force on 19 June 1986, the organisation assumed overall and financial control of the MOP Meteosat Operational Programme beginning on 12 January 1987. A special Council controls EUMETSAT Earth observation programmes and it represents the National Meteorological Services of 17 European states. Each European country makes a contribution to the programmes based on GNP (per cent): Austria 2.23, Belgium 2.70, Denmark 1.76, Finland 1.84, France 16.78, Germany 22.29, Greece 0.96, Ireland 0.54, Italy 15.46, Netherlands 4.03, Norway 1.47, Portugal 0.86, Spain 6.96, Sweden 3.20, Switzerland 3.33, Turkey 1.50 and UK 14.09.

In order to make a smooth transition from a research and technology programme to a fully operational system, ESA planned to develop its earth imaging resources in three phases: the Meteosat Operational Program (MOP), the Meteosat Transition Program (MTP) and the Meteosat Second Generation Program (MSG). Each phase uses distinct satellites and ground procedures. Meteosat 5 serves as the geostationary system's primary satellite, but utilises Meteosat 6 as a standby. Two older satellite's Meteosat 3/4 predated this pair but Eumetsat de-orbited both in 1995. Meteosat 7, the MTP Meteosat Transition Programme satellite, may join the others at geosynchronous in the second half of 1997 to maintain a continuous service until the MSG (Meteosat Second Generation) appears in 2000. Encryption of high resolution image data began 1 September 1995. On 1 December 1995, Eumetsat inaugurated a new facility dedicated to control the satellites at its MTP HQ of satellite launch and operations Primary Ground Station in Fucino, Italy. Before then, ESA provided these functions.

The Meteosat Operational Program (MOP) began 23 November 1983 under ESA's auspices and ended 30 November 1995. ESA transferred overall responsibility to EUMETSAT in January 1987 but in-orbit control remained with ESA. The Meteosat Operational Program receives funds from the Eumetsat member states in contributions of (per cent): Belgium 4.4, Denmark 0.58, Finland 0.35, France 25.60, Germany 26.39, Greece 0.30, Ireland 0.11, Italy 12.00, Netherlands 3.00, Norway 0.50, Portugal 0.30, Spain 5.24, Sweden 0.93, Switzerland 3.03, Turkey 0.50, UK 16.76 (0.01 per cent not covered). Various sources estimate the total cost for the three satellite series in 1982 terms at ECU378 million (revised to ECU721 million, or US$901 million, in 1995 terms).

Aerospatiale received an ECU139.1 million contract in May 1984 for three flight model MOP satellites and one spare. MOP had three primary missions: Earth imaging, dissemination of image and other meteorological data, and data collection/distribution, with two secondary objectives of meteorological processing and data archiving/retrieval. These MOP/Meteosat satellites transmit real-time image data (1,686.833 MHz) with up to 66 channels of DCP data (1,675.181-1675.381 MHz) at 333 kbit/s to EUMETSAT's Primary Ground Station in Fucino for relay to the Mission Control Centre in Darmstadt.

Eumetsat expected to put MOP 2 in service as the prime satellite in Summer 1991, replacing Meteosat 4. It received the first image on 3 April during commissioning at 4° west. However, an imaging anomaly required Eumetsat to make modifications to ground processing software. Then the satellite entered full service as prime in February 1994. ESA formally transferred the satellite to EUMETSAT 14 January 1992.

Then MOP 3 followed MOP 2 as part of the programme after it transmitted its first image (visible) on 29 November 1993. An onboard anomaly in the infra-red/water vapour imagery was detected in further testing, caused by an apparent movement of the cold optics within the radiometer. Additional ground processing software corrects the anomaly sufficiently to allow normal extraction of image products.

Darmstadt processes MOP images into a range of formats covering Europe only, or covering most of the full Earth disc with up to 24 frames of data. Final processed images include lat/long grids and coastlines which are added by transmission up to Meteosat (channel 1 2,101.5 MHz; channel 2 2,105.0 MHz) and relayed to users on two channels. Channel 1 (1,691.0 MHz) operates on conventional analogue WEFAX, and is compatible with other GEO metsats and NOAA's Automatic Picture Transmission service. This information is available to simpler Secondary Data User Stations. Channel 2 (1,694.5 MHz) provides high-resolution digital transmissions to Primary Data User Stations. Some of NOAA's GOES images are also available via Meteosat using the Lannion, northwest France relay station. Three other channels (1,695.725, 1,695.756, 1,695.787 MHz) primarily relay digitised facsimile and selected conventional met observations to

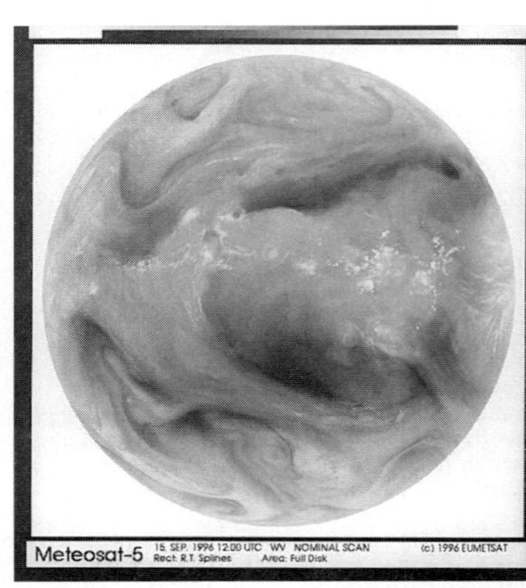

A three-channel Meteosat image is returned every 30 minutes: visible (left), infra-red (centre) and water vapour (EUMETSAT) ***1997***/0008086/7/8

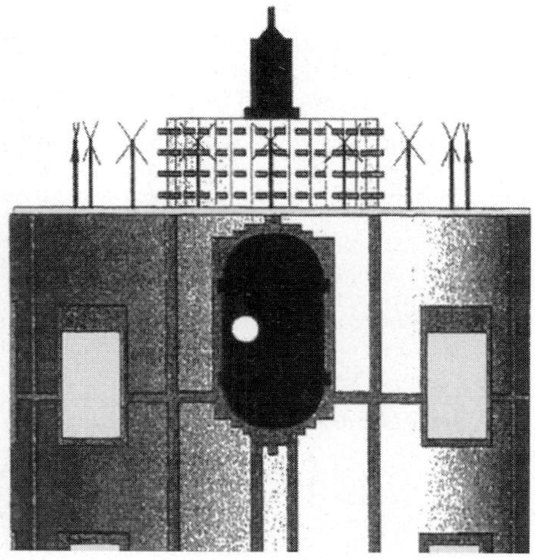

Meteosat Second Generation will retain the spin-stabilisation of the smaller current satellites ***1996***

Africa, as the Meteorological Data Dissemination (MDD) service.

In 1996, DCP messages received by Meteosat totalled 3,311,644, allowing a distribution of more than 1.1 million images to meteorological services around the world. The MARF Meteorological Archive and Retrieval Facility holds, as the single repository, all Meteosat image data and derived products acquired since 1978. Live data continuously enters the archive and Eumetsat copies historic data for later retrieval. All of the data are recorded on 6.6 Gbyte optical disks, with online access through the MARF's catalogue. During MOP, ESOC gathered some 40,000 tapes' worth of data, dating back to Meteosat 1.

The Meteosat Transition Programme began in 1991 when Eumetsat planned to orbit one Meteosat follow-on satellite, identical to MOP, to maintain a continuous service until MSG appears. The organisation signed FFr630 million contract in June 1993 with Aerospatiale to cover one satellite plus components for a second. ESA manages satellite procurement on EUMETSAT's behalf, but EUMETSAT has responsibility for the ground segment, launch and operations.

A new ground segment was developed by EUMETSAT to take full control of all satellites after 1 December 1995. It includes the MCC Mission Control Centre in Darmstadt and Primary/Secondary Ground Stations, replacing the system previously operated from ESOC. The PGS Primary Ground Station is at Fucino, Italy and includes facilities for operational support of two satellites. A high-speed link connecting PGS/MCC transmits data, telemetry and satellite commands. The back-up station is at Weilheim, Germany.

EUMETSAT budget
(ECUmillion)

	1994	1995	1996	1997
MOP	30.5	27.3	–	–
MTP	59.4	55.4	31.0	34.916
MSG	42.0	57.2	86.6	132.507
EPS polar	13.8	11.3	10.8	63.657
General budget	15.0	15.6	22.4	12.738
Total	**160.7**	**166.8**	**150.8**	**243.818**

1993 131.916; 1992 91.720; 1991 64.847; 1990 50.841; 1989 42.8; 1988 59.25; 1987 62.763. Staff total 141 by end-1996.

The Meteosat Second Generation (MSG) Phase C/D plan began in July 1995. Eumetsat planned to launch its first satellite under this programme in the fourth quarter of 2000. Eventually the organisation plans to launch three satellites, the second after 18 months and the third

The first satellite in the second generation of meteosats, (MSG-I), is shown here in Aerospatiale's facilities in Cannes during instrument integration (Aerospatiale) ***1998***/0038496

as needed. Arianespace signed an Ariane 5 launch contract for all three, worth more than ECU200 million on 27 July 1995. ESA maintains an active role in the programme as the developer of the first, prototype satellite. In addition to a financial contribution to that satellite, EUMETSAT will fund the three launchers, the next two satellites, the procurement of which has been entrusted to ESA, develop the ground facilities and provide operations until 2012.

MSG's mission is to provide basic multispectral imagery. Improved spatial, temporal and spectral resolution, for weather forecasting. It must also provide high resolution imagery. With AVHRR-type spatial resolution in visible band, for mesoscale convective monitoring over Europe. The information provided includes air mass analysis for Water/CO_2 absorption and meteorological product extraction for improved wind, and temperature data. Eumetsat intends to emphasise support of climate and environment monitoring, continuity of data collection and dissemination and the relay of search and rescue distress signals and observation of the global Earth radiation budget.

Metop, the Meteorology Operational Polar satellite, provides the European component of a joint NOAA/ EUMETSAT polar system that aims to provide continuous meteorological global observation from Sun-synchronous LEO. NOAA will launch the morning and afternoon (local time) orbits and EUMETSAT will eventually become responsible for the morning element.

Metop 1, after it is launched in 2001, will be the first in a series of operational satellites: Metop 2 follows in 2007 and Metop 3 in 2011. The preparatory programme, which prepared the proposals for the full EPS programme of ECU1.5 billion (1994 rates), began in January 1994.

Eumetsat implemented a technical means of controlling access to Meteosat HRI High-Resolution Image data on 4 September 1995. Now the satellite encrypts the data normally received by PDUSs, while the analogue (WEFAX) transmissions remains unencrypted. Over the past decade, a rapid increase in value-added commercial activity by the private sector has developed, based on meteorological forecasts and data provided by meteorological services. Simultaneously, government satellite organisations have brought increasing pressure on space resources to recover their costs by charging for services. The encryption of data began an effort to register all users of the high-resolution Meteosat data. Regular test transmission of encrypted data began in 1994. Decryption requires an MKU Meteosat Key Unit from EUMETSAT and some adaptation to the PDUS for MKU interfacing. Some exceptions exist. For example, HRI images at 00/06/12/18 GMT remain unencrypted, and the 3 hourly encrypted data are free to all National Meteorological Services. In general, the NMS of countries with a GNP per capita less than US$2,000 have free access to hourly and half-hourly data for internal use; wealthier countries pay according to their GNPs. Educational and science programmes have free access to all data. At the beginning of 1996, there were more than 300 PDUS and almost 2,000 SDUS.

Specifications

MOP-2/Meteosat 5

Launched: 2 March 1991 by Ariane V42 from Kourou, French Guiana (lifetime 5 years)
Mass: 681 kg at launch, 316 kg after firing of Mage apogee kick motor (including 39 kg hydrazine thruster propellant, with 2 kg conserved for removal from GEO at EOL)
Contractors: Aerospatiale (prime), MMS (radiometer, AOCS), Bosch Telecom (transponders), Alenia (TT&C), ETCA (power supply), MBB (structure, thermal)
Resolution: 2.5 km visible, 5 km water vapour/TIR
Orbit: 0° west geostationary
Spacecraft: 2.1 m diameter, 3.195 m high stepped cylindrical body with solar cells on six main body panels providing 300 W at beginning of 5 year life (200 W EOL). Spin-stabilised at 100 rpm around main axis aligned almost parallel to Earth's axis, with spin regulated by two small hydrazine thrusters. Two pairs of larger thrusters provide spin axis precession/inclination control and east-west station-keeping, respectively. East-west drifting of ±1° about 0° longitude is permitted for the prime satellite. Attitude information provided by pairs of Earth horizon + Sun-slit sensors. Meteosat's smaller cylinder carries radiating dipole antenna elements activated sequentially (that is, electronically despun) for 333 kbit/s S-band (1.670-2.300 GHz) image transmissions and TT&C operations (telecommand 2.0980 GHz; telemetry 1.675928 GHz); the protruding antenna provides toroidal-pattern S-band and low UHF links. Meteosat's other pole houses the radiometer's passive cooler open to space; the radiometer telescope itself scans along the satellite's equator
Sensors: Single Imaging Radiometer operating in three visible-infra-red bands and providing a full-disc Earth image in 25 min, followed by a 5 min retrace/ stabilisation period. Each infra-red image comprises 2,500 lines (each of 2,500 pixels) but the two

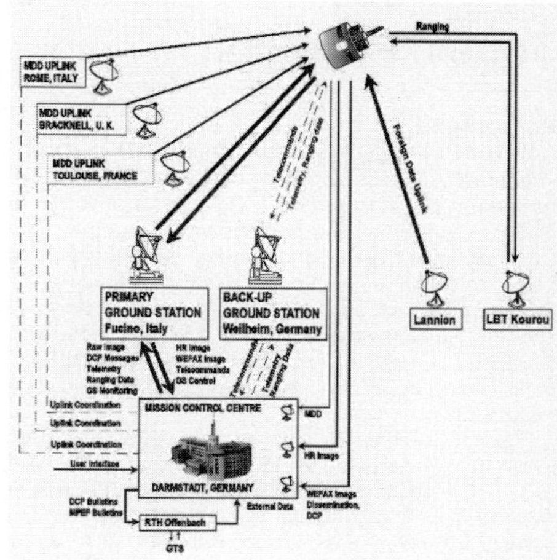

EUMETSAT ground segment from December 1995 (EUMETSAT)

simultaneous visible detectors provide 5,000-line images, corresponding to a resolution of 2.5 km. All four detectors are redundant. The two visible Si photodiode detectors cover 0.4– 0.9 μm, the thermal-infra-red HgCdTe detector 10.5-12.5 μm, and the water-vapour/ infra-red HgCdTe 5.7-7.1 μm, continuing the pre-operational Meteosat data
Optics: Ritchey-Chrétien 40 cm primary aperture, 365 cm fl telescope mounted on two flexible plate pivots attached to a step motor by a high precision jack screw via a gear box. The telescope is stepped 0.125 mrad every 100 rpm satellite rotation so that Earth's surface is scanned at 5 km intervals south to north, covering a total angular distance of 18°.

MOP-3/Meteosat 6

Launched: 20 November 1993 by Ariane V61 from Kourou, French Guiana (lifetime 5 years)
Mass: 704 kg at launch, 316 kg on-station BOL
Orbit: 10° west geostationary as standby
Other specifications are the same as Meteosat 5.

Meteosat Second Generation

Launch: MSG 1 planned for fourth quarter 2000 by Ariane-5 from Kourou; MSG 2 second quarter 2002 by Ariane-5, MSG 3 as required
Mass: about 1,750 kg at launch
Orbit: 0° geostationary
Contractors: The satellite is developed by an industrial consortium led by Aerospatiale, with MMS responsible for the SEVIRI instrument
Resolution: 1 km visible, 3 km infra-red
Spacecraft: the satellites will continue 100 rpm spin stabilisation, though considerably larger: 3.2 m diameter, 3.7 m high. Unified bipropellant system. 750 W BOL from Si panels. 7 year life. The communications package's three channels will downlink 3.2 Mbit/s raw images, housekeeping and relayed DCP data to the primary ground station, and retransmit processed imagery and other data to users in high-/low-rate data schemes (up to 1 Mbit/s without compression). MSG's data volume will be an order of a magnitude greater than the current system. Accomodation of the Search & Rescue package has been approved by EUMETSAT
Sensors: SEVIRI Spinning Enhanced Visible/IR Imager, scanning Earth's disc every 15 min in 12 channels. The 0.50-0.90 μm HRV High-Resolution Visible channel will provide 1 km resolution. The other 11 will return images of max 3,750 lines of 3,750 samples, corresponding to 3 km resolution visible 0.56-0.71, 0.74-0.88, 1.50-1.78; window 3.40-4.20, 8.3-9.1, 9.8-11.8, 11.0-13.00; water vapour 5.35-7.15, 6.85-7.85; ozone 9.38-9.94; CO_2 12.40-14.40 μm. Typical radiometric resolution 0.25K window channels, 0.75 k water vapour and 1.50 k ozone/1.80 k CO_2.

GERB Global Earth Radiation Budget instrument has been approved by EUMETSAT, provided by a UK-led consortium: Natural Environment Research Council; Rutherford Appleton Lab (technical management, data processing, archiving, distribution); Imperial College of Science, Technology and Medicine (PI Prof John Harries, science, calibration); Royal Meteorological Institute of Belgium (optics unit). 89% of the cost has been subscribed; EUMETSAT and the European Commission are expected to fund the remainder. Together with SEVIRI, GERB will enable study of water vapour and cloud forcing feedback, two of the most important (and poorly understood) feedback processes in climate prediction. It will measure Earth's long-wave and short-wave radiation every 15 min to an absolute accuracy of about 1 W/m^2. Bands: 0.35-4.0, 0.35-30 μm (4.0-30 μm by subtraction), 48 km nadir pixel, absolute accuracy 1% SW/0.5% LW, S/N 1,500 SW/LW, dynamic range 0-450/0-130 Wm^{2}SW/LW, <30 kg, <70 kbit/s, <35/40 W avg/peak.

Topex/Poseidon

Background

NASA/JPL's combined its Topex (The Ocean Topography Experiment) with CNES' similar Poseidon payload in FY87 to provide long-term observation of global ocean circulation and surface topography. The satellite carries two Ku-band altimeters to determine sea height to within 2 to 5 cm. NASA's dual frequency (C/Ku) altimeter built on experience from NASA Goddard and the Applied Physics Lab at The Johns Hopkins University of 1978's Seasat instrument. The CNES Ku-band device is a new design employing solid-state technology.

Topex was originally planned as a US$270 million NASA mission, targeted for 1989 launch, as a successor to Seasat. In 1985, NASA merged Topex with Poseidon. In April 1984, JPL awarded an eight month, US$1 million satellite definition contract to Fairchild, RCA Astro-Electronics and Rockwell based on their respective Solar Max, Tiros/DMSP and Navstar platforms. But Congress refused to grant new money in FY86 for the FY87 budget. NASA then issued requests for a proposal to the same three companies, in July 1986 with the intention of selecting a single satellite contractor in December. Fairchild won in January 1987, basing their design on the company's successful Solar Max multimission. Topex became the first complete spacecraft built by Fairchild. The total satellite cost ended 42 per cent over the US$121 million initial budget, largely due to NASA imposing more stringent requirements.

NASA projected total costs including launch and tracking at US$400 million in 1985 dollars and US$125 million for CNES. France's independently estimated its costs in 1991 to be about FFr900 million.

Topex/Poseidon flies as part of the World Ocean Circulation Experiment (WOCE), a major oceanographic field programme under the auspices of the World Climate Research Program. WCRP combines satellite-based data with traditional observations to generate global 3-D ocean current structure model. The operational orbit is phased to overfly a NASA calibration site on the Harvest oil platform near Pt Conception (California) and a CNES calibration site on Lampione Rock near Lampedusa in the Mediterranean (the sites provide independent measures of satellite height and sea level). A three year extended mission began in August 1995 after the US$30 million annual operations costs were reduced to less than US$20 million.

NASA anticipated at first that the inaccuracies in orbital determination would reduce certainty relative to the geoid to 13.4 cm. But the orbit restituted to less than 5 cm (radial) by the DORIS system, orbital determination is no longer the dominant item in the error budget; sea surface height accuracy has proved to be less than 5 cm rms.

Topex/Poseidon: built by NASA and launched by France, carrying sensors from both ***1995***

Specifications

Topex / Poseidon

Launch: 10 August 1992 by Ariane 42P from Kourou. Operational mission began late February 1993

Mass: 2,380 kg in-orbit BOL

Contractors: JPL (project management, Topex payload), Fairchild Space Co (bus), Alcatel Espace (Poseidon altimeter), Honeywell Satellite Systems Div (antenna pointing system, RW assemblies), Dassault Electronique (DORIS receiver)

Altitude precision: 2-3 cm Topex, 2-5 cm Poseidon

Orbit: 1,331 × 1,332 km, 66.05°, repeating within 1 km every 9.9 days (127 revs). This is a 'frozen' orbit, holding a fixed eccentricity and argument of perigee to minimise manoeuvres to maintain the ground track. The selected altitude was a trade-off between gravity anomalies and atmosphere introducing unacceptable perturbations versus increased altimeter power for higher resolution in a higher orbit

Spacecraft: based on NASA/Fairchild standard 3-axis Multi-Mission Spacecraft (MMS) bus designed for 3 to 5 year lifetime and 3.38 kW BOL solar power (2.14 kW after 5 years). 5.5 m length, 11.5 m span, 6.6 m high. Communications via TDRS. Power supply supported by 3 × 50 Ah batteries. Attitude control (nadir pointing) by RWs + magnetorquers; attitude determination by Earth/Sun sensors, star trackers, magnetometers, 3-axis gyros. Orbit control by 1/22 N hydrazine thrusters providing total ΔV 176 m/s (sufficient for 12 years)

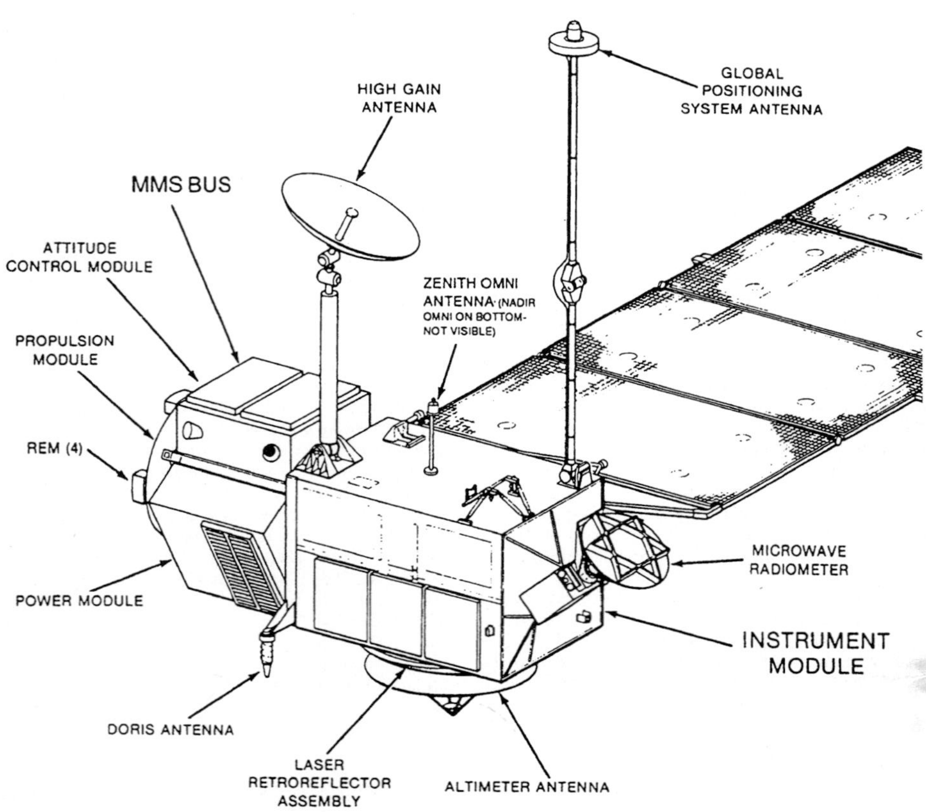

Principal Topex/Poseidon features

Sensors: Topex/Poseidon carries two altimetry systems, a microwave radiometer to correct for atmospheric effects, and three tracking systems:

Topex (US) Radar altimeter: operating with a prime channel at 13.6 GHz (Ku-band) and a secondary at 5.3 GHz (C-band), the two frequencies permitting corrections for ionospheric effects. The US/French altimeters share one 1.5 m dia dish, providing a 3 dB beamwidth of 2.7/1.1° at C/Ku. US radiated power 2/9 W C/Ku. Topex mass 206 kg, consumed power 237 W. Microwave radiometer: 18/21/37 GHz to correct altimetry data for tropospheric water vapour effects. Orbit determination: operationally by TDRS, precision (science) by laser reflectors (13 cm radial accuracy) and experimental NASA/Motorola GPS Demonstration Receiver operating at 1.2276/1.5754 GHz, using a new technique of GPS differential tracking for accuracy of a few cm.

Poseidon (France) Radar altimeter: solid-state instrument operating on a time-share basis with US version through same 1.5 m dia dish at single Ku-band centred on 13.65 GHz, bandwidth 330 MHz, pulse duration 100 μs, pulse repetition frequency 1,700 Hz. Radiated power 5 W, power consumption 49 W. An 80C86 microprocessor controls the instrument. Microwave unit size 36.5 × 28.8 × 23.2 cm; 13 kg. Processing unit size 35.5 × 25.5 × 20.3 cm; 12 kg. Orbit determination: Doppler ranging by DORIS payload operating on 2,036/401 MHz uplink (10 cm radial accuracy).

TRMM

Current status

The co-operative Tropical Rainfall Measuring Mission, launched 27 November 1997, entered orbit at 367 × 385 km and 35° inclination.

Background

TRMM plans to investigate the interactions of water in all three of its physical phases to determine the contribution it makes to the Earth's process. Water substantially alters incoming and outgoing radiation and affects global air motions and heat fluxes through condensation and evaporation. Moreover, the latent heat released by tropical rain provides 75 per cent of the energy the atmosphere receives, thus playing a key role in driving global atmospheric circulation. Yet scientists know the amount of tropical rainfall to within only a factor of two over the oceans and little better, if any, over the arid continents and jungles. As a result, five of the leading climate models differ widely in their estimates of rainfall under identical boundary conditions – they differ by factors of two to three for the mean annual rain in the equatorial belt and even more widely for rainfall on a regional scale.

TRMM will provide the first comprehensive precipitation data on a global scale. The rainfall package comprises the PR (the first weather radar to fly in space), the multichannel TMI microwave radiometer and VIRS. TMI improves rain remote sensing of its SSM/I Special Sensing Microwave/Imager predecessor, carried by the military DMSP metsats, by adding a 10 GHz channel. TRMM's data system will operate in collaboration with EOSDIS, which will perform its first collection/archiving of very large space/ground validation data with this mission. CERES and LIS, funded by EOS (LIS was deleted from EOS in 1992), were added later to improve investigation of global change. CERES will measure upwelling cloud radiation, helping to identify how clouds warm/cool the planet. LIS will clarify the relation between cloud electrification and rain processes, and why it apparently differs over oceans and land masses.

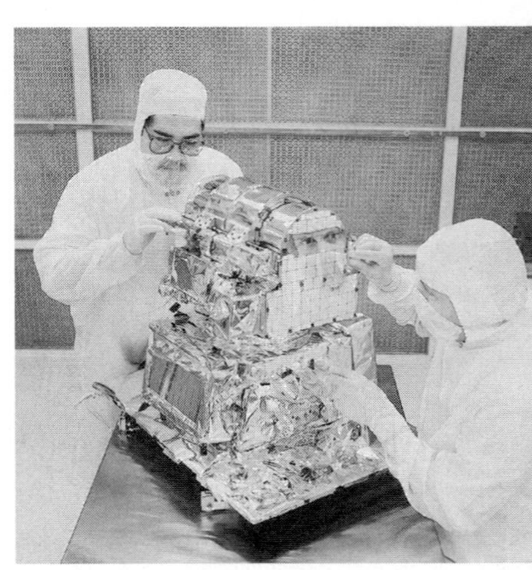

CERES will help to identify how clouds affect Earth's temperature (TRW) ***1996***

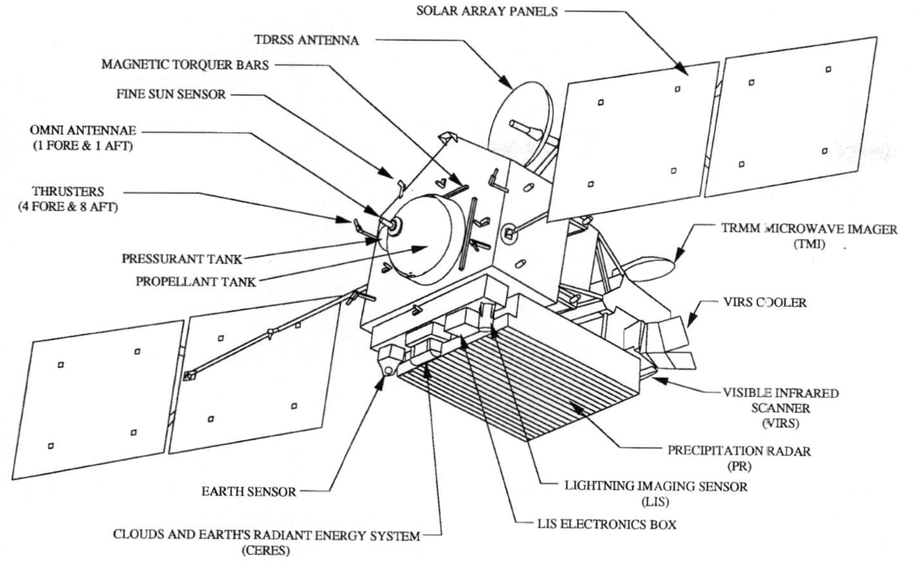

TRMM principal features (NASA) **1995**

TRMM will provide the first comprehensive precipitation data on a global scale. It is the largest spacecraft ever built in-house by NASA Goddard (NASDA) **1994**

TRMM will provide monitoring of tropical rainfall over three years, yielding monthly averages over 5 × 5° (500 × 500 km) cells. The US provided the spacecraft and four instruments (project management at NASA Goddard as an Earth Probe) and Japan launched the satellite aboard an H-2 rocket from Tanegashima. In addition, Japan added one instrument to the payload. NASA approved TRMM in 1990 following a 1986-88 Phase A evaluation programme. NASA put forward FY96 funding of US$23.1 million. (Previous years: FY95 48.1; FY94 63.1; FY93 51.5). NASDA for its part put up FY96 ¥3,630 million. (Previous years: FY95 4,964; FY94 2,537.5; FY93 919.1; FY92 874.) NASA has proposed a similar follow-on mission is being studied, flying at 55° to cover all the continental US.

Specifications

TRMM Satellite

Launch: 27 November, 1997 by H2 from Tanegashima, Japan
Mass: 3,620 kg (725 kg hydrazine to maintain altitude to ±1.25 km)
Orbit: planned 350 km circular, 35°
Spacecraft: 3-axis (0.2° knowledge) by Earth sensor (primary attitude reference), coarse/digital Sun sensors, gyros, four Ithaco 75 Nms RWs, magnetorquers, hydrazine thrusters, 1.1 kW min from twin GaAs solar wings + 2 × 50 Ah Ni/Cd batteries, S-band via TDRSS: data 32 kbit/s real-time; 2.048 Mbit/s playback. Instrument data rate 166.4 kbit/s. Two 264 Mbyte SSRs
Sensors: PR Precipitation Radar (NASDA, Japan; Toshiba prime). 0.7 mm/h measurable rainfall. 13.796/13.802 GHz, pulse width 1.67 μs, pulse repetition frequency 2.776 Gz, peak power 578 W, beam width 0.71°, cross track scan angle ±17°, 93.5 kbit/s, 128 slotted waveguide antennas (using SSPAs) in 2.3 × 2.3 m planar array, mass 470 kg, power 250 W.

VIRS Visible Infra-Red Scanner (NASA). Channels (μm): 1 0.63, 2 1.6, 3 3.75, 4 10.8, 5 12. IFOV ±0.210° at 350 km, ±45° cross track. 2 km res at 10 kbit/s/channel. Mass 49 kg, power 36 W.

TMI TRMM Microwave Imager (NASA). 9 channels (GHz): 10.65, 19.35, 21.3, 37, 85.5, mass 53 kg, power 43 W, data rate 8.5 kbit/s

CERES Cloud/Earth Radiant Energy System (NASA EOS; TRW). Channels (μm): 1 0.3-3.5, 2 8.0-12.0, 3 0.3->50, scan angle ±42.5° in 6.6 s cycle, 8 kbit/s, IFOV 25 km nadir, mass 45 kg, power 45 W.

LIS Lightning Imaging Sensor (NASA MSFC in-house). 0.7774 μm, scan angle ±42.5° cross track, res 5 km nadir, 0.5 kbit/s (6 kbit/s at full background), mass 20 kg, power 24 W.

JAPAN

ADEOS/MIDORI series

Background

NASDA's 1996 Advanced Earth Observation Satellite (renamed Midori, 'Green', and once in orbit) is a modular platform upon which the ADEOS 2 will be launched as well in ADEOS 2 in 2000. As part of its science and remote sensing research effort NASDA set aside ¥542 million for ADEOS 1 in FY88 to initiate the programme. Phase C/D began August 1990, following sensor selection in August 1989. Following a critical design review in 1991, integration and testing of the Engineering Model and the Structural Thermal Model began 1992.

MELCO, Nippon Electric and Toshiba submitted bids by the 14 April 1990 deadline in time for NASDA to announce the prime contractor in the summer. The Japanese government issued a shared contract in the aftermath of the CS-4 programme cancellation. NASA offered its Scatterometer to provide accurate measurements of ocean surface winds following NASA's successful demonstration of the technique by Seasat during 1978. ADEOS also carries NASA's TOMS Total Ozone Mapping Spectrometer, extending observations from Nimbus 7 and Russian Meteor 3, and 1996's Earth Probe TOMS. The satellite failed in orbit 30 June 1997 due to problems with the solar panel system used to generate electrical power.

The ¥276 million request to start ADEOS 2 was denied for FY92/FY93, but approved by the SAC in 1993 for FY94. Japan allocated ¥7.667 billion FY95, ¥9.896 billion FY96. NASA is developing SeaWinds (similar to NSCAT but with a mechanically scanning 1 m diameter antenna) to avoid a break in scatterometer data until its own EOS appears 2002

Specifications

ADEOS 1

Launch: 17 August 1996 by H2 from Tanegashima
Mass: 3,560 kg in-orbit BOL (1,300 kg instruments)
Contractors: Mitsubishi Electric, Nippon Electric and Toshiba sharing development contract; Odetics (DDS-6000EC tape recorders), Sodern (STD-16 Earth sensors) BAe Space Systems (IMG cryogenic coolers)
Resolution: from 8 m visible to 700 m thermal IR
Orbit: planned 797 km 98.59° Sun-synchronous, repeating every 41 days (585 revs), descending node 10.30 local time
Spacecraft: modular, providing minimum 3 year lifetime. 3-axis control (within 0.3° + 0.003°/s per axis) provided by zero momentum strapdown system derived from ETS-6; four RWs, two magnetorquers, inertial reference unit, 4 × 20 N + two sets 8 × 1 N RCS hydrazine thrusters supplied from three 55 cm diameter spheres. Single 3 × 24 m GaAs flexible paddle generates 4.5 kW EOL, supported by 5 × 35 Ah Ni/Cd batteries (max depth of discharge 20%). OBC stores 7-day batches of commands; 500 bit/s command uplink and 4,096 bit/s housekeeping down on 2.22 GHz S-band. High-rate data (AVNIR, OCTS, POLDER, IMG, and ILAS) stored at 3/6/60 Mbit/s (playback 60 Mbit/s) on three tape recorders. Low rate (NSCAT, TOMS, and TEDA) at 4,096 bit/s (playback 517 kbit/s) on two recorders (backed up by single housekeeping recorder, record 4,096 bit/s, and playback 33 kbit/s). Data transmitted on three 8.150-8.350 GHz X-band links: X1/X2 each 40 W SSPA 60 Mbit/s; X3 5 W SSPA 6 Mbit/s. ADEOS will also demonstrate IOL links, up to 1126 Mbit/s through COMETS, using 2.2875 GHz S-band and 25.850 GHz Ka-band. Low resolution OCTS real-time data transmitted on 465.00 MHz at 23 kbit/s to local users such as fishing vessels
Sensors: Advanced Visible/Near-Infra-red Radiometer (AVNIR). NASDA core instrument, prime contractor MELCO. Three visible and single near-infra-red band complemented by PAN band. 230 kg, 300 W, 60 Mbit/s. Resolutions: 16 m XS + 8 m PAN for 80 km swath. Si CCD array, electronic scanning. 5.7° FOV, can point 40° either side of ground track. Bands (μm): 0.42-0.50, 0.52-0.60, 0.61-0.69, 0.76-0.89; 0.52-0.69 PAN.

Ocean Colour & Temperature Scanner (OCTS). NASDA core instrument, NEC prime contractor. Three thermal-infra-red, one middle-infra-red, two near-infra-red and six visible channels. 360 kg, 315 W, 3 Mbit/s. Rotating scan mirror can tilt 20° forward/back along ground track to minimise Sun glitter. Resolution 700 m, swath 1,400 km. Bands (μm): 0.402-0.422, 0.433-0.453, 0.479- 0.500, 0.511-0.529, 0.555-0.575, 0.660-0.680, 0.745-0.785, 0.845-0.885, 3.55-3.88, 8.25-8.80, 10.3-11.4, 11.4-12.7.

Scatterometer (NSCAT). NASA/JPL active microwave radar for near-surface wind speed/direction over ocean accurate to 2 m/s + 20°. 13.995 GHz. 300 kg, 275 W, 3,178 bit/s. Spatial resolution 25 m. Radar cross section measured at 3 azimuth angles in both 600 km swaths.

Total Ozone, Mapping Spectrometer (TOMS). NASA Goddard mapping UV polychromator for total O_3 and SO_2 monitoring. 35 kg, 25 W, 700 bit/s. 3° resolution, swath 2,795 km. Wavelengths (μm): 0.3086, 0.3125, 0.3175, 0.3312, 0.3398, 0.3600.

Polarisation/Directionality of Earth's Reflectance (POLDER). CNES Toulouse. CCD detector behind rotating wheel of filters/polarisers. 33 kg, 40 W, 900 kbit/s. Swath 1,440 × 1,920 km. Wavelengths (μm): 0.443, 0.670, 0.865 at three polarisations; 0.443, 0.490, 0.565, 0.763, 0.765, 0.910.

Interferometric Monitor for Greenhouse Gases (IMG). MITI (Toshiba prime). Michelson interferometer for observation of CO_2, CH_4, NO and other greenhouse gases. 150 kg, 150 W, 600 kbit/s. 8 km resolution on 20 km swath. 3.3-4.3, 4.0-5.0, 5.0-14.0 μm.

Improved Limb Atmospheric Spectrometer (ILAS). Environment Agency. Trace constituents over 10-60 km (2 km resolution) at high latitudes measured by separate packages observing sunrise/sets. Each houses 12 cm diameter telescope with 44 pyroelectric detectors for 6.21-11.77 μm and 3 cm telescope with photodiode array for 0.753-0.784 μm. 130 kg, 80 W, 500 kbit/s.

Retroreflector In Space (RIS). Environment Agency. 50 cm diameter corner cube laser reflector to monitor O_3, fluorocarbon, CO and other constituents using 0.3-14 μm ground laser. 45 kg.

Technical Engineering Data Acquisition (TEDA) payload monitors the space environment and its effect on the spacecraft with Heavy Ion Telescope, Single Event Upset Monitor, Dosimeter, Contamination Monitor, Potential Monitor.

ADEOS 2

Launch: 2000 by H2 from Tanegashima

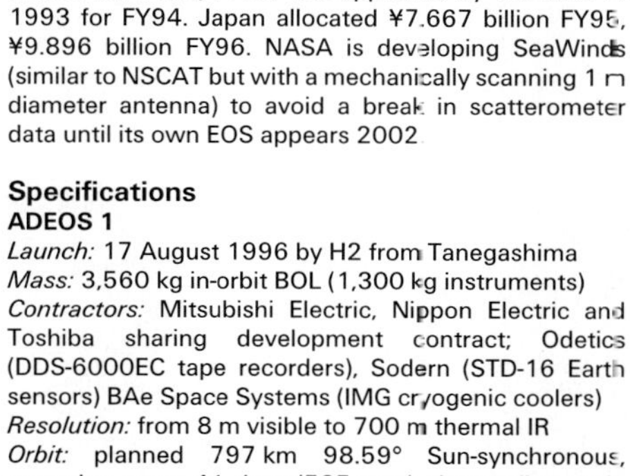

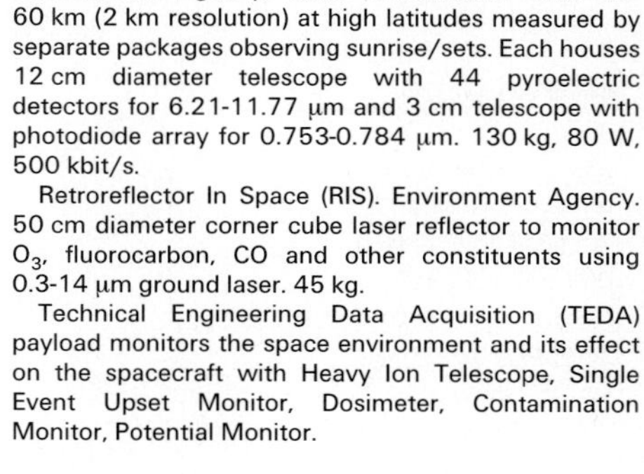

ADEOS instruments and principal features. 1 NASA Scatterometer, 2 DRTS antenna, 3 UHF antenna, 4 OCTS, 5 RIS, 6 NASA TOMS, 7 IMG, 8 AVNIR, 9 X-band antenna, 10 solar array, 11 mission data-processing unit, 12 ILAS, 13 control and data-handling unit, 14 S-band antenna, 15 POLDER, 16 direct transmit unit (NASDA)

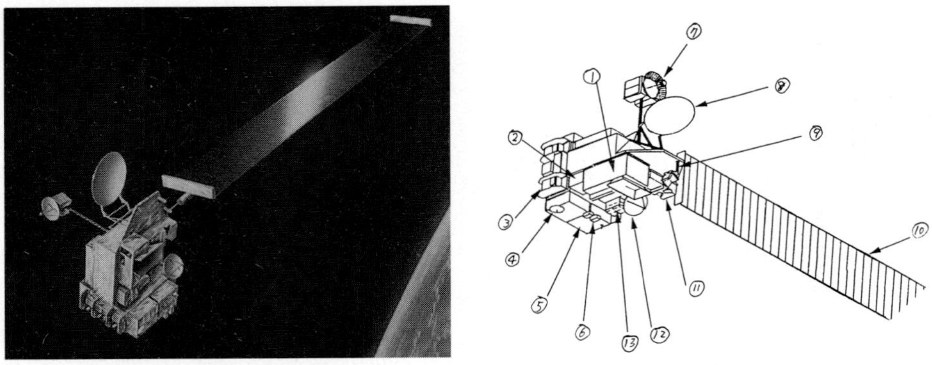

ADEOS 2 instruments and principal features. 1 Global Imager, 2 SeaWinds, 3 mission data processing unit, 4 direct transmit unit, 5 communications/data handling unit, 6 Earth sensor, 7 inter-orbit communications antenna, 8 AMSR scanning antenna, 9 UHF antenna, 10 solar array, 11 ILAS 2, 12 SeaWinds 1 m antenna, 13 POLDER (NASDA) ***1995***

Mass: 3.5 t at launch, 1.2 t payload
Orbit: 803 km 98.6° Sun-synchronous
Spacecraft: modular, providing minimum 3 years (5 year goal) lifetime. 3-axis control, single Si solar paddle providing 5.0 kW (1.2 kW required for instruments)
Sensors: Advanced Microwave Scanning Radiometer (AMSR). NASDA. Advanced version of MOS 1 MSR. Water vapour content, sea surface T + wind, precipitation, sea ice. 8 bands (6.9, 10.65, 18.7, 23.8, 36.6, 50.3, 52.8, 89.0 GHz), 50/50/25/25/15/10/10/5 km + 0.3-2K resolution on 1,600 km swath.

Global Imager (GLI). NASDA. Improved ADEOS 1 OCTS, 36-band visible to thermal-IR passive sensor for water colour, sea T, vegetation index, clouds, aerosols. 1 km/250 m resolution over 1,600 km swath.

SeaWinds. NASA/JPL. Modified NSCAT at 13.402 GHz for sea wind speed/direction. Resolution 2 m/s speed, 20° direction, 50 km spatial.

POLDER. CNES Toulouse. Same as ADEOS 1

Improved Limb Atmospheric Spectrometer 2 (ILAS 2). JEA. As ILAS, with resolution improved to 1 km.

Data Collection System (DCS): Argos compatible (adding platform command capability). CNES/NASDA.

GMS series

Background

The Geostationary Meteorological Satellite (GMS) system provides coverage of the Pacific region from Hawaii to India with a single primary satellite positioned over 140° east. It is Japan's contribution to the Global Observing System of the World Weather Watch. In addition to transmitting a full-disc visible and infra-red Earth image every 30 minutes, the satellite relays meteorological data from ships, aircraft, buoys and other weather stations, and relays ground-processed facsimile images to users.

Based on Hughes Aircraft's GOES design, the series began in 1977 with a Delta launch from Florida but the current satellites are launched from Japan. NASDA procures and launches the satellites according to requirements and funding from the Japan Meteorological Agency (JMA).

GMS-4 became operational at 140° east on 14 December 1989 following in-orbit testing at 160° east. When GMS-4 became operational NASDA moved GMS-3 to 120° east. Then GMS-5 subsequently replaced GMS-4 at the prime position in mid-1995.

GMS-5 acted as the last satellite in the series which the Ministry of Transport expects to replace with the MT-Sat in 1999. On GMS-5, the single infra-red channel of its predecessors is divided into two to improve temperature measurement accuracy, and NASDA added a third for atmospheric water vapour measurements, but the SEM detectors carried aboard GMS 1-4 were deleted. GMS-5 also carries an experimental payload to relay Sarsat/Cospas emergency signals. The Japanese government allocated ¥979 million in FY91, ¥1.138 billion FY92, ¥6.0585 billion FY93, ¥4.5068 billion FY94, ¥93 million FY95.

NASDA controls GMS satellites through the end of their in-orbit checkout phases from its Tsukuba Space Centre facilities and associated tracking stations. Operational control is exercised by JMA's Meteorological Satellite Centre, although station-keeping manoeuvres are commanded on JMA's behalf by NASDA (2.28/2.10 GHz up/down). Raw imagery, data collection platform signals and spacecraft telemetry are received on two 18 m Cassegrain dishes by JMA's Command & Data Acquisition Station at Hatoyama 56 km northwest of Tokyo. They are then relayed by 7.5 GHz microwave link to the Data Processing Centre (DPC). The DPC maintains four mainframe computers: two M-1600/2 (64 Mbyte each) for spacecraft control/operations, and three M-1600/10 (128 Mbyte each) for image processing and subsequent extraction of parameters such as sea T and cloud top height. The smaller machines can assume some of the 1600's processing duties in the event of failure. The Stretched-VISSR and WEFAX data are passed to the CDAS for uplinking and broadcast to users: Medium Data Utilisation Stations (MDUS) can receive S-VISSR but the Smaller Data Utilisation Stations (SDUS) receive the lower-resolution (7 to 8.5 km) WEFAX. Ranging data obtained from CDAS and two Turn-Around Ranging Stations (TARS) on Ishigaki Island (Japan) and Crib Point (Australia) are used for orbit determination.

GMS 5 provides Pacific region coverage from above 140° east via a scanning radiometer
(Hughes Space & Communications)

Specifications

GMS-1 (Himawari 1)
Launch: 14 July 1977 by McDonnell Douglas Delta from Cape Canaveral. 670 kg at launch, 315 kg on-station. Operational from April 1978; replaced by number 2 21 December 1981 and moved to 160° east. Shut down 30 June 1989.

GMS-2 (Himawari 2)
Launch: 10 August 1981 by N2 from Tanegashima. Design life 5 years, mission life 3 years. Operational life ended 27 September 1984; shut down 20 November 1987.

GMS-3 (Himawari 3)
Launch: 3 August 1984 by N2 from Tanegashima. 682 kg at launch, 304 kg on-station BOL. Design life 5 years, replaced as GMS prime satellite at 140° east 14 December 1989 and moved to 120° east. Possibility of moving it to contribute to the US GOES system, then with only a single operational satellite, was considered in 1989. Removed from GEO and shut down 23 June 1995 after GMS 5 became prime satellite for the series.

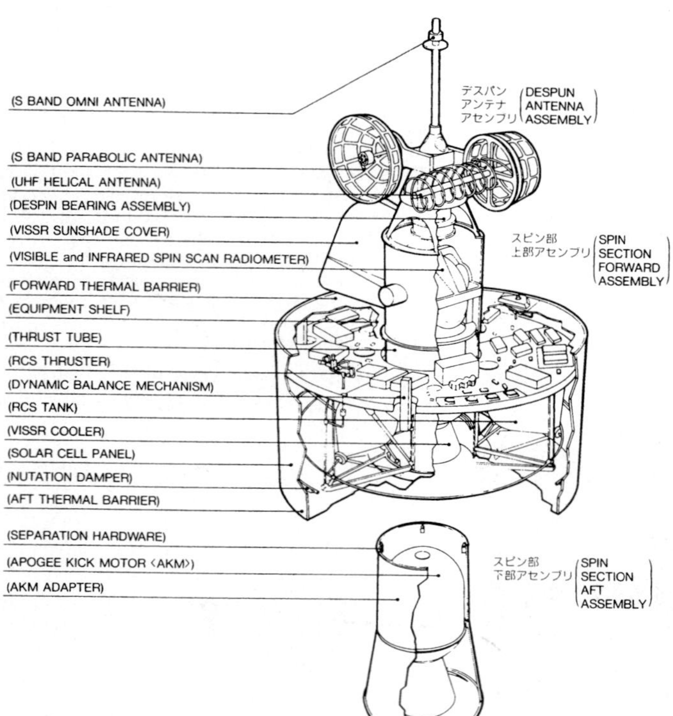

GMS is based on Hughes' spin-stabilised GOES design (NEC Corp)

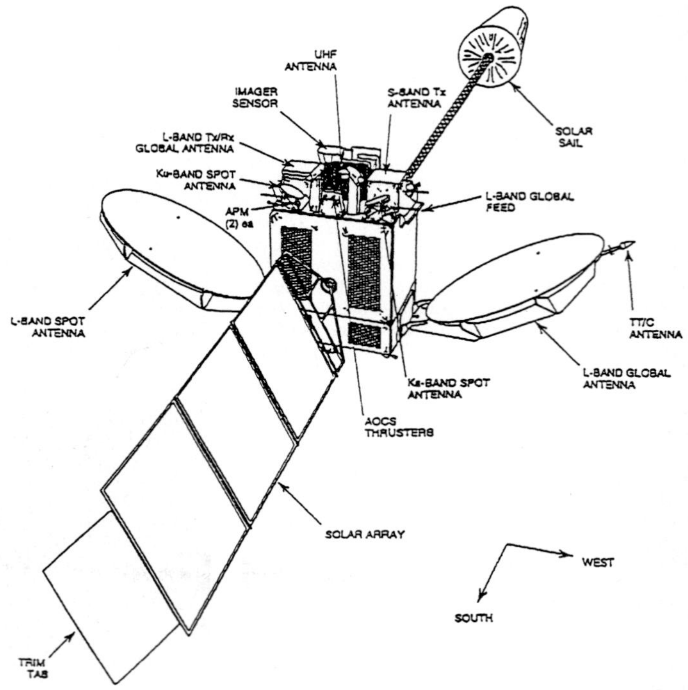

MT-Sat will replace GMS and provide aeronautical communications and navigation services (JRANSA) ***1996***

GMS-4 (Himawari 4)
Launched: 6 September 1989 by H1 from Tanegashima (design life 5 years)
Mass: 725 kg at launch, 325 kg on-station BOL
Orbit: 120° east GEO from July 1995; 140° east December 1989 to June 1995
Spacecraft: Design based on Hughes' US GOES. Reliability is greater than 0.5 after 5 years. 214.6 cm diameter, 345.1 cm high cylinder (after AKM separation, 444.1 cm before) with solar cells on main body generating 300 W BOL (265 W Summer solstice EOL). The cells comprise 4,588,000-7 2.2 × 6.2 cm cells for power generation, and 740 2 × 2 cm cells for charging the two 5.5 Ah Ni/Cd batteries (50 per cent max discharge). 1.7 GHz relays spacecraft imagery at 14 Mbit/s and SEM/telemetry at 250 bit/s PCM and IRIG-B FM-PM to CDAS main control centre. CDAS uplinks commands and processed imagery on 2.03 GHz for broadcast to users on 1.687-1.691 GHz; UHF (402/468 MHz up/down) for data collection platform relay
Sensors: Visible/Infra-red Spin Scan Radiometer (VISSR) with redundant detectors operating in one visible (0.5-0.75 μm) + one infra-red (10.5-12.5 μm) bands. Space Environment Monitor of five Si detectors for routine coverage of solar proton (1-500 MeV), electron (up to 2 MeV) and alpha particle (8-390 MeV) radiation levels in GEO.
Other specifications are the same as for GMS-5.

GMS-5 (Himawari 5)
Launched: 18 March 1995 by H2 from Tanegashima (design life 5 years)
Mass: 746 kg at launch, 345 kg on-station BOL
Contractors: Hughes Aircraft Co (spacecraft prime), NEC Corp (Japanese prime), Hughes Santa Barbara Research Center (VISSR)
Resolution: 1.25 km visible, 5 km infra-red
Orbit: 140° east geostationary by 15 June 1995; tested at 160° east (arrived 30 March)
Spacecraft: HS-378 design based on Hughes' US GOES. Reliability is >0.5 after 5 years. 214.6 cm diameter, 345.1 cm high cylinder (after AKM separation, 444 cm before) with solar cells on main body generating 300 W BOL (291 W summer solstice EOL, 22 W margin). Two 4.8 Ah Ni/Cd batteries. Spin-stabilised at 100 ±1 rpm around main axis, accurate to 0.5°, aligned parallel to Earth's axis with spin regulated by hydrazine thrusters. Two axial + four radial 4.45 N thrusters draw on a 40.8 kg supply of hydrazine. Attitude information provided by Earth horizon + Sun sensors, providing east-west antenna pointing accuracy of 0.39° (Sun sensor)/0.56° (Earth). GEO injection by Thiokol Star 27 solid AKM (providing 1.8 km/s δV) and then separated. GEO position held to within ±1.0° north-south/±0.5° east-west. UHF helical and S-band omni plus parabolic antenna carried on de-spun platform. 1.7 GHz relays spacecraft imagery/telemetry to CDAS main control centre at 14 Mbit/s. CDAS uplinks commands (omni) and processed imagery on 2.026-22.035 GHz for broadcast to users on 1.682-1.698 GHz. UHF (402/469 MHz up/down) for data collection platform relay
Sensors: Visible/Infra-Red Spin Scan Radiometer (VISSR) with redundant detectors operating in one visible (0.5-0.75 μm) + three IR (10.5-11.5, 11.5-12.5, 6.5-7.5 μm) bands, providing full-disc Earth image in 25 minutes followed by 5 minutes retrace/stabilisation. The visible channel comprises four Si photodiodes (+4 back-up), covering 1.1 km, scanning simultaneously. GMS spin provides east-west scanning, motor-actuated scanning mirror covers north-south in 2,500 steps. The HgCdTe pairs of redundant infra-red detectors are radiation cooled to 90K.

JERS satellite

Background
The Japan Earth Resources Satellite is the second domestic remote sensing satellite and the first to operate at all-weather radar wavelengths. It carries a Synthetic Aperture Radar accompanied by an optical sensor package. Japan began a study of this new ERS in 1980, and started preliminary design work on the new satellite in April 1986. The critical design review concluded in July 1989, allowing NASD to proceed from a prototype model to complete assembly during first half of 1990. NASDA funded the spacecraft and MITI the instruments.

Japan's Earth Observation Centre at Hatoyama manages all aspects of the mission control and acts as the primary processing, archiving and distribution centre. NASDA's Tsukuba provides tracking and spacecraft control stations (including an 11 m S-band transportable unit at Sweden's Kiruna). Only EOC and the University of Alaska's Fairbanks station (under a 1988 NASA agreement) receive stored data, while other stations handle only real time. Other stations are EOC, Kumamoto (Japan), Bangkok, Showa Base (Antarctica), Fucino, Kiruna, Maspalomas, Tromsoe, Gatineau, Prince Albert, Beijing and Fairbanks.

The first OPS image was recorded 12 March 1992 over the Sea of Okhotsk. Unfortunately, the two wings of the SAR antenna could not be deployed until 8 April so the first SAR image could not be returned until 21 April 1992. Full operations began formally 1 June 1992. By March 1995, JERS had transmitted more than 140,000 SAR and 90,000 optical scenes.

Japan did not release imagery to the general user community until 1993 because of striping problems on some radar and radiometer images and blurring on the SWIR radiometer. The SAR striping probably developed due to ground radar interference over Japan. VNIR is marred by horizontal striping, which Japanese ground stations now remove by data processing algorithms. SWIR's vertical striping was expected to remain, but the instrument was lost anyway 3 December 1993 because of a cooling fault. It is not expected to return to service.

JERS 3-D reconstruction of Mount Fuji from the high-resolution stereoscopic imagery acquired by NEC's optical sensor. The colour original clearly shows the effect of land development on the natural vegetation. JERS is the first satellite to provide continuous stereo imaging (NEC) **1995**

Specifications
JERS 1 (Fuyo 1)
Launch: 11 February 1992 by two-stage H1
Mass: 1,340 kg in-orbit at beginning of 2 year life; propellant mass 115 kg, instrument payload 497 kg
Contractors: Mitsubishi Electric (prime) and Nippon Electric, Toshiba, IHI, instruments from JAROS Japanese Resources Observation Systems Organisation, Odetics (DDS-6000EC tape recorder)
Resolution: 18 m SAR, 18 m OPS in 75 km swathes
Orbit: 567 × 569 km 97.7° Sun-synchronous, crossing equator at 10.30-11am local time, with 44-day repeat cycle (low orbit requires weekly adjustment)
Spacecraft: 3-axis 0.93 × 1.83 × 3.16 m box-like bus of CFRP struts/frames supporting nine aluminium honeycomb panels. Single 56.4 kg 7.03 × 3.46 m CFRP Solar Array Paddle (SAP) of six panels providing 2,053 W EOL from 22,344 2 × 4 cm Si cells and 4 × 30 Ah Ni/Cd batteries. 11.9 × 2.5 m SAR planar antenna. Zero momentum attitude control system four skewed 20 Nm RWs and magnetorquers, supported by three gyros, Earth/Sun sensors, provide 0.3° accuracy (design 0.1° roll, 0.18° pitch, 0.10° yaw). Redundant RCS: 22 × 1 N catalytic thrusters drawing on 115 kg hydrazine. Heat rejection via 3.6 m² radiator area (21 per cent total surface); 573 W/rev generated when SAR/OPS active. 72 Gbit data recorder holds up to 20 min imaging; 16 Mbit bubble memory for up to 102 minutes telemetry. Transmission at 64 Mbit/s on 8.15 + 8.35 GHz X-band, 50 MHz bandwidth, via 20 W TWTA
Sensors: L-band SAR, 1.275 GHz 1.3 kW peak power, 35 μs pulses, 18 m resolution, 75 km scan width, 64 Mbit/s data rate.

OPS. Optical Sensor covering visible-near infra-red at 18 m resolution (compared to 50 m resolution of MESSR on MOS 1). Seven bands (μm): B1 0.52-0.60, B2 0.63-0.69, B3/B4 0.76-0.86, B5 1.60-1.71, B6 2.01-2.12, B7 2.13-2.25, B8 2.27-2.40. The VNIR Visible/Near-IR Radiometer covers bands 1-4, and 5-8 by the SWIR Short Wavelength IR Radiometer, its detectors cooled to 80K (SWIR bands lost December 1993). Both employ 4,096-element CCDs. Nadir-viewing B3 and forward-looking (15.3°) B4 for continuous stereoscopy.

Japan's ERS: the primary instrument is the 18 m-resolution Synthetic Aperture Radar (NASDA) **1989**

MOS series

Background
The Marine Observation Satellite was Japan's first domestic Earth resources satellite. NASDA launched and operated it to monitor atmospheric water vapour, ocean currents, sea surface temperature, ice floe distribution and chlorophyll content, in addition to supporting land applications and acting as a relay for Data Collection Platforms. Originally NASDA planned to operate the satellite for just two years, with MOS 1B extending coverage following its launch in February 1990. The two could be operated simultaneously and independently.

Japan's Space Activities Commission approved MOS in 1978 as an element of the space development policy in order to acquire expertise for operational systems in the 1990s. NASDA completed MOS 1A's conceptual design and preliminary systems in 1978 and 1979, respectively, followed by the basic design freeze in July 1981. The last N2 booster carried the satellite into orbit in February 1987 but on 28 July 1987 transmissions ceased owing, most likely, to a failure of a pointing sensor. A back-up system activated successfully and the satellite resumed normal operations 4 August 1987.

Data from the Multi-Spectrum Electronic Self-Scanning Radiometer (MESSR), Visible/Thermal Infra-red Radiometer (VTIR) and Microwave Scanning Radiometer linked in real time to a network of seven ground stations worldwide. The primary station was the Earth Observation Centre at Hatoyama north of Tokyo. The other stations were operated by Tokai University (Kumamoto, Japan), National Institute of Polar Research (Showa Base, Antarctica), National Research Council of Thailand (Bangkok), CSIRO (Alice Springs, Australia) and Canada Centre for Remote Sensing (Prince Albert and Gatineau, Canada). RESTEC provides data to commercial users.

MOS 1A ended data operations in April 1995 and was shut down 29 November 1995. The second MOS was fabricated from 1a's prototype model, incorporating minor modifications gained from operational experience. It was shut down 25 April 1996 after battery failure.

Specifications
MOS 1A (Momo 1)
Launched: 18 February 1987 by N2 from Tanegashima
Mass: 738 kg in-orbit at beginning of life
Orbit: 909 km circular in 99.1°.

MOS 1B (Momo 1B)
Launched: 7 February 1990 by H1 from Tanegashima
Contractors: NEC Corp prime/MESSR sensor, MELCO attitude control system + MSR sensor, Fujitsu Corp VTIR sensor, Toshiba Corp solar array, IHI Corp attitude control thruster system
Resolution: 50 m MESSR, 0.9 km VTIR, 2.7 km VTIR thermal, 23 km (31 GHz), 32 km (23.8 GHz) MSR
Orbit: 912 × 920 km, 99.14° Sun-synchronous, crossing equator at 10.00-11.00 local time, with 17-day repeat cycle

MOS 1a was launched in February 1987, followed three years later by MOS 1b (NASDA)

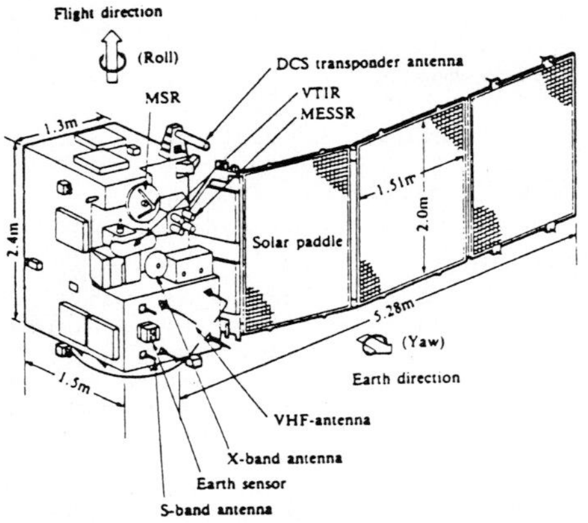

Principal features of Japan's MOS remote-sensing satellite

Spacecraft: 1.26 × 1.48 × 2.4 m high box-shaped satellite of aluminium honeycomb construction with single 2.0 m wide, 5.28 m solar wing of three 1.51 m-wide panels. 3-axis controlled by MWs + 4 × 1 N hydrazine thrusters. TT&C performed by Tsukuba Space Centre on 2 GHz, with MESSR/VTIR data returned in real time on 8 GHz and MSR data on 2 GHz. The Data Collection System Transponder relaying data from ground platforms on a random access basis operates in the 400 MHz band, transmitting to EOC at 1.7 GHz

Sensors: Multispectral Electronic Self-Scanning Radiometer (MESSR). MOS carries two MESSRs, each covering two wavelengths with 2,048-element CCDs having a total swath width of 200 km. The data rate of each 70 kg, 40 W instrument is 8 Mbit/s. Spectral ranges (in μm) for the four bands are: Band 1 0.51-0.59; Band 2 0.61-0.69; Band 3 0.72-0.80; Band 4 0.80-1.1. Applications include water turbidity of coastal areas, red tide, ice distribution, water resources, land use, vegetation, snow cover distribution, volcanic ash, geological structure.

Visible/Thermal Infra-red Radiometer (VTIR). A mechanical scanning radiometer with Ritchey-Chretien optics; the mirror scan rate is 7.3 rev/s. Visible band sensing is performed by Si photodiodes, with HgCdTe elements used for the three IR-bands. Data rate of the 25 kg, 35 W instrument is 800 kbit/s. Swath width is 1,500 km. Spectral ranges (in μm) with associated applications are: Band 1 0.5-0.7 (snow cover, ice, cloud); Band 2 6-7 (upper atmosphere vapour, cirrus clouds); Band 3 10.5-11.5 (snow cover, ice, cloud, ocean currents, sea surface T); Band 4 11.5-12.5 (applications as Band 3). Swath width 1,500 km.

Microwave Scanning Radiometer (MSR). Dicke-type radiometer operating at 23.8 GHz (resolution 32 km) + 31.4 GHz (23 km resolution) with the offset parabolic reflector rotating at 18 rpm providing the scanning action. Data rate of the 54 kg, 60 W instrument is 2 kbit/s; swath width 317 km. Spectral ranges (in GHz) with applications are: Band 1 23.8±0.2 (water vapour, rainfall area, snowfall) and Band 2 31.4±0.25 (ice, cloud water, snowfall). MSR is no longer operational.

RUSSIAN FEDERATION

Almaz

Background

The Soviet Union introduced a new class of Almaz ('Diamond') remote sensing satellites in 1987 with the launch of the 18 ½ tonne Kosmos 1870. With this test satellite the country intended to demonstrate the satellite's potential, in particular, techniques for oil pollution and ice monitoring. It carried a synthetic aperture radar which provided 30 m resolution all-weather coverage from the slotted waveguide antenna. The first satellite in the series, Almaz 1 appeared March 1991 featuring several improvements over older generation Earth monitoring platforms. The new SAR doubled resolution to less than 15 m and it transmitted data digitally through satellite links (C1870 employed optical processing). After launch the port antenna did not fully deploy, making the production of stereo pairs between 40° north and south difficult.

The Soviet Union originally designed Almaz as a military reconnaissance craft and reportedly intended to launch it in 1981. It was developed by Vladimir Chelomei's bureau, which was also responsible for Proton, the military Salyuts and the Heavy Kosmos modules, and predecessor to NPO Machinostroenye. The Almaz derives directly from the manned military Almaz (flown under the Salyut designation). A disagreement between chief designer Chelomei and Minister of Defence Dmitri Ustinov resulted in cancellation of the programme three days before the planned July 1981 launch and all work halted the following December. Both men died in December 1984 before the programme resumed.

Kosmos 1870 launched 25 July 1987 into 237 × 249 km, 71.9°, preceded the Almaz as a prototype. It required corrections every 12 days to combat orbital decay. In September 1988, ground controllers raised the height to 270 km and corrections switched to a 24-day cycle. The platform was commanded to re-enter 29 July 1989 after an apparently successful demonstration mission. It appears there was also a 1986 Almaz launch, possibly using 1981's vehicle, but the Proton failed. A TV interview with Machinostroenye officials in October 1991 noted that the 1981 Almaz was left at Tyuratam after the cancellation, untouched because it still carried destruct charges, until work restarted in January 1985.

As the Soviet Union broke up in late 1990, Space Commerce Corporation of the US sought exclusive western marketing rights to the data under a joint venture agreement with GlavKosmos and spacecraft manufacturer NPO Machinostroenye. These entities formed the Almaz Corporation to handle the marketing (NPO Machinostroenye had its own Almaz foreign trade company for marketing).

To prepare an Almaz 1 image, SSC had data transmitted via two articulated phased array panels to a Luch GEO satellite for relay to the Data Processing & Customer Support Centre in Moscow, which could handle 100 scenes daily. Imagery cost US$1,600 per 40 × 40 km scene on magnetic tape or floppy disk. Hughes STX signed a further agreement with NPO Machinostroenye for the distribution, processing and analysis of Almaz SAR data. Products included mosaics, annotated mapsheets, custom colourisations, enhancements, interpretation packages and georeferencing.

In December 1992, these companies halved the cost of a 40 × 40 km scene to US$800. By end of 1991, some 60 orders had been received, insufficient to break even, since the company had targeted a goal of US$2 million in sales in 1992. By the time of Almaz's re-entry, only a few hundred images had been sold. Hughes STX had sold about 175 for US$250,000 revenue. In addition to the commercial marketing, primary users could also receive data directly from Almaz using 2 m dishes, although no open source primary users ever appeared.

Russian officials claimed the Almaz 1 programme cost Rb300-400 million. At launch, its supporters sought a minimum 30-month operational life, but increased solar activity required frequent orbit raising, depleting propellant and requiring a controlled re-entry over the Pacific 17 October 1992. Financial difficulties prevented the launch of a second Almaz immediately. The bus for the second satellite is 60 per cent complete and is stored at Khrunichev awaiting funding. A new entity, the joint US-Russian SAR Corp (Sokol Almaz Radar) seeks financing, equivalent to US$126 million. SAR is 49 per cent owned by the US Sokol Group Inc and 47 per cent by Machinostroenye.

Specifications

Almaz 1

Launched: 31 March 1991 by Proton from Tyuratam
Mass: 18.55 tonne on-orbit (4 tonne instrument payload)
Resolution: 15-30 m
Orbit: typically 275 km, 72.7°, revisit 1-3 days. orbit adjusted typically every 24 days
Spacecraft: derived from Mir/Salyut core, about 15 m long, 4.15 m max diameter, two solar arrays total 86 m^2 (average power 2.4 kW), providing up to 10 kW for 20 minutes from the batteries. 3-axis control. Interior pressurised by N_2 and regulated to 5-35 ±1°C. Attitude stability provided by 6 rpm 3 m diameter RW
Sensors: 3.1 GHz synthetic aperture radar using two 1.5 × 15 m slotted waveguide antennas, horizontal polarisation, 190 W peak/ 80 W average pulse power, 0.07/0.1 ms pulses, 3,000 Hz repetition, 3-5 dB pixel accuracy, 30 km beam width in two 350 km swaths. Recorded in 20-240 km lengths. 2-channel radiometer (8 mm, 5 cm), 600 km swath, 0.3°C accuracy, spatial resolution 10-30 km. Carried backup 30 m res radar of C1870 type. Instrumentation for magnetic field mapping was also considered.

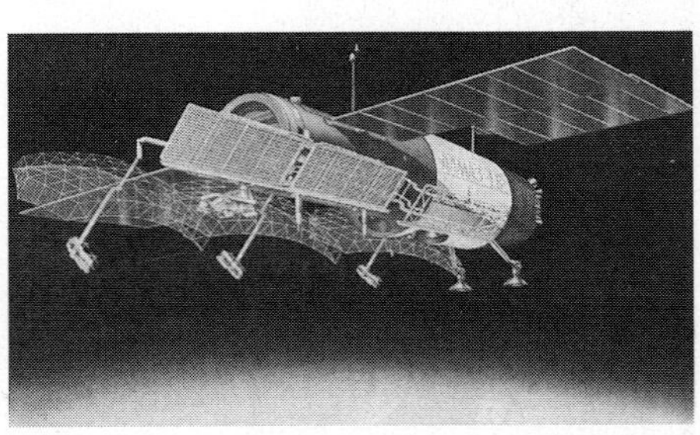

Almaz 1B is planned to carry a wider range of instrumentation. The two articulated white panels at far right are phased array antennas for data relay through GEO satellites (NPO Machinostroenye, courtesy Space magazine) ***1995***

Almaz 1B

Launch: uncertain
Mass: 18.55 t on-orbit (4.5 t instrument payload)
Resolution: 5-7, 15-40, 15-60 m (SAR)
Orbit: 350-400 km, 73°
Spacecraft: as Almaz 1. Users could receive data directly at 8.2 GHz: at 122.8 Mbit/s for main stations or 0.960/3 Mbit/s for small/mobile stations
Sensors: SAR-10 (based on Almaz 1 SAR). 3.125 GHz, 320 km swath, 25-51° off-nadir viewing. Starboard antenna provides two modes: observing 15-40 m res, VV, 100-150 km swath; intermediate 15 m res, V/HV, H/HV, 60-100 km swath. Port provides high-resolution 5-7 m res, HH, 25-50 km swath.

SAR-70 (carried on starboard side). 0.429 GHz, 15-60 km res, H/VH, V/VH, 100-150 km swath, 25-46° off-nadir viewing.

4-channel optical stereo imager, 0.5-0.8 μm, 2.5-4 m res, 70 m swath.

Two MSU-E scanners, channels 0.5-0.6, 0.6-0.7, 0.8-0.9 μm, 10 m res, 24 km swath (paired for 48 km).

MSU-SK visible/infra-red scanning radiometer. Two visible channels 0.53-0.59, 0.61-0.69 μm 80 m res; two near-infra-red 0.7-0.8, 0.9-1.0 μm 100 m res; one infra-red 10.4- 12.6 μm 300 m res. 300 km swath.

Sea surface radiometer, 0.1K res. 11 channels 0.4-12.5 μm, 600 m nadir res, 2 × 1,100 km swath.

Balkan 2 lidar to measure upper cloud altitude, 5,320 Å, 40 arcsec beamwidth, 1 s intervals, 3- 10 m vertical res, ±10° off-nadir swath.

Goms/Elektro

Background

Russia's first GEO meteorlogical satellite finally appeared in 1994 after years of delay. Plans for a Geostationary Orbit Meteorological Satellite system were announced in 1975 as part of the Global Atmospheric Research Programme, for which Europe, the US and Japan have long provided satellites. The delayed programme was acknowledged during 1988 as still active, with the possibility of the first launch during 1989. VNII Elektromekaniki leads development, in addition to its Meteor and Resurs-O activities. In late 1989, Chief Designer Yuri Trifonov noted the first launch was scheduled for 1991. Elektro 1 was completed that year but there were further delays. A ground station was built in Tashkent, Uzbekistan but the USSR's break-up in 1991 required a new one outside of Moscow. Elektro 1 was reportedly then firmly scheduled for July1993 but was postponed because of May 1993's Proton failure. Elektro is controlled by the VKS Military Space Forces for NPO Planeta, part of Roskomgidromet, the Federal Service for Hydrometeorology & Environmental Monitoring. Data are relayed to NPO Planeta and other Roskomgidromet centres. Slots are reserved at 14° west, 18° west, 166° east and 76° east and 40 to 50°

Elektro 1 at Baikonur during final launch processing (Theo Pirard) ***1995***

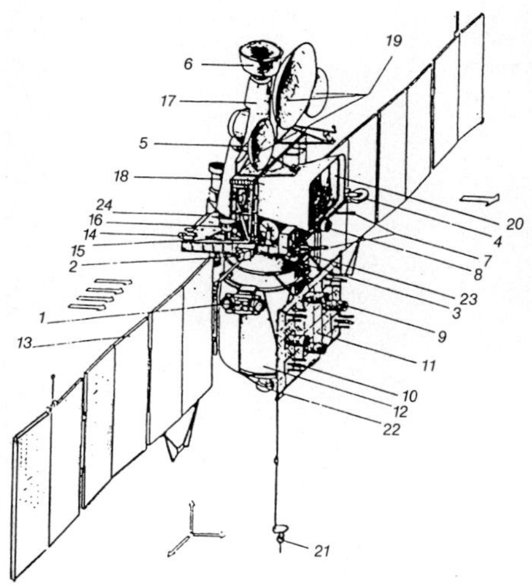

Elektro's key features. 1 arcjet thruster, 2 Sun sensor, 3 instrument platform, 4 reaction wheel, 5 2-axis antenna drive, 6 imaging system radiator cooler, 7 antennas, 8 nadir sensor (failed), 9 antenna for data collection platforms, 10 broadcast antennas, 11 antenna platform, 12 pressurised body, 13 solar array, 14 solar UV sensor, 15 coarse Sun sensor, 16 x-ray sensor, 17 imaging system, 18 polar star tracker, 19 main antennas (outer two are apparently dummies), 20 imaging system shade, 21 magnetometer, 22 thermal screen, 23 proton/electron sensor, 24 low-energy particle spectrometer

inclination satellites have been proposed for improved northern coverage.

Elektro 1 experienced initial control problems. The Earth sensor never operated and controllers instead provided orientation by ground monitoring the broadcast system's signal strength (it has also been claimed that instead, the coarse Sun sensor was used to aim the solar panels and the resulting power output indicated attitude). Elektro was injected into GEO at 90° east and was planned to reach 76° east 20 November but it overshot 3° because it could not be oriented properly while the plasma thrusters fired to unload the reaction wheels. The unplanned attitudes also caused thermal problems. Brought under initial control, it arrived at 76° east 6 December 1994. It was then planned to use the main parabolic antenna for more precise control for the pole star sensor to lock on; Elektro could then operate normally. Only infra-red images are returned; the visible channel has never worked. GOMS was also intended to carry a military telecom payload, but it was deleted when the military withdrew. This probably accounts for the two dummy parabolic antennas.

Specifications

Elektro 1 Satellite

Launch: 31 October 1994 by Proton SL-12
Mass: Elektro 1 2,580 kg, payload 700 kg
Resolution: 1.5 km visible, 6.5 km infra-red
Orbit: 76° east geostationary
Spacecraft: 3-axis controlled to 2 arcmin roll/pitch, 5 arcmin yaw. 16 DEN-16 electrothermal thrusters from NII Machinostroenye. Payload includes environmental platform data collection system (100 bit/s; 401-403/1,697 or 7,482 MHz up/down). Design life 2-3 years. 1,500 W at 28 V DC from 14.70 m-span 30 m² solar wings. OBC provides autonomous operation for 18 days
Sensors: Visible/infra-red TV system operating at 0.46-0.7 μm (8,000 lines) and 10.5-12.5 μm (1,400 lines), but visible inoperable. Elektro 2 will add 6-7 μm channel for water vapour. Global image returned every 1 h at 2.56 Mbit/s on 1.685 or 7.465 GHz to Moscow, Novosibirsk and Khabarovsk. WEFAX-standard imagery received at 2.115 or 8.195 GHz for broadcast at 1.691 GHz. Monitored are 0.2-2.5 MeV electrons, 0.2-500 keV protons, 2.0-120 MeVα particles, 2-10 E solar X-rays, 100-1,300 E solar UV and magnetic fields.

Meteor

Background

Similar to the US NOAA series, Meteor has launched more than 50 spacecraft over three generations. After three years of testing under the Kosmos label, the first Meteor 1 appeared in 1969, followed by the more capable Meteor 2 in July 1975 and Meteor 3 in November 1984. The Soviet Union retired the last of the 31 Meteor 1s in June 1984. The SL-14 Tsyklon began to replace SL-3 in 1982 as the launch vehicle for putting the Meteors in orbit. Meteor 1/2 typically flew at 800 to 900 km, but when the Soviet Union switched to the SL-14 the satellites increased their altitude to 950 km for Meteor 2 and to 1,200 km for Meteor 3. This provided a more comprehensive coverage. The new generation 3M may fly at 925 km. Four Meteor 1s entered into Sun-synchronous orbits from Tyuratam during 1977-81 as testbeds for the Resurs-O programme. All other Meteors have departed from Plesetsk since 1967. Meteor 3-05 broke new ground in 1991 when it carried NASA's TOMS.

Meteor 2-21 installation on its Tsyklon launcher. Italy's 32 kg Temisat data collection satellite (bottom left) was released by Meteor in orbit (Kayser-Threde GmbH)

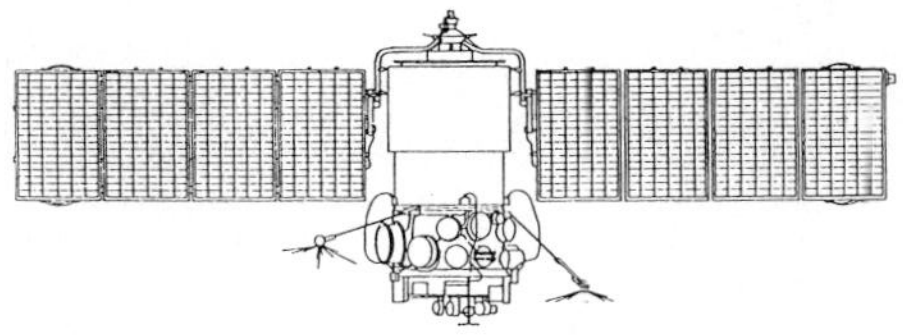

Meteor 2 second-generation weather satellite

In conjunction with the normal launch some Meteors have released third party satellites. Italy's Temisat entered space on a Meteor 2 and Germany's Tubsat B on a Meteor 3. That Meteor 3 included France's Scarab Earth radiation budget instrument and DARA's PRARE unit.

The Russian Federation planned to continue the successful series, but funding problems have hampered progress. The new generation Meteor 3M is based on as much of the old platform as possible. NASA's SAGE III instrument will fly on 3M-1 possibly as early as August 1998, and France is negotiating for Scarab. 3M-2 in August 2000 will carry NASA's Total Ozone Mapping Spectrometer.

Meteors provide a daily weather review for more than two-thirds of the globe, on clouds, ice cover, atmospheric radiation, weather fronts and jet streams. Kosmos 122, launched 25 June 1966 from Tyuratam, 589 × 643 km, 65°, later became the first dedicated meteorlogical satellite identified by the Soviets as meteorological. President de Gaulle of France witnessed the launch, the first Westerner to visit a Soviet launch site. By 1978, according to Moscow Radio, Meteors had transmitted one million images. These satellites cut the sailing time of Soviet ships because the satellites indicated the best courses to avoid storms, winds and ice. This maritime contribution alone was officially estimated to be worth more than R1 million. Meteors also monitor clear-air turbulence so that the airliner crews can be warned of potential hazards. In 1980, Meteor satellite made their first hail warnings, a prediction of particular value in Soviet Georgia, with its 600,000 hectares of vineyards.

Meteor 1

Launches, beginning with Meteor 1 on 26 March 1969 from Plesetsk (644 × 713 km, 81°), began at the rate of three to four annually so that Meteor 2 and 3 were always operational and providing a continuous survey of atmospheric conditions from pole to pole in a swath up to 1,500 km wide. The information dumped to ground stations in passes lasting only a few minutes. Three sites received these transmissions, the USSR Hydrometeorological Service in Moscow, Novosibirsk (Siberia) and Khabarovsk (Pacific Coast). It took about 1 h 30 minutes for the Soviets to process the data and transmit it to users. From Meteor 1-10, launched on 30 December 1971, the Meteor 1s used an 81° orbit situated at about 890 km. This satellite carried the first Soviet meteorological automatic picture transmission compatible with Western receivers. The bus had a small Fakel plasma thruster for orbital adjustments.

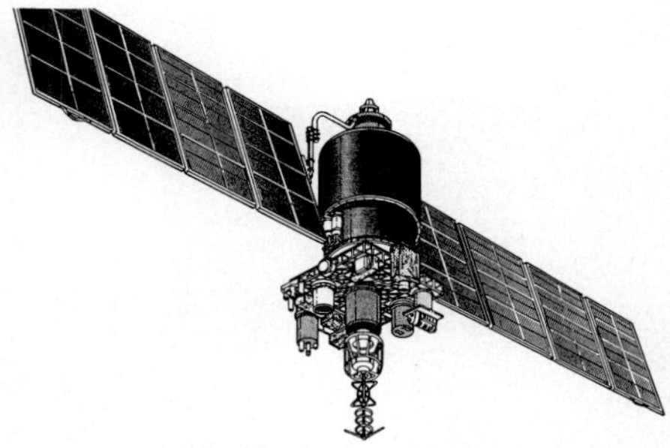

Meteor 3 can accommodate experimental instruments on its 1.6 × 1.8 m truss (Teledyne Brown Engineering)

From Meteor 1-28, launched 29 June 1977 (re-entered 28 August 1993), orbits were retrograde, Sun-synchronous and circular, at 630 km, 98°. This was the first of four Meteor 1s flown as testbeds for Resurs-O, which became operational in 1988 with Kosmos 1939. The last Meteor 1 was 1-31, launched 10 July 1981.

Meteor 1-31 also known as Priroda ('Nature') went into space 10 July 1981 aboard a SL-3 from Tyuratam into 630 km, 98° Sun-synchronous orbit.

The Meteor 1 had a main sensor consisting of a 32-channel scanning radiometer and 33-channel microwave radiometer for surface observations. The Bulgaria 1300 programme provided some of the equipment. At the end of 1983, this equipment was being used to determine snow depths on all Soviet agricultural regions to assist in planning the spring work season and forecasting the 1984 harvest. Transmissions ceased probably in early 1983.

Meteor 2

Introduced experimentally, while the Meteor 1 programme continued, Meteor 2-01 (11 July 1975), established the new operational orbits of 950 km, 82.5° orbits. The satellites have longer life, and include scanning radiometers. Various combinations of orbital plane have been tried, usually with two to three satellites giving passes at 6 to 8 h intervals. Since 1981, the Soviets maintained two or three in operation at any one time, although the Russians let the two remaining satellites fall silent in Summer 1992. August 1993's new launch remains the sole satellite. Automatic picture transmissions for visible imagery are switched off when, for the satellite, the Earth is in shadow.

Meteor 2-19/2-20 apparently replaced 2-16/2-18, respectively, but all five of 16-20 still operated at the beginning of 1991, probably explaining why there were no 1991 departures. 2-19/2-20 fell silent summer 1992.

Meteor 3

The first official Meteor 3 launch occurred on 24 October 1985 from Plesetsk using SL-14. Although employing the same 82.6° inclination as the SL-14 Meteor 2s, Meteor 3-01 was inserted into a much higher 1,227 × 1,251 km. In retrospect, the launch of Kosmos 1612 on 27 November 1984 (135 × 1,230 km)

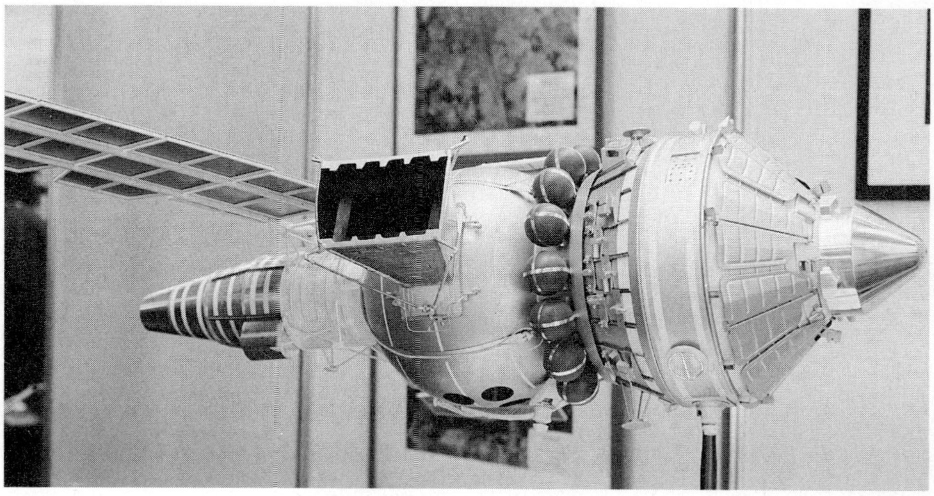

Russia's new Meteor 3 m weather satellite (Theo Pirard) ***1999***/0054350

Meteor 3 weather satellite first launched in 1983 ***1999***/0054352

appears to have been a precursor Meteor 3. This launch failed when Tsyklon's upper stage did not re-ignite. Meteor 3-01 incorporated plasma thrusters, which were used to lower its orbit to a mean altitude of 1,200 km in November-December 1985. In late June 1986, the Soviets lowered the orbit by another 4 km. No new 3s were launched in 1986 or 1987, but 3-02 appeared 26 July 1988, followed by 3-03 24 October 1989. Both satellites operated in early 1991, but were replaced during the year by 3-04 (24 April) and 3-05 (15 August). 3-05 carried NASA Goddard's Total Ozone Mapping Spectrometer to extend Nimbus 7 observations, which discovered Antarctica's ozone hole in 1987. The instrument failed December 1994, well beyond its two year design life. Data went via NASA's Wallops Flight Facility to the Russian Obninsk station.

Meteor 3-06 carries France's Scarab Earth radiation balance instrument. The Scarab (scanner for radiation budget) project began in 1987, made up of three 40 kg units built at FFr65 million. The second scanner is scheduled aboard the Resurs-O1 N4 and may fly on 3M-1. Data from Scarab go first to Obninsk and Medvezhi Ozera for relay to NPO Planeta in Moscow and then to CNES Toulouse. The instrument monitors four bands: visible 0.5-0.7, solar 0.2-4, total 0.2-50, infra-red window 10.5-12.5 μm. The US NOAA and ERBS satellites correlate the data.

The All-Russian Research Institute for Electromechanics (VNII Elektromekaniki), designed and built the Meteor/Resurs-O satellites. Meteor 3 and Resurs- O employ the same platform.

Specifications

Meteor 2/3

Launched: typically annually by Tsyklon from Plesetsk; 3M on Zenit 2

Mass: Meteor 2 1,300 kg on-orbit; Meteor 3 2,215 kg on-orbit, payload 500-700 kg; Meteor 3M 2,500 kg on-orbit, payload 900 kg

Resolution: 0.7-2 km visible, 3-42 km infra-red

Orbit: 950 km 82.5° Meteor 2, 1,200 km 82.5° Meteor 3, 925 km 82.5° Meteor 3M (3M-1 will satisfy SAGE requirements with: 1,020 km, 99.5°, 09.15 ascending node equatorial crossing)

Spacecraft: 3-axis stabilisation (0.5° accuracy; 3M will be improved), 4.2 m long 1.4 m diameter cylindrical body with twin 10 m span 2 m-wide solar wings providing 500 W to the payload (3M: 1 kW). Design life 2 years (3M, 3 years). Meteor 3 adjusts orbit by Fakel SPT plasma thrusters (these thrusters are now offered commercially; see the International Space Technology joint venture entry in the Launchers section). 3M will use GPS/Glonass for precise orbit determination. Transmissions directed at three primary sites (Moscow, Novosibirsk in Siberia and Khabarovsk on the Pacific coast; Tashkent was lost with independence; new sites were planned for Murmansk and Petropavlovsk-Kamchatski) in conjunction with >80 smaller ground stations. Internationally-compatible APT transmissions are made on 137-138 MHz to 15,000 CIS terminals; there is also a 10 W analogue store/forward mode at 466.5±0.125 MHz (3M will use 1.69-1.71 GHz digital). Data are processed within 90 minutes by the Hydrometeorological Centre

Sensors: Meteor 2. 0.5-0.7 μm 2,100 km swath medium resolution (2 km), 0.5-0.7 μm 2,400 km swath high resolution (1 km), 8-12 μm global scanning infra-red radiometer (2,600 km swath, 8 km resolution), and 11/ 10- 18/70 μm scanning infra-red radiometer (1,000 km swath, 30 km resolution). Two satellites can cover 80 per cent of the Earth's surface in 6 hours.

Meteor 3. 0.5-0.8 μm 3,100 km swath 0.7-1.4 km resolution, 10.5-12.5 μm infra-red radiometer 3,100 km swath 3 km resolution (1°C resolution), 9.4-19.68 μm 10-channel infra-red spectrometer 1,000 km swath 42 km resolution, 0.25-0.380 μm 8-channel ultra-violet backscatter spectrometer 200 km swath 3-5 km altitude resolution, Ozon-M 0.25– 0.29, 0.37-0.39, 0.6-0.64, 0.99-1.03 μm 4-channel UV ozone monitor 2 km altitude resolution, plus 0.17-600 MeV electron/proton detectors. 1991's Meteor 3-05 carried NASA's TOMS; second is planned for 3M-2 in 2000. 3M-1 will carry SAGE in 1998. Meteor 3-06 carries France's Scarab Earth radiation balance instrument (see below for details), DARA's PRARE unit (see ERS 1 for background), and released Germany's Tubsat B.

Resurs-F

Background

In the mid-1970s, the Russians started to fly recoverable remote sensing satellites within the Kosmos programme, these satellites being based upon the contemporary Zenit-4 class of photo-reconnaissance satellites. Over the years the majority of the launches were to orbits with inclinations of 81.2-82.5° and they were initially announced as simply undertaking remote sensing work 'for the national economy' or undertaking work for the 'Priroda Centre'. The first of these groups of recoverable remote sensing satellites was Kosmos 742, a rather isolated mission because it has been linked with four satellites in the Gektor-Priroda programme which were launched in 1979. A larger group called Fram debuted in September 1975 with the launch of Kosmos 771 and launches in this series continued through until 1984 (Kosmos 1681 was the last Fram satellite).

Resurs-F employs the Vostok-based photo-reconnaissance spacecraft, with more than 800 missions to its credit, flying for the Priroda centre of Roskartografia (the Federal Service of Geodesy & Cartography). Resurs-F began flying with C1127 in

Resurs-F1 recovery. The two KFA-1000 cameras sit above three Kate-200s. The camera section cover lies on the ground, with the Kate viewpoints just visible
(Resurs-F WorldMap) ***1994***

The nature and schedule of KB Photon's new Resurs-Spektr satellites are uncertain. They appear to be derived from the fifth-generation digital photo-reconnaissance satellites. Spektr-V would provide 3 to 5 m visible stereo resolution, complemented by microwave sensors. Spektr-R (shown here) would provide radar imagery
(Theo Pirard/Space Information Centre) ***1996***

September 1979. The older F1 type flew missions of typically two to three weeks (although active for only up to 14 days) with Kate-200 and KFA-1000 cameras. F2 undertook four-week missions using the MK-4 camera system, although lifetimes of up to 45 days are possible. F3 missions were identified in 1993, flying occasionally with two KFA-3000s for 2 m resolution. The film camera systems are returned in the 2.2 m diameter spherical descent capsules, which are reused an average of three times. The camera systems are also refurbished and reflown. F2's longer duration missions (see table) are supported by solar panels supplementing the standard battery power supply system. F3's appearance is identical externally to F1. The improved F1M had been planned to debut in 1994 but no Resurs-F missions flew that year because of the lack of launch vehicles. The satellite is completed but still lacks launch funding. F1M extends the active period to 19 days (+6 days inactive) and carries three KFA-3000s. F1M can also fly down to 190 km for improved resolution. The F2M is planned for 1997, carrying an improved film for higher resolution. Some Russian reports suggest that budget limitations have led to the indefinite delay or even cancellation of the Resurs-F1M and -F2M series.

JEBCO Information Services (UK) and Priroda, the primary non-military remote-sensing acquisition and processing organisation in Russia created the Resurs-F WorldMap consortium in 1993. Priroda's archives contain two million XS/PAN scenes at 2 to 20 m

Resurs-F2 in flight configuration. The solar panels allow longer missions than the battery-powered F1/F3

Resurs-F2 final flight preparations ***1995***

Resurs-F missions (1986 onwards)[1]		Launch	Orbit (km)	Days
C1746	F1	28 May 1986	259 × 273, 82.3°	14
C1757	F3	11 June 1986	159 × 391, 82.3°	14
C1762	F1	10 July 1986	259 × 273, 82.6°	14
C1768	F1	2 August 1986	259 × 273, 82.6°	14
C1789	F1	31 October 1986	322 × 342, 82.6°	14
C1846	F1	21 May 1987	323 × 342, 82.4°	14
–[3]	F1	18 June 1987	–	–
C1882	F1	15 September 1987	259 × 275, 82.3°	21
C1906[2]	F2	26 December 1987	257 × 277, 82.6°	–
C1920	F1	18 February 1988	330 × 334, 82.6°	20
C1951	F1	31 May 1988	259 × 275, 82.3°	14
C1956	F3	23 June 88	333 × 368, 82.4°	14
C1957	F1	7 July 1988	260 × 275, 82.6°	14
–[3]	F1	21 July 1988	–	–
C1965	F2	23 August 1988	257 × 277, 82.3°	30
C1968	F1	9 September 1988	260 × 275, 82.3°	14
C1990	F2	12 January 1989	256 × 269, 82.6°	30
C2000	F3	10 February 1989	343 × 395, 82.4°	20
Res 1	F1	25 May 1989	255 × 272, 82.3°	23
Res 2	F1	27 June 1989	260 × 274, 82.6°	14
C2029	F3	5 July 1989	338 × 362, 82.4°	14
Res 3	F1	18 July 1989	260 × 275, 82.6°	21
Res 4	F2	15 August 1989	259 × 268, 82.3°	30
Res 5	F1	6 September 1989	262 × 273, 82.3°	16
Res 6	F1	29 May 1990	259 × 273, 83.3°	16
Res 7	F2	17 July 1990	261 × 269, 82.3°	30
Res 8	F1	16 August 1990	259 × 271, 82.3°	16
Res 9	F1	7 September 1990	261 × 275, 82.6°	14
Res 10[4]	F2	21 May 1991	229 × 237, 82.3°	30
Res 11	F1	28 June 1991	259 × 273, 82.3°	23
Res 12	F1	23 July 1991	258 × 272, 82.3°	16
Res 13[4]	F2	21 August 1991	231 × 234, 82.3°	30
Res 14[4]	F2	29 April 1992	233 × 245, 82.1°	30
Res 15[4]	F1	23 June 1992	229 × 233, 82.3°	16
Res 16[4]	F1	19 August 1992	221 × 239, 82.6°	16
Res 17[4]	F2	21 May 1993	235 × 245, 82.6°	30
Res 18[4]	F1	25 June 1993	223 × 241, 82.6°	17
Res19[4]	F1	24 August 1993	234 × 239, 82.6°	17
Res 20[4]	F2	26 September 1995	231 × 235, 82.3°	30
Res 21	FIM	18 November 1997	238 × 270, 89.6°	25

Notes: 1) all launches made by Soyuz from Plesetsk; first Resurs-F mission was C1127 5 September 1979 (28 missions performed pre-1986). 2) detonated 31 January 1988 following spacecraft malfunction. 3) 1987 failure not identified until 1995; 1988 failure not identified until 1993; no further details. 4) new low-altitude regime to increase resolution (but Res 20 uniquely then rose to the old standard 256 × 280 km 22 October 1995). Res 20 image coverage was 20.7 million km².

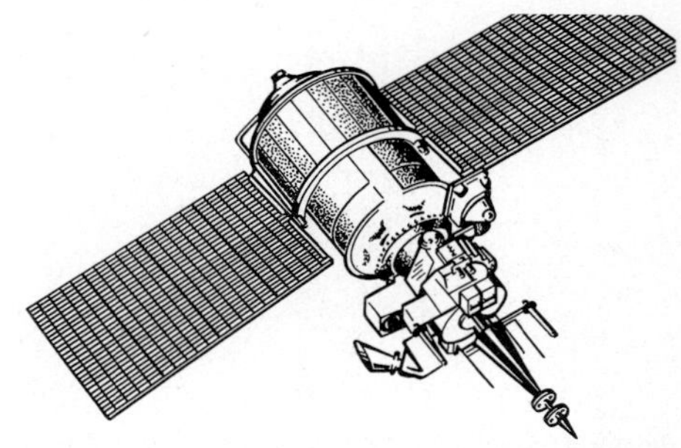

The Resurs-O1 digital Earth resources satellites are derived from their Meteor-Priroda predecessors

Resurs-O1 N3 ready for launch (OHB-System) ***1995***

resolution covering most of the globe. The consortium accepts requests for targets for future flights, and would like to reduce lead time. Usually 40 to 60 per cent of the full mission capability is specifically targeted, while the remainder of the film is acquired over targets of opportunity. 12 hour operating plans are uplinked twice daily. F1/F1M/F3 can power down for 11/11/6 days to wait for more suitable target conditions.

It has been reported that Resurs-F will be replaced eventually by KB Photon's Nika-K, although recent reports of funding problems make delays likely. It will provide more propellant, more film, improved orientation and 45 day flights. It is apparently derived from the Generation 6 Kuban photo-reconnnaissance satellite that is also being modified as the Nika-T μg craft to replace Photon.

Kosmos 1906, which was deliberately destroyed in orbit because of a control malfunction, was acknowledged at the time to be the first of a new type of craft carrying multispectral cameras, and later identified as the first F2. The first named Resurs-F satellite was launched 25 May 1989, accompanied by two small Pion

Resurs-F3 recovery, showing the paired KFA-3000 cameras. Above the cameras are fans to maintain convection for cooling (Resurs-F WorldMap)

passive upper atmosphere density satellites. The F2/MK-4 system completed its test flight programme in 1989, complementing the Kate/KFA cameras of Resurs-F1. Small μg payloads are also flown aboard Resurs: Intospace's Cosima 2 was the first, in September 1989, followed by number 3 on 29 May 1990 and the first Casimir unit in September 1990. August 1992's F1 carried the US Naval Research Laboratory's Be-7 Induced Radiation Experiment (aluminium foil on an extending arm) and released two further Pions. It was the first CIS/USSR satellite to carry a DoD experiment, under Space Test Program sponsorship. The arm failed to extend; there was an identical failure in mid-1993

Specifications

Resolution: 5-8 m visible for MK-4, 5 m visible KFA-1000, 2 m visible KFA-3000
Launched: typically 3-5 years, all by SL-4 Soyuz from Plesetsk, until funding slowed rate
Mass: about 6.3 t on-orbit, 2.3 t descent sphere
Orbit: typically 260-275 km, 82.5° for F1/2 majority (high 330-365 km for Kosmos 1283/1284/1472). Res 10 introduced new low of 230-235 km in 1991 for improved resolution. F3 now 275 km. 14 days active and 11 days quiescent for F1/F3, F1M 19/6 days; 30 days F2 (all active, none quiescent) initial orbit typically 180 × 30 km, then raised. F1M expected to employ 190 km operationally for improved resolution. F3 planned for 180 km to provide 1.6 m resolution
Spacecraft: Vostok-based
Sensors: Three Kate-200 and two KFA-1000 (F1); one Kate-200 + three KFA-1000 (F1M); MK-4 (F2); two KFA-3000 (F3) film cameras

Stellar camera (SA-33 for F1 and SA-3R for F2) provides pointing reconstruction.

Resurs-O

Background

The Resurs-O1 system, the multispectral digital equivalent of the US Landsat, became operational in 1988 with the launch of the series' fifth, Kosmos 1939. The Russians expected to replace it in June 1993, but the satellite's longevity deferred Resurs-O1 N3 until November 1994. The launch marked the first civil orbit for the Zenit booster. It will be superseded by the larger Resurs-O1 N4 and in 1999 by O1 N5. 1994's O1 carries Germany's SAFIR-R1 data relay demonstration attached package.

The All-Russian Research Institute for Electromechanics (VNII Elektromekaniki) builds the Resurs-O satellites and the Meteor 3 and Resurs-O1 employ the same platform modified for operations at different altitudes and offering three to five year lifetimes. The history of the two platforms has been intermixed throughout their lifetimes. Kosmos1939 (20 April 1988) followed Meteor 1-30 (deactivated 18 June 1988 after eight years' observations), Meteor 1-31, Kosmos 1484 and Kosmos 1689. All of these satellites used the same basic design.

Resurs control commands come from the VKS Military Space Forces using its Moscow centre. The imagery product from the satellites goes directly through Moscow, Novosibirsk and Khabarovsk. Commercial imagery for Western purchasers passes through Sweden's SSC Satellitbild, by an arrangement begun in 1995. The first SSC imagery came from MSU-SK downlinked from the November 1994 Resurs-O1. Sweden interest targeted agricultural and coastal water images.

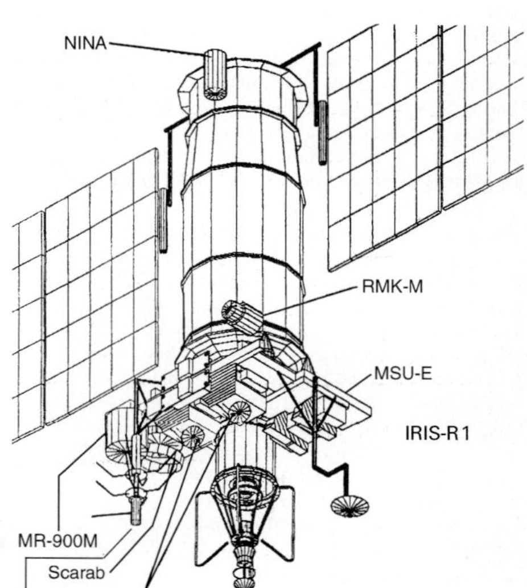

The upgraded Resurs-O1 N4 ***1996***

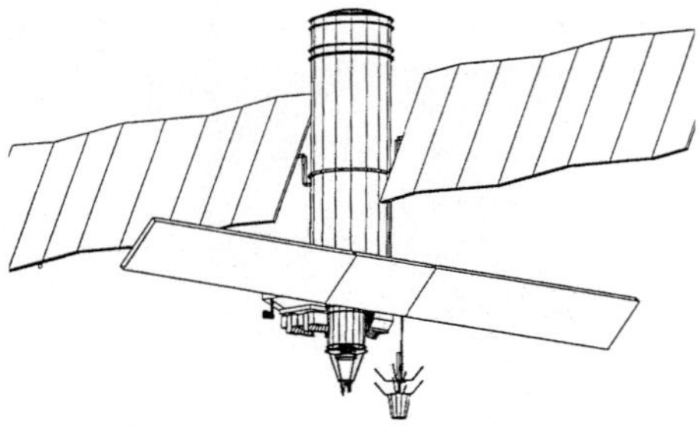

Resurs-O1 N5 will add a radar for ice patrol imagery to aid shipping (VNIIEM) ***1996***

SSC deals with the Sovzond consortium of VNII Elektromekaniki, NPO Planeta and the NIIKP Institute of Space Device Engineering (which builds the MSU sensors). Direct mode images began 3 July 1995, and recorded images arrived by the end of 1995. SSC and its partners had archived 1,000 images by the end of 1995. Up to 100 can be browsed on Eurimage's EINET web page (http: //www.eurimage.it/einet/einet – home.html). Archived images cost ECU1,100 path-oriented or ECU1,800 map-oriented.

VNIIEM has developed the USSP Universal Small Space Platform for release from Resurs-O. The 60 kg satellite employs a gravity gradient boom, reaction wheel and magnetorquers for attitude control. The bus is a hexagonal prism.

The first of an improved series of remote sensing satellites, Resurs-02/1 was launched on 10 July 1998.

Specifications

Resurs-O1 N3

Launch: 4 November 1994 by Zenit from Tyuratam
Mass: 1,907 kg (500 kg payload)
Resolution: 45 m visible, 170 m infra-red, 600 m thermal infra-red
Orbit: 663 × 691 km, 98.05° Sun-synchronous (16 day repeat)
Spacecraft: based on Meteor 3 platform (qv); 3-axis stabilised, about 5 m high, 10 m across solar panels. Real-time data rate 7.68 Mbit/s on 8.192 GHz. Power 500 W average (1.2 kW peak)
Sensors: Multiple multispectral package covering visible to thermal infra-red

MSU-SK conical scanner 0.5-0.6, 0.6-0.7, 0.7– 0.8, 0.8-1.1, 10.4-12.6 μm offers 600 km swath width, 170 m resolution for first four bands and 600 m for band 5.

MSU-E push-broom CCD imagers cover 0.5-0.6, 0.6-0.7, 0.8– 0.9 μm, providing 45 m resolution over 90 km total swath width of both instruments. 1,000-pixel linear CCD.

UKRAINE

Okean-O1/Okean-O/Sich

Background

The Okean-O1 all-weather radar, ice and oceanographic satellite system, became operational in 1988 with the launch of Okean 1. Kosmos 1076/1151 flew as precursor satellites, but 1983's Kosmos 1500 prototype was the first to incorporate 1.5 km resolution side-looking radar utilising a 12 m deployable antenna. Data return through the three primary Meteor ground stations. Ice forecasts are uplinked four times weekly to vessels via the Ekran geostationary satellites. Kosmos 1500 in December 1983 helped free 70 Soviet ships trapped in heavy Arctic ice near Wrangel Island. NPO Yuzhnoye in Dniepropetrovsk, Ukraine, built and designed the Okean and the VKS Military Space Forces

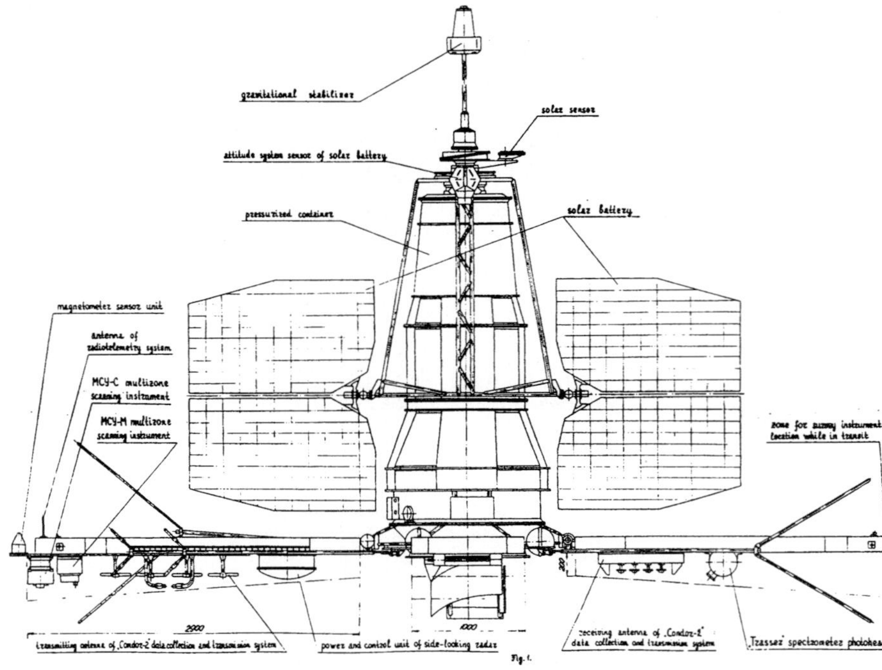

Okean-O1 configuration; the large base radar antenna is not visible

operate the system for NPO Planeta. Possibly the last of the 1.9 tonne Okean-O1 designs departed in August 1995 on Tsyklon as a joint Russian/Ukrainian project, designated Sich 1. The Ukraine took over full control from 11 October 1995 using Yevpatoria.

Sich is the Ukraine space agency's main satellite programme and the Okean-O platform will be used for the Ukraine-only Sich 2, carrying a large SAR. Sich 3 will carry more scanning radiometers.

As with so many former Soviet programmes, financial constraints have significantly reduced these plans.

Specifications

Okean-O1

Resolution: 1.5 km radar, 0.35-1.5 km visible/infra-red, 6-15 km radiometer
Design life: six months
Mass: 1,950 kg (550-650 kg payload)
Orbit: typically 630-660 km, 82.5°
Spacecraft: stepped cylindrical pressurised bus 3 m high, maximum 1.4 m diameter, 3-axis stabilisation (base nadir pointing, accuracy 0.75-1.0°, determination to 2 arc/min), aided by gravity gradient boom. Twin solar arrays spanning 4.82 m (each an irregular 2 m high, 1.6 m wide); power consumption 1.1 kW operating, 110-270 W daily average. Sensors and single 11.8 m radar antenna mounted on four 2.9 m panels. APT transmissions on 137.4 MHz. Condor 2 digital data collection from small automatic Condor 1 stations at 500 bit/s on 1533.4 MHz, relay at 2 kbit/s on 460.03 MHz). No orbit or yaw control
Sensors: RLS-BO side-looking radar operating at 3.15 cm and 450 km swath width to left of ground track. Power limits operations to 10 min sessions.

MSU-S visible/near-infra-red scanning radiometer (channels 0.5-0.7, 0.7-0.9 μm) providing 350 m resolution over 1,380 km swath width, nadir pointing.

MSU-M multispectral visible/near-infra-red scanner, 0.5-0.6, 0.6-0.7, 0.7-0.8, 0.8-1.1 μm, 1.5 km resolution over 1,930 km swath, nadir pointing.

RM-08 8 mm-wavelength scanning radiometer, 550 km swath, to left of track. C1500/1602/1766 carried the MSU-K circular scanning radiometer, 0.5-0.9 μm single band, 900 km swath, 500 m resolution, 6.5 kg.

Okean-O

Resolution: 1.5 km radar, 25-200 m visible, 100-600 m infra-red, 16– 130 km radiometer

The first 6.3 tonne Okean-O is expected in 1997 (Theo Pirard) ***1996***

Design life: 3 years (1 year Okean-O number 1)
Mass: up to 6.5 t (up to 2 t payload); Okean-O number 1 6.3/1.55 t
Orbit: planned 670 km, 98°
Spacecraft: total length 10.6 m, 1.8 m diameter 6 m pressurised bus, aluminium-magnesium alloy structure. 3-axis stabilisation (base nadir pointing, accuracy 5 arcmin, determination to 2 arc/min). orbit adjust/momentum dumping provided by base propulsion system: 10 × 30 N UDMH/NTO thrusters. Attitude control by RWs and magnetorquers. Single articulated solar wing; power consumption 5 kW operating, 1.5 W daily average (Okean-O number 1 3.5/0.80 kW). Sensors and radar antenna mounted on sides/base of bus and forward instrument platform. Condor 2 digital data collection and transmission system (receive at 500 bit/s on 1.5334 GHz; transmit at 2 kbit/s on 460.03 MHz). Data transmitted at 15.36-122.88 Mbit/s (Okean-O number 1 up to 64 kbit/s) on 8.2 GHz; APT transmitted on 137.4 MHz. 19 Gbit memory (Okean-O number 1 5.5 Gbit)
Sensors: RLS-BO side-looking radar operating at 3.15 cm and 450 km swath width to left of ground track

MSU-M multispectral visible/near-infra-red scanner operating at 0.5-0.6, 0.6-0.7, 0.7-0.8, 0.8-1.1 μm providing 1-1.7 km resolution over 1,930 km swath.

MSU-A 3-channel 0.5-0.9 μm scanner providing 30 m resolution over ±300 km swath.

MSU-SK visible/infra-red conical scanning radiometer. Two visible channels 0.53-0.59/0.61-0.69 μm 200 m resolution; two near-infra-red 0.7-0.8/0.9-1.0 μm 200 m resolution; one infra-red 10.4– 12.6 μm 600 m resolution. 600 km swath.

Okean-O1 radar programme launches[1]

	Launch	Orbit	End Ops
E1/C1076[2]	12 February 1979	593 × 620 km, 82.5°	31 March 1980
E2/C1151[2]	23 January 1980	613 × 640 km, 82.5°	13 October 1981
OE1/C1500	28 September 1983	643 × 679 km, 82.6°	16 July 1986
OE2/C1602	28 September 1984	629 × 664 km, 82.5°	5 December 1986
O1/C1766	28 July 1986	631 × 662 km, 82.5°	24 October 1988
O2/C1869[3]	16 July 1987	634 × 667 km, 82.5°	3 May 1989
O3/Okean 1	5 July 1988	635 × 666 km, 82.5°	14 June 1990
O4[4]	9 September 1989	-	-
O5/Okean 2	28 February 1990	639 × 666 km, 82.5°	18 July 1991
O6/Okean 3[5]	4 June 1991	634 × 666 km, 82.5°	4 January 1994
O7/Okean 4	11 October 1994	632 × 666 km, 82.6°	
O8/Sich 1[6]	31 August 1995	632 × 669km, 82.5°	

Notes: 1) all launches made by SL-14 Tsyklon 3 from Plesetsk; 2) experimental, did not carry SLR; 3) radar antenna failed to deploy but other sensors operational; 4) stage 3 failure; 5) plane 90° from number 2's; 6) Chile's 50 kg FASat-Alfa failed to separate.

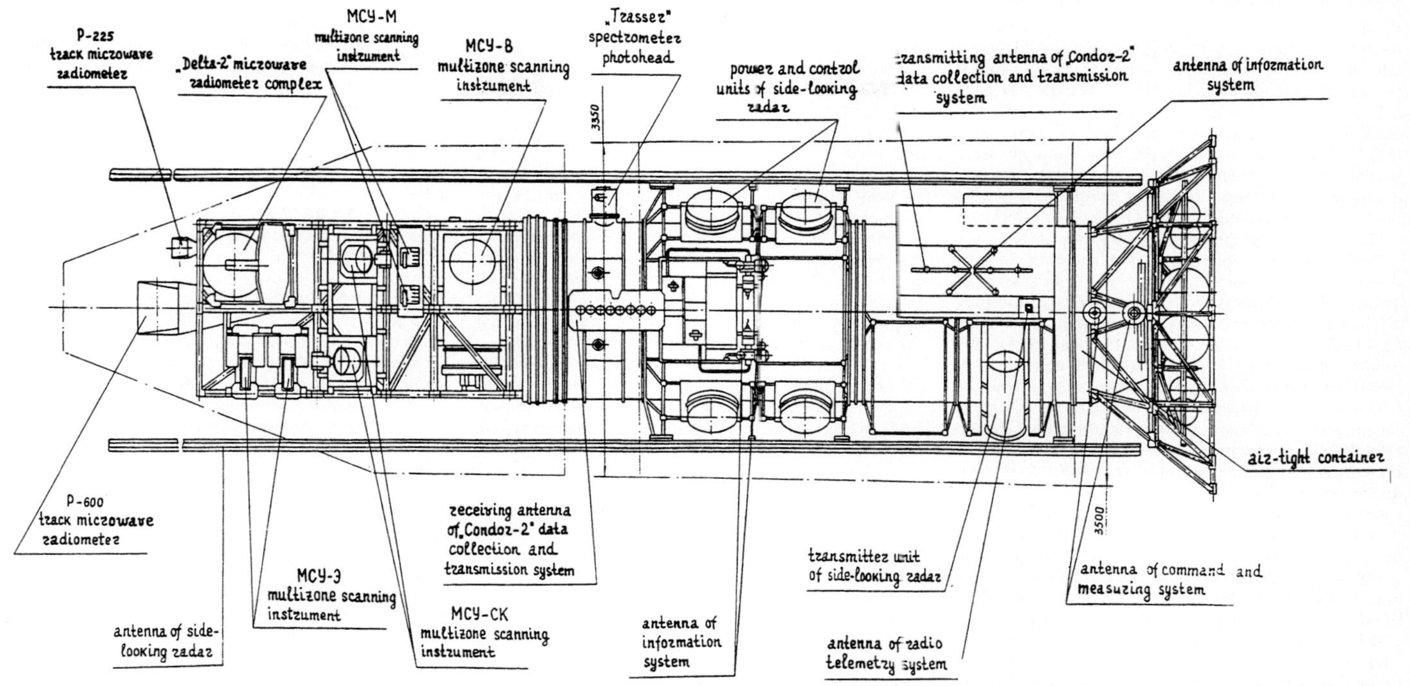

Earth-pointing face of 6.3 tonne Okean-O with potential sensor payload. The SLR antennas run along both sides (only the lower one will be carried); the propulsion system is at far right. The dashed line indicates the Zenit payload envelope

The Okean-O1 ocean radar satellite was first displayed at 1985's Paris Salon. The SLR antenna is seen diagonally at bottom left (Graham Turnill)

MSU-V conical scanner. Eight channels 0.45-0.52 (50 m resolution), 0.52-0.62 (50 m), 0.62-0.74 (50 m), 0.76-0.9 (50 m), 0.9-1.1 (50 m), 1.55-1.75 (100 m resolution), 2.1-2.35 (250 m), 10.0-12.0 (100 m resolution) μm. 180 km swath, 1.28-5.12 Mbit/s. Ice/snow imaging, land use.

Trasser-O spectroradiometer. 62 channels over 0.430-0.800 μm, <1 n mile resolution.

Land Delta 2 scanning radiometer, 7.0/13.0/22.5/36.5 GHz, 2,600 km swath.

R600 microwave radiometer. 6 cm/4.9 GHz, two polarisations, 130 km resolution, 1 kbit/s.

R225 microwave radiometer. 2.25 cm/13.3 GHz, two polarisations, 130 km resolution, 2 kbit/s.

Sich 2

Resolution: 10/50 m SAR, 1.5 km SLR, 30 m visible
Launch: undetermined
Design life: 3 years
Mass: up to 6.5 t (up to 2 t payload)
Orbit: planned 670 km, 98°
Spacecraft: see Okean-O
Sensors: RLBO double side-looking radar (with scatterometer mode), 12 m antenna, 3.2 cm wavelength, 700 km swath each side ground track.

2.4 × 15 m SAR, 23 cm wavelength, 85-122 km swath provides 50 m resolution, 36-57 km swath 10 m resolution.

MSU 3-channel visible/near-infra-red scanner, 0.5-1.0 μm, 48 km scan with 700 km for 30 m resolution.

Sich 3

Resolution: 3-10 m visible, 20 m infra-red, 3 km microwave
Launch: undetermined
Design life: 3 years
Mass: up to 6.5 t (up to 2 t payload)
Orbit: planned 670 km, 98°
Spacecraft: see Okean-O
Sensors: 11-channel spectroradiometer, 0.4-2.4 μm, 10-40 m res, 160-200 km swath within 600 km width, 32-64 Mbit/s. Ice/snow, land use.

3-5 m high resolution 0.4-0.8 μm camera, 30-50 km scan within 150 km width.

3-channel IR spectroradiometer, 0.3-13 μm, 20-60 m res, 50-150 km swath within 600 km width, 16-32 Mbit/s.

10-26-channel scanning radiometer, 5-120 GHz, 3-4 km res, 1,000 km swath, 100 kbit/s.

UNITED STATES OF AMERICA

Landsat series

Background

The United States floated the first proposals for a civil remote sensing satellite in the late 1960s following Earth observations from the manned Gemini and Apollo. Serious study began when the Department of Interior (DOI) in 1968 requested funding for an EROS Earth Resources Observation Satellite. Congress denied DOI the funding, because it felt the Department was not authorised to undertake space projects. Nevertheless DOI built a co-operative programme with NASA and funds appeared in 1969.

General Electric built Landsat 1-3. They were enhanced versions of the Nimbus weather and research satellites, originally known as the Earth Resources Technology Satellites (ERTS). NASA planned to launch two in consecutive years, 1972-73, as an experiment in systematically surveying the Earth's surface to study the health of its crops and the potential use and development of its land and oceans. 300 investigators in the 50 participating countries made use of the dramatic and vivid Landsat 1 images. Nations quickly realised these images were directly relevant to the management of the world's food, energy and environment and NASA laid plans for Landsat 2.

Before the Landsat 1 launch, NASA decreed that all ERTS data, including 9,000 weekly images, should be unclassified and made available to the public. The satellite downlinked the pictures through NASA ground stations to the US Department of Interior's EROS Data Center in Sioux Falls, South Dakota where they could be bought for prices starting at US$1.25. Unexpectedly, news agencies and private individuals studied detailed pictures of Soviet- and Chinese-launch centres and missile sites. The immediate success of the project caused the processing and distribution systems to be swamped with requests. As a result, a number of countries subsequently installed their own receiving stations, including South Africa, Canada, Brazil, Italy, Sweden, Japan, Thailand, India, Australia, Argentina, Zaire, Iran and China. They initially paid construction costs of US$4 to 7 million, US$1 to 2 million annual operational costs and an annual service fee of US$200,000.

The first three Landsats carried combinations of Return Beam Vidicon (RBV) and Multi-Spectral Scanner (MSS) imagers. Technical problems resulted in data from the RBVs of Landsat half being seldom used. For Landsat 3, even though RBV resolution was improved from 80 m to 40 m, the spectrally and radiometrically superior MSS was still preferred. RBVs were deleted from later spacecraft.

Despite its success, by 1982 the long-term Landsat programme had run into both technical and financial troubles. Landsat 2 and 3 had surpassed their design lives and NASA expected them to fail at any time. GE, whose prime contract for Landsat D/D-Prime (as Landsat 4 and its back-up were known before launch) had trouble selling its US$505 million contract in Congress. Hughes also seemed to be behind in the development of the advanced Thematic Mapper.

Finally, NASA managed to solve the technical and bureaucratic problems and got Landsat 4 into space on 16 July 1982. Today, Landsat 4 retains little operational capability. It was already suffering from multiple malfunctions by February 1983. First, the power cabling from the solar panel broke up under the day/night thermal cycling, leaving the satellite with only 50 per cent power capacity. Then the TM X-band link failed after transmitting 6,000 images. NOAA, faced with a monthly revenue loss of US$600,000, because of these failures, requested the US$20.4 million launch of the back-up Landsat D-prime, undertaken in March 1984. NASA studied a Landsat rescue mission to retrieve the US$60 million satellite in June 1986 using the second Vandenberg Shuttle mission. The end to Shuttle preparations at Vandenberg shelved the idea. Instead, EOSAT maintained full operations with the MSS and resumed TM transmissions via the TDRS satellite link in October 1987 after reprogramming the satellite to reactivate the back-up Ku-band.

L5 was built as L4's identical back-up (then modified to prevent L4-type failures) and launched when the prime vehicle suffered failure of its redundant power and transmitter systems. Capability began declining in 1992, as expected, but engineering tests indicate that transmission direct to ground stations is expected to continue through 1997.

The first three Landsats cost NASA a total of US$251 million, including US$149 million for the spacecraft and US$14 million for the launchers. By July 1981, it had become clear that the series would represent a total US investment of US$1 billion. The Reagan Administration decided at that stage that the remote sensing programme must either end, or be absorbed by the private sector. COMSAT, Hughes and several other companies expressed interest subject to government guarantees of a minimum level of data purchase.

The Landsat 4/5 system is operated by the Earth Observation Satellite Co (EOSAT), a partnership formed between Hughes Aircraft (subsequently purchased by GM) and RCA Corporation in May 1984. Under the Landsat Remote Sensing Commercialization Act of 1984, the Department of Commerce (DoC) awarded EOSAT a contract to operate the system for 10 years and to build two new satellites, plus ground systems. It granted EOSAT exclusive marketing and distribution rights to existing Landsat data until July 1994 and to new data for 10 years after collection. The government agreed to provide US$250 million for spacecraft development and to pay for the launches. In 1986, however, DoC eliminated funding for Landsat commercialisation from its FY87 budget request, beginning a series of battles with Congress ultimately accepting a scaled-back commercialisation plan. EOSAT/DoC, on 31 March 1988, signed modifications to the 1985 contract to bring it into line with the compromise reached between Congress and the Reagan Administration, Congress having finally released US$62.5 million for Landsat 6 programme start.

President Bush signed the Land Remote Sensing Policy Act of 1992 on 28 October 1992. This repealed 1984's Act, and transferred the satellites to the joint NASA-DoD Landsat Programme Management from NOAA/DoC. It also provided authorisation for the funding of Landsat 7 as a functional equivalent of Landsat 6. In response to public criticism of data release policy especially the close involvement of the military in the programme, the Act called for negotiations between LPM/EOSAT to modify data policy for Landsat 4 to 6 to reduce data price for US government agencies and their affiliated users.

Following Landsat 6's loss, EOSAT and NASA reached a preliminary agreement in April 1994 giving the company a further five years of rights to archived imagery and exclusive rights to new Landsat 4/5 scenes (and for five years after the satellites' deaths). In return, EOSAT would charge government users reduced rates: US$3,500 per scene for the rest of 1994, and US$2,500 from 1995, instead of the commercial US$4,400 rate. Once bought, these scenes could be copied for other government users at no extra cost. However, the DoC, in July 1994, decided to open the contract for bidding; EOSAT's contract was extended through 30 June 1995 while the matter was pursued; EOSAT was successful.

The 1992 Landsat Act also outlined a proposed Landsat 7 data policy to stimulate the use of data through sales at the cost of fulfilling user requests by the government and/or the competitive marketing of unenhanced data. Landsat 7 will be operated by NOAA and the imagery handled by the US Geological Survey. EOSAT will thus be left as any other value-added company. Landsat 6's launch loss in October 1993 was a severe blow to the system and the US must continue to rely on the ageing Landsat 4/5. There is a high chance that both will fail before Landsat 7 appears in 1998.

The National Space Council established a joint DoD/NASA Landsat 7 programme in early 1992. The NSC designated the air force for the development of the spacecraft, instruments and launch services, and NASA for the ground system. Because of severe budgetary pressures and coupled with Landsat 6's loss, the National Science and Technology Council reassessed the purpose of Landsat 7. Congress appropriated US$84 million for DoD in FY93 for the procurement of L7, allowing launch then planned for 1997. In October 1992, the USAF awarded the US$398 million contract to Martin Marietta Astro Space. DoD agreed to build Landsat 7 in the expectation that it would receive 5 m imagery from the new US$207 million High-Resolution Multispectral Stereo Imager but NASA continued to resist funding the high costs for the required ground system. The Clinton Administration decided, in February 1994, that NASA should develop Landsat 7 for 1998 launch.

This produced Presidential Decision Directive/NSTC-3, dated 5 May 1994, changing it to a joint NASA, NOAA and US Geological Survey programme, dropping the military component. The Landsat 7 programme aims to maintain data continuity and minimise any gap as well as make L7 data available to all users at the cost of fulfilling requests.

NOAA is responsible for operations and data capture, processing, archiving and distribution. USGS will be NOAA's operational arm for the data element, and will continue the operation and maintenance of the National Satellite Land Remote Sensing Data Archive for the long-term preservation of Landsat and other satellite land remote sensing data.

The new Newcastle ground station site at Norman, Oklahoma was developed in late 1991 for integrated TT&C and data reception of Landsat 6-era operations, adding 158 m^2 to an existing 279 m^2 building. The mid-continent location allows real-time reception of data covering the contiguous 48 states. The station was formally dedicated 8 May 1992, discontinuing the datalink through NASA Goddard. The data are then transferred to EOSAT's Lanham headquarters for processing and distribution as CCTs (6,250/1,600 bpi), 8 mm Exabyte tapes and photo products. Swedish Space Corporation's Kiruna site provides back-up TT&C. Control is maintained from HQ's LSOC Landsat Spacecraft Operations Center via Newcastle or Kiruna.

More than 2 million digital images are archived at EOSAT in addition to those held in countries with Landsat ground stations operating under annual US$600,000 fees plus royalty payments: Australia, Brazil, Canada, China, Ecuador, India, Japan, Pakistan, Saudi Arabia, South Africa and Thailand, plus two stations for Europe's Eurimage network.

Specifications

Landsat 1

Launched 23 July 1972 by Delta from Vandenberg. 891 kg; 901 × 920 km, 99°. Operations ended 6 January 1978. By 30 March 1973, when faults occurred in the tape recorder, the North American continent had been covered 10 times; the total of >34,000 images included all the world's major landmasses at least once. The recorder fault meant that images could no longer be stored for transmission, though when the satellite was over one of NASA's three ground receiving stations in California, Alaska and Maryland it was still possible to transmit real-time imagery of the entire North American continent. By the time L1 had to be retired in 1978, it had returned 300,000 images and demonstrated the potential of remote sensing

Sensors: Return Beam Vidicon (RBV). Landsat 1/2 each carried three 80 m instruments sensitive to Band 1 0.475 to 0.575 μm (blue-green), Band 2 0.580 to 0.680 μm (yellow-red); and Band 3 0.690 to 0.830 μm (red-near-infra-red). Simultaneous views were provided of 185 km^2 areas. Landsat 3 carried two single-band 40 m RBVs providing adjacent 98 km^2 images, covering 0.505 to 0.750 μm. TM replaced the RBV on later satellites.

Multi-Spectral Scanner (MSS). All three included 80 m resolution four-band sensors, with L3 adding a 5th, thermal 120 m band. Band 4 0.50-0.60 (green), Band 5 0.60 to 0.70 (red), Band 6 0.70 to 0.80 (red-near infra-red), Band 7 0.80 to 1.10 μm (near-infra-red). L3 only: Band 8 10.40 to 12.50 (thermal-infra-red). Coverage was identical to RBVs and subsequent Landsat MSS.

Landsat 2

Launched 22 January 1975 by Delta from Vandenberg. 816 kg; 907 × 918 km, 99°. Operations ended 25 February 1982. This Landsat incorporated a 14 kg digital computer with 4,096-word memory able to handle 55 separate commands from ground stations and carry out routine operations for up to 24 h – a vital function as it was out of tracking station range for 80 per cent of the time. L2 was intended to supplement and later replace L1, and the two were able to work together until 1978, providing repeat coverage every 9 instead of 18 days. Retired on its fifth birthday in 1980 because of attitude control problems, L2 was revived six months later by Goddard engineers, who developed a magnetic compensation system to provide stabilisation. Though its tape recorders had failed, it still provided real-time imagery in 1981, supplementing the limited activities of L3

Sensors: Same as Landsat 1.

Landsat 3

Launched 5 March 1978 by Delta from Vandenberg. 960 kg; 900 × 918 km, 99°. Operations switched to standby 31 March 1983. Further sensor refinements included the addition of a thermal band to the MSS, able to detect T differences in vegetation, bodies of water and urban areas by day/night; and an improved RBV offering a 50 per cent resolution increase and enabling areas as small as half an acre to be identified. The three year Large Area Crop Inventory Experiment (LACIE), concluded in November 1978, showed that Landsat's multispectral scanners revealed the Soviet Union's wheat crop to be 91.4 million tonnes, less than 1 per cent below the official Soviet figure. There was less success with predictions of wheat yields in Canada and the US, where the long, narrow fields can easily be confused with adjoining spring-planted crops. Improvements in L4's instruments were intended to remedy this, making it impossible for countries such as the Soviet Union to repeat past successes in buying up world wheat surpluses cheaply when anticipating a poor crop of its own. As wheat is the world's most important grain crop, Australia, China, Brazil and Argentina were also included in LACIE. As a follow-on, it was decided to monitor wheat, corn, rice, barley, sorghum, soya bean and sunflower crops in the Soviet Union, China, Brazil, Mexico, Australia, Canada and US. In January 1981, after two years of operations, L3's MSS malfunctioned but the RBV continued to complement L2's operational MSS

Sensors: Same as Landsat 1.

Landsat 4

Launched: 16 July 1982 by Delta 3920 from VAFB

Mass: 1,941 kg at beginning of life

Contractors: General Electric (now Lockheed Martin Astro Space) prime, Fairchild for Multimission Spacecraft, Hughes Santa Barbara Research Center for sensors

Resolution: 30 m Thematic Mapper (120 m thermal-infra-red), 80 m Multi-Spectral Scanner

Orbit: 705 km circular in 98.2° Sun-synchronous, crossing equator at 09.45 local time, with 16 day repeat, 2,760 km ground-track separation at equator between revolutions, 172 km between tracks in completed coverage

Spacecraft: about 4 m long and is based upon the NASA/Fairchild Multimission Modular Spacecraft (MMS), as used initially by Solar Max, attached to the instrument/solar panel/antenna forward section. The MMS is essentially a triangular polyhedron with a propulsion module on one end and three 1.2 × 1.2 × 0.3 m box-like side modules handling power, AOCS and data/communication links. 3-axis control with pointing accuracy of 0.01° provided by MWs in conjunction with an inertial reference system using attitude updates from two star trackers. The aft propulsion module houses 11 N + 0.5 N hydrazine thrusters for maintaining the 16-day repeat cycle and MW unloading. Real-time TM data transmitted at 84.9 Mbit/s on X-band and MSS data at 15 Mbit/s on S-band; an S/Ku-band dish on a 4 m mast accesses the TDRS network. No data recorders. The single 4-panel solar wing was designed to provide 990/814 W BOL/EOL; this failing power system prompted the launch of Landsat 5

Sensors: Thematic Mapper (TM). Landsat 4 was the first to incorporate the TM, with its greater number of spectral bands and enhanced resolution. The 245 kg 1.1 × 0.7 × 2.0 m TM draws 345 W. The spectral ranges (μm) and physical characteristics measured by the bands are: Band 1 0.45-0.52 (chlorophyll and carotenoid for soil/vegetation and deciduous/coniferous differentiation), Band 2 0.52-0.60 (green reflectance by healthy vegetation), Band 3 0.63-0.69 (chlorophyll absorption for plant species differentiation), Band 4 0.76-0.90 (near-infra-red reflectance of healthy vegetation for biomass surveys), Band 5 1.55-1.75 (vegetation moisture and snow/cloud reflectance differences), Band 6 10.40-12.50 (thermal mapping), Band 7 2.08-2.35 (vegetation moisture and hydroxyl ions in soils). Band 6 exhibits 120 m resolution; the remainder are 30 m. Each uses an oscillating mirror to scan the Earth's surface in the cross-track direction as Landsat's motion provides along-track coverage. The 170 km long 185 km wide images have a 14 per cent side overlap at the equator, increasing with latitude.

Multi-Spectral Scanner. The Landsat 4/5 MSS are similar to those flown on Landsat 1-3, with the four bands offering 80 m resolution and the same coverage as the TM. The 58 kg 35 × 40 × 90 cm MSS draws 81 W. Spectral ranges (μm): Band 1 0.50– 0.60 (green); Band 2 0.60-0.70 (red); Band 3 0.70-0.80 (near-infra-red); Band 4 0.80-1.10 (near-infra-red). Bands 1-3 in the Ritchey-Chrétien telescope system utilise six photomultiplier tubes; band 4 uses four silicon

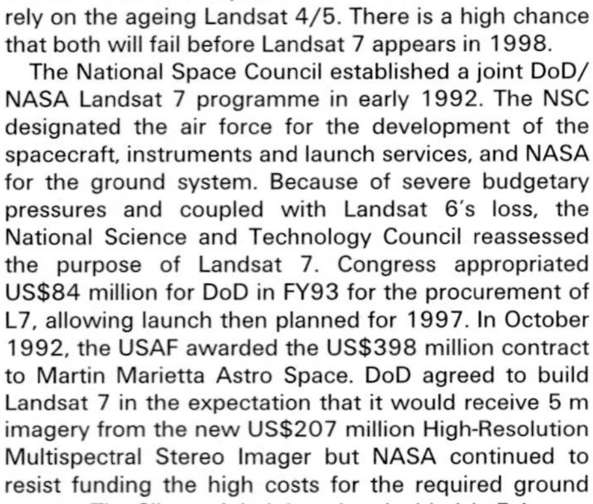

Principal Landsat 4/5 features (NASA)

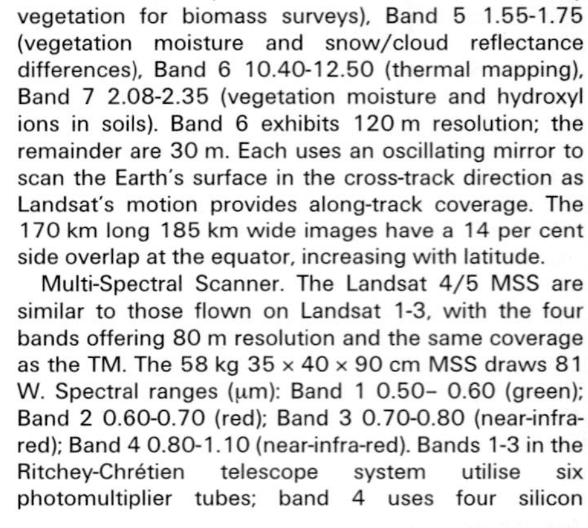

Landsat 5 checkout; Fairchild's MMS section is at bottom, with the instrument section and folded TDRS antenna above (EOSAT)

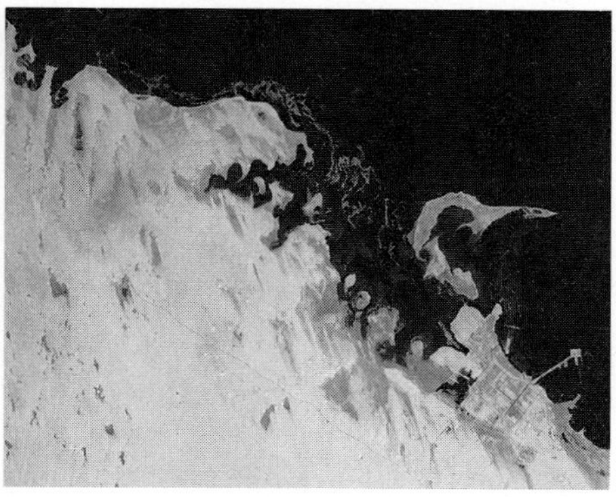

Landsat 5 image including an oil slick off Saudi Arabia (EOSAT) ***1995***

photodiodes. Data quantised to 6 bits (64 digital counts); transmission rate (real time only) 15 Mbit/s
Optics: The prime focal plane assembly of the Ritchey-Chrétien telescope with its 40 × 53 cm scan mirror contains four sets of 16 Si photodiodes for the VNIR bands 1-4. The cooled focal plane assembly contains two arrays of 16 InSb photodiodes for bands 5/7, and four HgCdTe detectors for thermal band 6. The TM also offers improved radiometric sensitivity over the MSS, permitting the data to be quantised to 8 bits (256 digital counts), thus providing a four-fold increase in the grey scale. Data rate (real-time only) 84.9 Mbit/s.

Landsat 5
Contractors: Same as Landsat 4
Resolution: Same as Landsat 4
Launched: 1 March 1984 by Delta 3910 from VAFB
Mass: 1,941 kg BOL (249 kg hydrazine, 150 kg remained May 1995)
Orbit: Same as Landsat 4
Spacecraft: Same as Landsat 4, but modified to prevent repetition of solar array power cabling and X-band transmitter failures
Sensors: Same as Landsat 4.

Landsat 6
Launched 5 October 1993 by Titan 2 from VAFB. It appears not to have attained orbit, although Titan apparently performed as expected. The converted ICBM released it with a 724.7 km (target 724.5 km) apogee and 98.0107° (target 98.006) inclination. L6's Thiokol Star 37XFP solid apogee motor was to have ignited at 819 seconds to establish the near-final 704 km 98.2° circular orbit. However, Landsat's attitude control thrusters were useless because their hydrazine tank had ruptured, leaving it without control during the kick motor's burn and leading to it burning up over the Pacific. The cause was difficult to identify as Landsat telemetry was switched off to avoid interference with Titan's stage 2 transmitter. L6 was almost identical to L7, except that ETM's thermal-IR band provided only 120 m resolution. Launch mass was about 2,750 kg, with 1,740 kg on-orbit BOL. Design life five years. EOSAT signed L6's rescoped development contract with the US Department of Commerce 31 March 1988, Congress having released US$62.5 million for programme start. Total cost was US$268.9 million, including US$36.5 million for launch, US$220 million for the space/ground segments, with US$10.8 million

Landsat 7 is almost identical to Landsat 6 (shown here) (Lockheed Martin) ***1995***

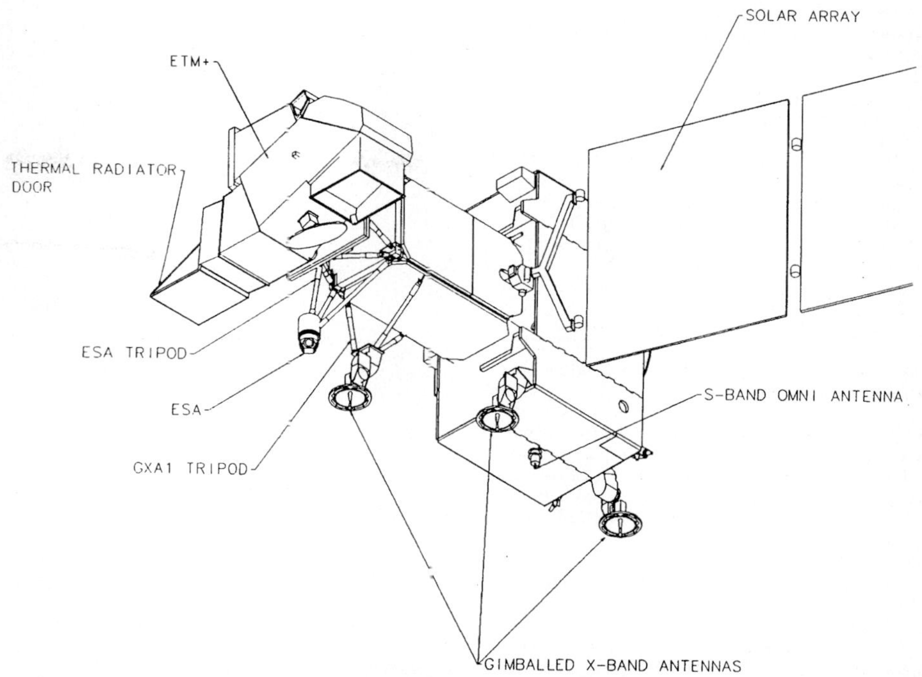

Landsat 7 principal features. ESA is the Earth Sensor Assembly. Velocity vector is to the left (NASA) ***1995***

contributed by EOSAT. L6 was to continue its predecessors' TM products, while adding a co-registered 15 m PAN band to the US$92 million 390 kg ETM. See p400 of the 1994/95 edition of *Jane's Space Directory* for the full L6 entry.

Landsat 7
Contractors: Lockheed Martin Astro Space (satellite, flight systems integration), Hughes Santa Barbara Research Center (ETM+)
Resolution: ETM+ 30 m (6 bands), 15 m panchromatic, 60 m thermal-infra-red
Launch: 15 April 1999 by Delta 27920 from VAFB
Mass: About 2,170 kg at launch, about 2,020 kg on-orbit BOL
Orbit: 705 km circular in 98.2° Sun-synchronous, descending node at 10.00±00.15 local time, with 16-day repeat cycle (233 revolutions) and 172 km ground track separation at equator
Spacecraft: About 4.3 m long and based on Tiros/DMSP meteorological satellite bus but with extensive subsystem modifications for Landsat-unique requirements and to improve mission reliability. Sensor mounted on the forward module. Three steerable high-gain X-band antennas operating at 8.0825, 8.2125 or 8.2434 GHz. Each provides two 75 Mbit/s streams (I/Q) of ETM+ data, with bands 1-6 on one channel and 7-8 on other. 375 Gbit SSR can store about 100 full ETM+ scenes. Stored and real-time data acquired at USGS EROS Data Center in Sioux Falls. Real-time data available to international stations on X-band. Command uplink 2.1064 GHz; telemetry downlink 2.2875 GHz. 1,550 W generated by a 4-panel single solar wing driven at orbital rate to follow the Sun. Eclipse power by 2 × 50 Ah Ni/H_2 batteries. 3-axis attitude control by RWs and magnetorquers, accuracy 180 arcsec; attitude information from star sensors, gyros Earth sensors and magnetometer. Ground track held within 5 km by 12 × 4.45 N hydrazine thrusters on lower module
Sensors: Enhanced Thematic Mapper + ETM+ is a nadir-viewing cross-track scanning 8-band multispectral radiometer, developed by Hughes SBRC. It provides continuity of TM data with Landsat 4/5 and adds a 15 m PAN band, improved resolution (60 m) for thermal, and improved radiometric accuracy (5 per cent). All other bands remain at 30 m. ETM+ can image >525 scenes daily, with a 185 km swath width and 170 km along-track length. Spectral ranges (µm): Band 1 0.45-0.52, Band 2 0.52-0.60; Band 3 0.63-0.69; Band 4 0.76-0.90; Band 5 1.55-1.75; Band 6 10.4-12.5 (thermal mapping); Band 7 2.08-2.35; Band 8 (PAN) 0.50-0.90. Imagery can be collected in low- or high-gain modes; high gain doubles the sensitivity.

Lewis and Clark satellites

Background
NASA's Small Spacecraft Technology Initiative plans to demonstrate new techniques, to speed the development and reduce the costs of next-generation satellites. The Office let two contracts on 8 June 1994 for Earth observation satellites with unprecedented sensor technology to be developed and delivered on-orbit within 24 months. From final announcement to contract signing consumed 70 days instead of the usual six months to one year.

TRW received US$59 million to build Lewis – the first hyper-spectral imaging system. Lewis' 384-channel HSI Hyper Spectral Imager has commercial applications in forestry, agriculture, water, land-use management and environmental monitoring. The high spectral resolution can distinguish between tree varieties, for example, and their states of health.

CTA's US$49 million Clark offers high resolution with stereo imaging. A key new capability of both is cloud editing, which 'remembers' cloud locations for later replacement with clear data. Both carry additional instruments for space physics and for global atmospheric pollution dynamics information. The high resolution locates utility pipelines and cables, and help town planners with, for example, assessments of transport needs and construction sites. The same imager will fly on EarthWatch's commercial satellite, which CTA Space Systems is building. CTA at one stage proposed to buy Clark back from NASA after one year of operations in order to use it as an EarthWatch back-up.

The Lewis and Clark have introduced key spacecraft technologies which include GPS attitude determination, a fibre optic data bus, CPV common pressure vessel, Ni/H_2 battery, composite structures, miniature star trackers, 32-bit processors, 2 Gbit solid-state recorders and multijunction GaAs solar arrays. After launch in August 1997, Lewis was found to be spinning at an unacceptable 2 rev/min and as this could not be stopped the satellite was inoperable. Subsequently, Clark was cancelled.

Specifications
Lewis
Launch: 23 August 1997 by Lockheed Martin LMLV 1 from VAFB SLC-6
Design life: 5 years
Mass: 385.6 kg at launch
Contractors: TRW Space & Electronics Group, Harris Corp, AlliedSignal Aerospace, Gulton, Ithaco, Hughes Danbury, Jackson & Tull, NASA Goddard, NASA Lewis, NASA Stennis, NASA Langley satellite built in only 24 months from the contract being awarded
Resolution: 5 m panchromatic, 30 m hyper-spectral
Orbit: planned 523 km circular, 97.4° Sun-synchronous
Spacecraft: bus 149.86 cm diameter hexagonal prism, 203.2 cm long. Zero momentum bias 3-axis stabilised using magnetically suspended RWs, GPS attitude

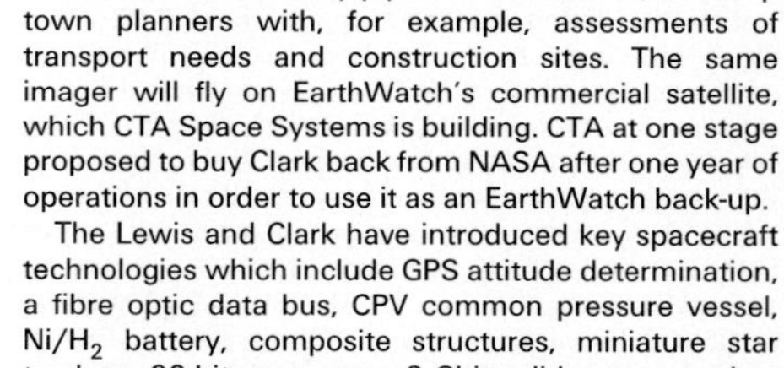

TRW's Lewis carries a 384-channel imager ***1996***

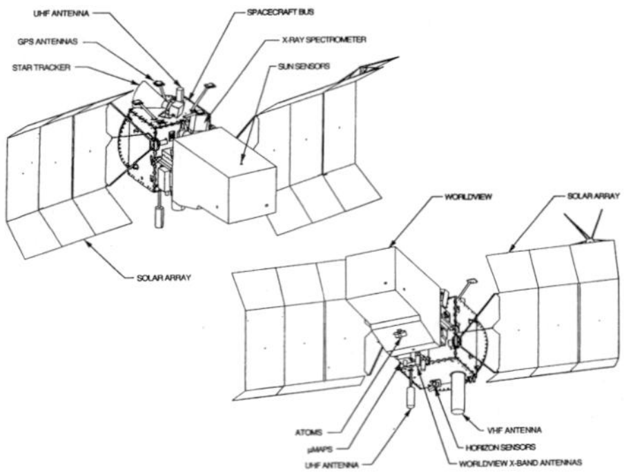

CTA's Clark carries a 3 m-resolution imager. WorldView is now named EarthWatch (CTA) ***1995***

determination + 20° FOV star tracker (6 star capability). Wheel unloading + orbit adjust by four pairs of 4.45 N hydrazine thrusters. 740 W from dual deployed solar wings, including multijunction GaAs and Amorphous Si. R3000 processor, fibre optic bus, 4 Gbit SSR, S-band STDN-compatible

Sensors/payload: HSI Hyper Spectral Imager, 384 0.01 μm channels over 0.4-2.5 μm, 30 m XS resolution, 5 m PAN.

LEISA Linear Etalon Imaging Spectral Array, 256 channels, 300 m resolution, map reflectance spectra of surface and atmospheric features.

UCB Ultraviolet Cosmic Background, 580- 950 Å grating spectrometer, map astrophysical component of EUV cosmic background

Components: TRW pulse tube cryocooler.

Clark

Launch: originally planned for November 1996 by Lockheed Martin LMLV 1 from VAFB SLC-6
Design life: 3 years
Mass: 278 kg at launch, including 94 kg payload and 12 kg hydrazine
Contractors: CTA Space Systems, Lockheed Martin Astronautics, Signal Corp, Futron, Odetics, WJ Schafer, CH2M Hill, NASA Goddard, NASA Langley, NASA Stennis, MIT, Stanford University
Resolution: 3 m panchromatic, 15 m multispectral
Orbit: planned 476 km circular, 97.3° Sun-synchronous, 11.15 descending node
Spacecraft: 73.9 × 83.8 × 61.1 cm bus; 163.7 cm high imager on 61.1 cm high bus. 3-axis zero momentum stabilisation (pitch bias back-up) by three 0.025 Nm RWs + three 17 Am^2 magnetorquers (0.15° attitude determination by GPS, 0.1° roll/pitch Earth horizon sensors, 5 arcsec star tracker, 0.015°/h gyro). Orbit correction/drag makeup by two 0.89 N thrusters. 584 W from cells on two wings totalling 3.09 m^2; 0.14 mm-thick GaAs cells on primary panels, CIS and multijunction cells on secondary (reflectors increase output), supported by two 15 Ah CPV Ni/H_2 batteries. RHC3000 processor with 8 Mbit × 40 bit DRAM memory, fibre optic bus, 16 Gbit 160 Mbit/s Odetics SSR. Imagery transmitted on X-band. Redundant 10 W UHF 2.4/19.2 kbit/s up/down; UHF for TT&C, VHF for store/forward. Shape memory devices replace pyrotechnics
Sensors/payload: PAN. 3 m plus 15 m XS imagery. Step-stare for 3 m collects 6 × 6 km images over 30 × 30 km by gimballed steering mirror. Mirror allows ±30° along/cross track, allowing stereo pairs with 5 m height resolution. Images transmitted at 160 Mbit/s X-band.

μMAPS Measurement of Air Pollution from Satellites is a miniaturised version of the NASA Langley MAPS gas filter radiometer flown on Shuttle. Measures CO mixing ratios in three altitude regions; experimentally uses N_2O capability to estimate image sensor cloud coverage.

Atmospheric Tomography. Proof of concept demonstration of 2-D mapping of atmospheric constituent gases by using ground laser (at air force Maui AMOS station) and satellite corner reflector to map CO using the 4.7 μm absorption band.

XRS X-ray spectrometer, 2-20 keV by eight 8 × 8 mm Si Avalanche PhotoDiodes (APD) and 10-80 keV by 12 8 × 8 mm CdZnTe detectors. Primary goal is to test these new semiconductor detectors, all operating at or near room temperature. Secondary is to determine spectra of solar flares and cosmic γ bursts. The ms timing of γ bursts will help to locate their origin by timing with similar measurements from other satellites. NASA Goddard.

MicroLab 1 satellite

OSC/NASA Marshall in August 1993 signed an all-inclusive contract for the provision of atmospheric research data from NASA's payload aboard OSC's MicroLab 1. The Optical Transient Detector is studying

MicroLab 1 attached to the Pegasus final stage. The optical sensor is covered by the box at top. OSC's first two Orbcomm satellites were then stacked forward of MicroLab (OSC) ***1995***

MicroLab 1 is returning lightning data (OSC) ***1995***

the spatial/temporal distribution of global lightning to better understand and predict major atmospheric storm systems and climate changes. It is a precursor to the LIS Lightning Imaging Sensor on 1997's NASA/Japanese TRMM mission. 500 images/s are recorded and the real-time event processor handles 10 million pixels/s to extract lightning signals from the bright background. Microlab data are downlinked twice daily to OSC's tracking station in Fairmont, West Virginia. The data are analysed in NASA Marshall's Global Hydrology & Climate Center and archived/distributed by the EOSDIS Earth Observing System Data Information System.

One useful finding of the satellite has resulted from observations of a 17 April 1995 storm in Oklahoma. Results suggested that increased intracloud lightning presages tornadoes, a possible method to perform tornado predictions and provide population alerts from space.

MicroLab's GPS meteorological experiment uses an onboard eight-channel TurboStar GPS receiver (donated by Allen Osborne Associates) daily recording 500 occultations of GPS signals to map atmospheric temperature and water vapour content. The experiment is managed by the UCAR (University Corporation for Atmospheric Research) for the National Science Foundation, which paid some US$500,000 to OSC. The total price paid to OSC for the two year mission is more than US$7 million. MicroLab is based on OSC's MicroStar small standard platform, developed for the Orbcomm satellite messaging system.

Specifications

Mass: 68 kg
Launch: 3 April 1995
Orbit: 733 × 749 km, 70.0°.

NOAA series (TIROS/TOS/ITOS/NOAA/GOES)

Current status

NOAA launched NOAA-K on 13 May 1998.

The GOES-West/East satellites, at 135° and 75° west GEO, provide synoptic visible/infra-red imaging and infra-red/thermal sounding for atmospheric temperature profiles. GOES 7 remains in orbit, left as the last representative of the earlier technology, after a bulb failed in GOES 6's scanning encoding system January 1989. The replacement new-generation GOES 8 did not occupy the 75° west slot until February 1995; GOES 9 took over at 135° west in January 1996.

Background

The United States has launched 35 civil Tiros/north polar meteorological satellites (and suffered three launch failures) in three distinct programmes since the beginning of the space age. NASA named the first series, launched during 1960-65, Tiros. Then it followed the Tiros programme with nine more satellites over the 1966-1969 period. It designated these the Tiros Operational Satellite (TOS) and sometimes called them the ESSA (Environmental Science Services Administration) satellites after the organisation that managed the programme during that period. NOAA introduced the second-generation Improved TOS (ITOS) in 1970. In order to enhance the ITOS satellites, NOAA built a prototype Synchronous Meteorological Satellites (SMS) in 1974, to provide synaptic coverage of short-term and other weather features from GEO. These geosynchronous satellites set the stage for a permanent weather station in geosynchronous and acted as the testbed and forerunners of GOES, which now provides continuous coverage of North and South America and its oceans. NOAA polar spacecraft distribute their data to Earth stations in 120 nations as they pass overhead, and thousands of schools, private individuals and others receive NOAA imagery.

TIROS

The Television & Infra-red Observation Satellite programme began as a joint NASA/Department of Defence project to develop meteorological satellites. Tiros 1 was launched in April 1960, and orbited at 692 × 740 km. The satellite demonstrated the value of an orbital viewpoint for meteorological applications. During its 78 day battery life, it returned 22,952 cloud cover images. The original Tiros satellites were basically cylindrical 18-sided prisms, 1.07 m in diameter and 0.55 m high. Solar cells covered the sides and top, with apertures for two TV cameras on opposite sides of the spacecraft body. Each camera captured 16 images a revolution at 128 second intervals. The image system passed its information to two tape recorders capable of storing up to 48 scenes when ground stations were out of range. Tiros 1's 119 kg increased to 138 kg over the course of design changes to Tiros 9/10.

By the time Tiros 10 had been launched, NASA had already begun to phase in the more advanced Nimbus and ESSA satellites. The first eight Tiros', all operating in similar orbits, transmitted several hundred thousand images, together with information on the Earth's heat budget. Tiros 9, marked the first time that NASA attempted to reach polar orbit from Cape Canaveral. It used a series of three Delta dogleg manoeuvres to reach the 82° inclination angle, but a stage 2 overburn meant that the orbit, instead of being 644 km circular, was 700 × 2,578 km.

Tiros 10 transmitted the first photomosaic of the entire world's cloud cover in 450 high-quality images. By the time Tiros 10 shut down on 3 July 1967, 500,000 cloud-cover pictures had been returned.

TOS/ESSA

TOS/ESSA satellites, similar but more advanced, carried two Automatic Picture Transmission (APT) cameras able to image a 3,000 km wide strip with 3 km resolution at the centre of the picture. The ESSA satellites transmitted Images every 352 seconds allowing a typical APT station to receive 8 to10 daily. They began operating on 3 February 1966 from a 702 × 845 km orbit. The series ended with ESSA 9 launched 26 February 1969 into a 1,427 × 1,508 km orbit. By that time, various countries had set up 400 receiving stations and the weather services of 45 countries, as well as 26 universities, and up to 30 US TV stations as well as an unknown number of private citizens relying on home-built receivers, used the imagery routinely. In 1969, an ESSA 7 scene made history by revealing that the snow cover over America's Midwest, in Minnesota and the Dakotas, was three times thicker than normal. Measurements showed that it was equivalent to 15 to 25 cm of water covering thousands of square miles. A disaster area was declared before the event, and when the floods came much had been done to control the situation.

ITOS

The Improved Tiros Operational System (ITOS) added IR sensors to permit night observations. NOAA superseded ESSA during the period that the agency launched the satellites and so they were designated ITOS before launch and once in orbit given a new series of NOAA designations. Thus NOAA 1, the second-generation prototype, and launched 11 December 1970 from Vandenberg into a 1,429 × 1,472 km, 101° inclination orbit began the series. ITOS-B since it failed to achieve orbit never received a NOAA designation. And the NOAA designation resumed with NOAA 3/4 launched in November 1973 and 1974 respectively.

NOAA

The US National Oceanic and Atmospheric Administration (NOAA) operates the polar-orbiting NOAA and geostationary GOES environmental systems of two satellites each, procured for NOAA by NASA through its Goddard operations centre. Launched into near-polar Sun-synchronous orbits, each NOAA can

Some NOAA polar satellites also operate as part of the Cospas/Sarsat international search and rescue system, relaying distress calls from ships and aircraft (Lockheed Martin Astro Space)

view almost all of Earth's surface twice every 24 hours. In addition to returning weather imagery, they monitor atmospheric temperature and humidity, sea surface temperature to 1.6°C accuracy, snow/ice cover, and total ozone content. near-Earth proton/electron fluxes atmospheric aerosols and radiation fluxes, and relay Argos surface platform environmental data and Cospas/Sarsat distress signals. NOAA designs its satellites for a nominal two year lifetime.

NOAA satellites carry two basic sensors (AVHRR/TOVS), and afternoon satellites such as NOAA 11/14 include the Solar Backscatter UV Spectral Radiometer, Mod 2 (SBUV/2, non-scanning 1,600-4,000 Å) for determining ozone levels to 1 per cent absolute accuracy. It also carries a Cospas/Sarsat distress relay, Argos Data Collection & Location System transponder, and a Space Environment Monitor.

NOAA 8 was the first of RCA's Advanced Tiros-N (ATN) satellites operating as part of the international Cospas-Sarsat search and rescue system. In June 1984 it began tumbling following failure of the master timing system, and the refusal of the back-up system to activate. All attempts to save the US$30 million satellite failed, and it had to be abandoned, presumed dead, leaving NASA with no search/rescue capability.

NOAA 9 introduced an ozone mapper being flown on subsequent satellites with near-noon equator crossings. NOAA launched the morning orbit satellite counterpart NOAA 10 in September 1986 as NOAA-G, which it replaced, by NOAA-D/12 in 1991. NOAA 9, after a series of weather and technical delays, restored both weather and rescue services from its afternoon orbit. It also carried NASA's Earth Radiation Budget Experiment scanner and non-scanner instruments. The morning satellite NOAA-10 replaced the degraded NOAA 6, which was brought back into use following the failure of NOAA 8. NOAA 10's Sarsat equipment restored the ability of the US to contribute to the international Cospas/Sarsat search/rescue system. Within six days of becoming operational, it had picked up its first distress signal, which led to the rescue of four Canadians whose aircraft had crashed in a remote area of Ontario.

One of the three attitude control gyros failed on NOAA 11 in August 1989 and the single back-up unit was brought on line. Software for operating with only two gyros was uplinked 8 August 1990 a month before the expected second failure. Software has been developed for operations with no gyros. AVHRR failed 13 September 1994 after operating since 8 November 1988. Placed on standby 1995 after being replaced by NOAA 14. The satellite cost US$53.5 million.

NOAA intended to use NOAA 13 as the prime afternoon satellite. However, failure of the power link from the solar panel lost contact 21 August during checkout. The September 1994 review board report identified a short-circuit in the battery charge assembly that prevented the solar arrays from charging the batteries. The short was probably caused by a 31 mm screw penetrating insulation and contacting a radiator plate. Inspection of NOAA-J (NOAA 14) showed 10 of the 12 screws to be in danger of causing the same fault.

NOAA plans to continue launches, on average, at 16-month intervals for the foreseeable future. It launched NOAA 13 (I) 9 August 1993 to replace NOAA 11 but this new satellite suffered a power failure within days. As a replacement NOAA converted NOAA-J, originally planned to replace NOAA 12, into an afternoon role and launched it in December 1994. In the coming decade NOAA plans to drop its morning operational satellites and turn meteorological monitoring into a co-operative effort with Europe's Metop.

President Clinton, in May 1994, approved a long awaited plan to combine NOAA's polar system with the military DMSP, halving the number of operational satellites to two. NOAA had planned a new series of Tiros-Next satellites (designated O-Q) for 2005, 2008 and 2011. The first joint military/civilian satellites may appear in 2007, controlled by NOAA from its Suitland, Maryland centre. To prepare, Suitland will take over DMSP control beginning in 1998, and DoD will phase out its own station retaining only one as back-up. The plan projects a US$300 million saving through 1999, by eliminating duplications and then at least US$1 bill on over the next 15 years.

NASA placed NOAA-11 on standby 1995 after being replaced by NOAA 14, crossing the equator in the afternoon. The Centre de Meteorologie Spatiale in Lannion, France can relay stored/real-time data. Afternoon satellites such as NOAA 11/ 14 include the Solar Backscatter UV Spectral Radiometer, Mod 2 (SBUV/2, non-scanning 1,6004,000 A) for determining ozone levels to 1 per cent absolute accuracy.

NOAA planned to place NOAA 14 in the morning orbit as a replacement for NOAA 12, but then the loss of NOAA 13 forced the agency to reconfigure the satellite for NOAA 11's afternoon role. This required replacing the Naval Research Lab's RAIDS Remote Atmosphere Ionospheric Detection System with SBUV. RAIDS is a precursor to two instruments planned for DMSP 5D3 that will improve prediction of ionospheric disruptions to communications and Over-the-Horizon radar systems. It probably will never fly since the next generation of satellites cannot carry it.

NASA built a new fourth-generation prototype for NOAA and named it the NOAA-N, and launched it into an 850 × 866 km orbit at 102° inclination. NOAA 6 joined it on 27 June 1979. Developed, like the earlier generations, by RCA, these were the start of a series based on the Block 5D bus developed for the USAF DMSP. NOAA-B failed to achieve the correct orbit, but by the time NOAA 7 was in orbit (launched 23 June 1981), NASA claimed another success for the programme. Fishermen in California, Oregon, Washington and Alaska noted improved catches as a result of following the NOAA sea surface temperature charts. One 60-vessel towing and transportation company announced fuel savings of 20 to 40 per cent as a result of choosing routes with the help of NOAA's stream and loop current information. NOAA 7 cost US$15 million plus US$7.5 million launch. NOAA-D was the last Tiros-N, launched out of sequence in 1991 as NOAA 12.

Beginning with NOAA-M, the US will fly only afternoon satellites; and Europe will be responsible for the morning orbit making NOAA-K the last US morning satellite. This satellite includes a channel for humidity profiles, delineating sea ice and open water, and monitoring snow thickness and soil moisture.

SMS/GOES

While the NOAA satellites added a significant improvement to the weather gathering capabilities of meteorologists, NOAA sought a continuous scene generator, which required a move to geosynchronous orbit. NASA built two prototype Synchronous Meteorological Satellites (SMS) to answer this need which it later released to NOAA as part of the Geostationary Operational Environmental Satellite system. NASA remains responsible for spacecraft procurement. SMS/GEOS satellites have a Visible/Infra-red Spin Scan Radiometer (VISSR) to differentiate between water land and clouds. By GOES 4 the more advanced VAS added data on atmospheric temperature and water vapour content. NASA stationed SMS 1 initially over the east Atlantic, and then moved it to 75° west over Bogota, to give the first day/night storm watch and alert capability, with full disc pictures of the Western hemisphere every 30 minutes.

As part of the Global Atmospheric Research Program (GARP), SMS-1 provided the first continuous coverage of a major hurricane, designated Carmen, in September 1974. SMS 2 joined the first satellites when NASA stationed it over 115° west, just east of Hawaii, completing total coverage for the western hemisphere. NOAA assigned it the tasks of watching California's forest areas to give warning within 90 minutes of fire outbreaks.

During the Global Weather Experiment of 1977-78, NOAA positioned GOES 1 over the Indian Ocean at 60° east, with ESA's ground station in Spain processing data. The satellite later moved to 116° west and 135° west.

GOES 2 performed as expected until a VISSR primary encoder failed 26 January 1979 and the back-up encoder failed 18 December 1978, ending this satellite's imaging function after returning 20,591 full disc images.

On GOES 3, the VISSR problem continued to plague the GOES system and this satellite the replacement for GOES 1 operated in a degraded mode from 14 September 1979 until the encoder finally failed 6 May 1981 after returning 36,190 full disc images. After its failure, NASA moved the satellite to 176° west from 130° west in June 1990 for use by Pacific island nations as a communication satellites through the Peacesat Pan-Pacific Educational and Cultural Experiments. But perhaps the most successful of all of the meteorological satellites, in 764 operational days, the GOES 4 provided warning of many Pacific storms, transmitting 28,500 scenes.

NASA intended to use the GOES 6 as the east coast satellite, but following GOES 5 failure, the Agency moved it to 98° west to cover both US coasts; then to 108° west to monitor winter storms in the Pacific off the US northwest coast. For almost three years following the failure of GOES 5, GOES 6 had to do the work of two satellites by being moved with the seasons between 98 and 108° west.

At the beginning of 1991, NOAA planned that it would follow the mid-1992 launch of GOES-I, eight months later by GOES-J and by the 1996 debut of GOES-K. However, instrument technical problems surfaced during the year, including the use of incorrect wiring, prompted the delay of GOES-I to at least December 1993. At one time during the correction process, NOAA considered launching at end of 1992 with below specification instruments. Congress set aside an additional US$110 million to ensure the programme continued, if required. The Department of Commerce decided instead to press for a full-performance GOES-I while buying time with Europe's Meteosat.

GOES 7 began at the GOES position, but later moved west following the GOES 6 failure, to 108° west for the winter season, then to 98° west in the Spring for observing hurricanes over the Atlantic and returning further west in the autumn. This GOES carries the Space Environment Monitor, a Data Collection System. To improve station-keeping lifetime, NASA included a small conical sail on a 17 m boom, on GOES 8 to balance solar radiation torque and reduce station-keeping propellant use; a trim tab on solar wing provides fine control.

GOES-L/M will take the series into the next century, with launches tentatively set for 2000/2004. NOAA began work on the series beyond GOES-M during 1988, holding meetings in early 1989 to define requirements. These were passed on to NASA in June 1989 for Goddard Space Flight Center to undertake Phase A, completed during 1990.

NOAA revealed in February 1995, in its FY96 budget request, that it intends to order three more similar satellites, plus an option on a fourth, before beginning Phase A of the GOES-R next generation in 1998 for first launch around 2008. FY96's request includes funding to begin procurement of GOES-N/O. The contract will be awarded in late 1997 (for 2002 first delivery) after the competition is opened by end-1996. It is expected that sounder resolution will be improved. Adding star trackers will improve location identification to 1 km from the current 4 to 6 km, improving forecasting.

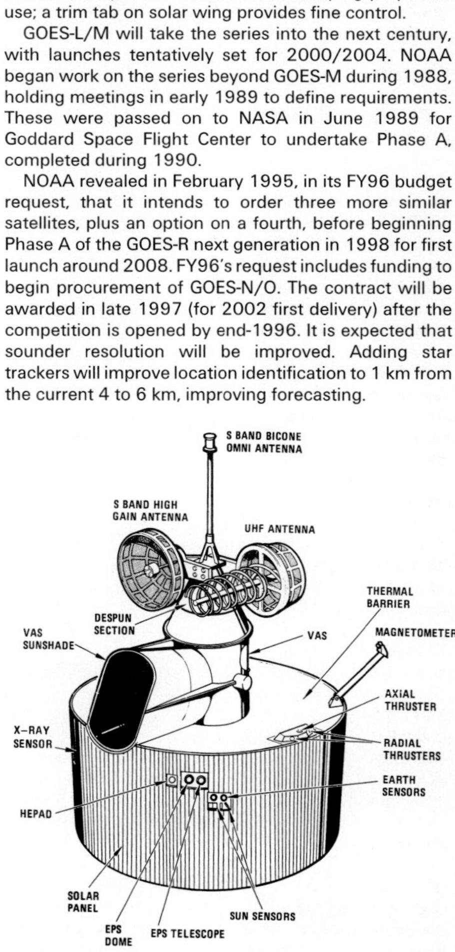

Principal GOES 7 features (Hughes)

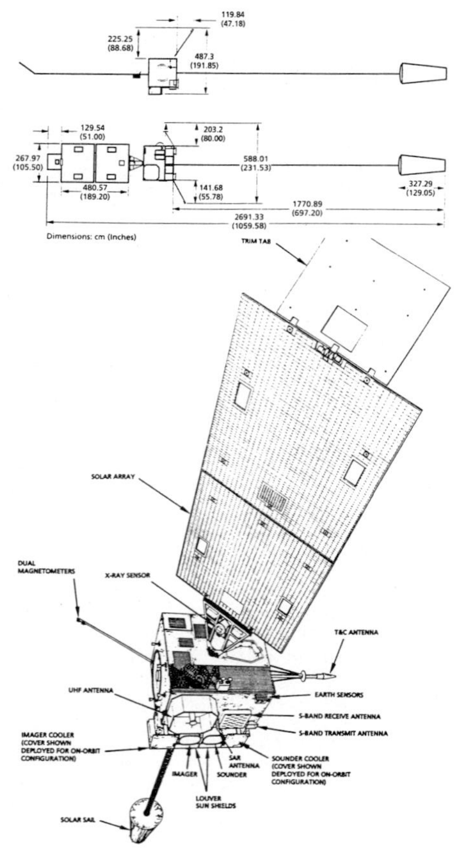

GOES 8-12 principal features (NOAA) ***1994***

Studies may show that this payload should be divided between several smaller carriers. Similar to the previous polar orbiters and built by Lockheed Martin Astro Space, this new generation carries different instruments. Three were initially ordered at US$160 million (excluding instruments and launches); N/N-prime were added in 1993 (NASA made the US$151 million award January 1995) to allow more time to prepare the next series. Beginning with NOAA-M, the US will fly only afternoon satellites; Europe will be responsible for the morning orbit. DSB has been eliminated, although sounder data are available through HRPT broadcasts. An eight-channel ocean colour instrument for VNIR sensing was considered.

Specifications

Tiros 1
Launched: 1 April 1960
Orbit: 692 × 740 km.

TOS
Launched: as ESSA 1 on 3 February 1966 (702 × 845 km) and ended with ESSA 9 (launched 26 February 1969, 1,427 × 1,508 km).

NOAA 1
Launched: 11 December 1970 from Vandenberg
Orbit: 1,429 × 1,472 km, 101°.

NOAA 2 (ITOS-B)
Launched: 21 October 1971
Comment: This launch was never allocated an NOAA designation because it failed to achieve a satisfactory orbit.

NOAA 3/4
Launched: November 1973/74.

NOAA 5
Mass: 340 kg
Launched: July 1976.

Tiros N/NOAA-A
Launched: 13 October 1978 (first of the series known as Tiros N – the NASA prototype)
Mass: 734 kg
Orbit: 850 × 866 km 102°
Comment: The complementary NOAA 6 joined the satellite on 27 June 1979. Developed, like the earlier generations, by RCA, these were the start of a series based on the Block 5D bus developed for the USAF DMSP.

NOAA-B
Launched: 29 May 1980
Comment: Failed to achieve the correct orbit, but by the time NOAA 7 was in orbit (launched 23 June 1981) it was possible to claim that fishermen in California, Oregon, Washington and Alaska were improving their catches by using the sea surface temperature charts. One 60-vessel towing and transportation company announced fuel savings of 20-40 per cent as a result of choosing routes with the help of NOAA's stream and loop current information.

NOAA 7 transmitted this image of Hurricane Lili over the north Atlantic on 20 December 1984, only the third recorded hurricane in that region (the others were in 1887 and 1954)

NOAA 7
Cost: US$15 million plus US$7.5 million launch.

NOAA 8
Launched: 28 March 1983 by Atlas from Vandenberg
Mass: 1,712 kg
Orbit: 833 km circular at 98.3°
Comment: First of RCA's Advanced Tiros-N (ATN) operating as part of the international Cospas-Sarsat search/rescue system. High-resolution instruments measured both surface and vertical temperature; a UK stratospheric sounding unit monitored upper atmosphere temperature and the French Argos system relayed data from balloons, buoys and remote weather stations. Attitude control problems began once in orbit. In June 1984, the satellite began tumbling following failure of the master timing system, and the refusal of the back-up system to activate. All attempts to save the US$30 million satellite failed, and it had to be abandoned, presumed dead, leaving NASA with no search/rescue capability. NOAA 6, placed on standby following a malfunction of its AVHRR, was reactivated to fulfil NOAA 8's other functions. Then, in June 1985, NOAA 8's back-up attitude control system unexpectedly came back online and its use was fully recovered. However, an overcharged battery burst in December 1985, some contact was restored, but NNOAA finally shut it down on 8 January 1986.

NOAA 9
Launched: 12 December 1984 by Atlas from Vandenberg
Comment: After a series of weather and technical delays, restored both weather and rescue services from its afternoon orbit. Carried NASA's Earth Radiation Budget Experiment scanner and non-scanner instruments (scanner failed 20 January 1987, non-scanner remains operational); see the ERBS satellite entry in the main national US section. Also the first to carry the SBUV/2 ozone mapping instrument. Satellite cost was US$43.5 million plus US$11.4 million launch. Placed on standby following the launch of NOAA 11. Its MSU channel 2 failed 8 March 1987, number 3 7 May 1987, prompting launch of NOAA 11. Collection of SBUV and ERBE non-scanner data continues. S&R function decommissioned March 1995. AVHRR remains usable, although 3.7 μm channel out of specification. The power system is marginal due to array shunt failures and battery problems.

NOAA 10
Launched: 17 September 1986 by Atlas E from Vandenberg after 16 delays (planned for August 1985)
Orbit: Morning 808 × 826 km, 98.8°
Comment: This satellite replaced the degraded NOAA 6, which was brought back into use following the failure of NOAA 8. NOAA 10's Sarsat equipment restored the ability of the US to contribute to the international Cospas/Sarsat search/rescue system. Within six days of becoming operational, it had picked up its first distress signal, which led to the rescue of four Canadians whose aircraft had crashed in a remote area of Ontario
Sensors: Same as NOAA 11 but does not carry SBUV/2 and incorporated Earth Radiation Budget Experiment scanner and non-scanner for NASA Langley. ERBE scanner failed 22 May 1989 (non-scanner solar monitor shutter stuck open, but continues to provide usable data). The Sarsat processor receiver failed 8 September 1988. Replaced by NOAA 12 in May 1991. NOAA 10 remains on standby, but has some degraded performance in the AVHRR, SEM and power systems.

NOAA 11
Launched: 24 September 1988 by Atlas E from Vandenberg AFB; two year design lifetime
Mass: about 1,700 kg BOL
Cost: US$53.5 million
Contractors: GE Astro Space (prime), ITT (HIRS, AVHRR), NASA/JPL (microwave sounding unit)
Resolution: 1.1 km AVHRR, 20 km HIRS/2, 147 km SSU, 105 km MSU
Orbit: 849 × 865 km, 98.9° Sun-synchronous, crossing equator in afternoon (originally 13.40; drifted to 16.02 by May 1993) on ascending node with daily repeat cycle
Spacecraft: Astro Space's Advanced Tiros-N platform, 1.80 m wide, 4.18 m long with single 2.37 × 4.91 m 8-panel 11.6 m² solar wing deployed, providing 1,500/1,400 W BOL/EOL; 3 × 26.5 Ah Ni/Cd batteries. Injection provided by Thiokol Star 37S solid. ±0.2° 3-axis control by RWs using magnetic unloading; 8 GN_2 thrusters (4.23 kg supply) provide control during solid burn and until wheels take over (also auxiliary wheel unloading); orbit adjust by 4 hydrazine (28.4 kg supply). Payload 386 kg. Real-time transmissions made for Automatic Picture Transmission (APT, 4 km resolution, 137.5 or 137.62 MHz), High Resolution Picture Transmission (HRPT, full resolution, 1.698 or 1.707 GHz) services and DBS Direct Broadcast Sounder (full resolution sounder data on 1.698, 1.7025 or 1.707 GHz). Spacecraft command/control at 148.56 MHz provided by NOAA's CDA stations in Wallops Island, Virginia and Fairbanks, Alaska; Satellite Operations Control Center is in Suitland, Maryland. The Centre de Meteoroloogie Spatiale in Lannion, France can relay stored/real-time data
Sensors: Advanced Very High-Resolution Radiometer (AVHRR). The instrument operates in five channels (μm): 0.58-0.68 (Si), 0.725-1.0 (Si), 3.55-3.93 (InSb), 10.3-11.3 (HgCdTe) and 11.4– 12.4 (HgCdTe), primarily for assisting weather forecasting, snow/ice monitoring and sea surface temperature, but also applied to marine oil pollution mapping, volcanic eruption monitoring, assessment of vegetation vigour on international scales and estimating atmospheric aerosols for climate monitoring. 1.3 mrad IFOV. 20 cm diameter Cassegrain telescope, beryllium scan mirror, ±55° scan angle.

Tiros Operational Vertical Sounder (TOVS) is a three-instrument system consisting of a four-channel Microwave Sounding Unit (MSU) for tropospheric temperature soundings in cloudy regions, a three-channel Stratospheric Sounding Unit (SSU) for 25-50 km stratospheric temperature probing, and the 20-channel High-Resolution Infra-red Sounder (HIRS/2) for vertical temperature profile, water vapour and total ozone content to 40 km. JPL's MSU, 50.30, 53.74, 54.96, 57.95 GHz, 220 MHz bandwidth, ±47.4° scan width from nadir, two scanning reflector antennas (9.5° steps through 360°), 0-350K dynamic range, 0.3K NEδT. Matra Marconi Space's SSU, pressure modulated CO_2 cell in each optical path for 14.926, 14.934, 14.940 μm, 10° IFOV, 0.25 NEδT at 273K, no collecting optics (5 cm aperture), ±40° scan width from nadir. ITT's HIRS/2 channels 1-12 6.72-14.95 μm HgCdTe, numbers 13-19 3.76-4.57 μm InSb, number 20 0.69 μm Si. 24 mrad IFOV, ±49.5° scan width from nadir, 15 cm diameter Cassegrain telescope, 1.8° step scanner covers 56 steps then retraces.

Space Environment Monitor records radiation levels to determine the energy deposited by solar particles in the upper atmosphere and to provide a solar storm warnings. TED cylindrical electrostatic analyzer and spiraltron: 0.3– 20 keV protons/electrons in 11 bands. MEPED solid state telescopes and omnidetectors: 30-20 keV protons, 11 bands; >30->300 keV electrons, 3 bands; >6 MeV ions; >16, >36, >80 MeV omniprotons.

NOAA 12 (NOAA-D)
Launched: 15.52 GMT 14 May 1991 by Atlas 50E from Vandenberg AFB
Orbit: 807 × 826 km, 98.7° Sun-synchronous in morning orbit. Replaced NOAA 10 as the prime morning satellite
Sensors: Same as NOAA 11 but does not carry SBUV/2, the S&R payloads or the SSU instruments. MSU has degraded because of gain changes. Other specifications as for NOAA 11
Comment: The last Tiros-N. First known as NOAA-D, but launched out of sequence in 1991 as NOAA 12.

NOAA 13
Launched: 9 August 1993 by Atlas 34E from Vandenberg AFB to replace NOAA 11
Orbit: Afternoon satellite
Comment: Failure of the power link from the solar panel lost contact 21 August during checkout. The September 1994 review board report identified a short-circuit in the battery charge assembly that prevented the solar arrays charging the batteries. A 31 mm screw penetrating insulation and contacting a radiator plate probably caused the short. Inspection of NOAA-J (NOAA 14) showed 10 of the 12 screws to be in danger of causing the same fault. The same design had flown on 16 NOAA and DMSP satellites

NOAA 14 became the prime satellite in 1995
(Lockheed Martin Astro Space) ***1995***

Sensors: Same as NOAA 11 plus the experimental MAXIE and EHIC sensors, sponsored by the Office of Naval Research. The Magnetospheric Atmospheric X-ray Imaging Experiment, provided by Lockheed, assisted by the Aerospace Corp and Norway's University of Bergen, was to map the intensities and energy spectra of X- rays generated by electrons in the upper atmosphere, and the associated auroral and substorm imaging. Good data were returned until the power failure. The Energetic Heavy Ion Composition Experiment, provided by the University of Chicago and Canada's NAC HIA, was to measure the chemical and isotopic composition of energetic particles between hydrogen and nickel over 0.5-200 MeV/nucleon.

NOAA 14 (NOAA-J)
Launch: 30 December 1994 by Atlas 11E from Vandenberg AFB. Control transferred from NASA to NOAA 3 January 1995
Mass: 1,712 kg at launch, 1,030 kg BOL on-orbit
Orbit: 848 × 863 km, 98.9° afternoon Sun-synchronous to replace NOAA 11.

NOAA-K to -N-prime
Launches: First launch (NOAA-15) on 13 May 1998 with subsequent launches in 1999, 2000, 2003; Titan 2G
Mass: 2,234 kg at launch, 1,454 kg BOL on-orbit
Power system: as NOAA 11 but 11-panel 17 m^2 solar wing
Sensors: AVHRR/3. Same as NOAA 11 but 6th time-shared channel (3A: 1.58-1.64 μm InGaAs) improves snow/cloud discrimination and changes in several channels improve calculation of worldwide vegetation levels and estimation solar and terrestrial radiation levels previously monitored by the two ERBEs. 33 kg, 28.5 W, ITT Aerospace/Communications Division.

HIRS/2 20-channel High-Resolution IR Sounder for vertical T profile, water vapour and total ozone content to 40 km. Channels 1-12 6.52-14.95 μm HgCdTe, numbers 13-19 3.76-4.57 μm InSb, number 20 0.69 μm Si. 24 mrad IFOV, ±49.5° scan width from nadir, 15 cm diameter Cassegrain telescope, 1.8° step scanner covers 56 steps then retrace. 34 kg, 24 W, ITT Aerospace/Communications Division.

AMSU-A 15-channel advanced microwave sounding unit for atmospheric T soundings from surface to 45 km at 23.8, 31.4, 50.3-57.3 and 89 GHz. 50 km nadir resolution, 2,200 km swath, 2.2 kbit/s.

AMSU-B 5-channel on NOAA-K/L/M for humidity profiles, delineating sea ice and open water, and monitoring snow thickness and soil moisture. 89, 157 and 183 GHz, 15 km nadir resolution, 2,200 km swath, 4 kbit/s. UK Met Office instrument; Matra Marconi Space was awarded a £9.5 million contract in March 1989.

MHS 5-channel microwave humidity sounder on NOAA-N/N-prime for humidity profiles up to 42 km. 89, 150 and 183 GHz, 15.4 km nadir resolution, 2,348 km swath, 3.95 kbit/s, 66 kg, 100 W. EUMETSAT (also on Metop). Argos system capacity is quadrupled, able to handle eight messages simultaneously at 2,560 bit/s over 80 kHz (currently 24 kHz) bandwidth

Cospas-Sarsat S&R. Same as NOAA 11

SEM. Same as NOAA 11

SBUV/2. Same as NOAA 11; not on NOAA-L.

SMS 1
Launched: 17 May 1974 by Delta from Cape Canaveral
Mass: 627 kg
Orbit: Initially over the East Atlantic, and then at 75° west over Bogota
Comment: This satellite provided the first day/night stormwatch, with full disc pictures of the western hemisphere every 30 min. As part of the Global Atmospheric Research Programme (GARP), it provided the first continuous coverage of a major hurricane, designated Carmen, in September 1974.

SMS 2
Launched: 6 February 1975
Orbit: 115° west, east of Hawaii
Comment: Positioned so that it and SMS 1 could cover the western hemisphere. One of its tasks was to keep watch on California's forest areas to give warning within 90 min of fire outbreaks.

GOES 1
Launched: 16 October 1975 by Delta from Cape Canaveral. 293 kg
Mass: 293 kg
Orbit: During the Global Weather Experiment of 1977-78, it was positioned over the Indian Ocean at 60° east, with ESA's ground station in Spain processing data. Later moved to 116° west and 135° west
Comment: Following GOES 4 failure in November 1982, it was functioning only in the visible band, but it was reactivated to transmit images via GOES 4.

GOES 2
Launched: 16 June 1977
Orbit: 105° west. At beginning of 1992, positioned over 60° west (±1.0°), inclined at 9.4° for relay of Mode AAA (Stretched VAS) from GOES 7, relay of East WEFAX and for East Data Collection Platform Interrogation (time-code). These functions were principally for western European and African users, who could not see GOES 7. It was moved in 1992 to 135° west to replace GOES 6 supporting all non-imaging functions of the west spacecraft (WEFAX, DCS, and SEM)
Comment: VISSR primary encoder failed 26 January 1979 (redundant encoder failed 18 December 1978), ending its imaging function after returning 20,591 full disc images and 6,838 partial. Its functions were replaced by GOES 7 in January 1995 and it retired

GOES 3
Launched: 16 June 1978, replacing GOES 1
Orbit: Moved to 176° west from 130° west in June 1990. It was held at 176° west, inclined at 10.6°, but it has been drifting since 1995 although still usable
Comment: The move to a new location occurred under an agreement with NASA and DoC's National Telecommunications & Information Administration for use by Pacific island nations through the Peacesat Pan-Pacific Educational & Cultural Experiments by Satellite project to re-establish the communications service inaugurated in 1971 by ATS 6. This provides 2-way voice/data services to small Earth stations at more than 100 sites in 22 countries. Peacesat is controlled for NOAA by NASA from the Kokee Park Geophysical Observatory on Kauai, Hawaii; NTIA co-ordinates user operations.

GOES 4
Launched: 9 September 1980
Orbit: 135° west
Comment: In 764 operational days provided warning of many Pacific storms, transmitting 28,500 scenes. Imaging system failed in November 1982, after which it was used as a standby transponder relaying data. In May 1985, it was loaned to ESA as temporary substitute for Meteosat 1 and was moved 4° daily to place it over the Atlantic at 43° west by June. This was in return for Meteosat 2's help to NOAA in 1984 following the GOES 5 failure. As GOES 4 was incompatible with ESA's ground equipment, NOAA operated GOES 4 for ESA from Suitland, Maryland and Wallops, Virginia. It was to be boosted out of GEO at the end of the loan. In August 1985 however, unidentified UHF interference at 401.9 MHz rendered its data collection capability useless as it was being moved. It was therefore returned to 43° west. It was boosted above GEO in November 1988 and deactivated.

GOES 5
Launched: 22 May 1981
Orbit: 75° west. Hydrazine was depleted November 1989 and it began drifting westwards. It will drift to about 145° west, then return east towards its original starting point of 60° west. This pattern will continue, with outer limits decreasing, until it settles around 105° west
Comment: Intended to monitor US east coast and tropical storm formations in the west Atlantic Ocean. VAS failed 29 July 1984 due to two tungsten filament lamp failures; then used to relay data from GOES 6 and then GOES 7. Relay of GOES 7 imagery to Europe was transferred to GOES 2 in May 1990. In July 1990 all other functions ended and it was deactivated.

GOES 6
Launched: 28 April 1983
Orbit: 135° west, moved to 98° west to cover both US coasts; then to 108° west to monitor winter storms in the Pacific off US northwest coast. Following the failure of GOES 5, GOES 6 had to do the work of two satellites by being moved with the seasons between 98/108° west. Into 1992, it was at 135.6° west (inclined 3.6°)
Comment: Intended to cover western US. With one of its four encoder lamps burned out, there was concern that the other three might fail; otherwise there was sufficient propellant to keep it operational until late 1988 or early 1989. Meanwhile, it could provide coverage for the main US continent, but could not provide hurricane coverage for some Atlantic islands, the Caribbean and areas west or south of Hawaii. The last lamp failed 21 January 1989 and GOES 6 was reduced to acting as a data relay, supporting all non-imaging functions of the west spacecraft (WEFAX, DCS, and SEM). It was low on hydrazine and each station-keeping manoeuvre could have depleted the supply. This happened in 1992 and it began slowly drifting, following a pattern similar to GOES 5's. GOES 2 took over its relay role, but return of SEM data continues.

GOES 7
Launched: 26 February 1987 by Delta
Mass: 835 kg at launch, 399 kg on-station BOL
Orbit: Initially GOES-East satellite, nominally at 75° west GEO. Following GOES 6 failure, it was moved to 108° west for the Winter season, then to 98° west in the spring for observing hurricanes over the Atlantic, returning further west in the Autumn. At 112° west from 1992. Arrived 135° west 27 January 1995 after GOES 8 began operations; replaced 11 January 1996 at 135° west by GOES 9. Started moving 26 January to 96° west storage position (arriving 10 July). The inclination is being allowed to increase (3.0° by April 1996, increasing at 0.9°/year)
Contractor: Hughes Aircraft Co (prime)
Resolution: 900 m visible, 6.9/13.8 km infra-red
Spacecraft: 2.16 m diameter 3.53 m high cylindrical spin-stabilised satellite; body-mounted solar array provides 450/330 W BOL/EOL. Two Ni/Cd batteries provide TT&C eclipse protection. Injected into GEO by solid AKM
Sensors: Visible Infra-red Spin-Scan Radiometric Atmospheric Sounder (VAS). Scans the full Earth disc in 1,820 successive steps by combination of spacecraft spin and stepping optics. VAS carries single visible channel (eight photomultiplier tube detectors in push-broom array) providing 900 m resolution, and six infra-red channels: two 6.7-14.7 μm with 6.9 km resolution primarily for imaging, two 3.7-7.3 μm and two 7.7-14.7 μm. The satellite also carries a Space Environment Monitor, a Data Collection System and was the first GEO satellite equipped for Cospas/Sarsat trials (see GOES-I to -M). DCS signals are relayed to a Command & Data Acquisition station, which performs error checks and routes them to NOAA offices at the World Weather Building in Camp Springs, Maryland for distribution to users. WEFAX broadcasts are made in the 10 minute intervals between VAS readouts.

GOES 8
Launched: 13 April 1994 by Atlas 1 from Cape Canaveral; lifetime five years (propellant sufficient for 8 years). Formally transferred from NASA to NOAA 26 October 1994; declared fully operational 9 June 1995
Mass: 2,105 kg at launch, 1,140 kg on-station BOL
Orbit: arrived 75° west GEO (GOES-East) 27 February 1995 after testing at 90° west (departed 20 January 1995)

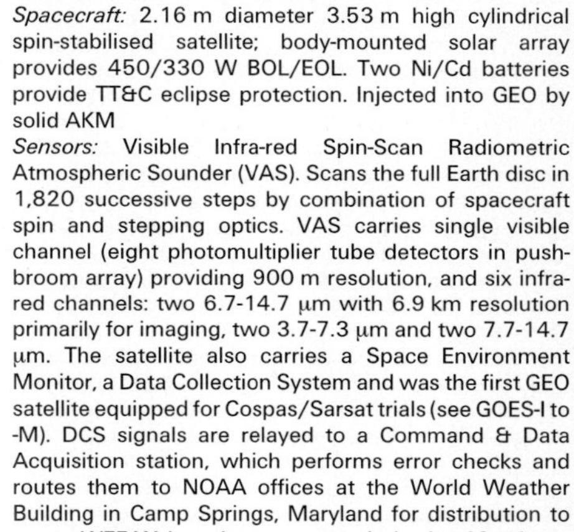

GOES 8's first visible engineering test image, 9 May 1994
(NOAA) ***1995***

GOES 9 began operations at 135° west in January 1996 (Space Systems/Loral) ***1996***

Contractor: Space Systems/Loral (prime), ITT Aerospace/Communications Division (sensors)
Resolution: 1 km visible, 4/8 km infra-red
Spacecraft: Space Systems/Loral's 3-axis stabilisation box-shaped platform, with single two-panel Si solar wing on south side and associated solar sail on north, allowing passive north-facing Imager/Sounder coolers to view cold space. Conical sail on 17 m boom balances solar radiation torque; a trim tab on solar wing provides fine control. Total span 26.91 m. Attitude control by two 51 Nms MWs, one 2.1 Nms yaw RW + two magnetorquers; during orbital manoeuvres by two sets of 6 × 22 N MMH/NTO thrusters. GEO insertion provided by three firings of 490 N Kaiser Marquardt R-4D. 431 kg MMH + 694 kg NTO each in single 103 cm diameter titanium sphere, pressurised by helium. Payload pointing accuracy: ±9.1 μrad roll; ±9.4 μrad pitch; ±73.3 μrad yaw (in 90 min). 268 × 481 cm solar wing can provide 1,057 W EOL Summer solstice; supported by 2 × 12 Ah Ni/Cd batteries; 42.0 V DC (30.0 V DC min in eclipse). Command/control provided via NOAA's Command & Data Acquisition station on Wallops Island, Virginia; the Satellite Operations Control Center is in Suitland, Maryland. Command transmitted 2.03420 GHz. Data are processed at the CDA and uplinked for broadcast
Sensors/payload: Imager. Imaging radiometer sweeping 8 km high swath along east-west/west-east path at 20°/s via servo-driven 2- axis gimballed mirror and 31.1 cm diameter aperture Cassegrain telescope. 120 kg, 119 W daily average. Channel 1: 0.55-0.75 μm, 8 Si detectors, nadir instantaneous FOV 1 km each, cloud cover; 2: 3.80- 4.00 μm, 4 InSb detectors, nadir IFOV 4 km each, night cloud cover; 3: 6.50-7.00 μm, 2 HgCdTe detectors, nadir IFOV 8 km each, water vapour; 4: 10.20-11.20 μm, 4 HgCdTe detectors, nadir IFOV 4 km each, surface temperature; 5: 11.50-12.50 μm, 4 HgCdTe detectors, nadir IFOV 4 km each, sea surface temperature and water vapour. Full Earth scan in <26 minutes; 3,000 × 3,000 km in 3 minutes; 1,000 × 1,000 km in 41 seconds. Data rate 2.621 kbit/s. Imaging stability 42 μrad for noon ±8 h, 70 μrad for midnight ±4 h. Passive cooler maintains infra-red detectors at 94/101K winter/summer. Full Disk Mode repeatedly scans the full disc every 30 minutes, except once every 6 hours when it performs an extended northern hemisphere scan to allow 10 minutes for housekeeping work. Routine Mode is a repeated 3 hour sequence. It begins with a 30 minute full disc scan, followed by a 30 minute sequence (performed five times) of a northern hemisphere scan, Conus scan and southern hemisphere scan. The last southern scan is omitted every 6 hours for 10 minutes housekeeping. Star looks (for precise attitude registration) and blackbody calibrations occur every 30 minutes. Rapid Scan Mode is similar but it emphasises Conus scans.

Sounder. 19-channel radiometer for atmospheric vertical temperature and moisture profiles, surface/cloud top temperature, and ozone distribution. 127 kg, 106 W daily average, 31.1 cm aperture. Optics and scan patterns similar to Imager. Wheel carries 18 filters in three concentric rings, one for each infra-red detector group. Longwave channels: 1-7, 14.71-12.02 μm, 4 HgCdTe detectors, and circular IFOV 242 μrad. Midwave channels: 8-12, 11.03-6.51 μm, 4 HgCdTe detectors, circular IFOV 242 μrad. Shortwave channels 13-18, 4.57-3.74 μm, 4 InSb detectors, circular IFOV 242 μrad; visible channel 19, 0.70 μm, 4 Si detectors, circular IFOV 242 μrad. Data rate 40 kbit/s. Scans in 10 km steps east-west/west-east, moving 40 km north-south at end of scan line. Each step 100/200/400 m/s, controlled by filter wheel.

Space Environment Monitor. Of four instruments for GEO particle radiation and solar X-ray output. EPS Energetic Particles Sensor: 0.8-500 MeV protons, 3.2-400 MeV α, 0.55-4.0 MeV electrons. HEPAD High Energy Proton & Alpha Detector: 370-970 MeV protons, 640->850 MeV α. XRS x-ray Sensor: 0.5-3.0 and 1.0-8.0 A real-time solar flux; mounted on solar wing yoke. Magnetometers: redundant 3-axis fluxgate magnetometers on 3 m boom, ±1,000 nT (±4 nT accuracy). GOES-M and successors will include a Solar X-ray Imager to aid prediction of geomagnetic storms. SEM data downlinked on 1.694000 GHz to Environmental Research Lab, Boulder, Colorado.

Cospas/Sarsat Distress Relay (406 MHz beacon relay only, no location service). Downlink at 1.544500 GHz.

WEFAX transponder. Processed imagery is returned to GOES at 2.03300 GHz for broadcast to users at 1.6910 GHz at 11 W to users
Data Collection System: Up to 233 100 bit/s signals received from DCP Data Collection Platforms on 401.9 or 402.2 MHz. Downlinked at 1.69450 or 1.69480 GHz to CDA and users. DCP interrogated by request uplinked to GOES from CDA on 2034.9000 or 2034.9125 MHz; DCP interrogated by GOES at 468.8250 or 468.8375 MHz. >12,000 DCPs active in 1995 (increases 15% yearly).

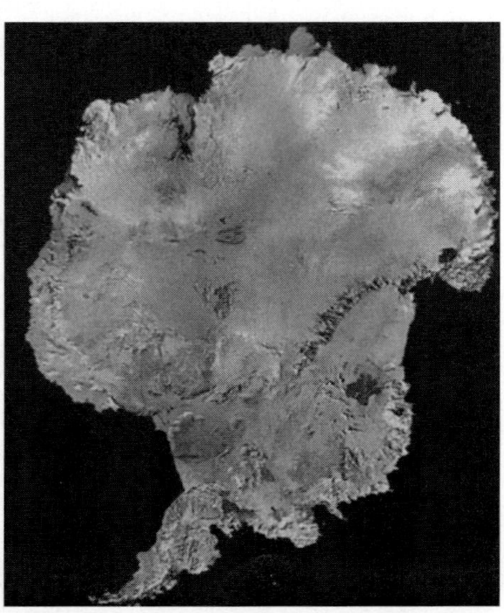

Mosaic of 28 NOAA AVHRR images covering Antarctica produced in 1988 in a joint venture between NOAA and the UK's National Remote Sensing Centre (NRSC)

GOES 9
Launched: 23 May 1995 by Atlas 1 from Cape Canaveral. NOAA assumed control from NASA 31 October 1995
Orbit: GEO at 135° west (GOES-West) from 11 January 1996, replacing GOES 7 after testing at 90° west. First (visible) image received 12 June 1995
Other specifications are the same as for GOES 8.

GOES 10
Launched: 25 April 1997 from Cape Canaveral using an Atlas-1 (final Atlas-1 launch)
Orbit: GEO over 106° west (early June 1997)
Other specifications as for GOES 8.

GOES-G
Comment: This US$57.5 million satellite should have become GOES 7 at the 70° west GOES-East position over the Atlantic Ocean, but was lost on 3 May 1986 when its Delta launcher failed. It was to have been the first satellite to operate Sarsat in GEO.

GOES-K to -M
Launches: GOES-K planned for 1999 by Atlas 1; GOES-L March 2000 by Atlas 2A; GOES-M 2004 by Atlas 2A
Orbit: GEO at 75° west (GOES-East) and 135° west (GOES-West) locations
Other specifications as for GOES 8.

OrbView series

Current status

Orbimage launched the SeaStar on 1 August 1997 from Vandenberg AFB, California using the Pegasus XL booster.

Background

OrbView 1 (MicroLab 1)
Orbital Sciences Corp's Eyeglass International, was formed in 1994, to build and operate the Eyeglass satellite system and to market imagery products and services. Orbimage, in 1995, decided instead to pursue the smaller, lower cost OrbView satellite, derived from its MicroLab design. The cost of deploying one satellite and establishing the ground system has been estimated at more than US$100 million. OSC's MacDonald Dettwiler Associates is managing the ground segment requirements. Saudi Arabia's EIRAD Co-signed June 1994 to become a major investor (operating a receive station).

OSC's SeaStar will provide Earth colour data

The United States' Department of Commerce approved the project 5 June 1995. EIRAD decided to take a 20 per cent stake in Orbimage since US regulations impose a 25 per cent limit. A second generation OrbView is planned for Taurus launch in 2001 into 700 km. It would add SAR to 15 km swath 1 m PAN/4 m XS.

OrbView 2 (SeaStar)
NASA Goddard selected OSC in March 1991 for a US$43.5 million contract to deliver multispectral ocean colour data to NASA investigators for five years as part of the agency's Earth Probes programme. This is the first time that the US government has purchased global environmental data from a privately-designed and operated remote sensing satellite. SeaStar, based on OSC's PegaStar bus, carries the SeaWiFS Sea-viewing Wide Field Sensor of Hughes' Santa Barbara Research Center. The instrument, originally planned for Landsat 6, is a next-generation design of Nimbus 7's Coastal Zone Colour Scanner.

SeaWiFS measures ocean surface-level productivity of phytoplankton and chlorophyll for ocean dynamics and marine life research. It will contribute to understanding the global carbon cycle and its effect on global warming. Orbimage retains exclusive worldwide rights to imagery for commercial and operational uses. It expects to find markets in ocean fishing, coastal monitoring, land management and the military. SeaStar was originally designed for ocean colour but SeaWiFS gains were changed during development to add higher land radiance dynamic range. It will provide a more environmentally stable Vegetation Index than NOAA's AVHRR, which is inaccurate under hazy atmospheric conditions because of its single visible and near-infra-red channels. Orbimage will sell to both end-users and value

The 1-metre resolution remote sensing satellite from Orbimage (Theo Pirard) ***1999***/0054351

added resellers at a price point of US$500 per image set.

A delay to mid-1994 was announced August 1993 for instrument modifications to cure a problem with stray light. The onboard computer then required redesign (the original computer from OSC's APEX was found to have insufficient throughput to process SeaWiFS data) and 1994's Pegasus XL failure resulted in further delays.

Specifications

OrbView 1 (MicroLab-1)

Launch: 3 April 1995 by Pegasus XL
Resolution: 1/2 m PAN (4/8 km swath), 4 m XS (8 km swath)
Lifetime: 3 years
Mass: 146 kg
Orbit: planned 460 km polar, 10.30 equatorial crossing, revisit <3 days
Spacecraft: 96.5 cm diameter, about 60 cm ht disc. Although there will be some onboard storage, most imagery will be downlinked in real time
Sensors: PAN. Bands: (μm): 0.50-0.90; XS 0.45-0.52, 0.52– 0.60, 0.63-0.69, 0.76-0.90.

OrbView 2 (SeaStar)

Launch: 1 August 1997 by Pegasus XL
Mass: 309 kg on-orbit (SeaWiFS 50 kg)
Orbit: 705 km, 98.2° equator crossing at noon
Resolution: 1.13 km Local Area Coverage and 4.5 km Global Area Coverage. L-band transmits encrypted continuous LAC data when SeaStar is visible to standard HRPT stations equipped with OSC's licensed decoder. GAC data are stored and downlinked every 12 h on S-band to central location
Spacecraft: 213 cm high, 112 cm diameter OSC PegaStar bus. 5-year design life; 10 year goal. Redundant 3-axis momentum biased ACS (0.5°; 1.23 mrad knowledge) of momentum wheels, magnetorquers, horizon sensors, Sun sensors, and 2-axis magnetometers. Redundant GPS receiver provides precise real-time position data to within 100 m. 200 W BOL orbit average from four 55.9 × 152.4 cm rigid solar panels supported by 10 Ah Ni/H_2 battery. TT&C and 0.665/2.0 Mbit/s data via L/S-band downlinks; 1.25 Gbit recorder. Four 22.2 N hydrazine thrusters to raise/maintain mission orbit
Sensors: SeaWiFS. The main telescope rotates 360° about a pivot axis to scan each scene, thereby avoiding the use of a scan mirror and its associated polarisation effects. Specular Sun reflection is avoided by tilting the telescope in the plane perpendicular to the scan plane to one of three positions: +20°, 0° or –20°. The continuous 360° scan permits reference sources to be viewed during the non-scene portion, in addition to including a deep space scan for zero referencing before the scan begins. The spread of observing bands addresses both chlorophyll and pigment absorption values, as well as water optical properties and suspended sediment measurements. Spectral band centres, (in μm, with bandwidth and application in brackets): Band 1 0.412 (0.020, gelbstoffe), Band 2 0.443 (0.20, chlorophyll), Band 3 0.490 (0.020, pigment), Band 4 0.510 (0.020, chlorophyll), Band 5 0.555 (0.20, suspended sediments), Band 6 0.670 (0.020, atmospheric aerosols), Band 7 0.765 (0.40; atmospheric aerosols), Band 8 0.865 (0.044, atmospheric aerosols). The bands for atmospheric aerosols provide atmospheric corrections.

Space Radar Lab

Background

NASA plans an 11-day SRTM Shuttle Radar Topography Mission in 2000 to help the defence Mapping Agency create a global digital elevation map with 16 m height accuracy at 30 m horizontal intervals. SRTM will use SRL's SIR-C antenna working interferometrically with another on a 60 m mast. SRL 1 flew aboard STS-59 in April 1994 and SRL 2 in September to October 1994 aboard STS-68. A third mission was, until 1993, planned for mid-1996. JPL is NASA's lead centre. A Spacelab pallet accommodated JPL's L/C-band SIR-C (Spaceborne Imaging Radar-C) and the German/Italian X-SAR X-band synthetic aperture radar. Novel mission features are multi-frequency radar imaging observations, dual polarisation (L/C) and 15 to 60° variable incidence angles. SRL returned imagery at up to 10 m resolution. JPL/Ball Communication Systems Division provided SIR-C's antenna systems. X-SAR's industrial team was headed by Dornier/Alenia Spazio; the 12 × 0.4 m planar slotted waveguide antenna comprises three leaves on a CFRP structure. Alenia Spazio was responsible for the RF subsystems, antenna electrical design and the mission operations ground segment. X-SAR represents the most extensive Earth observation co-operation yet between Germany/Italy.

SRL 1 returned imagery for 133 h covering 25 per cent of the globe's land surface and demonstrated its effectiveness for the monitoring of environmental parameters such as vegetation type, coverage, moisture distribution and energy transfer mechanisms. Individually, SIR-C covered 99.67 million km² in 90 h and X-SAR 47.89 million km² in 43 h. Some imagery was returned in real-time. The crew took >14.000 supporting photographs and 19 'supersites' were monitored intensively by ground teams. SRL 2's trajectory was carefully managed during its last four days to repeat within 800 m of some targets so that SRL interferometery would generate high-resolution topgraphical maps. Combining the data with SRL 1's revealed movements since April 1994 with cm accuracy. If used by a free-flying satellite, the technique would allow the minute movements of glaciers, volcanoes and Earthquake zones to be monitored routinely.

SRL also demonstrated an ocean wave data processor from Johns Hopkins' Applied Physics Laboratory. The data were relayed through TDRS and then processed in near-real-time at NASA's SAR Processor Facility at JPL.

Specifications

X-SAR

Frequency/polarisation: 9.6 GHz (X-band)/VV
Resolution: less than or equal to 30 m azimuth, 10-20 m slant
Swath width: 15-45 km (off-nadir angle 20-55°)
Peak transmit power: 3.2 kW; pulse length 40 μs; max DC power 1,060 W
Beamwidth (3 dB): 5.7° elevation, 0.4° azimuth
Antenna gain: 43 dB
Mass: 210 kg
Data rate: 45 Mbit/s.

SIR-C

Frequency/polarisation: 1.25/5.3 GHz (L/C-band); HH, HV, VH, VV
Resolution: <=30 m azimuth, 10-20 m slant
Swath width: 70 km
Data rate: 50-100 Mbit/s

TOMS-EP satellite

Background

The Total Ozone Mapping Spectrometer Earth Probe (TOMS-EP) extends the work of Nimbus 7's similar instrument in monitoring atmospheric ozone levels. It helps establish the continuity in the gathered data so a long-term picture of the Earth's ozone layer can be formed. NASA originally selected TOMS in 1989 as the fourth Small Explorer but then dropped its funding before reviving it as the first Earth Probe in NASA's Mission to Planet Earth.

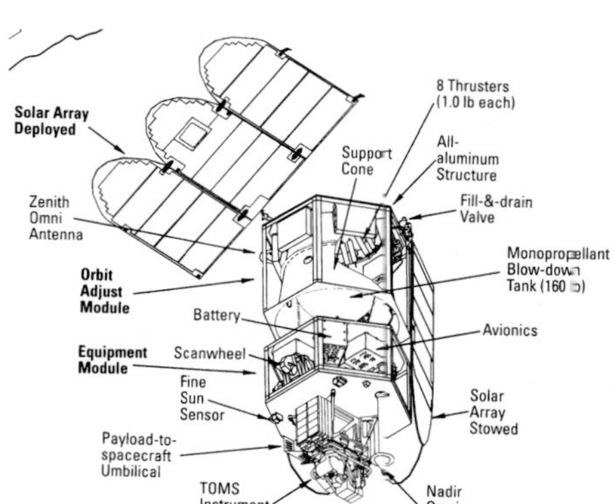

TOMS-EP principal features (TRW) ***1995***

TOMS-EP is mapping global ozone levels (TRW) ***1995***

NASA Goddard released a request for spacecraft proposals in late 1990; and the TRW's Eagle bus won out over its competitors for the June 1991 contract of US$29.3 million (now US$57 million). Mission cost, including launch but excluding operations has amounted to about US$80 million for NASA. TRW built the spacecraft and integrated the instrument.

In operation, the satellite stores up to 24 h of data stored on 16 Mbyte solid-state recorder and downlinks it via S-band transponder at 202.5, 50.6 or 1 kbit/s. The Total Ozone Mapping Spectrometer is based on 3,086-3,600 A Fastie-Ebert monochrometer, for 1 per cent/decade ozone trend determination in six wavelength bands.

Specifications

Launch: 2 July 1996 by Pegasus XL from Lockheed L1011 out of Vandenberg AFB
Mass: 294 kg (including 54 kg hydrazine)
Orbit: planned 500 km circular 97.4° Sun-synchronous at 11.00-11.30 ascending node
Contractors: TRW (bus), Orbital Sciences Corp (TOMS instrument)
Resolution: 50 × 50 km
Lifetime: minimum 2 years, goal 3 years
Spacecraft: TRW Eagle platform, two compartments, hexagonal prism of three equipment decks. Bus 99 cm diameter, 168 cm long without TOMS; stowed/deployed 112/388 cm max width, 178/243 cm long with TOMS. Two 3-panel dual-sided solar wings of composite material over honeycomb core providing 6.25 m² total cell area, mounted at 45° cant to body, for 233 W EOL (β=20°). 200 μm Si cells protected by 150 μm cover glass. 9 Ah Super Ni/Cd battery provides eclipse power. 3-axis control to 0.5° pitch/roll (0.25° knowledge, 3) pitch/roll, <1.0°/0.25° 3 yaw by momentum bias RW system utilising Earth/Sun sensors + gyros. 4 N dual-seat hydrazine thrusters for orbit insertion, trim and station-keeping. Up to 24 h TOMS data stored on 16 Mbyte solid-state recorder and downlinked via 2 W S-band transponder to DSN & Wallops stations at 202.5, 50.6 or 1 kbit/s. Satellite designed for 24 h autonomous operations. Combined Mission Operations Center and Science Operations Center at NASA Goddard
Sensors: Total Ozone Mapping Spectrometer. Based on 3,086-3,600 Å Fastie-Ebert monochrometer, for 1%/decade ozone trend determination in six wavelength bands. 3.0 × 3.0° FOV crosstrack scanning capability allows daily high spatial resolution (50 × 50 km) global O_3 maps. SO_2 monitoring also undertaken. Sensor mass is 32 kg, 1,525 cm³, 25 W average power consumption (including heaters); onboard solar and electronic calibration capabilities, plus reflectance calibration mode not previously flown. Microprocessor-based electronics provide flexibility in control and data formatting; data rate up to 736 bit/s.

ENGINEERING TEST

FRANCE

Proteus satellite

Current status

CNES plans the TPFO Topex-Poseidon Follow-On mission for launch in 1999 as the first of the 200 to 500 kg Proteus (Platforme Reconfigurable pour l'Observation, les Télécommunications et les Usages Scientifiques) multipurpose LEO platforms. France has shown its determination to make the programme cost effective by establishing a capital cap on the programme of FFr100 million for the platform and payload accommodation. Contractors studied the programme goals in the Phase A investigation, which ran from November 1994 to April 1995. Then CNES selected Aerospatiale as the Phase B prime contractor in May 1996. A final contract calls for one satellite plus one option. NASA plans to launch the TPFO. Candidates for subsequent missions are Tropiques (interaction between radiation, aerosols, cloud, water and atmospheric circulation in the tropics), Vagsat (conical scanning radar) and Irsute (biosphere 40 m resolution thermal-infra-red).

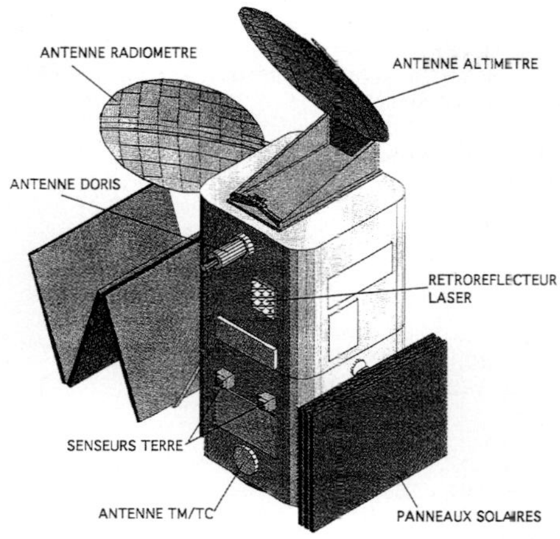

The Topex-Poseidon Follow-On mission will be the first to use the Proteus platform (CNES) ***1996***

SRETS series

Background

At the invitation of the Soviet government, France became one of the first Western governments to participate in Soviet space research. After a meeting with the Soviet premier, Charles de Gaulle approved a small satellite programme to investigate the effects of the space environment on operating satellites. The SRET programme, which resulted, sought to measure the decay in solar array performance due to radiation exposure. The French chose to launch their satellite in conjunction with, and in the same orbit as, a Soviet Molniya satellite. The Molniya semi-synchronous orbit carries a satellite through the van Allen radiation belts twice in one twelve-hour circuit of the Earth. By measuring the decay after each pass through the belts, and knowing the radiation exposure levels, the French could predict the exact decay any satellite would experience in space under different circumstances.

In the second experiment, SRET-2, the French-tested new cooling technology that it expected eventually to incorporate in its Meteosat bus, then being planned for a launch in conjunction with ESA.

Specifications

SRET 1
Launched: 4 April 1972 piggyback with Molniya 1-20 on SL-6 from Plesetsk
Mass: 15.4 kg
Orbit: 460 × 39,248 km
Re-entered: 26 February 1974.

SRET 2
Launched: 5 June 1975 piggyback with Molniya 1-30 on SL-6 from Plesetsk
Mass: 29.6 kg
Orbit: 513 × 40,825 km, 62.8°.

GERMANY

MIRCA satellite

Current status

DARA initiated a joint project between Kayser-Threde and DASA in 1992 using Jena-Optronik to develop and fly the 1 m diameter MIRCA Micro Re-entry Capsule for re-entry investigations. KT and Russia's Central specialised Design Bureau signed a contract in November 1994 to launch MIRCA attached to a Photon in September 1996. To date, the launch has not occurred.

MIRCA will be mounted on the forward face of Photon's own sphere, 12° below the centreline until it is ejected at 0.8-1.6 m/s separation after retrofire. KT's design uniquely uses the heatshield as the main load-carrying structure, comprising hemispheres bolted at the equator. The shield consists of a 1 mm carbon fibre composite core, 20/30 mm ablator for the upper/lower hemisphere and a 1.35 mm C/SiC outer layer. The heatshield has integrated temperature and pressure sensors and a pyrometer in the lower shield. Given the nose tip shape and other re-entry conditions, the Germans predict the shield experiences a 2,100°C stagnation point.

The mass budget is 150.0 kg total (71.0 kg heatshield, 16.7 kg mechanical support structure, 16.6 kg service system, 6.6 kg battery, 16.3 kg experiments, 14.3 kg recovery system, 8.0 kg balance mass) in a payload volume of 170 litres. The satellite will have 20 Ah, 28 V DC from $LiMnO_2$ cells of power which can be expanded to 50 Ah. The planned instrumentation includes HEATIN Heatshield Instrumentation: temperature, pressure and heat flux distribution within two meridian planes 90° apart. Also temperature at different depths in material and pressure caused by off-gassing. RAFLEX RArefied FLow EXperiment: pressure at stagnation point and 60° behind. PYREX Pyrometer Experiment: touchless temperature measurement of C/SiC layer behind shield at stagnation point. Upon completion, the experiment will also space qualify this mini-pyrometer. In flight the system has a command data, transputer-based, containing 16 Mbytes on EPROMs for control/data handling.

A Photon command at 105 seconds before separation activates the system. Three-axis body rates up to ±100°/s and ±64 *g* can be recorded. The payload will be recovered by pushing the cover clear with two 1,700 N solids burning for 80 ms. This action pulls out a conical ribbon chute at about 7 km, which is also activated redundantly by barometric switch and timer after peak deceleration. A reefed cross chute is released 25 seconds later, fully open 10 seconds later. Shortwave beacon (antennas on chute lines) is active for 10 minutes during descent; and a VHF transmitter and strobe remains activated on the ground for 48 hours.

A follow-on tether mission was studied, beginning in 1991 but cancelled by DARA in 1994. Project Rapunzel (Rope Attached Piggyback Unit Zooming on Environment Data at Low Cost) would have reeled out MIRCA on a 50 km tether from Photon to de-orbit when the line was cut. GPS receivers in each module would have provided relative position information. Before then, the first Rapunzel test was planned for a Resurs-F mission carrying a 45 kg 50 cm re-entry sphere developed by the Samara State Aerospace University. Later, Rapunzel studies concentrated on a MIRCA tethered to a Progress-M as it left Mir in 1997.

INDIA

Apple (Ariane Passenger Payload Experiment) satellite

Background

Apple was India's first indigenous GEO test communications satellite. The French launched the satellite to give ISRO experience in attaining its indigenous satellite goals before the completion of its own GSLV space launch booster. After launch, it relayed TV programmes and educational teleconferences despite a jammed solar panel. ISRO de-activated Apple in September 1983.

Specifications

Launched: 19 June 1981 by Ariane from Kourou
Mass: 670 kg
Orbit: 102°E GEO.

ISRAEL

Gurwin 1 (TechSat) satellite

Background

A collaboration between the Faculty of Aerospace Engineering at Technion Institute (under the direction of Professor Chaim Eshed), ISA, Israeli industry and Amsat-IL resulted in the design of a simple, low-cost, low-power platform. The consortium completed the Flight Model in October 1994. At launch, TechSat was lost in Start's launch failure. Techsat 1B, a near duplicate, is currently being funded. It will include the Netherlands' TNO-TPD wide field star sensor.

Technion proposes a 250-300 kg second-generation TechSat as a multipurpose platform, created by adding an AOCS unit and GPS navigation to the basic TechSat 1. The 100 kg AOCS module would carry Rafael 1 N liquid propellant and N_2 thrusters. TechSat 2A would carry a high-resolution Earth observation camera, gimballed on command and relaying imagery via a MMIC Ku-band phased-array antenna. Techsat 2B will reportedly be used for data relay and 2C for data and voice communication. The Israeli government has declined all requests for technical information on TechSat 1B.

Specifications

Launch: 28 March 1995 on Russian Start from Plesetsk
Orbit: planned orbit was 670 km 75.4°
Contractors/institutes: Technion (management, integration, ground station, ozone sensor, X-ray detector), MLM (solar panels), Rafael (antennas, structure), Tadiran (transceivers), TAMAM (wheel, magnetometer, magnetorquers), El-Op (horizon sensor, CCD camera), Elbit (OBC), Elisra (image compression system), IAI-MBT (battery, power control unit)
Mass: 52 kg
Configuration: box-shaped 460 mm high, 430 mm across (six Al plates, 2.5 mm thick). Earth-pointed face carries antennas, horizon sensor, and camera. 80C186EC-based CPU with two 512 kbyte independent RAM memories. Controlled from Technion station; command reception at 145.910/1,269.85 MHz
Power: solar cells on four faces (1,000 2 × 4 cm Si, 17 W average, 9 W required for housekeeping, Ni/Cd battery)
Attitude control: 3-axis Earth-pointing control (5° coarse, 0.1-0.5° fine) by 2.5 kg 3,600 rpm 4 Nms MW 3 × 1 Am^2 magnetorquers, 1 kg 1 W static horizon sensor (0.05° accuracy), 0.22 kg 0.8 W ±0.6 gauss 3-axis magnetometer
Payload: Amateur digital store/forward 1/3 W transponder. Transmit at 435.225 plus back-up 435.325 MHz band. Receive at 145.850, 145.890, 145.930, 1,269.700, 1,269.800, 1,269.900 MHz at 1,200/9,600 bit/s.

Elisra Earth-observing 350 × 570 pixel CCD camera. Compressed image data broadcast using mm-wave link experiment (also on UHF).

Technion ultra-violet CsTe spectroradiometer to monitor Earth's albedo at 0.24-0.30 μm (0.01 μm/2 mrad resolution), 15° FOV. The spectral profile is a function of the vertical ozone profile. Goals are investigating the long-term stratospheric impact of the Kuwaiti oil fires and the extent of the region's ozone depletion.

Technion X-ray detectors (up to 200 keV). Principal purpose is to prove that CdTe crystals can be used for X-ray imaging.

Ofeq series

Background

The original Ofeq ('Horizon') satellite was a demonstration testbed developed by IAI/MBT for ISA.

Shavit's Ofeq 3 launch (IAI) **1996**

It is controlled from MBT's ground station. The satellite itself resembles an irregular octagonal prism of conventional aluminium construction, measuring 2.3 m high, and 1.2 m at the base, shrinking to a top diameter of 0.7 m. It weighs 156 kg, broken down by: structure 33 kg, power system 58 kg, computer 7 kg and communications package 12 kg. The thermal control weighs 5 kg, the wiring 9 kg and the instrumentation 32 kg. To maintain control, the Israelis spin-stabilised the satellite at the rate of 400°/s. Sixteen solar panels provide 246 W through a 7 Ah Ni/Cd battery. The satellite consumes 53 W on average. It transmits 2.5 kbit/s S-band from 128 kbyte memory.

The Israelis refuse to disclose the full payload, but do say it included magnetometers and housekeeping monitoring. Attitude sensing results from the output of a three-axis rate gyro, magnetometer and Sun sensor. During its short lifetime, the only malfunction was in telemetry memory when the computer switched to back-up. Originally, the Israelis expected the satellite to last only a few weeks, but it did not re-enter until 14 January 1989 because of higher than expected orbit.

Several months later and almost identical to Ofeq 1, Ofeq 2 carried an improved computer, telemetry system back-up, thermal protection, gyros, and magnetometer. It could also execute some commands, whereas Ofeq 1 followed a pre-set program. The second satellite had a mass budget of: structure 34 kg, power system 59 kg, computer 7 kg, communications system 14 kg, thermal control 5 kg, wiring 10 kg and instrumentation/balance masses 31 kg. The orbit differed from Ofeq 1's to provide new perspective on radiation belts and to test Shavit's injection precision. It re-entered 9 July 1990 with an orbital life 40 days longer than expected because of reduced solar activity.

Some observers have speculated that the Israelis attempted another launch shortly after the launch of Ofeq 2 but that it failed.

By the time the Israelis were ready to proceed to their second-generation system, they had built a prototype 36 kg satellite. They described it in 1994 as a demonstration satellite with no science or other payload, but post-launch reports suggested a 2 m resolution Earth imager, supported by the orbit providing suitable lighting conditions over the region. Elop, in 1995, confirmed there was a camera similar to its ERMS Earth Resources Monitoring System.

The second-generation Ofeq has power provided by deployed twin three-panel solar wings totalling 3.6 m^2. The Sun sensors were developed by the Netherlands' TNO Institute of Applied Physics. Rafael provided the blowdown Hydrazine Propulsion Module for de-spin from 65 rpm, final orbit positioning, station-keeping and MW unloading. The thrusters are mounted on the top platform supported by eight composite struts, above the single 420 mm diameter positive expulsion sphere carrying 30 kg of hydrazine pressurised at 22 atmospheres. There are two independent branches of six LT-5N 5 N plus one HT-25N 25 N thrusters (HT-25N is principally for final positioning and orbit control). This propulsion system, with two tanks, appears to have been planned for South Africa's Greensat. Photographs show them to be identical.

Ofeq 3 is almost identical to South Africa's cancelled Greensat. Rafael's hydrazine propulsion module is visible at top (IAI) **1994**

Specifications

Ofeq 1
Launched: 09.32 GMT 19 September 1988 by Shavit from Palmachim Air Force Base south of Tel Aviv
Orbit: 248 × 1,170 km, 142.86°, 98.79' retrograde.

Ofeq 2
Launched: 12.02 GMT 3 April 1990 by Shavit
Orbit: 208 × 1,584 km, 143.23°
Mass: 160 kg.

Ofeq 3
Launch: 11.16 GMT 5 April 1995
Orbit: 247 × 732 km, 143.4° and raised itself at first apogee to 369 × 730 km, 143.4°. The goal was about 500 km circular but no further changes were made
Mass: 225 kg at launch
Design life: 2 years
AOCS: 3-axis control system providing 0.1° astronomical-class pointing accuracy
Data rate: 15 kbit/s with a command rate of 5 kbit/s.

Ofeq 4
Failed in launch attempt on 22 January 1998 due to a problem with the second stage of the Shavit launcher.

JAPAN

DEBUT (Orizuru) satellite

Background

The Japanese National Aerospace Laboratory's Deployable Boom and Umbrella Test repeatedly deployed a 15 cm diameter 1.2 m coilable lattice boom 34 times and a 24-panel 89.8 cm diameter umbrella 52 times during the 10 day life of the satellite. A 95 Ah lithium battery powered it. The behaviour mirrored the expectations the Japanese published prior to the test.

Researchers claim that the umbrella device could be utilised in future μg processing capsules released from space stations, using atmospheric drag to manoeuvre back to their orbital facility following 10^{-7} *g* operations.

Specifications

Launched: 7 February 1990, piggyback with MOS-1B
Mass: 50.3 kg
Orbit: 917 × 1,753 km, 99.05°.

ETS (Japanese Communications Engineering Test) series

Current status

The COMmunications and Broadcasting Engineering Test Satellite (COMETS) failed to achieve the correct orbit after lift-off on 21 February 1998. Japanese investigators traced the failure to a premature shutdown of the H2's, LE-5A second-stage engine. NASDA immediately announced a plan to rescue the satellite and change the orbit to prevent passage through the van Allen belt where it was stranded.

Background

NASDA still conducts communication experiments on its own set of Engineering Test Satellites. It launched four between 1975 and 1982, before privatising its space network. NASDA demonstrated the fruits of its research in August 1987 when it launched the ETS 5, the first Japanese three-axis stabilised satellite in geosynchronous orbit and the first to use an indigenous apogee kick motor.

ETS 6 appeared in 1994 incorporating intersatellite data relay and mobile communications packages. NASDA had intended to demonstrate a new ion propulsion engine to maintain north-south station-keeping on this two-ton platform. The launch vehicle left it short of geosynchronous orbit however, after launch in August 1994. The first of three planned firings by the new ETS 2 kN LAPS (Liquid Apogee Propulsion System) ended when controllers found the engine attained only 10 per cent of full thrust, probably because a fuel line valve had failed to open. The tanks ran out of oxidiser supply on the fifth attempt and ground controllers separated the LAPS on August 31 by command via NASA Goldstone, leaving ETS 6 in 7,796 × 38,707 km, 13.1° inclination orbit.

NASDA planned to launch the COMETS Communications & Broadcasting ETS in 1997 to develop new technologies, preceding the operational DRTS Data Relay & Test Satellites in 2000. The OICETS Optical Inter-orbit Communications Engineering Test Satellite should follow in 1998 to test laser links with ESA's Artemis. ETS 8 in 2002 will demonstrate a single-hop cellular phone system and a CD-quality radio broadcast service.

The Ministry of Posts & Telecommunications established the Partners (Pan-Pacific Regional Telecommunications Network Research Satellite) project in 1992, using Kiku 5 to transmit educational teleconferences (64 kbit/s) and material to less developed Asia and Pacific sites, such as Fiji and Papua New Guinea.

The Japanese will use ETS-7 for automatic and tele-controlled rendezvous and docking with a small target satellite. A GPS, laser, radar and optical sensor will simultaneously guide the satellite to its final rendezvous and docking. It will also demonstrate the capabilities of the 20 kg Orbital Replacement Unit exchange using a 2 m 6 DoF telerobotic arm. MITI's separate Advanced Robot Hand will include NAL's truss-building experiment, Atomic Oxygen Monitor. The work began on this project in FY91 when the Japanese expended ¥85 million; (FY92 ¥1,046, FY93 ¥2,709, FY94 ¥8,743, FY95 ¥8,659, FY96 ¥9,817).

Specifications

ETS-1 (Kiku, 'Chrysanthemum')
Launched: 9 September 1975 by N1 from Tanegashima
Mass: 83 kg
Orbit: 980 × 1,098 km, 47°
Comment: Test of launcher, tracking and antenna extension. Shut down 28 April 1982.

ETS-2 (Kiku 2)
Launched: 23 February 1977 by N1 from Tanegashima
Mass: 130 kg
Comment: First Japanese GEO satellite, manoeuvred to 130° east. Shut down 14 December 1990.

ETS-3 (Kiku 4)
Launched: 3 September 1982 by the last N1 from Tanegashima
Orbit: 968 × 1,229 km, 45°
Mass: 385 kg
Comment: Designed for 1 year of testing 3-axis control, deployable solar paddles and ion thrusters.

ETS-4 (Kiku 3)
Launched: 11 February 1981 by N2 from Tanegashima
Mass: 638 kg
Orbit: 220 × 35,082 km, 28.5°
Comment: Shut down 24 December 1984.

ETS-5 (Kiku 5)
Launched: 27 August 1987 on 3-stage test flight by H1 from Tanegashima
Mass: 1,096 kg (550 kg on station)
Orbit: 150° east, GEO
Comment: Japan's first 3-axis stabilised satellite. L-band mobile communications package. Still on-station and active, but hydrazine depletion imminent.

ETS-6
Launched: 28 August 1994 by H2 from Tanegashima
Comment: The first 2 t GEO satellite (3.8 t at launch) built and launched by Japan. The satellite provided a demonstration of ion propulsion, mobile communications and other technologies for future projects such as a Data Relay and Test Satellite system. Included the first laser communications demonstration. However, propulsion problem left it in low orbit, allowing only a truncated programme, completed 12 January 1996.

ETS-7
Launch: With TRMM on 28 November 1997
Orbit: 550 km, 35°
Mass: 2,370 kg chaser, 403 kg target
Lifetime: 1.5 years.

ETS-8
Launch: January/February 2002 by H2 from Tanegashima
Locations: 135° east, 129° east or 146° east GEO
Design life: 3 years
Contractors: to be determined
Transponders: S-band for mobile-mobile connections
BSS mobile broadcasting of CD-quality radio

Configuration: to be determined, but with two 10 m-class deployable antenna
AOCS: to be determined
Power: to be determined.

COMETS
Launch: 21 February 1998 by H2 from Tanegashima
Location: planned 121° east geostationary
Design life: 3 years
Contractors: NEC (system integrator)
Transponders: S-band 2.0-2.1/2.2-2.3 GHz forward/return inter-orbit link, working through 3.6 m antenna (steerable ±10°). Redundant 50 W 20 MHz SSPA/3 W 10 MHz SSPA forward/return, maximum EIRP 45.0 dBW. Developed by NASDA.

Ka-band 23/26 GHz forward/return inter-orbit link (working through same 3.6 m antenna). Redundant 3 W 30 MHz SSPA/30 W 150 MHz TWTA forward/return, EIRP greater than 56.0 dBW. Developed by NASDA.

Ka-band 30/20 GHz + O-band 50/40 GHz for high-quality mobile communication via 2.0 m feeder link antenna. Developed by Communication Research Lab.

2 (plus 1 back-up, 2-for-1) Ka-band 21 GHz 200 W TWTA 120 MHz high-quality DBS, covering Japan in two beams (Tokyo area + Kyushu Island) via 2.3 m antenna. Developed by CRL/NASDA
Technical Engineering Data Acquisition: TEDA payload will monitor the space environment and its effect on the spacecraft with Single Event Upset Monitor, Dosimeter, Magnetometer, Potential Monitor, Solar Cell Monitor
Principal applications: development of technology for inter-orbit data relay and tracking services, mobile communications and high-definition DBS
Configuration: 2 t class, similar to ETS 6. Solar array span 33.2 m, bus 1.96 × 2.96 × 2.8 m
Mass: 3.9 t at launch, 2 t on-station BOL
AOCS: highly accurate ACS (<±0.05° pitch/roll, <±0.15° yaw) required to control large antenna pointing. MELCO's 20-30 mN ion thrusters (35 kg Xe fuel) paired on E/W panels, provide NSSK. GEO insertion and attitude/orbit control by UPS Unified Propulsion System derived from ETS 6's LAPS: single 1.7 kN bipropellant Apogee Kick Engine plus 4 × 50 N and redundant 8 × 0.86 N hydrazine catalytic thrusters
Power: twin solar wings 2.93 × 15.6 m (rolled up for apogee motor firings) provide 4.7 kW EOL; Ni/H_2 battery provides 100% eclipse protection.

OICETS
Launch: August/September 1998 by J1 from Tanegashima
Orbit: planned 550 km circular, 35°
Design life: 1 year mission planned
Contractors: NEC (system integrator)
Transponders: 2.04425/2.22000 GHz S-band forward/return inter-orbit links (COMETS, Artemis); EIRP 4.4 dBW, G/T -30.6 dB/K.

2.04425/2.22000 GHz S-band up/down link, EIRP 12.3 dBW, G/T -47.5 dB/K.

Laser communications demonstration: 8,010 Å beacon, 8,190 Å receive (2 Mbit/s), 8,470 Å transmit (50 Mbit/s) 200 mW (CW) AlGaAs diode; 26 cm diameter telescope
Principal applications: demonstration of optical inter-satellite communications
Configuration: 3-axis stabilised box-shaped 78 × 110 × 150 cm bus, 9.36 m span across twin solar wings, 2.5 m high from base to top of optical assembly. Coarse pointing assembly provides 0-120° elevation, ±190° azimuth range
Mass: about 550 kg in orbit
Power system: twin 3-panel solar wings spanning 9.36 m, 1.75 m wide; 13 Ah NiMH battery.

Japanese launch monitoring test series

Specifications

Ohsumi
Launched: 11 February 1970 by Lambda 4S-5 from Kagoshima
Mass: 24 kg
Orbit: 340 × 5,150 km, 31°
Comment: Test satellite was 4th stage. Transmitted 17 h.

Tansei ('Light Blue')
Launched: 16 February 1971 by M-4S-2 from Kagoshima
Mass: 63 kg
Orbit: 990 × 1,110 km, 30°
Comment: Demonstrated launcher capability. Operated for full 10 days. Named after Tokyo University's school colour ('light blue').

Tansei 2
Launched: 16 February 1974 by M-3C-1 from Kagoshima
Mass: 56 kg
Orbit: 290 × 3,240 km, 31°
Comment: Tested M-3C performance and satellite attitude control. Decayed 22 January 1983.

Tansei 3
Launched: 19 February 1977 by M-3H-1 from Kagoshima
Mass: 130 kg
Orbit: 790 × 3,810 km, 66°
Comment: Tested launcher performance and satellite attitude control. Shut down March 1977.

Tansei 4
Launched: 17 February 1980 by M-3S-1 from Kagoshima
Comment: Tested M-3S performance and future science satellite subsystems. Decayed 13 May 1983.

VEP (Myojo) Vehicle Evaluation Payload
Launched: 3 February 1994 on H2 number 1
Mass: 2.4 tonnes
Orbit: 449 × 36,261 km 28.6° GTO
Payload: Box-shaped 3 × 2 × 0.8 m bus above ETS-6 type apogee propulsion system. The tanks carried 1,470 kg water for practice tests of pressurisation and depressurisation. Battery-powered from 4, 35 Ah Ni/Cd batteries
Comment: Payload was used to monitor H2 and range performance. The satellite had no attitude contro but attitude and position knowledge was derived within less than 15° from solar cell sensors. The satellite transmitted real-time acceleration, vibration and distortion data. Satellite operated for 100 h/10 orbits.

Japanese spaceplane test series

Background

NASDA, NAL and ISAS designed and are jointly researching technologies for aerospace vehicles. To review Japanese standing in the space transportation market, STA's Spaceplane Committee reported in 1987 that development of a manned space transportation system might be important but strongly recommended that unmanned cargo transports, based on existing technology, should be separated from manned craft, which require greater safety and reliability. As with all major policy decisions, the Space Advisory Committee reviewed the advice, and in mid-1993, released a report recommending development of a winged reusable vehicle launched from the H2.

A fully reusable spaceplane would take off and land horizontally, powered by air-breathing engines such as ramjets, Liquefied Air Cycle Engines (LACE) and scramjets. NASDA initiated Phase A studies of this HOPE unmanned spaceplane in 1986. Japanese government and industry spent some US$150 million in FY82-90 on air-breathing aerospace plane activities, and projected US$3.4 billion FY90-98 if HOPE, the prototype vehicle, moved off the drawing boards. The National Aerospace Laboratory is working with a 350 tonne baseline vehicle, capable of carrying a crew of four into a 500 km orbit. Landing mass would be 101 tonnes. Preliminary HOPE designs have six ramjets, six scramjets and three rocket engines, working with 80 tonnes liquid hydrogen (slush hydrogen is favoured) and 164 tonnes of liquid oxygen. Some economists claim the plane will cost Japan more than ¥20 billion. The first, uncooled, engine ran for 6 seconds in May 1994.

The OREX re-entry vehicle carried by H2 number 1 in February 1994 tested thermal protection and navigation as a preliminary step towards spaceplane re-entry. Then ALFLEX tested automatic approach and landing techniques by releasing test bodies from a helicopter in a test series. Japan conducted the series during 1996 at Woomera in Australia to demonstrate GN&C for the proposed HOPE Spaceplane's automatic landing. The first J1 in February 1996 carried the HYFLEX suborbital hypersonic flight experiment spaceplane model. The full-scale HOPE-X demonstrator is planned for launch by a single-stage H2A in 2000 on a suborbital flight. Thus far, Japan has expended ¥10.5 (NASDA) and ¥0.5 (NAL) billion in FY97 to permit Phase C to begin.

The HYFLEX was Japan's first hypersonic lifting body experience. HYFLEX's slender shape and flight profile was dictated by J1's fairing and test requirements. At the conclusion of the test, the vehicle should have had a parachute splashdown 1,300 km downrange for recovery 250 km northeast of the Ogasawara Islands. The chute deployed at 694 s and the flotation bag inflated, but HYFLEX sank in 6 km deep water after apparently breaking free of the bag. The thermal protection system could not be examined directly, but 12 of the 14 test objectives and the demonstration of the critical technologies for HOPE were achieved.

Following the successful experiments of OREX, HYFLEX and AFLEX in 1997, the SAC made the decision to proceed to Phase C. In 1997, STA proposed a pre-study of fully reusable launch vehicles and a spaceplane. NASDA started Phase C of HOPE-X in 1997 as a NASDA/NAL joint project, intending to provide servicing missions in the early 2000s to Japan's Experimental Module attached to the Space Station.

In parallel, ISAS is working on the HIMES highly Manoeuvrable Experimental Space unmanned suborbital delta-wing boost-glide vehicle, again as a technology demonstrator. The organisation plans to conduct a launch of the 14 tonne, 9.33 m wingspan 13.6 m long HIMES on a reusable LOX/LH_2 sounding rocket. ISAS began low-speed gliding tests from helicopters in 1986. In October 1987, it experimented with a 2 m long, 1.5 m wingspan 100 kg model released from 1 km and 2 km altitude over the Sea of Japan. ISAS further attempted a Mach 4 re-entry gliding test, but the 185 kg 2 m model and its solid rocket were lost 21 September 1988 at sea because of carrier balloon problems after ascent from Kagoshima.

ISAS and IHI are also pursuing development of the ATREX-500 air turbo-ramjet, which could power the flyback booster of a 2-stage spaceplane. The hydrogen expander ATREX would accelerate the vehicle to Mach 6 and carry it up to 30 km. ISAS used the 5.4 kN ATREX-500 ¼-scale version, 2.2 m long, to conduct 40 static sea-level runs during 1990-95 at ISAS' Noshiro Testing Centre. Efforts now centre on a 1.4 t flying test-bed launched from Japan and parachuting into the sea. A jet engine will take the FTB to Mach 0.5 from a horizontal rail for the 370 kg ATREX-500 with 190 kg of LH_2 to attain 25-35 km altitude 200 km downrange. The initial metal turbo machinery will achieve Mach 4.5, with Mach 6 targeted for the carbon-carbon composite design.

Specifications

OREX (Ryusei, 'Shooting Star') Orbital Re-entry Experiment
Launched: 3 February 1994 on H2 number 1
Cost: US$20 million
Vehicle: 3.4 m diameter 1.46 m high paraboloid vehicle re-entered at HOPE angle/speed
Mass: 865 kg launch, 761 kg re-entry. See p68 1995-96 edition of *Jane's Space Directory* for full description
Comment: Testing thermal protection and GPS navigation for HOPE spaceplane. Splashed down 460 km south of Christmas Island after 133 minutes 2 seconds. The recovery was not planned.

HYFLEX
Launched: 11 February 1996 with the HYFLEX hypersonic flight experiment model of HOPE by 2-stage J1
Mass: 1,040 kg
Test Description: 238 second flight at 110 km 3.9 km/s on a suborbital path to enter the atmosphere at Mach 14.4
Vehicle: Length 4.40 m, span 1.358 m, height 1.04 m, and planform area 4.27 m^2. Aluminium alloy primary structure, protected by ceramic tile underbody, flexible blankets on the upper body and carbon-carbon nose cap and elevons. Pitch and roll control during Mach 14.4 glide is performed by pair of elevons with six 40 N GN_2 thrusters for 3-axis exo- and endo-atmosphere control. Strapdown RLG and a C-band transponder for radar tracking perform autonomous navigation. T+P measured around body. 49° angle of attach for entry, 30-35° for later descent.

UME (Apricot) series

NASDA, under the direction of the Communications Research Laboratory (Ministry of Posts and Telecommunications) launched UME-1 to act as an ionosphere sounding satellite. NASDA used the fourth N-2 launch vehicle to put the satellite into orbit. The Space Agency expected to monitor radio waves in the ionosphere and use the results to forecast transmission conditions for short-wave communications. About one month after it was launched UME-1 failed. The satellite used spin-stabilisation for control, and NASDA hoped it would have a lifetime with a targeted residual rate of over 70 per cent of one and a half years. It decayed from orbit and impacted the Earth on 23 February 1983.

NASDA expected to use Ume-2 as an improved back-up for UME-1 (ISS-1) and pressed it into service shortly after the loss of UME-1. It performed well during its originally scheduled mission, and Japan continued to gather data from the satellite after its official lifetime had ended.

Specifications
UME-1
Launched: 29 February 1976 by N1 from Tanegashima
Mass: 135 kg
Orbit: 994 × 1,013 km, 70°
Comment: This ionospheric sounder experienced a power failure after 1 month but the back-up UME-2 compensated.

UME-2
Launched: 16 February 1978 by N1 from Tanegashima
Mass: 141 kg
Orbit: 980 × 1,220 km, 69°
Comment: The Japanese planned to use this UME-1 back-up satellite to study ionospheric radio propagation. It was shut down 23 February 1983.

RUSSIAN FEDERATION

Soviet spaceplane test series

Background

The Soviet Union began investigating the potential of small reusable aerodynamic spaceplanes for short duration manned missions to low Earth orbit in the early 1960s. Efforts became serious with the inauguration of the Mikoyan Design Bureau's 50/50 Spiral project, headed by Gleb Lozino-Lozinsky, in 1965. Mikoyan proposed a small reusable spaceplane carried aloft by a hypersonic aircraft. In the original design, the plane had a total 140 tonne take-off mass and carried one man on the 10.3 tonne 8 m long, 7.4 m span spaceplane. The carrier dropped the plane at high altitude and around M5.5-6.0.

Early tests produced the BOR 2/3 one-third and one-half scale model prototypes which the Soviets tested in wind tunnels and on suborbital ballistic trials using the Cosmos-3. The first BOR launch occurred 15 July 1969 on Cosmos-3. It used a wooden BOR 1 model that burned up at 60-70 km. The Politburo cancelled Spiral in 1969, following this test, but then revived it in 1972, after NASA proposed the Space Shuttle.

Mikoyan continued with a full-scale spaceplane model, constructed for runway trials in May 1976 and equipped with a turbojet. Aviard Fastovets piloted the first test. Igor Volk numbered among the first BOR astronaut corps (although he never got to fly the vehicle), but Volk made the first Buran approach and landing tests and was expected to command the first manned orbital mission. A Tu-95 conducted the first drop test 27 October 1977, followed by a further five piloted tests by Petr Ostapenko in 1978. At that point, the Soviets terminated Spiral altogether. The test vehicle is now displayed at the Monino museum near Moscow.

Chelomei's bureau began designing a new spaceplane in 1975. It was a 25 tonne vehicle, which would be able to carry 4-5 tonnes into LEO atop Proton.

Recovery of Cosmos 1445 in March 1983. Contemporary speculation suggested it was a one-third model of a manned military vehicle (Australian Department of Defence)

Pre-flight view of the BOR-4 shuttle testbed; illustration released 1989

The Mikoyan full-scale spaceplane model was flown from a Tu-95 in 1977/78. Photograph released 1989

A half-size mockup was built and the bureau presented its design to a government commission in 1980, claiming that development would require four years. However, the commission suspended work and accelerated Energia-Buran work.

The Soviet Union had begun the Buran Space Shuttle effort beginning in the 1970s, and it encompassed some of the same spaceplane technology work as a testbed for the other programme. As the technology matured, the Soviets conducted six suborbital flights of the BOR-5 one-eighth scale Buran between 4 July 1983 and 24 June 1988. Each test used an SL-8 Cosmos-3M launcher from Kapustin Yar. These tests verified the overall configuration.

Simultaneously, the Soviets also built a separate (military) spaceplane. Sources refer to this programme as the BOR-4 and indicate the Soviet Union adopted it to counteract the military applications of the US Space Shuttle. The 3.4 m long, 2.6 m span, 1.5 tonnes BOR-4 lifting body bore a strong resemblance to the 1960s' US DynaSoar. Subscale tests flew four orbital missions between 1982-84 of more than 2 h each, apparently using the same vehicle each time. (On 5 December 1980 and 20 October 1987, the Soviets also conducted suborbital tests.)

BOR-4 orbital test flights began in 1982 with Kosmos 1374, followed by the identical Kosmos 1445 (apparently the same vehicle), in March 1983. Both spaceplanes de-orbited and descended by parachute. Orion aircraft of the Royal Australian Air Force monitored these tests. When the Australians released the photos, the US Air Force made an unsuccessful attempt to classify them.

In the third test, a Soviet North Atlantic tracking ship monitored the retrofire of Kosmos 1517 (the test's international designation) instead of the usual tracking conducted from the main Soviet tracking station in the Crimea. A Soviet announcement reported a controlled descent and splashdown in the Black Sea. The smaller body of water indicated increased confidence and provided greater security from Western observation. The fourth subscale spaceplane test, Kosmos 1614, also re-entered into the Black Sea. By then it was believed to be one-third the actual size of the manned vehicle being developed. In 1987, the US Department of Defence suggested that full-scale production of the spaceplane might be already under way.

The Soviets reiterated their claim, in 1989, that this military craft, using the SL-16 Zenit medium-lift booster, was not under development. Western Intelligence services, nonetheless, suspected that a three-man 10 tonne design for quick-reaction military missions was being pursued. It was further claimed, in early 1991, that there was a manned programme, with cosmonauts selected in 1980 and 1983, and that it was ready to fly on Zenit in 1987. It later emerged that the 'Uragan' MiG craft, equipped with a kinetic weapon, was intended to intercept the US Shuttle when it operated out of Vandenberg. The Soviets cancelled this programme immediately when the US dropped the California site as a Shuttle spaceport. Programmatic inertia, however, meant that two Zenit launches in August 1987, of Cosmos 1871/1873, flew full-scale Uragan mockups.

NPO Molniya advertised BOR-4 in 1990 as available for foreign commercial flights to test re-entry materials and other hypersonic systems. A BOR-6 exists, designed to investigate radio transmission during re-entry, but remains grounded because of financial constraints.

The Soviets disclosed proposals in 1989 for several new spaceplane concepts involving air launches from the super-heavy lift An-225 or a dedicated hypersonic aircraft. Departing from the Antonov, a spaceplane accommodating two cosmonauts and 5 to 7 tonne payload could attain LEO, while the hypersonic platform approach could increase the payload to 25 tonnes.

Specifications

Bor-4 Test 1
Launched: 3 June 1982 by Kosmos-3M from Kapustin Yar
Orbit: 158 × 204 km, 50.7°
Comment: Recovered after 109 min 560 km south of the Cocos Islands in the Indian Ocean by a 7-ship Soviet task force.

Bor-4 Test 2
Launched: 15 March 1983 by Kosmos-3M from Kapustin Yar
Orbit: 158 × 204 km, 50.7°
Comment: Recovered 556 km south of the Cocos.

Bor-4 Test 3
Launched: 27 December 1983 by Kosmos-3M from Kapustin Yar
Orbit: 158 × 204 km, 50.7°
Comment: Recovered in the Black Sea.

Bor-4 Test 4
Launched: 19 December 1984 by Zenit from Tyuratam
Orbit: 158 × 204 km, 50.7°
Comment: Recovered in the Black Sea.

SPAIN

Intasat satellite

Background

Spain, through its INTA space and science agency, developed the Intasat to act as an ionospheric sounding beacon with a two-year operational life. Spain planned to use the results of the test to develop information for later communications satellites. The spacecraft body was a 12-sided polyhedron.

Specifications

Intasat 1
Launched: 15 November 1974 piggyback with NOAA 4/Oscar 7 by Delta from Vandenberg AFB
Orbit: 1,442 × 1,462 km, 115°
Mass: 20 kg.

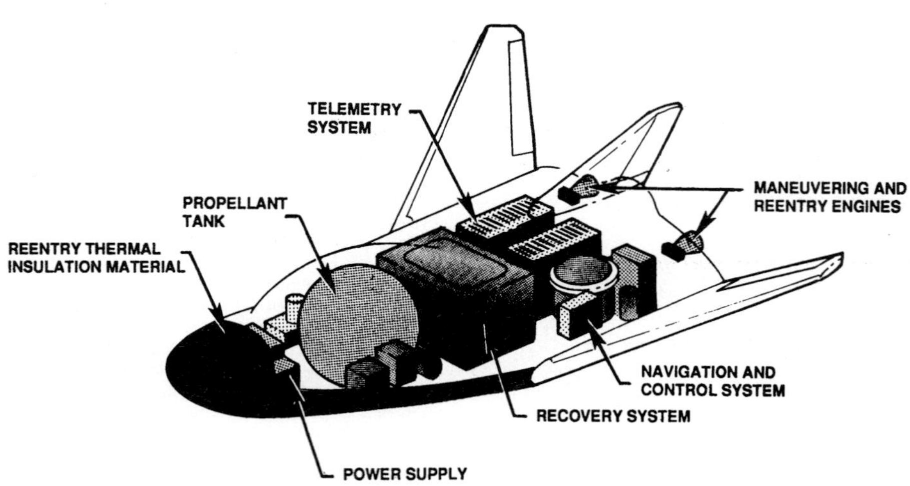

BOR-4 was claimed as available for re-entry and hypersonic trials (Teledyne Brown Engineering)

SWEDEN

Viking/Astrid-Freja engineering test series

Background

The Swedish government funded spaceborne scientific investigations primarily to gain experience in system development/management. It intended to follow up Sweden's sounding rocket studies of the interaction of the solar wind and Earth's magnetosphere. These tests would expand the investigations to 2 Earth Radii (RE) to gain insight into the behaviour of the Aurora Borealis. Since Sweden did not plan to develop an indigenous launcher capability, the government sought a satellite that could ride aboard existing launch vehicles dedicated to other missions. After an initial design phase, the Viking vehicle resulted from the Swedish effort.

Sweden funded its effort through the Swedish Space Corporation. SSC awarded a prime contract to Saab-Scania in September 1980 to begin work on a small precursor satellite. The resulting small satellite platform became known as Mesa. It was based on the USAF Small Scientific Satellites programme and NASA's Atmospheric Explorers. Given the space available on the Ariane launcher at the time of the first design review, the satellite met the objective of fitting below the main payload on Ariane launches. Swedish technical organisations and universities provided the first payload consisting of electric/magnetic field experiments, particle and magnetospheric wave experiments, and two Canadian UV imaging cameras.

As the satellite neared its launch date of 22 February 1986, the Swedish government officially named it the Viking. All of the experiments on the satellite performed well, until it ceased operations as planned 12 May 1987 after 444 days (goal: 8 months). The only reason the satellite finally shut down was due to a degrading power supply. It cost the Swedes SKr238 million (1991 rate) for the satellite and its launch.

By September 1986, the Canadian imaging cameras had returned more than 20,000 images showing the complete auroral oval and previously unobtainable sequences of the full cycle of auroral activity. The cameras were recording sequential shots at 20 second intervals and global views every 80 seconds, for real-time transmission to Kiruna. They were then relayed by tape or telemetry to Canada's Centre for Space Science and the University of Calgary.

SSC followed the programme with the Freja-C ('Compact') platform, which is a scaled-down version of the original Freja. Astrid became the first application for the Freja bus. In keeping with a tradition started by the British with their satellite names drawn from Shakespeare, the Swedes named the satellite after author Astrid Lindgren and the instruments after her characters. SSC states the first Astrid launch cost SKr9 million, including launch. Three instruments successfully monitored magnetosphere neutral particles and electrons and imaged aurora in ultra-violet, but their power supply failed 1 March 1995. Ground controllers had the last Astrid contact on 27 September 1995.

The Freja version, which SSC now offers to clients, is 45 × 45 × 39 cm and weighs 20 kg. The bus can carry 10 kg payload and it has 0.3 mm aluminium facesheets glued to aluminium honeycomb core. SSC promises attitude determination within 2° by analogue Sun sensors, coupled with two ±60 μT two-axis fluxgate magnetometers. A 12 Am^2 magnetorquer gives precession control, along with a 7 Am^2 for spin. In operation, the bus spins up and down by 10-20 Ns HTPB solids space-qualified motors. Power comes from deployable Si solar panels that generate 80 W BOL. Users communicate with the satellite by a 5 W downlink, 512 kbit/s, 2.20-2.30 GHz S-band antenna.

Specifications

Viking
Launched: 22 February 1986 piggyback with Spot 1 by Ariane V16 from Kourou
Mass: 286 kg (535 kg launch)
Orbit: 811 × 13,536 km, 98.8°
Size: 1.9 m diameter, height 50 cm.

Astrid 1
Launched: 24 January 1995 by Cosmos from Plesetsk
Orbit: 965 × 1,026 km, 82.93°
Comment: The first application of the Freja-C generic micro platform
Mass: 28 kg.

Astrid 2
Launch: 10 December 1998 on a Cosmos-3M launcher from Plesetsk
Contractors: prime Swedish Space Corporation on behalf of Swedish National Space Board
Mission life: 1 year
Mass: 30 kg
Orbit: planned 979 × 1,1013 km, 105.04°
Cost: budgeted at SKr10 million, including development of S-band ground station
Mission: electric and magnetic field polar phenomena investigations, electron density studies and a continuation of Astrid 1's neutral/charged particle imaging (secondary)
Spacecraft: Astrid 1 mechanical design slightly modified for increased solar cell area to provide more than 55 W peak and deployable booms. Dimensions with stowed solar panels: 40 × 45 × 95 cm. Payload power 15 W continuously. Sun-pointing spin axis is under automatic attitude control. A star imager will probably be test flown giving accurate attitude knowledge. 128 kbit/s 5 W S-band downlink to a 1.8 m ground station antenna.

UNITED KINGDOM

R Series/X-Shakespearean test series

Background

The United Kingdom began its own satellite programme in the early 1960s shortly after the United States and the Soviet Union had orbited satellites. Principally the UK wanted to build an indigenous European launch and satellite support service with the same capabilities as the superpowers' programmes. As a first demonstration platform, it built the R series beginning with the R-0 which it planned to launch from its co-operative sounding rocket range in Woomera Australia. In the first test of the R series, the UK planned to demonstrate stage 1/2 firings and a dummy stage 3 separation. The launch vehicle had to be destroyed at 64 seconds because it began tumbling, but R2 was successful as a demonstration of stage 1 and 2 engine firings, and a test spin up of the satellite. The second stage, as well as the payload, impacted in the Indian Ocean according to plan after 15 minutes, about 3,050 km northwest of Woomera. In the next test, R3, the British intended to use a live third stage and inject a satellite directly into orbit.

Originally intended as part of the Black Arrow launch vehicle programme, the British Parliament cancelled the X-series programme in July 1971. The controversy centred on whether the Europeans had the economic resources to build a launcher that was competitive in cost with the United States. Despite the cancellation, because some of the satellites had been built, the RAE continued with its launch programme. The X-4 (renamed Miranda in orbit, in the Shakespearean tradition of the UK series), was launched 9 March 1974 by Scout from Vandenberg. Despite minor control problems, good data were obtained. Total cost was more than US$8 million, plus US$2.35 million launch costs.

In anticipation of the project success, test officials named the mission the X-2 development payload, including a 13 kg Orba gold-plated drag sphere. As the test progressed, the second-stage shut down 15 seconds too early. In the wake of this failure, the British continued to fund the programme while Birmingham University prepared a new science payload to count micrometeoroids. The satellite had a one year design life. As the final preparations for the launch got under way the British government stopped the Black Arrow programme, but decided to let the one last test occur. The Royal Aircraft Establishment named it the Prospero.

Following a successful launch on 28 October 1971, it made 812 passes over the Woomera test range during a total of 4,960 orbits in the first year. The tape recorder replayed 697 times. It finally failed 24 May 1973 after 730 replays. RAE Lasham contacted the satellite annually until 1989, when the site's VHF system closed down. However, various stations reported continued reception of the 137.56 MHz beacon as late as February 1993. Unfortunately, Prospero's success did not come soon enough to save the Black Arrow programme, which was cancelled July 1971. The programme had sufficient funds available to continue with the X-4 Miranda satellite, which humiliatingly was launched by NASA Scout. Its intended launcher, R-4, is displayed at the Science Museum, London.

Specifications

R-0
Launched: 27 June 1969 from Woomera, Australia on the Black Arrow.

R-1
Launched: 4 March 1970.

R-2
Launched: 2 September 1970.

R-3 (X-3)
Launched: 28 October 1971 from Woomera
Mass: 66 kg
Description: 109 cm diameter
Comment: X-3 Prospero technology demonstration satellite
Orbit: 557 × 1,598 km, 82°.

X-4
Mass: 93 kg
Orbit: 703 × 912 km, 98.81°
Comment: UK-built by Hawker Siddeley for DTI, its principal goal was to demonstrate 3-axis ACS to replace the spin-stabilisation of previous UK satellites.

STRV series

Background

The DRA Space Department designed and built the STRV 1A/1B satellites. They are intended to test new technologies such as solar cells, microelectronics, electrostatic charging alleviation, structural materials and onboard data handling in a harsh environment. The Molniya-type orbit causes the satellites to traverse the van Allen belts four times daily over a year. This provides the approximate equivalent of a 10 year GEO radiation dose. STRV also provided advances in the use of carbon-PEEK composites, radiation-resistant microelectronics (SOS processors) and advanced systems architectures.

After launch, ground controllers lost contact with 1A temporarily during August 1994 because of a power system problem. The anomaly initially prevented full operation of the Xe plasma neutraliser, because of its high power requirement, but battery reconditioning overcame the problem and the tests of plasma discharges progressed satisfactorily. All other experiments have operated routinely but, on 1B, a hardware problem with the solar cell experiment has prevented operation in its primary mode.

With the nominal mission completed in June 1995, the accumulated radiation dose was several times lower than models had predicted. Combined with the high quality of data, this has allowed mission extension, with successful transfer of the day-to-day orbital maintenance of both satellites in September 1996 to the University of Colorado, under NASA funding. In the experiment, the satellite has not seen any transient system anomalies uniquely correlated with the radiation environment. Surprisingly dynamic behaviour in the van Allen belts, particularly the electron belt, has been observed, however.

The project definition phase of follow-on STRV 1C/1D satellites was completed in December 1996 with the aim of a paired GTO launch in 1999 as Ariane-5 auxiliary passengers.

Specifications

STRV 1A
Launch: 17 June 1994 Ariane V64 with Intelsat 702 from Kourou, French Guiana
Orbit: 1A 282 × 35,822 km, 7.04°, 10.56 h GTO
Orbit: 1B 284 × 35,899 km, 7.04°, 10.58 h GTO
Configuration: 40 × 50 × 50 cm box carrying no deployed structures. First space use of carbon-PEEK composite panels for main structural elements.

Final preparations before STRV 1B was attached to Ariane's carrier (DRA Space Department) ***1995***

Mass: 50 kg 1A, payload 12 kg; 53 kg 1B payload 14 kg
AOCS: 5 rpm (provided by Ariane), axis 90 ±5° from Sun line (for power/thermal control). Sun/Earth sensors. Three systems, one nutation damper and magnetorquer (to control precession and spin axis during perigee) perform attitude control. Each has two separate windings of aluminium wire (for redundancy). Three Xe thrusters exist on 1A only, two for spin up and one for precession in case of magnetorquer failure. The system uses cold gas from electrostatic discharge experiment
Power: GaAs cells on X/Y faces to provide 30.5/32.5 W 1A/1B BOL (normal to Sun), supported by 64 Ah Ni/Cd battery. Only 10 per cent array loss by May 1996
Communications: redundant 2 W transmit at 2.200-2.300 GHz at 1 kbit/s, 2.025-2.120 GHz receive at 125 bit/s. ESA/CCSDS packet telemetry/telecommand (custom chips) and ADA software. First space use of UK MIL-STD 1750/MAS 281 SOS 16-bit SOS μp set, 96/128 k word 1A/1B memories. Control room at DRA Farnborough working through 12 m antenna at DRA Lasham (passes up to 10 h). NASA's Deep Space Network routinely complements Lasham. Each satellite downlinks 60 kbyte experiment data daily
Payload: Charge alleviation package. Xe plasma emitted to discharge electrostatic buildup. First space test of the hollow cathode from the UK-10 ion thruster. Monitored by electrostatic charge detector (a new type of solid-state opto-electronic device to measure surface charge) and cold ion detector.

Atomic oxygen measurement package. Erosion rates of eight coatings (including carbon, Kevlar, Mylar, Kapton, FEP Teflon and SiO_2) by atomic oxygen around perigee. Measured by monitoring increasing resistance of a coated silver strip as it oxidises.

CREDO Cosmic Ray Effects and Dosimeter. Cosmic ray dosage at 5 minute intervals on 3 cm^2 300 μm diodes; total dose of all ionising radiation from RadFET dosimeters spread around STRV.

Battery Recharging Experiment. Temperature fluctuations to detect end of charge more accurately.

STRV 1B

Orbit: 1B 284 × 35,899 km, 7.04°, 10.58 h GTO
Configuration: 40 × 50 × 50 cm box carrying no deployed structures. First space use of carbon-PEEK composite panels for main structural elements
Mass: 53 kg 1B payload 14 kg
Other Specifications: Same as STRV 1A
Payload: SCTE Solar Cell Technology Experiment: New types of cells cover glasses, coatings and manufacturing techniques to determine radiation resistance. Includes Si, high-efficiency Si, amorphous Si, GaAs, GaAs/Ge, ultrathin GaAs, InP, CIS and CdTe.

Vibration suppression: Piezo-electric elements to suppress vibration of Stirling cycle cooler cold finger.

RADMON: Total radiation dose plus proton belt profiles, 4 kbit SRAM devices, results compared with CREDO.

HIP Heterojunction Internal Photo-emission: Novel Infra-Red (IR) detector studying radiation effects by measuring electrical characteristics.

Neural Net. Radiation fault tolerance of analogue neural network VLSI devices.

REM Radiation Environment Monitor (ESA). Trapped electrons, protons and some heavy ions for CREDO comparison. During an astronaut EVA an identical REM was installed on Mir's exterior 9 September 1994.

UoSat series

Current status

A new SSTL 100 kg enhanced microsatellite based on the heritage of the 50 kg microsatellite has been launched to demonstrate new technologies for Enhanced Small Satellite users.

Background

UoSAT-12 was launched by Dnepr on 21 April 1999 from Baikonur. Surrey University built UoSat-12 as a

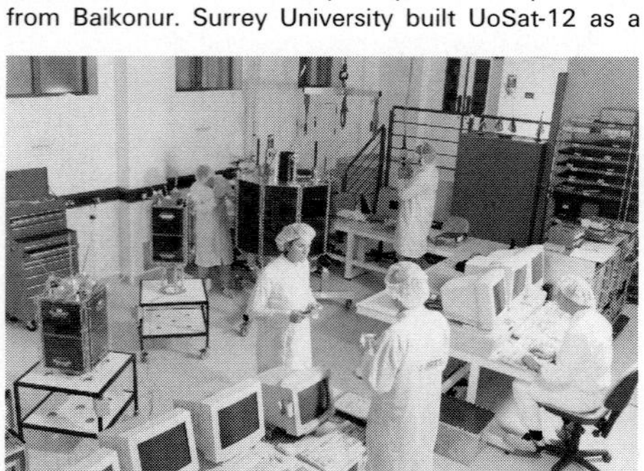

Surrey Space Centre's new mini satellite UoSat-12 was launched on 21 April 1999 on a Dnepr (SSC)
1999/0054359

Surrey University's series of successful 50 kg microsatellites will be joined in 1997 by the 300 kg UoSAT 12 minisat
(Theo Pirard/Space Information Centre) ***1996***

modular multipurpose bus which will provide S-band communications and enhanced OBDH for 180 kg payloads. The first payload carries RWs, a 40 m resolution Earth camera and propulsion system for orbit control. The bus is known as Bus Alpha and it weighs 23 kg and measures 600 × 600 × 500 mm, accomodates four body-mounted GaAs solar panels providing 38W average power. A full 3-axis altitude control system allows payload pointing accuracy better than 0.2°.

Bristol University emphasised simplification by making the satellite transmit-only and stripping the bus of attitude or orbit control. The 2 kg single main experiment can be varied for each satellite. On the first satellite a new type of battery cell will be tested. The other experiments examine GaAs solar cell degradation by exposing 14 experimental cells on the X dimensional face panel, radiation environment measurement, and monitoring the satellite's nutation and spin behaviour. The SSTL enhanced satellite can carry 5 CCD remote-sensing cameras providing 3-band multispectral imaging with 50 metres ground resolution and panchromatic imaging with 20 metres ground resolution.

SSTl built UoSat-12 to demonstrate advanced leo communications and Earth observation payloads (SSTL)
1999/0054360

UNITED STATES OF AMERICA

ASUSat satellite

Background

ASUSat 1 is the result of a challenge made to Arizona State University students in October 1993 by Orbital Sciences Corp's co-founder Scott Webster, to design and build a 4.5 kg nanosat to perform meaningful science. Total project cost is US$200,000 maximum and has involved 150 students in the Aerospace Research Center. The contractors are Arizona State University which is also building the control station, OSC which will donate the launch services on a Pegasus XL, Marshall Electronics building the cameras, Trimble Navigation providing a GPS receiver, Applied Solar Energy Corporation (GaAs cells), SAFT (batteries), Intel (80C188 processors).

The satellite has an expected design life of 2 years and it weighs only 4.5 kg. To support its primary objective of earth imaging it will be launched into a planned 550 km circular 97.6° Sun-synchronous orbit. In configuration it consists of a 14-sided prism, 24 cm high (dynamic envelope 26 cm height 31 cm diameter set by Pegasus avionics section), and has a structural weight of 1.1 kg. To reduce weight the student project uses carbon fibre in 954-2A epoxy resin, with a 0.8 mm wall. A gravity gradient boom and gravity anchor, which is a ball free to rotate in a fluid-filled spherical casing, and weighted to assume local vertical, provides attitude control by viscous drag to damp satellite. The orbit position determination comes from a commercial GPS receiver and spacecraft attitude (±10° error) from diodes (one Sun, three Earth) on each of 14 sides, supplemented by two Sun diodes on each end face.

Four VVL-1070 CMOS cameras (one side pointing, three Earth reference) provide 2 km/pixel resolution. Each orbit the satellite photographs up to eight 25-kbyte images as power permits. The Intel 80C188EC CPU has a 2 kbyte PROM, 256 kbyte EPROM and 1 Mbyte RAM. A GaAs solar array (30 18.5 per cent efficiency 2 × 2 cm cells on each of 14 sides plus 90 on top) powers the satellite along with two six-pack Ni/Cd switched batteries which provide 8.5-10 W. The satellite has an allocated amateur voice repeater with J-mode frequencies at 436.5 MHz for transmission, and 145.820 for receive. It is controlled from ASU on 2 m band.

ATS series

Current status

Despite their ages, ATS 1 and 3 remain active in inclined geosynchronous orbits. ATS 1 has exhausted its station-keeping propellant and is drifting slowly eastwards inclined at 14.6°, completing a circuit every two years. NASA Lewis says it can still be used for educational exchanges and search and rescue work. The VHF translator operates for about 12 hours each day. The signals from ATS 3 have grown weak but discernible. ATS 3 is employed and funded equally by NASA and the National Science Foundation at about US$180,000 annually for relaying data from Antarctica to the Space Physics Analysis Network. It also provides voice circuits to many islands in the Pacific basin under the Peacesat programme (Pan Pacific Education & Communications Experiments by Satellite).

Background

The Applications Technology Satellite series was conceived to demonstrate techniques for communications, meteorological and navigation satellites from geosynchronous altitude.

ATS 1 began operations in 1967 by providing the first transmissions of full-Earth, cloud-cover images from GEO, and the first real-time TV from the Pacific with coverage of Apollo 4's splashdown. In 1967, it provided emergency communications during Alaska's floods. In 1968, it demonstrated two-way links with commercial airliners, and in 1971 linked US/Soviet scientists during an atmospheric, sea and ice conditions experiment in the Bering Sea. Under the Peacesat project, ATS 1 transmitted educational, health, research, technology and community services to 23 autonomous terminals in Hawaii, Cook Islands, the Mariana and Caroline Islands, West and American Samoa, the Marshall Islands, Melanesia, New Zealand and Australia.

In 1973, ATS 1 pioneered transmission of electrocardiographs from Hawaii to New Zealand, and from Alaska to the University of Washington. Then, during 1971-78, it carried 12-nation medical conferences over

ATS 3 antenna pattern test at Hughes Aircraft in 1967. ATS 3 operations continue to this day

the Peacesat network. NASA announced in March 1985 that it had failed to respond to commands to correct its eastward drift from its 162° east Pacific position over the Gilbert Islands, and would drift out of a useful position within six months.

ATS 3 covers most of the Atlantic, East Pacific and US as far as Hawaii. It is supported equally by NASA/NSF and controlled from ATSOCC at the University of Miami as the node for data and telephone networks. Several Antarctica stations use ATS 3 as their primary link for voice, data and message traffic with the rest of the world. South Pole Antarctica uses it for about 5 h daily, while Palmer Station uses it 3 h and Seal Island 1 to 2 h. It also supports datalinks with research ships between 24° and 180° west and Arctic Ocean to Antarctica, and is still used about 5 hours a day by Peacesat for voice circuits to many Pacific basin islands.

The most notable of many ATS 3 achievements was the transmission of the first colour Earth imagery from space; it also obtained many cloud images and monitored severe storms. On 21 November 1967, it provided the first link over the Atlantic during a Pan American flight. ATS 3 then supported the Apollo Moon landings. In a series of maritime experiments covering ship location and ship-to-shore communications, it demonstrated that satellites could bring major improvements to the management of shipping fleets, presaging the Inmarsat system. At the end of 1974, ATS 3, as part of an NOAA programme, predicted and measured rainfall in remote areas by the comparison of satellite images of the brightness and size of clouds with radar density scans. During the 1980 Mount St Helens eruption, it provided links through a ground vehicle at the disaster site. In 1983, stationed at 105° west, it was being used for remote site communications by users of a two-suitcase terminal. Within 24 hours of 1985's Mexican earthquake, ATS 3 had been brought into use to provide critical support for the international relief effort.

ATS 6 used its 9 m diameter high-power antenna to relay demonstration TV to small receivers, including a year of educational programming over India. Other ATS 6 achievements included the first aircraft-to-ship relay via satellite; the first direct flight control of an aircraft by an ocean air traffic controller using a satellite; and the first direction of a search/rescue operation, albeit simulated, by satellite.

Specifications

ATS 1
Launched: 7 December 1966 by Atlas Agena D from Cape Canaveral
Mass: 352 kg
Size: 1.42 m diameter, 1.34 m high
Power: 23,870 solar cells
Orbit: In GEO over the Pacific
Lifetime: 3 years planned life

ATS 3
Launched: 6 November 1967 by Atlas Agena D from Cape Canaveral
Mass: 365 kg after apogee motor burnout
Orbit: Initially at 86° west GEO over the Atlantic, it continues transmission from 105.4° west inclined at 14.9°.

ATS 6
Launched: 30 May 1974
Mass: 1,402 kg.

LDEF satellite

Background

NASA Langley's 9,705 kg Long Duration Exposure Facility (LDEF) was deployed by Shuttle STS-41-C 7 April 1984 and retrieved by shuttle STS-32 on 12 January 1990. LDEF's 57 experiments covered materials, coatings and thermal systems, power and propulsion, science, electronics and optics. The satellite was a passively, free-flying unit with no central power or data systems or communications and command links. It was placed in a gravity gradient attitude upon release. A database by Langley's LDEF Material Special Investigation Group is maintained to be the most extensive open collection of information on materials behaviour in the space environment. It is being expanded to include material from other sources and is publicly available via Internet.

LDEF's emptied structure was placed in protected storage while the possibilities for further missions were clarified. It is now displayed at Kennedy Space Center.

SEDS Satellite (SEDSSat)

Background

SEDS began in 1983 as a Small Business Innovative Research programme and, through April 1997, the programme had flown two tests at a cost of US$10 million. March 1993's Delta carried NASA Marshall's 41 kg SEDS to observe the dynamics of a long tether system. It unreeled the 26 kg instrumented end mass on a 0.75 mm diameter polyethylene tether to 20 km in 77 minutes, allowing it to swing like a pendulum for 16 minutes before being detached to burn up. Observations corroborated modelling of tether dynamics.

The 41 kg SEDS 2 flew 10 March 1994 and this time remained attached to stage 2 on its 19.7 km tether. The 25.8 kg 20 × 33 × 41 cm payload included Langley's accelerometer, magnetometer and tensionometer to monitor its motion. Despite its slim diameter, the tether was visible to naked-eye ground observers.

GEODETIC

FRANCE

Stella satellite

Background
Stella is almost identical to 1975's Starlette geodetic laser ranging satellite. In appearance it is a 35.74 kg icosahedron core of depleted U_{-238} alloy, surrounded by 20 AlMg alloy spherical segments each carrying three reflectors of 3.29 cm diameter entry pupil. The satellite spins at 5 rpm imparted by Ariane stage 3. CNES built the satellite as an in-house effort and CEA (the atomic energy authority) at Bruyères-le-Châtel provided the structure and Aerospatiale the reflectors. The satellite was released piggyback by Spot 3's Ariane.

Specifications
Launched: 26 September 1993
Orbit: 718 × 805 km, 98.68°
Mass: 47.978 kg
Size: 24.000 ±0.002 cm diameter.

Stella joined the network of laser geodetic satellites in 1993 (CNES) ***1993***

GERMANY

GFZ series

Background
The GeoForschungsZentrum (GFZ, National Centre of Earth Sciences) in Potsdam signed a contract 31 May 1994 with Kayser-Threde to build a laser-reflecting geodetic satellite. Various government sources estimate the project cost DM1 million. GFZ's orbit is the lowest of any laser geodetic satellite yet, improving the modelling of Earth's gravitational field. The National Research Institute for Geology operates a laser station in Potsdam and a joint facility in Santiago de Cuba with the Cuban Centro Nacional de Investigaciones Sismologicas. NII KP provided the 60 laser reflectors.

Specifications
Launched: 9 April 1995 aboard Progress-M 27 to Russia's Mir and ejected from the station 19.12 GMT 19 April
Mass: 20 kg
Size: 21.5 cm diameter
Orbit: 384 × 394 km 51.65°.

GFZ 1 was released from Mir in April 1995 (Kayser-Threde) ***1995***

JAPAN

EGS (AJISAI) satellite

NASDA and Kawasaki Heavy Industries developed the Experimental Geodetic Satellite, (EGS or Ajisai, which means Hydrangea). EGS is a simple 685 kg, 2.15 m sphere carrying 318 mirrors and 120 laser reflector assemblies (1,436 corner cube reflectors) for precise tracking to provide Earth crustal movement and other data. Stage 2 also carried NAL's 295 kg MABES Magnetic Bearing Flywheel Experimental Satellite (although it was not designed for release) to test a two-axis actively controlled flywheel. A 2-D CCD star sensor and a linear CCD Sun sensor measure the attitude. The battery provided a three day life for the satellite.

Specifications
Launched: 13 August 1986
Orbit: 1,483 × 1,497 km, 49.98° by first H1.

EGS, carrying 120 laser reflector assemblies and 318 mirrors, is a passive geodetic satellite. The time-lapse exposure was recorded at Tsukuba Space Centre soon after launch (NASDA)

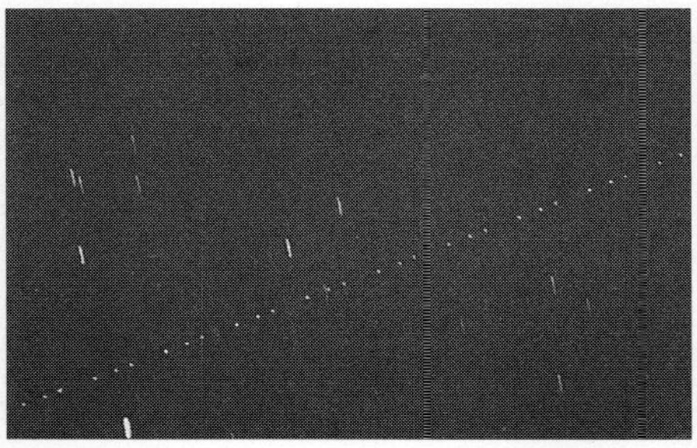

EGS, carrying 120 laser reflector assemblies and 318 mirrors, is a passive geodetic satellite. The time-lapse exposure was recorded at Tsukuba Space Centre soon after launch (NASDA)

UNITED STATES OF AMERICA

LAGEOS series

Background
NASA designed the Laser Geodynamics Satellite as a tool to aid geodetic research, including earthquake prediction. It was NASA's first effort devoted wholly to laser ranging. With a high orbit and resulting long life, LAGEOS 1 acts as a permanent reference point to measure distances between ground stations. From this information scientists derive knowledge of the Earth's crustal motion, Earth's rotational variations, solid Earth and ocean tides and movement of the polar axis. By paying particular attention to earthquake areas such as the San Andreas Fault in California, LAGEOS may ultimately provide information on impending earthquakes. Station co-ordinates are now determined to within 1 to 3 cm and relative motions of 5 to10 mm per year can be measured. NASA Goddard, through its Laboratory for Terrestrial Physics manages the programme.

LAGEOS allows researchers to determine position by bouncing LASER light off of its 426 3.8 cm diameter corner cube reflectors. The four germanium and 422 fused silica reflectors return laser beams to their source, regardless of the incident angle. To be effective, the satellite had to accommodate a large number of reflectors but at the same time be small enough to minimise solar radiation pressure and atmospheric drag effects. By the same arguments, NASA built the satellite from a combination of aluminium and brass. A pure aluminium construction would have been too light and brass too heavy, but their combination yielded a sufficiently high mass to surface area ratio to meet the objectives. NASA had to consider the interaction between the satellite and Earth's magnetic field in its selection of construction materials as well.

The LAGEOS carries a symbolic mission in addition to its scientific ones. Each end of the bolt connecting the hemispheres carries a copy of a message prepared by Dr Carl Sagan. It includes three maps of Earth's surface. One shows the period 225 million years ago when, it is believed, the landmasses were one super-continent (sometimes called 'Pangaea'). The second shows the position of the continents as they are now, and the third their estimated position 8.4 million years from now when the satellite, so solid it will survive re-entry, falls back to Earth.

LAGEOS 1's first four years were devoted to determining its precise orbit and establishing a global network of 14 permanent Satellite Laser Ranging (SLR) stations supported by mobile truck-mounted SLR systems. By accurately measuring the round trip time for a laser pulse, the position of the laser system could be determined to within about 10 cm. LAGEOS 2 improved this accuracy by a factor of 2.

As part of its Geodynamics Program, NASA in 1979 initiated the international Crustal Dynamics Project (CDP) to measure contemporary tectonic plate velocities, regional crustal deformation and various Earth orientation and rotation parameters. CDP was superseded in 1992 by the DOSE Dynamics Of Solid Earth programme. During the 1980s, 36 permanent SLR stations were operated by 18 countries (Australia,

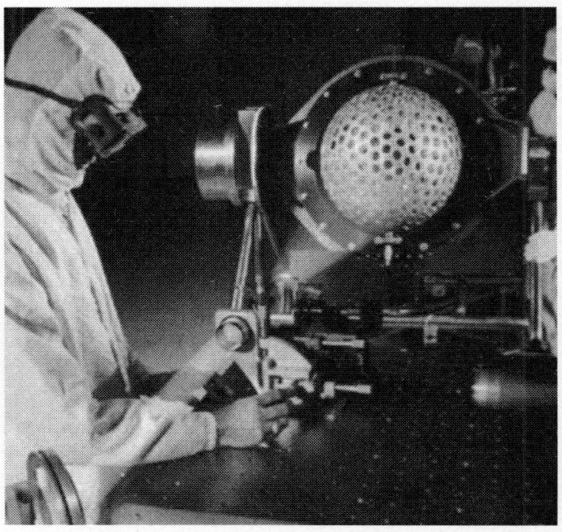

Lageos carries 426 laser reflectors embedded in its aluminium shell (NASA)

Austria, Chile, China, Czechoslovakia,Egypt, France, Germany, Israel, Italy, Japan, Mexico, Netherlands, Peru, Poland, UK, US, USSR), making repeated measurements with LAGEOS and providing their data to the CDP. NASA currently has 10 stations. Three sites (Texas, Hawaii and France) also range to reflectors on the Moon placed by Apollo 11/14/15 and the unmanned Lunokhod 1/2.

Typical velocities between tectonic plates vary over 1 to 20 cm per year, and the contemporary plate motions measured by space geodetic techniques are in good general agreement with the long-term average motion indicated by the geological record. Models derived from LAGEOS provided the first evidence of gravity field temporal variations, believed to result from continuing crustal relaxation following the last ice age.

Originally due for Shuttle launch, with the IUS upper stage in 1987, Italy's identical LAGEOS 2 was released by STS-52 23 October 1992. The altitude is the same as LAGEOS 1's but the 52° inclination improves coverage of the seismically active Mediterranean Basin and should enable investigators to identify the source of irregularities noted in LAGEOS 1's orbit. Some 16 mobile laser ranging facilities supplement the 10 fixed stations around the world that track both satellites. A complementary LAGEOS 3 is under consideration.

Specifications

LAGEOS 1

Launched: 4 May 1976 by Delta 2913 from Vandenberg
Orbit: 5,858 × 5,958 km 109.8°
Mass: 405 kg
Lifetime: Operational life of 50 years. Orbital life of 8 million years
Size: 60 cm diameter, 407 kg
Description: Two aluminium hemispherical shells, totalling 117 kg, surrounding a 31.76 cm diameter 27.5 cm long cylindrical brass core
Payload: 426 3.8 cm diameter laser reflectors for ground stations to make precise distance measurements.

LAGEOS 2

Released: From the Shuttle 23 October 1992
Orbit: 5,610 × 5,950 km, 52.84°
Other specifications: Same as LAGEOS 1.

MICROGRAVITY AND MATERIALS SCIENCE

CHINA, PEOPLE'S REPUBLIC

FSW μGravity (Jianbing) series

Current status

China refers to its recoverable orbital payloads as Jianbing (Progress). China became the third nation after the US and the USSR to recover an object it placed in space when the FSW-0 1 test reconnaissance platform returned to Earth in December 1975.

Development of China's recoverable system began in 1966, but the first eight launches, all made by CZ-2C from Jiuquan into 57 70° orbits carried a mix of Earth resources and reconnaissance imaging payloads. But then China used the same configuration for a μg payload the first time in August 1987. The payload itself consisted of a 15 kg demonstration package from France's Matra. China followed this flight with the FSW-1 design later in 1987. This design had the same external configuration but a completely reworked internal cargo space to improve capacity, duration and control. The FSW-2 debuted in August 1992 aboard the maiden CZ-2D, adding a pressurized equipment module and again improving the avionics. FSW-1 has presumably retired. As of May 1999, the last FSW launch had been that of FSW-2 3 on 20 October 1996.

Background

Overall, FSW-1 was 3.144 m high and 2.20 m across the base with an additional 200 kg of mass over the FSW-0's 1,800 kg. Chinese engineers deactivated the attitude control in orbit. The 2.4-3.1 tonne FSW-2 model can return up to 350 kg on dedicated missions. It comprises essentially an FSW-1 with added 1.50 m-high pressurized base equipment module in an overall length of 4.644 m and a diameter of 2.200 m. The later version improves on FSW-1's attitude control and avionics by using a triply redundant computer. It can also adjust its orbit using a hydrazine mono system. The Chinese noted in late 1989 that they plan to build a new spacecraft employing a lifting configuration for improved re-entry loads and landing accuracy.

The satellite initiates its re-entry over China by an unspecified solid-propellant motor housed in a section immediately below the capsule. The Ian Satellite Control Center at Ian City in Shane province transmits the de-orbit command through two mobile stations, while the third located the re-entry modules during descent. To deorbit, the satellite must generate 650 m/s if it uses the nose forward re-entry position, but only 150 m/s required for against-orbit returns. The higher velocity puts the satellite immediately down inside China. The dramatic descent results from China's lack of space partners to spread a tracking network outside of Hina's borders. Consequently the re-entry must followed immediately over China after the deorbit signal is sent. This maneuver produces a 20 *g* re-entry load, working against the goal of a microgravity environment.

Fifteen FSW capsules had been recovered by the beginning of 1996 **1995**

The Chinese built their heat shielding from oak wood using a blunt planform to slow the capsule. Then a parachute system takes over to reduce the speed to a terminal velocity of 10 to 15 km/h. FSW's can even survive a water landing since the landing site is a central area of the Sichuan Basin, where there are numerous rivers and lakes. A transponder and two beacons identify the position of the returned craft and aid recovery operations.

Some 275 μg experiments were returned on the six flights between 1987 and 1992, including guinea pigs in October 1990 and several GaAs, HgCdTe and protein crystal growth experiments. The Matra mission landed in central Sichuan Province in southwest China and the Chinese returned the sealed package to the owners 2 days after it was retrieved. August 1988's mission carried Intospace's 20 kg Cosima package of 104 protein crystal growth ampoules. All performance characteristics of the mission followed the flight plan, but large numbers of crystals became damaged during re-entry and landing. October 1992's mission included Japan's 710°C semiconductor furnace.

Launch schedules

Chinese recoverable missions

	Launch	Days	Payload
1	26 November 1975	3	FSW-0 1 imaging
2	7 December 1976	3	FSW-0 2 imaging
3	26 January 1978	3	FSW-0 3 imaging
4	9 September 1982	5	FSW-0 4 imaging
5	19 August 1983	5	FSW-0 5 imaging
6	12 September 1984	5	FSW-0 6 imaging
7	21 October 1985	5	FSW-0 7 imaging
8	6 October 1986	5	FSW-0 8 imaging
9	5 August 1987	5	FSW-0 9μg/imaging
10	9 September 1987	8	FSW-1 1μg/imaging
11	5 August 1988	8	FSW-1 2μg/imaging
12	5 October 1990	8	FSW-1 3μg/imaging
13	9 August 1992	16	FSW-2 1 μg/imaging
14	6 October 1992	7	FSW-1 4μg/imaging/Freja
15*	8 October 1993	-	FSW-1 5 μg/imaging
16	3 July 1994	15	FSW-2 2 μg/imaging
17	20 October 1996	15	FSW-2 3 μg/imaging

*indicates failure. All FSW-1 launched by CZ-2C (FSW-2 by CZ-2D) from Jiuquan. 100 per cent recovery rate until October 1993. The CZ-2A failed test launch of 5 November 1974 is believed to have carried an FSW. The long gap after 1978's test was used to increase orbital operations to 5 days, reduce structural mass and improve control. FSW-2 in 1992 was the first to manoeuvre: its adjustments every three days for the first nine days probably indicate Earth imaging; the manoeuvring system then failed and it was recovered 1 day late. First known FSW recovery failure was October 1993. Capsule was to be recovered 16 October but apparent attitude problem resulted in retro raising the orbit; decayed 12 March 1996 over South Atlantic.

The FSW-1 descent capsule, mass about 1.2 tonnes, incorporates oak for re-entry protection (CAST)

Chinese sources quote a price of US$45,000-50,000/kg for the small packages, and say the dedicated mission costs are negotiable. China can launch 2 to 3 missions a year, on a 12 month lead time for small packages and 24 months for dedicated missions.

Specifications

Mass: FSW-1 payload: recoverable 20 kg in two 10 kg, 200 × 200 × 250 mm packages; non-recoverable 50 kg in discarded unit. On dedicated missions total of 380 kg permissible, with 150-180 kg recoverable
FSW-2 payload: recoverable 150 kg in 200 × 200 × 250 mm volume; non-recoverable 200 kg in discarded unit. On dedicated missions total of 750 kg possible, with 310-400 kg recoverable
Microgravity level: 10^{-4}-10^{-5} *g*, duration unknown. Micro-accelerometer measurements during 1994's FSW-2 revealed 4-28 μ*g* during quiet periods; atmospheric drag reached 7 μ*g* during daytime perigee
Flight duration: 7-10 days for FSW-1; 15-18 days for FSW-2
Power supply: 1,300 Ah 27 V silver zinc batteries. CAST claims that all power requirements can be satisfied
Ambient internal conditions: <=50°C, 10^{-3}– 10^{6} torr
Experiment control and telemetry: unspecified but multiple programmed and ground commands available. Telemetry can be transmitted real-time or stored for subsequent dumping according to user requirements. Self-contained experiments yield lower costs
Attitude control: 3-axis but assumed de-activated during μg operations. Employs redundant digital computer (double redundancy on FSW-1; triply on FSW-2), inertial measurement unit, Sun/Earth sensors and cold gas thrusters. FSW-2 carries three-1 DOF fluid floated rate integrating gyros, two IR conical scanning Earth sensors, 120 × 128° FOV digital Sun sensors. Control accuracy: FSW-0 1° pitch/roll, 2.2° yaw; FSW-1 0.7°, 1°; FSW-2 0.5°, 0.7°, angular rate to within ±0.02°/s for fine Earth pointing, 3-axes to ±3°, to within ±0.3°/s for maneuver mode
Acceleration levels: payloads must be tested for 11 *g* longitudinal acceleration, 2 *g* lateral during launch, with 20 *g* for re-entry
Vibration/shock: payloads must be tested to 5-10 *g* sine wave and 70 *g*/6-10 ms shocks for launch; and 2.5 *g* sine wave, 70 *g*/6-8 ms for descent. Landing speed is typically 14 m/s
Acoustic loads: payloads must be tested to 150 dB for 120 s for launch.

GERMANY

Express satellite

Background

Germany used Japan's last M-3SII booster to launch its microgravity research satellite the Express from

Kagoshima on 15 January 1995. The payload consisted of the 765 kg Express 1, including its 405 kg reusable capsule with 128 kg of μg/technology experiments. Japan provided 80 per cent of the US$93.5 million funding. Originally Germany wanted to recover the satellite on the ground at Woomera, Australia after 5.5 days, but a stage 2 attitude control problem resulted in an approximate orbit of 110 × 250 km, 31° instead of the planned 210 × 398 km, resulting in re-entry after 2.5 orbits. When the satellite lost contact with the ground after 2 orbits, controllers believed it had been lost in the Pacific. Instead it descended safely in a remote region near Gushiago in northeast Ghana. DARA did not respond to local press reports of an unknown capsule with Russian labeling until November 1995. During this period the satellite lay untouched at Tamale airport until it was returned to Germany on 20 February 1996.

SPAS series

Background

Developed as a commercial venture under BMFT sponsorship, SPAS has a modular experiment pod built with a carbon fibre tubular structure connected by titanium elements. Subsystems and experiment modules sit within a standard 0.7 × 0.7 × 0.7 m cube assembled according to mission requirements. Experimenters may configure the basic structure in a variety of ways within a total spacecraft budget of 1,500 to 3,000 kg. The original SPAS-01 version was 1,500 kg, with 900 kg dedicated to payload. The 3,600 kg Astro-SPAS version contained the four approved Orfeus/Crista missions holding telescopes up to 3.9 m long, 1.2 m diameter with improved attitude control system for a few arcsec pointing accuracy.

Shuttle Mission STS-7 released the Shuttle Pallet Satellite in 1983 to make the platform the first retrievable satellite. NASA reflew the SPAS aboard STS-41B in 1984 and STS-39 in 1991. After this successful series of tests SDIO (now BMDO) bought the SPAS from ESA for US$16 million. SDIO wanted to fly it as SPAS 3 on STS-64 in 1994, but Congress cut funding in FY94. Unlike Eureca, SPAS cannot be left in space to be picked up by a subsequent Shuttle. It is therefore constrained to Shuttle-duration missions.

Specifications

Mass: 1,500 kg total, 900 kg payload
Microgravity levels: same as Eureca
Flight duration: typically 4-5 days
Power supply: 28 VDC direct to Shuttle, two AgZn batteries of 70 Ah each in free-flight
Telemetry/telecommand: two S-band links to orbiter (2.0603 GHz transmit, 2.2375 GHz receive)
Attitude control: 3-axis rate integrating gyro package, 12 N_2 cold gas 1 N thrusters drawing on four 10 l tanks
Thermal control: passive
Data processing/control: DASA's Modular Digital Universal System
Size: 4.8 m wide, 4.4 m high and 1.5 m deep.

INTERNATIONAL

Eureca satellite

Background

Eureca or, the European Retrievable Carrier is a Spacelab follow-on program initiated by ESA in April 1982. It is the first spacecraft designed specifically as a μg carrier and extends the capabilities of SPAS. Eureca 1 and its 15 experiments entered orbit on 2 August 1992 after release from the STS-46. It attained a 508 km circular orbit after a week and began full operations 18 August. After a highly successful mission, STS-57 recovered it in June 1993. Eureca is designed for five missions, each of six to nine months duration spread over a 10 year period. At about 4.4 tonne total mass, it incorporates a hydrazine propulsion system to take it to its 28.5°, 510 km operational altitude and then return it to 315 km for Shuttle recovery. The carrier is stored in Bremen, its future uncertain. ESA's October 1995 ministerial meeting rejected further funding and a Daimler-led commercial consortium took over responsibility for a commercial flight between 1998 and 2000.

Specifications

Payload mass: 1,000 kg in 8.5 m^3
Microgravity level: worst-case 10^{-5} *g* <1 Hz, 10^{-3} *g* >100 Hz; 5 x 10^{-7} *g* if solar activity/air drag at minimum and payload close to centre of gravity
Availability: <1.5 year turnaround
Flight duration: 6 months operational, 3 months dormant
Power supply: 1 kW average available payload, 1.5 kW peak (solar array output 5 kW); 28 VDC
Thermal control: liquid freon loop provides 1 kW heat rejection; multilayer insulation
Data management: 256 kbit/s high, 2 kbit/s low, 128 Mbit memory capacity (1.5 kbit/s) for average payload; S-band downlink
Attitude pointing accuracy: 1° at 3σ
Orbit: typically 525 km, 28.5°.

JAPAN

Express

Background

Japan's last M-3SII was launched from Kagoshima 15 January 1995 carrying the joint 765 kg Express 1, including its 405 kg reusable capsule with 128 kg of μg and technology experiments. Japan provided 80 per cent of the US$93.5 million funding. It was intended for recovery in Australia after 5.5 days but a stage 2 attitude control problem resulted in an approximate orbit of 110 × 250 km, 31° and contact was lost after two orbits. It was presumed lost in the Pacific, but it landed safely in Ghana and was returned to Germany in February 1996. The failure was attributed to insufficient modelling of the control system dynamics for such an unusually heavy payload.

Space Flyer Unit

Background

Japan launched the SFU-1 free-flying orbital platform on its third manufactured H2 on 18 March 1995. Then in a cooperative effort with the United States of America, Japan retrieved it on 13 January 1996 by the Space Shuttle STS-72. The H2 used three burns to achieve the 486 km 28.46° circular operational orbit by 23 March 1995. Japan planned to lower the orbit to 310 to 318 km for retrieval but thruster failures left it in 463 × 478 km. NASDA's astronaut Koichi Wakata performed the retrieval. As we worked to bring the SUF back aboard the Shuttle, the solar wings would not latch down after retracting and had to be pyrotechnically jettisoned.

ISAS leads the project, in collaboration with NASDA and the USEF Institute for Unmanned Space Experiment Free Flyer. SFU-1 had a microgravity quality of 10^{-5} *g*. In 1991, ISAS sources indicated the project cost about US$430 million or ¥60 billion. ISAS wanted to fly at least five missions.

Specifications

Contractors: MELCO (integrator)
Mass: 4.0 t launch, payload 1 t, 3.56 t retrieval
Orbit: 322 × 336 km 28.45°
Configuration: based on a 4.46 m diameter, 2.8 m high octagonal bus of aluminium alloy truss structure supporting six 150 kg box-like payload modules. Carries one keel and four longeron trunnions for Shuttle interface
AOCS: 3-axis control of 1° accuracy; Sun-pointing maintained. GPS and IMU provide 50 m, 0.1 m/s accuracy. RCS by two redundant sets of 6 × 3 N and 2 × 23 N N_2H_4 thrusters, total capacity 100 kg; orbit adjust by two redundant sets of 4 × 23 N N_2H_4, total 650 kg capacity. Attitude control maintained by RWs and 3 N thrusters
TT&C: ISAS Sagamihara Operations Center working via Kagoshima Space Center. S-band 2.0844/2.2636 GHz up/down; housekeeping down at 1 kbit/s; data down at 16 kbit/s packet and 128 kbit/s playback of 80 Mbit and 4 Mbit bubble memory recorders. Includes still b/w TV
Power: two 2.40 m wide retractable Solar Array Paddles spanning 24.42 m deployed generate 2.7 kW BOL, including 850 W for the experiments. Each carries 28,000 2 × 4 cm Si cells. Supported by 4 × 19 Ah Ni/Cd batteries. SAPs retracted before Shuttle grappling, but latching failure required both to be pyrotechnically jettisoned.

RUSSIAN FEDERATION

Nika-T series

Background

KB Photon is developing the enhanced Nika-T for μg missions although there are funding difficulties. It is possibly derived from the Kuban Generation 6/7 photo reconnaissance satellite, which will also be modified as the Nika-K to replace Resurs- F and as Nika-B to replace Bion. Capable of carrying more than 2.5 tonnes of payload, 1,200 kg of it recoverable, on missions of up to 120 days, Nika-T will be launched by Zenit 2. Small satellites of up to 100 kg 0.1 m^3 can also be deployed. Nika retains the 2.214 m diameter Vostok descent sphere adds non-recoverable but pressurized forward/aft payload modules, two large solar arrays, a 16.5 m^2 radiator and an updated service/control module. Capsule access is provided by 100 and 170 cm diameter hatches. The Zona 8 and Konstanta 4 furnaces are under development for Nika.

Specifications

Flight duration: 90-120 days
Spacecraft mass: about 9 t
Length: 9.27 m
Max diameter: 2.72 m
Orbit: 300-400 km circular Sun-synchronous or 200-300 × 300-500 km elliptical, both at 62.8-70.4°
Payload mass/volume: 900-1,200 kg recoverable in

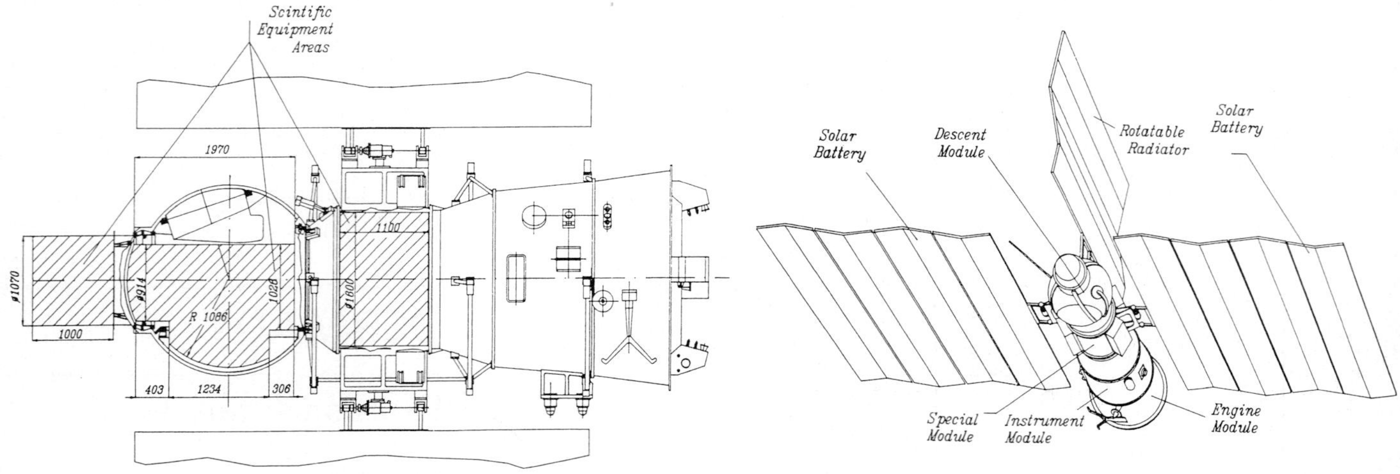

The Nika spacecraft is planned for g missions of up to 120 days. The hatched areas designate pressurised payload accommodation

4.5 m^3 capsule, 1,500-2,600 kg in and around non-recoverable forward pressure container and service module.
Power supply: 4.5 kW daily average at 23-32 V provided to payload from 6 kW generated by two deployed solar panels (totalling 72 m^2).
Ambient internal conditions: 0.46-1.5 atmosphere, 15-25°C (increasing to 30°C for high power operations) in recovery sphere
Experiment control: up to 80 commands transmitted in a single pass, up to 75 commands provided from the onboard computer (maximum of 70 in a day)
Telemetry: 50 channels totalling 3,000 measurements/s for recovery module, 280 channels providing 20,000 measurements/s for non-recoverable forward module (which incorporates a 60 Mbit memory). Depending on data rate, 108 min or 432 min of data can be dumped in 5.1 min pass
AOCS: 3-axis to within 1.5°.

Photon series

Background

The Soviets began marketing space facilities and services to the international µg community in 1987, notably at that year's Paris Salon. The USSR first offered temporary positions aboard the Mir/Kvant orbital complex, with equipment delivered by the Soyuz manned or Progress unmanned ferries, and accompanied by a specialist if required. The Russians have offered the Photon capsule, essentially the mature Vostok craft, for dedicated bookings on 14 to 30 day missions. Russian and foreign researchers also place small µg payloads (typically 33 kg, 60 × 40 × 25 cm) aboard the Resurs-F film camera Earth observation satellites.

Photon is based on the well-proven Vostok spacecraft, introduced in 1960 for the manned programme. In retrospect, it is clear that Photon-type materials processing missions were conducted by Kosmos 1645 (April 1985) and Kosmos 1744 (May 1986), followed by the acknowledged 14 day Kosmos 1841 of April 1987. The last's three primary payloads were the Zona 1 and Splav 2 electric furnaces and the Kashtan electrophoresis unit, all offered commercially.

The first named Photon materials processing spacecraft flew a 14-day mission during April 1988. Photon 2 followed on a 14-day mission beginning 27 April 1989. Its 405 kg payload included a CNES/Matra experiment investigating fluid behaviour. The Russians claimed the total mission cost was R6.5 million. Photon 3 in October 1991 carried CNES' Crocodile (Croissance de Cristaux Organiques par Diffusion Liquide dans l'Espace) equipment to grow NPP organic crystals for non-linear optics applications. The four cells achieved 12 days' crystal growth. CNES' 15 kg Sedex (Synthèse Enzymatique de Dextrane) flew on Photon 4 in October 1991 after the agency signed a 10 year agreement in June of 1991 with Glavcosmos and KB Photon for future Photon and Nika-T flights. Exercising the contract, CNES flew the Gezon experiment aboard Photon 6 in 1994 using Zona 4 to investigate the effects of magnetic fields on the growth of germanium crystals. Ibis flew on Photon 7 in February 1995 with a biology payload under a FFr8 million contract. It will refly on Photon 8.

Photon integration with its carrier ***1995***

The Photon materials processing return sphere was first displayed at June 1987's Paris Salon (Graham Turnill)

Eight German experiments flew on Photon 4-6. Photon 5 included four German crystal growth experiments using Zona 4 (continuing the CdTe growth on Photon 4). Then ESA conducted the first Biopan experiment, which it mounted externally to expose micro-organisms to free space (although this test flight carried only technology experiments). DARA used the Photon for a C/SiC test sample at the capsule's stagnation point to determine if this non-ablative material had heatshield applications. Kayser-Threde developed the Biopan biological experiment carrier for the exterior of Photon and Resurs descent capsules under ESA contract. ESA approved two further flights; first was Photon 6 carrying six biology experiments and another C/SiC sample. The next was Photon 8, which carried Germany's 1 m diameter 150 kg MIRCA re-entry capsule, and ESA's Biobox.

Photon 7 performed seven more German experiments under a DM1 million contract. The experiment tested five samples of AlSi for directional solidification in a Zona 4M and two glass interdiffusion experiments in a Konstanta 2M. ESA's Biobox 2 investigated cells responsible for bone mineralisation, flying under an ECU700,000 contract. Unfortunately, Photon 7 was destroyed after a safe landing when it was dropped by its recovery helicopter (Photon 8 is its reflight).

Commercially, the mission of September 1989 carried Intospace's Cosima 2 package (Cosima stands for Crystallisation of Organic Substances in Microgravity for Applied Research), followed by Cosima 3 May 1990 and the first 33 kg Casimir unit September 1990. Germany's Kayser Threde signed an agreement with Glavcosmos in December 1987 for options on Photon and Resurs missions, acting as marketing agent. Four missions flew between 1989 and 1991, carrying Intospace's Cosima 2-4, Casimir and LZZ experiments.

A Soyuz booster launches Photons from Plesetsk into 220-250 km perigee, 300-400 km apogee, 62.8° orbits for missions of 14 to 30 days. So far only a two week capability has been demonstrated. The mass is 6.2 tonnes. Russian trackers determine the spacecraft position to within 17.57 km along orbital track, 0.63 km transverse and 0.95 km in altitude. The 2.21 m diameter return sphere masses around 2.4 tonnes and is probably constructed from aluminium; thermal protection was originally provided by a >=5 cm thick layer of asbestos-type material embedded in a resin but the current technique is unknown. During the Photon 2 mission, it was acknowledged for the first time that a Soviet spacecraft was being reused: the descent sphere had already flown on C1841 and Photon 1. An attitude control system incorporating N_2 jets and Earth horizon sensors align Photon in preparation for the rear section's retroburn of around 90 m/s over Africa, some 8,000 km downrange. Predecessor Vostok utilised a 15.83 kN TDU-1 liquid-propellant retro; the current liquid engine is unknown. The rear section provides attitude control although marketing specifications indicate this is not available for Photon, possibly because of the need for conservation of limited resources for the re-entry sequence following a lengthy mission. Pyrotechnics release the four tension restraining straps and an offset centre of gravity generates the only attitude control during entry at peak 16 *g*, creating a tendency to wallow that was noted by the early cosmonauts. The three main 27 m diameter parachutes are deployed at 2½ km (following 4.8 m diameter main drogue), leading to a 7.5-8.5 m/s landing, cushioned by solid retros in the shrouds. Reserve chutes are 5.2 m diameter.

Specifications

User cost: ECU20,000/kg quoted in 1995
Availability: 12-18 months on signature of contract
Payload mass/volume: up to 500 kg in a volume of 4.7 m^3 internally; volume consists of two stacked cylinders 1.8 m dia/0.7 m high + 0.6 m dia/0.4 m high. There is also accommodation externally for =~0.06 m^3 non-recoverable
Microgravity level/duration: less than 10^{-5} *g* for unknown duration
Power supply: 400 W average daily; 700 W max for 90 min/day; 27±4 VDC; 25 A steady, 50 A for 200 ms. Battery supply accommodated in front-mounted unit ejected before re-entry; known as the Nauka module on science missions, 1.8 m diameter, 47 cm deep cylindrical section capped by domed end and mounted on four legs
Processing units available: Splav, Zona and Konstanta electric furnaces, Kashtan electrophoresis unit. Specifications provided under separate entries. If these are to be employed, information on material composition, behaviour and safety, experiment sequencing, T gradients and cooling rates is required. If the equipment is user-provided, dimensions, energy requirements, sequencing and telemetry requirement information must be provided
Ambient internal conditions: 0-40°C; 0.46-1.5 atmosphere; 0.055 rad/day radiation background plus 50 rad/flight from solar flares. During testing, the inert spacecraft and contents are subjected to a pressure differential of <=0.5 atmosphere by an air-5 per cent helium mixture in a 10^{-9} atmosphere chamber for 10 hours. Re-entry could generate internal temperatures up to 60°C; internal pressure drops to 0.053 atmosphere during parachute descent. External payloads must be able to withstand −150/120°C, 10^{-12} atm; 51.0 rad/day radiation background, 37 krad/flight from solar flares
Storage conditions: 5 to 30°C, 30 to 80 per cent relative humidity for 2½ year (maximum 35°C, 90 per cent humidity for 60 days/year). Prior to installation: −40/50°C, 20 to 80 per cent for 3 months at the launch site
Experiment control: 25 single ground commands, four preprogrammed commands from Photon's control system
Telemetry: 48 channels available. T measured (3.5 per cent error of measurement range) every 10.56 s with real-time transmission, or 42.24 s if stored; data transmission limited to 4 min/pass. The payload can command start/stop of the spacecraft's data storage system
Attitude control: none available (activated for retrofire sequence)
Payload mockups: customers must provide unit mockups for spacecraft fitting, accurate to within 0.5 kg, 5 mm for centre of masses and identical attachment points and means
Acceleration levels: 10 *g* for 600 s during launch; 0.1 *g* for 1,000 s during manoeuvres in-orbit, 0.1 *g*/1,000 s and 8 *g*/100 s during entry. Five 40 *g*/5-10 ms shocks for parachute deployment sequence, ended by a single 40 *g*/40-50 ms shock on landing
Vibration: unit resonance should occur >=25 Hz. Payloads are tested for 9 *g*/5-10 ms shocks at <=2 Hz,

Photon microgravity missions

	Launch	Orbit	Life (days)
C1645	16 April 1985	215 × 378km, 62.8°	13
C1744	21 May 1986	218 × 371km, 62.8°	14
C1841[1]	24 April 1987	218 × 381km, 62.8°	14
Photon 1[1]	14 April 1988	217 × 375km, 62.8°	14
Photon 2[1]	26 April 1989	216 × 380km, 62.8°	15
Photon 3	11 April 1990	225 × 398km, 62.8°	16
Photon 4	4 October 1991	216 × 396km, 62.8°	16
Photon 5	8 October 1992	221 × 359km, 62.8°	16
Photon 6	14 June 1994	221 × 364km, 62.8°	17 ½
Photon 7[2]	16 February 1995	220 × 369km, 62.8°	14 ½
Photon 8[3]	9 October 1997	218 × 375 km, 90.45°	13 ¾

Notes 1) reused descent capsule; subsequent reuse unknown. 2) capsule dropped by helicopter after safe landing; most of experiments destroyed or unusable. The Photon 1 listed here was the first named Photon launch and its successors were announced as 2-5. However, what should have been Photon 6 was announced as Photon 9 as a result of the three Cosmos predecessors being retrospectively referred to as Photon 1-3. The original designation system is employed here. 3) Photon 8 appears to be largely funded by foreign payloads: Biopan, Biobox, MIRCA, Crocodile 2, Ibis and Cosima 5.

and 1 *g* vibration over 2-10/10-20 Hz for 2 min each, and 0.5 *g* for 4 min over 10-20 Hz. Acoustic loads are 140 dB for internal equipment and 150 dB for external, both for 120 s over 20-4,000 Hz.

UNITED STATES OF AMERICA

COMET (Meteor) satellite

Background

As the Reagan Administration moved toward more commercial applications for space research, NASA chose to fund projects in other organisations. The Space Agency's Office of Advanced Concepts & Technology, explored the possibility of building a commercially viable re-entry module known as the COMET (Commercial Experiment Transporter) with both the Department of Transportation and commercial sources. NASA wanted the recoverable satellite to provide university and industry consortia with easy access to space.

Originally, COMET should have debuted in the summer 1994 carrying 11 experiments from Wallops Island into 460 km 40° for 30 days, but as the programme approached this deadline NASA chose in May 1994 to refuse additional funding unless the programme had a solid commercial basis. Nonetheless, NASA signed the launch contract 28 March 1995 with system owner EER Systems, who renamed it Meteor (Multiple Experiment Transporter to Earth Orbit & Return). The new flight plan called for the return capsule to carry four experiments, with eight more on the bus to continue operations for 1 to 2 years.

COMET's (Meteor's) mission ended after 45 seconds on 23 October 1995 when the maiden Conestoga that should have carried it into space self-destructed. NASA spent US$73 million on the programme. EER still takes bookings for Meteor 2, but few customers have shown an interest. A prospective customer has to pay about US$30 million to ensure a dedicated mission.

The CSTAR Center for Space Transportation & Applied Research at The University of Tennessee Space Institute headed the original three-flight programme. Westinghouse Electric Corporation provided the service module and systems engineering, and EER Systems supplied the Conestoga launcher. The launcher contract alone came to about US$45 million. Space Industries built the recovery capsule for about US$15 million, including orbital operations and payload integration.

After NASA lost interest, EER assumed responsibility for the program in late 1994. Because NASA's FY94 budget only allocated US$14.5 million, the entire system still needed about US$6 million in other money to get to the launch pad. Commercial customers provided the remainder, obligating NASA to release the US$14.5 million by 28 March 1995. For its money, NASA received half of the capacity of Meteor.

Originally, NASA wanted to fly three COMETs, but the budget squeeze required the Agency to modify the three into a single mission costing US$65.8 million. The first budget projected that the three separate missions would cost more than US$158 million.

At mission's end, the RM spins up to 73 rpm, separates and fires its Thiokol Star 13A solid retro motor for re-entry. The drogue and main parachutes are complemented by an airbag inflating just before landing in the Atlantic. Westinghouse hoped to land the first test at Utah Test & Training Range, with subsequent missions ending at White Sands.

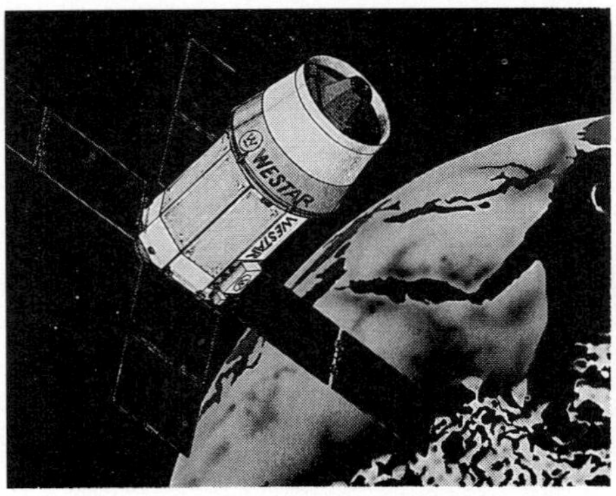

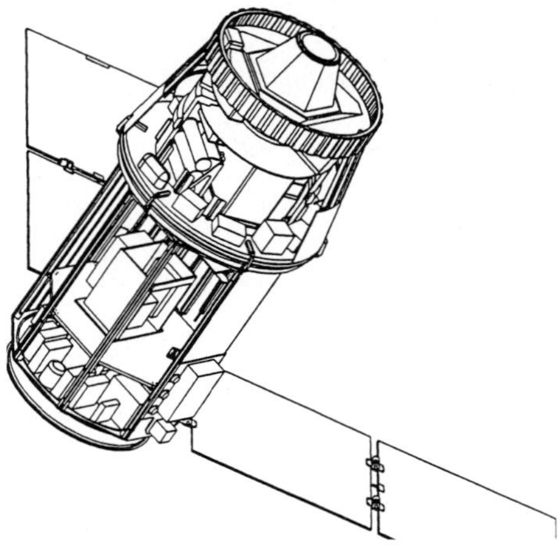

Meteor's first mission failed to reach orbit in October 1995

Specifications

Launcher: EER Systems' Conestoga 1620
Payload mass/volume: 136 kg/0.227 m^3 recoverable; 68 kg/0.426 m^3 non-recoverable
Payload access: samples can be loaded until 6 hours before launch
Microgravity level: better than 10^{-5} *g* (10^2 *g* during manoeuvres)
Mission duration: 30-100 days recoverable and 180 or more days non-recoverable
Power supply: 350 W continuous, 400 W peak for 200 h, 28±4 VDC, both recoverable/non-recoverable on-orbit from four 2-panel Si wings and Ni/Cd batteries (recovery capsule carries battery for descent, 5 W for 4 hours)
Ambient internal conditions: payload compartment pressurised to 1 atmosphere with dry air in recoverable capsule; unpressurized non-recoverable. Payload mounting plate maintained at 40±2.8°C
Communications: 9.6 kbit/s command uplink (2.075 GHz), 250 kbit/s down (2.315 GHz). 24 Mbit storage. Both event-driven and time-tagged experiment commands can be stored for autonomous execution
Pointing accuracy: roll ±1°, pitch/yaw ±0.1°. Provided by SM cold gas thrusters + RWs
Acceleration levels: <12 *g* ascent, 8 *g* re-entry, 6 *g* parachute, <10 *g* landing.

Wake Shield Facility series

Background

The Space Vacuum Epitaxy Center at the University of Houston built the Wake Shield Facility (WSF) under a grant from the NASA Office of Advanced Concepts & Technology and Space Industries Inc built the satellite. NASA sought a free-flying satellite it could deploy and retrieve from the Shuttle in order to conduct super vacuum and microgravity studies. The Shuttle launched it the first time in February 1994 (STS-60).

WSF flies in the wake that the Shuttle creates in the solar wind and residual atmosphere to minimise the drag forces that this small amount of material produces in other satellites at that altitude. The wake also establishes a higher vacuum of about 10^{-14} torr known as an ultra vacuum. NASA hoped to investigate the commercial possibilities of producing new electronic, magnetic and superconducting thin-film materials and devices.

It expected to use the first 52-hour mission to grow seven 6 to7 μm-thick GaAs wafers. However, a series of technical problems forced the crew to hold WSF on the Shuttle arm resulting in a vacuum of about 10^{-10} torr (about the best achievable on Earth).

WSF-02 did fly free during STS-69 in September 1995, achieving its primary objectives in five runs. WSF-03 flew aboard STS-80 and in free-flight in November 1996, growing 100 wafers. The WSF-3 flight achieved an extended duration of 180 hours. By NASA decision, the WSF-04 requires full commercial operation and industrial funding has not been forthcoming.

Specifications

Size: 3.66 m diameter, 1.81 m high, 1,700 kg deployed
Power provision: 50 by three AgZn batteries (WSF 01/02); 1 kW solar array (WSF-03/04)
Orbit: 305 km, 28.5°
AOCS: two MWs, three magnetorquers, conical Earth sensor, 3-axis magnetometer
Vacuum quality: 10^{-14} torr
Microgravity quality: 10^{-6} *g* for 45 hours
Payload: MBE/CBE equipment, mass spectrometers, retarding potential analyser, total pressure gauges, thermocouples, accelerometers and materials exposure experiments.

SVEC's Wake Shield Facility was released in September 1995. Its next flight is planned for November 1996

NAVIGATION

INTERNATIONAL

Nadezhda (COSPAS-SARSAT) series

Background

Canada, France, the US and the former Soviet Union (membership now assumed by Russia) established COSPAS-SARSAT in 1979 to provide an international satellite-based system for search and rescue. The system operates through a constellation of four polar satellites, which detect distress beacons. It locates an activated beacon to within 2 km and the founders of COSPAS-SARSAT credit it with saving more than 5,500 lives by June 1995 in 1,800 search and rescue operations. A Memorandum of Understanding between the signatories governed the actions of the body from 1984 until the International COSPAS-SARSAT Programme Agreement was signed by the four states on 1 July 1988 in Paris.

When a user employs the signal beacon, the distress message arrives via the satellites at a Local User Terminal (LUT) to determine beacon location. The information is passed to a Mission Control Centre to alert the rescue authorities. Globally COSPAS-SARSAT has provided some 550,000 121.5 MHz beacons and made them operational (600,000 are projected by 2000). Because the current system restricted usage to real-time relay by the satellite, a LUT must be within range (about 2,500 km) for position determination to within 10-15 km. The 406.025 MHz units include user identification codes in the message. Even with these

Russia's Nadezhda civil navigation satellites also carry the COSPAS payload to locate search/rescue beacons
(AKO Polyot) ***1995***

COSPAS-SARSAT satellite launches

	Launch	Orbit	Payload
C1383[1]	29 June 1982	989 × 1,028 km, 83°	COS 1
C1447[2]	24 March 1983	959 × 1,013 km, 83°	COS 2
NOAA 8[3]	28 March 1983	803 × 825 km, 98.7°	SAR 1
C1574[4]	21 June 1984	965 × 1,005 km, 83°	COS 3
NOAA 9[5]	12 December 1984	841 × 862 km, 99°	SAR 2
NOAA 10[6]	17 September 1986	808 × 826 km, 98.7°	SAR 3
NOAA 11	24 September 1988	849 × 865 km, 98.9°	SAR 4
Nadezhda 1[7]	4 July 1989	960 × 1,014 km, 83.0°	COS 4
Nadezhda 2[12]	27 February 1990	956 × 1,021 km, 83.0°	COS 5
Nadezhda 3[8]	12 March 1991	958 × 1,018 km, 82.9°	COS 6
NOAA 13[9]	9 August 1993	850 × 863 km, 98.9°	SAR 5
Nadezhda 4[10]	14 July 1994	954 × 1,005 km, 82.9°	COS 7
NOAA 14[11]	30 December 1994	848 × 863 km, 98.9°	SAR 6
NOAA 15	13 May 1998	808 × 823 km, 98.32°	SAR 7

GOES 7 (May 1986), 8 (April 1994) and 9 (May 1995), Insat 2A (July 1992) and 2B (July 1993), Cosmos 2054 Luch (December 1989), Luch 1 (December 1994) and GMS 5 (March 1995) carried 406 MHz relays into GEO and remain in experimental use.

Key: COS = COSPAS, SAR = SARSAT, C = COSMOS

Notes: 1) decommissioned March 88, 2) decommissioned December 89, 3) decommissioned December 1985, 4) decommissioned June 90, 5) 406 MHz local/global processor mode operational, global mode reactivated October 1991, 6) 406 MHz global/local modes not operating since September 88, 7) limited availability in Southern hemisphere, 8) started Southern hemisphere operations December 1991, 9) satellite contact lost 21 August 1993 while S&R testing under way, 10) acting as back-up to Nadezhda 3 in same orbit, 11) global mode failed soon after launch, 12) decommissioned February 1996.

systems, though, further localisation results from a measurement of the Doppler shift of the transmitted signal. Mathematically, two-signal localisation results in two possible positions, but the slight change due to the rotational velocity of the earth resolves the ambiguity. If the beacon's frequency stability is sufficient, as with the 406 MHz devices, the satellite can resolve true position on the second pass over the distress beacon. This system, therefore, provides a global service with an average waiting time of 44 minutes, with 94 per cent of signals detected in 90 minutes (results from 1990 exercise).

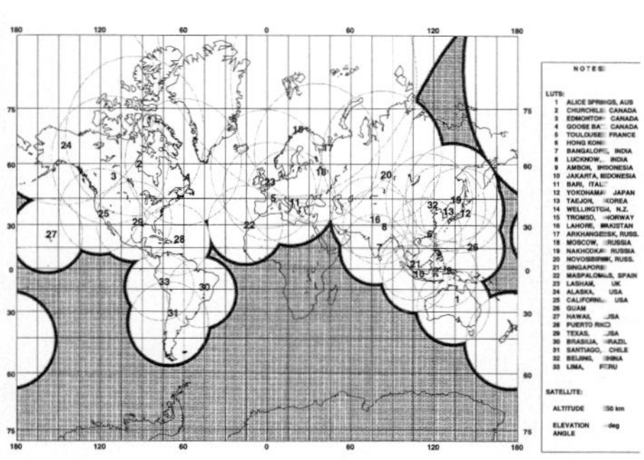

Local User Terminal 121.5 MHz coverage (406 MHz is global) ***1996***

All satellite downlinks operate at 1.5445 GHz. By end of 1995, 119,000 406 MHz beacons had been produced and distributed in 130 countries. Since the 406 MHz beacons emit a 5 W RF burst of about 500 ms duration every 50 seconds this frequency assures more precise location accuracy (90 per cent within 5 km), while the high peak power increases the probability of detection on a single pass to more than 98 per cent. Typical delay before detection is one to two hours near the equator. Position information from systems such as GPS and Glonass will be encoded in the messages of new 406 MHz beacons after January 1997.

The COSPAS organisation establishes Mission Control Centres in each country operating at least one LUT to disseminate information to the appropriate Rescue Co-ordination Centres. By end of 1995, there were 33 LUTs in 21 countries with others planned. The USSR began deploying the space segment with the launch of Kosmos 1383 in 1982. Designated COSPAS 1 ('Space System for Search of Vessels in Distress'), the 121.5 MHz signal remained operational until March 1988. Kosmos 1447 (1983), Kosmos 1574 (1984), Nadezhda 1 (1989), Nadezhda 2 (1990), Nadezhda 3 (1991) and Nadezhda 4 (1994, in the same orbit as COSPAS 5 as a spare) adopted the roles of COSPAS 2-7, with one more satellite on the ground and in preparation for launch if needed.

NOAA adds a data relay system with each of its meteorological satellites. SARSAT 1 (S&R Satellite-aided Tracking) flew with NOAA 8 in 1983. NOAA 9-11 brought the system to full strength.

SCIENTIFIC

ARGENTINA

MuSat (Victor) satellite

Background

Known as MuSat as it was built, the Argentines changed the name to Victor 1 when the Russians launched it on 29 August 1996. The Centro de Investigaciones Aplicades developed it as a technology demonstration satellite. The launch occurred aboard a Russian Molniya-M from Plesetsk. Victor weighed 32 kg and entered a 62.8° orbit at 235-1,170 km. The space craft itself measured 33 × 33 cm and was 43 cm high.

Argentina sought to develop space platform technology based upon low-cost engineering, to provide scientists and engineers with fast access to space for experiments. They expected the first experiments to include radio communications between Earth stations at a low cost, the demonstration of two CCD cameras to undertake colour photographs of Earth using low resolution (2 km/pixel) and high resolution (0.7 km/pixel) focal plane arrays. The image, sized 700 × 500 pixels, is immediately transmitted to Earth. The science agency also wanted to stimulate youth interest in space activities by promotion of participation by schools and universities.

SAC series

Background

Argentina emphasises applications and space science, in its modest space programme. Until May 1991, the air force executed all space research projects. In that year, the CNIE National Commission for Space Research was transformed into the civil CONAE National Commission for Space Activities, under direct supervision of the country's President. CONAE's proposed its first strategic programme covering 1995-2006 and President Menem approved it 28 November 1994. CONAR projected an expenditure of US$701 million from 1995-2006, although this includes significant contributions from unspecified third parties. The Argentine government reviews the plan every two years.

CONAE and its predecessor have worked on the SAC Satelite de Aplicaciones Cientificas (Scientific Applications Satellite) since 1981. Their research culminated in the SAC programme, first proposed in 1988, in response to NASA's request for Scout Small Explorer payloads. SAC-A would have been placed in a 500 km, 97.4° orbit to monitor solar output and its terrestrial effects during a solar maximum. Its instruments included a Fourier Imaging x-ray Spectrometer (FIXS), soft x-ray Bragg Crystal Locator (BCL), Hard x-ray Spectrometer (HXRS), Broad Band Photometer (BBP) array and a charged particle detector to monitor passage through the van Allen belts.

NASA ranked the SAC proposal in the top five, but it only had funds for the first three. Nonetheless, NASA officials maintained enough enthusiasm for the SAC to conduct a joint six-month study ending April 1991 on future co-operative missions. Then Argentina and NASA signed an MoU in August 1991, agreeing that Argentina would provide the spacecraft, the solar HXRS instrument, and the particle detector and mission operations. NASA provided the launch, other science instruments and back-up tracking coverage. Italy later agreed to provide one instrument (by IFSI) and the solar array (by FIAR). INVAP built the SAC-B as prime contractor INVAP. INVAP principal research prior to this was nuclear studies, in San Carlos de Bariloche.

When the satellite was ready, Brazil's INPE space agency performed all system qualification testing in its Sao Paulo facilities in return for participation in the science. INPE completed testing in 1995 and placed SAC in storage as it waited for the Pegasus programme to restart. Argentina's space segment cost is estimated at US$9 million and NASA's overall cost at US$13.4 million (including US$10.8 million for the Pegasus launch).

A Pegasus launched SAC-B on 4 November 1996, but it and the accompanying HETE (United States) satellite remained attached to the third stage of the Pegasus launch vehicle. NASA attempted to deploy its solar panels and return some experimental data. Although the solar panels emerged, the assembly tumbled, preventing the batteries from charging.

Argentina wants to launch two further SAC missions with NASA, the SAC-C Earth observation (multispectral medium resolution) satellite in 1999, possibly using a Delta vehicle and the SAC-D remote sensing, space physics and astrophysics satellite in 2001.

SAC-C candidate payload includes the Multispectral Medium Resolution Scanner, magnetometer and GPS receiver. Denmark is studying the addition of its Ørsted magnetometer payload to SAC-C.

Specifications

SAC-B

Launch: 4 November 1996 by Pegasus-XL from Wallops Island

Mass: 182.7 kg (bus 134.7 kg; experiments 48.0 kg)

Orbit: 38°, 490-555 km

Spacecraft: 72 cm³ 80 cm high. 3-axis control ±5° of Sun-pointing axis and 2° over 2 days X-axis for CUBIC; two 3-axis magnetometers plus two 2.0 Nms Ithaco RWs. Four 60.5 × 78.5 cm solar panels (224 cm span) using 1,032 30 × 41 mm GaAs cells provide 22 1/204 W BOL/EOL average (49 W payload requirement) at 28 VDC; 2 × 10 Ah Ni/Cd batteries. Redundant 2 Mbit memories, downlink 2.25550 GHz S-band at 100 kbit/s to CONAE's San Miguel centre

Payload: HXRS High Energy x-ray Spectrometer

IAFE, 20-320 keV solar x-ray and γ bursts, 7.6 kg

CUBIC Cosmic Unresolved Background Instrument with CCDs, The Penn State University, 0.1-10 keV, diffuse X-ray background, 30.0 kg.

GXRE Goddard x-ray Experiment, NASA Goddard, 4.8 kg, comprising: SoXS Soft X-ray Spectrometer (2-20 keV) plus GRaBS Gamma Ray Burst Spectrometer (>30 keV).

ISENA Imaging Particle Spectrometer for Energetic Neutral Atoms, Italy (IFSI Instituto di Fisica dello Spazio Interplanetario, CNR, Frascati), 5-200 keV, 5.6 kg.

AUSTRALIA

ARIES 1 satellite

Background

In October 1996, Australia announced it would build a remote-sensing satellite, with the first launch planned for 1999. The Australian Resource Information and Environment Satellites (ARIES) project is a joint programme between CSIRO, Auspace Ltd, the Australian Centre for Remote Sensing (ACRES) and has additional support from Earth Resource Mapping PTA Ltd, Geoimage PTA Ltd and Technical and Field Surveys PTA Ltd. The initial feasibility study was planned to take six months and cost in excess of A$1.2 million, with the Australian government sponsoring the study with a grant of A$300,000 and 10 international industry and agency sponsors have made commitments of A$75,000 each. The ARIES-1 board appointed the Macquarie Bank as financial advisers and believe that the financial package of A$70-100 million required for the launch of an operational satellite and the associated receiving and processing facilities can be achieved by 2000.

The design of the satellite is not finalised, but it is expected to have a mass of around 400 kg and would operate in a near-polar 450 km orbit.

Wresat satellite

Background

Australia joined the ranks of spacefaring nations on 29 November 1967, when it launched the Wresat, a test payload developed by the Weapons Research Establishment out of the Woomera range. The satellite weighed 45 kg and entered a 106 × 777 km, 83° inclination orbit. A modified US Redstone rocket, remaining from a re-entry test programme launched Australia's first satellite. The modifications included a solid second- and third-stage. Wresat transmitted data for five days and re-entered after 42 days. It carried several science payloads including particle radiation detectors, x-ray counter, ozone sensor, magnetometer, Lyman alpha telescope and solar aspect sensor to study effects of solar radiation on upper atmosphere.

BRAZIL

SACI 1

Background

INPE's 60 kg Satélite Científico will piggyback on CBERS and ride into a 778 km, 98.5°, using a platform designed for multiple missions. This is an independent effort for Brazil and not part of the MECB. The satellite is a 400 × 400 × 600 mm box-shaped body with four deployed solar panels and sun-pointing spin stabilisation. The entire package only costs about US$4.6 million, including US$800,000 for the experiments. The Brazilian Academy of Sciences selected four experiments for the project. They are: (1) ORCAS Solar/Anomalous Cosmic Rays Observation in the Magnetosphere, (2) the Plasma Bubbles study; development and decay over Brazil, to understand the strong influence of bubbles and their associated plasma turbulence on trans-ionosphere communications, space geodesy, radar remote sensing, (3) the Airglow Photometer measuring oxygen OI 557.7 nm, OI 630.0 nm and OH (8,3) and terrestrial airglow, particularly the equatorial ionospheric anomaly and the South Atlantic Anomaly, and (4) the Geomagnetic Experiment. The latter gives simultaneous measurements of the geomagnetic field.

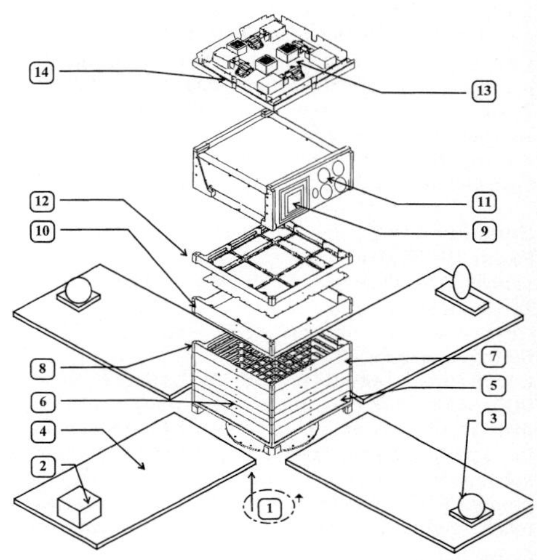

SACI 1 principal features. 1 Sun-facing solar panels, 2/12 magnetometer, 3 plasma sensor, 4 solar panel, 5 communications box, 6 OBC, 7 battery/power box, 8 plasma bubble expt, 9/10 cosmic ray detector, 11 airglow photometer, 13 antenna + open devices, 14 anti-Sun plate (INPE) ***1996***

CANADA

Alouette/ISIS

Background

Canada's first satellite, Alouette 1, was launched 29 September 1962 by Thor Agena B from Vandenberg into 997 × 1,026 km 80° orbit. Canada sought a joint agreement with NASA in order to fly their ionosphere sounder in order to explore the effects of the auroral regions on telecommunications. Canada had identified telecommunications to its vast northern regions as an application of spaceflight early in the space age. They built two satellites and when Alouette 1 performed as expected, they launched the back-up Alouette 2, on 29 November 1965, into a 505 × 2,987 km 80° orbit, as the first of three satellites in the newly created joint Canadian-NASA International Satellites for Ionospheric Studies (ISIS) programme.

After the first data from these satellites became widely available and showed some promise in predicting communication and navigation disruptions due to ionospheric turbulence, NASA and the CSA developed the ISIS which was first launched 30 January 1969 by Delta into 578 × 3,526 km, 88° orbit. The satellite weighed 241 kg. Both agencies agreed to extend the programme by ISIS 2, launched on 1 April 1971 by Delta into 1,358 × 1,429 km, 88° orbit. ISIS 2 weighed slightly more than its predecessor at 264 kg.

CHILE

FASat series

Background

Chile conducts its space affairs under the auspices of the CAE Committee for Space Matters, which it created in 1980, as part of the Ministry of Defence. The MOD's Space Division of the Fuerza Aérea de Chile (FACH, Chilean Air Force) launched the 50 kg FA Sat-Alfa as a passenger payload, in 1995, into polar orbit aboard a Ukrainian/Russian Tsyklon. Chile purchased from the UK's Surrey Satellite Technology Ltd, in May 1994, the satellite, ground station and on-the-job training for eight FACH engineers (launch cost was additional). A Chilean team built the actual satellite at Surrey working alongside SSTL staff. The programme helped students to gain Masters' Degrees on Surrey University's satellite engineering course.

Unfortunately, at launch, a pyrotechnic problem left the FASat-Alfa attached to the Russian Sich. FACH thus signed a £1.3 million contract with SSTL for the identical FASat-Bravo for 1997 launch. Educational activities using the satellite will help to create space awareness in the country. FACH plans the similar FASat-Beta (built in Chile) for 1997. Two others to form a constellation providing LEO communications services will follow it. Work is also under way on the FASat-Gamma mini-satellite for remote sensing from 2003. The expertise derived from these projects is expected to lead to a GEO satellite around 2010.

Specifications

FASat-Alfa/Bravo

Launch: Alfa 31 August 1995 piggyback on Sich 1 by Tsyklon from Plesetsk

Orbit: Alfa 632 × 669 km, 82.5° attached to Sich 1; Bravo planned for 650 km, 82.5°

Configuration: based on SSTL's UoSat 5 design. 600 × 345 × 345 mm box-shaped body with solar panels on four sides. 6 m boom deployed from top face for gravity gradient stabilisation. Each housekeeping subsystem is housed in a standard module box mass-produced by computer-controlled process (most were manufactured by the ENAER Chilean aerospace company)

Mass: 50 kg on-orbit; 53 kg including launcher fittings

Power supply: four GaAs panels (each 36 W) providing 20 W orbital average processed power; 6 Ah Ni/Cd battery supports eclipse operations

OBCs: OBC386 based on an 80C386EX with 387, 800 kbyte Error Detected/Corrected program memory and 16 Mbyte EDAC mass memory complemented by standard SSTL OBC186-Enhanced with 512k EDAC program and 16 Mbyte file store. The transputer system based on two transputers associated with the imaging payloads provides for high-performance data processing; the 256 Mbyte solid-state memory accommodates large onboard storage

Attitude control: Earth pointing maintained by combination of passive gravity gradient and active magnetorquing. 1-axis RW used for yaw rate control in conjunction with imaging payloads. Sun sensors in azimuth and elevation angle plus 3-axis fluxgate magnetometers (±2 nT resolution over ±64 nT range) provide attitude information. Magnetorquing provided by coils in each axis

Communications: data rates of 9.6, 38.4 or 76.8 kbit/s on 400 MHz using redundant UHF synthesised transmitters 2-10 W variable. 150 MHz VHF up (two communications + one command). 4-element VHF + UHF antenna arrays

TT&C: from ECM-Santiago control station at the Los Cerrillos Air Force Base

Payload: OLME Ozone Layer Monitoring Experiment. Two UV CCD cameras plus four UV photodiodes to monitor the ozone layer in southern Chile by measuring backscattered solar UV at 3,000 Å.

DTE Data Transfer Experiment. Provides enhanced store and forward communications through two synthesised VHF receivers under control of the dual Digital Signal Processing (DSP) computers. They interface with both up/down links for DTE receiver control, up/down modulation experiments and speech processing.

FASat-Alfa Flight Model; Bravo is identical. One of the Earth face's four blocks carried 5 g of Chilean soil; the others housed Sun sensors (2) and the 'Earth Underneath Detector'. At bottom is the tip mass of the compacted 6 m boom (FACH) ***1996***

EIS Earth Imaging System. Two monochromatic 578 × 576 pixel CCD cameras provide 100 m/1 km resolution. Images recorded on schedule controlled by the transputer data processing equipment. The transputers recover the images and provide onboard processing before passing to the OBC file system for distribution on the downlink.

GPS navigation experiment. GPS receiver controlled by the transputer data processing experiment. Keplerian elements are generated onboard to assist in image targeting and for downlink distribution. Bravo's receiver, developed by SSTL under ESA contract, will also provide attitude data.

SSDRE Solid State Data Recorder Experiment. The unit was developed by Sanders (Lockheed Martin) in collaboration with the FASat-Alfa project, providing 256 Mbyte storage via the CAN bus.

CHINA, PEOPLE'S REPUBLIC

Shi Jian series

Background

The majority of Chinese satellite launches in the domestic programme have been dedicated to practical uses rather than propaganda type missions. While there have been some science missions flown, these have been few and far between. Most domestic launches have been for either remote sensing (including meteorological) or communications.

Shi Jian 1 embodied the first true science payload, acting in a role of exploration and inquiry. It carried a magnetometer, cosmic ray and solar x-ray detectors. Apparently, the Chinese used the same bus for most of their mission applications and the Shi Jian 1 duplicated the DFH-1 configuration. It flew about a year after the successful launch of China's first satellite. The Chinese obviously learned from their earlier experience. This time the satellite flew solar cells for power and controlled the temperature of the platform by using louvres on the satellite. Then the Chinese launched Shi Jian 2, their first triple payload, made up of SJ-2, 2A and 2B. 2B apparently studied atmospheric density since the satellite, a balloon attached to a heavy mass – quickly decayed from orbit.

Shi Jian 3 is a designator used for an abandoned early-1980s meteorological satellite programme and was never used for an orbital mission. Shi Jian 4 – along with KF 1 – provided a scientific reason for the maiden flight of the CZ-3A launch vehicle. KF 1 (the acronym has never been explained by the Chinese) appears to have been an active satellite, but its purpose has never been justified. The Chinese are close-lipped about any successor such as a Shi Jian 5 satellite.

CZECH REPUBLIC

Magion series

Background

When it still formed part of the Soviet bloc, Czechoslovakia belonged to Interkosmos. Both the Czech and Slovak republics moved away from that organisation after they became separate powers on 1 January 1993. The Czech Republic continued on as a significant contributor to international space science missions.

Within its former borders, Czechoslovakia designed and built a series of small magneto/ionospheric (hence the name) research satellites. They came from a small group in the Upper Atmosphere Department of the Institute of Atmospheric Physics (until April 1994 the Ionosphere Department of the Geophysical Institute).

Interkosmos 18 released Magion 1, the first in the series and the first Czech satellite, shortly after launch on 24 October 1978 into a 404 × 772 km, 83° orbit. The satellite only weighed 14.5 kg, and it was 370 mm in diameter. The Institute planned to study low frequency fields between the pair of slowly diverging satellites, but a power malfunction degraded data acquisition. Even in its reduced capacity Magion 1 remained operational until entry 11 September 1981.

Then Magion 2, launched with Interkosmos 24 on 28 September 1989 into 500 × 2,492 km, 82.6° orbit continued the series by conducting propagation studies of VLF waves (from the parent) and their effects on the van Allen belts. The parent and the small sub-satellite separated by spring action at 0.2 m/s upon command from a ground station near Prague on 3 October while Interkosmos was nadir-pointing. Soon after release, the temperature fell to –40°C, which compromised the deployment performance since the satellite design only called for –20°C. Only one boom and one solar panel deployed, generating insufficient power for telemetry. Some telemetry did arrive on 21 December when the orbit had shifted into sunlight. Deployment commands retransmitted from the ground deployed all six booms and four solar panels on 24 December. The science programme began 21 January 1990, but power levels remained marginal, probably because of solar array degradation from intense solar activity. The last telemetry contact occurred 20 November 1990.

Magion 2 had a 26-face polyhedron, 560 mm diameter and weighed 52.18 kg. Within that weight budget a 9 kg AOCS package stabilised attitude control by a 30 Am^2 permanent magnet, using attitude information from magnetometer and Sun sensors. All of the science experiments used the 12 W of power from Si solar panels, supported by two Ni/Cd batteries (4 Ah each). Originally, ground controllers expected to adjust the orbit for varying separation from parent, by 150 atmospheres compressed air through 0.2 N thruster (total impulse 300 Ns). A cable failure prevented this from following the design. Telemetry arrived over a transmit carrier of 137/400 MHz, at the rate of 40.96 kbit/s max, using a 4 Mbit memory, 256 command combinations, 196 housekeeping parameters.

To perform its science investigations, Magion 2 carried a 3-axis fluxgate magnetometer, resolution 2-16 nT (Romania, Czech), a wave experiment with search coil magnetometers operating between 10 Hz-40 kHz, double probe electric component sensors

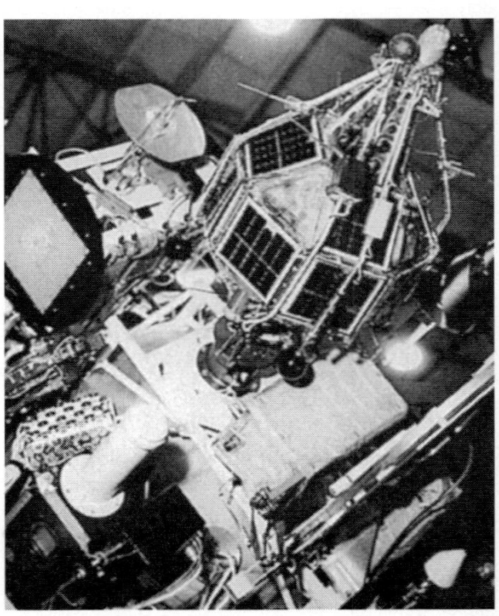

Magion 4/5 appear almost identical to Magion 2

Magion 4 Flight Model viewed from the solar direction (Dr Pavel Triska) ***1996***

0.1 Hz/120 kHz, frequency analyser 1-120 kHz, filter banks 20 Hz/15 kHz (Czech, Bulgaria), a radiospectrometer 0.01-10 MHz (Poland, Czech). The cold plasma data came from a Langmuir probe and RF probe (East Germany, Czech), and an ion trap (Bulgaria, Czech). The hot plasma experiments examined 0.01-20 keV electrons/protons at 16 energy levels (Bulgaria, USSR, Czech), and a 0.20-1 MeV solid-state detectors at 8 energy levels (Czech). In the visible/near visible wavebands Bulgaria provided a 6,300/5,777 Å photometer and 17 3 cm^2 laser reflectors on lower surface.

Magion 3, launched 18 December 1991 with Interkosmos 25 into an initial 438 × 3,071 km, 82.56° orbit monitored the propagation of electron beams and plasma injected by the parent. Magion separated at 02.03 GMT 28 December, drifting from its parent because of differential drag. It used its small thruster 9 January 1992 for the first time to close in again by 30 January. Similar manoeuvres followed on 6 May and 11-26 June 1992. The last approach was within 100 m with relative velocity almost zero. The sub-satellite operated with no problems until 20 August 1992 when it suddenly suffered a power shortage, probably caused by radiation damage to the battery charge system after a period of enhanced solar activity. Final telemetry downlinked on 9 September 1992. Magion 3 and its payload were almost identical to Magion 2.

Interball 1 provided the parent craft for Magion 4, launched 2 August 1995, into a 793 × 191,900 km 62.9° orbit to study mechanisms transporting solar wind energy into the magnetosphere. It weighed in at 58.7 kg. The design resembled Magion 2 except the onboard memory increased to 32 Mbit, and the design group added a 1.5 GHz band for additional telemetry. Its control improved too, to 3°/s spin-stabilised control around a Sun-pointing axis (within 15°). Magion 4 carried a science payload consisting of a 3-axis fluxgate magnetometer (0.015-8 nT resolution over ±128-65,535 nT), 3-axis search coils for 10-1,000 Hz ULF/ELF waves and 10 or 30 Hz ULF waves, a 1-axis search coil for 0.50-40 kHz VLF waves, and one for 0.04-20 Hz and 0.02-400 kHz electric fields and waves. An x-ray photometer searches for 1.5-500 keV solar x-rays.

Magion 5 went into orbit following launch with the Interball 2 on 29 August 1996. It consisted of an auroral probe on a 750 × 20,000 km 63° inclination orbit. Like Magion 4, the research craft studied the mechanisms transporting solar wind energy into magnetosphere, in conjunction with Magion 4, and it weighed 62 kg. It had a cold plasma detector in place of ion flux and solar x-ray instruments. Problems with the solar panel deployment ended after its release.

FRANCE

Cerise satellite

Background

Helios 1A's Ariane carried the 50 kg Cerise (Caractérisation de l'Environnement Radioélectrique par un Instrument Spatial Embarqué) to characterise the Earth's radio environment in research that could lead to a 'Zenon' electronic intelligence satellite in the next decade. Alcatel Espace acted as Cerise prime contractor for the DGA, using the University of Surrey's UoSAT bus. Cerise should have lasted 2½ years. France sought industry proposals in 1993 for 1998's Clementine Cerise follow-on to investigate different frequencies regimes in the ionosphere. Cerise became the first proven casualty of space debris when it collided with a piece of debris from the Ariane-1 third stage, which had orbited SPOT 1 in 1986. The collision with the debris 1986-019RF was at 09.48 GMT on 21 July 1996 and the upper portion of the Cerise gravity gradient boom was broken off. Control of Cerise was regained following the collision.

Cerise is contributing to the development of an electronic intelligence satellite (SSTL) ***1996***

French science satellite programme series

Background

France created the CNES in 1962 with a staff of 27 to execute the national space programme. Since April 1993 it has been under the regulating authority of three ministries (Defence; Industry, Posts, Telecommunications & Foreign Trade; Higher Education & Research). A President heads CNES and a Director General sets the science programme agenda. Today an 18-member Conseil d'Administration, with six members elected by agency employees, consult and advise the direction of space priorities.

In the beginning, France sought to separate its military and civilian space programmes, but in the interest of efficiency the country consolidated the two in 1993, in co-ordination with the Ministry of Defence. Now CNES has 2,488 personnel. Over the past 35 years, CNES has conducted a vigorous launch programme.

Historical French science satellites

French launches

A-1 (Asterix)
Launched: 26 November 1965 by Diamant from Hammaguir
Mass: 41.7 kg
Orbit: 528 × 1,758 km, 34°
Comment: First French satellite. Transmitted data for 2 days.

D-1A (Diapason)
Launched: 17 February 1966 by Diamant from Hammaguir
Mass: 20 kg
Orbit: 504 × 2,753 km, 34°
Comment: Third French satellite and the second national launch. Geodetic research.

D-1C (Diademe 1)
Launched: 8 February 1967 by Diamant from Hammaguir
Mass: 22.6 kg
Orbit: 580 × 1,340 km, 40°
Comment: Geodesy; operational in spite of low apogee.

D-1D (Diademe 2)
Launched: 15 February 1967 by Diamant from Hammaguir
Mass: 22.6 kg
Orbit: 592 × 1,886 km, 39°
Comment: Laser/Doppler ranging data to 5 April.

Wika/Mika
Launched: 10 March 1970 by Diamant B from Kourou
Orbit: 307 × 1,700/1,746 km, 5°
Comment: West German Wika was first foreign payload orbited by France. It provided 30 days of geocorona and upper atmosphere investigations. The satellite re-entered 5 October 1978. Mika monitored Diamant performance. Its re-entry occurred 9 September 1974.

Peole
Launched: 12 December 1970 by Diamant B from Kourou
Mass: 69.7 kg
Orbit: 580 × 747 km, 15°
Comment: Test flight to qualify Eole meteorology data relay satellite, launched by Scout B from US Wallops 16 August 1971 in collaboration with NASA to track hundreds of 4 m meteorology balloons. Re-entry occurred 16 June 1980.

D-2A (Tournesol)
Launched: 15 April 1971 by Diamant B from Kourou
Mass: 90 kg
Orbit: 456 × 703 km, 46°
Comment: Five instruments studied solar and UV radiation, including hydrogen content of geocorona. First French satellite to utilise active stabilisation (thrusters). Re-entered 28 January 1980.

D-2A (Polaire)
Launched: 5 December 1971 by Diamant B from Kourou
Mass: 97 kg
Comment: Tournesol's back-up; intended for higher inclination (75°) for complementary and simultaneous measurements, but launcher stage 2 failed.

D-5A/5B (Castor/Pollux)
Launched: 21 May 1973 by Diamant BP-4 from Kourou
Comment: Castor/Pollux fell into Atlantic when the Diamant failed to produce sufficient thrust.

Starlette
Launched: 6 February 1975 by Diamant B from Kourou
Mass: 48 kg
Orbit: 804 × 1,138 km, 50°
Comment: 24 cm sphere carrying 60 laser reflectors for geodetic studies. Observations continue. Similar 1993 Stella satellite added polar coverage.

D-5A (Pollux)
Launched: 17 May 1975 by Diamant BP-4 from Kourou
Orbit: 270 × 1,270 km, 30°
Mass: 36 kg
Comment: Intended to space-qualify SEP's 1.5-4 N thruster (7 kg hydrazine). Performed 11 sequences totalling 3,000 pulses of 0.125-100 s.

D-5B (Castor)
Launched: 17 May 1975 by Diamant BP-4 from Kourou
Orbit: 270 × 1,270 km, 30°
Mass: 76 kg
Comment: 800 mm diameter, 26 faces carrying 26 laser reflectors and solar cells to generate 20 W. Intended to space-qualify Onera's 10^{-5}-10^{-9} *g* microaccelerometer, which was also used for air density studies over 300-800 km (in conjunction with laser tracking) and registering micrometeoroid impacts. Re-entered 5 August 1975/18 February 1979.

D-2B (Aura)
Launched: 27 September 1975 by Diamant BP-4 from Kourou
Mass: 110 kg
Orbit: 499 × 723 km, 97°
Comment: Solar/stellar UV observations. Re-entered 30 September 1982.

European launches

SARA
Launched: 17 July 1991
Orbit: 770 × 777 km, 98.54°
Mass: 26.6 kg
Comment: 34 cm^3, intended to be the first French science amateur satellite. Total project cost FFr3 million, including FFr1.2 million launch fee. Planned to observe Jupiter's 2-15 MHz emissions but electronics failure prevented useful data, although SARA is otherwise in good condition. Switched to sleep mode in May 1992.

S80/T
Launched: 10 August 1992 by Ariane V52
Orbit: 1,302 × 1,329 km, 66.1°
Mass: 50 kg
Comment: Demonstrated technologies for proposed S80 TAOS 2-way messaging/localisation system for mobile users. MMS France's 1991 FFr25 million contract from CNES included a mobile station for tests. The S80/T employs Surrey University's subcontracted UoSAT platform and UK control station. Communications are tested through stations in Toulouse and Lanham. The 7 kg payload includes Dassault Electronique repeater, up/down 148-149.9/137-138 MHz, 2-3 year life. Stabilised by gravity gradient and magnetorquer. Released from ASAP carrier with Korea's Kitsat. 1993 activities included multipath observations and single event upsets. 1994-96 work

includes analysis of the 148-149.9 MHz band in several world regions and platform technology experiments. CNES planned to contract for four or five satellites in 1994, but in March 1994 decided not to pursue the project. TAOS would have formed the core of the Starsys system. The full system would have been operational by 1996/7, with users paying FFr1,500 for a hand-held terminal, a FFr500 annual fee and FFr5 per message.

USSR launches

Aureole 1
Launched: 27 December 1971 by SL-8 from Plesetsk
Mass: 300 kg
Orbit: 410 × 2,500 km, 74°
Comment: First Soviet-French satellite, launched under international 'Arcad' (Arctic auroral density) to investigate aurora borealis and ionosphere. Instruments provided by Moscow's Space Research Institute and France's Toulouse Centre for Study of Space Radiation.

Aureole 2
Launched: 26 December 1973 by SL-8 from Plesetsk
Mass: 400 kg
Orbit: 400 × 1,975 km, 74°.

Aureole 3
Launched: 21 September 1981 from Plesetsk
Mass: 1,000 kg
Orbit: 380 × 1,920 km, 82.6°
Comment: Third Arcad and the first Soviet satellite to be integrated at least partially in the West. It carried four French instruments and France also supplied the onboard programming and telemetry unit. Both countries received data directly.

SRET 1
Launched: 4 April 1972 piggyback with Molniya 1-20 on SL-6 from Plesetsk
Mass: 15.4 kg
Orbit: 460 × 39,248 km
Comment: Meant to study radiation effects on solar cells. Re-entered 26 February 1974.

SRET 2
Launched: 5 June 1975 piggyback with Molniya 1-30 on SL-6 from Plesetsk
Mass: 29.6 kg
Orbit: 513 × 40,825 km, 62.8°
Comment: Tested Meteosat cooling technology.

Signe 3
Launched: 17 June 1977 by SL-8 from Kapustin Yar
Mass: 102 kg
Orbit: 459 × 519 km, 50.6°
Comment: Monitored cosmic γ-rays for 733 days. Re-entry occurred 20 June 1979.

United States launches

FR-1
Launched: 6 December 1965 by Scout from Vandenberg
Orbit: 717 × 728 km, 75.9°
Comment: Meant to study ionosphere and VLF wave propagation by monitoring 16/25 kHz signals from St Assise (France) and Balboa (Panama) stations. 62 kg.

Eole
Launched: 16 August 1971 by Scout from Wallops
Orbit: 664 × 870 km, 50.1°
Mass: 84 kg
Comment: Data relay satellite interrogating 500 met balloons drifting in atmosphere at 12,000 m; returned data for 5 months.

Symphonie 1/2
Launched: 19 December 1974/27 August 1975 by Delta from Cape Canaveral into GEO at 11.5° west
Mass: 237 kg on-station
Comment: Franco-German telecom project, each providing two TV channels or 300 duplex phone. Aerospatiale built number 1 and MBB built number 2. Direct result was TV-Sat/TDF project.

GERMANY

Astro-SPAS (ORFEUS and CRISTA) series

Background
MBB began development of the SPAS carrier in 1986 (in preference to Dornier's Robus) into a free-flying

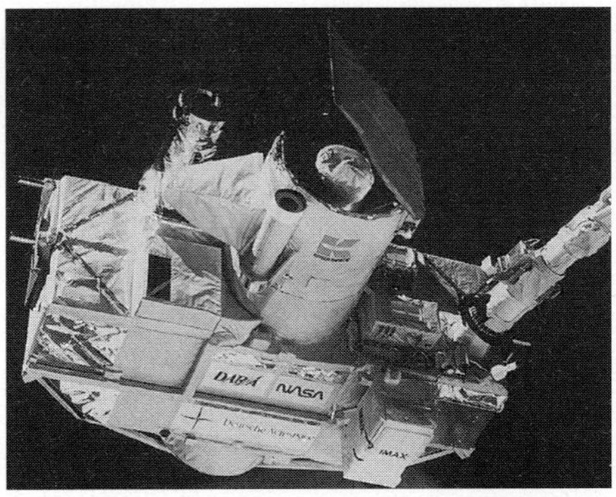

Germany's Astro-SPAS carries astronomical payloads: here it accommodates the Orfeus ultraviolet telescope. This 1993 configuration will be repeated in late 1996 (NASA) ***1995***

astronomical platform. In total, the DARA/NASA agreement calls for four co-operative science missions, with DARA providing the satellite, NASA the Shuttle launch, deployment/retrieval services and the two parties sharing the science instruments. Germany set aside DM115 million, for the first mission falling to DM40 million for the second as the carrier was reused. Germany's costs for the next two are projected at DM20 million each. NASA is providing the Shuttle free of charge, in return for access to data and inclusion of US experiments.

ORFEUS, the Orbiting and Retrievable Far/Extreme UV Spectrometer measuring radiation between 400-1,280 Å flew on STS-51 in September 1993 and was released at 14.06 GMT 13 September, retrieved 11.50 GMT 19 September. Science contributions came from the University of Tübingen, Sternwarte Heidelberg, University of Berkeley and Princeton University (IMPAS). ORFEUS' telescope was fabricated by Kayser-Threde in Germany; France's REOSC provided the 1 m f/2.5 mirror. The separate 950-1,150 Å IMAPS Interstellar Medium Absorption Profile Spectrograph added to the observations of hot galactic objects and the interstellar medium at high spectral resolution (240,000). Other payloads were DLR's Surface Effective Sample Monitor and Canada's IMAX camera. STS-80's mission will compare combined GPS/star tracker attitude determination against the gyro system.

CRISTA, the Cryogenic Infrared Spectrometers/Telescopes for the Atmosphere, developed by the University of Wuppertal, flew on the STS-66. It was released 12.50 GMT 4 November, and retrieved 13.05 GMT 12 November 1994. It carried as a Secondary instrument the US Naval Research Lab's MAHRSI Middle Atmosphere High Resolution Spectrograph Investigation for nitric oxide and OH chemistry measurements. CRISTA's primary goal is the first study of global atmospheric variability using limb sounding. 51,000 vertical scans were acquired during STS-66, measuring 15 trace gases every second. As this requires precise SPAS position information, it carries a GPS receiver qualified by the Orfeus mission.

Bremsat

Background
ZARM (University of Bremen), in co-operation with OHB-System and HTG Hypersonic Technology Goettingen, launched Bremsat aboard the Shuttle STS-60 flight in February 1994. It was released 9 February into 340 × 357 km, 57.0° orbit. The 12-sided 63 kg (30 kg science payload), 480 mm diameter 520 mm high aluminium prism completed two experiments over 48 h before ejection from its GAS canister. The first experiment tested thermal conductivity of a fluid (4 kg) and measured Shuttle's residual acceleration by a 3-axis micro-accelerometer (accurate to 10^{-6} *g*). After release, the satellite concentrated on measuring the near-Earth dust/micrometeoroid population to investigate the growth of manmade pollution. The same type of 0.8 kg detector flew aboard Japan's Hiten satellite.

This orbital phase also measured atomic oxygen concentrations. Once Bremsat had decayed to 180 km, the atmospheric force on a solar panel was measured to determine the accommodation coefficients of solar cells. For the entry phase, the MW was ejected to remove all satellite momentum. Beginning at 120 km (remaining life about 40 minutes before destruction at 60 km), temperature and pressure data were collected. Of particular interest was the transition from free molecular flow to continuum flow.

Bremsat re-entered 12 February 1995; the last data were collected at 16.48 GMT at 110 km altitude, 15 minutes before destruction.

Bremsat 2 has been proposed to DARA. Using a modified Bremsat 1 bus, it would carry seven experiments. In-Orbit: materials science; two technical (GPS orbit determination and new antenna design); active radiator thermal control tile; high resolution measurement of Earth's magnetic field. The main experiment is re-entry, using a 2,375 mm diameter parashield and recovery.

German science satellite programme series

Background
German space projects have always formed part of wider European and international activities – the country has never pursued an independent programme. When the Bundestadt created DARA, the German Space Agency, created in 1989, it put it in full charge of implementing the space programme, except for the hypersonic technology programme, which remains embedded within the BMFT Ministry for Research & Technology's aeronautics activities.

Despite the lack of a focused national programme, the Federal government has been supporting space research and technology since 1962. In February 1976 the government decided that there would be no further national or even bilateral research satellite projects such as Symphonie and two-thirds of the 1976-79 space budget of DM2.2 billion would be allocated to ESA.

Specifications
Azur
Launched: 8 November 1969 by Scout from Vandenberg
Mass: 71 kg
Orbit: 387 × 3,150 km, 103°
Comment: Seven instruments monitored radiation belts and solar particle flux in co-operation with the US to augment data from US Explorer and OGO.

Dial
Launched: 10 March 1970 by Diamant B from Kourou
Mass: 63 kg
Orbit: 301 × 1,631 km
Comment: France's first foreign launch, to study the hydrogen geocorona and ionosphere but excessive vibration during launch made data evaluation difficult. Re-entered 5 October 1988.

Aeros 1
Launched: 16 December 1972 by Scout from Vandenberg
*Mass:*127 kg
Orbit: 223 × 867 km, 97°
Comment: Five experiments (four German, one US); dipped into upper atmosphere 3,844 times for density measurements while monitoring solar UV output. Re-entered 22 August 1973.

Aeros 2
Launched: 16 July 1974 by Scout from Vandenberg
*Mass:*127 kg
Orbit: 224 × 869 km, 97°
Comment: Extended upper atmosphere research. Re-entered 25 September 1975.

Helios 1/2
Launched: 10 December 1974/15 January 1976 by Titan Centaur in collaboration with NASA
Mass: 370 kg
Comment: Solar probes passed within record 48/45 million km of Sun.

Firewheel
Comment: Designed by Max Planck Institute, with ESA/NASA support. The parent craft and all four sub-satellites were lost in Ariane L-02 failure of 23 May 1980. Magnetospheric investigations planned with injections of barium and lithium ions. The AMPTE programme continued the research.

AMPTE
Launched: 16 August 1984 by Delta from Cape Canaveral
Mass: 695 kg
Comment: 3-satellite Active Magnetospheric Particle Tracer Explorer mission. The German Ion Release Module, developed by Max Planck Institute, made seven lithium and barium releases for detection by US/UK satellites.

MIRCA

Background
DARA, in 1991, initiated a joint project between Kayser-Threde and DASA Jena-Optronik to develop and fly the

Germany's MIRCA re-entry capsule will be released from a Russian Photon craft. The central cylinder is the parachute container (Kayser-Threde) ***1996***

1 m diameter MIRCA Micro Re-entry Capsule for re-entry investigations. KT and Russia's Central Specialised Design Bureau signed the contract November 1994 for launching MIRCA attached to Photon in September 1996, however, the Russian launch is long-delayed and the satellite has not yet flown. MIRCA will be mounted on the forward face of Photon's own sphere, 12° below the centreline until it is ejected at 0.8-1.6 m/s separation after retrofire. Unusually, the heatshield is also the main load carrying structure, comprising hemispheres bolted at the equator. The shield consists of a 1 mm carbon fibre composite core, 20/30 mm ablator for the upper/lower hemisphere and a 1.35 mm C/SiC outer layer. Temperature and pressure sensors and a pyrometer are integrated in the lower shield.

A follow-on tether mission was studied, beginning in 1991, but cancelled by DARA in 1994. Project Rapunzel (Rope Attached Piggy-back Unit Zooming on Environment Data at Low Cost) would have reeled out MIRCA on a 50 km tether from Photon to de-orbit when the line was cut. GPS receivers in each module would have provided relative position information. Before then, the first Rapunzel test was planned for a Resurs-F mission carrying a 45 kg 50 cm re-entry sphere developed by the Samara State Aerospace University. Later Rapunzel studies concentrated on a MIRCA tethered to a Progress-M as it left Mir in 1997.

Specifications

Size: 1,000 mm-diameter sphere
Mass: 150.0 kg total (71.0 heatshield, 16.7 mechanical support structure, 16.6 service system, 6.6 battery, 16.8 experiments, 14.3 recovery system, 8.0 balance mass)
Payload volume: 110 litres
Payload power: 20 Ah 28 V DC from $LiMnO_2$ cells; can be expanded to 50 Ah
Command/data: transputer-based for control/data handling. System activated by Photon command 105 s before separation. 16 Mbyte on EEPROMs. 3-axis body rates up to ±100°/s and ±64 *g* recorded
Thermal: −20/70°C internal range specified; heater activated below −20°C
Recovery system: cover pushed clear by two 1,700 N solids burning for 80 ms will pull out conical ribbon chute at about 7 km, activated redundantly by barometric switch and timer after peak deceleration. Reefed cross chute released 25 s later, fully open 10 s later. Shortwave beacon (antennas on chute lines) active for 10 min descent; VHF transmit plus strobe activated on ground for 48 h

Payload: HEATIN Heatshield Instrumentation: T, P and heat flux distribution within two meridian planes 90° apart. Also T at different depths in material and P caused by off-gassing.

RAFLEX Rarefied Flow Experiment: P at stagnation point and 60° behind.

PYREX Pyrometer Experiment: touchless temperature measurement of C/SiC layer behind shield at stagnation point. Will also qualify this mini-pyrometer for other space applications.

Germany's MIRCA re-entry capsule will be released from a Russian Photon craft. The central cylinder is the parachute container (Kayser-Threde) ***1996***

TubSat

Background

The Technical University of Berlin's first satellite was released into polar orbit from ERS 1's Ariane launch on 17 July 1991. The 35 kg, 38 cm³ carries an L-band store/forward payload to demonstrate electronic mail applications, GaAs test cells and a 3-axis star sensor.

Following the successful first launch, TUB put the 40 kg TubSat B technology demonstrator into orbit, 25 January 1994, on Russia's Meteor 3-06. Ground controllers lost contact after 5 March. The 35 kg DLR-TubSat is under construction for 8 m-resolution Earth imaging following a possible launch on Russia's SL-8 Cosmos.

INDIA

Indian science satellite programme series

Specifications

Indian launches

Rohini 1B
Launched: 18 July 1980 by SLV-3 from Sriharikota
Mass: 35 kg
Orbit: 305 × 919 km, 44.7°
Comment: India's first successful launch, following the failure of Rohini 1A on 10 August 1979. Re-entered 20 May 1981.

Rohini 2
Launched: 31 May 1981 by SLV-3 from Sriharikota
Mass: 38 kg
Orbit: 186 × 418 km instead of planned 296 × 834 km, 46°, leading to re-entry after 8 days.

Rohini 3
Launched: 17 April 1983 by SLV-3 from Sriharikota
Mass: 41.5 kg
Orbit: 371 × 861 km, 46.6°
Comment: Carried two cameras and L-band beacon, and returned more than 5,000 Earth images before de-activated 24 September 1984. Re-entered 19 April 1990.

USSR/RFAS launches

Aryabhata
Launched: 19 April 1975 by SL-8 from Kapustin Yar
Mass: 360 kg
Orbit: 563 × 619 km
Comment: Named after the 5th-century astronomer/mathematician. Designed for 6 months' x-ray astronomy. Control was to have been handed over to Indian scientists on day 3, but the Indian-designed transformer failed and only a few days' useful data were obtained. Decayed 10 February 1992.

Bhaskara 1
Launched: 7 June 1979 from Kapustin Yar by SL-8
Mass: 442 kg
Orbit: 519 × 541 km, 50.6°
Comment: Named after 7th-century Indian astronomer. With two TV cameras and microwave radiometers, it cost Rs65 million and was to spend 1 year studying India's resources. Useful ocean/land data received but the cameras malfunctioned. Nonetheless housekeeping telemetry was received until re-entry 17 February 1989.

Bhaskara 2
Launched: 20 November 1981 from Kapustin Yar by SL-8
Mass: 444 kg
Orbit: 541 × 557 km, 50.7°
Comment: Declared operational after receipt of more than 300 TV images of Indian sub-continent. Housekeeping telemetry was still being received 1991; re-entered 30 November 1991.

European launches

APPLE (Ariane Passenger Pay Load Experiment)
Launched: 19 June 1981 by Ariane from Kourou
Mass: 670 kg
Orbit: 102° east GEO
Comment: India's first indigenous GEO test communications satellite. Indian TV programmes and educational teleconferences were relayed despite a jammed solar panel. De-activated September 1983.

SROSS series

Background

India took its second step towards an autonomous space programme with the first attempted launch of the Stretched Rohini on 24 March 1987. SROSS 1 disappeared in the vehicle failure of first ASLV. It represented the first in the series of low-cost 150 kg satellites designed to carry scientific and remote-sensing payloads of 15-35 kg on both three-axis and spin-stabilised missions. ISRO postponed the first attempt at launching SROSS 1, from late 1985 to early 1987 because of ASLV stage 1 problems. The Space Agency wanted to use the satellite carrying a battery of inertial measurement instruments to monitor launcher performance. SROSS 1 had an octagonal body and was powered by solar cells on each face and deployed panels, and also carried two retro-reflectors for laser tracking and ISRO's 30-1,000 keV Gamma-ray Burst Experiment.

SROSS 2 launched 13 July 1988, also disappeared in a launch failure, this time the second ASLV. The 150 kg, 85 cm diameter, 110 cm high satellite, with eight deployable solar wings feeding 100 W through a 12 Ah Ni/Cd battery, carried two instruments. West Germany provided the first instrument, a Monocular Electro Optical Stereo Scanner (MEOSS), a prototype for later earth resources instruments. The second instrument was ISRO's own 20-3,000 keV Gamma-ray Burst Experiment.

SROSS C had the honour to be the first fully successful launch in the series when it was launched 20 May 1992 by ASLV-D3 from Sriharikota. Lighter than its predecessors, at 106 kg, it carried the National Physics Laboratory's Retarding Potential Analyser for low-latitude ionosphere/thermosphere investigations and ISRO's 20-3,000 keV Gamma-ray Burst Experiment (two CsI(Na) scintillation detectors; 37/76 mm diameter, 12.5 mm thick).

During flight, the Stage 4 spun up to only 80 rpm instead of 180 rpm, yielding low perigee and a short life, but the satellite still met mission specification before re-entry 14 July 1992.

On 4 May 1994, the SROSS C2 took off aboard an ASLV-D4 from Sriharikota into a 433 × 922 km, 46.0° orbit. India planned a four-year design life for the payload, which was identical to SROSS C. Ground controllers activated the RPA on 9 May. The GRB detected 12 bursts of gamma rays from a potentially celestial origin in the first year.

This SROSS adjusted its orbit by lowering to an intended 429 × 608 km using RCS over 1-7 July 1994, giving India the potential to develop multiple deployment capabilities. The bus had 1.3 kg of fuel remaining of 5 kg fuel load at the end of the satellite lifetime.

INTERNATIONAL

European Space Agency science series

Background

ESA's contribution to the Inter-Agency Solar Terrestrial Physics programme forms the first cornerstone of the agency's Horizon 2000 space science plan. ESA's cost to completion was to be ECU761 million (1995 rate). The four Cluster satellites were lost in Ariane 5's failed debut in June 1996. Following the loss of the Cluster satellites, consideration was initially given to flying a ground spare satellite as a solo mission, the name being given as Phoenix. However, in April 1997, it was announced that this ground spare together with three other newly-built satellites would be launched as the Cluster-2 mission in 2000. Two launches of Russian Soyuz launch vehicles - each with an added upper stage - would each place a pair of Cluster-2 satellites into orbit.

The X-ray Multi-Mirror (XMM) observatory forms the second, High-Throughput X-Ray Astrophysics, cornerstone of ESA's Horizon 2000 space science programme. ESA's Science Program Committee approved the Science Management Plan and Schedule on 21 June 1988, with a launch scheduled for 1999, to provide astrophysicists with a wide-band, high sensitivity instrument for up to 10 years of data gathering. Two competing mirror technologies were tested: the baseline monolithic nickel approach of Italy's Media Lario company and epoxy replication on CFRP carriers at Zeiss. The Mirror Development Model using four of these shells underwent X-ray testing in May-August 1993 with breadboard instruments. CFRP did not meet requirements because of moisture sensitivity; the nickel approach was chosen end-1993. This safer route means that XMM's mass has increased from 3,300 to 3,900 kg at launch, requiring a switch from Ariane 4 to Ariane 5. One nickel mirror module will weigh 340 kg, instead of 137 kg for CFRP.

Using a delayed ignition of the Ariane-5 upper stage, XMM can achieve a 48 h orbit with apogee of 114,000 km over the northern hemisphere. If this option becomes unavailable, the apogee of the eccentric orbit would be in the southern hemisphere, with Perth and Kourou stations being used for data return.

ESA Science Satellites 1967-1989

ESRO-2A: 29 May 1967, 74 kg, solar/cosmic rays, launch fail
ESRO-2B (Iris): 17 May 1968, 164 kg, as 2A
ESRO-1A (Aurorae): 3 October 1968, 84 kg, ionosphere/auroras
HEOS 1: 5 December 1968, 108 kg, interplanetary magnetic fields and solar particles
ESRO-1B (Boreas): 1 October 1969, 80 kg, as 1A
HEOS 2: 31 January 1972, 117 kg, as number 1
TD-1A: 12 March 1972, 472 kg, UV sky survey
ESRO-4: 22 November 1972, 130 kg, ionosphere/solar particles
COS-B: 9 August 1975, 275 kg, γ-ray astronomy
GEOS 1: 20 April 1977, 573 kg, Earth's magnetosphere
ISEE 2: 22 October 1977, 166 kg, magnetosphere
IUE: 26 January 1978, 445 kg, UV spectroscopy
GEOS 2: 14 July 1978, 573 kg, as number 1
Exosat: 26 May 1983, 510 kg, x-ray astronomy
Giotto: 2 July 1985, 960 kg, comet fly-bys
Hipparcos: 8 August 1989, 1,140 kg, astrometric survey.

HIPPARCOS (ESA science series) satellite

Background

As its name implies, the High Precision Parallax Collecting Satellite (HIPPARCOS) launched 8 August 1989 by Ariane from Kourou, undertook the first space-based stellar position/parallax survey. Scientific investigations continued successfully, despite the failure of the launch craft's apogee motor. Stranded in GTO as it was and passing through the van Allen belts twice each orbit, the high radiation environment took its toll, and ESA terminated operations early on 15 August 1993.

HIPPARCOS solved the previous errors on positions, parallaxes and proper motions by taking measurements typically below the 0.002 arcsec target at ninth magnitude for 120,000 stars. The Tycho element provided data on more than one million objects. Final HIPPARCOS results offer parameters 10-100 times more accurate than those from ground observations. It was completed in December 1995 and released to participating scientists in March 1996.

ESA made the HIPPARCOS and Tycho catalogues publicly available in March 1997. Project cost to completion about ECU400 million (1990 rate).

Infra-red Space Observatory (ESA science series) satellite

Background

The cryogenically-cooled Infra-red Space Observatory is observing 2.5-200 μm IR radiation as a follow-up to the pioneering all-sky survey undertaken by IRAS (8-120 μm) in 1983. ISO has a sensitivity up to 1,500 times greater and ESA operates it as an observatory, studying specific targets, with about two-thirds of its observing time available to the general astronomical community. Among other objects, ISO studies selected galaxies and star formations within relatively cool clouds of gas and dust (typically 15-300,000). Within the solar system, it observes asteroids, comets and the atmospheres of the giant planets and moon Titan.

Estimated cost-to-completion to ESA of the specified 18-month in-orbit phase, excluding science instruments, is ECU750 million (1995 rate). Separately funded international consortia developed the four science instruments (delivered 1993). ESA/Aerospatiale signed the FFr1,500 million prime contract 17 September 1990. Two major technical problems in 1992 delayed launch by at least eight months to early 1995 and added 28.6 MAU to the cost. The first problem was leakage from the cryostat's liquid helium valves. The second problem was contamination of the primary mirror's gold coating. ISO's spare mirror became the primary after being coated and the flight telescope was delivered February 1993.

The flight model instruments and telescope were integrated into the FM cryostat by end-1993. Isocam returned its first image (galaxy M51) 28 November; Isophot observed star Gamma Draconis 29 November; LWS observed the S106 region 30 November.

The payload consists of a Ritchey-Chretien telescope, with an effective 60 cm aperture and providing a 20 arcmin unvignetted FOV. It is suspended in the cryostat. Four prime focus instruments are mounted below the optical support, each occupying an 80° segment and receiving a three arcmin unvignetted FOV, providing photometric and imaging capabilities over 2.5-200 μm and medium/high resolution spectroscopy over 2.5-180 μm. In general, only one instrument is operated at a time, although the camera can acquire extra data when one of the others is in use. Also, the long wavelength channel of the photometer is active when the satellite is being slewed, yielding a partial 200 μm sky survey (IRAS operated <120 μm).

Specifications

Launch: 19 November 1995 on Ariane 44P from Kourou, French Guiana
Contractors: Aerospatiale (prime), DASA (cryostat payload module), Fokker (attitude control subsystem)
Mass: 2,498 kg at launch, 2,418 kg BOL, 1,515 kg dry
Orbit: 1,038 × 70,578 km, 5.2°, 24 h; initial 500 × 71,850 km, 5.25°. ISO Control Centre is at Villafranca, Spain, allowing about 13 h of useful observing time over each 24 h rev (no onboard recorder). NASA's Goldstone extends observations to almost 17 h daily (in return, NASA has 0.5 h observing time daily)
Spacecraft: 5.3 m high, 2.3 m wide, 3-axis controlled to within a few arcsec. The cylindrical payload module carries a conical sunshade and two star trackers; the service module provides basic spacecraft functions. Attitude control for stable observations of up to 10 h incorporates Earth/Sun sensors, star trackers, four rate integrating gyros, four skewed RWs and two branches of 8 × 2 N hydrazine thrusters. Real-time downlink (no onboard recorders) 33 kbit/s, 24 kbit/s of which is devoted to science data. The payload module is essentially a large cryostat. Inside the vacuum vessel is a toroidal tank holding 2,250 litres of superfluid helium at mission start to provide at least 18 months' observations (two years is now projected). Some of the IR detectors are cooled to 2K by direct connection to the tank; the boil-off gas cools other units. This is first routed through the optical support structure, where it cools the telescope and science instruments to about 3K; it then passes along the baffles and radiation shields before being vented to space. A small toroidal auxiliary tank near the cryostat's neck containing about 60 litres of normal liquid helium provided cooling on the pad of the last 100 h before launch. On 5 September 1996 a measurement of the remaining mass of liquid helium suggested that the operational lifetime of ISO should be around two years rather than eighteen months which was the design lifetime. This would mean that the satellite should operate through to the end of 1997.
Payload: Isocam: 2.5-17 μm camera/polarimeter, 1.5/3/6/12 arcsec resolutions, two channels each with 32 × 32 element arrays. PI: Catherine Cesarsky, CEN-Saclay, France.

Isophot: 2.5-200 μm imaging photopolarimeter, operating in three modes as 30-200 μm far-IR camera, 2.5-12 μm spectrophotometer and 3-110 μm multi-band multi-aperture photopolarimeter. PI: Dietrich Lemke, MPI für Astronomie, Germany.

SWS: 2.5-45 μm Short Wavelength Spectrometer incorporating two gratings and two Fabry-Perot interferometers, 10 × 20 arcsec and 20 × 30 arcsec res. PI: Thijs de Graauw, SRON, Netherlands.

LWS: 45-180 μm Long Wavelength Spectrometer incorporating grating and two Fabry-Perot interferometers, 1.65 arcmin resolution. PI: Peter Clegg, Queen Mary & Westfield College, UK.

IASTP series

Background

The Inter-Agency Solar Terrestrial Physics programme is a 12-satellite project, involving NASA, ESA, ISAS and Russia's Institute of Space Research (IKI). IASTP plans and executes detailed investigations of the Sun, Earth's space environment and Sun-Earth interaction. The origins of the Solar-Terrestrial science programme trace back to a decision to investigate new subjects made by the Inter-Agency Consultative Group (IACG) following the highly successful Comet Halley collaboration. IASTP, started in 1977, with planned completion in 2001, involves more than 100 universities, research labs and major contractors in 16 countries.

NASA budgeted US$18.6 million in FY88 to start development of Wind/Polar (rising to 64.4 FY89, 57.6 FY90, 96.6 FY91, 75.3 FY92, 72.6 FY93, 21.6 FY94, 40.0 FY95, 0 FY96) and provide instruments for Geotail, Soho and Cluster. Russia's KONUS γ-burst detector was delivered 1992 to fly aboard Wind, the first CIS instrument on a US satellite bus. GE Astro Space built the Polar/Wind. Total NASA development cost for GGS, Geotail, Soho and Cluster is estimated at US$732 million. Wind/Polar design/construction difficulties resulted in launch delays, leaving Geotail as the first in orbit.

Cluster (IASTP series) satellites

Background

Cluster had four identical spacecraft scheduled on board the first Ariane 5, for field and plasma measurements of the magnetosphere. Cluster planned to map the 3-D extent and dynamic behaviour of small scale structures (from a few hundred km to a few thousand km) in the Earth's plasma. They would observe how solar particles interact with Earth's magnetic field and study the interactions of solar-terrestrial magnetic/electrical fields by direct measurement. Their loss on Ariane 5's debut was a major scientific blow. Only a month later at the SPC meeting held in London on 2-3 July 1996, it was agreed that a Cluster-2/Phoenix mission should be readied as soon as possible, with a final decision coming at the November 1996 meeting.

Studies to determine the right replacement included launching the ground spare which was shelved because the Cluster concept required a group of satellites in orbit, not just one. The next most plausible solution called for ESA to build three new satellites to accompany the ground spare, thus duplicating the planned 1996 mission, or developing a set of four new smaller satellites. At the November 1996 meeting, the ESA Board chose the option of flying the existing ground spare with three new identical satellites, on condition that the cost could be capped at 210 MAU. In April 1997, ESA announced that it had chosen to launch the Cluster-2 mission (as it is now called) using two Russian Soyuz-U launch vehicles, each carrying two Cluster-2 satellites. The launches would come in 2000.

Since the Soyuz-U launch vehicle is designed for launches to low Earth orbit only, the Russians may need to add a new upper stage to launch the two pairs of satellites into their highly eccentric orbits. Whether this will be the Soyuz-U/Ikar launch vehicle being marketed by STARSEM or another Soyuz-U variant has still to be clarified.

Specifications

Contractors: Dornier (prime), Alcatel (TT&C), British Aerospace (AOCS, RCS), Contraves (structure), FIAR (power), Fokker (thermal control), Laben (OBDH), DASA/MBB (solar array, thrusters), Saab (harness, separation system), Sener (booms), Odetics (tape recorder), Cubic (HPA)
Launch: 4 June 1996 by Ariane 501 from Kourou
Mass: FM 1 1,183/531 kg launch/dry; FM 2 1,169/518 kg; FM 3 1,171/520 kg; FM 4 1,184/533 kg. 650 kg propellant, 72 kg payload
Orbit: initial 10° GTO, rising to planned 25,513 × 140,318 km, 90°, 66 h in five manoeuvres (numbers 1-3 each 240 m/s; number 4 630 m/s; number 5 370 m/s for plane change). Tuned so that the four satellites would fly in tetrahedral formation, passing in/out of Earth's magnetic field, crossing cusp, magnetopause, bow shock, magnetotail and other features. Separations adjusted between 200 km in cusp and 18,000 km in plasma sheet
Lifetime: 2 years nominal
Spacecraft: 2.9 m diameter 1.3 m high cylinders, with conductive surfaces, spin stabilised at 15 rpm. Magnetic field experiments mounted on two 5 m radial booms; two pairs of 100 m tip to tip wire antennas for electric field measurements. Solar array to provide 224 W,

The four Clusters were launched in two double stacks (ESA) ***1996***

including 47 W for the payload. Science data rate 16.9 kbit/s normal operation and 105 kbit/s burst mode; payload data outside of ground coverage stored in 1 Gbit tape recorder or 2.25 Gbit SSR for subsequent dumping. Telemetry downlink 2-262 kbit/s. DASA's S400 400 N MON/MMH thruster to establish the operational orbits, including the inclination increase in five firings from Ariane's 10° GTO, using 500 kg of propellant. RCS by eight DASA S10 MkII 10 N MON/MMH thrusters. Attitude determination <0.25° by star mapper and Sun sensor. Data from four satellites synchronised via highly stable onboard clock and time stamping at ground stations. Operated from ESOC via Odenwald (Germany) and Redu (Belgium). Science operations co-ordinated through a Joint Science Operations Centre in the UK; data distributed via Cluster Science Data System, using ESANET to transfer it to centres in Austria, France, Germany, Hungary, Scandinavia, UK, China and US

Payload: STAFF Spatio-Temporal Analysis of Field Fluctuations (AC magnetometer; France).

EFW Electric Fields & Waves (utilising long wire antennas; Sweden). WHISPER Waves of High Frequency and Sounder for Probing of Density by Relaxation (active density measurement of solar plasma; France).

WBD Wide Band Data (high frequency electric fields of several 100 kHz; US), DWP Digital Wave Processor (controls STAFF, EFW, WHISPER, WBD wave consortium experiments; UK).

FGM Fluxgate Magnetometer (DC magnetometer; UK).

EDI Electron Drift Instrument (measurement of electric field by firing electron beam in circular path for many tens of km around satellite, detector on other side picks up return beam; FRG).

CIS Cluster Ion Spectrometry (composition/dynamics of slowest ions; France).

PEACE Plasma Electron/Current Analyser (distribution, direction, flow and energy distribution of low/medium-energy electrons; UK).

RAPID Research with Active Particle Imaging Detectors (essentially a pinhole camera to measure high energy electrons/ions; FRG).

ASPOC Active Spacecraft Potential Control (removal of satellite excessive charge by emitting indium ions; Austria).

Geotail (IASTP series) satellite

Background

Geotail investigates the energy mechanisms in the magnetosphere tail and the physical processes in other important regions, such as the dayside magnetopause and bow shock. Scientists are interested in the geomagnetic field lines on the dayside, in a hemisphere of some 10 Earth radii, because they are compressed by the solar wind. The nightside lines are correspondingly stretched into a tail extending beyond 2,000R_E. A substantial fraction of the solar wind energy imparted to the magnetosphere is stored in those stretched lines, and enhanced activity arises in the inner magnetosphere and polar ionosphere when this energy is released.

NEC developed the Geotail under ISAS direction. EPIC, CPI, the Multichannel Analyser and the Inboard Magnetometer on the Magnetic Field Measurement Investigation are funded by NASA Goddard.

Specifications

Launch: 24 July 1992 by Delta 6925 from Cape Canaveral

Mass: 1,009 kg (360 kg propellant; 105 kg science payload)

Orbit: for distant tail Sun-synchronous Moon orbit plane double lunar swing-by, max apogee 200R_E; for near tail 7.5° ecliptic inclination, 8 × 30R_E. Time spent in Earth's

Japan's Geotail was the first IASTP craft (ISAS)

shadow was 1.7 h max, including penumbra. In the first part of the mission, lasting 1.75 years, its apogees were held in the magnetotail (on Earth's night side) by repeated paired lunar swing-bys. After 4½ Earth orbits, the first lunar swing-by was 13,657 km at 15.34 GMT 8 September 1992, producing a 133R_E apogee in the magnetotail 18 days later. Farthest apogee was 199R_E on 25 April 1994. Four δV manoeuvres totalling 154.2 m/s in November 1994 reduced apogee to 50R_E and then six more totalling 161.5 m/s in February 1995 reduced it to 30R_E to study the neutral sheet current. No further δV manoeuvres are planned; it will remain in that orbit to support other IASTP missions

Life: 1.75 years in the distant tail and 1.45 years in the near tail (nominal mission completed July 1996)

Spacecraft: 20 rpm spin-stabilised cylinder 2.2 m diameter 1.6 m height with mechanically despun antennas. 340 W EOL provided by Si N-on-P BSFR cells + 3 × 19 Ah Ni/Cd batteries. Data at 16.384, 65.536 or 131.072 kbit/s real-time or playback (two 450 Mbit Odetics tape recorders) to NASA DSN, Usuda, Kagoshima (Japan). Operation/control from SOCC Sagamihara Operation Control Centre

Payload: Electric Field (EFD). Comprising the Spherical Probe, Wire Antenna and Electron Boomerang.

Magnetic Field (MGF). Fluxgate and Search Coil.

Plasma (LEP). Made up of Ion/Electron Energy Analyser, Solar Wind Ion Analyzer and Ion Mass Spectrometer.

Plasma (CPI). Consisting of the Hot Plasma Analyser, Solar Wind Ion Analyser and Ion Composition Analyser.

Energetic Particles (HEP). Constituted by the Low Energy Detector, Burst Detector and Isotope Telescopes.

Energetic Particles (EPIC). Composed of the Supra-Thermal Ion-Composition Spectrometer and the Ion Composition Subsystem.

The Plasma Waves (PWI). Made up of the Multichannel Analyser, the Sweep Frequency Analyser and the Wave Form Capture (both E/B fields).

Polar (IASTP series) satellite

Background

Polar will determine the ionosphere's role in substorm phenomena and in overall magnetospheric energy balance. It will measure plasma energy input through the dayside cusp, and determine the characteristics of ionospheric plasma outflow and energised plasma inflow to the atmosphere. Like the Wind, it will also study characteristics of the auroral plasma acceleration regions, and provide multispectral auroral images of the footprint of the magnetospheric energy disposition into the ionosphere/upper atmosphere.

Lockheed Martin Astro Space built the Polar satellite. It carries 11 instruments all with high data compression capabilities.

Lockheed Martin Astro Space was Polar and Wind prime contractor

Specifications

Launch: 24 February 1996 by Delta 7925-10 from VAFB

Mass: 1,300 kg (301 kg propellant, 226 kg science payload, 999 kg dry)

Orbit: 5,141 × 50,605 km, 85.9°

Life: 3 years nominal, extended mission under study

Spacecraft: 2.7 m diameter 2.5 m height cylinder (stowed envelope) with conductive surfaces, 10 rpm spin-stabilised with a 180° precession manoeuvre every 6 months. Despun platform for imagers inertially stabilised to 0.2°. Two deployable 6 m lanyard booms for magnetic field instruments, two deployable 5.5 m Z-axis booms and four radial wire antenna (two 50 m, two 80 m) for electric field measurements. 14.9 m² solar array provides 511 W BOL, including 186 W for payload; 3 × 26.5 Ah Ni/Cd batteries. Four Olin MR-111 2.2 N spin, four MR-111 4.4 N spin axis precession + four MR-50 22 N δV thrusters. Telemetry received by DSN over four 45 min nominal daily contacts. Science data 55.5 kbit/s real-time and up to 512 kbit/s playback. Payload data outside of ground coverage stored on 1.3 Gbit digital tape recorder. Mission operations/data processing from GSFC

Payload: MFE Magnetic Fields Experiment (University of California at LA).

EFI Electric Fields Instrument (University of California, Berkeley).

PWI Plasma Wave Instrument (University of Iowa).

TIDE/PSI Thermal Ion Dynamics Experiment/Plasma Source Instrument (NASA Marshall).

HYDRA Fast Plasma Analyser Experiment (University of Iowa).

TIMAS Toroidal Imaging Mass-Angle Spectrograph (Lockheed Palo Alto Research Lab, California).

CEPPAD/SEPS Comprehensive Energetic Particle Pitch Angle Distribution/Source Loss Cone Energetic Particle Spectrometer (Aerospace Corp, California).

CAMMICE Charge & Mass Magnetospheric Ion Composition Experiment (Boston University).

PIXIE Polar Ionospheric X-ray Imaging Experiment (Lockheed Palo Alto Research Lab, California).

VIS Visible Imaging System (University of Iowa).

UVI Ultraviolet Imager (University of Washington).

SOHO (IASTP series) satellite

Current status

During June 1998, ground controllers at the NASA Goddard facility lost contact with the SOHO satellite. Using the Deep Space Network, Goddard engineers managed to locate the satellite one month later. Apparently operator fault turned the satellite antenna away from ground stations. It appears it will be difficult to recover the mission.

Background

Until its loss, the Solar Heliospheric Observatory made continuous observations of the solar surface, corona and wind, from a vantage point in the L1 Lagrangian libration point, about 1.5 million km out. ESA and the international scientific community seeks the data to investigate physical processes that form and heat the corona, maintain it and give rise to the expanding solar wind, and to investigate internal structure by helioseismology and solar irradiance variations.

Matra Marconi Space France is the prime contractor, with help from BAe (AOCS), MMS UK (payload module, propulsion subsystem), Alenia (structure, harness), CASA (thermal control), Saab Ericsson (communications, data handling, software), FIAR (power), SAFT (batteries), Odetics (tape recorder), Cubic (HPA).

Specifications

Launch: 2 December 1995 by Atlas 2AS from Cape Canaveral. First solar image recorded 19 December 1995 and commissioning formally completed 16 April 1996

Mass: 1,864 kg at launch (655 kg payload)

Orbit: circling L1 libration point since 14 February 1996, 1.5 million km out on Earth-Sun line

Lifetime: 2 years nominal, consumables sized for 6 years (accurate launch left 20 years of propellants)

Spacecraft: 3.6 m height, 3 m diameter, 3-axis stabilised Sun-pointing with 10 arcsec accuracy and a pointing stability of 1 arcsec per 15 min (10 arcsec per 6 months). Instruments accommodated in payload module, with a high degree of decoupling from the service module. Solar panels generate about 1,400 W (payload requires 440 W). S-band telemetry received by NASA's Deep Space Network over 3 × 1.3 h + 1 × 8 h periods each day. Continuous science data stream 40 kbit/s, increased to 246 kbit/s when MDI solar oscillation imaging instrument is operating in high rate mode; 1 Gbit tape recorder + 2 Gbit SSR. EOF

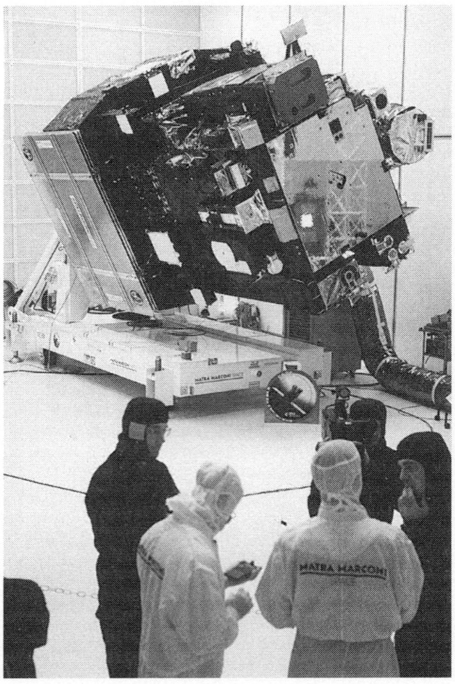
Soho final flight processing at Canaveral (Theo Pirard/Space Information Centre) ***1996***

Experiment Operations Facility at NASA Goddard co-ordinate/plan science operations. Its main tasks are to organise the payload's real-time operation and to control the imaging/spectrometric instruments during the daily 8 h contacts. MDI's solar oscillations imaging data requires a specialised data facility at University of Stanford, Palo Alto

Payload: SUMER Solar Ultraviolet Emitted Radiation (FRG)
CDS Coronal Diagnostic Spectrometer (UK)
EIT Extreme UV Imaging Telescope (France)
UVCS UV Chronograph Spectrometer (US)
LASCO White Light/Spectrometric Chronograph (US)
SWAN Solar Wind Anisotropies (France)
CELIAS Charge, Element/Isotope Analysis (FRG)
COSTEP Suprathermal/Energetic Particle Analyser (FRG)
ERNE Energetic Particle Analyser (Finland; COSTEP/ERNE are combined in the single CEPAC instrument)
GOLF Global Oscillations at Low Frequencies (France)
VIRGO Variability of Solar Irradiance (Switzerland)
MDI Michelson Doppler Imager (US).

Wind (IASTP series) satellite

Background

The Wind was built to investigate sources, acceleration mechanisms and propagation processes of energetic particles and the solar wind. IATSP also wants it to provide complete plasma, energetic particle and magnetic field input data for magnetospheric and ionospheric studies, determine the magnetospheric energy output to interplanetary space in the upstream region, and provide baseline ecliptic plane observations for heliospheric studies.

Lockheed Martin Astro Space built the satellite.

Specifications

Launch: 1 November 1994 by Delta 7925-10 from Cape Canaveral
Mass: 1,250 kg (300 kg propellant, 198 kg science payload, 752 kg dry)
Orbit: lunar swing-by (first 27 December 1994) to 250 R_E in first 2 years; L1 halo (3 × 10^5 km) from November 1996
Life: 3 years nominal, extended mission plans under study
Spacecraft: 2.6 m diameter 2.5 m height cylinder with conductive surfaces, spin-stabilised at 20 rpm. Two deployable 12 m lanyard booms for magnetic field instruments, two deployable 5.28 m Z-axis booms plus four wire antennas (two 7.5 m, two 50 m) for electric field measurements. 11.5 m^2 solar array provides 472 W BOL, including 144 W for payload; 3 × 26.5 Ah Ni/Cd batteries. Four Olin MR-111 2.2 N spin plus eight MR-50 22 N δV thrusters. Telemetry received by DSN over 2 h nominal daily contact. Science data rates 5.85/11.3 kbit/s real-time and up to 128 kbit/s playback. Payload data outside of ground coverage stored on 1.3 Gbit digital tape recorder. Mission operations/data processing from NASA Goddard

Wind is installed on Delta (NASA) ***1996***

Payload: MFI Magnetic Fields Investigation (GSFC)
WAVES Radio & Plasma Wave Experiment (Observatoire de Meudon, France)
SWE Solar Wind Experiment (GSFC)
EPACT Energetic Particles Acceleration, Composition, Transport (GSFC)
TGRS Transient Gamma Ray Spectrometer (GSFC)
SMS SWICS / Mass STICS (Solar Wind Mass / Suprathermal Ion Composition Studies) (University of Maryland)
3D PLASMA Energetic Particles & Three Dimensional Plasma Analyser (University of California, Berkeley)
KONUS Gamma Ray Burst Investigation (Ioffe Physical Technical Institute, St Petersburg).

Ibiza/Impact satellite

Background

The International Investigation BIsatellitaire des Zones Aurorales (Impact: Investigation of Magnetospheric Particle Acceleration and Turbulence) dual satellite mission will pick up where Explorers S3-3 and DE-1, Viking, Akebono and Freja ended their investigations. The former satellite's coverage revealed that high-time resolution using two co-ordinated satellites is essential for a deep understanding of particle acceleration phenomena in the ionosphere.

The steering group, set up in January 1996, issued an Invitation to Propose experiments in late March 1996 with a 30 September 1996 due date. Cost to completion is estimated at FFr230 million. 75 per cent will be spread between Germany (DARA, with Max Planck Institute for Aeronomy the major participant), France (CNES) and Sweden (Swedish National Space Board). Canada, UK and US may be contributors.

Once in space, the satellite will investigate, with unprecedented spatial and temporal resolution, the auroral acceleration region at 1,000-10,000 km altitude, where electrons/ions are accelerated by energy from the magnetosphere. Twin satellites will cross the same magnetic field lines simultaneously, but tens to hundreds of km apart.

Specifications

Launch: planned for 2001 aboard a vehicle to be selected, but an Ariane polar mission is a strong candidate
Design life: 2 years, covering the Northern hemisphere twice + Southern hemisphere once
Mass: each 320 kg at launch
Orbit: about 3,000 × 10,000 km, 97.2°. Dual launch with Earth observation satellite into initial 800 km, 98.7° is assumed, then joint solid motor establishes required orbit. Natural rotation of the line of apsides will fully rotate the apogee over 16 months, allowing coverage of the Northern, then Southern, then Northern hemispheres
Spacecraft: 15 rpm spin-stabilised perpendicular to orbital plane (180° switch every 8 months). 650 kbit/s max S-band downlink to 9 m antenna at Kiruna (Antarctica station required when apogee over Southern hemisphere); 1.6 Gbit science data/rev (1 Mbit/s max, typically 100 kbit/s), 3 Gbit memory. Two 50 m wire booms. 140 W required. Onboard GPS orbit determination. 75 m/s δV capacity

Candidate payload: fluxgate magnetometer (2.4 kg, 2.5 W, 8 kbit/s), search coil magnetometer (2.5 kg, 1 W, 8 kbit/s), electric field/waves (9 kg, 22 W, 250 kbit/s), electron drift (7.2 kg, 6.6 W, 2 kbit/s), electron analysers (6.8 kg, 9 W, 250 kbit/s joint for particles analysers + correlator), ion analysers (9.0 kg, 14 W), correlator/FPA (2.0 kg, 3 W), cold plasma (5 kg, 8 W, 30 kbit/s), potential control device (1.5 kg, 4 W, 3 kbit/s), intersatellite propagation experiment (2 kg, 2 W, 10 kbit/s), UV imager (one satellite only; 8.5 kg, 7 W, 120 kbit/s), electron gun (one satellite only; 5 kg, 10 W); two 50 m wire booms (14 kg), radial booms + HF antenna (9 kg).

Interkosmos series

Current status

Interkosmos, formally the Council for International Co-operation in the Studies and Uses of Outer Space, built collaborative science projects among its 10 member states and with other nations, until it closed in 1994. The projects continued under IKI, IZMIRAN and other institutes. Current activities include the Aktivny, APEX, Interball and Relikt science satellite projects. After it closed, funds for joint Interkosmos projects evaporated. Russia offered its technology and launchers free of charge and each country bore its own costs. No new missions have recently been announced and existing projects continue to suffer delays.

Background

The first Interkosmos launch, Cosmos 261 in 1968 from Plesetsk, studied air density and polar auroras. Each member financed its own contributions, with the CIS/USSR supplying the basic satellite and the launch costs. Scientific results became the common property of all member nations.

The first programme ran from 1967 to 1980, with Czechoslovakia a noted contributor to every unmanned mission. This culminated in the debut of the first Magion

RFAS/USSR-led International Earth-orbit Science Missions

Project	Launch date	Spacecraft	Orbit	Objective
Aktivny-IK	September 1989	AUOS parent	Low Earth Orbit	Study the effects of ULF waves on the magnetosphere;
		+ Magion sub-sat		Interkosmos project; parent still operational
Granat	December 1989	Astron-type	Prognoz-type	Study of gamma/x-ray radiation with France/Denmark; still operational
Gamma	July 1990	Progress-type	Low Earth Orbit	Study of gamma/x-ray radiation in co-operation with France
APEX	December 1991	AUOS parent	Low Earth orbit	Electron and plasma probing of magnetosphere;
		+ Magion sub-sat		Interkosmos project; parent still operational
Coronas-I	March 1994	AUOS-SM	Low Earth Orbit	Helioseismology and solar activity; Interkosmos project, still operational
Interball	August 1995	Prognoz-M2	Prognoz + intermediate	Study of magnetosphere and plasmasphere in 14-nation
	August 1996	+ Magion sub-sats	Molniya-type	co-operation; one Prognoz/Magion pair in each orbit type
Priroda	April 1996	Mir module	Low Earth Orbit	Mir remote sensing module, including France's Alissa radar
Bion 11	December 1996	Vostok-type	Low Earth Orbit	Life sciences research on living organisms, including 2 monkeys; NASA involvement

satellite on Interkosmos 18, the first non-Soviet satellite in the series.

Interkosmos' second programme, lasting from 1980 to 1985, saw a move towards Earth surface studies, with the ocean dynamics investigations of Interkosmos 20/21. By October 1984, when the 17th Interkosmos conference was held in Berlin, 11 high-altitude Vertikal research rockets and 22 Interkosmos satellites had been launched, the last in August 1981.

By the time for the third programme in 1985, the programme plan stated an economic justification based on the results already achieved, but made no mention of any new launches. However, as a member of the Inter-Agency Consultative Group, Interkosmos participated in the Vega missions and the international collaboration on the 1985/86 studies of Halley's Comet. It also made major contributions to the 1988/89's Phobos mission. In April 1985, The Soviets renamed Prognoz 10 (Intershock), one of the solar radiation series, Interkosmos 23. Aktivny, APEX and Interball subsequently appeared as major projects in October 1987. Interkosmos announced proposals for two Coronas observatories that would monitor the Sun during its period of maximum activity in the early 1990s. The SAS aeronomy satellites were described in 1989 (but later cancelled) and agreement was near on the Regatta contribution (now postponed indefinitely) to the Inter-Agency Solar-Terrestrial Physics (IASTP) programme.

With the exception of Interkosmos 6, a Vostok-type vehicle, and Interkosmos 15, an AUOS Automatic Unified Orbital Station prototype, Interkosmos 1-16 all were light research satellites. Weighing 320-375 kg, they were based around a standard ellipsoidal pressurised shell. Czech scientists made contributions to all these satellites, and provided the orientation system for the more advanced 422 kg AUOS, introduced in operational form with Interkosmos 17. Features included onboard data processing and provision for a sub-satellite. APEX is flying the last AUOS-Z model (Z means Seemly, or Earth, denoting 3-axis pointing about a z-axis pointing vertically from the centre of the earth). Coronas-I introduced the AUOS-SM in 1994; the Prognoz-M2 debuted on Interball in 1995.

Aktivny-IK/Interkosmos 24

Background

After a two-year delay, Aktivny was launched in 1989 under the Interkosmos programme, releasing Czechoslovakia's Magion 2 to monitor the magnetospheric propagation of VLF waves and their effects on the van Allen belts. The parent craft was also designated Interkosmos 24. It was the first dual-craft active probing of the deep magnetosphere. Both Magion 2 and the Interkosmos 24 craft suffered initial problems. First, the large ORU-20 antenna failed to deploy as expected and Magion could not begin its science programme until 21 January 1990. Then, the failure of the small satellite's thruster prevented control of the separation from its parent. The Interkosmos received its last Magion contact on 20 November 1990. The main satellite is operable but is not used regularly, although some data is still used in support of other missions.

The development of Aktivny cost the Soviets a reported Rb4.8 million. Its payload consisted of a VLF generator, that was a 5 kW 9.6±0.2 kHz transmitter, radiating via 20 m diameter ORA-20 loop antenna. The antenna unfurled from a reel of 1 mm-thick aluminium (deployment problem meant that in-space inductivity was 25 per cent of expected). The Plasmas generator: modulated plasma environment around antenna. A neutral xenon injector pumped xenon around satellite for monitoring of ionisation of neutral gas flow and its propagation effects. A Low-frequency Wave System (LWS) was made up of a 0.30-22 kHz receiver. The Plasma radio spectrometer was a 0.10-10 MHz receiver. Proton/Electron Spectrometer (PES) monitored in 30-700 keV electron plus 40-700 keV proton flux. The Soft Electron Detector (SED) measured 0.01-10 keV electron flux. KM-6 and Langmuir probed plasma electron/ion temperature and number densities. The NAM-5: measured plasma mass composition.

The Institute of Terrestrial Magnetism, Ionosphere & Radio Wave Propagation (Izmiran) support the Interkosmos project, with IKI leading the effort, with contributions from Bulgaria, Former East Germany, Cuba, Hungary, Poland, Czechoslovakia. Soviet (now Russian), Czech, Cuban, US, Japanese, Canadian and Brazilian ground stations provided complementary observations.

Specifications

Launch: 28 September 1989 by Tsyklon from Plesetsk
Mass: about 450 kg parent craft, 51.2 kg sub-sat
Orbit: 500 × 2,493 km, 82.6°, permitting injection at auroral latitudes. Sub-sat separation was planned to vary from 100 m to 100 km but Magion thruster failed
Spacecraft: Interkosmos AUOS 3-based parent craft.

APEX/Interkosmos

Background

The Soviet Union launched APEX (Active Plasma Experiment) under the Interkosmos programme, releasing Czechoslovakia's Magion 3 sub-satellite to monitor the propagation of electron beams and plasma injected into the magnetosphere by the parent craft, as their separation varied from tens of m to 1,000-2,000 km. It was the first CIS/Soviet mission of this type, but closely complemented Aktivny's investigations. Simultaneous ground, balloon and sounding rocket observations allowed the spacecraft release and separation to be measured through a variety of paths. It was the last flight of this AUOS-Z model. The AUOS-SM design was introduced in 1994 on the Coronas mission.

Magion separated and began operating on 28 December 1991. It drifted from its parent because of differential drag. A small thruster fired onboard on 9 January 1992 for the first time to bring the two craft back together again by 30 January. Similar manoeuvres on May and 11-26 June 1992 accomplished the same objective. The last approach was within 100 m with relative velocity almost zero.

About 200 plasma injections lasting 4-20 minutes performed the same task that normally would have required an estimated 200 sounding rocket flights. Magion detected all of these events. The small sub satellite operated with no problems until 20 August 1992 when it suddenly suffered a power shortage, probably caused by radiation damage to the battery charge system after a period of enhanced solar activity. A ground station received the last telemetry 9 September 1992.

Specifications

Launch: 18 December 1991 by Tsyklon from Plesetsk
Mass: 1,000 kg parent satellite; sub-satellite 52 kg
Orbit: initially 440 × 3,083 km, 82.5°. Separation varied from <100 m to 1,000-2,000 km, using Magion's single 0.2 N compressed air thruster
Spacecraft: Interkosmos AUOS-Z based parent craft; 56 cm diameter sub-sat almost identical to Magion 2. AUOS was introduced on the Interkosmos 15 mission.

APEX launch processing at Plesetsk on the SL-14 Tsyklon. Magion 3 is at far left. The Aktivny mission configuration was identical (Dr Pavel Triska) ***1995***

Interball /Interkosmos

Background

Interball is a 20-nation collaborative project to investigate the interaction of solar wind with Earth's magnetosphere and the movement of solar wind's particles through the auroral oval. Interkosmos launches Interball satellites into tow types of orbits using the new Prognoz-M2 design. The first orbital type monitors magnetotail processes in conjunction with its S2-T sub-sat, and the second set operating at a higher inclination observe auroral phenomena with its S2-A companion, a Magion sub-satellite provided by the Czech Institute of Atmospheric Physics' Upper Atmosphere Department S2-Y sub-satellites have the same Canadian UV imager first flown aboard Sweden's Viking auroral research satellite in 1986.

All four Interball craft carry fields and particle instrumentation. Sweden provided its Promix-3 ion composition spectrometers. France's CESR supplied two types of particles detector, one for the lower craft only to study auroral acceleration mechanisms. LPCE contributed to electric and magnetic field monitoring on the lower pair, and CRPE contributed an ion mass spectrometer. The auroral probe carries the University of Calgary's paired UV cameras, built by Canadian Astronautics. The independent cameras cover different UV bands and each comprise a photocathode, multi-channel plate, fibre optic transfer bundle and two dimensional CCD. Images, taken via an f/1 inverse Cassegrain wide field telescope, are motion-compensated by the radiation hardened read-out electronics and then processed before transmission. The imager provides sequential shots at 20 second intervals and global views in 80 seconds.

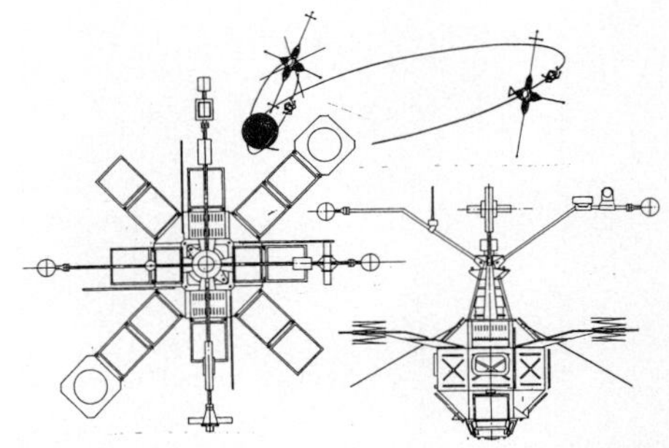

Magion subsatellite configuration and Interball orbital arrangement

Specifications

Institutions/contractors: 13 nations collaborating with RFAS (Austria, Bulgaria, Canada, Cuba, Czech Rep, ESA, Finland, France, former GDR, Greece, Hungary, Poland, Romania, Sweden), Canadian Astronautics Ltd was hardware contractor for University of Calgary's ultra-violet auroral imager
Launches: tail pair 2 August 1995; auroral pair 29 August 1996
Mass: Interball 1 (tail parent) 1,270 kg, Magion 4 sub-sat 58 kg; Interball 2 (auroral parent) 1,400 kg, Magion 5 sub-sat 62 kg
Orbit: tail probe 793 × 191,900 km, 62.9°; auroral probe planned 770-19,200 km, 63° (passing 5,000-15,000 km over the auroral oval). Magion thruster varies separation 10-10,000 km from parent
Spacecraft: 0.5 rpm spin-stabilised (about Sun axis, accurate to 1°) Prognoz-M2 parent craft incorporating pressurised (0.8-1.2 atmospheres, 20°C) quasi-cylindrical equipment bay and four tapered solar panels totalling 8.34 m² cell area; 36 Ah Ni/H_2 battery (tail probe also has 135 Ah Li battery for eclipses). ACS by 0.7 N N_2 thrusters (eight tanks holding 8 l/18.6 kg LN_2. Data stored in 30 Mbyte memory, dumped at 16 kbit/s (65 kbit/s
Tail probe: 270 kg science payload. Nine plasma instruments; three energetic particles (0.01-150 MeV) and solar x-rays; 3-axis magnetometer and wave analysers; 0.1-2 MHz radio detector
Auroral probe: 400 kg science payload. Seven plasma instruments (four hot plasma 0.02-25 keV, three thermal 0.1-100 eV), ion emitter for spacecraft potential control; 3-axis magnetometer; wave analysers and 0.1-2 MHz auroral radio detector; energetic particle detectors; paired auroral oval imagers.

Oscar

Background

The Orbiting Satellites Carrying Amateur Radio (Oscar) series of small satellites was initiated for radio amateurs to experience satellite tracking and participate in radio propagation experiments. The World Administrative

Japan's Fuji 3 provides worldwide amateur radio communications ***1996***

Oscar 13 was launched by Ariane 4 in June 1988 and incorporated a liquid thruster to attain a high-inclination elliptical orbit (Amsat-DL)

UoSAT 3/4 and the four Microsats integrated on their Ariane carrier. Webersat is at the rear, slightly taller than the other Microsats. DOVE is at far right (University of Surrey) ***1990***

Radio Conference allocated frequencies for the Amateur Satellite Service, including 29 MHz (10 m), 145 MHz (2 m), 435 MHz (70 cm), 1,270 MHz (24 cm) and 2,400 MHz (13 cm).

Transmitting low-powered signals, initially battery-operated and offering short lives, the satellites have become increasingly sophisticated. They have served school science groups, provided emergency communications for disaster relief, acted as technology demonstrators and transmitted Earth imagery. The UK's UoSAT series alone has involved hundreds of schools and thousands of groups worldwide in activities using simple antennas, receivers and personal computers. The work ranges from telemetry transmission only, through amateur radio communications, to advanced experiments such as testing indium phosphide solar cells and transputers, and monitoring radiation effects on electronics – satellite engineering for a fraction of the time and cost of the more advanced conventional satellites. Somewhat confusingly, US military Transit satellites also carry Oscar designations.

Israel's TechSat, launched 28 March 1995 on a Russian Start, included a store/forward transponder operating at amateur frequencies, but the launch failed. Mexico's Unamsat was lost at the same time. Italy's Itamsat and Portugal's PoSAT include amateur services. Similarly, SA Amsat is providing a voice and digital package to South Africa's Sunsat. Arizona State University's ASUSat will provide a J-mode voice repeater. Chile's Amsat-CE is planning a Microsat-type satellite providing a digital voice transponder. Malaysia's 1997 micro-satellite includes amateur services.

Amsat-P3DL has built the 500 kg Phase 3-D satellite for operation in a 4,000 × 47,000 km, 63.4° Molniya-type orbit following launch by Ariane 502. Project cost is US$3.5 million, including DM1.3 million Ariane contract. It is the first to cover 10/24 GHz. Three-axis control (three MWs with magnetic bearings) point the antennas at Earth, along with two color CCD cameras (narrow/wide-angle) provided by Amsat-Japan. 12 GPS antennas allow it to derive its own Kepler elements. The CEDEX experiment monitors particle radiation. The bus is hexagonal prism; 67.50 cm height, 2.23 m diameter. Solar array provides 870 W BOL. 100 mN 600-700 W ATOS ammonia plasma station-keeping thruster from the University of Stuttgart's Institute of Space Systems. It is released into a 500 × 35,000 km, 10° GTO, where a 400 N burn will raise apogee to 47,000 km. A major burn will then lift perigee to 4,000 km while cranking inclination to 60°. Total 400 N burn duration is 20 minutes. Finally, ATOS will create the required 63.4° over two years. Some 250 kg of 3D's initial mass is propellant.

Analogue uplink is (MHz): 21.210-21.250, 145.840-145.990, 435.550-435.800, 1,269.250-1,269.500, 1,268.325-1,268.575, 2,400.350-2,400.600, 2,446.450-2,446.700, 5,668.550-5,668.800. Analogue downlink: 145.805-145.955, 435.475-435.725, 2,400.225-2,400.475, 10,451.025-10,451.275, 24,048.025-24,048.275. Digital uplink: 145.800-145.840, 435.300-435.550, 1,269.000-1,269.250, 1,268.075-1,268.325, 2,400.100-2,400.350, 2,446.200-2,446.450, 5,668.300-5,668.550. Digital downlink: 29.325-29.335, 145.955-145.990, 435.900-436.200, 2,400.650-2,400.950, 10,451.450-10,451.750, 24,048.450-24,048.750. The Rudak digital packet unit allows up to 256 kbit/s.

Specifications

Oscar 1 was launched 12 December 1961 with Discoverer 36 by Thor Agena. A group of enthusiasts in California persuaded the USAF to replace ballast on the Agena upper stage with the 4.5 kg Oscar package. 570 amateurs in 28 countries reported receiving its simple 'Hi-hi' morse code signals on the 2 m band before it re-entered after 312 revolutions. It led to the creation of Amsat in 1969.

Oscar 2 launched on 2 June 1962 and Oscar 3 on 9 March 1965, were DoD piggybacks. Oscar 3 was the first true amateur satellite, relaying voice contacts at 145 MHz.

Oscar 4 was launched 21 December 1965 by Titan 3C provided the first US-Soviet amateur link.

Oscar 5 was launched 23 January 1970 with ITOS 1. 18 kg. 1,435 × 1,481 km, 102°. Built by Melbourne University, Australia. Battery-powered; transmitted telemetry for 46 days. The university compiled tracking reports from hundreds of stations in 27 countries.

Oscar 6 was launched 15 October 1972 with NOAA 2. 16 kg. 1,450 × 1,459 km, 101.7°. First Phase 2 satellite. Solar panels powering Ni/Cd battery permitted operations until mid-1977. Provided store/forward for morse and teletype; up 145.90-146 MHz, down 29.450-29.550 MHz at 1 W.

Oscar 7 was launched 15 November 1974 with NOAA 4. 29 kg. 1,444 × 1,462 km, 101.7°. Phase 2 similar to Oscar 6, built by German, Canadian and Australian radio hams under direction of US Amsat. Two 2 W repeaters provided store/forward morse & teletype. Ended operations 1981.

Oscar 8 was launched 5 March 1975 with Landsat 3. 27 kg; 33 cm high, 38 × 38 cm. 903 × 917 km, 99°. Last Phase 2 providing similar store/forward service (145.90 up, 29.40 MHz 1-2 W down) of number 7 but also educational applications. Japan transponder 145.90-146.00 up, 435.10-435.20 MHz 2 W down (telemetry 435.095 MHz). Design life 3 years; served to mid-1983.

Oscar 9 (UoSAT 1) was launched 6 October 1981 with SME. Re-entered 13 October 1989.

Oscar 10 First successful Phase 3 satellite, built by Amsat-DL, launched 16 June 1983 by Ariane L6. The first was lost on the failed Ariane L02 of 23 May 1980. Oscar 10 was struck twice by stage 3 soon after separation, damaging an antenna. Its 400 N liquid thruster also fired 80 seconds too long, so that the planned 1,500 × 35,800 km, 55° 11 hours orbit ended as 3,952 × 35,510 km at 25.9°. 140 kg at launch, 90 kg on-station mass, spin-stabilised and three-point star structure (see Oscar 13). It was to have provided emergency communications for disaster relief, transmissions of education programmes direct to classrooms, and two-way long-distance communications for 17 hours daily but its computer failed, through particle radiation, although its transponders are usable. See Oscar 13 for transponder details. It is operational only periodically, depending on the Sun's angle.

Oscar 11 (UoSAT 2) was launched 1 March 1984 with Landsat 5. 60 kg; Sun-synchronous at 700 km. Built by Surrey University, UK. With three year design life, it remains operational providing Earth imaging and store/forward. The DCE Digital Communications Experiment pioneered store/forward mailbox links. The Digitalker Speech Synthesiser is used for school demonstrations (in 1988 it supported a Canadian/USSR expedition to the North Pole). Down: 145.826/436.025MHz 1,200/4,800bit/s AFSK-FM.

Oscar 12 (Fuji 1, JAS 1) was launched 13 August 1986 piggyback with EGP into 1,448 km, 50°orbit 50 kg, including 11 kg dummy mass. Japan Amateur Satellite was 26-faced polyhedron, 40 × 40 × 47 cm, powered by 6.9 W/14 V average from 979 2 × 2 cm Si cells and 6 Ah Ni/Cd battery. The satellite carries two transponders: 145.9-146.0/435.8-435.9 MHz up/down analogue (1 W transmit), and 145.85-145.91/435.91 MHz up/down digital (1 W transmit). Developed by Japan Amateur Radio League, with system design/integration by NEC. Number 1 used Flight Model 2; FM 1 was launched 7 February 1990 as Oscar 20. Number 1 was taken out of service November 1989 because of battery failure.

Oscar 13 was launched 15 June 1988 by first Ariane 4. Built by Amsat-DL and similar to Oscar 10. Released into GTO and fired 400 N liquid bipropellant thruster twice (51 seconds 22 June; 341 seconds 6 July) to raise orbit to 2,545 × 36,265 km, 57.7°. Re-entry came 6 December 1996 resulting from orbital perturbations lowering perigee. 140 kg initial mass decreased to 90 kg at beginning of six-year operational life. 24,000 solar cells on three-point star structure provide 40 W (25 W after five years). Payload included the Rudak digital packet communications unit and a microprocessor for managing/monitoring the satellite's operation. Three transponders: 50 W 435.423-435.573/145.825-145.975 MHz up/down mode B; 50 W 1,269.641-1,269.351/435.715-436.005 MHz mode L, 144.423-144.473/435.990-435.940 MHz up/down mode J, Rudak 400 bit/s digital 1,269.710/

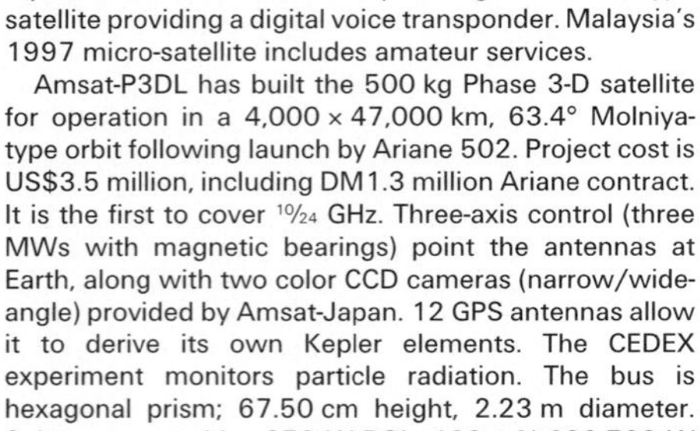

Oscar 13 was launched by Ariane 4 in June 1988 and incorporated a liquid thruster to attain a high-inclination elliptical orbit (Amsat-DL)

Spot 3's Ariane carried six passengers on its ASAP platform. Clockwise, from far left: Itamsat, Kitsat 2, Stella (apex), PoSAT 1, Eyesat and HealthSat 2

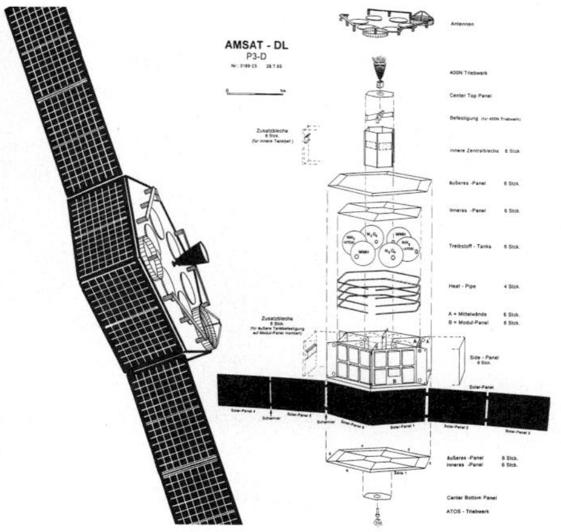

Principal features of Amsat-DL's Phase 3-D satellite (Amsat-DL) ***1994***

435.677 MHz up/down; 1 W 435.603-435.639/ 2,400.711-2,400.747 MHz up/down mode S. Rudak failed to work; a second version was launched 29 January 1991 as the Oscar 21 part of the Soviet Informator 1 RS-144 package.

Oscar 14/15 (UoSAT 3/4) was launched 22 January 1990 by Ariane 4 into 789 × 804 km/790 × 805 km, 98.7° orbit. 45kg. First use of Ariane ASAP rack to orbit six payloads simultaneously (see also Oscars 16-19). Number 3 carries store/forward digital transponder and cosmic particle detector; number 4 included CCD Earth imaging camera, Transputer data processing experiment and solar cell technology experiment. UoSAT 3 operations continue but contact with number 4 was lost after 30hours. UoSAT 5 reflew the payload plus an enhanced Packet Communications experiment. All non-amateur links were switched to UoSAT 3 in early 1992 and all amateur services moved to UoSAT 5.

Oscar 16-19 (Microsat 1-4) was launched 22 January 1990 jointly with numbers 14/15. All four developed jointly by Amsat-NA and Weber State University's Center for AeroSpace Technology. Based on 228 mm cubic bus powered by solar cells (Webersat's height extended to 31.7 cm). Orbits about 787 × 805 km, 98.7°. Numbers 1/4 (10/12 kg) are Pacsat and Lusat packet radio satellites for Amsat-NA and Amsat-Argentina, respectively, number 2 (12 kg) relays messages using a synthesised voice on 145.825 MHz at 4 W for Project DOVE (Digital Orbiting Voice Encoder) for Amsat-Brazil, and number 3, Webersat, (12.25 kg) carries a colour Canon CCD camera (400 × 700 pixel) for Weber State University, Utah, returning 2-3 km resolution Earth imagery. Webersat features visual light spectrometer, flux gate magnetometer, video flash digitiser, horizon sensor, 3.175 × 15.24 cm micrometeoroid impact detector (from Brighton High School in Sandy, Utah), L-band (1.265 GHz) AM receiver for amateur video store/forward, amateur radio packet store/forward, and sensors for TT, voltages and currents throughout the satellite. Solar panels plus Ni/Cd batteries provide 6 W orbit average. All four satellites operational.

Oscar 20 (Fuji 2, JAS 1b) was launched 7 February 1990, piggyback with MOS-1b. 50 kg; 912 × 1,746 km, 99.1° orbit. Identical to JAS 1. Operational but failure of degrading solar array is imminent. JAS 2 replacement.

Oscar 21, Amsat-DL's Rudak II package was launched 29 January 1991 aboard the Soviet Informator satellite in co-operation with Russia's Amsat-U-Orbita. Operations continued until the satellite was deactivated in September 1994 by its VKS Military Space Forces controllers.

Oscar 22 (UoSAT 5) was launched 17 July 1991 on ERS 1's Ariane. 47.5kg. Reflight of UoSAT 4 payload plus enhanced Packet Communications Transponder for US SatelLife medical information services in developing countries; it is also used by the US National Science Foundation for Antarctic links. The satellite is both amateur/non-amateur to enable use within the Amateur Satellite Service while supporting the payloads (which paid for the project). Operations continue; all non-amateur links were switched to UoSAT 3 in early 1992 and all amateur services moved to UoSAT 5.

Amsat-DL's 500 kg Phase 3-D satellite will raise itself into a Molniya-type path (Amsat-DL) ***1995***

Oscar 23 (Kitsat 1) was launched 10 August 1992 with Topex on Ariane into 1,301 × 1,402 km circular, 66.1° orbit.

Oscar 24 (Arsene) was launched 12 May 1993 by Ariane into 17,699 × 37,094 km, 1.10° orbit. 150.6 kg; 96 kg on-station. Ariane Radio amateur Satellite pour l'ENseignement de l'Espace conceived by CNES radio amateurs. Hexagonal prism, height 618 mm, diameter 785 mm; first satellite completely powered by European GaAs array. After three months of amateur operations, falling power levels forced Arsene into dormancy; the failed command system prevented revival.

Oscar 25 (Kitsat 2) was launched 26 September 1993 with Spot 3 by Ariane into 795 × 805 km 98.7° orbit. Released from Ariane's 4th ASAP platform.

Oscar 26 (Itamsat) was launched 26 September 1993 with Spot 3 by Ariane into 796 × 803 km 98.7° orbit. Italy's first amateur satellite, built by Amsat-Italy from Amsat-NA's Microsat design, for store/forward digital communications (started 24 October). 9.7 kg, 23 cm cube carrying no deployables. Five aluminium frame modules bolted together. 264 2 × 4 cm GaAs cells provide 8-10 W orbit average, supported by 6.5 Ah Ni/Cd battery. Passive magnetic stabilisation with solar spin (1.7 rpm); magnets maintain alignment with local geomagnetic field. CMOS V-40 microprocessor, 10 Mbyte RAM, 256 k RAM EDAC memory. 435.867 MHz (primary, 1,200 baud PSK) + 435.825 MHz (secondary, 300-9,600 baud AFSK) downlink, output of each controllable up to 4 W; omni four-element turnstile antenna. 145.875, 145.900, 145.925, 145.950 MHz (300-9,600 baud) user uplink, omni whip antenna; command channel 1 in 144-146 MHz band.

Oscar 27 (Eyesat 1) was launched 26 September 1993 with Spot 3 by Ariane into 791 × 806 km 98.6° orbit. Built by Interferometrics of US to demonstrate commercial store/forward digital communications. 10.5 kg, 23 cm cube carrying no deployables. Eyesat includes the AO-27 experimental amateur payload, built by the AMRAD Amateur Research & Development Corp of Vienna (Virginia), supporting digital communications at 300-9,600 baud. One UHF transmit (436.7975 MHz), output controllable to 0.5 or 2.5 W; omni whip antenna. two-channel VHF rx (145.85 MHz), omni whip antenna. The receiver's audio output can be switched directly to the transmitter output as a single analogue channel.

Oscar 28 (PoSAT 1) was launched 26 September 1993 with Spot 3 by Ariane into 794 × 806 km 98.7° orbit. Built in co-operation with Surrey Satellite Technology Ltd; similar to UoSAT 5.

Fuji 3 (JAS 2) was launched 17 August 1996 piggyback with ADEOS into 98.6°, 800-1,325 km orbit to replace JAS 1b. Design is basically as JAS 1/1b but added are attitude control, 9,600 bit/s packet system and Digitalker for hand-held receivers with simple aerials. 50 kg, 26-faced polyhedron, 44 × 44 × 47 cm, powered by 700 2 × 2 cm and 748 1 × 2 cm GaAs cells and 6 Ah Ni/Cd battery. Spin axis held at 90° to orbit plane by magnetorquer, Sun sensor and geomagnetic aspectometer. Transponders: 145.9-146.0/435.8-435.9 MHz 1 W up/down analogue voice and Morse; 145.85-145.91/435.91 MHz 1 W digital 1,200 or 9,600 bit/s; Digitalker (activated occasionally by ground control) provides 435.91 MHz 1 W 25 second message for FM receivers. Developed by Japan Amateur Radio League, system design/integration by NEC.

Oscar (ASUSat 1). Arizona State University's 4.5 kg student-built satellite will provide a J-mode voice repeater at 145.820/436.5 MHz up/down. See the separate ASUSat entry in the US National section for further information.

ROSAT

Background

Roentgensatellit (ROSAT) carries the largest imaging x-ray telescope flown to date. It is a joint German/UK/NASA project and follows the Uhuru, HEAO and Exosat projects. After a six week calibration and verification phase immediately after launch on 1 June 1990, on 30 July 1990 it began its six months all-sky survey in the course of which it constructed the first all-sky x-ray map at soft x-ray energies obtained with an imaging telescope. These images of the entire sky provided the first high-resolution map of the diffuse but structured x-ray background emission in addition to containing more than 80,000 point-like x-ray sources. A simultaneously all-sky survey at XUV wavelengths with the Wide Field Camera instrument resulted in 500 XUV sources. Since February 1991, more than 6,000 pointed observations have been carried out with the x-ray telescope's focal plane cameras (the PSPC and the HRI) adding more than another 70,000 sources.

ROSAT's principal instruments are housed within the body; the UK Wide Field Camera at top right was a later addition (Dornier)

The ROSAT has paid in science dividend with the detection of soft x-ray emission from SN 1987A (in 1991), the detection of x-ray emission from SN 1993J only a few days after an optical outburst, the discovery of x-ray emission from the Moon, from Comet Hyakutake and other comets, the detection of x-ray emission from the impact of comet Shoemaker-Levy on Jupiter, the x-ray emission from neutron stars embedded in supernova remnants as well as isolated thermally cooling neutron stars and reheated neutron stars. X-ray shadows produced by optically thick interstellar matter have been discovered high-lighting the structure of the interstellar medium.

In the Andromeda Nebula M31, 560 discrete sources have been resolved for the first time, using ROSAT's camera. Now scientists know that the clusters of galaxies from the merging of subclusters appears a common phenomenon and the radial surface brightness profile of clusters of galaxies has revealed the distribution of hot and dark matter.

X-ray emission from quasars could be traced to a redshift of z = 4.33. The ultra-deep surveys have resolved at least 60 per cent of the so-called extragalactic x-ray background emission as a superposition of faint point sources. The ROSAT sources include many thousands of previously unknown or not studied objects including stars, supernova remnants, galaxies, clusters of galaxies and Active Galactic Nuclei (AGN), including Seyfert galaxies and quasars. More than 2,400 publications in scientific journals have been the harvest of ROSAT's first six years of operation.

The mission is controlled from the German Space Operations Centre at Oberpfaffenhofen via the Weilheim ground station. NASA provided launch services and ground station support from its Deep Space Network. The only consumable on board, the PSPC gas, was exhausted in June 1994, two years after the nominal mission lifetime. Since then x-ray observations have been carried out with the ROSAT HRI only. One of the three star trackers on board failed in late 1990, but the two remaining star trackers are fully sufficient for attitude control and attitude reconstruction (required for the science data). Two of the four gyros have failed, the Y-gyro in May 1991, and the X-gyro in November 1993. After each failure, the onboard software was successfully reconfigured. After the Y-gyro failure, nominal operations could be resumed in November 1991 with a reduced pointing capability being available between May 1991 and November 1991.

In the period between November 1993 and February 1994 there was also only a reduced pointing capability available. The X-gyro failure in November 1993 necessitated the use of magnetometer and Sun sensor data for target acquisition. Since then new targets can be acquired only on the Sunlit side of the ROSAT orbit which led to an overall efficiency loss of about 20 per cent compared to the nominal system performance but otherwise satisfactory observations.

The BMFT German Federal Ministry for Research & Technology provided DM160 million funding. NASA's total ROSAT expenditure is $106 million under a 1982 agreement providing for the launch and HRI, first flown aboard HEAO 2 in 1978. Observing time for the mission's pointed phase is divided 50 per cent US, 38 per cent Germany and 12 per cent UK.

Specifications

Launch: 21.48 GMT 1 June 1990 by Delta 6920-10 from Cape Canaveral complex 17

Contractors: Dornier (prime, telescope structure, thermal control, power supply, data handling, ground support equipment), MPE Max Planck Institut for Extraterrestrial Physics (science management, PSPC, German data analysis centre), SAO Smithsonian Astrophysical Observatory (HRI), MBB (40%, covering AOCS, data transmission system, satellite structure),

Carl Zeiss (mirror assembly), Telefunken Systemtechnik (solar arrays), Sira Ltd. (star trackers)
Mass: 2,426 kg at launch, including 1,555 kg payload
Orbit: 580 × 584 km, 53.0°
Spacecraft: box-shaped bus, 2.40 × 2.15 × 4.50 m at launch, 2.30 × 4.70 × 8.90 m with solar array deployed on-orbit. Array provides 1 kW EOL to satisfy 905 W total consumption, including 340 W for charging 24 Ah battery. 3-axis stabilisation by four RWs and three 0.014N m magnetic coils; attitude measurement by two CCD star trackers (WFC also carries its own), four rate integrating gyros, three coarse Sun sensors and 3-axis fluxgate magnetometer. Real-time data transmitted at 8 kbit/s, stored at 1 Mbit/s; max data volume 700 Mbit per 21 h. Uplink on 2.096/2.771 GHz; downlink on 2.2765 GHz. Controlled from GSOC via Weilheim (Bavaria); German science centre is at MPE Garching
Payload: an 83.5 cm aperture 50 cm long telescope of four nested Wolter I grazing incidence mirrors (gold coated Zerodur), range 6-100 Å, rms roughness 3 Å, focal length 2,400 mm, 5 arcsec on-axis resolution feeds three focal plane instruments mounted on a carousel for operation as required: two 0.1-2 keV Position Sensitive Proportional Counters (MPE) and a High Resolution Imager (3 arcsec; SAO), providing source positional accuracy to within 10 arcsec during the observatory phase. The PSPCs undertook the 6-month survey, scanning the sky in 2° wide great circles perpendicular to the Earth-Sun line.

The UK's 135 kg Wide Field Camera was subsequently added to perform the first all-sky survey at 60-300 Å EUV, concurrent with the x-ray survey. The three nested Wolter Schwartzschild I gold coated aluminium grazing incidence mirrors provide a 475 cm^2 collecting area for a 3.5 arcmin resolution (1 arcmin on-axis). Max aperture 576 mm, focal length 525 mm, 5° FOV, drawing 78 W when fully operating. Focal plane instruments: two microchannel plate detectors. Sponsored by the UK's Science & Engineering Research Council, built by a consortium led by Leicester University.

SAX

Background

The 'Satellite Astronomia raggi-X' (SAX) Italian-Dutch bilateral project is undertaking 0.1-300 keV x-ray observations at moderate angular resolution (1 arcmin) with emphasis on spectral/time variability measurements of x-ray binaries, pulsars, transients, supernovae remnants, stellar coronae, active galactic nuclei and normal galaxies. 80 per cent of observing time is devoted to a set core programme, with the remainder available for opportunity and guest observations.

Shuttle uncertainties slowed the development time of SAX and made its developers decide to make it ELV-compatible, delaying completion of the phase into mid-1988. The consortium selected an Atlas launch in 1988; the contract was signed 31 August 1990 in Rome.

ASI, in December 1992, quoted SAX's budget as Lit408 billion; the cost was reported in January 1994 at Lit614 billion, including launch. The Operations Control Centre and a Science Data Centre are sited at Fucino, Italy. Data are returned via a Kenyan station (required because of the low inclination), with Kourou as back-up. The NIVR Netherlands Agency for Aerospace Programmes is responsible for Dutch involvement. SRON Space Research Organisation Netherlands contributed two X-ray Wide Field Cameras and Fokker Space provided the AOCS and solar arrays. The bilateral MoU, worth NLG65 million to Dutch companies, was signed 29 June 1990.

Italian instruments are a concentrator/spectrometer, 0.1-10 keV 1 arcmin resolution (three concentrators constitute the 1-10 keV Medium Energy Concentrator Spectrometer, a 0.10-10 keV Low Energy Concentrator Spectrometer (provided by ESA's Space Science Department), a High Pressure Gas Scintillation Proportional Counter, 3-120 keV, and a Phoswich Detector System 15-300 keV (also operating as 100-600 keV γ-burst monitor).

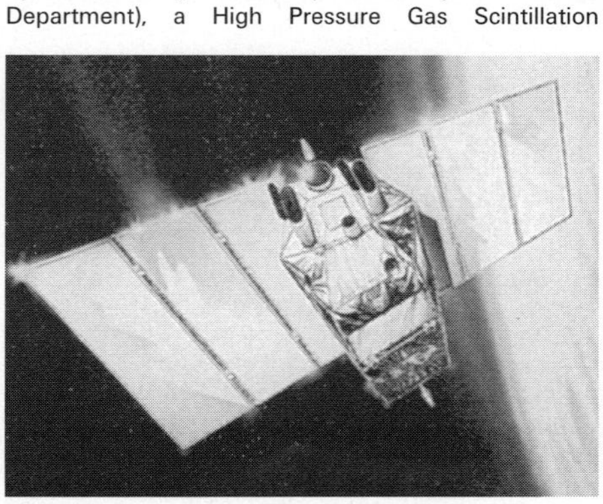

The Italian-Dutch SAX is making x-ray observations

Two SRON Wide Field Cameras viewing in opposite directions (each 42 kg, 14.5 W 1.8-30 keV, 5 arcmin resolution over 20 × 20° FOV; coded mask 70 cm from 25 × 25 cm multiwire proportional counter filled with 2.1 atmosphere Xe). As the NFI Narrow Field Instrument group studies a source, the 3-axis SAX rotates about its pointing axis for the perpendicular WFCs to scan the sky for simultaneous observations.

Specifications

Launch: 30 April 1996 by Atlas 1 from Cape Canaveral
Contractors: Alenia Spazio (satellite, system, assembly, integration/test; structure, thermal control, harness subsystems), Alenia Spazio/Fokker (AOCS, TT&C), Fokker (solar arrays), FIAR (power distribution), Laben (OBDH, NFI Narrow Field Instruments), BPD (RCS), Teldix (four RWs), Nuova Telespazio (ground segment)
Mass: 1,400 kg at launch (490 kg payload)
Orbit: 583 × 603 km, 3.96°
Spacecraft: 3.623 m high, 2.718 m diameter at launch, 26-month min life (up to 4 year projected); 3-axis stabilised, 1.5/0.5 arcmin accuracy/stability. Power (2.42 kW EOL) from two fixed deployable wings of three 2.58 × 1.15 m panels (spanning 8.97 m) each plus a single body-mounted panel; supported by 2 × 30 Ah Ni/Cd batteries (650 W mean over eclipses of up to 37 min). Data production 50 kbit/s mean, 100 kbit/s max, 300 Mbit per orbit. Downlinked at 2.2455 GHz S-band from 450 Mbit redundant tape recorder memory at 917 kbit/s (high) or 16 kbit/s (low) or 131 kbit/s real-time over Malindi (Kenya) during 5-8 min contact period each pass. 2.0675 GHz uplink at 2 kbit/s. RCS by 12 × 10 N thrusters. Attitude control by four RWs, three magnetorquers, gyro package, magnetometer, three star trackers, four Sun acquisition sensors (requirement 1.5 arcmin error for 10^5 s observation).

Tethered Satellite System (TSS)

Background

The Tethered Satellite System (TSS), designed to continue Italy's atmospheric studies, was a joint project with NASA. Alenia built the satellite to trail out behind the Shuttle. The two tests cost ASI/NASA $404 million.

A joint US/Italian project, agreed to in 1984, the Tethered Satellite System comprised an ASI/Alenia Spazio aluminium spherical satellite capable of being reeled upwards or downwards from the Shuttle cargo bay on a 2.5 mm diameter Kevlar tether up to 100 km long. Martin Marietta developed the deployment system for the STS-46 first mission in 1992, involving a 20 km upward deployment, with an insulated copper wire incorporated in the tether, to study electrodynamic and other phenomena. Downward deployment with the help of N_2 thrusters on the satellite could 'troll' the Earth's atmosphere at heights of 130-150 km – normally inaccessible because of drag – for upper atmosphere research at 7.65 km/s.

TSS 1 attained only 256 m (insufficient for significant science returns) before a protruding bolt fouled the U.S. developed winding mechanism. The first TSS deployment should have allowed about 20 km of tether to be played out. The Italians made a second attempt with the replacement satellite TSS 1R in February 1996 on STS-75. This second try succeeded until the tether snapped at the 19.7 km mark. Up to 4 kW of power should have flowed along the core down to the Shuttle,

Final processing at Cape Canaveral for installing the Tethered Satellite aboard Shuttle (Alenia Spazio) ***1997***/0006595

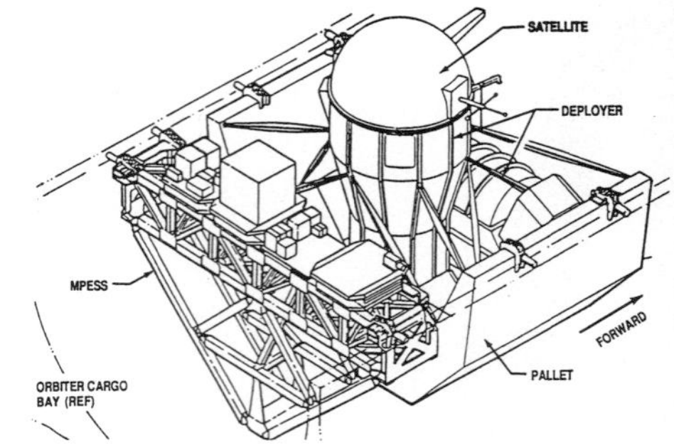

The tethered satellite is deployed from a modified Spacelab pallet

but the cable generated only 40 V at 15 mA during the STS-75.

NASA/ASI are jointly defining requirements for TSS follow-on missions, though none is funded. An atmospheric verification mission would possibly employ the TSS deployer and a NASA reusable satellite. An all-up TSS atmospheric science mission would lower Italy's satellite and its 44 kg science payload to around 130 km altitude on a 100 km-long non-conducting heat resistant tether. The configuration would be essentially as TSS 1/1R but a simple aerodynamic surface and thermal protection would add 22 kg. Battery power supply would be 2; data rate 54 kbit/s. Consideration was given to modifying TSS 1's structural model as a non-recoverable TSS 2. A TSS 3 might extend TSS 1's investigations but be based at the Space Station.

Specifications

Contractors: Alenia Spazio (satellite), Lockheed Martin (deployer system, contract value US$68 million)
Mass: satellite 518 kg
Orbit: up to 20.7 km above Shuttle for electrodynamics TSS 1/1R mission; up to 100 km below for atmospheric mission
Spacecraft: 1.6 m diameter spherical shell of conventional Al construction. Allowable payload mass 70 kg, volume 0.43 m^3. Power provided by 2 batteries accessible through four equatorial doors. Attitude control exercised by 12 N_2 thrusters: 4 × 1 N thrust along line direction to aid deployment and maintain tension, 4 × 2.5 N at 90° intervals along equator for in-plane and perpendicular-to-plane thrusting, and 4 × 2.5 N yaw jets in two clusters 180° apart on equator. Attitude information by Earth sensors and four Sun sensors. TT&C via S-band link: 16 kbit/s data rate, 2kbit/s command rate. TSS 1/1R tether 5-layer: Nomex core, copper conductor, insulation, Kevlar strength member and outer Nomex braid; outer diameter 2.5 mm, mass 8.2 kg/km, capable of carrying 1 A current, 10 kV, resistance 0.12 ohm/m. Satellite housed in deployer support structure mounted on enhanced Spacelab pallet, deployed by electric reel mechanism from extended 12 m boom.

ISRAEL

Gurwin

Background

The Technion Institute of Technology proposed building three or four small satellites for continuous early warning coverage. The Gurwin 1 TechSat civil satellite may have been its first attempt to provide technology demonstrations of future imaging systems. The satellite came about as a result of a collaboration between the Faculty of Aerospace Engineering at Technion Institute (under the direction of Professor Chaim Eshed), ISA, industry and Amateur satellite-IL to develop a simple, low-cost, low-power platform. The Flight Model was completed October 1994, but the first TechSat was lost in Start's launch failure. The TechSat 1B near-duplicate is being funded. It will include the Netherlands' TNO-TPD wide field star sensor. All requests for 1B information were declined.

The TechSat 2 is apparently a 250-300 kg second-generation TechSat proposed by Technion as a multipurpose platform, created by adding an AOCS unit and GPS navigation to the basic TechSat 1. The 100 kg AOCS module would carry Rafael 1 N liquid propellant and N_2 thrusters. TechSat 2A would carry a high-resolution Earth observation camera, gimballed on command and relaying imagery via an MMIC Ku-band phased array antenna.

Specifications

Launch: 28 March 1995 on Russian Start from Plesetsk; planned orbit was 670 km 75.4°

Mass: 52 kg
Configuration: box-shaped 460 mm height, 430 mm across (six Al plates, 2.5 mm thick). Earth-pointed face carries antennas, horizon sensor, and camera. 80C186EC-based CPU with two 512 kbyte independent RAM memories. Controlled from Technion station; command reception at 145.910/1,269.85 MHz
Power: solar cells on four faces (1,000 2 × 4 cm Si, 17 W average, 9 W required for housekeeping, Ni/Cd battery)
Attitude control: 3-axis Earth-pointing control (5° coarse, 0.1-0.5° fine) by 2.5 kg 3,600 rpm 4 Nms MW 3 × 1 Am^2 magnetorquers, 1 kg 1 W static horizon sensor (0.05° accuracy), 0.22 kg 0.8 W ±0.6 gauss 3-axis magnetometer (see GN&C Space Industry entries for detailed specifications)
Payload: Elisra Earth-observing 350 × 570 pixel CCD camera, compressed image data broadcast using millimetre-wave link experiment (also on UHF).

Technion UV CsTe spectroradiometer to monitor Earth's albedo at 0.24-0.30 μm (0.01 μm/2 mrad resolution); 15° FOV. The spectral profile is a function of the vertical ozone profile. Goals are investigating the long-term stratospheric impact of the Kuwaiti oil fires and the extent of the region's ozone depletion.

Technion x-ray detectors (up to 200 keV). Principal purpose is to prove that CdTe crystals can be used for x-ray imaging.

ITALY

San Marco series

Background

Italy became a force in international space science with the introduction of the San Marco, a small but effective University of Rome programme built around Italy's international platform in the Indian Ocean off San Marco, Kenya. University of Rome scientists began their investigations in 1964 and continued the programme through 1974 with a small revival in the mid-1980s. The primary purpose of the San Marco satellites were atmospheric density tests.

Specifications

San Marco 1
Launched: 15 December 1964 by Scout from Wallops Island
Mass: 24 kg
Orbit: 194 × 697 km, 38°
Decay date: 13 September 1965.

San Marco 2
Launched: 26 April 1967 by Scout from San Marco
Mass: 129 kg
Orbit: 185 × 211 km, 3°
Comment: First launch from the platform and first in a series to measure atmospheric density: two concentric spheres connected by flexible instrumented arms measured drag
Decay date: 14 October 1967.

San Marco 3
Launched: 24 April 1971 by Scout from San Marco
Mass: 164 kg
Orbit: 222 × 718 km, 3°
Decay date: 29 November 1971.

San Marco 4
Launched: 18 February 1974 by Scout from San Marco
Mass: 164 kg
Orbit: 231 × 910 km, 3°
Comment: With Explorer 51, sampled the auroral zone and SM4 over the equator. The satellite detected

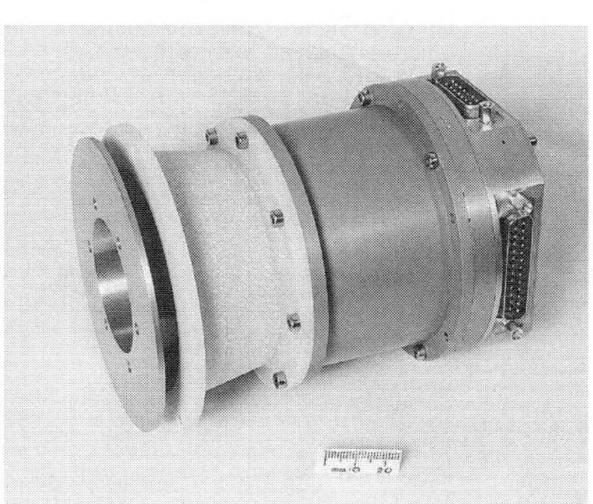

The UK's cloud camera will fly on Badr 2 (Rutherford Appleton Lab) ***1996***

variations in atmospheric density due to differences in solar heating. Transmitted 806 days
Decay date: 4 May 1976.

San Marco 5
Launched: 25 March 1988 by Scout from San Marco
Mass: 237 kg
Orbit: 262 × 619 km, 2.89°
Decay date: 6 December 1988
Comment: Drag on 1 m sphere measured from spring arms supporting external 3 kg 6 mm-thick shell. Covered 500 km to re-entry at 141 km. Also monitored charged particles, neutral winds, electric fields, UV solar fluxes and UV airglow.

Sirio series

Background

Italy began its first investigations of experimental space communication when it launched the Sirio 1 satellite on 25 August 1977 by Delta from Cape Canaveral. The satellite, weighing 398 kg and 220 kg on-station at 15° west GEO, carried two 9 W transponders operating at 12/18 GHz. As an all-Italian project, the ground control operations centred at Fucino. Italy prepared for an eventual communications system for the entire country by investigating SHF propagation in rain, snow and fog. When the experiments were complete, Italy turned the results over to ESA's to build the Sirio 2 follow-on, using Sirio 1's back-up bus. Italy provided 74 per cent of funding for Sirio 2 which was lost in Ariane L5 failure September 1982. Following the 1983 space co-operation agreement with China, Sirio 1 was moved to 65° east to assist China in development of 18 GHz GEO communications.

NETHERLANDS

ANS satellite

Background

The first Dutch satellite was the ANS astronomical platform launched in 1974, followed by significant participation in IRAS and a 10 per cent interest in the SAX satellite. The ANS was launched 30 August 1974 by Scout from Vandenberg. It weighed 129 kg and entered a 258 × 1,173 km, 98° orbit. A Dutch industrial consortium designed and built the Netherlands' first satellite. Scientists in the Netherlands and other investigators wanted a near-polar, Sun-synchronous circular orbit of 500 km for studies of the UV spectra of young, hot stars and soft/hard x-rays from cosmic sources.

In appearance, the ANS sits 1 12 m high and has a depth of 70 cm. The solar array spans 1.5 m (totalling 0.95 m^2, 2,000 cells). It has a controls system with 3-axis 1 arcmin pointing accuracy by RWs/ magnetorquers. A computer could hold data from a 12-hour observation programme. The science package had three instruments: 5-band UV photometer 150-350 nm (Groningen); soft x-ray detectors 2-4, 4-12, 27-35, 44-55 Å (Utrecht); 2-40 keV hard x-ray telescope (MIT, US).

ANS re-entered 14 June 1977.

PAKISTAN

BADR (Full Moon) series

Background

Pakistan has plans to develop an independent remote-sensing, science and communications space presence. The country currently uses Landsat images from a ground station it installed as part of a cooperative international agreement. To pursue its goal of space science, the Pakistanis built the BADR-1 in conjunction with the Chinese. BADR-1 passed over the country about five times a day to give ground controllers experience in handling spacecraft operations and preparing for an eventual store/forward communications capability

After a successful launch on 16 July 1990 by CZ-2E from Xichang, the satellite entered into a 205 × 983 km, 28.49° orbit. This demonstrator satellite weighed 52 kg, and had a 483 mm diameter 26-sided polyhedron. It carried a UHF/VHF store/forward payload. Chinese officials quoted the launch cost to Pakistan as US$300,000. The orbit provided four or five usable passes daily for its controllers, each lasting as much as 20 minutes, instead of the two pairs separated by at least 10 h available from a polar orbit. Suparco's Karachi campus provided TT&C, supported by the Sparc station at Lahore. The sites also operated two low-cost tracking and receiving stations. Unfortunately, contact was lost after 35 days, seriously affecting the communications experiments. BADR re-entered 8 December 1990 after 145 days.

The Pakistanis still plan for BADR-2 to be launched on a vehicle yet to be selected, but possibly on Russian Cosmos-3M, into 1,000 km 83° (originally planned for Ariane ASAP into 800 km). The new satellite will weigh 50 kg, and have a design life of one year. Gravity gradient stabilisation using 4 kg tip mass on a 6 m boom will maintain the transponder orientation. BADR-2 will carry a CCD camera for cloud monitoring, provided by the UK's RAL and University of Leicester. Other experiments are SAFE Store/Forward Experiment (Suparco), RadFET Compact Dosimeter (SIL/ESA) and Battery End of Charge Detector (SIL/ESA). S-band/UHF will provide 128/9.6 kbit/s data rates, respectively. RAL's Chilton will act as Lahore's back-up. The UK's Satellites International Ltd is providing the entire attitude control system consisting of one MFM-3L fluxgate magnetometer, two MTR-25 magnetorquers, two DSS-256 Digital Sun Sensors (each a pair) and attitude control electronics. It is also providing the S-band transceiver, 4 Ah Ni/Cd battery, OBDH and power conditioning system. The UK's EEV Ltd is providing the GaAs solar panels. CCD camera: 0.840-0.890 μm, 2.5 kg, 7 W, 17° FOV, 72 mm fl, f/6, EEV 576 × 770-pixel CCD for cloud monitoring with 310 m resolution (from 1,000 km). The ATSR Along Track Scanning Radiometers on ESA's ERS 1/2 measure sea surface T but have to reject pixels covering clouds. However, pixels contaminated by clouds below their 1 km resolution can still be accepted, biasing the readings. BADR's camera will thus quantify the sub-1 km clouds and may demonstrate the need for such monitors on future ATSRs. The camera will be carried free of charge for RAL in return for Suparco using it for land imaging.

PORTUGAL

PoSat series

Background

The first Portuguese satellite, PoSat 1, was launched in 1993, built by a Portuguese team within a technology transfer programme between the UK and Portugal. The primary objective is to stimulate a national space industry and generate a nucleus of engineers with first-hand expertise for possible future satellite projects. The PoSat consortium is led by INETI (National Institute of Engineering & Industrial Technology; see below) and includes EFACEC, Alcatel Portugal, OGMA Oficinas Gerais de Material Aeronautico, CPRM Companhia Portuguesa de Rádio Marconi, IST Instituto Superior Técnico, UBI Universidade de Beira Interior and CEDINTEC. Overall cost was about £5 million, funded by the Ministry of Industry through the PEDIP programme, plus each consortium member supported its own manpower. The programme included on the job training for 11 engineers at Surrey Satellite Technology Ltd. in the UK, who also provided the ground station in Portugal.

Specifications

PoSat 1
Launched: 26 September 1993 as an Ariane ASAP passenger
Orbit: 794 × 806 km, 98.68°
Mass: 47.5 kg. Based on SSTL's UoSAT 5 design
Configuration: Same as UoSAT 5
Power supply: four GaAs panels (each 34 W) totalling 1,344 cells providing 20 W (orbit average processed); 6 Ah Ni/Cd battery supporting eclipse operations
OBC: 80C186 processor with 512k memory and associated 16 Mbyte Ramdisk bulk data store. The payload transputers provide high performance data processing and attitude control
Attitude control: Earth-pointing maintained by combination of gravity gradient stabilisation and active magnetorquing; Sun/Earth horizon sensors and 3-axis fluxgate magnetometer with ±2 nT resolution over ±64 μT provide attitude information. Gravity gradient supplied by 3 kg tip mass on 6 m boom deployed from top face, magnetorquing by six 150-turn Al wire coils (one on each PoSAT facet)
TT&C: control from CPRM's site at Sintra, near Lisbon
Communications: data rate of 9.6 or 38.4 kbit/s, 3-channel VHF receiver (2 communications, 1

command), redundant UHF transmit (four bands) switchable to 2/10 W RF output, 4-element turnstile VHF antenna, one monopole UHF antenna. 148.585 + 148.285 MHz up; 429.950 + 429.450 MHz down (switchable to amateur service on 435.250 + 435.275 MHz down and 145.925 + 145.975 MHz up)
Payload: EIS Earth Imaging System. Two monochromatic CCD 578 × 576-pixel cameras, 2 km/200 m resolution. Image read from each by TDPE Transputer Data Processing Experiment and, after processing, passed to OBC for storage on 16 Mbyte Ramdisk.

DSPE Digital Signal Processing Experiment. Two Texas Instruments C25 + C30 TMS320 processors for relaying speech.

SAF Store/Forward Packet Communications. Stored on OBC's Ramdisk for downloading by other stations.

GPS Navigation Experiment. Based on Trimble TANSII receiver to provide position, velocity and onboard time reference. Allows PoSat to generate its own orbital element sets and ground stations with GPS receiver to experiment with real-time differential applications.

CRE Cosmic Ray Experiment particle radiation detector to monitor environment and effects on electronics.

SS Star Sensor. Patterned after EIS camera but with stellar optics, analysed by TDPE and data passed to OBC. Allows 1 arcsec attitude measurement.

RUSSIAN FEDERATION/ UKRAINE

Coronas I/F

Background

These are the first Russian/Ukraine observatories dedicated to helioseismology (solar oscillations) and identification of coronal/chromospheric acceleration mechanisms. They serve additionally as solar system 'weather' forecasters. Coronas stands for Comprehensive Orbital near-Earth Observations of the Active Sun.

The satellites are placed in Prognoz-type highly elliptical orbits at 500 kms. They monitor solar events such as flares to evaluate their effects on near-Earth space. This facility will be particularly useful for warning crews aboard permanent space stations of potentially hazardous solar activity. The Soviet Union first revealed the missions in 1987 and stated it planned to begin the launches in 1990 and through to 1992.

The Institute of Terrestrial Magnetism, Ionosphere & Radio Wave Propagation (Izmiran, =I), and by the Lebedev Physical Institute (=F) manage the project. German and Polish scientists also collaborate.

Coronas I marks the first flight of NPO Yuzhnoye's AUOS-SM (S=Sun, M=modified) Sun-pointing bus, derived from the AUOS design last employed by Apex. It can accommodate a science payload of 700 kg, providing daily-averaged 300 W of power, and point at the Sun with 10 arcmin accuracy. Electronics are housed in the bus pressurised at 0.77-1.35 atmosphere with nitrogen at 0-40°C.

Each Coronas carries solar-observing and near-Earth particle radiation instrumentation for detection of solar events and resulting terrestrial interaction. The solar instruments include x-ray telescope (10-300 Å), coronograph, heliometer (10-30 keV), Thomson polarimeter, x-ray and visible photometers, 0.03-30 MHz radio detector. The particle radiation monitoring includes 0.1-100 MeV gamma-rays, a monitor for 30 MeV or less neutrons, one for 1 MeV or less protons/nuclei and 0.05 MeV or greater electrons. Yuzhnoye is investigating power production techniques with four 64 × 53 × 3 cm panels carrying different solar concentrators and cell types, each panel contributing a maximum 50 W.

Specifications

Coronas satellite series
Launch: Coronas I 2 March 1994 by Tsyklon from Plesetsk; Coronas F planned 1997-1998
Orbit: Coronas I 488 × 527 km, 82.49°
Design life: 5 years
Mass: 2,160 kg (700 kg payload)

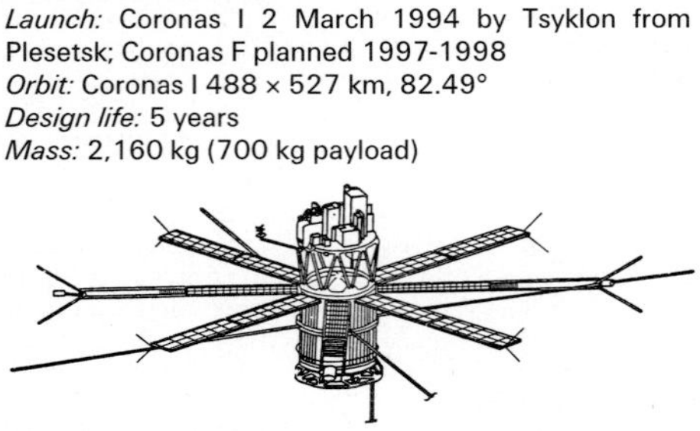

Coronas I began solar observations in 1994. The arrays are also testing sample solar panels

Bion missions employ a Vostok-derived spacecraft, carrying specimens in the recoverable sphere. The two monkey housings are visible through the hatch. The same basic vehicle is offered commercially as the Photon microgravity facility (qv) (Graham Turnill)

TT&C: provided by VKS centre at Golitsyno-2; science data returned to Izmiran's Troitsk station.

RUSSIAN FEDERATION

Bion/Biokosmos

Current status

Bion 12, the last in the long-running series of CIS/Soviet biological satellites has yet to be launched. It had been planned to fly two Rhesus monkeys on the mission, but this was thrown into doubt when the United States announced in April 1997 that they were pulling out of the monkey research following pressure from animal rights groups – pressure which increased following the loss of one of the monkeys launched on the Bion 11 mission. If monkeys are not flown on Bion 12, then other animals such as rodents and insects, which have flown in previous Bion/Kosmos flights, can be carried. However, Illyin has noted that experiments and equipment planned for the mission would have to be re-designed at the cost of a significant mission delay.

Background

The Bion programme is operated by the Institute of Biomedical Problems in Moscow and is headed by Professor Yevgeni Illyin. Illyin had undergone preliminary cosmonaut training in the mid-1960s as the science officer for a Voskhod mission conducting biomedical research. The Biokosmos programme utilises a modification of the Vostok spacecraft similar to Photon microgravity capsule.

Ten international biosatellites had been launched by

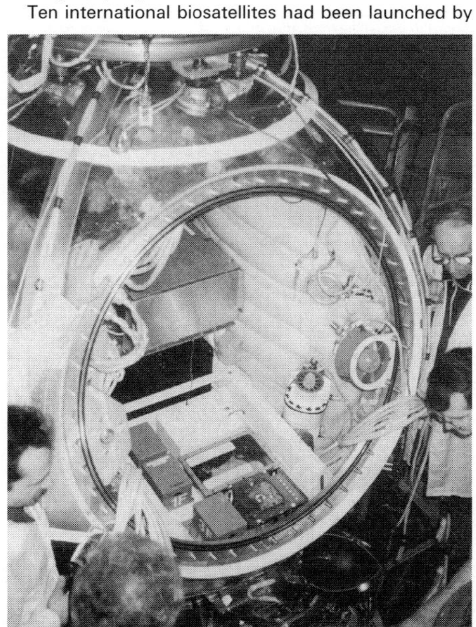

Bion 10 preparations. The silvery unit in the rear of the compartment is ESA's Biobox (ESA)

the CIS/USSR by end 1995 in the Biokosmos series, with the US participating in the last eight. US interest began after the collapse of the US's own Biosatellite programme in 1969. CIS/USSR and US scientists have used more than 50 biological specimens in this series, ranging from viruses to mammals, and including monkeys, rabbits, guinea pigs, rats, mice, tortoises, higher plants, fish, birds and insects. At least nine countries have participated, with ESA contributing for the first time to Kosmos 1887.

Bion capsules enter space aboard Soyuz-U from Plesetsk into about 220 × 400 km at 62.8° or 82.3°. The capsules weigh about 6 tonnes, with 700 kg internal cargo and 200 kg external. Zonds 5/6/8 and Soyuz 20 also carried biological specimens such as turtles, flies, plants and seeds. The two most recent Bion missions were Kosmos 2229, launched 29 December 1992 by Soyuz-U from Plesetsk into 226 × 397 km, 62.8° orbit, again carrying two Rhesus monkeys (Krosh/Ivasha), and Bion 11.

The Russians recovered Bion 10 on 10 January 1993 at 04.16 GMT in a forest some 100 km north of Karaganda, 800 km from the pre-mission site near Kustanai and two days early because of high internal temperatures. 7 of 15 salamanders died from the heat, one monkey went without food for two to three days and two of the four external pans failed to close for re-entry, damaging the experiments. NASA provided some flight hardware for the primate experiments in addition to participating in several primate experiments. NASA also conducted bone and muscle biopsies, neurovestibular pre/post-flight measurements, thermoregulation and immunology experiments.

ESA's 40 kg Biobox debuted with six experiments, investigating bone weakening in μg and 1 *g* centrifuge conditions. ESA hoped this and later flights would settle the question of bone sensitivity to gravity. The incubator's temperature is controlled to 4-37°C and the conditions recorded on a 1 Mbyte memory card. Biobox can also be used for other experiments such as protein crystallisation, plant metabolism and genetics.

ESA/IBMP shared five other experiments: three μg flights for fruit flies and unicellular algae, and two radiation exposure flights for plants and seeds and to test dosimetry. The algae experiment was planned for the last days so was truncated by the early return.

Bion 11 was launched from Plesetsk, 24 December 1996 using a Soyuz-U vehicle. The Russians used the Bion name openly for the first time on this flight – previous flights had always carried a Kosmos number. The spacecraft entered a 62.8 °, 217-379 km orbit. On board were two Macaque monkeys (Lapik and Multik), together with newts, snails, drosophila flies and other insects and bacteria. Both Russian and United States researchers carried out experiments. The United States had paid about US$15 million towards the flight.

The spacecraft landed 130 km north of Kustani 7 January 1997. The following day Multik died, resulting in overseas protests about the flight from animal rights sympathisers. Yevgeni Illyin said that the monkey's death had nothing to do with the two weeks just spent in Orbit: while coming around from a general anaesthetic Multik's heart simply stopped, while Lapik who had undergone the same medical procedure came through without any trouble.

Elektron

Background

The Soviet Union became interested in the van Allen belts, principally because its proposed series of communications satellites, the Molniyas, had to traverse the belts twice a day exposing them to a potentially lethal dose of ionised particles. They developed a civilian satellite research programme known as the Elektron to investigate the environment of the van Allen belts.

Elektron 1 and 2 went into space on the same booster, an SL-3 Vostok, launched on 30 January 1964 from Tyuratam. The two satellites weighed 350 and 445 kg respectively, the greater weight on the second being used for propellant to put the satellite in a more energetic orbit than its sibling. Elektron 1 had a 394 × 7,126 km orbit and Elektron 2 a 441 × 67,988 km orbit. Both were inclined 61° to the equator. This launch marked the first Soviet dual launch. The Soviets announced that the flight would gather information on radiation for protection of manned spacecraft. By the time, however, the needs for manned shielding for one or two brief passages through the van Allen belts was fairly well established and long term exposure to the belts had a higher investigative priority.

Elektron 3 and 4 on similar missions left Earth on 11 July 1964 by SL-3 Vostok from Tyuratam. Both had similar masses, orbits and missions as Elektron 1 and 2. Elektron 4 decayed 12 October 1983. During the height of the Cold War, many space buffs assigned an early warning role to this mission, but no proof from the Soviets has ever been revealed.

Granat ('Garnet')

Background

Russia's answer to the Hubble Space Telescope, the gamma and x-ray observatory Granat continues operations into its ninth year. It was the second flight of the Astron astrophysical platform. This was the last flight built around the Venera planetary spacecraft; the Spektrum bus, designed specifically for astronomical missions is superseding it. Its main mission was to conduct observations of 3-2,000 keV γ/x-ray sources, including imaging in 3-500 keV, source pinpointing, recording temporal/spectral characteristics of burst-type and transient sources, monitoring of bright, permanent x-ray sources, and investigation of the afterglow of γ bursts in the soft x-ray region

Granat observations began in January 1990 following in-orbit checkout, which included instrument testing beginning 11 December. Russian planners wanted a mission duration of about 18 months and delayed the launch 1½ years while they checked out payload development problems. In the first nine months of operations it completed more than 140 observing sessions, some lasting up to 46 h. By early 1992, about 400 sessions typically of 24 h had been logged. 1992 provided 163 sessions and 143 in 1993.

A truly international collaboration of space programmes contributed to building the Granat. CNES provided the primary 1 tonne Sigma (Système d'Imagerie Gamma Masque Aléatoire, or random mask imaging gamma-ray system) telescope at a cost of FFr150 million. NASA and the Russian Space Agency collaborated on observations in September 1991, collating data from the Granat and NASA's Compton Observatory. As its attitude control propellant approached exhaustion, Granat switched on 1 October 1994 from pointing mode to a 30°/min slow yaw, providing its instruments with an all-sky scan every six months. This provided 5,500 h of observations until 9 September 1995, mapping the Galaxy's diffuse gamma emission. For the rest of September 1995, Granat returned to its original pointing mode for 110 h of observations in the hope (unrealised) of confirming pulsars in the galactic centre. It began slow yawing again by month's end, before further galactic centre pointing in May 1996.

Granat marks several major findings among its accomplishments, among them, correlating the diffuse gamma emission with the interstellar gas column density. CNES reported at end-1992 that Granat had identified eight black hole candidates. Granat's engineering model was sold at Sotheby's December 1993 Russian space auction in New York.

Specifications

Granat satellite

Contractors/Institutions: NPO Lavotchin/Babakin Engineering Research Centre (spacecraft), IKI/CNES Toulouse (science lead), CESR of Toulouse/CEA Service d'Astrophysique near Paris (Sigma telescope), Sodern (SED 12 star trackers), Denmark (Watch X-ray detector), Bulgarian Space Research Institute (Podsolnukh), Sagem (Sigma's magnetic bubble memory), Sextant Avionique (Sigma's 8 kg onboard management unit, 8 Mbit 5.5 kg buffer memory, 90 W 8.5 kg power converter)
Launch: 1 December 1989 by Proton from Tyuratam
Mass: 4,402 kg, payload 2,146 kg (including 1 t Sigma)
Orbit: Prognoz-type 4-day, 1,760 × 202,480 km, 51.86°, permitting 56 h observation sessions above radiation belts beginning 60,000 km

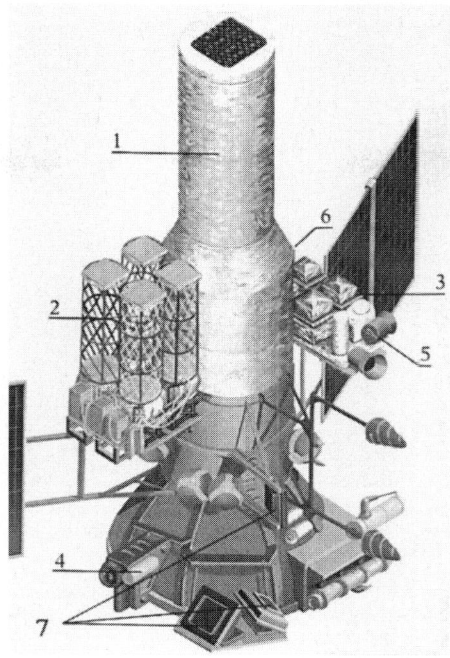

Granat: last of the Venera series of spacecraft. 1 Sigma (France), main gamma telescope; 2 ART-P (USSR), imaging x-ray telescope; 3 ART-S (USSR), x-ray detector; 4 Phebus (France), gamma/x-ray burst detector; 5 Konus (USSR), gamma/x-ray burst detector; 6 Podsolnukh (USSR), detection of burst sources; 7 Watch (Denmark), transient and burst detection (CNES)

Spacecraft: based on Venera/Astron bus, with 2.35 m diameter pressurised base torus housing electronics. 6.5 m high. 3-axis controlled, capable of 20 arcmin pointing accuracy over 24 h. Soviet star sensors complemented by CNES' SED 12 trackers for Sigma pointing. Real-time data transmissions at 3,072 bit/s or 65,546 bit/s playback from 150 Mbit memory. Power (400 W allocated to science instruments) derived from twin double-segment solar wings spanning 8.5 m
Payload: Sigma Mass: 1,050 kg, 3.5 m high, 1.2 m base diameter topped by 49 × 53-element 1.5 cm-thick tungsten coded mask permitting 16 arcmin resolution imaging with 57 cm diameter 1.25cm thick Na (Tl) crystal over 0.5-2 MeV. The detector lies 2.5 m below the mask with effective area 800 cm². Full FOV 10 × 10°; 16 arcmin resolution available over 4.2 × 4.2°. Pointing information provided by two bore-sighted Sodern CCD SED 12 star trackers; one provides visible imaging of γ bursts identified by the main instrument. Sigma is the most sensitive γ burst detector aboard Granat and can additionally use its 19,200 cm² CsI (Tl) anti-coincidence detectors for 0.10-7.5 MeV monitoring. Sigma's 28 kg 128 Mbit Sagem magnetic bubble mass memory unit also used for data processing and as a telemetry buffer for data downlinks.

ART-P: four identical modules with 41 × 43-element coded masks for 5 arcmin resolution imaging over 3-100 keV. Effective area 4 × 600 cm², time resolution 1 ms, 4 × 64 kbit storage.

ART-S: two pairs of 3-150 keV non-imaging detectors, each pair alternately observing source/background via a rocking collimator. FOV 2 × 2° FWHM, time resolution 100 μs, effective area 4 × 600 cm², 4 × 64 kbit memory.

Konus: detection of 0.20-4 MeV bursts with seven NaI (Tl) detectors 20 cm dia/5 cm thick. Location accuracy 1-5°, 40-channel spectrum, time resolution 1/64-1/256 s.

Phebus: detection of 0.10-100 MeV bursts with six bismuth germanate crystals. FOV 4π, 1/128 s time resolution, energies measured over 102 bands over 0.10-10.0 MeV and 14 bands over 10.0-100 MeV. Instrument provided by France.

Watch: detection of 5-120 keV x-ray transient sources and γ bursts with four Danish 45 cm² all-sky 4π detectors. Accuracy of burst localisation about 0.1°. Each detector is a mosaic of NaI (Tl) + CsI (Tl) scintillators 2 mm thick. The rotation modulation collimator permits real-time localisation of rapidly varying x-ray sources.

Podsolnukh: two 2-25 keV proportional counters mounted on a rotational platform steered from Konus data. Instrument provided by Bulgaria.

Pion

Background

In conjunction with a Resurs-F film return mission, the Soviet Union released pairs of 45-47 kg, 33 cm diameter Pion spheres for upper atmosphere studies. The photo mapping satellites released the spheres, launched 25 May and 18 July 1989, a few days into each mission. Built by the student design bureau at the Korolev Aviation Institute in Samara, they carried only simple radio transmitters to aid orbit determination at a time of enhanced solar activity. Pion 1 and 2 began their mission 9 June 1989, and re-entered 23 and 24 July. Pion 1 has been described as a 45 kg 'glass sphere' with an internal corner reflector. Pion 2 was a 45 kg sphere with a 'magnesium aluminium alloy surface'. The second two separated from the Resurs-F carrier vehicle on 7 August and decayed 18 and 19 September 1989. Pion 3 matched Pion 2, but with a mass of 47 kg, and Pion 4 matched Pion 1, but with a mass 47 kg.

The Soviets felt the experiment was successful enough to release another pair, Pion-Hermes from a Resurs-F on 1 and 2 September 1992. The Resurs was launched on 19 August. KB Photon (responsible for Resurs-F) built the new satellites under contract to the Russian-US Hermes company, to provide upper atmosphere data. The Russian Defence Ministry invited US Space Command to provide joint tracking, but the US military snubbed the request. Earlier reports described them as 78 kg solid metal spheres. PH-1 decayed 25 September and PH-2 on 24 September 1992.

Prognoz ('Forecast')

Background

The Prognoz series specialises in solar radiation and its interaction with Earth's magnetosphere. Spheroidal, with four cruciform solar panels and weighing 900-1,000 kg, they enter highly eccentric orbits ranging from 500 × 200,000 km to 400 × 720,000 km at 65°. SL-6 Molniya vehicles from Tyuratam launch them.

Prognoz orbits provide comparative measurements from near-Earth space and the upper atmosphere to out beyond the magnetosphere. France, Sweden, Czechoslovakia, Hungary and Poland have provided instruments and conducted experiments in parallel observations undertaken from ground stations.

The Soviets placed the satellites at opposite orbital extremes during their lifetimes. Prognoz 1, 14 April 1972, was at its apogee while Prognoz 2, 29 June 1972, was at perigee. Several launches came in rapid succession. Prognoz 3 15 February 1973, Prognoz 4 22 December 1975, Prognoz 5 25 November 1976, Prognoz 6 22 September 1977, Prognoz 7 30 October 1978 and Prognoz 8, 25 December 1980. All tested instruments to be used by Vega to study VLF plasma waves. Prognoz 9, introducing a second generation of satellite, lifted off 1 July 1983, weighing 930 kg, and adopting a 380 × 720,000 km orbit (taking it beyond Moon). It carried equipment from Czechoslovakia and France.

Prognoz 10/Intercosmos 23, launched 26 April 1985, weighed an identical 930 kg and was placed in a 421 × 200,320 km orbit. It was developed under Project Intershock, with Czechoslovakia, to study the structure of shock waves generated from solar wind plasma striking the magnetosphere. Czech scientists developed the computer required to record hundreds of measurements in the few seconds, about half way between Earth/Moon, during which the satellite crossed the front of the near-Earth shock wave. Soviet scientists developed the main instrumentation for measuring plasma waves and particles.

The Soviets began the Interball programme with a first launch of a 1,200 kg Prognoz-M2 satellite August 1995. It entered a highly elliptical orbit. The second followed in 1996. A third Prognoz-M2 will be placed in a halo orbit around Earth's L2 libration point with the Relikt 2 payload for mapping the Universe's microwave background; its plasma wave and particle detectors will contribute to Interball.

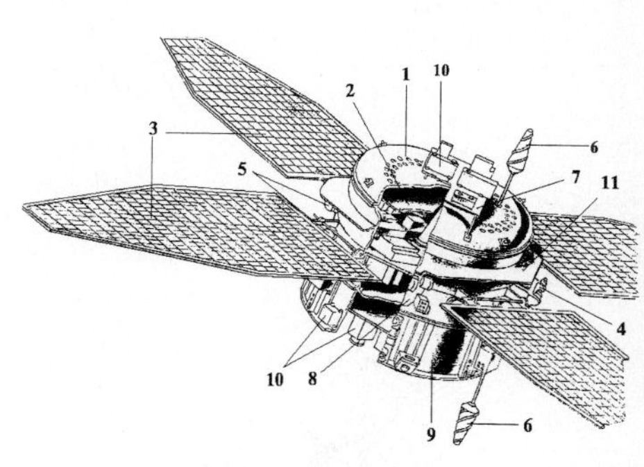

Prognoz-M2 principal features. 1 pressurised bus, 2 platforms, 3 solar panels, 4 Sun sensors, 5 ACS propellant tanks, 6 telemetry antennas, 7 upper experiment platform, 8 side frames, 9 cylindrical frame (Magion released from here), 10 science payload, 11 thermal shield ***1996***

Relikt/Libris

Background

Relikt 1 was launched as part of Prognoz 9's payload in 1983 to search for variations in the Universe's microwave background at 8 mm/37 GHz. Two horn antennas, one anti-solar and its perpendicular twin scanned the entire celestial sphere every six months as Prognoz rotated at 1-2 rpm. Throughout their lifetimes, they had detected no unevenness at the 0.5mK/5.8° level at mission's end in February 1984. The Soviets subsequently revealed, however, that thermal radiation from the Earth and the Moon had disturbed the instruments.

Relikt 2 (apparently also called Libris) is thus planned for a 180-day halo orbit around the L2 libration point beyond the Moon to avoid this interference. Launch is currently scheduled for 1999. Four radiometers with angular resolution 7, 3 and 1.5° will scan at 22, 35, 59 and 72 GHz. Relikt will also carry four plasma instruments to monitor geotail processes. As the Prognoz spacecraft has limited δV propulsion capacity, the halo orbit lies outside the core geotail, compromising the plasma detectors.

Investigators are interested primarily in large-scale (more than 5°) inhomogeneous regions in the Big Bang and medium-scale (0.5-5°) variations. Relikt will be the third mission of the Prognoz-M2 platform.

Libris/Relikt 2 will orbit around the L2 libration point to scan the Universe's microwave background

Spektrum-R/Radioastron

Background

The Astro Space Centre of the Lebedev Physics Institute is the lead centre; Professor Nikolai Kardashev is project manager. High resolution Very Long Baseline Interferometry (VLBI) radio mapping of active galaxies, black holes, neutron stars and other exotic objects will be undertaken in conjunction with 70 m dishes at Yevpatoria (the control centre), Usserisk and a new antenna in Uzbekistan on the Suffa plateau. The Australia Telescope National Facility will play a critical role in the southern hemisphere observations of both Radioastron and Japan's VSOP. Tracking accuracy required is 5-15 cm range and 0.3 mm/s range-rate.

The US announced in March 1990 that it would participate in data processing, provide four data recorders on extended loan, and support tracking operations with its Deep Space Network (doubling coverage to 90 per cent). Total cost to the US was estimated at $20 million over five to six years. Finland is providing the 22 GHz ground/space receivers

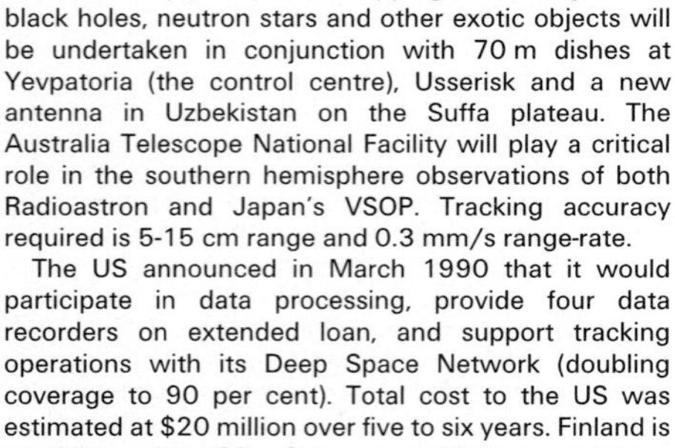

Spektrum-RG is the first of a new generation of astronomical observatories. The assembly at top right is the Sodart telescope, folded out 180° from its launch position (Danish Space Research Institute)

Radioastron's 10 m dish will operate in conjunction with ground telescopes and Japan's VSOP satellite for VLBI studies

(designed/manufactured by Juhana Ylinen Electronics Co). BAe Australia built the 1.66 GHz L-band receivers and amplifiers. India made the 0.327 GHz unit and Europe the 4.83 GHz.

IKI issued a report in September 1987, outlining a 20 year plan of space-based radio astronomy missions. What had been believed to be a single Radioastron launch of 1990 was revealed to be a series of up to six craft, the first three of which would be based on Spektrum. Radioastron-cm comprised two satellites for cm-wavelength observations beginning in 1994/5, followed by at least one Radioastron-mm after 1996. The ultimate three-craft network of 30 m-dish Radioastron-KK beyond 2001 in GEO, 27-day and L2 halo orbits appeared to require a follow-on generation to Spektrum - payload mass was described as 20 tonnes.

Spektrum-UV is planned for launch around 2002 and it is intended as an ultra-violet observatory, which the Russians see as complementing the United States' Hubble Space Telescope. The observatory is to be built around a primary mirror (designated T-170) with a diameter of 1.7 m.

Specifications

Spektrum RG

Planned launch: undetermined
Contractors/Institutes: Lavotchkin/Babakin Engineering Research Centre (bus), IKI (payload; PI. Driving force behind project is Academician R Sunyaev, Head of Dept of High Energy Astrophysics).
Mass: 6.0 t at launch, science payload 2,750 kg
Orbit: planned 500 × 200,000 km, 51.6-65°, 96 h, launch by Proton-K from Tyuratam with apogee over northern hemisphere
Spacecraft: sized for accommodation within Proton's 3 m diameter × 5 m payload envelope. Central pressurised cylinder about 2 m diameter houses most electronic systems. 3-axis control provided by Sun/star/ Earth sensors, RWs and microthrusters yielding 2-2.5 arcmin pointing accuracy (30-40 arcsec stability) for up to 30 h and 0.0001°/s angular stability. Twin Sun-following 3-panel solar wings provide 500 W to the science payload at 27±1 VDC. Data transmitted at 16-65 kbit/s or 18-128 Mbit/s through articulated 90 cm parabolic dish. Daily total 2-2.5 Gbyte for 4-5 h as each instrument in turn dumps its data load. Thermal control provided by radiator below cylindrical equipment bay. Design life 2 years
Payload: SODART two multimirror (145 shells, 60 cm outer diameter) grazing incidence telescopes packages, one with Bragg crystal, each with focal plane detectors selected by a sliding mechanism. Telescope focal length 8.0 m, 1° FOV. Instrumentation covers 0.1-20 keV, 2 arcmin resolution, 30 arcmin FOV. For launch, SODART's upper section, housing the mirror modules, star/optical monitors and Bragg crystal, folded against the lower unit. Deployed by electric motor through 180°. SODART total mass 1,570 kg. Telescope A, with Bragg crystal: two soft x-ray XSPECT position-sensitive proportional counters (Denmark/Finland) for 0.5-3 keV and 2-25 keV, 2 arcmin resolution. Sealed position-sensitive gas proportional counter (CIS), 1.5-15 keV, 2 arcmin resolution. Two SXRP stellar x-ray polarimeters (USA, Italy) 0.26-20 keV, with 0.3 arcmin resolution are included. Telescope B carries two XSPECT counters as Telescope A, a solid state Si(Li) spectrometer (Finland/ CIS) radiatively cooled to 100K, 00.3-20 keV, 3 arcmin resolution. Part of SODART is the 33.5 kg TAUVEX (Israel), providing simultaneous UV photometry via six filters covering 1,400-3,000 Å. It also provides offline aspect solutions for the x-ray experiment and is part of Spektrum's acquisition/pointing system to improve pointing to tens of arcsec.

Spektrum-ultra-violet may carry a 1.7 m telescope

JET-X 551 kg grazing incidence multimirror dual 3.5 m fl telescope with UK, Italian and Russian participation, 0.1-10 keV, 10-30 arcsec resolution, 30 × 40 arcmin FOV, optical star tracker provides pointing, optical monitor images faint sources to magnitude +23 for correlation with x-ray imagery.

MART coded mask x-ray telescope (Italy, CIS), 4-100 keV, 6 arcmin resolution, 8 × 8° FOV.

Euvita six EUVE telescopes (CIS, Switzerland) 0.031-0.12 keV, 10 arcsec resolution, 2 × 2° FOV, 10 cm^2 effective area.

Spin two γ-burst and scintillation spectrometers (CIS) 0.010-10 MeV, 30 arcmin resolution, 4π steradian FOV.

MOXE four coded mask x-ray burst monitors (USA) 3-12 keV, 15 X 600 arcmin resolution, 4π steradian FOV.

Spektrum-R Radioastron satellite

Contractors/institutes: Lavotchkin/Babakin Engineering Research Centre (spacecraft), Astro Space Centre (payload)
Mass: 5.22 t, payload 2.5 t
Orbit: 2,000 × 80,000 km, 51.5°, 24 h
Spacecraft: as Spektrum platform description but payload reduced at 1.5 t. Pointing accuracy required 60 arcsec
Payload: 10 m parabolic radio telescope (27 panels around 3 m central disc, surface accuracy rms. 0.5 mm) operating at 0.327, 1.66, 4.83, 22.2 GHz. Resolution 2.1, 0.4, 0.14, 0.031 marcsec, respectively.

Spektrum-UV satellite

Contractors/Institutes: Lavotchkin/Babakin Engineering Research Centre (spacecraft), Crimean Astrophysical Observatory (science payload)
Mass: 5.87 t, payload 2.5 t
Orbit: 51.5°, 500-300,000 km, period - 7 days.

Spektrum series

Background

Soviet scientists introduced a new class of astronomical observatory platform design to supersede the Venera-based Astron/Granat type. Until recently, sources indicated that the Spektrum would appear in late 1995, but RKA financing problems have delayed the series and the first launch date has still not arrived. The Spektrum class bus provides 128 Mbit/s data rates, 30-40 arcsec pointing stability and 500 W power for a 2.7 tonne payload. While these features seemed impressive, in fact, it became apparent in 1993 that Lavotchkin's bus is already in service for the Prognoz GEO early warning satellites, the first of which flew as Kosmos 1940 in April 1988.

The first design in the new series, Spektrum-RG, includes extensive international x-ray instrumentation. The second derivative, Spektrum-R accommodates the Radioastron VLBI payload due to fly in 2000, followed by the Spektrum-UV observatory in 2002. The cryogenically cooled Spektrum-IR for sub-millimetre observations appears to have been cancelled. Despite the turmoil, sufficient funding is at least allowing the projects to survive.

As a follow-up to 1989's Granat x-ray observatory, the Spektrum-RG will cost Russia an estimated Rb3 billion (1992 Roubles). Russia spent Rb500 million through 1992. When the Spektrum series was first described in 1987 the cost per mission was estimated at Rb480 million. Denmark, Italy, UK and the US are contributing major x-ray and UV telescopes and instruments. The UK proposed to fly the Euvita UV instrument in October 1986, but UK participation was subsequently cancelled and a consortium led by Leicester University and RAL, and including Birmingham University and the Mullard Space Science Lab, together with Italian, Russian and

German groups, was formed to develop Jet-X. The planned Dutch/UK TTM coded mask telescope was subsequently replaced by the Soviet's own Art-SP (now also cancelled). US MOXE is slated to replace the UK's Fourpi (4π) all-sky burst monitor. Some instruments were cancelled in 1991 when the scan platform was dropped because of mass problems (which continued).

Spektrum is planned to fly in a highly eccentric orbit reaching above Earth's radiation belts which permits protracted observations of x-ray sources and monitoring for γ-burst events. Up to 30,000 x-ray sources are expected to be mapped, including 3,000 active galactic nuclei.

Tenuous references to a subsequent complementary Spektrum-gamma mission in a 400 km orbit were made but no firm details have emerged and the Spektrum-X payload was extended to include a γ-ray capability.

Znamya (Banner) Reflector satellite

Background

In 1984, the Moscow Institute of Aviation proposed an experimental 200 kg satellite reflector with a 110 m^2 working area capable of reflecting sunlight to illuminate 9 km^2 regions in different time zones at night. Operational reflectors, the Institute thought, would provide seven times as much light as a full moon over areas of 10 km diameter, and should be able to prolong daylight for several hours in cities like Moscow. The first designs used a metallised polymer film. An early systems study of the scheme said the cost would be repaid in four to five years by electricity savings. The reflected light could be switched from one group of towns to another at short notice, providing a useful facility when night work was needed. When they launched Energia, in May 1987, Soviet officials linked it to this idea and future projects were designed to provide both illumination and power to the Earth from satellites. Radio Moscow reported in April 1989 that work on the Institute's experimental satellite had been halted, partly because of concerns over ecological effects of illuminating large areas of the Earth and its atmosphere.

The first flight test of a solar reflector was achieved 4 February 1993 when Progress-M 15 undocked from Mir and deployed a rotating 20 m 5 μm thick aluminised sail, directing sunlight to the ground. Rotation maintained the shape since there was no framework. The Znamya (Banner) Reflector remained visible after it was jettisoned from M15's nose after four orbital revs. The reflector weighed 40 kg, including the 4.2 kg sail. RKK Energia proposed a 200 m full-scale sail for late 1993 or early 1994 to travel beyond the Moon as part of the international race to celebrate Christopher Columbus' 1492 voyage. RKK Energia also proposed using it to illuminate the Earth's night side, followed by a plan to launch four or five reflectors by 1995. Funding has not been forthcoming.

SOUTH AFRICA

Sunsat

Current status

Sunsat, a micro-satellite built by post-graduate engineering students at the University of Stellenbosch.

Background

Sunsat was announced in June 1991, to promote development of industrial and educational space engineering. It is part of a programme to establish an internationally recognised satellite engineering activity at Stellenbosch, started with the Grinaker company's sponsorship of a Chair of Satellite Systems. The control centre will be sited at the university over the five-year life. Stellenbosch already has an Oscar ground station and a 4 m antenna. Students are supported by major electronics companies: Altech and Siemens of Johannesburg, Grinaker Electronics of Pretoria, Plessey SA of Cape Town, Analysis Management & Services of Stellenbosch, and First National Bank of Johannesburg. Cost is estimated at US$800,000. NASA provides launch free in exchange for the GPS receiver and laser reflector payloads. Sunsat was originally planned for Ariane's Helios 1A launch but the US$400,000 cost was prohibitive. Instead, it replaces the balance mass for Denmark's Ørsted on Delta's Argos launch. Breadboard prototypes were completed January 1992 to March 1993; the engineering was completed April 1994.

Specifications

Sunsat satellite

Orbit: planned 450 × 840 km, 98.7°
Mass: 61 kg at release
Configuration: 450 × 450 mm 600 mm-high box-shaped body with solar panels on four sides, 4 m boom deployed from top face for gravity gradient stabilisation
Power supply: 20 W orbital average
OBC & Data Handling: 80C188EC + 80C386EX redundantly handle flight management tasks and digital communications, and access a 64 Mbyte RAM disk
Attitude control: 2 m boom with 5 kg tip mass provides gravity gradient stabilisation. Magnetorquers damp libration and provide slow yaw spin for thermal control. Redundant attitude determination using Kalman filtered 3-axis magnetometer data, linear CCD Sun/horizon sensors (0.5 mrad) and CCD star camera. T800 transputer used as a processor; back-up by flight computer. When imaging, RW stops satellite spin and re-orients Sunsat
Communications: 10 W transmit allows 1,200/9,600 baud. 5-10 W 2.3 GHz wideband downlink for imager. Adding 1.2 GHz receiver creates a wideband transponder capable of 1 Mbyte/s to 2 m dishes
Payload: Stereo 3-colour Earth imager providing 15 m resolution. Linear 3,500-pixel CCD plus 10 cm aperture optics (by CSIR in Pretoria). Real-time downlink to Stellenbosch. Swath width 50 km. Images up to 50 × 50 km stored in RAM for downloading by amateur operators. For stereo imaging, after 50-500 km strip recorded, camera attitude reversed for second view.

Voice/digital communications package at amateur frequencies.

SA Amateur satellite project. 2 m up/down + 2/0.70 m up/down both used. 'Parrot' mode for novice users uplinks speech on 2 m and re-transmits on same frequency. 2 m also distributes SA Radio League's weekly bulletin. SA Amateur satellite also provided linear 2/10 m Mode A transponder.

Precision receiver and laser retro-reflectors. Intended to reveal fine orbital perturbations for gravity field mapping in conjunction with Ørsted.

SPAIN

Minisat series

Background

INTA manages a programme under the Interministerial Commission for Science & Technology (CICYT) to produce a small, low cost, rapid-delivery platform for science, communications and Earth observation missions. When the first bus design was complete, INTA signed a US$10.2 million launch services contract with Orbital Sciences Corp on 26 October 1994 for the L-1011/Pegasus XL to depart from Torrejon Air Base in September 1996. It actually flew on 21 April 1997 from Gando, making it the first orbital mission to depart from western Europe.

CASA is responsible for Minisat's platform and subsystems, Sener for systems/mission analysis, and Inisel for the Earth segment. Buying the unit as subsystem components (a technical assistance agreement was signed with Ball Aerospace 9 October 1992) held down development costs and by using one of Spain's existing three sites for the ground station.

Minisat 01, the first in the series, was launched 21 April 1997 by a Pegasus XL from Canary Islands. Although early plans suggested Minisat 01 would enter a 28.5°, 600 km orbit, the one actually used was 151°, 560-581 km. It is possible that early reports of the planned inclination mis-interpreted the true value

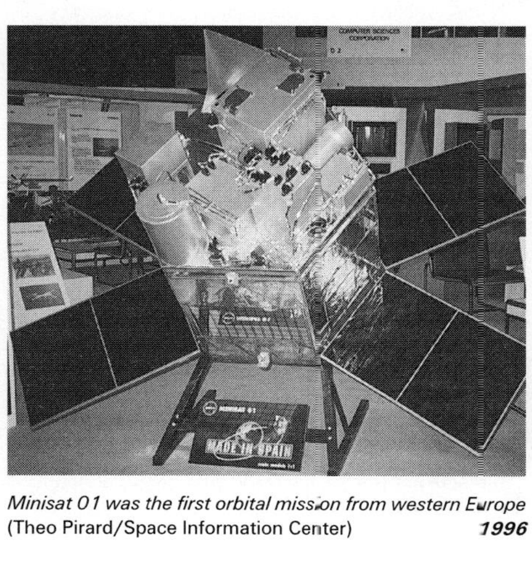

Minisat 01 *was the first orbital mission from western Europe* (Theo Pirard/Space Information Center) ***1996***

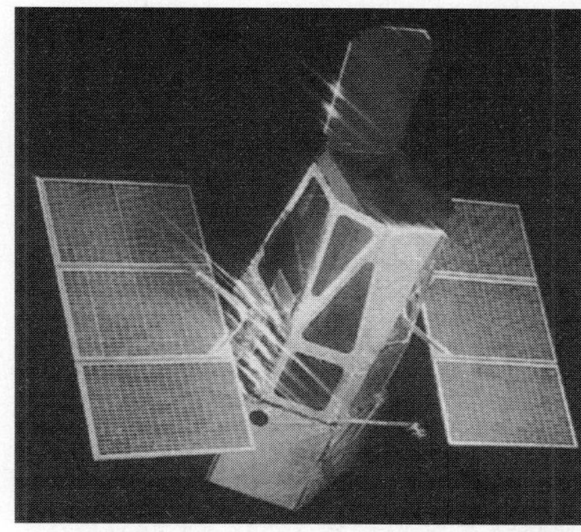

Minisat 01 is planned for 1998 to return 1 m-resolution imagery (INTA)

(180 – 28.5° = 151.5°). It carried three science payloads, the EURD Extreme Ultraviolet Radiation Detector diffuse background study (INTA/University of California at Berkeley), the CPLM to study behaviour of a liquid bridge in μg (also on UPM-Sat 1; Polytechnic University of Madrid) and the LEGRI 10-100 keV Low Energy Gamma Ray Imager (INTA, University of Valencia, Spain's CIEMAT Lab, UK's Rutherford Appleton Lab). Mass 201 kg (platform 100 kg; payload 100 kg); orbital life one to two years first mission; power 160 W EOL (100 W required, 40 W by payload); 32 Mbyte storage; 1 Mbit/s down by 5 W S-band; 3-axis stabilisation (±3° accuracy).

The 550 kg Minisat 01 will use the platform to its maximum capability for a 1998 remote sensing mission (1 m + 20 m resolution). 330 kg payload, 480 W, 150 Mbit storage, 3-axis control (0.1° pointing error), N_2/N_2H_4 100 m/s propulsion.

UPM-Sat 1 satellite

Background

This science and educational micro-satellite was designed and integrated by the CIDE/UPM Centro de Investigación y Desarrollo Espacial de la Universidad Politécnica de Madrid for launch as an Ariane ASAP passenger. Another satellite project is proposed, with several universities and research institutes from Europe and Latin America.

An Ariane 4 launched the satellite on 7 July 1995 from Kourou, French Guiana into a 665 × 676 km, 98.07° orbit. It weighed 47 kg, including separation system, packed into a 530 × 450 × 450 mm box-shaped body with solar panels on four sides. The main structural elements are four aluminium 7075 alloy shelves joined by four L-section beams. Four closing panels, supporting the solar panels, are screwed to the beams. Equipment is divided into three modular layers. For attitude control, the satellite uses active magnetorquing from 3-axis fluxgate magnetometer for attitude data, to provide spin for thermal control and to hold spin axis parallel to geomagnetic dipole axis.

UPM-Sat 1 is the first satellite built in Spain for more than 20 years (CIDE/UPM) ***1995***

The communications and datalinks provide 9.6 kbit/s downlink on single 400 MHz 10 W UHF dipole antenna. Data goes to a ground station at CIDE (Madrid). To supply the system with power, one GaAs experimental panel and three Si panels provide 20 W orbit average processed from 696 cells. Two sets of 14 Ni/Cd 6 Ah cells support eclipse operations.

In the payload bay, experimenters gather data on fluid behaviour in space with the LBE Liquid Bridge Experiment, at liquid mass held by surface tension forces between two solid supports. The ALE Al interconnector solar panel Experiment (with DASA/ESTEC) allows research in the efficient connection of solar cells with a panel of Si cells.

UNITED KINGDOM

United Kingdom science satellite programme

Background

By early 1999, the UK had flown 21 indigenously-built national satellites. This total comprised one launched by Black Arrow, 13 by the USA and seven by Ariane. The total includes the five UoSATs of the University of Surrey. These are in addition to craft produced for international and commercial programmes, such as Giotto and Inmarsat. Prospero remains the only UK satellite launched on a UK vehicle. Like the original USA/USSR programmes, the national space effort depended initially on the conversion of a ballistic missile into a launcher. This was the Hawker Siddeley Blue Streak, finally cancelled as a missile in 1960 though continued until 1973 as first stage for ELDO's Europa. A more modest attempt at a national launch vehicle was announced in September 1964, developing a small three-stage vehicle from the successful Westland Black Knight research rocket. Construction of three Black Arrows was ordered in March 1967. Success in October 1971 with only the second complete vehicle was contrasted by the programme's previous cancellation in July 1971. With Prospero, the UK became the sixth nation to achieve orbital capability – after the USSR, USA, France, Japan and China.

The Prospero technology demonstration satellite was placed in 557 × 1,598 km, 82° orbit. Its sole science payload was Birmingham University's micrometeoroid counter. Prospero made 812 passes, which were monitored, out of the 4,960 orbits in first year. The science portion ended when the tape player failed, after replaying 730 times, on 24 May 1973. RAE Lasham contacted the satellite annually until 1989, when the site's VHF system closed down. The Black Arrow programme was cancelled July 1971 before Prospero's success. NASA Scout launched X-4 Miranda. Its intended launcher, R-4, is displayed at The Science Museum, London.

The UK also provided a 74 kg, 1 m diameter sub-satellite, known as the UKS, to the AMPTE project in a joint effort with the US and West Germany to study how energy carried by the solar wind is transferred to Earth's radiation belts and magnetosphere out to 100,000 km. The three-satellite mission began with a Delta launch in August 1984. The Rutherford Appleton Lab collected data for 4 h on each 44 h orbit for five months until UKS failed unexpectedly 16 January 1985.

The UK science ambitions suffered a serious setback in June 1996, with the uninsured loss of the Cluster solar monitoring missions on Ariane 501. Nonetheless, the country presses ahead by participating in current international missions such as ESA's Ulysses, ISO and SOHO, the ESA-NASA Hubble Space Telescope and the US/German/UK ROSAT and future missions with a UK contribution such as Russia's Spektrum-RG, and ESA's Rosetta cometary probe. The loss of Ariane 501, carrying Cluster, was immediately subjected to an ESA investigation, from which it was clear that no UK components were at fault.

Specifications

Ariel 1
Launched: 26 April 1962 by Delta from Cape Canaveral
Mass: 60 kg
Orbit: 389 × 1,214 km, 54°
Comment: Transmitted ionosphere and x-ray data until November 1964. First international satellite, evolving from 1959 NASA offer on payloads of mutual scientific interest
Re-entered: 24 May 1976.

Ariel 2
Launched: 27 March 1964 by Scout from Wallops Island
Mass: 60 kg
Orbit: 285 × 1,362 km, 52°
Comment: Similar mission to Ariel 1. Returned data until November 1964
Re-entered: 18 December 1967.

Ariel 3
Launched: 5 May 1967 by Scout from Vandenberg
Mass: 90 kg
Orbit: 498 × 606 km, 80°
Comment: First all-UK satellite. Observed radio noise from lightning and galactic sources at frequencies too low to penetrate the atmosphere. Spin-stabilised, operated for more than two years
Re-entered: December 1970.

Ariel 4
Launched: 11 December 1971 by Scout from Vandenberg
Mass: 100 kg
Orbit: 485 × 599 km, 83°
Comment: Electron density studies, radio noise and other monitoring. Mission ended April 1973, but the satellite is re-activated occasionally to support sounding rocket experiments
Re-entered: 12 December 1978.

Ariel 5
Launched: 15 October 1974 by Scout from San Marco
Mass: 129 kg
Orbit: 504 × 549 km, 3°
Comment: Six instruments to study x-ray sources. Working with Dutch ANS and NASA Explorer 42 it identified 12 rapid x-ray bursters. Operated for 5 years
Re-entered: 14 March 1980.

Ariel 6
Launched: 2 June 1979 by Scout D from Wallops Island
Mass: 154 kg
Orbit: 600 × 654 km, 55°
Comment: Observations ended in February 1982 on exhaustion of ACS gas after having studied 30 x-ray objects in detail, including several black hole candidates
Re-entered: 23 September 1990.

UNITED STATES OF AMERICA

Discovery series

Background

NASA's strategic plan, formulated in 1991 for solar system exploration, included the Discovery programme of near-Earth missions. These are rapid turnaround missions each costing less than US$150 million (US$75 million in any one year) at 1992 rates for spacecraft development; total cost must be less than US$245 million. Launch must be within three years of approval. Costs are minimised by emphasising focused science return from mature instrument and spacecraft technology providing simplicity and reliability. Only flight proven launchers (no larger than Delta 2) are employed. Each proposal requires a consortium of a NASA centre (or other federally funded research centre), industry and university partners.

The first two missions are the NEAR (Near-Earth Asteroid Rendezvous) in February 1996 and the Mars Pathfinder in December 1996. US$66.6 million and US$60.8 million was approved by Congress for FY94 to begin NEAR and Mars Pathfinder, respectively. US$129.7 million was provided FY95; US$102.2 million FY96; US$74.8 million requested FY97.

NASA received 28 proposals for round two by October 1994 in response to August 1994's request. NASA announced February 1995 that mission 3 would be the US$59 million Lunar Prospector, to be launched October 1997 to map the Moon's chemical composition and possibly identify polar water ice. The proposal was sufficiently mature to proceed to development after final technical definition. See the separate entry. Three other missions were selected for studies before the final choice in November 1995. Stardust will be launched February 1999, to fly through the coma of active comet Wild 2 in January 2004, returning comet and interstellar dust samples on aerogel plates to Earth January 2006. Total cost to NASA will be US$199.6 million. See the separate entry. Unsuccessful were Venus Multiprobe Mission (16 23 kg capsules into Venusian atmosphere to study complex circulation) and Suess-Urey (collect solar wind samples outside of Earth's magnetosphere).

The next round for Discovery missions started in June 1996 and, in April the following year, NASA announced the five proposals for further detailed studies as part of the series. Feasibility studies for the five missions are due for submission by 15 August 1997, and one or two will be selected the following October for full development in the Discovery programme. The five proposals selected in April 1997 were:

Aladdin, a mission to gather samples of the small Martian satellites Phobos and Deimos, plans to fire four projectiles into the satellites' surface and gather the ejecta during slow fly-bys, and then return the samples to Earth for detailed study;

The Comet Nucleus Tour (CONTOUR), a mission to take images and comparative spectral maps of at least three comet nuclei and analyse the dust flowing from them;

Genesis, a mission to collect a sample of the solar wind and return it to Earth for detailed analysis. It had been one of the finalists in the fourth round of Discovery programme selections under the name Seuss-Urey;

The Mercury Surface, Space Environment, Geochemistry and Ranging mission, or MESSENGER, an orbiter spacecraft carrying seven instruments to image and study the closest planet to the Sun;

The Venus Environmental Satellite (VESAT), an orbiter spacecraft to study the atmospheric chemistry and meteorology of Earth's cloud-covered neighbour using an imager, near-infra-red spectrograph, a temperature mapper and an X-band radar.

The spacecraft must be ready for launch no later than 30 September 2002, and meet the Discovery Programme's development cost (launch plus 30 days) cap of US$183 million in FY97 dollars over 36 months.

NEAR Discovery satellite

Current status

After launch on 17 February 1996, NEAR performed a fly-by of the asteroid Mathilde on 27 June 1997 coming within 1,212 km of its surface. NEAR performed an Earth fly-by on 23 January 1998, passing 540 km above south-western Iran. An attempted engine burn on 20 December 1998 to align the trajectories for the encounter with Eros failed and the spacecraft was placed in 'safe' mode. In a revised mission plan, NEAR performed a fly-by of Eros on 23 December 1998 coming to within 3,830 km and taking 222 pictures. An engine burn on 3 January 1999 ensured that NEAR will return to Eros for the planned orbit insertion burn on 14 February 2000, about a year later than originally planned.

Background

The Near-Earth Asteroid Rendezvous mission is the first of the Discovery programme. NEAR's first target was the S-type 13 × 14 × 35 km asteroid catalogued as 433 Eros. It is the second largest of the near-Earth asteroids (only three are larger than 10 km). NEAR was expected to orbit the asteroid as a satellite. Scientists assigned NEAR the mission of determining the asteroid's physical and geological properties as well as its chemical and mineralogical composition.

NEAR integrated on Delta's PAM kick stage. The science instruments are mounted on the bottom face and the nozzle of the single 445 N thruster protrudes between the solar wings (APL/JHU) ***1996***

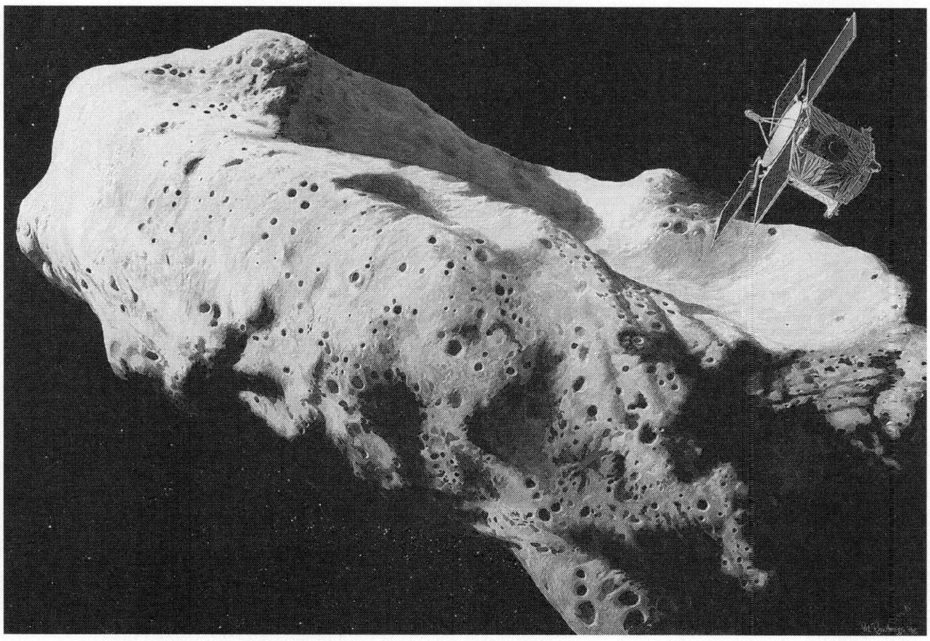

NEAR entered orbit around asteroid Eros in February 1996 (Applied Physics Lab/Johns Hopkins) ***1996***

As the first asteroid orbiter, NEAR was designed to evaluate the environment for orbital missions during its lifetime. It was to approach to within 15 km during the one year investigation, and photograph the surface using a camera with about 1 m resolution. NEAR promises to throw light on the nature of planetesimals (the building blocks for the terrestrial planets), the origin of objects impacting Earth, and the relationship between asteroids, comets and meteorites. The data will significantly enhance interpretation of Galileo data from the Gaspra and Ida S-type fly-bys.

En route, NEAR was to fly by the 61 km diameter C-class main belt asteroid Mathilde.

The Applied Physics Laboratory of The Johns Hopkins University was selected in 1992 to build/control NEAR, with the design reflecting simplicity and reliability, and the payload addressing the highest priority science issues. It is the first NASA planetary mission to be conducted by a non-NASA centre. APL is conducting mission and science operations from its campus.

Congress capped NASA's cost ceiling at US$150 million (1992 value) for construction, launch and 30 days' operations. When the programme was finished NASA actually paid only US$122 million in real-year dollars. Development began December 1993.

It was launched 17 February 1996 by Delta 7925-8 from Cape Canaveral at 20.43.27 GMT into a two-year Venus Earth Gravity Assist (VEGA) trajectory. By using the Venus fly-by, NEAR only needs a launch energy of 25.9 instead of 42 km^2/s. The 1,200 km 9.9 km/s 253 Mathilde fly-by was scheduled to occur on 27 June 1997. This was to be followed by a 279 m/s deep space manoeuvre 3 July 1997 (in two segments) and a 478 km Earth swing-by 22 January 1998 – cranking the orbital inclination to 10.8°. Then, the mid-space manoeuvres were to take over to nudge NEAR to its final orbital entry around Eros. These manoeuvres, totalling 949 m/s were to start 9 January 1999 to slow approach speed to 5 m/s and enter 500 km Eros orbit 6 February 1999 (2.5 AU from Earth) The Eros orbit is 1.13 × 1.78 AU, at 10.8°. Data gathering was to begin from the initial 200 × 400 km orbit, slowly reduced to 35 to 50 km circular by mid-March 1999. NASA was to shut the mission down on 31 December 1999.

See 'Current status' above for in-flight changes to this pre-mission plan.

Specifications

Mass: 805 kg at launch, 312 kg propellants, 437 kg dry, 56 kg science payload

Spacecraft: 1.5 m dia hexagonal bus, 3-axis stabilised (1.5 mrad control, 50 µrad knowledge), fixed HGA and solar panels. Four fixed GaAs panels (NEAR's only deployable system, and the first solar cells beyond Mars) provide 1,800 W at 1 AU (81 W required by science instruments). Fixed 1.5 m HGA returns 6-27 kbit/s at X-band; two 512 Mbit SSRs. 445 N Leros NTO/MMH + 12 × 22 N MMH thrusters provide 1,500 m/s total δV. Passive thermal design

Payload: CCD visible imager (95 × 161 µrad resolution, 2.25 × 2.9° FOV, eight-position filter wheel, 10 kg, 15 W, APL, MSX heritage).

X/γ-ray spectrometer (1-10 keV solar monitor, 0.3-10 MeV; Al, Mg, Si, Fe, Ti, Ca, U, Th, K; 26 kg, 10 W, Goddard).

Near-infra-red spectrograph (0.8-2.6 µm, 2-3% spectral resolution, 18 kg, 15 W, APL, DMSP heritage).

Magnetometer (4-65,536 nT, 16-bit resolution, 1 kg, 1 W, Goddard).

Laser altimeter (6 m ht resolution at 50 km, part of attitude/navigation system).

Radio science (2-way Doppler to 0.1 mm/s, part of engineering payload).

ERBS

Background

The 2,240 kg Earth Radiation Budget Satellite (ERBS), launched 5 October 1984 by Shuttle 41G into 610 km, 57° orbit remains operational. Ball Aerospace as part of the three-satellite Earth Radiation Budget Experiment managed by NASA Goddard built ERBS; NASA Langley was responsible for instrument development and data reduction. Earth's radiation budget is probably the weather's principal driver. Energy balances for the entire planet resolve around the energy exchange among the poles, the tropics and outer space. The tropics absorb more energy than they emit, and seasonally the polar regions receive no solar radiation though they continue to emit. The difference in absorbed and emitted energy causes the atmosphere and oceans to circulate and transfer energy, creating weather systems.

ERBS's instruments were supplemented by similar packages on NOAA 9/10 in near-polar orbits to monitor thermal and solar radiation of the entire Earth at some time each day, and most of the planet at all times of day during every month. All three scanning instruments have failed: NOAA 9's on 20 January 1987; NOAA 10's on 22 May 1989 and ERBS' on 28 February 1990. The eight radiometers of each satellite were divided into two separate instruments: a scanning and a non-scanning radiometer (wide/medium FOV). They monitor the incoming solar ultra-violet/visible and subtract the component reflected back into space (principally by clouds/snow) and the infra-red/thermal infra-red emission from the heated atmosphere and surface. The result is Earth's net radiative heating. The observations are indicating variations in the Sun's output, the veiling of Earth by volcanic dust, increases in atmospheric CO_2 from fossil fuel burning.

ERBS demonstrated for the first time that cloud cover is responsible for a net global cooling, depressing surface temperature by an average of 11°C during April 1985. ERBS observations also provided the first direct evidence that a volcanic eruption can lead to significant changes in the global radiation balance. A 3.8 per cent increase in reflected sunlight was detected for several months following Mt Pinatubo's 1991 eruption.

ERBS was released by Challenger's manipulator. Sally Ride had to use the arm to shake the satellite before the 7 m solar array would deploy. Its thrusters later fired for over 10 hours to attain a 610 km orbit

In separate experiments, ERBS carries Langley's SAGE (Stratospheric Aerosol/Gas Experiment) to monitor the vertical distribution of stratospheric aerosols, O_3 and N_2O_4. SAGE and Nimbus 7 showed that Antarctica's 1987 ozone hole, which disappeared during 1988, re-appeared almost as deeply in 1989 and worse in each subsequent year.

In order to deploy ERBS, NASA launched the Shuttle into a 57° inclination for only the second time. It was released by the remote manipulator and employed an integral propulsion system to reach the required orbit. 3-axis control is normally maintained to within 1° by a magnetorquing system, backed up by the hydrazine thrusters, employed primarily for orbit adjust and yaw turns. ERBS must be yawed 180° every four to five weeks to maintain solar array illumination. It requires 150 real-time and 550 stored commands each day for normal operations. The solar arrays (470 W average load) are supported by a 50 Ah Ni/Cd battery (2 of 22 cells failed; the other battery failed). Commands and telemetry are routed through TDRSS by the electronically steerable spherical array antenna, backed up by two omni antennas via TDRSS and ground stations. Data are recorded at 1.6/12.8 kbit/s and dumped at 128 or 32 kbit/s on alternate orbits; normal real-time rates are 1.6/12.8 kbit/s.

Spacecraft systems and the SAGE and ERBE non-scanner instruments continue to operate normally except for a failed battery and some failed gyros. The gyro losses do not affect operations apart from the yaw turns, which are performed by timed thruster firings. As of 5 February 1996, after 4,110 days on orbit, ERBS had completed 61,775 orbits, 114 manoeuvres and more than 120,000 SAGE science events.

Explorer series

Background

Explorer is the oldest US satellite series and will continue to fly science payloads in Earth orbit into the next century. 74 Explorers have been launched since Explorer 1 became the first US satellite in February 1958. The Advanced Composition Explorer (ACE) is the last of the Delta-class Explorers, before FUSE appears as the pathfinder for the smaller and cheaper MIDEX Medium Explorer line. The first two MIDEX missions were selected in April 1996 for definition studies leading to 1999 and 2000 Med-Lite launches: MAP Microwave Anistropy Probe and IMAGE Imager for Magnetopause to Auroral Global Exploration. The SMEX Small Explorer programme is providing a return to lower cost, rapid turnaround LEO missions emphasising space physics and astrophysics. Three are flying under the first cycle: Solar, Anomalous & Magnetospheric Particle Explorer (SAMPEX, launched 1992), Fast Auroral Snapshot Explorer (FAST, 1996) and Sub-millimetre Wave Astronomy Satellite (SWAS, 1997). Two, Transition Region & Coronal Explorer (TRACE, 1997) and Wide-Field Infra-red Explorer (WIRE, 1999), were selected in 1994 as the second set.

NASA also plans the UNEX University Explorer series, giving the PI authority and management responsibility for less complex missions than SMEX. It will be managed under a fixed cost ceiling, with NASA retaining oversight authority sufficient only to discharge its responsibility for handling public money. The SNOE and Terriers missions

SWAS will investigate the chemistry of interstellar clouds (Ball Aerospace) ***1995***

of the USRA Universities Space Research Association are UNEX precursors.

The Explorer programme is headed by NASA's Goddard Space Flight Center for the Office of Space Science. FY95 development funding US$120.4 million, US$132.2 million FY96, US$135.0 million requested for FY97.

Advanced Composition Explorer (ACE)

Background

NASA selected the Advanced Composition Explorer in October 1989 for Phase B, as an intermediate class Explorer, to study the composition of energetic particles. It has a collecting power 10 to 1,000 times greater than past missions. NOAA agreed in 1994 to pay US$450,000 for three instruments to return data continuously, instead of NASA's planned 4 h daily, to act as a Real-Time Solar Wind Monitor. NOAA wanted this capability to provide 1 hour warnings of solar storms to satellite operators and power companies.

Specifications

Contractors: Goddard (project management, ground system, mission operations, orbit determination), Johns Hopkins Applied Physics Lab (spacecraft, including integration and launch support), California Institute of Technology (science payload management), Seakr Engineering (SSRs), Olin Aerospace (propulsion), Ball Aerospace (star tracker), Adcole (Sun sensors), SAFT (battery), Tecstar (solar arrays)
Launch: 25 August 1997 by Delta 2 from Cape Canaveral
Mission life: 2 years (5 year goal)
Mass: 765 kg (including 189 kg hydrazine)
Orbit: 150,000 × 300,000 km halo orbit around the L1 Sun-Earth Lagrangian point (arriving 3 months after launch)
Spacecraft: octagonal prism 160 cm diameter main body with four 86 × 152 cm deployed solar panels, providing 500 W; 12 Ah Ni/Cd battery. Two panels carry magnetometer booms. 5 rpm spin-stabilised, Earth-pointing spin axis. AOCS by 10 hydrazine 4.5 N thrusters, solid-state star tracker + digital Sun sensors. Solid-state storage capacity is >80 h; data stored for 24 h at 7 kbit/s interleaved with real-time data and returned to Deep Space Network via 66 cm parabolic antenna for up to 4 h daily at 77 kbit/s (but RTSW data returned continuously)
Payload: CRIS Cosmic Ray Isotope Spectrometer. Four identical Si solid-state telescopes with common scintillating optical fibre hodoscope. Few × 100 MeV/nucleon. CIT, GSFC, JPL, Washington University (St Louis).

SIS Solar Isotope Spectrometer. Two identical solid-state telescopes, 10-100 MeV/nucleon. CIT, GSFC, JPL.

ULEIS Ultra Low Energy Isotope Spectrometer. Time of flight *vs.* energy spectrometer. Few × 100 keV/nucleon. University of Maryland (UMD)/APL.

SEPICA Solar Energetic Particle Ionic Charge Analyser. Electrostatic deflection *vs.* energy loss charge state analyser. 1 MeV/nucleon. University of New Hampshire/Max Planck Institute.

SWIMS Solar Wind Ion Mass Spectrometer. Time of flight spectrometer. Few × 100 km/s. UMD/University of Bern.

SWICS Solar Wind Ionic Charge Spectrometer. Electrostatic deflection, time of flight and energy. Few × 100 keV. UMD, University of Bern and Max Planck Institute.

EPAM Electron, Proton and Alpha Particle Monitor. Two telescope assemblies have five apertures. Particle fluxes from 50 keV to several MeV. APL.

SWEPAM Solar Wind Electron, Proton and Alpha Particle Monitor. Separate electron (1-1,450 eV) + ion (260 eV-35 keV) electrostatic analysers. Los Alamos National Lab/Sandia National Lab.

Magnetometer. Two triaxial fluxgate magnetometers. 0.004-65,000 nT. Bartol Research Institute University of Delaware/GSFC

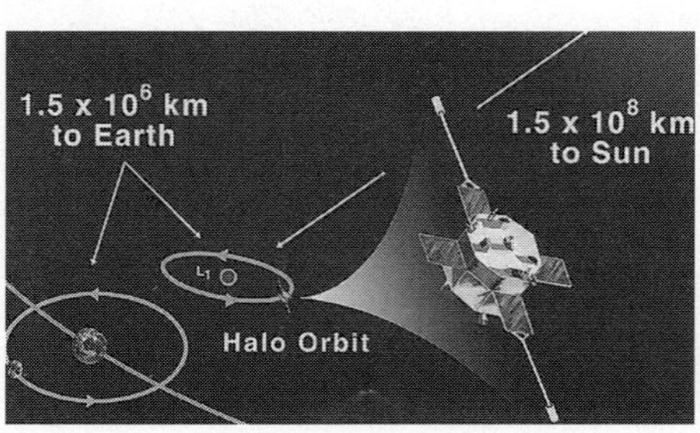

ACE will study charged particles ranging from the solar wind to galactic matter (NASA) ***1995***

Secondary payload: RTSW Real-Time Solar Wind monitor for solar storm prediction. EPAM, SWEPAM and MAG data broadcast 20 h/day via a 434 bit/s selectable telemetry format to NOAA/USAF-provided ground stations. Data taken during 4 h ACE ground contact sent to NOAA in real time.

Cosmic Background Explorer (COBE)

Background

Cosmic Background Explorer (COBE) set out to answer fundamental questions about the universe. Its microwave/infra-red sky maps at 100 wavelengths are helping to determine if the universe is rotating, is distributed homogeneously and is uniformly expanding. In addition, present interstellar and interplanetary dust clouds were mapped. COBE was conceived in the mid-60s after the Big Bang theory of creation received a strong verification in 1964 when Arno Penzias and Robert Wilson serendipitously discovered the remnant radiation resulting from the bang.

COBE carried three main instruments. The Differential Microwave Radiometer (DMR) mapped the sky at 3.3/5.7/9.6 mm, where the galaxy is more transparent, using standard microwave hardware to compare the emissions from two patches of sky, 60° apart, at a time. If the early universe was thermally and structurally uniform the background would be equal around the sky. If not, the map will highlight the variations. The detectors for the FIRAS Far Infra-Red Absolute Spectrometer were cooled to less than 2K by liquid helium inside a cryostat and protected against Sun/Earth radiation by the conical shield. The 95.7 kg of cryogen limited the far-infra-red mission to a 10-month life, sufficient for 1.6 surveys.

Operating with a sensitivity two orders of magnitude greater than previous balloon/rocket instruments, FIRAS collected radiation with a horn antenna viewing along the spin axis. Like DMR, it provided an angular resolution of 7°. An electrically heated calibrator was inserted into the horn and warmed until its emission matched that of the sky in the field of view for measurements with an accuracy of 0.001°C over 100 μm-1 cm. The Diffuse Infra-Red Background Experiment (DIRBE) searched for the remnant infra-red radiation, previously undetected, from the initial burst of radiation from the primordial galaxies as they formed. Ten filters studied selected bands over 1-300 μm, using cooled bolometers, photoconductors of Ge and Si, and photovoltaic detectors of In and Sb; observations at the shorter wavelengths were still possible after cryogen exhaustion.

FIRAS, measured the background temperature at 2.728±0.004K, varying 0.01 per cent from the blackbody spectrum and indicating that 99.99 per cent of the universe's early radiant energy was released within one year. FIRAS' results mean that the Big Bang theory has been tested against observation to a fine degree of precision. Another instrument, DIRBE mapped the radiation from the interstellar medium, solar system dust and stars; it is hoped that extensive modelling now under way to extract the foreground infra-red will reveal the glow from the universe's first stars and galaxies. DMR's momentous discovery of the predicted background ripples in support of the Big Bang theory was announced April 1992. These 30 × 10^{-6}°C variations, required for the appearance of galaxies, date back to when the was only 10^6 years old. They also support the theory that most of the universe is composed of dark matter. FIRAS mapped interstellar

COBE: the Cosmic Background Explorer was the first of the new Explorer series (NASA)

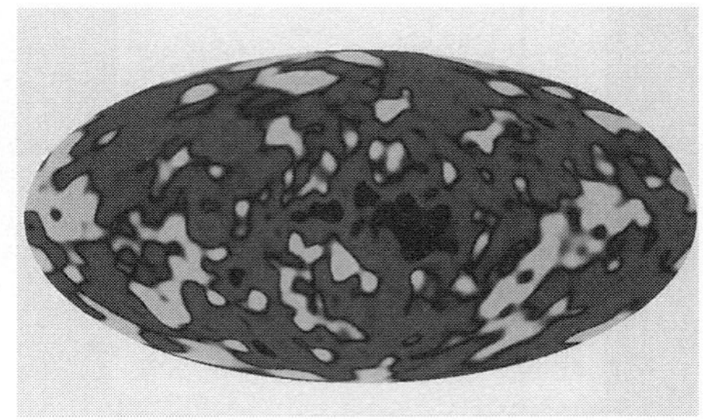

All-sky microwave map generated by COBE's DMR instrument. Our Galaxy's radiation has been processed out, to leave the 0.02 per cent variations in the 2.73K background. Detailed analyses show that fainter cosmic signals are present from the earliest clumping of material in the Universe ***1995***

ionised nitrogen (the first detection at 205 μm), carbon and dust for the first time, atomic emissions that control the cooling of interstellar gas. DIRBE's continuing survey at shorter wavelengths, although six times less sensitive than originally, was primarily aimed at solar system dust/gas mapping. Repeated coverage could show temporal variations (the analysis has yet to be performed).

COBE completed a full-sky survey in mid-June 1990 and was performing its second when the cryogen was exhausted 21 September 1990. FIRAS' observations therefore ended, but the un-cooled DMR continued and DIRBE could still operate at its shorter IR wavelengths with the cryostat's interior warmed to 72K. Science operations ended 23 December 1993 and EOL engineering tests were completed in January 1994 before COBE was handed over to the Wallops Flight Facility for testing WFF's ground tracking system two to three times monthly, evaluating telemetry reception, command transmission and the angle and Doppler tracking capabilities. TDRSS and JPL similarly access it when WFF work allows. Commanding is limited to basic transponder functions and playback of tape data recorded when COBE was last fully operational. No other housekeeping functions are exercised but, with the high power margin, COBE should continue for some years. It is commanded locally by any of WFF's stations. The original flight engineers are occasionally given data for a quick checkup.

Mission cost was approximately US$230 million.

Specifications

Launched: 18 November 1989 by Delta 5920 from VAFB
Lifetime: 1 year nominal
Mass: 2,205 kg BOL, 2,109 kg EOL
Orbit: 887 × 898 km 99.0° Sun-synchronous
Spacecraft: hexagonal bus spin stabilised at 0.8 rpm normal to Sun by RWs (supported by Sun sensors, magnetometers and three scanning infra-red Earth sensors) for scanning the instruments, 95.7 kg/600 litre liquid helium initially carried in cryostat, protection against solar/Earth radiation provided by conical shield. Three 3-segment solar wings totalling 31 m^2 and spanning 8.5 m generate 1,050 W BOL/905 W EOL (615 W EOL eclipse). TT&C provided through NASA's TDRS system but data dumped to Wallops in 10 min each day for relay to the COBE Science Data Center at Goddard.

Extreme Ultra-Violet Explorer (EUVE)

Background

EUVE carries four ultra-violet telescopes covering 60-800Å for the first detailed all-sky observations at these wavelengths. The initial all-sky mapping required six months, 24 July 1992 to 21 January 1993, pinpointing sources with 0.1° accuracy for more detailed subsequent observations over the following year. Some 700 bright EUV sources appeared in EUVE's instruments and more than 40 extragalactic sources have been detected.

The observatory phase began 22 January 1993, looking at one to four targets weekly. Most EUV objects discovered to date have been stars at advanced stages of evolution; white dwarfs and stars with active coronas, although 30 per cent have not been identified with an optical counterpart.

The Challenger accident delayed and forestalled plans for the EUVE and in October 1986 Goddard began procurement of an Explorer Platform, launchable by either Shuttle or Delta, that would permit these payloads to be exchanged in orbit. It was decided, in 1991, that launches would be on expendables to maximise science return. EUVE is the first payload flown on the new Explorer Platform, based on the NASA/Fairchild Multi-

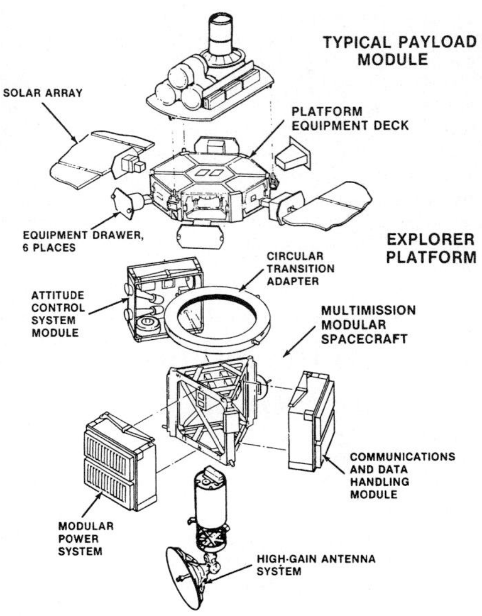

EUVE is undertaking detailed extreme Ultra-Violet observations

Mission Spacecraft (MMS). The EUVE's four instruments are 3 to 188 kg grazing incidence telescopes and a 323 kg deep survey/EUV telescope/spectrometer. The fourth instrument took measurements in the anti-Sun direction along the ecliptic, taking advantage of the low background to complete a survey in two bands in 60-400 Å. Its three spectrometers spanning 60-760 Å to study the brightest sources with a resolving power of 350. Each instrument incorporates 40 cm diameter Wolter Schwarzschild Ni-coated aluminium mirrors: two Type I covering 60-400 Å (both for the sky survey) and two Type II for 400-800 Å (one for the deep survey and spectrometer). Detectors consist of Z stacks of micro-channel plates mounted in windowless assemblies that are read out in an array of 1,024 × 1,024 pixels.

EUVE is managed from Goddard's POCC Payload Operations Control Center and the science payload from the Center for EUV Astrophysics at the University of California.

Specifications

Contractors: Goddard (project management; development/integration of payload module), Fairchild Space Corp (prime; communications/data handling MMS module), General Electric (attitude control system MMS module), McDonnell Douglas (power systems MMS module), University of California Space Sciences Lab (instruments, Science Operations Center)
Launch: 7 June 1992 by Delta 2 from Cape Canaveral
Lifetime: 42 months nominal mission; 20 month/US$10 million extended mission funded from January 1996. No degradation of spacecraft or instruments has occurred
Mass: 3,256 kg at launch
Orbit: 514 × 529 km, 28.5°
Spacecraft: based on NASA/Fairchild MMS, with Platform Equipment Deck mated with the Explorer Platform by motorised connectors. MMS carries modular power system (working with two solar wings providing about 300 W to payload from >1 kW BOL total), command/data handling module, the Modular Attitude Control System (MACS) and the signal conditioning unit. 3-axis stabilisation. Data rate of 16 kbit/s shared by the four telescopes, recorded with engineering data at 32 kbit/s on a 10E9 tape recorder for dumping to ground at 512 kbit/s through TDRS system. Platform designed for 10-year life; degrading solar arrays can be replaced in orbit (but not planned).

Far Ultra-violet Spectroscopic Explorer (FUSE)

Background

NASA initially studied FUSE as a 1 m telescope for very high-resolution spectroscopy over 900-1,200 Å, expanding the preliminary investigations made by OAO 3 and complementing HST and the x-ray missions. FUSE was planned for launch by Delta in 2000 in a total US$254 million programme, but in 1994, Congress reduced its total budget. NASA reduced the budget again in 1998 and renamed it Med-Lite. As with the Hubble Space Telescope, NASA decided to save money by lowering the apogee of the orbit at the expense of the scientific mission. With the FUSE, the Agency lowered

FUSE will study ultra-violet spectra below IUE's range ***1996***

the original 71,000 km apogee to LEO, reducing the number of observations but not affecting the basic science.

Instead of NASA Goddard managing the programme and building the spacecraft for the PI's payload, PI Warren Moos of JHU's Center for Astrophysical Sciences has full responsibility for the mission, selecting Orbital Sciences Corp in August 1995 for the US$37 million bus contract. McDonnell Douglas is providing the Delta 7320 at the same cost as planned for Delta-Lite, which it now does not plan to develop.

One FUSE goal is studying trace species in interstellar and intergalactic gas using absorption spectroscopy of distant faint sources. Scientists believe the primordial deuterium/hydrogen ratio is a critical parameter in understanding the first seconds of the Big Bang. In order to conduct its investigations four coaligned mirror telescopes, gather spectra on an individual focal plane assembly carrying four slits: 1.5 × 20 arcsec (highest resolution), 4 × 20 (extended objects), 30 × 30, 0.2 diameter pinhole (bright objects that would overwhelm the detector). Each telescope illuminates a holographic diffraction grating and the output from all four is detected by two 10 × 179 mm microchannel plates.

Specifications

Launch: planned 1999 by Delta 7320 from Cape Canaveral
Lifetime: 3 year nominal mission
Mass: 1,360 kg (bus 500 kg, payload 780 kg, no propellant)
Orbit: 800 km, 25°
Spacecraft: 3-axis 0.9 × 0.9 × 1.3 m box-shaped bus with simple mechanical/thermal links to payload. 0.5 arcsec pointing (1σ) with FES (2° without) using four MWs (16.6 Nms at 5,100 rpm), 3 × 100 Am^2 torque rods, Sun sensors and magnetometers. No propulsion. 1 Mbit/s on 5 W S-band downlink (2 kbit/s command link) from 240 Mbyte memory. Twin GaAs 3.5 m^2 solar wings provide 500 W EOL (payload requires 340 W); 40 Ah Ni/Cd battery.

Fast Auroral Snapshot Explorer (FAST)

Background

The Fast Auroral Snapshot Explorer (FAST) is investigating plasma phenomena in auroral processes discovered in previous satellite and sounding rocket missions. Its orbit at around 100 km takes it through the source region for much of the energy that appears as auroral light. A transportable 8 m antenna is positioned in Alaska to collect real-time data as FAST passes through the northern aurora. These observations are being complemented by data from higher altitude satellites, placing FAST's data in a global context.

FAST had difficulty making it into space because August 1994/5's six-week launch windows were both missed because of June 1994/5's Pegasus failures. FAST was placed in storage waiting its mid-1996 launch in order to make it synchronous with peak northern auroral activity.

Specifications

Contractors: GSFC provided the spacecraft for the UCB University of California at Berkeley's payload
Launch: 21 August 1996 by Pegasus XL/Lockheed L-1011 from VAFB
Lifetime: 1 year minimum
Mass: 180 kg (51 kg science)
Orbit: planned 83°, 350-4,165 km
Spacecraft: bus irregular octagonal prism. 50 W orbit average provided by GaAs body panels (5.6 m^2 total) + one 9 Ah Ni/Cd battery (science instruments require 15 W orbit average, 31 W operating, so frequently turned off). 1 Gbit solid-state memory, 0.90/1.5/2.25 Mbit/s

FAST is investigating how electrical and magnetic fields accelerate charged particles in the auroral regions. It is shown during spin balancing at Goddard (NASA) ***1995***

rate via NASA standard S-band 5 W transponder. 2 kbit/s commands up, 4 kbit/s housekeeping down. 12 rpm spin stabilisation (normal to orbit) by two magnetorquers, spin Sun sensor, horizon crossing indicator, magnetometer. Attitude post-knowledge 1°. 86 cm diameter × 1.5 m length launch envelope. Four 30 m radial wire booms, 2 axial booms, 2 magnetometer booms
Payload: EESA. Quadrispherical Electrostatic Electron Analyser (UCB, 11.2 kg, on AMPTE, Giotto, Mars Observer, Wind, Cluster).

TEAMS Time-of-Flight Energy Angle Mass Spectrograph (University of New Hampshire + Lockheed Palo Alto Research Lab, 6.4 kg, on Cluster).

EFPE Electric Field Plasma Experiment (UCB, 13.6 kg, on ISEE 1, CRRES, Polar, Cluster).

Magnetometer (UCLA, 3.7 kg, on Pioneer, ISEE 1/2, Polar, Cluster).

IMP Explorers

Background

Ten Interplanetary Monitoring Platforms monitored Earth's radiation environment during a complete 11-year cycle of solar activity, defining the nature and extent of the magnetosphere. IMPs provided continuous warning of solar flare radiation during the Apollo/Skylab missions, and provided baselines for Pioneer 10/11 and Mariner 10. IMP 8 continues to return data. In December 1992, it provided key data as Galileo passed through Earth's magnetotail. IMP 8's battery is now long since dead (array output 4.40 A) and ranging cannot be performed because the 136.800 MHz transmitter has failed, and insufficient gas precludes manoeuvres. Even so, it remains a principal source of solar wind data and is funded to 2002.

IMP 1, launched 16 November 1963 as Explorer 8 is a 16-sided drum, 135.6 cm diameter, 157 cm high. It has a 22 rpm spin-stabilised control (no AOCS remaining).

Explorer 50 (IMP 8) continues to return fields and particle data from high Earth orbit after 23 years (NASA)

Data returns from the LAP 2-D electrostatic analyser (4-20 keV electrons and 0.070-20 keV protons), MAP Faraday cup (0.017-7 keV electrons, 0.050-7 keV protons), IOE electrostatic analyser (0.5 eV-50 keV), MAE electrostatic analyser, APP solid-state telescope (fluxes/spectra greater than 220 keV electrons, greater than 300 keV protons, greater than 640 keV alpha particles, and fluxes of more than 1 MeV/n particles), GME cosmic ray telescopes (0.150-15 MeV electrons, 4-80 MeV isotopes), GNF fluxgate sensors, CHE cosmic ray telescopes (0.3-10 MeV electrons; 0.5-1,200 MeV nucleons), CAI multi-element telescope (0.16-2.8 MeV electrons; 0.5-40 MeV isotopes), GWP energetic particles telescopes (30-450 keV electron fluxes and 0.030-8.6 MeV proton fluxes).

Its orbit is 125,000 × 202,000 km, with an on-orbit weight of 62 kg.

Explorer 35, launched 19 July 1967, entered lunar orbit to study the Moon's magnetic field and radiation belt, in addition to Earth's magnetotail.

IMP 8 (Explorer 50), launched 26 October 1973 had an initial orbit of 204,200 × 236,000 km, 28.8°, which has decayed to a now near circular one at 223,000 km, 50°. Its mass at launch was 298 kg and provided 120/42 W to spacecraft and experiments.

ISEE Explorer (ICE)

Background

The International Sun-Earth Explorers, part of the International Magnetosphere Study, followed up the IMP series. NASA built ISEE 1 and 3 and ESA provided 2. ISEE 1 and 2 were launched jointly on 22 October 1977. ISEE 3 followed on 12 August 1978 by Delta from Cape Canaveral (469 kg). ISEE 3 had the distinction of attaining the first halo orbit around the L1 libration point 1.5 million km out on the Earth-Sun line. Eight annual hydrazine thruster firings maintain the orbit and it oscillated 150,000 km about the ecliptic annually so that solar radio emissions did not interfere with tracking and data acquisition.

The ISEE 3 provided 1 hour warning of solar radiation before events reached ISEE 1 and 2. NASA diverted ISEE 3, due to remain as a solar flare sentinel for 10 years, towards Earth in late 1982 so that it could be directed into a series of lunar swing-bys (the last, in December 1983, within 100 km) for the first ever comet encounter. Its speed was increased from 4,679 km/h to 8,278 km/h by sweeping behind the Moon on 22 December 1983, surviving 28 minutes without solar power (its battery failed in 1983).

After the swing-by, ISEE became the ICE, or International Cometary Explorer. On 11 September 1985 it passed 7,860 km behind Giacobini-Zinner's 2.4 km diameter nucleus at a closing speed of 75,300 km/h. Goddard controllers commanding the 422 kg 1.8 m diameter vehicle, built by Fairchild, had feared cometary dust could coat its solar cells and damage its 91 m-span antennas. Throughout the 20 minute coma passage its 5 W transmitter and small antenna, designed to provide data from no further than 1.6 million km, transmitted 70.8 million km to Earth. NASA's Deep Space Network had six large dishes trained on it (Goldstone, Madrid and Canberra), with additional facilities in Japan and Arecibo (Puerto Rico). Though lacking an imaging system, it returned data on plasma densities, flow speeds, temperature and heavy ions in the tail. Initial analysis revealed water ice to be the primary constituent, with cold slow-moving plasma, and water and carbon monoxide ions inside the tail, supporting Whipple's dirty snowball model.

ICE achieved the first comet interception. It will be recaptured by the Earth-Moon system in 2014 (NASA)

Unexpectedly, ICE began encountering atmospheric ions 1.6 million km out, and the 24,000 km-wide tail was much wider than expected. The first instrument to detect the comet was a European energetic proton instrument. ICE also passed within 40.2 million km of Halley's Comet on 28 March 1986. It was then manoeuvred into an Earth-like orbit with a period of 355 days so that, on 10 August 2014, it will return for capture by the Earth-Moon system, when it could be retrieved for analysis of its cometary dust coating. Ownership has already been formally transferred to the National Air & Space Museum.

Daily data return at 64 bit/s to Goddard continued until December 1995 from the: magnetometer, plasma composition, solar wind plasma, plasma waves, energetic protons, low/medium/high energy cosmic rays and cosmic ray electrons instruments. The telemetry transponders were turned off 19 December for JPL to begin a radio science mission. The low bit rates means that AOCS cannot be commanded until about 2010. Hydrazine tank 1 is empty (leak), number 2 holds 135 kg; solar array is 6 A. ICE is expected to remain active until at least 2014's return.

IUE Explorer

Background

The International Ultraviolet Explorer followed a modified geosynchronous orbit at 30,221 × 41,349 and 34.9° of inclination, making it the first astronomical satellite in geosynchronous orbit. A joint venture by NASA/ESA/UK, it operated until October 1995 from its 26 January 1978 launch by Delta from Cape Canaveral. At launch it weighed 1,479 kg (445 kg in orbit).

IUE provided astronomers with a unique tool and requests for observing time remained three times greater than could be accommodated. Despite the appearance of the Hubble Space Telescope, IUE's 1,150-3,200 Å spectral band covered an entire region not accessible simultaneously to HST's high resolution spectrographs. The fast response time of less than 1 hour provided an unparalleled flexibility in scheduling targets of opportunity. Within one year, IUE had provided 200 investigators from 17 countries with more than 9,000 ultra-violet spectra of objects unobtainable from ground-based instruments. 113,300 spectra had been obtained by the end of 1995 for 2,000 astronomers and 3,300 papers using IUE data had been published in referred journals.

ESA joined NASA and independent researchers from the UK. As part of its programme, to enhance the ESA science contribution to space exploration, European approval for IUE came in 1973, to build the provision of the spacecraft's deployable solar array and a ground station at Villafranca near Madrid. Having passed 18 years of operations, it is responsible for generating more scientific papers than any other astronomical telescope.

Hydrazine propellant is plentiful, with 17.45 kg (75 per cent) remaining by end of 1995; and an annual consumption of only 0.57 kg. On 17 August 1985, a fourth attitude control gyro failed, but software for the remaining two had been prepared in anticipation and only a months' observations were lost. Solar cell degradation had been about 3 to 4 per cent annually (3.31 per cent average 1978-95), with 148 W is required for normal operations.

Still, NASA hinted in January 1994, that it would pull the plug on funding the operations beyond September 1994, and instead dedicate the US$3.5 million annual saving to other science missions. It had to back down from this position, because of European opposition, on 1 October 1995, and since then, science operations have been performed only during the 16 hours low radiation part of the orbit, run from Villafranca. Goddard supplies eight hours of satellite management and housekeeping, while NASA concentrates on the IUE Final Archive.

After the 1994 decision all parties agreed to halt operations on 31 December 1997, but ESA decided in February 1996 that it would shut IUE down by September 1996 at the latest to save ECU5 million annually. The satellite was finally shut down at 18:42 GMT on 30 September 1996: the day before shutdown the orbit was 35.9°, 29,990-41,620 km.

IUE was shut down in 1996 because of funding problems

ESA ran science operations from 1 October 1995 from Villafranca, until the 1996 deactivation. IUE has had a profound impact on the way astrophysical studies are done through the highly flexible operations mode and satellite design, which required continuous contact but allowed for a very efficient science observation schedule. Many programmes of co-ordinated observations with other satellites and ground sites were often performed.

Transition Region And Coronal Explorer (TRACE)

Background

The Transition Region And Coronal Explorer studies the connections between fine scale magnetic fields in the solar surface and features in the photosphere, chromosphere, transition region and corona. Scientists want to use it to explore the relation between diffusion of the surface magnetic fields and the changes in heating and structure throughout the solar transition region and corona. The simultaneous ultra-violet films of the 6,000-10,000,000,000 regions at 1 arcsec spatial and 1 second time resolutions will reveal the rate of change of magnetic topology. The launch coincides with the onset of a new solar cycle, so the emerging magnetic flux will be observed in a relatively uncomplicated atmosphere.

TRACE's payload consists of a 30 cm diameter ultra-violet/extreme ultra-violet telescope with image motion compensation. Four quadrant normal-incidence coatings on the primary/secondary mirrors form identically-sized and perfectly co-aligned images. A 1,024 × 1,024 CCD collects images over 8.5 × 8.5 arcmin FOV. TRACE's image processor can continuously monitor the data stream and adapt the observing programme and instrument pointing. Guide telescope is integral part of satellite/instrument pointing system.

TRACE and SOHO will co-ordinate their daily observing programme. TRACE is an 'open' mission, in which the PI and co-investigators have waived all priority data rights. The PI will make data available online within 24 hours and science-level data will be released to the PI and general community simultaneously.

Specifications

Contractors: GSFC is providing the spacecraft for the SLISR Stanford Lockheed Institute for Space Research's payload
Launch: 2 April 1997 by Pegasus XL/Lockheed L-1011 from VAFB
Lifetime: 1 year minimum
Mass: 250 kg (59 kg science)
Orbit: 599 × 641 km, Sun synchronous
Spacecraft: bus irregular octagonal prism, 86 cm diameter × 1.5 m length launch envelope. 220 W orbit average provided by four deployed GaAs wings + one 9 Ah Super Ni/Cd battery (science instruments require 83 W). 80C86 instrument processor, 300 Mbyte solid-state memory (66 used by error detection/correction), 2 kbit/s uplink, 0.0234/1.125/2.25 Mbit/s rate via NASA standard S-band 5 W transponder. Four to six daily passes at Wallops/Poker Flat; 320-480 Mbyte science down daily. 3-axis solar 20 arcsec pointing momentum biased around Sun line (stabilised by secondary mirror to <1 arcsec), 3-axis magnetometer, six coarse Sun sensors, bright object sensor, digital Sun sensor, three gyros, guide telescope (fine Sun sensor on instrument), three torque rods, four RWs.

Rossi X-ray Timing Explorer (RXTE)

Background

RXTE carries three instruments for intensive studies of x-ray source luminosity variations, for the first time ranging from μs to years, to provide information on processes and structures of white dwarfs, black holes, neutron stars and other exotic objects. The large effective area (0.8 m^2) and broadband sensitivity (2-200 keV) are valuable for timing intensity variations and determination of spectra from high energy sources.

The PCA Proportional Counter Array and HEXTE (High Energy X-ray Timing Experiment) form a single telescope, viewing a single source in their common 1° FOV. The large areas and low backgrounds provide high sensitivity to weak sources. The ASM All-Sky Monitor scans 70 per cent of the sky every 1.5 hours to monitor

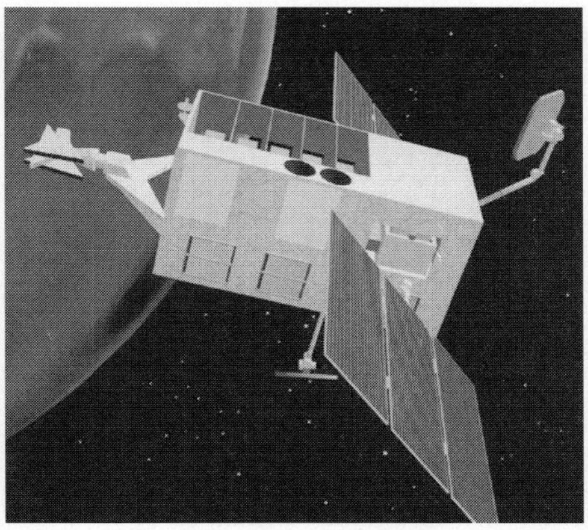

RXTE is aimed at monitoring luminosity variations of x-ray sources (NASA) ***1996***

about 75 of the sky's brightest sources. It thus measures long-term variations and identifies changes for PCA/HEXTE to study at short notice. The microprocessor EDS Experiment Data System provides onboard processing/compression of PCA/ASM data from 500,000 bit/s to the telemetry rate.

XTE was renamed in February 1996 after x-ray astronomy pioneer Professor Bruno B Rossi. Satellite cost, including ground equipment, was US$195 million; annual operations cost is US$11 million.

Specifications

Contractors: Goddard (project management; development/integration of observatory, development of PCA), MIT (ASM, Experiment Data System), UCSD (HEXTE)
Launch: 30 December 1995 by Delta 7920-10 from Cape Canaveral
Lifetime: 24 months nominal mission (up to 5 years possible)
Mass: 3,045 kg at launch
Orbit: 580 km, 23°
Spacecraft: >6°/min manoeuvres via four 75 Nms Ithaco RWs, pointing accuracy <0.1° (knowledge from star trackers + gyros to 60 arcsec). PCA/HEXTE FOV can be pointed to any position if Sun angle >30°. Command/datalinks through TDRS: 32 kbit/s (26 kbit/s science) plus 256 kbit/s for 30 minutes daily. Science Operations Center at Goddard. Two pointable high gain antennas maintain TDRS link while XTE pointed at any target. Twin 3-panel rotating solar wings (total 17.86 m^2) provides 800 W; 2 × 50 Ah batteries
Payload: PCA. Five Xe proportional counters, 6,250 cm^2, 2-60 keV, 1 × 1° FWHM, 1 μs resolution, 0.1 mCrab sensitivity (10 min), 18 kbit/s. Smaller version flown on HEAO 1. NASA Goddard.

HEXTE. Two rocking (every 15 s) clusters of four NaI/CsI detectors, 1,600 cm^2, 20-200 keV, 1 × 1° FWHM (rocking), 10 μs resolution, 1 mCrab sensitivity (10^5 s), 5 kbit/s. University of California at San Diego.

ASM. Three scanning shadow cameras on rotating boom, 90 cm^2, 2-10 keV in 3 channels, 0.2 × 1° FOV, 1.5 h resolution, 30 mCrab sensitivity (1.5 h), 3 kbit/s. MIT.

Solar Anomalous and Magnetospheric Particle Explorer (SAMPEX)

Background

SAMPEX is detecting solar/interplanetary charged particles and galactic cosmic rays of energies from 0.4 MeV to hundreds of MeV while over the poles, and observing magnetospheric electrons before they interact with the middle atmosphere. Of the instruments on board, LEICA and HILT were built for GAS (Getaway Special) Shuttle missions but not flown because of delays. Another instrument, MAST/PET, was partially fabricated to fly on the US probe half of Ulysses.

SAMPEX science operations began 10 July 1992. The LEICA initially had power problems, which later resolved themselves. MAST confirmed a belt of anomalous cosmic rays (particles resulting from the Sun's interaction with the interstellar medium) at 600 km within the inner van Allen belt. A US/Russian team in 1991 found the first clear evidence from Cosmos satellite data, but could not determine its location. Because of its orbit, SAMPEX can sample interstellar gas directly from this belt.

Specifications

Contractors: GSFC provided the satellite for the University of Maryland's payload
Launch: 3 July 1992 by Scout from Vandenberg AFB
Lifetime: 3 years nominal

SAMPEX was the first Small Explorer ***1994***

Mass: 158 kg (41 kg science)
Orbit: 550 × 675 km, 81.67°
Spacecraft: 100 W orbit average provided by four GaAs solar panels (3.4 m^2 total) + one 9 Ah Ni/Cd battery (science instruments require 22 W). 32 Mbyte memory, telemetry by NASA standard 5 W S-band transponder: data down at 900 kbit/s for 10 minutes twice daily. Attitude control by MW, three magnetorquers, two-axis fine Sun sensor, four coarse Sun sensors, 3-axis magnetometer. 86 cm diameter × 1.5 m length launch envelope
Payload: LEICA Low-Energy Ion Composition Analyser (8.0 kg, 0.16-25 MeV, University of Maryland).

HILT Heavy Ion Large Telescope (21.0 kg, 3.9-177 MeV, Max Planck Institute Germany).

MAST Mass Spectrometer Telescope (7.5 kg, 7-470 MeV, California Institute of Technology).

PET Proton Electron Telescope (part of MAST, 1-500 MeV).

Sub-millimeter Wave Astronomy Explorer Satellite (SWAS)

Background:

The Sub-millimeter Wave Astronomy Explorer Satellite (SWAS) will observe dense galactic molecular clouds, particularly at water, molecular oxygen, neutral carbon and carbon monoxide wavelengths for information on chemistry, energy balance and structure. Ball Aerospace's Electro-Optics/Cryogenics Division received a US$14.5 million NASA contract in the autumn of 1989 for the primary SWAS instrument.

Specifications

Contractors: GSFC provided the spacecraft for the Smithsonian Astrophysical Observatory payload; Ball Aerospace (antenna, star tracker, instrument integration); Millitech Corp (sub-mm heterodyne receiver)
Launch: 6 December 1998 by Pegasus XL/Lockheed L-1011 from VAFB
Lifetime: nominally two years
Mass: 283 kg (science 102 kg)
Orbit: planned 637 × 653 km, 70°
Spacecraft: 97 cm diameter. Payload attaches to top of bus as single module. 3-axis stellar pointing to 38 arcsec (<19 arcsec jitter), typically three to five targets per orbit. Nods up to 3° off-target in 15 s increments every 40 s for background reading. Control by three magnetorquers, digital Sun sensor, five coarse Sun sensors, three MWs, magnetometer, three inertial gyros, high accuracy CCD star tracker. 200 W orbit average provided by four deployed fixed + one body panel (3.4 m^2 total GaAs) plus one 21 Ah Ni/Cd battery (science instruments require 50 W orbit average). 100 Mbyte bulk memory, 1.8 Mbit/s rate via NASA standard S-band 5 W transponder. 2 kbit/s commands up
Payload: 55 × 71 cm diameter off-axis f/4.8 Cassegrain antenna feeding two sub-mm receivers (cooled to 150K) for University of Cologne's acousto-optical spectrometer (1.4 GHz bandwidth, 1,400 channels, 2.2 GHz centre frequency).

Gravity Probe

Background

Following launch by Delta in October 2000, GP-B will compare the observed precession of gyroscopes in Earth orbit with that predicted by Einstein's theory of

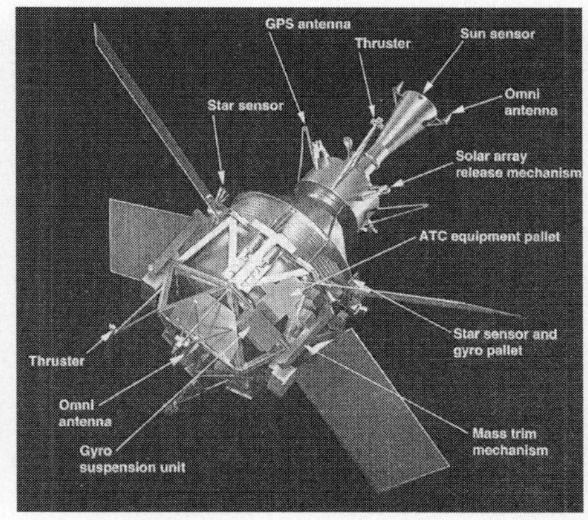

GP-B will test Einstein's theory of general relativity by measuring the precesion of gyroscopes in Earth orbit (NASA Marshall) ***1995***

general relativity. Stanford University is leading the experiment for NASA; to put angular inertial measuring instruments on a free-flying 650 km polar satellite to look for the expected 6.6 arcsec/year precession perpendicular to Earth's rotational axis and the 0.042 arcsec/year parallel to it.

In order to distinguish between fine grained effects, the 3,300 kg symmetric spacecraft will slowly roll about the line of sight to the guide star. The spin axis will be measured to 0.1 arcsec by superconducting readout of the gyro's magnetic 'London moment' and compared to the line of sight observations from the reference telescope. Maintaining the proper conditions requires near absolute zero temperatures. All four gyros will be cooled to 1.8K by 2,000 litres of superfluid helium, which should be sufficient to keep them cooled for up to 18 months.

Each gyro is electrically suspended and will spin at more than 100 Hz and operate in a 10^{-14}atm vacuum and a less than 10^{-7} gauss magnetic field. To reduce translational contamination, a drag compensation system will reduce the mean gyro acceleration to less than 10^{-10} *g* employing thrusters utilising the Helium boil-off from the dewar. Up to 600 W of power comes from the GaAs wings, spanning 8.8 m and supported by two 35 Ah batteries.

Stanford researchers built a ground-based prototype SIA Science Instrument Assembly with two gyros operating at 1.8K. The second configuration began tests in February 1995, using a flight prototype SIA with four gyros supported by the flight back-up hardware and the engineering development dewar. The third configuration in 1996 tested a protoflight SIA with the flight back-up support hardware and the flight dewar. Ground test of the full flight science payload in 1998 will verify the payload for flight. A Shuttle Test Unit was planned to establish the characteristics of the suspension system in μg, but it was cancelled in early 1994 in a major programme reorganisation.

NASA's cost estimate is US$280 million from FY96. US$10.3 million was expended during FY88 and US$17.9 million in FY89, but GP-B was then deleted from FY90's budget request by the agency. However, congressional review re-instated the programme with US$25 million FY92, 1.9 FY93; 42.4 FY94, 50 FY95, 51.5 FY96 and 59.6 requested FY97. Stanford is supported by two major subcontracts with Lockheed Martin Missiles & Space Co for the spacecraft, dewars and experiment support hardware.

Gravity Probe-A started the series in 1976, when a Scout suborbital package launched it to 10,000 km altitude, on a 116 minute mission over the Atlantic on 18 June 1976, carrying a hydrogen maser clock for telemetric comparison with a duplicate terrestrial version to test the equivalence principle of general relativity.

CHANDRA (AXAF)

Current status

NASA renamed the Advanced X-ray Astrophysics Facility (AXAF) after the renowned astrophysicist and Nobel-Laureate Subrahmanyan Chandrasekhar, giving it the name CHANDRA in late 1998. CHANDRA was launched by STS-93 on 23 July 1999.

Background

The Advanced X-ray Astrophysics Facility (CHANDRA) constitutes the third in NASA's Great Observatories Series, following the Hubble Space Telescope and the Compton Observatory. AXAF-I (Advanced X-Ray Astrophysics Facility) will resume the work started by Uhuru in 1970 and continued by HEAO 2 until re-entry in 1981. AXAF-I's four pairs of nested mirrors for focusing 0.1-10 keV high-energy x-rays will offer about 10 times the spatial resolution and 100 times the sensitivity of HEAO 2. With 1.2 m outer diameter, the 10 m focal length mirrors will provide 1,100 cm^2 collecting area (instead of the 1,700 cm^2 of the original six pairs).

In March 1985, NASA announced 11 investigators who will provide scientific and technical guidance from initial design to on-orbit operation; nine were US, one Netherlands and one UK. To initiate the programme, Congress gave NASA Marshall US$29 million start-up funding in FY89. In subsequent years NASA allocated US$64 million in FY90, US$113 million in FY91, US$141 million in FY92, US$158 million in FY93. The major contracting work began with the authorisation of US$224 million for AXAF-I in FY94 (reduced from the original request of US$260 million for both AXAF-I/AXAF-S), 234.3 FY95, 237.6 FY96, 178.6 FY97, 92.2 FY98. Congress terminated funding for AXAF-S in November 1993 as part of a far-reaching NASA cost-reduction exercise. The US$60 million shortfall in FY92's request, and Congress's declaration of no FY93 increase, slipped the launch a year to 23 July 1999.

NASA approved a facility design in 1991 for the US$86.7 million contract to design and operate the AXAF Science Center in Cambridge, Massachusetts. Under a US$40.7 million contract to the MIT Center for Space Research, in support of the Principal Investigator at PSU, Martin Marietta was selected in 1986 as subcontractor for the CCD Imaging Spectrometer; the company's Space Systems division was awarded the US$13.1 million contract in mid-1990. The HRC instrument is primarily an in-house effort at SAO, utilising a few small subcontracts. Ball Aerospace Systems Group in Broomfield, Colorado was awarded the US$24 million contract in September 1990 by GSFC for the XRS 490 l helium dewar to cool the instrument to 1.3K. Fokker Space is providing the solar arrays.

Preliminary plans called for the US$1.4 billion 5,200 kg AXAF-I (imaging) to be launched by the Shuttle with an unspecified upper stage in September 1998 into a 10,000 × 140,000 km, 28.5° orbit. The US$320 million 2,700 kg AXAF-S (spectroscopy) would follow in December 1999 aboard Delta into polar orbit. NASA wanted both to last at least five years in orbit. AXAF-I consists of a four-mirror pair-shell grazing incidence telescope (instead of the original six parabola/hyperbola pairs), two gratings and two focal plane instruments. Boeing's agreed to provide an IUS upper stage in May 1994 under a US$25 million contract to establish the final orbit. AXAF-S changed the replicated optics telescope with the supercooled x-ray spectrometer in the focal plane.

NASA adopted a phased programme to avoid HST-type cost overruns. Hughes Danbury Optical Systems Inc began grinding the largest parabola/hyperbola pair (P1/H1) of the four paired nested grazing incidence mirrors in December 1989 under TRW's US$158.1 million contract. Congress, as a measure to check NASA's repeated history of cost escalation, required the agency to make a semi-annual progress report through the completion of testing on 1 October 1991. Testing on P1/H1 was completed that September in MSFC's new US$17 million X-Ray Calibration Facility (XRCF).

All eight flight mirrors were completed in early 1995, followed by coating at Optical Coating Laboratories, Inc (H1 was the last, in January 1996) and delivered to Eastman Kodak. Kodak provides the optical bench and assembles the eight mirror elements (HRMA High Resolution Mirror Assembly) and the telescope. Ball Aerospace Electro-Optics/Cryogenics Div will supply a visible wavelength aspect camera for attitude reference. Ball will also provide the SIM Science Instrument Module, which supports the focal plane instruments, provides focus control and translates the instrument in/out of the focal position as required.

Specifications

Launch: 23 July 1999 by shuttle mission STS-93 from Kennedy Space Centre using IUS as boost stage
Contractors: TRW Space and Electronics Group
Mass: 4,790 kg
Orbit: 9,977 × 138,400 km, 64 hr orbit.
Spacecraft: 13.8 m long, 19.5 m span across deployed solar panels. Mission life >5 years. Power provided by 3-panel, silicon solar arrays (2.35 KW) and 3 × 40 Ah metal/hydrogen batteries for eclipse power. Data recording from solid-state recorder with 1.8 giga-bytes (16.8 hr) capability
Payload: AXAF CCD Imaging Spectrometer, sub-arcsec angular resolution (ACIS, Pennsylvania State University/MIT).

High-Resolution Camera, microchannel plate imager with less than 0.5 arcsec angular resolution (HRC, Smithsonian Astrophysical Observatory).

High-Energy Transmission Grating Spectrometer, 0.4-8.0 keV, resolution 800 at 1.0 keV.

ACIS, images spectral dispersion of grating (HETGS, MIT).

Low-Energy Transmission Grating Spectrometer, 0.1-3.0 keV, resolution 750 at low energies, HRC images spectral dispersion of grating (LETGS, Utrecht Lab for Space Research).

Payload weight 954 kg. Focal length 10m, outer diameter 1.2m

Compton Gamma Ray Observatory (Great Observatories series)

Background

Before launch, this satellite was simply known as the Gamma Ray Observatory, but shortly after launch, it was re-named the Compton Gamma Ray Observatory (CGRO) in honour of Arthur H Compton, the US joint-Nobel laureate in physics. One of the largest astronomical satellites ever launched, Compton is providing a deeper understanding of high-energy processes within supernovas, pulsars, quasars and radio galaxies, illuminating the character of the Universe shortly after its creation, and indicating the possible existence of anti-matter in the Universe.

As its scientific priority, the Compton seeks to explain the typically 1 second gamma bursts that sweep across the solar system. BATSE had detected 1,500 by the end of 1995. The brightest occurred on 31 January 1993. The first scientists expected they would concentrate in the galactic disc, since they felt the most likely origin was detonations on the surfaces of neutron stars. However, BATSE has found them uniformly over the sky. This indicates they originate either from some exotic small objects near the Solar System or from extremely powerful distant objects well beyond the galaxy. The latter source is more likely, requiring a new type of celestial body generating unprecedented quantities of energy from previously unknown processes. Such an object would have to channel many times the energy of a supernova into a 1 second gamma-only flash.

In another discovery, Compton's EGRET instrument found that quasar 3C-279 emits 10^8 times the galaxy's gamma energy, a step towards explaining the nature of these exotic objects. In early 1993, Germany's COMPTEL made the first detection of a Ti-44 γ line, in the Cas-A supernova remnant. As Ti-44 has a half-life of only 70 years, this line can now be used to look for nearby but obscured supernovas in our galaxy. COMPTEL also found Al-26 emissions from the 10,000 year-old Vela SuperNova (SN). Al-26's 10^6 year half-life means it can be used to look for traces of very old nearby SNs.

COMPTEL's third major 1993 discovery was to identify the Orion nebula as a source of cosmic rays. Such particles filling the galaxy were discovered in 1911 but their origin was a mystery. Following a 17 February 1994 gamma burst, EGRET detected high energy gamma rays for more than 90 minutes. OSSE has detected a cyclotron spectral line from the galactic neutron star A0535+26, indicating a surface magnetic field of more than 5 × 10^{12} gauss.

Specifications

Launch: 5 April 1991 by Shuttle from KSC; deployed 7 April 1991
Contractors: TRW Space & Communications Co
Mass: 15,623 kg at release
Orbit: 445 × 459 km, 28.45° at release; decayed to 345 km by mid-1993; reboosted to 441 × 452 km October to December 1993; decayed to 386 km by March 1996
Spacecraft: 7.6 m long, 3.8 m diameter, solar array span 21.3 m. 3-axis stabilised. Based on 6.30 m long, 3.15 m wide, 61 cm deep platform of aluminium sill/keel beams. Mission life 2¼ years but four tanks carry 1,814 kg hydrazine sufficient for 10 years AOCS via 4 × 445 N orbit adjust + 8 × 22.2 N attitude control thrusters (although not planned, can be refuelled in orbit; total impulse 4.03 MNs). 500 kg hydrazine is reserved for controlled re-entry at mission's end. Control torques and momentum storage provided by four reaction wheel assemblies. Inertial Reference Unit is a self-contained strapdown three-axis dual redundant attitude rate sensing unit, providing reference for high rate (±2°/s) manoeuvring and low rate (±400 arcsec/s) for pointing; associated with fixed head star trackers, fine/coarse Sun sensor assemblies and 3-axis magnetometer. Two solar wings of four 137 × 336 cm panels each (total of 40,488 Si cells) provide 4,500 W BOL/3,980 W EOL to two modular power systems, each supported by three 50 Ah Ni/Cd batteries (one battery has failed; three are adequate). Recorded data were returned at 256/512 kbit/s via 2.2875 GHz S-band 152 cm diameter high gain antenna on 442 cm boom working through TDRSS; both tape recorders have failed (the first in 1991). TDRS 1 was moved to 85° east February 1994 to provide a dedicated 32 kbit/s real-time link through Australia's Tidbinbilla for relay to White Sands via Intelsat, increasing coverage from 65 to 90%. TDRS 3 took over in June 1995. Command rate 1 kbit/s, backed up by 0.125 kbit/s through two omni antennas at 2.1064 GHz
Payload: Oriented Scintillation Spectrometer Experiment (OSSE). Four identical 33 cm diameter NaI (Tl)-CsI(Tl) phoswich detectors for 0.05-10 MeV measurements. Operated in pairs, they can be turned through 180° for background and secondary target monitoring. Tungsten collimator provides 3.8 × 11.4° FOV (resolution). Naval Research Lab; Ball Aerospace.

Imaging Compton Telescope (COMPTEL). 2° resolution over 1.0-30 MeV using two planes of scintillators 1.5 m apart. Compton's most sensitive instrument. Max Planck Institute (Germany); MBB prime.

Energetic Gamma-Ray Experiment Telescope (EGRET). Operating at top end of Compton's energy range, 20-30,000 MeV, using 6,400 cm^2 spark chamber and providing best resolution (0.5° at 2 GeV). NASA Goddard.

Burst and Transient Source Experiment (BATSE). Eight wide field detectors covering the whole celestial sphere (4π radians) each with 50.8 cm diameter, 1.3 cm thick NaI(Tl) scintillation crystal to register 0.20-100 MeV transient events of a few μs and locating positions to within a few degrees.

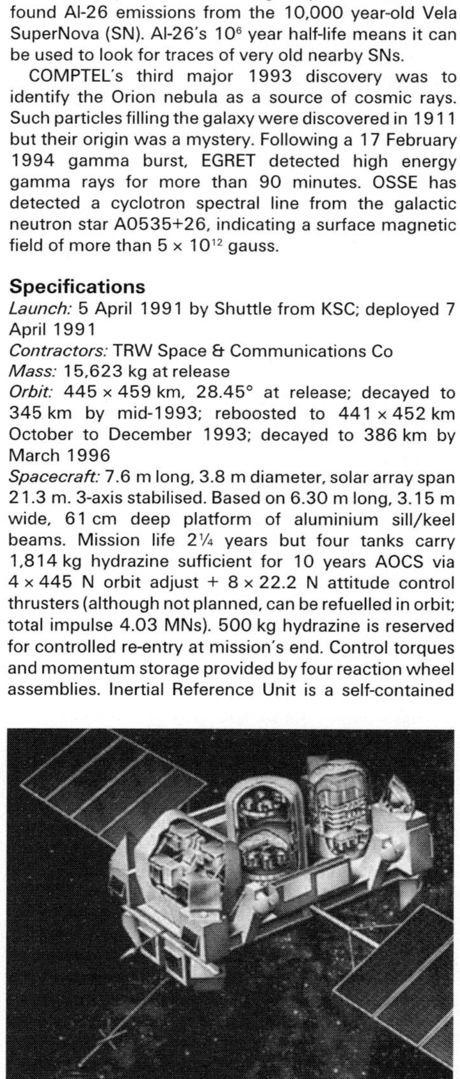

NASA's Compton Observatory is probing the Universe's higher energy processes. COMPTEL is the central cylinder, EGRET is to the right and OSSE is housed in the unit at left. The eight BATSE burst detectors are mounted on the platform's corners (NASA)

Hubble Space Telescope (Great Observatories series)

Background

Renamed the Edwin P Hubble Space Telescope, after the eminent American astronomer in October 1983, this much-postponed project was described by NASA as the most important scientific instrument ever flown. NASA expected the Hubble to resolve objects 50 times fainter and seven times more distant than those visible from Earth-based telescopes. In the most distant views, astronomers would be observing the universe as it was around 14 billion years ago.

NASA began studies of a Large Space Telescope of 3 m aperture in 1971 and 1972. The studies followed closely a realignment of NASA priorities with those of the air force in order to build a space plane according to air force dictates. At the time, the air force planned to launch a new generation of 'Keyhole' close-look spy satellites and, instead of a scientific instrument, the definition studies followed closely the specifications for these satellites.

Astronomers originally expected to launch the Telescope on an expendable booster into an efficient sun synchronous orbit and thereby produce more of a scientific payoff. However, in its final programme review before manufacturing began, NASA shrank the allowable aperture for the Telescope to its current size in May 1975.

ESA agreed to make a 15 per cent contribution in October 1976. In 1976 terms the full ESA cost was projected to be US$595 million, but delays and technical problems caused a rise to more than US$1 billion. NASA promised a return of 15 per cent of observation time for European astronomers. In fact, observing allocation continues to be 20 per cent in open competition.

NASA awarded the US$69.4 million contract for the mirrors and fine guidance sensors to Perkin-Elmer in 1977 following a competition with Kodak Federal System Division of Rochester, New York. Corning Glass, New York cast the 900 kg primary mirror in October 1977. Perkin-Elmer began nine months of fine polishing in August 1980, and aluminised the completed mirror in December 1981. Kodak produced a back-up mirror, although never fully polished it. The Kodak mirror remains in storage.

The main elements came together during 1985 for the first verification programme to begin, although NASA never asked for an end-to-end optical test. BAe supplied its two solar array wings in 1985 under a 1979 £13 million contract for the 1986 launch. Following the Challenger accident in January 1986 however, the

Hubble's first servicing mission (NASA) ***1995***

arrays were returned to BAe for a £2 million modification programme. BAe added features to facilitate EVA exchange with new arrays, and the new arrays arrived at Lockheed in April 1989 for installation.

Following the Shuttle accident, NASA delayed Hubble's launch from October 1986 until a possible window in 1988 or 1989. Many astronomers doubted whether the Hubble would have been ready for the 1986 launch anyway. One principal stumbling block was the data transmission of collected images from Hubble to the ground. The Hubble, like its KH-9 counterparts had to use relay satellites or foreign ground stations to receive information. KH-8 satellites used ground stations at remote locations and relays in common orbits with the spy satellites. The TDRS constellation was designed to replace these expensive links. Unfortunately, the air force decided, after the TDRS programme started, that the links needed to be hardened against radio frequency and laser attack from the Soviet Union. The air force argued that the Soviet Union could blind all US intelligence gathering capability with such attacks. In order to make TDRS attack-hardened, the transponders had to be retrofitted at a large cost over-run to the programme and about a two year delay in deployment of the full network. This similarly delayed the relays for the Hubble and without the national security argument behind it, it was unclear how Hubble could use a ground-based data relay network for its image production, let alone do it in a timely fashion. Hence, the Telescope would have remained under-used in orbit while the ground system was brought to full operational capability.

As NASA grappled with this problem, the guide star catalogue for steering the Shuttle had to be finished as well. ST ScI completed the initial Guide Star Catalogue in 1989 after four years of work. When finished, it constituted the largest-ever sky survey. A total of 1,477, 6.4 × 6.4° plates exposed by the US Palomar Schmidt and UK Australian Schmidt telescopes had to be digitised to generate position and magnitude data on 18,819,291 objects for HST's three guidance sensors. Some 15 million objects in the catalogue are stars, most of the remainder are galaxies.

HST arrived at Kennedy Space Center 11 October 1989 aboard a USAF C-5 aircraft for its eventual Shuttle launch. Ground technicians powered it up for the first time on 28 October by remote command from Sunnyvale. Then HST underwent an 11 day GST-8

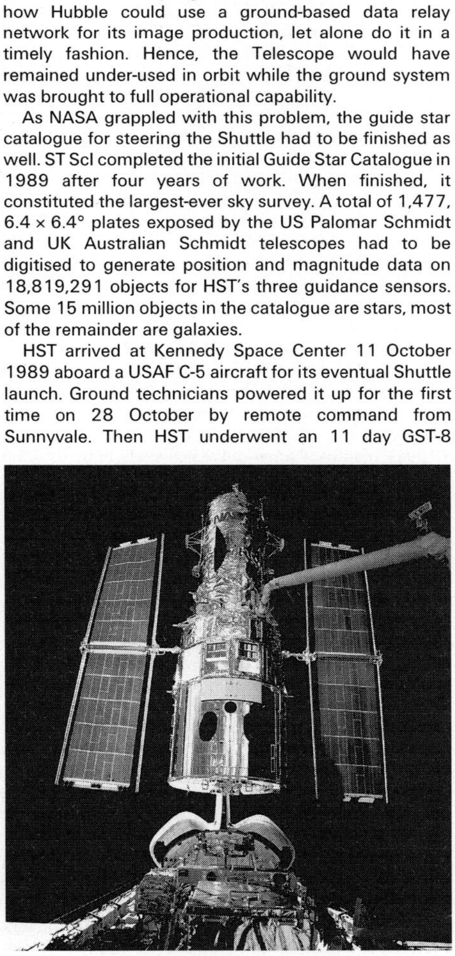

Hubble deployment by STS-82. Photographed by an aft-mounted wide-angle IMAX camera (NASA) ***1997***/0006567

ground systems test that concluded 8 December 1989, including simulation of early orbit operations and safemode recovery.

After a launch on the Shuttle Discovery, Hubble entered orbit at 20.39 GMT 25 April 1990 619 km over the east Pacific. NASA planned that Orbital Verification (OV) would take three months, followed by five months' Science Verification (SV) in which the instruments would be checked out and calibrated, while allowing some of the Guaranteed Time Observer programmes. Instead the OV took over 19 months.

Once in its lower and less energetic non-sun synchronous orbit, HST passed out of the earth's shadow. Differential heating caused the telescope to ring like a bell for nearly half of the orbit. This ringing prevented precise tracking of the stars that is required for gathering a high resolution image. Lockheed attempted several software 'patches' to correct the problem, but all of them failed to fully alleviate the problem. Later the air force admitted it had known about the problem in advance of the launch, because of similar problems experienced on its KH-9 spy satellites. The air force had refused to alert NASA to the problem, however, for national security reasons. This decision cost the Hubble more than half of its available data gathering time.

The air force nearly cost the United States the use of the HST at another point as well. After release from the Shuttle, Hubble entered its deployment phase in which the solar panels are unfurled. Many problems had developed as the panels unfurled, but the crew, which could have assisted, had moved on to deliver a classified payload to a different orbit.

Cycle 1 of the full observational schedule was to begin towards the end of 1990. The first image, of Carina's NGC 3532 open star cluster, chosen not for scientific reasons, but so that NASA could release a publicity photo, was returned by WF/PC on 20 May 1990. Resolution was 0.7 arcsec, as against the expected 1.5 arcsec, but an improvement by a factor of seven was still expected. This first light observation showed light, which appeared to ground-based telescopes as a single star, was, in fact, composed of two stars. Astronomers had long suspected the fact, but let NASA use the information anyway, for a press release.

After first light was achieved 6 June, it soon became apparent that a large flaw existed in the optics. The first images showed that HST suffers from classical spherical aberration, caused by a 2 μm grinding error around the primary's edge. The mirror's 24 actuators could not compensate and would complicate the 'pure' aberration (the final compromise focus was not decided on until November 1991). Specifications required 70 per cent of the incoming light to fall within a 0.2 arcsec area, and measurements showed 1.4 arcsec. Some 15 per cent covered 0.1 arcsec, with the remainder spread across 1 arcsec (equivalent to ground-based performance). Computer processing for objects seen against a dark background could achieve the target 0.1 arcsec resolution. Low sky background at shorter wavelengths meant that ultra-violet observations were largely unaffected – the visible work suffered most. WF/PC was the most serious loss: it was to have taken 40 per cent of the observing time, but occupied only about 10 per cent.

Hubble was launched with several principal scientific objectives. One key objective was the measurement of the Universe's expansion rate (Hubble constant) by observing the brightness of magnitude 24 and fainter stars (Cepheid variables) in distant galaxies with WF/PC. With the flaws in the optics definitive observations had to wait for WF/PC-II arrived later. ESA's reduced its FOC work by about half after visible observations offered no improvement over ground systems.

The independent six-man HST Optical Systems Board of Investigation established 2 July under JPL Director Dr Lew Allen reported in late November 1990 that the primary mirror had been well figured but to the wrong shape. Perkin-Elmer's Reflective Null Corrector laser interferometer, used to check the figure around 1981, apparently had a lens 1.3 mm out of position. Perkin-Elmer had some indication of the problem, but disregarded the test results. Kodak's bid for the 1977 contract included US$10 million for an end-to-end test, which would have immediately identified the problem.

The Board concluded that the defect could not be solved by a global repair in-orbit but individual instruments could be corrected optically. NASA planned anyway, that WF/PC would be replaced by a second-generation version on SM1. FOC, HRS and FOS are corrected by replacing HSP with the 290 kg 220 cm long 87.6 cm². COSTAR Correcting Optics Space Telescope Axial Replacement, carrying a boom to insert 10 mirrors into the light paths (one picks off the light for a second to provide the correction) before they reach the instruments. The coatings call for 56 per cent throughput of 1,216 Å UV. HSP was lost but it was the least-used instrument. Ball Aerospace began work on COSTAR under a February 1991 letter contract, finalised October 1991 at US$30.4 million. ESA considered an on-orbit FOC internal modification and flying a new instrument, although at an estimated US$180 million it proved too costly.

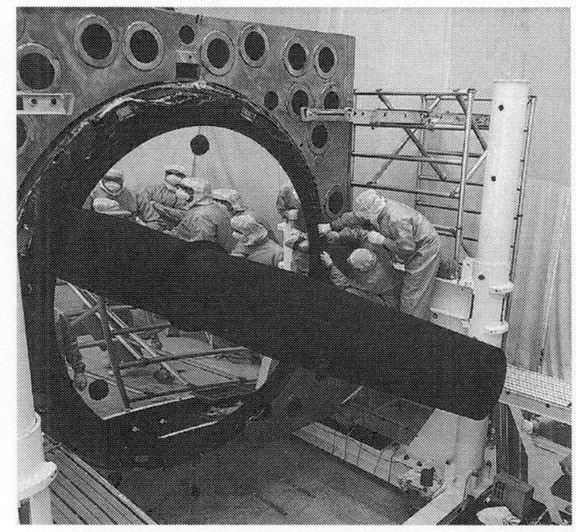

HST's primary 2.4 m, 830 kg mirror with central light baffle installed. The surface is so precise that if it were expanded to cover the entire US its largest bumps would be only a few cm high (Perkin-Elmer)

Early observations by the FOC showed that the specified 0.1 arcsec performance was still possible in some circumstances. HST then gathered images of Supernova 87A's ring, Pluto/Charon and of the R Aquarii red giant/white dwarf double star nova. Scientists then made a major step towards determination of the Hubble constant by accurately measuring the distance to the Large Magellanic Cloud. HST/IUE combined to measure the diameter of SN87A's ring at 1.37 light years, translating to an LMC distance of 169,000 light years ±5 per cent (replacing the previous ±30 per cent spread).

As the Telescope settled into its observation schedule, in 1991 it used the HRS to measure the Universe's deuterium content and found the amount supports the Big Bang and ever-expanding cosmos concepts. HRS observed evidence of planetary formation around star Beta Pictoris and HRS discovered huge hydrogen clouds that, theoretically, should have evolved into galaxies long ago. The WF/PC instrument found evidence of a black hole 2.6 billion times the Sun's mass in a region 1-10 times the size of the solar system at the centre of galaxy M87.

In 1992, the HST uncovered strong evidence that discs of planet-forming material surround many stars. WF/PC found 15 in the Orion Nebula; only four were known before HST. Work continued on refining the Hubble constant, but it really required improved optics.

SM1, the first telescope maintenance mission solved a second major problem in December 1993. It corrected the pointing flaws due to the thermal ringing. The ringing produced two modes: a 0.1 Hz end-to-end movement lasting up to 10 minutes after terminator crossing, and a lesser 0.6 Hz transverse action during daylight. The solar array tips flexed up to 1 m under worst conditions. The wobble prevented long exposures. Previous software fixes took up too much of HST antiquated computer memory. SM-1 put in place replacement solar arrays (STSA-2) and an aluminium layer supported by 900 plastic discs in an accordion-like structure covered each modified boom. This reduced thermal gradients by a factor of 20. Then BAe also built a system for countering expansion at the array tips with frictionless springs and immobilised the storage drums by an electrical brake. They were found to reduce jittering by a factor of 10. These arrays should be sufficient until about 2000.

SM1 also replaced two gyro pair packages. The number 6 rate gyro failed 3 December 1990, but one of two back-ups assumed its duties. Then number 4 failed July 1991, possibly from radiation damage of the hybrid

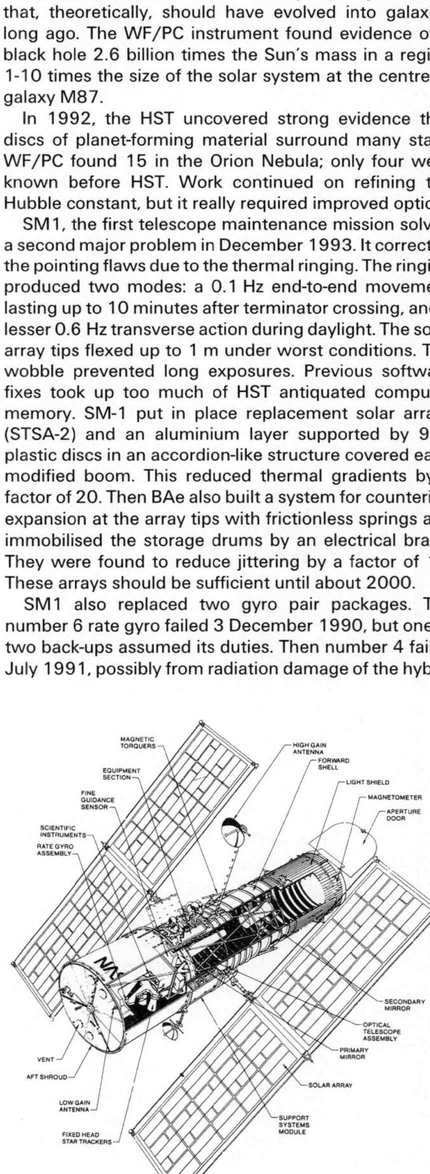

Hubble Space Technology

circuits generated by high solar activity. Number 5, too, began showing signs of distress, possibly because of dirt in its rotor mechanism. Number 3 failed in November 1992, probably because of a random electronic part failure in the control unit. If they had all failed simultaneously before replacement, HST would have entered a magnetic torque mode to hold safe attitude.

SM1 achieved all objectives in December 1993 during five EVAs totalling more than 35 hours by four crew members.

EVA-1: Replaced Rate Sensing Units numbers 2/3 (each with two gyros) and control units for numbers 1/3, and eight of 12 3 A fuse plugs that protect HST's electrical circuits with 5 A versions.

EVA-2: replaced starboard solar array discarded into orbit. Port wing removed/stowed for return to Earth; two new wings attached.

EVA-3: replaced WF/PC with WF/PC-II. Two new magnetometers installed on top of failed pair at the aperture end.

EVA-4: replaced HSP with COSTAR. Added 386 coprocessor to DF-224 computer.

EVA-5: replaced Solar Array Drive Electronics. Both wings unfurled by 10.35 GMT 9 December; HST switched to array power 15.30 GMT. Installed HRS redundancy kit.

By 1994, NASA had corrected HST's defective optics and observations with the improved optics provided seemingly conclusive evidence for this black hole. WF/PC also began a long-term programme of Mars observations to characterise its climatic processes and surface changes at a resolution of 50 km. 1994 observations showed that Mars' average temperature had dropped 20°C since Viking, but the climate now appears more stable.

In 1994, WF/PC-II observations of 20 Cepheid variables in M100 placed the galaxy at 56±6 million light-years, yielding a Hubble constant equivalent to an age of 8 to 12 billion years for the Universe, instead of the generally accepted 15 billion years. If confirmed, it would require a major reworking of cosmology. However, many more galaxies must be measured because those around the Virgo cluster are mutually perturbed by their concentration. The FOC in 1994 provided the first image of star G1623b, the red dwarf companion in a double star system in Hercules and one of the smallest stars known in our Galaxy. It has only 10 per cent of the Sun's mass and is 60,000 times fainter – it would appear only eight times brighter than the full Moon from 1 AU. Its presence was known from astrometric measurements of its parent but ground-based telescopes were unable to resolve it. Red dwarfs were believed to be the most abundant stars but Hubble's observations show them to be surprisingly rare. They cannot therefore account for much of the Universe's missing mass.

From a study of very young galaxies, HST provided the surprising result that elliptical galaxies formed very early on, but spirals took much longer and most were torn apart. FOC observations of a distant quasar revealed the type and quantity of helium predicted by the Big Bang theory. FOC and WF/PC II observations of 525 km asteroid 4 Vesta in November/December 1994 showed the solar system's oldest terrain at 80 km resolution. The gouged surface revealed details below the crust, lava flows and impact basins. The once-molten interior shows that Vesta should be regarded more as a mini-planet, rather than an inert lump of rock.

As the data gathering entered its full programme, astronomers became disenchanted with the programme as the consequences of working with NASA became apparent. At first, NASA promised to allow scientists unfettered access to the Telescope through the TDRS datalinking network. In fact, NASA had an agreement with the US Air Force to allow the military to have priority linkage through the data relays for nuclear attack planning. NASA also promised the Hubble could be repaired on orbit at least every 30 months. The solar arrays were to be retracted, first every five years, then every three years, and the whole vehicle brought to Earth for refurbishment and relaunch about a year later. By 1987, NASA announced it would avoid Earth return if at all possible and make in-orbit servicing more infrequent.

WF/PC-II observations in 1995 confirmed the existence of the postulated Kuiper Belt of comets 40-500 AU out, beginning just beyond Neptune and the source of comets that orbit the Sun in periods of greater than 200 years. The observations indicate the belt hosts at least 200 million comets that have remained essentially unchanged since the solar system formed. Other 1995 observations strongly suggest that volcanoes are still erupting on Jupiter's moon Io and that Ganymede has a tenuous oxygen atmosphere, created from the bombardment of water ice by Jupiter's charged particles. The first unambiguous brown dwarf was confirmed: Gliese 229B is the faintest object seen orbiting another star. At 20-50 Jupiter masses, it is too large to be classed as a planet but too small and cool (720°C) to shine as a star.

The second HST servicing missions began with a Shuttle launch on 11 February 1997 as the STS-82 flight of Discovery.

When the Shuttle had captured HST and the astronauts were able to look in detail at the telescope's thermal insulation cover, it was found that it had degraded more than expected during the three and a half years since its first servicing mission. As a result, an extra EVA was scheduled so that some running repairs could be done. During the servicing:

EVA-1 (Lee/Smith): removed the High-Resolution Spectrograph and the Faint Object Spectrograph and replaced these with the Space Telescope Imaging Spectrograph and the Near Infra-red Camera and Multi-Object Spectrometer.

EVA-2 (Harbaugh/Tanner): replaced the degraded Fine Guidance Sensor 1 and Engineering/Science Tape Recorder 2 with new models: installed Optical Control Electronics Enhancement Kit. It was during this EVA that the insulation degradation was first noticed.

EVA-3 (Lee/Smith): removed and replaced Data Interface Unit 2 and a reel-to-reel ESTR with a new digital Solid-State Recorder that permits simultaneous recording and playback of data.

EVA-4 (Harbaugh/Tanner): replace Solar Array Drive Electronics 2, replace the covers of the Magnetic Sensing System and replaced some of the thermal sets on two areas around the light shield of the telescope.

1 Scientific Instrument Control and Data Handling
2 Faint Object Camera
3 Wide Field/Planetary Camera
4 Data Interface Unit for Optical Telescope Assembly
5 High Gain Antenna (2)*
6 Data Management Unit
7 DF-224 Computer
8 Fine Guidance Electronics (3)
9 Actuator Control Electronics*
10 Battery (6)
11 Fixed Head Star Tracker (3)
12 Rate Sensor Unit (3)
13 Charge Current Controller (3)
14 Data Interface Unit for Support Systems Module
15 Electrical Power Thermal Control Electronics
16 Power Control Unit
17 Power Distribution Unit
18 Fuses (14)
19 Multiple Access Transponder (2)
20 Single Access Transmitter (2)
21 Faint Object Spectrograph
22 Fine Guidance Sensor (3)
23 Reaction Wheel Assembly (4)
24 Solar Array (2)
25 Low Gain Antenna
26 Sun Sensor

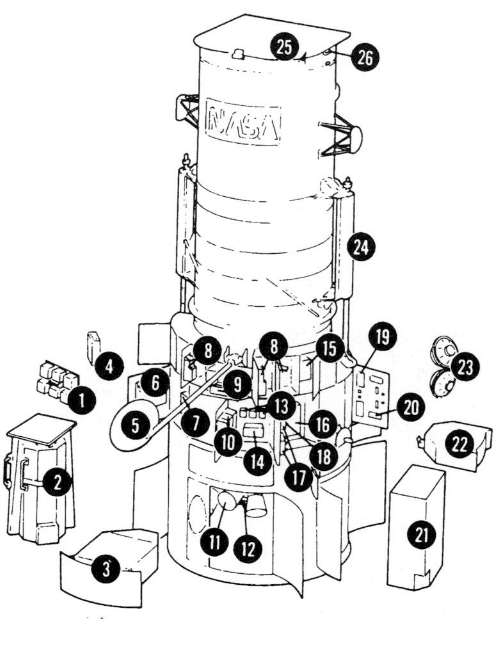

Hubble Space Technology

Hubble Space Technology

EVA-5 (Lee/Smith): attached replacement thermal coverings on three equipment compartments at the top of the Support System Module.

While attached to Discovery, the orbiter was used to re-boost the altitude of HST: before being captured it had been in a 590-599 km orbit and when released it was in a slightly higher 599 to 620 km orbit.

Degradation of several key systems and the failure of two crucial gyroscopes out of sync, plus intermittent performance from a fourth, stimulated the HST team to request an interim servicing mission before the planned visit in late 2000. Dubbed mission 3A, the interim visit has been assigned to Shuttle Discovery (OV-103) on mission STS-103 scheduled for launch on 14 October 1999. In addition to replacing all six control moment gyroscopes, EVA astronauts will install a new HST 486 class computer to dramatically increase computing power, speed and storage capability. They will charge out one of the Fire Guidance Sensors and replace a tape recorder with a new Solid-State Recorder. The EVA crew will also install a new 5-band transmitter and voltage improvement kits for the nickel-cadmium batteries. They will finally carry out repairs to the multilayer insulation on the outer surface of the telescope. Three spacewalks are planned for mission 3A with a fourth as a contingency. Astronauts on servicing mission 3B scheduled for late 2000 will install the Advanced Camera for surveys offering 10 × the resolution of on-board cameras. In addition, new high-efficiency solar arrays from Lockheed Martin Missiles and Space will generate enough power for simultaneous operation of several science instruments. SM-4, the fifth servicing mission planned for late 2002 will upgrade the HST to a state-of-the-art facility.

Observing proposals are still running at five times the level that can be accommodated. Total cost to the end of 15 years of operations was projected in 1989 at US$2,800 million, inclusive of US$1,450 million for development and activities up to launch. In 1990, NASA added US$600 million to the estimate. SM1 was priced at US$674 million, including US$378 million for Shuttle's element. Cost of repairs and correcting for the mirror defects was included at $86 million.

As launch was originally planned for 1983 and full operations were not possible until 1994, advances in the technology of ground-based astronomy have removed some of HST's anticipated unique advantages. It does maintain superiority, however, with its 0.01 arcsec angular resolution (ground telescopes can achieve 0.3 arcsec and then only briefly) and its ability to work with the ultra-violet wavelengths blocked by the atmosphere.

Since launch in 1990 the HST has looked at more than 10,000 astronomical objects, taken more than 120,000 exposures, produced 4.5 trillion bytes of science data and given rise to over 1,700 scientific papers and research reports by astronomers from 35 countries.

Specifications

Launched: 24 April 1990 by Shuttle STS-31 from Kennedy Space Center, released 20.39 GMT 25 April. Serviced 4-10 December 1993

Contractors: NASA Marshall (lead NASA centre), NASA Goddard (instrument development management and vehicle TT&C), Lockheed Missiles & Space Co (vehicle integration, Support Systems Module development, ST Operations Control Center), Perkin-Elmer Corp (Optical Telescope Assembly, Fine Guidance Sensors; company now Hughes Danbury Optical Systems), Ball Aerospace (HRS, NICMOS, STIS, HACE instruments), Dornier/Matra (Faint Object Camera), Martin Marietta (Faint Object Spectrograph), British Aerospace (solar arrays, FOC photon detector assembly), Eagle Picher (batteries),

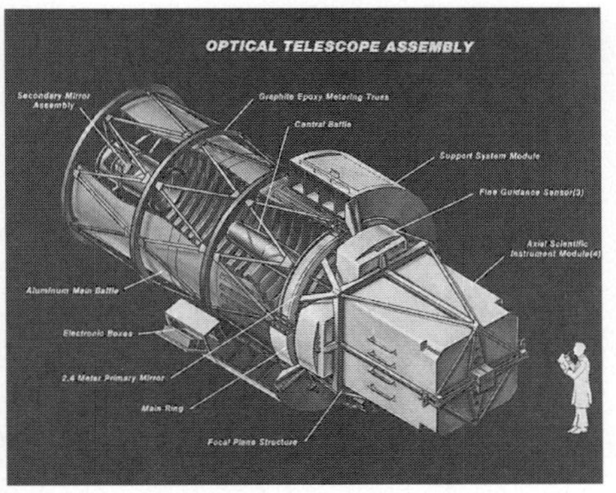

Hubble Space Technology

Boeing (mirror metering truss), Fairchild Space Co (science instrument command/data handling computer), Honeywell Satellite Systems Operations (RW assemblies), Odetics (three DDS-3100 digital tape recorders, to be replaced by SM2)

Mass: initially 10,843 kg on-station; 10,960 kg after SM1

Orbit: released at 613 × 620 km, 28.5° (altitude to minimise atmospheric drag), 586 × 596 km by time of SM1 capture, raised to 592 × 601 km for SM1 release

Lifetime: 15 years, with 3 to 5 years revisits by Shuttle for maintenance, replacement of instruments and reboost. SM1 revisit December 1993; SM2 planned for STS-82 February 1997, SM3 November 1999, SM4 November 2002, SM5 November 2005

Configuration: HST is 13.1 m long (without aperture door), 4.27 to 4.7 m diameter carrying a 2.40 m diameter primary mirror and optical system held rigidly in an Optical Telescope Assembly (OTA) relaying light to five (four after SM1) aft-mounted focal plane instruments. The Support Systems Module (SSM) encircling the OTA's base and protecting the instruments provides housekeeping functions. 426 kg OTA is required to maintain the 5 m mirror-to-mirror spacing to within 3 μm/0.003 arcsec over 24 h using the 5.08 m long graphite epoxy metering truss. The primary is held in a main support ring inside a large square tooling fixture. Inside the truss is a 3 m diameter aluminium light baffle. Externally, the OTA is covered forward with the SSM light shield and end-aperture door. The twin 33 m^2 solar arrays and primary communications antenna are mounted on the SSM forward shell. The SSM equipment section encircles the metering truss base, accommodating power, control, communications, RWs, rate gyros and other subsystems in its 10 bays, accessible externally for servicing

AOCS: HST avoids thruster exhaust contamination by employing four 3,000 rpm RWs for attitude control (turn rate of 90° in 20 min), with angular momentum offloading by magnetorquers. Periodic reboosting during Shuttle servicing missions combats orbital decay. Six rate gyros (four prime plus two back-up; two pairs replaced by SM1) and three fixed head star trackers in the SSM equipment section provide overall attitude information; three Sun sensors mounted on the aft face prevent pointing within 50° of the Sun (aperture door available for protection). Two magnetometers at the front end sense attitude relative to Earth's magnetic field. Pointing stability requirement is 0.007 arcsec for up to 10 hours, using three interferometric Fine Guidance Sensors on the field of view's periphery. Any two are sufficient for guidance; magnitude +13 was detection requirement, but +15.5 appeared feasible pre-mission. FGS are also operated as a science instrument (see Payload section). The Guidance Star Catalogue, providing position and magnitude data on 18,819,291 objects, was completed in 1989 but continues to be expanded, including proper motion and colour information

Power: two ESA/British Aerospace solar arrays provide 5.0 kW BOL, 4.5 kW after two years and 4.3 kW after five years at 34 V; 5 per cent degradation measured after one year. Each 150 kg 2.83 × 11.8 m wing carries 24,380 Si cells, rolled into a 20 cm diameter drum for launch. Unfurling achieved by Bi-stem actuator units at either end unwinding and interleaving two spring steel elements to form 2.2 cm diameter booms either side of the thermal covers. The arrays can be restowed or jettisoned, and all functions can be achieved manually. The original arrays were replaced by SM1 because incorrect design allowed excessive flexing during day/night cycling. The new arrays reduced jittering by × 10. Each component is qualified for 30,000 thermal cycles of -100/100°C over a five year life. Payload requirement is 2.4 kW. The six 90 Ah Eagle Picher Ni/H_2 batteries, in three modules, can be replaced on each servicing mission

TT&C: 1 Mbit/s data via NASA's TDRS satellite system to White Sands for relay to the ST Operations Control Center at Goddard and to the ST Science Institute at Johns Hopkins. Goddard exercises vehicle control, ST ScI manages the observations. HST is controlled by a triply redundant 50 kg Rockwell Autonetics DF-224 computer (48,000 word total memory; two of six memories have failed, three required for full operation; a co-processor was added by SM1 to increase memory/speed) and Lockheed's 37.7 kg Data Management Unit. All incoming commands are decoded/relayed by the DMU and it receives the data from the Scientific Instrument Control & Data Handling computer, which relays DMU's commands and formats the science data

Thermal: heaters/coverings maintain mirror at 21°C. Elements such as batteries requiring greater heat dissipation are positioned on the anti-solar side

Payload: Ritchey-Chretien f/24 Cassegrain Optical System comprises an 829 kg 2.40 m diameter primary mirror of ultra-low expansion titanium silicate glass and a 30 cm diameter Zerodur secondary 5 m distant. Effective focal length is 57.6 m. The primary is constructed of 2.5 cm-thick face and rear glass plates separated by a honeycomb layer, and provides l/20 rms at 6,330 Å. Both hyperboloid mirrors are polished to less than 0.05 μm, and coated with a 0.064 μm layer of aluminium and a 0.025 μm layer of protective magnesium fluoride. That outer layer is required for adequate ultra-violet reflectance, although it reduces the aluminium's visible light reflectivity to 85 per cent. Reflection at 1,216 Å is 70 per cent; spectral range is 1,150 Å-1 mm. The secondary directs the light cone through the primary's 60 cm diameter central aperture to a focus 1.5 m behind the face plate for dispersion to the science instruments and FGS (three instruments now receive their light via COSTAR). Optical adjustment by 24 actuators attached to the primary's rear plate and six on the secondary's support structure. The specification required 70 per cent of the incoming light to be focused in a 0.2 arcsec diameter core. The primary mirror spherical aberration discovered in-orbit produces 15% within 0.1 arcsec. COSTAR improved the performance to less than 70 per cent. Four instruments are axially aligned (COSTAR replaced one), the fifth (Wide Field/Planetary Camera) and guidance sensors are in radial bays. All are latched for in-orbit replacement.

Wide Field/Planetary Camera (WF/PC) (designed/built by JPL; US$69 million) for general astronomical far-ultra-violet to near-infra-red applications over 1,200-11,000 Å, including astrometric searches for planetary bodies about other stars (prevented by mirror defect), solar system and galactic/extra-galactic observations. 280 kg. Took light from HST's central FOV for greatest quality. Operated in two modes: f/12.9 2.7 × 2.7 arcmin FOV Wide Field camera (resolution 0.1 arcsec per pixel), and f/30 1.2 × 1.2 arcmin Planetary Camera (resolution 0.043 arcsec per pixel). Light from the f/24 pick-off mirror passed through a wheel of 48 filters, polarisers and gratings, and was divided into four beams by a pyramid mirror for relay to the required camera, selected by pyramid rotation. Each camera employed four CCDs of 800 × 800 pixels (1,515 μm) each. Exposures from 0.110 second to 28 hours. Planetary mode planned for 400 km resolution on Jupiter. WF/PC was most affected by HST's optical defect: was to have taken 40 per cent of observing time but reduced to 10 per cent. Replaced on SM1 by US$101 million 278 kg WF/PC-II: integral optics to correct HST's defect, revised filter set (including far-ultra-violet), updated CCDs reaching into far-ultra-violet, articulation of the pick-off mirror and 3 of the 4 fold mirrors to ensure accurate pointing required by optical correction. 1991 decision reduced detectors from eight to four (one PC plus three WF) on cost grounds.

Faint Object Camera (FOC), provided by ESA (Dornier prime), uses HST's full resolution capabilities for faint objects/structures (magnitude 24-29) using cumulative exposures. 320 kg; 150 W required operating, 75 W standby. Length 2 m, cross section 1 × 1 m. Range 1,150-6,500 Å; most sensitive at blue visible/ultra-violet. Operates in four principal modes: direct imaging at f/48 (22 × 22 arcsec FOV, × 2 magnification), f/96 (11 × 11 arcsec, × 4 magnitude) and f/288 (4 × 4 arcsec, × 12 magnitude), and as a long-slit spectrograph. Four filters provide 12 filters each. Employs two separate imaging photon-counting detectors, each with a three-stage intensifier tube coupled to a TV camera. HST optical defect rendered FOC's visible capability similar to ground-based instruments, but ultra-violet work remained attractive; observing programme was reduced to half. Following SM1, FOC's chains are fed corrected but magnified images: f/96 became f/150 (7 arcsec FOV), f/48 became f/75 (14 arcsec FOV), f/288 became f/4500.

Faint Object Spectrograph (FOS), built by Martin Marietta for University of California. 1,150-8,500 Å; operates in low or intermediate resolution modes plus offers spectropolarimetry capability. Incorporates 1-D Digicon detector.

High-Resolution Spectrometer, built by Ball Aerospace for NASA Goddard. Similar to FOS but optimised for ultra-violet (1,100-3,200 Å, so relatively unaffected by HST optical defect) and high resolution (2,000, 20,000, 120,000) observations. Employs six interchangeable gratings. Began suffering serious power supply fluctuations July 1991; switching to back-up permanently would reduce effectiveness by 30 per cent. SM1 installed relay to allow either detector channel to use either power supply.

High-Speed Photometer (HSP), designed/built by University of Wisconsin. Four photon-counting image dissector tubes, a photomultiplier and >50 focal plane/aperture combinations could detect temporal variations down to 16 μs. 1,150-7,000 Å. Least affected by HST optical defect, but could not begin data return until end-1991 because of HST problems. Sacrificed for COSTAR optics correction module.

Fine Guidance Sensors (FGS) are primarily used for pointing system. Only two are required for 0.007 arcsec pointing goal, but third added for astrometric observations to 0.002 arcsec. Could not begin astrometry until end-1991 because of HST problems.

First results after Hubble's corrective optics were installed. At top are pre/post-COSTAR images of a star. Below is M100 seen by WF/PC-I and (right) the corrected WF/PC-II (ESA/ NASA) ***1994***

HETE

Background

MIT's Center for Space Research proposed the High Energy Transient Experiment in 1987 to observe the bright, brief bursts from astronomical objects over UV to gamma energies. NASA suggested in mid-1989 combining HETE's science goals with a proposal from Argentina's space agency for a joint satellite project. That approach was dropped in 1990 and HETE became an MIT project managed from NASA HQ.

A nuclear test monitor series, called the Vela, first detected gamma bursts in 1969 and they have yet to be precisely mapped. NASA's Compton Observatory is currently providing 10° accuracy of their location, but HETE will locate about 100 to within 10 arcsec. The gamma instruments are derived from the sensors used on the French-Soviet Phobos Mars mission of 1988-89.

HETE went into space along with Argentina's SAC-B on 4 November 1996 on a Pegasus-XL with the L-1011 aircraft taking off from Wallops Island – the first Pegasus launch from there. Although orbit was achieved, neither satellite separated from the Pegasus third stage. Since HETE was contained with the launch vehicle's Dual Payload Attach Fixture there was no opportunity for it to deploy its solar panels and try to obtain some experimental results, and thus it died due to a lack of power.

Specifications

Contractors/Institutes: NASA HQ (lead), MIT Center for Space Research (science lead, UV instruments), CESR (France, gamma instrument), RIKEN (Institute of Physical & Chemical Research, Japan, X-ray instrument, 2-way ground station), ASI (Italy, primary 2-way ground station), Los Alamos National Lab (X-ray coded mask aperture, software), AeroAstro (bus, payload integration/test, ground stations), Sextant Avionique (GPS receiver)

Launch: 4 November 1996 by Pegasus-XL

Mass: 118 kg (55 kg payload)

Orbit: 38°, 490-555 km (550 km circular planned)

Spacecraft: 66 cm diameter 89 cm high (inclination payload 34 cm high). Momentum biased; ±0.5° attitude

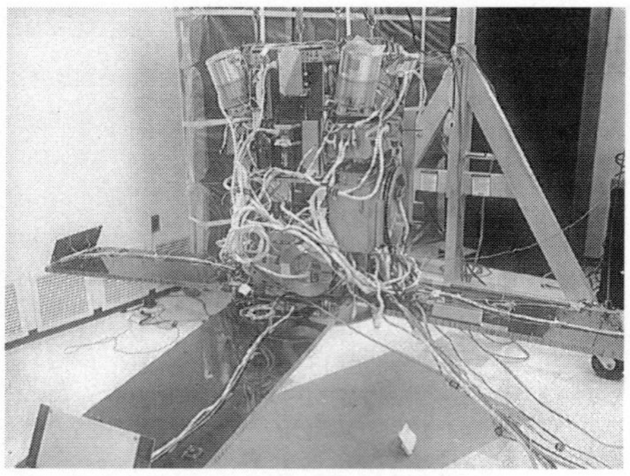

HETE will return near-realtime data on transient astronomical events to observatories worldwide. AeroAstro provided 23 ground stations as part of the turnkey system (Peter Gridley) ***1996***

control. GPS receiver for position determination. Four 31.5 × 83.2 cm solar wings provide 70 W average. 96 Mbyte mass memory, downlink S-band at 230 kbit/s to three master stations (KPNO Arizona, Sicily and Japan; 7.5 kbit/s uplink) + 20 science receive-only stations worldwide (including Russia). Mission life 1 year, goal 3 years; spacecraft reliability 95 per cent after 18 months
Payload: CESR Omnidirectional Gamma-Ray Spectrometer, CESR, 6 keV-1 MeV.

Wide-Field X-ray Transient Monitor, RIKEN/LANL, 2-25 keV. X-ray emissions of gamma bursts will provide localisation to 10 arcmin.

UV Transient Camera Array, MIT, 4-7 eV (1,800-3,100 Å). Ultra-violet emissions of gamma bursts will provide localisation to 6 arcsec.

Mission to Planet Earth series – UARS Satellite

Background

NASA's Upper Atmosphere Research Satellite is investigating the mechanisms controlling the structure and variability of the upper atmosphere. It is the first Mission to Planet Earth satellite and the largest ever flown for atmospheric research. Four instruments (CLAES, ISAMS, MLS, HALOE) provide atmospheric temperature profiles and concentrations of ozone, methane, water vapour and key trace species such as chlorofluorocarbons. Using Doppler shift measurements, two other instruments, HRDI and WINDII, map upper atmosphere wind fields, which shape the global distribution of chemical species. Four other instruments (ACRIM, SUSIM, SOLSTICE, PEM) are monitoring energy inputs from the Sun, which strongly influence chemistry and dynamics.

After gathering data for 4½ years, UARS confirmed that man-made chlorofluorocarbons are responsible. To verify the satellite's findings, mission scientists compare and co-ordinate data, with results from the Solar Backscatter Ultra-Violet (SBUV) spectrometer flying on NOAA meteorological satellites and the Shuttle.

NASA estimates that the total cost through 4½ years of operations and data processing and analysis was US$710 million, with about US$730 million projected through September 1997. The baseline programme authorised 18 months' operations covering two northern winters. Current plans extended those operations through September 1997. Contractors and NASA designers established the 18 month operational window based on the volume constraints in CLAES' coolant supply (exhausted 5 May 1993). All the other instruments had a design life of three years.

The other instruments began failing later in the mission. The ISAMS has not been operational since the mid-1992 failure of the chopper wheel. The MLS 183 GHz radiometer, measuring water vapour, failed April 1993. MLS's vertical scanning activator began to show signs of ageing in 1994. The instrument's operational mode now uses a reduced scan range and duty cycle. MLS focuses its observations on lower stratospheric ClO and ozone, particularly during polar winter. A solar array drive clutch problem halted full science operations 2 June to 20 July 1992. Operations with the degraded primary drive were maintained until October 1993, when the back-up took over. Problems in March 1995 with both drives prompted the May 1995 decision to park the solar array in its safety zone and operate the eight remaining instruments in power-sharing mode. The batteries have given some cause for concern but periodic recharging and revised charge control procedures are allowing full operations.

UARS is based on Fairchild's Multi-mission Modular Spacecraft and employs the refurbished Modular Attitude Control Subsystem package returned from the Solar Max satellite by Shuttle 41C in April 1984. General Electric's Valley Forge centre received the US$221 million prime contract in March 1985 from NASA Goddard. Fairchild's US$41 million contract for the MMS was awarded August 1985.

Specifications

Contractors: GE Astro Space (prime), Fairchild (multimission modular spacecraft)
Launched: 12 September 1991 by Shuttle STS-48
Mass: 6,526 kg on-orbit BOL, 2,700 kg science payload; 7,474 kg in Shuttle bay (including support equipment)
Orbit: 571 × 596 km, 57.0°, permitting sampling of all local solar times about every 33 days
Spacecraft: 10.36 m long, 4.57 m diameter, 3-axis control, based on Fairchild's MMS. OBC is 64 k NASA Standard Spacecraft Computer (NSSC-1) employing software derived from Landsat
AOCS: 3-axis control provided by inertial reference unit, star trackers, Sun sensors, Earth sensors, RWs, magnetorquers and four monopropellant blowdown hydrazine engine modules, each accommodating 3 × 0.9 + 1 × 22.2 N thrusters; total 204 kg hydrazine. Control to within 36 arcsec, determination to within 20 arcsec
Communications: control is exercised by GSFC at 0.125/1.2 kbit/s on 2.1064 GHz, data return on 5 W 2.2875 GHz 32 kbit/s real-time, 512 kbit/s playback (two NASA Standard 450 Mbit tape recorders); all links via TDRS-based Space Network
Power system: 1.6 kW orbit average provided by single 3.25 × 9.25 m 6-panel Si array and 3 × 50 Ah Ni/Cd batteries
Payload: SUSIM Solar ultra-violet Spectral Irradiance Monitor.

SOLSTICE Solar Stellar Irradiance Comparison Experiment (solar ultra-violet intensity and spectral distribution).

PEM Particle Environment Monitor (magnetospheric energy input to atmosphere from protons/electrons, plus x-ray imaging spectrometer and vector magnetometer).

ACRIM Active Cavity Radiometer Irradiance Monitor (solar output at all wavelengths for long-term measurement of solar constant, continuing work of Solar Max).

CLAES Cryogenic Limb Array Etalon Spectrometer (cryogen depleted).

ISAMS Improved Stratospheric and Mesospheric Sounder (chopper wheel drive system failed).

MLS Microwave Limb Sounder (63/183/205 GHz molecular thermal emission from atmosphere; 183 GHz failed).

HALOE Halogen Occultation Experiment (vertical distribution of key trace species).

HRDI High Resolution Doppler Imager + WINDII Wind Imaging Interferometer (wind speeds in upper atmosphere).

New Millennium series

Current status

Deep Space 1 (DS 1) was launched by Delta 2 on 24 October 1998 from CCAFS Florida, to an intercept trajectory with asteroid 1992KD in July 1999. Early tests with the xenon thrusters were confounded with random shutdown before a series of successful runs were achieved in late November. During December, tests with the astronomers navigation system were successfully carried out.

JPL/NASA Lewis are developing this 30 cm 92 mN Xe thruster under the NSTAR (NASA SEP Technology Application Readiness) project to validate low-power ion propulsion technology. It was carried on DS 1. The thruster is shown before a 2,000 h test at LeRC; a 7,000 h test at JPL began April 1996 (NASA) ***1996***

New Millennium's first Earth Observation mission will carry an advanced land imager ***1996***

Background

NASA's New Millennium programme covers the identification, development and flight validation of key technologies for incorporation into 21st century science missions without the risks inherent in their first use. NASA projects frequent and affordable science missions with highly focused objectives: numerous micro-spacecraft carrying advanced miniaturised instruments returning a continuous information flow. NM will demonstrate these revolutionary technologies and architectures with Deep Space (DS) and Earth Orbiting (EO) validation flights on a semi-annual basis from 1998 to 2000. These flights also provide opportunities for meaningful science.

The DS 1-2 and EO 1 missions have been selected, and the DS 3 candidate optical interferometer is in the concept phase. It is projected that spacecraft development time will be reduced from the eight-year average of the early 1990s to two years a decade later. This will help to reduce spacecraft development cost from US$590 million to US$105 million (FY96 rates). DS FY96 budget is US$30 million, with US$30 million planned FY97 and US$50 million for each succeeding year. The EO spacecraft development budget from the Office of Mission to Planet Earth is projected at US$15 million FY96 and US$40 million/year from FY97, plus US$6 million/year for launches.

JPL manages the programme, and the advanced technologies for flight validation are identified and developed by integrated product development teams of representatives from NASA, industry, academia and other government centres.

DS 1, the first New Millenium mission, is a comet/asteroid fly-by using solar electric propulsion. The industrial partner for mission definition and implementation is Spectrum Astro, Inc. It will fly by the small asteroid McAuliffe followed roughly a after launch by an approach of Comet West-Kohoutek-Ikemura. NASA may decide to extend it to include a visit to another object within four years of launch. DS-1 weighs 365 kg at launch, including 125 kg advanced technologies, 23 kg hydrazine, 58 kg Xe. Its design incorporates a fixed 1.5 m X/Ka-band HGA, two LGAs, two star trackers and hydrazine ACS into an aluminium bus, 96 × 112 × 79 cm with a hexagonal spaceframe. Autonomous remote software would plan and execute activities such as autonomous optical navigation. The optical system will image asteroids against star background for position and trajectory information as well as imaging target for fly-by corrections. Primary propulsion is a single JPL/NASA Lewis SEPS 30 cm 20-90 mN, 2.5 kW, SI 3,260 s. Two 152 × 533 cm solar concentrator arrays will generate 2.6 kW.

It carries a payload of the SEPS, solar concentrator array (BMDO), miniature integrated camera/spectrometer, small deep space transponder, autonomous optical navigation, integrated space physics package, beacon mode operations, composite HGA, 3D stack processor.

DS 2 will fly two microprobes piggyback on 1999's Mars Surveyor Lander to demonstrate key technologies that will enable network science and *in situ* and subsurface science data acquisition analysis. Within its non-ablative aeroshell, it has a micro-telecommunications subsystem with programmable transceiver, power microelectronics with mixed digital and analogue ASIC, ultra-low temperature lithium battery, microcontroller, flexible interconnects for system cabling, meteorological high-*g* pressure sensor, sample/H_2O experiment soil conductivity and a high-*g* temperature sensor.

SEDS/SEDSat

Background

SEDS began in 1983 as a Small Business Innovative Research programme. March 1993's Delta carried NASA Marshall's 41 kg SEDS to observe the dynamics of a long tether system. It unreeled the 26 kg instrumented end mass on a 0.75 mm diameter polyethylene tether to 20 km in 77 minutes, allowing it to swing like a pendulum for 16 minutes before being detached to burn up. Observations corroborated modelling of the tether dynamics, which introduced a new mathematical description invented by Professor Thomas Kane of Stanford University. The 41 kg SEDS 2 flew 10 March 1994 and this time remained attached to stage 2 on its 19.7 km tether. The 25.8 kg 20 × 33 × 41 cm payload included Langley's accelerometer, magnetometer and tensionometer to monitor its motion. Despite its slim diameter, the tether was visible to naked eye ground observers.

NASA Marshall planned a third SEDS flight aboard STS-85 in July 1997, releasing the University of Alabama in Huntsville's SEDSat 1. The SEDS will be mounted on top of a Goddard Hitch-hiker carrier in the cargo bay to deploy SEDSat 20 km upwards over 80 minutes from Shuttle's 300 km orbit. The tether will be cut at Shuttle's end first and then at SEDSat's half an orbit later. SEDSat will end up in a 330 × 525 km orbit, while the tether will enter the atmosphere after about one day. SEDSat includes transponders for amateur communications: mode A 145.915-145.975/29.350-29.410 MHz up/down; mode L 1,270/435 MHz up/down at 9.6 or 57.6 kbit/s.

The two flown SEDS cost US$10 million.

Spartan

Background

The Shuttle Pointed Autonomous Research Tool for Astronomy was developed by NASA Goddard as a series of short duration low cost free-flyers, meant to extend sounding rocket experiments by using the Shuttle. Spartan 101 used the 100-class carrier, a rectangular 3.2 × 1 × 1.2 m structure, deployed the first time 20 June 1985 by Shuttle 51G with a 136 kg payload and retrieved 45 h later. NRL's x-ray instrument studied the Perseus cluster of galaxies and the central core of the Milky Way galaxy for about 24 hours during 16 orbits at a maximum distance of 192 km from Discovery. It shut down at 40.5 hours, sensing low levels of consumables and battery power. In reality it had used about half its 2,800 m of tape.

Following 101, Goddard developed the 200 carrier for future missions. Three flights were planned for 1986, but the programme was placed on hold after Spartan-Halley, carrying a ultra-violet spectrometer to study Halley's Comet, was lost with Challenger. The Spartan 200 carriers are built from milled aluminium plates bolted together. No welding is used anywhere on the frame. The structure comprises the Instrument Carrier, Upper Structure and Service Module (power, data handling, thermal control, attitude control). The tape recorder were replaced FY96 by a solid-state recorder.

Spartan is deployed by Shuttle's RMS and all subsequent events are pre-programmed with the gathered data stored on the tape recorder. Autonomous free-flying duration is typically 40 hours.

Specifications

Spartan 201-02 (general)
Mass: 900 kg Spartan, 500 kg payload (1,000 kg Shuttle support structure); 201-01=1,289 kg deployed
Pointing: 3-axis inertial. Pointing accuracy ±3 arcmin; 0.1 arcsec/s gyro drift; cold gas attitude control
Power: 30 kWh AgZn batteries (28 V DC).

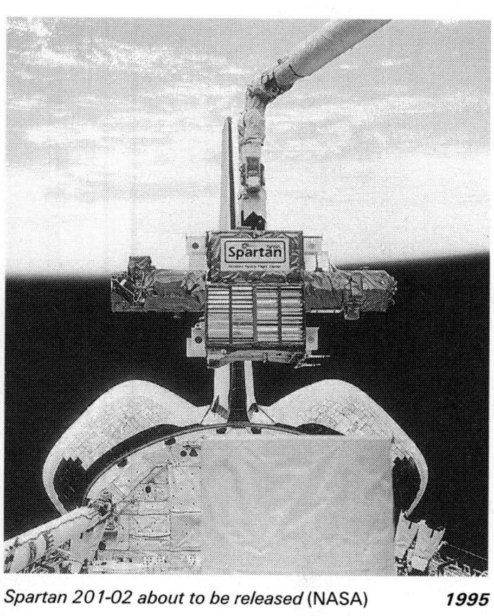

Spartan 201-02 about to be released (NASA) ***1995***

Spartan 201-01
Released: STS-56 April 1993
Mass: 1,289 kg
Comment: Two solar physics instruments: ultra-violet coronal spectrograph and white light coronograph to probe the physics of solar wind acceleration.

Spartan 201-02
Released: STS-64 September 1994
Comment: Reflight of 201-01 timed to coincide with Ulysses south polar pass.

Spartan 204
Released: STS-63 January 1995
Comment: Naval Research Lab's FUVIS far-ultra-violet imaging spectrograph observed Shuttle surface glow and thruster firings while attached to RMS, and celestial targets as a free-flyer. Used as large handling mass by EVA astronauts.

Spartan 201-03
Released: STS-69 August 1995
Comment: Reflight of 201-01 timed to coincide with Ulysses north polar pass.

Spartan 206/OAST-Flyer
Released: STS-72 January 1996
Comment: Platform for technology experiments. Goddard's Return Flux Experiment characterised return flux contamination to validate computer modelling. Johnson's laser-initiated pyrotechnic characterised space environment effects on reliability. Goddard's Attitude Determination and Control experiment used GPS receivers for navigation and pointing; it was the first satellite to use GPS attitude control. A packet radio system was also demonstrated.

Spartan 207
Released: STS-77 May 1996
Comment: The Inflatable Antenna Experiment deployed JPL's 12.2 m diameter 30 m long aluminised Mylar parabolic antenna structure. It was jettisoned after its surface accuracy was measured, and Spartan was recovered.

Spartan 201-04
Released: STS-87 October 1997
Comment: Reflight of 201-01.

STEDI series

Background

The USRA Universities Space Research Association is managing the STEDI Student Explorer Demonstration Initiative to launch three small satellites under a US$24 million NASA grant. The first two were selected in February 1995 from 66 proposals for missions costing within US$4.4 million each, excluding launch costs. NASA, in November 1994, awarded OSC the 'Ultralight' launch services contract for two firm Pegasus XL launches plus eight options. The third was selected as a result of intense lobbying by advocates. Cost to USRA is about US$6.7 million per launch, but OSC is offering secondary payload places for US$3.5 million to US$5 million. CATSAT has been pursued at a low level as back-up, but USRA has formally requested full funding for it as the third STEDI.

CATSAT (STEDI series)

Background

The University of New Hampshire's Co-operative Astrophysics and Technology Satellite will investigate the distance and polarisation of gamma bursts. Though high density bursts occur daily, their source remains a mystery. It is not even known if they originate from near our own galaxy or from cosmological distances, leading to an uncertainty in their source energy of 26 orders of magnitude. CATSAT will measure the 500 eV to 10 MeV burst energies to derive the photoelectric absorption along the line of sight. The point of cut-off below a few keV in this absorption depends on the column depth of material traversed. This depth will be compared with known line of sight column densities from radio surveys of galactic neutral hydrogen, thus yielding the burster distance. In its secondary science role, CATSAT will be the only all-sky monitor in the critical 500 eV to 2 keV range, directly complementing MOXE, HETE and RXTE by filling the gaps at 500 eV-2 keV and 1-10 MeV. NASA has scheduled a launch in 2002.

Specifications

Contractors/Institutes: UNH, Weber State University (satellite), University of Leicester (UK)
Launch: planned for fourth quarter of 1998 by Pegasus XL/Lockheed L-1011 from Vandenberg AFB
Planned life: 1 year
Orbit: planned 550 km, 97° Sun-synchronous
Mass: 115 kg
Spacecraft: 53 × 53 × 56 cm. 3-axis control, nadir pointing to ±5°. 100 W nominal power. 1.5 Mbaud S-band science data down to UNH; VHF/UHF up/down command at WSU
Payload: soft x-ray spectrometer, hard x-ray spectrometer, directional γ-ray spectrometer, x-ray albedo polarimeter.

SNOE (STEDI series)

Background

The University of Colorado's Lab for Atmospheric and Space Physics will use the Student Nitric Oxide (NO) Explorer (SNOE) to test possible mechanisms for NO density variations in the lower thermosphere, and to explain why low latitude changes are independent of polar changes. Although only a trace constituent, NO is a major player in many atmospheric processes, mainly ionospheric charge exchange reactions, thermosphere radiative cooling, ozone destruction in the mesosphere and stratosphere.

The compound's density varies with altitude and latitude but nature of the variations and their transport mechanisms are not clearly understood. SNOE will investigate two hypotheses on the energy sources for NO formation. First that solar soft x-rays at low latitude provide the active energy or secondly that secondary auroral electrons in polar regions, where the highest NO levels are found, supply energy. SNOE will simultaneously measure NO density, soft solar x-ray and auroral electron flux.

Specifications

Contractors/Institutes: built inhouse at LASP, collaborating with Ball Aerospace and National Center for Atmospheric Research/High Altitude Observatory
Launch: 26 February 1998 by Pegasus XL/Lockheed L-1011 from Vandenberg AFB
Planned life: one year
Orbit: planned 580 × 550 km, 97.75° Sun-synchronous
Mass: 132 kg
Spacecraft: Compact hexagonal structure 91.4 cm high and 99 cm across its widest dimension. Power consumption 35W (average). Data rate: 6 MB/day
Payload: ultra-violet spectrometer, auroral photometer, solar x-ray photometer.

TERRIERS (STEDI series)

Background

Boston University's Center for Space Physics will use the Tomographic Experiment using Radiative Recombination Ionospheric EUV and Radio Sources satellite to create a 2-D, and eventually a 3-D, model of the ionosphere's electron density and photo-emissive components. Terriers will use CAT Computer-Assisted Tomography similar to that developed for medical imaging. Highly directional radiation detectors will measure the total radiation in a given spectral band from a particular direction as Terriers spins about a horizontal axis. As Terriers spins and moves along its orbit, the same points in space will be viewed from many different directions, allowing the emission from any one point to be calculated. Visible emission will provide data on chemical sources, such as oxygen ions and nitrogen molecules; EUV emissions from radiative recombinations will be directly related to electron density. Measurements will be co-ordinated with ground-based observations.

Specifications

Contractors/Institutes: Boston University collaborating with Naval Research Lab, MIT Haystack Observatory, University of Illinois, AeroAstro (satellite)
Launch: planned for May 1999 by Pegasus XL/Lockheed L-1011 from Vandenberg AFB
Planned life: about 1 year
Orbit: planned 550 km, 97° Sun-synchronous
Mass: 132 kg
Payload: Tomographic EUV spectrographs, two visible photometers, gas ionisation solar spectral monitors, radio beacon.

PLANETARY AND SPACE SCIENCE

INTERNATIONAL

Cassini

Current status
Cassini was launched at 08:43 UT on 15 October 1997. In late July 1998, Cassini was traveling faster than 107,000 km/h (66,000 mph) having travelled more than 769 million km (478 million miles) since it left Earth last October. It remained in good health. The spacecraft functioned well earlier in its flight during test and housekeeping chores including lubricating the reaction wheel assembly mechanisms.

Background
NASA pegged Cassini as the fourth of the 'Core' set of planetary missions sought by its Solar System Exploration Committee in 1983. Cassini was supposed to save funds by employing the sophisticated Mariner Mark II spacecraft, designed to undertake the more demanding deep space missions, by accommodating scan platforms, nuclear power supplies, high data rates and decade-long flight times.

NASA chose Titan as the target for Cassini, partly because the organic processes taking place on Saturn's largest moon provide the only planetary-scale laboratory for studies of a pre-life terrestrial atmosphere. Titan has a dense, predominantly nitrogen atmosphere and the mission will permit intensive scrutiny. Then, suffering from budget constraints due to the Shuttle problems, NASA asked ESA to participate in the programme as well. Following a joint NASA/ESA study, the agencies agreed in June 1985 that NASA would supply a 1,540 kg (dry mass) Saturn orbiter and ESA a 190 kg Titan atmospheric probe.

Originally, NASA set a schedule for a Shuttle/Centaur launch in May 1994 giving the craft the opportunity to study asteroids in fly-bys before Cassini's trajectory returned it to Earth for a swing-by to provide the energy for the 4.5 year journey to Saturn. The mission slipped in 1987 at NASA's insistence and because of the new need to carry Centaur upper stages – then the only possible vehicle for the flight – on expendable launchers.

ESA began to have doubts about the mission in 1988, but at the request of NASA pushed the programme into the phase B programme plan that according to ESA's charter almost ensures a launch date. ESA's Council approved the mission during its June 1990 meeting, planning to begin Phase B in January 1991, although the MoU had still to be signed because of delays in the preceding Soho/Cluster MoU. Aerospatiale received the prime contract (signed with ESA 20 September 1993). Sole bidder General Electric (now Lockheed Martin) was selected by the Department of Energy in mid-1990 to provide the RTGs for both CRAF/Cassini. The mission remained essentially unchanged (calling for Cassini launch 6 April 1996 for 6 December 2002 Saturn

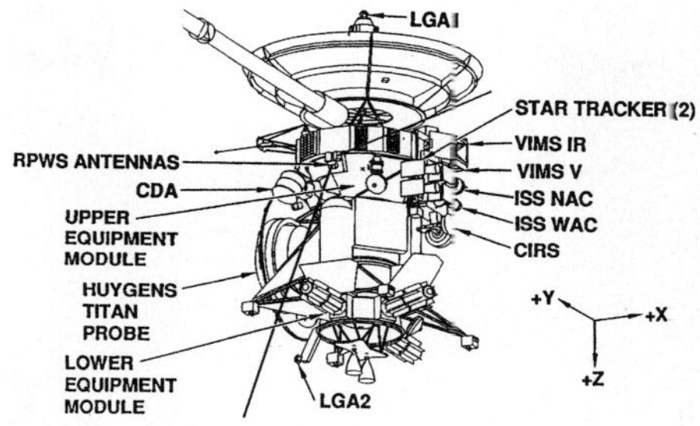

Cassini cruise configuration (NASA/JPL)

arrival) until early 1991 when the CRAF/Cassini launches were swapped and Venus gravity assists added.

The increased mass capacity allowed the addition, for example, of a third RTG, increasing margins and removing the need for batteries. Cassini's 25 November 1995 departure would have produced Saturn arrival 15 May 2004 and allowed an 18 November 1998 fly-by of 40 km diameter F-class asteroid Clarissa in the main belt. CRAF was cancelled in January 1992 and Cassini was slipped to spread out NASA's funding obligations. This two-year delay, which cost ESA contract funds as well, still increased total Cassini funding beyond the original US$1.6 billion Congressional ceiling.

To meet other funding obligations, NASA dropped the scan platforms in 1992. Coupled with a policy of no pre-Saturn science acquisition (an asteroid fly-by would cost 70 m/s delta-V, the equivalent of 7-8 Titan fly-bys), this reduced development cost to US$1.4 billion.

After launch on 15 October 1997, Cassini entered into a Venus/Venus/Earth/Jupiter gravity assist multi-planet swing-by path, thereby reducing launch energy from 80 km^2/s^2 to 22 km^2/s^2. The 3 December 1998 450 m/s deep space manoeuvre at aphelion sets up the second Venus gravity assist for 24 June 1999 at 600 km. Then a 1,166 km Earth fly-by of 16 August 1999 provides the final energy to reach Jupiter. A 115-140 R_J Jupiter swing-by on 30 December 2000 sets up Saturn arrival on 1 July 2004.

Cassini will pass through the asteroid belt December 1999 to April 2000, but will not gather data because the science committee will not turn on their instruments again until two years before Saturn arrival (apart from a 40-day gravitational wave experiment about December 2001).

Cassini also plans a 52,000 km Phoebe fly-by on 11 July 2004, where a 96-minute 621 m/s braking burn establishes an initial 147-day, 1.3 × 176 R_S 17° inclination orbit. This is the closest approach to Saturn during the entire mission and the insertion burn precedes its optimal point (centred on periapsis) by 48 minutes for science observations. A 332 m/s burn 2 September 2004 will raise periapse to 8.2 R_S for Europe's Titan probe to spin up and release itself on 6 November 2004. A 57 m/s deflection burn follows two days later to fly past Titan at 1,500 km during 27 November while the probe descends. The orbiter will fly by Titan on 33 of the planned 63 orbits over the four-year mission, sometimes approaching to within 950 km and eventually raising the orbital inclination to 85° (Voyager photographed the planet from an inclination of less than 30°). Plans allow the Cassini to perform radar mapping during the encounters with the rings if power is available. Fly-bys of Iapetus, Enceladus, Dione and Rhea will also be made, plus more distant encounters with Tethys, Mimas and Hyperion, before the mission formally ends on 1 July 2008.

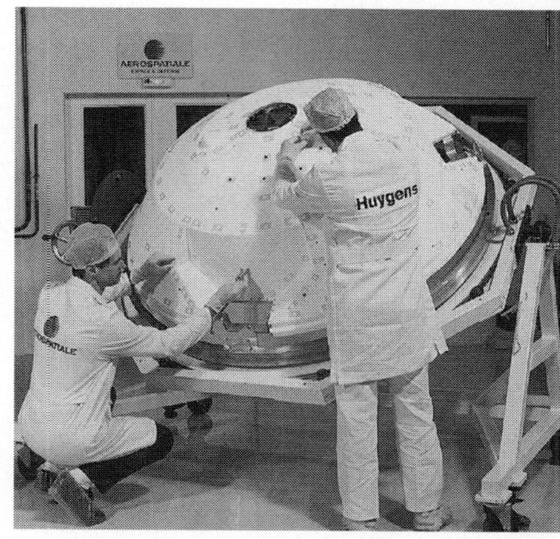

Huygens heat shield and aft cover (ESA/Aerospatiale) ***1995***

During the Titan fly-by, Europe's Huygens will take 2.5 h to descend during 27 November 2004. The probe will enter the moon's atmosphere at 6.15 km/s over Titan's day side at about 18° north/200-220° longitude (where 0° is defined as the sub-Saturn point; recent Hubble observations confirmed that Titan is tidally locked to Saturn). ESA plans to use the probe to determine the atmosphere's chemical composition and measure winds, temperature and pressure profiles from 170 km down to the surface. It is possible that Huygens could continue returning data after touchdown, but planners speculate that, at the very least, accelerometer data will reveal if it has splashed into an alien sea.

As the probe descends, Huygens will jettison a canister when M1.5 has been attained at around 170 km. Then a mortar-deployed 2.6 m diameter pilot parachute will pull off the aft cover to allow deployment of the 8.3 m diameter first main parachute. Following stabilisation, the heat shield will be ejected to free the instruments' inlet ports. After a 15-minute descent, a 3.0 m diameter stabiliser will take over for the final descent. $LiSO_2$ batteries will provide 350 W for 3 h, sufficient for an 8 kbit/s data rate on each of two redundant probe to orbiter S-band links. All data are stored aboard the orbiter for later relay to Earth.

Specifications

Cassini satellite

Launch: 15 October 1997 by Titan 4 Centaur from Canaveral
Mass: 2,150 kg dry orbiter, 3,132 kg propellants, 373 kg Huygens (335 kg probe), 165 kg launch adapter. Total injected mass 5,820 kg
Design life: 11 years
Communications: 4 m diameter HGA and GHz X-band 20 W TWTA will return 20 bit/s – 169 kbit/s. Data stored on two 1.8 Gbit SSRs, providing up to 403.2 kbit/s record/playback
Power: Three RTGs provide 815 W BOL (638 W EOL; no batteries)
AOCS: 3-axis attitude control (2 mrad pointing accuracy, 6.98 mrad/s max slewing) maintained via SRU, four gyros, RWs and four clusters of four 1.0-0.6 N hydrazine blowdown thrusters on propulsion module. (primary: two redundant 445 N MMH/NTO bipropellant R-4D thrusters will be used for up to 200 manoeuvres totalling 2,360 m/s)
Orbiter payload: VIMS Visual Infra-red Mapping Spectrometer (320 channels; 40.0 kg, facility instrument)

Cassini releases Huygens into Titan's atmosphere

Huygen's final descent sequence was demonstrated with a drop from 36 km over Kiruna, Sweden (Fokker Space/ESA)

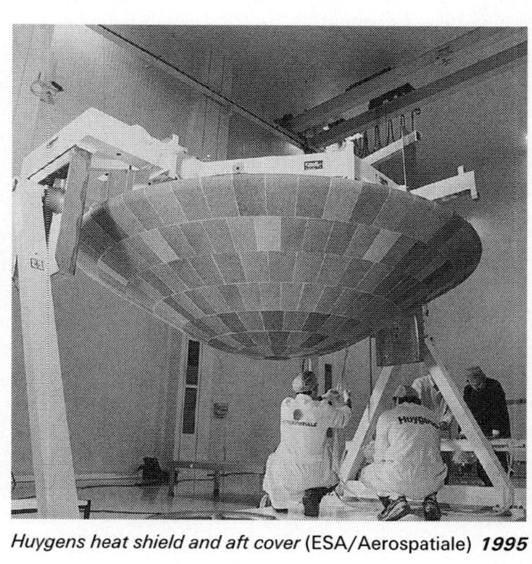

Huygens heat shield and aft cover (ESA/Aerospatiale) ***1995***

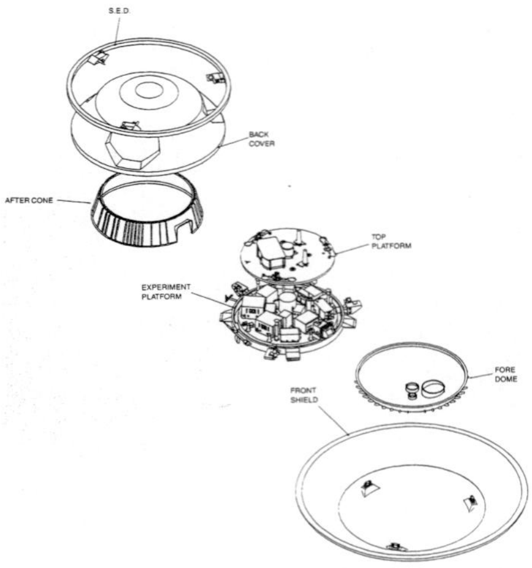

Huygens exploded view. SED is the spin and ejection device (ESA) ***1995***

Huygens approaches the surface of Titan (ESA) ***1996***

ISS Imaging (65.1 kg, 250 mm fl wide angle, 2,000 mm narrow angle, 1,024 × 1,024 element CCDs; facility instrument)

RADAR (58.0 kg, 13.8 GHz, 120 W raw power, employs HGA; facility instrument)

Ion Neutral Mass Spectrometer (10.8 kg facility instrument)

Cosmic Dust Analyser (15.1 kg, Max Planck Institute für Kernphysik, Germany), Plasma/Radio Wave Spectrometer (37.9 kg, University of Iowa), Plasma Spectrometer (20.1 kg, Southwest Research Institute)

UV Spectrometer/Imager (14.9 kg, University of Colorado)

Magnetospheric Imaging Instrument (24.9 kg, The Johns Hopkins University)

Dual Technique Magnetometer (8.7 kg, Imperial College, UK)

RF Instrument Subsystem (13.2 kg)

Composite infra-red Spectrometer (36.2 kg, NASA Goddard)

Huygens payload: HASI Huygens Atmospheric Structure Instrument: 6.7 kg, temperature and pressure of Titan's atmosphere, winds and turbulence, atmosphere electricity. (University of Rome)

GCMS Gas Chromatograph Neutral Mass Spectrometer: 19.5 kg, atmosphere composition. (NASA Goddard)

ACP Aerosol Collector/Pyrolyser: 6.7 kg, atmosphere aerosol composition. (France)

DISR Descent Imager/Spectral Radiometer: 8.5 kg, image clouds/surface and UV-NIR spectroscopy. (University of Arizona)

SSP Surface Science Package: 4.2 kg, surface state after touchdown. (University of Kent, UK)

DWE Doppler Wind Experiment 2.1 kg, high accuracy zonal wind characteristics. (University of Bonn, Germany)

Huygens probe

Mass: 335 kg probe containing an 87 kg 2.7 m diameter decelerator. Science instrumentation: 48 kg.

Payload: 15.4/15.8 GHz phased array radar to show altitude and surface reflectivity data from 10 km to 150 m.

Ulysses (ISPM)

Current status

Ulysses made a Jupiter's gravity assist manoeuvre in early 1992 to swing south of the ecliptic towards the Sun, becoming the first probe to undertake detailed investigations of the heliosphere's third dimension. On 9 June 1993, it became the first spacecraft to exceed 32°, surpassing Voyager 1's record. The trajectory brought Ulysses over the Sun's south pole during 1994, with 132 days spent above 70° solar latitude between 26 June and 6 November. A maximum 80.2° was reached on 13 September 1994, at a distance of 2.3 AU from the Sun. The equator was crossed on 5 March 1995. Passage over the north pole came during 19 June to 30 September 1995; 80.2° was attained on 31 July. Ulysses arrived during a period of minimum solar activity, instead of the original schedule's maximum.

Ulysses continues to monitor activity (ESA) ***1995***

Ulysses is tracked 8 hr per day throughout the mission by the 34 m dishes of NASA's Deep Space Network. A joint NASA/ESA team at JPL's Mission Control and Computing Center conducts operations.

Ulysses will make a second pass near the next solar maximum, returning in 2000-2001 after reaching 5.4 AU aphelion in April 1998. Its second southern polar pass above 70° will occupy September 2000 to January 2001, followed by perihelion in May 2001 and the northern pass September-December 2001. Power-saving measures are necessary. This extension, beginning 1 October 1995, is formally known as the Second Solar Orbit and is costing ECU14 million through December 2001. The original mission, ended 30 September 1995, cost ECU168 million (1995 rate).

Background

Until 1981, the International Solar Polar Mission was a two-craft project to be launched by a single Shuttle and three-stage IUS in February 1983. ESA had the responsibility to provide one spacecraft and about half of the 16 experiments, while NASA promised to provide the other craft, the rest of the experiments, RTGs and Shuttle launch. The two agencies signed the MoU 29 March 1978.

The original plan called for the two spacecraft to fly for 15 months to Jupiter, one passing above and the other below the planet at distances of 450,000 km, causing both to enter high inclination, but opposite (posigrade and retrograde solar) orbits. Four years after launch, they would have passed over opposite solar poles at 250 million km, through the ecliptic plane and then over the other poles. The five-year mission would have been the first to investigate the Sun in three dimensions, providing an accurate assessment of the solar magnetic field and solar wind and studying solar flares, coronal holes and loss of mass/energy.

Then NASA's decision in 1980 not to proceed with the three-stage IUS raised doubts about the feasibility of the dual launch plan. The project switched to the cryogenic liquid Centaur upper stage, but US budget cuts in February 1981 forced NASA to save US$250 million to US$300 million on ISPM by cancelling its craft. Shuttle delays and other problems had already delayed the launch to 1985. It was finally agreed that NASA would continue to provide launch/tracking services and an RTG. With ESA's stepped up prominence as the majority partner in the project, it was renamed Ulysses (from ISPM), at ESA's request to honour both Homer's mythological hero, and a reference in Dante's *Inferno* of Ulysses' urge to explore 'an uninhabited world behind the Sun'.

Doubts again crept into the programme plan when controversy arose about Shuttle/Centaur threatening the planned May 1986 launch even before the Challenger accident in January 1986. After the accident, the earliest possible launch window became September/October 1989, with subsequent windows available only every 13 months because of Jupiter's motion. Following NASA's June 1986 decision to prohibit the use of the liquid Centaur in the Shuttle, the ESA and NASA agreed to move the craft to an IUS/PAM-S combination. Each had been employed separately from the Shuttle, but never in combination.

There was also concern over launch/injection loads and Ulysses' RTG. A Centaur launch would have restricted peak loads to 5.8 *g*, whereas the substituted solid rocket upper stages could have imposed up to 11 *g*. However, Ulysses was designed to withstand 11.5 *g* because of uncertainty over upper stages in the programme's early days.

Following the closure of the US Department of Energy plant that produced the RTG fuel, 1989 marked the last year a launch could be made, without reductions in the power available for the experiments. These issues added complexity to the competition for Shuttle launch slots with Galileo and Magellan, all requiring the October 1989 window. In the end, the ESA agreement with NASA for an October 1990 launch was signed April 1987, Galileo took the Autumn 1989 slot and Magellan departed the previous May.

Following Challenger's loss, Ulysses was returned from Kennedy Space Center to ESA's ESTEC facility in The Netherlands on 16 June 1986. ESA removed its instruments and returned the majority to the Principal Investigators. The spacecraft itself spent 1987 in storage at prime contractor Dornier in Friedrichshafen, with some units, such as the tape recorders and axial boom, which required special treatment, removed.

By February 1989, a new schedule was issued after Ulysses was brought out of storage in October 1988. It called for it to be reintegrated with its instruments starting in May 1989, tested during July 1989 to February 1990, and delivered to Cape Canaveral in May 1990 for mating with its upper stages in July 1990 in preparation for installation in Shuttle's cargo bay in September 1990. Some 200 microprocessor memory chips were replaced in the decoder, data handling and experiment units beginning April 1988 following the discovery of a potential generic fault.

NASA launched the Ulysses by Shuttle during the 5-23 October 1990 Jupiter window, 13 years after ESA's Science Program Committee approved the mission. The Inertial Upper Stage and PAM-S motor accelerated it to a record 15.4 km/s, exceeding the velocities of NASA's Pioneers and Voyagers. After launch, the probe had to arrive at Jupiter within a 160 km corridor, forcing an IUS guidance system in orbit from Discovery's more accurate unit. Some 45 minutes after release, at 18.53 GMT, IUS stage 1 ignited for a 148-second burn. The 125-second coast was followed by stage 2's 108-second firing. The IUS provided final pointing and separated, leaving four small Atlantic Research Corp solids around PAM's waist to spin the craft up to 70 rpm. They were released and PAM fired for about 88 seconds to add 4.3 km/s for a record 15.4 km/s departure speed, outstripping the four Pioneers and Voyagers and crossing the Moon's orbit in about 8 hr. Two 0.9 kg masses on 12 m wires were freed to reduce the spin to 6.8 rpm and Ulysses separated about 10 seconds after PAM's burn. The deployment of Ulysses' 5.55 m experiment boom on 7 October cut the spin to 4.7 rpm. The RTG was found to be providing 283 W, 2 W less than expected.

The Ulysses deployed its 7.5 m boom on 4 November and nutation peaked after 11 days at 6.5°, before decreasing to about 3°. Such motion must be <0.75° for the narrow-beam X-band link to lock on and when Ulysses' automatic Conscan system was activated on 12 December (designed to cope with up to 1.5° nutation) it fell to 0.5°. The cause was solar heating of the 7.5 m boom as Ulysses rotated. The effect disappeared as the craft moved further from the Sun.

Ulysses' outbound path would have taken it to 17 AU, almost to Uranus's orbit, if Jupiter had been avoided. A 0.29 m/s tweak 8 July 1991 set up the Jupiter encounter. After travelling 993 million km and already collecting important information on the interplanetary medium, Ulysses began the Jupiter encounter by detecting the bowshock crossing at 17.33 GMT 2 February 1992, some 113 R_J out. This was more distant than expected: Pioneer's was 100 R_J and Voyager's 60 R_J. It approached through the late morning region of the magnetosphere at about 30° north, came within 378,400 km of the cloud tops (5.3 R_J) at 12.02 GMT 8 February and then exited unscathed on the previously unexplored evening side, at high southern latitudes, having spent more than a week in the magnetosphere.

Just before closest approach, Ulysses entered the magnetosphere's polar cap. The close 13.5 km/s fly-by was necessary to produce the high ecliptic inclination – IUS/PAM could have reached only 23° without the

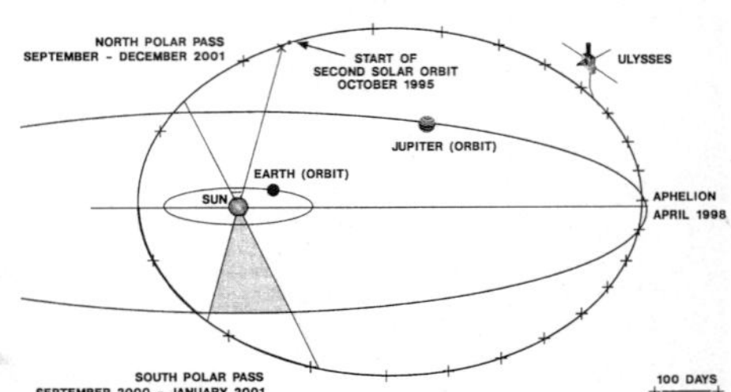

Ulysses' flight profile will again take it over both solar poles. It will never pass within Earth's orbit (ESA) ***1996***

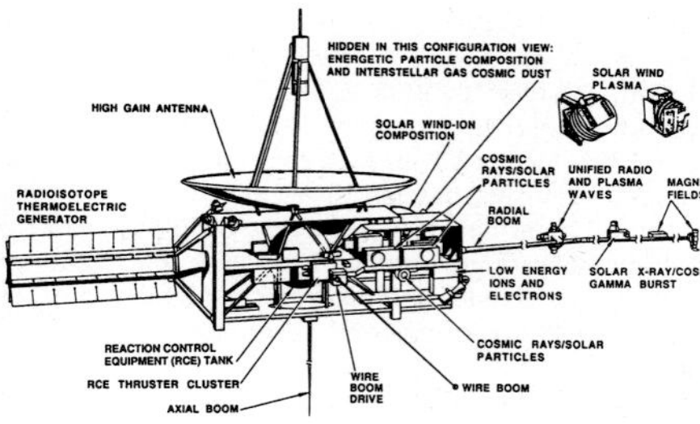

Ulysses principal features (ESA)

assist. Ulysses found that Jupiter's intense radiation belts reach only 40° latitude, whereas Earth's extend to 70°. Ulysses then passed directly through the Io plasma torus (only Voyager 1 preceded it), which plays a key role in refuelling the magnetosphere with plasma. The overall density was about as expected, implying injection from Io has not changed dramatically since Voyager, but significant azimuthal asymmetries were found.

At end-February 1992, the Sun, Earth and Ulysses were in direct opposition, the best time to detect gravitational waves by observing a minute shift in the craft's position. Although no scientist could positively identify gravity waves, this radio science experiment, set new upper limits with a factor of 20 improvement in sensitivity on their value. Other important Ulysses' observations included the first direct detection of ionised O, N and He (and neutral He) atoms arriving from interstellar space and the measurement of μm-sized dust grains from interstellar space.

By mid-1993, Ulysses was permanently immersed in the region of space dominated by the Sun's southern pole. This could be seen in the consistently negative polarity measured by the magnetometer since April 1993. Increasing latitude also reduced the intensity of charged particles.

Ulysses monitored radio science events as Jupiter was impacted 16-22 July 1994 by the 21 large objects from the disrupted Comet Shoemaker-Levy 9. No evidence for radio emission from the impacts was found.

Now in high inclination heliocentric orbit, Ulysses has passed over the solar south pole and conducted the first long-term in situ observations of high speed solar wind flowing from the large coronal hole that covers that polar cap at solar minimum (the Sun's current phase). Ulysses measurements of ionised interstellar neutralised gas particles, so-called interstellar pick-up ions, have led to a major advance in our understanding of the processes affecting this component of our heliospheric particle population.

Unexpectedly, the magnetometers detected a wide variety of fluctuations at many spatial scales in the Sun's high latitude field. They are believed to represent relatively unevolved turbulence originating at the southern polar coronal hole. Another important finding was that the magnetic field radial component varies relatively little with latitude. This is contrary to the expectation that the imprint of the underlying dipole-like magnetic field at the Sun's surface, showing an increase in the radial field at the poles, would be detected at Ulysses' position. In simple terms, a south magnetic pole was absent at high latitudes. Undoubtedly related to the presence of the large-scale fluctuations in the polar magnetic field was the detection of an unexpectedly small increase in the measured influx of cosmic ray particles over the poles compared with model predictions.

Moving back towards the equator, Ulysses left the fast solar wind (750 km/s) in which it had been immersed for more than 18 months, crossed briefly into the equatorial region dominated by slow wind (400 km/s) and once again became immersed in the fast wind, this time from the North pole. The boundaries were strikingly symmetric: 22° south/21° north. The magnetic field changed from negative (directed inwards) to positive (outwards), reflecting the current configuration of the Sun's dipole-like surface field. The radial component's strength continues to show no change with latitude. With the exception of the north-south solar wind asymmetry and cosmic ray intensity, the north pass was generally similar to the S pass.

Observations as Ulysses heads slowly south are now being correlated with those from Soho. During 1997-98, it will remain at almost constant distance from the Sun during aphelion, allowing temporal effects to be separated from spatial.

Specifications

Ulysses satellite

Mission: fields/particles and cosmic dust investigations of the heliosphere up to high solar latitudes.

Launch: 6 October 1990 aboard Shuttle Discovery from Kennedy Space Center.

Contractors: Dornier (prime), BAe (AOCS, HGA), Fokker (thermal, nutation damper), FIAR (power), Officine Galileo (Sun sensors), Laben (data handling), Thomson-CSF (telecommand), MBB (thrusters).

Mass: 371 kg at launch; 366.6 kg at PAM separation. Science payload 55 kg.

Orbit: heliocentric, including Jupiter swing-by, perihelion 1.34 AU.

Description: spin-stabilised at 4.99 rpm with the 1.65 m diameter HGA pointed continuously towards Earth. Powered by a single DoE RTG providing 284 W BOM and 255 W EOM October 1995 (4.5 kW heat). Double-hinged 5.55 m radial boom carries four sensors for HED, STO and HUS. 7.41 m axial boom as monopole for STO. two wire booms 72.5 m tip-to-tip as dipole for STO.

Communications: X-band (20 W, 8,408.209 MHz, 2° 3 dB beamwidth); 5 W 2 GHz S-band 2,111.607/ 2,293.148 MHz up/down is used for dual frequency radio science investigations and early manoeuvres. During contact periods, real-time data at 1,024 bit/s are interleaved with playback of stored data. Two redundant tape recorders can each hold 45.8 Mbit, stored at 128/256/512 bit/s.

Propulsion: 33.5 kg high purity hydrazine in single bladder tank with dual manifold feeds two clusters of four 2 N catalytic thrusters (total delta-V 185 m/s)

Payload: BAM solar wind plasma (two electrostatic analysers with channel electron multipliers; US)

GLG solar wind ion composition (electrostatic analyser with time-of-flight/energy measurement; Switzerland/US)

HED magnetic fields (triaxial vector helium and fluxgate magnetometers; UK)

KEP energetic particle composition/neutral gas (four solid state detectors; Germany)

LAN low energy charged particle composition/ anisotropy (two sensor heads with five solid state detectors: US)

SIM cosmic rays and solar particles (five solid state detectors, US)

STO radio/plasma waves (72.5 m radial dipole plus 7.55 m axial monopole antennas with 2-axis search coil; US)

HUS solar X-rays and cosmic γ bursts (two silicon plus two cesium iodide detectors; France/Germany)

GRU cosmic dust (multi-coincidence impact detector with channeltron: Germany).

JAPAN

Lunar-A (SS-17) satellite

Background

ISAS selected a lunar orbiter/penetrator mission in 1990 for its launch vehicle M-5's second mission. It then chose NEC to act as system integrator. As it has laid out the programme to date, ISAS plans for an orbiter to fire three 13.6 kg, 13.8 cm diameter, 71.4 cm-long (1.44 m with motor attached) penetrators carrying seismometers and heat flow probes 1-3 m into the surface. The probes will penetrate both the far side and the near side of the moon. Japan has designed the seismic data experiments to clarify long-standing questions on the far side crustal and deep mantle structure and existence of the lunar core. The orbiter will act to relay data to Earth, fleshing out the Apollo programmes near side measurements. In order to test its plans, ISAS has set up an N_2-powered gas gun at ISAS' Noshiro facility to fire penetrator designs into a simulated lunar surface. Lower speed drop tests began in August 1986. A one-stage S-520 sounding rocket, launched on 17 September 1995, also tested penetrator separation and attitude control plans but during the test the ground station lost telemetry after 25 seconds.

Physically, Lunar-A's penetrator is a 540-kg spin-stabilised unit, 2.2 m in diameter and 2.2 m high vehicle. The three-stage M-5 plus two solid kick stages will inject it into a Geotail-type highly eccentric Earth orbit with an apogee of about 500,000 km. N_2O_4/N_2H_4 thrusters account for 189 kg of the weight budget and the penetrator modules about 124 kg. Six 20 N and four 1 N/N_2H_4 thrusters augment the main 500 N bipropellant engines.

In order to reach the final trajectory, Lunar-A will fly-by the moon to obtain an apogee of 1,189,726 km and reduce the delta-V to 550 m/s for the main engine to establish a 200 × 200 km, 20° orbit during the second encounter.

Preparing to insert the RTG's fuel elements on the Shuttle pad (NASA)

1936

The Lunar-A 1997 lunar orbiter/penetrator mission is selected for the second M-5 (ISAS)

For each penetrator release, the orbiter will descend to 45 × 200 km before returning to its higher orbit to use its LIC Lunar Imaging Camera (with an image compression chip provided by CNES). The orbiter will spin up to 120 rpm for release from its normal 30 rpm. A 30 cm diameter spherical solid motor (20 second burn, 5 kN max thrust) will brake the penetrator out of orbit. Its N_2 attitude control system will ensure a head-on impact at 8° and more than 300 m/s. One penetrator is planned for the Apollo 12 and 14 region, one into the far side and one into the far/near side boundary. The orbiter will then rise to a 300 × 300 km orbit for data relay.

Planet-B (Nozomi)

Current status

Japan launched the Planet-B spacecraft on 3 July 1998 aboard an M-5 rocket built by ISAS. After lift-off, ISAS renamed Planet-B Nozomi, or 'Hope'. Nozomi performed two lunar fly-bys on 24 September and 18 December (2,809 km) and then a first Earth fly-by (1,003 km) on 20 December 1998 for a gravity assist to trans-Mars trajectory. However, a perigee burn planned at this time provided insufficient velocity and an extra burn a day later used more propellant than planned. Instead of reaching Mars on 11 October 1999 and entering a 300 × 47,500 km orbit, Nozomi will remain in solar orbit until December 2003 when the Mars orbit injection burn will take place. Japan launched Nozomi on 3 July specifically to acknowledge the first anniversary of the Mars Pathfinder mission. Japan has spent approximately ¥11 billion (US$80 million) for the orbiter, which will examine the planet's atmosphere for details of the Aerian atmosphere's interaction with the solar wind and the escape trajectories of oxygen molecules from the thin atmosphere of Mars.

Japan is only the third country to launch a planetary mission.

Background

Planet-B, operating for at least a full Mars year, will provide synoptic data on turbulence, motion and seasonal variation of the upper atmosphere, interaction with the solar wind and map Mars' magnetic field. ISAS has overall responsibility for the spacecraft and it contracted with NEC to perform the systems integration and MHI to build the 500 N thruster. The spacecraft has a mass of 536 kg at launch of which 258 kg is the dry mass carrying a 33 kg payload. After launch, Planet-B enters a 7000 × 400,000 km Earth orbit for two months before a lunar swing-by now scheduled for the middle of October. When the spacecraft completes 4.5 orbits, then a 420 m/s burn, it will be on its planned trajectory for Mars. Upon arrival it will then perform a 1,257 m/s Mars injection burn originally planned for 11 October 1999 to create an initial 300 × 47,500 km 170° inclination (ecliptic) orbit further refined and adjusted to 150 × 47,500 km (at times lowered to 130 km), but see current status.

Specifications

Planet B

Description: 2 m diameter irregular octagonal prism body, 7.5 rpm spin

Mass: 536 kg; 258 kg dry

Telemetry: fixed 1.6 m diameter Earth-pointing X/S-band HGA. LGA provides S-band link during non-Earth pointing, 1.7 m boom and four 25 m sounder antennas deployed in Mars orbit. 2-32 kbit/s X-band + 64-1024 bit/s S-band, depending on Earth distance (2 kbit/s X-band minimum), 16 + 125 bit/s command uplink; 128 Mbit SSR (1 kg).

Power: 2 fixed solar panels provide 200 W min from 20 × 40 mm Si cells (18 per cent efficiency). 2 × 15 Ah Ni-MH batteries. 5 m magnetometer boom

Propulsion: Trans-Mars injection and Mars orbit insertion by 500 N NTO/N_2H_4 thruster (10 rpm spin for stability). RCS by 4 axial, 2 radial, 4 tangential 2.3 N N_2H_4 thrusters.

AOCS: Attitude data from V-slit star scanner, spin Sun aspect sensor (170° FOV), and rate gyros for active nutation control in launch phase.

Payload: MIC visible camera (with data compression chips provided by CNES)

MGF magnetometer (3-axis 0.1 nT accuracy)

ESA energetic electrons (5 eV-22 keV)

ISA energetic ions (10 eV-20 keV/q)

IMI energetic ion mass (0.5 eV-40 keV/q; Swedish Institute of Space Physics)

EIS high-energy particles (40-500 keV)

TPA thermal ion drift (0.1-100 eV; Canadian Space Agency)

PET electron T, UVS UV spectrometer (H, O, CO, CO_2, imaging)

PWS sounder/HF waves (electron density profile and HF waves)

LFA plasma waves (VLF/ELF waves)

NMS neutral gas mass spectrometer (NASA)

MDC dust counter (Munich Technical University, Germany)

XUV EUV spectrometer (helium)

USO (ultra-stable oscillator for radio science, NASA).

RUSSIAN FEDERATION

Mars series

Current status

Russia abandoned its ambitious planetary exploration programme after the failure of the Mars -96 mission on 16 November 1996. IKI still discusses the Mars Together programme with JPL Russian Space Agency, although funding for any planetary mission remains tight and the budget is being drawn away from the exploration programme to finance the Russian contribution to the International Space Station.

Background

Russia finally launched Mars-96 on 16 November 1996 at 20.48 GMT from pad 200L at Tyuratam/Baikonur aboard a four-stage Proton-K using a Block D-2 fourth stage. The Block D-2 (11S824F) stage had previously been used for the two 1988 Phobos launches. Initially, the launch went well with the Block D-2 performing its first burn to place itself and the attached Mars probe – re-named Mars 8 after launch – into an initial 51.5°, 139-155 km parking orbit. The flight plan called for the Block D-2 to fire again about an hour after launch to place itself and Mars 8 into a highly eccentric Earth orbit. (Block D2 should have provided a delta-v of 2.850 km/s). After this firing Mars 8 would fire the Fregat-class propulsion module for the final burn to a heliocentric, trans-Mars orbit.

At the time of the Block D-2 re-ignition, an incorrectly aligned assembly caused the engine to shut down after 20 seconds on command from Mars 8. Stranded in a low Earth orbit, the Mars 8 still separated from the Block D-2. Some reports suggest that the Fregat module then burned as planned, placing the Mars 8 spacecraft on an orbit with a perigee of about 87 km and apogee of 1,500 km or more and the low perigee quickly brought the stage back to Earth. Initially the Russians thought that it had re-entered over the southern Pacific Ocean, but later debris tracks indicated it survived long enough to possibly land in Chile or Bolivia, raising public health concerns over the spacecraft's plutonium power source. Sweeps by the Chilean Air Force failed to turn up any sign of radioactive contamination. Meanwhile, the Block D-2 continued in orbit until 18 November. The Russians stated that the stage re-entered at 01.13 GMT with debris falling into the southern Pacific Ocean close to 50.9° south, 191.9° east. USSPACECOM challenged this, stating the decay time was actually 01.32 GMT over 32° south, 264° east.

Mars-96

Mars-96 consisted of a single spacecraft based on the Phobos design. It was meant to follow a Type II trajectory to arrive at the Red Planet in September 1997 after 315 days. At launch, the spacecraft assembly had a mass of 6.2 tonnes. If it succeeded, the orbiter would have remained in its initial 12 h orbit for three to four weeks for regular data contact with surface elements. Then it would have released two small landers before rising to the mapping path that allows surface contact only every three to four days. The mapping was planned to last a full seasonal cycle or roughly two Earth years from 250 × 18,000 km 101°.

The imaging platform known as ARGUS had many development problems at VNII TransMash as Mars 8 neared completion. The problems prompted Russia to propose ARGUS' deletion, but DARA/DLR strongly resisted the move because body-mounting the instruments instead would cost at least 60 per cent of the science return and compromise the project goal of 10 m resolution (from 300 km) stereo colour imagery.

MARS-96's orbiter would have sent the two autonomous stations on a trajectory to the surface three to five days after achieving a final orbit. The small stations would have followed a descent path with a 5.6 km/s entry. Just before landing the re-entry capsule would have ejected a 65 cm diameter heatshield leaving the lander suspended on 130 m line and protected by a 1.35 m inflated shell. Upon touching down, the shell would have split around the equator and cut the parachute and the carrier free. Russia expected the final landing to impact at between 15-20 m/s activating four petals à la Luna 9 through 13. After a final checkout, the lander would have uplinked data from a 32 Mbit memory at 8 kbit/s on 401 MHz. The power supply could provide 2 W for 20 minutes daily initially, then every three days. A 9 kg payload stereo panoramic camera provided by Russia would use most of the telemetry band. NASA and RKA signed a contract in April 1993 for the US to fly a duplicate Mars Oxidiser experiment on two landers. The MOX would measure the soil to determine if it is strongly oxidising. It would have exposed bundles of optical fibers with coatings sensitive to peroxides and superoxides to the air and soil and then measured changes in the light transmission characteristics.

Russia based its designs for the independent surface penetrators on those planned for the Vesta mission. As envisioned, the 75 kg penetrators would directly enter the Martian surface, by spinning up to 95 rpm in orbit and upon release, igniting a four-nozzle solid propellant motor to provide 4.6 km/s entry. The 45 kg 1.5 m long 11 cm diameter titanium body would impact at 76 m/s, reaching 6-8 m depth. The deceleration peak of 500 *gs* breaks a forebody free to penetrate deeper. Fore and aft bodies are connected by cable. Once in place, a small isotope generator provides one year of life. The penetrators and surface stations will create a seismic network.

Specifications

Mars-96 (Mars-8)

AOCS: 3-axis controlled to 1° per axis

Data rate 64 kbit/s when Earth distance is greater than 200 million km, 128 kbit/s when less. 0.5 Gbit/s daily downlink capacity

Mass (science payload): 250 kg

Mass (dry): 1,750 kg

Orbiter sensors (Plasma science): ASPERA-S energy-mass (Sweden).

FONEMA omni non-scanning energy-mass ion analyzer (UK)

DYMIO omni ionosphere energy mass spectrometer (France)

MARIPROB ionosphere plasma spectrometers (Russia)

MAREMF electron analyser/magnetometer (Russia)

ELISMA wave complex (France)

SLED-2 low energy charged particle spectrometer (Ireland)

Orbiter sensors (Cruise science): PGS precision γ spectrometer (Russia)

LILAS-2 spectrometer (cosmic/solar ν bursts, France)

EVRIS photometer (stellar seismology, France, on PAIS 2-axis platform)

SOYA photometer (part of SPICAM, solar oscillations, France)

RADIUS-MD dosimeter (Bulgaria)

MORION-S science data acquisition system (Russia), 2 Gbit mass memory (ESTEC)

Imaging Sensors: ARGUS 3-axis scan platform with four imaging instruments

WAOSS Wide Angle Opto-electronic Stereo Scanner (DLR Germany). WAOSS: focal length 21.7 mm, f/5.6, 0.4-0.76 μm, each push-broom linear CCD 5,184 elements, swath 80°, viewing nadir + 20° back/forward, 500 kbit/s; resolution 100 m.

HRSC High Resolution Stereo Camera (DLR Germany). HRSC: fl 175.0 mm, f/5.6, nine push-broom linear CCDs, 3,950-9,250 Å, nine channels (4 XS, 10 m resolution PAN, 2 stereo PAN, 2 photometric), swath 11°, 1.6 Mbit/s. HRSC/WAOSS total 47.7 kg, 75 W; data stored in 1 Gbit mass memory shared with Omega.

Omega/VIMS 0.5-5.2 μm visible/IR mapping spectrometer (CNES; Italy contributing to visible channel)

NK navigation camera (Russia). 4.0 kg, 520 × 580 CCD, 100 mm fl, f/3.5, 5 × 8° FOV.

Other planetology instruments: TERMOSCAN mapping radiometer (Russia)

SVAT mapping spectrometer (Russia)

SPICAM multi-channel spectrometer (on PAIS platform, Russia)

UVS-M UV spectrometer (Russia)

LWR longwave radar (Russia), Photon ν spectrometer (on PAIS platform, Russia)

NEUTRON-S neutron spectrometer (Russia), MAK quadropole mass spectrometer (Russia).

SAS small autonomous stations

Mass: 90 kg each (34-kg surface mass).

SAS sensors: DESCAM descent camera (France, 200 images during descent from 10 km at 1 s intervals, 544 × 240-pixel CCD, 10 mm fl)

Optimism vertical axis seismometer (France, 0.350 kg, 9 cm cube, 0.01-4 Hz, 10^{-8} m/s² sensitivity), magnetometer (on deployed arm, France)

Metegg meteorology package (FMI Finland, on deployed arm)

Mars Oxident Experiment (MOx, NASA), carrying (a) a neutron-proton spectrometer (on deployed arm, Germany) (b) optical depth analyser (France). 1 year surface life provided by two 150 mW_e RTGs and Ni/Cd battery (four RTGs in total generate 34 W heat) (c) Oxidiser/superoxidiser experiment

Penetrator sensors: TV cameras (1.2 kg Kassei + 0.15 kg Mikrotel; RFAS, Japan, Czech Republic)
Meteorology unit (0.4 kg Mekom; RFAS, Finland)
Gamma-ray spectrometer (1.2 kg Pegas, with 3 cm tungsten shield against RTG; RFAS, Czech Republic)
X-ray spectrometer (0.3 kg Angstrem; RFAS, Germany),
Alpha-particle spectrometer (0.3 kg Alpha; RFAS, Germany)
Neutron detector (0.2 kg Neutron-P; RFAS, Germany)
Accelerometer (0.2 kg Grunt; RFAS, UK)
Thermometer (0.3 kg Termozond; RFAS, UK)
Seismometer (0.3 kg Kamerton; RFAS, UK)
Conductometer (0.1 kg Epsilon; RFAS, UK)
Magnetometer (<0.1 nT resolution, 0.2 kg IMAP-6; RFAS, Bulgaria).

Zond

Background

The Zond missions can be split into two classes, the original series which comprised Zonds 1-3 (plus the launch failure identified as Cosmos 27) and the second series was initiated by Cosmos 146 and Cosmos 154, launched in 1967 as test flights of the L-1 circumlunar spacecraft (called L-1P) derived from Soyuz.

All first series Zonds were dedicated at specific solar system research and planetary studies, but the Zond name covered the general target planet because of the poor performance of previous planetary spacecraft. Zond 1 aimed at Venus, Zond 2 at Mars and Zond 3 – rumoured to have been a delayed Mars probe – flew past the Moon into heliocentric orbit, taking photographs of the lunar far side on its way out of the Earth-Moon system.

As the moon race with the Americans heated up, the Russians chose to concentrate on lunar exploration missions with the second series of Zonds. After launch failures in September and November 1967, Zond 4 flew away from the Moon in March 1968 for a communications and high-velocity re-entry test. The spacecraft possibly reached out to approximately the lunar orbit, but in a different direction. A further launch failure occurred in April 1968 and the Soviets cancelled a July 1968 launch which may have been manned if tests had gone according to plan after a Proton-K Block D propellant tank ruptured before launch.

The Soviets achieved a first with Zond 5, when they recovered it after a circumlunar mission, albeit unmanned, in September 1968, beating NASA's manned Apollo 8 mission by several months. Zond 6 tried to repeat the success but crashed on landing. Zond 7 flew August 1969 and appears to have been the only fully successful mission in this series. Zond 8 closed down the flights in October 1970.

The Soyuz-based Zond 4-8 flew in preparation for a Soviet manned circumlunar programme

Zond 5 awaiting recovery in the Indian Ocean. It is clearly a Soyuz descent module

In their desperation to compete with the Americans, the Soviets cannibalised the last two L-1 spacecraft, intended for manned missions around the Moon for the November 1969 and December 1970 L-1E missions. These were primarily Block D propulsion tests with the spacecraft heavily instrumented for launch vehicle monitoring. The first launch failed and the second was named Kosmos 382 indicative of a failure. The Zond/L-1 spacecraft – called L-1S – went aloft on the first two N-1 lunar boosters, launched February and July 1969, but neither reached orbit.

UNITED STATES OF AMERICA

Clementine satellite

Background

Clementine, also known as the Deep Space Program Science Experiment, demonstrated that useful payloads could cost less than US$6 million. The project began 24 March 1992, taking only 19 months to full integration. Total cost came out to be about US$75 million, including US$17 million paid to USAF for a Titan 2 launch. The Naval Research Lab built/integrated the spacecraft; form off-the-shelf hardware and Lawrence Livermore National Lab provided the sensor suite. NASA supplied Deep Space Network tracking. Each component acted autonomously and the bus only provided power and communication services.

The InterStage Adapter (ISA) remained in the 189 × 126,990 km 66.77° inclination 51.2 hr Earth orbit with its package of radiation detectors while the satellite entered moon orbit to conduct a radar mapping mission. With Clementine 1 spinning at 60 rpm, the Thiokol Star 37 FM solid ignited 06 29 GMT 3 February 1992 during a north bound pass over the Indian Ocean for a 2,959.7 m/s delta V burn. The Star/ISA dropped off and during perigee 1, at 10.01 GMT on 5 February, the 490 N motor added 213.5 m/s to stretch the

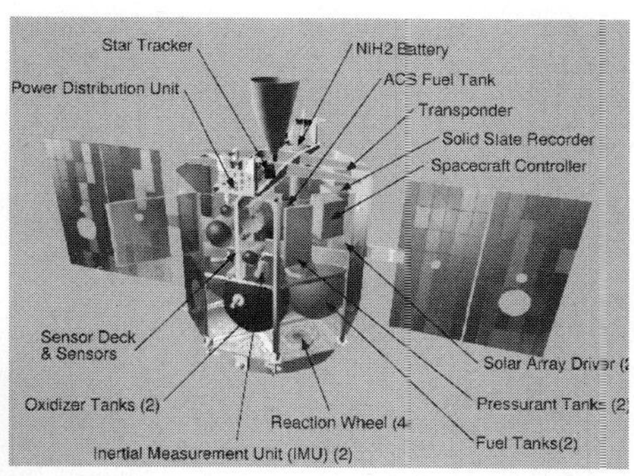

Clementine 1 in February 1993 became the first mapping lunar orbiter since Luna 22 in 1974. It was the first US lunar mission since 1972. Note the instrumented interstage adapter below the main craft (Naval Research Lab)

apogee out to the Moon, with a 31.9 m/s trim at 10.00 6 February at the Moon. A few other trim manoeuvres occurred and the spacecraft performed a 460.3 m/s burn at 12.52 GMT 19 February, 103.5 m/s at 12.16 21 February and a 5.3 m/s at 12.17 22 February to establish a 382 × 2,971 km, 89.3° inclination lunar orbit. This provided global mapping over two lunar months; while data were taken at 15 hours around perigee and the rest of the orbit was used to relay the data to Earth. Clementine returned a total of 1.8 million images from lunar orbit.

Two major burns on 26 March to change the argument of perigee carried Clementine up to 1,009 × 3,560 km and back. When the mission concluded another major burn on 4 May, it started the return to high Earth orbit with a stop to image 1,620 Geographos. A computer problem 7 May exhausted all the hydrazine attitude control propellant and created an 80 rpm spin. Engineers determined that spin could be reduced to 30 rpm – still too high for worthwhile asteroid data return. Instead, the main engine on 19 May made the first of three burns to establish a 44,769 × 388,197 km orbit for prolonged radiation exposure studies. BMDO attempted to contact the satellite in February 1995 but it had insufficient power for any significant telemetry. However, since the solar array Sun angle improved daily Clementine responded to commands 10 April 1995.

Galileo

Current status

Galileo is close to completing one of the most successful planetary missions of all time. In four years since December 1995, it will have performed 25 fly-by encounters with Jupiter and four of its moons. Designed to perform at Jupiter for two years, Galileo has returned data for twice that time, extending the number of encounters from 11 to 25. Its last encounter will be a close fly-by of Io on 26 November 1999 where intense radiation is expected to seriously affect the spacecraft's electronics.

Background

Plans to follow up the Voyager fly-bys with a Jupiter orbiter and probe started in 1978 and by 1981, NASA had invested US$300 million in the project. In November 1980, Hughes Aircraft received a US$40 million contract to develop the probe carrier, to be launched by Shuttle in March 1984 on a trajectory to reach the planet 1,200 days later in July 1987.

As NASA envisioned the project, the probe would separate 100 days before planet arrival and enter Jupiter's atmosphere on the sunlit side at 160,000 km/h. After slowing in a few seconds to a few hundred km/h, a parachute would open to allow the instruments to measure the vertical profile of the atmosphere. The carrier would relay the probe's data back to Earth, from an orbit similar to that of Ganymede, at about 1,002,400 km from Jupiter. It would use that body's gravity to skew its orbit at each near approach so that in 11 passes in 20 months it would cover all regions of the Planet. Powered by two RTG nuclear sources used on previous outer planet missions, Galileo would incorporate a 5 m diameter furlable antenna to relay 30-50 m resolution imagery of the moons.

As NASA faced financial restrictions, the project suffered many delays and changes in the original plan. The abandonment of a three-stage IUS in favour of a Centaur upper stage modified for the Shuttle caused one delay and then NASA's cancellation of this wide-body Centaur in favour of a two-stage IUS delayed the programme yet again. By the end of 1982, the wide-body Centaur had been reinstated for a two year journey beginning May 1986.

By 1983, NASA set the total cost already expended at US$630 million, but said that spare parts already acquired could be used for a US$100 million Galileo Saturn orbiter. This would have been launched in 1987-88, using Earth gravity assist for an eight-year transit.

There were growing concerns even before the Shuttle accident of January 1986 over safety considerations with the Centaur in the payload bay. When the Shuttle version of Centaur was finally cancelled in June 1986, some Galileo proponents proposed dumping a Shuttle launch altogether in favour of a Titan 4 Centaur G-prime combination in 1991 as an alternative to competing with Ulysses for a Shuttle launch. NASA refused to entertain the notion even at the risk of delaying the project until it might be ultimately cancelled. Ultimately, a Shuttle launch combined with IUS #19, previously assigned to a DoD mission, in conjunction with a Venus double-Earth swing-by (VEEGA: Venus-Earth-Earth Gravity Assist), emerged as the most promising solution to the launch dilemma.

Because a Shuttle/IUS combination is insufficient to fling Galileo into a direct Jupiter trajectory, JPL's mission

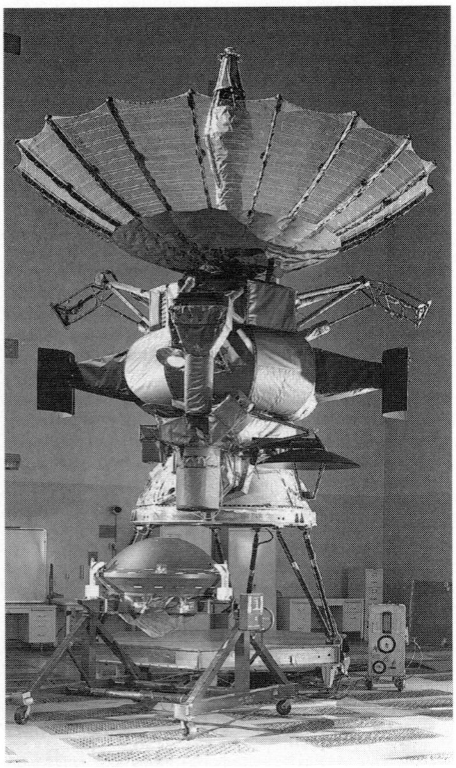

Galileo in JPL's Spacecraft Assembly Facility prior to modifications for the VEEGA trajectory. Germany provided the central Retro Propulsion Module. The conical probe, on stand in foreground, was carried below the RPM for deployment 147 days before arrival. Galileo is topped by the 4.8 m High Gain Antenna, which has not deployed because of three stuck ribs (NASA)

design section identified a multiple gravity assist scheme leading to arrival in December 1995. JPL identified no additional benefit with a Mars swing-by, but it was found that the six-year VEEGA route would generate the required energy. On that basis, NASA set the launch of the 2,561 kg craft (including 925 kg propellant and 339 kg probe) for the window of 8 October-24 November 1989, with 30-minute daily windows. The lengthy period within Earth's orbit presented the greatest problem and Galileo underwent modifications to combat the almost doubled solar flux and the effect it would have on delicate instrumentation. The modified spacecraft, even with an additional 100 kg of shielding and equipment, emerged slightly lighter. Modifications included the addition of a large thermal disc above the bus and an LGA on the RTG boom for Earth communications during the Venus/Earth fly-bys.

By the end of 1987, delays and modifications had increased the total cost of development, launch and 30 days of space operations from US$673.5 million to US$893 million, with the rest of the mission rising from US$200 million to US$500 million (assuming a 7 per cent inflation rate). Added to the fact that Galileo's design would represent 20-year-old technology when it reached Jupiter, this generated speculation during 1987 that the mission would be cancelled. Galileo competed with Ulysses for a Shuttle launch in a Jupiter window that recurs only at 13-month intervals. ESA and NASA finally agreed in April 1987 that Galileo would be launched in the 45-day October-November 1989 window, with Ulysses following in October 1990. There was to be no repetition of the 1986 plan (foiled by Challenger) of attempting both during the same window.

Many observers doubted that departure would occur at all, because the spacecraft suffered a legal challenge prompted by concerns that the 22 kg of plutonium dioxide carried by Galileo's two RTGs (the first new-model RTGs referred to as GPHS General Purpose Heat Source RTG) might spill in an explosion. The RTGs were not inserted until 3 October. Atlantis began its fifth mission at 16.53.40 UT 18 October 1989.

Checkout of the Galileo/IUS combination began 2 h into the mission and after 5 h 42 minutes the crew reported the IUS table was raised to 29°, followed 10 minutes later by the required 58°. Mission Specialist Shannon Lucid commanded spring deployment at 6 h 21 minutes 23 seconds. The IUS automatically ignited its first stage 60 minutes later over Borneo, following the 149 second burn with stage 2's 105 seconds. At this stage, the vehicle was being monitored by the USAF Consolidated Space Test Center at Onizuka AS, California via their ARIA Advanced Range Instrumentation Aircraft. IUS then pointed Galileo within 9.5° of the Sun for thermal shading, spun it up to 2.8 rpm and deployed the craft's magnetometer and RTG booms, before separating at 01.05 GMT 19 October.

The probe and its instruments underwent 6 h of tests on 26 October 1989 before being closed down until its next annual test during December 1990's Earth fly-by. The orbiter's instruments were also checked and the Heavy Ion Counter and magnetometer left on to return cruise data. The HIC was conceived in 1984 to answer concerns over Single Event Upsets that were being found on Earth-orbit satellites. The test model of Voyager's Cosmic Ray Science instrument was reworked to monitor the rich heavy ion environment found around Jupiter. Ironically, HIC's older components are more SEU-resistant than Galileo's more modern elements. Four large solar events occurred soon after launch but the detected impacts caused no damage.

The first mid-course correction sequence TCM-1 was executed over 9-11 November, yielding a 17 m/s delta-V. 10 N thrusters on the spin section fired >5,500 times in typically 1 second pulses. TCM-2 on 22 December provided a 0.8 m/s trim for a 16,000 km Venus fly-by. TCM-3 was not required. During its passage over Venus' morning terminator, Galileo employed its NIMS infra-red mapper to observe cloud features on the dark side for the first time and map airglow at 2.7 μm. The instrument provided the first views of small scale features over the mid-latitude and equatorial regions, in addition to performing spectroscopy of the middle/deep layers of the thick atmosphere. Features matching those on radar images suggest the surface may be visible in the 1 μm carbon dioxide window. The plasma wave instrument confirmed the presence of lightning. A closest approach of 16,106 km was made over 41° south at 05.58.48 GMT on 10 February 1990, gaining 8,030 km/h. For a day before and for up to seven days after, the charged particles, dust and magnetic fields, infra-red/ultra-violet instruments were active but most of the data were recorded for playback mid-November shortly before the first Earth fly-by. The link went via one of the LGAs (thermal considerations required the HGA to remain furled until April 1991). However, three TV images and some NIMS mapping were returned. As planned, 81 TV images were recorded but a software timing problem led to the system shuttering an additional 452 times (without recording) in the 5 hr after closest approach. Final imaging took place 17 February.

Following the fly-by, Galileo's heliocentric orbit was 0.699 × 1.28 AU, heading towards first perihelion 25 February and first aphelion 23 August 1990. Galileo performed its largest course adjustment before Jupiter in April/May 1990, to patch its post-Venus path into the required Earth fly-by trajectory. The VEEGA route initially identified in July 1986 required a November 1989 launch but the insertion of this large adjustment expanded the window into October. TCM-4's first set of pulses, over 9-12 April, yielded a 24 m/s (24.8 m/s planned) change, followed by 11.3 m/s on 11/12 May, using the lateral thrusters.

After the Venus fly-by, mission planners approved 1991's encounter with asteroid Gaspra. Propellant margins had eased with IUS' accurate injection, further trajectory calculations and the discovery that Galileo's thrusters were providing 2 per cent greater efficiency than expected all encouraged confidence in the mission.

Observations were made of Comet Austin during May 1990 and the following August the Extreme UV instrument registered hydrogen in the coma and tails of Comet Levy. Only small corrections were required after TCM-4 to adjust the trajectory for the first Earth fly-by: TCM-5 provided 0.9 m/s on 17 July; TCM-6 0.5 m/s on 9 October, TCM-7 1.3 m/s on 13 November and TCM-8 0.05 m/s on 28 November. The tape recorders' stored Venus data dumped their data to earth station on the way by on 19-21 November.

Passage through the Earth-Moon system provided the opportunity to certify the instruments and undertake the first terrestrial observations from a planetary craft.

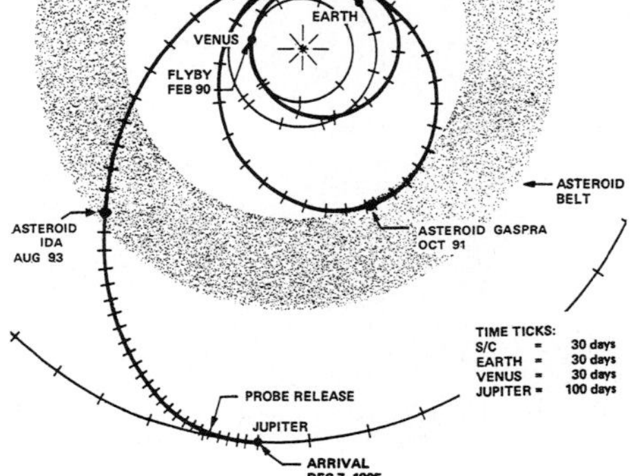

Galileo's unique Venus-Earth-Earth swingby path enhanced the Inertial Upper Stage's Jupiter capability and permitted dual asteroid encounters

Galileo was the first to make in situ observations of an outer planet's atmosphere (Hughes)

Galileo flew within 960 km (8 km above target) of Earth at 20.34.34 GMT 8 December 1990, detecting clear evidence of life in atmospheric trace constituents. The cameras made one exposure each minute for a day for a 1,500-frame film sequence to be assembled, showing Antarctica from a unique perspective. Lunar observations were aimed at composition of the far side crust, Mare Orientale's structure and imaging little-known regions around the south pole. Galileo confirmed the existence of the suspected (from Apollo 15 data) 1,940 km diameter Aitken Basin in the south, underlying the younger Orientale impact feature. The entire Earth/Moon encounter returned 2,675 images. It also set up October 1991's Gaspra fly-by, by increasing Galileo's heliocentric period from one to two years and increasing ecliptic inclination to 4.5°. The TCMs before Gaspra fine-tuned this gravity assist.

Previously furled because of thermal requirements, the 4.8 m diameter HGA was commanded to open 11 April 1991 but jammed less than half way, with three of the 18 ribs still slotted in their launch positions on the central mast. Analysis and tests with the spare flight model suggested that insufficient lubricant, aggravated by chafing during road transport caused the problem. JPL turned Galileo in late May 1991 to use solar heating induced thermal expansion to unjam the mast. Then JPL engineers realised contraction and not expansion would be the best solution. The mast was held in shadow for 32 hr in July and then for 50 hr in August while Galileo was 2.0 AU from the Sun. These attempts failed. The HGA had been scheduled for the Gaspra fly-by, but the data had to be recorded instead for replay during the December 1992 Earth fly-by. The fifth and most aggressive cold soak was tried in December 1991 2.2 AU out around aphelion but the ribs remained stuck. As Galileo approached 1 AU during 1992 for the Earth fly-by, hot/cold cycling was attempting to 'walk' the ribs out. Rotating the craft cost 4 kg of valuable propellant.

Galileo, meanwhile, achieved the first close encounter with a known asteroid during 29 October 1991. Following Earth's December 1990 gravity assist, five TCMs fine-tuned the approach to 951 Gaspra. Ground-based observations had located Gaspra to within 200 km and Galileo's own navigation images during September and October 1991 refined the ephemeris to 50 km. To avoid any debris cloud, the probe approached no closer than 1,600 km (actual 1,604 km) during the 8.0 km/s fly-by at 22.37 GMT, some 410 million km from Earth. In fact, no dust impacts were registered. Even with the relatively large separation, the gravitational deflection of Galileo's trajectory was sufficient to yield a good estimate of Gaspra's mass and hence, average density. Field reversals detected by the magnetometer indicate a strong magnetic field, requiring Gaspra to be iron or nickel.

A total of 150 images were recorded during the encounter, 126 of them during the last hour before closest approach. From 35-29 minutes before closest approach, a mosaic of nine colour frames (that is 27 frames through three filters) was acquired to ensure Gaspra was seen. Similarly, a 49-frame black/white mosaic was recorded 29-16 minutes before closest approach. No frames were taken at closest approach because of the angular speed. The 800 × 800 element Si CCD camera has a broader spectral range and faster response than the vidicon tubes aboard Voyager. The Near-IR Mapping Spectrometer (NIMS) is a powerful diagnostic tool never flown before, capable of classifying the surface mineralogy through the solar reflection

Inserting the probe's descent module into the heatshield (Hughes) ***1995***

spectrum, highlighting silicates, carbonates and nitrates, for example. The photopolarimeter radiometer studied the photometric and thermal surface properties. If there is an associated dust field, the German dust detector would have measured the size distribution, spatial extent and trajectories of its sub-μm particles. With a sensitive area of 1,000 cm^2, particles are registered after impact with a gold target which generates a small plasma cloud. All of the data were recorded for playback when the HGA was released or during end-1992 Earth's fly-by, whichever came sooner.

In fact, mankind's first close-up view of a known asteroid came on 7 November 1991 when controllers retrieved the second-highest resolution image (comprising four frames for full colour), providing 160 m/pixel from 16,000 km, at 40 bit/s via the LGA. The 17 × 12 × 11 km S-type Gaspra was surprisingly rounded, suggesting a considerable dust layer held in place by the 0.0005 *g* gravity. Craters ranged up to around 2 km diameter. Grooves similar to those seen on Phobos were at the edge of resolution. The highest resolution image (50 m/pixel) was retrieved 18 May – 5 June 1992, and all the rest in late November 1992. The International Astronomical Union in 1994 approved names for Gaspra's features based on the world's spas and resorts, as the asteroid itself was named after a resort on the Crimean Peninsula.

November 1991's TCM-13 was not required but four sequences were executed for the second Earth/Moon fly-by: TCM-14 21.0 m/s 4 August, -15 0.7 m/s 9 October, -16 0.9 m/s 14 November and -17 0.03 m/s 28 November. Mass was 2,415 kg after TCM-14. The probe was put through its first Mission Sequence Test (MST) during 20 November, taking it through its actual mission sequence. Performing observations similar to those on the first passage, Galileo flew past the Moon at 110,000 km at 03.58 GMT 8 December 1992 and then Earth at 303.1 km (303.8 km planned) at 15.09.25 GMT over 34° south/6° west, adding 3.7 km/s to enter the direct Jupiter route. The passage permitted high resolution (and the first multi-spectral) observations of the lunar north pole. UVS discovered that Earth's geocorona is far more extensive than previously believed. As last time, time lapse films were made, including the first-ever showing of a full 24 hr rotation of Earth.

After thermal walking failed to release the HGA by July 1992, engineers switched to a 'hammering' technique, pulsing the two electric motors to drive the ballscrew mechanism further. They waited until December 1992 around perihelion when the antenna was warmed to its assembly temperature. The first attempt 29-30 December did move rib No 2 by 8° but the 5-11/14-19 January 1993 sequences failed. After 13,320 pulses, they gave up hope of deploying the HGA on this first Jupiter orbiter. A final attempt by spinning Galileo up to 10.5 rpm 11 March 1993 while hammering, also failed. 312 pulses of a 10 N thruster were required for the spin up from 2.89 rpm and 313 pulses for the spin down. It was next hammered 1,080 times again 29 August 1994, as a last try before arriving at Jupiter, although there was little hope of it working. It was hammered 1,080 times for the last time 22 March 1996.

The LGA officially became the baseline mission in March 1993, although software and ground hardware improvements should yield around 70 per cent of the originally planned science return (80 per cent atmosphere, 100 per cent probe, 70 per cent satellites and 60 per cent magnetosphere). 1,500-2,000 images will be returned instead of the 50,000 expected. The 40 bit/s has been improved to an equivalent of more than 1 kbit/s (160 bit/s are actually returned, but data compression multiplies the effective rate). By using data compression and new error correction algorithms, arraying Earth antennas and increasing their sensitivities (particularly of the Canberra station, the most important, as Jupiter is then in the southern sky).

Galileo made history again, when in conjunction with the Mars Observer and Ulysses, JPL used the craft as along baseline gravity wave detector between 21 March and 11 April 1993. The results of this experiment were negative.

Approval for the asteroid Ida fly-by on 28 August 1993 was required before the Earth-2 TCMs. Although it cost 30 kg propellant and earlier in the mission looked likely to be rejected (principally because the six-day launch delay imposed a 58 kg penalty), Galileo's performance and the science return from Gaspra prompted approval from NASA HQ 25 June 1992. Earth-2 fly-by was so accurate that TCM-18 13 days later was not required. TCM-19 (2.1 m/s) on 19 March 1993 provided the bulk of retargeting to Ida and the 0.6 m/s TCM-20 trim 13 August cancelled the need for TCM-21 26 August. Galileo followed a Gaspra-type sequence and made its closest approach of 2,410 km at 16.51.59 0 GMT 28 August 1993. A potentially disastrous event hit Galileo 256 minutes out: for some reason, it turned off its gyros and returned to cruise mode. Engineers determined the encounter could still be successful in this mode by pointing the scan platform using the actuators' encoders. However, the mode had stowed the platform, so a precisely timed command was sent 198 minutes before arrival to restore proper pointing. 150 image frames were recorded, about half of them to ensure capture. 13 colour and 8 black and white images were stored, to be transmitted over 1,300 h at 40 bit/s via the LGA September 1993 and 16 February – 26 June 1994 when Earth was close enough. The highest resolution five-frame mosaic was recovered over about 30 h by 22 September 1993 (the last day before Earth was too distant), showing the 52 km-long Ida at 25 m resolution. The battered surface shows it to be an old body. Ann Harch of the SSI team discovered a 1.2 × 1.4 × 1.6 km diameter moon, subsequently named Dactyl, while examining image slices on Galileo's recorder 17 February 1994. It was found in the NIMS data 23 February. The image return completed 26 June 1994 showed it to be as battered, and thus as old, as Ida. They are similar bodies but NIMS revealed different compositions, suggesting that Dactyl was captured and did not break off Ida. The pair appeared together in 47 SSI images, allowing Dactyl's orbit and hence Ida's density to be modelled. The orbit could not be calculated with certainty, but gave an Ida density range of 1.9-3 2 g/cm^3, suggesting a light and/or porous structure.

The 38.6 m/s TCM-22 was performed 4-8 October 1993 using 35.2 kg propellant, for the first time slotting Galileo into the Jupiter atmosphere entry corridor. The almost 5,000 pulses on each of the two lateral thrusters provided the most accurate burn yet – only a 0.15 per cent underburn. The 10 N thrusters will not be used for such a large delta-V again. 15 February 1994's 0.1 m/s TCM-22A corrected the error. Five instruments provided astronomers' only direct observations of the disrupted Comet Shoemaker-Levy 9 impacting Jupiter 16-22 July 1994. Data return was completed 29 January 1995.

TCM-23 (12 April) was perfect, so TCM-24 (23 June) was cancelled. The probe was released on 13 July 1995, 147 days and 80 million km from Jupiter. The umbilical was cut 11 July, meaning that it could never again receive commands and would not transmit telemetry until after its parachute opened. Galileo assumed the release attitude, spun up to 10.5 rpm and fired the three explosive nuts on the probe's tripod assembly. Springs ensured a 30 cm/s separation. Galileo avoided following the probe into the planet by making the 62 m/s 308 second TCM-25/ODM Orbiter Deflection Burn 27 July, firing its 400 N engine for the first time. It also set up unique inbound fly-bys of Europa and Io, the reason for selecting arrival 7 December. The 10 N thrusters could have made the ODM, but the larger engine's performance was demonstrated for Jupiter Orbit Insertion. Galileo's approach 6° below the satellites' orbital planes provided the mission's only views of Europa's south pole region, from 32,500 km (six times closer than Voyager) within 9 h of perijove. Then, 4 h before perijove, came Galileo's only close (900 km) fly-by of Io because of the harsh radiation environment, providing 20-30 m resolution. The encounter reduced Galileo's JOI delta-V by 175 m/s and provided passage through Io's plasma torus. However, the tape recorder jammed 11 October 1995 and all of the inbound remote sensing was cancelled. Instead of rewinding after recording an image, the tape continued slipping on a capstan for 15 h before a ground command shut it down. Fearing the tape was weakened, that spot was hidden under 25 turns on the takeup reel. As the tape was essential for recording the probe's unique descent data, it was deemed safer to cancel further remote sensing until April 1996 after the last probe data had been retrieved. In fact, 3 h of high-resolution fields/particles data was recorded during Io plasma torus passage.

As Galileo hit perijove 130,000 km above the clouds at 21.53 GMT 7 December 1995, the probe entered the atmosphere at 6.5° north/4.4° west at 22.04 GMT and transmitted data for 57 minutes. the descent lasted 156 km after its parachute opened 53 seconds/26 km late (which meant data began below the cloud tops. The cause was a faulty *g*-switch). Timers activated the probe 6 h before entry. It was expected to survive down to 20 Earth atmospheres following the 47 km/s, 250 *g* entry (designed to withstand 400 *g*), but it achieved 22 atmospheres and 425K. The full data were stored on Galileo's tape recorder and in partial back-up form in its solid-state Command & Data Subsystem. The back-up data were read out at intervals in December/January and the tape January-April 1996. Preliminary findings showed an intense radiation belt 50,000 km above the clouds, a sun-like helium to hydrogen ratio, few organic compounds, one cloud layer (three expected) and 640 m/s winds below the clouds (powered by internal heat instead of the Sun).

Regardless of the probe's survival at 75 minutes, the orbiter had to break off to prepare for JOI. The 643 m/s, 48.98 minutes 400 N burn starting 00.27 GMT 8 December 1995 and consuming 377 kg propellant established Galileo in a 198-day closed path about Jupiter – the first man-made satellite of the giant planet. Perijove was raised from 4 R_J to 10 R_J at first apojove on 14 March 1996 in a 377 m/s burn to avoid the radiation hazards. The planned 11 tours over 23 months with Europa, Callisto and Ganymede close encounters thus began. The 400 N engine is not needed after this third firing. The 185 kg consumed left 105 ±5 kg, leaving a 5 kg margin to complete the mission with 90 per cent confidence (13 kg is still held aside for the project manager to allocate). Galileo's total delta-V capacity is 1.6 km/s; the tour would require 6 km/s without gravity assist, whereas 146 m/s is allocated.

On 15 April, the play-back of the tape-recorded science data from the Atmospheric Probe was completed. Other experiments on the spacecraft were gathering and transmitting data about the jovian magnetic field during the period leading up to the first Ganymede encounter. On 3 May, an orbital trim manoeuvre was completed to adjust the arrival time and geometry for the encounter. On 3 June, the spacecraft began to transmit data which had been recorded during its fly-by of the satellite Io six months earlier on 7 December. Images showed how much the satellite's geography had changed due to volcanism since the jovian fly-by visits of the Voyager spacecraft.

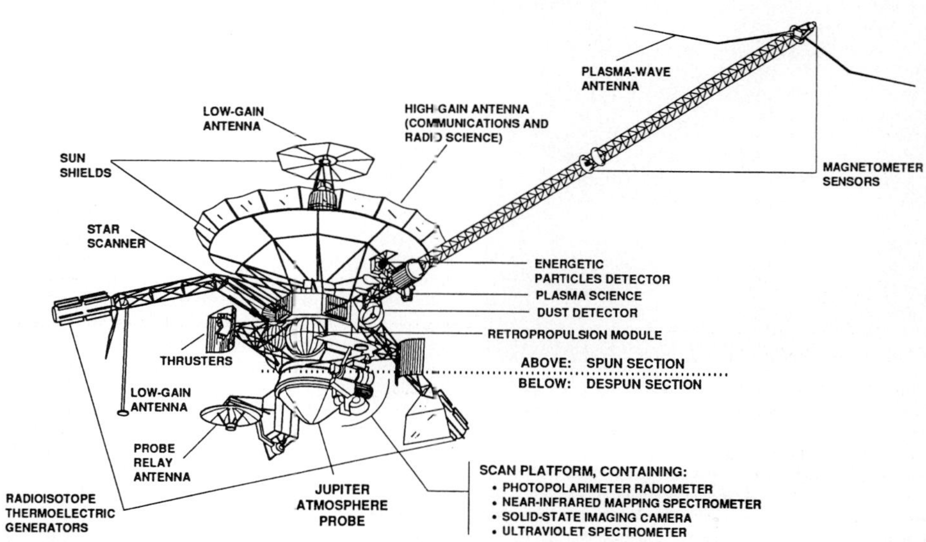

Galileo in planned cruise configuration before release of its atmospheric probe (NASA/JPL)

A final orbital trim manoeuvre took place on 24 June and the first fly-by on Ganymede came at 09.29 GMT on 27 June, when Galileo passed 844 km above the satellite's surface. On 24 August, Galileo unexpectedly put itself into a safe mode due to processing time allocation being exceeded in a central computer on the spacecraft: control was switched to a back-up computer which performed basic engineering functions. Four days later, the main on-board computer was brought back on-line. In the meantime, on 27 August the spacecraft had fired its thrusters once more to fine-tune its trajectory for the second Ganymede fly-by. Encounter came at about 19.00 on 6 September, this time the minimum distance being 262 km. Once more, instruments were not only taking photographs of the satellite's surface, but also probing the magnetic and particle environment and gravity characteristics.

The first Galileo fly-by of the satellite Callisto took place at 13.34 GMT on 4 November, the minimum distance being 1,104 km. On 26 November, a trajectory trim was performed in advance of the first encounter with Europa. The encounter took place at 06.53 on 19 December at a minimum distance of 692 km. There was a lull in Galileo's encounters with the Jovian moons while Jupiter was experiencing solar conjunction, so the revolution 5 encounter was with the planet Jupiter itself, closest approach of 646,000 km coming at 09:36 GMT on 19 January 1997. Then it was back to the moons of Jupiter beginning again on revolution 6. On 20 February 1997 at 17.06 GMT, the second encounter with Europa took place with the minimum distance being within 1 km of the planned 586 km. Further trim manoeuvres took place on 13 and 31 March in advance of the next satellite encounter, the third with Ganymede. This fly-by came at 07.11 on 5 April, the distance being 3,102 km.

Galileo began its investigation of the Jovian moons with two encounters at Ganymede on successive orbital passes, each elliptical swing around Jupiter taking several weeks while the spacecraft drifted several million kilometres out to the apojove of its orbit ensuring protection from the planet's intense radiation belts. Revolutions 7 and 8 also provided two successive encounters with Ganymede in April and May respectively, the last time this Jovian moon would be visited by Galileo. Following this, two close passes of Callisto where Galileo swooped to within 415 kilometres of the surface preceded a third encounter with Europa to close out the 11 fly-by encounters that formed the primary mission for which the spacecraft had been launched. Successful performance in the Jupiter system and equally successful lobbying from the Galileo team gained approval for a two-year extension to be known as the Galileo Europa Mission (GEM).

GEM would double the science time at Jupiter, extending the mission from two to four years and extend the number of encounters from 11 to 25. Originally scheduled to end on 7 December 1997, the mission would continue to the end of 1999. Divided into three tightly focused periods of observation known as 'Ice', 'Water' and 'Fire', GEM would build upon findings from encounters at Europa to investigate key aspects of the moon and its interaction with Jupiter. The 'Ice' phase would require eight successive passes of Europa (December 1997 to February 1999) during which the spacecraft would search for further evidence of a salty ocean beneath an ice crusted surface. It would look for volcanic activity to indicate a geologically active interior and obtain images with a resolution of just six metres during a close pass of just 200 km.

The second, or 'Water' phase would call for four successive swings round Callisto (May, June, August, September 1999) to halve the distance of closest approach to Jupiter (perijove reduction) for investigations into water circulation in cloud tops that surround the giant planet. On each encounter, Galileo would fly through the ring-shaped torus of charged particles that links innermost major moon Io with Jupiter itself, mapping the density of sulphur spewing from the surface.

The most dramatic phase, 'Fire', provides a unique opportunity on the last two GEM passes (October and November 1999) for a close encounter with Io. On the last one, the spacecraft dives down to a distance of just 300 km as Galileo swoops through the active plume of frozen sulphur from Io's Pillan Patera volcano taking pictures with a resolution of six metres while intense radiation peppers the camera's light detector with flashes visible in the images. GEM formally ends on the last day of 1999.

The following table identifies all the encounters by Galileo accomplished to date and those scheduled for the remainder of the mission.

Orbit	Moon (sequence)	Date	Time (UT)	Distance from objective (km)
Primary Mission:				
1	Ganymede (1)	27 June 1996	06:29	835
2	Ganymede (2)	6 September 1996	19:00	262
3	Callisto (1)	4 November 1996	13:34	1,118
4	Europa (1)	19 December 1996	06:53	698
5	Jupiter (1)	19 January 1997	09:36	646,000
6	Europa (2)	20 February 1998	17:06	586
7	Ganymede (3)	5 April 1998	07:10	3,102
8	Ganymede (4)	8 May 1998	15:57	1,596
9	Callisto (2)	25 June 1997	13:48	415
10	Callisto (3)	17 September 1997	00:19	538
11	Europa (3)	6 November 1997	20:32	2,042
Galileo Europa Mission:				
12	Europa (4)	16 December 1997	12:05	200
13	Europa (5)	10 February 1998	17:57	3,562
14	Europa (6)	29 March 1998	13:23	1,649
15	Europa (7)	31 May 1998	21:12	2,521
16	Europa (8)	21 July 1998	05:07	1,837
17	Europa (9)	26 September 1998	03:54	3,582
18	Europa (10)	22 November 1998	11:47	2,281
19	Europa (11)	1 February 1999	02:10	1,495
20	Callisto (4)	5 May 1999	14:06	1,311
GEM encounter schedule:				
21	Callisto (5)	30 June 1999	07:46	1,050
22	Callisto (6)	14 August 1999	08:39	2,288
23	Callisto (7)	16 September 1999	17:35	1,053
24	Io (1)	11 October 1999	04:36	500
25	Io (2)	26 November 1999	04:00	300

Galileo flew past asteroid Ida in August 1993, but the first asteroid moon was not revealed until February 1994. The bottom (discovery) image shows Ida and Dactyl from 10,870 km, 14 minutes before closest approach. Inserted is the best image of Dactyl, 10 minutes later, from 3,900 km. At the top is the highest resolution achieved at Ida, some 25 m/pixel 46 s after closest approach. The images remained stored until 1994 when Earth was close enough again for the 40 bit/s relay (NASA/JPL) ***1995***

Orbiter description

The 2,222 kg orbiter is divided into an upper section incorporating an HGA, an 8-sided bus and retro-propulsion module (RPM) spun at 3.15 rpm (increased to 10.5 rpm for probe deployment). This was also used March 1993 in an attempt to free the HGA, and a despun lower section of probe, its data relay antenna and scan platform (0-1°/s slew rate, 0.2° pointing accuracy) with the four remote sensing instruments. The five fields/particles instruments are mounted on or near a single 11 m boom. Science payload is 118 kg.

Two new-model RTGs on 5 m booms provided 570 W at launch, projected to decay to 480 W in 1997 (496 W, 31 W margin, as of September 1995); each carries 10.9 kg of plutonium dioxide. Germany's 1,120 kg RPM provides all attitude control and retro capability from two independent networks of six 10 N thrusters, each mounted on two 2 m arms and central 400 N motor blocked by the probe until release. A usable total of 925 kg of MMH/NTO is divided between four tanks, with two associated helium pressurant spheres. The three major burns (planet avoidance after probe release, orbit insertion and orbit raising) were performed by the large motor with the entire spacecraft spinning at 10 rpm, but four of the 10 N thrusters are positioned as back-ups. With all of the tankage emptied the probe has a total delta V capability of 1,600 m/s.

The 4.8 m diameter TDRS-type HGA, comprising gold-plated molybdenum wire mesh stretched between 18 graphite epoxy ribs, is designed for X-band (8.415 GHz) 115.2 kbit/s real-time and 134 kbit/s playback datalinks at Jupiter. The LGA gives a 40 bit/s in S-band at 2.296 GHz. Uplinking occurs at 34 kbit/s on 2.115 GHz S-band. Probe data were to be relayed at 28.8 kbit/s in real-time, received by the dedicated despun dish, and also recorded on the single Odetics DDS-3100 900 Mbit digital tape recorder, able to hold the equivalent of about 200 images. Instead, the LGA required 3 months to transmit the stored probe data between January and April 1996.

Probe description

The 339 kg probe, built by Hughes Aircraft under NASA Ames' management, was activated by timer 6 hr before entry to follow a pre-programmed descent with sequences triggered by accelerometer cues. Pre-programmed system testing was undertaken annually during coast and before release. The Probe contained two main elements: the external 220 kg Deceleration Module, and the Descent Module uncovered after entry.

Some 90 kg of the 145 kg carbon phenolic heatshield was lost during the 43 kW/cm^2 peak heating entry; a phenolic nylon shield protected the upper surface. A 2.5 m Dacron parachute pulled the Descent Module free. The DM was not a pressure vessel (as were Pioneer Venus' capsules) but the 28 kg of science instrumentation, mounted on a 66 cm diameter aluminium honeycomb shelf, was protected inside individual housings expected to survive up to 20 atmospheres. Data were transmitted on 1.3870/1.3871 GHz L-band at 28.8 kbit/s for orbiter storage. A 21 Ah $LiSO_2$ battery provided power, with 72 W peak allocated for science instruments.

NASA did not sterilise the probe since Jupiter is a Category 2 planet under NASA's 1984 quarantine guidelines.

Specifications

Orbiter payload

Solid-State Imaging: 28 kg, 1,500 mm, f/8.5, 800 × 800 CCD, map Galilean moons at 1 km resolution and monitor atmospheric circulation, 115 kbits/s.

Near-IR Mapping Spectrometer: 18 kg, 0.7-5.2 μm, satellite surface composition, atmosphere T/composition.

UV Spectrometer: 4 kg, 1,150-4,300 Å, atmospheric gases/aerosols. 544-1,280 Å Extreme UV detector added later.

Photopolarimeter-radiometer: 5 kg, visible/near-IR bands, radiometry to greater than 42 μm, atmospheric particles, thermal radiation.

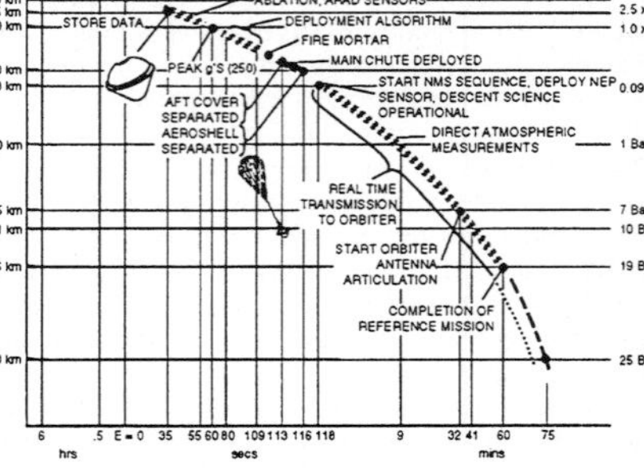

Galileo probe's descent sequence was pre-programmed, with events triggered by accelerometer cues (JPL) ***1995***

Magnetometer: 7 kg, 32-16,384 v, magnetic fields.

Energetic Particle Detector: 9 kg, 0.02-55 MeV ions, 0.015-11 MeV electrons, and high energy ions in magnetosphere.

Plasma Detector: 12 kg, 1 eV-50 keV in 64 bands, energy composition/distribution of low energy ions.

Plasma Wave: 6 kg, 6-31 Hz, 50 Hz-200 kHz, 0.1-5.65 MHz, electromagnetic waves and wave/particle interactions. Dust Detector: 4 kg, 10^{-16}-10^{-6} g, 2-50 km/s, measures γ and charge.

Heavy Ion Counter: modified from Voyager Cosmic Ray Science detector, added in 1985 to monitor spacecraft exposure to damaging radiation in addition to observing solar flares and Jupiter's environs. Five single Si crystal wafers 30-2,000 μm thick, 6-200 MeV.

Radio system: determine planet/moon masses and radio/atmospheric structures.

Probe Payload

Atmospheric Structure Instrument: 4 kg, 0-540K, 0-28 atmosphere, temperature and pressure, density and molecular weight with altitude

Neutral Mass Spectrometer: 11 kg, 1-150 amu, atmospheric chemical composition

Helium Abundance Detector: 1 kg, 1% accuracy

Net Flux Radiometer: 3 kg, 0.3 to more than 100 μm, ambient thermal and solar energy.

Nephelometer: 5 kg, 0.2-20 μm cloud particles more than 3/cm^3

Lightning/Energetic Particles: 2 kg, fisheye lens, 1 Hz-100 kHz, lightning, measure energetic particles.

Magellan

Current status

One of the most successful planetary missions in history ended in October 1994 when the NASA/JPL Magellan Venus orbiter fell silent and then burned up in the Venusian atmosphere. Having completed the planet's first detailed radar and gravity mapping, Magellan lowered itself into the atmosphere to generate unique aerodynamic data for future aerobraking missions. The US$800 million mission completed the first 243-day cycle of radar mapping Venus on 15 May 1991, providing the first clear views of 83.7 per cent of the surface at 120-300 m resolution. The formal project goal of 70 per cent coverage passed as a milestone on 2 April 1991. Magellan returned 1,200 Gbit of data and far surpassed the 900 Gbit from *all* of NASA's other planetary missions.

Mapping cycle 2 ended 15 January 1992, increasing coverage to 96 per cent, particularly adding high southern latitudes by changing attitude to look right. Cycle 3, ending 13 September 1992, concentrated on stereo imaging (achieving 21 per cent coverage and increased mapping to 98 per cent. No further radar mapping was performed because of communications problems. Cycle 4, ending 25 May 1993, emphasised gravity data between ±30° latitudes. The first ever extended aerobraking sequence, 25 May to 3 August 1993, then lowered the orbit. Cycles 5/6, covering 3 August 1993 to 10 October 1994, thus collected 200-400 km high resolution gravimetry data over 95 per cent of the planet with 1.5 milligals sensitivity – better than so far acquired for Earth.

Mars Global Surveyor

Background

NASA began a new initiative, the Mars Surveyor programme in the wake of the Mars Observer loss in 1993. In effect the administration intended to demonstrate a new commitment towards cheaper, faster and more frequent missions. Rather than flying a duplicate mission of the lost Mars Observer, NASA shook up its approach by including US$78.4 million in FY95's request (receiving $59.4 million) to begin the Mars Surveyor series of small orbiters/landers. Congress capped the programme to US$100 million annually, and contrary to NASA's track record the budget cap seems to be holding.

The 1 tonne first orbiter, Mars Global Surveyor, carries five of Mars Observer's seven experiments. The Mars Pathfinder orbiter carries one of MO's two remaining instruments and acts as a relay for the separate lander. 2001's mission will carry MO's last instrument. NASA hopes to return 2 kg of samples in 2005, while at the same time remaining within the US$200 million ceiling, and in situ propellant production may be required to meet the mass target.

Lockheed Martin Astronautics built MGS, employing Mars Observer's spare thrusters and HGA but with a graphite composite bus instead of aluminium, because of mass constraints. Major elements draw on MO spares: most major electronic assemblies are MO spares, retrofitted to eliminate MO-identified discrepancies. The dual-mode propulsion system draws on Cassini designs.

NASA launched MGS at 17.00.50 GMT on 7 November 1996 and initially the Delta-2 launch vehicle placed the satellite and PAM-D third stage into a near-circular orbit at 28.5°, 185 km. At 17.41, the second stage of the Delta-2 re-ignited to place the assembly into an orbit reaching out to 4,700 km and then separated from the combined PAM-D/MGS. PAM-D ignition at 17.44.30 put the spacecraft into a heliocentric, trans-Mars orbit.

At about 17.52, MGS spread out its solar array, but telemetry indicated that one of the solar panels (-Y) did not fully open, tilting instead about 20° from the expected angle relative to the spacecraft's main bus. Both solar panels were receiving sunlight, while still fully charging the batteries. The orientation of the (-Y) solar panel is expected to have little effect during the cruise phase to Mars.

On 10 November, the spacecraft used its star sensor to establish its orientation in space and later that day it was put into an 'array normal spin' orientation which was used for the rest of the trans-Mars coast. On 21 November, MGS fired its main engine and performed the first of its four planned flight path correction manoeuvres. The 44 s burn achieved a change in spacecraft velocity of about 27 m/s. By performing the burn, the third stage of the Delta-2 launch vehicle is steered away from a Mars collision.

On 22 January 1997, the spacecraft's flight computer activated a 53 W heater in the Mars Orbiter Camera to begin 'baking' the instrument and removing residual moisture. Without this 14-day bakeout period, the moisture in the camera's tube-like structure would leak into space at a slow rate and cause a gradual shift in the camera's focus. The bakeout will remove all of the moisture at once and stabilise the focus of the camera. Flight controllers also performed a series of very slight manoeuvres on 22, 23 and 24 January to attempt to manipulate the MGS solar array and characterise the exact condition of the debris that is preventing the panel from fully deploying.

A 26 second burn on 20 March 1997, refined MGS's flight path to Mars and achieved a change in spacecraft velocity of about 3.87 m/s. The burn was performed in two stages, in which flight controllers first commanded the spacecraft to fire its small thrusters for 20 seconds, then to fire its main engine for another 6 seconds. During April the spacecraft was in a quiet phase and conducted a search campaign to detect gravity waves.

MGS arrived at the Red Planet on 11 September 1997, and made a 973.03 m/s delta-V burn of 22 minutes 39 seconds to put it into an initial 250 × 54,000 km, 48 hour orbit. Aerobraking over a five-month period was to have changed the initial, elliptical path to a near circular 378 km orbit with an inclination of 92.9°. This would have provided an orbital period of 117.65 minutes and a 2 pm equatorial sun-synchronous repeat cycle of 7 days (88 revolutions) so that full mapping could begin in March 1998. On 1 October 1997, when the spacecraft descended through periapsis of 110 km, engineers observed movement in the –Y solar cell array that had given trouble shortly after launch. This generated concerns about possible damage caused by successive aerobraking passes through the tenuous outer layers of the Mars atmosphere and the periapsis altitude was temporarily raised back up to 121 km while an engineering analysis was performed. On 11 October, it was decided to raise periapsis to 171 km and on 27 October a management meeting was held to discuss options; at this date apoapsis was about 45,135 km and the orbital period 35.4 h. Periapsis reduction recommenced 7 November in a burn that put the low point of the orbit down at 134.8 km but aerobraking would be performed at a much slower rate than originally planned, enabling engineers to carefully monitor any stress on the –Y panel. Engineers worked

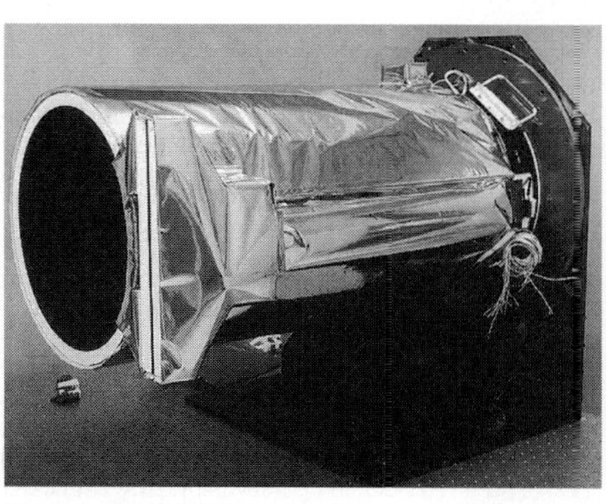

The completed MGS camera. The narrow-angle camera views through the main 35 cm aperture; the two wide-angle cameras (red/blue filters) view through the narrow slit (Malin Space Science Systems) ***1996***

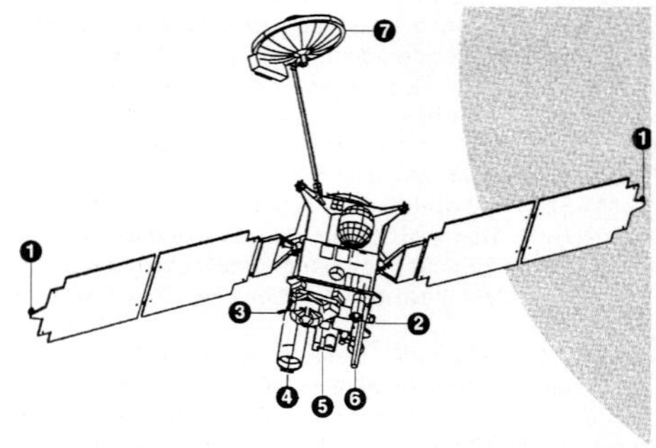

Mars Global Surveyor. 1) magnetometer, 2) electron reflectometer, 3) laser altimeter, 4) camera, 5) thermal spectrometer, 6) Mars Relay, 7) HGA (NASA/JPL) ***1995***

out that during initial deployment, a damper arm had fractured resulting in structural damage to one end of the solar cell frame.

The revised aerobraking plan reduced the periapsis pressure exerted on the solar panels to 0.2 N/m^2, one-third the level assumed in the original plan. Global Surveyor would reach a circular mapping orbit in March 1999 after a total of more than 900 orbits. To achieve that, aerobraking continued until 23 March 1998 when it was suspended for four months during the period covered by solar conjunction. Four days later, the science instruments were turned on for a period of observation and imaging. On 23 September, Global Surveyor performed the first of three propulsive manoeuvres, completed two days later, totalling 11.62 m/sec to step back into the atmosphere, lower periapsis to 121 km and settling it into a pressure regime of 0.14 N/m^2. On 26 and 27 September, further manoeuvres of 0.18 m/sec lowered periapsis to 116 km but engineers recorded peak aerobraking pressures of 0.25 N/m^2. As the spacecraft continued to reduce its orbital period it experienced peak pressures of 0.28 N/m^2. At 12:40 UT on 12 December it passed within 300 km of the moon Phobos.

On 29 January 1999, with apoapsis altitude down to 1,000 km and periapsis of 103 km, the spacecraft conducted the first of four tiny 'walk-out' manoeuvres, each less than 1 m/sec, allowing it to step out of the atmosphere and stabilise in the desired orbit. Having slowed the spacecraft by more than 1,200 m/sec, aerobraking was completed at 20:11 UT on 4 February when a 61.9 m/sec burn of the main propulsion engine raised periapsis and put the spacecraft in a 1.97 hour orbit with a local solar mean time of 2:04 pm. At 22:00 UT on 19 February, a further burn marginally lowered the newly raised periapsis from 405 to 367 km establishing a fully sun-synchronous orbit. On 9 March the mapping mission began and on 28 March the high-gain antenna was successfully deployed for optimised data return to Earth. On April 15, the spacecraft put itself into an automatic contingency mode when the azimuth hinge on the high-gain antenna failed to point the antenna toward Earth. The entire spacecraft was turned to align the antenna with Earth and normal mapping operations resumed 5 May. Having completed an 88-orbit repeat pattern, a propulsive manoeuvre of 3.5 m/sec on 7 May started the spacecraft 'walking' round the planet by 58 km on each successive revolution for the next mapping phase. The spacecraft was expected to continue its mapping operations until 31 January 2001.

Specifications

Mars Orbiter

Mass: 1,062.1 kg at launch (388.3 kg propellant, 75.8 kg science, 598 kg bus); body approx 1.5 m square by 3 m high.

Attitude Control: 3-axis pointing to 10 mrad (3 mrad knowledge, stability 1 mrad for 0.5 s and 3 mrad for 12 s) using 4 RWs, Sun sensors, IRU, star sensor, Mars Horizon sensor. 596 N bipropellant thruster for large delta-Vs; 12 x 4.45 N for momentum unloading, small delta-Vs and TVC.

Communications: 2 × 1.5 Gbit SSRs. 25 W X-band transmitter provides max 85.3 kbit/s via 1.5 m parabolic articulated boom-mounted HGA. Command uplink 7.8-500 bit/s.

Power: Twin 2-panel solar wings (12 m top-to-top extended) provide 940/660 W aphelion/perihelion (GaAs inner panel, Si outer); 2 × 20 Ah Ni/H_2 batteries

Payload: (all surface instruments are nadir body-mounted.)

MOC Mars Orbital Camera. CCD line-scan camera for wide angle (f/6.5, 11.3 mm focal length, 140° FOV, 3,456 CCD array length, resolution 7.5 km/240 m) and narrow angle (f/10, 350 cm fl, 2,048 CCD array length, resolution 1.4 m) imaging to monitor surface and atmospheric changes. 20.5 kg, 6 W. Imaging systems include 96 Mbit solid-state memory. (California Institute of Technology)

MOLA Mars Orbital Laser Altimeter. 10 Hz pulses, surface topography within 30 m (1-10 m relative) and surface reflectance at 1.06 μm. 160 m diameter footprint. NASA Goddard

TES Thermal Emission Spectrometer. Mapping of rock mineral composition, frost coverage and cloud composition at up to 3 km resolution across 9 km FOV using 6.25-50 μm Michelson interferometer

MAG/ER Magnetometer/Electron Reflectometer. Two NASA Goddard triaxial fluxgate magnetometers and French CNES ER to map magnetic field and monitor solar wind-Mars plasma interactions; resolution ±0.004 to ±16 nT

RS Radio Science. Uses radio system's X-band transponder + ultrastable oscillator to measure vertical profile of atmospheric refractive index (indicating pressure, temperature, density) by monitoring Earth-received signal after atmospheric passage; map gravity field by spacecraft Doppler tracking. JPL

MR Mars Relay. Relays data from NASA Mars Surveyor landers. Data stored in MOC memory. Downlink beacon activates data uplink.

Mars Pathfinder

Current status

Scientists at the JPL in Pasadena, California lost contact with the Mars Pathfinder in October 1997. During the four-month life of the mission, Pathfinder transmitted 2.3 billion bits of data from the six-wheeled Sojourner rover and the Pathfinder lander. First impressions from the scientific team sifting through the data indicated that all the rocks studied by Pathfinder's alpha proton X-ray spectrometer, contain a high-silicon content and resemble volcanic rock on Earth known as andesite.

Background

NASA approved the Mars Pathfinder in 1993 as one of the first two Discovery missions, which feature low cost planetary missions. To ensure a cost effective mission, NASA set a US$150 million FY92 development cost cap with fast development schedules of less than three years and focused science objectives.

NASA wants Pathfinder to demonstrate a low cost cruise, entry, descent and landing system for exploration of the Martian surface, with the additional objectives of the deployment and operation of three science instruments and operation of the small rover (Sojourner Truth, named after a pre-Civil War slavery abolitionist).

The spacecraft and instruments cost US$171 million (excluding the separately funded US$25 million MFEX). It cost NASA US$61 million to launch the craft and US$14 million for one year of tracking and mission operations.

Mars Pathfinder was originally proposed as a technology demonstration for the MESUR Mars Environmental Survey Network mission, but after the loss of the Mars Observer, NASA shelved MESUR and kept Mars Pathfinder as the flagship mission for the Mars Surveyor and Discovery programmes.

Sojourner is a NASA Office of Advanced Concepts & Technology experiment for mobile vehicle technologies, whose primary mission is to determine microrover performance in the poorly understood Martian terrain. The total rover budget was US$25 million including development and operation.

To save time and operational costs, the spacecraft performed no science experiments on the entry, which was a direct hyperbolic approach. The rover landed about 850 km southeast of Viking 1, in the Ares/Tiu Valles. This site elicited special interest among project planners because of its outflow of a catastrophic flood system.

Mars Pathfinder will demonstrate a low-cost approach to delivering payloads to the surface of Mars. The 9 kg Sojourner rover will probe the elemental composition of rocks and soil. The flood plain should provide a grab bag of rocks washed down from the highlands (NASA/JPL) ***1996***

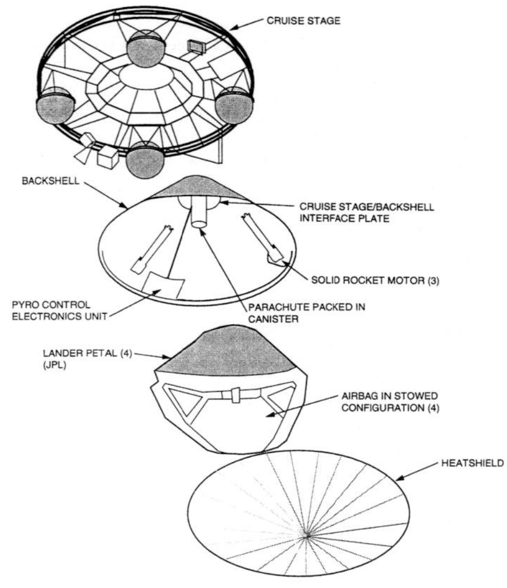

Mars Pathfinder exploded view (JPL) ***1995***

JPL built the spacecraft and handled the mission operations. NASA launched the Pathfinder on a McDonnell Douglas Delta. Lockheed Martin contributed the aeroshell and the IMP optics, Loral Federal Systems the flight computer, Motorola the transponder, ILC Dover the airbags, Pioneer Aerospace the parachute, and Thiokol the Rocket Assisted Deceleration (RAD) subsystem, airbag gas generators.

The Mars Pathfinder was launched at 06:58:06 on 4 December 1996. Unlike the Mars Global Surveyor, the solar panels on MPF deployed successfully and performed some 10 per cent better than predicted. On 11 December, the spacecraft spin rate was reduced from 12.3 rpm down to just 2 rpm and the star scanner was activated to establish full three-axis knowledge of orientation in space. On 17 December, the Sojourner rover, encased within the MPF spacecraft, awakened on a ground signal and performed an internal health check.

MPF underwent its first trajectory correction manoeuvre (TCM) on 10 January 1997 with two of the spacecraft's eight 0.4 kg thrusters firing continuously for 90 minutes, changing the spacecraft's velocity by 31 m/s. With the spacecraft 19 million km from Earth, the second TCM was performed on 3 February, a two-part manoeuvre designed to null out dispersions from TCM-1 and shift the trajectory closer to a planetary encounter. For quarantine reasons, Mars Pathfinder would not be placed on a direct Mars intercept path until TCM-3. The first component of TCM-2 involved the two small forward-firing thrusters operating for five minutes, changing velocity by 1.5 m/sec. The second burn required all four thrusters on one side to fire for five seconds and change velocity by 0.5 m/sec. Conducted on 6 May, TCM-3 was a three-component manoeuvre consisting of a 0.4 m/sec lateral burn away from Mars, a 0.1 m/sec axial burn to correct arrival time and a 0.5 m/sec burn back toward Mars. About 17 million km from Mars, activity during the last week in May focused on charging up the flight battery. About 22 Ah had been taken out of the 56 Ah capacity through degradation and by re-charging the level was returned above the 40 Ah level felt necessary for the remainder of the mission.

TCM-4, the last scheduled course adjustment, was accomplished on 25 June with the spacecraft 180 million km from Earth, about 3.5 million km from Mars and just nine days to go before direct entry. The first component involved firing four thrusters on one side of the vehicle for 1.6 seconds while the second burn involved firing two thrusters close to the heat shield for 2.2 seconds. The combined velocity change was a mere 0.018 m/sec and aligned the trajectory for a direct entry on target to a landing in the Ares Vallis flood plain on 4 July 1997. On 30 June, controllers turned the spacecraft into its proper orientation for entry and switched to the 'entry, descent and landing' (EDL) computer sequence. An optional TCM-5 at either 12 or six hours prior to entry was not necessary, tracking indicating the spacecraft was aligned with the centre of the pre-planned entry path and at an angle of 14.06° instead of 13.9°.

The EDL phase began about 35 minutes prior to landing (L−35) with cruise stage separation followed by entry at L−5 at an altitude of 130 km, a velocity of 7,470 m/sec and peak deceleration of 25 *g*. At L−134 sec and 9.4 km, with velocity reduced to 370 m/sec, the parachute deployment was triggered by an accelerometer sensing 16 *g* and diminishing. Heatshield release occurred 20 seconds later followed by the lander separating, so that it was suspended beneath the backshell by a 20 m kevlar bridle, 20 seconds after that (L−94 sec). Ground radar acquisition came at L−28.7 sec at a height of 1.6 km and a descent rate of 68 m/sec followed at L−10.1 sec (355 m altitude) by inflation of the airbag. Just four seconds later (L−6.1 sec) and at an altitude of 98 m with the spacecraft falling at 61.2 m/sec, the descent rate was dramatically arrested by ignition of the rockets in the backshell which fired for 2.5 seconds. At L−3.8 seconds and at a height of 21.5 m, the bridle holding the spacecraft to the backshell was cut. Encapsulated by the multi-lobe airbag, in free-fall the lander struck the surface at 16:56:55 UT event time, 17:07:25 UT Earth receive time, on 4 July 1997, at a true solar time of 2:58 am. Impact velocity of 10.5 m/sec and 18 *g* was sufficient to cause Pathfinder to bounce about 15 m off the surface with a second bounce of 9 *g* and about 7 m. This was followed by a third impact of 11 *g*, probably caused by the airbag/lander combination spinning up as it bounced. The spacecraft bounced at least 15 times before rolling up a slight incline and some way back down again before coming to a stop. Landing co-ordinates are 19.30° N by 33.52° W, 19 km south-west of the planned spot with the spacecraft at a slight tilt of 2.5°.

Named the Sagan Memorial Station after the late and eminent planetologist and astronomer Carl Sagan, the Mars Pathfinder site was explored by the Sojourner rover deployed to the surface down extended metal tapes serving as runners for the rover wheels. Before that the airbag was partly reinflated and recollapsed in an effort to rearrange the folds from obstructing the rover's traverse down to the surface. An extensive panorama of images secured the preliminary survey of local topography and pictures taken of Sojourner on the lander helped verify the condition of the 10.5 kg rover. Minor software problems between Pathfinder and Sojourner were cleared up and the rover moved halfway down the ramps, one of which was inclined 20°, before taking pictures. Resuming its movement down to the surface, which took four minutes, the first TV pictures showing all six wheels on Mars were received at 05:59 UT on 6 July. It was the first vehicle to roam the surface of another planet. The APXS designed to analyse surface materials for their chemical content was activated on sol 2 (Mars day 2 of the mission). Data assessment at the end of the mission indicated that local rocks resemble volcanic types with high silicon content, a rock known as andesite on Earth, covered with a fine layer of dust.

Mars Pathfinder was as much a proof-of-concept mission as a scientific exploration of Mars and proved that low-cost technology and innovation can dramatically reduce the price of planetary missions. It demonstrated that the small rover is an ideal mobile vehicle for placing instruments up against rocks and soil samples and for conveying a camera around the surface for topographic, geologic and morphologic studies of surface features, in both mono and stereo images. The last signals from the lander were received on 27 September 1997 after which 2.6 Gbits of data had been transmitted including more than 16,000 images from the lander camera and 550 from the rover. Approximately 8.5 million temperature, pressure and wind measurements had been made. The rover travelled a total 100 m in 230 commanded manoeuvres, exploring about 250 m² of the local area. The lander had been in contact for 83 days, nearly three times its design life of 30 days and the Sojourner rover operated for 12 times its design life of seven days.

Specifications

Mass: 870 kg at launch; 785 kg dry, 85 kg hydrazine. 556 kg at entry, 325 kg lander mass includes 25 kg payload

Cruise stage

Description: 2.65m diameter, 2 rpm spin stabilised during cruise, attitude control knowledge to less than 0.5° using Magellan-heritage star scanner and Mars Observer-heritage Sun sensors.

Telemetry: Minimum command rate (from DSN 34 m High Efficiency net) 125 bit/s; min telemetry rate 40 bit/s to DSN 34 m net.

Propulsion: Four hydrazine spheres with total 100 kg capacity feed propulsion system for TCM manoeuvres and attitude control by eight 4.45 N thrusters. TCMs in

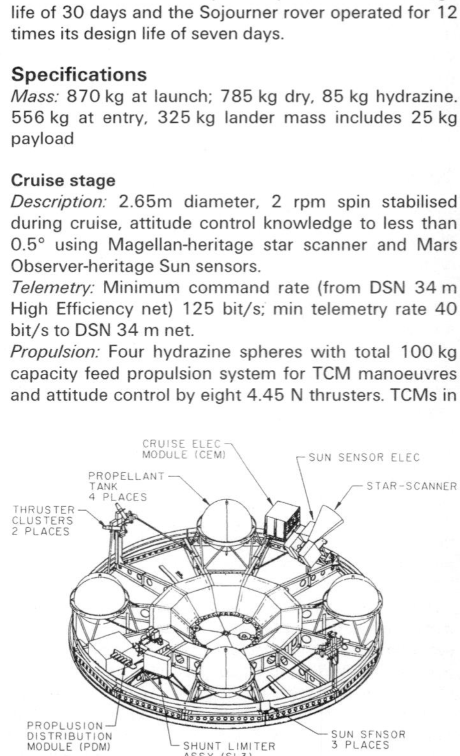

Mars Pathfinder cruise stage. Separated 30 min before entry, it will impact the surface (JPL) ***1995***

Mars Pathfinder airbag testing (NASA) ***1996***

Earth-line vector mode or turn/burn.
Power: 4.4 m² GaAs solar array covering the top face provides power for cruise ops.

Entry vehicle
Description: 2.65 m diameter aeroshell, parachute (based on Viking design), radar altimeter (operates from 1,500 m), RAD solid retros and 20 m³ double-layer airbags. Aeroshell, based on Viking-heritage, employs SLA-561 ablative material. 11.5 m diameter parachute deployed through backshell; three modified Star 5 solid retros fired to reduce landing speed. Airbag mounted on each face of tetrahedral lander limits lander deceleration to 50 *g*. Communication is performed via lander RF equipment and backshell LGA. Telemetry rate to DSN 70 m antenna 40 bit/s.

Lander
Description: 1.1 m isosceles tetrahedron with deployable sides for self-righting.
Power: 2.9 m² GaAs array on petal provides power during day. 40 Ah AgZn secondary battery allows night operations.
Computation: A single board Loral R6000 computer (20 Mips max). Data storage by 128 Mb DRAM plus 4 Mb EEPROM. Flight software encoded in 'C' using commercial operating system (VxWorks). Computer and associated command/data handling hardware use standard VME architecture.
Communication: Cassini-heritage deep space transponder plus 13 W X-band SSPA for telecoms (including 5 W X-band back-up). 125 bit/s command rate via HGA from DSN 34 m HEF net. 5.0 kbit/s telemetry to DSN 70 m net. HGA is a mechanically steered slotted plate in thermal enclosure. Command rate through lander LGA (whip antenna on bottom, used during final descent) is 7.8 bit/s, telemetry 40 bit/s.
Lander payload: IMP Imager for Mars Pathfinder. Stereo colour imager mounted on 80 cm deployable mast. Stepper motors provide camera head steering, ±178° azimuth, −72° to +83° elevation. Loral CCD mounted at foci of two optical paths divided into two 256 × 256 square frames; stereo optical paths separated by 120 mm. f/10, stepped down to f/18 with 23 mm effective fl, 14.4° FOV, 0.5-32,000 ms exposures. Pixel IFOV 1 mrad. Fused silica window at each path entrance prevents dust intrusion. Each path's 0.5-1.05 μm filter wheel has 12 positions: eight for geology, two for water vapour, blue for atmospheric dust, and a broadband for stereo. IMP also carries magnets for magnetic properties experiment (Denmark; infra-red imaging may identify magnetic minerals) and wind socks for 5-50 m/s wind direction observations. 5.20 kg, 2.6 W, US$6 million cost cap.

APXS Alpha/proton/X-ray Spectrometer. Alpha-particle source/detector for backscattered alpha, protons and X-rays to determine sample elemental composition. APXS electronics mounted in rover WEB; sensor head mounted on rover deployment mechanism. 0.74 kg, 0.8 W. (Max Planck Institute/University of Chicago)

ASI/MET Atmospheric Structure/Meteorology Package. JPL facility instrument assesses atmospheric structure and surface meteorology. Surface meteorology hardware includes 1 m weather boom with a set of temperature and pressure sensors and wind speed/direction sensors. Acceleration data acquired throughout entry using range switchable accelerometers. temperature and pressure measurements after heatshield separation, using surface temperature and pressure sensors. 2.04 kg, 3.2 W.

Sojourner (MFEX Microrover Flight Experiment). 9 kg free-ranging rover for autonomous ops in lander vicinity; nominal life 7 sols but >30 sols possible. 63 cm long, 48 cm wide, 28 cm deployed height. Six 13 cm diameter wheels, independently powered (front/rear independently steerable), on rocker bogie suspension; top speed 0.4 m/min. Power provided by 0.22 m² GaAs top face array plus 150 Wh $LiSOCl_2$ primary battery. Peak solar panel power production 15 W; mobility requires 10 W. Sensitive electronics housed in insulated WEB warm electronics box; no night heating. Control by integrated set of computing/power distribution electronics. 80C85 computer rate at 100 Kips; memory includes 176 kbyte PROM plus 576 kbyte RAM. Motion control partially autonomous using feedback from onboard sensors: three CCD monochrome cameras (stereo at front; rear for 1 mm resolution of APXS sites), laser stripers, accelerometers, potentiometers. Distance travelled and heading from averaged odometry and gyro. Deployed from lander over ramps, deployable metal reels providing path over airbags. Communicates to lander via UHF modems at up to 9,600 bit/s.

Mars Climate Orbiter

Current Status

Just one day late, Mars Climate Orbiter was launched by Delta 7425 at 18:45:51 UT on 11 December 1998 from launch pad 17A at the Cape Canaveral Air Force Station, Florida. Before the flight, engineers discovered a potential problem with a charge control unit that regulates the flow of current from the solar cell arrays to the battery. If it had failed in flight the battery would have overcharged and been rendered useless. Just 63 seconds after lift-off, the four solid rocket boosters burned out and were jettisoned, two at a time, with first stage cut-off at approximately 4 minutes 24 seconds. Second stage burn lasted about 11 minutes 22 seconds and put the spacecraft in a low Earth orbit at about 189 km. Second stage restart came at 34 minutes and lasted 19 seconds before third stage ignition at 40 minutes 30 seconds for 1 minute 28 seconds pushing the MCO out of Earth orbit and on its way to Mars. This was followed by a de-spin operation using extended lines and yo-yo weights with final dispersions nulled by spacecraft thrusters. MCO's first trajectory correction manoeuvre at 21:33 UT on 21 December to remove an intentional bias placed in the launch vehicle's guidance steering to prevent the unsterilised third stage of Delta from following on to Mars and entering the atmosphere. The spacecraft's thrusters were fired for 2 minutes 48 seconds effecting a 19.1 m/sec change in velocity. Two days later, engineers successfully conducted a health check on the MCO's two science instruments. On 4 March 1999, the MCO performed its second course correction manoeuvre with a burn involving its four thrusters for 8.2 seconds (0.86 m/sec) 17 million km from Earth. The next course correction is scheduled for 23 September just two months before it arrives at Mars.

Background

NASA developed the Mars Surveyor programme in 1994 as a commitment to sending low-cost spacecraft to Mars at every launch window, which take place roughly every 26 months. The first mission in this series was Mars Global Surveyor launched in 1996 and under the mission opportunity for 1998, Mars Climate Orbiter (MCO) and Mars Polar Lander (MPL) were developed collectively as the next two missions under the Mars 98 umbrella. All of NASA's Mars missions planned for the first decade of the 21st century come under the Surveyor programme and in efforts to cut costs, all will involve a single industrial partner with the Jet Propulsion Laboratory, Lockheed Martin. In support of this, JPL set up the Mars Surveyor Operations Project. Along with Mars Polar Lander, MCO is one of two Mars 98 missions and, like Mars Global Surveyor, will be designed for aerobraking into the optimised mission orbit following initial capture by retro-propulsion. Synchronised from a staggered launch with MCO being sent off first, both spacecraft were scheduled to begin 65 days of simultaneous operations in orbit and at the surface in December 1999.

Excluding launch vehicle costs, the Mars 98 Surveyor orbiter and landing funding was capped at US$193.1 million for all project management and spacecraft, instrument and mission system development activities from the authority to proceed (November 1995) through lander launch +30 days (February 1999). This sum included a US$24.7 million contingency fund. In addition, launch costs were budgeted at US$91.7 million and missions operations costs at US$42.8 million, excluding the Deep Space 2 costs for Mars Polar Lander. The schedule required a preliminary design review in March 1996, critical design review in January 1997, MCO integration and test in May, instrument delivery in August and delivery to the Cape Canaveral Air Force Station in September 1998 for the launch in December 1998. MCO and MPL shared 70 per cent in hardware and 89 per cent in software modules.

While Mars Global Surveyor is dedicated to topographic mapping and geomorphologic studies, Mars Climate Orbiter is dedicated to better understanding of the atmosphere and climate. Lockheed Martin was selected for Phase B definition studies in March 1995 as the successful respondent to an industry-wide competitive solicitation for the Mars 98 Surveyors. MCO was designed for launch by Delta 2 (7425) launch vehicle, a more powerful version of the rocket used to send Near East Asteroid Rendezvous (NEAR) on its way. The primary launch window was 10-17 December 1998 with a secondary window 18-25 December. Launch during the initial period insured a higher probability that the MCO would complete aerobraking before Mars Polar Lander, the second Mars 98 mission, arrived at the planet. Launch during the secondary period would result in a higher capture orbit at arrival requiring more aerobraking to bring the spacecraft into its final science orbit. Launch after 25 December was possible but would result in further compromise to the capture orbit.

MCO would take an estimated 286 days from launch to encounter, reaching Mars on 23 September 1999. Between two and four mid-course corrections can be made to refine the trajectory during interplanetary cruise, the first of these to be carried out 10 days after launch. This manoeuvre is expected to be the largest and longest and would be used to correct injection errors and to adjust the Mars arrival aimpoint. The remaining manoeuvres would be used to direct the spacecraft to the proper aimpoint and are scheduled at 45 days after launch and 60 days and 10 days before arrival at Mars. During cruise the science instruments are turned on, calibrated and tested, the Pressure Modulator Infra-red Radiometer and Mars Colour Imager will be calibrated during a week-long checkout 80 days after launch. Eighteen days before arrival, the imager will take pictures of Mars that will be transmitted to Earth over the following three days.

Mars orbit insertion is scheduled for 23 September 1999. As the spacecraft nears its closest point to the planet, coming in over the northern hemisphere, it will fire its 640 N thrust main motor for 16-17 minutes to brake into an elliptical capture orbit, reducing speed from 5.9 km/sec to 4.7 km/sec. In an initial orbit of 160 × 39,000 km, the spacecraft will loop around Mars roughly once every 29 hours, but about 22 minutes after completing the burn the spacecraft will be turned to point its high-gain antenna at Earth. Continuous communication with the Deep Space Network's 70 and 34 m antennas will be maintained for two days after orbit insertion and based on the vehicle's state vector the spacecraft will fire its thrusters during periapsis to lower the orbit and reduce the period by 2-4 h. The aerobraking phase, similar to that planned for Mars Global Observer, will begin when periapsis is lowered to 100 km and last 200 aeropasses or 57 days. At the end of this phase, on 22 November 1999, MCO will be in an orbit of 90 × 405 km and periapsis will have shifted north from 34° latitude to 89° latitude, almost directly over the planet's north pole. Small thruster firings will keep the MCO at the desired level for heating and dynamic pressure, a maximum 0.46 W/cm².

The operating orbit will be one in which MCO passes over Mars' equator at 4:40 pm local time over the sunlit side and 4:30 am on the night side. It is scheduled to achieve this when the MCO manoeuvres to a 405 km near circular orbit, a Sun-synchronous path, on 1 December and the high-gain antenna is deployed. For the first three months, the MCO will act as a radio relay for the Mars Polar Lander, relaying commands from Earth to the lander as well as sending data from the lander to the Earth via the UHF two-way relay link. The orbiter will overfly the lander typically 5-6 minutes 10 times a day and in these periods communicate with the lander at 128 kbits/s with X-band telecommunications systems. This support is scheduled to take place for 10 hours each day through to the end of February 2000. From 3 March 2000 until 15 January 2002, the MCO will perform its primary mission, making systematic observations of the atmosphere and the surface of Mars using its two science instruments, the Pressure Modulator Infra-red Radiometer and the Mars Colour Imager. The duration of the science mission is one Mars year (687 Earth days) to capture a full cycle of seasonal changes in Martian weather. After this the MCO will drift in a low-maintenance orbit awaiting the arrival of the Mars 2001 spacecraft in January 2002 when it will once again operate as an orbital relay station.

Specifications

Description: The Mars Climate Orbiter is a three-axis stabilised spacecraft comprising a bus 2.1 m tall, 1.6 m wide and 2 m deep. Framework is made of combined graphite composite facesheets on aluminium honeycomb with a gusset plate construction propulsion module (scaled down from the Mars Global Surveyor design) and a truss construction equipment module.
Mass: 629 kg at launch; 338 kg dry plus 291 kg of fuel.

Computation: RAD6000 32-bit processor with Mars Pathfinder heritage in the Orbiter Command and Data Handling subsystem, provides a central processing capability for all spacecraft systems including payload elements. It can be switched between clock speeds of 5, 10 or 20 MHz and includes 128 Mbytes of RAM of which about 20 per cent is used for running spacecraft programmes. The rest of the space is used for science and for storing data which will be relayed to Earth or the Mars Polar Lander. Unlike many other spacecraft it does not have an on-board tape recorder or solid-state data recorder, but instead uses its RAM for transmission to Earth. The computer also has 18 Mbytes of flash memory which will be used to store data even when the computer is off, of which 8 Mbytes is for storing triplicate copies of high priority data. When the computer accesses this triplicate data, it checks all three copies of each byte to ensure there has been no corruption.

Telecommunications: Primary communication between the spacecraft and the Earth is via the Deep Space Network's X-band (up/down) link and the orbiter's deep space transponder developed for the Cassini spacecraft. MCO has 15 W RF solid-state power amplifiers and a 1.3 m diameter articulated, two-axis, high gain antenna, one transmit only medium gain antenna and one receive-only low gain antenna. A two-way 10 W RF UHF link is provided for on-orbit communication with the lander while it is on Mars.

Power: The two-axis articulated, gallium arsenide, three-panel solar array (7.4 m^2 cell area; 11 m^2 total wing area), 5.5 m tip-to-tip, provides electrical power and serves as the most significant drag brake during aerobraking. Nickel-hydrogen (NiH_2) 16 Ah common pressure vessel batteries provide power during eclipses and during peak power needs. The electrical power control electronics are derived primarily from the Small Spacecraft Technology Initiative (SSTI).

Thermal control: The TC system uses passive methods and louvres to control the temperature of the batteries and the solid-state power amplifiers. Passive coatings as well as multi-layer insulation blankets are used to control temperatures along with thermostatically controlled and computer controlled heater circuits. Where needed, radiators are used to take the excess heat out of the spacecraft components to maintain proper operating temperatures.

Attitude control: The orbiter is three-axis stabilised in all mission phases except following separation from the launch vehicle. Primary attitude determination is via star camera derived from the Clementine spacecraft star camera and an inertial measurement unit with analogue star sensors as back up. Reaction wheels provide primary attitude control during most mission phases, desaturated via RCS thrusters. Because of lifetime concerns the IMUs are switched off during significant portions of cruise and mapping and the vehicle operated in an all-stellar mode except during manoeuvres. RCS thrusters also provide attitude control during trajectory correction manoeuvres, orbit insertion, aerobraking drag passes, orbital trim manoeuvres and safe mode. The MCO has four 22.24 N thrusters used for trajectory correction and for pitch and yaw control and four 0.89 N thrusters for roll control.

Propulsion: Derived from the propulsion subsystem for Mars Global Surveyor the MCO has a LEROS 640 N main engine operating with hydrazine and nitrogen tetroxide bi-propellants for Mars orbit insertion. Following this burn the oxidiser is gone and the attitude thrusters use the remaining fuel (hydrazine) as a monopropellant.

Payload: Consists of a rebuilt version of the Mars Observer Pressure Modulated Infra-red Radiometer (PMIR) and the Mars Colour Imaging (MARCI) system. The PMIR weighs 42 kg and uses 41 W of power in an instrument box 23 × 30 × 74 cm. A smaller cooling unit 58 × 65 × 30 cm is attached to the box. PMIR has nine channels, visible light + IR wavelengths 6-50 microns. These will construct vertical profiles of the atmosphere from 80 km to the surface. Bands of water vapour and CO_2 will be detectable at a vertical resolution of 5 km. The two MARCI wide-angle and medium-angle cameras weigh 2 kg and each measures 6 × 6 × 12 cm and use 4 W power. The wide-angle camera detects light in the visible (5 × 425-750 nm bands) and UV (2 × 250-350 nm) with a resolution of 1 km when large numbers of pictures are being sent or 7.2 km at other times when averaging adjacent pixels. The camera can also be used to image the planet's limb for cloud structure and haze patterns at a resolution of 4 km. The medium-angle camera detects light in eight spectral bands from violet to near-IR (425-930 nm) with a maximum resolution of about 40 m across a 6° Field of View (FoV) covering 40 km.

Mars Polar Lander

Current status

Mars Polar Lander was launched by Delta 2 (7425) at 20:21:10 UT on 3 January 1999 from Launch Complex 17B at the Cape Canaveral Air Force Station, Florida. The four solid rocket boosters were jettisoned at 60 seconds followed by first stage cut-off 3 minutes 24 seconds later and separation 8 seconds after that. The second stage lit up 5.5 seconds after first stage separation and burned for 6 minutes 44 seconds placing the vehicle in a low Earth orbit of 119 km. After a 23-minute coast over the Indian Ocean, the second stage fired briefly a second time followed by the third stage which fired for 88 seconds to inject the MPL to a Type 2 Mars transfer trajectory. Shortly thereafter the spacecraft separated from the third stage and deployed the cruise solar panels. The spacecraft signal was acquired by the DSN station at Canberra, Australia. The first trajectory correction manoeuvre was performed on 21 January when the four manoeuvring engines fired for almost 10 seconds to change velocity by 0.89 m/s. The second correction manoeuvre was performed on 15 March, a burn using the same engines and with the same parameters. The spacecraft continues in health with only minor irregularities in performance and is scheduled to land on Mars on 3 December 1999.

Background

Mars Polar Lander (MPL) is one of the Surveyor missions instituted by NASA in 1994 as a series of small spacecraft for the sustained exploration of Mars. MPL is one of two spacecraft developed for the Mars 98 mission opportunity. The other is Mars Climate Orbiter (MCO). See Mars Climate Orbiter (this section) for general Mars 98 background and costs. MPL fills the scientific need for further information about the planet's past and present water resources and will seek answers by probing layered terrain near Mars' south pole. It will deploy two penetrators before impact and a 2 m robotic arm on the main spacecraft will dig into the top of the terrain. This is the first time a lander has examined the surface of Mars at such a high latitude and it is the second lander mission in the new strategy of sending spacecraft to this planet at every launch window approximately 26 months apart, the first being Mars Pathfinder. Developed and built by Lockheed Martin, the MPL will also carry two Deep Space 2 microprobes, released for ballistic descent to the surface shortly before the lander performs touch down. Scientific objectives specifically include a search for ancient and modern ground and surface water environments.

The mission plan required MPL to be launched less than one month after Mars Climate Orbiter by Delta 2 (7425) identical to the one used to send MCO on its way and similar to the one used to launch NEAR. The trajectory selected for MPL was a Type 2 path where the spacecraft would travel more than 180° round the Sun, launching marginally inward towards the Sun before spiralling out towards Mars. This enables MPL to reach the appropriate landing zone selected close to the south pole. Because it flew a Type 1 trajectory (travelling less than 180° round the Sun before encounter) Mars Climate Orbiter would reach Mars more than two months ahead of Polar Lander. In the intervening period it will use aerobraking to achieve its operational (near circular) orbit by 1 December 1999, two days ahead of the date Polar Lander is scheduled to reach the surface. This would configure the relative position of MCO to operate as a relay to the Lander for approximately three months. The primary launch window of 3-10 January 1999 provided the best opportunity for MPL to reach the surface of Mars on 3 December 1999 in the target area 73-77° south latitude. The secondary launch window of 11-16 January would incur a northward drift in the landing site to about 71° south and extend the landing date to no later than 17 December. Launch times within both periods were almost instantaneous to the second allowing no room for unexpected holds in the countdown.

The lander portion of the MPL is encapsulated in a protective aeroshell to the backshell of which is attached the cruise stage providing power, propulsion and communications and housekeeping duties until shortly before entry into the Martian atmosphere. Less than one hour after launch with the spacecraft on its Type 2 trajectory, the cruise stage solar panels are deployed and the vehicle orientated towards the Sun at which point the 34 m DSN antenna at Canberra, Australia, acquires it. Interplanetary cruise lasts 11 months during which communication is maintained via X-band link and the medium gain antenna on the cruise stage. Up to six course correction manoeuvres are built in to the cruise plan, the first conducted 15 days after launch to adjust the Mars arrival aimpoint. The remainder are scheduled at 45 days after launch and at 60 days and 10 days before Mars arrival with two manoeuvres possible between 3 days and 7 hours before arrival. Tracking will increase 45 days before encounter and the precise aimpoint calculated while science instruments that gathered data during the cruise phase will be transmitted to the 70 m DSN antennas on Earth. Selection of the precise aimpoint within the preferred landing zone will be made between June and August 1999 using images of the region transmitted to Earth by Mars Global Surveyor. Mission managers expect to finally designate that spot no later than four months before arrival. Unlike the Viking landers that touched down in 1976 the MPL will conduct a ballistic entry rather than first going into Mars orbit but like the Viking landers and unlike the Mars Pathfinder of 1997 the MPL will perform a controlled descent using retro-motors almost to touchdown.

Entry events commence about 14 hours before encounter when the final four-hour tracking phase begins, the final opportunity for navigation data which could signal the need for a trajectory correction manoeuvre. A final 30-minute tracking session starts at 7 hours 25 minutes before entry, at which point the final trajectory correction command can be issued. At E–5 hours, a one-hour tracking session provides spacecraft health data and status checks following any last course correction and at E–4 hours 40 minutes a series of valves open to vent the descent motors. Final contact with the spacecraft begins at E–25 minutes and lasts 15 minutes. At E–20 minutes the lander's descent motor heaters are turned on five minutes before the software controlling the Mars descent imager is initialised. At E–10 minutes the MPL switches to inertial navigation computations obtaining data from gyroscopes and accelerometers and at E–6 minutes the spacecraft fires its thrusters for 80 seconds to put the vehicle in the correct orientation. At E–5 minutes (10 minutes to touchdown) with the MPL travelling at 6,200 m/s and about 2,300 km from the surface of Mars, the cruise stage and solar panels are jettisoned followed 18 seconds later by the two Deep Space 2 microprobes. Five minutes later, at 6,900 m/s and an altitude of 125 km, the aeroshell encounters the atmosphere with just 4 minutes 33 seconds to touchdown. At maximum deceleration, the aeroshell will experience 12 *g* and a temperature of 1,650° C

About two minutes before landing when the MPL is descending at 493 m/s and just 8.8 km from the surface, the parachute will be released by mortar followed 10 seconds later by release of the heat shield from beneath the lander, which is suspended under the backshell. At this point the descent imager will have been switched on and the first picture taken 0.3 seconds before shield separation, the first of 10 pictures taken during descent. The first pictures will show a 9 × 9 km area at 7.5 m resolution while the final picture will show a 9 × 9 m area with a resolution of 9 mm per pixel. Between about 70 and 100 seconds before touchdown, the landing gear will deploy followed 1.5 seconds later by activation of the doppler velocity landing radar. This will determine the lander's altitude approximately 44 seconds later about 2.5 km above the surface descending at 80 m/s with 30 seconds to touchdown. At 1.4 km the manoeuvre thrusters will be turned off and lander will separate from the backshell and parachute assembly. Just 0.5 seconds later the descent engines are turned on for deceleration, attitude and roll control to turn the descent path vertical and put the lander in the proper orientation so that it sits on the surface with the solar panels facing the Sun as it moves across the sky. The radar is turned off at a height of 40 m and the lander relies on inertial navigation. At 12 m or a descent rate of 2.4 m/s, whichever comes first, the lander will lower itself at a constant velocity (<1 m/s horizontal velocity) and the descent engines will be turned off when pad sensors touch the surface.

The lander is expected to contact the surface at 5:00 am local time, 21:03 UT, on 3 December 1999. About two minutes later, the solar panels will deploy and three minutes later the medium-gain antenna will align with the Earth. Two-way communication with Earth will begin about 20 minutes after landing and much of that Mars morning will be spent diagnosing spacecraft health and condition, sending meteorological data, some descent imager pictures and some from the surface stereo imager via the high-data rate UHF transmitter. When the lander is ready to begin science operations, the robotic arm will dig a shallow trench a few centimetres deep, probably taking the sunlight hours of two Mars days (sols) to complete the initial task. A sample will be scooped up and dropped into the Thermal and Evolved Gas Analyser where a low temperature cooking sequence heats it to 27° C followed by a high-temperature cycle to 1,027° C a day later. Each day the MPL will have 3-4 opportunities to send engineering and science data to Earth via the Mars Climate Orbiter, when the latter is at least 20° above the horizon.

Meanwhile, the two Deep Space 2 probes released from the MPL about 5 minutes before entry, encapsulated in an aeroshell to protect them from the 2,000° C of re-entry, head towards impact at 200 m/s about 50-85 seconds before the lander touches down 100 km distant. The penetrator comprises a main surface body and a probe connected by a flexible line protected by the aeroshell during descent. When it strikes the surface the aeroshell shatters and the shock of sudden deceleration (30,000 *g* on the probe and 60,000 *g* on the main surface body) drives the probe into the surface still connected to the main surface body, which remains above ground. The two probes, up to 100 km apart, will each obtain a small sample of subsurface soil using a small electric drill extending sideways from the probe pulling less than 100 mg into a small cup. The sample is then sealed and heated turning any water ice into vapour through which a tiny laser beam is shone to record the quantity. The probes also measure the rate at which heat dissipates into the surrounding soil. The primary mission of the probes is expected to last no more than 36 hours although they will continue to obtain data for as long as their batteries last. The prime mission of the lander is due to finish on 1 March 2000, releasing the Mars Climate Orbiter from its role as relay enabling it to begin systematic observations of the atmosphere and the surface.

Specifications

Description: The Mars Polar Lander stands 1.06 m tall from the ground to the tip of the science deck and measures 3.6 m wide. The spacecraft is constructed of a composite material with honeycomb aluminium core and graphite epoxy facesheets bonded to each side of the bus. A thin aluminium sheet is bonded to the composite to provide a Faraday cage around the thermal enclosure. The landing legs are made of aluminium and equipped with compression springs to deploy them from the stowed position. Tapered, crushable, aluminium honeycomb inserts in each leg provide shock absorption at landing. The central enclosure houses the onboard computer, power distribution, the 16 Ah, 12-cell, nickel-hydrogen battery, a battery charger and radio equipment. A separate thermally-isolated component deck outside of the central electronics enclosure contains gyroscopes, electronics to fire pyrotechnic devices for instrument deploy and radar equipment for entry, descent and landing (EDL). Embedded within the component deck are the evaporators, which transfer heat from the aluminium facesheets to the radiators outside the thermal enclosure.

The lander flight system consists of a separable cruise stage with a MDAC V-band launch vehicle separation interface and propulsive lander/entry assembly. The cruise stage is jettisoned before EDL providing a clean aerodynamic shape for entry and a reduced ballistic coefficient of 58-62 kg/m^2 across the 2.4 m ablator heat shield. Cruise stage operational components include star cameras and sun sensors for attitude determination, two solar array wings (2.6 m^2 area) for power generation, X-band medium gain transmit/receive horn antenna and one low gain receive patch antenna and a redundant pair of SSPAs for cruise telecom. Three-axis attitude control is achieved using redundant IMUs and four cruise reaction engine modules (REMs) on the lander, each REM containing one aft-facing 22.2 N TCM thruster and one 4.45 N RCS thruster (canted 20° out, 15° aft). The EDL stage consists of a 2.4 m diameter ablator aeroshell utilising tooling originally developed for Mars Pathfinder and sharing the same nose radius and cone angle. The parachute is also from the Pathfinder design to lower cost and is deployed based on an on-board navigation velocity estimate eliminating the long range radar used on the Viking landers in 1976.

Mass: 576 kg at launch; lander 290 kg; cruise stage 82 kg; aeroshell/heatshield 140 kg; propellant 64 kg.

Computation: The lander employs the same SSPA and RAD6000 32-bit processor as the orbiter (see Mars Climate Orbiter).

Telecommunications: The primary telecom link for science data relay and spacecraft commands is the UHF relay through the Mars Climate Orbiter. A UHF link to Mars Global Surveyor is also available for data relay only. The maximum duration of an X-band transmission is limited to one hour by the capability of the Loop Heat Pipe to transfer heat energy from the X-band SSPA out of the thermal enclosure. The number of UHF and X-band cycles is limited by the amount of daytime power available. The lander can communicate direct with Earth via X-band at 5.7 kbits/s through the DSN's 70 m antennas or at 1.4 kbits/s via the 34 m antennas. Data relayed via Mars Climate Orbiter or Mars Global Surveyor can be transmitted at 128 kbits/s.

Power and thermal: For the landing footprint of 73-77° South latitude, the Sun does not go below the normal horizon for the season of the prime mission. However, a 10° terrain mask is assumed for power analysis and results in a defined day and night interval when the Sun goes below that horizon mask. The solar array consists of six panels (four fixed and two deployed) to provide daytime power for payload operations and for charging the 16 Ah batteries which provide night-time heater power for the thermal enclosure. The two deployed auxiliary solar panels were necessary because the fixed panels constrained in size by the launch vehicle payload shroud were of insufficient area to provide the required amount of electrical power. The C&DH and Power Distribution electronics have a low-power sleep mode to reduce night-time energy consumption. The lander lifetime is limited by the size of the batteries, as are daytime operations, and by the amount of power available to recharge them. At time of landing the lander can perform 8-9 hours of daytime operations but this decreases as the sun gets lower in the sky. The power system is designed for full operation in a dusty atmosphere of opacity (Tau) = 0.5. The rechargeable NiH_2 battery will keep temperatures in the central electronics enclosure above −30° C during −90° C night-time temperatures at the south pole.

Propulsion: The lander is equipped with eight thrusters in four clusters. Each cluster consists of one 22 N thruster and one 4.4 N thruster. The lander's final descent is controlled by 12 pulse-modulated, 266 N thrust, descent engines (retro-rockets) arranged in three groups of four on the underside of the lander. Two spherical diaphragm propellant tanks underneath the lander's solar arrays carry 64 kg of hydrazine for all 20 thrusters and retro-rockets. This is a pressure-regulated system with serially-redundant pressure regulators utilising helium gas as pressurant. For final descent doppler radar provides accurate altitude and 3D velocity estimates. As the spacecraft descends to within 12 m of the surface, the spacecraft control system begins the 2.4 m/s terminal velocity descent phase. The AACS subsystem controls the orientation of the vehicle on landing, placing the X-axis within 5° of the desired azimuth (45° west of north) to maximise solar array efficiently and minimise direct-to-Earth antenna blockage.

Payload: The science payload for the Mars Polar Lander consists of six instruments: the four instruments comprising the Mars Volatile And Climate Surveyor (MVACS) suite, the Mars Descent Imager (MDI) and the LIght Detection And Ranging (LIDAR) instrument. The MVACS consists of the Stereo Surface Imager (SSI), the robotic arm, the meteorology package and the Thermal and Evolved Gas Analyser (TEGA). The SSI is mounted at the top of a 1.5 m mast that will deploy upward from the lander deck and is identical to the imager on the Mars Pathfinder lander. It has a spectral range from violet to IR (400-1,100 nanometer) and two sets of lenses mounted slightly wider apart than a normal pair of human eyes so as to capture stereo images. Two sets of optics share a single CCD detector and two filter wheels allow the camera to take images in selectable spectral ranges. The 2 m robotic arm is attached to the lander deck and has an articulated member on the other end with a digging scoop, camera and temperature probe. The scoop will dig trenches and deliver soil samples to the TEGA. The temperature probe will measure the ambient temperature and thermal conductivity of the soil. The meteorology package includes a 1.2 m mast with wind speed and direction sensor, temperature sensors and tunable diode lasers that detect water vapour and isotopes of water and CO_2. A 0.9 m submast with a wind speed sensor and two temperature sensors points downward from the lander deck and is designed to study atmospheric effects in the zone 10-15 cm above the surface to determine the threshold wind speed required for dust storms to start. The TEGA heats soil samples and analyses them to determine concentrations of volatiles such as water and CO_2, whether present as ice or in volatile-bearing minerals. The arm deposits 0.1 g soil samples at a time in a receptacle which is mated with a cover to form a small oven. Heater wires heat the sample across a range of temperatures up to 1,027° C and a tunable laser emits beams which pass through the gases to a detector which measures the amount of light to determine the presence of vapour. The instrument can perform eight soil analyses. The MDI consists of a single camera head with an electronically shuttered CCD to capture black and white images in 1,000 × 1,000 pixel format. The LIDAR uses a gallium aluminium arsenide laser diode to emit 2,500 pulses of near-IR energy per second upward from the lander deck. A detector shows how long it takes the pulses to return following reflection from ice and dust haze in the lower part of the atmnosphere (2-3 km maximum altitude). provided by Dr V S Linkin of IKI, this is the first Russian experiment flown on a US planetary spacecraft

Pioneer series

Current status

Pioneer Venus Orbiter burned up in October 1992 and project funding ended September 1994, but the Venus Data Analysis Program continued until December 1995. Pioneers 6 to 8 continue their occasional operations.

Tracking of Pioneers 6, 7, and 8 takes place once or twice a year for an hour, mainly out of curiosity about system functionality. The solar arrays continue to deteriorate and the transmitters can turn on only around perihelion where there is enough solar flux to provide sufficient power. If power is adequate, NASA activates some instruments, although the scientific value is insignificant. Three of Pioneer 6's six instruments, two Plasma Analysers and the Cosmic Ray Detector, can still return real-time data. It was tracked 29 July and 15 December 1995 (after 30 years), returning 1 h science data each time.

Pioneer 10 is travelling at about a constant 12.5 km/s relative to the Sun and round-trip communications time

The flight of Pioneer 10. Its firsts include: Asteroid Belt crossing January 1973, Jupiter fly-by December 1974, crossing of Saturn's orbit February 1976, crossing of Uranus' orbit August 1979, crossing of Neptune's orbit and leaving the solar system June 1983 (NASA)

with Earth required 17 hours 27 minutes at the end of 1995. It is heading towards Aldebaran, but it will take two million years to reach that point and by then the star will have moved on. Pioneer 10 should encounter another system about once every million years.

NASA Ames officially ended routine contact attempts with Pioneer 11 on 30 September 1995. A few minutes of healthy engineering data was received 24 November 1995. This was the last communication from Pioneer 11 and shortly afterwards the Earth moved out of view of the spacecraft's antenna.

Background

The relatively simple spin-stabilised Pioneer craft, managed by NASA Ames, paved the way for more sophisticated subsequent explorations of the Sun, Jupiter, Saturn and Venus. Pioneer 1, launched in October 1958, was nominally the first NASA deep space probe, having been transferred by the USAF. Pioneers 1-4 failed to attain their lunar target but probed Earth's radiation belts. Three more Pioneer launches by Atlas Able in November 1959, September 1960 and December 1960 and what should have been Pioneer 10, launched by Delta in August 1969, were failures.

NASA Ames' deep space and interplanetary Pioneers began with number 5, which, with its successors, monitored solar particle radiation and provided up to 15 days' warning of solar flares. Originally designed for six months' operation, Pioneers 6 through 9 provided coverage from widely separated points during an entire solar cycle of 11 years and were all still predicting solar storms at the end of 1981. Pioneer 9 was the first to fail, in 1983. Pioneer 6 marked a quarter century of operations in December 1990, and could still be re-activated again at any time after 30 years in orbit.

Pioneers were the first spacecraft to detect the Earth's magnetotail, measuring some 5.6 million km on the anti-solar side.

Beginning in 1972 and continuing through 1974, Pioneer first began to fly to Jupiter. These flights introduced a new generation of planetary probes designed as precursors to more sophisticated vehicles such as Voyager. Pioneer 10 placed on a sidereal trajectory became the first man-made object to exit the solar system, the first to fly beyond Mars, the first to enter the asteroid belt and the first to reach Jupiter. It proved that the asteroid belt contains fewer dispersed particles than at one time believed and showed that traversing vehicles faced little hazard from collisions.

When Pioneer 10 retired after 25 years of deep space exploration, only a few weeks' of data have been lost due to spacecraft problems. It became the first spacecraft to reach 50 AU from the Sun, on 22 September 1990.

To date, Pioneer 10 flies on a predictable trajectory indicating that no gravitational effects from a 10th planet occur in its present vicinity. However, there is an unaccounted constant acceleration of 8×10^{-10} m/s^2 towards the Sun acting on both craft. It is not caused by a 10th planet, a dark star companion to the Sun or an unseen disc of outer solar system material; alternative causes are under investigation. In view of its potential importance, The Aerospace Corp is using independent software to check the discovery.

Pioneer 10 continues to move outward at about 2.7 AU annually in a direction generally away from the Galactic centre and nearly opposite the direction of the basic solar system motion with respect to the nearby stars. Pioneer 11 continues to depart at about 2.48 AU per year, towards the Galactic centre in the general direction of Sagittarius.

The two Pioneers have shown the heliospheric boundary with the interstellar medium to be significantly more distant than expected. Measurements of the solar wind indicate the boundary of the heliosphere may lie some 90-120 AU from the Sun. Inference on the heliosphere's extent is also provided by monitoring the solar modulation of cosmic ray intensities and from UV observations of changing atomic interactions with sunlight that result from incursion of extra-solar particles into the heliosphere.

Pioneer 10 continues to observe the solar wind and its interactions with planets and their magnetosheaths and the gradient in cosmic particle density with solar distance. In October 1990, Pioneer 11's primary receiver developed a problem that required very high power commands and prevented measurement of spin axis orientation. The radio science search for extra sources of gravitation ended with that failure.

Pioneer Progress: distance from Sun in Astronomical Units on 1 January each year

Year	Pioneer 10	Pioneer 11
1987	40.0	22.4
1988	42.7	24.9
1989	45.4	27.3
1990	48.1	29.8
1991	50.7	32.2
1992	53.4	34.8
1993	56.0	37.3
1994	58.6	39.8
1995	61.2	42.2
1996	63.8	44.7
1997	66.5	47.2
1998	69.1	49.6
1999	71.7	52.0
2000	74.3	54.5
2001	76.9	57.1
2002	79.5	59.5

Pioneers 10 and 11 ventured into the outer solar system by design as the first man made objects to ever leave the Solar System. NASA built them to last for seven years of data transmission. In their 258 kg of mass, these Pioneers carried 30 kg of science instruments and 27 kg of propellant (18 kg remained in January 1996, the only drain on this supply is an insignificant periodic precession manoeuvres).

Six 4N thrusters each provide control and mid-course corrections and using the original propellant supply can produce a total delta-V of 670 km/h. The Pioneers retain their attitude with respect to the fixed stars by two Sun and one star sensors mounted on a spin-stabilised platform having a full-circle scan rate of about four times a minute. (The spin rate is now decreasing slowly; the cause is unknown but has been theorised as a build-up of electrostatic charge interacting with the interplanetary magnetic field). Four nuclear units, held on 2.7 m booms to avoid interference with the science experiments provide power.

The heart of the communications system is the fixed 2.7 m diameter dish antenna and redundant 8 W S-band transmitter (2,110/2,292 MHz up/down through NASA's Deep Space Network).

Pioneer's instruments have enjoyed a remarkable and unpredicted longevity and continue returning data on the solar atmosphere from the outer reaches of the solar system. Rather than instrument death, it now appears operations will fail due to the decreasing power, which diminishes by 81 mA annually. About 140 mA was available for science as of 17 January 1996 (the actual level is unknown as the telemetry provides 50 mA steps). Only those instruments with the lowest power consumption can now be operated and even then not all simultaneously.

The Plasma Analyser (179 mA required) could be operated until 9 September 1995. The Cathode Ray Telescope (100 mA) was dropped in 1996 and the Charged Particle Instrument was turned off on 30

Pioneer 6 is the first spacecraft to operate for more than a quarter century. This is the similar Pioneer 8 undergoing final balance checkout at the Cape

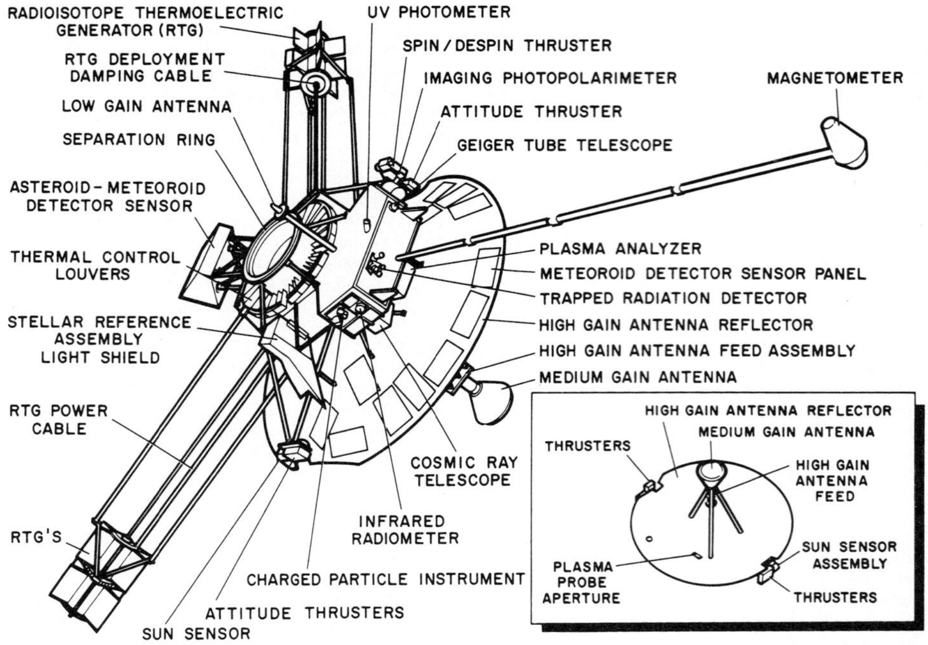

Pioneer 10/11 principal features (NASA)

December 1996, leaving just the Geiger Tube Telescope and the Ultraviolet Photometer operating as 1997 opened. On 31 March 1997 – more than 25 years after launch and nearly 10 billion km from Earth, Pioneer 10 retired and fell silent.

At 20.19 GMT on 22 September 1990, Pioneer 10 became the first spacecraft to penetrate to 50 AU, although it had still not reached the heliosphere's boundary (nor had it by the beginning of 1996).

Each craft carries a 15 × 23 cm plaque indicating its origin should it be recovered by an intelligent species.

Specifications

Pioneer 6
Launched: 16 December 1965 by Delta from Cape Canaveral
Orbit: 0.814 × 0.985 AU solar orbit.
Characteristics: Mass: 63 kg cylinder, 1 m diameter, 0.9 m high.
Mission Description: Pioneer 6 returned the first detailed data on the tenuous solar atmosphere and recorded passage of Comet Kohoutek's tail. By December 1990, having circled the Sun 29 times and covered 24,800 million km, it had been operational for 25 years – a record for an interplanetary spacecraft – compared with planned six months. With Pioneers 7-9 it formed a network of solar weather stations spaced along Earth's orbit. On 26 November 1988, it made its closest approach to Earth since launch, the 18.66 million km encounter increasing its heliocentric orbital period from 311 to 317 days.

Pioneer 7
Launched: 17 August 1966 by Delta from Cape Canaveral
Orbit: 1.010 × 1.125 AU solar orbit
Characteristics: 63 kg.
Mission Description: In 1977, Pioneer 7 registered Earth's magnetic tail 19.3 million km out, three times deeper into space than before. Pioneer 7 was tracked 17 April 1994 but insufficient power prevented the transmitter operation. On 31 March 1995, 2 hours of contact was made and the Plasma Analyser turned on.

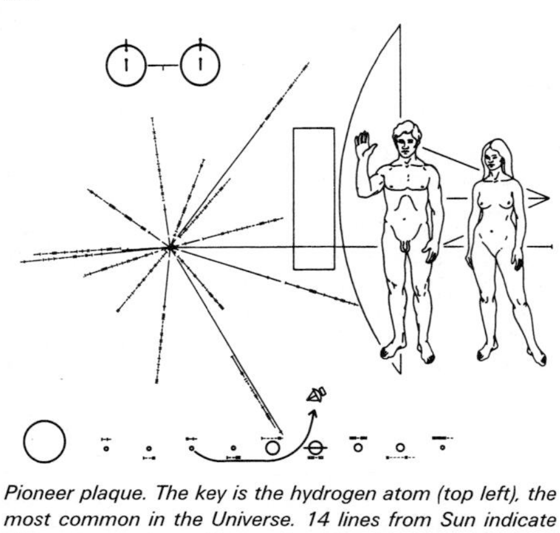

Pioneer plaque. The key is the hydrogen atom (top left), the most common in the Universe. 14 lines from Sun indicate distance to pulsars. Human figures, with the man's hand raised in goodwill gesture, are compared with spacecraft to give scale. Bottom, plan of solar system, showing Pioneer originating from Earth

Pioneer 8
Launched: 13 December 1967 by Delta from Cape Canaveral
Orbit: 1.0 × 1.1 AU solar orbit
Mission Description: Pioneer 8 defined Earth's magnetotail. NASA tracked it on 23 July 1995 but could not turn on the transmitter, either because of spacecraft failure or too great a Sun distance for the solar cells.

Pioneer 9
Launched: 8 November 1968 by Delta from Cape Canaveral.
Orbit: 0.75 × 1.0 AU solar orbit.
Characteristics: 66.6 kg
Mission Description: This spacecraft was the fourth of what was to be a series of five solar monitors (Pioneer E, intended to be Pioneer 10, failed to orbit 27 August 1969). NASA contacted Pioneer 9 last on 18 May 1983, but it was not officially declared inactive until March 1987 after attempts to detect it using SETI equipment with a very low signal detection threshold failed on 3 March.

Pioneer 10
Launched: 3 March 1972 by Atlas Centaur from Cape Canaveral.
Mission Description: After launch Pioneer 10 had an initial 51,800 km/h velocity, faster than any previous man-made object. Its trajectory took it within 130,300 km of Jupiter's cloud tops on 4 December 1973. On 13 June 1983, it crossed Neptune's orbit; further from the Sun than any known planet and effectively it left the solar system at that point.

Pioneer 11
Launched: 5 April 1973 by Atlas Centaur from Cape Canaveral.
Mission Description: NASA added a second magnetometer to measure Jupiter's intense fields. With the precedent of Pioneer 10, NASA knew how to retarget Pioneer 11 to approach three times closer to Jupiter and then proceed to the first encounter with Saturn. At 17.22 GMT 3 December 1974, travelling faster than any previous man-made object, 171,000 km/h, Pioneer 11 passed 42,940 km below the Saturian South pole. Its closest approach to Saturn, occurred on 1 September 1979 after a six-year and 3,200 million km journey. At that time it was only 20,800 km from the cloud tops and travelling at 114,500 km/h. Pioneer 11 crossed the orbit of Neptune, then the most distant planet, during 23 February 1990, becoming the fourth spacecraft to exit the solar system. By 1995, Pioneer 11 could support only two science instruments, the Geiger Tube Telescope and UV Instrument. JPL turned off the Charged Particle Instrument by the end of 1994.

Voyager

Current status

Instruments no longer collecting data and non-essential heaters on Voyager 2's scan platform were turned off in November 1998 as a power conservation move. Five instruments continue to provide data. Voyager 2 is now more than 8.6 billion km (8 light hours) from Earth travelling at 16 km/s. The Voyager 1 spacecraft scan

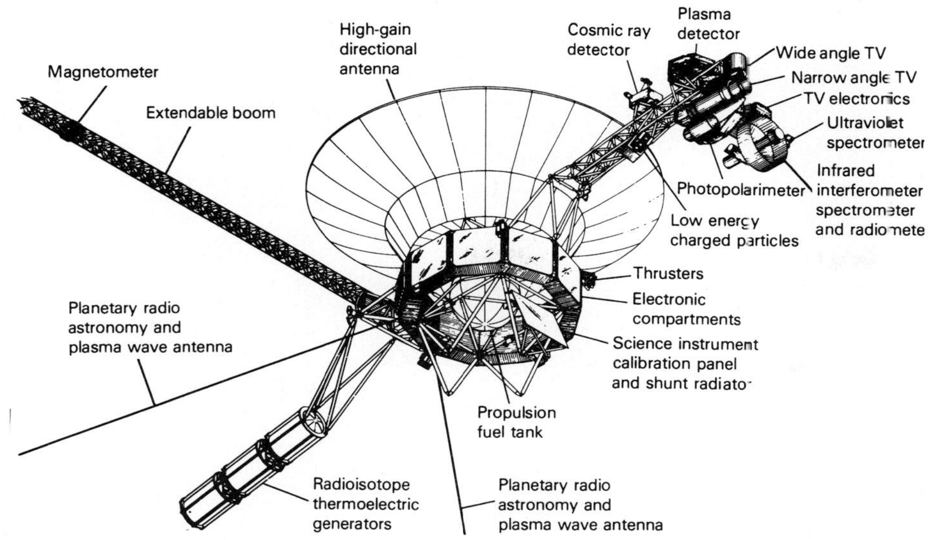

Voyager principal features

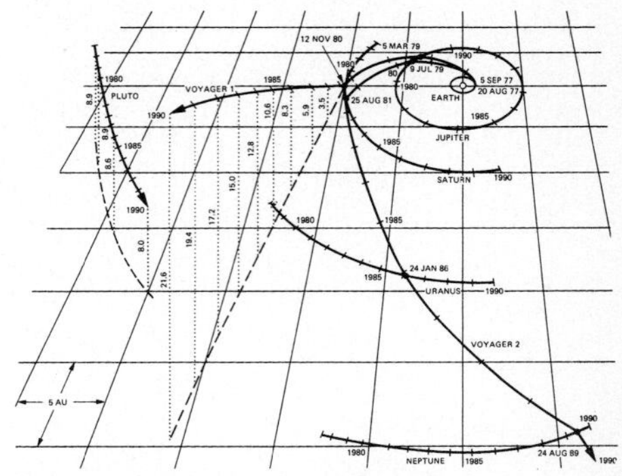

Voyager trajectories out of the solar system

platform will be turned off in mid-2000 to conserve power. The most distant object made by humans, Voyager 1, is 11 billion km (10 light hours) or 73 AU from Earth travelling at 17.3 km/s. Both spacecraft confirmed the earlier Pioneer finding that the heliopause exists further than this earlier estimate of its position. Based on the rate of change of solar wind material, the Voyager science data sets the current estimate for heliopause at 80-90 AU. The satellites scan the path ahead for low energy cosmic rays penetrating the outer reaches of the solar system from nearby supernovae remnants. The plasma wave receivers have been recording intense 2-3 kHz emissions since August 1992, believed to originate at the heliopause where the solar wind interacts with the cold interstellar plasma.

JPL's Deep Space Network continues to receive daily ultra-violet and fields/particles data. The Laboratory's scientists turned off the imaging cameras, photopolarimeter and IR spectrometer on both craft after Voyager 2's Neptune fly-by as they are of no further scientific use. The remaining seven instruments (UV spectrometer, cosmic ray telescope, low energy charged particle experiment, magnetometer, plasma and plasma wave subsystems and the planetary radio astronomy instrument) continue to return data at a combined rate of 160 bit/s. Their data return and communications could be maintained to around 2017, when power levels reduce to the minimum 230 W required for critical life functions. Voyager 1 and 2 will then be 138/113 AU out. Electrical power outputs at end-1995 were 339/345 W (480 W at launch). With sufficient power, contact would be lost with their 20 W transmitters around 2033 because of the distances (29,000 million km for Voyager 1). Attitude control propellant is not a limiting consumable: the 34.1/36.4 kg of hydrazine remaining at end-1995 (104 kg at launch) is sufficient for operations well beyond 2020.

Background

Dr Gary Flandro of the University of Utah, first noticed that a rare alignment of the planets late in the 1970s allowed a single spacecraft to visit all of the major outer planets of the solar system. Working with NASA's JPL, he and JPL engineers conceived of the 'Grand Tour' in the 1960s. In the mission profile, gravitational forces of the outer planets served to swing one spacecraft in turn from Jupiter to Saturn, Uranus and Neptune and another from Jupiter to Saturn and Pluto. Financial cuts at NASA required to pay for the burgeoning budget of the Shuttle pared the project to a mere extension of the Mariner programme, which had explored the inner planets. Work began in January 1972 but the project underwent many changes. For example, NASA lacked the funds for Jupiter orbiters and postponed a visit until Galileo. The craft were designed, built and operated by JPL.

As the flight date neared, JPL prepared the craft for launch. But following the four-day road journey from JPL to Cape Canaveral, engineers discovered faults in the attitude and articulation control and flight-data subsystems of Voyager 2, due to depart first because its trajectory included the Uranus option. The two Voyagers thus had to be interchanged, with consequent swapping of equipment.

The overall project cost about US$320 million not including launchers and flight support activities and NASA sources indicated the entire programme cost after 1986's Uranus encounter was $600 million. Voyager 1 left on a Titan-Centaur 3E on 5 September 1977 and coming within 278,000 km of Jupiter on 5 March 1979 and 124,000 km of Saturn 12 November 1980. Voyager 2 left on 20 August 1977, coming within 650,000 km of Jupiter 9 July 1979, 101,300 km of Saturn 25 August 1981, 71,000 km of Uranus 24 January 1986 and 5,016 km of Neptune 25 August 1989.

Voyager 2's fly-past of Neptune formally ended 2 October 1989, the sixth and last planetary encounter of the two-craft programme. When this happened, NASA formally renamed the project Voyager Interstellar Mission and estimated the programme had cost US$863 million. Under its new designation, VIM cost US$70 million through 1995, exclusive of tracking. Voyager's two Jupiter, two Saturn and the first Uranus and Neptune fly-bys sustained NASA's planetary programme during an 11-year dearth of launches. Voyager 1 captured the last of some 67,000 images in February 1990 to create a mosaic portrait of the solar system.

The Voyagers and two Pioneers are exiting the Sun's system in different directions, heading towards the hypothetical heliopause, where the solar wind is turned back by the interstellar wind. Voyager 1 has been heading 35° north of the ecliptic since its November 1980 Saturn encounter. Viking 2 on a 48° south trajectory since Neptune.

Specifications

Mass: 2,016 kg launch mass, Voyager 792 kg.
Communications: 3.7 m antenna dish, normally Earth-pointing (largest flown on a planetary mission). Transmitter power is 20 W.
Power: 3 RTGs, each 39 kg, encased in beryllium on a deployed boom. (To avoid their radiation, the most sensitive science instruments are mounted on an opposite 2.3 m boom).
Computation: 536 Mbit tape recorder memory (allowed the storage of the equivalent of 100 full-resolution images.) Six computers (three back-up), total 32 kbit capacity, reprogrammable to meet the needs of each encounter.
Propulsion: Spherical propellant tank at the centre of the basic 10-sided aluminium bus supplies hydrazine (104 kg loaded) for the 12 attitude control and four trajectory correction thrusters.
Payloads: TV: 200 mm f/3 and 1,500 mm f/8.5 cameras.

IR spectrometer: 51 cm diameter gold-plated telescope covering 2,000 wavelengths over 4-50 μm ultra-violet spectrometer: 128 wavelengths over 500-1,700 Å.

Photopolarimeter: 15 cm Cassegrain telescope.

Cosmic ray telescope: covering 0.5-500 MeV.

Low energy charged particle experiment: 0.01-11 MeV electrons and 0.015-150 MeV protons/ions in the planets' magnetospheres.

Planetary radio astronomy: radio emissions.

Plasma particles: 10-6,000 eV charged particles in the solar wind and the planet/moon magnetospheres.

Plasma waves: 0.01-56 kHz oscillations in surrounding plasma to determine temperature and density.

Magnetometers: magnetic fields around planets/moons and in interplanetary space; $20^{-2} \times 10^{-8}$ gauss.

Comment: Each carries a 30 cm copper 'Sounds of Earth' record containing greetings in 60 languages, sounds of birds, whales and other animals, samples of music and other information in case they are ever retrieved by intelligent beings. The record's aluminium jackets carry playing instructions using the cartridge and needle provided.

HUMAN SPACE FLIGHT

Human Space Flight
Resupply Vehicles
Space Stations

HUMAN SPACE FLIGHT

INTERNATIONAL

Spacelab

Current status
The last Spacelab flight was launched on 17 April 1998 aboard Columbia on STS-90. In 22 flights, the pressure module has accumulated 244 days in space over more than 14 years of operations. The 10 pallet-only flights between November 1981 and February 1996 accumulated 89 days of mission time.

Background
Spacelab was Europe's major contribution to the US Space Shuttle programme, and ESA's most important long-term project for the nine-year span between 1974 and the first flight in November 1983, about four years behind schedule because of delays in the Shuttle programme. NASA and ESA jointly ran the programme until the third Spacelab flight (STS-51F/Spacelab 2) in 1985. Later that year, ESA delivered to NASA the second Spacelab, including IPS, for which NASA paid about 200 MAU, formally ending the joint programme. Europe's US$1 billion development of Spacelab at no cost to NASA did not provide ESA with continuing right of access to or use of the laboratory once the joint programme ended. By the end of 1985, four Spacelab missions had been completed, the fourth being the German-funded/controlled D1 mission.

Discussions that eventually led to Spacelab, began in 1969, when NASA invited Europe to participate in the post-Apollo programme. Agreements on Spacelab between NASA/ESA and between the US and nine European governments followed in September 1973. ESA, basing its plans upon American claims of space launch affordability believed US would put the Spacelab into orbit at least 50 times during its lifetime. To date, the flight rate is less than half that.

NASA also promised the Europeans that Spacelab would be a first step towards internationalising the US manned programme, however NASA would control all of its activities. NASA would permit European flight crew opportunities in conjunction with flight projects sponsored by ESA or by governments participating in the Spacelab programme. West Germany contributed the most in percentage terms. The breakdown of contributions is: West Germany 54.94 per cent, Italy 15.57 per cent, France 10.29 per cent, UK 6.51 per cent, Belgium 4.32 per cent, Spain 2.88 per cent, Netherlands 2.16 per cent, Denmark 1.54 per cent, Switzerland 1.00 per cent and Austria 0.79 per cent.

An initial six-year contract worth US$226 million and covering Spacelab design/development was awarded in June 1974 to VFW-Fokker/ERNO of Bremen, whose major subcontractors were Bell Telephone in Belgium, Aeritalia (pressure shell) in Italy and BAe Dynamics (Pallets). In all, the industrial consortium finally embraced 50 companies in 10 European countries. TRW provided technical assistance to ERNO.

During 1979, the NASA/ESA partnership nearly dissolved. Spacelab integration tests brought to light numerous mismatches of hardware and incompatibilities between hardware and software. New cost projections to fix the problems showed the project would break the 120 per cent ceiling and more funds had to be sought from the participating countries. By then, the original estimated cost of US$515.7 million had grown to US$850 million.

With additional funding, however, in November 1980, after more than six years of development, the EM was delivered to NASA at the Cape, where it was used for interface verification and training. Then, US Vice-President Bush formally took delivery of the first flight version of Spacelab on 5 February 1982, together with an EM, ground support equipment and some computer software. ESA supplied a Long Module, five Pallets, one igloo, an Instrument Pointing System and two sets each of EGSE/MGSE.

In 1980, NASA ordered a second Spacelab worth US$144 million and subsequently a second IPS worth US$18.4 million for delivery in 1983. NASA would have preferred a better deal under which it would have provided flight facilities in return for the hardware. NASA dashed ESA hopes of continuing the project by dropping an option for a third Spacelab.

Beginning with 1990's Astro mission, NASA took over control, with ESA participation, of all science activities from Marshall's Spacelab Mission Operations Control Center, replacing Johnson's Payload Operations Control Center. McDonnell Douglas, the American aerospace company, provides Spacelab processing and support. NASA extended the contract in 1992 by four years at US$163 million, plus a four-year US$120 million option. Germany dropped out of Spacelab citing the soaring costs related to American financial management of the overall programme. In 1993, ESA dropped plans for its own Spacelab-E flight on cost grounds.

The Spacelab long module continues to be used for science missions (NASA) ***1996***

Spacelab flies in the cargo bay of the Shuttle and relies on the Orbiter for life support during the 9-16-day missions. Columbia accommodates most Spacelab flights and, from 1992, its Extended Duration Orbiter capability has permitted 13-day missions. In 1995, NASA adopted the Mir Shuttle visits, greatly reducing the need for Spacelab sorties.

Spacelab comprises three principal units: a 75 m^3 Pressure Module, Pallets for mounting external payloads and an 880 kg tunnel linking the PM to the Orbiter airlock. In preparing the Spacelab for flight, the ESA sets a landing mass and centre of gravity restriction that require aftward mounting in the cargo bay of the Shuttle, but a five-Pallet train can be accommodated along the full length. The total load limit on the Spacelab upper mass limit is 14.5 tonnes. British Aerospace constructs each 725 kg 3 m long 4 m wide Pallet with conventional aircraft-type designs and aluminium skin over an aluminium frame. The final pallet configuration is capable of carrying up to 3 tonnes of payload. Data, power and cooling links are routed through the PM. The first Pallet, an engineering model, flew on STS-2. If the PM is absent, a 640 kg 2.2 m^3 2.4 m high 1.1 m diameter aluminium igloo cylinder at the front provides a pressurised environment for the avionics.

The Pressure Module consists of a single shell of 0.16-0.35 cm thick aluminium capped at both ends by an Al cone. Each 4.06 m diameter 2.7 m long segment consists of eight butt-welded panels ending in 24 cm wide 4.10 cm outer diameter aluminium load-bearing rings. A two-segment Long Module is created by bolting two units together, with O-rings providing the seal; a single segment Short Module configuration is also available (but has not yet flown). Two longerons fastened to the aluminium rings attach Spacelab to the Orbiter's longerons via a single 82 mm diameter Inconel pin on each side and a keel point underneath.

The 78.2 cm deep 4.11 m diameter cones are bolted to the cylinder only when the internal equipment is in place and thereafter provide access through central 1.3 m circular openings. ESA removes the rear hole for flight and uses the forward opening for the tunnel. The LM's overall length is about 7 m. Each cone includes three 40 cm holes: two at the base for Orbiter supplies and Pallet connections and one near the top. On the rear cone, the latter is always a 30 cm diameter 2.6 cm thick window for viewing the Pallets, whereas the forward version carries pressure relief valves (the structure is stressed for a maximum pressure differential of 1.1 atmosphere).

Dual-triangular aluminium beams run longitudinally inside each PM segment and are attached to the end rings as support for the floor and experiment racks. Prior to flight, experimenters place their packages and integrate them into the Spacelab offline and then the racks are slid with the floor into place on the beams. The aluminium racks are designed to accept standard lab trays 48.3 cm wide: the 42 kg Single Rack can hold 0.9 m^3/290 kg of equipment and the Double Rack 1.75 m^3/580 kg. Each Spacelab segment can hold two DR and one SR on both sides, creating a central aisle similar in size to Mir. Typically, a rack is devoted to a specific discipline or reserved as a storage unit. The racks are attached to two overhead Al beams that also support up to eight storage containers. The Al honeycomb floor panels can be lifted to expose additional space in the rear segment and equipment can be bolted down on the 60 cm wide central aisle. This was employed, for example, during Spacelab D1 for the Sled chair/track experiment.

The two forward DRs house connections for standard functions, even on a Short Module flight. The port side workbench rack carries a work surface, tools, high-quality lighting and writing equipment. The starboard unit controls Spacelab's systems. Three identical IBM AP-101SL (replacing the Matra 125/MS 64 kbit) computers handle either life support or experiments, retaining one as a back-up. The Orbiter's flight deck hosts a duplicate terminal and a third can be added to the second Spacelab segment. The same rack holds the High Data Rate Recorder for storing data at up to 32 Mbit/s. Data can be transmitted in real time through NASA's TDRS satellite system at up to 50 Mbit/s via the steerable Ku-band dish on the Orbiter's forward cargo bay starboard side. The recorder is employed when TDRS is unavailable.

The roof of each segment incorporates a 1.3 m diameter opening for windows or, for the rear opening only, a science airlock. The window unit contains a general viewport and a high-quality optical 4.1 cm single thickness 40.9 × 55.2 cm remnant Skylab pane. ESA's large format Metric Camera employed this port during Spacelab 1. An external aluminium cover provides protection against radiation, contamination and

Spacelab 1 in Columbia's payload bay. The Z-shaped docking tunnel at the far end (top right) matches the levels of the Orbiter and Spacelab hatches and allows longitudinal movements during launch and re-entry to reduce stress (NASA)

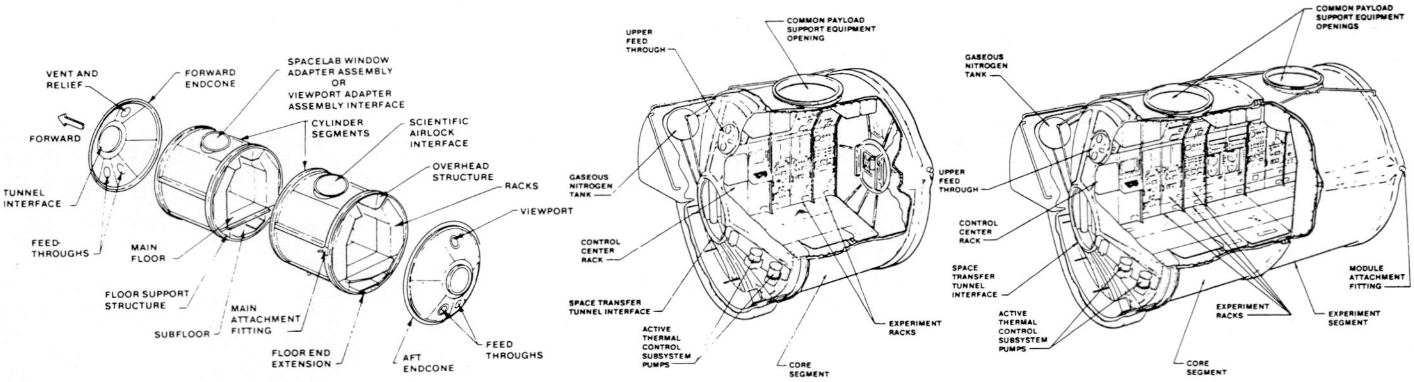

The Pressurised Module can be flown in either long or short form. It has always flown as a Long Module so far

Spacelab's life support system is carried under the forward segment's floor. The twin holes each side in the main floor beams blow cooling air up behind the equipment racks. The racks and second segment await mating in the background

Spacelab Mission History

Date	Configuration	Mission
November 1983	LM+1P	Spacelab 1[1]
April 1985	LM+MPESS	Spacelab 3[2]
July 1985	IG+3P	Spacelab 2[3]
October 1985	LM+MPESS	Spacelab D1[4]
December 1990	IG+2P	Astro-1 + BBXRT[5]
June 1991	LM	SLS-1[6]
January 1992	LM	IML-1[7]
March 1992	IG+2P	Atlas-1[8]
June 1992	LM	USML-1 (1st EDO)[9]
September 1992	LM	Spacelab-J (Japan)[10]
April 1993	LM	Spacelab-D2 (Germany)[11]
April 1993	IG+P	Atlas-2[12]
October 1993	LM	SLS-2 (EDO)[13]
July 1994	LM	IML-02 (EDO)[14]
November 1994	IG+P	Atlas-3[12]
March 1995	IG+P	Astro-2 (EDO)[15]
June 1995	LM	SL-M Mir visit
October 1995	LM	USML-2 (EDO)[16]
June 1996	LM	LMS (EDO)
April 1997	LM	MSL-1 (EDO)[17]
July 1997	LM	MSL-1 (EDO)[17]
April 1998	LM	Neurolab (EDO)

Key:
EDO: Extended Duration Orbiter; IG: Igloo; LM: Long Module (pressurised);
LMS: Life & Microgravity Spacelab;
MPESS: Mission Peculiar Experiment Support Structure; P: Pallet;
Atlas: Atmospheric Lab for Applications & Science; SLS: Spacelab Life Sciences;
IML: International Microgravity Lab; MSL: Microgravity Sciences Lab; USML: US Microgravity Lab.
The first eight PM missions (through D2) involved 387 experiments covering 323 PIs from 148 institutes in 26 countries.
1 38 NASA/ESA experiments (16 Pallet/22 PM), to demonstrate basic Spacelab capabilities.
2 10 µg experiments in LM, including Drop Dynamics Module and animal holding units with 2 monkeys & 24 rats, 2 on MPESS (Atmos for monitoring upper atmosphere trace constituents and Indian Ions for studying cosmic rays).
3 carried IPS with four solar telescopes, plus x-ray, cosmic ray & IR telescopes and retrievable Plasma Diagnostics Package first flown on SM3.
4 first 8-man crew, 76 µg experiments: life sciences, Biorack, Fluid Physics Module, Sled, materials processing, plus Materials Experiment Assembly on MPESS. Sponsored by FRG BMFT & managed by DFVLR.
5 7,836 kg Astro telescopes on ESA/Dornier IPS (HUT Hopkins UV Telescope, The Johns Hopkins University; UIT UV Imaging Telescope, GSFC; WUPPE Wisconsin UV Photopolarimeter Experiment, University of Wisconsin). 3,294 kg BBXRT BroadBand X-Ray Telescope, GSFC.
6 most intensive life sciences research mission flown, first three-woman crew, 29 rats and 2,478 jellyfish, plus 12-can GAS Bridge. **7** 55 materials and life sciences experiments, plus 10-can GAS Bridge. **8** 11 instruments on 2 pallets for atmosphere, solar, space plasma and astronomy.
9 31 experiments in four broad areas: materials, fluid physics, combustion and biotechnology.
10 43 materials and life sciences experiments (34 from Japan).
11 90 experiments including Anthrorack, Advanced Fluid Physics Module, Medea, Holop, Biolabor, MOMS-02 Earth observations, Gauss galactic plane UV film imaging. Funded by BMFT, managed and payload controlled by DLR.
12 six instruments on one pallet plus seventh in two side cans, for atmosphere and solar studies.
13 14 experiments continuing SLS-1 life sciences studies.
14 >80 experiments, including Large Isothermal Furnace, Critical Point Facility, Advanced Protein Crystallisation Facility, Free-Flow Electrophoresis Unit, Biorack, Aquatic Animal Experiment Unit, Bubble Drop & Particle Unit.
15 as Astro-1 (note 5), except no BBXRT.
16 followed on and repeated USML 1 (note 9).
17 Mission terminated early and to be reflown July 1997.

micrometeoroids. A simple viewport can replace this unit as required. D1 used the simple window.

Dornier's Instrument Pointing System (IPS) provides three-axis control for a payload of up to 3 tonnes, although with modifications the system can handle 7 tonnes. It carries all inertial sensors, data and power electronics and the dedicated software for control via the Spacelab computers in a unit that weighs 1,184 kg. The nominal payload is 2 tonnes, on a 0.5 to 3 m diameter. Pointing accuracy (arcsec) for a 2 tonnes payload is 0.4 lateral/11.2 roll under star tracker (3) control; 0.5 lateral 41.0 roll in the Sun mode. Pointing stability is sub-arc/sec. The available control torque is 30 Nm/axis. Slewing rate is 1°/s maximum in inertial mode, power required during pointing is 500 W. Payload power provided is 1.25 kW. Data rate 16 Mbit/s on six channels. The control system has flown on Spacelab 2 and Astro 1/2.

The 1 m long 1.05 m diameter airlock can accommodate 98 cm diameter 99 cm long packages bolted to a side-mounting table, which is then hand-cranked 96 cm out into open space. When astronauts require easy access they may remove the inner hatch. D1 deleted the airlock, for instance. It carries an 18 cm diameter off-centre porthole for observing experiment progress. Spacelab 1/3 used it to place an ultra-violet Very Wide Field Camera outside the main structure for celestial imaging. Spacelab carries sufficient N_2 for up to seven airlock repressurisations on a normal mission. Should the outer hatch jam open, preventing the Orbiter's door from closing, it can be ejected or removed during EVA.

Astronauts perform EVAs as needed through a top hatch on the transfer tunnel adjacent to the Orbiter's airlock. The 1.3 m diameter aluminium tunnel, built by McDonnell Douglas, provides a 1.02 m internal diameter for passage of a suited astronaut and accommodates the 1.07 m height difference between the Shuttle rear hatch and the Spacelab central opening. Only the 5.75 m long six-segment tunnel has been employed so far but a 2.66 m five-segment version is available for the Short Module (which can be carried further forward). Flexible sections at both ends allow some movement in the bay.

The Orbiter and Spacelab work in concert to perform life support activities and to maintain a 1.013 atmosphere 18 to 27°C 30 to 70 per cent humidity environment. A ventilation unit incorporating an old Skylab fan at the bottom of the tunnel bend draws Orbiter cabin air into Spacelab through a floor conduit and returns it via the main volume. A charcoal scrubber converts CO to CO_2 and removes other trace materials. When the end hatch is closed (the Spacelab end does not carry a door) during ascent and landing, the system continues to ventilate the tunnel/lab. The air is cooled in heat exchangers before being blown through roof-mounted diffusers. Air is also forced up inside the racks for cooling and drawn off at several points inside the curved wall.

The payload is limited to about 8 kW (28 V DC) from the fuel cells because of the Orbiter's heat rejection capacity. To maintain thermal control some experiments and avionics are mounted on cold plates using the Orbiter's radiator system. The whole module carries an external jacket of 39 interleaved layers of Dacron and gold-covered Kapton topped by Teflon-coated beta cloth, the whole held in place by studs.

Before a launch, researchers normally insert their experiments while Spacelab is horizontal and the whole configuration is slotted into the Cargo Integration Test Equipment (CITE) at KSC for Orbiter simulations. Then the NASA ground crew installs Spacelab in the Orbiter Processing Facility before the spaceplane is stacked and transferred to the pad.

RUSSIAN FEDERATION

Soyuz

Current status

An improved Soyuz-TM will act as the initial Assured Crew Return Vehicle (ACRV) for the international Space Station. NASA signed an agreement with RKK Energia in December 1992, when the Phase A study ended, affirming that Soyuz is technically suitable. Plans call for two docked three-man Soyuz-TMs to be used as the emergency escape vehicle in much the same role as the

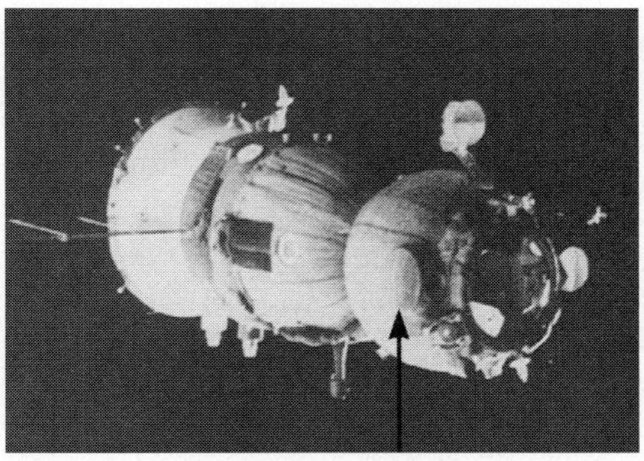

TM16 displays the forward-looking cupola (arrowed) and androgynous docking system ***1995***

spacecraft is now used aboard the Mir for American astronauts. Each Soyuz costs US$30 million, plus enhancements such as increasing service life from six months to three years. The vehicle must be able to land within 20 km of a designated landing spot. Crews will not wear pressure suits. The current ferry will also be modified to accommodate the larger torso size range of US astronauts. For example, in 1995, Scott Parazynski stood down from May 1997's Mir visit because of his torso length. Although he was to fly up and down on Shuttle, he had to be able to return in Soyuz in an emergency.

Background

Soyuz is the most-utilised manned spacecraft and is now into its third decade with a flight total comparable to the aggregate for Mercury, Gemini, Apollo and Shuttle. There have also been dozens of unmanned orbital tests and it probably has a version that houses a photo-reconnaissance satellite. Despite its age, Soyuz continues on as a Mir ferry. RKK Energia in Kaliningrad is responsible for design and production. Russian planners have long floated plans for an upgraded version of the Soyuz but, in 1991, Energia halted design work on the 13 tonne Zarya super-Soyuz, carrying five to six crew to be launched from a Zenit booster, because of funding difficulties. The overall form of the descent module was similar to that of Soyuz, but with a 3.7 m base diameter. The manned version would have included 1.5 tonnes of cargo and the unmanned 3.8 tonnes.

Originally, Chief Designer Sergei Korolev intended to build several versions of Soyuz. His ideas called for a full three-module craft for Earth-orbit operations, a stripped-down two-module Zond for demonstrating lunar mission techniques, and a lunar-landing orbital version. As flown, Soyuz has had three distinct mission categories. Soyuz 1-9 (1967-69) carried three-man LEO solo missions of durations up to 2½ weeks. Then when it became evident to the Soviets that they had lost the moon race, design work began on the Salyut space station in late 1969 and Soyuz 10 and 11 functioned as crew ferries with integral transfer tunnels. After the Soyuz 11 failed, killing its crew during descent, the Soviets initiated a thorough redesign and Soyuz became a two-man station ferry capable of only 2½ days' independent flight. At this time, the Soviet space program evolved towards a more automated program than its American counterpart and as a result, the unmanned Progress cargo version of the Soyuz capsule came about.

The post-Soyuz 11 version of the craft was 24 times in its manned configuration when Soyuz 40 retired the model in May 1981. After Soyuz 40, the Soviet Union introduced the Soyuz-T with its updated systems and three-man crews. This capsule took over for Salyut 7 operations until a further reworking produced the 'TM' Mir ferry. The 'TM' designates all Russian craft as modified transports. Soyuz has thus been transformed from a spacecraft with lunar missions to a dedicated space station ferry, through several generations of space exploration objectives.

The Soyuz Orbital Module (OM) provides space for cargo, food, waste management services, orbital experimentation on solo missions and EVA operations through the side hatch which is employed primarily for crew boarding on the pad. The forward hatch houses the docking apparatus of a central cone and eight metal blades inside a 1.3 m diameter docking collar that has eight mechanical latches (Soyuz 16/19/TM16 carried androgynous docking units). Two crew use sleeping bags in the OM, while the third sleeps in the Descent Module (DM) for safety (a rule not always adhered to). The toilet swings out from behind a panel, using pressurised airflow as a gravity substitute.

When delivering payloads to space, the Soyuz uses built- in connections for electrical, communications and hydraulics links. The original Soyuz design included four OM waist portholes, reduced to two on Soyuz-T and one on TM. Since 1991 and possibly before, TM has carried at least one front-viewing cupola in the OM, with controls for flyaround operations using direct vision. The OM is jettisoned by 12 pyrotechnic charges after retrofire. Until the TM5 failure of September 1988, cosmonauts ejected the OM before retrofire. This sequence saved about 10 per cent of the propellant. Failure investigations proved that crew safety required the post-retrofire jettisoning operation.

The Soyuz Descent Module (DM) accommodates a crew of two or three, with the Cdr occupying the central couch, the flight engineer the port side and the cosmonaut-researcher the starboard side. Alternatively, a 100 kg cargo pod can replace one crew member. These Kazbek-Y couches are supported on shock-absorbing struts and can be lowered to provide increased space. Ten minutes before landing, compressed N_2 raises the couch 10 cm to help absorb the shock. Each seat liner is customised for the occupant and when cosmonauts return in different craft from their delivery craft, then the liners must be changed.

The commander provides manual thruster control via translational and rotational handles to his left and right, respectively, performing rendezvous and docking operations using the range and range-rate radar data and the 15° Field of View (FoV) VSK-4 periscope below the centre of the main control panel. From the introduction of Soyuz-T, the onboard computer has been capable of performing a fully automatic docking. The DM's separate 67 N H_2O_2 thruster network is activated only for re-entry: two nozzles near the apex some 60° apart provide roll control, with pitch control from a pair located between, and two at the base for yaw.

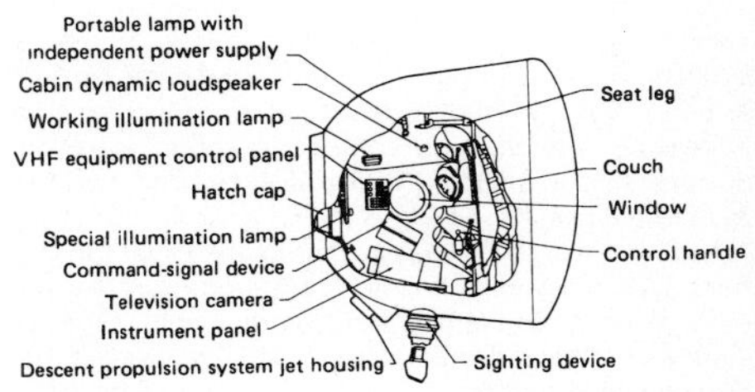

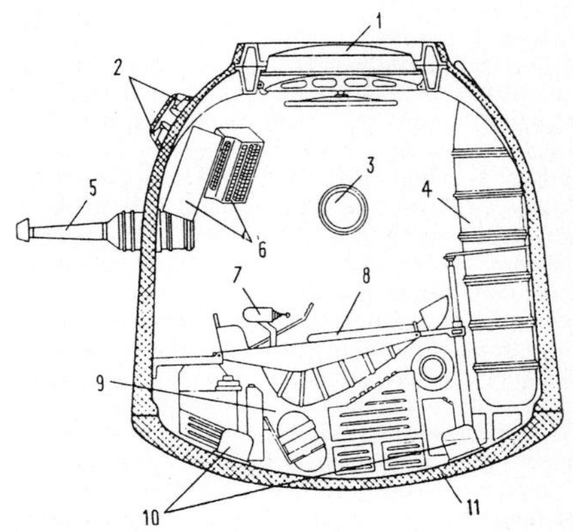

Soyuz descent module. Key: 1 crew hatch, 2 pitch thrusters, 3 porthole (one of two), 4 'chute container (one of two), 5 periscope, 6 control panels, 7 hand controller, 8 couch, 9 life-support equipment, 10 soft-landing thrusters, 11 detachable heatshield

Soyuz-TM: the current version of the Soyuz manned craft is used for Mir duties. This full-scale representation at 1987's Paris Salon was the first public appearance. Note the SM's two Earth horizon sensors, the DM's periscope and umbilical panel and the OM's hatch for on-pad access (Andrew Wilson)

The service module completes the vehicle, supporting the DM on its forward face. The Service Module (SM) carries the propulsion unit. In the old Soyuz, the propulsion system employed a KTDU-35 assembly at the rear of the SM. Pumps fed propellant and oxidisers to a single main engine, which fired for 2½ minutes to initiate re-entry. A single start twin-nozzle engine served as a back-up unit. AOCS thrusters distributed around the SM's forward and aft faces provided guidance and control of the vehicle. Soyuz-T introduced the ODU unified propulsion system in which all engines were supplied from the same tank. Soyuz T's smaller AOCS provided a back-up re-entry capability. The Russians have revealed less about the Soyuz-TM system. Jane's believes the primary engine provides 3.1 kN of thrust. Mir incorporates similar engines and the timeframe of the development programme, the contractors and the history of Russian design practices suggest the two engines would be used for both projects.

As a safety measure, the propulsion system tanks now use metallic membranes to avoid pressurising N_2 which would have leaked through the previously elastic barrier. Some 150 kg of the propellants are reserved for rendezvous and docking and the retrofire operation consumes an additional 200 kg (400 kg allocated) for Earth return.

In the event of an emergency, an escape tower can fire the capsule away from the immediate area. From the top of the launcher on the pad, the capsule would reach an altitude of 1 km and land some 2½ km away, using the 27 m diameter reserve parachute. On a normal mission, descent is initiated by a 3 to 4 minute ~155 m/s retroburn by the main engine over the south Atlantic, followed by OM and then SM ejection. Re-entry occurs over Africa at an angle of attack of about 30°; a lift to drag ratio between 0.25-0.30. A 0° bank (heads-up) generates maximum lift and a deceleration of 3 *g*, 60° produces 4 *g* and the no-lift and a 90° bank leads to 8 *g*. The Soyuz TM can land cross-range of ±65 km, with a landing accuracy of 20 × 30 km ellipse. During re-entry, the external temperature reaches 3,000°C, but internal temperature maintains a 25-30°C constraint.

When landing, a pressure switch, activated at 9 to 11 km altitude and 850 km/h, unfurls two sequential stabilising drogues releasing the 4.25 m braking parachute from the port compartment. The primary canopy is deployed reefed at 8 km some 40 to 45 minutes after retrofire. It is freed to its full 35.5 m diameter at a descent rate of around 35 m/s to reduce sink rate to 8 m/s. Soyuz-TM retains the same size parachutes but their lighter material accounts for 110 kg of the craft's increased lift capacity. The 90 kg reserve system yields 10 m/s descent rate with its 25 m parachute, activated at 6 km. The heatshield is dropped at about 3 km some 5 minutes before landing to clear the base retromotors for a soft landing.

In the final burn, four solids initiate upon command from a radar altimeter about 2 m above the ground to cushion the impact. Soyuz-TM carries an improved system that has reduced contact speed from 3 to 4 m/s to 1 to 2 m/s. Following the Soviet design, the Soyuz generally touches down on land but the craft is equipped for water landings. Coping with splashdown is a routine element of cosmonaut training.

Each capsule carries standard Granat-6 survival kits of Forel nylon flotation suit, Neva self-inflating life preserver, TZK cold weather suit, dried food, medical supplies, flares, radio, machete and canteen. The equipment is designed to sustain the crew for up to 3 days in the capsule in severe conditions and up to 12 hours in 2°C water with −10°C ambient air. A recovery beacon automatically activates on touchdown from behind a small ejectable cover to the left of the parachute compartment. In the event of a water landing, a balloon inflates from the same location. A radioactive source also acts as a beacon, something the recovery crew has to avoid after landing.

TM3's descent module. A rare view of the capsule's underside. Note the scorch marks from the soft-landing rockets

Cosmonauts Malyshev (left) and Aksynov train in Star Town's simulator. The attitude controller can be seen by Malyshev's right knee, the periscope below panel centre, with oxygen controls above, and the Globus Earth position mechanical display at top right. Globus' main function is to show the landing site at any instant if retrofire begins immediately; pushing a button turns the globe from the current orbital position to the landing co-ordinates. Rods are used for pressing buttons difficult to reach in the cramped conditions. The commander holds a flimsy column bearing an abort button during launch. Note the hatch at top and the parachute container bulging between the cosmonauts' positions

Specifications

Cargo/Manned Unit

First manned launch: 23 April 1967 Soyuz; 5 June 1980 Soyuz-T; 5 February 1987 Soyuz-TM
Number manned launches: 37 Soyuz, 14 Soyuz-T, 25 Soyuz-TM (to mid-1997)
Principal applications: space station ferry, LEO solo operations
Cost: Rb95 million (plus Rb56 million launcher) quoted in November 1992. Rb5 million quoted during Bulgarian TM5 mission of 1988, plus Rb4 million for Soyuz-uluz launcher.
Availability: typically 2-4 launched annually.
Performance: TM can accommodate a crew of three and 50 kg cargo for Mir-type operations, returning a full crew and 50 kg to Earth (no crew allows 450 kg up and 300 kg down).
Flight longevity: Independent 3.2 days and attached to Mir 180 days. Capabilities are constrained partially by launch escape system.
Cargo size: 45 × 60 × 100 cm up and 30 × 40 × 50 cm down.
Length: 6.98 m with docking probe withdrawn (on old Soyuz retraction was 43 cm)
Mass: 7,070 kg at launch (Soyuz-T 6,850 kg). Typically 6,790 kg at docking. Soyuz 19 mass breakdown was 1,224 kg OM, 2,802 kg DM, 2,654 kg SM.
Habitable volume: ~10 m³ (T/TM), previously ~9 m³
Landing mass: about 3,000 kg

Soyuz Orbital Module

Mass: 1.2 t
Size: 2.2 m diameter spheroidal
Volume: 6 m³
Hatches: 65 cm diameter side hatch, 80 cm forward hatch

Soyuz Descent Module

Size: 2.2 m base diameter, 2.2 m high
Mass: 2.85 t re-entry capsule
Usable volume: ~4 m³
Standing height limit: 1.86 m, 94 cm when seated. Some cosmonauts have exceeded the height limit while still within the seat range.
Hatches: A single 70 cm diameter overhead hatch, opened from either side and two 70 cm diameter side hatches for the primary and reserve parachute systems. The hatch lips extending from the 7° slope wall add to the DM's lift during re-entry and help to flip the hatches away in the airstream. The DM provides occupants with two 20 cm diameter portholes, one either side of the control panels; the discoloured outer pane is ejected after entry to restore vision. It is possible the aluminium airframe has been reworked and lightened.

Soyuz Service Module

Size: 2.3 m long, 2.2 m diameter (flaring to an aft 2.72 m)
Power systems: Twin Si solar wings (spanning 10.6 m).
Thermal Control: ~8 m² of thermal radiators.

KTDU-35

Fuel/Oxidiser: ~500 kg of UDMH/nitric acid in four 80 cm diameter spheres.
Engine: A single 4.09 kN main engine with back-up of a single start twin-nozzle engine generating 4.03 kN,
AOCS thrusters: 14 with 4 back-up providing 98 N and 8 with 4 back-up providing 9.8 N, supplied by H_2O_2 from spheres in the SM structure below the DM.

Soyuz-T

Fuel/Oxidiser: ~700 kg of UDMH/NTO (150 kg reserved for retrofire)
AOCS thrusters: 14 providing 137 N and 12 providing 24.5 N. Capable of acting as retrofire back-up.

Soyuz-TM

Thrust: 3.1 kN
Fuel Oxidiser: 800 kg supply of UDMH/NTO. 150 kg reserve for rendezvous/docking.

Power

Soyuz-T

Type: Twin 4-segment solar wings 1.4 × 4.4 m with batteries in the service module.

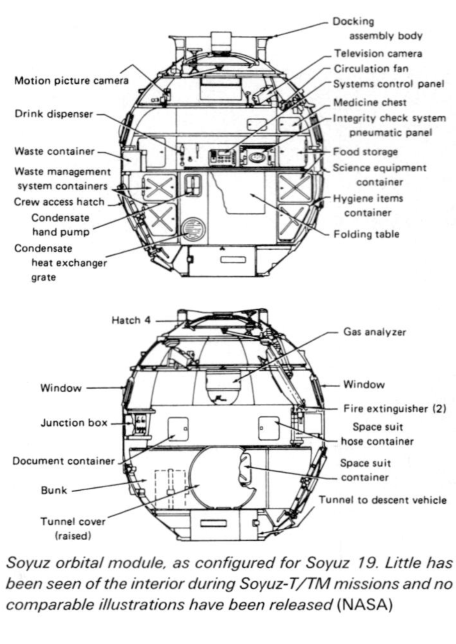

Soyuz orbital module, as configured for Soyuz 19. Little has been seen of the interior during Soyuz-T/TM missions and no comparable illustrations have been released (NASA)

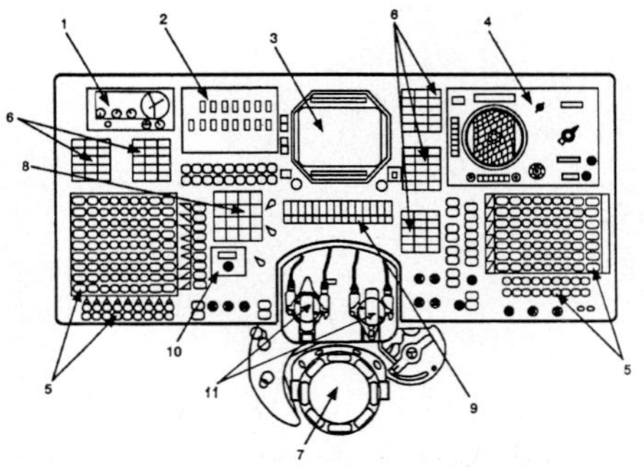

The Soyuz-TM main control panel in Star City's primary TM trainer. The commander's abort column is at bottom left 1 clock, 2 computer keyboard, 3 computer display, 4 Globus Earth position display, 5 control panels and status lights (81 on left, 62 right), 6 systems status lights, 7 periscope screen, 8 ignition panel (protected), 9 keyboard for information selection, 10 propellant supply indicator, 11 oxygen supply valves ***1996***

Output ~1.3 kW, with a significant proportion lost in cabling/processing.

Soyuz-TM
Type: Solar cells with a span of 10.6 m. Soyuz locks on to the Sun with a sensor and slowly cartwheels around the Sun axis. DM includes a small battery supply for power after SM separation.

Thermal protection
Soyuz T
Type: Hand-cut asbestos sheets impregnated with ED-5 or ED-20 epoxy resin laid on to the cleaned aluminium structure. The base was made separately and attached mechanically before the whole module was baked at 200°C and then machined to the required dimensions on a large lathe. From 1969, a 1-2 cm thick titanium honeycomb layer injected with an asbestos fibre/binder mixture provided base protection.

Soyuz-TM
Type: 8 m² of SM radiators with the rest of the surface area covered by green thermal blanketing.

Descent Module
Type: Eight blankets held by apex and base rings released when the other modules separate.
Comment: Three of these blankets broke free of the base on 1990's TM9 at shroud separation, requiring EVA repairs.

Life support systems
Soyuz -TM
Pressure: Hermetically sealed on the pad under sea level conditions at 184-263 mbar partial O_2 pressure.
Regeneration: Potassium superoxide and LiOH cylinders.
Suits: Sokol-KV2 (Falcon) pressure suits. Wearer has to be minimum of 1.64 m tall.
Contractors: NPO Zvezda is responsible for all Russian life support systems and suits.
Comment: The bulk of the system is housed in the OM but the DM carries a smaller, independent unit under the couches. O_2 bottles provide emergency protection. Waste management and eating is handled entirely in OM; DM offers air sufficient for 48 hours and only food/water packs for landing emergencies.

Few details are available of Soyuz-TM's propulsion system. This mockup at 1987's Paris Salon suggests a single main engine, with protective cover and four peripheral thrusters. The large backup retro of previous designs appears to be absent

Avionics/control
Soyuz
Type: Ground controlled by command and/or pre-programmed analogue sequencers activated from switch panels either side of the main display.

Soyuz-T
Type: 16 kbyte capacity Argon computer, capable of completing a fully automatic rendezvous and docking. (Cdrs previously assumed manual control 180 m from the target). Its Russian acronym was BTsVK, 'onboard numerical calculation complex'.
Attitude: Provided by infra-red Earth horizon plus Sun sensors in conjunction with an improved strapped-down inertial platform. Range/range-rate information comes from radar deployed on a 2.5 m pylon.

Soyuz-TM
Docking type: S-band Kurs ('course') docking system. Allows approach to Mir from any angle without the entire station having to rotate, as under the previous Igla ('needle') system.
Sensors: 15° FoV VSK-4 periscope for final approach. Two external 625-line 25 pictures/s TV cameras. The OM's waist carries VHF TV, radio and telemetry/command aerials, T-shaped telemetry/command antennas encircle its base.
Communication channels: Direct voice communications on 121.75 MHz. Data relay performed by the Rassvet system working via Mir through GEO satellites providing simultaneous dual-voice channels.

Landing/escape system
Soyuz/SL-4 launcher
Contractor: NPO Iskra's
Type: Tractor escape motor system providing protection for the first 115 s (the fairing's own separation rockets can then separate with Soyuz attached up to 165 s).

Soyuz TM
Type: 1,971 kg solid propellant tower, 6.680 m long, 1.415 m diameter.
Operation: Pyrotechnics separate below the DM and halfway down the shroud for the main 441 to 715.4 kN solid (1,607 kN total impulse) to pull the DM/OM combination free.
Peak acceleration: about 14 *g*. The motor attains 70 per cent thrust in less than 0.07 seconds. The main escape motor has two components, a larger motor with four nozzles at its apex, plus an upper smaller motor with four smaller nozzles at its base. Both fire for aborts up to 20 seconds, but from 20 seconds only the larger one operates. The firing sequence lasts for about 5 seconds and then four shroud panels pivot outwards to slow the ascent. Throughout the escape, a small apex unit with four nozzles curves the vehicle onto its recovery path.
Engines/Motors: 150 kg, 98 to 171.5 kN (98 kN total impulse) 12 nozzle tower separation motor.

Recent missions

Soyuz-TM 21
Launched 06.11 GMT 14 March 1995 with the eighteenth main crew and NASA astronaut Dr Norman Thagard with a primary interest in life sciences. Commander Vladimir N Dezhurov and engineer Gennadi Strekalov (the first Russian to six launches). This was

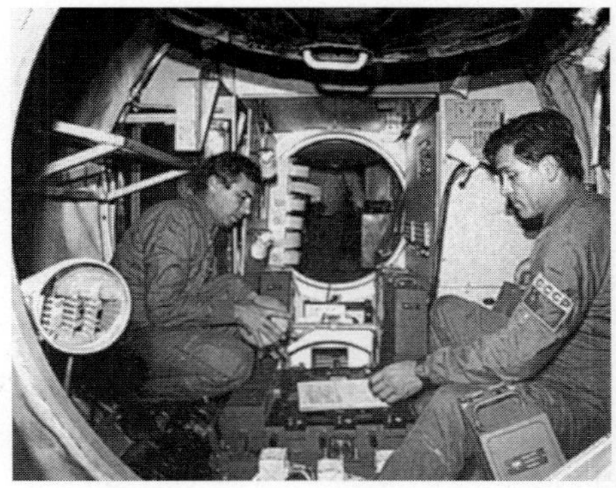

TM8's Viktorenko (left) and Serebrov with the MKF-6MA camera in the Kvant 2 trainer at Zvezdny Gorodok, seen through the 1 m EVA hatch

the 100th space launch of a US astronaut; all planned to return aboard NASA Shuttle Atlantis. The launch created a record (13) for the number of people in space at one time. Docked automatically with Kvant 1 07.46 GMT 16 March at a final rate of 27 cm/s. Thagard began his experiments 17 March with body mass measurements but his full programme had to wait for Spektr's delivery of his equipment. Progress-M 27 docked 21.01 GMT 11 April, bringing Germany's 20 kg 21.5 cm diameter GFZ 1 laser-reflecting geodetic satellite for release 19 April from Mir's core garbage airlock (see German entry for other details). 48 Japanese quail eggs, fertilised just before launch, also arrived. Thagard 'fixed' them at various points in their development, the fourth and last time on 29 April at full term, for return on Atlantis. Dezhurov/Strekalov began 18 April removing Kvant 2's unsuccessful shower unit (unused as a shower for two years) to make space for more gyrodynes, breaking it up to fit into Progress-M 27 for disposal. M27 undocked 23.40 GMT 22 May (leaving the axial port clear for Spektr) and de-orbited 03.17 23 May.

Dezhurov/Strekalov made three EVAs to move Kristall's solar wings before Shuttle's arrival. EVA-1: 12 May, 6 hours 15 minutes, Kvant 2 hatch opened 04.20 GMT; planned to transfer one wing to Kvant 1, but retraction problems. EVA-2: 17 May, 6 hours 52 minutes, exited Kvant 2 02.38. Moved wing to Kvant 1 using Strela, but unable to complete connections in time. EVA-3: 22 May, 5 hours 15 minutes, Kvant 2 hatch open 00.10. Completed wing-1 installation and extension. Wing-2 was partially retracted to give clearance while still helping the power shortage.

Mir's 19.6 tonne Spektr module was launched 03.33 GMT 20 May on Proton, although Mir was not yet ready to receive it. Dezhurov commanded the node's Lyappa arm to start swinging Kristall from its -Y/Earth-pointing position at 23.28 26 May to the -X/axial face, taking about 30 minutes. A 28 May internal EVA in the depressurised node moved the docking cone to the -Z/starboard port; Kristall moved there 30 May. Spektr docked 00.58 GMT 1 June – the first time a station module had arrived at the first attempt. It included 755 kg of NASA equipment for Thagard and his NASA successors. Spektr added 62 m³ of habitable volume. Its two forward solar wings remained stowed in transit, but one deployed only partially 5 June; a second attempt

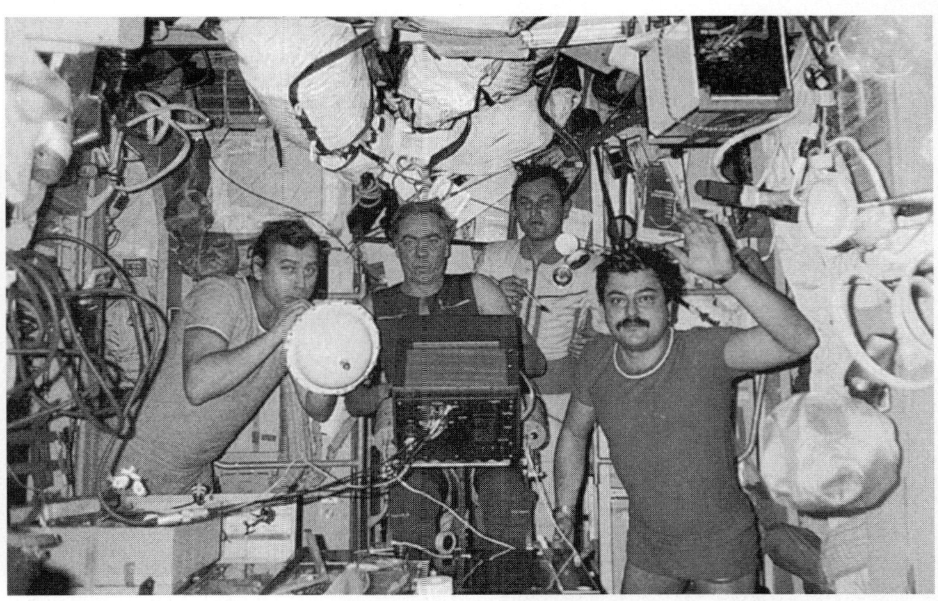

Alexander Alexandrov (left) aboard Mir June 1988 with, from left, Levchenko (on veloergometer), Titov and Manarov. Compare the cluttered appearance with the photograph of the same main living area in the Mir space station section. Titov and Manarov became the first to spend a year in space. Levchenko was a candidate for the first manned shuttle test but died of a brain tumour soon after

TM8 commander Viktorenko tests the new manoeuvring backpack 5 February 1990. Note Kvant 2's 1 m hatch, the first outward-opening hatch on a Soviet craft. It was stored on Kvant 2's exterior 8 February 1996 to move it out of the way

was unsuccessful 8 June. It was believed that a launch restraint was blocking it, so a July EVA was planned to cut it free using one of two cutters (one Russian and one US) brought up by Atlantis. Another array was to be checked because it had problems tracking the Sun.

Another internal node EVA, 1 June, starting 22.05 GMT moved the cone to -Y for Spektr to swing there 19.52-21.50 GMT 2 June. With Spektr in its final location, Kristall returned to the axial face 10 June to await Atlantis, which docked at 13.00 GMT 29 June 395 km over Lake Baikal, creating a ~223 tonne complex. Dezhurov and Shuttle commander Gibson shook hands some 2 hours later and the record 10-man team collected in Mir's core for photographs. Mir was formally handed over to the fresh cosmonauts. Most of the joint mission was devoted to investigating the long-term physiological effects on the Mir 18 crew in Spacelab. Some 400 saliva, urine and blood samples were returned from Mir/Spacelab. Spacelab's treadmill and Lower Body Negative Pressure unit helped to prepare for Earth return. 200 kg of food, clothing and science equipment and >500 kg of water were delivered, and Mir's pressure was raised from 0.92 atmosphere to 1.02 atmosphere. Shuttle provided most of the attitude control, holding the complex for optimuum power generation during working hours. It would then return to the local vertical, with its belly pointing at Earth, as had been planned before Mir's power shortage occurred.

Kristall's hatch was closed at 19.32 GMT 3 July and Shuttle's at 19.48. Solovyov/Budarin undocked from Kvant 1 in Soyuz-TM 21 at 10.55 4 July to photograph Shuttle's 11.10 undocking and flyaround. However, Mir began drifting when its main computer failed so TM21 redocked 5 minutes early at 11.39. Mir remained drifting until they could replace the hardware. Gibson fired his separation thrusters at 12.35 4 July to end the joint phase. Thagard reported that his greatest challenge had been cultural isolation – sometimes going days without talking to a native English-speaker. He broke the US endurance record of 84 days 1 hour at 04.28 6 June, eventually achieving 115 days 8 hours.

Soyuz-TM 22

Launched 09.00.23 GMT 3 September 1995 as the 20th main crew; commander Yuri R Gidzenko and engineer Sergei Avdeyev, with ESA researcher Thomas Reiter (back-up Christer Fuglesang) for the EuroMir-95 mission. For the first time, a visitor trained (3,500 hr) as a flight engineer, working with Mir systems such as thermal control for 2 h daily and qualified for EVA. Fuglesang remained at TsPK until mid-1996 to qualify for landing a Soyuz for the Space Station era; he could also fly EuroMir-97/98. TM22 docked automatically at Mir's forward node 10.29.54 GMT 5 September, and the hatches were opened at 12.01. Most of ESA's equipment was delivered by Progress-M 28, but number 29 arrived at 20.33 GMT 10 October at Kvant 1 with another 85 kg (including Reiter's EVA cassette). 10 kg went up with TM22. M29's hatch was opened 21.57 GMT.

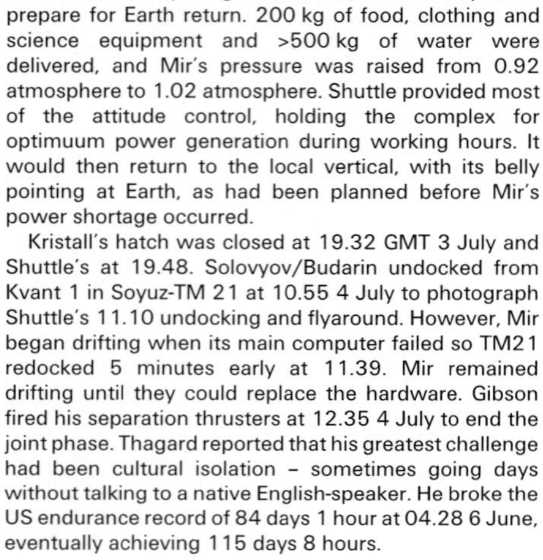

Soyuz-TM 14 training: test pilot Klaus-Dietrich Flade (left) and physicist Dr Reinhold Ewald (right), with Kazakh candidate Talgat Musabayev (flew 1994 aboard TM19) (DLR)

It became clear in late September that TM23's Soyuz booster would be delayed because of funding problems. It was confirmed with TM22's crew 17 October that they would not return until 29 February 1996, instead of 16 January. The extension had been an option before launch and allowed Reiter a second EVA and increased science return. It meant that Progress-M 30 in December 1995 needed to carry 58 kg of ESA science consumables for the extension (ESA paid US$900,000 for the cargo plus US$700,000 for EVA 2). Before the extension, 41 experiments were planned, mostly in Spektr and Mir's core: 18 life sciences, 10 technology, 8 materials science and 5 astrophysics, totalling 450 h, or 4.5 h daily. Germany's TITUS (Tubular furnace with Integrated Thermal analysis Under Space conditions), developed for EuroMir from CSK-1 experience, was used for solidification experiments (see EuroMir-94 mission for CSK-1 failure). Germany's Oberpfaffenhofen facility provided EuroMir operations control.

ESA's first EVA began 11.50 GMT 20 October when Avdeyev/Reiter opened Kvant 2's hatch with the main goal of installing a dust collection experiment on Spektr's exterior. This European Space Exposure Facility (ESEF) was launched attached to Spektr and Reiter added four boxes (delivered by Progress-M 29) opened by remote control: two to collect dust particles (UK's University of Kent), one for environment monitoring and the fourth for electronics. Reiter and the cassettes rode to the site on the Strela arm, operated by Avdeyev, who then climbed along it. ESEF's four covers were removed and retained as they were experiments themselves, having collected radiation data since launch (they were brought back by STS-74). The pair then exchanged cassettes on the nearby Russian-Swiss Komza experiment, finishing at 15.23 GMT, and returned to Kvant 2, closing the hatch at 17.06 GMT. The dust collectors were soon opened for Leonid micrometeoroids. One remained open for most of the time until it was retrieved 8 February 1996 (see below), closing briefly to avoid contamination from approaching and departing craft; the other was opened only during passage through cometary dust streams.

Kvant 1's Vozdukh CO_2-scrubbing molecular sieve was discovered to be leaking glycol coolant 1 November. Kvant 2 carries another, but Mir otherwise then had to rely on the 25 to 30 day supply of chemical canisters aboard the core. Kvant 1's leak was plugged and tested satisfactorily 3 November. Nevertheless, STS-74 still delivered 20 LiOH canisters and an adaptor to plug them into the foreign system. By coincidence, ESA's Biokin scrubber was tested for a week starting 9 November, operating it in an enclosed bag to see if microbes could clean out a preset level of noxious gases. The membrane bioreactor was then frozen and returned on STS-74 for analysis.

STS-74 made Mir's first radial docking at 06.27.39 GMT 15 November 1995, attaching Energia's 2.2 m diameter 4.7 m long 4,087 kg docking module to Kristall so that future dockings will adequately clear Mir's solar wings. Without it, Kristall would have to be shifted to the forward face each time. The hatches were opened at about 09.00 when the pressures had equalised. Atlantis delivered 275 items totalling 968 kg and took away 195 items (370 kg). Delivered were 400 kg of water in containers, 65 kg of air, experimental equipment, fresh food (including vegetables, steak and ice cream), gifts of flowers and sweets and a collapsible classical guitar. It departed with 20 kg of ESA cargo from Reiter's activities: frozen medical samples, furnace material samples and science data on videotape and computer disk. Left behind on the DM were two stowed solar wings, which were planned for deployment by EVAs during 1996 on Kvant 1. Atlantis undocked 08.16 GMT 18 November to circle Mir at 130 m for 45 minutes while the crew took detailed engineering photographs.

An internal 'EVA' by Gidzenko/Avdeyev 19.23-19.52 GMT 8 December in Mir's node moved a docking cone to prepare for Priroda's April 1996 arrival. Progress-M29 departed 09.15 19 December, leaving Kvant 1's port free for M30's 16.30 20 December docking. A 182 minute EVA starting 14.03 GMT 8 February 1996 from Kvant 2 with Gidzenko allowed Reiter to retrieve the two ESEF dust collectors and install a third. They began by removing the old manoeuvring backpack, unused since 1990 and cluttering up the airlock, to an outside storage position on Kvant 2. The crew rode back aboard TM22, which undocked 07.20 GMT 29 February, began its 261.6 second 115.2 m/s retro burn 09.47 and landed 105 km northeast of Arkalyk at 10.42.08, only 2 km off target.

Soyuz-TM 23

Launched 12.34.05 GMT 21 February 1996 with the 21st main crew, Commander Yuri Onufrienko and flight engineer Yuri Usachev, docking with the Kvant 1 port 14.20.36 23 February: Progress-M 30 had vacated this port at 07.26.47 22 February. Soyuz-TM 23 had been planned for 25 December 1995, but funding problems and the unavailability of a Soyuz-U2 launch vehicle required an extension of the TM 22 crew's occupation of Mir. The 351 minute EVA, starting 01.04 15 March, installed a new core crane for moving the new Docking Module solar wing packages to Kvant 1.

The UK's Helen Sharman aboard Mir with (from left) Afanasyev, Artsebarsky, Manarov and Krikalev

Third docking of Atlantis to Mir (STS-76 mission) brought NASA astronaut Shannon Lucid to join the main crew until the STS-79 visit in August. Atlantis docked at 02.34 24 March, attaching to the Docking Module delivered during the previous visit of Atlantis. An EVA lasting for 362 minutes by Clifford and Goodwin exiting from the Shuttle's tunnel, began 6.34 27 March (the first US EVA outside Mir) attached the MEEP to Mir to gather space debris and materials data (to be retrieved by an EVA during the STS-86 mission) and also tested common US/Russian tether hooks and foot restraints. Some 640 kg of water was transferred from the Shuttle to Mir, along with 880 kg of Russian supplies (including gyrodynes and batteries), 740 kg US science equipment and 93 kg of other items. More than 60 per cent of the Spacehab's cargo of 2,010 kg comprised supplies for Mir. Of these supplies, 450 kg of material was returned to Earth aboard Atlantis, including 110 kg science payload. Mir's radio and electric field environment was mapped. Atlantis undocked 001.08 29 March.

The final module planned for Mir finally arrived at the station following launch on 23 April. The 19.7-tonne Priroda remote sensing module carrying a tonne of US equipment docked with the forward port of Mir 12.43 26 April and the following day it was relocated to the portside position. The battery-powered module was the first large module not to carry solar panels (the smaller Kvant 1 had not carried its own panels either), although early in its design a single large solar panel had been planned for the module.

Progress-M 31 was launched 07.04 5 May and docked at the front port 08.54 two days later. Cargo freighter carried 1,140 kg of propellant for Mir plus 1,700 kg of other supplies.

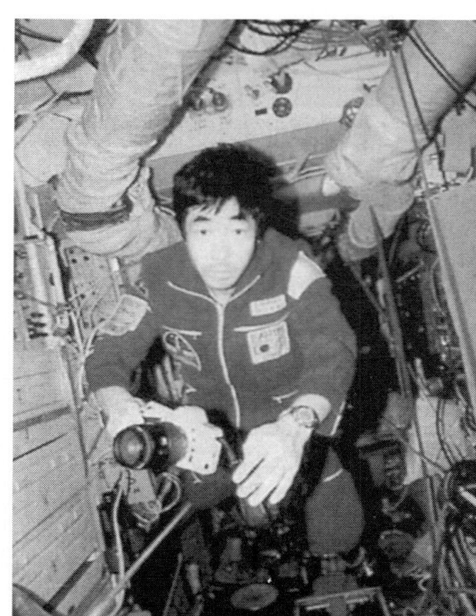

Tokyo Broadcasting Service paid ¥12 million for its journalist, Toyohiro Akiyama, to travel to Mir aboard Soyuz-TM 11

Tognini, Kaleri and Viktorenko (from left to right) aboard Mir during Soyuz-TM 15 (CNES)

On 20 May, Onufrienko and Usachev began a series of EVAs: beginning at 22.50, the cosmonauts exited via the Kvant 2 airlock. They used the telescopic lifting device installed during their earlier EVA to transport a folded solar battery (manufactured in the USA) to outside Kvant 1: the crew also filmed an advertisement for the Pepsi-Cola soft drink company while outside Mir, using a larger-than-life mockup of a newly designed can. The EVA lasted for 320 minutes. Three days later at 20.47 the two men were outside Mir again for 347 minutes, during which they completed the installation of the solar batteries on Kvant 1. At 18.20 on 30 May, the fourth EVA began, during which Onufrienko and Usachev mounted the MOMS-2P multispectral scanner (intended for geophysical studies) outside the Priroda module. They then installed a new handrail outside Priroda, before returning inside Mir after spending 260 minutes outside. The mission's fifth EVA began on 6 June at 16.56 and the cosmonauts used the telescopic crane to move to the Spektr module. They replaced equipment for a Swiss-Russian experiment to measure interstellar gas and then returned to Kvant 2. They installed two US experiments for the detection of micrometeorites and a Russian experiment to measure the effects of the space environment on various materials. The EVA lasted for 214 minutes. The sixth and final EVA of the mission began at 12.45 on 13 June when the prime objective would be to replace a truss on Kvant 1 with a new 5.9 m model. The old truss (named Rapana) would later be attached to the new one, allowing additional experiments to be deployed outside the station. Onufrienko and Usachev also filmed the second part of the soft drink commercial.

On 21 June, it was announced that Onufrienko and Usachev would remain on board Mir until around 30 August, 40 days longer than planned, since the Soyuz-TM 24 mission was being delayed. On 2 and 4 July, two orbital manoeuvres reduced the orbital altitude of Mir by 11 km, to a level at which Soyuz-TM 24 could dock with the station after being launched using a Soyuz-U vehicle. Onufrienko and Usachev continued working with Lucid on Mir: in June they assembled a new Crater-5M furnace in the Kristall module, this being planned for use in producing improved-quality semiconducting materials.

The launch of the Progress-M 32 cargo freighter was cancelled three times in July before it was launched, first on 22 July, then 25 July and finally on 27 July. Launch took place at 20.00 on 31 July and the freighter docked at the front port of Mir at 22.04 2 August. In the meantime, Progress-M 31 had undocked 16.45 on 1 August and had been de-orbited later that day. The 2.4 tonnes of supplies carried to Mir included 400 kg of propellant, 300 kg of food and 340 kg of equipment for the forthcoming French Cassiopeia mission. At 09.34 on 18 August, Progress-M undocked from Mir to free the port for the new Soyuz-TM, but it remained in orbit to redock after the return of Soyuz-TM 23.

After the arrival of Soyuz-TM 24 on 19 August, Onufrienko and Usachev worked with the new crew on the French Cassiopeia experiments. Soyuz-TM 23 undocked from Mir at 04.20 2 September with Onufrienko, Usachev and Andre-Deshays (who had been launched aboard Soyuz-TM 24) and landed south west of Akmola at 07.41 the same day.

Soyuz-TM 24 (Cassiopeia)

Russian Space Agency informed CNES that the planned launch of Soyuz-TM 24 with French spationaut Claudie Andre-Deshays (for the Cassiopeia mission), originally planned for June 1996 and then delayed until 6 to 22 July would be further delayed until 14 to 30 August. A further change to the mission was announced on 7 August when the prime Russian crew of Gennady Manakov and Pavel Vinogradov would be grounded because doctors had found minor heart troubles with Manakov. Rather than split the Russian team, RSA decided to ground the prime crew and fly with the back-ups, Valery Korzun and Aleksandr Kaleri. The prime spationaut crew assignment would not be changed. At the same time, it was announced that the Soyuz-TM 24 mission would begin on 17 August.

At 13.18 GMT on 17 August 1996, Soyuz-TM 24 was finally launched using a Soyuz-U vehicle to begin the 22nd residency aboard Mir. The spacecraft was manually docked with Mir at the front port two days later at 14.50.

On board Mir the French Cassiopeia research programme included five main experiments. Cognilab was a laboratory which is dedicated to neuroscience and robotics and was used to study the mechanisms used by the central nervous system and to evaluate the adaptive capabilities of the brain while in a microgravity environment. Fertile was a biology experiment, devoted to the study of the development of embryos of amphibians under microgravity conditions. Castor was a technical experiment, intended to measure the vibrations of the Mir orbital station complex, analyse its modes of deformation and study the dynamic behaviour in the microgravity environment of a trellis equipped with shock absorbing systems. Physiolab was designed to study the cardiovascular system with two planned goals. First, the study of the fundamental mechanisms that regulate blood pressure and second, to be used as a tool for medical monitoring. Finally, ALICE-2 (Analyse des LIquides Critiques dans l'Espace) covered scientific research into the behaviour of fluids in microgravity. The French had paid FFr82 million for the mission.

With the completion of her part of the mission, Andre-Deshays returned to Earth with Onufrienko and Usachev aboard Soyuz-TM 23 on 2 September (see Soyuz-TM 23 above).

On 3 September at 09.35, Progress-M 32 redocked with Mir, this time at the rear port, which had been vacated by Soyuz-TM 23.

The fourth visit of Atlantis to Mir for the STS-79 mission began on 19 September, with the Orbiter docking at 03.13. Atlantis carried more than 2.1 tonnes of supplies to Mir. Shannon Lucid ended her residency aboard Mir and her place was taken by John Blaha. Atlantis undocked from Mir 23.33 23 September and returned to the Kennedy Space Center three days later at 12.13.

Following the departure of Atlantis, the Mir cosmonauts and astronauts settled into the routine of space station operations. They undertook experiments using the astrophysical equipment aboard Kvant 1, the Mariya magnetic spectrometer was used for measurements of high-energy charged particles and Earth photography using the Priroda-5 equipment was undertaken.

The launch of Progress-M 33 had been scheduled for 15 October, but because there was no Soyuz-U launch vehicle available for that date it was delayed until November. Launch finally came at 23.21 on 19 November and docking at the rear Mir port took place at 01.01 22 November. In the meantime, Progress-M 32 had undocked from Mir 19.44 21 November and was de-orbited later that day. Progress-M 33 brought about 2.4 tonnes of supplies to the station.

On 26 November, preparations began for the first EVA during this residency on board Mir as Korzun and Kaleri checked the EVA suits. On 2 December at 15.54, the Kvant 2 airlock hatch opened and the two cosmonauts began their EVA to install a new solar battery: the EVA lasted for nearly 6 h. A second EVA took place on 9 December, lasting for more than 6 h (an hour longer than planned). During the EVA, the cosmonauts completed the connections of the new solar battery and installed a new Kurs antenna on the docking Module, intended to ease future dockings of the Atlantis shuttle orbiter.

The New Year of 1997 saw the Mir crew looking forward to the arrival of Atlantis for its fifth visit to the station. Launched 12 January on the STS-81 mission, Atlantis docked with the station at 03.55 15 January. While the Russian and American crews were on board Mir they undertook joint biomedical experiments, assessing adaptation to the space environment and radiation studies. A further crew rotation took place, as Blaha prepared to return to Earth aboard Atlantis and Jerry Linenger prepared for four months in orbit. Atlantis undocked from Mir 02.16 20 January and landed at the Kennedy Space Center two days later at 14.22.

In preparation for the forthcoming Soyuz-TM 25 launch, Progress-M 33 undocked from Mir 12.14 6 February and remained in orbit for a planned redocking after the return of Soyuz-TM 24. At 16.28 on 7 February, Soyuz-TM 24 containing Korzun, Kaleri and Linenger undocked from the front port of Mir and redocked at the rear port 23 minutes later.

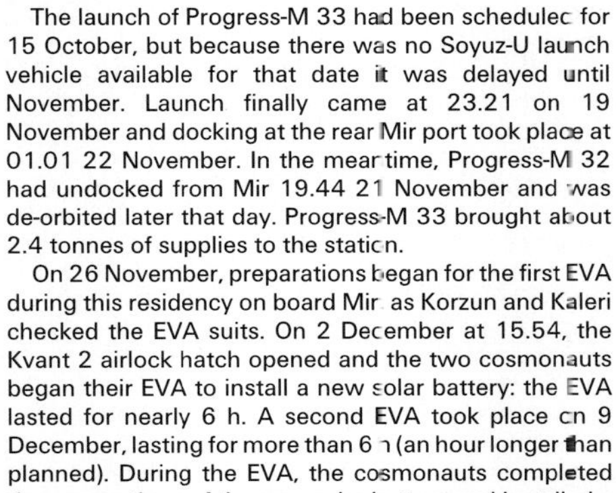

Valeri Poliakov takes a blood sample from Ulf Merbold (ESA) ***1995***

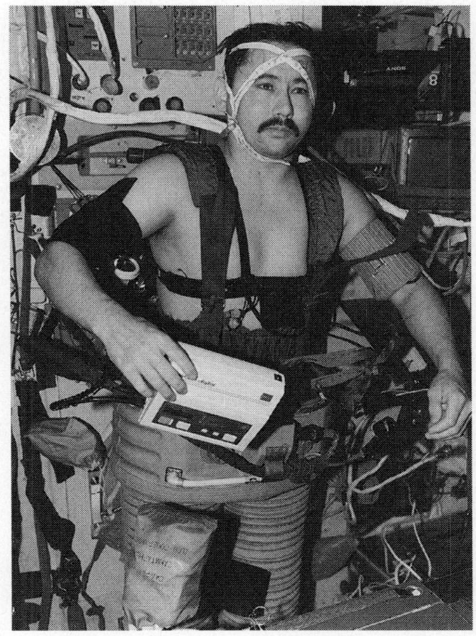

Preparing for return to Earth, cosmonauts spend increasingly long periods in the Chibis lower body negative pressure suit. Shown is TM19's Talgat Musabayev ***1995***

With the arrival of Soyuz-TM 25 on 12 February, the Soyuz-TM 24 residency drew to a close. Joint work with the new Russian crew and the German researcher Ewald was undertaken before the return to Earth. Soyuz-TM 24 with Korzun, Kaleri and Ewald undocked from Mir 03.25 2 March and landed near Arkelyk at 06.44.

Soyuz-TM 25 (Mir-97)

The launch of Soyuz-TM 25 with a Russian-German crew had originally been planned for 15 December 1996, but delays in the delivery of a Soyuz-U launch vehicle for the flight meant that on 7 November that year, it was announced that the flight would be delayed until February 1997. Soyuz-TM 25 was actually launched 14.09 GMT 10 February 1997 with two Russian cosmonauts, Vasily Tsibilev (Cdr), Aleksandr Lazutkin (flight engineer) and German cosmonaut Reinhold Ewald.

Docking at the front port of Mir came two days later at 15.51. Once more the docking had to be conducted manually after the automatic Kurs system failed. Russia noted when reporting the docking that there would be only one further flight using Kurs, presumably Soyuz-TM 26.

During the handover period, there was a fire on board Mir. While there are more than three people on Mir, the oxygen supply is maintained by the burning of oxygen generators using lithium cartridges: when combusting, the cartridges release oxygen into the station's atmosphere, a process that is accompanied by a great deal of heat. During the evening of 23 February, the fourth and final cartridge of the day was burning. Without warning, the cartridge exploded and some of the panels inside Kvant 1 caught fire. The crew used three fire extinguishers to fight the fire, working in gas masks for 15 minutes. When the fire had been put out, the crew dismantled the air line and retreated to their respective Soyuz-TM spacecraft, while ground controllers considered the emergency evacuation of the station. It was decided that the mission could continue, although the cosmonauts continued to wear breathing masks until 25 February as they conducted repairs. The Russians noted that this was the second time that a lithium cartridge had caught fire: Valery Poliakov had to extinguish a similar fire in October 1994.

Ewald returned to Earth on 2 March aboard Soyuz-TM 24, leaving Tsibilev and Lazutkin in orbit with Linenger. On 4 March, the attempt to redock Progress-M 33 at the rear port of the station failed and the spacecraft had too little propellant available for further docking attempts: the spacecraft was de-orbited on 12 March. On 19 March, the primary angular rate sensor in the Spektr module on Mir failed and 3 minutes later, the motion control system computer switched to a backup. In the interim, the gyrodynes started manoeuvring Mir in all three axes at rates beyond the capacity of the gyrodynes to compensate. The oven switched off the attitude control system relaxing the station into 'free drift' and used thrusters to stabilise it. With Mir in gravity gradient attitude the crew installed a spare Omega attitude sensor in the Kristall module and its sensor served as a backup.

On 22 March, the United States announced that they

From left: Malenchenko, Musabayev, Viktorenko and Kondakova, during EuroMir 94. At bottom right is the Chibis suit (ESA) ***1995***

had achieved a year-long presence in space, thanks to astronauts working on board Mir: however, since Mir had been permanently occupied by Russians since September 1989, the American record paled into insignificance.

As April approached it became clear that Mir was showing its age: launched more than eleven years earlier and expected to remain operational only until around 1992, the core module's life support system was causing serious problems. In November 1996, there had been problems reported with the recycling of water, as 'industrial' water from the waste management system had begun to enter the station's cooling system. The two Elektron oxygen supply systems broke down in March 1997, meaning that the crew had to rely on burning the lithium cartridges to maintain an oxygen supply.

The launch of Progress-M 34 came at 16.04 on 6 April and it docked at the rear port of the station at 17.30 two days later. Among the supplies being taken to the station were spare parts for the repair of the two Elektrons, a fresh supply of lithium cartridges, which could supply oxygen in the meantime, and 50 kg of fresh oxygen. The day following the docking, the Elektron spare parts had been unloaded from Progress-M and the crew began their repairs of the Elektrons. By the end of the month the Elektrons were once more operating.

One of the flight tasks conducted by NASA astronaut Linenger was participation in a year-long sleep study involving three experiments put together between Pittsburgh University, Harvard University and the University of Toronto together with Moscow's Institute of Bio-Medical Problems. The North American work was sponsored by the Canadian Space Agency and co-ordinated between NASA and the Russian Space Agency. The results of the research were said to be important for the clinical understanding of several disorders such as stress and depression which it appeared to the experimenters could be linked to altered immune functions brought about by sleep disorder. The difficulty astronauts face in matching the circadian rhythm cycle of living things on Earth seriously affects their immune systems. In attempts to understand this, Linenger was to be assessed at 81 and 141 days and again when he returned to Earth and the same process would be applied to Russian cosmonauts Lazutkin and Tsibilev as well as their backups, Musabayev and Budarin.

Linenger was one of the first in space to have use of the new Orlan M pressure suit, brought by Progress M 34, which weighs 80 kg and is designed to afford easier access through the rear. Linenger used this suit to conduct an EVA beginning 05:10 UT, 29 April, and ending 4 hours 58 minutes later. During this, Linenger collected the Partial Impact Experiment and the Mir Sample Experiment from their respective locations on the outside of the Kvant-2 module where they had been placed in 1996 and installed the Optical Properties Monitor Experiment on the Docking Module at the end of the Kristall module. Linenger was moved between locations by the Strela telescopic boom and worked with Tsibilev to place the Benton radiation dosimeter on Kvant-2. In the first two weeks of May, the crew tidied Mir and stowed equipment ready for the arrival of Shuttle Atlantis (STS-84) and its seven-person crew including Michael Foale, replacement for Linenger. While the Elektron oxygen generators were working satisfactorily, the 50 kg of oxygen brought up by Progress-M 34 was to be used first. There was still a problem with fumes getting into the station atmosphere.

Launched on 15 May 1997, Atlantis docked with Mir for the sixth time at 14:33 UT two days later, bringing more than 2,800 kg of food, water and experiments in addition to a new Elektron unit installed as a backup to the main device on Kvant-2. During joint operations involving the 10 Mir-Shuttle crewmembers, the KOB-2 cooling loop was reactivated although a leak in a supplementary cooling loop, the VGK, awaited repair. With Linenger in Atlantis and Foale left aboard Mir, the Shuttle undocked at 01:04 UT 22 May and tested a European laser docking sensor from a distance of about 900 m, returning to Earth two days later. Foale integrated quickly with Lazutkin and Tsibilev aboard Mir and helped track down various coolant leaks and check out experiments before packing redundant equipment in Progress-M 34 prior to its departure from the rear docking port.

M 34 undocked from the Kvant port at 10:22 UT 24 June under a plan to have Progress perform a one-day autonomous flight for a manual re-docking test. Next day, out of radio contact with the ground, Tsibilev attempted to re-dock the Progress using the dual-band controllers of the TORU system and a TV link when the cosmonaut noticed that the vehicle was deviating from its intended course. It missed the docking port, travelled down the side of Kvant and the main Mir station block and struck one of the four solar arrays on the Spektr module causing it to tilt and collide with one of the radiator panels sending a pronounced shock throughout the complex. Physical damage breached the hull causing a pressure leak audibly detectable to the crew, who quickly severed electrical cables across the open hatch to Spektr and sealed the module to secure the leak. Damage to the solar panel cut Mir electrical power by 50 per cent but this was partially restored later that day through manual tracking of the arrays. However, the Elektron in Kvant-2 was shut down and the Vozdukh CO_2 removal system turned off. For the time being, CO_2 levels would be kept to acceptable values by lithium hydroxide canisters brought up by STS-84. Meanwhile, discussions began about entering Spektr wearing space suits to connect electrical cables from the solar arrays to batteries elsewhere on Mir and restore power.

Early on 27 June, the Kvant-2 batteries ran down and thrusters aboard Soyuz TM-25 were used to adjust station attitude and by building battery power back up it was hoped that the gyrodones could be brought back on the following day. Next day with power being slowly restored, the Vozdukh CO2 scrubber was turned back on and the crew activated the new Elektron unit brought aboard by the STS-84 crew. By early July, plans were made to bring special repair equipment to Mir in Progress-M 35 and for the two Russian cosmonauts aboard Mir to perform the suited inspection of Spektr with access to all other modules sealed and Foale in Soyuz TM-25. Progress-M 34 was de-orbited on 2 July and its successor was launched at 04:11:54 UT on 5 July, carrying 2.4 tonnes of supplies and equipment. M 35 docked at the Kvant port at 05:59:24 UT 7 July, a day in which the Mir crew succeeded in switching on 10 of the 11 operational gyrodones (one had not been working for some time). Most of 8-10 July was spent unloading the Progress cargo-tanker but conditions in the Kristall and Priroda modules was getting worse. The crew used towels to mop up condensation caused by high humidity and Foale abandoned his traditional sleep location in the docking node and moved into Kvant-2.

On 13 July, Tsibilev displayed a cardiac arrhythmia, not unknown among Mir crewmembers (including NASA astronauts) but one which under the stress of the circumstances caused concern. On 16 July, during lengthy cable disconnection procedures, a line to the Omega attitude control system computer was inadvertently removed causing the station to slowly lose orientation on the Sun and most of the power from the solar arrays. When the crew woke next day to find the situation they manually rotated the Kvant-2 module arrays and gradually re-charged the batteries. Once again the crew had to go through lengthy power-down procedures similar to those following the collision three weeks earlier and the internal inspection of the crippled Spektr module was postponed. On 19 July, the crew were informed that the suited activity would be performed by Soyuz TM-26 cosmonauts Solvyev and Vinogradov, due to replace them in early August.

Soyuz TM-26

Launched at 15:36 UT on 5 August 1997, Solovyov and Vinogradov carried 96.5 kg of extra equipment for Mir, including 40 kg of oxygen along with tools and EVA equipment, instead of French astronaut Leopold Eyharts who would fly to the space station on Soyuz TM-27. Next day, at 11:46 UT, Progress-M 35 was undocked from the aft port and removed to a safe station-keeping distance to make room for the unmanned vehicle. At 17:02 UT on 7 August, Solovyov and Vinogradov steered Soyuz TM-26 to a manual docking at the aft Kvant port. Cosmonauts Tsibilev and Lazutkin returned to Earth in Soyuz TM-25, touching down 168 km from Dzezkagan, Kazakhstan, at 12:17 UT on 14 August, leaving Foale to support Solovyov and Vinogradov in repairs to the Mir complex. Next day, the Mir crew conducted a fly-around inspection of the station exterior from Soyuz TM-26, undocking at 13:30 UT and re-docking at 14:14 UT. Progress-M 35 was re-docked to the aft port at 12:53 UT where it would remain as a refuse bin until shortly before the arrival of Progress-M 36 in October.

The much delayed suited internal inspection of Spektr began with the opening of the hatch at 11:10 UT on 22 August with Foale in the Soyuz. Retrieving a vacuum cleaner, Foale's personal ditty bag and effects and after having video documented the state of the interior and installed the power line patch, Solovyov and Vinogradov returned to the node and at 14:30 UT, closed the hatch on the unpressurised module. With more power flowing, systems were turned on, sequentially over the next several days, in Kristall, Kvant-2 and Prioroda modules respectively. For the next two weeks the station complex was slowly restored to its former power levels and plans were made for a Spektr inspection EVA involving Foale in attempts to locate the rupture which had caused depressurisation after the impact.

The inspection EVA commenced at 01:07 UT on 6 September when Foale opened Kvant-2's airlock hatch. In the early part of the excursion, Foale assisted Solovyov as he was manoeuvred by the Strela crane into position over Spektr. Cutting tools were used to remove thermal blankets to inspect the hull before moving to the damaged solar arrays where they installed handrails to aid later cosmonauts on full-scale repair tasks. Moving them to undamaged Spektr arrays they eased the panels into a new orientation to increase solar exposure and improve the electrical supply. Foale removed the Benton dosimeter left by Linenger before the EVA ended after exactly 6 hours, Solovyov's eleventh space walk.

On 8 September, the crew were busy correcting a major failure with the Mir main command computer, vital for station orientation and maintain the correct

The 8 kg Sokol-KV pressure suit flew for the first time on Soyuz-T 2. It was improved to the KV2 for Soyuz-TM

Solovyov and Budarin close Kristall's hatch before STS-71's departure. Note the docking system petals at top/bottom, and the docking target (NASA) ***1996***

attitude alignment for optimum solar energy, but general maintenance was now taking increasing slices of work time. A new fluid unit for the back-up Elektron brought up by STS-84 was installed in Kvant-2, the old one being returned aboard Atlantis after its late-September visit and water purification and elements on the condensate recovery system were replaced. On 14 September, the main computer failed again and power requirements were cut. In an operation similar to that when the computer failed in 1995, the cosmonauts stripped two failed machines to make one good working computer, taking the central data processing unit from the failed unit and the central processing unit from the back-up. Two days later, the solar power was back to normal and the gyrodones were switched back on, one by one. However, the planned launch of Progress-M 36 on 20 September was postponed to allow time for a new computer to be produced and delivered by the cargo-tanker in October. The third computer failure on 22 September and again the gyrodynes stopped and systems were cut until the crew had repaired the faulty item and begun to restore the station once more.

Shuttle Atlantis was launched to Mir at 02:34:19 UT on 26 September at the start of STS-86 during which Russian cosmonaut Titov and NASA astronaut David Wolf would be left at Mir and Foale would be returned to Earth. Docking took place at 19:28 UT on 28 September and over the next two days, the crew, including veteran visitor Jean-Loup Chretien, offloaded 3,160 kg of supplies and equipment to the Mir complex. A late addition to the cargo manifest was a new interim computer for Mir mounted to the aft bulkhead on Spacehab via the Module Vertical Access kit. The computer was installed by Solovyov and Vinogradov while Parazynski and Titov performed a 5 hour 1 minute EVA from Atlantis to retrieve four Mir Environmental Effects Packages and test the Simplified Aid for EVA Rescue (SAFER). They also installed a 121 kg Solar Array Cap to the Docking Module which cosmonauts could use on a future EVA to seal the leak on Spektre. With astronaut Foale on board, Atlantis undocked from Mir at 17:28 UT on 3 October and returned to a landing at 21:55 UT.

At 15:09 UT on 5 October, the Progress-M 36 supply ship was launched toward Mir and, after Progress-M 35 was undocked at 12:04 UT two days later (following some unexpected technical difficulties), a docking was achieved at 17:08 UT on 8 October. Progress had 2,500 kg of equipment including a back-up Motion Control Computer, 1,000 kg of propellant and 100 litres of water which the crew, including Wolf, offloaded to Mir. On 20 October, Solovyov and Vinogradov performed a second suited inspection of Spektr's interior and in an operation lasting 6 hours 38 minutes routed cables to partially restore orientation control for the remaining healthy solar arrays. Two more maintenance EVAs to install a new solar array brought up aboard STS-84 were conducted over the next few weeks.

On 3 November, Solovyov and Vinogradov went outside to remove the old array from Kvant-2 with the aid of the Strela device. While outside they released a small model of Sputnik 1. On trying to return through the hatch, they found it buckled, requiring them to enter through the second Kvant-2 module and use its inner hatch as the pressure seal. In all the EVA lasted in excess of 6 hours. The second EVA 3 days later successfully installed the replacement arrays to Kvant-2 but they had to be manually deployed. After installing 10 clamps around the failed airlock hatch, the 6 hour 17 minute EVA ended.

On 14 November, the Mir complex sustained another major setback when ground controllers failed to command rotation of the Kvant-2 solar arrays to prevent a power surge. This caused the main computer to fail again, once more requiring the crew to switch off all non-essential systems including the Vozhdukh and Elektron systems. To supplement diminished power, the crew transferred charged batteries from the Kristall module to the main core module to restore power to five gyrodynes which spun down after the failure. Following yet another failure on 21 November, the crew began to replace the interim computer brought up in Atlantis with the completely new computer delivered by Progress-M 36 and the recently removed Kristall batteries ensured power levels prevented the gyrodynes spinning down as they had on earlier occasions. Despite frequent problems with systems the crew were able to conduct some science tasks and plans were laid for two space walks in January 1998.

On 17 December at 06:02 UT, Progress-M 36 was undocked at the start of an exercise to deploy a small German Inspector sub-satellite designed as a test precursor to a fly-around TV inspection satellite for use with the International Space Station. Inspector was released from the cargo section of Progress at 07:37 UT but a software problem prevented it responding to commands from a laptop computer on Mir although it did take a few TV shots of the complex. Next day, the crew reported a leak of freon gas from the air conditioning unit and concerns were expressed over the quantity of water from condensation gathering because of temperature differences between the modules and the base block. Progress-M 36 was commanded to destructive re-entry over the Pacific Ocean at 13:20 UT 19 December and at 08:45 UT next morning Progress-M 37 was launched from Baikonur. It docked with the aft Kvant-1 port at 10:22 UT on 22 December.

The first day of 1998 was marked by another main computer failure triggering the now routine power-down procedure, reversed when a new central exchange unit was hooked up. Solovyov and Vinogradov conducted an EVA lasting little more than 4 hours during which they retrieved the OPM experiment and examined the outer airlock hatch before successfully locking all latches. By 12 January it was found to be slowly leaking again. On 14 January, Wolf and Solovyov conducted a 3 hours 52 minutes EVA to use the NASA Space Portable Spectroreflectometer, a device for measuring the effects of the space environment on several materials in different places.

The seventh and penultimate Shuttle-Mir docking began with the launch of STS-89 on 23 January carrying Andrew Thomas, the last NASA astronaut to serve aboard Mir. Endeavour docked with Mir at 20:14 UT next day. More than 3.6 tonnes of equipment and supplies were transferred to Mir over 5 days of joint activities, returning with Wolf after separating from the Docking Module at 16:57 UT 29 January leaving Thomas to await the arrival of a new crew Baikonur To vacate the aft docking port, Progress-M 37 was undocked at 12:53 UT on 30 January and pulled back to a safe station-keeping distance for re-docking after the next manned visit.

Soyuz TM-27

The 34th manned spacecraft to dock with the Mir space station, Soyuz TM-27 was launched at 16:33:20 UT on 29 January 1998 and mated to the Kvant-1 port at 17:54 UT two days later. The mission was a repeat of the Cassiopeia flight of 1996 with Pegasus experiments occupying the full attentions of CNES astronaut Leopold Eyharts using cosmonauts Talgat Musabayev and Nikolai Budarin as test subjects when time permitted. A dubious 'first' of a questionable nature, Solovyov and Vinogradov appeared live on TV selling space pens through a US cable shopping channel before returning home, with Eyharts, in the Soyuz TM-26 spacecraft.

After undocking from the front port of the Mir complex at 05:52:50 UT on 19 February, the three men landed at 09:10 UT, 30 km south-east of Arkalykh in the middle of a blizzard. Next day Musabayev, Mudarin and Thomas boarded TM-27 and, at 08:47 UT, undocked from the aft port and backed away 60 m while Mir rotated 180° facilitating a manual docking to the forward port at 09:32 UT. Using the Kurs automated system Progress-M 37 re-docked to the aft port at 09:42:27 on 23 February.

In an EVA attempt on 3 March, Musabayev and Budarin were unable to release all 10 manual latches on the troublesome Kvant-2 airlock module hatch because the tool they needed could not be found and the attempt was aborted. Science was interspersed with maintenance duties although on 13 March, a problem with the air conditioning unit cleared up when new parts brought by STS-89 were fitted. Science tasks included operations with the MOMS-2P multispectral scanner on the Priroda module. At 22:45:55 on 14 March, the Progress-M 38 cargo tanker was launched from Baikonur. At 19:16 UT next day, the Progress-M 37 ship was undocked from the aft port allowing its successor to dock at 00:31:17 on 17 March. Musabayev took over manual control with the TORU system after the automatic Kurs unit failed.

On 27 March, the two cosmonauts spent several hours in the Kvant-2 airlock replacing broken latches on the troublesome hatch and freeing up others in preparation for several major EVAs to repair the Spektr solar arrays. They would also search for possible leaks on the module itself and change a VDU thruster module on the outside of the station. The first EVA began at 13:35 on 1 April and lasted 6 hours 40 minutes during which Musabayev and Budarin worked at the base of the array using the Strela access device. A second EVA took place on 6 April but it ended after 4 hours 23 minutes when Mir appeared to ground controllers to be drifting off alignment with the Sun, and jets on the Priroda module had to be used to restore attitude. The third EVA started at 09:55 UT on 11 April when Musabayev and Budarin exited the Kvant-2 module, moved up the Sofora girder to remove the VDU block and pushed it away into space from where its independent orbit would decay within a year. They returned to Kvant-2 after 6 hours 25 minutes but again experienced problems with the hatch latches.

The fourth EVA in this work series started at 07:40 UT on 17 April and in 6 hours 33 minutes set up the Sofora girder for attachment of the new VDU which was raised by ground command from its special mounting on the modified Progress cargo ship. A science support girder, Rapana, which the crew had originally planned to jettison was stowed for possible later use and the VDU was locked in place at an angle to facilitate further work on a later EVA. The fifth EVA got underway at 05:34 UT on 22 April and the new VDU was attached to the Sofora plate with several cables. It took several hours to complete the assembly, 14 m long, with the VDU block on top of the Sofora in its working position. Inside Mir, Thomas activated heaters to prepare the VDU for operations. The 6 hour 21 minute EVA brought the total amount of work time outside the complex during April 1998 to 30 hours 22 minutes.

About this time, space officials began to plan the demise of Mir and with it the historic sequence of Soyuz missions to the orbiting complex that had kept it permanently manned since 1989. NASA wanted the Russians to end the Mir programme to ease the pressure on limited funds for their contribution to the International Space Station (ISS). Efforts to find commercial sponsors began in the hope of maintaining a human presence aboard the station until the ISS could be permanently manned early in the next decade. The final Shuttle visit to Mir – STS-91 – would complete what NASA called Phase 1 of the ISS, but the Russians wanted NASA to help them maintain Mir by using the Shuttle for two additional visits. NASA refused to consider this and the Russians postponed plans to lower Mir in May 1998, essential for a controlled de-orbit to an unpopulated part of the Pacific Ocean in 1999.

Progress-M 39 was launched at 22:13 UT on 14 May carrying 1.5 tonnes of water, food, scientific equipment, instruments and personal items for the crew. Progress-M 38 was undocked from the aft axial port at 18:41 UT next day clearing the way for its successor to dock at 23:51 UT on 16 May. Routine servicing and technical checks followed the transfer of cargo and equipment from Progress but at the end of the month the main core computer once again failed, Mir lost Sun-lock and the manoeuvring engines on Soyuz TM-27 were used to control the complex. This occurred three days before the launch of STS-91, the last Shuttle visit to Mir, coming to deliver supplies and to collect NASA astronaut Andrew Thomas. On board with five NASA astronauts was the veteran cosmonaut and head of the Russian management team on the Shuttle-Mir programme, Valeri Ryumin, who would conduct a full inspection of Mir and report on its potential future use.

STS-91 was launched at 22:06:18 on 2 June when mission managers were assured that the computer problem aboard Mir had been solved and attitude control, an essential prerequisite for docking, was assured. After a nominal rendezvous, Discovery docked

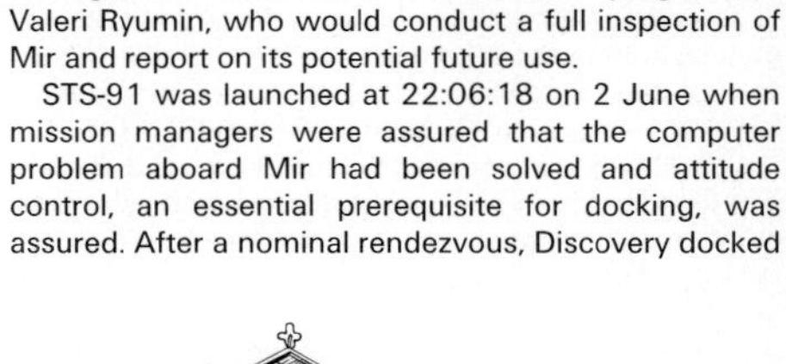

Soyuz-T spacecraft cutaway. TM's appearance is almost identical

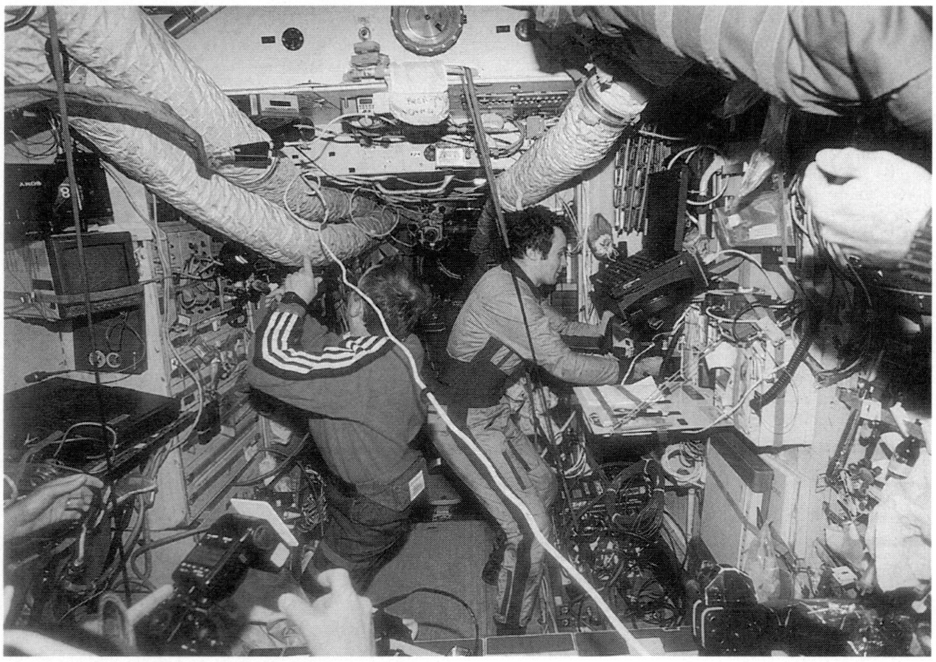

Solovyov (left) and Dezhurov work inside Mir's core (NASA) ***1996***

with the DM at 16:58 UT two days later carrying 500 kg of water and 2,100 kg of cargo experiments and supplies. The visit was a busy one with removal from Mir and stowage in Discovery of equipment supporting the eight Shuttle visits, which had begun with STS-71 in September 1995. Ryumin performed his inspection and videoed the interior, claiming that the station was undermanned and that it needed at least six cosmonauts on permanent duty to cope with all the work required. Leaving Musabayev and Budarin aboard Mir, Thomas said his farewells and joined Ryumin and the other five NASA astronauts aboard Discovery for the ride home, undocking at 16:01 UT on 8 June, landing at KSC on 12 June.

Three days later, mission control formally announced that the orbit of Mir had been lowered by 6 km to start the process whereby a specially configured Progress vehicle could, at some date in 1999, perform a retrograde burn to destruction in the atmosphere. While the crew set about the busy scientific and engineering tasks aboard Mir, the fate of the complex was discussed at length and resistance to its demise stiffened as potential backers supported the possibility of a commercial purpose to its continued operation in orbit. This even extended to the Menatep bank which had been sending souvenirs up to Mir for some potential commercial gain in the future. Severe funding problems were causing difficulties with mission life planning. After a representative from the Russian Finance Ministry stepped in, the desire to keep Mir aloft until the end of 1999 was summarily cut by six months.

This, plus difficulties with paying wages and utilities bills at the Baikonur cosmodrome in Kazakhstan, delayed the launch of the next Progress and, in turn, the next crew to Mir. Finally, it was agreed that the next Soyuz crew would be launched in August 1998 with the last crew going aboard in February 1999 before their return, and de-orbit of the station, in June 1999. The launch of Progress-M 40 would be delayed by the inability of the space agency to buy a launch vehicle until later in the year.

Soyuz TM-28

In preparation for the next Mir visit, Progress-M 39 was undocked from the aft port at 09:28 UT on 12 August and displaced several kilometres to a safe station-keeping position. On 13 August 1998 at 09:43 UT, the Soyuz TM-28 carrying Padalka, Avdeyev and Baturin was launched from Baikonur and only a relatively minor problem with the Kurs docking system marred a smooth link-up, which took place under Padalka's control at 10:56:52 two days later. Once a presidential aide, Yuri Baturin would conduct physics, technology and medical experiments during his relatively brief stay aboard Mir. During the 10 days of changeover, the three cosmonauts transferred several new items including a new furnace for materials processing activities. With five crewmembers, it was possible to perform critical maintenance tasks, especially on the thermal control system and the Vozdukh CO_2 removal equipment. Musabayev, Budarin and Baturin undocked the Soyuz TM-27 from the forward port at 02:05 UT on 25 August and landed 40 km from Dzhezkazgan 3 hours 17 minutes later.

At 16:44 UT 27 August, Padalka and Avdeyev withdrew from Mir in Soyuz TM-28 while the station complex rotated 180° to allow the spacecraft to dock at the forward port 23 minutes later. Finally, two days later, Progress-M 39 was returned to a docking with the aft port at 05:35 UT where it would remain until its successor came up with fresh cargo and equipment. At 20:00 UT 15 September, Padalka and Avdeyev performed a suited inspection inside the depressurised Spektr module and within minutes had successfully reconfigured cables to allow orientation of three operable solar panels on the crippled module; the fourth had been damaged beyond repair.

Plans to send Progress-M 40 to Mir were seriously delayed by funding difficulties and activities aboard Mir were severely curtailed by the absence of the supply ship. Most of October 1998 was spent trading maintenance time with repeating experiments already carried out. Signs of friction with senior space agency managers flared when Padalka and Avdeyev sympathised with striking science and engineering workers in Moscow but declared they would continue working as usual, much of which was devoted to repairs and continued tinkering with station systems.

Progress-M 40 was finally launched at 04:17:57 UT on 25 October carrying the second Znamya 'space-mirror' device similar to the one carried in 1993. The plan was to unfurl the 25 m reflector after Progress had undocked following the transfer of cargo and fresh supplies to Mir. In preparation for its arrival, on the same day at 23:03 UT, Progress-M 39 was undocked from the aft port and the cosmonauts rehearsed the manoeuvres they would apply to M 40 during the Znamya experiment. M 40 docked with the aft port at 05:43:42 on 27 October and two days later, M 39 was de-orbited over the Pacific Ocean.

To retrieve particles from the Leonid meteor shower the two cosmonauts performed an EVA starting at 19:24 UT 10 November, during which they attached sets of equipment to the exterior of the complex including a US-built experimental solar array, a commercial experiment paid for by the development company. One of their first tasks was to release a small miniature of Sputnik 1 built by French students and Energia similar to the one released one year and one week earlier. The EVA ended after 5 hours 54 minutes. With mixed emotions on 20 November, the crew acknowledged the launch of Zarya, Russia's contribution to the International Space Station and the first element sent into space. The ISS would remove all need to maintain Mir in orbit but Energia claimed potential investors were again discussing commercial uses for the station.

On 24 December, station perigee was raised by 14 km to preserve it for the last scheduled visit to Mir, placing it in a 373 × 361 km orbit. Early in January 1999 it was announced that the Soyuz TM-29 flight would be launched on 20 February, the 13th anniversary of the launch of the Mir core block. This last scheduled crew would include the first Slovak, Ivan Bella and the French cosmonaut J-P Hagniere flying as the first non-Russian flight engineer on Soyuz. If investors could be found, the mission would last until 21 August anticipating further visits but if talks failed, the crew was to return on 1 June. Three weeks later, Prime Minister Primakov declared that if investors could be found, Mir would continue in orbit for a further three years.

On 27 January, Progress-M 40 was used again to raise the orbit of Mir and on 2 February at 09:59:32 UT, it undocked from the aft port and backed away 400 m from the station in preparation for the Znamya tests which were to reflect sunlight to specified locations on Earth. At 11:33 UT, commands were sent to deploy the mirror assembly, but as it began to unfurl, a portion of the reflective fabric snagged a Kurs docking antenna which controllers had unwisely commanded to deploy. The material failed to break free despite thruster firings to jerk it away from the antenna and Progress-M 40 was commanded to de-orbit at 10:16 UT next day. At 11:23 UT on 8 February, Soyuz TM-28 carrying Padalka and Avdeyev backed away from Mir as it rotated for a re-docking at the aft port, completed 16 minutes later.

Just three days later, it was announced that commercial investors had gone away and that the Soyuz TM-29 crew would be the last aboard Mir. Now the plan was to use the propulsion system on Progress-M 41 to de-orbit the station on 28 August just five days after its crew returned home, little more than one week short of a record 10 years of continuous human occupation.

UNITED STATES OF AMERICA

Apollo

Current status

Use of the laser reflectors deposited by Apollos 11/14/15 in 1969-72 continues for measurement of the Earth-Moon distance to within 2.5 cm. Laser beams are routinely fired through optical telescopes sited at McDonald Observatory in Texas and near Grasse in southern France at the 100 fused silica half cubes of each reflector. The research has identified that the Moon is receding at 3.8 cm/year, has measured the combined Earth-Moon mass within one part in 100 million, determined the character of small scale variations in lunar rotation and identified variations of 0.001 s/year in the length of a terrestrial day. Lunar ranging has also yielded a large improvement in knowledge of the Moon's orbit.

Background

NASA protects and scientifically investigates Apollo's lunar samples in NASA Johnson's Lunar Sample Building (number 31N), built in 1979 specifically to provide permanent storage for the moon rocks. In all, Apollo astronauts returned with a total of 381.7 kg, comprising 2,196 individual specimens from the Moon. NASA scientists and other researchers have processed the samples into more than 97,000 individually catalogued samples.

Four of the 54 4 cm diameter 40 cm long core tube sections collected at 27 localities remain to be opened and analysed. As of 27 October 1995, specimens were widely distributed. JSC held 298 kg, Brooks AFB remote vault 52 kg, 18.8 kg was out with researchers, on display, being used for educational purposes or given as gifts, while 13.2 kg had been consumed during research. Only 33 g ended up classified as 'lost'.

Two vaults store the samples: the 'pristine' vault houses those that have never been outside the facility (270 kg as of October 1995) and a second accommodates those that have been returned (27.6 kg by October 1995) by researchers. More than 60 laboratories worldwide actively pursue sample studies and some 1,100 samples are sent out to researchers annually. They are prepared in stainless steel environmental cabinets purged by high-purity nitrogen

Apollo's lunar samples are stored and processed in NASA Johnson's Lunar Sample Building. They are handled in cabinets filled with high-purity nitrogen (NASA) ***1995***

that is continuously monitored to keep oxygen and moisture contents within 6 ppm. Samples that are not consumed in analysis are returned and can be recycled to other users, but are stored separately from the pristine samples. The facility also includes rooms to support sample examinations and experiments by visiting scientists. The building features steel-lined storage vaults with thick reinforced concrete walls and stands above projected storm-surge sea level heights. Bank-type vault doors remain closed except for sample transfer from storage to work areas. All pipes and openings into the vault close automatically in response to fires or attempted intrusions. Other parts of the building, including separate laboratories where samples are processed, are also heavily alarmed. All materials used in constructing and equipping the building, including floor coverings, walls, plumbing, light fixtures and paint, were screened to exclude chemical elements posing sample contamination threats (particularly lead and gold).

Columbia approaches touchdown at the conclusion of STS-73 **1996**

Space Shuttle

Current status

If Shuttle schedules stay on course, this reusable launch vehicle, logistics freighter and manned spacecraft since 1981 will log its 100th flight into space some time during the year 2000. The transition of the Shuttle from the reusable, piloted, replacement for all US ELVs, aspired to by NASA during the first half of the 1980s and negated by the 'Challenger' disaster of January 1986, is complete. In December 1998, it became the space station logistics, supply and crew delivery vehicle it was once supposed to be with the launch of 'Unity', the first US element for the International Space Station, aboard Endeavour on STS-88. This was followed in May 1999 by the first cargo mission when STS-96 delivered an international crew.

In 1999, the Shuttle will display every aspect of its revised operational role. As scheduled in mid-June, in late July Columbia launches the last of the NASA 'Great Observatories' (Chandra), in September Endeavour conducts an Earth science mission, in October Discovery performs the third servicing mission on the Hubble Space Telescope and in December Atlantis carries cargo up to the International Space Station. If these flights take place as planned, by the end of 1999 NASA will have launched 97 Shuttles and put all but one into orbit.

Preparation of the Shuttle for service until at least 2020 is confirmed with a series of modifications and upgrades that began with the installation of a 'glass cockpit' in Atlantis, bringing the orbiter to current standards required by the commercial airline industry but lacking on this ageing vehicle. The concept of the Shuttle is more than 35 years old and the technology used in the orbiters is of early 1970s vintage. There is no certainty that candidate replacements such as the Lockheed Martin Venture Star will achieve their design objectives and justify the investment necessary for development. The four NASA orbiters may yet be only halfway through their operational lives.

Background

Shuttle manifest changes

The Space Shuttle has clearly returned to full operational status following Challenger's loss in January 1986 and NASA is implementing a programme to make Shuttle processing and turnaround time match the original goals announced for the programme in 1972 when the United States committed itself to reusable boosters.

The recovery programme cost the United States an estimated US$2.4 billion, with an additional US$1.8 billion spent on the replacement vehicle Endeavour. Rockwell delivered the Orbiter in 1991 for its 1992 debut. NASA currently aims to maintain an annual flight rate of seven Shuttle launches, spending no more than 88 days processing each Orbiter after return from orbit. In this 88 day objective, NASA budgets 23 days on the pad.

In order to make operations run more smoothly, NASA has removed all commercial satellites from Shuttle manifests, leaving only a backlog of science and defence missions and beginning in November 1998, Space Station assembly and utilisation launches. STS-53 in December 1992 entered history as the last dedicated military mission. The Department of Defense has dramatically reduced its Shuttle requirements and transferred most of its launches to new Atlas, Titan and Delta launchers. One of the last DoD entries on the Shuttle mission docket is the September 1999 Shuttle Radar Topography Mission, added in July 1996 after struggling for years to find a slot.

While the Shuttle was able to maintain most of its target launch dates (apart from minor delays generally due to the weather), NASA has announced repeated delays in the first missions dedicated to the International Space Station. NASA tried to blame these delays on Russian contributions to the Station, but later NASA's own contractors admitted they were behind schedule, as well, and over budget. Initially, NASA delayed Shuttle missions starting with STS-88 by about eight months. After the May 1998 release of a study that showed that NASA had provided unrealistic estimates of the Station's cost to the American public, all Shuttle missions related to the Space Station's construction were rescheduled.

When conceived in the mid-1960s, the Space Transportation System was to be fully recoverable, but the concept was abandoned because of its high development cost in favour of the present system. This involves the Solid Rocket Boosters parachuting into the sea for recovery and the External Tank being irrecoverably jettisoned. This compromise, it was intended, would still enable the system to reduce overall programme costs by ⅚ compared with the then available expendable launch vehicles, which were to be phased out. This was not achieved because manpower requirements and turnaround times were not reduced to the expected levels.

Shuttle components

Shuttle Orbiters

The Orbiters are about the size of a DC-9 airliner. Long-term initial planning was based on production of five, each capable of at least 100 flights. Before being named, what was intended to be the structural test article Orbiter Vehicle was designated STA-099 and the five flight-standard vehicles OV-101 to OV-105. Shortly before the Approach & Landing Tests (ALT) began in 1977, NASA intended to name OV-101 Constitution, but a letter campaign to President Ford produced a change to Enterprise after the starship in the 'Star Trek' TV series. However, it was decided shortly afterwards to upgrade STA-099 as the second flight vehicle, replacing it with Enterprise as the ground test vehicle after the ALT programme. The first four operational Orbiters were then named after pioneer sailing ships. Columbia (OV-102) is named after one of the first navy frigates to circumnavigate the world (it was also the name of the Apollo 11 Command Module). Challenger (OV-099) is named after the ship that explored the Atlantic/Pacific 1872-76 (also the name of the Apollo 17 Lunar Module). Discovery (OV-103) was named after two ships, the vessel that discovered Hudson's Bay in 1610-11 and the one in which Captain Cook discovered the Hawaiian Islands, and Atlantis (OV-104), after the ketch that performed oceanographic research 1930-66. Challenger's OV-105 replacement was named in May 1989 as Endeavour after Cook's first command; it was also used by Apollo 15's Command Module.

About 210 significant hardware changes were made to each Orbiter between missions 51L & STS-26, in addition to 100 software and 35 SSME changes. The system now contains some 800 Criticality 1 items – that is, equipment without back-up and creating significant mission impact should they fail. Improvements included RCS rewiring so that thrusters cannot fail into the 'on' position, replacement of the limited-life APUs (the retrofitted models offer 75 h lives), and rerouteing of fuel cell waste water outlets so that freezing cannot disable the whole power production system. The main axles of the landing gear were thickened and the capacity for nosewheel steering, introduced on Columbia for mission 61C, was added to the other two vehicles. Higher temperature thermal protection was added in the wing elevon cove areas, along with RCC protection between the nose cap and landing gear door. Columbia's low-T thermal protection along its mid-fuselage, payload doors and tail was replaced by Advanced Flexible Reusable Surface Insulation

OV-103/104 now have strengthened wings after analyses showed that the weight reduction programme during construction left them vulnerable to flight loads higher than predicted. Closure of the flapper valves on the 43 cm External Tank propellant lines during SSME burn was a major concern before Challenger's loss because of the fatal consequences, but latching mechanisms were added to ensure they cannot close while liquid is passing through to the engines.

Typical Space Shuttle launch. Note the 'white room' (centre bottom), which provides crew access and the only pad escape route (NASA) **1996**

Enterprise (OV-101)

Used for the Approach & Landing Tests (ALT) of 1977-78, Enterprise subsequently served as a ground test vehicle, ending at Vandenberg in 1985. It was used there for preflight test matings on the SLC-6 complex, later closed. It now belongs to the Smithsonian's National Air & Space Museum at Washington DC's Dulles Airport. In June 1987, NASA borrowed 101 for arrester net tests at Dulles. The need for such nets at emergency landing sites was under consideration before the Challenger accident and particularly at Dakar (Senegal) and the Pacific Ocean runways on Easter Island and Hoa. The net is based on those already in use at military airfields.

Columbia (OV-102)

As the heaviest flying Orbiter, Columbia's utilisation potential was enhanced by the increase in permitted landing weights. It was modified for longer duration missions of nine to twelve days, instead of the normal five to seven days, and flew most of the Spacelab missions. It was taken out of service after STS-40, in June 1991 for about a year, for conversion in Rockwell's Orbiter final assembly facility as the first Extended Duration Orbiter to accommodate missions of up to 16 days (plus two days' contingency); dry mass is 82,267 kg. Additionally, improved APUs, GPCs, carbon brakes and a drag parachute were fitted and a major structural inspection recertified the airframe for a further three years. Obsolete development instrumentation was removed and provision made for later installation of the new Collins' TACAN landing navigation system. Its third inspection/recertification took place October 1994 – April 1995.

As the first Orbiter in space, Columbia had made six flights by STS-9 in November 1983, had flown 18 astronauts and launched two commercial satellites.

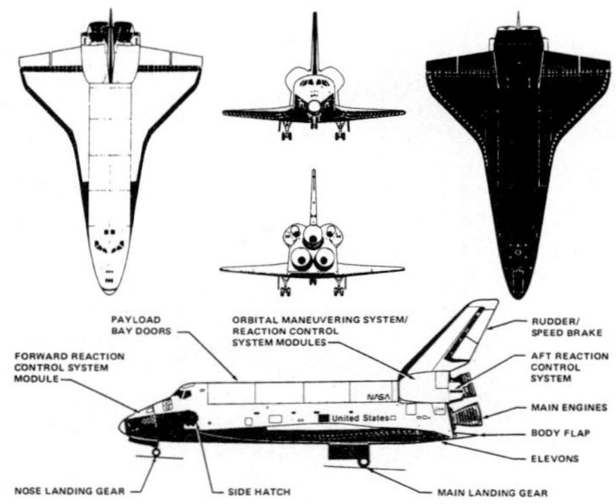

Diagram of Space Shuttle

Columbia returned to service for STS-61C in January 1986 after 18 months' major modifications to bring it to an operational configuration. It should have returned to Palmdale for removal of its ejection seats and overhead escape windows after STS-5 in late 1982, but DoD insisted that two Orbiters should be available in case of loss, so the work was delayed until after STS-9/ Spacelab 1. When it was returned to Palmdale, the planned 24 modifications became >240. Because of the greater dry mass, monitoring instrumentation and extended mission facilities such as additional cryogen storage, Columbia's maximum payload capacity for a 204 km, 28.5° orbit was 3,800 kg less than that of OV-103/104. Work included structural strengthening, to enable it to withstand higher aerodynamic loading, wiring changes and avionics updating and a reinforced carbon-carbon nose cap, incorporating pressure sensor inlets as part of the programme to collect data on dynamic pressure during re-entries. A pod was also installed on the vertical stabiliser to obtain detailed infra-red (IR) imagery of upper surface heating of the port wing and fuselage during re-entry. A plan to take Columbia to Vandenberg for Shuttle handling tests before the facilities were shut down, was abandoned in October 1986. NASA estimated that each of Columbia's first six flights cost US$300 million.

Challenger (OV-099)
The second Orbiter to fly, in April 1983. So much was learned during construction that it was upgraded from the Structural Test Article and replaced by Enterprise in that role. By November 1985, it had made nine flights, flown 45 astronauts and launched five commercial satellites. Its milestones included the first 7/8-member crews, the first EVA (including the first untethered), and the first capture, repair and redeployment of a disabled satellite. Following STS-41G in 1984, 13,984 thermal tiles had to be replaced in an emergency operation involving 200 Rockwell tile specialists being sent to KSC, after 2,500 new tiles had been manufactured at Palmdale. Challenger was lost in the launch explosion of 28 January 1986; the debris is sealed in two Minuteman silos at Cape Canaveral. Average cost of each of its eight flights was estimated at US$290 million.

Discovery (OV-103)
The third Orbiter to fly, OV-103 undertook STS-26, the first mission of the new post-Challenger series. It debuted in August 1984 and by November 1985, it had made six flights, flown 35 astronauts and launched 13 commercial satellites. Its firsts included the first capture/return to Earth of two disabled satellites, the first orbital test of a laser tracking system, the first trial of a 32 m solar array and the first classified military flight. Before the Challenger accident it was to have been based permanently at Vandenberg from its seventh flight for polar operations, but its delivery from Kennedy Space Center had been delayed by problems with the new lightweight SRBs, followed by delays with completion of the Vandenberg facilities. Between the accident and the cancellation of Shuttle/Centaur, Discovery was modified, at a cost of US$5 million, to enable it to carry the Centaur payloads in place of Challenger. Average cost of each of Discovery's first six flights was estimated at US$170 million. Capacity of the 78,700 kg inert-mass Orbiter into a 204 km, 28.5° orbit is 24,950 kg. Discovery was the first to use carbon-carbon brakes, on STS-31 during 1990. It was removed from service after STS-42 in January 1992 for extensive modification (the 78 major changes included addition of a drag' chute and nosewheel steering) and inspection at Kennedy Space Center, returning for STS-53 in December 1992. It returned to Palmdale following the July 1995 STS-70 mission for its regular three year major service, this lasting from September 1995 to June 1996. During the servicing a complete structural inspection was made of the orbiter, which found no major problems.

More than 750 kg of performance-enhancement weight savings were achieved to reach the requirements for International Space Station missions, the first of which was then planned to be STS-94 in November 1998 (this mission has now been cancelled). AFT ballast was added, as well as a fifth cryogenic tank which set provisions for additional on-orbit capabilities, additional nitrogen supply, installation of lightweight seats and a reinforcement of the crew module's floor under seats 1, 2 and 5 to meet the requirements for a 20 *g* crash overload. Modifications in support of the International Space Station and Shuttle/Mir missions included air-cooling provisions for mid-deck locker payloads, installation of a UHF space station communications system, orbiter provisions for ISS assembly power conversion units, removal of the internal airlock and the installation of a new airlock which provides a Mir docking capability. Design improvements included a redesign of the main propulsion system gaseous hydrogen flow control valves and a wing doubler modification. Four next-generation toughened single-piece fibrous insulation tiles were installed on Discovery's side hatch. Discovery returned to flight for the STS-82 second Hubble Space Telescope servicing mission in February 1997.

The 1995-1996 Discovery servicing marked the first use of a new vehicle automated power-on checkout system at the Palmdale facility. This demonstrated programme savings which reduced the number of data storage tapes from 1,060 to 10, cutting the number of checkout support staff from 21 to 3 and cutting the start-up time from four hours to five minutes.

Atlantis (OV-104)
OV-104 performed STS-27, the second mission of the post-Challenger series, reportedly deploying the new Lacrosse imaging radar reconnaissance satellite. Its first mission, the classified STS-51J for DoD in September 1985, was claimed to be the most trouble-free of all Shuttle flights until then. It had been used for only one more, STS-61B in November 1985, before the 2 year shutdown. The scheduled December 1984 delivery to Kennedy Space Center was delayed until 13 April 1985 for modifications to enable it to carry the Centaur upper stage in the payload bay. These modifications included extra plumbing to load/vent cryogenic propellants and controls in the aft flight station for loading, sensing and monitoring the cryogenics. When delivered, Atlantis' inert weight was 76,658 kg – slightly above Discovery's, which had not undergone wing spar modifications. Atlantis was protected with 21,801 high-temperature reusable insulation tiles and 1,977 thermal protection tiles. Final assembly began at Palmdale in 1983, with experience of earlier Orbiters enabling it to be completed with a 49.5 per cent reduction in man-hours. In October 1986, it was rolled out to Pad 39B and fully stacked with ET/SRBs for tests of a new weather protection system, emergency evacuation and slide wire tests. Following the post-Challenger modifications, its dry mass was 78,255 kg, with a 24,950 kg capacity into a 204 km, 28.5° orbit; dry mass is now 78,372 kg. Atlantis was the first to fly the upgraded GPCs, on STS-37 in 1991. It was returned to Palmdale following return from STS-46 in August 1992 for conversion as the second EDO (although it has no scheduled EDO missions) and modifications to allow it to carry an RKK Energia docking unit for the flight to Mir in June 1995. It returned to KSC 29 May 1994.

Endeavour (OV-105)
Work began 1 August 1987, at Rockwell's Palmdale and Downey facilities, on construction of the Challenger replacement. Final assembly was performed in Palmdale Building 150. NASA awarded a US$1.3 billion contract to Rockwell (total cost about US$1.8 billion), which at peak production employed 2,000. OV-105 was rolled out, on schedule, 25 April 1991 and formally handed over to NASA. It arrived at Kennedy Space Center 6 May minus its main engines, forward RCS module and OMS pods. The 20 seconds on-pad Flight Readiness Firing (FRF) standard for all new Orbiters was made 6 April 1992. While the FRF was considered successful, all three of the new engines were replaced because of minor irregularities. Rockwell completed mating of the major structural elements, apart from the OMS pods, in January 1991. Onboard systems were powered up for the first time 6 July 1990. Endeavour made its debut as an Extended Duration Orbiter in March 1995 for STS-67. If NASA ever decides on a 28-day Orbiter, then OV-104/ 105 will provide the capacity.

While duplicating the original OV-103/104 configurations in most respects, post-Challenger modifications included the new GPCs with higher-density memories. They also included quick disconnect valves from ET to Orbiter; carbon brakes to resolve the overheating problems experienced on most landings; modified RCS valves to prevent them sticking open and operating continuously or dumping fuel overboard; and crew escape hatch. A manually deployed 12.2 m diameter nylon/Kevlar drag parachute on a 26 m line is carried at the tail base to shorten rollout by 760 m and cut nosewheel contact speed by 37 km/h to 260 km/h, reducing tyre/brake wear. It also improves handling characteristics, particularly in crosswinds and on wet runways. The system has been retrofitted to the other Orbiters. Eight 'chute trials were conducted July to October 1990 at the Dryden Flight Research Center by NASA's B52, employed the previous April to launch OSC's orbital Pegasus vehicle. OV-105 tested the 'chute on its six initial missions under a US$33 million Rockwell contract.

OV-105 carries the same thermal tile patterns as 103/ 104 but employs a new type of gap filler: Ceramic Ames Gap Filler (CAGF), developed by Rockwell. The older rubber-impregnated quartz fabric leaves a whitish residue on the tiles as it burns away. CAGF avoids that build-up and can undertake at least 30 missions. It will gradually appear on the other Orbiters as attrition occurs.

Five assemblies (mid-fuselage, body flap, wing gloves, vertical stabiliser and main landing gear doors) were already available, having been built as operational spares under a 1983 NASA contract. Structural assemblies remaining to be built included wings, forward/aft fuselage sections, crew module, payload bay doors and the forward RCS module. Endeavour's waste management system is based on the Space Station design. Previous Orbiters carry nine computer-controlled 23 × 56 cm vent doors along each fuselage side to manage pressure variations during launch/ landing, but assemblies 4/7 were eliminated on OV-105 because mission experience showed the original design to be conservative. Following the STS-77 mission in May 1996, Endeavour was returned to Palmdale for its first servicing since being introduced to flights in 1992. OV-105 returned to flight with the STS-89 mission to the Mir space station in January 1998 and flew the first US International Space Station mission (STS-88) in December 1998.

OV-106
The Augustine committee report of December 1990 was against a sixth Orbiter, preferring the funding for a heavy-lift unmanned vehicle. President Bush signed a domestic launch policy in Summer 1991 that rejected any new Orbiters. As a follow-on to Endeavour's August 1987 contract, NASA and Rockwell began discussions on replacing the spares used for 105. Although for use primarily as spares to replace possible damage on existing Orbiters, they could also be used for construction of a new Orbiter. Rockwell was awarded a 4.5-year, US$375 million contract in late 1989. The contract included US$35 million for FY90, covering fabrication of major structural spares: aft fuselage, crew compartment, lower/upper forward fuselage, mid-fuselage, wings (including elevons), vertical stabiliser (including rudder and speed brake), body flap, forward RCS system and one set of OMS pods. The spares were to be of the same configuration as Endeavour. However, NASA decided, in January 1994, to cancel the work on cost grounds. In Summer 1989, NASA was planning to include, in its FY91 budget request, funding for two further Orbiters, partly to support demands on the fleet from Space Station assembly and operations (the funding was not requested). It was then considered for the FY92 request, aiming for 1997 delivery, but again was not included.

External Tank (ET)
The largest Shuttle component (46.88 m high, 8.40 m diameter), the ET dominates the cluster and carries typically 730 tonnes of propellant for the Orbiter's three main engines. It remains the only part of the stack not recovered. As its 29,930 kg dry mass equates with that of the Shuttle's original theoretical maximum payload, there are continuing proposals for it to be taken into orbit. It is jettisoned shortly before orbital speed to ensure a controlled re-entry. Releasing it at the most economic moment would leave it tumbling in low orbit, to re-enter within days.

The ET provides the cluster's backbone at lift-off, absorbing the SRB/SSME thrust. It incorporates three units, the LOX tank forward, containing 625,850 kg at −183°C, the intertank, with instrumentation, and aft, the LH_2 tank, containing 104,308 kg at −253°C. Total loaded mass, which varies for each mission, is about 760 tonnes. External insulation is provided by a 2.5 cm multilayered polyurethane foam coating to withstand the extreme thermal differential. At launch, the propellants are pressure-fed at a combined rate of 3,900 litres/s (1,407 kg/s) through 43.2 cm diameter

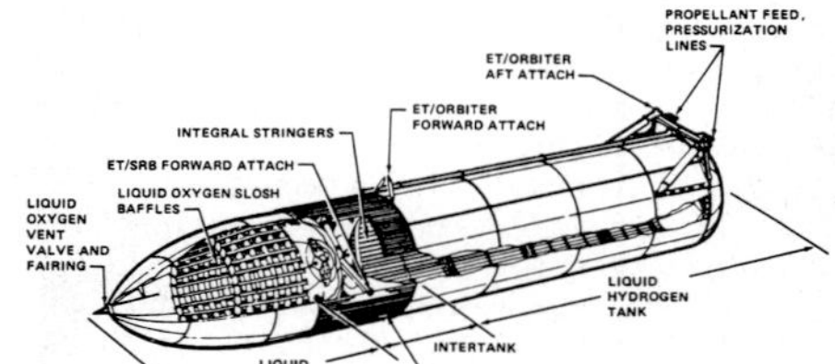

External Tank principal features (Rockwell)

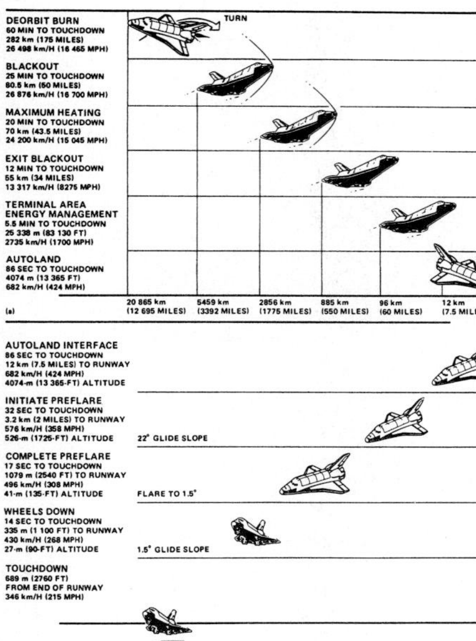

Space Shuttle re-entry and landing sequence

feedlines to the three SSMEs. The tanks are self-pressurised: 3.0 atmosphere LH_2; 1.43 atmosphere LOX. The ET is jettisoned after 8.5 minutes, at about 114 km altitude. Venting residual LOX was used to begin a slow tumble, preventing skipping off the atmosphere and promoting break-up over a remote ocean area. The valve has been omitted since STS-65 as the tank tumbled anyway when venting was deliberately not used on STS-31. A 'beanie cap' on the launch tower was added, to siphon off vented oxygen from the apex during countdown, to prevent ice formation and possible launch damage to the thermal tiles. It was first fully implemented for STS-2. The ET carries typically 770 kg of residual propellants at separation, intended as a safeguard against premature depletion before main engine cut-off. Early exhaustion of LH_2 would be particularly serious as running the SSMEs on an oxygen-rich mixture would cause erosion.

NASA decided, in 1994, to proceed with development of an SLWT Super LightWeight Tank version, awarding Lockheed Martin Manned Space Systems US$172.5 million for the work. Replacing 2219 aluminium with the Weldalite 2195 aluminium-lithium alloy, changing to an orthogrid design and refining the foam insulation spraying technique will save 3,600 kg. The alloy is 30 per cent stronger and 5 per cent less dense than the current material and comprises 1 per cent Li, 4 per cent Cu, 0.4 per cent Ag, 0.4 per cent Mg and 94.2 per cent Al. The saving trades directly to payload capacity, necessary for Space Station missions. A 12.2 m long full-diameter version, essentially a segment of the hydrogen tank with a LOX tank dome at one end, was delivered to NASA Marshall 1 February

A completed ET leaves NASA's Michoud plant **1996**

ET hydrogen tank (Lockheed Martin)

1996 for 6 months' dynamic and pressure testing to verify the structural design. A one-piece graphite phenolic nose cap that does not require the spray-on foam is already under development for first delivery and flight in 1997. Studies showed it was possible for the insulation to break off and impact an Orbiter window, with fatal consequences. The new design also eliminates 961 fasteners, requiring only about 50. It was used for the first time on the STS-91 flight launched in June 1998.

The major post-Challenger modification involved the installation of a mechanical latch on the Orbiter side of the 43 cm feed lines to prevent the fatal closing of either flapper valve before separation. These 43 cm disconnect mechanisms were replaced by simplified 35.6 cm versions in 1993, with flapper valves folded against the walls (instead of opening sideways). This cleaner design, using 51 per cent fewer parts, permits the same propellant flow; the rest of the 43 cm plumbing remains. NASA also plans to redesign the systems tunnel. The projected cost of an ET rose from US$1.8 million in 1971 to US$10.1 million 10 years later. A September 1983 contract for 26 ETs plus material for another 21 (following the existing contract for 15 ETs) was worth US$505 million. ETs are manufactured by Lockheed Martin Manned Space Systems at NASA's Michoud Assembly Facility in New Orleans for NASA Marshall and shipped by ocean-going barge to Kennedy Space Center. After the 1986 accident, production of the 59 contracted tanks was halted with seven available, with a 1985 contract for a further 60 held in abeyance. The US$1,797 million contract was finally awarded in August 1989; the first was completed in late 1991 and the 60th is planned for 2000.

Solid Rocket Boosters (SRB)

The SRBs produce their thrust rapidly upon ignition and burnout in about 130 seconds, having provided the main thrust to lift the Shuttle cluster up to about 45 km. Thiokol is the motor prime contractor, selected in 1974. Other contractors include McDonnell Douglas for the structures and USBI for SRB checkout, assembly, launch and refurbishment (USBI's contract was most recently extended by two years through September 1999. Each booster generates a sea level thrust of 14,678 kN at launch and together they provide 71.4 per cent of the thrust at lift-off and during first stage ascent. After separation, they coast upwards to about 65.9 km. They then fall in a ballistic trajectory for four minutes; at 4,700 m, a barometric switch causes the 3.5 m drogue first and then three 41.4 m parachutes to deploy; the SRBs splashdown at about 95 km/h for recovery in the Atlantic about 227 km downrange. Each SRB is 45.46 m long, 3.7 m diameter and (from STS-6) weighs about 589,670 kg, of which about 503,950 kg is propellant – a mixture of ammonium perchlorate (oxidiser, 69.6 per cent by mass), aluminium (fuel, 16 per cent), iron oxide (a catalyst, 0.4 per cent), a polymer binder and an epoxy curing agent (1.96 per cent). It is configured to provide high thrust at ignition, reducing by about one-third after 50 seconds to avoid overstressing the vehicle during maximum dynamic pressure. The SRBs are interchangeable matched pairs, each comprising four motor segments. SRB ignition can occur only when the ground crew manually removes a lock pin from the Safe & Arm device. The computer launch sequence starts the three SSMEs in sequence (each within ¼ second), at −6.6 second; only when they have all attained 90 per cent, rising to 104 per cent, will the four onboard computers command SRB ignition. Two Orbiter-type APUs provide hydraulic power from T−28 seconds until separation for TVC. Each APU is fuelled by 10 kg of hydrazine.

On 1 February 1987, a year after the Challenger accident, NASA reported that the PDR of the Redesigned Solid Rocket Motor (called the Reusable SRM since 1995) was complete; CDR followed in October 1987. The field joints now incorporate a tang capture feature to reduce joint deflection, a third O-ring, changes to the design of the joint insulation, sealing without putty and external heaters. The heaters maintain seal temperature at greater than or equal to 24°C in all weather conditions and a weather band prevents water penetration. The nozzle/casing joint, which showed evidence of erosion on some flights, was similarly redesigned. Five full-scale RSRM firings were made, culminating in the PVM-1 (Product Verification Motor) test at Thiokol's Brigham City facility on 18 August 1988. At the time of Challenger's loss, 11 motors remained to be used. Thiokol fired these over November 1988 to late 1991 to empty them for conversion to the current design; the firings also generated further engineering data. NASA announced July 1991 a US$2.6 billion Buy 3 contract extension to Thiokol for 67 further flight pairs to 1999/STS-105, plus eight ground test motors.

Proposals by contractors for a second-generation SRM design led to the award, in August 1987 by NASA Marshall of study contracts to five companies for an Advanced SRM, providing the Shuttle with payload

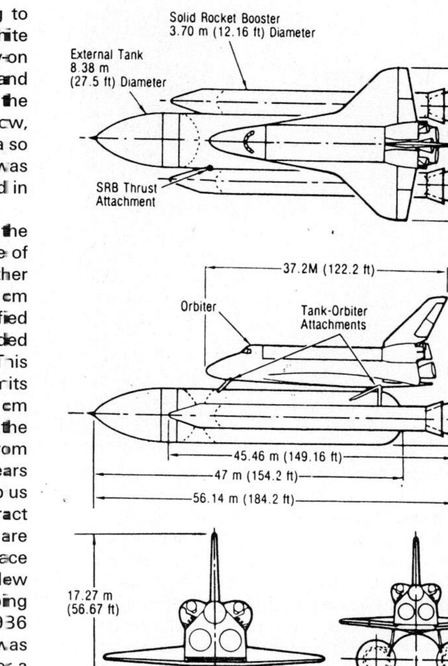

Space Shuttle

increases of up to 5,440 kg. This almost makes the Shuttle deliver as much payload as NASA originally promised, being close to the original 29.5 tonnes (65,000 lb). Capacity was to be 27.67 tonnes into a 280 km orbit and raised Space Station 28.5° orbit capability from 20.9 tonnes to 26.3 tonnes, essential for station assembly. Thrust profile was designed to greatly reduce Orbiter main engine throttling during maximum dynamic pressure, removing some 175 SSME and APU critical failure modes associated with the potential inability to throttle up. Lockheed Missile Systems Division, teamed with Aerojet Space Booster Co, was selected April 1989 to develop ASRM. An associated 3½-year US$550 million contract required construction of a NASA production facility at Yellow Creek in Iuka, Mississippi capable of handling 32 motors annually. Programme cost was estimated at US$3.9 billion, but Congress cancelled it in October 1993 after about US$2 billion had been expended. Cost of cancellation is estimated at US$240 million; and Lockheed/Aerojet handed over the Yellow Creek facility to NASA in December 1994. Thiokol then planned to transfer its RSRM nozzle manufacturing there, but was directed 2 May 1995, by NASA, to discontinue work.

Studies have also been conducted into liquid boosters for future Shuttle and other launch vehicle use. They would increase Shuttle payload capability by 5.5 to 9 tonnes and add an abort capability during early ascent, but would require US$5 billion and eight years to develop, plus US$500 million for pad modifications. NASA's initial studies ended in Spring 1989, with General Dynamics and Martin Marietta completing their studies in December 1989.

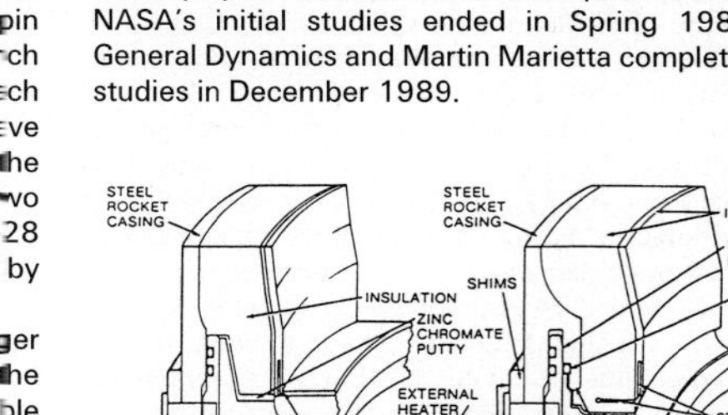

The post-Challenger SRB field joint includes a tang capture feature, with a third O-ring, to reduce joint deformation and a combined weather seal/heating band to exclude water and maintain joint temperature at 24°C

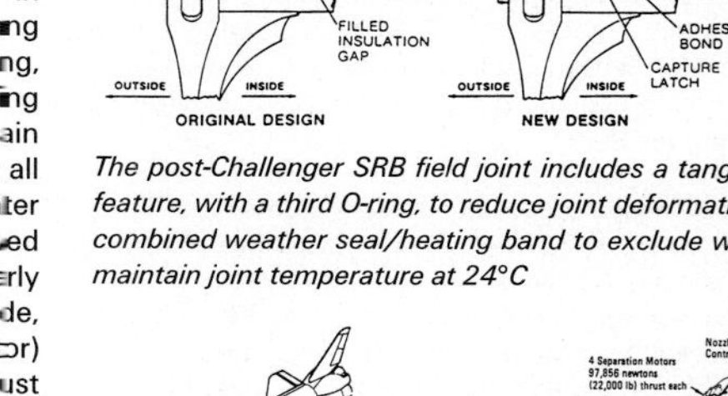

Space Shuttle Solid Rocket Booster (NASA)

The first Remote Manipulator System, being assembled in Toronto (Spar Aerospace)

The escape system retrofitted on each Orbiter consists of a telescopic pole extended from the side hatch. The astronauts use a lanyard to slide along it to clear the Orbiter's wings and tail. This picture was taken from a C-141 aircraft as a Navy parachutist demonstrated the technique

The intention was to order five SRB pairs, at about US$50 million per pair, for each Orbiter and reuse them 20 times. However, one pair was lost on STS-4 as a result of excessive water impact speeds, leading to enlargement of the main parachutes from the original 35 m. A second pair was lost with Challenger on STS-51L. After recovery, each is broken down into 14 major components for refurbishing, with the sections being reused separately. Nine is the record held to date for reuse. Total mass is being steadily reduced – for example, reducing wall thickness in some areas produced a 1.8 tonne saving. A proposal was also made to use the SRBs as sounding rockets on each mission, since each could carry 90 kg of instruments without loss of performance. A US$46 million programme was planned (but cancelled when Orbiter weight reduction proved cheaper) to reduce the weight of each SRB by 2,720 kg, adding 540 kg capacity into the planned 51.6° Space Station orbit. A further 1,025 kg capacity would be created if the 'chutes and various recovery structural items were omitted, saving 5,125 kg per booster.

Space Shuttle Main Engines (SSME)

The SSMEs have been undergoing upgrades since 1983 under a 10-year US$1 billion improvement programme. Improved turbopumps were introduced for STS-26. In the 18 months before that mission, SSME firing trials accumulated the equivalent of more than 40 missions. On 15 November 1991, total testing and flight time passed 500,000 seconds, the equivalent of 320 missions; it exceeded 600,000 seconds in December 1993. New Pratt & Whitney high-pressure turbopumps, allowing 7.5 hr of firing (55 mission equivalents) between major overhauls are replacing Rocketdyne's versions. The older pumps require refurbishment after 33 minutes use (three flights). Problems increased development cost to US$504 million, so NASA halted fuel pump work from December 1991 in order to focus on the more critical oxygen unit for the Block 1 upgrade.

It restarted work May 1994 for 1997 debut as part of the Block 2 upgrade once the LOX pump was on schedule for its July 1995 flight debut. The Block 1 upgrade includes a single coil heat exchanger (eliminating seven weld joints) and a two-duct hot gas manifold, which allows the fuel pump to run at lower temperatures and pressures. A single Block 1 upgrade engine (number 2036) flew successfully on STS-70 in July 1995. All three engines were upgraded for STS-77 in May 1996. Testing of the Block 2 upgrade began at NASA Stennis on number 0521, once Block 1 certification was completed 26 May 1995. Block 2 includes P&W's fuel pump and a 10 per cent diameter larger throat combustion chamber that will reduce wear, reducing pump discharge pressure by 4.5 to 6 per cent and pump operating temperature by 7 to 12 per cent. All three engines on STS-86 in September 1997 introduced the Block 2. STS-85 will be the last mission to fly with an old engine; Block 1 or 2 will be used on all subsequent missions.

Computer and automated control systems

Each Orbiter's five IBM AP-101 General Purpose Computers of 1972 vintage are being replaced by IBM AP-101S models designed in 1984 and offering 2.5 times the capacity and three times the speed for half the size and mass. At peak, they will be used at more than 50 per cent of their capacities, contrasted with 80 per cent for the older models. The new GPCs first flew aboard Atlantis on 1991's STS-37. New Block 2 Honeywell SSME controllers were initially installed on OV-105 and then retrofitted to the others, providing four times the capacity with the first space-qualified 68000 series microprocessors. They also eliminate a separate 13.6 kg vibration monitoring box and add 20 engine parameter monitoring channels to the previous 120. The last of 30 ordered was delivered April 1993. The KT-70 Inertial Measurement Units are being augmented and eventually replaced by the High Accuracy Inertial Navigation System, of similar configuration but with increased accuracy and reliability (the first one flew on Atlantis during STS-44 in 1991). A new TACAN system is also due. Endeavour was delivered with the new items installed. Replacing the copper wire payload links in the cargo bay with a fibre optic system would save 545 kg.

Remote Manipulator System (RMS)

Spar Aerospace developed the RMS as Canada's principal Shuttle contribution. 38 cm diameter, 15.3 m long, mass of 408 kg, capable of handling 29.5 tonne payload. Shoulder, elbow and wrist joints are driven by DC electric motors and can be controlled in five different modes, ranging from manual hand controllers for the crew, using direct observation and TV cameras on the elbow/wrist, to full computer control. Stowed on the bay's port side and latched into three cradle pedestals, it can be restowed by EVA if it jams in use or, in the last resort, jettisoned by pyrotechnics. Each RMS is designed for a 10 year, 100 mission life. The first, the number 201 DDT&E arm and associated ground equipment, was delivered April 1981 at a cost to Canada of C$100 million. The operational 301/302/303 models were delivered May 1982, 14 December 1983 and 29 March 1985, respectively. 302 were lost on Challenger, but were not replaced until the refurbished 202 qualification model debuted on STS-66 in November 1994. RMS' highly successful use as a mobile crew work platform, in addition to its primary role as a manipulator, established it as an indispensable tool. A servicing centre based on RMS derivatives is Canada's primary Space Station contribution. Spar was contracted in 1991 to upgrade the arms to handle up to 265 tonne for berthing the Orbiter to Space Station. This includes a C$30 million redesign of arm control electronics, followed by a refurbishment and overhaul of various critical components completed in 1996.

Modifications to the RMS will be certified to handle up to 265 tonnes and a remotely operated power/data connector will be carried on one side of the cargo bay for early Station flights until the pressurised docking system can be used. The existing internal airlock will be refurbished and mounted outside, attached to the cabin by a short flexible tunnel. On top will be an ASTP-type androgynous docking unit, with integral power/datalinks, and crew hatch. Three docking units will be built and carried as required.

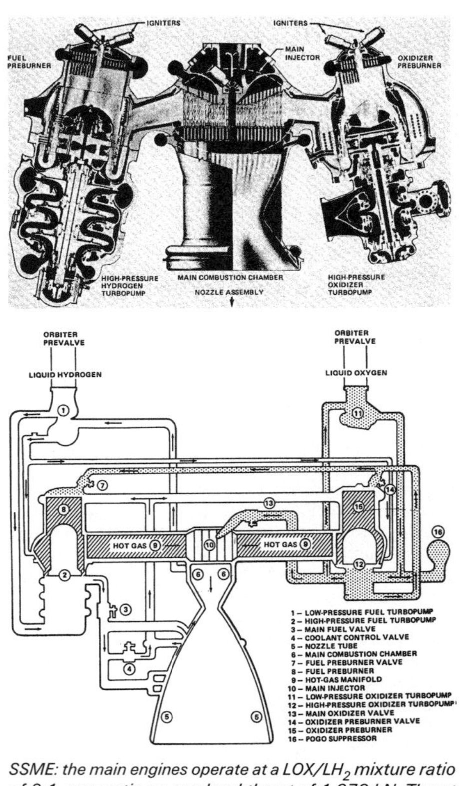

SSME: the main engines operate at a LOX/LH_2 mixture ratio of 6:1, generating a sea level thrust of 1,670 kN. Thrust can be throttled over 65 to 109 per cent, allowing thrust reduction to limit maximum dynamic pressure and ensure that astronauts and equipment endure no more than 3 g. Following Challenger, the upper limit was established at 104 per cent to reduce engine wear, although 109 per cent would be used in the event of a trans-Atlantic abort. Improvements will allow Space Station assembly missions currently beyond Shuttle's capability to fly at 106 per cent. Total engine testing exceeds 650,000 s

Environmental Mobility Unit (Spacesuit)

EVAs have played a major role in Shuttle activities. Through February 1996, 35 different astronauts had conducted 60 EVAs totalling 378.2 hours, had repaired and redeployed four satellites and brought back to Earth two more (see EVA table at front of Directory). Crew members must be able to leave the Orbiter for routine work and emergencies. Three of the 14, 117 kg, suits are normally carried on each EVA mission (four were carried for the first time on STS-49 May 1992 and again STS-61 December 1993) and two for emergencies on non-EVA missions. They are stowed in the airlock, where they are donned/doffed, their size adjusted to fit the crew members trained for EVA on the particular mission. The crew uses 30.5 m tethers when working in the payload bay, unless they are working in the MMU or on the RMS foot restraint.

Cheaper and more flexible than Apollo's lunar suits, donning the suit begins with a zip-on one-piece Spandex liquid cooling/ventilation garment. Weighing 3 kg and with a projected life of 15 years, it includes urine collection facilities for up to 950 ml and a drink bag containing 0.6 litres of water. A Snoopy cap with headphones and microphone fits over the head and chin, and also provides caution and warning tones. Over this fits the two-piece spacesuit itself, comprising: hard upper torso, made in five sizes with hard waist ring. The lower torso comes in various sizes with hard waist ring; the helmet (one size only), and the visor assembly, which protects against micrometeoroids and the Sun's UV/IR radiation fits over the helmet. A more dextrous glove was tested during STS-37 in 1991. Expected life is 15 years for the hard sections and six years for the softer portions. There are no zippers on the suit; the components connect with hard snap-rings. The suit is composed of several bonded layers, beginning with a polyurethane-on-nylon pressure bladder, followed by Kevlar layers with folded/tucked joints (for mobility) and ending with a Kevlar, Teflon and Dacron anti-abrasion layer. The hard upper torso has a glass fibre shell. The materials are designed to prevent fungal or bacterial growth: unlike the individual Apollo suits, they are used by different astronauts and must be cleaned and dried after each use. Each suit can be used up to 25 times during a mission.

The Portable Life-Support System (PLSS), attached to the back of the upper torso, provides sufficient oxygen and water for 7 h inside the suit, including 6 h for EVA and a 30-minute reserve. There is also a secondary pack to supply O_2 and maintain suit pressure for 30 minutes if the primary system fails. Astronauts can plug in to the Orbiter airlock for fresh supplies of O_2 and water. A liquid crystal display/control module on the front of the upper

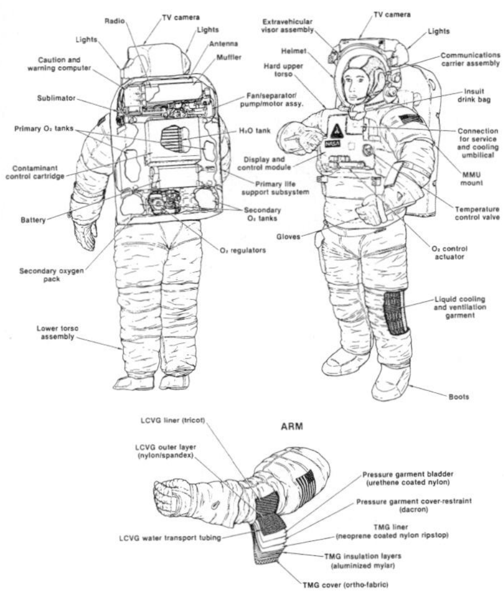

Shuttle EVA Mobility Unit (Hamilton Standard)

torso indicates remaining EVA capacity, and warns of excessive consumption or problems with the water pressure and T in the cooling garment. A suit vents 0.7 kg of water/O_2 hourly, a potential source of contamination. Hamilton Standard was awarded a US$97.3 million contract in 1991 for two further PLSS (adding to the 15 built) and continued engineering support through September 1997, bringing contract total value to US$212 million. Astronaut training and suit maintenance costs NASA some US$100 million annually.

It was planned to provide suitless crew members with a Personal Rescue Enclosure, an 86 cm diameter inflatable sphere made of suit material. Sitting inside with a portable oxygen system, the occupant would have had a communications system and viewing port, and sufficient pressurising gas for a 1 hr rescue. If an Orbiter became disabled, suited crew members could carry the others across to a rescue Orbiter; alternative methods included use of the RMS and passing them along a line as in a ship-to-ship rescue. The PREs were not produced.

The initial disadvantage of Shuttle's suit was that, as it is pressurised to only 0.29 atmosphere (4.3 psi) of pure O_2, the wearer had to prebreathe pure O_2 for 3.5 hr prior to EVA to prevent N_2 bubbles forming in the joints. Pre-breathing now begins with an hour while cabin pressure is reduced from 14.7 psi to 10.2 psi. After 12 hr, the EMU is donned and, during checkout, another 40 minutes is spent breathing pure O_2 before depressurizing the airlock. To overcome the prebreathing necessity, which might delay an EVA to meet an emergency, NASA/Hamilton Standard began work on a zero prebreathe high-pressure hard suit. NASA decided later because of the cost (estimated in 1995 at US$700 million) to use Shuttle-type suits on Space Station. The Station will use 1 atmosphere, requiring the full 3.5 h prebreathe. Hamilton Standard has proposed a US$10 million minor upgrade that would operate the suit at 0.4 atmosphere, the same as Russia's Orlan suit, which requires only 30 minutes prebreathing. Thermal modifications are under way as Station EVA work will be away from Shuttle's warm cargo bay. STS-69 in September 1995 tested fingertip heaters and the body cooling loop could be shut off.

Manned Manoeuvring Unit (MMU)

The 153 kg Manned Manoeuvring Unit, which snapped on to the back of the PLSS, enabled an astronaut to operate free of tethers. Two hand controllers, operating 24 × 7.6 N N_2 thrusters, provided 6 DoF and a total 6,190 Ns impulse. Two were carried in the forward cargo bay when required. Martin Marietta built only three at US$10 million each. On missions 10/11/14 they were used nine times, with an accumulated flight time of 10 hr 22 minutes. NASA began a US$600,000 study in mid-1990 to determine the cost of restoring them to flight readiness, possibly for a flight in late 1991 carrying an IMAX camera and other detectors to observe Orbiter thruster firings and water dumps. A total of about US$5 million was projected. However, NASA management ordered them returned to storage in 1991 as they were developed for demonstration purposes and are too limited for effective deployment.

The 37.6 kg SAFER Simplified Aid for EVA Rescue was tested September 1994 during STS-64 to show its value in an emergency, such as when an astronaut comes adrift from Space Station. At the moment, astronauts have to work in pairs, carefully tethered, and with the Orbiter ready to provide rescue. A drifting astronaut would activate SAFER, to stop any tumbling, using its automatic attitude hold mode and then jet back to safety. SAFER is attached below the PLSS, providing a 3 m/s total impulse from a 1.4 kg pressurised N_2 supply via 24 3.5 N thrusters. It was first used operationally during March 1996's STS-76 EVA at Mir, as it would have taken Atlantis 20 minutes to undock for a rescue.

Shuttle support equipment, maintenance and reliability

Shuttle carrier aircraft

Boeing 747-123 N-9668, purchased by NASA from American Airlines in June 1974 after logging 8,999 hr, was modified (and registered as N-905NA) to ferry Orbiters between Shuttle facilities and to carry Orbiter *Enterprise* for the original ALT programme. For ferry flights, a tail cone fairing is installed on the Orbiter to reduce aerodynamic drag and buffet and aerosurface control locks are added to the Orbiter's elevons. The Orbiter is carried unmanned with its systems inert. A bailout system is carried for the 747 crew. With use of the Kennedy runway limited, the aged lone 747 was a vital element of the Shuttle system. Had it been damaged or lost, the Shuttle fleet would have been grounded. In September 1987, NASA signed a letter of agreement with Boeing for a second aircraft. An ex-Japan Air Lines 747-SR was delivered to Boeing's Wichita plant 17 April 1989 under a US$55 million contract to undergo conversion; NASA accepted it in November 1990. Designed to carry 109 tonnes (22 tonnes more than the older aircraft), N-910NA has a 1,700 km range and is designed for 265 ferry flights over 15 years. The aircraft may be used for carrying the X34 hypersonic test vehicle.

The USAF's 433rd Military Airlift Wing at Kelly AFB, Texas, operates two Lockheed C5As modified by the company under an October 1985 US$133 million contract for carrying a loaded Shuttle cargo container and its towing tractor. These C5A SCM (Shuttle Cargo Modification) aircraft, delivered November 1988/Oct 1989, can handle a full cargo bay load such as the Hubble Space Telescope.

Shuttle operations

Processing/turnaround

NASA plans an annual flight rate of seven for the full four-Orbiter fleet through the Space Station era. The processing goal for each mission is 88 days, covering 60 days in the Orbiter Processing Facility, 5 days in the VAB and 23 days on the pad (Endeavour set the current 52-day OPF record preparing for STS-57 in 1993, beating the previous best, of 1993's STS-45, by 3 days). Platforms and equipment from the mothballed Vandenberg site are installed in KSC's Orbiter Maintenance & Refurbishment Facility, creating a third OPF position from late 1991 and allowing simultaneous processing of three Orbiters.

In September 1983, it was announced that Lockheed Space Operations Co (now Lockheed Martin Space Operations), a consortium including Grumman, Thiokol and Pan American World Services, had won the NASA/USAF Shuttle processing contract, effective for three years from 1 October 1983, with four three-year options. NASA exercised the US$1.60 billion second option in September 1989, extending the contract to 30 September 1992. The third was exercised in 1992, worth US$1.830 billion through September 1995, bringing the total for 1983-95 to US$6.3 billion. A one year US$638 million extension began 1 October 1995, in preparation for the new approach of awarding Shuttle operations to a single contractor. LMSOC assumed all the work from October 1995. Before then, Lockheed Martin's share as consortium leader was 75 per cent. Grumman, responsible for launch processing, had a 13 per cent stake. Thiokol's task of processing the ET and SRB (including recovery after launch) was worth 11 per cent, and Johnson Controls World Services (who bought out Pan American), responsible for operations analysis, process planning/control and logistics engineering, received 1 per cent.

NASA's policy was to hand over responsibility for the whole STS processing operation at the Cape and Vandenberg, thus freeing itself for development work on Space Station. It was also hoped that the single processing contract would improve safety and mission effectiveness while reducing costs and turnaround time. Turnaround time was down to 69 days between STS-6/7. Processing had taken 649 days for STS-1, compared with 198, 114, 80 and 120 days for the next four missions. With the near achievement of 10 flights in 1985, turnaround time was down to an average of 36 days.

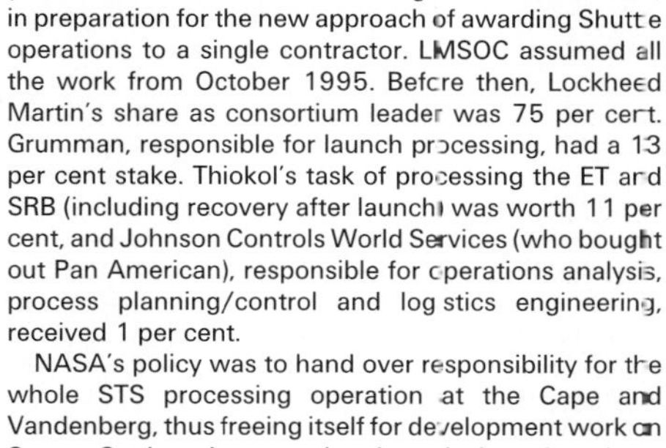

Four Gulfstream IIs, powered by Rolls-Royce engines, are fitted out as Shuttle flight decks for pilot training. They simulate the powerless handling characteristics and steep descent during approach and landing (Grumman)

NASA's STS Operations Contract (STSOC), a four-year award worth an estimated initial US$685 million, was awarded to Rockwell's Shuttle Operations Co in September 1985, effective 1 January 1986. A US$2.3 billion extension through December 2000 was granted Spring 1991, bringing total contract value to US$4.8 billion. RSOC became responsible for managing 22 Shuttle operations functions previously performed by 17 contractors, involving mission planning, flight design, mission data production, flight crew/ground controller training, and ground facility engineering. The work complements Rockwell's Florida logistics contract, extended in October 1991 to 1994 at a cost of US$453 million. The 25,000 m² NASA Shuttle Logistics Depot (NSLD) became fully operational in 1993, certified to handle >3,200 of Shuttle's 4,000 Line Replaceable Units.

Orbiter pilot training

Pilot training includes at least 30 flights on Shuttle Training Aircraft (STA), with about 900 practice landing approaches. Some 10 training sorties are made weekly at the White Sands Space Harbour. The aircraft are housed in the NASA hangar at El Paso International Airport, Texas. STAs are Gulfstream II executive jets, with Rolls-Royce engines and extensive structural, wiring and cockpit changes and with installation of direct lift control systems and thrust reversers to simulate Orbiter flight characteristics. In early 1985, NASA ordered a third STA from Grumman, the modifications costing US$14 million. A fourth was being considered before the Challenger accident to support a launch rate of about 10 per year. It became operational in April 1991. Honeywell's glass cockpit displays will be fitted from 1997 as they are installed on the Orbiters. The two pilots during STS-58 in 1993 introduced a new training aid designed to refresh landing skills on long missions: the PILOT Pilot In-flight Landing Operations Training unit is a portable computer with an Orbiter-type hand controller. It is mounted on the console in front of the pilot's flight deck seat and the controller attached to the actual Orbiter controller.

Shuttle maintenance and reliability

All of the Orbiters return to Palmdale every three years for a major inspection, recertification and maintenance. Realistic analyses of the chance of a Shuttle loss resulting from a 1988 NASA analysis, state the chance of loss at a 1 in 78 median probability of loss during ascent. In 1989, NASA recalculated the values at 1 in 94. Then in 1995, NASA again upgraded safety projections for any Shuttle mission as 1 in 145 (estimates ranged from 1 in 76 to 1 in 230), made up in part of a factor of 1 in 248 for ascent only. The main engines are the major contributors with a mean probability of failure of 1 in 410. With the upgraded engines, the chance will level off at less than 1 in 200 for the whole mission and less than 1 in 300 for ascent. The first estimates that NASA provided to the American public in 1975 indicated that the loss rate would be less than 1 in 1,000,000.

Shuttle orbiter modifications

Through the first several years of launch operations, Shuttle in-orbit missions usually lasted less than 10 days. Columbia and Atlantis have undergone major modifications to provide additional fuel cell cryogenic supplies and improved life support systems for 16-day missions and, as a result, are now referred to as Extended Duration Orbiters (EDO). The Columbia retrofit was completed in 1992. The 13-day US Microgravity Lab STS-50, in June 1992, used the EDO for the first time. *Atlantis* underwent the same conversion following return from STS-46 in August 1992 (although it has no scheduled EDO missions yet) and was also equipped for docking with Mir in June 1995. Endeavour was built with the required plumbing and electrical connections and flew its first EDO mission as STS-67 in March 1995.

In order to make the modifications, Rockwell added a cryogenic pallet in the aft cargo bay to extend fuel cell capacity and enlarged the storage space in the cabin for additional food and clothing. In addition, they installed an improved waste collection system (first flown operationally STS-65 July 1994) and a reusable CO_2 scrubber to replace the previous lithium hydroxide canisters. Atlantis and Endeavour are capable of carrying EDO kits allowing up to 28-day missions, although none is yet planned.

During the preliminary planning for the Shuttle, NASA announced that the spaceplane would have a mission duration of up to 30 days. When it actually flew, NASA could not promise an on-orbit duration of more than eight days and returned to Congress to request additional funds for modification. In November 1987, NASA provided Congress with a report on methods of extending the in-orbit capability of at least one Orbiter to 16 days. This extended the time from the pre-Challenger

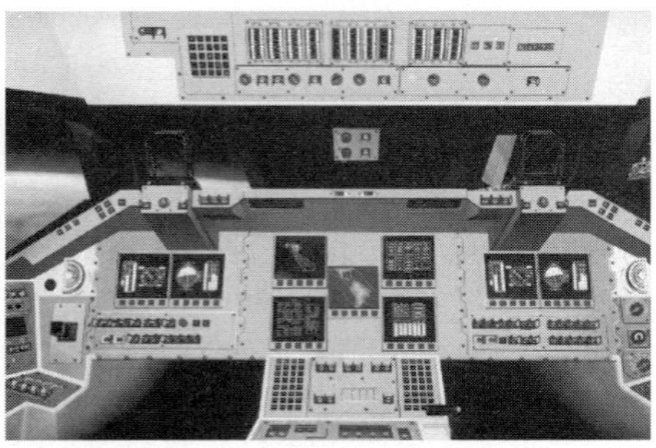

Flight Deck view of the 1,800 switches, push buttons, circuit breakers and other controls. A 'glass cockpit' Multifunction Electronic Display Subsystem incorporating nine 17 × 17 cm flat LCD colour screens will be retrofitted 1997-98 (below; the aft panels will have another two). They can also show video. Other functions can be added because of the increased computing power: one candidate is landing simulation to refresh pilot training during longer missions. See Rockwell's Shuttle Orbiter entry in the Launchers section for further details (Honeywell)

limitation by fuel-cell life of 9-10 days. In May 1983, Rockwell proposed a two-stage plan to extend stay-time to 15 to 18 days by 1985 and to 45 days by 1988-89, but NASA did not proceed with it. The first stage, comparable to the current improvements, was projected to cost US$200 million. The second stage, costing US$200-250 million, would have required an orbiting solar array and battery system with which the Shuttle could dock. This required relocation of the airlock and mid-deck modifications and gained some DoD support as a cheaper alternative to space station development. NASA agreed to Congressional suggestions in early 1989 of providing OV-105 with a 28 day mission capability at a FY90 cost of US$52 million.

Rockwell agreed in March 1990 to contribute US$15 million to the US$53 million cryogenic pallet element under a semi-commercial agreement. If Rockwell signs up a non-government user, the company will receive US$1.85 million for each day beyond an 8-day mission, thereby recouping the investment in a single 16-day flight. Similar missions up to end-1997 would provide profit. NASA management indicated in January 1990 that a 28 day modification is unlikely because of the associated risks.

For EDO, the 2,926 kg cryo pallet plumbs four tank sets to the standard five, each capable of supporting two days' activities. The pallet could carry four further sets on its aft side but Columbia is too heavy to be converted to a 28 day Orbiter. Hamilton Standard's 147 kg Regenerative CO_2 Removal System, forward of the mid-deck lockers, operates two beds in a 15-minute cycle, one absorbing the gas while the other is exposed to vacuum. A similar system was flown on the Skylab space station. Some N_2 is also recovered for addition to the two extra tanks carried under the payload bay against the aft bulkhead. STS-50, the first EDO flight, still carried 50 LiOH canisters as back-up (astronauts had to repair the RCRS in flight). Although RCRS is now fully qualified, EDO missions continue to carry sufficient canisters for four days.

The EDO 22 kg trash compactor was demonstrated aboard 1990's STS-35, carried in place of a mid-deck locker. Rubbish is loaded in a 0.014 m^3 plastic bag, equivalent to one person's daily waste and compacted by pistons at 41 N/cm^2 into quarter the volume. The bag is then stored under the mid-deck floor. Some items, such as the rehydratable food package, have been redesigned to allow crushing. STS-35 also demonstrated the new Waste Containment System. The old toilet has to be removed/cleaned after each flight, but Hamilton Standard's newer design collects the waste in bags, which are then compacted into storage tubes with odour/bacterial filters. Using removable bags means the toilet has unlimited capacity. Development versions were carried by Endeavour on STS-54/57 in 1993, allowing the then record-breaking STS-65 July 1994 to be the first operational mission.

NASA has studied the Longer Duration Orbiter (LDO) modification for longer missions. The preliminary results indicated that 30 days are possible and 60 days appear feasible if the Shuttle remains attached to the Space Station for mutual support. Power (fuel cells) is the limiting factor beyond 30 days. A solar array module has been proposed, which would also make it independent of Space Station. LDO was deleted from FY94's budget to save US$43 million, particularly as long missions are now available aboard Mir.

Beginning in 1994, NASA and its contractors modified Atlantis to fly seven Mir docking missions between 1995 and 1997 and Discovery for two missions. Engineers believe the Orbiter structures are good for more than 100 missions, allowing flights until around 2030. The thrust structures are the principal limiting factor on a Shuttle lifetime. Discovery, Atlantis and Endeavour are modified for Space Station operations. Orbiter Atlantis stood down for 10 months while engineers and technicians replaced electromechanical cockpit instruments with 11 colour flat-panel screens, improving pilot efficiency and reducing costs of eliminating the need to maintain outdated equipment. Atlantis arrived at Boeing's orbiter processing facility at Palmdale, California, on 14 November 1997 and returned with more than 130 major modifications on 27 September 1998.

Space Shuttle Orbiter specifications

Shuttle payload capacity

Structural analyses and reviews of landing stresses following Challenger's loss showed that landing weight could be increased by 8,619 kg to 104,328 kg. Abort landings, which constrain launch mass, are limited to 108,860 kg. Discovery, Atlantis and Endeavour can deliver 24,990 kg into 204 km orbits; they are referred to as 55,000 Orbiters because of their 55,100 lb capacities into NASA's standard 120 n mile (220 km) orbit. Originally NASA promised the Shuttle would deliver 65,000 to the standard orbit. Columbia can handle 3,152-3,840 kg less because of its greater dry mass. As a result, it will not be used for Station assembly. Advanced solid boosters, available from 1999 at the end of a development cycle cancelled in October 1993, would have added 5,440 kg capacity. Some Station assembly flights might require SSME levels of 106 per cent, allowed by pump improvements and other changes.

Shuttle costs

NASA's own figures for the first 20 flights made by Columbia, Challenger and Discovery showed an average cost of US$257 million. The improved Discovery, new Atlantis and modified Columbia were expected to reduce the average, but not to the US$74 million per 'dedicated flight' (use of the whole payload capacity) which NASA intended to charge after 1988. NASA quoted a cost per mission of US$286 million in 1990, but charges only US$130 million at 1988 rates. True amortised rates suggest that the real cost of Shuttle missions is probably in excess of about US$350 million (1986 rates). The true cost is debatable, depending on what costs are included. NASA in 1992 estimated the *marginal* cost (that is, the saving by reducing the flight rate from eight to seven) at only about US$40 million. But NASA's Shuttle operations budget produces a figure of US$450 million for FY95. Adding other programme costs such as TDRS, ASRM and civil service salaries puts the figure at more than US$500 million in 1992 terms.

NASA estimates the Shuttle's original development at US$5,150 million (FY71 rate). The actual cost was US$10.1 billion (real year cost), or US$6,744 million (FY71 rate). The 1971 estimate of US$250 million for each Orbiter rose to about US$2 billion in 1983 dollars, though this included, for the first time, engines and equipment, Remote Manipulator System, galley and the closed-circuit TV. The major Shuttle contracts, with their values in 1982 US$ million were:

Rockwell (Shuttle Orbiter)	3,560
Grumman (wings)	45
McDonnell Douglas (OMS/RCS pods)	85
McDonnell Douglas (Support)	52
Rocketdyne (Main Engines)	1,546
Thiokol (Solid Rocket Boosters)	206
USBI (SRB assembly/retrieval)	89
Martin Marietta (External Tank)	529

Following the Challenger accident, NASA submitted, in February 1987, a series of proposed actions in response to the Rogers Commission recommendations, with a total estimated cost of US$1,379 million. The estimated total expenditures to meet this goal were actually US$2.4 billion at the time of STS-26.

Other costs may be estimated from subsequent actual mission costs. As an example, President Reagan's 15 point policy, issued in March 1988, was designed to create opportunities for US commercial space interests. The proposals emphasised the Shuttle-launched Industrial Space Facility as a man-tended microgravity platform and the use of Spacehab inside the payload bay. It was projected that the former price of US$74 million for the use of the whole cargo bay could go as high as US$255 million by the early 1990s. As the 1990s began, NASA cited a new price of around US$180 million, which represented full cost recovery if there were 14 flights annually. NASA reached a *quid pro quo* agreement with the DoD in October 1987 citing US$115 million (FY89 rate) for a full cargo bay but, in 1996, NASA quoted US$200 million to US$250 million for DoD.

President Reagan's August 1986 edict, prohibiting NASA from using the Shuttle for commercial satellite launches, made his earlier Shuttle pricing policy. The intention was to charge a minimum of US$74 million (1982 US$) for a full payload bay and to offer three such missions by auction two years in advance for commercial and foreign use. At less than half the cost of a Shuttle mission, NASA was hoping that demand would increase the price.

NASA is consolidating more than 80 Shuttle processing and operations contracts, saving US$770 million annually, into a single prime contractor. The final contract, lasting for six years and worth US$7 billion, was awarded to United Space Alliance. This consolidated ground processing and in-flight operations within a single company. United Space Alliance will suffer penalties if it fails to meet cost-cutting targets, but it will be awarded 35 per cent of additional costs saved after an initial target of US$400 million has been reached. NASA retains the responsibility for safety, scheduling and budgeting. 1995's operating budget was already 25 per cent lower than 1992's. US$3.149 billion was allocated to Shuttle operations and hardware for FY96; US$3.151 billion was requested for FY97. For other years the requests were FY95 US$3.155 billion, FY94 US$3.549 billion, FY93 US$4.061 billion, FY92 US$4.271 billion, FY91 US$4.066 billion, FY90 US$3.682 billion.

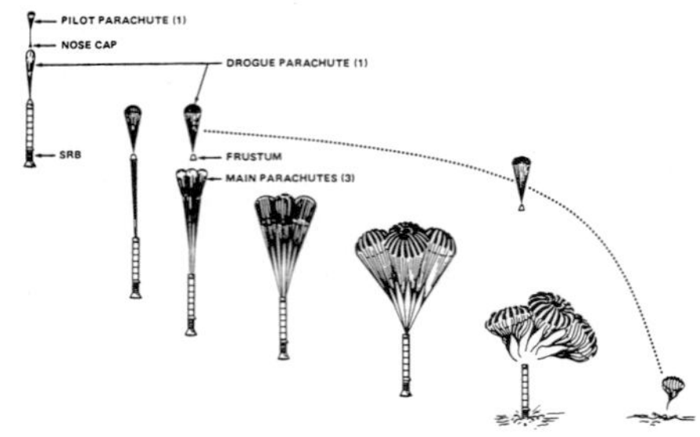

SRB recovery: after separation, the motors follow a 4-minute ballistic path before the recovery sequence begins

Aborts

Four basic abort modes are available to Shuttle controllers, the precise scheduling of each varying with the mission ascent profile. Return To Launch Site (RTLS) would be used in the event of a main engine failure in the first 260 seconds. The remaining engines and RCS thrusters would achieve a pitch-around, enabling the Orbiter to jettison the ET 45 km from the coastline and glide back to the KSC runway. RTLS is considered to be a very risky manoeuvre.

The Transatlantic Landing (TAL) abort is required when an SSME fails after the last opportunity for RTLS and before an Abort to Orbit is possible. There are four TAL sites where an Orbiter would land about 25 minutes after launch: Moron Air Base (Spain), Yundum International Airport (Banjul, Gambia), Zaragoza Air Base (Spain, prime site for high-inclination flights) and Ben

Harris (left) and Foale prepare to leave the airlock for STS-63's EVA. The new electronic checklist can be seen on Foale's right arm; it failed, probably because of the cold conditions, but has since been successful. This was Shuttle's first EVA to exit from a hatch in the tunnel leading to Spacehab (NASA) ***1995***

Stacked SRBs await ET mating in the Vehicle Assembly Building at Kennedy Space Center

Guerir Airfield (Morocco). All sites provide Shuttle visual and navigational aids (principally TACAN and a Microwave Landing System); two or three are activated for each ascent and one (usually Ben Guerir) remains active throughout in case of emergency de-orbit. There are also two ACLS (Augmented Contingency Landing Site), at Hickham AFB (Honolulu) and Andersen AFB (Guam). They are similar to TAL sites in facilities and equipment but used instead for contingency de-orbits and not ascent aborts.

Abort Once Around (AOA) would be used from about 2 minutes after SRB separation, again in the event of a main engine failure. The procedure relies on two OMS firings after ET jettison to place the Orbiter in a sub-orbital coast and free return orbit for re-entry and glide back to the runway at KSC, Edwards or White Sands. Abort To Orbit (ATO) is available in the event of a main engine failure after passing the AOA point. Again the procedure relies upon two OMS firings, one for orbit insertion and the other to circularise the path. An alternative mission might then be possible, depending on mission rules. An ATO occurred on STS-51F in July 1985, but most of that mission was completed. Each mission has its emergency landing sites, which vary with the launch profile, but they carry no Shuttle equipment or aids. In the case of STS-51F, a landing could have been made at Zaragoza, although it would have taken at least 5 weeks to recover Challenger. The AOA site was White Sands Space Harbour, New Mexico.

One abort site safety device is a 13.7 m high arresting net attached to an energy-absorbing tape drum, capable of stopping an Orbiter travelling at 185 km/h. Tests of this Shuttle Orbiter Arresting System (SOAS) were made with Enterprise in June 1987 at Washington DC Dulles airport. Nets were installed at the four TAL sites and at Hickham AFB. Similar abort procedures were devised for Vandenberg launches and arrangements were made for emergency landings on several Pacific Islands, including Chile's Easter Island.

ESA's STS-66 mission specialist Clervoy tested a new reclining seat for returning long-duration occupants from Mir. Behind him is the airlock hatch; to his right are sleep cubicles (NASA) ***1995***

Crew escape system

Following Challenger's loss, each Orbiter was retrofitted with a telescoping pole escape system to permit evacuation while the craft is in a controlled glide descent. As such, it provides very limited cover and would primarily be called upon in the event of a water ditching, which would probably be non-survivable for a crew remaining with the craft. The system comprises a 109 kg, 7.5 cm diameter aluminium/steel pole that would be extended 2.67 m through the jettisoned hatch to guide evacuating crew 45° down and 15° back to avoid the port wing. The abort decision must be made below 18,300 m, with the Cdr and GPCs achieving a controlled glide descent before the designated jumpmaster, occupying the mid-deck seat closest to the door, manually depressurises the cabin and blows the hatch pyrotechnically at 6,710 m. Bale-out begins at 6,100 m, with each of up to eight crew members attaching the suit lanyard to a hook to slide down the pole. The main parachute opens automatically within 6 seconds. All crew members now wear 32 kg high-altitude pressure suits carrying emergency oxygen, liferaft and other survival equipment.

The transcript of onboard crew conversation, plus the fact that at least two of the air-breathing packs were activated before Challenger fell into the Atlantic, convinced the astronauts that even though it was a 'worst case' accident, at least some crew members could have survived if there had been an escape system.

Recent missions

STS-66 ATLANTIS

Orbiter flight: 13
Mobile Launch Platform: 3
Launch pad: KSC 39B
Launch date/time: 3 November 1994 at 16.59.43 GMT
Landing date/time: 14 November 1994 at 15.33.45 GMT
Mission duration: 10 days 22 hours 34 minutes 2 seconds
Landing site: EAFB
Main engines: 1: 2030 2: 2034 3: 2017
Launch mass: =~2,045.0 t
Landing mass: 95,184 kg
Crew: Donald R McMonagle (Commander), Curtis L Brown Jr (Pilot), Ellen Ochoa (Payload Commander/ Mission Specialist -1), Joseph R Tanner (Mission Specialist 2/EV2), Jean-François Clervoy (Mission Specialist 3/ESA), Scott E Parazynski (Mission Specialist 4/EV1).
Payload (Primary):
Atlas 3 Spacelab (3,759 kg)
Crista-SPAS (3,263 kg)
Payload (Secondary): cargo bay,
ESCAPE II Experiment of the Sun for Complementing the Atlas Payload and for Education
Payload (Secondary): mid-deck,
PCG Protein Crystal Growth
HPP 2 Heat Pipe Performance
PARE R1 Physiological & Anatomical Experiment – Rodents (see STS-56)
SAMS Shuttle Acceleration Measurement System (supporting PCG)
STL Space Tissue Loss (flew STS-45/53/56/59)

Mission summary

The third Atmospheric Lab for Applications and Science (Atlas) Spacelab carried six instruments on a Spacelab pallet, plus a seventh in two sidewall cans, to study the atmosphere's chemical composition over 15-600 km, measure sunlight's total energy and variation and how these factors affect ozone levels. Atlas was intended to fly every one to two years throughout the Sun's 11-year activity cycle, complementing UARS observations and providing a baseline for discovering changes by flying highly calibrated instruments. NASA has dropped the original schedule and Atlas 3 was the last manifested. Six of the instruments (ATMOS, MAS, SSBUV, ACRIM, SOLCON, SOLSPEC) flew on STS-45/56 Atlas 1/2. SUSIM, added to measure the Sun's UV spectrum, also flew on Atlas 2.

Although MAS failed on day 2, Atlas 3 allowed investigators to study the northern hemisphere atmosphere as it moved from its relatively quiet Summer state to its more active Winter conditions. It also observed the Antarctic ozone hole near its maximum for comparison with the March and April data of Atlas 1/2.

Germany's retrievable Crista-SPAS complemented Atlas with the first study of global atmospheric variability by limb sounding. It returned 51,000 vertical scans, measuring 15 trace gases every one second. A GPS receiver provided the required position information. A secondary SPAS payload was the US Naval Research Lab's MAHRSI Middle Atmosphere High-Resolution Spectrograph Investigation for nitric oxide and hydroxyl measurements. OH is a key compound in ozone's destruction. MAHRSI returned four global maps of its distribution and concentration. ESA's Clervoy released SPAS from the robot arm at 12.50 GMT 4 November for it to be controlled by a German team at KSC.

SPAS' recovery demonstrated a new rendezvous technique for Shuttle missions to Mir. The Orbiter approached from below, firing thrusters away from its target. Ochoa retrieved the free-flyer at 13.05 GMT 12 November. Crista-SPAS was planned for reflight aboard STS-85 in July 1997.

ESCAPE obtained digital images of the Sun at 1,216 Å to help correlate how the Sun's extreme UV affects the temperature and chemical composition of Earth's upper atmosphere. PARE carried 10 pregnant rats in two Animal Enclosure Modules for NASA/National Institute of Health studies into µg effects on living organisms.

Clervoy also tested a special reclining seat for returning Mir's long-duration crews. The landing was 1994's fourth diverted to Edwards because of bad weather at KSC. Each diversion costs about US$1 million to return the Orbiter to Florida.

STS-63 DISCOVERY

Orbiter flight: 20
Mobile Launch Platform: 2
Launch pad: KSC 39B
Launch date/time: 3 February 1995 at 05.22.04 GMT
Landing date/time: 11 February 1995 at 11.50.19 GMT
Mission duration: 8 days 6 hours 28 minutes 15 seconds
Landing site: KSC
Main engines: 1: 2035 2: 2109 3: 2029
Launch mass: =~2,046.4 t
Landing mass: 95,835 kg
Crew: James B Wetherbee (Commander), Eileen M Collins (Pilot), Bernard A Harris Jr (Payload Commander/ Mission Specialist 1/EV2), C Michael Foale (Mission Specialist 2/EV1), Janice E Voss (Mission Specialist 3), Vladimir G Titov (Mission Specialist 4, Russia).
Payload (Primary):
Spacehab 3
Spartan 204 *Payload (Secondary): cargo bay,*
CSE Cryo Systems Experiment
GLO 2 (flew STS-53)
ODERACS 2 Orbital Debris Radar Calibration Spheres (flew STS-60)
IMAX
Payload (Secondary): mid-deck,

Mark Lee tests the SAFER rescue propulsion pack during STS-64. SAFER sits below his suit's backpack, with the controls at front. Behind him resting along the bay's sill is SPIFEX (NASA) ***1995***

Thomas (left) and Cabana at the conclusion of STS-65 (NASA) ***1995***

SSCE Solid Surface Combustion Experiment (flew STS-41/40/43/50/47/54/64)
AMOS AF Maui Optical Site

Mission summary
This was Shuttle's first launch into 51.6° and with the five minute window that will be standard for Alpha missions. NASA had to fly the high-inclination orbit with its dogleg ascent to complete the Mir rendezvous and flyaround as a final dress rehearsal for the series of dockings beginning with STS-71. Discovery approached to within 10 m of Mir for 10 minutes without disturbing the delicate solar wings. Russian controllers evinced some concern during the three-day rendezvous because an aft Mir RCS thruster began leaking oxidiser soon after launch. The problem was solved by turning off the entire group of three leaving no redundancy in the event of an emergency. A nose thruster began leaking 4 February but the problem was cured by warming it in the Sun (a standard technique that failed on the aft unit).

Shuttle astronauts Wetherbee and Collins took over manual control from 600 m out using the aft flight deck, slowing the closing rate down at 100 m to 30 cm/s. A camera in Spacehab's top window provided a direct view. Foale operated two laser ranging systems, namely the cargo bay Trajectory Control System and a hand-held unit. Discovery halted within 10 m at about 19.20 GMT 6 February. 10 minutes later it began backing off to 150 m and spent 45 minutes photographing Mir in detail. Discovery finally departed 21.13 GMT.

Spartan 204 carried the Naval Research Lab's FUVIS Far-UV Imaging Spectrograph. Titov used it on day 2 on the RMS to observe oxygen glow on Discovery's tail and observed thruster firings. He released it at 12.26 GMT 7 February for autonomous astronomical observations. As Voss recovered it at 11.53 GMT 9 February, Harris and Foale waited in the airlock to begin their EVA. They began their 278 minute EVA on the end of the RMS away from the warm cargo bay to test suit modifications. One modification disconnected the arm tubes in their liquid-cooled undergarment. While they were holding Spartan during a night pass to investigate the handling of large masses, they reported they were both very cold and the EVA was ended. Sensors showed Foale's glove fingers had fallen to 4°C. New electronic cuff checklists failed, apparently because of the cold.

The small SAFER (seen below the main backpack, with chest controls) carries rescue thrusters should the wearer drift free of Mir or Alpha (NASA) ***1995***

Spacehab 3 carried 20 primarily μg experiments (some were actually housed in the mid-deck). McDonnell Douglas' robotic box-like Charlotte moved along cables extending from each of its eight corners to Spacehab attach points. Reels allowed it to move anywhere inside Spacehab, demonstrating remote control and observation of experiments. Its arm, equipped with a simple gripper, can flip switches and exchange samples and data cartridges.

Goddard's Hitchhiker bridge carried four experiments plus the IMAX camera. Among the experiments the CSE evaluated an integrated Stirling 65K cryocooler and oxygen diode heat pipe, the GLO 2 made glow observations of Orbiter surfaces, airglow and contaminating events at 1,150-11,590 Å. The Hitchhiker also contained two SAAMD stand-alone acceleration measurement devices and ODERACS 2.

On day two, ODERACS ejected 5, 10 and 15 cm diameter spheres and three dipoles (two 13.3 cm long and one 4.42 cm) to help calibrate radar and optical tracking systems for debris monitoring, particularly the Haystack Radar.

At the end of the mission, the Shuttle performed the first KSC landing since the runway was resurfaced in September 1994 to decrease tyre wear and increase crosswind tolerances.

STS-67 ENDEAVOUR
Orbiter flight: 8 (5th Extended Duration Orbiter)
Mobile Launch Platform: 1
Launch pad: KSC 39A
Launch date/time: 2 March 1995 at 06.38.13 GMT
Landing date/time: 18 March 1995 at 21.47.47 GMT
Mission duration: 16 days 15 hours 9 minutes 34 seconds
Landing site: EAFB
Main engines: 1: 2012 2: 2033 3: 2031
Launch mass: =~2,050.6 t
Landing mass: 98,879 kg
Crew: Stephen S Oswald (Commander), William G Gregory (Pilot), John Grunsfeld (Mission Specialist 1/EV2), Wendy B Lawrence (Mission Specialist 2), Tamara E Jernigan (Payload Commander/Mission Specialist 3/EV1), Ronald A Parise (PS1), Samuel T Durrance (PS2).
Payload (Primary):
Astro 2
Payload (Secondary): cargo bay,
Endeavour UV telescope (flew STS-42)
Payload (Secondary): mid-deck,
MACE Mid-deck Active Control Experiment
PCG Protein Crystal Growth (flew STS-66)
CMIX 3 Commercial Materials
ITA Experiment (flew STS-52/56)
SAREX-II Shuttle Amateur Radio Experiment

Mission summary
Astro's set of three broadband ultra-violet telescopes, designed to observe astrophysical targets such as quasars and active galactic nuclei, first flew aboard SM38/STS-35 in December 1990. Astro 2 was cancelled in January 1990 to save US$10 million to US$20 million and the telescopes were stored at KSC.

Endeavour configuration by late 1989

However, NASA announced in May 1991 that an Astro 2 mission would be funded, although Goddard's complementary Broad Band X-ray Telescope (BBXRT) was not re-flown because Japan's 1993 Asuka satellite carries a similar telescope.

Carried by ESA's Instrument Pointing System (IPS) on two Spacelab pallets, Astro's payload carried three instruments. The most important is the Hopkins UV Telescope (HUT), The Johns Hopkins University, 90 cm f/2 silicon carbide-coated paraboloid mirror optimised for 900-1,200 Å but extending to 425-1,850 Å. It has a prime focus Rowland Circle spectrograph, 3.0 Å spectral resolution and weighs 789 kg. Next is the UIT UV Imaging Telescope and film cameras, a 38 cm f/9 Ritchey-Chretien optics, 1,200-3,200 Å, 40 arcmin FOV, 2 arcsec resolution, device built by NASA Goddard. It can see down to a magnitude limit of +25 with two image intensifiers and 70 mm film. The cameras can take 1,000 frames each, with exposures up to 30 minutes. It weighs 474 kg. A University of Wisconsin research group built the Wisconsin UV Photo-Polarimeter Experiment (WUPPE), with 50 cm f/10 Cassegrain optics, 1,400-3,200 Å, and detector spectropolarimeter with dual electronic diode array detectors. The device studies stars (particularly white dwarfs), interstellar medium, asteroids, comets, nebulas and galaxies with 3.3 × 4.4 arcmin resolution and a magnitude limit of +16, and spectral resolution of 6 Å. It weighs 446 kg.

With the UIT/WUPPE Spacelab's IPS stabilisation improves to less than 1 arcsec by a gyro-based image motion compensation system adjusting the secondary mirrors. IPS activation began 200 minutes after launch.

MACE was a 1.5 m long flexible beam with scanning and pointing payloads at each end and a central control system with a rate gyro and reaction wheels to test active damping techniques.

Endeavour's first EDO mission (and the 99th US manned launch) set a Shuttle duration record.

STS-71 ATLANTIS
Orbiter flight: 14
Mobile Launch Platform: 3
Launch pad: KSC 39A
Launch date/time: 27 June 1995 at 19.32.19 GMT
Landing date/time: 7 July 1995 at 14.54.34 GMT
Mission duration: 9 days 19 hours 22 minutes 15 seconds
Landing site: KSC
Main engines: 1: 2028 2: 2034 3: 2032
Launch mass: =~2,046. 4 t
Landing mass: 97,447 kg
Crew: Robert L Gibson (Commander), Charles J Precourt (Pilot), Ellen S Baker (Mission Specialist 1/EV2), Gregory J Harbaugh (Mission Specialist 2/EV1), Bonnie J Dunbar (Mission Specialist 3), Anatoli Y Solovyov (Russia, Mir 19; up only), Nikolai M Budarin (Russia, Mir 19; up only), Vladimir N Dezhurov (Russia, Mir 18; down only), Gennady M Strekalov (Russia, Mir 18; down only), Norman E Thagard (Mir 18; down only).
Payload (Primary):
Spacelab
Payload (Secondary): cargo bay,
ODS Orbiter Docking System
Payload (Secondary): mid-deck,
SAREX-II Shuttle Amateur Radio Experiment
IMAX In-Cabin Camera

Mission summary
Atlantis departed at the beginning of a 10-minute window on the 100th US manned flight and the first of seven planned dockings with Mir in preparation for the International Space Station. After launch, the pilot Precourt halted 350 m directly underneath Mir by 10.50

TSS in Alenia Spazio's anechoic chamber during final tests in Turin before delivery to Kennedy Space Center ***1992***

GMT 29 June and then used a mean distance navigation strategy for final approach. This allowed thruster firings *away* from the passive Mir to avoid damage to the delicate solar wings, which remained in a feathered configuration as a further precaution.

Precourt station-kept at 82 m until Mir control gave permission to proceed. He used a camera inside the ODS to assist with alignment by aiming it through the hatch window at Kristall's hatch target. He again halted at 10 m until the final approach at 12.55 GMT. The ODS docking module completed capture at 13.00.15 GMT at 395 km altitude over Lake Baikal. The ODS capture ring carried three guide petals, each with two latches for soft docking. The ring was held 33 cm out from the ODS base ring by six 6-DoF shock absorbers. At contact, the closing rate was 3.26 cm/s (5.1 cm/s allowed), displacement error 1.8 cm, angular error 0.4° (4° allowed). At that point, Precourt triggered a 2.46 second computer sequence firing Shuttle's aft thrusters 14 times and the forward thrusters six times to push the two vehicles together. Harbaugh retracted the ring for the 12 base latches to hook on to Kristall, creating a 223 tonne space complex.

Precourt & Dezhurov shook hands some 2 h later and the record 10-man team collected in Mir's core for photographs. Mir was formally handed over to the fresh cosmonauts. The Shuttle delivered 200 kg of food, clothing and science equipment and more than 500 kg of water were delivered and Mir's pressure was raised from 0.92 atmosphere to 1.02 atmosphere. Shuttle provided most of the attitude control, holding the complex for optimum power generation during working hours.

Most of the rest of the joint mission investigated the long-term physiological effects on the Mir 18 crew, aboard Mir since March 1995. Some 400 saliva, urine and blood samples were returned from Mir/Spacelab. Spacelab's treadmill and Lower Body Negative Pressure unit helped to prepare the cosmonauts for return to Earth.

Kristall's hatch was closed at 19.32 GMT 3 July and Shuttle's at 19.48 GMT. Solovyov and Budarin undocked from Kvant 1 in Soyuz-TM 21 at 10.55 GMT 4 July purely to photograph Shuttle's 11.10 GMT undocking and flyaround. However, Mir began drifting when its main computer failed, so TM21 redocked 5 minutes early at 11.39 GMT. Mir remained drifting until they could replace the hardware. Precourt fired his separation thrusters at 12.35 GMT 4 July to end the joint phase. The three Mir 18 returnees occupied reclining

STS-48 astronauts Brown (left) and Buchli assemble a Space Station truss model to study its vibration characteristics. An improved version flew on STS-62 in March 1994 (NASA) ***1991***

Endeavour's cockpit, minus seats (Rockwell)

mid-deck seats designed to help their adaptation to 1 *g*. US record-holder Thagard insisted on standing up 10 minutes after landing, while his Russian colleagues were stretched out to three weeks of biomedical tests. Thagard reported that his physical condition was excellent, but his greatest challenge had been cultural isolation – sometimes going days without talking to a native English speaker.

STS-70 DISCOVERY

Orbiter flight: 21
Mobile Launch Platform: 2
Launch pad: KSC 39B
Launch date/time: 13 July 1995 at 13.41.55 GMT
Landing date/time: 22 July 1995 at 12.02.00 GMT
Mission duration: 8 days 22 hours 20 minutes 5 seconds
Landing site: Kennedy Space Center
Main engines: 1: 2036/B1 2: 2019 3: 2017
Launch mass: =~2,051.1 t
Landing mass: 88,411 kg
Crew: Terrence T Henricks (Commander), Kevin R Kregel (Pilot), Donald A Thomas (Mission Specialist 1/EV2), Nancy J Currie (Mission Specialist 2), Mary E Weber (Mission Specialist 3/EV1).
Payload (Primary):
TDRS 7
Payload (Secondary): mid-deck,
BRIC Biological Research in Canisters (flew STS-64/68)
BDS Bioreactor Development System
CPCG Commercial Protein Crystal Growth block III
PARE R2 Physiological & Anatomical Experiment -Rodents
STL Space Tissue Loss (flew STS-45/53/56/59/63)
HERCULES Hand-held Earth-oriented Real-time Co-operative User-friendly Location-targeting & Environmental System (flew STS-53/56)
MIS Microcapsules in Space (flew STS-53)
MAST Military Application of Ship Tracks (flew STS-65/64/68)
RME-III Radiation Monitoring Experiment (flew STS-48/44/42/45/53/56/51/64)
SAREX-II Shuttle Amateur Radio Experiment
WINDEX Window Experiment
VFT Visual Function Tester 4 (flew STS-27/28/36/44/45/53/59)

Mission summary

NASA wanted to put STS-70 in space on 8 June but cancelled it on 2 June after a woodpecker attacked the ET's insulation. It left more than 100 holes of up to 4 cm, which NASA thought, would delay the tight schedule of STS-71. So STS-70 was rolled back to the VAB on 8 June and STS-71 became the next mission. Only six days separated the launch of 70 and 71, a NASA record.

NASA described the debut of the Block 1 improved SSME (#2036) as flawless. After launch, STS-70 released TDRS-7 for orbital insertion at 19.55 GMT 13 July. Houston's new-generation Flight Control Room handled, albeit partially (the orbital phase after IUS deployment). NRL's HERCULES camera, which first flew on STS-53 to photograph Earth features and simultaneously record their latitude and longitude to within 3.7 km added the capability of transmitting images and location data to the ground for analysis during flight with STS-56. The STS-70 version added a 10 m resolution multispectral night vision array package, although there were IMU alignment problems. HERCULES also transmitted an image of a debris crater discovered in one window. Following STS-70, OV-103 stood down from service before its nine-month maintenance period at Palmdale from October 1995.

STS-69 ENDEAVOUR

Orbiter flight: 9
Mobile Launch Platform: 1
Launch pad: KSC 39A
Launch date/time: 7 September 1995 at 15.09.00 GMT

Crew module for replacement Orbiter Endeavour

Landing date/time: 18 September 1995 at 11.37.56 GMT
Mission duration: 10 days 20 hours 28 minutes 56 seconds
Landing site: KSC
Main engines: 1: 2035 2: 2109 3: 2029
Launch mass: =~2,050.3 t
Landing mass: 99,740 kg
Crew: David M Walker (Commander), Kenneth D Cockrell (Pilot), James S Voss (Payload Commander/Mission Specialist 1/EV1), James H Newman (Mission Specialist 2), Michael L Gernhardt (Mission Specialist 3/EV2).
Payload (Primary):
Wake Shield Facility. 2 × 1,950 kg Spartan. 1,288 kg *Payload (Secondary): cargo bay,*
GAS Bridge Assembly including CAPL Capillary Pumped Loop 2 (flew STS-60, ESA)
IEH International Extreme UV Hitchhiker
ITEPC Inter-Mars Tissue-Equivalent Proportional Counter
Payload (Secondary): mid-deck,
APE-B Auroral Photography Experiment (flew STS-43/51/60/62)
BRIC Biological Research in Canisters (flew STS-64/68/70)
CGBA Commercial Generic Bioprocessing Apparatus (flew STS-50/54/60/62/63)
CMIX 4 Commercial Materials
ITA Experiment (flew STS-52/56/67)
EPICS Electrolysis Performance Improvement Concept Study
STL Space Tissue Loss (flew STS-45/53/56/59/66/70)

Mission summary

Engineers delayed STS-69 from its 5 August launch when they discovered hot gases had damaged O-rings on STS-70 and 71's SRB nozzle joints. Thiokol reinsulated STS-69's.

Spartan 201-03 was a reflight of STS-56's 201-01 and 201-02 on STS-64. Gernhardt released it from the RMS 15.42 GMT 8 September, leaving Endeavour to move

STS-49 astronauts Hieb (left), Akers (centre) and Thuot (right) capture Intelsat 603 during the first-ever three-man EVA (NASA) ***1992***

Endeavour makes the 16th Shuttle landing in Florida to conclude STS-57 (NASA) ***1993***

105 km behind. The RMS retrieved the satellite at 15.02 GMT 10 September. This time the observatory viewed the Sun while ESA's Ulysses made its northern polar pass.

WSF, developed by the Space Vacuum Epitaxy Center under a NASA grant, investigated the commercial possibilities of producing new electronic, magnetic and superconducting thin-film materials and devices in the 10^{-14} torr ultravacuum of space. In orbit, however, technical problems prevented the first WSF from being released from the Shuttle during STS-60. Still it grew five 7 cm diameter GaAs wafers under 10^{-10} torr (about the best achievable on Earth). Endeavour's astronauts released WSF 2 at 11.25 GMT 11 September. The satellite used its own 0.6 N N_2 thruster to move 25 km behind Endeavour (making it the first payload to perform its own separation manoeuvre). It shut down during the fifth of seven planned runs when its batteries depleted. Despite cooling, attitude control and communications problems, SVEC declared that WSF had achieved its primary objectives. Newman grappled it 13.59 GMT 14 September after the approaching Orbiter deliberately hit it with thruster exhaust to test plume effects on rendezvous targets. During day 9, Newman hoisted it to various positions to map Endeavour's plasma wake for 5 h.

Voss and Gernhardt performed a 406 minute generic EVA during 16 September to test new glove heaters as they worked on typical station tasks. Newman lifted them away from the cargo bay's warmth and they could also shut off their body cooling loops to examine extreme conditions.

STS-73 COLUMBIA

Orbiter flight: 18 (6th Extended Duration Orbiter)
Mobile Launch Platform: 3
Launch pad: KSC 39B
Launch date/time: 20 October 1995 at 13.53.00 GMT
Landing date/time: 5 November 1995 at 11.45.21 GMT
Mission duration: 15 days 21 hours 52 minutes 21 seconds
Landing site: KSC
Main engines: 1: 2037/B1 2: 2031 3: 2038/B1
Launch mass: =~2,051.5 t
Landing mass: 104.400 kg (heaviest landing to date)
Crew: Kenneth D Bowersox (Commander), Kent V Rominger (Pilot), Catherine G Coleman (Mission Specialist 1/EV2), Michael E Lopez-Alegria (Mission Specialist2/EV1), Kathryn C Thornton (Payload Commander/Mission Specialist 3), Fred W Leslie (PS1), Albert Sacco Jr (PS2).
Payload (Primary):
US Microgravity Lab-2 Spacelab 10,308 kg of experiments composed of Crystal Growth Furnace-4 experiment, Drop Physics Module, Surface Tension Driven Convection Experiment, Geophysical Fluid Flow Cell, ESA Advanced Protein Crystallisation Facility, ESA Glovebox - 7 experiment, SAMS 07 Space Acceleration Measurement System, 3-D µg Accelerometer, STABLE Suppression of Transient Accelerations by Levitation Evaluation.
Payload (Primary): mid-deck,
Astroculture
CGBA Commercial Generic Bioprocessing Apparatus
CPCG Commercial Protein Crystal Growth
ZCG Zeolite Crystal Growth
SPCG Single Locker Protein Crystal Growth
Payload (Secondary):
OARE Orbital Acceleration Research Experiment

Mission summary

USML 2 followed up and repeated experiments flown aboard STS-50/USML 1, conducting 16 experiments in materials, fluid physics, combustion and biotechnology. The crew divided into two shifts to pursue the experiments 24 hours a day, laying the foundations for µg research aboard the International Space Station. It was Columbia's first flight since its most recent inspection at Palmdale after STS-64 and the first to carry two Block 1 engines. The Orbiter should have been the first to fly the full set of new engines, but #2039 was replaced by the old #2031 when technicians feared that a seal in the upgraded LOX pump was improperly seated. It took six attempts to get the Columbia launched.

STS-74 ATLANTIS

Orbiter flight: 15
Mobile Launch Platform: 2

Russian Vladimir Titov handles samples for Spacehab 3's Commercial Generic Bioprocessing Apparatus during STS-63 (NASA) ***1995***

STS-71 cargo bay configuration for Mir docking (NASA) ***1996***

Launch pad: KSC 39A
Launch date/time: 12 November 1995 at 12.30.43 GMT
Landing date/time: 20 November 1995 at 17.01.27 GMT
Mission duration: 8 days 4 hours 30 minutes 44 seconds
Landing site: KSC
Main engines: 1: 2012 2: 2026 3: 2032
Launch mass: =~2,046.5 t
Landing mass: 92,700 kg
Crew: Kenneth D Cameron (Commander), James D Halsell Jr. (Pilot), Chris A Hadfield (Mission Specialist 1, Canada), Jerry L Ross (Mission Specialist 2/EV1), William S McArthur Jr (Mission Specialist 3/EV2)
Payload (Secondary): cargo bay,
IMAX camera
GLO 4
PASDE Photogrammetric Appendage Structural Dynamics Experiment (measured motion of Mir solar arrays)
ITEPC Inter-Mars Tissue Equivalent Proportional Counter (flew STS-69)
Payload (Secondary): mid-deck,
SAREX-II Shuttle Amateur Radio Experiment

Mission summary

RSA and NASA planned to use this second docking mission to Mir to attach RKK Energia's 4,087 kg docking adaptor and to perform two days of life sciences experiments. After installation, the 2.2 m diameter 4.7 m long androgynous docking module attaches to Kristall at Mir's starboard position so that future dockings will adequately clear Mir's solar wings. Without it, Kristall has to be shifted to the forward face during each Shuttle visit. In order to perform the final docking, four DM retroreflectors guide the Shuttle's Laser Trajectory Control Sensor to generate 6 Degrees of Freedom (DoF) information about position between the two vehicles.

During 14 November, Hadfield unberthed the DM from the aft cargo bay and held it 12 cm above 104's Orbiter Docking System, reused from STS-71. With RMS in 'limp mode', Cameron thrusted Atlantis into contact for retraction at 07.17 GMT. Atlantis docked at 06.27.39 GMT 15 November and astronauts opened the seven hatches at 09.00 GMT when the pressures had equalised.

STS-74 delivered 275 items totalling 968 kg and took away 195 items (370 kg). 400 kg of water in containers, 65 kg of air, experimental equipment, fresh food (including vegetables, steak and ice cream), gifts of flowers and sweets, and a collapsible classical guitar were among the goods brought aboard. The Shuttle

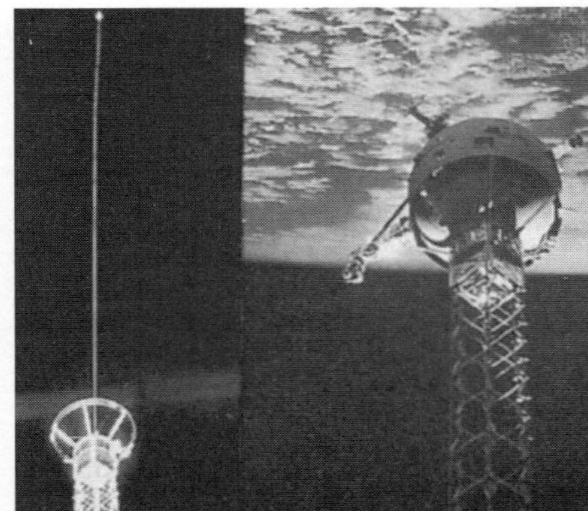

Beginning deployment of Italy's tethered satellite during STS-46. The 2.54 mm tether repeatedly jammed in the reel mechanism (the box unit at the boom's apex). At left, the sphere approaches its maximum altitude (NASA and Alenia Spazio) ***1992***

Training with the HERCULES camera, which includes a ring laser gyro for accurate pointing information ***1993***

departed with 20 kg of ESA cargo from Reiter's activities, including frozen medical samples, furnace material samples and science data on videotape and computer disk.

Atlantis undocked 08.16 GMT 18 November to circle Mir at 130 m for 45 minutes while the crew took detailed engineering photographs. STS-74 carried the 200th individual to fly in the Shuttle.

STS-72 ENDEAVOUR
Orbiter flight: 10
Mobile Launch Platform: 1
Launch pad: KSC 39B
Launch date/time: 11 January 1996 at 09.41.00 GMT
Landing date/time: 20 January 1996 at 07.41.41 GMT
Mission duration: 8 days 22 hours 0 minutes 41 seconds
Landing site: KSC
Main engines: 1: 2028 2: 2039/B1 3: 2036/B1
Launch mass: =~2,048.0 t
Landing mass: 98,553 kg
Crew: Brian K Duffy (Commander), Brett W Jett Jr (Pilot), Leroy Chiao (Mission Specialist 1/EV1), Winston E Scott (Mission Specialist 2/EV3), Koichi Wakata (Mission Specialist 3, Japan), Daniel T Barry (Mission Specialist 4/EV2).
Payload (Primary):
Spartan 206 OAST-Flyer
Payload (Secondary): cargo bay,
SSBUV/A Shuttle Solar Backscatter UV
SLA Shuttle Laser Altimeter
5 GAS
Payload (Secondary): mid-deck,
PARE R3 Physiological & Anatomical Rodent Experiment (three female rats, each with 10 newborn)
PCG/STES 4 Protein Crystal Growth Single Locker Thermal Enclosure System
CPCG Commercial Protein Crystal Growth
STL Space Tissue Loss (flew STS-45/53/56/59/66/70/69)

Mission summary
NASDA's astronaut Koichi Wakata captured Japan's SFU (Space Flyer Unit) with the RMS at 10.57 GMT 13 January after one orbit caused by the free-flyer's solar wings refusal to latch down after retraction. They had to be pyrotechnically jettisoned. Wakata released Goddard's OAST-Flyer from the RMS 11.32 GMT 14 January for two days of separate operations, regrappling it 09.47 GMT 16 January. The satellite conducted four experiments, namely: surface contamination investigations; the first attitude control of a satellite using GPS data; a laser-triggered pyrotechnic device firing and amateur packet radio.

STS-69 prepares to release the Wake Shield Facility (NASA) ***1996***

Astronauts Chiao and Barry performed a 369-minute EVA 1 15 January, to test generic Space Station operations such as moving equipment, the strength of a flexible foot restraint and connecting simulated fluid lines. Chiao and Scott followed with a 414-minute EVA on 17 January performing similar duties and including Scott remaining still for 30 minutes in the shadowed cargo bay to test suit modifications. The test proved the suit could cope with −60°C conditions during Station work.

SLA began a series of four flights sampling land topography and vegetation data from a spaceborne laser altimeter.

STS-75 COLUMBIA
Orbiter flight: 19 (7th Extended Duration Orbiter)
Mobile Launch Platform: 3
Launch pad: KSC 39B
Launch date/time: 22 February 1996 at 20.18.00 GMT
Landing date/time: 9 March 1996 at 13.58.21 GMT
Mission duration: 15 days 17 hours 40 minutes 21 seconds
Landing site: KSC
Main engines: 1: 2029 2: 2034 3: 2017
Launch mass: =~2,053.2 t
Landing mass: 103,160 kg
Crew: Andrew M Allen (Commander), Scott J Horowitz (Pilot), Jeffrey A Hoffman (Mission Specialist 1), Maurizio Cheli (Mission Specialist 2/ESA), Claudie Nicollier (Mission Specialist 3/ESA/EV2), Franklin R Chang-Diaz (Payload Commander/Mission Specialist 4/EV1), Umberto Guidoni (PS1, Italy).
Payload (Primary):
TSS 1R Tethered Satellite System (system mass 5,506 kg)
USMP 3 US Microgravity Payload
Payload (Secondary):
OARE Orbital Acceleration Research Experiment
Payload (Secondary): mid-deck,
Glovebox (combustion studies)
CPCG 09 Commercial Protein Crystal Growth

Mission summary
The joint Italian and US TSS 1R reflew STS-46's tethered satellite system. In the first flight the tether jammed at only 256 m instead of the planned 20 km. When fully deployed, the Italian designers expected that 2.3 kW should have flowed along the 2.5 mm tether's copper core down to the Shuttle. In this test, the researchers expected to deploy the 1.6 m diameter sphere from a 12 m boom to 20.7 km. They planned for the deployment to take 5.5 h with no stops. Then, after 22 h on station, the satellite retrieval would require 17 h, including a stop at 3.2 km for experiment operation.

Unreeling began 20.45 GMT 25 February. Unfortunately, the tether snapped near the top of the boom without warning at 01.30 GMT 26 February while TSS was 19.7 km out, leaving the sphere in a 316 × 414 km, 28.47° orbit. Since the original orbit began at 294 × 304 km, the Italians at least demonstrated orbit raising without propellant. At the time of failure, the cable experienced only 70 N of tens on against a testing breaking strength of 1.78 kN. NASA Marshall controllers continued taking TSS data for 4 days until the batteries were exhausted early 1 March. Ground observers clearly witnessed the stable tether before the 00.55 GMT re-entry 19 March over the Atlantic.

The two missions cost US$404 million, but no more are planned. A failure review board concluded in June 1996 that the cable insulating layer had been penetrated, possibly by debris squeezed in while it was still wound on the reel. Electric arcing began while it was in the lower tether control mechanism, sputtering intermittently for 9 seconds as the moving tether passed through the deployer mechanisms and into the boom area. The remaining Kevlar strands could not withstand the small deployment forces.

STS-76 ATLANTIS
Orbiter flight: 16
Mobile Launch Platform: 2
Launch pad: KSC 39B
Launch date/time: 22 March 1996 at 08.13.04 GMT
Landing date/time: 31 March 1996 at 13.28.57 GMT
Mission duration: 9 days 5 hours 15 minutes 53 seconds
Landing site: EAFB
Main engines: 1: 2035 2: 2109/B1 3: 2019
Launch mass: =~2,045.5 t
Landing mass: 95,400 kg
Crew: Kevin P Chilton (Commander), Richard A Searfoss (Pilot), Ronald M Sega (Mission Specialist 1), Michael R Clifford (Mission Specialist 2/EV2), Linda M Godwin (Mission Specialist 3/EV1), Shannon W Lucid (Mission Specialist 4, up only).
Payload (Secondary): cargo bay,
MEEP Mir Environmental Effects Programme
TRIS Trapped Ions in Space
Payload (Secondary): mid-deck,
Kidsat
SAREX-II Shuttle Amateur Radio Experiment

STS-57's Wisoff on the RMS watches David Low working on Eureca's antenna. Spacehab is in the foreground (NASA) ***1993***

Mission summary
NASA intended to use this flight to deliver Shannon Lucid to the Mir in its third docking with the space station, to perform an EVA and deliver supplies. Atlantis docked at 02.34 GMT 24 March with Kristall's adaptor module (attached by STS-74). Clifford and Goodwin's 363-minute EVA, beginning 06.34 GMT 27 March from Shuttle's tunnel hatch, attached the MEEP to Mir to gather space debris and materials data. It also tested common US/Russian tether hooks and foot restraints. The Shuttle transferred some 640 kg of water along with 880 kg of Russian supplies (such as one gyrodyne and three batteries returned by STS-71), 740 kg US science equipment and 93 kg other items. About 60 per cent of Spacehab's 2,010 kg cargo were Russian supplies. 450 kg returned, with the Shuttle including 110 kg science payload. One experiment mapped Mir's radio and electric field environment. Then the Shuttle undocked at 01.08 GMT 29 March. Because of the orbit inclination, the more stringent landing conditions resulted in a switch to EAFB for the return.

STS-77 ENDEAVOUR
Orbiter flight: 11
Mobile Launch Platform: 1
Launch pad: KSC 39B
Launch date/time: 19 May 1996 at 10.30.00 GMT
Landing date/time: 29 May 1996 at 11.09.18 GMT
Mission duration: 10 days 0 hours 39 minutes 18 seconds
Landing site: KSC
Main engines: 1: 2037/B1 2: 2040/B1 3: 2038/B1
Launch mass: =~2,049.8 t
Landing mass: 100,420 kg
Crew: John H Casper (Commander), Curtis L Brown Jr (Pilot), Andrew S W Thomas (Mission Specialist 1), Daniel W Bursch (Mission Specialist 2/EV2), Mario Runco Jr (Mission Specialist 3/EV1), Marc Garneau (Mission Specialist 4, Canada)
Payload (Primary):
Spacehab 4 Spacehab carried commercial space product development payloads made up of the Advanced Separation Process for Organic Materials, CGBA Commercial Generic Bioprocessing Apparatus, Plant GBA, Fluids GBA 2, Hand-Held Diffusion Test Cell, Immune 03, CPCG Commercial Protein Crystal Growth, GPPM Gas-Permeable Polymeric Membrane, Commercial Float Zone Furnace, Space Experiment Facility, Space Tissue Loss/NIH-Cells 07.

STS-79 was the first to carry the Spacehab Double Module, providing 2.7 t cargo capacity (Spacehab) ***1996***

The new Multifunction Electronic Display Subsystem (MEDS) is shown in the cockpit of Orbiter Atlantis ***1999***/0054448

Spartan 207/IAE The Spartan 207 Inflatable Antenna Experiment deployed JPL's 12.2 m diameter 60 kg aluminised mylar parabolic antenna structure on three 28 m inflatable struts. It was jettisoned after its surface accuracy was measured and Spartan was recovered.

TEAMS 01 Technology Experiments Advancing Missions in Space. TEAMS' PAMS released a small passive satellite from its Hitchhiker canister, illuminating it with a laser to observe its attitude as it was stabilised aerodynamically and magnetically.

GANES GPS Attitude & Navigation Experiment

Liquid Metal Test Experiment

PAMS Passive Aerodynamically Stabilised Magnetically Damped Satellite

VTRE Vented Tank Resupply Experiment

Payload (Secondary): cargo bay,

BETSCE Brilliant Eyes Ten-Kelvin Sorption Cryocooler Experiment,

GAS Bridge (12 cans)

RFTPCE Reduced-Fill Tank Pressure Control Experiment

Payload (Secondary): mid-deck:

ARF Aquatic Research Facility. The Canadian Space Agency/NASA ARF debuted as a modular facility that can support different biological experiments in two standard mid-deck lockers. This flight carried mussels, starfish and sea urchins.

BRIC 07 Biological Research in Canisters (flew STS-64/68/70/69)

STS-78 COLUMBIA

Orbiter flight: 20 (8th Extended Duration Orbiter)
Mobile Launch Platform: 3
Launch pad: KSC 39B
Launch date/time: 20 June 1996 at 14.49.00 GMT
Landing date/time: 7 July 1996 at 12.37.45 GMT
Mission duration: 16 days 21 hours 47 minutes 45 seconds
Landing site: KSC
Main engines: 1: 2041/B1 2: 2039/B1 3: 2036/B1
Launch mass: =~2,049.3 t
Landing mass: 103,424 kg
Crew: Terrence T Hendricks (Commander), Kevin R Kregel (Pilot), Richard M Linnenhan (Mission Specialist 1/EV2), Susan J Helms (Payload Commander/Mission Specialist 2/EV1), Charles E Brady Jr (Mission Specialist 3), Jean-Jacques Favier (PS1, France), Robert Brent Thirsk (PS2, Canada)
Payload (Primary):

LMS 1 Life and Microgravity Sciences.

Payload (Secondary): payload bay,

Orbital Acceleration Research Experiment

Payload (Secondary): mid-deck,

SAREX-2 (Shuttle Amateur Radio Experiment). Flown in configuration C which operates in either voice or data mode in communications with amateur stations within line of sight of the orbiter).

BRIC-08 (Biological Research in Canisters, block 2 configuration)

NASA's ACTS communications satellite begins the journey out of Shuttle's cargo bay (NASA) ***1993***

Mission summary

After a successful launch, the crew of the 78th Shuttle mission successfully activated the Spacelab, which housed more than 40 life science and microgravity experiments in the Shuttle's cargo bay. The experiments measured the physiological changes that take place in the human body during space flight and also on utilising the unique microgravity environment to study materials processing techniques. On the first day in orbit, the flight controllers dumped the back-up flight software that had experienced intermittent transients during Columbia's launch. The day after launch, Henricks worked with the Bubble Drop Facility while Kregel laboured over a computer workstation to assess human behaviour changes during long-duration space flight. The other crewmembers worked with neurological and cardiovascular experiments. Human physiology tests on the flight included the Direct Measurement of the Initial Bone Response to Space Flight investigation, which scientists used to explore ways to counter this detrimental effect of space flight.

On 22 June, flight controllers assisted the astronauts in a procedure to flush an ice blockage out of the Flash Evaporator System (FES), used to dissipate heat from the shuttle. The procedure, which ran warm freon through the FES, succeeded in making the system once again fully operational. To supplement the FES, flight controllers also deployed one of two sets of radiators mounted on the Shuttle's payload bay doors to further radiate heat from the orbiter. Henricks and Kregel tested a device designed to permit voice control of the Shuttle's closed circuit television system. The voice command system should allow astronauts to command the CCTV system verbally while performing other tasks. Kregel reported that the system responded to at least 90 per cent of his verbal commands. On 24 June, the STS-78 crew members contacted friends and family members on the ground with the SAREX.

Three days later, Brady held an unlit Olympic Torch aloft while exercising on the bicycle ergometer in the Spacelab. The other astronauts on board joined him in that informal salute to the 1996 Olympics, which opened in Atlanta on 19 July.

On 28 June, the crew performed a second successful flush of a small buildup of ice from orbiter's FES after the ice formed during a dump of the supply water tanks. The procedure took about 20 minutes to complete. Canadian payload specialist Bob Thirsk observed Canada Day 1 July as the other crewmembers spent their twelfth day in orbit. The Americans then had the opportunity to observe their own 4 July holiday three days later.

After a one-day extension from the ground, on 6 July the astronauts conducted routine firings of the orbiter's reaction control system jets and checked out its flight control systems and aero surfaces in anticipation of the forthcoming landing at the Kennedy Space Center. Henricks and Kregel pulsed Columbia's vernier reaction control system jets to gently raise Columbia's altitude. These firings tested the manoeuvres needed by Discovery's next mission (STS-82) to service the Hubble Space Telescope. The vernier jet firings must be gentle enough to raise the orbit without disturbing any payloads on board, or in the case of the Hubble Space Telescope, without placing any force on the telescope's fragile solar arrays.

The return on 7 July ended the longest flight to that date in the Shuttle programme.

STS-79 ATLANTIS

Orbiter flight: 17
Mobile Launch Platform: 1
Launch pad: KSC 39A
Launch date/time: 16 September 1996 at 08.54.49 GMT
Landing date/time: 26 September 1996 at 12.13.15 GMT
Mission duration: 10 days 3 hours 18 minutes 26 seconds
Landing site: KSC
Main engines: 1: 2012 2: 2031 3: 2033
Launch mass: =~2,045.9 t
Landing mass: 97,603 kg
Crew: William F Readdy (Commander), Terrence W Wilcutt (Pilot), Jay Apt (Mission Specialist 1/EV1), Thomas D Akers (Mission Specialist 2), Carl E Walz (Mission Specialist 3/EV1), John E Blaha (Mission Specialist 4, up only), Shannon W Lucid (Mission Specialist 4, down only)
Payload (Primary):

SPACEHAB 05 Double module, orbiter docking system.

Payload (Secondary): mid-deck,

SMM-04

SAREX-2 (configuration M)

IMAX in-cabin camera

MSX Mid-course Space Experiment

Mission summary

On 1 July 1996, NASA transferred Atlantis from the VAB to pad 39A. As a precautionary measure, NASA, on 10 July, rolled the Shuttle stack back to the VAB to avoid any damage from the approach of Hurricane Bertha. Once retrieved, NASA decided to delay the launch until September, as a result of the discovery that hot gases had seeped into the field joints of the STS-78 SRBs. The ground crew destacked Atlantis from its SRBs and re-stacked it with new solids. Then NASA rolled the stack back to the pad on 20 August. Hurricane Fran forced NASA to return the stack to the VAB on 4 September as the launch awaited a one day delay. Launch finally occurred on 16 September.

After reaching its high-inclination orbit, Atlantis docked with Mir at the Docking Module attached to the Kristall on 19 September at 03.13 GMT. After two hours of pressure and leak checks, astronauts opened the hatches between the two spacecraft at 05.40 GMT. The two crews greeted one another to begin five days of joint operations. Soon after the crew members completed their welcoming ceremony, they hauled bags of water and other supplies from the Shuttle's SPACEHAB module into the Mir, eventually transferring more than two tonnes of equipment and logistical supplies to Mir.

During the first night that Atlantis was docked with Mir, scientists on the ground continued work with the Active Rack Isolation System in the SPACEHAB module, completing several successful tests. The system experimented with dampening crew motion and engine firings to protect the microgravity environment on the International Space Station.

During the first full day of joint experiments, the Biotechnology System, an investigation to study the long-term development of cartilage cells in microgravity, moved from the Shuttle to Mir. During his stay on Mir, John Blaha took weekly samples of the culture, to determine the feasibility of engineering cartilage cells for possible use in transplantation.

On 20 September, Carl Walz worked with the Mechanics of Granular Materials experiment in the double SPACEHAB module while Mission Specialist Shannon Lucid had a workout on the Mir treadmill, continuing physical conditioning exercises designed to prepare her body for the return to Earth's gravity after six months in orbit. Jay Apt monitored the ARIS experiment as Readdy and Korzun fired manoeuvring jets on their spacecraft to test the ability of the International Space Station prototype experiment rack to dampen the resulting vibrations.

Lucid's stay on Mir ended on 23 September at 23.33 GMT when Atlantis undocked from the orbital station to begin the return home, leaving John Blaha behind. After undocking, Atlantis flew around Mir 1½ times at a distance of over 100 m, before the orbiter performed its final separation manoeuvre. When Atlantis returned to KSC on 26 September, Lucid claimed the record for the longest flight by a non-Russian (non-Soviet), having been in orbit for 188 days 4h.

STS-80 COLUMBIA
Orbiter flight: 21 (9th Extended Duration Orbiter)
Mobile Launch Platform: 3
Launch pad: KSC 39B
Launch date/time: 19 November 1996 at 19.55.47 GMT
Landing date/time: 7 December 1996 at 11.49.05 GMT
Mission duration: 17 days 15 hours 53 minutes 18 seconds
Landing site: KSC
Main engines: 1: 2032 2: 2026 3: 2029
Launch mass: =~2,052.7 t
Landing mass: 103,365 kg
Crew: Kenneth D Cockrell (Commander), Kent V Rominger (Pilot), Tamara E Jernigan (Mission Specialist 1/EV1), Thomas D Jones (Mission Specialist 2/EV2), F Story Musgrave (Mission Specialist 3)
Payload (Primary):
WSF 3 (Wake Shield Facility)
ORFEUS-SPAS 2 (Orbiting and Retrievable Far and Extreme Ultraviolet Spectrograph - Shuttle Pallet Satellite). 2 free-flying spacecraft
Payload (Secondary): payload bay,
ITEPC Inter-Mars Tissue Equivalent Proportional Counter
SEM 01 Space Experiment Module/Getaway Special
Payload (Secondary): mid-deck,
PARE Physiological and Anatomical Rodent Experiment
MDA Materials Dispersions Apparatus
VIEW-CPL Visualisation in an Experimental Water Capillary Pumped Loop
CCM Cell Culture Module
BRIC 09 Biological Research in Canisters

Mission summary
NASA employed the Mobile Transporter to move The STS-80 stack from the VAB to launch pad 39B on 16 October 1996, ready for launch the following month. The space agency had to delay the launch on 4 November until no earlier than 15 November to allow additional time to study unexpected SRB nozzle erosion which was observed following the STS-79 launch. Then on 13 November, NASA announced that the launch needed an additional five-day delay until 19 November because of inclement weather. The final launch had a two-minute hold to monitor the hydrogen concentration in the orbiter's aft compartment.

Columbia released its first satellite, the ORFEUS-SPAS 2, from the orbiter's RMS at 04.11 GMT 20 November. The release occurred just over an hour late as a result of a longer-than-planned predeployment checkout. During its mission, the science satellite gathered data on the origin and makeup of stars.

Two days later, Columbia released WSF 3 by the Shuttle's robot arm at 00.38 GMT 22 November. Within forty-eight hours, ground controllers feared that ORFEUS-SPAS might be closing in on WSF at a higher rate than planned and they decided to retrieve the latter satellite about a day earlier than the original plan. The remote manipulator latched on to the satellite at 01.01 GMT on 26 November and reberthed the satellite in the orbiter's payload bay. WSF scientists reported they completed seven thin film growths of semiconductor materials, the maximum capability for the satellite. At 08.23 GMT the Shuttle retrieved ORFEUS-SPAS as well.

NASA cancelled a planned EVA activity for STS-80 on 29 November. Depressurisation had already begun and Jernigan and Jones had to leave Columbia's airlock because, despite repeated attempts, they were unable to open the airlock due to a stuck handle. The EVAs had been planned as tests of equipment and procedures for International Space Station operations.

Bad weather cancelled the first planned landings on 5 and 6 December, and Columbia finally returned to KSC on 7 December, setting a new Shuttle duration record.

STS-81 ATLANTIS
Orbiter flight: 18
Mobile Launch Platform: 2
Launch pad: KSC 39B
Launch date/time: 12 January 1997 at 09.27.23 GMT
Landing date/time: 22 January 1997 at 14.22.55 GMT
Mission duration: 10 days 4 hours 55 minutes 32 seconds
Landing site: KSC
Main engines: 1: 2041/B1 2: 2034 3: 2042/B1
Launch mass: =~2,046.1 t
Landing mass: 97,275 kg
Crew: Michael A Baker (Commander), Brett W Jett Jr (Pilot), Peter J K Wilsoff (Mission Specialist 1), John M Grunsfeld (Mission Specialist 2), Marsha Ivins (Mission Specialist 3), Jerry M Linenger (Mission Specialist 4, up only), John E Blaha (Mission Specialist 4, down only)
Payloads (Primary): mid-deck,
SMM-05
SPACEHAB 06 Double module, orbiter docking system.

Shuttle Discovery lifts off on 27 May 1999 to deliver supplies and cargo to the ISS ***1999***/0054452

Payloads (Secondary): mid-deck,
CREAM Cosmic Radiation Effects and Activation Monitor
KidSat

Mission summary
On 10 December 1996, the ground crew moved the STS-81 stack from the VAB to pad 39B. After launch, one month later, the orbiter docked with the Docking Module attached to the Kristall module of Mir to begin joint operations as part of the fifth Shuttle-Mir Mission. The docking occurred at 03.55 GMT on 15 January. When the Americans entered Mir the handover between John Blaha and Jerry Linenger as a resident Mir crewmember took place. The Shuttle carried three tonnes of supplies to Mir, including equipment to be used by Linenger for experiments.

In the morning on 17 January, one of three Inertial Measurement Units (IMU) which provide navigation information to the Shuttle's general purpose computers showed significant drifting and Mission Control put it in standby mode. The IMU returned to service for use during Atlantis' landing.

Undocking from Mir came at 02.16 GMT on 20 January and Blaha returned to Earth with the remaining STS-81 crew two days later. Linenger stayed in orbit, awaiting collection by the STS-84 (SMM-06) mission due in May 1997.

STS-82 DISCOVERY
Orbiter flight: 22
Mobile Launch Platform: 1
Launch pad: KSC 39A
Launch date/time: 11 February 1997 at 08.55.17 GMT
Landing date/time: 21 February 1997 at 08.32.37 GMT
Mission duration: 9 days 23 hours 37 minutes 20 seconds
Landing site: KSC
Main engines: 1: 2037/B1 2: 2040/B1 3: 2038/B1
Launch mass: =~2,047.8 t
Landing mass: 97,076 kg
Crew: Kenneth D Bowersox (Commander), Scott J Horowitz (Pilot), Joseph R Tanner (Mission Specialist 1/EV4), Steven A Hawley (Mission Specialist 2), Gregory J Harbaugh (Mission Specialist 3/EV3), Mark C Lee (Mission Specialist 4/EV1), Steven L Smith (Mission Specialist 5/EV2)

Mission summary
On 17 January 1997, NASA transported the STS-82 stack with the Shuttle orbiter *Discovery* from the VAB to pad 39A for launch the following month. NASA dedicated this flight to the second servicing mission for the Hubble Space Telescope, the installation of Near-Infra-red Camera and Multi-Object Spectrometer (NICMOS), the Solid-State Recorder (SSR) 1 and the Space Telescope Imaging Spectrograph (STIS). The crew also replaced the Fine Guidance Sensor (FGS) 1, Reaction Wheel Assembly (RWA) 1. As a secondary objective, the crew planned to install the Optical Control Electronics Enhancement Kit (CCE-EK), Magnetic Sensing System (MSS) covers and to replace the Data Interface Unit (DIU) 2, the Engineering/Science Tape Recorder (ESTR) 2 and the Solar Array Drive Electronics (SADE) 2.

The launch occurred on time. Two days after launch, at 08.34 GMT on 13 February Steven Hawley – who had first deployed HST during the STS-31 mission – used the orbiter's robotic arm to capture the telescope. Within half an hour it was berthed in Discovery's payload bay. The first EVA to service HST began at 04.34 GMT on 14 February when astronauts Lee and Smith left the Shuttle proper for 402 minutes. Once outside, the astronauts opened a hatch on the aft side of HST and removed the Goddard High-Resolution Spectrograph and the Faint Object Spectrograph and replaced them with the STIS and the NICMOS.

On 15 February, the second EVA took place. Astronauts Harbaugh and Tanner began their EVA at 01.25 GMT and it lasted for 442 minutes. They replaced a degraded FGS and a failed ESTR with new models. They also installed the OCE-EK, which will increase the capability of the new FGS. During the EVA activities the astronauts and flight controllers noticed that the thermal insulation showed signs of cracking and general wear and suggested an additional EVA to allow extra unplanned repair work to be completed. As the EVA drew to a close, Discovery's RCS fired to slightly raise the orbit of the assembly, in advance of a larger re-boost later in the mission. On 16 February, NASA decided that a fifth EVA would be added to the four planned for the STS-82 mission.

EVA number three started at 01.53 GMT 16 February. During this effort, astronauts Lee and Smith removed and replaced a DIU and a reel-to-reel style ESTR with a new digital SSR that will allow the simultaneous recording and playback of data. Then they replaced an RWA, and ended the EVA after 431 minutes.

The fourth EVA began at 03.45 GMT on 17 February with Harbaugh and Tanner outside Discovery. They first replaced a SADE used to control the positioning of HST's solar arrays. Next, the astronauts moved to the top of the telescope and replaced the covers of the MSS. They then placed thermal protection over two areas of degraded insulation around the light shield part of the telescope. Harbaugh and Tanner returned inside Discovery after an EVA lasting for 394 minutes.

The fifth (unplanned) and final EVA began at 03.15 GMT on 18 February, when astronauts Lee and Smith left Discovery for the third time. They attached several thermal insulation patches to three equipment compartments at the top of the Support Systems Module of HST which contain key data processing, electronics and scientific instrument telemetry packages. Following the completion of that work, Lee and Smith briefly returned to the airlock while flight controllers evaluated a possible problem with one of four RWA units used to manoeuvre the telescope for its scientific observations. After determining that further analysis of the Reaction Wheel Assembly would be required, ground control directed the astronauts to end their spacewalk and re-enter the airlock for the final time.

With the HST servicing complete, Discovery raised the orbit of the assembly in anticipation of the re-deployment of the telescope the following morning. Additional orbital 'tweaking' resulted from an orbital prediction that HST might collide with a piece of space debris (1994-029HU). Before being captured, HST orbited at 590-599 km and when it had been released, its new orbit was 599-620 km. The Shuttle released the HST on 19 February at 06.41 GMT. Within minutes, thrusters on Discovery had fired to begin the retreat from the telescope. Two days later, Discovery returned to Earth.

With the mission complete, testing of the new HST instruments showed that one of the three cameras on the NICMOS is not focusing properly. Controllers hope that this problem might cure itself with time.

STS-83 COLUMBIA
Orbiter flight: 22 (10th Extended Duration Orbiter)
Mobile Launch Platform: 3
Launch pad: KSC 39A
Launch date/time: 4 April 1997 at 19.20.32 GMT
Landing date/time: 8 April 1997 at 18.33.23 GMT
Mission duration: 3 days 23 hours 12 minutes 51 seconds
Landing site: KSC
Main engines: 1: 2012/B2 2: 2109/B2 3: 2019/B2
Launch mass: =~2,051.7 t
Landing mass: 96,643 kg
Crew: James D Halsell Jr (Commander), Susan L Still (Pilot), Janice S Voss (Payload Commander/Mission Specialist 1), Michael L Gernhardt (Mission Specialist 2/EV1), Donald A Thomas (Mission Specialist 3/EV2), Roger K Crouch (PS1), Gregory T Linteris (PS2)
Payload (Primary): payload bay,
MSL 1 Microgravity Science Laboratory
Payload (Secondary): payload bay,
CRYOFD Cryogenic Flexible Diode Experiment
OARE Orbital Acceleration Research Experiment
Payloads (Secondary): mid-deck,
SAREX 2 Shuttle Amateur Radio Experiment

Mission summary
The ground crew moved the stack for STS-83 from the VAB to pad 39A on 11 March 1997 in advance of the launch planned for the following month. Launch came

The Unity connecting module is moved into the payload bay of Orbiter Endeavour prior to STS-88 ***1999***/0054449

without any delays on 4 April. Two hours after reaching orbit, astronauts Voss and Crouch entered SPACELAB for the first time.

Shortly after the mission began, Mission Control discovered that one of the three fuel cells suffered some degradation. The loss of a fuel cell usually prompts a flight termination, but initially controllers decided to simply monitor the problem. Finally, on 6 April, Shuttle programme managers decided that the fuel cell problem meant that the mission would have to end. Astronauts shut down the faulty fuel cell that afternoon but, with careful management of the available electricity, the astronauts continued to conduct some experiments as part of the MSL 1 programme. The science activities ended on 7 April as preparations for the early return continued. The following day Columbia landed at KSC.

Once on the ground, NASA considered the delays in the Shuttle flights related to the International Space Station programme and decided to immediately refurbish Columbia for relaunch around 1 July for a mission designated STS-94 (originally the reflight was tentatively designated STS-83R).

STS-84 ATLANTIS

Orbiter flight: 19
Launch pad: KSC 39A
Launch date/time: 15 May 1997 4:07:48.62 EST
Landing date/time: 24 May 1997 09:27:44 EDT
Mission duration: 9 days, 5 hours, 19 minutes 56 seconds
Landing site: KSC, Runway 33
Main engines: SSME-1: SN-2032, SSME-2: SN-2031,SSME-3: SN-2029
Launch mass: =~2,052.7 t
Landing mass: 103,365 kg
Orbit: Altitude,184 statute miles. Inclination: 51.6°
Orbits: 145
Crew: Charles J Precourt (Commander), Eileen M Collins (Pilot), C Michael Foale (Mission Specialist-1), Carlos I Noriega (Mission Specialist-2), Edward T. Lu, (Mission Specialist-3), Jean-François Clervoy (Mission Specialist-4), Elena V. Kondakova (Mission Specialist -5)
Payload (Primary):
SpaceHab-DM
LME
SAMS
CGEL

Mission summary

The STS-84 mission was the 6th Shuttle/Mir docking mission and is part of the NASA/Mir programme, which consists of nine Shuttle-Mir dockings and seven long duration flights of US astronauts aboard the Russian space station. STS-84 transferred 7,314 lb of water and other material to and from the Mir. During the docked phase, 1,025 lb of water, 844.9 lb of US science equipment, 2,576.4 lbs of Russian logistics along with 392.7 lb of miscellaneous material passed between the vehicles. 897.4 lb of US science material, 1,171.2 lb of Russian logistics, 30 lb of ESA material and 376.4 lb of miscellaneous material returned to Earth aboard the Shuttle. C Michael Foale stayed aboard Mir, replacing Jerry M Linenger who will have arrived on Mir from STS-81.

The Shuttle missed the first landing opportunity due to low cloud cover over the KSC. At 08:19 EDT, the STS-84 crew received approval for the de-orbit burn of 3 minutes 7 seconds on the second KSC landing opportunity for 24 May 1997. The burn occurred at 08:24 EDT. This allowed Atlantis to enter the atmosphere over the Pacific Ocean and travel in a northwest to southeast path toward the Kennedy Space Center.

STS-85 DISCOVERY

Orbiter: Discovery (OV-103), 23rd flight
Launch date: 7 August 1997
Launch pad: KSC 39-A
Launch time: 10:41 EDT
Launch Window: 1 hour, 39 minutes
Orbit Altitude/Inclination: 160 n miles, 57°
Mission duration: 10 days, 20 hours, 24 minutes
Landing date: 18 August 1997
Landing time: 07:05 EDT
Primary landing site: Kennedy Space Center, Florida
Launch mass: =~2,028.4 tonnes
Landing mass: 97,919 kg
Crew: Curt Brown, (Commander), Kent Rominger (Pilot), January Davis, (Payload Commander/ Mission Specialist -1), Robert Curbeam (Mission Specialist -2), Steve Robinson (Mission Specialist -3), Bjarni Tryggvason (Payload Specialist-1)
Payload (Primary): cargo bay,
CRISTA-SPAS-02. Cryogenic Infra-red Spectrometers and Telescopes for the Atmosphere-Shuttle Pallet Satellite-2
TAS-01 Technology Applications and Science-01
IEH-02 International Extreme Ultraviolet Hitchhiker-02
MFD
Payload (Secondary): crew cabin,
BDS-03 Bioreactor Demonstration System-3
BRIC
PCG-STES-05
ACIS
MSX
SIMPLEX
SWUIS Southwest Ultraviolet Imaging System
SSCE

Mission summary

STS-85 had the prime objective of launching the CRISTA-SPAS-2 which is making its second flight on the Space Shuttle and is the fourth mission in a co-operative venture between the German Space Agency (DARA) and NASA. The RMS will deploy the satellite and then the CRISTA-SPAS system will free-fly for over 200 hours and will use three telescopes and four spectrometers to measure infra-red radiation emitted by the Earth's middle atmosphere. Data gathered will help investigators from 15 countries to understand how small-scale tracer 'filaments' in the stratosphere contribute to transport of ozone and chemical compounds that affect the distribution of ozone.

NASDA, the Japanese Space Agency, plans to test the RMS with the MFD series of experiments. MFD consists of three separate experiments located on a support truss in the payload bay. It is designed to demonstrate applications of a mechanical arm for possible use on the Japanese Experiment Module of the future International Space Station.

The TAS holds seven separate experiments that provide data on the Earth's topography and atmosphere study the sun's energy and test new thermal control devices. It is joined by the four experiments comprising the IEH payload. NASA intends to study ultraviolet radiation from the stars, the Sun and other sources in the Solar System with the IEH.

Payload and experiments flying in the crew cabin area include the (SWUIS), a 7 in imaging telescope that will be pointed at the Hale-Bopp comet. The BDS-3 payload performs cell biology experiments under controlled conditions on small samples of material.

STS-85 carries several payloads to continue NASA's study of natural and human-induced changes in Earth's environment as part of the Mission to Planet Earth programme. Mission to Planet Earth programme investigations include three experiments – SOlar CONstant Experiment (SOLCON), Infra-red Spectral Imaging Radiometer (ISIR) and Shuttle Laser Altimeter (SLA), that are a part of the Technology Applications and Science-1 (TAS-1) payload.

The Middle Atmosphere High-Resolution Spectrograph Investigation (MAHRSI) is another NASA payload to obtain data on the natural environment. During STS-85 MAHRSI will obtain new vertical profile data on the distribution of hydroxyl (OH) in the mesosphere and upper stratosphere under different conditions from its previous flight on STS-66. At the end of the flight, NASA plans to develop global maps of the vertical distribution of OH and measurements of nitric oxide in the middle atmosphere.

STS-86 ATLANTIS

Orbiter: Atlantis, 20th flight
Launch date: 25 September 1997
Launch pad: KSC 39-A
Launch time: 22:34 EDT
Launch window: 10 minutes
Orbit altitude/Inclination: 184 statute miles, 51.6°
Orbits: 169
Mission duration: 10 days 19 hours 22 minutes 12 seconds.
Landing date: 6 October 1997
Engines: SSME-1: SN-2012, SSME-2: SN-2040, SSME-3: SN-2019
Landing time: 17:55 EDT
Primary landing site: Kennedy Space Center, Florida, Runway 15
Crew: James D Wetherbee (Commander), Michael J Bloomfield (Pilot),Vladimar G Titov (Mission Specialist), Scott E Parazynski (Mission Specialist), Jean-Loup J M Chretien (Mission Specialist), Wendy B Lawrence (Mission Specialist), David A Wolf (Mission Specialist)

Russian cosmonaut Sergei Krikalev and NASA astronaut James Newman (right) sit inside Orbiter Endeavour during a terminal countdown demonstration test ***1999***/0054450

Payload (Primary): cargo bay,
SpaceHab-DM
MEEP-R
EDFT-06
SEEDS-II
CCM-07
MSX-09
CREAM-09
KidSat-03
RME-III-21
SIMPLEX-02
Payload (Secondary): crew cabin,
GAS(G-036)

Mission summary

After receiving final clearance to launch from NASA Administrator Daniel Goldin earlier in the day, Atlantis' astronauts lifted off from the Kennedy Space Center at 21:34 Central time to begin the seventh docking mission to the Mir Space Station.

Shortly after reaching orbit, the astronauts began to activate Shuttle and Spacehab systems and planned a minor firing of the orbital manoeuvring system engines to refine Atlantis' path to the Mir, which was flying over southern Europe at launch time at an altitude of about 215 n miles.

On 27 September 1997, 14:58 EDT, for the seventh time in two years, the Shuttle Atlantis gently docked with the Russian Mir Space Station. Operating from Atlantis' aft flight deck, Cdr Jim Wetherbee guided Atlantis to a smooth docking with the Mir at 14:58 Central time as the two vehicles flew north of the Caspian Sea east of the Russian-Kazakh border at an altitude of about 215 n miles.

Astronauts Scott E Parazynski and Vladimir G Titov performed an EVA outside the Atlantis-Mir complex to retrieve four Mir Environmental Effects Packages which had been mounted on the docking module last March by astronauts Linda Goodwin and Rich Clifford. Parazynski and Titov tethered the Spektr Solar Array Cap to the docking module. The Solar Array Cap may be used in a future Mir spacewalk to seal any hole found in the hull of the damaged Spektr module. The EVA began on 1 October at 12:29 Central Time.

Atlantis' landing with astronaut Mike Foale was postponed for 24 hours, after clouds at the Kennedy

Endeavour lifts off into a night sky for the start of mission STS-88, 4 December 1998 ***1999***/0054451

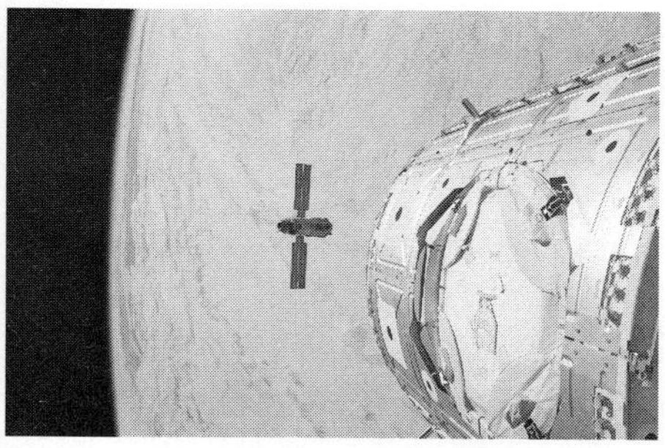

Russian-built FGB approaches Shuttle Endeavour and Unity module (NASA) ***1999***/0062166

Space Center in Florida prevented a landing on the two opportunities available at the Florida spaceport. The Space Shuttle Atlantis finally landed on Kennedy Space Center, Florida at 16:55 CDT 6 October.

STS-87 COLUMBIA
Orbiter flight: 24 (Extended Duration Orbiter)
Launch pad: KSC LC-39B
Launch date/time: 19 November 1997/14:46:00 EST
Launch window: 2 hours 30 minutes
Main engines: SSME-1, 2031; SSME-2, 2039; SSME-3, 2037
Lift-off weight: 2,051,833 kg
Orbiter weight at lift-off: 118,298 kg
Orbit altitude/inclination: 278 km/28.45°
Mission duration: 15 days, 16 hours, 34 minutes 04 seconds
Landing site: KSC (41) Runway 33
Landing date/time: 5 December 1997/07:20:04 EST
Rollout distance: 2,440 m
Rollout time: 57 seconds
Crew: Kevin R Kregel (Commander), Steven W Lindsey (Pilot), Kalpona Chawla (Mission Specialist), Winston E Scott (Mission Specialist), Takao Doi, NASDA (Mission Specialist), Leonid Kadenyuk, National Space Agency of Ukraine (Payload Specialist)
Payload (Primary):
United States Microgravity Payload-4 (USMP-4)
Spartan 201-04
Payload (Secondary):
Shuttle Ionospheric Modification With Pulsed Local Exhaust (SIMPLEX)
EVA Demonstration Flight Test 05 (EDFT)
Shuttle Ozone Limb Sounding Experiment (SOLSE)
Loop Heat Pipe (LHP)
Sodium Sulphur Battery Experiment (NaSBE)
Turbulent Gas Jet Diffusion Flames (TGDF)
GAS 036
Collaborative Ukraine Experiment (CUE)
Autonomous EVA Robotic (AER) Camera/Sprint (RME 1323)
Midcourse Space Experiment
Payload weight: 10,038 kg

Mission summary
OPF: 17 July 1997
VAB: 24 October 1997
Pad: 29 October 1997
Following a near flawless countdown, STS-87 was launched on time and, for the first time, on to a new attitude profile for ascent. Approximately 6 minutes after launch, the orbiter/ET combination was rolled from 'heads-down' to 'heads-up' allowing the antennas to communicate through the TDRSS network of geosynchronous satellites about 2 minutes 30 seconds earlier than would otherwise be possible. This profile will be flown on all subsequent low-inclination trajectories to eliminate the need for the Bermuda tracking station.

The microgravity science and technology research aboard Columbia performed well throughout the mission. Experiments conducted aboard USMP-4 included: Advanced Automated Directional Solidification Furnace; Confined Helium Experiment; Isothermal Dendritic Growth Experiment; Material for the Study of Interesting Phenomena of Solidification on Earth and in Orbit (MEPHISTO); Microgravity Glovebox Facility with several unique experiments; Enclosed Laminar Flames; Wetting Characteristics of Immiscibles; Particle Engulfment; Pushing by a Solid/Liquid Interface; Space Acceleration Measurement System; Orbital Acceleration Research Experiment.

The plan to deploy the Spartan was not a success. Following deployment at 16:04 EST on 21 November Spartan executed a pirouette manoeuvre, indicating it might have had an attitude control problem. On attempting to re-capture the grapple fixture, Chawla failed to obtain a proper attachment and on trying to re-grapple the platform inadvertently imparted a 2°/s spin to Spartan. An attempt to synchronise the rotation of the orbiter with that of the platform was called off by the Flight Director. During the first EVA on 24 November, Scott and Doi managed to grab the platform and stow it aboard the payload bay. The EVA ended after 7 hours 43 minutes. The second EVA lasted 4 hours 59 minutes during which Scott and Doi completed unfinished tasks from the first EVA.

Columbia landed at KSC's runway 33 on orbit 252 at 07:20:04 EST.

STS-89 ENDEAVOUR
Orbiter flight: 12
Launch pad: KSC LC-39A
Launch date/time: 22 January 1998/21:48:15 EST
Launch window: ~10 minutes
Main engines: SSME-1, 2043; SSME-2, 2044; SSME-3, 2045 (all Block II engines)
Lift-off weight: 2,046,919 kg
Orbiter weight at lift-off: 114,167 kg
Orbit altitude/inclination: 394 km/51.6°
Mission duration: 8 days 19 hours 46 minutes 54 seconds
Landing site: KSC (42) Runway 15
Landing date/time: 31 January 1998/17:35:09 EST
Crew: Terrence W Wilcutt (Commander), Joe Frank Edwards, Jr (Pilot), James F Reilly II (Mission Specialist), Michael P Anderson (Mission Specialist), Bonnie J Dunbar (Mission Specialist (Payload Commander)), Salizhan Shakirovich Sharipov (Mission Specialist), Andrew S W Thomas (Mission Specialist (up)), David Wolf (Mission Specialist (down))
Payload (Primary):
Spacehab 08-DM, transfer tunnel, transfer tunnel extension, orbiter docking system
Payload (Secondary):
GAS 093, 141, 145 and 432
CREAM
SIMPLEX
EarthKAM (Kidsat)
MPNE
HP
CEBAS
Payload weight: 9,952 kg (up); 8,858 kg (down)

Mission summary
OPF-1: 28 March 1997
VAB: 8 April 1997 (temporary stowage)
OPF-3: 21 April 1997
VAB: 23 May 1997 (temporary stowage)
OPF-1: 4 June 1997 (begin preflight processing)
VAB: 12 December 1997
Pad: 19 December 1997
Mission managers announced 22 May 1997 orbiter Endeavour would fly STS-89 instead of Discovery and that it would become the first orbiter other than Atlantis to dock with Mir. Launch was originally scheduled for 15 January but this was changed first to 20 January and then to 22 January at the request of the Russians who needed extra time in orbit to prepare for the visit. Following the retirement of veteran launch director Jim Harrington, this was the first launch overseen by Dave King, one of two rotational launch directors.

Following nominal ascent to orbit, the crew activated secondary experiments. Nominal rendezvous following standard ascent to orbit with docking at Mir at 15:14 EST 24 January. The hatches were opened at 17:25 EST at the start of five days of joint activities. Endeavour carried 3,629 kg of scientific equipment, logistical hardware and water for the Russian space station. Astronaut Thomas was switching positions with astronaut Wolf and would remain aboard Mir as the last NASA astronaut to conduct a long duration stay aboard the Russian space station. Wolf had spent 119 days aboard Mir and a total on-orbit time of 128 days.

The Shuttle-Mir science programme focused on 27 areas of study: advanced technology, in which commercial development of research tasks was mooted as a test bed for the ISS; Earth sciences, including research in ocean biochemistry, land surface, hydrology and meteorology; human life sciences, focusing on crewmembers' adaptation to zero-*g*; and ISS risk mitigation, involving crew health and safety.

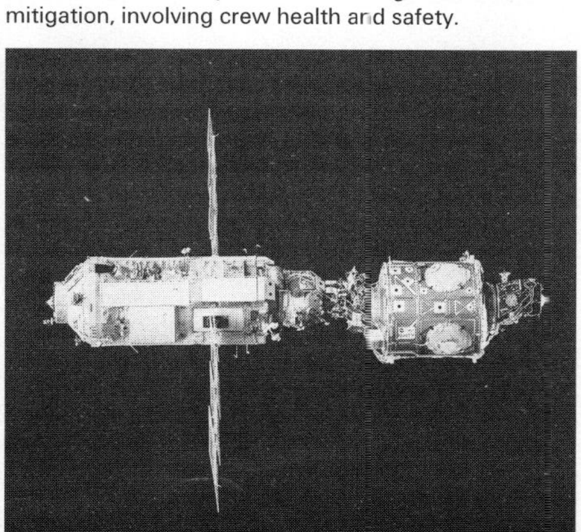

Zarya and Unity modules viewed from Endeavour (STS-88) (NASA) ***1999***/0062168

Musgrave and Hoffman ride the RMS near the end of 35 hr of EVA servicing of the Hubble telescope in Endeavour's cargo bay (NASA) ***1994***

Hatches between Mir and the Shuttle were closed 17:34 EST on 28 January and the two spacecraft undocked at 11:57 EST the following day. Orbiter landed on orbit 139 at 17:35:08, 31 January, on runway 15 at the Kennedy Space Center, the 13th time in succession and the 20th time in the previous 21 landings, having travelled 5.8 million km.

STS-90 COLUMBIA
Orbiter flight: 25
Launch pad: KSC LC-39B
Launch date/time: 17 April 1998/14:19:00 EDT
Launch window: ~2 hours 30 minutes
Main engines: SSME-1, 2041; SSME-2, 2032; SSME-3, 2012
Lift-off weight: 2,051,982 kg
Orbiter weight at lift-off: 119,005 kg
Orbit altitude/inclination: 278 km/39°
Mission duration: 15 days 21 hours 49 minutes 59 seconds
Landing site: KSC (43) Runway 33
Landing date/time: 3 May 1998/12:08:59 EDT
Rollout distance: 3,047 m
Rollout time: 58 seconds
Crew: Richard A Searfoss (Commander), Scott D Altman (Pilot), Kathryn 'Kay' Hire (Missions Specialist), Dafydd (Dave) Rhys Williams, Canadian Space Agency (Mission Specialist), Dr Jay C Buckey (Payload Specialist), Dr James A Pawelczyk (Payload Specialist)
Payload (Primary):
Spacelab LM Neurolab
EDO cryogenic pallet (5 full and 4 extended duration cryo tank sets)
Payload (Secondary):
GAS 197, 744 and 722
Shuttle Vibration Forces
Bioreactor Demonstration System 04
Payload weight: 11,862 kg

Mission summary
OPF: 5 December 1997
VAB: 16 March 1998
Pad: 23 March 1998
Launch originally planned for 16 April 1998 but delayed for 24 hours because of difficulties with Network signal processor No 2, one of two processors which format data and communications between the ground and the Shuttle and which was subsequently replaced. Lift-off occurred on time a day later and the direct insertion ascent was nominal with only one OMS thrusting manoeuvre to reach the operational orbit. Neurolab mission was the last flight of the Spacelab pressurised modules to fly.

Mission was to combine biomedical techniques with new and sophisticated measurement and recording devices to advance neuroscience as NASA's contribution to the Decade of the Brain designated by President Bush and Congress. The experiments were to examine the effects of space flight and microgravity on the nervous system. Neurolab's 26 experiments targeted basic research in attempts to understand the role played by gravity in the development of the nervous systems in humans and animals. Indications are strong that critical early development periods shape the way the brain interprets stimuli in later life. These tests were conducted with the astronauts and with rats, mice, crickets, snails and two kinds of fish.

STS-95 crew commanded by Curtis L Brown Jr (right, seated) and including John Glenn (right, standing) (NASA)
1999/0062170

About one week into the mission, the crew had to jury-rig a bypass system for a suspect regulator valve in the CO_2 removal system, preventing the need for an earlier return to Earth than planned. Consumables resources permitted consideration of a one-day extension but this was not adopted because of a forecast for deteriorating weather at KSC. Columbia landed on orbit 256 after logging almost 10.26 million km making the 14th consecutive touchdown at KSC.

STS-91 DISCOVERY
Orbiter flight: 24
Launch pad: KSC LC-39A
Launch date/time: 2 June 1998/18:06:24 EDT
Launch window: ~10 minutes
Main engines: SSME-1, 2047; SSME-2, 2040; SSME-3, 2042
Lift-off weight: 2,047,782 kg
Orbiter weight at lift-off: 117,861 kg
Orbit altitude/inclination: 320 km/51.6°
Mission duration: 9 days 19 hours 53 minutes 57 seconds
Landing site: KSC (44) Runway 15
Landing date/time: 12 June 1998/14:00:18 EDT
Rollout distance: 3,576 m
Rollout time: 1 minute 4 seconds
Crew: Charles (Charlie) J Precourt (Commander), Dominic (Dom) L Gorie (Pilot), Franklin Chang-Diaz (Mission Specialist), Wendy Lawrence (Mission Specialist), Janet Kavandi (Missions Specialist), Valeriy Ryumin, Russian Space Agency (Mission Specialist), Andrew (Andy) S W Thomas (down (Mission Specialist))
Payload (Primary):
Spacehab 09-SM, transfer tunnel, transfer tunnel extension, orbiter docking system
Payload (Secondary):
GAS 090, 743, 765, 648, 2 × SEM, 2 × Phase 1 Program Support Packages
Alpha Magnetic Spectrometer
Phase 1 Requirements
Commercial Protein Crystal Growth
Solid Surface Combustion Experiment
Shuttle Ionospheric Modification with Pulsed Local Exhaust
SIMPLEX
Payload weight: 11,782 kg (up); 11,843 kg (down)

Mission summary
OPF-3: 19 August 1997 (temporary storage)
OPF-2: 1 October 1997 (temporary storage)
OPF-2: 30 October 1997 (begin preflight processing)
VAB: 21 April 1998
Pad: 2 May 1998
With few exceptions, the countdown proceeded smoothly and final launch time was determined at the T–9 minutes built-in hold to synchronise the orbit insertion point with the Mir orbital state vector. This was the first time Discovery had been used on a Shuttle-Mir mission and the docking took place on 4 June at 12:58 EDT at an altitude of 334 km. At hatch opening (14:34 EDT), NASA astronaut Andy Thomas officially became a member of the STS-91 crew, ending 130 days of living and working aboard Mir and bringing to 907 days the total amount of time spent on Mir by seven US astronauts as long-duration crewmembers.

During the next four days, the crews transferred more than 499 kg of water and 2,132 kg of supplies between the two spacecraft. Long-term US experiments were transferred to the mid-deck locker area and into the Spacehab single module in the orbiter payload bay including the Space Acceleration Measurement System and the tissue engineering co-culture investigations, as well as two crystal growth experiments. The crews also conducted Risk Mitigation Experiments and Human Life Sciences investigations. Carried for the first time, the Alpha Magnetic Spectrometer was designed to look for 'dark matter' in the universe and was switched on during Flight day 1 but data that should have been sent to the ground via the Ku-band antenna was recorded onboard because of a problem with that antenna. On 3 June, the crew were able to set up a bypass system allowing AMS data to be downlinked through the S-band/FM system. The Ku-band failure prevented the transmission of TV from Discovery and a fault between ground stations and mission control in Moscow prevented TV from being broadcast out of Mir.

Hatches between Mir and Discovery closed at 09:07 on 8 June 1998 and the two undocked at 12:01 EDT ending Phase 1 of the ISS programme. Discovery landed on orbit 155 having logged 6.1 million km making the 15th consecutive landing at that facility and the 22nd in the last 23 missions.

STS-95 DISCOVERY
Orbiter flight: 25
Launch pad: KSC LC-39A
Launch date/time: 29 October 1998/14:20:19 EST
Launch window: 2 hours 30 minutes
Main engines: SSME-1, 2048; SSME-2, 2043; SSME-4, 2045
Lift-off weight: 2,051,142 kg
Orbiter weight at lift-off: 103,061 kg
Orbit altitude/inclination: 471 km/28.45°
Mission duration: 8 days 21 hours 43 minutes 57 seconds
Landing site: KSC (45) Runway 33
Landing date/time: 7 November 1998/12:04:16 EST
Rollout distance: 2,898 m
Rollout time: 59 seconds
Crew: Curtis L Brow, Jr (Commander), Steven W Lindsey (Pilot), Scott E Parazynski (Missions Specialist), Stephen K Robinson (Missions Specialist), Pedro Duque (NSDA, Japan (Mission Specialist)), Chiaki Mukai (Payload Specialist), Senator John H Glenn (Payload Specialist)
Payload (Primary):
Spacehab
Spartan 201-5
Hubble Space Telescope Orbital Systems Test Platform
International Extreme Ultraviolet Hitchhiker
Payload (Secondary):
Cryogenic Thermal Storage unit
Space Experiment Module
GAS
Biological Research in Canisters
Electronic Nose
Payload weight: 12,937 kg (up); 12,867 kg (down)

Mission summary
OPF: 15 June 1998
VAB: 14 September 1998
Pad: 21 September 1998
The launch was scheduled for 14:00 EST but an unscheduled hold of 8 minutes 30 seconds at the T–9 minute built-in hold period was made in order to investigate a master alarm during cabin leak checks after hatch closure. The Range Safety Officer requested a further hold at T–5 minutes when an unidentified aircraft entered the restricted airspace around KSC. After main engine ignition but before solid rocket booster ignition, the drag chute compartment door came off without any effect on the lift-off or ascent to orbit. This was the first time veteran Mercury MA-6 astronaut John Glenn had been into space since 20 February 1962 when he became the first American to orbit the Earth 36 years 8 months and 9 days earlier.

The nine days in space were filled with concentrated scientific activity in a variety of disciplines and involved the independent flight of the free-flying Spartan on its fifth flight into space. During STS-95 it would perform a repeat of the failed Spartan 201-4 mission on STS-87 when it had to be retrieved by a contingency EVA (see STS-87 Columbia). Some of the on-board science supported studies into ageing and the 77-year old Glenn was a test subject required to deliver 10 blood and 16 urine samples over the period of the mission. On 30 October, the crew released into orbit the 70 kg Petite Amateur Naval SATellite (PANSAT), a small telecommunications satellite developed by the Naval Postgraduate School in Monterey, California. Next day Glenn began giving the cycle of blood and urine samples.

Spartan was released from the remote manipulator arm at 13:59 EST on 1 November and for about 9 h, the Shuttle remained 9.6-16 km away before moving to a distance of around 48 km. Spartan 201-5 carried instruments to study the Sun and the solar wind and

John Glenn returns to space aboard Shuttle STS-95 (NASA)
1999/0062169

NASA astronaut Newman during EVA at Unity module (NASA) ***1999***/0062167

after a series of rendezvous manoeuvres, it was back on the remote manipulator at 15:45 CST on 3 November and berthed back in the payload bay shortly thereafter. A further use of Spartan on 4 November extended it above the Shuttle for additional measurements. In all, STS-95 completed some 80 experiments.

Preparations for the return to Earth included closing the payload bay doors at 08:17 EST on 7 November followed by the 4 minute 40 seconds firing of the deorbit engines on orbit 135 starting at 10:53 EST. Entry began at 11:30 EST followed by touchdown at 12:04 EST, the 16th consecutive landing at KSC. The drag chute was not deployed.

STS-88 ENDEAVOUR
Orbiter flight: 13
Launch pad: KSC LC-39A
Launch date/time: 4 December 1998/03:35:34 EST
Launch window: 5-10 minutes
Main engines: SSME-1, 2043; SSME-2, 2044; SSME-3, 2045
Lift-off weight: 2,049,542 kg
Orbiter weight at lift-off: 119,717 kg
Orbit altitude/inclination: 320 km/51.6°
Mission duration: 11 days 19 hours 17 minutes 55 seconds
Landing site: KSC (46) Runway 15
Landing date/time: 15 December 1998/22:53:29 EST
Rollout distance: 2,543 m
Rollout time: 44 seconds
Crew: Robert D Cabana (Commander), Frederick (Rick) Sturckow (Pilot), Nancy Currie (Mission Specialist), Jerry Ross (Mission Specialist), Jim Newman (Mission Specialist), Sergei Krikalev, Russian Space Agency (Mission Specialist)
Payload (Primary):
Unity Module (Node 1/PMA1/2)
Imax Cargo Bay Camera
Argentinian Scientific Applications Satellite-C
Payload (Secondary):
MightySat 1 Hitchhiker
Space Experiment Module-07
GAS 093
Payload weight: 11,612 kg (Unity)

Mission summary
OPF: 1 February 1998
VAB: 13 October 1998
Pad: 21 October 1998

STS-88 was the long-awaited first US assembly flight in support of the ISS, the culmination of almost 15 years of design, redesign and preparation in an effort pledged by President Ronald Reagan in his state of the union address on 25 January 1984. The ISS was re-scoped by President Bill Clinton during the mid-1990s to include the Russians as well as the Europeans, the Japanese and the Canadians. STS-88 followed the launch of the first element, the Russian-built Zarya (or SUV, Space Utility Vehicle), on 20 November 1998 in ISS Flight 1A/R. The STS-88 mission was known as Flight 2A.

The first launch attempt aimed for a lift-off at 03:58 EST on 3 December 1998 but at T–4 minutes 24 seconds, a master alarm halted the clock 24 seconds later. The alarm had been triggered by pressure in the orbiter's hydraulic system #1 temporarily registering a low of 19,306 kPa. The countdown was held at T–31 seconds to assess the situation, but engineers ran out of time before constraints governed by LOX drain-back redlines closed the launch window. Next day the countdown proceeded without major incident and Endeavour was launched at 03:35 EST. Mission objectives focused on attaching the Unity (Node 1) module and PMA attachments to Zarya, the first (Russian-built) element in the ISS launched 20 November 1998.

Following a trouble-free ascent and during the early stages of rendezvous, Endeavour was reconfigured. Using the remote manipulator, the crew manoeuvred the Unity module from its cradle at the rear of the payload bay to a vertical position and placed it on top of the external docking module and airlock assembly in the forward part of the payload bay. Mating occurred at 18:45 EST on 5 December. Next day Endeavour rendezvoused with Zarya and, using TV cameras and an Orbiter Space Vision System tested on earlier flights, drew Zarya down on the PMA#1 docking ring at 21:07. Initial attempts at a docking were unsuccessful and only after Currie ungrappled the module and tried again was the Shuttle assembly docked to the Russian element, at 21:48 UT on 6 December.

The first of two EVAs was conducted by Ross and Newman starting at 17:10 EST on 7 December and lasted 7 hours 21 minutes during which the two astronauts hooked up external power and data cables, installed handrails to be used later during assembly flights and checked out the exterior. Power was applied to Unity at 22:49 EST to verify the integrity of the connectors before the crew completed their EVA. Newman was moved by the manipulator arm to inspect Zarya's two TORU rendezvous antennas which did not fully deploy after launch. Shortly before the EVA ended, Ross broke the record for the most cumulative EVA time (29 hours 41 minutes) previously held by astronaut Akers during five EVAs on STS-49 and STS-61. He ended his EVA with 30 hours 8 minutes accumulated time.

On 8 December, the orbiter propulsion system was used in a sequence of pulsed firings lasting 22 minutes to raise the orbit of the docked configuration giving it greater orbital life before the next visit in 1999. The second EVA began at 15:33 EST 9 December with Ross and Newman setting out to install two systems health monitoring boxes on the exterior of Unity which double as a video-conferencing system. Then they removed restraint pins on the four Unity hatchways which will be used by future crews during the assembly phase. Newman was raised to one of the Zarya TORU antennas which he freed with a grappling hook, ending the EVA at 7 hours 2 minutes.

Cabana and Krikalev were first to begin the occupation of the new complex and moved into the pressurised Unity module at 14:54 EST on 10 December, the rest of the crew following. Turning on lights, unstowing gear and checking equipment, Cabana and Krikalev opened the hatch to Zarya at 16:12 EST and were followed inside by the remainder of the crew. Several work tasks were completed before the ensuing rest period. Following further work sessions, Cabana and Krikalev closed the hatch on Zarya at 17:41 EST on 11 December and shut the hatch on Unity at 19:26 EST completing the first 28 hours 32 minutes visit to the docked assembly.

The third and final EVA began at 15:33 EST on 12 December during which Ross and Newman disconnected umbilicals, stowed an exterior tool work bag for later astronauts and took an extensive series of photographs to document the exterior configuration. Ross then freed the second jammed TORU antenna and completed a total 21 hours 22 minutes EVA on this mission. Ross has now completed seven EVAs totalling 44 hours 9 minutes. Endeavour was undocked from the PMA at 15:25 EST on 13 December for a complete fly-around and photograph exercise after which, at 16 49 EST, the final separation burn was made. At 22:31 EST the crew deployed the Argentinian satellite.

Gibson (foreground) and Dezhurov shake hands across the Mir-Shuttle docking plane. At bottom is one of the androgynous docking unit petals. The mission marked the first US-Russian docking since Apollo-Soyuz in 1975 (NASA) ***1996***

Endeavour executed the de-orbit burn at 21:48 EST, 15 December and landed at 22:53 EST making the 10th night landing, the 5th night landing at KSC and the 18th consecutive landing in Florida.

STS-96 DISCOVERY
Orbiter flight: 26
Launch pad: KSC LC-39B
Launch date/time: 27 May 1999/06:49:42 EDT
Launch window: 5-10 minutes
Main engines: SME-1, 2047; SSME-2, 2051; SSME-3, 2049
Lift-off weight: 2,047,756 kg
Orbiter weight at lift-off: 118,859 kg
Orbiter altitude/inclination: 320 km/51.6°
Mission duration: 9 days 19 hours 13 minutes 1 second
Landing site: KSC (47) Runway 15
Landing date/time: 6 June 1999/02:02:43 EDT
Rollout distance: ?
Rollout time: 56 seconds
Crew: Kent V. Rominger (Commander), Rick D Husband (Pilot), Tamara E Jernigan (Mission Specialist), Ellen Ochoa (Mission Specialist), Daniel T Barry (Mission Specialist), Julie Payette, Canadian Space Agency (Mission Specialist), Valery Tokarev, Russian Space Agency (Mission Specialist)
Payload:
Spacehab DM
Integrated Cargo Carrier
STARSHINE
Shuttle Vibration Forces (experiment)
Integrated Vehicle Health Monitoring
Payload weight: 9,047.4 kg

The STS-71/Mir crew, creating a record for the largest crew of a single complex, mark the accomplishment inside Spacelab (NASA) ***1996***

Mission summary
VAB: 15 April 1999
Pad: 23 April 1999
VAB: 16 May 1999
Pad: 21 May 1999
STS-96 was the second manned mission in the assembly sequence for the ISS and was known as Flight 2.A1. Following a countdown marred only by the potential hold waiting for a sailboat to clear restricted waterways, STS-96 was launched on time at 06:49 EDT and cleared to orbit at an initial altitude of 315 km. At orbit insertion, the ISS was in a near circular orbit of 389 km and over the next two days, the two vehicles converged through a sequence of rendezvous manoeuvres actioned by Discovery. Contact between the orbiter and the docking unit on PMA-1 attached to the Unity module launch by STS-88 came at 00:24 EDT on 29 May with docking 15 minutes later.

The object of the mission was to transfer 1,618 kg of supplies and equipment in 115 separate itemised units from Discovery to the interior of the ISS modules. The crew would also move 18 items weighing 89 kg from the ISS to Discovery and remove 300 kg of equipment from the integrated Cargo Carrier and attach it to the outside of the ISS. The first EVA began at 06:51 EDT on 30 May. Jernigan and Barry retrieved operator posts and grapple fixture plates for the Strela crane which will be assembled at the station on later flights and placed on the PMA-2 adapter. The EVA ended after 7 hours 55 minutes, the 45th Shuttle-based EVA but the fourth ISS EVA raising that total to 29 hours 17 minutes.

At 21:14 EDT on 30 May, Jernigan and Tokarev opened the hatch into Unity and at 22:07 EDT moved across into Zarya followed by the rest of the crew for more than three days of busy work. They replaced 12 of 18 MIRTS battery recharge controllers, three to each of the six batteries which had shown signs of under-charge during the months since Zarya's launch six months earlier. On 31 May, the crew installed sound mufflers inside Zarya and completed the transfer of 318 litres of water for the use of the first resident crew. The hatch to Zarya was finally closed at 02:40 EDT on 3 June followed by the hatch to Unity at 03:12 EDT and the hatch to PMA-2 at 04:44 EDT completing a 79 hour 30 minute period of occupation.

A series of 17 pulsed firings over 37 minutes raised the orbit of the docked configuration to 396 × 388 km. By the time the Service Module (Zvezda) is launched late in 1999, this orbit will have degraded to about 357 km. Undocking came at 18:39 EDT on 3 June after 5 days 18 hours 13 minutes in the docked configuration. A photographic fly-around was performed before the separation burn at 20:53 EDT that day. At 03:31 EDT on 5 June, the STARSHINE satellite was ejected to independent orbit. Designed and built by the Naval Research Laboratory, the 160 kg satellite carries 878 reflective mirrors polished by school children in 18 countries enabling them to track it for calculations on the atmospheric density. Discovery landed at 02:03 EDT on 6 June performing the 11th Shuttle night landing, the 6th at KSC.

RESUPPLY VEHICLES

RUSSIAN FEDERATION

Progress series

Background

The constant demand by Salyut and its Mir successor for consumables sparked the Soviet Union to build an automatic resupply vessel. RKK Energia designed and constructed the Progress vessel based on its manned Soyuz vehicle. In the new configuration, the Soviets replaced the manned descent module with a compartment for transporting up to 940 kg of propellants, other liquids and compressed gases. The new design features a fully autonomous docking port with pipes to carry the consumables into the station.

The Soyuz orbital module accommodates up to 1.8 tonnes of removable cargo. It has been used to carry experimental equipment, food, film and air regeneration cylinders. For the return trip, the crew fills this module with unwanted materials. A Progress may also raise the host station's orbit to combat atmospheric drag. Ground controllers command the ferry to undock and retrofire and then target it to re-enter that atmosphere over Russia for destructive re-entry along a Pacific corridor. Cargo accounts for about 35 per cent of a Progress launch mass, the high proportion made possible by the absence of escape systems and re-entry thermal protection.

All 43 vehicles in the series successfully docked with their intended targets, delivering a total of 99 tonnes. Progress delivered a total of 24.97 tonnes to Salyut 7 by 12 named flights. Kosmos 1669 added 2,254 kg to this total. Mir's first 17 deliveries totalled more than 40 tonnes, or double the station's original mass. The first design lacked solar panels, which limited it to an independent flight time of at least three days, although the capsule itself remained docked to Mir for two or three months. The vehicle stretches to a slightly greater length than Soyuz because RKK Energia added an additional pressurised instrument section to the rear unit.

Progress-M, the most current version of the design, flew for the first time in August 1989. It incorporates Soyuz-TM's Kurs rendezvous/docking system, computer, propulsion unit and solar arrays. Kurs allows forward port docking, whereas the previous Progress could only approach a space station through the aft docking unit. The Russians at first planned to build an improved version for delivering supply modules to the ISS, but Russia now expects to improve the current capacity by 200 kg and use a Proton-launched modified FGB for large deliveries. A Zenit 2 booster would have launched the 13.3 tonne Progress-M2 and might have delivered as much as 2.3 tonnes of propellant. Four of these loads would satisfy the Station's projected annual requirement.

Progress-M 18 attached to Mir's forward port **1995**

Launch and re-entry dates up to the time of going to press were as follows:

Salyut 6
1 20 January to 7 February 1978
2 7 July to August 1978
3 8 August to 24 August 1978
4 4 October to 26 October 1978
5 12 March to 5 April 1979
6 13 May to 9 June 1979
7 28 June to 18 July 1979
8 27 March to 26 April 1980
9 27 April to 22 May 1980
10 29 June to 19 July 1980
11 29 September to 11 December 1980
12 24 January to 19 March 1981

Salyut 7
13 23 May to 6 June 1982
14 10 July to 11 August 1982
15 18 September to 16 October 1982
16 31 October to 13 December 1982
17 17 August to 18 September 1983
18 20 October to 16 November 1983
19 21 February to 2 April 84
20 15 April to 7 May 1984
21 8 May to 26 May 84
22 28 May to 15 July 1984
23 14 August to 26 August 1984
24 21 June to 15 July 1985
C1669 19 July to 30 August 1985

Mir
25 19 March to 21 April 1986
26 23 April to 23 June1986
M5* 27 September to 28 November 1990
27 16 January to 25 February 1987
M6 14 January to 16 March 1991
28 3 March to 28 March 1987
M7* 19 March to 7 May 1991
29 21 April to 11 May 1987
M8 30 May to 15 August 1991
30 19 May to 19 July 1987
31 3 August to 22 September 1987
32 23 September to 19 November 1987
33 20 November to 19 December 1987
34 20 January to 2 March 1988
35 25 March to 5 May 1988
36 13 May to 5 June 1988
37 18 July to 12 August 1988
38 9 September to 23 November 1988
39 25 December 1988 to 7 February 1989
40 10 February to 5 March 1989
41 16 March to 25 April 1989
M1 23 August to 1 December 1989
M2 20 December 1989 to 9 February 1990
M3 28 February to 28 April 1990
42 5 May to 27 May 1990
M4 15 August to 20 September 1990
M6 14 January to 16 March 1991
M7* 19 March to 7 May 1991
M8 30 May to 15 August 1991
M9* 20 Aug-30 September 1991
M10* 17 October to 20 January 1992
M11 25 January to 13 March 1992
M12 19 April to 28 June 1992
M13 30 June to 24 July 1992
M14* 15 August to 21 October 1992
M15 27 October to 7 February 1993
M16 21 February to 27 March 1993
M17 31 March to 1993 to 3 March 1994
M18* 22 May to 3 July 1993
M19* 13 August to 13 October 1993
M20* 11 October to 21 November 1993
M21 30 January to 23 March 1994
M22 22 March to 23 May 1994
M23* 22 May to 2 July 1994
M24 25 August to 4 October 1994
M25* 11 November to 16 February 1995
M26 15 February to 15 March 1995
M27 9 April to 23 May 1995
M28 20 July to 4 September 1995
M29 8 October to 19 December 1995
M30 18 December to 22 February 1996
M31 5 May to 1 August 1996
M32 31 July to 21 November 1996
M33 19 November 1996 to 12 March 1997
M34 6 April to 2 July 1997
M35 5 July 1997 to 7 October 1997
M36 5 October 1997 to 19 December 1997
M37 20 December 1997 to 15 March 1998
M38 14 March 1998 to 15 May 1998
M39 14 May 1998 to 29 October 1998
M40 25 October 1998 to 3 February 1999
M41 2 April 1999

M17 undocked 11 August 1993 but was held in orbit for longevity tests. * return capsule carried. Progress Mir cargo (kg): 25 2,482, 26 2,405, 27 2,406, 28 2,084, 29 2,227, 30 1,856, 31 2,441, 32 2,341, 33 2,082, 34 2,324, 35 2,283, 36 2,337, 37 2,305, 38 2,282, 39 2,242, 40 1,993, 41 2,238, M1 2,682, M2 ?, M3 2,643, 42 2,409, M4 2,689, M5 2,594, M6 2,546, M7 2,542, M8 2,693, M9 2,730, M10 2,623.5, M11 2,576, M12 2,748.5, M13 2,752, M14 2,532, M15 2,558, M16 2,598

The last original Progress spacecraft flew between 5 and 27 May 1990. Progress-M 7 was the first to suffer docking failure, in March 1991, because of a broken antenna on Mir itself. Consequently, it became the first to take propellant from a station for orbit raising. Progress-M 4 carried an open cargo carrier in place of the refuelling mid-section for the first time.

Russian space science has made great strides in docking compared to the West. Manakov Poleshchuk demonstrated this when he undocked Progress M16 on 26 March 1993, took it out to 70 m and brought it back in for docking. In September 1994, TM19's Malenchenko brought in Progress-M 24 (with its malfunctioning Kurs) manually from 150 m using TV signals from the unmanned craft. Remote-control docking from the ground may be added, with a cosmonaut in the Flight Control Center controlling Progress using Soyuz-type controls and a TV monitor.

Some Progress's now deliver a coni-cylindrical re-entry capsule (VBK: ballistic return capsule) among their cargo. Loaded with 80-150 kg of film and material samples, the capsule mounted in Progress's hatch by the crew, releases itself after retrofire at about 120 km for recovery by parachute (initially in Kazakhstan's normal manned landing area but beginning with M18 within Russia to avoid political problems). M5 carried the first of these configurations and landed with at least some crystals among its 113 kg payload. Unfortunately, the M7's Raduga and 94 kg payload was lost.

After separation, the capsule retains Mir's 1 atmosphere pressure and the outside nose temperature reaches 2,200°C during entry. The braking chute deploys barometrically at 15.5±2.5 km (Mach 1.5±0.5), the main chute spreads to a diameter of 5.6 m. Antennas on 'chute lines and clouds of dipole reflectors released during the descent pinpoint the location. The reflectors are the same ones used as penetration aids for Russia's nuclear re-entry vehicles. A balloon provides buoyancy for water landings.

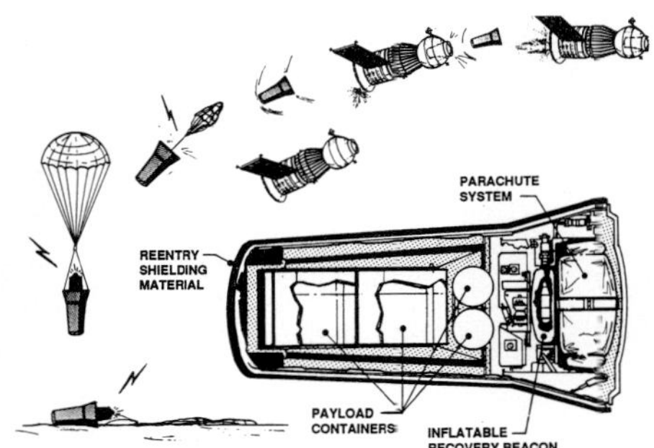

The Raduga return capsule is mounted in Progress' docking hatch and de-orbited by its host for a ballistic re-entry (Teledyne Brown Engineering)

Specifications

First launch: 20 January 1978; 19 July 1985 for original Mir design; 23 August 1989 for Progress-M (27 September 1990 for M with return capsule)
Launch sites: Tyuratam
Launch vehicle: Soyuz SL-4 (without escape system)

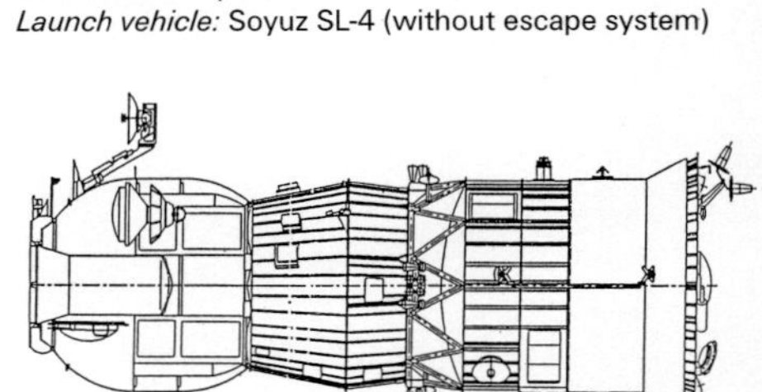

Progress-M side view, configured for undocking with the Raduga return capsule positioned for ejection (RKK Energia)

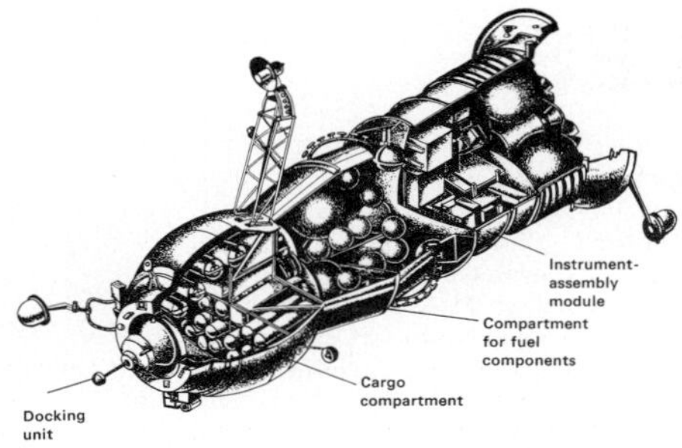

Progress space station cargo vehicle. The current M version incorporates solar panels

Principal applications: unmanned LEO space station ferry
Cost: reported at about Rb10 million/mission
Availability: typically launched every 2 to 3 months during station occupation
Performance: up to 2,750 kg total. 1,500-1,800 kg in crew-accessible 7.0 m^3 Orbital Module and 940 kg in mid-section (870 kg UDMH/NTO in two spheres each; 50 kg O_2). Progress-M can also transfer its own excess propellant. The preceding design carried up to 2,480 kg (up to 1,400 kg in OM + up to 1,200 kg refuelling section). Raduga, when carried, can return up to 150 kg. Independent flight time 30 days (3 days without solar arrays), attached to Mir max 180 days
Total length: 7.23 m; 2.2 m typical diameter
Launch mass: 7.3 t; 7,240 kg in original Mir design, 7,020 kg w/2,300 kg cargo for Soyuz.
Propulsion: derived from ODU unified propulsion system of Soyuz-T/TM. Hypergolic UDMH/NTO is employed with the 3.1 kN main engine and a network of 14 × 98 N (four back-up) plus 12 × 9.8 N (four back-up) thrusters. It is not known if a back-up to the main engine is carried (the smaller thrusters provide de-orbit capability). The system is used for Mir orbit raising to combat drag effects before Progress departs. Progress-M can transfer its excess propellant to Mir (M7, because of its docking problems, became the first to take propellant from Mir for orbit raising)
Power: TM-type solar arrays generating 1.3 kW from 10 m^2; batteries carried in pressurised instrument compartment. Augments Mir supply when docked
Avionics/control: Progress M docks with Mir/Kvant's forward or aft port, employing the Kurs ('course') approach system. Two external TV cameras are carried and range/range-rate data are transmitted to both mission control and to the station crew, who can assist in the docking operation if necessary. Attitude reference provided by Earth horizon sensors + gyro package.

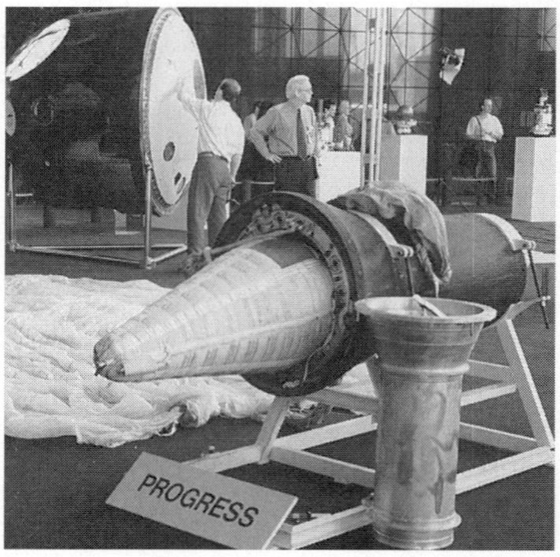

Raduga displayed with its payload container (Theo Pirard/Space Information Center) ***1996***

Soyuz carries its avionics in its manned descent module, for Progress they are repackaged into the instrument unit. Navigational computations are performed by mission control and uplinked for execution.

SPACE STATIONS

Mir

Background

Mir ('Peace'), a close cousin of the Salyut 7, replaced that programme in 1986. It was configured as the core habitation module of a 135 tonne multivehicle complex. The Soviets launched Mir with only 20 days' provisions on board and Progress 25 had to resupply the consumable by Progress 25. The Soyuz-T15 cosmonauts occupied it first during 1986, Vladimir Solovyev having the honour of entering the spacecraft as the first person aboard. Despite this inauguration, science module delays deferred the beginning of routine occupation until February 1987. Resupply vessels routinely visit Mir every two to three months when it is occupied.

TM2's Yuri Romanenko established a space endurance record of 326 days on his return to Earth 29 December 1987, surpassed by the 365.94 days of TM4's Titov/Manarov during 1988 and then the 437.7 days of TM18's Poliakov in 1995. Manarov's TM11 mission 1990-91 made him the first person to exceed 500 days total in space.

Mir hosted three international visiting crews during 1988 (Bulgarian, Afghani and French), but continuing module delays and the ground test failure of Soyuz-TM 8's service module resulted in it being left unmanned from April to August 1989. Kvant 1's arrival in April 1987 allowed the beginning of detailed astrophysical observations, but without other modules Mir was limited to basic medical, Earth observation and μg research.

Official Soviet sources stated that the Mir programme up to the return of Soyuz-TM7 27 in April 1989 cost Rb1,471 million, including the development, launch and operating costs of the core, Soyuz, Kvant 1 and Progress, and the development costs of the first two awaited modules. Kvant 1/2, Kristall and all Progress through M16 delivered 87 tonnes of cargo.

Recycling reduced the cargo needs of the Progress vehicles. Water and urine recovery has proved a valuable method of reducing cargo deliveries. Through mid 1993, some 7,500 litres of water had been recovered from the atmosphere. Since a Progress can deliver a maximum of 420 litres of water, this saved 18 Progress flights alone. Shuttle crews now carry across their waste water, which would otherwise be dumped, in plastic bags holding up to 43 kg. Kvant 2 had recycled 2,500 litres of urine by electrolysis, providing enough oxygen for 750 days and saving 4 tonnes of Progress cargo capacity. The molecular sieves of Kvant 1/2 for scrubbing out CO_2 had saved the delivery of 450 chemical canisters (3,600 kg cargo. By comparison a Progress can handle 40 canisters.

NASA Administrator Dan Goldin and RKA General Director Yuri Koptev signed a $400 million protocol on 16 December 1993 that includes $305 million for up to 10 American astronaut visits to Mir. The Spektr and Priroda modules carry some US equipment for a broad programme of materials science, biotechnology, life sciences, Earth observations and space technology. The agreement covers upgrading and maintaining Mir to ensure it remains operational through late 1999.

NASA/RKA, in 1993, began joint development of a solar power dynamic power system, for testing aboard Shuttle-Mir in 1997/98 and environmental control and life support systems. NASA has since dropped the plans for the ISS. When NASA signed the agreement, Mir managers told NASA they could operate their vehicle for at least another five years, and possibly 10 years, after space station occupation begins. NASA asked them to refrain from planning to avoid competition with its space station.

First aboard Mir was Dr Norman Thagard, delivered for a three and a half-month stay by Soyuz-TM 21 in March 1995 (Sergei Krikalev flew on STS-60 in February 1994 and Vladimir Titov was aboard STS-63 in February 1995 when it flew around Mir to demonstrate rendezvous and docking procedures). STS-71 docked June 1995, with Spacelab carried aboard Atlantis, delivering the new Russian crew and taking home TM21's three-man team. STS-74 in November 1995 delivered a docking module for Kristall, to hold Shuttle away from Mir's solar arrays and two new solar wings. It also brought 967 kg of cargo and returned with 370 kg of mainly science hardware and results.

Four radial science modules completed attachment in 1996, after the core had already exceeded its planned life. The first, Kvant 2, houses gyrodynes to help control the initially asymmetric complex, water electrolysis equipment, crew shower (removed in April 1995) and manned manoeuvring unit (disassembled July 1991) for use in conjunction with a large EVA hatch of at least 1 m diameter. This general purpose unit was followed in 1990 by the technology Kristall module, providing materials processing and biotechnology facilities. The Spektr remote sensing module appeared May 1995 and the Priroda Earth observation unit April 1996.

Soviet space planners saw Mir as a transition element to an even larger space station, which they would build in the 1993 to 1995 time frame Much as the United States had, they wanted to use their Shuttle, the Buran to build this new station. After the first flight of the Buran, the Soviets realised that building a space station by Shuttle launch was a highly inefficient process and dropped any subsequent plans. By that time they had a capability that the United States has still not been able to match in the Progress and Merkur space tugs.

Such a grandiose and costly Mir 2 was finally scrapped in early 1991. The new RSA evolving out of the space activities of the former Soviet Union wanted instead a more modest new core installed in 1997 by Proton to dock with Mir, for Buran's manipulator to transfer the newer Spektr and Priroda modules before the old units were de-orbited. In 1994, plans indicated a new Mir one-size core might be launched late 1996 for immediate manning, awaiting a second core and resource, docking and user modules. A truss beam would accommodate solar arrays and possibly solar dynamic collectors. RSA even planned to move the launch site away from Tyuratam and use instead a Proton KM vehicle from Plesetsk into 65° orbit.

Paradoxically, although RKK Energia leads the Mir programme, the core, heavy modules and their Proton launchers come from the old Chelomei bureaus. Station designer Konstantin Feoktistov commented in 1993 that the original Mir design called for 7 tonne (Progress-sized) add-on modules but 'higher authority' imposed 20 tonne units.

Mir began hosting its 21st long-stay crew in February 1996, with crew exchanges at four to six monthly intervals allowing visits by foreign passengers. CNES/Energia signed a memorandum 29 July 1992 for four missions, in July 1993, August 1996 (Cassiopeia), 1998 and 2000 to exploit Mir's facilities. They paid FFr126 million for the 1993/96 missions, plus equipment. The 2000 Mir visit cannot now occur, but 2000's visit may, in fact, be made to Russia's portion of Alpha. RKA/DARA signed an agreement December 1995 for a second German visit, aboard TM25 in February 1997. ESA's two visits were the 30 day EuroMir-94 from October 1994 and the six-month EuroMir-95 from September 1995, paying ECU45 million for the pair. RKA acknowledged in February 1996 that South Africa and South Korea were interested in visits.

By 1997, NASA began to raise serious questions about the fitness of Mir to continue in orbit. With the fire in February 1997 and the failure of the oxygen-generating equipment the following month, the cosmonauts spent more and more time simply doing repair work to keep the station operating. Then the catastrophic collision late in the Summer caused an almost daily media assault on the station.

On 22 April, the first authoritative Russian comments appeared which raised the possibility of abandoning Mir. Viktor Blagov at the Mission Control in Korolyov said that at the moment there was no reason to evacuate Mir, but added: "The team is continuing its repairs. It's a difficult task and if it cannot be solved Mir might be abandoned by the cosmonauts". Blagov, the deputy head of the flight programme, said the cosmonauts had repaired two leaks in the cooling system with a special glue but had yet to find a third through which antifreeze was escaping into Mir's air supply. "It's on the limit of what is acceptable," He told a news conference, adding that it was not clear how dangerous the concentration of antifreeze vapour was. He said an urgent evacuation of the cosmonauts could be ordered from ground control or by the space station commander. "That could happen if there's a big fire on the station or if there is decompression," Blagov said, adding that the men could reach Earth in an hour using the Soyuz-TMM spacecraft docked with Mir.

If Mir had to be left unmanned for any length of time then it is questionable whether it could ever be used

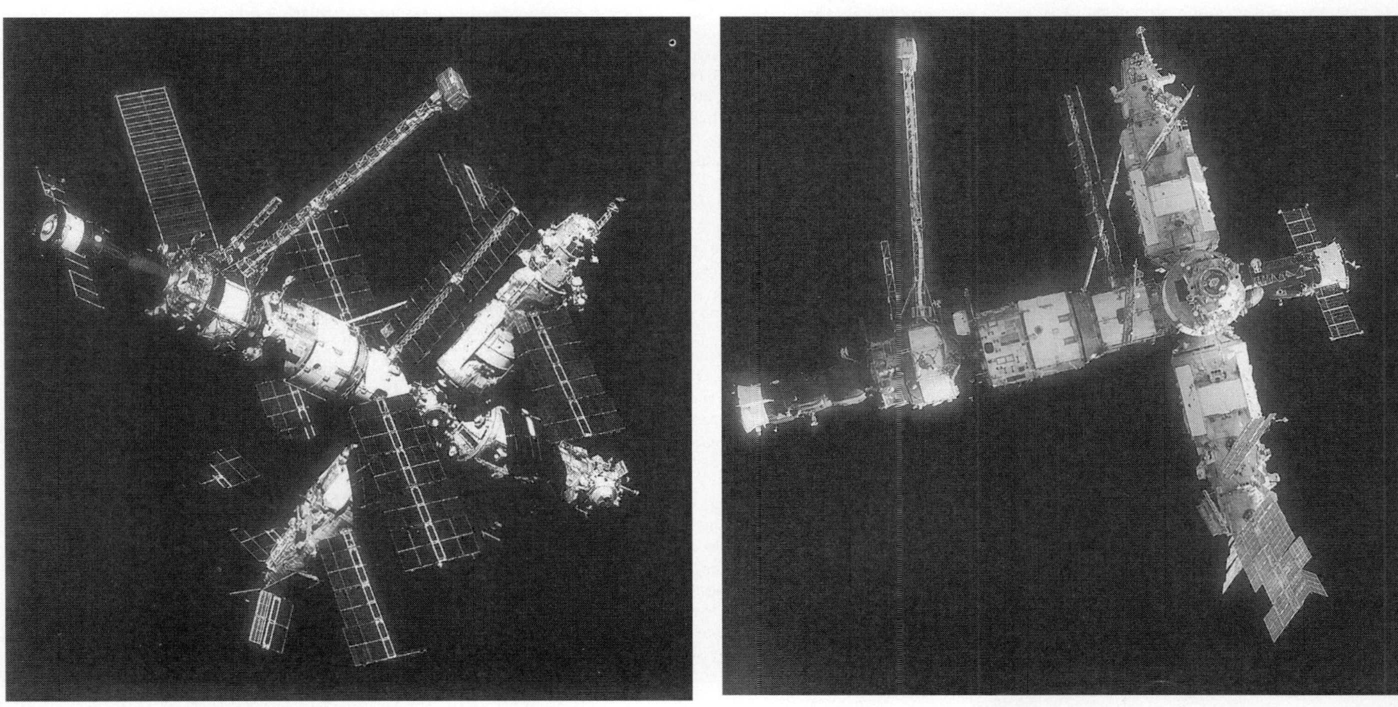

Mir configuration as STS-74 (left) and STS-71 (right) approached for docking. The main difference is that STS-71 docked with Kristall at Mir's axial node; STS-74 brought a docking module that allows Kristall to remain permanently at a radial face. Kvant 2 is at top and Spektr at bottom. Protruding from Kvant 1 is the 14 m Sofora tower carrying the roll control thruster package. Kvant 1 also carries the single solar wing transferred from Kristall (NASA)

1995

Mir's main living area, looking towards the rear tunnel leading to Kvant 1. At bottom is the main table, at right are the refrigerator (white circle), individual cabin and food store

again. In 1985, the Soyuz-T 13 crew brought the Salyut 7 out of mothballs establishing a precedent for such activities. That station was nowhere near as old as is the core module of Mir. At present, it is reported that the cosmonauts on Mir are spending at least half of their working days doing repairs to keep the station operational. If the station is left untended for even a short period of time, it is thought likely that it would rapidly degrade to a point where it could not be revived for habitation.

If Mir has to be abandoned, then the chances are that it will re-enter the atmosphere without any control over where debris might land. With the radial modules breaking off and coming down independently along the same ground track the chances of serious damage at ground level cannot be ignored. Accordingly, plans emerged during 1998 for a Progress cargo-tanker to de-orbit the station several days after its last crew leaves in August or September 1999. In this way it can be brought down in a controlled manner far from populated regions of Earth.

Specifications

Mir

Launched: 19 February 1986 by Proton from Tyuratam
Mass: 20.9 t at launch; about 51 t for Soyuz/Mir/Kvant 1/Progress; 72.8 t with Kvant 2 added; 90.0 t with Kristall added; up to 135 t fully assembled with four radial modules
Orbit: typically 300-400 km, 51.6°
Crew size: two-six (two long stay typical)
Length: 13.13 m core, about 32.9 m for Soyuz/Mir/Kvant 1/Progress
Diameter: 4.15 m max cylindrical diameter
Pressurised volume: 90 m^3 for core
Configuration: derived from Salyut 6/7 but the interior reworked as a habitation core vehicle and a five-port docking unit replaced the previous single-port forward unit. The 2.2 m diameter node can accommodate a Soyuz-TM/Progress ferry or large module (axial face) and four science modules at the radial positions. Each module docks first at the axial port drogue, inserts a Lyappa manipulator arm into one of two node sockets and swings around mechanically to its allocated site. A conventional docking approach from the side would impose unacceptable loads on the structure (STS-74 made first radial docking 15 November 1995, its 2 cm/s contact within rates).

Valeri Poliakov observes the 6 February 1995 rendezvous by STS-63 Atlantis from Mir's core (NASA) ***1995***

All ports incorporate the standard range of electrical, mechanical and hydraulic connections of all the 80 cm diameter hatches. Limited space meant that only two docking drogues were included, the others using blank doors. Swapping doors with drogues to accommodate module movements has required some 'internal' EVAs (11 January 1990, 28 May 1995, 2 June 1995, 8 December 1995). Length of the sphere and its frustum collar is 2.84 m. A hatch leads into the first, 2.9 m diameter, habitation cylinder, which houses the primary station control facilities. The wall units either side of the main control panel carry the air regeneration, water recovery and thermal control systems; this section's exterior provides a mounting for the thermal radiators and the three solar arrays. The habitation section flares to a 4.15 m diameter towards the rear.

The instrumentation carried by every Salyut, which viewed Earth through a large underbody aperture, was removed and floor lockers added. There is also a small airlock in the floor under the wardroom table for experiments. This large table protrudes from the starboard wall with two seats either side to create a dining/recreation area. Heating and drinking facilities are provided, including aft food lockers and a 50 litre portside refrigerator (there is no freezer – ESA had to bring its own). Two personal cubicles are provided behind the table on either side, each with a sleeping bag, porthole and communications panel. On the floor between them is the treadmill aligned with the longitudinal axis. Starboard is a toilet cubicle that includes a spherical vessel for washing hands in running water. Aft of the toilet, a hatch completes the 7.67 m long main cylindrical sections. A 2.26 m long, 4.15 m diameter propulsion unit that provides access to aft-docked vehicles through a 1.67 m long, 2 m diameter tunnel caps Mir.

Power: 9 kW was generated by two Sun-following 38 m^2 4-segment 2 × 2 cm-cell Si solar wings with a rigid central spine; the two outer segments are slightly larger. Span is 29.73 m. The top Salyut attachment point was retained and a 340 kg third array, delivered by Kvant 1 in April 1987, was erected to bring the total area to 98 m^2 and supply to 10.1 kW. Voltage is 28.5 ±0.5 V DC. 12 74 kg Ni/Cd batteries provide storage. Appears that Kvant 2 added to the power supply but some recabling was performed during 1991 to allow pooling of all modules' power. With Soyuz/Progress attached, total array area 250 m^2. This would provide 28 kW BOL, but shading/ageing reduced it to 10 kW by mid-1992. Power shortage is a persistent Mir problem; ageing reduces the supply by 5% annually. STS-74 in November 1995 delivered two new solar arrays for attaching to Kvant 1 in two 1996 EVAs to add 12 kW. Both built by Energia, but one carries 84 Lockheed Martin panels of 80 Si cells each, totalling 42 m^2 and 6.72 kW.

Propulsion: twin aft-mounted 2.94 kN main engines gimballed ±5° and 32 (in two independent networks of 16 each) 137 N attitude control thrusters. Pressure-fed with hypergolic UDMH/NTO from four tanks incorporating metallic membranes and offering 800 kg capacity. System derived from Soyuz-T/TM. Refuelled by Progress ferries via lines around Kvant; the primary engines cannot be used while Kvant 1 remains attached. Progress' own engines are employed for orbit raising/changing before departure (M7, because of its docking problems, became the first to take propellant *from* Mir). AOCS requirements were reduced by adoption of control moment gyros (see avionics/control). 700 kg thruster package added to 14 m Kvant 1 Sofora mast September 1992 to reduce roll control consumption eight-fold.

Life support: pressure maintained at 1,053-1,276 mb, T 18-28°C, humidity 20-70%, O_2/N_2 mixture with O_2 max content 40%, CO_2 3%. Pressurised air delivered by Progress + Shuttle (20 litres stored for atmospheric makeup), CO_2 scrubbed chemically by LiOH (now used as back-up to molecular sieves on Kvant 1/2; 15 cartridges retained for 3.3 man-days capacity). 40 cartridges of O_2-generating $NaClO_3$ sodium chlorate retained as back-up to Kvant 1/2's electrolytic system and used when >3 people aboard. A 1,000-1,100K charge ignites the reaction. Other modules provide additional life support capabilities. Salyut's systems were extensively reworked to accommodate crews of up to six. Long-stay cosmonauts remain aboard for typically 6-12 months (more recently six months), punctuated by 8-30 day visiting missions and working 8.5 h days with weekends off. It was discovered that a three-man crew accomplishes little more work than a two-man. Up to 90 min are spent exercising daily. General equivalent protection inside Mir against radiation is typically 1 g/cm^2. Background noise level is 75 dB; Manarov reportedly required treatment for deafness caused by Mir's high levels. Mir's life support system is the responsibility of NPO Zvezda.

Inside Mir: looking forward from the living section across the main control panels towards the docking node. TV and computer screens face the crew. The pictures were taken from the same perspective, but November 1994's (right) shows how cluttered Mir has become. Work is affected because of the lack of a proper inventory; crews sometimes have to radio previous occupants to find stored items (Graham Turnill/ESA)

Avionics/control: ODCC On-board Digital Computer Complex in conjunction with PINS Platform-less Inertial Navigation System, offering 11/2° automatic/15 arcmin manual pointing accuracy. Eight computers have been referred to aboard Mir's core but the two primary units are: a Soyuz-T Argon 16B derivative capable of performing 200,000 short ops/s and the more powerful Salyut 5B computer (500,000 ops/s) that is employed now as the Mir complex develops. Limited memory means that activities software has to be loaded daily. As modules are added, new control software has to be integrated to accommodate the configuration change; the L-shaped assembly formed after Kvant 2's arrival proved more difficult to control than expected. Salyut 5B was brought into operation during 1990 but difficulties with it delayed the launch of Kristall. There are two CRTs in Mir's forward command position; Kvant 1 includes a third, permitting station control during observing

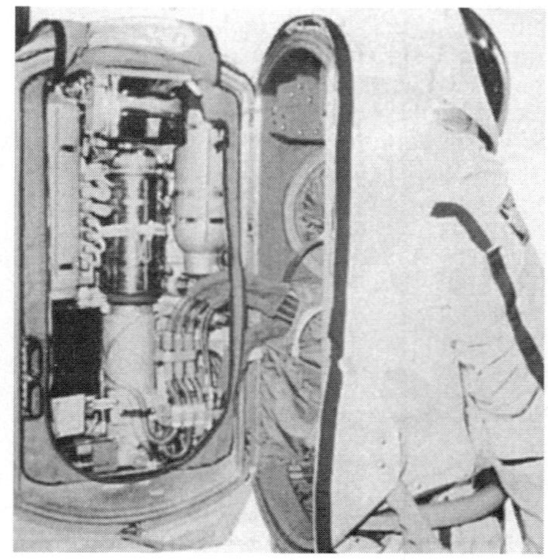

Entry into the Mir EVA suit is via the open backpack. Note the two 1 kg oxygen bottles

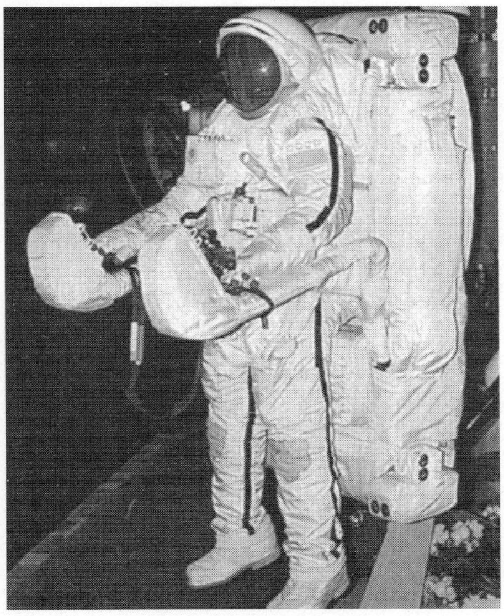

Mir EVA/manoeuvring unit. Note the paired thrusters on the corner units

sessions. Attitude/navigation information is provided by: IR Earth horizon scanning sensors (local vertical), solar sensors, star sensors, Sun-presence sensors (4π field of view, signalling Earth shadow entry/exit), automatic/manual star sextant, magnetometer, gyros and linear accelerometers. The Kvant 1/2 modules each carry six gyrodyne control moment gyros, providing 15 arcmin pointing accuracy during astronomical observations. Each 10,000 rpm, 1,000 Nms/200 Nm unit requires 90 W and masses 165 kg with its electronics packages. Mir carries the S-band Kurs ('course') approach/docking system forward/aft, and the older Salyut Igla ('needle') only at the rear. Kurs holds Mir stationary, thus conserving propellant, while the approaching vehicle conducts all the manoeuvres. Soyuz-TM, Progress-M and the new science modules employ Kurs, while the older Progress used Igla. Both generate range, closing rate, line-of-sight angular velocity, perpendicular deviation from the required approach and, from 200 m out, roll deviation. Unmanned vessels are docked under ground control; crews originally occupied their descent craft during such operations on safety grounds. Malenchenko docked progress-M 24 from Mir using TV relayed from the ferry. Mir carries an aft steerable 11-14 GHz antenna for continuous links through GEO telecom satellites at 95/200/344° east (Cosmos 2054 at 344° east, Luch 1 at 95° east and Luch-1 1 at 77° east were in position at the beginning of 1996). C1700 at 95° east permitted 40 min contact/rev during Mir's early life, in contrast to Salyut's 15-20 min, although the satellite subsequently drifted off station.

Mir components

EVA suit/manoeuvring unit

Description: The Zvezda Factory under the direction of Chief Designer Gai Severin is responsible for EVA equipment and Mir's life support system. October 1988's EVA of Titov/Manarov introduced a modified Orlan-DM ('Bald Eagle') suit that did not require power or communications umbilicals; a simple tether was provided for safety. The suits were delivered by Progress 37 in July 1988. Improved Orlan-DMA versions were delivered by Kvant 2 with the manoeuvring unit. Standard duration is limited to 6.5 h by the capacity of 2.2 kg LiOH 120 mm diameter cylinders for scrubbing out CO_2, but can cope with further 90 min. However, the suit can be linked to Mir for up to a week in emergencies. Adjusting via cables/pulleys accommodates different arm/leg lengths. There are two glove sizes. The cosmonaut wearing a liquid cooling garment climbs into the suit from the rear, through the open backpack and inserts his head into the integral helmet. The rear door is closed by pulling a chest-mounted wire and sealed by a lever on the right waist. The aluminium hard upper torso carries a display/control panel at upper right visible from the helmet; bottom left are temperature regulation and emergency oxygen controls with reversed lettering read by a left-wrist mirror. The rear hatch prevents there being a waist bearing (unlike Shuttle's), making shoulder twisting difficult. No food, drink or urine collection is provided. Each suit can undertake 10 EVAs over four years.

Suit Mass: 105 kg loaded

Oxygen system: 1.0/1.0 kg O_2 stored primary/back-up. Primary 0.206-0.255 kg/h maintains 0.4 atmosphere pressure (30 min pre-breathing required); back-up 2 kg/h activated manually or automatically if pressure drops to 0.22 atmosphere, routeing some flow directly to the helmet. If a glove leak is responsible, cuffs inflate to provide 15 mins' grace (unlike the rest of the suit, the gloves do not have a back-up bladder). Mir can also supply 2-3 kg/h through umbilical

Cooling garment: 3.6 kg water circulated at 60-180 kg/h at 16-22°C, removal capacity 300/600 W average/max. Also available from Mir, in addition to 0.15-0.25 m³/min cooling gas.

Power: 42 W required. Can be provided by Mir or the manoeuvring pack, along the same cable that relays suit parameters to the pack's telemetry unit.

Communications: 100.125 MHz voice transmit; 23 1/247 MHz suit parameters.

Comment: The new Orlan-M was delivered to Mir in 1996 and will be used in the Space Station era. 110 kg loaded, warranty for 12 sorties in 4 years. 54 W consumed power. It provides easier donning/doffing, improved visibility, increased reliability and partial interoperability with Shuttle EVA suits. The entry hatch is raised 6 cm to eliminate the awkward head insertion into the helmet. The torso provides increased volume. Shortening the hard upper torso and making the longer soft lower torso adjustable (the DMA allows only leg/arm adjustment) accommodates the larger range of US/Western European wearers. New bearings between the shoulder and elbow and above the ankle provide greater flexibility. A small top visor allows overhead viewing. A 'hot' back-up water pump in series increases cooling reliability. The 0.22 atmosphere back-up pressure mode has been eliminated – it has never been used intentionally (but has been activated accidentally) and would be of limited value. The 1 kg emergency oxygen supply cuts in at 0.27 atmosphere (instead of 0.22 atmosphere) and the redundant bladder comes into effect at 0.32 atmosphere instead of 0.24 atmosphere. The LiOH cartridge provides 9 hr operation, allowing EVAs from Shuttle's airlock (the current 6.5 hr is acceptable because Mir provides scrubbing while airlocking).

The Icarus (Ikar) equivalent to NASA's Manned Manoeuvring Unit, known generically as the YMK, was introduced during Serebrov's 1 February 1990 EVA The original version, incorporating 14 air thrusters and some *solid* motors, was completed in 1968 but cancelled because of lack of missions. The new unit (Zvezda built two) was housed in Kvant 2 and could be used only through that module's 1 m diameter hatch. It was apparently developed for Buran tile inspection/repair and after 1990's two demonstrations was disassembled into six pieces and stored inside Kvant 2, with no plans for further use; it was attached to Kvant 2's exterior in the EVA of 8 February 1996 to move it out of the way. A rigid waist belt, that also provides attachment to the parent spacecraft and payload mounting, holds in the occupant. Two adjustable arms provide consumables status reports and rotational (right) and translational (left) thruster controllers. Four T-shaped corner units each incorporate eight thrusters in redundant pairs powered by compressed air from two titanium tanks; total 30 m/s V (NASA's MMU 20.1 m/s). Fresh tanks delivered by Progress ferries. Two independent gyro systems provide rotational reference and an automatic stabilisation to ±2°. Range for Mir operations limited to 60 m by the safety tether, carried because, unlike the Shuttle, there is no immediate rescue capability; 100 m is the limit without it. Serebrov's initial demonstration involved only translational movements 5 m out from Mir; Viktorenko's four days later included 3-axis variations. Both were tethered: an electric winch on the chest took up slack and could provide propulsion back to the station. Potential improvements include a navigation system, return of an incapacitated occupant by remote control from the station and an uprated propulsion system.

Mass: 218 kg loaded, about 400 kg with suited occupant.

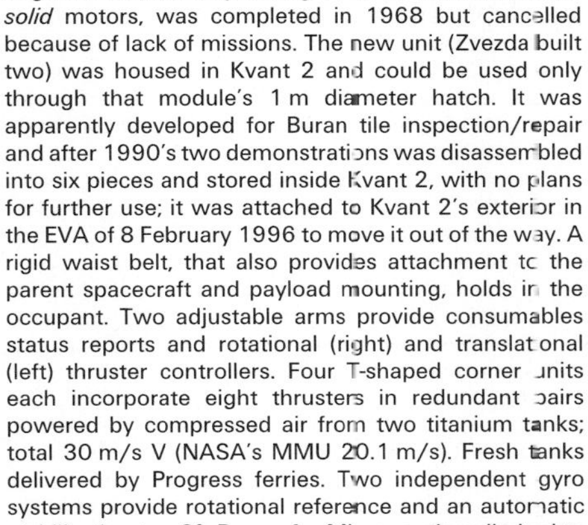

The Kvant 1 astronomical module. 1 main compartment, 2 Igla antennas, 3 docking unit, 4 Mir cargo, 5 Svetlana electrophoresis unit, 6 Kurs antennas, 7 drogue docking unit, 8 science compartment, 9 HEXE x-ray telescope, 10 Sirene 2 spectrometer, 11 Glazar ultra-violet telescope, 12 TTM x-ray telescope, 13 Pulsar X-1 x-ray telescope, 14 optical sight, 15 star tracker, 16 main control panel, 17 life support units

Kvant 2, featuring its ASPG scan platform at top left (NASA)
1996

Power: single 18.0 Ah primary + 8.5 Ah back-up 27 V DC AgZn batteries. Power can be provided to suit.

Communications: about 100 unit + suit parameters transmitted automatically. Audio warnings provided in the cosmonaut's helmet.

Total impulse: 12 kNs (δV 30 m/s)

Propulsion system: two independent systems of 16 × 5.0 N thrusters each, mounted on four corner units in redundant pairs. Each system has two titanium tanks holding 28 litres of air compressed at 345 atmospheres, reduced by valves to 12.3 atmospheres. One system utilised until tank pressure drops to 108.5 atmospheres; system number 2 then used to exhaustion, leaving number 1's as an emergency reserve. Crossover lines allow each tank to serve the other system. In economy mode, for proximity operations, the translational thrusters fire for 1 s and rotational provides 3°/s. In boosted mode, values are 4 s and 8°/s, respectively.

Kvant 1

Description: Mir's first additional module was Kvant ('quantum'), comprising a laboratory bay and transfer compartments totalling 40 m³ and an unpressurised aft payload bay accommodating a large array of high energy telescopes. The principal bay houses the control systems, life support systems, including T and air composition regulators, plus radio, TV and command/display units for Mir's computer system. Six gyrodynes, electrically powered by Mir's solar panels, help to orient the station economically. This bay carries two viewports: one, 43 cm diameter, is intended for a mounted optical device and the other, 22.8 cm diameter, for a visual star tracker. The aft transfer compartment has two 8 cm diameter portholes for visual observations; it also includes an airlock for servicing the Glazar UV telescope. The suite of astronomical instruments can be operated from the ground in untended mode; Glazar's exposed films are delivered to Earth by returning crews. Glycol coolant was discovered to be leaking 1 November 1995 from the Vozdukh CO_2 scrubbing system. It was plugged and tested satisfactorily 3 November.

Launched: 31 March 1987, Mir docking 9 April 1987. Jettisoned service module re-entered 25 August 1988.

Principal application: manned/automatic astronomical observatory, expansion of Mir experimental facilities.

Length: 5.8 m (total length with service module attached about 12 m).

Diameter: maximum 4.15 m (as Mir core).

Mass: 20.6 t at launch with service module, 11,050 kg docked to Mir (including 1.5 t science payload and 2.6 t Mir equipment such as solar panel).

Pressurised volume: about 40 m³

Propulsion: none initially, rendezvous/docking operations were performed by the jettisoned service module. Kvant's exterior incorporates plumbing to permit refuelling of Mir's tanks by Progress ferries. 700 kg self-contained thruster package was attached to 14 m Sofora mast by 11 September 1992 EVA to reduce the complex's roll control consumption by up to eight times. Sofora is inclined 11° forward to be closer to centre of gravity.

Power: derived from Mir core. One Kristall wing was transferred by EVAs in May 1995. Two 6 kW wings delivered by STS-74 November 1995 are planned to replace it in 1996.

Life support systems: Kvant was hermetically sealed on the pad under sea level conditions. Mir maintains a pressure of 1,053-1,276 mb, T 18-28°C, with max O_2 content 40%, CO_2 up to 3%. Kvant carries experimental unit for O_2 production by electrolytic decomposition of water. 150 kg Vozdukh molecular sieve scrubs CO_2 (280 W daily average requirement).

Avionics/control: derived from Mir (from service module before docking), but Kvant was first to incorporate six gyrodyne control moment gyros, providing 10 arcsec orientation during astronomical observations.

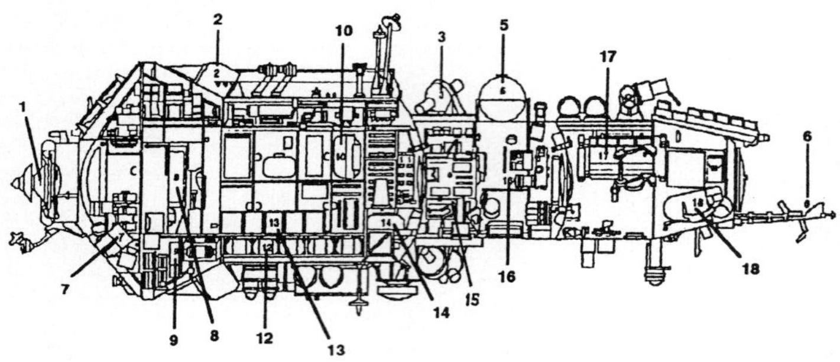

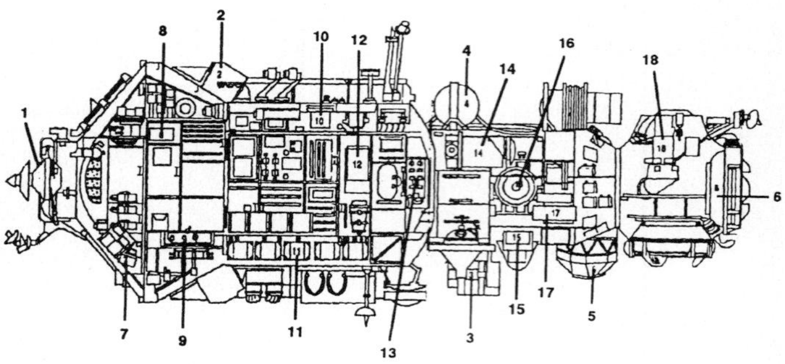

Kvant 2 (left) and Kristall cutaways. Kvant 2 key: 1 docking unit, 2 thruster packages, 3 gyrodynes, 4 star camera, 5 Rodnik water tanks, 6 EVA hatch, 7 control station, 8 shower/hygiene unit, 9 water regeneration unit, 10 Progress water units, 11 attitude control system, 12 power system, 13 food/supply containers, 14 payload container, 15 MKF-6M camera, 16 VIKOD equipment, 17 EVA manoeuvring backpack (stored outside since February 1996), 18 EVA suit. Kristall key: 1 docking unit, 2 thruster packages, 3 Glazar 2 ultra-violet telescope, 4 Rodnik water system, 5 Marina γ-ray telescope, 6 androgynous docking unit, 7 control station, 8 Krater 5 control panel (unit is on opposite wall), 9 running track, 10 attitude control system, 11 power system, 12 SVET plan cultivation unit, 13 Rodnik system control panel, 14 freezer, 15 Ksenia device, 16 reusable solar wing drive, 17 Kristallisator CSK-1 furnace, 18 Priroda 5 Earth camera assembly. The Krater 5, Optizon 1 and Zona 2/3 units are on the starboard wall (The Cosmos Center) ***1995***

Each 10,000 rpm, 1,000 Nms/ 200 Nm unit requires 90 W and with its electronics packages masses 165 kg. Five were reported to be still operational after five years. Kvant also carries two IR Earth horizon sensors, two star sensors, star tracker (two more were added during the EVA of 8 January 1990), Sun sensors and optical sight to contribute to the station's control system; attitude control can be commanded manually from the module during observing sessions. See under propulsion heading above for added roll control thrusters.

Science payload: the unpressurised instrument bay girdles the aft transfer compartment and accommodates the 800 kg 'Roentgen' battery of X-ray and UV telescopes developed in co-operation with the UK, Netherlands, West Germany and ESA. Kvant also houses the Svetlana automated electrophoresis plant in its pressurised volume for biotechnology investigations. Roentgen's detector array comprises: TTM Coded Mask Imaging Spectrometer (Netherlands/UK, 2-30 keV, 7° FOV, 2 arcmin resolution), Sirene 2 Gas Scintillation Proportional Spectrometer (ESA, 2-100 keV, 3° FOV), HEXE High Energy X-ray Experiment (West Germany, 15-250 keV, 1.6° FOV), Pulsar X-1 (USSR, 0-1,300 keV, 3° FOV) and Glazar UV Spectrometer (1,150-1,350 Å, 1-2 Å resolution, 1.3° FOV). Titov/Manarov undertook two EVAs June/October 1988 to replace TTM's failed 60 kg detector package.

Following launch, Kvant underwent a five-day propellant-efficient rendezvous with Mir and on 5 April 1987, locked on for final approach from 17 km. Mir occupants Romanenko/Laveikin were sealed inside Soyuz-TM2 for safety as Kvant closed in from 500 m but the approach was aborted 200 m out because of control difficulties. The module docked 01.36 GMT 9 April but telemetry indicated that hard dock had not been achieved and the cosmonauts had to perform Mir's first EVA, removing fabric left in Kvant's docking probe area during ground processing. The service module was cast off 13 April to expose the Roentgen observatory and the aft docking unit for visiting Progress/Soyuz. The TM8 cosmonauts attached two 80 kg star trackers during 8 January 1990's EVA to improve the pointing accuracy of Mir's attitude control system. The Sofora roll control thruster package was added September 1992.

Kvant 2

Description: Mir's first radial module, also originally referred to as Module D, incorporates a gyrodyne attitude control system, two water regeneration units, Elektron/Vika oxygen production electrolysis equipment (using recovered urine), shower, MKF-6MA Earth resources camera, 1 m diameter outward-opening airlock (plus suits; Icarus EVA manoeuvring unit was housed here until July 1991), in addition to water (300 litres in external tanks of the Rodnik system), food and propellant supplies. The shower was first used by TM9's crew in February 1990 but was not successful. Air flow substituted for gravity, but the cubicle shape needs to be improved. It was dismantled in April 1995 and disposed of in Progress-M 27; a new design was expected to be delivered. Two portholes in Kvant's central section are used for camera, science instrument and visual observations. Kvant 2's operational life was quoted as three years.

Launch: 26 November 1989 by Proton SL-13 from Tyuratam; docked with Mir's forward axial port 6 December, transferring to radial port 8 December using the Lyappa arm.

Principal application: Mir extension unit, EVA activities.

Length: 13.73 m

Maximum diameter: about 4.35 m

Mass: 18,500 kg

Pressurised volume: 61.3 m^3 in three units: airlock; central science/equipment; base service and cargo.

Propulsion: 2 × 3.9 kN KB Khimautomatiki main engines plus clusters of 400 N KB Khimmach thrusters. Carried sufficient NTO/UDMH propellant for one year of free-flying operations.

Power: 6.9 kW provided by two 26.6 m^2 126 kg solar arrays (40 × 48 mm Si cells). Ni/Cd batteries providing 360 Ah total capacity.

Avionics/control: Kurs approach/docking system. Six external gyrodyne control moment gyros (see Kvant 1) assist complex orientation and pointing (four malfunctioned within three years; two replaced internally in July 1992; more added March 1993). Includes DASA Jena-Optronik Astro CCD star attitude sensor.

Payload: Icarus EVA unit (dismantled 1991; stored outside 8 February 1996), East German MKF-6MA 6-band Earth resources film camera, KAP-350 topographic camera, the 110 kg Czech ASPG-M platform (derived from Vega's scan platform; capable of 3°/s) carrying 115 kg of equipment (ITS-7D IR spectrometer, ARIZ X-ray spectrometer, MKS-M2 optical spectrometer + TV cameras; the Gamma 2 TV/ spectrometer package appears to have been added in EVA of 26 January 1990), Sprut 5 charged particle spectrometer installed during EVA of 23 January 1991, Phaza AFM-2 telespectrometer, Spektr-256 spectrometer, 250 kg Volna 2 propellant tank demonstration, Inkubator 2 for hatching birds' eggs, cosmic dust detectors.

Launch was postponed first in early 1989 because of construction delays and again in mid-September when faults were discovered in electronic components of the Kurs rendezvous/docking system. Telemetry indicated after attaining orbit that the starboard solar array's inboard panel had not deployed, leaving the three outer sections free to move during manoeuvres. Full extension/locking was achieved 30 November by rotating the drive mechanism during a vehicle roll. During the final approach to Mir on 2 December for a planned 14.25 GMT docking, Kvant stopped 19 km out, apparently because Kurs detected an excessive closing rate. Docking at Mir's forward axial port came at 12.22 GMT 6 December, bringing the complex's mass to 65.8 t and length to 40 m. On 8 December, Kvant slotted an arm into a socket on Mir's node and took 60 min to swing around to a radial port, the first such procedure in space. Viktorenko/Serebrov then began activating the new section. The first EVA conducted from Kvant 2, beginning with hatch opening at 12.09 GMT on 26 January 1990, prepared for the 1 February demonstration of the manoeuvring pack, followed by a second excursion four days later. Kvant had delivered the suits. Solovyov/Balandin damaged the hatch on 17 July 1990 when they opened it before the compartment was depressurised, preventing its closure at the end of the EVA. They were forced to use the central chamber as an emergency airlock as their suits' capacities were exceeded. The manoeuvring backpack's volume requirements mean that Kvant 2 is the first Russian manned vehicle to carry an outward-opening hatch. The door is equipped with a stop to hold a 1 mm gap for internal pressure to bleed off before full opening, but it was apparently removed prematurely. The hatch was repaired on a 26 July EVA from inside Kvant 2 but the damage was sufficiently serious to warrant further work. Manakov/Strekalov failed to repair the hinge in their EVA of 29 October 1990, leaving Afanasyev/Manarov to replace the hinge unit 7 January 1991. The unused backpack was stored outside on 8 February 1996 EVA.

Kristall

Description: Launch of the technology Module T was delayed from 30 March 1990 because of problems with Mir's Salyut 5B control system computer. Equipment includes materials processing furnaces and biotechnology units, initially claimed capable of generating up to 100 kg of material annually. The 2.2 m diameter node carries two APAS-89 Androgynous Peripheral Assembling System docking ports (89=1989; 1975's Apollo-Soyuz carried APAS-75) to accommodate Shuttle-type visits (first used by TM16 in January 1993); a third face houses the Priroda 5 Earth resources cameras. Buran was to have used the axial port; the other was reserved for a 1 t X-ray module (housing an uprated version of Kvant 1's Pulsar phoswich telescope) carried up by the spaceplane. Because of Shuttle requirements and as the Spektr Earth observation module is designed to operate opposite Kvant 2, Kristall faced four moves in 1995. On 26 May, it swung from the bottom face to the axial and then to starboard on 30 May, allowing Spektr to move directly to the bottom after docking. On 10 June, it swung around to the axial face, where STS-71 could dock safely away from Mir's solar wings. After STS-71's departure, it moved to its final starboard position 17 July, where STS-74 in November 1995 left it with a 4,087 kg 2.2 m diameter 4.7 m-long docking module, lengthening it for future Shuttles to avoid Mir contact.

Launch: 31 May 1990 by SL-13 Proton from Tyuratam; docked with Mir's forward axial port 10 June, transferring to radial port 11 June using the Lyappa arm.

Principal application: materials processing units.

Length: 13.73 m

Maximum diameter: about 4.35 m

Mass: 19.64 t in orbit (including 7 t cargo)

Pressurised volume: 60.8 m^3 in two volumes separated by 80 cm diameter hatch

Propulsion: 2 × 3.9 kN KB Khimautomatiki main engines plus clusters of 400 N KB Khimmach thrusters. Secondary system activated for final approach following problems with primary.

Power: 5.5-8.4 kW was provided by two concertina arrays spanning about 36 m, totalling 72 m^2 of 50 × 50 cm Si cells. Each 36-panel wing and drive unit >500 kg and the first designed to be redeployable – the first wing was removed in May 1995 and attached to less shaded positions on Kvant 1. Each wing can be deployed to 5/10/15 m, attaining full extension only after Mir docking. Ni/Cd batteries totalling 360 Ah. Daily average power consumption 0.5-1 kW; peak 3-7 kW.

Avionics/control: Kurs approach/docking system.

Payload: Krater 5, 5 Optizon 1 and 5 CSK-1/Kristallizator units for semi-conductor production (Krater initially used with zinc oxide, Optizon for growing silicon crystals via zonal heating); Zona 2/3 furnaces for other materials processing. The 500 kg of equipment was expected to generate up to 100 kg of materials annually, but output reportedly affected by equipment reliability and power constraint. Ainur electrophoresis unit (also on Priroda module). Glazar 2 UV telescope to extend Kvant 1's Glazar observations. Priroda 5 package of two KFA-1000 film cameras for Earth photography from third face of docking node. Bulgarian Svet plant cultivation unit (inclination cultivating lettuce and radishes). Mariya magnetic spectrometer, Marina γ telescope, Buket γ spectrometer and Granat astrophysical spectrometer.

Mir's third module finally lifted off aboard an SL-13 from Tyuratam at 10.33 GMT 31 May 1990 into an economical seven-day transfer orbit, planning to dock 11.36 GMT 6 June. However, only 2 h before docking, Kristall's onboard computer detected a thruster problem during the last burn for final approach and abandoned the attempt. The back-up thruster system was activated, initiating final approach with a burn at 08.04 GMT 6 June, followed by docking at Mir's forward axial port 10.47 GMT. The complex's mass was now reported to be 83 t. Restoration of the complex's symmetrical configuration reduced the attitude control requirements. Transfer using the Lyappa arm system to the face opposite Kvant 2's took an hour the next day before the crew entered and began activating the module.

Spektr

Description: The module's remote sensing instruments are particularly aimed at atmospheric studies. A two-year minimum life is planned. Spektr and Priroda also carry US equipment for a broad programme of materials science, biotechnology, life sciences, Earth observations and space technology. Spektr is equipped with a small external manipulator which can take 'Pelican' packages of up to 33 × 50 cm from a small airlock for positioning

Kvant 2 processing at Tyuratam. The EVA hatch is at right

at 11 attach points with control, power and telemetry links. The arm can also deploy small satellites.
Launch: 20 May 1995 by SL-13 Proton from
Principal application: remote sensing module, emphasising atmospheric studies.
Length: 14.44 m
Maximum diameter: 4.10 m
Mass: 19,640 kg at launch (11.5 t dry), including 1,260 kg cargo/consumables, 3,240 kg Mir equipment, 2,150 kg science equipment, 700 kg US equipment.
Pressurised volume: 61.9 m^3
Propulsion: 2 × 3.9 kN KB Khimautomatiki NTO/UDMH main engines plus clusters of 400 N KB Khimmach thrusters.
Power: 6.9 kW provided by two 26.6 m^2 Kvant 2-type Si solar wings spanning 23.30 m plus two 38 m^2 angled Si wings adding an estimated 9.3 kW. Ni/Cd batteries providing 360 Ah total capacity.
Avionics/control: Kurs approach/docking system
Payload: Balkan 1 lidar to measure upper cloud altitude, 5,320 Å, 0.15 mJ pulse, 5.5 s intervals, 4.5 m vertical resolution.

Phaza spectrometer for surface studies: 0.340-285 μm, 200-250 km resolution.

Astra 2 atmospheric trace constituents, Mir environment; Taurus + Grif monitors Mir's induced X/γ-ray background.

KOMZA interstellar gas detector.

286 K binocular radiometer.

VRIZ UV spectroradiometer.

EFO 2 photometer.

MIRAS Mir IR Atmosphere Spectrometer monitoring neutral atmospheric composition over at least 1 year using 10 of 32 daily solar occultations for infra-red absorption spectrometry. The Belgian Institute for Space Aeronomy's 205 kg grille spectrometer was launched stowed in two parts inside Spektr, assembled and then installed July 1995 by EVA. 30 cm diameter Cassegrain telescope, 6 m fl. Observing 15 key chemical species using InSb (2.5-5 μm) + HgCdTe (2.5-10.5 μm) detectors, cooled to 77K by Stirling-cycle cooler. Science data stored in 16 Mbyte memory, downlinked at 256 kbit/s in 30 minute daily period to FCC, then transferred to participants via Internet. 230 W power.

Priroda
Description: Mir's fourth module emphasises remote sensing of land, oceans and atmosphere. Bulgaria, Germany, Poland, Romania and Czech Republic have made instrument contributions an Interkosmos project. Spektr/Priroda also carry US equipment for a broad programme of materials science, biotechnology, life sciences, Earth observations and space technology. A two-year minimum life is planned.
Launch: 23 April 1996 by SL-13 Proton from Tyuratam; docked with Mir's forward axial port 12.43 GMT 26 April, transferred to portside position 27 April using the Lyappa arm.
Length: about 13 m
Maximum diameter: about 4.35 m
Mass: 19.7 t in orbit (including 2 t remote sensing payload).
Pressurised volume: 66.2 m^3
Propulsion: 2 × 3.9 kN KB Khimautomatiki NTO/UDMH main engines plus clusters of 400 N KB Khimmach thrusters.
Power: no solar arrays, battery power only.
Avionics/control: Kurs approach/docking system.
Payload: about 2 t of remote sensing instrumentation.

Ikar-N nadir microwave radiometers: R-30 (0.3 μm), R-80 (0.8), R-135 (1.35), R-225F (2.25, polarimeter), R-600 (6.0). Each 60 km/0.15K resolution.

Ikar-D scanning microwave radiometer system viewing 40° off-track in 400 km swath. R-400 (4.0), resolution 30 km/0.15K. Delta system: R-30 (0.3; 5 km resolution), R-80 (0.8; 8 km resolution), R-135 (1.35; 15 km resolution). 8 kbit/s. Bulgaria provided R-400.

Ikar-P panoramic microwave radiometers viewing 40° off-track. Two sets of three RP-225 (2.25) and RP-600 (6.0), all 750 km swath, resolution 75 km/0.15K. 1 kbit/s.

Istok 1 64-channel IR spectroradiometer, 4.0-16.0 μm, 7 km swath, resolution 0.7 × 2.8 km, 0-90° viewing angle. 8 kbit/s. Also used as 0.4-0.75 μm TV system, resolution 6.5 km. Contributions from Czech Republic, Poland and Romania.

Ozon-M spectrometer for ozone/aerosol profiles, 0.257-1.155 μm, 160 channels, 1 km altitude resolution, 50 kbit/s.

Travers SAR, 1.28/3.28 GHz, 50 km swath, 38° look angle, 50-150 m resolution, 16 Mbit/s each channel.

Alissa cloud height, structure and optical properties lidar (second priority: strato/tropo aerosols), four 5,320 Å Nd-Yag Russian lasers, 40 mJ (4 × 10 mJ) pulses at 50 Hz, width 5-15 ns, 150 m vertical resolution, 1 km horizontal resolution, 40 cm diameter 160 cm fl Cassegrain telescope receiver (110 kg in weight). Laser life (25 h or 4.5 million pulses) and 3 kW total consumption limit use. Operates in conjunction with other Priroda instruments and Meteosat. In co-operation with France. Proposed by CNES in 1985 to fly on Salyut (hence Alissa: l'Atmosphere par Lidar Sur Saliout).

The Spektr module (bottom) emphasises atmospheric remote sensing instruments. Note the jammed solar array. The Kristall docking port faces the camera at top for STS-71's docking (NASA) ***1996***

MOS-Obzor 17-channel spectrometer, 0.750-1.01 μm, 80 km swath, 700 m resolution, nadir viewing. Contributed by Institute of Space Research of former East Germany. 256 kbit/s. MOS-A: four high resolution (14 Å) channels 7,570-7,664 Å measuring aerosol profile (data also required to correct MOS-B). MOS-B: 13 medium resolution channels 4,080-10,100 Å sampling ocean reflectance signature to identify, for example, chlorophyll and sediments.

MSU-E2 3-channel high resolution (10 m nadir) optical scanner, 0.5-0.6/0.6-0.7/0.8-0.9 μm, 2 × 24.5 km swath, nadir viewing, 11.5 Mbit/s.

DOPI 2.4-20 μm interferometer studying gases/aerosols. With DARA.

Greben ocean altimeter (10 cm height resolution), 13.76 GHz, 2.5 km swath, nadir viewing, 6.5 kbit/s.

Centaur 400 MHz receiver to collect geophysical buoy data.

MOMS-02P 0.440-0.810 μm electro-optical Earth imager with four channels (three stereo). Operates in three modes: multi-spectral, stereo or high-resolution: 37 or 78 km swath, 6 or 18 m res. Provided by Germany, built by DASA Jena-Optronik GmbH; Kayser-Threde responsible for orbit/attitude determination package. MOMS-02 flew on Spacelab D2 Shuttle mission in 1993, returning 5 h data (>6,000 images) at 4.4 m res. The unit was refurbished and added to Priroda (P = Priroda) by EVA 30 May 1996 in a joint venture with RKK Energia. MOMS is the first single unit capable of generating both high-resolution panchromatic and thematic mapping XS images. Imagery transmitted via an intermediate Mir tape recorder.

Mir chronology

1992
20 January: Progress-M 10 undocks 07.14 GMT.
27 January: Progress-M 11 docks at Mir's forward axial port at 09.31 GMT.
20 February: Volkov/Krikalev perform 252 min EVA, curtailed after Volkov suit problem.
13 March: Progress-M 11 undocks 08.44 GMT.
14 March: Volkov/Krikalev undock in TM13 at 11.43 GMT from Kvant 1 port, docking at forward port after 27 min.
19 March: Soyuz-TM 14 carrying Viktorenko/Kaleri/Flade docks at Kvant 1 at 12.33 GMT.
25 March: Soyuz-TM 13 carrying Volkov/Krikalev/Flade undocks from forward port.
21 April: Progress-M 12 docks at Mir's forward axial port at 23.22 GMT.
28 June: Progress-M 12 undocks.
4 July: Progress-M 13 docks at Mir's forward axial port at 16.55 GMT.
8 July: Viktorenko/Kaleri perform 123 min EVA to install two new gyrodynes on Kvant 2.
24 July: Progress-M 13 undocks 08.49 GMT.
29 July: Soyuz-TM 15 carrying Solovyov/Avdeyev/Tognini docks at Mir's forward axial port at 07.51 GMT.
9 August: Soyuz-TM 14 carrying Viktorenko/Kaleri/Tognini undocks from Kvant 1 aft port.
18 August: Progress-M 14 docks at Mir/Kvant 1's aft port at 00.21 GMT carrying 5th return capsule and Mir Sofora roll control thruster package; it undocks 16.46 GMT 21 October.
3 September: Solovyov/Avdeyev perform 236 min EVA via Kvant 2 to prepare for Sofora thruster package installation.
7 September: Solovyov/Avdeyev perform 308 min EVA via Kvant 2 to attach control cabling to Sofora in preparation for thruster package installation.
11 September: Solovyov/Avdeyev perform 344 min EVA via Kvant 2 to complete Sofora's thruster package installation.
15 September: Solovyov/Avdeyev perform 213 min EVA via Kvant 2 to attach Kurs docking antenna to Kristall.
29 October: Progress-M 15 docks at Mir/Kvant 1's aft port at 19.06 GMT. It undocks 00.45 GMT 4 February 1993 and deploys 20 m test solar reflector.
20 November: Mak 2 satellite released from Mir at 09.00 GMT.

1993
26 January: Soyuz-TM 16 carrying Manakov/Poleshchuk docks 07.31 GMT at Kristall's APAS-89 androgynous axial port.
1 February: Soyuz-TM 15 carrying Solovyov/Avdeyev undocks at 03.00 GMT from Mir forward axial port.
23 February: Progress-M 16 docks at Mir/Kvant 1's aft port at 20.18 GMT. Manakov/Poleshchuk undock it 06.50 GMT 26 March to demonstrate remote control, taking it 70 m and then redocking it. Undocked 04.21 GMT 27 March.
2 April: Progress-M 17 docks at Mir/Kvant 1's aft port at 05.16 GMT. Undocked 15.36 GMT 11 August.

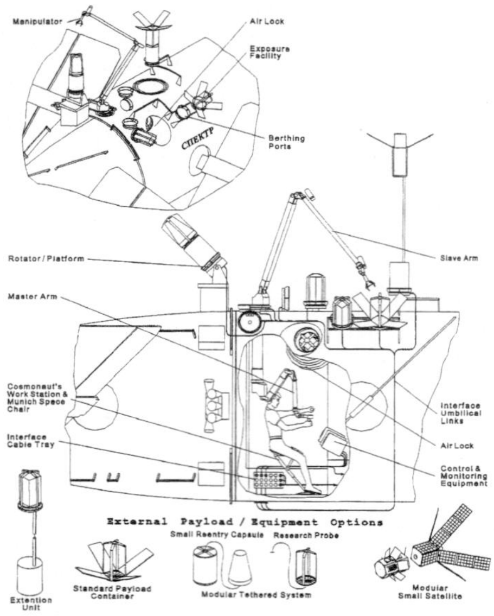

Spektr's manipulator can extract small payloads from an airlock for attachment to the exterior or release into orbit (RKK Energia) ***1996***

19 April: Manakov/Poleshchuk perform 325 min EVA to prepare for removal of solar arrays from Kristall to Kvant 1. Handle lost from transport boom.
24 May: Progress-M 18 docks at Mir's forward axial port at 08.25 GMT, carrying 6th return capsule. Undocked 3 July. First occasion with two docked Progress'.
18 June: Manakov/Poleshchuk perform 273 min EVA to prepare for removal of solar arrays from Kristall to Kvant 1. New handle installed on transport boom.
3 July: Soyuz-TM 17 carrying Tsebliyev, Serebrov and French researcher Haignere docks 16.24 GMT at Mir's forward axial port.
22 July: Soyuz-TM 16 carrying Manakov/Poleshchuk/ Haignere undocks from Kristall axial port.
13 August: Progress-M 19 docks at Kvant 1 at 00.00 GMT, carrying 7th return capsule. Undocked 17.59 GMT 12 October.
16 September: Tsebliyev/Serebrov perform 258 min EVA 1 to install Rapana truss package on Kvant 1.
19 September: Tsebliyev/Serebrov perform 193 min EVA 2 to deploy Rapana truss to 5 m on Kvant 1.
28 September: Tsebliyev/Serebrov perform 112 min EVA 3 to video Mir's exterior to assess condition.
22 October: Tsebliyev/Serebrov perform 38 min EVA 4 to video Mir's exterior to assess condition.
29 October: Tsebliyev/Serebrov perform 252 min EVA 5 to video Mir's exterior to assess condition.
14 October: Progress-M 20 docks at Kvant 1 at 23.25 GMT, carrying 8th return capsule. Undocked 02.36 GMT 21 November.

1994
10 January: Soyuz-TM 18 carrying Afanasyev/ Usachev/Poliakov docks at Kvant 1 at 11.50 GMT. Undock 03.12 GMT 24 January and redock Mir forward axial port 05.01 GMT 24 January.
14 January: Soyuz-TM 17 carrying Tsebliyev/Serebrov undocks at 04.37 GMT from Mir forward axial port.
30 January: Progress-M 21 docks at Kvant 1 at 03.56 GMT. Undocked 01.20 GMT 23 March.
24 March: Progress-M 22 docks at Kvant 1 at 06.40 GMT. Undocked 00.57 GMT 23 May.

Backside of damaged Spektr solar panel (NASA) ***1999***/0062163

24 May: Progress-M 23 docks at Kvant 1 at 06.00 GMT, carrying 9th return capsule. Undocked 08.47 GMT 2 July.
3 July: Soyuz-TM 19 carrying Malenchenko/Musabayev docks at Kvant 1 at 13.55 GMT.
9 July: Soyuz-TM 18 carrying Afanasyev/Usachev undocks from Mir forward axial port.
2 September: Progress-M 24 docks at Mir forward axial port. Attempts 27/30 August failed because of Progress Kurs fault. Malenchenko controlled M24 manually from inside Mir for number 3 attempt. Undocked 18.56 GMT 4 October.
9 September: Malenchenko/Musabeyev perform 304 min EVA to examine damage from Progress-M 24 and Soyuz-TM 17 bumps.
13 September: Malenchenko/Musabeyev perform 361 min EVA to continue inspection, including Sofora mast.
6 October: Soyuz-TM 20 carrying Viktorenko/ Kondakova/Merbold docks at Mir forward axial port 00.28 GMT.
2 November: Soyuz-TM 19 carrying Malenchenko/ Musabayev/Merbold undocks from Kvant 1 10.30 GMT, redocking 11.05 GMT to test automatic Kurs system.
4 November: Soyuz-TM 19 carrying Malenchenko/ Musabayev/Merbold undocks from Kvant 1 at 07.29 GMT.
13 November: Progress-M 25 docks at Kvant 1 at 09.04 GMT, carrying 10th return capsule. Undocked 13.05 GMT 16 February 1995.

1995
11 January: Soyuz-TM 20 carrying Viktorenko/ Kondakova/Poliakov undocks from Mir's forward axial port 08.57 GMT, redocking 09.23 GMT to test automatic Kurs system.
17 February: Progress-M 26 docks at Kvant 1 at 18.22 GMT. Cargo included 100 kg equipment for Thagard visit. Undocked 02.27 GMT 15 March.
16 March: Soyuz-TM 21 carrying Dezhurov/Strekalov/ Thagard docks at Kvant 1 at 07.46 GMT. Undocks 03.31 GMT 11 September 1995, carrying Solovyov/Budarin from STS-71.
22 March: Soyuz-TM 20 carrying Viktorenko/ Kondakova/Poliakov undocks from Mir forward axial port 00.40 GMT.
11 April: Progress-M 27 docks at Mir forward axial port at 21.01 GMT, carrying Germany's GFZ 1 laser geodetic satellite. Undocked 23.40 GMT 22 May.
19 April: Germany's GFZ 1 laser geodetic satellite is ejected from Mir at 19.12 GMT.
12 May: 375 min EVA-1 by Dezhurov/Strekalov retracts first Kristall solar wing for move to Kvant 1.
17 May: 412 min EVA-2 by Dezhurov/Strekalov transfer first Kristall solar wing to Kvant 1 portside.
22 May: 315 min EVA-3 by Dezhurov/Strekalov completes connection of first Kristall solar wing to Kvant 1.
28 May: 'internal' EVA by Dezhurov/Strekalov moves docking cone in Mir node for Kristall to move to starboard face 30 May.
1 June: Spektr module docks at Mir's forward axial port 00.58 GMT.
1 June: 'internal' EVA by Dezhurov/Strekalov moves docking cone in Mir node for Spektr to move to lower face 2 June.
10 June: Kristall swings from Mir's starboard face to the forward axial port, to allow Shuttle docking.
29 June: Shuttle STS-71 docks with Kristall androgynous port at 13.00 GMT, delivering 19th main crew Solovyov/ Budarin.
4 July: Soyuz-TM 21 with Solovyov/Budarin undocks from Kvant 1 at 10.55 GMT to photograph STS-71 departure; redocks 11.39 GMT.
4 July: Shuttle STS-71 undocks from Kristall at 11.10 GMT, returning 18th main crew Dezhurov/Strekalov/ Thagard.
14 July: 334-min EVA-1 by Solovyov/Budarin partially deploys Spektr's jammed solar wing.
17 July: Kristall is moved from Mir's forward axial port to the starboard port.
19 July: planned 300-min EVA-2 by Solovyov/ Budarin to install MIRAS spectrometer is shortened to 188 min because of Solovyov suit cooling system problem.
21 July: 335-min EVA-3 by Solovyov/Budarin installs MIRAS spectrometer.
22 July: Progress-M 28 docks at Mir forward axial port at 04.40 GMT, carrying 2,400 kg cargo. Undocked 05.10 GMT 4 September.
5 September: Soyuz-TM 22 carrying Gidzenko/Avdeyev/ Reiter docks at Mir's forward axial port at 10.30 GMT.
11 September: Soyuz-TM 21 carrying Solovyov/Budarin undocks from Mir's Kvant 1 port 03.31 GMT.
10 October: Progress-M 29 docks at Kvant 1 at 20.33 GMT. Undocked 09.15 GMT 19 December.
20 October: 316-min EVA by Avdeyev/Reiter installs ESA experiments on Spektr.
15 November: Shuttle STS-74 docks with Kristall androgynous port at 06.28 GMT, using 4.1 t docking module, left attached when Shuttle undocks 08.16 GMT 18 November. Mir's first radial docking – Kristall is at starboard port.

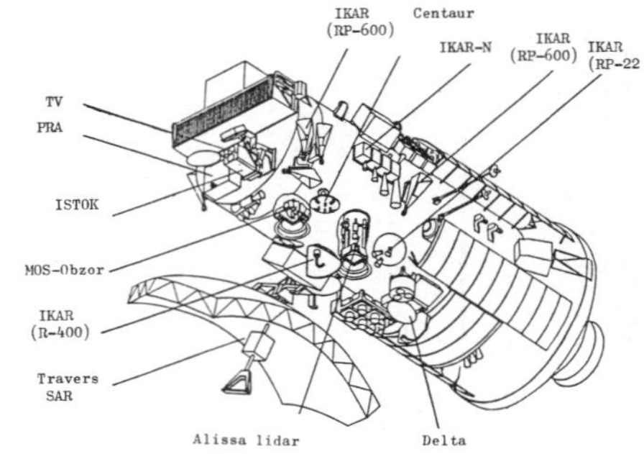

The Priroda module's suite of remote sensing instruments began operations in 1996. Germany's MOMS-02P was added by an EVA

8 December: 19.23-19.52 GMT 'internal' EVA by Gidzenko/Avdeyev moves docking cone in Mir node for Priroda's docking.
20 December: Progress-M 30 docks at Kvant 1 at 16.30 GMT. Undocked 07.26.47 GMT 22 February.

1996
8 February: 182-min EVA by Gidzenko/Reiter recovers ESA experiments from Spektr.
19 February: Mir marks 10 years in space, completing 56,150 revs.
23 February: Soyuz-TM 23 carrying Onufrienko/ Usachev docks at Kvant 1's port at 14.20.36 GMT.
29 February: Soyuz-TM 22 carrying Gidzenko/ Avdeyev/Reiter undocks from Mir's forward axial port at 07.20 GMT.
15 March: 351-min EVA 1 by Onufrienko/Usachev installs core crane for moving STS-74 docking module solar wing packages to Kvant 1.
24 March: Shuttle STS-76 docks with Kristall adapter at 02.34 GMT, undocks 01.08 GMT 29 March. Lucid remains with main crew.
27 March: 362-min EVA by Clifford/Godwin from Shuttle hatch attaches Mir experiments and tests common EVA equipment.
26 April: Priroda module docks at Mir's forward axial port 12.43 GMT.
27 April: Priroda module swings around to portside position. Mir assembly is now complete.
7 May: Progress-M 31 docks at Mir's forward port.
21 May: Onufrienko and Usachev conduct EVA lasting for 320 min to install a solar panel on Kvant 1. During EVA, the crew photographed a larger-than-life mockup of a soft drinks can to be used by the United States drink manufacturer as part of an advertising campaign.
24 May: Onufrienko and Usachev perform EVA lasting approximately six hours to complete installation of a solar panel to Kvant 1.
30 May: Onufrienko and Usachev conduct fourth EVA lasting for 260 min to mount the optical unit of MOMS-2P multispectral scanner outside the Priroda module. They also installed an additional handrail on Priroda.
6 June: Fifth EVA by Onufrienko and Usachev, lasting for 214 min to remove cassettes from the Swiss-Russian

NASA astronaut Foale operates video camera in Mir's base block (NASA) ***1999***/0062159

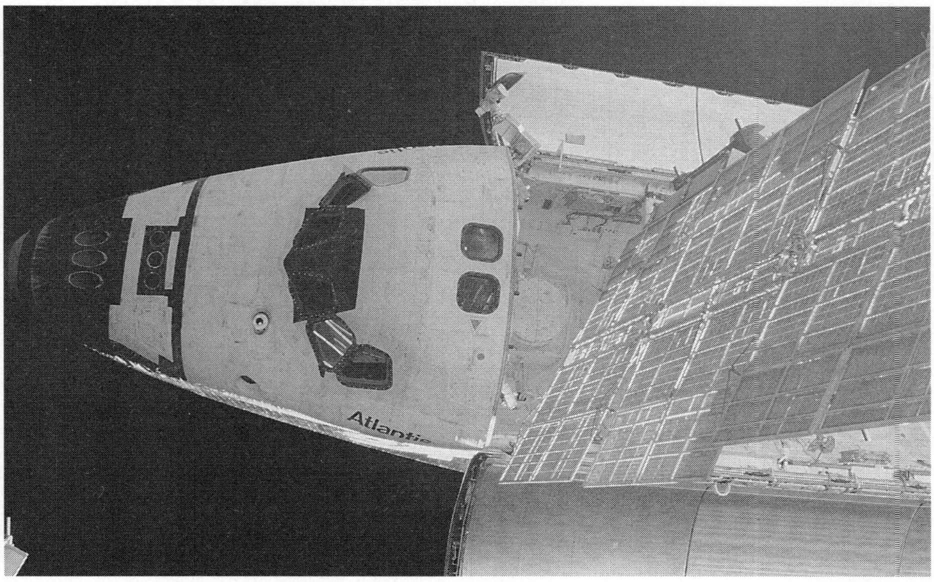

Forward section of Atlantis from a port on Kvant-2 module (NASA) ***1999***/0062160

interstellar gas detector on Spektr and installed two US micrometeorite detectors and a Russian experiment for testing the effects of the space environment on various materials.
13 June: EVA number 6 conducted by Onufrienko and Usachev lasting for approximately 330 min: they filmed the second part of the soft drinks commercial and replaced a truss on Kvant 1 with a newer model.
21 June: Russian Space Agency announces that the stay of Onufrienko and Usachev aboard Mir has been extended by 40 days.
2/4 July: Orbit of Mir is slightly lowered to permit the less-powerful Soyuz-U launch vehicle to launch Soyuz-TM 24, due mid-August.
1 August: Progress-M 31 undocks from Mir, de-orbited the same day.
3 August: Progress-M 32 docks with Mir at the front port.
7 August: Announced that back-up Russian crew will fly the next Soyuz-TM mission: prime crew Cdr Manakov is grounded because of heart trouble.
18 August: Progress-M 32 undocks from Mir: remains in orbit.
19 August: Soyuz-TM 24 crew of Korzun, Kaleri and French spationaut Andre-Deshays arrive on Mir: French Cassiopeia mission.
2 September: Soyuz-TM 23 with Onufrienko, Usachev and Andre-Deshays returns to Earth.
3 September: Progress-M 32 re-docks with Mir, using the rear port vacated by Soyuz-TM 23.
19 September: Atlantis docks with module attached to Kristall for the STS-79 mission: undocks 23 Sep, leaving Blaha on Mir and returning Lucid to Earth.
5 November: Russia announces that the Kurs automatic rendezvous system will be dropped due to its unreliability.
21 November: Progress-M 32 undocks from Mir, de-orbited later in the day.
24 November: Progress-M 33 docks with Mir at the rear port.
2 December: Korzun and Kaleri perform EVA lasting for 357 min, installing a new solar panel.
9 December: Second EVA by Korzun and Kaleri lasting more than 6 h installing an antenna on a docking unit to ease docking by Atlantis and also connecting the new solar battery.

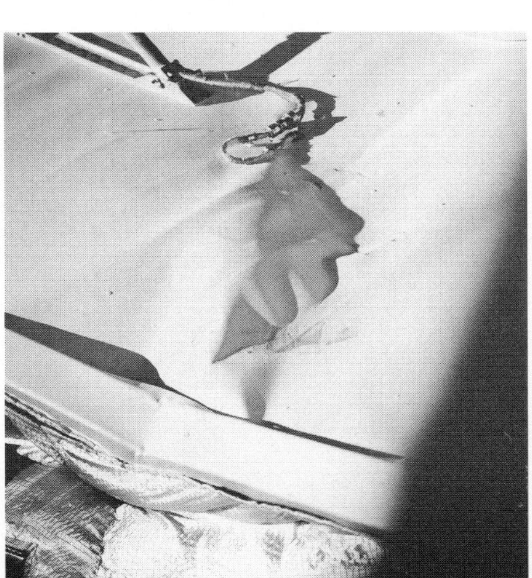

Radiator damage to Mir's Spektr module (NASA) ***1999***/0062161

1997
15 January: Atlantis docks with Kristall's docking module for STS-81 mission. Undocked 20 January, returning Blaha to Earth and leaving Linenger aboard Mir.
6 February: Progress-M 33 undocks from Mir re-docking planned following the return of Soyuz-TM 24 to Earth.
7 February: Soyuz-TM 24 undocks from front port of Mir and re-docks at the rear.
12 February: Soyuz-TM 25 docks at front of Mir, carrying Tsibilev, Lazutkin and German researcher Ewald flying MIR-97 mission. Docked manually at second attempt after automatic system failed.
23 February: Fire breaks out on Mir: lithium cartridge exploded, but fire soon extinguished. Crew has to wear breathing apparatus until gases vented.
2 March: Soyuz-TM 24 with Korzun, Kaleri and Ewald undocks from Mir and returns to Earth.
4 March: Progress-M 33 fails in re-docking attempt: no further attempt planned and spacecraft is de-orbited 12 March.

Spektr solar array damage (NASA) ***1999***/0062162

8 April: Progress-M 34 docks at rear of Mir carrying supplies and equipment to repair Elektron oxygen-producing systems.
29 April: Linenger performs 4 h 58 min EVA to retrieve/install equipment on exterior of station.
17 May: Atlantis docks with Mir bringing new Elektron unit, collecting Linenger and leaving Foale. Undocked 22 May.
24 June: Progress-M 34 undocks from Mir to perform autonomous operation. Next day during attempted re-docking collides with Spektr module damaging a solar array wing and causing module depressurisation.
27 June: Kvant-2 batteries run down, thrusters on Soyuz TM-25 used for orbit adjust.
7 July: Progress-M 35 docks at Kvant-2 port bringing supplies and equipment.
13 July: Cosmonaut Tsibilev displays cardiac arrhythmia and mission managers decide to control workload to prevent increase in stress.
19 July: Crew given approval to perform a suited entry to depressurised Spektr for situation assessment.
6 August: Progress-M 35 undocked from aft port and removed to a safe distance.
7 August: Soyuz TM-26 docks to aft port brining Solovyov and Vinogradov to replace Tsibilev and Lazutkin.
14 August: Tsibilev and Lazutkin return to Earth in Soyuz TM-25 leaving Foale with Solovyov and Vinogradov.
15 August: Mir crew perform fly-around inspection of complex exterior in Soyuz TM-26 and re-dock to forward port. Progress-M 35 re-docks to aft port.
22 August: Leaving Foale in Soyuz, Solovyov and Vinogradov perform six-hour suited inspection of Spektr.
8 September: Crew replace old back-up Elektron following major power failure.
14 September: Main computer fails again and crew power down while repairs are made. Launch of Progress-M 36 delayed to await new computer it will deliver to Mir.

Spektr station module showing collision damage to solar panels (NASA) ***1999***/0062164

NASA astronaut David Wolff sorts microbial samples aboard Mir (NASA) ***1999***/0062165

Three generations of Soviet stations

	Sal 1	**Sal 6**	**Mir**
Year of launch	1971	1977	1986
Original mass (t)	19	19	21
Body length (m)	15	15	13.1
Max diameter (m)	4.15	4.15	4.15
Primary crew size	3	2-3	2-6
Design life (month)	3-4	24-36	120?
Docking ports	1	2	6
Main engines	1	2	2
Main thrust (kN)	4.09	2 × 2.94	2 × 2.94
Refuellable?	no	yes	yes
Solar span (m)	11	17	29.73
Panel area (m^2)	<28	51	98
Power (kW)	<2	4	10.1

28 September: Shuttle Atlantis docks with Mir bringing 3.16 tonnes of supplies and effecting an astronaut change.
3 October: Atlantis undocks taking Foale, leaving Wolf.
7 October: Progress-M 35 undocked from Kvant-1 aft port.
8 October: Progress-M 36 docks to Kvant-1 aft port bringing 2.5 tonnes of supplies and equipment.
20 October: Solovyov and Vinogradov perform second suited inspection of Spektr module in 6 h 38 min operation to jury-rig electrical cables for solar arrays.
3 November: Solovyov and Vinogradov perform 6 h EVA to remove old solar arrays from Kvant-2 module and to release a small model of Sputnik as an independent satellite. Hatch locking problem on attempting ingress diverts them to Kvant-2 hatch for entry.
6 November: Solovyov and Vinogradov perform 6 h 17 min EVA to install replacement arrays on Kvant-2. Failed to mend problem hatch.
14 November: Kvant-2 solar array orientation system breaks down necessitating another power-down.
21 November: Following another computer failure crew replace computer with interim unit brought up by Atlantis.
17 December: Progress-M 36 undocked and removed for release of German subsatellite designed to act as precursor TV inspection satellite for International Space Station but suffers software failure.
19 December: Progress-M 36 de-orbited over Pacific Ocean.
22 December: Progress-M 37 docks to aft Kvant-1 port bringing new computer and supplies and equipment for ailing Mir.

1998
14 January: Solovyov and Wolf perform 3 h 52 min EVA to conduct scientific investigations including a NASA environment monitoring device.
24 January: Shuttle Endeavour docked with Mir bringing 3.6 tonnes of supplies. Five days on joint activities.
29 January: Shuttle Endeavour undocks from Mir bringing astronaut Foale back to Earth and leaving Thomas for the final long stay by NASA astronauts aboard the station complex.
30 January: Progress-M 37 undocked from the aft Kvant-1 port and was backed away to a safe station-keeping distance.
31 January: Soyuz TM-27 docks to the aft Mir port and cosmonauts Musabayev, Budarin and French cosmonaut Eyharts to join Solovyov, Vinogradov and Thomas for nearly three weeks of joint activity.
19 February: Solovyov, Vinogradov and Eyharts return to Earth in Soyuz TM-26 leaving Musabayev, Budarin and Thomas aboard Mir.
23 February: Using the Kurs automated system, Progress-M 37 is re-docked to the aft port at the Lvant-1 module.
3 March: Musabayev and Budarin perform an EVA in a futile attempt to repair the sticking hatch on Kvant-2 airlock module.
15 March: Progress-M 37 undocked from Mir aft port and removed to a safe station-keeping distance.
17 March: Progress-M 38 docks to aft Kvant-1 port after Musabayev took over manual control when the Kurs automatic system failed.

Mir space station viewed from Shuttle discovery (NASA) ***1999***/0062156

27 March: Musabayev and Budarin spend several hours attempting to repair the Kvant-2 airlock module hatch, replacing some latches and freeing up others.
1 April: Musabayev and Budarin perform a 6 h 40 min EVA to repair Spektr module solar arrays and worked at the base of the array using the Strela crane for access.
6 April: Musabayev and Budarin cut short an EVA after 4 min 23 s when ground controllers notice Mir drifting off solar alignment with consequent loss of electrical power.
11 April: Musabayev and Budarin complete third EVA lasting 6 h 25 min working to restore power to Spektr solar arrays and change a VDU thruster module.
17 April: Musabayev and Budarin perform fourth EVA lasting 6 h 33 min and set up Sofora girder for VDU thruster module replacement brought up by Progress-M 38.
22 April: Fifth EVA by Musabayev and Budarin lasts 6 h 21 min and brings month's total to 30 h 22 min by installing VDU block and preparing several experiments for future use.
16 May: Progress-M 38 undocked from Kvant-1 port and backed away to a safe distance and Progress-M 39 docks at the same port 5 h later.
4 June: Shuttle Discovery makes the last docking in the Shuttle-Mir programme bringing 2.6 tonnes of water and equipment and for four days of joint activity as well as a full survey of the station conducted by cosmonaut Ryumin brought up by Discovery.
8 June: Discovery departs the Mir space station carrying astronaut Thomas back from his four-month stay aboard the station complex.
15 June: Moscow announces that the orbit of Mir has been lowered by 6 km so that a specially prepared Progress vehicle launched some time in 1999 can de-orbit the station over the Pacific Ocean.
12 August: Progress-M 39 undocked from Kvant-1 aft port and removed several kilometres to a safe distance.
15 August: Cosmonauts Padalka, Avdeyev and Baturin aboard Soyuz TM-28 dock to the aft port on the Kvant-1 module for nine days of joint activity with Musbayev and Budarin.
25 August: Soyuz TM-27 departs Mir and returns to Earth with Musabayev, Budarin and Baturin leaving Padalka and Avdeyev aboard the space station.
27 August: Padalka and Avdeyev board Soyuz TM-28 and transfer their spacecraft from aft to forward docking ports.
29 August: Progress-M 39 was returned to a docking with the aft Kvant-1 port.
15 September: Padalka and Avdeyev conduct a suited inspection of the Spektr module and reconfigure cables to restore attitude orientation control to the operable solar arrays.
25 October: Progress-M 39 was undocked from the aft port and removed, clearing the way for this replacement and was de-orbited over the Pacific Ocean on 29 October.
27 October: Progress-M 40 docked to the aft Mir port bringing supplies and equipment as well as a repeat Znamaya space mirror experiment designed to reflect sunlight to areas on Earth.
10 November: Padalka and Avdeyev conduct a 5 hr 54 min EVA to attach experiments to the exterior, to catch microscopic particles associated with the Leonid meteor shower and to release a small replica of Sputnik.
24 December: Mir perigee raised by 14 km to preserve its orbit for the last human crew scheduled for launch in February 1999.

1999
27 January: Progress-M 40 engines fired for 1 min 40 s to raise the orbit of the Mir complex to preserve the orbit under the effect of a denser upper atmosphere due to solar activity.
2 February: Progress-M 40 undocks and backs away 400 m from Mir to conduct the Znamya mirror reflector experiment but the thin foil array fouls on a rendezvous antenna.
8 February: Padalka and Avdeyev transfer their Soyuz TM-28 spacecraft from the forward to the aft docking ports clearing the forward port for Soyuz TM-29.
22 February: Carrying Afanasyev, French cosmonaut

J-P Hagniere (the first non-Russian Flight Engineer) and Slovakian researcher Bella, Soyuz TM-29 docks with Mir's forward port as the last Mir crew to visit the space complex.

28 February: Padalka and Bella undock from Mir's aft port in the Soyuz TM-28 spacecraft and return to Earth leaving cosmonauts Afanasayev, Avdeyev and Hagniere as the last crew aboard the station.

4 April: Progress-M 41 docks at the aft Kvant-1 port bringing supplies and equipment for the three-man crew aboard Mir scheduled to return to Earth in late August 1999 before the Mir de-orbit burn expected to take place over the Pacific Ocean.

Salyut/Almaz

Background

Salyut development began in the 1960s, but details remained hidden because of the resemblance to a military programme that was in competition with the United States' own version. Korolev's successor, Vasili Mishin, published an article about the station in the 1990s. In this article, Mishin described the differences between the military space stations known as Almaz, and the civilian versions known as Salyut.

The manned programme emphasised long-term occupancy from the beginning when the first Salyut appeared in April 1971. That station was smaller and less capable than NASA's solo Skylab but it had the significant advantage of quantity production with improvements to successive models leading to the current Mir. The first six orbited Salyuts (one was cloaked under a Cosmos designation) were only moderately successful and none was occupied for more than a total of 100 days.

The parallel military programmes, Almaz, unfolded in 51.6° orbit. The boundaries between civilian and military were sometimes indistinct. The programmes eventually merged and matured with Salyut 6. This single station hosted 16 crews of 27 individuals for a total occupancy of almost 2 years. It demonstrated that a Salyut-type module could support a manned Mars mission. Salyut 7 expanded the record with 812 days' occupation by 19 individuals between May 1982 and June 1986. Its capabilities were also enhanced by the docked Cosmos 1443/1686 modules.

Chelomei's bureau worked on both the military manned station, Almaz, and the civilian station, Salyut. At the time, Korolev's bureau was under the direction of Defence Minister Ustinov in late 1969. Chelomei designed and built the first DOS or 'Permanent Orbital Station' within 18 months just as the United States moved to shut down its own efforts on the MOL or Manned Orbiting Laboratory. Chelomei's failure in the manned moon programme, prompted the Soviet Defence Ministry to move the Almaz to Korolev's bureau. Mishin took over two Almaz vessels. The Salyut programmes remained reasonably autonomous at Chelomei's bureau, although they both used Chelomei's three-stage Proton. NPO Machinostroenye still apparently has four manned Almaz craft in storage, along with five small return capsules (400 kg including 120 kg 0.9 m^3 payload, 85 cm diameter, spin-stabilised).

A slightly different history, however, claims that Almaz was almost completed by Chelomei – with 10 being built at Khrunichev Design Bureau. As the vessels moved down the assembly line the control system lagged behind in development, and the Politburo passed the programme on to a reluctant Mishin, who had no interest in Earth orbital stations. His bureau added the Soyuz elements (particularly the control system, solar panels and propulsion unit) to create Salyut. The programme was known as Zarya but changed shortly before launch as a salute to the 10th anniversary of Gagarin's flight.

Mishin put a kinetic energy interceptor device on his Almaz for defence against US anti-satellite weapons and inspection, although apparently it was never implemented.

Salyut 3 (Almaz 2) and 5 (Almaz 3), with their all-military crews and lower orbits probably originated as military stations. The Soviets may have changed the designation as an answer to the United States' Skylab civilian programme. In contrast, Salyut 4 (DOS 4) and 6 (DOS 5) were unquestionably civil.

Salyut 3 included a 10 m focal length camera, able to resolve Earth surface objects down to 30 cm. Before Popovich/Artyukhin left after their 15-day stay, they prepared Salyut 3 for automatic control and for a possible further manned visit, but Soyuz 15 failed to dock. The return capsule ejected on orbit 1,451 from 253 × 275 km and recovered. The flight ended 24 January 1975 following a re-entry manoeuvre. Salyut had operated for seven months, more than twice as long as originally planned.

The Soyuz 17 crew docked with Salyut 4, 17 days after the launch of the space station and remained aboard until 9 February 1975. It was to have been boarded two months later by the Soyuz 18A crew, whose launch was aborted. It therefore continued under automatic control until the back-up 18B crew arrived for a record 63-day stay from 23 May to 26 July. Much of the subsequent success came from the personal repairs which the Soyuz 17/18B crews accomplished on the solar telescope and cosmic ray equipment, as well as to more fundamental items such as air conditioning. There was also a series of failures of both Soyuz and Salyut equipment between the aborted Soyuz 18A launch and the Soyuz 25 docking attempt, bringing the total success rate to five out of 13 space station missions. Four months after crew 2's departure, the unmanned Soyuz 20 was automatically docked, remaining for 91 days before returning to Earth. It not only demonstrated a resupply capability later employed by Progress, but also the possibility of using an unmanned Soyuz to collect sick crewmembers. On 3 February 1977, only 1.5 days before natural decay would have resulted in re-entry, Salyut 4 entered a re-entry trajectory over the Pacific. It had completed more than 12,000 orbits and two crews had occupied it for a total of 93 days.

The two-man Soyuz 21 crew boarded Salyut 5 14 days after launch, but an acrid odour from the life support system drove them home. Their successors bled off the atmosphere and replaced it from tanks of compressed air. Indications were that, like Salyut 3, the Soviets wanted the vehicle for mainly military purposes. For photo-reconnaissance, the Almaz carried the 6 m focal length Agat camera system, producing 50 × 50 cm negatives. Salyut 5's chief cosmonaut, Gorbatko claimed in 1994 that the manned crew once managed to snap the photographs, develop them on-board and dispatch them to Earth in a capsule all within 30 minutes.

Following the failure of the Soyuz 23 docking attempt, Salyut 5's orbit steadily decayed and re-entry was predicted for March 1977. However, corrections on 14 and 18 January raised it to 256 × 275 km. Then, on 7 February, came the successful boarding and 17 day stay by the Soyuz 24 crew. One day after their departure, the automatic return capsule separated and brought back with it film cassettes and other material loaded via an internal airlock. This capsule was sold for US$48,875 at Sotheby's December 1993 New York Russian space auction. It was the first public sight of the design. On 8 August 1977, after 6,630 orbits, Salyut 5's engines placed it in a descent trajectory over the Pacific.

The steady development of the Progress cargo ferry, also gave the Soviets a unique space station capability

Salyut 7: the 5-year-old station, with T14 at its aft docking port, was photographed by the T13 crew. Note the solar panels added during EVAs

still not possessed by the US as it builds the International Space Station. The Salyuts had one major advantage over NASA's Skylab, which descended almost uncontrolled in July 1979. They were equipped with orbital manoeuvring engines that permitted re-entry to be commanded safely over the Pacific, with surviving debris falling harmlessly into the sea (Salyut 7 did not). The manoeuvring motors resulted from a need to initially put the stations in a low orbit to compensate for the limited lifting power of Soviet rockets and, until the T series, for the restricted manoeuvring capability of the Soyuz spacecraft.

Apart from the 'random' Soyuz 25/33 failures, the Soviets realised consistent success with Salyut 6. What appeared to be a routine launch on 29 September 1977 was actually a significant step towards a permanently manned space station. Over 44 months, 30 craft successfully docked. 16 carried a total of 33 crew members and the Progress came into its own with 12 docking. Five long-term missions of 96, 140, 175, 185, 75 days, plus 11 short visits (by crews including nine cosmonauts from Interkosmos countries), meant that Salyut 6 was occupied for a total of 676 days, far surpassing the 171 days' occupation of the US Skylab in 1973-4. Salyut 6 became the home of more than 200 experiments in the material science furnaces and 48 million km^2 of Earth's surface was photographed. The Progress delivered more than 20 tonnes of supplies, enabling crews to replace more than a quarter of the station's equipment and to add to the range of its scientific apparatus.

The station's career began inauspiciously with the failure of Soyuz 25's docking. The conjectured existence of a second port was confirmed when Soyuz 26 docked without difficulty on 19 December at the aft end. Following the Soyuz 33 failure in April 1979, there was a one year gap in manned launchings, but the cosmonauts were so successful in replacing and repairing components that a new lease of life began in April 1980. This phase continued through 1981, when four missions were flown, beginning with Progress 12 in January with supplies for the last long-stay crew in Soyuz-T 4. Then came the final docking, on 19 June 1981. Salyut 6 was in a 335 × 377 km, 51.6° orbit and

Cosmos 1443/Salyut 7/Soyuz-T: about the same length as NASA's 1973 Skylab but providing less than half the living volume. The TKS/Heavy Cosmos craft (left) was originally intended as a three-man ferry and habitation module for the military Salyuts. The Merkur descent module was derived from Chelomei's small UR500/LK-1 lunar fly-by and later UR700/LK-700 lunar lander designs. C1443's descent module was sold for US$552,500 to an anonymous US collector at Sotheby's in December 1993 space auction in New York (Teledyne Brown Engineering)

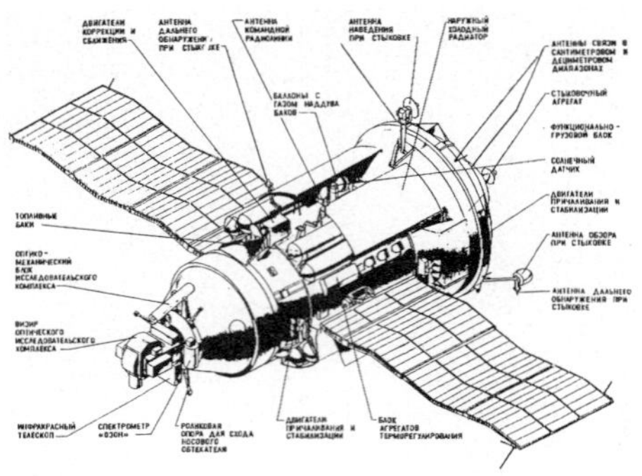

Cosmos 1686 carried a heavily modified Merkur capsule. The Earth observation instruments included an Ozon spectrometer (NASA Johnson, JSC 26770) ***1995***

the docking craft was the 15.1 tonne Kosmos 1267, the first use of the TKS/Heavy Kosmos. After manoeuvres on 28 July 1982, Kosmos' 1267 engines were fired the following day to nudge the two vehicles – still joined – into a controlled re-entry over the Pacific.

In its orbit of about 220 × 270 km, periodically raised to about 350 km, Salyut 6 could lose 142 m of altitude every 16 orbits (24 h). This decay depended heavily on solar activity and station attitude. For a long period, Salyut 6's main engines were unusable, but the Soviets compensated for this by showing they could raise the orbit of the whole complex by using the Progress's engines. Soon, spare propellant on Soyuz spacecraft refuelled the stations.

The Soviets expected to follow-on to the Progress with a better vehicle called the Merkur. It became clear in 1991 that the Soviets built the Merkur, otherwise known as the Heavy Kosmos craft (called TKS or 'Logistics transport spacecraft') as a three-man ferry and habitation module for Almaz. However, the Almaz as flown employed only Soyuz and a small payload return capsule ejected from the docking collar.

The Merkur should have made its debut on another military Almaz in 1979-80, to be occupied by a three-man crew flying the TKS/Heavy Kosmos in flight trials. In support of this, Heavy Cosmos Merkur ('Mercury') descent modules were flown two at a time on Proton re-entry tests cloaked under the identifiers Kosmos 881 and Kosmos 882 on 15 December 1976, Kosmos 997/998 on 30 March 1978, and Kosmos 1100/1101 on 22 May 1979. There was also a launch failure 4 August 1977, reusing the capsules from the first two. The tower rescued one but the other was lost. Western reports indicated these were supposed to be small spaceplane trials. Merkur provided three hatches, one on the side for pad ingress, another on top for exit after landing and a base hatch that required removal of the central seat for access to the spacecraft. Kosmos 1443, although it flew unmanned, was equipped as a manned vehicle in one trial.

Salyut 7 was launched amid Western expectations that it would carry three docking ports and form the basis for a 6- to 12-man station. Experience from Salyut 6's career allowed the Soviets to declare a four to five year design life, although major systems failures almost produced an early demise. Ten crews totalling 19 individuals (four paying two visits and Dzhanibekov three) spent a total of 812 days aboard during five periods of occupancy. Repairs, modifications and tests resulted in 13 EVAs and the T13/14 cosmonauts became the first to effect an in-orbit crew exchange. 13 Progress's and two Heavy Kosmos (1443/1686) delivered 37.65 tonnes of cargo.

Upgrades included a docking unit strengthened by an additional collar to handle the torques imposed by 20 tonne Heavy Cosmos modules. The portholes included movable external covers to combat the micrometeoroid and UV degradation suffered during Salyut 6. The MKF-6M and Kate 140 Earth resources cameras were again mounted on the floor and allowed each of the first two long-stay crews to return more than 20,000 images. French visitor Chrétien left his Piramig and PCN cameras in July 1982 for sensitive Earth and astronomical imaging at visible/IR wavelengths. The astronomical instrumentation viewing through the large body aperture included the RT-4M X-ray telescope (first flown on Salyut 4), the SKR-02M X-ray spectrometer and the Yelena γ-ray detector (carried by Salyut 6). Materials processing again formed a large portion of Salyut's work, with the Kristall, Magma-F and 136 kg Korund furnaces. Korund produced semiconductor crystals of cadmium selenide and indium arsenide. Evaporator-M used the ShK airlock to deposit metal coatings on surfaces in vacuo and samples were also exposed to study plasma charging effects. The airlock was also employed to launch the 28 kg Iskra 2/3 relay satellites in May/November 1982 on behalf of students at a Moscow engineering institute. A Soviet craft carried electrophoresis equipment for the first time to purify biological materials in a buffer solution subjected to an electric field. Salyut reportedly played a significant role in developing a new flu vaccine. The first run generated material 10-15 times purer than possible under normal gravity.

Some time during late August/September 1983, power was lost from one main solar panel and the life support system struggled to cope under the reduced supply. Heaters maintained 18°C but humidity rose to 100 per cent and condensation dripped off the walls. Fortunately, C1443 had already delivered at least two sets of smaller 1/2 kW panels to satisfy increased loads of the main experiments. Lyakhov/Alexandrov completed two EVAs 1 to 3 November 1983 and attached a 1½ × 5 m GaAs panel either side of the central length. Kizim/Solovyev attached a second pair on 18 May 1984 and brought the supply up to 4 kW, indicating that one main panel was still not contributing. Dzhanibekov/Savinykh added the third pair following delivery by Progress.

Salyut 7's career was further threatened in early September 1983 when oxidiser leaked from the propulsion system during Progress 17's refuelling operations. Lyakhov/Alexandrov retreated to Soyuz in preparation for an emergency departure but were allowed back when it became apparent there was no large breach. The plumbing was shut down, cutting off half of the 32 attitude thrusters and the back-up route to the main engines; the curtailment of attitude manoeuvres seriously affected Earth and astronomical observations. T9's crews could do little but Progress 20 delivered tools and other equipment, and T10's Cdr/engineer arrived trained for a full repair. Kizim/Solovyov undertook four EVAs in April/May 1984 to isolate the leaking pipe and bypass it with new valves and plumbing. They did not complete the work until 8 August after Dzhanibekov arrived with new tools and a videotape illustrating how to pinch off the offending pipe. The duo eventually accounted for eight of Salyut 7's 13 EVAs.

Salyut came closest to orbital death following T10's departure in October 1984. A fault in the radio command system broke contact with ground control and the onboard systems were unable to maintain Sun-lock for the solar panels. The eight Ni/Cd batteries discharged and the station could no longer maintain its vital functions. Experienced cosmonauts Dzhanibekov/Savinykh were dispatched aboard T13 to assess the situation and docked with the inert vehicle using range-speed data from an optical instrument. Salyut's interior was covered in frost and icicles hung from burst pipes, but they discovered that six batteries were serviceable and power could be restored from one panel. Equipment damaged by ice, including eight ventilation fans, was replaced and P24 delivered three new batteries 23 June 1985 so that, within two weeks of boarding, the crew had revived Salyut.

The station saw the beginning of a long occupation by the T14 crew (Vasyutin's illness precipitated an early return) and a brief visit by Kizim/Solovyev from Mir. Salyut was clearly ageing but it and the docked C1686 were still fired into a 474 km circular storage orbit providing at least eight year's life in the hope of a future inspection. During 1988, the Soviets noted that their Shuttle might retrieve it for detailed inspection. However, increased solar activity triggered rapid decay to 445 km by 1 May 1989. Altitude was 413 km by 16 February 1990, 371 × 390 km, 51.61° by 17 April 1990 and dropping by about 1 km per week, leading to a prediction of re-entry by April 1991. It was down to 275 × 305 km by 29 December 1990 and to 176 × 190 km at 12.30 GMT 5 February 1991.

A de-orbiting mission by Progress or Soyuz was mooted during 1989, as a Shuttle retrieval would be both very demanding and too early in the spaceplane's delayed schedule, but it was rejected apparently on cost grounds. The control system had been inactive for two years and only 70 kg propellant remained (500 kg was required for a controlled de-orbit burn). An attitude manoeuvre was attempted the day before entry to direct it by drag modulation into the Atlantic, but it was apparently incomplete. The Soviets predicted that up to 2 tonnes could survive, including 250 lumps of several kg; compensation would be paid for damage or injury. The two largest recovered chunks were reportedly 4 and 8 kg. Several Soviet news releases claimed that C1686 carried a 2.5 tonne 3 m diameter spherical entry capsule; these were the first references to a descent module. Some Soviet reports claimed it had been planned to be separated in January 1986 but Vasyutin's illness during T14 prevented it. Others said Earth return was never planned and the solid retros were not even loaded with propellant. It now appears it was a heavily modified Merkur, carrying Earth observation instruments at its apex instead of the propulsion section. The 40 tonne combination re-entered at about 04.00 GMT 7 February 1991 over 34.9° south/63.8° west, scattering debris over Argentina and Chile as it headed north-east.

Salyut 7 had three habitable compartments totalling 100 m³, compared with Skylab's 357 m³. For Salyut 6 and later Mishokin redesigned the original engine system and replaced the single nozzle with twin nozzles on each side of the second docking port added at the rear. He also added more than 20 portholes for experiments, visual observations and photography.

Specifications

Salyut (General)

Forward transfer compartment: The front end, 3 m long, 2 m diameter, was basically an access tunnel with an inner bulkhead enabling it to be sealed off for EVAs from the side hatch; two suits stored here were for the shared use of all visiting crews. There were seven portholes, some carrying astro-orientation sensors.

Working/operations compartment: This main living/working area comprised two cylindrical shells connected by a conical section. The smaller first section was 2.9 m diameter, 3.5 m long, the conical section 1.2 m long and the second section 4.15 m diameter, 2.7 m long. Equipment was arranged in standard racks around the interior's sides. Salyut's central control position in the lower part of the first section had two seats. Further back, a table offered hot food and hot/cold water. On one side was the system that regenerated water from the atmosphere and on the other, the onboard computer (which according to US sources had no back-up). To the outside of the first section were three solar panels, two horizontal and one vertical, spanning 17 m. They rotated 340° to follow the Sun and produce 4 kW of power.

On Salyut 1-3 there were two smaller pairs of horizontal panels, with the Soyuz providing supplementary power from a similar pair. From Salyut 4 onwards the Soyuz began to fly without solar panels as they were able to recharge internal chemical batteries from Salyut's power supplies. This limited Soyuz flying time to 48 h in the event of failure to dock, which happened so often that the solar panels were later restored.

In the rear of the larger-diameter section, on opposite sides, were the crew's bunks and food storage. Aft were a toilet and two small airlocks for ejecting waste. The shower was at the forward end of this section. In the lower centre of the larger-diameter section, the MKF-6M Zeiss camera occupied one of two portholes. Much cosmonaut time was spent nearby on the treadmill, bicycle ergometer and lower-body vacuum suit.

Rear transfer compartment: On the early Salyuts this was an unpressurised service module 2.17 m long, 2.2 m diameter. On Salyut 6, it was widened to the same 4.15 m diameter as the working compartment to take the second docking port. Here the short-term Interkosmos visitors were received when they docked, as were the unmanned Progress ferries, which plugged into the propellant resupply lines. The main engine and thrusters used the same propellants, and refuelling was almost automatic. However, the 1 kW refuelling compressor used so much solar-generated power that refuelling had to be spread over six shifts so that sufficient power remained for other vital systems.

More than 2 tonnes of equipment was used for materials processing, biology, astronomy and Earth resources surveys. The 650 kg BST-1M sub-mm telescope and the MKF-6M multispectral camera dominated Salyut 6's work compartment. The two furnaces (Splav 'Alloy' for smelting metals and Kristall for crystal growth) were delivered by Progress 1/2.

Salyut 1/DOS-1
Launched: 19 April 1971
Orbit: initial 200 × 222 km
De-orbited: 11 October 1971 after about 2,800 orbits
Comment: Following the death of the Soyuz 11 crew on 29 July, the Salyut continuously maintained the temperature and pressure at life-supporting levels. It seemed likely that at least one more Soyuz docking was intended, until the inquiry into the Soyuz failure showed that major modifications would be needed.

Salyut A/ DOS-2
Launched: 29 July 1972
Comment: Proton failure

Salyut 2/Almaz 1
Launched: 3 April 1973
Orbit: 215 × 260 km, 51°
Comment: After 130 orbits, the station was said to be stable and the orbit had been raised to 261 × 296 km

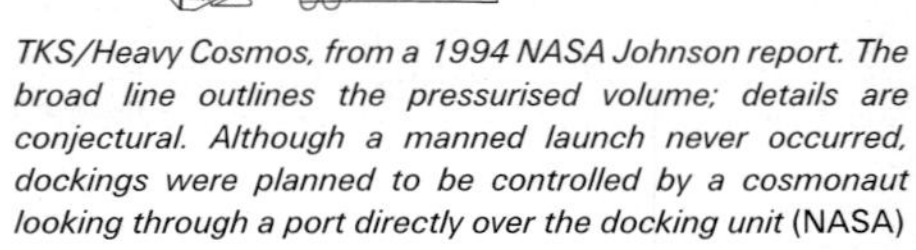

TKS/Heavy Cosmos, from a 1994 NASA Johnson report. The broad line outlines the pressurised volume; details are conjectural. Although a manned launch never occurred, dockings were planned to be controlled by a cosmonaut looking through a port directly over the docking unit (NASA) ***1995***

(now known that main work compartment lost pressure after burn). Shortly after, Western observers reported that it had become unstable. On 14 April, it apparently broke up into 25 fragments; decayed 28 May.

Kosmos 557/DOS-3
Launched: 11 May 1973
Orbit: initial 214 × 243 km
Comment: Initial orbit indicated a propulsion or command system failure. Decayed naturally after 11 days. Telemetry suggested civil station. Carried three solar panels, like Salyut 4.

Salyut 3/Almaz 2
Launched: 25 June 1974
Orbit: 270 × 219 km
Comment: Modifications included three solar arrays, automatically rotating through 180° for tracking the Sun, instead of the four fixed small arrays on Salyut 1.

Salyut 4
Launched: 26 December 1974 by SL-13 from Tyuratam
Mass: 18,900 kg
Orbit: initial 219 × 270 km, 51.6°, raised 11 days later to 343 × 355 km

Salyut 5 /Almaz 3
Launched: 22 June 1976 by SL-13 from Tyuratam
Mass: 19,000 kg
Orbit: 219 × 260 km, 51.6°

Salyut 6/DOS 5
Launch: 29 September 1977
Orbit: an initial 219 × 275 km, 51.6°
Mass: 19,825 kg

Salyut 7
Launched: 19 April 1982
Mass: 19.92 tonnes
Orbit: 212 × 260 km, 51.6°

UNITED STATES OF AMERICA

International Space Station (ISS)

Current status

Following years of delay and postponements, most recently caused by problems with the delivery of Russian-built hardware, assembly of the International Space Station began on 20 November 1998 with the launch of the FGB (Zarya) by a Proton launch vehicle from Baikonur. Known as Flight 1A/R, it put the first element of the ISS into space. It came almost 15 years after President Reagan announced the goal of operating a permanently manned space station, 12 years after the launch of Russia's Mir station and nine years after the Russians began permanent habitation of their laboratory in orbit. The second ISS flight, 2A, was launched on 3 December 1998 when the Shuttle Endeavour (STS-88) carried the Unity module (Node 1) to a docking with Zarya. The third mission, flight 2A.1, was launched on 27 May 1999 when Shuttle Discovery (STS-96) took several tonnes of supplies to the dual complex in preparation for later visits.

The third element, Russia's Service Module, is modelled after the Mir core module and is to be launched by Proton from Baikonur late in 1999. It will dock to the aft end of the FGB, which will serve as a connecting and propellant module between Unity and the Service Module. Unity will form the connecting link to US, Japanese and European elements while the Service Module will be the strongback for Russian

Space Station configuration as presented to President Reagan in December 1983; he used this model at the London Economic summit in June 1984

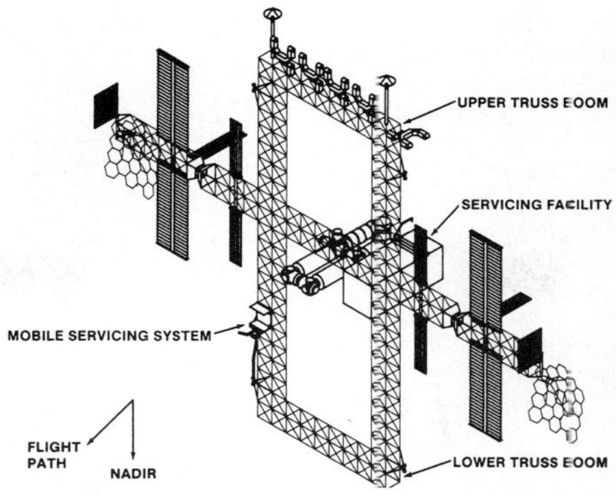

The Dual Keel assembly was descoped to create the Phase 1 design, based on its central boom

elements. As of June 1999, the fifth ISS flight, 2A.2 (STS-101), will be launched 2 December 1999 when Shuttle Atlantis carries additional equipment and a crew to prepare the evolving ISS for later assembly flights and the first occupation crew known as Expedition 1.

Background

The evolution, final configuration and Russian involvement in the US International Space Station all came to a conclusion in 1994. NASA had named the assembly Alpha until then, but now simply calls it the International Space Station (ISS). 'Alpha' normally meaning first follows by a decade the permanent manned presence in space that the Soviet Union pioneered.

President Reagan directed NASA to begin development of a permanent Space Station, crewed by six to eight astronauts, in his January 1984 State of the Union Address. Echoing Kennedy's May 1961 Moon landing pronouncement, he proposed that it should be in orbit 'within a decade' and should constitute a bold and imaginative programme to maintain US space leadership into the 21st century. It began as an eight-year, US$8 billion project requiring possibly seven Shuttle flights for initial assembly. NASA administrator James Beggs was requested to discuss participation with European nations, Canada and Japan.

Almost immediately, NASA's various centres suggested methods of building the Station. JSC's Space Station Office advocated an evolutionary facility with 'modular add-ons'. Tasks would include servicing of satellites in low orbits and refuelling of Orbital Transfer Vehicles for GEO satellites. JSC also advocated design of the station as a stepping stone towards a lunar base for mining, military surveillance and communications, and perhaps also for hazardous industrial activities. NASA emissaries seeking international support in Europe in 1983 noted that both polar and LEO stations would be required by 2000, by which time returns from the stations would be worth US$2.4 billion annually.

In its first incarnation, the Station would be placed in a 500 km, 28.5° orbit, with five modules, each requiring three to four hours work per day by crew members. A two-year trial by two crewmen was envisaged in the

Space Station after completion in 2002. Russia's elements are clustered to the right; the only point of contact with the rest of Alpha is the node's docking adaptor (John Frassanito & Associates/NASA) ***1995***

1990s, to validate closed-cycle life support systems, ultimately enabling ground supply requirements to be much reduced. The plan specified that another 10 modules would be needed by 1996, with station mass rising from 36 to 94 tonnes by 2000. Included in the US$8 billion price tag were two free-flying platforms. One would accompany the Station, undertaking astrophysics observations, materials processing and other research activities which, if based in the main Station, could be disturbed by crew movements. The second, Shuttle-serviced in a 98° Sun-synchronous polar orbit, would perform Earth observation.

CW6.8Grumman, under a contractor study, advocated a Power Tower concept, which was eventually accepted by NASA because of the advantages provided by assembling the Station around a 120 m-long central tower. By keeping one end nadir pointing, Earth viewing occurred simultaneously with celestial-viewing. Grumman placed the power generation system between the two sets of instruments, so that it did not obscure the view of either. A massive cross-arm supported either eight solar panels on gimballed joints so that they faced the Sun for about 60 per cent of each orbit; or alternatively, in the solar dynamic concept, four dish reflectors. Further down, above the modules, a pair of large thermal radiators constantly orientated the system to avoid solar illumination as they discharged waste heat. Tether experiments could be simply accommodated, because both Tower and tether orientation stay approximately constant in relation to Earth.

Furthermost from the power system, the Tower offered a large clear area to accommodate the approach

Space Station configuration in November 1998 after the US Lab module and Canada's RMS are attached (at right). Although occupation was possible from May 1998, the main solar wing was not due to be attached to the US node until September 1998. This wing will be moved later to become one of the four truss-mounted arrays of the completed station (NASA) ***1995***

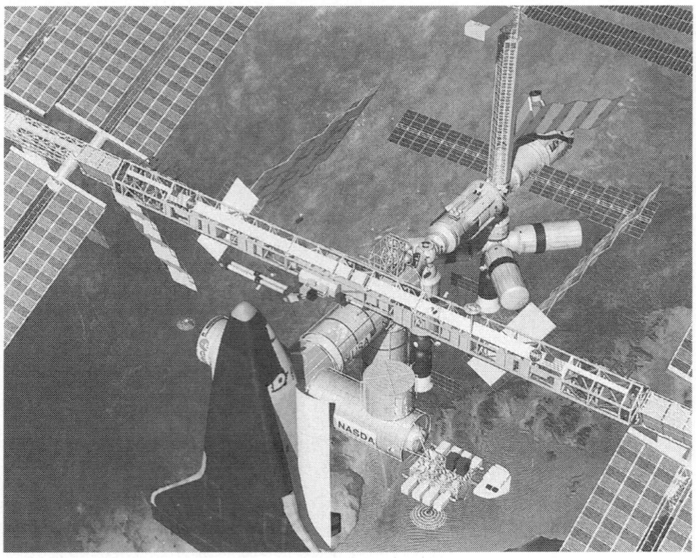

Space Station after completion in 2002 (NASA) ***1995***

and departure of Shuttle Orbiters or OMVs and their payloads. OMVs flew in and out of the lower end for removal and servicing by the RMS. The tower's 2.4 m^2 truss structure provides a convenient track for RMS, as well as supplying a versatile storage area. Finally, the Tower configuration was attractive because it was not sensitive to mass changes from additional modules. By using gravity gradient stabilisation, it inherently accounted for mass changes and therefore did not place tight constraints on the locations of payloads or upon the direction of growth. Attitude control gyros could easily compensate for any stability loss.

Then NASA abandoned the Power Tower approach in October 1985 in favour of the Dual Keel concept. Largely based on Lockheed and McDonnell Douglas designs, it provided the users with a better μg environment (10^{-5} g for all modules) by housing the modules centrally. The plan also increased usable area on the structure for attaching external payloads, allowed better pointing accuracy due to the stiffer structure, and reduced traffic through the lab modules. This radically changed plan, announced on 14 May 1986 (and modified again, four months later), was the result of criticisms of the Power Tower expressed in both an internal report and by a group of external engineers and retired astronauts. According to them, the Power Tower design paid insufficient attention to the needs of science and commercial users; in addition, the arrangement of the crew modules, mounted end-to-end in a racetrack pattern, offered poor safety. Individual modules could not be sealed off in emergency – for instance, if punctured by debris. The new design had the modules placed in parallel, and connected with external airlock hubs, enabling any one module to be sealed off without affecting the others. It also provided more room inside the modules.

The design was based on two vertical keels 110 m long, connected by upper and lower horizontal booms 44.5 m long. The central transverse boom upon which the modules would be mounted was 153.3 m long and the truss would consist of graphite/epoxy tubes 5.4 cm in outer diameter, forming a 5 m^2 open structure, designed to provide maximum stiffness for stability during manoeuvres. Astronauts using assisting fixtures during EVAs would erect it. The upper/lower booms were designed for mounting astronomical instruments at the top, and Earth sensors at the bottom, with the central transverse boom supporting the pressurised modules. The outer ends of the transverse boom would support the power generation and heat rejection systems, and Canada's mobile RMS would traverse the boom to reach required points of operation.

The new station needed two power generation systems. One was the well-proven Si photovoltaic system, providing 37.5 kW of the total 87.5 kW required. The other 50 kW would come from the newer, and still experimental, solar dynamic system. Large mirrors would concentrate the Sun's energy at a receiver adjacent to the generator, with energy being stored thermally in a molten salt. The solar dynamic collectors must be kept pointing towards the Sun within about 0.1°, whereas solar panels do not have to track the Sun so precisely. In September 1986, the surrounding truss was stripped away to cut costs, leaving the central beam and the configuration in basically the current form.

Before the project was even three years old, NASA began to experience its first real cost overruns. That year NASA spokesmen told Congress the Station cost to completion would be US$21.5 billion, with US$14.5 billion paid by the US (excluding launch costs). By 1989, Station development and initial operations costs by one estimation would be US$24.7 billion at 1989 rates; with launch costs added, the total was around US$30 billion. NASA projected total US cost through FY99 in 1990 at US$38.3 billion. NASA dropped this figure to US$30 billion for its restructured programme unveiled in March 1991. The General Accounting Office estimated in April 1991 US$40 billion, followed by US$78 billion up to 2027.

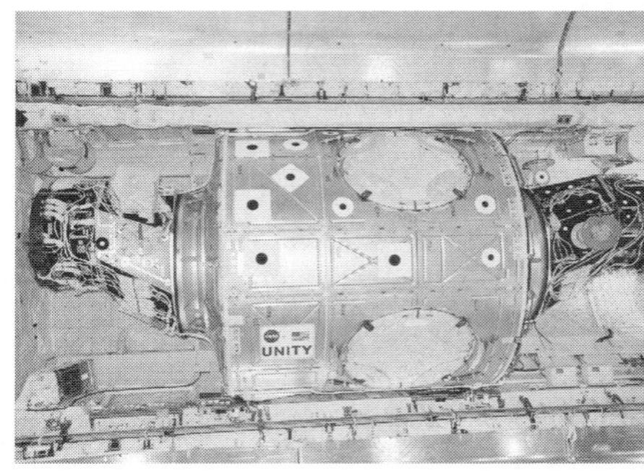

Unity connecting module fitted to payload bay of Shuttle Orbiter Endeavour (NASA) ***1999***/0054453

A major programme review held in the Summer of 1989, under threats from Congress of budget cuts, stated the first element would be in place by March 1995. The Man Tended Configuration (MTC) milestone slipped by five months to April 1996, Permanent Manned Configuration (PMC) by seven months to July 1997 and assembly completion after 29 Shuttle launches by one and a half years to August 1999. NASA issued a new baseline assembly sequence in November 1989, requiring 387 elements to be delivered over four and a half years.

System modifications made as a result of 1989's review included changing from a hydrogen/oxygen propulsion system, involving 4 kW electrolysis of Shuttle waste water, to a modular hydrazine approach. The hybrid AC/DC system was switched to an all-DC network. Development of hard high pressure EVA suits was postponed indefinitely in favour of current Shuttle models. It was decided that various subsystems, such as the closed loop environmental control system, data management, communications and tracking would be incrementally brought to full capability as assembly progressed. It was expected that only 37.5 kW would be available by the time PMC was attained in July 1997. The active thermal control system of fluid/gas heat pipes was replaced by a passive system of body-mounted radiators.

The mass ceiling set for the November 1989 assembly sequence called for a 232 tonne station

US Laboratory module 'Destiny' in space station processing facility at NASA's Kennedy Space Center (NASA) ***1999***/0054454

Alpha's FGB, built by Khrunichev under Boeing contract, will provide initial control and power (Theo Pirard/Space Information Centre) ***1996***

Laboratory module 'Destiny' undergoes checkout at KSC's space station processing facility (NASA) ***1999***/0054455

Interior of US Laboratory module 'Destiny' (NASA) ***1999***/0054456

(including 87 tonne modules, 58 tonne framework, 39.5 tonne power system), with the basic power requirement of 45 kW leaving 30 kW for experimenters. The mass projection peaked at 298 tonnes in June 1990 (and power at 61 kW), before changing the storage tanks' stainless steel to Inconel alloy yielded 283 tonnes the following month. Other measures reduced the power level to 56 kW.

In Autumn 1990, Congress ordered a restructuring of the programme that would reduce expenditure by US$5.7 billion over FY92-96 and peak annual spending from US$4 billion to US$2.6 billion. NASA's 90-day review began in November 1990, encompassing reductions already under consideration. The main US modules were shortened by 40 per cent, and the hexagonal bolted aluminium alloy truss by 50 per cent. In this design, the Shuttle would carry the prices into orbit fully assembled. Assembly elements were reduced from 122 to 17 (an 18th was later added for a node carrying the 2.5 m life sciences centrifuge).

The goal was to reduce in-orbit activity to at most one EVA weekly and preferably monthly. At PMC in late 1999, two years later than planned previously, it would accommodate a permanent crew of four (instead of eight). Mass would be 281 t. Data system capacity was reduced from 300 to 50 Mbit/s and now employing a Shuttle-type Ku-band antenna. The simplified assembly sequence resulted in cancellation of the FTS Flight Telerobotic Servicer, a multi-armed robot for station assembly/maintenance.

An operator viewing via four cameras would control the two manipulators, supported by a leg clamped on the nearest structure. It was to be available from the first mission, mounted initially on Shuttle's RMS and then the station's Canadian Mobile Servicing System. The project was transferred to NASA's Office of Aeronautics, Exploration & Technology and a late 1993 Shuttle test planned, but it was cancelled in the FY92 budget.

Some modules, such as the US lab/hab, Japan's JEM and ESA's Columbus, required the use of the more powerful Advanced Solid Rocket Motor (ASRM) on Shuttle. It was cancelled from NASA's FY93 budget request in January 1992 on cost grounds. Some US$1.2 billion was allocated through September 1992, with a further US$300-400 million required in termination costs. If it had remained cancelled, then Station's 26 assembly and utilisation flights before PMC would have required an additional two assembly and one utilisation flight using the current Shuttle boosters. ASRM was re-instated September 1992 by Congress, but they finally cancelled it in October 1993.

NASA and Italy's ASI signed an MoU 6 December 1991, for Alenia to supply three Mini-Pressurised Logistics Modules. The MoU included a Mini Laboratory, initially dedicated to life sciences with a 2.5 m tilting centrifuge and three racks. In return for these modules, an Italian astronaut would serve a 90-day tour every six years and one user rack would be allocated 90 days annually (in addition to Italy's ESA entitlement).

FY93's budget allocation US$150 million shortfall in September 1992 prompted further schedule slips: March 1996 first launch (four-month delay), June 1997 MTC (six months) and June 2000 PMC (nine months). The programme and its management were overhauled and by the end of 1993, Russia had agreed to become a fully-fledged international partner. The arrival of a new administration and concern over costs resulted in President Clinton in February 1993 ordering a restructuring to cut cost to completion to US$9 billion (US$9 billion has already been expended), capping annual spending at US$2.1 billion. After years of struggling to survive annual budget procedures, the possibility of cancellation could still not be ruled out. The restructuring aimed to reduce life from 30 years to 10 years (extendable to 15 years) and permanent occupancy was no longer guaranteed. Much simpler schemes, possibly employing Shuttle– and Spacelab-based stations or co-operating on Russia's Mir, were also examined. The 45-member Redesign Team began work 10 March 1993 and submitted its final report 7 June 1993.

The team proposed three options: *Option A* using existing flight proven hardware and cost-effective *Freedom* systems, *Option B* making maximum use of existing *Freedom* designs, *Option C* using a Shuttle-derived vehicle placing a relatively simple station in orbit on a single launch. The White House announced on 17 June 1993, the President's selection of Option A. The plan based on Option A was submitted 7 September 1993 as the 'Alpha Station'. Cost and schedule projections were added 20 September (the US$9 billion goal could not be achieved, but the administration stuck with the US$17.4 billion projection NASA offered for work from 1994 to 2002 (excluding launch costs).

The heads of all the involved space agencies met in Montreal 7 November 1993 to review the outcome of Station's redesign and discuss Russia's participation. The international partners on 7 December formally invited Russia to join the project. Russia's funding

The US Lab module pressure vessel after completion of welding (Boeing) ***1996***

STS-79 was the first to carry the Spacehab Double Module, providing 2.7 tonne cargo capacity (Spacehab) ***1997***/0007086

The Power Tower concept was abandoned in late 1985

problems prompted suggestions in 1995 that Mir should, particularly with the new Spektr and Priroda modules, form the basis for ISS. NASA declined this late suggestion, insisting that any changes must not alter ISS's cost or schedule. Instead, RKA/Energia now plan to keep Mir operational at least through 1999 as a much-needed source of revenue, and NASA has added two 1998 Shuttle missions for Mir supplies. Shuttle will also deliver most of Russia's SPP power platform in 1999, saving three Zenit flights.

From 1985 to 1993, NASA spent US$10.1 billion on the station. The US$17.4 billion comprises: US$9.1 billion development, US$4.6 billion operations, US$1.0 billion utilisation plus US$2.7 billion for payloads and Mir support. Operations costs 2003-2012 are projected at US$13.0 billion. The United States plans to pay Russia more than US$1 billion for its contributions, including the US$400 million already agreed on for FY94-97, largely encompassing the Shuttle-Mir missions. Upon completion, NASA will own the FGB 'functional cargo block' under a US$190 million agreement signed by Boeing Defens e & Space Group with Khrunichev 15 August 1995 for in-orbit delivery.

Boeing Defense & Space Group received the prime contract for the revised station on 17 August 1993, with the award being US$5.63 billion. Boeing is responsible for delivering the full-up vehicle and for co-ordinating and integrating the US portion with international elements, as well as the design, development, physical and analytical integration, test, delivery and launch of the vehicle. Under the arrangement, Boeing also handles the first one year of sustaining engineering following launch of each package, including spares.

As prime contractor, Boeing oversees three Product Group subcontractors: first, the McDonnell Douglas, truss, external thermal control, command/data handling, communications and tracking, node/cupola integration, GN&C, pressurised mating adaptor components; second, the Rocketdyne, solar arrays, Ni/H_2 batteries, power management/distribution systems; and lastly, Boeing's own, US lab/hab modules, node/cupola primary structures, lab integration, life support systems, international thermal control, internal audio/video, secondary power subsystem.

Assembly will require about 36 Shuttle and nine Russian launches. From the beginning, NASA has had a troublesome time managing the disparate elements of the programme. The FGB was supposed to be on-orbit in November 1997 with the initial Shuttle assembly flight following the next month. However, the Service Module,

Astronauts take part in a Node 1 systems verification inspection (NASA) ***1999***/0054457

ISS Assembly Sequence: June 1999 Planning Reference

Date	Flight	Launch vehicle	Element(s)
20 November 1998	1A/R	Russian Proton	Zarya Control Module (Functional Cargo Block – FGB)
3 December 1998	2A	US Orbiter STS-88	Unity Node (One Stowage Rack) Two Pressurised Mating Adaptors attached to Unity
27 May 1999	2A.1	US Orbiter STS-96	Spacehab Double Cargo Module
November 1999	1R	Russian Proton	Service Module
3 December 1999	2A.2	US Orbiter STS-101	Spacehab Double Cargo Module
February 2000	3A	US Orbiter STS-92	Integrated Truss Structure (ITS) Z1 PMA-3 Ku-band Communications System Control Moment Gyros (CMGs)
March 2000	2R	Russian Soyuz	Soyuz Expedition 1 Crew
March 2000	4A	US Orbiter STS-97	Integrated Truss Structure P6 Photovoltaic Module Radiators
April 2000	5A	US Orbiter STS-98	US Laboratory Module
June 2000	5A.1	US Orbiter STS-102	Logistics and Resupply; Lab Outfitting Multi-Purpose Logistics Module (MPLM) carries equipment racks
July 2000	6A	US Orbiter STS-100	Leonardo MPLM (US Lab outfitting) Ultra High-Frequency (UHF) antenna Space Station Remote Manipulating System (SSRMS)
August 2000	7A	US Orbiter STS-104	High Pressure Gas Assembly
September 2000	4R	Russian Soyuz	Docking Compartment Module-1 (DCM-1)
November 2000	7A.1	US Orbiter STS-105	MPLM
January 2001	UF-1	US Orbiter STS-106	MPLM PV Module batteries Spares Pallet (spares warehouse)
March 2001	8A	US Orbiter STS-108	Central Truss Segment (ITS SO) Mobile Transporter (MT)
May 2001	UF-2	US Orbiter STS-109	MPLM with payload racks Mobile Bass System (MBS)
July 2001	9A	US Orbiter STS-111	First Starboard truss segment (ITS S1) with radiators Crew & Equipment Translation Aid (CETA) Cart A
August 2001	11A	US Orbiter STS-112	First port truss segment (ITS P1) Crew & Equipment Translation Aid (CETA) Cart B
November 2001	9A.1	US Orbiter STS-114	Russian provided Science Power Platform (SPP) with four solar arrays
January 2002	12A	US Orbiter STS-115	Second port truss segment (ITS P3/P4) Solar array and batteries
March 2002	12A.1	US Orbiter STS-117	Third port truss segment (ITS P5) Multi-Purpose Logistics Module (MPLM)
May 2002	13A	US Orbiter STS-118	Second starboard tress segment (ITS S3/S4) Solar array set and batteries (Photovoltaic Module)
June 2002	3R	Russian Proton	Universal Docking Module
July 2002	5R	Russian Soyuz	Docking Compartment 2 (DC2)
July 2002	10A	US Orbiter STS-120	Node 2
October 2002	1J/A	US Orbiter STS-123	Logistics Module (JEM ELM PS) Science Power Platform (SSP) solar arrays
January 2003	1J	US Orbiter STS-124	Japanese Experiment Module (JEM) Japanese Remote Manipulator System (JEM RMS)
February 2003	UF-3	US Orbiter STS-125	Multi-Purpose Logistics Module (MPLM) Express Pallet
May 2003	UF-4	US Orbiter STS-127	Express Pallet SPACELAB Pallet carrying 'Canada Hand' (Special Purpose Dexterous Manipulator) Alpha Magnetic Spectrometer
June 2003	2J/A	US Orbiter STS-128	Japanese Experiment Module Exposed Facility (JEM EF) Solar Array Batteries
July 2003	9R	Russian Proton	Docking and Stowage Module (MPLM)
August 2003	14A	US Orbiter STS-130	Cupola Science Power Platform (SPP) Solar Arrays Service Module Micrometeroid and Orbital Debris Shield (SMMOD)
September 2003	UF-5	US Orbiter STS-131	Multi-Purpose Logistics Module (MPLM) Express Pallet
January 2004	20A	US Orbiter STS-133	Node 3
February 2004	1E	US Orbiter STS-134	European Laboratory – Columbus Orbital Facility (COF)
March 2004	17A	US Orbiter STS-135	MultiPurpose Logistics Module (MPLM) US Lab racks for Node 3
May 2004	8R	Russian Soyuz	Research Module 1
May 2004	18A	US Orbiter STS-136	US Crew Return Vehicle (CRV)
June 2004	19A	US Orbiter STS-137	MultiPurpose Logistics Module (MPLM)
July 2004	15A	US Orbiter STS-138	Module S6
August 2004	10R	Russian Soyuz	Research Module 2
August 2004	UF-7	US Orbiter STS-139	Centrifuge Accommodation Module (CAM)
September 2004	UF-6	US Orbiter STS-140	Multi-Purpose Logistics Module (MPLM) Batteries
November 2004	16A	US Orbiter STS-141	US Habitation Module

Notes: Additional Progress, Soyuz, H-II Transfer Vehicle and Automated Transfer Vehicle flights for crew transport, logistics and resupply are not listed.

Unity module pressurised mating adapter (NASA)
1999/0054458

scheduled to follow FGB by about five to six months, has not been finished. NASA has paid Russia for the manufacturing of the Service Module, but those funds had not been reaching the Khrunichev factory home of the Service Module construction. Because of the lack of funds the work on the Service Module fell behind schedule and it became clear that it would be impossible to meet the scheduled launch in April 1998. As a result it was decided to delay all of the launches for ISS.

Finances and launch delays were not the only problems that ISS suffered during 1996. The first resident crew to ISS was named as A Solovyov and S Krikalev from Russia and W Shepherd from the United States – the latter being backed-up by D Thomas. Both Russian cosmonauts had gained a great deal of experience on multiple visits to the Mir space station, while only Thomas has any space station flight experience. While Solovyov was not questioned as the Soyuz-TM commander, the United States insisted that Shepherd should become the commander once the crew was on board the fledgling ISS. The Russians became reluctant to cede the commander's duties to anybody without space station experience, a matter eventually taken up by the Russian Duma.

Legal precedents were sought out to back up the Russian claim that Solovyov and Krikalev should not be under the command of a foreigner. Of course, the United States is the major financial partner in the ISS programme, but at the time of the flight there would be more Russian units comprising ISS than American ones.

Under pressure from NASA, the RSA removed Solovyov and replaced him with the less experienced Y Gidzenko, while Gidzenko's place on the Soyuz-TM 26 mission to Mir was taken by Solovyov.

In December 1996, ITAR-TASS announced that there were nine Russians in the training group for flights to ISS. From the Russian Air Force, there were Y Gidzenko, Y Malenchenko, Y Onufrienko, V Dezhurov, V Korzun and S Sharipov: the Energiya bureau had supplied N Kuzhelnaya, S Krikalev and M Tyurin. At the time of the report, eight of the nine people were in training at Zvezdny Gorodok for ISS missions: Korzun would join the group at a later date since at that time he was in orbit aboard Mir. Despite having all of these people in training, the Russians have not scheduled any visits beyond Soyuz-TM 29.

A NASA flight director in Houston will control ISS, even when only Soyuz has access in the early phase. The physical control in that Russian-only phase will be from Kaliningrad, but under direction from Houston. NASA will assume full control when the US Lab is attached. English is the working language.

Three Shuttle Orbiters will be modified for station operations. The RMS is undergoing recertification for loads up to 265 tonnes and a remotely-operated power/data connector will be carried on one side of the cargo bay for early flights until the pressurised docking system can be used. The existing internal airlock will be refurbished and mounted outside, attached to the cabin by a short flexible tunnel. On top will be an ASTP-type androgynous docking unit, with integral power/datalinks, and crew hatch. Three docking units will be built and carried when required.

Specifications

International Space Station

Mass: 420 t at completion
Crew size: six at completion, including four science (three continuous crew possible from mid-1998). Crews serve six-month tours; typically two will be Russian.
Total pressurised volume: 908 m³ at completion
Orbit: 426 km circular at 51.6° attained late in assembly sequence; assembled in 355 km, allowing each Shuttle to carry 1.8 t more cargo. The inclination overflies 85% of Earth's landmass and 95% of population.
Microgravity: 180 days quiet µg annually
Configuration: pressurised modules in T-arrangement with transverse 3.7 × 4.9 m hexagonal truss carrying four end-mounted solar arrays for total 110 m length. US will provide two 8.36 × 4.42 m modules outfitted as separate habitation and laboratory facilities, and two connecting nodes (one for storage, other for power control); each 4.42 m long 4.82 m diameter node has six 2 m diameter attachment ports. Node 1 has an attached EVA airlock and a 1.4 m high 2.2 m diameter cupola with seven windows. The 15.5 t 8.364 m long 4.420/4.216 m outer/inner diameter lab (USL) will carry 13 1.054 m wide 2.019 m height 1.50 m³ usable volume (1.97 m³ total) standard payload racks. USL hatch 1.3 m², 8 latches, central 20 cm diameter window 60 mm thick. 16.1 t hab module includes galley (oven, fridge/freezer, trash compactor, hand washer, water supply) and wardroom (table, 2 × 51 cm windows at one end and bathroom plus shower in mid section. US modules are principally 2219 aluminium alloy with skin panels featuring external waffle grid.

Interior of Russian built FGB (NASA) ***1999***/0062158

Zarya and Unity modules viewed from STS-88
1999/0062157

Canada's Remote Manipulator System will move along truss on mobile transporter carrying 17.6 m long seven-jointed manipulator with 265 t load capacity, tactile feedback, a machine vision system and, if required, Special Purpose Dexterous Manipulator (SPDM) with two 2 m arms. RMS applications include station assembly/servicing, Orbiter docking/undocking, payload manipulation and EVA assistance. Russia will provide 19.3 t FGB, 21 t Mir Service Module; two modified Soyuz-TM as interim Assured Crew Return Vehicles; Science Power Platform to provide 20 kW (US solar cells), attitude control (gyrodynes) and thermal control for early Station; science modules; Progress M and further FGBs to deliver propellant. FGB will provide initial AOCS, power and control; after SM docks, its primary function is as propellant depot/distribution. It also forms a link in Alpha's overall power and data distribution system. Japan's JEM is a 9.7 m long 4.2 m internal diameter pressurised module with attached remote manipulator and exposed platform; a 4 m long pressurised Experiment Logistics Module will be delivered once annually.

ESA's descoped COF will carry 2 t payload in four double racks for materials sciences, fluid physics and life science research. As its contribution to operations costs, ESA will provide servicing and reboost using its ATV Automated Transfer Vehicle launched on Ariane 5; it receives 5.3% of Alpha's resources in return. Italy is providing three 16-rack (including up to five powered refrigerators/freezers) Shuttle-launched Mini Pressurised Logistics Modules: 9.1 t payload, 4.7 t empty, 6.17 m long, 4.5 m diameter. ESA is providing COF's ECLSS. Italy might also provide a Mini Laboratory initially dedicated to life sciences with 2.5 m tilting centrifuge plus 3 racks.
Power: 110 kW at completion, including 46 kW for science. Single US array will provide 23 kW during initial assembly, in addition to the small Russian arrays. At completion, NASA's four photovoltaic Si solar arrays of 32,800 cells each will provide 92 kW. Each thermal cover 9.8 × 29.3 m. Supported by six Ni/H_2 batteries on each array module. Russian SPP Science Power Platform will add 20 kW. All DC system: 160 V primary, 120 V secondary.
AOCS: initially by FGB then by SM, plus ATV reboost. FGB stores 6.14 t UDMH/NTO in eight external cylinders, refuelled via SM. Attitude determination by US, principally by GPS receivers.
Life support: SM supports three crew, US Hab four crew.
Communications: data 43 Mbit/s Ku-band downlink; 70 kbit/s S-band uplink (300 Mbit/s down projected for future).
Thermal: US will provide six 1,045 kg 22.7 m-long liquid ammonia radiators.
Command/control: from NASA Johnson's 9,500 m² five-storey SSCC Space Station Control Center. Working language is English.

MILITARY SATELLITES, SATELLITE SERIES AND SATELLITE CONSTELLATIONS

ASAT
Ballistic missile defence
Communications
Data relay
Early warning
Elint
FOBS
Geodetic
Meteorological
Minor military
Navigation
Nuclear surveillance
Ocean surveilllance
Photo reconnaissance
Radar calibration
Radar imaging
Science and engineering test
Unknown

ASAT

RUSSIAN FEDERATION

Polyot series

Background

In 1992, some Russian scientists admitted the Polyot manoeuvring vehicle tests conducted in November 1963 and April 1964 tested ASAT propulsion systems. They were planned for launch on Chelomei's UR200 but booster delays forced the Soviets to switch the tests to an SL-3 Vostok minus its upper stage. With cancellation of UR200 in late 1964, the ASATs were switched to the SL-11 Tsyklon derivative of the R36/SS-9 Scarp missile. TsNPO Kometa had overall responsibility for ASAT development; and Chelomei (now NPO Machinostroenye) provided the bus.

At the time, the Soviets suggested that manoeuvres supported their manned space programme objective of rendezvous and docking. It emerged in 1992 that these missions were tests of the ASAT programme's propulsion systems. They were planned for launch on Chelomei's UR200 but booster delays (and then cancellation in late 1964), forced a switch to Vostok minus its upper stage. The 1,950 kg P1, placed in initial 339 × 592 km, manoeuvred several times, ending in 343 × 1,437 km. It decayed on 16 October 1982. P2 performed several similar manoeuvres and also changed its inclination from 58 to 60°, decaying finally on 6 June 1966.

Soviet ASAT series

Current status

Inactive.

Background

In 1982, the Soviet Union built and deployed a satellite system which destroyed other satellite systems in space by drawing near and detonating a conventional warhead. Contemporary administration officials and particularly the Department of Defence, often cited this system to claim that the RFAS possesses the world's only operational ASAT system. During the Reagan administration various officials cited this fact while seeking public funding for a new American ASAT system. But the claim still has merit. The RFAS Tsyklon SL-11 booster, basis for the RFAS ASAT system, still serves as an active part of the RFAS launch vehicle inventory. Sources within the RFAS military also point out that the SH-11 Gorgon (Russian name: Topol) ABM, deployed as part of Moscow's ABM defences, has enough energy to launch a LEO ASAT weapon.

The Soviet Union introduced its first ASAT without specifically identifying the purpose in October 1968. The country conducted a series of tests by launching

Russia's ASAT vehicle was publicly revealed in 1993. The main radar antenna is at right

target satellites first from Tyuratam and then from Plesetsk, followed by the interceptor from Tyuratam. The Soviet Union conducted 20 tests in 1982 using radar/ optical sensors, to track and close on the target. Sources later revealed that only the radar devices succeeded. The pattern of orbits indicated that the interceptor satellites were expected to kill the targets within one or two orbits. Only two of the four single orbit tests succeeded. Overall, the interceptor hit the target nine of the twelve times. No interceptor moved specifically toward the target if it missed, and observers differ on whether the ASAT has this capability.

ASAT interceptions tended to take place over East Europe or the Soviet Union in the early hours, local time, apparently to allow optical tracking of the ASAT and the target during the closing manoeuvres. Western observers studied the test between 20 and 30 October 1970 by observing Kosmos 373-375. Kosmos 373 acted as the target.

Approximately 3.9 m long and 2.1 m in diameter, it probably had sensors to determine miss distance. Kosmos 374 trailed it three days later, passed very close on its second orbit and then exploded into more than 100 tracked pieces. Kosmos 375 followed on 30 October and also passed very close to Kosmos 373 on the second orbit, 230 minutes after launch. It exploded into almost 50 pieces at its closest approach.

The US DoD estimates the Soviets had an operational system for intercepting and destroying enemy satellites up to altitudes of about 5,000 km since 1972. In 1985 the Soviets publicly announced their system in part as a response to the SDI. As part of the statement, the Soviet Union also announced it had a direct ascent ASAT which would provide a launch on warning capability attack posture towards the SDI satellites the USA expected to deploy.

The last Soviet test of the series of 20, in June 1982, preceded the 'unilateral moratorium' on tests, declared by President Andropov in August 1983. The Soviet Union called for a negotiated ban on space weapons in June 1984. The USA responded with a congressional ban on USAF tests of the ASAT system in effect until October 1988. General Herres, commander in chief of the US Space Command, objected to the congressional ban and warned that almost all near-Earth payloads were within Soviet ASAT range. A publication of the DoD, Soviet Military Power 1985, widely dismissed as a propaganda publication of the Pentagon, noted the

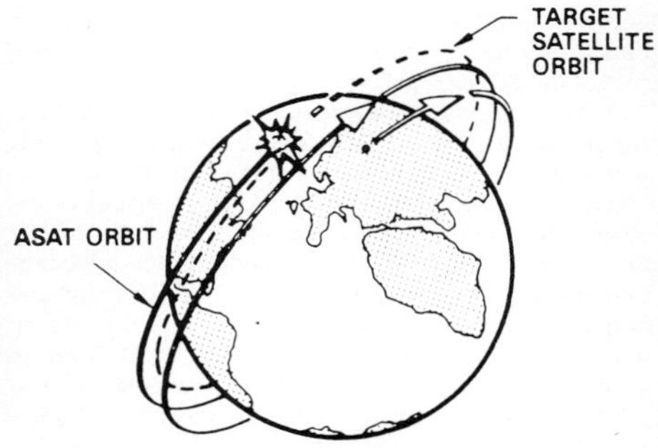

Russia's orbital ASAT typically requires about 3½ hours (2 revolutions) to intercept its prey (Teledyne Brown Engineering)

Soviets could reach satellite targets orbiting at more than 5,000 km.

The highest observed Soviet era interception occurred at 1,575 km. US estimates suggested the ASAT could reach about 20 US surveillance, weather and navigation satellites known to orbit at or below 1,000 km. However, because of the launch azimuths available, satellites below 45° latitude are probably not vulnerable. NASA pointed out that with the usual 28.5° inclination orbit of the Shuttle, it is not at risk. (Military Shuttle missions have flown up to 62° inclination). Early warning, communication and spy satellites in geosynchronous orbits remain out of RFAS/Soviet ASAT reach; a larger booster, such as Proton, could be used, but would be easily detectable and become a vulnerable target in itself.

The Soviet Union conducted its seventh interception test by the end of 1971 (of which five are believed to have been successful). It concluded this initial testing period with Kosmos 404 launched 4 April 1971, which successfully engaged Kosmos 400 at 1,005 km altitude. Kosmos 462, launched 3 December 1971, did equally well against target satellite Kosmos 459 at only 230 km altitude. Then a four year gap followed these successes until Kosmos 803 went into orbit against Kosmos 462 in February 1976. Some observers have stated the Soviets intentionally conducted this test to coincide with the Salt talks. In April 1980, the test resumed when Kosmos 1171 was launched. The interceptor missed its target by 60 km.

Kosmos 1174 apparently used optical-thermal detection of targets in place of the older radar systems in two tests, one in 1981 and one in 1982. After the collapse of the Soviet Union, researchers in to the Soviet era archives discovered in 1992 that the Polyot missions of November 1963/April 1964 carried ASAT propulsion systems. They were scheduled to fly aboard Chelomei's UR200 but booster delays forced a switch to an SL-3 Vostok. When the Soviet Union cancelled UR200 in late 1964, the ASATs were switched to the SL-11 Tsyklon derivative (first launch 1966) of the R36/SS-9 Scarp missile.

Besides direct hit or kinetic kill ASATs, both the Soviet Union and the United States expended considerable sums on ground-based lasers as a means to kill satellites. For years, Western observers have held that the Sary Shagan and Dushanbe lasers may be capable of damaging passing satellites' sensitive components. Although weather and atmospheric beam dispersion may limit such ground-based lasers, they have the advantage of being able to refire and thus disable several satellites. The large-scale ballistic missile defence research into laser, particle beam, RF and kinetic energy technologies could also provide significant advances in space-based ASATs.

In 1986, some reports suggested the Soviets used the Dushanbe laser, north of the Afghanistan border to disable some US satellites. Secretary of Defense Caspar Weinberger refused to confirm the story; the Soviets made no specific refutation. Interestingly, cosmonauts Kizim and Solovyov mentioned Dushanbe earlier in the year in connection with a 'geophysical survey' they conducted from the Mir space station. The high elevation around Dushanbe permits its laser beam to be particularly effective as an ASAT due to low atmospheric attenuation of the beam strength. The longitudinal spacing of what could well be ASAT facilities at Tyuratam, Dushanbe, Sary Shagan and Semipalatinsk make it possible to attack low-altitude satellites on at least two passes daily.

The Swedish Space Media Network, using the French Spot satellite in October 1987, produced high-resolution photographs that pinpointed new construction work at Nurek, near Dushanbe. Purportedly, the Soviets were building four domed laser installations, with direct access to the 2,700 MW Brezhnev hydroelectric plant.

The Dushanbe facility was described as a space tracking centre but its elevation is ideal for siting directed energy weapons (Soviet Military Power)

The Soviets had once described a separate facility nearby as an acquisition/tracking system that could operate in conjunction with the lasers. In response to accusations of SDI research at Dushanbe, Soviet officials said that facilities near both Sary Shagan and Dushanbe were used for satellite tracking, but not SDI-related activities. On 2 January 1988, *Pravda* issued a specific denial that the Sanglok observatory, near Nurek, was part of the SDI-type programme.

A US Congressional delegation received a tour from the Soviets in 1989 of Sary Shagan. According to this delegation, a single 1 m diameter beam director supported two different systems: a very low-power ruby laser and a more powerful CO_2 laser. The 0.7 μm ruby laser housed in the main building adjacent to the beam director consists of 19 individual 5 W lasers whose outputs were merged into a single beam. In a nearby building, a 20 kW CO_2 laser operating at 10.6 μm is linked via a tunnel to the director for a final output of 1 to 2 kW. The Soviet guides claimed the devices had been employed only for tracking, with the most recent exercise in August 1988.

UNITED STATES OF AMERICA

Army ASAT

Current status

In 1997, the US Army spent US$50 million to develop an anti-satellite weapon, up from US$30 million last year. The principal target of the new army ASAT is foreign photo-reconnaissance satellites. In May 1998, the Pentagon withdrew a funding request to continue the programme.

The army programme bases its kill strategy on a direct ascent kinetic vehicle, launched from Vandenberg AFB, which hits the enemy satellite in a battle space within the immediate area, rather than as a co-orbital or partial orbital system. In an attempt to reduce space debris, the army ASAT releases a 6 m² Mylar bag to trap debris created by the impact. The plan has earned the programme the nickname "Green Wars".

United States ASAT series

Background

Just shortly after the Space Age dawned, the US services began plans to shoot satellites out of space. In 1960, one military proposal gained funding and the first test of a system framework came out of Project Bold Orion where a US air-launched missile flew from the atmospheres up to the orbital altitude of NASA's Explorer 6, purposely missing by the satellite by just a few miles. Then the US Air Force began a project named SAINT, variously reported to be an acronym for SAtellite INspector, SAtellite INTerceptor, or SAtellite Inspection & NegaTion. Congress cancelled the programme in 1962 without any flight tests, but the US Air Force followed up quickly with a proposal to knock satellites out of the sky with nuclear-armed Thor missiles fired directly into orbit. The air force conducted tests and began plans to build a base on Johnston atoll and President Johnson declared the system operational in 1964. The Nixon administration dismantled it in 1975, although by then Soviet ASAT tests had been under way for seven years.

Development of the USAF ASM-135 system began under President Ford following concern that the Soviet system was by then operational. But ASM-135 first had to win a bureaucratic and then a technical battle to stay alive. The USAF did get a missile into the air for the first time when it tested the two-stage hit-to-kill ASM-135 from a modified McDonnell Douglas F-15 fighter on 21 January 1984 over the Western Test Range. It was aimed at a point in space, not at a target. The USAF called the second test, in November 1984, 'a partial success'.

On 13 September 1985, the USAF conducted its third test. An F-15 aircraft, piloted by Major Wilbert Pearson, director of the ASAT test force at Edwards AFB, launched an ASM-135 from a position 320 km off the Californian coast, at the DARPA P78-1 Solwind gamma ray spectrometer satellite. The attack followed a 60 to 65° climb out at just below M1.0, and was aimed at 55 m to the right and 25 m above the planned altitude of 11,582 m within a 12.8 km geographic window. The target, launched 24 February 1979, was then 8,600 km distant at 525 km altitude. The ASAT hit it as planned and the impact broke it into at least 285 catalogued pieces (more than 10 pieces are still in orbit). Solwind, it was subsequently revealed amid controversy, had been operational at the time it was hit and was still returning scientific data, returning images of the Sun and corona; it had also recorded comets diving into the Sun.

In the mid-1980s, as the Soviet Rorsats and Eorsats began to orbit, the US Space Command justified funding to Congress on the basis that these satellites could neutralise the US Navy in a time of hostilities. In July 1985, the USAF postponed the launch of two ITVs – the 1.98 m diameter balloons target vehicle which the ASAT would hit – because of reported problems with the ITV's miss-distance signalling system. While the maker, Avco Systems of Wilmington, Massachusetts, was studying the problem, the Summit talks got under way in Geneva. At the talks, the Soviet Union made a dramatic proposal to halt all ASAT testing. However, US$65 million had by then been provided by Congress to start ASAT production, and 58 missiles had been ordered for delivery by 1989.

Space Command continued to argue vigorously for the ASAT but the Congressional Budget Office in February 1987 proposed cancellation. The office contended the move would save US$2.6 billion over five years and would not affect US security because direct ascent technology was rapidly becoming obsolete in light of emerging SDI technologies.

Some details of the ASAT itself became available in September 1981 when a full-size mockup was fitted to an F-15. The interceptor, based on Boeing's SRAM missile, was 5.4 m long, 50.8 cm in diameter and weighed 1,179 kg. A simple infra-red seeker spinning at 1,200 rpm and guided by a laser gyro roll sensor indicated the intercept trajectory. The USAF wanted a direct ascent attack to allow less time for the Soviets to move their satellite out of the way.

Space Command claims that, in addition to ocean reconnaissance satellites, the F-15 ASAT could attack Russian reconnaissance, weather and navigation satellites, in addition to manned space stations and associated spacecraft. In order to reach the Glonass satellites it would have to fit on to a Minuteman missile. The development and procurement of the F-15 low-altitude system is reported to have cost US$3.850 billion.

The air force laid plans to base F-15 ASAT squadrons at Langley AFB (Virginia) and McChord AFB (Washington) because of the orbital inclinations of Soviet target spacecraft and the need for booster debris from tests to fall into the Atlantic or Pacific Oceans. They conducted ASAT command and control functions from a facility developed since 1979 in Space Command's Cheyenne Mountain complex. Full cancellation of the USAF aircraft-based ASAT activities came in March 1988.

The Congressional ban imposed in year-long blocks since September 1985 against ASAT testing on space targets, automatically ended on 30 September 1987. At this time the USAF hoped to launch three intercept demonstrations during 1988 and it launched the ITVs – (Instrumented Target Vehicle) by Scout in December 1985. They re-entered August 1987 and May 1989 respectively. NASA's October 1987 manifest showed that further Scout/ITVs were planned for May in each year between 1988 and 1991. The Congressional ban was extended, however, and all the launches were deleted by the time March 1988's ELV listing was released. The USAF's initial projection of a fully operational system by 1989, and then by the early 1990s, was untenable and the project's cancellation was announced in March 1988. Five F-15 ASM-135 interceptors are now stored in Dallas.

The USAF argued in 1988 that Congress should fund the resumption of tests because the target vehicles would soon be lost. The service argued the instrumented target vehicle balloons cost US$20 million and the opportunity should not be lost to continue the multi-billion dollar programme. DoD followed up the US$183 million FY86 funding and requested US$322 million FY87 funding, with requests for US$448.7 million in FY88 and US$784.2 million in FY89 for its 'restructured' ASAT programme.

BALLISTIC MISSILE DEFENCE

UNITED STATES OF AMERICA

Introduction

Current status
Inactive space component, but the MSX satellite remains operational.

Delta 180 satellite

Background
Delta 180, conceived by the SDIO, planned to demonstrate that small kinetic energy weapons could destroy ballistic missiles during the launch phase, before the release of multiple warheads. The complex exercise, costing US$150 million, lasted only 205 minutes, and involved six aircraft, 38 radars, 31 satellite communications links and the co-ordination of the White Sands and Kwajalein missile ranges with the Cape and Vandenberg test ranges. The orbiting Delta 3920 second stage was equipped with four forward-looking sensors to assess the target's rocket plumes and four others pointing aft to assess its own. From its initial 222 × 220 km, 28.5°, stage 2 released a satellite mounted on a new McDonnell Douglas Payload Assist System (PAS) powered by an Aerojet 201 engine. The SDI satellite carried not only advanced infra-red and other sensors but also, mounted on a mast, the first laser radar flown in space. After stage 2 had manoeuvred to a separation of about 200 km, they crisscrossed and obtained data simulating observations of ICBMs against various Earth horizon and terrain backgrounds. At 92 minutes a Minuteman stage 2, fired from White Sands, was detected by the infra-red system (modified from the front end of a Maverick anti-tank missile). Finally, the satellite used its radar (modified from a Phoenix air-to-air missile) to impact the burning stage 2 at 10,450 km/h over Kwajalein. TV images relayed by aircraft showed two plumes merging into a flash as a conventional warhead on one exploded to ensure complete destruction of the classified sensor packages; impact was within 75 cm of dead centre. SDI director General Abrahamson claimed it as the most complex command and control mission undertaken by the USA.

Delta 181 satellite

Background
In this second satellite of the SDIO series, the Office demonstrated that space-based trackers could identify hostile objects and provide the information to homing seekers. At a cost of US$250 million, the SDIO reported that the Delta 181 mission met all of its scientific objectives. In the first step of the test, the 1.8 tonne payload module ejected 14 objects for seven sensors (infra-red, visible, ultra-violet and a single LIDAR) to track against the cold background of deep space, then against the more cluttered infra-red background of Earth's surface and finally against the horizon. During the test, four of the target objects also ignited solid-propellant motors to simulate missile signatures. The test concluded with the sensor observation of a Strypi sounding rocket launch.

SDIO states the mission succeeded in tracking cold objects against a variety of backgrounds and demonstrated that the laser radar's ability could discriminate between warheads and decoys by determining the shapes and masses of potential targets for the first time.

Delta Star satellite

Background
A Delta 3920 launched SDI's third major orbital mission from Canaveral 24 March 1989 into 482 × 504 km, 47.7° inclination orbit. The 2,720 kg craft had a 698 kg package of eight visible, infra-red, ultra-violet and laser radar sensors that were intended to collect signature

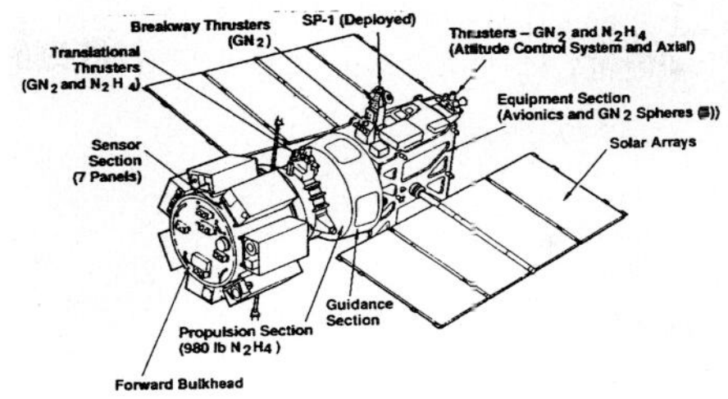

Delta Star principal features (McDonnell Douglas)

Delta Star principal features (McDonnell Douglas)

data on hot rocket motors and other targets against a variety of space and terrestrial backgrounds. In the second primary set of experiments, the payload returned real-time information on materials degradation in space. The US$140 million vehicle failed to observe Delta's stage 2 separation because of a slow opening sensor door.

Mission planners wanted to observe the Delta, Titan and Trident launches from the Cape in order to establish the feasibility of identifying boosters from their plumes. The sensors would also have trained on at least six Black Brant sounding rockets from a variety of US sites. McDonnell Douglas provided the bus accommodating the Johns Hopkins Applied Physics Lab sensors. Apparently, the manoeuvres to observe these launches did not occur as planned.

The USAF controlled the mission from its Consolidated Space Test Centre in Sunnyvale. SDIO announced in March 1990 that Delta Star shut down on command in December 1989 after propellant exhaustion and 4,253 orbits. Orbital decay was predicted after six years.

The mission did succeed in returning 110 observations, including the return of 1,000 minutes of video spectral data at 30 frames/s.

Losat-X satellite

Background
SDIO funded Ball Aerospace to create the 'Low-Altitude Satellite Experiment' to demonstrate small satellite and complementary sensor technologies. Losat carried an integrated avionics suite built around two 80C86 processors, a 0.25 Gbit mass memory, Ball reaction wheels and new wide field of view camera. Losat's mission included collection of multispectral data on rocket launches and the space environment. It was originally planned to fly on the RME/LACE Delta but was not ready in time. In order to reach orbit the Delta 7925 ejected the satellite from stage 2 adaptor after release of primary GPS payload. Losat-X weighed 75.3 kg.

MSTI series

Background
BMDO started the Miniature Sensor Technology Integration programme to provide realistic tests of miniature sensors on low-cost and rapidly-built satellites in LEO. MSTI-1 reached orbit a within a year of the programme start. The Office wanted to launch two to three satellites annually over five years following what they called a 'build a little, test a little' approach. BMDO transferred MSTI to the AF Space and Missile Systems Center 1 October 1994. MSTI 1 cost US$15 million plus an additional US$3 million for initial costs such as ground equipment. Each follow on flight cost about US$10 million each.

MSTI 1 started the test series to track and guide a homing vehicle against a detected track. Rockwell's 4.4 to 4.8 micrometer mid-infra-red 256 × 256 element HgCdTe staring array detected the target and provided 1.4 arc/min resolution with the 100 mm fl telescope. Observations included a static solid motor firing. MSTI spun up after two leaking thrusters lost all the nitrogen propellant on day one, some imaging still occurred. Imaging continued until April 1993. AF Philips Lab directed the project; JPL was responsible for structure, power, thermal and attitude control systems.

MSTI 2 examined Loral Electro Optical Systems' 2.7 to 3.0 micrometer platinum silicide camera and Amber Engineering 3 to 6 micrometer InSb instrument with six filters. The new technology introduced on MSTI 2 included a Ni/H_2 CPV battery, lightweight GaAs solar array, high-performance Sun/horizon sensors and a lightweight low power erasable disk magnetic memory recorder. More capable than MSTI 1, the two-camera system with a gimbaled tracking mirror could follow both fixed targets and ballistic missiles. MSTI 2 acquired and tracked Minuteman 3 from Vandenberg, observed two Sergeant launches from Wallops and made multiple tracking observations of ground test objects. In addition to its primary TMD space-based sensor demonstration role, it achieved connectivity with US Navy shipboard assets in a theatre space-based queuing demonstration.

MSTI 3 carried a SWIR/MWIR 256 × 256 pixel InSb array to observe ground and atmospheric features to determine their variations with aspect, time of day and season. BMDO intended to use it to test future early warning infra-red instruments. The visible wedge 499 × 768 pixel CCD spectrometer provides comparison data for a variety of target signatures. Payload provided by SAIC Science Applications International Corporation under contract to AF Phillips Laboratory.

MSX satellite

Current status
The Midcourse Space Experiment Program successfully completed demonsrations of discrimination, tracking and surveillance from space, as well as data collection to characterise stressing backgrounds against which SBIRS must operate. During the lifetime of the primary infra-red sensor (April 1996 to February 1997), MSX collected approximately two terabytes of data. MSX science teams are in the process of analysing this data in support of the SBIRS and GBI programmes.

Targets tracked include a dedicated STARS rocket and two sounding rockets with a variety of deployed objects; co-operative target launches with the UK (Red Tigress II), Australia (DUNDEE) and Russia (AGRE) and co-operative tracking demonstrations of US, UK and Australian aircraft Aurora, airglow, clouds and other phenomena were observed; stellar occulation, a measurement technique for determining atmospheric composition by observing stars as they go below the

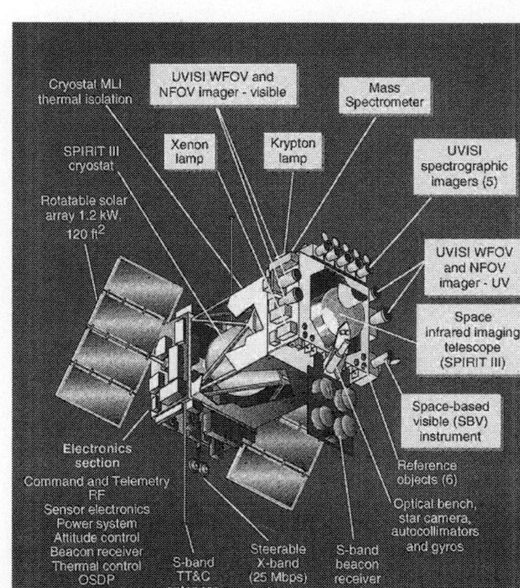

MSX principal features (APL/JHU) **1995**

The completed MSX at Johns Hopkins' Applied Physics Laboratory before it was shipped to Vandenberg ***1995***

horizon, has been successfully demonstrated. Space-based space surveillance was successfully demonstrated.

BMDO data collection with MSX has been completed, although the spacecraft and remaining sensors have three years of life remaining. BMDO is maintaining the capability to collect data and pursuing co-operative data collection efforts with other agencies.

Based on the successful space surveillance demonstration, BMDO has entered into a joint Advanced Concept Technology Development (ACTD) programme with Air Force Space Command. AFSPC will use the MSX Space-based Visible (SBV) sensor as a pathfinder for operational space-based space surveillance. The ACTD is a three-year, jointly funded effort.

Background

The US$866 million Midcourse Sensor Experiment mission is demonstrating SMTS-type sensors for discriminating and tracking mid-course missile warheads. Johns Hopkins' Applied Physics Lab directed the construction of the satellite. The main instrument was projected to last for 20 months, only limited by the cryogenic coolant supply. All other sensors will operate for five years. MSX is also gathering global environmental data, such as ozone levels, in

MSX is demonstrating SMTS-type sensors for discriminating/tracking mid-course targets (APL/JHU)

co-operation with NASA. The APL was originally scheduled for launch on 18 November 1994 but a Dewar vacuum leak allowed the frozen hydrogen to boil off. The cause (incompatible aluminium types attaching equipment to the Dewar developing microcracks) took months to determine.

To help improve orbital debris models, MSX will track along the orbits of Kosmos 2227 in LEO, Kosmos 1278 in GTO and a Titan transtage in GEO looking for debris from their fragmentation. Aurora, airglow, clouds and other phenomena are being observed; key trace constituents will be measured to help establish a database for climate change research. MSX will gather data for two years before NASA's first Earth Observing System (EOS) will take its place. The satellite also performs many astronomy experiments including the UVISI instrument, which looks for photons emitted by decaying neutrinos in a search for the Universe's 'missing mass'. MSX released MIT's 2.0 cm reference spheres for instrument calibration.

Specifications

Launch: 24 April 1996 by Delta 7920 from Vandenberg AFB
Orbit: 897 × 906 km, 99.4° Sun-synchronous
Mass: 2,680 kg at launch
Configuration: 510 cm high bus with three 150 × 150 cm cross section modules consisting of electronics, Dewar support module and instrumentation (11 optical sensors precisely aligned). Attitude control by four RWs plus three magnetorquers to avoid thrusters contaminating the sensors' environment: 0.1° real-time pointing accuracy (post-processing knowledge 9 μrad). Attitude sensors include two three-axis RLGs, star camera, two horizon sensors, five digital Sun sensors plus 3-axis magnetometer. Line of sight jitter held to ±9 μrad over instrument integration periods of 1 s. Background data collected for 20 min/rev, plus 35 min for target tracking. Data recorded at 5/25 Mbit/s on two 54 Gbit Odetics tape recorders and later dumped at 25 Mbit/s X-band to the primary ground site at Johns Hopkins. 1,200 W provided from twin 11.16 m^2 solar wings, 28 VDC; supported by 50 Ah Ni/H_2 battery. The 200 cm long mid-section graphite epoxy truss supports the Dewar, thermally isolating the 200K outer surface from the much warmer bus
Sensors: SPIRIT III (Space Infra-Red Imaging Telescope) scanning radiometer and interferometer/spectrometer: five radiometer bands collect data over 4.2-26 μm and six interferometer channels over 2.5-28 μm. 33.7 cm aperture telescope, 1 × 3° FOV, 90 μrad resolution. 967 kg, 334 W. Nominal life 18-20 months, limited by the solid hydrogen (8.5°K) supply. Utah State University. Space Dynamics Laboratories.

UVISI (Ultra-violet Visible Imagers & Spectrographic Imagers). Five 1,100-9,000 Å 1.0 × 1.0° spectrographic imagers plus four 1,100-3,000/3,000-9,000 Å ultra-violet/visible narrow (1.6 x 1.3°) + wide (10.5 × 13.1°) FOV imagers. 207 kg, 105 W. APL

SBV (Space-Based Visible) sensor. 15.2 cm 4,000-10,000 Å visible telescope. 1.4 × 6.6° FOV, 60 μrad resolution. 78 kg, 68 W. MIT Lincoln Labs
Experiments: CE Contamination Experiment: mass spectrometer, quartz crystal microbalance, and Xe/Kr flashlamps to measure contamination around MSX and on optical surfaces. Separate balances measure mass deposition on MSX's external surface and on Spirit's main mirror.

Skipper satellite

Background

This 230 kg satellite flew 28 December 1995 as a secondary payload on a Russian SL-6 Molniya from Tyuratam to investigate aerothermochemistry and aerobraking by dipping into the atmosphere. Unfortunately, the solar arrays had a wiring problem and the battery was drained within a day. The SDL/USU Space Dynamics Labs of Utah State University meant Skipper to conclude the three bow shock studies flights conducted. Bow Shock 1 and 2 sounding rocket missions investigated ultra-violet emissions from bodies at 3.5 to 5.1 km/s and 40 to 110 km. These new data prompted Skipper to look at the 7 km/s region. The elliptical orbit had a perigee of 130 to 150 km for aerodynamic studies.

Skipper's payload included two scanning spectrometers and 18 photometers. The spectrometers when powered, scan every 0.25 second over 0.200 to 0.420 micrometers in 10 angstrom steps and the photometers view ultra-violet and VUV, with emphasis on atomic oxygen, Lyman-alpha, NO, OH and N_2.

COMMUNICATIONS

NATO

NATO series

Background

Before dedicated international geosynchronous satellites appeared, NATO paid the US government about US$60,000 annually to utilise IDSCS. However, beginning in the late 1960s, NATO began to assemble its own satellite communication network, generally referred to as the second-generation system. Philco-Ford built the NATO 2A and 2B, 243 kg GEO communications satellites, which operate about 5,950 km apart at 18° west and 26° west. These two initial satellites covered the northern hemisphere from Ankara (Turkey) to Virginia in the US, providing 57 voice and 100 telegraph point-to-point circuits through a dozen 12.8 m diameter fixed Earth stations. NATO retired 2A in May 1972 and 2B June 1976, after a total operational expenditure of US$15 million. NATO Communications and Information Systems Agency (NACISA) in Brussels manages NATO communication traffic and has had this responsibility since the inception of the second-generation system.

In February 1973, Ford Aerospace received a US$27.7 million contract to build three NATO 3 satellites and continue the programme into its third generation. Ford's new satellite weighed 720 kg and had a seven-year life. Each could handle hundreds of voice, telegraph, facsimile and wideband datalinks simultaneously on its three X-band channels. NATO manoeuvred NATO 3A into position, following its launch on 28 April 1976 by Delta from Canaveral, and stationed it between Africa and South America. The satellite retired December 1982. NATO 3B followed shortly after the NATO 3A launch on 28 January 1977 where it took up position off the US West Coast above the east Pacific. This satellite also filled in for the traffic that began overwhelming the US DSCS system after 1979. It retired from service in October 1986 and was boosted above GEO from 60° west in July 1993. NATO 3C joined 3A on 19 November 1978 in the slot midway between Africa and South America as an in-orbit spare.

Because NATO 3B continued to perform flawlessly beyond its expected lifetime, NATO 3C remained in orbital storage for seven and a half years, the longest time on record. In October 1986, 3C emerged from storage to begin communication on command from USAF's Sunnyvale facility, relayed through a New Hampshire link. It remained in service until retirement in October 1991, ending its life at 21° west. A small boost command lifted it above GEO in mid-1992. NATO 3D, launched 13 November 1984 by Delta from Canaveral completed the series. 3D cost US$80 million, including US$40 million for the satellite.

At the end of the NATO 3D lifetime, NATO began considering a more advanced, follow-on satellite, but deferred the plans because it normally needs only one satellite, and was owed two years' service from the US DSCS system in return for its use of NATO satellites. NATO 3D moved from 30° west to 21° west in mid-1993, where it remains as a back-up. At the beginning of 1996, the satellite had drifted into an inclination of 3.70°. NATO 3 operational control transferred on 11 July 1991 from the USAF CSTC Consolidated Space Test Center in Sunnyvale to the 3rd Satellite Control Squadron at the CSOC Consolidated Space Operations Center in Colorado, now referred to as the 3rd Space Operations Squadron of Space Command's 50th Space Wing. The control transferred again in 1994 to 5th SOPS.

NATO ordered two NATO 4 satellites, almost identical to the UK's Skynet 4 in February 1987 at a total cost of about £150 million. The total programme will ultimately cost NATO US$364 million at 1987 rates. 4A entered orbit from a Delta 7925 from Canaveral on 7 January 1991 for positioning at 18° west, where it remains as the principal satellite. Like Skynet, it has three 40 W SHF transponders to provide four channels of 60-135 MHz traffic and two UHF transponders which each carry a single 25 kHz channel on a full-Earth helix antenna. The payload includes signal processing and anti-jamming features that are normally associated with electronic intelligence satellites. 4B departed 8 December 1993, also aboard a Delta 7925 and NATO located it at 6° east, where it serves as a hot standby for 4A. The United Kingdom manages the NATO 4 programme on behalf of NACISA.

British Aerospace and Marconi Space Systems act as prime and payload contractors respectively, while the operational control emanates from No 10001 Signals Unit at RAF Oakhanger in the UK. Since the Skynet 4 programme also originates at that facility, speculation is rife that the Zircon capability also exists on the NATO 4 satellites.

The NATO IV satellite employs the same basic configuration as the Skynet 4 and the NATO IV satellites for military communications (Matra-Marconi Space/BAe) ***1998***/0019684

RUSSIAN FEDERATION

Molniya (Lightning) constellation

Background

Molniya-1

The Soviet Union built a continuous communication constellation in a sub-geosynchronous orbit by placing eight satellites in orbital planes separated by 45°. Though the Soviet Union never commented publicly on the satellites and the new openness in Russia has added little new information, most observers believe Molniya-1 served primarily for government and military communications. Gorbachev's government made a substantial financial investment and replaced half of the satellites, which have operating lifetimes of around two to three years, in 1988. In 1989, two more launches continued the replacement programme, three in 1990, three in 1991, two in 1992, three in 1993, one in 1994, none in 1995 and one in 1996. The Molniya-M launch vehicle places the Molniya-1 satellites in orbit mainly from the launch field from Plesetsk but very occasionally from Tyuratam. Tyuratam's last launch, Molniya-1, occurred in February 1989.

Raduga (Rainbow) series

Background

Raduga satellites provide domestic telecommunications links and include an X-band military and government transponder for the Gals service dating to the 1970s. The L-band Volna maritime/aeronautical module and Ku-band Luch transponders also occupy spots on the Raduga satellites. Confusion arose in Western circles when Soviets introduced the Gals and Luch satellites, since they seemed at first to be identical to the supplementary payloads carried on Statsionar system satellites.

Raduga satellites have used the Statsionar geosynchronous slots as one of the three types of satellites to operate in these positions. (The other satellite types are Ekran and Gorizont. The Russians have refused to issue any illustrations that can be definitely identified as Raduga satellites, although some pictures have been released claiming to be Raduga. Careful examination of the photographs indicates that they have been Ekran or Gorizont payloads, which have been misidentified. Thirty-six Raduga satellites have orbited with the first orbit starting in December 1975. 35 conventional Radugas have reached their intended GEO orbits, one (Raduga 33) reached geosynchronous transfer orbit but an explosion of the Block DM-2 fourth stage destroyed the satellite at geosynchronous orbit injection. The Soviets lost only one Raduga to a launch failure on 24 December 1982. Russian statements released with Raduga's 1-3 launch in February 1994 indicated that this was an improved type of satellite intended to replace the existing Raduga satellites with Raduga-1 class payloads. Press releases accompanying the three subsequent launches identified these satellites as the Raduga variant, not Raduga-1.

In February 1992, Italy joined Russia in the joint Astelit venture between Italy's Telespazio and Russia's NPO Astra, which has held responsibility previously for most of the Soviet ground segment, including the Moskva and Orbita stations. Astelit formed to provide Russia's first reliable telephone links into the international telephone network. Ground stations in Moscow and St. Petersburg uplink calls via a single transponder on a Raduga to Fucino in Italy and from there on to the network.

Specifications

Launches: typically one to two annually using a four-stage Proton-K from Tyuratam (Baikonur)
Design life: two to three years, although operational lifetimes well in excess of this have been demonstrated
Contractors: NPO PM (spacecraft), NPO Radio (payload)
Transponders: 6 × 15 W TWTA, 5.75-6.25/3.42-3.92 GHz up/down C-band global, zonal and spot beams, 34 Mhz bandwidth, edge EIRP 26 dBW global, 35 dBW zonal, 45 dBW spot circular polarisation . Single 7.9-8.4/ 7.25-7.75 GHz up/down X-band beam.
Configuration: unknown, but probably similar to Ekran and Gorizont
Mass: 1,965 kg
AOCS: three-axis, east-west station-keeping to within 0.5° by liquid thrusters

Strela (Arrow) series

Current status

There have been no launches in the Strela-1 octuple of tactical communications satellites since the flight of Kosmos 2187-2194 in June 1992 and this series has clearly been terminated, especially since the Strela-3/ Kosmos launches of a second-generation system are under way. Similarly, the series of singleton Strela-2 store/dump communications satellites terminated with the last launch, named Kosmos 2298 in December 1994.

Background

Strela 1 series

Soviet direct communications between ships, aircraft and bases used to be passed over a constellation of eight 61 kg 100 cm diameter Strela 1 ('Arrow') satellites put in orbit by a single SL-8 Kosmos launched from Plesetsk into 1,500 km circular orbits at 74° with 115-minute periods. Strela 1s offered medium-range links between forces in the field, and possibly at sea, over VHF/UHF bands. The Soviets inaugurated the series with the Kosmos 336-343 satellites launched 25 April 1970. The 45th and latest missions occurred on Kosmos 2187-94, launched 3 June 1992.

At any one time, the Soviets had to keep about 24 satellites operating in order to ensure global coverage. Each had an operational life of about two years, but as their orbital life is about 10,000 years, with a 20,000-year life for the rocket stages, the orbits had to accommodate more than 350 satellites and stages. These orbits are now largely avoided because of the debris left behind from the Strela-1.

Strela 2 series

The Soviets conducted longer-range military communications on a constellation of three 875 kg Strela 2 satellites, built by AKO Polyot for NPO PM. The practice remains in place with the RFAS. An SL-8 Kosmos places them in 786-810 km orbits at 74° inclination and separated in spacing on the orbital plane by 120°. They apparently receive information from low-power spy and other clandestine transmitters around the world, and store the information until a suitable pass enables it to be dumped to RFAS receiving stations. Their ground tracks repeat approximately every 85 revolutions. Two or more satellites have been operated in recent years in each plane of the system, which began with Kosmos 372, 16 October 1970. K2298 (20 December 1994) was the 48th.

Strela 3 series

Analyses of a more recent system, which carries six 225 kg Strela 3 satellites on each launch to orbits of 1,400 km at 82.6° inclination, suggest it is a second-generation replacement for the octets. AKO Polyot builds the satellites for NPO PM. The first test launch of the Strela 3 occurred in 1986 (Kosmos 1617-1622) followed by another later that year (Kosmos 1690-95). On 15 October 1986, a Tsyklon failure prevented the Soviets from completing an operational constellation, which did not occur until Kosmos 1827-32 and Kosmos 1875-80 in 1987. Each year except 1995, the Soviets, and later the RFAS, launched at least one Strela-3 load of satellites. In annual terms, the Russian military has launched Kosmos 1909-14 in 1988, Kosmos 1994-99 and Kosmos 2038-43 in 1989, Kosmos 2090-95 and Kosmos 2114-19 in 1990, Kosmos 2143-95, Kosmos 2157-62 and Kosmos 2165-70 in 1991. During 1992, they launched Kosmos 2197-2202, and Kosmos 2211-16, Kosmos 2245-50 and Kosmos 2252-57 in 1993, Kosmos 2268-73 and Kosmos 2299-2304 in 1994, and Kosmos 2328-2330 in 1996. In 1990, NKO PM offered this sextet system, without the military transponders, commercially to foreign buyers interested in establishing their own store/dump communications network.

UNITED KINGDOM

Skynet series

Current status

The two most recent satellites in this series were Skynet 4D launched by Delta 7925 on 10 January 1998 and Skynet 4E launched by Ariane 44L on 26 February 1999.

Background

The UK began planning to build a network of communication satellites for its far-flung outposts even before the country had begun to withdraw from Asia and the Middle East. Britain needed secure voice, telegraph and fax links for strategic and tactical communications between UK military headquarters and ships and bases around the globe. However, the first attempts to obtain this capability proved unsatisfactory and the UK began to lose interest.

Skynet 1 was launched on a Delta launcher in November 1969 from Cape Canaveral almost as soon as geosynchronous orbits became accessible. But Skynet 1's TWTA failed within a year. Skynet 1B followed shortly, but it failed to reach the correct orbit and remained stranded in a highly elliptical orbit. The first three satellites in this series offered two UHF global channels for submarine and other mobile users, four SHF channels ranging from global to 3° spot beam coverage, and an experimental EHF uplink channel for propagation studies.

Skynet's 14 m Earth stations at RAF Oakhanger. The site also controls NATO 4. Siemens Plessey provided the turnkey station under a £40 million contract; a £60 million contract covers a second large anchor station, a Satellite Network Control Centre and two large transportable terminals (Siemens Plessey Defence Systems)

For its second-generation satellites the UK sought to build an indigenous capability. Ironically, the UK, with the most sophisticated electronics industry in the world, contracted with Philco-Ford to build the first Skynet. With help from Philco, the UK's Marconi Space & Defence Systems built the first communications satellite produced outside the US/USSR. A launch failure set the Skynet 2A tumbling through space until it plunged back into the atmosphere six days after being launched. Skynet 2B succeeded and broke the string of failures for the UK. It linked the UK directly with Western Australia for the first time in history.

The original Skynet system included 17 ground stations, with a 12.8 m master terminal at RAF Oakhanger, air-transportable terminals for use on land and shipborne terminals. By the time the system became operational, successive defence cuts and British withdrawals from Asia and the Near East made it seem like an unnecessary luxury. The British quietly cancelled the planned Skynet 3 series in 1975 and replaced it by leasing capacity from US/NATO satellites.

In the decade before the Falklands conflict of 1982, the UK's military leased channels from the US and the plan worked well for all UK defence needs. Nonetheless, the Falklands conflict itself ironically proved the difficulty of routeing military communications through the good offices of a foreign provider, in this case the US DSCS circuits. The UK revived attempts shortly thereafter to establish an independent and secure communications network for the Royal Navy and other UK forces. The idea passed in Parliament and the budget included funds for the Skynet 4 programme.

Skynet 4: the series of UK milcom satellites will continue beyond 2000 (MMS) **1996**

Skynet 2B was the first communications satellite built outside of the US/USSR. Launched in 1974, it can still be used (MMS)

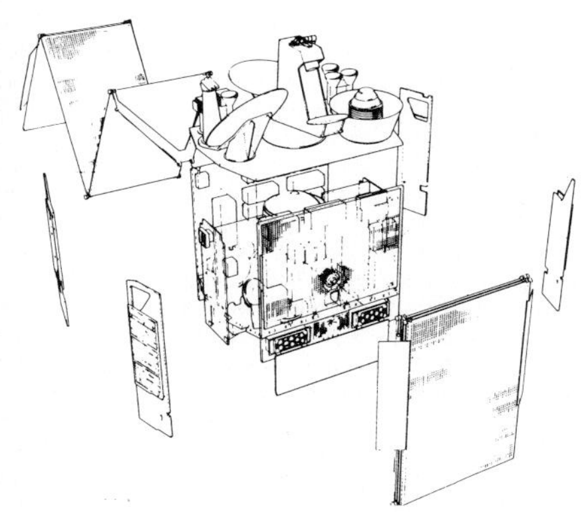

The Skynet 4 series is based on the ECS family of civil communications satellites of the 1970s (BAe)

After several delays, the UK finally scheduled Skynet 4A and 4B for launches in June and December 1986, both by Shuttle despite strong French pressure to use Ariane. The UK finally agreed to the Shuttle launches because of NASA's standing offer to let a British astronaut go along and the hope that the US would help the UK make the Skynet 4 the standard NATO system. By early 1988, the Skynet 4B/C had berths on the Ariane 4 and in late 1988 and early 1990, and 4A had a Shuttle booking.

The continuing problems with the Shuttle moved Skynet 4A to a Titan 3 launch in August 1989 at a reported cost of US$55 million. With the launch vehicles chosen, the UK government ordered Skynet 4C from BAe in May 1985.

The programme, popular in Parliament, seemed to have assured backing but, in January 1987, a journalist reported that the Skynet actually cloaked an electronics intelligence programme known as Zircon. In a secret intelligence sharing deal with the United States, the UK would provide all the message traffic that its Skynets snatched from their Asian orbital slots in return for US information. The United States benefited from this by averting suspicion from electronic intelligence satellites that it needed to have in the region through the mechanism of the false flag. The satellite eavesdropped on other electronic communication at geosynchronous. In the published article, the author cast suspicion on the location registered for the satellite, which was far from any conceivable orbit for communications. The Ministry of Defence eventually admitted its role in Project Zircon and confessed to spending £70 million on the programme.

In operation it would resemble the US Rhyolites and the later Magnum signal intelligence satellites. Reports indicate that the UK would share intelligence with the US via the UK government's monitoring centre GCHQ Cheltenham. Zircon was abandoned following disclosure amid concerns over its ageing technology. By the time Zircon was ready to go though, the ageing satellites, delayed after years of waiting for a Shuttle ride, were simply obsolete.

Skynet suffered from another American problem. Originally, the British military projected Zircon would cost about £400-500 million to build, launch and operate two to three satellites to provide Elint reportedly from GEO at 53° east. The UK Parliamentary Space Committee received a report in June 1990 that the entire Skynet series had cost almost £1 billion, even before Skynet 4D-4F came along.

The Skynet 4 generation of telecom satellites assures continuity of UK tactical and strategic defence communications to at least 2006. The fourth-generation Skynet adds a considerable capability to the communication system with steerable antennas for SHF spot beams, increased power, improved anti-jamming and a fully tunable UHF system for increased flexibility.

Specifications

Skynet 1
Launched: 22 November 1969 from Cape Canaveral by Delta
Launch mass: 243 kg, 129 kg on-station. Payload of two 3 W X-band TWTAs. Cylindrical bus 137 × 81 cm high (157 cm with solid AKM), 90 rpm spin-stabilised with a despun antenna. 7,236 solar cells supported by two 16 Ah Ni/Cd batteries. Prime contractor Philco-Ford; same design as NATO 2A/B. Positioned in GEO over the Indian Ocean but its TWTAs failed within a year

Skynet 1B
Launched: 19 August 1970 by Delta from Cape Canaveral. Abandoned in 270 × 36,058 km GTO after AKM failure. Other details as Skynet 1

Skynet 2A
Launched: 19 January 1974 by Delta from Cape Canaveral. Launch mass 435 kg; 235 kg on-station. Built by the UK's Marconi Space & Defence Systems (with Philco-Ford providing technical direction) as the first communications satellite produced outside US/USSR. Payload of two 16 W X-band TWTAs provided redundant 2 + 20 MHz channels. Cylindrical bus 191 × 208 cm high (with solid AKM), 90 rpm spin-stabilised with despun antenna. Launch failure left it tumbling in 120 × 1,857 km orbit; re-entered after six days

Skynet 2B
Launched: 23 November 1974 by Delta from Cape Canaveral, placed in GEO over Indian Ocean. With a design life of at least three years, it provided links from the UK to W Australia. Other details as 2A. It began drifting in late 1992, indicating its propellant was exhausted. Tests in 1994 showed that its two 16 W TWTAs still met specifications. It remains usable, inclined at >13° and slowly drifting. The original Skynet system included 17 ground stations, with 12.8 m master terminal at RAF Oakhanger, air-transportable terminals for use on land and 1 m shipborne terminals. By the time the system was operational, successive defence cuts and British withdrawals from Asia and the Near East had rendered it largely unnecessary

Skynet 4A
Launched: 1 January 1990 by Titan 3 from Cape Canaveral
Location: 34° east geostationary (6° east until 1992), inclination 2.2°
Other specifications as for Skynet 4B, except McDonnell Douglas PAM-D2 provided perigee kick

Skynet 4B
Launched: 11 December 1988 by Ariane V27 from Kourou, French Guiana, declared operational 22 February 1989
Location: 53° east geostationary (1° west until 1992), inclination 3.0°
Design life: 7 years
Contractors: BAe (prime), Marconi Space Systems (payload), and CAL Ltd. (UHF antenna)
Transponders: 3 × 7.250-8.400 GHz X-Band 40 W TWTAs providing four SHF channels made up of channel 1, 31 dBW EIRP 17.0 dB gain global 135 MHz bandwidth; channel 2, 34 dBW EIRP 24.7 dB gain narrow (Europe) 85 MHz beam; channel 3, 35 dBW EIRP 21.7 dB gain wide (hemispherical) 60 MHz beam; channel 4, 39 dBW EIRP 29.5 dB gain 3° spot (central Europe) 60 MHz beam

2 × UHF 305-315/250-260 MHz up/down global 40 W transponders, 25 kHz bandwidth, 26 dBW EIRP, working through single helical antenna. With EMP hardened anti-jamming and ECM interference features
Principal applications: UK Ministry of Defence. Submarines, aircraft, maritime patrol, tankers, AWACs, transportable Earth terminals, 1.7 m mobile terminals, manpacks
Configuration: based on MMS ECS/OTS bus, 2.1 m high, 1.9 m deep, 1.4 m wide; 16.0 m span with solar wings deployed. The Communications Module is a U-shaped structure sitting on the Service Module; major structural elements are predominantly aluminium alloy honeycomb with aluminium or CFRP skins. Payload components are mounted on both sides of the upper floor and on the sidewalls. The central 2.2 m diameter thrust cone is CFRP honeycomb with aluminium support struts
Mass: 1,433 kg at launch, 790 kg after AKM firing, 770 kg on-station BOL, 655 kg dry
AOCS: 3-axis control with ±0.1° east-west station-keeping incorporating IR Earth sensors (Officine Galileo, accuracy ±0.04°) and Sun acquisition sensors (TNO/TDP, accuracy ±2° coarse) maintained by two 25 Nms and one 16 Nms Teldix MWs and 20 catalytic hydrazine thrusters. (0.5 N roll/yaw, 2 N east-west station-keeping, 5 N pitch, 20 N active nutation damping). The thrusters draw on 70 kg propellant in two N_2-pressurised tanks. Pointing accuracy 0.07° roll/pitch, 0.35° yaw
TT&C: through RAF Oakhanger in Hampshire, UK, which also acts as the primary communications node using 14 m dish. A dedicated Skynet 4 in-orbit checkout facility is also maintained at Oakhanger. The TT&C subsystem, working through the SHF receivers with S-band utilised during launch and as back-up, can handle 512 8-bit words about every 16 s; a total of 750 telecommands is available
Power system: twin 228 × 730 cm 3-segment (each 128 × 205 cm) Si solar wings and auxiliary body panels, providing 1,200 W at regulated 42 V in sunlight; eclipse power provided by two banks of 14-cell 35 Ah Ni/Cd batteries with 30-37 V unregulated output. Eclipse power requirement 940 W, maximum requirement 1,083 W (equinox). Degradation of AEG back surface reflector cells 9% over 7 years
Thermal control: heaters plus passive radiators, including Secondary Surface Mirrors and 10-15 layers
Apogee kick motor: Thiokol Star 30E solid propellant, burn time 49 s, specific impulse 290 s

Skynet 4C
Launch: 30 August 1990 by Ariane V38 44LP from Kourou, French Guiana. AKM fired 2 September, 69 h after launch; solar wings deployed 3 September; UHF antenna deployed 4 September
Location: 1° west geostationary (moved from storage position at 53° east in 1992, where it was ideally placed to provide links during the 1990-91 Gulf conflict); inclination 1.1°

Skynet 4D
Launch: 10 January 1998 by Delta 7925 from Cape Canaveral
Location: 35/1°W or 6/53°E
Life: 7 years
Platform: as 4A-C but reworked TT&C (including increased autonomy); improved power system; improved reaction control system
Transponders: 3 × 7.250-7.725 GHz X-Band 50 W TWTAs providing four SHF channels: channel 1, more than 31 dBW EIRP 17.0 dB gain global 125 MHz bandwidth; channel 2, >34 dBW EIRP 26 dB gain narrow (Europe) 75 MHz beam; channel 3, >36 dBW EIRP 21.5 dB gain wide (hemispherical) 75 MHz beam; channel 4, >40 dBW EIRP 30 dB gain 3° steerable spot (central Europe) 60 MHz beam; two UHF 295-318/254-258 MHz up/down global 50 W transponders, 25 kHz bandwidth, >25 dBW EIRP, working through single helical antenna (up/down frequency selection independently tunable). With EMP-hardened anti-jamming and ECM interference features
Mass: 1,510 kg at launch, 868 kg after AKM firing, 851 kg on-station BOL, 256 kg payload
Power system: as Skynet 4A-C, except 2 × 29 Ah Ni/Cd batteries
Other specifications as Skynet 4A-C

Skynet 4E
Launch: 26 February 1999 by Ariane 44L from Kourou, French Guiana
Location: 35/1° west or 6/53° east
Mass: 1,510 kg at launch, 873 kg after AKM firing, 851 kg on-station BOL, 256 kg payload
Other specifications as Skynet 4A-C

Skynet 4F
Launch: planned for 1999 by Ariane from Kourou, French Guiana
Location: 35/1° west or 6/53° east
Mass: 1,510 kg at launch, 851 kg on-station BOL, 256 kg payload
Other specifications as Skynet 4A-C.

UNITED STATES OF AMERICA

DSCS II series

Background

The IDSCS, often referred to as phase I of military defence communication of the defence Satellite Communications System, emphasised the Cold War needs of nuclear hardening and anti-jamming protection. In phase II, building a network made up of 16 satellites, the military emphasised an increased capacity and system flexibility that reflects the importance of the initial system. The satellite system can handle 1,300 voice channels or 100 Mbit/s data. Each satellite carries two steerable narrowbeam antennas to work through small ground terminals. Self-contained and mobile, these small antenna enable US Army units to operate full communications through DSCS within 20 minutes of arriving at a new site. The satellite can redirect its beams to different locations within minutes, and the army can move the satellite within a few days to a new GEO position to meet defence contingencies in any part of the world. Antenna nodding reduces north-south station-keeping, a measure that reduces the propellant requirements and extends useful lifetimes.

DSCS 1 and 2, launched 3 November 1971, ceased operations by 1978, but DSCS 3 and 4, launched 14 December 1973, continued to operate over the west Pacific while 1 and 2 shut down. Number 4 was raised above GEO from 63° west in March 1993 and de-activated 13 December 1993. DSCS 5 and 6, launched 20 May 1975, would have completed world coverage except for polar areas, but they fell into the Pacific after six days, due to a Titan guidance failure. A second set of six satellites offering five-year design lives, began with DSCS 7 and 8, launched 12 May 1977. The DSCS system still did not reach completion as projected when DSCS 9 and 10 were lost 25 March 1978 following Titan stage 2 failure. During this period NATO provided a 'loaner', repayable in similar services when needed, of the use of its own system.

DSCS-2 15 was launched with the first DSCS-3 satellite on October 19 1982 and number 16 (the last one) with the third DSCS-3. In March-April 1997, the orbital period of DSCS-2 15 was raised from 1,436 minutes to 1,451 minutes and the satellite began to drift off-station: it is unclear whether the satellite is being moved to a new location or – more likely – being retired. Orbital data for DSCS-2 16 has been classified and therefore its location history is not known.

A full complement of four operational satellites occurred when the successful launch of DSCS 11 and 12 on 14 December 1978 placed them over the east and west Pacific. They still remain operational 10 years later. DCSC 11 boosted itself above GEO in mid-1993 after being loosely stationed around 90° west. The army deactivated DSCS 11 and 12 on 13 December 1993. DSCS 13 and 14 took up position on 21 November 1979. DSCS 13 left service and boosted itself above GEO from 180° east June 1993. DSCS 14 retired in April 1995 after drifting for at least nine months, ending the phase II programme.

Specifications

Contractors: TRW Space & Defence Group
Transponders: numbers1-12. 2 × 20 W TWTA 7/8 GHz up/down X-band 410 MHz bandwidth circular polarisation transponders employable as 28 dBW global (18° beamwidth), 32 dBW regional (6.5° beamwidth, 4,100 km diameter coverage) and 43 dBW spot (2.5° beamwidth, 1,600 km diameter coverage). numbers13-16 increased TWTA to 40 W
Principal applications: 1,300 duplex voice channels or up to 100 Mbit/s data
Configuration: 2.7 m diameter, 3.95 m deployed height, spin-stabilised cylindrical bus with de-spun platform incorporating 2 global coverage horns and two narrow coverage steerable (±10°) dishes
Mass: 522 kg numbers 1-5, 567 kg numbers 7-12, 612 kg numbers 13-16 fuelled

DSCS Phase III: these models are the primary user of the Atlas 2 launcher. At bottom is the large 61-beam receive array; the 19-beam transmit antennas are at top, below the single steerable spot. The two UHF antennas are central. At left are the global horns (Lockheed Martin Astro Space)

AOCS: spin-stabilised using infra-red Earth horizon sensors and hydrazine thrusters; small solid thrusters at base of drum provided initial spin-up. Inserted into GEO by launcher final stage. NSSK not provided.
TT&C: via base S-band biconical horn antenna with 32° toroidal beamwidth
Power system: eight solar array panels on cylindrical bus provide 535 W BOL and 358 W after five years; 235 W required by communications payload. 3 × 12 Ah Ni/Cd batteries provide full eclipse protection for minimum of five years.

DSCS III series

Current status

The DoD launched a 1,040 kg DSCS 3 from Cape Canaveral on 25 October 1997. The DoD code-named the satellite USA 134.

Background

DSCS Phase III satellites have 10-year lives and are three-axis stabilised. The DoD ordered 14 in total when the programme began in 1975. Their multiple-beam antennas employ advanced encryption to protect their data and 1,300 duplex voice links for both tactical and strategic users. In addition to nuclear hardening, sensors can detect a jamming attempt, report it and wait for ground control to plot the location of the jammer and dispatch instructions on how to use its steering capacity to null the jammer. The nuclear hardening cannot protect against direct attack, but enables it to withstand low-level flash effects, X-rays, gamma rays and electromagnetic pulses (EMPs). The system provided more than 80 per cent of all communications during Operation Desert Storm/Shield in 1990 and 1991.

A complete DSCS III ground terminal can be transported by a C-5 Galaxy. The first operational satellite in this series weighed 1,042 kg. and orbited with II-15 from the same Titan 34D/IUS on 30 October 1982 from Canaveral. The entire Phase II/III system had a network of six satellites, four active and two spares, which operated at 7/8 GHz X-band. They provided secure strategic and tactical voice and data transmission, military command and control and ground mobile communications. In the event of a nuclear war, the National Command Authority, the White House Communications Agency and the Diplomatic Telecommunications Service all had priority usage for the DSCS.

The constellation had seven satellites by end-1985, (the latter ones launched cost US$150 million each). Four DSCS IIIs were required on-station in GEO to provide full coverage; and the first two (B-4/B-5) left aboard the single IUS from STS-51J on 4 October 1985. The Shuttle Atlantis had to fly to an extremely high orbit of 515 × 475 km, to help the IUS deliver its heavy cargo into GEO. The next DSCS III number 4 (A-2), departed 4 September 1989. The third IIIB created, for the first time, a five-satellite DSCS III network.

The operational system requires four active satellites plus two in-orbit spares, and the army acknowledged that five primary and four back-up satellites were in place by September 1995. The ban on releasing orbital elements for the post-1982 satellites prevents a comprehensive identification of these satellites. The available elements indicated the army still actively controlled one of the last three IIs (II-16 is post-1982) and the first IIIs were still controlled at the beginning of 1996. A tentative identification locates these satellites at 64° east (II-15), 130° west (III-1). Space observers identified II-16 as still operational in September 1995. II-13 retired in mid-1993 and boosted itself above GEO. II-4, II-11 and II-12 left active service in December 1993 and the army took II-14 off-line in mid-1995. One source indicated in 1992 that II-13 and II-14 were severely degraded and barely usable, as was III-1.

The primary DSCS III positions are 12° west Atlantic, 136° west (east Pacific), 175° east (west Pacific), 60° east (Indian Ocean). The army has also registered slots with the ITU for 130° west, 52.5° west, 42.5° west, 57° east, l80° east.

The new US Army Space Command, created in 1988, collocated at Colorado Springs with Space Command/NORAD, accepted full authority for the system in the summer of 1988. Operational control was transferred 11 July 1991 from the Consolidated Space Test Center at Sunnyvale to the 3rd Satellite Control Squadron at the Consolidated Space Operations Center in Colorado (now referred to as the 3rd Space Operations Squadron of Space Command's 50th Space Wing).

The air force planned to award Lockheed Martin a US$100 million to US$125 million contract in March 1996 to upgrade at least four of the five satellites in storage in order to prolong full system operability until 2008. The IIIs, despite a design goal of 10 years, only average a seven year lifetime. Improved thrusters will stretch the station-keeping propellant from 9 to 11½ years, and replacing the Si cells on the two inboard panels with GaAs will combat degradation. The 40 W TWTAs of channels 1/2 will be replaced with either SSPAs or improved TWTAs. The company has already replaced amplifiers with upgraded versions capable of handling 50 per cent more traffic under a US$9.6 million contract in early 1990.

1995's DCSC IIIB B-7 carries STP's CHARGECON-GEO (Satellite Charge Control Experiment at GEO) to test the ability to detect and discharge an electrostatic build-up while still small.

Specifications

Contractors: Lockheed Martin Astro Space
Transponders: 6 × SHF 7.900-8.400/7.250-7.750 GHz up/down channels: 1 40 W 50 MHz ; 2 40 W 75 MHz; 3/4 10 W 85 MHz; 5 10 W 60 MHz; 6 10 W 50 MHz. Two (plus two back-up) 40 W TWTAs + four (plus two back-up in 1-for-2) 10 W SSPAs.
Working through: One high gain (30.2 dB min) steerable 3.0° spot dish for channel 4 (EIRP 37.5 dBW) + channels 1/2 (EIRP 44 dBW).

Two 19-beam waveguide lens Multi-Beam Antennas (MBAs) for ch1-4 with beam-forming networks to produce selected antenna patterns accommodating the ground receiver network. EIRP: Earth coverage channels 1/2 29.0 dBW, channels 3/4 23.0 dBW; 1° narrow coverage channel 1/2 40.0 dBW channels 3/4 34.0 dBW. Receive through 61-beam waveguide lens MBA, with associated beam-forming network to provide selective coverage and jamming protection.

Lockheed Martin produces the third-generation Defense Satellite Communications System (DSCS III). It features high-data rate and around the globe, secure communications. Each satellite has six independent SHF transponders with a 500 MHz bandwidth (Lockheed Martin) ***1998***/0022567

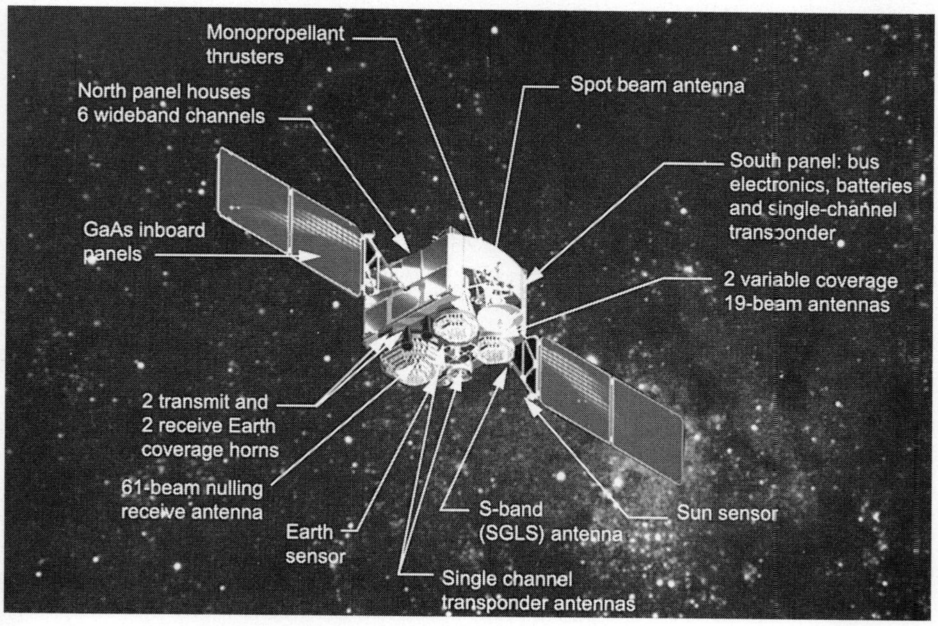

All of the major features of the DSCS III are shown in this drawing ***1998***/0022568

Two global horns receive and two transmit for channels 3-6 (EIRP 25.0 dBW)

UHF bow tie 300-400 MHz receive and cross dipole 225-260 MHz SSPA transmit for Single Channel Transponder to support AFSATCOM. AFSATCOM II signals handled at either UHF or SHF (channels 1)

Configuration: 206 cm long body (279 cm across antennas), 193 cm wide, and 196 cm deep to antenna tips. Transponder amplifiers and associated components are mounted on the north panel. Modular structure, 3-axis stabilised

Mass: about 1,170 kg at launch

AOCS: 3-axis, pointing accuracy 0.09° pitch/roll, 1.0° yaw, 16 × 4.45 N thrusters, 276 kg capacity hydrazine system utilising four spheres. Inserted into near-GEO by IUS stage 2 (III on Shuttle) or IABS (IIIB)

TT&C: via S-band (1,807.764 and 1,823.779 up/2,257.5 and 2,277.5 MHz down) and SHF (channels 1/5 up/7,600.0 and 7,604.70588 MHz down). The SHF link's primary function is to control channel and antenna configuration in real time through the defence Information Systems Agency's SCCE Satellite Configuration Control Elements

Power system: 2-panel Si dual solar wings spanning 11.62 m, totalling 11.72 m², providing 1,240 W BOL, 980 W after 10 years, supported by 3 × 35 Ah Ni/Cd batteries. 190 W provided by stowed array during spin transfer phase.

FltSatCom/AFSATCOM series

Current status

Four US Navy FltSatCom continued operations. Some of these navy vehicles are being required to double their seven-year design lifetimes but attitude control/station-keeping propellant depletion will soon take them out of service. Controllers are permitting them to drift 10° beyond the equatorial arc in order to conserve propellant.

Background

When it was the backbone of navy communications, FltSatCom handled 90 per cent of the USN's communications and provided two-way UHF links for the Strategic Command and the US Army/Marine components of the Rapid Deployment Force. The navy placed its satellites at 110° west, 23° west, 72.5° east and 172° east, although the service has also recently used 100° west, 23° west, 15° west and 171° east. The four operational satellites will begin to drop out of service between 1997 and 2000 as the last of the onboard propellant is exhausted, but they will still exceed their planned seven-year lifetimes by a good measure. The NAVSOC Naval Satellite Operations Center at Point Mugu in California operates the system after taking over from the USAF June 1996.

During its operational lifetime, FltSatCom provided high-priority UHF communications between the US President and commanders in the field. The system began full operations in January 1981, providing global links for command and control of US nuclear forces. It handled between 900 ships, submarines and aircraft of the navy and related fleet ground stations, while also supplementing communication between 1,000 USAF aircraft and air-to-ground terminals for Strategic Command and other air force operations. The coverage extended into the high latitudes and only the polar regions remained without service.

Comprehensive coverage up to 70° latitude required four satellites and generally five were in orbit because the navy scheduled one in-orbit spare. By 1988, five of the seven placed in orbit up to that time still remained operational and Fl had passed its tenth anniversary in space. F8 completed the series on 25 September 1989. FltSatCom experienced a spate of launch failures between 1987 and 1988 when on 27 March 1987 an Atlas Centaur carrying FltSatCom F6 suffered a lightning strike and had to be destroyed by the range safety officer 51 seconds after lift-off from Canaveral.

FSC has 1,005 kg on-station mass and 13 to 19 m span with solar wings extended. It carries enough propellant to remain operational for 10 years. Two redundant networks of eight (2 roll, 2 pitch, 4 orthogonal) thrusters maintain the satellite position and attitude. FltSatCom's solar panels generate more than 1,400 W. The antennae consist of a deployable mesh dish 4.9 m in diameter with a 3.4 m long helical receive antenna. Each provides nine 25 kHz bandwidth channels and 12 × 5 kHz channels for small mobile users, plus a 25 kHz fleet broadcast channel and a 500 kHz National Command Authority channel.

Two onboard FltSatCom processors, one each for the air force and the navy, protect against uplink UHF/SHF signal jamming. The older satellites provide 75 bit/s UHF down with limited anti-jamming capability. AFSATCOM-II provides 150 bit/s at UHF/SHF. The system would handle nuclear weapon release authorisation codes, for example. Air force Space Command assumed AFSATCOM responsibility 1 June 1992 from Strategic Command at Offutt AFB, Nebraska. During 1990, the Single Channel Transponder on DSCS III entered AFSATCOM service.

FltSatComs provide global communications for command and control of US nuclear forces. They will remain operational into the late 1990s (TRW)

Specifications

FSC-1

Launched: 9 February 1978, took up a position at 100° west over Midway Island for coverage across the US to the Azores and Atlantic. It was moved in October 1992 over a 25-day period to provide more capacity over the Atlantic at 15° west near FSC-8, which is inclined at 12.5°. The navy manoeuvred FltSatCom-1 off-station in December 1996 over approximately 344-345°E and the satellite had its location restabilised over 71-72°E at the end of January 1997.

FSC-2

Launched: 4 May 1979, filled the slot at 71.5° east to cover the Indian Ocean from Africa to the Philippines. By 1992, it was clearly out of service, and drifting rapidly.

FSC-3

Launched: 17 January 1980 into 23° west, covered the Atlantic and Mediterranean; it has been drifting slowly since 1991.

FSC-4

Launched: 30 October 1980 to 172° east over the Pacific, provided coverage from East Asia to the US west coast. It remained there in an inclination of 10.6°.

FSC-5

Launched: 6 August 1981, had been scheduled for 93° west near FSC-1, the busiest region, to act as an instantly available spare, but was damaged by implosion of the Centaur's payload fairing during launch.

FSC-6

Launched: 27 March 1987 and disappeared when lightning struck its Atlas Centaur. The satellite was reported to be worth US$83 million, and the launcher US$78 million.

FSC-7

Launched: 4 December 1986 by Atlas Centaur from Canaveral after eight delays from the original 22 May date. The more advanced FSC-7, equipped with the jam-resistant EHF, actually preceded FSC-6 when the latter needed to have suspect components replaced. It was positioned at 105° west and later moved to 100° west January 1996, in an inclined orbit of 2.55'.

FSC-8

Launch scheduled: 24 July 1987, got bumped on 13 July when its Centaur upper stage, the last of its type, was irreparably punctured by a pad work platform. The mission departed on 25 September 1989 instead. By January 1996, it held the 23° west slot, inclined at 2.0°.

IDSCS series

Background

Lack of launcher power, together with political and economic arguments, delayed the first deployment of a robust communication satellite architecture for the US military until 16 June 1966, when a Titan 3C deployed eight satellites, including the first seven of the Initial Defense Satellite Communications System (IDSCS). Built by Ford Aerospace, these 45 kg, 26-sided 86 cm diameter polygons served as signal repeaters. When the Titan released its payload, the individual satellites dispensed themselves over six hours at slightly different orbital velocities to provide global coverage at points just below GEO altitude, at 33,915 km. Because they were not perfectly geosynchronous, the satellites drifted about 30 minutes relative to Earth daily, but stayed in view of an equatorial station for four and a half days so that even when one malfunctioned another drifted over the horizon to take up the traffic. The satellites maintained attitude control by spin-stabilisation. Over the years the network of IDSCS satellites slowly grew to total 26 satellites on 13 June 1968.

The system provided five high-quality voice, 1,550 Teletype or 11 tactical voice channels using 20 MHz bandwidth in the 7.266-7.286 GHz X-band. They linked users 16,000 km apart and, from 1967, provided the military with direct communications between South Vietnam and Hawaii, and Hawaii to Washington. The military intelligence services in Vietnam transmitted high quality reconnaissance photographs. Designed to last 18 months, five were still operational after eight years and two after 10 years. New studies of cost-effective communication satellites indicate IDSCS-type satellites may appear again since it has been estimated that 200 could be launched at the same cost as a single Milstar.

Leasat/Syncom 4 series

Current status

Three Hughes/US Navy Leasats continued operations. Some of these navy vehicles are being required to double their seven-year design lifetimes but attitude control/station-keeping propellant depletion will soon take them out of service. Controllers are permitting them to drift 10° beyond the equatorial arc in order to conserve propellant.

Background

Also known as Syncom IV, Leasats took full advantage of the Shuttle's wide payload bay, the first satellites to require the full capacity of the Shuttle. A Leasat launch uses the unique 'Frisbee' or rollout method. The satellite has a built-in upper stage that remains in place on station. Each satellite sits 6.1 m high with deployed antennas and has a 4.2 m diameter. Two large helical UHF antennas provide receive and transmit signals at 240 to 400 MHz. The principal Fleet Broadcast function includes an SHF uplink and SHF/UHF downlinks.

Owned and operated by Hughes Communications Inc, the system has five HS-381 satellites (which included one ground spare) which DoD leases for worldwide UHF links between ships, aircraft and fixed facilities. DoD and the navy specified that US$83.75 million would be paid annually for each satellite once it was operational in its GEO position – but not before. This clause served the US Navy well when Leasat 3 became stranded in the wrong orbit for eight months and Leasat 4 failed completely after attaining GEO.

Leasat's original design kept it unpowered in the payload bay. Unlike other Shuttle-launched satellites, astronauts could not perform a final electrical checkout before release. As they left the bay, an arming lever automatically initiated a timing sequence which first raised the 2.4 m omni antenna after 80 seconds, then at 6.5 minutes commanded 22 N thrusters to increase the spin rate to 2 to 33 rpm. At 45 minutes the sequencer fired the 131 kN Minuteman 3 stage 3 solid motor for 64 seconds to raise apogee to 15,450 km. Three additional firings by two 448 N liquid thrusters then took them into GEO. After the failure of Leasat 3 however, NASA required a redesign of this automatic arming feature.

Leasats occupied four operational slots at 177° west, 105° west, 15° west and 75° east. The ground segment operates from Hughes Operations Control Center in Los Angeles working through 10 m antennas in Italy (NATO base at Lago Patira, near Naples), California (Stockton), Hawaii (Wahiawa), Guam (Finegayan) and Norfolk, Virginia. Guam and Norfolk also have transportable stations originally used during Leasat deployment but now held as back-ups. Although controlled from Hughes in Los Angeles, each station can assume control of the whole fleet if required. Operations are co-ordinated via dedicated leased terrestrial lines to the US Naval Space Command Operations Center in Dahlgren, Virginia.

Leasats supplement the FltSatCom network

TRW technicians integrate the LDR payload on to DFS 2 in Lockheed Martin's facilities in Sunnyvale. Prominent is the 60 GHz inter-satellite antenna at top (TRW) ***1996***

Leasat Launch History

Satellite	Launch Date	Vehicle	Position	Comments
Leasat 1	8 November 1984	STS-51A	105° west /15° west	In October 1985, the only fully operational Leasat. Moved to 33,093 × 39,532 km early 1993.
Leasat 2	30 August 1984	STS-41D	177° west	Operational until September 1985, when its wideband channel began to malfunction
Leasat 3	12 April 1985	STS-51D	107° west	Automatic timing sequence failed to activate after deployment. Hughes claimed US$85 million insurance and paid NASA US$8.5 million for a repair operation. NASA left the satellite in low orbit until 27 October to warm up the solid motor.
Leasat 4	29 August 1985	STS-51I	177° east	The primary communications system failed eight days after deployment. Hughes claimed the US$85 million insurance.
Leasat 5	9 January 1990	STS-32	75° east	

LES/Tacsat series

Current status

Powered by RTGS, both satellites LES 8 and LES 9 remain usable. LES 9 provided links with a polar exploration team in 1995 while LES 9's AFSATCOM transponder carried Gulf War communications.

Background

Tactical communication satellites require much higher power levels than conventional satellites because they must work through small ground terminals. Lincoln Laboratories acted as the prime contractor for the US military to study methods to bring down the costs of these communication satellites.

LES 5 (Lincoln Experiment Satellite) orbited 1 July 1967 with a six-satellite payload, which included IDSCS 16/17/18. It employed a 225-400 MHz military UHF band. Two days after manoeuvring into a 33,360 km near-synchronous orbit, the satellite conducted the first communications between US aircraft, a US Navy submarine and surface vessel and army ground units.

LES 6 continued these investigations after September 1968. With the success of the two LES satellites the US DoD launched the follow-on Tacsat 1 (LES 7) on 9 February 1969. A 7.6 m tall, 2.3 m diameter satellite weighing 726 kg, it maintained its attitude with spin stabilisation on a cylinder bearing a despun platform. It targeted 30 cm diameter UHF/X-band land-based antennas using about 40 coded UHF voice channels for mobile antennas; and in X-band either the same or up to 700 teletype channels which it transmitted to 1 m diameter dishes.

LES 8 and 9 launched together on 15 March 1976, took up positions in 25° GEO-altitude orbits to conduct inter-satellite K-band links at ranges up to thousands of km, plus UHF downlinks. LES 8/9 appear to be stationed over 103° and 105° west respectively now inclined at 15.5°.

MACSAT series

Background

MACSAT, the Multiple Access Communications Satellites, M1 and M2, built by Defence Systems Inc, demonstrated tactical UHF data store/forward techniques. NAVSOC operated the satellites from its operations centre at Naval Satellite Operations Center Pt Mugu, California until the satellites failed in 1994.

A single Scout carried the two 68 kg satellites into space on 9 May 1990 to a 613 × 769 km, 89.90° inclination and a 614 × 767 km, 89.9° inclination orbit respectively. DARPA provided the initial seed money, to build the satellites, through its 'lightsat programme'. Each satellite has a diameter of 61 cm, and sits 35.6 cm high. MACSAT incorporated two 80C86 processors, each with 1.2 Mbyte RAM providing 8.64 kbyte mailboxes in eight, 192 kbyte increments. A 64 kbyte (32 page) document could be relayed at 2.4 kbit/s from a given footprint during one pass. 54 blocks of solar cells generated 14 W orbital average power to supply the one 10 W and one 60 W transmitters. The design called for the MACSAT to maintain stabilisation by deploying two gravity gradient booms. Radar images indicated on boom remained deployed.

Originally designed for demonstration purposes only, the DoD pressed M2 into service by US Persian Gulf forces from 20 August 1990 until 5 April 1991. MACSAT also provided links between US Navy personnel in Antarctica, New Zealand and California. RADM John Weaver, Commander of USN Space & Naval Warfare Systems Command noted, during 1990, that USN plans included several (described in 1991 as a network of six) polar UHF 'Arcticsat' relays based on MACSAT. Apart from improving high-latitude communications, they would be available for rapid launch and replacement in times of need. Arcticsat failed to acquire the US$37 million requested for FY92 and was cancelled from FY93.

The NAVSOC Naval Satellite Operations Center at Point Mugu in California operated them until their failures in 1994.

Milstar series

Background

The troubled Milstar programme, providing secure tactical links via hardened satellites, is now operational. The first of six authorised satellites left for orbit on 7 February 1994 aboard the first Titan 4 Centaur. At first DFS 1 (Development Flight Satellite) took up a position over the 90° west test slot until the air force moved it about September 1995 to 120° west . DFS2 appeared 5 November 1995 for testing at 90° west, before the air force put it in the planned operational slot at 4° east, inclined at 4°. The air force continues to plan to launch 3 in 1999. The DoD has not concluded plans for the final constellation.

The air force initiated the Milstar programme in 1983 as an all-service, jam-resistant and survivable system to become operational in the second half of that decade. The 4.67-tonne satellites offer tactical communications via manpacks that access the 44.5 GHz uplink and 20.7 GHz downlink frequencies. The EHF band ensures communications continue during nuclear conflicts and for the first time provides interservice links. Frequency-hopping techniques eliminate jamming to some degree and EHF inter-satellite transmissions permit users to communicate around the globe without routeing via intermediate ground stations.

Each satellite has a 0.5 Mbit/s total capacity, which is limited because of the nuclear countermeasures, spread over 192 channels at 752,400 bit/s, with onboard processing and five agile antennas covering 185 service areas. The first two satellites, Milstar 1A, carry only this LDR Low Data Rate payload. Tactical forces will carry 1,050 terminals, strategic forces have 279 and the USAF EC-135 aircraft has 138 installed terminals. The services had built some 200 army, USN and air force terminals by early 1996.

DoD notified the International Telecommunications Union it would use 148° west, 120° west, 90° west, 68° west, 16° west and 9° west. In the eastern hemisphere it would use 14° east, 19° east, 30° east, 55° east, 90° east, 133° east, 150° east, 152° east, 177.5° east.

Congress ordered a restructuring of DoD's Military Strategic &Tactical Relay programme in 1990 to reduce cost and refocus the satellite communication needs from strategic applications during and after nuclear wars to tactical uses. The 4th Space Operations Squadron (4th SOPS) of Space Command's 50th Space Wing, operates Milstar after commissioning by 3rd SOPS. The Congress announced that the total cost between 1991 and 2011 for development, launch, 20 years of operations and user terminals is projected at $17 billion. Each Milstar 2 unit costs $800 million. The helical antennas provide US Navy UHF links (larger for receive, smaller transmit) and the two dishes are the source of the EHF/SHF spot beams.

At one time, provisions on Milstar allowed it to relay encrypted KH11 imagery, but that apparently has been

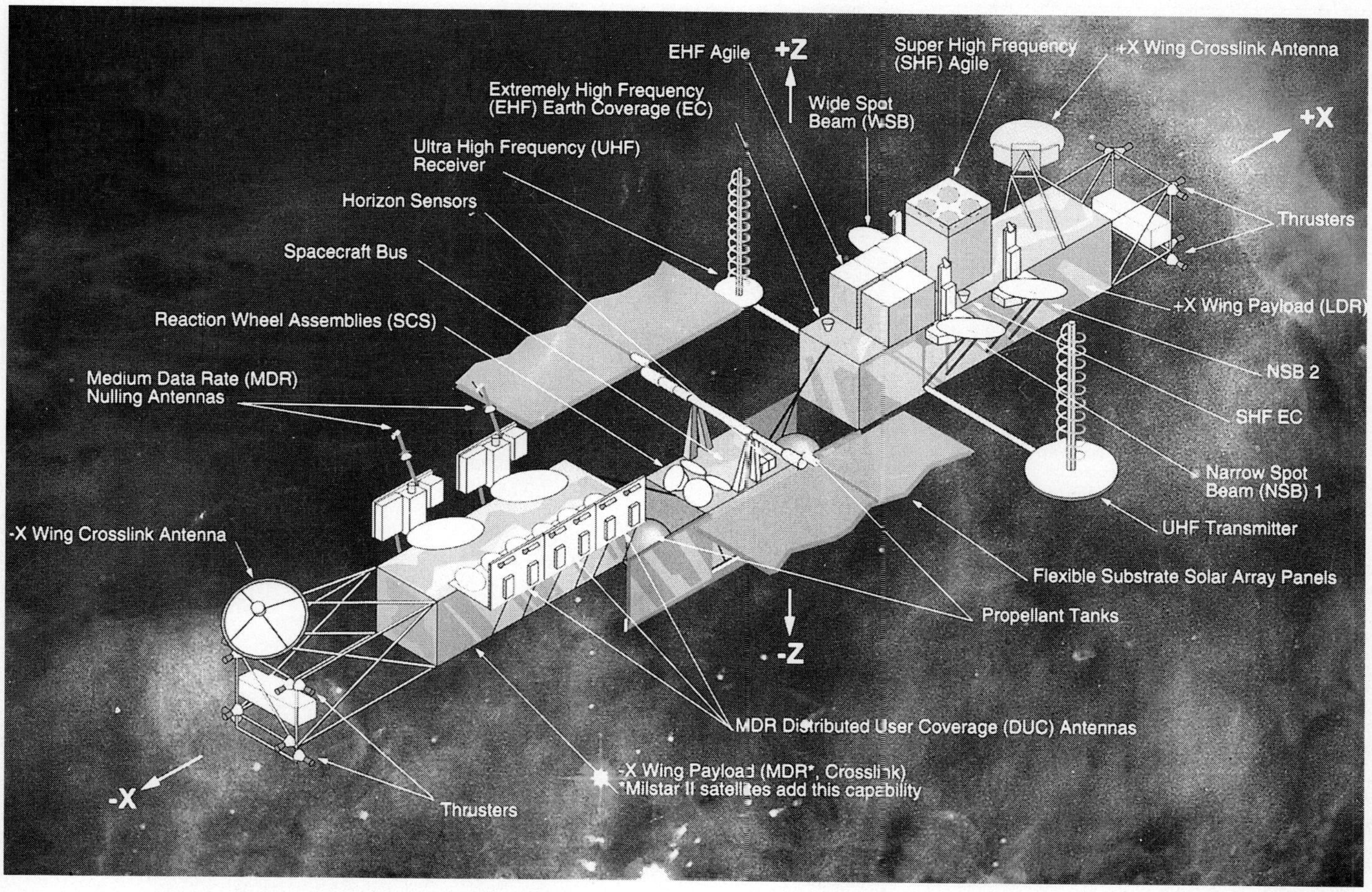

A drawing showing the major features of the Milstar satellite (Lockheed Martin)

1998/0022566

dropped. Milstar's space testing phase began with the launch of FltSatCom 7 in December 1986 which carried its highly jam-resistant 44 GHz EHF transmitter. The US headquarters in Riyadh, Saudi Arabia used a prototype Milstar terminal during 1991's Desert Storm to provide a demonstration link with the Pentagon.

Milstar is the first system to allow all three armed services to communicate with each other on the same network. The extensive onboard signal processing and resource management allows it to automatically set up, maintain and reconfigure voice/data networks in real time. It is the first payload to operate at EHF using unique frequency hopping to avoid interception and interference. Crosslinking between satellites eliminates the need for costly, politically sensitive and vulnerable ground stations.

In his testimony in 1988, Air Force Secretary Edward Aldridge admitted that technical problems had raised the cost for each satellite and its Titan Centaur launch from Cape Canaveral to US$1 billion, but in 1992, other air force sources quoted a price of US$515 million per satellite. As it debated a funding bill in 1990, the Senate Armed Services committee estimated that the entire system would cost US$35-40 billion by the time it was complete. The Congress gave the air force US$345 million in FY90, with US$956 million in FY91, US$1.6 billion FY92, US$1.14 billion FY93, US$918 million FY94 and US$647 million FY95. Milstar's redirection reduced the number of satellites from 10 to four and cut the numbers of ground stations and user terminals.

Specifications

EHF/SHF Services: EHF 44.5 GHz (2 GHz bandwidth) up/ SHF 20.7 GHz (1 GHz bandwidth) down
Data rates: LDR 75-2,400 bit/s; MDR 4.8-1,544 kbit/s
Channels: LDR 192 (100 at 2,400 bit/s); MDR 32
Antenna coverage: LDR 1 up/down (Earth), 5 up/1 down agile, 2 up/down narrow spot, 1 up/down wide spot; MDR 2 up nulling spots (allowing operation in presence of in-beam jammers), 2 coincidental down spots, 6 up/down spots (distributed user coverage)
UHF Services: 4 transmit/rx 75 bit/s AFSATCOM IIR channels (Earth coverage) and 1 transmit-only 1.2 kbit/s Fleet Broadcast channel (Earth coverage)
Crosslinks: (60 GHz) 2 per satellite (1 each direction), compatible with LDR/MDR

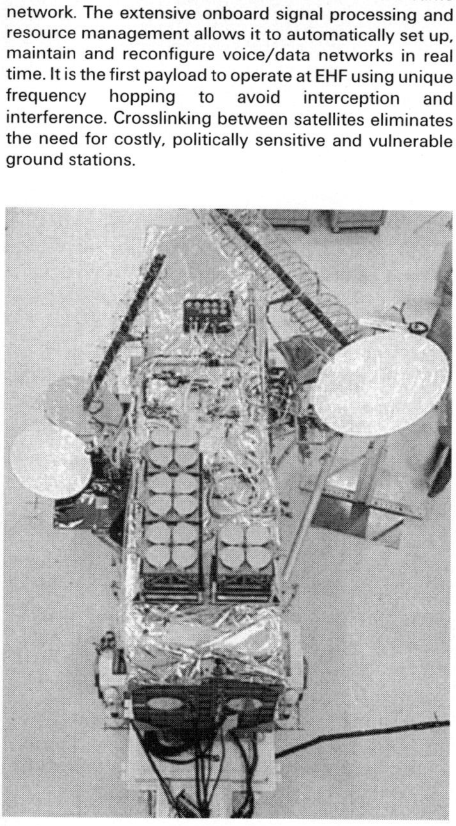

Milstar 1 LDR payload final test/integration at TRW during 1989. The helical antennas provide US Navy UHF links (larger for receive, smaller transmit) and the two dishes provide EHF/SHF spot beams. The smaller arrays provide agile EHF links; missing at centre is the SHF agile unit. Milstar's other payload wing, originally planned for the Science Prime package to relay KH11 imagery, will carry the MDR package from number 3

The USAF established the Milstar programme in 1983 to provide military communication before and after a massive nuclear exchange. Each satellite carries 13 assorted antennas. The US Air Force paid a reported US$1.5 billion for the first Milstar. A total constellation requires at least four satellites (TRW)

1998/0022565

Polar Bear satellite

Background

The Polar Beacon Experiment and Auroral Research satellite was a joint USAF/defence Nuclear Agency mission. The air force launched the 122 kg craft by Scout on 13 November 1986 from Vandenberg into a 962 × 1,019 km at 89.5° inclination. It was also known as STP P87-1, and the air force intended to use it to study communications interference caused by solar flares and increased auroral activity. DAN contributed to the effort in order to estimate the effects of nuclear bursts on communication signals. The USAF recycled an existing Transit from the navy programme and saved US$2 million on the cost of the US$13 million satellite. The satellite had originally been displayed from 1976 to 1984 at the National Air and Space Museum, Washington DC. Johns Hopkins University discovered that only one solar cell out of thousands was marginal and required replacement. The Auroral Imaging Remote Sensor, sponsored by the Air Force Geophysics Laboratory, imaged the aurora borealis; the Beacon Experiment, sponsored by the DNA, monitored ionosphere propagation over the poles. The satellite lost most of its battery capacity in early October 1992, leaving insufficient power for the remaining SIDEX Signal Identification Experiment. The air force deactivated Polar Bear 2 November 1992.

UHF Follow-On series

Current status

FO-9: Launched 20 October 1998 by Atlas-2A as UFO 9 to geosynchronous location at 186° E.

FO-8: The navy had an 80-minute launch window on 16 March 1998 to put the UFO FO-8 satellite in orbit. Hughes contracted with Lockheed Martin for launch on board an Atlas II rocket. The satellite is a commercial-based HS 601 spacecraft and is the first in the fleet to carry Global Broadcast Service (GBS). GBS uses high-power satellite transponders to provide high-speed wideband, simplex broadcast signals. The GBS payload utilises four 130 W, 24 Mbit/s military Ka-band transponders operating in the 30/20 GHz frequency. When the navy's third GBS satellite, UHF FO, is launched in 1999, the DoD will have near-global GBS coverage. UBF FS, F9, and F10 will carry the GBS payload, in addition to the existing IW and EHF payloads. UFO FO-8 will be stationed at 172° east longitude over the Pacific Ocean. The UBF Follow-On satellites have replaced the Fleet Satellite Communications (FltSatCom) and the Hughes-built Leasat spacecraft. In March 1996, the navy ordered the special GBS payloads for FS, F9, and F10, bringing the total contract value to US$1.85 billion.

The US Navy has contracted with Hughes Space and Communications Company to add GBS capabilities to the UHF Follow-On (UHF F/0) satellites 8, 9, and 10, the last three in a series it is building under a total US$1.9 billion contract. Hughes derives the space segment GBS package from its commercial system. The GBS payload will replace the current SHF X-band payload with four 130 W, 24 Mbit/s military Ka-band (30/20 GHz) transponders. This modification results in a 96 Mbit/s capability per satellite.

Data uplinks to the satellite pass through a fixed receive antenna from a broadcast management centre and a steerable receive antenna from theatre injection points. Each of the four transponders can be accessed through either of the receive paths, configured by ground command. Data is transmitted on three steerable spot beam antennas per spacecraft into 22 in receive antennas. Each of two spot beams covers an area of 500 n miles in diameter at the sub-satellite point. The third downlink spot beam covers an area of 2,000 n miles in diameter at the sub-satellite point,

To integrate the GBS payload requires several spacecraft modifications. The payload's high-power amplifiers are integrated on to a larger south radiator panel with heat pipes added to accommodate the increased thermal dissipation. The GBS fixed receive antenna together with the forward SGLS omni antenna will be mounted on the structure supporting the current SHF antennas. The GBS steerable receive antenna will be mounted on a deployable boom. A new pallet structure will be added to integrate the three GBS transmit spot beam antennas. The power subsystem requires modification to provide additional capacity to support the high-power GBS payload.

More efficient Gallium Arsenide (Ga As) solar cells will replace silicon-type cells on four solar wing panels. The battery grows in capacity to support eclipse operations. All the spacecraft bus changes can be made with commercially proven space hardware.

The US Navy has built the UHF Follow-On, shown in this artist's conception, to replace and upgrade its UHF (Ultra High Frequency) satellite communications network. Navy communications needs have prompted it to order 10 satellites in the series through January 1994 (Hughes Space and Communications Division) ***1998***/0022569

Spacecraft	Blk	Weight (lbs)*	Power (w)**	L (ft)	W (ft)	Payload
F1-F3	I	2,600	2,500	60	23	UHF/SHF
F4-F7	II	3,000	2,800	60	23	UHF/SHF/EHF
F8-F10	III	3,400	3,800	86	22	UHF/EHF/GBS

*Begining of life **End of life

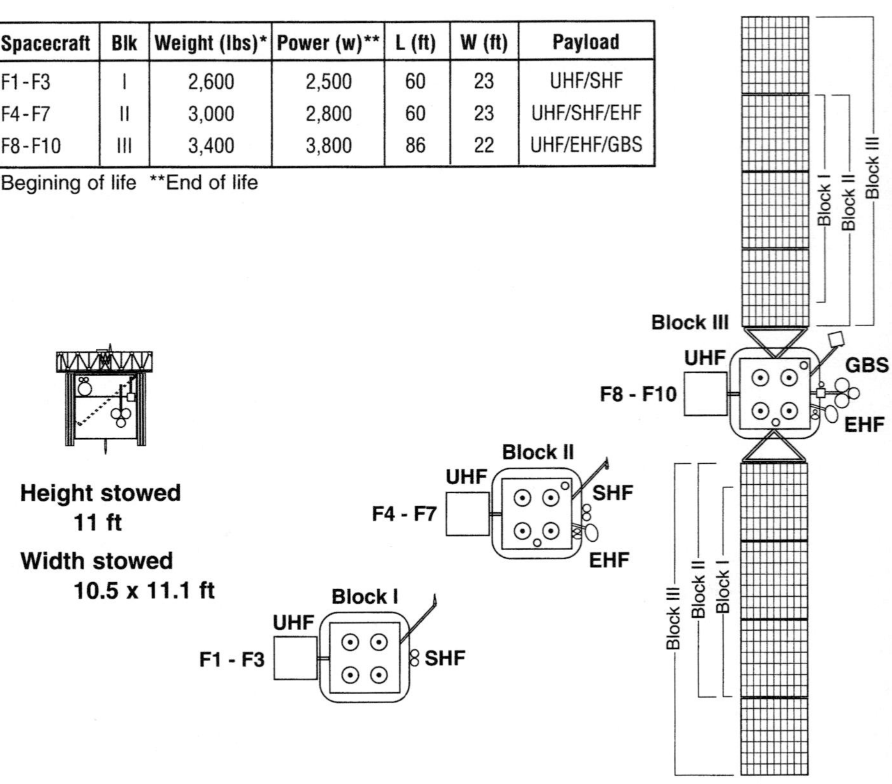

Major features of the UHF Follow-On satellite are shown in this diagram (Hughes Space and Communications Division) ***1998***/0022570

Background

UFO offers double the communications capacity over FltSatCom for about the same amount of weight in orbit. An EHF package provides 11 channels on some satellites and is distributed between a 5° spot beam and global coverage.

The first of FltSatCom's UFO follow-ons, with 14-year lifetimes, departed March 1993, but its Atlas 1 launcher left it in an unusable low orbit. UFO 2 departed successfully in September 1993 and the USN formally accepted command authority on 2 December. The satellite occupies a position over the Indian Ocean, collocated with Leasat 5. Two more UFOs departed successfully by March 1995 and a further six are scheduled through 1998.

UFO Satellites numbers 1 through 3 cost US$138 million and each of the successors will cost US$198 million. Hughes builds the initial 1 tonne satellites, using its HS-601 three-axis design. They have 21 narrowband (5 kHz), 17 relay (25 kHz) and one fleet broadcast (25 kHz) channel working through 11 SSPAs using TDMA. UFO features enhanced anti-jamming and EMP hardening. It can function for up to 30 days without ground contact of any kind. The 3rd Space Operations Squadron of the 50th Space Wing has command of the satellites for the Naval Space Command.

UFO Satellite Launch History

Satellite	Launch Date	Position	Comments
UFO 1	March 1993		In an unusable low orbit because of an Atlas 1 failure. Hughes was required to pay the navy a US$137 million penalty.
UFO 2	3 September 1993	174° west/ 71.5° east	Navy and Hughes originally tested it at 174° west. Hughes formally handed it over 2 December 1993.
UFO 3	24 June 1994	15° west	
UFO 4	29 January 1995	161° west / 177° west	
UFO 5	31 May 1995	17.1° east / 71.5° east	
UFO 6	22 October 1995	106° west	Inclination of 4 to 5°.
UFO 7	25 July 1996	337° east	
UFO 8	16 March 1998	188° east	
UFO 9	20 October 1998	186° east	

The US government made the contract require a first launch before 29 July 1992 and the military would levy fines against Hughes up to US$80,000 daily for late launch, up to a maximum of US$16 million. The USN had a problem with a computer it planned to supply to the project and the first departed March 1993 but was left in an unusable low orbit by its Atlas 1. Hughes still had to pay the navy a US$137 million penalty.

The initial 1 tonne satellite, based on Hughes' HS-601 three-axis design, previously selected only for Australia's Aussat 2, replaces ageing Leasat/FltSatCom families with 21 narrow band (5 kHz), 17 relay (25 kHz) and one fleet broadcast (25 kHz) channel working through 11 SSPAs using TDMA.

DATA RELAY

RUSSIAN FEDERATION

Geizer/Potok series

Current status

As of 1999, a single satellite occupies each of the formerly occupied Potok slots at geosynchronous, Kosmos 2319 over 80° east and Kosmos 2291 over 346.5° east.

Background

Izvestia noted in a 5 February 1992 story that the Russian armed forces employ about 10 satellites in geosynchronous orbits, but these satellites utilise only 30 per cent of their available 1,500 telephone and 1,000 telegraph channels. The remainder lie idle, except when the Russian military conducts exercises. Then the network usage swells to 60 per cent for 60 days annually. Izvestia suggested the military turn this capacity over to more productive, non-military applications.

Within this network the Russian military has registered three geosynchronous orbit locations for the Potok military data relay system. Potok uses NPO PM's Geizer satellites launched within the Kosmos programme. Potok 1 resides at 346.5° east, Potok 2 at 80° east, and Potok 3 at 192° east, all potential useful orbits for civilian communication systems or data relay networks. The Russians have never occupied the third location Each Potok employs C-band transponders at frequencies up and down of 4.40-4.68 and 3.95-4.00 GHz. NPO ELAS adopted Potok's phased-array antenna system for the commercial Kupon satellite, but the Russians have thus far refused to publish an illustration of the satellite.

Intelligence sources report that the Geizer satellites permit communications with the fifth-generation photo-reconnaissance satellites using a ground-space-space link to provide near real-time imagery. Launch dates correspond to this theory since the first Geizer satellite went into space in May 1982 and the first fifth-generation launch came in December of that year. Amateur satellite observers have been puzzled for years by the lack of transmissions from fifth– generation photo-reconnaissance payloads and an uplinked transmission to geosynchronous would explain the absences of direct links to a ground station. In this vein, speculation also holds that the Geizer satellites may perform the same function for continuous communications with the EORSATs.

The relay network has no fixed boundary in space. During their operational lifetime Kosmos 1888, Kosmos 1961 and Kosmos 2291 have each relocated once and frequently more than one satellite has been located over a particular longitude. For example, from March 1984 to November 1987, both Kosmos 1366 and Kosmos 1540 operated at 80° east. When Kosmos 1366 died, Kosmos 1888 replaced it and operated with Kosmos 1540 from October 1987 to December 1988 when Kosmos 1540 died. In July 1990, Kosmos 2085 replaced Kosmos 1888 and, from April 1992 to mid-1993, Kosmos 1961 operated with Kosmos 2085 until the former satellite died. During September 1994 and continuing until February 1995, Kosmos 2085 operated in tandem with Kosmos 2291, until Kosmos 2085 died. Finally in September 1995, Kosmos 2319 replaced Kosmos 2291.

A similar pattern occurred at the other Potok slot, 346.5° east. From August 1988 to August 1989, Kosmos 1738 operated with Kosmos 1961 until the older satellite died. From September 1990 to March 1992, Kosmos 1888 and Kosmos 1961 operated together and during December 1991 to March 1992, Kosmos 2172 joined the two older satellites, so that for a few months there were three satellites operating at this position. Kosmos 1888 and Kosmos 2172 continued to operate together, after the departure of Kosmos 1961 (which manoeuvred to 80° east), until December 1994 when Kosmos 1888 died. And from November 1995 to March 1996, Kosmos 2172 and Kosmos 2291 operated together until Kosmos 2172 died.

UNITED STATES OF AMERICA

Glomr series

Background

Defence System Incorporated built the 68 kg Global Low Orbiting Message Relay satellite, a 62-sided polyhedron, designed for the defence Advanced Research Projects Agency. The satellite launched itself on a spring ejection from a Getaway Special canister in the payload bay of STS-61A on 30 October 1985. It was the second launch attempt for the satellite after a failure to deploy it on STS-51B in April 1985 because of a battery problem. Glomr had a one-year life and DARPA wanted to use it to demonstrate techniques for locating, commanding and relaying data from oceanographic and other sensors such as submarine detectors dispersed on Arctic pack ice. A follow-up satellite was launched aboard the first Pegasus vehicle in April 1990.

SDS series

Background

The Satellite Data System craft operate in highly elliptical orbits similar to those of the Russian Molniya, relaying imagery from KH 11 and other reconnaissance satellites, to ground stations in the United States. They also provide AFSATCOM communications links with US nuclear and other forces in the polar regions that have spotty coverage from GEO satellites. The indistinguishable Jumpseat versions act as communications intelligence satellites. DoD refuses to release any details about the programme or its associations with the KH-11 reconnaissance satellites.

Trade reports in 1994 claimed that the 'Heritage' sensor being proposed for DSP's follow-on has long flown aboard the Jumpseat and SDS to augment the DSP system. Each satellite weighs about 700 kg. Hughes builds them at a cost to the DoD of about $100 million each. Until the Shuttle became available the satellites had to be launched by Titan 3B Agena D from Vandenberg into 63.4° inclination orbits. In association with each new reconnaissance satellite the probable launches are SDS-1 2 June 1976, SDS-2 6 August 1976, SDS-3 5 August 1978, SDS-4 24 April 1981, SDS-5 31 July 1983, SDS-6 28 August 1984, SDS-7 11 February 1987 and SDS-8 2 July 1996. Sources within DoD claim the programme ended in 1984 and the last launch occurred in 1984. A review of the Shuttle manifest and other orbital details indicates that Shuttle STS-28 in August 1989 and STS-53 in December 1992 both released new versions of the Jumpseat. The first two Advanced KH 11s (KH 12 in some sources), appearing February 1990 and November 1992, and the launches of the advanced Jumpseat coincide with these departures.

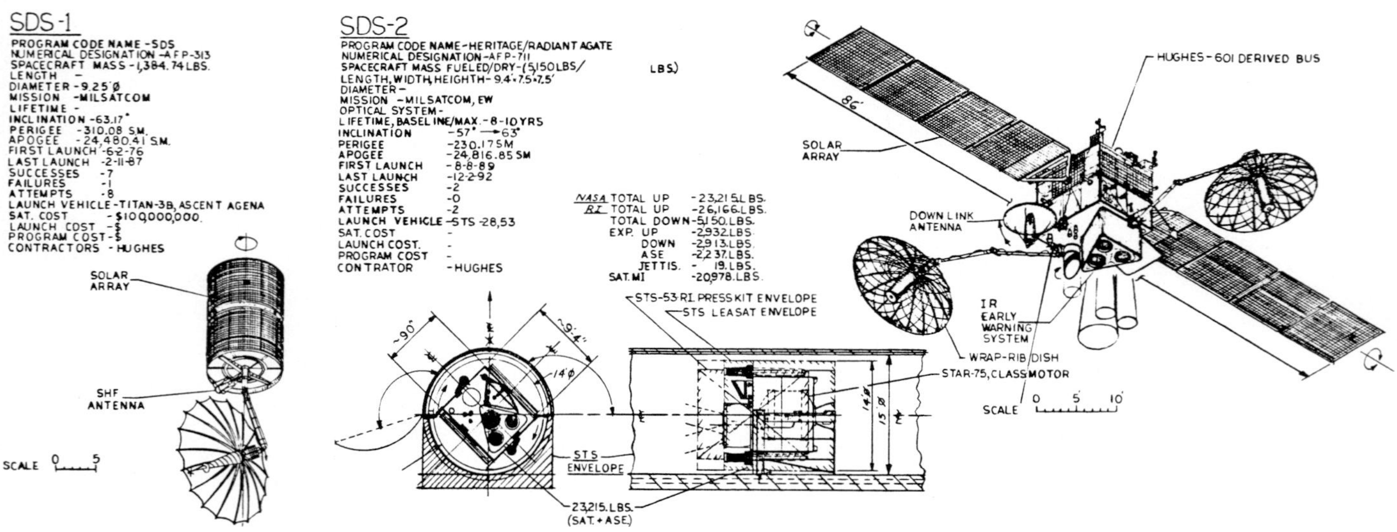

SDS conceptual studies using information from open sources. The current version, suggested as being launched by STS-28/53, is at right. Trade reports in 1994 claimed that the 'Heritage' sensor being proposed for DSP's follow-on has long flown on Jumpseat/SDS to augment the DSP system. Dimensions are in feet/inches (C P Vick) ***1995***

EARLY WARNING

RUSSIAN FEDERATION

Oko (Eye) constellation

Background
The system of nine Oko satellites roughly corresponds to the US DSP system. The Soviets intended for the system to produce an early warning of a ballistic missile attack. It took 15 difficult years for the Soviets to place the system and get it fully operational in 1987. US military observers claim that 'it is capable of providing about 30 minutes warning of a US attack and of determining the general area from which it originated'. Okos probably also monitor nuclear tests, as do the US systems.

The Soviets began a four-year test programme for the first-generation system in 1972, and operationally deployed the system starting in 1976. An SL-6 Molniya, launched from Plesetsk, places each Oko in a highly elliptical (about 600 × 40,000 km) semi-synchronous orbit at 40° intervals. A full system requires nine satellites operating 160 minutes apart. With apogees over Western Europe and the Pacific and overlapping coverage, each views ICBM sites in the US for 5 to 6 h/revolution (12 hours daily) and transmits data to RFAS bases at the same time. Oko also monitor routine launches from the US and other countries.

When Kosmos 1849 reached orbit on 4 June 1987, the Soviets filled all their requested operational slots for the first time. The RFAS keeps one or two older satellites on a more westerly ground track, possibly for better viewing of the western US. Despite its age and the lessening of tension with the United States, Russia shows no sign of abandoning the system and, during the last days of the Soviet Union, Oko established a new orbital plane, in late 1990, by Kosmos 2105 (20 November 1990). Western analysts still have not determined the reason for this new plane. The Soviets also created full-time global coverage by launching four unmodified Okos in geosynchronous orbit at 335° east to provide Pacific Ocean coverage: Kosmos 775 (1975), Kosmos 1546 (1984), Kosmos 1629 (1985) and Kosmos 1894 (1987). The last one finally shut down in January 1992 when Prognoz took over.

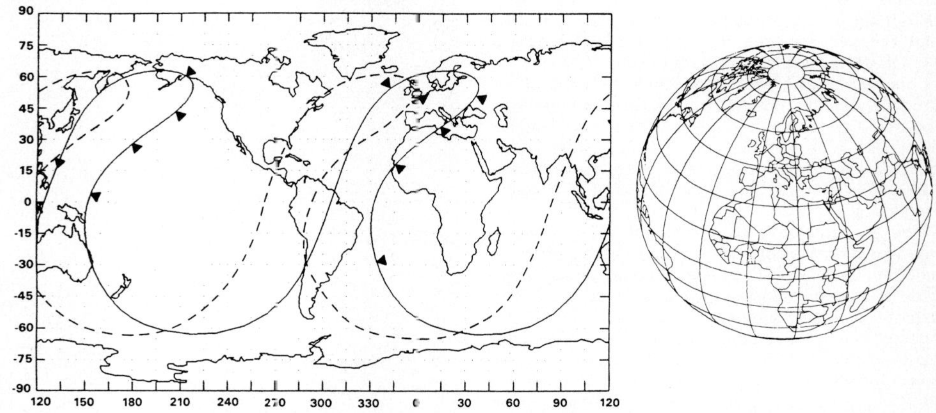

The Oko early warning satellites typically operate with apogees that provide surveillance of the US while remaining in contact with the RFAS/USSR. The globe illustrates the view from one apogee. The solid groundtrack depicts the daily path of the primary satellite, while one or two satellites are usually maintained in the more westerly dotted track (Teledyne Brown Engineering)

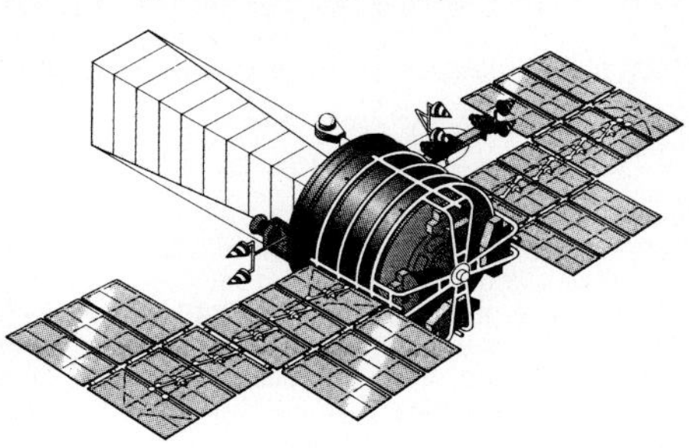

The original Oko early warning satellite (Europe and Asia in Space, 1991-92; USAF Phillips Lab) ***1995***

In the late 1990s, a decline in launches has caused the constellation to fall below its full population. In 1990, there were six launches to replenish the system, then none in 1991, four in 1992, three in 1993, one in 1994, one in 1995 and none in 1996. The Russians launched a new Oko satellite in April 1997 with the new Kosmos 2340. Kosmos 2222 apparently died in the second half of 1996 and a standard manoeuvre reduced the orbital period by 0.4 minute. Through April 1997, no further manoeuvres appeared, indicating that the satellite is no longer operational.

Interfax News Agency initially stated that the planned launch of a Kosmos satellite aboard a Molniya-M on 18 December 1996 had been delayed without giving a specific reason. After the Russians rescheduled the launch attempt for 10 January 1997, it was again postponed 'for technical reasons'. After a launch on 9 April 1997, the Russians named the satellite Kosmos 2340 once in orbit.

Russian drawings have depicted the Oko satellites, showing a cylindrical drum about 2 m diameter and 2 m long, plus two sets of solar panels and a telescopic housing for the instruments. They appear to weigh about 1,250 kg. Oko is the responsibility of TsNPO Kometa, with NPO Lavochkin responsible for the satellite bus.

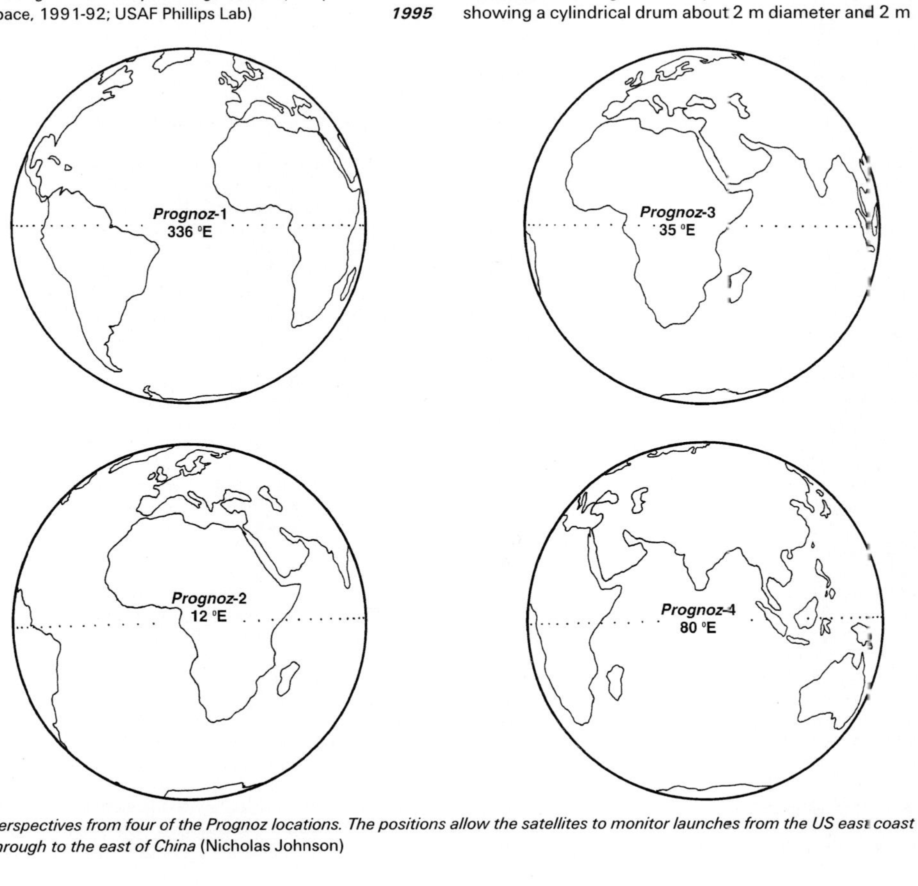

Perspectives from four of the Prognoz locations. The positions allow the satellites to monitor launches from the US east coast through to the east of China (Nicholas Johnson)

Prognoz (Forecast) series

Current status
Kosmos 2224 is the sole operating Prognoz satellite.

Background
As a follow-on plan to Oko, the Soviets requested seven GEO positions in 1981 and named the satellites Prognoz. The Soviets claimed the Prognoz would study the Earth's oceans and natural resources beginning in 1982. The Soviets requested slots at: 12° east, 35° east, 80° east, 130° east. 166° east, 159° west, and 24° west. The first Prognoz satellite took up its position 26 April 1988 named Kosmos 1940. An SL-12 Proton carried it into orbit, but the flight plan left it in GTO for several weeks prior to transfer.

For 14 years before this, the Soviets inserted all their GEO satellites close to 90° east and then shifted them into final operating positions. Kosmos 1940 instead remained in its low parking orbit for an additional 6 h and then transferred directly into 15° west before moving to 24° west. The Soviets moved it to 12° east during July/August 1988 to support Buran's maiden flight. It failed on 14 September 1988. In 1993, the Russians finally identified Prognoz as an early warning system.

Kosmos 2133 (14 February 1991) established the series as the second vehicle launched under the Prognoz programme. It occupied two slots first at 80° east and Western observers claimed it handled Potok communications at the time. Then it manoeuvred first to 35°, 12°, 336° and 80° east – each of the Prognoz locations. In August 1995, it drifted off its assigned longitude and appears to be no longer operational.

Prognoz 2 was joined at 336° east by Kosmos 2209, launched 10 September 1992 then Kosmos 2224 17 December 1992 at 12° east, and later moved to 336° east. It returned to 12° east in April 1994. Kosmos 2282 appeared 6 July 1994 and settled at 335° east; but began drifting in October 1995.

The satellite is made up of a 2 m diameter main instrument section, two large solar wings and a large

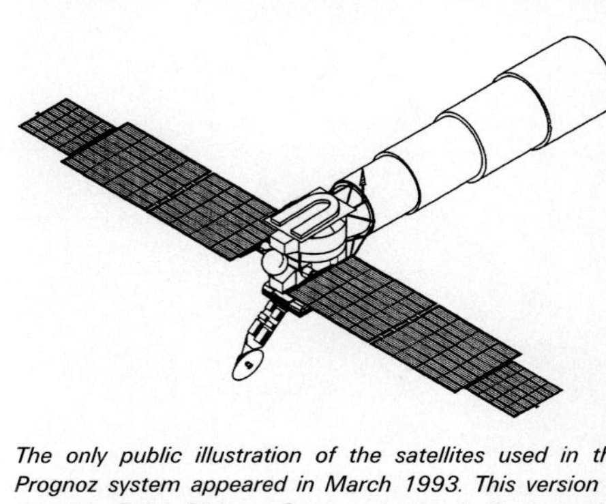

The only public illustration of the satellites used in the Prognoz system appeared in March 1993. This version is drawn by Ralph Gibbons. Compare it with the Spektrum-UV artwork in the Spektrum section ***1994***

telescope tube housing a beryllium mirror. The PbS linear CCD can apparently detect aircraft on afterburners. The same bus will be used for the Spektrum science mission in 1998; and the sole published illustration closely reflects the Spektrum-UV artwork the Russians have released. Proton GEO performance indicates a mass of about 2.5 tonnes. Oko and Prognoz are both the responsibility of TsNPO Kernels; NPO Lavochkin provides the bus.

The Russians hinted at the close association of the Prognoz and Oko programmes because, though they registered the Prognoz system in 1981, no Prognoz launches took place until 1988. The three geosynchronous Oko satellites launched during 1984-1987 all took up orbital slots registered at Prognoz longitudes.

The first Prognoz satellite apparently failed because, after one orbital relocation, during July of that year, the satellite was boosted off-station in September and never had its geosynchronous longitude restabilised. The satellite probably suffered some failure during the attempt to relocate it around the geosynchronous orbit band. Kosmos 2155 appears to have been another failed mission in terms of operating time since the Russians manoeuvred the satellite off-station nine months after launch.

The launches in 1992 turned out to be two of the most successful Prognoz satellites. Kosmos 2209 operated for just over four years, while Kosmos 2224 is still operational. The latest Prognoz launch, Kosmos 2282 in July 1994, only operated for 15 months, drifting off-station in October 1995.

India's latest IRS satellite operates with a special resolution of better than 20 m and has four spectral bands. It resembles the Landsat 4 in performance characteristics (Radarsat) ***1998***/0022563

UNITED STATES OF AMERICA

Defense Support Program (DSP) series

Current status

A total of 18 satellites has been launched and TRW is under contract to build a total of 23. DSP-18 was launched on 24 February 1997.

Background

The Defense Support Program (DSP) is the air force catch-all phrase invented in 1966 to hide the placement of geosynchronous satellite monitors of Soviet ballistic missile launches. The air force did not acknowledge the existence of the programme until the launch of DSP 16 aboard the Shuttle. It still refuses to reveal which DSP satellites remain operational. Other sources indicate that DSP 17 is operational at 37° west Atlantic, DSP 16 at 69° east, DSP 15 at 10° east, and DSP 14 at 152° west. DSP 13 is probably still operational in a limited capacity at the 110° east back-up position, to cover Russian ICBM/SLBM launch areas. DSP 16 probably replaced number 14. Another five satellites sit in storage at the manufacturer and the air force cancelled the manufacture of DSP 24 and 25 to help pay for the next generation. This next generation known as the SBIRS Space-Based Infra-Red System will eventually cost $22 billion, but its overlappping coverage will provide stereo views in order to show three dimensional track files for ballistic missiles thereby improving warnings of theatre launches. The more agile system will work in conjunction with the JTAGS vehicle to provide launch site and target data directly to tactical commanders and ballistic missile defence batteries.

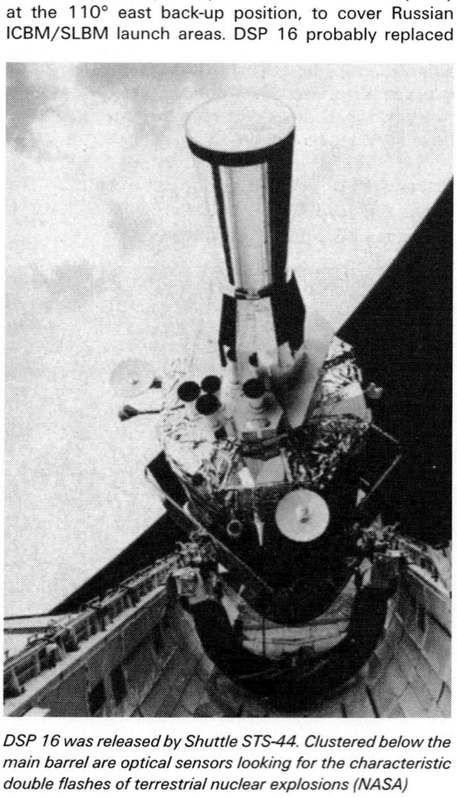

DSP 16 was released by Shuttle STS-44. Clustered below the main barrel are optical sensors looking for the characteristic double flashes of terrestrial nuclear explosions (NASA)

The Space Command's 50th Space Wing at Falcon AFB controls the DSP for its user, the 21st Space Wing at Peterson AFB, working through stations in Guam, Pine Gap and Nurrangar. Loral Federal Systems received a five-year, US$150 million contract July 1995 to provide software and engineering support for DSP ground stations. Observational data are received and processed at Buckley Air National Guard Base in Aurora, Colorado, Nurrangar and Kapuan (Germany). The United States has plans to close down the two foreign sites and consolidate operations, from 1999, at a new central processing station at Buckley. Pine Gap, 19 km southwest of Alice Springs in Australia, controls all military satellites when they are in the southern hemisphere. The 21st Space Wing's 5th Space Warning Squadron (jointly with the Australian 1st Joint Communications Squadron) acts as the control authority. When the United States publicly renewed the leases on the facilities in November 1988, the public knew the functions of the bases for the first time.

Nurrangar acts as the principal data and imaging centre and it became operational in early 1971 to take over the early warning satellite system from Pine Gap. The United States had to upgrade the facility in order to use the current Block 14 satellites, and it can now receive, process and re-encrypt data at 100 Mbit/s.

Some concern occurred after the loss of the Shuttle Challenger that the new generation of DSP would be stranded until a new launch vehicle could be found. As the older second-generation satellites began to go offline there seemed to be no interim solution to keep an early warning system in place for the United States. Instead, the USAF launched the last of the second-generation satellites in November 1987and the Shuttle became available again to handle the new Block 14 DSPs. The first Block 14 orbited aboard the new Titan 4 launch vehicle on 14 June 1989. The third Block 14, DSP 16, appeared aboard Shuttle mission STS-44 in November 1991. This was the first advance notice of a DSP departure, given by NASA in August 1990 on USAF clearance. These enhanced spacecraft incorporate more sensitive IR and Vela-type nuclear detectors for improved discrimination. They also employ two infra-red wavelengths to combat laser jamming.

Acquisition of DSPs continues apace. TRW received a US$743 million contract by USAF's Space and Missile Systems Center in mid-1987 to build a second batch of third-generation satellites, numbers 18-22, to follow on from 14-17. Then the USAF ordered numbers 23 to 25 in June 1993 under an eight-year US$724 million contract to TRW, plus US$501 million to Aerojet for the payload.

The DSPs monitored 166 exchanges of Soviet SS-1 Scud medium-range missiles between Iraq and Iran during the missile exchange known as the 'War of the Cities'. During 1991's allied Gulf War against Iraq, the last two second-generation DSPs – the first to carry the 6,000 element detectors – moved their orbits from their ordinary positions to give stereo coverage. The 69° east slot provided an ideal position to view the launch field of the H2 airfield in Iraq. Reports from field commanders indicate the satellites provided launch information within two minutes of the launch and therefore could give the Patriot batteries information for targeting well before the end of the Scud's seven-minute flight. Target field identification was said to be less than 6 km of uncertainty.

Prompted by the war, two programmes have developed ground stations for tactical commanders to receive DSP data directly. A prototype of the army/Aerojet Tactical Surveillance Demonstration station began operating in Germany in 1993 and was transferred to South Korea in 1994 as a precaution against North Korean missiles. The station merges data from two or more satellites. The army plans to acquire six operational JTAGS Joint Tactical Ground Stations. The USAF built its own version despite the 'joint' claim and called it Talon Shield. It became operational in October 1994. The 21st Space Wing's 11th Space Warning Squadron at Falcon AFB provides theatre commanders with continuous tactical surveillance and alert information.

Although DSP provides launch location, some partial information about missile type and approximate azimuth, contrary to popular reports it cannot pinpoint

The second-generation DSP was first launched and entered orbit in 1975; launches continued until 1987 (TRW)

First of the new DSP early warning satellites was launched by Titan 4 in June 1989 (TRW)

accurate range information. After a launch signal is passed to the ground station a BMEWS or ground-based radars must calculate the range with supplementary information. DSPs have been used formally since 1993 to log asteroid atmospheric impacts. There are 10-30 annually, equivalent to 1 kT explosions.

The second-generation DSP busses extend 2.78 m to house the optics on a 2.91 m high cylindrical bus. The bus provides power from its solar cells on its outer surface supplemented by four small panels, generating a total of 680 W. The air force hardened the satellites against laser jamming and nuclear blasts.

DSP 3s have a 3.63 m long Schmidt telescope with 92 cm aperture, to collect IR radiation and permit target discrimination by ground controllers within 1.20 seconds. The satellite spins at 5.7 rpm to maintain orientation. The telescope's optical axis points at 7.5° off the spin axis to generate a conical scanning pattern as the vehicle rotates. From GEO, each of the Aerojet IR array's 2,000 lead sulphide cells views a region on Earth about 3 km in diameter. The USAF introduced a new-generation sensor and retrofitted the last two in the primary series – DSP 12 and 13 – with this new sensor. Each new sensor has about 6,000 elements commensurately increasing the resolution. DSP 12 carries HgCdTe detectors as well. DSP 12 and 13 provided the primary coverage of the Gulf to detect and warn Patriot batteries of Scud launches.

The earliest DSP satellites weighed 2,000 lbs, had 400 W of power, 2,000 infra-red detectors and a design life of one and a quarter years. In the 1970s, the satellites were upgraded. The satellite weight grew to 3,690 lbs, the power to 680 W, and the number of detectors increased to 6,000 and the design life to three years.

Late in the decade, TRW upgraded the sensor to improve its survivability against blinding attacks from the Soviet Union. The company added a second infra-red colour to protect against attack from lasers and particle beams. The second colour or medium-wavelength infra-red resulted, for a focal plan array made up of mercury cadmium telluride, the first application of the material in a space system.

Evolutionary growth has continued to improve satellite capability, survivability and life expectancy without major redesign. TRW is currently building DSP spacecraft that accommodate 6,000 detectors. Today's DSP satellite weighs 5,200 lbs and uses 1,274 W of power. At its peak, in the late 1980s, DSP employed more than 800 people at TRW.

Early in the programme, engineers used a so-called 'zero momentum' approach that allows for spacecraft attitude control with a minimum of fuel expenditure. The DSP satellite is spun about its Earth-pointing axis to provide a scanning motion for the infrared sensor. To reduce the satellite spin momentum to a nominal value of zero, engineers introduced a reaction wheel that achieved an equal and opposite momentum. To maintain the satellite's pointing accuracy throughout its lifetime, the DSP has an Inertial Properties Adjustment Device (IPAD) which balances the changes in moments of inertia as fuel and other consumables are depleted.

DSP Block 1

In 1966, TRW and Aerojet began a joint programme to build the first DSP satellites. Initially designated Program 266, then Program 949 and finally, in June 1969, Program 647 to hide its true purpose. The design used a 400 W panels spanning about 7 m with 3-year design life. Titan 3s placed the satellites in orbit. Each focal plane array carried 2,000-element infra-red detectors. The original specification for the satellites called for them to include nuclear test detection among their missions, thus making the Velas obsolete. The first DSP (DSP 1) went into orbit 6 November 1970 from Cape Canaveral. It weighed 820 kg and entered a 26,070 × 36,050 km 7.8° inclination orbit. By March 1974, it had exhausted all of its propellant. DSP 2 and 3 flew 5 May 1971 and 1 March 1972 respectively into synchronous orbits at 0.1 and 0.2° inclinations The USAF placed DSP number 2 over the Panama Canal to watch for submarine missiles. Number 3 sat over the Indian Ocean to warn of an attack by Soviet land-based missiles in addition to providing information on Soviet and Chinese missile tests. DSP 4 followed on 12 June 1973. The four satellites detected 1,014 missile launches by 30 June 1973.

DSP Block 2

The second-generation satellites completed their launch programme in 1987. DSP 5 entered orbit on 14 December 1975, but its hydrazine system ruptured four days after launch. DSP 6 followed on 26 June 1976 and replaced number 3 over the Atlantic. The USAF moved it March 1981 to 75° east on standby; and removed it from GEO March 1985. Then DSP 7 flew on 6 February 1977, DSP 8 on 10 June 1979 (retired by the USAF August 1986), DSP 9 on 16 March 1981 (removed from GEO April 1988), DSP 10 on 6 March 1982. The air force added DSP 11 on 15 April 1984 (retired 1992 from 110° east), DSP 12 on 22 December 1984 (operational since May 1985 at 152° west, probably retired 1995) and, concluding the series, DSP 13 on 29 November 1987 (operational at 37° west).

The USAF planned to give the second- and third-generation satellites the capability to fly in Molniya-type orbits if required. When launched, they expected the satellite to last three years, but hoped for five years. The loss of the Challenger forced the air force to keep them in service much longer.

DSP Block 3

By the late 1980s, the air force exhibited some anxiety about getting the DSP third-generation satellites into space. It had scheduled with NASA the first launch of the new series in December 1986, but the Challenger disaster ended the plan. Then discussions of a modified Titan launch changed when a Titan, flying from Vandenberg, failed in April of that year. The first DSP had to wait until the beginning of 1989, to be launched aboard the maiden Titan 4 in March 1989. The first Shuttle launch didn't follow until a year later. The air force will eventually launch 10 new DSPs, in intervals of roughly 18 months. The Titan 4 launch on 14 June 1989 from Canaveral's pad 41 to 152° west, replaced DSP 10.

Each third-generation DSP has a seven- to nine-year lifetime and senses targets at two IR wavelengths (with HgCdTe and PbS detectors) to avoid laser jamming and improve discrimination. The air force claims this new two-colour system has the sensitivity to detect aircraft operating on afterburners. The satellites also carry nuclear explosion detectors for the Department of Energy. Since each one between numbers 14 and 23 carries a Magnetic Plasma Analyser and a Synchronous Orbit Particle Analyser they can monitor spacecraft charging and radiation environments. They thus coincidentally provide unique 3-D observations of Earth's near-equatorial inner magnetosphere. The air force makes data available to qualified researchers through the IASTP Inter-Agency Solar Terrestrial Physics programme. Los Alamos National Laboratory makes the instrument on funding from the Department of Energy's Office of Intelligence and National Security. The first result to come from this scientific investigation came from Los Alamos when it announced in 1994 that electron levels observed back to 1979 exhibit a solar-like 11-year cycle but, surprisingly, peak one year before the Sun's minimum activity.

The third-generation DSP weighs 2,360 kg and is 10.0 m long by 6.74 m across the four deployed solar panels. The solar arrays produce 1,274 W of power, almost double the preceding generations. Impact detectors monitor physical interference and provide the option for manoeuvring away from the source in case a kinetic kill vehicle attacks the satellite.

The first spacecraft (number 14) carried a piggyback SDI visible/ultra-violet sensor to characterise targets against bright backgrounds such as Earth's limb. Number 15 launched by Titan 4/IUS 13 November 1990 from Canaveral to 100° east, replaced #11. DSP 16 left for geosynchronous aboard Shuttle mission STS-44, and probably replaced number 12 at 69° east. NASA announced this launch the first ever for the classified programme on USAF clearance. The USAF named this DSP Liberty and the NASA launch cost was quoted at US$300 million in contrast to the standard US$180 million. DSP 17 appeared 22 December 1994

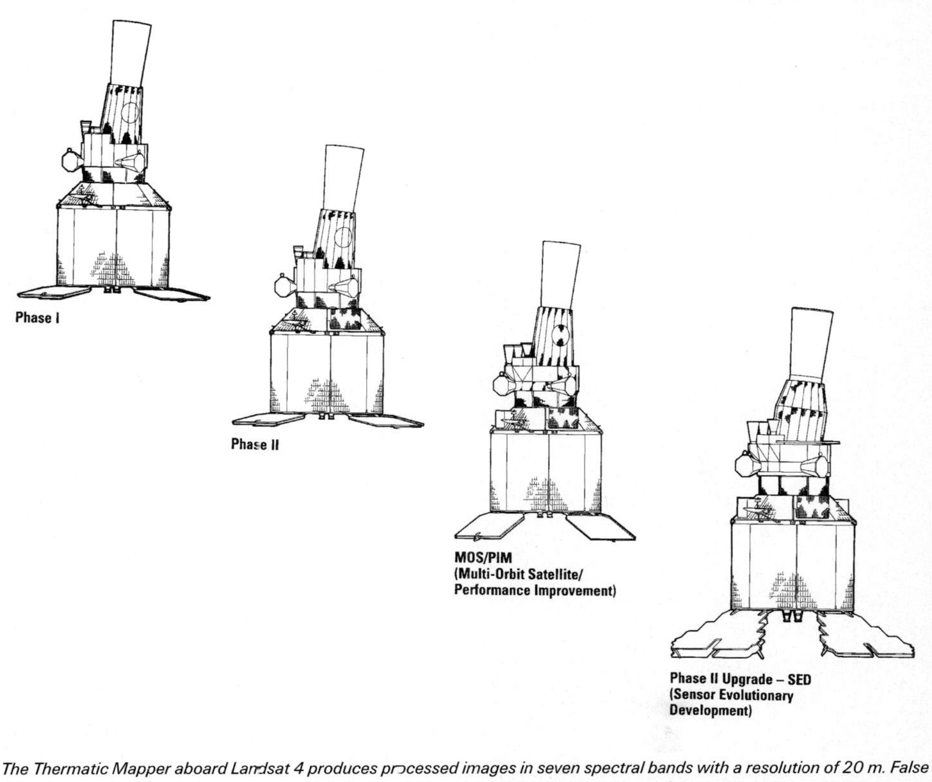

The Thermatic Mapper aboard Landsat 4 produces processed images in seven spectral bands with a resolution of 20 m. False colour images reveal topographical details of the mountain range traversing White Sands Missile Range, New Mexico
1998/0022562

on Titan 4, probably replacing DSP 13 at 37° west. Many observers say DSP 17 is the satellite that witnessed the 10 December 1995 re-entry, some 300 km north of the Falkland Islands, of Russia's Kosmos 398 lunar module test vehicle, launched 26 February 1971.

At the beginning of 1994, both DSP 14-16 remained operational but only #16 offered full capabilities. DSP 14 and 15 had both suffered failures in the thermal control of their focal plane arrays because of blown fuses, reducing their sensitivity to smaller missiles.

Midas series

Background

As both sides in the Cold War built ballistic missiles, the United States felt an acute need to augment its DEWS radar tracking stations with a space-based missile detection system. The USAF called the first design for this system the MIDAS. Each MIDAS satellite used IR sensors capable of detecting the hot exhaust plume of a ballistic missile against the background of the Earth.

Ground-based BMEWS had to wait until missiles rose above the horizon and could only provide the US with a 15-minute attack warning. 12-15 Midas satellites spaced equally around a polar orbit detected an attack soon enough to provide the US with 30 minutes warning. The USAF launched 12 satellites in three series.

Midas 4, launched 21 October 1961 aroused an international furore when details of Project Westford became known. Westford ejected a 36 kg canister containing 350 million hair-like copper dipoles each 21 mm long. After separation, the spinning canister should have dispensed the dipoles in a belt 3,220 km high, 8 km wide and 40 km deep to act as passive reflectors to relay military communication messages. Radio astronomers protested the possibility of interference with their work. Midas 4 ejected its canister but the dipoles failed to disperse.

Midas suffered from a large number of false alarms, interpreting Sun reflections off cloud tops as missile launches. The USAF dropped the name Midas and the satellites became simply Program 461. However, by 1963 President Johnson, apparently referring to the 7 and 9 launches claimed that Midas detected both liquid and solid ICBM launches.

Midas Launch History

Satellite	Launch date	Outcome
Midas 1	25 February 1960	Failed to reach orbit
Midas 2	24 May 1960	484 × 511 km orbit. Telemetry link failure
Midas 3	12 July 1961	3,428 km × 91° inclination
Midas 4	21 October 1961	Ejected its canister but the dipoles failed to disperse
Midas 6	17 December 1962	Failed
Midas 8	9 May 1963	Successfully released a Westford canister
Midas 9	13 June 1963	Failed
Midas 10	9 June 1966	Failed

SBIR T series

Current status

The SBIR programme will cost US$13 billion for the high satellites, plus US$9 billion for SMTS if it is approved in 2000.

Background

In 1992, the USAF Space & Missiles Systems Center awarded US$240 million 24-month contracts for the next generation of early warning satellites. The USAF projected its expenditures at US$1.6 billion through the end of 1994 and asked Congress to provide US$84 million FY92, US$251 million FY93, and US$215 million requested FY94. However, the Congress cancelled the FEWS (Follow-On Early Warning System), programme in December 1993. Instead, the air force proposed the ALARM (Alert Locate And Report Missiles) approach for deployment beginning in 2004. To accompany the proposal, the air force requested US$150 million for FY95 and projected US$1 billion for 1995-99. The air force planned to reduce the cost of FEWS by reduced onboard processing and tactical links in the first years of the programme and then gradually build to FEWS capabilities in the out years.

Many observers expected the air force to cancel DSPs 24/25 to help pay for ALARM, and DSP 23 production slowed. As the Ballistic Missile Defense Office sought to incorporate its own planning into the programme ALARM disappeared into the broader-based SBIR Space-Based Infra-Red programme in December 1994. SBIR requested US$283 million FY96, US$319 million FY97, with US$2.6 billion projected between FY98-01.

As proposed by BMDO, SBIR will initially cover four GEO satellites (the first 2002) and two highly inclined elliptical satellites. Trade reports in 1994 claimed SBIRs would use the 'Heritage' sensor now included or potentially included on the Jumpseat/SDS surveillance satellites. BMDO will add LEO satellites beginning in 2006, based on the Brilliant Eyes (now called the Space and Missile Tracking System or SMTS). The air force picked up the programme in 1994 from BMDO and selected two teams headed by Lockheed Martin and Hughes/TRW in August 1995 for 15-month US$80 million pre-engineering and manufacturing development contracts.

If SMTS is cancelled, then SBIR will exclude tracking after burnout, manoeuvring, decoy discrimination, accurate impact prediction and cueing ground/space defenses. TRW/Aerojet proposed adding a small field of view sensor (developed for a cancelled classified mission) in place of Trailblazer on a current DSP to provide coverage of theatre-sized areas, significantly improving tactical value.

The USAF was planning to spend a total of US$2.6 billion between 1998 and 2001, dependent upon Congressional approval of the deployment of SBIRS (Boeing) ***1998***/0022578

ELINT

CHINA, PEOPLE'S REPUBLIC

JSSW series

Background
The series of Ji Shu Shiyan Weixing satellites has never been discussed in Chinese literature, but Western sources claim they are a series of Elint satellites which the Chinese dropped as an operational programme. Since Mao Zedong backed the Shanghai Space Activities Programme (which was responsible for this programme) and interest faded when he died, it is likely that his was the sole political support for this technology.

RUSSIAN FEDERATION

Tselina Constellations

Current status
Launches of the Tselina-2 satellites are continuing at a rate of one to two per year.

Background

Tselina/Tselina-O
The Soviet Union apparently named its electronic intelligence programme Tselina and initiated it with some Elint payloads on the small Kosmos-1/2 class launch vehicle. Starting in the late 1960s, the Kosmos-3M booster launched the Tselina-O family of Elint satellites which appear to be the payloads primarily launched into 74°, 540 km orbits. These satellites, like all Elints, pinpoint to within about 10 km, the sources of radar and radio emissions and make it possible to identify command and control centres, forward battle elements, air defence units and, even more importantly, reveal movements and changing tactics.

Tselina 2
The Soviets built a variety of electronic intelligence-gathering satellites over the years, but the Russians now use only two varieties: Tselina D and Tselina 2, both from NPO Yuzhnoye. The new Tselina 2, 3.2 tonne Elint type appeared as Kosmos 1603, launched 28 September 1984, and Kosmos 1656, launched 30 May 1985, by Protons from Tyuratam. Both performed very extensive manoeuvres. Starting from an initial 198 by 186 km orbit in an inclination of 51.6° ; the satellite and its upper stage remain in this parking orbit for only about 30 minutes until it uses the booster to raise the apogee to 860 km. Some 50 minutes later, another manoeuvre raises the perigee to 850 km and changes the inclination to 66.6°.

On 22 October 1985, the Soviets launched the third satellite in the series, Kosmos 1697. It left no debris in the 51.6° inclination and 66.6° inclination orbits it used as it adopted its final position. A re-examination of the programme led Western observers to conclude that, for the first time, the new SL-16 Zenit launcher carried this satellite into orbit. A second similar satellite followed on 28 December 1985, but this time the Zenit upper stage failed to re-ignite and the Tselina re-entered in February 1986.

Flights resumed in 1987 with Kosmos 1833 (18 March) and Kosmos 1844 (13 May), and two more in 1988: Kosmos 1943 (15 May) and Kosmos 1980 (23 November). Kosmos 2082 appeared next on 22 May 1990. The Soviet attempt to complete a constellation of four satellites in orbital planes spaced 45° apart failed 4 October 1990 when the Zenit exploded within seconds of launch destroying both the payload and pad. The Russians experienced another major setback 30 August 1991 when the next Zenit exploded after launch from a new pad. The launch authority issued a statement saying: it was a 'military-technical satellite to verify the fulfilment of disarmament treaty commitments.' A third Zenit failure followed on 5 February 1992, but the Zenit of 17 November 1992 carried Kosmos 2219 into an orbit almost coplanar with 1988's Kosmos 1943. The second Zenit in 6 weeks delivered Kosmos 2227 on 25 December 1992, into an orbital plane 90° separated from Kosmos 2219's. Kosmos 2237 appeared 26 March 1993, apparently replacing Kosmos 1980. Kosmos 2263 appeared 16 September 1993. These most recent additions form a 3-satellite system with planes 120° degrees apart. Kosmos 2278 appeared 23 April 1994 with its plane 40° west of Kosmos 2263's and Kosmos 2297 appeared 24 November 1994 with its plane 40° west of Kosmos 2278's. Kosmos 2322 appeared 31 October 1995. A study of the orbits indicated in 1995 that Tselina 2's mass is only 3.2 tonnes, rather than the 9 tonnes previously believed, from Zenit's capacity in that orbit.

The tests using the Proton-K launch vehicle were possibly a result of delays in flying the new Zenit-2 and when that vehicle started to fly problems arose. The first Zenit-2 launch in April 1985 carried a mockup Tselina-2 intended for a 64.4°, 850 km orbit, but for unknown reasons the launch failed. In June the second test flight failed when a second stage vernier engine exploded 511 seconds after launch. Three pieces of debris reached short-lived orbits with inclinations of 64.4°. The launch of Kosmos 1714 marked the third Zenit-2 failure in four launches, (Kosmos 1697 being a success), when the payload shroud failed to separate from the second stage and a second stage vernier engine exploded. The combined second stage shroud and encapsulated satellite reached an intermediate orbit with an apogee of 850 km without circularisation.

Unlike the earlier Tselina constellations, Tselina-2 series show no regular orbital plane spacing indicative of a constellation.

Tselina D
Another Soviet era operational electronic intelligence system, consists of a constellation of six 2.5 tonne satellites, spaced at 60° intervals in roughly 630 km, 81° inclination orbits. The Soviets used these satellites to identify and pinpoint military radio/radar stations. An SL-3 carried the first of these Elints into space on Kosmos 389, launched 18 December 1970 from Plesetsk (642 × 687 km). The constellation continued to be upgraded at a launch rate of approximately two each year until 1976, when the launch rate suddenly increased to four per year. Each had an average operational life of 18 months. Western observers who studied the orbits, believe the orbits suggest that the satellite is based on the Meteor bus configuration and when the series began, many initially believed them to be Meteor failures.

In 1978, the SL-14 Tsyklon 3 began launching Kosmos satellites into 635 × 665 km, 82.5° inclination orbits. The Soviets named the first in the series the Kosmos 1025 (although this might be Okean-related). Various statements by Soviet era officials called a few of these satellites (Kosmos 1076, Kosmos 1151, Kosmos 1500, Kosmos 1602) oceanographic satellites. But their true mission as an Elint became apparent in 1984, when a complete constellation of six satellites spaced in planes 60° apart, emerged just as the Soviets announced the termination of the SL-3 programme. The coincidence of these events indicated that the SL-3 assembly line remained open just to launch the last Elint Kosmos 1441 (launched 16 February 1983), the 35th in the series. Built by NPO Yuzhnoye, the Soviets call the Tsyklon ELINTS the Tselina D.

The last of 1986's six Tsyklon Elints (Kosmos 1805, 10 December, 635 × 662 km, 97.7° inclination) entered a plane exactly between two established planes. These clues indicated that the constellation experienced an enlargement or reconfiguration, although a supplementary satellite did not arrive until Kosmos 2242 in 1993 (between Kosmos 2221 and Kosmos 2228). 1987's five Elints routinely replaced existing satellites and this pattern continued through the four missions of 1988. Kosmos 1805 transmitted Elint-type signals throughout 1988. No new Tselina's appeared during 1989. Then the single Kosmos 2058 orbited on 30 January 1990, replacing Kosmos 1812. Another long break occurred until Kosmos 2151 appeared on 13 June 1991, replacing Kosmos 1908. Kosmos 2221 and Kosmos 2228 appeared in 1992, replacing Kosmos 1842 and Kosmos 2058, respectively. Then 1993's single Tselina D Kosmos 2242, slipped into a new plane midway between them. A launch failure on 25 May 1994 may have been a Tselina-D and that is the last recorded launch of the series.

UNITED STATES OF AMERICA

Introduction

Elints record radio and radar transmissions from areas of military activity. When replayed to ground stations the radar signatures (characteristics such as pulse repetition frequency, pulsewidth, transmitter frequency and modulation) enable the likely function and method of operation of a particular centre to be identified. The number and type of electronic systems at a particular site, and subsequent changes in the signals, provide a valuable indication of its purpose and capability. The ability to intercept and decode satellite and ground communications, and to interfere with them by rival satellite activities, is likely to be a decisive factor in any large scale hostilities. In times of hostilities, even if the contents of the messages are not decoded in time, their interception can indicate the locations and identities of the communicators. Together with the movements of the mobile transmitters, an Electronic Order of Battle (EOB) can be constructed. It is continued monitoring of large ground radars such as ballistic missile early warning systems, with their specific ranges, viewing angles and other operational restraints, that enable details of their modernisation and upgrading to be logged and analysed.

Canyon (Program 827) series

Background
The air force built Canyon to monitor radio communications. They began to go into orbit, with the Atlas Agena D launch of 6 August 1968, in a 10° inclination orbit that traced a figure-eight over each hemisphere improving northern coverage. Each satellite weighed about 230 kg. The other six launches appear to be 12 April 1969, 31 August 1970, 4 December 1971 (launch failure), 20 December 1972, 18 June 1975 and 23 May 1977. The orbits reached near to, but did not use, geosynchronous conditions.

Chalet/Vortex (Program 366) series

Background
Chalet succeeded Canyon and weighed more than 1 tonne. It first flew on 10 June 1978 following the US loss of ground stations in Iran. Five more followed on Titan 3s: 1 October 1979, 31 October 1981, 31 January 1984, 2 September 1988 and 10 May 1989. The 1988 launch apparently suffered a Transtage failure. Sources hinted that the 15th and last Titan 34D launch, of 4 September 1989, may have been a Chalet mission but it was subsequently confirmed as a DSCS/IUS flight. USA 67 on STS-38, in November 1990, has also been suggested as a Vortex, although trade reports at the time claimed it was a Magnum. Information released following 1988's failure indicated that Chalet had been renamed Vortex and that the two first craft remained operational at 45° east and 115° east. An advanced Vortex is claimed to have debuted on the Titan 4 Centaur of 14 May 1995. The USA 118 launch of 24 April 1996 was similar.

Heavy Ferret series

Background
The United States achieved its first significant Elint capability starting 18 June 1962 when a Thor Agena B launch carried a Heavy Ferret from Vandenberg into 370 × 411 km 82° inclination orbit. Eventually the United States launched 15 satellites of about 900 kg, at

a rate of two to three annually. The Elint programme apparently began with test packages aboard at least the early Samos satellites. As they were developed, the Thrust Augmented Thors and Agena Ds then later the Long Tank Thrust-Augmented Thors and Agena Ds, carried these satellites into orbit. The NRO launched its last Heavy Ferret apparently on 16 July 1971 into 488 × 508 km 94.5° orbit. The programme eventually became obsolete as it was succeeded by Canyon.

Jumpseat series

Background

Jumpseat record radio and radar transmissions from areas of military activity. When replayed to ground stations the radar signatures (characteristics such as pulse repetition frequency, pulsewidth, transmitter frequency and modulation) enable the likely function and method of operation of a particular centre to be identified. The number and type of electronic systems at a particular site, and subsequent changes in the signals, provide a valuable indication of its purpose and capability. The ability to intercept and decode satellite and ground communications, and to interfere with them by rival satellite activities, will most likely help determine the course of future wars. In times of hostilities, even if the contents of the messages are not decoded in time, their interception can indicate the locations and identities of the communicators. Together with the movements of the mobile transmitters, an Electronic Order of Battle (EOB) can be constructed. It is continued monitoring of large ground radars such as ballistic missile early warning systems, with their specific ranges, viewing angles and other operational restraints, that enable details of radar modernisation and upgrading to be logged and analysed.

Jumpseat have a 12-hour, 63.4° inclination Molniya-type orbit to provide polar coverage. They use almost the same orbits as SDS and use the same Titan 3B launcher. Jumpseat probably began to be manufactured in the late 1960s by either TRW or Hughes under the air force, reportedly to monitor Moscow ABM site radar emissions. Other targets include space tracking radars, SAM sites, airbases and SLBM submarine bases. The air force ordered six satellites. They were launched as number 1 on 21 March 1971, number 2, 21 August 1973, number 3, 10 March 1975, number 4, 25 February 1978, number 5, 13 December 1980. Number 6 may have departed 8 February 1985 (on the circumstantial evidence that SDS appearances were usually associated with KH 11 launches). Advanced Jumpseat have followed. The Titan failure of 16 February 1992 probably carried a Jumpseat. The first Titan 4 Centaur launch into an apparently Molniya-type orbit, 3 May 1994, has been suggested as carrying a Jumpseat successor. The Titan 4 Centaur of 10 July 1995 was apparently similar. Shuttle STS-28 and STS 53 flew similar profiles with IUS, but their payloads have generally been attributed to advanced SDS.

Magnum/Orion/Intruder series

Background

The CIA initiated development of these advanced Ferrets to replace Rhyolite following the complete loss of Iranian ground stations by early 1979, but they suffered deployment delays because of launcher problems. Designed to take advantage of Shuttle launches, and pick up signals across the whole radio frequency range, the CIA had to pull together plans to the Titan/IUS launch as the Shuttle fell behind schedule. Although NASA finally scheduled the first Magnum for Shuttle/IUS launch STS 10, the mission did not take place until STS-51C on 25 January 1985. The satellite started at a 28.5° GTO orbit. It attained GEO despite the IUS' small liquid thrusters being required to make up for I's delta velocity shortfall of some 1.6 m/s. The Magnum, many believe used STS-33 to travel to LEO during 23 November 1989. The same speculation surrounds STS-38's payload a year later. Magnums weigh 2.7 tonnes. Each spacecraft cost US$300 million. The Titan 4 Centaur of 14 May 1995 headed due east from Canaveral, indicating a 28.5° orbit, preparatory to departure for GEO. This may have been the first Intruder, which combines the functions of Magnum and Vortex. The new satellite weighs about 4.5 tonnes and has a 100 m diameter antenna. Extensive building work at Menwith Hill supports the debut of a new vehicle. The 24 April 1996 launch of USA 118 probably included the same type of satellite. A new advanced Orion was launched on 9 May 1998 on a Titan 4B.

The satellites are controlled from Pine Gap, which was expanded 1983-85 for Magnum operations.

The Magnum/Orion type tend to specialise in datalinks, such as missile telemetry transmitted during tests, while Vortex's voice communications speciality may have been added to Magnum's successor. Chalet was originally tasked with communications intelligence but from the second vehicle it was modified to encompass signals intelligence. Jumpseat is targeted at radar installations.

Rhyolite/Aquacade series

Background

Electronic intelligence satellites, also known as Ferrets, matured with Rhyolite in the 1970s. Earlier Elints were usually launched into circular 500 km orbits. The first Rhyolite was launched 6 March 1973, by Atlas Agena D into geosynchronous orbit at 0 to 2° inclination. It only weighed about 275 kg, but it became the first operational GEO Ferret. Some test versions may have premiered from 1970 until 1972 (the first may have been 19 June 1970). Rhyolites pick up telemetry from Soviet missile tests, monitor weapons development and check compliance with arms control treaties. Built by TRW for the CIA, two more were launched: 11 December 1977 and 7 April 1978. The CIA originally funded and then dropped project Argus, a more sophisticated follow-on, in 1976.

The GEO satellites return their data to Pine Gap (Australia), Bad Aibling (Germany), Menwith Hill (UK), and Ft Meade, Maryland for DSCS relay to Buckley Air National Guard Base, Aurora, Colorado. Pine Gap, near Alice Springs, is a 2.25 km^2 base under CIA/National Range Operations Control. Operational from 1970 in time for Rhyolite, two new radomes were added during the expansion of 1983-85 for Magnum operations, just as others were added in the early 1970s for Rhyolite. There were 11 radomes by 1991, operated by almost 700 personnel. Data tapes are usually airlifted out of Alice Springs to avoid interception.

Subsatellite Ferret series

Background

A series of small subsatellites, launched piggyback with photo-reconnaissance satellites, identifies possible targets for detailed examination by the heavy Ferrets. The first of these, weighing 50 kg, left Vandenberg on 29 August 1963 from Vandenberg aboard a Thor Agena D. It adopted a 310 × 431 km 82° inclination orbit and had a 30 day life. By 8 June 1975, 35 had been identified; Big Bird took over as their launcher from 20 January 1972. There were interesting variations on 12 December 1968, 5 February 1969, 10 October 1972, 10 November 1973, 8 June 1975, 18 June 1980 and 20 June 1983, when the Ferrets went as high as 1,400 km. They possibly monitored the Sary Shagan centre at the time of Soviet ABM tests.

Titan 2 Ferret series

Background

Russia's Tass news agency reported on 22 April 1992 that the USA planned to launch a third new type of Elint satellite on a Titan 2. The first was USA 32 on 5 September 1988 in 185 × 292 km 85° inclination. Titan 2 had been expected to continue the Whitecloud series but these satellites clearly did not fit the Whitecloud mould. The second in the series, USA 45 on 5 September 1989, suffered technical failures according to Tass. USAF claimed a successful launch, indicating the fault lay with the payload. USA 81 ascended 25 April 1992 into 784 × 805 km, 85.1° inclination orbit and appeared to continue the series.

Trumpet series

Current status

The NRO launched a Trumpet class electronic intelligence satellite on 8 November 1997, from Cape Canaveral aboard a Titan 4A. The Intelligence Agency code-named the satellites USA 136. The launch was the third in the series for Trumpet satellites, following USA 103 and USA 112. The three satellites are all operational providing continuous coverage of the Russian signal spectrum.

Background

The Trumpet is a massive Hughes electronic intelligence satellite with the more specific objective of signal intelligence (signals intelligence). To eavesdrop on foreign military signals it unfolds an antenna about the size of a football field. The spacecraft cost roughly a billion dollars, and cost an additional US$200 million to US$300 million to launch.

It carries a special phased-array broadband eavesdropping antenna that measures roughly 300 ft across when it is unfolded. In order to keep the satellite placed over the proper point for the longest period of time, the spacecraft flies a Molniya orbit with an apogee of 23,000 miles, and a perigee of a few hundred miles, at an orbital inclination of 63°. The spacecraft uses its broadband capability to monitor more than 2,000 ground sources simultaneously, including communications between Russian generals and the country's nuclear submarine fleet.

The antenna demonstrates a new technology of construction that differs from the Magnum, Mentor and Jumpseat, 'wraprib' umbrella type under girding structure. The Trumpet has a mesh antenna and small pulleys and tiny motors unfurl it when it reaches its assigned orbit. Like the TDRS and Jumpseat systems it carries an Extremely High-Frequency (EHF) relay system.

In order to achieve final orbit the TITAN / CENTAUR must fire four separate times, two to achieve the inclination change and two to change the eccentricity of the orbit.

FOBS

RUSSIAN FEDERATION

Fractional Orbital Bombardment System (FOBS) series

Background

The Soviets provided their answer to the United States deployment of a missile tracking system on the northern perimeter of the western hemisphere when they launched The Fractional Orbit Bombardment System (FOBS). They demonstrated its potential in 18 tests conducted between 1966 and 1971. The purpose of FOBS is to fly a nuclear warhead into the United States by using a southern trajectory, essentially coming into the western hemisphere over the south pole instead of the north pole.

FOBS are fired into an orbit of 160 km and then braked by retro for re-entry, their nuclear warheads (though none was ever carried during the tests) falling on the target before completion of the first revolution (hence the name fractional orbit). In these series of tests, the Soviets purposely launched the satellites on a track to miss the United States, with the test warhead being called down on to Soviet territory shortly before the end of the first orbit. From the Soviet point of view, FOBS forced the West to invest in more complicated defences such as Over-The-Horizon radar (OTH). But they paid a price in reduced accuracy and payload capacity.

All FOBS tests used a SL-11 Tsyklon (SS-9 Scarp) from Tyuratam, with a retro stage fitted to the warhead. The 1966 tests 'unannounced' by the Soviet Union and allocated a 'Kosmos-U' designation by RAE Farnborough, initiated the series and the response in the West prompted the Soviets to continue with nine in 1967: Kosmos 139, 160, 169, 170, 171, 178, 179, 183, 187. Then the Soviets continued with two in each of the following three years, and the last, C433, came on 8 August 1971. Opinion in intelligence services is divided. Some feel the lack of tests indicate the system is operational and others feel it means the system was abandoned. Under Article VII of the unratified Salt 2 Treaty and the accompanying MoU, all FOBS must be dismantled. Specifically, the acknowledged 18 FOBS launchers at Tyuratam were to be either destroyed or converted.

GEODETIC

RUSSIAN FEDERATION

Etalon series

Background

Etalon 1, launched 10 January 1989 as Kosmos 1989 and Etalon 2 launched 31 May 1989 as Kosmos 2024 are 1,415 kg, 1,294 mm diameter geodetic satellites. Etalon reflects laser light in a passive role. The first was placed in 19,102 × 19,149 km at 64.9° along with two Glonass satellites. Etalon carries 306 laser retro-reflector assemblies for international geodetic studies; six germanium reflectors may be used for future IR interferometric measurements. Etalons supplement Glonass information, aiming to improve the navigation system's accuracy. Three laser stations at Ternopol, Yevpatoria and Maydanak provide 25 cm ranging accuracy, and were later joined by 10 foreign sites.

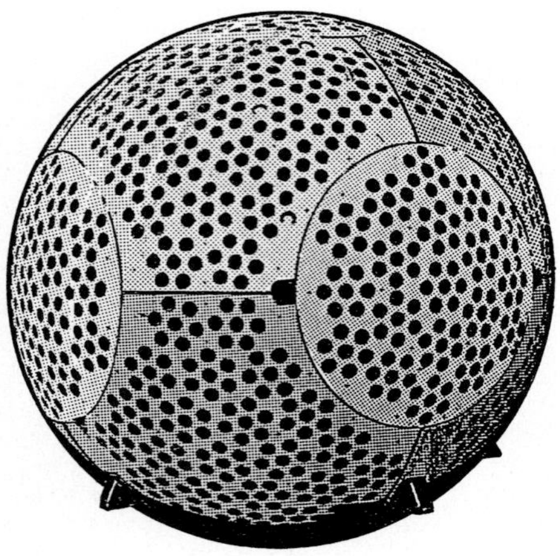

The Etalon laser-reflecting geodetic satellite complements the Lageos, Starlette and EGP spheres (Teledyne Brown Engineering)

Geodetic Type 1 series

Background

Three satellites have been launched using the Kosmos-3M into the normal Parus/Tsikada navigation satellite orbital regime but their orbital planes and transmissions indicated that they were not part of the navigation satellite system. Speculation centres upon navigation calibrations as their ultimate mission. Kosmos 842 and Kosmos 911 had orbital planes 180° apart.

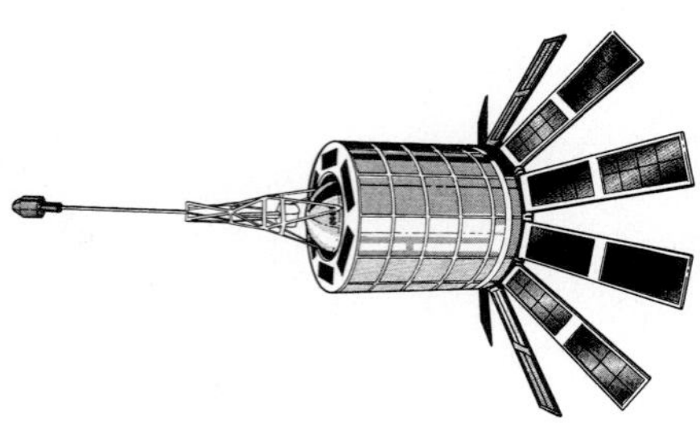

The GEO-IK class of geodetic satellites, derived from NPO PM's series of low altitude navigation satellites, was acknowledged in 1989. 1994's was actually named GEO-IK 1 (Teledyne Brown Engineering) ***1991***

Kosmos 2285 remains an isolated mission. Although it shared the orbital altitude of the navigation satellites, the orbital inclination of 74° had not been used in that programme since 1972. This mission has not been explained and geodesy is perhaps the most likely explanation.

There is also a small group of satellites which are apparently geodetic but their orbital parameters differ from the other acknowledged geodetic missions: these might be part of the Sfera series or a different type of satellite.

GEO-IK Musson (Monsoon) series

Background

The Soviets publicly discussed this series for the first time in 1989. They stated it used a design based on the Parus/Tsikada military/civil navigation satellites. NPO PM builds the satellite bus, and the NIIKP Institute of Space Device Engineering provides the payload. Soviet era reference call them the 'Musson' (Monsoon). Solar cells power the instrumentation through 10 deployed petals, working with Ni/H_2 batteries. The satellite takes measurements five days a week using a 150 to 400 MHz Doppler transmitter which operates for up to 12 hours daily, providing 1-3 cm/s accuracy. Musson uses 9.4 GHz radar to measure height over sea with 3 to 5 m accuracy, and a 5.7/3.4 GHz C-band up/down transponder measures satellite range within 3 to 5 m. Laser retro-reflectors calibrate the range measurement to within 1.5 m. A flashlight system emits a sequence of nine 800 to 1,200 Joule flashes up to 55 times daily, limiting positioning errors to 15 m for a satellite at 1,500 km.

GEO-1K 1 became the first satellite in the series to carry a name and not a Kosmos serial number. It included an Elekon payload to demonstrate a location and data relay service for Russian cargo transport services. A joint venture between NPO PM and Germany's Elbe Space and Technology (Dresden) GmbH, the operational Elekon system will comprise payloads on seven Tsikada navigation satellites.

Launches of GEO-IK Geodetic satellites

Launch date	Cosmos	Incl deg	Period min	Perigee km	Apogee km
23 January 1981	Failed to reach orbit				
23 September 1981	1312	82.6	116.0	1,493	1,505
24 September 1982	1410	82.6	116.0	1,495	1,503
24 November 1983	1510	73.6	116.1	1,481	1,526
8 August 1984	1589	82.6	116.0	1,496	1,504
14 June 1985	1660	73.6	116.1	1,482	1,526
11 February 1986	1732	73.6	116.1	1,480	1,526
2 December 1986	1803	82.6	116.0	1,498	1,503
20 February 1987	1823	73.6	116.0	1,479	1,526
30 May 1988	1950	73.6	116.1	1,485	1,522
28 August 1989	2037	73.6	116.1	1,484	1,525
30 July 1990	2088	73.6	116.1	1,483	1,525
22 December 1992	2226	73.6	116.0	1,482	1,526
29 November 1994	GEO-IK 1	73.6	116.1	1,480	1,527

Sfera (Sphere) series

Background

The Soviet Union built the preponderance of geodetic satellites and launched them as part of their Kosmos series. They called the first generation the Sfera geodetic satellites and put them into three orbital inclination slots and two orbital altitude slots. Typically, The Soviets launched two each year during the period 1968-78. No public photo or drawing of the satellites has appeared, but they may have resembled the Parus/Tsikada navigation satellites. NPO Prikladnoi Mekhaniki built them.

Launches of Sfera Geodetic satellites

Launch date	Cosmos	Incl deg	Period min	Perigee km	Apogee km
20 February 1968	203	74.1	109.2	1,178	1,208
4 June 1968	Failed to reach orbit				
30 November 1968	256	74.1	109.5	1,175	1,227
17 March 1969	272	74.0	109.4	1,181	1,211
24 November 1969	312	74.0	108.6	1,144	1,179
28 April 1971	409	74.0	109.4	1,177	1,216
20 November 1971	457	74.0	109.5	1,185	1,221
25 March 1972	480	83.0	109.2	1,175	1,203
21 December 1972	539	74.0	113.0	1,343	1,383
8 September 1973	585	74.0	113.6	1,368	1,416
29 April 1974	650	74.0	113.5	1,369	1,402
29 August 1974	675	74.0	113.7	1,365	1,426
12 February 1975	708	69.2	113.6	1,369	1,413
24 September 1975	770	82.9	109.2	1,169	1,210
24 November 1977	963	82.9	109.4	1,182	1,210
26 December 1978	1,067	83.0	109.1	1,158	1,208

METEOROLOGICAL

UNITED STATES OF AMERICA

Defense Meteorology Satellite Program (DMSP) series

Background

The US military launched 37 military meteorological satellites beginning in 1965 through 1995. The military uses these satellites to forecast, identify and track developing patterns of severe weather, hurricanes and typhoons and to prevent imaging reconnaissance satellites and aerial surveys from wasting film and flying time on targets that are covered by clouds. In recent years, the USAF has also begun to build three dimensional cloud profiles to optimise aircraft route planning, and during the Cold War and since, these same satellites have watched the atmosphere for telltale signs of nuclear blast effects. The military provides its information about the atmosphere using two satellites in near-polar Sun-synchronous orbits to both ground and shipborne stations to give warning of thunderstorms, typhoons and hurricanes.

After the first reconnaissance satellites returned many useless photos because of cloud cover at the target, the air force proposed the DMSP system to sample the target ahead of photo-reconnaissance satellites and only use the film if the conditions at the target warranted it. The mission requires two satellites in Sun-synchronous orbits. The first crosses the equator at 6 am local time and the second at 10.30 am. When the satellites go off line a replacement can wait no longer than 90 days.

When the Thor Burner booster became available in 1965 the air force pressed ahead with its plans for the Block 4A DMSP. The first of the satellites appeared in 1965 on Thor Burner. The air force launched 10 satellites on 18 January 1965, 17 March 1965, 20 May 1965, 9 September 1965, 6 January 1966, 30 March 1966, 15 September 1966, 8 February 1967, 22 August 1967, and 11 October 1967. As the satellites improved, the next generation appeared. Known as the Block 5A, the first left for orbit on 22 May 1968, and the remainder on 22 October 1968, 22 July 1969, 11 February 1970, 2 September 1970, and 16 February 1971. As the satellite continued to improve and grow heavier, the next generation, the Block 5B, switched to a more powerful Thor Burner combination: 14 October 1971, 24 March 1972, 9 November 1972, 16 August 1973, 16 March 1974, and 8 August 1974. Two 5C models were launched: 23 May 1975 and18 February 1976.

The five Block 5D-1, Advanced Meteorological Satellites, placed in 850 km, 98° inclinations with 1½-year operational lifetimes had a spotty record: one failed after launch and another never achieved full operational status. The air force launched AMS-1, 11 September 1976; AMS-2, 5 June 1977; AMS-3, 1 May 1978; AMS-4, 6 June 1979; AMS-5, 14 July 1980 but this last one failed to reach orbit. When Block 5D-2 appeared it offered additional redundant systems and three-year operational life, adding 155 kg to the 613 kg mass. This last mass addition required the air force to switch to the more powerful Atlas. The first 5D-2 (2-01 SI) orbited on 20 December 1982 by Atlas E from Vandenberg into a 724 km Sun-synchronous orbit; the second a 751 kg. 2-02 S2 on 18 November 1983 by Atlas F from Vandenberg, and the next also a 751 kg satellite went into a 816 × 833 km, 98.7° orbit. Its primary sensor failed after 47 months.

DMSP has grown in complexity and sophistication in several stages. The air force called the next version of the satellites Block 5D-2. The air force purchased nine (S6-14) of these satellites and one remains to be launched. Both of the original DMSP 5D-2s are now well beyond their three-year design lives, but an upgraded 823 kg DMSP 5D-2 (S8) replaced the original satellites after being launched by Atlas E from Vandenberg SLC-3W in June 1987. S8 and S9 still operated well when S10 joined them in December 1990, guaranteeing DMSP services during Operation Desert Storm in the Gulf. S9's appeared about to fail during this period and the air force hoped to use S10 to maintain services, but S10's eccentric orbit reduced its effectiveness. S8 failed at the end of 1991 and the replacement S11 arrived promptly thereafter. S12 appeared in August 1994 and S13 in March 1995, on the last Atlas E in the inventory. Both remained fully operational at the end of 1995, supported by three on standby with degraded capabilitiies.

In May 1994, President Clinton approved a long debated plan to combine DMSP and the civil NOAA system, halving the number of operational satellites to two. The first joint satellite is expected in 2005, controlled by NOAA from its Suitland, Maryland centre. Suitland will take over DMSP control from 1998. DoD will close its own stations, but retain one as backup.

The USA instituted an Integrated Program Office on 1 October 1994, headed by James T Mannen to set up the joint operations. The plan projects a US$300 million saving through 1999, and then at least US$1 billion over the next 15 years. Until March 1996, it was planned that two US$100 million Phase 1 contracts for system demonstration, risk reduction and prototyping would begin August 1996, but the programme is being restructured as a result of budget constraints and performance of current generations. Five or six satellites may be ordered, launched on Delta 2-class vehicles and offering 5½-year lives. Phase 2, starting in 1999, was to cover two satellites, and Phase 3 three or four. The USA will also fly instruments on Europe's Metop satellites, due 2001 and 2006.

DMSP 5D-2, the current model. Among other applications, DMSPs prevent reconnaissance satellites wasting resources when targets are cloud-covered (Lockheed Martin Astro Space)

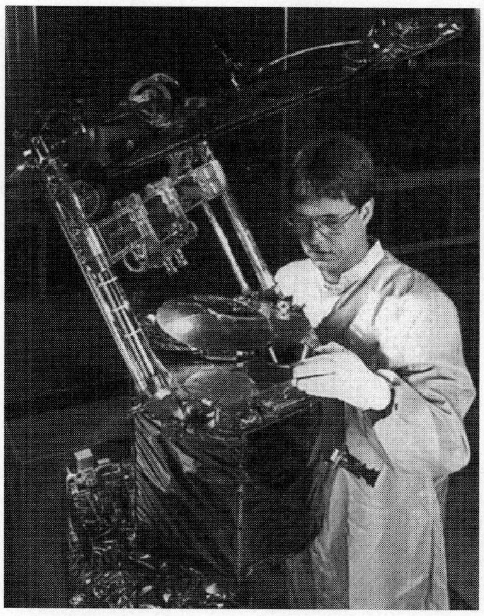

Hughes' DMSP microwave imaging sensor. NASA's Tropical Rainfall Mapping Mission has an enhanced version

Lockheed Martin Astro Space received the contract for the new generation of satellites valued at US$228.2 million in July 1989 for five satellites (S15-19). They are expected to feature increased onboard processing and anti-jamming protection. Similar in appearance to their predecessors, some US$40 million of the US$60 million unit cost is attributed to the instrument package, which Aerojet will provide for US$62.1 million for three SSMIS Special Sensor Microwave Imager Sounders. The sounders combine the functions of the current Barnes Engineering humidity profiler, Aerojet's own temperature sounder and Hughes' microwave imager. The air force US$7.9 million option was exercised in January 1990 for a fourth sensor, the US$8.8 million fifth followed in August 1991. Each 57 kg, 70° west sensor will provide 19 microwave channels over 19 to 183 GHz for observations from 30 km to ground level, including rainfall to an accuracy of 5 mm/h and ocean wind speeds to within 2 m/s. Temperature profiles up to 73 km will also be generated.

S15 will carry STP's POGS II Polar Orbiting Geomagnetic Survey magnetometer on a 5 m boom to update geomagnetic maps. S16-19 will also carry two ionosphere instruments for observing UV emissions for improved prediction of disruptions to communications and Over-the-Horizon radar systems. The Space Forecast Center at Falcon AFB will be responsible for processing the data into warnings. Johns Hopkins University's Applied Physics Lab is providing the USAF instrument for nadir viewing, while the USN's own Naval Research Lab is building the complementary horizon-viewing unit. The NRL's RAIDS Remote Atmospheric & Ionospheric Detection System test version was planned for NOAA 13 in 1994 but it had to be removed and may now never fly.

The air force is expected to issue a Block 6 contract in 1998 for first spacecraft delivery in 2004-05. Martin Marietta Astro Space and Lockheed Missiles & Space were awarded US$8.2 million and US$8.3 million contracts, respectively, July 1991 for sensor risk reduction experiments. A multispectral instrument was under consideration in addition to meteorological sensors. A radar altimeter would have continued the ocean topography work of the GFO Geosat Follow-On. Block 6 is now submerged within the joint DMSP/NOAA programme.

The air force, through the USAF Space Command, manages the defence Meteorological Satellite Program. The AF Space Command's 6th Space Operations Squadron (part of 50th Space Wing, Falcon AFB, performs TT&C at Offutt AFB's Satellite Operations Center. Each satellite uses a visible/IR OLS Operational Linescan System. An 85.5 GHz Microwave Imaging Sensor (an enhanced version is on NASA's TRMM) measures ocean surface wind speed, ice coverage/age, areas and intensity of precipitation, cloud water content and land surface moisture; the microwave temperature sounder returns cloud temperature profile data. The precipitating electron spectrometer monitors electron and proton densities at different energy levels, and the electrostatic ionosonde locates auroral boundaries to aid radar and long-range radio operations. There is also a Gamma/X-ray Spectrometer and Magnetometer on board.

Data are provided in real-time to tactical ground stations and navy ships worldwide. They can be

DMSP Block 5D3 will incorporate the first Special Sensor Microwave Imager Sounder instrument

encrypted at times of hostility. They are also stored onboard for transfer via US readout stations and then commercial telecommunication satellites to the US Air Force Global Weather Center at Offutt AFB, Nebraska and the US Navy's Fleet Numerical Meteorology and Oceanography Center in Monterey, California. The command readout stations are sited at Fairchild AFB (Washington), Keene Pt (Hawaii) and Loring AFB (Maine; Loring's TT&TC facility closed FY89); Thule AFB in Greenland provides TTETC support. DMSP information is also made available for civil use through NOAA, and much of it is kept on file at NOAA's depository at Colorado University. Forces equipped with Mark 4 truck-mounted terminals receive the imagery.

Ground forces used six mobile receiver units during the Gulf War but their bulk (1.3 tonnes) required 75 per cent of a C-141 cargo transport plane and so slowed their introduction into the theatre. In an ironic twist, tactical users in the forward lines of the war carried commercial receivers for civil satellites.

Specifications

DMSP 2-03 (S8)

Carries a microwave imaging sensor intended to identify tropical storms earlier than previous systems. The sensor can penetrate clouds to measure storm intensity, instead of measuring only the cloud tops; the system can also determine wind speeds over the ocean, measure ground and snowpack moisture content, determine ice coverage and indicate whether rain is falling over land.

DMSP 2-04 (S9)

Launched: 3 February 1988 by Atlas E from Vandenberg to replace 2-02. 823 kg; 820 × 829 km, 98.77°. Was to be replaced by 2-05 but operations continued. See 2-03 for comments.

DMSP 2-05 (S10)

Launched: 1 December 1990 by Atlas E from Vandenberg SLC-3W at 15.57 GMT into 732 × 849 km, 98.9°. Low-perigee (almost 50 trackable fragments indicate solid motor explosion) reduced effectiveness. Intended to replace 2-04, guaranteeing services during Operation Desert Storm in the Gulf.

DMSP 2-06 (S11)

Launched: 28 November 1991 by Atlas E from Vandenberg SLC-3W into 840 × 857 km, 98.9°. Replaced 2-03.

DMSP 2-07 (S12)

Launched: 29 August 1994 by Atlas E from Vandenberg SLC-3W at 17.38 GMT into 843 × 859 km, 98.9°. Replaced 2-04/2-05.

DMSP 2-08 (S13)

Launched: 24 March 1995 by Atlas E from Vandenberg SLC-3W at 14.05 GMT into 847 × 854 km, 98.83°. Replaced 2-06.

DMSP 2-09 (S14)

Launched: 4 April 1997 by Titan 239 from Vandenberg SLC-3W into 844 × 855 km, 98.94° orbit.

Contractors: Lockheed Martin Astro Space (prime), Westinghouse Electric Corp (Operational Linescan imaging system), Hughes Aircraft (microwave imager), Aerojet (microwave temperature sounder), Barnes Engineering (infra-red temperature/moisture sounder)
Resolution: typically 0.5 km
Design life: 48 months
Mass: 823 kg at launch, including 180 kg sensor payload
Orbit: typically 820-860 km Sun-synchronous at 98.8°, providing repeated passes at either 6am or 10.30am local time. Scan width 2,960 km, creating continuous coverage at equator and global coverage in 12 h
Spacecraft: based on Astro Space's NOAA Tiros bus, 1.22 m wide bus, 6.4 m long with single solar wing (1 kW from 8 panels of 12,500 Si cells) deployed. 127 kg structure provides 5.4 m^2 of payload mounting area. Zero momentum 3-axis control by MWs plus magnetorquers; 4 hydrazine (15.9 kg propellant) plus 8 N_2 (2.3 kg propellant) thrusters. Earth pointing accuracy of 0.01° provided by strapdown star sensor and gyros; 0.12° accuracy backup system of static Earth sensor, Sun sensor, gyros. Final orbit insertion provided by Thiokol Star 37S solid. TT&C assigned to AF Space Command's 6th Space Operations Squadron (part of 50th Space Wing, Falcon AFB, which can also provide control) at Offutt AFB's Satellite Operations Center. Data downlink on 2,207.5, 2,252.5, 2,267.5 MHz; telemetry downlink 2,237.5 MHz; command uplink 1,792 MHz at 2 kbit/s (minimum of 300 stored commands); data stored on Odetics DDS-5000 tape recorder. Passive thermal control provided by finishes and radiators; active control by electrically-controlled louvres and heaters (thermal subsystem mass 24 kg)
Sensors: primary meteorological imaging is provided by the visible/infra-red OLS Operational Linescan System. The system includes an 85.5 GHz Microwave Imaging Sensor (SSM/I, an enhanced version rides on NASA's TRMM) measures ocean surface wind speed, ice coverage/age, areas/intensity of precipitation, cloud water content and land surface moisture, the microwave temperature sounder (SSM/T-1) returns cloud temperature profile data, the SSM/T-2 microwave water vapour sounder will be added later for atmospheric humidity monitoring to improve global forecasting. The precipitating electron spectrometer (SSJ/4) monitors electron plus proton densities at different energy levels, and the electrostatic ionosonde locates auroral boundaries to aid radar and long-range radio operations. Gamma/X-ray Spectrometer (SSB/X2) and SSM Magnetometer also carried.

DMSP 2-10 (S15)

Contractors: Lockheed Martin Astro Space (prime), Westinghouse Electric Corp (Operational Linescan imaging system), Aerojet (Special Sensor Microwave Imager Sounder)
Resolution: typically 0.5 km
Design life: 60 months
Mass: 1,154 kg. on-station
Orbit: as 5D2
Spacecraft: based on 5D2 bus, 1.22 m wide bus, 6.4 m long with single solar wing (10 panels of Si cells) deployed. 187 kg structure provides 6.5 m^2 of payload mounting area. Zero momentum 3-axis control by MWs + magnetorquers; 4 hydrazine (21.8 kg propellant) plus 8 N_2 (3.6 kg propellant) thrusters. Earth pointing accuracy of 0.01° provided by strapdown star sensor plus gyros; 0.12° accuracy backup system of static Earth sensor, Sun sensor, gyros. Final orbit insertion provided by Thiokol Star 37FP solid (19 per cent offload). TT&C and frequency details as 5D2. Passive thermal control provided by finishes and radiators; active control by electrically-controlled louvres and heaters (thermal subsystem mass 29.5 kg)

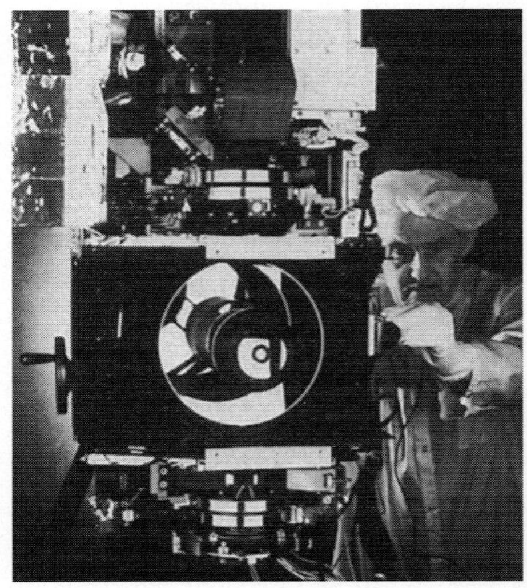

DMSP's primary imager is the Operational Linescan System (Westinghouse Electric Corporation)

MINOR MILITARY

RUSSIAN FEDERATION

Introduction

Current status

Russian literature has now identified three classes of satellites within the 'minor military' series: Yug (south), Vektor (vector) and Romb (rhombus). All three appear to have been designed by a laboratory at the Mozhaisky Military Engineering Space Academy (VIKA) in St. Petersburg. After subtracting the Yug, Vector and Romb satellites from the list of satellites generally classified as 'minor military' or 'radar calibration' in the west, there is still a large number of satellites not identified with these groups. None of these satellites produced the additional sub-satellites which are associated with Romb missions although in many cases the satellites were flown to Romb-type orbital altitudes.

Background

Type 1

The late Dr Charles Sheldon coined the term 'minor military' to describe families of Soviet satellites for which no other information had leaked to the west, but which appeared to have a military purpose. Dr Sheldon filed his reports with the US Library of Congress, and so the coinage appears in most official US documents describing the Soviet era space programme. Most minor military satellites fly at low altitude and their purpose has been suggested (but not confirmed) to include radar calibration, atmospheric drag measurement and spacecraft technology experiments. In 1993 one Soviet era official acknowledged that some dedicated missions each carry a 2 m sphere for passive atmospheric monitoring. Others carry 1 m spheres as sub-satellites on other missions.

The categorisation of Dr Sheldon covers four orbital inclinations with two subclasses. The Soviets placed the first subclass in eccentric orbits about twice a year. Between 1980 and 1984 they launched six by the Kosmos booster into 300 to 400 × 1,600 to 2,000 km orbits. Then a hiatus occurred until the next such satellites appeared on 28 August 1990 as Kosmos 2098 entered a 396 × 1,977 km, 83.0° inclination orbit. Kosmos 2265 (26 October 1993) released a 2 m sphere into 291 × 1,574 km, 82.9° orbit for passive atmospheric measurements, and the follow on satellites Kosmos 2292 (27 September 1994) and Kosmos 2332 (24 April 1996) performed similar functions. The Soviets tinkered with the basic design before launching a new variant in 1989, on Kosmos 2002 (14 February 1989) placed in 186 × 2,307 km, 65 to 85° orbit. Kosmos 2002 for the first time, released 10 objects shortly after launch and then itself decayed naturally during 15 October 1989. Kosmos 2059, 6 February 1990 performed a similar operation.

Type 2

The second sub-class has less eccentric, more circular orbits having 92-95 minute periods. These satellites decay naturally in only a few years but many operate simultaneously. Some minor military satellites – C 1501 (30 September 1983, 468 by 614 km. 82.94 degree inclination), Kosmos 1601 (27 September 1984, 475 × 516 km, 65 to 84° inclination), and Kosmos 1662 seem to be cylinders about 4 m long and 2 m in diameter. They periodically jettison small fragments about 0.1m^2 in radar cross-section. Originally, they dropped 22 to 24 objects, but after Kosmos 1601 the number rose to 28. Kosmos 2075, launched 25 April 1990 appears to be the last of this series.

The released objects decay much more rapidly than the parent body and may simulate MIRVs (Multiple Independent Re-entry Vehicles) to give the Russians radar detection system practice for engaging possible US nuclear warheads. Kosmos 2306 (2 March 1995, 469 × 517 km, 65.9° inclination) immediately released four sub-satellites. Three minor military launches occurred in 1991, all using SL-8 Kosmos. Kosmos 2137 entered 449 × 495 km, 65.8° inclination orbit on 19 March and Kosmos 2164 entered 286 × 708 km, 74° inclination orbit. The launch of 25 June should have placed its payload into 485 × 515 km, 74° inclination orbit but stage 2 failed to re-ignite and prevented orbital injection.

Type 3

A new larger type of minor military satellite, possibly an Elint, debuted in late 1988. The launch on 23 December had the name Kosmos 1985. It had a 622 × 633 km, 73.6° inclination and was the first minor military satellite ever launched by SL-14 Tsyklon. Four small sub-satellites soon appeared from the main bus and two more emerged in late January 1989. By 1990, a total of 36 objects had appeared on radar tracks. The satellite Kosmos 2053 (27 December 1989) appears to be the second of this type ever launched. The third, Kosmos 2106, entered a higher inclination (82.5°) 28 November 1990.

Launch Date	Cosmos	Incl deg	Period min	Perigee km	Apogee km
24 July 1975	752	65.9	94.6	481	515
17 December 1976	885	65.8	94.4	467	512
2 February 1977	891	65.8	94.5	473	516
22 July 1977	933	65.8	92.5	384	408
12 May 1978	1006	65.8	92.4	382	408
8 February 1979	1075	65.8	94.5	473	516
4 December 1979	1146	65.8	94.0	444	494
27 March 1980	1169	65.8	94.5	477	515
23 September 1981	1310	65.8	94.5	477	518
21 October 1982	1418	50.7	92.3	371	413
29 December 1982	1427	65.8	94.0	445	499
25 January 1983	Launch failure				
6 April 1983	1450	65.9	94.4	472	512
5 October 1983	1502	65.8	92.3	369	412
26 January1984	1534	65.8	94.5	469	517
20 December 1984	1615	65.8	94.0	441	501
27 February 1985	1631	65.8	94.5	472	512
27 October 1986	1788	65.8	94.5	468	517
15 December 1987	1902	65.8	92.4	368	411
14 July 1988	1958	65.8	92.4	371	413
14 June 1989	2027	65.8	94.6	482	513
19 March 1991	2137	65.8	94.0	449	495

NAVIGATION

RUSSIAN FEDERATION

Introduction

Current status

Russia maintains three navigational satellite programmes, one low altitude military and one low altitude civil and the Russian equivalent of GPS known as Glonass. The first two employ satellites in 1,000 km, 83° inclination orbit and give positions to within 80 to 100 m, or roughly Transit-type positioning, for navy/merchant marine vessels equipped with Shkhuna terminals.

Both the civil and military systems use 810 kg, 2 m in diameter gravity gradient satellites which fly from Plesetsk on SL-8 Kosmos vehicles. A receiver establishes its position from the Doppler shift of various satellites in the constellation employing the same method as Transit and uses similar VHF frequencies (150 to 400 MHz) as Transit. More than 1,200 ships around the world have terminals to receive both systems' signals. Extra accuracy comes from the military's six satellite constellation as opposed to the four satellite civil equivalent.

Glonass series

Background

The GPS equivalent Global Navigation Satellite System, Globalnaya Navigatsionnaya Sputnikovaya Sistema (Glonass) system has been under deployment since 1982 and became fully operational in December 1995. Glonass follows in a long tradition of Soviet navigation satellites going back to 1967's Kosmos 192 'Tsyklon'. The Soviet Union refused to publicly acknowledge it had a navigation satellite system until the launch of Kosmos 1000 in 1978. Then the Soviet Union informed the International Telecommunications Union it would inaugurate the Glonass service in 1982. On 12 October 1982, an SL-12 Proton launched the first three satellites on a single booster from Tyuratam and identified them as Kosmos 1413-1415. In order to achieve the constellation spacing the satellites first entered a low altitude parking orbit at 51-60 kms. Then the upper stage performed two manoeuvres to place the three payloads into their required orbits. The Soviets repeated this pattern on their next five launches during the next three years. Curiously, each trio had only one or two Glonass pre-operational satellites and the remaining seven were dummies, although still called Kosmos and given Kosmos numbers. The 10 operating satellites weighed 1,260 kg and had one-year design lives. But they achieved an average of 14 months. During this pre-operational testing phase the Soviets only had six satellites simultaneously operating at any one time. The last launch of the pre-operational phase included the first Block 2a, transmitter with improved clocks. Six Block 2a satellites eventually appeared and each had an average life of 17 months.

With the pre-operational phase complete, the Soviets entered into a new phase and launched the satellites on a more direct profile, to distinguish it from the earlier phase. Each Proton carried the new satellites directly into an initial inclination of almost 65°. This track surprised long time space observers because it was the first time the Soviets had used it in almost 20 years. Again, two manoeuvres by the upper stage spaced the satellites properly, but no orbital plane change occurred during this manoeuvre. All subsequent missions have used this more direct route.

With September 1986's launch, the Soviets put up the first payload in which all three satellites were live. But a launch failure marred the debut of the Block 2b, which the Soviets claimed had a 24-month design life. The Soviets launched 12 Block 2Bs between 1987 and 1988, but only half reached orbit. These satellites achieved a 22-month average life. The 1,415 kg Block 2v first appeared in September 1988. The new satellite has a reported three-year lifespan and is the basis of the current Glonass system.

Two 1990 missions raised the number of operational craft to 11 and 1992's first launch maintained the active number at 11. There were 13 active satellites by September 1992. Lifetimes began to improve significantly following the launch on April 1994, which had a radiation, hardened satellite. This launch also filled every slot in planes one and three at some time. Then August 1994's launch began the process of filling plane two.

Glonass planners wanted to have a Phase 1 system of 10 to 12 operational satellites ready by the end of 1990, but didn't realise that goal until 11 March 1992. From this original deployment the Russians planned to grow the system to a full operational capability of 21 craft, plus three in-orbit spares by the end of 1995. Still, by end of 1993 the Russians had 13 active satellites (the oldest from 1990) plus one reserve. The system continued to grow in 1994, building to 16 satellites before the end of that year. When the Russians announced a full deployment in December 1995 they had thus made up for schedule slippage and achieved a considerable success. At least part of the urgency centred on keeping pace with the west in having a commercial navigational system in place to compete with GPS.

Russia's own Institute of Radionavigation & Time estimated in 1993 that a scant 200 to 500 civil Glonass receivers had made it into widespread use compared with the hundreds of thousands of contemporary GPS receivers. As the Soviets originally conceived the system, maritime and aeronautical users would be the principal consumers of signals allowing positional accuracy within 100 m surface location and 150 m altitude and 15 cm/s velocity readings. Most civil users now report the actual figures are about 20 m of accuracy, better than original expectations primarily because phantom signal ghosting from multiple paths has not proven a serious problem. Unlike GPS, Glonass cannot degrade with selective availability. The P-code yields accuracy about twice as good as the civil code.

Glonass satellites	Launch		Plane	Slot	Block
1	12 October 1982	Cosmos 1414	1	1	1
2	10 August 1983	Cosmos 1490	1	3	1
3		Cosmos 1491	1	2	1
4	29 December 1983	Cosmos 1519	3	18	1
5		Cosmos 1520	3	17	1
6	19 May 1984	Cosmos 1554	3	19	1
7		Cosmos 1555	3	18	1
8	4 September 1984	Cosmos 1593	1	2	1
9		Cosmos 1595	1	3	1
10	17 May 1985	Cosmos 1650	1	1	1
11		Cosmos 1651	1	2	2a
12	24 December 1985	Cosmos 1710	3	18	2a
13		Cosmos 1711	3	17	2a
14	16 September 1986	Cosmos 1778	1	2	2a
15		Cosmos 1779	1	3	2a
16		Cosmos 1780	1	8	2a
17*	24 April 1987	Cosmos 1838	3		2b
18*		Cosmos 1839	3		2b
19*		Cosmos 1840	3		2b
20	16 September 1987	Cosmos 1883	3	18	2b
21		Cosmos 1884	3	17	2b
22		Cosmos 1885	3	24	2b
23*	17 February 1988	Cosmos 1917	1		2b
24*		Cosmos 1918	1		2b
25*		Cosmos 1919	1		2b
26	21 May 1988	Cosmos 1946	1	8	2b
27		Cosmos 1947	1	7	2b
28		Cosmos 1948	1	1	2b
29	16 September 1988	Cosmos 1970	3	24	2v
30		Cosmos 1971	3	18	2v
31		Cosmos 1972	3	19	2v
32	10 January 1989	Cosmos 1987	1	2	2v
33		Cosmos 1988	1	3	2v
34	31 May 1989	Cosmos 2022	3	19	2v
35		Cosmos 2033	3	24	2v
36	19 May 1990	Cosmos 2079	3	17	2v
37		Cosmos 2080	3	19	2v
38		Cosmos 2081	3	20	2v
39	8 December 1990	Cosmos 2109	1	4	2v
40		Cosmos 2110	1	7	2v
41		Cosmos 2111	1	5	2v
42	4 April 1991	Cosmos 2139	3	21	2v
43		Cosmos 2140	3	22	2v
44		Cosmos 2141	3	24	2v
45	29 January 1992	Cosmos 2177	1	3	2v
46		Cosmos 2178	1	8	2v
47		Cosmos 2179	1	1	2v
48	30 July 1992	Cosmos 2204	3	24	2v
49		Cosmos 2205	3	18	2v
50		Cosmos 2206	3	21	2v
51	17 February 1993	Cosmos 2234	1	3	2v
52		Cosmos 2235	1	2	2v
53		Cosmos 2236	1	6	2v
54	11 April 1994	Cosmos 2275	3	17	2v
55		Cosmos 2276	3	23	2v
56		Cosmos 2277	3	18	2v
57	11 August 1994	Cosmos 2287	2	12	2v
58		Cosmos 2288	2	16	2v
59		Cosmos 2289	2	14	2v
60	20 November 1994	Cosmos 2294	1	3	2v
61		Cosmos 2295	1	6	2v
62		Cosmos 2296	1	4	2v
63	7 March 1995	Cosmos 2307	3	20	2v
64		Cosmos 2308	3	22	2v
65		Cosmos 2309	3	19	2v
66	24 July 1995	Cosmos 2316	2	15	2v
67		Cosmos 2317	2	10	2v
68		Cosmos 2318	2	11	2v
69	14 December 1995	Cosmos 2323	2	9	2v
71		Cosmos 2325	2	13	2v

* Launch failure. The slot listing is the initial position for each; some later transferred to other slots.

Note: 1: spare. **Active satellites** January 1996: C2111, C2178, C2179, C2204, C2205, C2235, C2236, C2275-77, C2287-89, C2294-96, C2307-09, C2316-18, C2323-25 (C2324 spare). Active satellites as of 15 December 1993 were: C2079, C2080, C2110, C2111, C2140, C2178, C2179, C2204, C2205, C2206, C2234, C2235, C2236; C1987 was held in reserve.

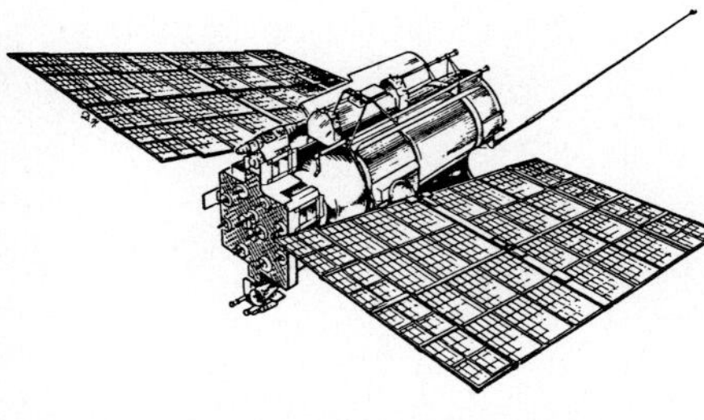

Glonass navigation satellites are launched by Proton, typically in triplets. Height is 7.840 m with magnetometer boom deployed; 7.230 m span across solar wings (Nicholas L Johnson/Kaman Sciences) ***1995***

Although the P-code encryption is being rewritten following the demonstrated ease with which Professor Peter Daly of Leeds University, England was able to crack it. Glonass now offers better coverage than GPS at higher latitudes, because of the constellation configuration but GPS is better nearer the equator.

The system has an operating philosophy that closely mirrors GPS. One difference arises from the geoid model upon which the two systems base their calculations. GPS uses the WGS-84 geodetic systems, Glonass uses SGS-85 (Soviet Geocentric System), adding an error of a few metres whenever a receiver averages between signals from the two systems. Satellites transmit in narrow bands centred around 1.250 and 1.6035 GHz. The higher frequency band creates complaints from radio astronomers that cite its interference with the hydroxyl line at 1.612 GHz. In response the Russians have begun changing the frequencies of some satellites beginning in September 1993. Glonass itself will soon suffer interference in its own upper L1 band from LEO satellite communication systems, such as Iridium, because these systems have been allocated 1.610-1.6265 GHz frequencies. Both problems force Glonass to use only 12 channels, instead of the planned 25, because satellite transmissions are doubled at the poles. By January 1996, nine pairs were doubling up on channels.

In terms of constellation the Glonass does differ from the US system, which uses circular semi-synchronous orbits near 20,000 km (718-minute period). Glonass satellites fly at a slightly lower altitude of 19,100 km (676 minute period) at inclinations of 64.8° which gives a precise retrace of the ground track every 17 revolutions (eight days). The VKS Military Space Forces launch and operate Glonass and monitors it through five principal monitoring and command stations: Moscow, St. Petersburg, Eniseyisk, Komsomolsk-on-Amur and Balkhash. The VKS performs control functions from a single location at Golitsyno-2 in the Moscow region, the core of VKS's command system.

NPO PM has responsibility to produce the satellites which are manufactured at AKO Polyot, using the NIIKP Institute Space Device Engineering's payloads. NIIKP has responsibility for manufacturing both the civil and military receivers. The Institute of Radionavigation and Time maintains the ground time/frequency standards. Individual satellites are called Uragan (hurricane).

The 1,480 kg Glonass M improves stability of the three cesium clocks from 5×10^{-13} daily to 1×10^{-13} and extending the series lifetimes to five to seven years per satellite. The new civilian band includes an ionospheric correction transmission, which allows for compensation based on the comparison of effects on signals of two different frequencies. Passage through the ionosphere generally creates the largest error in a receiver's position. Pending military review the entire system may be uprated again providing inter-satellite links and 60-day satellite autonomy.

Parus Sail series (Generation 1 and 2)

Current status

The latest Parus satellite entered service on 23 September 1997 as Kosmos 2346. It flies in the same plane as Kosmos 2334. It ended in a lower than expected orbit for Parus satellites.

Background

The Soviet Union built a low altitude military system similar to the US Transit system to locate and guide its ballistic missile submarines. The first trace of this system appeared on an SL-8 Kosmos launch in 1967. Following a two-phase test programme at 74° inclination between 1967 and 1972, the Soviets declared the system operational following the launch of Kosmos 514, which completed a three-satellite constellation at 83° inclination. Eventually the Soviets phased this system out in 1978 (the last launch, Kosmos 1027, occurred on 27 July 1978). In its place they established the current constellation of six Parus 'sail' satellites in orbital planes spaced 30° apart and inaugurated by Kosmos 700 (26 December 1974). Kosmos 2061 appeared on 20 March 1990, apparently replacing Kosmos 1959; and reconfiguring the constellation as Kosmos 2026, 1904, 2061, 2016, 2034, 2004 in order of the orbital planes. Kosmos 2074 appeared 20 April 1990, replacing Kosmos 1904, and Kosmos 2100 14 September 1990, replaced Kosmos 2016. In 1991, Kosmos 2135 (26 February) replaced Kosmos 2026, Kosmos 2142 (16 April) took up the duties for Kosmos 2034, Kosmos 2154 (22 August) for Kosmos 2061 and Kosmos 2173 (26 November) for 2100. In 1992, Kosmos 2180 (18 February) replaced Kosmos 2004, Kosmos 2184 (15 April) substituted for Kosmos 1904, Kosmos 2195 (1 July) for Kosmos 2026 and Kosmos 2218 (29 October) for Kosmos 2154. By the beginning of 1993, the operational system consisted of Kosmos 2142, Kosmos 2173 and 1992's four satellites. In 1993, Kosmos 2233 (9 February) replaced Kosmos 2141, Kosmos 2239 (1 April) replaced Kosmos 2173 and Kosmos 2266 (2 November) replaced Kosmos 2195. In 1994, Kosmos 2279 (26 April) replaced Kosmos 2180. There were no more launches through 1994, leaving the operational system at the beginning of 1995 as Kosmos 2184, Kosmos 2218, Kosmos 2233, Kosmos 2239, Kosmos 2226. Kosmos 2285 (2 August 1994) remains to be identified although it resembles an original generation Parus, the 74° inclination has been vacant since 1972. In 1995, Kosmos 2310 (22 March) replaced Kosmos 2184, but Kosmos 2321 (6 October) was stranded in 258 × 793 km and 82.94° inclination, because of a stage 2 sticking fuel valve. The orbital plane suggests that it was intended to replace 1993's Kosmos 2266. It was claimed to be usable, but Kosmos 2327 (16 January 1996) then replaced Kosmos 2266.

Launches of Parus Navigation satellites 1990 – 1997

Launch Date	Satellite	Node	Plane °E
20 March 1990	Cosmos 2061	90	3
20 April 1990	Cosmos 2074	60	2
14 September 1990	Cosmos 2100	120	4
16 April 1991	Cosmos 2142	150	5
22 August 1991	Cosmos 2154	90	3
27 November 1991	Cosmos 2173	120	4
17 February 1992	Cosmos 2180	180	6
15 April 1992	Cosmos 2184	60	2
1 July 1992	Cosmos 2195	30	1
29 October 1992	Cosmos 2218	90	3
9 February 1993	Cosmos 2233	150	5
1 April 1993	Cosmos 2239	120	4
2 November 1993	Cosmos 2266	30	1
26 April 1994	Cosmos 2279	180	6
22 March 1995	Cosmos 2310	60	2
6 October 1995	Cosmos 2321	[30]	[1]
16 January 1996	Cosmos 2327	30	1
5 September 1996	Cosmos 2334	30	1
20 December 1996	Cosmos 2336	120	4
17 April 1997	Cosmos 2341	60	2

The nodes are referred to an arbitrary zero point. Cosmos 2321 was left in its transfer orbit due to a malfunction with the second stage of the Cosmos-3M launch vehicle.

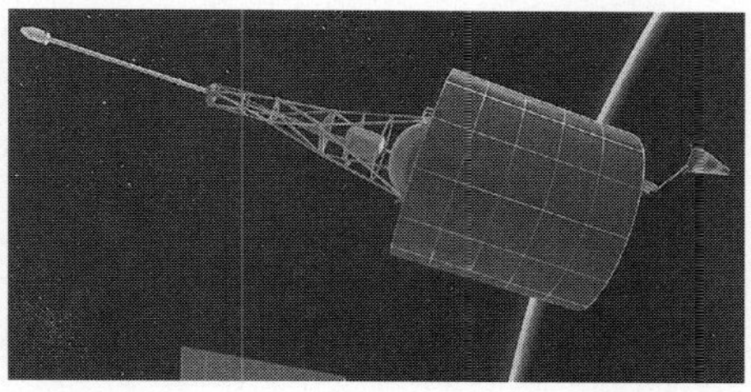

The Parus military navsats are probably almost identical to this Nadezhda civil version. See the Cospas-Sarsat entry for an illustration showing Nadezhda under construction ***1995***

Tsikada 'Cicada' series

Background

The system Soviet sources refer to as both Tsikada 'Cicada' and Nadezhda ('Hope', when applied to the Cospas-related satellites) provides location information and warning and distress signal identification. As the Russian republic inherited some of the existing Soviet launch infrastructure, the country has sought to maintain the Tsikada network. It still launches replacement satellites into each civil plane approximately every 26 months, and the payload sometimes includes a Cospas/Sarsat transponder. After the original constellation emerged no new satellites appeared in 1988 but a replacement took up an orbit at 960 × 1,014 km, 82.96° on 4 July 1989 with the Cospas 4 transponder. Nadezhda 2 (Cospas 5) appeared 27 February 1990 in 956 × 1,021 km, 83.0° inclination orbit. Nadezhda 3 (Cospas 6) followed on 12 March 1991 at 958 × 1,018 km, 82.9° inclination, then Kosmos 2181 on 9 March 1992 at 973 × 1,014 km, 83.0° inclination replaced Naddezhda 1. Kosmos 2230 on 12 January 1993 at 973 × 1,007 km, 82.9° inclination replaced Kosmos 2181. Nadezhda 4 (Cospas 7) on 14 July 1994, replaced Nadezhda 2. The first named Tsikada appeared in 1995 when the Russians launched Tsikada 1, on 24 January 1995, probably replacing Kosmos 2123. Kosmos 2315 appeared 5 July 1995, replacing Kosmos 2230, and including the Kurs marine location/control system.

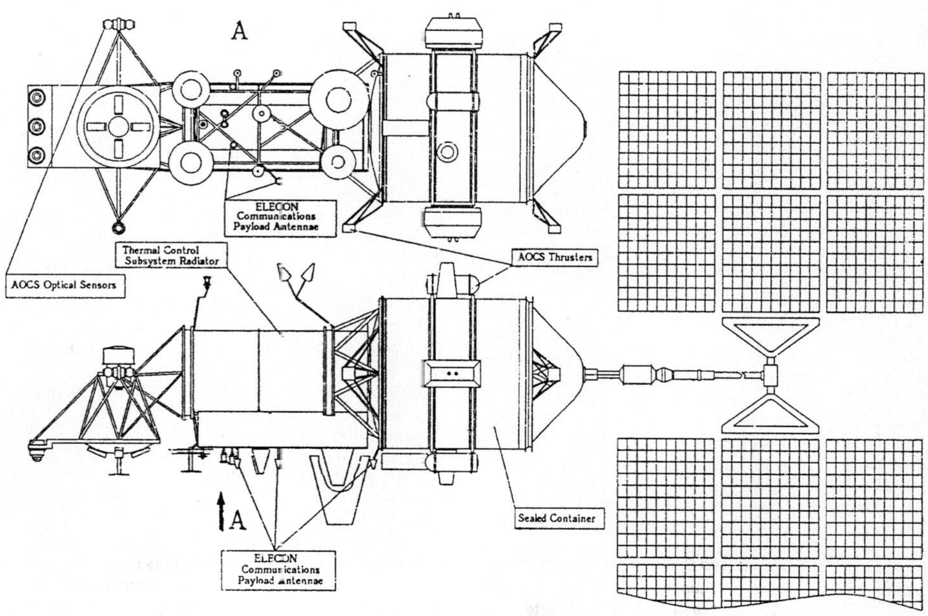

The improved Tsikada includes the commercial Elekon store/forward package (Elbe Space & Technology Dresden) ***1996***

Launch date	Satellite	Node deg E	Plane
27 February 1990	Nadezhda 2	0	14
5 February 1991	Cosmos 2123	315	13
26 February 1991	Cosmos 2135	0	14
12 March 1991	Nadezhda 3	270	12
9 March 1992	Cosmos 2181	235	11
12 January 1993	Cosmos 2230	235	11
14 July 1994	Nadezhda 4	0	14
24 January 1995	Tsikada 1	315	13
5 July 1995	Cosmos 2135	235	11

The nodes are referred to an arbitrary zero point.

Guided by hints from Russia and press reports, space analysts expect that the Russians will eventually launch seven Tsikada satellites, the first two in 1996, carry the Elekon payload to provide location and data relay for global store/forward services.

Tsyklon (Cyclone) series

Background

Some observers believe the original Soviet navigation programme began with this series of satellites operated under the name Tsyklon (Cyclone). Flights started with the launch of Kosmos 158 (15 May 1967) into a 74°, 740 × 820 km orbit: the series continued with Kosmos 192 (1967), 220 (1968), 292, 304 (both 1969), 332 and 371 (both 1970), all using a slightly lower orbital altitude. Starting with Kosmos 385, flights used a higher orbital regime, but the same inclination – 74°, 980 × 1,020 km: later flights were Kosmos 422, 465 (both 1971), 475 and 489 (both 1972). Flights then switched to 83°, using the same altitude regime: Kosmos 514 (1972), 574 (preceded by a launch failure on 25 May), 586, 627 (all 1973), 628, 663, 689 (all 1974), 729 (1975), 800, 823, 846 (all 1976), 890, 962 (both 1977), 994 and 1027 (both 1978). By the time that the last Tsyklon series navigation satellite had been launched the Parus and Tsikada launches were underway.

UNITED STATES OF AMERICA

GPS constellation

Background

The Navstar Global Positioning System (GPS) is a space-based radio positioning, navigation and time transfer system operating at L-band frequencies, providing 16 m, 0.1 m/s and 0.1 μs accuracy to military users. The operational space segment consists of 24 Navstar (Navigation System using Timing & Ranging) satellites, including three hot in-orbit spares, in six orbital planes (A-F). Each carries four atomic clocks, two caesium with a stability of 1 second in 300,000 years and two rubidium with about half that stability (Block 1 satellites carry only rubidium clocks). The satellites fly in circular 20,182 km, 55° inclination, 11.967 hour (half a sidereal day) orbits, spaced so that a minimum of four appear to a receiver worldwide. They transmit information on L1 (1,575.42 MHz) and L2 (1,227.60 MHz), phase modulated by the data they carry and by spread spectrum pseudorandom number codes.

The GPS programme began in 1973 as a navy initiative to replace the 200 m accuracy of the Transit system. Successful as the Transit had been it did not allow the navy to build accurate SLBMs for its submarine fleet. The navy awarded Rockwell the prime contract in 1974 to build the first 12 Navstars, but then delayed the programme by deciding to harden the satellites and to re-designate the 11 already launched by Atlas as prototypes. So number 12 became a ground qualification unit to prepare for the Block 2 production series. As part of the Cold War the navy then added funding in 1979 to develop an onboard sensor able to report whether the satellite had been illuminated by laser energy or interfered with by another spacecraft. The design review completed in mid-1982 resulted in the Block 2 being one-third larger, with lifetime increased from 5 years to 7 years 6 months, larger solar panels increasing power supply from 400 to 700 W, and the adoption of improved atomic clocks.

An L1 carrier provides the C/A coarse/acquisition code at 1.023 MHz and both Ll/L2 are modulated by the P precise code at 10.23 MHz. These codes contain a message at 50 bit/s giving details about the satellite ephemeris, atmospheric propagation correction and clock bias information. By measuring the time interval between transmission and reception, the receiver calculates the distance to a satellite. Using the distance to at least three satellites in an algorithm provides the position fix. A minimum of three satellites signals provides latitude/longitude; the fourth adds altitude.

With the C/A code Standard Positioning Service (SPS) yields about 100 m civil accuracy and the P-code the PPS Precise Positioning Service, intended to provide 16 m accuracy on 50 per cent of occasions. Commercial C/A SPS receivers originally provided 20 to 40 m accuracy, but the C/A code is now deliberately degraded by SA Selective Availability to 100 m by broadcasting slightly erroneous clock and ephemeris information. Differential GPS (DGPS) overcomes SA and provides increased overall accuracy. DGPS places a transmitter at a known location and its position information is used to calculate corrections communicated to other receivers in the area. With DGPS, civilian users get real-time accuracy of 10 to 15 m, with 3 m possible for post-processing calculations in static positions.

GPS has an additional A-S (anti-spoofing) signal to prevent friendly forces being deceived by an enemy deliberately transmitting inaccurate signals. A-S encrypts the P-code to become the Y-code, which is accessed by an auxiliary output chip in the receiver. The military classified early P-code receivers once they had been loaded with the cryptographic material. The new generation incorporate a PPS-SM Precision Positioning Service-Security Module which isolates the security aspects into a sealed tamperproof module so that the receiver as a whole is unclassified when keyed. Current military GPS receivers require a hand-held, hand-inserted cryptographic code. With thousands of users, handling that cryptographic key becomes very cumbersome and expensive. So the United States has adopted a national policy of an electronic key management system – which transmits the code to a user without involving paper.

Four Monitor Stations on Hawaii, Kwajalein (Pacific), Diego Garcia (Indian Ocean) and Ascension Island (Atlantic) continually track the health of the system and route data through to the Master Control Station of the 2nd Space Operations Squadron of the 50th Space Wing, Falcon AFS, Colorado Springs.

The DoD has ordered about 27,000 receiver sets, ranging from man-packs to F-16 navigation systems. The military had to rush 10,000 hand-held units to the Gulf in 1990 and 1991. Currently, the military has outstanding orders for 13,999 receivers and accessories, with options that could bring the total purchase to 93,999 units through 1998. The 50,000th unit was delivered October 1995.

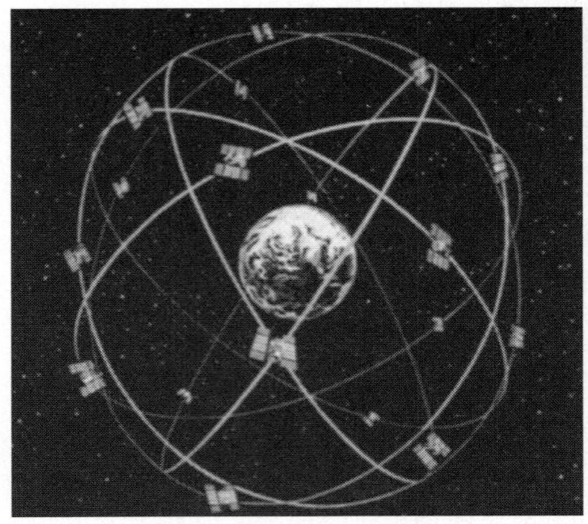

The operational GPS network comprises 24 satellites in six orbital planes. The system can fix the positions of military users to within 16 m and offers a downgraded civil facility (Rockwell)

Each Block 2 incorporates the Integrated Operational Nuclear Detection System (IONDS) devices to detect nuclear explosions for assessment of nuclear attacks. Detonations can be pinpointed with 100 m accuracy, with every point on the globe within view of at least four satellites. The Dept of Energy's Sandia National Labs provide the Earth-facing 'bhangmeter' optical sensor that detects the characteristic double flash of an Earthbound nuclear explosion, while the Los Alamos National Lab's X-ray/ neutron detectors are spread around each satellite to monitor for space events. Rockwell received the US$1.2 billion production contract in 1983 for 28 satellites of these Block 2s. The navy planned to deliver them to orbit aboard the Shuttle (10 planned for 1987 alone) in conjunction with McDonnell Douglas' PAM-D2 solid upper stage. These plans had to be dropped after the Challenger accident and the satellites had to be retrofitted for the Delta 2 expendable launcher.

In practice the GPS Navstar system has been operating at full strength (24 Block 2/2A satellites, including three hot spares) since March 1994. One ground spare remains in the military's inventory to be launched as an emergency replacement. Another can be built with two months' notice.

The Department of Transportation formally became part of the GPS management system in 1994 to accommodate civil interests. The DoD is thus officially operating the system in part for the civil community and must ensure that further testing and development has minimum impact on those users. With DoT on board pilots have now been given authority to use GPS as primary guidance for non-precision approaches without monitoring the ground navigation aid, providing they have autonomous integrity monitoring. Improvements in military cryptography may lead to at least partial civil access to the military PPS signal, improving accuracy in the near future.

President Clinton, on 29 March 1996, declared his intention to halting civil signal degradation within 10 years. March 1994's launch brought the system up to full Block 2/2A strength for the first time when it became operational on 28 March, but he left the USAF Space Systems Division to the USAF Space Command in

GPS: the first Navstar Block 2 satellite (Navstar 14) undergoes solar panel deployment testing at Rockwell's Seal Beach facility. It was delivered to the US Air Force 27 April 1987 as a Pathfinder in the Navstar Processing Facility at Cape Canaveral

Colorado with operational control of the system. The DoD is thus officially operating the system in part for the civil community and it must seek Presidential approval and then notify the FAA and Coast Guard (USCG) at least 48 hours before any planned reduction in satellite numbers. GPS became integrated 17 February 1994 into the US National Airspace System and the FAA plans the US$500 million Air Traffic Management system to be operational in late 1997, using GPS and 24 ground stations for aircraft.

In FY97, Congress gave the USAF a GPS budget of US$278 million (which supports procurement of first three Block 2F satellites). Other budget requests have been FY96 US$201 million, FY95 US$226.5 million, FY94 US$94.2 million, FY92 US$66.41 million.

Specifications

Navstar Block 1

Navstar 1-6 used the Atlas F from Vandenberg to achieve a 20,000 km circular orbit (all planes A through C) at 63° inclination. Number 3 suffered a clock problem. Number 6 took over the detection of nuclear explosions from the Vela Hotel satellites. Number 3 had to be taken out of service at the height of the Gulf Conflict on 11 December 1990 because a reaction wheel problem affected attitude control. The satellite returned to partial service 16 January 1991. Number 7 burned up in the Atlas launch failure. Navstar 8 was deactivated in May 1993, because of battery and attitude problems. Number 31 replaced it. Number 9 had a similar battery problem and left service in February 1994. Number 10 retired in 1996 as the last in-service Block 1 satellite. Number 11 had already retired from service February 1994 because of clock problems and removed from the orbit 14 April.

Signal strength: +23.8 dBW P signal, +26.8 dBW C/A signal (+28.8 dBW in high power mode)
Atomic clocks: three rubidium, one caesium
Mass: 760 kg at launcher separation, 455 kg initial on-orbit
Span: 5.34 m across deployed solar panels
Design life: 5 years (ACS propellant sufficient for 7 years)
Structure: aluminium honeycomb panels
AOCS: spin-stabilised. 18 hydrazine thrusters in two modules: 2 × 22 N + 16 × 4.5 N. Orbital injection by 28.8 kN maximum Thiokol TE-M-616 solid (25 kg burnt-out motor remains attached)
Power: 410 W EOL from 5.02 m^2 array of two Si cell panels; 3 × 15 Ah Ni/Cd batteries

Navstar Block 2

Navstars 13-21 weigh about 840 kg in orbit and each unit cost US$38.8 million in FY87 dollars. All operated throughout 1996, but the aging GPS 16 and GPS 13 had already been replaced by GPS 33 and GPS 40 respectively.

Navstar Block 2A

Navstars 22-40 have enhanced survivability in the event of a nuclear attack and weigh about 1,881 kg at launch. Navstar 40, the last of the contracted 28 satellites, was delivered in 1992 and launched by Delta 7925. The first (Navstar 22 or GPS 23) appeared in November 1990, increasing the operational system to 16 satellites. However, a problem developed in the circuit board of one solar array's control system 12 December just two days after launch. This glitch turned out to be a design fault and delayed the launch of Navstar 23 (GPS 24) from February to July 1991 while it was corrected. Then number 24 (GPS 25) experienced a further delay from August 1991 to February 1992 after a reaction wheel failed in a ground test because its extended storage had affected lubrication. Navstar 40 was launched 12 September 1996.

Navstar Block 2/2A

Signal spread: −160 dBW C/A & −163.0 dBW P on L1; −166.0 dBW C/A or P on L2, −165.2 dBW C/A on L3
Atomic clocks: two rubidium and two caesium clocks
Mass: 843/930 kg on-orbit BOL (1,667/1,881 kg launch)
Span: 5.34 m across deployed solar panels
Design life: 7.5 years
Structure: aluminium honeycomb panels, 1.95 m^2 radiating array
AOCS: spin-stabilised, 22 hydrazine thrusters in two modules with 60 kg supply sufficient for >10 years of normal operations. Orbital injection by 47.1 kN (2,097 kN total impulse) Thiokol Star 37XFP
Power: 710 W EOL from 7.26 m^2 array of two Si cell panels; 3 × 35 Ah Ni/Cd batteries.

Navstar Block 2R

Navstar 41-60 became the first round of replenishment satellites to maintain the fully occupied constellation. Each has been improved to provide six months of autonomous satellite operations without ground control corrections, increased service life and improved survivability. The atomic clocks, normally updated daily, offer five times the stability of their predecessors. The design phase contract cost US$119 million: the first 20 satellites will then cost US$575 million. First delivery was planned for October 1995 for launch March 1996, although payload development problems delayed first launch to 23 January 1997. First Block 2R satellite was launched aboard a Delta-2 (7925) 17 January 1997, but following a fault with one of the GEM strap-on boosters the launch vehicle self-destructed 21 s after launch, resulting in the loss of the satellite. The satellite had been intended to fill the vacant F5 slot. The second satellite (No 44) was launched on 6 November 1997. GAO estimated during 1990 that each wil cost US$200,000 annually to store. Launch will be on USAF's MLV-3 medium launch vehicle: Delta 7925 was selected April 1993. Cost of design and development was acknowledged in March 1993 to have risen from the planned US$99.1 million to US$119 million, primarily because of the requirement that the satellites should remain in continuous service throughout a nuclear war.
Mass: 2,032 kg at launch; 1,075 kg BOL on-station
Deployed dimensions: 152 × 193 × 191 cm body, 19.3 m span across solar arrays
Design life: 10 years minimum
Structure: equipment is mounted on six aluminium honeycomb structural panels mounted on central aluminium core. Communications equipment and some housekeeping units are mounted on the N/S faces; antennas and Earth sensors are on the Earth-facing panel
Power: twin 2-panel solar wings (totalling 13.4 m^2) using N-on-P Si cells providing 1,136 W EOL (970 W requirement); Ni/H^2 batteries provide 100 per cent eclipse protection
Pointing: not available
TT&C: at 2,227.5 MHz S-band

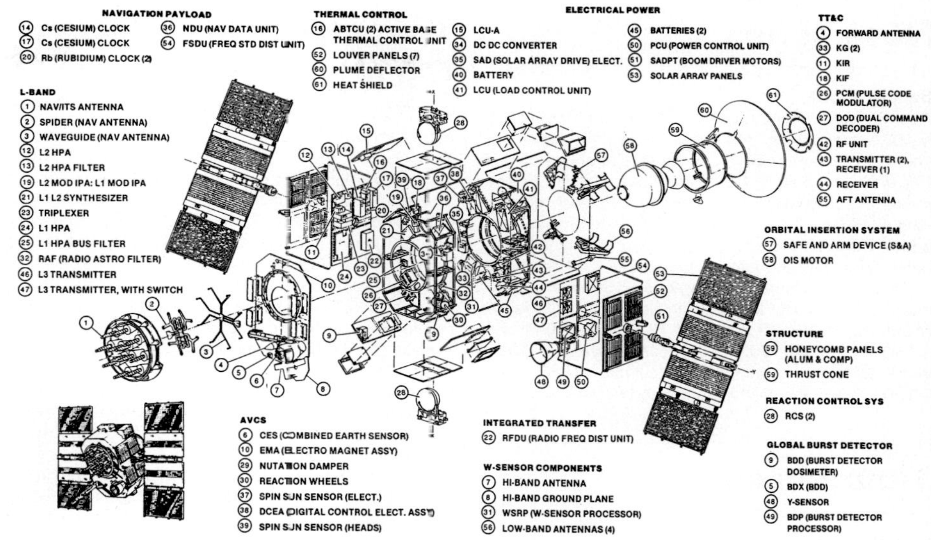

Navstar Block 2A satellite configuration; Block 2 was very similar (Rockwell)

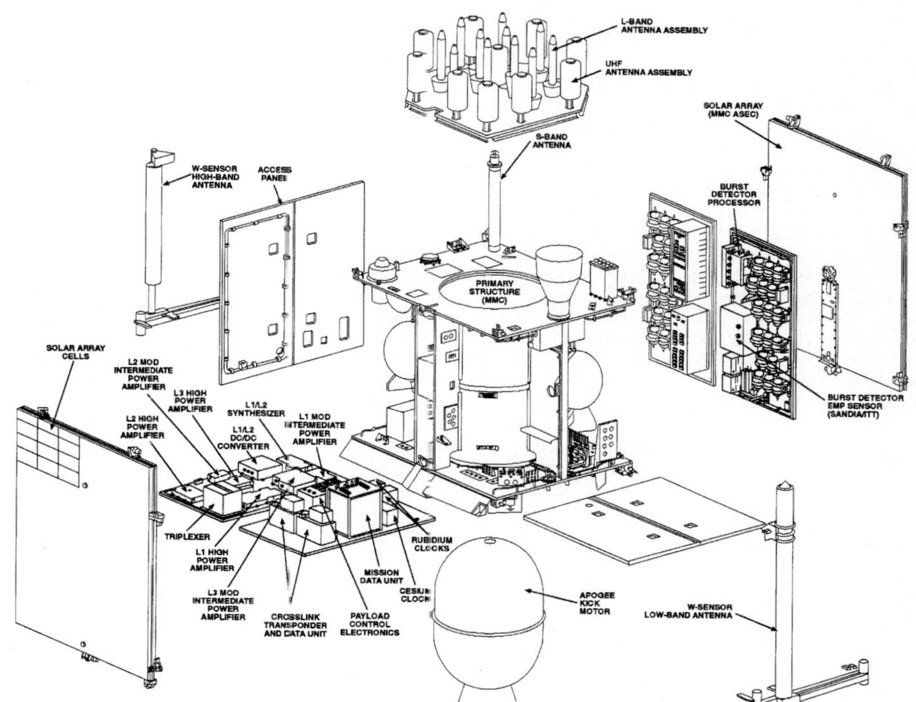

GPS Block 2R principal features (ITT)

1996

AOCS: 3-axis by momentum wheel, magnetorquing and 16 blowdown monopropellant hydrazine thrusters. The thrusters are divided into two redundant sets of eight: six 0.89 N catalytic Reaction Engine Assemblies (REAs) accompanied by two 22.2 N Electrothermal Hydrazine Thrusters (EHT) for increased performance NSSK, drawing on four tanks (94.8 kg total hydrazine). Insertion by Thiokol Star 37FM solid
Thermal systems: passive and heaters.

Navstar Block 2F

Navstar 61-93 have got underway with the initial US$382 million award for the next generation of satellites made on 22 April 1996 to Rockwell, the single team aiming for first delivery April 2001 and launch third quarter of 2002 on the AF Evolved ELV. Total programme cost over its life cycle is US$3.5 billion. The first award for six satellites will be followed by a contract in 1998 for 15 and in 2002 for 12, totalling US$1.3 billion. The original intention was for 51 satellites, but that was reduced to 33 in June 1995. DoT is funding a civil addition to provide the L5 signal for improving civil accuracy. The satellite network will be all-2F by 2010. Individual minimum life will be 10 years; design life is 15 years. Rockwell teamed with Computer Sciences Corp to bid; Lockheed Martin Astro Space again teamed with ITT Aerospace/Communications and Loral Federal Systems; Hughes' team covered Hughes Information Technology Co, Space Applications Corp and Stanford Telecom. Hughes had not bid on GPS contracts before. The Nuclear Detection System is expected to include new technology demonstrated by the FORTE satellite launched in 1997. It will look for RF pulses from low technology explosions, such as might result from a nuclear test by a proliferant nation.

Nova series

Background

After the commercial success of the original Transit (US Navy Oscar) series, the navy introduced Nova, an improved version of Oscar, designed by Johns Hopkins University and built by RCA. Each Nova includes a more powerful transmitter, improved reference clock, greater computer capacity, an onboard ability to compensate for orbital disturbances and a 7.6 m long gravity gradient boom for stabilisation. Nova 1 launched 14 May 1981 by Scout from Vandenberg; weighed 136 kg and went into a 1,170 × l,187 km, 109° inclination orbit. It failed March 1991. Nova 3 launched 12 October 1984 into 1,150 × 1,198 km, 90.1° orbit failed November 1993 and Nova 2 launched 16 June 1988 into 1,150 × 1,199 km, 90° inclination orbit is still active.

Timation series

Background

Time Navigation, later called Timation, began in 1964 as a Naval Research Lab project to demonstrate the principle of navigation by ranging using synchronised space and ground clocks as opposed to Doppler navigation used by Transit. The 25-kg Timation 1 appeared 31 May 1967 as a piggyback satellite released by Thor Agena D into 906 × 915 km 70° inclination orbit. Timation 2 followed on 30 September 1969 transmitting at 150 to 400 MHz. The USAF had its own version of time ranging Project 621B, calling for 12 satellites in eccentric 40,800 km orbits. The DoD combined the two programmes in 1973 creating GPS Navstar. NRL's 293 kg Timation 3, built before the merger, appeared in orbit after the merger on 14 July 1974 and received the name NTS 1 (Navigation Technology Satellite). The 430 kg NTS 2 launched 23 June 1977 into 20,181 × 20,181 km, 63.3° inclination orbit actually served as the first Navstar verifying the launcher, onboard systems, ground control software and tracking techniques. It carried the first caesium standard in space, now flying on Block 2/2A GPS.

Transit/NNSS (US Navy Oscar) series

Background

The Applied Physics Laboratory of Johns Hopkins University originally proposed and designed Transit to update the inertial navigation systems it had helped build on Polaris submarines. Eisenhower approved initial project funding in December 1958, but the first Polaris navigation fix did not occur until January 1964. After this success the US Navy opened up the sitemaps to civil users in 1967. Transit can be used in any situation where dead reckoning can be used for about an hour between updates. The navy did not shut the system down until 1996.

Even though more than 400 military users relied on the system, the commercial system found the greatest audience by far, with an estimated 80,000 commercial sets available at one time.

Transit operates by measuring the Doppler shift in the carrier signals from the satellite as it passes overhead. Each Doppler sample requires 10 to 15 minutes for proper accuracy. The satellites transmit timing marks and orbital information on two stable coherent frequencies (149.99/399.97MHz) that permit corrections for ionospheric refraction.

The navy created the Naval Satellite Operations Center (NAVSOC) in 1962 to operate Transit from Point Mugu, California. It has TT&C stations at Prospect Harbor (Maine), Rosemount (Minnesota), Laguna Peak (California) and Wahiawa (Hawaii). Wahiawa closed in September 1993 as a step towards phasing out Transit.

Using data from these stations the Pt Mugu computing centre refines the orbital model of each satellite every 24 hours and uplinks new ephemeredes typically every 12 hours. Sample tests indicate Transit locates a receiver to within an RMS accuracy of 15 to 25 m utilising a dual-frequency receiver. Single frequency reception yields 80 to 100 m. Because the speed of the vessel affects the signal, an error of 370 m is introduced for each 0.5 m/s of unknown vessel speed. As the current constellation has reached the end of its useful life, intervals between fixes now vary over 35 to 100 minutes. For stationary users, such as surveying and oil platform location, integration of measurements from 25 passes yields 5 m RMS accuracy. If a second receiver can be located several hundred km distant, then simultaneous tracking over 10 passes generates 0.5 m RMS accuracy.

After an initial launch failure, Transit 1B, launched 13 April 1960 into a 373 × 745 km, 51° inclination orbit inaugurated a test for the service by transmitting for three months. Over the next six years, 22 more Transit satellites joined the constellations. Most operated for about one year. However, many far exceeded their projected three-year lives. The most long-lived satellites included number 12 151 months, number 13 260 months and number 14 199 months.

In September 1983, the navy announced that of 15 satellites built by RCA (now Lockheed Martin Astro Space) in the 1970s, 12 remained in store because the three launched (2 March 1968, 27 August 1970, 30 October 1973) had far exceeded their design lives. RCA received a US$9.9 million contract to modify eight for dual launches by Scout from 1985 under the Stacked-Oscar-on-Scout (SOOS) programme. The navy intended to store these satellites in orbit rather than on the ground. The first pair, Oscar 24 and Oscar 30 went into orbit 3 August 1985 on a Scout.

At 55 kg, the Scout could put them into 1,005 × 1,263 km, 89.8° inclination orbits. The 59 kg Oscar 27/29 pair joined the predecessors on 16 September 1987 from Vandenberg in a 1,017 × 1,183 km, 90.3° inclination orbit. They brought the total of Transits in orbit at that time to nine. They were joined by the SOOS-3 pair (Oscar 23/32) on 26 April 1988 and then SOOS-4 (Oscar 25/31) on 25 August 1988.

The Defense Nuclear Agency purposely destroyed one of the four remaining satellites – Oscar 22 – on the ground by a 1.50 g aluminium pellet travelling at 6 km/s to simulate a space debris strike. The navy gave Oscar 26 to The Naval Post-graduate School in Monterey, California. The other two (Oscar 21/28) remain in storage at APL.

The US Navy shut Transit down in December 1996 now that GPS is in place. Three Oscars and one Nova were active at end-1995 in the system, plus three Oscars were stored on-orbit.

NUCLEAR SURVEILLANCE

UNITED STATES OF AMERICA

Alexis (STP P891-B)

Current status

Launched 25 April 1993, the US$17 million Department of Energy-funded satellite has lasted well beyond its nominal one-year mission to demonstrate its telescope and radio-receiver technology for nonproliferation applications and past a three-year lifetime engineering estimates gave it. After five years it still has most of its capabilities intact, even though the satellite is starting to show signs of wear and tear. The Blackbeard experiment has been superseded by Los Alamos' FORTE satellite.

A broken solar panel at launch imparted a wobble to the spin-stabilised spacecraft. Los Alamos scientists modelled the wobble and mapped the actual positions of the photons captured by the six telescopes to their originating location in the sky. The final computer software for decoding the telescope data was recently completed and used to process all 49 months' worth of archived data, stored on more than 150 CDs.

Background

LANL's Alexis (Array of Low-Energy X-ray Imaging Sensors) satellite performs an all-sky survey of the Universe's soft X-ray background glow at 62/72/93 electron Volts. The X-rays, possibly generated by interstellar gas, are detectable by Alexis's three telescope pairs, each with 30° fields of view. Los Alamos National Laboratory designed the project to demonstrate that low cost complex sensors can be built quickly for verification of future arms control agreements. Although Alexis's telescopes perform an astrophysical survey, they really monitor space for nuclear flashes as revealed by the fact that the Dept of Energy's Office of Intelligence and National Security funded the US$17 million programme.

In order to put the satellite in space the USAF provided launch services through its Space Test Programme and called the satellite P89-IB. AeroAstro provided the platform and the Apple Macintosh-based ground station. The launcher's avionics deck carries the VHF Spectrum Utilisation Measurement Experiment (VSUME) to research VHF usage and densities in specified military frequencies for the Naval Space Systems Activity. Alexis also carries the Blackbeard radio propagation technology development experiment for ionospheric research, Blackbeard can take 150 million measurements/second and detect radio noise from everyday sources such as car ignitions and toy walkie-talkies. This survey could lead to detection by future satellites of RF pulses from nuclear detonations.

It took the ground station several months to gain full control of Alexis because one solar panel broke loose during launch. Blackbeard began operating 11 July and the first X-ray telescope 26 July. Blackbeard has discovered bright and rapid near-surface RF flashes: hundreds of transionospheric pulse pairs. Although associated with thunderstorms, they are not regular lightning, but they are similar to nuclear pulses and so must be characterised. They were explained in 1995 as resulting from a runaway process creating an upward avalanche of electrons in the strong electric fields above thunderstorms. The initial electrons are provided by cosmic gamma rays colliding with air molecules. The second pulse arises from the weakening electric field still being strong enough in the very thin air at high altitude to generate another avalanche. The theory also explains the blue and red flashes discovered shooting up from thunderstorm clouds: the electrons cause air molecules to fluoresce, blue in thicker air and red in thinner.

Alexis carries an X-ray telescope array and radio propagation experiment (AeroAstro)

Specifications

Launch: 25 April 1993 by Pegasus
Mass: 109 kg (bus 45 kg, payload 64 kg)
Orbit: 747 by 836 km, 69.93°
Spacecraft: 61 cm diameter 104 cm high. Spun around Sun-pointing axis (to within 2°) at 3 rpm; 0.25° attitude determination. Four solar panels provide 50 W average to payload. 96 Mbyte mass memory, downlink S-band at 750 kbit/s to 1.8 m Los Alamos station daily, uplink 9.6 kbit/s. Mission life 1 year, goal 3 years; spacecraft reliability 95% after 18 months.

FORTE (STP P94-1)

Current status

Los Alamos National Laboratory's FORTE satellite has detected many thousands more radio bursts from lightning strikes and other phenomena than previously reported. Los Alamos scientists are using another major FORTE instrument, an event classifier, to distinguish the inherent structures of radio frequency signals in the 30 to 300 MHz range, which includes commercial television, FM radio, aircraft navigation and communication bands. Los Alamos claims a library of these sources will assist it in sorting out suspected nuclear detonations from ordinary events.

Background

Sandia National Laboratories launched the Fast On-orbit Recording of Transient Events satellite on a Pegasus XL on 29 August 1997 into 800 km 68° to test the detection of nuclear detonations by RF pulses. The mission is sponsored by the Department of Energy and jointly executed by the Los Alamos and Sandia National Laboratories. The purpose of the mission is to detect and distinguish between electromagnetic pulses caused by events such as low-technology nuclear detonations and lightning-strikes. This characterisation will be aided by optical sensors in order to correlate the optical signals with the RF signals. The satellite will also provide a proof of concept for an advanced composite bus made of graphite epoxy and cut into shape with a water jet. The advanced structure allows the satellite to carry an additional 23 kg of payload. The satellite is three axis stabilised, and most of the exterior panels are covered with solar cells to provide 160 W of power.

Forte will study RF pulses from nuclear detonations (Los Alamos National Laboratory) ***1995***

The satellite has an expected lifetime of at least one year (three-year goal). SNL supplied an optical monitor to eliminate lightning flashes. The satellite serves as a prototype for the GPS 2F-satellite generation, which the air force intends to use to look for the pulses from low technology explosions. The 10.7 m antenna, built by Astro Aerospace and comprising two log-periodic orthogonal arrays, is held pointing Earthwards by magnetorquers. Sandia is contributing the data to a database of lightning's global distribution. The optical monitor uses a 10 km² resolution wide field of view imager to locate lightning and a 50 kHz sampling rate to record separate flashes and trigger RF recording. Forte will test if it can sort out lightning automatically.

Ionds sensors

Background

The United States monitors the world for nuclear detonations represented by X- and gamma ray wavelengths with the Integrated Operational Nuclear Detection System (IONDS) packages aboard the DSP early warning and GPS navigation satellites. Detonations can be pinpointed with 100 m accuracy now that the Navstar system is completed, with every point on the globe within view of at least four satellites. The Dept of Energy's Sandia National Labs provide the Earth-facing 'bhangmeter' optical sensor that detects the characteristic double flash of an Earthbound nuclear explosion, while the Los Alamos National Lab's X/neutron detectors are spread around each satellite to monitor for space events. Rockwell equipment also looks for electromagnetic pulses. USAF in 1993 decided to delete detectors from the FEWS follow-on to DSP in order to save mass and money (FEWS was cancelled anyway later that year), relying instead on GPS after DSP has expired. LANL/SNL are flying a series off three satellites to demonstrate techniques for monitoring of nuclear detonations.

OCEAN SURVEILLANCE

RUSSIAN FEDERATION

Eorsat series

Current status

The Russian Federation launched a new Eorsat, code named Kosmos 2347 on 9 December 1997 at 7.17 pm from Baikonur aboard a Tsyklon rocket. The Russian launch announcement at first indicated it was a Yantar photo-reconnaissance satellite, but this was later proven to be false. The Eorsat is working in conjunction with Rorsat Kosmos 2335.

Background

Eorsats appeared for the first time at the end of 1974. They differ from the actively transmitting Rorsats and instead have the passive task of identifying the type and role of ships of the Western navies by noting transmission frequencies of their radio and radar. Consequently Eorsats do not have the same huge power requirements that the Rorsats have. As they gather information they manoeuvre extensively in low orbits and consistently maintain an exact orbital period of 93.3 minutes in circular orbits at 65°.

During the Falklands War, Eorsats monitored the progress of the British Task Force's journey to the Southern Hemisphere. A British commentator asked one commander how he avoided being tracked by Soviet satellites. He replied that the British Navy had countermeasures against the satellite monitoring. In 1982, the Soviets had three Eorsats in orbit, Kosmos 1286 (4 August 1981, 432 × 434 km, 93° inclination); Kosmos 1306 (14 September 1981, 409 × 462 km, 93° inclination) and Kosmos 1337 (11 February 1982, 429 × 447 km, 93° inclination). The Soviets did not have a paired Rorsat in orbit at the time of the movement of the British fleet, but did launch one when the fleet arrived in the Falklands.

Following their testing phase, the Soviets since 1979 usually used Eorsats in pairs, like the Rorsats. The Eorsats enjoy longer lifetimes than the Rorsats however, with some having operational lives of more than a year. Curiously, Eorsats usually exploded just before re-entry. The Soviets refused to announce whether the explosions were intentional or not, but characteristically ceased after 1987, when the space debris issue became a widespread concern. In 1985, the Soviets had three Eorsats active simultaneously for the first time, Kosmos 1567, Kosmos 1588 and Kosmos 1646 between 18 April and 14 July, and then again with Kosmos 1567, Kosmos 1646 and Kosmos 1682 beginning 19 September 1985. But the simultaneous operation ended 19 November 1985 when Kosmos 1567 left its tightly controlled orbit after a record 538 days on station.

The Eorsat programme began to evolve in 1986 when the first Eorsat of the year, Kosmos 1735 (27 February), entered a slightly lower orbit with a three day repeating pattern instead of the standard four day repetition. When the second satellite of the series that year, Kosmos 1737 (25 March) it followed the new orbital pattern, but at the higher inclination of 73.4°. All previous Rorsats and Eorsats had flown only at 65° inclination. But more importantly, Kosmos 1735 and Kosmos 1737 had exactly the same ascending node indicating a connection between the information gathered by the satellites. The two flew together until the Soviet ground control commanded Kosmos 1737 to

Eorsat bus (Kaman Sciences Corporation) ***1994***

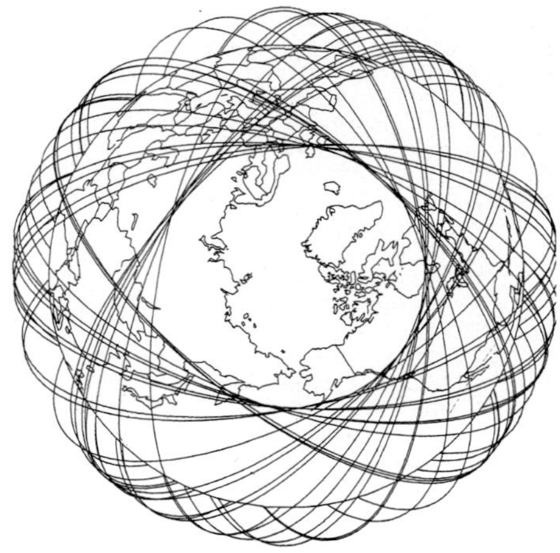

A considerable number of spent nuclear reactors and fuel cores are now in long-lived orbits about the Earth (Teledyne Brown Engineering)

re-enter. The ocean surveillance programme had never before witnessed an intentional re-entry.

By 1988, all Eorsats manoeuvred to achieve the three day repeating pattern. Those that had a 65° inclination however remained at that inclination with no repeat of the Kosmos 1737 profile. The first launch of 1988, Kosmos 1949 (28 May), established the first complete constellation of evenly-spaced Eorsats. The Soviets maintained that the constellation Kosmos 1834 terminated its mission on 29 August. A new constellation formed when Kosmos 1979 joined the other two in November, but by 8 December Kosmos 1890 completed its mission, leaving only the two 1988 satellites still operating. In a surprise move, Kosmos 1834 and Kosmos 1890 were both commanded to lower their orbits to accelerate natural decay. The two satellites re-entered without incident, signalling a departure from earlier fragmentation practices.

By 1989, the Soviet Union decided to expand the highly successful Eorsat programme and add new operational characteristics. Three Eorsats appeared as the year progressed. Kosmos 2033 went up on the 24 July, Kosmos 2046 on the 27 September, and Kosmos 2051 on 24 November enlarged the constellation to four active craft and for the first time inaugurated a multiple plane system to permit faster revisits to ocean areas of interest. Kosmos 2033 completed the plane started by Kosmos 1949 and Kosmos 1979. When Kosmos 2046 orbited it occupied a position exactly 172° behind the earlier plane. The orbit is significant in that its plane allows it to repeat exactly Kosmos 2033's ground track just 11 hours later. Kosmos 2051 joined Kosmos 2046 in the new plane but immediately experienced a propulsion problem. Several weeks of small manoeuvres saved the mission, although Kosmos 1979 had come to the end of its own useful life in the meantime. The replacement philosophy already in place repeated itself in March 1990 when Kosmos 2060 (14 March 1990) completed the plane started by Kosmos 2046 and Kosmos 2051, while Kosmos 1949 finally failed after a record mission of 22 months.

The Eorsats programme continued in 1990 with three new launches: Kosmos 2096 (23 August), Kosmos 2103 (14 November) and Kosmos 2107 (4 December), creating a record network of six satellites. Kosmos 2103 failed during the first few days of 1991, but Kosmos 2122 (18 January 1991) quickly replaced it. The CIS allowed the Soviet network to gradually deplete itself during 1991 as satellites ended their careers and no replacements appeared: Kosmos 2046 re-entered in April, and Kosmos 2060 retired in August and re-entered during September. At the end of 1991, only Kosmos 2096, Kosmos 2107 and Kosmos 2122 still operated. Kosmos 2107 manoeuvred off-station 13 March 1992 and decayed 6 April. Kosmos 2096 ended operations 2 August 1992 and re-entered 30 August, leaving only Kosmos 2122 at the beginning of 1993. Even this satellite re-entered 28 March 1993 after operating for 776 days.

Then Kosmos 2238 appeared 30 March 1993, and was joined by Kosmos 2244 on the 28 April 1993, in a co-planar orbit but trailing 120° behind. Kosmos 2258 appeared 7 July, coplanar with the previous two but 120° ahead of Kosmos 2238 and 120° behind Kosmos 2244. Kosmos 2264 started a new triplet 17 September 1993 with its plane 145° away and Kosmos 2293 (2 November 1994) opened up a third plane.

1995 began then with the Eorsat slowly rebuilding a full constellation having Kosmos 2244 and Kosmos 2258 in plane 1, Kosmos 2264 in plane 2 and Kosmos 2293 in plane 3. Kosmos 2244 moved off-station about 14 February 1995 and decayed 18 March. Kosmos 2258 moved off-station about 2 March 1995 (decaying 8 June), leaving only Kosmos 2264 and Kosmos 2293. Kosmos 2264 then deactivated 4 April and decayed 7 August 1995. Kosmos 2313 appeared 8 June 1995, joined on 20 December 1995 by Kosmos 2326. So at the beginning of 1996 the CIS had Kosmos 2293 Kosmos 2313 and Kosmos 2326 operating. Kosmos 2293 decayed May 1996. The role of Eorsats expanded when the Ioffee Physics Institute of St. Petersburg included a 131 kg Konus, a gamma ray burst detector from Kosmos 2326. The Granat satellites also carry the Konus.

Okean series

Background

During the summer of 1985, US Intelligence agencies reported that the Soviets had recently deployed a radar carrying satellite system. The satellite according to Soviet sources mapped ice formations in polar regions, and thereby enhanced the ability of the Soviet Navy to operate in icebound areas. The system aided navigation of northern sea routes, in order to assist in moving naval ships from construction yards in the western Soviet Union to new ports in the Pacific.

Radar mapping of ice floes certainly appeared to explain those satellites belonging to the Kosmos 1500 type, with its side-looking radar, and later emerging in a civilian guise as the Okean satellites. The system uses a wavelength of 3.2 cm, a swath width of 460 km, and a resolution of 1.5 km. Since 1985, the satellites transmit their data to Moscow, which uplinks it to ships via the Ekran geosynchronous satellites.

Kosmos 1602 followed by Kosmos 1500 on 28 September 1984, in the traditional 636 × 669, 82.5° inclination orbit. The third oceanographic satellite, Kosmos 1766, appeared on 28 July 1986 from Plesetsk into 635 × 666 km, 82.5° inclination orbit. The orbital plane for 1766 was 60° away from Kosmos 1602 and gave the impression that the Soviets were in the process of establishing a fixed constellation.

The next oceanographic satellite, Kosmos 1869, launched 16 July 1987, assumed a position 60° from its predecessors. The satellite failed to deploy its antenna correctly rendering the SLR unusable. In 1988, the full programme became operational with the 5 July launch of Okean 1.

Rorsat series

Current status

The Russian launch of an Eorsat (Kosmos 2347) on 9 December to work in conjunction with Kosmos 2335 indicates this Rorsat is still operating normally, following its launch in late 1996.

Background

During the Cold War the Soviet Union sought to monitor US aircraft carriers in the proximity of the USSR and to maintain the capability 'to oppose them in the event of war'. To carry out this plan they invested in a suite of ocean surveillance satellites. The US Navy viewed the development of this capability as a severe threat to its interests and has justified the existence of the US ASAT programme based on the need to eliminate the Soviet monitoring network. The satellites as configured by the Soviets and carried on by the Russians apparently relay the position, heading and identification of Western naval forces to RFAS forces in real time.

Each nuclear-powered Rorsats (radar ocean reconnaissance satellites) or Eorsats (Elint ocean reconnaissance satellites) works as part of a pair and the corresponding interferometric radar return allows the RFAS to build up a comprehensive view of surface activity. The satellites maintain the required separation accuracy and height by very low thrust engines. If a vessel attempts to use electronic countermeasures against the low altitude (255 km) Rorsats the radio activity probably exposes the existence of the ship to the accompanying Eorsat, orbiting above at about 435 km. Significant as this threat is to US Naval operations, the Russians are apparently phasing the programme out and the country has launched no identifiable Rorsat launches since 1988.

Viadimir Vasilkovsky, Deputy Chief of Department for nuclear reactors of the Russian Atomic Energy Ministry, said in January 1994 that the Russians have no plans to launch nuclear power sources in the next few years.

The Soviet Union launched two Eorsats and two Rorsats in 1987. Kosmos 1834 (April 1987) joined two similar Eorsats, Kosmos 1735 (27 February 1986) and Kosmos 1769 (4 August 1986), at the same time that five Soviet Victor 111-class nuclear attack submarines conducted simulated attacks against New London and Norfolk. Another Eorsat, Kosmos 1890 (10 October 1987) soon followed Kosmos 1735 after Kosmos 1769 ceased functioning. Kosmos 1769 and another Eorsat, Kosmos 1646 (18 April 1985), broke up in September/November, respectively, continuing a pattern set with the very first Eorsat, Kosmos 699 (1974).

Of the two Rorsats launched in 1987, the first lasted for only a brief period. The second had an unusual orbit. Kosmos 1860 (18 June 1987) flew shortly after the Soviet Union publicly complained of US violations of Soviet territorial waters and roughly coincident with the build-up of US naval forces in the Persian Gulf after the Iraqi attack on the USS Stark. During the 1980s, most Rorsats lasted in orbit about 80 days (Kosmos 1365 lasted 136 days, the maximum of any in the series), but Kosmos 1860 ended its mission after only 40 days when its nuclear reactor separated and manoeuvred into a higher altitude storage orbit. Kosmos 1900 (12 December 1987) deviated from the 20-year-old orbit unique to the radar satellites when it entered an orbit about 8 km higher than its predecessors.

In 1988, Kosmos 1900 experienced a control malfunction and almost re-entered with its nuclear power supply. Soviet ground controllers attempted to change its orbit to no avail and then on 30 September 1988, only five days before re-entry a back-up system propelled the reactor to a safe storage orbit. The Soviet's launched their last mission Kosmos 1932 on 14 March 1988. Lincoln Lab's Haystack radar reports the Rorsats appear to be leaking coolant, according to orbital debris research.

While the Rorsat programme has been inactive since Kosmos 1900, the Eorsat operations continued to expand into 1991. The number of operational craft in working constellations has grown from an average two to three operational craft in 1987. The constellation grew steadily until a record six satellites operated simultaneously in late 1990. Perhaps to compensate for the absence of Rorsats, the CIS launched five Eorsats between March 1990 and January 1991. Three remained operational at the end of 1991 and only Kosmos 2122 still operated at the end of 1992. That re-entered 28 March 1993 after operating for 776 days, leaving no operational satellite for the first time in a decade.

The powerful pulsed X-band radar signals, used to locate naval ships in all weathers with resolution of tens of metres, easily indicate the purpose of the Rorsat satellites. These satellites can identify formations of ships and their headings/speeds. Then they downlink targeting date within one revolution to ships or other weapons platforms carrying long range anti-ship missiles. Ships of the western powers therefore have less than 90 minutes to take evasive action or otherwise prepare for attack.

The paired satellites flew out of Tyuratam on SL-11 into 270 km circular orbits. Each satellite is about 10 m long, 1.3 m in diameter. A 2 kW Bouk multi-cell thermionic reactor built by Kraznaya Zvezda containing 31.1 kg of 90 per cent enriched U235 divided into 37 fuel elements in beryllium reflectors powers each of the X-band phased array radars. The USAF uses its visible/infra-red telescopes, such as those at the Maui Optical Station on Hawaii, to determine if the satellite's reactor remains active by its glow. The Soviets and then the RFAS normally separates the reactor and boosts it to a 900 km orbit with a 500- to 600-year lifetime at the end of its 60- to 70-day mission.

The Soviet Union placed 33 Rorsats in orbit (plus two launch failures), although some early flights might have been reactor tests only with no radar. Kosmos 198, launched 27 December 1967 (249 × 270 km, 66° inclination), began the series. In the 1970s, a Rorsat created an international furor when it failed and its nuclear reactor began an unpredictable re-entry sequence. The satellite Kosmos 964 launched 18 September 1977 into 251 × 265 km, 64.98° inclination orbit malfunctioned and started to decay on 1 November. The Soviet controllers were unable to boost its nuclear power pack to a safe high orbit, and the world awaited with grave speculation wondering which country the radioactive debris would impact. 92 days after launch, on 24 January 1978, Canada became the site and main debris fell west and south of Yellowknife along an 800 km strip of the Northwest Territories. In one of the first international applications of the US Nuclear Emergency Search Team the American nuclear specialist set up operations known as Operation Morning Light to recover the debris.

Still, the US withheld protesting too vigorously because its own nuclear powered US Navy satellite launched from Vandenberg and powered by a Snap-9 nuclear generator with plutonium 238 had failed to reach orbit, broken up and released radioactive material over the Indian Ocean. In the end the Canadian government estimated that the Kosmos 954 search had cost C$6 million, the Soviet Union contributing about C$3 million.

In April 1979, the Soviet Union used a pair of non-nuclear satellites for ocean survey, but then quickly resumed the nuclear powered ones a few months later. The US State Department issued a 'statement of regret' to protest the resumption of Rorsat nuclear power when Kosmos 1176 orbited on 29 April 1980. Despite this, the programme continued in 1981 with Kosmos 1249, launched 5 March into 252 × 265 km. At the end of its life the Soviet Union boosted its nuclear power pack to 898 × 985 km. Kosmos 1266, launched 2 April 1981, operated for only eight days before its manoeuvring rocket and platform detached and manoeuvred to an 891 × 965 km with 600-year life. Apparently the satellite malfunctioned causing the Soviets to abandon it early and a similar incident prematurely ended the life of Kosmos 1299, launched 24 August 1981 into 248 × 269 km and boosted to 910 × 984 km after only 12 days.

For the first time since the Kosmos 954 drama, a pair of Rorsats, Kosmos 1365 and Kosmos 1372, operated as planned for 136 and 71 days respectively before being boosted into higher orbits. They were thus able to work together during June, July and early August, surveying the South Atlantic and, presumably, giving the Soviets information about the Falklands war.

Two more satellites followed, identified as Kosmos 1402 and Kosmos 1412, launched 30 August and 2 October. They operated for 118 and 39 days respectively. They too monitored events in the South Atlantic together during October and early November. Then, exactly five years after Kosmos 964, another Rorsat Kosmos 1402 created the same heart-stopping drama when the Kettering Group reported that between revolutions 1,925 and 1,926 the separation and boost ng events failed. The nuclear reactor would remain stuck in a lower orbit that would eventually decay.

The Soviets expressed confidence that their new design would ensure that the reactor would burn up upon re-entry. The safety measure succeeded: the less radioactive reactor vessel and instrument section re-entered over the Indian Ocean in January 1983, and the core disintegrated over the South Atlantic on 7 February. No increase in radioactivity was detected in either area.

Other Soviet end of life operations proceeded successfully and the booster put Kosmos 1412 in the proper end of life orbit at 901 km. The Soviets may have again modified the units after the Kosmos 1402 incident, because no new Rorsats entered service in 1983, but two, Kosmos 1579 and Kosmos 1607, orbited in the second half of 1984. Both operated for about three months and then their nuclear power supplies separated and entered higher orbits.

In 1985, the Soviets launched two more Rorsats Kosmos 1670 (1 August) and Kosmos 1677 (23 August). The timing of these launches as two other Eorsats already operated in orbit, appeared to coincide with the large NATO naval exercise Ocean Safari 85. Both reactors separated during 22-23 October 1985 after missions of 82 and 61 days respectively. They then manoeuvred into the standard storage orbits. The 1986 Rorsats, Kosmos 1736 (21 March) and Kosmos 1771 (20 August), completed separate missions of 92 and 56 days. At the end of 1987, only Kosmos 1900 still operated. During the second week of April, Kosmos 1900 failed to respond to radio commands and began to decay slowly from its 265 km mean attitude. On 13 May, Tass acknowledged the satellite was in trouble and predicted a re-entry in August or September. Meanwhile, the spacecraft's nuclear reactor and attitude control system continued to operate which prevented any of the backup safety systems from being activated and sending the reactor to a higher orbit. On 30 September the satellite apparently lost all stabilisation when it used the last of its propellant supply. The depletion of propellant triggered a stored computer command to separate the reactor and propel it to a higher storage orbit.

In 1988, the Soviets followed up on the programme with only one Rorsat launch, Kosmos 1932, which operated for 66 days. Because of the malfunction of Kosmos 1900 and a growing worldwide concern about nuclear reactors in near Earth orbit, the Rorsat programme apparently ended with this launch. Nonetheless, Kraznaya Zvezda, which developed the 5 to 6 kW Topaz multi-cell thermionic reactor, has discussed sales to the United States and provided prototype reactors to the US Department of Energy.

UNITED STATES OF AMERICA

Geosat FO series

Current status

A Taurus rocket successfully deployed the US Navy GEOSAT Follow-On (GFO) satellite on 9 February 1998 at 5.20 am Pacific time. The launch from Vandenberg placed the satellite released from the Taurus's fourth stage into a 775 km circular, 108° inclination orbit. The GFO has a final configuration of 813 lb. Ball Aerospace and Technologies Corporation built it for the US Navy's Space and Naval Warfare Systems Command. The satellite will use a radar altimeter and a water vapour radiometer to measure differences in sea surface height associated with ocean currents and eddies.

Background

The US Navy awarded a US$46 million contract to Ball Corp by their Space & Naval Warfare Systems Command in August 1992 to build the Geosat Follow-On. This satellite is a radar altimeter, which provides accurate measurement of ocean topography for the Naval Meteorology & Oceanographic Command in Bay St. Louis, Missouri. By measuring sea height the satellite provides the navy with information on sea floor topography and water density, surface roughness and by association, an index of wind speed. The navy will not discuss altitude precision for reasons of classification, but the similar civil Topex/Poseidon routinely achieves 2 to 5 cm wave height resolutions. GFO when properly interpreted forms a vital link to the scientific understanding and accuracy of global ocean circulation models under development for the navy's new class of supercomputers. It also identifies warm and cool water masses, which is important knowledge for anti-submarine warfare. For the first time, the navy may also

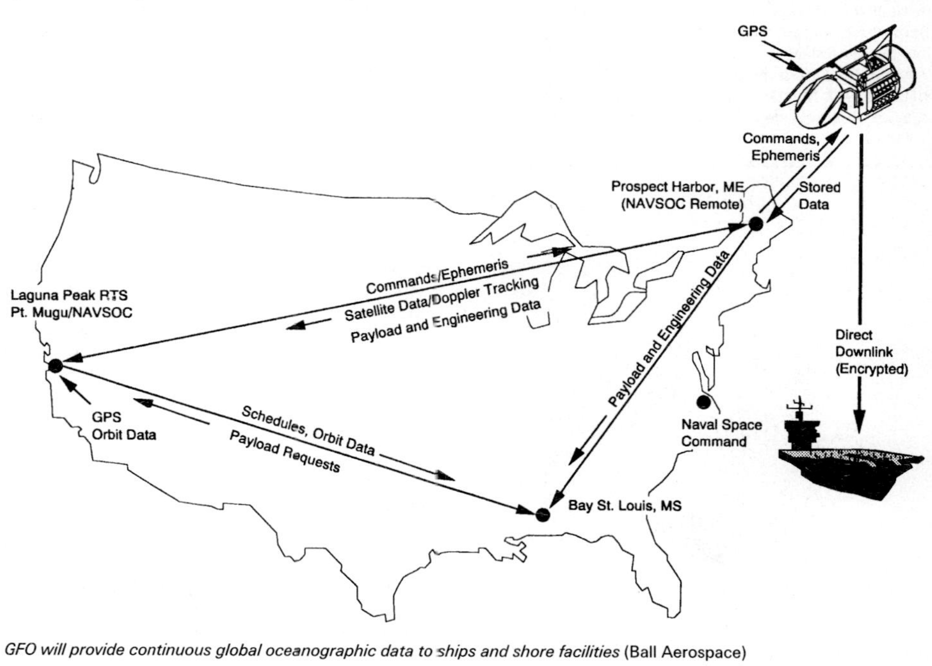

GFO will provide continuous global oceanographic data to ships and shore facilities (Ball Aerospace)

use the satellite for real-time direct encrypted readout to 65 ship/shore terminals. GFO has an important role for NOAA as well as filling in gaps in the data for climate and long range weather forecasting and global change predictions.

Specifications

Launch: 9 February 1998 by Taurus from VAFB
Orbit: 775 km 108° polar, 17 day repeat
Design life: 8-year minimum
Mass: 347 kg at launch
Payload: Pulse radar altimeter, measure sea surface height with <3 cm accuracy
Passive radiometer to correct altimetry data for tropospheric water vapour effects
AOCS: 3-axis control. GPS will provide precise orbit determination (altitude precision <5 cm). Eight hydrazine thrusters drawing on two tanks
TT&C: by NAVSOC Naval Satellite Operations Center at Pt Mugu, California.

Geosat series

Background

The Navy launched Geosat on 13 March 1985 by Atlas F from Vandenberg into 760 × 817 km, 108.05° inclination orbit. Currently the Johns Hopkins Applied Physics Laboratory, developers of the satellite, operate it for the US Navy. Geosat houses a radar altimeter to measure sea surface height and map wave action. The navy has been largely without such data of 3 to 4 m accuracy since Seasat's failure in 1978. Radar altimetry and the instruments to conduct it are also carried by the navy NROSS programme.

When it was launched, the navy hoped to get three years out of Geosat, but RADM Richard Pittenger, Oceanographer of the Navy, indicated in April 1989 that it remained operational and would probably remain so for several years more. However, its onboard recording facility failed in October 1989 and the altimeter failed 5 January 1990. In June 1992, the navy declassified all Geosat data covering south of 30° south and followed in July 1995 by declassifying the remainder. NOAA is responsible for public distribution.

NOSS

Background

The advent of NOSS-1, launched 30 April 1976 by Atlas from Vandenberg into 1,092 × 1,128 km, 63.5° inclination orbit, marked a new capability in US monitoring programmes because of the much larger main satellite. This programme became known as Navy Ocean Surveillance Satellite programme or NOSS. In orbit NOSS-1 and its cylindrical successors released three box-shaped subsatellites into similar orbits. 1993's Russian article claimed the satellites flew in a triangular formation around the parent, separated by 50 to 240 km. Actually, the new system flies with 30 to 110 km separations. NOSS locates targets by passive radar interferometry (hence the subsatellites), and is able to track surface ships in all weathers, with RF facilities for listening to their radar and communications. It may incorporate infra-red sensors to track submerged nuclear submarines, detecting their warm water wakes. Some have suggested the infra-red detectors also find low flying missiles.

NOSS-2 was launched 8 December 1977, followed by NOSS-3 on 3 March 1980, suggesting a three- to five-year operational life. A launch failure of 9 December

Geosat Follow-On will provide accurate sea height measurements. It will directly support naval operations such as ship routeing, anti-submarine warfare and amphibious operations (Ball Aerospace)

1980 was followed by NOSS-4, on 9 February 1983 by Atlas from Vandenberg into 1,063 × 1,186 km, 63.4° inclination orbit. It released four box-shaped subsatellites, designated SS (for surveillance satellite) A-D into similar orbits. NOSS-5, launched 10 June 1983, released three subsets designated GB1-3; and NOSS-6, launched 5 February 1984, released three subsets designated JD1-3 into similar orbits. NOSS-7 followed 9 February 1986 and NOSS-8 concluded the series of five Atlas H launches on 15 May 1987, heading into 1,050 × 1,170 km 63.40° orbits. The Titan 2 launch of 5 September 1988 was initially believed to have carried NOSS-9 but the lack of subsets and the subsequently announced orbit of 185 × 292 km, 85.0° seemed incompatible with the programme. No orbit is known for the second Titan 2 launch, of 5 September 1989, but again no subsatellites appeared. These are now believed to have been Elints.

PARCAE/Whitecloud series

Background

As Soviet Naval power grew during the 1960s, the United States Navy responded by developing a Naval surface ship detection programme to track and locate the Soviet fleet. Satellites offered the only hope because US aircraft had become woefully inadequate at tracking and monitoring the newly expanded fleet. The resulting programme called Whitecloud in unclassified documents is also known by the classified name PARCAE. The name Whitecloud surfaced in October 1984, when it was reported that postal covers carrying the name and some details were on sale at NASA Johnson Space Center's gift shop.

Whitecloud resembles the Russian Rorsat programme in that it passively measures signals emitted from naval ships. The Naval Research Lab developed and built the first two prototype satellites, and the navy launched the first one on 14 December 1971 by Thor Agena. Then contractor Martin Marietta took over manufacture along with E-Systems of Dallas, which built the payload itself. These satellites provided the basis for all US surveillance of Russian shipping until about 1976 when other programmes took their place.

At one time the navy planned for a programme equivalent to the Russian Eorsat called Clipper Bow. Congress killed the programme in 1980, because among other reasons the navy could not offer a sufficient justification for duplicating the capability already available on the LaCrosse. Some reports still indicate that Clipper Bow is alive and the navy will deploy it sometime in the 1990s.

When the US Navy launched the second Titan 4 vehicle off the assembly line into an unusual orbit, space observers speculated that the navy had finally completed its Clipper Bow programme. The launch on 8 June 1990, originally put the satellite into a 448 km, 61° inclination orbit. The satellite, a single USA-59 surveillance satellite went into this orbit, but the principal satellite then manoeuvred into a 1,116 km 63.4° inclination orbit and released three payloads (USA 60, USA 61 and USA62). Then Vandenberg's 8 November 1991 Titan 4 launch of USA 72 followed a similar pattern with objects (USA 74, USA 76 and 77) released into 1,053 × 1,165 km 63.4° inclination orbits. The Titan 4 failure of 2 August 1993 probably carried the same platforms. Titan's 5 December 1995 launch of USA 116 may have been the first replacement for the original payload, but the navy did not give this satellite a USA designation which is generally the surest indication of the nature of the payload. The May 1996 launch carried the pattern forward with the first object being identified as USA 119 and three subsets as USA 120, USA 121 and USA 122.

An article published in July 1993's Foreign Military Review by Russia's Ministry of Defence, provided the first generally available information that these satellites are the advanced version of the US Whitecloud programme satellites. According to the article, the main satellite weighs 7 tonnes, compared to the initial Whitecloud which weighed only 600 kg. A fully operational Whitecloud uses four clusters of satellites, one transmitter and three interferometric receivers. The detection of ocean going vehicles requires 1.5 m of resolution or better a level of radar imaging that has only been available on the Vortex satellites to date.

In May 1987, the navy launched its tenth Whitecloud and the last (fifth) by Atlas H.

Whitecloud/NOSS: a rare sight of the US Navy's Ocean Surveillance Satellite. Three subsatellites cluster around the parent vehicle before release (Aviation Week)

PHOTO RECONNAISSANCE

FRANCE

Helios 1A satellite

Current status
In March 1998, Germany announced it was withdrawing from the Helios 2 partnership and France began to re-evaluate its continuation of the programme.

Background
France joined the rank of nations capable of performing autonomous military reconnaissance from space when it launched the Helios 1 surveillance satellite in July 1995 aboard an Ariane. On a platform based on the civil Spot Earth resources observation satellite, Helios provides the fourth independent military surveillance capability after those of the USA, RFAS and China. During the 1987 parliamentary season, France announced it had paid FFr7.6 billion for the programme, and had saved FFr300 million by joint development with Spot 4. France claims the satellite allows it autonomy from the US decision-making process in the western alliance, by giving it an independent source of corroboration for US Intelligence claims. France justified the decision at the time by saying it needed to upgrade its data gathering ability to prepare for the expected improvements in Soviet strike capability in response to the US Strategic defence Initiative. Germany for a while agreed to participate in the programme but then later backed out, citing a desire to build a radar imaging system rather than a visible spectrum satellite. After this, CNES proposed that the Helios and Spot 4 production programmes be merged to reduce costs. The world recognised France's sophistication in this new field when it first published Spot photographs of 10 m resolution, showing Soviet naval, aircraft and nuclear weapon storage facilities at Murmansk and Severemorsk in 1987. The USA and the CIA immediately challenged France's right to publish such data at the time. The White House Senior Interagency Group for Intelligence had to file a report on the national security aspects of civil remote-sensing programmes.

The Italians and Spanish share part of the financial burden and the image products with France. Helios 1 is controlled from the CMP (Centre de Maintien Poste) at Toulouse, and has a back-up control facility at Francazal Air Base. The new Fresnel building at CNES Toulouse handles both Helios 1 and Spot 4 control and schedule uplinking. The images are downlinked to three CRI (Centre de Reception d'Images): Colmar in northeast France, Lecce in southeast Italy and Maspalomas in Spain. Each of the three capitals has an individual processing centre. Controllers can select onboard encryption so that imagery is restricted to one or more of the three principal nations or is more widespread. Initial reports declared Helios 1A imagery to be excellent. Helios I initially flew into a 673 × 676 km orbit, but the French adjusted it 14 July 1995 to 680 × 682 km, 98° inclination Sun-synchronous orbit that has a spot revisit time of 48 hours for one satellite.

Helios 1A platform testing at Intespace. See France's National entry for an artist's impression of the satellite in orbit (Ministère de la Défense) ***1996***

The Western European Union may become a partner after deciding in 1995 against its own system. Aerospatiale is working on the IHR Instrument Haute Resolution, a second-generation imaging system, increasing resolution to 50 cm and adding an infra-red capability.

Specifications
Helios IA
Launch: 1A 7 July 1995, by Ariane 4 from Kourou, French Guiana
Lifetime: 5 years
Resolution: 1 m
Lifetime: 4 years
Mass: 2,537 kg.
Comment: Italy is providing 14.1% funding and Spain 7%. Based on the extensive experience derived from the civil Spot, Helios provides the 4th independent military surveillance capability after those of the USA, RFAS/ USSR and China. Programme cost about FFr10 billion. Helios 1B is planned for launch by Ariane to back-up and complement 1A operations; orbits will be phased to provide 24-hour revisit periods. The FFr11.5 billion Helios 2 programme, in parallel with Spot 5 development, will deliver the first of two satellites in orbit in 2001. Germany is providing 20%; Italy and Spain are expected to participate.
Contributors: Délégation Générale pour l'Armement/ Direction des Missiles et de l'Espace (DGA/DME, executive programme management under AGEX Agence Exécutive), CNES (system/spacecraft architect), Aerospatiale (system industrial architect, EPV imaging system, GSR-3 solar array), Matra Marconi Space France (satellite, processing centre), SEP (processing centre in collaboration with Matra), Alenia/Inisel (ground stations), Alcatel Espace (payload electronics, command/communications systems), Sodern (DTA 04 cameras).
Orbit: 1A released into 673 × 676 km, adjusted by 14 July 1995 to 680 × 682 km, 98.07° Sun-synchronous; revisit time 48 hours for one satellite.
Comment: Spacecraft: based on Matra's Spot 4 bus 3-axis control: Spot 4's 0.05° RMS pointing accuracy improved to 0.005° RMS by addition of star tracker. 5-panel GSR-3 solar wing.
Sensors: EPV Ensemble de Prises de Vues imaging system using DTA 04 multispectral CCD cameras, and electronic intelligence equipment.

RUSSIAN FEDERATION

Introduction

Current status
The Russians launched a Yantar 4K class (Kobalt) close look photo-reconnaissance satellite from Plesetsk on 15 December 1997. The spacecraft carries two or possibly four externally mounted data return capsules which are ejected for earth return while the main payload remains in orbit.

Kosmos 2349a Yantar 1KFT/Kometa topographical mapping satellite orbited on 17 February 1998, aboard a Soyuz-U launched from Baikonur. The satellite had an expected lifetime of 45 days at launch.

Background
In the more than 35 years since the Soviet Union orbited a Zenit 2 photo-reconnaissance satellite aboard Kosmos 4, the country has launched more satellites for this purpose than any other. With about 770 launches having reached orbit, no other country even comes close to this dedication to photo surveillance.

But since the break-up of the Soviet Union and the introduction of new technology, the new Russian Republic has greatly decreased the number of its photo-reconnaissance launches. In part, simple economics influenced this trend, but another factor is that the third-generation satellites which remained in orbit only about two to three weeks are being replaced by the fourth- and fifth-generation satellites, with lifetimes of around two months and around a year respectively.

In 1987, photo-reconnaissance satellites reached a peak in orbital days of operation by completing more than 1,400 days in orbit. In the 1990s this has dropped to: 1990 - 962 days, 1991 - 896 days, 1992 - 808 days, 1993 - 840 days, 1994 - 940 days, 1995 - 745 days and 1996 - 361 days. Part of the lost operational time resulted from launch failures of fourth-generation satellites in May and June 1996. Following the de-orbiting of Kosmos 2320 on 28 September 1996 the Russians have not had a photo-reconnaissance satellite in orbit through April 1997.

The Russians launched only two of their new fourth-generation Yantar-2K satellites in each of 1994 and 1995, and so, if the two launch failures of 1996 each carried a Yantar-2K, the Russians may simply have depleted their inventory of satellites for the year. Normally the Russians also launch only one (sometimes two) Kometa satellites each year, but failed to put one in orbit during 1995.

What intrigued Western observers is the fact that the Russians did not replace the fifth generation Zenit satellite after the de-orbit of Kosmos 2320. The Russians did phase out the Soyuz-U2, so it could be simply that no method to put the satellite in orbit currently exists.

Zenit 2 (Generations 1-3) series

Background
The Soviets' first of the type of photo-reconnaissance satellite and only their 18th launch, Kosmos 4 in 1962, carried the 4,300 kg Zenit 2. After the fall of the Soviet Union, various commentators revealed the name and some operational details of the programme in 1990. More recent accounts in the 1995 book Corona fill in some of the details. Sergei Korolev developed the Zenit 2 at his design bureau OKB-1 and fitted the film capsule into the manned Vostok platform which could be retrieved from space automatically. The affiliate Central specialised Design Bureau took over the programme in 1964 and continued to develop new and better film canisters. As the programme became more sophisticated, the film canisters became able to carry more film and the vehicle itself could manoeuvre more in space. The satellites referred to in the West as Generations 3 through 5 grew in mass to about 6 to 7 tonnes. The Soviets launched roughly 30 to 35 a year through the 1980s, but cut back sharply to 21 in 1990 as the more advanced longer-lived models took over. 1991 and 1992 saw the Russians only launch 11 Zenits in each year. In 1993 and 1994, they cut back further to seven each, and in 1995 the number of launches fell to only three.

The Russians continue to rely on film returned from space at a time when the United States has long abandoned the practice preferring instead digital imaging. Film canisters offer higher resolution than the digital equivalent and some sources say the Russian photographic resolution is 20 cm. Electro-optical digital imaging provides sub metre resolution. The Russians have moved into digital processing however, to receive more timely the information than film. The increasingly-important Generation 5 Zenit returns digital data and the electronic emission form the Generation 6 and Generation 7 satellites indicate this is their method as well.

Initially, the Soviets snapped both high- and low-resolution pictures from orbits of about 200 × 250 km. They increased the image quality, beginning in 1974, by lowering the perigee of the satellites to about 170 km. The Generation 4 Zenits can lower their orbit to 150 km or lower over areas of military or geographical significance. They also co-ordinate these changes with information from weather satellites to check whether the target cloud cover would waste film.

The Soviets claim their satellite, Kosmos 932, launched 20 July 1977, 180 × 342 km, 65° inclination, warned the world that South Africa verged on detonating a nuclear device in the atmosphere. As a result of this the Western Alliance joined with the Eastern Bloc to urge the South Africans to desist.

Photo-reconnaissance satellites found other uses beyond their primary missions. The Soviets manoeuvred some Generation 3 low-resolution photo mappers into the unusual orbit of 350 × 420 km early in their lifetimes. From this altitude the satellites composed photo-mosaics of the earth in 11 days, covering the globe in strips of 2° of longitude. The Soviets launched their early Zenit satellites from Plesetsk at an inclination of 73° or of 70° from Tyuratam. By the fifth generation

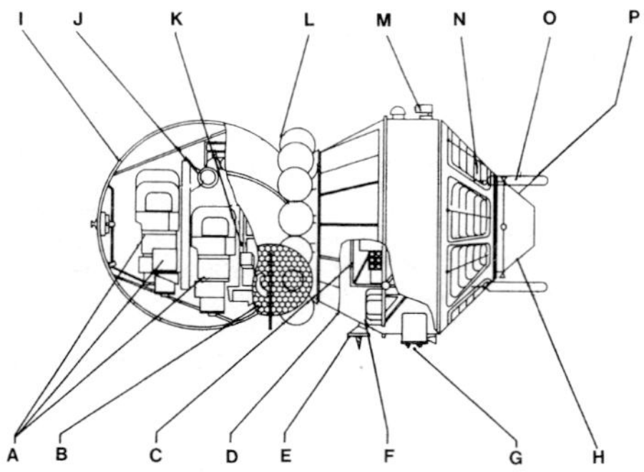

This first drawing of the Zenit 2 photo-reconnaissance satellite was released in 1993. It is very similar to the manned Vostok except principally for the cylindrical section in the service module. This carried IR/Sun sensors for the improved orientation required for imaging, and also possibly additional batteries. A cameras, B sigint antenna, C electronics, D service module, E radio antenna, F/K control system, G IR vertical sensor, H retrorocket, I parachute hatch, J self-destruct system, L N_2 spheres for cold gas attitude thrusters, M solar sensor, N thermal radiators, O/P radio antennas **1996**

all launched activity had switched exclusively to Tyuratam.

Several marketing firms have begun to sell old photographs with 5 m resolution to foreign customers. Even the US Air Force has become a potential customer. It has its own interest in buying Generation 5 satellite imagery. Priroda plans to distribute up to the minute digital 3 m resolution panchromatic imagery via their own ground stations from a military satellite, presumably a Generation 5 Zenit.

Operationally, some Russian and former Soviet spy satellites use a beam-splitter mirror to locate themselves extremely precisely by photographing a star background. This triangulation technique is often used to find exact locations of US ICBM silos and may even be used operationally to increase the accuracy of SLBMs. The Soviets adopted the practice of destroying their spy satellites at the end of their useful life in order to prevent them from falling into western hands if they did not completely disintegrate upon re-entry. The Russians apparently ended this practice after the detonation of Kosmos 2262, 102 days after its launch 7 September 1993.

The Soviets have co-ordinated launches and orbits with events unfolding internationally. When Iraq invaded Iran in 1980, the Soviets quickly launched Kosmos 1210, into a 187 × 320 km 82° inclination orbit. The satellite passed over the battlegrounds at noon each day and was recovered on the 14th day, 3 October Kosmos 1211 left for orbit before Kosmos 1210 returned and this new satellite followed the exact orbit of its predecessor.

The venerable Zenit-class photo-reconnaissance satellites ended their tenure as the workhorse of photo reconnaissance with Kosmos, launched in June 1994 – the last of the series.

Generation 4 (Yantar/ Kometa) series

Background

At the same time as the United States moved toward extreme close up photography, the Soviet Union had similar projects in the works. One was the Yantar (Amber) 'Close Look' satellites also known as Generation 4. The Series got underway with the launch of Kosmos 697, on 13 December 1974. At 6,700 kg, the satellite flew originally into orbits resembling the premier mission conditions of orbits of 174 × 326 km at 67° inclination. Some Soyuz hardware has found a place on the Yantar, but it carries at least two film capsules. At the end of its lifetime, the Soviets de-orbited the film canister module and often reused the hardware. In those cases where platform did not de-orbit, the Soviets commanded its destruction from the ground.

The Russians planned to operate one Yantar at all times but later abandoned the plans after 1994, when only two launches took place, an insufficient number to keep up continuous surveillance. But the lifetimes of Yantars increased from 58-59 days to 70 days and then to 89 days with the flight of Kosmos 2331. When that satellite returned on 11 June 1996, the Russians tried to put a replacement in orbit within nine days, but a failure of the payload shroud caused the launch to fail 50 seconds after take-off when the Soyuz-U vehicle's strap-on and core engines shut down.

The Soviets experienced some launch failures and acknowledge that vehicles carrying Yantars failed on 23 May 1974 from Plesetsk and 28 March 1981 from Tyuratam. Other failures carrying photo-reconnaissance satellites occurred at Tyuratam on 12 June 1982, 26 March 1986, 9 July 1988 and 11 November 1988. Speculation holds that these were probably third or fourth generation satellites

At the time of publication, the Russians have not attempted another in this particular series following this launch failure. Some Russian sources state that the country is out of Yantars in its inventories with no prospects for more because of budget cutbacks.

One subgroup of the Generation 4 Yantar performs topographic and mapping functions and is known in its civilian mode as Kometa. The first of these satellites appeared with Kosmos 1246 in February 1981. Western observers easily identified it as a new mission class because of its unusual orbit. Most Yantar-2K close look satellites orbit around 170 to km. Kosmos 1246 instead manoeuvred to a more circular 210 to 265 km orbit. Other similar satellite orbits soon followed, typically with operational perigees of around 210 km and apogees within the range 270 to 300 km. The character of the electronic emissions from these satellites identified them as Yantar – like and also resembling the transmissions from known to be mapping and surveying topography. Much later the Russians confirmed this deduction, identifying the name of the satellite series as Kometa (Comet). Some Russian reports indicate the original name was Yantar-1KFT, but this is slightly out of character since the Yantar-2 designator appeared six years prior to the flight of Kosmos 1246.

The Kometa launches occur less frequently than the Yantar-2K missions. Kometas go into space only one or two times a year and all have originated from the Tyuratam launch site. Western journalists witnessed the launch of Kosmos 1944 in May 1988 and their Soviet hosts described the mission as cartographic in nature.

Kosmos 2243, probably a Kometa mission exploded at orbital injection because the satellite's on-board self-destruct system inadvertently activated, causing the satellite to explode. 172 pieces of debris entered orbit, but the main object decayed after nine days. In its first commercial application, the Russians attempted to launch a Kometa in May 1996 to begin the SPIN-2 (Space Information – 2 m) imagery programme for a United States consortium under an agreement with Sovinformsputnik. The payload failed to reach orbit. The payload shroud failed 49 seconds after launch due to inferior bolts and materials having been used in its construction and debris from the launch crashed at Tyuratam 310 seconds after launch.

Sovinformsputnik (whose dominant share-holder is the Kosmosentral specialised Design Bureau) holds a licence to sell imagery from the Kometa's KVR-1000, TK-350 and DD-5 camera systems of military satellites. Orders can be directed through Space Liaison and Imaging Kosmosorp and the WorldMap consortium. The company has released details of the camera systems aboard the satellite. The KVR-1000 is a panoramic low-angle camera flying aboard Kometa since 1983, usually in conjunction with TK-350. The focal length is 1,000 mm, and has a ground resolution of 75 cm. Commercially the photos only have a resolution of 2 m. The frame size is 18 × 72 cm covering a swath of 37 × 165 km; cut into three portions and a scale 1:220,000. The image has been enlarged to 1:10,000 without loss of detail. The TK-350's topographic cameras have a focal length of 350 mm, with 10 m resolution, a frame size of 30 × 45 cm for a scale of 1:660,000 and 60 to 80 per cent stereo overlap. The digital DD-5 imaging system makes high-resolution data, which the Russians degrade, to 2 m resolution in the digitising process. DD-5 does not refer to a specific camera system but is a catch-all designation for digitised data from military satellites.

Generation 5 series

Background

The fifth generation photo-reconnaissance satellites remain unidentified as such and the title derives from Western observation of their behaviour based on a model of a digital imaging remote-sensing satellite identified at the 1995 Paris Air Show as Resurs-Spektr-V. Typically Generation 5 satellites fly at 230 km perigee and 280 to 320 km apogee. Kosmos 2320, which de-orbited on 28 September 1996 after 365 days in orbit, did not have a replacement. Unlike the budgetary problems alluded to for Generation 4 satellites, the likely reason the Generation 5s have stopped flying is the lack of a launch vehicle. When a factory involved with the production of SYNTIN fuel closed, Russia had to abandon the Soyuz U2 vehicle, which relied on the fuel. The series may resume when the Soyuz-2 (Russian Federation) launch vehicle begins flights.

Generation 6 series

Background

On 18 July 1989, Kosmos 2031 entered a rarely used orbital inclination of 50.6° from Tyuratam and during its lifetime it manoeuvred in a manner different from the fourth and fifth generation satellites. As planned, perigee would be maintained at 210 to 220 km while allowing apogee to be adjusted to a maximum 350 km. Ground controllers destroyed the satellite after 44 days in orbit. In October of the following year, Kosmos 2101 entered a routine 64.8° inclination slot, with apogee and perigee similar to Kosmos 2031. The Soviets destroyed this satellite too after 60 days. Whether coincidentally or by intention, these lifetimes matched the standard ones being seen for Kometa and Yantar-2K missions respectively. Three further missions took place at approximately yearly intervals between 1991 and 1993, with Kosmos 2262 remaining in orbit for 101 days before being destroyed.

Generation 7 series

Background

To all indications, the Russians introduced a new type of satellite on 26 August 1994 when a Zenit-2 flew from Tyuratam carrying Kosmos 2290. It had a mass of 10.6 tonnes and entered a 212 × 293 km, 64.81° inclination orbit. Extensive manoeuvres produced a 182 × 577 km orbit by the end of March 1995. The Russians commanded it to re-enter on 4 April 1995. No further missions had appeared through July 1997.

UNITED STATES OF AMERICA

Introduction

Current status

The National Reconnaissance Office set up the new National Imaging and Mapping Agency to facilitate the distribution of images from the Discoverer, Argon Corona, and Lanyard satellites. Gambit imagery remains classified. The Department of Commerce still places tight restrictions on the export of remote-sensing technology. DoC estimates that the current US$400 million annual market for remote-sensing data will grow to US$2 billion by 2000. The National Archives holds more than 800,000 images from some reconnaissance satellites flown between 1960 and 1972. In November 1997, the National Reconnaissance Office (NRO) released about 38,000 newly declassified records associated with early photosatellite programmes, plus historical videotape. The records involve the development, construction, launch and operation of the CORONA, ARGON and LANYARD photo-reconnaissance satellite systems. This release added no satellite imagery to that already available.

Background

President Eisenhower approved the first programme to build photo-reconnaissance satellites in the United States in 1958 and heartily endorsed the idea behind the programme with a policy initiative he referred to as the 'Open Skies'. The CIA brought Eisenhower's dream into action in developing the Discoverer satellites series by drawing on commercial industry in America. From the earliest days, the programme considered two technologies: 1) snapping a picture of the area of interest and then returning a film canister or 2) developing the picture on the satellite and scanning it for transmission to the earth. Both had strengths and liabilities. The returned film capsules offered, by far, the highest resolution and best quality, but it often took as much as two weeks to get the actual photograph back from space and into the hands of the President. The second technique offered immediate transmission so the President could follow unfolding events, but often the quality of the imagery gave little information.

To carry out Eisenhower's vision, the Secretary of Defense set up the National Reconnaissance Office on 25 August 1960. The NRO introduced the 'Keyhole' designation in 1962 and thereafter all American reconnaissance satellites carried the moniker 'KH'. The first officially named was KH-4, fourth in the series. The name came about not because of the satellite but because of the camera it carried, the fourth different type used in Corona/Discoverer by 1962. Although DoD did not officially acknowledge the NRO even existed until 18 September 1992, most Pentagon insiders found it by looking up the name of the long time director who had a cover title in the DoD phone book. NRO co-ordinated all overhead reconnaissance operations, including those from satellites and aeroplanes. Today it runs on a US$5 billion annual budget, reports to the Secretary of Defense but allows the CIA to target and control its assets.

Until 1992, the inter-agency Comirex (Committee on Imagery Requirements & Exploitation) met daily to review requests and direct targeting. But the DoD Central Imagery Office, created in June 1992 received authority to take over the obligations in the aftermath of the Gulf War, which revealed a clear need for more efficient dissemination of wartime imagery.

Imagery is relayed to Fort Belvoir in Virginia, from where it is passed to the National Photographic Interpretation Center in Washington DC established 1961 and run by the CIA for all of the intelligence community. Another major function of overhead imagery, defence mapping, became the responsibility of the new NIMA National Imaging & Mapping Agency (NIMA) created 1 October 1996 to co-ordinate imagery collection, processing, exploitation, analysis and distribution. NIMA consolidates all defence and intelligence imagery and mapping functions, absorbing the defence Mapping Agency, CIA, NPIC, imagery analytical support elements from the Defense Intelligence Agency, and dissemination and exploitation assets from the National Reconnaissance Program. NRO will continue to build the satellites.

KH-1-KH-3 (Corona/Discoverer) series

Background

After Eisenhower released the authority to build a spy satellite to the CIA, the Agency built an entire front programme, putatively to conduct important civilian scientific work. Ironically, Congress never explained to the American public why it had a separate scientific satellite programme in the Pentagon just as it was in the throes of setting up the new supposedly all-civilian agency NASA, which was to conduct all American space science. Corona began from the CIA offices as a high-risk programme funded from an almost unlimited intelligence budget to prove that a satellite could be built and controlled adequately enough for photography in the first place.

Even with the first disappointing results, Corona clearly had the potential to revolutionise the collection of intelligence. The United States had suffered a major embarrassment when the Soviet Union shot down CIA U-2 pilot Gary Powers on 1 May 1960. But flights like Powers proved that the Soviet missile build-up was largely posturing on the part of the Soviets who had maybe four satellites all together instead of the hundreds being hinted at in the Kennedy presidential campaign. After the loss of the U-2, for reasons of pure practicality, the United States had to continue with the Discoverer programme since without the spy plane the USA had no means to keep abreast of Soviet developments.

The CIA rushed the satellite through production and testing and launched the first in the series on 28 February 1959, only 13 months after the launch of Explorer 1, America's first satellite. For 18 months failure dogged the programme. The air force launched the first satellite into a polar orbit, but it tumbled and re-entered only five days later. Discoverer 2 carried the first RV but it re-entered at the wrong time and landed in Spitsbergen touching off a behind-the-scenes diplomatic flap as the CIA sought to recover the satellite before the Soviets found it. They did not succeed and the Soviets retrieved it ahead of the American search team.

Missions then suffered a continued series of failures, the most obvious occurring because the cold conditions in orbit led to the loss of capsule batteries and solid rocket motors. The acetate film also degraded in the

The pioneering Discoverer 14/Corona imaged its first target 18 August 1960, showing a military airfield near Mys Schmidta on the Chukchi Sea in NE Russia. It was a remarkable achievement; before then the best had been simple cloud cover images scanned by Tiros. Resolution is about 7 m. The airfield became a regular Corona target (CIA) ***1995***

vacuum of space and jammed the camera. When this problem appeared the CIA turned to Kodak to develop polyester film base. Discoverer 11 carried this new film for the first time, but that mission failed when the solid spin rockets exploded. Nonetheless, Kodak inaugurated a new civilian line of film based on the new substrate. To prevent failure of the solid rocket motors, Lockheed designers introduced new cold gas thrusters for attitude control. To test the concept, the CIA launched the next two Discoverers numbers 12 and 13 with no film canister on board, just a dummy mass. Because of this these two satellites have never carried the venerable KH designator. To improve the recovery procedure, which until then had a spotty record, the CIA used Discoverer 13 on 10 August 1960 to test new techniques. They retrieved it intact from the ocean. Even with progress clearly being made, Eisenhower sometimes lost hope for the satellite programme and considered cancelling before the end of his term.

Then with D14, all problems suddenly evaporated. Occurring immediately after the loss of the U-2, Discoverer 14 left Vandenberg on 18 August 1960, flew a perfect mission over the Soviet Union and returned on its 18th orbit. This one satellite provided more photographic coverage of the USSR than all the previous U-2 missions combined. It returned from space with 1,082 m of film covering 4.27 million km² of the Soviet land mass. It succeeded in imaging all Soviet MRBM, IRBM and ICBM complexes, revealing there to be no missile gap.

Discoverer 14 changed its recovery mode to an air capture when a C 119 transport snatched its parachute from mid-air over the Pacific Ocean. Discoverer 14 aimed its cameras at a military airfield near Mys Schmidta on the Chukchi Sea in Northeast Russia, which showed up clearly in the 1.5 m resolution photographs. The CIA declared Corona operational.

From 1961 onwards, Corona provided regular imagery of the USSR, and eventually led the Pentagon to make the public admission less than one month into Kennedy's administration that one of the issues that elected him – the missile gap – did not exist, nor had it ever. Corona imagery had brought the estimates of Soviet ICBMs down from hundreds to tens. Eventually Corona provided the United States with images of the Plesetsk test range, the Severodvinsk submarine construction facility and ABM deployments such as Galosh around Moscow, and their radars. Corona became almost the sole source of the defence Mapping Agency's data. Then casting the veil of secrecy back over the programme starting on 22 November 1961 the DoD classified all military missions.

In practice, a specially modified air force Thor IRBM put the satellite into orbit and then the Agena's 66.7 kN second stage engine fired to put the whole stage in a circular orbit. Once in orbit, the booster used its attitude control thrusters to turn the camera in the nose cone around to view the earth. This attitude manoeuvre often caused the re-entry capsule to enter at the wrong angle far overshooting the projected return zone. Sometimes the capsule even boosted itself into a higher orbit. One such incident resulted in a false alarm in February 1960, when the US Navy's early warning detection system found an unknown satellite, believed to be Soviet, in near-polar orbit. It was eventually established to be an errant Discoverer 5 re-entry vehicle. Depending on the quantity of film onboard, explosive bolts separated the capsule for re-entry on either orbit number 17 or orbit number 33. At 15,240 m a parachute pulled the recovery package clear of the heatshield, which fell into the sea, usually the Pacific near Hawaii. Beginning with Discoverer 30 in September 1961, the air force began using C-130s for mid-air recovery.

The 10 KH-1 missions used the C camera with an f/5 Tessar 61 cm focal length lens. D16 unsuccessfully introduced the C camera for the KH-2 missions, although it was little different from C/KH-1. But Discoverer 16 also flew aboard the more capable Agena B upper stage, allowing it to almost double the mission time by using an 18 kg film payload.

Discoverer 29 introduced the C/KH-3 camera with the f/3-5 Petzval 61 cm focal length lens. It should have doubled resolution to 4 to 6 m, but a design fault blurred all of Discoverer 29's exposures. Discoverer 38 in February 1962 introduced the radically new Mural camera system made up of two KH-3 cameras angled 15° fore/aft from the vertical for stereoscopic photography. These cameras flew successfully on 20 of the 26 KH-4 missions. Mural also added a wide angle index camera and a stellar location and triangulation system to help identify target areas more precisely. The next major upgrade came when the CIA put the KH-4A camera into service in August 1963. This camera designated the J1 retained Mural but the 73 kg film load split out between two RVs, so part of the secret data could be retrieved more quickly.

In public bland announcements from various front organisations said that the Discoverer made important discoveries every day about 'radiometric' conditions in space.

KH-4A/B Corona series

Background

As the spy satellite programme moved beyond its frontier days and matured into a significant technical and political asset, the CIA introduced the more sophisticated KH-4A/B cameras on their Agena carrier. At the very end of the Discoverer programme – on Discoverer 38 – the stereo KH-4 flew for the first time in February 1962 providing the most accurate mapping data ever returned from space.

The upgraded KH-4A took over as the programme lost the Discoverer designation in August 1963, with dual film capsules intended to double the mission duration. The programme experienced the same start-up glitches

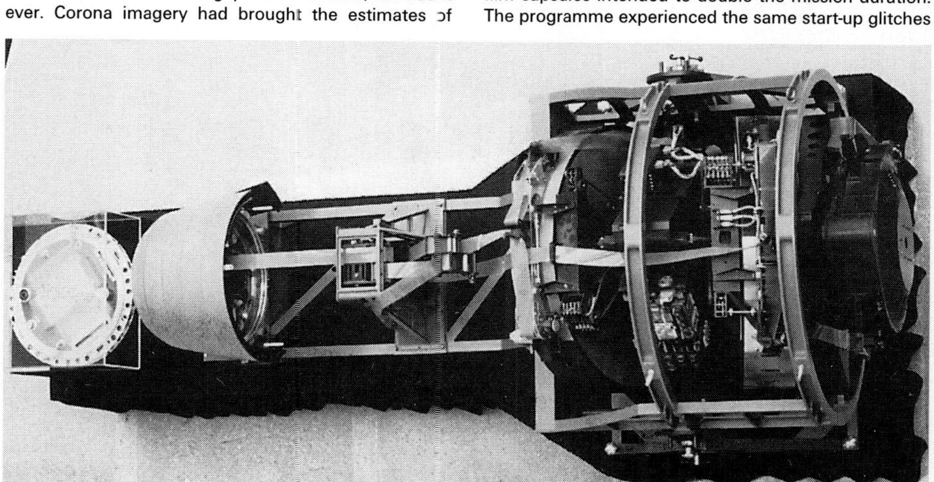

The KH-4B payload carried two return capsules (the central one is omitted here), fed film streams from the two rotating panoramic cameras. The film cassette is at far right (CIA) ***1996***

that beset the Discoverer in its earlier days. When the CIA first tried to retrieve the dual canister, they discovered that only one capsule could be recovered from the first two missions. The first true double recovery came on the third mission, launched 15 February 1964. Eventually, 50 of 52 launches returned RVs and most succeeded in delivering the double RVs. The KH-4A provided resolution of about 3 m from the satellite's altitude of around 175 km. Each satellite lifetime lasted about 1-4 weeks, complementing the new close look KH-7.

Once again the introduction of a new booster extended the amount of film the KHs carried when the Thorad Agena D launched a KH-4A on 9 August 1966. The KH4B camera, first launched in 1967, substantially improved resolution to 2 m and had selectable exposure and filters. It used the DISIC Dual Improved Stellar Index Camera for target location identification and attained orbit 16 out of 17 times. A more flexible control system also allowed the spacecraft to power down for up to 21 days.

102 of the 144 missions succeeded when the Corona programme formally ended by putting a last KH-4B camera in space in May 1972.

KH-5 Argon series

Background

As the Soviet Union deployed more missiles in hardened underground silos during the early 1960s, the United States' accuracy for its own missiles had to improve in order to destroy the silos. After a successful programme to advance missile guidance systems and the re-entry vehicle itself, the United States arrived at a point where its missiles had become more precise than its knowledge of the target locations. To support the new accuracy the air force needed more exact mapping co-ordinates. The defence Mapping Agency provided all targeting maps to the air force, and so the US Army devised its own overhead mapping project to be flown within Corona. It called the programme Argon and the new camera the KH-5. In all it had seven successful flights out of 12 attempts.

When the army finished its programme it could locate targets to within 300 m using Itek's 7.5 cm focal length camera which covered 550 × 550 km on each 11.4 × 11.4 cm frame at 140 m resolution. The first KH-5 orbited as Discoverer 20, but the camera failed. The first success came May 1962 and the last of the 12 Argons flew August 1964, when the entire Soviet Union had been mapped.

KH-6 Lanyard series

Background

Perhaps the least used of the cameras, Itek's 168 cm focal length panoramic camera inaugurated the KH-6 programme. The CIA originally intended to use this device on the Samos. It was designed to acquire 60 cm resolution images of targets such as the Tallin (Estonia) ABM site. But the first, on 18 March 1963, failed to achieve orbit; then the Agena upper stage failed on the second, launched 18 May 1963. The third attained orbit 31 July 1963 but this time the camera failed after 32 hours. Then when the first successful photographs appeared, investigators realised the lens had a serious focus problem, later traced to thermal factors. The programme ended after these three attempts because of KH-7's successful introduction.

KH-7 Gambit series

Background

The USAF built its own photo-reconnaissance system to provide detailed assessments of certain installations in order to determine the effectiveness of its targeting programme. It named the system the Gambit. Gambit returned 0.5 m resolution film by flying at a perigee as low as 137 km. Often flown in conjunction with a KH-4A, one system provided detailed images while the other provided a more panoramic view. Between 12 July 1963 and 4 June 1967 when the programme ended, 36 out of 38 KH-7s launches achieved orbit. Its principal contribution to intelligence lore was to photograph the new SS-7 and SS-8 ICBMs silos in the Soviet Union. Like KH-4A, film came back in two capsules and the satellite had an approximate lifetime of six days. No images have been released.

KH-8 Gambit Follow-on series

Background

As the KH-7 provided the test bed for new cameras, the KH-8 brought them into being. The first KH-8 flew on a Titan 3B Agena D on 29 July 1966. It had a longer life than its KH-7 predecessor and carried first the KH-4B and then KH-9 camera which had a resolution down to 15 cm. Agena's restartable engine enabled perigees as low as 135 km to be maintained for more than 50 days, so fewer of the expensive satellites had to be flown. In the early years from 1960 on, the CIA flew about nine spy flights a year. The KH-8 allowed them to reduce this number to two or three. Even though the Big Bird entered service in the mid-1970s, the CIA continued to fly the KH8 apparently to fill gaps in Big Bird launches. Finally in April 1984 the CIA depleted the stockpile of KH-8 cameras and the programme ended after 53 launches. The production stopped in 1980 after Congress became alarmed about cost overruns, reported at about US$400 million, in the Keyhole programmes.

KH-9 Hexagon (Big Bird) series

Background

By the mid-1960s, the air force desired its own completely independent method to observe Soviet ground installation. It had proceeded with the Manned Orbiting Laboratory programme, which had Earth observation as one of its primary objectives. To support the plan, the air force built the KH-10 camera. But when Congress cancelled the MOL in 1969, the air force still made its new camera available for imaging principally as the Big Bird. Instead of broad swaths of countryside the Big Bird is a 'close look' operation, intended to make detailed photos. Some programme managers refer to it by its service acronym as LASP (Low Altitude Surveillance Platform).

The orbiting capsule weighs more than 11 tonnes. Lockheed built it for the air force in a cylindrical canister 15.2 m long, 3.05 m in diameter and powered by two large solar arrays along with a 6 m diameter antenna. The Big Bird expanded the number of film canisters to four and it could eject them as required one at a time in times of crisis. Big Bird truly serviced two audiences, the camera combined KH-8's 45 cm resolution with KH-4B's 64 km swath width and achieved a 60 cm resolution across a 130 km swath. In addition to their primary mission, many Big Birds carried 60 kg Ferret subsatellites (some two), signals intelligence antennas and covert communications relays.

Lockheed enjoyed a boom in its launcher business and developed the Titan 3D specifically to carry the Big Bird into space. Most observers think that the air force tried to clandestinely launch the first Titan 3D without registering it. Each satellite entered a Sun-synchronous orbit to pass regularly over the targets at the same time of day. In this manner each target in subsequent frames should have had exactly the same size shadows, so if new construction occurred, an analyst could follow the progress by watching the shadow lengthen. Lockheed overcame the drag that such a vehicle incurred, normally enough to make it re-enter within 7 to 10 days, by periodic Agena firings. At the end of the satellite's lifetime, the air force de-orbited it directly into the ocean rather than allowing it to decay naturally. Big Bird production ended in 1981, with a number held in store for annual launches until later systems made them unnecessary.

After an initial start-up period the air force managed to get a Big Bird into orbit about once every 90 days. Only one six-month gap occurred between number 8, launched 10 April 1974 and number 9 launched 29 October 1974. The air force may have been using the time to modify the camera in order to counter Soviet efforts to camouflage missile silos and control centre construction. Number 9 provided surveillance at a time of increased tension in the Middle East and witnessed 16 Soviet ships unloading crated materials, believed to include SA-6 spare parts and components for Scud surface-to-surface missiles.

Big Bird 16 returned film over a period of 261 days, watching the Iran-Iraq war. Big Bird 17 launched 11 May 1982 watched the Falklands War. A Big Bird probably

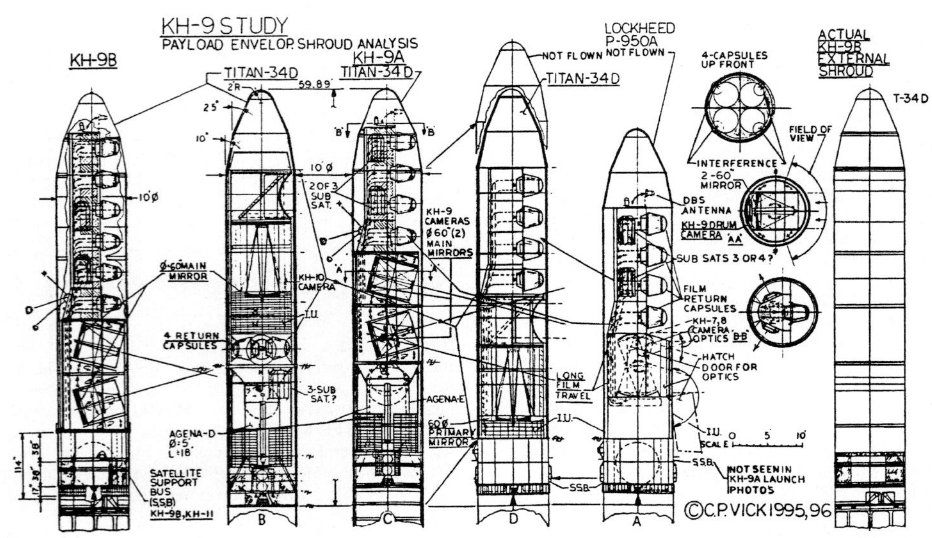

1997/0003433

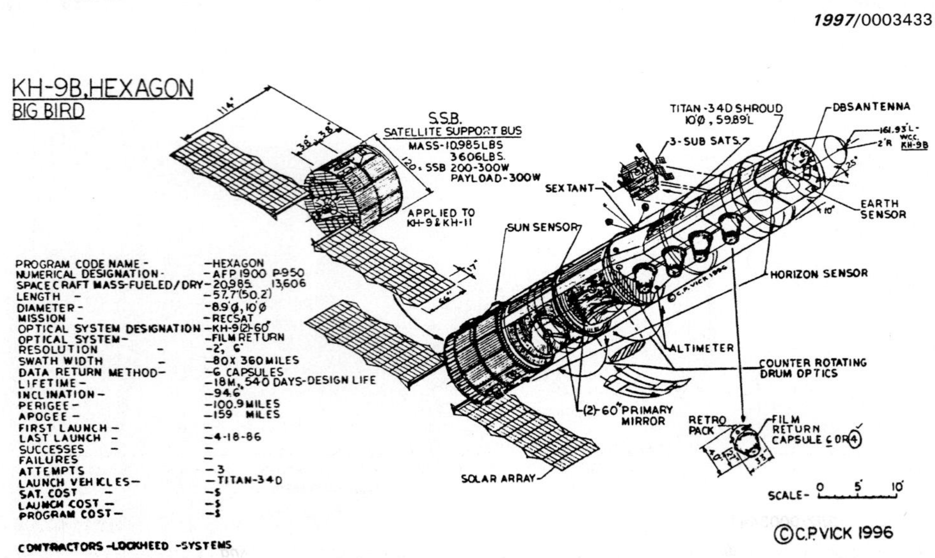

KH-9 conceptual studies. The last three KH-9s were retrofitted with the SSB Satellite Support Bus and two more return capsules. Dimensions are in feet (C P Vick) **1997**/0003434

identified the new Soviet missile radar at Abalakova, central Siberia, which became important to the Reagan administration's claims that the Soviets violated the ABM treaty.

KH-11 Kennan/Crystal series

Background

KH-11 introduced electro-optical imaging to the photo-reconnaissance inventory of the United States. Instead of returning a film canister several days to weeks after a launch an electro-optical imaging system takes the equivalent of a high resolution television picture of the earth which is then transmitted to a ground station. KH-11, despite a lower resolution than its film bearing predecessors, became more practical for routine use, and the 1970s semi-annual launch rate for espionage satellites with film aboard dropped to a few a decade.

KH-11s also occupied a higher orbit than the Big Bird because their lengthy mission duration did not allow for the frequent orbit adjustments that atmosphere skimming satellites require. A usual KH-11 orbit had a perigee of about 300 km and an apogee of 500 km with some as high as 1,000 km. Each KH-11 weighs more than 13 tonnes, is about 19.5 m long and 3 m in diameter. TRW builds the satellites for the USAF who uses the Titan 3D/34D to place them in 97° inclination orbits. Even at the higher altitude the KH-11 must raise its altitude about every three months during a three-year lifetime. Its real-time sensing systems (8-12 images per minute) and high resolution CCD cameras enable it to distinguish military from civilian personnel on the ground. The infra-red and multi-spectral sensing devices of Advanced KH-11 can locate mobile missiles, trains and launchers by day or night, and distinguish camouflage and artificial vegetation from real plants and trees. Each satellite has a side-looking capability, doubling the number of daily passes over a site from two to four.

An SDS helper satellite relays the data to Fort Belvoir

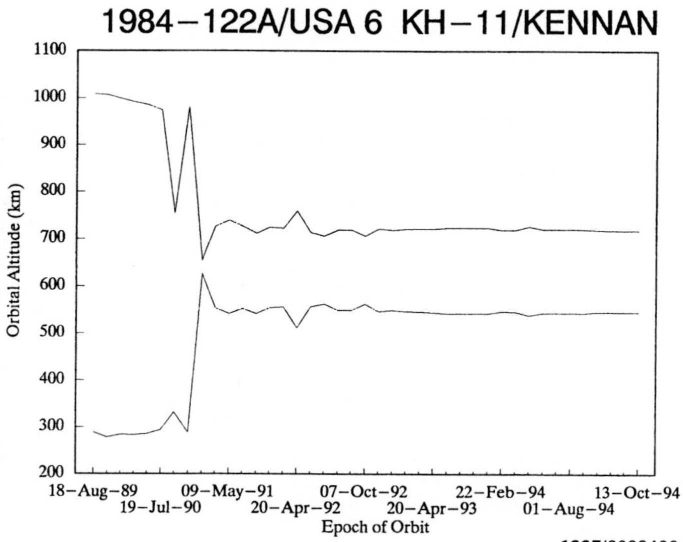

1997/0003400

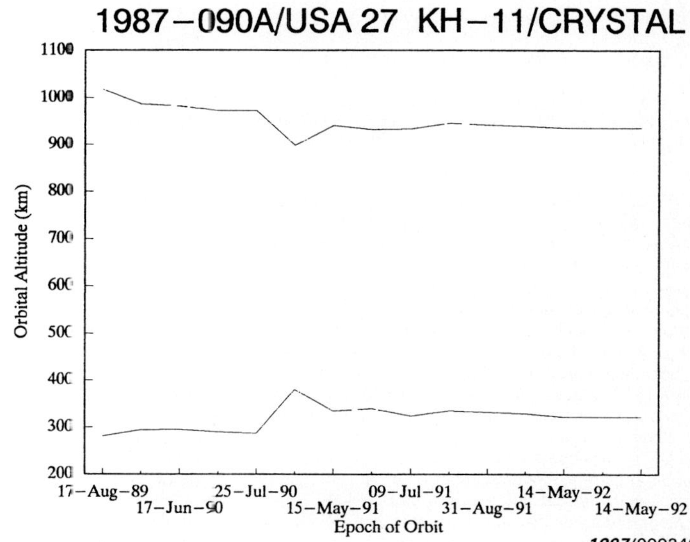

1997/0003401

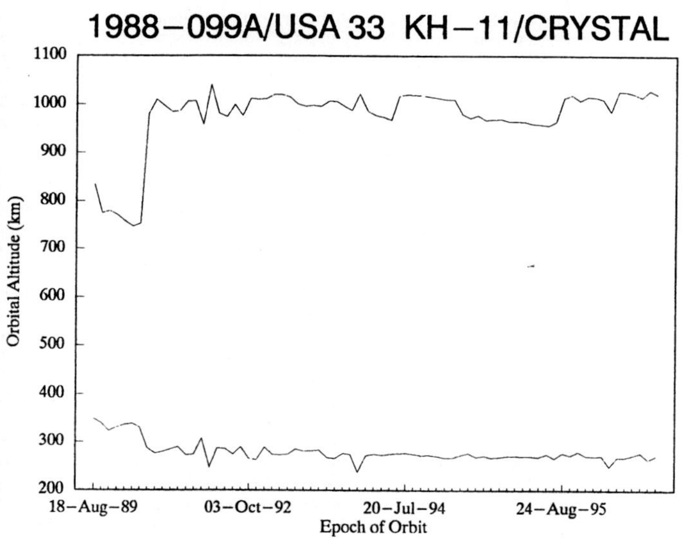

1997/0003439

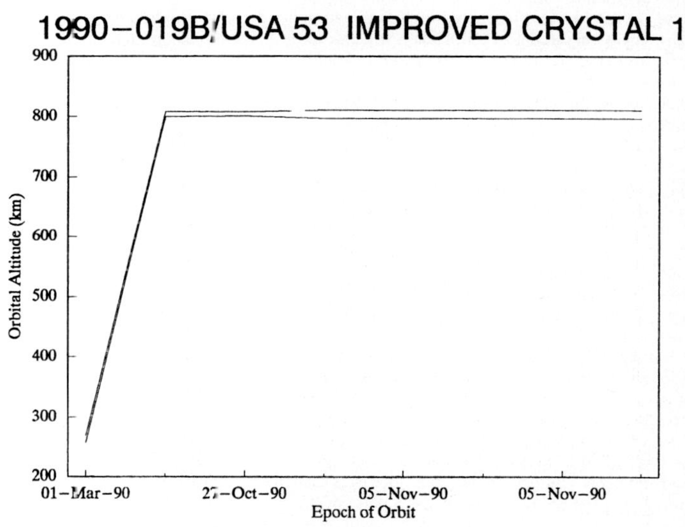

1997/0003440

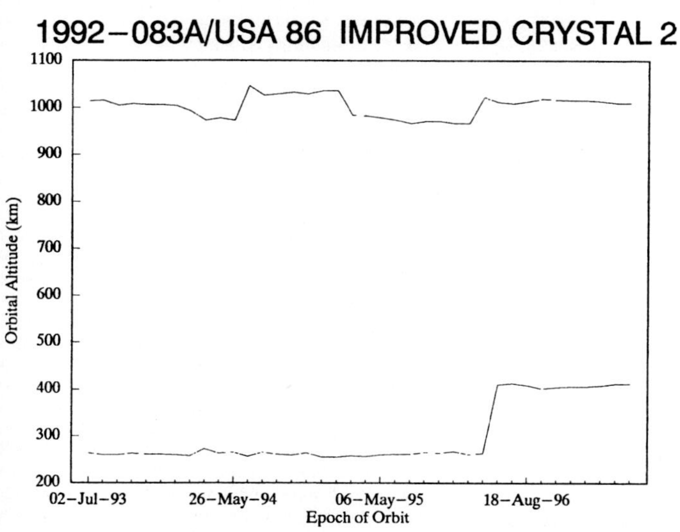

1997/0003441

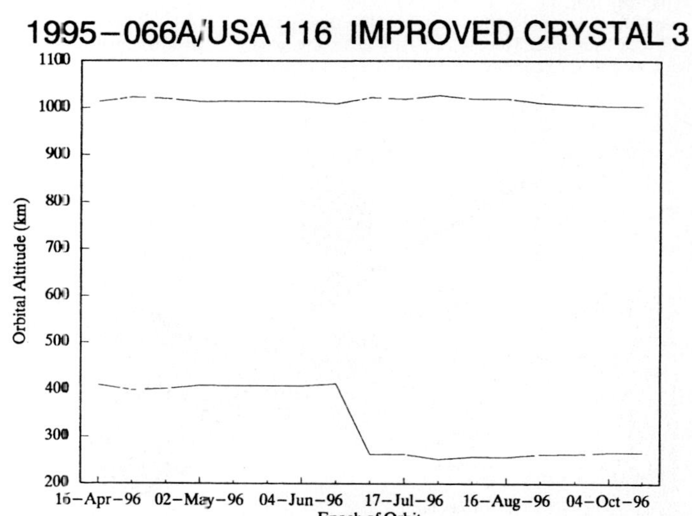

1997/0003442

Graphs showing the orbital altitude regimes for photo-reconnaissance missions operating during the 1990s. USA 6 was a KH-11/Kennan satellite, USA 27 and USA 33 were KH-11/Crystal satellites and USA 53, USA 86, USA 116 and USA 129 were Improved Crystal satellites. These orbits are based upon amateur observations of the satellites, since the actual operating orbits are classified

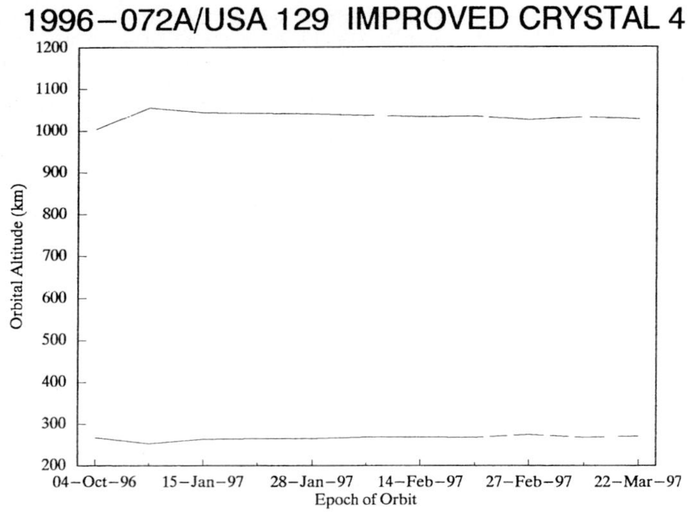

Graph showing the orbital altitude regimes for photo-reconnaissance missions operating during the 1990s. USA 129 was an Improved Crystal satellite. These orbits are based upon amateur observations of the satellites, since the actual operating orbits are classified

1997/0003443

in Virginia as the KH-11 snaps each picture. As configured by the USAF, the KH-11 operational system consists of a constellation of two satellites in planes 48.7° apart so that one covers a target in mid-morning while the second follows mid-afternoon. Each satellite repeats its track every four days.

There had been no public reference to the KH-11 designation on the part of the CIA until a CIA agent sold an operating manual to a KGB agent in Greece in 1978. The KH-11 scored its largest intelligence coup by monitoring a shed in early 1980 at the Severodvinsk naval yard on the White Sea. A submarine larger than the US Trident boats finally emerged from this shed a few months later. In September 1980 the Pentagon purposefully leaked information that a KH-11 had photographed side by side emplacements of SS-20 and SS-16 missiles. The Pentagon claimed the Soviets planned to use their own reconnaissance satellites as surrogate KH-11s to determine how much camouflage each missile required to be mistaken for the other. In short order the Pentagon cited this as evidence that Soviet deception would not allow the US to verify adherence to the Salt 2 agreement on ballistic missile limitation.

KH-11-1, launched 19 December 1976, and KH-11-2, launched 14 June 1978, worked in pairs during part of their lifetimes. Both made their observational passes in the morning. After KH 11-3 joined them on 7 February 1980, the pairs alternated observations in mornings and afternoons to watch the early stages of the Iran-Iraq war in conjunction with Big Bird.

KH-11-4, launched 3 September 1981, set an operational record at 1,177 days, KH-11-5, launched 7 November 1982, and later joined by KH-11-6 on 4 December 1984, worked as a pair until 13 August 1985, when KH-11-5 re-entered. Two weeks later, on 28 August 1985, its KH-11-7 replacement disappeared when the Titan 34D launcher exploded shortly after lift-off, seriously damaging the Vandenberg pad. Because of the loss of the Shuttle Challenger and the Titan explosion, the United States had lost a significant portion of its photo-reconnaissance ability and heroic measures prolonged KH 11-6's life by using it to only relay images of the highest priority intelligence targets.

The satellite executed few manoeuvres in order to conserve fuel, but amateur observations revealed a manoeuvre in mid-1989 into a still unexplained 350 x 900 km orbit. The CIA allowed the orbit to decay back to 336 x 732 km when its perigee was raised to 559 km in November 1990, possibly to support Persian Gulf War operations. No further manoeuvres were seen before it apparently de-orbited in late 1994 after an impressive 10-year life. KH-11-8 was joined by KH-11-6 on 26 October 1987 in an afternoon orbit. This satellite was probably the last one in the USAF inventory, and may itself have only been a modified ground demonstration model. The Titan 34D launch of 6 November 1988 has all the earmarks of a KH-11, however, arriving at the morning slot in unison with KH-11-8.

Even with the lengthened lifetimes, many in Congress talked of killing the KH programme. Total cost overruns in the KH-11 had come to more than US$1 billion and the Soviets had thwarted it by introducing an elaborate camouflage system. To overcome the objection of cost the air force at one time planned to refuel each satellite in orbit to stretch its three-year lifetime. Some observers have suggested that military astronauts made a practice refuelling attempt using the Space Shuttle. Others have indicated this was just a plan that was never tried.

Advanced KH-11 (KH-12) Improved Crystal series

Background

In 1989, the CIA had clearly dropped the KH-11, but while some observers called the new series the KH-12 others claimed it is really an upgraded KH-11. Since few references had been made to KH-12 by the military or intelligence community it is not now clear if any satellite will ever carry this designation. In August 1989, the Shuttle placed a payload called USA-40, into a 303 x 315 km, 57° orbit, which later raised itself to 460 km, although observations by amateur astronomers indicated initial tumbling problems. This new satellite obviously differed from the KH-11 since it had a mass of only 9,350 kg, which included a large propellant load for manoeuvring.

The high orbit requires improved optics to maintain the resolution of earlier KH-11s, and some circumstantial evidence exists for such a change. As part of its promotion for using the 'Bus I' service module for the International Space Station, Lockheed claims the bus is operational on classified missions. Advanced KH- 11 is a prime candidate. Its 4.04 m diameter and 2.67 m length accommodates a larger mirror than KH-11's. (Compare this to the 2.40 m diameter mirror on the Hubble Space Telescope and NASA's claim that HST has the largest mirror in space is suspect.)

Some observers suggest USA-40 is an improved SDS satellite intended to relay data from a Molniya type orbit. If true, Shuttle STS-36 actually deployed the first upgraded KH-11 when it released the 16.9 tonne AFP-731 into a 204 km 62° orbit on 28 February 1990. The fact that this new satellite is significantly larger than KH-11, possibly up to 18 tonnes fully fuelled, lends credence to the belief USA-40 is actually an SDS. In order to launch this satellite from Vandenberg the Shuttle would have had to use 109 per cent main engine thrust – a condition no longer permitted. So one of the principal justifications for the original Shuttle has proved to be false. During 1986, the lack of appropriate photo-reconnaissance satellites prompted some discussion of an emergency Shuttle flight to orbit an advanced KH-11 (KH-12). Since the flight would have been unmanned, the military astronaut corps fought it, because it would have indicated they were not needed to pilot the spaceplane. The first advanced KH-11 should have flown on Shuttle 62C – the first scheduled flight from VAFB in September 1986 – but when the air force mothballed the west coast's US$3 billion Shuttle facilities it had to wait until a 1989 Shuttle departure from the Kennedy Space Center.

Each advanced KH-11 carries both digital imaging sensors and signals intelligence receivers covering a wide range of transmissions over the Soviet Union, from telephone to TV. Each also supposedly carries a laser altimeter and the ICMS Improved Crystal Metric System for the defence Mapping Agency, adding fiducial marks to imagery.

Publication of commercial imagery of Chernobyl and Soviet military bases brought the subject of space-based observations into the open so heavily by April 1986 that Keyhole, and even KH-12, were for the first time discussed in the US House of Representatives. Representative George Brown called for relaxation of the existing limit of 10 m resolution on civilian remote sensing satellites (which President Reagan subsequently agreed to). Earth resources satellites being developed by Europe, Japan and China do not adhere to such limits.

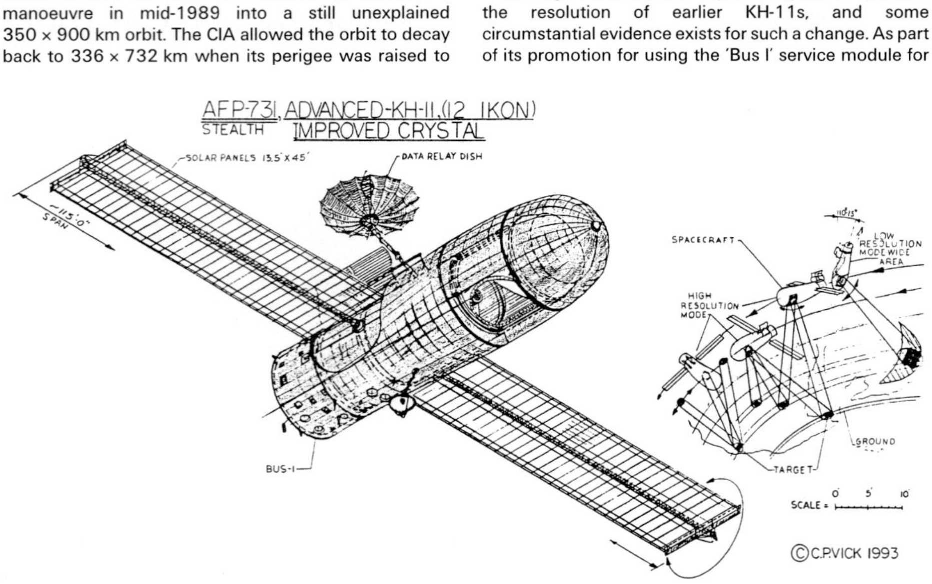

Advanced KH-11/Improved Crystal conceptual study. Dimensions are in feet (C P Vick) ***1997***/0003438

Samos series

Background

As the Corona/Discoverer programme developed, a competing philosophy struggled to find a voice in the United States bureaucracy. The USAF had long proposed and funded background studies in direct transmission of data from space in an existing RAND programme named WS-117L. But James Killian and Edwin Land, two of Eisenhower's advisors on photographic intelligence, still recommended recovering a film canister from space. The air force refused to let the direct readout transmission idea die and continued funding WS-117L on a parallel track with Corona. Eisenhower's National Security team actually encouraged this programme as a public shield for the Corona programme. From its press releases the public believed the air force programme constituted the sum of the United States investigations into satellite reconnaissance. Eventually the programme became known as Sentry. Many historians feel the name SAMOS stood for Space and Missile Observation Satellite, but in fact the name was chosen to honour the Greek island where Aristarchus lived.

The Polaroid corporation provided the onboard film processing procedure, which NASA adapted for the Lunar Orbiters of 1966-68. Samos itself though was apparently a dismal failure and had to be cancelled in 1962, although there have been references to missions beyond the five admitted to at the end of 1961.

The publicly acknowledged missions are Samos 1 launched 11 October 1960 by Atlas Agena A from Vandenberg, which failed to achieve orbit. Samos 2 launched 31 January 1961 made it into a 476 × 554 km 97° orbit, but Samos 3 exploded on the pad 9 September 1961. Samos 4 also failed to attain orbit, 22 November 1961. Samos 5 succeeded 22 December 1961. Some observers hold that 7 March 1962 may have been a Samos. One author who has studied the programme claims that the entire series returned only a single set of images, with resolution about 30 m, and that analysts were unable to decide if it showed China or the USSR. A second Samos, E-6, series apparently flew five times in 1962 (26 April, 17 June, 18 July, 5 August, 11 November), returning film aboard capsules and using Agena to initiate re-entry.

RADAR CALIBRATION

RUSSIAN FEDERATION

Romb series

Background
The third class of satellite is the largest group to be identified: during the period 1980-1990, 20 launches of the Romb series are acknowledged, although there might have been other satellites launched during this period which were failures. Additionally, there appear to have been launches outside this specified period. The outer shell of the satellite is described as having several spherical objects, which are ejected in orbit as sub-satellites perhaps to calibrate discrimination capabilities of radars for detecting re-entry vehicle decoys.

Launch date	Cosmos	Incl (degrees)	Period	Perigee	Apogee
28 April 1976	816	65.8	94.6	481	515
27 April 1977	906*	50.6	94.3	463	515
30 May 1977	913	74.0	94.5	472	520
19 July 1977	930*	74.0	94.6	481	514
8 December 1977	965	74.0	94.4	465	516
22 December 1978	1065	50.7	93.4	344	548
6 July 1979	1112**	50.7	93.4	344	542
6 June 1980	1186	74.0	94.5	473	519
31 July 1980	1204	50.7	93.3	345	538
28 September 1981	1311	83.0	94.5	463	519
29 January 1982	1335	74.1	94.6	482	518
21 April 1982	1351	50.7	93.5	348	548
29 July 1982	1397	50.7	93.4	345	540
19 April 1983	1453	74.0	94.4	463	517
26 May 1983	1465	50.7	93.4	349	543
31 August 1983	1494	50.7	93.4	345	550
3 September 1983	1501	82.9	94.5	468	514
27 September 1984	1601	65.8	94.5	475	516
19 June 1985	1662	65.8	94.5	476	513
2 October 1985	1688***	50.7	93.5	347	548
3 September 1986		74.0	94.6	474	516
22 January 1987	1815***	50.7	93.5	345	550
28 July 1988	1960	65.8	94.5	473	513
23 December 1988	1985	73.6	95.2	522	533
15 February 1989	2002	65.8	110.3	186	2,299
27 February 1989	2053	73.5	95.1	518	526
6 February 1990	2059	65.8	110.2	189	2,280
24 April 1990	2075	74.0	94.6	484	514
28 November 1990	2106	82.5	95.2	518	537
25 June 1991		74.0	Launch failure		
2 March 1995	2306	65.8	94.5	469	517

The inclusion of some satellites in this listing is based upon their orbital parameters rather than the existence of sub-satellites deployed in orbit. The launch failure in June 1991 might not have been a Romb mission, but this seems to be the most likely satellite group for the failure.

* Cosmos 906 and Cosmos 930 remained attached to their respective Cosmos-3M launch vehicle second stages, so could have been Romb failures (no additional sub-satellites were tracked).

** The Cosmos-3M second stage from the Cosmos 1112 mission exploded in orbit and a determination of any Romb mission-related sub-satellites cannot be confirmed.

*** These two satellites are the only ones to have used this particular orbital regime during 1980 – 1990 which are not identified as Romb missions: it is possible that they were mission failures (or a totally different class of satellite).

Vektor series

Background
Vektor satellites fly to allow ground-based radars perform experiments for the determination of a satellite's characteristics. Unlike Yug, Vektor is an active satellite and carries solar cells over its 2 m diameter spherical surface, as well as four antennas. Two classes of Vektor satellites have appeared. The first group started with Kosmos 660 and flew into approximately 400 – 2,000 km orbits while the second group started with Kosmos 687 and flew into approximately 300 – 700 km orbits. The Russians have not revealed whether satellites of slightly different designs fly the two different orbital regimes or whether all of the Vektor satellites were identical.

Launch date	Cosmos	Incl	Period	Perigee	Apogee
18 June 1974	660	83.0	109.1	397	1,972
12 March 1976	807	83.0	109.1	398	1,973
16 January 1981	1238	83.0	109.0	406	1,958
9 April 1981	1263	83.0	109.1	397	1,970
11 November 1983	1508	82.9	109.1	400	1,966
28 August 1990	2098	83.0	109.2	396	1,977
27 September 1994	2292	83.0	108.9	400	1,954
Vektor Series: Class 2 Missions (Cosmos 687 Type)					
11 October 1974	687	74.0	94.5	286	698
28 May 1976	822	74.0	94.5	280	711
14 July 1987	1868	74.0	94.6	284	709
10 October 1991	2164	74.0	94.5	285	708

Yug series

Background
According to Russian literature the Yug satellites are simple spheres with a diameter of 2 m and no appendages. They are used for passive work specifically connected with the calibration and operation of ground-based radars. Kosmos 2265 appears to be a model for these satellites launched by the Soviets and Russians since 1979. In fact the first use of the Kosmos 2265 type of orbital altitude came with the launch of Kosmos 1463 in May 1983, although it is possible that Kosmos 893 in a higher-apogee orbit and launched in 1977 might also have been in this series. No definitive tally of the number of Yug satellites exists.

Launch date	Cosmos	Incl (degrees)	Period	Perigee	Apogee
15 February 1977	893*	74.0	105.2	332	1,680
19 May 1983	1463	82.9	103.6	301	1,551
28 June 1984	1578	50.7	104.4	295	1,641
26 October 1993	2265	82.9	103.6	291	1,574
24 April 1996	2332	83.0	103.6	295	1,565

* Cosmos 893 might not be a Yug satellite, although the orbital altitudes are a close match for such a mission.

RADAR IMAGING

UNITED STATES OF AMERICA

Lacrosse/Vega series

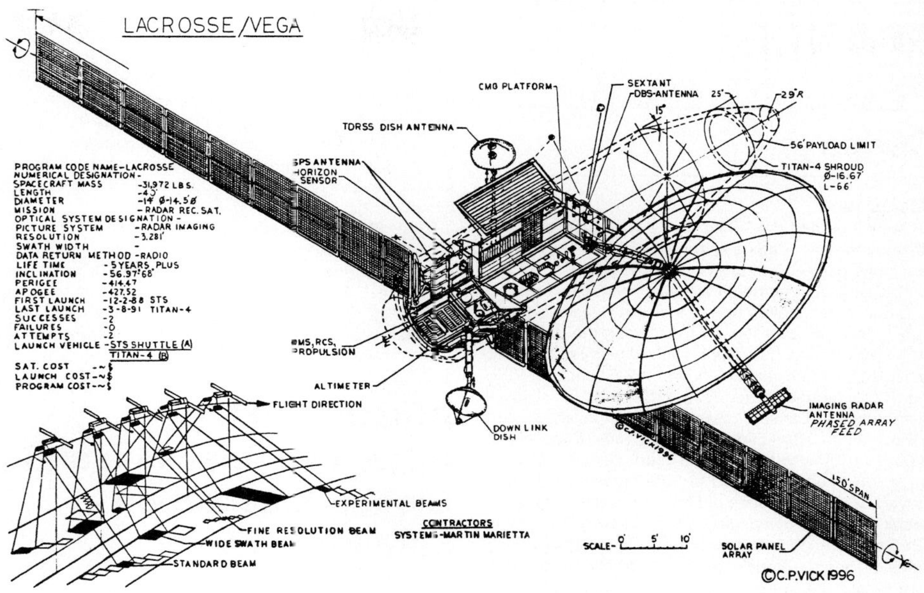

Lacrosse conceptual study. Dimensions are in feet (C P Vick) ***1997***/0003435

Current status

The Central Intelligence Agency launched its third Lacrosse/Vega high resolution imaging satellite on 25 October 1997. The Agency code-named the satellite USA 133. It arrived in a higher than expected 674 km orbit at 57° inclination. The new Lacrosse has a higher resolution than the 1 m capability of its two predecessors. This Lacrosse had to go into the 57° orbit because the CIA planned to operate it in conjunction with an earlier version to obtain stereo coverage. Because the CIA chose to launch the earlier satellite on the Shuttle in 1988, the mission could not achieve a higher inclination.

Background

The first high resolution imaging radar satellite and the first US military active radar craft of any type left the Shuttle cargo bay as USA-34 from Atlantis on 2 December 1988 during the STS-27 mission. The air force used it to provide all weather day/night coverage of Soviet-bloc armour movements. To keep track of armour requires that the satellite images at 1 m resolution or better. By comparison NASA's Shuttle Imaging Radar (SIR) has a 10-20 m resolution. Some sources feel that the Lacrosse also carries optical sensors. If true the sensors probably only identify targets optically for the Radar to scan. The US military probably proposed that Lacrosse act to detect Soviet mobile missiles, but the end of the Cold war required the DoD to dream up another mission to keep the programme afloat. One candidate often cited by sources is to verify adherence to mobile ICBM treaties. Lacrosse returns data via TDRS to the White Sands, New Mexico ground station.

Lockheed Martin Astronautics in Denver acted as the prime contractor for the Vega/Lacrosse satellite, which reportedly cost in excess of US$500 million. The CIA originally named the programme Indigo. In its present version it somewhat resembles the nuclear-powered RORSAT, utilised by the Soviet Union for monitoring the world's navies.

Atlantis's remote manipulator arm released the Lacrosse during orbit number 5 some 7 hours after launch but the solar panels initially failed to deploy to their 45.7 m span. The satellite subsequently raised its orbit by some 240 km to 668 × 703 km at 57° inclination. The Lacrosse cost the US government more than US$500 million. Some sources say it was originally called Indigo.

Vandenberg's first Titan 4 launch on 8 March 1991 probably carried a Lacrosse. Amateur observations of USA-69 during March 1991 identified a 683 km 68.0° orbit, indicative of a radar satellite. Titan 4's 8 November 1991 mission might have carried a Lacrosse as well, but was more likely an advanced Whitecloud. USA-116 of 5 December 1996 may have been a Lacrosse. It has been claimed that a Titan 3B Agena mission of 21 January 1982 was launched into 553 × 646 km, 97.3° orbit in order to demonstrate the Lacrosse technology. Orbit manoeuvres on the part of several nearby satellites indicate they acted as calibration targets for the radar satellite.

This orbit was maintained through to the end of 1996. In March 1997 amateur satellite observers found that USA-34 had disappeared from orbit, presumably having been de-orbited that month.

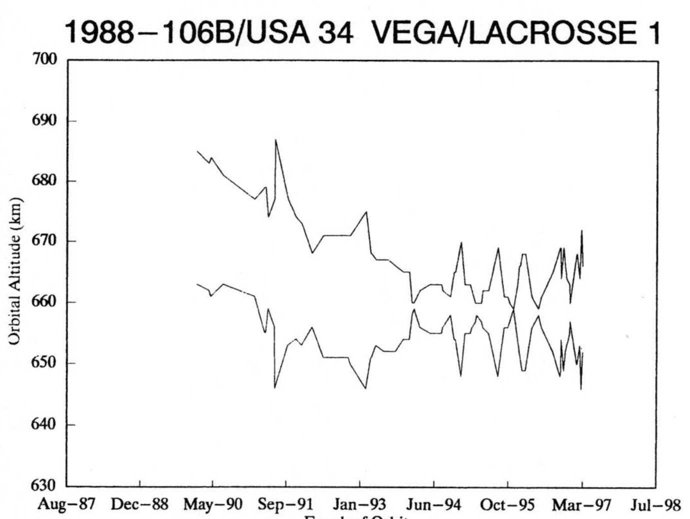

1997/0003436

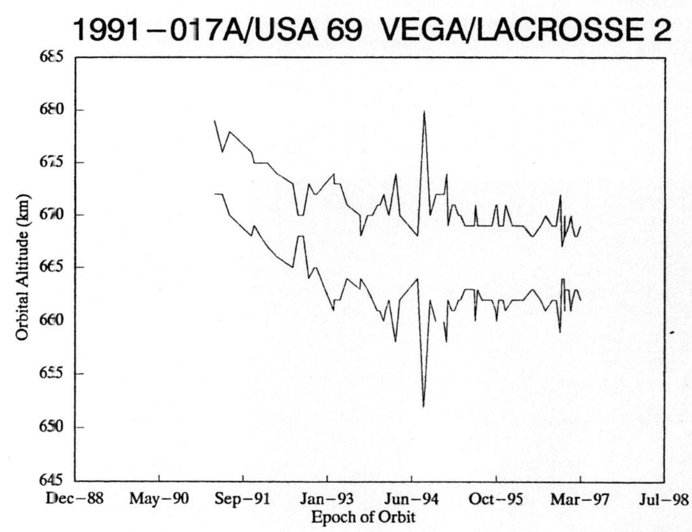

1997/0003437

The orbital altitudes of the two Lacrosse satellites which operated during the 1990s. It will be noted that each satellite started life in an orbit slightly higher than the final one and was allowed to slowly drift down until the altitude was stabilised at around 660 to 670 km. USA 34/Lacrosse 1 disappeared from orbit in March 1997. These orbits are based upon amateur observations of the satellites, since the actual operating orbits are classified

SCIENCE AND ENGINEERING TEST

FRANCE

Cerise

Background
An Ariane carried the 50 kg Cerise (Caractérisation de l'Environnement Radioélectrique par un Instrument Spatial Embarqué) into orbit in order to characterise the Earth's radio environment in research that could lead to a 'Zenon' Elint satellite during the next decade. Alcatel Espace built the satellites for DGA, using the University of Surrey's UoSAT bus. The French Ministry of Defence expects to use the satellite for about 2.5 years. In1993, the MoD requested industry proposals for 1998's Clementine Cerise follow-on to investigate environmental interaction at different radio frequencies.

Cerise collided with a piece of debris from Ariane third stage, which orbited Spot 1 in 1986, making it the first proven casualty of space debris. The collision with the debris 1986-019RF occurred at 09.48 GMT on 21 July 1996 breaking off the upper portion of the Cerise gravity gradient boom. French ground controllers regained control over the satellite following the collision.

UNITED STATES OF AMERICA

APEX (P90-6 Advanced Photovoltaic and Electronics Experiment)

Background
A Pegasus launched the 261 kg APEX (Advanced Photovoltaic & Electronics Experiment) 3 August 1994 to determine the long-term effects of harsh radiation and plasma environments. The satellite achieved a 361 × 2,528 km, 70° inclination orbit that passed repeatedly through Earth's radiation belts at apogee and through the upper atmosphere at perigee.

The USAF awarded Orbital Sciences Corporation a US$9.6 million contract on 18 March 1991 to build the APEX P90-6 mission in the STP series of experiments. Orbital Sciences built the test bed that housed the many experiments on board. They included the PASP Plus (Photovoltaic Array Space Power Plus Diagnostics) which tested 12 solar cell samples with Fresnel lens and Cassegrain concentrators, the CRUX (Cosmic Ray Upset Experiment) which studied microelectronics single event upsets, and the radiation degradation of ferro-electric devices with the FERRO (Thin Film Ferro-electric Experiment).

PASP Plus ceased operating 11 August 1995 after collecting 10 times more data that all such previous space experiments combined. CRUX/FERRO remains operational. The total mission cost to the air force is about US$30 million, including about US$12 million for the spacecraft and US$7.8 million for the Pegasus.

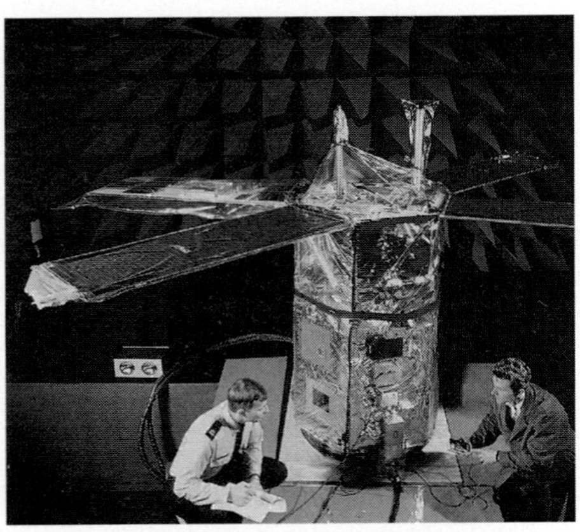

APEX is the first application of OSC's PegaStar bus (OSC)
1995

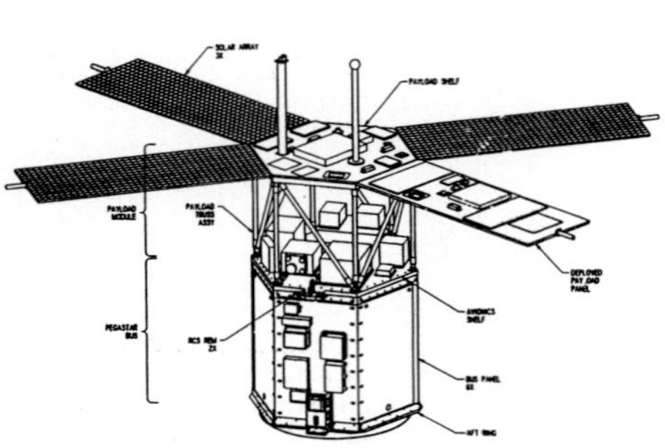

APEX principal features (Orbital Sciences Corporation)

ARGOS (P91-1, Advanced Research and Global Observation Satellite)

Background
The ARGOS mission is to fly and operate joint service payloads which include two technology demonstrators and seven experiment payloads for global and celestial observation. ARGOS weighs approximately 2,700 kg and has a goal of three years of on-orbit operations. The ARGOS mission is scheduled to launch from the Vandenberg Air Force Base Space Launch Complex-2 in autumn 1997. ARGOS will be launched on a Delta-2 launch vehicle into a Sun-synchronous 835 km, 98° orbit. ARGOS incorporates an accurate attitude determination and control system with an embedded Global Positioning System (GPS) receiver for position determination and time reference and a solid-state recorder, providing 2.6 Gbits of onboard data storage. ARGOS also includes an automated mission planning system, developed by the University of Colorado's Laboratory of Atmospheric and Space Physics, to optimise ground contacts as well as onboard power and data storage requirements. The automated mission planning tool adds tremendous capability to manage the ARGOS downlink of 9.6 Gbits each day. On board experiments are:

The High Temperature Superconductivity Space Experiment, or HTSSE 2, developed by the Naval Research Laboratory (NRL), Naval Center for Space Technology. HTSSE 2 will space qualify super-conducting digital and RF subsystems to demonstrate an operational space capability. Expected performance of super-conducting components include factors of 100-1,000 in power reduction, more than 10 times higher speed and similar weight reduction than today's silicon or gallium arsenide (GaAs) based electronics.

The Extreme Ultra-Violet Imaging Photometer (EUVIP) payload is sponsored by the US Army Space and Strategic Defense Command and built by the University of California at Berkeley Extreme Ultra-Violet Astrophysics Laboratory. EUVIP will establish the behaviour of the upper atmosphere and plasmasphere as needed for Army RF secure communication systems design, prediction of magnetic storms, and characterisation of the aurora.

The Naval Research Laboratory, Space Science Division, sponsors the Unconventional Stellar Aspect (USA) payload. ARGOS will be one of the first research satellites to fly an embedded Global Positioning System (GPS) receiver, while characterising astronomical x-ray sources for potential use as autonomous position, attitude and time-keeping references for military space systems. The augmentation of a proven x-ray detector design with advanced computers will help to demonstrate the feasibility of autonomous satellite navigation using x-ray pulsars in place of GPS timing and navigation signals.

Electric Propulsion Space Experiment or ESEX, developed at TRW's Space and Electronics Group facilities will demonstrate reliable high powered arcjet thruster operations in space. ESEX will also execute orbit transfers and verify compatibility with spacecraft subsystem controls. Electric propulsion is projected to double the payload-to-orbit capability of current space propulsion systems. The managing organisation for this payload is the Air Force Phillips Laboratory, Electric Propulsion Laboratory.

The Space Dust Experiment of SPADUS will provide a 3-D survey map of the present dust distribution in low Earth polar orbit and allow prediction of orbital debris 'showers' which could affect spacecraft such as the Space Shuttle. SPADUS will also obtain early flight experience for sensors and electronics, which are planned for International Space Station and the Cassini mission to Saturn in 1998. This payload is sponsored by the Office of Naval Research, Electronics Division and built by the University of Chicago Enrico Fermi Institute, Laboratory for Astrophysics and Space Research.

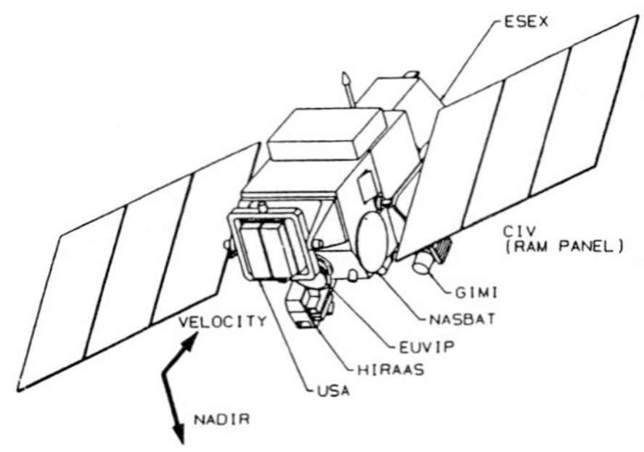

Argos: the artist's impression highlights the operating ammonia arcjet (TRW)

The Critical Ionisation Velocity experiment of CIV will release xenon and carbon dioxide to study ionisation caused by plasma and collision processes in the upper atmosphere. The results will be used to help identify plumes and atmospheric wakes of launch and orbital vehicles. CIV is developed and sponsored by the Air Force Phillips Laboratory, Satellite Assessment Division.

The High Resolution Airglow Aura Spectroscope (HIRAAS) payload, developed and sponsored by the Naval Research Laboratory, Space Science Division, consists of three high spectral resolution ultra-violet spectrographs designed to measure the naturally occurring thermospheric and ionospheric airglow. Measurements of the ionosphere can improve a number of DoD systems that depend on radio and microwave propagation through the upper atmosphere and ionosphere. Neutral density measurements also support NORAD operational requirements for satellite drag forecasting and the ability to predict orbital life and re-entry impact locations.

Global Imaging Monitor of the Ionosphere or GIMI will demonstrate operational charged coupled device (CCD) sensor technology for environmental monitoring of upper atmospheric perturbation due to meteors, rocket exhausts, and aurora. Use of wide-field sensors in three separate wavelengths will enable GIMI to continuously image 560 m^2 of the Earth's limb. GIMI is developed and sponsored by the Naval Research Laboratory, Space Science Division.

The Coherent Radio Topography Experiment Payload (CERTO) consists of a three-frequency radio beacon and radiating antenna mounted on the space vehicle. Receivers on the ground use the differential phase techniques to derive the integrated electron density. Characterising CERTO data plays a critical role in assessing impacts on navigational accuracy, communication systems, and remote sensing by radars. CERTO is managed by the Naval Research Laboratory Plasma Physics Division.

Co-manifested on the Delta-2 launch vehicle are two secondary payloads from Denmark and South Africa. Assembled is a coalition of international test teams to support space test programmes from the United States, Denmark and South Africa. The Danish satellite ORSTED will map the Earth's magnetic field and the South African satellite SUNSAT will map the Earth's gravitational fields. The integration and development for these micro-satellites are managed by the NASA Orbital Launch Services Project.

JAWSAT (P98-1, Joint Academy, Weber State University Satellite)

Background

STP continues to prepare for the JAWSAT (P98-1), Joint Academy/Weber State University Satellite. JAWSAT, along with the Air Force Academy's FalconSat, was due to be launched on the first Orbital/Suborbital Program (OSP) launch vehicle in September 1999. The OSP launch vehicle will place the satellites into a 600-700 km Sun-synchronous orbit, at approximately 98° inclination. JAWSAT will carry three experiments: a Thermospheric Temperature and Nitric Oxide Spectrograph, an Ionospheric occultation Experiment, and the Coherent Electromagnetic Radio Tomography Probe (CERTO) series experiment.

MightySat satellite

Background

The Space Experiments Directorate of AF Phillips Lab is developing a series of MightySat micro-satellites carrying experimental payloads. Phillips' Program Office was created in January 1994 to provide a standard small platform for testing new technologies without placing larger missions at risk. Phase 1 covers the refurbishment by CTA under a US$2 million contract of an existing Space & Missiles Systems Center bus from an earlier programme. Launch took place from Shuttle STS-87 in July 1998. Spectrum Astro was awarded the contract 22 November 1995 for the five-year Phase 2, for at least three micro-satellites, with options for two more, worth up to US$23.5 million. First launch was planned for March 1998 and then at one-year intervals. Payload includes pulsed plasma thrusters, Kestrel Corp's Fourier transform hyperspectral imager (256-512 channels, 1.7 nanometer resolution, 0.350-1.050 μm), and NRL/Space Instruments' solar output Total and Ultra-Violet Irradiance Radiometer.

The air force will use Phase 1 to prove the concept

MightySat 1 composite mockup, with four 45° antennas for the UHF half duplex TT&C system (Phillips Laboratory) ***1996***

MightySat 2 will carry a hyperspectral imager (Spectrum Astro Inc) ***1996***

effort, retrofitting the bus with an advanced composite structure and high-efficiency, multijunction solar cells. It will carry an experimental release mechanism using shape memory devices and a reliability experiment space qualifying a new set of advanced electronics and electronic packaging techniques.

MTI (P97-3, Multispectral Thermal Imager)

Background

The Multispectral Thermal Imager is a Department of Energy (Sandia National Laboratory, Los Alamos National Laboratory, and Savannah River Technology Center) and STP sponsored mission. MTI will be placed into a 575 km circular, 97.5° inclination Sun-synchronous orbit in a launch planned for August 1999 on a Taurus launch vehicle. The 1,610 kg spacecraft is 3-axis stabilised, provides an average of 540 W of power, and is designed for a 14-month mission life with a three-year goal. The objective of the MTI mission is to demonstrate and evaluate multispectral and thermal imaging technology of nuclear proliferation monitoring and for other military and civiliar applications. The MTI sensor will provide multispectral data with high ground resolution and radiometric accuracy. Ball Aerospace is building the MTI spacecraft bus and Hughes Santa Barbara Research Center, Hughes Danbury Optical Systems, TRW, and Sandia National Laboratory are building the sensor.

OPREX (P95-C) Optical Reflection Experiment satellite

Background

The Optical Reflection Experiment was planned for the fourth quarter of 1997 launch into 834 km, 82 to 92°, to study the ionospheric attenuation of a CO_2 laser link. An RF link would be functionally similar to SIDEX.

PICOSat (P97-1)

Background

Surrey Satellite Technology Ltd (SSTL) is building PICOSat. It will provide a test of an off-the-shelf micro-satellite to determine if cost effective spaceflight of DoD experiments can be achieved in this manner. The four experiments to be flown on the satellite are: the Polymer Battery Experiment, the Ionospheric Occultation Experiment (IOX) to demonstrate the use of occultation of GPS signals to characterise the ionosphere, Coherent Electromagnetic Radio Tomography Ionospheric (CERTO), and Optical Precision Platform Experiment (OPPEX). P13EX is an all plastic battery that has virtually limitless charge/discharge cycles and other space-favourable characteristics. IOX and CERTO are experiments measuring electron content and densities in the ionosphere. OPPEX is an experimental platform that demonstrates active and passive vibration isolation for future precision instrument applications. The 68 kg spacecraft is scheduled to launch in 1999 into a circular orbit above 625 km, with a 40 to 70° inclination.

RADCAL (P92-1 Radar Calibration Satellite)

Background

A Scout booster launched the RADCAL 25 June 1993 from Vandenberg into 759 × 888 km, 89.5° orbit. It provides radar calibration via two-C band transponders for more than 13 agencies. The Doppler transmitter permits use of ground processed ephemeris (by Naval Surface Warfare Center) as a standard reference for the radars. Two onboard GPS receivers using 4 patch antennas demonstrate a new technique for generating ephemeredes using ground processing. The USAF Phillips Laboratory Small Satellite Power System Regulator (SSPSR) experiment is also testing improved methods of battery charging. The RADCAL experiments are housed on a modified X-Sat bus. Defense Systems Inc (then CTA Space Systems, now Orbital Sciences

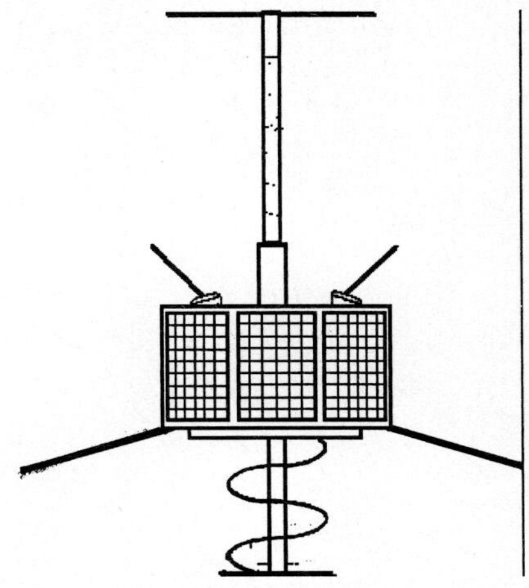

Radcal is providing calibration for >70 C-band radars

Corporation) was prime contractor for the 87 kg spacecraft at a total cost of less than US$4 million.

Future Payloads

POGS 2 (S92-1, Polar Orbiting Geomagnetic Survey 2)
The experiment uses a Special Sensor Magnetometer to collect data, which would be used to update the geomagnetic maps of the Earth. POGS 2 was due to be carried onboard defence Meteorological Satellite Program spacecraft S-15, which had a predicted launch need date of 1998. The magnetometer was to be placed on an extendible 5 m boom in order to reduce the effects of electromagnetic noise from the spacecraft, planned to fly on DMSP's S15-S20. First was planned for the second quarter of 1997.

CERTO PLUS (S97-2, Coherent Electromagnetic Radio Tomography/Profiling the Limb with Ultra-violet Sensors)
CERTO PLUS will provide measurements of the integrated electron density of the ionosphere in the satellite orbit plane. It will also provide a database for global models of the ionosphere and will test tomographic algorithms for reconstruction of ionospheric densities and irregularities. CERTO PLUS is scheduled to fly on STRV ID, a spin stabilised, 100 kg class spacecraft to be launched into a geosynchronous transfer orbit by an Ariane 5 launch vehicle in July 1999.

REX II (P94-2) satellite

Background

STP's follow-on mission to 1991's REX (PS9-1) for research into the physics of electron density irregularities that cause disruptive scintillation effects on radio signals. The payloads include a Rome Laboratory Communications experiment and an OPS Attitude Determination and Control System (ADACS) experiment to perform on-orbit attitude determination and direct feedback to the control system using the Global Positioning System.

The air force successfully launched the spacecraft by a hybrid Pegasus on 8 March 1996 from Vandenberg. Built by CTA Space Systems, the 113 kg, gravity gradient stabilised with three-axis magnetotorque coils and a momentum wheel for pitch momentum bias, with four deployable solar arrays and gravity gradient boom. A failure occurred in the REX payload in July 1996, which prevents normal operations, but efforts continue to identify and compensate for the cause of the failure. The spacecraft and ADACS payload continue to operate nominally. The air force paid less than US$14 million for REX, including US$9 million for Pegasus.

STEP series

Background

TRW signed a US$5.48 million contract 17 April 1990 for STP's Space Test Experiments Platform (STEP). The STEP designs and flies satellites below 450 kg, with lifetimes of one to three years, which communicate with small ground stations. The bus upon which the air force places its experiments provides up to 300 W power and is 3-axis, or gravity gradient stabilised. The TRW contract allows for options on a further 11 satellites. OSC's Pegasus became the prime launch vehicle after selection in April 1991 for one firm (APEX) and 39 optional launches under the AFSLV Air Force Small Launch Vehicle Program.

TAOS is the first STEP mission (TRW) ***1994***

STEP MO (P90-5)
STEP MO carried a payload of ten TAOS Technology for Autonomous Operational Survivability experiments, including seven computers. It weighed 503 kg at launch on 13 March 1994 into 539 to 560 km, 105° on Taurus with the classified DARPAsat. The air force declared it fully operational on 31 March 1994. The satellite package cost US$56 million, including US$16 million for launch. It demonstrated Phillips' Lab project for satellite autonomy from ground controllers using a 6-channel GPS receiver, horizon scanner and high speed computer for 400 m and 0.03° accuracy position determination. It had a design life of 18 months and two articulated solar panels, generating 195 W (100 W available to payload).

MO suffered a serious blow on 20 July 1994 when the IMU failed and full capability was not restored with new software circumventing the problem until 28 February 1995. Responsibility for operating the MO spacecraft has now been transferred to Air Force Space Command.

STEP M1 (P90-1)
The air force originally planned for a 500 km 90° orbit on a Pegasus-XL to investigate HF radio propagation in the ionosphere below 250 km, atmospheric drag over 195-400 km, and plasma instabilities and electric fields. The satellite weighed 348 kg, had 3-axis control and 105 W of power from deployed solar panels. It carried payloads of Ducted HF Propagation; Satellite Electrostatic Triaxial Accelerometer (SETA); Absolute Density Mass Spectrometer (ADMS); Composition and Density Sensor (CADS); Coordinated Heating and Modification Process in the Ionosphere (CHAMPION), Plasma Environment Analyser (PEA). The satellite was lost in the 27 June 1994 Pegasus launch failure.

STEP M2 (P91-2)
The air force launched the 180 kg SIDEX (Signal Identification Experiment) on 19 May 1994 for the USAF Rome Laboratory Air Development Center into 603-821 km, 81.96° orbit (830 km circular planned) aboard Pegasus. The low orbit degraded data collection.

STEP's first four missions: MO TAOS (bottom), M1 (centre), M2 SIDEX (top left) and M3 (top right) (TRW) ***1995***

The STEP M2 spacecraft failed December 1995 and investigators never determined the cause. Full operations of the spacecraft did not begin until October 1994 because of several anomalies, and then suffered power problems because of overheating. The satellite payload consisted of a narrow beam parabolic gimbaled antenna; wide beam fixed helix antenna, and a GPS antenna. The experiments were designed to evaluate communications in dense signal environments. Controllers flipped the M2 180° in yaw 22 December 1994 to position the battery on the anti-Sun side to resume normal operations after the over-heating.

STEP M3 (P92-2)
The fourth STEP was lost 22 June 1995 due to failure of the Pegasus-XL launch vehicle. It carried five experiments: ACTEX advanced controls technology experiment (BMDO), EDMM erasable disk mass memory, SAMMES space active modular materials experiment (BMDO), SAWAFE strategic attack warning and assessment flight experiment ('smart skin', BMDO), SQUOD space qualified optical disk. The mission cost the air force about US$40 million. M3 was completed in August 1994 but then had to be stored because of Pegasus-XL's 1994 failure.

STEP M4 (P95-1)
The fifth and final STEP Mission is a 402 kg spacecraft launched on 22 October 1997 aboard a Pegasus XI, into a 430 × 511 km parking orbit at 45°. Ground radar confirms that the solar arrays are not deployed, leaving the satellite in a deep under-voltage condition. STEP M4 hosts three experiments. They are an Orbiting Ozone and Aerosol Measurement (OOAM) experiment, which measures global ozone depletion and water vapour in the middle atmosphere to help gauge infra-red sensor performance, the Electromagnetic Propagation Experiment (EMPE), which characterises ionospheric effects and the Digital Ion Driftmeter (DIDM), which measures the flow of ionospheric ions. Mission requirements demand that the attitude control system operate in both spin and 3-axis modes. In July 1996, CTA Space Systems delivered the spacecraft to TRW S&EG in Chantilly, Virginia for final integration and test. The mission cost the air force an estimated US$80 million.

STP (Space Test Programme) payloads

Current status
Last year STP launched or operated four dedicated payloads and twelve experiments on other satellites.

Background
The Air Force Space and Missile Systems Center, created in May 1965, runs a programme of science and engineering test payloads known as the Space Test Program. The Department of Defense relies on the STP (formerly called STEP) programme to test new concepts and gather information about the space environment. STP put its first payload in space in 1967 and since that time had either launched or participated in the launches of 415 different experiments. DoD provided US$38.3 million for STP operations and experiments this year. In past years the budget has been FY97 US$43.1 million, FY96 US$43.8 million, FY95 US$53.l million, FY94 US$73.1 million, FY93 US$77.9 million, FY92 US$69 million, FY91 US$62.4 million. STP participates in space experiments in three ways: free-flyers, Shuttle pallets and piggyback flights on other experimental packages. For experiments with unique orbital requirements that can best be met by free-flying spacecraft, STP contracts for spacecraft development, experiment integration, and launch service. A recent example is TRW's Space Test Experiments Platform. These experiments usually have a 'P' designation and carry a numerical identifier closely related to the fiscal year of their completion. STP also flies experiments as piggyback payloads on spacecraft of various agencies of the US (NASA and DoD) and other countries (CNES-France, DRA-UK, Russia), usually carrying an 'S' designation. STP also makes use of the Space Shuttle mid-deck lockers, the Hitchhiker, the Getaway Specials, Spartan and other low cost available space with the NASA programme.

Current and recent payloads

STS-83, Columbia (OV-102)
STS-83 launched on 4 April 1997 carried the Cryogenic Flexible Diode (CRYOFD), a DoD Cargo Bay payload. CRYOFD is designed to determine the behaviour of Cryogenic 2-phase thermal control components in microgravity, demonstrate oxygen and methane heat pipe startups from a super-critical condition, demonstrate operations, verify analytical performance models, and establish the correlation between one *g* and microgravity thermal performance. A secondary objective is to validate the performance of an American Loop Heat Pipe with Ammonia (ALPHA).

STS-84, Atlantis (OV-104)
STS-84 launched on 15 May 1997 as the sixth Shuttle/Mir docking mission. Atlantis carried two DoD payloads, which were the first DoD sponsored experiments to operate inside the Mir Space Station. The Radiation Monitoring Equipment – RME III – attempted the first ever direct measurement of the East-West effect, a theory that predicts higher radiation levels (by a factor of 2 to 4) on the West facing side of the Mir. Data was collected throughout the docked phase of the mission and also from inside the Spacehab on the way back to Earth. The Cosmic Radiation Effects and Activation Monitor is composed of passive and active radiation detectors that work together to better understand the effects of cosmic radiation on both the crew and electronics. CREAM recorded over 151 hours of data.

STS-94, Columbia (OV-102)
STS-94, launched on 1 July 1997, again carried a Cryogenic Flexible Diode (CRYOFD). It performed successfully.

STS-85, Discovery (OV-103)
STS-85 was launched from KSC on 7 August 1997 carrying the COOLLAR Flight Experiment activated nine hours into the mission aboard the TAS-01 bridge. CFE completed all of its planned activities and was able to run for an additional 18 hours to gather secondary data.

STS-86, Atlantis (OV-104)
STS-86 launched on 25 September 1997 carried three DoD STP payloads, the Cell Culture Module (CCM), Cosmic Radiation Effects and Activation Monitor, and Shuttle Ionospheric Modification with Pulsed Localised Exhaust (SIMPLEX). CCM performed very well throughout the mission. CREAM was deployed and activated inside the Mir Space Station on 28 September. The experiment ran flawlessly until its deactivation and partial stowage in the Shuttle on 2 October. The remainder of CREAM's components (all passive) remained deployed in the Mir module, Kvant-II module, and Kristall module until STS-89 (January 98). NASA supported SIMPLEX with a pass over the Jicamarca, Peru ground station at approximately 15:30 CDT on 4 October 1997. The pass consisted of a 10-second OMS burn producing an exhaust plume northward. The orbiter aft cameras focused on the starboard OMS engine throughout the burn which produced excellent video.

STS-87, Columbia (OV-102)
STS-87 lifted off from KSC on 19 November 1997 with the Midcourse Space Experiment (MSX) and Shuttle Ionospheric Modification with Pulsed Localised Exhaust (SIMPLEX). A primary objective of this mission was the deployment and retrieval of a Spartan free-flyer. Due to problems with the Spartan deploy and retrieve operations, propellant was at a premium for this mission causing SIMPLEX to 'stand down' halfway through the mission. MSX was able to observe a 10-second, 2-engine, RAM burn (from the Orbiter's rear engines). The ram burn was visible using the LTV narrowband sensor, and the spectrographic imagers. The result was excellent data, making the experiment a success.

STS-89, Endeavour (OV-105)
STS-89 launched on 22 January 98 with the Cosmic Radiation Effects and Activation Monitor (CREAM) active monitors on board the Spacelab Payload Processing Facility (SPPF). Some modifications took place in order to extend the data collection capacity during the Mir phase of the flight. CREAM was successfully deployed and activated inside the Mir Space Station on 25 January. CREAM's active and passive components were deployed in the Mir, Kvant-II module, and Kristall module. All of the CREAM components from STS-86 aboard the Mir were returned to Endeavour at the same time that the STS-89 components were deployed.

Argos was STP's next major satellite, planned for 1997 (Naval Research Laboratory) ***1995***

CHARGECON-GEO (SP90-3)
The Satellite Charge Control Experiment at GEO flies on the DSCS 3B B-7 communications satellite. The satellite tests the build-up of an electrostatic charge on the spacecraft and methods to then dissipate the charge by releasing an ionised xenon plasma. The experiment started gathering data in August 1995 and through July 1996 the experiment had completed 20 automatic charge dissipation operations. The experiment continues to run.

SWIM (S91-4)
The Solar Wind Interplanetary Measurements instrument on NASA's Wind satellite, launched 1 November 1994.

POAM 2 (S88-1)
NRL's 25 kg POAM Polar Ozone and Aerosol Measurement package orbits attached to France's SPOT 3 launched 26 September 1993. Thermo Trex of La Jolla, California built the instrumentation. Nine UV-IR channels observe sunsets/rises over the poles to provide vertical profiles of polar ozone, aerosols, water vapour and nitrogen tetroxide, and stratosphere and upper troposphere density and temperature. POAM 2 characterises and clarifies the intense turbulence in the middle atmosphere that adaptive optics in systems such as laser communications would have to accommodate. Two of POAM's channels are at important laser wavelengths: 0.353 and 1.059 μm. The other frequencies are 447.4, 452.5, 550.0, 600.0, 763.4, 780.0 and 945.0 μm. When NASA's TOMS instrument aboard Russia's Meteor-3 failed in December 1994, POAM became the only operating US ozone sensor, returning data in unprecedented detail on the ozone hole. The French announced the end of the SPOT 3 mission in December 1996 after an unexplained failure rendered the spacecraft inoperative.

POAM 3 (S95-A)
The Naval Research Laboratory Polar Ozone and Aerosol Measurement 3 is a congressionally mandated follow-on to the POAM 2 project in support of an integrated NRL effort to globally monitor environmental parameters. NRL contracted with the Centre National d'Etudes Spatial (CNES) to integrate, test and launch POAM 3 on the SPOT 4 spacecraft. An Ariane 5 launch vehicle boosted POAM 3/SPOT 4 to an 833 km Sun-synchronous orbit from Kourou, French Guiana.

BINRAD (S93-1, Beryllum-7 Induced Radiation Experiment)
STP conducted a third reflight of the experiment from 14 March 1996 to 15 June 1996. The experiment first flew with Kosmos 2331, launched on a Soyuz-U launch vehicle. The analysis of the foil showed a Be-7 level only 1 per cent of the expected level and analysts are still struggling to explain these results. The experiment followed-up on NASA's LDEF to help explain how LDEF accumulated Be-7 at a rate several thousand times greater than modelled for that altitude. Some researchers have theorised a heretofore unknown transport mechanism from a lower altitude exists. Two previous reflight attempts in 1992 and 1993 on Resurs-F spacecraft failed when the aluminium foil remained unexposed because of improper release from the experiment containment system.

MPTB (S96-1)
The Naval Research Laboratory Microelectronics & Photonics Test Bed is a piggyback experiment that was launched into a high radiation orbit in Autumn 1997. MPTB is presently returning data vital to predicting the performance of advanced microelectronic and photonic devices. Once proven, these off-the-shelf devices will be used in military SPO procurements. The experiment contains 24 individual experiment boards and a charged particle telescope. The decline of the Cold War means that a full complement of radiation hardened parts is no longer available for space systems, leaving commercial unhardened devices to be employed.

TSX-5 (P95-2) Tri-Service Experiments Mission-Mission 5

Background
The air force plans to launch TSX-5 on a Pegasus XL launch vehicle. It will be placed in a 410 × 1,750 km orbit at a 69° inclination and will be controlled by the Research and Development, Test and Evaluation Support Complex (RSC) at Kirkland AFB NM. TSX-5 will carry the Space Technology Research Vehicle II (SRRV 2) experiment package and the USAF Research Laboratory's Compact Environment Anomaly Sensor (CEASE). STRV-2 is jointly sponsored by the Ballistic Missile defence Organization and the United Kingdom's Defence Research Agency (DRA) and will evaluate several space technologies.

TSX-6 (P97-x) Tri-Service Experiments Mission-Mission 6

Background
Tri-Service Experiment Mission 6 is expected to carry the Optical Reflection Experiment Spacecraft Evaluation Model (OSEM) and other, yet to be determined, experiments into an 835 km orbit at an inclination of 82-98°. Planned for launch in August 1999, a contract for TSX-6 has not been awarded.

UNKNOWN

RUSSIAN FEDERATION

Kosmos series

Background

More than 90 small Kosmos launches have taken place since 1964 whose purpose has not been identified. They share the common feature of an SL-7 launch. The last in the series seems to be Kosmos 919 in 1977. Western observers have classed them as military because they emit no characteristic communications signals and no scientific results have been published at international scientific conferences that reference these launches.

One series, with 62 total launches, used Kapustin Yar as a spaceport and had an orbital parameter of 280 × 500 km, 49° inclination. These satellites lasted until 1967-68. When Plesetsk opened, the satellites came from that spaceport and flew into 71° inclination orbits. A second series, adopting an orbit of 200 × 1,100-1,500 km and 49° inclination, also flew from Kapustin Yar beginning in 1967, and then switched to Plesetsk in 1968. A third series also left for orbit from Kapustin Yar into 260 × 2,100 km at 49° inclinations. The fourth series used a 280 × 840 km, 71° inclination orbit from Plesetsk. The constant replacement of one satellite by another suggested that the Soviets were making continuous observations. Another 40 Kosmos satellites used similar orbits, but these could be identified as scientific payloads from publication of their findings.

The Soviets conveniently announced launches under the Kosmos title, accompanied by serial numbers, orbital parameters and a vague description of purpose from the time the series began on 16 March 1962. Kosmos designations have also been used to cloak inevitable failures during the development of a new series. Interplanetary probes, to cite one example, received a name in the Venera, Mars or other series only after successful injection. Otherwise they were allocated the next Kosmos number and a routine announcement released. A typical launch failure disguised by the Kosmos label was Kosmos 1612 in November 1984. Orbital analysis suggests that this satellite, stranded in an unusable transfer orbit when its upper stage failed to re-ignite, was the first of the new-generation Meteor 3 weather satellites. Similarly, a geosynchronous launch failure in January 1987 received the designation of Kosmos 1817 before it re-entered the following day.

CONTRACTORS

Civil and commercial space organisations
Military space organisations
Communication satellite systems and services
Earth observation satellite systems and services
Navigation satellite systems and services
Civil and commercial ground facilities
Space industrial providers

CIVIL AND COMMERCIAL SPACE ORGANISATIONS

ARGENTINA

Asociacion Argentina de Tecnologia Espacial (AATE)
C.C. 142-Suc 28, 1428 Buenos Aires
Tel: (+54 1) 555 60 30
WWW: http://www.aate.org.ar
President: Pablo Gabriel de Leon

Centro de Investigaciones Aplicades
Avenida Fuerza Aérea Argentina Km 5. 1/2,
5103 Córdoba
Tel: (+54 51) 69 06 66
Telex: 51965 ancorar
Fax: (+54 51) 69 06 31
Director: Brig Raul Juan Carlos Camussi

Comision Nacional de Actividades Espaciales
Paseo Colon 751, 1063 Buenos Aires
Tel: (+54 1) 331 54 47
WWW: http://www.conae.gov.ar
President: Conrado F Varotto
SAC-B Project Manager: Daniel Caruso
SAC-B Project Scientist: Marcos E Machado

Comision Nacional de Telecomunicaciones
Peru 103, piso 11, 1067 Buenos Aires
Tel: (+54 1) 343 00 75/343 92 19/342 16 27/
342 99 01
Telex: 21706 cntel ar
Fax: (+54 1) 334 49 35
President: Rinaldo Colome
Director, Asuntos Satelitales: Carlos Alberto Killian

Instituto de Astronomia y Fisica del Espacio
Postal: Casilla de Correo 67, Sucursal 28,
1428 Buenos Aires
Cuidad Universitaria, Núñez, Buenos Aires
Tel: (+54 1) 783 26 42/781 67 55
Fax: (+54 1) 786 81 14
Director: Marta Rovira

AUSTRALIA

Commonwealth Scientific & Industrial Research Organisation (CSIRO)
PO Box 53, Parkville, Victoria 3052
Tel: (+61 3) 662 71 11
Fax: (+61 3) 662 71 96

CSIRO Office of Space Science & Applications
PO Box 3023, Canberra ACT 2601
Tel: (+61 6) 216 72 00
Fax: (+61 6) 216 72 22
Director, COSSA/EOC: Dr Brian J J Embleton
Manager, Science/Applications: Jeff Kingwell
Manager, Technology Development: Chris Graham
Business Manager: Jill Huntington
EOC Science Program Leader: Dr David Jupp

Space Policy Unit
Department of Industry, Science and Tourism
20 Allara Street, Canberra, ACT 2601
Postal: GPO Box 9839, Canberra, ACT 2601
Tel: (+61 6) 213 72 46
Telex: AA62654
Fax: (+61 6) 213 72 49
WWW: http://www.dist.gov
Director: Elio Grohavaz
Assistant Director: Mike Fowler
Projects Officer: Michael Hettinger

AUSTRIA

Austrian Space Agency
Garnisongasse 7, A-1090 Vienna
Tel: (+43 1) 403 81 77
Fax: (+43 1) 405 82 28
Managing Director: Prof Dr Johannes Ortner

Austrospace
Garnisongasse 7, A-1096 Vienna
Tel: (+43 1) 403 81 77
Fax: (+43 1) 405 82 28
Contact: Rudolf Oezelt

Joanneum Research
Inffeldgasse 12, A-8010 Graz
Tel: (+43 316) 437 21
Telex: 31121 TUGRZ A
Fax: (+43 316) 46 36 97
Director: Prof Dr Willibald Riedler

Space Research Institute
Austrian Academy of Sciences
Inffeldgasse 12, A-8010 Graz
Tel: (+43 316) 46 36 96
Telex: 31121 TUGRZ A
Fax: (+43 316) 46 36 97
Head, Experimental Space Research:
Prof Dr Willibald Riedler
Head, Satellite Geodesy: Prof Hans Sünkel
Head, Near-Earth Physics: Prof Siegfried Bauer

BELGIUM

Belgacom
2 Rue des Palais, B-1210 Brussels
Tel: (+32 2) 213 42 67/42 82
Telex: 29154 gentl b
Fax: (+32 2) 218 52 62/223 15 39

Belgian Institute of Space Aeronomy
3 Ringlaan, B-1180 Brussels
Tel: (+32 2) 374 27 28
Telex: 21563
Fax: (+32 2) 374 84 23
Director: Dr Marcel Ackerman

Office for Science, Technology & Cultural Affairs
Rue de la Science 8, B-1040 Brussels
Tel: (+32 2) 238 34 11
Telex: 24501 prosci b
Fax: (+32 2) 230 59 12
Secretary General: Jacques Wautrequin
Space Research/Technology Manager:
Monique Wagner

Royal Meteorological Institute of Belgium
3 Ringlaan, B-1180 Brussels
Tel: (+32 2) 373 06 00
Telex: 21563
Fax: (+32 2) 374 67 88
e-mail: dcr@radio.oma.be
Head, Aeronomy Department:
Dr Dominique Crommelynck

BRAZIL

Agencia Espacial Brasileira (AEB)
Palacio do Planalto, anexo 11 sala 135,
70150-900 Brasilia DF
Tel: (+55 61) 211 17 37/224 77 69/322 52 41
Fax: (+55 61) 223 36 64
President: Dr Luiz Gylvan Meira Filho
Director General: Ajax Barros de Melo

Centro Tecnico Aeroespacial (CTA)
Instituto de Aeronáutica e Espaço
Praça Marechal Eduardo Gomes 50, Villa das Acácias,
12228-904 São José dos Campos, São Paulo
Tel: (+55 11) 41 46 11
Telex: 1133393
Fax: (+55 11) 41 25 22
Director: Col Tiago da Silva Ribeiro
VLS Project Manager: Dr Jayme Boscov

Instituto Nacional de Pesquisas Espaciais
Av dos Astronautas 1758,
CP 515, 12227-010 São José dos Campos, São Paulo
Tel: (+55 11) 41 89 77
Telex: 11 3530
Fax: (+55 11) 41 25 22
Director: Marcio Nogueira Barbosa
Co-ordinator for Space Engineering and Technology: Oscar Pereira Dias Júnior
Co-ordinator for Atmospheric and Space Sciences: José Humberto Andrade Sobral
Co-ordinator for Earth Observation:
Luiz Alberto Vieira Dias
Co-ordinator, Institutional Relations:
Dr Roberto Pereira de Cunha
Head of Staff: Sérgio Sobral de Oliveira
Head, Satellite Tracking and Control Centre: Pawel Rozenfeld

BULGARIA

Bulgarian Aerospace Agency (BASA)
James Boucher Boulevard, Block 122, Apartment 29,
BG-1407 Sofia
Tel/Fax: (+359 2) 68 30 94
President: Prof Dr Boris Bonev
Science Programmes: Dt Tania Ivanova
International Relations: Pavel Bonev

Navigation Maritime Bulgare Ltd
1 Primorski Boulevard, BG-9000 Varna
Tel: (+359 52) 22 24 74/23 55 91
Telex: 77351
Fax: (+359 52) 22 24 91/60 03 60
Managing Director: Ivan Borisov
Manager, Communications Department:
Vassil Dimitrov

Space Research Institute (SRI-BAS)
6 Moskovska Str, BG-1000 Sofia
Tel: (+359 2) 88 35 03
Fax: (+359 2) 80 13 47
Director: Prof Dr Boris Bonev
Vice-Director: Dr Dimitar V Iordanov
Science Secretary: Dr Stefan Chapkanov

CANADA

Canadian Space Agency
6767 Route de l'Aéroport,
St Hubert, Quebec J3Y 8Y9
Tel: (+1 514) 926 48 00
Fax: (+1 514) 926 48 78
President: William MacDonald (Mac) Evans
Executive Vice-President: Alain-F Desfossés
Vice-President Corporate Development:
Dr Garry Lindberg
Vice-President Programs: Dr Karl H Doetsch
Vice-President Engineering: Joseph L McNally
Vice-President Operations: to be announced
Executive Secretary: Michel Giroux
Director General, Radarsat Program:
Shabeer Ahmed (acting)
Director General, Space Station: Alain Poirier
Director General, Space Technology:
Virendra Jha (acting)
Director General, Space Science Program:
Barry Wetter (acting)
Director General, Astronaut Program:
Steve MacLean (acting)
Director General, David Florida Laboratory:
Dr Rolf Mamen

CHILE

Centro de Estudios Espaciales
Universidad de Chile,
A. Prat 1171, Casilla 411-3, Santiago de Chile
Tel: (+56 2) 555 34 00
Fax: (+56 2) 844 10 03
Executive Director: Eduardo Dias Araya

Comite de Asuntos Espaciales (CAE)
Zenteno 45, Santiago
Tel: (+56 2) 671 12 13
Fax: (+56 2) 696 45 81

President: Gral Máximo Venegas
Secretary: Col Ricardo Ortega

Empressa Nacional de Telecomunicaciones
Casilla 4254, Santiago de Chile
Tel: (+56 2) 690 21 21/25 48
Telex: 240683/240543 entel cl
Fax: (+56 2) 699 34 24

Fuerza Aérea de Chile División Espacial
Zenteno 45, Santiago
Tel: (+56 2) 671 12 13
Fax: (+56 2) 696 45 81
Space Division Head: Col Ricardo Ortega
FASat-Alfa Project Manager: Lt Col Fernando Mujica
FASat-Bravo Project Manager: Lt Col Mario Arancibia

CHINA, PEOPLE'S REPUBLIC

Beijing Marine Communications & Navigation Company
10 Fu-Xing Lu, 100845 Beijing
Tel: (+86 10) 64 22 53 58
Telex: 22462 comct cn
Fax: (+86 10) 64 21 35 09
President: Mrs Guiyu Tang
Chief Engineer, Satcoms: Yumin Zhao
Inmarsat Marketing: Jing Dong Du, Tang Shaohua

China Academy of Launch Vehicle Technology
PO Box 9200-1, No.1 Nandahong Men Road,
Fengtai District, 100076 Beijing
Tel: (+86 10) 68 38 10 35/34
Fax: (+86 10) 68 38 10 34
President: Li Jianzhong
Vice Chief Engineer: Liu Erxun
Vice-President Quality Assurance: Han Jingquan
Chief Designer, CZ-3B: Long Lehao
Chief Designers, circa 1995: C2-2C Li Zhankui, C2-2E Wang Dechen, C2-3 Fan Shihe, C2- 3A/B/C Long Lehao

China Academy of Space Electronic Technology (CASET)
PO Box 2421-9, 8 Chedaogou, Zizhuyuan Road,
Haidian, 100081 Beijing
Tel/Fax: (+86 10) 68 42 21 19
President: Tao Jiaqu

China Academy of Space Technology
PO Box 2417, 31 Bai Shi Qiao Lu, 100081 Beijing
Tel: (+86 10) 68 37 82 30/94 39
Telex: 22473 ccsc cn
Fax: (+86 10) 68 37 91 63/82 37
President: Xu Fuxiang
CBERS Project Manager: Chen Yiyuan

China Chang Feng Mechanics & Electronics Technology Academy
PO Box 3905, 50 Yongding Road, 100854 Beijing
Tel: (+86 10) 68 38 54 17
Telex: 222278 cfic cn
Fax: (+86 10) 68 21 36 32
President: Chen Dingchang

China Great Wall Industry Corporation
30 South Haidian Road, Haidian District, 100080 Beijing
Tel: (+86 10) 68 74 88 10/08 88
Fax: (+86 10) 68 74 88 65/88 76
Chairman: Guo Guozheng
President: Zhang Tong
Vice-Presidents: Yu Fusheng, Xu Yinfu, Liu Zhixiong

China Great Wall Industry Corporation
Space Division
22 Fucheng Road, Haidian Qu, 100036 Beijing
Tel: (+86 10) 68 37 86 21/36 24
Fax: (+86 10) 68 37 36 20/842 92 17
General Manager: Zhang Jianye

China Great Wall Industry Corporation
GW Aerospace, Inc
21515 Hawthorne Boulevard No 1065, Torrance, California 90503, USA
Tel: (+1 310) 540 77 06
Telex: 910 250-4940
Fax: (+310) 540 34 75
President: Zuoyi Huang
Vice-President: Guodong Tian

China Great Wall Industry Corporation
GWIC German office
Emil-Riedel Str 18, D-80538 München, Germany
Tel: (+49 89) 22 66 56/57
Fax: (+49 89) 22 66 59
Chief Representative: Jin Haoxing

China Great Wall Industry Corporation
GWIC Washington DC office:
3400 International Drive NW, Suite 2J-300, Washington DC 20008, USA
Tel: (+1 202) 364 85 75/85 80
Fax: (+1 202) 364 85 63
Chief Representative: Chen Baosheng
Representative: Jun Wang, Ms Minzhi Li

China Haiying
Electro-Mechanical Technology Academy
PO Box 7200-1, 100074 Beijing
Tel: (+86 10) 68 37 55 86
Telex: 222832 ahy cn
Fax: (+86 10) 68 37 53 78
President: Wang Jianmin

China Jiangnan Space Group
36 Beijing Road, 563003 Zunyi City, Guizhou Province
Tel: (+86 852) 22 61 16
Telex: 0201 zunyi
Fax: (+86 852) 22 32 82
General Manager: Li Guozhong

China National Space Administration
8 Fucheng Road, Haidian District,
PO Box 848, 100830 Beijing
Tel: (+86 10) 68 37 08 11/08 12/01 56/22 21
Fax: (+86 10) 68 37 08 49
President: Liu Jiyuan
Vice-Presidents: Wang Liheng, Luan Enjie, Bai Baier, Xia Guohong
Director, General Office: Zhang Baoqian
Director, Programmes and Planning: Wang Guolin
Director, Department of International Co-operation: Wang Xiuting
Director, Department of Foreign Affairs: Bai Jingwu

China Sanjiang Space Group
Fuyuan Alley, 1st Qianjin Road, 430022 Wuhan, Hubei
Tel/Fax: (+86 27) 586 40 37
President: Cao Lijia

China Space Civil & Building Engineering Design & Research Academy (CSCBA)
PO Box 2964, 93 Fengtai Road, 100071 Beijing
Tel: (+86 10) 63 82 28 31
Fax: (+86 10) 68 37 08 65

Chinese Society of Astronautics
PO Box 838, 100830 Beijing
Tel: (+86 10) 68 37 20 81/21 77
Fax: (+86 10) 68 37 20 81
President: Liu Jiyuan

The 4th Academy of CASC
PO Box 159, 710000 Xian City
Tel: (+86 29) 721 45 69
President: Ye Dingyou

Institute of Space Medico-Engineering
PO Box 5104, 1 Yuanmingyuan West Road,
100094 Beijing
Tel: (+86 10) 62 58 22 27
President: Min Guirong
Director: Yuan Wang

Shanghai Academy of Space Electronics Technology (SASET)
222 Caoxi Road (Spaceflight Tower), Shanghai 200233
Tel: (+86 21) 472 08 00
President: Zhang Wenzhong

Shanghai Academy of Space Technology
805 Research Institute, 819 Jiang Chuan Road,
Min Hang District, 202240 Shanghai
Tel: (+86 21) 430 12 51
Telex: 30246
Fax: (+86 21) 430 22 54
President: Zhang Wenzhong

Shanghai Institute of Satellite Engineering
251 Huayin Road, Shanghai 200240
Director: Prof Meng Zhizhong
Vice-Director: Dr Lu Bili

Sichuan Aerospace Industry Corporation
PO Box 1, 636450 Wanyuan City, Sichuan Province
Tel: (+86 811) 881 39 13/39 11
General Manager: Yu Ruihua

Xian Aerospace Corporation
PO Box 15 Yuhang Street, Aerospace Town,
710000 Xian, Shaanxi Province
Tel: (+86 29) 525 41 61/41 65
President: Hu Hongfu

Xian Institute of Space Radio Technology
PO Box 165 Xian, Shaanxi Province
Vice-Director: Wei Zhang

CZECH REPUBLIC

Astronomical Institute
Turkovicka 278, CZ-251 65 Ondrejov
Tel: (+420 2) 04 64 92 01
Telex: 121579 astro
Fax: (+420 2) 88 16 11
Director: Dr Jan Palous

Commission on Space Activity
Academy of Science Narodni Trida 3, 111 42 Prague 1
Tel: (+420 2) 24 34 41/9
Fax: (+420 2) 26 56 71
Commission Chairman: Dr Ladislav Sehnal

Institute of Atmospheric Physics
Bocni II, CZ-141 31 Prague 4
Tel: (+420 2) 76 25 48
Telex: 121546 ionp c
Fax: (+420 2) 76 25 28
e-mail: ptr@ufa.cas.cz
Director: Dr Josef Stekl
Upper Atmosphere Department (Magion): Dr Pavel Triska

DENMARK

Danish Research Advisory Board for Space
The Ministry of Research and Information Technology
Bredgade 43, DK-1260 Copenhagen K
Tel: (+45) 33 92 97 00
Telex: 15089 fd dk
Fax: (+45) 33 92 35 01
Secretary: Charlotte Bloch (head of section)

FINLAND

Academy of Finland
PO Box 99, FIN-00501 Helsinki
Tel: (+358 9) 77 48 82 66
Fax: (+358 9) 77 48 83 93
e-mail: Mirja.Vihma-Kaurinkoski@aka.fi
Scientific Secretary: Mirja Vihma-Kaurinkoski MSc

Finnish Space Committee
Secretariat: PO Box 69, FIN-00101 Helsinki
Tel: (+358 0) 69 36 78 54
Fax: (+358 0) 69 36 77 95
e-mail: petrie.peltonen@tekes.fi
WWW: http://www.tekes.fi
Chairman: Dr Martti Mäenpää
Secretary: Petrie Peltonen

Technology Development Centre (TEKES)
Malminkatu 34, PO Box 69, FIN-00101 Helsinki
Tel: (+358 10) 521 51
Fax: (+358 9) 694 91 96/(+358 10) 521 59 05 Space Technology
Director General: Dr Martti Mäenpää
Director (Space Technology): Martin Maklin
Project Manager, Earth Observation: Esa Panula-Ontto
Research Manager: Petri Peltonen
Space 2000 Co-ordinator: Kimmo Ahola
Technology transfer and launchers: Haken Sandell

FRANCE

Centre de Lancements de Ballons
F-40800 Aire-sur-l'Adour
Tel: (+33 5) 58 71 61 82
Telex: 560951
Fax: (+33 5) 58 71 67 44

Centre National d'Études Spatiales
2 place Maurice Quentin, F-75039 Paris Cedex 01
Tel: (+33 1) 44 76 75 00
Telex: 214674
Fax: (+33 1) 44 76 76 76
Chairman: Alain Bensoussan
Director General: to be announced
Assistant Director General: Jean-Jacques Sussel

Inspector General: Pierre Bescond
Special Advisor to President: Daniel Sacotte
Director General Advisor: Jacques-Emile Blamont
Secretary General: Denis Soubeyran
Director, Programs/Industrial Policy: Gérard Brachet
Director, International Relations: Isaac Revah
Director, Strategic Planning: Jean-Yves Le Gall
Director, Astronauts: Jean-Loup Chrétien
Director, Quality Assurance: Pierre Marx
Press Contacts: Daniel Metzlé, Philippe Collot
Tel: (+33 1) 44 76 76 88; Fax: (+33 1) 44 76 78 16

Centre Spatial de Toulouse (CST)
18 avenue Edouard-Belin, F-31055 Toulouse Cedex
Tel: (+33 5) 61 27 31 31
Telex: 531081
Fax: (+33 5) 61 27 31 79
Director, Toulouse Centre: Michel Courtois
Deputy Director: Yves Trempat
Technical Inspector General: Bernard Estadieu
Press Contact: Eliane Moreaux Tel: (+33 61) 27 33 44; Fax: (+33 61) 28 21 47

Centre Spatial d'Evry
Rond-Point de l'Espace, F-91023 Evry Cedex
Tel: (+33 1) 60 87 71 11
Telex: 604701
Fax: (+33 1) 60 87 70 47/8
Director, Evry Centre and Launchers: Daniel Mugnier
Ariane 5 Project Manager: Guy Laslandes
Press Contact: Danielle Charvet
Tel: (+33 1) 60 87 70 29; Fax: (+33 1) 60 87 74 43

Centre Spatial Guyanais
BP 726, F-97387 Kourou Cedex
Tel: (+33 5) 94 33 51 11
Telex: 910 516
Fax: (+33 5) 94 33 47 66
Director: Michel Mignot
Tel: (+33 5) 94 33 55 53; Fax: (+33 5) 94 33 47 19
Press Contact: Hélène Michel-Donadieu

DARA/DLR Bureau Paris
17 avenue de Saxe, F-75007 Paris
Tel: (+33 1) 42 19 94 26
Fax: (+33 1) 42 19 96 29
Head of Office: Manfred Gerstenlauer

Etat Major des Armées
Bureau Espace
231 boulevard St Germain, F-00456 Armées
Tel: (+33 1) 42 19 30 11
Fax: (+33 1) 42 19 43 56
Chief: RADM de Bazelaire

ISRO Technical Liaison Unit
25 rue d'Argenteuil, F-75001 Paris
Tel: (+33 1) 42 96 63 35
Telex: 214 213
Fax: (+33 1) 42 61 24 71
Liaison Officer: M Y S Prasad

Ministère de la Défense
14 rue Saint-Dominique, F-00456 Paris Armées
Tel: (+33 1) 45 55 95 20/44 39 59
Telex: 270003 defnat f, 202681 infodet f
President, Space Co-ordination Group:
GAA Jean-Philippe Douin

GERMANY

Bundesminister für Forschung und Technologie (BMFT)
Postfach 20 02 40, Heinemannstrasse 2,
D-53175 Bonn
Tel: (+49 59) 35 16
Telex: 885 674
Fax: (+49 59) 36 01

Deutsche Agentur für Raumfahrtan-gelegenheiten (DARA) GmbH
Koenigswinterer Str 522-524,
D-53227 Bonn-Oberkassel.
Postal: Postfach 30 03 64, D-53183 Bonn
Tel: (+49 228) 44 70
Fax: (+49 228) 44 77 00
Director General: Dr Jan B Mennicken
Managing Director: Dipl-Ing Klaus D Berge
Earth Observation/Telecoms: Heinz Seipel
Space Science/Infrastructure Systems:
Manfred Otterbein
Industrial Policy/Technology: Paul-Jürgen Weber
Budget/Strategic Planning: Dr Ulrich Krafft
Administration: Dr Thomas Wehdeking

International/National Co-operation:
Prof Peter Oesberg
Public Relations: to be announced

Deutsche Forschungsanstalt für Luft- und Raumfahrt (DLR)
Executive Headquarters
Linder Höhe, D-51147 Köln
Tel: (+49 2203) 60 10
Telex: 8810-0 dlr d
Fax: (+49 2203) 673 10
WWW: http://www.dlr.de
Chairman: Prof Dr Walter Kröll
Vice-Chairman: Dr Jürgen Blum

DLR/DARA Office Washington
1627 Eye Street NW, Suite 540,
Washington DC 20006-4020, USA
Tel: (+1 202) 785 44 11
Telex: 023 248652 gacc
Fax: (+1 202) 785 44 10
Head of Office: Dr Dietmar Wurzel (DLR)
Dr R Densing (DARA)

Energia Deutschland GmbH
Wolfratshauser Str 44-48, D-81379 Munich
Tel: (+49 89) 72 49 50
Telex: 529749
Fax: (+49 89) 72 49 52 91
Managing Director: Reiner Klett

GREECE

Hellenic National Space Committee
PO Box 14631 GR-11510 Athens
Tel: (+30 1) 771 43 79
Fax: (+30 1) 771 38 10
President: Prof Dr L N Mauridis

HUNGARY

Antenna Hungaria
Hungarian Radiocommunications Corp
Petzvál J.u. 31-33, H-1119 Budapest
Tel: (+36 1) 203 60 87
Fax: (+36 1) 203 60 91

Hungarian Satellite
Communications Association (HUNSAT)
Városmajor u. 12-14, H-1122 Budapest
Tel: (+36 1) 212 50 15
Fax: (+36 1) 155 26 57
Director: Istvan Hazay

Hungarian Space Office
PO Box 565, H-1374 Budapest
Tel: (+36 1) 117 59 00
Telex: 224233 omfb h
Fax: (+36 1) 118 79 98
Director: Dr Gyula Tófalvi

KFKI Research Institute for Particle and Nuclear Physics (RIPNP)
PO Box 49, H-1525 Budapest
Tel: (+36 1) 395 92 89
Telex: 224722 kfki h
Fax: (+36 1) 395 91 51
Director: Dr Károly Szegö
e-mail: szego@rmki.kfki.hu
Department of Space Physics: András Varga
Department of Space Technology: Dr Sándor Szalai
e-mail: szalai@rmk530.rmki.kfki.hu

INDIA

Indian Space Research Organisation
Antariksha Bhavan, New BEL Rd, Bangalore 560 094
Tel: (+91 80) 333 44 74
Telex: 0845-2499
Fax: (+91 80) 333 42 29
Chairman: Dr Krishnaswamy Kasturirangan
Scientific Secretary: M G Chandrasekhar
Director, Earth Observation: M G Chandrasekhar
Director, Satellite Communications: K Narayanan
Director, Launch Vehicles: D Narayanamoorthi
Director, Technology Transfer: N Sampath
Deputy Director, Computers: Dr S K Sinha
Director, Publications/Public Relations:
S Krishnamurthy

Physical Research Laboratory (PRL)
Navrangpura, Ahmedabad 462029
Tel: (+91 79) 656 85 50
Telex: 0121 397
Fax: (+91 79) 656 05 02
Director: Dr G S Agrawal

Space Commission/Department of Space
Antariksha Bhavan, New BEL Rd, Bangalore 560 094
Tel: (+91 80) 333 44 74
Telex: 0845-2499
Fax: (+91 80) 333 42 29
Chairman: Dr Krishnaswamy Kasturirangan

Videsh Sanchar Nigam Ltd
287 Videsh Sanchar Bhavan,
Mahatma Gandhi Road, Bombay 400 001
Tel: (+91 22) 262 40 20/204 23 64
Telex: 1182429 vsnh in
Fax: (+91 22) 262 44 66/95 43 21
Chairman/Managing Director: B K Syngal
Director (Operations): Chilmakuru V Naidu

INDONESIA

National Institute of Aeronautics & Space
JLN Pemuda, Persil No 1,
PO Box 20/JAT, Jakarta 13220
Tel: (+62 21) 489 49 41/489 28 02
Telex: 49175
Fax: (+62 21) 489 48 15
Chairman: Maj Gen Z B Palaguna
Secretary: Ir J Soegijo
Director, Remote Sensing Technology Centre: Ir Mahsun Irsyam

INTERNATIONAL

European Space Agency
8-10 rue Mario-Nikis,
F-75738 Paris Cedex 15, France
Tel: (+33 1) 53 69 76 54
Telex: 202746 esa f
Fax: (+33 1) 53 69 75 60/61/62
WWW: http://www.esrin.esa.it
Director General: Jean-Marie Luton
Council Chairman: P G Winters
Head, Director General's Cabinet: Karl-Egon Reuter
Associate Director for Strategy, Planning and International Policy: Jean-Jacques Dordain
Associate Director, Information System: Brian Walker
Inspector General: Prof Massimo Trella
Director, Administration: Adalbert Plattenteich
Director, Science: Dr Roger M Bonnet
Director, Earth Observation and Its Environment: Lanfranco Emiliani
Director, Telecommunications: René Collette
Director, Launchers: Dr Frederik Engström
Director, Manned Spaceflight and Microgravity: Jörg Feustel-Büechl
Director, ESTEC: Vacant
Director, ESOC: Felix Garcia-Castañer
Head, ESRIN: Francis Roscian
Head, Public Relations: Roger Elaerts
Tel: (+33 1) 53 69 71 55; Fax: (+33 1) 53 69 76 90

Washington Bureau
955 L'Enfant Plaza, suite 7800,
Washington DC 20024, USA
Tel: (+1 202) 488 41 58
Fax: (+1 202) 488 49 30
Head of Office: Ian Pryke

Brussels Bureau
Rue de la Loi 223, Boite 4, B-1040 Bruxelles, Belgium.
Tel: (+32 2) 230 90 39
Fax: (+32 2) 230 90 81
Head of Office: Eric Slachmuylders

Moscow Bureau
Sretensky Bd 6 – A.44, 103045 Moscow, Russia
Tel: (+7 095) 928 75 29
Fax: (+7 095) 928 53 52
Head of Office: Alain Fournier-Sicre

Kourou Bureau
BP 816, F-97388 Kourou, French Guiana
Tel: (+594 33) 71 59
Telex: 910 302
Fax: (+594 33) 47 54

NETHERLANDS

National Aerospace Laboratory NLR
Postal: PO Box 90502, NL-1006 BM Amsterdam
Anthony Fokkerweg 2, NL-1059 CM Amsterdam
Tel: (+31 20) 511 31 13
Fax: (+31 20) 511 32 10
WWW: http://www.nlr.nl
General Director: Dr B M Spee
Space Division: B J P van der Peet
Co-ordinator Space Projects: Dr J C Venema

Netherlands Agency for Aerospace Programs
Postal: PO Box 35, NL-2600 AA Delft
Kluyverweg 4, NL-2629 HT Delft
Tel: (+31 15) 278 80 25
Fax: (+31 15) 262 30 96
Director: Dr R Roos

Netherlands Industrial Space Organisation
PO Box 32070, NL-2303 DB Leiden
Tel: (+31 71) 524 51 24
Fax: (+31 71) 524 51 25

Observatory for Satellite Geodesy
PO Box 581, NL-7300 AN Apeldoorn
Tel: (+31 57) 693 41
Telex: 36442 satko nl
Fax: (+31 57) 693 44
Contact: Prof R Rummel

Space Research Organisation Netherlands
Sorbonnelaan 2, NL-3584 CA Utrecht
Tel: (+31 30) 253 56 00
Telex: 47224 astro
Fax: (+31 30) 254 08 60
WWW: http://www.sron.ruu.nl
e-mail: secr@sron.ruu.nl
Chairman of the Board: Dr A P M Baede
General Director: Prof J A M Bleeke
Deputy Director: Prof T de Jong

NIGERIA

Nigeria Telecommunications Ltd (Nitel)
Corporate Administration
3/5 Tafawa Balewa Square, Lagos
Tel: (+234 1) 61 35 42
Telex: 11001 netad ng/28112 cirtel ng
Fax: (+234 1) 262 28 45
Chief, International Relations: Dr E B Ojeba

NORWAY

Norsk Romsenter
(Norwegian Space Centre)
Drammensveien 165, PO Box 113, Skoyen, N-0212 Oslo
Tel: (+47 22) 51 18 00
Fax: (+47 22) 51 18 01
Managing Director: Pal Sorensen
Director, Applications: Jostein Rosenberg
Director, Finance: Knut Myrvang
Director, Industrial Programmes: Geir Hovmork
Director, Information: Per Torbo
Director, Space Science: Bo N Andersen

PAKISTAN

Space & Atmospheric Research Centre
PO Box 8402, Karachi 75270
Tel: (+92 21) 47 10 01 05
Telex: 25720 SPACE PK
Fax: (+92 21) 496 05 53
Member, Space Research: vacant

Space and Upper Atmosphere
Research Commission (Suparco)
Sector 28, Gulzar-e-Hijri,
PO Box 8402, Off University Road, Karachi 75270
Tel: (+92 21) 47 26 30/47 01 58
Telex: 25720 SPACE PK
Fax: (+92 21) 496 05 53
Chairman: Sikander Zaman
Member, Space Technology: Sikandar Zaman
Member, Space Research: vacant
Member, Space Electronics: S Akram H Abidi
Member, Finance: Javid Ahmed Khan
Secretary: M Zafrul Hasan
Director, International Affairs: M Nasim Shah

POLAND

Copernicus Astronomical Centre
Ul. Bartycka 18, 00716 Warsaw
Tel: (+48 22) 41 00 41/10 86
Telex: 813 978 cbk pl
Fax: (+48 22) 41 08 28
Director: Prof Jozef I Smak

Institute of Aviation
Aleja Krakowska 110/114, 02256 Warsaw
Tel: (+48 22) 46 00 11/46 08 01
Telex: 813 537
Fax: (+48 22) 46 44 32
Director: Dr Roman Czerwinski

Polspace Ltd
Ul. Bartycka 18A, 00716 Warsaw
Tel: (+48 40) 30 35
Fax: (+48 40) 30 35
WWW: http://www.korpo.pol.pl/polspace
President: Prof Janusz B Zielinski
Executive Vice-President/Research and Development Director: Dr Zygmunt Krawczyk

Space Research Centre
Ul. Bartycka 18A, 00716 Warsaw
Tel: (+48 22) 40 37 66/ 40 31 31
Telex: 825 670 cbk pl
Fax: (+48 39) 12 12 73
e-mail: CBKPAN@PLEARN.BITNET
WWW: http://www.cbk.waw.pl
Director: Prof Zbigniew Klos

Telekomunikacja Polska SA
Biuro Systemow Satelitarnych
Ul. Twarda 14, 00105 Warsaw
Tel: (+48 22) 620 38 87
Telex: 813 348 pl
Fax: (+48 22) 26 36 65
Director: Zbigniew Zubik

PORTUGAL

Instituto Nacional de Engenharia e Tecnologia Industrial
Azinhaga dos Lameiros à Estrada do Paço do Lumiar, 1699 Lisboa, Codex
Tel: (+351 1) 716 51 81
Fax: (+351 1) 716 30 48
Space Activities: Dr José Manuel Rebordão

ROMANIA

Ministry of Communications
14 Aleea Libertatii, Sector 5, 70060 Bucharest
Tel: (+40 1) 400 15 75
Telex: 11372 r
Fax: (+40 1) 400 12 30
Minister of Communications: Adrian Turicu
Director General for Regulations: Cantemir Ionescu

RUSSIAN FEDERATION

AKO Polyot
226 Bogdana Khmelnitskogo Street, 644021 Omsk 21
Tel: (+7 8) 381 12 39 74 32/57 92 00
Fax: (+7 8) 381 12 57 92 00/57 91 00
Director General: Stanislav R Osmulski
Chief Designer: Alexander I Ilyin
Design/Test Complex: Dr Alexei I Ilyin
Marketing Director: Vasili I Gorlov

Association for the Advancement of Space Science and Technology (ASCONT)
84/32 Profsoyuznaya St, 117810 Moscow
Tel: (+7 095) 333 24 45
Telex: 411498 star su
Fax: (+7 095) 330 12 00
e-mail: ykrasnop@iki3.bitnet
aberezni@esoc1.bitnet
Director General: Guenrikh A Avenesov
Deputy Director General: Yan L Ziman
Executive Director: Ivan V Polyansky

Astelit
Apt 32-33, 1 Mytnaya St, 121069 Moscow
Tel: (+7 095) 291 08 63
Fax: (+7 095) 291 09 25
President: Semyon V Lubman
Director General: Luigi Alberto Ciavoli Cortelli

Central Specialised Design Bureau
18 Pskovskaya Street, 443009 Samara
Tel: (+7 8462) 22 28 14
Telex: 214 153 nika su
Fax: (+7 8462) 27 20 70
General Designer: Dmitri I Kozlov
First Deputy/General Designer: Gennadi P Anshakov
Foreign Trade Director: Vladimir M Saigak

Centre for Space Biotechnology
16/10 Miklucho-Maklay ul, 117871 Moscow
Tel: (+7 095) 336 56 44
Telex: 411982
Chief: Dr Viktor Glotov

ESKOS SA
20 rue Troyon, F-92310 Sèvres, France
Tel: (+33 1) 46 23 85 00
Fax: (+33 1) 46 23 81 26
e-mail: 100073.2774@compuserve.com
President: Viktor Kouznetsov
General Manager: Anatole Volikov
Development Director: Yuri Ivchtchenko

Bureaux 602-603, Cosmonaut Volkov Street, 125299 Moscow
Tel/Fax: (+7 095) 159 11 15
General Manager: Mr Pitchkhadze

Glavcosmos
9 Krasnoproletarskaya ul, 103030 Moscow
Tel: (+7 095) 258 22 20
Telex: 411879 start
Fax: (+7 095) 288 95 83
Director: Alexander I Dunayev
Deputy Director: Vyacheslav I Dukov
International Relations: Oleg Firsyuk
Commercial Department: Yevgeny Sherstobitov

Global Information Systems, Inc
Prostornaya ul. 7, 107392 Moscow
Tel: (+7 095) 963 91 83
Fax: (+7 095) 168 31 15
President: Vladimir A Andrianov
Vice-President: Mikhail Y Senatorov
General Marketing Manager: Valeri Koutoukov

Informcosmos
6 Kazakova Street, 103064 Moscow
Tel: (+7 095) 267 80 21/261 15 27
Fax: (+7 095) 263 18 94/261 00 90
Director General: Igor S Tsirline
Commercial Director: Boris Chirkov

Informcosmosconsult LLC
9 Krasnoproletarskaya Street, suite 3, 103030 Moscow
Tel: (+7 095) 978 85 54/85 65/85 69
Fax: (+7 095) 978 85 38
e-mail: iavmos@sovam.com

201 North Mill Street, suite 201, Aspen, Colorado 81611, USA
Tel: (+1 303) 925 65 20/63 00
Fax: (+1 303) 925 78 86/11 81
e-mail: informcosmos@igc.apc.org
Managing Director: B Joseph Krabacher

4217 McCain Ct, Kensington, Maryland 20895, USA
Tel: (+1 301) 946 46 86
Fax: (+1 301) 946 74 66
e-mail: iav@ari.net

Institute of Biomedical Problems
Khoroshevskoye Shosse 78, 123007 Moscow D7
Tel: (+7 095) 195 68 71
Telex: 417508 vual su
Fax: (+7 095) 195 22 53
Director: Dr Anatoli I Grigoriev
First Deputy Director: Dr Viktor M Baranov
Deputy Directors: Drs Valeri V Poliakov, Yevgeni Ilyin

Intersputnik
2nd Smolensky Lane, 1/4, 121099 Moscow
Tel: (+7 095) 244 03 33
Telex: 411288/411289 DISK SU

Fax: (+7 095) 253 99 06
e-mail: dir@intersputnik.msk.ru
Director General: Gennadi G Kudryavtsev
Marketing Manager: Neil Bakmann (244 06 98)
Director, Legal/Financial Department: Istvan Kovacs
Director, Technical Development Department: Mario Fernandez
Director, Operations Department: Joseph Dolecki
US contact: Jeffrey Minsky
Tel: (+1 212) 247 38 46
Fax: (+1 212) 977 81 22
Italy contact: Moshe Lerner
Tel: (+39 2) 80 40 19

Izmiran
Troitsk, 142092 Moscow oblast
Tel: (+7 095) 334 02 93
Telex: 412623 scstp su
Fax: (+7 095) 334 01 24
Director: Dr Viktor Orayevski
First Deputy: Dr Valeri Fomichev
Head, Astrophysics Research: Dr Vladimir Ptuskin

KB Arsenal
1/3 Komsomola Ulitsa, 195009 St Petersburg
Tel: (+7 812) 542 29 73
Telex: 121608 ars su
Fax: (+7 812) 542 20 60
Director/Chief Designer: Boris I Poletaev

KB Khimautomatiki
22 Vorochilova Ul, 394055 Voronezh
Tel: (+7 0732) 33 36 80/57 25 58
Telex: 153212 ravor su
Fax: (+7 0732) 33 36 80/33 41 22
General Director: Dr Vladimir S Rashuk
Marketing: Dr Gennadi I Zavizion
Production: Anatoli I Solianik

KB Khimmach
2 Lyesoparkovyj Tupic, 141070 Kaliningrad
Tel: (+7 095) 516 38 20/30 64/30 54
Telex: 346757
Fax: (+7 095) 516 05 85/06 62
Director: Dr Nikolai I Leontyev
Foreign Relations: Yevgeni P Selyeznyov

KB Kolomna
Okski Prospekt 42, 140402 Kolomna
Tel: (+7 096) 613 75 78/75 05
Fax: (+7 096) 613 76 99/30 64
General Designer: Sergei P Nepobedimiy

KB Machinostroenye
1 Turgoyakskoe Shosse, 456300 Miass
Tel: (+7 35) 13 52 39 70
Telex: 124858 rosa su
Fax: (+7 35) 13 52 60 47
General Director: Dr Igor I Vyelichko

KB Photon
151 Molodogvardeiskaya St,
443001 Samara (Kuibyshev)
Tel: (+7 8462) 33 26 32
Telex: 214 153 nika su
Fax: (+7 8462) 27 20 70
Director: Vladimir D Kozlov
Deputy Director: Valeri I Abrashkin

Keldysh Institute of Applied Mathematics
4 Miusskaya pl, 125047 Moscow
Tel: (+7 095) 972 37 14
Fax: (+7 095) 972 07 37
Director: Dr Sergei P Kurdyumov
Head, Space Division: Dr Ephraim L Akim
Head, Spacecraft Attitude Control: Vasili A Sarychev

Keldysh Research Center
8 Onezhskaya, 125438 Moscow
Tel: (+7 095) 456 46 08
Fax: (+7 095) 456 82 28

Khrunichev State Research and Production Space Centre
18 Novozavodskaya, 121309 Moscow
Tel: (+7 095) 145 80 36/83 43
Fax: (+7 095) 142 59 00
Telex: 412427 su
e-mail: proton@sovam.com
General Director: Anatoli I Kiselev
First Deputy General Director: Dr Alexander A Medvedev
Deputy General Director for International Co-operation: Alexander V Lebedev
Deputy General Director, KB Salyut General Designer: Anatoli K Nedaivoda
Deputy General Director, Director Rocket/Space Technologies Plant: Anatoli A Kalinin
Deputy General Director, Director Test/Rocket Technical Ops Plant: Igor S Dodin
Director, Rockot Programme: Andrei G Novikov
Director, Iridium Programme: Alexander M Seregin
Director, Alpha/FGB Programme: Sergei K Shaevich
Proton Designer: Oleg I Davidov
Rockot Designer: Vladimir F Gusev

Lebedev Physics Institute Astro Space Centre
84/32 Profsoyuznaya St, 117485 Moscow
Tel: (+7 095) 333 21 89
Fax: (+7 095) 333 23 78
Radioastron Project Manager: Dr Nikolai Kardashev

Morsviasputnik
1/4 Rozhdestvenka Street, 103759 Moscow
Tel: (+7 095) 978 70 45/44 77/44 78
Telex: 871-411197 mmf su
Fax: (+7 095) 253 99 10
Inmarsat A fax: (+873) 140 21 62
Director General: Valeri Bogdanov
Deputy Director General: L Pcheliakov

Moscow Aviation Institute
4 Volokolamskoe shosse, 125871 Moscow
Tel: (+7 095) 158 13 73/43 33
Fax: (+7 095) 158 29 77
e-mail: aet@tk.mainet.msk.su
Rector: Prof Alexander Matveenko
First Deputy Rector (education): Igor Prokhorov
Deputy Rector (research): Viktor Veretennikov

Moscow Institute Of Thermal Technology
Otryadnii Proyezd 97, 127976 Moscow
Tel: (+7 095) 189 03 17
General Director/Chief Designer: Boris N Lagutin
First Deputy General Director: Grigori Yasinski
Deputy General Designer/Deputy Chief Designer: Dr Yuri S Solomonov

NII Argon
25 Varshavskoye Street, 113405 Moscow
Tel: (+7 095) 319 78 67
Telex: 411218
Fax: (+7 095) 315 40 83
Director: Maxim A Tereschenko

NII Grafit
2 Elektrodnaya Street, 111524 Moscow
Tel: (+7 095) 176 13 06
Telex: 412280 inkar su
Director: Valeri I Kostikov

NII Khimicheskogo Machinistroenye
Sergeyev Posad, 141300 Moscow
Tel: (+7 09654) 633 21
Telex: 846246 agat
Fax: (+7 09654) 676 98
Director: Dr Alexander A Makarov
Deputy Director: M N Demidov
International Relations: Dr Konstantin P Denisov

NII Kosmisheskovo Priborostroenye
Institute of Space Device Engineering
53 Aviamotornaya Street, 111024 Moscow
Tel: (+7 095) 273 47 19
Telex: 412176 intec su
Fax: (+7 095) 273 19 37
Director General/Chief Designer: Leonid I Gusev
Research and Development Director: Dr Nicolai E Ivanov

NII Machinostroenye
72 Stroitelei Street, Nizhnyaya Salda,
624610 Yekaterinburg oblast
Tel: (+7 34345) 217 03
Telex: 348172 omul
Fax: (+7 34345) 222 58
Director: Boris A Nekrasov
Sales/Marketing: Yuri P Bondar

NII Parachutostroenye
2 Irkutskaya ul, 107241 Moscow
Tel: (+7 095) 462 13 19
Telex: 411700
Fax: (+7 095) 462 52 33
Director: O V Rysev

NII PM (Prikladnoi Mekhaniki)
53 Aviamotornaya Street, 111023 Moscow
Tel: (+7 095) 273 38 76
Fax: (+7 095) 273 38 76
Director: Yuri S Sarimov

NII Radio
16 Kazakova St, 103064 Moscow
Tel: (+7 095) 261 36 94
Fax: (+7 095) 261 00 90
Director: Yuri B Zubarev

NPO AP
1 Vvedenskogo Street, 117342 Moscow
Tel: (+7 095) 330 65 70
Fax: (+7 095) 334 83 84
General Director: Prof Vladimir L Lapigin
Chief Engineer: Yuri Novikov

NPO Astra
Bolshoi Vyzovskyi per 2, 109028 Moscow
Tel: (+7 095) 297 33 43
Fax: (+7 095) 973 20 26
Director General: A P Radimov

NPO Cryogenmash
67 Lenin Prospekt, Balashiha 7,
143900 Moscow region
Tel: (+7 095) 521 57 33
Telex: 412561 cryo su
General Director: Yuri M Broshko

NPO ELAS
Zelenograd, 103482 Moscow
Tel: (+7 095) 531 17 49
Fax: (+7 095) 531 32 13/84 03
Chief Designer: Dr Gennady J Guskov

NPO Energomash
1 Budrenko Street,
Khimky-1, 141400 Moscow oblast
Tel: (+7 095) 572 22 00/573 00 38
Telex: 911649 almaz su
Fax: (+7 095) 573 12 01/251 75 04
Director General/Chief Designer: Dr Boris I Katorgin
First Deputy General Director/General Designer: Vladimir Chvanov
Director, Design Bureau: Vladimir Trofimov
Engine Department Chief/RD-170 Chief Designer: Felix J Chelkis (57301 10)
Deputy Director General, External Affairs: Viktor A Sigayev

NPO Geofizika
SKB GeoKos
11/17 Irkutskaya Street, 107497 Moscow
Tel: (+7 095) 462 03 43
Fax: (+7 095) 462 15 06/06 84
Director General: Alexei A Fomin
Chief Designer/First Deputy Director General Science Activities/Chief SKB GeoKos: Vladimir S Kouzmin
Deputy Chief Designer, International Co-operation: Alexandre P Dmitriev
Tel: (+7 095) 462 21 68

NPO Iskra
28 Vedeneyev Street, 614038 Perm
Tel: (+7 3422) 72 80 00
Telex: 134162 nota
Fax: (+7 3422) 72 84 68
Director General/General Designer: Prof Lyev N Lavrov
Chief Designer: Mikhail I Sokolovsky
Marketing: Yuri L Makarevitch

NPO Istochnik
10 Dahlia Street, 197376 St Petersburg
Tel: (+7 812) 234 90 30
Telex: 12140 su
Fax: (+7 812) 234 90 26

NPO Istok
Fryazina Ulitsina,
141120 Fryazino, Moscow oblast
Tel: (+7 095) 465 86 90/86 83
Fax: (+7 095) 465 86 86
General Director: Alexander Korolev
Deputy General Director: Sergei A Zaitsev

NPO Kompozit
4 Pionerskaya Street,
Kaliningrad, 141070 Moscow oblast
Tel: (+7 095) 516 81 72
Telex: 411204 nex su
Fax: (+7 095) 187 49 22
Director General: Prof Stanislav P Polovnikov
Deputy Director General: Vitaly A Tatarchenko

NPO Kraznaya Zvezda
Elektrolitniy Proezd 1a, 115230 Moscow
Tel: (+7 095) 113 23 09
Telex: 412748 space
Fax: (+7 095) 113 34 88
General Director: Georgi M Gryaznov

NPO Kvant
Kulakov per, 129626 Moscow
Tel: (+7 095) 287 97 58
Fax: (+7 095) 286 89 22

NPO Lavotchkin
Babakin Engineering Research Centre
24 Leningradskaya Street,
Khimky-2, 141400 Khimki, Moscow oblast
Tel: (+7 095) 573 90 56/55 65/86 92
Telex: 911 721 irbis
Fax: (+7 095) 573 35 95/27 86
Director General: Anatoli M Baklounov
Chief Designer: Vladislav Soloviey
Director General for International Affairs: M I Zytsev
Babakin Chief Desiger: Roald S Kremnev

Babakin, Deputy Chief Designer: Konstantin G Sukhanov
Babakin, Mars Designer: Garry Rogovsky
Babakin, Head Ballistics: O V Papkov
Babakin, Head Mission Control: I D Tseryonin

Lunokhod
Tel: (+7 095) 573 00 75
Fax: (+7 095) 571 23 45
Director General: Sergei E Prokopchuk

NPO Machinostroenye
33 Gagarin Street,
Reutov, 143952 Moscow oblast
Tel: (+7 095) 302 11 85
Fax: (+7 095) 302 20 01
General Director: Dr Herbert A Yefremov
Vice General Director: Anatoli V Chromushkin
Chief Designer: Igor U Postnikov
First Deputy General Designer: Vladimir Viter

NPO Molniya
4 Novoposelkova Street,
Tushino, 123459 Moscow
Tel: (+7 095) 493 92 55
Fax: (+7 095) 493 43 88/93 71
General Director: Alexander Bashilov
Chief Designer: Dr Genady P Dementyev

NPO Orion
Ul Plekhanova 2/46, 111123 Moscow
Tel: (+7 095) 176 16 39
Fax: (+7 095) 176 72 21
General Director: Vladimir Khryapov
Marketing Manager: Alexandr Chapkevich

NPO Planeta
1 Pervomaiskaya Street,
Dogoprudni, 141700 Moscow oblast
Tel: (+7 095) 483 41 18
Telex: 846672 zenit su
Director General: Vladimir Aksenov

NPO PM (Prikladnoi Mekhaniki)
660026 Krasnoyarsk-26
Tel: (+7 391) 973 20 32
Telex: 288 794 more
Fax: (+7 391) 972 26 35
Deputy Director General: Albert Kozlov
Head, Patents/Information: Roman Turkenich

NPO Precision Instruments
1 Yurlovski Proyezd, 127490 Moscow
Tel: (+7 095) 404 78 81
Telex: 411700
Fax: (+7 095) 404 79 66
Director General: Vladimir A Gorkovoi

NPO Saturn
13 Kasatkina Street, 129301 Moscow
Tel: (+7 095) 283 09 13/94 93
Telex: 2867 566 su
Fax: (+7 095) 286 75 66
General Director: Viktor M Chepkin
Head, Foreign Relations: Viktor G Nesterov

NPO Scientific Centre (NPO-NTs)
Zelenograd, nr Moscow

NPO Soyuz
Ul Sovetskaya 6, 140056 Dzerzhinsk
Tel: (+7 095) 172 26 92
Fax: (+7 095) 175 24 94
General Director: Dr Zinovi P Pak

NPO Start
Nizhnyaya Salda, 620082 Yekaterinburg
Tel: (+7 3482) 26 01 03
Telex: 221534 start su
Fax: (+7 3432) 26 03 55

NPO Stekloplastik
Kaliningrad

NPO Tekhnomash
3rd Proyezd Marinoi Roschi 40,
127018 Moscow
Tel: (+7 095) 289 74 33
Telex: 411351 tent su
Fax: (+7 095) 289 73 45
Director General: Dr Vyacheslav Bulavkin
Vice General Director for External Economic Activities: Viktor V Kuznetzov

NPO Trud
Sergeya Lazo St, 443026 Samara
Tel: (+7 8462) 50 02 28
Telex: 214 172 zelj
Fax: (+7 8462) 50 12 11
General Director/General Designer: Yevgeni A Gritsenko
Chief Designer: Valentin S Ansimov

NPO VEGA-M
34 Kutuzovsky Prospekt, 121170 Moscow
Tel: (+7 095) 249 76 10
Telex: 412268 vega su
Fax: (+7 095) 148 79 96

OKB Fakel
Moskovsky Prospekt 181, 236001 Kaliningrad
Tel: (+7 0112) 46 16 16/19 64
Telex: 262116 rtv su
Fax: (+7 0112) 46 17 62
Chief Designer: Alexandr Bober
Director: Vjacheslov M Murashko
Foreign Relations: Dr Nikolai A Maslennikov

OKB Moskovskaya Energetitsheski Institut
Moscow

PO Mayak
Ul. Lenina 31,
454058 Ozersk/Chelyabinsk-65
Tel: (+7 35151) 316 55
Telex: 124846 atom su
Fax: (+7 35151) 338 26
Director: Viktor Fetisov

PO Samara Frunze Engine Building
Zavodskoye shosse 29, 443009 Samara
Tel: (+7 8462) 51 12 12
Telex: 214157 ktex su
Fax: (+7 8462) 27 16 70
General Director: Igor L Shitarev

PPO Motorostroitel
Komsomolski Prospekt 93, 614600 Perm
Tel: (+7 3422) 45 11 87
Telex: 134812 don su
Fax: (+7 3422) 45 13 20
General Director: Eduard I Cherkashin
Commercial Director: Vladimir I Goryainov

Priroda
45 Volgogradski Prospekt, 109125 Moscow
Tel: (+7 095) 177 80 53
Fax: (+7 095) 177 65 74
Deputy Director: Anatoli Rubakha
Operations Manager: Yuri Karionov

Progress
18 Pskovskaya Street, 443009 Samara
Tel: (+7 84) 62 58 47 72
Telex: 224869 zalp
Fax: (+7 84) 62 58 46 11
General Director: Anatoli A Chizhov

RD&PE Zvezda
39 Gogol Street,
Tomilino, 140070 Moscow oblast
Tel: (+7 095) 557 30 65
Fax: (+7 095) 557 33 88
Director General: Prof Guy I Severin

Research Institute of Applied Electromechanics
Director: Garry Popov

RKK Energia
4a Lenin Street,
Kaliningrad, 141070 Moscow oblast
Tel: (+7 095) 513 30 65
Telex: 911518 buran su
Fax: (+7 095) 513 86 20
Chairman/President: Dr Yuri P Semyenov
Vice-President Manufacturing and Production: Nicholai I Zelenshchikov
Vice-President International Space Station Programme: Dr Viktor P Legostayev
Head, International Commercial Programmes: Alexander Derechin
Head, Mir Space Station Program: Dr Leonid Gorshkov
Head, Mission Controller: Victor Blagov
Head, Space Biotechnology Lab: Oleg Mitichkin
Head, Control, Propulsion and Navigation Systems: Dr Vladimir Branets
Head, Aerospace Capsules: Viktor Minenko
Head, Development of Electromechanical and Large Deployable Structures: Prof Vladimir S Syromiatnikov
Head, Remote Sensing: Oleg Lebedev
Director, Soyuz-TM Operations: Dr Vladimir Timchenko
Head, Life Support and Thermal Regulation: Edward Grigorov
Head, Power Department: Vyacheslav Khorounov

Russian Academy of Sciences (RAN)
Leninsky Prospekt 14, 117901 Moscow
Tel: (+7 095) 237 98 49
Telex: 411964 ran
Fax: (+7 095) 230 26 30
President: Yuri S Osipov
Vice-President: Nikolai P Laverov
Deputy: Vladimir A Kotelnikov

Russian Space Agency
42 Shchepkin Street, 129857 Moscow
Tel: (+7 095) 288 99 05/971 80 21
Fax: (+7 095) 251 87 02/883 56 22
Director General: Dr Yuri Nikolaivich Koptev
First Deputy Director General: Valeri V Alaverdov
Deputy Director General, International: Alexander I Medvedchikov
Deputy Director General, Space Science: Yuri G Milov
Deputy General Director: Boris D Ostroumov

Science & Technology International of Russia
1st Schipovski per.20, 113093 Moscow
Tel/Fax: (+7 095) 237 73 51

SKAS (Moscow Institute of Radio Links)
Moscow
Tel: (+7 095) 362 56 52/75 54
Telex: 417715 su

Smolsat
2a Moskvoretskaya nab, 109240 Moscow
Tel: (+7 095) 297 89 63
President: Alexander A Kisselev

Sovinformsputnik
47 Leningradski Prospekt, 125167 Moscow
Tel: (+7 095) 943 97 57
Fax: (+7 095) 943 05 85
General Director: Mikhail M Fomchenko

Sovzond
53 Aviamotornaya Street, 111250 Moscow
Tel: (+7 095) 273 21 42
Fax: (+7 503) 232 92 03
Director General: Dr Vladimir I Mikhailov

Soyuzkarta
45 Volgogradsky Prospekt, 109125 Moscow
Tel: (+7 095) 177 40 50
Telex: 411942 REN SU
Fax: (+7 095) 177 51 11
President: Dr Vyacheslav A Piskulin
Vice-President: Leonid G Suslonov
Deputy Director, Kosmokarta: Marina Rezepova

Space Research Institute (IKI)
84/32 Profsoyuznaya St, 117810 Moscow
Tel: (+7 095) 333 52 12/14 66
Telex: 411498 star su
Fax: (+7 095) 310 70 23/51 78
Director: Dr Albert A Galeyev
Deputy Directors: Gennadi M Tamkovich, Vyacheslav M Balebanov
Director, High Energy Astrophysics: Dr Rashid Sunyaev

Splav Technical Centre
9 Baikalskaya Street, 107497 Moscow
Tel: (+7 095) 462 46 88
Fax: (+7 095) 462 40 35
Director: Alexandr V Egorov

STC Complex
Berjozovaya Alleya ul 10/1, 127276 Moscow
Tel: (+7 095) 402 73 21/58 53
Fax: (+7 095) 402 82 29
Director: Dr Yuri S Solomonov

Tekhnoinvest
2 Mantulinskaya Street, 123100 Moscow
Tel: (+7 095) 289 95 86
Telex: 412 115 infas su
Fax: (+7 095) 205 76 81

TsAGI
(Tsentralny Aero-Girodinamichesky Institut)
1 Zhukovsky Street,
Zhukovsky-3, 140160 Moscow oblast
Tel: (+7 095) 556 44 12/41 53
Telex: 412573 zakat su
Fax: (+7 095) 271 00 19
Director General: Prof Gherman I Zagainov

TsNII Mashinostroenye
4 Pionerskaya Street,
Kaliningrad, 141070 Moscow oblast
Tel: (+7 095) 581 91 55
Telex: 411952 mcc su
Fax: (+7 095) 274 00 25
Director: Vladimir F Utkin
First Deputy Director, Science: Nikolai A Anfimov
First Deputy Director, General: Yevgeny N Shepelsky
Director General, TsNIIMash Export: Reshetin Andreyevitch

TsNPO Kometa
Velozavodskaya 5, 109280 Moscow
Tel: (+7 095) 275 15 33
General Director/Designer: Dr Anatoli I Savin

Ural Electrochemical Integrated Plant
Dzerzhinsky Street 2, Novouralsk,
624130 Yekaterinburg oblast
Tel: (+7 34370) 225 77
Telex: 721742 rikon su
Fax: (+7 34370) 241 41
Director: V Kornilov
Director, Electrochemical Converters: Dr Alfred N Arshinov
Chief Designer, Electrochemical Converters: Boris S Pospelov
Head, Fuel Cell Department: Dr Anatoli Ovchinikov

VNII Elektromekaniki
4 Khoromni Tupik, 107817 Moscow
Postal: PO Box 496, 101000 Moscow,
Tel: (+7 095) 208 84 67/928 34 54/924 44 39
Telex: 111807 alit su, 113530 ustup, 411390 metp
Fax: (+7 095) 207 49 62
Director General: Dr Stepan A Stoma
Deputy Director General/Chief Designer: Dr Yuri V Trifonov
Deputy Chief Designer: Georgi Arkopov

VNII TransMash
2 Zarechnaya ul, 198323 St Petersburg
Tel: (+7 812) 135 98 50
Fax: (+7 812) 135 98 37
Director: Eduard K Potyomkin
Planetary rovers: Valeri Gromov

Vympel
4 Ul 8 Marta 3, 125319 Moscow
Tel: (+7 095) 152 94 95/95 95
Fax: (+7 095) 152 93 34
General Director: Nikolai V Mikhailov

SOUTH AFRICA

University of Stellenbosch
Dept of Electrical Engineering, 7600 Stellenbosch
Tel: (+27 21) 808 45 25
Fax: (+27 21) 808 49 81
e-mail: milne@firga.sun.ac.za
WWW: http://esl.ee.sun.ac.za
Dean of Engineering: Prof P W van der Walt
Sunsat Project Leader: Prof Garth Milne

SPAIN

CDTI
Paseo de la Castellana 141, E-28046 Madrid
Tel: (+34 1) 581 55 00
Telex: 23121 cdti e
Fax: (+34 1) 581 55 84
Director General, Head of Spanish Delegation to ESA: Vicente Gómez
Director of Strategic Programmes: Emilia Buergo
Head of ESA Programmes Department: J Perez-Echague
Head of Space Technology and Applications Department: J M Leceta
Directorate of Space Programmes, comprising ESA Programmes and Technology and Applications Departments:
Tel: (+34 1) 581 55 63/57/11
Fax: (+34 1) 581 55 84/70

INTA
Carretera de Ajalvir km. 4,
E-28850 Torrejón de Ardoz Madrid
Tel: (+34 1) 627 03 00
Telex: 22026 inta e
Fax: (+34 1) 627 03 10/03 84
Director General: Dr Alvaro Girninez Cañeti
International Programmes: Eduardo Mezquida
Director Space: Dr Jose Maria Hoyos
Director Tracking: Luis Ruiz de Gopegui
Minisat Programme Manager: Miguel A Garcia Primo
Capricornio Programme Manager: Ricardo Dorado Gutierrez

Telefónica de Espana SA
Satellite Communications Systems Department
Plaza de Espana 4, E-28008 Madrid
Tel: (+34 1) 542 47 33
Telex: 42109 tsns e
Fax: (+34 1) 559 23 22
Deputy Director: Carlos Herrera de la Rosa

SWITZERLAND

Federal Space Affairs Commission
Federal Department of Foreign Affairs (EDA)
Political Directorate
Eigerplatz 1, CH-3003 Berne
Tel: (+41 31) 324 10 65
Fax: (+41 31) 324 10 73
President: Anton Cottier
Secretary: Dr Marc Bertschi

General Directorate of PTT
Telecom Department
Viktoriastrasse 21, CH-3030 Berne
Tel: (+41 31) 338 40 10
Telex: 911 025 rtv ch
Fax: (+41 31) 338 78 95
Director General: Karl Wuhrmann

International Space Science Institute
Hallerstrasse 6, CH-3012 Bern
Tel: (+41 31) 631 48 96
Tel: (+41 31) 631 48 97
WWW: http://ubeclu.unibe.ch/issi/
Executive Director: Prof Johannes Geiss

TAIWAN

National Space Program Office
8F, No 9 Prosperity 1st Road
Science-Based Industrial Park, 30077 Hsin-Chu City
Tel: (+886 35) 78 42 08
Fax: (+886 35) 78 42 10
e-mail: jiaming@nspo.gov.tw
Director: Dr Jia-Ming Shyu

THAILAND

Space Development Agency
Ministry of Transport & Communications
Ratchadamnoen Nok Avenue, Bangkok 10100
Tel: (+66 2) 283 30 09
Fax: (+66 2) 280 17 12/12 37
Director: Dr Khanef Chamarakkul

UKRAINE

Marlin Yug Ltd
2 Kapitanskaya Street, 335000 Sevastopol
Tel/Fax: (+380 0692) 52 04 50
e-mail: ocean@mhi2.sebastopol.ua
Director: Dr S V Motyzhov

NPO Khartron/Elektropribor
1 Akademik Proskura Street, 310070 Kharkov
Tel: (+7 0572) 43 31 31/44 56 1'
Telex: 115242 glava uv
Fax: (+7 0572) 43 16 77
General Director: Yakov Aizenberg
Commercial Director: Vladimir Sosunov

NPO Musson
18 Vakulinchuka Street, 335053 Sevastopol
Tel: (+7 69) 23 34 90
Telex: 187115
Fax: (+7 69) 23 80 81
Deputy Technical Director: Nikolai I Belikov
Inmarsat contact: I Kalyuzhniy

NPO Yuzhnoye
3 Krivorozhskaya Street,
320059 Dniepropetrovsk
Tel: (+7 0562) 42 00 23
Telex: 143404 orekh su
Fax: (+7 0562) 92 50 41
General Designer: Stanislav N Konyukhov
Deputy General Designer: M V Fomenko
Director General, Yuzhmash: Yuri S Alexeyev
SL-16 Chief Designer: Prof Yuri A Smetanin
SL-11/14 Chief Designer: Stanislav N Konyukhov
Chief Designer, Satellites: Vladimir Dranovsky
Deputy Chief Designer, Satellites: Sergei S Kavelin

PO Monolith
310644 Kharkov
Tel: (+7 0572) 27 58 80
Telex: 311027 rita
Fax: (+7 0572) 22 47 23

Ukrainian Space Agency
Commission for Space Research
Academy of Sciences
252-127 Kiev 127
Tel: (+7 044) 266 31 10
Telex: 131 406 sky
Fax: (+7 044) 266 47 58
Director General: Alexander Nehoda
Director, Space Complexes and Transportation Systems: Oleg S Uruski
International Department: Kostantin Yartsev
Business Manager: Edward I Kuznetsov
Commission Secretary: Sergei A Silich

Yuzhkosmos
3 Krivorozhskaya Street,
320059 Dnieprepetrovsk
Tel: (+7 0562) 42 00 23
Telex: 143404 orekh su
Fax: (+7 0562) 92 50 41
General Director: Prof Vyacheslav S Fomenko
Vice-Director: Igor K Butbayev
Technical Director: Yevgeny I Busheyev
Director, Foreign Trade Service: Yevgeny N Pisarev

UNITED KINGDOM

British National Space Centre
Bridge Place, 88/89 Eccleston Square,
London SW1V 1PT
Tel: (+44 171) 215 08 08
Fax: (+44 171) 821 53 87
WWW: http://www.open.gov.uk/bnsc/bnschome.htm
Director General: Derek Davis
Deputy Director General: Dr David Leadbeater
Director, Policy/Finance: Hugh Evans
Director, Space Applications: Dr David Lumley
Director, Science/Microgravity: Dr Paul Murdin
Director, Earth Observation: to be announced
Information Unit: Mike Blackwell
e-mail: Mike-Blackwell@BNSC-HQ.ccMail.Compuserve.com

DERA – Space Department
Farnborough, Hampshire GU14 6TD
Tel: (+44 1252) 39 39 44
Fax: (+44 1252) 39 63 35
WWW: http://www.dera.gov.uk
Manager: Graham Davison
Business Development: Alan Haskell

Particle Physics and Astronomy Research Council
Polaris House, North Star Avenue, Swindon,
Wiltshire SN2 1SZ
Tel: (+44 1793) 44 20 00
Fax: (+44 1793) 44 20 02
e-mail: firstname_surname@pparc.ac.uk
WWW: http://www.pparc.ac.uk
Chairman: Dr Peter Williams (44 20 89)
Chief Executive: Prof Ken Pounds (44 20 9)
Director, Observatories: Prof Alec Boksenberg
(+44 1223) 37 40 00; e-mail: boksy@uk.ac.cam.ast
Director, Science: Dr Ian Corbett (44 20 28)
Head, Astronomy Division: Dr Paul Murdin (44 20 75)
Head, Corporate Affairs: Geoffrey Findlay (44 20 33)
Press Officer: Andrew Bennett (44 20 12)

Strategic Communications & Networks Department
Defford, Worcestershire WR8 9DU
Tel: (+44 1386) 75 77 00
Fax: (+44 1386) 75 79 99
Head: Richard L Harris (75 77 30)
Civil Satcoms Contact: Dr John H Crowther (75 77 64)

UNITED STATES OF AMERICA

Department of Commerce
Office of Space Commerce, room 7060,
Washington DC 20230
Tel: (+1 202) 482 61 25
Telex: 892536 usdoc wash dc
Fax: (+1 202) 482 51 73
Director: Keith Calhoun-Senghor
Program Director, International: Kathryn Sullivan

Department of Energy
Forrestal Building, 1000 Independence Ave SW,
Washington DC 20585
Tel: (+1 202) 586 58 06
Fax: (+1 202) 586 05 53

Department of Transportation
Room 5415, 400 Seventh Street SW,
Washington DC 20590
Tel: (+1 202) 366 57 70
Fax: (+1 202) 366 72 56
Associate Administrator, Office of Commercial Space Transportation: Frank Weaver
Deputy Associate Administrator/OCST: Patti Grace Smith

Energia Ltd
631 South Washington Street,
Alexandria, Virginia 22314
Tel: (+1 703) 836 19 99
Fax: (+1 703) 836 19 95
Managing Director: Jeffrey Manber
Deputy Managing Director: Christopher J Faranetta

Federal Communications Commission
1919 M Street, Washington DC 20554
Chairman: Reed E Hundt Tel: (+1 202) 418 10 00
Chief, International Bureau: Donald Gips
Tel: (+1 202) 418 04 20; Fax: (+1 202) 418 28 18
Chief Engineer: Richard Smith

NASA Johnson Liaison Office
Room 476, Building 1, Mail Code KT,
NASA Johnson Space Center,
Houston, Texas 77058
Tel: (+1 713) 483 78 70
Fax: (+1 713) 483 39 14

National Aeronautics & Space Administration
Two Independence Square Building
300 E Street SW, Washington DC 20546
Tel: (+1 202) 358 00 00
Telex: 4979843 nasawsh
Fax: (+1 202) 358 00 37
Administrator: Daniel S Goldin
Deputy Administrator: John R Dailey (acting)
Chief Engineer: Dr Daniel R Mulville
Chief Scientist: Dr France Ann Cordova
Associate Administrators: Wilbur C Trafton (Space Flight), Dr John E Mansfield (Space Access and Technology), Dr Robert Whitehead (Aeronautics), Dr Wesley Huntress (Space Science), to be announced (Mission to Planet Earth), Dr Arnauld Nicogossian (Life/Microgravity Sciences and Applications), Frederick D Gregory (Safety/Mission Assurance), to be announced (Communications), Alan Ladwig (Policy/Plans), John Schumacher (External Relations)
Directors: Steve Oswald (Shuttle Program Co-ordinator), Michael Hawes (acting Space Station; deputy Robert W Moorehead), to be announced (Launch Vehicles)

Office of Science & Technology Policy
The White House, Washington DC 20506
Tel: (+1 202) 456 71 16
Director: Dr John H Gibbons
Assistant Director, Aeronautics and Space: Richard DalBello
Associate Director, Science: Ernie Moniz
Associate Director: Lionel (Skip) Johns

Office of Space and Defense Power Systems
Tel: (+1 301) 903 43 06
Fax: (+1 301) 903 28 85

MILITARY SPACE ORGANISATIONS

INTERNATIONAL

Western European Union
4 rue de la Regence, B-1000 Brussels
Tel: (+32 2) 500 44 00/11
Fax: (+32 2) 511 32 70
Information Officer: Patrick Raes

WEU Satellite Centre
Apdo de Correos 511, Edificio 457, Avenida de Cádiz, Base Aérea de Torrejón,
E-28850 Torrejón de Ardoz (Madrid), Spain
Tel: (+34 1) 677 77 29
Fax: (+34 1) 677 72 28
Director: Col Bernard Molard
Head of Implementation: Peter Lingk
Head of Operations/Training: Francis Bonnet
Head of Scientific Support: Giulio Prisco
Head of Informatics: Willy Tack
Head of Administration/Personnel: Robert Hagen

UNITED STATES OF AMERICA

14th Air Force
Vandenberg AFB, California 93437-6268
Tel: (+1 805) 734 82 32; ext 61400
Commander: Maj Gen Gerald F Perryman

21st Space Wing
Peterson AFB, Colorado 80914-1294
Tel: (+1 719) 556 46 96
Fax: (+1 719) 556 78 48
Commander: Brig Gen Franklin J Blaisdell

30th Space Wing
Vandenberg AFB, California 93437-6261
Tel: (+1 805) 734 82 32
Fax: (+1 805) 866 32 32
e-mail: Kanode@vafb3.af.mil
Commander: Col Claude R Kehler
Commander, 30th Ops Gp: Col Ronald J Haeckel
Commander, 2nd SLS: Lt Col Richard Hornbeak
Commander, 4th SLS: Lt Col Edwin Zehner
Director, Public Affairs: Maj L Brad Peck

45th Space Wing
Patrick AFB, Florida 32925-5002
Tel: (+1 407) 494 11 10
Fax: (+1 407) 494 7302
Commander: Brig Gen Randy Starbuck
Commander, 45th Range Squadron: Lt Col Richard Reynolds
Commander, 45th Weather Squadron: Lt Col Thomas Adang

50th Space Wing
Falcon AFB, Colorado 809 12-3024
Tel: (+1 719) 550 50 00
Fax: (+1 719) 550 64 59
Commander: Col Elwood Torcuit
Vice-Commander: Col John Harris
Commander 750th Space Group: Col Michael A Hamel
Commander 1 SOPS: Lt Col Evan Hoapili
Commander 2 SOPS: Lt Col Roger Hunter
Commander 3 SOPS: Lt Col Mark Owen
Commander 4 SOPS: Lt Col Michael R Mantz
Commander 6 SOPS: Lt Col Neil Wyss
Commander 7 SOPS: Lt Col Frank Casserino

AF Maui Optical Station (AMOS)
535 Lipoa Parkway, Suite 200, Kihei, Hawaii 96753
Tel: (+1 808) 874 15 41
Fax: (+1 808) 874 16 40
Chief Scientist: Dr Paul W Kervin

AF Space & Missile Systems Center
2430 East El Segundo Boulevard, Suite 4049, Los Angeles AFB, California 90245-4687
Tel: (+1 310) 363 00 30
Fax: (+1 310) 363 25 49
Commander: Lt Gen Roger G DeKok
Vice-Commander: Mr William Maikisch
Milsatcom JPO System Program Director: Brig Gen Leonard F Kwiatkowski
Navstar GPS JPO System Program Director: Col John Clay
AF BMD Programs Systems Program Director: Col Howard J Mitchell
Space-based Infra-red Systems Director: Col Craig Weston
DMSP System Program Director: Col John A Goyette
Launch Programs System Program Director: Col William S Files
Satellite Control and Handling Systems Program Director: Col Terry W Graham
EELV Program Manager: Robert Steele
Atlas Program Manager: Lt Col M E Lacaillade
Upper Stages Program Director: Col Norman H Buchanan
Chief, Media Relations: Howard B Antelis
e-mail: antelishb@m105x3.laafb.af.mil

Air Force Space Command
Peterson AFB, Colorado 80914-3010
Tel: (+1 719) 554 30 01/33 02
Commander: Lt Gen Howell M Estes III
Vice-Commander: Lt Gen Patrick P Caruana

Arnold Engineering Development Center
100 Kindel Drive, Arnold AFB, Tennessee 37389-5000
Tel: (+1 615) 454 52 01
Fax: (+1 615) 454 64 15
Commander: Col Michael P Wiedemer
Vice-Commander: Col L Griffin

Ballistic Missile Defense Organisation
7100 Defense, Pentagon,
Washington DC 20301-7100
Tel: (+1 703) 695 87 43
Fax: (+1 703) 614 70 59
WWW: http://www.acq.osd.mil/bmdo/bmdolink.html
Director: Lt Gen Lester Lyles
Deputy Director: Richard D West
Chief of Staff: William C Smith
Deputy, Acquisition/Theater Missile Defense (Acting): BG Curtis H Emery II
Deputy, Technology Readiness: Dr James D Carlson
Deputy, Program Operations: Robert Snyder
Deputy, Strategic Relations: Dr J David Martin
Contracts Directorate: Sallie H Wake
Management Operations Directorate: Edward L Gray
Security, Intelligence/Countermeasures Directorate: Dr Thomas G Ward Jr

Consolidated Space Test Center (CSTC)
Onizuka AS, Sunnyvale, California 90045-2960
Tel: (+1 213) 643 02 54
Commander: Col J L Grogan III
Vice-Commander: Col Michael Harvey

Defense Advanced Research Projects Agency
3701 North Fairfax Drive, Arlington,
Virginia 22203-1714
Tel: (+1 703) 696 24 44
Fax: (+1 703) 696 22 07
Director: Larry Lynn
Director, Tactical Technology: Dr David Whelan

Defense Information Systems Agency
701 South Court House Road, Arlington,
Virginia 22204-2199
Tel: (+1 703) 607 69 00 (Public Affairs)
Fax: (+1 703) 607 40 81
Director: Lt Gen Albert J Edmonds,
USAF Deputy Director: MG David J Kelley, USA

Department Of The Air Force
The Pentagon, Washington DC 20330-1000
Tel: (+1 202) 695 53 82
Fax: (+1 703) 614 74 86 (media desk)
Office of Assistant Secretary for Acquisition
Director, Space: Maj Gen Robert S Dickman
Office of Assistant Secretary for Space
Assistant Secretary: Keith Hall (acting)
Space Plans/Policy: Richard McCormick
Director, Space Systems: Brig Gen Thomas J Scarlan
AF Program Executive Offices
Program Executive Officer for Space: Maj Gen Garry A Schnelzer

Department Of The Navy
The Pentagon, Washington DC 20350-2000
Tel: (+1 703) 695 64 79/78
Fax: (+1 703) 614 29 45
e-mail: acebrows@is1.js.mil
Director, C^4: RADM Arthur Cebrowski

Florida Spaceport Authority
150 Cocoa Isles Boulevard, Suite 401,
Cocoa Beach, Florida 32931
Tel: (+1 407) 868 69 83
Fax: (+1 407) 868 69 87
Executive Director: Edward O'Connor, Jr
Operations Director: Edward Ellegood
LC46 Project Manager: Albert Thomas

National Reconnaissance Office
14675 Lee Road, Chantilly, Virginia 22021
Tel: (+1 703) 808 10 17 Media
Tel: (+1 703) 808 10 15 Public/Employee Information
Fax: (+1 703) 808 10 24
WWW: http://www.nro.odci.gov
Acting Director: Keith R Hall (Acting USAF Assistant Secretary, Space)

Naval Satellite Operations Center
661 13th Street, Point Mugu, California 93042-5003
Tel: (+1 805) 989 42 04
Fax: (+1 805) 989 42 80
e-mail: centerc@mugu.navy.mil
Commanding Officer: Capt William W Wittmann
Executive Officer: CDR M A Klee
Technical Director: Gary C Kennedy

Naval Space Command
5280 Fourth Street, Dahlgren, Virginia 22448-5300
Tel: (+1 540) 653 61 00
Fax: (+1 540) 653 61 08
e-mail: gwagner@nsc.navy.mil
Commander: RADM Katharine Laughton
Deputy Commander: Col Victor Simpson
Technical Director: Pieter J Traas

Navstar GPS Joint Program Office
US Air Force Space & Missile Systems Center
PO Box 92960, Worldway Postal Center,
Los Angeles, California 90009
Program Director: Col John Clay

Office of Naval Research
800 North Quincy Street, Arlington, Virginia 22217
Tel: (+1 703) 696 40 00
Fax: (+1 703) 696 59 40
Chief Naval Research: RADM Paul G Gaffney
Assistant Chief: Capt R A Burnett

Naval Research Laboratory
4555 Overlook Avenue SW,
Washington DC 20375-5000
Tel: (+1 202) 767 25 41
Fax: (+1 202) 767 69 91
Commanding Officer: Capt Bruce W Buckley
Director of Research: Dr Timothy Coffey
Naval Center for Space Technology: Peter G Wilhelm
Remote Sensing Division: Dr Philip Schwartz (acting)
Space Science Division: Dr Herbert Gursky

Phillips Laboratory
3550 Aberdeen Avenue SE,
Kirtland AFB, New Mexico 87117-5776
Tel: (+1 505) 846 19 11
Fax: (+1 505) 846 04 23

Geophysics Directorate
29 Randolph Road,
Hanscom AFB, Massachusetts 01731-3010
Tel: (+1 617) 377 36 01
Fax: (+1 617) 377 56 88

Propulsion Directorate
5 Pollux Drive,
Edwards AFB, California 93524-7048
Tel: (+1 805) 275 54 65
Fax: (+1 805) 275 51 44
Commander: Col Michael L Heil
Vice-Commander: Col James Ledbetter
Executive Director: Dr Earl Good
Director, Advanced Weapons & Survivability: Col William Heckathorn
Director, Lasers & Imaging: Col Lanny Larson
Director, Space & Missiles Technology: Christine Anderson
Director, Geophysics: Dr Harold Roth
Director, Propulsion: Col Robert J Karner
Director, Space Experiments: Col Michael Havey

Space And Naval Warfare Systems Command
2451 Crystal Drive, Arlington, Virginia 22245-5200
Tel: (+1 703) 602 30 06
Fax: (+1 703) 602 75 47
e-mail: wagnerg@smtp-gw.spawar.navy.mil
Commander: RADM George F A Wagner
Deputy Commander: Dr Dale Uhler

US Army Space & Strategic Defense Command (USASDC)
PO Box 15280, Arlington, Virginia 22215-0280

Tel: (+1 703) 607 19 38
Fax: (+1 703) 607 14 91
WWW: http://www.ssdc.army.mil
Commander: Lt Gen Edward G Anderson III

US Army Space Command (Forward)
Colorado Springs, Colorado 80916-2749
Tel: (+1 719) 554 88 84/88 85
Commander: Col E Paul Semmens

US Space Command North American Aerospace Defense Command
Peterson AFB, Colorado 80914-3010
Tel: (+1 719) 554 30 01/30 02/30 05
Commander in Chief, NORAD/US Space Command:
Lt Gen Howell M Estes

COMMUNICATION SATELLITE SYSTEMS AND SERVICES

ARGENTINA

Nahuelsat SA
Lavalle 472 3° Piso (1047) Capital Federal,
Buenos Aires
Tel: (+54 1) 327 22 92; 22 95
Fax: (+54 1) 327 23 08
Vice-President/General Manager: Eckart Schober

AUSTRALIA

Codan Pty Ltd
PO Box 96, Campbelltown, South Australia 5074
Tel: (+61 8) 305 03 11
Fax: (+61 8) 305 04 11
Manager, Satellite Communications: Allan Gobolos

10359 Bear Creek Drive,
Manassas, Virginia 22111, USA
Tel: (+1 703) 361 27 21
Fax: (+1 703) 361 38 12
Contact: Gregory Walker

Defence Communications Research Centre
Institute for Telecommunications Research
University of South Australia
The Levels, South Australia 5095
Tel: (+61 8) 302 38 97
Fax: (+61 8) 302 38 73
Director, DCRC: Prof Ken Lever
Deputy Director: Dr Les Sabel

Mitec Limited
21-23 Hasp Street, Seventeen Mile Rocks, Brisbane,
Queensland 4073
Tel: (+61 7) 376 55 88
Fax: (+61 7) 376 37 72
Managing Director: Dr John Ness
Marketing Director: Kevin Berridge
Chief Engineer: Peter Bradley

Optus Communications Pty Ltd
Optus Centre, 101 Miller Street,
North Sydney, New South Wales 2060
Postal: PO Box 1, North Sydney, New South Wales 2060
Tel: (+61 2) 93 42 78 00
Fax: (+61 2) 93 42 71 00
WWW: http://www.optus.net
Chairman: Russell Fynmore
CEO: Dr 'Ziggy' Switzowsky
Chief Operating Officer: Phil Jacobs
Manager, satellite planning: Gordon Pike

2 Challenger Drive, Belrose, New South Wales 2085
Postal: PO Box 235, Frenchs Forest, New South Wales 2086
Tel: (+61 29) 342 31 00
Fax: (+61 29) 342 31 11
Manager, Satellite Operations: Duncan Henry

Satellite Communications Research Centre
Institute for Telecommunications Research
University of South Australia
The Levels, South Australia 5095
Tel: (+61 8) 302 33 36
Fax: (+61 8) 302 38 73
Director, SCRC: Prof Michael Miller
Business Manager: Dr Graham Gurr

Space Industry Centre For Microwave Technology
Griffith University, Nathan, Queensland 4111
Tel: (+61 7) 875 71 11
Fax: (+61 7) 875 51 98
Director: Prof Barry Harrison

Telstra Corp Ltd
GPO Box 7000, Sydney, New South Wales 2001
Tel: (+61 2) 92 87 50 00
Fax: (+61 2) 92 87 52 87
Chief Executive: W Frank Blount
National GM, Network Operations: Phil Hastings

Tunra Ltd
Aviation Building, University of Newcastle,
Callaghan, New South Wales 2308
Tel: (+61 49) 67 18 11
Fax: (+61 49) 67 49 46
Chief Executive Officer: Dr Soozy J Smith
TIED Manager: Ian Dick
TIED Technical Director: Dr Richard Middleton
Senior Engineer: David J L Brown

BELGIUM

Alcatel Bell NV
(Alcatel Telecom/Space Business Line)
Francis Wellesplein 1, B-2018 Antwerp
Tel: (+32 3) 240 40 11
Telex: 72128 bella b
Fax: (+32 3) 240 99 99
WWW: http://www.alcatelecom.be

Space Systems
Berkenrodelei 33, B-2660 Hoboken
Tel: (+32 3) 829 50 11
Fax: (+32 3) 829 55 89
e-mail: jacquetp@btmaa.bel.alcatel.be
President/CEO: J De Wilde
Director Radio, Space and Defence Division: Noel Parmentier
Space Systems Marketing: J Geenen

Newtec CY NV
Space Telecommunications
Mechelsesteenweg 27, B-2018 Antwerp
Tel: (+32 32) 32 97 92
Fax: (+32 32) 33 76 90

Sirius Communications NV
Wingeparke 51, B-3110 Rotselaar
Tel: (+32 16) 44 44 02
Fax: (+32 16) 44 54 81
WWW: http://www.sirius.be

BRAZIL

Avibras Fibras Oticas E Telecomunicacoes S/A
Rua Ricardo Hausen, 100 – PO Box 229,
12227-820 São José dos Campos, São Paulo
Tel: (+55 123) 21 74 33
Telex: 55 123-3844/3845
Fax: (+55 123) 21 70 14
Director: Fausto Carlos de Madureira Pará

Empresa Brasileira de Telecomunicacoes SA
Avenida Presidente Vargas 1012,
20177-700 Rio de Janeiro
Tel: (+55 21) 253 44 38
Fax: (+55 21) 233 64 46
President: Dílio Sergio Penedo
National Operations Director: Luis Francisco Perrone
International Operations Director: Edson Soffiatti
Development Director:
Francisco dos Santos Pires Albuquerque
Head, Satellite Systems: Dr Elisabete Trachez do Couto

CANADA

Andrew Canada Inc
606 Beech Street, Whitby, Ontario L1N 5S2
Tel: (+1 905) 668 33 48
Fax: (+1 905) 668 85 90
President: T E Charlton
Chief Engineer: R Boyko
Vice-President Government Antenna Systems: G Tong

CAL Corporation
Director, Communications Systems: Gerry Frappier
Aeronautical Satcom Program: Gary Nash
Sales Manager, Aeronautical Satcom Program: John Menard

Calian Communications Systems Ltd
300 Legget Drive, Kanata, Ontario K2K 1Y5
Tel: (+1 613) 592 86 00
Fax: (+1 613) 592 33 78
Director of Technology: John Bodie
Business Development: Esme Kwong

Canadian Marconi Co
600 Dr Frederik Philips Boulevard,
Ville Saint-Laurent, Quebec H4M 2S9
Tel: (+1 514) 748 31 48
Telex: 05-827822
Fax: (+1 514) 748 30 55

PO Box 13330, 415 Legget Drive,
Kanata, Ontario K2K 2B2
Tel: (+1 613) 592 65 00
Telex: 05-34805
Fax: (+1 613) 592 74 27
President/CEO: Carmen L Lloyd
Vice-President Aerospace: S Rauch
General Manager Commercial Avionics: D W Bailey
Chief Engineer: S Mariano
Marketing Manager, Air Transport: E Ford
Manager Marketing Services: Roger Hart

Com Dev Canada
155 Sheldon Drive, Cambridge, Ontario N1R 7H6
Tel: (+1 519) 622 23 00
Fax: (+1 519) 622 16 91
Chairman/CEO: M Val O'Donovan
President: William R Fitzgerald (acting)
Executive Vice-President: Anita Davis
Vice-President Business Development: John Stuart

Sed Systems Inc
PO Box 1464, 18 Innovation Boulevard,
Saskatoon, Saskatchewan S7K 3P7
Tel: (+1 306) 931 34 25
Telex: 074-2495
Fax: (+1 306) 933 14 86
Director, Business Development: Kent McKerlie

Techwest Data Systems Ltd
8176 Winston Street,
Richmond, British Columbia V5A 2H5
Tel: (+1 604) 421 73 22
Fax: (+1 604) 421 73 66
e-mail: tds@mindlink.bc.ca
President: Alan McLeod
General Manager: Jeff Payne

Teleglobe Canada Inc
1000 de La Gauchetière West,
Montreal, Quebec H3B 4X5
Tel: (+1 514) 868 79 91
Fax: (+1 514) 868 76 89
Manager, Satellite Systems: Jag Wanchu

Telesat Canada
1601 Telesat Court, Gloucester, Ontario K1B 5P4
Tel: (+1 613) 748 01 23
Fax: (+1 613) 748 87 12
e-mail: info@telesat.ca
WWW: http://www.telesat.ca
President/CEO: Laurry J Boisvert
Vice-President Space Systems: Dr Leonard G Stass
Vice-President Marketing/Sales: Dennis G Billard
Vice-P Engineering: Donovan E Weese
Vice-PresidentNetwork Services: J Gordon Fraser
Public Affairs: Stuart MacMillan
e-mail: s.macmillan@telesat.ca

TMI Communications
PO Box 9826, 1601 Telesat Court,
Ottawa, Ontario K1G 5M2
Tel: (+1 613) 742 00 00
Fax: (+1 613) 742 41 00
WWW: http://www.tmi.ca
President/CEO: John H Farrell
Vice-President: Peter E Vivian
Vice-President Sales/Marketing: T Kent Elliott
Vice-President Business/Service Development: Lynda Partner
Vice-President Service Delivery: John Forsey
Vice-President Finance/Administration: Jim Hone

CHINA, PEOPLE'S REPUBLIC

Apt Satellite Co Ltd
One Pacific Place, 31/F room 3111-3112,
88 Queensway, Hong Kong
Tel: (+852) 526 22 81
Fax: (+852) 522 04 19

Chairman: Xie Gaojue
President: He Kerang
Chief Engineer: Bao Miao Qin
Project Manager: Hu Qianming
Director, Marketing: Yang Xue Ming

Asia Satellite Telecommunications Co Ltd
23-24/F East Exchange Tower,
38-40 Leighton Road, Causeway Bay, Hong Kong
Tel: (+852) 28 05 66 66
Telex: 68345 ASAT HX
Fax: (+852) 25 76 41 11
WWW: http://www.asiasat.com.hk
Chief Executive Officer: Peter Jackson
Deputy Chief Executive Officer: William Wade
General Manager, Marketing: Sabrina Cubbon
General Manager, China: Chen Liqun
General Manager, Engineering: David J Whalen
General Manager, Operations: Ya Chiu
General Manager, Finance/Administration: Denis Lau

Shijiazhuang Communications Labs
PO Box 174, Shijiazhuang, Hebie
Tel: (+86 311) 333 30
Contact: Luo Zheng Bin

DENMARK

Philips Radio Communications Systems
Prags Boulevard 80, DK-2300 Copenhagen
Tel: (+45) 32 88 22 22
Fax: (+45) 31 88 39 30
Contact: Finn Christensen

S.P. Radio A/S
Porsvej 2, DK-9200 Aalborg SV
Tel: (+45) 98 18 09 99
Telex: 69789
Fax: (+45) 98 18 67 17
Inmarsat contact: Peter Andersen

Skanti A/S
Kirke Vaerlosevej 34, DK-3500 Vaelose
Tel: (+45) 42 48 25 44
Fax: (+45) 42 48 45 09
Managing Director: Jorgen F Mikkelson

Thrane & Thrane A/S
Tobaksvejen 23, DK-2860 Soborg
Tel: (+45) 31 56 41 11
Telex: 055-19298 thrane dk
Fax: (+45) 31 56 21 40
Sales Manager: Jens Christian Andersen

Ticra
Kronprinsensgade 13,
DK-1114 Copenhagen K
Tel: (+45) 33 12 45 72
Fax: (+45) 33 12 08 80
General Manager: Dr Niels Christian Albertsen
Executive Secretary: Mads Bue Johnsen

FRANCE

Aerospatiale
Espace & Défense
66 route de Verneuil, F-78130 Les Mureaux, Cedex
Postal: BP2, F-78133 Les Mureaux Cedex.
Tel: (+33 1) 34 92 12 34
Telex: 695 988 F
Fax: (+33 1) 34 92 12 54
President, Space & Defence: Michel Delaye
Senior Vice-Presidents: André Motet (Technical) and Jean-Marie Mir (Operations)
Vice-President Satellite Affairs: Emmanuel Sartorius
Managing Director Space Operations: Francois Calaque

Satellite Business Unit
100 Boulevard du Midi, BP99,
F-06322 Cannes-La-Bocca Cedex
Tel: (+33 1) 92 92 70 00
Telex: 470 902 F
Fax: (+33 1) 92 92 71 90
Managing Director Satellite Operations: Henri Sala

Alcatel Espace (part of Alcatel-Alsthom group)
Headquarters
5 rue Noël Pons, F-92734 Nanterre
Tel: (+33 1) 46 52 62 00
Fax: (+33 1) 46 52 62 50

Industrial Site
26 avenue Jean-François Champollion, BP1187,
F-31037 Toulouse Cedex
Tel: (+33 5) 61 19 50 50
Telex: atesp 521 355
Fax: (+33 5) 61 44 49 90
President/CEO: Jean-Claude Husson
Executive Vice-President: Bernard Deloffre, Lionel Fleury
Senior Vice-Presidents: Pierre de Bayser, Henry Paul Brochet
Director Strategy: Pascale Sourisse
Director Corporate Relations and Advertising: Thierry Deloye
Military Advisor: Christian Fontaine

Applications Des Techniques Nouvelles En Electrique (ATNE)
7 avenue de l'Atlantique, Les Ulis,
F-91955 Courtaboeuf Cedex
Tel: (+33 1) 69 28 76 72
Fax: (+33 1) 69 28 40 08
General Manager: Christian Petrigny
Programme Manager: Jean-Marie Auré

Ateliers de Constructions du Centre
Rue du Pré-la-Reine, BP93,
F-63016 Clermont-Ferrand Cedex 1
Tel: (+33 2) 73 98 38 38
Telex: 990835 acc
Fax: (+33 2) 73 98 38 20
Space & Sales Manager: Christian Regnier

Compagnie de Signaux et d'equipements Électroniques
Le Banville, 153 rue de Courcelles,
F-75817 Paris Cedex 17
Tel: (+33 1) 44 15 61 61
Fax: (+33 1) 44 15 94 14
Chairman: Yazid Sabeg

CSEE-Défense
ZA de Courtaboeuf, 6 Avenue des Tropiques, BP80,
F-91943 Les Ulis Cedex A
Tel: (+33 1) 69 86 85 00
Fax: (+33 1) 69 07 03 70
General Manager: Claude Bras

Dateno
ZAC Es-Passants – BP 123,
Rue Amiral Bérenger
35801 Dinard Cedex
Tel: (+33 2) 99 46 24 75
Fax: (+33 2) 99 46 47 27
President: Alain Laporte
Vice-President/General Manager: J Y L'Honnen
Marketing/Sales Manager: Gérard Barrere

Elecma
13-15 rue Pagès, BP305, F-92156 Suresnes Cedex
Tel: (+33 1) 40 99 81 84
Telex: 620385
Fax: (+33 1) 40 99 82 18
Commercial Manager: Jean-Louis Sauvetre

France Telecom
6 place d'Alleray, F-75740 Paris Cedex 15
Tel: (+33 1) 44 44 22 22
Telex: 202715
Fax: (+33 1) 45 31 52 11
President: Marcel Roulet
Director General: Charles Rozmaryn
Senior Vice-President, International: Michel Hirsch
Head, International Space Ops: Pierre Godinaux

IN-SNEC
2 rue de Caen, BP7, F-14740 Bretteville-l'Orgueilleuse
Tel: (+33 2) 31 80 71 22
Telex: 171 813
Fax: (+33 2) 31 80 65 49
Inmarsat project leader: Frank Meuleman

Avenue des Andes, BP101, ZA de Courtaboeuf,
F-91943 Les Ulis Cedex
Tel: (+33 1) 69 82 72 00
Telex: 604179
Fax: (+33 1) 69 07 01 70
Marketing/Sales Manager: Jean-Claude Dardelet

Radiall
101 rue Philibert Hoffman, F-93116 Rosny-sous-Bois
Tel: (+33 1) 49 35 35 35
Telex: 235220
Fax: (+33 1) 48 54 63 63
Military/Space Business Development: Gérard Vallet
Vice-President Sales: Michel Molles

Satel Conseil
5 rue Louis Lejeune, F-92128 Montrouge
Tel: (+33 1) 46 57 75 65
Fax: (+33 1) 46 57 70 34
President: Michel Popot
Director General: Robert Blanc
Technical Director: Claude Chalendard

Starec
12/14 avenue Carnot, F-91349 Massy Cedex
Tel: (+33 1) 69 20 13 30
Telex: 604353
Fax: (+33 1) 60 13 02 38
President: M Faure
Sales Executive: D Marquis
Production Director: J Lafargue

Télédiffusion De France
BP518, 21-27 rue Barbès, F-92542 Montrouge Cedex
Tel: (+33 1) 49 65 10 00
Telex: 250738
Fax: (+33 1) 46 57 48 50
Chairman: Bruno Chetaille
Managing Director: Philippe Levrier

Thomson Tubes Electroniques
18 avenue du Maréchal Juin,
F-92366 Meudon La Foret Cedex
Tel: (+33 1) 30 70 35 00
Telex: thomtub 699 407 f
Fax: (+33 1) 30 70 35 35
Chairman/CEO: Henri Magnan
Vice-President Marketing/Sales: Pierre Menes
Space Director: Georges Fleury
Marketing Manager, Space TWTs: Paul Barker

Thomson-CSF Communications
66 rue du Fossé Blanc, BP156,
F-92231 Gennevilliers Cedex
Tel: (+33 1) 46 13 20 00
Telex: TCSFRGS 620 840 F
Fax: (+33 1) 46 13 21 63

GERMANY

EurasSpace GmbH
Munich, Germany
Tel: (+89 49) 60 72 27 83
Fax: (+89 49) 60 72 35 22
Manager: Klaus Ninke
Programme Manager: Ludwig Baur

AEG Elektronische Röhren GmbH (AERG)
Söflingerstr 100, D-89077 Ulm
Tel: (+49 731) 933 12 10
Telex: 712 601 aegr d
Fax: (+49 731) 933 17 95
Managing Director: Dieter Deml
Satellite Sales Manager: Dr Günter Palz

Bosch Telecom GmbH
Space Communications Product Group
Gerberstr 33, D-71522 Backnang
Tel: (+49 7191) 130
Fax: (+49 7191) 13 34 22
Managing Director: Rolf Huebner
General Manager Business Development/Programmes: Reinhard Schnabel

Detecon
Oberkasselerstrasse 2, D-53227 Bonn
Tel: (+49 228) 70 00
Fax: (+49 228) 700 37 89
Satellite Services: Dipl Ing Rolf Menzel

Deutsche Telekom AG
Zentrale, Godesberger Allee 117, Bonn
Postal: Postfach 2000, D-53105 Bonn
Tel: (+49 228) 181 49 49
Telex: 886668 gdt d
Fax: (+49 228) 181 89 41
CEO: Dr Ron Sommer
Director, Satellites: Axel Jörns

MAN Technologie AG
System Electronics Division
Postfach 1347, D-85751 Karlsfeld
Tel: (+49 8131) 891 3 54
Fax: (+49 8131) 891 3 17
Marketing/Sales, Mobile Data Acquisition & Mobile Comms: Armin Keller

OHB-System GmbH
Universitätsallee 27, D-28359 Bremen
Tel: (+49 421) 202 08
Telex: 245316 ohb
Fax: (+49 421) 202 06 10
Managing Directors: Prof Dipl-Ing Manfred Fuchs, Christa Fuchs, Marco Fuchs (VP)
Directors: Wolfgang Wienss, Alfred Tegtmeier, Dr Ingo Engeln, Dr Peter Junk, Dr Reinhardt Schlitt, Dr Bernd Hofer
Head of Satellite Department: Amnon Ginati

Siemens AG
Public Communication Networks
Hofmannstrasse 51, OEN MA 12, D-81359 Munich
Tel: (+49 89) 72 23 89 27
Fax: (+49 89) 72 24 10 72
Head of PCN: P Pribilla

STN Atlas Elektronik
Sebaldsbruecker Heerstrasse 235, D-28305 Bremen
Tel: (+49 421) 457 35 35
Fax: (+49 421) 457 39 35
Contact: Juergen Michel

STN Atlas Australia
39 Herbert Street, St Leonards, New South Wales 2065, Australia
Tel: (+61 2) 437 45 77
Fax: (+61 2) 439 75 76
Contact: Brian Groves

VSAT Satellitenkommunikation GmbH
Anzingerstrasse 5, D-85560 Ebersberg
Tel: (+49 80) 92 82 58 33
Fax: (+49 80) 928 28 11
President: Dipl Ing Josef AH Geiger

INDIA

Afro-Asian Satellite Communications Ltd (ASC)
135 Dr Annie Besant Rd, Worli, Bombay 400018
Tel: (+91 22) 493 96 85
Fax: (+91 22) 493 51 88
Chairman/CEO: Subhash Chandra
Marketing Manager: Raj Mitta
Chief Engineer: David Greenwood

UK Office
The Communications Building,
48 Leicester Square, London WC2H 7DA, UK
Tel: (+44 171) 930 20 00
Fax: (+44 171) 930 12 34
Director: Stephen Moss
Chief Engineer: David Greenwood
Contact: Georgia Gordon

Bharat Electronics Limited
Trade Centre, 116/2 Race Course Road,
Bangalore 560 001
Tel: (+91 80) 225 73 22
Telex: 0845-8650 be in
Fax: (+91 80) 225 56 57; 84 10
Chairman/Managing Director: RADM J J Baxi
General Manager International Marketing: Col H S Shankar
Deputy General Manager International Marketing: Col A Govindarajan

Electronics Corporation Of India Limited
Industrial Development Area, Cherlapalli, PB No.5
Hyderabad 500762
Tel: (+91 40) 851272

Indian National Satellite System
Department of Space
Antariksha Bhavan, New BEL Rd, Bangalore 560 094
Tel: (+91 80) 333 44 74
Telex: 8452499
Fax: (+91 80) 333 42 29
Programme Director: K Narayanan

Karthik Engineering
6, (NP) Developed Plot,
Ekkaduthangal, Madras 600097
Tel: (+91 44) 43 37 72

Prabhakar Products
C-2 Industrial Estate, Guindy, Madras 600032
Tel: (+91 44) 43 15 71
Telex: 041-26064 PCO IN
Managing Partner: T Babu

INDONESIA

PT Compact Microwave Indonesia
PO Box 7606, Bandung 40076
Tel: (+62 22) 31 11 33
Telex: 28716 cmibdo ia
Fax: (+62 22) 31 11 11

PT Indosat
Jalan Merdeka Barat 21, Jakarta 10110
Tel: (+62 21) 384 27 70; 386 91 00
Telex: 44064/44383 INDSAT IA
Fax: (+62 21) 344 04 84; 385 77 42
President Director: Tjahjono Soerjocibroto
General Manager: Budi Prasetyo
General Manager Satcom Division: Djoko Prajitno
Operation Director: Sumitro Roestam
Director, Development/Construction: Habimono Koesoebjono
Finance/Administration Director: Johanes Djajadi

Earth Station
Stasiun Bumi Satelit
Jatiluhur, Purwakarta, Jawa Barat
Manager: Djoko Ismujono

PT Mediacitra Indostar
Gedung Elektrindo, 7th & 8th Floors
J1. Kuningan Barat 8
Jakarta 12710
Tel: (+62 21) 526 80 00
Chairman: Peter Gontha
Indostar Programme Manager: Arief Juwanto
Indostar Technical Managers: Dr Chen Chao, Doopy Irwan

PT Pasifik Satelit Nusantara
Mulia Center, suite 1201,
Jalan HR Rasuna Said,
Kav. X6 No.8, Jakarta 12940
Tel: (+62 21) 522 92 92
Fax: (+62 21) 522 92 93
President: Adi Adiwoso
Vice-President Service Development: Kevin Smyth
Director: W Louk Jurgens

PT TELKOM (PT Telekomunikasi Indonesia)
J1 Japati No. 1,
Bandung 40133, West Java
Tel: (+62 22) 452 71 10
Fax: (+62 22) 44 03 13
WWW: http://www.telkom.co.id
President/CEO: A A Nasution
Directors: Dadad Kustiwa, Harry Supangkat, John Welly, Andi Sisuaka Faisal

Head Office: Jalan Jend. Gatot Subroto 52,
Jakarta 12150
Tel: (+62 21) 520 33 21
Telex: 60751
Fax: (+62 21) 520 33 22

INTERNATIONAL

Alcatel Qualcomm
Mannheimer Strasse 22B,
D-75179 Pforzheim, Germany
Tel: (+49 7231) 39 71 11
Fax: (+49 7231) 39 71 01
President: Patrick A Buhannic
Commercial Manager: Charles J Doll
e-mail: g-cdoll@qualcomm.com

Arabsat
PO Box 1038, Riyadh 11431, Saudi Arabia
Tel: (+966 1) 464 66 66
Telex: 201300
Fax: (+966 1) 465 69 83
Chairman: HE Eng Mohammed Jameel Al- Molla
Director General: Eng Saad bin Abdul-Aziz al-Bidna
Assistant Director General: Saeed Al-Qahtani
Acting Director, Ops & Marketing: Nabil A Shanti

Aseta
PO Box 17-11-06042, Quito, Ecuador
Tel: (+593 2) 56 38 12; 50 98 21; 50 98 22
Telex: 22860
Fax: (+593 2) 56 24 99
Secretary General: Ing Marcelo López Arjona

European Satellite Consulting Organisation
5 rue Louis-Lejeune,
F-92128 Montrouge Cedex, France
Tel: (+33 1) 46 57 75 65
Telex: 632352 f
Fax: (+33 1) 46 57 70 34
Executive Secretary: Robert Blanc

Eutelsat
70 rue Balard, F-75015 Paris, France
Tel: (+33 1) 53 98 47 47
Telex: 203823
Fax: (+33 1) 53 98 37 00
WWW: http://www.eutelsat.org
Director General: Jean Grenier
Director, Finance/Administration: David G Hardman
Technical Director: Andreas Langemeyer
Operations Director: Jesus Domingo Laborda
Commercial Director: Giuliano Berretta

Ico Global Communications
1 Queen Caroline Street,
Hammersmith, London W6 9BN, UK
Tel: (+44 181) 600 10 00
Fax: (+44 181) 600 11 99
Chief Executive Officer: Olof Lundberg
Chief Financial Officer: Mike Whiddett
Executive Vice-President/Programme Director: Jai Singh
Executive Vice-President/Chief Commercial Officer: Peter Cummings

Inmarsat
99 City Road, London EC1Y 1AX, UK
Tel: (+44 171) 728 10 00
Telex: 297201
Fax: (+44 171) 728 10 44
Director-General: Warren Grace
Executive Vice-Presidents: Gene Jilg, George Symeonidis
Vice-Presidents: John Angus, Paul Branch, George Novelli
Chief Financial Officer: Ramin Khadem
Head, Media Relations: Elizabeth Hess
e-mail: elizabeth_hess@Inmarsat.org

China Resident Office
PO Box 9788, Block E suite 2404, 100101 Beijing
Tel: (+86 1) 499 16 39
Fax: (+86 1) 493 10 67
Manager: Francis Man-Hong Choi

Intelsat
3400 International Drive NW,
Washington DC 20008-3098, USA
Tel: (+1 202) 944 75 00
Telex: 89-2707
Fax: (+1 202) 944 78 60
e-mail: firstname.lastname@intelsat.int
WWW: http://www.intelsat.int/
Director-General/CEO: Irving Goldstein
Vice-President, Chief Financial Officer: Andrew Browne
Vice-President, Engineering: Claude Burgio
Vice-President, Operations: Conny Kullman
Vice-President, Sales/Marketing: Gary Smith
CIO and Director: Ramu Potarazu
Director, Procurement: David Meltzer
Director, External Relations: Len Dooley
Director, Corporate Communications: Tony A Trujillo

International Telecommunications Union
Place des Nations, CH-1211 Geneva 20, Switzerland
Tel: (+41 22) 730 51 11
Telex: 421 000 uit ch
Fax: (+41 22) 733 72 56
e-mail: itumail@itu.ch
WWW: http://www.itu.ch
Secretary-General: Dr Pekka Tarjanne
RRB Chairman: M Miura
Director, Telecommunication Standardisation Bureau: Theodor Irmer
Space Systems Co-ordination: Peter Korobenkov
Press Officer: Francine Lambert

Odyssey Telecommunications International Inc
One Space Park, E2/5000
Redondo Beach, California 90278, USA
Tel: (+1 310) 814 88 88
Fax: (+1 310) 814 14 00

1000 Rue de la Gauchetiere West
Montreal, Quebec, Canada H3B 4X5
Tel: (+1 514) 868 80 00
Fax: (+1 514) 868 80 80

Orion Atlantic LP
Orion Satellite Corp
2440 Research Boulevard, Suite 400,
Rockville, Maryland 20850, USA
Tel: (+1 301) 258 81 01
Fax: (+1 301) 258 32 22
President/CEO, Orion Network Systems: W Neil Bauer
Senior Vice-President/General Manager Facilities: Dr Denis J Curtin
Senior Vice-President/General Manager Satellite Services: Jack Albert
Vice-President Operations: David J Nelson
Vice-President Marketing, OrionSat: Steven B Salamoff
Vice-President/General Manager Network Services: Richard Grecco
Director, Sales, Transponder Services: Jerry Porter
Director, Sales, Satellite Services: William Houtz
Director, Media Services: David Gardner
Manager, Satellite Sales: Ed Wright

Rascom
PO Box 3628, Avenue Thomasset 01,
Abidjan, Ivory Coast
Tel: (+225) 22 36 83
Fax: (+225) 22 36 79
Director General: Gounde Desire Adadja

Satellite Aircom Consortium
France Telecom
16 Boulevard du Mont d'Est,
F-93160 Noisy le Grand, Paris, France
Tel: (+33 1) 48 15 73 20
Fax: (+33 1) 48 15 73 08
Satellite Aircom contact: Gilles Negrel

IDB-Aeronautical
10525 West Washington Boulevard,
Culver City, California 90232-3380, USA
Tel: (+1 213) 240 37 83
Fax: (+1 213) 240 39 01
Satellite Aircom contact: Jim Jancso

SITA192 avenue Charles de Gaulle,
F-92522 Neuilly sur Seine, France
Tel: (+33 1) 46 41 19 08
Fax: (+33 1) 46 41 19 43
Satellite Aircom Product Manager: Gérard Collin

Teleglobe Canada Inc
Aeronautical Mobile Satellite Services
1000 rue de La Gauchetière Street West,
Montreal, Quebec, Canada H3B 4X5
Tel: (+1 514) 868 81 25
Fax: (+1 514) 868 77 64
Satellite Aircom contact: Stephane Brisebois

Telstra Corp Limited
Mobile Satellite & Radio Services
79 St Hilliers Road, Auburn, New South Wales 2144
Australia
Tel: (+61 2) 202 41 30
Fax: (+61 2) 737 95 45
Satellite Aircom contact: David Brooks

Skyphone Consortium
BT
43 Bartholomew Close, London EC1A 7HP, UK
Tel: (+44 181) 754 85 52
Telex: 916554 btimr g
Fax: (+44 171) 492 46 33
Service Use Development Manager, Skyphone Services: Paul Murton

Telenor
Mobile Satellite Services
PO Box 6701 St Olavs Plass, N-0130 Oslo 1, Norway
Tel: (+47) 22 48 71 82
Telex: 56 72666 trm n
Fax: (+47) 22 77 71 78
Manager Sales/Marketing: Odd Hollung

Singapore Telecom
International Mobile Division Marketing
Telephone House, 15 Hill Street, No 05-00,
Singapore 0617
Tel: (+65) 331 67 79
Telex: +87 20213 RS BTS
Fax: (+65) 334 61 10
Marketing Manager: Chong Keng Yong

Skyways Alliance
COMSAT Aeronautical Services
6560 Rock Spring Drive, Bethesda,
Maryland 20817, USA
Tel: (+1 301) 214 31 50
Fax: (+1 301) 214 72 05
Vice-President/General Manager: Christopher J Leber

Kokusai Denshin Denwa Co Ltd
PO Box 1, 3-2 Nishi-Shinjuku 2-chome,
Shinjuku-ku, Tokyo 163-03, Japan
Tel: (+81 3) 33 47 74 35
Fax: (+81 3) 33 47 71 99
Director, Mobile: Takaomi Yamamoto

SovCan Star Satellite Communications Inc
155 Sheldon Drive,
Cambridge, Ontario N1R 7H6, Canada
Tel: (+1 519) 622 65 77
Fax: (+1 519) 622 16 91

Project office: 21025 Trans Canada Highway,
Montreal, Quebec H9X 3R2, Canada
Tel: (+1 514) 457 21 50
Fax: (+1 514) 425 30 45
President: Mel Gafner
Vice-President Engineering: Gareth Lewis

IRAN

Telecommunications Company of Iran
Dr Shiriati Avenue, Tehran
Tel: (+98 21) 860 10 41
Telex: 210400/213425 or ir
Fax: (+98 21) 85 85 66
Director General: Mohammad H Harati
Managing Director: M Naraghi
Deputy Managing Director: Aliakbar Sanati
Satcoms Director: Hossein Sheikhattar

ISRAEL

Carcom/Rafael
PO Box 2250 (8K) IL-31021 Haifa
Tel: (+972 4) 79 48 92
Fax: (+972 4) 79 40 37
President: David Eshet
Contact: Dr Adam Live

Elisra Electronic Systems Ltd
48 Mivtza Kadesh Street, IL-51203 Bene Beraq
Tel: (+972 3) 754 51 11
Telex: 03-3553
Fax: (+972 3) 754 54 68
President/CEO: Avner Raz
Vice-President Marketing: Meir Bar-Tov
Head, Satcom Department: Daniel N Nissani

Gilat Satellite Networks Ltd
24A Habarzel Street, IL-69710 Tel Aviv
Tel: (+972 3) 645 42 22
Fax: (+972 3) 648 74 29
e-mail: name@gilat.mhs.compuserve.com
Chairman/CEO: Yoel Gat
President/COO: Amiram Levinberg
Vice-President Marketing/Business Development: Joshua Levinberg
Vice-President Technology: Gideon Kaplan

Gilat Satellite Networks Inc
1700 Old Meadow Rd, McLean, Virginia 22102, USA
Tel: (+1 703) 734 94 01
Fax: (+1 703) 734 94 49
Vice-President/General Manager: Erez Antebi

Gilat Satellite Networks (Europe) SA
Tour Maine Montparnasse, 33 avenue du Maine,
F-75755 Paris Cedex 15, France
Tel: (+33 1) 45 38 89 90
Fax: (+33 1) 45 38 89 99
Vice-President/General Manager: Nissan Leviathan

Gilat Satellite Networks (HK) Ltd
902 Hutchison House, 10 Harcourt Road,
Central Hong Kong
Tel: (+852) 28 69 06 18
Fax: (+852) 25 26 29 69
Vice-President Sales/Marketing: Raymond Swanson

Israel Aircraft Industries Ltd
MBT Systems & Space Technology
Industrial Zone, IL-56000 Yehud
Tel: (+972 3) 531 41 84
Fax: (+972 3) 531 41 80
Manager, Space Technologies: Dr Moshe Bar-Lev
Director, Communications Satellites and Amos Program Manager: Amitsur Rosenfeld

Israel Telecommunications Corp (Bezeq)
9 Ahad Haam Street, PO Box 29555, IL-61290 Tel Aviv
Tel: (+972 3) 515 15 71
Fax: (+972 3) 515 13 18
Head, International Section: Yoram Grushkevitz

SpaceCom Satellite
Communication Services Ltd
Atidim Advanced Technology Park, Building 12
PO Box 58156, IL-61581 Tel Aviv
Tel: (+972 3) 49 04 84
Fax: (+972 3) 49 04 85
Chief Executive Officer: Shlomo Tirosh
Chief Operations Officer: Noam Fink
Managing Director: David Pollack
Chief Financial Officer: Dror Chevion
Director of Marketing: Avraham Ziv-Tal

ITALY

Alcatel Telettra
Via Trento 30, I-20059 Vimercate (MI)
Tel: (+39 039) 68 61
Fax: (+39 039) 608 14 83

Italtel
Via A di Tocqueville 13, I-20154 Milan
Tel: (+39 02) 43 88 51 50
Fax: (+39 02) 43 88 51 79
President/CEO: Salvatore Randi
COOs: Fausto Plebani, Giovanni Barbieri
Space Sales Manager: Camillo Enrico

Nuova Telespazio SpA
Via Tiburtina 965, I-00156 Rome
Tel: (+39 06) 407 91
Telex: 620424
Fax: (+39 06) 40 79 36 31
Managing Director: Dr Andrea Pucci
Public Relations/External Relations: Dr Alfonso Gregorio
Tel: (+39 06) 40 79 33 38
Fax: (+39 06) 40 79 36 31

Centro Spaziale 'Piero Fanti' del Fucino ('Piero Fanti' Fucino Space Centre)
I-67051 Avezzano (L'Aquila)
Tel: (+39 0863) 55 01
Fax: (+39 0863) 550 24 01
Manager: Aldo Bonetto

Stazione del Lario
I-22050 Colico (Como)
Tel: (+39 0344) 931 11
Fax: (+39 0344) 932 72
Manager: Ercole Marello Stazione dello Scanzano

I-90037 Piana Degli Albanesi (Palermo)
Tel: (+39 091) 845 11 11
Fax: (+39 091) 846 46 77
Manager: Piero Coffaro

Page Europa SpA
Via Simone Martini 127/129, I-00142 Rome
Tel: (+39 06) 50 39 53 77
Telex: 610540 pageur i
Fax: (+39 06) 50 39 52 11/12
President/General Manager: Paolo Capperi
Director, Business Development: Fabrizio Giulianini

JAPAN

Broadcasting Satellite System Corporation
Parkside-Yamamotokan 2F,
16-4 Tomigaya 1-chome, Shibuya-ku, Tokyo 151
Tel: (+81 3) 54 53 65 21
Fax: (+81 3) 54 53 65 24
President: Toshio Kataoka
Director, Business Division: Katsuji Saito

Furuno Electric Co Ltd
9-52 Ashihara-cho, Nishinomiya City 662
Tel: (+81 798) 65 21 11
Telex: 5644-325
Fax: (+81 798) 65 42 00
General Manager, International Marketing: Hiroaki Komatsu

Furuno (UK) Ltd
Ocean House, Parklands Business Estate,
Denmead, Hampshire PO7 6XP, UK
Tel: (+44 1705) 23 03 03

Japan Radio Company
Akasaka Twin Tower (Main), 17-22, Akasaka 2-chome,
Minato-ku, Tokyo 107
Tel: (+81 3) 35 84 88 36
Telex: 2425420 jrctok j
Fax: (+81 3) 35 84 88 78
President: Akeyoshi Sumita
Executive Managing Director: Hiroshi Yokomizo
General Manager, International Sales: T Sasahara

Japan Satellite Systems, Inc
Toranomon 17 Mori Building, 5F,
1-26-5 Toranomon, Minato-ku, Tokyo 105
Tel: (+81 3) 55 11 77 45
Fax: (+81 3) 35 97 71 80
President: Takuya Yoshida
Executive Vice-Presidents: Masayoshi Marumo, Yasuharu Inashima
General Manager, Engineering: Masaru Nomura
Manager, International Marketing: Kazuo Tomimura
General Manager, Corporate Comms: Yasuo Okuyama

Kokusai Denshin Denwa Co Ltd
3-2 Nishi Shinjuku 2-chome, Shinjuku-ku, Tokyo 163-03
Tel: (+81 3) 33 47 69 60
Telex: 72 22500 kdd tokyo
Fax: (+81 3) 33 47 63 06
Director, Transmission Systems Engineering Department: Dr Y Hirata
Manager, Satellite Communications: Muneo Abe
Director, Inmarsat Relations: Rikuo Koike

KDD UK Ltd
Bush House, Suites 126/132A,
Aldwych, London WC2B 4PA, UK
Tel: (+44 171) 379 78 78
Telex: 888052 kdd ldn g
Fax: (+44 171) 379 73 47

Ministry of Posts & Telecommunications
Communications Policy Bureau
3-2 Kasumigaseki, l-chome,
Chiyoda-ku, Tokyo 100-90
Tel: (+81 3) 35 04 49 60
Fax: (+81 3) 35 91 35 20
WWW: http://www.mpt.go.jp
Director, Space Communications Policy Division: Masao Matsumoto (Communications Research)

Laboratory (CRL)
2-1 Nukui-Kitamachi 4-chome, Koganei, Tokyo 184
Tel: (+81 423) 27 75 01
Fax: (+81 423) 27 66 98
Director, Space Communications Div: Shingo Ohmori
Global Environment Division: Kenichi Okamoto

NHK Nippon Hoso Kyokai
2-2-1 Jinnan, Shibuya-ku, Tokyo 150-01
Tel: (+81 3) 34 65 11 11
Telex: j34778 nhkhq
Fax: (+81 3) 34 69 81 10
President: Mikio Kawaguchi
General Managing Director/Executive Director General Engineering: Shuichi Morikawa
Deputy Director, Public Relations: Yasuo Onuki

Nippon Telegraph & Telephone Corp
1-6 Uchi Saiwai-cho 1-chome, Tokyo 100
Tel: (+81 3) 35 09 51 11
Wireless Systems Laboratories
1-2356 Take, Yokosuka-shi, Kanagawa-ken 238-03
Tel: (+81 468) 59 34 00
Fax: (+81 468) 59 33 51
Director of Satellite Communications Laboratory: Dr Eng Shuichi Samejima

Space Communications Corporation of Japan
2-8 Higashi-shinagawa 2-chome,
Shinagawa-ku, Tokyo 140
Tel: (+81 3) 54 62 13 66
Fax: (+81 3) 54 62 13 90
President: Yoshio Taniguchi
Executive Vice-Presidents: Yasuo Otaki, Teruhiko Ena
Senior Managing Director: Masakazu Takeuchi
Managing Director/General Manager, Spacecraft Engineering and Operating Division: Yoshihito Kamiya
General Manager, Communication Network Division: Takeshi Hashimoto
Manager, Marketing/Sales Dept: Kazunobu Iijima
Manager, Corporate Planning: Junta Fujikawa
Manager, General Affairs Department: Kunio Seki
Manager, Satellite Control Centre: Yoshiharu Shii

Telecommunications Advancement Organisation of Japan
31-19 Shiba 2-chome, Minato-ku, Tokyo 105
Tel: (+81 3) 37 69 68 11
Fax: (+81 3) 34 52 76 00
President: Moriya Koyama
Managing Directors: Toshiaki Kaneko, Misao Matsushita, Ken Fujta
Director, General Affairs: Shin-ichi Watanabe

Kimitsu Satellite Control Centre
1743-5 Sasa, Aza-Eishi, Kimitsu Chiba 292-05
Tel: (+81 439) 39 21 41
Fax: (+81 439) 39 33 06
Director: Minoru Yamamoto

Toshiba Corporation
1-1 Shibaura 1-chome, Minato-ku, Tokyo 105-01
Tel: (+81 3) 34 57 30 89
Telex: 22587 toshiba j
Fax: (+81 3) 54 44 93 33
Senior Manager, Telecommunication Systems International Business Department, Electronics & Telecommunication Systems Group: Yasushi Hamano
Manager: Tadashi Watanabe

Toyo Communication Equipment Company Ltd
12-32 Konan 2-chome, Minato-ku, Tokyo 108
Tel: (+81 3) 54 62 96 10
Fax: (+81 3) 54 62 96 25
Chairman: Sukehiro Ito
President: Shinichi Itoh
Executive Vice-President: Toshio Soejima
Senior Vice-Presidents: Yutaka Kominato, Yasuji Yamaguchi
Director/General Manager, International Operations Division: Hiroshi Kakuya
Senior Manager, Radio Applications Division: Toshikiyo Hirata
Contact, International Operations Division: Hiroshi Araki

KOREA, SOUTH

Korea Telecom
Satellite Business Group
680-63 Jayang-Dong, Sungdong-Gu, Seoul
Tel: (+82 2) 750 37 00
Fax: (+82 2) 750 37 22
President: Dr Han Hwangbo
Managing Director: Mike Yongsun Hion
Director, Satellite Technology Division: Myung-Soo Rhee

Electronics & Telecommunications Research Inst
PO Box 106, Yuseong Post Office, Taejeon 305-600
Tel: (+82 42) 860 68 30
Fax: (+82 42) 860 64 30
Vice-President Satellite Comms Division: Dr Seon Jong Chung

LUXEMBOURG

Astra/Société Européenne Des Satellites (SES)
Château de Betzdorf, L-6815 Château de Betzdorf
Tel: (+352) 710 72 51
Telex: 60625 astra lu
Fax: (+352) 710 72 53 24; 227
e-mail: 78:DGS 1011
Chairman: Pierre Werner
Director General: Romain Bausch
Technical Director: Aldis Grinbergs
Financial Director: Jürgen Schulte
Marketing Director: Harold Melzer
Public Relations Manager: Yves Feltes

MALAYSIA

Binariang SDN BHD
Ground Floor, Block B, Wisma Semantan,
12 Jalan Gelenggang, Bukit Damansara,
50490 Kuala Lumpur
Tel: (+60 3) 252 20 00
Fax: (+60 3) 252 19 20/1
e-mail: webmaster@gateway.Maxisnet.com.my
Chief Operating Officer: Duffy Swan
MEASAT Programme Manager: Ali Ebadi

MEXICO

Secrétaria De Comunicaciones Y Transportes (SCT)
San Francisco 1626-3ER Piso,
Colonia Del Valle, Delegacion Benito Juarez,
Codigo Postal 03100 Mexico City DF
Tel: (+52 5) 538 04 50/519 89 20
Telex: 170902 upeme

Telecomunicaciones De Mexico
Eje Central Lazaro Cardenas 567,
03020 Mexico City DF
Tel: (+525) 629 11 00
Director General: Carlos Mier y Teran Ordiales
Tel: (+525) 629 11 66
Fax: (+525) 559 98 12
Director, Satellites/Teleports: Juan Zamudio
Tel: (+ 525) 629 11 73
Fax: (+525) 538 42 26
Director of Planning: Enrique Cervantes Martinez
Tel: (+525) 629 11 70
Fax: (+525) 709 96 38
Commercial Director: Jose M Rivas
Tel: (+525) 629 11 23
Fax: (+525) 519 86 24

NETHERLANDS

CHL Netherlands BV
PO Box 3072, NL-2220 CB Katwijk
Tel: (+31 71) 402 55 14
Fax: (+31 71) 402 50 78
Senior Sales: Robbert Lens
e-mail: marketing@chl.nl

PTT Telecom Netherlands
PO Box 30150, NL-2500 GD The Hague
Tel: (+31 70) 343 78 65
Telex: 31255 ITEL NL
Fax: (+31 70) 343 27 23
Manager, Satcom Strategy: Ing Dick S Hoefsloot

NORWAY

Nera A/S
PO Box 10, N-5061 Kokstad, Bergen
Tel: (+47) 55 22 51 00
Telex: 42113
Fax: (+47) 55 22 52 99
President/Managing Director: Asbjørn Birkeland
Vice-President/General Manager Satcom Division: Bjorn Ove Skeie

PO Box 91, N-1361 Billingstad
Tel: (+47) 66 84 47 00
Telex: 71721 umeb n
Fax: (+47) 66 84 46 21
e-mail: knud.klein@nera.co
Vice-President/General Manager Inmarsat LES: Håkon Otterlei
Vice-President/General Manager Satcom Marine: Ottar Bjåstad
Sales Manager: Knud Klein

Telenor Satellite Services AS
PO Box 6914, St Olav's Plass,
N-0130 Oslo 1
Tel: (+47) 22 77 86 84
Telex: 71203 gentl n
Fax: (+47) 22 77 79 80
Head, Telenor International - Satellites: Jostein Rønneberg
Thor Project Manager: Hans Fjøsne
Press Officer: Dag Melgaard
Tel: (+47) 22 77 85 90

TSAT A/S
PO Box 50, Manglerud, N-0612 Oslo
Tel: (+47) 23 18 05 00
Telex: 0056-77140 nm n
Fax: (+47) 23 18 02 19
Managing Director: Leiu Erik Sverning
R&D Director: Slein Harstad

POLAND

Institute Of Telecommunications
Szachowa 1, 04894 Warsaw
Tel: (+48 22) 12 90 28
Telex: 813 630
Fax: (+48 22) 12 90 07
Director: Prof Andrzej Zielinski

Gascom
4a Lenin Street, Kaliningrad,
141070 Moscow oblast
Tel: (+7 095) 513 66 28
Fax: (+7 095) 513 69 02
General Director: Nikolai Sevastyanov
1st Dep General Director: Andrei V Shestakov

Global InformationSystems, Inc
Prostornaya ulitsa 7, 107392 Moscow
Tel: (+7 095) 963 91 83
Fax: (+7 095) 168 31 15
President: Vladimir A Andrianov
Vice-President: Mikhail Y Senatorov
General Marketing Manager: Valeri Koutoukov

Intersputnik
2nd Smolensky Lane, 1/4, Moscow 121099
Tel: (+7 95) 244 03 33
Telex: 411288/411289 DISK SU
Fax: (+7 95) 253 99 06
e-mail: dir@intersputnik.msk.ru
Director General: Gennadi G Kudryavtsev
Marketing Manager: Neil Bakmann (244 06 98)
Director, Legal/Financial Department: Istvan Kovacs
Director, Technical Development Dept: Mario Fernandez
Director, Operations Department: Joseph Dolecki

US contact: Jeffrey Minsky
Tel: (+1 212) 247 38 46
Fax: (+1 212) 977 81 22

Italy contact: Moshe Lerner
Tel: (+39 2) 80 40 19

Ministry For Posts & Telecommunications
7 Gorky Street, Moscow K-375
Tel: (+7 095) 295 51 08
Minister: Vladimir Bulgak

Morsviasputnik
1/4 Rozhdestvenka Street,
Moscow 103759
Tel: (+7 095) 258 70 45
Telex: 871-411197 mmf su
Fax: (+7 095) 253 99 10
Inmarsat A Fax: (+873) 140 21 62
Director General: Valeri Bogdanov
Deputy Director General: L Pcheliakov

SINGAPORE

Singapore Telecommunications Ltd
31 Exeter Road, #28-00 Comcentre, Singapore 0923
Tel: (+65 838) 26 02
Fax: (+65 732) 06 73/222 10 20
President: Wong Hung Khim

SOUTH AFRICA

Grinaker Telecom
185 Witch Hazel Avenue, PO Box 8492,
0046 Verwoerdburg (PTA)
Tel: (+27 12) 672 60 00
Fax: (+27 12) 672 62 22
e-mail: jlandsberg@avitronics.co.za
Contact: Johan Landsberg

SPAIN

Alcatel Espacio SA
(Alcatel Telecom/Space Business Line)
C/Einstein 7, P.T.M. Tres Cantos, E-28760 Madrid
Tel: (+34 1) 807 47 10
Fax: (+34 1) 804 00 16
President/General Manager: Luis García Echegoyen
Commercial/Marketing Director:
Javier Hernández Laille
Industrial Operations Director: Luis Cervera González

Hispasat SA
C/Gobelas 41-45, planta 2, La Florida, E-28023 Madrid
Postal: APDO 95000, E-28080 Madrid
Tel: (+34 1) 372 90 00
Telex: 47911 HSAT E
Fax: (+34 1) 372 89 41
President: José Villar Uribarri
General Director: Gabriel Barrasa
General Secretary: Josa Antonio Ferrer-Sama
Programme Director: Pedro Pintó
Exploitation Director: Pascual Menendez

Telefonica Sistemas De Satelites SA
C/Ramirez de Arellano 19, E-28043 Madrid
Tel: (+34 1) 337 31 01
Telex: 41008 tssat e
Fax: (+34 1) 337 56 72
President: Juan Carlos Martinez
Director General: Luís M García
Commercial Director: Domingo Díaz
Marketing Manager: J Gerardo Muros

SWEDEN

Ericsson Microwave Systems AB
SE-431 84 Mölndal
Tel: (+46 31) 747 00 00
Fax: (+46 31) 87 66 39
President: Jan-Ake Kark
Marketing Director: Bertil Hellstrom
Production Manager: Reimar Borgljung

Swedish Space Corporation
PO Box 4207, Albygatan 107
Tel: (+46 8) 627 62 00
Telex: 17128 spaceco s
Fax: (+46 8) 98 70 69
General Manager, Telecom Division: Per Zetterquist

Telia AB
SE-123 86 Farsta
Tel: (+46 8) 713 10 00
Fax: (+46 8) 713 33 33
Eutelsat contact: Stefan Carlsson
Intelsat contact: Leif Blomquist

Telia Mobitel AB
SE-136 80 Haninge
Tel: (+46 8) 707 45 00
Fax: (+46 8) 707 49 07
Inmarsat contact: Torsten Smedberg

TAIWAN

JONSA
TV Chung Enterprise Co Ltd
11 Lane 160, Nan Shing Road, Nantou City
Tel: (+886 49) 31 71 11
Fax: (+886 49) 35 08 39
Sales Manager: John Wu

THAILAND

Asian Broadcasting and Communications Network Co Ltd
SM Tower, 979/66-70 Phaholyothin Road, Phayathai, Bangkok 10400
Tel: (+66 2) 298 00 44
Fax: (+66 2) 271 30 78 77
Chairman/CEO: Sondhi Limthongkul
President: Suradet Mukyangkoon
Chief Technical Officer: Jrarat Pingclasai

Post & Telegraph Department
87 Phaholyothin 8, Phayathai, Bangkok 10440
Tel: (+66 2) 271 35 12
Telex: 82503 deposthe th
Fax: (+66 2) 271 35 12
Director General: Dr Kitti Yupho

Shinawatra Satellite Public Co Ltd
Thaicom Satellite Station
41/103 Rattanathibet Road, Nonthaburi 11000
Tel: (+66 2) 591 07 36 to 07 49
Fax: (+66 2) 591 07 06
Chairman: Dr Taksin Shinawatra
President: Dr Dumrong Kasemset
Vice-President: Dr Nongluck Phinainitisart
General Manager: Paiboon Panuwattanawong
Senior Manager, Commercial Department: Avudh Ploysongsang

TONGA

Friendly Islands Satellite Communications, Ltd (TONGASAT)
c/o W L Pritchard & Co, Inc
7315 Wisconsin Avenue, Suite 520E,
Bethesda, Maryland 20814, USA
Tel: (+1 301) 654 11 44
Fax: (+1 301) 654 18 14
Managing Director: Wilbur L Pritchard

TURKEY

Turkish Telekom
TR-06103 Ankara
Tel: (+90 312) 313 11 30
Fax: (+90 312) 313 15 97

UNITED KINGDOM

Advent Communications Ltd
Preston Hill House, Nashleigh Hill,
Chesham, Buckinghamshire HP5 3HE
Tel: (+44 1494) 77 44 00
Fax: (+44 1494) 79 11 27
Directors: Steve McGuinness, David Garrood,
Roger Crawshaw

Andrew Limited
Ilex Building, Mulberry Business Park,
Fishponds Road, Wokingham, Berkshire RG40 2GY
Tel: (+44 118) 977 68 86
Telex: 849650 andant g
Fax: (+44 118) 979 40 05
WWW: http://www.andrew.com
Marketing: Steven Gibson
Sales Manager: A Walp

BT (British Telecommunications plc)
Holborn Centre, 120 Holborn, London EC1N 2TE
Tel: (+44 171) 492 40 10
Telex: 21601 bti g
Fax: (+44 171) 492 44 90
Head, International Consortia & Organisations:
Richard Vos
Regional Satellite Policy Manager: Jeremy Simons
Mobile Satellite Investment Manager: Derek Simmons

COM DEV EUROPE LTD
Triangle Business Park, Stoke Mandeville,
Aylesbury, Buckinghamshire HP22 5SX
Tel: (+44 1296) 61 44 44
Fax: (+44 1296) 61 44 00
Managing Director: B N F Eddleston
Technical Director: R J Cameron
Business Development Manager: Colin McLaren

ERA Technology Limited
Cleeve Road, Leatherhead, Surrey KT22 7SA
Tel: (+44 1372) 36 70 00
Fax: (+44 1372) 36 70 99
e-mail: antenna@era.co.uk
WWW: http://www.era.co.uk
Business Development: Dr Barry Claydon

Gec-Marconi Communications Ltd
Marconi House, New Street,
Chelmsford, Essex CM1 1PL
Tel: (+44 1245) 35 32 21
Telex: 99201 marcon g
Fax: (+44 1245) 28 71 25
Managing Director: Dr Giorgio Ghiglione
Marketing Manager, Civil Communications:
Gary Dawson

GO-SAT International
24 Stagbury Avenue, Chipstead, Surrey CR5 3PD
Partners: Peter Moss
Tel: (+44 1737) 55 50 42
Fax: (+44 1737) 55 03 54
John Price
Tel: (+44 1202) 86 10 90
Fax: (+44 1202) 87 67 61

Kudos Technology Ltd
41 Rylett Crescent,
Shepherd's Bush, London W12 9RP
Tel: (+44 181) 746 19 95
Fax: (+44 181) 746 11 62
e-mail: 101350.1412@compuserve.com
Managing Director: Dr Michael J S Quigley

M/A-COM Limited
Humphrys Road, Woodside Estate,
Dunstable, Bedfordshire LU5 4SX
Tel: (+44 1582) 47 12 00
Telex: 82295
Fax: (+44 1582) 47 22 77
Engineering Director: Peter Volckman

Magnavox Systems Limited
Lancaster House, 7 Church Street,
Slough, Berkshire SL1 1TL
Tel: (+44 1753) 53 80 25
Telex: 847583 magvox g
Fax: (+44 1753) 82 46 33
Product Support Manager: John Wood

Marconi Marine
Marconi House, New Street,
Chelmsford, Essex CM1 1PL
Tel: (+44 1245) 35 32 21
Telex: 99228
Fax: (+44 1245) 35 87 76
Divisional Manager: David Evans
Sales/Marketing Manager: Roger Branscomb
Divisional Sales: Malcolm Bain

Marine Technology International Ltd
50 Barwell Business Park, Leatherhead Road,
Chessington, Surrey KT9 2NY
Tel: (+44 181) 391 60 34
Fax: (+44 181) 391 92 25
Chief Executive: Rodger Perks
Sales/Marketing: James Green
Technical Director: Michael Rambaut
Operations Director: Alan Todd

Multipoint Communications Ltd
Satellite House, Eastways Industrial Park,
Witham, Essex CM8 3YQ
Tel: (+44 1376) 51 08 81
Telex: 987313 mpoint
Fax: (+44 1376) 50 22 33

Nera Limited
Mobile Products & Systems Division
20 Imperial Way, Croydon, Surrey CR0 4RR
Tel: (+44 181) 686 57 01
Telex: 946384
Fax: (+44 181) 686 68 11
Chief Executive: M L Braime
Divisional Managing Director: Robert J Chewter
Director, Satellite Products & Systems: S J O'Neill
Chief Engineer: Andrew J Blaydes
Defence Sales Manager: John A Boys
Manager, Customer Services: P A Lanaste

Telecommunications Division
171 Camford Way, Sundon Park,
Luton, Bedfordshire LU3 3AN
Tel: (+44 1582) 49 11 49
Fax: (+44 1582) 58 18 73
Managing Director: Andy Lockwood
Sales & Marketing Managers:
Peter Moss (SatCom Component Division)
Barry Dalton (Radio Link Division)

Queen Mary & Westfield College
Microwave Antenna Research Group
Electronic Engineering Department,
Mile End Road, London E1 4NS
Tel: (+44 171) 975 53 39
Telex: 94018094
Fax: (+44 171) 981 02 59
WWW: http://www.elec.qmc.ac.uk/antenna
Contacts: Dr Clive Parini and Prof David Olver

Racal Antennas Ltd
First Avenue, Millbrook Trading Estate,
Southampton, Hampshire SO1 0LJ
Tel: (+44 1703) 70 57 05
Telex: 47476
Fax: (+44 1703) 70 11 22
Managing Director: J M Hirst
Commercial Director: Derek C Henderson

Racal Avionics Ltd
88 Bushey Road, London SW20 0JW
Tel: (+44 181) 946 80 11
Telex: 928592
Fax: (+44 181) 946 75 30
Managing Director: Sean O'Malley
Marketing Director: P Leo Gallagher
Marketing Manager: Simon Glenister

Siemens Plessey Systems
Defence Systems Division
Grange Road, Christchurch, Dorset BH23 4JE
Tel: (+44 1202) 48 63 44
Telex: 418417
Fax: (+44 1202) 40 49 68
Managing Director: Clive Dolan
Managing Directors, Defence Systems Division:
David Barnes
Satcom Marketing: David Ward

Racal Research Ltd
Satellite Systems Division
Worton Drive, Worton Grange Industrial Estate,
Reading, Berkshire RG2 0SB
Tel: (+44 1734) 86 86 01
Fax: (+44 1734) 75 23 00
Director: C P Ash
Technical Manager: P D Britten

Racal Communications Systems Ltd
Western Road, Bracknell, Berkshire RG12 1RG
Tel: (+44 1344) 48 32 44
Telex: 848166
Fax: (+44 1344) 48 13 67
Managing Director: Dr S Wilder
Technical Director: A P Graham

UNITED STATES OF AMERICA

American Mobile Satellite Corp
10802 Parkridge Boulevard, Reston, Virginia 22091
Tel: (+1 703) 758 60 00
Fax: (+1 703) 758 61 11
President/CEO: Gary Parsons
Executive Vice-President and Chief Operating Officer:
Robert L Goldsmith

Andrew Corporation
10500 W 153rd Street, Orland Park, Illinois 60462
Tel: (+1 708) 349 33 00
Telex: 25-3897
Fax: (+1 708) 349 59 43
President/CEO: Dr Floyd L English
Vice-President Government Products: William Shockley
Director Marketing: Jerry Carter
Manager Business Development: Matt Melester

Antenna Technology Communications Inc
1128 East Greenway, Mesa, Arizona 85203
Tel: (+1 602) 264 72 75
Telex: 187217
Fax: (+1 602) 898 76 67
International Sales/Marketing: Gary S Hatch
Director of Engineering: Kevin Hatch
Director of Marketing: Charles M Willman

Astroguide International
1403 Fifth Street, LaSalle, Illinois 61301
Tel: (+1 815) 224 27 00
Fax: (+1 815) 224 27 01
President: A S Kline
e-mail: antenna@era.co.uk
Vice-President/Marketing Director: Alan Howarter

AT&T Alascom
210 East Bluff Drive, Anchorage, Alaska 99501
Tel: (+1 907) 264 70 00
Fax: (+1 907) 276 59 37
President: Jerry DeFrancisco
Vice-President Network: John Higginbottom
Vice-President Sales/Customer Service: Jim Colson
External Affairs Director: Laurie Herman

AT&T Tridom
840 Franklin Court, Marietta, Georgia 30067
Tel: (+1 404) 426 42 61
Fax: (+1 404) 514 17 37
President: Eileen McNamara-Raisch
Vice-President Engineering: Michael Smith
Vice-President Sales: Ruth Seydel

Aydin Corporation
700 Dresher Road, PO Box 349,
Horsham, Pennsylvania 19044
Tel: (+1 215) 657 75 10
Fax: (+1 215) 657 38 30
Chairman/CEO: I Gary Bard
President: Dr Donald S Taylor
Vice-President Sales/Marketing: Manfred Hirt
Vice-President/President, Aydin Telecom:
Thomas LoCasale

Ball Aerospace & Technologies Group
Telecommunications Products Division
PO Box 1235, Broomfield, Colorado 80038
Tel: (+1 303) 939 40 00
Fax: (+1 303) 442 48 12
Telex: 45-605
TWX: 910-940-3241
e-mail: airlink@ball.com
WWW: http://www.ball.com
President: Don Vanlandingham
Vice-President and General Manager,
Telecommunications Products Division:
Patrick Dummigan

California Microwave, Inc
555 Twin Dolphin Drive, Suite 650,
Redwood City, California 94065
Tel: (+1 415) 596 90 00
Telex: 352016
Fax: (+1 415) 596 66 00
WWW: http://www.cmi-ged.com/corp.htm
Chairman/CEO: Philip Otto
Vice-President Corporate Communications:
Stephanie M Day
Director Government Affairs: Charles Feldmayer

Satellite Transmission Systems
125 Kennedy Drive, Hauppauge, New York 11788
Tel: (+1 516) 272 56 00
Fax: (+1 516) 272 55 00
WWW: http://www.cmi-sts.com
President/General Manager: Brian Maloney
Vice-President Marketing: Philip A Arcona

Government Communication Systems
10820 Guilford Road, Suite 202,
Annapolis Junction, Maryland 20701
Tel: (+1 410) 792 22 25
Fax: (+1 410) 792 42 45
e-mail: gcs@aol.com
WWW: http://www.cmi-ged.com/gcs.htm
General Manager: Nelson Ward
Vice-President Business Development: Richard McClure

EFData
2105 W 5th Place, Tempe, Arizona 85281
Tel: (+1 602) 968 04 47
Fax: (+1 602) 921 90 12
WWW: http://www.efdata.com
President: Don Anderson
Vice-President Government Programs: Charles Stillings

Sunnyvale Products
1143 Borregas Avenue, Sunnyvale, California 94089
Tel: (+1 408) 752 89 00
Fax: (+1 408) 743 47 70
WWW: http://www.cmi-ged.com/spd.htm
President: E Russel Hardy
Vice-President Business Development: Carol Flaherty

Center for Commercial Space Communications
Virginia Tech, Blacksburg, Virginia 24061-0111
Tel: (+1 703) 231 44 61
TWX: 910 333 1861 vpi bks
Fax: (+1 703) 231 33 55
e-mail (Bitnet): bowen@VTVM1
Executive Director: Prof Charles W Bostian
Associate Director, Technology
Research: Prof Timothy Pratt
Associate Director, Business
Research: Prof George E Morgan

Center for Satellite & Hybrid Communication Networks
University of Maryland, Institute for Systems Research
A V Williams Building, College Park, Maryland 20742
Tel: (+1 301) 405 79 00
Fax: (+1 301) 314 85 86
Director: Dr John S Baras

Collins Air Transport Division (CATD)
400 Collins Road NE, Cedar Rapids, Iowa 52498
Tel: (+1 319) 395 21 23
Fax: (+1 319) 395 43 33
Vice-President/General Manager: Clayton Jones
Vice-President Marketing: S M Murchison
Satellite Comms Marketing: Thomas L McDonald

Collins Avionics & Communications Division (CACD)
350 Collins Road NE, Cedar Rapids, Iowa 52498
Tel: (+1 319) 295 51 00
Fax: (+1 319) 295 47 77
Vice-President Business Development:
Michael K McDonald
Contact: Rhonda Edwards, MS 120-130

Communications & Power Industries, Inc
607 Hansen Way, Palo Alto, California 94304-1130
Tel: (+1 415) 424 53 36
Fax: (+1 415) 424 62 76
President/CEO: Al Wilunowski
National Sales Manager: John Beighley

CPI Satcom Division
3200 Patrick Henry Drive,
Santa Clara, California 95054
Tel: (+1 408) 496 62 73
Fax: (+1 408) 496 62 35
President: Jim Commendatore
Vice-President Business Development: Frank Blaha
Vice-President Product Development: Al Drewke

Communications Center
2723 Green Valley Road,
Clarksburg, Maryland 20871-8599
Tel: (+1 301) 831 67 00
Fax: (+1 301) 865 55 77

COMSAT Corporation
22300 COMSAT Drive, Clarksburg, Maryland 20871
Tel: (+1 301) 428 40 00
Telex: 197800
Fax: (+1 301) 428 77 47
President/CEO: Betty C Alewine
Senior Vice-President Business Development:
Ronald J Mario
President, COMSAT World Systems: Betty C Alewine
President, COMSAT Labotatories: John V Evans
President, COMSAT Video Enterprises: Charles Lyons
President, COMSAT RSI: Richard Thomas
Vice-President/General Manager Mobile Comms:
Thomas Collins
Vice-President/General Manager Personal Comms:
Christopher J Leber
Vice-President Corporate Affairs: Arthur R Sando

COMSAT RSI
1501 Moran Road, Sterling, Virginia 20166
Tel: (+1 703) 450 56 80
Telex: 899156
Fax: (+1 703) 450 47 06
President: Richard E Thomas
Vice-President Advanced Systems: William Thomas
Vice-President Communications Systems:
Marvin Shoemake
Vice-President Wireless Networks: John Morris

Comsearch
2002 Edmund Halley Drive, Reston, Virginia 20191
Tel: (+1 703) 620 63 00
Fax: (+1 703) 476 27 20
WWW: http://www.comsearch.com
President and CEO: Michael K Morin
Vice-President: Timothy P Eckersley

COMSTREAM
10180 Barnes Canyon Road,
San Diego, California 92121
Tel: (+1 619) 657 54 16
Fax: (+1 619) 657 54 04
President: Ronald Derry
Vice-President Operations: John Wadsworth
Vice-President Rural Networks: Preston Vorlicek
Vice-President HiMedia: Stephen A Blake
Vice-President Sales: Sal Benti
Vice-President R&D: Dr Itzhak Gurantz

Comtech Antenna Systems Inc
3100 Communications Road, St Cloud, Florida 34769
Tel: (+1 407) 892 61 11
Telex: 810 870-0220
Fax: (+1 407) 957 34 02
Vice-President Marketing: Tom Christy

Constellation Communications, Inc
10530 Rosehaven Street, Suite 200,
Fairfax, Virginia 22030
Tel: (+1 703) 352 17 33
Fax: (+1 703) 352 92 79

Cynetics Corporation
PO Box 2422, 3824 Jet Drive,
Rapid City, South Dakota 57709
Tel: (+1 605) 394 64 30
Fax: (+1 605) 394 64 56
President: Don K Lefevre
Vice-President: Dan Mulally

Datron Systems Inc
200 West Los Angeles Avenue,
Simi Valley, California 93065-1650
Tel: (+1 805) 584 17 17
Telex: 910-494-2787
Fax: (+1 805) 526 08 85
President: Donald Rondeau
Vice-President Marketing: John Ricardi

Decom Systems Inc
1939 Palomar Oaks Way, Carlsbad, California 92009
Tel: (+1 619) 431 19 45
Fax: (+1 619) 431 19 46
President: John A Robinson
Vice-President Manufacturing: Richard Carrine

DirecTV, Inc
2230 East Imperial Highway,
El Segundo, California 90245
Tel: (+1 310) 535 50 00
Fax: (+1 310) 535 52 25
President: Eddy W Hartenstein
President, DirecTV International: Kevin N McGrath
Public Relations: Linda Brill

Dorne & Margolin Inc
2950 Veterans Memorial Highway,
Bohemia, New York 11716
Tel: (+1 516) 585 40 00
Fax: (+1 516) 585 48 10
Vice-President/Marketing and Sales:
Timothy F Ursprising

Echostar Satellite Corporation
90 Inverness Circle East, Englewood, Colorado 80112
Tel: (+1 303) 799 82 22
Fax: (+1 303) 799 62 22
Chairman: Charles Ergen
President/COO: Carl Vogel
Vice-President Marketing/Sales: David Carlson

Electromagnetic Sciences, Inc
660 Engineering Dr, Norcross, Georgia 30091-7700
Tel: (+1 404) 263 92 00
Fax: (+1 404) 263 92 07
CEO: Dr Thomas E Sharon

Electronic Space Systems Corporation
Old Powder Mill Road,
Concord, Massachusetts 01742-4697
Tel: (+1 508) 369 72 00
Fax: (+1 508) 369 76 41
CEO/President: Albert Cohen
President: Steven R Scheff
Executive Vice-President: J Leo Clark
Vice-President Engineering: J B Sangiolo
Director, International Marketing: Edwin Tuarog
WWW: http://www.es.scoradomes.com

Ireland
ESSCO Collins Limited
Kilkishen, County Clare, Ireland
Tel: (+353 61) 36 72 44
Fax: (+353 61) 31 10 44
Operations Manager: Liam Brown

Ellipsat International Inc
1120 19th St NW, Suite 480, Washington DC 20036
Tel: (+1 202) 466 44 88
Fax: (+1 202) 466 44 93
President/CEO: David Castiel
Executive Vice-President: Neal Howard

E-Systems Inc
PO Box 660248, 6250 LBJ Freeway,
Dallas, Texas 75266-0248
Tel: (+1 214) 661 10 00
Telex: 703365
Fax: (+1 214) 661 85 08
Chairman/CEO: A Lowell Lawson
Senior Vice-Presidents:
Dr T W Heil, P A Marino, B D Cullen
Vice-President International Operations:
David Matthews

1901 N Moore Street, Suite 609,
Arlington, Virginia 22209-1772
Tel: (+1 703) 524 23 10
Fax: (+1 703) 524 55 75
Vice-President Government/International Operations:
William McClain

ECI Division
PO Box 12248, 1501 72nd St N,
St Petersburg, Florida 33733-2248
Tel: (+1 813) 381 20 00
Fax: (+1 813) 343 12 95
Vice-President/General Manager: Dr James C Garrett
Vice-President Communications Systems:
Joe L Brumbaugh
Space Systems Business Development: Greg Seymour
Space Systems Engineering Manager: William Mosley

Greenville Division
PO Box 6506, Greenville, Texas 75403-6056
Tel: (+1 903) 455 34 50
Fax: (+1 903) 457 44 13
Vice-President/General Manager: Alan J Doshier
Manager, Satcom Products: John Parker

Garland Division
PO Box 660023, 1200 Jupiter Road,
Dallas, Texas 75266-0023
Tel: (+1 214) 272 05 15
Fax: (+1 214) 272 81 44
Vice-President/General Manager: P A Marino
Manager, Commercial Remote Sensing:
Robert Clemons

Fairchild Data
350 North Hayden Road, Scottsdale, Arizona 85257
Tel: (+1 602) 949 11 55
Fax: (+1 602) 941 00 23
President/General Manager: Tom Breeden
Vice-President Operations: Rich LaMoure

FEI Microwave, Inc
825 Stewart Drive, Sunnyvale, California 94086-4514
Tel: (+1 408) 732 08 80
Fax: (+1 408) 730 16 22
President: Peter O Clark
Executive Vice-President: Leonard J Martire
Manager Marketing, Satellite Products: Mickey Elwood
Vice-President Engineering: Matias Pardo

Final Analysis, Inc
7500 Greenway Center, Suite 1240,
Greenbelt, Maryland 20770
Tel: (+1 301) 474 01 11
Fax: (+1 301) 474 32 28
e-mail: fai@facs.com
President: Nader Modanlo
Executive Vice-President: Michael Ahan
Director, Business Development: Don Erat

GE Americom
Four Research Way,
Princeton, New Jersey 08540-6684
Tel: (+1 609) 987 40 00
Fax: (+1 609) 987 44 50
Chairman/CEO: John F Connelly
President, GE Spacenet Services: Gino Picasso
Vice-President Satellite Services: Andreas M Georghiou
Senior Vice-President/General Manager, Technical and Government Operations: Walter H Braun
Vice-President Technology: Rick Langhans
Vice-President Business Development and International Marketing: George Monaster
Vice-President Business Development and Quality Programs: Yousef Javadi

Globalstar Limited Partnership
3200 Zanker Road, PO Box 640670,
San Jose, California 95164-0670
Tel: (+1 408) 473 55 50
Fax: (+1 408) 473 55 48

Washington Operations
1725 Jefferson Davis Highway, Suite 1007,
Arlington, Virginia 22202-3501
Tel: (+1 703) 414 10 45
Fax: (+1 703) 414 10 75
Chairman/CEO: Bernard Schwartz
President: Douglas G Dwyer
Executive Vice-President, Business Development:
Anthony J Navarra
Vice-President Communications: Jeanette Cloran
General Manager, Washington Operations:
Marc Newman
Director, Business Development Operations:
Reid Stephenson
Manager, Public Relations: John Cunningham

Glocom, Inc
1803 Research Boulevard, Suite 401,
Rockville, Maryland 20850
Tel: (+1 301) 309 19 40
Fax: (+1 301) 309 19 32
Inmarsat Products: Steven Wooldridge

Harris Corporation
Government Communications Systems Division
PO Box 91000, 1380 Harris Road,
Melbourne, Florida 32902
Tel: (+1 407) 727 40 00
Fax: (+1 407) 727 63 04
Vice-President/General Manager: Harry L Deffebach

Hughes Communications, Inc
PO Box 9712, Long Beach, California 90810
Tel: (+1 310) 525 52 11
Fax: (+1 310) 525 52 20
WWW: http://www.hcisat.com
President: Jerald F Farrell
Executive Vice-President: Hal McDonnell
Senior Vice-President Sales, Marketing and Network Operations, Galaxy Satellite Services: Carl Brown
Vice-President Operations and Government Services:
Floyd Stuart
Vice-President Galaxy/Spaceway: Edward Fitzpatrick
Vice-President Galaxy International: Gary Bertonneau
Director, Engineering and Network Operations: Bernard Vecerek

Hughes Electron Dynamics Division
PO Box 2999, Torrance, California 90509-2999
Tel: (+1 310) 517 64 63
Fax: (+1 310) 517 64 51
Division Manager: Dr Timothy Fong
Division Marketing Manager: Levon Thorose
Space Business Area Manager: Stan Sosa

Hughes Magnaphone
3100 W Lomita Boulevard, Torrance, California 90505
Tel: (+1 310) 517 68 00
Fax: (+1 310) 517 61 10
Manager, Commercial Satellite Communications:
William C Euler

Hughes Network Systems Inc
11717 Exploration Lane,
Germantown, Maryland 20876
Tel: (+1 301) 428 55 00
TWX: 710 828 0541
Fax: (+1 301) 428 18 68/28 30
President/COO: Pradman Kaul
Chairman/CEO: Jack A Shaw
President, HNS International: Andrew Werth
Senior Vice-President VSAT Domestic Sales/Marketing: Dennis Conti
Vice-President VSAT International Marketing:
Paul Gaske
Vice-President /General Manager Satellite Networks Division: Bahram Pourmand
Director, Marketing Communicationss: Judy Blake

Hughes Space & Communications Company
PO Box 92919, Los Angeles, California 90009
Tel: (+1 310) 364 60 00
Fax: (+1 310) 334 49 11
President: Donald L Cromer
Principal Deputy: Jeffrey E Grant
General Manager Commercial Business Dev:
William Pomevanz
General Manager Government Business Dev:
William Moore
Communications/Customer Relations: Don O'Neal
Chief, Media Relations: Fran Slimmer
e-mail: fmslimmer@ccgate.hac.com

Integral Systems, Inc
5000 Philadelphia Way, Suite A,
Lanham, Maryland 20706-4417
Tel: (+1 301) 731 42 33
Fax: (+1 301) 731 96 06
Contact: Steve Carchedi

Iridium, Inc
1401 H Street NW, Washington DC 20005
Tel: (+1 202) 326 56 00
Fax: (+1 202) 842 00 06
Chairman/CEO: Robert W Kinzie
President/COO: Jerrold D Adams

Vice-President/CFO: Paul Daverio
Vice-President Government Affairs: Leo Mondale
Vice-President Telecommunications Policy: Candace Johnson
Director, Space Segment Engineering: Peter Brunt

Keiser Engineering, Inc
2046 Carrhill Road, Vienna, Virginia 22181
Tel/Fax: (+1 703) 281 95 82
President: Bernhard E Keiser

LNR Communications Inc
180 Marcus Boulevard,
Hauppauge, New York 11788-3795
Tel: (+1 516) 273 71 11
Telex: 221620
Fax: (+1 516) 273 71 19
President: Seymour Okwit
Vice-President Marketing: Frank Arams
Director of Marketing: Frank Kelly

Lockheed Martin Astro Space
PO Box 800, Princeton, New Jersey 08543-0800
Tel: (+1 609) 490 34 00
Telex: 843458
Fax: (+1 609) 490 38 18

Spacecraft Operations Valley Forge
PO Box 8555, Philadelphia, Pennsylvania 19101
Tel: (+1 215) 354 10 00
Telex: 843597
Fax: (+1 609) 354 53 35
President: Alexander Horvath
President, Astro Space Commercial: Russ McFall
Vice-President Valley Forge Defense Programs: Edward M Morgan
Vice-President Valley Forge Earth Remote Sensing Programs: K Michael Henshaw
Vice-President Commercial Programs: Russ McFall
Vice-President E Windsor Government Programs: Ted Gavrilis
Vice-President New Procurements: J Hemingway
Vice-President Mission Success and Product Assurance: Nathan Lindsay
Vice-President Business Operations: Bill Firman
Vice-President Human Resources: J Gill
Director, Public Affairs: J Tierney

Lockheed Martin Conic-Terracom
9020 Balboa Avenue,
San Diego, California 92123-3507
Tel: (+1 619) 279 04 11
Telex: 910 335 1183
Fax: (+1 619) 27 06 93

Loral Skynet
Room 2A115W, 900 Route 202/206 North, Bedminster, New Jersey 07921-0752
Tel: (+1 800) 847 93 35
Fax: (+1 908) 234 79 40
President: Terry Hart

Loral Space & Communications Ltd
600 Third Avenue, New York, New York 1006
Tel: (+1 212) 697 11 05
Fax: (+1 212) 661 89 88
Chairman/CEO: Bernard L Schwartz
President/COO: Michael B Targoff

MCL Incorporated
501 South Woodcreek Road,
Bolingbrook, Illinois 60440-4999
Tel: (+1 708) 759 95 00
Telex: 910-683-18 99
Fax: (+1 708) 759 50 18
President: Frank P Morgan
Vice-President Marketing: Ronald Adamson

Microdyne Corporation
Aerospace Telemetry Division
491 Oak Road, Ocala, Florida 34472-0213
Tel: (+1 904) 687 46 33
Fax: (+1 904) 687 33 92
e-mail: mcdy@ocala.com
Division Vice-President: Christian Spitz
Vice-President Operations: Rosendo Leon
Vice-President Engineering: William Lennox
Marketing/Sales Manager: Thomas H Kidd

MITEQ, Inc
100 Davids Drive, Hauppauge, New York 11788-2034
Tel: (+1 516) 436 74 00
Fax: (+1 516) 436 74 30
WWW: http://www.miteq.com
President: Arthur Faverio
Director, Corporate Communications: Helle M Kiiss

Mobile Telesystems Inc
300 Professional Drive,
Gaithersburg, Maryland 20879
Tel: (+1 301) 590 85 00
Telex: 901885
Fax: (+1 301) 590 85 58
WWW: http://www.mti-usa.com
President/CEO: William Bearry
Vice-President, Marketing and Sales: Shafiq Chaudhuri
Video Products: Steven Geweny
Portable Satellite Products (US Govt): Robert Demers
Portable Satellite Products (International): Madhav Kadambi

Motorola
Space & Systems Technology Group
Radio Systems Operations
PO Box 9040, 8220 East Roosevelt Road,
Scottsdale, Arizona 85252-9040
Tel: (+1 602) 441 40 79
Fax: (+1 602) 441 23 68
Corp Vice-President /General Manager: Earl Cobb
Sales: Robert Boyle (Tel: (+1 602) 441 40 79)

Space and Systems Technology Group
Satellite Communications Group,
Government Space Systems Division
8201 East McDowell Road,
PO Box 1417, Scottsdale, Arizona 85252-1417
Tel: (+1 602) 441 67 38
Fax: (+1 602) 441 67 77

Motorola, Inc
Satellite Communications
2501 South Price Road,
Chandler, Arizona 85248-2899
Tel: (+1 602) 732 20 00
Fax: (+1 602) 732 21 48
Senior Vice-President/General Manager: Durrell Hillis
Assistant General Manager: Bary R Bertiger
Systems Engineering Director: Ray J Leopold
Iridium Manufacturing Manager: Larry Casey
Business Manager: James Foley
Director of Marketing: Mark Gercenstein

NASA Goddard Space Flight Center
Greenbelt, Maryland 20771
Tel: (+1 301) 286 80 34
Fax: (+1 301) 286 17 21
TDRSS Project Manager: Anthony Comberiate
TDRSS Deputy Project Manager: Vacant

NASA Lewis Research Center
21000 Brookpark Road, Cleveland, Ohio 44135
ACTS Project Manager: Rodney M Knight
ACTS Archived Data Requests: Joanne R Poe
ACTS Spacecraft Operations: Louis R Ignaczak
ACTS System Engineering: Sina Javidi

O'Gara Satellite Networks
1 Brandywine Drive, Deer Park, New York 11729
Tel: (+1 516) 586 51 00
Fax: (+1 516) 586 55 31
President: Neil Saldin

Orbcomm Global LP
21700 Atlantic Boulevard, Dulles, Virginia 20166
Tel: (+1 703) 406 60 00
Fax: (+1 703) 406 35 04
President: Alan L Parker
Executive Vice-President Marketing/Business Development: Robert J Pizzimenti
e-mail: rpizz@hqgate.orbital.com
Vice-President International Business Development: Martin Deckett
Program Manager: David Steffy

Orbitron
351 South Peterson Street,
Spring Green, Wisconsin 53588
Tel: (+1 608) 588 29 23
Fax: (+1 608) 588 22 57

Panamsat Corporation
One Pickwick Plaza, suite 270,
Greenwich, Connecticut 06830
Tel: (+1 203) 622 66 64
Fax: (+1 203) 622 91 63
WWW: http://www.panamsat.com
President/CEO: Frederick A Landman
Executive Vice-President: Lourdes Saralegui
Senior Vice-President Engineering/Ops: Robert Bednarek
Senior Vice-President, Broadcast Sales: David P Berman
Vice-President Broadcast Services: Michael Antonovich
Vice-President Asia: Andrew Jordan
Vice-President Business Development: Ammar Hanafi
Vice-President Corporate Communications: Daniel J Marcus

Qualcomm Inc
6455 Lusk Boulevard,
San Diego, California 92121-2779
Tel: (+1 619) 587 11 21
Fax: (+1 619) 658 15 78
Chairman/CEO: Dr Irwin M Jacobs
Chief Operating Officer: Richard Sulpizio
President: Harvey P White
Chief Technical Officer: Dr Andrew J Viterbi
Vice-President OmniTracs General Manager/Senior: John Sarto

Raytheon Company
Equipment Division
1001 Boston Post Road,
Marlborough, Massachusetts 01752
Tel: (+1 508) 490 10 00
Telex: 92-3455
Fax: (+1 508) 490 28 22
Vice-President/Division Manager: C Dale Reis
Division Marketing: Steve S Teel

Raytheon E-Systems Richardson
PO Box 831359
Richardson, Texas 75083-1359
Tel: (+1 972) 470 21 07
Fax: (+1 972) 470 24 66
Vice-President: George T Spiva

Satellife
1360 Soldiers Field Road,
Boston, Massachusetts 02153
Tel: (+1 617) 789 54 55
Fax: (+1 617) 789 47 71
e-mail: hnet@usa.healthnet.org
WWW: http://www.healthnet.org
Chairman/Founder: Dr Bernard Lown
Executive Director: John Mullaney
Technical Director: Alexander Fraser
Development Director: Melissa Boudreault

Scientific-Atlanta, Inc
Corporate Headquarters
PO Box 105600, One Technology Parkway South,
Atlanta, Georgia 30348
President and CEO: James F McDonald

Network Systems Group
4356 Communications Drive,
Norcross, Georgia 30093
Tel: (+1 770) 903 61 70
Telex: 4611804 sciatl
Fax: (+1 770) 903 62 45
e-mail: @sciatl.com
President: Jack W Simpson
Vice-President Mobile Satellite Systems: Charles R Smith
Vice-President Marketing, Satellite Networks: B Scott Fogle
Sales/Marketing, Mobile Satellite: Jack W Bright

Sea Tel, Inc
1035 Shary Court, Concord, California 94518
Tel: (+1 510) 798 79 79
Fax: (+1 510) 798 79 86
WWW: http://www@seatel.com
President: J Patrick Matthews
Vice-President Engineering: R J Matthews
Vice-President Technical Sales: R C Matthews
Marketing Director: Richard D Stallbaum

Sierracom
99 South Street, Hopkinton, Massachusetts 01748
Tel: (+1 508) 435 24 00
Fax: (+1 508) 435 20 22/24 12
President: Richard Gawlik
Vice-President Marketing: Ramunas A Kalvaitis

Space Communications Technology Centre
Florida Atlantic University
500 NW 20th Street, S&E building 43/486,
Boca Raton, Florida 33431-0991
Tel: (+1 407) 367 23 43
Fax: (+1 407) 367 34 18
Director: Dr William E Glenn

Space Systems/Loral
3825 Fabian Way, Palo Alto, California 94303-4697
Tel: (+1 415) 852 40 00
Telex: +1 34-8481
Fax: (+1 415) 852 47 88
Chairman: Bernard L Schwartz
President: Robert E Berry
Executive Vice-President Operations: Dr Alexis M Livanos
Executive Vice-President Globalstar: Dr John M Klineberg
Senior Vice-President, Strategic Ventures: Dr Ronald Maehl
Vice-President, Marketing and Sales: Daniel E Collins
Vice-President, Systems Planning: Troy D Ellington
Vice-President Systems Engineering, Integration and Test: Christopher F Hoeber
Executive Director, Govt Programs: Cor Westerhof
Executive Director, Intelsat Program: John Celli
Executive Director, Military Satcoms: Raymond Barker
Director, Strategic Partnership Management: Duncan L Reynard

SSE Technologies Inc
47823 Westinghouse Drive,
Fremont, California 94539-7437
WWW: http://www.sset.com
Tel: (+1 510) 657 75 52
Telex: 710-824-0096
Fax: (+1 510) 490-85 01
President: Fred Toombs
Vice-President Engineering: Claudio Mariotta
Vice-President Sales: Fred Kelly
Manager, Corporate Marketing: Mary Ellen Hannon

SSE Telecom Inc
President: Fred Toombs
Vice-President Sales/Marketing: Fred Kelly

ST Systron Donner Corporation
Microwave Products Division
13100 Telfair Avenue, Sylmar, California 91342
Tel: (+1 818) 364 72 73
Fax: (+1 818) 367 47 09
President/CEO: Dick Nabozny
Marketing Manager: Bruce A Carpenter

Stanford Telecommunications Inc
1221 Crossman Avenue,
Sunnyvale, California 94089
Tel: (+1 408) 745 08 19
Fax: (+1 408) 745 77 56
President/CEO: Dr Val P Peline
Chief Engineer: John Ohlson

Tecom Industries, Inc
9324 Topanga Canyon Boulevard,
Chatsworth, California 91311-5795
Tel: (+1 818) 341 40 10
Fax: (+1 818) 718 14 02
President: Vito P Minerva
Vice-President Engineering: Bruce Bishop
Director, Satcom Products: Dick Radawicz

TELSYS Inc
4407A Forbes Boulevard, Lanham, Maryland 20706
Tel: (+1 301) 459 27 27
Fax: (+1 301) 459 30 86
WWW: http://www.tsi-telsys.com
President: Jim Chesney

Tempo DBS, Inc
5619 DTC Parkway, Engelwood, Colorado 80111
Tel: (+1 303) 267 55 00
Fax: (+1 303) 488 32 25
President, TCI: John Malone
President, Tempo DBS: David Beddow

TIW Systems, Inc
2211 Lawson Lane, Santa Clara, California 95054
Tel: (+1 408) 654 56 00
Fax: (+1 408) 654 56 13/14
President/CEO: Dr Rein Luik
Executive Vice-President: Louis Becker
Marketing Director: Rob Wellins

W L Pritchard & Co, Inc
7315 Wisconsin Avenue, Suite 520E
Bethesda, Maryland 20814
Tel: (+1 301) 654 11 44
Fax: (+1 301) 654 18 14
President: Wilbur L Pritchard
Vice-Presidents: Ellen Hoff, Jack Dicks

Worldspace, Inc
11 Dupont Circle NW, 9th Floor,
Washington DC 20036
Tel: (+1 202) 884 78 00
Fax: (+1 202) 884 79 00
e-mail: *name*@worldspace.com
WWW: http://www.worldspace.com
Chairman/CEO: Noah A Samara
CFO: Heinz Stubblefield
Senior Vice-President Engineering/Operations: Pierre Madon

EARTH OBSERVATION SATELLITE SYSTEMS AND SERVICES

ARGENTINA

Aeroterra SA
Gorostiaga 2465, 1426 Buenos Aires
Tel: (+54 1) 771 58 81/775 62 05
Fax: (+54 1) 774 61 83
Executive Vice-President: Carlos M Viola

AUSTRALIA

Australian Centre For Remote Sensing
Postal: PO Box 28, Belconnen, ACT 2616
Dunlop Court, Fern Hill Park, Bruce, ACT 2617
Tel: (+61 6) 201 42 01
Telex: 61510
Fax: (+61 6) 251 63 26
WWW: http://www.auslig.gov.au
Manager: Paul Trezise
Chief Engineer: Robert Denize
Business Manager: Donna Scott
Production Director: Paul Wise
Production Manager: Mike Pasfield
Director, Marketing: Jim Mollison

Centre For Remote Sensing & GIS
The University of New South Wales
New South Wales 2052
Tel: (+61 2) 93 85 41 72
Fax: (+61 2) 93 13 74 93
WWW: http://www.gmat.unsw.edu.au
Director: Prof Bruce Forster

CSIRO Earth Observation Centre
PO Box 3023, Canberra ACT 2601
Tel: (+61 6) 216 72 00
Fax: (+61 6) 216 72 22
e-mail: murray.wilson@eoc.csiro.au

Earth Resource Mapping Pty Ltd
87 Colin Street, Level 2,
West Perth, Western Australia 6005
Tel: (+61 9) 388 29 00
Fax: (+61 9) 388 29 01
e-mail: firstname@erm.oz.au
Managing Director: Stuart Nixon
Corporate Marketing Manager: Heather Aquilina

Geoimage Pty Ltd
13/180 Moggill Road, Taringa, Queensland 4068
Tel: (+61 7) 38 71 00 88
Fax: (+61 7) 38 71 00 42
WWW: http://www.geoimage.com.au
Contacts: Bob Walker and Sylvia Michael

Ground Floor, CML Building, 59 Smith Street,
Darwin, Northern Territory 0800
Tel: (+61 8) 89 41 36 77
Fax: (+61 8) 89 41 36 70
Contact: Bernie Fitzpatrick

Land Information Centre
Department of Conservation & Land Management
Panorama Avenue, PO Box 143
Bathurst, New South Wales 2795
Tel: (+61 63) 32 84 19
Fax: (+61 63) 31 80 95

Moonraker Technology Pty Ltd
Tasmania Technopark, Dowsing Point, Tasmania 7010
Tel: (+61 3) 62 73 15 33
Fax: (+61 3) 62 73 17 49
Managing Director: C W Edwards
Engineering: J Serpanos
Marketing Manager: Geraldine Edwards

Remote Sensing Applications Centre
Department of Land Administration (DOLA)
Leeuwin Centre for Earth Sensing Technologies,
65 Brockway Road, Floreat, Western Australia 6014
Postal: PO Box 471, Wembley, Western Australia 6014
Tel: (+61 9) 340 93 30
Fax: (+61 9) 383 71 42
Manager: Dr Richard Smith

Spot Imaging Services Pty Ltd
PO Box 197, 156 Pacific Highway,
St Leonards, New South Wales 2065
Tel: (+61 2) 99 06 17 33
Fax: (+61 2) 99 06 51 09
Managing Director: Carl McMaster
Business Development: Damian Carroll

AUSTRIA

Geospace
Beckel-Satellitenbilddaten GmbH
Jakob-Haringer-Str 1, A-5020 Salzburg
Tel: (+43 662) 45 81 15
Fax: (+43 662) 458 11 54
e-mail: beckel@geospace.co.at
WWW: http://geospace.co.at
Director: Dr Lothar Beckel

BELGIUM

CLEO
Leuvensesteenweg 510, Bux 45, B-1930 Zaventem
Tel: (+32 2) 773 94 44
Fax: (+32 3) 773 79 10
Manager: Win Van Diest

Eurosense
Eurosense Belfotop NV
Nerviërslaan 54, B-1780 Wemmel
Tel: (+32 2) 460 70 00
Telex: 26687
Fax: (+32 2) 460 49 58
Managing Director: Emile Maes

Eurosense BV
PO Box 4923, NL-4803 EX Breda, The Netherlands
Tel: (+31 76) 565 88 50
Fax: (+31 76) 565 59 99
Deputy Managing Director: Fred A Hagman

Eurosense SA
8 rue du Mal de Lattre de Tassigny, F-59800, Lille, France
Tel: (+33 3) 20 57 30 27
Fax: (+33 3) 20 54 03 24
Deputy Managing Director: Didier Mendel

Eurosense GmbH
Unnauer Weg 17, D-50767 Cologne, Germany
Tel: (+49 221) 79 80 31
Fax: (+49 221) 79 66 88
Deputy Managing Director: Hubert Minten

Eurosense GmbH
Markgrafendamm 24, D-10245 Berlin, Germany
Tel: (+49 30) 23 24 68 17
Fax: (+49 30) 23 14 68 26

Eurosense SRO
Nevädzová 5, SK-82101 Bratislava, Slovakia
Tel: (+421 7) 57 87
Fax: (+421 7) 57 87
Deputy Managing Director: Robert Barca

Eurosense Kft
Ulloi ut 200, H-1191 Budapest, Hungary
Tel: (+36 1) 282 20 19
Fax: (+36 1) 282 95 74
Deputy Managing Director: Miklós Gross

Eurosense Sp z o.o
ul Lipowa 27/2, Pl-05-830 Nadarzyn, Poland
Tel: (+48 90) 20 73 56
Fax: (+48 22) 729 86 56
Deputy Managing Director: Boguslav Okupny

Eurosense Beijing Ltd
WanQuan Land House 5, ZhiXing DongLu Haidian, Beijing 100083, People's Republic of China
Tel: (+86 10) 62 05 44 99
Fax: (+86 10) 62 05 44 99

Walphot SA
Rue van Opré 97, B-5100 Namur (Jambes)
Tel: (+32 81) 30 24 01
Fax: (+32 81) 30 41 67
Director: A Jadot
Project Manager: Th Engels

CANADA

Alliance For Marine Remote Sensing
Sun Tower, Suite 620, 1550 Bedford Highway
Bedford, Nova Scotia B4A 1E6
Tel: (+1 902) 835 22 09
Fax: (+1 902) 835 20 70
e-mail: brian@orbit.acrso.ns.ca
WWW: http://www.acrso.ns.ca
Executive Director: Brian Whitehouse

Array Systems Computing Inc
401 Magnetic Drive, Suite 24,
Downsview, Ontario M3J 3H9
Tel: (+1 416) 736 09 00
Fax: (+1 416) 736 47 15
President: Stuart Berkowitz
Product Manager: Robert Bruce

Barringer Research Ltd
1730 Aimco Boulevard, Mississauga, Ontario L4W 1V1
Tel: (+1 905) 238 88 37
Fax: (+1 905) 238 30 18
President: John H Davies
Vice-President: Ludo Daubner
Sales Manager: David Martinak

Canada Centre For Remote Sensing
Geomatics Canada
Department of Energy, Mines and Resources
588 Booth Street, 3rd floor, Ottawa, Ontario K1A 0Y7
Tel: (+1 613) 947 12 14
Fax: (+1 613) 947 13 82
WWW: http://www.ccrs.nrcan.gc.ca
Director General: Edryd Shaw
Director, Applications Division: Florian Guertin
Director, Methods/Systems Div: Dr Robert O'Neil
Director, Data Acquisition Division: Dr Susan Till
Chief, Industrial Co-operation/ Communication: Dr Robert A Ryerson

Dendron Resource Surveys Inc
880 Lady Ellen Place, Suite 206,
Ottawa, Ontario K1Z 5L9
Tel: (+1 613) 725 29 71
Fax: (+1 613) 715 17 16
e-mail: dendron@dendron.com
President: Dr Udo Nielson
Marketing Manager: Andrew Welch

Eidetic Digital Imaging Ltd
1210 Marin Park Drive,
Brentwood Bay, British Columbia V8M 1G7
Tel: (+1 250) 652 93 26
Fax: (+1 250) 652 52 69
President: Dr Fred G Peet

Gregory Geoscience Ltd
Kanata Square, Suite 504,
260 Hearst Way, Kanata, Ontario K2L 3H1
Tel: (+1 613) 599 74 65
Fax: (+1 613) 592 91 35
President: Harold D Moore
Sales Manager: Elizabeth Moore

Horler Information Inc
9th floor, 275 Slater St, Ottawa, Ontario K1P 5H9
Tel: (+1 613) 233 48 83
Fax: (+1 819) 243 12 85
e-mail: horler@istar.ca
President: Dr David N H Horler

Intermap Technologies Corporation
2 Gurdwara Road, Suite 200, Nepean, Ontario K2E 1A2
Tel: (+1 613) 226 54 42
Fax: (+1 613) 226 55 29
Vice-President GEO Technologies: Garth Lawrence

Macdonald Dettwiler
13800 Commerce Parkway,
Richmond, British Columbia V6V 2J3
Tel: (+1 604) 278 34 11
Fax: (+1 604) 278 12 85
e-mail: sales@mda.ca
CEO: Daniel Friedmann
Vice-President, Space/Defence Business: David Caddey

Provincial Remote Sensing Office
Ministry of Natural Resources,
90 Sheppard Avenue East, 4th Floor,
North York, Ontario M2N 3A1
Tel: (+1 416) 314 13 12
Fax: (+1 416) 314 13 37
Manager: Christopher I Goddard

Radarsat International, Inc
3851 Shell Road, Building D Suite 200,
Richmond, British Columbia V6X 2W2
Tel: +1 (604) 231 50 00
Fax: +1 (604) 231 49 00
e-mail: info@rsi.ca
Corporate WWW: http://www.rsi.ca
Education WWW: http://www.radarsatinaction.com
Client services: Tel: (+1 604) 244 04 00
Fax: (+1 604) 244 04 04
President: Robert E Tack
Director, Operations: John Hornsby
Director, Worldwide Sales: Pierre Engel
Director, Finance: Roland Knight
Director, Market Development: Dennis Nazarenko
Manager, Communications: Cory Aspden
Manager, Client Services: E. Patricia Connor

Ottawa, Ontario, Canada
Tel: (+1 613) 238 54 24
Fax: (+1 613) 238 54 25
Gatineau, Quebec, Canada
Tel: (+1 819) 827 30 01
Fax: (+1 819) 827 19 55
Farnborough, United Kingdom
Tel: (+44 1252) 39 48 84
Fax: (+44 1252) 39 26 59

SATLANTIC
Pier 9, Richmond Terminals,
3295 Barrington St, Halifax, Nova Scotia B3K 5X8
Tel: (+1 902) 492 47 80
Fax: (+1 902) 492 47 81
Vice-President Information Systems Division:
Richard Olsen

Seimac Limited
271 Brownlow Avenue,
Dartmouth, Nova Scotia B3B 1W6
Tel: (+1 902) 468 30 07
Fax: (+1 902) 468 30 09
e-mail: jsteeves@seimac.com
Marketing Director: Jennifer S Steeves

The Bercha Group
1220 Kensington Road NW, Suite 250,
Calgary, Alberta T2N 3P5
Tel: (+1 403) 270 22 21
Telex: 03-827666
Fax: (+1 403) 270 20 14
Managing Director: Dr Frank Bercha
Vice-President: Brian J Griffin

CHINA, PEOPLE'S REPUBLIC

Chinese Academy of Space Technology
PO Box 2417, 31 Bai Shi Qiao Lu, 100081 Beijing
Tel: (+86 1) 837 82 30/94 39
Telex: 22473 ccsc cn
Fax: (+86 1) 837 91 63/82 37
President: Qi Fayen

CZECH REPUBLIC

GISAT
Charkovská 7, CZ-101 00 Prague 10
Tel: (+42 2) 74 83 90/48 34 03
Fax: (+42 2) 74 83 90
e-mail: rsat@gisat.anet.cz
Director: Dr Jan Kolár
Project Manager: Marie Háková
GIS Applications: Tomás Soukup

DENMARK

Kampsax Geoplan
PO Box 1138, Stamholmen 112, DK-2650 Hvidovre
Tel: (+45) 36 39 09 00
Fax: (+45) 36 77 24 21
Managing Director: Peter Normann Hansen
Contact: Søren Buch

ECUADOR

Clirsen
Apartado 17-08-8216,
Edificio del Instituto Geográfico Militar, P.4, Quito
Tel: (+593 2) 52 12 59/52 65 08 (Quito)
Tel: (+593 2) 58 24 47/58 10 64 (Cotopaxi)
Fax: (+593 2) 58 10 66/54 94 72
Executive Director: CRNL TEC ING Filemón Valencia

EGYPT

Eight M Ltd
Environmental & Remote Sensing Services
5 El Ahram Street, Heliopolis, Cairo
PO Box 1041 Heliopolis Bahari
Tel: (+20 2) 67 07 18/291 83 30/291 63 44
Fax: (+20 2) 290 69 16
Executive Vice-President: Dr Hossam Fawzy
Executive Vice President: Ihab EL Ghazzani MSc

FINLAND

Finnish Geodetic Institute
Geodeetinrinne 2, FIN-02430 Masala
Tel: (+358 0) 29 55 50
Fax: (+358 0) 29 55 52 00
Contact: Prof Risto Kuittinen

Finnish Meteorological Institute
PO Box 503, FIN-00101 Helsinki
Tel: (+358 9) 192 91
Telex: 124436 efkl fi
Fax: (+358 9) 19 29 22 03
Director General: Prof Erkki Jatila
Director, Geophysical Research: Prof Risto Pellinen
Director, Operations: Prof Jorma Riissanen
International Relations: Lea Leskinen

Helsinki University of Technology
Space Technology Laboratory
Otakaari 5A, FIN-02150 Espoo
Tel: (+358 0) 451 23 71
Fax: (+358 0) 451 28 98
STL Director: Prof Martti Halikainen

FRANCE

Fleximage
43 rue Brèche aux Loups, F-75012 Paris
Tel: (+33 1) 53 02 99 00
Fax: (+33 1) 43 44 10 05
Managing Director: Dr Francois Louange
Computer-Aided Photo-Interpretation: P Flament
Geographic Databases: P Barbier

GDTA
18 avenue Edouard Belin, F-31055 Toulouse Cedex
Tel: (+33 5) 61 27 42 76
Telex: 531081
Fax: (+33 5) 61 28 14 98
e-mail: daniel@venus.gdta.cnes.fr
Director: Jean-Luc Bessis
Deputy Director: Jean-Pierre Paris
Secretary General: Alain Gusthiot
Director, Education: Marc Bernard

GEOSYS
3 rue Hermès,
Parc Technologique de Toulouse-Ramonville,
F-31526 Ramonville Cedex
Tel: (+33 5) 61 75 12 10
Telex: 530 955 F (box B61)
Fax: (+33 5) 61 75 87 22
CEO: Damien Lepoutre
Commercial Director: Alain Killmayer

ISTAR
Espace Beethoven, route des Lucioles-Sophia Antipolis,
F-06560 Valbonne
Tel: (+33 4) 93 95 72 30
Fax: (+33 4) 93 95 83 29
President: Hervé Le Roux

Matra Cap Systemes
6 rue Dewoitine, BP14, F-78142 Vélizy-Villacoublay
CEDEX
Tel: (+33 1) 34 63 70 00
Fax: (+33 1) 34 63 70 70
General Manager: Bernard Plano
Space Activities: Michel Mariton
Sales Manager: Michel Coustere
Communications Manager: Christian Badé

Meteo-France
1 quai Branly, F-75340 Paris Cedex 07
Tel: (+33 1) 45 56 71 71
Telex: Metfran 202876
Fax: (+33 1) 45 56 70 05
International Relations: Denis Lambergeon

Service Central de la Communication
et de la Commercialisation
Tel: (+33 1) 45 56 71 71
Fax: (+33 1) 45 56 71 11
Head, Communications: Patrick Tchang
Head, Commercial Affairs: Charles Dupuy

Centre de Météorologie Spatiale
SCEM-CMS
BP147, F-22302 Lannion Cedex.
Tel: (+33 2) 96 05 67 00
Telex: 950 256 lannsat
Fax: (+33 2) 96 05 67 37
Head, CMS: André Bellocq

Novespace
15 rue des Halles, F-75001 Paris
Tel: (+33 1) 42 33 41 41
Fax: (+33 1) 40 26 08 60
Chairman/CEO: Jean-Pierre Fouquet
Spacelink contact: Eric Brun

Scot Conseil
1 rue Hermès, Parc du Canal,
F-31526 Ramonville Saint Agne Cedex
Tel: (+33 5) 61 39 46 00
Telex: 531752 argos f
Fax: (+33 5) 61 39 46 10
President/General Manager: Jean-Claude Cazaux

Spot Image SA
5 rue des Satellites, BP4359, F-31030 Toulouse Cedex
Tel: (+33 5) 62 19 40 40
Telex: 532079
Fax: (+33 5) 62 19 40 11
WWW: http://www.spotimage.fr/
Chairman: Jacques Mouysset
General Manager: Philippe Renault
Commercial Director: David Balducci
Production/Technical Services: Philippe Delclaux
Development Director: Philippe Munier

GERMANY

ACCESS eV
Intzestrasse 5, D-52072 Aachen
Tel: (+49 241) 80 67 21
Telex: 832704
Fax: (+49 241) 385 78
Board Chairman: Peter R Sahm
Managing Directors: Robert Guntlin, Stephan Rex

Daimler-Benz Aerospace AG
Space Infrastructure Division
PO Box 10 59 09,
Hunefeldstrasse 1-5, D-28059 Bremen
Tel: (+49 421) 53 91
Telex: 245 548 a erno d
Fax: (+49 421) 539 50 74
Microgravity: Dr Peter Vits
Texus Manager: Dr Berkhard Franke
Marketing: Chris R Gilbert

DLR Microgravity User Support Centre
Linder Höhe, D-51147 Köln
Tel: (+49 2203) 601 21 74
Telex: 8874 410 dfvd
Fax: (+49 2203) 614 71
MUSC Manager: Dr K Wittmann
Life Science Support: D Padeken
Materials Science Support: M Schuber, R Willnecker

Energia Deutschland GmbH
Wolfratshauser Str 44-48, D-81379 Munich
Tel: (+49 89) 72 49 50
Telex: 529749
Fax: (+49 89) 72 49 52 91
Managing Director: Reiner Klett

GAF mbH
Leonrodstrasse 68, D-80636 Munich
Tel: (+49 89) 121 52 80
Telex: 5218713 gafm d
Fax: (+49 89) 123 31 48
Managing Director: Dr Rupert Haydn

INTOSPACE GmbH
Sophienstrasse 6, D-30159 Hannover
Tel: (+49 511) 30 10 90
Telex: 9218233
Fax: (+49 511) 301 09 28
Managing Director: Jürgen K von der Lippe
Commercial Manager: Carsten H Möhlenhof
Head, Materials Science: Dr Heinz J Sprenger

Leybold Durferrit AG
Rückinger Strasse 12, D-6455 Erlensee
Tel: (+49 6181) 34 32 70
Telex: 415206-12 ld d
Fax: (+49 6181) 34 32 99
High Vacuum Furnace Dept: Fernand Kuhlmann

OHB-System GmbH
Universitätsallee 27, D-28359 Bremen
Tel: (+49 421) 202 08
Telex: 245316 ohb
Fax: (+49 421) 202 06 10
Managing Directors: Professor Dipl Ing Manfred Fuchs, Christa Fuchs, Marco Fuchs (Vice-President)
Directors: Wolfgang Wienss, Alfred Tegtmeier, Dr Ingo Engeln, Dr Peter Junk, Dr Reinhardt Schlitt, Dr Bernd Hofer
Head of Satellite Department: Amnon Ginati

VCS Nachrichtentechnik GmbH
Borgmannstrasse 2, D-44894 Bochum
Tel: (+49 234) 925 80
Telex: 825537 vcsbo d
Fax: (+49 234) 925 81 90
Managing Director: Dr Klaus-G Meng
Marketing Manager: Klaus M Heidrich
Manager, Met Systems: Dr Peter Scheidgen
Account Manager, Satellite Ground Systems: Ulli Leibnitz

ZARM
Centre of Applied Space Technology & Microgravity
University of Bremen, Am Fallturm, D-28359 Bremen
Tel: (+49 421) 218 27 52
Telex: 245 811 unibr d
Fax: (+49 421) 218 25 21
e-mail: rath@zarm.uni-bremen.de
Director: Professor Dr Ing Hans J Rath
Microgravity Applications: Dr Hendrik C Kuhlmann, Dr Michael Dreyer
Research Systems: Dr Hansjörg Dittus
Space Technology: Dipl Ing Hans J Königsmann, Dipl Ing Holger W Oelze
Hypersonic Technology: Dipl Ing Holger W Oelze, Dr Ing Georg Koppenwallner
Drop Tower Operations: Dipl Ing Christian Eigenbrod

HUNGARY

Remote Sensing Centre Of The Institute For Geodesy, Cartography & Remote Sensing (FÖMI RSC)
PO Box 546, H-1373 Hungary
Tel: (+36 1) 163 66 69
Fax: (+36 1) 252 79 18
e-mail: h3874win@ella.hu
Head: Péter Winkler

INDIA

Indian Institute Of Remote Sensing
No 4 Kalidas Road, Dehra Dun 248 001
Tel: (+91 135) 245 83
Telex: 0595-224
Head: S K Bhan

National Remote Sensing Agency
Balanagar, Hyderabad 500 037
Tel: (+91 842) 27 95 72 77
Telex: 0425-6522
Fax: (+91 842) 27 86 48
Director: Professor B L Deekshatulu

Regional Remote Sensing Services Centres/Natural Resources Management System
ISRO, Department of Space,
Antariksh Bhavan, New BEL Rd, Bangalore 560 094
Tel: (+91 80) 33 26 77
Telex: 0845-8867
Fax: (+91 80) 33 42 29
Director: K Radhakrishna

Speck Systems PVT Ltd
B-49 Electronics Complex,
Kushaiguda, Hyderabad 500 762
Tel: (+91 40) 62 18 12/62 53 05
Telex: 0425-7065 spek in
Fax: (+91 40) 62 07 57
Directors: N Chandrasekhar, KCM Kumar

INDONESIA

PT Indica Dharma Consulting Services
Golden Plaza Blok 43-44, Fatmawati no 15, Jakarta
Tel: (+62 21) 750 89 86; 63 61
Fax: (+62 21) 750 89 85/63 50
e-mail: indica@server.indo.net.id
Director: Mohammed Isnaini

INTERNATIONAL

CLS Argos
18 avenue Edouard Belin,
F-31055 Toulouse Cedex, France
Tel: (+33) 61 39 47 00
Telex: 531752
Fax: (+33) 61 75 10 14
Chairman: Michel Cazenave
General Manager: Michel Taillade
Head, Data Processing: Philippe Schwab
Marketing Contact: Antoine Monsaingeon

Service Argos Inc (SA Inc)
1801 McCormick Drive, Suite 10,
Landover, Maryland 20785, USA
Tel: (+1 301) 925 44 11
Telex: 898146
Fax: (+1 301) 925 89 95
WWW: http://www.argosinc.com
President/CEO, Service Argos Inc: Archie Shaw
Vice-President: Jeff Wingenroth

North American CLS (NACLS)
9200 Basil Court, Suite 306,
Landover, Maryland 20785, USA
Tel: (+1 301) 341 18 14
Fax: (+1 301) 341 21 30
CEO: Christophe Vassal
Business Development Manager: Peter Griffith
North American CLS Contact: Peter Griffith

CLS Argos Australasia
Satellite Information Technology
c/o Bureau of Meteorology
GPO Box 1289 K, 150 Lonsdale Street,
Melbourne, Victoria 3001, Australia
Tel: (+61 3) 669 46 50
Telex: aa 30434
Fax: (+61 3) 669 46 75
Director: Shlomit Shachori

CLS Argos Japan
Cubic I Limited
2-9-7-312 Nishi Gotanda,
Shinagawa Ku, Tokyo 141, Japan
Tel: (+81 3) 37 79 55 06
Fax: (+81 3) 37 79 57 83
President: Takashi Shibata

Committee On Earth Observation Satellites
NASA/NOAA
300 E Street SW, Washington DC 20546, USA
Tel: (+1 202) 358 07 93
Fax: (+1 202) 358 27 98
e-mail: lshaffer@codei.hq.nasa.gov
Secretariat: Dr Lisa Shaffer

European Space Agency
8-10 rue Mario-Nikis, F-75738 Paris Cedex 15, France
Tel: (+33 1) 53 69 71 31
Fax: (+33 1) 53 69 75 60
Secretariat: Dr Huw Hopkins

NASDA/STA
2-2-1 Kasumigaseki, Chiyoda-ku, Tokyo 100, Japan
Tel: (+81 3) 54 01 86 90
Fax: (+81 3) 54 01 87 02
Secretariat: Mr Chu Ishida

Eumetstat GmbH
Am Kavalleriesand 31, D-64294 Darmstadt, Germany
Tel: (+49 6151) 80 77
Fax: (+49 6151) 80 75 55
Director General: Dr Tillmann Mohr
Council Chairman: Dr Jorma Riissanen
Head, Technical Department: Alain Datiev
Head, Administration Department: Dr Volker Thiem
Geostationary Prog Manager: Dr Anthony Harrison
EPS Programme Manager: Michel Langevin

Eurimage S.c.r.l.
Via Edoardo D'Onofrio 212, I-00155 Rome, Italy
Tel: (+39 6) 40 69 41
Fax: (+39 6) 40 69 42 32
WWW: http://www.eurimage.it
Chairman: Lars Bjerkesjö (SSC Satellitbild)
Deputy Chairman: Stephen La Pensee (NRSC)
Director: Alfred Setzer (Dornier)
Managing Director: Marcello Maranesi (Telespazio)
Marketing Director: Fabrizio Lombardi
ERS Customer Service:
Tel: (+39 6) 94 18 07 41
Fax: (+39 6) 94 18 06 52
Multimission Customer Service:
Tel: (+39 6) 94 18 07 52
Fax: (+39 6) 94 16 77 2

European Association of Remote Sensing Companies
PO Box 66, B-1020 Bruxelles 2, Belgium
Tel: (+44 1705) 67 44 68
Chairman: F Geoff Sawyer

Resurs-F Worldmap Joint Venture
JEBCO Information Services
22 Armoury Way, London SW18 1EZ, UK
Tel: (+44 181) 874 01 23
Fax: (+44 181) 877 99 81
Marketing Manager: Mark Paton

10260 Westheimer, Suite 400,
Houston, Texas 77042, USA
Tel: (+1 713) 975 02 02
Fax: (+1 713) 975 92 93
Vice-President: Grant Lichtman

State Centre PRIRODA
45 Volgogradski Prospekt, 109125 Moscow, Russia
Tel: (+7 095) 177 80 53
Fax: (+7 095) 177 65 74
e-mail: ADM@PRIRODA.MSK.SU
Deputy Director: Anatoli Rubakha
Operations Manager: Yuri Karionov

Small State Enterprise ROSVNESHGEO
Novocheryomushkinskaya 69B,
117418 Moscow, Russia
Tel: (+7 095) 254 80 74
Fax: (+7 095) 943 00 13
Director: Yevgeni A Popov

ISRAEL

Interdisciplinary Centre For Technological Analysis And Forecasting (ICTAF)
Tel-Aviv University, Ramat Aviv, Tel Aviv 69978
Tel: (+972 3) 641 28 82
Fax: (+972 3) 641 01 93
Director: Professor Mordechai Sokolov
Head, Space & Remote Sensing Division: Abraham Tal

ITALY

Mars Centre
Via Comunale Tavernola-Capodichino, I-80144 Napoli
Tel: (+39 81) 234 45 80
Fax: (+39 81) 234 71 00
President: Professor Rodolfo Monti
General Manager: Dr Francesco Laccetti

JAPAN

Institute For Unmanned Space Experiment Free Flyer (USEF)
Shinko Building,
2-12 Ogawamachi, Kanda, Chiyoda-ku, Tokyo 101
Tel: (+81 3) 32 94 48 34
Fax: (+81 3) 32 94 71 63
President: Takashi Kitaoka
Executive Managing Director: Hirotaka Kawamoto
Managing Director: Noboru Wakasugi

Earth Observation Centre
1401 Ohashi, Hatoyama-machi,
Hikigun, Saitama 350-03
Tel: (+81 492) 98 12 00
Telex: 2952841
Fax: (+81 492) 96 02 17
Director: Hideo Sato

Japan Meteorological Agency
3-4 Ote-machi 1-chome, Chiyoda-ku, Tokyo 100
Tel: (+81 3) 32 12 83 41
Telex: 02222163 mettok j
Fax: (+81 3) 32 11 20 32
Director General: T Nitta
Deputy Director: S Mochizuki
Head Aviation Weather Management Division: T Furukawa
Senior Assistant for International Affairs: Koichi Nagasaka

Meteorological Satellite Centre
3-235 Nakakiyoto, Kiyose, Tokyo 204
Tel: (+81 424) 93 11 11
Fax: (+81 424) 92 24 33

Japan Microgravity Centre
Tokyo Tatemono Sapporo Building, 7F,
20 Nishi 2-chome, Kita 7-Jo, Kita-ku,
Sapporo 060, Hokkaido
Tel: (+81 11) 757 71 11
Fax: (+81 11) 757 77 11
President/Director: Takeyoshi Morihana
Vice-President: Tetsuhiro Nakanomori
Senior Managing Director: Shoji Sakai

JAMIC Tokyo Office
3-30-16 Nishiwaseda, Shinjuku-ku, Tokyo 169
Tel: (+81 3) 52 72 72 76
Fax: (+81 3) 52 73 06 07
Sales Manager: Masaaki Hirota

Japan Space Utilisation Promotion Centre (JSUP)
3-30-16 Nishiwaseda, Shinjuku-ku, Tokyo 169
Tel: (+81 3) 52 73 24 41
Fax: (+81 3) 52 73 98 47
President: Tadahiro Sekimoto
Vice-Presidents: S Suzuki, K Inaba, Y Iida, K Imai
Executive Director: Yasuhiro Morishita
Director, Survey/Research Dept: Ryuichi Okumura

Nissho Iwai Corporation
4-5 Akasaka 2-chome, Minato-ku, Tokyo 107
Tel: (+81 3) 35 88 26 24
Telex: 22233 nictk aj
Fax: (+81 3) 35 88 47 40
Associate Director; GM Space Project Office; Landsat Data Sales: Takeshi Miyazaki
Spot Data Sales: Toshiaki Futagawa

RESTEC
Roppongi First Building, 12F,
1-9-9 Roppongi, Minato-ku, Tokyo 106
Tel: (+81 3) 55 61 97 71
Telex: 02-426780 restec j
Fax: (+81 3) 55 61 95 40
President: Hiroyuki Osawa
Executive Senior Managing Director: Yoshio Yorimizu
Executive Directors:
Yukio Mukai, Takeshi Osugi, Tetsuo Sakashita

Space Technology Corporation
c/o IHI Space Development Division, Mizuho Plant
229 Tonogaya, Mizuho-machi,
Nishitama-gun, Tokyo 190-12
Tel: (+81 425) 68 71 99
Fax: (+81 425) 68 75 75
Counsellor: Kazuo Ishida

KENYA

Teleos International BV
c/o RCSSMRS, PO Box 18118, Rasarani Road, Nairobi
Tel: (+254 2) 86 16 73
Fax: (+254 2) 86 16 74
Director, East Africa Ops: Ken Ferguson

KOREA, SOUTH

Chang Woo, Inc
Yeo Eui Do, PO Box 747, Seoul
Tel: (+82 2) 782 90 56
Fax: (+82 2) 782 90 58
Director, Systems Division: Hae Uk Chung
Contact: Jong Hun Kim

LIBYA

Libyan Centre For Remote Sensing & Space Science (LCRSSS)
Secretariat of Scientific Research,
PO Box 397, Tripoli
Fax: (+218 21) 90 00 71

MOROCCO

Centre Royal De Télédétection Spatiale Du Maroc (CRTS)
16 bis avenue de France, Agdal, Rabat
Tel: (+212 7) 77 63 05
Fax: (+212 7) 77 63 00
Director: Dr Mouslim Kabbaj

NETHERLANDS

Bradford Engineering
Bradfordstraat 1-2, NL-4645 HE Putte
Postal: Postbus 120, NL-4645 ZJ Putte
Tel: (+31 164) 60 35 50
Fax: (+31 164) 60 21 77
Director: Eduard C J Voeten
Contact for microgravity: Raoul G H M Voeten

Environmental Analysis And Remote Sensing BV (EARS)
Kanaalweg 1, NL-2628 EB Delft
Postal: Postbus 449, NL-2600 AK Delft
Tel: (+31 15) 56 24 04
Telex: 26401
Fax: (+31 15) 62 38 57
General Manager: Ir J H Bijleveld
Remote Sensing: Ir A Rosema

Netherlands Remote Sensing Board
Kanaalweg 3b, PO Box 5023, NL-2600 GA Delft
Tel: (+31 15) 269 11 11
Fax: (+31 15) 261 89 62
e-mail: p.b.bcrs@mdi.rws.minvenw.nl
Head, Programme Bureau: Dr N J J Bunnik

NEW ZEALAND

DSIR Division Of Physical Sciences
Information Technology Group
PO Box 31311, Lower Hutt
Tel: (+64 4) 66 69 19
Fax: (+64 4) 69 00 67
Group Manager: Mr E A King

Manaaki Whenua Landcare Research New Zealand Ltd
Private Bag 11 052, Palmerston North
Tel: (+64 6) 356 71 54
Fax: (+64 6) 355 92 30
e-mail: stephensp@landcare.cri.nz
WWW: http://www.landcare.cri.nz
Contact: Peter Stephens

Meteorological Service Of New Zealand Ltd
30 Salamanca Road, PO Box 722, Wellington 1
Tel: (+64 4) 472 93 79
Fax: (+64 4) 473 52 31
Chief Executive: John Lumsden
General Manager, National Weather Services: Dr Neil Gordon
Marketing Manager: Tom Sutherland

Sirtrack Limited
Private Bag 1403, Havelock North
Tel: (+64 6) 877 77 36
Fax: (+64 6) 877 54 22
Manager: Dave Ward

NORWAY

Norut Information Technology Ltd
N-9005 Tromsø
Tel: (+47) 77 62 94 00
Fax: (+47) 77 62 94 01
Managing Director: Arne-Wilhelm Theodorsen
Information Officer: Sigurd Sjursen

Norwegian Meteorological Institute
PO Box 43 Blindern, N-0313 Oslo 3
Tel: (+47) 22 60 30 00
Telex: 21564
Fax: (+47) 22 96 30 50

POLAND

Military Institute Of Aviation Medicine
Krasinskiego 54, PL-01-755 Warsaw
Tel: (+48 22) 633 41 54
Telex: 817 135
Fax: (+48 22) 633 70 68
Director: Professor Stanislaw Baranski

Institute of Geodesy & Cartography
Jasna 2/4, PL-00-950 Warsaw
Tel: (+48 22) 27 03 28
Telex: 812 355
Fax: (+48 22) 27 03 28
Director: Dr Adam Linsenbarth

Institute Of Meteorology And Water Management
Podlesna 61, PL-01-673 Warsaw
Tel: (+48 22) 34 16 51
Telex: 814 331
Fax: (+48 22) 34 54 66
Director: Professor Jan Zielinski

Satellite Data Receiving/Processing Centre
Ul. Borowego 14, PL-30-215 Krakow
Tel: (+48 12) 11 38 44; 11 67 98
Telex: 325514 hm pl
Fax: (+48 12) 11 67 98

SINGAPORE

Centre For Remote Imaging, Sensing And Processing (CRISP)
National University of Singapore,
Faculty of Science, Lower Kent Ridge Road,
Singapore 119260
Tel: (+65 772) 32 20
Fax: (65 775) 77 17
e-mail: crisp@nus.sg
WWW: http://www.crisp.nus.sg

Spot Asia Pte Ltd
73 Amoy Street, Singapore 069892
Tel: (+65 227) 55 82/74 95
Fax: (+65 227) 62 31
General Manager: Yves Béchacq
Sales Manager: Christer Colliander
Sales Executive: Weng Kee Foo

SOUTH AFRICA

CSIR
PO Box 395, Pretoria 0001
Tel: (+27 12) 841 29 11
Fax: (+27 12) 349 11 53
President: Dr Geoff Garrett

Satellite Applications Centre
PO Box 395, Pretoria 0001
Tel: (+27 12) 326 47 00
Fax: (+27 11) 642 24 46
WWW: http://www.sac.co.za
Programme Manager: Renier Balt
Business Development: Piet van der Westhuizen
e-mail: piet.vdwesthuizen@csir.co.za

SPAIN

Aurensa
San Francisco de Sales 38, 1 Izq, E-28003 Madrid
Tel: (+34 1) 553 38 65
Fax: (+34 1) 554 47 80
Managing Director: Recaredo del Potro

Ibersat SA
Araquil 11, E-28023 Madrid
Tel: (+34 1) 357 18 60
Fax: (+34 1) 357 31 92
e-mail: ibersat@box.eunet.es
President: Alfredo Sánchez de Zavala
General Manager: Carlos Sánchez Irle
Director, New Business: Bernard Denore

SWEDEN

Centre For Environmental Satellite Data
c/o Swedish Space Corp,
PO Box 4207, SE-171 04 Solna
Tel: (+46 8) 627 62 00
Fax: (+46 8) 98 70 69
Chairman: Bengt Hultqvist
ETC/LC Programme Manager: Ulf von Sydow

SSC Satellitbild
PO Box 816, SE-981 28 Kiruna
Tel: (+46 980) 121 40
Telex: 8761 sbild s
Fax: (+46 980) 160 44
Managing Director: Lars Bjerkesjö
Marketing Manager: Jörgen Hartnor
Production Manager: Jörgen Forsgren
Administration Manager: Martin Hedlin

TAIWAN

Centre For Space And Remote Sensing Research
National Central University
Chung-Li, Taiwan 320
Tel: (+886 3) 425 72 32
Fax: (+886 3) 425 55 35

THAILAND

Thailand Remote Sensing Centre
196 Phahonyothin Road, Chatuchak, Bangkok 10900
Tel: (+66 2) 57 91 37 09/03 45
Telex: 82213 narecou th
Fax: (+66 2) 561 30 35/326 91 54
Deputy Secretary-General for Natural Sciences: Dr Suvit Vibulsresth
Director: Dr Paibul Ruangsiri
Head, Applications: Dr Darasri Dowreang
Head, Earth Station: Nares Chamboonrot
Head, User Service: Mrs Praneet Ditsariyakul

TURKEY

ISLEM Group
Kader Sok. 11/3 Gaziosmanpasa, TR-06700 Ankara
Tel: (+90 312) 468 08 30
Fax: (+90 312) 467 41 48
WWW: http://www.islem.com.tr
President: Dr Halil Sogut
General Manager: Mehmet Tankut

ISLEM Information Systems GmbH
Flughafen Strasse 15, D-1000 Berlin 44, Germany
Tel: (+49 30) 624 92 01
Fax: (+49 30) 623 89 02

UNITED KINGDOM

British Association Of Remote Sensing Companies
PO Box 85, Farnborough, Hampshire GU14 6TQ
Chairman: Gordon Oswald
Tel: (+44 1223) 42 00 24
Executive Secretary: Keith Muirhead
Tel: (+44 181) 843 24 11

Data Sciences (UK) Ltd
Pinehurst House, Farnborough Road,
Farnborough, Hampshire GU14 7NB
Tel: (+44 1252) 54 43 21
Telex: 858228
Fax: (+44 1252) 51 37 39/54 67 12
Business Development Manager: John Palfreyman

Dundee Centre for Coastal Zone Research
Dundee University, Dundee DD1 4HN
Tel: (+44 1382) 231 81
Fax: (+44 1382) 20 28 30
Contact: Dr Robert Vaughan

Cray Systems Space & Defence Division,
DAS House, Bristol, Avon BS1 6NH
Tel: (+44 117) 927 78 54
Fax: (+44 117) 929 09 17
Company Directors: Professor Arthur Cracknell, Dr Dave Sloggett

Earth Observation Sciences Ltd
Broadmede, Farnham Business Park,
Farnham, Surrey GU9 8QT
Tel: (+44 1252) 72 14 44
Fax: (+44 1252) 71 25 52
e-mail: paulh@eos.co.uk
Managing Director: Dr Paul Howard
Business Operations Director: Dr Neil McIntyre

Earth Resource Mapping
Blenheim House, Crabtree Office Village
Eversley Way, Egham, Surrey TW20 8RY
Tel: (+44 1784) 43 06 91
Fax: (+44 1784) 43 06 92
Director: Jason Reeve

Feedback Instruments Ltd
Park Road, Crowborough, East Sussex TN6 2QR
Tel: (+44 1892) 65 33 22
Telex: 95255
Fax: (+44 1892) 66 37 19
Chairman/Managing Director: R F Barnett
Weather Satellite Systems (Sales): C S Jacobson

Hunting Technical Services Ltd
Thamesfield House, Boundary Way,
Hemel Hempstead, Hertfordshire HP2 7SR
Tel: (+44 1442) 23 18 00
Fax: (+44 1442) 21 98 86
Executive Chairman: R Hugh Hunting
Managing Director: D H Potten
Director, Remote Sensing: Graham C Deane
Associate Director: Michael Barr

National Remote Sensing Centre Ltd
Delta House, Southwood Crescent,
Southwood, Farnborough, Hampshire GU14 0NL
Tel: (+44 1252) 54 14 64
Fax: (+44 1252) 37 50 16
e-mail: data-services@nrsc.co.uk
WWW: http://www.nrsc.co.uk
Managing Director: David Bartlett
Marketing Director: Tristram Cary

Nigel Press Associates Ltd
1 Fircroft Way, Edenbridge, Kent TN8 6HS
Tel: (+44 1732) 86 50 23
Fax: (+44 1732) 86 65 21
Chairman: Nigel Press

Space Technology Systems Ltd
Amery House, Steeple Drive,
Alton, Hampshire GU34 1TN
Tel: (+44 1420) 886 83
Telex: 858028 metsat g
Fax: (+44 1420) 891 90
e-mail: 101473.2250@compuserve.com
Managing Director: Dewi Evans

The Meteorological Office
Remote Sensing Branch
Building Y70, Defence Research Agency,
Farnborough, Hampshire GU14 6TD
Tel: (+44 1252) 39 57 18
Fax: (+44 1252) 51 55 23
WWW: http://www.meto.gov.uk
Chief Executive: Professor Julian C R Hunt
Director (Observations): Dr Jim Caughey
Head, Remote Sensing: Dr Peter D Curtis
Manager, Metstar Consultants: Dr Malcolm Hough
Fax: (+44 1344) 85 40 24

Timestep Weather Systems Ltd
PO Box 2001, Newmarket, Suffolk CB8 8XB
Tel: (+44 1440) 82 00 40
Fax: (+44 1440) 82 02 81
Managing Director: David J Cawley
Software Consultant: Peter M Arnold

University of Dundee
Satellite Station, Dundee DD1 4HN
Tel: (+44 1382) 34 44 06
Fax: (+44 1382) 20 25 75
e-mail: peb@uk.nerc-dundee.unixa
Director: Peter E Baylis

UNITED STATES OF AMERICA

Bioserve Space Technologies
Campus Box 429,
Department of Aerospace Engineering Sciences,
University of Colorado, Boulder, Colorado 80309
Tel: (+1 303) 492 10 05
Fax: (+1 303) 492 88 83
Center Director: Dr Marvin W Luttges
Associate Director Technical Affairs: Dr Louis Stodieck
Associate Director Engineering: Dr Michael Robinson
Associate Director External Affairs: John Berryman

Aerojet
Electronic Systems Plant
1100 W Hollyvale Street,
Azusa, California 91702-0296
Tel: (+1 818) 812 80 00
Fax: (+1 818) 969 90 10
Vice-President: Walter Chin

Alden Electronics, Inc
40 Washington Street,
Westborough, Massachusetts 01581-0500
Tel: (+1 508) 366 88 51
Fax: (+1 508) 898 24 27
e-mail: info@alden.com
WWW: http://www.alden.com
President/CEO: Arnold A Kraft

Center For Cell Research
117 Research Office Building,
The Pennsylvania State University,
University Park, Pennsylvania 16802
Tel: (+1 814) 865 24 07
Fax: (+1 814) 865 24 13
Director: Dr Wesley C Hymer
Associate Directors: Dr William Kraemer, Dr Daniel Deaver
Program Manager: Joseph T Baker

Center For Macromolecular Crystallography
Box 79 THT, UAB Station,
University of Alabama at Birmingham,
Alabama 35294-0005
Tel: (+1 205) 934 53 29
Fax: (+1 205) 934 04 80
WWW: http://www.cmc.uab.edu
Director: Dr Lawrence J DeLucas

Center For Mapping
The Ohio State University
1216 Kinnear Road, Columbus, Ohio 43212
Tel: (+1 614) 292 16 12
Fax: (+1 614) 292 80 62
Director: Dr John Bossler
Associate Director: Frank Jelinek

Consortium For Materials Development In Space
University of Alabama in Huntsville,
Research Institute, Room M-65,
301 Sparkman Drive,
Huntsville, Alabama 35899
Tel: (+1 205) 895 66 20 (public affairs 890 64 14)
Fax: (+1 205) 895 67 91
Director: Dr Charles Lundquist
Associate Director: Dr Francis Wessling
Public Affairs: Ray Garner

Earth Data Analysis Center
2500 Yale Boulevard SE, Suite 100
University of New Mexico,
Albuquerque, New Mexico 87131-6031
Tel: (+1 505) 277 36 22
Fax: (+1 505) 277 36 14
e-mail: edac@spock.unm.edu
Director: Dr Stanley A Morain
Associate Director: Michael Inglis

Earth Resource Mapping Pty Ltd
4370 La Jolla Village Drive, Suite 900,
San Diego, California 92122-1253
Tel: (+1 619) 558 47 09
Fax: (+1 619) 558 26 57
President: Andrew Nickerson

Earth Satellite Corporation
6011 Executive Boulevard, Suite 400,
Rockville, Maryland 20851
Tel: (+1 301) 231 06 60
Fax: (+1 301) 231 50 20
e-mail: rmitchel@earthsat.com
President/COO: Leonard Jaffe
Vice-President Program Development: Roger Mitchell

EarthWatch, Inc
1900 Pike Road, Longmont, Colorado 80501
Tel: (+1 303) 682 38 00/440 03 74
Fax: (+1 303) 440 03 10
President/COO: Douglas Gerull
CEO: Richard N Herring
Vice-President Government Operations: James Frelk

EOSAT
4300 Forbes Boulevard,
Lanham, Maryland 20706-9954
Tel: (+1 301) 552 05 00
Telex: 277685 lsat ur
Fax: (+1 301) 552 05 07
WWW: http://www.eosat.com
President/CEO: Dr Arturo Silvestrini

Executive Vice-President: David T Edwards
Vice-President Programs: Allen H Watkins
Chief Scientist: Dr David Fischel
Director, NASA Programs: Jim Love
Director, Strategic Marketing: Dr Tina Cary
Director, Business Development and Applications: Terry Lehman
Director, Marketing: Timothy Puckorius
Director, International Sales: Shawana P Johnson
Director, External Affairs: Richard Mroczynski
Director, Public Affairs: Richard Keating

Via E D'Onofrio 212, I-00155 Rome, Italy
Tel: (+39 6) 40 69 42 20
Fax: (+39 6) 40 69 42 22

Eros Data Center
Mundt Federal Building,
Sioux Falls, South Dakota 57198
Tel: (+1 605) 594 65 11
Fax: (+1 605) 594 65 89
Chief: Donald T Lauer
Chief, Satellite Data Systems Office: R J Thompson

Instrumentation Technology Associates, Inc
Administration Office:
Box 871, Exton, Pennsylvania 19341
Tel: (+1 610) 524 19 88
President: John M Cassanto
Systems Engineering Manager: Ulises Alvarado
Project Engineer: William Dunfee

Integration Facility:
35 E Uwchlan Avenue, Suite 300,
Exton, Pennsylvania 19341
Tel: (+1 610) 363 83 43
Fax: (+1 610) 363 85 69

Intergraph Corporation
Huntsville, Alabama 35894-0001
Tel: (+1 205) 730 20 00
Fax: (+1 205) 730 67 50
Marketing Representative: Michael Baker
Media Relations: David Joffrion

Lockheed Martin IR Imaging Systems
2 Forbes Road,
Lexington, Massachusetts 02173-7393
Tel: (+1 617) 862 62 22
Fax: (+1 617) 863 34 96
President: John S Dehne
Vice-President Marketing/Contracts: P A Gaita
Manager, Marketing Comms: David W Plummer

Microwave Telemetry, Inc
10280 Old Columbia Road, Suite 260,
Columbia, Maryland 21046
Tel: (+1 410) 290 86 72
Fax: (+1 410) 290 88 47
President: Dr Paul W Howey

NASA Goddard Space Flight Center
Shuttle Small Payloads Project,
Special Payloads Division, Code 740.3,
Greenbelt, Maryland 20771
Tel: (+1 301) 286 87 99/67 60
Fax: (+1 301) 286 16 94
e-mail: sssp@sssp.gsfc.nasa.gov
Acting Project Manager, Shuttle Small Payloads Project: S Chris Dunker
GAS Mission Manager: Susan Olden
Customer Support: Fran Mosier

NASA Stennis Space Center
Mississippi 39529
Tel: (+1 601) 688 33 41
Fax: (+1 601) 688 10 94
Director: Roy S Estess

ITD Space Remote Sensing Center
Building 1103, Suite 118
Tel: (+1 601) 688 25 09
Fax: (+1 601) 688 28 61
Director: Dr George May

National Oceanic & Atmospheric Administration
Herbert C Hoover Building, Room 5128,
14th & Constitution Avenue NW,
Washington DC 20230.
Tel: (+1 202) 482 34 36
Fax: (+1 202) 408 96 74

Department of Commerce
Under Secretary for Oceans & Atmosphere: D James Baker
Deputy Under Secretary: Diana Josephson

National Environmental Satellite,
Data, and Information Service
Federal Office Building #4, Room 2069,
Washington DC 20233
Tel: (+1 301) 457 51 15
Fax: (+1 301) 457 52 76
Assistant Administrator for Satellite and Information Services: Robert S Winokur
Director, Office of Systems Dev: W John Hussey
Director, Office of Satellite Data Processing and Distribution: Helen M Wood
Director, Satellite Operations: Gary Davis
GOES Program Manager: Gerald J Dittberner
POES Program Manager: Gregory Mandt

National Snow & Ice Data Center
University of Colorado,
Boulder, Colorado 80309-0449
Tel: (+1 303) 492 51 71
Fax: (+1 303) 492 24 68
e-mail: nsidc@kryos.colorado.edu
WWW: http://www-nsidc.colorado.edu
Director: Dr Roger Barry

OrbImage / Orbital Imaging Corporation
21700 Atlantic Boulevard, Dulles, Virginia 20166
Tel: (+1 703) 406 54 36
Fax: (+1 703) 406 55 52
President: Gilbert D Rye
Director, Business Development: Mark Pastrone

Payload Systems Inc
247 Third Street, Cambridge, Massachusetts 02141
Tel: (+1 617) 868 80 86
Telex: 650 295 7096 mciuw
Fax: (+1 617) 868 66 82
WWW: http://www.payload.com
President: Dr Javier de Luis
Principals: Vinit Nijhawan, Dr Anthony Arrott

Santa Barbara Remote Sensing
75 Coromar Drive, Goleta, California 93117
Tel: (+1 805) 562 72 00
Fax: (+1 805) 562 77 67
SBRS General Manager: Arthur F Napolitano
SBRS Business Development: Dominick M DellaValle

Seaspace Corporation
9240 Trade Place, #100,
San Diego, California 92126-6314
Tel: (+1 619) 578 40 10
Fax: (+1 619) 578 36 25
e-mail: info@seaspace.com
WWW: http://www.seaspace.com/

Space Environment Center
Environmental Research Laboratories
National Centers for Environmental Prediction
325 Broadway (R/E/SE),
Boulder, Colorado 80303-3328
Tel: (+1 303) 497 33 11
Fax: (+1 303) 497 36 45
e-mail: ehildner@sec.noaa.gov
Director, SEC: Dr Ernest Hildner
Chief, Services Division: Gary Heckman
Chief, Research Division: Ron Zwickl

Space Imaging Inc
9351 Grant Street, Suite 500,
Thornton, Colorado 80229
Tel: (+1 303) 254 20 00
Fax: (+1 303) 254 22 15
WWW: http://www.spaceimage.com
Chief Executive Officer: John R Copple
President: Jeffrey K Harris
Chief Technical Officer: John Neer
Executive Vice-President/CFO: Ronald A Forkel
Vice-President Marketing/Sales: William Folchi
Tel: (+1 303) 254 21 00
Vice-President Systems Development/Operations: Robert Iwai
Vice-President Product Research & Production: Tish Vajta-Williams
Director, Federal Programs: Ralph Mason
Director, Systems Engineering: Gene Steiger
Director, Space Segment: Scott Smith
Director, Ground Segment: James Snow
Director, Product Marketing: Jolyon Thurgood
Director, Customer Services: Larry F Konty
Media contacts:
Brian Webster
Tel: (+1 303) 254 21 04
e-mail: bwebster@spaceimage.com
Linda Turner
Tel: (+1 303) 254 21 06
email: lturner@spaceimage.com

Space Instruments, Inc
4403 Manchester Avenue, Suite 203,
Encinitas, California 92024
Tel: (+1 619) 944 70 01
Fax: (+1 619) 944 70 56
Technical Director: James W Hoffman

Space Liaison & Imaging Corporation
11440 West Bernardo Court, Suite 206,
San Diego, California 92127
Tel: (+1 619) 66 18 17 49
Fax: (+1 619) 618 18 73
President: John M Papazian
Vice-President Marketing: Brook Watts

Space Vacuum Epitaxy Center
University of Houston, Bldg SR-1, Room 724,
4800 Calhoun Road, Houston, Texas 77204-5507
Tel: (+1 713) 743 36 21
Fax: (+1 713) 747 77 24
Director: Dr Alex Ignatiev
WSF Program Manager: Mark Sterling

Spacehab, Inc
1215 Jefferson Davis Highway, Suite 1501,
Arlington, Virginia 22202
Tel: (+1 703) 414 81 00
Fax: (+1 703) 414 81 07
e-mail: darossi@aol.com
President/CEO: Richard P Hora
Chief Financial Officer: Margaret E Grayson
Vice-President Operations: Chester M Lee
Senior Vice-President Business Development: David A Rossi
Vice-President Mir Program Manager: John M Lounge
Vice-President Engineering/Integration: M Dale Steffey
Director, Government Relations: Nick Fuhrman
Director, Corp Communications: Rebecca B Gray

620 Magellan Road, Cape Canaveral, Florida 32920
Tel: (+1 407) 868 74 04
Fax: (+1 407) 868 74 21

13100 Space Center Boulevard, Room 2100,
Houston, Texas 77059.
Tel: (+1 713) 244 46 65
Fax: (+1 713) 244 46 72
Vice-President Operations: Dr Bernard A Harris

Spot Image Corporation
1897 Preston White Drive,
Reston, Virginia 22091-4368
Tel: (+1 703) 715 31 00
TWX: 703-499-30 73
Fax: (+1 703) 648 18 13
e-mail: name@spot.com
President: Theodore Nanz
Vice-President Technical Operations: Ron Hodge
Vice-President Sales: Dan Carayiannis
New Products Manager: Mark Whalley
Marketing/Communications Manager: Clark Nelson

Telonics Inc
932 E Impala Avenue, Mesa, Arizona 85204-6699
Tel: (+1 602) 892 44 44
Fax: (+1 602) 892 91 39
Telemetry Systems Manager: Brenda Milam

Terra-Mar
1710 South Amphlett Blvd, Suite 215,
San Mateo, California 94402
Tel: (+1 415) 655 25 65
Fax: (+1 415) 655 25 66
President/CEO: James D Nichols
International Sales: Steve Hurst

The Geosat Committee, Inc
PO Box 1762, Norman, Oklahoma 73070-1762
Tel: (+1 405) 799 15 15
Fax: (+1 405) 799 15 65
Chairman: Dr H R Hopkins
Administrator: Lisa McArthur

Wirin & Associates
555 Indian Trail, Palm Springs, California 92264
Tel: (+1 619) 320 49 66
Fax: (+1 619) 328 66 61
Managing Director: William B Wirin

VENEZUELA

Centre For Digital Image Processing (CDIP)
1040-A Caracas
Tel: (+58 2) 962 11 24
Telex: 21865
Fax: (+58 2) 962 10 25
Director: Gustavo Ruiz

ZIMBABWE

Realtime Computers (PVT) Ltd
Realtime House, Third Street/Central Avenue,
PO Box 4567, Harare
Tel: (+263 4) 73 96 55
Telex: 0907-26123 zw
Fax: (+263 4) 79 54 96
Managing Director: Ron F Hyslop

NAVIGATION SATELLITE SYSTEMS AND SERVICES

AUSTRALIA

Commonwealth GPS Group
PO Box 2 Belconnen, ACT 2616
Tel: (+61 6) 201 42 92
Fax: (+61 6) 201 43 66
Secretary: Bob Irwin

Sigtec
PO Box 304, Seaford, Victoria 3198
Tel: (+61 3) 97 86 00 77
Fax: (+61 3) 97 86 00 63

Space Centre For Satellite Navigation
GPO Box 2434, Brisbane, Queensland 4001
Tel: (+61 7) 864 14 03
Fax: (+61 7) 864 15 17
Director: Professor Miles Moody
Business Manager: Mark Kinsella

FRANCE

Informatique Electronique Sécurité Maritime
Zone Industrielle des Cinq Chemins, F-56520 Guidel
Tel: (+33 2) 97 02 49 49
Telex: 950535 F
Fax: (+33 2) 97 65 00 20
President: James Audren
Export Manager: Reine Petch
Aviation Department: Florence Crossay

INTERNATIONAL

Cospas-Sarsat
Cospas-Sarsat Secretariat
Inmarsat, 99 City Road,
London EC1Y 1AX
Tel: (+44 171) 728 13 91
Telex: 297201 inmsat g
Fax: (+44 171) 728 11 70
e-mail: cospas_sarsat@inmarsat.org
WWW: http://www.cospas-sarsat.org
Head: Daniel Levesque
Deputy: Jim King

NORWAY

Jotron Electronics AS
PO Box 85, N-3280 Tjodalyng
Tel: (+47) 33 12 45 77
Fax: (+47) 33 12 67 80
Chairman: Olaf Stavik
Managing Director: Steinar Saeter

Kongsberg Navigation AS
PO Box 98, N-5034 Ytre Laksevåg
Tel: (+47) 55 34 77 10
Telex: 40580 koben n
Fax: (+47) 55 34 79 65
Managing Director: Torunn Økland

UNITED KINGDOM

GEC-Marconi Radar & Defence Systems
Electronic Systems Division
Browns Lane, The Airport,
Portsmouth, Hampshire PO3 5PH
Tel: (+44 1705) 67 52 36
Fax: (+44 1705) 67 40 41
Division Director: D J Pardoe

Lokata Limited
New North Road, Hainault, Ilford, Essex IG6 2UR
Tel: (+44 181) 500 10 20
Telex: 896401 kelhue g
Fax: (+44 181) 559 85 22
Marketing Services: Barry Morgan

Magnavox Systems Limited
Lancaster House, 7 Church Street,
Slough, Berkshire SL1 1TL
Tel: (+44 1753) 380 25
Telex: 847583 magvox g
Fax: (+44 1753) 82 46 33
Managing Director: Robin Guenier
Product Support Manager: John Wood

Navstar Systems Ltd
Mansard Close, Westgate Industrial Estate,
Northampton NN5 5DL
Tel: (+44 1604) 58 55 88
Fax: (+44 1604) 58 55 99
Sales Manager: Mark Sampson
Sales and Business Development Manager:
Dr Clive de la Fuente

Nortel Limited
Brixham Road, Paignton, Devon TC4 7BE
Tel: (+44 1803) 66 28 74
Fax: (+44 1803) 66 28 09
Contact: Mark E Wilson
GPS Simulator Project Manager: Peter Boulton

Northern Telecom Inc
475 North Martingale Road,
Shaumburg, Illinois 60173, USA
Tel: (+1 708) 706 86 00
Fax: (+1 708) 706 86 06
Contact: Charles Fay

Nova Marine Systems Ltd
Rodney Road, Fratton,
Portsmouth, Hampshire PO4 8SG
Tel: (+44 1705) 77 50 00
Fax: (+44 1705) 81 90 87
General Manager: Alan Higgins
International Sales Manager: Richard Peacock

Satellite Navigation Group
The Royal Institute of Navigation
1 Kensington Gore, London SW7 2AT
Tel: (+44 171) 591 31 30
Fax: (+44 171) 591 31 30
Chairman: M A Savill
Secretary: Dr Terry Moore

UNITED STATES OF AMERICA

3S Navigation
23141 Plaza Pointe Drive,
Laguna Hills, California 92653
Tel: (+1 714) 830 37 77
Fax: (+1 714) 830 84 11
e-mail: nav3s@aol.com
WWW: http://www.3snavigation.com
Director, Business Development: F Earl Weile

Allen Osborne Associates, Inc
756 Lakefield Road,
Westlake Village, California 9136‑-2624
Tel: (+1 805) 495 84 20
Fax: (+1 805) 373 60 67
Vice-President Business Development:
Dr Robert W Snow

Ashtech, Inc
1170 Kifer Road, Sunnyvale, California 94086
Tel: (+1 408) 524 14 00
Fax: (+1 408) 524 15 00
Chairman: Bill Bottoms
President/CEO: Chuck Boesenberg
Vice-President Sales: Thomas Hunter
Vice-President Marketing: Jonathan Ladd
Vice-President New Product Development/New Business Development: Todd Townsend

Ashtech Europe Ltd
Blenheim Office Park, Lower Road, Long Hanborough,
Oxfordshire OX8 8LN, UK
Tel: (+44 1993) 88 35 33
Fax: (+44 1993) 88 39 77

Ashtech
Park Place Moscow, 113 Leninski Prospekt,
Moscow 117198, Russian Federation
Tel: (+7 502) 256 54 00
Fax: (+7 502) 256 53 60

Cast
5450 Katella Avenue, Los Alamitos, California 90720
Tel: (+1 310) 594 88 83
Fax: (+1 310) 594 46 94

Civil GPS Service
Navigation Information Service
7323 Telegraph Rd, Alexandria, Virginia 22315-3998
Tel: (+1 703) 313 59 00
Fax: (+1 703) 313 59 20
Fax: (+1 703) 313 59 31/59 32 provides system status
BBS: (+1 703) 313 59 10 all baud rates
WWW: http://www.navcen.uscg.mil

Datum Incorporated
Timing Division
1363 South State College Boulevard
Anaheim, California 92806-5790
Tel: (+1 714) 533 63 33
TWX: 910 592 12 89
Fax: (+1 714) 533 63 45
Vice-President Marketing: Charles Neumann
Vice-President Engineering: Ronald H Schlund
Vice-President Manufacturing: Kenneth B Weatherly

Del Norte Technology, Inc
1100 Pamela Drive,
Euless, Texas 76040
Tel: (+1 817) 267 35 41
Fax: (+1 817) 354 57 62
WWW: http://www.onramp.net/del_norte
Vice-President Marketing/Sales: John Pointon

Garmin International, Inc
1200 E 151st Street, Olathe, Kansas 66062
Tel: (+1 913) 397 82 00
Fax: (+1 913) 397 82 82
President: Gary Burrell
Vice-President Engineering: Min H Kao

Garmin/Europe Ltd
Unit 5, The Quadrangle, Abbey Park, Romsey, Hampshire SO51 9AQ, UK
Tel: (+44 1794) 51 99 44
Fax: (+44 1794) 51 99 22

Garmin Corporation
4th Floor, No. 1, Lane 45, Pao-Hsing Road, Hsin Tien, Taipai, Taiwan (ROC)
Tel: (+886 02) 917 41 07
Fax: (+886 02) 917 17 58

GE Starsys
4400 Forbes Boulevard,
Lanham, Maryland 20706-4392
Tel: (+1 301) 459 88 32
Fax: (+1 301) 794 71 06
Program Manager: Alan B Renshaw
Business Development: Steve Morgan

Honeywell Inc
Air Transport Systems
PO Box 21111, 21111 N 19th Avenue,
Phoenix, Arizona 85036-1111
Tel: (+1 602) 436 23 11
Telex: 667405
Fax: (+1 602) 436 92 14

Interstate Electronics Corp
1001 East Ball Road, PO Box 3117,
Anaheim, California 92803
Tel: (+1 714) 758 05 00
Fax: (+1 714) 758 41 48
President: Larry A LaCotti
Vice-President Operations: Michael J Patrick
Vice-President Business Development:
C Steven Jungers
Vice-President Commercial Communication Systems:
Faris Gaffney
Vice-President Sales & Marketing, Satcoms:
Robert Levin
Director, GPS Marketing: James Grace
Director, Government Comms Marketing: Fred Gould
Director, Business Development, GPS Systems:
James Grace

Leica, Inc
Navigation and Positioning Division
23868 Hawthorne Boulevard,
Torrance, California 90505
Tel: (+1 310) 791 53 00
Fax: (+1 310) 791 61 08
General Manager: Sharon Jones

Litton Aero Products
21050 Burbank Boulevard,
Woodland Hills, California 91367
Tel: (+1 818) 226 20 00
Telex: 662619
Fax: (+1 818) 226 27 55
President: L A Webrand
Director, Advanced Programs: Victor F Strachan
Vice-President Business Development: J W Winchester
Marketing Director: Larry Hancey

Magellan Systems Corp
960 Overland Court, San Dimas, California 91773
Tel: (+1 909) 394 50 00
Fax: (+1 909) 394 70 50
President/CEO: Randy D Hoffman
Vice-President Marketing/Sales: Richard R Sill
Director, Consumer Products Marketing:
Carol Ann Gorden

Micrologic
9174 Deering Ave,
Chatsworth, California 91311-5801
Tel: (+1 818) 998 12 16
Fax: (+1 818) 709 36 58
President: David Caldwell
Vice-President Sales/Marketing: Michael Deveso

Unit 7, Minister Court, Tuscam Way,
Camberley, Surrey GU15 3YY, UK
Tel: (+44 1276) 67 70 78
Fax: (+44 1276) 68 61 12/294 71
CEO Europe: Colin Saunders
Sales/Marketing Engineer: Patrick Woolcocks

NavSymm
85 West Tasman Drive,
San Jose, California 95134
Tel: (+1 888) 367 79 66
Fax: (+1 408) 428 79 98
e-mail: navsymm@telecom.com
USA Business Development Manager: Joe Blanda

Rockwell International Corporation
Autonetics & Missile Systems Division
3370 Miraloma Avenue, PO Box 3105,
Anaheim, California 92803-3105
Tel: (+1 714) 762 81 11
Fax: (+1 714) 762 41 27
President: Bob Paster
Vice-President, Operations/Product Support:
James E Lake
Vice-President Integrated INS/GPS:
Edward P Barry, Jr

Collins Avionics & Communications Division
350 Collins Road NE, Cedar Rapids, Iowa 52498
Tel: (+1 319) 395 10 00
Telex: 464421
Fax: (+1 319) 395 16 42
WWW: http://www.cacd.rockwell.com
President: John D Cosgrove
Executive Vice-President: Robert P Marovich
Vice-President Engineering: Brian T Wright
Vice-President Program Management:
Robert M Chiusano
Vice-President Business Development:
Michael K McDonald
Director, International Marketing: Rod N Thorpe
GPS User Equipment: H W Pomije

SCI Systems, Inc
PO Box 1000, Huntsville, Alabama 35807
Tel: (+1 205) 882 41 99
Fax: (+1 205) 882 46 52
President: Eugene Sapp
Senior Vice-President, Government Division:
Richard Holloway
Space Systems Program Manager: Tom Runner
Marketing Director: Ray Minter

Sokkia Corporation
9111 Barton, PO Box 2934,
Overland Park, Kansas 66201
Tel: (+1 913) 492 49 00
Fax: (+1 913) 492 01 88
President: Herbert C Johnson, III

Sony Electronics Inc
Sony Drive, Park Ridge, New Jersey 07656-8003
Tel: (+1 201) 930 74 16
Fax: (+1 201) 930 79 09
Marketing Manager, Navigation Systems:
Kenneth Turner

Trimble Navigation Limited
645 North Mary Avenue,
Sunnyvale, California 94086-3642
Tel: (+1 408) 481 80 00
Telex: 6713973
Fax: (+1 408) 481 20 00
WWW: http://warrior.com/trimble/index.html
President/CEO: Charles Trimble
Executive Vice-President Worldwide Sales, Field Ops:
Charles Joseph
Executive Vice-President Surveying/Mapping:
Jim Sorden
Executive Vice-President Tracking/Communications:
David Vaughn
Vice-President/GM Avionics: Bruce Alspach
Vice-President Corporate Development: Doug Gerhart
Vice-President Business Development:
Ralph Eschenbach

Trimble Navigation Europe Ltd
Trimble House, Meridian Office Park,
Osborn Way, Hook, Hampshire RG27 9HX
Tel: (+44 1256) 76 01 50
Fax: (+44 1256) 76 01 48
Managing Director: Mike Coles

Truetime, Inc
2835 Duke Court, Santa Rosa, California 95407
Tel: (+1 707) 528 12 30
Telex: 176687
Fax: (+1 707) 527 66 40
e-mail: truetime@nbn.com
Director, International Sales: Gregory J Kret

United States GPS Industry Council
1100 Connecticut Avenue NW, Suite 535,
Washington DC 20036
Tel: (+1 202) 296 16 53
Fax: (+1 202) 862 39 29
Contact: Mike Swiek

Universal Navigation Corp
3260 East Lerdo Road, Tucson, Arizona 85706
Tel: (+1 602) 295 23 00
Telex: 705806
Fax: (+1 602) 295 2395
Executive Vice-President: Charles H Edmondson
Vice-President Marketing: Donald D Berlin

Welnavigate, Inc
510 Spectrum Circle, Oxnard, California 93030
Tel: (+1 805) 485 59 08
Fax: (+1 805) 485 76 87
President: Saburo Ifune
Vice-President: Frank Bletzacker
Director of Marketing: Dean Ifune

CIVIL AND COMMERCIAL GROUND FACILITIES

BELGIUM

Centre Spatial De Liege
Parc Scientifique du Sart Tilman,
Avenue du Pré-Aily, B-4031 Angleur-Liège
Tel: (+32 41) 67 66 68
Telex: 41320 ial sp b
Fax: (+32 41) 67 56 13
e-mail: cslulg@vm1.ulg.ac.be
Director: Dr Claude Jamar
Deputy Director: Dr J P Macau

CANADA

Akjuit Aerospace Incorporated
Suite 800, One Lombard Place,
Winnipeg, Manitoba R3B 0X3
Tel: (+1 204) 934 56 36
Fax: (+1 204) 947 03 76
WWW: http://www.spaceport.ca/spaceport
Chairman: George T Richardson
President/CEO: Siobhan Mullen
Chief Financial Officer and Chief Operating Officer: Rees M Jones
Vice-President, International Launch Systems: Dr Andriy Zhalko-Tytarenko
Vice-President, International Relations and Government Affairs: Lawrence G Prout

Canadian Space Agency
David Florida Laboratory
PO Box 11490, Station H,
3701 Carling Avenue, Ottawa, Ontario K2H 8S2
Tel: (+1 613) 998 23 83
Fax: (+1 613) 993 61 03
WWW: http://www.dfl.doc.ca
Director General: Dr Rolf Mamen
Planning/Communications Officer: Charly Geier

DENMARK

Danish Space Research Institute
Juliane Maries Vej 30, DK-2100 Copenhagen
Tel: (+45) 35 32 58 30
Fax: (+45) 35 36 24 75
e-mail: internet user@danru.dk
Telex: SPAN:DSRISN::USER
Director: Prof Herbert W Schnopper

FRANCE

Intespace
18 avenue Edouard Belin, BP4356,
F-31029 Toulouse Cedex
Tel: (+33 5) 61 28 11 11
Fax: (+33 5) 61 28 11 12
Chairman/CEO: Jean-Louis Marcé
Administration/Finance: Gérard Boures
Marketing/Sales: André Mouton
Operations: Jean-Claude Pasquet
Engineering: Joseph Morlef
Research and Development: Louis-Patrice Bugeat
Head of Communications: Michel Leotard

Onera
29 Avenue de la Division Leclerc, F-92322 Châtillon
Tel: (+33 1) 46 73 40 40
Telex: ONERA 632186
Fax: (+33 1) 46 73 41 41
WWW: http://www.onera.fr
President: Michel Scheller
Secretary General: Michel Belva
General Technical Director: Gérard Dorey
General Scientific Director: Georges Duvaut
Director, Military/Space: Philippe Cazin
Director, Aerodynamics Department: Bernard Monnerie
Director, Systems Department: Jean Fave
Director, Energetics Department: Paul Kuentzmann
Director, Structures Department: Roger Labourdette
Director, Materials Department: Paul Costa
Director, Physics Department: Daniel Lépine
Director, Large Testing Facilities: Xavier Bouis
Director, Toulouse Research Centre: François Jouaillec
Director, Lille Fluid Mechanics Institute: Francis Dupriez
Press/Public Relations Officer: Nicole Landré

FRENCH GUIANA

Centre Spatial Guyanais
BP726, F-97387 Kourou Cedex, French Guiana
Tel: (+594 33) 51 11
Telex: 910516 csg fg
Fax: (+594 33) 47 66
Director: Michel Mignot
Public Affairs Contact: Hélène Michel-Donadieu
Tel: (+594 33) 55 53; Fax: (+594 33) 47 19

GERMANY

German Space Operations Centre
Oberpfaffenhofen, D-82230 Wessling/Obb
Tel: (+49 8153) 28 27 00
Telex: 527420 soc d
Fax: (+49 8153) 24 68
Head: Dr Franz Schlude

Industrieanlagen- Betriebsgesellschaft GmbH
Einsteinstrasse 20, D-85521 Ottobrunn
Tel: (+49 89) 60 88 22 27
Fax: (+49 89) 60 88 39 03
Executive Board: Dr William Sweeney, Dr Rudolf Schwarz
Head, Space Test Centre: Karl Ahrensdorf
Marketing Manager: Gerhard Vodermeier

INDIA

Development & Educational Communications Unit (DECU)
Jodhpur Tekra, Ahmedabad 390 053
Tel: (+91 79) 674 39 54
Telex: 0121-6261
Fax: (+91 79) 656 80 73
Director: B S Bhatia

ISRO Satellite Centre (ISAC)
Airport Road, Vimanapura Post Office,
Bangalore 560 017
Tel: (+91 80) 56 62 51
Telex: 0845-2325/2769
Fax: (+91 80) 56 61 97
Director: R Aravamuthan
Deputy Director (Special Projects): Dr A S Prakash Rao
Deputy Director (Mechanical): A V Patki
Deputy Director (Controls/Mission): P S Goel
Deputy Director (Electronics): P Ramachandran
Director (LEOS): T K Alex
Director, Programme Planning and Evaluation Group: V A Thomas

ISRO Telemetry Tracking And Command Network (ISTRAC)
A1-6, Peenya Industrial Estate,
Bangalore 560 058
Tel: (+91 80) 839 45 15
Telex: 0845-5066/5072
Fax: (+91 80) 839 42 49
Director: K V Venkatachary
Deputy Director: R Nandakumar

Shar Centre
Sriharikota Range
Andhra Pradesh 524124
Tel: (+91 8) 623 50 60
Telex: 041-6723/7353
Fax: (+91 4) 456 85 94
Director: Dr S Vasantha
Deputy Director, ISRO Range/Test Entity: K Narayana
Deputy Director, PRT: M C Uttam

Space Applications Centre (SAC)
Jodhpur Tekra, Ahmedabad 390 053
Tel: (+91 79) 656 40 43
Telex: 0121-6261
Fax: (+91 79) 656 80 73
Director: Dr George Joseph
Deputy Director (Remote Sensing): Dr Baldev Sahai
Deputy Director (SCA): O P N Calla

Vikram Sarabhai Space Centre
Trivandrum 695022
Tel: (+91 471) 56 24 44/56 25 55
Telex: 0435-201/202
Fax: (+91 471) 797 95
Director: Dr S Srinivasan
Deputy Director (SPC): Rajaram Nagappa
Deputy Director (PPC): Dr V N Krishnamoorthy
Deputy Director (SRG): M S R Dev

INTERNATIONAL

European Space Operations Centre
Robert-Bosch-Strasse 5,
D-64293 Darmstadt, Germany
Tel: (+49 6151) 900
Telex: 419453 eu d
Fax: (+49 6151) 904 95
WWW: http://www.esoc.esa.de
Director, Technical and Operational Support: David Dale
Head, Mission Operations Department: Georges Mazza
Head, Ground Systems Engineering Department: Carlo Mazza

European Space Research Institute
Via Galileo Galilei, CP 64,
I-00044 Frascati (Rome), Italy
Tel: (+39 6) 94 18 01
Telex: 610637 esrin i
Fax: (+39 6) 94 18 02 80
WWW: http://www.esrin.esa.it
Director: Francis Roscian
Head Exploitation Department: Willy Jensen
Head Projects and Engineering Department: Livio Marelli

European Astronauts Centre
Porz-Wahnheide, Linder Höhe, D-51147 Köln, Germany
Postal: PO Box 90 60 96, D-51127 Köln, Germany
Tel: (+49 2203) 600 10
Fax: (+49 2203) 60 01 66
WWW: www/esrin.esa.nl/spaceflight/astronaut
Head: Heinz Oser
Training: Klaus Damian
Astronauts: Franco Rossitto
Co-ordination: Walter Peters

European Space Research And Technology Centre
Postbus 299, NL-2200 AG Noordwijk zh,
Netherlands
Tel: (+31 71) 65 65 65
Telex: 39098
Fax: (+31 71) 65 60 40
WWW: http://www.estec.esa.nl
Director: Marius Le Fèvre
Systems engineering: Niels Jensen
Mechanical systems: Erhard Classen
Electrical systems: Mario Lopriore
Automation/Informatics: Michel Guerin
Product Assurance/Safety: Alex Soons
Testing: Peter Brinkmann

JAPAN

Kagoshima Space Centre
Uchinoura, Kimotsuki-gun, Kagoshima 893-14
Tel: (+81 99) 467 22 11
Fax: (+81 99) 467 38 11
Director: Professor Yasunori Matogawa

Kakuda Propulsion Centre
1 Koukuzo, Jinjiro, Kakuda-shi, Miyagi 981-15
Tel: (+81 224) 68 32 11
Fax: (+81 224) 67 10 32
Director: Hirokazu Nagai

Tanegashima Space Centre
Minamitane-machi, Kumage-gun, Kagoshima 891-37
Tel: (+81 9) 97 26 21 11
Fax: (+81 9) 97 24 40 04
Director: Takehiro Iida
Deputies: Masafumi Sasahara, Takane Kobayashi

Tsukuba Space Centre
1-1, Sengen 2-chome, Tsukuba-shi, Ibaraki 305
Tel: (+81 2) 98 52 22 11
Fax: (+81 2) 98 52 23 84
Director: Toshihiko Kikuyama

NORWAY

Andøya Rocket Range
PO Box 54, N-8480 Andenes
Tel: (+47) 76 14 16 44
Fax: (+47) 76 14 18 57
e-mail: range@arr.nsc.no
SPAN: 21813::SKYTEFELT
Range Director: Kolbjørn Adolfsen
Head of Operation: Ivar Nyheim
Head of Quality/Safety: Kjell Bøen

Tromsø Satellite Station
N-9005 Tromsø
Tel: (+47) 77 68 48 17
Fax: (+47) 77 65 78 68
e-mail: adm@tss.no
WWW: http://www.tss.no
Managing Director: Einar Ellingsen
Marketing Director: Dr Jan Petter Pedersen

RUSSIAN FEDERATION

Plesetsk
Mirny-12, Arkhangelskaya oblast, 164170
Tel: (+7 095) 330 91 90
Fax: (+7 095) 330 91 90
Commander: Gen Anatoli P Perminov
Chief, Space Vehicle Testing: Maj Gen Anatoli F Ovchinnikov

Tsentr Upravlenyia Polyotom
4 Pionerskaya Street,
Kaliningrad, 141070 Moscow oblast
Tel: (+7 095) 187 06 66
Telex: 411952 mcc su
Fax: (+7 095) 274 00 25
Director: Vladimir Lobachev
Deputy Director: Viktor Blagov
Mir Director: Vladimir Solovyov
Chief, Public Information: Vsevolod P Latyshev

Zvezdny Gorodok
141160 Moscow
Tel: (+7 095) 526 28 20/83 42
Fax: (+7 095) 526 26 12
Director: Lt Gen Pyotr I Klimuk
Head of Training: Maj Gen Yuri Glazkov

SOUTH AFRICA

OTB
Private Bag X12, Bredasdorp 7280, Western Cape
Tel: (+27 2847) 590 10
Fax: (+27 2847) 590 67
WWW: http://www.denel.co.za
Test Range Manager: Dr Jan G Malan
Senior Manager, Programmes: Jacob Venter
Range Systems Engineer: Phillip S van der Westhuizen

SWEDEN

Esrange
PO Box 802, S-981 28 Kiruna
Tel: (+46 980) 720 00
Telex: 8744 esrange s
Fax: (+46 980) 128 90
General Manager, Esrange Division: Jan Englund
Head, Sounding Rockets and Balloon Department, Esrange Division: Stig Kemi
Head, Science Satellites, Esrange: Lennart Marcus
Head, Satellite Station, Earth Observation Division: Lars Alm
Head, Satellite Control, Telecom Division: Jonny Järnmark

UNITED KINGDOM

Central Laboratory of the Research Councils
Rutherford Appleton Laboratory
Chilton, Didcot, Oxfordshire OX11 0QX
Tel: (+44 1235) 82 19 00
Telex: 83159
Fax: (+44 1235) 44 58 48
WWW: http://www.ssd.rl.ac.uk
Chief Executive, CCL: Dr Paul R Williams
Director, Research and Development: Dr Gordon Walker
Head, Daresbury Laboratory: Dr Ron Newport
Director Space Science/Department Head: Dr Eric Dunford (+44 1235 44 66 75)
Head of Marketing, Space Science Department: Jeremy Curtis

UNITED STATES OF AMERICA

Alaska Aerospace Development Corporation
4300 B Street, Suite 101, Anchorage, Alaska 99503
Tel: (+1 907) 561 33 38
Fax: (+1 907) 561 33 39
e-mail: aadc@alaska.net
Executive Director: Pat Ladner
Assistant Director: David A Sadlowski

Astrotech Space Operations, Inc
12510 Prosperity Drive, Suite 100,
Silver Spring, Maryland 20904
Tel: (+1 301) 622 58 04
Telex: 380143 astrotech ud
Fax: (+1 301) 622 29 37

1515 Chaffee Drive, Titusville, Florida 32780
Tel: (+1 407) 268 38 30
Fax: (+1 407) 268 38 34

PO Box 5097, Vandenberg AFB, California 93437
Tel: (+1 805) 734 11 02
Fax: (+1 805) 734 25 51
President: James Grohowski
Vice-President Marketing/Operations: George D Baker

California Commercial Spaceport, Inc
3769-C Constellation Road, Lompoc, California 93436
Tel: (+1 805) 733 73 70
Fax: (+1 805) 733 73 72
e-mail: dbarry@calspace.com
President/CEO: Orlando Severo
Director, Business Operations: Dominick Barry

Los Alamos National Laboratory
Los Alamos, New Mexico 87545
Tel: (+1 505) 667 70 00
Fax: (+1 505) 665 55 52
Director: Dr Siegfried Hecker
Head, Nonproliferation and International Security Division: Don Cobb

NASA Ames Research Center
Moffett Field, California 94035-1000
Tel: (+1 415) 604 90 00
Fax: (+1 415) 604 39 53
Director: Dr Henry McDonald
Deputy Director: William E Dean
Director, Space Research: William Berry
Director, Aeronautics: John S Burks
Director, Information Systems: Dr Henry Lum
Director, Center Operations: Jana M Coleman
Comptroller: Ralph H Robinson
Chief, Space Science: Dr David Morrison
Pioneer Project Manager: Manfred N Wirth
Biocosmos Project Manager: Michael Skidmore
Chief, External Affairs: Michael Marlaire
Special Assistant, Public Affairs: Ann Hutchison

NASA Dryden Flight Research Center
PO Box 273, Edwards, California 93523
Tel: (+1 805) 258 34 49
Fax: (+1 805) 258 35 66
WWW: http://www.dfrc.nasa.gov
Director: Kenneth J Szalai
Chief, Space Shuttle Support Office: Vance D Brand

NASA Goddard Space Flight Center
Greenbelt, Maryland 20771
Tel: (+1 301) 286 20 00
Fax: (+1 301) 286 17 07 (Public Affairs)
WWW: http://pao.gsfc.nasa.gov/gsfc.html
Director: Joseph Rothenberg
Assistant Director: Alphonso V Diaz
Director, Management Operations: Sharon Foster
Acting Director, Flight Assurance: Wentworth O Duncan
Director, Flight Projects: Vernon Weyers
Director, Mission/Data Systems Operations: Arthur Fuchs
Director, Space Sciences: Stephen Holt
Director, Engineering: Dr Allan Sherman
Director, Suborbital Projects and Operations: Arnold Torres
Director, Earth Sciences: Vincent Salomonson
Director, Mission to Planet Earth: Dr Bob Price

NASA Jet Propulsion Laboratory
4800 Oak Grove Drive, Pasadena, California 91109
Tel: (+1 818) 354 50 11
Fax: (+1 818) 354 45 37
Director: Dr Edward C Stone Jr
Deputy Director: Larry N Dumas
Chief Scientist: Moustafa T Chahine
Chief Engineer: John R Casani
Associate Director: Kirk Dawson
Director, Engineering/Science: James King
Director, Business Operations: John Heie
Director, Space and Earth Science Programs: Charles Elachi
Director, Technology and Applications Programs: James A Evans
Director, Telecommunications and Mission Operations: Norman R Haynes

NASA Johnson Space Center
Houston, Texas 77058
Tel: (+1 713) 483 51 11
Fax: (+1 713) 483 20 00
Director: George W S Abbey
Deputy Director: James Wetherbee
Director, Space Shuttle Operations: Tommy Holloway
Director, Flight Crew Operations: David Leestma
Director, Mission Operations: John W O'Neill
Manager, Orbiter/GFE Projects: Daniel Germany
Director, Engineering: Leonard S Nicholson
Deputy Director, Engineering: Max Engert
Director, Center Operations: James A Hickmon
Manager, Space Station Program Office: Randy Brinkley
Chief Engineer, Space Station: Chester Vaughan
Director (acting), Space/Life Sciences: Don Robbins
Chief Astronaut: Robert D Cabana
EVA Project Office: Donald R McMonagle
Director (acting), Public Affairs: Jeffrey E Carr

NASA Kennedy Space Center
Florida 32899
Tel: (+1 407) 867 24 68
Fax: (+1 407) 867 26 92
WWW: http://www-pao.ksc.nasa.gov/kscpao
Director: Roy Bridges
Deputy Director: Jim Jennings (acting)
Associate Director: vacant
Director, Shuttle Processing: Robert B Sieck
Deputy Director, Shuttle Processing and Launch Director: James F Harrington
Director, Engineering Development: Sterling Walker
Director, Safety/Mission Assurance: Thomas Breakfield
Director, Payload Processing: Bobby Bruckner
Acting Director, ELVs: Floyd A Curington
Director, Payload Flight Systems: Shannon Bartell
Chief, Space Station Launch Site Support Office: Stephen M Francois
Director, Space Station Hardware Integration Office: John Talone

NASA Langley Research Center
Hampton, Virginia 23681-0001
Tel: (+1 804) 864 61 24
Fax: (+1 804) 864 63 33
WWW: http://www.larc.nasa.gov
Director: Dr Jeremiah F Creedon
Deputy Director: Dr H Lee Beach Jr
Associate Director: Ming Tang
Director, Space and Atmospheric Sciences Program Group: Darrell R Branscome
Chief Scientist: Dennis M Bushnell

NASA Lewis Research Center
21000 Brookpark Road, Cleveland, Ohio 44135
Tel: (+1 216) 433 89 20
Fax: (+1 216) 433 81 43
Director: Donald J Campbell
Deputy Director: Martin P Kress
Director, Aeronautics: Dr Carol J Russo
Chief, High Speed Research Propulsion Project Office: Dr Robert J Shaw
Director, Aerospace Technology: Fred Teren
MMSL Manager: Leslie A Greenbauer-Seng

Director, Space Flight Systems: Wilson F Ford
ACTS Project Manager: Rodney M Knight
Associate Manager, Launch Vehicle Project: Eugene P Simmons

NASA Marshall Space Flight Center
Huntsville, Alabama 35812
Tel: (+1 205) 544 00 34
Fax: (+1 205) 544 58 52
WWW: http://www.msfc.nasa.gov
Director: Dr J Wayne Littles
Director, Science & Engineering: G Sherman Jobe
Director, Program Development: Axel Roth
Payload Projects Manager: Richard A Marman
Shuttle Projects Manager: Alexander A McCool
Manager, SSME Project: George Hopson
Manager, RSRM Project: Victor K Henson
Manager, External Tank Project: Parker Counts
Director, Public Affairs: John B Taylor

Michoud Assembly Facility
PO Box 29300, New Orleans, Louisiana 70189
Tel: (+1 504) 257 26 01
Fax: (+1 504) 257 26 06
Manager: John White

NASA Space Telescope Science Institute
Johns Hopkins Homewood Campus
3700 San Martin Drive, Baltimore, Maryland 21218
Tel: (+1 410) 338 47 00
Fax: (+1 410) 338 47 67
Director: Dr Robert E Williams

NASA Stennis Space Center
Mississippi 39529-6000
Tel: (+1 601) 688 33 41
Fax: (+1 601) 688 10 94
WWW: http://www.ssc.nasa.gov
Director: Roy S Estess
Deputy Director: Mark Craig
Director, Center Ops: Florence Kailiwai-Barnett
Director, Propulsion Test Directorate: Leslie 'Boyce' Mix
Chief, Earth Systems Office: Dr Richard Miller

NASA Wallops Flight Facility
Wallops Island, Virginia 23337
Tel: (+1 804) 824 15 79
Fax: (+1 804) 824 19 71
WWW: http://www.wff.nasa.gov
Director, Suborbital Projects and Operations: Dr Arnold L Torres
Head, Vehicle Assembly and Launch: Claude Linton

NASA White Sands Test Facility
PO Drawer MM, Las Cruces, New Mexico 88004
Tel: (+1 505) 524 50 11
Fax: (+1 505) 524 57 98
Manager: Grady E McCright
Deputy Manager: Joseph Fries
Chief, Propulsion Test Office: Pleddie M Baker

Poker Flat Research Range
Geophysical Institute, University of Alaska Fairbanks,
903 Koyukuk Drive, PO Box 757320,
Fairbanks, Alaska 99775-7320
Tel: (+1 907) 474 74 13 Improvement Office
Tel: (+1 907) 474 70 15 Range
Telex: 090-354140
Fax: (+1 907) 474 57 05
WWW: http://www.gi.alaska.edu/
Scientific Director: Dr Charles S Deehr
Assistant Director: Stanley R Schwafel
Range Manager: Ronald Pierce
Launch Officer: Ray Martinez
Operations Controller: Kathe Rich

Spaceport Systems International
3769-E Constellation Road, Lompoc, California 93436
Tel: (+1 805) 733 73 70
Fax: (+1 805) 733 73 72
e-mail: cwaller@calspace.com
CEO: Orlando Severo
Business Development: Cheryl Waller

The Johns Hopkins University Applied Physics Laboratory
Johns Hopkins Road, Laurel, Maryland 20723-6099
Tel: (+1 301) 953 50 00
Telex: 511303
Fax: (+1 301) 953 10 93
WWW: http://sd-www.jhuapl.edu
Director, APL: Dr Gary L Smith
Head, APL Space Department: Dr Stamatios M Krimigis
Assistant Department Head, Business: Larry J Crawford
Assistant Department Head, Operations: Robert J Danchik
Chief Engineer, Space Department: Eric J Hoffman
Programmes Manager: Tom Coughlin

Western Commercial Space Center, Inc
3765-A Constellation Road, Lompoc, California 93436
Tel: (+1 805) 733 47 00
Fax: (+1 805) 733 07 17
e-mail: wcsc41110267@www1.utech.net
Executive Director: Gary Stanley
Director, External Affairs: Jana Dawn Bott

White Sands Missile Range
Las Cruces, New Mexico 88002-5032
Tel: (+1 505) 678 11 34
Fax: (+1 505) 678 71 74
WWW: http://www.wsmr.army.mil
Commander: Brig Gen Jerry L Laus

SPACE INDUSTRIAL PROVIDERS

AUSTRALIA

Asia Pacific Aerospace Consultants Pty Ltd
PO Box 98, Hawker, Australian Capital Territory 2614
Tel: (+61 6) 255 26 67
Fax: (+61 6) 255 22 31

Auspace Limited
50 Hoskins Street, Mitchell, Australian Capital Territory 2911
Postal: PO Box 17 Mitchell, Australian Capital Territory 2911
Tel: (+61 6) 242 26 11
Fax: (+61 6) 241 66 64
Managing Director: Ted Stapinski
Major Projects Manager: Ed Roberts

Australian Space Insurance Group
15 Queen St, Level 5, Melbourne, Victoria 3000
Tel: (+61 3) 96 14 33 37
Fax: (+61 3) 96 29 42 95
Manager: Warwick D Jones

British Aerospace Australia
14 Park Way, Technology Park,
The Levels, South Australia 5095
Tel: (+61 8) 290 88 88
Fax: (+61 8) 290 88 33
e-mail: enquiries@baea.com.au
WWW: http://www.baea.com.au
Chairman: A E Harris AC
Group Chief Executive: Robin Southwell
Managing Director: Barry Murphy
Head of International Marketing: Tony Ennis
Head of Corporate Affairs and Communications: Julie Shearn

Electro Optic Systems Pty Ltd
Queanbeyan, New South Wales
Tel: (+61 6) 299 24 70
Fax: (+61 6) 299 76 87
Chief Executive Officer: Dr Ben Greene
Marketing Director: R Quodling

Hawker De Havilland Ltd
361 Milperra Road,
Bankstown, New South Wales 2200
Tel: (+61 2) 97 72 81 11
Telex: aa 20719
Fax: (+61 2) 97 92 36 04
Marketing Manager: F J Makin

Kel Aerospace Pty Ltd
231 High Street, Ashburton, Victoria 3147
Tel: (+61 3) 889 00 22
Fax: (+61 3) 889 00 06
WWW: http://www.kel.com.au
Managing Director: Terry D Kelly
Information Systems Manager: Garry Elliott

AUSTRIA

Österreichische Raumfahrt-Und Systemtechnik GmbH (ORS)
Operngasse 20b, A-1040 Vienna
Tel: (+43 1) 58 81 40
Fax: (+43 1) 58 81 42 21
Managing Director: Dipl Ing U C F Hansson

ACT Hochleistungskunststofftechnik GmbH
Advanced Composites Technology
Hauptstrasse 2, A-2630 Ternitz
Tel: (+43 2630) 314
Fax: (+43 2630) 314 65
Managing Director: August Fenz

Andritz AG
Stattegger Strasse 18, A-8045 Graz
Tel: (+43 316) 69 02 24 39
Telex: 311313 andz
Fax: (+43 316) 690 24 02
Contact: Dr Hansjörg Weinberger

Austrian Research Centre Seibersdorf
Österreichisches Forschungszentrum
Seibersdorf GmbH
A-2444 Seibersdorf
Tel: (+43 2254) 78 00
Telex: 14353 fzs a
Fax: (+43 2254) 740 60
Managing Director: Dr Franz Leberl
Space Activities: Prof Dr Friedrich Rüdenauer
Space Materials and Technology: Dipl Ing Dr Ernst Semerad

Böhler Edelstahl GmbH
POB 96, Mariazeller Strasse 25, A-8605 Kapfenberg
Tel: (+43 3862) 20 71 81
Fax: (+43 3862) 20 74 60

Fischer – FACC
POB 192, Fischerstrasse 9, A-4910 Ried Im Innkreis
Tel: (+43 7752) 909
Telex: 27734
Fax: (+43 7752) 837 98
General Manager, Sales/R&D: Walter Stephan
Technical Director: Manfred Neuboeck

Plansee AG
Postfach 74, A-6600 Reutte/Tirol
Tel: (+43 5672) 60 00
Telex: 55505 plan a
Fax: (+43 5672) 60 05 00/10
President: Dr Michael Schwarzkopf
Marketing: Alfred Troy
Base Products: Josef Tinzl
Finished Products: Fritz Simander
Technology Centre: Dr Günther Kneringer

Steyr-Daimler-Puch Fahrzeugtechnik AG Co KG
Space Technology
Puchstrasse 85, A-8020 Graz
Tel: (+43 316) 404 31 28/60
Fax: (+43 316) 404 38 83
General Manager Space Technology: Dr H-P Gryksa
Product Manager: Helmut Malnig

Schrack Aerospace GmbH
106-108 Breitenfurter Strasse, A-1120 Vienna
Tel: (+43 1) 80 19 90
Fax: (+43 1) 80 19 96 95
Managing Directors: Mario Rohracher, Ulf Hansson
Marketing Manager: Bernhard Eichinger

Siemens AG Österreich
Gudrunstrasse 11, A-1100 Vienna
Tel: (+43 1) 170 74 52 10
Fax: (+43 1) 170 75 56 89
e-mail: oskar.beckmann@siemens.co.at
Space Business: Oskar W Beckmann

BELGIUM

Advanced Mechanical and Optical Systems SA
Rue des Chasseurs Ardennais, B-4031 Angleur Liège
Tel: (+32 43) 61 40 40
Fax: (+32 43) 67 20 07
General Manager: B Collin
Technical Manager: C Flebus

Advanced Products NV
Industrieterrein Krekelenberg,
Rupelweg 9, B-2850 Boom
Tel: (+32 3) 880 81 50
Fax: (+32 3) 888 48 62
Managing Director: M Nedee

Alcatel ETCA
(Alcatel Telecom/Space Business Line)
101 Rue Chapelle-Beaussart,
B-6100 Mont-sur-Marchienne
Tel: (+32 71) 44 28 61
Telex: 51766 belsdt b
Fax: (+32 71) 43 57 78
Managing Director: R Hannon
Strategy: M Praet
Advanced Programmes, Onboard: N Limbourg
Advanced Programmes, Ground: M Melotte
Programme Control/Co-ordination: A Poncin
Sales/Contracts: G Lacroix

Barco NV
Theodoor Sevenslaan 106, B-8500 Kortrijk
Tel: (+32 56) 23 32 11
Telex: 85842
Fax: (+32 56) 23 34 62
Vice-President: Luc Vandenbroucke
Product Managers: Guy Putman, Geert Callens

Belgospace
Rue des Drapiers 21, B-1050 Brussels
Tel: (+32 2) 510 25 20
Telex: 21078
Fax: (+32 2) 510 24 56
President: Noél Parmentier
Director: Roger Godechoul

CISET International SpA
Rue St Jacques 350, B-5500 Dinant
Tel: (+32 82) 22 29 38
Fax: (+32 82) 22 68 27
Managing Director: Antonio Moscatelli
Business Development: Gian Carlo Coletta

Delft Sensor Systems OIP NV/SA
Westerring 21, B-9700 Oudenaarde
Tel: (+32 55) 33 38 11
Fax: (+32 55) 31 68 95
WWW: http://www.oip.be
Managing Director: Freddy Versluys
Scientific/Technical Advisor: G Brandt

IMEC vzw
Kapeldreef 75, B-3001 Leuven
Tel: (+32 16) 28 12 11
Fax: (+32 16) 22 94 00
WWW: http://www.imec.be
President: Prof Roger Van Overstraeten
Director, Research Contracts: Johan Van Helleputte

Logica SA/NV
Rue Colonel Bourgstraat 105, B-1040 Brussels
Tel: (+32 2) 745 07 45
Fax: (+32 2) 705 51 93
e-mail: dwl@brussels.logica.com
WWW: http://www.logica.com
General Manager: Michel Correz
Business Development Manager, Space: David W Lytton

NV Bekaert SA
Bekaert Fibre Technologies
Bekaertstraat 2, B-8550 Zwevegem
Tel: (+32 56) 76 69 83
Fax: (+32 56) 76 79 66
Marketing Manager: Dr Roger De Bruyne

NV Fabrisys SA
Heideveld 1, B-1654 Huizingen
Tel: (+32 23) 63 17 11
Fax: (+32 23) 60 25 85
Managing Director: Marc Mertens

Pedeo Techniek NV
Martijn Van Torhoutstraat 15, B-9700 Oudenaarde
Tel: (+32 55) 31 35 61
Fax: (+32 55) 31 26 85
Space Co-ordinator: A Lafaut

SABCA
1470 Chausée de Haecht, B-1130 Brussels
Tel: (+32 2) 729 55 11
Telex: 21237 sabca b
Fax: (+32 2) 729 58 96
General Manager: Remo Pellichero
Commercial Manager: Marc Humblet
Marketing Manager: Ranbir Bhatti

SETTAS SA
Allée Centrale – Zone Industrielle, B-6040 Jumet
Tel: (+32 71) 34 44 88
Telex: 51418
Fax: (+32 71) 34 56 11
General Manager: Francois Hauzeur
Sales Manager: Larry I J Lambert

SONACA SA
Parc Industriel, Route National Cinq, B-6041 Gosselies
Tel: (+32 71) 25 51 11
Fax: (+32 71) 34 40 35
R&D/Space Contracts Manager: Pierre Taquet

Space Applications Services SA/NV
Leuvensesteenweg 325, B-1932 Zaventem
Tel: (+32 2) 721 54 84
Fax: (+32 2) 721 54 44
e-mail: sas@sas.be
Managing Director: R Aked

Space Information Centre
(Centre d'Information Spatiale)
Route de la Croix-Maga, 54 B-4860 Pepinster
Tel: (+32 87) 46 11 39
Fax: (+32 87) 46 91 13
Managing Director: Theo Pirard

Spacebel
Parc Scientifique du Sart Tilman
Avenue Pré Aily, B-4031 Angleur Liège
Tel: (+32 41) 66 66 11
Fax: (+32 41) 66 04 33
Managing Director: Bernard Stevenot
General Manager: Francis Vanderhaeghen

Trasys Space SA
Leuvensesteenweg 510, bus 43, B-1930 Zaventem
Tel: (+32 2) 773 86 37
Telex: 21852 tract b
Fax: (+32 2) 773 79 10
Managing Director: Philippe Cacheux
Trasys Space Manager: Bruno Lemaire
Sales Manager: Wilfried van Welden

Verhaert Design & Development NV
Hogen Akkerhoekstraat 9, B-9150 Kruibeke
Tel: (+32 3) 250 14 14
Fax: (+32 3) 253 14 64
e-mail: 100331.3220@compuserve.com
Space Division: Piet Holbrouck
Marketing Manager: Katrien Verelst

BRAZIL

Aeroeletronica Industria De Componentes Avionicos SA
Av Sertório 4400, 91040-620 Porto Alegre RS
Tel: (+55 51) 361 12 22
Fax: (+55 51) 361 27 73
President: Claudio Miguel Barreto Viana
Engineering Technical Director: Joáo Claudio Jotz

Andrade Gutierrez Quimica Ltda
Rua Dos Pampas, 484 Sala 63 Prado,
30410-580 Belo Horizonte Mg
Tel: (+55 31) 290 67 77/67 78
Fax: (+55 31) 290 67 75
Directors: Sérgio Andrade, Roberto Gutierrez
General Manager: Fernando B Dias
Plant Manager: Miguel C Dos Santos

Bernardini SA
Rua Hipólito Soares 79,
04201-090 Ipiranga, São Paulo, SP
Tel: (+55 11) 274 80 33
Telex: 11-21605 bsaibr
Fax: (+55 11) 274 85 67
President: Ricardo Bernardini Cury
Vice-President: Flavio Bernardini

Elebra Sistemas de Defesa e Controles Ltda
Rua Bogaert 326, 04298-020 São Paulo, SP
Tel: (+55 11) 969 16 64
Fax: (+55 11) 969 19 66
President: Alexandre de Carvalho
Aerospace Manager: Giacomo Rondani

Esca Engenharia de Sistemas de Controle e Automação SA
Alameda Araguaia 1142 – Alphaville,
06455-940 Barueri, São Paulo
Tel: (+55 11) 725 41 00
Fax: (+55 11) 725 25 45
President: Steve A Ortiz
Engineering Division Director: Pedro Cury
Marketing Director: Antonio Almeida Filho

Promon Eletronica Ltda
Av Pres Juscelino Kubitscheck 1830,
04543-900 São Paulo, SP
Tel: (+55 11) 827 44 11
Telex: 11 37875 prom br
Fax: (+55 11) 827 04 72
President: Carlos M Siffert
Technology Centre Manager: Américo Richieri

CANADA

Bristol Aerospace Limited
PO Box 874, 660 Berry Street,
Winnipeg, Manitoba R3C 2S4
Tel: (+1 204) 775 83 31
Telex: 07-57774/57804
Fax: (+1 204) 885 31 95
President: Keith F Burrows
Vice-President Rockets and Space: Dr Fred Christie
Director of Space Engineering: B R Payne
Marketing Manager, Propulsion and Space Systems: David T O'Connor

CAE Electronics Ltd
CP 1800, Saint-Laurent, Quebec H4L 4X4
Tel: (+1 514) 341 67 80
Telex: 05-824856
Fax: (+1 514) 341 76 99
President: J E Caldwell
Vice-President, Sales/Marketing: A Morris

CAL Corporation
Space Systems Group
1050 Morrison Drive, Ottawa, Ontario K2H 8K7
Tel: (+1 613) 820 82 80
Telex: 053-3937
Fax: (+1 613) 820 29 94
President/COO: Dr Neil Mackay
Vice-President, Space Systems Group: John Graham
Director, Business Development, Space Systems: Tony Raab
Director, Marketing/Sales: Chris Gay

EDO Canada Ltd
1940 Centre Avenue NE, Calgary, Alberta T2E 0A7
Tel: (+1 403) 569 54 00
Fax: (+1 403) 569 54 99
President/CEO: J D Moore

Electronics & Space Corporation Of Canada Ltd
1 Nicholas Street, Suite 1102,
Ottawa, Ontario K1N 7B7
Tel: (+1 613) 241 11 10
Fax: (+1 613) 241 26 81
President: David Roberts

Fell-Fab Products
2343 Barton, Station E,
Hamilton, Ontario L8V 5V8
Tel: (+1 905) 560 92 30
Fax: (+1 905) 560 98 46
President: Glen Fell

Fleet Industries
1011 Gilmore Road, Fort Erie, Ontario L2A 5N3
Tel: (+1 905) 871 21 00
Fax: (+1 905) 871 27 22
President/General Manager: Ernie G Dueck
Marketing/Sales Manager: Brian M Oakley

IMAX Corporation
45 Charles Street East, Toronto, Ontario M4Y 1S2
Tel: (+1 416) 960 85 09
Fax: (+1 416) 960 85 96
Chairman/CEO: Bradley J Wechsler
Vice-Chair: Richard Gelfond, Douglas Trumbull
Producers: Graeme Ferguson and June Myers

MPB Technologies Inc
1725 North Service Road, Trans-Canada Highway,
Dorval, Quebec H9P 1J1
Tel: (+1 514) 683 14 90
Fax: (+1 514) 683 17 27
President: Dr M P Bachynski
Project Co-ordinator: Jane Bachynski

Novatronics Inc
PO Box 610, 677 Erie Street,
Stratford, Ontario N5A 6V6
Tel: (+1 519) 271 38 80
Fax: (+1 519) 271 97 81
President: Donald J McDougall
General Manager: Peter VanDrunen
Sales Manager: Paul Smith

Spar Aerospace Limited
Corporate Office
5090 Explorer Drive, Suite 900,
Mississauga, Ontario L4W 4X6
Tel: (+1 905) 629 77 27
Telex: 069-60106
Fax: (+1 905) 629 08 54
President/CEO: Colin D Watson
Senior Vice-President Business Development: David F Masotti
Director, Public Affairs: Deborah M Allan

Spar Space Systems
21025 Trans Canada Highway,
Ste Anne de Bellevue, Quebec H9X 3R2
Tel: (+1 514) 457 21 50
Telex: 05-822792
Fax: (+1 514) 457 27 24

9445 Airport Road, Brampton, Ontario L6S 4J3
Tel: (+1 905) 790 28 00
Telex: 065-27360
Fax: (+1 905) 790 44 00
Senior Vice-President/General Manager: J Cliff Mackay
Vice-President/Assistant General Manager: Steve J Sorocky

Telespace Ltd
505 Eglinton Avenue West, Suite 403
Toronto, Ontario M5N 1B1
Tel: (+1 416) 481 70 05
Fax: (+1 416) 481 47 38
President: Uriel Domb

DENMARK

Alcatel Danmark A/S
(Alcatel Telecom/Space Business Line)
Alcatel Denmark Space
Lautrupvang 2, DK-2750 Ballerup
Tel: (+45) 44 86 75 00
Fax: (+45) 44 86 75 01
Managing Director: Niels Oberg, Alcatel Danmark A/S
Division Director: Michael Peytz, Alcatel Danmark A/S
Commercial Marketing/Sales: Aksel Møller Christensen
ESA Marketing/Sales: Chresten Overbeck

DELTA
4 Venlighedsvej, DK-2970 Hoersholm
Tel: (+45) 42 86 77 22
Fax: (+45) 42 86 58 98
Managing Director: Torben Grønning
Electronics Engineering: Ole Steensen

Per Udsen Co Aircraft Industry A/S
Fabrikvej 1, DK-8500 Grenaa
Tel: (+45) 86 32 19 88
Fax: (+45) 86 32 14 48
Space Manager: Henrik Steen Larsen
Marketing Manager: John Rygaard

Terma Elektronik A/S
Hovmarken 4, DK-8520 Lystrup
Tel: (+45) 86 22 20 00
Telex: 68106 terma dk
Fax: (+45) 86 22 27 99
President: Johannes Jacobsen
Vice-President Aerospace: Borge Witthoft
Manager, Space Systems: Lars Sejr Kristensen

Computer Resources International A/S
Bregnerødvej 144, DK-3460 Birkerød
Tel: (+45) 45 94 94 00
Fax: (+45) 45 94 94 01
WWW: http://www.cri.dk
President: Steen Hansen
Head of Space Division: Jens Langeland-Knudsen
Head of Commercial Affairs: Jan Lauridsen
Head of Business Development: Jens Guldberg

FINLAND

Finnyards Ltd Electronics
Naulakatu 3, FIN-33100 Tampere
Tel: (+358 3) 12 45 01 11
Fax: (+358 3) 12 13 01 88
Space Technology Manager: Osmo Laaksonen

Metorex International
PO Box 85, FIN-02201 Espoo.
Tel: (+358) 42 14 51
Fax: (+358) 421 35 53
President: Tuula Lukander
Executive Vice-President: Heikki Sipilä

Nokia Research Center
PO Box 45, Heikkiläntie 7, FIN-00211 Helsinki
Tel: (+358 0) 437 61
Fax: (+358 0) 43 76 65 33
Director: Dr Juhani Kuusi
Manager, R&D Co-operation: Markku Ranta
R&D Manager: Pekka Ojanen

Space Systems Finland Ltd
Keilaranta 8, FIN-02151 Espoo
Tel: (+358 9) 61 32 86 00
Fax: (+358 9) 61 32 86 99
e-mail: seppo.korpela@ssf.fi
WWW: http://www.ssf.fi
Managing Director: Dr Seppo Korpela

Technical Research Centre Of Finland
PO Box 1000, FIN-02044 VTT
Postal: Vuorimiehentie 5, Espoo
Tel: (+358) 45 61
Telex: 122972 vttha sf
Fax: (+358) 456 70 00
Director General: Markku Mannerkoski
Space Instruments: Prof Väinö Kelhä
Satellite Communications: Dr Jouko Aurinsalo
Remote Sensing: Dr Tuomas Häme

University Of Helsinki
PO Box 14 (Tähtitorninmäki), FIN-00014
Tel: (+358) 191 29 10
Fax: (+358) 191 29 52
e-mail: kari.lumme@helsinki.fi
Head, Observatory: Prof Kari Lumme

Vaisala Ltd
PO Box 26, FIN-00421 Helsinki
Tel: (+358) 894 92 05
Fax: (+358) 894 92 06
Director, Business Development: Pekka J Kostamo

Ylinen Electronics Co
Teollisuustie 9A, FIN-02700 Kaunianen
Tel: (+358) 505 52 11
Fax: (+358) 505 55 47
Managing Director: Juhana Ylinen
Business Development Director: Nicholas Hughes

FRANCE

Aerospatiale
Espace & Défense
66 route de Verneuil, F-78130 Les Mureaux
BP2, F-78133 Les Mureaux Cedex
Tel: (+33 1) 34 92 12 34
Telex: 695 988 F
Fax: (+33 1) 34 92 12 54
President, Espace & Défense: Michel Delaye
Senior Vice-President, Operations: Jean-Marie Mir
Senior Vice-President, Technical: André Motet
Vice-President, Satellite Affairs: Emmanuel Sartorius
Managing Director, Space Operations: François Calaque
Managing Director, Satellite Operations: Henri Sala
Military Advisor: Gen Jean-Marie Menu
Manned Spaceflight Advisor: Patrick Baudry

Angénieux SA
Aerospace Department
F-42570 Saint-Héand
Tel: (+33 4) 77 90 78 00
Telex: 330695
Fax: (+33 4) 77 90 78 05
General Manager: Denis Suveran
Sales, Defence/Aerospace: Pierre Duranton

Bertin & Cie
59 rue Pierre Curie, BP3, F-78373 Plaisir Cedex
Tel: (+33 1) 34 81 85 00
Telex: 696231
Fax: (+33 1) 30 54 04 14
President: Hervé Hamon
Chief Engineer/Space Contact: Jean Louis Lafon

Brochier SA
33 avenue Franklin Roosevelt, F-69153 Décines Cedex
Tel: (+33 4) 72 05 39 39
Telex: 340829
Fax: (+33 4) 72 02 07 59

CAP Gemini
8 rue Mesplé, BP1155, F-31036 Toulouse Cedex
Tel: (+33 5) 61 31 52 00
Fax: (+33 5) 61 31 53 85
Aerospace Contact: Jean-Pierre Gleyse

CISI
Aerospace Division
13 rue André Villet, BP4042, F-31029 Toulouse
Tel: (+33 5) 61 17 66 66
Fax: (+33 5) 61 54 13 39
Director: Claude Scavazza
Space Systems: Christian Labezin

Credit Lyonnais
Telecommunications & Space Division
1/3 rue des Italiens, F-75009 Paris
Tel: (+33 1) 42 95 54 61
Telex: 285 043 F
Fax: (+33 1) 42 95 03 82
Head of Division: Jean-Pascal Orcel
Senior Manager: Gilles Gantois

Cryospace Gie
route de Verneuil, BP2, F-78133 Les Mureaux Cedex
Tel: (+33 1) 34 92 11 73
Telex: cryospa 695 243
Fax: (+33 1) 34 92 10 52
CEO: Pascal Mauberger

Dactem SA
1545, chemin sous St-Etienne, F-30319 Alès Cedex
Tel: (+33 4) 66 54 21 00
Fax: (+33 4) 66 54 21 01
Director General: Alix Reinhalter

Dassault Aviation
Direction Générale de l'Espace
78 quai Marcel Dassault
F-92552 Saint-Cloud Cedex 300
Tel: (+33 1) 47 11 40 00
Telex: 270929
Fax: (+33 1) 47 11 31 95
Vice-President Research/Engineer:
Bruno Revellin-Falcoz
Director General, Space: Jean Gérard Roussel
Director, Development/Communications: Michel Aubert

Dassault Electronique (Group)
55 quai Marcel Dassault, F-92214 Saint-Cloud
Tel: (+33 1) 49 11 80 00
Telex: 633 299
Fax: (+33 1) 46 02 57 58
Chairman/CEO: Bertrand Daugny
Vice-Chairman/Managing Director: Jacques Mijonnet
Vice-Chairman/Managing Director:
Jean-Etienne Connet
Director, Space Activities: Michel Guionnet
Space Sales Manager: Françoise Gatinois
Manager, Marketing/Sales (Satcoms):
Christophe Courtade

Epsilon Ingenierie
BP653 Voie 5, F-31319 Labège
Tel: (+33 5) 62 24 31 07
Fax: (+33 5) 61 39 99 16
e-mail: epsilon@epsilon.fr
Managing Director: Bruno Desaunettes

Faugere & Jutheau SA
La Sémaphore, 54 quai Michelet,
F-92681 Levallois-Perret Cedex
Tel: (+33 1) 41 34 50 55
Telex: 616 826 elysu
Fax: (+33 1) 41 34 55 00
Contacts: Stanislas Chapron,
Jean Fournier and José Da Costa

Filotex
140-146 rue Eugène-Delacroix, F-91210 Draveil Cedex
Tel: (+33 1) 69 83 78 00
Telex: 601666
Fax: (+33 1) 69 42 05 70
Space Manager: Jean-Paul Bouyer
Commercial Director: Claude Lemaire

Fortech Airforge
BP45, F-63502 Issoire Cedex
Tel: (+33 4) 73 55 54 00
Fax: (+33 4) 73 55 54 01
Space Activities: M Dupont
Marketing/Sales Director: Y C Ricci

Framatome Connectors France
145 rue Yves Le Coz, F-78035 Versailles Cedex
Tel: (+33 1) 39 49 21 83
Telex: 695506
Fax: (+33 1) 39 49 20 00

Space Division
3 avenue Maréchal Devaux, F-91550 Paray-Vieille Poste
Tel: (+33 1) 46 87 23 23
Fax: (+33 1) 45 60 58 69
Contact: Didier Chabaud

FRB-Connectron SA
BP251, F-92602 Asnières
Tel: (+33 1) 46 13 36 36
Telex: 620764
Fax: (+33 1) 47 90 05 73
Sales Manager: P Courteille

L'Air Liquide
75 quai d'Orsay, F-75321 Paris Cedex 07
Tel: (+33 1) 40 62 55 55
Telex: 270716 alss
Fax: (+33 1) 45 55 58 76
Chairman: Alain Joly
CEO: Gérard Levy
Executive Vice-President/Company Secretary:
Michel Bourgarel

Advanced Technology Division
PO Box 15, F-38360 Sassenage
Tel: (+33 4) 76 43 60 30
Fax: (+33 4) 76 43 60 98
General Manager: Robert Claveyrolas

LanceXport
84 boulevard de la République, F-92420 Vaucressor
Tel: (+33 1) 47 95 17 01
Fax: (+33 1) 47 41 25 20
Senior Partner: Louis Moussy

Latecoere
135 rue de Périole, F-31079 Toulouse Cedex
Tel: (+33 5) 61 58 77 00
Telex: 531714
Fax: (+33 5) 61 26 28 41
Vice-President, Engineering/Design: Michel Carles

Le Blanc De Nicolay
134 rue Danton, BP149,
F-92304 Levallois-Perret Cedex
Tel: (+33 1) 47 59 11 11
Telex: 612430 lbnp
Fax: (+33 1) 47 59 93 82
Space Manager: Thierry Mangot

Lucas Aerospace
Flight Control Paris
7-9 avenue de l'Eguillette, BP7186,
F-95056 Cergy-Pontoise Cedex
Tel: (+33 5) 34 32 63 00
Telex: luca fcs 606330 f
Fax: (+33 5) 34 32 63 10
Sales Manager: Yvon Allain

Progespace Ingenierie
113 avenue Jean-Baptiste Clément, F-92100 Boulogne
Tel: (+33 1) 46 05 66 11
Fax: (+33 1) 46 05 58 38
President/CEO: Daniel Philippe Erard
General Manager: Vincent Rouaix
International Affairs: Carmel Conroy

Prospace
25 rue du Pont Neuf, F-75039 Paris Cedex 01
Tel: (+33 1) 44 76 77 70
Telex: 214 674
Fax: (+33 1) 44 76 78 47
President: Jean Jaques Sussel
Executive Manager: Alain Pernot
Deputy Executive Manager: Eric Cerf-Mayer

Pyromeca
chemin Charles Battezzati, Quartier Lagoubran,
BP2148, F-83063 Toulon
Tel: (+33 4) 94 22 86 86
Fax: (+33 4) 94 22 86 99

Pyrospace
139 route de Verneuil, BP52, F-78130 Les Mureaux
Tel: (+33 1) 34 92 44 22
Fax: (+33 1) 34 74 30 93
Commercial Manager: Serge Morlan

Realix Technologies
Innoparc 2, rue de la Découverte,
BP146, F-31676 Labège Cedex
Tel: (+33 5) 62 24 99 44
Fax: (+33 5) 61 39 96 59
Contact: Christian Gayraud

Relais Electroniques Deutsch
22 rue des Chaises, F-45140 St Jean de la Ruelle
Tel: (+33 2) 38 70 45 19
Telex: 783509
Fax: (+33 2) 38 70 45 72
Space Product Manager: G Dormeuil

REOSC – Group SFIM
avenue de la Tour Maury,
F-91280 Saint Pierre du Perray
Tel: (+33 1) 69 89 72 00
Fax: (+33 1) 69 89 72 20
Chairman/CEO: Michel Faure
Vice-President, Operations Optics: Roland Geyl

Rosemount Aerospace
2 bis, rue Marcel Doret, F-31700 Blagnac
Tel: (+33 5) 61 30 01 61
Fax: (+33 5) 61 30 00 31
Aero Marketing/Sales Manager: Joél Belon

RTC-Compelec
117 quai du President Roosevelt,
F-92134 Issy-les-Moulineaux Cedex
Tel: (+33 1) 40 93 80 00
Telex: 680495
Fax: (+33 1) 40 93 81 27
Sales Director: Alain le Corvec

SAFT
Defence and Space Division
156 avenue de Metz, F-93230 Romainville
Tel: (+33 1) 49 15 36 00
Telex: 235 566
Fax: (+33 1) 49 15 34 00
Director: Jean Barbeau
Marketing/Sales (Europe, Asia, Africa): Anne Sennett
Marketing/Sales (US, S America): Barbara Clark
Tel: (+1 410) 771 32 00; Fax: (+1 410) 771 11 44

Sagem SA
6 avenue d'Iéna, F-75783 Paris Cedex 16
Tel: (+33 1) 40 70 63 63
Fax: (+33 1) 40 70 66 78
Chairman: Pierre Faurre
Vice-President/Managing Director, Defence & Security Division: Daniel Dupuy
Managing Director, Navigation/Missile Guidance Department: Christian Jaeger
Space Products Manager: Jean-Claude Dubuc

SAMM
chemin de la Malmaison, F-91570 Bièvres
Tel: (+33 1) 69 35 80 07
Telex: 604 085
Fax: (+33 1) 69 35 81 98
Chairman: Daniel Riviere
Director of Marketing: Alain Baudry

SAT
41 rue Cantagrel, BP389, F-75626 Paris Cedex 13
Tel: (+33 1) 45 82 34 24
Fax: (+33 1) 45 82 32 28
Space Manager: Michel Royer

Schlumberger Industries
Data Acquisition & Recording Division
1 rue Nieuport, BP54,
F-78140 Vélizy-Villacoublay Cedex
Tel: (+33 1) 30 70 30 70
Telex: enervil 698 201 f
Fax: (+33 1) 30 70 30 80
General Manager: Jean-Paul Vernhes
Director, Marketing: Gilles Danan

SEMA
Group Headquarters
16 rue Barbès, F-92126 Montrouge Cedex
Tel: (+33 1) 40 92 40 92
Fax: (+33 1) 46 56 96 53
Space Manager, Europe: Jean Michel

56 rue Roger Salengro,
F-94126 Fontenay sous Bois
Tel: (+33 1) 43 94 57 10
Space Manager: Bernard Quiquemelle

Sextant Avionique
Aèrodrome de Villacoublay, Zone Aéronautique Louis Brèguet, BP200, F-78141 Velizy-Villacoublay Cedex
Tel: (+33 1) 46 29 70 00
Telex: sxt 631 240 f
Fax: (+33 1) 40 94 02 51
WWW: http://www.sextant-avionique.com

Space Division
25 rue Jules-Védrines, F-26027 Valence Cedex
Tel: (+33 4) 75 79 87 31
Telex: 345 807
Fax: (+33 4) 75 79 86 60
President/CEO: Jean-Robert Martin
Strategy/Communications Manager: Gérard Delalande
Space Division Manager: Henry F Col
Press Contact: Florence Fayolle

SNPE
Le Bouchet Research Centre
9 rue Lavoisier, BP2, F-91710 Vert le Petit
Tel: (+33 1) 64 99 12 34
Fax: (+33 1) 64 99 13 55
WWW: http://www.snpe.fr
Chairman/CEO: Jean Faure
Advanced Technologies & Propulsion: Jacques Cardin
Space Propulsion Programmes: David Quancard

Space Propulsion Programmes
BP57, F-33166 Saint-Médard en Jalles Cedex
Tel: (+33 5) 56 70 50 16
Fax: (+33 5) 56 70 50 52

Société Européenne De Propulsion
24 rue Salomon de Rothschild,
BP303, F-92150 Suresnes Cedex
Tel: (+33 1) 47 28 65 00
Telex: 630 906
Fax: (+33 1) 47 28 05 36
Chairman/CEO: Roger Vignelles
Executive Vice-President: Pierre Betin
Executive Vice-President, Operations: Jacques Rossignol
Director, External Communications: Claude Jarry

Large Liquid Propulsion Division
BP 802, F-27207 Vernon
Tel: (+33 2) 32 21 72 00
Telex: 770 019
Fax: (+33 2) 32 21 27 01
Vice-President/General Manager: Michel Laroche
Ariane 5/Vulcain Programme Manager: Jacques Pasquier
Vulcain Chief Project Engineer: P Signol

Solid Propulsion & Composites Division
Les Cinq Chemins, Le Haillan,
F-33165 St-Medard-en-Jalles
Tel: (+33 5) 56 55 30 00
Telex: 560 678
Fax: (+33 5) 56 55 30 01
Vice-President/General Manager: Guy Corteel

Small Propulsion & Equipment Division
F-77550 Moissy-Cramayel
Tel: (+33 1) 64 71 46 97
Fax: (+33 1) 64 37 68 27
General Manager: Bernard Portejoie

SODERN
20 avenue Descartes,
F-94451 Limeil-Brévannes Cedex
Tel: (+33 1) 45 95 70 00
Telex: 265543
Fax: (+33 1) 45 69 14 02
President/General Manager: Henri Dugré
Director, Space Activity: Philippe Lugherini
Director, Optical Activity: Claude Babolat
Public Relations: Jean-Luc Guillaume

Sopemea
Zone Aéronautique Louis-Bréguet, BP48,
F-78142 Vélizy-Villacoublay Cedex
Tel: (+33 1) 46 30 22 74
Telex: 634 137 f
Fax: (+33 1) 46 30 54 06
Managing Director: J Paul

Starsem SA
33 rue Fernand Forest
F-95150 Suresnes
Tel: (+33 1) 46 25 03 60
Fax: (+33 1) 45 06 05 78
President: François Calaque (Managing Director of Aerospatiale's Space Business Unit)
Director General: Viktor Kuznetsov (Russian Space Agency)
Commercial Operations: Jean-Charles Vincent (Arianespace)

Steel Technologies
F-31260 Mazères-sur-Salat
Tel: (+33 5) 61 98 45 45
Telex: 521 132
Fax: (+33 5) 61 97 45 60
Technical Manager: Jean-Marie Ribeyrol

Thomson-CSF
173 boulevard Haussmann, F-75415 Paris Cedex 08
Tel: (+33 1) 53 77 84 78
Fax: (+33 1) 53 77 86 05
Chairman/CEO: A Gomez
Space Director: Michel Traizet
Marketing: Bernard Exartier

Thomson-CSF Semiconducteurs Specifiques
route Départementale 128, F-91401 Orsay Cedex
Tel: (+33 1) 69 33 00 00
Fax: (+33 1) 69 33 03 45
CEO: Michel Roger
Technical Manager: Jean-Michel Brice

GERMANY

Alcatel ANS
Lorenzstrasse 10, Postfach 40 07 49,
D-7000 Stuttgart 40
Tel: (+49 711) 821 20 11
Fax: (+49 711) 821 46 66

Amsat-Deutschland eV
Holderstrauch 10, D-35041 Marburg/Lahn
Tel: (+49 6421) 28 35 50/51
Fax: (+49 6421) 28 56 65
e-mail: ga950@mailer.uni-marburg.de
President: Dr Karl Meinzer
Vice-President: Werner Haas

Anite Systems GmbH
Im Leuschnerpark 4, D-64347 Griesheim
Tel: (+49 6155) 871 70
Fax: (+49 6155) 87 17 87
Managing Director: Bob Rouse

Autoflug GmbH & Co
Industriestrasse 1, D-25462 Rellingen
Tel: (+49 4101) 30 72 15
Fax: (+49 4101) 30 72 07
Managing Director: Andreas Sedlmayr

Bank Reuschel & Co
Maximiliansplatz 13, D-80333 Munich
Tel: (+49 89) 239 50
Telex: 523 690 reuex d
Fax: (+49 89) 33 11 45
Space Activities: Rainer Willner

Carl Zeiss
D-73446 Oberkochen
Tel: (+49 7364) 200
Fax: (+49 7364) 20 43 43
Space Marketing: Wolfgang Hasselmann

Computer Anwendung für Management GmbH
Oberföhringerstrasse 2, D-81679 Munich
Tel: (+49 89) 92 80 40
Fax: (+49 89) 928 04 27
Managing Director: Frank H Beck

Daimler-Benz Aerospace AG (DASA)
Postfach 80 11 09, D-81663 Munich
Tel: (+49 89) 60 70
Telex: 5287-0 dasam d
Fax: (+49 89) 60 72 64 81
President/CEO: Dr Manfred Bischoff
President, Space Systems Group: Werner Heinzmann
Executive Vice-President: Andreas Sperl
Vice-President, Communications: Christian Poppe

Daimler-Benz Aerospace/Space Infrastructure Division
PO Box 10 59 09,
Hunefeldstrasse 1-5, D-28059 Bremen
Tel: (+49 421) 539 53 26
Telex: 245 548 a erno d
Fax: (+49 421) 539 45 34
Vice-President /General Manager: Josef Kind
Vice-President, Technical Operations: Dr Stefan Graul
Vice-President, Programmes: Werner Inden

Daimler Jena-Optronik GmbH
Prüssingstrasse 41, D-07745 Jena-Göschwitz
Tel: (+49 3641) 20 01 10
Fax: (+49 3641) 20 02 22
Managing Director: Helmut Kappel
Technical Director: Dr Waldemar Skarus
Commercial Director: Dr Elke Schmiedeknecht

Dornier Satellitensysteme GmbH
D-88039 Friedrichshafen
Tel: (+49 7545) 801
Fax: (+49 7545) 844 11
President: Dr Klaus Ensslin

Drägerwerk AG
Aerospace Division
Moislinger Allee 53/55, Postfach 1339,
D-23542 Lübeck
Tel: (+49 451) 882 37 35
Telex: 26807-0 dwl d
Fax: (+49 451) 882 41 60
Space Technical Manager: Ulrich von Czapiewski

Elbe Space & Technology Dresden GmbH & Co KG
Universitatallee 5, D-28359 Bremen
Tel: (+49 421) 20 13 40
Fax: (+49 421) 201 34 56
Senior Project Manager: Peter W Kunigk

Elektronik-und Luftfahrtgeräte GmbH
Justus-von-Liebig Strasse 18, D-53121 Bonn
Tel: (+49 6151) 90 26 34
Fax: (+49 6151) 90 20 19
Space Operations Manager: Dr Gerhard Gritner

ESG Elektronik System GmbH
Einsteinstrasse 174, D-81675 Munich
Tel: (+49 89) 921 60
Telex: 522594
Fax: (+49 89) 92 16 26 31
Space Manager: Axel Popella

FRIWO Silberkraft
Gesellschaft für Batterietechnik mbH
Meidericher Strasse 6-8,
PO Box 10 05 52, D-47005 Duisburg
Tel: (+49 203) 300 20
Telex: 855543 friwo
Fax: (+49 203) 30 02 240/185
General Managers: Dr Albrecht Leuschner, Helmut Ebert, Johannes Cornelissen
Director, Application Engineering: Johannes Cornelissen
Space Product Manager: Heribert Kickel
Director, Sales & Marketing: Helmut Ebert

Institut für Raumfahrtsysteme
Universität Stuttgart
Pfaffenwaldring 31, D-70550 Stuttgart
Tel: (+49 711) 685 23 75
Telex: 7255445 univ
Fax: (+49 711) 685 35 96
WWW: http://www.irs.uni-stuttgart.de
Director: Prof Dr Ernst W Messerschmid
Deputies: Prof Dr-Ing habil Monika Auweter-Kurtz (Space Transportation Technology), Dr Ulrich Schöttle

Kayser-Threde GmbH
Wolfratshauserstrasse 44-48, D-81379 Munich
Tel: (+49 89) 72 49 50
Telex: 529749
Fax: (+49 89) 72 49 52 91
WWW: http://www.kayser-threde.de
Managing Directors: Reiner Klett, Axel Schmalz

Kayser-Threde Moscow
Splav Technical Centre
9 Baikalskaya Street, 107497 Moscow, Russia
Tel: (+7 095) 462 46 88
Fax: (+7 095) 462 40 35

Kayser Italia srl
Piazza 185 Regg.art.Folgore 1, I-57128 Livorno, Italy
Tel: (+39 0586) 50 02 33
Fax: (+39 0586) 50 07 30
Manager: Dr Valfredo Zolesi

VRS GmbH
Verkehr Raumfahrt Systemtechnik
Jüdengasse 14-16, D-04600 Altenburg
Tel: (+49 34) 47 56 56 30
Fax: (+49 34) 47 56 56 56
Managing Director: Günter Buchmüller

Medialario srl
Localita Pascolo, I-22040 Bosisio, Parini (Como), Italy
Tel: (+39 031) 86 71 11
Fax: (+39 031) 87 65 95
Manager: Dr Arnaldo Valenzuela

Krupp VDM GmbH
PO Box 1820, D-58778 Werdohl
Tel: (+49 2392) 550
Telex: 8264330 vmd
Fax: (+49 2392) 55 22 17
Marketing Manager: W R Herda

Linde AG
Dr Carl von Linde Strasse 6-14,
D-82049 Hoellriegelskreuth
Tel: (+49 89) 744 50
Telex: 528 327-0 li d
Fax: (+49 89) 74 45 49 29
Managing Director: Dr Aldo Belloni
Space Cryogenics Department: Dr Joachim Wolf

MAN Gutehoffnungshutte AG
(Mechanical Structures and Systems Department)
Ginsheimer Strasse 1, D-65462 Ginsheim-Gustavsburg
Tel: (+49 6134) 552 19
Telex: 4 182 058 man d
Fax: (+49 6134) 552 04

MAN Technologie AG
Space Systems Division
Liebigstrasse 5a, D-85757 Karlsfeld
Postal: Postfach 1347, D-85751 Karlsfeld
Tel: (+49 8131) 89 01
Fax: (+49 8131) 89 19 00
Director, Space Systems Division: Klaus Kaiba

Munich Reinsurance Company
Aviation & Space Division
Königinstrasse 107, Postfach 40 13 20,
D-80791 Munich
Tel: (+49 89) 389 10
Telex: 5215233-0 mr d
Fax: (+49 89) 38 91 36 22
WWW: http://www.munichre.com
Executive Management: Hartmut Hesse

Racal-Heim GmbH
Huttenstrasse 38, D-51469 Bergisch Gladbach
Tel: (+49 2202) 93 68 30
Fax: (+49 2202) 93 68 35
Managing Director: D A Poole
Operations Director: D H Nutton

Rockot
Hünefeldstrasse 1-5,
Postfach 10 59 46, D-28199 Bremen
Tel: (+49 421) 539 65 01/65 10
Fax: (+49 421) 539 65 00
CEO: Günther Stamerjohanns
Mission Manager: Dr Mark A Kinnersley

Satellite Operational Services GmbH
Postfach 1449,
Gutenbergstrasse 9,
D-82199 Gilching
Tel: (+49 8105) 384 30
Fax: (+49 8105) 38 43 12
Managing Directors: Carl E Johnson
and Tonia F Johnson van Werven
General Manager: Georg Foraita
Mission Operations: Rolf Burgdorf
Tel/Fax: (+49 8153) 25 80
Flight Dynamics: Dr Jacobus Herman
Tel: (+49 8153) 28 12 92
Fax: (+49 8153) 25 80

Technical University of Berlin
Institute for Aeronautics & Astronautics, F6
Marchstrasse 12, D-10587 Berlin
Tel: (+49 30) 31 42 23 08
Telex: 184262 tubln d
Fax: (+49 30) 31 42 28 66
Space Department Managers:
Professors Udo Renner, R Lo

Teldix GmbH
PO Box 10 56 08, Grenzhöfer Weg 36,
D-69046 Heidelberg
Tel: (+49 6221) 51 20
Fax: (+49 6221) 51 23 05
Director: Richard Hopman (Rainer Götting, deputy)
Head of Sales, Space Products: Uco A Merk
Head of Product Planning/Marketing: Albrecht Wand
Head of Space Products: Gerhard Röckl

W L Gore & Associates GmbH
Nordring 1, D-91785 Pleinfeld
Tel: (+49 9144) 60 10
Telex: 624 689
Fax: (+49 9144) 60 14 53
General Managers: Dr Heinrich Flik, Ralf Tillmanns
Space Products (Co-ordination & Sales): Michael Wahl

Zeppelin Technologie GmbH
Leutholdstrasse 108, D-88045 Friedrichshafen
Tel: (+49 7541) 20 23 60
Fax: (+49 7541) 20 24 80

INDIA

Alfa-Laval (India) Ltd
Bombay-Pune Road, Dapodi, Pune 411 012
Tel: (+91 212) 77 63 21
Telex: 0146-219/259
Fax: (+91 212) 864 79
Managing Director: Mrs L F Poonawalla
General Manager, Marketing: H Kaushal

Antrix Corporation Ltd
Antariksh Complex, Near New Bel Road,
Bangalore 560 094
Tel: (+91 80) 333 44 74
Telex: 0845-2499/2326
Fax: (+91 80) 333 42 76/42 29
Chairman: Dr Krishnaswamy Kasturirangan
Executive Director: N Sampath

Anup Engineering PVT Ltd
Near Anil Starch, Anil Road, Odhav,
Ahmedabad 380025
Tel: (+91 272) 36 46 21
Telex: 0120-327
Managing Director: Arun P Sheth

Development Consultants PVT Ltd
24B Park Street, Calcutta 700016
Tel: (+91 33) 29 76 01
Fax: (+91 33) 29 28 97
Deputy Managing Director: G C Nundy

Godrej & Boyce Manufacturing Co PVT Ltd
Plant 18 Pirojshanagar, Bombay 400079 Vikhroli
Tel: (+91 22) 517 22 30
Telex: 011-71502
Fax: (+91 22) 517 09 00

HBL Ltd
6-3-871 Snehalata, Greenlands Road, Begumpet,
Hyderabad 500 016
Tel: (+91 40) 31 14 67/8/9
Telex: 0425-6993 NIFE IN
Fax: (+91 40) 31 35 08
Chairman/Managing Director: Dr A J Prasad
Technical Director: Ashok Nagarkatti

Hindustan Aeronautics Limited
Aerospace Division
PO Box 7502, Bangalore 560 075
Tel: (+91 80) 558 14 03
Telex: 0845-2908 pslv in
Fax: (+91 80) 558 16 51
General Manager: L M Bhardwaj
Deputy Manager (Marketing): Jitendra Bhatia

London Liaison Office
India House, High Commission of India
Aldwych, London WC2B 4NA, UK
Tel: (+44 171) 497 23 60
Telex: 264616 halldn g
Fax: (+44 171) 497 83 98
Resident Manager: Mr S C V Rathnam

Hindustan Organic Chemicals Ltd
Harchandrai House, 81 Mahashri Karve Road,
Bombay 400 002

ISRO Inertial Systems Unit
ISRO Post Office, Trivandrum 695 022
Tel: (+91 471) 56 24 44
Telex: 0435-201
Fax: (+91 471) 797 75
Director: N Vedachalam

Laboratory for Electro-Optic Systems (LEOS)
ISRO Satellite Centre (ISAC)
A 5-6 Peenya Industrial Estate,
Peenya, Bangalore 560 058
Tel: (+91 80) 39 64 70/39 42 19
Telex: 0845-5066
Fax: (+91 80) 56 76 21
Director: T K Alex

Larsen & Toubro Ltd
Powai Works, Saki Vihar Road,
PO Box 8901, Bombay 400 072
Tel: (+91 22) 578 14 01/11
Tel: (+91 22) 579 19 14
Telex: 11-71693/71694/71695/72152
Fax: (+91 22) 578 45 94
Vice-President, Operations: A M Naik
Deputy General Manager, Marketing: N Nigam

Liquid Propulsion Systems Centre
Valiamala, Trivandrum 695547
Tel: (+91 471) 57 22 57
Telex: 0435-396 LPSCIN
Fax: (+91 471) 46 26 86
Director: G Madhavan Nair

Siva Vasu Electronics Works
Kowdiar PO, Trivandrum
Tel: (+91 471) 623 73

Solid Propulsion Group
Vikram Sarabhai Space Centre
Trivandrum 695022
Tel: (+91 471) 56 24 44/ 56 25 55
Telex: 0435-201, 0884-202
Deputy Director: R Nagappa

INTERNATIONAL

Arianespace
boulevard de l'Europe, BP177,
F-91006 Evry Cedex, France
Tel: (+33 1) 60 87 60 00
Telex: aresp 602392 f
Fax: (+33 1) 60 87 62 47
Chairman/CEO: Jean-Marie Luton
Executive Vice-President/COO: Francis Avanzi
Secretary General: Francoise Bouzitat
Senior Vice-President, Industrial/
Technical Affairs: Claude Quievre
Senior Vice-President, Marketing/
International Affairs: Ralph-Werner Jaeger
Senior Vice-President, Finance/Admin: Hervé Loiseau
Vice-President Comms/International Relations:
Patrice Albrecht
Operations Director: Jean-Marcel Agasse
Marketing Service: Patrice Larcher
Industrial Director: Edouard Perrez

Arianespace Inc
700 13th Street NW, Washington DC 20005, USA
Tel: (+1 202) 628 39 36
Telex: ariane wsh 897006
Fax: (+1 202) 628 39 49
President: Douglas Heydon

Arianespace Tokyo
Kasumigaseki Building, 31F, 3-2-5, Kasumigaseki,
Chiyoda-ku Tokyo 100, Japan
Tel: (+81 3) 35 92-27 66
Fax: (+81 3) 35 92-27 68
Manager: Jean Louis Claudon
Assistant Manager: Jacques Roelandts

Arianespace Kourou
BP809, 97388 Kourou Cedex, French Guiana
Tel: (+594) 33 47 07
Telex: aresp 910387 fg
Fax: (+594) 33 69 13

Boeing Commercial Space Co
Sea Launch Program, PO Box 3999, MS 8X-57,
Seattle, Washington 98124-2499, USA
Tel: (+1 206) 477 57 70
Fax: (+1 206) 393 02 50
WWW: http://www.boeing.com

Cospar
51 boulevard de Montmorency, F-75016 Paris, France
Tel: (+33 1) 45 25 06 79
Telex: 645554 COSPAR F
Fax: (+33 1) 40 50 98 27
e-mail: COSPAR@paris7.jussieu.fr
President: Prof Gerhard Haerendel
Executive Director: Prof S Grzedzielski

ESTL (European Space Tribology Laboratory)
AEA Technology PLC, Risley, Warrington, Cheshire WA3 6AT, UK
Tel: (+44 1925) 25 36 50
Fax: (+44 1925) 25 24 15
Head: Dr Robert A Rowntree (25 36 54)
Head of Research: Dr E W Roberts (25 30 15)
Head of Engineering: S Gill (25 21 14)
Marketing and Sales: Simon Griffin (25 31 03)

Eucosat
6 rue Galilée, F-75116 Paris, France
Tel: (+33 1) 47 20 41 58
Fax: (+33 1) 47 20 41 41
President: Jean-Pierre Fourré

European Centre for Space Law
8-10 rue Mario-Nikis, F-75738 Paris Cedex, France
Tel: (+33 1) 53 69 76 05
Fax: (+33 1) 53 69 75 60
e-mail: rroeland@esa.bitnet
President: Gabriel Lafferranderie
Secretary: Rafaël Roelandt

EUROPEAN COMMISSION Directorate General 12 on Science
Research and Development
Rue de la Loi 200, B-1049 Brussels, Belgium
Tel: (+32 2) 235 11 11
Fax: (+32 3) 235 01 22/01 23/01 24
Head of Space Division: Michel Paillon

European Space Agency
ESTEC
PO Box 299, NL-2200 AG Noordwijk zh, Netherlands
Tel: (+31 71) 565 39 84
Fax: (+31 79) 565 54 21
e-mail: hschoeye@vmprofs.estec.esa.nl
ESA Senior Propulsion Engineer Special Assignments: Herman F R Schöyer

European Space Agency
8-10 rue Mario-Nikis,
F-75738 Paris Cedex, France
Tel: (+33 1) 53 69 76 54
Fax: (+33 1) 53 69 75 60
Director, Launchers: Fredrik Engström
Head, Ariane Department: Raymond Orye
Programme Manager, Ariane 5: Jacques H Durand

Europropulsion
11 rue Salomon de Rothschild,
F-92150 Suresnes, France
Tel: (+33 1) 40 99 10 41
Telex: eurprop 612436 f
Fax: (+33 1) 40 99 10 48
Chairman: Erder Mingoli
Deputy General Manager: Jean Pierre Ledey

Eurospace
16 rue Hamelin, F-75116 Paris, France
Tel: (+33 1) 47 55 83 00
Fax: (+33 1) 47 55 63 30
President: Professor Ernesto Vallerani
Secretary General: Alain Gaubert
Director of Studies: Marcel Toussaint
Public Relations: Rosy Plet

International Launch Services (ILS)
101 West Broadway, Suite 2000,
San Diego, California 92101, USA
Tel: (+1 619) 645 64 00
Fax: (+1 619) 645 65 00
President, Proton: Charles H Lloyd
President, Atlas: Michael R Wash
Vice-President Sales and Marketing: Gregory Gilmore
Vice-President Programmes and Mission Management: Dennis Dunbar
Strategic Analysis Director: Victoria Kuick
Atlas Customer Services: Joan Underwood
Proton Customer Services: Tammy McDonald
Public Affairs Director: Julie Andrews

Washington Office
1725 Jefferson Davis Highway, Suite 300,
Arlington, Virginia 22202-4127, USA.
Director, Washington Office: Greg Gilmore

Denver Operations
PO Box 179, Denver, Colorado 80201-0179, USA
Tel: (+1 303) 977 52 50
Fax: (+1 303) 971 24 72

International Space University
Parc d'Innovation, Strasbourg Central Campus,
boulevard Gonthier d'Andernach,
F-67400 Illkirch, France
Tel: (+33) 88 65 54 30
Fax: (+33) 88 65 54 47
WWW: http://www.isunet.edu
President: Dr Roland Doré
Admissions: Fabienne Muller
Tel (+33 3) 88 65 54 46
e-mail: admissions@isu.isunet.edu
Public Relations/Press: Caroline Vignard

3400 International Drive NW, 4M-Suite 400,
Washington DC 20008-3098, USA
Tel: (+1 202) 237 19 87
Fax: (+1 202) 237 83 36
Contact: Greg Richey

Lockheed-Khrunichev-Energia (LKE) International, Inc
President: Charles H Lloyd
Vice-President: Arthur Woods
Senior Director: Larry Whitfield
Chief Engineer: Eric Laursen

Matra Marconi Space
37 avenue Louis Bréguet, BP1,
F-78146 Vélizy-Villacoublay Cedex, France
Tel: (+33 1) 34 88 30 00
Telex: matra 698077 f
Fax: (+33 1) 34 88 43 43
Chairman/CEO: Louis-Armand Carlier
Deputy CEO/Managing Director: Nicholas Frank
Marketing/Strategy: Patrick Jourdan
Launchers/Manned Flight: Glaude Guionnet
Civil Communications: André Hardy
Science/Optical Observation: Jean-Michel Aubertin
Science/Radar Observation: Dave Link
Military Space: Philippe Aubay, Paul Down
Ground Systems: Jon Sayer
Public Relations: Daniel Berthelot

31 rue des Cosmonautes, ZI du Palays,
F-31077 Toulouse, France
Tel: (+33 5) 61 39 61 39
Telex: matra 530980 f
Fax: (+33 5) 61 54 57 10
Senior Director: Bernard Plano

Anchorage Road,
Portsmouth, Hampshire PO3 5PU, UK
Tel: (+44 1705) 70 57 05
Telex: 86666 marsat g
Fax: (+44 1705) 70 57 06/7
Senior Director: Bob Waterhouse

Gunnels Wood Road,
Stevenage, Hertfordshire SG1 2AS, UK
Tel: (+44 1438) 31 34 56
Fax: (+44 1438) 73 66 37
Senior Director: Paul M Down

PO Box 16, Filton, Bristol BS12 7YB, UK
Tel: (+44 117) 900 69 00
Fax: (+44 117) 900 68 49/50
Senior Director: Dave Link

Sea Launch Co, LDC
Strandveien 37, N-1324 Lysaker, Norway
Tel: (+47) 67 52 63 50
Fax: (+47) 67 52 63 99
President: Ron Olson

IRELAND

Adtec Teoranta Timoney Technology
Gibbstown, Navan, County Meath
Tel: (+353 46) 541 83
Telex: 43588 til ei
Fax: (+353 46) 544 24
Chief Executive: E P Timoney
Aerospace Manager: C Byrne

Cara Data Processing Ltd
Palmerston House, Fenian Street, Dublin 2
Tel: (+353 1) 661 90 66
Fax: (+353 1) 671 61 13
Managing Director: Patrick J McNamara
General Manager, CARA Software/Services (International) Ltd: Jim Fitzgerald

Devtec Limited
20-21 Viscount Avenue,
Airways Industrial Estate, Dublin 17
Tel: (+353 1) 42 66 88
Telex: 31998
Fax: (+353 1) 42 60 20
Managing Director: Ciaran Mulhall
Technical Marketing Manager: Bryan Reid

Farran Technology Ltd
Ballincollig, Cork
Tel: (+353 21) 87 28 14
Fax: (+353 21) 87 38 92
WWW: http://www.iol.ie/farran
Managing Director: Dr Gerry T Wrixon
Marketing Manager: David Vizard

Forbairt
Glasnevin, Dublin 9
Tel: (+353 1) 37 01 01
Telex: 32501 olas ei
Fax: (+353 1) 37 96 20
Chief Executive: Dan Flinter
Science/Technology: Peter D Coyle
Industrial: Tony McDonald
Scientific: Brian O'Donnell

Space Technology (Ireland) Ltd
Industrial Park, St Patrick's College,
Maynooth, County Kildare
Tel: (+353 1) 628 67 88
Fax: (+353 1) 628 64 70
e-mail: stil@vax1.may.ie
Managing Director: Professor Susan McKenna-Lawlor
Marketing Director: Daniel Gleeson
Technical Director: Peter Rusznyak

Team Aer Lingus
Dublin Airport, MD 108, Dublin
Tel: (+353 1) 705 33 06
Telex: 33556
Fax: (+353 1) 705 64 73
Manager, Space Department: Michael Kelly

ISRAEL

Amsat-Israel
1 Hahartsit Street, IL-76568 Rehovot
IARU Satellite Co-ordinator: Shlomo Menuhin

Elbit Ltd
Advanced Technology Centre
PO Box 539 Hof Hacamel, IL-31053 Haifa
Tel: (+972 4) 31 53 15
Fax: (+972 4) 55 00 02
President/CEO: Emmanuel Gill
Executive Vice-President/Business Development: Shlomo Haruvi

Elop Electro-Optics Industries Ltd
Advanced Technology Park
Kiryat Weizmann, PO Box 1165, IL-76111 Rehovot
Tel: (+972 8) 38 62 11
Telex: 381328
Fax: (+972 8) 38 62 37
President/CEO: J Toren
Corporate Marketing/Sales: J Alon
Manager, Space Product Line: Avi Blasberger
Manager, Remote Sensing Marketing and Business Development: Eytan Reis
Marketing Communications Manager: Aviva Haleva

Elta Electronics Industries Ltd
PO Box 330, IL-77102 Ashdod
Tel: (+972 8) 57 23 33
Telex: 381807/381874 elta il
Fax: (+972 8) 56 39 30/56 39 33
Managing Director: Yehuda Frankel
Sales/Marketing Director: Dr Avia Kafri

Israel Aircraft Industries Ltd
Ben-Gurion International Airport, IL-70100 Tel Aviv
Tel: (+972 3) 935 31 11
Telex: 371133/371102/371114
Fax: (+972 3) 935 22 90/31 31
President/CEO: Moshe Keret
Executive Vice-President, Planning/Business Development: Dr M Dvir
Director, Corporate Advertising: Doron Suslik
Advertising/Foreign Press: Ilana Sternfeld

Israel Military Industries Ltd (IMI)
Rocket Systems Division
PO Box 1044/66-04, IL-47100 Ramat Hasharon
Tel: (+972 8) 924 26 84/927 74 47
Fax: (+972 8) 925 28 96
President/CEO: Shlomo Milo
Marketing Director: Israel Weinreb

MBT Systems & Space Technology
Industrial Zone, PO Box 105, IL-56000 Yehud
Tel: (+972 3) 531 55 55
Telex: 341450 mbt il
Fax: (+972 3) 536 52 05/33 76
Corporate Vice-President/General Manager: Shmuel Alkon
MBT General Manager: A Herzug
MBT Deputy Manager: Uri Manor
Manager, Space Technologies Directorate: Dr Moshe Bar-Lev
Deputy Manager, Space Technologies Directorate: Alex Kachko

MLM System Engineering & Integration
PO Box 45, IL-70350 Beer Yaakov
Tel: (+972 8) 27 25 11
Fax: (+972 8) 27 28 90
Corporate Vice-President/General Manager: O Harari
NEXT Program Manager: Professor Claude Oiknine
Marketing Manager: A Har-Even

Rafael Manor
Propulsion and Explosives Systems
PO Box 2250, IL-31021 Haifa
Tel: (+972 4) 79 24 51/79 39 79
Fax: (+972 4) 79 43 92/79 47 16
WWW: http://www.rafael.co.il
Marketing: Siegfried Wenkert

Rotem Industries Ltd
PO Box 9046, IL-84190 Beer Sheva
Tel: (+972 7) 56 83 00/56 85 57
Telex: 5375 rtmkg il
Fax: (+972 7) 55 45 02
CEO: Gideon Shavit
Chief Scientist: Emanuel Riklis
Export Manager: Oded Kaplan

Soreq Nuclear Research Centre
IL-81800 Yavne
Tel: (+972 8) 943 47 36/41 25
Fax: (+972 8) 943 44 03
Head, Space Technology Section: Dr Yeshayahu Lifshitz

Tadiran Spectralink Ltd
29 Hamerkava Street, PO Box 150, IL-58101 Holon
Tel: (+972 3) 557 31 02
Fax: (+972 3) 557 73 84
Marketing Manager: Ovadia Cohen

US Office
Tadiran Electronic Industries Inc
1215 Jefferson Davis Highway,
Arlington, Virginia 22202, USA
Tel: (+1 703) 416 91 43
Fax: (+1 703) 416 22 13

Tamam Precision Instrument Industries
Yehud Industrial Zone, PO Box 75, IL-56100 Yehud
Tel: (+972 3) 531 50 03
Fax: (+972 2) 531 51 40
Director, Technologies Department: Ehud Aviran

ITALY

Aerocosmos srl
Corso Vittorio Emanuele 24, I-20122 Milano
Tel: (+39 02) 76 00 18 47
Fax: (+39 02) 78 31 47
Chairman: Dr Luciano Zanotti
Director General: Dr Amilcare Bosso
New Project Manager: Dr Alberto Ortona

Alenia Difesa
(A Finmeccanica Company)
Officine Galileo Industrial Unit
Via A Einstein 35, I-50013 Campi Bisenzio, Florence
Tel: (+39 055) 895 01
Telex: 570126 galile i
Fax: (+39 055) 895 06 05
Space Manager: Dr Ivo Varano
Engineering Manager: Ing Armando Buccheri
Marketing Manager: Ing Guglielmo Cecchi

Alenia Spazio SpA
HQ: Via Saccomuro 24, I-00131 Rome
Tel: (+39 06) 415 11
Telex: 614337/614047
Fax: (+39 06) 419 06 75/419 07 73
Chairman: Prof Ernesto Vallerani
Managing Director: Ing Antonio Rodotà
General Manager: Giuseppe Viriglio

Corso Marche 41, I-10146 Turin
Tel: (+39 11) 718 01
Telex: alspat i
Fax: (+39 11) 72 33 07/72 48 07

Amsat-Italy
c/o Meridian Microtech srl,
Viale Piave 7, I-20129 Milano
Tel: (+39 02) 76 02 17 35
Fax: (+39 02) 76 02 22 91
e-mail: 71360.3467 (CompuServe)
President: Paolo Pitacco

Assicurazioni Generali SpA
Piazza Duca degli Abruzzi 2, I-34132 Trieste
Tel: (+39 040) 67 16 52 (direct Space Dept)
Fax: (+39 040) 67 18 72 (direct Space Dept)
Chairman: Antoine Bernheim
Deputy General Manager: Benito Pagnanelli
Head, Space Department: Giovanni Gobbo

BPD Defence & Space
Corso Garibaldi 22, I-00034 Colleferro
Tel: (+39 06) 972 91
Fax: (+39 06) 97 29 22 05
CEO: Ing Davide Maccagnani
Space Transportation Business Line: Antonio Fabrizi
Marketing Manager: André Van Gaver
External Relations Director: Roberto Bonetti

Cap Gemini SpA
Via dei Berio 91, I-00155 Rome
Tel: (+39 06) 22 59 31
Fax: (+39 06) 228 66 49
General Manager: Dr Giulio Chiarini
Space Manager: Gianfranco Corini

Carlo Gavazzi Space SpA
Via Gallarate 139, I-20151 Milano
Tel: (+39 02) 33 40 00 38
Telex: 335066 cgait i
Fax: (+39 02) 308 64 58
e-mail: carlo.gavazzi.space@galact ca.it
Chairman/Managing Director: Manfred Fuchs
Marketing Manager: Lanfranco Zucconi

Cisi Aid
Piazza della Repubblica 32, I-20124 Milano
Tel: (+39 02) 670 55 12
Fax: (+39 02) 670 41 69
Managing Director: Antonio Mazzarelli
Space Marketing: Guido Albertini

CNUCE
Via Santa Maria 36, I-56126 Pisa
Tel: (+39 050) 59 31 11
Fax: (+39 050) 90 40 52 (G3)
Fax: (+39 050) 90 40 51 (G4/ISDN)
WWW: http://www.cnuce.pi.cnr.it (CNUCE Institute)
WWW: http://apollo.cnuce.cnr.it
(Spaceflight Dynamics Section)
CNUCE Director: Professor. Luca Simoncini
Head, Space & Structure Mechanics Dept:
Dr Andrea Pagni
Co-ordinator, Spaceflight Dynamics Section:
Dr Andrea Cardillo

Dataspazio SpA
Via Elio Vittorini 129, I-00144 Rome
Tel: (+39 06) 502 22 43 44
Telex: 613436 dtmt i
Fax: (+39 06) 502 23 31
President: Ing Mario De Leo
Managing Director: Ing E Cuturi
General Manager: Ing Sabatino Stornelli
Technical Manager: Ing L Magliozzi

Elettronica SpA
Via Tiburtina Km 13.7, I-00131 Rome
Tel: (+39 06) 415 41
Telex: 626527
Fax: (+39 06) 419 28 69

Fiar SpA
Via Montefeltro 8, I-20156 Milan
Tel: (+39 02) 35 79 01
Telex: 331140 fiarmo i
Fax: (+39 02) 33 40 09 81
President: Raffaele Esposito
General Manager: Giancarlo Grasso
General Manager Space Group: Luigi Giardino
Space Division Marketing: Marialberto Mensa

FiatAvio SpA
312 Via Nizza, I-10127 Turin
Head Office: 112 Corso Ferrucci, I-10138 Turin
Tel: (+39 011) 685 81 11
Fax: (+39 011) 685 84 86
Space Commercial Manager: Pier G Lasagni

Intecs Sistemi SpA
Via Zoe Fontana, Tecnocitta, I-00131 Rome
Tel: (+39 06) 41 88 61
Fax: (+39 06) 419 16 67
Managing Director: S Ciarrocca
Space Marketing: Dr Sergio Ferri

Laben Proel Tecnologie Division
Viale Machiavelli 31, I-50125 Florence
Tel: (+39 055) 22 87 31
Telex: 574 152 proel i
Fax: (+39 055) 229 82 28
Laben General Manager: M Pascucci

Laben SpA
SS Padana Superiore 290,
I-20090 Vimodrone (Milan)
Tel: (+39 02) 25 07 51
Telex: 312616
Fax: (+39 02) 250 55 15
Chairman: Dr Ing Antonio Zibellini
Managing Director: Dr Ing Alessandro Bellini
General Manager: Dr Ing Marco Pascucci
Technical Director: Dr Francesco di Tolle

Microtecnica SpA
Via Madama Cristina 147, I-10100 Turin
Tel: (+39 011) 693 21
Telex: 221190 micro i
Fax: (+39 011) 693 23 11
President: Giorgio Lapidari
Space Department Manager: Marino Ferrara

Space Software Italia
Viale del Lavoro 101, Quartiere Paolo VI,
I-74100 Taranto
Tel: (+39 099) 470 11 01
Fax: (+39 099) 470 17 77
General Manager: Giuseppe Corona
Technical Director: Piero La Vopa
Marketing Manager: Sabino Cacucci

Tecnodata Italia srl
Via Luciano Manara 5, I-00044 Frascati (Rome)
Tel: (+39 06) 941 98 01/2/3
Fax: (+39 06) 941 94 26
Managing Director: Alan Brown
Director, Sales/Marketing: John Bowley
Business Development Managers:
Fred Bonus (N Europe), John Harland (S Europe)

Tecnospazio SpA
Via Montefeltro 8, I-20156 Milan
Tel: (+39 02) 38 00 50 49
Fax: (+39 02) 38 00 44 69
General Manager: Vittorio Venturini
Marketing Manager: Mauro Fenzi

Vitrociset SpA
Via Salaria 1027, I-00138 Rome
Tel: (+39 06) 88 17 01
Telex: 611422 ciset i
Fax: (+39 06) 88 17 02 52
President/Managing Director: Girolamo Cartia
General Manager: Alberico Pellacchia
Marketing/Sales Manager: Vittorio Vergara
Space Division Director: Riccardo Grazi

JAPAN

Daicel Chemical Industries Ltd
Propulsion Systems Division
1 Teppo-cho, Sakai, Osaka 590
Tel: (+81 722) 27 31 11
Telex: 522 2987 daicel j
Fax: (+81 722) 27 30 00
General Manager: Takehiko Takao
Propulsion Systems Division: Mineo Moriike

Fuji Heavy Industries Limited
Subaru Building, 7-2 1-chome Nishi-shinjuku,
Shinjuku-ku, Tokyo 160
Tel: (+81 3) 33 47 25 25
Telex: 0 232-2268 fuji j
Fax: (+81 3) 33 47 25 88
Managing Director/General Manager Aerospace Division: Yasuyuki Kogure
General Manager Commercial Business: Kisaburo Wani

Fujitsu Limited
Space Technology Development Group
1015 Kamikodanaka Nakahara-ku, Kawasaki 211
Tel: (+81 44) 777 11 11
President: Tadashi Sekizawa
General Manager, Space Technology Development Group: Tomoyoshi Yamashita

High-Reliability Components Corp
Shuwa No 3 Shiba-Park Building,
2-10-12 Shiba-Daimon, Minato-ku, Tokyo 105
Tel: (+81 3) 34 35 95 91
Fax: (+81 3) 34 35 95 92
President: Masayuki Shimodaira

IHI Company Ltd
Space Development Division
Aero-Engines & Space Operations
2-1 Ohtemachi 2-chome, Chiyoda-ku, Tokyo 100
Tel: (+81 3) 32 44 57 79
Telex: ihihqt j 22232
Fax: (+81 3) 32 44 56 68
Division Director: Yukio Kitamura
Deputy Division Director: Masato Manago
General Manager Marketing/Sales:
Takashi Shiota
General Manager Space Experiment Project Office:
Yoshiaki Ano
General Manager Propulsion Sys Engineering:
Masakuni Nakamura

Itochu Corporation
5-1 Kita-Aoyama 2-chome, Minato-ku, Tokyo 107-77
Tel: (+81 3) 34 97 31 86
Fax: (+81 3) 34 97 31 77

President/CEO: Minoru Murofushi
Division President: Shogo Honma
Communications Business Dept: Yoshitada Sonoda
Group Manager, Satellite and Visual Media Business Section: Chiaki Fujino

Iwatsu Electric Company, Ltd
1-7-41 Kugayama, Suginami-ku, Tokyo
Tel: (+81 3) 33 34 11 11
Fax: (+81 3) 33 34 60 53
Space Systems: Hirosi Isshiki

Japan Aircraft Manufacturing Co Ltd
3175 Showa-machi, Kanazawa-ku, Yokohama 236
Tel: (+81 45) 773 51 18
Fax: (+81 45) 771 12 53
President: Kanji Sonoda
Managing Director/General Manager Sales Division: Masanao Horie
Vice-President: Tomihisa Mori
Director/General Manager Sales Division: Motonobu Akiyama
Managing/General Manager Space Equipment Dept: Kenschi Yuyama
Associate Director/General Manager Engineering Div: Dr Akira Obata

Japan Aviation Electronics Industry Ltd
21-6 Dogenzaka 1-chome, Shibuya-ku, Tokyo 150
Tel: (+81 3) 34 61 33 06
Telex: 29197
Fax: (+81 3) 37 80 29 45
Aerospace Manager: Yukihiko Muraishi
Aerospace Sales: Yoshioki Asakura

Japan Manned Space Systems Corp
Central Building, 1-29-6 Hamamatsu-cho, Minato-ku, Tokyo 105
Tel: (+81 3) 34 36 45 91
Fax: (+81 3) 34 36 45 15
Contact: Tsuneo Miyagawa

Tsukuba office
Yasuda Building, 1-1-26 Kawaguchi, Tsuchiura-shi 300
Tel: (+81 298) 26 87 51
Fax: (+81 298) 26 87 61

Kyokuto Boeki Kaisha, Ltd
CPO Box 330, 2-1 Otemachi 2-chome, Chiyoda-ku, Tokyo 100-91
Tel: (+81 3) 32 44 35 11
Telex: 22440
Fax: (+81 3) 32 46 21 48
President: Motoo Imakita
General Manager, Aerospace and Defense Systems Department: Hiroshi Inoue
Business Manager, General Affairs Department: Isao Maekawa

Mitsubishi Corporation
3-1 Marunouchi 2-chome, Chiyoda-ku, Tokyo 100-86
Tel: (+81 3) 31 20 45 85
Fax: (+81 3) 32 10 42 95
President: Minoru Makihara
General Manager Aerospace Department: Toru Ariga

Mitsubishi Electric Corp
2-2-3 Marunouchi, Chiyoda-ku, Tokyo 100
Tel: (+81 3) 32 18 33 96
Fax: (+81 3) 32 18 33 14
President: Takashi Kitaoka
Manager, Space Programs and Satellite Communications Department: Youkichi Hirota
Electronics Systems Marketing: H Kamei

Mitsubishi Heavy Industries Ltd
Space Systems Department
5-1 Marunouchi 2-chome, Chiyoda-ku, Tokyo 100
Tel: (+81 3) 32 12 31 11
Fax: (+81 3) 32 18 28 95
President: Nobuyuki Masuda
General Manager, Space Systems Department: Keiji Yasuki

Launcher production:
Nagoya Aerospace Systems Works, 10 Oye-cho, Minato-ku, Nagoya 455
Tel: (+81 52) 611 21 11
Telex: 446561
General Manager: Masanori Hirose

Engine production:
Nagoya Guidance & Propulsion Systems Works, 1200 Higashi-Tanaka, Komaki-shi, Aichi 485
Tel: (+81 568) 79 21 11
Telex: 4485850 hishiju j
General Manager: Yasuyuki Oka

Mitsubishi Precision Co Ltd
Mita No 43 Mori Building, 13-6 Mita 3-chome, Minato-ku, Tokyo 108
Tel: (+81 3) 34 53 64 21
Fax: (+81 3) 34 53 64 34
President: Enkoh Itoh
Vice-President/General Manager Group Business Development Electronics Systems: Norihide Eguchi
Deputy General Manager Group Business Development Electronics Systems: Hirosuke Takahashi

Mitsubishi Space Software Co Ltd
World Trade Centre Building, 4-1-2 Hamamatsu-cho, Minato-ku, Tokyo 105
Tel: (+81 3) 34 35 47 26
President: Kazuaki Wakata

Mitsui & Co Ltd
Aerospace Systems Division
2-1 Ohtemachi, 1-chome, Chiyoda-ku, Tokyo 100-91
Tel: (+81 3) 32 85 44 91
Telex: 22253 mitsui j
Fax: (+81 3) 32 85 99 27
General Manager, Aerospace Systems Division: Tomoo Abe

NEC Corporation
7-1, Shiba 5-chome, Minato-ku, Tokyo 108-01
Tel: (+81 3) 34 54 11 11
Telex: nectok j22686
Fax: (+81 3) 37 98 15 10 to 19

Space Systems Division
4035 Ikebe-cho, Tsuzuki-ku, Yokohama 224
Tel: (+81 45) 932 11 11
Fax: (+81 45) 934 21 69
General Manager Space Systems Development Division: Kiyoshi Murata
General Manager Guidance/Electro-Optics Division: Katsutoshi Nakada
General Manager Overseas Microwave and Satellite Communications Systems Division: Motohiro Shichiri

Nippon Propellant Industry Co Ltd
5-9 Nishibenzai 2-chome, Asaka-city, Saitama 351
Tel: (+81 484) 64 14 21
Fax: (+81 484) 64 14 10
General Manager: Shigeki Yamamoto

Production Plant
5-2 Shiraishi-cho, Kawasaki-city, Kanagawa 210
General Manager: Teio Shinya

Nippon Sanso Corporation
Space and Cryogenic Equipment Division
6-2 Kojima-cho, Kawasaki-ku, Kawasaki City, Kanagawa 210
Tel: (+81 44) 288 69 37
Fax: (+81 44) 299 41 08
Marketing: Kizou Sasaki

Nissan Motor Company Ltd
Aerospace Division
3-5-1 Momoi, Suginami-ku, Tokyo 167
Tel: (+81 3) 33 01 66 10
Fax: (+81 3) 33 01 67 14
General Manager: Akio Shinohara
Administration General Manager: Shigeyuki Ogawa
Sales/Marketing General Manager: Masaharu Hama
Space Systems General Manager: Tatsuo Asai

NOF Corporation
Yurakucho Building, 10-1 Yuraku-cho 1-chome, Chiyoda-ku, Tokyo 100
Tel: (+81 3) 32 83 72 77 (Space Dev Group)
Fax: (+81 3) 32 83 72 64
General Manager Explosive/Propulsion Division: Hayami Takahashi
Leader, Space Development Group: Hiroshi Oka

OKI Electric Industry Co Ltd
7-12 Toranomon 1-chome, Minato-ku, Tokyo 105
Tel: (+81 3) 35 01 31 11
Telex: 02226324 okiden j
Managing Director: Yoshio Masuda

Rocket Systems Corporation
Hamamatsu-cho Central Building, 1-29-6 Hamamatsu-cho, Minato-ku, Tokyo 105
Tel: (+81 3) 54 70 79 00
Fax: (+81 3) 54 70 79 50
Chairman: Masato Yamano
President: Takaaki Yamada
Executive Vice-President: Hiroshi Imamura

Sharp Corporation
Photovoltaics Division
282-1 Hajikami, Shinjo-cho, Kitakatsuragi-gun, Nara 639-21
Tel: (+81 7456) 511 61
Fax: (+81 7456) 282 53
Division General Manager: Akio Suzuki

Electronic Components Sales Department
International Sales/Marketing Group
IC/Electronic Components
22-22 Nagaike-cho, Abeno-ku, Osaka 545
Tel: (+81 6) 621 12 21
Fax: (+81 6) 624 01 63
Vice Group General Manager: Nobukazu Yagi and Tadashi Inoue

Showa Denko KK
13-9 Shiba Daimon 1-chome, Minato-ku, Tokyo 105
Tel: (+81 3) 34 38 27 64
Telex: shodenko j 26232
Fax: (+81 3) 34 35 77 06
Chief Manager, Space Sales: Takeshi Adachi

Sumitomo Heavy Industries Ltd
2-1 Ohtemachi 2-chome, Chiyoda-ku, Tokyo
Tel: (+81 3) 32 45 43 21
Telex: 22264
Manager Engineering: Akira Noguchi

Taisei Corporation
Space Technology Section, Technology Development Department
3-25-1 Hyakunin-cho, Shinjuku-ku, Tokyo 169
Tel: (+81 3) 53 86 75 65
Fax: (+81 3) 53 86 75 78
Manager, Technology Development Department: Tetuya Hanamura
Manager, Space Technology Section: Masahito Mori

Toshiba Corporation
1-1 Shibaura 1-chome, Minato-ku, Tokyo 105-01
Tel: (+81 3) 34 57 30 77
Telex: 22587 toshiba j
Fax: (+81 3) 54 76 40 57
Senior Vice-President Electronics and Telecom Systems Group: Shiro Matsuo
General Manager Space Programs Division: Johji Sakuragi
Space Marketing: Minoru Homma

The Japan Amateur Radio League, Inc
PO Box 377, Tokyo

Kawasaki Heavy Industries Ltd
4-1, Hamamatsu-cho 2-chome, Minato-ku, Tokyo 105
Tel: (+81 3) 34 35 22 20
Telex: 242 4371 kawaju
Fax: (+81 3) 34 35 23 54
President: Hiroshi Ohba
Senior General Manager, Aerospace Group: Ryozo Tsutsui
General Manager, Space Systems Dept: Kazunori Sugimura

LIECHTENSTEIN

Balzers AG
FL-9496 Balzers
Tel: (+41 75) 388 41 11
Fax: (+41 75) 388 54 25

NETHERLANDS

Aerospace Propulsion Products BV
PO Box 697, NL-4600 AR Bergen op Zoom
Tel: (+31 1640) 24 59 50
Fax: (+31 1640) 24 59 60
Managing Director: F W M Zee
Marketing Director: Paul A O G Korting

Bradford Engineering
Bradfordstraat 1-2, NL-4645 HE Putte
Postal: Postbus 120, NL-4645 ZJ Putte
Tel: (+31 164) 60 35 50
Fax: (+31 164) 60 21 77
e-mail: Bradford@unisys.iaf.nl
Director: Eduard C J Voeten
Contact, Propulsion Components: Patrick van Put
Contact, Electrical Propulsion Launcher Analysis: Berry Sanders

Brandt Fijnmechanische Industrie BV
H J E Wenckebachweg 117, NL-1096 AM Amsterdam
Tel: (+31 20) 668 12 81
Fax: (+31 20) 692 83 54
Managing Directors: M H Vlastra/T H Dijk
Sales Engineer: E de Jong

CCM
PO Box 12, NL-5670 AA Nuenen
Tel: (+31 40) 83 44 05
Fax: (+31 40) 83 71 35
Space Contact: Ir H P Willemsen

DEP (BV Delft Electronische Producten)
Postbus 60, NL-9300 AB Roden
Tel: (+31 50) 501 88 08
Fax: (+31 50) 501 35 10

Dinfa BV
PO Box 45, NL-2690 AA Gravenzande
Tel: (+31 1748) 144 41
Fax: (+31 1748) 201 00
Managing Director: P G Blom

Dreves Engineering BV
PO Box 1117, NL-6801 BC Arnhem
Tel: (+31 85) 45 46 18
Fax: (+31 85) 51 01 43
Space Department: L M E Verstraeten

Drukker International BV
Beversestraat 20, NL-5431 SH Cuijk
Tel: (+31 485) 39 57 00
Fax: (+31 485) 31 61 04
Contact: Dr R Colston

Fokker Space BV
PO Box 32070, NL-2303 DB Leiden
Tel: (+31 71) 524 50 00
Fax: (+31 71) 524 59 99
e-mail: info@fokkerspace.nl
President: Jan P P Groen
Vice-President Technology: Boudewijn Baud
Vice-President Marketing: B Paul van Ijsselstein
Manager, Business Development: Geert Mennenga
Solar arrays: Koos van der Straaten
Science projects: Hans Werner
Simulation: Erik-Jan Hammega
Launchers: Michiel Meijer
Robotics: Wart Mandersloot

Genius Klinkenberg BV
Noordeinde 71, PO Box 49, NL-1520 AA Wormerveer
Tel: (+31 75) 28 23 41
Fax: (+31 75) 28 71 25
Aerospace Division: S J N M Bakker

Hymec BV
Postbus 336, NL-6130 AH Sittard
Tel: (+31 46) 451 32 20
Fax: (+31 46) 452 68 28
General Manager/Sales: M H Th ten Wolde

ICT Aerospace Group
Keulenstraat 7, PO Box 701, NL-7400 AS Deventer
Tel: (+31 5706) 938 88
Fax: (+31 0570) 62 13 62
Contact: N Koopman

Mechanical Engineering & Consulting (MECON) BV
Postbus 679, NL-7000 AR Doetinchem
Tel: (+31 314) 36 02 11
Fax: (+31 314) 36 04 50
Contact: R A Wiersma

Netherlands Aerospace Group
Boerhaavelaan 40,
PO Box 190, NL-2700 AD Zoetermeer
Tel: (+31 79) 353 13 56
Fax: (+31 79) 353 13 65
General Manager: Ir ing Jan JC Bartels
Office Manager: Erika Roodzant

Origin/Nieuwegein BV
PO Box 1444, NL-3430 BK Nieuwegein
Tel: (+31 30) 608 88 88
Fax: (+31 30) 606 05 77
Managing Director: R Rijke

Polymarin BV
PO Box 151, NL-1670 AD Medemblik
Tel: (+31 227) 54 30 44
Fax: (+31 227) 54 38 60
Managing Director: J J W Opdam
Head of R&D Department: B Roorda

Prins Maurits Laboratory TNO
PO Box 45, NL-2280 AA Rijswijk
Tel: (+31 15) 284 28 42
Fax: (+31 15) 284 39 91

Ramaer Connection Technology BV
PO Box 32, Vossenbeemd 101, NL-5700 AA Helmond
Tel: (+31 4920) 849 11
Fax: (+31 4920) 509 75
Contact: A H M Bakermans

RDM Technology BV
PO Box 1039, NL-3000 BA Rotterdam
Tel: (+31 10) 487 21 98
Fax: (+31 10) 487 22 99

Satellite Services BV
Scheepmakerstraat 40, NL-2222 AC Katwijk
Tel: (+31 71) 402 81 20
Fax: (+31 71) 402 79 34
Managing Director: G S Sims
Technical Director: P van Duijn

Stork Engineers & Contractors BV
Department High Technology/Space Technology
PO Box 58026, NL-1040 HA Amsterdam
Tel: (+31 20) 580 71 11
Telex: 11542
Fax: (+31 20) 580 70 50
Manager, Environmental, Safety and High Technology Division: Ir A Valk
Co-ordinator High Technology/Space Technology Department: Dr RJLH Breukers

Stork Product Engineering BV
Czaar Peterstraat 229,
PO Box 379, NL-1000 AJ Amsterdam
Tel: (+31 20) 556 34 44
Fax: (+31 20) 556 35 56
Managing Director: Ir C R H Bredt
Program Manager, Space Projects: Ir C A L Kemper
Manager Engineering: Ir G F J Janssen

Technisch Physische Dienst TNO-TU Delft
Space Instrumentation Department
Stieltjesweg 1, NL-2628 CK Delft
Postal: PO Box 155, NL-2600 AD Delft
Tel: (+31 15) 269 20 00
Telex: 38091 TPDDT NL
Fax: (+31 15) 269 21 11
Department Manager: Dr Roel Hoekstra
Commercial Manager: ING M Rien van der Kraan
Manager Attitude Sensors: Ir Kees de Boom

NORWAY

AME Space AS
(Alcatel Telecom/Space Business Line)
PO Box 138, Knudsroun 7, N-3191 Horten
Tel: (+47) 3303 27 00
Fax: (+47) 3303 28 00
Managing Director/Space Activities: Sverre Bisgaard
Technical Manager: Jørgen Galteland

CAP Gemini Norway AS
Space Skill Centre Trondheim
Ingvald Ystgaards vei 15,
N-7005 Trondheim
Tel: (+47) 73 82 72 00
Fax: (+47) 73 82 72 01
General Manager: Arne Oen
Space Skill Centre Manager: Vegard Stuan
Space Sales Manager: Eirik Vefsnmo

DET Norske Veritas Industry A/S
PO Box 300, N-1322 Høvik
Tel: (+47) 67 57 75 74
Telex: 0056 76192 verit n
Fax: (+47) 67 57 99 85
Vice-President/Space Manager: Per O Araldsen

Eidel (Eidsvoll Electronics AS)
Gruemyra, N-2080 Eidsvoll
Tel: (+47) 63 96 42 30
Fax: (+47) 63 96 20 48
General Manager: Tore Havstein
Technical Manager: Øystein Sandholt
Product Manager: Ivar Nordby

Kongsberg Gruppen AS
Aerospace
PO Box 1003, N-3601 Kongsberg
Tel: (+47) 32 73 82 00
Fax: (+47) 32 73 93 13 (space)
President: Jan T Jørgensen
Vice-President/General Manager Kongsberg Aerospace: Tom Gerhardsen
Space Programme Director: Gunnar K Berge
Space Programme Marketing Manager: Jon E Kvistedal

Kongsberg Spacetec A/S
Prestvannveien 38, N-9005 Tromsø
Tel: (+47) 77 68 45 00
Fax: (+47) 77 65 58 59
Managing Director: Bjorn Arne Kanck
Marketing Manager: Jan Børre Rydningen

Konsberg Infomasjonskontroll A/S
PO Box 265, Solbråveien 32, N-1371 Asker
Tel: (+47) 66 79 06 60
Fax: (+47) 66 90 03 45
Managing Director: Ingmar Nyheim
Programme Manager: Morten Toverud

Nordic VLSI
PO Box 79, N-7079 Flatåsen
Tel: (+47) 72 58 62 11
Fax: (+47) 72 58 60 05
Managing Director: Oddbjørn Troøyen
Technical Director: Trond Saether
Managing Director: Vegard Wollan

Norsk Elektro Optikk AS
PO Box 17, N-2001 Lillestrøm
Tel: (+47) 63 81 92 00
Fax: (+47) 63 81 90 80
Managing Director: Tycho Jaeger
Director: Peter Kaspersen

Norwegian Defence Research Establishment
PO Box 25, N-2007 Kjeller
Tel: (+47) 63 80 70 00
Fax: (+47) 63 80 72 12
Managing Director: Nils Holme
Chief Scientist, Head of Electronics Division: Henry Kjell Johansen
NISSE Project Co-ordinator: Dr Bjorn T Narheim

Norwegian Industrial Forum for Space Activities (NIFRO)
Middelthunsgate 27,
Postal: PO Box 5250 Majorstua,
N-0303 Oslo
Tel: (+47) 22 96 50 00
Fax: (+47) 22 96 53 71
Manager: Hans Fr Dahle

Prototech A/S
Christian Michelsen Institute
PO Box 12, Fantoftvegen 38, N-5036 Fantoft, Bergen
Tel: (+47) 55 57 41 10
Telex: 40006 cmi n
Fax: (+47) 55 57 41 14
Managing Director: Morten Meibom
Project Managers, Space Technology:
Asbjørn Søreide (Sounding Rockets), Arild Vik (Space)

Raufoss Technology A/S
PO Box 77, N-2831 Raufoss
Tel: (+47) 61 15 26 04
Fax: (+47) 61 15 29 47
Space Manager: Per Eger

Sintef Delab
N-7034 Trondheim
Tel: (+47) 73 59 26 00
Telex: 55 552 delab n
Fax: (+47) 73 59 43 02
Managing Director: Aage J Thunem
Assistant Director: Odd Helge Longva
Branch Head, Satellite Systems: Inge Melhus

Sintef Group
N-7034 Trondheim
Tel: (+47) 73 59 30 00
Telex: 55 620 sintf n
Fax: (+47) 73 59 24 80
President: Roar Arntzen

Ticon Plastindustrier A/S
PO Box 26, N-3001 Drammen
Tel: (+47) 3 84 08 00
Fax: (+47) 3 84 06 95
General Manager: Svein Landaas

RUSSIAN FEDERATION

KB Machinostroenye
1 Turgoyakskoe Shosse,
456300 Miass
Tel: (+7 3513) 52 39 70/52 61 05/52 61 08
Telex: 124858 rosa su
Fax: (+7 3513) 52 60 47
General Director: Dr Igor I Vyelichko
Foreign Relations: Vyacheslav Danilkin

SOUTH AFRICA

Denel (Pty) Ltd
PO Box 8322, Hennopsmeer 0046
Tel: (+27 12) 428 09 11
Fax: (+27 12) 347 01 09
Chairman: Dr John Maree
Managing Director: Johan PL Alberts
Aerospace Group Executive: B Kruger
Corporate Communications: Eben Mouton

SPAIN

Canava Electronica, SA
Adorves Bajos 2, E-23001 Jaén
Tel: (+34 953) 27 20 24
Fax: (+34 953) 27 27 62
President: Jesus Hermoso Pobes
Commercial Director: Alfredo Serrano Chacon

Construcciones Aeronauticas SA (CASA)
Space Division
Avenida Aragón 404, E-28022 Madrid
Tel: (+34 91) 586 37 00
Telex: 48540 casa e
Fax: (+34 91) 747 47 99
Director: Antonio Fuentes
Business Development Manager: Miguel Angel Llorca
Programmes Manager: V Badenes
Engineering Manager: J Iglesias
Production Manager: P Méndez
Ariane Production: F Martínez

Cray Sistemas S.A.
Plaza de Matias, Gutierrez 4, Cobena, E-28863 Madrid
Tel: (+34 91) 620 65 43
Fax: (+34 91) 620 87 79
Managing Director: Petr Bares

CRISA
Computadoras, Redes e Ingeniería SA
Parque Tecnológica de Madrid – PTM,
C/Torres Quevedo 9, E-28760 Tres Cantos – Madrid
Tel: (+34 91) 803 27 28
Fax: (+34 91) 803 65 28
General Manager: Victor Rodrigo
Commercial Director: Enrique Domingo

GMV, S.A.
C/Isaac Newton 11, PTM Tres Cantos, E-28760 Madrid
Tel: (+34 91) 807 21 00
Telex: 48487 gmev e
Fax: (+34 91) 807 21 99
President: Professor JJ Martínez García
Technical Department Manager: Dr C Bachmaier
Marketing Group: LA Mayo, LF Alvarez, JJ Galan

Iberespacio
Glorieta de Quevedo 9, E-28015 Madrid
Tel: (+34 91) 594 48 11
Fax: (+34 91) 445 17 64
Chairman: Manuel Castillo Rubio
Managing Director: Adolfo Garcia Rodriguez
General Manager: Juan Francisco Nebrera Garcia
Technical Manager: Hervé Bigand
Marketing Manager:
Jesús Sánchez Alvarez-Campana

Indra Espacio, SA
Mar Egeo, S/N, Polígono Industrial, no1,
E-28830 San Fernando de Henares (Madrid)
Tel: (+34 91) 396 39 11
Fax: (+34 91) 396 39 12
Managing Director: Jose Maria Marti Fluxá
Business Development Director: Vicente Ruiz

INSA
Orense 4, 9a Planta, E-28020 Madrid
Tel: (+34 91) 556 14 18
Fax: (+34 91) 597 21 81
WWW: http://www.insa.es
Chairman: A Gimenez
Managing Director/CEO: José Ma Dorado
General Secretary: José Ma Parga

NTE SA
Can Malé s/n, E-08186 Lliça d'Amunt (Barcelona)
Tel: (+34 93) 841 42 50
Fax: (+34 93) 841 40 51
e-mail: francesc@NTE.ES
General Manager: Miguel Pastor
Technical Director: Albert Tomas
Marketing Manager: F Gallart

Sener Ingenieria Y Sistemas, SA
Avda de Zugazarte 56, E-48930 Las Arenas (Vizcaya)
Tel: (+34 94) 481 75 00
Fax: (+34 94) 464 12 60
C/Severo Ochoa, s/n,
Parque Tecnológico de Madrid (PTM),
E-28760 Tres Cantos (Madrid)
Tel: (+34 91) 807 70 00
Fax: (+34 91) 807 72 01
Manager Space Department: Alvaro Azcárraga

Tecsidel SA
Avenida República Argentina 6-5, E-08023 Barcelona
Tel: (+34 93) 237 29 80
Telex: 51416 tdel e
Fax: (+34 93) 237 27 61
General Manager: Ramon Recio
Space Manager: Jordi Galera

Universidad Politecnica de Madrid
Centro de Investigación y Dessarrollo Espacial
CIDE/UPM, ETSI Aeronáuticos, E-28040 Madrid
Tel: (+34 91) 336 63 53
Fax: (+34 91) 336 63 63
e-mail: meseguer@sat.cide.upm.es
Director: Dr José Meseguer
Space Activities: Prof Angel Sanz

SWEDEN

ABB HAFO AB
PO Box 520, Bruttovägen 1, SE-17526 Järfälla
Tel: (+46 758) 245 00
Telex: 17262 hafo
Fax: (+46 758) 319 52
President: Christer Ovrén
Space Group Manager: Jan Wassénius
ASICs Sales Manager: Philip R Todd

ACR Electronics AB
PO Box 99, SE-61900 Trosa
Tel: (+46 156) 191 77
Fax: (+46 156) 191 88

Celsiustech Electronics AB
SE-175 88 Järfälla
Tel: (+46 8) 58 08 40 00
Fax: (+46 8) 58 03 22 44
Sales Representative: Stig Gnauder

FFV Aerotech AB
Avionics Division
Kungörsvägen 60, SE-732 32 Arboga
Tel: (+46 589) 800 00
Telex: 73246 ffva s
Fax: (+46 589) 116 52
President: Jan Eiborn
Division Manager: Bengt Bardvall
Manager, Space Activities: Stig Nilsson

SAAB Ericsson Space
SE-405 15 Gothenburg
Tel: (+46 31) 35 00 00
Fax: (+46 31) 35 95 20
President: Ivan Öfverholm
Executive Vice-President: Ulf Berg
Commercial Director: Per Gunnar Lundh
Technical Director: Peter Möller
Manager, Separation and Guidance Systems:
Ingvar Bengtson
Manager, Communications and Public Affairs: Irene Svensson

Technology Trade International
PO Box 14057, SE-104 40 Stockholm
Tel: (+46 8) 660 29 98
Fax: (+46 8) 662 3430
Managing Director: Johan Lundberg

Volvo Aero Corporation
SE-461 81 Trollhättan
Tel: (+46 520) 940 00
Telex: 42040
Fax: (+46 520) 985 50
President/CEO: Arne Wittlöv
Space Division Manager: Bjôrn Sjiblom
Marketing Managers: Ulf Palmnäs and Ike Holm

SWITZERLAND

Aerospace Engineering Office
Nelkenstrasse 7, CH-8033 Zurich
Tel: (+41 1) 363 40 12
Fax: (+41 1) 364 01 04
Managing Director: Kaspar J Zimmermann

APCO Technologies SA
PO Box 59, Avenue de Corsier 1, CH-1800 Vevey 1
Tel: (+41 21) 922 11 22
Fax: (+41 21) 922 11 26
President: André Pugin
Sales Manager: M Flückiger

Compagnie Industrielle Radioelectrique SA (CIR)
Neuenburgstrasse 7, CH-2076 Gals/Berne
Tel: (+41 32) 338 98 00
Fax: (+41 32) 338 98 92
Director: Alain Maurissen

Contraves
Oerlikon-Contraves AG
Schaffhauserstrasse 580, CH-8052 Zurich
Tel: (+41 1) 306 22 11
Telex: 823402
Fax: (+41 1) 306 33 84
Director Space Department: Hanspeter Schneiter

Cryophysics SA
39 Rue Rothschild, CH-1202 Geneva
Tel: (+41 22) 732 95 20
Fax: (+41 22) 738 52 46
WWW: http://ourworld.compuserve.com/homepages/cryophysicsch
President: K A Geiger
General Manager (UK): John Graham
General Manager (France): Philippe Benoist
General Manager (Switzerland/Italy): R Koch
Sales Manager (Germany): Detlef Cieslikowski

CSEM SA
PO Box 41, Jaquet-Droz 1, CH-2007 Neuchâtel
Tel: (+41 327) 20 51 11
Fax: (+41 327) 20 57 20
Chairman: P Arnold
Director General: Dr T Hinderling
Aerospace Programme Manager: Dr M Roulet

ETEL SA
Rue de la Gare 13, CH-2112 Môtiers
Tel: (+41 328) 62 01 23
Fax: (+41 328) 62 01 01
WWW: http://www.etel.ch
Chief Executive Officer: Dr N Wavre
Marketing Manager: Claude-E Brunner

LEICA plc
CH-9435 Heerbrugg
Tel: (+41 71) 70 31 31
Telex: 881222 wi ch
Fax: (+41 71) 72 18 65
Special Products: Hans Hess

Mecanex SA
Vuarpillière 29, CH-1260 Nyon
Tel: (+41 22) 361 77 33
Fax: (+41 22) 361 67 52
General Management: Nicola Thibaudeau
Commercial Manager: Pierre Salzmann
Marketing Manager: Jean-Frédéric Berthoud
Technical Manager: M-Thérèse Ivorra

Oscilloquartz SA
CH-2002 Neuchâtel 2
Tel: (+41 38) 25 85 01
Fax: (+41 38) 25 85 08
President: E Sutcliffe
Sales/Marketing Manager: Alain Jendly

Reusser AG
Gruebstrasse 25, CH-8706 Meilen/Zurich
Tel: (+41 1) 923 29 33
Fax: (+41 1) 923 08 35
General Manager: Peter Reusser

RST Raumfahrt Systemtechnik AG
Gaiserwaldstrasse 14, CH-9015 St Gallen
Tel: (+41 71) 311 28 75
Fax: (+41 71) 311 28 76
Manager: Dr Hans Martin Braun

Swiss Aircraft & Systems Company
CH-6032 Emmen
Tel: (+41 41) 268 41 11
Fax: (+41 41) 260 25 88
WWW: http://www.sfaerospace.ch
Managing Director: Dr W Glanzmann
Commercial Manager: P Engelhard
Technical Manager: H Surbe

Vibro-Meter SA
Route de Moncor 4, CH-1701 Fribourg
Tel: (+41 26) 407 11 11
Telex: 942292 vibr ch
Fax: (+41 26) 402 36 62
Space Manager: Roger Blaser
Space Engineer: Claude Pittet

THAILAND

Thai Microsatellite Company Ltd
Makonakorn University of Technology
51 Chenm-Sampan Road, Nong Chok, Bangkok 10530
Tel: (+66 2) 988 36 88
Fax: (+66 2) 988 36 87
Chairman: Professor Sitthichai Pookiayaudom

UNITED KINGDOM

Advanced Products (Seals & Gaskets) Ltd
No 1 Industrial Estate,
Consett, County Durham DH8 6SR
Tel: (+44 1207) 50 03 17
Fax: (+44 1207) 50 12 10

AEA Technology
Space Systems, Business Dev Manager:
John Andrews
Tel: (+44 1235) 46 35 40; Fax: (+44 1235) 46 42 51
Space Sales Manager: Simon Griffin
Tel: (+44 1235) 46 42 71; Fax: (+44 1235) 46 42 51
e-mail: simon.griffin@aeat.co.uk

Culham Laboratory
Abingdon, Oxfordshire OX14 3DB
Tel: (+44 1235) 52 18 40
Fax: (+44 1235) 46 36 49

Harwell Laboratory
Didcot, Oxfordshire OX11 0RA
Tel: (+44 1235) 43 55 55
Fax: (+44 1235) 43 28 59

Agema Infrared Systems Ltd
Arden House, West Street,
Leighton Buzzard, Bedfordshire LU7 7DD
Tel: (+44 1525) 37 56 60
Fax: (+44 1525) 37 92 71
WWW: http://www.agema.com
Contact: Jenny Jeffreys

Air Products plc
Hersham Place, Molesey Road,
Walton-on-Thames, Surrey KT12 4RZ
Tel: (+44 1932) 24 92 00
Telex: 917243
Fax: (+44 1932) 24 95 65
WWW: http://www.airproducts.con

Alcoa International Ltd
PO Box 6, Marmion House,
Copenhagen Street, Worcester WR1 2EL
Tel: (+44 1905) 61 35 51
Fax: (+44 1905) 251 61
Contact: Darol Baker

Ampex (GB) Ltd
Ampex House, Beechwood, Chineham Business Park,
Chineham, Basingstoke, Hampshire RG24 8WA
Tel: (+44 1256) 81 44 10
Fax: (+44 1256) 81 44 74
Managing Director: Ken Baxter
European Marketing Manager: Willy Björklund
UK Sales Manager: Paul Drew

AMSAT-UK
94 Herongate Road, Wanstead Park, London E12 5EQ
Tel: (+44 181) 989 67 41
Fax: (+44 181) 989 34 30
e-mail: r.broadbent@ee.surrey.ac.uk
WWW: http://www.amsat.org
Honorary Chairman: Professor Martin S Sweeting
Honorary Secretary/Treasurer: Ronald J C Broadbent

Analytical Graphics
2 Hambeldon Place, Dulwich Common,
London SE21 7EY
Tel: (+44 181) 299 38 36
Fax: (+44 181) 693 56 79
WWW: http://www.stk.com (main US site)
UK/European Account Executive: Joseph J Cunningham

Anvil Technology Ltd
4 Wigmores South,
Welwyn Garden City, Hertfordshire AL8 6PL
Tel: (+44 1707) 39 30 99
Fax: (+44 1707) 39 30 76
e-mail: anvil@anvtech.co.uk
Contact: Mark Bentley

Argos Electronics Ltd
Units 1-3, Victory Trading Estate, Kiln Road,
Portsmouth, Hampshire PO3 5LB
Tel: (+44 1705) 69 06 11
Fax: (+44 1705) 69 85 13
Managing Director: Derek H Lloyd
Quality Assurance: John Cornish

Asdor Filters
Division of Howden Aircontrol Ltd
Hawthorne Road, Staines, Middlesex TW18 3AY
Tel: (+44 1784) 46 22 60
Fax: (+44 1784) 46 58 94
Sales Manager: Ron Monger

Astos
109 Chestnut Lane,
Amersham, Buckinghamshire HP6 6DZ
Tel/Fax: (+44 1494) 72 12 10
Chairman: Peter Milne
Secretary: Stan W Tomalin

Avica Equipment
Boundary Way,
Hemel Hempstead, Hertfordshire HP2 7SL
Tel: (+44 1442) 647 11
Telex: 82256
Fax: (+44 1442) 23 00 35
Managing Director: Steve Dowling
Sales/Commercial Director: Roger Tyson
Sales Manager: Mark Bolton

BP Chemicals Ltd
Advanced Materials Division, Aerospace Composites
Avonmouth Road, Avonmouth, Bristol, Avon BS11 9DU
Tel: (+44 117) 982 48 21
Fax: (+44 117) 982 46 40
Business Development Manager: John Janke

British Aerospace Defence Company Ltd
Royal Ordnance Rocket Motors Division
Summerfield, Kidderminster, Worcestershire DY11 7RZ
Tel: (+44 1562) 82 40 61
Fax: (+44 1562) 82 81 26
Managing Director: David Singleton
Division Sales Manager: Malcolm Lassan
Technical Director: Peter Penny

Westcott, Aylesbury, Buckinghamshire HP18 0NZ
Tel: (+44 1296) 65 20 00
Fax: (+44 1296) 65 20 39
Liquid Engine Business: Jeff Curtis

C T Bowring Space Projects Ltd
The Bowring Building, Tower Place, London EC3P 3BE
Tel: (+44 171) 357 10 00
Telex: 882191
Fax: (+44 171) 357 52 78
Chairman: Brian G Moore
Managing Director: J Mc L Munro
Technical Director: B R Yeomans

Commercial Space Technologies Ltd
67 Shakespeare Road, Hanwell, London W7 1LU
Tel: (+44 181) 840 10 82
Fax: (+44 181) 840 77 76
e-mail: james@cstltd.demon.co.uk
Chairman: Gerry Webb
Director, London: James Keravala
Director, North America: Derek Thorpe
Director, Moscow: Nina Pestmal
(Fax: +7 095) 415 77 32

Control Techniques Dynamics Ltd
South Way, Walworth Industrial Estate,
Andover, Hampshire SP10 5AB
Tel: (+44 1264) 32 41 55
Fax: (+44 1264) 35 65 61
Managing Director: M H Boughtwood

Cossor Electronics Limited
The Pinnacles, Elizabeth Way,
Harlow, Essex CM19 5BB
Tel: (+44 1279) 42 68 62
Telex: 81228
Fax: (+44 1279) 41 04 13
Managing Director: T Ligon

Cray Systems Ltd
Space and Defence Division
DAS House, Quayside, Temple Back
Bristol, Avon BS1 6NH
Tel: (+44 117) 927 78 54
Fax: (+44 117) 929 09 17
Managing Director: Bill Black
Sales/Marketing Director: Philip Bostock
Business Development Manager, Earth Observation:
Mark Protheroe

Dowty Aerospace Wolverhampton
Wobaston Road,
Wolverhampton, West Midlands WV9 5EW
Tel: (+44 1902) 39 77 00
Telex: 338220
Fax: (+44 1902) 39 43 94
Chief Executive: Tony Edwards
Space Projects General Manager: Michael Horton
Space Projects Production Manager: Julian Chetter
Space Projects Engineering Manager: Alan Lawrie

Easams Limited
Waters Edge, Riverside Way (Watchmoor Park),
Camberley, Surrey GU15 3PD
Tel: (+44 1276) 68 67 77
Fax: (+44 1276) 68 66 23
Managing Director: Keith Wilman
Space Applications: Bill Summers

EDS Ltd
1/3 Bartley Heath Business Park,
Bartley Way, Hook, Hampshire RG27 9XA
Tel: (+44 1256) 74 20 00
Fax: (+44 1256) 74 23 49
UK SBU President: David Thorpe
Managing Director Defence: Roger Gilbert
Sales Director Defence: Peter Appleby

EEV Ltd
106 Waterhouse Lane, Chelmsford, Essex CM1 2QU
Tel: (+44 1245) 49 34 93
Fax: (+44 1245) 49 24 92
Manager, Solar Power Systems: Dr Trevor Cross
Manager, Special CCDs: Dr Ralph Holtom
Strategic Marketing: Dr John Ashton

EG&G LIMITED
Sealol, Wright Components and
Pressure Science Divisions
Coronation Road, Cressex Industrial Estate,
High Wycombe, Buckinghamshire HP12 3TP
Tel: (+44 1494) 45 16 61
Fax: (+44 1494) 45 24 25
Commercial Manager: John Moore
European Sales Manager: Malcolm Ward

Electron Tubes Ltd
Bury Street, Ruislip, Middlesex HA4 7TA
Tel: (+44 1895) 63 07 71
Fax: (+44 1895) 63 59 53
WWW: http://www.electrotech.demon.co.uk
Managing Director: Jan P Frederiksen
Marketing Director: Anthony G Wright
Technical Director: Ronald McAlpine
Production Director: Clifford Wade

Empcom Systems Ltd
7 Robinia Close, Waterlooville, Hampshire PO7 8HF
Tel: (+44 1705) 26 71 90
Fax: (+44 1705) 26 71 90
Managing Director: Graham Packer
Business Development Manager: Anne Packer

Encore Computer (UK) Ltd
Marlborough House, Mole Business Park,
Leatherhead, Surrey KT22 7BA
Tel: (+44 1372) 81 42 00
Telex: 947707
Fax: (+44 1372) 81 42 01
General Manager: David Rhys-Davies
Storage Sales Manager: Glen Robinson
Marketing Manager: Christopher Atkins

ESYS Limited
Berkeley House, London Square, Cross Lanes,
Guildford, Surrey GU1 1UE
Tel: (+44 1483) 30 45 45
Fax: (+44 1483) 30 38 78
e-mail: mdillon@esys.co.uk
Managing Director: Michael Dillon

Fairey Microfiltrex Ltd
Fareham Industrial Park,
Fareham, Hampshire PO16 8XG
Tel: (+44 1329) 28 56 16
Fax: (+44 1329) 82 24 42
Managing Director: Adrian Ryder
General Sales Manager: L Lee

Firth-Rixson plc
Firth Rixson House, 25 Carbrook Hall Road,
Sheffield, South Yorkshire S9 2EJ
Tel: (+44 114) 261 22 22
Fax: (+44 114) 261 22 23
Chief Executive: David J Hall
Marketing Manager: Mike Wolstenholme

GEC Plessey Semiconductors
High Reliability and Space Facility
Lincoln Industrial Park,
Doddington Road, Lincoln LN6 3LF
Tel: (+44 1522) 50 05 00
Fax: (+44 1522) 50 05 50
e-mail: trefor.davies@lincoln.gpsemi.com
Managing Director: Tom Urwin
Marketing Manager: Trefor Davies

GEC-Marconi Materials Technology Limited (GMMT)
Caswell, Towcester, Northamptonshire NN12 8EQ
Tel: (+44 1327) 35 05 81
Fax: (+44 1327) 35 67 75
Managing Director: Dave Richardson
Commercial Director: Alan Marshall
Sales Manager: Ken Hollands

Wood Burcote
Woods Burcote Way, Burcote Road,
Towcester, Northamptonshire NN12 7JS
Tel: (+44 1327) 35 05 81
Fax: (+44 1327) 35 69 96

GEC-Marconi Radar & Defence Systems Ltd
Navigation & Electro-Optic Systems Division
Silverknowes, Ferry Road,
Edinburgh EH4 4AD
Tel: (+44 131) 332 24 11
Fax: (+44 131) 343 50 50
Managing Director (NESD): L Butler
Marketing Manager: H Spoor

Graseby Dynamics Ltd
Park Avenue, Bushey,
Watford, Hertfordshire WD2 2BW
Tel: (+44 1923) 22 85 66
Fax: (+44 1923) 22 83 20
Technical Director: Dr Robert B Turner
Manager, R&D Business Development: Alan Rhodes

Hunterskil Howard plc
110/113 Tottenham Court Road, London W1A 1BX
Tel: (+44 171) 388 45 55
Fax: (+44 171) 388 04 55
Chairman: Ian Humphries
Technical Executive, Space: Kathy Freeman

IGG Component Technology Ltd
Waterside House, Waterside Gardens,
Fareham, Hampshire PO16 8RR
Tel: (+44 1329) 82 93 11
Telex: 86727
Fax: (+44 1329) 82 93 12
Managing Director: William H Manuel
Project Director: Richard G Matthews
Columbus Director: Geoffrey K Penhaligon
Sales/Marketing Director: Robert Smith

Spartan Microwave
Divisional Director: Richard G Matthews

IMI Titanium Ltd
PO Box 704, Witton, Birmingham B6 7UR
Tel: (+44 121) 356 11 55
Telex: 336771 IMI KYN G
Fax: (+44 121) 356 54 13
Managing Director: John Tamberlin
Technical Director: A C Barber
Sales/Marketing Director: A J Bacon

International Space Brokers Ltd
8 Lloyds Avenue, London EC3N 3HD
Tel: (+44 171) 488 14 14
Telex: 8956151 warcry
Fax: (+44 171) 265 14 53
Chairman: John H Howes
Director, Space Technical Services: Clive P G Smith

Irvin Group Ltd
Reddings Wood, Ampthill, Bedfordshire MK45 2HD
Tel: (+44 1525) 84 11 48
Telex: 82105
Fax: (+44 1525) 40 58 61

US Office
Irvin Aerospace Inc
Santa Ana, California, USA
Fax: (+1 714) 662 15 86 (Phil De Lurgio)

UK Office
Irvin Aerospace Ltd
Letchworth, Hertfordshire, UK
Fax: (+44 1462) 48 20 07 (Mike Woollard)

Italian Office
Irvin Aerospace SpA
Aprilia, Italy
Fax: (+39 06) 92 72 71 65 (Attilio Ciuffi)

Canadian Office
Irvin Aerospace Canada Ltd
Fort Erie, Ontario, Canada
Fax: (+1 416) 871 65 34 (David Wright)

J S Chinn Engineering Co Ltd
Faraday Road, Harrowbrook Industrial Estate,
Hinckley, Leicestershire LE10 3DE
Tel: (+44 1455) 23 83 33
Telex: 31303
Fax: (+44 1455) 25 12 83/89 05 85
Managing Director: T Latimer

Johnson & Higgins Ltd
Aldgate House,
33 Aldgate High Street, London EC3N 1AQ
Tel: (+44 171) 945 74 40
Fax: (+44 171) 945 74 91
Director, Aviation/Space: Timothy J Wright

JRA Aerospace & Technology Ltd
CED House, Taylors Close,
Marlow, Buckinghamshire SL7 1PR
Tel: (+44 1628) 89 11 05
Fax: (+44 1628) 89 05 19
e-mail: jratech@leapfrog.almac.co.uk
Managing Director: John G Rootes
Technology Consultant: Jim Beatty

Kalgard (UK) Ltd
Canalwood Industrial Estate, Chirk, Clwyd LL14 5RL
Tel: (+44 1691) 77 80 70
Fax: (+44 1691) 77 83 03
Chairman: Dr P Reasbeck
Managing Director: A D Thompson
Marketing: D Barlow

KG Coating Limited
Chairman: H Westgaard
Managing Director: A D Thompson
Marketing: D Barlow

Kinesix (Europe) Ltd
5 Bracknell Beeches, Old Bracknell Lane,
Bracknell, Berkshire RG12 7BW
Tel: (+44 1344) 74 23 50
Fax: (+44 1344) 74 23 51
e-mail: sammi@quaduk.co.uk
Marketing Executive: Helen Slaughter

Ling Dynamic Systems
Heath Works, Baldock Road,
Royston, Hertfordshire SG8 5BQ
Tel: (+44 1763) 24 24 24
Fax: (+44 1763) 24 97 15
Marketing Director: Christopher Williams
Sales Manager: Graham Turgoose

Logica UK Ltd
Space Division
74 Portsmouth Road, Cobham, Surrey KT11 3LG
Tel: (+44 171) 637 91 11
Telex: 918096
Fax: (+44 1932) 86 91 12
WWW: http://www.logica.com
Space Director: Walter Cormack
Business Development Manager: Pat Norris

LRE Relays + Electronics Ltd
Signal House, Jacklyns Lane,
Alresford, Hampshire SO24 9JJ
Tel: (+44 1962) 73 44 33
Telex: LRE UK 477516
Fax: (+44 1962) 73 54 38
Senior Sales Executive: Peter Searle

Marshall Aerospace
The Airport, Cambridge CB5 8RX
Tel: (+44 1223) 37 37 37
Telex: 81208
Fax: (+44 1223) 32 10 32
Director of Marketing: Mick E Milne

Martin-Baker Aircraft Co Ltd
Lower Road, Higher Denham,
Uxbridge, Middlesex UB9 5AJ
Tel: (+44 1895) 83 65 92
Telex: 23617 ejects g
Fax: (+44 1895) 83 25 87
Chairman: D J Burrell
Managing Directors: John S Martin, James W Martin
Head of Marketing/Public Relations: Brian A Miller

Matra Marconi Space
PO Box 16, Filton, Bristol, Avon BS12 7YB
Tel: (+44 117) 900 69 00
Fax: (+44 117) 900 68 49/50
Project Manager Sounding Rocket:
Hugh Whitfield (65 84)

Mitsubishi Electric UK Ltd
Electronics Division
Travellers Lane, Hatfield, Hertfordshire AL10 8XB
Tel: (+44 1707) 27 61 00
Telex: 296196
Fax: (+44 1707) 27 86 93
Joint Managing Director: Takeo Iinuma
Project Manager: John F Pacey

Molniya Space Consultancy
25 Redfern Avenue, Whitton, Middlesex TW4 5NA
Tel/Fax: (+44 181) 898 45 71
e-mail: psclark@dircon.co.uk
Contact: Phillip S Clark

Moreton Hall Associates
Morar House, Altwood Close,
Maidenhead, Berkshire SL6 4PP
Tel: (+44 1628) 78 34 55
Fax: (+44 1628) 375 86
Managing Director: Geoffrey E Hall

Mullard Space Science Laboratory
Holmbury St Mary, Dorking, Surrey RH5 6NT
Tel: (+44 1483) 27 41 11
Fax: (+44 1483) 27 83 12
Director: Professor J L Culhane

Normalair-Garrett Ltd
Yeovil, Somerset BA20 2YD
Tel: (+44 1935) 47 51 81
Telex: 46132
Fax: (+44 1935) 42 76 00
Head of Marketing: Mike P Bednall

Parker Hannifin plc
Seal Group
Star House, 69/71 Clarendon Road,
Watford, Hertfordshire WD1 1LP
Tel: (+44 1923) 23 90 09
Fax: (+44 1923) 23 88 33
UK Sales Manager: Don Champken
Sales Office Manager: Jeff Panovic

Penny & Giles Data Systems Ltd
The Mill, Wookey Hole, Wells, Somerset BA5 1BB
Tel: (+44 1749) 67 54 54
Fax: (+44 1749) 67 10 92
WWW: http://www.penny-giles.co.uk
Managing Director: Adrian Gurling
Sales Manager: Richard Ludlow
Product Manager: Edwin Kayes

Pilkington Space Technology
Unit 2, Kinmel Park Industrial Estate,
Bodelwyddan, Rhyl, Clwyd LL18 5TY
Tel: (+44 1745) 58 45 84
Telex: 61532
Fax: (+44 1745) 58 42 08
General Manager: Dr Cedric Kitchen
Sales Manager: David E Jones

Plowshare Technology Ltd
PO Box 9, Horsham, West Sussex RH12 4FA
Tel: (+44 1403) 21 04 94
Fax: (+44 1403) 21 04 94
Directors: Dr Alan Johnstone, Svetlana Yegorova

Quality CNC Engineering Ltd
46-48 Azura Close, Woolsbridge Industrial Park,
Three Legged Cross, Wimborne, Dorset BH21 6SZ
Tel: (+44 1202) 82 75 17
Fax: (+44 1202) 82 39 58
Sales Director: Allen Davis

Queensgate Instruments Ltd
Waterside Park, Bracknell, Berkshire RG12 1RB
Tel: (+44 1344) 48 41 11
Fax: (+44 1344) 48 41 15
e-mail: space@queensgate.co.uk
Managing Director: Dr Paul D Atherton
Projects Director: Chris M Shannon

1760 Grand Avenue, Merrick, New York 11566, USA
Tel: (+1 516) 623 97 25
Fax: (+1 516) 546 70 31

Reaction Engines Ltd
34, Hunters Field, Stanford in the Vale,
Oxfordshire SU7 8LX
Tel/Fax: (+44 13677) 188 80
Managing Director: Alan Bond
Technical Director: Richard Varvill
Engineering Director: John Scott-Scott

Rose Bearings Ltd
Doddington Road, Lincoln LN6 3RA
Tel: (+44 1522) 50 09 33
Fax:(+44 1522) 50 09 75
Managing Director: Ron Chapman

Rosemount Ltd
Heath Place, Bognor Regis, West Sussex PO22 9SH
Tel: (+44 1243) 86 31 21
Telex: 86218
Fax: (+44 1243) 86 30 47
Managing Director: John Pollock
Director, Aerospace Division: Mel White

Rotadata Limited
Bateman Street, Derby, Derbyshire DE23 8JQ
Tel: (+44 1332) 34 80 08
Fax: (+44 1332) 33 10 23
Managing Director: Jack Taylor
Sales Director: Derek J Taylor

Saft Nife Limited
Station Road, Hampton, Middlesex TW12 2BY
Tel: (+44 181) 979 77 55
Fax: (+44 181) 783 04 94
Managing Director: Ken Klinker
Deputy Managing Director: J Searle
Technical Director: Dr J Thompson

Science Systems (Space) Ltd
23 Clothier Road, Brislington, Bristol, Avon BS4 5SS
Tel: (+44 117) 971 72 51
Fax: (+44 117) 971 11 25
Managing Director: Dr John Haynes
Sales Director: John Auburn

Sedgwick Space Services
Sedgwick House, The Sedgwick Centre,
London E1 8DX
Tel: (+44 171) 377 30 78
Fax: (+44 171) 377 30 77
Chairman: David P Trezies
Managing Director: H Ben Poole

Serco Europe Limited
Serco House, Hayes Rd, Southall, Middlesex UB2 5NJ
Tel: (+44 181) 843 24 11
Telex: 24114
Fax: (+44 181) 843 31 70
Managing Director: John B Wyke
Business Development Director: Mark Logan

Sira Limited
South Hill, Chislehurst, Kent BR7 5EH
Tel: (+44 181) 467 26 36
Telex: 896649 sira g
Fax: (+44 181) 467 65 15
e-mail: QUESTQ04843
Managing Director: Richard A Brook
Space Managers: Dr Mike A Cutter, David J Purll
Sales/Marketing Manager: John A Hother

Smith System Engineering Ltd
Surrey Research Park, Guildford, Surrey GU2 5YP
Tel: (+44 1483) 44 20 00
Fax: (+44 1483) 44 21 44
WWW: http://www.smithsys.co.uk
Chairman: Dr Bruce G Smith, OBE
Managing Director: Dr Tom Black
Marketing Director: R J M Sweet

Space Innovations Ltd
The Paddock, Hambridge Road,
Newbury, Berkshire RG14 5TQ
Tel: (+44 1635) 462 54
Telex: 846188 satint g
Fax: (+44 1635) 387 85
WWW: http://www.sil.com
Managing Director: A Kim Ward
Technical Director: James Barrington-Brown
Marketing Commercial Director: Dennis R Brindley

Spacesense Consultants
19c Kew Gardens Road, Richmond, Surrey TW9 3HD
Tel: (+44 181) 287 84 98
Fax: (+44 181) 287 93 09
Principal Consultant: Dr Adrian Lloyd-Lawrence

Spectrogon UK Ltd
Kinnoull Street, Dundee DD2 3TN
Tel: (+44 1382) 82 83 03
Fax: (+44 1382) 82 89 09
General Manager: Graeme Robb

Spur Electron Ltd
Asgard House, Hayward Business Centre,
New Lane, Havant, Hampshire PO9 2NL
Tel: (+44 1705) 45 55 64
Fax: (+44 1705) 47 08 74
Engineering Director: Robert Hymers
Business Development: Malcolm Snowdon

Surrey Satellite Technology Ltd
Centre for Satellite Engineering Research
University of Surrey, Guildford, Surrey GU2 5XH
Tel: (+44 1483) 25 92 78
Telex: 859331
Fax: (+44 1483) 25 95 03
e-mail: m.sweeting@ee.surrey.ac.uk
CEO/Managing Director: Prof Martin Sweeting
Technical Director: Dr Jeff Ward
General Manager: Ed Milton (1483 25 98 72)

Systems Engineering & Assessment Ltd
Beckington Castle, PO Box 800, Bath, Avon BA3 6TB
Tel: (+44 1373) 83 18 00
Fax: (+44 1373) 83 11 33
e-mail: pcj@sea.co.uk
Chairman/Managing Director: Ian Dale-Staples
Business Director, Space Systems: Paul Phillips
Business Development Director: Philip James

T S Space Systems
Unit V, Rose Business Estate,
Marlow Bottom, Marlow, Buckinghamshire SL7 3ND
Tel: (+44 1628) 47 40 40
Fax: (+44 1628) 47 78 00
Consultant Physicist: Dr William G Williams
Project Manager: D C Green

UK Industrial Space Committee
Secretariat: PO Box 14,
Wisbech, Cambridgeshire PE13 1JZ
Tel: (+44 1945) 46 49 75
Fax: (+44 1945) 46 19 88
WWW: http://www.ukisc.com/
Chairman: Pat Norris
Secretary General: Alan G Hicks

University Of Kent
Unit for Space Sciences and Astrophysics
Physics Laboratory, The University,
Canterbury, Kent CT2 7NR
Tel: (+44 1227) 45 96 16
Telex: 965449 ukclibg
Fax: (+44 1227) 76 26 16
e-mail: jcz@ukc.ac.uk
WWW: http://wwwspace.ukc.ac.uk
Unit Head: Prof JAM (Tony) McDonnell
Staff: Dr John Zarnecki, Dr Simon Green,
Dr Mark Burchell, Dr Laurie Kay

Vega Space Systems Engineering Ltd
2 Falcon Way, Shire Park,
Welwyn Garden City, Hertfordshire AL7 1TW
Tel: (+44 1707) 39 19 99
Fax: (+44 1707) 39 39 09
e-mail: sales@vega.co.uk
WWW: http://www.vega.co.uk
Chairman: Robert Drummond
UK Operations Director: Tony Constable
German Operations Director: John Lewis
Business Development Manager: Graham Peters

Willis Corroon Inspace
10 Trinity Square, London EC3P 3AX
Tel: (+44 171) 488 87 08
Telex: 882141
Fax: (+44 171) 860 91 41
Space Group Managing Director: Jerry Flaxman

UNITED STATES OF AMERICA

Adcole Aerospace Products
669 Forest Street, PO Box 39,
Marlborough, Massachusetts 01752
Tel: (+1 508) 485 91 00
Fax: (+1 508) 481 61 42
Vice-President for Aerospace: Richard S Warner

Advanced Products Co
33 Defco Park Road, Box 396,
North Haven, Connecticut 06473
Tel: (+1 203) 239 33 41
Fax: (+1 203) 234 72 33
President: Nancy Nicholson
Vice-President Sales/Marketing: Peter G Amos

AEC-Able Engineering Co, Inc
93 Castillian Drive, Goleta, California 93117
Postal: PO Box 588, Goleta, California 93116-0588
Tel: (+1 805) 685 22 62
Fax: (+1 805) 685 13 69
President: Max D Benton
Vice-President Engineering: Robert F Crawford
Vice-Presidents: Allister F Fraser, William M Robbins

Aeroastro Corporation
520W Huntmar Park Drive, Herndon, Virginia 22070
Tel: (+1 703) 709 22 40
Fax: (+1 703) 709 07 90
e-mail: info@aeroastro.com
WWW: http://www.newspace.com/aeroastro
President: Dr Rick Fleeter
Vice-President Chief Technical Officer: Richard Warner
Vice-President Advanced Programs: François Martel
Director, Analog Systems: Greg Huffman
Project Manager: Peter Goldberg
Director, Operations: Lisa Feldman

Aerojet
PO Box 13222, Sacramento,
California 95813-6000
Tel: (+1 916) 355 10 00
Telex: 377409
Fax: (+1 916) 355 24 48
President: Roger I Ramseier
Vice-President Strategic & Space Product Sector:
Robert Davis
Titan Program Director: Terry M Jenkins
Delta Program Director: Jim H Kurtz
Director, Strategic & Space Propulsion Systems Business Development: Marc T Constantine
Vice-President Business Development: Tom Peoples
Vice-President Communications: Edie Cartwright
(351 85 01)

Aerospace Industries Association Of America Inc
1250 Eye Street NW, Washington DC 20005-3922
Tel: (+1 202) 371 84 00
Fax: (+1 202) 371 85 73
President: Don Fuqua
Vice-President, International: Joel L Johnson
Vice-President, Procurement/Finance: LeRoy J Haugh
Vice-President, Technical/Operations: Stanley N Siegel
Executive Director, Research Center: Virginia C Lopez

Air Products & Chemicals, Inc
7201 Hamilton Boulevard,
Allentown, Pennsylvania 18195-1501
Tel: (+1 610) 481 62 13
Telex: 847416
Fax: (+1 610) 481 25 76
Government Systems: Charles Anderson

Alliant Techsystems, Inc
Bacchus Works
PO Box 98, Magna, Utah 84044-0098
Tel: (+1 801) 250 59 11
Fax: (+1 801) 251 20 09
Vice-President/General Manager, Space & Strategic Propulsion: Paul Ross
Vice-President Engineering, Technology: Ed Hikida
Vice-President Titan Projects: Jeff Foote
Director MLV/ILV: Travis Campbell
Director, Composite Structures: John Weidner
Director, Business Development: Brent Carroll
Manager, Public Affairs: David Nicponski

AlliedSignal Aerospace Co
2525 West 190th Street, PO Box 2960,
Torrance, California 90509-2960
Tel: (+1 310) 323 95 00/321 50 00
Telex: 674490
Fax: (+1 310) 512 22 21
President: Daniel Burnham
Senior Vice-President Marketing/Sales:
Joseph P Leonard
Senior Vice-President Planning Development:
Ernest Linneman
Vice-President Public Affairs: Dennis Signorovitch

AlliedSignal Aerospace Co
Aerospace Equipment Systems
Power Management and Generation Systems (PMGS)
1300 W Warner Road, Tempe, Arizona 85284-2896
Tel: (+1 602) 893 44 21
Fax: (+1 602) 893 49 22
Director, Space Systems: Len H Grove
Manager, Business Development: Dave Conrad

AlliedSignal Aerospace Co
Government Electronic Systems
46 Williams Avenue, Teterboro, New Jersey 07608
Tel: (+1 201) 228 20 00
Telex: 134414 bendix ttbr
Fax: (+1 201) 393 66 10
President: David A Tawfik
Vice-President General Manager:
Christopher C Bernhardt
Vice-President Space Systems: Stanley Seeb
Space Systems Business Development: Karl S Bogdan

AlliedSignal Technical Services Corporation
One Bendix Road, Columbia, Maryland 21045-1897
Tel: (+1 410) 964 70 00
Telex: 87860 benfld colb
Fax: (+1 410) 730 67 75
President: Ivan Stern
Vice-President/Director, Military Space: Peter R Dachel
Vice-President Space Operations: Philip H Johnson
Senior Engineering Advisor: Richard Hieb

AMP
PO Box 3608, Harrisburg, Pennsylvania 17105-3608
Tel: (+1 717) 564 01 00
Telex: 510-657 4110
Fax: (+1 717) 986 75 75
President: William J Hudson
Vice-President Aerospace/Government Systems Sector:
A P Kastel

AMSAT-NA
Radio Amateur Satellite Corp
850 Sligo Avenue, Suite 600,
Silver Spring, Maryland 20910
Tel:(+1 301) 589 60 62/73 90

Fax: (+1 301) 608 34 10
e-mail: wtynan@amsat.org
WWW: http://www.amsat.org
President: William A Tynan
Executive Vice-President: Keith Baker
Vice-President Engineering: Dick Jansson

Analytical Graphics, Inc
660 American Avenue,
King of Prussia, Pennsylvania 19406
Tel: (+1 610) 337 30 55
Fax: (+1 610) 337 30 58
e-mail: info@stk.com
WWW: http://www.stk.com
President/CEO: Paul L Graziani
Vice-President Product Development: Doug Claffey
Vice-President Sales: George Palmer
Director of Marketing: Donna T Milewski

Anser, Inc
1215 Jefferson Davis Highway, Suite 800,
Arlington, Virginia 22202-3251
Tel: (+1 703) 416 20 00
Fax: (+1 703) 416 84 40
e-mail: anserciac@anser.org
WWW: http://www.anser.org/
President/CEO: Dr John M Fabian
Vice-President Aerospace Systems: Thomas L Moser
CIAC Director: Richard Kohrs
CIAC Deputy Director: Stephen M Hopkins

16/2 Tverskaya Ul, Bldg 312,
103009 Moscow, Russian Federation
Tel: (+7 095) 935 70 57
Fax: (+7 095) 935 70 56
e-mail: 76110.342@compuserve.com
Director, Moscow Office: Susan Kieev

Applied Technologies Associates
1320 Villa Street, Mountain View, California 94041
Tel: (+1 415) 965 71 90
Fax: (+1 415) 965 29 89
President: John Ng
Senior Vice-President: Warren T Fujii
Vice-President, Strategic Planning: Gerald Hansen

ARAC
23 N Main Street, Suite C, Franklin, Indiana 46131
Tel: (+1 317) 738 38 86
Fax: (+1 317) 738 39 80
e-mail: t2s@iquest.net
WWW: http://www.advicom.net/~arac
President, Specialists in Technology Transfer: Dr F Timothy Janis

Ardé Inc
500 Walnut Street, Norwood, New Jersey 07648
Tel: (+1 201) 784 98 80
Fax: (+1 201) 784 97 10
President: Reginald Wood
Chief Engineering: Kirk Sneddon
Director, Operations: Jeff Ehalt
Marketing: Bob Holland

Ardak Corporation
6849 Old Dominion Drive, Suite 370,
McLean, Virginia 22101-3705
Tel: (+1 703) 893 37 02
Fax: (+1 703) 893 40 10
President: Mac McCutchan
Executive Vice-President: Walter Kennedy
Vice-President: Gary Brosch

Arrowhead Products
4411 Katella Avenue, Los Alamitos, California 90720
Tel: (+1 714) 828 77 70
Fax: (+1 714) 995 34 52
President: Dale T Shanahan
Vice-President Marketing: Steven Schwarz
Vice-President Engineering: Dale Carey
Vice-President Quality: Andy Mead

Assured Space Access, Inc
301 East Chilton Drive,
Chandler, Arizona 85225
Tel: (+1 602) 813 59 20
Fax: (+1 602) 813 39 50
e-mail: asaisl8@aol.com
President: Michael Rendine
Corporate Vice-President: Larry Foor
Director, Launch Services: Tracy Kennedy

Astro Aerospace Corp
6384 Via Real, Carpinteria, California 93013-2920
Tel: (+1 805) 684 66 41
Fax: (+1 805) 684 33 72
President: James Conlan
Antenna development: Mark Thomson
Solar array development: Brian Spence
Structures: Geoff Marks

Atlantic Research Corporation
5945 Wellington Road,
Gainesville, Virginia 22065-1699
Tel: (+1 703) 754 50 00
Telex: 456241
Fax: (+1 703) 754 53 16
President/CEO: Antonio L Savoca
Group President, Aerospace Group: James R Sides
Vice-President Advanced Projects: Harry Crosby
Vice-President Propulsion Marketing: A D Mattox

ARC Liquid Propulsion Division
PO Box 300 LPO, 6686 Walmore Road,
Niagara Falls, New York 14304
Tel: (+1 716) 731 60 00
Fax: (+1 716) 731 62 81
Vice-*President/General Manager:* Gerald L Greene
Marketing Director: Ruben Canny

Virginia Propulsion Division
5945 Wellington Road, Gainesville, Virginia 22065
Tel: (+1 703) 754 50 00
Telex: 456241
Fax: (+1 703) 754 53 16
Vice-President/General Manager: B Frank Rohrback

Arkansas Propulsion Division
PO Box 1036 Camden, Arkansas 71701
Tel: (+1 501) 574 06 10
Fax: (+1 501) 574 25 09
Vice-President/General Manager: Keith B Mullin

Advanced Materials Division
5945 Wellington Road,
Gainesville, Virginia 22065
Tel: (+1 703) 754 50 00
Fax: (+1 703) 754 56 38
Vice-President/General Manager: Jerry Fields

Autometric, Inc
5301 Shawnee Road, Alexandria, Virginia 22312-2333
Tel: (+1 703) 658 40 00
e-mail: omni@cs.autometric.com
WWW: http://www.autometric.com
Marketing Manager: Chris Gamble
Omni Engineering Manager: Lynn Mattie

Avibank Mfg, Inc
210 South Victory Boulevard,
PO Box 391, Burbank, California 91503-0391
Tel: (+1 818) 843 43 30
Fax: (+1 818) 846 18 72
Vice-President Sales: D P Arnold
Sales Office/Marketing Manager: David Ladin
General Sales Manager: Mike A Yates

Ball Aerospace & Technologies Corporation
Ball Space Systems Division
PO Box 1062, Boulder, Colorado 80306-1062
Tel: (+1 303) 939 40 00
Telex: 45-605
TWX: 910-940-3241
Fax: (+1 303) 442 48 12
President: Donovan Hicks
President, Space Systems Division: Jesse W Moore

Ball Aerospace Systems Group
Ball Electro-Optics/Cryogenics Division
PO Box 1062, Boulder, Colorado 80306-1062
Tel: (+1 303) 939 60 02
Fax: (+1 303) 939 44 63
Group Vice-President: Donovan B Hicks
Ball EO/C Division President: Dr Ed L Van de Noord

Barnes Engineering Division
PO Box 867, 88 Long Hill Cross Road,
Shelton, Connecticut 06484-0867
Tel: (+1 203) 926 17 77
Telex: 6819115
Fax: (+1 203) 926 10 30/92 85
Vice-President/General Manager: Harry Elmendorf
Director, Detector Operations: Grant Albright
Director, Business Development, Space Products: Robert Barnes

BDM International, Inc
409 Third Street NW, Suite 340,
Washington DC 20024.
Tel: (+1 202) 863 29 50
Fax: (+1 202) 479 52 69

BDM Federal, Inc
1501 BDM Way, McLean, Virginia 22102-3204
Tel: (+1 703) 848 51 15
Fax: (+1 703) 848 52 21
President/CEO: Philip Odeen
Group Vice-President/National Security/Defense: John J Todd
Vice-President Advanced Programs, Space Systems/Technology: Robert A Kennedy

Bergen Cable Technologies Inc
PO Box 1300, 170 Gregg Street,
Lodi, New Jersey 07644
Tel: (+1 201) 487 35 21
Fax: (+1 201) 368 05 32
President: Thomas E Carr

BFGoodrich Aerospace
PO Box 5501, 250 N Cleveland-Massillon Road,
Akron, Ohio 44334-0501
Tel: (+1 216) 374 22 00
Telex: 910 240-5438
Fax: (+1 216) 374 32 76
President: Marshall O Larsen
Director of Communications: Brayton Harris

Aircraft Wheels/Brakes
PO Box 340, Waco Street, Troy, Ohio 45373
Tel: (+1 513) 339 38 11
Fax: (+1 513) 339 38 13
General Manager Commercial: William Walthall
General Manager Aviation/Military: Mike Piscatua

Aircraft Integrated Systems
Panton Road, Vergennes, Vermont 05491
Tel: (+1 802) 877 44 22
Fax: (+1 802) 877 41 13
Group General Manager: Ronald W Hodges
General Manager Commercial Systems: Joe Wilson
General Manager Military Systems: Al Daniels

Carbon Products
50 William White Boulevard, Pueblo, Colorado 81001
Tel: (+1 719) 948 35 00
Fax: (+1 719) 948 34 22

Super-Temp
11120 South Norwalk Boulevard,
Sante Fe Springs, California 90670
Tel: (+1 310) 944 62 44
Fax: (+1 310) 944 24 85

Boeing (Rockwell SOC)
600 Gemini Avenue, Houston, Texas 77058
Tel: (+1 713) 282 20 00
Fax: (+1 713) 282 45 75
President: Paul Smith
Vice-President Programme Manager: Glynn S Lunney

Boeing Aerospace Operations, Inc
PO Box 58747, 16840 Buccaneer,
Houston, Texas 77258
Tel: (+1 713) 280 12 02
Fax: (+1 713) 280 00 71
President/General Manager: Frank Smith
Director: George Davis

Boeing Commercial Space Company
PO Box 3999, M/S 8X-56, Seattle, Washington 98124
Tel: (+1 206) 393 11 18
Fax: (+1 206) 393 10 50
President: Jim Noblitt
Executive Vice-President: Scott Carson
Vice-President Business Development: Allen Ashby
Vice-President/Sea Launch Program Manager: Ronald C Olson

Boeing Defense & Space Group
PO Box 3999, Seattle, Washington 98124-2499
Tel: (+1 206) 773 21 21
Telex: 329430
Fax: (+1 206) 773 21 11
President: Allan Mulally
Vice-President Missiles & Space Division: Thomas Gubala
Director, Space Transportation: Hal DiRamio
IUS Program Manager: Paul Bay

Boeing Defense & Space Group
PO Box 3999, Seattle, Washington 98124-2499
Tel: (+1 206) 773 21 21
Telex: 329430
Fax: (+1 206) 773 42 61
President: C G Jerry King
Vice-President Missiles/Space: Jim Noblitt
Vice-President Electronic Systems: John Dempster
Vice-President Engineering: Larry Winslow

Missiles/Space Division
PO Box 240002, 499 Boeing Boulevard,
Huntsville, Alabama 35824-6402
Tel: (+1 205) 461 28 05
Fax: (+1 205) 461 35 92
Vice-President/General Manager: Jim Noblitt
Vice-President Space Station Laboratory: John Winch
Vice-President Missile Systems: Ron Olson
Director, Space Transportation: Hal DiRamio
Director, Space Systems: Allen Ashby

Space Station Program Office
PO Box 58747, 1045 Gemini,

Houston, Texas 77258-8747.
Tel: (+1 713) 283 48 00
Fax: (+1 713) 480 98 87
Vice-President Space Station Program: Doug Stone

Boeing North American Inc
12214 Lakewood Boulevard,
Downey, California 90241
Tel: (+1 310) 922 20 66/18 56
Fax: (+1 310) 922 20 32
President: Robert G Minor
Vice-President Advanced Programmes/Business Development: Henry J Dinenno
Chief Engineer, Shuttle Orbiter: Donald H Emero

PO Box 3644,
2600 Westminster Boulevard SK09,
Seal Beach, California 90740-7644
Tel: (1 310) 797 33 11
Fax: (+1 310) 797 10 51
GPS Program Director: Richard L Horner

Houston Operations
555 Gemini Avenue, Houston, Texas 77058
Tel: (+1 713) 282 20 30
Fax: (+1 713) 282 51 35
Vice-President /General Manager: Glynn S Lunney

Huntsville Operations
555 Discovery Drive, Huntsville, Alabama 35806
Tel: (+1 205) 971 32 32
Fax: (+1 205) 971 33 33
Site Director: Harry R Wiener

Florida Operations
PO Box 21105, Kennedy Space Center,
Florida 32815
Tel: (+1 407) 867 41 53
Fax: (+1 407) 867 30 29

NASA Shuttle Logistics
Depot Building 1, 8550 Astronaut Boulevard,
Cape Canaveral, Florida 32920
Tel: (+1 407) 799 72 00
Fax: (+1 407) 799 72 43
Vice-President/General Manager: Leroy D Solid

Boeing Rockwell Division
Rockwell International Corp
6633 Canoga Avenue,
Canoga Park, California 91303
Tel: (+1 818) 586 63 00
Telex: 698478
TWX: 910-494 12 19
Fax: (+1 818) 586 44 65
President: Paul B Smith
Executive Vice-President: Michael L Skolnick
Vice-President Engineering/Test: Byron K Wood
Vice-President Controller Finance: John W Bradley
Vice-President Propulsion Systems: Paul N Fuller
Vice-President /Director, Advanced Programs Div: Hank Wieseneck
Vice-President /Manager, SSME Flight Engines: Alan L Hallden
Vice-President Production Operations: Jim Albaugh
Vice-President /Manager, Space Station: Royce Mitchell
Manager, Program Development for Propulsion Systems: Henry Minami
Programme Director, ELV Propulsion Systems: Bob Smith
Director, Advanced Power Systems: Bob Mucica
Director/Programme Manager RLV Programs: Steve Bouley
XLR-132 Program Manager: Ron Pauckert
Business Development Manager, Directed Energy and Imaging: Brent Brentnall
Director, KSC Launch Operations: John Plowden
Director, SSME Launch Operations: Larry Connell
Director, Huntsville Operations: Charley Kraus

Booz, Allen & Hamilton Inc
Space Systems Division
8283/8251 Greensboro Drive,
McLean, Virginia 22102
Tel: (+1 703) 902 50 00
Fax: (+1 703) 902 33 33/32 66

1725 Jefferson Davis Highway, Suite 1100,
Arlington, Virginia 22202
Tel: (+1 703) 769 77 00
Fax: (+1 703) 685 65 08

1050 S Academy Boulevard, Suite 148,
Colorado Springs, Colorado 80910
Tel: (+1 719) 570 31 00
Fax: (+1 719) 597 81 31

2525 Bay Area Boulevard, Suite 290,
Houston, Texas 77058
Tel: (+1 713) 488 67 50
Fax: (+1 713) 488 56 18

7404 Executive Place, Suite 500,
Seabrook, Maryland 20706
Tel: (+1 301) 805 54 00
Fax: (+1 301) 805 95 35

Rust Envreugdlaan 2,
NL-2243 AS Wassenaar, Netherlands
Tel: (+31 1751) 794 41
Fax: (+31 1751) 188 02

112 Avenue Kleber, F-75116 Paris, France
Tel: (+33 1) 44 34 31 31
Fax: (+33 1) 44 34 30 00
Vice-President: William Brooks
Principal, Houston Manned Space Operations: Karol Bobko

Brunswick Defense
3333 Harbor Boulevard,
Costa Mesa, California 92628-2009
Tel: (+1 714) 546 80 30 ext 6385
TWX: 910 595 1527
Fax: (+1 714) 434 74 92
Vice-President/General Manager: D Parsons
Vice-President Business Development: August H Bickel

4300 Industrial Avenue,
Lincoln, Nebraska 68504
Tel: (+1 402) 464 82 11
TWX: 910 621-0769
Fax: (+1 402) 464 22 47

Brush Wellman Applications Development Center
34325 Ardenwood Boulevard,
Fremont, California 94555
Tel: (+1 415) 795 11 00
TWX: 910 373-1106
Fax: (+1 415) 792 11 97
Aerospace Manager: Phillip Arnold

California Space Institute
CalSpace A-021, University of California at San Diego,
9500 Gilman Drive, La Jolla, California 92093-0221
Tel: (+1 619) 534 63 81
Fax: (+1 619) 534 74 52
Director: Dr Sally Ride
Associate Director: David Rogers

Calspan Corporation
PO Box 400, Buffalo, New York 14225
Tel: (+1 716) 632 75 00
Fax: (+1 716) 631 41 31
President: Dr H Robert Leland

Calspan Operations
4455 Genesee Street, Cheektowaga, New York 14225
Tel: (+1 716) 631 67 97
Fax: (+1 716) 631 69 56
Senior Vice-President/General Manager: Jack Wagner

Calspan Service Contracts Division
110 Mitchell Boulevard, Tullahoma, Tennessee 37388
Tel: (+1 615) 455 47 59
Fax: (+1 615) 454 95 10
Senior Vice-President/General Manager: Francis L Shill

Calspan Systems Research Laboratories
200 Indian Ripple Road, Dayton, Ohio 45440-3696
Tel: (+1 937) 427 72 00
Fax: (+1 937) 429 27 77
Senior Vice-President/General Manager: John W Pitts

Calspan Srl Corporation
Suite 1111, 800 Connecticut Avenue NW,
Washington DC 20006
Tel: (+1 202) 887 47 00
Fax: (+1 202) 861 03 21
Chairman: James K Baker
President: David H Langstaff
Executive Vice-President/COO: Dr H Robert Leland
Vice-President Contracts: Eli Polner

Celestis Inc
Suite 260, 244 Times Boulevard, Houston,
Texas 77005-3253
Tel: (+1 713) 522 72 82
Fax: (+1 713) 522 73 80
WWW: http://www.celestis.com
Vice-President: Charles Chafer

Center for EUV Astrophysics
University of California at Berkeley
2150 Kittredge Street, Berkeley, California 94720
Tel: (+1 510) 642 16 48
Fax: (+1 510) 643 83 03
Director: Professor C Stuart Bowyer
Executive Director: Dr Roger F Malina

Center for Space Power
Texas Engineering Experiment Station,
The Texas A&M University System,
223 Wisenbaker Engineering Research Center,
College Station, Texas 77843-3118
Tel: (+1 409) 845 87 68
Fax: (+1 409) 847 88 57
e-mail: DRB6624@ZEUS.TAMU.EDU
Director: Dr Frederick R Best
Deputy Director: Dr David R Boyle
Associate Director: Dr Frank E Little
Chairman, CSP Board: Donald M Ernst

Cincinnati Electronics Corp
Space Electronics Business Unit
7500 Innovation Way, Mason, Ohio 45040-9699
Tel: (+1 513) 573 62 82
Fax: (+1 513) 573 67 67
e-mail: blampe@CinEle.com
Space Marketing Manager: William R Lampe

Circle Seal Controls, Inc
2301 Wardlow Circle, Corona, CA 91720
Postal: PO Box 3300, Corona, CA 91718
Tel: (+1 909) 270 62 00
Fax: (+1 909) 270 62 01

Commercial Launch Services
PO Box 179, Denver, Colorado 80201
Tel: (+1 303) 977 51 91
Fax: (+1 303) 977 61 77
President: Michael Wash
Vice-President Programs/Technical Operations Dennis R Dunbar
Vice-President Marketing: Ed Ward
Atlas 2AR Program Manager: John Karas

Computer Sciences Corporation
2100 East Grand Avenue, El Segundo, California 90245
Tel: (+1 213) 615 03 11
TWX: 910 348-6286
Fax: (+1 213) 64 02 64 80
CEO: Van B Honeycutt
Vice-President/President Consulting Group: James Champy
Vice-President/President Systems Group: Milton E Cooper
Vice-President/President Industry Services Group: Dale Elam
Vice-President/President CSC Europe: Ron Mackintosh
President, Technology Management Group: Edward Boykin

Applied Technology Division
6565 Arlington Boulevard,
Falls Church, Virginia 22046
Tel: (+1 203) 237 20 00
Fax: (+1 703) 533 95 60
President: Jerome B Hilmes

Conatec, Incorporated
1511 Ritchie Highway, Suite 302, Arnold,
Maryland 21012-2465
Tel: (+1 410) 349 87 93
Fax: (+1 410) 349 16 38
President: Eugene F Kadar
Executive Vice-President: Wayne H Montag

Condor Pacific Industries, Inc
31829 La Tienda Drive,
Westlake Village, California 91362
Tel: (+1 818) 865 30 00
Fax: (+1 818) 865 30 10
President: Ernest J Smith

Cosmic
The University of Georgia
382 East Broad Street, Athens, Georgia 30602-4272
Tel: (+1 706) 542 32 65
Fax: (+1 706) 542 48 07
e-mail: service@cossack.cosmic.uga.edu
WWW: http://www.cosmic.uga.edu

CTA International
1521 Westbranch Drive, McLean, Virginia 22102-3201
Tel: (+1 703) 883 10 50
Fax: (+1 703) 883 96 56
President: Dr Ali Atia
Vice-President/General Manager: Thomas van der Heyden
Vice-President: Jason O'Neill

CTA Space Systems
1521 Westbranch Drive,
McLean, Virginia 22102-3201
Tel: (+1 703) 883 10 00/14 71
Telex: 70972
Fax: (+1 703) 883 96 56
President: Dr George Sebestyen
Director of Marketing: Christopher Roberts

CTA, Incorporated
6116 Executive Boulevard, Suite 800,
Rockville, Maryland 20852
Tel: (+1 301) 816 12 00
Fax: (+1 301) 816 14 16
Chairman and CEO: Dr C E (Tom) Velez
President, CTA Space and Telecommunications Systems: Ricardo de Bastos
President, CTA Information Technology Services: Raymond V McMillan
President, CTA International: Dr Ali E Atia

Dallas Remote Imaging Group
PO Box 118053, Carrollton, Texas 75011-8053
Tel: (+1 214) 394 73 25
Fax: (+1 214) 492 77 47
e-mail: jwallach@drig.com
BBS: (+1 214) 394 74 38
WWW: http://www.drig.com/
President: Dr Jeff Wallach

Datatape Incorporated
605 East Huntington Drive,
Monrovia, California 91016
Tel: (+1 818) 358 95 00
Fax: (+1 818) 358 91 00
Chairman/CEO: Dominick Saccacio
Chief Financial Officer: David Patterson
Vice-President, Business Development: Bruce Peters
Director, Sales: Jerry Norton

Decca Valves Corporation
1631 Tenth Street, Santa Monica, California 90404
Tel: (+1 213) 450 64 61
Fax: (+1 213) 450 45 98
Executive Vice-President: Michael Figoff

Delco Systems Operations
6767 Hollister Avenue, Goleta, California 93117
Tel: (+1 805) 961 53 70
Fax: (+1 805) 961 72 26
President: Michael G Johnston
Business Development: Christine Doyle
Sales/Marketing: Jack Easterbrook
Public Affairs Manager: Teri Conrad

Derlan Inc
2040 East Dyer Road,
Santa Ana, California 92705-5777
Tel: (+1 714) 250 31 23
Fax: (+1 714) 261 20 13
Director of Marketing: Les Jennings

Deskin Research Group, Inc
2270 Agnew Road, Santa Clara, California 95054
Tel: (+1 408) 496 90 15
Fax: (+1 408) 496 53 54/53 33
Director of Marketing: Rokki Leonard

Draper Laboratory
The Charles Stark Draper Laboratory
555 Technology Square,
Cambridge, Massachusetts 02139-3563
Tel: (+1 617) 258 10 00
Fax: (+1 617) 258 30 50

Dynamic Engineering Inc
703 Middle Ground Boulevard,
Newport News, Virginia 23606
Tel: (+1 804) 873 13 44
Fax: (+1 804) 873 21 83
President/CEO: David H Mullins
Vice-President Marketing: Richard L Best
Vice-President Engineering: James R Brunges
Vice-President Manufacturing: Rodney F Kramer

Eagle Aerospace, Inc
16915 El Camino Real, Suite 200,
Houston, Texas 77058
Tel: (+1 713) 283 60 00
Fax: (+1 713) 283 60 20
CEO: Robert F Thompson
President: Bass Redd
Vice-President Space Division: Tom Milton
Vice-President Business Development: Larry Griffin
Vice-President Corporate Development: John Watson

Eagle-Picher Industries, Inc
Electronics Division
PO Box 47, Joplin, Missouri 64802-0047
Tel: (+1 417) 623 80 00
Fax: (+1 417) 623 13 08
Vice-President: L Rex Erisman
Director of Marketing: Lee Miller
Operations Manager: Ron Repplinger
Li Ion Business Manager: Steve Westfall
Li Ion Operations Manager: Robert Higgins

Eastman Kodak Company
Commercial/Government Systems
1447 St Paul Street, Rochester, New York 14653-7004
Tel: (+1 716) 253 69 92
Fax: (+1 716) 253 69 88
President: T Christopher Fitzsimmons
Manager, Planning/Administration: Robert J Leonard

Eaton Corporation
Pressure Sensors Division
15 Durant Avenue, PO Box 504,
Bethel, Connecticut 06801
Tel: (+1 203) 743 67 21/796 60 00
Telex: 6819305
Fax: (+1 203) 796 63 13
General Manager: John K Engelberger
Electromechanical Products Manager: Gary Werner
New Business Manager: Donald J Rupar

EER Systems Inc
10289 Aerospace Road, Seabrook, Maryland 20706
Tel: (+1 301) 577 89 00
e-mail: lastname@eer.com
President/CEO: Dr Jai N Gupta
Vice-President Space Systems Group: Jim Hengle
Director, Corporate Communications: Michael Bryant

EG&G Florida, Inc
PO Box 21267, Kennedy Space Center,
Florida 32815
Tel: (+1 407) 867 72 95
Fax: (+1 407) 867 11 99
President/General Manager: Richard H Jolley
Manager, Public Affairs: Judyth S Casper

EG&G Wright Components, Inc
Route 96, PO Box 160,
Phelps, New York 14532
Tel: (+1 315) 548 95 01
Fax: (+1 315) 548 20 03
General Manager: Robert Barrett
Business Unit Director: Patrick Rousseau
Marketing Manager: Lawrence J Smith

Encore Computer Corporation
4061 Powder Mill Road, Suite 400,
Calverton, Maryland 20705
Tel: (+1 301) 595 98 96
Fax: (+1 301) 595 92 65
CEO: Kenneth G Fisher
Range/Telemetry Markets: Bill Edmonston

Engineered Magnetics, Inc
18435 Susana Road,
Rancho Dominguez, California 90221
Tel: (+1 310) 635 95 55
Fax: (+1 310) 631 03 13
President: Irwin D Miller
Senior Vice-President: Harry E Wert
Vice-President Marketing: Steven J Miller

Engineered Systems (ESCO)
2550 Market Street, Aston, Pennsylvania 19014
Tel: (+1 610) 494 80 00
Fax: (+1 610) 494 89 89
Vice-President Marketing: G Greer MacMasters
Regional Director: Kevin Mulligan

Entran Devices, Inc
10 Washington Avenue, Fairfield, New Jersey 07004
Tel: (+1 201) 227 10 02
Fax: (+1 201) 227 68 65
WWW: http://www.entran.com
President: William F Caplan
Executive Vice-President: Daniel M Slavik
Vice-President, Sales: Robert L Monell

Environmental Control Systems (ECS)
2525 West 190th Street, Torrance, California 90509
Tel: (+1 310) 323 95 00
Telex: 674490
Fax: (+1 310) 512 22 21
President: Ron Hoge
Vice-President: John Hamilton
Space Marketing Manager: Scott Manatt

E'Prime Aerospace Corporation
PO Box 6472, Titusville, Florida 32782
Tel: (+1 407) 269 09 00
Fax: (+1 407) 269 09 59
President/CEO: Robert G Davis
Executive Vice-President/COO: Larry Post
Executive Vice-President, Washington Operations: Wayne Hartke
Director, Launch Operations: Frank Zaccagnino
Programs: Charles Wofford

Essex Corporation
9150 Guilford Road, Columbia, Maryland 21046-1891
Tel: (+1 301) 939 70 60
Fax: (+1 410) 953 78 80
WWW: http://www.essexcorp.com
Chairman/CEO/President: Dr Harry Letaw, Jr
Vice-President Corporate Development: Anthony L Ward

Fiberite, Inc
2055 East Technology Circle, Tempe, Arizona 85284
Tel: (+1 602) 730 20 10
Fax: (+1 602) 730 23 90
Sales Director: John McKnight

Furon
Aerospace Components Division
3711 South Hudson Street,
Seattle, Washington 98118
Tel: (+1 206) 723 56 00
Fax: (+1 206) 723 68 69
President: Terrence A Noonan
Customer Development: Douglas A Swoope

G&H Technology, Inc
Special Products Division
1649 17th Street,
Santa Monica, California 90404-3893
Tel: (+1 310) 314 31 05
Fax: (+1 310) 452 54 78
Marketing Director: Bill Weaver

General Atomics
3550 General Atomics Court,
PO Box 85608, San Diego, California 92186-5608
Tel: (+1 619) 455 26 51
Fax: (+1 619) 455 36 79
Senior Vice-President: David Roberts
Director, Space Power Program: Dr Thomas A Lechlenberg

General Research Corporation
1900 Gallows Road, Vienna, Virginia 22182
Tel: (+1 703) 506 50 00
Fax: (+1 703) 506 01 79
President/CEO: James Roth
Vice-President/Director, Decision Technologies Division: Thomas E Maultsby

Gulton Data Systems
6600 Gulton Court NE,
Albuquerque, New Mexico 87109
Tel: (+1 505) 345 90 31
Fax: (+1 505) 344 98 79
Vice-President Marketing: David Thomas

Harris Corporation
Government Aerospace Systems Division
PO Box 37 Melbourne, Florida 32902
Tel: (+1 407) 727 51 15
Fax: (+1 407) 729 76 75
Vice-President/General Manager: Charles J Herbert
Vice-President Business Development: Bill Braselton
Director, Business Development: Dr B C Tankersley
Electronic Systems Sector President: Albert E Smith

Harris Space Systems Corporation
PO Box 37 Melbourne, Florida 32902
Tel: (+1 407) 633 38 00
Fax: (+1 407) 633 39 00
Vice-President/Core Project Manager: Bill McCaslin
Manager, Business Development: Bob Garceau

Hernandez Engineering Inc
17625 El Camino Real, Suite 200,
Houston, Texas 77058
Tel: (+1 713) 280 51 59
Telex: 774289
Fax: (+1 713) 480 75 25
President: Miguel A Hernandez Jr
COO: Dave Franz
Vice-President: Teresita Z Hernandez
Business Development Director: William Holmberg
Director of Engineering: Jerry Hammack

Hernandez Engineering BV
Van Panhuysstraat 2A,
NL-2203 JR Noordwijk, Netherlands
Tel: (+31 7136) 489 84
Fax: (+31 7136) 489 85

Honeywell Inc
Space Systems Division
13350 US Highway 19 North,
Clearwater, Florida 34624-7290
Tel: (+1 813) 539 40 00
Fax: (+1 813) 539 34 47
WWW: http://www.honeywell.com
Vice-President/General Manager: Bill R Poe
Vice-President Business Development: Rudy Oswald

Satellite Systems Operations
PO Box 52199, Phoenix, Arizona 85072-2199
Tel: (+1 602) 561 30 00
Telex: 71-187184
Fax: (+1 602) 561 32 00
e-mail: ssocommu@space.honeywell.com
Vice-President/Location Manager: W Jay Lovelace
Director, Program Development: Andy Engelhardt

Space/Strategic Systems Operations
13350 US Highway 19 North,
Clearwater, Florida 34624-7290
Tel: (+1 813) 539 40 00
Fax: (+1 813) 539 34 47
e-mail: info-fl51@space.honeywell.com
Vice-President Space and Strategic Systems: Larry Speight
Director, Houston Space Engineering Center: Robert McAfoos
Director, Business Development: Scott Starrett

Hughes Danbury Optical Systems Inc
100 Wooster Heights Road,
Danbury, Connecticut 06810-7589
Tel: (+1 203) 797 52 66
Fax: (+1 203) 797 51 14
e-mail: info@hdos.hac.com
WWW: http://www.hac.com
President: Dr John C Rich
Executive Director, Reconnaissance Systems: Victor Abraham
Executive Director, Space Programs: Thomas Pitts
Manager, Star Tracker Business Development: Les Schlom

Hughes Research Laboratories
3011 Malibu Canyon Road,
Malibu, California 90265
Tel: (+1 310) 317 50 00
Manager, Plasma Sources Department: Dr Robert Beattie

Hughes STX Corporation
4400 Forbes Boulevard,
Lanham, Maryland 20706-4392
Tel: (+1 301) 794 50 00
Telex: 317630
Fax: (+1 301) 794 71 06
President: Ashok Kaveeshwar

Humphrey Incorporated
9212 Balboa Avenue,
San Diego, California 92123-1590
Tel: (+1 619) 565 66 31
Fax: (+1 619) 565 68 73
President: Errol Ekaireb
Vice-President/General Manager: H H Kries

Hydro Fitting Manufacturing Corporation
733 East Edna Place, PO Box 1558,
Covina, California 91722
Tel: (+1 818) 967 51 51
Telex: 670439
Fax: (+1 818) 339 44 54
President: Arnold C Schwartz
Vice-President/General Manager: Anthony Samartina

ICI Explosives
Valley Forge Corporate Center
PO Box 819, Valley Forge,
Pennsylvania 19482
Tel: (+1 215) 666 86 00
Fax: (+1 215) 666 86 25
Product Group Manager: Jim Donahue

ILC Dover, Inc
One Moonwalker Road,
Frederica, Delaware 19946
Tel: (+1 302) 335 39 11
Fax: (+1 302) 335 07 62
WWW: http://www.ilcdover.com
Director of Progammes: Bradford Walters

Inframetrics, Inc
16 Esquire Road,
North Billerica, Massachusetts 01862-2598
Tel: (+1 508) 670 55 55
TWX: 710 326-0659
Fax: (+1 508) 667 27 02
President: Brian O'Donohue
Contact: Robert F Bruno

Interferometrics Inc
14120 Parke Long Ct, Chantilly, Virginia 20151
Tel: (+1 703) 222 58 00
Fax: (+1 703) 968 88 08
President: Dennis Fecteau
Director of Engineering: Rick Grenwis
e-mail: grenwis@interf.com

International Space Brokers
1616 North Fort Myer Drive, Suite 1220,
Rosslyn, Virginia 22209
Tel: (+1 703) 841 13 34
Fax: (+1 703) 841 05 25
President/COO: John W Vinter
Senior Vice-Presidents: Jean-Michel Eid, Robert T Johnson
Vice-President: Richard E Rankin

International Space Technology, Inc
3825 Fabian Way, G79,
Palo Alto, California 94303-4604
Tel: (+1 415) 852 52 35
Fax: (+1 415) 852 41 48

International Technology Underwriters
4800 Montgomery Lane, 11th floor,
Bethesda, Maryland 20814
Tel: (+1 301) 654 85 85
Fax: (+1 301) 654 75 69
President/CEO: Frederick H Hauck
Senior Vice-President, Space Underwriting: Patrick Rivalan
Director, Space Systems Engineering: Russell Fillers

Intraspace Corporation
421 West 900 North, Salt Lake City, Utah 84054
Tel: (+1 801) 292 04 40
Fax: (+1 801) 292 63 41
President: Robert D'Ausilio
Vice-President: Frank Williams

ISSI
500 Wynn Drive, No 306,
Huntsville, Alabama 35816-3429
Tel: (+1 205) 890 11 78
Fax: (+1 205) 890 11 74
President: Gerald Wittenstein

ITHACO Incorporated
PO Box 6437, Ithaca, New York 14851-6437
Tel: (+1 607) 272 76 40
Fax: (+1 607) 272 08 04
e-mail: ithaco@lightlink.com
WWW: http://www.ithaco.com
President/CEO: David Burns
Senior Vice-President, General Manager: Joe Maddock
Senior Vice-President, Finance: Leo McGrattan
Senior Vice-President, Marketing/Sales: David Burns

ITT Defense & Electronics, Inc
1000 Wilson Boulevard, Arlington, Virginia 22209
Tel: (+1 703) 276 83 00
Fax: (+1 703) 276 97 04
President: Louis J Giuliano
Vice-President Government Relations: E George Riedel
Vice-President International Marketing: W Frank Parker

ITT Aerospace/Communications Division
PO Box 3700, 1919 W Cook Road,
Fort Wayne, Indiana 46801
Tel: (+1 219) 487 63 84
TWX: 810 332-1413
Fax: (+1 219) 487 60 03
President/General Manager: Marvin R Sambur
Vice-President Business Development: Barry M Abzug
Vice-President/Director of Operations: Edmund Stuczynski
Vice-President/Director Contract Administration: James J Vogt
Senior Manager, Space Systems Product Line Manager: Steven G Johnson
Manager, Public Affairs: Ray Crockett

John Crane Belfab
305 Fentress Boulevard,
Daytona Beach, Florida 32114
Tel: (+1 904) 253 06 28
Fax: (+1 904) 257 01 22
President: Douglas D Fockler
Marketing Director: Dick Idtensohn

Johnson & Higgins
Space & Telecom Group
1401 Eye Street NW, Suite 400, Washington DC 20005
Tel: (+1 202) 408 61 99
Fax: (+1 202) 789 25 74
Vice-President/Chairman STG: A Michael Hewins

Kaiser Electro-Optics, Inc
2752 Loker Avenue West, Carlsbad, California 92008
Tel: (+1 619) 438 92 55
Fax: (+1 619) 438 68 75
President: Jerry Carollo
Vice-President Operations: Bill Maffucci
Chief Scientist: Dr Warren Smith
Business Director: Benedict J Mall

Kaiser Marquardt
16555 Saticoy St, Van Nuys,
California 91406 1739
Tel: (+1 818) 989 64 00
TWX: 910 495-1710
Fax: (+1 818) 989 66 23
e-mail: kaiser@marquardt.com
President: Ron McMahon
Director, Business Development: R Carl Stechman

Kaman Aerospace Corporation
Old Windsor Road, Bloomfield, Connecticut 06002
Tel: (+1 860) 243 44 81
Telex: 99326 kaman acft blfd
Fax: (+1 860) 243 75 14
Chairman: Charles H Kaman
President: Walter R Kozlow

Electro-Optics Development Center
3480 East Britannia Drive, Suite 120,
Tucson, Arizona 85706-5007
Tel: (+1 602) 889 70 00
Fax: (+1 602) 889 02 11
Vice-President: Dr Bobby L Ulich

Kaman Sciences Corporation
1500 Garden of the Gods Road,
Colorado Springs, Colorado 80907
Postal: PO Box 7463, Colorado Springs, Colorado 80933-7463
Tel: (+1 719) 599 15 00
Fax: (+1 719) 599 19 42
Chairman: Charles H Kaman
President: J Lamar Allen
Vice-President Operations: Charles A Eklund

Kamatics Corporation
1330 Blue Hills Avenue, PO Box 3,
Bloomfield, Connecticut 06002-0003
Tel: (+1 860) 243 79 04
Fax: (+1 860) 243 79 98
Chairman: Charles H Kaman
President: Alan A Whitfield
Vice-President Marketing: John E Reed

Kearfott Guidance & Navigation Corporation
150 Totowa Road, Wayne, New Jersey 07470-0946
Tel: (+1 201) 785 60 00
Telex: 13-3440
Fax: (+1 201) 785 60 25
CEO: Dr Ronald E Zelazo
Vice-President Engineering: Thomas Shanahan
Vice-President Business Development/Management: Harvey Truppi
Director, Strategic and Space Programs: Richard A Kunzier

Kelly Space & Technology
294 S Leland Norton Way,
San Bernardino, California 92408
Tel: (+1 909) 382 20 10
Fax: (+1 909) 382 20 12
WWW: http://www.kellyspace.com
President & CEO: Michael S Kelly
Executive Vice-President and Chief Operating Officer: Michael J Gallo (+1 909) 382 20 13
Vice-President Contracts: Vaughn M Best
Chief Engineer: Donald P Anctil

Ketema Programmed Composites
380 Cliffwood Park, Brea, California 92621
Tel: (+1 714) 671 38 90
Fax: (+1 714) 671 16 25
General Manager: John C Van Doren

Kinesix Corporation
9800 Richmond Avenue, Suite 750,
Houston, Texas 77042
Tel: (+1 713) 953 83 00
Fax: (+1 713) 953 83 06
e-mail: sammi@kinesix.com

Kistler Aerospace Corporation
10877 Wilshire Boulevard, Suite 805,
Los Angeles, California 90024
Tel: (+1 310) 208 20 01
Fax: (+1 310) 208 04 30
Chairman: Robert Wang
CEO: Dr George E Mueller
CFO/Vice Chairman: Jeffrey I Scheinrock
Executive Vice-President: Dan Brandenstein

KKI
1011 Chapel Road, Suite 100,
Middletown, Virginia 22645
Tel: (+1 703) 247 43 33
Fax: (+1 540) 869 05 54
President: Paul F Kisak

KPMG Peat Marwick
Commercial Space Group
2001 M Street, NW, Washington DC 20036
Tel: (+1 202) 467 30 85
Telex: 440477 PMMDCU1
Fax: (+1 202) 293 54 37
Director: Larry Janski
Senior Managers, Washington DC: Pete Stahl, Bob Morris
Senior Manager, Colorado: Chip Houlihan

Launchspace, Inc
7235 1/2 Arlington Boulevard,
Falls Church, Virginia 22042
Tel: (+1 703) 560 95 22
Fax: (+1 703) 698 02 11
e-mail: mhk@launchspace.com
Chairman: Dr Marshall H Kaplan

L'Garde, Inc
15181 Woodlawn Avenue,
Tustin, California 92680-6487
Tel: (+1 714) 259 07 71
Fax: (+1 714) 259 78 22
President: Gayle D Bilyeu

Litton Applied Technology
4747 Hellyer Avenue,
San Jose, California 95150-7012
Tel: (+1 408) 365 47 47
TWX: 910-339-9271
Fax: (+1 408) 365 40 40
President: Clayton Williams
Vice-President, Business Development: Harry H Pursel, Jr

Litton Guidance & Control Systems
5500 Canoga Avenue,
Woodland Hills, California 91367-6698
Tel: (+1 818) 712 20 59
Fax: (+1 818) 712 70 19
President: Darwin D Beckel
Director, Space/Strategic Systems: Ben Tipps

Lockheed Martin Astronautics
PO Box 179, Denver,
Colorado 80201
Tel: (+1 303) 977 30 00
TWX: 910 935-0718
Fax: (+1 303) 971 49 02
President: James W McAnally
Vice-President/General Manager, Space Launch Systems: Frederick H Hudoff
Vice-President, Advanced Launch Systems: Gareth D Flora
Vice-President, Defense Systems: Virgil F Young
Vice-President, Flight Systems: Noel W Hinners
Vice-President, Ground Systems: Grover W Hall
Vice-President, Special Programs: Robert L Boyce
Vice-President, Business Development: Raymond C Colladay
Vice-President, Production Operations: Edwin M Squires
Vice-President, Launch Operations: Forrest S McCartney
Vice-President, Business Management: Chester F Wheeler
Vice-President/Chief Engineer, Mars Surveyor: Parker Stafford
Director, Advanced Remote Sensing: Joel S Porter
Director, MSLS: Roman Mathern
LMLV Program Director: Richard Strauss
LMLV Business Development Director: Jerry L Cobb
Director, Public Affairs: Linda Strine Shahan

Space Launch Systems
Vice-President/General Manager: John Adamoli
Titan 4 Program Director: Frederick H Hudoff
Titan 4 Business Development: Richard Blakley
Titan 2/3 Program Manager: Mel Hughes
Titan 2/3 Business Development: Larry Price
Director, Canaveral Operations: William Murphy
Director, Vandenberg Operations: Andrew Hall

Lockheed Martin Corporation
6801 Rockledge Drive, Bethesda, Maryland 20817
Tel: (+1 301) 897 63 87
Fax: (+1 301) 897 60 28
Chairman: Norman R Augustine
CEO: Vance D Coffman
President/COO: Peter B Teets
Executive Vice-President, Business Strategy: Frank C Lanza
Executive Vice-Presidents: Thomas Young, Vincent N Marafino
Executive Vice-President/CFO: Marcus C Bennett
President, Space & Strategic Missiles Sector: Mel R Brashears
President, Electronics Sector: Thomas A Corcoran
President, Aeronautics Sector: James A Blackwell
Vice-President, Washington Operations: David S Osterhout

Lockheed Martin Data Systems
PO Box 3041, Sarasota, Florida 34230
Tel: (+1 813) 371 08 11
Fax: (+1 813) 378 69 95
Vice-President, Instrumentation Recorders: Ray Shuford
Contact: Jim Horvath (Tel: (+1 813) 378 67 11)

Lockheed Martin Electro-Optical Systems
600 East Bonita Avenue, Pomona,
California 91769-2737
Tel: (+1 909) 624 80 21
Fax: (+1 909) 447 22 30
President: Bob Mueller
Director, Space Systems Business Development: Bob McBride

Lockheed Martin Engineering & Science Services
2625 Bay Area Boulevard, Houston, Texas 77058
Tel: (+1 713) 283 44 00
Fax: (+1 713) 283 46 60
Executive Vice-President/General Manager: James C Adamson
Vice-President, Engineering/Sciences Programs: Dr Moe M Miller
Vice-President Business Development: Kenneth Asbury
Vice-President, Science Programs: Richard S Sade
Vice-President, Houston Operations: Richard Y Lyons

Lockheed Martin Federal Systems
6600 Rockledge Drive, Bethesda, Maryland 20817
Tel: (+1 301) 493 81 00
Fax: (+1 301) 493 17 46

3700 Bay Area Boulevard,
Houston, Texas 77058-1199
Tel: (+1 713) 282 78 18
Fax: (+1 713) 282 80 72
Group Vice-President: Arthur E Johnson
Vice-President, Engineering/Technology: Donald Del Giorno
Vice-President, Business Development: David Leitner
Marketing Manager: E 'Sol' Solomon

Lockheed Martin Manned Space Systems
PO Box 29304, New Orleans,
Louisiana 70189
Tel: (+1 504) 257 33 11
Fax: (+1 504) 257 44 31
President: G Thomas Marsh
Vice-President, External Tank: Dennis R Deel
Vice-President, Technical Operations: Richard Foll
Vice-President, Production Operations: Robert Hieter
Director, SLWT: Terry Hibbard
Deputy Director, SLWT: Larry Knauer
Director, Business Operations: Brent Clayburn
Director, Facilities and Environmental Operations: Reynold Abadie
Director, Public Relations: Thomas L Branigan (+1 504) 257 37 88

Lockheed Martin Missiles & Space Company
Orgn 74-02, Building 159,
PO Box 3504, 1111 Lockheed Way,
Sunnyvale, California 94089-3504
Tel: (+1 408) 756 52 16
Telex: 346409
Fax: (+1 408) 756 86 36
President: Minoru S (Sam) Araki
Vice-President, Assistant General Manager Space Systems Division: David Klinger
Vice-President, Military Satcoms: Myles Crandall
Vice-President, NASA Programs: Michael Henshaw
Vice-President, Commercial Space Programs: Arnold Aldrich
Vice-President, Commercial Business Development: Nicholas Ordon
Director, New Business Planning: J E Hatton
Director, New Business Development NASA: Carol Lane
Director, International Development: Jon Salmon
Director, Scientific Space Programs: Dr Franklin Martin
Director, Iridium Program: Tom Tadano
Manager, Hubble Program: Tom Dougherty

Advanced Technology Center (ATC)
3251 Hanover St, Palo Alto, California 94304-1191
Tel: (+1 415) 424 30 00
Fax: (+1 415) 424 35 48
Vice-President: Dr Ronald A Andrews
Director, Space Physics: Dr Richard R Vondrak
Manager, Cryogenic Cooler: Ted Nast

Lockheed Martin Skunk Works
1011 Lockheed Way,
Palmdale, California 93599
Tel: (+1 805) 572 29 74
Fax: (+1 805) 572 41 63

Lockheed Martin Tactical Systems Sector
Washington Operations
1725 Jefferson Davis Highway, Suite 900,
Arlington, Virginia 22202
Tel: (+1 703) 416 55 00
Vice-President: Jimmie V Adams
Vice-President, Communications: Joanne Hvala

Lockheed Martin Command & Control Systems
9970 Federal Drive, Colorado Springs,
Colorado 80921-3697
Tel: (+1 719) 594 10 00
Fax: (+1 719) 594 11 58
President: Jack Thornell
Vice-President, Business Development: Joseph Salute

Lockheed Martin Space Mission Systems
7375 Executive Place, Seabrook, Maryland 20706
Tel: (+1 301) 805 03 00
Fax: (+1 301) 805 05 17
President: David Blanchard

Lockheed Martin Space Information Systems
1322 Space Park Drive, Houston, Texas 77058
Tel: (+1 713) 335 60 00
Fax: (+1 713) 335 63 01
President: Clint H Denny
Vice-President, Business Development: Jon Guthrie
Director, Advanced Programs and Technical Planning: Michael L Coats

Lockheed Martin Space & Range Systems
1260 Crossman Avenue,
Sunnyvale, California 94089-1198
Tel: (+1 408) 734 54 00
Fax: (+1 408) 734 64 44
President: William Blanchard
Vice-President, Space Applications Operations: William Schuster
Vice-Presidents, Marketing: Robert S Laskin

Lockheed Martin Fairchild Imaging Sensors
1801 McCarthy Boulevard, Milpitas, California 95035
Tel: (+1 408) 433 25 00
Fax: (+1 408) 433 26 04
Vice-President/General Manager: Bob Drew
Vice-President, Sales: Bill Walker

Lockheed Martin Vought Systems
PO Box 650003, Dallas, Texas 75265-0003
Tel: (+1 214) 603 10 00
Fax: (+1 214) 603 05 93
President: Jay A Musselman
Tel: (+1 214) 603 92 46
Director, Space Programs: Ray French
Tel: (+1 214) 603 94 09

Loral Space & Communications Ltd
600 Third Avenue, New York, New York 10016
Tel: (+1 212) 697 11 05
Fax: (+1 212) 661 89 88
Chairman/CEO: Bernard L Schwartz
President/COO: Michael B Targoff

Space Systems/Loral
3825 Fabian Way, Palo Alto, California 94303
Tel: (+1 415) 852 40 00
Fax: (+1 415) 852 47 88
President: Robert E Berry

Loral Skynet
900 Route 202/206 North Room 2A115W,
Bedminister, New Jersey 07921-0752
Tel: (+1 800) 847 93 35
Fax: (+1 908) 234 79 40
President: Terry Hart

Globalstar
3200 Zanker Road,
San Jose, California 95164-0670
Tel: (+1 408) 473 55 50
Fax: (+1 408) 473 55 48
President: Douglas G Dwyre

Lord Mechanical Products
1952 West Grandview Boulevard,
PO Box 10040, Erie, Pennsylvania 16514-0040
Tel: (+1 814) 868 54 24
Telex: 291935
Fax: (+1 814) 868 31 09
Corporate Senior Vice-President/General Manager: T J Loftus
Manager, International Business Development: John S Fisher

Lucas Control System Products
Schaevitz Products
7905 North Route 130,
Pennsauken, New Jersey 08110-1489
Tel: (+1 609) 662 80 00
Fax: (+1 609) 662 62 81
Plant Manager: Bob Starita
Vice-President, Marketing/Sales: Joe Urbanek

Malin Space Science Systems
PO Box 910148, San Diego, California 92191-0148
Tel: (+1 619) 552 26 50
Fax: (+1 619) 458 05 03
e-mail: info@msss.com
WWW: http://www.msss.com
President/Chief Scientist: Dr Michael C Malin

Marsh & McLennan Space Projects
1166 Avenue of the Americas,
New York, New York 10036-2774
Tel: (+1 212) 345 60 00
Fax: (+1 212) 345 34 13
Contacts: Mary Ann Curnan, Philip K Crawford, Ruth I Freeman

Matra Aerospace Inc
550 Highland Street, Frederick, Maryland 21701
Tel: (+1 301) 624 11 39
Fax: (+1 301) 682 68 85
Acting Vice-President: Philippe Balducchi

McDonnell Douglas Aerospace
5301 Bolsa Avenue,
Huntington Beach, California 92647
Tel: (+1 714) 896 34 11
Telex: 678426
Fax: (+1 714) 896 13 13
President: John P Capellupo
Deputy President: Herbert J Lanese
Vice-President/General Manager, Space/Defense Systems: Willard P Olson
Vice-President/General Manager, Space Flight Programs: R Gale Schluter
Vice-President/General Manager, Space Transportation: Tom D Burson
Vice-President/General Manager, New Space/Defense Products: Charles A Ordahl

Kennedy Space Center Division
PO Box 21233, Kennedy Space Center, Florida 32815
Tel: (+1 407) 867 26 30
Fax: (+1 407) 867 30 89
Vice-President/General Manager: George R Faenza
Vice-President/Deputy General Manager: Dallas K Gillespie

Houston/Space Station Division
Tower Building, 16055 Space Center Boulevard,
Houston, Texas 77062-6208
Tel: (+1 713) 280 15 00
Fax: (+1 713) 280 15 83
Vice-President/General Manager, Space Flight Programs: Rodney M Linford

Huntsville Division
689 Discovery Drive, Huntsville, Alabama 35806
Tel: (+1 205) 922 66 00
Fax: (+1 205) 922 72 15
Vice-President/General Manager: Charles F Marvin

Meda, Inc
485 Spring Park Place, Suite 350,
Herndon, Virginia 22070
Tel: (+1 703) 471 14 45
Fax: (+1 703) 471 91 30

Metalmart, Inc
12225 Coast Drive, Whittier, California 90601
Tel: (+1 562) 692 90 81
Fax: (+1 562) 699 68 68
WWW: http://www.metal-mart.com
President: Randall Biggs
Vice-President: Derek Lafavor

Microcosm, Inc
2377 Crenshaw Boulevard, Suite 350,
Torrance, California 90501
Tel: (+1 310) 320 05 55
Fax: (+1 310) 320 02 52
e-mail: microcosm@smad.com
WWW: http://www.smad.com/
President: Dr James R Wertz
Vice-President: Robert E Conger

Moog Inc
Space Products Division
East Aurora, New York 14052-0018
Tel: (+1 716) 652 20 00
Fax: (+1 716) 687 44 66
President/CEO: Robert T Brady
General Manager: Richard C Sherrill
Director, Sales/Programs: Jay Hennig
Manager, International Buisness Development: Michelle Milanostri
Tel: (+1 716) 687 44 20; Fax: (+1 716) 687 44 66

MTS Systems Corporation
14000 Technology Drive,
Eden Prairie, Minnesota 55344-2290
Tel: (+1 612) 937 45 55
Telex: 29-0521 MTS SYS ENPE
Fax: (+1 612) 937 40 95
WWW: http://www.mts.com
President: Donald M Sullivan
Executive Vice-President: Keith D Zell
Vice-President Aerospace Business Group: Kenneth E Floren

Nerac, Inc
One Technology Drive,
Tolland, Connecticut 06084
Tel: (+1 860) 872 70 00
Fax: (+1 860) 875 17 49
President: Dr Daniel U Wilde
Vice-President: Kevin A Bouley
Manager, Communications: Nan R Cooper

Northrop Grumman Corporation
Electronic Sensors & Systems Division
PO Box 17319, Baltimore, Maryland 21203-7319
Tel: (+1 410) 765 10 00
TWX: 710 862-0100
Fax: (+1 410) 993 87 71
WWW: http://sensor.northgrum.com
Corporate Vice-President/General Manager: James G Roche
Vice-President, International: Lloyd C Carpenter
Vice-President, Business Development: David J Beck
Director Domestic Marketing: Dallas L Knudson
Vice-President, Space Division: Robert P Iorizzo
Vice-President, Airborne Surveillance Systems: William R Adams

Northrop Grumman Corporation
Electronics & Systems Integration Division
1111 Stewart Avenue,
Bethpage, New York 11714-3580
Tel: (+1 516) 575 51 19
Fax: (+1 516) 575 36 91
Vice-President/General Manager: John E Harrison
Vice-President/Deputy General Manager: Max T Weiss

Oceaneering Space Systems
16665 Space Center Boulevard,
Houston, Texas 77058-2268
Tel: (+1 713) 488 90 80
Fax: (+1 713) 488 64 85
e-mail: rklusend@oss.oceaneering.com
Vice-President/General Manager: Arthur G Stephenson
Vice-President, Operations: Ronald L Welch
Director of Marketing/Sales: Roy K usendorf

Odetics, Inc
Communications Division
1585 South Manchester Avenue,
Anaheim, California 92802-2907
Tel: (+1 714) 758 01 00
Telex: 372-0423
Fax: (+1 714) 780 76 49
CEO: Joel Slutzky
President/General Manager: Crandall Gudmundson
Vice-President Marketing: Jerry Muench
Manager, Business Development, Space Data Storage Products: Bob Tripp
Marketing Manager, Advanced Products: Ellie Kurrasch
Public Relations Manager: Pam Rigdon

Olin Aerospace Company
11441 Willows Road NE,
PO Box 97009, Redmond, Washington 98073-9709
Tel: (+1 206) 885 50 00
Fax: (+1 206) 882 57 47
President: William W Smith
Marketing Vice-President: J J Galbreath
Manager, Business Development: James H Bartron

Olin Chemicals
350 Knotter Drive, Cheshire,
Connecticut 06410-0586
Tel: (+1 203) 271 41 57
Fax: (+1 203) 271 40 50
Director, Aerospace/Defense: Peter Danna
Marketing Manager: Stu Grossberg

Optical Coating Laboratory, Inc
2789 Northpoint Parkway,
Santa Rosa, California 95407-7397
Tel: (+1 707) 545 64 40
Fax: (+1 707) 525 70 18
President/CEO: Herbert M Dwight

Orbital Sciences Corporation
21700 Atlantic Boulevard, Dulles, Virginia 20166
Tel: (+1 703) 406 50 00
Fax: (+1 703) 406 35 02
President/Chairman/CEO: David W Thompson
Executive Vice-President/General Manager, Communications and Information Systems Group: Bruce W Ferguson
Executive Vice-President/General Manager, Launch Systems Group: James R Thompson
Executive Vice-President/General Manager, Space and Electronics Systems Group: Jack A Frohbieter
Executive Vice-President/General Manager, Advanced Systems Group: Dr Michael D Griffin
Executive Vice-President/General Manager, Systems Integration and Software Group: Daniel E Friedman
Senior Vice-President/Chief Technology Officer: Dr Antonio L Elias
Senior Vice-President/General Counsel: Leslie C Seeman
Senior Vice-President/Corporate Strategy: John H Mehoves
President, Orbimage: Gilbert D Rye
President, Orbcomm: Alan L Parker
Vice-President Government Relations: Mark Bitterman
Vice-President/General Manager, Intelligent Transportation Systems: Michael P Kushner
Taurus Program Manager: Joseph F Padavano
Pegasus Program Manager: W Jean Floyd

Chandler Operations
3380 South Price Road, Chandler, Arizona 85248
Tel: (+1 602) 899 60 00
Fax: (+1 602) 899 88 00
Senior Vice-President, Technical Operations: David L Cooper

Germantown Operations
20301 Century Boulevard,
Germantown, Maryland 20874
Tel: (+1 301) 428 60 00
Senior Vice-President/General Manager: Larry Yermack

Pomona Operations
2771 North Garey Avenue, Pomona, California 91767
Tel: (+1 909) 593 35 81
Fax: (+1 909) 596 23 01
Senior Vice-President: Albert F Tillman

Oremet Titanium
PO Box 580, 530 34th Avenue SW,
Albany, Oregon 97321-0177
Tel: (+1 541) 967 90 00
Fax: (+1 541) 917 06 47
Vice-President, Commercial: David G Floyd

Para-Flite Inc
5800 Magnolia Avenue,
Pennsauken, New Jersey 08109-1399
Tel: (+1 609) 663 12 75
Fax: (+1 609) 663 30 28
President: Elek Puskas
Executive Vice-President: William F Bruno

Parker Bertea Aerospace
18321 Jamboree Road, Irvine,
California 92715
Tel: (+1 714) 833 30 00
Telex: 678383
Fax: (+1 714) 851 32 77
President: Stephen L Hayes
General Manager, Air/Space Division: Jim Sabin

Control Systems Division
14300 Alton Parkway, Irvine, California 92718
Tel: (+1 714) 833 30 00
Telex: 678427
Fax: (+1 714) 586 84 56
Vice-President/General Manager: Robert Barker
Vice-President, Marketing: Jim Ryder

1425 West 2675 North, Ogden,
Utah 84404
Tel: (+1 801) 782 31 00
Telex: 388424
Fax: (+1 801) 786 30 45
Plant Manager: Mike Romito

Penny & Giles Data Systems, Inc
1250 Northmeadow Parkway, Suite 114,
Roswell, Atlanta, Georgia 30076
Tel: (+1 404) 740 04 15
Fax: (+1 404) 740 04 20
President: Phil Lancaster
General Manager: Ed Shea

Precision Rolled Products, Inc
PO Box 60010, 14255 Mt Bismark Street,
Reno, Nevada 89506
Tel: (+1 702) 972 02 72
Fax: (+1 702) 972 07 22
President: Manfred Wenzlaff

Specialty Alloy Melt Division
306 Columbia Road, Florham Park, New Jersey 07932
Tel: (+1 201) 822 91 00
Fax: (+1 201) 822 09 32

Pressure Systems, Inc
6033 East Bandini Boulevard,
City of Commerce, California 90040
Postal: PO Box 6999,
Los Angeles, California 90022
Tel: (+1 213) 722 02 22
Fax: (+1 213) 721 60 02
President/CEO: Alex Roberts
Marketing Manager: Michael Hersh

Princeton Synergetics, Inc
900 State Road, Princeton, New Jersey 08540
Tel: (+1 609) 924 20 20
Fax: (+1 609) 924 69 42
President: Joel S Greenberg

Puroflow Corporation
1631 Tenth Street, Santa Monica,
California 90404
Tel: (+1 310) 450 64 61
Fax: (+1 310) 450 45 98
Executive Vice-President: Michael Figoff

Quantic Industries, Inc
2751 San Juan Road, Hollister,
California 95023
Tel: (+1 408) 637 58 51
Fax: (+1 408) 637 10 13
President: Robert Valenti
Marketing Manager: Kenneth E Willis
Contracts Manager: Michael Donaldson

Questar Corporation
PO Box 59, New Hope, Pennsylvania 18938
Tel: (+1 215) 862 52 77
Fax: (+1 215) 862 05 12
President: Douglas M Knight
Vice-President: Douglas Knight Jr

R-Cubed Composites Inc
3392 West 8600 South,
West Jordan, Utah 84088-9706
Tel: (+1 801) 569 04 01
Fax: (+1 801) 569 08 17
WWW: http://www.r-cubed.com
Sales Manager: Mark Edwards

Sanders
65 Spit Brook Road,
Nashua, New Hampshire 03061-0868
Tel: (+1 603) 885 60 22
Fax: (+1 603) 885 72 64
President: John R Kreick
Executive Vice-President: Dr Hugo Poza
Vice-President/General Manager: Ehtisham U A Siddiqui
Vice-President, Operations: Steven J Walker
Vice-President, Business Development:
Joseph Giacoponello
Vice-President, Washington Operations:
Burlie A Brunson

Sandia National Laboratories
Albuquerque, New Mexico 87185
Tel: (+1 505) 844 66 01
Fax: (+1 505) 844 63 67
Media Relations: Rod Geer

Scaled Composites, Inc
1624 Flight Line, Mojave, California 93501-1663
Tel: (+1 805) 824 45 41
Fax: (+1 805) 824 41 74
e-mail: info@scaled.com
President/CEO: Burt Rutan
CFO: Patricia L Storch
Vice-President/General Manager: Michael W Melvill
Vice-President, Business Development: Douglas B Shane

Schaeffer Magnetics Inc
9175 Deering Avenue,
Chatsworth, California 91311-5897
Tel: (+1 818) 341 51 56
Fax: (+1 818) 341 38 84
President/COO: Keith DeConde
CEO: Ernie Schaeffer
Executive Vice-President: Max Jacobson
Vice-President, Technology Development:
Jim C Hammond
Deputy Program Manager: Kuai Lok Tsui
Marketing Manager: Peter E Chase
Production Manager: R Andolsek

Schonstedt Instrument Co
1775 Wiehle Avenue, Reston, Virginia 22090-5199
Tel: (+1 703) 471 10 50
Fax: (+1 703) 471 17 95
WWW: http://www.schonstedt.com
President/CEO: Rawling G Davenport, Jr
Vice-President/COO: Joseph L Andrews, Jr
Marketing Manager: Wilma Merrick
Vice-President Engineering/Chief Engineer:
John B Rippingale

Schuller International
PO Box 5108, Denver, Colorado 80217-5108
Tel: (+1 303) 978 52 50
Fax: (+1 303) 978 27 00
Product Manager: Charles Lostak

Science Applications International Corporation
10260 Campus Point Drive,
San Diego, California 92121
Tel: (+1 619) 535 60 00
Fax: (+1 619) 546 66 34
WWW: http://www.saic.com
Chairman/CEO: J R Beyster
President/COO: William A Owens
Executive Vice-President: John H Warner Jr
Vice-President Commercial/International: John Glancy
Corporate Vice-President Civil Space Systems:
Neil Hutchinson
Director, Space Systems Business Development:
Jules McNeff
Director, Corporate Development: Jim Idell

SEAKR Engineering, Inc
4030 Spencer Street, #108,
Torrance, California 90503
Tel: (+1 310) 542 93 02
Fax: (+1 310) 542 32 07

Sedgwick Space Services
3 Becker Farm Road,
Roseland, New Jersey 07068-3786
Tel: (+1 201) 533 47 42
Fax: (+1 201) 535 05 93
Chairman: David P Trezies

Semiconductor Sector
PO Box 883 Melbourne, Florida 32902
Tel: (+1 407) 724 40 00
Fax: (+1 407) 724 39 37
Vice-President/General Manager, Military/Aerospace:
Jay A Davis

Servo Corporation of America
123 Frost St, Westbury, New York 11590
Tel: (+1 516) 938 97 00
Fax: (+1 516) 938 96 44
WWW: http://www.servo.com
President: Stephen Barre
Senior Vice-President: Kent Martin
Director of Engineering: Alan Doctor
Marketing Manager: Richard G Burton

SETI Institute
2035 Landings Drive,
Mountain View, California 94043
Executive Director: Tom Pierson
Project Scientist: Dr Jill Tarter

Smiths Industries Aerospace
Florham Park, 7-9 Vreeland Road,
Florham Park, New Jersey 07932
Tel: (+1 201) 822 13 00
Fax: (+1 201) 822 16 25

4141 Eastern Avenue, SE,
Grand Rapids, Michigan 49508
Tel: (+1 616) 241 70 00
Fax: (+1 616) 241 73 18
President: Robert Ehr
Vice-President, Business Development: James Sweeney
Vice-President/General Manager, Engineering:
Angelo Dimitriou
Vice-President, Manufacturing: Steve Marcus
Director, Strategic Development: Peter Howells

Space Analysis & Research, Inc
PO Box 49446, Colorado Springs,
Colorado 80949-9446
Tel: (+1 719) 260 05 00
e-mail: jburger@oldcolo.com
President: Joseph J Burger

Space Applications Corporation
901 Follin Lane, Suite 400,
Vienna, Virginia 22180
Tel: (+1 703) 255 52 00
Fax: (+1 703) 255 36 67
e-mail: stolarik@spaceapp.com
President/CEO: Michael G Stolarik
Vice-President, Advanced Programs: Gary L Barnes

Space Commerce Corporation
5718 Westheimer, Suite 1515,
Houston, Texas 77057
Tel: (+1 713) 977 96 91
Fax: (+1 713) 977 15 43
e-mail: ART@DULA.COM
President: Arthur M Dula

Space Electronics Inc
4031 Sorrento Valley Boulevard,
San Diego, California 92121-1404
Tel: (+1 619) 452 41 67
Fax: (+1 619) 452 54 99
WWW: http://www.newspace.com/spaceelec
CEO: Robert Czajkowski
CEO: David J Strobel
Space Product Manager: David Czajkowski
Sales Manager: Edward Li

Space Industries, Inc
101 Courageous Drive, League City, Texas 77573
Tel: (+1 281) 538 60 00
Fax: (+1 281) 334 40 10
General Manager: Robert Hammond
Manager, Engineering: Dr Michael F Lembeck

Space Machine Advisors, Inc
289 Greenwich Avenue,
Greenwich, Connecticut 06830
Tel: (+1 203) 661 02 02
Fax: (+1 203) 661 52 43
CEO: Alden M Richards
CFO: Richard L Guest

Space Policy Institute
Elliott School of International Affairs
The George Washington University
2013 G Street, NW, Suite 201, Washington DC 20052
Tel: (+1 202) 994 72 92
Fax: (+1 202) 994 16 39
WWW: http://www.gwu.edu/~spi
Director: Professor John M Logsdon

Space Power Institute
Auburn University, 231 Leach Center,
Auburn, Alabama 36849-5320
Tel: (+1 205) 844 58 94
Fax: (+1 205) 844 59 00
CCDS Director: Dr Henry W Brandhorst Jr

Space Studies Institute
PO Box 82 Princeton, New Jersey 08540
Tel: (+1 609) 921 03 77
Fax: (+1 609) 921 08 39
e-mail: ssi@ssi.org
President: Dr Freeman J Dyson
Executive Director: Bettie Greber

Space Vector Corporation
9223 Deering Avenue,
Chatsworth, California 91311-6900
Tel: (+1 818) 886 65 00
Fax: (+1 818) 886 12 76
e-mail: SpaceVectr@aol.com
President: Joseph H Jerger
Vice-President Operations: Pat Golden
Manager, Marketing/Program Development:
Joel M Bridges
Program Administration: Ray Tensfeldt

17330 Brookhurst Street, Suite 150,
Fountain Valley, California 92708
Tel: (+1 714) 963 88 00
Fax: (+1 714) 963 22 51

Sparta Inc
23041 Avenue de la Carlota, Suite 400,
Laguna Hills, California 92653
Tel: (+1 714) 768 81 61
Director, Space Technology: Richard Quartararo

Specialty Steel & Forge
26 Law Drive, Fairfield, New Jersey 07004-3293
Tel: (+1 201) 808 83 00
Fax: (+1 201) 808 44 88
Director, International Sales: Stef Branden

Spectrolab Inc
12500 Gladstone Avenue,
Sylmar, California 91342-5373
Tel: (+1 818) 365 46 11
Fax: (+1 818) 361 51 02
Manager, Advanced Programs: Terry Cavicchi
Business Manager: Ron Diamond

Spectrum Astro, Inc
1440 North Fiesta Boulevard,
Gilbert, Arizona 85233-8366
Tel: (+1 602) 892 82 00
Fax: (+1 602) 892 29 49
WWW: http://spectrumastro.com
Chairman/CEO: Martha S Martin
President: W David Thompson
Senior Vice-President: Stanley Dubyn
Vice-President Advanced Systems: Jerry Chicoine
Manager, Special Projects: Greg Moody

Sundstrand Aerospace
PO Box 7002, 4747 Harrison Avenue,
Rockford, Illinois 61125-7002
Tel: (+1 815) 226 60 00
Telex: 257440
TWX: 910 631 42 55
Fax: (+1 815) 226 74 88
Executive Vice-President/COO: Ronald F McKenna
Vice-President Aerospace Programs: Victor D Marone
Vice-President/General Manager, Sundstrand Power Systems: Omar Winter
Vice-President, Operations: Kenneth G Eitenmiller
Vice-President, Aerospace Business Development:
James A Smith
Vice-President, Aerospace Engineering:
Robert J Wetherel

Sundstrand Power Systems
PO Box 85757, 4400 Ruffin Road,
San Diego, California 92138-5757
Tel: (+1 619) 569 40 00
TWX: 910 335 1502
Fax: (+1 619) 569 46 41
Vice-President/General Manager: Omar Winter

Systron Donner
Inertial Division
2700 Systron Drive, Concord, California 94518-1399

Tel (switchboard): (+1 510) 682 61 61
Tel (sales): (+1 510) 671 64 00
Telex: 910 481-9478
Fax: (+1 510) 671 65 90
President: Richmond B Terrell
Vice-President/COO: Charles A Long
Director, Business Development: Brad B Sage
Chief Scientist: Harold D Morris
Director, Engineering: Sheridan W Hammons

Taber Industries
455 Bryant Street, North Tonawanda, New York 14120
Tel: (+1 716) 694 40 00
Fax: (+1 716) 694 14 50
Senior Vice-President: Bernie W Daube
Sales Engineer: Matt Randazzo

Tayco Engineering, Inc
10874 Hope Street, PO Box 6034,
Cypress, California 90630
Tel: (+1 714) 952 22 40

Technical Services
1250 Grumman Place, Titusville, Florida 32780-7900
Tel: (+1 407) 264 45 22
Fax: (+1 407) 268 83 57

Tecstar Inc
15251 Don Julian Road,
City of Industry, California 91745-1002
Tel: (+1 818) 968 65 81
Fax: (+1 818) 336 86 94
e-mail: BusMgnt@appliedsolar.com
WWW: http://www.appliedsolar.com
President/CEO: David M Van Buren
Vice-President Solar Cell Operations: Frank Ho
Vice-President Advanced Programs/Technology: Patrick Park

Teledyne Brown Engineering
Cummings Research Park, PO Box 070007,
300 Sparkman Drive NW,
Huntsville, Alabama 35807-7007
Tel: (+1 205) 726 50 50
Fax: (+1 205) 726 50 90
President: James McGovern
Vice-President Space Programs: Marty T Runkle
Space Business Development: Thomas M Perrin
Tel: (+1 205) 726 57 24
e-mail: tom.perrin@pobox.tbe.com

Teledyne Electronic Systems
19601 Nordhoff Street, Northridge, California 91324
Tel: (+1 818) 886 22 11
Fax: (+1 818) 886 58 69
President: S C Warrick
Senior Vice-President, Business Development: Joseph T Gallagher
Director, Business Development Administration: Andrew S Frohlich

Telegenix/Grim Corporation
PO Box 5550, 26 Olney Avenue,
Cherry Hill, New Jersey 08034
Tel: (+1 609) 424 52 20
Fax: (+1 609) 424 20 22
Sales Manager: Al Henry

Textron Inc
Textron Defense Systems
201 Lowell Street, Wilmington, Massachusetts 01867
Tel: (+1 617) 657 51 11
Fax: (+1 617) 657 20 60
President: Harold K McCard
Vice-President Communications/Planning: Edward Josephson

Textron Aerostructures
PO Box 210, Nashville, Tennessee 37202
Tel: (+1 615) 361 20 00
Telex: 554421
Fax: (+1 615) 360 57 79
President: Dick Wells
Executive Vice-President: John R Murphey
Marketing Director: Brian Fairhurst

Textron Specialty Materials
2 Industrial Avenue, Lowell, Massachusetts 01851
Tel: (+1 508) 452 89 61
TWX: 710 343 1284
Fax: (+1 508) 454 56 19
President: Paul R Hoffman
Vice-President Operations: Michael Rinaldi
Marketing Director: Michael Dorf

The Aerospace Corporation
2350 East El Segundo Boulevard,
El Segundo, California 90245-4691
Postal: PO Box 92957, Los Angeles,
California 90009-2957
Tel: (+1 310) 336 50 00
Fax: (+1 310) 336 70 55
WWW: http://www.aero.org
Chairman: Dr Ruth M Davis
Vice-Chairman: Gen Lawrence A Skantze
President/CEO: Edward C Aldridge, Jr
Executive Vice-President: Dr George A Paulikas
National Systems Group:
Group Vice-President: Jon H Bryson
Vice-President, Orbital Systems Operations: Dr Rodney C Gibson
Vice-President, Space Technology Applications: Ronald K Sable
Engineering/Technology Group:
Group Vice-President: Dr Michael J Daugherty
Vice-President, Technology Operations: Dr Hugo R Rugge
Space Systems Group:
Group Vice-President: Richard M Allman
Vice-President, Space Launch Operations: Chester L Whitehair
Vice-President, Space Programme Operations: John R Parsons

Washington DC Office
1000 Wilson Boulevard, Suite 2600,
Arlington, Virginia 22209-3988
Tel: (+1 703) 812 06 00
Fax: (+1 703) 812 94 15
Vice-President, Washington Region: Ronald K Sable
General Manager, Business Development, Space Technology Applications: Vincent Boles
Director, Government Operations: Gary Pulliam

Eastern Range Office
PO Box 21205, Kennedy Space Center,
Florida 32815-0205
Tel: (+1 407) 853 66 66
Fax: (+1 407) 853 36 90
Principal Director: Michael R Spence

Herndon Office
The Hallmark Building,
13873 Park Center Road, Suite 187,
Herndon, Virginia 22071
Tel: (+1 703) 318 16 50
Fax: (+1 703) 318 54 46
Vice-President, National Systems Group: Jon H Bryson
General Manager, Advanced Programmes Operations: Dr James P Campbell
Tel: (+1 703) 808 22 61

Western Range Office
PO Box 5068 Vandenberg Air Force Base,
California 93437-0068
Tel: (+1 805) 866 20 18
Fax: (+1 805) 866 20 03
Principal Director: Donald L Nichols

Houston Office
Mail code ZR-12, Johnson Space Center,
Houston, Texas 77058
Tel: (+1 713) 483 22 71
Fax: (+1 713) 483 35 16
Project Engineer: Larry Dungan

Sunnyvale Office
1320 Orleans Drive,
Sunnyvale, California 94089-1135
Tel: (+1 408) 744 64 00
Fax: (+1 408) 744 66 68
Systems Director: Richard R Kistler

Albuquerque Office
Kirtland AFB, PO Box 9045,
Albuquerque, New Mexico 87119
Tel: (+1 505) 846 02 22
Fax: (+1 505) 846 02 11
Principal Director: Dr Alfred N Sorensen

Colorado Springs Office
1150 Academy Park Loop, Suite 109,
Colorado Springs, Colorado 80910
Tel: (+1 719) 573 24 00
Fax: (+1 719) 573 24 97
General Manager: Dr James R LaFrieda

The Mitre Corporation
7525 Colshire Drive, McLean, Virginia 22102-3481
Tel: (+1 703) 883 60 00
Fax: (+1 703) 883 54 68
Senior Vice-President/General Manager, Center for Environment, Resources and Space: Dr Lydia W Thomas

Thiokol Corporation
2475 Washington Boulevard, Ogden,
Utah 84401- 2398
Tel: (+1 801) 629 22 70
Fax: (+1 801) 629 24 20
Chairman/President/CEO: James R Wilson
Vice-President/CFO: Richard L Corbin
Vice-President/Controller: Michael R Ayers
Vice-President, Defense and Launch Vehicles (DLV) Division: Winston N Brundige
Director, Communications: Gilbert R Barley

Space Operations
PO Box 707, Brigham City, Utah 84302-0707
Tel: (+1 801) 863 35 11
Fax: (+1 801) 863 22 34
Vice-President/General Manager: Joseph A Lombardo
Vice-President RSRM Program Manager: George C Alford

Utah Defense & Launch Vehicles Operations
PO Box 689, Brigham City, Utah 84302-0689
Tel: (+1 801) 863 35 11
Fax: (+1 801) 863 88 06
Vice-President/General Manager: Winston N Brundige
Vice-President/Operations: Robert T Kirkham
Vice-President/Programs & Business Development: Bruce H Marshall

Elkton DLV Operations
PO Box 241, Elkton,
Maryland 21922-0241
Tel: (+1 410) 392 14 29
Fax: (+1 410) 392 12 05
Division Manager: W Lloyd McMillan
Director, Programs: Don E Shaffer

Ordnance Operations
PO Box 1029, Marshall, Texas 7567
Tel: (+1 903) 679 31 81
Fax: (+1 903) 679 23 24
Vice-President/General Manager: Luther C Johnson
Director, Marketing: Jim E Hawley

Huck International, Inc
6 Thomas, Irvine, California 92718
Tel: (+1 914) 855 90 00
Fax: (+1 914) 855 85 37
Contact: Richard L Corbin

Thomas Electronics Inc
100 Riverview Drive, Wayne, New Jersey 07470
Tel: (+1 201) 696 52 00
Fax: (+1 201) 696 82 98
President: David Ketchum
Vice-President, Sales: Bruce Piaget
Vice-President Engineering: Denis Young

TRW Space & Electronics Group
One Space Park, Redondo Beach, California 90278
Tel: (+1 310) 812 43 21
Fax: (+1 310 812 71 11
Vice-President/General Manager: Timothy W Hannemann

Defense Systems Division
Vice-President/General Manager: Donald C Winter
Deputy General Manager: Joanne M Maguire
Chief Scientist: Joseph M Denney
Mission/System Engineering Manager: Richard A Croxall
Ground Development/Operations Manager: Gerald Steinker
DSP Program Manager: Elliot I Bailis
SMTS Program Manager: James W Burnett
Advanced Technology Programs: Dale P Hoffman
Remote Sensing Center Manager: Richard W Sullivan

Civil/International Systems Division
Vice-President/General Manager: Paul Y Sasaki
Advanced Programs/Systems Engineering Manager: Thomas E Hill
SSTI Program Manager: George M Reppucci
AXAF Manager: Edgar G Wheeler
EOS PM/Chemistry Manager: Brooks S Vogt
TOMS-EP Program Manager: Alan W Fleming
Rocsat Program Manager: Robert E Sharples
Kompsat Program Manager: Jerry Gruninger

Electronic Systems/Technology Division
Vice-President/General Manager: David B Vandervoet
Deputy General Manager: Alexis C Livanos
Milstar Payload Manager: Paul S Borzcik
Odyssey Communications System: Martin Melnick

Space/Technology Division
Vice-President/General Manager: Joanne M Maguire
Deputy General Manager: Steven L Korn
Major Spacecraft Programs Manager: Ronald I Gilje
Major Technology Programs Manager: Grant A Hosack
Advanced Bus Technology Programs Manager: Garrett R Shier
Sensors, Lasers and Research Center Manager: Thomas E Romesser
Propulsion and Combustion Center Manager: Robert L Sackheim
Spacecraft System Design and Integration Center Manager: Robert E Patterson
Avionics Systems Center, Manager: Mary Ann Chory

Odyssey System Program
Vice-President/Program Manager: Edward J
Manager, Cryogenics Program: Manny Tward

TRW Components International Inc
Tel: (+1 310) 214 55 00
Fax: (+1 310) 214 56 92
Vice-President/General Manager: Douglas E McCormac

Unisys Space & Information Systems (USIS)
595 Gemini Avenue, Houston, Texas 77058
Tel: (+1 713) 283 66 62
Fax: (+1 713) 488 13 82
Vice-President: Jay W Evers
Director Business Development: Frank Perez
Deputy Program Manager, Spaceflight Simulation and Training Software: Richard Covey
Communications Manager: Edward C Conaway

United Technologies
USBI Co
PO Box 21212, Kennedy Space Center,
Florida 32815
Tel: (+1 407) 867 79 00
Fax: (+1 407) 867 23 39
Executive Vice-President/General Manager: Joseph P Zimonis
Senior Vice-President SRB Programs: Guy Perry

United Technologies
Pratt & Whitney Space Propulsion Operations Chemical Systems Division
PO Box 49028, San Jose, California 95161-9028
Tel: (+1 408) 779 91 21
Fax: (+1 408) 778 45 99
Vice-President SPO/CSD Solid Programs: Charles A Sinclair
Director, Business Development: Frank O Verlot
Manager Business Development, Space Programs: Charles A Chase
Public Relations: Charles L Vecchio, Jr

United Technologies Corp
Hamilton Standard Space Systems International, Inc
One Hamilton Road,
Windsor Locks, Connecticut 06096-1010
Tel: (+1 860) 654 42 82
Telex: 99-4439
Fax: (+1 860) 654 26 73
President, Hamilton Standard Space Systems International: Harry Garfinkel
Vice-President Business Development: Deborah Barnhart
Manager, Environmental Control and Life Support Marketing: Merlin A Shuey
Manager, Commercial Business Development: Larry McNamara

United Technologies Pratt & Whitney
Government Engines & Space Propulsion
PO Box 109600, West Palm Beach,
Florida 33410-9600
Tel: (+1 407) 796 67 96
Fax: (+1 407) 796 72 58
President, P&W GESP: John A Balaguer
Public Relations: Patrick W Louden

P&W Space Propulsion Operations
PO Box 49028, San Jose,
California 95161-9028
Tel: (+1 408) 779 91 21
Fax: (+1 408) 778 45 99
Executive Vice-President/General Manager: Douglas A North
Vice-President Liquid Space Programs: Joseph P Zimonis
RL10 Program Director: Peter G Scharf
SSME/ATD Program Manager: John L Price

Universal Propulsion Company, Inc
25401 North Central Avenue,
Arizona 85027-7899
Tel: (+1 602) 516 33 40
Fax: (+1 602) 516 33 64
President: Harold G Watson
Vice-President Business Development: R F Sadler

Universities Space Research Association
The American City Building, Suite 212,
Columbia, Maryland 21044
Tel: (+1 410) 730 26 56
Fax: (+1 410) 730 34 96
President: Dr Paul J Coleman, Jr
Executive Director: Dr W David Cummings

Washington DC Office
300 D Street, SW, Suite 801, Washington DC 20024
Tel: (+1 202) 479 26 09
Fax: (+1 202) 479 26 13

Valcor Engineering Corp
2 Lawrence Road, Springfield, New Jersey 07081
Tel: (+1 201) 467 84 00
Fax: (+1 201) 467 83 82
President: George G Landberg

Veda Incorporated
1800 North Beauregard St, Alexandria, Virginia 22311
Tel: (+1 703) 998-83 32
Fax: (+1 703) 845-02 23
President: Gary L Palonen
Vice-President: Charles M Mitchell
Program Director for Civil and Military Space: Dr Daniel J Heimerdinger
Program Director for Commercial Space: Dr Robert A Nelson

Virginia's Center
2214 Rock Hill Road, Suite 600,
Herndon, Virginia 20170
Tel: (+1 703) 689 30 00
Fax: (+1 703) 689 30 41
WWW: http://www.cit.org
President: Dr Robert G Templin Jr

Wear-Cote International
PO Box 4177, 101 Tenth Street,
Rock Island, Illinois 61204-4177
Tel: (+1 309) 793 12 50
Fax: (+1 309) 786 65 58
Chief Executive Officer: James R Henry
Chief Operating Officer: Mark Henry

UK Office
Wear-Cote 1980, Ltd
Unit 32 Cambridge Trading Estate,
Hanwell, London W7 3UP, UK
Tel: (+44 181) 567 19 11
Managing Director: David James

Weber State University
Center for AeroSpace Technology (CAST)
1805 University Circle, Ogden, Utah 84408-1805
Tel: (+1 801) 626 72 72
Fax: (+1 801) 626 79 51
e-mail: kreister@weber.edu
WWW: http://cast.weber.edu
Director: Dr Kermit W Reister

WESGO Metals
555 Harbor Boulevard, Belmont, California 94002
Tel: (+1 415) 598 33 55
Fax: (+1 415) 498 32 97
President: Daniel Apfelblum

Western Electrochemical Co
PO Box 629, 10622 West 6400 North,
Cedar City, Utah 84721
Tel: (+1 801) 865 50 00
Fax: (+1 801) 865 50 05
Chairman/CEO: Fred D Gibson Jr
President: James J Peveler
Plant Manager: Dave A Thayer
Customer Services: Laurie A Finnegan

Willis Corroon Inspace
Wall Street Plaza, 88 Pine St, New York 10005-1843
Tel: (+1 212) 612 75 73
Fax: (+1 212) 363 42 13
Executive Vice-President: Bill Lloyd

Wyle Laboratories
PO Box 77777, 7800 Highway 20 West,
Huntsville, Alabama 35807-7777
Tel: (+1 205) 837 44 11
TWX: 810 726-2225
Fax: (+1 205) 830 26 89
President/COO: Gus Yiakas
Vice-President: Sherwyn Hyten
Director, Advanced Programs/Marketing: Don McAvin

Yardney Technical Products, Inc
82 Mechanic Street, Pawcatuck, Connecticut 06379
Tel: (+1 860) 599 11 00
Fax: (+1 860) 599 39 03
President/CEO: Richard M Scibelli
Senior Vice-President Corporation Development/ Marketing: Peter A Karpinski

Zigzag Venture Group, Inc
Olympic Tower East,
645 5th Avenue, 7th Floor,
New York, NY 10022
Tel: (+1 212) 725 67 00
Fax: (+1 212) 725 69 15
WWW: http://www.military-parade.com
e-mail: zigzag@mail.idt.net
President/CEO: Thomas J Langan

Moscow Office Contact and Address:
George Lyssenko
Publishing House Passport International
80 Leningradsky Prospekt, Bldg. 17,
Moscow 125178, Russian Federation

SPACE LOGS

Last updated in June 1999

1996-2000 CHRONOLOGY

1996

11 Jan: the year's 1st Shuttle mission, STS-72, is launched. It retrieved Japan's Space Flyer Unit and deployed/retrieved Spartan. See the Shuttle entry for mission description.
12 Jan: Ariane 44L (V82) is launched from Kourou's ELA-2 pad at 23:10 GMT into 200 × 35,992 km 6.99° GTO carrying PanAmSat's 2,918 kg PAS 3R and Malaysia's 1,450 kg MEASAT 1 GEO telecom satellites. 36 primary satellites remain contracted for Ariane launches. PAS 3R is the Hughes-built replacement for PAS 3, lost in Ariane's 1 Dec 1994 failure, to provide Latin America DTH from 43°W. Malaysia's first satellite was commercially provided by Hughes, but a national microsatellite built with India's assistance is in development for 1997 launch. See 5 Feb.
14 Jan: Korea's second national telecom satellite, Korea Telecom's 1,459 kg Mugunghwa 2/Koreasat 2, is launched by McDonnell Douglas Delta 7925 from Cape Canaveral pad 17B at 11:10 GMT towards 116°E GEO. The orbit achieved after stage 3 firing was 1,358 × 35,421 km, 21.0° GTO. Koreasat's apogee kick motor was fired 16 Jan to attain GEO altitude. Matra Marconi Space delivered the payload Mar 1995 fully tested and ready for integration into its Lockheed Martin 3000-series bus at the Princeton, NJ facility. Koreasat 1 was left in a low orbit 5 Aug 1995 when one of Delta's strap-ons failed to separate.
16 Jan: Cosmos 2327 is launched by SL-8 Cosmos 3M from Plesetsk at 15:34 GMT into 952 × 1,021 km, 82.98° GTO. The orbital plane suggests that it is a Parus military navigation satellite, intended to replace 1993's C2266. The original replacement, C2321 (6 Oct 1995), was stranded in a low orbit.
25 Jan: Gorizont 31 is placed by a Proton SL-12 into geosynchronous orbit (1.5° inc) following launch from Tyuratam at 09:55 GMT. The satellite is positioned at 40°E 2 Feb, replacing G22. The launch announcement indicated there would be no more Gorizonts, but see 25 May.
26 Jan: Mikhail F Reshetnev, Director General and General Designer of NPO Prikladnoi Mekhaniki, dies. NPO PM is responsible for almost 900 satellites, including the Express, Gals, Ekran, Gorizont, Raduga and most other major telecom satellites; also the Strela 3 military sextets, Glonass and Cosmos navigation and GEO-IK geodetic satellites. Reshetnev founded NPO PM in 1959 as a design bureau under Sergei Korolev, specialising in missiles, before becoming independent in the mid-1960s.

1 Feb: Indonesia's 2,989 kg Palapa C1 telecom satellite is launched by Atlas 2AS AC-126 at 01:15:01 GMT from Cape Canaveral pad 36B towards 113°E GEO, replacing Palapa B2P (launched Mar 1987).
5 Feb: Ariane 44P+3 (V83) is launched from Kourou's ELA-2 pad at 07:19:38 GMT into 200 × 35,952 km 7.00° GTO carrying NTT's 3,420 kg NStar b to provide domestic Japanese telecom services from 136°E. 39 primary satellites remain contracted for Ariane. See 14 Mar.
8 Feb: a 182 min EVA by Mir's Gidzenko/Reiter recovers ESA experiments from Spektr.
11 Feb: NASDA's J1 launcher successfully debuts at 23:00 GMT from Tanegashima. The 2-stage suborbital version carried the 1,040 kg Hyflex hypersonic lifting body demonstrator to 110 km for a M14.4 re-entry. The chute deployed at 694 s and the flotation bag inflated, but Hyflex sank in 6 km-deep water after apparently breaking free of the bag. The thermal protection system could not be examined, but 12 of the 14 test objectives were achieved. The planned HOPE unmanned spaceplane will draw on Hyflex technology.
14 Feb: China's maiden CZ-3B is launched at 19:01 GMT from Xichang, carrying Intelsat's 4,576 kg Intelsat 708 towards 50°W. However, the inertial platform veered after 2 s and CZ crashed after 25 s into a hillside 1.5 km from the pad. The fault was determined to be caused by poor workmanship. The official death toll was six plus 57 injured, although unofficially it was claimed as up to 100. 250 homes were damaged or destroyed. The launch was reportedly insured for US$204.7 million at 13 per cent (the low rate reflecting its part in a 10-launch package).
17 Feb: NASA's first Discovery mission, the 805 kg NEAR (Near-Earth Asteroid Rendezvous), is launched by Delta 7925-8 at 20.43 GMT from Cape Canaveral's pad 17. NEAR, built by Johns Hopkins Univ's Applied Physics Lab, was due to fly out to the main asteroid belt (approaching to within 1,200 km of Mathilde Jun 1997) returning to Earth for a 415 km swing by 22 Jan 1998, setting up arrival at Eros Jan 1999 (entering orbit 6 Feb). NEAR is the first asteroid orbiter, coming within 24 km during its 1 yr of primary investigations.
19 Feb: three 225 kg Strela 3 Cosmos satellites, 2328-2330, and three Gonets-D are launched aboard a single SL-14 Tsyklon 3 from Plesetsk at 00:58 GMT into orbits around 1,410 km, 82.6°. Strela are apparently part of a military tactical comms system, coplanar with 1994's C2299-2304, suggesting replacement. The Smolsat consortium plans a constellation of 36 Gonets satellites for E-mail store/forward and realtime bulk data relay.
19 Feb: Raduga 36 is launched by a Proton SL-12 from Tyuratam at 08:18 GMT towards geosynchronous orbit but is left in GTO when Block-DM2 stage 4 explodes, producing about 200 fragments. DM's burn 2 should have raised GTO into GSO but ignition failed; Raduga was released automatically soon after into 242 × 36,502 km, 48.60° GTO. The review board concluded that the Triethyl Aluminium ignition hypergol had not reached the gas generator or main chamber because of reduced pressure from a leaking joint in the pipe from the TEA bottle. The fitting nut had a broken lockwire, probably from poor installation, allowing it to back off during burn 1. Corrective actions include adding a second lockwire.
21 Feb: Soyuz-TM 23 is launched at 12:34 GMT by SL-4 Soyuz-U2 from Tyuratam site 2 carrying Commander Yuri Onufrienko and engineer Yuri Usachev as Mir's 21st main crew. Launch had been planned for 25 Dec 1995 but funding problems meant that its launcher was not ready in time, requiring extension of TM22's occupation. TM23 docked with Kvant 1's port 14:21 GMT 23 Feb.
22 Feb: the year's 2nd Shuttle mission, STS-75, is launched carrying USMP 3 and TSS 1R. The tether broke 26 Feb at about 19 km, leaving TSS as a separate satellite. See the Shuttle entry for mission description.
24 Feb: NASA's 1,258 kg Polar ionospheric research satellite is launched by Delta 7925-10 from Vandenberg pad 2W at 11:24 GMT into initial 186 × 50,495 km, 85.98°; perigee was raised 9 Mar to 5,141 km.

9 Mar: Orbital Sciences Corp's 10th Pegasus launcher is released from its Lockheed L-1011 carrier aircraft at 01:33 GMT at 11.94 km altitude, some 105 km out over the Pacific after taking off from Vandenberg AFB at 00:38 GMT. Stage 2 separated at 608 s instead of the expected 529 s as a result of high headwinds at the drop site; the autopilot ordered the delay to gain height before stage 3 ignition. Pegasus achieved 805 × 836 km 89.96° (target 821 × 834 km), releasing the 113 kg USAF REX II satellite to investigate radio propagation. This was the first successful Pegasus XL after two failures. REX was planned for a standard Pegasus but it switched to requalify XL; USAF still paid only US$6 million.
14 Mar: Intelsat's 4,175 kg 707 satellite, built by Space Systems/Loral, is launched by Ariane 44LP (V84) from Kourou's ELA-2 pad at 07:11 GMT into 200 × 35,913 km, 7.04° GTO. 707 entered service at 1°W GEO. It can handle up to 90,000 telephone circuits and three TV channels. Stationkeeping propellant should be sufficient for 15 yr. 38 primary satellites remain contracted for Ariane. See 20 Apr.
14 Mar: the Cosmos 2331 4th gen photorecon satellite is launched by SL-4 Soyuz-U from Plesetsk pad 43 at 17:40 GMT into 164 × 358 km, 67.1° for a Yantar close look mission. The descent capsule carried the USAF Space Test Program's deployable Birrad experiment to collect Be-7 samples. Returned to Earth 11 Jun after record 89 days.
15 Mar: 351 min EVA 1 starting 01:04 GMT by Mir cosmonauts Onufrienko/Usachev installs a crane on Mir's core module for moving the solar wing packages delivered Nov 1995 by STS-74 to Kvant 1. See 20 May.
21 Mar: India's 922 kg IRS P3 Earth observation satellite is launched by PSLV-D3 from Sriharikota at 04:53 GMT into 819 × 821 km, 98.8°. This was the second successful PSLV. Germany's stereo scanner (first activated 22 Mar on rev 15) is optimised for ocean studies and India's WiFS wide field sensor (first activated 23 Mar on rev 29) for vegetation dynamics. P3 also carries the XAP X-ray astronomy payload. India's space agency now plans to commercialise PSLV. Programme cost was Rs4,150 million through the first two launches; unit vehicle cost is Rs450 million.
22 Mar: the year's 3rd Shuttle mission, STS-76, is launched. It docks with Mir 25 Mar, delivering Shannon Lucid to join the station's main crew until STS-79's return Sep 1996. See the Shuttle entry for mission description.
22 Mar: former NASA astronaut Bob Overmyer is killed while testing an experimental Cirrus VK-30 aircraft near Duluth International Airport. Overmyer retired from NASA in 1986 after flying the STS-5/51B missions. He was originally selected as an astronaut for the AF Manned Orbiting Laboratory.
28 Mar: Rockwell's 16th GPS Block 2A satellite, Navstar 2-25 (USA 117; GPS 38), is launched by McDonnell Douglas Delta 7925 from Cape Canaveral complex 17 at 00:21 GMT into the system's C2 slot. The Thiokol Star 37F solid propellant apogee motor was fired some 44 h later to establish the final orbit, joining 24 other operational Block 2/2A satellites. A further three 2As are available for launches at 2 months' notice to guarantee the service until Block 2R appears in 1997.

3 Apr: Inmarsat's 2,068 kg initial third-generation satellite is launched by Atlas Centaur 2A (AC-122) from Cape Canaveral's pad 36A at 23:01 GMT. The payload was activated 13 Apr for testing at 28°E from Fucino; it replaced Inmarsat 2-F1 at 64°E in mid-May. 2-F1 moved to 65°E as backup. 3-F2 is planned for Aug aboard Proton.
8 Apr: Luxembourg's 3,010 kg Astra 1F DTH satellite is launched at 23:09 GMT from Tyuratam by Proton SL-12 towards 19.2°E. It is Proton's first commercial launch. Block DM injected Astra into a GTO of 12,100 × 36,000 km, 7.0° using two burns. Astra separated after 6 h 41 min. It is Hughes' 29th HS-601 model to be launched.
20 Apr: TMI Communication's 2,855 kg MSAT 1 is launched by Ariane 42P (V85) from Kourou's ELA-2 pad at 22:36:00 GMT into 199.5 × 35,923 km, 7.00° GTO. MSAT began initial services at 106.5°W providing mobile phone services over N America. It is Hughes' 30th HS-601 model to be launched. The identical US AMSC 1 appeared 7 Apr 1995. 47 primary satellites remain contracted for Ariane. Ariane has now successfully launched 112 primary and 26 auxiliary payloads.
23 Apr: Russia's 19.7 t Priroda remote sensing module is launched by Proton from Tyuratam at 11:48 GMT to complete the Mir station. It docked with Mir's main axial face 12.43 GMT 26 Apr and transferred to the portside position 27 Apr. Cargo included 1 t of US equipment. Without solar wings, Priroda relied on battery power.
24 Apr: BMDO's US$800 million MSX Midcourse Sensor Expt mission is launched at 12:27 GMT by Delta 7920-10 from Vandenberg's pad 2W into 897 × 906 km, 99.4° Sun-synchronous. The 2,680 kg MSX is demonstrating early warning sensors for discriminating and tracking mid-course targets. The main instrument's nominal life is up to 20 months, limited by the cryogenic coolant supply, but the other sensors will operate for 5 yr. MSX is also gathering global environmental data, such as ozone levels, in co-operation with NASA.
24 Apr: Cosmos 2332 is launched at 13:03 GMT by SL-8 Cosmos 3M from Plesetsk into 295 × 1,565 km, 82.96°. A 2 m ∅ sphere (500 kg?) for tracking for radar calibration and passive monitoring of upper atmosphere.
24 Apr: the 7th Titan 4 Centaur is launched from Canaveral's ETR 41 at 23:37:01 GMT carrying the classified USA 118 payload. Titan's core was #K16. Observers saw Titan heading due east (100.4° aximuth), indicating a 28.5° orbit preparatory to departure for GEO. Centaur's employment and the lack of other information suggest a GEO elint satellite for the National Reconnaissance Office. See 14 May 1995 and 12 May 1996.
30 Apr: the 1.4 t Italian-Dutch SAX X-ray satellite is launched by Atlas 1 from Canaveral's pad 36B at 04:31 GMT into 583 × 603 km, 3.96°. This was the 100th Atlas Centaur mission and the penultimate Atlas 1.

5 May: Progress-M 31 is launched by Soyuz-U from Tyuratam site 2 at 07:04 GMT, docking with Mir's forward port 08.54 GMT 7 May.
12 May: a Titan 4 Centaur is launched from Vandenberg's SLC 4E at 21:32 GMT carrying the classified USA 119 payload. The mission was identified as ocean surveillance when three subsatellites were designated (USA-120/121/122) and, uniquely, a 53 kg

tether experiment was released 20 Jun into 1,023 km, 63.4°. The TiPS Tether Physics and Survivability experiment was provided by the US Naval Center for Space Technology, with the US$4 million total cost (including US$1.9 million for 2 yr of tracking) funded by the National Reconnaissance Office. The 4.0 km tether and its two end masses (9 kg + 41 kg) unreeled in 42 min; it became passive along the gravity gradient after its battery was exhausted. TiPS is being tracked by radar and laser to model its motion and longevity. Titan vehicle processing required 112 days; Titan's core was #K22. See 24 Apr.

14 May: a Kometa 4th generation topographic/ mapping satellite is lost when the payload fairing of its Soyuz-U launcher fails after 49 s. Launch was at 08:55 GMT from Tyuratam site 31. In addition to its military mission, Kometa was to return 2/10 m-resolution film imagery after 45 days under Sovinformsputnik's SPIN-2 (Space Information 2 m) agreement with three US companies (Aerial Images, Inc; Lambda Tech International; Central Trading Systems). The commercial element was insured with Moscow's Megaruss Co for US$2.7 million. A second SPIN-2 mission is reportedly planned for early 1997. Sovinformsputnik planned to fill the current image orders using archived material and other military satellites.

16 May: Ariane 44L (V86) is launched from Kourou's ELA-2 pad at 00:56:29 GMT into 299 × 35,934 km, 4.00° GTO carrying Israel's 996 kg Amos 1 and Indonesia's 2,989 kg Palapa C2. Amos will provide mainly TV distribution from 4°E, and PT Satelindo's Palapa will replace Palapa B2R at 108°E. It is Hughes' 31st HS-601 model to be launched. 45 primary satellites remain contracted for Ariane. Ariane has now successfully launched 114 primary and 26 auxiliary payloads.

17 May: the 211 kg BMDO/USAF MSTI 3 is launched aboard Orbital Sciences Corp's 11th Pegasus launcher, released from its Lockheed L-1011 carrier aircraft at 02:44 GMT at 11.59 km altitude, some 96 km out over the Pacific after taking off from Vandenberg AFB. Pegasus achieved 296 × 361 km 97°; MSTI then manoeuvred into 425 km. MSTI is mapping Earth's IR background for a full season to help design future early warning sensors. This was the last standard Pegasus launch; future versions will be the extended XL.

19 May: the year's 4th Shuttle mission, STS-77, is launched. It deploys a 12 m inflatable antenna and carries Spacehab 4. See the Shuttle entry for mission description.

20 May: 320 min EVA 2, beginning 22:50 GMT, by Mir cosmonauts Onufrienko/Usachev attaches to Kvant 1 the Russian/US solar wing package delivered Nov 1995 by STS-74. They also filmed footage of a large inflated Pepsi Cola can for a reported fee of >US$1 million. See 24 May.

24 May: Hughes Communications' 1,397 kg Galaxy 9 is launched towards 123°W by a Delta 7925 from Canaveral's pad 17B at 01:10 GMT. Galaxy will primarily distribute cable TV programming.

24 May: 343 min EVA 3, beginning 20:47 GMT, by Mir cosmonauts Onufrienko/Usachev handcranks out the solar wing installed during 24 May EVA, adding 6 kW to Mir's power supply. Space Station Alpha will use the same US solar cells. See 30 May.

25 May: Gorizont 32 is placed by a Proton SL-12 into geosynchronous orbit (1.5° inc) following launch from Tyuratam at 02:05 GMT. The satellite is heading for 52°E.

30 May: 260 min EVA 4, beginning 18:20 GMT, by Mir cosmonauts Onufrienko/Usachev mounts Germany's MOMS-02P multi-spectral scanner on Mir's Priroda module.

4 Jun: ESA's maiden Ariane 5 breaks up after a software problem loses attitude control. The main Vulcain engine ignited at 12:33:59 GMT on Kourou's ELA 3 pad for launch 7.5 s later when the P230 solid boosters ignited. Flight was normal until 30 s after launch, at M0.7 and 3,500 m altitude, when both booster nozzles swivelled within 2 s to their extreme positions, followed by Vulcain. The intense aerodynamic loads broke up the vehicle and the self-destruct system was triggered by rupture of the electrical links between the boosters and core. An independent inquiry board reported 23 Jul that both inertial reference systems had failed simultaneously because of software specification and design errors. Flight 502 was planned for Sep/Oct 1996, but is now probable for mid-1997, followed by a new test flight 4Q 1997. 501 was to release four 1.17 t Cluster satellites into 10° GTO; they would have then manoeuvred into 25,513 × 140,318 km, 90° to study the 3D structure of Earth's plasma environment. ESA decided in Jul to build the single Phoenix replacement satellite from spare Cluster hardware, and will decide in Nov 1996 whether to fund another three.

6 Jun: Onifrienko and Usachyov began EVA 5 at 16:56 GMT during which they spent 3 h 34 min replacing experiment cassettes and installing American-made micrometeoroid devices.

13 Jun: Russian cosmonauts Onifrienko and Usachyov began their mission's EVA 6 at 12:45 GMT which lasted 5 h 42 min to assemble and erect a four-section truss to the side of Kvant module and deploy a large antenna on the Priroda module.

15 Jun: Intelsat's 3,420 kg 709 satellite, built by Space Systems/Loral, is launched by Ariane 44P (V87) from Kourou's ELA-2 pad at 06:55:09 GMT into 200 × 35,955 km, 6.99 GTO towards 18°W GEO. It can handle up to 90,000 telephone circuits and three TV channels. Stationkeeping propellant should be sufficient for 15 yr. 44 primary satellites remain contracted for Ariane. Ariane has now successfully launched 115 primary and 26 auxiliary payloads.

27 Jun: NASA's Galileo spacecraft conducted a fly-by of Ganymede when it passed within 835 km.

27 Jun: Canada's Industry Minister John Manly announced far reaching expansion of his country's participation in the International Space Station. The Canadian Space Agency reached agreement with NASA for three new positions for Canadian astronauts involving Julie Payette, Steve MacLean and Bjarni Tryggvason.

2 Jul: At 07:48 GMT Pegasus-XL F-12 dropped from a converted L-1011 aircraft and launched the 295 kg TOMS-EP (Total Ozone Mapping Spectrometer-Earth Probe) from WTR. This is the first satellite dedicated to mapping the Earth's ozone layer and is instrumented to also monitor sulphur dioxide levels emitted by volcanic eruptions. TOMS-EP was placed in a 494 × 511 km, 97.44° orbit with a period of 94.67 min.

3 Jul: At 00:31 GMT Titan 4 405 K-2 (45H-1) launched USA-125, the 700 kg SDS-B4 (SDS-2 series) satellite from LC-40 at the ETR. The satellite was placed in a 292 × 319 km, 54.99° orbit with a period of 90.64 min.

3 Jul: At 10:47 GMT APStar 1A, a satellite for the Asia-Pacific Telecommunications Satellite Co Ltd based in Hong Kong was launched by CZ3-10 from China's Xi Chang launch site to a 35,785 × 35,791 km, 0.07°, geostationary orbit. The 1,400 kg satellite was located at 133.9°E to replace the APStar 2 satellite lost during launch in January 1995.

9 Jul: At 22:24 GMT Ariane 44L V89 launched the 2,617 kg Arabsat 2A and 1,743 kg Turksat 1C from Kourou, French Guiana. Arabsat 2A is a telecommunications and direct broadcast satellite for the Arab League group of countries and was placed in a 35,782 × 35,794 km, 0.08°, geostationary orbit. BOL mass is 1,570 kg and the Spacebus 3000 satellite was located over 25° E. A telecommunications and direct broadcast satellite, the Turksat 1C Spacebus 2000 satellite was launched to a 35,781 × 35,791 km, 0.02°, geostationary orbit at 42° E with a BOL mass of 1,078 kg. Turksat 1C was initially located over 31° E and moved later to its present location.

10 Jul: Canada signed agreements with the European Space Agency extending co-operation into the fields of remote sensing and telecommunications.

12 Jul: NASA decides to replace STS-79's solid rocket boosters with new ones when examination of the boosters recovered from STS-78 indicated gas penetration down into the joint where the O-rings seal the pressure and prevent blow-through.

16 Jul: At 00:50 GMT Delta 7925A (237) launched from LC17A, ETR, carried the 1,881 kg (930 kg dry) Navstar 26 (SVN40) to a 20,138 × 20,224 km, 55.04°, orbit with a period of 717.93 min. Satellite also known as USA-126 was to be operated in plane E slot 3 of the GPS network

21 Jul: The first public acknowledgement of an accidental collision between two objects in space was made when a piece of debris from the Ariane 1 launcher that carried SPOT 1 into orbit in 1986 collided with CERISE launched July 1995, breaking off a section of boom making it a third object in orbit.

22 Jul: Rockwell and NPO Yuzhnoye concluded an agreement under which Rockwell's Space Systems Division would market commercial launch services for the Ukraine's Cyclone rocket. Derived from the two-stage SS-9 ICBM, with a third stage Cyclone can lift 3,600 kg to low Earth orbit.

25 Jul: At 12:42 GMT an Atlas 2 (AC-125 IABS/ Centaur II) launched the 3,020 kg UHF F7 (UHF FO F-7-EHF) satellite from LC36A, ETR, for the DoD. Also known as UFO 7, it is an HS-601 communications and was placed in a geosynchronous orbit of 34,579 × 36,562 km with an inclination of 5.07° and a period of 1,425.08 min.

31 Jul: Progress M-32 was launched by Soyuz U (11A511U) from LC1, Baikonur, at 20:06 GMT to a 293 × 333 km, 51.65° orbit with a period of 90.78 min. This was the first successful launch of a Soyuz U vehicle in the last three attempts. It docked with the Mir forward port (-X) at 22:03:40 GMT 2 Aug, undocked at 09:33:45 GMT Aug 18 and redocked to the rear docking port (+X) at 09:35:00 GMT Sep 3. M-32 finally undocked at 19:51:20 GMT 20 Nov and broke up on re-entry at 22:42:25 GMT. Total docked time was 93.91 days, free time 2.2 days. Total mission duration was 96.11 days.

31 Jul: The McDonnell Douglas Clipper Graham VTOL test launcher was destroyed when it toppled and burst into flames after a landing leg failed to deploy on touchdown at the end of its fourth flight this summer.

6 Aug: NASA announced that a team of scientists from the Johnson Space Center had concluded that a meteorite (ALH84001) found in Antarctica in 1984 probably contained the tracks of primitive life forms and that isotopic analysis of the 4.5 billion year old rock indicated that it came from Mars.

6 Aug: Starsem S.A. is founded as a joint Russian-European venture to market industrial and commercial opportunities for the Soyuz-Icare series of launchers capable of lifting up to 5 tonnes to low orbit.

8 Aug: Ariane 44L V90 launched from ELA2 at Kourou at 22:49 GMT and carried the 1,990 kg Italsat 2 and 2,260 kg Telecom 2D satellites to geostationary orbits. Italsat 2 is a telecommunications satellite used for data transmission and digital TV with ESA's EMS package for mobile communications. It was placed in a preliminary orbit of 34,911 × 35,791 km with a period of 1,413.9 min drifting west to its planned position at 10.2° E. BOL mass was about 1,200 kg. The Eurostar 3000 Telecom 2D is used for telephone and TV communications and French government communications. It was placed in a 35,768 × 35,795 km geostationary orbit at 3° E. BOL mass was 1,400 kg.

14 Aug: Molniya-M launcher 8K78M/ML lifted off from LC43/3, Plesetsk, at 22:21 GMT carrying the 1,600 kg Molniya-1T/89 telecommunications satellite, replacing Molniya-1T/85.

17 Aug: Japan's ADEOS (ADvanced Earth Observing Satellite) and Fuji 3 amateur radio satellite were launched from Tanegashima LC-Y by H-2 (H-II-4F) at 01:53 GMT. The 3,560 kg ADEOS was placed in a 797 × 799 km, 98.62° orbit with a period of 100.83 min and is designed to monitor the Earth's surface, atmosphere and oceans to measure environmental changes. Designated JAS 2 prior to launch, Fuji 3 was placed in a 802 × 1,323 km, 98.58° orbit with a period of 106.46 min.

17 Aug: The Soyuz TM-24 manned spacecraft (TM 11F732 s/n 73) was launched at 13:18 GMT from LC1 at Baikonur by a Soyuz U (11A511U) launch vehicle. Designated Mir Expedition EO-22 the Soyuz TM-24 was placed in a 375 × 390, 51.65° orbit with a period of 92.2 min. It carried V. G. Korzun (Commander), A. Y. Kaleri (Flight Engineer) and C. Andre-Deshays, the first French woman in space. Soyuz TM-24 docked to the front longitudinal port (+X) on Mir at 14:50:21 GMT 19 Aug, allowing cosmonauts Korzun, Kaleri and Andre-Deshays to enter the space station.

18 Aug: At 10:27 GMT a CZ-3 (CZ3-11) launched the 1,200 kg Zhongxing 7 satellite from the LC1 complex at the Xi Chang launch site. An HS-376 telecommunications satellite for domestic internal use, it was stranded in a 200 × 17,229 km, 27.25° orbit with a period of 307.53 min when the second stage engine stopped firing 48 s prior to the planned time.

20 Aug: Maritime satellite Marecs A was retired from service.

21 Aug: A Pegasus-XL (s/n F13) launched at 09:47 GMT carried the FAST (Fast Auroral Snapshot Explorer) to an orbit of 351 × 4,165 km with an inclination of 82.98° and a period of 133.12 min. FAST is a 180 kg scientific satellite developed by NASA-GSFC for investigating plasma phenomena in aurora and general magnetospheric phenomena.

29 Aug: Launched at 05:22 GMT by Molniya-M (8K78M/2BL) from LC43/3 at the Plesetsk launch facility, the Interbal 2, Magion 5 and Musat science satellites were placed in elliptical orbits. Provided by the Instituto Aeronautico de Cordoba, Argentina, to demonstrate low-cost technologies, the 30 kg Musat was placed in a 236 × 1,171 km, 62.79° orbit with a period of 98.84 min. The 62 kg Magion 5 (Magnetosphere — Ionosphere) was provided by the Czech Republic to operate with Magion 4 launched in 1995 and Interball satellites to study energy transport mechanisms between the solar wind and the Earth's magnetosphere. It was placed in a 791 × 19,196 km, 62.8° orbit with a period of 347.46 min. Interball 2, also called Auroral Probe, is the second flight of the Prognoz-M2 science bus, weighs about 1,300 kg and was placed in a 769 × 19,211 km, 62.77°, orbit with a period of 347.34 min.

2 Sep: At 04:20:00 GMT the Soyuz TM-23 spacecraft carrying cosmonauts Y. Onufrienko, Y. Usachoyov and C. Andre-Deshays undocked from Mir and performed a small separation burn at 04:24:40 prior to the deorbit burn at 06:47:20 GMT. The modules separated at 07:14:36 with parachute deployment at 07:26 GMT followed by a landing at 07:41:40 100 km southwest of Akmola in Kazakhstan.

4 Sep: At 09:01 GMT the 3,250 kg Cosmos 2333 was carried by Zenit-2 from LC45L, Baikonur, to a 849 × 852 km, 71.01° 101.94 min, orbit. Cosmos 2333 was a Tselina-2 ELINT satellite.

5 Sep: A Parus military navigation satellite, Cosmos 2334 was launched from LC132/1, Plesetsk, at 12:47

GMT by Cosmos-3M (Kosmos 11K65M) launcher to a 970 × 1,000 km, 82.94° 104.8 min orbit. The navigation satellite was positioned in plane 1 of the constellation and operated at radio frequencies of 150.03 MHz and 400.08 MHz. Also launched with Cosmos 2334 was UNAMSAT 2, a science satellite from the Autonomous University of Mexico designed to study micrometeorites. It replaces the satellite of this type lost on the first flight of the Start launcher on 25 March 1995 and was placed in a 968 × 1,011 km, 82.93° 104.89 min, orbit.

6 Sep: At 17:37 GMT a Proton-K (8K82K/DM-1; s/n 375-01/DM-1 s/n 1L) launched Inmarsat-3 F2 into a 35,958 × 36,136 km, 2.65° orbit from LC81L, Baikonur. This was the second commercial launch for the Proton-K but the second Inmarsat-3 satellite did not need an apogee kick motor. The 1,144 kg (launch mass) satellite was temporarily positioned at 28° E on 18 Sep, but on 27 Sep it was moved further west.

6 Sep: The NASA Jupiter spacecraft Galileo performed a fly by of Ganymede at a distance of just 262 km.

8 Sep: An Atlas 2A (IIA (1N); s/n AC-123) launched from LC36B carried the 2,764 kg GE 1 telecommunications satellite to a 35,788 × 35,792 km supersynchronous orbit inclined 0.01° with a period of 1,436.27 min. Launched for GE American Communications (Americom), the satellite was located over 257° E.

11 Sep: An Ariane 42P (V91) launched from ELA2, Kourou, at 00:00 GMT carried the 2,865 kg (BOL 1,600 kg) EchoStar 2 direct broadcast satellite to a 35,679 × 35,787 km, 0.04° orbit for the EchoStar Communications Corporation. The satellite design was based on the Aerospatiale AS 7000 bus.

12 Sep: At 08:49 GMT a Delta 7925 (7925A; s/n 238) carried the 1,881 kg (launch mass) Navstar 27 (SVN30) navigation satellite from LC17A, ETR, to a 20,057 × 20,307 km, 54.72° orbit with a period of 717.98 min.

16 Sep: STS-79 (Atlantis) was launched from LC-39A at 08:54 GMT, the fourth Shuttle-Mir mission, carrying six astronauts and more than 2,100 kg of supplies to the Russian space station. Astronaut Blaha replaced Shannon Lucid as the resident NASA astronaut aboard Mir. See Shuttle log and the Human Space Flight entry for details.

26 Sep: At 17:50 GMT a Proton-K (8K82K; s/n 375-01) carried the 2,500 kg Ekspress telecommunications satellite from LC200L, Baikonur, to a 35 m 838 × 35,908 km, 0.21° orbit with a period of 1,440.53 min. This was the second launch of the new generation Ekspress and the satellite was stationed at 80° E.

28 Sep: With the de-orbiting of Cosmos 2320, launched 29 Sep 1996, there was no Russian military reconnaissance satellite currently in orbit.

20 Oct: At 07:20 GMT a Chinese CZ-2D (s/n CZ2D-3) launched from Jiuquan complex carried the 2,600 kg FSW2-3 (Fanhui Shi Weixing) military reconnaissance satellite to a 171 × 342 km, 63.04° orbit with a period of 89.64 min. The recoverable module was returned to Earth 4 Nov and landed in the Sichuan Province.

24 Oct: At 11:37 GMT a Molniya-M (8K78M) launch vehicle carried the 1,750 kg Molniya-3 48 telecommunications satellite from LC43/4 Plesetsk to a 610 × 39,768 km, 62.83° orbit with a period of 718.27 min coplanar with Molniya-3 36.

4 Nov: Pegasus XL (s/n F14) was launched from the L-1011 carrier aircraft at 17:08 GMT carrying the SAC-B (Satellite de Aplicaciones Cientificas) and HETE (High Energy Transient Experiment) satellites. The 183 kg SAC-B was built by INVAP of Argentina with both Argentinian and NASA instruments. The 118 kg HETE was a US science satellite. The Pegasus XL stage performed as planned but the satellite separation system malfunctioned and failed to release either payload. The solar panels on SAC-B were deployed but the combined third stage (176 kg) and satellites caused it to tumble and the batteries could not be recharged. The solar panels on HETE could not be deployed because it remained encapsulated. Initial orbit of the assembly was 488 × 556 km at 37.97° with a period of 95.07 min.

4 Nov: NASA's Galileo spacecraft conducted a fly-by of Jupiter's moon Callisto, passing within 1,104 km of the surface.

7 Nov: At 17:00 GMT a Delta 7925A (s/n 239) carried the 1,060 kg Mars Global Surveyor (MGS) spacecraft from LC-17A to a transplanetary trajectory via heliocentric orbit. Spacecraft carried a camera, laser altimeter, electron reflectometer, thermal emission spectrometer, Mars relay radio system and magnetometer. MGS was scheduled to arrive at Mars 11 September 1997 for a planned two years of operations from Mars orbiter, achieved through aerobraking techniques.

13 Nov: The 2,661 kg Arabsat 2B and the 1,512 kg MEASAT 2 telecommunications satellites were launched by Ariane 44L (s/n V92) from ELA-2 at French Guiana. Arabsat 2B was placed in a geostationary orbit over 32° E and MEASAT, launched for Binariang (Kuala Lumpur, Malaysia), operated from 148° E.

16 Nov: At 20:48 GMT a Proton K launcher (8K82K s/n 392-02) carried the 6,825 kg Mars 96 spacecraft to an initial Earth orbit of 160 km at 51.53° inclination. The Block D-2 fourth stage was then to have put Mars 96 in an elliptical 160 × 100,000 km Earth orbit. After separating from the D-2 stage, Mars 96 should have executed an escape manoeuvre to a heliocentric Mars transfer orbit. An attitude error put the D-2/Mars 96 combination in the wrong attitude for elliptical transfer and a single burn of two D-2 ullage motors resulted in a 146 × 171 km Earth orbit. Controlled by timer, the spacecraft separated and performed its own burn, putting it in a 87 × >1,500 km orbit from where it decayed early 17 November followed by the D-2 next day. The spacecraft carried two 90 kg surface landers to have been deployed before Mars orbit insertion and two 75 kg surface penetrometers released after orbit insertion.

19 Nov: Shuttle Columbia was launched at 19:55 GMT from LC-39B at the start of the STS-80 carrying five astronauts including Mission Specialist Story Musgrave, at 61 years the oldest astronaut to date. Problems with a stuck hatch prevented planned EVAs. See Shuttle flight log and Human Space Flight section for full details.

19 Nov: Progress M-33 was launched from Baikonur complex LC1 at 23:20 GMT and placed in a 371 × 390 km, 51.65° orbit with a period of 92.16 min. The 7,190 kg cargo tanker was launched by Soyuz U (11A511U) and docked with Mir (-X port) at 01:01:30 GMT on 22 November, undocking at 12:13:53 GMT on 6 February 1997 after 76.47 days. Continued in independent flight in 377 × 395 km orbit for 35.7 days until destroyed on re-entry at 03:23:37 GMT 12 March 1997 following a failed re-docking attempt 4 March.

20 Nov: NASA astronauts aboard the Shuttle Columbia on the STS-80 mission released the 3,573 kg ORFEUS-SPAS 2 (Orbiting Retrievable Far and Extreme Ultraviolet Spectrometer – Shuttle PAllet Satellite) to a 349 × 356 km orbit. From there it conducted studies of the magnetosphere and energy from very hot and very cold matter in the universe. It was retrieved by Columbia on 4 December and returned to Earth.

21 Nov: An Atlas 2A (AC-124) carrying the 2,912 kg Hot Bird 2 telecommunications satellite for Eutelsat was launched from LC-36A, ETR, at 20:47 GMT. The satellite was placed in a geostationary orbit at 13° E.

22 Nov: Astronauts aboard the Shuttle Columbia released the 2,109 kg Wake Shield Facility into a 347 × 359 km orbit from where it conducted microgravity experiments. The WSF was retrieved and re-berthed aboard Columbia on 26 November.

2 Dec: Russian cosmonauts Korzun and Kaleri perform a 5 h 57 min EVA to begin the installation of solar array cables and connectors and a new solar panel to augment power supplies.

4 Dec: A Delta 7925 (s/n 240) lifted off from LC-17B, ETR, at 06:58 GMT carrying the 890 kg Mars Pathfinder spacecraft to a heliocentric Mars transfer orbit. The lander would use a new means of safely reaching the surface, impact energy cushioned by inflated spheres. The lander spacecraft encapsulated a rover, Sojourner, designed to operate independently on the surface.

5 Dec: In a transaction valued at around US$8.1 billion, Boeing acquired the aerospace and defence interests of Rockwell, including Rocketdyne, while moving to acquire McDonnell Douglas.

9 Dec: Russian cosmonauts Korzun and Kaleri performed an EVA lasting more than 6 hours to install more electrical power cables for the new solar battery and attach an antenna on the docking unit to improve the final approach for Shuttle Atlantis.

11 Dec: A Tsyklon-2 launch vehicle carrying the 3,150 kg Cosmos 2335 EORSAT (ELINT Ocean Reconnaissance SATellite) was launched at 12:00 GMT from the Baikonur LC90L complex carrying its payload to a 403 × 419 km, 65.04° orbit with a period of 92.79 min. Cosmos 2335 operates with Cosmos 2313 and Cosmos 2326. It is virtually coplanar with Cosmos 2313, crossing the equator 120° radially ahead of it.

11 Dec: Equatorial Guinea joined Intelsat becoming its 140th member with an investment of 0.05 per cent, more than 11 years after it began using Intelsat services.

18 Dec: An Atlas 2A (AC-129) carrying the 2,074 kg Inmarsat-3 maritime communications satellite operated by Inmarsat was launched at 01:57 GMT from LC-36A to a 35,698 × 35,877 km geostationary orbit over 157° E. The AS4000 satellite has a BOL mass of 1,100 kg and a dry mass of 860 kg.

20 Dec: A Cosmos-3M (11K65M) launched at 06:43 GMT from LC132/1, Plesetsk, carried the 825 kg Cosmos 2336 Parus military navigation satellite to a 979 × 1,012 kg, 82.95°, 105.03 min, orbit in plane 4 of the constellation. Parus operated at 149.97 MHz and 399.92 MHz.

20 Dec: Titan 4 404A (s/n K-13) launched at 18L04 GMT carried the 19,600 kg USA 129 payload comprising a KH-12 (No. 4) Advanced Crystal military reconnaissance satellite to a 153 × 949 km, 97.9° orbit from SLC4E, WTR.

23 Dec: US DoD communications satellite FltSatCom 1 was retired from duty and moved off station to 344° E.

24 Dec: A Soyuz-U (11A511U) carrying the 6,000 kg Bion 11 biological satellite to a 217 × 379 km, 62.80°, 90.48 min, orbit was launched at 13:50 GMT from LC43/4, Plesetsk. Carrying two macaque monkeys, newts, snails, drosophilia flies and a variety of insects and bacteria, this was the 12th in the series. Bion 11 was returned to Earth, 130 km north of Kustani, on 7 January 1997.

1997

9 Jan: OSC announces that it will use a Taurus launch vehicle to place 420 kg Korean KOMPSAT into orbit during mid-1999.

10 Jan: Surge in electromagnetic radiation following an event on the Sun on 6 Jan resulted in serious damage to the Telstar 401 satellite four days later when that radiation reached the Earth's environment. Electronics on board the satellite appear to have short-circuited.

12 Jan: Launch of shuttle orbiter Atlantis on STS-81/SMM-5 mission from KSC: docks with Mir 15 Jan, undocked 20 Jan, landed KSC 22 Jan. Returned J Blaha from Mir, left J Linenger aboard the station.

14 Jan: Reported that GE purchases 17.25 per cent stake in Argentina's Nahuel-1 satellite.

16 Jan: India announces that two advanced remote sensing satellites, IRS-P4 and IRS-P5 are to be launched aboard PSLVs before the end of the century. Approval also given for R&D of a third generation earth observation satellite series.

17 Jan: First launch failure of a Delta-2 (7925); SRB anomaly just after launch from Cape Canaveral led to the vehicle self-destructing 21 seconds into the flight. Navstar 28 payload was the first Block 2R satellite to be launched.

22 Jan: Delta-2 second stage from the MSX launch re-entered the atmosphere and was almost intact until impact in Georgetown TX (US).

23 Jan: Chinasat authorises SS/Loral to build Chinasat 8 with 16 Ku-band and 32 C-band channels. Launch planned for late 1998.

30 Jan: Ariane-44L launched from Kourou; successfully places GE 2 (US) and Nahuel 1A (Argentina) comsat into orbit.

SS/Loral signs US$100 million contract with Kistler Aerospace for 10 COMSAT launches aboard the K-1 vehicle between 1999 and 2002.

JCSAT selects Atlas-2AS for the launch of JCSAT 6 in mid-1998.

3 Feb: An Australian businessman from Korea and several South Korean companies will put up A$160 million from a required A$200 million for a proposed launch site at Temple Bay, Queensland. Russian launch vehicles could start orbital missions from there around 2000.

6 Feb: Progress-M 33 undocked from rear port of Mir: planned re-docking on 4 Mar failed and the spacecraft was de-orbited 12 Mar.

7 Feb: Soyuz-TM 24 undocked from the front longitudinal port of the Mir Complex and re-docked at the back, in advance of the TM25 launch.

10 Feb: Soyuz-TM 25 launched from Baikonur (previously known a Tyuratam) using a Soyuz-U: carried two Russians and German astronauts to Mir: German returned to Earth aboard TM24 (see blow).

11 Feb: Discovery shuttle orbiter launched from KSC on the second mission to service Hubble Space Telescope. HST captured 13 Feb and two pairs of astronauts complete a total of five EVAs, undertaking repairs and replacing new equipment for old; released 19 Feb and Discovery landed KSC 21 Feb.

12 Feb: Japanese Haruka (Muses-B) launched using M-5 from Kagoshima: radio telescope with an 8 m diameter dish.

14 Feb: Russia launches three Gonets-D and three Strela-3/Cosmos communications satellites from Plesetsk using a three-stage Tsyklon vehicle.

SS/Loral books satellite launches aboard five Delta-3 vehicles; this gives 18 satellites booked to fly aboard the new launch vehicle, due to debut in 1998.

17 Feb: Japanese JCSAT 4 comsat launched from Cape Canaveral using an Atlas-2AS.

20 Feb: Galileo performs fly-by of jovian satellite Europa, minimum distance 587 km.

Reports that Malaysia is to fly Tiung (= mynah bird) satellite — built by Surrey Satellite Technology — aboard Russian launcher.

Wall Street Journal reports that Teledesic communications satellite system could become operational around 2002 with only 288 satellites operating in 12 orbital planes: full system is planned to have 840 satellites in orbit.

23 Feb: A solid-fuel oxygen-generating candle exploded in the Kvant 1 module, causing a fire which was brought under control by the crew on board Mir.

24 Feb: Maiden flight of the Titan-4B using an IUS upper stage assembly launches DSP 18 (USA 130) early warning satellite. Delay of six months in the ISS launch schedule is confirmed. First launch — Russian FGB module aboard a three-stage Proton-K — now scheduled for around June 1998. Delay caused by slippage of work on second Russian module, the Service Module, which cannot be launched until around November 1998.

1 Mar: INTELSAT 801 comsat launched by Ariane-44P from Kourou.
2 Mar: Soyuz-TM 24 undocks from Mir, returning to Earth with the original crew plus German astronaut launched aboard TM25.
Pioneer 10 celebrates 25 years in space; the spacecraft was the first to penetrate the asteroid belt and perform a fly-by of Jupiter.
Reports that Iran plans to manufacture communications satellites in the country's 'new year' which begins 21 March. First to fly would be an educational satellite in around three years.
4 Mar: Maiden launch from new Russian cosmodrome at Svobodny using a Start-1, places small Zeya satellite into orbit.
8 Mar: US Tempo 2 comsat launched from Cape Canaveral using Atlas-2A.
10 Mar: Carlos Varotto, president of Argentinean space agency, stated that a launch vehicle capable of placing around one tonne into a 1,000 km orbit is currently being designed with the backing of Argentina and Brazil. The programme is to replace Condor-2 which is to be cancelled.
Indonesia plans the launch of a replacement for Palapa-B 2R before the end of 1998.
12 Mar: NASDA selects Thiokol (US) for the supply of the solid-propellant strap-on boosters for the H-2A launch vehicle. Thiokol will supply Castor 4A-XL strap-ons to Mitsubishi Heavy Industries.
14 Mar: Russian government spokesman told ITAR-TASS that the first launch of the Sea Launch programme will take place in late 1998; a Zenit-2 launch vehicle will be used, although geosynchronous orbit missions require the still-unflown Zenit-3 version, carrying a Block DM third stage based upon the fourth stages used by the Proton-K launcher.
15 Mar: Mars Pathfinder overtakes Mars Global Surveyor en route to Mars.
16 Mar: Indonesia wants Lockheed Martin to build Telkon-1 COMSAT for launch aboard an Ariane in 1998-1999. Satellite will be of the same generation as Palapa-B.
18 Mar: Korea Telecom signs US$108 million contract with Lockheed Martin for Koreasat 3 (aka Mugunghwa), to be launched in August 1999 aboard an Ariane.
20 Mar: SS/Loral to build two INTELSAT 'follow-on series 2' comsats, for launch in Summer and Autumn 2000.
22 Mar: NASA celebrates having astronauts in orbit for a full 12 months, due to the visits to the Mir station. Russia has had a permanent presence in space since September 1989.
25 Mar: Official at KB Polyot indicates that the Kapustin Yar launch site will return to orbital launches around February 1999 with the launch of German Abrixas satellite aboard a Cosmos-3M launcher; last orbital launch from the site was in 1987.
31 Mar: Final signals received from Pioneer 10 after more than 20 years of communications.

2 Apr: MARISAT 1 manoeuvred off-station in geosynchronous orbit and retired after 21 years of communications service.
4 Apr: US DMSP-2 9 (USA 131) military metsat launched from Western Range (Vandenberg) using Titan-23G.
Columbia shuttle orbiter launched on STS-83 Microgravity Science Laboratory mission: planned 16-day mission curtailed due to fuel cell problems and shuttle landed KSC 8 Apr. Mission reflown in July as STS-94.
5 Apr: Galileo performs fly-by of jovian satellite Ganymede, minimum distance 3,102 km.
6 Apr: Launch of Progress-M 34 cargo freighter from Baikonur using Soyuz-U; docked with Mir 8 Apr, undocked 24 Jun and following a collision with Spektr module (see 25 Jun below) was de-orbited 2 Jul.
9 Apr: Launch of Cosmos 2340, Oko early warning satellite, from Plesetsk using Molniya-M.
16 Apr: Ariane-44LP from Kourou successfully launches THAICOM 3 (Thailand) and B-SAT 1A (Japan) comsats.
17 Apr: Cosmos 2341, Parus military navigation satellite, launched from Plesetsk using a Cosmos-3M. Planned launch of US FaiSat 2 aboard the same vehicle was cancelled, reportedly because documentation was not complete.
18 Apr: India plans to fly orbital mission using two-stage liquid oxygen/kerosene launcher by 2000.
21 Apr: Spanish MINISAT 01 launched from Gando AFB on the Canary Islands using US L-1011/Pegasus-XL system. Pegasus-XL third stage also carried Celestis 1 payload, comprising partial cremated ashes of 24 personalities.
22-23 Apr: Cosmos 2313 ELINT Ocean Reconnaissance Satellite performs end-of-life manoeuvre and begins to decay from orbit. Satellite disintegrates in orbit 26 Jun, marking the first EORSAT break-up since Novemeber 1987.
24 Apr: NASA withdraws from planned primate experiments aboard Russian Bion 12 because of unacceptable mortality risk to the primates. Criticism of primates aboard Bion missions ignores the fact, that primates were originally introduced to the programme at NASA's request.
Five candidates for the next one or two Discovery missions were approved; missions included flights to Mercury, Venus, the martian satellites, comets or the Sun. Detailed plans to be submitted by 15 Aug, launches to take place no later than 30 Sept 2002 and cost no more than US$183 million (1997 rate) over three years.
25 Apr: Final Atlas-1 launched from Cape Canaveral, carrying GOES 10 (GOES K before launch), meteorological satellite.
26 Apr: Xinhua news agency reports that India is to build a launch facility near the town of Irian Jaya in Indonesia: agreement with LAPAN calls for launches of Indonesian rockets, training from Indian personnel and TTC facilities.
Reports that Taiwan's first commercial satellite, ST-1, costing US$100 million, is to be launched towards geosynchronous orbit aboard an Ariane-5 in February 1998.
29 Apr: Russian Vasily Tsibliyev and NASA astronaut Jerry Linenger complete EVA outside Mir to install equipment outside the station.
RKA head Yuri Koptev said that a consortium of banks would offer an Rb80 billion (US$140 million) loan to permit work on the country's ISS commitment to continue.
30 Apr: Israel selects a shuttle astronaut candidate — identified only as 'Lt Col A' — to fly a mission during 1998 to mark the country's 50th anniversary. Since serving Israeli officers cannot be named under the country's security laws, the fighter pilot and electrical engineer will have to 'resign' from the military forces before his flight takes place.
Loral to build two digital radio comsats for CD Radio, launches due in 1999. Agreement signed in July calls for the satellites to be launched on separate Ariane-5 missions.

5 May: Return to flight for Delta-2 launch vehicle from Western Range when 7920 variant places first cluster of five Iridium satellites into orbit; full Iridium system calls for 60 operating satellites and six spare satellites spread around six orbital planes.
7 May: Galileo performs fly-by of Jovian satellite Ganymede, minimum distance 1,580 km.
9 May: London Daily Telegraph reports that launch platform (a re-fitted oil rig) for the Sea Launch programme has completed testing at Stavanger, Norway and is to process to Russia for further installations.
11 May: China launches second DFH-3 comsat from Xi Chang using CZ-3A launcher; first DFH-3 launched in 1993 suffered attitude control problems and had to be written off.
13 May: Amateur 'Rockoon' launched by Huntsville L5, a chapter of the National Space Society in the US, reached a peak altitude of 70 km. High Altitude Lift-Off Mission carried to an altitude of 15 km by a balloon and then a hybrid propulsion system fired to reach a higher altitude. Orbital mission planned by later version of the launch system.
14 May: Launch of Cosmos 2342, Oko early warning satellite, from Plesetsk using Molniya-M.
Reports that NASA and ISAS (Japan) are to co-operate on the Muses-C mission to collect asteroid samples. Due for launch using an M-5 in Jan 2002. Spacecraft to land on asteroid Nereus Sep 2003 and samples should return to Earth Jan 2006.
15 May: Launch of shuttle orbiter Atlantis on STS-84/SMM-6 mission from KSC; docks with Mir 17 May, undocked 22 May, landed KSC 24 May. Returned J Linenger from Mir, left M Foale aboard the station.
Russia launches Cosmos 2343 photo-reconnaissance satellite from Baikonur using Soyuz-U. This marks the first Russian photorecon satellite in orbit since 28 Sep 1996 when Cosmos 2320 was de-orbited — an unprecedented break in the programme since the first photorecon satellite was orbited in Apr 1962.
17 May: NASA confirms that Ukrainian cosmonaut Leonid Kadenyuk will fly a shuttle mission in October 1997.
20 May: Zenit-2 explodes seconds after launch from Baikonur; Tselina-2 ELINT satellite carried. Norwegian Thor 2 comsat launched from Cape Canaveral using Delta-2 (7925).
23 May: Kelly Space and Technology reported that it has been granted patent protection for the Eclipse tow-launch technique.
24 May: Russian four-stage Proton-K successfully launches US Telstar 5 comsat from Baikonur.
25 May: Inaccurate press reports suggest that a Chinese launch had placed a military communications satellite into orbit 'last week'; these are garbled reports of the successful second DFH-3 launch on 11 May.

3 Jun: Successful launch of INMARSAT-3 4 (UK) and INSAT 2D (India) comsats from Kourou using Ariane-44L.
6 Jun: Russian Proton-K launched from Baikonur carrying Cosmos 2344, said to be a new class of optical intelligence-gathering satellite. Orbit of 63.4°, 1,509-2,732 km is unlike any other known optical reconnaissance satellite. ESA considering a plan for the launch of a Mars sample-return mission in 2003 using an Ariane-5.
10 Jun: China successfully launches second Feng Yun-2 #1 geosynchronous orbit metsat from Xi Chang using CZ-3; first FY-2 lost in 1994 pre-launch accident.
13 Jun: Arianespace announces that it has signed four new launch contracts: Sirius 3 will fly aboard an Ariane-4 in 1998, Telstar 6 aboard an Ariane-5 in late 1998 and two launches for an undisclosed customer.
18 Jun: Second cluster of Iridium satellites launched on four-stage Proton-K from Baikonur; seven satellites launched on this flight.
19 Jun: PANAMSAT selects Arianespace for the launch of its second HS-702 high-power satellite in 1999. Sydney Morning Herald reports that a deal has been signed with Starsem for the launch of Soyuz-class vehicles from Cape York by 2000.
21 Jun: China Great Wall Industry Corporation signs a deal with Hughes Space & Communications for the launch of 10 Hughes satellites aboard Chinese launchers during the period late 1998 to the end of 2006.
25 Jun: INTELSAT 802 comsat launched from Kourou using Ariane-44P.
Galileo performs fly-by of Jovian satellite Callisto, minimum distance 415 km.
Because of the possibility that ERS 1 might collide with Cosmos 614, the orbit of ERS 1 was raised slightly to ensure that a miss would take place and then it was lowered within 24 hours to the original altitude.
Manual re-docking of Progress-M 34 with the Mir Complex fails when the cargo freighter missed the rear of the station and performed a series of collisions with the Spektr module, damaging some of the solar panels and puncturing the module at least once. Two cosmonauts and NASA astronaut on Mir shut the hatch connecting Spektr to the Mir core module as the module loses pressure. Mir is powered down to conserve power while repairs are planned.
27 Jun: NEAR spacecraft performs a fly-by of asteroid 253 Mathilde: small manoeuvre 3 Jul put spacecraft on an Earth-fly-by trajectory (fly-by due Jan 1998) in advance of encounter with asteroid 433 Eros in early 1999.
30 Jun: Following a power loss over a period of a week, contact was lost with Japanese Midori (ADEOS 1).

1 Jul: Columbia shuttle orbiter launched on STS-94 Microgravity Science Laboratory mission — reflight of curtailed STS-83 mission; first time that a complete crew has been re-cycled to repeat a mission and the fastest return-to-orbit for the crew. Landed KSC 17 Jul. Arianespace signs an agreement for the launch of Mexico's Morelos 3 comsat; HS-601 satellite scheduled for launch at the end of 1998.
3 Jul: Gyrodynes on board Mir fail, causing the station to lose its lock on the Sun. Attitude control was later restored.
4 Jul: Mars Pathfinder spacecraft successfully lands on Mars in the Ares Vallis region. Spacecraft renamed Carl Sagan Memorial Station in honour of astronomer who died in Dec 1996. Sojourner rover deployed on martian surface 6 Jul. Both base station and rover still operating at end of August, far in excess of the planned mission.
5 Jul: Progress-M 35 cargo freighter launched to Mir Baikonur using Soyuz-U. Docked with Mir 7 Jul.
9 Jul: Third launch of Iridium satellite cluster; five satellites launched from Western Range using Delta-2 (7920); one of the satellites fails after orbital injection.
10 Jul: Indian government approves proposal for the development and launch of two advanced remote sensing satellites with an estimated cost of Rs3,900 million (US$120.4 million). CartoSat 1, a cartography satellite with a 2.5 metres ground resolution, and ResourceSat will both be launched using the PSLV.
14 Jul: OSC will pay US$12 million cash and refinance about US$25 million of debts to acquire the satellite manufacturing and communications service business of CTA Inc.
16 Jul: Cosmonaut aboard Mir inadvertently disconnects power/computer cable, resulting in the station losing its lock on the Sun. Systems restored within 24 hours.
21 Jul: Russia announces that the next Soyuz-TM launch to the Mir Complex will carry only the two Russians planned: the planned flight of a French spationaut is delayed until the following Soyuz-TM launch, due in early 1998.
22 Jul: Japan cancels the HOPE unmanned spaceplane programme in favour of a cheaper reusable 'rocket plane'.

23 Jul: Navstar satellite launched from Cape Canaveral using Delta-2 (7925); first Block 2R satellite to reach orbit. Japan delays the launch of COMETS aboard H-2 pending the results of the investigation into the Midori loss; launch now scheduled for early 1998.
24 Jul: Xinhua news agency reports that China will launch its first 'ocean exploration satellite' in 1999.
28 Jul: Japan's Superbird C comsat launched from Cape Canaveral using Atlas-2AS.
Norway's HS-376 Thor 3 comsat will be launched using a Delta-2 (7925) in July 1998.

1 Aug: L-1011/Pegasus-XL launches OrbView 2 remote sensing satellite from Vandenberg.
5 Aug: Two-man Soyuz-TM 26 launched from Baikonur using Soyuz-U: docks with Mir 7 Aug. Undocked 15 Aug from the rear port and re-docked at the front port in anticipation of the Progress-M 35 re-docking.
OSC will act as the prime contractor for 17 satellites which will form the Ellipso system.
7 Aug: Discovery shuttle orbiter launched from KSC for the STS-85 mission. CRISTA SPAS remote sensing payload deployed and re-captured before shuttle's return to Earth 19 Aug, landing KSC.
8 Aug: US PAS 6 comsat launched from Kourou using Ariane-44P.
14 Aug: Soyuz-TM 25 with original two Russian cosmonauts undocks from Mir and makes a successful landing, marred by the solid propellant cushioning-rockets not firing, thus making the landing rougher than normal.
Cosmos 2345 launched from Baikonur using four-stage Proton-K; Prognoz early warning satellite.
18 Aug: Progress-M 35 re-docks with Mir after a delay of 24 hours. Just before docking the Mir computer system crashed, causing the Progress-M docking system to shut down; docking completed manually, but Mir lost its alignment with the Sun for about a day.
19 Aug: Chinese CZ-3B (currently the most powerful Chinese launch vehicle) launched from Xi Chang and successfully places Philippines Agila 2 comsat into orbit. South Africa plans to develop Overberg, Western Cape Province as a satellite launch site.
21 Aug: Fourth launch of Iridium satellite cluster: five satellites launched from Western Range using Delta-2 (7920). First Delta-2 launch since McDonnell Douglas merges with Boeing.
22 Aug: Mir cosmonauts Solovyov and Vinogradov perform IVA (intra-vehicular activity) using their EVA suits while NASA astronaut Foale waited in the Soyuz-TM 26 spacecraft. Russians isolated Spektr from the remaining Mir modules and working in the depressurised multiple docking adapter were able to reconnect cables from the Spektr module's solar panels via a new Spektr hatch attachment to the Mir power system. Vinogradov tried without success to spot where Spektr had been punctured.
23 Aug: LMLV-1 (Lockheed-Martin Launch Vehicle) successfully launches Lewis remote sensing satellite, but satellite is found to be tumbling after orbital injection.
24 Aug: Belgian company Verhaert Design and Development proposes to build 60 kg PROBA satellite for ESA for remote sensing and educational work; could be launched as a piggy-back payload aboard PSLV in mid-1999.
25 Aug: NASA's Advanced Composition Explorer (ACE) launched from Cape Canaveral using Delta-2 (7920). Mir briefly suffers problems with primary and back-up oxygen systems.
28 Aug: US PAS 5 comsat launched from Baikonur using four-stage Proton-K.
29 Aug: FORTE (Fast On-orbit Recording of Transient Events) launched from Western Range using L-1011/Pegasus-XL.

1 Sep: At 14:00 GMT two 689 kg Iridium Mass Frequency Simulator satellites were by a CZ-2C/SD launcher (s/n CZ2C-15) from the T'ai Yuan site in China to a 623 x 632 km, 86.34° orbit with a period of 97.26 min. These dummy satellites used the new CZ-2C Smart Dispenser launcher combination with MFS-1 in ascending node 271.9° and MFS-2 in ascending node 272°, both in plane 5 of the projected constellation.
2 Sep: At 22:21 GMT an Ariane 44LP (V99) carried the 2,915 kg (1,715 kg BOL) Hot Bird 3 communications satellite and the 703 kg (281 kg BOL) Meteosat 7 meterology satellite. They were launched from the ELA-2 launch pad at Kourou to geostationary orbits over 13.3° E and 349° E, respectively. Meteosat 7 was eventually located over the prime meridian.
4 Sep: From LC-36A, ETR, at 12:03 GMT an Atlas 2AS (AC-146) launched the 2,750 kg GE 3 telecommunications satellite to a geostationary orbit at 276° E, moved later to 272° E. Owned by GE American Communications, GE 3 is a Lockheed Martin AS2100 satellite.
14 Sep: A Proton-K (8K82K) launcher carried seven 689 kg Iridium satellites from complex LC91L, Baikonur, to medium orbit, the fifth Iridium cluster launch and the second to use the four-stage Proton-K. Launch came at 01:36 GMT. One (Iridium 27) was placed in a 542 × 553 km, 86.7° orbit and failed but six (Iridium 23-33 inclusive) were placed in a 776 × 779 km, 86.4° orbit conforming to plane 3 of the planned constellation
23 Sep: At 16:44 GMT a Cosmos-3M (11K65M) launcher carried Cosmos 2346, a 795 kg Parus military navigation satellite from complex LC132, Plesetsk, to a 939 × 996 km, 82.92° orbit with a period of 104.42 min putting the satellite co-planar with Cosmos 2334. Also carried aloft was the 120 kg Faisat 2V store-forward communications satellite for Fina Analysis, Inc., placed in a similar orbit.
23 Sep: Launched at 23:58 GMT, a Ariane 42L (V100) carried the 3,455 kg Intelsat 803 telecommunications satellite from the ELA-2 complex at Kourou to a geostationary orbit over 333.5° E. Satellite is built on a Lockheed Martin AS7000 bus.
24 Sep: At 21:32 GMT a Molniya-M launcher (8K78M) carried the 1,600 kg Molniya-1 90 telecommunications satellite from complex LC43/4 at Plesetsk to a 449 × 39,906 km 62.84° orbit with a period of 717.78 min.
26 Sep: At 02:34 GMT the Shuttle Atlantis was launched from LC-39A at the start of STS-86, the last NASA mission to the Mir space station carrying seven astronauts (including one cosmonaut) and NASA crewmember Wolff who replaced Foale aboard the Russian facility. See Shuttle Log and Human Space Flight sections for further details.
27 Sep: At 01:23 GMT a Delta 7925 carried four Iridium satellites (19 and 34-37) from SLC2W at the WTR to an orbit of 775 × 790 km, inclination 86.4° with a period of 100.4 min. This was the sixth launch of Iridium series satellites, the fourth to use Delta.
28 Sep: At 04:47 GMT a PSLV (PSLVC-1) launched from India's Sriharikota launch site carried the 1,250 kg IRS 1D (Indian Remote Sensing 1D) satellite to a 308 × 822 km, 98.64° orbit with a period of 95.96 min. Following a malfunction of the fourth stage IRS 1D was left in an eccentric rather than circular (820 km) orbit. Onboard propulsion was used to change this orbit to a more useable 739 × 826 km orbit with the resultant reduction in lifetime to less than three years.

5 Oct: At 15:08 GMT a Soyuz-U (11A511U) carried the 7,195 kg Progress M-36 (s/n 237) from complex LC1 to a 382 × 391 km, 51.65° orbit with a period of 92.29 min. Unmanned supply vehicle docked with Mir -X port at 17:07:09 GMT, 8 Oct, undocked at 06:01:53 GMT 17 Dec after 69.54 days and was destroyed on re-entry at 13:20:01 GMT 19 Dec. Total free time 5.39 days.
5 Oct: The 2,900 kg (1,700 kg EOL) EchoStar 3 direct broadcast telecommunications satellite was launched at 21:01 GMT by Atlas 2AS from LC-36B, ETR, to a geostationary orbit at 298.6° E.
9 Oct: At 18:00 GMT a Soyuz-U (11A511U) launcher carried the 5,800 kg Foton 8 (F 11) satellite from LC43/3 at Baikonur to a 218 × 375 km, 62.81° orbit with a period of 90.45 min. Foton F11 carried microgravity experiments. The descent module landed 23 October 170 km northeast of Orsk in the Orenberg Region followed by the 150 kg MIRCA (Micro Re-entry Capsule) built by Kayser-Threde which had been attached to the exterior until after the de-orbit burn. It landed 110 km south-east of Orsk
15 Oct: The 5,712 kg Cassini-Huygens spacecraft was launched at 08:43 GMT from LC-40, ETR, by direct ascent on a Titan 4B (401B/Centaur s/n 4B-33) to heliocentric orbit designed to facilitate two fly-bys of Venus, one fly-by of Earth and an encounter with Saturn in November 2004 where it will conduct observations from orbit and deploy the 320 kg Huygens probe to the surface of Titan. See Cassini under Planetary and Space Science section.
16 Oct: A CZ-3B (s/n CZ3B-3) launched from LC2 at the Xichang complex at 19:13 GMT carried the 3,745 kg APStar 2R telecommunications satellite to a geostationary orbit over 76° E. Launched as a replacement for APStar 2 lost shortly after lift off on 25 January 1995 the satellite is based on the FS-1300 bus from Space Systems/Loral.
22 Oct: The 395 kg STEP 4 (Space Test Experiment Platform 4) satellite was launched from a Pegasus XL (s/n F18) dropped from a Lockheed L-1011 carrier at 14:13 GMT. Placed in a 434 × 501 km, 44.96° orbit with a period of 93.95 min, STEP 4 carried a Digital Ion Drift Meter (DIDM), an ElectroMagnetic Propagation Experiment (EMPE) and a Orbiting Ozone and Aerosol Measurement (OOAM) instrument.
24 Oct: Under the designation USA 133, at 02:32 GMT a Titan 4 (403A s/n 4A-18) carried the 14,500 kg Lacrosse 3 high-resolution radar imaging satellite from SLC4E at the WTR into a 666 × 679 km, 57° orbit for the National Reconnaissance Office. Lacrosse 1 was de-orbited in March 1997 but Lacrosse 2 was still in orbit at this date.
25 Oct: At 00:46 GMT an Atlas 2A (AC-131) carried the 1,040 kg USA 134, DSCS-3 (B13) military communications satellite from LC-36A to a geostationary orbit and deployed the USAF Falcon Gold technology satellite to a 151 × 34,457 km, 26.2° orbit with a period of 185.8 min. Falcon Gold was designed to demonstrate the use of GPS navigation technique in geosynchronous orbit.
30 Oct: The second, and first successful, Ariane 5 (s/n 502-V101) launched two satellites from ELA3 at Kourou at 13:43 GMT. The 2,650 kg MAQSAT-H/Teamsat satellite was placed in a 533 × 26,635 km, 7.79° orbit with a period of 467.8 min. The 1,400 kg MAQSAT-B/EPS satellite remained attached to the second stage and was placed in a 533 × 26,569 km, 7.78° orbit with a period of 466.64 min. The 60 kg Young Engineers Satellite (YES) was to have been unreeled from Teamsat but the tether was deleted and it became an independent satellite in a 540 × 26,626 km, 7.79° orbit with a period of 467.77 min. A malfunction in the second stage prevented the two MAQSAT satellites from being placed in elliptical orbit with apogee at geosynchronous altitude, about 9,000 km lower than planned.

2 Nov: Launched at 12:25 GMT from the Alacantara facility in Brazil, the first VLS launcher failed to reach orbit and was destroyed 65 s after liftoff when one of the four strap-on boosters while carrying the 115 kg SCD 2A data collection satellite.
6 Nov: At 00:30 GMT a Delta 7925 (s/n 249) carried the 1,881 kg Navstar 38 satellite from LC-17A, ETR, to a 19,910 × 20,449 km, 54.9° orbit with a period of 718 min. This was the last launch of the Block 2A series GPS series.
8 Nov: At 02:05 GMT a Titan 4 (401A/Centaur s/n 4A-17) carried the 5,000 kg Trumpet 3 Elint satellite from LC-41 at the ETR to a 1,100 × 39,060 km, 63.6° orbit with a period of 713.8 min. Under the designation USA 136, Trumpet 3 is the third such satellite after USA 103 (1994) and USA 112 (1995) launched for the NRO and the NSA.
9 Nov: a Delta 7925 (7920-10C s/n 250) from SLC2W at the WTR carried five 689 kg Iridium satellites (38-41 and 43) to a 775 × 779 km, 86.4° orbit with a period of 100.3 min. This is the seventh Iridium cluster to launch and was coplanar with the third sequence launched in July 1997.
12 Nov: At 17:00 GMT a Proton-K (8K82K) launched from LC200L at Baikonur carried the 2,500 kg Kupon satellite to a geostationary orbit at 55° E. Kupon (K95K) is the first in a series of four satellites built by NPO Lavochkin for the Russian Federation Central Bank. The series will relay information throughout the Bankir system.
12 Nov: An Ariane 44L (V102) launched at 21:48 GMT carried two communications satellites from ELA2 at the Kourou site to geostationary orbits. The 2,920 kg Sirius 2 telecommunications satellite launched for NSAB, Stockholm, was positioned over 5° E and the 1,385 kg (802 kg BOL).
18 Nov: A Soyuz U (11A511U) launched from Plesetsk carried the first 6,300 kg Resurs F-1M remote sensing satellite to a 180 × 236 km, 82.33° orbit with a period of 88.66 min. Derived from the old Zenit/Vostok design, the F-1M changed to a 238 × 270 km orbit with a period of 89.59 min.
19 Nov: At 19:46 GMT the Shuttle Columbia was launched from LC-39B, KSC, to a 281 × 286 km, 28.46° orbit with a period of 90.18 min at the start of the STS-87 mission to deploy a Spartan free-flying platform and conduct an EVA. For more details see Shuttle Log and the appropriate Human Space Flight section.
27 Nov: A Japanese H-2 (s/n H-II-6F) launched at 21.27 GMT carried the 3,620 kg TRMM (Tropical Rainfall Measuring Mission) satellite from complex Y at Tanegashima to a 367 × 385 km, 35° orbit with a period of 92.07 min. Manufactured by NASA's Goddard Space Flight Center, TRMM is designed to monitor rainfall conditions in the tropics. It also deployed the 2,900 kg Kiku-7 two-part satellite designed to demonstrate automated rendezvous and docking sequences.

2 Dec: At 22:52 GMT an Ariane 44P (V103) carried the 2,982 kg (1,819 kg BOL) JCSAT telecommunications satellite from ELA2 at the Kourou site to a geostationary orbit over 139° E and the 230 kg satellite Equator-S science satellite. Operated by the Max Planck Institute fur Extraterrestische Physik in Germany it was placed in a 496 × 67,232 km, 3.98° orbit with a period of 1,339.02 min. JCSAT was moved later to 150° E.
2 Dec: At 23:10 a Proton-K (9K82K) launched the 3,300 kg Astra 1G (HS 601) direct broadcast satellite for SES from complex LC81L at Baikonur to a geostationary position over 23-24°, moved later to 19.2° E.
8 Dec: A CZ-2C/SD (CZ2C-16) carried two 689 kg Iridium satellites (42 and 44) from China's T'ain Yuan launch site to a 775 × 779 km, 86.39° orbit with a period of 100.4 min. This was the first operational use of this launcher following a successful mass simulator launch on 1 Sep. They joined the cluster 3 orbit plane.
8 Dec: An Atlas 2AS (AC-149) launched from LC-36B at the ETR at 23:52 GMT carried the 3,000 kg class

Galaxy 8I (HS 601) telecommunications satellite to a geostationary position above 280° E. Galaxy 8I was originally known as Galaxy 3R.

9 Dec: A Russian ELINT satellite, US-P EORSAT (Cosmos 2347) was carried to a 403 × 429 km, 65.04° orbit with a period of 92.79 min after launch at 07:17 GMT on a Tsyklon-M (Tysklon 2) from LC90 at the Baikonur complex.

15 Dec: Russia launched a Yantar-4K, fourth generation, military reconnaissance satellite on a Soyuz U (11A511U) at 15:40 GMT from the Plesetsk complex. Designated Cosmos 2348, it was placed in a 165 × 345 km, 67.15° orbit with a period of 89.6 min.

20 Dec: The Progress M-37 logistics vehicle was launched by Soyuz U (11A511U) at 08:45 GMT from complex LC1 at Baikonur. It docked with the -X port on Mir's Kvant 1 module at 10:22:20 GMT 22 Dec, undocked at 12:00:00 GMT 30 Jan 1998, re-docked at 09:42:28 GMT 23 Feb, undocked again at 19:16:01 15 Mar and was destroyed on re-entry at 23:04:00 the same day. Total docked time 59.47 days.

20 Dec: At 13:16 GMT a Delta 7925 (s/n 251) launched from SLC2W at the WTR carried five Iridium satellites (45-49) to a 775 × 779 km, 86.4° orbit, the ninth such cluster launched and the sixth by Delta 2.

22 Dec: The 3,455 kg (2,079 kg BOL) Intelsat 804 telecommunications satellite was launched at 00:16 GMT by Ariane 42L (V104) from the ELA 2 complex at Kourou to a geostationary orbit over 46° E, later moved to 64° E.

23 Dec: The first eight of 24 Orbcomm satellites (FM5-FM12 or A1 to A8 in orbit) were launched at 19:11 GMT by a Pegasus XL-HAPS (s/n F19). The 43 kg satellites were placed in closely spaced orbits of approximately 815 × 823 km, incln 45° with an orbital period of 101 min. Three 45° orbital planes of eight satellites each will be supplemented by 2-4 satellites in 70° orbits to cover high latitude requirements.

24 Dec: At 13:32 GMT a Start 1 launcher carried the 300 kg EarlyBird 1 commercial remote sensing satellite from the Svobodny launch site to a 480 × 488 km, 97.3° orbit with a period of 94.28 min. The satellite had been manufactured for EarthWatch of Longmont, USA, by Orbital Sciences Corp.

24 Dec: At 23:19 GMT the 3,410 kg (2,534 kg BOL) AsiaSat 3 was launched by Proton-K (8K82K/DM3) from LC81L at Baikonur to a geosynchronous transfer orbit of 361 × 35,999 km with an inclination of 51.1°. Asiasat 3 is a Hughes HS601 launched for the Asia Satellite Telecommunications Co of Hong Kong. The Proton fourth stage (DM3) failed 1 s into a planned second burn of 110 s and was unable to place the satellite in the planned orbit of 10,000 × 36,000 km, incln 12° leaving it stranded. Following an insurance settlement the satellite reverted to Hughes and a complex series of on-board manoeuvres was made to carry it around the moon and bring it to geostationary orbit.

1998 (orbital details can be found in the Satellite Launches 1996-1998 log)

7 Jan: NASA's Lunar Prospector, the third Discovery mission) is launched toward the moon at 02:28 GMT with lunar orbit injection four days later into a 100 km orbit. Prospector supported a geochemical study of the moon and exercised the first use of Athena 2 launcher.

10 Jan: The UK's Skynet 4D military communications satellite was launched by Delta 2 at 00:32 GMT. BOL mass is 850 kg.

22 Jan: Israel launched its fifth Ofeq military surveillance satellite (Ofeq 4) at 12:56 GMT but a malfunction in the Shavit launcher prevented it from reaching orbit.

23 Jan: NASA launched Shuttle Endeavour on STS-89, the eighth Mir visit and the first by this Orbiter to the space station, at 02:48 GMT returning to Earth 31 Jan after leaving astronaut Thomas and returning Wolff to Earth.

28 Jan: Russia launches Soyuz TM-27 with three cosmonauts on board including French cosmonaut L. Eyharts.

29 Jan: Under the designation USA 137, National Reconnaissance Office flies the Capricorn intelligence data relay satellite to a highly eccentric orbit following a launch at 18:37 GMT from the ETR.

4 Feb: At 23:29 GMT an Ariane 44LP (V105) launched the Brasilsat B3 telecommunications satellite located over 295° E (BOL mass is 1,052 kg) and the Inmarsat-3 F5 maritime mobile communications satellite (BOL mass 1,149 kg) over 25° E.

10 Feb: At 13:20 GMT the first launch of the full Taurus launcher from the WTR carried the GFO (Geosat Follow-On) and two Orbcomm satellites for Orbcomm Global LP into orbit.

14 Feb: The first four satellites in the planned constellation of 54 for Globalstar Ltd Partnership were launched by Delta 2 from the ETR at 14:34 GMT. The complete network would comprise six satellites and one spare in eight orbit planes. This was the first launch of the MedLite Delta 7420 with four solid strap-ons.

17 Feb: Following launch at 10:30 GMT a Soyuz U launcher (11A511U) carried Cosmos 2349 into orbit, a Yantar-1KFT/Kometa military topographic and mapping satellite with a planned lifetime of 45 days. The descent module was recovered late 2 April.

18 Feb: Four Iridium satellites (50-54) were launched at 13:58 GMT by Delta 2 (7920).

21 Feb: An H-2 launched from Tanegashima at 07:55 GMT placed the Kakehashi communications and broadcast engineering test satellite into orbit. Premature shutdown of the second stage prevented the satellite from reaching geosynchronous transfer orbit.

26 Feb: SNOE (Student Nitric Oxide Explorer), a science satellite operated by the University of Colorado, was launched at 07:07 GMT by Pegasus XL. It is designed to measure the density of nitric acid as a function of altitude and was the first satellite to be launched under the STudent Explorer Demonstration Initiative (STEDI). The final stage of Pegasus carried the second cluster of capsules in the Celestis series carrying human remains into space. A second satellite, Teledisc T-1 was released to a separate orbit where it would test micro-comsat technology.

27 Feb: Ariane 42P V106 launched at 22:38 GMT carrying the Hot Bird 4 telecommunications satellite for Eutelsat, placed at 13° E. Built on the Matra Marconi Eurostar 2000 satellite bus, it had a BOL mass of 1,770 kg.

28 Feb: Based on the Lockheed Martin (formerly RCA) AS7000 bus, the Intelsat 806 telecommunications satellite was launched by Atlas 2AS from LC-36B at the ETR at 00:21 GMT to a geostationary orbit over 40.5° W.

5 Mar: During a press conference at NASA's Ames Research Centre scientists from the Lunar Prospector programme announced that instruments aboard the moon probe had evidence for the existence of water ice on the lunar surface.

10 Mar: Operations with NASA's Mars Pathfinder ceased when the lander refused to acknowledge signals from the Deep Space Network's 34 m antenna at Goldstone, Calif.

12 Mar: The X-38 precursor Crew Return Vehicle for possible utilisation with the International Space Station made its first free-flight descent from a B-52 launched out of Edwards AFB, California. The origin of the X-38 design owes much to the evolution of M2, HL-10 and X-24 series lifting bodies.

14 Mar: The Progress M-38 cargo-tanker was launched at 22:46 GMT to a docking with the Mir space station just over two days later. It carried to orbit the Vynosnaya Dvigatel'naya Ustanovka (VDU) external engine unit to supplement attitude control equipment aboard Mir. M-38 was de-orbited 15 May.

16 Mar: Designated USA 138, the UFO 8 (UHF Follow-On) US Navy communications satellite was launched to geostationary orbit at 188° E by Atlas 2 at 21:32 GMT.

24 Mar: The 2,755 kg SPOT 4 remote sensing satellite was launched at 01:46 GMT by Ariane 40 (V107).

25 Mar: Two Iridium satellites (51 and 61) were launched by China's CZ-2C/SD into plane A of the system at 17:01 GMT.

30 Mar: Five Iridium satellites were launched into plane D of the system by Delta 2 (7920) which lifted off at 06:02 GMT.

2 Apr: At 02:42 GMT a Pegasus XL launched NASA's TRACE (Transition Region And Coronal Explorer) to study ultraviolet radiation from the Sun as a means of discriminating between the solar surface and features in the photosphere, the chromosphere and the corona. TRACE was the third in NASA's Small Explorer series.

7 Apr: Seven Iridium satellites (62-68) were launched by Russia's Proton-K at 02:13 GMT for the sixth and final orbit plane. It was the last cluster to be launched by this vehicle.

17 Apr: NASA's Shuttle Columbia was launched from KSC at 18:19 on the STS-90 mission to study a variety of neurological effects on living things including humans. See Shuttle Log and the Human Space Flight section. Columbia returned to Earth on 3 May.

24 Apr: Four Globalstar satellites (5-8) were launched by Delta 2 (7420) at 22:38 GMT to an orbit plane of 52° displaced approximately 49° to that of the first cluster.

26 Apr: NASA's Saturn-bound Cassini spacecraft performed a fly-by of Venus at a distance of only 284 km. It was expected to fly past Venus again 24 June 1999.

28 Apr: At 22:53 GMT Nilesat 1 became the first satellite to be launched for an African country when Ariane 44P (V108) carried Egypt's first telecommunication satellite into geostationary orbit over 353° E. Also launched was B-SAT 1B for the B-SAT Corporation, Japan, placed over 110° E.

29 Apr: Russia launched Cosmos 2350 at 04:37 GMT, a Geizer data relay satellite in the Potok system, to a geostationary orbit over 73° E.

2 May: Two Iridium satellites (69 and 71) were launched by China's CZ-2C/SD at 09:16 GMT and placed into plane B of the constellation.

7 May: Under the guise of Cosmos 2351 an Oko early warning satellite was launched at 08:53 GMT for the Russian Ministry of Defence intio an orbit plane 80° east of its companion Cosmos 2340.

7 May: At 23:45 GMT a Proton-K from Baikonur launched the EchoStar 4 direct broadcast satellite to a geostationary orbit at 233° E. Operated by the EchoStar Communications Company, the satellite was built around the Lockheed Martin AS2100 bus.

8 May: Renamed HGS 1 (Hughes Global Services), the AsiaSat 3 satellite launch 24 December 1997 was boosted toward the Moon using its onboard propulsion. It passed within 6,250 km of the Moon on 13 May and performed a fly-by of Earth at 36,000 km. A second lunar fly-by was made on 6 June at 34,354 km. HGS arrived in geosychronous drift orbit on 14 June with an inclination of 8.7° over 207° E.

9 May: USA 139, an Advanced Orion 2 ELINT satellite for the National Reconnaissance Office was launched at 01:38 GMT to a geostationary orbit by Titan 4B/Centaur.

13 May: NOAA 15 was carried into orbit by a Titan 23G launched at 15:52 GMT, the first of a new generation of meteorological satellite for the National Oceanic and Atmospheric Administration. The satellite carried a Thiokol Star-37XFP solid rocket motor which fired to take over from the two Titan stages and place it in orbit.

14 May: Progress M-39 was launched at 22:12 GMT for a rendezvous with Mir and a docking at the -X port two days later. M-39 de-orbited 29 Oct.

16 May: Technicians at the European Space Agency's ground station at Villafranca, Spain, turned off the Infra-red Space Observatory following more than two years of outstanding science studying the sources of infra-red energy in the universe.

17 May: Five Iridium satellites (70 and 72-75) were launched at 21:16 GMT by Delta 2 to plane A of the constellation.

30 May: The Zhongwei 1 (China Star) telecommunications satellite (a Lockheed Martin AS2100) was launched by CZ-3B at 10:00 GMT. It was located over 87.5° E. The satellite and liquid hydrogen upper stage were initially placed in a 216 × 85,035 km supersynchronous orbit.

2 Jun: At 22:06 GMT the Shuttle Discovery was launched from LC-39A at the start of STS-91, the ninth and last Shuttle-Mir mission. Six astronauts were launched but seven came back when Thomas was returned to Earth from Mir. Discovery landed on 12 Jun.

10 Jun: The Thor 3 (HS 376) Norwegian telecommunications satellite was launched at 00:35 GMT by Delta 2 and through a unique sequence of six burns by three motors was propelled to geostationary orbit. The first burn of the second stage placed the assembly in low orbit, the second burn raised apogee to 1,400 km, the third burn circularised the orbit prior to separation of the stage from the satellite. The fourth burn of the second stage de-orbited it into the atmosphere and the third stage fired to place the satellite on a geosynchronous transfer path with a 1,400 km perigee. The satellite fired its own Star 30 apogee kick motor for geostationary orbit over 0.8° W.

15 Jun: Six Strela 3 store-dump military communications satellites (Cosmos 2352-2357) were launched at 22:58 GMT by Tsyklon. They were delivered to the same orbital plane as the previous cluster of Gonets D1 and Strela 3 satellites in early 1997. The final stage of Tsyklon, however, failed to place them in the correct orbit and they gradually drifted out of phase with the previous clusters.

18 Jun: An Atlas 2AS launched at 22:48 GMT carried the Intelsat 805 satellite from LC-36A at the ETR to a geosynchronous orbit above 304.5° E.

24 Jun: A two-day ESA Council meeting in Brussels concludes with approval for the long-term programme establishing the initial step toward a Global Navigation Satellite System (GNSS), a vigorous Earth observation programme, first steps toward a more powerful Ariane 5 and development of the small VEGA launcher.

24 Jun: A Soyuz U launched at 18:30 GMT carried a fourth generation Yantar-4K2 Kobalt close-look military reconnaissance satellite (Cosmos 2358) to orbit.

25 Jun: A Yantar fifth generation photo-reconnaissance satellite was launched by Soyuz U at 14:00 GMT.

1 Jul: Molniya 3-49 was launched at 00:48 GMT from Plesetsk into a co-planar orbit with Molniya 3-39.

3 Jul: Japan's Mars probe Planet-B (Nozomi after launch) was placed on an interim Earth orbit of 146 × 417 km followed by ignition of the fourth (KM-V1) stage putting the spacecraft in a highly elliptical 359 × 401,491 km circumlunar orbit performing multiple Earth fly-bys and lunar fly-bys on 24 Sep and 18 Dec.

6 Jul: One component of Japan's technology satellite Kiku 7 satellite, called Orihime, launched 27 Nov 1997 separated from the other component, called Hikoboshi, and separated by 2 metres prior to re-docking after 21 min of free flight.

7 Jul: In the first orbital launch from a submerged submarine, Tubsat N and N1 were launched from the Delfin class submarine Novomoskovsk attached to Russia's Northern Fleet operating in the Barent Sea at 69.3° N × 35.3° E. The Shtil-1 launcher is a modified R-29RM ICBM known to NATO as the SS-N-23 Skiff.
10 Jul: At 06:30 GMT a Zenit-2 launched six separate technology and science satellites for Russia, Chile, Thailand and Australia.
18 Jul: A telecommunications satellites for the SINO Satellite Communications Company, SINOSAT 1 was launched at 09:20 from the Xi Chang launch complex on a CZ-3B. The Alcatel Spacebus 3000 satellite had been built in Cannes and was jointly owned by Daimler-Benz Aerospace and the Chinma Aerospace Corporation. Located at 110° E the satellite was handed over to its operators at the end of a checkout session.
28 Jul: A Zenit 2 launched Cosmos 2360, Tselina 2 ELINT satellite at 09:15 GMT from Baikonur.

2 Aug: Eight Orbocomm satellites (13-20) were launched by a Pegasus XL at 16:24 GMT in a cluster 120° to the east of the cluster orbited 23 Dec 1997.
6 Aug: Second undocking of the two elements of Japan's Kiku 7 satellite when Hikobashi separated from Orihime to a distance of 525 metres. Following several separate re-rendezvous and docking attempts during which maximum separation became 5 km, re-docking was achieved on 27 Aug.
12 Aug: A Titan 4A/Centaur launched the advanced Vortex Mercury ELINT 3 satellite from the ETR at 11:30 GMT but a power interrupt pitched the launcher into the sea at 40 s.
13 Aug: Soyuz TM-28 was launched by Soyuz U from Baikonur at 09:43 GMT, docked with Mir 12 Aug on a crew exchange mission, the spacecraft returning to earth 28 Feb 1999.
19 Aug: Two Iridium satellites (3 and 76) were orbited by CZ-2C/SD, launched at 23:01 GMT, the first mission to replace in-orbit failures.
25 Aug: An Ariane 44P (V109) launched at 23:07 GMT carried the Singapore-Taiwan ST 1 telecommunications satellite to a geostationary orbit at 88° E.
27 Aug: Galaxy 10, launched for the PanAmSat Corporation by the first Delta 3 lifted off at 01:17 GMT but was destroyed when a guidance failure caused erratic flight and the guidance officer destroyed the vehicle at 75 s.
30 Aug: A Proton-K4 launched at 00:31 GMT carried the Astra 2A direct broadcast satellite to a geostationary orbit at 28.2° E where it was to be operated by SES.
31 Aug: Although initially calling it a success, North Korea suffered a failure with its first orbital attempt using the Taepo Dong 1 launcher which lifted off at 03:07 GMT.

8 Sep: Five Iridium satellites (77 and 79-82) were launched by Delta 2 as replacement satellites for plane F of the constellation.
9 Sep: A cluster of 12 Globalstar satellites were lost when the Zenit-2 that launched them at 20:29 GMT was destroyed after a failure in the second stage guidance at 272 s.
16 Sep: An Ariane 44LP launched the PAS 7 direct broadcast satellite at 06:31 GMT. Satellite was eventually located over 68.5° E from where it was operated by PanAmSat.
23 Sep: Eight Orbcomm satellites (22-28) were launched by Pegasus XL at 05:06 GMT, the third cluster to be oribited.
28 Sep: Molniya 1T telecommunications satellites was sent into the same plane as Molniya 1-86 by a Molniya-M launched at 23:41 GMT.

3 Oct: Sponsored by the US National Reconnaissance Office, the STEX (Space Technology Experiment) satellite was launched on a Taurus at 10:04 GMT. STEX incorporated a TAL-D55 plasma engine and a tether experiment which partly failed in flight and had to be jettisoned.
5 Oct: The Eutelsat W2 and Sirius 3 telecommunications satellites were launched by Ariane 44L (V111) at 22:51 GMT. Eutelsat W2 was the first of the new widebeam series and was located over 1° E. Sirius 3 was launched for the Nordic Television Company and was placed initially at 23° E prior to taking up its operating position at 16° E.
9 Oct: Hot Bird 5 was launched at 22:50 GMT by an Atlas 2A and placed over 13° E.
20 Oct: Launched as USA 140, the UFO 9 (HS 601) military communications satellite lifted off at 07:19 GMT and was located over 186° E.
21 Oct: The third qualification flight of the Ariane 5 launcher (s/n 503) began at 16:37 GMT with the launch of the Atmospheric Re-entry Demonstrator and the MAQSAT 3, a Eutelsat W mockup.
23 Oct: At 00:02 GMT the second SCD (Satelite de Coleta de Dados/SCD Data Collection Satellite) was launched by Pegasus to a medium orbit where it would be used as a Brazilian environmental data relay satellite.
24 Oct: NASA's Deep Space 1 technology spacecraft was launched at 12:08 GMT on a Delta 7326 from the ETR. The prime objective was to test new technologies for deep-space and interplanetary spacecraft, including an ion propulsion motor which would perform trajectory adjustment allowing encounters with an asteroid and a comet. As a secondary payload, SEDSAT 1 built by the Students foir the Exploration and Development of Space was released from the second stage of the Delta 7326 which was also making its first flight.
25 Oct: Progress M-40 was launched by Soyuz U at 04:14 GMT, docking at the Mir -X port two days later. It was de-orbited 5 Feb.
28 Oct: AfriStar, a broadcast communications satellite operated by the WorldSpace Corporation, and GE 5, a direct broadcast satellite for GE American Communications (Americom) were launched at 22:16 GMT. AfriStar was the first WorldSpace satellite and would broadcast 24-96 digital radio channels to hand-held L-band receivers throughout Africa and the Middle East from a position over 21° E. Originally the backup for Argentina's Nahuelsat, GE 5 was an Alcatel Euostar 2000 bus.
29 Oct: Shuttle Discovery was launched at 19:19 GMT carrying seven crewmembers including the first US astronaut to orbit the Earth, John Glenn, on a life sciences flight. Designated STS-95 payloads included a small telecommunications satellite called PANSAT and the Spartan free-flying platform released on 1 Nov and retrieved two days later prior to re-entry 7 Nov.

4 Nov: PAS 8, a direct broadcast satellite for the PanAmSat Corporation was launched by Proton-K from Baikonur at 05:12 GMT.
6 Nov: A Delta 2 (7920) launched from the WTR at 13:37 GMT placed five Iridium satellites (2 and 83-86) into plane E of the Iridium constellation.
20 Nov: The first element in the much delayed and re-designed International Space Station was launched from Baikonur at 06:40 GMT when a Proton-K carried the 20,700 kg Zarya FGB to Earth orbit. Funded by NASA and built by Khrunichev in Moscow under contract to Boeing, Zarya carried a multiple docking unit at one end and incorporated a pressurised section, a propulsion and instrument section and a rear docking port.
23 Nov: Owned by a Russian media consortium, the Bonum 1 telecommunciations satellite was launched at 23:54 GMT by Delta 2 (7925) from the ETR to geostationary orbit at 36° E. A HS 376, Bonum I is capable of broadcasting 50 channels to western Russia.

4 Dec: NASA's Shuttle Endeavour was launched at 08:35 GMT with a crew of six astronauts on the STS-88 mission to begin assembly of the International Space Station by attaching the Unity module to the multiple docking adaptor on Zarya. The mission was a complete success and the Shuttle returned to Earth 16 Dec.
6 Dec: An Ariane 42L (V114) launched the SATMEX 5 telecommunications satellite at 00:43 GMT to a geostationary position over 243.2° E from where it would serve Mexico.
6 Dec: Another in the NASA Small Explorer series, the SWAS (Submillimeter Wave Astronomy Satellite) was launched from the WTR by Pegasus at 00:57 GMT.
10 Dec: At 11:57 GMT a Cosmos 3M launcher carried the civilian navigation satellite Nadezhda 5 and the Swedish particles and fields research satellite Astrid 2 into orbit from Plesetsk.
11 Dec: NASA launched the Mars Climate Orbiter to a heliocentric trajectory at 18:45 GMT. First of the two Mars '98 missions it is expected to arrive at Mars 23 Sep 1999.
19 Dec: Two Iridium satellites (11A and 20A) were launched by CZ-2C/SD from China's T'ai Yuan launch site at 11:30 GMT.
22 Dec: The PAS 6B telecommunications satellite was launched on an Ariane 42L at 01:08 GMT from Kourou, French Guiana.
24 Dec: Cosmos 2361, a navigation satellite in the Parus series, was launched at 20:03 GMT into a coplanar orbit with Cosmos 2233.
29 Dec: Launched 3 Jul, Japan's Nozomi Mars probe passed within 1,000 km of Earth at the end of multiple Earth-Moon fly-bys and performed a 7 min propulsive burn. The combined energies were to have placed the spacecraft on a heliocentric orbit for encounter with Mars on Oct 11, 1999. Course corrections on 21 December left Nozomi with too little propellant to reach Mars and the spacecraft was placed on a trajectory for Mars encounter Dec 2003.
30 Dec: Three Uragan navigation satellites in the Glonass series (Cosmos 2362-2364) were launched by Proton-K from the Baikonur LC200L complex.

1999

3 Jan: Second of the two Mars '98 planetary spacecraft scheduled for this launch window, Mars Polar Lander was launched at 20:21 GMT on a Delta 2 (7425) from the ETR. The spacecraft, also carrying two microprobes which will be released from the cruise stage just prior to entry into the Mars atmosphere on 3 Dec.
27 Jan: An Athena 1 launched from the ETR at 00:34 GMT placed Taiwan's first satellite (Chunghua or ROCSAT — Republic of China satellite) in orbit. It carries an ocean colour imager for fisheries and Earth resource evaluation.

7 Feb: Fourth mission in NASA's Discovery programme, the Stardust spacecraft was launched by Delta 2 (7426) at 21:04 GMT into a heliocentric orbit to fly by the comet Wild 2 in Jan 2004, capture primordial comet material and return it to Earth in Jan 2006.
9 Feb: Four Globalstar communications satellites (23, 36, 38 and 40) were launched by Soyuz U/Ikar at 03:54 GMT marking the first flight of this vehicle.
15 Feb: The Telstar 6 telecommunications satellite was launched at 05:12 GMT by Proton-K and placed in a geostationary orbit at 267° E.
16 Feb: An Atlas 2AS launched the JC-SAT 6 telecommunications satellite at 01:45 GMT for Japan Satellite Systems and placed it in geostationary orbit over 120° E.
20 Feb: Three cosmonauts from Russia, France and Slovakia were launched aboard the Soyuz TM-29 spacecraft at 04:18 GMT for a rendezvous with the +X Mir port two days later. The Slovak cosmonaut Bella returned with Padalka in Soyuz TM-28 on 28 Feb, but the Russian Afanasyev and the Frenchman Hagnere remained aboard Mir with cosmonaut Avdeyev. They were scheduled to return in late August as the last Mir crew.
23 Feb: A Delta 2 (7920) launched at 10:30 GMT carried three satellites to medium orbit: the joint-services ARGOS (Advanced Research and Global Observation Satellite) for near-Earth physics experiments; the Danish Ørstad microsatellite operated by the Danmarks Meterologiske Institut for mapping Earth's magnetic and particle environment; and Sunsat, South Africa's first satellite for Earth imaging and digital communications.
26 Feb: An Ariane 44L launched from Kourou at 22:44 GMT carried the ARABSAT 3A telecommunications satellite to a geostationary orbit at 26° E and the UK Skynet 4E military communications satellite to a position over the Gulf of Guinea at 6° E.
28 Feb: A Proton-4 launched at 04:00 GMT carried the Raduga-1 4 Globus telecommunications satellite into geostationary orbit.

5 Mar: At 02:56 GMT a Pegasus rocket launched the fifth of NASA Small Explorer satellites, the WIRE (Wide-Field Infra-Red Explorer) satellite designed to examine the origin and evolution of galaxies but technical problems prevented the equipment from obtaining data when an instrument cover was ejected prematurely.
15 Mar: Four Globalstar satellites (22, 37, 41 and 46) were launched at 03:06 GMT, the second flight of the Soyuz U/Ikar.
21 Mar: AsiaSat 3S, a HS 601P telecommunications satellite for the Asia Satellite Telecommunications Co Ltd of Hong Kong, was launched by Proton-K at 00:09 GMT and placed in geostationary transfer ellipse. From there, using its onboard R4D motor (the same as that used for attitude control on the Apollo spacecraft) it would be positioned for operations over 105.5° E. Asiasat 3S is the replacement for Asiasat 3 (now HGS), lost when the DM-SL stage of an earlier Proton failed dumping it in the wrong orbit.
28 Mar: A Zenit-3SL performed the first flight in the Sea Launch programme when, at 01:30 GMT it lifted off from the mobile launch platform named Odyssey located at 0° latitude by 154° W carrying a dynamic simulator model of a Hughes HS 702 satellite to geosynchronous transfer orbit. The DM-SL stage fired itself out of orbit.

2 Apr: The Progress M-41 cargo-tanker was launched at 11:28 GMT for a rendezvous with the Mir station two days later. Its propulsion system was also used to boost the orbit of Mir to inhibit premature orbital decay.
2 Apr: Ariane 44P (V117) was launched from the ELA-2 complex at Kourou carrying the Insat 2E communications and meteorology satellite.
2 Apr: RKA, the Russian Space Agency, announced that it had arranged finance to keep Mir in operation beyond August 1999 and NASA announced it would pay the Russians US$100 million to buy an additional Soyuz TMA to operate as an interim rescue vehicle.
9 Apr: A failure in the Inertial Upper Stage (IUS) of a Titan 4B/IUS left the DSP-19 early warning satellite in a useless orbit of 188 × 38,000 km. Launched at 17:01 GMT the first two stages of Titan placed the IUS/DSP-19 assembly in low Earth orbit. The first stage of the IUS fired as planned for geosynchronous transfer ellipse. However one or more connectors failed to disengage and prevented total separation of the lower stage and when the upper stage fired it caused the assembly to tumble.
12 Apr: An Atlas 2AS launched at 22:50 GMT carried Eutelsat W3, a Spacebus 3000B2 class telecommunications satellite, to a geostationary position over 7° E.

15 Apr: Four L-band Globalstar satellites (19, 42, 44 and 45) built by Space Systems/Loral were launched at 00:46 GMT by a Soyuz U/Ikar from Baikonur.
15 Apr: The Landsat 7 remote sensing satellite was launched by Delta 2 (7920) at 18:32 GMT into a 668 × 698 km orbit. It will be operated by NASA's Goddard Space Flight Center until control is transferred to the US Geological Survey in October 2000.
16 Apr: At 10:30 GMT an OSCAR amateur radio satellite was released from the Mir space station and designated Sputnik-99.
21 Apr: UOSAT-12 was launch at 04:59 GMT aboard the first Dnepr launch vehicle adapted from the R-36M2 ICBM. The third stage manoeuvering bus was used to propel UOSAT-12 into its proper orbit.
27 Apr: Ikonos 1, the first high-resolution civilian Earth observation satellite was launched at 18:22 GMT but failed to reach orbit when the nose fairing failed to separate four minutes into flight. The added weight prevented the terminal stage on the Athena 2 launcher from reaching orbital velocity or position and the assembly re-entered over the Pacific.
28 Apr: A Cosmos 3M launched at 20:30 GMT carried the Abrixas X-ray astronomy satellite and the 35 kg Megsat 0 communications satellite into medium altitude orbits of 544 × 603 km.
30 Apr: The first in a series of upgraded Milstar 2 (F1) military communications satellites was stranded in a useless orbit when computer software in the Titan 4B/Centaur that launched it (at 16:30 GMT) commanded engine firings at the wrong times. Instead of three sequential Centaur burns over 6 h placing Milstar in parking orbit, geosynchronous transfer orbit and final geostationary orbit the three burns took place on the first orbit leaving the satellite in a 740 × 5,000 km path. Milstar was moved to a 953 × 4,475 km orbit where all propellant was depleted and it was declared dead.
30 Apr: Orbital Sciences Corporation rolled out the first of three X-34 spaceplane technology demonstrators. The X-34 would first fly a series of demonstration drops from a B-52 but eventually reach altitudes of 23,225 m and speeds of around Mach 8.

5 May: Following launch at 01:00 GMT from LC-17B at the ETR, the second Delta 3 failed to achieve its objective when the nozzle extender failed to operate and the engine burned for only 1 s, leaving the Orion 3 telecommunications satellite in a useless orbit of 162 × 1,378 km.
10 May: At 01:33 GMT a CZ-4A launched from China's T'ai Yuan complex carried the Fen Yung 1C meteorological satellite and the Shi Jian 5 magnetospheric satellite into orbit.
13 May: The US House of Representative's Science Committee voted to cancel the Triana project inspired by Vice President Al Gore conceived as a means of placing an imaging satellite between the Earth and Moon to provide continuous images of the sunlit Earth for access through the Internet.
13 May: AeroAstro Inc announces it has been authorised to build the first commercial interplanetary spacecraft designed to carry pictures, messages and human DNA from several million people worldwide on a gravity assist slingshot at Jupiter out of the Solar system following launch aboard an Ariane 5 in late 2002.
16 May: Canada's first contribution to the International Space Station, the Space Station Remote Manipulator System (SSRMS) arrived at NASA's Kennedy Space Center ready for launch in July 2000 aboard the STS-100 mission.
18 May: Launched at 05:10 GMT, a Pegasus XL carried two small satellites to orbit. One, the Terriers (Tomographic Experiment using Radiative Recombinative Ionospheric and Extreme UV and Radio Sources), was the second STEDI (Student Explorer Demonstration Initiative) and would study the Earth's upper atmosphere and ionosphere. The second, MUBLCOM (Multiple Path Beyond Line of Sight Communications) was operated by the Defense Advanced Research Projects Agency (DARPA) to research digital voice and data communications between handheld terminals without using any other terrestrial infrastructure.
20 May: Telesat Canada's first direct broadcast satellite, with 32 Ku-band transponders, the Lockheed Martin A2100 Nimiq, was launched by Proton-K from the Baikonur cosmodrome to a geostationary orbit over 91° W.
22 May: A Titan 4B launched from the ETR carried a classified military payload to orbit for the National Reconnaissance Office.
26 May: At 06:22 GMT In India's first commercial flight a PSLV-C2 launcher carried the 1,050 kg Oceansat 1 (IRS-P4) satellite into orbit along with the 107 kg Kitsat 3 for South Korea's Advanced Institute of Science and Technology and the 45 kg Tubsat-DLR for Germany's space agency. This was also the first time India had carried three satellites on one launcher.
27 May: NASA's Shuttle Discovery was launched at 10:49 GMT at the start of the STS-96 mission to carry about 1,800 kg of supplies and materials to the two docked International Space Station modules and perform three EVA operations to install equipment and prepare the assembly for later visits. Discovery returned to Earth 6 June.

1 Jun: Scheduled to fly in February 2000, the LE-7A engine for Japan's new H-IIA launcher performed its first static firing which had to be curtailed when inlet pressure fell and the gimbal system failed.
7 Jun: Sea Launch announced that its first commercial launch from the special equatorial ocean launch platform will take place in August or September. A Zenit 3SL will lift the Hughes built DirecTV-1R broadcast satellite to a geostationary position at 101° W replacing the 5.5 year old DBS-1 which began to fail last year.
10 Jun: A Delta 2 (7920) carried four Space Systems/Loral Globalstar communication satellites (25, 47, 49 and 52) from the ETR to orbit.
11 Jun: Two Iridium satellites (14A and 21A) were launched by China's CZ-2C/SD from the T'ai Yuan launch complex.
16 Jun: Sea Launch announced four more launch contracts for the period 2002-2003 bringing firm launch commitments to 19 satellites. A Zenit 3SL costs US$70-90 million.
18 Jun: A Proton-K/DM-3 launched from Baikonur at 01:49 GMT carried the Hughes Astra 1H (HS 601P) direct broadcast satellite into a geostationary orbit at 19.2° E for SES.
18 Jun: In a major challenge to the US Navstar GPS navigation programme, the 15 EU ministers of transport agreed in Brussels to initiate funding for the Galileo programme of European navigation satellites, approving an initial payment of US$42 million. ESA has already spent US$62 million on the project.
20 Jun: A Titan 23G launched at 02:15 GMT from the WTR carried Nasa's Ball Aerospace QuickScat remote sensing satellite into a near-polar, 800 km, orbit. The satellite will provide detailed data on ocean winds to study their effects on oceans, weather and climate.
21 Jun: Iridium announced massive price cuts for its global mobile telephone service and shed 15 per cent of its 500 employee base in efforts to increase the number of subscribers. Company stock fell to US$5.375 a share compared with nearly US$71 a share in May 1998. Iridium had until 30 Jun to hit a total 27,000 subscribers.
24 Jun: A Delta 2 (the first 7320) launched from the ETR at 15:44 GMT carried NASA's 1,350 kg FUSE (Far Ultraviolet Spectroscopic Explorer) astronomy satellite, built by the Johns Hopkins Applied Physics Laboratory into orbit.
24 Jun: NASA's Cassini spacecraft, bound for Saturn, completed the second gravity-assist fly-by of Venus at 20:30 UT passing within 600 km of the planet. A 1,166 km fly-by of Earth 18 Aug will be followed by a fly-by of Jupiter on 30 December 2000 and a rendezvous with Saturn 1 Jul 2004.
29 Jun: NASA cancelled the Champollion mission designed to land a probe on the nucleus of comet P/Tempel 2 in late 2005.

NOTES: 1. All times and dates are in GMT. **2.** Satellite masses are based on available information; estimates should be within 10% of the actual figure. **Launch sites: CC** Cape Canaveral Air Station (US); **EAFB** Edwards Air Force Base (US); **IS** Yevne (Israel); **JQ** Jiuquan (People's Republic of China; US Space Command designator is Shuang Cheng Tzu); **KA** Kagoshima (Japan); **KO** Kourou (French Guiana); **KY** Kapustin Yar (Russia); **KSC** Kennedy Space Center (US); **PL** Plesetsk (Russia); **SM** San Marco platform (Indian Ocean); **SR** Sriharikota (India); **TA** Tanegashima (Japan); **TT** Tyuratam/Baikonur (USSR/Kazakhstan); **TY** Taiyuan (People's Republic of China; US Space Command designator is Wuzhai); **WR** Vandenberg AFB (US); **XI** Xichang (People's Republic of China).

SATELLITE LAUNCHES 1995/June 1999

Desig	*Name*	*Date*	*Site*	*Vehicle*	*Mass (kg)*	*Perigee (km)*	*Apogee (km)*	*Period (min)*	*Inc (deg)*	*Notes*
1995										
1A	Intelsat 704	10 Jan	CC	Atlas 2AS	3,661	geostationary above 66° east				International telecommunications
–	Express	15 Jan	KA	M-3SII	765	110	250	88.1	31	German/Japanese μg craft but entered after 2½ revs over Ghana because of low orbit from stage 2 control problem. Returned to Germany 20 Feb 1996. It was tracked by US Space Command but beaurocratic error meant that it did not receive orbital designation
2A	Tsikada 1	24 Jan	PL	Cosmos-3M	810	965	1,021	105.0	83.0	Tsikada civil navigation, replacing C2123
2B	Astrid 1				28	965	1,026	105.0	82.9	Swedish magnetospheric science satellite. Instruments' power supply failed after 5 weeks; last contact 27 Sep 1995
2C	Faisat 1				114	968	1,021	105.0	82.9	US commercial store/forward demonstrator
–	APStar 2	25 Jan	XI	CZ-2E						failed to achieve orbit; vehicle exploded at 51 s
3A	UFO 4	29 Jan	CC	Atlas 2	3,023	geostationary above 177° west				US Navy communications (USA 108)
4A	*Discovery* (STS-63)	3 Feb	KSC	Shuttle	95,853	312	342	91.1	51.6	Wetherbee, Collins, Harris, Foale, Voss, Titov
						408	410	92.8	51.7	(Russia) rendezvoused with Mir, operated Spacehab 3 and deployed/retrieved Spartan; returned KSC 11 Feb
4B	Spartan 204		Shuttle		1,167	388	388	92.3	51.7	Autonomous astronomy payload retrieved after 2 days
4C-F	Oderacs 2A-F		Shuttle		0.0005-5	320	340	91.1	51.6	Spheres + dipoles for radar calibration
5A	Progress-M 26	15 Feb	TT	Soyuz-U	7,250	188	224	88.6	51.6	Mir supply ferry, docked 17 Feb, undocked/
						391	397	92.5	51.7	de-orbited 15 Mar 1995
6A	Photon 7	16 Feb	PL	Soyuz-U	6,200	220	369	90.4	62.8	Microgravity experiments; down 3 Mar
7A	*Endeavour* (STS-67)	2 Mar	KSC	Shuttle	98,740	347	358	91.6	28.5	Oswald, Gregory, Grunsfeld, Lawrence, Jernigan, Durrance, Parise operated Astro 2 astrophysics payload; returned EAFB 18 Mar
8A	Cosmos 2306	2 Mar	PL	Cosmos-3M	1,000?	469	517	94.5	65.9	Radar calibration; four sub-satellites deployed shortly after launch; six further appeared in Apr 1997
9A	Cosmos 2307	7 Mar	TT	Proton-K + DM-2	1,400	19,113	19,149	675.8	64.8	Uragan/Glonass maritime/aeronautical
9B	Cosmos 2308				1,400	19,101	19,159	675.7	64.8	navigation satellites
9C	Cosmos 2309				1,400	19,117	19,142	675.7	64.8	
10A	Soyuz-TM 21	14 Mar	TT	Soyuz-U	7,150	191	221	88.6	51.7	Dezhurov, Strekalov, NASA's Thagard to Mir,
						391	396	92.4	51.6	docked 16 Mar. Crew returned aboard US Shuttle 7 Jul 1995; TM21 returned 11 Sep 1995

Desig	*Name*	*Date*	*Site*	*Vehicle*	*Mass (kg)*	*Perigee (km)*	*Apogee (km)*	*Period (min)*	*Inc (deg)*	*Notes*
11A	SFU	18 Mar	TA	H2	4,000	313	341	91.1	28.5	Space Flyer Unit with μg experiments,
						467	496	94.2	28.5	retrieved by STS-72 in Jan 1996
11B	GMS 5				746	geosynchronous above 140° east				Japanese meteorology
12A	Cosmos 2310	22 Mar	PL	Cosmos-3M	825	980	1,011	105.0	82.9	Parus military navigation, replacing C2184
13A	Intelsat 705	22 Mar	CC	Atlas 2AS	3,660	geostationary above 50° west				International telecommunications
14A	Cosmos 2311	22 Mar	PL	Soyuz-U	6,500	168	336	89.5	67.2	Yantar 4th gen photoreconn; down 31 May
15A	DMSP 2-08	24 Mar	WR	Atlas E	823	847	854	101.9	98.8	US military meteorology (USA 109)
–	TechSat 1	28 Mar	PL	Start	52	failed to achieve orbit				Israel technology demonstrator
	Unamsat				12	stage 4/5 failure				Mexican store/forward + meteor detector
	EKA 2				200	Start debut				Russian dummy payload
16A	Brasilsat B1	28 Mar	KO	Ar 44LP	1,780	geostationary above 65° west				Brazilian telecommunications
16B	Hot Bird 1				1,798	geostationary above 13° east				European TV distribution
17A	Orbcomm 1	3 Apr	VAFB	L-1011/	40	736	749	99.7	70.0	Commercial messaging/position determination
17B	Orbcomm 2			Pegasus	40	734	747	99.6	70.0	Commercial messaging/position determination
17C	MicroLab 1				68	733	749	99.6	70.0	NASA lightning observations
18A	Ofeq 3	5 Apr	IS	Shavit	225	247	732	94.4	143.4	Technology demonstrator; Earth imaging
						368	730	95.6	143.4	
19A	AMSC 1	7 Apr	CC	Atlas 2A	2,700	geostationary above 101° west				US commercial mobile communications (initial voice services began Dec 1995)
20A	Progress-M 27	9 Apr	TT	Soyuz-U	7,250	187	221	88.6	51.7	Mir supply ferry, docked 11 Apr, undocked 22 May,
						389	396	92.4	51.7	de-orbited 23 May 1995. Delivered GFZ 1
1986-17JE	GFZ 1				20	384	394	92.3	51.7	German laser geodetic satellite, delivered to Mir by M27, ejected from Mir 19 Apr
21A	ERS 2	21 Apr	KO	Ar 40	2,516	783	785	100.5	98.6	ESA radar imaging/microwave sounding
22A	USA 110	14 May	CC	Titan 401	?	geosynchronous?				Advanced Orion sigint?
23A	Intelsat 706	17 May	KO	Ar 44LP	3,693	geostationary above 53° west				International telecommunications
24A	Spektr	20 May	TT	Proton-K	19,500	216	317	89.8	51.7	Fourth Mir expansion module, docking 1 Jun,
						393	400	92.5	51.7	rotating to final position 2 Jun
25A	GOES 9	23 May	CC	Atlas 1	2,105	geostationary above 135° west				NOAA meteorological imaging
26A	Cosmos 2312	24 May	PL	Molniya-M	1,900	604	39 745	717.7	62.9	Oko missile early warning, replacing C2063
27A	UFO 5	31 May	CC	Atlas 2	3,020	geostationary above 71.5° east				US Navy communications (USA 111)
28A	Cosmos 2313	8 Jun	TT	Tsyklon-M	3,000	403	418	92.8	65.0	Eorsat elint ocean reconn; end-of-life manoeuvre during 22-23 Apr 1997, then disintegrated 26 Jun 1997
29A	DBS 3	10 Jun	KO	Ar 42P	2,934	geostationary above 100.8° west				US Hughes domestic direct TV
–	STEP M3	22 Jun	WR	L-1011/ Pegasus XL	268					failed to achieve orbit; interstage 1/2 remained attached
30A	*Atlantis* (STS-71)	27 Jun	KSC	Shuttle	97,387	158	303	89.0	51.6	Gibson, Precourt, Baker, Harbaugh, Dunbar,
						392	400	92.5	51.6	Solovyov/Budarin docked with Mir 13.00 GMT 29 Jun, creating orbital complex of ~223 t. Solovyov/Budarin swapped with TM21's Dezhurov/Strekalov/Thagard. *Atlantis* undocked 4 Jul; landed KSC 7 Jul
31A	Cosmos 2314	28 Jun	PL	Soyuz-U	6,500	166	340	89.6	67.1	Yantar 4th gen photoreconn; down 6 Sep
32A	Cosmos 2315	5 Jul	PL	Cosmos-3M	810	970	1,014	104.9	82.9	Tsikada civil navigation, replacing C2230
33A	Helios 1A	7 Jul	KO	Ar 40	2,537	680	682	98.4	98.1	First French reconnaissance satellite
33B	Cerise				50	667	675	98.2	98.1	Monitoring RF to support elint study
33C	UPM-Sat 1				47	665	676	98.2	98.1	Spanish science and technology research
34A	USA 112	10 Jul	CC	Titan 401	8,000?	no orbital data issued				Second Titan 4 Centaur launch into apparently Molniya-type orbit (profile similar to 3 May 1994 launch); Advanced Jumpseat elint?
35A	*Discovery* (STS-70)	13 Jul	KSC	Shuttle	88,500	316	331	91.0	28.5	Henricks, Kregel, Thomas, Currie, Weber
						288	315	90.6	28.5	deployed TDRS 7; landed KSC 22 Jul
35B	TDRS 7				2,225	geostationary above 150° west				NASA data relay
36A	Progress-M 28	20 Jul	TT	Soyuz-U	7,250	293	351	91.0	51.6	Mir supply ferry, docked 22 Jul,
						394	398	92.5	51.7	undocked/de-orbited 4 Sep
37A	Cosmos 2316	24 Jul	TT	Proton-K + DM-2	1,400	19,084	19,176	675.7	64.9	Uragan/Glonass maritime/aeronautical
37B	Cosmos 2317				1,400	19,037	19,223	675.7	64.9	navigation satellites
37C	Cosmos 2318				1,400	19,037	19,223	675.7	64.8	
38A	DSCS 3B-7	31 Jul	CC	Atlas 2A	1,040	geosynchronous				US military communications (USA 113)
39A	Interball 1	2 Aug	PL	Molniya-M	1,270	793	191,900	5,458	62.9	Russian-led magnetospheric science
39B	Magion 4				59	793	191,900	5,458	62.9	Interball's Czech subsat
40A	PAS 4	3 Aug	KO	Ar 42L	3,043	geostationary above 69° east				Commercial PanAmSat Indian Ocean telecom services
41A	Koreasat 1	5 Aug	CC	Delta-2 (7925)	1,459	geosynchronous above 116° east				Korean domestic telecom services. One stage 1 strap-on failed to separate, leaving Koreasat in GTO with apogee some 6,000 km below GEO altitude. Koreasat had to use its own propellant to attain GEO, reducing useful life to <5 yr
42A	Molniya-3 47	9 Aug	PL	Molniya-M	1,750	427	39,369	718.6	62.8	Orbita 2 system communications, replacing 3-31
–	GEMStar 1	15 Aug	WR	LMLV 1	127	failed to reach orbit				Maiden LMLV veered off course, destruct by range safety
43A	JCSat 3	29 Aug	CC	Atlas 2AS	3,100	geostationary above 128° east				Commercial Pacific Rim telecommunications
44A	NStar a	29 Aug	KO	Ar 44P	3,410	geostationary above 132° east				Domestic Japan telecommunications
45A	Cosmos 2319	30 Aug	TT	Proton-K + DM-2	2,300	geosynchronous above 80° east				Geizer satellite for CIS Potok government/military data communications
46A	Sich 1/FASat-Alfa	31 Aug	PL	Tsyklon	1,560?	632	669	97.7	82.5	Ukraine/Russian oceanography radar satellite. Chile's 50 kg FASat failed to separate
47A	Soyuz-TM 22	3 Sep	TT	Soyuz-U	7,150	197	218	88.6	51.6	Gidzenko, Avdeyev, ESA's Reiter to Mir,
						393	398	92.5	51.6	docked 5 Sep. Returned 29 Feb 1996
48A	*Endeavour* (STS-69)	7 Sep	KSC	Shuttle	99,663	368	377	92.0	28.5	Walker, Cockrell, Voss, Newman, Gernhardt
						338	346	92.0	28.5	deployed/retrieved Spartan 201-03 and Wake Shield Facility 2; landed KSC 18 Sep
48B	Spartan 201-03				1,289	368	377	92.0	28.5	Autonomous astronomy payload retrieved after 47.3 h
48C	WSF 2				1,979	396	408	92.6	28.5	Autonomous μg payload retrieved after 3.1 d
49A	Telstar 402R	24 Sep	KO	Ar 42L	3,410	geostationary above 95° west				AT&T US domestic telecom services
50A	Resurs-F2 7	26 Sep	PL	Soyuz-U	6,300	181	248	88.8	82.3	Earth resources photography mission;
						256	278	89.9	82.3	recovered 26 Oct
51A	Cosmos 2320	29 Sep	TT	Soyuz-U	7,000	242	302	90.0	64.9	5th gen photoreconn, de-orbited 28 Sep 1996
52A	Cosmos 2321	6 Oct	PL	Cosmos-3M	825	258	793	95.1	82.9	Parus military navigation, to replace C2266, but stage 2 failure resulted in low orbit
53A	Progress-M 29	8 Oct	TT	Soyuz-U	7,250	188	223	88.6	51.7	Mir supply ferry, docked 10 Oct,
						393	396	92.5	51.7	undocked 19 Dec, de-orbited 20 Dec
54A	Luch-1 1	11 Oct	TT	Proton-K	2,400	geosynchronous above 77° east				First improved Altair; CIS civil data comms
55A	Astra 1E	19 Oct	KO	Ar 42L	2,924	geostationary above 19.2° east				European direct TV
56A	*Columbia* (STS-73)	20 Oct	KSC	Shuttle	104,400	267	278	90.0	39.0	Bowersox, Rominger, Thornton, Coleman,
						246	301	90.0	39.0	Lopez-Alegria, Leslie, Sacco on US Microgravity Lab 2 (USML 2); landed KSC 5 Nov
57A	UFO 6	22 Oct	CC	Atlas 2	3,017	geostationary above 106° west				US Navy communications (USA 114)
–	Meteor 1	23 Oct	WI	Conestoga	127	failed to reach orbit				Maiden Conestoga veered off course, destruct by range safety
58A	Cosmos 2322	31 Oct	TT	Zenit-2	3,250	849	852	101.9	71.0	Tselina 2 elint
59A	Radarsat 1	4 Nov	WR	Delta-2 (7920)	2,749	788	794	100.7	98.6	Canadian radar imaging
60A	Milstar DFS 2	6 Nov	CC	Titan 401	4,670	geosynchronous above 4° east				Milstar milcoms (USA 115)

Desig	Name	Date	Site	Vehicle	Mass (kg)	Perigee (km)	Apogee (km)	Period (min)	Inc (deg)	Notes
61A	*Atlantis* (STS-74)	12 Nov	KSC	Shuttle	93,000	294	306	90.5	51.6	Cameron, Halsell, Hadfield, Ross, McArthur,
						337	349	91.4	51.6	docked with Mir 15 Nov, undocked 18 Nov (leaving Docking Module attached to Kristall); landed KSC 20 Nov
62A	ISO	17 Nov	KO	Ar 40	2,498	1,038	70,569	1,437	5.2	ESA IR celestial observations
63A	Gals 2	17 Nov	TT	Proton-K + DM-2	2,500	geosynchronous above 70° east				Second successor to Ekran direct TV
64A	AsiaSat 2	28 Nov	XI	CZ-2E	3,379	geostationary above 100.5° east				Commercial Hong Kong telecoms
65A	Soho	2 Dec	CC	Atlas 2AS	1,875	L1 halo orbit				ESA solar science observations
66A	USA 116	5 Dec	WR	Titan 403						US military Lacrosse or Advanced KH11?
67A	Telecom 2C	6 Dec	KO	Ar 44L	2,283	geostationary above 3° east				French domestic telecommunications
67B	Insat 2C				2,050	geostationary above 93.5° east				Indian telecommunications
68A	Cosmos 2323	14 Dec	TT	Proton-K + DM-2	1,400	19,086	19,175	675.7	64.9	Uragan/Glonass maritime/aeronautical
68B	Cosmos 2324				1,400	19,114	19,142	675.7	64.8	navigation satellites
68C	Cosmos 2325				1,400	19,113	19,147	675.7	64.8	
69A	Galaxy 3R	15 Dec	CC	Atlas 2A	2,860	geostationary above 95° west				US Hughes DTH Latin America service
70A	Progress-M 30	18 Dec	TT	Soyuz-U	7,250	194	316	89.6	51.6	Mir supply ferry, docked 20 Dec,
						391	399	92.5	51.7	undocked/de-orbited 22 Feb 1996
71A	Cosmos 2326	20 Dec	TT	Tsyklon-M	3,150	407	415	92.8	65.0	Eorsat elint ocean reconn
72A	IRS 1C	28 Dec	TT	Molniya-M	1,350	819	821	101.3	98.7	Indian Earth observation
72B	Skipper				230	804	813	101.1	98.6	US BMDO aerothermochemistry + aerobraking, intended to lower perigee to 150 km and then by 10 km steps until decay but wrongly connected solar array drained battery in <1 day
73A	EchoStar 1	28 Dec	XI	CZ-2E	3,288	geostationary above 119° west				US direct TV
74A	RXTE	30 Dec	CC	Delta-2 (7920)	3,045	565	585	96.2	23.0	NASA celestial X-ray observatory
1996										
1A	*Endeavour* (STS-72)	11 Jan	KSC	Shuttle	98,430	459	466	93.8	28.4	Duffy, Jett, Chiao, Scott, Wakata, Barry
						302	311	90.7	28.5	retrieved Japan's Space Flyer Unit 13 Jan and deployed/ retrieved Spartan 206/OAST-Flyer; landed KSC 20 Jan
1B	Spartan 206		Shuttle		1,198	303	312	90.7	28.5	Autonomous technology payload retrieved after 2 days
2A	PAS 3R	12 Jan	KO	Ar 44L	2,918	geostationary above 43° west				Atlantic telecommunications
2B	Measat 1				1,450	geostationary above 91.5° east				Malaysian telecommunications
3A	Koreasat 2	14 Jan	CC	Delta-2 (7925)	1,459	geosynchronous above 116° east				Korean domestic telecommunications
4A	Cosmos 2327	16 Jan	PL	Cosmos-3M	825	952	1,021	104.8	83.0	Parus military navigation, replacing C2266
5A	Gorizont 31	25 Jan	TT	Proton-K + DM-2	2,125	geosynchronous above 40° east				CIS civil telecommunications
6A	Palapa C1	1 Feb	CC	Atlas 2AS	2,989	geostationary above 113° east				Indonesian telecommunications
7A	NStar b	5 Feb	KO	Ar 44P	3,420	geostationary above 136° east				Domestic Japan telecommunications
–	Hyflex	11 Feb	TA	J1	1,040	suborbital				Spaceplane test; J1 debut (2-stage)
–	Intelsat 708	14 Feb	XI	CZ-3B	4,576	failed to achieve orbit				Inertial platform failed within seconds
8A	NEAR	17 Feb	CC	Delta-2 (7925)	805	heliocentric				NASA Near-Earth Asteroid Rendezvous flew by 253 Mathilde 27 Jun 1997, due to encounter 433 Eros early 1999
9A	Gonets-D1	19 Feb	PL	Tsyklon	225	1,400	1,414	113.9	82.6	Gonets civil and Strela 3 military
9B	Gonets-D2					1,407	1,414	114.0	82.6	tactical communications
9C	Gonets-D3					1,410	1,417	114.1	82.6	
9D	Cosmos 2328					1,411	1,415	114.1	82.6	
9E	Cosmos 2329					1,412	1,422	114.2	82.6	
9F	Cosmos 2330					1,412	1,428	114.2	82.6	
10A	Raduga 36	19 Feb	TT	Proton-K + DM-2	2,000					released into GTO after stage 4 failed; CIS government/military comms
11A	Soyuz-TM 23	21 Feb	TT	Soyuz-U	7,150	200	234	88.8	51.6	Onufrienko/Usachov to Mir,
						391	398	92.5	51.6	docked 23 Feb. Undocked and landed 2 Sep, also carrying André-Deshays launched on TM24
12A	*Columbia* (STS-75)	22 Feb	KSC	Shuttle	103,370	297	303	90.5	28.5	Allen, Horowitz, Hoffman, Cheli, Nicollier,
						279	322	90.5	28.5	Chang-Diaz, Guidoni deployed/lost TSS 1R and operated USMP 3 μg payload. Returned KSC 9 Mar
12B	TSS 1R				518	316	414	91.9	28.5	Italy's tethered satellite lost when tether snapped 26 Feb; re-entered 19 Mar
13A	Polar	1 Nov	WR	Delta-2 (7925)	1,258	186	50,495	937.0	86.0	NASA fields/particles in near-Earth space
						5,141	50,605	1,051.4	86.0	
14A	REX 2	9 Mar	VAFB	L-1011/ Pegasus XL	113	804	832	101.3	90.0	USAF investigation of radio propagation; first successful Pegasus XL (third attempt)
15A	Intelsat 707	14 Mar	KO	Ar 44LP	4,175	geostationary above 1° west				International telecommunications
16A	Cosmos 2331	14 Mar	PL	Soyuz	6,500	164	358	89.7	67.1	Yantar 4th gen photoreconnaissance; recovered 11 Jun after record 89 days. Carried USAF Space Test Program Binrad experiment
17A	IRS P3	21 Mar	SR	PSLV D3	922	818	848	101.4	98.8	Indian Earth observation
18A	*Atlantis* (STS-76)	22 Mar	KSC	Shuttle	111,740	158	293	89.0	51.6	Chilton, Searfoss, Sega, Clifford, Godwin,
						337	349	91.4	51.6	Lucid docked with Mir 24 Mar, undocked 29 Mar (leaving behind Lucid to return STS-79 Aug 1996); landed EAFB 31 Mar
19A	Navstar 2-25	28 Mar	CC	Delta-2 (7925)	930	20,070	20,285	717.8	54.7	GPS Block 2A navigation satellite (USA 117)
20A	Inmarsat 3 F1	3 Apr	CC	Atlas 2A	2,068	geostationary above 64° east				International mobile telecommunications
21A	Astra 1F	8 Apr	TT	Proton-K + DM-3	3,010	geosynchronous above 19.2° east				European direct TV
22A	MSAT 1	20 Apr	KO	Ar 42P	2,855	geostationary above 106.5° west				N American mobile telecommunications
23A	Priroda	23 Apr	TT	Proton-K	19,700	214	328	89.9	51.7	Mir remote sensing module, docked 26 Apr,
						391	396	92.4	51.7	completed Mir complex after 10 yr
24A	MSX	24 Apr	WR	Delta-2 (7920)	2,680	897	906	103.0	99.4	US BMDO Midcourse Space Experiment
25A	Cosmos 2332	24 Apr	PL	Cosmos-3M	500	295	1,565	103.6	83.0	2 m ∅ sphere for radar calibration and passive atmospheric monitoring
26A	USA 118	24 Apr	CC	Titan 401	?	geosynchronous?				Advanced Orion sigint?
27A	SAX	30 Apr	CC	Atlas 1	1,400	583	603	96.5	4.0	Italian-Dutch X-ray observatory
28A	Progress-M 31	5 May	TT	Soyuz-U	7,250	186	227	88.6	51.6	Mir supply ferry, docked 7 May, undocked and
						391	396	92.4	51.6	de-orbited 1 Aug
29A	USA 119	12 May	WR	Titan 4		1,023	1,023		63.4	probably PARCAE ocean surveillance. Three subsats designated (USA 120/121/122) and 53 kg TiPS Tether Physics & Survivability expt released 20 Jun into orbit typical for PARCAE
–	Cosmos	14 May	TT	Soyuz-U	6,500	failed to reach orbit				Kometa 4th gen topographic/mapping photoreconn, including commercial SPIN-2 for US companies, but shroud failed at 49 s
30A	Palapa C2	16 May	KO	Ar 44L	2,989	geostationary above 108° east				Indonesian telecommunications
30B	Amos 1				996	geostationary above 4° west				Israel/Middle East telecommunications
31A	MSTI 3	17 May	VAFB	L-1011/ Pegasus	211	291	365	91.1	97.0	BMDO sensor technology demonstration;
						422	435	93.1	97.1	last standard Pegasus
32A	*Endeavour* (STS-77)	19 May	KSC	Shuttle	115,600	280	290	90.2	39.0	Casper, Brown, Thomas, Bursch, Runco, Garneau operated Spacehab 4; deployed/retrieved Spartan 207/Inflatable Antenna Expt; deployed PAMS; landed KSC 29 May
32B	Spartan 207/IAE				1,296	282	294	90.3	39.0	Autonomous payload deployed 20 May, inflated 14 m ∅ antenna (released after 90 min), Spartan retrieved 21 May

Desig	Name	Date	Site	Vehicle	Mass (kg)	Perigee (km)	Apogee (km)	Period (min)	Inc (deg)	Notes
32D	PAMS				52	278	288	90.2	39.0	Passive Aerodynamically Stabilised Magnetically-Damped Satellite, ejected 22 May
33A	Galaxy 9	24 May	CC	Delta-2 (7925)	1,397	geosynchronous above 123° west				US Hughes commercial communications
34A	Gorizont 32	25 May	TT	Proton-K	2,125	geosynchronous above 52° east				CIS civil telecommunications
–	Cluster F1	4 Jun	KO	Ar 5	1,183	failed to reach orbit				Maiden flight of ESA's Ariane 5 ended after control software problem caused severe attitude failure from 37 s and vehicle breakup. Satellites were to be released into 10° GTO before raising themselves to 25,513 x 140,318 km, 90° to study 3D structure of Earth's plasma environment
	Cluster F2				1,169					
	Cluster F3				1,171					
	Cluster F4				1,184					
35A	Intelsat 709	15 Jun	KO	Ar 44P	3,420	geostationary above 18° west				International telecommunications
36A	*Columbia* (STS-78)	20 Jun	KSC	Shuttle	116,200	273	286	90.1	39.0	Henricks, Kregel, Helms, Linnehan, Brady, Favier
						268	278	90.0	39.0	Thirsk undertook record-duration LMS Life & Microgravity Sciences Spacelab mission landed KSC, 7 Jul
–	Cosmos	20 Jun	PL	Soyuz-U	6,500	failed to reach orbit				Yantar 4th gen photoreconnaissance, but Soyuz shroud failed after 49 s
37A	TOMS-EP	2 Jul	VAFB	L-1011/ Pegasus XL	294	345	953		97.4	NASA's Total Ozone Mapping Spectrometer
						502	502		97.4	
38A	USA 125	3 Jul	CC	Titan 4	?	orbital data not available; inc >55°				DoD classified; trade reports suggest advanced SDS Satellite Data System (orbit about 400 x 40,000 km, 63°) for relaying imagery from photorecon satellites. Amateur observations showed manoeuvres into a high orbit
39A	APStar 1A	3 Jul	XI	CZ-3	1,400	geostationary above 77° east				Hong Kong regional telecommunications
40A	ARABSAT 2A	9 Jul	KO	Ar 44L	2,617	geosatationary above 26° east				Arab League communications
40B	TURKSAT 1C				1,743	geostationary above 42° east				Turkish Telecom communications
41A	Navstar-2 26	16 Jul	CC	Delta-2 (7925)	1,881	20,138	20,224	717.93	55.0	GPS Block 2A navigation: USA 126
42A	UFO 7	25 Jul	CC	Atlas-2	3,020	geostaionary above 23° west				US Navy communications: USA 127
43A	Progress-M 32	31 Jul	TT	Soyuz-U	7,250?	186	229	88.7	51.6	Mir supply ferry: docked 2 Aug, undocked 18 Aug,
						375	391	92.2	51.6	redocked 3 Sep, undocked and de-orbited 21 Nov
44A	ITALSAT 2	8 Aug	KO	Ar 44L	1,990	geostationary above 10° east				Italian communications, carries EMS package
44B	Telecom 2D				2,260	geostationary above 5° west				French civil and military communications
45A	Molniya-1 89	14 Aug	PL	Molniya-M	1,600?	464	39,887	717.7	62.8	Orbita system communications
46A	Midori 1	17 Aug	TA	H-2	3,560?	797	799	100.8	98.6	Japanese remote sensing; ADEOS 1 before launch; power failure June 1997
47A	Soyuz-TM 24	17 Aug	TT	Soyuz-U	7,150?	233	292	89.8	51.6	Korzun, Kaleri, André-Deshays (CNES); docked with Mir
						375	390	92.2	51.6	19 Aug; André-Deshays returned to Earth aboard TM23 2 Sep; Korzun, Kaleri and Linenger undocked from Mir 7 Feb 1997 and re-docked at the opposite end of Mir the same day; Korzun, Kaleri and Ewald undocked and returned to Earth 2 Mar 1997
48A	Zhongxing 7	18 Aug	XI	CZ-3	1,200?	200	17,229	307.5	27.3	China Star communications; third stage shut down early, mission abandoned
49A	FAST	21 Aug	WR	L-1011/ Pegasus-XL	180	351	4,165	133.1	83.0	Fast Auroral Snapshot Explorer, science
50A	MUSAT	29 Aug	PL	Molniya-M	30	236	1,171	98.8	62.8	Argentine technology satellite
50B	MAGION 5				62	791	19,196	347.5	62.8	Czech Republic solar/magnetosphere interaction research
50C	Interball 2				1,300?	769	19,211	347.3	62.8	Auroral Probe, international magnetosphere, plasmasphere research satellite
51A	Cosmos 2333	4 Sep	TT	Zenit-2	3,250?	849	852	101.9	71.0	Tselina-2 ELINT
52A	Cosmos 2334	5 Sep	PL	Cosmos-3M	825?	972	1,001	104.8	82.9	Parus navigation; co-planar with cosmos 2327
52B	UNAMSAT 2				17	968	1,011	104.9	82.9	Mexican micrometeorite research satellite
53A	INMARSAT-3 2	6 Sep	TT	Proton-K + DM-1	1,144	geostationary above 15° west				Maritime communications
54A	GE 1	8 Sep	CC	Atlas-2A	2,764	geostationary above 103° west				GE Americom communications
55A	EchoStar 2	11 Sep	KO	Ar 42P	2,865	geostationary above 119° west				EchoStar Communications Corporation communications
56A	Navstar 27	12 Sep	CC	Delta-2 (7925)	1,881	20,057	20,307	718.0	54.7	GPS Block 2A navigation: USA 128
57A	*Atlantis* (STS-79)	16 Sep	KSC	Shuttle	113,105	372	378	92.1	51.6	Readdy, Wilcutt, Apt, Akers, Walz and Blaha on fourth Shuttle-Mir mission: Blaha remained on Mir until STS-81, Lucid returned aboard Shuttle. Docked with Mir 19 Sep, undocked 23 Sep, landed KSC 26 Sep
58A	Ekspress 2	26 Sep	TT	Proton-K + DM-2	2,500	geostationary above 80° east				Communications
59A	FSW-2 3	20 Oct	JQ	CZ-2D	2,600?	171	342	89.6	63.0	Microgravity, remote sensing; capsule recovered 4 Nov, bus decayed
60A	Molniya-3 48	24 Oct	PL	Molniya-M	1,750?	610	39,768	718.3	62.8	Orbita system communications
61A	SAC-B + HETE + Pegasus 3rd stage	4 Nov	WI	L-1011/ Pegasus-XL	480?	488	556	95.1	38.0	SAC-B Argentine, HETE US: both satellites failed to separate from Pegasus-XL final stage once in orbit
62A	Mars Global Surveyor	7 Nov	CC	Delta-2 (7925)	1,060	heliocentric orbit				Planned to reach Mars orbit 11 Sep 1997
63A	ARABSAT 2B	13 Nov	KO	Ar 44L	2,661	geostationary above 30° east				Arab League communications
63B	MEASAT 2				1,512	geostationary above 148° east				Binariang, Malaysia, communications
64	Mars 8 (Mars-96)	16 Nov	TT	Proton-K + D-2	6,825	139	155	87.4	51.5	Mars probe, failed to leave Earth orbit; first orbit
						87?	1,500?	100.7?	51.6?	is for the fourth stage, second one for the spacecraft (approximate orbit); spacecraft not assigned an International Designator
65A	*Columbia* (STS-80)	19 Nov	KSC	Shuttle	102,946	347	358	91.6	28.5	Cockrell, Rominger, Jernigan, Jones and Musgrave; landed KSC 7 Dec
65B	ORFEUS-SPAS 2				3,573	349	356	91.6	28.5	Astrophysics payload, deployed 20 Nov, recaptured 4 Dec
65C	WSF 3				2,109	347	359	91.6	28.5	Wake Shield Facility microgravity payload, deployed 22 Nov, re-captured 26 Nov
66A	Progress-M 33	19 Nov	TT	Soyuz-U	7,150?	250	309	90.1	51.6	Cargo freighter to Mir, docked 22 Nov, undocked 6 Feb 1997,
						371	390	92.2	51.6	re-docking 4 Mar 1997 failed and was de-orbited 12 Mar 1997
67A	Hot Bird 2	21 Nov	CC	Atlas-2A	2,912	geostationary above 13° east				Eutelsat communications
68A	Mars Pathfinder	4 Dec	CC	Delta-2 (7925)	890	heliocentric orbit				Landed on Mars 4 Jul 1997, deployed Sojourner rover
69A	Cosmos 2335	11 Dec	TT	Tsyklon-M	3,000?	403	419	92.8	54.0	EORSAT
70A	INMARSAT-3 3	18 Dec	CC	Atlas-2A	2,074	geostationary above 178° east				Maritime communications
71A	Cosmos 2336	20 Dec	PL	Cosmos-3M	825?	979	1,012	105.0	83.0	Parus navigation: co-planar with Cosmos 2173
72A	USA 129	20 Dec	WR	Titan-4	13,000?	orbital data classified				Advanced Crystal photoreconnaissance
73A	Bion 11	24 Dec	PL	Soyuz-U	6,000?	217	379	90.5	62.8	Biological research with two monkeys: US experiments carried: recovered 7 Jan 1997
1997										
1A	*Atlantis* (STS-81)	12 Jan	KSC	Shuttle	113,370	370	390	92.2	51.6	Baker, Jett, Wisoff, Grunsfeld, Ivins and Linenger on fifth Shuttle-Mir mission: Linenger remained on Mir until STS-84, Blaha returned to Earth aboard the Shuttle. Docked with Mir 15 Jan, undocked 20 Jan, landed at KSC 22 Jan
–	Navstar 28	17 Jan	CC	Delta-2 (7925)	2,032	failed to reach orbit				First launch of a GPS Block 2R: launch vehicle self-destructed 21 seconds after launch (first Delta-2 to fail to reach orbit)
2A	GE 2	30 Jan	KO	Ar 44L	2,469	geostationary above 85° west				GE Americom communications
2B	Nahuel 1A				1,790	geostationary above 72° west				Nahuelsat, Argentina, communications
3A	Soyuz-TM 25	10 Feb	BA	Soyuz-U	7,150?	188	231	88.7	51.6	Tsibliev, Lazutkin and Ewald (Germany): docked with Mir
						378	394	92.3	51.6	12 Feb: Ewald returned to Earth aboard TM24, Tsibliev and Lazutkin aboard TM25 in August

Desig	Name	Date	Site	Vehicle	Mass (kg)	Perigee (km)	Apogee (km)	Period (min)	Inc (deg)	Notes
4A	*Discovery* (STS-82)	11 Feb	KSC	Shuttle	96,758	589	594	96.5	28.5	Bowersox, Horowitz, Tanner, Hawley, Harbaugh, Lee and Smith: second Hubble Space Telescope servicing mission. HST captured 13 Feb, released 19 Feb: shuttle returned to KSC 21 Feb
5A	Haruka	12 Feb	KA	M-5	823	247	21,409	374.4	31.3	Radio telescope; space segment of VSOP
	(MUSES-B)					573	21,402	379.6	31.3	
6A	Gonets-D 4	14 Feb	PL	Tsyklon	225?	1,413	1,423	114.2	82.6	Civil communications; it is not certain which of the six satellites
6B	Gonets-D 5				225?	1,413	1,429	114.2	82.6	are the Gonets payloads and which are the Cosmos/
6C	Gonets-D 6				225?	1,409	1,414	114.0	82.6	Strela-3 payloads
6D	Cosmos 2337				225?	1,402	1,414	114.0	82.6	Strela-3 military communications
6E	Cosmos 2338				225?	1,413	1,415	114.1	82.6	
6F	Cosmos 2339				225?	1,412	1,415	114.1	82.6	
7A	JCSAT 4	17 Feb	CC	Atlas-2AS	3,105	geostationary above 124° east				Japan Satellite Systems Inc communications
8A	DSP 18	24 Feb	CC	Titan-4B + IUS	2,400?	geostationary orbit				Early warning satellite; orbital data classified; maiden flight of Titan-4B; USA 130
9A	INTELSAT 801	1 Mar	KO	Ar 44P	3,420	geostationary above 62° east				International telecommunications
10A	Zeya	4 Mar	SV	Start-1	87	467	480	94.1	97.3	Geodetic satellite; maiden launch from Svobodny site
11A	Tempo 2	8 Mar	CC	Atlas-2A	3,561	geostationary above 119° west				Tele-Communications Inc communications
12A	DMSP-2 9	4 Apr	WR	Titan-23G	823	844	855	101.9	98.9	Block 5D-2 military meteorological satellite
13A	*Columbia* (STS-83)	4 Apr	KSC	Shuttle	96,643	299	304	90.5	28.5	Halsell, Still, Voss, Gernhardt, Thomas, Crouch and Linteris; first Microgravity Science Laboratory mission; flight curtailed because of fuel cell problems and re-flown as STS-94 in July; landed KSC 8 Apr
14A	Progress-M 34	6 Apr	BA	Soyuz-U	7,156	187	227	88.6	51.6	Cargo freighter; docked with Mir 8 Apr, undocked 24 Jun: re-docking attempt 25 Jun failed and spacecraft collided with Spektr module, the module's vanes of solar cells and puncturing its shell; Progress de-orbited 2 Jul
15A	Cosmos 2340	9 Apr	PL	Molniya-M	1,250?	541	39,815	717.8	62.9	Oko early warning
16A	THAICOM 3	16 Apr	KO	Ar 44LP	2,650	geostationary above 78.5° east				Shinawatra Satellite, Thailand, communications
16B	B-SAT 1A				1,236	geostationary above 110° east				B-SAT, Japan, communications
17A	Cosmos 2341	17 Apr	PL	Cosmos-3M	825?	978	1,014	105.0	82.9	Parus navigation, co-planar with Cosmos 2310
18A	MINISAT 01	21 Apr	GA	L-1011/ Pegasus-XL	209	563	582	96.1	151.0	Spanish technology satellite: Pegasus third stage carried Celestis 1 payloads, carrying cremated ashes of 24 people: first launch from Gando AFB, Canary Islands
19A	GOES 10	25 Apr	CC	Atlas-1	2,105	geostationary above 105° west				NOAA meterological satellite: final launch of Atlas-1
20A	Iridium 8	5 May	WR	Delta-2 (7920)	657	773	783	100.4	86.4	First launch within the Iridium communications constellation; Iridium 1-3 were non-flight vehicles
20B	Iridium 7				657	773	782	100.4	86.4	
20C	Iridium 6				657	771	783	100.4	86.4	
20D	Iridium 5				657	769	787	100.4	86.4	
20E	Iridium 4				657	771	784	100.4	86.4	
21A	Zhongxing 8	11 May	XI	CZ-3A	2,230?	geostationary above 125° east				Second DFH-3 communications
22A	Cosmos 2342	14 May	PL	Molniya-M	1,250?	525	39,825	717.7	62.8	Oko early warning
23A	*Atlantis* (STS-84)	15 May	KSC	Shuttle	100,284	371	396	92.2	51.6	Precourt, Collins, Clervoy, Noriega, Lu, Kondakova and Foale on sixth Shuttle-Mir mission; Foale remained on board Mir until STS-86, Linenger returned to Earth aboard the Shuttle; docked with Mir 17 May, undocked 22 May, landed KSC 24 May
24A	Cosmos 2343	15 May	BA	Soyuz-U	7,000?	197	292	89.4	64.9	Photoreconnaissance, probably sixth generation type
						203	343	90.0	64.9	
–	Cosmos	20 May	BA	Zenit-2	3,250	Failed to reach orbit				Tselina-2 ELINT: first stage engine shut down 85 seconds after launch
25A	Thor 2	21 May	CC	Delta-2 (7925)	1,300?	geostationary above 1° west				Telenor Satellite Services AS, Norway, communications
26A	Telstar 5	24 May	BA	Proton-K + DM4-1L	3,650	geostationary above 97° west				Loral Skynet communications
27A	INMARSAT-3 4	3 Jun	KO	Ar 44L	1,999	geostationary above 54° west				Maritime communications
27B	INSAT 2D				2,079	geostationary above 74° east				ISRO communications
28A	Cosmos 2344	6 Jun	BA	Proton-K + DM-5	5,000?	1,509	2,732	130.0	63.4	New class of optical reconnaissance platform
29A	Feng Yun-2 1R	10 Jun	XI	CZ-3	1,380	geostationary above 105° east				First Chinese geosynchronous meteorological satellite: first FY-2 1 lost in pre-launch accident, Apr 1994
30A	Iridium 14	18 Jun	BA	Proton-K + DM-5	657	555	560	95.8	86.4	Second launch of Iridium cluster, first to use Proton-K: orbits are as at the end of Jun 1997
30B	Iridium 12				657	608	616	96.9	86.4	
30C	Iridium 10				657	555	558	95.8	86.4	
30D	Iridium 9				657	560	561	95.9	86.4	
30E	Iridium 13				657	580	583	96.3	86.4	
30F	Iridium 16				657	585	591	96.4	86.4	
30G	Iridium 11				657	670	676	98.2	86.4	
31A	INTELSAT 802	25 Jun	KO	Ar 44P	3,435	geostationary above 174° east				International communications
32A	*Columbia* (STS-94)	1 Jul	KSC	Shuttle	117,801	300	304	90.6	28.5	Halsell, Still, Voss, Gernhardt, Thomas, Crouch and Linteris refly Microgravity Sciences Laboratory returned to earth early on STS-8 in Apr 1997. Landed 17 Jul
33A	Progress-M 35	5 Jul	TT	Soyuz-U	7,250	386	392	92.3	51.6	Mir supply
34A	Iridium 15	9 Jul	VAFB	Delta 2	657	774	780	100.4	86.4	Third Iridium cluster launch
34B	Iridium 17				657	775	780	100.4	86.4	
34C	Iridium 18				657	774	779	100.4	86.4	
34D	Iridium 20				657	772	782	100.4	86.4	
34E	Iridium 21				657	629	643	97.4	86.4	Failed to reach op orbit
35A	USA 132	23 Jul	CC	Delta 2	2,032	20,124	20,247	718.1	54.9	First successful Navstar Block 2R
36A	Superbird C	28 Jul	CC	Atlas AS2	3,130	geostationary above 143° east				HS 601 for Space Communication Corp of Japan
37A	Seastar	1 Aug	VAFB	L-1011/ Pegasus-XL	309	297	313	90.6	98.3	Remote sensing ocean satellite with
						707	709	98.9	98.2	SeaWifs sensor
38A	Soyuz-TM 26	5 Aug	TT	Soyuz-U	7,150	386	392	92.3	51.6	Solovyov and Vinogradov to Mir, relieving Tsibliev and Lazutkin who returned to earth in Soyuz-TM 25 leaving Foale on board
39A	*Discovery* (STS-85)	7 Aug	KSC	Shuttle	98,953	290	301	90.4	57.0	Brown, Reminger, Davis, Curbean, Robinson, Tryggvanson on earth science and ISS support mission. Landed 19 Aug
39B	CHRISTA-SPAS 2				3,504	291	301	90.4	57.0	Platform with cryogenic IR spectrometer deployed 7 Aug recaptured 16 Aug
40A	PAS 6	8 Aug	KO	Ar44P	3,420	geostationary above 316.5° east				DBS for PanAmSat
41A	Cosmos 2345	14 Aug	TT	Proton-K	2,500	geostationary above 336° east				Prognoz early warning
42A	Mabuhay Agila	19 Aug	XI	CZ-3B	2,560	geostationary above 147° east				Communications sat for Mabuhay Philippine Satellite Corporation
43A	Iridium 26	21 Aug	VAFB	Delta 2	657	765	763	100.2	86.4	Fourth Iridium cluster launched
43B	Iridium 25				657	775	780	100.4	86.4	
43C	Iridium 24				657	774	781	100.4	86.4	
43D	Iridium 23				657	775	780	100.4	86.4	
43E	Iridium 22				657	775	780	100.4	86.4	
44A	Lewis (SSTI)	23 Aug	VAFB	LMLV-1	386	298	320	90.7	97.6	Experimental TRW Small Spacecraft Technology Initiative satellite

Desig	Name	Date	Site	Vehicle	Mass (kg)	Perigee (km)	Apogee (km)	Period (min)	Inc (deg)	Notes
45A	ACE	25 Aug	CC	Delta 2	752	177	1,367,769	95,140	28.7	Advanced Composition Explorer to measure heliospheric matter
46A	PAS 5	28 Aug	TT	Proton-K	3,720	geostationary over 302° east				HS 601P for PanAmSat
47A	FORTE (P94-1)	29 Aug	VAFB	L-1011/ Pegasus-XL	215	800	834	101.2	79.0	Detects and measures radio frequency pulses for US Dept of Energy
48A	Iridium MFS 1	1 Sep	TY	CZ-2C/SD	650	623	632	97.3	86.4	Test flight of new upper stage
48B	Iridium MFS 2				650	623	633	97.3	86.4	
49A	Hot Bird 3	2 Sep	KO	Ar44LP	2,915	geostationary over 13° east				DBS satellite for Eutelsat
49B	METEOSAT 7				703	geostationary over 349° east				Eumetsat satellite
50A	GE 3	4 Sep	CC	Atlas 2AS	2,750	geostationary over 276° east				For GE American Communications
51A	Iridium 29	14 Sep	TT	Proton-K	657	776	779	100.4	86.4	Fifth Iridium cluster
51B	Iridium 32				657	775	779	100.4	86.4	
51C	Iridium 33				657	776	779	100.4	86.4	
51D	Iridium 27				657	554	558	95.8	86.6	
51E	Iridium 28				657	775	780	100.4	86.4	
51F	Iridium 30				657	774	780	100.4	86.4	
51G	Iridium 31				657	776	779	100.4	86.4	
52A	Cosmos 2346	23 Sep	PL	Cosmos-3M	795	939	996	104.4	82.9	Parus military navigation satellite
52B	FAISAT 2V				120	939	996	104.4	82.9	Store-forward communication satellite
53A	Intelsat 803	24 Sep	KO	Ar42L		geostationary over 333.5° east				Intelsat telecommunications satellite
54A	Molniya-1 90	24 Sep	PL	Molniya-M	3,455	449	39,906	717.8	0.1	Molniya-IT series; coplanar with Molniya-1 84
55A	*Atlantis* (STS-86)	26 Sep	KSC	Shuttle	114,184	256	298	90.1	51.7	Wetherbee, Bloomfield, Titov Parazynski, Chretien, Lawrence, Wolf on 7th Mir visit. Wolf replaced Foale on Mir who returned with the STS-86 crew 6 Oct
56A	Iridium 19	27 Sep	VAFB	Delta 2	657	775	779	100.4	86.4	Sixth Iridium launch
56B	Iridium 37				657	774	779	100.4	86.4	
56C	Iridium 36				657	641	652	97.7	86.4	
56D	Iridium 35				657	774	781	100.4	86.4	
56E	Iridium 34				657	775	780	100.4	86.4	
57A	IRS 1D	28 Sep	SR	PSLV	1,250	308	822	95.9	98.6	ISRO remote sensing but spacecraft left in wrong orbit
						739	826	100.5		due to stage 4 malfunction. Perigee raised using satellite propellant as shown
58A	Progress-M 36	4 Oct	TT	Soyuz-U	7,250	382	392	92.3	51.6	Mir cargo supply flight
58C	Sputnik model				50	385	389	92.3	51.6	Sputnik model launched from Progress
59A	EchoStar 3	4 Oct	CC	Atlas 2AS	2,900	geostationary over 298° east				DBS for EchoStar Communications Corp
60A	Foton 8 (F11)	9 Oct	PL	Soyuz-U	5,800	218	375	90.4	62.8	Microgravity research. Capsule with samples landed 23 Oct
61A	Cassini/Huygens	15 Oct	KSC	Titan 4B/ Centaur	5,712	Heliocentric				Saturn orbiter by way of two fly-bys of Venus and one fly-by of Earth, Jupiter and Phoebe arriving Nov 2004
62A	APStar 2R	16 Oct	XI	CZ-3B	3,747	geostationary over 76° e				Replaced APStar 2 for APT Satellite Co.
63A	STEP M4	22 Oct	WI	L-1011/ Pegasus XL	395	434	501	93.5	45.0	Near-earth space properties satellite. Failed when solar panels did not deploy
64A	USA 133	24 Oct	VAFB	Titan 4A	14,500	673	674	98.2	57.0	Third Lacrosse/Vega launch. Orbit shown is unofficial
65A	USA 134	25 Oct	CC	Atlas 2A	1,040	geostationary				DSCS-3B 5, improved type
66A	MAQSAT-H/TEAMSAT	30 Oct	KO	Ar 5 (502)	2,650	533	26,635	467.8	7.8	Technology and debris monitor but prevented from achieving 36,000 km apogee by second stage malfunction
66B	MAQSAT-B/EPS				4,100	533	26,569	466.6	7.8	Remained attached to second stage
66C	YES				60	540	26,626	467.8	7.8	Young Engineers Satellite planned for tether deploy to become independent satellite
67A	USA 135	6 Nov	CC	Delta 2	1,881	19,912	20,449	358.3	54.9	NAVSTAR 38
68A	USA 136	8 Nov	CC	Titan 4A/ Centaur	5,000	1,100	39,060	713.8	63.6	ELINT Trumpet satellite
69A	Iridium 43	9 Nov	VAFB	Delta 2	657	774	780	100.4	86.4	Seventh cluster launch
69B	Iridium 41				657	772	782	100.4	86.4	
69C	Iridium 40				657	768	770	100.2	86.4	
69D	Iridium 39				657	773	780	100.4	86.4	
69E	Iridium 38				657	768	771	100.2	86.4	
70A	Kupon 1	12 Nov	TT	Proton-K	2,500	geostationary above 55° east				First satellite for Bankir system
71A	Sirius 2	12 Nov	KO	Ar44L	2,920	geostationary above 5° east				Communications satellite for NSAB, Stockholm
71B	Cakrawarta				1,385	geostationary above 107.7° east				DBS for Media Citra, Indonesia
72A	Resurs-F 21	18 Nov	PL	Soyuz-U	6,300	180	236	88.7	82.3	First flight of improved F-1M
						238	270	88.6	82.3	
73A	*Columbia* (STS-87)	20 Nov	KSC	Shuttle	102,717	281	286	90.2	28.5	USMP-4 microgravity payload with Kregel, Lindsey, Chawla, Scott, Doi and Kadenyuk. Landed 5 Dec
73B	Spartan 201-04				1,352	280	285	90.2	28.5	Deployed 21 Nov, retrieved 22 Nov
74A	TRMM	27 Nov	TA	H-2	3,620	367	385	92.7	35.0	NASA Tropical Rainfall Mission
74B	Hikobashi/Orihame				2,900	379	538	93.8	35.0	Kiku-7 (ETS 7 before launch)
75A	JCSAT 5	2 Dec	KO	AR44P	2,982	geostationary over 139° east				Telecommunications satellite for Japan Satellite Systems Inc. To operate from 150° east
75B	Equator-S				230	496	67,232	1,339	4.0	Science satellite from Max Planck Inst, Germany
76A	Astra 1G	2 Dec	TT	Proton-K	3,300	geostationary over 19.2° east				DBS satellite for SES, Luxembourg
77A	Iridium 42	8 Dec	TY	CZ-2C/SD	657	775	779	100.4	86.4	Eighth Iridium cluster
77B	Iridium 44				657	775	779	100.4	86.4	
78A	Galaxy 8I	8 Dec	CC	Atlas 2AS	3,000	geostationary over 280° east				Telecommunications satellite operated by Hughes Communications Inc
79A	Cosmos 2347	9 Dec	TT	Tsyklon-M	3,000	403	419	92.8	65.0	ELINT ocean reconnaissance satellite
80A	Cosmos 2348	15 Dec	PL	Soyuz-U	6,600	165	345	89.6	67.1	Yantar-4K fourth-generation reconnaissance satellite
81A	Progress-M 37	20 Dec	TT	Soyuz-U	7,195	378	389	92.3	51.7	Mir cargo-tanker supply ship, docked 22 Dec. Re-entered 15 Mar 1998
82A	Iridium 45	20 Dec	VAFB	Delta 2	657	775	778	100.4	86.4	Ninth Iridium cluster launch
82B	Iridium 46				657	775	780	100.4	86.4	
82C	Iridium 47				657	776	778	100.4	86.4	
82D	Iridium 48				657	775	779	100.4	86.4	
82E	Iridium 49				657	716	720	100.4	86.4	
83A	Intelsat 804	22 Dec	Ko	Ar42L	3,455	geostationary over 64° east				Telecommunications satellite for Intelsat
84A	Orbcomm FM 8	23 Dec	WI	L-1011/Pegasus	43	815	823	101.3	45	First cluster for Orbcomm Global LP
84B	Orbcomm FM 10				43	814	823	101.3	45	
84C	Orbcomm FM 11				43	815	823	101.3	45	
84D	Orbcomm FM 12				43	815	823	101.3	45	
84E	Orbcomm FM 9				43	815	823	101.3	45	
84F	Orbcomm FM 5				43	816	827	101.3	45	
84G	Orbcomm FM 6				43	815	826	101.3	45	
84H	Orbcomm FM 7				43	816	826	101.3	45	
85A	Early Bird 1	24 Dec	SV	Start-1	300	480	488	101.3	97.3	Commercial remote sensing satellite for EarthWatch Inc.
86A	AsiaSat 3	24 Dec	TT	Proton-K	3,401	301	35,999	638.3	51.1	Telecommunications satellite for Asia Satellite Telecommunications Co Ltd, Hong Kong
1998										
01A	Lunar Prospector	7 Jan	CC	Athena-2	233	translunar/heliocentric				Geochemical lunar orbiter

Desig	Name	Date	Site	Vehicle	Mass (kg)	Perigee (km)	Apogee (km)	Period (min)	Inc (deg)	Notes
02A	Skynet 4D	10 Jan	CC	Delta 2	1,493	geostationary over 6° east				UK military communications satellite
03A	*Endeavour* (STS-89)	23 Jan	KSC	Shuttle	114,130	378	387	92.2	51.7	Wilcutt, Edwards, Reilly, Anderson, Dunbar, Sharipov and Thomas, replacing Wolff on Mir. Landed 31 Jan
04A	Soyuz-TM 27	28 Jan	TT	Soyuz-U	7,150	379	385	92.2	51.7	Musabayev, Budarin and Eyharts, the latter returning with Solovyov and Vinogradov in Soyuz-TM 26
05A	USA 137	29 Jan	CC	Atlas 2A	2,500	200	39,500		63.0	Capricorn classified successor to the DSP data relay satellites for the National Reconnaissance Office
06A	Brasilsat B3	4 Feb	KO	Ar44LP	1,780	geostationary over 295° east				Telecommunications satellite for EBRATEL (Brazil)
06B	Inmarsat-3 5				2,000	geostationary over 25° east				Mobile communications satellite for Inmarsat
07A	GFO	10 Feb	VAFB	Taurus	347	780	878	101.5	108.0	Geosat Follow-On radar altimeter satellite operated by Naval Meteorology and Oceanographic Command
07B	Orbcomm FM 3				43	781	876	101.5	108.0	Orbcomm operated by Orbcomm Global LP
07C	Orbcomm FM 4				43	781	874	101.5	108.0	
08A	Globalstar 1	14 Feb	CC	Delta 2	450	1,244	1,259	110.5	52.0	First launch for Globalstar Ltd Partnership
08B	Globalstar 2				450	1,237	1,254	110.4	52.0	
08C	Globalstar 3				450	1,240	1,260	110.5	52.0	
08D	Globalstar 4				450	1,244	1,256	110.5	52.0	
09A	Cosmos 2349	17 Feb	TT	Soyuz-U	6,600	212	278	89.4	70.4	Yantar-1KFT/Kometa topographic and mapping photo-reconnaissance satellite
10A	Iridium 50	18 Feb	VAFB	Delta 2	657	723	726	99.3	86.5	Tenth Iridium cluster
10B	Iridium 51				657	768	773	100.2	86.4	
10C	Iridium 52				657	699	694	98.6	86.5	
10D	Iridium 53				657	766	771	100.2	86.4	
10E	Iridium 54				657	774	780	100.4	86.4	
11A	Kakehashi	21 Feb	TA	H-2	3,960	247	1,883	106.5	30.1	Broadcast research and development test satellite
12A	SNOE	26 Feb	VAFB	Pegasus-XL	132	535	581	95.8	97.8	Student Nitric Oxide Explorer operated by University of Colorado
12B	Teledisc T-1				45	535	580	95.8	97.7	Test satellite for Teledisc communications satellite system
13A	Hot Bird 4	27 Feb	KO	Ar42P	2,885	geostationary orbit at 13° east				Telecommunications satellite for Eutelsat
14A	Intelsat 806	28 Feb	CC	Atlas 2AS	3,400	geostationary orbit at 304° east				Telecommunications satellite for Intelsat
15A	Progress-M 38	14 Mar	TT	Soyuz-U	7,200	376	383	92.1	51.7	Cargo freighter to Mir
16A	USA 138	16 Mar	CC	Atlas 2	3,000	geostationary orbit at 188° east				UHF Follow-On military communications satellite
17A	SPOT 4	24 Mar	KO	Ar40	2,755	824	826	101.4	98.7	CNES remote sensing satellite
18A	Iridium 51	25 Mar	TY	CZ-2C/SD	657	775	780	100.4	86.4	Eleventh Iridium cluster launch
18B	Iridium 61				657	687	692	98.6	86.5	
19A	Iridium 55	30 Mar	VAFB	Delta 2	657	728	731	99.4	86.5	Twelfth Iridium cluster launch
19B	Iridium 57				657	665	668	98.1	86.6	
19C	Iridium 58				657	666	668	98.1	86.6	
19D	Iridium 59				657	687	696	98.6	86.6	
19E	Iridium 60				657	675	676	98.3	86.6	
20A	TRACE	2 Apr	VAFB	Pegasus-XL	250	599	641	97.1	97.8	Transition Regional & Coronal Explorer
21A	Iridium 62	7 Apr	TT	Proton-K	657	776	779	100.4	86.4	Thirteenth Iridium cluster of satellites
21B	Iridium 63				657	776	779	100.4	86.4	
21C	Iridium 64				657	775	779	100.4	86.4	
21D	Iridium 65				657	775	778	100.4	86.4	
21E	Iridium 66				657	772	782	100.4	86.4	
21F	Iridium 67				657	774	780	100.4	86.4	
21G	Iridum 68				657	775	778	100.4	86.4	
22A	*Columbia* (STS-90)	17 Apr	KSC	Shuttle	105,500	257	286	89.5	39.0	Searfoss, Altman, Linnehan, Hire, Williams, Buckey and Pawelczyk on Neurolab mission. Landed 3 May
23A	Globalstar 5	24 Apr	CC	Delta 2	450	1,258	1,265	110.7	52.0	Second launch of Globalstar cluster
23B	Globalstar 6				450	1,256	1,270	110.8	52.0	
23C	Globalstar 7				450	1,260	1,263	110.8	52.0	
23D	Globalstar 8				450	1,259	1,269	110.8	52.0	
24A	Nilesat 101	29 Apr	KO	Ar44P	1,840	geostationary orbit at 353° east				First satellite to be launched for an African country, Nilesat of Egypt
24B	B-Sat 1B				1,230	geostationary orbit at 110° east				Telecommunications satellite for B-SAT Corporation of Japan
25A	Cosmos 2350	29 Apr	TT	Proton-K	2,500	geostationary orbit at 80° east				Data relay satellite
26A	Iridium 69	2 May	TY	CZ-2C/SD	657	625	641	97.4	86.4	Fourteenth cluster of Iridium satellites
26B	Iridium 71				657	774	780	100.4	86.4	
27A	Cosmos 2351	7 May	PL	Molniya-M	1,250	525	39,829	717.7	63.0	Oko early warning satellite
28A	EchoStar 4	7 May	TT	Proton-K	2,900	geostationary orbit at 233° east				DBS for EchoStar Communications Corp. Operated over 212° east
29A	USA 139	9 May	CC	Titan 4B/ Centaur	4,500	geostationary orbit				Advanced Orion ELINT satellite
30A	Noaa 15	13 May	VAFB	Titan 23G	2,234	808	823	101.2	98.7	First of new generation meteorological satellite for NOAA
31A	Progress-M 39	14 May	TT	Soyuz-U	7,135	371	379	92.1	51.7	Mir cargo freighter
32A	Iridium 70	17 May	VAFB	Delta 2	657	775	780	100.4	86.4	Fifteenth cluster of Iridium satellites
32B	Iridium 72				657	773	780	100.4	86.4	
32C	Iridium 73				657	775	779	100.4	86.4	
32D	Iridium 74				657	774	778	100.4	86.4	
32E	Iridium 75				657	692	695	98.6	86.4	
33A	Zhongwei 1	30 May	XI	CZ-3B	2,984	35,761	84,991	2,860.7	1.3	China Star telecommunications satellite operated for China Orient Telecom Satellite Company. Eventually operated in eccentric geosynchronous orbit at 87.5° east
34A	*Discovery* (STS-91)	2 Jun	KSC	Shuttle	117,860	370	378	92.0	51.7	Precourt, Gorie, Chang-Diaz, Lawrence, Kavandi and Ryumin on ninth and last Mir visit. Returned 12 Jun
35A	Thor 3	10 Jun	CC	Delta 2	1,300	geostationary orbit at 358° east				Telecommunications satellite for Telenor, Norway
36A	Cosmos 2352	15 Jun	PL	Tsyklon	225	1,310	1,875	118.0	82.6	First of six Strela military store-dump
36B	Cosmos 2353				225	1,300	1,870	117.9	82.6	communications satellites. Failure in final
36C	Cosmos 2354				225	1,307	1,872	118.0	82.6	stage attitude system left satellites in eccentric
36D	Cosmos 2355				225	1,302	1,868	117.9	82.6	rather than near-circular orbits of 1,400-1,415 km
36E	Cosmos 2366				225	1,298	1,867	117.8	82.6	Strela 3 satellites launched into same orbit
36F	Cosmos 2367				225	1,294	1,863	117.7	82.6	plane as Gonets-D1, 4-6 and Cosmos 2337-2339 cluster
37A	Intelsat 805	18 Jun	CC	Atlas 2AS	3,524	geostationary at 304.5° east				Telecommunications satellite for Intelsat
38A	Cosmos 2358	24 Jun	PL	Soyuz-U	6,600	167	334	89.5	67.1	Yantar-4K2 Kobalt fourth-generation military reconnaissance satellite
39A	Cosmos 2359	25 Jun	TT	Soyuz-U	7,000	240	303	89.9	64.9	Yantar fifth-generation photo-reconnaissance satellite
40A	Molniya-3 49	1 Jul	P	Molniya-M	1,750	432	39,943	718.2	62.8	Telecommunications satellite co-planar with Molniya-3 39
41A	Nozomi	3 Jul	KA	M-5	536	1,421	516,130	22,646.8	25.4	Planet-B Mars spacecraft in earth-moon orbit with lunar fly-bys
42A	Tubsat N	7 Jul	BS	Shtil-1	8.5	401	777	96.5	78.9	Sensor and transceiver test platform. First launch from a submarine, located in Barents Sea at 69.3° north x 35.3° east
42B	Tubsat N 1				3	400	771	96.4	78.9	Store-forward transceiver communications
43A	Resurs-02 1	10 Jul	TT	Zenit 2	2,800	817	818	101.2	98.8	First Resurs-02 class. Belgian/German IRIS
43B	Fasat-Bravo				50	817	819	101.2	98.8	Chilean Air Force to replace Sich 1, 1995-46
43C	Tmsat				50	817	818	101.2	98.8	Thai remote sensing satellite sensor payloads
43D	Techsat 1B				50	816	818	101.2	98.8	Replaces failed Techsat 1B, March 1995
43E	Westpac				24	817	819	101.2	98.8	Australian-Russian geodetic satellite

Desig	Name	Date	Site	Vehicle	Mass (kg)	Perigee (km)	Apogee (km)	Period (min)	Inc (deg)	Notes
43F	Safir 2				60	815	819	101.2	98.8	Successor to Safir R1, 1994-97
44A	Sinosat 1	18 Jul	XI	CZ-3B	2,840	geostationary at 87.5° east				Telecommunications satellite for Sino Satellite Communications Company
45A	Cosmos 2360	28 Jul	TT	Zenit 2	3,250	848	854	101.9	71.0	Tselina-2 ELINT satellite
46A	Orbcomm FM 13	2 Aug	WI	L-1011/Pegasus XL + HAPS	43	813	825	101.3	45.0	Orbcomm Global LP communications
46B	Orbcomm FM 14				43	813	826	101.3	45.0	
46C	Orbcomm FM 15				43	813	826	101.3	45.0	
46D	Orbcomm FM 16				43	813	827	101.3	45.0	
46E	Orbcomm FM 20				43	816	828	101.3	45.0	
46F	Orbcomm FM 19				43	816	828	101.3	45.0	
46G	Orbcomm FM 18				43	815	828	101.3	45.0	
46H	Orbcomm FM 17				43	816	828	101.3	45.0	
47A	Soyuz-TM 28	13 Aug	TT	Soyuz-U	7,150	364	374	91.9	51.7	Padalka, Avdeyev, who remained aboard Mir, and Baturin, who returned to earth in Soyuz-TM 27
48A	Iridium 3	19 Aug	TY	CZ-2C/SD	657	776	780	100.4	86.4	Sixteenth Iridium cluster launched
48B	Iridium 76				657	775	779	100.4	86.4	
49A	ST 1	25 Aug	KO	AR44P	3,255	geostationary orbit at 88° east				Telecommunications satellite for Singapore Telecom and Chunghwa Telecom International
50A	Astra 2A	30 Aug	TT	Proton-K	3,500	geostationary orbit at 28° east				Broadcast satellite for SES, Luxembourg
51A	Iridium 82	8 Sep	VAFB	Delta 2	657	703	707	98.9	86.5	Seventeenth Iridium cluster launch
51B	Iridium 81				657	776	779	100.4	86.4	
51C	Iridium 80				657	776	779	100.4	86.4	
51D	Iridium 79				657	704	708	98.9	86.4	
51E	Iridium 77				657	704	708	98.9	86.4	
52A	PAS 7	16 Sep	KO	Ar44LP	3,838	geostationary at 68.5° east				Broadcast satellite for PanAmSat
53A	Orbcomm-FM 21	23 Sep	WI	L-1011/Pegasus XL + HAPS	43	811	822	101.2	45.0	Third cluster launch for Orbcomm Global LP
53B	Orbcomm-FM 22				43	810	818	101.2	45.0	
53C	Orbcomm-FM 23				43	812	821	101.2	45.0	
53D	Orbcomm-FM 24				43	812	821	101.2	45.0	
53E	Orbcomm-FM 25				43	809	819	101.2	45.0	
53F	Orbcomm-FM 26				43	809	819	101.2	45.0	
53G	Orbcomm-FM 27				43	809	819	101.2	45.0	
53H	Orbcomm-FM 28				43	810	819	101.2	45.0	
54A	Molniya-1 91	28 Sep	PL	Molniya-M	1,600	420	40,657	732.5	62.8	Telecommunications satellite launched to the same orbital plane as Molniya-1 86
55A	Stex	3 Oct	VAFB	Taurus	540	697	716	98.9	85.0	Space Technology Experiment satellite for National Reconnaissance Office
56A	Eutelsat W2	5 Oct	KO	Ar44L	2,950	geostationary orbit at 16° east				First of new widebeam Eutelsat telecommunications satellite
56B	Sirius 3				1,465	geostationary orbit at 5° east				Broadcast satellite for Nordic Television Co in Sweden
57A	Hot Bird 5	9 Oct	CC	Atlas 2A	2,900	geostationary orbit at 13° east				Telecommunications satellite for Eutelsat
58A	USA 140	20 Oct	CC	Atlas 2A	3,208	geostationary orbit at 186° east				UHF Follow-On (UFO) 9 for US Navy
59A	Maqsat 3	21 Oct	KO	Ar5	6,000	1,003	35 494	641.0	7.0	Eutelsat W communications satellite mock-up
60A	SCD 2	23 Oct	CC	Pegasus	115	744	768	99.9	25.0	Brazilian data collection satellite
61A	Deep Space 1	24 Oct	CC	Delta 2	486	heliocentric orbit				First of NASA's New Millenium Programme satellites; flew by asteroid 9969 Braille on 28 Jul 1999 at 15 km distance
61B	Sedsat 1				40	547	1,079	101.1	31.4	Student microsatellite with imaging camera
62A	Progress-M 40	25 Oct	TT	Soyuz-U	7,150	355	365	91.7	51.7	Cargo freighter to Mir
63A	AfriStar	28 Oct	KO	Ar44L	2,739	geostationary orbit at 21° east				Radio broadcast satellite for WorldSpaceCorp
63B	GE 5				1,698	geostationary orbit 281° east				Broadcast satellite for GE Americom
64A	*Discovery* (STS-95)	29 Oct	KSC	Shuttle	130,973	552	561	95.8	28.5	Brown, Lindsey, Robinson, Parazynski, Duque, Mukai, Glenn (the first US astronaut to orbit the earth) on life sciences mission. Returned 7 Nov
64B	Pansat				70	554	558	95.8	28.5	US Navy student telecommunications satellite
64C	Spartan 201-5				1,269	550	560	95.7	28.5	Reflight of Spartan 201-4. Retrieved 3 Nov
65A	PAS 8	4 Nov	TT	Proton-K	3,800	geostationary orbit 166° east				Broadcast satellite for PanAmSat
66A	Iridium 2	6 Nov	VAFB	Delta 2	657	566	579	96.1	86.1	Eighteenth Iridium cluster launch
66B	Iridium 86				657	517	533	96.1	86.1	
66C	Iridium 85				657	514	534	95.3	86.1	
66D	Iridium 84				657	702	706	98.8	86.1	
66E	Iridium 83				657	515	533	95.1	86.1	
67A	Zarya	20 Nov	TT	Proton-K	20,700	388	401	92.4	51.6	First element in the orbital assembly of the International Space Station
68A	Bonum 1	23 Nov	CC	Delta 2	1,425	geostationary orbit at 36° east				Telecommunications satellite owned by Russian Bonum-1 news group
69A	*Endeavour* (STS-88)	4 Dec	KSC	Shuttle	119,717	383	395	92.3	51.6	Cabana, Sturckow, Ross, Currie, Newman and Krikalev. Second element in the orbital assembly of the International Space Station, Unity module attached to Zarya. Landed 16 Dec
69B	SAC A				268	383	400	92.4	51.6	Argentine satellite
69C	MightySat 1				320	382	399	92.4	51.6	Developed by USAF Phillips Laboratory
69D	Slidewire				3.9	382	394	92.3	51.6	Carrier lost overboard from Endeavour
69E	Socket				1.5	382	394	92.3	51.6	Lost overboard during mission
69F	Unity				11,600	383	395	92.3	51.6	Interconnecting ISS module
70A	Satmex 5	6 Dec	KO	Ar42L	4,144	geostationary at 243.2° east				Telecommunications satellite owned by Satmex S.S. de C.V., Mexico
71A	Swas	6 Dec	VAFB	L-1011/ Pegasus-XL	283	637	653	97.6	69.9	Submillimetre Wave Astronomy Satellite
72A	Nadezhda 5	10 Dec	PL	Cosmos-3M	825	977	1,013	105.0	82.9	Tsikada system civilian navigation satellite
72B	Astrid 2				30	979	1,013	105.0	82.9	Swedish science payload
73A	MCO	11 Dec	CC	Delta 2	629	heliocentric orbit				NASA Mars Climate Orbiter due to arrive in Mars orbit 23 Sep 1999
74A	Iridium 11A	19 Dec	TY	CZ-2C/SD	657	688	713	98.8	86.5	Nineteenth Iridium cluster launch
74B	Iridium 20A				657	686	713	98.8	86.5	
75A	PAS 6B	22 Dec	KO	Ar42L	3,600	geostationary at 313° east				Broadcast telecommunications satellite for PanAmSat. Operated from 317° east
76A	Cosmos 2361	24 Dec	PL	Cosmos-3M	795	969	1,013	104.9	82.9	Parus military navigation satellite
77A	Cosmos 2362	30 Dec	TT	Proton-K	1,300	19,125	19,129	675.6	64.8	Three Uragan navigation satellites
77B	Cosmos 2362				1,300	19,126	19,129	675.6	64.8	
77C	Cosmos 2363				1,300	19,066	19,200	675.8	64.8	
1999										
01A	MPL + DS2	3 Jan	CC	Delta 2	583	heliocentric orbit				Mars Polar Lander + 2 DS penetrators, due to arrive at Mars 3 Dec
02A	Chunghua 1	27 Jan	CC	Athena-1	410	595	605	96.7	35.0	First satellite launched for Taiwan, a marine resources satellite
03A	Stardust	7 Feb	CC	Delta 2	385	heliocentric orbit				Fourth in NASA's Discovery series, to retrieve cometary material following 100 km fly-by of P/Comet Wild 2, returning to earth Jan 2006

Desig	Name	Date	Site	Vehicle	Mass (kg)	Perigee (km)	Apogee (km)	Period (min)	Inc (deg)	Notes
04A	Globalstar M36	9 Feb	TT	Soyuz-U/Ikar	450	1,337	1,356	112.6	52.0	Fourth cluster launch for Globalstar and first
04B	Globalstar M23				450	1,323	1,338	112.2	52.0	use of Soyuz-U/Ikar launcher
04C	Globalstar M38				450	1,343	1,404	113.2	52.0	
04D	Globalstar M40				450	1,005	1,032	105.5	52.0	
05A	Telstar 6	15 Feb	TT	Proton-K	3,650	geostationary orbit at 267° east				Loral Skynet high power telecommunications satellite
06A	JC-Sat	16 Feb	CC	Atlas-2AS	3,000	geostationary orbit at 120° east				Telecommunications satellite owned by Japan Satellite Systems
07A	Soyuz-TM 29	20 Feb	TT	Soyuz-U	7,120	346	364	91.6	51.7	Afanasyev, Hagnere and Bella to dock with Mir. Bella returned to earth with Padalka in Soyuz-TM 28
08A	Argos	23 Feb	VAFB	Delta 2	2,700	831	838	101.7	98.7	Joint-services advanced research and global observation satellite
08B	Ørsted				62	647	864	99.9	96.5	Danish magnetic and particle science satellite
08C	Sunsat				61	647	865	99.9	96.5	South Africa's first satellite carries earth imager and voice/ digital communications
09A	Arabsat 3A	26 Feb	KO	Ar44L	2,708	geostationary at 26° east				First of new generation Arab broadcast and telecommunications satellite
09B	Skynet 4E				1,490	geostationary at 6° east				UK military communications satellite
10A	Raduga-1 4	28 Feb	TT	Proton-K	2,000	geostationary				Globus military and governmental communications satellite
11A	Wire	5 Mar	VAFB	L-1011/ Pegasus-XL	270	539	594	96.0	97.5	Wide-Field Infra-Red Explorer to study the origin and evolution of galaxies
12A	Globalstar M22	15 Mar	TT	Soyuz-U/Ikar	450	1,409	1,417	114.1	52.0	Fifth cluster launch for Globalstar Ltd
12B	Globalstar M41				450	1,413	1,414	114.1	52.0	
12C	Globalstar M46				450	1,411	1,416	114.1	52.0	
12D	Globalstar M37				450	1,413	1,415	114.1	52.0	
13A	AsiaSat 3S	21 Mar	TT	Proton-K	3,465	geostationary orbit at 98° east				Telecommunications satellite for Asia Satellite Telecommunications Co, Hong Kong
14A	Sea Launch	28 Mar	PO	Zenit-3SL	4,500	639	36,064	645.0	1.2	HS 702 mass model in first test of Sea Launch Odyssey platform at 0° north × 154° west
15A	Progress-M 41	2 Apr	TT	Soyuz-U	7,180	339	355	91.5	51.7	Unmanned freighter carrying 2,438 kg cargo to Mir
15B	Sputnik 99				3	338	352	91.4	51.7	Amateur radio satellite released by Mir crew
16A	Insat 2E	2 Apr	KO	Ar42P	2,600	geostationary orbit at 83° east				Indian telecommunications and meteorology satellite
17A	USA 142	9 Apr	CC	Titan-4B/IUS	2,400	188	38,000			Defense Support Program19 satellite stranded in GTO by IUS burn timing failure
18A	Eutelsat-W 3	12 Apr	CC	Atlas-2AS	3,183	geostationary orbit at 7° east				Second wideband Eutelsat satellite
19A	Globalstar M19	15 Apr	TT	Soyuz-U/Ikar	450	902	944	103.5	51.9	Sixth cluster launch for Globalstar
19B	Globalstar M42				450	904	946	103.5	51.9	Partrnership Ltd
19C	Globalstar M44				450	931	977	103.5	51.9	
19D	Globalstar M45				450	1,412	1,415	114.1	52.0	
20A	Landsat 7	15 Apr	VAFB	Delta 2	2,170	685	702	98.6	98.2	NASA, NOAA, USGS remote sensing satellite
21B	Uosat 12	21 Apr	TT	Dniepr	320	649	652	97.7	64.6	Surrey Satellite Technology Ltd satellite with mobile radio experiment, GPS receiver and imaging cameras. Intl designation is correct following an officially incorrect designation of Sputnik 99 as 21A
22A	Abrixas	28 Apr	KY	Cosmos-3M	470	554	603	96.2	48.4	Broadband X-ray survey astronomy satellite
22B	Megsat-0				35	549	603	96.2	48.4	Italian technology development satellite
23A	USA 143	30 Apr	CC	Titan-4B/ Centaur	4,700	147?	740?	147.5	28.0	Milstar-2 DoD communications satellite stranded in wrong orbit by Centaur
24A	Orion 3	5 May	CC	Delta 3	4,300	422	1,317	102.3	29.1	Loral Orion Inc telecommunications satellite stranded in useless orbit by Delta second stage
25A	Fen Young-1 3	10 May	TY	CZ-4B	958	849	868	102.1	98.8	Third launch of FY-1 meteorological satellite
25B	Shi Jian 5				298	846	865	102.0	98.8	Test of new satellite bus
26A	Terriers	18 May	VAFB	Pegasus-XL + HSPS	123	542	554	95.6	97.7	NASA Student Explorer Demonstration Initiative satellite
26B	Mublcom				45	773	776	100.3	97.7	US Army mobile communications test satellite
27A	Nimiq 1	20 May	TT	Proton-K	3,646	geostationary at 269° east				Broadcast satellite for Telesat Canada
28A	USA 144	22 May	VAFB	Titan-4B						Classified payload
29A	Kitsat 3	26 May	SR	PSLV	107	711	737	99.3	98.4	Second operational PSLV launch with earth imaging satellite for South Korea
29B	DLR-Tubsat				45	716	737	99.3	98.4	High resolution earth observation satellite for DLR and University of Berlin
29C	Oceansat				1,048	718	739	99.4	98.4	ISRO ocean resources satellite (IRS-P4)
30A	*Discovery* (STS-96)	27 May	KSC	Shuttle	100,236	324	342	91.2	51.6	Rominger, Husband, Jernigan, Ochoa, Barry, Payette, and Tokarev on second Shuttle flight to ISS Unity/Zarya modules
30B	Starshine 1				160	380	397	97.3	51.6	US Navy/University of Utah atmospheric
31A	Globalstar M52	10 Jun	CC	Delta 2	450	1,360	1,380	113.1	52.0	Seventh cluster for Globalstar Partnership
31B	Globalstar M49				450	1,413	1,414	114.1	52.0	
31C	Globalstar M25				450	1,413	1,414	114.1	52.0	
31D	Globalstar M47				450	1,411	1,416	114.1	52.0	
32A	Iridium 14A	11 Jun	TY	CZ-2C/SD	667	709	713	99.0	86.5	Twentieth Iridium cluster launch
32B	Iridium 21A				667	746	752	99.8	86.4	
33A	Astra 1H	18 Jun	TT	Proton-K	3,728	geostationary at 19.2° east				Broadcast satellite for SES, Luxembourg
34A	Quikscat	20 Jun	VAFB	Titan-23G	870	280	813	95.6	98.6	NASA remote sensing technology satellite
						792	821	101.0	98.6	with in-orbit perigee boost for circularisation
35A	Fuse	24 Jun	CC	Delta 2	1,335	754	770	100.1	25.0	NASA Far Ultraviolet Spectroscopic Explorer

US Manned Flights (Mercury to Apollo-Soyuz)

Mission	Launch Date	Crew	Revs	Duration (d, h, m)
Mercury MR-3	05.05.61	Alan Shepard	Sub	00.00.15

1st American in space; Freedom 7 attained suborbital 187.5 km, 8,336 km/h

Mission	Launch Date	Crew	Revs	Duration (d, h, m)
Mercury MR-4	21.07.61	Virgil Grissom	Sub	00.00.16

Second suborbital flight, attained 190.4 km. Liberty Bell 7 sank

Mission	Launch Date	Crew	Revs	Duration (d, h, m)
Mercury MA-6	20.02.62	John Glenn	3	00.04.55

1st American in orbit, Friendship 7 attained 161 × 261 km. Telemetry falsely indicated heatshield unlatched

Mission	Launch Date	Crew	Revs	Duration (d, h, m)
Mercury MA-7	24.05.62	M Scott Carpenter	3	00.04.56

MA-6 repeat, Aurora 7 overshot landing site by 420 km

Mission	Launch Date	Crew	Revs	Duration (d, h, m)
Mercury MA-8	03.10.62	Walter Schirra	6	00.09.13

Engineering evaluation, conserving propellant. Sigma 7 landed 7.24 km from target

Mission	Launch Date	Crew	Revs	Duration (d, h, m)
Mercury MA-9	15.05.63	L Gordon Cooper	22	01.10.20

1st US 1-day mission; performed manual re-entry in Faith 7 after automatic stabilisation & control system failed, landing 6.44 km from recovery carrier. In-orbit mass 1,373 kg, carrying additional supplies for long flight

Mission	Launch Date	Crew	Revs	Duration (d, h, m)
Gemini 3	23.03.65	Virgil Grissom John Young	3	00.04.53

Test flight; first manned orbit changes; first computer, allowing onboard calculation of manoeuvres

Mission	Launch Date	Crew	Revs	Duration (d, h, m)
Gemini 4	03.06.65	James McDivitt Edward White	62	04.01.56

White made 1st US EVA (21 min); 1st US 4-day flight. Manual entry made after computer failed

Mission	Launch Date	Crew	Revs	Duration (d, h, m)
Gemini 5	21.08.65	L Gordon Cooper Charles Conrad	120	07.22.56

Endurance record (sufficient for return lunar mission)

Mission	Launch Date	Crew	Revs	Duration (d, h, m)
Gemini 7	04.12.65	Frank Borman James Lovell	206	13.18.35
Longest US flight for 8 yr, record until Soyuz 9. See G6				
Gemini 6	15.12.65	Walter Schirra Thomas Stafford	16	01.01.51
1st manned rendezvous, to within 1.8 m of Gemini 7 (as planned Agena was lost)				
Gemini 8	16.03.66	Neil Armstrong David Scott	6	00.10.41
1st space docking (with Agena); emergency re-entry after Gemini thruster jammed on and re-entry thrusters had to be activated to regain control				
Gemini 9	03.06.66	Thomas Stafford Eugene Cernan	45	03.00.21
127 min EVA by Cernan, rendezvous but no docking with target. Landed 800 m from recovery ship				
Gemini 10	18.07.66	John Young Michael Collins	43	02.22.47
Docked with Agena 10 and used engine to attain record 763 km; rendezvous with Agena 8, Collins retrieved package in 39 min EVA				
Gemini 11	12.09.66	Charles Conrad Richard Gordon	44	02.23.17
Agena engine used to attain record 1,369 km; Gordon 163 min SEVA/EVA. Gordon connected Gemini and Agena by tether; when undocked, Gemini initiated cartwheel motion, generating 0.00015 g. First automatic, computer-guided re-entry				
Gemini 12	11.11.66	James Lovell Eugene Aldrin	59	03.22.34
Agena docking; SEVA/EVA totalling 330 min by Aldrin. Automatic computer-guided re-entry				
Apollo 7	11.10.68	Walter Schirra Donn Eisele Walter Cunningham	163	10.20.09
CSM demonstration, 1st manned Apollo, 1st US 3-man flight				
Apollo 8	21.12.68	Frank Borman James Lovell William Anders	10 lunar	06.03.00
1st manned lunar mission, 1st manned Saturn 5				
Apollo 9 *Gumdrop* *Spider*	03.03.69	James McDivitt David Scott Russell Schweickart	151	10.01.01
1st manned Lunar Module (Earth orbit); 56 min Schweickart EVA tested lunar suit				
Apollo 10 *Charlie Brown* *Snoopy*	18.05.69	Thomas Stafford Eugene Cernan John Young	31 lunar	08.00.03
Descent to within 14 km of Moon in landing dress rehearsal. Holds manned speed record: 11.0825 km/s at atmospheric entry				
Apollo 11 *Columbia* *Eagle*	16.07.69	Neil Armstrong Eugene Aldrin Michael Collins	31 lunar	08.03.18
1st manned lunar landing; 151 min EVA, 21 kg samples, 21 h 36 min on Moon				
Apollo 12 *Yankee Clipper* *Intrepid*	14.11.69	Charles Conrad Alan Bean Richard Gordon	49 lunar	10.04.36
2 lunar EVAs totalling 465 min; 34 kg samples; 31 h 31 min on Moon				
Apollo 13 *Odyssey* *Aquarius*	11.04.70	James Lovell John Swigert Fred Haise		05.22.55
Mission aborted following Service Module oxygen tank explosion; circumlunar return. Holds manned altitude record: 400,187 km above Earth's surface				
Apollo 14 *Kitty Hawk* *Antares*	31.01.71	Alan Shepard Edgar Mitchell Stuart Roosa	34 lunar	09.00.42
2 lunar EVAs, total 563 min; 43 kg samples; 33 h 31 min on Moon				
Apollo 15 *Endeavour* *Falcon*	26.07.71	David Scott James Irwin Alfred Worden	74 lunar	12.07.12
3 lunar EVAs, total 19 h 8 min; 77 kg samples; 1st lunar rover; 66 h 54 min on Moon. 38 min Worden EVA. Subsatellite released				
Apollo 16 *Casper* *Orion*	16.04.72	John Young Charles Duke Tom Mattingly	64 lunar	11.01.51
3 lunar EVAs, total 20 h 14 min; 94 kg samples; 71 h 14 min on Moon. Subsatellite released				
Apollo 17 *America* *Challenger*	07.12.72	Eugene Cernan Jack Schmitt Ron Evans	75 lunar	12.13.51
3 lunar EVAs, total 22 h 4 min; 110 kg samples; 74 h 59 min on Moon. 66 min Evans EVA				
Skylab 1	14.05.73	Unmanned	34,981	
Space station, suffered launch damage. Hosted 3 crews. Re-entered 11.07.79				
Skylab 2	25.05.73	Charles Conrad Paul Weitz Joe Kerwin	404	28.00.50
Endurance record. Repaired Skylab, deploying 6.7 × 7.3 m parasol through 20 cm science airlock on 7.6 m boom and releasing jammed solar wing. 81 h of solar observations and 11 Earth resources passes (9,850 images); 2 EVAs + SEVA				
Skylab 3	28.07.73	Alan Bean Owen Garriott Jack Lousma	858	59.11.09
Endurance record. Erected new sunshade, tested EVA manoeuvring units, 305 h solar observations (24,940 frames), 39 Earth resources passes, welding & materials processing experiments; 3 EVAs				
Skylab 4	16.11.73	Gerald Carr Edward Gibson William Pogue	1214	84.01.15
Endurance record. Observed Comet Kohoutek (2,500 images) and returned 73,400 solar + 19,400 Earth images; 4 EVAs				
ASTP	15.07.75	Thomas Stafford Vance Brand Deke Slayton	138	09.01.28
1st US-Soviet joint flight, 46 h 44 min docked over two periods for activities with Leonov/Kubasov in Soyuz 19. 28 US expts (including 5 with Soyuz). Docked mass 1,500 kg (Apollo 12,700 kg). US cost $218 million				

The above manned flights totalled 189,985,890 km (118,077,569 miles).

US Space Shuttle Flights

Flight	Launch date Orbiter	Crew	Duration (d, h, m) Incl
01 **STS-1**	12.04.81 Columbia	Young, Crippen	02.06.21 40.4°
Shuttle debut mission; some thermal tiles lost			
02 **STS-2**	12.11.81 Columbia	Engle, Truly	02.06.13 38.0°
Tested RMS arm; 5-day mission halved by fuel cell fault			
03 **STS-3**	22.03.82 Columbia	Lousma, Fullerton	08.00.05 38.0°
Third test; White Sands landing delayed 1 day by storm			
04 **STS-4**	27.06.82 Columbia	Mattingly, Hartsfield	07.01.10 28.45°
Final test; first concrete runway landing. SRBs lost. Cirris DoD payload failure			
05 **STS-5**	11.11.82 Columbia	Brand, Overmyer, Allen, Lenoir	05.02.14 28.45°
First operational flight. SBS 3/Anik C3 deployed. EVA cancelled			
06 **STS-6**	05.04.83 Challenger	Weitz, Bobko, Peterson, Musgrave	05.00.23 28.45°
Challenger debut. TDRS 1 deployed on IUS. First Shuttle EVA			
07 **STS-7**	18.06.83 Challenger	Crippen, Hauck, Fabian, Ride, Thagard	06.02.24 28.45°
First US woman (Ride). First deploy/retrieve (SPAS). Anik C2/Palapa B deployed			
08 **STS-8**	30.08.83 Challenger	Truly, Brandenstein, Bluford, Gardner, W Thornton	06.01.07 28.45°
First night launch/landing. India's Insat 1B telecom sat deployed			
09 **STS-9**	28.11.83 Columbia	Young, Shaw, Garriott, Parker, Lichtenberg, Merbold	10.07.47 57.0°
Spacelab 1 demonstration; 90% science achieved. First 6-man flight			
10 **41B**	03.02.84 Challenger	Brand, Gibson, McCandless, Stewart, McNair	07.23.17 28.45°
First MMU tests + SMM EVA repair rehearsal. Westar 6/Palapa B2 lost. 1st KSC landing			
11 **41C**	06.04.84 Challenger	Crippen, Scobee, G Nelson, van Hoften, Hart	06.23.40 28.45°
LDEF deploy. First satellite retrieval/repair (Solar Max)			
12 **41D**	30.08.84 Discovery	Hartsfield, Coats, Resnik, Hawley, Mullane, C Walker	06.00.56 28.45°
Discovery debut. SBS 4/Leasat 2/Telstar 3C deploy. Solar panel deploy test. First commercial Payload Specialist			
13 **41G**	05.10.84 Challenger	Crippen, McBride, Sullivan, Ride, Leestma, Garneau, Scully-Power	08.05.23 57.0°
OSTA-3, ERBS, Large Format Camera. Firsts: 7 crew, US woman EVA, 2-woman flight. 2nd KSC landing			
14 **51A**	08.11.84 Discovery	Hauck, D Walker, A Fisher, D Gardner, Allen	07.23.45 28.45°
First sat recovery/return: Palapa B2/Westar 6. Anik D/Leasat 1 deployed. 3rd KSC landing			

Flight	Launch date	Orbiter	Crew	Duration (d, h, m)	Incl	Notes
15 51C	24.01.85	Discovery	Mattingly, Shriver, Onizuka, Buchli, Payton	03.01.33	28.45°	*First dedicated DoD flight: Magnum elint on IUS. 100th manned orbital mission. 4th KSC landing*
16 51D	12.04.85	Discovery	Bobko, Williams, Seddon, Hoffman, Griggs, C Walker, Garn	06.23.56	28.45°	*Anik C/Leasat 3 deployed; EVA failed to repair Leasat 3. 5th KSC landing*
17 51B	29.04.85	Challenger	Overmyer, Gregory, Lind, Thagard, W Thornton, van den Berg, Wang	07.00.08	57.0°	*Spacelab 3 science mission; released small Nusat satellite*
18 51G	17.06.85	Discovery	Brandenstein, Creighton, Lucid, Fabian, Nagel, Baudry, Al Saud	07.01.38	28.45°	*Spartan deploy/retrieve; Morelos/Arabsat/Telstar 3D deployments*
19 51F	29.07.85	Challenger	Fullerton, Bridges, Musgrave, Henize, England, Acton, Bartoe	07.22.45	49.5°	*Spacelab 2 astronomy/science mission (no pressurised module)*
20 51I	27.08.85	Discovery	Engle, Covey, van Hoften, Lounge, W Fisher	07.02.18	28.45°	*MSL; Aussat/ASC/Leasat 4 deployed; Leasat 3 EVA repair*
21 51J	03.10.85	Atlantis	Bobko, Grabe, Hilmers, Stewart, Pailes	04.01.44	28.5°	*DoD DSCS 3 pair on IUS. Atlantis debut*
22 61A	30.10.85	Challenger	Hartfield, Nagel, Dunbar, Buchli, Bluford, Messerschmid, Furrer, Ockels	07.00.44	57.0°	*Spacelab D1: First foreign (West German) Shuttle flight; First 8-man crew*
23 61B	27.11.85	Atlantis	Shaw, O'Connor, Cleave, Ross, Spring, C Walker, Vela	06.21.05	28.45°	*Morelos B/Satcom K2/Aussat 2 deploy, Ease/Access First Space Station assembly demo*
24 61C	12.01.86	Columbia	Gibson, Bolden, G Nelson, Hawley, Chang-Diaz, Cenker, W Nelson	06.02.04	28.45°	*Satcom K1 deploy; SDI experiments*
25 51L	28.01.86	Challenger	Scobee, Smith, Resnik, McNair, Onizuka, McAuliffe, Jarvis	00.00.01		*Vehicle exploded after 73 s; crew killed*
26 STS-26	29.09.88	Discovery	Hauck, Covey, Hilmers, Lounge, Nelson	04.01.00	28.5°	*TDRS 3 deployed on IUS stage; 11 mid-deck secondary payloads*
27 STS-27	02.12.88	Atlantis	R Gibson, G Gardner, Mullane, Ross, Shepherd	04.09.06	57.0°	*First DoD Lacrosse imaging radar satellite for all-weather day/night reconnaissance*
28 STS-29	13.03.89	Discovery	Coats, Blaha, Buchli, Springer, Bagian	04.23.39	28.45°	*TDRS 4 on IUS; five mid-decK + two cargo bay secondary payloads*
29 STS-30	04.05.89	Atlantis	D Walker, Grabe, Thagard, Cleave, Lee	04.00.58	28.85°	*Magellan Venus orbiter on IUS stage; arrived Venus Aug 1990*
30 STS-28	08.08.89	Columbia	Shaw, Richards, Leestma, Adamson, Brown	05.01.00	57.00°	*Deployed military satellite into Molniya-type orbit for relaying imaging reconnaissance satellite imagery? Flight similar to STS-53*
31 STS-34	18.10.89	Atlantis	Williams, McCulley, Lucid, Baker, Chang-Diaz	04.23.39	34.30°	*Galileo Jupiter orbiter on IUS stage; arrive Jupiter Dec 1995*
32 STS-33	23.11.89	Discovery	Gregory, Blaha, Musgrave, K Thornton, Carter	05.00.07	28.45°	*DoD mission; deployed second Magnum signal intelligence satellite?*
33 STS-32	09.01.90	Columbia	Brandenstein, Wetherbee, Dunbar, Low, Ivins	10.21.01	28.50°	*Deployed USN Leasat 5 telecom satellite, retrieved Long Duration Exposure Facility*
34 STS-36	28.02.90	Atlantis	Creighton, Casper, Hilmers, Mullane, Thuot	04.10.18	62.00°	*DoD mission; deployed AFP-731 reconnaissance satellite*
35 STS-31	24.04.90	Discovery	Shriver, Bolden, Hawley, McCandless, Sullivan	05.01.16	28.45°	*Deployed Hubble Space Telescope; Shuttle altitude record (619 km)*
36 STS-41	06.10.90	Discovery	Richards, Cabana, Shepherd, Melnick, Akers	04.02.10	28.45°	*Ulysses solar probe on IUS stage*
37 STS-38	15.11.90	Atlantis	Covey, Culbertson, Springer, Meade, Gemar	04.21.54	28.45°	*DoD mission; deployed third Magnum signal intelligence satellite? 6th KSC landing*
38 STS-35	02.12.90	Columbia	Brand, G Gardner, Lounge, Hoffman, Parker, Parise, Durrance	08.23.05	28.45°	*Carried Astro-1/BBXRT-1 ultraviolet/X-ray telescopes*
39 STS-37	05.04.91	Atlantis	Nagel, Cameron, Ross, Godwin, Apt	05.23.33	28.45°	*Deployed Gamma Ray Observatory; 2 EVAs (first since 1985)*
40 STS-39	28.04.91	Discovery	Coats, Hammond, Harbaugh, McMonagle, Bluford, Veach, Hieb	08.07.22	57.0°	*DoD AFP-675/IBSS IR Background Signature Survey platforms. 7th KSC landing*
41 STS-40	05.06.91	Columbia	O'Connor, Gutierrez, Bagian, Jernigan, Seddon, Gaffney, Hughes-Fulford	09.02.14	39.0°	*Spacelab Life Sciences (SLS-1); first dedicated life sciences research; 12 GAS canisters*
42 STS-43	02.08.91	Atlantis	Blaha, Baker, Adamson, Low, Lucid	08.21.21	28.45°	*Deployment of TDRS 5 on IUS stage; Lucid first woman to make 3 flights. 8th KSC landing*
43 STS-48	12.09.91	Discovery	Creighton, Reightler, Buchli, Brown, Gemar	05.08.28	57.0°	*Deployment of UARS Upper Atmosphere Research Satellite*
44 STS-44	24.11.91	Atlantis	Gregory, Henricks, Jim Voss, Musgrave, Runco, Hennen	06.22.51	28.5°	*Unclassified DoD deployment of DSP early warning satellite on IUS*
45 STS-42	22.01.92	Discovery	Grabe, Oswald, Thagard, Readdy, Hilmers, Bondar, Merbold	08.01.15	57.0°	*Spacelab International Microgravity Lab (IML-1)*
46 STS-45	24.03.92	Atlantis	Bolden, Duffy, Foale, Leestma, Sullivan, Lichtenberg, Frimout	08.22.09	57.0°	*Atmospheric Lab for Applications & Science (Atlas-1). 9th KSC landing*
47 STS-49	07.05.92	Endeavour	Brandenstein, Chilton, Melnick, Thuot, Hieb, K Thornton, Akers	08.21.18	28.35°	*Endeavour debut; Intelsat 603 recovery/repair. 4 EVAs by 4 crew, total 60.1 man-h (including 1st 3-man)*
48 STS-50	25.06.92	Columbia	Richards, Bowersox, Dunbar, Baker, Meade, DeLucas, Trinh	13.19.31	28.5°	*US Microgravity Lab (USML 1) Spacelab. 1st EDO; Shuttle duration record. 10th KSC landing*
49 STS-46	31.07.92	Atlantis	Shriver, Allen, Chang-Diaz, Ivins, Hoffman, Nicollier, Malerba	07.23.15	28.5°	*Deployment of ESA's Eureca 1; deployment/retrieval of Italy's Tethered Satellite. 11th KSC landing*
50 STS-47	12.09.92	Endeavour	Gibson, Brown, Lee, Apt, Davis, Jemison, Mohri	07.22.30	57.0°	*Japan-sponsored Spacelab-J: 43 materials/life sciences expts. 12th KSC landing*
51 STS-52	22.10.92	Columbia	Wetherbee, Baker, Veach, Shepherd, Jernigan, MacLean	09.20.56	28.45°	*Deployment of Italy's Lageos 2. USMP 1 US Microgravity Payload. 13th KSC landing*
52 STS-53	02.12.92	Discovery	D Walker, Cabana, Bluford, Jim Voss, Clifford	07.07.19	57.00°	*Deployment of DoD 1 satellite, purpose unknown, into Molniya-type orbit possibly to relay reconnaissance satellite imagery. Flight similar to STS-28*
53 STS-54	13.01.93	Endeavour	Casper, McMonagle, Runco, Harbaugh, Helms	05.23.38	28.45°	*Deployment of TDRS 6 on IUS. 268 min EVA. 14th KSC landing*
54 STS-56	08.04.93	Discovery	Cameron, Oswald, Foale, Cockrell, Ochoa	09.06.08	57.00°	*Atmospheric Lab for Applications & Science (Atlas 2); Spartan 201 deploy/retrieve. 15th KSC landing*
55 STS-55	26.04.93	Columbia	Nagel, Henricks, Ross, Precourt, Harris, Walter, Schlegel	09.23.40	28.46°	*Spacelab D2: 90 materials/life sciences, Earth observation & astronomy expts. Shuttle programme exceeded 1 yr aggregate flight time*
56 STS-57	21.06.93	Endeavour	Grabe, Duffy, Low, Jan Voss, Sherlock, Wisoff	09.23.45	28.45°	*Eureca 1 retrieval; Spacehab 1 with 21 expts. Generic EVA. 16th KSC landing*
57 STS-51	12.09.93	Discovery	Culbertson, Readdy, Bursch, Newman, Walz	09.20.11	28.45°	*ACTS deploy, German Orfeus-SPAS deploy/retrieve, generic EVA. 17th KSC landing (1st night)*
58 STS-58	18.10.93	Columbia	Blaha, Searfoss, Seddon, Lucid, McArthur, Wolf, Fettman	14.00.13	39.0°	*Spacelab Life Sciences (SLS-2) dedicated life sciences research; 2nd EDO (record duration)*

Flight	Launch date Orbiter	Crew	Duration (d, h, m) Incl	Notes
59 **STS-61**	02.12.93 Endeavour	Covey, Bowersox, Musgrave, Akers, Hoffman, K Thornton, Nicollier	10.19.59 28.47°	*First Hubble servicing mission; 5 EVAs for 4 crew totalled 70.93 man-h. 18th KSC landing*
60 **STS-60**	03.02.94 Discovery	Bolden, Reightler, Davis, Sega, Chang-Diaz, Krikalev	08.07.10 57.0°	*Spacehab 2; employed Wake Shield Facility; first Russian aboard US craft. 19th KSC landing*
61 **STS-62**	04.03.94 Columbia	Casper, Allen, Thuot, Gemar, Ivins	13.23.17 39.0°	*USMP 2 US Microgravity Payload; OAST 2; 3rd EDO. 20th KSC landing*
62 **STS-59**	09.04.94 Endeavour	Gutierrez, Chilton, Godwin, Jones, Apt, Clifford	11.05.49 57.0°	*Space Radar Lab 1 mapped 20% Earth, 133 h data. Endeavour made record 412 manoeuvres*
63 **STS-65**	8.07.94 Columbia	Cabana, Halsell, Hieb, Chiao, D Thomas, Walz, Mukai	14.17.55 28.5°	*Spacelab International Microgravity Lab IML-2; 4th EDO (record duration). 21st KSC landing*
64 **STS-64**	09.09.94 Discovery	Richards, Hammond, Helms, Lee, Linenger, Meade	10.22.50 57.0°	*LITE-1 Lidar In-space Technology Expt; deploy/retrieve Spartan 201-02; EVA*
65 **STS-68**	30.09.94 Endeavour	Baker, Wilcutt, Jones, Bursch, Smith, Wisoff	11.05.46 57.0°	*Space Radar Lab SRL-2*
66 **STS-66**	03.11.94 Atlantis	McMonagle, Brown, Ochoa, Parazynski, Tanner, Clervoy	10.22.34 57.0°	*Atmospheric Lab for Applications & Science (Atlas-3); deploy/retrieve Germany's Crista-SPAS 1*
67 **STS-63**	03.02.95 Discovery	Wetherbee, Collins, Foale, Harris, Jan Voss, Titov	08.06.28 51.6°	*Rendezvous with Mir; Spacehab 3; deploy/retrieve Spartan 204. 1st female pilot (Collins); 2nd Russian aboard US craft (Titov); generic EVA. 22nd KSC landing*
68 **STS-67**	02.03.95 Endeavour	Oswald, W Gregory, Jernigan, Grunsfeld, Lawrence, Durrance, Parise	16.15.10 28.5°	*Astro-2 UV telescope array; 5th EDO (record duration)*
69 **STS-71**	27.06.95 Atlantis	Gibson, Precourt, Baker, Harbaugh, Dunbar, Solovyov, Budarin	09.19.22 51.6°	*1st docking with Mir, delivering station's 19th main crew (Solovyov/Budarin), returning with 18th crew (Dezhurov/Strekalov) + Soyuz-TM 21's Dr Norman Thagard (aboard since Mar 1995). Spacelab carried. 23rd KSC landing*
70 **STS-70**	13.07.95 Discovery	Henricks, Kregel, Sherlock, D Thomas, Weber	08.22.20 28.5°	*Deploy TDRS 7 on IUS. 24th KSC landing*
71 **STS-69**	7.09.95 Endeavour	D Walker, Cockrell, Jim Voss, Newman, Gernhardt	10.20.29 28.45°	*Deploy/retrieve WSF 2 Wake Shield Facility; deploy/retrieve Spartan 201-03; generic EVA. 25th KSC landing*
72 **STS-73**	20.10.95 Columbia	Bowersox, Rominger, K Thornton, Coleman, Lopez-Alegria, Leslie, Sacco	15.21.52 39.0°	*Spacelab USML 2 US Microgravity Lab; 6th EDO; 26th KSC landing*
73 **STS-74**	12.11.95 Atlantis	Cameron, Halsell, McArthur, Ross, Hadfield	08.04.31 51.64°	*#2 Mir docking. Delivered docking unit for future Shuttle missions and new solar arrays. 27th KSC landing*
74 **STS-72**	11.01.96 Endeavour	Duffy, Jett, Barry, Chiao, Scott, Wakata	08.22.01 28.45°	*Retrieve Japan's SFU Space Flyer Unit; deploy/retrieve Spartan OAST-Flyer; 2 generic EVAs. 28th KSC landing*
75 **STS-75**	22.02.96 Columbia	Allen, Horowitz, Chang-Diaz, Cheli, Hoffman, Nicollier, Guidoni	15.17.40 28.45°	*USMP 3 US Microgravity Payload; deploy/retrieve Italy's TSS-1R tethered satellite (TSS lost: tether snapped); 7th EDO. 29th KSC landing*
76 **STS-76**	22.03.96 Atlantis	Chilton, Searfoss, Lucid, Godwin, Clifford, Sega	09.05.16 51.6°	*#3 Mir docking, with Spacehab short module. Delivered logistics and Lucid for Mir stay (returning STS-79). Godwin/Clifford EVA mounted expts on Mir's docking module*
77 **STS-77**	19.05.96 Endeavour	Casper, Brown, Bursch, Runco, A Thomas, Garneau	10.00.39 28.5°	*Spacehab 4; deploy/retrieve Spartan 207 (inflatable antenna expt); deploy PAMS Passive Aerodynamically Stabilised Magnetically Damped Satellite. 30th KSC landing*
78 **STS-78**	20.06.96 Columbia	Henricks, Kregel, Helms, Linnehan, Brady, Favier, Thirsk	16.21.48 39.0°	*Life & Microgravity Science Spacelab; 8th EDO. 31st KSC landing*
79 **STS-79**	16.09.96 Endeavour	Readdy, Wilcutt, Apt, Akers Walz, Blaha (up), Lucid (down)	10.03.18 51.6°	*#4 Mir docking with Spacehab double module. Delivered supplies to Mir, left Blaha on board for medium-length stay, returned with Lucid (mission duraction 188.04.00) from STS-76*
80 **STS-80**	19.11.96 Columbia	Cockrell, Rominger, Jernigan, Jones, Musgrave	17.15.53 28.5°	*Deployment and retrievel of Wake Shield Facility 3 and ORFEUS-SPAS 2; 9th EDO; planned EVAs by Jernigan and Jones cancelled*
81 **STS-81**	12.01.97 Atlantis	Baker, Jett, Wilsoff, Grunsfeld, Ivins, Linenger (up), Blaha (down)	10.04.56 51.6°	*#5 Mir docking with Spacehab double module. Delivered supplies to Mir, left Linenger on board for medium-length stay, returned with Blaha from STS-79 (mission duration 128.05.27)*
82 **STS-82**	11.02.97 Discovery	Bowersox, Horowitz, Tanner Hawley, Harbaugh, Lee, Smith	9.23.37 28.5°	*Second servicing mission for the Hubble Space Telescope: new instruments installed, older ones returned to Earth, repairs to insulation; five EVAs completed, Lee/Smith 3, Harbaugh/Tanner 2*
83 **STS-83**	04.04.97 Columbia	Halsell, Still, Voss, Gernhardt Thomas, Crounch, Linteris	3.23.13 28.5°	*Microgravity Science Laboratory 1 primary payload for 10th EDO: mission curtailed due to fuel cell problems and re-flown as STS-94*
84 **STS-84**	15.05.97 Atlantis	Precourt, Collins, Clervoy, Noriega Lu, Kondakova, Foale (up), Linenger (down)	9.05.21 51.6°	*#6 Mir docking mission with Spacehab-DM module. Delivered supplies to Mir, left Foale on board for medium-length stay, returned with Linenger from STS-81 (mission duration 132.03.50)*
85 **STS-94**	01.07.97 Columbia	Halsell, Still, Voss, Gernhardt, Thomas, Crouch, Linteris	15.16.45 28.5°	*Reflight of STS-83, Microgravity Science Laboratory 1 primary payload for 11th EDO: fastest return-to-orbit for the crewmembers*
86 **STS-85**	07.08.97 Discovery	Brown, Rominger, Davis, Curbeam, Robinson, Tryggvason	11.20.28 57.0°	*CRSTA-SPAS deployment, test of Japanese remote manipulator system*
87 **STS-86**	25.09.97 Atlantis	Wetherbee, Bloomfield, Titov, Parazynski, Chretien, Lawrence, Wolf (up), Foale (down)	10.19.22 51.6°	*#7 Mir docking; Lawrence originally planned to remain aboard Mir but replaced by Wolf because her height precludes use of Russian EVA suit. EVA with Parazynski and Titov, first to involve non-US crewmember*
88 **STS-87**	19.11.97 Columbia	Kregel, Lindsey, Chawla, Scott, Doi , Kadenyuk	15.16.35 29.45°	*US microgravity payload #4, deployment and retrieval of SPARTAN 201-04. 12th EDO. Scott and Doi performed two EVAs, 1 to manually capture SPARTAN*
89 **STS-89**	22.01-98 Endeavour	Wilcutt, Edwards, Reilly, Anderson, Dunbar, Sharipov, Thomas (up) and Wolf (down)	08.19.48 51.6°	*#8 Mir docking. Thomas replaced Wolf*
90 **STS-90**	17.04.98 Columbia	Searfoss, Altman, Hire, Linnehan, Williams, Buckey and Pawelczyk	15.21.51 39.0°	*US Neurolab mission. 13th and last planned EDO before ISS operations for long duration missions*
91 **STS-91**	02.06.98 Discovery	Precourt, Gorie, Chang-Diaz, Lawrence, Kavandi, Ryumin and Thomas (down)	09.19.55 51.6°	*#9 and last Mir docking, first with Discovery. Cosmonaut Ryumin conducted Mir inspection to determine suitability for sustained operations. Thomas returned to Earth*
92 **STS-95**	29.10.98 Discovery	Brown, Lindsey, Parazynski, Robinson, Duque, Mukai and Glenn	08.21.45 28.45°	*Life sciences mission including world's oldest astronaut, John Glenn age 77. Deployed and retrieved SPARTAN 201-05 and tested Hubble Space Telescope orbital systems test platform*
93 **STS-88**	04.12.98 Endeavour	Cabana, Sturckow, Ross, Currie, Newman and Krikalev	11.19.53 51.6°	*Carried Unity module to Zarya in first docking to begin assembly of ISS. Ross and Newman made three EVAs to attach cables and deploy fixtures to ISS exterior*
94 **STS-96**	27.05.99 Discovery	Rominger, Husband, Jernigan, Ochoa, Barry, Payette, Tokarev	09.19.14 51.6°	*First Shuttle ISS servicing mission to docked Zarya/Unity modules. Released Starshine satellite*

FUTURE MISSIONS (as of June 1999)

Flight	Launch date Orbiter	Crew	Duration (d, h, m) Incl	Notes
95 **STS-93**	20.07.99 Columbia	Collins, Ashby, Hawley, Coleman, Tognini	04.23.00 28.45°	*Launch of Chandra X-ray observatory*
96 **STS-99**	16.09.99 Endeavour	Kregel, Gorie, Kavandi, Voss, Mohri, Thiele	11.04.00 57.0°	*Shuttle Radar Topography Mission*
97 **STS-103**	14.10.99 Discovery	Brown, Kelly, Smith, Foale, Grunsfeld, Nicollier, Clervoy	09.21.00 28.45°	*Third Hubble Space Telescope servicing mission*
98 **STS-101**	02.12.99 Atlantis	Halsell, Horowitz, Weber, Tsang Lu, Williams, Malenchenko, Morukov	11.00.00 51.6°	*Shuttle's third ISS Flight (2A.2) bringing logistical supplies. First crew to enter Zvezda Service Module scheduled for Russian launch November 1999*
99 **STS-92**	24.02.00 Discovery	Duffy, Melroy, Wakata, Chiao, Wisoff, Lopez-Algeria, McArthur	10.00.00 51.6°	*Shuttle's fourth ISS flight (3A) carrying integrated truss structure (ITS) Z1, pressurised mating adapter — 3, Ku-band communications system and control moment gyros (CMGs)*

Flight	Launch date Orbiter	Crew	Duration (d, h, m) Incl
100 **STS-97**	23.03.00 Endeavour	Jett, Bloomfield, Tanner, Noriega, Garneau	09.00.00 51.6°
Shuttle's fifth ISS flight (4A) carrying integrated truss structure P6, photovoltaic module and radiators			
101 **STS-98**	20.04.00 Atlantis	Cockrell, Polansky, Lee, Jones, Ivins	10.00.00 51.6°
Shuttle's sixth ISS flight (5A) carrying US Laboratory Module Destiny to orbit			
102 **STS-102**	29.06.00 Discovery	Hadfield, Curbeam, Voss, Helms, Usachev, Shepherd, Gidzenko, Krikalev	09.00.00 51.6°
Shuttle's seventh ISS flight (5A.1) on logistics and resupply mission carrying lab outfitting equipment, Leonardo Multi-Purpose Logistics Module and equipment racks			

*rev on which landing occurred (equals completed revs for Kennedy Space Center landings). Durations are in days/hours/minutes. **Shuttle Orbiters:** *Challenger* 099, *Enterprise* 101, *Columbia* 102, *Discovery* 103, *Atlantis* 104, *Endeavour* 105. EDO = Extended Duration Orbiter

CIS/USSR Manned Flights

Mission	Launch Date	Crew	Duration (d, h, m)
Vostok 1	12.04.61	Yuri Gagarin	00.01.48
1st manned spaceflight, single rev			
Vostok 2	06.08.61	Gherman Titov	01.01.18
1st full day in space			
Vostok 3	11.08.62	Andrian Nikolayev	03.22.22
1st double flight (with Vostok 4)			
Vostok 4	12.08.62	Pavel Popovich	02.22.57
1st double flight (with Vostok 3)			
Vostok 5	14.06.63	Valeri Bykovsky	04.23.06
2nd double flight (with Vostok 6)			
Vostok 6	16.06.63	Valentina Tereshkova	02.22.50
1st woman in space; joint flight with Vostok 5			
Voskhod 1	12.10.64	Vladimir Komarov Konstantin Feoktistov Boris Yegorov	01.00.17
Modified Vostok; 1st 3-man flight			
Voskhod 2	18.03.65	Pavel Belyayev Alexei Leonov	01.02.02
Modified Vostok; 1st EVA (Leonov) via inflatable airlock			
Soyuz 1	23.04.67	Vladimir Komarov	01.02.48
Komarov killed after descent parachute snarled			
Soyuz 3	26.10.68	Georgi Beregovoi	03.22.51
Rendezvous with unmanned Soyuz 2			
Soyuz 4	14.01.69	Vladimir Shatalov	02.23.21
1st docking of 2 manned craft; Khrunov/Yeliseyev returned in Soyuz 4			
Soyuz 5	15.01.69	Boris Volynov Yevgeni Khrunov Alexei Yeliseyev	03.00.54
1st manned docking; Khrunov/Yeliseyev transferred by EVA to Soyuz 4			
Soyuz 6	11.10.69	Georgi Shonin Valeri Kubasov	04.22.42
Rendezvous with Soyuz 7/8 (intended Soyuz 7 docking); welding experiment			
Soyuz 7	12.10.69	Anatoli Filipchenko Vladislav Volkov Viktor Gorbatko	04.22.41
Triple rendezvous with Soyuz 6/8 (intended Soyuz 6 docking)			
Soyuz 8	13.10.69	Vladimir Shatalov Alexei Yeliseyev	04.22.51
Triple rendezvous with Soyuz 6/7 (remains only triple flight)			
Soyuz 9	01.06.70	Andrian Nikolayev Vitali Sevastyanov	17.16.59
Endurance record for solo craft remains			
Salyut 1	19.04.71	Unmanned space station	–
First space station; re-entered 11 Oct 1971			
Soyuz 10	23.04.71	Vladimir Shatalov Alexei Yeliseyev Nikolai Rukavishnikov	01.23.46
Docked twice with Salyut 1 but no boarding made			
Soyuz 11	06.06.71	Georgi Dobrovolsky Vladislav Volkov Viktor Patsayev	23.18.22
23 days aboard Salyut 1; Soyuz pressure loss killed crew before re-entry			
Soyuz 12	27.09.73	Vasili Lazarev Oleg Makarov	01.23.16
Tested post-Soyuz 11 modifications			
Soyuz 13	18.12.73	Pyotr Klimuk Valentin Lebedev	07.20.55
Solo flight with Orion ultraviolet telescope			
Salyut 3	25.06.74	Unmanned space station	–
2nd station, occupied by Soyuz 14; re-entered 24 Jan 1975			
Soyuz 14	03.07.74	Pavel Popovich Yuri Artyukhin	15.17.30
Sole occupation of Salyut 3			
Soyuz 15	26.08.74	Gennadi Sarafanov Lev Demin	02.00.12
Failed to dock with Salyut 3 because of Soyuz systems failures			
Soyuz 16	02.12.74	Anatoli Filipchenko Nikolai Rukavishnikov	05.22.24
Solo rehearsal for joint Apollo-Soyuz mission			
Salyut 4	26.12.74	Unmanned space station	–
3rd station; occupied by 2 crews for 93 days; re-entered 3 Feb 1977			
Soyuz 17	11.01.75	Alexei Gubarev Georgi Grechko	29.13.20
1st Salyut 4 occupation; Soviet endurance record			
Soyuz 18A	05.04.75	Vasili Lazarev Oleg Makarov	00.00.21
Launch to Salyut 4 aborted when stage 1 failed to separate; 14 g re-entry			
Soyuz 18B	24.05.75	Pyotr Klimuk Vitali Sevastyanov	62.23.20
2nd Salyut 4 occupation; Soviet endurance record			
Soyuz 19	15.07.75	Alexei Leonov Valeri Kubasov	05.22.31
1st joint US/USSR flight, docked with Apollo for 2 days			
Soyuz 20	17.11.75	Unmanned test	–
Resupply ferry rehearsal to Salyut 4			
Salyut 5	22.06.76	Unmanned space station	–
Occupied by 2 crews for 65 days; re-entered 8 Aug 1977			
Soyuz 21	06.07.76	Boris Volynov Vitali Zholobov	49.06.23
1st Salyut 5 occupation (48 days), acrid fumes forced return			
Soyuz 22	15.09.76	Valeri Bykovsky Vladimir Aksyonov	07.21.52
Solo Earth observation mission using backup ASTP craft			
Soyuz 23	14.10.76	Vyacheslav Zudov Valeri Rozhdestvensky	02.00.06
Salyut 5 docking cancelled after launch error (almost causing ascent abort) and Igla automatic approach system failure 200 m out left only enough propellant for one docking attempt and one de-orbit burn; mission rules required enough for two de-orbits. 1st Soviet splashdown after emergency return, into Lake Tengiz at night with −22°C outside. Frost was forming inside when rescuers found them after 9 h (presuming them dead because of no radio contact) and helicopter pulled them to shore			
Soyuz 24	07.02.77	Viktor Gorbatko Yuri Glazkov	17.17.26
2nd Salyut 5 occupation			
Salyut 6	29.09.77	Unmanned space station	–
Occupied for 676 days by 5 long stay + 11 visiting crews. Re-entered 28 Jul 1982 guided by C1267			
Soyuz 25	09.10.77	Vladimir Kovalyonok Valeri Ryumin	02.00.45
Failed to convert soft docking with Salyut 6 into hard docking (see S26)			
Soyuz 26	10.12.77	Yuri Romanenko Georgi Grechko	96.10.00
1st Salyut 6 occupation, endurance record, returned in Soyuz 27. Grechko EVA examined forward docking unit for reason for S25 docking failure, none apparent			
Soyuz 27	10.01.78	Vladimir Dzhanibekov Oleg Makarov	05.22.59
1st station dual occupancy; returned in Soyuz 26			
Soyuz 28	02.03.78	Alexei Gubarev Vladimir Remek	07.22.16
1st international flight (Czech), to Salyut 6			

Mission	Launch Date	Crew	Duration (d, h, m)
Soyuz 29	15.06.78	Vladimir Kovalyonok Alexander Ivanchenkov	139.14.18
1st 100+ day flight, Salyut 6; returned in Soyuz 31			
Soyuz 30	27.06.78	Pyotr Klimuk Miroslaw Hermaszewski	07.22.03
Polish Intercosmos visit to Salyut 6			
Soyuz 31	26.08.78	Valeri Bykovsky Sigmund Jahn	07.20.49
E German Intercosmos Salyut 6 visit; returned in Soyuz 29			
Soyuz 32	25.02.79	Vladimir Lyakhov Valeri Ryumin	175.00.36
Endurance record, Salyut 6; returned in Soyuz 34			
Soyuz 33	10.04.79	Nikolai Rukavishnikov Georgi Ivanov	01.23.01
Bulgarian Salyut 6 visit; failed to dock after engine failure			
Soyuz 34	06.06.79	Unmanned	73.18.24
Delivered for Soyuz 32 crew return to compensate for Soyuz 33 failure			
Soyuz-T 1	16.12.79	Unmanned	100.09.20
T test, docked with Salyut 6			
Soyuz 35	09.04.80	Leonid Popov Valeri Ryumin	184.20.11
Endurance record, 4th Salyut 6 long stay, returned in Soyuz 37			
Soyuz 36	26.05.80	Valeri Kubasov Bertalan Farkas	07.20.46
Hungarian Salyut 6 visit; returned in Soyuz 35			
Soyuz-T 2	05.06.80	Yuri Malyshev Vladimir Aksyonov	03.22.19
1st manned T flight, Salyut 6 visit			
Soyuz 37	23.07.80	Viktor Gorbatko Pham Tuan	07.20.42
N Vietnamese Intercosmos Salyut 6 visit; returned in Soyuz 36			
Soyuz 38	18.09.80	Yuri Romanenko Arnaldo Tamayo Mendez	07.20.43
Cuban Intercosmos visit to Salyut 6			
Soyuz-T 3	27.11.80	Leonid Kizim Oleg Makarov Gennadi Strekalov	12.19.08
Resumption of 3-man flights; Salyut 6 repair work			
Soyuz-T 4	12.03.81	Vladimir Kovalyonok Viktor Savinykh	74.17.38
Last Salyut 6 long-stay; 50th Soviet/100th spaceman			
Soyuz 39	22.03.81	Vladimir Dzhanibekov Jugderdemidiyn Gurragcha	07.20.43
Mongolian Intercosmos visit to Salyut 6			
Soyuz 40	14.05.81	Leonid Popov Dumitru Prunariu	07.20.42
Romanian Intercosmos Salyut 6 visit; last flight of old Soyuz design			
Salyut 7	19.04.82	Unmanned space station	–
Occupied for 812 days by 10 crews; 13 EVAs. Re-entered 7 Feb 1991 with C1686 attached			
Soyuz-T 5	13.05.82	Anatoli Berezovoi Valentin Lebedev	211.09.05
1st Salyut 7 occupation + long stay; returned in T7			
Soyuz-T 6	24.06.82	Vladimir Dzhanibekov Alexander Ivanchenkov Jean-Loup Chrétien	07.21.51
French visiting mission to Salyut 7			
Soyuz-T 7	19.08.82	Leonid Popov Alexander Serebrov Svetlana Savitskaya	07.21.52
Salyut 7 visit, 2nd Soviet woman, returned in T5			
Soyuz-T 8	20.04.83	Vladimir Titov Gennadi Strekalov Alexander Serebrov	02.00.18
Failed to dock with Salyut 7 for long stay. Radar failed (possibly torn off during launch) and manual approach aborted			
Soyuz-T 9	27.06.83	Vladimir Lyakhov Alexander Alexandrov	149.10.46
Salyut 7 long stay; 2 EVAs added 2 solar panels			
Soyuz-T 10A	26.09.83	Vladimir Titov Gennadi Strekalov	
Launchpad fire, 1st use of escape tower. Descent module re-used by T15			
Soyuz-T 10B	08.02.84	Leonid Kizim Vladimir Solovyev Oleg Atkov	236.22.50
1st long stay triple crew, Salyut 7; 6 EVAs totalling 22 h 56 min			
Soyuz-T 11	03.04.84	Yuri Malyshev Gennadi Strekalov Rakesh Sharma	07.21.40
Indian visit to Salyut 7; returned in Soyuz-T10B			
Soyuz-T 12	17.07.84	Vladimir Dzhanibekov Svetlana Savitskaya Igor Volk	11.19.15
Salyut 7 visit, 1st EVA by a woman			
Soyuz-T 13	06.06.85	Vladimir Dzhanibekov Viktor Savinykh	112.03.12 168.03.51
Reactivation of Salyut 7; Savinykh stayed with T14 crew			
Soyuz-T 14	17.09.85	Vladimir Vasyutin Georgi Grechko Alexander Volkov	64.21.52 08.21.13 64.21.52
Salyut 7, 1st mission to be ended by illness (Vasyutin); Grechko returned with T13			
Mir	20.02.86	Unmanned space station	–
New-generation space station with multiple ports			
Soyuz-T 15	13.03.86	Leonid Kizim Vladimir Solovyev	125.00.01
1st Mir occupation; excursion to Salyut 7 5 May – 26 Jun. 2 Salyut EVAs 8 h 50 min. Re-used descent module from T10A abort			
Soyuz-TM 1	21.05.86	Unmanned test	08.22.27
TM demonstration; docked to Mir for 6 days			
Soyuz-TM 2	05.02.87	Yuri Romanenko Alexander Laveikin	326.11.18 174.03.26
2nd Mir long stay; Romanenko new endurance record; Romanenko returned TM3			
Kvant 1	31.03.87	Unmanned Mir module	–
Astrophysics module docked with Mir 09.04.87			
Soyuz-TM 3	22.07.87	Alexander Viktorenko Alexander Alexandrov Mohammed Faris	07.23.05 160.07.17 07.23.05
Syrian Mir visit, Alexandrov replacing Laveikin			
Soyuz-TM 4	21.12.87	Vladimir Titov Musa Manarov Anatoli Levchenko	365.22.39 365.22.39 07.21.58
Titov/Manarov swapped with Romanenko/Alexandrov for Mir's 3rd long stay occupation: endurance record, 3 EVAs totalling 13 h 40 min, returned in TM6			
Soyuz-TM 5	07.06.88	Anatoli Solovyov Viktor Savinykh Alexander Alexandrov	09.20.10
Bulgarian Mir visit, all returning in Soyuz-TM 4			
Soyuz-TM 6	29.03.88	Vladimir Lyakhov Valeri Poliakov Abdol Mohmand	08.20.27 240.22.36 08.20.27
Afghani Mir visit, physician Poliakov remained with Titov/Manarov			
Soyuz-TM 7	26.11.88	Alexander Volkov Sergei Krikalev Jean-Loup Chrétien	151.11.09 151.11.09 24.18.07
French Mir visit, Volkov/Krikalev swapped with Titov/Manarov. Chrétien EVA, returned in TM6, Volkov/Krikalev in TM7			
Soyuz-TM 8	05.09.89	Alexander Viktorenko Alexander Serebrov	166.06.58
5th Mir long stay: received Kvant 2. 5 EVAs including 2 with manoeuvring backpack			
Kvant 2	26.11.89	Unmanned Mir module	–
Radial expansion module, docked with Mir 06.12.89			
Soyuz-TM 9	11.02.90	Anatoli Solovyov Alexander Balandin	179.01.19
6th Mir long stay: received Kristall, 2 EVAs			
Kristall	31.05.90	Unmanned Mir module	–
2nd radial expansion (technology) module, docked with Mir 10.06.90			
Soyuz-TM 10	01.08.90	Gennadi Manakov Gennadi Strekalov	130.20.36
7th Mir long stay: emphasised Kristall materials processing, 1 EVA			
Soyuz-TM 11	02.12.90	Viktor Afanasyev Musa Manarov Toyohiro Akiyama	175.01.52 175.01.52 07.21.15
8th Mir long stay: 4 EVAs; Akiyama (Japan) returned in Soyuz-TM 10			
Soyuz-TM 12	18.05.91	Anatoli Artsebarsky Sergei Krikalev Helen Sharman	144.15.22 311.20.01 07.21.15
9th Mir long stay; 1st UK cosmonaut (Sharman) returned in Soyuz-TM 11, Artsebarsky in TM12, Krikalev in TM13. 6 EVAs totalling 31 h 58 min			
Soyuz-TM 13	02.10.91	Alexander Volkov Toktar Aubakirov Franz Viehböch	175.02.52 07.22.13 07.22.13
10th Mir long stay; 1st Austrian- + Kazakh-sponsored cosmonauts (1st mission with 2 researchers), returned in TM12, Volkov in TM13. 1 EVA			

Mission	Launch Date	Crew	Duration (d, h, m)
Soyuz-TM 14	17.03.92	Alexander Viktorenko	145.14.11
		Alexander Kaleri	145.14.11
		Klaus-Dietrich Flade	07.21.57
11th Mir long stay; 1st German cosmonaut, returned in TM13, Viktorenko/Kaleri in TM14. 123 min EVA			
Soyuz-TM 15	27.07.92	Anatoli Solovyov	188.21.40
		Sergei Avdeyev	188.21.40
		Michel Tognini	13.18.57
12th Mir long stay; 2nd French cosmonaut (3rd French visit mission), returned in TM14; Solovyov/Avdeyev in TM15. 4 EVAs totalling 18 h 21 min			
Soyuz-TM 16	24.01.93	Gennadi Manakov	179.00.44
		Alexander Poleshchuk	179.00.44
13th Mir long stay; 1st docking with Kristall androgynous port. 2 EVAs to prepare for move of Kristall solar arrays to Kvant 1			
Soyuz-TM 17	01.07.93	Vasili Tsebliyev	196.17.45
		Alexander Serebrov	196.17.45
		Jean-Pierre Haignere	20.16.09
14th Mir long stay; 3rd French cosmonaut (4th French visit mission), returned in TM16; Tsebliyev/Serebrov in TM17. 5 EVAs to test new support truss and monitor ageing Mir			
Soyuz-TM 18	08.01.94	Viktor Afanasyev	182.00.27
		Yuri Usachev	182.00.27
		Valeri Poliakov	437.17.58
15th Mir long stay; physician Poliakov remained aboard for record duration, returning with TM20 (record unlikely to be broken for many years)			
Soyuz-TM 19	01.07.94	Yuri Malenchenko	125.22.53
		Talgat Musabayev	125.22.53
16th Mir long stay. 2 EVAs inspected damage from TM17/Progress-M 24 impacts			
Soyuz-TM 20	03.10.94	Alexander Viktorenko	169.05.22
		Yelena Kondakova	169.05.22
		Ulf Merbold	31.12.36
17th Mir long stay; ESA researcher Merbold returned with TM19			
Soyuz-TM 21	03.03.95	Vladimir Dezhurov	115.08.44
		Gennadi Strekalov	115.08.44
		Norman Thagard	115.08.44
18th Mir long stay; NASA researcher Thagard; all crew returning on NASA Shuttle STS-71 7 Jul 1995. 3 EVAs transferred one Kristall solar array to Kvant 1			
Spektr	20.05.95	Unmanned Mir module	–
3rd radial module (remote sensing), docked with Mir 01.06.95			
NASA Shuttle STS-71 launch 27 Jun 1995 to deliver 19th Mir main crew, Cdr Anatoli Solovyov + engineer Nikolai Budarin; return TM21 crew. Solovyov/Budarin returned in TM21. 3 EVAs partially deployed jammed Spektr solar wing and installed MIRAS spectrometer			
Soyuz-TM 22	03.09.95	Yuri Gidzenko	179.01.42
		Sergei Avdeyev	179.01.42
		Thomas Reiter	179.01.42
20th Mir long stay; ESA researcher Reiter performed 2 EVAs			
NASA Shuttle STS-74 launch 12 Nov 1995. Attached docking module to Kristall, delivered solar arrays to *increase Mir's power, delivered supplies and returned with experimental results*			

Mission	Launch Date	Crew	Duration (d, h, m)
Soyuz-TM 23	21.02.96	Yuri Onufrienko	193.19.08
		Yuri Usachev	193.19.08
21st Mir long stay			
NASA Shuttle STS-76 launch 22 Mar 1996. Shannon Lucid remained on Mir, returning aboard STS-79 in Sep 1996. NASA's Clifford/Godwin exited docking adapter hatch to mount three experiments on Mir's docking module			
Priroda	23.04.96	Unmanned Mir module	–
4th radial module (remote sensing), docked with Mir 26 Apr			
Planned:			
Soyuz-TM 24	17.08.96	Valeri Korzun	196.17.26
		Alexandr Kaleri	196.17.26
		Claudie André-Deshays	15.18.24
22nd long-stay Russian mission to Mir; French 'Cassiopeia' mission (5th French visit)			
NASA Shuttle STS-79 launched 16.09.96 delivered Blaha to Mir, returned Lucid (188.04.00)			
NASA Shuttle STS-81 launched 22.12.97 delivered Linenger to Mir, returned with Blaha (128.05.27)			
Soyuz-TM 25	10.02.97	Vasily Tsibilev	184.22.07
		Aleksandr Lazutkin	184.22.07
		Reinhold Ewald	19.16.35
23rd long-stay Russian mission to Mir; second German visit to the station. Accident-prone residency saw fire aboard Mir during the hand-over period with the Soyuz-TM 24 crew, collison of Progress-M 34 with Spektr module and the brief accidental disconnection of the computer system			
NASA Shuttle STS-84 launched 15.05.97 delivered Foale to Mir, returned Linenger (132.03.59)			
Soyuz-TM 26	05.08.97	Anatoli Solovyov	197.17.34
		Pavel Vinogradov	197.17.34
24th long-stay Russian mission to Mir: planned flight of a French spationaut delayed until Soyuz-TM 27 because of power problems on Mir following the Progress-M 34 collision with Spektr			
NASA Shuttle STS-86 launch 26.09.97 delivered Wolf to Mir, returned with Linenger (144.13.48)			
NASA Shuttle STS-89 launch 23.01.98 delivered Thomas, final US astronaut to Mir, returned Wolf (127.10.01)			
Soyuz-TM 27	29.01.98	Talgat Musabeyev	207.12.50
		Nikolai Budarin	207.12.50
		Leopold Eyharts (CNES)	20.16.37
25th long-stay aboard Mir carrying French cosmonaut Eyharts			
NASA Shuttle STS-91 launch 02.06.98 returned Thomas, last US astronaut to Mir (140.15.12)			
Soyuz-TM 28	13.08.98	Gennady Padalka	198.04.31
		Sergei Avdeyev	?
		Yuri Baturin	11.07.40
26th long-stay aboard Mir and with seats on the next and last, manned flight to Mir pre-sold to Slovak and French cosmonauts, their flight plan was being constantly revised			
Soyuz-TM 29	20.02.99	Afanasyev	?
		Haignere	?
		Bella	07.21.56
27th long-stay aboard Mir with Slovak and French cosmonauts and Avdeyev the sole Russian on board. Planned return to Earth for Avdeyev, Afanasyev and Haigere in late August 1999 as final Mir crew			

NASA Astronauts in order of First Flight

No	Name	Total Flts	Total Duration (d, h, m)	Missions
1	A Shepard	2	09.00.17	M3, A14
2	V Grissom	2	00.05.08	M4, G3
3	J Glenn	2	08.26.40	M6, SM92
4	S Carpenter	1	00.04.56	M7
5	W Schirra	3	12.07.13	M8, G6, A7
6	G Cooper	2	09.09.16	M9, G5
7	J Young	6	34.19.41	G3, G10, A10, A16, SM1, SM9
8	J McDivitt	2	14.02.57	G4, A9
9	E White	1	04.01.56	G4
10	C Conrad	4	49.03.38	G5, G11, A12, SL2
11	T Stafford	4	21.03.43	G6, G9, A10, ASTP
12	F Borman	2	19.21.36	G7, A8
13	J Lovell	4	29.19.05	G7, G12, A8, A13
14	N Armstrong	2	08.14.00	G8, A11
15	D Scott	3	22.18.54	G8, A9, A15
16	E Cernan	3	23.14.15	G9, A10, A17
17	M Collins	2	11.02.05	G10, A11
18	R Gordon	2	13.03.53	G11, A12
19	E Aldrin	2	12.01.53	G12, A11
20	D Eisele	1	10.20.09	A7
21	W Cunningham	1	10.20.09	A7
22	W Anders	1	06.03.00	A8
23	R Schweickart	1	10.01.01	A9
24	A Bean	2	69.15.45	A12, SL3
25	J Swigert	1	05.22.55	A13
26	F Haise	1	05.22.55	A13
27	S Roosa	1	09.00.42	A14
28	E Mitchell	1	09.00.42	A14
29	J Irwin	1	12.07.12	A15
30	A Worden	1	12.07.12	A15
31	T Mattingly	3	21.04.34	A16, SM4, SM15
32	C Duke	1	11.01.51	A16
33	R Evans	1	12.13.51	A17

No	Name	Total Flts	Total Duration (d, h, m)	Missions
34	J Schmitt	1	12.13.51	A17
35	J Kerwin	1	28.00.50	SL2
36	P Weitz	2	33.01.13	SL2, SM6
37	O Garriott	2	69.18.56	SL3, SM9
38	J Lousma	2	67.11.14	SL3, SM3
39	G Carr	1	84.01.15	SL4
40	E Gibson	1	84.01.15	SL4
41	W Pogue	1	84.01.15	SL4
42	V Brand	4	31.02.04	ASTP, SM5, SM10, SM38
43	D Slayton	1	09.01.28	ASTP
44	R Crippen	4	23.23.48	SM1, SM7, SM11, SM13
45	J Engle	2	09.08.31	SM2, SM20
46	R Truly	2	08.07.22	SM2, SM8
47	C Fullerton	2	15.22.50	SM3, SM19
48	H Hartsfield	3	20.11.40	SM4, SM12, SM22
49	R Overmyer	2	12.02.22	SM5, SM17
50	J Allen	2	13.01.58	SM5, SM14
51	W Lenoir	1	05.02.15	SM5
52	K Bobko	3	16.02.03	SM6, SM16, SM21
53	D Peterson	1	05.00.23	SM6
54	S Musgrave	6	53.09.59	SM6, SM19, SM31, SM44, SM59, SM80
55	F Hauck	3	18.03.08	SM7, SM14, SM26
56	S Ride	2	14.07.47	SM7, SM13
57	J Fabian	2	13.04.02	SM7, SM18
58	N Thagard	5	140.13.29	SM7, SM17, SM29, SM45, TM21/ SM69
59	D Brandenstein	4	32.21.06	SM8, SM18, SM33, SM47
60	D Gardner	2	14.00.53	SM8, SM14

No	Name	Total Flts	Total Duration (d, h, m)	Missions
61	G Bluford	4	29.10.34	SM8, SM22, SM40, SM52
62	W Thornton	2	13.01.17	SM8, SM17
63	B Shaw	3	21.05.53	SM9, SM23, SM30
64	R Parker	2	19.06.52	SM9, SM38
65	R Gibson	5	36.22.19	SM10, SM24, SM27, SM50, SM69
66	B McCandless	2	13.00.33	SM10, SM35
67	R Stewart	2	12.01.01	SM10, SM21
68	R McNair	2	07.23.18	SM10, SM25
69	F Scobee	2	06.23.41	SM11, SM25
70	T Hart	1	06.23.40	SM11
71	J van Hoften	2	14.01.58	SM11, SM20
72	G Nelson	3	17.02.44	SM11, SM24, SM26
73	M Coats	3	19.07.57	SM12, SM28, SM40
74	J Resnik	2	06.00.57	SM12, SM25
75	S Hawley	4	27.21.56	SM12, SM24, SM35, SM82
76	R Mullane	3	14.20.20	SM12, SM27, SM34
77	J McBride	1	08.05.23	SM13
78	K Sullivan	3	22.04.48	SM13, SM35, SM46
79	D Leestma	3	22.04.32	SM13, SM30, SM46
80	D Walker	4	30.14.30	SM14, SM29, SM52, SM71
81	A Fisher	1	07.23.44	SM14
82	L Shriver	3	16.02.04	SM15, SM35, SM49
83	E Onizuka	2	03.01.34	SM15, SM25

No	Name	Total Flts	Total Duration (d, h, m)	Missions
84	J Buchli	4	20.18.24	SM15, SM22, SM28, SM43
85	D Williams	2	11.23.35	SM16, SM31
86	R Seddon	3	31.02.23	SM16, SM41, SM58
87	J Hoffman	5	50.11.55	SM16, SM38, SM49, SM59, SM75
88	D Griggs	1	06.23.56	SM16
89	F Gregory	3	18.23.06	SM17, SM32, SM44
90	D Lind	1	07.00.08	SM17
91	J Creighton	3	16.20.24	SM18, SM34, SM43
92	S Lucid	5	223.02.51	SM18, SM31, SM42, SM58, SM76*
93	S Nagel	4	30.09.35	SM18, SM22, SM39, SM55
94	R Bridges	1	07.22.45	SM19
95	K Henize	1	07.22.45	SM19
96	A England	1	07.22.45	SM19
97	R Covey	4	26.21.11	SM20, SM26, SM37, SM59
98	J Lounge	3	20.02.23	SM20, SM26, SM38
99	W Fisher	1	07.02.18	SM20
100	R Grabe	3	26.03.42	SM21, SM29, SM45, SM56
101	D Hilmers	4	20.14.17	SM21, SM26, SM34, SM45
102	B Dunbar	5	50.08.26	SM22, SM33, SM48, SM69, SM89
103	B O'Connor	2	15.23.20	SM23, SM41
104	S Spring	1	06.21.06	SM23
105	M Cleave	2	10.22.04	SM23, SM29
106	J Ross	6	47.06.14	SM23, SM27, SM39, SM55, SM73, SM93
107	C Bolden	4	29.02.39	SM24, SM35, SM46, SM60
108	F Chang-Diaz	6	53.15.43	SM24, SM31, SM49, SM60, SM75, SM91
109	M Smith	1	00.00.01	SM25
110	G Gardner	2	13.08.11	SM27, SM38
111	W Shepherd	3	18.08.12	SM27, SM36, SM51
112	J Blaha	5	161.02.47	SM28, SM32, SM42, SM58, SM79*
113	J Bagian	2	14.01.53	SM28, SM41
114	R Springer	2	09.21.33	SM28, SM37
115	M Lee	4	32.21.56	SM29, SM50, SM64, SM82
116	R Richards	4	33.21.31	SM30, SM36, SM48, SM64
117	J Adamson	2	13.22.21	SM30, SM42
118	M Brown	2	10.09.28	SM30, SM43
119	M McCulley	1	04.23.39	SM31
120	E Baker	3	28.14.32	SM31, SM48, SM69
121	K Thornton	4	50.15.16	SM32, SM47, SM59, SM72
122	M Carter	1	05.00.07	SM32
123	J Wetherbee	4	39.19.47	SM33, SM51, SM67, SM87
124	D Low	3	29.18.17	SM33, SM42, SM56
125	M Ivins	4	43.00.29	SM33, SM49, SM61, SM81
126	J Casper	4	34.09.51	SM34, SM53, SM61, SM77
127	P Thuot	3	27.06.52	SM34, SM47, SM61
128	R Cabana	4	37.22.42	SM36, SM52, SM63, SM93
129	B Melnick	2	12.23.28	SM36, SM47
130	T Akers	4	33.22.46	SM36, SM47, SM59
131	F Culbertson	2	14.18.05	SM37, SM57
132	C Meade	3	29.16.15	SM37, SM48, SM64
133	C Gemar	3	24.05.38	SM37, SM43, SM61
134	K Cameron	3	23.10.12	SM39, SM54, SM73
135	L Godwin	3	26.10.38	SM39, SM62, SM76
136	J Apt	4	35.07.11	SM39, SM50, SM62. SM78
137	B Hammond	2	19.06.12	SM40, SM64
138	G Harbaugh	3	34.02.00	SM40, SM53, SM69, SM82
139	D McMonagle	3	25.05.34	SM40, SM53, SM66
140	C Veach	2	18.04.18	SM40, SM51
141	R Hieb	3	31.22.35	SM40, SM47, SM63
142	S Gutierrez	2	20.08.03	SM41, SM62
143	T Jernigan	5	63.01.28	SM41, SM51, SM68, SM80, SM94
144	M Baker	4	40.05.00	SM42, SM51, SM65, SM81
145	K Reightler	2	13.15.38	SM43, SM60
146	T Henricks	4	42.22.40	SM44, SM55, SM70, SM78
147	M Runco	3	22.23.08	SM44, SM53, SM77
148	Jim Voss	4	40.19.24	SM44, SM52, SM71, SM85
149	S Oswald	3	33.22.33	SM45, SM54, SM68
150	W Readdy	3	27.00.46	SM45, SM57, SM79
151	B Duffy	3	27.19.55	SM46, SM56, SM74
152	M Foale	5***	229.14.57	SM46, SM54, SM67, SM84*, SM87
153	K Chilton	3	29.08.23	SM47, SM62, SM76
154	K Bowersox	4	50.12.00	SM48, SM59, SM72. SM82
155	A Allen	3	37.16.11	SM49, SM61, SM75
156	C Brown	5	48.03.56	SM50, SM66, SM77, SM86, SM92
157	J Davis	3	28.06.08	SM50, SM60, SM86
158	M Jemison	1	07.22.30	SM50
159	M Clifford	3	27.18.24	SM52, SM62, SM76
160	S Helms	3	33.20.16	SM53, SM64, SM78
161	K Cockrell	3	37.18.32	SM54, SM71, SM80
162	E Ochoa	3	29.23.56	SM54, SM66, SM94
163	C Precourt	4	38.20.18	SM55, SM69, SM84, SM91
164	B Harris	2	18.06.08	SM55, SM67
165	N Currie	3	30.17.23	SM56, SM70, SM93
166	J Wisoff	3	31.10.28	SM56, SM65, SM81
167	Janice Voss	4	37.23.12	SM56, SM67, SM83, SM85**
168	J Newman	3	32.11.58	SM57, SM71, SM93
169	D Bursch	3	31.01.36	SM57, SM65, SM77
170	C Walz	3	34.17.25	SM57, SM63, SM79
171	R Searfoss	3	38.26.20	SM58, SM76, SM91
172	W McArthur	2	22.04.44	SM58, SM73
173	D Wolf	3	141.20.14	SM58, SM87 (up), SM89 (down)
174	R Sega	2	17.12.26	SM60, SM76
175	T Jones	3	40.03.29	SM62, SM65, SM80
176	J Halsell	4	42.18.25	SM63, SM73, SM83, SM85
177	L Chiao	2	23.15.16	SM63, SM74
178	D Thomas	4	43.08.14	SM63, SM70, SM83, SM85**
179	J Linenger	2	143.02.49	SM64, SM81*
180	T Wilcutt	3	30.04.53	SM65, SM79, SM89
181	S Smith	2	21.05.24	SM65, SM82
182	J Tanner	2	20.22.12	SM66, SM82
183	S Parazynski	3	29.03.41	SM66, SM87, SM92
184	E Collins	2	17.11.49	SM67, SM84
185	W Gregory	1	16.15.10	SM68
186	J Grunsfeld	2	26.20.07	SM68, SM81
187	W Lawrence	3	37.14.27	SM68, SM87, SM91
188	K Kregel	3	40.36.43	SM70, SM78, SM88
189	M Weber	1	08.22.20	SM70
190	M Gernhardt	3	30.12.28	SM71, SM83, SM85**
191	K Rominger	4	55.23.28	SM72, SM80, SM86, SM94
192	C Coleman	1	15.21.52	SM72
193	M Lopez-Alegria	1	15.21.52	SM72
194	B Jett	2	19.02.58	SM74, SM81
195	D Barry	2	18.17.15	SM74, SM94
196	W Scott	2	24.14.36	SM74, SM88
197	S Horowitz	2	25.17.18	SM75, SM82
198	A Thomas	2	122.15.51	SM77, SM89 (up), SM91 (down)
199	C Brady	1	16.21.48	SM78
200	R Linnehan	2	32.19.39	SM78, SM90
201	S Still	2	19.15.59	SM83, SM85**
202	C Noriega	1	09.05.21	SM84
203	E Lu	1	09.05.21	SM84
204	R Curbeam	1	11.20.28	SM86
205	S Robinson	2	19.42.13	SM86, SM92
206	M Bloomfield	1	10.19.22	SM87
207	S Lindsey	2	24.14.20	SM88, SM92
208	K Chawla	1	15.16.35	SM88
209	J Edwards	1	08.19.48	SM89
210	M Anderson	1	08.19.48	SM89
211	J Reilly II	1	08.19.48	SM89
212	S Altman	1	15.21.51	SM90
213	K Hine	1	15.21.51	SM90
214	D Gorie	1	09.19.55	SM91
215	J Kavandi	1	09.19.55	SM91
216	P Duque	1	08.21.45	SM92
217	F Sturckow	1	11.19.18	SM93
218	R Husband	1	09.19.14	SM94

Note The table excludes Payload Specialists, Observers and non-US Mission Specialists listed separately. Abbr: M = Mercury, G = Gemini, A = Apollo, ASTP = Apollo/Soyuz, SL = Skylab, SM = Shuttle Mission. Astronaut biographies can be found on http://www.jsc.nasa.gov/Bios/

*This mission left the astronaut aboard the Mir Complex for an extended flight. **SM85 (STS-94) was a re-flight of SM83 (STS-83) with the same crew. ***Foale was in orbit aboard the Mir Complex at the end of August 1997; flight time is calculated to midnight, 31 August 1997 (107.15.52).

List of non-career Cosmonauts and Astronauts

Name	Country	Year of First Flight	Number of Flights	Total Flight Time	Missions
V Remek	Czechoslovakia	1978	1	07.22.17	S28
M Hermaszewki	Poland	1978	1	07.22.04	S30
S Jahn	GDR	1978	1	07.20.49	S31
G Ivanov	Bulgaria	1979	1	01.23.01	S33
B Farkas	Hungary	1980	1	07.20.46	S36
Pham Tuan	North Vietnam	1980	1	07.21.42	S37
A Mendez	Cuba	1980	1	07.20.34	S38
J Gurragcha	Mongolia	1981	1	07.20.43	S39
D Prunariu	Romania	1981	1	07.20.38	S40
B Lichtenberg	US	1981	2	19.05.06	STS-9, STS-45
U Merbold	Germany/ESA	1991	3	49.21.38	STS-9, STS-42, TM20
J-L Chretien	France	1982	3	43.11.20	T6, TM7, STS-86
R Sharma	India	1984	1	07.21.51	T11
C Walker	US	1984	3	19.21.58	STS-41D, STS-51D, STS-61B
M Garneau	Canada	1984	2	18.06.02	STS-41G, STS-77
P Scully-Power	Australia	1984	1	08.05.23	STS-41G
G Payton	US	1985	1	03.00.26	STS-51C
J Garn	US	1985	1	06.23.56	STS-51D
L van den Berg	Netherlands	1985	1	07.00.08	STS-51B
T Wang	US	1985	1	07.00.08	STS-51B
S Al-Saud	Saudi Arabia	1985	1	07.01.38	STS-51G
P Baudry	France	1985	1	07.01.38	STS-51G
L Acton	US	1985	1	07.22.45	STS-51F
D Bartoe	US	1985	1	07.22.45	STS-51F
W Pailes	US	1985	1	04.00.44	STS-51J
R Furrer	Germany	1985	1	07.08.44	STS-61A
E Messerschmidt	Germany	1985	1	07.08.44	STS-61A
W Ockels	Netherlands	1985	1	07.08.44	STS-61A
R Vela	Mexico	1985	1	06.21.06	STS-61B
R Cenker	US	1986	1	06.20.04	STS-61C
W Nelson	US	1986	1	06.20.04	STS-61C
G Jarvis	US	1986	1	00.00.01	STS-51L
C McAuliffe	US	1986	1	00.00.01	STS-51L
M Faris	Syria	1987	1	07.23.05	TM3
A Alexandrov	Bulgaria	1988	1	09.20.10	TM5
A Mohmand	Afghanistan	1988	1	08.20.27	TM6
S Durrance	US	1990	2	25.14.15	STS-35, STS-67
R Parise	US	1990	2	25.14.15	STS-35, STS-67
T Akiyama	Japan	1990	1	07.21.55	TM11
H Sharman	UK	1991	1	07.21.15	TM12
D Gaffney	US	1991	1	09.02.14	STS-40
M Hughes-Fulford	US	1991	1	09.02.14	STS-40
F Viehboch	Austria	1991	1	07.22.13	TM13
T Aubakirov	Kazakhstan	1991	1	07.22.13	TM13
T Hennen	US	1991	1	06.22.51	STS-44
R Bondar	US	1992	1	08.01.15	STS-42
K-D Flade	Germany	1992	1	07.21.57	TM14
D Frimout	Belgium	1992	1	08.22.06	STS-45
L DeLucas	US	1992	1	13.19.31	STS-50
E Trinh	US	1992	1	13.19.31	STS-50
M Tognini	France	1992	1	13.18.57	TM15
F Malerba	Italy	1992	1	07.23.15	STS-46
C Nicollier	Switzerland/ESA	1992	3	34.12.53	STS-46, STS-61, STS-75
M Mohri	Japan/NASDA	1992	1	07.22.30	STS-47
S MacLean	Canada	1992	1	09.20.56	STS-52
H Schlegel	Germany	1993	1	09.23.40	STS-55
U Walter	Germany	1993	1	09.23.40	STS-55
J-P Haignere	France	1993	1	20.16.09	TM17
M Fettman	US	1993	1	14.00.13	STS-58
S Krikalev*	Russia	1994	1	08.07.10	STS-60
C Naito-Mukai	Japan	1994	1	14.17.56	STS-65
J-F Clervoy	France/ESA	1994	2	20.03.55	STS-66, STS-84
V Titov*	Russia	1995	2	19.01.50	STS-63, STS-86
N Thagard*	US	1995	1	115.08.44	TM21
T Reiter	Germany/ESA	1995	1	179.01.42	TM22
F Leslie	US	1995	1	15.21.52	STS-73
A Sacco	US	1995	1	15.21.52	STS-73
C Hadfield	Canada	1995	1	08.04.31	STS-74
K Wakata	Japan/NASDA	1996	1	08.22.01	STS-72
M Chelli	Italy/ESA	1996	1	15.17.40	STS-75
U Guidoni	Italy	1996	1	15.17.40	STS-75
J-J Favier	France	1996	1	16.21.48	STS-78
R Thirsk	Canada	1996	1	16.21.48	STS-78
C André-Deshays	France	1996	1	15.18.24	TM24
R Ewald	Germany	1997	1	19.16.35	TM25
R Crouch	US	1997	2	20.15.59	STS-83. STS-94
G Linteris	US	1997	2	20.15.59	STS-83, STS-94
Y Kondakova*	Russia	1997	1	09.05.21	STS-84
B Tryggvason	Canada	1997	1	11.20.28	STS-85
T Doi	Japan	1997	1	15.16.35	STS-87
L Kadenyuk	Ukraine	1997	1	15.16.35	STS-87
S Sharipov	Russia	1998	1	08.19.48	STS-89
L Eyharts	France	1998	1	20.16.37	TM27
D Williams	Canada	1998	1	15.21.51	STS-90
J Buckey	US	1998	1	15.21.51	STS-90
J Pawelczyk	US	1998	1	15.21.51	STS-90
Y Baturin	Russia	1998	1	11.19.40	TM28
C Mukai	Japan	1998	1	08.21.45	STS-95
I Bella	Slovakia	1999	1	07.21.56	TM29
J-P Hagniere**	France	1999	1	161.19.42	TM29
T Payette	Canada	1999	1	09.19.14	STS-96
V Tokarev	Russia	1999	1	09.19.14	STS-96

Both the former Soviet Union and the United States have flown 'guests' as part of their piloted missions. In the Soviet Union the programme was initially conducted as part of the Intercosmos collaborative space programme with cosmonauts from the Soviet-Bloc countries flying missions to Salyut 6 (1978-1981); other countries have had guests fly to Salyut 7 and the Mir Complex, the missions starting in 1990 being on a commercial basis. Crews for the United States Space Shuttle have included non-career astronauts, fulfilling the roles of payload specialists. In this listing there are also a few Russian flight-experienced career cosmonauts who have flown aboard some shuttle missions and one career astronaut has been launched aboard a Soyuz-TM; these people are marked * and the time spent in orbit only reflects launches on 'foreign' spacecraft. Missions are identified as S – Soyuz, STS – shuttle mission designator, T – Soyuz-T and TM – Soyuz-TM; although not officially 'STS numbers', missions like 41D are shown as 'STS-41D', STS-94 was a reflight of STS-83 with the same crew.

** Still aboard Mir on 31 July and flight time is calculated up to midnight on that day; Hagniere is expected to return to Earth in late August 1999.

US EVA Record (excluding lunar)

Mission	EVA date	Astronaut	h, min
Gemini 4	03.06.65	White	00.36
Gemini 9	05.06.66	Cernan	02.08
Gemini 10	19.07.66	Collins (SEVA)	00.49
	20.07.66	Collins	00.39
Gemini 11	13.09.66	Gordon	00.33
	14.09.66	Gordon (SEVA)	02.08
Gemini 12	12.11.66	Aldrin (SEVA)	02.29
	13.11.66	Aldrin	02.06
	14.11.66	Aldrin (SEVA)	00.55
Apollo 9	06.03.69	Schweickart	00.37
	06.03.69	Scott (SEVA)	00.46
Apollo 15	05.08.71	Worden	00.38
	05.08.71	Irwin (SEVA)	00.38
Apollo 16	25.04.72	Mattingly	01.13
	25.04.72	Duke (SEVA)	01.13
Apollo 17	17.12.72	Evans	01.06
	17.12.72	Schmitt (SEVA)	01.06
Skylab 2	25.05.73	Weitz (SEVA)	01.15
	07.06.73	Conrad	03.30
	07.06.73	Kerwin	03.30
	19.06.73	Conrad	01.44
	19.06.73	Weitz	01.44
Skylab 3	06.08.73	Garriott	06.31
	06.08.73	Lousma	06.31
	24.08.73	Garriott	04.30
	24.08.73	Lousma	04.30
	22.09.73	Bean	02.45
	22.09.73	Garriott	02.45
Skylab 4	22.11.73	Gibson	06.33
	22.11.73	Pogue	06.33
	25.12.73	Carr	07.01
	25.12.73	Pogue	07.01
	29.12.73	Carr	03.28
	29.12.73	Gibson	03.28
	03.02.74	Carr	05.19
	03.02.74	Gibson	05.19
STS-6	07.04.83	Musgrave	03.54
	07.04.83	Peterson	03.54
STS-41B	08.02.84	McCandless	05.35*
	08.02.84	Stewart	05.35
	10.02.84	McCandless	06.02
	10.02.84	Stewart	06.02*
STS-41C	09.04.84	Nelson	02.59*
	09.04.84	van Hoften	02.59
	11.04.84	Nelson	07.07
	11.04.84	van Hoften	07.07*
STS-41G	12.10.84	Sullivan	03.29
	12.10.84	Leestma	03.29
STS-51A	13.11.84	Allen	06.13*
	13.11.84	Gardner	06.13
	15.11.84	Allen	06.01
	15.11.84	Gardner	06.01*
STS-51D	16.04.85	Hoffman	03.10
	16.04.85	Griggs	03.10
STS-51I	31.08.85	van Hoften	07.20
	31.08.85	W. Fisher	07.20
	01.09.85	van Hoften	04.31
	01.09.85	W. Fisher	04.31
STS-61B	29.11.85	Spring	05.34
	29.11.85	Ross	05.34
	01.12.85	Spring	06.46
	01.12.85	Ross	06.46
STS-37	07.04.91	Ross	04.31
	07.04.91	Apt	04.31
	08.04.91	Ross	05.58
	08.04.91	Apt	05.58
STS-49	10.05.92	Thuot	03.53
	10.05.92	Hieb	03.53
	11.05.92	Thuot	05.38
	11.05.92	Hieb	05.38
	13.05.92	Thuot	08.30
	13.05.92	Hieb	08.30
	13.05.92	Akers	08.30
	14.05.92	Thornton	07.47
	14.05.92	Akers	07.47
STS-54	17.01.93	Runco	04.28
	17.01.93	Harbaugh	04.28
STS-57	25.06.93	Low	05.50
	25.06.93	Wisoff	05.50
STS-51	16.09.93	Walz	07.05
	16.09.93	Newman	07.05
STS-61	05.12.93	Musgrave	07.54
	05.12.93	Hoffman	07.54
	06.12.93	Akers	06.36
	06.12.93	Thornton	06.36
	07.12.93	Musgrave	06.47
	07.12.93	Hoffman	06.47

Mission	EVA date	Astronaut	h, min
	08.12.93	Akers	06.50
	08.12.93	Thornton	06.50
	09.12.93	Musgrave	07.21
	09.12.93	Hoffman	07.21
STS-64	16.09.94	Lee	06.51
	16.09.94	Meade	06.51
STS-63	09.02.95	Foale	04.38
	09.02.95	Harris	04.38
STS-69	16.09.95	Voss	06.46
	16.09.95	Gernhardt	06.46
STS-72	15.01.96	Chiao	06.09
	15.01.96	Barry	06.09
	17.01.96	Chiao	06.54
	17.01.96	Scott	06.54
STS-76	27.03.96	Clifford	06.03
	27.03.96	Godwin	06.03
STS-82	14.02.97	Lee	06.42
	14.02.97	Smith	06.42
	15.02.97	Harbaugh	07.27
	15.02.97	Tanner	07.27
	16.02.97	Lee	07.11
	16.02.97	Smith	07.11
	17.02.97	Harbaugh	06.34
	17.02.97	Tanner	06.34
	18.02.97	Lee	05.17
	18.02.97	Smith	05.17
STS-86	01.10.97	Parazynski	05.01
STS-87	24.11.97	Scott	07.43
	24.11.97	Doi	07.43
	02.12.97	Scott	05.00
	02.12.97	Doi	05.00
STS-88	07.12.98	Ross	07.21
	07.12.98	Newman	07.21
	09.12.98	Ross	07.02
	09.12.98	Newman	07.02
	12.12.98	Ross	06.59
	12.12.98	Newman	06.59
STS-96	30.05.99	Jernigan	07.55
	30.05.99	Barry	07.55

*= includes untethered flight with Manned Manoeuvring Unit (total 10.22); SEVA = Stand-Up EVA, ie astronaut stood in hatchway but did not leave craft. Times given are depressurisation of spacecraft; actual time spent in space will be less. Planned EVAs for STS-80 by Jernigan and Jones were cancelled because of faulty hatch.

US Lunar EVA Record

Mission	EVA date	Astronaut	h, min
Apollo 11	21.07.69	Armstrong	02.31
	21.07.69	Aldrin	02.31
Apollo 12	19.11.69	Conrad	03.56
	19.11.69	Bean	03.56
	20.11.69	Conrad	03.49
	20.11.69	Bean	03.49
Apollo 14	05.02.71	Shepard	04.48
	05.02.71	Mitchell	04.48
	06.02.71	Shepard	04.35
	06.02.71	Mitchell	04.35
Apollo 15	30.07.71	Scott (SEVA)	00.33
	31.07.71	Scott	06.33
	31.07.71	Irwin	06.33
	01.08.71	Scott	07.12
	01.08.71	Irwin	07.12
	02.08.71	Scott	04.50
	02.08.71	Irwin	04.50
Apollo 16	21.04.72	Young	07.11
	21.04.72	Duke	07.11
	22.04.72	Young	07.23
	22.04.72	Duke	07.23
	23.04.72	Young	05.40
	23.04.72	Duke	05.40
Apollo 17	11.12.72	Cernan	07.12
	11.12.72	Schmitt	07.12
	12.12.72	Cernan	07.37
	12.12.72	Schmitt	07.37
	13.12.72	Cernan	07.16
	13.12.72	Schmitt	07.16
Totals	**EVA**	**28**	**161.06**
	SEVA	**1**	**00.33**

CIS/USSR Cosmonauts in order of First Flight

No	Name	Total Flts	Total Duration (d, h, m)	Missions
1	Y Gagarin	1	00.01.48	V1
2	G Titov	1	01.01.18	V2
3	A Nikolayev	2	21.15.25	V3, S9
4	P Popovich	2	18.16.29	V4, S14
5	V Bykovsky	3	20.17.49	V5, S22, S31
6	V Tereshkova	1	02.22.50	V6
7	V Komarov	2	02.02.54	Vd1, S1
8	K Feoktistov	1	01.00.17	Vd1
9	B Yegorov	1	01.00.17	Vd1
10	P Belyayev	1	01.02.02	Vd2
11	A Leonov	2	07.00.33	Vd2, S19
12	G Beregovoi	1	03.22.51	S3
13	V Shatalov	3	09.21.59	S4, S8, S10
14	B Volynov	2	52.07.20	S5, S21
15	A Yeliseyev	3	08.22.24	S5, S8, S10
16	Y Khrunov	1	01.23.49	S5
17	G Shonin	1	04.22.42	S6
18	V Kubasov	3	18.17.19	S6, S19, S36
19	A Filipchenko	2	10.21.05	S7, S16
20	V Volkov	2	28.17.03	S7, S11
21	V Gorbatko	3	30.12.31	S7, S24, S37
22	V Sevastyanov	2	80.16.19	S9, S18B
23	N Rukavishnikov	3	09.22.10	S10, S16, S33
24	G Dobrovolsky	1	23.18.22	S11
25	V Patsayev	1	23.18.22	S11
26	V Lazarev	2	01.23.16	S12, S18A
27	O Makarov	4	20.18.28	S12, S18A, S27, T3
28	P Klimuk	3	78.18.19	S13, S18B, S30
29	V Lebedev	2	219.05.00	S13, T5
30	Y Artyukhin	1	15.17.30	S14
31	G Sarafanov	1	02.00.12	S15
32	L Demin	1	02.00.12	S15
33	A Gubarev	2	37.11.37	S17, S28
34	G Grechko	3	133.20.33	S17, S26, T14
35	V Zholobov	1	49.06.24	S21
36	V Aksyonov	2	11.20.35	S22, T2
37	V Zudov	1	02.00.06	S23
38	V Rozhdestvensky	1	02.00.06	S23
39	Y Glazkov	1	17.17.23	S24
40	V Kovalyonok	3	216.10.17	S25, S29, T4
41	V Ryumin	4	371.12.19	S25, S32, S35, STS-91
42	Y Romanenko	3	430.18.12	S26, S38, TM2
43	V Dzhanibekov	5	147.15.59	S27, S39, T6, T12, T13
44	A Ivanchenkov	2	147.12.39	S29, T6
45	V Lyakhov	3	333.06.49	S32, T9, TM6
46	L Popov	3	199.01.32	S35, S30, T7
47	Y Malyshev	2	11.20.32	T2, T11
48	L Kizim	3	374.17.59	T3, T10B, T15
49	G Strekalov	6	268.22.37	T3, T8, T10A, T11, TM10, TM21
50	V Savinykh	3	196.17.00	T4, T13, TM5
51	A Berezovoi	1	211.08.05	T5
52	A Serebrov	4	372.22.53	T7, T8, TM8, TM17
53	S Savitskaya	2	19.17.07	T7, T12
54	V Titov	4	375.05.25	T8, T10A, TM4, STS-63
55	A Alexandrov	2	309.17.03	T9, TM3
56	V Solovyev	2	360.22.51	T10B, T15
57	O Atkov	1	236.22.50	T10B
58	I Volk	1	11.19.14	T12
59	V Vasyutin	1	64.21.52	T14
60	A Volkov	3	391.11.53	T14, TM7, TM13
61	A Laveikin	1	174.03.26	TM2
62	A Viktorenko	4	489.01.36	TM3, TM8, TM14, TM20
63	M Manarov	2	541.00.31	TM4, TM11
64	A Levchenko	1	07.21.58	TM4
65	A Solovyov	5	651.00.04	TM5, TM9, TM15, STS-71, TM26
66	V Poliakov	2	678.16.34	TM6, TM18
67	S Krikalev	4	487.06.55	TM7, TM12, STS-60, STS-88
68	A Balandin	1	179.01.19	TM9
69	G Manakov	2	309.21.20	TM10, TM16
70	V Afanasyev	3*	518.22.01	TM11, TM18, TM29
71	A Artsebarsky	1	144.15.22	TM12
72	A Kaleri	2	342.07.37	TM14, TM24
73	S Avdeyev	3*	720.13.39	TM15, TM22, TM28
74	A Poleshchuk	1	179.00.44	TM16
75	V Tsibliyev	2	381.15.52	TM17, TM25
76	Y Usachev	2	376.19.35	TM18, TM23
77	Y Malenchenko	1	125.22.53	TM19
78	T Musabayev	2	333.11.43	TM19, TM27
79	Y Kondakova	2	178.10.43	TM20, STS-84
80	V Dezhurov	1	115.08.44	TM21
81	N Budarin	2	283.00.11	STS-71, TM27
82	Y Gidzenko	1	179.01.42	TM22
83	Y Onufrienko	1	194.19.08	TM23
84	V Korzun	1	196.17.26	TM24
85	A Lazutkin	1	184.22.07	TM25
86	P Vinogradov	1	197.17.34	TM26
87	G Padalka	1	198.16.31	TM26

Notes Some cosmonauts have flown aboard shuttle missions. Dezhurov and Strekalov were launched aboard Soyuz-TM 21 but returned to Earth aboard Atlantis/STS-71; A Solovyov and Budarin were launched on Atlantis/STS-71 and returned to Earth aboard Soyuz-TM 21. Other cosmonaut shuttle missions did not involve a switch between launch and landing craft. T10A was a launch pad abort and is not included in the totals. V = Vostok, Vd = Voskhod, S = Soyuz, T = Soyuz-T, TM = Soyuz-TM.

*Afanasyev and Avdeyev were still aboard Mir on 31 Jul 1999 and the accumulated time is calculated to midnight on that day; the two crewmembers were expected to return to Earth in late August 1999.

CIS/USSR EVA Record

Mission	*EVA date*	*Cosmonaut*	*h, min*
Voskhod 2	18.03.65	Leonov	00.24
Soyuz 4	16.01.69	Yeliseyev	01.00
Soyuz 5	16.01.69	Khrunov	01.00
Soyuz 26	20.12.77	Grechko	01.28
		Romanenko*	
Soyuz 29	29.07.78	Kovalyonok	02.05
		Ivanchenkov	02.05
Soyuz 32	15.08.79	Lyakhov	01.23
		Ryumin	01.23
Soyuz-T 5	30.07.82	Berezovoi	02.33
		Lebedev	02.33
Soyuz-T 9	01.11.83	Lyakhov	02.50
		Alexandrov	02.50
Soyuz-T 9	03.11.83	Lyakhov	02.55
		Alexandrov	02.55
Soyuz-T 10B	23.04.84	Kizim	04.15
		Solovyev	04.15
Soyuz-T 10B	26.04.84	Kizim	05.00
		Solovyev	05.00
Soyuz-T 10B	29.04.84	Kizim	02.45
		Solovyev	02.45
Soyuz-T 10B	04.05.84	Kizim	02.45
		Solovyev	02.45
Soyuz-T 10B	18.05.84	Kizim	03.05
		Solovyev	03.05
Soyuz-T 12	25.07.84	Dzhanibekov	03.35
		Savitskaya	03.35
Soyuz-T 10B	08.08.84	Kizim	05.00
		Solovyev	05.00
Soyuz-T 13	02.08.85	Dzhanibekov	05.00+
		Savinykh	05.00+
Soyuz-T 15	28.05.86	Kizim	03.50
		Solovyev	03.50
Soyuz-T 15	31.05.86	Kizim	05.00
		Solovyev	05.00
Soyuz-TM 2	12.04.87	Romanenko	03.40
		Laveikin	03.40
Soyuz-TM 2	12.06.87	Romanenko	01.53
		Laveikin	01.53
Soyuz-TM 2	16.06.87	Romanenko	03.15
		Laveikin	03.15
Soyuz-TM 4	26.02.88	Titov	04.25
		Manarov	04.25
Soyuz-TM 4	30.06.88	Titov	05.10
		Manarov	05.10
Soyuz-TM 4	20.10.88	Titov	04.12
		Manarov	04.12
Soyuz-TM 7	09.12.88	Volkov	05.57
		Chrétien	05.57
Soyuz-TM 8	08.01.90	Viktorenko	02.56
		Serebrov	02.56
Soyuz-TM 8	11.01.90	Viktorenko	02.54
		Serebrov	02.54
Soyuz-TM 8	26.01.90	Viktorenko	03.02
		Serebrov	03.02
Soyuz-TM 8	01.02.90	Viktorenko	04.59
		Serebrov[1]	04.59
Soyuz-TM 8	05.02.90	Viktorenko[2]	03.45
		Serebrov	03.45
Soyuz-TM 9	17.07.90	Solovyov	07.16
		Balandin	07.16
Soyuz-TM 9	26.07.90	Solovyov	03.31
		Balandin	03.31
Soyuz-TM 10	29.10.90	Manakov	03.45
		Strekalov	03.45
Soyuz-TM 11	07.01.91	Afanasyev	05.18
		Manarov	05.18
Soyuz-TM 11	23.01.91	Afanasyev	05.33
		Manarov	05.33
Soyuz-TM 11	26.01.91	Afanasyev	06.20
		Manarov	06.20
Soyuz-TM 11	25.04.91	Afanasyev	03.34
		Manarov	03.34
Soyuz-TM 12	24.06.91	Artsebarsky	04.58
		Krikalev	04.58
Soyuz-TM 12	28.06.91	Artsebarsky	03.24
		Krikalev	03.24
Soyuz-TM 12	15.07.91	Artsebarsky	05.56
		Krikalev	05.56
Soyuz-TM 12	19.07.91	Artsebarsky	05.28
		Krikalev	05.28
Soyuz-TM 12	23.07.91	Artsebarsky	05.34
		Krikalev	05.34
Soyuz-TM 12	27.07.91	Artsebarsky	06.49
		Krikalev	06.49
Soyuz-TM 13	20.02.92	Volkov	04.12
		Krikalev	04.12
Soyuz-TM 14	08.07.92	Viktorenko	02.03
		Kaleri	02.03
Soyuz-TM 15	03.09.92	Solovyov	03.56
		Avdeyev	03.56
Soyuz-TM 15	07.09.92	Solovyov	05.08
		Avdeyev	05.08
Soyuz-TM 15	11.09.92	Solovyov	05.44
		Avdeyev	05.44
Soyuz-TM 15	15.09.92	Solovyov	03.33
		Avdeyev	03.33
Soyuz-TM 16	19.04.93	Manakov	05.25
		Poleshchuk	05.25
Soyuz-TM 16	18.06.93	Manakov	04.33
		Poleshchuk	04.33
Soyuz-TM 17	16.09.93	Tsebliyev	04.18
		Serebrov	04.18
Soyuz-TM 17	20.09.93	Tsebliyev	03.13
		Serebrov	03.13
Soyuz-TM 17	28.09.93	Tsebliyev	01.52
		Serebrov	01.52
Soyuz-TM 17	22.10.93	Tsebliyev	00.38
		Serebrov	00.38
Soyuz-TM 17	29.10.93	Tsebliyev	04.12
		Serebrov	04.12
Soyuz-TM 19	09.09.94	Malenchenko	05.04
		Musabayev	05.04
Soyuz-TM 19	13.09.94	Malenchenko	06.01
		Musabayev	06.01
Soyuz-TM 21	12.05.95	Dezhurov	06.15
		Strekalov	06.15
Soyuz-TM 21	17.05.95	Dezhurov	06.52
		Strekalov	06.52
Soyuz-TM 21	22.05.95	Dezhurov	05.15
		Strekalov	05.15
'STS 71'	14.07.95	Solovyov	05.34
		Budarin	05.34
'STS 71'	19.07.95	Solovyov	03.08
		Budarin	03.08
'STS 71'	21.07.95	Solovyov	05.35
		Budarin	05.35
Soyuz-TM 22	20.10.95	Avdeyev	05.16
		Reiter	05.16
Soyuz-TM 22	08.02.96	Gidzenko	03.02
		Reiter	03.02
Soyuz-TM 23	15.03.96	Onufrienko	05.51
		Usachev	05.51
Soyuz-TM 23	20.05.96	Onufrienko	05.20
		Usachev	05.20
Soyuz-TM 23	24.05.96	Onufrienko	05.43
		Usachev	05.43
Soyuz-TM 23	30.05.96	Onufrienko	04.20
		Usachev	04.20
Soyuz-TM 23	06.06.96	Onufrienko	03.34
		Usachev	03.34
Soyuz-TM 24	02.12.96	Korzun	05.57
		Kaleri	05.57
Soyuz-TM 24	09.12.96	Korzun	06.36
		Kaleri	06.36
Soyuz-TM 25	29.04.97	Tsibilev	04.58
		Linenger	04.58
Soyuz-TM 26	22.08.97	Solovyov	06.00
		Vinogradov	06.00
Soyuz-TM 26	20.10.97	Solovyov	06.38
		Vinogradov	06.38
Soyuz-TM 26	06.11.97	Solovyov	06.17
		Vinogradov	06.17
Soyuz-TM 26	14.01.98	Solovyov	03.52
		Wolf	03.52
Soyuz-TM 27	01.04.98	Musabayev	06.40
		Budarin	06.40
Soyuz-TM 27	06.04.98	Musabayev	04.23
		Budarin	04.23
Soyuz-TM 27	11.04.98	Musabayev	06.25
		Budarin	06.25
Soyuz-TM 27	17.04.98	Musabayev	06.33
		Budarin	06.33
Soyuz-TM 27	22.04.98	Musabayev	06.21
		Budarin	06.21
Soyuz-TM 28	10.11.98	Padalka	05.54
		Avdeyev	05.54

*Unauthorised, duration unknown. + Approximate time. [1] first test of manoeuvring unit, [2] second test of manoeuvring unit. Linenger was the visiting NASA astronaut at the time.

SOLAR SYSTEM EXPLORATION CHRONOLOGY

The following table presents a chronological listing of known and, in some cases, surmised planetary missions. Speculative material on early Soviet failures abounds; the flights included here are the more probable ones. Beginning in 1964, Cosmos numbers were assigned to Soviet payloads stranded in Earth parking orbit. US Apollo manned lunar and Helios solar missions are also included.

	Launch	*Nation*	*Launcher*	*Mission Type*	*Comments*
1958					
Pioneer 0	Aug 17	US	Thor-Able	Lunar orbiter	Stage 1 explosion
Luna-1958A	Sep 23	USSR	Vostok	Lunar impact	Structural failure T+92 s
Pioneer 1	Oct 11	US	Thor-Able	Lunar orbiter	Failed to reach Earth escape velocity; attained 113,854 km during 43 h 17 m flight. 38 kg
Luna-1958B	Oct 12	USSR	Vostok	Lunar impact	Launcher exploded T+100 s
Pioneer 2	Nov 8	US	Thor-Able	Lunar orbiter	Stage 3 failure. 39.5 kg
Luna-1958C	Dec 4	USSR	Vostok	Lunar impact	Stage 1 failure T+245 s
Pioneer 3	Dec 6	US	Juno II	Lunar flyby	Failed to reach Earth escape velocity, re-entered from 102,333 km apogee. 5.9 kg
1959					
Luna 1	Jan 2	USSR	Vostok	Lunar impact	First lunar flyby (5,995 km)
Pioneer 4	Mar 3	US	Juno II	Lunar flyby	Partial success; 59,983 km flyby. 5.9 kg
Luna-1959A	Jul 18	USSR	Vostok	Lunar impact	Failed to reach Earth orbit
Luna 2	Sep 12	USSR	Vostok	Lunar impact	First lunar impact
Luna 3	Oct 4	USSR	Vostok	Lunar flyby	First far side images (barycentric)
Atlas-Able 4	Nov 26	US	Atlas-Able	Lunar orbiter	Launcher failure
1960					
Pioneer 5	Mar 11	US	Thor-Able	Solar monitor	0.81 × 0.99 AU solar orbit; returned solar wind data to 26 Jun 1960. 43 kg
Luna-1960A	Apr 15	USSR	Vostok	Lunar flyby	Failed to reach Earth orbit (premature stage 2 cutoff); Luna 3-type
Luna-1960B	Apr 16	USSR	Vostok	Lunar flyby	Strap-on failed to ignite, broke up near pad; Luna 3-type
Atlas-Able 5A	Sep 25	US	Atlas-Able	Lunar orbiter	Launcher failure
Mars-1960A	Oct 10	USSR	Molniya	Mars flyby	Failed to reach Earth orbit
Mars-1960B	Oct 14	USSR	Molniya	Mars flyby	Failed to reach Earth orbit
Atlas-Able 5B	Dec 15	US	Atlas-Able	Lunar orbiter	Launcher failure

	Launch	*Nation*	*Launcher*	*Mission Type*	*Comments*
1961					
Venera-1961A	Feb 4	USSR	Molniya	Venus impact	Failed to leave Earth orbit
Venera 1	Feb 12	USSR	Molniya	Venus impact	Contact lost en route; 100,000 km flyby. 644 kg
Ranger 1	Aug 23	US	Atlas-Agena B	Lunar test	High-altitude lunar test; failed to leave low orbit
Ranger 2	Nov 18	US	Atlas-Agena B	Lunar test	As Ranger 1
1962					
Ranger 3	Jan 26	US	Atlas-Agena B	Lunar lander	Launch error, lunar flyby
Ranger 4	Apr 23	US	Atlas-Agena B	Lunar lander	Spacecraft failure; first US lunar impact
Mariner 1	Jul 22	US	Atlas-Agena B	Venus flyby	Launcher failure; error in flight guidance equations
Venera-1962A	Aug 25	USSR	Molniya	Venus capsule	Failed to leave Earth orbit
Mariner 2	Aug 27	US	Atlas-Agena B	Venus flyby	First successful Venus flyby; 34,830 km after 109 days. Surface T far higher than expected
Venera-1962B	Sep 1	USSR	Molniya	Venus capsule	Failed to leave Earth orbit
Venera-1962C	Sep 12	USSR	Molniya	Venus flyby	Failed to leave Earth orbit
Ranger 5	Oct 18	US	Atlas-Agena B	Lunar lander	Spacecraft failure; close lunar flyby
Mars-1962A	Oct 24	USSR	Molniya	Mars flyby	Final stage exploded in Earth orbit. 893.5 kg
Mars 1	Nov 1	USSR	Molniya	Mars flyby	Contact lost 21 Mar 1963; 200,000 km flyby
Mars-1962B	Nov 4	USSR	Molniya	Mars lander?	Failed to leave Earth orbit
1963					
Luna-1963A	Jan 4	USSR	Molniya	Lunar lander	Failed to leave Earth orbit
Luna-1963B	Feb 2	USSR	Molniya	Lunar lander	Failed to reach Earth orbit
Luna 4	Apr 2	USSR	Molniya	Lunar lander	Lunar miss (8,529 km)
Cosmos 21	Nov 11	USSR	Molniya	Venus test	Failed to leave Earth orbit
1964					
Ranger 6	Jan 30	US	Atlas-Agena B	Lunar impact	Cameras failed
Venera-1964A	Feb 19	USSR	Molniya-M	Venus flyby	Failed to reach Earth orbit
Venera-1964B	Mar 1	USSR	Molniya-M	Venus flyby	Tentative identification; launcher failure?
Venera-1964C	Mar 21	USSR	Molniya-M	Venus flyby	Failed to reach Earth orbit
Cosmos 27	Mar 27	USSR	Molniya-M	Venus flyby	Failed to leave Earth orbit
Zond 1	Apr 2	USSR	Molniya-M	Venus flyby	Contact lost May; 100,000 km flyby 19 Jul. 825 kg
Luna-1964B	Apr 20	USSR	Molniya-M	Lunar lander	Failed to reach Earth orbit
Ranger 7	Jul 28	US	Atlas-Agena B	Lunar impact	Successful image return
Mariner 3	Nov 5	US	Atlas-Agena D	Mars flyby	Payload shroud failed to separate. Intended 13,840 km flyby
Mariner 4	Nov 28	US	Atlas-Agena D	Mars flyby	First successful Mars flyby (14 Jul 1965; 9,844 km) + imaging. 21 TV images (1% surface) showed arid, lunar-like surface
Zond 2	Nov 30	USSR	Molniya	Mars flyby	Contact lost May 1965; 1,500 km flyby 6 Aug 1965. 1,145 kg
1965					
Ranger 8	Feb 17	US	Atlas-Agena B	Lunar impact	Successful image return
Cosmos 60	Mar 12	USSR	Molniya	Lunar lander	Failed to leave Earth orbit
Ranger 9	Mar 21	US	Atlas-Agena B	Lunar impact	Successful image return
Luna-1965A	Apr 10	US	Molniya	Lunar lander	Failed to reach Earth orbit
Luna 5	May 9	USSR	Molniya-M	Lunar lander	Retro failed; lunar impact
Luna 6	Jun 8	USSR	Molniya-M	Lunar lander	Missed Moon (161,000 km)
Zond 3	Jul 18	USSR	Molniya	Lunar flyby	Photographed lunar far side; Mars probe test. 1,145 kg
Luna 7	Oct 4	USSR	Molniya	Lunar lander	Failed landing sequence, crashed
Venera 2	Nov 12	USSR	Molniya-M	Venus flyby	Contact lost en route; 23,810 km flyby. 963 kg
Venera 3	Nov 16	USSR	Molniya	Venus capsule	Contact lost before entry. 960 kg
Cosmos 96	Nov 23	USSR	Molniya-M	Venus capsule	Failed to leave Earth orbit
Luna 8	Dec 3	USSR	Molniya-M	Lunar lander	Retro delay; lunar impact
Pioneer 6	Dec 16	US	Delta	Solar orbiter	0.81 × 0.985 AU orbit; still operable for brief periods
1966					
Luna 9	Jan 31	USSR	Molniya-M	Lunar lander	First semi-soft landing; first surface images
Cosmos 111	Mar 1	USSR	Molniya-M	Lunar orbiter?	Failed to leave Earth orbit
Luna 10	Mar 31	USSR	Molniya-M	Lunar orbiter	First lunar orbiter; in contact 56 days
Luna-1966A	Apr 30	USSR	Molniya-M	Lunar orbiter	Failed to reach Earth orbit
Surveyor 1	May 30	US	Atlas-Centaur	Lunar lander	First true soft landing
Explorer 33	Jul 1	US	Delta	Lunar orbiter	Entered high Earth orbit
Lunar Orbiter 1	Aug 10	US	Atlas-Agena D	Lunar orbiter	Second lunar orbiter; photographic survey
Pioneer 7	Aug 17	US	Delta	Solar monitor	1.012 × 1.125 AU orbit; still operable for brief periods
Luna 11	Aug 24	USSR	Molniya-M	Lunar orbiter	Third lunar orbiter; possibly carried failed TV system
Surveyor 2	Sep 20	US	Atlas-Centaur	Lunar lander	Impacted Moon
Luna 12	Oct 22	USSR	Molniya-M	Lunar orbiter	Fourth lunar orbiter; TV imaging
Lunar Orbiter 2	Nov 6	US	Atlas-Agena D	Lunar orbiter	Fifth lunar orbiter; photographic survey
Luna 13	Dec 21	USSR	Molniya-M	Lunar lander	Third lunar landing
1967					
Lunar Orbiter 3	Feb 5	US	Atlas-Agena D	Lunar orbiter	Sixth lunar orbiter; photographic survey
Surveyor 3	Apr 17	US	Atlas-Centaur	Lunar lander	Fourth lunar lander
Lunar Orbiter 4	May 4	US	Atlas-Agena D	Lunar orbiter	Seventh lunar orbiter; photographic survey
Venera 4	Jun 12	USSR	Molniya-M	Venus capsule	First successful Venus atmospheric probe (18 Oct). 1,106 kg (383 kg capsule); tx to 27 km altitude
Mariner 5	Jun 14	US	Atlas-Agena D	Venus flyby	Second successful Venus flyby; 3,990 km 19 Oct
Cosmos 167	Jun 17	USSR	Molniya-M	Venus capsule	Failed to leave Earth orbit
Surveyor 4	Jul 14	US	Atlas-Centaur	Lunar lander	Soft-landed? Contact lost
Explorer 35	Jul 19	US	Delta	Lunar orbiter	Eighth lunar orbiter
Lunar Orbiter 5	Aug 1	US	Atlas-Agena D	Lunar orbiter	Ninth lunar orbiter; photographic survey
Surveyor 5	Sep 8	US	Atlas-Centaur	Lunar lander	Fifth lunar lander
Zond-1967A	Sep 28	USSR	Proton-K + Block D	Lunar test	Stage 1 failure; Zond 4-type mission
Surveyor 6	Nov 7	US	Atlas-Centaur	Lunar lander	Sixth lunar lander
Zond-1967B	Nov 22	USSR	Proton-K + Block D	Lunar test	Stage 2 engine failure; Zond 4-type mission
Pioneer 8	Dec 13	US	Delta	Solar monitor	0.990 × 1.087 AU orbit; still operable for brief periods
1968					
Surveyor 7	Jan 7	US	Atlas-Centaur	Lunar lander	Seventh lunar lander
Luna 1968A	Feb 7	USSR	Molniya-M	Lunar orbiter	Failed to reach Earth orbit
Zond 4	Mar 2	USSR	Proton-K + Block D	Lunar distance	Test of manned craft
Luna 14	Apr 7	USSR	Molniya-M	Lunar orbiter	Tenth lunar orbiter
Zond-1968A	Apr 23	USSR	Proton-K + Block D	Lunar test	Control system short circuit shut down stage 2 at 260 s
Zond 5	Sep 14	USSR	Proton-K + Block D	Lunar flyby	Earth return; test of manned craft
Pioneer 9	Nov 8	US	Delta	Solar monitor	Successful; 0.756 × 0.990 AU orbit
Zond 6	Nov 10	USSR	Proton-K + Block D	Lunar flyby	Earth return; test of manned craft
Apollo 8	Dec 21	US	Saturn 5	Lunar orbiter	First manned lunar orbiter
1969					
Venera 5	Jan 5	USSR	Molniya-M	Venus capsule	Atmospheric entry (16 May), tx to 25 km altitude. 1,130 kg (405 kg capsule)
Venera 6	Jan 10	USSR	Molniya-M	Venus capsule	Atmospheric entry (17 May), tx to 12 km altitude. 1,130 kg (405 kg capsule)
Zond-1969A	Jan 20	USSR	Proton-K + Block D	Lunar test	One stage 2 engine shut down 25 s early
Luna-1969A	Feb 19	USSR	Proton-K + Block D	Lunar rover	Lunokhod. Exploded at 40 s
Zond L1S	Feb 21	USSR	SL-15/N-1	Circumlunar	Exploded T+70 s. N1 test but Zond cameras to image potential manned landing sites. Mockup lander
Mariner 6	Feb 25	US	Atlas-Centaur	Mars flyby	Imaging flyby; 3,412 km 31 Jul, 24 near-encounter images concentrating on equatorial region

	Launch	*Nation*	*Launcher*	*Mission Type*	*Comments*
Mariner 7	Mar 27	US	Atlas-Centaur	Mars flyby	Imaging flyby; 3,524 km 5 Aug, 33 near-encounter images. Confirmed ice cap mainly CO_2 + some H_2O ice
Mars-1969A	Mar 27	USSR	Proton-K + Block D	Mars lander?	Stage 3 exploded at 438 s
Mars-1969B	Apr 2	USSR	Proton-K + Block D	Mars lander?	Early stage 1 failure, crashed nearby pad
Apollo 10	May 18	US	Saturn 5	Lunar orbiter	Second manned lunar orbiter
Luna-1969B	Jun 14	USSR	Proton-K + Block D	Lunar sampler	Launcher failure
Zond L1S	Jul 3	USSR	SL-15/N-1	Circumlunar	Exploded T+5 s. N1 test but Zond cameras to image potential manned landing sites. Mockup lander
Luna 15	Jul 13	USSR	Proton-K + Block D	Lunar sampler	11th unmanned lunar orbiter; crashed landing attempt (Luna 16 type)
Apollo 11	Jul 16	US	Saturn 5	Lunar lander	First manned lunar landing (eighth in all)
Zond 7	Aug 7	USSR	Proton-K + Block D	Lunar flyby	Earth return; test of manned craft
Cosmos 300	Sep 23	USSR	Proton-K + Block D	Lunar sampler	Failed to leave Earth orbit
Cosmos 305	Oct 22	USSR	Proton-K + Block D	Lunar sampler	Failed to leave Earth orbit
Apollo 12	Nov 14	US	Saturn 5	Lunar lander	Second manned landing (ninth in all)
1970					
Luna-1970A	Feb 6	USSR	Proton-K + Block D	Lunar sampler	Failed to reach Earth orbit. Luna 16-type
Apollo 13	Apr 11	US	Saturn 5	Lunar lander	Manned Earth return/flyby after craft failure
Venera 7	Aug 17	USSR	Molniya-M	Venus capsule	23 min data after landing 15 Dec. 1,180 kg (500 kg capsule)
Cosmos 359	Aug 22	USSR	Molniya-M	Venus capsule	Failed to leave Earth orbit
Luna 16	Sep 12	USSR	Proton-K + Block D	Lunar sampler	13th unmanned lunar orbiter; first sample return (~100 g)
Zond 8	Oct 20	USSR	Proton-K + Block D	Lunar flyby	Earth return; test for manned craft
Luna 17	Nov 10	USSR	Proton-K + Block D	Lunar rover	14th unmanned orbiter; first rover: Lunokhod 1 operated for 11 months, travelled 10,540 m, returned >22,000 TV images
1971					
Apollo 14	Jan 31	US	Saturn 5	Lunar lander	Third manned landing, 12th in all
Mariner 8	May 8	US	Atlas-Centaur	Mars flyby	Booster failure; Centaur autopilot failure
Cosmos 419	May 10	USSR	Proton-K + Block D	Mars orbiter	Stage 4 failed to ignite to leave Earth orbit because of incorrect timer setting. Orbiter-only to beat Mariner 8/9 as first Mars orbiter
Mars 2	May 19	USSR	Proton-K + Block D	Mars orbit/land	Second in Mars orbit; 27 Nov landing failed during dust storm. 4,650 kg
Mars 3	May 28	USSR	Proton-K + Block D	Mars orbit/land	Third in Mars orbit; 2 Dec lander failed at 45°S/158°W. Claimed 20 s of TV signals received, but dust storm in progress; claimed in 1993 there were no signals – a propaganda lie to claim first 'successful' landing. Also carried a small tethered rover
Mariner 9	May 30	US	Atlas-Centaur	Mars orbiter	First in Mars orbit, 13 Nov; mapping survey (7,329 images) in 349 days' operations. Mapped most of surface at 1-2 km resolution (2% at 100-300 m)
Apollo 15	Jul 26	US	Saturn 5	Lunar lander	Fourth manned landing; 13th in all
Apollo 15 subsat	Aug 4	US	Apollo 15	Lunar orbiter	Released from Apollo 15
Luna 18	Sep 2	USSR	Proton-K + Block D	Lunar sampler	Failed landing attempt
Luna 19	Sep 28	USSR	Proton-K + Block D	Lunar orbiter	Returned images from orbit; contact >1 yr
1972					
Luna 20	Feb 14	USSR	Proton-K + Block D	Lunar sampler	Second successful unmanned sample return (~50 g?); 14th landing in all
Pioneer 10	Mar 2	US	Atlas-Centaur	Jupiter flyby	First jovian flyby; still operating (contact to 67 AU in 1997 projected)
Venera 8	Mar 27	USSR	Molniya-M	Venus capsule	Successful atmospheric probe; tx for 50 min after touchdown 22 Jul. 1,184 kg
Cosmos 482	Mar 31	USSR	Molniya-M	Venus capsule	Failed to leave Earth orbit
Apollo 16	Apr 16	US	Saturn 5	Lunar lander	Fifth manned landing; 15th successful in all
Apollo 16 subsat	Apr 24	US	Apollo	Lunar orbiter	Released from Apollo 16
Soyuz L3	Nov 23	USSR	SL-15/N-1	Lunar distance	Exploded T+107 s. N1 test away from Moon, with orbiter + model lander
Apollo 17	Dec 7	US	Saturn 5	Lunar lander	Sixth manned landing; 16th in all
1973					
Luna 21	Jan 8	USSR	Proton-K + Block D	Lunar rover	17th lunar landing; second rover, Lunokhod 2 covered 37 km, returned >80,000 TV images
Pioneer 11	Apr 5	US	Atlas-Centaur	Jup/Sat flyby	First Saturn, second Jupiter flyby; last contact 30 Sep 1995 at 42 AU
Explorer 49	Jun 10	US	Delta	Lunar orbiter	Successful
Mars 4	Jul 21	USSR	Proton-K + Block D	Mars orbiter	Retro failure; 2,200 km flyby 10 Feb 1974 (returned one image swath). 4,385 kg
Mars 5	Jul 25	USSR	Proton-K + Block D	Mars orbiter	Fourth Mars orbiter; operated 'a few days' returning S hemisphere images, 4,385 kg
Mars 6	Aug 5	USSR	Proton-K + Block D	Mars lander	Contact lost shortly before landing 12 Mar 1974 at 24°S/25°W; first atmosphere data
Mars 7	Aug 9	USSR	Proton-K + Block D	Mars lander	Missed planet 9 Mar 1974 by 1,300 km (attitude or retro failure)
Mariner 10	Nov 4	US	Atlas-Centaur	Merc/Venus flyby	First Mercury flyby; first Mercury/Venus imaging (6,800 images). Venus 5,760 km 5 Feb 1974. Mercury 271 km 29 Mar 1974 (2,300 images), 48,069 km 21 Sep 1974 (1,000), 319 km 16 Mar 1975 (349; 1 km res); in total covered 57%
1974					
Luna 22	May 29	USSR	Proton-K + Block D	Lunar orbiter	Imaging survey. Contact >15 months
Luna 23	Oct 28	USSR	Proton-K + Block D	Lunar sampler	18th successful landing; sampler damaged (abandoned after 3 days)
Helios 1	Dec 10	US/FRG	Titan-Centaur	Solar probe	Approached to within 47 million km of Sun
1975					
Venera 9	Jun 8	USSR	Proton-K + Block D	Venus orbit/lander	First Venus orbiter; first surface image; operated for 53 min after landing, 4,936 kg (660 kg lander)
Venera 10	Jun 14	USSR	Proton-K + Block D	Venus orbit/lander	Second Venus orbiter; second surface image; operated for 65 min after landing. 5,033 kg
Viking 1	Aug 20	US	Titan-Centaur	Mars orbit/lander	First Mars soft landing; first surface images; fifth orbiter (2,325 kg; operated to 7 Aug 1980). 1,067 kg lander 20 Jul 1976, operated to Nov 1982
Viking 2	Sep 9	US	Titan-Centaur	Mars orbit/lander	Second Mars soft landing; second surface images; sixth orbiter (2,325 kg; operated to 24 Jul 1980). 1,067 kg lander 3 Sep 1976, operated to 11 Apr 1980
Luna-1975A	Oct 16	USSR	Proton-K + Block D	Lunar sampler	Failed to reach Earth orbit
1976					
Helios 2	Jan 16	US/FRG	Titan-Centaur	Solar probe	Approached to within 43 million km of Sun
Luna 24	Aug 9	USSR	Proton-K + Block D	Lunar sampler	19th landing; ninth sample return in all
1977					
Voyager 2	Aug 20	US	Titan-Centaur	Multiple flyby	Fourth Jupiter, third Saturn, first Uranus/Neptune flybys; still operating
Voyager 1	Sep 5	US	Titan-Centaur	Jup/Sat flyby	Third Jupiter, second Saturn flybys; still operating
1978					
Pioneer Venus 1	May 20	US	Atlas-Centaur	Venus orbiter	Third Venus orbiter; op until burned up 8 Oct 1992
Pioneer Venus 2	Aug 8	US	Atlas-Centaur	Venus capsules	Five successful atmospheric probes
ISEE-3/ICE	Aug 12	US	Delta	Interplanetary monitor	First cometary encounter
Venera 11	Sep 9	USSR	Proton-K + Block D	Venus lander	No surface images, but op for 95 min after landing 23 Dec. 4,450 kg
Venera 12	Sep 14	USSR	Proton-K + Block D	Venus lander	No surface images, but op for 110 min after landing 21 Dec. 4,450 kg
1981					
Venera 13	Oct 30	USSR	Proton-K + Block D	Venus lander	Third surface images (first colour); first soil analysis. Op for 127 min after 1 Mar 1982 landing. 4,363 kg (760 kg lander)
Venera 14	Nov 4	USSR	Proton-K + Block D	Venus lander	Fourth surface images (second colour), second soil analysis. Op for 57 min after 3 Mar 1982 landing. 4,364 kg (760 kg lander)
1983					
Venera 15	Jun 2	USSR	Proton-K + Block D	Venus orbiter	Radar mapper (1-2 km res); fourth Venus orbiter 10 Oct 1983. 5,250 kg
Venera 16	Jun 7	USSR	Proton-K + Block D	Venus orbiter	Radar mapper (1-2 km res); fourth Venus orbiter 16 Oct 1983. 5,300 kg

	Launch	Nation	Launcher	Mission Type	Comments
1984					
Vega 1	Dec 15	USSR	Proton-K + Block D	Venus lander/ comet flyby	First Venus balloon 11 Jun 1985 (46 h flt), lander survived 56 min (no camera); first Halley flyby 6 Mar 1986, 8,890 km. 4,920 kg (21 kg balloon)
Vega 2	Dec 21	USSR	Proton-K + Block D	Venus lander/ comet flyby	Second Venus balloon 15 Jun 1985 (46½ h flt), lander survived 57 min (no camera); second Halley flyby 9 Mar 1986, 8,030 km. 4,920 kg (21 kg balloon)
1985					
Sakigake	Jan 7	Japan	M-3SII	Halley flyby	Distant flyby successful; last Earth approach of 4,000R_E Jul 1994; contact lost 15 Nov 1995
Giotto	Jul 2	Europe	Ariane 1	Halley flyby	First comet coma encounter, Mar 86; Earth swingby Jul 90; 2nd comet flyby Jul 92; Earth flyby due Jul 1999; dormant
Suisei	Aug 18	Japan	M-3SII	Halley flyby	Flyby successful; mission ended Feb 1991
1988					
Phobos 1	Jul 7	USSR	Proton-K + Block D	Mars orbiters,	Contact lost Sep 1988 en route
Phobos 2	Jul 12	USSR	Proton-K + Block D	Phobos flyby/ landings	Seventh Mars orbiter; contact lost Mar 1989
1989					
Magellan	May 4	US	Shuttle/IUS	Venus orbiter	Radar, altimeter, gravity mapper; sixth Venus orbiter (10 Aug 1990); last contact 12 Oct 1994, burned up 13 Oct
Galileo	Oct 18	US/FRG	Shuttle/IUS	Jupiter orbiter/ atmos probe	First Jupiter orbiter, Dec 1995; 16,000 km Venus flyby 10 Feb 90; Earth flyby 8 Dec 90; first asteroid (Gaspra) 29 Oct 91; Earth flyby 8 Dec 92; second asteroid (Ida) 28 Aug 93; 57 min probe data 7 Dec 1995; 1st orbiter 8 Dec 1995; began 2 yr orbital tour with 835 km Ganymede flyby 27 Jun 1996; still operating
1990					
Hiten/Hagomoro	Jan 24	Japan	M-3SII	Lunar flyby orbiter	Hagomoro subsat injected into lunar orbit 19 Mar during parent Hiten's geocentric flyby; Hiten also became lunar orbiter 15 Feb 1992, impacted 10 Apr 1993
Ulysses	Oct 6	ESA/US	Shuttle/IUS	High ecliptic/ solar polar	Jupiter swingby 8 Feb 1992 for solar polar overflights 1994/95 + 2000/01; still operating
1992					
Mars Observer	Sep 25	US	Shuttle/TOS	Mars orbiter	Mars arrival due 24 Aug 1993 for global survey but last contact 22 Aug; fate unknown
1994					
Clementine	Jan 25	US	Titan 2G	Lunar orbiter	In lunar orbit 19 Feb – 4 May; 31 Aug flyby of asteroid 1620 Geographos cancelled after attitude propellant exhausted 7 May while in Earth orbit; perturbed from Earth-Moon system early Sep 1994, entering 1.0243 × 1.0561 AU heliocentric orbit. Intermittent contact made in 1995
1996					
NEAR	Feb 17	US	Delta	Asteroid	Near Earth Asteroid Rendezvous planned as first asteroid orbiter; flew to within 1,200 km of C-class asteroid 253 Mathilde on 27 Jun 1997, Earth on 23 Jan 1998 and to within 3,830 km of 433 Eros on 23 Dec 1998 when 222 pictures were sent back. Scheduled to begin year-long orbit of 433 Eros in Feb 2000
MGS	Nov 7	US	Delta	Mars orbiter	Mars Global Surveyor arrived in Mars orbit 11 Sep 1997 but aerobraking extended by problem with solar panel strut and mapping operations only started 9 Mar 1999, to continue until 31 Jan 2001 at an average altitude of 400 km
Mars-8	Nov 16	Russia	Proton-K	Mars orbiter W/2 landers	Failed to leave Earth orbit
Mars Pathfinder	Dec 4	US	Delta	Mars lander	Landed 4 Jul 1997 and returned copious data and images from lander base and Sojourner roving vehicle for three months compared to predictions of one month and one week respectively
1997					
Cassini/Huygens	Oct 15	US	Titan 4	Saturn orbiter	Performed fly-bys of Venus on 26 Apr 1998 and 24 Jun 1999 and Earth on 18 Aug 1999. Will fly past Jupiter in Dec 2000 and reach orbit around Saturn in June 2004 releasing the Huygens probe to Titan six months later
1998					
Lunar Prospector	Jan 7	US	LMLV-2	Lunar orbiter	Began geochemical mapping operations from 100 km orbit from 15 Jan 1998 until start of extended mission at 30 km altitude from Jan 1999. Impacted on the moon 31 Jul 1999
Nozomi	Jul 3	Japan	M-5	Mars orbiter	Aeronomy orbiter given insufficient energy during multiple earth-moon swing-bys for planned arrival in Oct 1999 and will now reach Mars in December 2003
Deep Space 1	Oct 24	US	Delta	Asteroid flyby	Demonstrated 12 key technologies for future missions including ion propulsion. Passed within 15 km of asteroid 9969 Braille on 29 Jul 1999
MCO	Dec 11	US	Delta	Mars orbiter	Mars Climate Orbiter is second flight in Mars Surveyor programme expected to enter Mars polar orbit 23 Sep 1999 for aerobraking through 23 Nov prior to climate/weather studies
1999					
MPL	Jan 3	US	Delta	Mars lander	Carries two Deep Space 2 Microprobes released prior to landing on 3 Dec 1999 at polar site to search for frozen water
Stardust	Feb 7	US	Delta	Comet sampler	Due to fly within 100 km of comet 81P/Wild-2 in Jan 2004 and return samples to earth in Jan 2006

List of Chinese satellite launches, 1970-1998 (including known test flights)

Launch Date	Satellite	Mass kg	Vehicle	Site	Orbit Class
10 January 1970			CZ-1	J	Test flight
24 April 1970	DFH 1	173	CZ-1	J	E-LEO
3 March 1971	SJ 1	221	CZ-1	J	E-LEO
10 August 1972			FB-1	J	Test flight
18 September 1973	JSSW	1,100 ?	FB-1	J	Failure
14 July 1974	JSSW	1,100 ?	FB-1	J	Failure
4 November 1974	FSW-0	1,800 ?	CZ-2A	J	Failure
26 July 1975	JSSW 1	1,107	FB-1	J	LEO
26 November 1975	FSW-0 1	1,790	CZ-2C	J	LEO (3)
16 December 1975	JSSW 2	1,109	FB-1	J	LEO
30 August 1976	JSSW 3	1,108	FB-1	J	E-LEO
10 November 1976	JSSW	1,100 ?	FB-1	J	Failure
7 December 1976	FSW-0 2	1,812	CZ-2C	J	LEO (3)
14 September 1977			FB-1	J	Test flight
26 January 1978	FSW-0 3	1,810	CZ-2C	J	LEO (3)
16 April 1978			FB-1	J	Test flight
28 July 1979	SJ	770 ?	FB-1	J	Failure
19 September 1981	SJ 2	257	FB-1	J	E-LEO
	SJ 2A	483			E-LEO
	SJ 2B	28			E-LEO
9 September 1982	FSW-0 4	1,783	CZ-2C	J	LEO (5)
19 August 1983	FSW-0 5	1,842	CZ-2C	J	LEO (5)

Launch Date	*Satellite*	*Mass kg*	*Vehicle*	*Site*	*Orbit Class*
29 January 1984	SW (DFH-2)	1,000	CZ-3	XC	GEO fail
8 April 1984	STTW A (DFH-2)	1,000	CZ-3	XC	GEO
12 September 1984	FSW-0 6	1,809	CZ-2C	J	LEO (5)
21 October 1985	FSW-0 7	1,820	CZ-2C	J	LEO (5)
1 February 1986	STTW 1 (DFH-2)	1,000	CZ-3	XC	GEO
6 October 1986	FSW-0 8	1,820	CZ-2C	J	LEO (5)
5 August 1987	FSW-0 9	1,820	CZ-2C	J	LEO (5)
9 September 1987	FSW-1 1	2,100	CZ-2C	J	LEO (8)
7 March 1988	STTW 2 (DFH-2A)	1,025	CZ-3	XC	GEO
5 August 1988	FSW-1 2	2,100	CZ-2C	XC	LEO (8)
6 September 1988	FY-1 1	757	CZ-4A	T	SSO
22 December 1988	STTW 3 (DFH-2A)	1,025	CZ-3	XC	GEO
4 February 1990	STTW 4 (DFH-2A)	1,024	CZ-3	XC	GEO
7 April 1990	AsiaSat 1 (Hong Kong)	1,442	CZ-3	XC	GEO
16 July 1990	Badr 1 (Pakistan)	52	CZ-2E/PKM	XC	LEO
	AUSSAT-B (model only)	2,700 ?		XC	GTO fail
3 September 1990	FY-1 2	881	CZ-4A	T	SSO
	QQW 1	4 ?			SSO
	QQW 2	4 ?			SSO
5 October 1990	FSW-1 3	2,080	CZ-2C	J	LEO (8)
28 December 1991	STTW 5	1,025 ?	CZ-3	XC	GEO fail
9 August 1992	FSW-2 1	2,400	CZ-2D	J	LEO (16)
13 August 1992	Optus-B 1 (Australia)	3,000 ?	CZ-2E/STAR-63F	XC	GEO
6 October 1992	Freja (Sweden)	259	CZ-2C	J	E-LEO
	FSW-1 4	2,100			LEO (7)
21 December 1992	Optus-B 2 (Australia)	3,000 ?	CZ-2E/STAR-63F	XC	GEO fail (satellite exploded)
8 October 1993	FSW-15	2,099	CZ-2C	J	LEO
8 February 1994	SJ 4	410	CZ-3A	XC	GTO
	KF 1	1,600 ?			GTO
3 July 1994	FSW-2 2	2,600 ?	CZ-2D	J	LEO (15)
21 July 1994	APStar 1	1,383	CZ-3	XC	GEO
27 August 1994	OPTUS-B 3	3,000 ?	CZ-2E/STAR-63F	XC	GEO
29 November 1994	Zhongxing 6	2,230	CZ-3A	XC	GEO
25 January 1995	APStar 2 (Hong Kong)	3,000 ?	CZ-2E/STAR-63F	XC	Failure (satellite exploded ?)
29 May 1995		?	CZ-1D	J	Failure
25 November 1995	AsiaSat 2 (Hong Kong)	3,379	CZ-2E/EPKM	XC	GEO
28 December 1995	EchoStar 1 (USA)	3,288	CZ-2E/EPKM	XC	GEO
14 February 1996	INTELSAT 708 (USA)	4,576	CZ-3B	XC	Failure
4 July 1996	APStar 1A (Hong Kong)	1,385 ?	CZ-3	XC	GEO
18 August 1996	Zhongxing 7	1,400 ?	CZ-3	XC	GEO fail
20 October 1996	FSW-2 3	2,600 ?	CZ-2D	J	LEO (15)
11 May 1997	Zhongxing 8	2,230	CZ-3A	XC	GEO
10 June 1997	Feng Yun-2	1,380	CZ-3	XC	GEO
19 August 1997	Mabuhay Agila	2,560	CZ-3B	XC	GEO
1 September 1997	Iridium MFS-1	650	CZ-2C/SD	T	LEO
	MFS-2	650			
16 October 1997	ApStar 2R	3,747	CZ-3B	XC	GEO
8 December 1997	Iridium 42	657	CZ-2C/SD	T	LEO
	Iridium 44				
25 March 1998	Iridium 51	657	CZ-2C/SD	T	LEO
	Iridium 61				
2 May 1998	Iridium 69	657	CZ-2C/SD	T	LEO
	Iridium 71				
30 May 1998	Zhongwei 1	2,984	CZ-3B	XC	GEO
18 July 1998	Sinosat 1	2,840	CZ-3B	XC	GEO
19 August 1998	Iridium 3	657	CZ-2C/SD	T	LEO
	Iridium 76	657			
19 December 1998	Iridium 11A	657	CZ-2C/SD	T	LEO
	Iridium 20A	657			

Sub-orbital test flights are shown as 'Test flight'. Abbreviations of satellite names are: DFH – Dong Fang Hong (The East is Red), SJ – Shi Jian (Practice), JSSW – Ji Shu Shiyan Weixing (technical experimental satellite), FSW – Fanhui Shi Weixing (recoverable experimental satellite), SW – Shiyan Weixing (experimental satellite), STTW A – Shiyan Tongbu Tongxin Weixing – (experimental stationary communications satellite), STTW 1-5 – Shiyong Tongbu Tongxin Weixing (operational stationary communications satellite), FY – Feng Yun (wind and cloud), QQW – Qi Qui Weixing (atmospheric observation satellite). An asterisk indicates a non-Chinese satellite, with the owning country shown under the satellite's name. Launch sites are J – Jiuquan, XC – Xi Chang and T – T'ai Yuan. Orbits are indicated in general terms: LEO – Low Earth Orbit, E-LEO – Eccentric Low Earth Orbit, GEO – Geosynchronous Earth Orbit, GTO – Geosynchronous Transfer Orbit and SSO – Sun-Synchronous Orbit. 'Failure' indicates that nothing reached Orbit 'GEO fail' indicates a mission intended to reach GEO which failed after the initial orbital injection. For the recoverable FSW satellites the lifetime in days is shown in parentheses following 'LEO'.

Parking orbits used by CZ-2E when launching HS-601 satellites

Payload	*Inclination deg*	*Period min*	*Perigee km*	*Apogee km*
AUSSAT-B model	28.49	96.6	204	990
Optus-B 1	27.99	97.1	203	1,032
Optus-B 2	28.15	96.9	208	1,010
Optus-B 3	27.85	97.5	190	1,086
APStar 2	Failed to reach orbit			

AUSSAT-B was a model with a Chinese third stage, but was flown to simulate the stack for operational launches. From the data in this table, it would appear that the CZ-2E performed successfully in placing the third stage and payload into the planned parking orbit on all four orbital missions.

List of Chinese launch failures

Launch date	*Satellite*	*Comments*	
18 September 1973	Ji Shu Shiyan Weixing?	Rumoured launch failure	(FB-1)
14 July 1974	Ji Shu Shiyan Weixing?	Second stage engine lost thrust	(FB-1)
5 November 1974	Fanhui Shi Weixing -0	First stage control system failure	(CZ-2A)
10 November 1976	Ji Shu Shiyan Weixing?	Rumoured launch failure	(FB-1)
28 July 1979	Shi Jian	Second stage thrusters failure	(FB-1)
29 January 1984	Shiyan Weixing	Third stage failed to re-ignite after parking orbit injection	(CZ-3)
16 July 1990	Badr + AUSSAT model	Chinese solid-propellant PKM failed to place the AUSSAT model in an eccentric Orbit: Badr was deployed as planned	(CZ-2E/PKM)
28 December 1991	Shiyong Tongbu Tongxin Weixing 5	Third stage failed to re-ignite after parking orbit injection	(CZ-3)
21 December 1992	Optus-B 2	Explosion of the satellite shortly after launch	(CZ-2E/STAR-63F)
25 January 1995	APStar 2	Possible explosion of the satellite shortly after launch	(CZ-2E/STAR-63F)
29 May 1995			No details of failure (CZ-1D)
14 February 1996	INTELSAT 708	Attitude control lost within seconds of launch	(CZ-3B)
18 August 1996	Zhongxing 7	Third stage shutdown during burn to GTO	(CZ-3)

Notes: In April 1994 the first Feng Yun-2 geosynchronous orbit satellite was destroyed during propellant loading: since it had not been attached to its CZ-3 launch vehicle, this cannot be classed as a launch failure.

Geosynchronous satellite catalogue for 17 August 1999

Number of satellites: 721

Cat No	*International Designator*	*Common Name*	*Source*	*Apogee (km)*	*Perigee (km)*	*Inclination (°)*
553	1963-004A	SYNCOM 1	US	37015	34181	33.5
634	1963-031A	SYNCOM 2 (A 26)	US	35833	35753	24.9
858	1964-047A	SYNCOM 3	US	35858	35743	8.6
1317	1965-028A	INTELSAT 1-F1	ITSO	35794	35756	13.4
2207	1966-053A	OPS 9381 (GGTS)	US	33895	33629	11.4
2215	1966-053B	OPS 9311 (IDSCS 1)	US	33878	33668	10.1
2216	1966-053C	OPS 9312 (IDSCS 2)	US	33914	33661	11.5
2217	1966-053D	OPS 9313 (IDSCS 3)	US	33937	33685	11.5
2218	1966-053E	OPS 9314 (IDSCS 4)	US	34011	33694	11.6
2219	1966-053F	OPS 9315 (IDSCS 5)	US	34108	33693	11.7
2220	1966-053G	OPS 9316 (IDSCS 6)	US	34211	33711	11.8
2221	1966-053H	OPS 9317 (IDSCS 7)	US	34346	33718	11.9
2222	1966-053J	TITAN 3C TRANSTAGE R/B	US	34702	33439	10.7
2608	1966-110A	ATS 1	US	35788	35757	13.9
2639	1967-001A	INTELSAT 2-F2	ITSO	35880	35728	14.4
2645	1967-003A	OPS 9321 (IDSCS 8)	US	33809	33536	11.5
2649	1967-003B	OPS 9322 (IDSCS 9)	US	33783	33579	11.4
2650	1967-003C	OPS 9323 (IDSCS 10)	US	33837	33554	11.4
2651	1967-003D	OPS 9324 (IDSCS 11)	US	33840	33614	11.5
2652	1967-003E	OPS 9325 (IDSCS 12)	US	33901	33623	11.6
2653	1967-003F	OPS 9326 (IDSCS 13)	US	33993	33631	11.8
2654	1967-003G	OPS 9327 (IDSCS 14)	US	34097	33647	11.8
2655	1967-003H	OPS 9328 (IDSCS 15)	US	34238	33647	10.6
2660	1967-003J	TITAN 3C TRANSTAGE R/B	US	33796	33561	11.4
2717	1967-026A	INTELSAT 2-F3	ITSO	35909	35661	13.8
2862	1967-066A	OPS 9331 (IDSCS 16)	US	33532	33015	12.5
2863	1967-066B	OPS 9332 (IDSCS 17)	US	33549	33025	12.5
2864	1967-066C	OPS 9333 (IDSCS 18)	US	33550	33078	12.5
2865	1967-066D	OPS 9334 (IDSCS 19)	US	33571	33136	12.5
2866	1967-066E	LES 5	US	33616	33189	12.6
2867	1967-066F	DODGE 1	US	33666	33262	12.6
2868	1967-066G	TITAN 3C TRANSTAGE R/B	US	33834	33221	12.6
2969	1967-094A	INTELSAT 2-F4	ITSO	35867	35741	14.3
3029	1967-111A	ATS 3	US	35854	35718	14.5
3284	1968-050A	OPS 9341 (IDSCS 20)	US	33826	33747	11.4
3285	1968-050B	OPS 9342 (IDSCS 21)	US	33859	33741	11.9
3286	1968-050C	OPS 9343 (IDSCS 22)	US	33887	33734	12.0
3287	1968-050D	OPS 9344 (IDSCS 23)	US	33955	33725	12.0
3288	1968-050E	OPS 9345 (IDSCS 24)	US	34026	33734	12.1
3289	1968-050F	OPS 9346 (IDSCS 25)	US	34106	33750	11.9
3290	1968-050G	OPS 9347 (IDSCS 26)	US	34233	33749	12.2
3291	1968-050H	OPS 9348 (IDSCS 27)	US	34377	33741	12.3
3292	1968-050J	TITAN 3C TRANSTAGE R/B	US	35048	33666	11.3
3428	1968-081A	OV2-5	US	35773	35085	13.1
3430	1968-081C	ERS 21 (OV5-4)	US	35823	35765	13.6
3431	1968-081D	LES 6	US	35853	35761	13.6
3432	1968-081E	TITAN 3C TRANSTAGE R/B	US	35843	35039	13.2
3623	1968-116A	INTELSAT 3-F2	ITSO	37117	35976	15.5
3674	1969-011A	INTELSAT 3-F3	ITSO	35803	35767	6.1
3691	1969-013A	OPS 0757 (TACSAT)	US	35772	35766	7.0
3692	1969-013B	TITAN 3C TRANSTAGE R/B	US	35819	35662	14.2
4068	1969-069A	ATS 5	US	36018	35993	15.2
4250	1969-101A	SKYNET 1	UK	35858	35716	14.6
4297	1970-003A	INTELSAT 3-F6	ITSO	36542	36183	5.2
4353	1970-021A	NATO 2A	NATO	35787	35786	14.4
4376	1970-032A	INTELSAT 3-F7	ITSO	35791	35775	0.9
4478	1970-055A	INTELSAT 3-F8	ITSO	36599	33876	13.5
4881	1971-006A	INTELSAT 4-F2	ITSO	36260	36125	15.3
4902	1971-009A	NATO 2B	NATO	35805	35768	15.2
5587	1971-095A	OPS 9431 (DSCS II-1)	US	35804	35774	15.2
5588	1971-095B	OPS 9432 (DSCS II-2)	US	35786	35762	15.1
5589	1971-095C	TITAN 3C TRANSTAGE R/B	US	37368	35979	16.1
5709	1971-116A	INTELSAT 4-F3	ITSO	36031	35912	13.7
5775	1972-003A	INTELSAT 4-F4	ITSO	35934	35889	13.2
6052	1972-041A	INTELSAT 4-F5	ITSO	35843	35832	13.9
6278	1972-090A	ANIK A1 (TELESAT 1)	CA	36251	36143	13.9
6437	1973-023A	ANIK A2 (TELESAT 2)	CA	35977	35861	13.1

Cat No	*International Designator*	*Common Name*	*Source*	*Apogee (km)*	*Perigee (km)*	*Inclination (°)*
6796	1973-058A	INTELSAT 4-F7	ITSO	36130	36082	13.3
6973	1973-100A	OPS 9433 (DSCS II-3)	US	36659	36414	16.0
6974	1973-100B	OPS 9434 (DSCS II-4)	US	36879	36273	15.5
6976	1973-100D	TITAN 3C TRANSTAGE R/B	US	38530	36136	16.8
7229	1974-017A	COSMOS 637	CIS	35802	35484	14.8
7250	1974-022A	WESTAR 1	US	35920	35863	12.8
7298	1974-033A	SMS 1	US	36329	36186	17.3
7318	1974-039A	ATS 6	US	35436	35191	14.7
7324	1974-039C	TITAN 3C TRANSTAGE R/B	US	35776	35581	15.1
7392	1974-060A	MOLNIYA 1-S	CIS	35838	35772	15.2
7466	1974-075A	WESTAR 2	US	35915	35892	12.6
7544	1974-093A	INTELSAT 4-F8	ITSO	35948	35900	11.9
7547	1974-094A	SKYNET 2B	UK	35780	35726	14.2
7578	1974-101A	SYMPHONIE 1	FGER	35879	35869	14.7
7648	1975-011A	SMS 2	US	36046	35953	14.5
7790	1975-038A	ANIK A3 (TELESAT-3)	CA	35869	35834	12.0
7815	1975-042A	INTELSAT 4-F1	ITSO	36110	36035	11.9
8132	1975-077A	SYMPHONIE 2	FGER	35892	35849	14.8
8330	1975-091A	INTELSAT 4A-F1	ITSO	35896	35871	11.9
8357	1975-097A	COSMOS 775	CIS	35864	35748	15.2
8366	1975-100A	GOES 1 (SMS-C)	US	35807	35785	14.6
8476	1975-117A	SATCOM 1	US	36077	35880	12.0
8513	1975-123A	RADUGA 1	CIS	35808	35776	15.0
8585	1976-004A	CTS	CA	35827	35715	14.9
8620	1976-010A	INTELSAT 4A-F2	ITSO	35963	35939	12.0
8697	1976-017A	MARISAT 1	US	36126	36050	13.5
8746	1976-023A	LES 8	US	35834	35734	13.0
8747	1976-023B	LES 9	US	35879	35694	13.0
8751	1976-023F	TITAN 3C TRANSTAGE R/B	US	36928	35791	13.4
8774	1976-029A	SATCOM 2	US	36504	36006	11.8
8808	1976-035A	NATO 3A	NATO	36006	35811	13.0
8832	1976-023J	TITAN 3C DEB	US	36947	35773	13.4
8838	1976-042A	COMSTAR 1	US	35929	35897	11.9
8882	1976-053A	MARISAT 3	US	37606	36511	14.4
9009	1976-066A	PALAPA 1	INDO	35844	35829	11.7
9047	1976-073A	COMSTAR 2	US	35885	35859	11.7
9416	1976-092A	RADUGA 2	CIS	35881	35680	15.0
9478	1976-101A	MARISAT 2	US	35793	35780	12.7
9503	1976-107A	EKRAN 1	CIS	36055	35489	15.0
9785	1977-005A	NATO 3B	NATO	37448	37051	13.2
9852	1977-014A	KIKU 2 (ETS 2)	JPN	35877	35841	14.3
9862	1977-018A	PALAPA 2	INDO	35875	35823	10.9
10000	1977-034A	OPS 9437 (DSCS II-7)	US	36906	36747	15.2
10001	1977-034B	OPS 9438 (DSCS II-8)	US	37360	37044	14.9
10002	1977-034C	TITAN 3C TRANSTAGE R/B	US	38494	35828	15.5
10024	1977-041A	INTELSAT 4A-F4	ITSO	36062	35981	11.0
10061	1977-048A	GOES 2	US	35792	35770	13.5
10143	1977-065A	HIMAWARI 1 (GMS 1)	JPN	36135	36017	13.7
10159	1977-071A	RADUGA 3	CIS	35813	35755	14.8
10294	1977-080A	SIRIO	IT	35879	35717	12.5
10365	1977-092A	EKRAN 2	CIS	35941	35601	14.8
10489	1977-108A	METEOSAT 1	ESA	35855	35734	14.3
10516	1977-118A	SAKURA 1A (CS-1A)	JPN	36186	36155	13.4
10557	1978-002A	INTELSAT 4A-F3	ITSO	35897	35881	10.5
10637	1978-012A	IUE	US	42011	29683	37.4
10778	1978-035A	INTELSAT 4A-F6	ITSO	35825	35787	10.5
10792	1978-039A	YURI (BSE)	JPN	35837	35699	14.1
10855	1978-044A	OTS 2	ESA	36132	36081	12.3
10953	1978-062A	GOES 3	US	35806	35771	12.6
10975	1978-068A	COMSTAR 3	US	36173	36008	10.4
10981	1978-071A	GEOS 2	ESA	36058	36022	14.3
10987	1978-073A	RADUGA 4	CIS	35812	35718	14.5
11115	1978-106A	NATO 3C	NATO	36311	36277	10.8
11144	1978-113A	OPS 9441 (DSCS II-11)	US	37738	37525	13.0
11145	1978-113B	OPS 9442 (DSCS II-12)	US	36331	36284	12.8
11147	1978-113D	TITAN 3C TRANSTAGE R/B	US	38922	36514	14.9
11153	1978-116A	ANIK B1 (TELESAT-4)	CA	35926	35904	10.0
11158	1978-118A	GORIZONT 1	CIS	49736	21832	25.1
11256	1979-007A	SCATHA	US	43139	27736	14.3
11261	1979-009A	AYAME 1 (ECS 1)	JPN	37278	29407	5.6
11273	1979-015A	EKRAN 3	CIS	35945	35633	14.3
11343	1979-035A	RADUGA 5	CIS	35805	35761	14.3
11353	1979-038A	OPS 6392 (FLTSATCOM 2)	US	36361	36195	12.8
11440	1979-062A	GORIZONT 2	CIS	35794	35774	14.0
11484	1979-072A	WESTAR 3	US	35896	35853	8.9
11561	1979-087A	EKRAN 4	CIS	35810	35757	14.1
11567	1974-017F	SL-12 R/B(2)	CIS	35779	35386	14.6
11568	1975-123F	SL-12 R/B(2)	CIS	35785	35656	14.9
11569	1976-107F	SL-12 R/B(2)	CIS	35506	35406	14.7
11570	1977-071F	SL-12 R/B(2)	CIS	36571	36452	15.5
11571	1977-092G	SL-12 R/B(2)	CIS	35557	35453	14.5
11581	1977-092H	EKRAN 2 DEB	CIS	35847	35731	14.8
11621	1979-098A	OPS 9443 (DSCS II-13)	US	37196	37103	12.6
11622	1979-098B	OPS 9444 (DSCS II-14)	US	36358	36309	12.4
11623	1979-098C	TITAN 3C TRANSTAGE R/B	US	38615	35855	14.0
11648	1979-105A	GORIZONT 3	CIS	35809	35722	13.8
11669	1980-004A	OPS 6393 (FLTSATCOM 3)	US	35875	35695	11.7
11676	1975-097F	SL-12 R/B(2)	CIS	35938	35740	15.2
11684	1979-105E	SL-12 R/B(2)	CIS	36295	36179	14.1
11708	1980-016A	RADUGA 6	CIS	35782	35770	13.9
11715	1980-018A	AYAME 2 (ECS-2)	JPN	36839	32784	1.4
11728	1980-016D	SL-12 R/B(2)	CIS	36598	36494	14.4
11841	1980-049A	GORIZONT 4	CIS	36272	36237	13.7
11862	1980-049F	SL-12 R/B(2)	CIS	36575	36332	14.0
11890	1980-060A	EKRAN 5	CIS	35868	35723	11.3
11926	1978-118C	SL-12 R/B(2)	CIS	49072	21765	24.7
11941	1978-073E	SL-12 R/B(2)	CIS	36635	36487	15.2
11964	1980-074A	GOES 4	US	36235	35927	12.4
12003	1980-081A	RADUGA 7	CIS	35827	35796	13.6

Cat No	*International Designator*	*Common Name*	*Source*	*Apogee (km)*	*Perigee (km)*	*Inclination (°)*
12065	1980-091A	SBS 1	US	35929	35900	9.6
12089	1980-098A	INTELSAT 502	ITSO	36258	36103	8.2
12120	1980-104A	EKRAN 6	CIS	35832	35756	13.5
12309	1981-018A	COMSTAR 4	US	35789	35785	10.4
12351	1981-027A	RADUGA 8	CIS	36093	35400	13.5
12447	1980-081F	SL-12 R/B(2)	CIS	35904	35848	13.6
12471	1980-104E	SL-12 R/B(2)	CIS	35626	35353	13.3
12472	1981-049A	GOES 5	US	35791	35779	9.8
12474	1981-050A	INTELSAT 501	ITSO	36214	36164	8.7
12544	1981-057A	METEOSAT 2	ESA	36351	36100	9.8
12545	1981-057B	APPLE	IND	35942	35769	12.8
12564	1981-061A	EKRAN 7	CIS	35799	35744	13.2
12618	1981-069A	RADUGA 9	CIS	35805	35790	13.2
12635	1981-073A	FLTSATCOM 5	US	36297	36224	13.0
12677	1981-076A	HIMAWARI 2 (GMS 2)	JPN	36027	35949	12.1
12850	1981-069F	SL-12 R/B(2)	CIS	36614	36432	13.6
12851	1981-061F	SL-12 R/B(2)	CIS	35605	35560	13.1
12855	1981-096A	SBS 2	US	35832	35815	8.7
12897	1981-102A	RADUGA 10	CIS	35788	35771	13.0
12967	1981-114A	SATCOM 3R	US	35836	35814	6.5
12994	1981-119A	INTELSAT 503	ITSO	36103	35921	7.9
13010	1981-122A	MARECS A	ESA	37848	36792	9.6
13035	1982-004A	SATCOM 4	US	35986	35973	5.9
13056	1982-009A	EKRAN 8	CIS	35992	35760	12.9
13069	1982-014A	WESTAR 4	US	35941	35912	5.9
13083	1982-017A	INTELSAT 504	ITSO	36207	35955	7.9
13092	1982-020A	GORIZONT 5	CIS	36411	36152	13.0
13129	1982-031A	INSAT 1A	IND	36177	35394	0.4
13177	1982-044A	COSMOS 1366	CIS	35814	35752	12.8
13269	1982-058A	WESTAR 5	US	36152	36020	5.7
13431	1982-082A	ANIK D1 (TELESAT 6)	CA	35850	35795	6.2
13554	1982-093A	EKRAN 9	CIS	35904	35701	12.4
13595	1982-097A	INTELSAT 505	ITSO	36418	36203	7.4
13624	1982-103A	GORIZONT 6	CIS	35821	35808	12.0
13630	1982-103E	SL-12 R/B(2)	CIS	35806	35694	12.0
13631	1982-105A	SATCOM C5	US	35809	35764	6.3
13636	1982-106A	DSCS II-15	US	37308	37300	10.0
13643	1982-106D	TITAN 34D IUS R/B(2)	US	36232	35846	11.2
13651	1982-110B	SBS 3	US	35917	35849	6.0
13652	1982-110C	ANIK C3 (TELESAT-5)	CA	35905	35875	6.1
13669	1982-113A	RADUGA 11	CIS	36691	36350	11.8
13753	1976-023K	LES 8,9/SOL 11A,B DEB	US	35508	35471	13.8
13782	1983-006A	SAKURA 2A (CS-2A)	JPN	36067	35993	9.3
13878	1983-016A	EKRAN 10	CIS	37480	37167	13.4
13899	1982-020F	SL-12 R/B(2)	CIS	36360	36141	13.1
13900	1979-015D	SL-12 R/B(2)	CIS	35515	35464	14.1
13907	1977-108D	METEOSAT 1 AKM	ESA	36883	36103	15.7
13954	1982-113F	SL-12 R/B(2)	CIS	36647	36474	12.0
13969	1983-026B	TDRS 1	US	35793	35779	10.4
13974	1983-028A	RADUGA 12	CIS	35807	35770	11.5
13983	1983-028F	SL-12 R/B(2)	CIS	35948	35743	11.5
13984	1983-030A	SATCOM 1R	US	35943	35867	5.1
14005	1979-062D	SL-12 R/B(2)	CIS	36541	36523	14.5
14050	1983-041A	GOES 6	US	35803	35801	8.8
14077	1983-047A	INTELSAT 506	ITSO	36123	36074	6.5
14086	1983-016F	SL-12 R/B(2)	CIS	35602	35510	12.3
14114	1982-044F	SL-12 R/B(2)	CIS	35858	35719	12.8
14115	1982-093F	SL-12 R/B(2)	CIS	35523	35503	12.3
14117	1982-009F	SL-12 R/B(2)	CIS	35744	35431	12.8
14128	1983-058A	EUTELSAT 1-F1 (ECS 1)	EUTE	36213	36154	7.6
14133	1983-059B	NAHUEL I2 (ANIK C2)	ARGN	36277	35945	5.9
14134	1983-059C	PALAPA B1	INDO	35839	35795	7.0
14158	1983-065A	GALAXY 1	US	35836	35821	4.2
14160	1983-066A	GORIZONT 7	CIS	36381	36289	11.3
14193	1980-060F	SL-12 R/B(2)	CIS	35480	35351	13.5
14194	1981-027F	SL-12 R/B(2)	CIS	36603	36464	14.0
14195	1981-102F	SL-12 R/B(2)	CIS	35850	35751	13.0
14234	1983-077A	ARABSAT 1DR (TELSTAR 3A)	AB	36094	35902	4.4
14248	1983-081A	SAKURA 2B (CS-2B)	JPN	36223	36177	8.7
14307	1983-088A	RADUGA 13	CIS	36469	36301	11.4
14318	1983-089B	INSAT 1B	IND	35858	35777	7.5
14328	1983-094A	SATCOM 2R	US	36104	35952	3.5
14333	1983-088F	SL-12 R/B(2)	CIS	36623	36472	11.5
14365	1983-098A	GALAXY 2	US	35835	35775	4.1
14377	1983-100A	EKRAN 11	CIS	35785	35756	11.7
14394	1983-100F	SL-12 R/B(2)	CIS	35663	35478	11.6
14421	1983-105A	INTELSAT 507	ITSO	35980	35921	6.9
14532	1983-118A	GORIZONT 8	CIS	36466	36250	11.0
14548	1983-118F	SL-12 R/B(2)	CIS	35966	35594	10.7
14659	1984-005A	YURI 2A (BS-2A)	JPN	36180	36081	8.6
14725	1984-016A	RADUGA 14	CIS	35800	35762	10.8
14783	1984-022A	COSMOS 1540	CIS	35808	35757	11.8
14786	1984-023A	INTELSAT 508	ITSO	36724	36563	6.4
14821	1984-028A	EKRAN 12	CIS	37062	36958	12.4
14867	1984-031A	COSMOS 1546	CIS	35886	35694	10.7
14899	1984-035A	STTW-T2	PRC	35800	35735	9.4
14940	1984-041A	GORIZONT 9	CIS	35793	35791	10.5
14943	1984-041D	SL-12 R/B(2)	CIS	36307	36200	10.7
14948	1984-022F	SL-12 R/B(2)	CIS	36005	35793	11.8
14951	1984-031F	SL-12 R/B(2)	CIS	36078	35975	10.8
14985	1984-049A	CHINASAT 5 (SPACENET 1)	PRC	35793	35783	2.7
15057	1984-063A	RADUGA 15	CIS	35789	35754	10.5
15139	1984-028F	SL-12 R/B(2)	CIS	35552	35377	11.6
15141	1983-066F	SL-12 R/B(2)	CIS	36591	36503	11.5
15144	1984-078A	GORIZONT 10	CIS	35802	35776	10.2
15152	1984-080A	HIMAWARI 3 (GMS 3)	JPN	35944	35880	8.4
15158	1984-081A	EUTELSAT 1-F2 (ECS 2)	EUTE	36225	36184	6.9
15159	1984-081B	TELECOM 1A	FR	36477	36161	6.8
15181	1984-078F	SL-12 R/B(2)	CIS	35872	35703	10.2

Cat No	International Designator	Common Name	Source	Apogee (km)	Perigee (km)	Inclination (°)
15219	1984-090A	EKRAN 13	CIS	37076	36967	11.6
15235	1984-093B	SBS 4	US	35794	35781	4.9
15236	1984-093C	LEASAT 2	US	36723	36466	11.8
15237	1984-093D	TELSTAR 302	US	35942	35910	3.4
15308	1984-101A	GALAXY 3	US	35947	35873	3.5
15383	1984-113B	ARABSAT 1D (ANIK D2)	AB	36291	36057	4.7
15384	1984-113C	LEASAT 1	US	36407	36149	6.2
15385	1984-114A	SPACENET 2	US	35977	35886	1.8
15386	1984-114B	MARECS B2	ESA	35802	35771	8.2
15391	1984-115A	NATO 3D	NATO	35790	35783	5.9
15484	1985-007A	GORIZONT 11	CIS	35804	35784	9.8
15487	1985-007D	SL-12 R/B(2)	CIS	35090	34973	9.4
15560	1985-015A	ARABSAT 1A	AB	35772	35718	6.5
15561	1985-015B	BRAZILSAT 1	BRAZ	35791	35782	3.7
15574	1985-016A	COSMOS 1629	CIS	35811	35785	10.0
15581	1985-016F	SL-12 R/B(2)	CIS	36126	35936	10.1
15626	1985-024A	EKRAN 14	CIS	37477	37312	11.4
15629	1985-025A	INTELSAT 510	ITSO	36503	36203	5.3
15630	1985-024D	SL-12 R/B(2)	CIS	35568	35475	10.5
15642	1985-028B	NAHUEL I1 (ANIK C1)	ARGN	35793	35779	1.9
15643	1985-028C	LEASAT 3	US	37075	36406	13.1
15677	1985-035A	GSTAR 1	US	35816	35757	2.5
15678	1985-035B	TELECOM 1B	FR	35832	35808	8.8
15693	1984-063F	SL-12 R/B(2)	CIS	35022	34899	10.1
15824	1985-048B	MORELOS 1	MEX	36032	35996	4.2
15825	1985-048C	ARABSAT 1B	AB	35809	35688	5.8
15826	1985-048D	TELSTAR 303	US	35902	35890	3.2
15873	1985-055A	INTELSAT 511	ITSO	35801	35774	4.9
15946	1985-070A	RADUGA 16	CIS	35813	35800	9.6
15963	1985-070F	SL-12 R/B(2)	CIS	36542	36444	9.9
15993	1985-076B	OPTUS A1 (AUSSAT 1)	AUS	35947	35910	5.0
15994	1985-076C	ASC 1	US	35818	35767	3.8
15995	1985-076D	LEASAT 4	US	36522	36466	8.4
16101	1985-087A	INTELSAT 512	ITSO	36138	36089	4.1
16199	1985-102A	COSMOS 1700	CIS	35812	35784	9.3
16214	1985-102D	SL-12 R/B(2)	CIS	35775	35594	9.2
16250	1985-107A	RADUGA 17	CIS	35836	35784	9.4
16274	1985-109B	MORELOS 2	MEX	35795	35778	0.8
16275	1985-109C	OPTUS A2 (AUSSAT 2)	AUS	35798	35773	4.6
16276	1985-109D	SATCOM K2	US	35799	35776	1.8
16339	1985-107F	SL-12 R/B(2)	CIS	36676	36490	9.7
16482	1986-003B	SATCOM K1	US	35994	35986	2.1
16497	1986-007A	RADUGA 18	CIS	36491	35908	9.4
16526	1986-010A	STTW-1	PRC	35803	35785	8.3
16597	1986-016A	BS-2B (YURI 2B)	JPN	36130	35995	6.8
16649	1986-026A	GSTAR 2	US	35941	35906	3.3
16650	1986-026B	BRAZILSAT 2	BRAZ	35796	35778	2.3
16667	1986-027A	COSMOS 1738	CIS	35856	35711	9.3
16676	1986-027F	SL-12 R/B(2)	CIS	36689	36363	9.6
16729	1986-038A	EKRAN 15	CIS	36937	36792	10.3
16732	1986-038D	SL-12 R/B(2)	CIS	35573	35390	9.7
16769	1986-044A	GORIZONT 12	CIS	35810	35801	8.8
16797	1986-044F	SL-12 R/B(2)	CIS	36567	36496	9.1
16870	1986-007F	SL-12 R/B(2)	CIS	36620	36368	9.6
17046	1986-082A	RADUGA 19	CIS	36347	36257	8.7
17065	1986-082F	SL-12 R/B(2)	CIS	36669	36436	8.9
17083	1986-090A	GORIZONT 13	CIS	36877	36744	8.8
17125	1986-090D	SL-12 R/B(2)	CIS	35842	35749	8.5
17561	1987-022A	GOES 7	US	35819	35805	5.2
17611	1987-028A	RADUGA 20	CIS	37166	36912	9.2
17705	1987-028D	SL-12 R/B(2)	CIS	36027	35770	8.9
17706	1987-029A	AGILA 1 (PALAPA B2P)	RP	35913	35861	2.7
17872	1976-092F	SL-12 R/B(2)	CIS	35844	35702	15.0
17873	1979-035E	SL-12 R/B(2)	CIS	35917	35695	14.3
17874	1984-016F	SL-12 R/B(2)	CIS	35935	35663	10.7
17875	1984-090F	SL-12 R/B(2)	CIS	35588	35436	10.9
17939	1979-087C	SL-12 R/B(2)	CIS	35909	35556	14.0
17969	1987-040A	GORIZONT 14	CIS	36652	36419	10.3
17972	1987-040D	SL-12 R/B(2)	CIS	35088	34979	9.8
18316	1987-070A	KIKU 5 (ETS 5)	JPN	36083	36020	6.1
18328	1987-073A	EKRAN 16	CIS	36894	36871	9.1
18331	1987-073D	SL-12 R/B(2)	CIS	35558	35396	8.6
18350	1987-078A	OPTUS A3 (AUSSAT 3)	AUS	35792	35778	2.5
18351	1987-078B	EUTELSAT 1-F4 (ECS 4)	EUTE	35797	35772	4.7
18384	1987-084A	COSMOS 1888	CIS	35794	35773	7.7
18387	1987-084D	SL-12 R/B(2)	CIS	35958	35733	7.7
18443	1987-091A	COSMOS 1894	CIS	35795	35764	7.8
18446	1987-091D	SL-12 R/B(2)	CIS	35867	35709	7.8
18570	1987-095A	TVSAT 1	GER	36144	36071	8.1
18575	1987-096A	COSMOS 1897	CIS	35821	35773	7.6
18578	1987-096D	SL-12 R/B(2)	CIS	35788	35623	7.6
18631	1987-100A	RADUGA 21	CIS	35786	35771	7.9
18634	1987-100D	SL-12 R/B(2)	CIS	34987	34875	7.5
18715	1987-109A	EKRAN 17	CIS	37237	36890	7.7
18718	1987-109D	SL-12 R/B(2)	CIS	35900	35354	7.4
18877	1988-012A	CS 3A	JPN	36426	36365	2.5
18922	1988-014A	STTW-2	PRC	35803	35762	4.7
18951	1988-018A	SPACENET 3R	US	35806	35770	0.1
18952	1988-018B	TELECOM 1C	FR	36861	36042	3.2
19017	1988-028A	GORIZONT 15	CIS	36624	36347	7.7
19020	1988-028D	SL-12 R/B(2)	CIS	36597	36401	7.8
19073	1988-034A	COSMOS 1940	CIS	35752	35602	7.4
19076	1988-034D	SL-12 R/B(2)	CIS	35935	35741	7.5
19090	1988-036A	EKRAN 18	CIS	37347	37227	8.6
19094	1988-036E	SL-12 R/B(2)	CIS	35643	35458	8.1
19121	1988-040A	INTELSAT 513	ITSO	35806	35768	3.1
19215	1988-051A	METEOSAT 3	ESA	36776	36714	5.5
19217	1988-051C	PAS 1	US	35807	35766	0.1
19330	1988-063A	INSAT 1C	IND	35795	35764	7.4

Cat No	International Designator	Common Name	Source	Apogee (km)	Perigee (km)	Inclination (°)
19331	1988-063B	EUTELSAT 1-F5 (ECS 5)	EUTE	35804	35769	4.0
19344	1988-066A	COSMOS 1961	CIS	35814	35759	7.0
19347	1988-066D	SL-12 R/B(2)	CIS	36390	36098	7.2
19397	1988-071A	GORIZONT 16	CIS	35943	35802	7.1
19400	1988-071D	SL-12 R/B(2)	CIS	35778	35642	7.0
19483	1988-081A	GSTAR 3	US	35814	35759	9.3
19484	1988-081B	SBS 5	US	35792	35781	0.0
19508	1988-086A	CS 3B	JPN	36100	36063	1.6
19548	1988-091B	TDRS 3	US	35824	35759	4.7
19550	1988-091D	IUS R/B(2)	US	35802	35671	6.5
19596	1988-095A	RADUGA 22	CIS	35789	35736	6.9
19621	1988-098A	TDF 1	FR	36098	36086	2.5
19683	1988-108A	EKRAN 19	CIS	36949	36714	7.0
19686	1988-108D	SL-12 R/B(2)	CIS	35496	35385	6.7
19687	1988-109A	SKYNET 4B	UK	35971	35932	5.5
19688	1988-109B	ASTRA 1A	LUXE	35788	35785	0.1
19710	1988-111A	STTW-3	PRC	35802	35776	3.2
19765	1989-004A	GORIZONT 17	CIS	36211	36044	6.7
19772	1989-006A	INTELSAT 515	ITSO	35800	35776	1.6
19776	1989-004F	SL-12 R/B(2)	CIS	36547	36327	6.8
19777	1988-095F	SL-12 R/B(2)	CIS	36506	36402	7.1
19874	1989-020A	JCSAT 1	JPN	36005	35984	1.7
19876	1989-020B	METEOSAT 4 (MOP 1)	ESA	36771	36624	4.3
19883	1989-021B	TDRS 4	US	35800	35774	2.2
19913	1989-021D	IUS R/B(2)	US	35798	35591	9.0
19919	1989-027A	TELE-X	SWED	36118	36072	2.0
19928	1989-030A	RADUGA 23	CIS	35864	35712	6.5
19931	1989-030D	SL-12 R/B(2)	CIS	36520	36395	6.7
20040	1989-041A	SUPERBIRD A	JPN	35959	35910	6.6
20041	1989-041B	DFS 1	GER	35729	35621	2.8
20083	1989-048A	RADUGA 1-1	CIS	36306	36154	6.4
20086	1989-048D	SL-12 R/B(2)	CIS	36564	36371	6.5
20107	1989-052A	GORIZONT 18	CIS	36176	35888	6.3
20110	1989-052D	SL-12 R/B(2)	CIS	35143	34900	6.0
20122	1989-053A	OLYMPUS 1	ESA	35550	35415	5.5
20168	1989-062A	TVSAT 2	GER	35808	35766	0.9
20193	1989-067A	SIRIUS (MARCOPOLO 1)	SWED	35796	35775	0.2
20217	1989-070A	HIMAWARI 4 (GMS 4)	JPN	35791	35781	4.0
20263	1989-081A	GORIZONT 19	CIS	35797	35780	6.1
20266	1989-081D	SL-12 R/B(2)	CIS	35823	35567	6.1
20315	1989-087A	INTELSAT 602	ITSO	35793	35778	0.0
20317	1989-070C	HIMAWARI 4 AKM	JPN	37280	35148	5.9
20367	1989-098A	RADUGA 24	CIS	35798	35778	6.0
20370	1989-098D	SL-12 R/B(2)	CIS	36548	36408	6.2
20391	1989-101A	COSMOS 2054	CIS	35795	35779	5.9
20394	1989-101D	SL-12 R/B(2)	CIS	36398	36328	6.0
20401	1990-001A	SKYNET 4A	UK	35795	35780	3.9
20402	1990-001B	JCSAT 2	JPN	35793	35780	0.0
20410	1990-002B	LEASAT 5	US	35793	35781	3.8
20473	1990-011A	STTW-4	PRC	35801	35773	3.1
20499	1990-016A	RADUGA 25	CIS	35793	35762	5.8
20502	1990-016D	SL-12 R/B(2)	CIS	36017	35695	5.8
20523	1990-021A	INTELSAT 603	ITSO	35796	35778	0.0
20558	1990-030A	ASIASAT 1	AC	35785	35775	0.3
20570	1990-034A	PALAPA B2R	INDO	35789	35787	0.1
20643	1990-051A	INSAT 1D	IND	35819	35776	1.2
20659	1990-054A	GORIZONT 20	CIS	35818	35783	5.5
20662	1990-054D	SL-12 R/B(2)	CIS	35796	35649	5.5
20667	1990-056A	INTELSAT 604	ITSO	35802	35768	0.0
20693	1990-061A	COSMOS 2085	CIS	35810	35768	5.5
20696	1990-061D	SL-12 R/B(2)	CIS	35891	35629	5.4
20705	1990-063A	TDF 2	FR	36475	36045	0.6
20706	1990-063B	DFS 2	GER	35799	35774	0.0
20762	1990-074A	THOR 1 (MARCOPOLO 2)	NOR	35795	35778	0.0
20771	1990-077A	BS-3A (YURI 3A)	JPN	36249	36155	1.1
20776	1990-079A	SKYNET 4C	UK	35796	35778	2.3
20777	1990-079B	EUTELSAT 2-F1	EUTE	35808	35766	0.5
20799	1977-048G	GOES 2 AKM	US	36637	34773	14.5
20800	1989-020E	METEOSAT 4 AKM	ESA	36291	35174	6.6
20801	1978-062D	GOES 3 AKM	US	39916	32242	14.6
20835	1975-011F	SMS 2 AKM	US	36690	35840	15.5
20836	1974-060F	SL-12 R/B(2)	CIS	35866	35691	15.2
20837	1981-057F	METEOSAT 2 AKM (MAGE 1)	ESA	36343	35739	13.0
20872	1990-091A	SBS 6	US	35793	35782	0.0
20873	1990-091B	GALAXY 6	US	35793	35781	0.0
20918	1990-093A	INMARSAT 2-F1	IM	35799	35771	1.8
20923	1990-094A	GORIZONT 21	CIS	35793	35785	5.3
20926	1990-094D	SL-12 R/B(2)	CIS	35770	35472	5.2
20945	1990-100A	SATCOM C1	US	35791	35778	0.1
20946	1990-100B	GSTAR 4	US	35794	35779	0.0
20953	1990-102A	GORIZONT 22	CIS	35791	35782	5.2
20962	1975-100F	GOES 1 AKM	US	36504	34151	14.7
21016	1990-112A	RADUGA 26	CIS	35814	35747	5.1
21019	1990-112D	SL-12 R/B(2)	CIS	35959	35749	5.1
21038	1990-116A	RADUGA 1-2	CIS	35801	35757	5.1
21041	1990-116D	SL-12 R/B(2)	CIS	36610	36293	5.2
21046	1990-102D	SL-12 R/B(2)	CIS	36547	36401	5.4
21047	1991-001A	NATO 4A	NATO	35798	35776	2.5
21052	1969-069D	ATS 5 AKM	US	36926	35842	15.5
21055	1991-003A	ITALSAT 1	IT	35812	35758	1.9
21056	1991-003B	EUTELSAT 2-F2	EUTE	35791	35783	0.0
21111	1991-010A	COSMOS 2133	CIS	35802	35771	4.0
21129	1991-010F	SL-12 R/B(2)	CIS	35898	35769	4.1
21132	1991-014A	RADUGA 27	CIS	35805	35792	5.3
21135	1991-014D	SL-12 R/B(2)	CIS	35035	34810	5.1
21139	1991-015A	ASTRA 1B	LUXE	35795	35777	0.0
21140	1991-015B	METEOSAT 5 (MOP 2)	ESA	35822	35753	2.9
21149	1991-018A	INMARSAT 2-F2	IM	35799	35774	1.9
21222	1991-026A	ANIK E2	CA	35796	35777	0.0

Cat No	International Designator	Common Name	Source	Apogee (km)	Perigee (km)	Inclination (°)
21227	1991-028A	SPACENET 4 (ASC 2)	US	35789	35784	0.0
21392	1991-037A	AURORA 2	US	35794	35778	0.1
21533	1991-046A	GORIZONT 23	CIS	36201	36146	4.8
21536	1991-046D	SL-12 R/B(2)	CIS	35659	35545	4.6
21639	1991-054B	TDRS 5	US	35796	35774	1.3
21641	1991-054D	IUS R/B(2)	US	35939	35653	6.2
21648	1989-101G	COSMOS 2054 DEB	CIS	36623	36337	6.1
21653	1991-055A	INTELSAT 605	ITSO	35794	35780	0.0
21668	1991-060A	BS-3B (YURI 3B)	JPN	36215	36198	1.8
21702	1991-064A	COSMOS 2155	CIS	35810	35804	4.7
21703	1991-064B	SL-12 R/B(2)	CIS	35916	35871	4.7
21726	1991-067A	ANIK E1	CA	35795	35778	0.0
21759	1991-074A	GORIZONT 24	CIS	36384	36231	4.5
21762	1991-074D	SL-12 R/B(2)	CIS	35961	35933	4.5
21765	1991-075A	INTELSAT 601	ITSO	35791	35784	0.0
21789	1991-079A	COSMOS 2172	CIS	35775	35733	4.4
21792	1991-079D	SL-12 R/B(2)	CIS	36290	36224	4.5
21803	1991-083A	EUTELSAT 2-F3	EUTE	35803	35770	0.0
21813	1991-084A	TELECOM 2A	FR	35804	35769	0.0
21814	1991-084B	INMARSAT 2-F3	IM	35798	35774	0.9
21821	1991-087A	RADUGA 28	CIS	35793	35783	4.3
21824	1991-087D	SL-12 R/B(2)	CIS	36517	36342	4.4
21893	1992-010A	SUPERBIRD B1	JPN	35810	35763	0.0
21894	1992-010B	INSAT 2R (ARABSAT 1C)	IND	35808	35765	0.0
21904	1991-015E	METEOSAT 5 AKM	ESA	36487	35168	5.1
21906	1992-013A	GALAXY 5	US	35792	35780	0.0
21922	1992-017A	GORIZONT 25	CIS	35791	35782	4.1
21925	1992-017D	SL-12 R/B(2)	CIS	35657	35457	4.1
21939	1992-021A	TELECOM 2B	FR	35804	35768	0.1
21940	1992-021B	INMARSAT 2-F4	IM	35801	35775	2.6
21964	1992-027A	PALAPA B4	INDO	35788	35786	0.0
21989	1992-032A	INTELSAT K	ITSO	35799	35775	0.0
22027	1992-041A	INSAT 2A	IND	35850	35722	1.7
22028	1992-041B	EUTELSAT 2-F4	EUTE	35800	35774	0.0
22041	1992-043A	GORIZONT 26	CIS	35793	35779	3.8
22044	1992-043D	SL-12 R/B(2)	CIS	36581	36387	3.9
22087	1992-054A	OPTUS B1 (AUSSAT B1)	AUS	35802	35772	0.1
22096	1992-057A	SATCOM C4	US	35801	35770	0.0
22112	1992-059A	COSMOS 2209	CIS	35763	35746	3.9
22115	1992-059D	SL-12 R/B(2)	CIS	35969	35868	3.9
22116	1992-060A	HISPASAT 1A	SPN	35801	35774	0.0
22117	1992-060B	SATCOM C3	US	35792	35778	0.1
22175	1992-066A	DFS 3	GER	35798	35774	0.0
22205	1992-072A	GALAXY 7	US	35795	35778	0.0
22210	1992-074A	EKRAN 20	CIS	35795	35777	3.6
22213	1992-074D	SL-12 R/B(2)	CIS	35636	35451	3.5
22245	1992-082A	GORIZONT 27	CIS	35796	35776	3.6
22248	1992-082D	SL-12 R/B(2)	CIS	36477	36380	3.7
22253	1992-084A	SUPERBIRD A1	JPN	35806	35768	0.0
22266	1984-080E	N-2 R/B(3)	JPN	36493	35372	10.0
22269	1992-088A	COSMOS 2224	CIS	35801	35788	2.7
22272	1992-088D	SL-12 R/B(2)	CIS	35891	35809	2.6
22314	1993-003B	TDRS F6	US	35791	35782	0.6
22316	1993-003D	IUS R/B(2)	US	36133	35520	4.1
22557	1993-013A	RADUGA 29	CIS	35791	35774	3.3
22563	1993-015A	UFO 1 (USA 98)	US	36091	36059	23.1
22624	1993-013D	SL-12 R/B(2)	CIS	36506	36358	3.4
22653	1993-031A	ASTRA 1C	LUXE	35793	35780	0.1
22694	1993-039A	GALAXY 4	US	36089	35890	0.9
22723	1993-048A	HISPASAT 1B	SPN	35795	35780	0.0
22724	1993-048B	INSAT 2B	IND	35810	35766	0.0
22796	1993-058B	ACTS	US	35790	35783	0.8
22836	1993-062A	RADUGA 30	CIS	35785	35783	2.9
22839	1993-062D	SL-12 R/B(2)	CIS	35864	35702	2.9
22871	1993-066A	INTELSAT 701	ITSO	35796	35775	0.0
22880	1993-069A	GORIZONT 28	CIS	35793	35778	2.9
22883	1993-069D	SL-12 R/B(2)	CIS	35754	35692	2.9
22907	1993-072A	GORIZONT 29	CIS	35790	35781	2.8
22910	1993-072D	SL-12 R/B(2)	CIS	35070	34918	2.7
22911	1993-073A	SOLIDARIDAD 1	MEX	35796	35776	0.0
22912	1993-073B	METEOSAT 6	ESA	35789	35784	0.8
22921	1993-076A	NATO 4B	NATO	35801	35773	1.5
22927	1993-077A	TELSTAR 401	US	35802	35771	2.0
22930	1993-078A	DBS 1	US	35798	35776	0.0
22931	1993-078B	THAICOM 1	THAI	35791	35785	0.1
22963	1994-002A	GALS 1	CIS	35797	35777	1.2
22966	1994-002D	SL-12 R/B(2)	CIS	36190	35835	3.9
22981	1994-008A	RADUGA 1-3	CIS	35795	35778	2.6
22984	1994-008D	SL-12 R/B(2)	CIS	36547	36342	2.7
23010	1994-012A	RADUGA 31	CIS	35829	35771	2.6
23013	1994-012D	SL-12 R/B(2)	CIS	36557	36251	2.7
23016	1994-013A	GALAXY 1R	US	35795	35776	0.0
23051	1994-022A	GOES 8	US	35796	35772	0.1
23108	1994-030A	GORIZONT 30	CIS	35793	35778	2.6
23111	1994-030D	SL-12 R/B(2)	CIS	34989	34640	2.4
23118	1993-073E	METEOSAT 6 AKM	ESA	36152	35576	2.9
23124	1994-034A	INTELSAT 702	ITSO	35799	35773	0.0
23168	1994-038A	COSMOS 2282	CIS	35822	35740	1.5
23171	1994-038D	SL-12 R/B(2)	CIS	36126	35921	1.5
23175	1994-040A	PAS 2	US	35856	35717	0.0
23176	1994-040B	BS-3N	JPN	35794	35782	0.0
23185	1994-043A	APSTAR 1	PRC	35791	35783	0.0
23192	1994-047A	DBS 2	US	35797	35776	0.0
23199	1994-049A	BRAZILSAT B1	BRAZ	35792	35783	0.0
23200	1994-049B	TURKSAT 1B	TURK	35790	35781	0.1
23227	1994-055A	OPTUS B3	AUS	35799	35775	0.0
23267	1994-060A	COSMOS 2291	CIS	35796	35778	2.2
23270	1994-060D	SL-12 R/B(2)	CIS	35880	35811	2.2
23305	1994-064A	INTELSAT 703	ITSO	35937	35634	0.1

Cat No	International Designator	Common Name	Source	Apogee (km)	Perigee (km)	Inclination (°)
23313	1994-065A	SOLIDARIDAD 2	MEX	35796	35776	0.0
23314	1994-065B	THAICOM 2	THAI	35816	35758	0.0
23319	1994-067A	EXPRESS 1	CIS	35797	35778	0.0
23322	1994-067D	SL-12 R/B(2)	CIS	35811	35738	3.4
23327	1994-069A	ELEKTRO (GOMS)	CIS	35812	35750	2.4
23330	1994-069D	SL-12 R/B(2)	CIS	35929	35732	2.4
23331	1994-070A	ASTRA 1D	LUXE	35802	35770	0.0
23413	1994-079A	ORION 1	US	35796	35779	0.0
23415	1994-080A	DFH-3 1	PRC	35975	35212	3.2
23426	1994-082A	LUCH	CIS	35810	35795	1.1
23429	1994-082D	SL-12 R/B(2)	CIS	35760	35515	1.0
23448	1994-087A	RADUGA 32	CIS	35793	35785	2.0
23451	1994-087D	SL-12 R/B(2)	CIS	35883	35790	2.0
23461	1995-001A	INTELSAT 704	ITSO	35807	35765	0.0
23522	1995-011B	GMS 5	JPN	35792	35778	0.4
23524	1995-011D	GMS 5 AKM	JPN	36408	34620	2.2
23528	1995-013A	INTELSAT 705	ITSO	35800	35774	0.0
23536	1995-016A	BRAZILSAT B2	BRAZ	35800	35775	0.0
23537	1995-016B	HOT BIRD 1	EUTE	35809	35760	0.0
23553	1995-019A	AMSC 1	US	35797	35777	0.0
23571	1995-023A	INTELSAT 706	ITSO	35798	35777	0.0
23581	1995-025A	GOES 9	US	35809	35769	0.5
23598	1995-029A	DBS 3	US	35797	35776	0.0
23613	1995-035B	TDRS F7	US	35797	35774	3.0
23615	1995-035D	IUS R/B(2)	US	35766	35680	4.6
23636	1995-040A	PAS 4	US	35813	35759	0.0
23639	1995-041A	KOREASAT 1	KOR	35794	35783	0.1
23649	1995-043A	JCSAT 3	JPN	35795	35780	0.0
23651	1995-044A	NSTAR 1	JPN	35794	35780	0.0
23653	1995-045A	COSMOS 2319	CIS	35791	35785	1.4
23656	1995-045D	SL-12 R/B(2)	CIS	35918	35723	1.4
23670	1995-049A	TELSTAR 402R	US	35798	35776	0.0
23680	1995-054A	LUCH 1	CIS	35810	35766	0.5
23683	1995-054D	SL-12 R/B(2)	CIS	35890	35691	0.5
23686	1995-055A	ASTRA 1E	LUXE	35799	35773	0.1
23715	1995-062A	ISO	ESA	70354	494	5.2
23716	1995-062B	ARIANE 44P+3 R/B	ESA	73745	927	5.4
23717	1995-063A	GALS 2	CIS	35804	35771	0.4
23720	1995-063D	SL-12 R/B(2)	CIS	35970	35634	2.6
23723	1995-064A	ASIASAT 2	AC	35798	35777	0.0
23730	1995-067A	TELECOM 2C	FR	35801	35771	0.0
23731	1995-067B	INSAT 2C	IND	35812	35762	0.1
23741	1995-069A	GALAXY 3R	US	35797	35777	0.0
23754	1995-073A	ECHOSTAR 1	US	35791	35781	0.0
23764	1996-002A	PAS 3R	US	35792	35781	0.0
23765	1996-002B	MEASAT 1	MALA	35792	35783	0.0
23768	1996-003A	KOREASAT 2	KOR	35807	35766	0.2
23775	1996-005A	GORIZONT 31	CIS	35792	35779	1.2
23778	1996-005D	SL-12 R/B(2)	CIS	36574	36487	1.2
23779	1996-006A	PALAPA C1	INDO	35802	35771	0.4
23781	1996-007A	NSTAR 2	JPN	35800	35774	0.1
23816	1996-015A	INTELSAT 707	ITSO	35800	35773	0.0
23839	1996-020A	INMARSAT 3-F1	IM	35811	35759	0.2
23842	1996-021A	ASTRA 1F	LUXE	35792	35782	0.0
23846	1996-022A	MSAT M1	CA	35808	35765	0.0
23864	1996-030A	PALAPA C2	INDO	35794	35781	0.0
23865	1996-030B	AMOS	ISRA	35797	35776	0.1
23877	1996-033A	GALAXY 9	US	35793	35780	0.0
23880	1996-034A	GORIZONT 32	CIS	35796	35782	0.9
23883	1996-034D	SL-12 R/B(2)	CIS	36594	36307	1.0
23915	1996-035A	INTELSAT 709	ITSO	35797	35777	0.0
23943	1996-039A	APSTAR 1A	PRC	35790	35785	0.0
23948	1996-040A	ARABSAT 2A	AB	35799	35771	0.0
23949	1996-040B	TURKSAT 1C	TURK	35794	35778	0.1
24208	1996-044A	ITALSAT 2	IT	35793	35782	0.0
24209	1996-044B	TELECOM 2D	FR	35798	35774	0.0
24282	1996-048A	CHINASAT 7	PRC	46543	21631	27.5
24307	1996-053A	INMARSAT 3-F2	IM	35803	35770	0.6
24313	1996-055A	ECHOSTAR II	US	35794	35779	0.0
24315	1996-054A	GE 1	US	35796	35777	0.0
24435	1996-058A	EXPRESS 2	CIS	35796	35778	1.1
24438	1996-058D	SL-12 R/B(2)	CIS	35842	35759	1.9
24652	1996-063A	ARABSAT 2B	AB	35800	35772	0.1
24653	1996-063B	MEASAT 2	MALA	35790	35784	0.0
24665	1996-067A	HOT BIRD 2	EUTE	35815	35757	0.0
24674	1996-070A	INMARSAT 3-F3	IM	35804	35769	0.7
24700	1995-062C	ISO DEB	ESA	70428	1190	4.7
24713	1997-002A	GE 2	US	35801	35774	0.0
24714	1997-002B	NAHUEL 1A	ARGN	35802	35774	0.1
24732	1997-007A	JCSAT 4	JPN	35801	35775	0.0
24742	1997-009A	INTELSAT 801	ITSO	35814	35761	0.1
24748	1997-011A	TEMPO 2	US	35801	35770	0.1
24768	1997-016A	THAICOM 3	THAI	35815	35759	0.1
24769	1997-016B	B-SAT 1A	JPN	35791	35784	0.0
24786	1997-019A	GOES 10	US	35796	35775	0.3
24798	1997-021A	DFH-3 2	PRC	35799	35775	0.1
24808	1997-025A	THOR 2A	NOR	35791	35781	0.0
24812	1997-026A	TELSTAR 5	US	35795	35778	0.0
24819	1997-027A	INMARSAT 3-F4	IM	35804	35771	0.0
24820	1997-027B	INSAT 2D	IND	35958	33172	1.5
24834	1997-029A	FY-2	PRC	35787	35783	0.6
24846	1997-031A	INTELSAT 802	ITSO	35793	35779	0.0
24880	1997-036A	SUPERBIRD C	JPN	36432	35170	0.1
24891	1997-040A	PAS 6	US	35797	35777	0.0
24894	1997-041A	COSMOS 2345	CIS	36498	35067	0.2
24897	1997-041D	SL-12 R/B(2)	CIS	37385	34334	0.2
24901	1997-042A	AGILA 2	RP	35802	35772	0.0
24916	1997-046A	PAS 5	US	35793	35781	0.0
24931	1997-049A	HOT BIRD 3	EUTE	35821	35751	0.1

Cat No	*International Designator*	*Common Name*	*Source*	*Apogee (km)*	*Perigee (km)*	*Inclination (°)*
24932	1997-049B	METEOSAT 7	ESA	35792	35785	0.4
24936	1997-050A	GE 3	US	35791	35783	0.1
24957	1997-053A	INTELSAT 803	ITSO	35802	35772	0.0
25000	1968-081G	TITAN 3C TRANSTAGE DEB	US	36070	35077	13.2
25001	1968-081H	TITAN 3C TRANSTAGE DEB	US	35697	35077	13.1
25004	1997-059A	ECHOSTAR 3	US	35818	35757	0.0
25010	1997-062A	APSTAR 2R	PRC	35804	35769	0.0
25021	1997-065C	IABS R/B	US	35610	35520	1.3
25045	1997-070A	KUPON	CIS	35804	35791	1.1
25048	1997-070D	SL-12 R/B(2)	CIS	36029	35908	1.3
25049	1997-071A	SIRIUS 2	SWED	35792	35779	0.0
25050	1997-071B	INDOSTAR 1	INDO	35795	35777	0.1
25067	1997-075A	JCSAT 5	JPN	35795	35779	0.0
25068	1997-075B	EQUATOR-S	FGER	66985	700	6.5
25071	1997-076A	ASTRA 1G	LUXE	35791	35782	0.1
25086	1997-078A	GALAXY 8	US	35792	35781	0.0
25110	1997-083A	INTELSAT 804	ITSO	35854	35717	0.0
25126	1997-086A	HGS-1 (ASIASAT 3)	US	35978	35595	7.9
25134	1998-002A	SKYNET 4D	UK	35936	35786	3.3
25152	1998-006A	BRAZILSAT B3	BRAZ	35796	35779	0.1
25153	1998-006B	INMARSAT 3-F5	IM	35800	35771	1.9
25237	1998-013A	HOT BIRD 4	EUTE	35808	35765	0.0
25239	1998-014A	INTELSAT 806	ITSO	35803	35772	0.0
25311	1998-024A	NILESAT	EGYP	35797	35776	0.0
25312	1998-024B	B-SAT 1B	JPN	35791	35784	0.0
25315	1998-025A	COSMOS 2350	CIS	35800	35766	1.4
25318	1998-025D	SL-12 R/B(2)	CIS	35986	35724	1.4
25331	1998-028A	ECHOSTAR 4	US	35791	35783	0.1
25339	1996-053D	SL-12 R/B(2)	CIS	36627	36190	0.5
25353	1997-049E	METEOSAT 7 AKM	ESA	36003	35361	0.3
25354	1998-033A	CHINASTAR 1	PRC	35800	35773	0.1
25358	1998-035A	THOR 3	NOR	35793	35780	0.1
25371	1998-037A	INTELSAT 805	ITSO	35795	35779	0.0
25404	1998-044A	SINOSAT 1	PRC	35800	35774	0.0
25460	1998-049A	ST-1	STCT	35795	35779	0.0
25462	1998-050A	ASTRA 2A	LUXE	35802	35768	0.0
25473	1998-052A	PAS 7	US	35810	35763	0.1
25491	1998-056A	EUTELSAT W2	EUTE	35801	35770	0.0
25492	1998-056B	SIRIUS 3	SWED	35791	35780	0.0
25495	1998-057A	HOT BIRD 5	EUTE	35904	35666	0.1
25515	1998-063A	AFRISTAR	US	35804	35768	0.0
25516	1998-063B	GE 5	US	35801	35774	0.0
25522	1998-065A	PAS 8	US	35806	35767	0.0
25546	1998-068A	BONUM 1	CIS	35790	35784	0.0
25558	1998-070A	SAT MEX 5	MEX	35791	35782	0.0
25585	1998-075A	PAS 6B	US	35898	35677	0.1
25611	1997-029C	FY-2 AKM	PRC	36033	34109	0.5
25626	1999-005A	TELSTAR 6	US	35793	35781	0.0
25630	1999-006A	JCSAT 06	JPN	35959	35615	0.0
25638	1999-009A	ARABSAT 3A	AB	35798	35775	0.0
25639	1999-009B	SKYNET 4E	UK	35799	35776	3.5
25642	1999-010A	RADUGA 1-4	CIS	35810	35756	1.3
25645	1999-010D	SL-12 R/B(2)	CIS	36513	36332	1.3
25657	1999-013A	ASIASAT 3S	AC	35795	35780	0.0
25666	1999-016A	INSAT 2E	IND	35815	35758	0.1
25673	1999-018A	EUTELSAT W3	EUTE	35805	35767	0.0
25740	1999-027A	NIMIQ 1	CA	35794	35781	0.0
25785	1999-033A	ASTRA 1H	LUXE	35795	35777	0.0
25867	1999-040B	CXO	US	139136	5689	28.5
25869	1999-040D	CXO R/B(2)	US	72936	422	28.6

World orbital launch rates

(First digit indicates total known launches; second, the number of failures)

	1983	*1984*	*1985*	*1986*	*1987*	*1988*	*1989*	*1990*	*1991*	*1992*	*1993*	*1994*	*1995*	*1996*	*1997*	*1998*
BRAZIL																
VLS	–	–	–	–	–	–	–	–	–	–	–	–	–	–	1/1	
CHINA																
CZ-2C	1	1	1	1	2	1	–	1	–	1	1	–	–	–	2	4
CZ-2D	–	–	–	–	–	–	–	–	–	1	–	1	–	1		
CZ-2E	–	–	–	–	–	–	–	1	–	2*	–	1	3/1	–		
CZ-3	–	2/1*	–	1	–	2	–	2	1/1	–	–	1	–	2/1	1	
CZ-3A	–	–	–	–	–	–	–	–	–	–	–	2	–	–	1	
CZ-3B	–	–	–	–	–	–	–	–	–	–	–	–	–	1/1	2	2
CZ-3C	–	–	–	–	–	–	–	–	–	–	–	–	–	–		
CZ-4A	–	–	–	–	–	1	–	1	–	–	–	–	–	–		
* CZ-2E attained orbit, but without payload; CZ-3 partial failure																
EUROPE																
Ariane 1	2	2	1	1	–	–	–	–	–	–	–	–	–	–		
Ariane 2	–	–	–	1/1	1	3	2	–	–	–	–	–	–	–		
Ariane 3	–	2	3/1	1	1	2	1	–	–	–	–	–	–	–		
Ariane 40	–	–	–	–	–	–	–	1	1	–	1	–	2	–		1
42P	–	–	–	–	–	–	–	1	–	3	1	2/1	1	2		1
44P	–	–	–	–	–	–	–	–	2	–	–	–	2	2	4	2
42L	–	–	–	–	–	–	–	–	–	–	1	1	3	–	2	2
44LP	–	–	–	–	–	2	2	1	1	1	2	3/1	2	1	2	2
44L	–	–	–	–	–	–	2	3/1	4	3	2	2	1	5	3	2
Ariane 5	–	–	–	–	–	–	–	–	–	–	–	–	–	1/1	1	1
Ariane 5E	–	–	–	–	–	–	–	–	–	–	–	–	–	–		
INDIA																
SLV-3	1	–	–	–	–	–	–	–	–	–	–	–	–	–		

	1983	1984	1985	1986	1987	1988	1989	1990	1991	1992	1993	1994	1995	1996	1997	1998
ASLV	–	–	–	–	1/1	1/1	–	–	–	1	–	1	–	–		
PSLV	–	–	–	–	–	–	–	–	–	–	1/1	1	–	1	1/1	
GSLV	–	–	–	–	–	–	–	–	–	–	–	–	–	–		
ISRAEL																
Shavit	–	–	–	–	–	1	–	1	–	–	–	–	1	–		1/1
JAPAN																
J-1!	–	–	–	–	–	–	–	–	–	–	–	–	–	1		
M-3S	–	1	–	–	–	–	–	–	–	–	–	–	–	–		
M-3SII	–	–	2	–	1	–	1	1	1	–	1	–	1/1*	–		
M-5	–	–	–	–	–	–	–	–	–	–	–	–	–	–	1	1
N-2	2	2	–	1	1	–	–	–	–	–	–	–	–	–		
H-1	–	–	–	1	1	2	1	2	1	1	–	–	–	–		
H-2	–	–	–	–	–	–	–	–	–	–	–	2	1	1	1	1
H-2A	–	–	–	–	–	–	–	–	–	–	–	–	–	–		
* payload left in untenable low orbit.																
! J-1 launch was sub-orbital																
KOREA																
Taepo Dong			1/1													
UNITED STATES																
Atlas E	3	3	2	1	1	2	–	2	2	–	1	2	1	–		
Atlas H	2	1	–	1	1	–	–	–	–	–	–	–	–	–		
Atlas 1	1	1/1	3	1	1/1	–	1	1	1/1	2/1	2/1*	1	1	1		
Atlas 2	–	–	–	–	–	–	–	–	1	2	2	–	3	1		1
Atlas 2A	–	–	–	–	–	–	–	–	–	1	–	2	3	4	2	3
Atlas 2AS	–	–	–	–	–	–	–	–	–	–	1	1	4	1	5	2
Atlas 2AR	–	–	–	–	–	–	–	–	–	–	–	–	–	–		
Conestoga	–	–	–	–	–	–	–	–	–	–	–	–	1/1	–		
Delta 3910	1	–	–	–	–	–	–	–	–	–	–	–	–	–		
3914	2	1	1/1	–	–	–	–	–	–	–	–	–	–	–		
3920	–	1	1	–	2	1	1	–	–	–	–	–	–	–		
3920/PAM	3	1	–	–	–	–	–	–	–	–	–	–	–	–		
3924	2	1	–	–	–	–	–	–	–	–	–	–	–	–		
4925	–	–	–	–	–	–	1	1	–	–	–	–	–	–		
5920	–	–	–	–	–	–	1	–	–	–	–	–	–	–		
6920	–	–	–	–	–	–	–	2	–	1	–	–	–	–		
6925	–	–	–	–	–	–	5	7	1	1	–	–	–	–		
7326	–	–	–	–	–	–	–	–	–	–	–	–	–	–		1
7424	–	–	–	–	–	–	–	–	–	–	–	–	–	–		2
7920	–	–	–	–	–	–	–	–	–	–	–	–	2	–	7	5
7925	–	–	–	–	–	–	–	1	4	9	7	3	1	3	3/1	3
3	–	–	–	–	–	–	–	–	–	–	–	–	–	–		
LMLV 1 Athena 1	–	–	–	–	–	–	–	–	–	–	–	–	1/1	–	1	
LMLV 2 Athena 2	–	–	–	–	–	–	–	–	–	–	–	–	–	–		1
Pegasus	–	–	–	–	–	–	–	1	1	–	2	2	1	1	5	6
Pegasus XL	–	–	–	–	–	–	–	–	–	–	–	1/1	1/1	3		
Scout G-1	1	1	2	1	1	4	–	1	1	2	1	1	–	–		
Space Shuttle	4	5	9	2/1	–	2	5	6	6	8	7	7	7	7	8	5
Start 1															2	
Taurus	–	–	–	–	–	–	–	–	–	–	–	1	–	–		2
Titan 2G	–	–	–	–	–	1	1	–	–	1	1/1*	1	–	–	1	1
3B	1	1	–	–	–	–	–	–	–	–	–	–	–	–		
34B	1	1	1	–	1	–	–	–	–	–	–	–	–	–		
3D	1	–	–	–	–	–	–	–	–	–	–	–	–	–		
34D	–	5/1	1/1	1/1	2	2	2	–	–	–	–	–	–	–		
4	–	–	–	–	–	–	1	2	2	1	1/1	4	4	4	4	2
Titan 3	–	–	–	–	–	–	–	3/1	–	1	–	–	–	–		
* Atlas 1 failed to attain operational orbit; payload failed to provide orbital injection after separation from Titan 2G																
RFAS/USSR																
Vostok	4	–	1	–	–	2	–	–	1	–	–	–	–	–		
Soyuz-class	43/1	44	40	37/1	44/1	45/3	38	32/2	24	24	17	15	12	4/2	10	8
Molniya-m	11	11	16	14/1	4	11	6	12/1	5	8	8	3	4	3	3	3
Kosmos 3M	21/1	17	12/1	15	13	7	9	10	12/1	7	6	5	5/1	4	2	2
Tsyklon 2/m	2	4	5	5	6	3	3	4	1	–	4	1	2	1	1	
Proton-K*	11	13	9	7	11/2	13/2	10	10/1	8	8	6/1	13	6	9/2**	9	7
Proton-K	1	–	1	2/1	2	–	1	1	1	–	–	–	1	1		
Tsyklon	5	7/1	12	12/1	11	10	8/1	8	8	5	4	7/1	1	1	1	1
Zenit-2	–	–	4/3	2	5	2	–	2/1	1/1	3/1	2	4	1	1	1/1	3
Zenit-3	–	–	–	–	–	–	–	–	–	–	–	–	–	–		
Energia	–	–	–	–	1	1	–	–	–	–	–	–	–	–		
Shtil																1
Start-1	–	–	–	–	–	–	–	–	–	–	1	–	–	–		
Start	–	–	–	–	–	–	–	–	–	–	–	–	1/1	–		
Rockot	–	–	–	–	–	–	–	–	–	–	–	1	–	–		

Note: some failures might not be included in the Chinese and RFAS/USSR analyses. * Four stage variant.
** In-orbit failures

Company Index

Argentina
Aeroterra SA: 71
Nahuelsat SA: 293

Australia
Asia Pacific Aerospace Consultants Pty Ltd: 79
Auspace Ltd: 79
Australian Space Insurance Group: 79
British Aerospace Australia Ltd: 79
Earth Resource Mapping Pty Ltd: 71
Electro Optic Systems Pty Ltd: 79
Geoimage Pty Ltd: 71
Hawker de Havilland Ltd: 79
KEL Aerospace Pty Ltd: 79
Moonraker Technology Pty Ltd: 71
Satellite Aircom: 63
Spot Imaging Services Pty Ltd: 71

Austria
ACT Hochleistungskunststofftechnik GmbH: 79
Andritz AG: 79
Austrian Research Centre Seibersdorf: 79
Böhler Edelstahl GmbH: 79
Fischer FACC: 79
Geospace: 71
Österreichische Raumfahrt und Systemtechnik (ORS): 79
Plansee AG: 79
Schrack Aerospace GmbH: 79
Siemens AG Österreich: 80
Steyr-Daimler-Puch Fahrzeugtechnik: 80

Belgium
Alcatel Bell Telephone: 127
CLEO: 71
Eurosense NV: 71
Newtec CY NV: 127
SAIT Systems SA: 127
Sirius Communications NV: 63

Canada
Array Systems Computing, Inc: 71
Barringer Research Ltd: 71
Bercha Group: 72
Bristol Aerospace Ltd: 105
CAL Corp: 117
Canadian Marconi Co: 117
Dendron Resource Surveys Inc: 71
Eidetic Digital Imaging Ltd: 71
Gregory Geoscience Ltd: 71
Horler Information Inc: 71
Intermap Technologies Corp: 71
MacDonald Dettwiler: 72
Novatronics Inc: 121
Radarsat International: 72
Satlantic: 72
Seimac Ltd: 72

China
China Aerospace Corp: 105
China Great Wall Industry Corp: 105

CIS/USSR
see Russian Federation

Czech Republic
GISAT: 72

Denmark
Alcatel Danmark A/S: 121
Kampsax Geoplan: 72

Ecuador
CLIRSEN: 72

Egypt
Eight M Co: 72

France
Aerospatiale: 121, 133
Alcatel Espace: 141
Alcatel SAFT: 121
Fleximage: 72
FORTECH Airforge: 105
Geosys: 72
IESM: 117
ISTAR: 72
Matra Cap Systèmes: 72
Matra Marconi Space: 141
Novespace: 115
Pyrospace: 105
Sagem: 133
Schlumberger Industries: 141
SEP: 105, 264
Sextant Avionique: 117, 133
SNPE: 105, 265
Sodern SA: 133
Spot Image SA: 72
Starsem: 105, 265

Germany
Anite Systems GmbH: 80
Autoflug GmbH: 80
Bank Reuschel & Co: 80
Carl Zeiss: 80
Computer Anwendung für Management GmbH: 80
DaimlerChrysler Aerospace AG (DASA): 80, 106, 115, 121, 266
Daimler Jena-Optronik GmbH: 134
Dornier Satellitensysteme GmbH (DASA): 80
Drägerwerk AG: 80
Elbe Space & Technology Dresden GmbH & Co KG: 80
Elektronik- und Luftfahrtgeräte GmbH: 80
Energia Deutschland GmbH: 72
ESG Elektronik System GmbH: 80
FRIWO Silberkraft: 121
GAF mbH: 72
Institut für Raumfahrtsysteme, Stuttgart: 268
Kayser-Threde GmbH: 80
Krupp VDM GmbH: 81
Leybold Durferrit AG: 115
Linde AG: 121
MAN Gutehoffnungshutte AG: 81
MAN Technologie AG: 106, 117, 268
Munich Reinsurance Co: 81
OHB-System GmbH: 115
Racal-Heim GmbH: 141
Satellite Operational Services GmbH: 81
Teldix GmbH: 121
VCS Nachrichtentechnik GmbH: 73
W L Gore & Associates GmbH: 81
Zeppelin Technologie GmbH: 81

India
Afro-Asian Satellite Communications Ltd: 63
Alfa-Laval (India) Ltd: 106
ISRO Inertial Systems: 134
ISRO Propulsion Systems: 268
Laboratory for Electro-Optic Systems (LEOS): 134
Larsen & Toubro Ltd: 106
Speck Systems Pvt Ltd: 73

Indonesia
PT Indica Dharma: 73
PT MediaCitra Indostar: 63
PT Pasifik Satelit Nusantara: 63
PT Satelit Palapa Indonesia: 63
PT Telkom: 63

International
Alcatel Qualcomm: 127
CLS Argos: 73
COSPAR: 39
DASA/ERNO & SSC: 249
Eurimage: 73
Europropulsion: 106, 249
ICO Global Communications: 63
ISTI International Space Technology: 106
Lockheed Khrunichev Energia (LKE) International, Inc: 106
Matra Marconi Space: 81
Orion Atlantic LP: 64
Regulus: 106, 250
RESURS-F WorldMap: 73
Sea Launch joint venture: 106
Spacelink Europe: 82

Israel
Amsat-Israel: 82
Elbit Ltd: 82
Elop: 134
Elta Electronics Industries Ltd: 82
Israel Aircraft Industries: 64, 82
Israel Military Industries Ltd: 107
Rafael Propulsion Department (RPD): 107
Rotem Industries Ltd: 107
Soreq Nuclear Research Center: 82
Tadiran: 83
TAMAM Precision Instrument Industries: 135

Italy
Alenia Aerospazio SpA: 73, 107
Alenia Difesa (DIOS): 135
Alenia Spazio: 83, 115
Assicurazioni Generali SpA: 83
BPD Difesa e Spazio: 107
CAP Gemini SpA: 83
Carlo Gavazzi Space SpA: 83
CISI AID: 83
Dataspazio SpA: 83
Elettronica SpA: 83
FIAR SpA: 83, 121
FiatAvio SpA: 107
INTECS Sistemi SpA: 83
Laben SpA: 83
Laben Proel Tecnologie: 83
Microtecnica-Torino SpA: 83
Space Software Italia SpA: 83
Tecnodata Italia srl: 83
Tecnospazio SpA: 83
VITROCISET SpA: 83

Japan
Broadcasting Satellite System Corp: 64
Daicel Chemical Industries Ltd: 107
Fuji Heavy Industries Ltd: 84
Fujitsu Ltd: 84
High-Reliability Components Corp: 84
IHI Ishikawajima-Harima Heavy Industries Company Ltd: 84, 107, 270
ITOCHU Corp: 84
Iwatsu Electric Co Ltd: 107
Japan Aircraft Manufacturing Co Ltd: 84
Japan Amateur Radio League Inc: 85
Japan Aviation Electronics Industry Ltd: 135
Japan Manned Space Systems Corp: 84
Japan Radio Co Ltd: 117
Japan Satellite Systems, Inc: 64
Kawasaki Heavy Industries Ltd: 84
Kokusai Denshin Denwa Co Ltd: 64
Kyokuto Boeki Kaisha Ltd: 84
Mitsubishi Corp: 84
Mitsubishi Electric Corp (MELCO): 84
Mitsubishi Heavy Industries: 107, 270
Mitsubishi Precision Co Ltd: 84
Mitsubishi Space Software Co: 84
Mitsui & Co Ltd: 84
NEC Corporation: 74, 85, 135
Nippon Hoso Kyokai (NHK): 64
Nippon Propellant Industry Co Ltd: 108
Nippon Sanso KK: 85
Nissan Motor Company Ltd: 108, 271
Nissho Iwai Corp: 74
NOF Corp: 108
Oki Electric Industry Co Ltd: 85
Rocket Systems Corporation (RSC): 108
Sharp Corp: 121
Showa Denko KK: 108
Space Communications Corp: 64
Space Technology Corp: 115
Sumitomo Heavy Industries Ltd: 85
Taisei Corp: 85
Telecommunications Advancement Organisation of Japan (TAO): 64
Toshiba Corp: 85

Kenya
Teleos International BV: 74

Korea, South
Chang Woo Inc: 74

Liechtenstein
Balzers AG: 85

Luxembourg
Astra/SES: 65

Malaysia
Binariang SDN BHD: 65

Netherlands
Aerospace Propulsion Products BV: 108
Bradford Engineering BV: 115
Brandt Fijnmechanische Industrie BV: 85
CCM: 85
DEP: 85

Dinfa BV: 85
Dreves Engineering BV: 85
Drukker International BV: 85
Fokker Space: 85, 122
Genius Klinkenberg BV: 85
Hymec BV: 85
ICT Aerospace Group: 85
Mechanical Engineering & Consulting (MECON) BV: 86
Netherlands Aerospace Group (NAG): 86
Origin/Nieuwegein BV: 86
Polymarin BV: 86
Ramaer Connection Technology BV: 86
RDM Technology BV: 86
Satellite Services BV: 86
Stork Engineers & Contractors BV: 86
Stork Product Engineering BV: 86

New Zealand
Meteorological Service of New Zealand Ltd: 74
Sirtrack Ltd: 74

Norway
AME Space A/S: 86
CAP Gemini Norway A/S: 86
Det Norske Veritas A/S: 86
Eidel (Eidsvoll Electronics A/S): 86
Jotron Electronics A/S: 117
Kongsberg Gruppen A/S: 86
Kongsberg Navigation A/S:117
Nordic VLSI: 86
Norsk Elektro Optikk A/S: 86
NORUT Information Technology Ltd: 74
Norwegian Industrial Forum for Space Activities (NIFRO): 87
Prototech A/S: 87, 115
Raufoss Technology A/S: 87
SINTEF Group: 87
Telenor Satellite Services (TNSS): 65
Ticon Plastindustrie A/S: 87

Russian Federation
AKO Polyot: 46
Eurockot Launch Services: 108
KB Khimautomatiki: 47, 108, 257
KB Khimmach: 47, 108, 259
Keldysh Research Center: 256
Korolev Design Bureau: 108, 261
Kuznetsov Company SSC: 48, 108, 261
NII Machinostroenye (R&D IME): 48, 108, 262
NII Thermal Processes: 48, 108, 263
NPO Energomash: 49, 109, 252
NPO Saturn/Lyulka: 49, 109, 263
NPO Yuzhnoye: 263
OKB Fakel: 50, 109, 263

Singapore
Government sponsored organisations: 65
Singapore Telecom: 65

South Africa
Denel (Pty) Ltd: 87
Grinaker Telecom: 127

Spain
Alcatel Espacio SA: 127
Aurensa: 74
Canava Electronica SA: 87
Construcciones Aeronauticas SA (CASA): 87
Cray Sistemas SA: 87
CRISA: 87
GMV SA: 87
Iberica del Espacio AS: 87
INDRA Espacio SA: 87
INSA: 88
NTE SA: 88
SENER Ingenieria y Sistemas SA: 88
Tecsidel SA: 88
Telefonica Sistemas de Satelites SA: 127
Universidad Politecnica de Madrid: 88

Sweden
ABB Hafo AB: 88
CelsiusTech Electronics AG: 135
FFV Aerotech AB: 88
Saab Ericsson Space: 88
Swedish Space Corporation: 65, 116, 249
Technology Trade International: 88
Volvo Aero Corp: 109

Switzerland
Aerospace Engineering Office: 88
APCO Technologies SA: 88
Compagnie Industrielle Radioelectrique SA (CIR): 88
Contraves AG: 88
Cryophysics SA: 122
CSEM SA: 88
ETEL SA: 122
Leica: 89
Mecanex SA: 89
Oscilloquartz SA: 89
Reusser AG: 122
RST Raumfahrt Systemtechnik AG: 89
Swiss Aircraft & Systems Co: 89
Vibro-Meter SA: 89

Taiwan
JONSA: 127

Thailand
Asian Broadcasting & Communications Network Co Ltd: 66
Shinawatra Satellite Public Co Ltd: 66
Thai Micro-Satellite Co Ltd: 89

Ukraine
Marlin Yug Ltd: 52
NPO Elektropribor: 52
NPO Khatron: 52
NPO Musson: 52
NPO Yuzhnoye: 52, 109
PO Monolith: 52

United Kingdom
Advanced Products (Seals & Gaskets) Ltd: 89
AEA Technology: 89
AGEMA Infra-red Systems Ltd: 89
Air Products plc: 109
Alcoa International Ltd: 89
Ampex (GB) Ltd: 141
Analytical Graphics: 89
Andrew Ltd: 127
Anvil Technology Ltd: 89
Argos Electronics Ltd: 89
Asdor Filters: 89
Association of Specialist Technical Organisations for Space (ASTOS): 89
Avica Equipment: 90
BAJ Coatings Ltd: 90
BP Chemicals Ltd: 90
British Aerospace Defence: 272
C T Bowring Space Projects Ltd: 90
COM DEV Europe Ltd: 127
Commercial Space Technologies Ltd: 90
Control Techniques Dynamics Ltd: 122
Cossor Electronics Ltd: 90
Cray Systems Ltd: 90
Dowty Aerospace Wolverhampton: 109, 143
Earth Observation Sciences Ltd: 74
EASAMS Ltd: 90
EDS Ltd: 74, 90
EEV Ltd: 122
Electron Tubes Ltd: 90
Empcom Systems Ltd: 90
Encore Computer (UK) Ltd: 90
ESYS Ltd: 90
Fairey Microfiltrex Ltd: 109, 143
Feedback Instruments Ltd: 74
Firth-Rixson plc: 143
GEC-Marconi Communications Ltd: 127
GEC-Marconi Materials Technology Ltd: 90
GEC-Marconi Navigation & Electro-optics Systems: 135
GEC-Marconi Radar & Defence Systems: 117
GEC Plessey Semiconductors: 90
Go-Sat International: 128
Graseby Dynamics Ltd: 91
Hunterskil Howard plc: 91
Hunting Technical Services Ltd: 74
IGG Component Technology Ltd: 91
IMI Titanium Ltd: 91
International Space Brokers Ltd: 91
IRVIN Group Ltd: 91
J S Chinn Engineering Co Ltd: 91
Johnson & Higgins Ltd: 91
JRA Aerospace & Technology Ltd: 91
Kalgard (UK) Ltd: 91
KG Coating Ltd: 91
Kinesix (Europe) Ltd: 91
Kudos Technology Ltd: 128
Ling Dynamic Systems: 91
Logica UK Ltd: 91
Lokata Ltd: 117
LRE Relays and Electronics Ltd: 91
M/A-Com Ltd: 128
Marconi Marine: 128
Marine Technology International Ltd: 128
Marshall Aerospace: 91
Martin-Baker Aircraft Co Ltd: 91
Matra Marconi Space: 74, 122, 125
Mitsubishi Electric UK Ltd: 91
Molniya Space Consultancy: 91
Moreton Hall Associates: 92
Multipoint Communications Ltd: 128
Navstar Systems Ltd: 118
Nigel Press Associates Ltd: 74
Normalair-Garrett Ltd: 92
Parker Hannifin plc: 92
Penny & Giles Data Systems Ltd: 141
Pilkington Space Technology: 123
Plowshare Technology Ltd: 92
Quality CNC Engineering Ltd: 92
Queensgate Instruments Ltd: 92
Racal Antennas Ltd: 128
Racal Avionics Ltd: 128
Reaction Engines Ltd: 109
Rose Bearings Ltd: 92
Rosemount Ltd: 92
Rotadata Ltd: 109, 143
SAFT Nife Ltd: 123
Satellite Navigation Group: 118
Science Systems (Space) Ltd: 92
Sedgwick Space Services: 92
Serco Europe Ltd: 92
Sira Ltd: 136
Smith System Engineering Ltd: 92
Space Innovations Ltd: 92, 123, 136
Space Technology Systems Ltd: 75
Spacesense Consultants: 92
Spectrogon UK Ltd: 93
Spur Electron Ltd: 93
Surrey Satellite Technology Ltd: 93
Systems Engineering & Assessment Ltd: 93
T S Space Systems: 93
Timestep Weather Systems Ltd: 75
UK Industrial Space Committee (UKISC): 93
University of Dundee Centre for Coastal Zone Research: 54
Vega Space Systems Engineering Ltd: 93
Willis Corroon Inspace: 93

United States of America
3S Navigation: 118
Adcole Aerospace Products: 136
Advanced Products Co, Inc: 94
AEC-Able Engineering Co, Inc: 94
AeroAstro: 94
Aerojet: 75, 109, 273
Aerospace Corp: 94
Aerospace Industries Association of America Inc: 94
Air Products & Chemicals Inc: 109
Alden Electronics: 75
Allen Osborne Associates Inc: 118
Alliant Techsystems: 110, 275
AlliedSignal Aerospace Company: 94, 137
AlliedSignal Aerospace Environment Control Systems: 94
AlliedSignal Aerospace Power Management Systems: 94
AlliedSignal Technical Services Corp: 94
American Mobile Satellite Corp: 66
AMP Inc: 94
Analytical Graphics, Inc: 94
Andrew Corporation: 129
ANSER, Inc: 95
Antenna Technology Communications Inc (ATCi): 129
Applied Technology Associates: 95
ARAC: 95
Ardak Corp: 95
Ardé Inc: 110, 143
Arrowhead Products: 95
Ashtech, Inc: 118
Assured Space Access, Inc: 110
Astro Aerospace Corp: 95
Astroguide International: 129
Atlantic Research Corp: 110, 275
AT&T Alascom: 66
AT&T Tridom: 129
Autometric, Inc: 95
Avibank Mfg, Inc: 95
Aydin Corporation: 129
Ball Aerospace & Technologies Corp: 95, 129, 137
Barnes Engineering Division: 137
BDM International, Inc: 95
Bergen Cable Technologies Inc: 95
BFGoodrich Aerospace: 95
Boeing Aerospace Operations, Inc: 95
Boeing Commercial Space Co: 95
Boeing Rocketdyne: 110
Boeing Space Systems Group: 95, 110, 275
Booz, Allen & Hamilton Inc: 95
Brunswick Defense: 143
Brush Wellman Applications Development Center: 96
California Microwave Inc: 129
California Space Institute: 96
Calspan Corp: 96
CAST: 118
Celestis Inc: 96
Center for EUV Astrophysics (CEA): 96
Center for Satellite & Hybrid Communication Network: 129
Cincinnati Electronics Corporation: 96
Circle Seal Controls: 96
Collins Air Transport Division: 129
Collins Avionics & Communications Division: 129
Communications & Power Industries, Inc: 129
Computer Sciences Corp: 96
COMSAT RSI: 130
Comsearch: 130
ComStream: 130
Comtech Antenna Systems Inc: 130
Conatec Inc: 96
Condor Pacific Industries, Inc: 137
Constellation Communications Inc: 66
COSMIC: 96
CTA, Inc: 96
CTA Space Systems: 96
Cynetics Corp: 130
Dallas Remote Imaging Group: 96
Datron Systems Inc: 130
Datum Inc: 118
Decca Valves Corp: 96
Decom Systems Inc: 130
Del Norte Technology, Inc: 118
Delco Systems Operations: 138
Derlan Inc: 97
Deskin Research Group, Inc: 97, 130
DirecTV, Inc: 66
Dorne and Margolin Inc: 130
Draper Laboratory: 97
Dynamic Engineering Inc: 97
Eagle Aerospace, Inc: 97
Eagle-Picher Industries Inc: 123
Earth Satellite Corp (EarthSat): 75
Eastman Kodak Co: 97
Eaton Corp: 97
EchoStar Satellite Corp: 67
EER Systems Inc: 111
EG&G Florida, Inc: 97
EG&G Wright Components: 143
Electromagnetic Sciences Inc: 130
Ellipsat International Inc: 67
Encore Computer Corp: 97
Engineered Magnetics, Inc: 123
Engineered Systems: 97
Entran Devices Inc: 97
E'Prime Aerospace Corp: 110
ESSCO Collins: 130
Essex Corp: 97
E-Systems Inc: 130
FEI Microwave, Inc: 130
Fiberite, Inc: 97
Furon Aerospace Composites Division: 97
G&H Technology, Inc: 97
Garmin International, Inc: 118
GE Americom: 67
General Atomics: 123
General Research Corp: 97
Globalstar Ltd Partnership: 67
Glocom, Inc: 130
Gulton Data Systems: 97
Harris Corporation: 97, 130
Hernandez Engineering Inc: 98
Honeywell Inc: 98, 118, 138
Hughes Communications Inc: 67
Hughes Danbury Optical Systems Inc: 98, 138
Hughes Electron Dynamics Division: 130
Hughes Network Systems Inc: 131
Hughes Research Center: 111
Hughes Space & Communications Co: 67
Hughes STX Corp: 98
Humphrey Inc: 138
ICI Explosives: 111
ILC Dover: 98
Inframetrics, Inc: 98
Instrumentation Technology Associates Inc: 116
Integral Systems, Inc: 131
Interferometrics Inc: 98
Intergraph Corp: 75
International Space Brokers Inc: 98
International Space Technology Inc: 278
International Technology Underwriters: 98
Intraspace Corp: 98
Iridium, Inc: 68
ISSI: 98
Ithaco Inc: 138
ITT Defense & Electronics: 98
John Crane Belfab: 123
Johnson & Higgins: 98
Kaiser Electro-Optics Inc: 98
Kaiser Marquardt: 111, 143, 278
Kaman Aerospace Corp: 99
Kaman Sciences Corp: 99
Kamatics Corp: 99
Kearfott Guidance & Navigation Corp:

138
Keiser Engineering Inc: 131
Kelly Space & Technology Inc: 111
KETEMA Programmed Composites: 99
Kistler Aerospace Corp: 111
KKI: 99
KPMG Peat Marwick: 99
Launchspace, Inc: 99
Leica, Inc: 118
L'Garde, Inc: 99
Litton Aero Products: 118
Litton Applied Technology: 139
LNR Communications Inc: 131
Lockheed Martin Astro Space: 111, 125
Lockheed Martin Astronautics: 99, 111
Lockheed Martin Conic-Terracom: 131
Lockheed Martin Corporation: 99
Lockheed Martin Federal Systems: 99
Lockheed Martin Infra-red Imaging Systems: 75
Lockheed Martin Manned Space Systems: 112
Lockheed Martin Missiles and Space Company: 99, 123, 131, 139
Lockheed Martin Space Launch Systems: 112
Lockheed Martin Tactical Systems Sector: 100
Lockheed Martin Vought Systems: 123
Loral Space & Communications Ltd: 69, 100
Lord Mechanical Products: 100
Lucas Control System Products: 139
Magellan Systems Corp: 118
Malin Space Science Systems: 100
Marsh & McLennan Space Projects: 100
McDonnell Douglas Aerospace: 100
MCI Communications/News Corp: 69
MCL Inc: 131
Meda, Inc: 139
Metalmart Inc: 100
Microcosm, Inc: 100, 112
Microdyne Corp: 131
Micrologic: 119
Microwave Telemetry, Inc: 75
Miteq Inc: 131
Mobile Telesystems Inc: 131
Moog Inc: 143
Motorola Inc: 119
Motorola Space & Systems Technology: 131
MTS Systems Corp: 100
NERAC, Inc: 100
Northrop Grumman Corp: 101
Northrop Grumman Corp (Electronics & Systems Integration): 100
Northrop Grumman Corp (Technical Services): 101
Oceaneering Space Systems: 101
Olin Aerospace: 112, 279
Olin Chemicals: 112
Optical Coating Laboratory Inc: 101
Orbcomm Global: 69
Orbimage: 77
Orbital Sciences Corp: 101, 112
Orbital Sciences Corp/Orbimage: 77
Oremet Titanium: 101
PanAmSat Corp: 69
Para-Flite Inc: 112, 143
Parker Bertea Aerospace: 143, 281
Precision Rolled Products Inc: 101
Pressure Systems Inc: 112, 144, 281
Princeton Synergetics Inc: 101
Puroflow Corp: 144, 281
Qualcomm Inc: 86
Quantic Industries Inc: 281
Questar Corporation: 102
Raytheon Company: 131
Raytheon E-Systems Richardson: 131
R-Cubed Composites, Inc: 102
Rosemount Aerospace Inc: 102
Sandia National Laboratories: 112
Santa Barbara Remote Sensing: 75
Scaled Composites, Inc: 102
Schaeffer Magnetics Ltd: 124
Schonstedt Instrument Co: 102
Schuller International: 102
Science Applications International Corp: 102
Scientific-Atlanta Inc: 131
Sea Tel, Inc: 132
SeaSpace Corp: 75
Sedgwick Space Services: 102
Servo Corp of America: 139
SierraCom: 132
Smiths Industries Aerospace: 140
Sokkia Corp: 119
Sony Electronics Inc: 119
Space Analysis & Research, Inc: 102
Space Applications Corp: 102
Space Commerce Corp: 102
Space Communications Technology Center: 132
Space Electronics Inc: 102
Space Imaging, Inc: 77
Space Industries Inc: 102
Space Instruments, Inc: 75
Space Liaison & Imaging Corp: 75
Space Machine Advisors: 102
Space Systems/Loral: 70
Space Vector Corp: 113
Sparta Inc: 102
Specialty Steel & Forge: 102
Spectrolab Inc: 124
Spectrum Astro Inc: 102
Spot Image Corp: 75
SSE Technologies Inc: 132
SSI Inc: 113
ST Systron Donner: 132, 140
Stanford Telecommunications Inc: 119, 132
Sundstrand Aerospace: 124
Taber Industries: 103
Tayco Engineering, Inc: 124
TECOM Industries, Inc: 119, 132
Tecstar Inc: 124
Teledyne Brown Engineering: 103
Teledyne Electronics Systems: 103
Telegenix/Grim Corp: 103
Telonics Inc: 75
Telsys Inc: 132
Tempo DBS, Inc: 70
Terra-Mar Resource Information Services, Inc: 75
Textron Inc: 103
Thiokol Corp: 281
Thomas Electronics Inc: 103
TIW Systems Inc: 132
Trimble Navigation Ltd: 132
TRW: 113
TRW Space & Electronics Group: 103, 124, 283
Unisys Space & Information Systems: 103
United Technologies/CSD: 285
United Technologies/Hamilton Standard SSI: 103
United Technologies/Pratt & Whitney: 113, 287
United Technologies/USBI: 289
Universal Navigation Corp: 119
Universal Propulsion Co Inc: 113
Valcor Engineering Corp: 104
Veda Inc: 104
W L Pritchard & Co Inc: 132
Wear-Cote International: 104
Welnavigate, Inc: 119
WESGO Inc: 104
Western Electrochemical Co: 113, 289
Willis Corroon Inspace: 104
Wirin & Associates: 75
WorldSpace, Inc: 70
Wyle Laboratories: 104
Yardney Technical Products Inc: 124

Zimbabwe
Realtime Computers (Pvt) Ltd: 76

General Index

11th Space Wing (US): 29
21st Space Wing (US): 29
30th Space Wing (US): 29
45th Space Wing (US): 29
50th Space Wing (US): 29
750th Space Group (US): 29

Aachen Centre for Solidification in Space: 36
Academy of Finland: 34
ACTS: 329
ADEOS: 355
Advanced Composition Explorer (ACE): 408
AFSATCOM: 493
Agencia Espacial Brasileira (AEB, Brazil): 32
Agenzia Spaziale Italiana (ASI): 42
Aktivny-IK: 397
Alcantara launch centre: 147
Alexis: 515
Alliance for Marine Remote Sensing: 33
Almaz: 358, 475
Alouette: 389
Altair: 341
Ames Research Center: 56
Amos: 313
AMSC 1: 330
Andøya Rocket Range: 153
Andrade Gutierrez Quimica Ltda: 32
Angara: 205
Anik: 293
ANS: 401
APEX (Russia): 397
APEX (US): 528
Apollo: 446
Apple (Ariane Passenger Payload Experiment): 373
Applied Physics Lab, The Johns Hopkins University (US): 175
APstar: 295
Aquacade: 504
Arabsat: 38, 302
Argentina:
- Communications: 293
- Earth observation facilities: 71
- Government & institutional organisations: 31
- Government & NGO space programmes: 3
- Scientific: 389

Argon: 522
ARGOS: 38, 528
Ariane: 189
Ariane 4: 192
Ariane 5: 195
ARIES 1: 389
Army Space Command (US) ground stations: 169
Arnold Engineering Development Center (US): 175
Artemis (ESA telecom sat): 303
ASAT (RFAS): 485
ASAT (US): 486
ASCONT: 46
ASETA: 39
Asher Space Research Institute: 42
AsiaSat 1-3: 295
Asociacion Argentina de Tecnologia Espacial (AATE): 31
Astelit: 46
Astra: 319
Astrid 1/2: 377
Astro-SPAS: 392
ASUSat: 378
Athena: 234
Atlas launcher: 224
ATS series: 378
Aurora 2: 338
Aussat: 31
Australia:
- Communication satellites: 63
- Earth observation facilities: 71
- General services & suppliers: 79
- Government & institutional organisations: 31
- Government & NGO space programmes: 3
- Launch sites: 147
- Scientific: 389
- Telemetry, tracking & command: 167

Australian Centre for Remote Sensing: 31
Austria:
- Earth observation facilities: 71
- General services & suppliers: 79
- Government & institutional organisations: 32
- Government & NGO space programmes: 3

Austrian Space Agency: 32
Austrospace: 32

Badr: 401
Baikonur: 154
Balasore launch site: 151
Barreira do Inferno launch centre (Brazil): 147
Belgacom: 32
Belgian Institute of Space Aeronomy: 32
Belgium:
- Communication satellites: 63
- Earth observation facilities: 71
- Government & institutional organisations: 32
- Government & NGO space programmes: 4
- Research & development facilities: 171
- Satellite communications: 127

Biokosmos: 402
Bion: *see* Biokosmos
Bioserve Space Technologies: 59
Black Brant: 182
BOR-4: 376
Brazil:
- Data relay: 341
- Government & institutional organisations: 32
- Government & NGO space programmes: 4
- Launch sites: 147
- Launch vehicles: 181
- Scientific: 389

Bremsat: 392
British Association of Remote Sensing Companies: 53
British National Space Centre (BNSC): 53
British Telecommunications (BT): 53
BS/BSAT satellites: 314
Bulgaria:
- Government & institutional organisations: 33
- Government & NGO space programmes: 4

Bundesminister für Forschung: 36

Canada:
- Communications: 293
- Earth observation facilities: 71
- Earth observation & meteorological: 343
- Government & institutional organisations: 33
- Government & NGO space programmes: 4
- Launch sites: 147
- Launch site services: 105
- Launch vehicles: 182
- Navigation/SAR: 117
- Research & development facilities: 171
- Satellite actuation systems: 121
- Scientific: 389

Canada Centre for Remote Sensing: 33
Canadian Space Agency: 33, 171
Canyon: 503
Cape Canaveral Air Station (US): 157
Cape York: 147
Capricornio: 221
Cassini: 419
CATSAT: 417
Centaur upper stage: 227
Center for Cell Research: 59
Center for Commercial Space Communications (US): 54
Center for Macromolecular Crystallography (US): 59
Center for Mapping (US): 59
Center for Satellite & Hybrid Communication Networks (US): 54
Center for Space Transportation: 60
Central Laboratory of the Research Councils (UK): 53, 174
Central Specialised Design Bureau, Progress Division: 46
Centre for Digital Image Processing (Venezuela): 60
Centre for Environmental Satellite Data (Sweden): 51
Centre for Remote Imaging, Sensing & Processing (CRISP): 51
Centre for Remote Sensing & GIS (Australia): 31
Centre for Space Biotechnology: 46
Centre for Space & Remote Sensing Research (Taiwan): 52
Centre National d'Etudes Spatiales: 35
Centre Royal de Télédétection Spatiale du Maroc (CRTS): 45
Centre Spatial de Liège: 32, 171
Centre Spatial de Toulouse: 35
Centre Spatial d'Evry: 35
Centro de Investigaciones Aplicades (Argentina): 31
Centro para el Desarollo Tecnologico Industrial: 51
Centro Tecnico Aeroespacial (CTA) (Brazil): 32
CEOS: 39
Cerise: 391, 528
Chalet/Vortex: 503
Chandra (AXAF): 411
Chelomei Design Bureau: *see* KB Salyut & NPO Machinostroeniye
Chile:
- Government & NGO space programmes: 5
- Scientific: 390

China:
- Communications: 295
- Earth observation & meteorological: 343
- Elint: 503
- Government & institutional organisations: 34
- Government & NGO space programmes: 5
- Launch sites: 148
- Launch site services: 105
- Launch vehicle propulsion: 250
- Launch vehicles: 184
- Microgravity & materials: 383
- Scientific: 390
- Telemetry, tracking & command: 167

Chinasat: 296
China Telecommunications Broadcast Satellite Co: 34
Chinese Society of Astronautics: 34
Churchill launch site: 147
Civil Earth Observation Satellites (table): 351
Civil GPS Service (US): 54
Clark remote sensing satellite: 366
Clementine: 423
CLIRSEN Centro de Levantamientos Integrados: 34
Cluster satellites: 394
CNIE (Argentine Space Research Agency): *see* CONAE
Comet (microgravity carrier): 386
COMETS: 375
Comision Nacional de Actividades Espaciales (Argentina): *see* CONAE
Comision Nacional de Investigaciones Espaciales (CNIE): *see* CONAE
Comision Nacional de Telecomunicaciones (CNT): 31
Commission on Space Activity (Czech): 34
Commonwealth GPS Group (Australia): 31
Commonwealth Scientific & Industrial Research Organisation: *see* CSIRO
Compton Gamma Ray Observatory (CGRO): 412
COMSAT Corporation: 54
Comstar/SBS: 303
CONAE (Argentina space research agency): 31
Conestoga: 228
Consolidated Space Operations Center (CSOC, US): *see* 50th Space Wing
Consortium for Materials Development in Space: 59
Copernicus Astronomical Centre (Poland): 46
Corona: 521
Coronas: 402
Coscon: *see* Koskon
Cosmic Background Explorer (COBE): 408
COSPAR Committee of Space Research: 39
Cospas/Sarsat: 39
Coupon: *see* Kupon
Crista: 392
Crystal: 523 *see* KH
CS-3/JCSat: 315
CS-4/NStar: 317
CSIRO (Australia): 31
Czech Republic:
- Earth observation facilities: 72
- Government & institutional organisations: 34
- Government & NGO space programmes: 8
- Scientific: 390

Danish Research Advisory Board for Space: 34
Danish Space Research Institute: 34, 171
DARA (Deutsche Agentur für Raumfahrtangelegenheiten): 36
DARPA: *see* Defense Advanced Research Projects Agency
Darwin: 147
Data relay satellite: 341
DC-XA: 228
DEBUT: 374
Deep Space Network (DSN): 169
Defense Advanced Research Projects Agency (US): 55
Defense Information Systems Agency (DISA) (US): 55
Defense Support Program (DSP) series: 500
Delta 1: 229
Delta 2: 229
Delta 3: 232
Delta 4: 232
Delta 180/181: 487
Delta Clipper: 228
Delta Star: 487
Denmark:
- Earth observation facilities: 72
- Government & institutional organisations: 34
- Government & NGO space programmes: 8

Research & development facilities: 171
Satellite actuation systems: 121
DERA (Space Dept): 53
Deutsche Forschungsanstalt für Luft und Raumfahrt: 36
Deutsche Telekom AG: 36
DFS Kopernikus: 298
Discoverer: 521
Discovery solar system missions: 406
DLR (Germany): 37
DMSP (Defense Meteorological Satellite Program, USAF): 507
Dnepr launch vehicle: 206
Dryden Flight Research Center: 56
DSCS II series: 491
DSCS III series: 492

Early Bird: *see* Intelsat
Earth Data Analysis Center: 59
Earth Observation Centre (Japan): 43
Earth Observation Data Centre (UK): 53
Eastern Space & Missile Center: *see* 45th Space Wing
Eastern Test Range (US): 169
EchoStar 1/2/3/4: 330
ECS (European Communications Satellites): 304; *see also* Eutelsat
Ecuador:
Earth observation facilities: 72
Government & institutional organisations: 34
EGS: 381
Egypt:
Earth observation facilities: 72
Ekran: 323
Elekon: 323
Elektro: 358
Elektron: 402
Empresa Brasileira de Telecomunicacões: 33
Envisat: 350
Eole: 341
Eorsat (Elint Ocean Reconnaissance Satellite): 516
ERBS (Earth Radiation Budget Satellite): 407
EROS Data Center (US): 59
ERS (European Remote Sensing satellite): 348
ERS (Japan): *see* JERS
ERTS: *see* Landsat
ESA: *see* European Space Agency
ESA communications demonstration missions: 304
ESKOS SA: 47
Esrange: 156
Etalon: 506
Etat Major des Armées: 35
ETS 1-8: 317, 374
Eucosat: 39
EUMETSAT: 40, 353
Eureca (European Retrieval Carrier): 384
European Astronauts Centre: 165
European Center for Space Law: 40
European Commission: 40
European Space Agency (ESA): 11, 165
European Space Agency science series: 393
European Space Operations Centre (ESOC): 167
European Space Research & Technology Centre (ESTEC): 173
European Space Research Institute (ESRIN): 173
European Telecommunications Satellite Oranisation (Eutelsat): 40
European Union: *see* European Commission
Eurospace: 40
Eutelsat 1: 304
Eutelsat 2: 305
Explorer: 407
Express microgravity capsule: 383, 384
Express telecom sat: 323
Extreme Ultraviolet Explorer (EUVE): 408
Eyeglass: *see* OrbView

Faint Object Camera: *see* Hubble Space Telescope
FAISat: 331
Far Ultraviolet Spectroscopy Explorer: *see* FUSE
FASat-Alfa/Bravo: 390
FAST Explorer: 409
Federal Communications Commission (US): 55
Feng Yun: 343
Ferrets: *see* Elint states
Finland:
Government & institutional organisations: 34
Government & NGO space programmes: 9
Finnish Geodetic Institute: 34
Finnish Meteorological Institute: 34
Finnish Space Committee: 35
Fleetsatcom: 493
Florida Spaceport Authority: 56
Fltsatcom: *see* Fleetsatcom
Fluid Physics Module: *see* Spacelab
FOBS (Fractional Orbit Bombardment System): 505
FÖMI RSC: 37
FORTE: 515
France:
Communications: 297
Data relay: 341
Earth observation facilities: 72
Earth observation & meteorological: 344
Engineering test: 373
Geodetic: 381
Government & institutional organisations: 35
Government & NGO space programmes: 9
Launch sites: 149
Launch site services: 105
Launch vehicle propulsion: 264
Microgravity facilities: 115
Navigation/SAR: 117
Photo reconnaissance: 519
Research & development facilities: 171
Satellite actuation systems: 121
Satellite guidance & navigation: 133
Satellite information storage: 141
Science & engineering test: 528
Scientific: 391
French Guyana Space Centre: *see* Kourou
French science satellite programme: 391
FSW µgravity (Jianbing): 383
Fuji: *see* Oscar
FUSE Explorer: 409
FY: *see* Feng Yung

Gabon:
Government & NGO space programmes: 10
Gagarin Cosmonaut Training Centre: *see* Zvezdny Gorodok (Star Town)
Galaxy DBS: 332
Galileo (Jupiter probe): 423
Gals: 324
Gambit: 522 *see* KH
Gamma Ray Observatory (GRO): *see* Compton Observatory
Gascom: 47
GAS Dynamics Laboratory: *see* NPO Energomash
GDTA: 35
Geizer: 498
Gelikon: 324
Geodetic Type 1: 506
GEODSS: 177
GeoForschungsZentrum: *see* GFZ
GEO-IK: 506
Geosat: 518
Geosat Committee: 56
Geosat Follow-On: 517
Geosys: 35
Geotail: 395
German science satellite programme: 392
German Space Operations Centre (GSOC): 167
Germany:
Communications: 298
Data relay: 341
Earth observation facilities: 72
Engineering test: 373
General services & suppliers: 80
Geodetic: 381
Government & institutional organisations: 36
Government & NGO space programmes: 10
Launch site services: 106
Launch vehicle propulsion: 266
Microgravity facilities: 115
Microgravity & materials: 383
Navigation/SAR: 117
Research & development facilities: 172
Satellite actuation systems: 121
Satellite guidance & navigation: 134
Satellite information storage: 141
Scientific: 392
Telemetry, tracking & command: 167
GFZ 1: 381
Glavcosmos: 47
Glenn Research Center: 57
Global Positioning System: 512 *see* Navstar
GLOBALSTAR: 331
Glomr (Global Low Orbiting Message Relay): 498
Glonass (Global Navigation Satellite System): 510
GMS: 356
Goddard Space Flight Center: 56
GOES (Geostationary Operational Environmental Satellites): 367
Goldfinch motor: *see* Skylark
GOMS (Geostationary Orbit Meteorological Satellite): 358
Gonets: 324
Gorizont: 324
GPS: 512 *see also* Navstar
GPS Industry Council: 59
GPS Information Center: *see* Civil GPS Service
Granat: 403
Gravity Probe: 411
Greece:
Government & NGO space programmes: 11
GSLV: 186
GStar 1/2/3/4: 339
Gurwin 1 (TechSat): 373, 400

H1/H2 launchers: 201
Hainan/Harbin launch site: 148
Healthsat: 331
Heavy Ferret: 503
Heavy lift launch vehicles: 213
Helios 1A (France/Italy/Spain): 519
Helsinki University of Technology: 35
HETE (High Energy Transient Experiment): 415
Hexagon: 522 *see* KH
Hipparcos: 394
Hispasat: 326
HOPE: 203
HOTOL/SKYLON: 222
Hubble Space Telescope: 412
Hungarian Satellite HUNSAT: 37
Hungarian Space Office: 37
Hungary:
Government & institutional organisations: 37
Government & NGO space programmes: 11
Huygens: *see* Cassini

IABG Industrieanlagen-Betriebsgesellschaft mbH: 172
IAE Instituto de Aeronautica e Espacio: 33
IAFE (Instituto de Astronomia y Fisica del Espacio): 31
IASTP (Inter-Agency Solar Terrestrial Physics): 14, 394
Ibiza: 396
IDSCS: 493
Igla docking system: *see* Soyuz & Mir
Ikar: 216
IKI: 50 *see* Space Research Institute (Russia)
Impact: 396
IMP Explorer: 409
India:
Communications: 299
Communication satellites: 63
Earth observation facilities: 73
Earth observation & meteorological: 346
Engineering test: 373
Government & institutional organisations: 37
Government & NGO space programmes: 11
Launch sites: 151
Launch site services: 106
Launch vehicle propulsion: 268
Launch vehicles: 185
Research & development facilities: 173
Satellite guidance & navigation: 134
Scientific: 393
Telemetry, tracking & command: 167
Indian National Satellite System: 37
Indian Remote Sensing satellite: *see* IRS
Indian science satellite programme: 393
Indian Space Research Organisation *see* ISRO
Indonesia:
Communications: 300
Communication satellites: 63
Earth observation facilities: 73
Government & institutional organisations: 38
Government & NGO space programmes: 11
Launch site: 152
Indosat PT: 38
Indostar: 300
Inertial Upper Stage: *see* IUS
Informatr: 325
Infra-red Space Observatory: 394
Inmarsat Organisation: 40
Inmarsat 1: 306
Inmarsat 2/3: 307
INPE (National Institute for Space Research, Brazil): 33
Insat: 299, 346
Institute for Unmanned Space Experiments Free Flyer (USEF, Japan): 43
Institute of Aviation (Poland): 46
Institute of Biomedical Problems (Russia): 47
Institute of Geodesy & Cartography (Hungary): 37
Institute of Geodesy & Cartography (Poland): 46
Institute of Meteorology & Water Management (Poland): 46
Institute of Space and Astronautical Science (ISAS, Japan): 43
Institute of Space Devices Engineering (Russia): *see* NII Kosmisheskovo Priborostroenye
Instituto de Astronomia y Fisica del Espacio (Argentina): *see* IAFE
Instituto Nacional de Engenharia e Tecnologia Industrial (INETI, Portugal): 46
Instituto Nacional de Pesquisas Espaciais (Brazil): *see* INPE
Instrument Pointing System (IPS): *see* Spacelab
INTA: 51
Intasat 1: 376
Intelsat Organisation: 41
Intelsat: 309
Intelsat 7/8/9: 311
Intelsat K: 312
Interball: 14, 397
Interdisciplinary Centre for Technological Analysis and Forecasting (Israel): 42
Interkosmos: 14, 396
International:
Communications: 302
Communication satellites: 63
Data relay: 341
Earth observation facilities: 73
Earth observation & meteorological: 348
General services & suppliers: 81
Government & institutional organisations: 38
Government & NGO space programmes: 11
Human space flight: 437
Launch sites: 152
Launch site services: 106
Launch vehicle propulsion: 249
Launch vehicles: 189
Microgravity & materials: 384
Navigation/SAR: 387
Planetary science: 419
Research & development facilities: 173
Satellite communications: 127
Scientific: 393
Telemetry, tracking & command: 167
International Microgravity Laboratory: *see* Spacelab
International Space Science Institute: 52
International Space Station (ISS): 477
International Space University: 41
International Telecommunications Union: 41
Intersputnik: 41
Intespace (Ingénière Tests en Environnement Spatial): 171
Intruder: 504
IONDS sensors: 515
Iran:
Government & institutional organisations: 42
Government & NGO space programmes: 15
Iraq:
Government & NGO space programmes: 15
Launch vehicles: 198
Ireland:
Government & institutional organisations: 42
Government & NGO space programmes: 15
IRIDIUM: 334
IRS (Indian Remote Sensing satellite): 347
IRS (Institut für Raumfahrtsysteme): 37
ISEE (International Sun-Earth Explorers): 410
ISIS programme: 389
ISPM (International Solar Polar Mission): *see* Ulysses

Israel:
Communications: 313
Communication satellites: 64
Engineering test: 373
General services & suppliers: 82
Government & institutional organisations: 42
Government & NGO space programmes: 15
Launch sites: 152
Launch site services: 107
Launch vehicles: 198
Satellite guidance & navigation: 134
Scientific: 400
Israel Military Industries Ltd: 42
Israel Space Agency: 42
ISRO (Indian Space Research Organisation): 37
ISRO Inertial Systems Unit: 134
ISRO Satellite Centre (ISAC): 173
ISRO Telemetry Tracking & Command Network (ISTRAC): 167
ISTP (International Solar Terrestrial Physics programme): *see* IASTP
Italian Space Agency: *see* Agenzia Spaziale Italiana (ASI)
Italsat: 313
Italy:
Communications: 313
Earth observation facilities: 73
General services & suppliers: 83
Government & institutional organisations: 42
Government & NGO space programmes: 16
Launch sites: 152
Launch site services: 107
Launch vehicles: 199
Microgravity & materials: 115
Satellite actuation systems: 121
Satellite guidance & navigation: 135
Scientific: 401
ITOS (Improved Tiros Operational Satellite): *see* NOAA
ITU: *see* International Telecommunications Union
IUE Explorer: 410
IUS (Inertial Upper Stage) motors: 233
Izmiran: 47

J1 launcher (Japan): 204
Japan:
Communications: 314
Communication satellites: 64
Earth observation facilities: 74
Earth observation & meteorological: 355
Engineering test: 374
General services & suppliers: 84
Geodetic: 381
Government & institutional organisations: 43
Government & NGO space programmes: 16
Launch sites: 153
Launch site services: 107
Launch vehicle propulsion: 270
Launch vehicles: 201
Microgravity facilities: 115
Microgravity & materials: 384
Navigation/SAR: 117
Planetary science: 421
Research & development facilities: 174
Satellite actuation systems: 121
Satellite guidance & navigation: 135
Telemetry, tracking & command: 168
Japan Meteorological Agency: 43
Japan Microgravity Centre: 43
Japan Space Utilisation Promotion Centre (JSUP): 43
JAWSAT: 529
JCSat: 315
JERS 1: 357
Jet Propulsion Laboratory (JPL; US): 57
Jianbing: 383
Jiuquan launch centre: 148
Joanneum Research: 32
Johnson Space Center: 57
JSSW: 503
Jumpseat: 504

Kagoshima Space Centre: 153
Kapustin Yar launch centre: 155
KB Arsenal: 47
KB Khimautomatiki: 47
KB Khimmach: 47
KB Kolomna: 47
KB Machinostroenye: 47
KB Photon: 47
KB Salyut: 47
Keldysh Institute of Applied Mathematics: 47
Keldysh Research Centre: 47
Kennan: 523 *see* KH
Kennedy Space Center: 158
Kenya:
Earth observation facilities: 74
KFKI Research Institute for Particle & Nuclear Physics (RIPNP, Hungary): 37
KH-1/-3 Corona/Discoverer satellites: 521
KH-4A/B Corona satellites: 521
KH-5 Argon: 522
KH-6 Lanyard: 522
KH-7/8 Gambit: 522
KH-9 Hexagon: 522
KH-11 Kennan/Crystal: 523
KH-11/-12 improved Crystal: 524
Khrunichev State Research & Production Space Centre: 47
Kodiak launch complex: 160
Kometa: 520
Koreasat: 319
Korea, South:
Communications: 319
Earth observation facilities: 74
Government & institutional organisations: 44
Government & NGO space programmes: 17
Korea Telecom (KT): 44
Korolev Design Bureau: *see* RKK Energia
Kosberg Design Bureau: *see* KB Khimautomatiki
Koskon: 47
Kosmos: 531
Kosmos 3M: 206
Kourou Space Centre: 149
Kristall: 470
Kupon: 325
Kurs docking system: *see* Soyuz & Mir
Kuznetsov N D, Company: 48
Kvant 1: 469
Kvant 2: 470

Lacrosse: 527
Lageos (Laser Geodynamics Satellite): 381
Land Information Centre: 31
Landsat: 363
Langley Research Center: 57
Lanyard: 522 *see* KH
LAPAN: 38
Laser weapons: *see* ASAT
LDEF (Long Duration Exposure Facility): 379
LE-5 (rocket engine): 261; *see also* H2 launcher
LE-7 (rocket engine): 262; *see also* H2 launcher
Leasat: 494
Lebedev Physics Institute (Russia): 48
LES: 494
Lewis remote sensing satellite: 365
Libya:
Government & institutional organisations: 44
Libyan Centre for Remote Sensing & Space Science: 44
Liechtenstein:
General services & suppliers: 85
LMLV launcher: 234
Locsys: *see* Gonets
Long Duration Exposure Facility: *see* LDEF
Long March: 184
Los Alamos National Lab (US): 175
Losat-X: 487
LStar 1/2: 329
Luch: 341
Lunar-A: 421
Lunar Rover Vehicle: *see* Apollo
Lunokhod: 48
Luxembourg:
Communications: 319
Communication satellites: 65
Government & NGO space programmes: 17
Lyulka Engine Design Bureau: *see* NPO Saturn

M-5 launchers: 204
Maanmittauslaitos: 35
MACSAT: 494
Magellan: 427
Magion (Magnetospheric and Ionospheric): 390
Magnum: 504
Malaysia:
Communications: 320
Communication satellites: 65
Government & NGO space programmes: 17
Marecs: 306
Marine Observation Satellite: *see* MOS
Mariner: *see* tables at front of Directory
Marisat: 312
Marlin Yug Ltd: 52
Marots: *see* ECS
MARS Centre: 42
Mars Climate Orbiter: 429
Mars Global Surveyor: 427
Marshall Space Flight Center: 58
Mars Pathfinder: 428
Mars Polar Lander: 430
Mars series: 422
Maui space surveillance site: 177
McDonnell Douglas F-15 ASAT: *see* ASAT
MEASAT: 320
MESUR Mars mission: *see* Mars Pathfinder
Météo-France: 35
Meteor 1/2/3 (RFAS): 359
Meteorological Office (UK): 53
Meteosat: 351
Metop 1: 350
Mexico:
Communications: 321
Government & institutional organisations: 44
Government & NGO space programmes: 17
Microgravity Laboratory of Japan: 43
Microgravity User Support Centre (MUSC): 37
MicroLab: 366
Midas (Missile Defence Alarm System): 502
MIDORI: 355
MightySat: 529
Military Institute of Aviation Medicine (Poland): 29
Milstar: 494
Minisat 405
Mir: 467
MIRCA micro re-entry capsule: 373, 392
Mir chronology: 471
MITI: 43
Molniya-1: 325, 489
Molniya-M: 207
Morelos 2/3: 321
Morocco:
Government & institutional organisations: 45
Morsviasputnik: 48
MOS 1 357
Moscow Aviation Institute: 48
Moscow Institute of Thermal Technology: 48
MOTIF Maui Optical Tracking & Identification Facility: 177
MSAT 294
MSTI: 487
MSX: 487
MTI: 529
Mugunghua 1/2: 319
MURST (Directorate of International Affairs, Italy): 42
MuSat: *see* VICTOR 1

N1/2 launch vehicle: 205
Nadezhda: 387
Nahuel: 293
NASA: 78-80, 203, 369; *see* specific projects, especially Space Shuttle, and centres
National Aeronautics & Space Administration: 56
National Aerospace Laboratory (NLR; Netherlands): 45
National Aerospace Laboratory (NAL, Japan): 44
National Committee for Space Research (Israel): 42
National Imaging & Mapping Agency (NIMA): 55
National Land Survey of Finland: 35
National Reconnaissance Office (US): 55
National Remote Sensing Agency (India): 37
National Snow & Ice Data Center (US): 59
National Space Development Agency of Japan (NASDA): 44
NATO:
Communications: 489
Naval Research Laboratory (US): 175
NAVSPASUR: 177
Navstar: 512 *see also* Navigation section
Navy Ocean Surveillance Satellites: *see* NOSS
Near-Earth Asteroid Rendezvous (NEAR): 406
NESDIS: *see* NOAA
Netherlands:
General services & suppliers: 85
Government & institutional organisations: 45
Government & NGO space programmes: 17
Launch site services: 108
Microgravity & materials: 115
Satellite actuation systems: 122
Scientific: 401
Netherlands Agency for Aerospace Programs (NIVR): 45
Netherlands Industrial Space Organisation (NISO): 45
Netherlands Remote Sensing Board: 45
New Millennium: 416
New Zealand:
Earth observation facilities: 74
Government & NGO space programmes: 17
NEXT launcher: 198
Nigeria:
Government & NGO space programmes: 17
NII Argon: 48
NII Grafit: 48
NII Khimicheskogo Machinostroenye: 48
NII Kosmisheskovo Priborostroenye: 48
NII Machinostroenye (R&D IME): 48
NII Parachutostroenye: 48
NII PM (Prikladnoi Mekhaniki): 48
NII Radio: 48
NII TP (Thermal Processes): 48
Nika-T: 384
NOAA (National Oceanic & Atmospheric Administration): 58, 366
NORAD: 169
Norsk Romsenter: 45
Norway:
Communication satellites: 65
Communications: 322
Earth observation facilities: 74
General services & suppliers: 86
Government & institutional organisations: 45
Government & NGO space programmes: 18
Launch sites: 153
Microgravity & materials: 115
Navigation/SAR: 117
Telemetry, tracking & command: 168
Norwegian Meteorological Institute: 45
NOSS Navy Ocean Surveillance Satellite: 518
Nova: 514
NPO AP: 48
NPO Astra: 48
NPO Cryogenmash: 48
NPO Elas: 48
NPO Energomash: 49
NPO Geofizika: 49
NPO Iskra: 49
NPO Istochnik: 49
NPO Istok: 49
NPO Khartron/Elektropribor: 52
NPO Kompozit: 49
NPO Kraznaya Zvezda: 49
NPO Kvant: 49
NPO Lavotchkin: 49
NPO Machinostroenye: 49
NPO Molniya: 49
NPO Musson: 52
NPO Orion: 49
NPO Planeta: 49
NPO PM (Prikladnoi Mekhaniki): 49
NPO Precision Instruments: 49
NPO Radio: 49
NPO Saturn: 49
NPO Scientific Centre: 49
NPO Soyuz: 49
NPO Start: 49
NPO Stekloplastik: 49
NPO Tekhnomash: 49
NPO Tekhnologiya: 49
NPO Trud: 49
NPO Vega-M: 50
NPO Yuzhnoye: 52
NStar: 317
Nuova Telespazio: 43

Observatory for Satellite Geodesy (Netherlands): 45
Ofeq 1/2/3: 373
Office for Science, Technology & Cultural Affairs (Belgium): 32
Office of Naval Research (ONR, US): 55
Office of Science & Technology Policy (US): 59
Office of Space Acquisition & Technology: 55

Office of Space & Defense Power Systems: 55
Office of Space Commerce: 55
Office of Space Transportation: 55
Ohio State University Center for Mapping: 59
Ohsumi: 375
OICETS: 375
OKB Fakel: 50
OKB Moskovskaya: 50
Okean: 516
Okean-O1/Sich 1: 362
Oko: 499
Olympus: 312
ONERA (France): 172
Ontario Centre for Remote Sensing (OCRS, Canada): *see* Provincial Remote Sensing Office
OPREX: 529
Optical Inter-orbit Communications Engineering Test Satellite: *see* OICETS
ORBCOMM GLOBAL: 335
Orbita: *see* Molniya
Orbiters OV 099-106: 447
OrbView: 370
Orfeus: 392
Orion 1/2/3: 336
Orion/Orion derivatives: 235
OSCAR (Orbiting Satellite Carrying Amateur Radio): 15, 397
OTS (Orbital Test Satellite): 312
Overberg Test Range (OTB): 155

PA-X sounding rocket: 236
Pakistan:
  Government & institutional organisations: 46
  Government & NGO space programmes: 18
  Scientific: 401
Palapa: 301
Palmachim air force base: 152
PAM (Payload Assist Module): 236
PanAmSat: 37
Paneungpeuk launch site: 152
PARCAE/Whitecloud : 518
Particle Physics & Astronomy Research Council (UK): 53
Parus Sail: 511
Pegasus launcher: 236
Perumtel: *see* PT Telkom
Phillips Lab (US): 176
Photon spacecraft: 385
Photon Design Bureau: *see* KB Photon
Physical Research Lab (India): 38
PICOSat: 529
Pion: 403
Pioneer: 431
Planet B: 422
Plesetsk: 155
PO Mayak: 50
PO Monolith: 52
PO Samara Frunze Engine Building: 50
Poker Flat Research Range: 160
Poland:
  Government & institutional organisations: 29, 46
  Government & NGO space programmes: 18
Polar Bear: 496
Polar Platform: *see* Envisat
Polar satellite : 395
Polspace Ltd: 46
Polyot: 485
Polyot AKO: 46
Portugal:
  Government & institutional organisations: 46
  Government & NGO space programmes: 18
  Scientific: 401
PoSAT: 401
Poseidon: 354
Potok: 498
PPO Motorostroitel: 50
Priboi: 217
Prins Maurits Lab (PML): 45
Priroda module: 471
Priroda State Centre: 50
Prognoz: 403, 499
Progress: 465; *see also* individual Soyuz, Salyut and Mir missions
Proteus: 373
Proton-K launch vehicle: 209
Provincial Remote Sensing Office (Ontario, Canada): 33
PSDE: *see* Artemis
PSLV: 186
PTT Telecom Netherlands: 45

Queen Mary & Westfield College: 53

R series test craft: 377
Radarsat: 343
RADCAL: 529
Radioastron: 404
Raduga: 489
RASCOM: 41
Raven motor: *see* Skylark
RD&PE Zvezda: 50
Relikt/Libris: 404
Remote Manipulator System (RMS): *see* Space Shuttle
Remote Sensing Applications Centre (Australia): 31
Remote Sensing Centre (Thailand): 52
Remote Sensing Centre of the Institute for Geodesy, Cartography & Remote Sensing (Hungary): 37
Remote Sensing Lab (Israel): 42
RESTEC (Japan): 44
Resurs-F: 360
Resurs-O: 361
REX II: 529
Rhyolite: 504
RKK Energia: 50
RMIB (Royal Meteorological Institute of Belgium): 32
Rockot: 197
Roentgen: *see* Kvant
Rohini sounding rockets: 189
Romania:
  Government & NGO space programmes: 18
Romb: 526
Rorsat: 156
Rosat: 399
Rossi X-ray Timing Explorer: 410
Royal Aerospace Establishment (UK): *see* DERA
Royal Aircraft Establishment, Culham: 54
Royal Meteorological Institute of Belgium: 32
Royal Signals & Radar Establishment (UK): *see* DERA
RSAC (Remote Sensing Applications Centre, Australia): 31
Russian Academy of Sciences: 50
Russian Federation:
  ASAT: 485
  Communications: 323, 487
  Data relay: 341, 498
  Early Warning: 499
  Earth observation & meteorological: 358
  Elint: 503
  Engineering test: 376
  FOBS: 505
  Geodetic Satellites: 506
  Government & institutional organisations: 46
  Government & NGO space programmes: 18
  Human space flight: 438
  Launch site services: 108
  Launch sites: 154
  Launch vehicle propulsion: 251
  Launch vehicles: 205
  Microgravity & materials: 384
  Minor military satellites: 509
  Navigation: 510
  Ocean surveillance: 516
  Photo reconnaissance: 519
  Planetary science: 422
  Radar calibration: 526
  Resupply vehicles: 465
  Science & engineering test: 531
  Scientific: 402
  Space monitoring facilities: 177
  Space stations: 467
  Telemetry, tracking & command: 168
  Training facilities: 165
Russian Space Agency: 50

S-520 sounding rocket: 205
SAC-B: 389
SACI 1: 389
SAFIR: 341
Salyut: 475
Samos (Satellite & Missile Observation System): 524
SAMPEX Explorer: 411
San Marco launch site: 152
San Marco satellites: 401
San Marco Scout: *see* Vega
Sarsat: *see* Cospas/Sarsat
Satcom satellites: 338
Satellite Applications Centre (SAC): 51
Satellite Navigation Group: 54
SAX (astronomy satellite): 400
SBIR T: 502
SCD 1: 341
Science & Technology Agency (Japan): 44
SCOT Conseil: 36
SCRC (Satellite Communications Research Centre): 31
SCT: 44
SDRN: 341
SDS (Satellite Data System): 496
SEDS Satellite: 379, 416
SETI Institute (US): 102
Sfera: 506
Shar Centre: 151
Shavit: 198
Shi Jian satellites: 390
Shinsei: *see* ISAS
Shuttle Radar Lab: *see* Space Radar Lab
Shuttle Tethered Satellite: *see* Tethered Satellite System (TSS)
Sich 2/3: 363
Singapore:
  Communication satellites: 65
  Government & institutional organisations: 51
Sirio: 401
Sirius DBS satellite: 327
SKAS: 50
Skipper: 488
Skylark: 222
SKYLON: 222
Skynet: 490
Small Expendable Deployer System: *see* SEDS
Smolsat Consortium: 50
SMS (Synchronous Meteorological Satellites): *see* NOAA
SNOE: 417
SNPE: 36
Soho (Solar Heliospheric Observatory): 395
Solar Polar Mission: *see* Ulysses
Solid Propulsion Group, VSCC: 38
Solidaridad: 321
Sonda 1-4: 181
South Africa:
  General services & suppliers: 87
  Government & institutional organisations: 51
  Government & NGO space programmes: 20
  Launch site: 155
  Satellite communications: 127
  Scientific: 405
Sovinformsputnik: 50
Sovzond: 50
Soyuz-TM: 438
Soyuz-U: 214
Soyuz-U2: 216
Soyuz-2: 216
Soyuz/Ikar: 216
Soyuzkarta: 50
Space & Atmospheric Research Centre (Pakistan): 46
Space & Upper Atmosphere: 46
Space Activities Commission (SAC): 44
Space Activities Promotion Council (Japan): 44
Space Applications Centre (India): 173
Space Centre for Satellite Navigation (Australia): 32
Space Clipper: 216
Space Commission (India): 38
Space Flyer Unit: 384
Space Policy Institute (US): 102
Space Policy Unit (Australia): 32
Space Radar Lab: 371
Space Research Centre (Poland): 46
Space Research Institute (Austria): 32
Space Research Institute (Russia): 50
Space Research Organisation Netherlands: *see* SRON
Space Shuttle (US): 447
  Aborts: 452
  Carrier Aircraft: 451
  Computer & automated systems: 450
  Costs: 452
  Crew escape system: 453
  Environmental Mobility Unit: 450
  External Tank: 448
  Maintenance & reliability: 451
  Manned manoeuvring unit: 451
  Missions:
    STS-63: 453
    STS-66: 453
    STS-67: 454
    STS-69: 455
    STS-70: 455
    STS-71: 454
    STS-72: 457
    STS-73: 456
    STS-74: 456
    STS-75: 457
    STS-76: 457
    STS-77: 457
    STS-78: 458
    STS-79: 458
    STS-80: 459
    STS-81: 459
    STS-82: 459
    STS-83: 459
    STS-84: 460
    STS-85: 460
    STS-86: 460
    STS-87: 461
    STS-88: 463
    STS-89: 461
    STS-90: 461
    STS-91: 462
    STS-95: 462
    STS-96: 463
  Orbiter modifications: 451
  Orbiter pilot training: 451
  Orbiters (general description): 447
  Orbiters (*see also* individual missions)
    Atlantis: 448, 453, 454, 456, 457, 458, 459, 460
    Challenger: 448
    Columbia: 447, 456, 457, 458, 459, 461
    Discovery: 448, 453, 455, 459, 460, 462, 463
    Endeavour: 448, 454, 455, 457, 461, 463
    Enterprise: 447
  Payload capacity: 452
  Processing and turnround: 451
  Remote Manipulator System: 450
  Solid Rocket Boosters: 449
  SSME: 450
  Tethered Satellite: *see* Tethered Satellite System (TSS)
Space Station (RFAS): *see* Salyut and Mir
Space Studies Institute (US): 102
Space surveillance network/NORAD: 169
Space surveillance system (RFAS): 177
Space Telescope: *see* Hubble Space Telescope
Space Telescope Science Institute: 447; *also see* Hubble Space Telescope
Space Test Experiments Platform: *see* STEP
Space Test Program (USAF): *see* STP
Space Transportation System: 238
Space Vacuum Epitaxy Center: 59
Spacelab: 437
Spacenet 2-4: 340
Spaceplane (RFAS): 376
Spaceplanes (Japan): 375
Spaceport West: 161
Spain:
  Communications: 326
  Earth observation facilities: 74
  Engineering test: 376
  General services & suppliers: 87
  Government & institutional organisations: 51
  Government & NGO space programmes: 20
  Launch vehicles: 221
  Satellite communications: 127
  Scientific: 405
Spartan: 417
SPAS: 384
Spektr (Mir module): 470
Spektrum series: 404
Spektrum-R (satellite): 404
Spelda (Structure Porteuse Externe pour Lancement Double Ariane): *see* Ariane
Speltra: *see* Ariane
Spinning Solid Upper Stage: *see* PAM
Splav Technical Centre (Russia): 50
Spot (Satellite Probatoire d'Observation de la Terre): 344
SRET 1/2: 373
Sriharikota launch centre: *see* SHAR
SRON: 45
SROSS: 393
SS-17 (Lunar-A): 421
SS-18K (Dnepr): 206
SS-24 missile: *see* Space Clipper
SSRD: 341
Star City: *see* Zvezdny Gorodok
Starfire: 239
Starfire optical range: 177
Start launcher: 216
STEDI: 417
Stella: 381
Stellenbosch University: 51
Stentor: 297
STEP: 529
STP: 530
Strela: 490
STRV: 377
STTW: 296

Subsatellite Ferret: 504
Sunsat: 405
SUPARCO: 46
Super Arcas: 239
Superbird: 318
Support ships (RFAS/USSR): 168
Surf launcher: 217
Svobodny: 155
SWAS Explorer: 411
Sweden:
  Communication satellites: 65
  Communications: 327
  Engineering test: 377
  General services & suppliers: 88
  Government & institutional organisations: 51
  Government & NGO space programmes: 21
  Launch site services: 109
  Launch sites: 156
  Satellite guidance & navigation: 135
Swedish Institute of Space Physics: 51
Swedish National Space Board: 52
Swedish Space Corporation: 52
Switzerland:
  General services & suppliers: 88
  Government & institutional organisations: 52
  Government & NGO space programmes: 21
  Satellite actuation systems: 122
Syncom 4: 494 *see also* Leasat
Syria:
  Government & NGO space programmes: 21

Tacsat (Tactical Communications Satellite): 494
Taiwan:
  Government & institutional organisations: 52
  Government & NGO space programmes: 21
  Satellite communications: 127
Taiyuan launch centre: 148
Tamouz 1: 198
Tanegashima launch centre: 153
Tansei 1/2/3/4: 375
Taurus orbital launcher: 240
TDF: 36, 298
TDRS (Tracking & Data Relay Satellite): 341
Team Aer Lingus: 42
Technisch Physische Dienst (TPD): 45
Technology Application Center (US): *see* Earth Data Analysis Center
Technology Development Centre (TEKES; Finland): 35
Telecom (French satellites): 297
Telecommunication Company of Iran: 42
Telecommunications Advancement Organisation of Japan: 44
Telefonica de España SA: 51
Teleglobe Canada: 33
Telekomunikacja Polska SA: 46
Telenor Satellite Services AS: 45
Telesat Canada: 33
Tele-X: 327
Telia AB: 52
Telstar 5/6: 330
Telstar 301: 293
Telstar 302/3: 330
Telstar 401/2: 330
Telstra Corp Ltd: 32
Temisat: 313
Terrier-Malemute: 241
Terriers: 417
Tethered Satellite System (TSS): 400
Thaicom 1-3: 328
Thailand:
  Communication satellites: 65
  Communications: 328
  General services & suppliers: 89
  Government & institutional organisations: 52
  Government & NGO space programmes: 21
Thematic Mapper: *see* Landsat
Thor launcher: *see* Delta
Thor 1/2/3 satellites: 322
Thumba launch site: 152
Timation (Time Navigation): 514
Tiros (Television & Infra-red Observation Satellite): *see* NOAA
Titan 2: 241
Titan 2 Ferret: 504
Titan 3: 242
Titan 4: 244
TOMS-EP: 371
Tonga:
  Government & institutional organisations: 52
Tongasat: 52
Topex: 354
TOS (Tiros Operational Satellite): *see* NOAA
TRACE Explorer: 410
Transit/NNSS: 514
Transtage: *see* Titan
TRMM Tropical Rainfall Measuring Mission: 354
Tromsø satellite station: 168
Trumpet: 504
TsAGI (Central Aerohydrodynamics Institute): 50
TsDUC deep space network: 168
Tselina: 503
Tsentr Upravlenyia Polyotom (TsUP): 168
Tsentralny Aero-Girodinamichesky Insitut (TsAGI): 50
Tsikada: 511
TsNII Mashinostroenye: 51
TsNPO Kometa: 51
TsSKB: 46
Tsukuba Space Centre: 168, 174
TsUP (Tsentr Upravlenyia Polyotom): 168
TSX-5/-6: 531
Tsyklon: 512
Tsyklon-2/Tsyklon-M launcher: 218
Tsyklon-3: 218
TubSat: 393
Turkey:
  Communications: 329
  Government & institutional organisations: 52
  Government & NGO space programmes: 22
Turkish Telekom: 52
Turksat: 329
TV-Sat: 299
Tyuratam (Baikonur): 154

UARS (Upper Atmosphere Research Satellite): 416
UFO (UHF Follow-on): 496
Ukraine:
  Earth observation & meteorological: 362
  Government & institutional organisations: 52
  Government & NGO space programmes: 22
  Launch site services: 109
  Scientific: 402
Ukrainian Space Agency: 53
Ulysses: 420
Ume 1/2: 375
United Kingdom:
  Communications: 490
  Earth observation facilities: 74
  Engineering test: 377
  General services & suppliers: 89
  Government & institutional organisations: 53
  Government & NGO space programmes: 22
  Launch site services: 109
  Launch vehicle propulsion: 272
  Launch vehicles: 222
  Navigation/SAR: 117
  Research & development facilities: 174
  Satellite actuation systems: 122
  Satellite busses: 125
  Satellite communications: 127
  Satellite guidance & navigation: 135
  Satellite information storage: 141
  Satellite & propulsion support: 143
  Scientific: 406
United States:
  ASAT: 486
  Ballistic missile defence: 487
  Communication satellites: 65
  Communications: 329, 491
  Data relay: 341, 498
  Early warning: 500
  Earth observation & meteorological: 363
  Earth observation facilities: 75
  Earth observation satellites: 77
  Elint: 503
  Engineering test: 378
  General services & suppliers: 94
  Geodetic: 381
  Government & institutional organisations: 29, 54
  Government & NGO space programmes: 23
  Human space flight: 445
  Launch site services: 109
  Launch sites: 157
  Launch vehicle propulsion: 273
  Launch vehicles: 224
  Meteorology: 507
  Microgravity & materials: 386
  Navigation: 512
  Navigation/SAR: 118
  Nuclear surveillance: 515
  Ocean surveillance: 517
  Photo reconnaissance: 520
  Planetary science: 423
  Radar imaging: 527
  Research & development facilities: 175
  Satellite actuation systems: 123
  Satellite busses: 125
  Satellite communications: 129
  Satellite guidance & navigation: 136
  Satellite & propulsion support: 143
  Science & engineering test: 528
  Scientific: 406
  Space monitoring facilities: 177
  Space stations: 477
  Telemetry, tracking & command: 169
  Training facilities: 165
University of Dundee (Coastal Zone Research; UK): 54
University of Dundee (Satellite Laboratory; UK): 54
University of Surrey (UK) UoSAT: 378
UPM-Sat 1: 405
Uragan: *see* Glonass
Ural Electrochemical Integrated Plant (UEIP): 51
US Space Command: 29
USASDC US Army Space & Strategic Defense Command: 29
USSPACECOM space surveillance network: 178

Vandenberg Air Force base: 162
Vega: 199, 527
Vektor: 526
Venezuela:
  Government & institutional organisations: 60
VentureStar launch vehicle: 246
VEP (Vehicle Evaluation Payload): 375
VICTOR 1: 389
Videsh Sanchar Nigam Ltd (VSNL): 38
Viking (Sweden): 377
Viking engines: 243; *see also* Ariane
Vikram Sarabhai Space Centre (India): 173
VNII Elektromekhaniki: 51
VNII TransMash: 51
Vortex: 503
Voyager: 432
VSCC Liquid Propulsion Systems Centre: 38
VSCC Solid Propulsion Group: 38

Wake Shield facility: 368
Wallops Flight Facility: 58, 163
Westar: 340
Western Space & Missile Center: *see* 30th Space Wing
White Sands Missile Range: 164
White Sands Space Harbor: 165
Whitecloud: 518
Wind satellite: 396
Woomera launch site: 147
Woomera tracking & telemetry station: 167
Wresat: 389

X-33: 246
X-34: 246
Xian Satellite Control Centre: 167
Xichang Satellite Launch Centre (China): 148

Yantar: 520
YF rocket motors: 184
Yoshinobu Launch Complex: *see* Tanegashima Space Centre
Yug: 526
Yuzhnoye: *see* NPO Yuzhnoye

ZARM: 37
Zenit-2 (Gen 1-7): 519
Zenit-2/-3: 220
Zenit-3SL: 221
Zhongxing: 297
Zimbabwe:
  Earth observation facilities: 76
Znamya reflector: 405
Zond: 423
Zvezdny Gorodok: 165

NOTES

NOTES

NOTES